CELLULAR AND MOLECULAR NEUROPHYSIOLOGY

FOURTH EDITION

Companion Web Site:

http://booksite.elsevier.com/9780123970329

All the figures from the volume will be hosted on the site

Cellular and Molecular Neurophysiology, Fourth Edition

Constance Hammond, Author

Instructor Support Website:

http://textbooks.elsevier.com/web/Manuals.aspx?isbn=9780123970329

CELLULAR AND MOLECULAR NEUROPHYSIOLOGY

FOURTH EDITION

CONSTANCE HAMMOND

ELSEVIER

AMSTERDAM • BOSTON • HEIDELBERG • LONDON
NEW YORK • OXFORD • PARIS • SAN DIEGO
SAN FRANCISCO • SINGAPORE • SYDNEY • TOKYO
Academic Press is an Imprint of Elsevier

AP

Academic Press is an imprint of Elsevier
32 Jamestown Road, London NW1 7BY, UK
525 B Street, Suite 1800, San Diego, CA 92101-4495, USA
225 Wyman Street, Waltham, MA 02451, USA
The Boulevard, Langford Lane, Kidlington, Oxford OX5 1GB, UK

Cover figure: Ghost neuron.
Newborn rat hippocampal pyramidal neuron in a dissociated culture. Neurons were transfected with the green fluorescent protein (GFP) attached to a protein of the cell's cytoskeleton (MAP2β) located to the cell body and dendritic arbor. Artificially colored here to evoke a ghostly ambience. Supplied by Damien Guimond, Institut de Neurobiologie de la Méditerranée, INSERM UMR 901, Marseille, France.

British Library Cataloguing-in-Publication Data
A catalogue record for this book is available from the British Library

Library of Congress Cataloging-in-Publication Data
A catalog record for this book is available from the Library of Congress

ISBN: 978-0-12-397032-9

For information on all Academic Press publications
visit our website at http://store.elsevier.com/

Typeset by Thomson Digital

Table of Contents

Foreword

This excellent and highly acclaimed textbook by Hammond and co-authors, now in its fourth edition, remains faithful to its central philosophy that teaching science cannot simply rely on the presentation of facts, but must also include the intellectual journey that gives birth to key discoveries and results in the solution of long-standing puzzles. Science is in constant motion, and neuroscience, in particular, cannot be understood without appreciating how ground-breaking experiments were designed to test novel hypotheses using a combination of intuition, knowledge and experience. Readers will greatly appreciate this book for several reasons. Chief among them is that the text introduces them to the core scientific process, beyond simply presenting solutions to problems. A unique aspect of this textbook is that most figures are reproduced from the original papers that first demonstrated a given finding. This is an excellent didactic approach, because these original figures not only convey information concerning particular experimental arrangements and results, but they also introduce students to the history of neuroscience first hand. At the bottom of each figure legend, for example, readers will find the full reference including the authors, title, and journal for the paper that reported the particular discovery. There is no better way to teach students integrity and self-confidence than to introduce them to the original papers.

Because the current edition contains many updated chapters and appendices, students will learn about experiments performed by pioneering authors whom they can actually meet in person at scientific conferences. This is where the true power of the book lies, as this approach enables students to familiarize themselves with the ever-changing, flesh-and-blood frontlines of cutting-edge research, while they are still in the process of studying the key concepts of molecular and cellular neuroscience. The topics range from elementary properties of excitable cells to detailed discussions of ion channels, receptors, and synaptic transmission, all the way to dendritic integration and various forms of neuronal plasticity. This book is a concise, yet in-depth, highly informative text that will continue to inspire present and future practitioners of neuroscience.

Ivan Soltesz, PhD
University of California, Irvine

Acknowledgments

Constance Hammond would like to thank her colleagues who took part of their time to give their expertise: Agnès Baude for Chapters 1, 6, and 19 and Romain Nardou for Chapters 14 and 19.

And the following individuals who have contributed to previous editions of the book or who helped in carefully reading the proofs:

Andrea Nistri, International School for Advanced Studies (SISSA), Trieste, Italy and Aron Gutman (deceased) formerly of Kaunas Medical Academy, Kaunas, Lithuania (previous Chapter 4, now part of Chapter 3); Gautam Bhave and Robert Gereau, Baylor College of Medicine, Houston, TX, USA (Chapter 12); Yusuf Tan, Bogazici University, Istanbul, Turkey (Appendix 5.1); Charles Bourque; Diana Ferrari (Chapter 7); and Geneviève Chazal (Chapters 19 and 20).

PART I

NEURONS: EXCITABLE AND SECRETORY CELLS THAT ESTABLISH SYNAPSES

CHAPTER

1

Neurons

Constance Hammond

OUTLINE

By using the silver impregnation method developed by Golgi (1873), Ramon y Cajal studied neurons, and their connections, in the nervous system of numerous species. Based on his own work (1888) and that of others (e.g. Forel, His, Kölliker and Lenhossék), he proposed the concept that neurons are isolated units connected to each other by contacts formed by their processes: 'The terminal arborizations of neurons are free and are not joined to other terminal arborizations. They make contacts with the cell bodies and protoplasmic processes of other cellular elements.'

As proposed by Cajal, neurons are independent cells making specific contacts called *synapses*, with hundreds or thousands of other neurons sometimes greatly distant from their cell bodies. The neurons connected together form circuits, and so the nervous system is composed of neuronal networks which transmit and process information. In the nervous system, there is another class of cells, the glial cells, which surround the various parts of neurons and cooperate with them. Glial cells are discussed in Chapter 2.

Neurons are *excitable* cells. Depending on the information they receive, neurons generate electrical signals and propagate them along their processes. This capacity is due to the presence of particular proteins in their plasma membrane which allow the selective passage of ions: the ion channels.

Neurons are also *secretory* cells. Their secretory product is called a *neurotransmitter*. The release of a neurotransmitter occurs only in restricted regions, the synapses. The neurotransmitter is released in the extracellular space. The synaptic secretion is highly focalized and directed specifically on cell regions to which the neuron is connected. The synaptic secretion is then different (with only a few exceptions) from other secretory cells, such as from hormonal and exocrine cells which respectively release their secretory products into the general circulation (endocrine secretion) or the external environment (exocrine secretion). Synapses are discussed in Chapter 6.

Neurons are *quiescent* cells. When lesioned, most neurons cannot be replaced, since they are postmitotic cells. Thus, they renew their constituents during their entire life, involving the precise targeting of mRNAs and proteins to particular cytoplasmic domains or membrane areas.

part

Cellular and Molecular Neurophysiology. DOI: 10.1016/B978-0-12-397032-9.00001-7

1.1 NEURONS HAVE A CELL BODY FROM WHICH EMERGE TWO TYPES OF PROCESSES: THE DENDRITES AND THE AXON

Although neurons present varied morphologies, they all share features that identify them as neurons. The cell body or *soma* gives rise to processes which give the neuron the regionalization of its functions, its polarity and its capacity to connect to other neurons, to sensory cells or to effector cells.

1.1.1 The somatodendritic tree is the neuron's receptive pole

The soma of the neuron contains the nucleus and its surrounding cytoplasm (or *perikaryon*). Its shape is variable: pyramidal soma for pyramidal cells in the cerebral cortex and hippocampus; ovoid soma for Purkinje cells in the cerebellar cortex; granular soma for small multipolar cells in the cerebral cortex, cerebellar cortex and hippocampus; fusiform soma for neurons in the pallidal complex; and stellar or multipolar soma for motoneurons in the spinal cord (**Figure 1.1**).

One function of the soma is to ensure the synthesis of many of the components required for the structure and function of a neuron. Indeed, the soma contains all the organelles responsible for the synthesis of macromolecules. Most neurons in the central nervous system cannot further divide or regenerate after birth, and the cell body must maintain the structural integrity of the neuron throughout the individual's entire life. Moreover, the soma receives numerous synaptic contacts from other neurons and constitutes, with the dendrites, the main receptive area of neurons (see **Figure 1.5** and Section 6.2). The neurons have one or several processes emerging from the cell body and arborizing more or less profusely. The two types of neuronal processes are the dendrites and the axon (**Figures 1.1 and 1.3**). This division is based on morphological, ultrastructural, biochemical and functional criteria.

The dendrites, when they emerge from the soma, are simple perikaryal extensions, the primary dendrites. On average, between one and nine primary dendrites emerge from the soma and then divide successively to give a dendritic tree with specific characteristics (number of branches, volume, etc.) for each neuronal population (**Figures 1.1 and 1.2**). The dendrites are morphologically distinguishable from axons by their irregular outline, by their diameter, which decreases along their branchings, by the acute angles between the branches, and by their ultrastructural characteristics (**Figures 1.1, 1.3 and 1.7**). The irregular outline of dendrites is related to the presence of numerous appendices of various shapes and dimensions at their surface. The most frequently observed are the dendritic spines which are lateral expansions with ovoid heads binding to the dendritic branches by a peduncle that is variable in length (**Figure 1.3**). Some neurons are termed 'spiny' because there are between 40 000 and 100 000 spines on the surface of their dendrites (e.g. pyramidal neurons of the cerebral cortex and hippocampus, the medium-sized neurons of the striatum, and the Purkinje cells of the cerebellar cortex). However, other neurons with only a few spines on their dendritic surface are termed 'smooth' (e.g. neurons of the pallidal complex) (**Figure 1.1**). The transition from the cell body to proximal dendrites is gradual, and the cytoplasmic architectures of proximal dendrites and the cell body are similar. In particular, the endoplasmic reticulum and ribosomes are almost as abundant in the proximal dendrites as in the cell body. Moreover, even distal dendrites contain ribosomes and endoplasmic reticula.

Dendrites and soma receive numerous synaptic contacts from other neurons and constitute the main receptive area of neurons (see **Figure 1.5** and Section 6.2). In response to afferent information, they generate electrical signals such as postsynaptic potentials (EPSPs or IPSPs; see **Figure 1.5** left 1, 2) or calcium action potentials, and integrate the afferent information. Chapters 8–10 look at the mechanisms underlying the excitatory (EPSP) and inhibitory (IPSP) postsynaptic potentials generated in the postsynaptic membrane in response to transmitter release. Chapters 13–16 discuss how these postsynaptic responses are integrated along the somatodendritic tree. Although dendrites are generally a receptive zone, there are certain exceptions: some dendrites are connected with other dendrites and act as a transmitter area by releasing neurotransmitters (see **Figure 6.2d**).

1.1.2 The axon and its collaterals are the neuron's transmitter pole

The axon is morphologically distinct from dendrites in having a smooth appearance and a uniform diameter along its entire extent, and by its ultrastructural characteristics (see **Figures 1.3, 1.6 and 1.7**). Axons are narrow from their origin, and do not usually contain ribosomes or endoplasmic reticula. The transition from the cell body to axon is distinct; the region of the cell body from which an axon originates is called the axon hillock and it tapers off to the axonal initial segment, where action potentials begin. Although most parts of the cell body are rich in endoplasmic reticula, the axon hillock is not. At the axon initial segment, the plasma membrane has thick underlying structures, and there is a specialized bundle of microtubules. In some neurons, the axon emerges at the level of a primary dendrite.

The axon is not a single process; it is divided into one or several collaterals which form right-angles with the main axon. Some collaterals return toward the cell body area; these are recurrent axon collaterals. The axon and

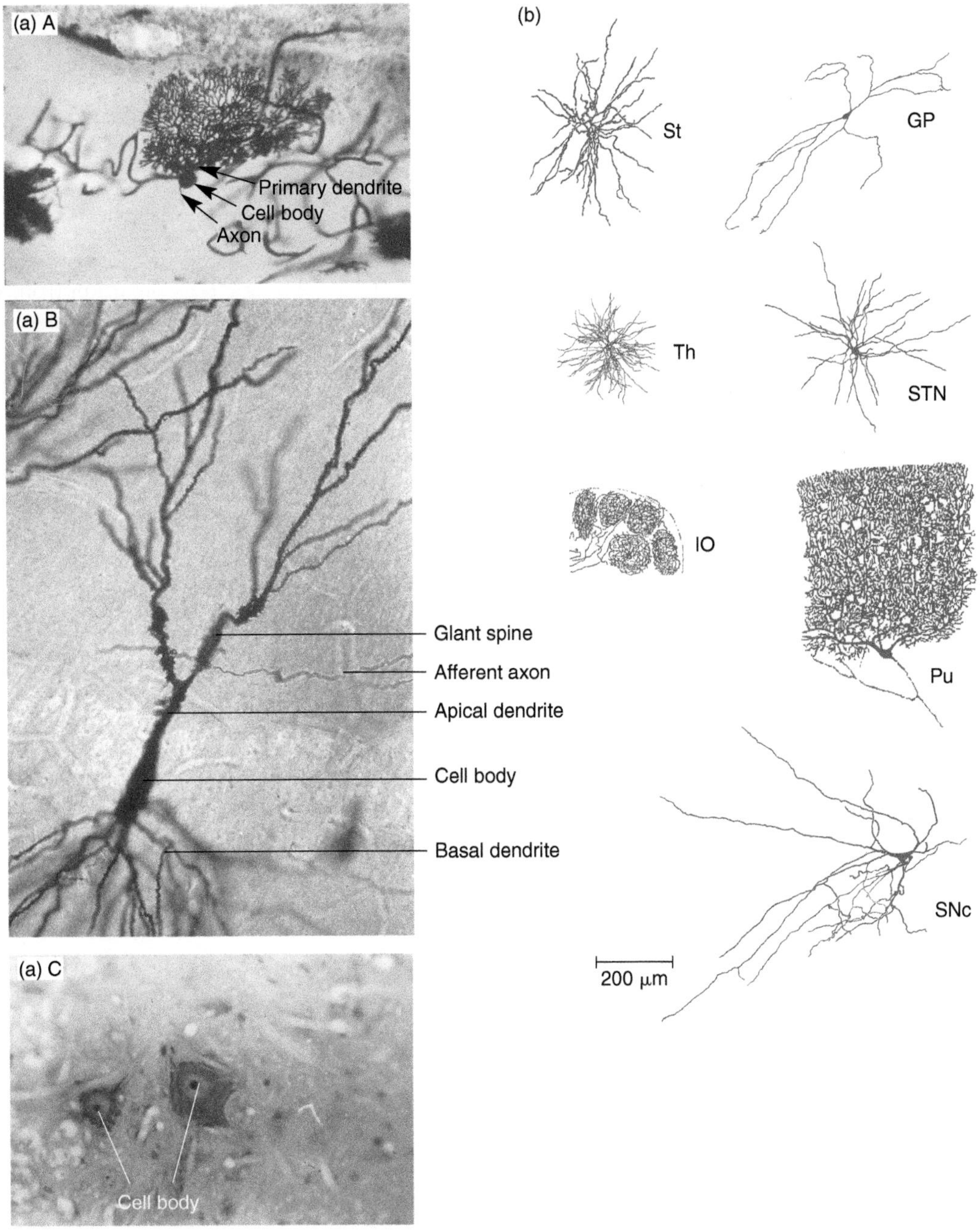

FIGURE 1.1 **The neurons of the central nervous system present different dendritic arborizations. (a)** Photomicrographs of neurons in the central nervous system as observed under the light microscope. A – Purkinje cell of the cerebellar cortex; B – pyramidal cell of the hippocampus; C – soma of a motoneuron of the spinal cord. Golgi (A and B) and Nissl (C) staining. The Golgi technique is a silver staining which allows observation of dendrites, somas and axon emergence. The Nissl staining is a basophile staining which displays neuronal regions (soma and primary dendrites) containing Nissl bodies (parts of the rough endoplasmic reticulum). **(b)** Camera lucida drawings of neurons in the central nervous system of primates, revealed by the Golgi silver impregnation technique and reconstructed from serial sections: St, medium spiny neuron of the striatum; GP, neuron of the globus pallidus; Th, thalamocortical neuron; STN, neuron of the subthalamic nucleus; IO, neurons of the inferior olivary complex; Pu, Purkinje cell of the cerebellar cortex; SNc, dopaminergic neuron of the substantia nigra pars compacta. All these neurons are illustrated at the same magnification. Photomicrographs by Francoise Condé (**a**A), Olivier Robain (**a**B) and Paul Derer (**a**C). Drawings by Jérôme Yelnik, except OL and PU by Ramon Y Cajal (1911).

its collaterals may be surrounded by a sheath, the myelin sheath. Myelin is formed by glial cells (see Sections 2.2 and 2.4). The length of an axon varies. Certain neurons in the central nervous system have axons that project to one or several structures of the central nervous system that are more or less distant from their cell bodies (**Figure 1.4**), whereas other neurons have short axons (a few microns in length) that are confined to the structure where their cell bodies are located; these are interneurons or local circuit neurons (see **Figure 1.13**).

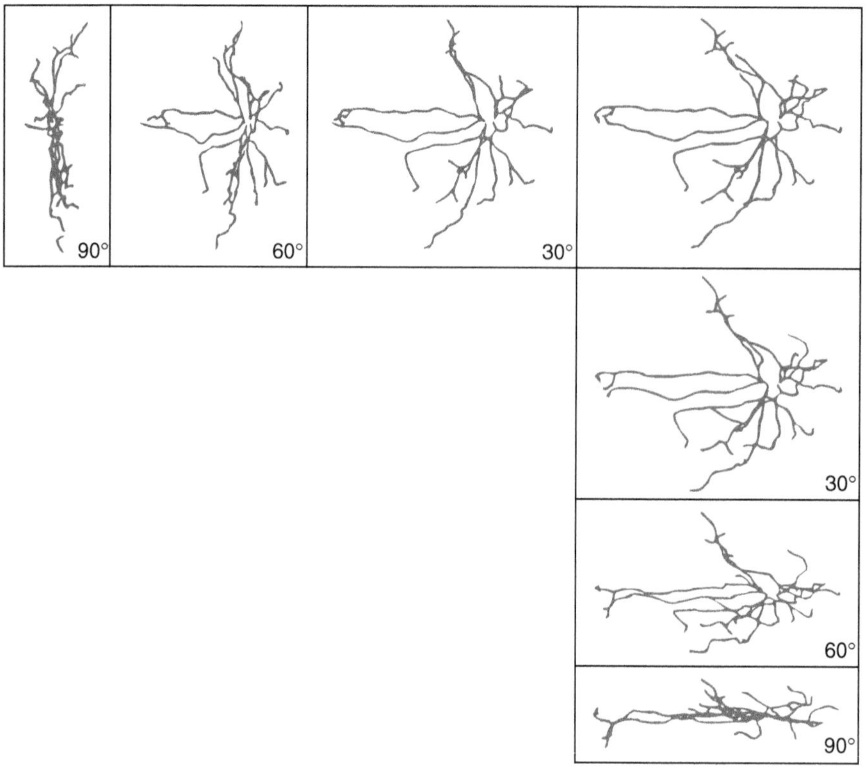

FIGURE 1.2 **Tridimensional illustration of a dendritic arborization.** Computer drawing of a neuron of the subthalamic nucleus injected intracellularly with horseradish peroxidase (HRP) and reconstructed in three dimensions from serial sections. At 0°, the dendritic arborization of this neuron is represented in its principal plane; i.e. in the plane where it has its largest surface. In this plane, the dendritic field is almost circular (859 μm long and 804 μm wide). 30°, 60° and 90° rotations from the principal plane around the horizontal (horizontal column) and vertical (vertical column) axis show that the dendritic field has a flattened ovoidal form (230 μm thick). From Hammond C, Yelnik J (1983) Intracellular labelling of rat subthalamic nucleus with horseradish peroxidase: computer analysis of dendrites and characterization of axon arborization. *Neuroscience* **8**, 781–790, with permission.

Thus projection (Golgi type I) neurons and local-circuit (Golgi type II) neurons can be differentiated. In Golgi type I neurons, the length of the axon is variable: certain projection neurons are directed to one structure only (e.g. corticothalamic neurons; see **Figure 1.14**) whereas other projection neurons have numerous axon collaterals which project to several cerebral structures (**Figure 1.4**).

The axon and axonal collaterals in certain neurons end in a terminal arborization, i.e. numerous thin branches whose extremities, the synaptic boutons, make synaptic contacts with target cells (see **Figure 6.3**). In other neurons, the axon and its collaterals have enlargements or varicosities which contact target cells along their way: these are 'boutons en passant' (see **Figures 6.14 and 6.15b**). It can be noted that both types of boutons are called *axon terminals*, although 'boutons en passant' are not the real endings of the axon.

The main characteristic of axons is their capacity to trigger sodium action potentials and to propagate them over considerable distances without any decrease in their amplitude (**Figure 1.5** left 3). Action potentials are generated at the initial segment level in response to synaptic information transmitted by the somatodendritic tree. Then they propagate along the axon and its collaterals toward the axon terminals (synaptic boutons or boutons en passant). When action potentials reach the axon terminals these trigger calcium action potentials (**Figure 1.5** left 4) which may cause the release of the neurotransmitter(s) contained in axon terminals in a specific compartment, the synaptic vesicles. This secretion is localized only at the synaptic contacts. Overall, the axon is considered as the transmitter pole of the neuron.

Chapter 4 discusses the mechanisms underlying the abrupt, large and transient depolarizations called (sodium) action potentials, and how they are triggered and propogated. Chapters 4, 5 and 7 look at how sodium action potentials trigger calcium action potentials, the entry of calcium in synaptic terminals and the secretion of transmitter molecules.

Certain regions – such as the initial segment, nodes of Ranvier (zones between two myelinated segments; see **Figure 1.5**) and axon terminals – can also be receptive areas (a postsynaptic element) of synaptic contacts from other neurons (see Section 6.2).

1.2 NEURONS ARE HIGHLY POLARIZED CELLS WITH A DIFFERENTIAL DISTRIBUTION OF ORGANELLES AND PROTEINS

The somatodendritic tree is the neuron's receptive pole, whereas the axon and its collaterals are the neuron's transmitter pole. Neurons are highly polarized cells. Cellular morphology and accurate organelles and protein distribution lay the basis to this polarization. The

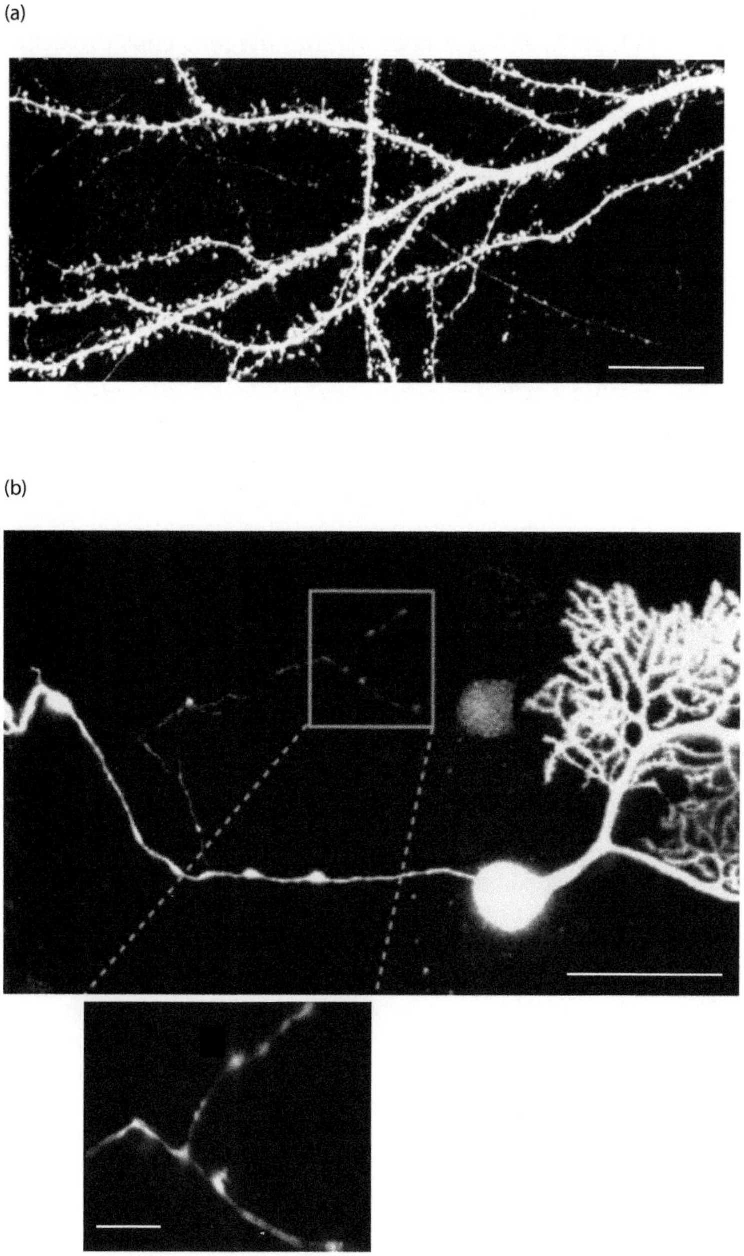

FIGURE 1.3 **Dendritic spines and axon collaterals.** Dendritic spines of a pyramidal neuron of the hippocampus after 19 days of culture *in vitro*. Scale bar: 20 μm. (**b**) Axonal arborization of a Purkinje cell and enlargement of axonal bifurcation. Scale bars 50 and 10 μm. Part (**b**) photo by Hartmut Schmidt.

organelles and cytoplasmic elements present in neurons are the same organelles found in other cell types. However, some elements such as cytoskeletal elements are more abundant in neurons. The non-homogeneous distribution of organelles in their soma and processes is one of the most distinguishing characteristics of neurons.

1.2.1 The soma is the main site of macromolecule synthesis

The soma contains the same organelles and cytoplasmic elements that exist in other cells: cellular nucleus, Golgi apparatus, mitochondria, polysomes, cytoskeletal elements and lysosomes. The soma is the main site of synthesis of macromolecules since it is the one compartment containing all the required organelles.

Compared with other types of cells, the neuron differs at the nuclear level and more specifically at the chromatin and nucleolus levels. The chromatin is light and sparsely distributed: the nucleus is in interphase. Indeed, in humans, most neurons cannot divide after birth since they are postmitotic cells. The nucleolus is the site of ribosomal synthesis and ribosomes are essential for translating messenger RNA (mRNA) into proteins. The large size of the nucleolus indicates a high level of protein synthesis in neurons.

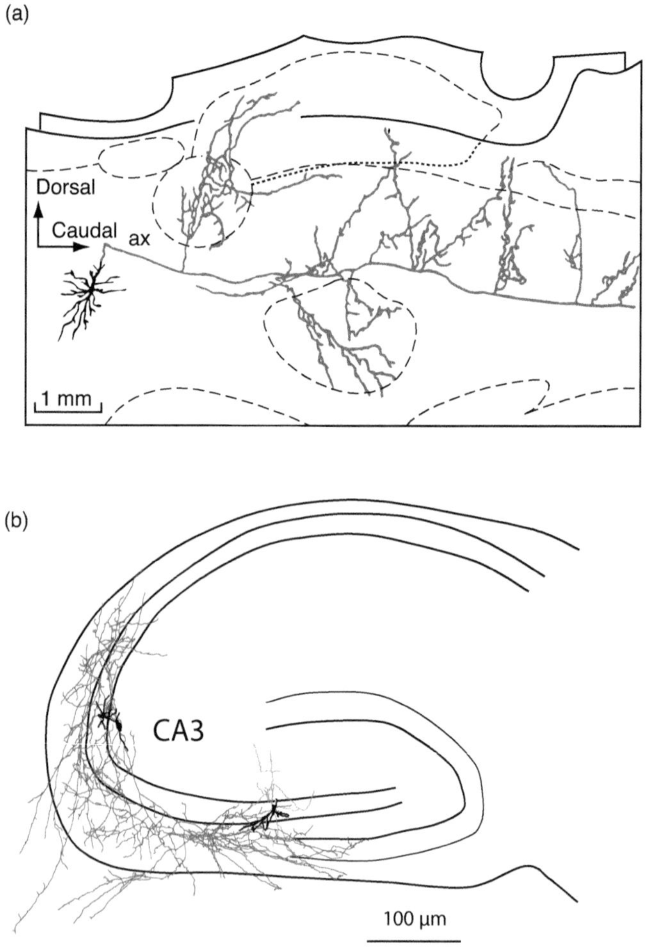

FIGURE 1.4 **Neurons showing complex axonal arborizations.** Drawing of a cat reticulospinal neuron has been stained by intracellular injection of peroxidase and drawn in a parasagittal plane obtained from serial sections. The axon (ax, black) gives off numerous collaterals along its rostrocaudal trajectory, making contacts with different neuronal populations (delimited by broken lines). Scale: 7 mm = 1 mm. **(b)** A rat GABAergic 'hub' neuron of the CA3 region of the hippocampus filled with neurobiotin during whole cell recording shows numerous axonal collaterals (blue) that expand inside and outside the hippocampus. In comparison a control GABAergic interneuron shows a restricted axonal arborization (green). Cells bodies and dendrites are in black. Part (**a**) from Grantyn A (1987) Reticulo-spinal neurons participating in the control of synergic eye and head movement during orienting in the cat. *Exp. Brain Res.* **66**, 355–377, with permission. Part (**b**) from Picardo MA, Guigue P *et al.* (2011) Pioneer GABA cells comprise a subpopulation of hub neurons in the developing hippocampus. *Neuron* **71**, 695–709, with permission.

1.2.2 The dendrites contain free ribosomes and synthesize some of their proteins

In dendrites can be found smooth endoplasmic reticulum, elongated mitochondria, free ribosomes or polysomes, and cytoskeletal elements including microtubules which are oriented parallel to the long axis of the dendrites (**Figure 1.6**).

By using the hook procedure, microtubules have been shown to have two orientations in proximal dendrites: half of them are oriented with the plus-ends distal to the cell body, and the other half has the plus-ends proximal to the cell body. This is very different from the orientation in distal dendrites and axons (**Figure 1.7**), which is uniform. Moreover, one microtubule-associated protein (MAP), the high-molecular-weight MAP2 protein and more precisely the MAP2A and MAP2B, are more common to dendrites than to axons. For this reason MAP2A or MAP2B antibodies coupled to fluorescent molecules are useful for labeling dendrites, particularly for dendrite identification in cell cultures.

mRNA trafficking and local protein synthesis in dendrites

The dendritic compartment contains many ribosomes whereas an axon has considerably fewer ribosomes. One particular feature of dendrites, compared with axons, is the presence of synapse-associated polyribosome complexes (SPRCs); these are clusters of polyribosomes and associated membranous cisterns that are selectively localized beneath synapses (more precisely, beneath postsynaptic sites), at the base of dendritic spines when spines are present.

What is the origin of this selective distribution of ribosomes in neurons? This question is particularly important since this compartmentalization leads to different properties of dendrites and axons: dendrites can locally synthesize some of their proteins, whereas axons would synthesize very few of them, if any.

Whereas most proteins destined for dendrites and dendritic spines are conveyed from the cell body, a subset of mRNAs is transported into dendrites to support local protein synthesis. Such a local dendritic protein synthesis requires that a particular subset of mRNAs synthesized in the nucleus is transported into the dendrites up to the polysomes where they are translated. How are dendritic mRNAs recognized and routed to dendrites and not axons? How are they transported to their final dendritic destination?

Chimeric constructs consisting of a reporter gene fused to putative *cis*-acting RNA localization elements were used to visually identify *cis*-acting sequences that direct mRNAs to their destination in dendrites near synapses. These experiments convincingly proved the existence of *cis*-acting elements commonly referred to as 'zipcodes' which allow mRNAs targeting to distinct subcellular compartments in order to regulate gene expression with temporal and spatial control.

In cultured hippocampal neurons, RNA labeled with tritiated uridine is shown to be transported at a rate of 250–500 μm per day. This transport is blocked by metabolic poisons and the RNA in transit appears to be bound to the cytoskeleton, since much of it remains following detergent extraction of the cells. Studies using video microscopy techniques and cell-permeant dyes which

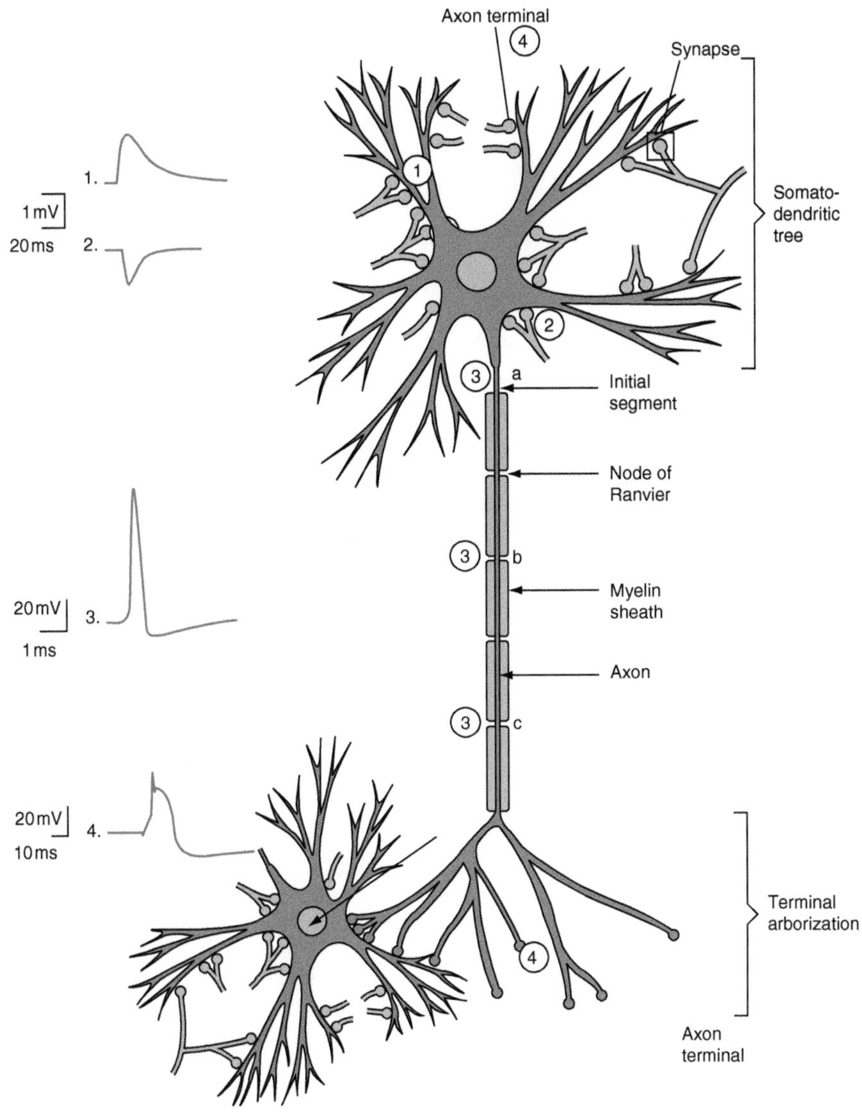

FIGURE 1.5 **Comprehensive schematic drawing of neuron polarity.** The somatodendritic compartment of a neuron receives a large amount of information from other neurons that establish synapses with it. At each synapse level, the neuron generates postsynaptic potentials in response to the released neurotransmitter (1, EPSP; 2, IPSP). These postsynaptic potentials propagate and summate in the somatodendritic compartment, then they propagate to the initial segment of the axon where they generate (or not) action potential(s) (3a). The action potentials propagate along the axon (3b, 3c) and its collaterals up to the axon terminals where they evoke (or not) the entry of calcium (4) and neurotransmitter release. Note the different voltage and time calibrations.

fluoresce on binding to nucleic acids have permitted observation of the movement of RNA-containing particles along microtubules in dendrites. These studies suggest that mRNAs are transported as part of a larger structure. The visualized RNA particles co-localize with poly(A) mRNA, the 60S ribosomal subunit, suggesting that the RNA-containing particles may represent translational units or complexes (**Figure 1.8**). This energy-dependent transport seems to be associated with the dendrite cytoskeleton as also shown by the delocalization of RNA-containing particles in response to colchicin (a drug which blocks microtubule polymerization).

To visualize mRNA translocation in live neurons, studies used nucleic acid stains and green fluorescent protein fused to RNA-binding proteins. It showed that mRNAs are transported in the form of large particles, the ribonucleoprotein particles, containing mRNAs, RNA-binding proteins, ribosomes, and translational factors (RNA-containing granules) in a rapid (average speed, 0.1 μm/s), bidirectional, and microtubule dependent manner. Ribonucleoprotein particles that are transported to dendrites bind to motor proteins (kinesin) as a large detergent resistant, RNAse-sensitive particle (see **Figure 1.11** and Section 1.3 for kinesin).

The mRNAs present in dendrites encode proteins with specialized synaptic functions. Among the mRNAs detected in dendrites by *in situ* hybridization (see Appendix 6.2) are mRNAs that encode certain cytoskeletal proteins (Arc, MAP2), kinases (the α-subunit of calcium/calmodulin-dependent protein kinase II), an integral membrane protein of the endoplasmic reticulum (the inositol trisphosphate receptor), calcium-binding proteins, certain subunits of neurotransmitter receptors (GluR1 and 2 of AMPA receptors, NR1 of NMDA receptors) as well as other proteins of unknown function. Moreover, within dendrites, different mRNAs are localized in different domains and different mRNAs are localized in the dendrites of different neuron types.

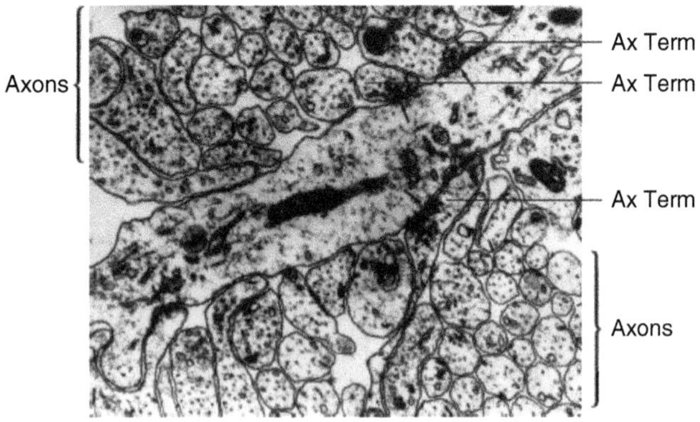

FIGURE 1.6 **Photomicrograph of a tissue section of the central nervous system at the hippocampal level.** This shows the ultrastructure of a dendrite, numerous axons and their synaptic contacts (observation under the electron microscope). The apical dendrite of a pyramidal neuron contains mitochondria, microtubules, ribosomes and smooth endoplasmic reticulum. It is surrounded by fascicles of unmyelinated axons with mitochondria and microtubules but no ribosomes. The axon's trajectory is perpendicular to the section plane. Three synaptic boutons (Ax Term) with synaptic vesicles make synaptic contacts (arrows) with the dendrite. Photomicrograph by Olivier Robain.

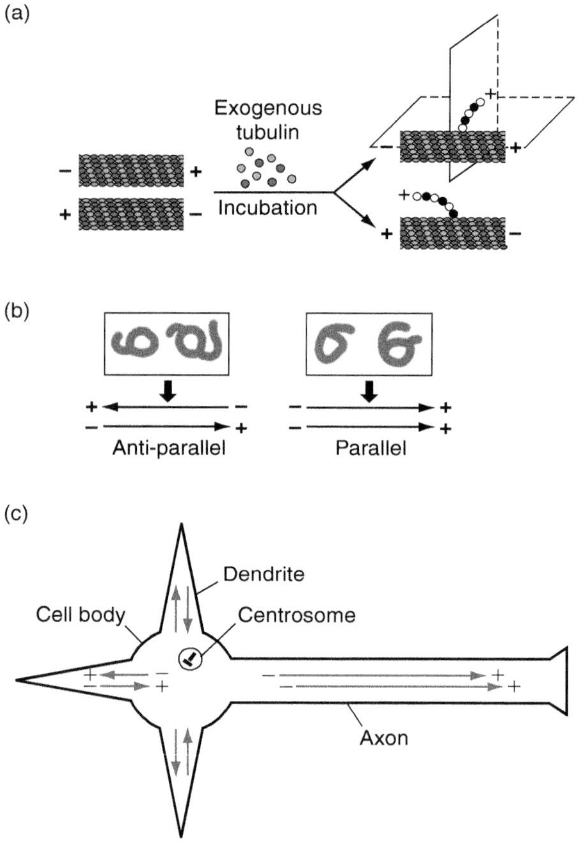

FIGURE 1.7 **Microtubule polarity in neuronal processes. (a)** The polarity of microtubules is defined by the hook procedure. Neurons in culture are permeabilized in the presence of taxol to stabilize microtubules. Monomers of tubulin, purified from brain extracts, are added in the extracellular medium. Several minutes after, transversal cuts are performed at the level of dendrites or at the level of an axon. Slices are treated for electron microscopy. Hook-like structures are observed. They result from exogenous tubulin polymerization at the surface of endogenous microtubules. Hooks are always oriented toward the plus-end of microtubules. **(b)** When hooks, at the electron microscopic level, have mixed orientations (clockwise and anticlockwise), this means that endogenous microtubules are antiparallel (left). Uniformly oriented hooks (right) indicate that endogenous microtubules are parallel. **(c)** Orientation of microtubules in dendrites and the axon. Drawing **(a)** by Lotfi Ferhat. Drawing **(b)** adapted from Sharp DJ, Wenqian Yu, Ferhat L *et al.* (1997) Identification of a microtubule-associated motor protein essential for dendrite differentiation. *J. Cell. Biol.* **138**, 833–843. Drawing **(c)** adapted from Baas PW, Deitch JS, Black MM, Banker GA (1988) Polarity orientation of microtubules in hippocampal neurons: uniformity in the axon and nonuniformity in the dendrite. *Proc. Natl Acad. Sci. USA* **85**, 8335–8339, with permission.

In summary, it has become clear that certain mRNAs are targeted to dendrites within ribonucleoprotein particles (>1000S). These particles are transported distally by a kinesin, which binds RNA-binding proteins by a recognition motif in its tail domain. mRNAs targeted to dendrites would travel to their final synaptic destination in a translationally repressed state due to the presence of various binding partners acting as translational regulators. Once an individual synapse is activated, these particles would partially disassemble and translational repression would be relieved locally.

The relatively large amount of RNA transported into dendrites raises the question of why neurons need this supply. Targeting of mRNAs to dendritic synthetic machinery located at the base of dendritic spines suggests that local translation could be regulated in an activity-dependent manner. For example, in response to synaptic information local protein synthesis is triggered to modulate synaptic transmission by changing, for example, the subunits or the number of receptors to the neurotransmitter in the postsynaptic membrane (this can occur during plasticity and may produce long-lasting changes in synaptic strength) (see Chapter 18). This in turn would allow an individual synapse of a given neuron to modify its function as well as its morphology, thereby providing a mechanism for synaptic plasticity and memory consolidation.

1.2.3 The axon, to a large extent, lacks the machinery for protein synthesis

The axoplasm contains thin elongated mitochondria, numerous cytoskeletal elements and transport vesicles. It is devoid of ribosomes associated to the reticulum but may contain ribonucleoprotein complexes especially during development. Nevertheless, axons cannot restore the vast majority of the macromolecules from which they are made; neither can they ensure alone the synthesis of the neurotransmitter(s) that they release since they are unable to synthesize proteins (such as enzymes). This problem is resolved by the existence of a continuous supply of macromolecules from the cell body to the axon through anterograde axonal transport (see Section 1.3).

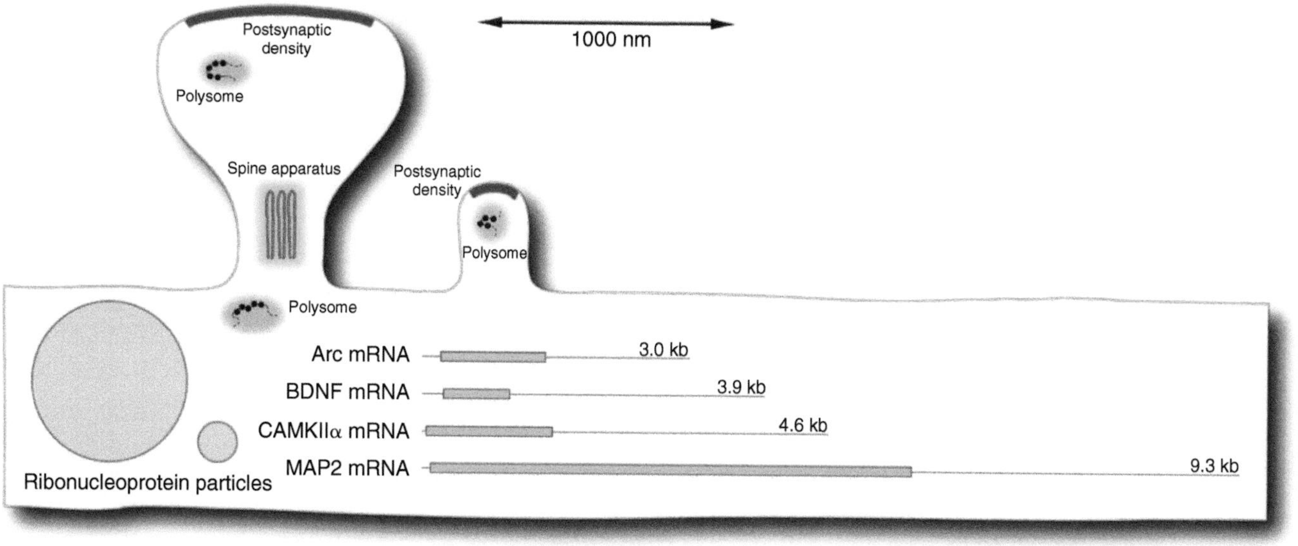

FIGURE 1.8 **Approximate sizes of representative dendritic mRNAs and translational elements at synaptic sites on dendrites.** The drawing illustrates the approximate size range of spine synapses that would be found in rat forebrain structures such as the hippocampus and cerebral cortex. The lines represent the approximate length of representative dendritic mRNAs if they were straightened out. Shading indicates the length and position of the coding region. Adapted from Schuman EM, Dynes JL, Steward O (2006) Synaptic regulation of dendritic mRNAs. *J. Neurosci.* **26**, 7143–7146, with permission.

Another major difference between dendrites and axons is the orientation of microtubules. By using the hook procedure (see **Figure 1.7**), it has been shown that the polarity of microtubules is uniform in the axon, meaning that all their plus-ends point away from the cell body, toward the axon terminals. The polarity of the microtubules is relevant for transport properties (see Section 1.3). Moreover, one MAP, the Tau protein, is more common to axons than to dendrites. Tau antibodies coupled to fluorescent molecules are useful for labeling axons, particularly for axon identification in cell cultures.

1.3 AXONAL TRANSPORT ALLOWS BIDIRECTIONAL COMMUNICATION BETWEEN THE CELL BODY AND THE AXON TERMINALS

Axonal transport is the movement of subcellular structures (such as vesicles, mitochondria, etc.) and proteins (like those of the cytoskeleton) from the cell body to axonal sites (nodes of Ranvier, presynaptic release sites, etc.) and from axon terminals to the cell body.

1.3.1 Demonstration of axonal transport

Weiss and Hiscoe (1948) first demonstrated the existence of material transport in growing axons (during development) as in mature axons. Their work consisted of placing a ligature on the chicken sciatic nerve, and then examining the change in diameter of the axons over several weeks. They showed that these neurons became enlarged in their proximal part and presented degenerative signs in their distal part (**Figure 1.9**). The authors suggested that material from the cell body had accumulated above the ligature and ensured the survival of the distal part.

Later, Lubinska *et al.* (1964) elaborated the concept of anterograde and retrograde transport. These authors placed two ligatures on a dog sciatic nerve, isolated part of the nerve and divided it into short segments in order to analyze their acetylcholinesterase content. This

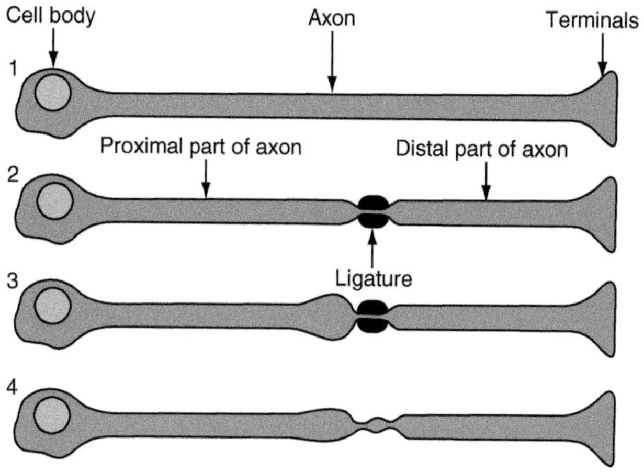

FIGURE 1.9 **Experiment by Weiss and others demonstrating anterograde axonal transport.** Schematic of a chicken motoneuron (1). When a ligature is placed on the axon (2) an enlargement of the axon's diameter above the ligature is noted after several weeks (3). When this ligature is removed, the enlargement progressively disappears (4). From Weiss P, Hiscoe HB (1948) Experiments on the mechanism of nerve growth. *J. Exp. Zool.* **107**, 315–396, with permission.

enzyme is responsible for acetylcholine degradation and was used here as a marker. They showed that it accumulates at the level of both ligatures. This result therefore suggested the existence of two types of transport: an anterograde transport (from cell body to terminals) and a retrograde transport (from terminals to cell body). Moreover, it appeared that both types of transport are distributed along the entire extent of the axon.

We presently know of three types of axonal transport: fast (anterograde and retrograde), slow (anterograde) and mitochondrial.

1.3.2 Fast anterograde axonal transport is responsible for the movement of membranous organelles from cell body towards axon terminals, and allows renewal of axonal proteins

Fast anterograde axonal transport consists in the movement of vesicles along the axonal microtubules at a rate of 200–400 mm per day (i.e. 2–5 μm/s). These transport vesicles, which are 40–60 nm in diameter, emerge from the Golgi apparatus in the cell body (**Figure 1.10a**). They transport, among other things, proteins required

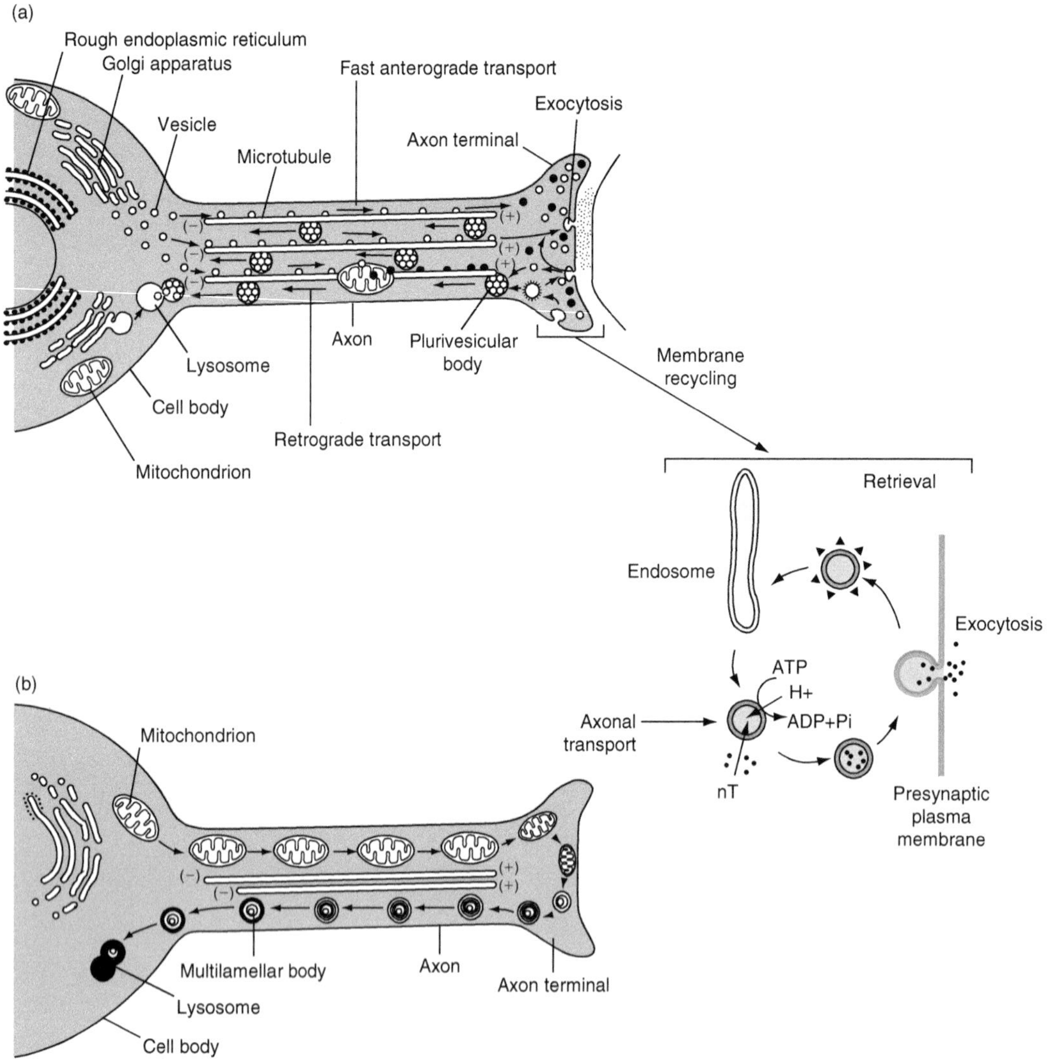

FIGURE 1.10 **Fast axonal transport. (a)** Schematic of fast anterograde axonal transport (anterograde movement of vesicles) and retrograde axonal transport (retrograde movement of plurivesicular bodies). These two transports use microtubules as substrate. The detail shows recycling of small synaptic vesicles. Vesicles synthetized in the cell body and transported to the axon terminals are loaded with cytoplasmic neurotransmitter and targeted to the presynaptic plasma membrane. In response to Ca^{2+} entry, they fuse with the plasma membrane and release their content into the synaptic cleft (exocytosis); then they are recycled via an endosomal compartment. **(b)** Schematic of mitochondrial transport. Note that the neuron representation is extremely schematic since axons do not give off *one* axon terminal. Drawing (**a**) adapted from Allen R (1987) Les trottoirs roulants de la cellule. *Pour la Science*, April, 52–66; and Südhof TC, Jahn R (1991) Proteins of synaptic vesicles involved in exocytosis and membrane recycling. *Neuron* **6**, 665–677, with permission. Drawing (**b**) adapted from Lasek RJ, Katz M (1987) Mechanisms at the axon tip regulate metabolic processes critical to axonal elongation. *Prog. Brain Res.* **71**, 49–60, with permission.

to renew plasma membrane and internal axonal membranes, neurotransmitter synthesis enzymes and neurotransmitter precursors when the neurotransmitter is a peptide. This transport is independent of the type of axon (central, peripheral, etc.).

The pioneer living preparation

The squid's giant axon is most commonly used for these observations since its axoplasm can easily be extruded and a translucent cylinder of axoplasm devoid of its membrane is thus obtained. This living extruded axon keeps its transport properties for several hours. The absence of plasma membrane allows precise control of the experimental conditions and entry into the axoplasm of several components that cannot usually pass through the membrane barrier *in vivo* (e.g. antibodies). The improvement of video techniques applied to light microscopy allowed the first observations of the movement of a multitude of small particles along the microtubules in a living extruded axon.

Identification of the moving organelles and their substrates

Analysis of the particles that accumulate on each side of the 1.0–1.5-mm long isolated frozen segments of the squid axon has permitted the identification of moving organelles in axons. Correlation between video and electron microscopy images of these axonal segments has shown that the particles moving anterogradely on video images are small vesicles. Indeed, when a purified fraction of small labeled vesicles (with fluorescent dyes) is placed in an extruded axon, these vesicles and also native vesicles are transported essentially in the anterograde direction.

Evidence demonstrating the implication of microtubules in fast anterograde transport came from experiments with antimitotic agents (colchicin, vinblastin) which prevent the elongation of microtubules and block this transport. Finally, video techniques have also demonstrated that the vesicles are associated to microtubules by arms of 25–30 nm length (**Figure 1.11**a).

The role of ATP and kinesin

By analogy with actin–myosin movements in muscle cells, scientists tried to isolate in neurons an ATPase (the enzyme responsible for the hydrolysis of ATP) associated with microtubules and able to generate the movement of vesicles. To demonstrate molecular components responsible for interactions between vesicles and microtubules, the vesicle–microtubule complex system has been reconstituted *in vitro*: isolated vesicles from squid giant axons are added to a preparation of purified microtubules and placed on a glass coverslip. These vesicles occasionally move in the presence of ATP. If an extract of solubilized axoplasm is then added to this system, the number of transported vesicles is considerably increased.

In order to determine the factor present in the solubilized fraction responsible for vesicle movement, a nonhydrolyzable ATP analog has been used: the 5′-adenylyl imidophosphate (AMP-PNP). In the presence of AMP-PNP, the vesicles associate with the microtubules but then stop. In these conditions, vesicles are bound to the microtubules and also, consequently, to the transport factor. When an overdose of ATP is added to this vesicle–microtubule complex isolated by centrifugation, the AMP-PNP is removed and so vesicles are released and the transport factor is solubilized. Kinesin has been thus isolated and purified. It is a soluble microtubule-associated ATPase (MAP) belonging to the family of mechanochemical ATPases that couple ATP hydrolysis to unidirectional movement of vesicles along the microtubule. As we have already seen, in axons, all microtubules are oriented, their plus-end being distally located from the cell body. It has been shown that kinesin, now named conventional kinesin (KIF5), moves vesicles in one direction only: from the minus-end toward the plus-end. All these results show that kinesin is responsible for anterograde transport.

In mammals, kinesin is a homodimer composed of two identical heavy chains associated with two light chains. These form an 80 nm rod-like molecule. Kinesin is composed of a motor domain (head, formed by the heavy chains), a stalk domain and a tail region (formed by the light chains). The motor domain binds to microtubules and moves on them by hydrolyzing ATP, whereas the tail regions recognize and bind to the cargo(s) (**Figure 1.11a,b,c**). The head binds to and dissociates from a microtubule through a cycle of ATP hydrolysis (**Figure 1.11c**). In proposed mechanism models, the arms observed between vesicles and microtubules *in vitro* would be kinesin.

The effects of mutations of the kinesin heavy-chain gene (*khc*) on the physiology and ultrastructure of *Drosophila* larval neurons have been studied. Motoneuron activity and corresponding synaptic (junctional) excitatory potentials of the muscle cells they innervate were recorded in control and mutant larvae in response to segmental nerve stimulation. The mutations dramatically reduced the evoked motoneuron activity and synaptic responses. The synaptic responses were reduced even when the terminals were directly stimulated. However, there was no apparent effect on the number of axons in the nerve bundle or the number of synaptic vesicles in the nerve terminal cytoplasm. These observations show that kinesin mutations impair the function of action potential propagation and neurotransmitter release at nerve terminals. Thus kinesin appears to be required for axonal transport of material other than synaptic vesicles: for example, vesicles containing ion channels such as Na^+ channels delivered to Ranvier nodes and Ca^{2+} channels delivered to presynaptic membranes. These vesicles,

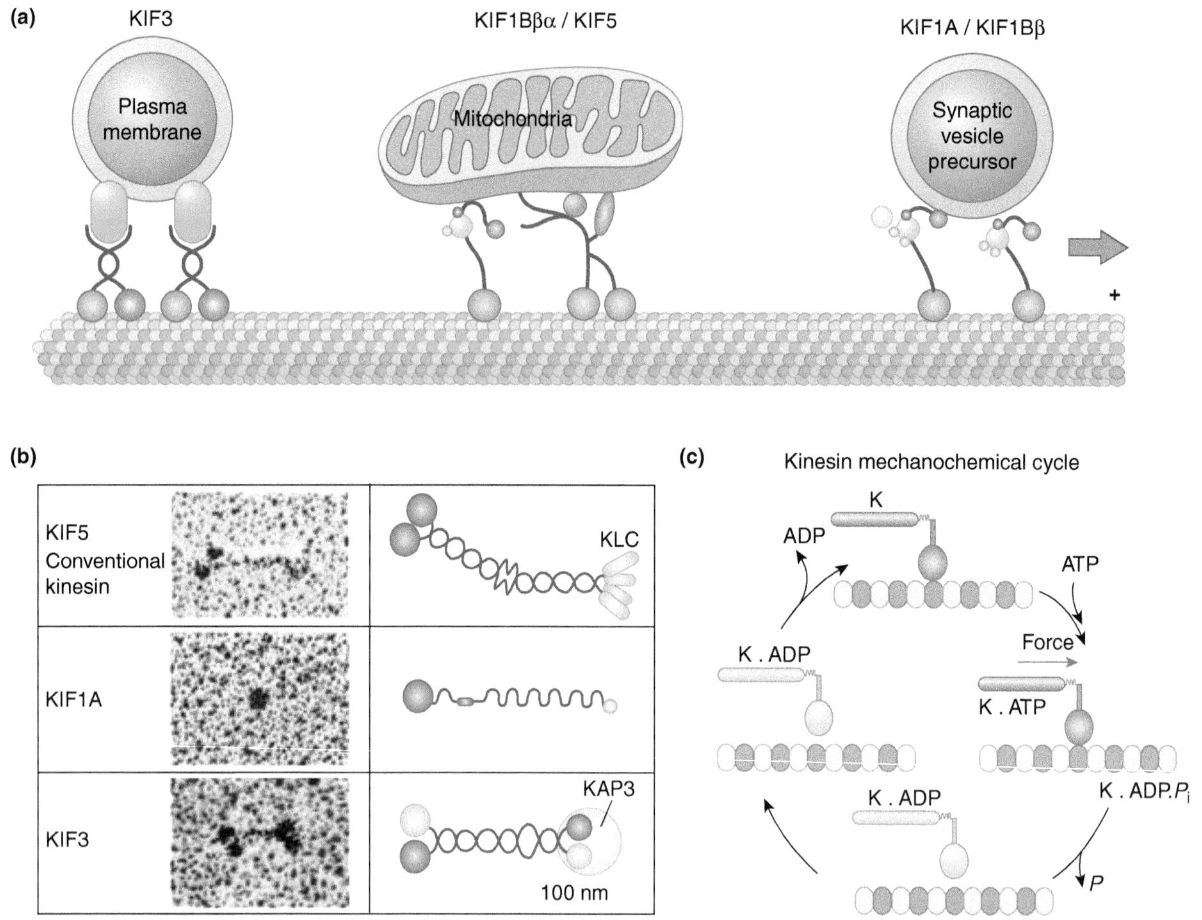

FIGURE 1.11 **The motors of fast anterograde transport. (a)** Kinesin motors (KIFs) carry cargoes (membranous organelles in the axon) anterogradely in axons along a unipolar array of microtubule towards the plus-ends. **(b)** Some members of kinesin superfamily proteins (KIFs) observed by low angle rotary shadowing (left column). Diagrams, constructed on the basis of electron microscopy or predicted from the analysis of their primary structures, are shown on the right (the larger orange ovals in each diagram indicate motor domains). Kinesin-1 consists of two KIF5s and two kinesin light chains (KLCs). The heavy chains form a homodimer that binds to the two KLCs to form a heterotetramer. KIF1A is a unique monomeric motor. KIF3 form a tetramer with two kinesin superfamily associated protein 3 (KAP3). **(c)** Simplified kinesin (K) mechanochemical cycle. ATP binding is thought to be the force producing step for kinesin motors. Part **(a)** and **(b** right) from Hirokawa N, Niwa S, Tanaka Y (2010) Molecular motors in neurons: transport mechanisms and roles in brain function, development, and disease. *Neuron* **68**, 610–638, with permission. Part (**b** left) adapted from Hirokawa N (1998) Kinesin and dynein superfamily proteins and the mechanism of organelle transport. *Science* **279**, 519–526. Part (**c**) adapted from Kull FJ, Endow SA (2013) Force generation by kinesin and myosin cytoskeletal motor proteins *J. Cell Sci.* **126**, 9–19, with permission.

called 'cargoes', are linked to kinesin. The observation that mutation of kinesin heavy chain had no effect on the number of synaptic vesicles within nerve terminals would obviously not be expected if conventional kinesin were the universal anterograde axonal transport motor.

Plus-end vesicle motors

Since the original discovery of kinesin, a large family of proteins (kinesin superfamily proteins or KIFs) with homology to kinesin's motor domain has been discovered. The kinesin superfamily is a large gene family of microtubule-dependent motors with more than 40 members identified in mice and humans. The murine and human KIF genes have been classified into three types on the basis of the positions of their motor domains: N-terminal motor domain KIFs (N-KIFs), middle motor domain KIFs (M-KIFs) and C-terminal motor domain KIFs (C-KIFs). All KIFs have a globular motor domain that shows high degrees of homology and contains a microtubule-binding sequence and an ATP-binding sequence but, outside the motor domain, each KIF has a unique sequence (**Figure 1.11a**). Of these, only N-terminal KIFs generally move towards microtubule plus-ends (from the soma to the axon terminal). Kinesins have either a monomeric (ex KIF1A), a homodimeric (ex KIF5) or a heterodimeric (ex KIF3) structure (**Figure 1.11b**). The 'classical kinesin' corresponds to KIF5. Many KIFs are expressed primarily in the nervous system, but KIFs are also expressed in other tissues and participate in various types of intracellular transport. The diversity of cargo-binding domains explains how KIFs can transport numerous different cargoes. The hypothesis is that each

KIF member is targeted to a specific cargo population, allowing the trafficking of the different neuronal compartments to be regulated independently. While there is some functional redundancy among members of the kinesin superfamily, there is also a remarkable degree of cargo specialization. For example, KIF1A/KIF1Bβ transport synaptic vesicle precursors, KIF5 transport presynaptic membrane or active zone vesicles and KIF1Bα/KIF5 transport mitochondria, all down the axon (**Figure 1.11a**).

1.3.3 Retrograde axonal transport returns old membrane constituents, trophic factors, exogenous material to the cell body

Retrograde axonal transport allows debris elimination and could represent a feedback mechanism for controlling the metabolic activity of the soma. The vesicles or cargoes transported retrogradely are larger (100–300 nm) than those transported anterogradely. Structurally, they are prelysosomal structures, multivesicular or multilamellar bodies (**Figure 1.10**). In the squid extruded axoplasm, vesicles move on each filament in both directions and frequently cross each other without apparent collisions or interactions.

Do filaments used for the fast transport of vesicles form a complex made up of several distinct filaments where certain filaments would be implicated in fast anterograde and others in retrograde transport? By using a monoclonal antibody raised against α-tubulin (a specific component of microtubules), it has been demonstrated that all the filaments implicated in anterograde or retrograde axonal transport contain α-tubulin. Moreover, by using a toxin-binding actin (and so consequently binding microfilaments), it was shown that filaments used for fast anterograde transport or retrograde transport were devoid of actin in their structure. Thus it appeared that filaments used for the movement of vesicles in both directions are microtubules.

The minus-end motor(s)

Morphometric analysis of the arms between retrograde vesicles (pluricellular bodies) and microtubules demonstrated that these are similar to arms between anterograde vesicles and microtubules. Studies looking to find a factor different from, but homologous to, kinesin and responsible for retrograde transport were undertaken. This factor present in axoplasm homogenate might be lost during kinesin purification procedures since no retrograde vesicles movement was observed *in vitro* with kinesin. Cytoplasmic dynein has been thus isolated. It is a microtubule-associated protein with an ATPase activity. Dynein belongs to the AAA+ (ATPase associated with diverse cellular activities) family of proteins.

Dynein is unique compared with kinesin and myosin because dynein molecules form large molecular complexes. Mammalian cytoplasmic dynein consists of two heavy chains, so termed because of their large molecular mass (each around 500 kDa), and several smaller subunits. Each heavy chain is divided into four domains: tail, linker, head, and stalk (**Figure 1.12a,b**). The tail is the cargo binding domain, the head is the motor domain (site of ATP hydrolysis), the linker is the mechanical amplifier, and the stalk is the microtubule binding domain. The divergent amino-terminal tail domain serves as a platform for the binding of several types of associated subunit which, in turn, mediate interactions with cargo either via direct binding or through the recruitment of adaptor proteins. Cytoplasmic dynein assembles around two identical heavy chains and is thus known as a two-headed motor.

The motor domain has been uncovered through sequence analysis, two-dimensional (2D) electron microscopy and, most recently, X-ray crystallography and three-dimensional (3D) cryo-electron microscopy (cryo-EM). Like conventional AAA+ ATPases, dynein has a ring of six AAA+ modules at its core. The AAA+ ring can be thought of as the engine of dynein as it converts the chemical energy from ATP hydrolysis into motion.

Functions of retrograde transport

The removal of misfolded or aggregated protein is a key problem in neurons. Cytoplasmic dynein's role as a retrograde motor makes it an ideal candidate for 'taking out the trash' in the cell, returning misfolded or degraded proteins from the cell periphery to the cell center for recycling and/or degradation. Evidence for such a role has come from the analysis of aggresome formation, in which the formation of perinuclear aggregates of misfolded protein was found to be dynein-dependent. The organelles trafficked by cytoplasmic dynein include endosomes, lysosomes, phagosomes, melanosomes, peroxisomes, mitochondria and vesicles from the endoplasmic reticulum (ER) destined for the Golgi. Retrograde axonal transport allows the return of membrane molecules to cell bodies, where they are degraded by acidic hydrolases found in lysosomes.

Retrograde axonal transport is not only a means of transporting cellular debris for their elimination, but is also a way of communicating information from the axon terminals to the soma. The retrogradely transported molecules would inform the cell body about activities taking place at the axon terminal level, or they may even have a neurotrophic action on the neuron. One key role for cytoplasmic dynein in neurons is in retrograde signaling, specifically the transport of neurotrophic factors from synapse to cell body. Neurotrophins are a family of small molecules, such as the nerve growth factor (NGF), the brain-derived neurotrophic factor (BDNF), and the

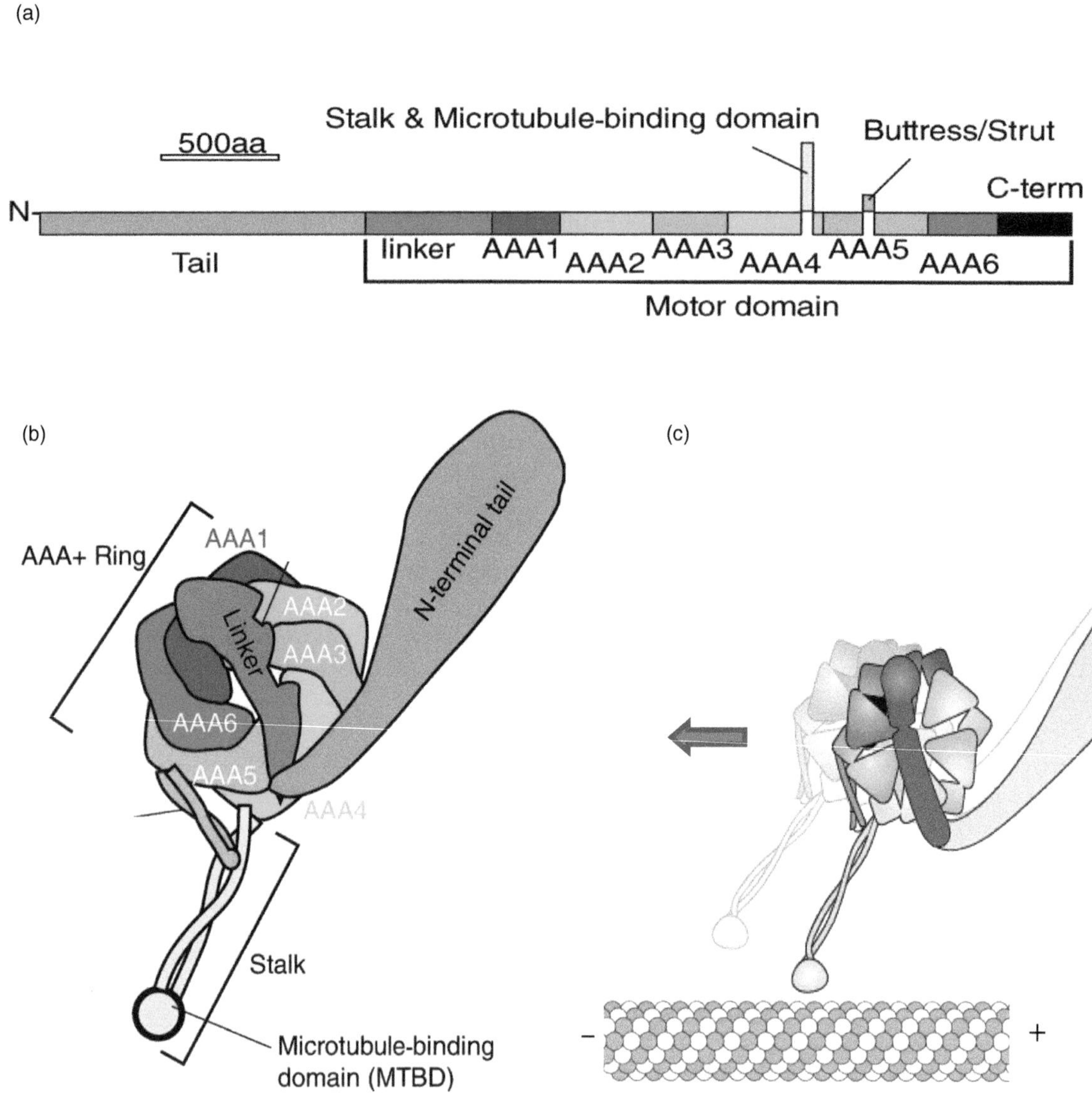

FIGURE 1.12 **Cytoplasmic dynein drives retrograde axonal transport in motor neurons.** (**a**) The primary structure of cytoplasmic dynein. (**b**) Schematic domain structure of dynein. (**c**) The cytoplasmic dynein complex contains a pair of identical heavy chains. Within each heavy chain, the six AAA+ modules fold into a ring. The stalk protrudes as an extension from the small subdomain of AAA4. The tail is connected to AAA1 by the linker domain, which arches over the AAA+ ring. The cytoplasmic dynein heavy chains assemble with up to five types of associated subunit, which are also dimers. Part (**a**) and (**b**) © Kikkawa 2013. Originally published in J. Cell Biology. 202:15–23. doi:10.1083/jcb.201304099. Part (**c**) adapted from Roberts AJ, Kon T, Knight PJ, Sutoh K, Burgess SA (2013) Functions and mechanics of dynein motor proteins. *Nat. Rev. Mol. Cell Biol.* **14**, 713–726, with permission.

neurotrophic factor NT3, which are secreted by target tissues, and then bind to receptor tyrosine kinases (Trk receptors) on the surface of the neuron. The neurotrophin/Trk receptor complex is then internalized (taken up by endocytosis) transported to the cell body where it initiates signaling cascades that regulate cell growth and survival.

Furthermore, viruses such as HIV, herpes virus and adenovirus have evolved mechanisms to hijack cytoplasmic dynein to reach the nucleus. Tetanus toxin or cholera toxin macromolecules are taken up by axon terminals and have a toxic effect on the cell body. These toxins, as well as horseradish peroxidase (HRP), an enzyme taken up by the axon terminals, are used in research studies for the retrograde labeling of neuronal pathways.

In conclusion, cargoes are transported in either the antero- or retrograde direction, depending on whether plus- or minus-end motors are active on their surface. Cargoes destined for the nerve terminal, such as synaptic vesicles or their precursors, are transported by plus-end motors; while cargoes targeted for the cell body, such as vesicles containing neurotrophin-receptor complexes, are transported by minus-end motors. In axons, oriented microtubules establish a 'road map' inside the neuron to motors that are linked to particular intracellular cargoes.

1.3.4 Slow anterograde axonal transport moves cytoskeletal proteins and cytosoluble proteins

The cytoskeleton (microtubules, neurofilaments and microfilaments) and cytosolic proteins (intermediate metabolic enzymes including glycolysis enzymes) are transported anterogradely along axons at a slow rate of about 0.002–0.1 μm/s (0.17–8.6 mm/day). In the elongating axon (i.e. during development or regeneration), the function of the slow transport is to supply axoplasm required for axonal growth. In mature neurons, its function is to renew continuously the total proteins present in the axon and axon terminals and to act as a substrate for the anterograde and retrograde axonal transport. To appreciate fully the structural achievement of this transport, one must put the size of cell bodies and axons into relation. The neuronal cell bodies (10–50 μm diameter) are connected by axons that can be over 1 m length (the axonal diameter is 1 to 25 μm). This is a factor of 100 000 difference. As there is relatively little protein synthesis in the axon, the proteins that comprise the microtubules, neurofilaments and microfilaments must be actively transported from the cell body into and down the length of the axon.

To understand the mechanisms involved in slow axonal transport, several questions can be raised: (i) in which state are cytoskeletal proteins transported in the axons: as soluble proteins or as polymers? (ii) in which axonal region(s) is the cytoskeleton (i.e. the complex network of filaments) assembled? The following are, in chronological order, the diverse hypotheses that have been proposed.

The different cytoskeletal elements are assembled and connected by bridges in the cell body

They then progress as a whole (a matrix) in the axon. However, studies have demonstrated that cross-bridges between the different cytoskeletal elements are weak and unstable. Moreover, numerous cytoskeletal discontinuities exist along the axon as seen in the nodes of Ranvier. Thus, the hypothesis of the continuous transport of a stable matrix of assembled cytoskeletal elements explaining the ultrastructure of the axon is now known to be false.

The cytoskeletal proteins are transported in a soluble form or as isolated fibrils and assembled during their progression

Lasek and his colleagues proposed that the microtubules and other cytoskeletal elements in slow transport are moved as polymers by sliding. When they are assembled some become stationary and would be renewed onsite. This hypothesis came from pulse-labeling studies and particularly those coupled with photobleaching experiments. Purified subunits of cytoskeletal proteins (tubulin or actin) coupled to a fluorescent dye molecule are introduced into living neurons in culture by injection into their soma. The observation with fluorescent microscopy shows that these labeled subunits are gradually incorporated into the polymer pool of the corresponding cytoskeletal proteins (microtubules and microfilaments) throughout the axon. A highly focused light source is then used to extinguish or bleach the fluorescence of the molecules contained within a discrete axonal segment (about 3 μm long). The fate of the bleached zone is followed over a period of hours. The bleached zone does not move along the axon or widen and recovers a low level of fluorescence within seconds. This latter effect is ascribed to the diffusion of free fluorescent subunits from the neighboring fluorescent regions into the bleached region. These observations suggested that microtubules and microfilaments are essentially stationary and are exchanging subunits.

The transport of microtubules and neurofilaments is bidirectional, intermittent, asynchronous, and occurs at the fast rate of known motors (stop and go model)

However, when Wang and Brown widened the parameters of the live-cell imaging paradigm, such that a much longer bleached zone (about 30 μm in length) was created, and the zone was imaged every several seconds rather than minutes, they found that the transport of microtubules is bidirectional, intermittent, highly asynchronous, and at the fast rates of known motors (average rates of 1 μm/s) such as cytoplasmic dynein and the kinesin superfamily. These observations indicate that microtubules are propelled along axons by fast motors. The average moving microtubule length is around 3 μm.

The rapid, infrequent, and highly asynchronous nature of the movement may explain why the axonal transport of tubulin has eluded detection in so many other studies. In addition, these results offer an explanation for the slow rate of tubulin transport documented in the early kinetic studies: it reflects an average rate of fast movements and non-movements (long pauses). The overall rate of microtubules movement is slow because the microtubules spend only a small proportion of their time moving.

Similarly, the initial studies of neurofilament transport using radiographic labeling suggested a velocity of 0.25–3 mm/day which is slower than any speed produced by known molecular motors. Recent studies using green-fluorescent-protein (GFP)-tagged neurofilament subunits and real-time confocal microscopy show more accurately that the conventional fast axonal transport also applies to neurofilaments. Peak velocities of 2 μm/s occur anterogradely and retrogradely and are interrupted by prolonged resting phases resulting in the overall slow transport originally described.

1.3.5 Axonal transport of mitochondria allows the turnover of mitochondria in axons and axon terminals

Mitochondria are prominent members of the cast of axonally transported organelles. They are essential for the function of all aerobic cells, including neurons. They produce ATP, buffer cytosolic calcium, are essential steps for the synthesis of certain neurotransmitters and sequester apoptotic factors. Like many other neuronal organelles, mitochondria are thought to arise mainly in the neuronal cell body, but their transport is distinctive. In postmitotic neurons, mitochondria are delivered to and remain in areas of the axon where metabolic demand is high, such as initial segments, nodes of Ranvier and synapses. How do mitochondria achieve these distributions in the axon?

The motility patterns of mitochondrial transport in neurons can be visualized by applying time-lapse imaging in live primary neurons. Mitochondria are labeled for example by expressing DsRed-Mito, a mitochondria-targeted fluorescent protein. The motility of mitochondria differs from that of other axonal organelles, in that it shows frequent pausing and halting at stationary residence sites throughout the axon and they frequently change direction. The mitochondria are transported anterogradely in axons up to axon terminals at a rate of 10–40 mm per day. A retrograde movement of mitochondria showing degenerative signs is also observed (see **Figure 1.10b**). Specific inhibition of kinesin-1 family stops most mitochondrial movement in *Drosophila melanogaster* motor axons. Also time-lapse imaging of GFP-tagged mitochondria in *Drosophila* axons has shown that kinesin-1 mutations cause a profound reduction in the transport of mitochondria. Similar approaches indicate that KIF1Bα/KIF5 anterogradely transport mitochondria (**Figure 1.11a**) whereas cytoplasmic dynein is the primary retrograde mitochondrial motor (**Figure 1.12**). In mature neurons, around 20–30% of axonal mitochondria are motile, some of which pass by presynaptic terminals while the remaining two thirds are stationary.

1.4 NEURONS CONNECTED BY SYNAPSES FORM NETWORKS OR CIRCUITS

1.4.1 The circuit of the withdrawal medullary reflex

Sensory stimuli (including visual, auditive, tactile, gustative, olfactive, proprioceptive, and nociceptive stimuli) are detected by specific sensory receptors and transmitted to the central nervous system (encephalon and spinal cord) by networks of neurons. These stimuli are analyzed at the encephalic level. They can also evoke movements such as motor reflexes on their way to higher central structures.

Thus, when a noxious stimulus (i.e. a stimulus provoking tissue damage, for example pricking or burning) is applied to the skin of the right foot, it induces a withdrawal reflex consisting of the removal of the affected foot (contraction of flexor muscles of the right inferior limb) to protect itself against this stimulus. The noxious stimulus activates nociceptive neurons which are the peripheral endings of primary sensory neurons whose cell bodies are located, in this case, where injury is located at the body level – in dorsal root ganglia. Action potentials are then generated (or not, if the intensity of the noxious stimulus is too small) in primary sensory neurons and propagate to the central nervous system (spinal cord). Local circuit neurons of the dorsal horn of the spinal cord (**Figure 1.13a**) relay the sensory information. Sensory information is thus transmitted to motoneurons (neurons innervating skeletal striated muscles and located in the ventral horn) through a complex network of local circuit neurons (Golgi type II neurons) which have either an excitatory or an inhibitory effect. It results on the stimulus side (ipsilateral side) in an activation of the flexor motoneurons (F) and an inhibition of the extensor motoneurons (E): the right inferior limb is being withdrawn (is in flexion). The opposite limb is extended to maintain posture.

This pathway illustrates peculiarities present in numerous other circuits.

- *Divergence of information.* Primary sensory information is distributed to several types of neurons in the medulla: local circuit neurons connected to motoneurons that innervate posterior limb muscles and also projection neurons that relay sensory informations to higher centers where they are analyzed.
- *Convergence of information.* Motoneurons receive sensory information via local circuit neurons and also descending motor information via descending neurons whose cell bodies are located in central motor regions (motor commands elaborated at the encephalic level) (**Figure 1.13a**).
- *Anterograde inhibition* (feedforward inhibition). A neuron inhibits another neuron by the activation of an inhibitory interneuron (**Figure 1.13b**).
- *Recurrent inhibition* (feedback inhibition). A neuron inhibits itself by a recurrent collateral of its own axon which synapses on an inhibitory interneuron. The inhibitory interneuron establishes synapses on the motoneuron (**Figure 1.13c**). This recurrent inhibition allows for rapid cessation of the motoneuron's activity.

The last two circuits described are also called *microcircuits*, since they are included in a larger circuit

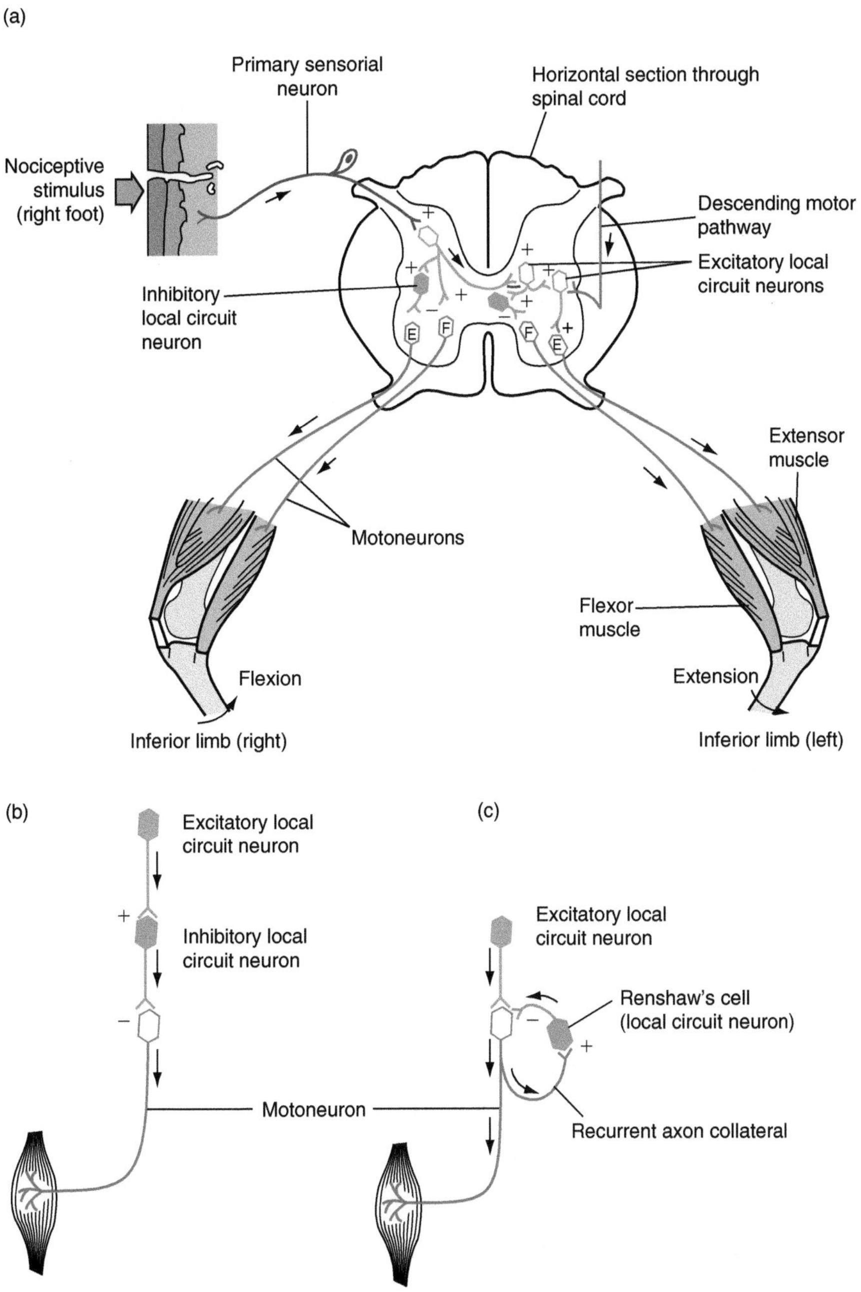

FIGURE 1.13 **Withdrawal medullary reflex pathway.** **(a)** Schematic of a horizontal section through the spinal cord and of connections between a primary nociceptive sensory neuron, medullary local circuit neurons and ipsi- and contralateral motoneurons innervating inferior limb muscles. See text for details. **(b)** Anterograde inhibitory circuit. **(c)** Recurrent inhibitory circuit. Arrows show the direction of action potential propagation.

or *macrocircuit*. In this selected example, all the neurons forming the microcircuit enable precise regulation of motoneuron activity.

1.4.2 The spinothalamic tract or anterolateral pathway is a somatosensory pathway

Noxious stimuli (temperature and sometimes touch) are detected at the skin level by nerve endings of nociceptive neurons, are transduced (or not) in action potentials and are conveyed to the somatosensory cortex via relay neurons. Information from the body first reaches the dorsal horn neurons of the spinal cord, and information from the face reaches the trigeminal nuclei in the brainstem, via primary sensory neurons (nociceptive neurons) whose cell bodies are located in dorsal root ganglia or cranial ganglia, respectively. They relay on projection neurons located in the dorsal horn of the spinal cord or in trigeminal nuclei which send axons to the thalamus. These axons cross the midline, form a tract in the anterolateral part of the white matter, and terminate in non-specific thalamic nuclei. Thalamic neurons then send the sensory information to cortical areas specializing in noxious perception (somatosensory cortex). At each level of synapses (dorsal horn or trigeminal nucleus, thalamus, cortex), the somatosensory information is not simply relayed, it is also processed through local microcircuits receiving afferent sensory information and descending information from higher centers which modulate incoming sensory information.

When superposing horizontal sections through the spinal cord (**Figures 1.13a and 1.14**), it becomes clear that a noxious stimulus applied to the skin of the right inferior limb is transduced in action potentials by nociceptive neurons and transmitted to motoneurons where it can evoke a withdrawal reflex. Action potentials evoked by a noxious stimulus also reach the somatosensory cortex where they are analyzed. The reflex is evoked before the consciousness of the stimulus because of the longer distance to brain areas than to the ventral horn of the spinal cord.

1.4.3 How are the different noxious stimuli analyzed? Functional dissection of pain circuits with targeted ablative or optogenetic methods

How do we know that the pain we feel is due to skin burn or skin pinch? Though nociceptive neurons are heterogeneous by a variety of molecular criteria, this heterogeneity may not be linked to a heterogeneity of functions. Electrophysiological recordings of unmyelinated primary afferent C fibers (axons of nociceptive neurons) from dorsal root ganglia (DRG) show that most

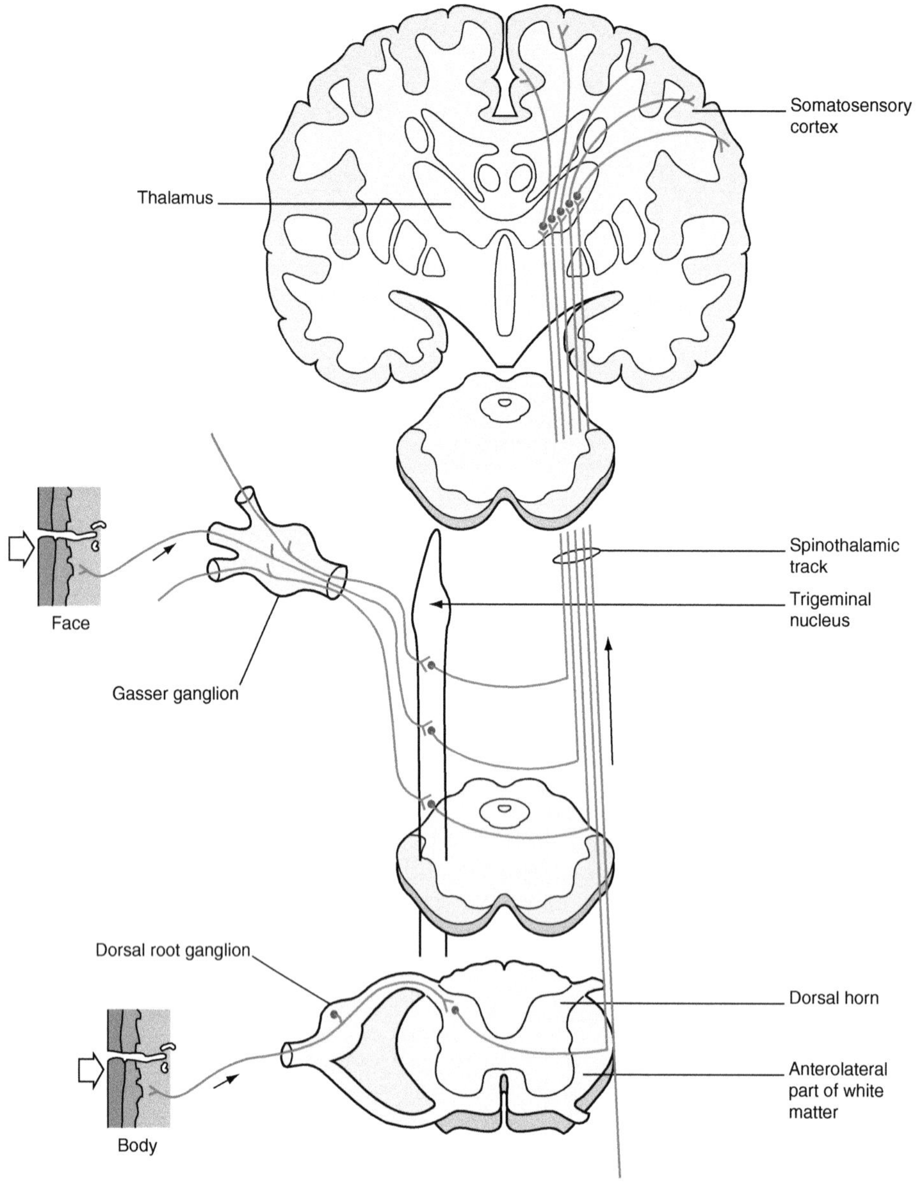

FIGURE 1.14 **The spinothalamic tract or anterolateral ascending sensory pathway.** This pathway integrates and conveys sensory information such as nociception, temperature and some touch. Bottom to top: horizontal sections through the spinal cord, the pons and frontal section through the diencephalon. See text for explanations.

(70%) nociceptive neurons are polymodal since they are activated by multiple types of painful stimuli, such as mechanical or thermal. This suggests that the brain discriminates thermal and mechanical stimuli thanks to spinal or brain neuronal circuits. To re-test whether discrimination of different noxious stimuli is attributable to primary nociceptive neurons, targeted ablation of neurons or optogenetic techniques are used. The result of such studies is highly dependent on the specificity of the targeting: only one type of nociceptive neuron must be ablated or transfected.

For this, specific markers have to be identified.

Mas-related G-protein-coupled receptor D (Mrgprd) is a marker for a subset of non-peptidergic nociceptive neurons in the rodent, while calcitonin gene-related peptide (CGRP) is a marker for a subset of peptidergic nociceptive neurons. Within the skin, Mrgprd$^+$ neurons innervate the most superficial layer of epidermis, the stratum granulosum, while peptidergic CGRP$^+$ nociceptive neurons innervate the underlying stratum spinosum. To identify the role of Mrgprd$^+$ nociceptive neurons their selective ablation is performed by inserting in the *Mrgprd* locus the gene coding for a receptor to a lethal toxin (the human diphtheria toxin receptor: DTR). DTR expression in the absence of the diphtheria toxin does not impair the survival of Mrgprd$^+$ nociceptive neurons. In contrast, injection of diphtheria toxin into adult *Mrgprd^{DTR}* mice produces a complete loss of Mrgprd$^+$ cell bodies in the dorsal root ganglion and selective loss of Mrgprd$^+$ axon terminals in the dorsal horn of the spinal cord (**Figure 1.15a**). This produces selective deficits in behavioral responses to mechanical stimuli (but not to noxious thermal or cold stimuli). Conversely, selective ablation of CGRP$^+$ nociceptive neurons produces selective deficits in behavioral responses to noxious thermal stimuli. This strongly suggests that behavioral discrimination between different noxious stimuli can occur at nociceptive neurons, the earliest stage of sensory processing.

Although it is known that Mrgprd$^+$ axons terminate largely in lamina II of the dorsal horn, the precise connectivity of this nociceptive neuronal subpopulation is

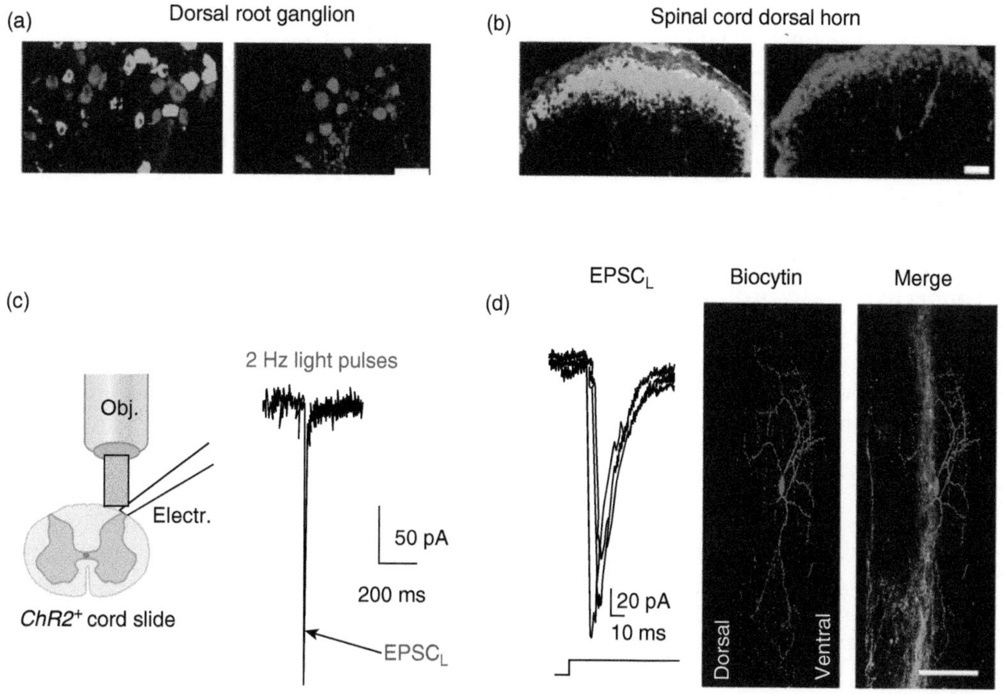

FIGURE 1.15 **The specific ablation or activation of Mrgprd+ nociceptive neurons in mice. (a,b)** The green fluorescent protein (GFP) is expressed in Mrgprd$^+$ nociceptive neurons whereas CGRP$^+$ cells are labeled with immunocytochemistry in red. Section of a dorsal root ganglion **(a)** and of the spinal cord **(b)** from *Mrgprd$^{EGFP/DTR}$* mice before (left) and after (right) diphtheria toxin injection. Note selective loss of green Mrgprd$^+$ somas (**a**) or axons (**b**). Scale bars: 50 μm. **(c)** Illumination of the dorsal region of the spinal cord in spinal cord slices from *Mrgprd-ChR2-Venus$^+$* mice selectively to activate Mrgprd$^+$ axon terminals (left). This generates a light-evoked response ($EPSC_L$) in postsynaptic spinal neurons (right). **(d)** Overlay of five successive light-evoked responses from a dorsal spinal cord neuron in spinal cord slices from *Mrgprd-ChR2-Venus$^+$* mice. Delay between light presentations is 1 min. The recorded neurons are filled with biocytin during the recording session. Following recording, slices are stained for biocytin (red) to identify the recorded spinal neurons and for Venus (green) to label Mrgprd-ChR2-Venus+ axon terminals. Slices are then imaged by confocal microscopy. Proximity of the recorded red neuron to Mrgprd-ChR2-Venus+ axon terminals is shown in the merged image. Scale bar: 50 μm. Parts (**a,b**) from Cavanaugh DJ, Lee H, Lo L *et al.* (2009) Distinct subsets of unmyelinated primary sensory fibers mediate behavioral responses to noxious thermal and mechanical stimuli. *Proc. Natl Acad. Sci. USA* **106**, 9075–9080, with permission. Parts (**c,d**) from Wang H, Zylka MJ (2009) Mrgprd-expressing polymodal nociceptive neurons innervate most known classes of substantia gelatinosa neurons. *J. Neurosci.* **29**, 13202–13209.

not known. To identify the target neurons of Mrgprd$^+$ nociceptive neurons in the dorsal horn of the spinal cord, optogenetic techniques worked out by K. Deisseroth are used. To summarize briefly this approach, researchers force certain neurons to express a light-activated channel isolated from green algae (opsins): for example channel rhodopsin 2 (ChR2). By using genetic engineering techniques, opsins are selectively inserted into any chosen subpopulation of neurons. Then, when the tissue is illuminated with blue light (480 nm) through fiberoptics or other light-guiding tools, only neurons that express ChR2 are activated. Why? Because ChR2 open in response to blue light and, due to their permeability to cations, depolarize the transfected neurons and only these neurons. The resultant effect is the precise activation of a certain class of neurons. When applied to pain pathways, this rendered possible activation of the axon terminals of *Mrgprd*$^+$ nociceptive neurons in isolation.

ChR2 expression is exclusively targeted in Mrgprd$^+$ nociceptive neurons (in their dendrites, soma, axon and axon terminals) and slices of the spinal cord are performed. In these slices, Mrgprd$^+$-ChR2$^+$ axon terminals are activated by illumination during electrophysiological recordings of postsynaptic lamina II (dorsal horn) spinal neurons from *Mrgprd-ChR2-Venus*$^{+/+}$ mice (**Figure 1.15b left)**. Illumination evokes monosynaptic excitatory postsynaptic currents (EPSC$_L$) in a quarter of the neurons recorded in lamina II (**Figure 1.15b right)**, suggesting that Mrgprd$^+$ nociceptive neurons are directly connected to a subset of lamina II neurons only. Further studies are needed to identify whether lamina II neurons not connected to Mrgprd$^+$ nociceptive neurons are innervated by other types of nociceptive neurons and convey other noxious modalities. A positive result would support the view that pain processing is a modality specific at both the peripheral and spinal levels.

1.5 SUMMARY: THE NEURON IS AN EXCITABLE AND SECRETORY CELL PRESENTING AN EXTREME FUNCTIONAL REGIONALIZATION

This chapter has described how the various functions of neurons, such as their metabolism, excitability and secretion, are localized to specific regions of the neuron. The main neuronal compartments are the dendrites (more precisely postsynaptic sites), soma, axon and axon terminals (more precisely presynaptic sites). These regions are sometimes located at great distances from each other, and so neurons have to resolve the problems of communication between these regions and harmonization of their activities.

Regionalization of metabolic functions

The essential synthesis activity of a neuron is localized in its cell body, since dendrites can synthesize only some of their proteins, and axons are able to synthesize only a few. In this cell, where the axon's volume represents up to a thousand times the volume of the cell body, the structural and functional integrity of the axon and its terminals requires an important and continuous supply of macromolecules. This supply is ensured by anterograde axonal transport. In dendrites, RNA transport from the cell body to the polysomes has been demonstrated and would allow the synthesis of some of their proteins.

The degradation of cellular metabolism debris and non-neuronal elements taken up from the external environment by endocytosis (e.g. uptake of viruses) takes place in the lysosomes of the cell body. They are transported from axon terminals to the cell body via the retrograde axonal transport. Finally, to coordinate synthesis activity in the cell body with the needs of the axon terminals, the existence of a feedback mechanism (from terminals to cell body) seems essential. This could take place through retrograde axonal transport.

Anterograde transport moves newly synthesized material outward from the cell body along the axon. Retrograde transport drives the movement of organelles, vesicles, and signaling complexes from the cell periphery and distal axon back to the cell center. Key motors for this transport include members of the kinesin superfamily and cytoplasmic dynein.

Regionalization of functions implicated in reception and transmission of electrical signals

The neuronal regions receiving synapses are mainly the dendritic (primary segments, branches and spines of dendrites) and somatic regions, but also some axonal regions. These receptive regions, called postsynaptic elements, have a restricted surface. They contain, within their plasma membrane, proteins specialized in the recognition of neurotransmitters: the neurotransmitter receptors (receptor channels and receptors coupled to G proteins). These proteins synthesized in the cell body are then transported toward the dendritic, somatic or axonal postsynaptic membranes to be incorporated. Similarly, the proteins specialized in the generation and propagation of action potentials (voltage-dependent channels) are synthesized in the soma and have to be transported and incorporated in the axonal membrane.

Regionalization of secretory function

This function is localized in regions making synaptic contacts and more generally in presynaptic regions such as axon terminals (and sometimes in dendritic and somatic regions). At the level of presynaptic structures, the neurotransmitter is stocked in synaptic vesicles and released. The secretory function implicates the presence of specific molecules and organelles in the presynaptic region: neurotransmitter synthesis enzymes, synaptic vesicles, microtubules and associated proteins, voltage-dependent channels, etc.

In conclusion, owing to its extreme regionalization and the extreme length and volume of its processes, the neuron has the challenge to deliver the proteins synthesized in the soma at the appropriate sites (targeting) at appropriate times.

Further reading

Aizawa, H., Sekine, Y., Takemura, R., Zhang, Z., Nangaku, M., Hirokawa, N., 1992. Kinesin family in murine central nervous system. J. Cell Biol. 119, 1287–1296.

Brady, S.T., Lasek, R.J., Allen, R.D., 1982. Fast axonal transport in extruded axoplasm from squid giant axon. Science 218, 1129–1131.

Brown, A., Yung, P., 2013. A critical reevaluation of the stationary axonal cytoskeleton hypothesis. Cytoskeleton 70, 1–11.

Davies, L., Burger, B., Banker, G.A., Steward, O., 1990. Dendritic transport: quantitative analysis of the time course of somatodendritic transport of recently synthetized RNA. J. Neurosci. 10, 3056–3068.

Doyle, M., Kiebler, M.A., 2011. Mechanisms of dendritic mRNA transport and its role in synaptic tagging. EMBO J. 30, 3540–3552.

Gho, M., McDonald, K., Ganetzky, B., Saxton, W.M., 1992. Effects of kinesin mutations on neuronal functions. Science 258, 313–316.

He, Y., Francis, F., Myers, K.A., Yu, W., Black, M.M., Baas, P.W., 2005. Role of cytoplasmic dynein in the axonal transport of microtubules and neurofilaments. J. Cell Biol. 168, 697–703.

Hirokawa, N., 2011. From electron microscopy to molecular cell biology, molecular genetics and structural biology: intracellular transport and kinesin superfamily proteins, KIFs: genes, structure, dynamics and functions. J. Electron Microsc. 60 (Suppl. 1), S63-S92.

Johnston, J.A., Illing, M.E., Kopito, R.R., 2002. Cytoplasmic dynein/dynactin mediates the assembly of aggresomes. Cell. Motil. Cytoskeleton 53, 26–38.

Kanai, Y., Dohmae, N., Hirokawa, N., 2004. Kinesin transports RNA: isolation and characterization of an RNA-transporting granule. Neuron 43, 513–525.

Kull, F.J., Sablin, E.P., Lau, R., Fletterick, R.J., Vale, R.D., 1996. Crystal structure of the kinesin motor domain reveals a structural similarity to myosin. Nature 380, 550–555.

Martin, K.C., Ephrussi, A., 2009. mRNA localization: gene expression in the spatial dimension. Cell 136, 719–730.

Mikami, A., Paschal, B.M., Mazumdar, M., Vallee, R., 1993. Molecular cloning of the retrograde transport motor cytoplasmic dynein (MAP1C). Neuron 10, 787–796.

Muresan, V., Godek, C.P., Reese, T.S., Schnapp, B.J., 1996. Plus-end motors override minus-end motors during transport of squid axon vesicles on microtubules. J. Cell Biol. 135, 383–397.

Nangaku, M., Sato-Yoshitake, R., Okada, Y., et al., 1994. KIF1B, a novel microtubule plus-end-directed monomeric motor protein for transport of mitochondria. Cell 79, 1209–1220.

Pilling, A.D., Horiuchi, D., Lively, C.M., Saxton, W.M., 2006. Kinesin-1 and dynein are the primary motors for fast transport of mitochondria in *Drosophila* motor axons. Mol. Biol. Cell 17, 2057–2068.

Roberts, A.J., Kon, T., Knight, P.J., Sutoh, K., Burgess, S.A., 2013. Functions and mechanics of dynein motor proteins. Nat. Rev. Mol. Cell Biol. 14, 713–726.

Saxton, W.M., Hollenbeck, P.J., 2012. The axonal transport of mitochondria. J. Cell Sci. 125, 2095–2104.

Schnapp, B.J., Reese, T.S., 1989. Dynein is the motor for retrograde axonal transport of organelles. Proc. Natl Acad. Sci. USA 86, 1548–1552.

Schnapp, B.J., Vale, R.D., Sheetz, M.P., Reese, T.S., 1985. Single microtubules from squid axoplasm support bidirectional movement of organelles. Cell 40, 455–462.

Scott, D.A., Das, U., Tang, Y., Roy, S., 2011. Mechanistic logic underlying the axonal transport of cytosolic proteins. Neuron 70, 441–454.

Steward, O., Levy, W.B., 1982. Preferential localization of ribosomes under the base of dendritic spines in granule cells of the dentate gyrus. J. Neurosci. 2, 284–291.

Tanaka, Y., Kanai, Y., Okada, Y., et al., 1998. Targeted disruption of mouse conventional kinesin heavy chain, kif5B, results in abnormal perinuclear clustering of mitochondria. Cell 93, 1147–1158.

Vale, R.D., Reese, T.S., Sheetz, M.P., 1985. Identification of a novel force-generating protein, kinesin, involved in microtubule-based mobility. Cell 42, 39–50.

Wang, D.O., Martin, K.C., Zukin, R.S., 2010. Spatially restricting gene expression by local translation at synapses. Trends Neurosci. 33, 173–182.

Williams, S.C.P., Deisseroth, K., 2013. Optogenetics. Proc. Natl Acad. Sci. USA 110, 16287.

CHAPTER

2

Neuron–glial cell cooperation

Constance Hammond, Myrian Cayre, Aude Panatier, Elena Avignone

OUTLINE

There are roughly twice as many glial cells as there are neurons in the central nervous system. They occupy the space between neurons and neuronal processes and separate neurons from blood vessels. As a result, the extracellular space between the plasma membranes of different cells is narrow, of the order of 15–20 nm.

Virchow (1846) was the first to propose the existence of non-neuronal tissue in the central nervous system. He named it 'neuroglia' (neural putty), because it appeared to stick the neurons together. Following this, among others, Deiters (1865) and Golgi (1885) identified glial cells and distinguished them from neurons.

There are several categories of glial cells. Depending on their anatomical position they are classed as follows:

- *Central glia* are found in the central nervous system, and comprise two groups: macroglia and microglia. This first group comprises astroglia, NG2-glial cells, and oligodendrocytes. Astroglia regroups radial glia (principally during embryonic and developmental stages), ependymal cells which form the epithelial surface covering the walls of the cerebral ventricles and of the central canal of the spinal cord, tanycytes, pituicytes, ependymocytes, choroid plexus cells, retinal pigment epithelial cells and astrocytes.
- *Peripheral glia* comprise non-myelinating perisynaptic Schwann cells, Schwann cells, olfactory ensheathing cells, satellite glial cells and enteric glia.

Glial cells, excluding microglia, have an ectodermal origin. Those of the central nervous system derive from the germinal neural epithelium (neural tube), while peripheral glia (Schwann cells) are derived from the neural crest. Microglia, in contrast, have a mesodermal origin.

While glial cells were considered for a long time only as a supporting tissue, each category has in fact its own roles which are essential for the integrity and processing of the nervous system. Glial cells have morphological as well as functional and metabolic characteristics that distinguish them from neurons, in particular, they do not generate or conduct action potentials.

We will explain in this chapter the roles of astrocytes, oligodendrocytes, microglia and Schwann cells.

2.1 ASTROCYTES FORM A VAST CELLULAR NETWORK OR SYNCYTIUM BETWEEN NEURONS, BLOOD VESSELS AND THE SURFACE OF THE BRAIN

In 1895, Michael Lenhossék introduced the term 'astrocyte' because these cells may have a 'star' appearance. Astrocytes lie between vessels and neurons, suggesting that they play important roles in adult brain metabolism and synaptic transmission. Astrocytes send out processes that end on the walls of blood vessels or beneath the pial surface of the brain and spinal cord. Two kinds

Cellular and Molecular Neurophysiology. DOI: 10.1016/B978-0-12-397032-9.00002-9

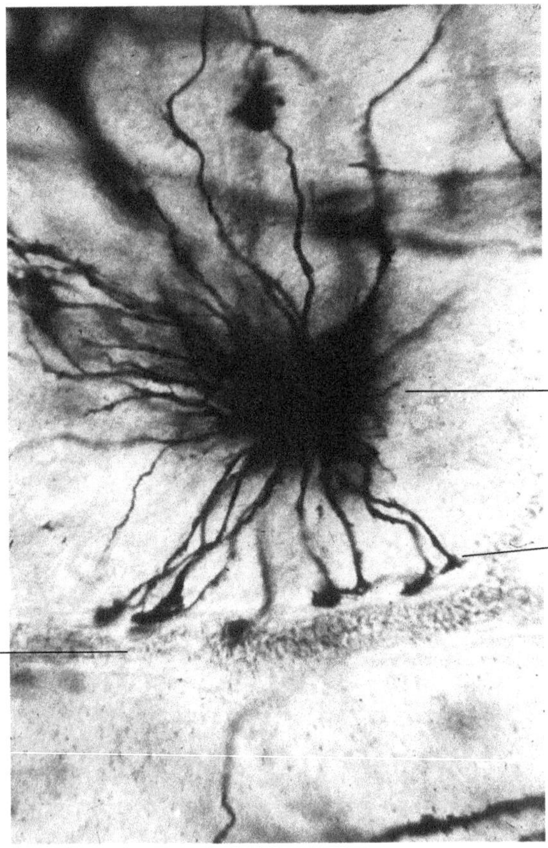

FIGURE 2.1 **Fibrillary astrocyte.** Micrograph of a fibrillary astrocyte stained with a Golgi stain observed through an optical microscope. The processes of this astrocyte make contact with a blood vessel: these are the terminal end feet. Photograph by Olivier Robain.

of astrocytes are recognized: fibrillary astrocytes in the white matter (**Figure 2.1**) with numerous radial processes, which are infrequently branched, and protoplasmic astrocytes, in the gray matter.

2.1.1 Astrocytes are star-shaped cells characterized by the presence of glial filaments in their cytoplasm

The principal characteristics of astrocytes are the glial filaments (≈8–10 μm diameter) and glycogen granules present in the cytoplasm of their somata and main processes. A glial filament commonly used to identify astrocytes is the glial fibrillary acidic protein (GFAP). In physiological conditions, it reveals only around 13% of the morphology of the cell and its expression varies with age and brain region.

Protoplasmic astrocytes are characterized by spongiform morphology, with one to several stubby main processes, from which emerge several ramifications with an irregular shape. Each astrocyte extends at least one process, called end feet, in contact with the wall of blood vessels (**Figure 2.2**), and a high number of processes, which are in close apposition with the two neuronal elements of the synapse. In contrast to fibrillary astrocytes, protoplasmic astrocytes have their own territory, called domain, in which they contact around 120 000 synapses. Despite their poor overlap (only 5% at the edge), these glial cells are coupled via gap junctions to form a syncytium. These gap junctions are permeable to small molecules (less than 1000 kDa) including Ca^{2+} ions.

2.1.2 Astrocytes form an active bridge between neurons and the blood–brain barrier

Astrocytes play an important role in the development, maintenance and regulation of the blood–brain barrier. The blood–brain barrier is a physical barrier isolating the brain cells from the blood. This is formed, in most regions of the central nervous system, by vascular endothelial cells joined together by tight junctions. It is surrounded by basal lamina and astrocytic end feet, which are not part of the blood–brain barrier. The blood–brain barrier supplies brain cells with essential nutrients and allows the efflux of large molecules and brain metabolites to maintain brain homeostasis.

The permeability of the blood–brain barrier is regulated by the expression of transporters and receptors mediating transcytosis, at the level of the plasma membrane

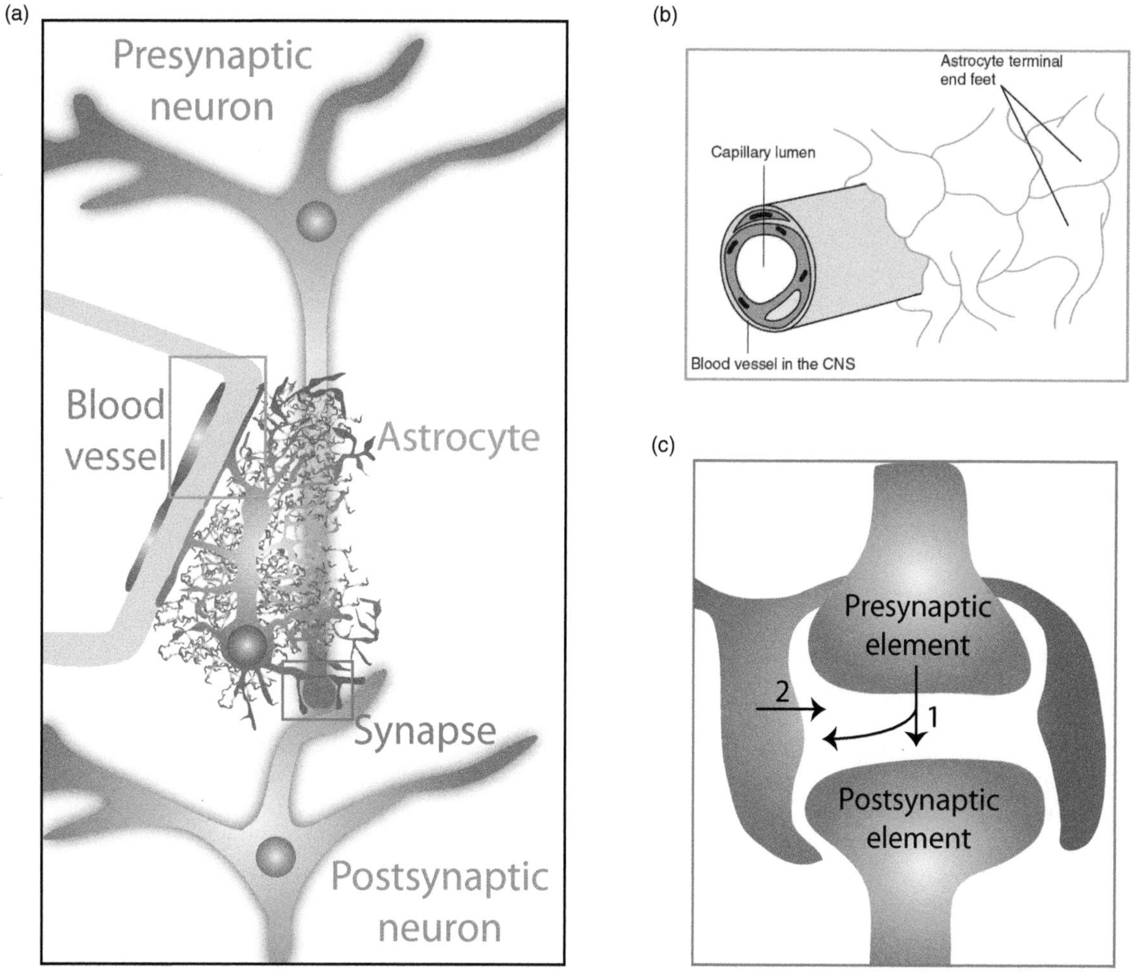

FIGURE 2.2 **Protoplasmic astrocytes are well positioned between neurons and vessels and organized in domains.** (**a**) Diagram representing the position of the astrocyte in between blood vessels and synapses. (**b**) Diagram of the covering formed by astrocyte end feet around a capillary. (**c**) Tripartite synapse model showing that astrocytes detect the synaptic signal (1) and in turn regulate its efficacy (2). Parts (**a, c**) drawing by Aude Panatier. Part (**b**) from Goldstein G, Betz L (1986) La barrière qui protège le cerveau. *Pour la Science*, November, 84–94, with permission.

of endothelial cells. Then, the exchange between endothelial cells and astrocytes is due to the presence of receptors, transporters and channels in end feet plasma membrane.

Depending on the metabolic demand, astrocytes control the local blood supply. To this end, astrocytes, together with neurons, microvessels, endothelial cells and probably microglia and oligodendrocytes, are organized into neurovascular units, a key element for the neurovascular coupling.

2.1.3 Astrocytes regulate the ionic composition of the extracellular fluid

Proper function of the brain depends on the regulation of the composition of extracellular and intracellular ions. Indeed, survival, function, excitability and communication of brain neurons are based on ionic equilibrium.

In 1965, Leif Hertz proposed that astrocytes play a key role in K^+ homeostasis. During action potential propagation, and more precisely during the hyperpolarizing phase, there are large fluxes of potassium ions into the extracellular space through voltage-activated potassium channels (see Sections 4.3 and 5.3). Extracellular K^+ concentration needs to be tightly regulated to a concentration of 3 mM: higher concentration would induce neuron depolarization leading to an increase of neuronal excitability followed by a decrease of action potential propagation. To limit this side effect, astrocytes uptake local K^+ ions through potassium channels and transporters ($K_{ir}4.1$, Na^+/K^+ pump and $Na^+/K^+/Cl^-$ co-transporter). K^+ ions are then redistributed in neighboring astrocytes through gap junctions, a mechanism called 'potassium spatial buffering'.

Beside their contribution in potassium clearance, astrocytes play a major role in Ca^{2+}, Cl^-, pH as well as water homeostasis.

2.1.4 Astrocytes regulate the efficacy of synaptic transmission

As mentioned above, astrocytes extend processes that are in close apposition with synapses. Astrocytes are intimate partners of neurons during synaptic transmission. First, astrocytes are required for the synthesis of transmitters and particularly glutamate and GABA. Thanks to the presence of glutamine synthetase in astrocytes (see **Figure 10.13**), glutamine is synthesized from glutamate. Glutamine is then uptaken by neurons and transformed back into glutamate.

Secondly, astrocytes play a fundamental role in removing neurotransmitters released in the synaptic cleft, avoiding steady high concentrations of transmitters that would interfere with synaptic transmission and induce toxicity triggered by long-lasting activation of receptors (particularly glutamate receptors). Most transmitters are transported into cells from the extracellular space by specialized carrier molecules in the cell membrane, called transporters (but acetylcholine is hydrolyzed; see **Figure 6.12**). Although both neurons and glia express such carrier proteins, uptake into astrocytes is of particular importance, especially for glutamate. Indeed, glutamate is primarily uptaken by astrocytes through the high-affinity glutamate transporter 1 (GLT-1) and glutamate aspartate transporter (GLAST).

Finally, although initially considered passive and non-excitable, these glial cells display a form of non-electrical excitability based on intracellular calcium variations. Indeed, they express several receptors through which they can detect the synaptic signal. Following their activation, astrocytes regulate the efficacy of synaptic transmission by releasing active substances called gliotransmitters, such as purines, D-serine and glutamate. Thus, from numerous works performed worldwide emerged the concept of the 'tripartite synapse' proposing that astrocytes are functional components of synapses.

2.2 OLIGODENDROCYTES FORM THE MYELIN SHEATHS OF AXONS IN THE CENTRAL NERVOUS SYSTEM AND ALLOW THE CLUSTERING OF Na^+ CHANNELS AT NODES OF RANVIER

Two types of oligodendrocytes are recognized: interfascicular or myelinating oligodendrocytes, found in the white matter where they make the sheaths of myelinated axons; and satellite oligodendrocytes which surround neuronal somata in the gray matter. We will deal with the former type in detail. By forming the myelin sheath, their major role is to electrically isolate segments of axons, induce the formation of clusters of Na^+ channels at nodes of Ranvier and therefore to allow the fast propagation of Na^+ action potentials (see Section 5.4). Oligodendrocytes also have other less characterized functions such as trophic support of long axons and rapid regulation of action potential propagation through axons.

2.2.1 Processes of interfascicular oligodendrocytes electrically isolate segments of central axons by forming the lipid-rich myelin sheath

The cell bodies of interfascicular oligodendrocytes are situated between bundles of axons

Interfascicular, or myelinating, oligodendrocytes have small spherical or polyhedral cell bodies of diameter 6–8 μm and few processes. They are called interfascicular because their cell bodies are aligned between bundles (fascicles) of axons. They are distinguished from astrocytes by the sites of termination of their processes: oligodendrocyte processes enwrap axons and make no contact with blood vessels.

Observed by electron microscopy, the nucleus and perikaryon of oligodendrocytes appear dark (**Figure 2.3**), there are no glial filaments, but there are many microtubules in the somatic and dendritic cytoplasm.

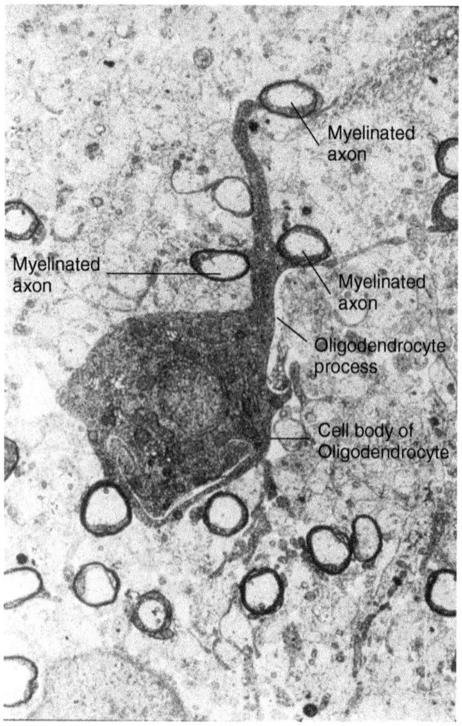

FIGURE 2.3 **Myelinating oligodendrocyte.** Electron micrograph of an oligodendrocyte. The cell body and one of its processes enwrapping several axons can be seen. Section taken through the spinal cord. Photograph by Olivier Robain.

Because of this, oligodendrocyte processes may be confused with fine dendrites, and it is by the absence of chemical synapses that the mature oligodendrocyte processes are identified.

Oligodendrocytes can be identified by immunohistochemistry. This is done using an anti-galactoceramide immune serum (anti-gal-C), galactoceramide being a glycolipid found exclusively in the membrane of processes of myelinizing oligodendrocytes.

The myelin sheath is a compact roll of the plasmalemma of an oligodendrocyte process: this glial membrane is rich in lipids

Myelinated axons are surrounded by a succession of myelin segments, each about 1 mm long. The covered regions of axons alternate with short exposed lengths where the axonal membrane (axolemma) is not covered. These unmyelinated regions (of the order of a micron) are called nodes of Ranvier (**Figures 2.4** and **2.5a**).

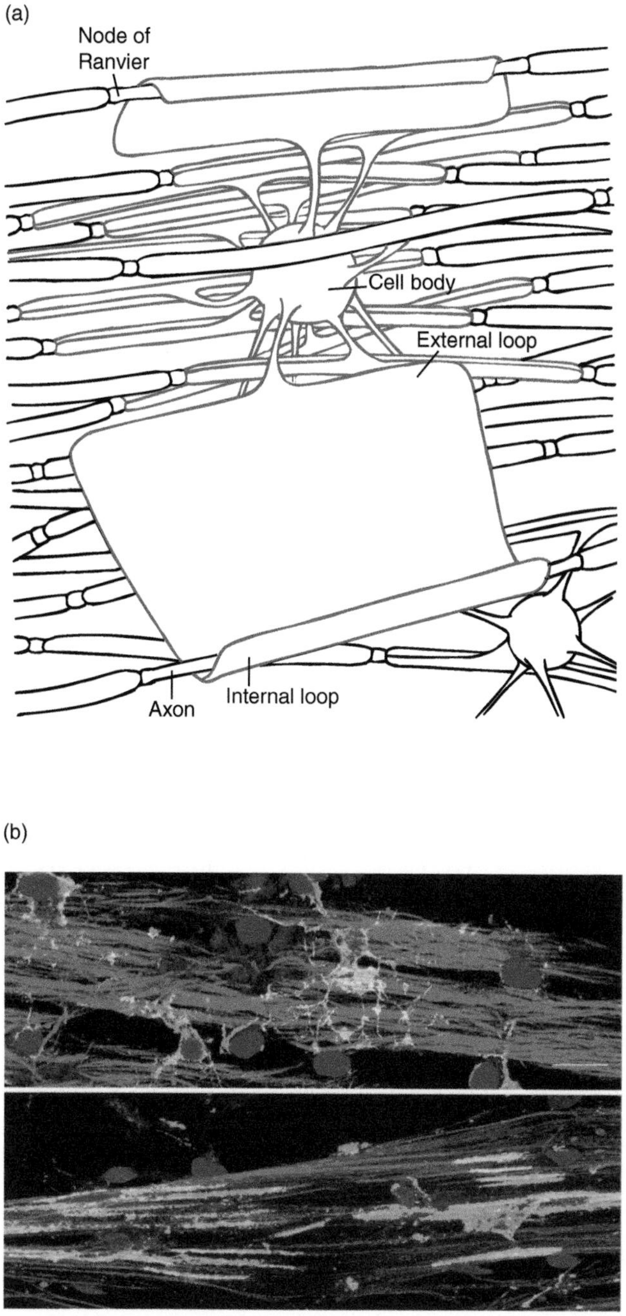

FIGURE 2.4 **Diagram and photomicrographs of myelinating oligodendrocytes and their numerous processes.** (**a**) Each oligodendrocyte process forms a segment of myelin around a different axon in the central nervous system. Two myelin segments are represented, one partially unrolled, the other completely unrolled. (**b**) Maturation of oligodendrocytes grown on axons of dorsal root ganglionic (DRG) neurons in culture. DRG axons are immunostained for neurofilament (red), and cell nuclei are labeled with DAPI (blue). Upper part: oligodendrocyte progenitor cells (revealed by NG2 labeling in green) plated onto DRG axons (red) after 2 days in co-culture. They extend many processes which contact multiple axons. Lower part: after 7 days in co-culture, oligodendrocyte precursor cells have differentiated into myelinating oligodendrocytes which form multiple segments of compact myelin associated with the axons (myelin is shown in green by myelin basic protein, a component of the myelin sheath, immunolabeling). Scale bar = 15 μm. Part (**a**) drawing by Tom Prentiss. In Morell P, Norton W (1980) La myéline et la sclérose en plaques. *Pour la Science* **33**, with permission. Part (**b**) from Lee PR, Fields RD. (2009) Regulation of myelin genes implicated in psychiatric disorders by functional activity in axons. *Front. Neuroanat.* **3**, 4, with permission.

A myelinated segment comprises the length of axon covered by an oligodendrocyte. One oligodendrocyte can form 20–70 myelin segments around different axons (**Figure 2.4b**). Thus the degeneration or dysfunction of a single oligodendrocyte leads to the disappearance of myelin segments on several different axons.

Origin of oligodendrocytes and myelination process

Myelination represents a crucial stage in the ontogenesis of the nervous system. In the human at birth, myelinization is only just beginning and, in some regions, is not complete even by the end of the second year of life. Oligodendrocytes are derived from oligodendrocyte progenitor cells (OPCs). These progenitors (sometimes called NG2 cells because they express the neural/glial antigen 2 proteoglycan (NG2) on their surface) represent a large and enigmatic population of glial cells. Because of their ability to generate all neural cell types *in vitro* (neurons, astrocytes and oligodendrocytes), NG2 cells have been proposed to be multipotent stem cells; however, studies using genetic lineage tracing with inducible Cre mouse lines demonstrate that NG2 cells only generate oligodendrocytes *in vivo*. Nevertheless, oligodendrocyte progenitor cells present remarkable properties: (i) they persist throughout life and represent approximately 5% of all cells in the CNS; (ii) they are the main population of slowly dividing cells in the postnatal brain; (iii) they are activated after insult of the central nervous system: they start proliferating actively, migrate toward the lesion and, finally, differentiate into oligodendrocytes thus contributing to spontaneous remyelination; and (iv) they are coupled to axons through synapses that use GABA and glutamate as neurotransmitters (half of these cells can even fire action potentials). Remarkably, synapses of oligodendrocyte progenitor cells disappear as soon as the cell differentiates into a mature oligodendrocyte. In fact, this terminal differentiation of oligodendrocyte progenitor cells is promoted by neuronal activity. Indeed, optogenetic stimulation of premotor cortex in awake mice stimulates proliferation of oligodendrocyte precursor cells and promotes myelination of axons within the deep layers of the premotor cortex and subcortical white matter.

Oligodendrocytes do not myelinate all axons in the CNS. Myelination starts with the initial turn of myelin around the axon which is rapidly formed (**Figure 2.4**). Myelin is then slowly deposited over a period which, in humans, can reach several months. Myelinization is responsible for a large part of the increase in weight of the central nervous system following the end of neurogenesis.

In order to form the compact spiral of myelin membrane, the oligodendrocyte process must roll itself around the axon many times (up to 40 turns) (**Figure 2.5**). It is the terminal portion of the process, called the inner loop, situated at the interior of the roll, which progressively spirals around the axon. This movement necessitates the sliding of myelin sheets which are not firmly attached. During this period, the oligodendrocyte synthesizes several times its own weight of myelin membrane each day.

Within the spiral, the cytoplasm disappears entirely (except at the internal and external loops). The internal leaflets of the plasma membranes can thus adhere to each other. This adhesion is so intimate that the internal leaflets virtually fuse, forming the period, or major, dense line of thickness of 3 nm (**Figure 2.5b**). The extracellular space between the different turns of membrane also disappears, and the external leaflets also stick to each other. This apposition is, however, less close and a small space remains between the external leaflets. The apposed external leaflets form the minor, or interperiod, dense line (**Figure 2.5b**).

Thus, a cross-section of a myelinated axon observed by electron microscopy shows alternating dark and light lines forming a spiral around the axon. The major dense line terminates where the internal leaflets separate to enclose the cytoplasm within the external loop. The interperiod dense line disappears at the surface of the sheath at the end of the spiral (**Figure 2.5b**).

In the central nervous system, there is no basal lamina around myelin segments, so myelin segments of adjacent axons may adhere to each other forming an interperiod dense line. The g ratio, determined by the axon diameter divided by axon plus myelin diameter, is often used as a myelination index.

Myelin

Myelin consists of a compact spiral (without intracellular or extracellular space) of glial plasma membrane of a very particular composition. Lipids make up about 70% of the dry weight of myelin and proteins only 30%. Compared with the membranes of other cells, this represents an inversion of the lipid:protein ratio.

This lipid-rich membrane is highly enriched in glycosphingolipids and cholesterol. The major glycosphingolipids in myelin are galactosylceramide and its sulfated derivative sulfatide (20% of lipid dry weight). There is also an unusually high proportion of ethanolamine phosphoglycerides in the plasmalogen form, which accounts for one-third of the phospholipids.

In myelin, a number of structural classes of proteins are present. Some proteins are extremely hydrophobic membrane-embedded polypeptides, some integral membrane proteins have a single transmembrane domain and clearly define extra- and intracellular domains, and some of the myelin proteins are cytosolic; however, they are often intimately associated with the myelin membrane.

Myelin basic protein (MBP) and the proteolipid proteins (PLP/DM20) are the two major myelin proteins in the CNS. Myelin basic proteins are found on the

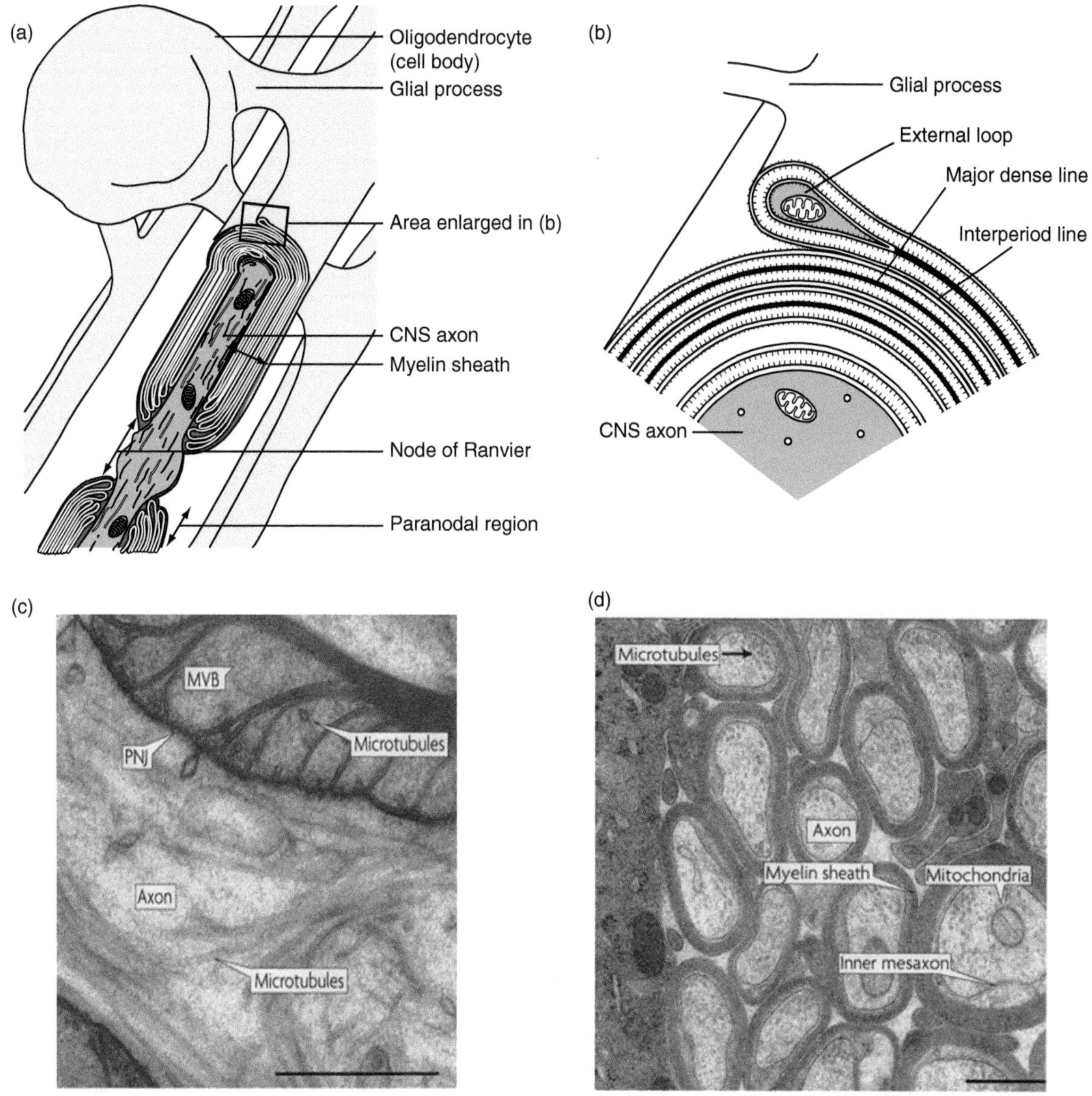

FIGURE 2.5 **Myelin sheath of central axons.** (**a**) Three-dimensional diagram of the myelin sheath of an axon in the central nervous system (CNS). The sheath is formed by a succession of compact rolls of glial processes from different oligodendrocytes. (**b**) Cross-section through a myelin sheath. The dark lines, or major dense lines, and clear bands (in the middle of which are found the interperiod lines) visible with electron microscopy are accounted for by the manner in which the myelin membrane surrounds the axon, and by the composition of the membrane. The dark lines represent the adhesion of the internal leaflets of the myelin membrane while the interperiod lines represent the adhesion of the external leaflets. The lines are formed by membrane proteins while the clear bands are formed by the lipid bilayer. (**c**,**d**) Electron micrographs of myelinated axons in sciatic (**c**) and optic (**d**) nerves. Structural details of myelin ensheathment at the paranodal level (longitudinal section, **c**). Cross-section of myelinated axons (**d**). Numerous microtubules and mitochondria are present in the axons. The innermost layer of myelin sheath is often non-compacted. MVBs: multivesicular bodies. PNJ: paranodal junction. Part (**a**) drawing from Bunge MB, Bunge RP, Ris H (1961) Ultrastructural study of remyelination in an experimental lesion in adult cat spinal cord. *J. Biophys. Biochem. Cytol.* 10, 67–94, with permission of Rockerfeller University Press. Part (**b**) drawing by Tom Prentiss. In Morell P, Norton W (1980) La myéline et la sclérose en plaques. *Pour la Science* 33, with permission. Parts (**c**,**d**) from Nave KA (2010) Myelination and the trophic support of long axons. *Nat. Rev. Neurosci.* 11, 275–283, with permission.

cytoplasmic side and play a role in the adhesion of the internal leaflets of the specialized oligodendroglial plasma membrane. Proteolipid proteins are integral membrane proteins. Though they are in high abundance (they represent around 50% of the total myelin protein in the central nervous system), their exact biological role has not yet been elucidated.

We have seen that myelin has an inverted lipid:protein ratio, while the cell body membrane of the oligodendrocyte has a ratio comparable to that of other cell membranes. As the myelin of the oligodendrocyte process is in continuity with the plasma membrane of the cell body, it is necessary to postulate gradients in the composition of lipids and proteins (in opposite directions to each other) between the cell body and the various processes. During the active phase of myelination, each oligodendrocyte must produce as much as 5–50 000 μm^2 of myelin membrane surface area per day.

Interestingly, myelination does not stop with the end of development but is maintained throughout life. Using transgenic mice with inducible Cre recombinase system to label adult oligodendrocyte progenitor cells and trace their progeny, it was demonstrated that almost 30% of total oligodendrocytes of the corpus callosum in an 8-month-old animal were generated during adulthood. The function of this constant production of myelinating oligodendrocytes and of myelin remodeling in the adult brain remains poorly understood but, together with the activity-dependent control of myelination, it highlights the remarkable plasticity of myelin.

Nodes of Ranvier

In the central nervous system, the nodes of Ranvier, regions between myelin segments, are relatively long (several microns) compared with those in the peripheral nervous system. Here the axolemma is exposed and an accumulation of dense material is seen on the cytoplasmic side. The myelin sheath does not terminate abruptly. Successive layers of myelin membrane terminate at regularly spaced intervals along the axon, the internal layers (close to the axon) terminating first. This staggered termination of the different layers of myelin constitutes the paranodal region (**Figure 2.5**).

2.2.2 Myelination enables rapid conduction of action potentials for two reasons

Isolation of internode axonal segments

The high lipid content and compact structure of the myelin sheath help make it impermeable to hydrophilic substances such as ions. It prevents transmembrane ion fluxes and acts as a good electrical insulator between the intracellular (i.e. intra-axonal) and extracellular media. Between the nodes of Ranvier the axon therefore behaves as an insulated cable. This permits rapid, saltatory conduction of action potentials along the axon (see Section 5.4).

Formation of Ranvier nodes with a high density of Na^+ channels

Na^+ channels are clustered in very high density within the nodal gap whereas voltage-dependent K^+ channels are segregated in juxta-paranodal regions, beneath overlying myelin (**Figure 2.5**). To test whether oligodendrocyte contact with axon influences Na^+ channel distribution, nodes of Ranvier in the brain of hypomyelinating mouse *Shiverer* are examined. *Shiverer* mice have oligodendrocytes that ensheath axons but do not form compact myelin and axoglial junctions. In these mutant mice, there are far fewer Na^+ channel clusters than in control littermates and aberrant locations of Na^+ channels are observed. If Na^+ channel clustering depended only on the presence of oligodendrocytes and were independent of myelin and oligodendroglial contact, one would expect to find normal Na^+ channel distribution along 'shiverer' axons.

Contribution of myelination to signal transmission efficiency

Myelination of axons by glial cells has appeared quite late in evolution. This innovation in ancient jawed vertebrates allowed speeding up nerve conduction more than 20-fold and the appearance of efficient signal transmission in large animals. Today, numerous correlative evidences suggest myelin involvement in cognition or in the development of fine motor skills.

2.2.3 Myelin-independent functions of oligodendrocytes

Facilitation of conduction velocity by depolarization of oligodendrocytes

Myelin increases the velocity of action potentials along axons but oligodendrocytes also contribute to optimize signal transmission via myelin-independent functions. Like neurons and astrocytes, oligodendrocytes express a large variety of ion channels and neurotransmitter receptors. They can thus participate in K^+ buffering, monitor neuronal activity, be depolarized and directly increase conduction velocity along the axon. The mechanisms involved in oligodendrocyte depolarization-induced conduction velocity are not completely deciphered but may involve structural changes leading to increased insulation in the paranodal and intermodal regions. Since one oligodendrocyte myelinates multiple axons, oligodendrocytes could also contribute to signal synchronization.

Oligodendrocytes support axonal integrity

Axonal support is probably a common and ancestral function of all axon-ensheathing glial cells. Paradoxically, because myelin completely insulates the axon, it also deprives it from metabolic support by the extracellular milieu. Thus, oligodendrocytes need to overcome this by supplying trophic and metabolic support to axons, thereby preserving axonal integrity. Diseases affecting oligodendrocytes often lead to axon degeneration. Part of this neuroprotection effect is independent of myelin's function. Loss of myelin proteins such as PLP or CNP, which are dispensable for myelination, leads to progressive axonal degeneration. Mechanisms underlying glial cell support are not completely understood yet.

2.2.4 Impact of myelin defects

Oligodendrocytes show the highest metabolic rates among all brain cells in order to produce but also to maintain very high volume of membranes which can represent up to 100 times the weight of the cell. This metabolic requirement makes them highly sensitive to pathological conditions such as ischemia, trauma and inflammation.

The consequences of massive oligodendrocyte death and demyelination have long been known from diseases such as multiple sclerosis, with functional loss depending of the areas of the CNS where demyelination occurs (blindness, incontinency, motor deficit…). More recently, it has been discovered that more subtle myelin abnormalities can contribute to a wide range of neuropsychatric disorders including depression and schizophrenia. White matter abnormalities also represent an early feature of neurodegenerative pathologies such as Huntington and Alzheimer diseases (far before neurofibrillary tangles and clinical symptoms) and might also contribute to the pathogenesis of the disease.

2.3 MICROGLIAL CELLS ARE THE MACROPHAGES OF THE CENTRAL NERVOUS SYSTEM

Microglial cells are unique since they are at the interface between the immune and nervous systems, which confers them remarkable properties. Microglia are equipped by a plethora of receptors allowing them to detect signals deriving from different origins, such as an immune challenge or activity from other brain cells, and readily react to them. As macrophages, they are professional phagocytes, but they can also release several mediators which can influence neuronal activity as well as the fate of surrounding cells. After a decade long debate about their origin, we know today that these cells arise early during development from progenitors in the embryonic yolk sac (YS) that seed the brain rudiment. During development, microglial cells colonize the entire CNS and they differentiate to assume a mature phenotype adapted to their function in the adult brain.

2.3.1 Microglial cells contribute to brain physiology

Microglia are highly dynamic cells

In the adult CNS, microglial density and morphology change among different CNS regions, likely to better fulfill their function. Their density varies from about 12% of total cells in substantia nigra to ≈5% in corpus callosum. Despite this variability, their distribution in space in a specific region is rather uniform with minimal overlap. In white matter, their morphology is characterized by an elongated soma and processes preferentially oriented along fiber tracts, while in the circumventricular organs, a region with a leaky blood–brain barrier, they are compact with a few short processes. In contrast, in gray matter, they present a small cell body with highly branched processes. This typical appearance, which is quite different from a classical macrophage, has been associated with microglial 'resting' state and, until recently, this state was considered quiescent, i.e. functionally dormant. However, recent *in vivo* imaging studies employing mice with EGFP-expressing microglia demonstrated that they constantly extend and retract their fine cellular processes at a rate of few micrometers per minute to scan the brain parenchyma. During movements, a single microglia maintains its cell body position as well as its overall size and symmetry of its territory, and avoids contacts with other microglial processes. Thus, microglia are able to monitor their extracellular space and cellular neighborhood, always ready to intervene and change their activity status if abnormal changes are detected such as infections, lesions and neurodegeneration, critical variation in neuronal activity...

Microglia sense and control neuronal activity

During their scanning movements microglial processes rapidly contact other CNS cells, like neurons, perivascular astroglial and other glial cells. In particular, they contact post- and presynaptic elements of synapses. Interestingly, the frequency of these contacts is regulated by the neuronal activity. Indeed, it is decreased by sensorial deprivation, pharmacological block of action potential, as well as a decrease of body temperature. Thus, microglial cells can sense and react to neuronal activity. In turn, they can locally influence neuronal excitability by releasing several factors such as cytokines, neurotrophic factors, and neurotransmitters. A study in the larvae of zebra fish recently showed that an increase of neuronal firing attracts microglia through the release of ATP, and this contact decreases the activity of the targeted neuron.

Microglia control adult neurogenesis

Another important role of microglia is the regulation of the adult neurogenesis. Two neurogenic niches, in the ventricles and in the dentate gyrus of the hippocampus (see Chapter 19), produce new neurons throughout life. However, only a few newborn cells are incorporated into the circuitry and the majority of them are presumed to die at immature neuronal stage. Microglia can control proliferation, migration and differentiation of stem cells, but the factors regulating different steps of neurogenesis are still unknown. Furthermore, they are responsible for the phagocytosis of apoptotic cells preventing them from undergoing secondary necrosis, with the consequent spillage of cellular contents leading to an inflammatory response, detrimental for neurogenesis.

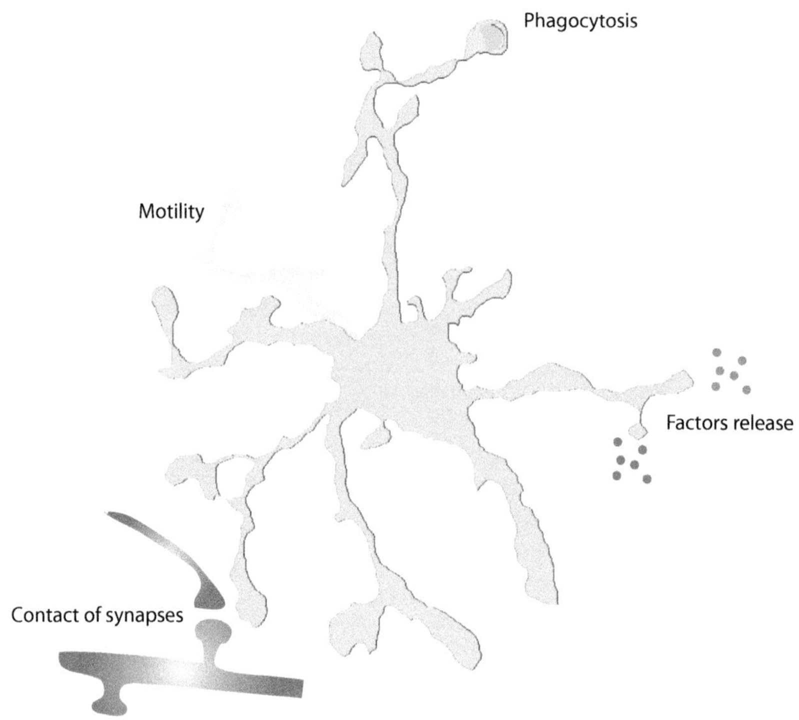

FIGURE 2.6 **Scheme of the different functions of microglia.** Microglia (green) constantly move their processes to scan the brain parenchyma. During their movements they contact synapses and neuronal dendrites (orange), as well other brain cells. They can control brain activity and surrounding cells' fate by releasing several factors. They phagocytose cells and neuronal debris, but also synaptic elements and newborn cells (orange), thus they participate in sculpting the neuronal circuits. Drawing by E. Avignone.

2.3.2 Microglia exert their activity through physical interaction and release of soluble factors

We can classify the main actions of microglia into two broad categories: physical interactions and release of soluble factors (**Figure 2.6**). These actions are triggered by changes in the environment which modify microglia properties.

Microglia with physical interactions sculpt neuronal circuits and isolate brain injuries

Three types of physical interactions are described: phagocytosis, synapse stripping, and formation of a barrier around a damaged area. Phagocytosis is a major feature of microglial cells. This phenomenon is particular important during pathologies involving cell death, where cellular debris has to be removed, as well as during development. In immature circuits, microglia phagocyte presynaptic terminals and dendritic spines (synapse pruning). They thus contribute to circuit formation.

Synapse stripping occurs when microglia processes form a barrier between the pre- and postsynaptic elements at the synaptic cleft. It is observed during lesions where there is an injury of innervating cells and a loss of synaptic boutons, as after facial nerve injury.

As previously seen, microglia constantly move their processes to scan the brain parenchyma. As soon as a problem is detected they can extend their processes in a few minutes towards sites of injury or substances that may represent a danger signal, like ATP or nitric oxide. These substances could be released by necrotic cells or by neurodegenerating axons. The ensemble of microglial processes then forms a barrier around the injured area in a time course from minutes to hours, thus limiting the spread of the damage of an acute lesion to the surrounding healthy tissue.

Microglia release soluble factors to control neuronal survival and activity

Pioneer studies in culture identified several soluble factors released by microglia, including a multitude of growth factors, cytokines, and neurotransmitters. Many released factors are known to affect neuronal activity, either directly or through other cells such as astrocytes. However, direct evidences of neuronal activity modulation by microglia in physiological conditions are still lacking. Indeed, most of these studies have been performed in culture, where microglia properties do not faithfully reflect those in the normal, non-pathological brain.

Microglia change their properties in pathological conditions to respond to injuries

Microglial cells are extremely sensitive to changes in their local environment. As a response to homeostasis perturbation, they adapt their phenotype in a multifaceted process called activation. Activation develops over

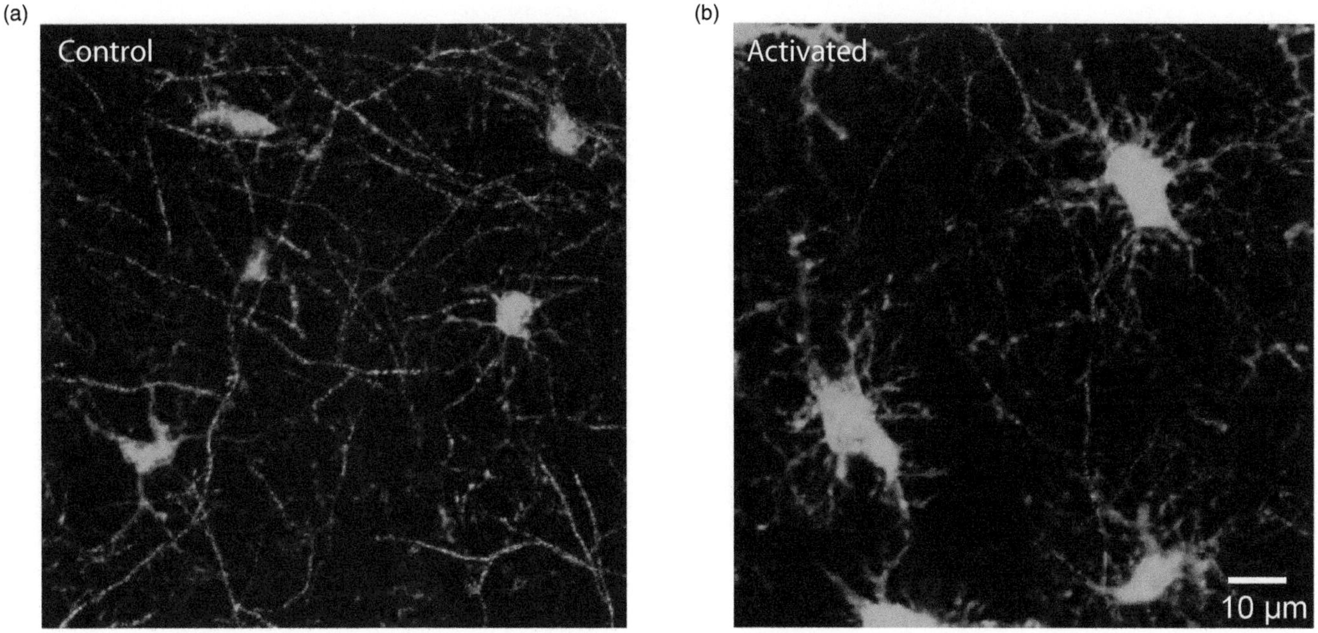

FIGURE 2.7 **Microglia change properties after activation.** The images show an example of morphological changes of microglia 48 hours after activation induced by *status epilepticus*. In control conditions (**a**) microglial cells have a small body with long and ramified processes. (**b**) In contrast, activated microglial cells have larger body with shorter and thicker processes. From Menteyne A, Levavasseur F, Audinat E, Avignone E (2009) Predominant functional expression of Kv1.3 by activated microglia of the hippocampus after status epilepticus. *PLoS One* 4, e6770, with permission.

periods ranging from hours to days, and it includes changes in morphology (**Figure 2.7**), electrical membrane properties and expression of a variety of markers. Activation triggers migration, proliferation and release of a variety of mediators, such as pro- and anti-inflammatory cytokines and neurotrophines.

Activation is not an all-or-none process. It is variable and adaptive and it depends on the nature of triggering factors. Thus, microglia can exist in a variety of activated states. The heterogeneity of activation phenotypes and functions caused a debate on the beneficial versus detrimental roles of microglia in pathology, since it can exacerbate damage or, on the contrary, protect neurons from degeneration. A major challenge today is to characterize several aspects of microglia activation in specific diseases, and develop new drugs that block only harmful facets of activation.

2.4 SCHWANN CELLS ARE THE GLIAL CELLS OF THE PERIPHERAL NERVOUS SYSTEM; THEY FORM THE MYELIN SHEATH OF AXONS OR ENCAPSULATE NEURONS

There are three types of Schwann cell:

- those forming the myelin sheath of peripheral myelinated axons (myelinating Schwann cells);
- those encapsulating non-myelinated peripheral axons (non-myelinating Schwann cells);
- those that encapsulate the bodies of ganglion cells (non-myelinating Schwann cells or satellite cells).

2.4.1 Myelinating Schwann cells make the myelin sheath of peripheral axons

Along an axon, several Schwann cells form successive segments of the myelin sheath. In contrast to oligodendrocytes, it is not a process that enwraps the peripheral axon to form the segment of myelin, but the whole Schwann cell (**Figure 2.8**). Each Schwann cell therefore forms only one myelin segment.

The composition of peripheral myelin differs from that of central myelin only in the proteins it contains. The principal protein constituents of peripheral myelin are: peripheral myelin protein 2 (P2), protein zero (P0) and myelin basic proteins (MBPs). The first two proteins are specific to peripheral myelin. MBP comprises a major part of the cytosolic protein of myelin and is present both in the CNS and peripheral nervous system (PNS). Protein zero is a glycoprotein that has adhesive properties and is located in the interperiod line. It functions, in part, as a homotypic adhesion molecule throughout the full thickness of the myelin sheath. It is a good marker for myelinating Schwann cells as it represents over 50% of total PNS myelin protein.

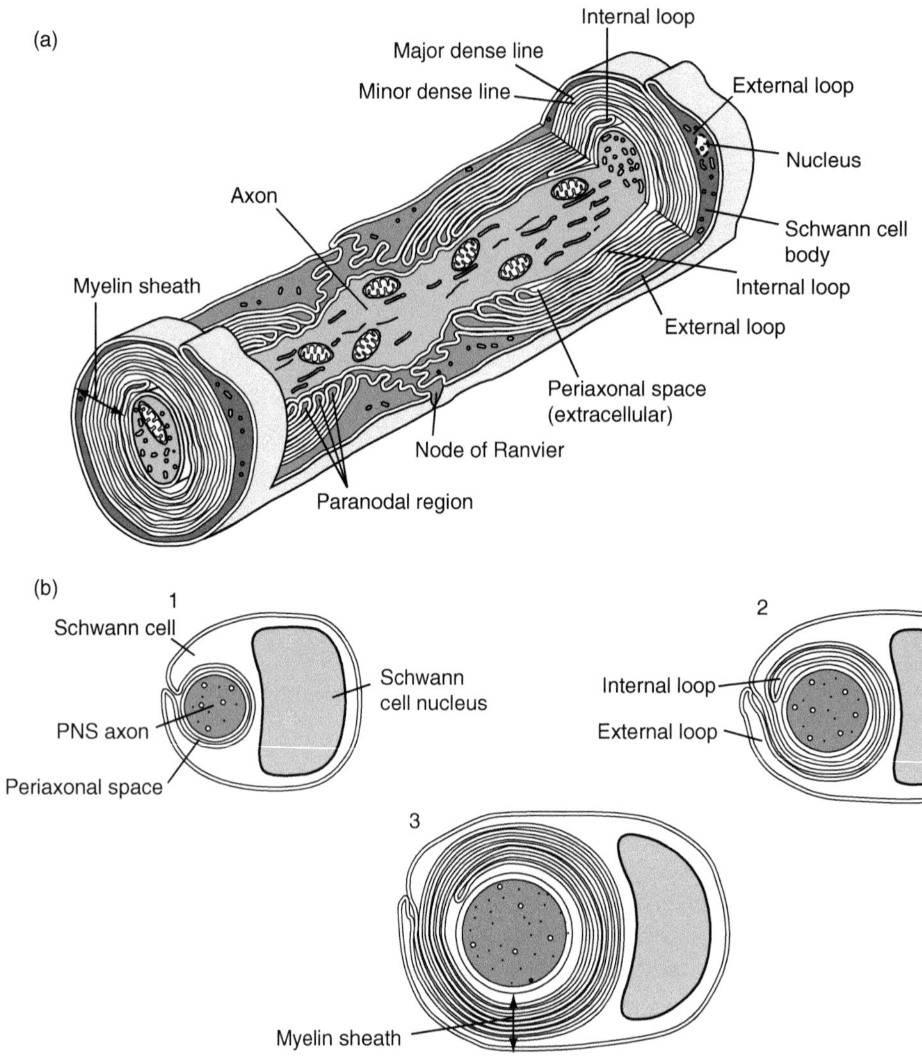

FIGURE 2.8 **Myelin sheath of a peripheral axon.** (**a**) Three-dimensional diagram of the myelin sheath of an axon of the peripheral nervous system (PNS). The sheath is formed by successive rolled Schwann cells. (**b**) Process of myelinization. The internal loop wraps around the axon several times. During this process, the axon grows and the myelin becomes compact. Contact between the Schwann cell and axon occurs only at the paranodal and nodal regions. Elsewhere an extracellular, or periaxonal, space always remains. Part (**a**) drawing adapted from Maillet M (1977) *Le Tissu Nerveux*, Paris: Vigot, with permission. Part (**b**) drawing by Tom Prentiss. In Morell P, Norton W (1980) La myéline et la sclérose en plaques. *Pour la Science* **33**, with permission.

2.4.2 Non-myelinating Schwann cells encapsulate the axons and cell bodies of peripheral neurons

Non-myelinated axons are not uncovered in the peripheral nervous system as they are in the central nervous system; they are encapsulated. A single non-myelinating Schwann cell surrounds several axons (about 5–20) for a distance of 200–500 μm in humans.

In addition, spinal and cranial ganglia contain a large number of Schwann cells that do not produce myelin. These Schwann cells cover the somata of the ganglionic cells, leaving an extracellular space of about 20 nm between themselves and the surface of the covered neuron.

The lipid and protein composition of the plasma membrane of non-myelinating Schwann cells is the same as that of other eukaryotic cells (30% lipid, 70% protein).

Apart from their role in the saltatory conduction of action potentials (myelinating Schwann cells), Schwann cells also play a role in maintenance of axonal integrity and survival. A remarkable property of Schwann cells is their ability to promote the regeneration of peripheral nerve cells. It has long been known that cut peripheral nerves can, within certain limits, regrow and reinnervate deafferented regions while central axons are not capable of this. The substantial regenerative capacity of the Schwann cells is driven by their ability to dedifferentiate following axonal injury, to divide and produce axonotrophic factors. This property of regeneration is due in large part to an enabling effect of Schwann cells on axon regrowth.

Further reading

Azevedo, F.A., Carvalho, L.R., Grinberg, L.T., et al., 2009. Equal numbers of neuronal and nonneuronal cells make the human brain an isometrically scaled-up primate brain. J. Comp. Neurol. 513, 532–541.

Bélanger, M., Allaman, I., Magistretti, P.J., 2011. Brain energy metabolism: focus on astrocyte-neuron metabolic cooperation. Cell Metab. 14, 724–738.

Bushong, E.A., Martone, M.E., Jones, Y.Z., Ellisman, M.H., 2002. Protoplasmic astrocytes in CA1 stratum radiatum occupy separate anatomical domains. J. Neurosci. 22, 183–192.

Butt, A.M., Kalsi, A., 2006. Inwardly rectifying potassium channels (Kir) in central nervous system glia: a special role for Kir4.1 in glial functions. J. Cell. Mol. Med. 10, 33–44.

Buttermore, E.D.L., Thaxton, C.L., Bhat, M.A., 2013. Organization and maintenance of molecular domains in myelinated axons. J. Neurosci. Res. 91, 603–622.

Dupree, J.L., Mason, J.L., Marcus, J.R., et al., 2005. Oligodendrocytes assist in the maintenance of sodium channel clusters independent of the myelin sheath. Neuron Glia Biol. 1, 1–14.

El Waly, B., Macchi, M., Cayre, M., Durbec, P., 2014. Oligodendrogenesis in normal and pathological brain. Front. Neurosci. 8, 145.

Farber, K., Kettenmann, H., 2005. Physiology of microglial cells. Brain Res. Brain Res. Rev. 48, 133–143.

Fields, R.D., Araque, A., Johansen-Berg, H., et al., 2014. Glial biology in learning and cognition. Neuroscientist 20, 426-431.

Freeman, M.R., Doherty, J., 2006. Glial cell biology in Drosophila and vertebrates. Trends Neurosci. 29, 82–90.

Gibson, E.M.L., Purger, D., Mount, C.W., et al., 2014. Neuronal activity promotes oligodendrogenesis and adaptive myelination in the mammalian brain. Science, Apr 11. [Epub ahead of print].

Greer, J.M., Lees, M.B., 2002. Myelin proteolipid protein – the first 50 years. Int. J. Biochem. Cell Biol. 34, 211–215.

Kettenmann, H., Kirchhoff, F., Verkhratsky, A., 2013. Microglia: new roles for the synaptic stripper. Neuron 77, 10–18.

Montague, P., McCallion, A.S., Davies, R.W., Griffiths, I.R., 2006. Myelin-associated oligodendrocytic basic protein: a family of abundant CNS myelin proteins in search of a function. Dev. Neurosci. 28, 479–487.

Nave, K.A., 2010. Myelination and support of axonal integrity by glia. Nature 468, 244–252.

Pannasch, U., Rouach, N., 2013. Emerging role for astroglial networks in information processing: from synapse to behavior. Trends Neurosci. 36, 405–417.

Parkhurst, C.N., Gan, W.-B., 2010. Microglia dynamics and function in the CNS. Curr. Opin. Neurobiol. 20, 595–600.

Parpura, V., Heneka, M.T., Montana, V., et al., 2012. Glial cells in (patho)physiology. J. Neurochem. 121, 4–27.

Peles, E., Salzer, J.L., 2000. Molecular domains of myelinated axons. Curr. Opin. Neurobiol. 10, 558–565.

Pfeiffer, S.E., Warrington, A.E., Bansal, R., 1993. The oligodendrocyte and its many cellular processes. Trends Cell Biol. 3, 191–197.

Popko, B., 2000. Myelin galactolipids: mediators of axon–glial interactions? Glia 29, 149–153.

Shapiro, L., Doyle, J.P., Hensley, P., Colman, D.R., Hendrickson, W.A., 1996. Crystal structure of the extracellular domain from P0, the major structural protein of peripheral nerve myelin. Neuron 17, 435–449.

Streit, W.J., 2000. Microglial response to brain injury: a brief synopsis. Toxicol. Pathol. 28, 28–30.

Tremblay, M.-É., 2011. The role of microglia at synapses in the healthy CNS: novel insights from recent imaging studies. Neuron Glia Biol. 7, 67–76.

Ueno, M., Yamashita, T., 2014. Bidirectional tuning of microglia in the developing brain: from neurogenesis to neural circuit formation. Curr. Opin. Neurobiol. 27C, 8–15.

Vandenberg, R.J., Ryan, R.M., 2013. Mechanisms of glutamate transport. Physiol. Rev. 93, 1621–1657.

Ventura, R., Harris, K.M., 1999. Three-dimensional relationships between hippocampal synapses and astrocytes. J. Neurosci. 19, 6897–6906.

Yamazaki, Y.L., Hozumi, Y., Kaneko, K., Fujii, S., Goto, K., Kato, H., 2010. Oligodendrocytes: facilitating axonal conduction by more than myelination. Neuroscientist 16, 11–18.

CHAPTER

3

Ionic gradients, membrane potential and ionic currents

Constance Hammond

OUTLINE

The neuronal plasma membrane delimits the whole neuron, cell body, dendrites, dendritic spines, axon and axon terminals. It is a barrier between the intracellular and extracellular environments. The general structure of the neuronal plasma membrane is similar to that of other plasma membranes. It is made up of proteins inserted in a lipid bilayer, forming as a whole a 'fluid mosaic' (**Figure 3.1**). However, insofar as there are functions that are exclusively neuronal, the neuronal membrane differs from other plasma membranes by the nature, density and spatial distribution of the proteins of which it is composed.

The presence of a large diversity of transmembrane proteins called *ionic channels* (or simply 'channels') characterizes the neuronal plasma membrane. They allow the passive movement of ions across membranes and thus electrical signaling in the nervous system. Among the ions present in the nervous system fluids, Na^+, K^+, Ca^{2+} and Cl^- ions seem to be responsible for almost all of the action.

3.1 THERE IS AN UNEQUAL DISTRIBUTION OF IONS ACROSS THE NEURONAL PLASMA MEMBRANE. THE NOTION OF CONCENTRATION GRADIENT

3.1.1 The plasma membrane separates two media of different ionic composition

Regardless of the animal's environment (seawater, freshwater or air), potassium (K^+) ions are the predominant cations in the intracellular fluid and sodium (Na^+) ions are the predominant cations in the extracellular fluid. The main anions of the intracellular fluid are organic molecules (P^-): negatively charged amino acids (glutamate and aspartate), proteins, nucleic acids, phosphates, etc., which have a large molecular weight. In the extracellular fluid, the predominant anions are chloride (Cl^-)

Cellular and Molecular Neurophysiology. DOI: 10.1016/B978-0-12-397032-9.00003-0

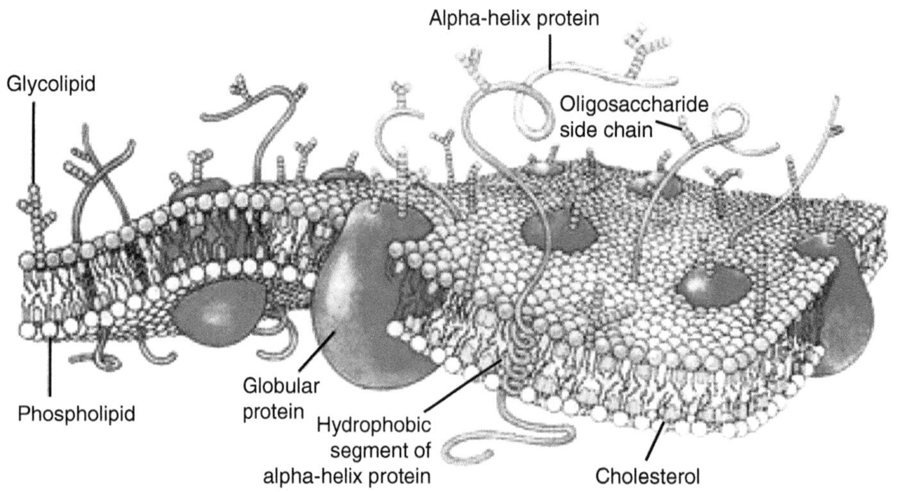

FIGURE 3.1 **Fluid mosaic.** Transmembrane proteins and lipids are kept together by non-covalent interactions (ionic and hydrophobic). From dictionary.laborlawtalk.com/Plasma_membrane.

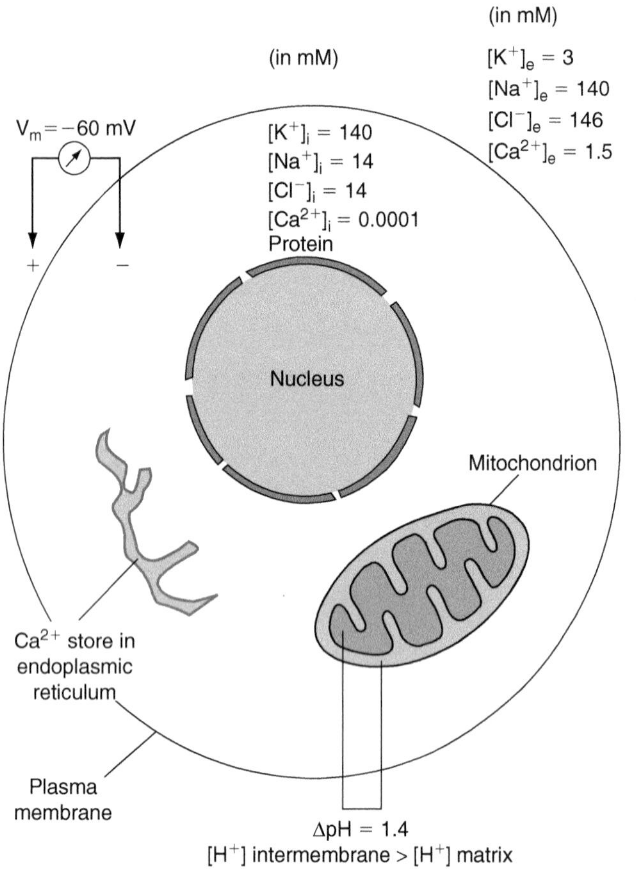

FIGURE 3.2 **There is an unequal distribution of ions across neuronal plasma membranes.** Idealized nerve cell (depicted as a sphere) with intra- and extracellular ionic concentrations. Membrane potential is the difference of potential (in mV) between the intracellular and extracellular faces of the plasma membrane.

ions. A marked difference between cytosolic and extracellular Ca^{2+} concentrations is also observed (**Figure 3.2**).

Spatial distribution of Ca^{2+} ions inside the cell deserves a more detailed description. Ca^{2+} ions are present in the cytosol as 'free' Ca^{2+} ions at a very low concentration (10^{-8} to 10^{-7} M) and as bound Ca^{2+} ions (bound to Ca^{2+}-binding proteins). They are also distributed in organelles able to sequester calcium, which include endoplasmic reticulum, calciosome and mitochondria, where they constitute the intracellular Ca^{2+} stores. Free intracellular Ca^{2+} ions present in the cytosol act as second messengers and transduce electrical activity in neurons into biochemical events such as exocytosis. Ca^{2+} ions bound to cytosolic proteins or present in organelle stores are not active Ca^{2+} ions; only 'free' Ca^{2+} ions have a role.

In spite of the unequal distribution of ions across the plasma membrane, intracellular and extracellular media are neutral ionic solutions: in each medium, the concentration of positive ions is equal to that of negative ions. According to **Figure 3.2**,

$$[Na^+]e + [K^+]e + 2[Ca^{2+}]e = 140 + 3 + (2 \times 1.5) = 146\,mM$$

$$[Cl^-]e = 146\,mM$$

$$[Na^+]_i + [K^+]_i + 2[Ca^{2+}]_i = 14 + 140 + 0.0002(2 \times 0.0001) = 154\,mM$$

$$\text{But } [Cl^-]_i = 14\text{ mM}$$

In the intracellular compartment, anions other than chloride ions are present and compensate for the positive charges. These anions are HCO_3^-, PO_4^{2-}, amino acids, proteins, nucleic acids, etc. Most of these anions are organic anions that do not cross the membrane.

3.1.2 The unequal distribution of ions across the neuronal plasma membrane is kept constant by active transport of ions

A difference of concentration between two compartments is called a *'concentration gradient'*. Measurements of Na^+, K^+, Ca^{2+} and Cl^- concentrations have shown that concentration gradients for ions are constant in the external and cytosolic compartments, at the macroscopic level, during the entire neuronal life.

At least two hypotheses can explain this constancy:

- Na^+, K^+, Ca^{2+} and Cl^- ions cannot cross the plasma membrane: plasma membrane is impermeable to these inorganic ions. In that case, concentration gradients need to be established only once in the lifetime.
- Plasma membrane is permeable to Na^+, K^+, Ca^{2+} and Cl^- ions but there are mechanisms that continuously re-establish the gradients and maintain constant the unequal distribution of ions.

This has been tested experimentally by measuring ionic fluxes. When proteins are absent from a synthetic lipid bilayer, no movements of ions occur across this purely lipidic membrane. Owing to its central hydrophobic region, the lipid bilayer has a low permeability to hydrophilic substances such as ions, water and polar molecules; i.e. the lipid bilayer is a barrier for the diffusion of ions and most polar molecules.

The first demonstrations of ionic fluxes across plasma membrane by Hodgkin and Keynes (1955) were based on the use of radioisotopes of K^+ or Na^+ ions. Experiments were conducted on the isolated squid giant axon. When this axon is immersed in a bath containing a control concentration of radioactive $^*Na^+$ ($^{24}Na^+$) instead of cold Na^+ ($^{22}Na^+$), $^*Na^+$ ions constantly appear in the cytoplasm. This $^*Na^+$ influx is not affected by dinitrophenol (DNP), a blocker of ATP synthesis in mitochondria. It does not require energy expenditure. This is *passive* transport. This result is in favor of the second hypothesis and leads to the following question: what are the mechanisms that maintain concentration gradients across neuronal membranes?

When the reverse experiment is conducted, the isolated squid giant axon is passively loaded with radioactive $^*Na^+$ by performing the above experiment, and is then transferred to a bath containing cold Na^+. Measuring the quantity of $^*Na^+$ that appears in the bath per unit of time (d^*Na^+/dt, expressed in counts per minute) allows quantification of the efflux of $^*Na^+$ (**Figure 3.3a**). In the presence of dinitrophenol (DNP) this $^*Na^+$ efflux quickly diminishes to nearly zero. The process can be started up again by intracellular injection of ATP. Therefore, the $^*Na^+$ efflux is *active* transport. The movement of Na^+ from the cytosol to the outside (efflux) can be switched off reversibly by the use of metabolic inhibitors.

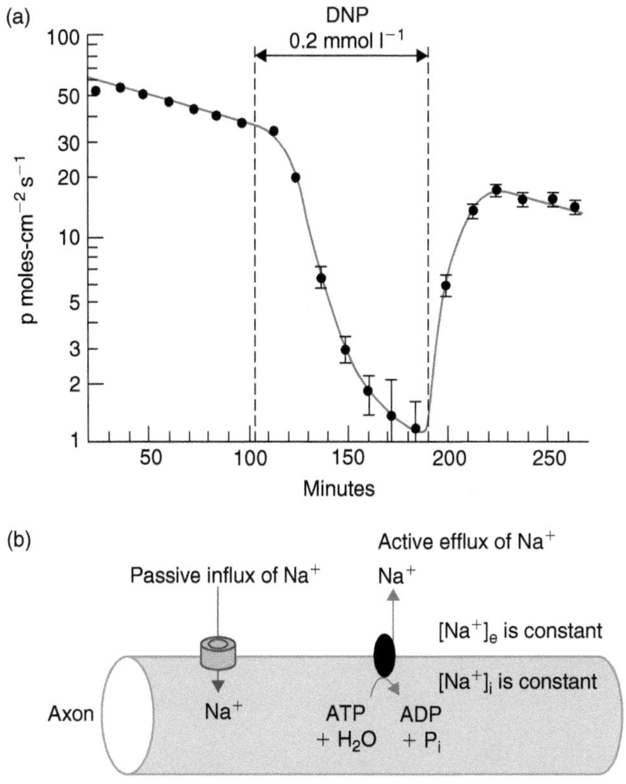

FIGURE 3.3 **Na^+ fluxes through the membrane of giant axons of sepia.** (**a**) Effect of dinitrophenol (DNP) on the outflux of $^*Na^+$ as a function of time. The axon is previously loaded with $^*Na^+$. At $t = 1$, the axon is transferred in a bath devoid of $^*Na^+$. The ordinate (logarithmic) axis is the quantity of $^*Na^+$ ions that appear in the bath (that leave the axon) as a function of time. At $t = 100$ min, DNP (0.2 mM) is added to the bath for 90 min. The efflux, which previously decreased linearly with time, is totally blocked after one hour of DNP. This blockade is reversible. (**b**) Passive and active Na^+ fluxes are in opposite directions. Plot (**a**) adapted from Hodgkin AL and Keynes RD (1955) Active transport of cations in giant axons from sepia and loligo. *J. Physiol. (Lond.)* **128**, 28-60, with permission.

This experiment demonstrates that cells maintain their ionic composition in the face of continuous passive exchange of all principal ions by active transport of these ions in the reverse direction. In other words, ionic composition of cytosol and extracellular compartments are maintained at the expense of a continuous basal metabolism that provides energy (ATP) utilized to actively transport ions and thus to compensate for their passive movements (see Appendix 3.1).

3.1.3 Na^+, K^+, Ca^{2+} and Cl^- ions passively cross the plasma membrane through a particular class of transmembrane proteins – the channels

Transmembrane proteins span the entire width of the lipid bilayer (see **Figure 3.1**). They have hydrophobic regions containing a high fraction of non-polar amino acids and hydrophilic regions containing a high fraction of polar amino acids. Certain hydrophobic regions organize themselves inside the bilayer as transmembrane α-helices while more hydrophilic regions are in contact with the aqueous intracellular and extracellular environments. Interaction energies are very high between hydrophobic regions of the protein and hydrophobic regions of the lipid bilayer, as well as between hydrophilic regions of the protein and the extracellular and intracellular environments. These interactions strongly stabilize transmembrane proteins within the bilayer, thus preventing their extracellular and cytoplasmic regions from flipping back and forth.

Ionic channels have a three-dimensional structure that delimits an aqueous pore through which certain ions can pass. They provide the ions with a passage through the membrane (see Appendix 3.2). Each channel may be regarded as an excitable molecule as it is specifically responsive to a stimulus and can be in at least two different states: closed and open. Channel opening, the switch from the closed to the open state, is tightly controlled (**Table 3.1**) by:

- a change in the membrane potential – these are voltage-gated channels;
- the binding of an extracellular ligand, such as a neurotransmitter – these are ligand-gated channels, also called receptor channels or ionotropic receptors;
- the binding of an intracellular ligand such as Ca^{2+} ions or a cyclic nucleotide;
- mechanical stimuli such as stretch – these are mechanoreceptors.

The channel's response to its specific stimuli, called gating, is a simple opening or closing of the pore. The pore has the important property of selective permeability, allowing some restricted class of small ions to flow passively down their electrochemical gradients (see Section 3.3). These gated ion fluxes through pores make signals for the nervous system.

3.2 THERE IS A DIFFERENCE OF POTENTIAL BETWEEN THE TWO FACES OF THE MEMBRANE, CALLED MEMBRANE POTENTIAL (V_m)

If a fine-tipped glass pipette (usually called a microelectrode), connected via a suitable amplifier to a recording system such as an oscilloscope, is pushed through the membrane of a living nerve cell to reach its cytoplasm, a potential difference is recorded between the cytoplasm and the extracellular compartment (**Figure 3.2**). In fact, the cell interior shows a negative potential (typically between −60 and −80 mV) with respect to the outside, which is taken as the zero reference potential. Membrane potential (V_m) is by convention the difference between the potential of the internal and external faces of the membrane ($V_m = V_i - V_o$). In the absence of ongoing electrical activity, this negative potential is termed the resting membrane potential (V_{rest}) (**Figure 3.4**).

We have seen above that in the intracellular and extracellular media the concentration of positive ions is equal to that of negative ions. However, there is a very small excess of positive and negative ions accumulated on each side of the membrane. At rest, for example, a small excess of negative ions is accumulated at the internal side of the membrane whereas a small excess of

TABLE 3.1 Examples of ionic channels

Channels	Voltage-gated	Ligand-gated			Mechanically gated
Opened by	Depolarization Hyperpolarization	Extracellular ligand	Intracellular ligand		Mechanical stimuli
Localization	Plasma membrane	Plasma membrane	Plasma membrane	Organelle membrane	Plasma membrane
Examples	Na^+ channels	nAChR iGluR	G protein-gated channels	IP_3-gated Ca^{2+} channel	Stretch-activated channels
	Ca^{2+} channels	5-HT_3	Ca^{2+}-gated channels	Ca^{2+}-gated Ca^{2+} channel	
	K^+ channels Cationic channels	$GABA_A$ GlyR	CNG channels ATP-gated channels		
Closed by	Inactivation Repolarization	Desensitization Ligand recapture or degradation			Adaptation End of stimulus
Roles	Na^+ and Ca^{2+}-dependent action potentials $[Ca^{2+}]_i$ increase	EPSP IPSP $[Ca^{2+}]_i$ increase	EPSP IPSP Action potential repolarization	$[Ca^{2+}]_i$ increase	Receptor potential

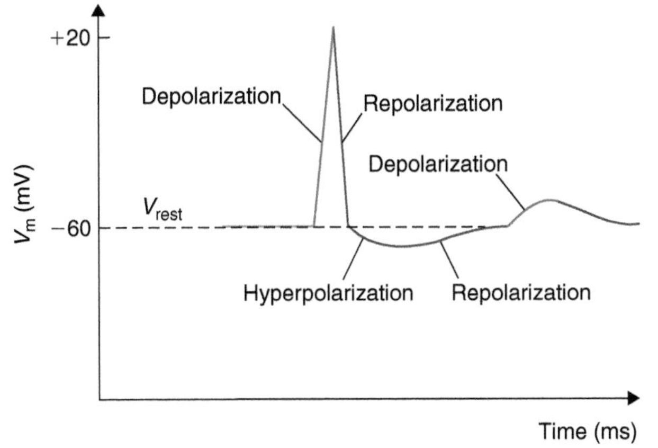

FIGURE 3.4 **Variations of the membrane potential of neurons (V_m).** When the membrane potential is less negative than resting membrane potential (V_{rest}), the membrane is said to be depolarized. In contrast, when the membrane potential is more negative than V_{rest}, the membrane is said to be hyperpolarized. When the membrane varies from a depolarized or hyperpolarized value back to rest, the membrane repolarizes.

positive ions is accumulated at the external side of the membrane (see Section 3.5). This creates a difference of potential between the two faces of the membrane: the external side is more positive than the internal side, which makes $V_m = V_i - V_o$ negative.

What is particular to the membrane of neurons (and of all excitable cells) is that V_m varies (**Figure 3.4**). It can be more negative or hyperpolarized or less negative (depolarized) or even positive (also depolarized, the internal face is positive compared to the external face). At rest, V_m is in the range −80/−50 mV depending on the neuronal type. But when neurons are active, V_m varies between the extreme values −90 mV and +30 mV. Since nerve cells communicate through rapid (milliseconds; ms) or slow (seconds; s) changes in their membrane potential, it is important to understand V_{rest} first.

3.3 CONCENTRATION GRADIENTS AND MEMBRANE POTENTIAL DETERMINE THE DIRECTION OF THE PASSIVE MOVEMENTS OF IONS THROUGH IONIC CHANNELS: THE ELECTROCHEMICAL GRADIENT

To predict the direction of the passive diffusion of ions through an open channel, both the concentration gradient of the ion and the membrane potential have to be known. The *resultant* of these two forces is called the electrochemical gradient. To understand what the electrochemical gradient is for a particular ion, the concentration gradient and the electrical gradient will first be explained separately.

3.3.1 Ions passively diffuse down their concentration gradient

The concentration gradient of a particular ion is the difference of concentration of this ion between the two sides of the plasma membrane. Ions passively move through open channels from the medium where their concentration is high to the medium where their concentration is lower. Suppose that membrane potential is null ($V_m = 0$ mV), there is no difference of potential between the two faces of the membrane, so ions will diffuse according to their concentration gradient only (**Figure 3.5a**). Since the extracellular concentrations of Na^+, Ca^{2+} and Cl^- are higher than the respective intracellular ones, these ions will diffuse passively towards the intracellular medium (when Na^+, Ca^{2+} or Cl^- permeable channels are open) as a result of their concentration gradient. In contrast, K^+ will move from the intracellular medium to the extracellular one (when K^+ permeable channels are open).

The force that makes ions move down their concentration gradient is *constant* for a given ion since it depends on the difference of concentration of this ion, which is itself continuously controlled to a constant value by active

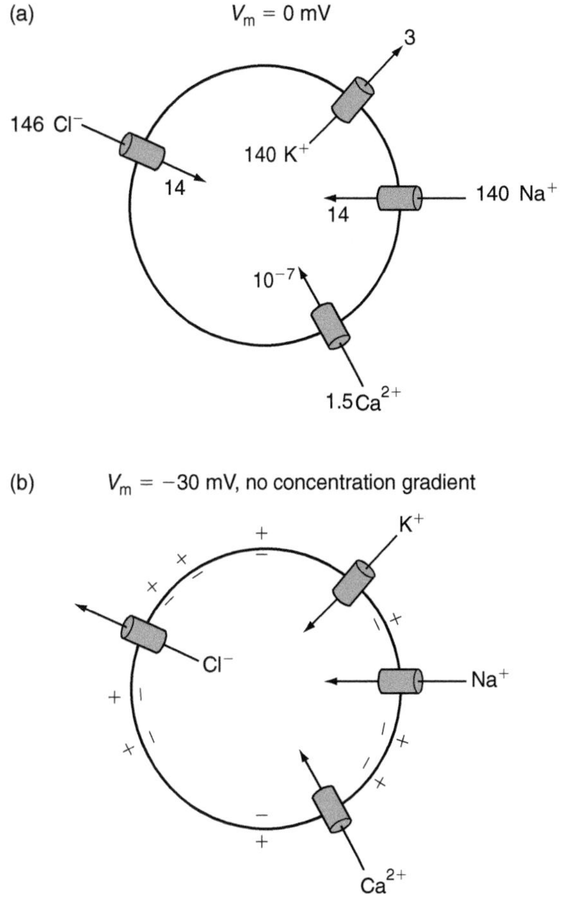

FIGURE 3.5 **Passive diffusion of ions.** Passive diffusion of ions according (**a**) their concentration gradient only, or (**b**) to membrane potential (electrical gradient) only (V_m = −30 mV).

transport (pumps and transporters). However, this is not always true; during intense neuronal activity, concentration of ions may change (K^+ concentration in particular) owing to the small volume of the external medium in physiological conditions. At the microscopic level, this is not true also; intracellular Ca^{2+} concentration, for example, can increase locally by a factor of between 100 and 1000 but stay stable in the entire cytosol. However, these increases of ion concentration do not change the direction of the concentration gradient for this ion since ionic gradients cannot reverse by themselves.

3.3.2 Ions passively diffuse according to membrane potential

Membrane potential is a potential gradient that forces ions to passively move in one direction: positive ions are attracted by the 'negative' side of the membrane and negative ions by the 'positive' one. If we suppose that there is no concentration gradient for any ions (there is the same concentration of each ion in the extracellular and intracellular media), ions will diffuse according to membrane potential only: at a membrane potential $V_m = -30$ mV (**Figure 3.5b**), positively charged ions, the cations Na^+, Ca^{2+} and K^+, will move from the extracellular medium to the intracellular one according to membrane potential. In contrast, anions (Cl^-) will move from the intracellular medium to the extracellular one.

3.3.3 In physiological conditions, ions passively diffuse according to the electrochemical gradient

In physiological conditions, both the concentration gradient and membrane potential determine the direction and amplitude of ion diffusion through an open channel. Since concentration gradient is constant for each ion, the direction and amplitude of diffusion varies with membrane potential. When comparing **Figure 3.5a** and **b** it appears that at a membrane potential of -30 mV, concentration gradient and membrane potential drive Na^+ and Ca^{2+} ions in the same direction, toward the intracellular medium, whereas they drive K^+ and Cl^- in reverse directions. The resultant of these two forces, concentration and potential gradients, is the electrochemical gradient. To know how to express the electrochemical gradient, the equilibrium potential must first be explained.

The equilibrium potential for a given ion, E_{ion}

All systems are moving toward equilibrium. The value of membrane potential where the concentration force that tends to move a particular ion in one direction is exactly balanced by the electrical force that tends to move the same ion in the reverse direction is called the 'equilibrium potential' of the ion (E_{ion}) or the reversal potential of the ion E_{rev}. The equilibrium potential for a particular ion is the value of V_m for which the net flux of this ion (f_{net}) through an open channel is null: when $V_m = E_{ion}$, $f_{net} = 0$ mol s^{-1}.

E_{ion} can be calculated using the Nernst equation (see Appendix 3.3):

$$E_{ion} = (RT/zF)\ln([\text{ion}]_e/[\text{ion}]_i),$$

where R is the constant of an ideal gas (8.314 VCK^{-1}mol^{-1}); T is the absolute temperature in kelvin (273.16 + the temperature in °C); F is the Faraday constant (96 500 C mol^{-1}); z is the valence of the ion; and [ion] is the concentration of the ion in the extracellular (e) or intracellular (i) medium. This gives:

$$E_{ion} = (58/z)\log_{10}([\text{ion}]_e/[\text{ion}]_i), \qquad (1)$$

From the equation and concentrations of **Figure 3.2**, the equilibrium potentials for each ion can be calculated:

$$E_{Na} = (58/1)\log_{10}(140/14) = +58\,\text{mV}$$

$$E_K = (58/1)\log_{10}(3/140) = -97\,\text{mV}$$

$$E_{Ca} = (58/2)\log_{10}(1.5/10^{-4}) = +121\,\text{mV}$$

$$E_{Cl} = (58/-1)\log_{10}(146/14) = -59\,\text{mV}.$$

These equations have the following meanings. If the channels open in a membrane where K^+ channels are the only channels open, the efflux of K^+ ions will hyperpolarize the membrane until $V_m = E_K = -97$ mV, a potential at which the net flux of K^+ is null since K^+ ions have exactly the same tendency to diffuse towards the intracellular medium according to their concentration gradient than to move in the reverse direction according to membrane potential. At that potential, the efflux of K^+ will be exactly compensated by the influx of K^+ and the membrane potential will stay stable at $V_m = E_K$ as long as K^+ channels stay open. Now, if only Na^+ channels are open, the membrane potential will move toward $V_m = +58$ mV, the potential at which the net flux of Na^+ is null. Similarly, when $V_m = E_{Cl} = -59$ mV, Cl^- ions have the same tendency to move down their concentration gradient than to move in the reverse direction according to membrane potential, the net flux of Cl^- is null. In contrast, when V_m is different from E_{Cl}, the net flux of Cl^- is not null. This holds true for all the other ions: when V_m is different from E_{ion} there is a net flux of this ion.

The electrochemical gradient

We have seen that when $V_m = E_{ion}$ (i.e. $V_m - E_{ion} = 0$), there is no diffusion of this particular ion ($f_{net} = 0$). In contrast, when V_m is different from E_{ion} there is a passive diffusion of this ion through an open channel. The difference ($V_m - E_{ion}$) is called the electrochemical gradient. It is the force that makes the ions move through an open channel.

3.4 THE PASSIVE DIFFUSION OF IONS THROUGH AN OPEN CHANNEL CREATES A CURRENT

To know the direction of passive diffusion of a particular ion and how many of these ions diffuse per unit of time, the direction and intensity of the net flux of ions (number of moles per second) through an open channel have to be measured. Usually the net flux (f_{net}) is not measured; the electrical counterpart of this net flux, the ionic current, is measured instead.

Passive diffusion of ions through an open channel is a movement of charges through a resistance (resistance here is a measure of the difficulty of ions moving through the channel pore). Movement of charges through a resistance is a current. Through a single channel the current is called 'single-channel current' or 'unitary current', i_{ion}. The relation between f_{net} and i_{ion} is:

$$i_{ion} = f_{net}\, zF$$

The amplitude of i_{ion} is expressed in amperes (A) which are coulombs per seconds ($C.s^{-1}$). F is the Faraday constant (96 500 $C.moles^{-1}$); z is the valence of the ion (+1 for Na^+ and K^+, −1 for Cl^-, +2 for Ca^{2+}); and f_{net} is the net flux of the ion in $mol.s^{-1}$.

In general, currents are expressed following Ohm's law: $U = RI$, where I is the current through a resistance R and U is the difference of potential between the two ends of the resistance. For currents carried by ions (and not by electrons as in copper wires), I is called i_{ion}, the current that passes through the resistance of the channel pore which has a resistance R (called r_{ion}). But what is U in biological systems? U is the force that makes ions move in a particular direction; it is the electrochemical gradient for the considered ion and is also called the driving force: $U = V_m - E_{ion}$ (**Figure 3.6**).

Unitary current, i_{ion}

According to Ohm's Law, the current i_{ion} through a single channel is derived from

$$(V_m - E_{ion}) = r_{ion} \cdot i_{ion}$$

So:

$$i_{ion} = (1/r_{ion})(V_m - E_{ion}) = \gamma_{ion}(V_m - E_{ion})$$

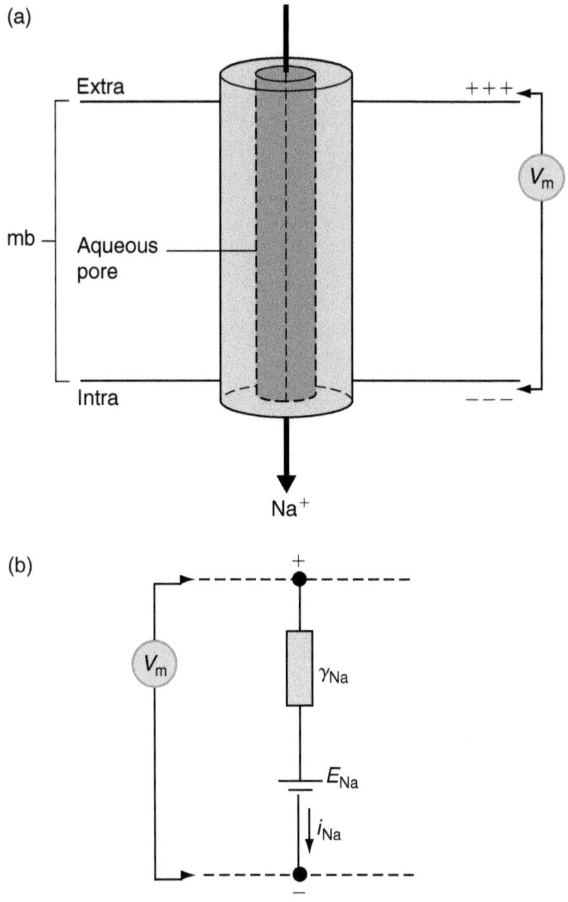

FIGURE 3.6 **The Na^+ channel.** (a) Schematic, and (b) its electrical equivalent.

γ_{ion} is the reciprocal of resistance; it is called the *conductance* of the channel, or unitary conductance (**Figure 3.6**). It is a measure of the ease of flow of ions (flow of current) through the channel pore. Whereas resistance is expressed in ohms (Ω), conductance is expressed in siemens (S). By convention i_{ion} is negative when it represents an inward flux of positive charges (cations) and i_{ion} is positive when it represents an outward flux of positive charges (**Figure 3.5c**). It is generally of the order of picoamperes (1 pA = 10^{-12} A). At physiological concentrations, γ_{ion} varies between 10 and 150 pico-siemens (pS), according to the channel type.

Total current, I_{ion}

In physiological conditions, several channels of the same type are open at the same time in the neuronal membrane. Suppose that only one type of channel is open in the membrane, for example Na^+ channels, the total current I_{Na} that crosses the membrane at time t is the sum of the unitary currents i_{Na} at time t:

$$I_{Na} = Np_o i_{Na}$$

where N is the number of Na^+ channels present in the membrane; p_o is the probability of Na^+ channels being open at time t (Np_o is therefore the number of open Na^+ channels in the membrane at time t); and i_{Na} is the unitary Na^+ current. More generally:

$$I_{ion} = Np_o i_{ion}$$

By analogy, the total conductance of the membrane for a particular ion is:

$$G_{ion} = Np_o \gamma_{ion}$$

and from $i_{ion} = \gamma_{ion}(V_m - E_{ion})$ above:

$$I_{ion} = G_{ion}(V_m - E_{ion})$$

I_{ion} and i_{ion} can be measured experimentally. The latter is the current measured from a patch of membrane where only one channel of a particular type is present. I_{ion} is the current measured from a whole cell membrane where N channels of the same type are present.

Roles of ionic currents

Ionic currents have two main functions:

- Ionic currents change the membrane potential: either they depolarize the membrane or repolarize it or hyperpolarize it, depending on the charge carrier. These terms are in reference to resting potential (**Figure 3.4**). Changes of membrane potential are signals. A depolarization can be an action potential (see Chapters 4 and 5) or a postsynaptic excitatory potential (EPSP; see Chapters 8 and 10). A hyperpolarization can be a postsynaptic inhibitory potential (IPSP; see Chapter 9). These changes of membrane potential are essential to neuronal communication.
- Ionic currents increase the concentration of a particular ion in the intracellular medium. Calcium current, for example, is always inward. It transiently and locally increases the intracellular concentration of Ca^{2+} ions and contributes to the triggering of Ca^{2+}-dependent events such as secretion or contraction.

3.5 A PARTICULAR MEMBRANE POTENTIAL, THE RESTING MEMBRANE POTENTIAL V_{rest}

In the absence of ongoing electrical activity (when the neuron is not excited or inhibited by the activation of its afferents), its membrane potential is termed the resting membrane potential (V_{rest}). For some neurons, V_{rest} is stable (silent neurons), for others it is not (pacemaker neurons, for example). In this section, we will consider stable V_{rest} only. To understand unstable V_{rest} many different channels must be known that are explained later in the book (see Chapter 14).

3.5.1 When most of the channels open at rest are K^+ channels, V_{rest} is close to E_K

It was Julius Bernstein (1902) who pioneered the theory of V_{rest} as due to selective permeability of the membrane to one ionic species only and that nerve excitation developed when such selectivity was transiently lost. According to this theory, under resting conditions, the cell membrane permeability is minimal to Na^+, Cl^- and Ca^{2+} while it is high to K^+. What is the membrane potential of a membrane permeable to K^+ ions only? This condition can be tested experimentally by measuring ionic fluxes with radioactive tracers through a plasma membrane where K^+ channels are the only open channels. K^+ moves outwards following its concentration gradient (the intracellular concentration of K^+ is around 50 times higher than the extracellular one): positive charges are thus subtracted from the intracellular medium and there is an accumulation of negative charges at the intracellular side of the membrane and positive charges at the external side of the membrane. These positive charges will oppose further outward movements of K^+ until an equilibrium is reached when the concentration gradient for K^+ cancels the drive exerted by the electrical gradient. This is by definition the *equilibrium potential* E_K. Hence, at $V_m = E_K$, although K^+ keeps moving in and out of the cell, there is no net change in its concentration across the membrane (**Figure 3.7**). In a physiological situation, the exact value of E_K is unknown since the exact $[K^+]_i$ is unknown. When $V_{rest} = -80/-70$ mV, though it seems close to E_K it may not be equal to E_K.

A way to test whether $V_{rest} = E_K$ is the following. Inspection of the Nernst equation applied to K^+ indicates that a 10-fold change in the concentration ratio should alter the membrane potential of a neuron by 58 mV. This relation can be tested in experiments in which the extracellular concentration of this ion is altered and the resulting membrane potential measured with a sharp or patch microelectrode. A semilog plot of the extracellular K^+ concentration (abscissa) against the membrane potential (ordinate) should thus have a slope of 58 mV per 10-fold change in K^+ (**Figure 3.8**); this condition is rarely encountered in neurons but it seems to be more common for glial cells (which sometimes are termed K^+ electrodes because their membrane potential is linearly dependent on K^+). In the case of neurons, non-linearity of this plot is frequently seen, particularly at low levels of extracellular K^+. These observations confirm that K^+ is a very important ion for setting the value of neuronal V_{rest} but that other ions must also play a significant role.

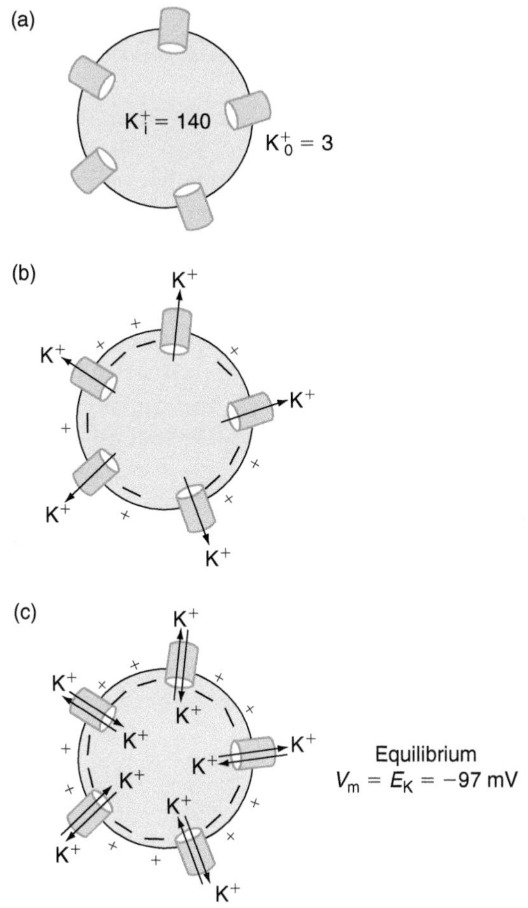

FIGURE 3.7 **Establishment of V_{rest} in a cell where most of the channels open are K^+ channels.** Suppose that at $t = 0$ and cell potential $V_m = 0$ mV (**a**), K^+ ions will move outwards due to their concentration gradient (**b**). Loss of intracellular K^+ induces a negative potential (V_m) as $V_m = E_K$ (**c**).

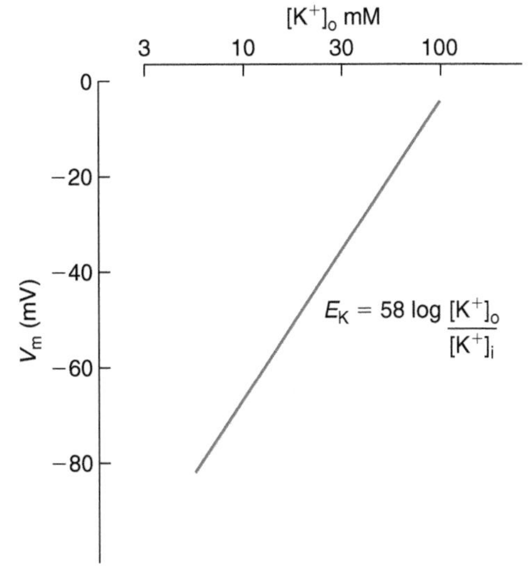

FIGURE 3.8 **Theoretical diagram of E_K versus the external concentration of K^+ ions ($[K^+]_o$).** $E_K = (RT/zF)\ 2.3 \times \log([K^+]_o/[K^+]_i)$.

3.5.2 In central neurons, K^+, Cl^- and Na^+ ion movements participate in resting membrane potential and V_{rest} is different from E_K: the Goldman–Hodgkin–Katz equation

Aside from K^+, which ions play a role in V_{rest}? Since the intracellular concentration of Na^+ is not negligible, this implies that this ionic species can accumulate inside the cytoplasm, presumably because of its rather positive E_{Na} (+58 mV) versus a very negative V_{rest} creates an electrochemical gradient extremely favorable to Na^+ entry. Equally, the asymmetric distribution of Cl^- suggests its possible role in determining V_{rest}. In order to take into account various ionic species it is useful to introduce what is commonly called the *Goldman–Hodgkin–Katz equation* (GHK), derived from the Nernst equation and named after the three physiologists responsible for its derivation:

$$V_{rest} = 58 \log \times \frac{p_K[K^+]_o + p_{Na}[Na^+]_o + p_{Cl}[Cl^-]_i}{p_K[K^+]_i + p_{Na}[Na^+]_i + p_{Cl}[Cl^-]_o} \quad (2)$$

where p is the permeability coefficient (cm s^{-1}) for each ionic species. The relative contribution of each ion species to the resting voltage is weighted by that ion's permeability.

Note that if the resting permeability to Na^+ and Cl^- is very low, the GHK equation closely resembles the Nernst equation for K^+.

In applying the GHK equation to nerve cells, the following assumptions must be made:

- The voltage gradient across the membrane is uniform in the sense that it changes linearly within the membrane. This assumption has led to the GHK equation being called the *constant field equation*.
- The overall net current flow across the membrane is zero as the currents generated by individual ionic species are balanced out.
- The membrane is in a steady state since there is no time-dependent change in ionic flux or channel density. This is obviously not applicable to non-steady state conditions of rapidly changing membrane potential as produced when a nerve cell fires action potentials.
- Any role of active transport mechanisms is ignored.
- The ionic species are monovalent cations or anions which do not interact among themselves or with water molecules. The first point does not hold true if there is a measurable permeability to divalent cations such as Ca^{2+}. Furthermore, it has been reported that ions can interact among themselves within the same channel.

- The role of membrane surface charges is ignored. This is a relatively major limitation because the cell membrane contains negative charges on its inner and outer layers (amino acid residues of membrane proteins which are typically negatively charged). The electric field generated by these charges is able to influence the kinetic properties of ionic channels (gating, activation and inactivation). Adding divalent cations such as Ca^{2+} or Mg^{2+} leads to screening of these charges and consequent changes in channel properties.
- The mobility of each ionic species and its diffusion coefficient (D) within the membrane of thickness (δ) is constant.
- The ions do not bind to specific sites in the membrane and their concentration (C) can be expressed by a linear partition coefficient ($\beta = C_{membrane}/C_{solution}$). However, there is evidence that ions can bind to sites inside channels and influence channel kinetics.
- The ionic activities (a) can be replaced by their concentrations.

3.6 A SIMPLE EQUIVALENT ELECTRICAL CIRCUIT FOR THE MEMBRANE AT REST

Since the plasma membrane does not allow the passage of all the ions at all times, it can be equated to an insulator separating two electrically conductive media (intracellular and extracellular electrolytes): it thus plays the role of a dielectric in a *capacitor* and it can be assigned an average capacity (C_m) value of 1 μF cm^{-2}.

In **Figure 3.9b**, instead of three parallel current sources for K^+, Na^+ and Cl^-, we have lumped them together into only one source with driving (electromotive) force E equal to V_{rest} and an inward conductance g_m equal to the sum of the specific ionic (channel) conductances $g_K + g_{Na} + g_{Cl}$. One may consider, instead of the absolute value of membrane potential, only its deviation from V_{rest}. In this case, the equivalent electromotive force becomes equal to zero and the equivalent scheme of the cell membrane simplifies to an *RC*-circuit (**Figure 3.9c**). If one includes more channel types, then the notion of resting current still holds true. The equivalent scheme of **Figure 3.9c** is

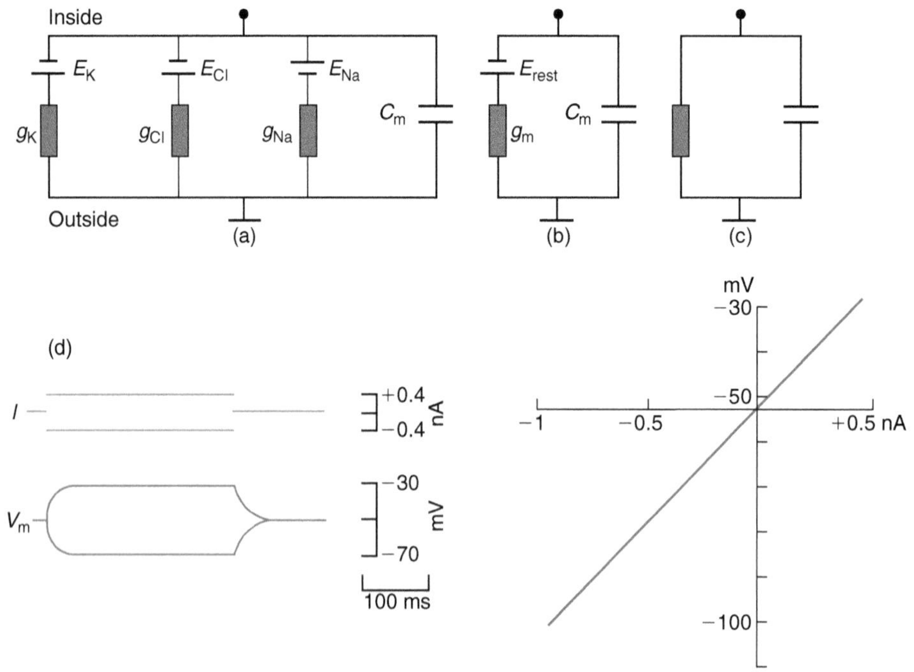

FIGURE 3.9 **Simplified equivalent scheme to account for membrane electrical characteristics near the resting potential and ohmic behavior of the membrane potential around the resting potential.** (**a**) Three main ionic current sources. Note: E_K and E_{Cl} are negative while E_{Na} is positive. (**b**) An equivalent current source for the resting potential. (**c**) Electrical scheme for below-threshold potential changes (passive de- and hyperpolarizations) relative to the resting potential. Battery symbols indicate electromotive forces, boxes represent conductances and parallel plates indicate membrane capacitors. (**d**) From top to bottom: Time-dependent responses to ±0.4 nA current injected for 300 ms; left: upper traces, current I; middle traces, membrane potential changes V_m; right: membrane potential at the end of the current pulse (i.e. at 300 ms) plotted against current intensity. From Adams PR, Brown DA, Constanti A (1982) M-currents and other potassium currents in bullfrog sympathetic neurones. *J. Physiol. (Lond.)* **330**, 537-572, with permission.

applicable only to depolarizations and hyperpolarizations characterized by linear (ohmic) current–voltage relations (**Figure 3.9d**). In standard excitable cells it means that these potential changes from V_{rest} are not activating voltage-gated currents; e.g. they are below the threshold for spike generation.

3.7 HOW TO EXPERIMENTALLY CHANGE V_{rest}

3.7.1 How to experimentally depolarize a neuronal membrane

The aim of the experiment is to lower the difference of potential between the two faces of the membrane and even to reverse it. There are at least three main ways of depolarizing a membrane: (a) by increasing the K^+ concentration in the external medium, (b) by applying a drug that opens cationic channels or (c) by injecting a positive current inside the neuron (**Figure 3.10**).

An *in vitro* preparation such as a neuronal culture or a brain slice is bathed in an extracellular solution of an ionic composition close to that of the extracellular medium. A recording electrode is implanted in a neuronal cell body. At rest the membrane potential is close to −70 mV. When the extracellular solution is changed to one containing a higher concentration of K^+ ions (30 mM instead of 3 mM) and a lower concentration of Na^+ ions (113 mM instead of 140 mM) to keep constant the extracellular concentration of positive ions, a depolarization is recorded. Since at rest most of the channels open are K^+ channels, V_m tends toward E_K which is now equal to −38 mV (E_K = 58 log 30/140) instead of −97 mV. The membrane depolarizes because E_K is more depolarized than V_{rest}.

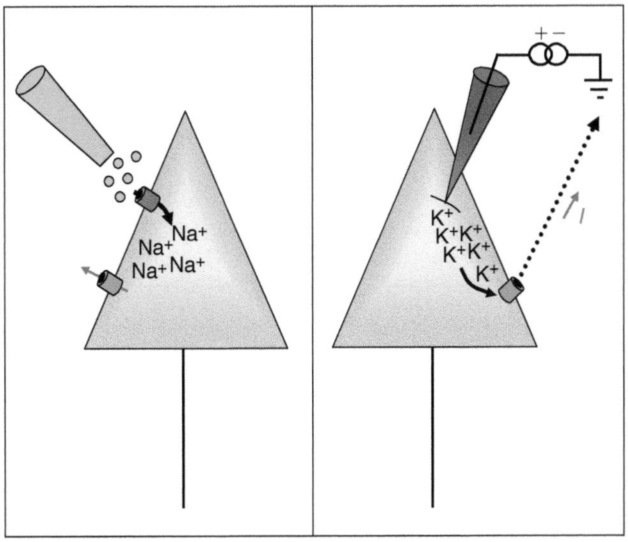

FIGURE 3.10 **V_m is depolarized by applying:** (**a**) a drug in the extracellular medium that opens Na^+ channels (veratridine) or (**b**) a positive current via an intracellular electrode.

In the same preparation bathed in control extracellular medium, veratridine is applied by pressure via a pipette located close to the recorded neuron. Veratridine induces a depolarization of the recorded membrane (**Figure 3.10a**). As this drug opens Na^+ channels, Na^+ ions enter the cell and create an inward current of positive charges. The electrical circuit is closed because + charges can go out of the cell via the K^+ channels open at rest. Since Na^+ channels now represent the major population of open channels, V_m tends toward E_{Na} (+58 mV) and the membrane depolarizes as long as veratridine is applied.

If now a positive current is applied through the recording pipette which contains a KCl solution, K^+ ions are expelled from the pipette. They create a current of positive charges that depolarizes the membrane (**Figure 3.10b**). The electrical circuit is closed because K^+ ions can go through the membrane via the K^+ channels open at rest. A depolarizing current pulse is a positive current injected via an intracellular electrode. One part of the stimulating current is used to load the capacity C_m of the neuronal membrane and the other part passes through the ion channels:

$$I_{stimulus} = C_m \, dV/dt + I_{ion}$$

$$dV/dt = [-I_{ion} + I_{stimulus}]/C_m$$

A positive stimulating current applied at the inside of a neuron (cell body, dendrite, axon) will cause a depolarization of V_m according to the above equation. Inversely, a negative current will hyperpolarize the membrane (see below). Once the membrane capacity is loaded (steady state) the injected current equals the current passing through the membrane via open channels.

In the case of a silver electrode inside the pipette, as a coat of AgCl is deposited on the silver metal, it provides a store of Ag^+ and Cl^- ions and mediates between electronic conduction in the metal ($Ag^+ + e^- \leftrightarrows Ag$) and ionic current due to Cl^- exchanges between precipitate (AgCl) and solution.

3.7.2 How to experimentally hyperpolarize a neuronal membrane

The aim of the experiment is to increase the difference of potential between the two faces of the membrane. There are at least two main ways of hyperpolarizing a membrane: (a) by applying a drug that opens K^+ channels or (b) by injecting a negative current inside the neuron.

An *in vitro* preparation such as a neuronal culture or a brain slice is bathed in a physiological saline of an ionic

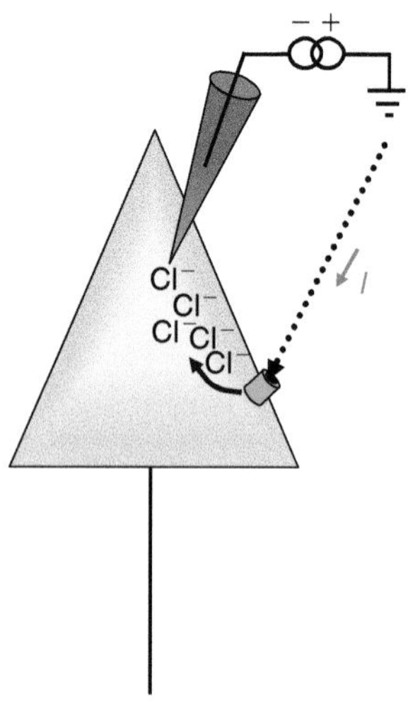

FIGURE 3.11 **V_m is hyperpolarized by applying a negative current via an intracellular electrode.**

composition close to that of the extracellular medium. A recording electrode is implanted in a neuronal cell body. A peptide that opens K^+ channels is applied by pressure via a pipette located close to the recorded neuron. This induces a hyperpolarization of the membrane, due to the outward flux of K^+ ions. As this drug opens K^+ channels (via metabotropic receptors such as $GABA_B$ receptors, see Chapter 11), K^+ ions exit the cell and create an outward current of positive charges. The electrical circuit is closed because ions can enter the membrane via the channels open at rest. As K^+ channels now represent the major population of open channels, V_m tends toward E_K (−97 mV) and the membrane hyperpolarizes as long as the peptide is applied.

If now a negative current is applied through the recording pipette which contains a KCl solution, Cl^- ions are expelled from the pipette. They hyperpolarize the membrane (**Figure 3.11**). The electrical circuit is closed because ions can go through the membrane via the channels open at rest.

3.8 SUMMARY

Passage of ions through the membrane is a regulated process and the flow of ions across the neuronal plasma membrane is not a simple and anarchic diffusion through a lipid bilayer. Instead, it is restricted through transmembrane proteins whose opening (channel proteins) or activation (pumps or transporters) are tightly controlled by different factors.

Where and how do ions passively cross the plasma membrane? (See also Appendices 3.2 and 3.3)

- Ions move passively across the plasma membrane through ionic channels that are specifically permeable to one or several ions of the same sign. They move down their electrochemical gradient. This passive movement of charges is a current that can be recorded. Through a single channel it is a unitary current i_{ion}, and through N channels it is a macroscopic current or total current I_{ion}.
- The type of ion that moves through an open channel (ionic selectivity of the channel pore) is determined by the structure of the channel itself. This ionic selectivity gives the name to the channel. For example, an Na^+ channel is permeable to Na^+ ions; a cationic channel is permeable to cations: Na^+, K^+ and sometimes also Ca^{2+}.
- The *direction* of ion diffusion through a single channel depends on the electrochemical gradient or driving force for this particular ion ($V_m - E_{ion}$).
- The *number* of charges that diffuse through an open channel per unit of time (i_{ion}) depends on the electrochemical gradient ($V_m - E_{ion}$) but also on how easily ions move through the pore of the channel (expressed as the conductance γ_{ion} of the channel): $i_{ion} = \gamma_{ion}(V_m - E_{ion})$.

How and where do ions actively cross the plasma membrane and thus compensate for the passive movements? (see also Appendix 3.1)

Active movements of Na^+, K^+, Ca^{2+} or Cl^- ions across the membrane occur through pumps or transporters. Pumps obtain energy from the hydrolysis of ATP, whereas transporters use the energy of an ionic gradient, for example the sodium driving force. These transports require energy since they operate against the electrochemical gradient of the transported ions or molecules. They maintain ionic concentrations at constant values in the extracellular and intracellular compartments despite the continuous passive movements of ions across the membrane.

What are the roles of electrochemical gradients and passive movements of ions?

The electrochemical gradients of ions are a reserve of energy: they allow the existence of ionic currents and drive some active transports. The large asymmetries in ion distribution imply a dynamic state through which cell-to-cell signaling is made possible. Ionic currents have two main functions: (i) they evoke transient changes of membrane potential which are electrical signals of the neuron (action potentials or postsynaptic potentials or sensory potentials) essential to neuronal communication; and (ii) they locally increase the concentration of a particular ion in the intracellular medium, for example Ca^{2+} ions, and thus trigger intracellular Ca^{2+}-dependent events such as secretion or contraction.

APPENDIX 3.1 THE ACTIVE TRANSPORT OF IONS BY PUMPS AND TRANSPORTERS MAINTAIN THE UNEQUAL DISTRIBUTION OF IONS

Passive movements of Na^+, K^+, Ca^{2+} or Cl^- ions across the membrane would finally cause concentration changes in the extracellular and intracellular compartments if they were not constantly regulated during the entire life of the neuron by transport of ions in the reverse direction, against passive diffusion; i.e. against electrochemical gradients. This type of transport is described as active since it requires energy in order to oppose the electrochemical gradient of the transported ions. Ions cross the membrane *actively* through specialized proteins known as pumps or transporters. Pumps obtain energy from the hydrolysis of ATP, whereas transporters use the energy of an ionic gradient, for example the sodium driving force (**Figure A3.1**). The energy is needed for the conformational changes that allow the pump or the transporter to change its affinity for the ion transported during the transport: the binding site(s) must have a high affinity when facing the medium where the transported ion is at a low concentration (in order to bind it) and must change to low affinity when facing the medium where the concentration of the transported ion is high in order to release it.

A3.1.1 Pumps are ATPases that actively transport ions

Pumps have ATPase activity (they hydrolyze ATP). This ATPase activity is generally the easiest way of identifying them. Pumps are membrane-embedded enzymes that couple the hydrolysis of ATP to active translocation

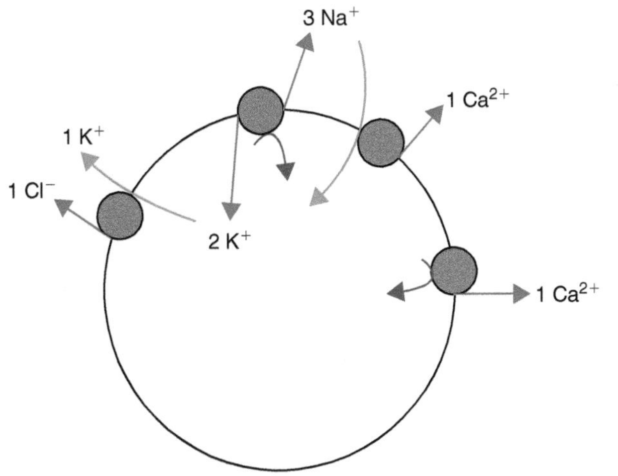

FIGURE A3.1
Transport against the electrochemical gradient of the ion
Transport along the electrochemical gradient of the ion
ATP hydrolysis

of ions across the membrane. The central issue of ion motive ATPases is to couple the hydrolysis of ATP (and their auto-phosphorylation) to the translocation of ions.

The Na/K-ATPase pump

Na/K-ATPases maintain the unequal distribution of Na^+ and K^+ ions across the membrane. Na^+ and K^+ ions cross the membrane through different Na^+ and K^+ permeable channels (voltage-sensitive Na^+ and K^+ channels plus receptor channels). This pump operates continuously at a rhythm of 100 ions per second (compared with 10^6–10^8 ions per second for a channel), adjusting its activity to the electrical activity of the neuron. It actively transports three Na^+ ions towards the extracellular space for each two K^+ ions that it carries into the cell.

The energy of ATP hydrolysis is needed for the conformational changes (they are energy dependent) that allow the pump to change its affinity for the ion transported, whether the binding sites are accessible from the cytoplasmic or the extracellular sides. For example, when the Na^+ binding sites are accessible from the cytoplasm, the protein is in a conformation with a high affinity ($K_A = 1$ mM) for intracellular Na^+ ions, and so Na^+ ions bind to the three sites. In contrast, when the three Na^+ have been translocated to the extracellular side, the protein is in a conformation with a low affinity for Na^+ ions so that the three Na^+ are released in the extracellular space.

The steady unequal distribution of Na^+ and K^+ ions constitutes a reserve of energy for a cell. The neuron uses this energy to produce electric signals (action potentials, synaptic potentials) as well as to actively transport other molecules.

The Ca-ATPase pump

The function of Ca-ATPases is to maintain (with the Na–Ca transporter) the intracellular Ca^{2+} concentration at very low levels by active expulsion of Ca^{2+}. In fact, the intracellular Ca^{2+} concentration is 10 000 times lower than the extracellular concentration despite the inflow of Ca^{2+} (through receptor channels and voltage-gated Ca^{2+} channels) and the intracellular release of Ca^{2+} from intracellular stores. Maintaining a low intracellular Ca^{2+} concentration is critical since Ca^{2+} ions control several intracellular reactions and are toxic at a high concentration. Ca-ATPases are located in the plasma membrane and in the membrane of the reticulum. The former extrude Ca^{2+} from the cytoplasm whereas the latter sequester Ca^{2+} inside the reticulum (see also **Figure 7.8**).

A3.1.2 Transporters use the energy stored in the transmembrane electrochemical gradient of Na^+, K^+, H^+ or other ions

When transporters carry Na^+, K^+ or H^+ ions (along their electrochemical gradient) in the same direction as the transported ion or molecule, the process is called *symport*. When the movements occur in opposite directions, the process is called *antiport*. We shall study only transporters implicated in the electrical or secretory activity of neurons.

The Na–Ca transporter

This transporter uses the energy of the Na^+ gradient to actively carry Ca^{2+} ions towards the extracellular environment. It is situated in the neuronal plasma membrane and operates in synergy with the Ca-ATPase and with transport mechanisms of the smooth sarcoplasmic reticulum to maintain the intracellular Ca^{2+} concentration at a very low level (see Section 7.2.4).

The K–Cl transporter KCC

Adult mammalian central neurons maintain a low intracellular Cl^- concentration. Cl^- extrusion is achieved by K^+–Cl^- cotransporters (KCC) fueled by K^+. As all transporters, it does not directly consume ATP but derives its energy from ionic gradients, here the K^+ gradient generated by the Na/K/ATPase.

Neurotransmitter transporters

Inactivation of most neurotransmitters present in the synaptic cleft is achieved by rapid reuptake into the presynaptic neural element and astrocytic glial cells. This is performed by specific neurotransmitter transporters, transmembrane proteins that couple neurotransmitter transport to the movement of ions down their concentration gradient. Certain neurotransmitter precursors are also taken up by this type of active transport (glutamine and choline, for instance). Once in the cytoplasm, neurotransmitters are concentrated inside synaptic vesicles by distinct transport systems driven by the H^+ concentration gradient (maintained by the vesicular H^+-ATPase) (see Section 7.4).

APPENDIX 3.2 THE PASSIVE DIFFUSION OF IONS THROUGH AN OPEN CHANNEL

It has been stated above that a channel is said to be in a closed state (C) when its ionic pore does not allow ions to pass. In contrast, when the channel is said to be in the open state (O), ions can diffuse through the ionic pore.

$$C \rightleftharpoons O$$

This diffusion of ions through an open channel is a passive transport since it does not require energy expenditure.

- Which type(s) of ions will move through a given open channel: cations, anions?
- In which direction will these ions move, from the external medium to the cytosol or the reverse?
- How many of these ions will move per unit of time?

The structure of the channel pore determines the type of ion(s) that diffuse passively through the channel

The pores of ion channels select their permeant ions. The structural basis for ion channel selectivity has been studied in a bacterial K^+ channel called the KcsA channel (it is a voltage-independent K^+ channel). All K^+ channels show a selectivity sequence $K^+ = Rb^+ > Cs^+$, whereas permeability for the smallest alkali metal ions Na^+ and Li^+ is extremely low. Potassium is at least 10 000 times more permeant than Na^+, a feature that is essential to the function of K^+ channels. Each subunit of the KcsA channel consists of an N-terminal cytoplasmic domain, followed by two transmembrane helices and a C-terminal globular domain in the cytoplasm. The P loop (P for pore) situated between transmembrane helices 1 and 2 is the region primarily responsible for ion selectivity.

The KcsA channel is overexpressed in bacteria and the three-dimensional structure of its pore investigated by the use of X-ray crystallography. The KcsA channel is a tetramer with fourfold symmetry around a central pore (**Figure A3.2**). The pore is constructed of an inverted teepee with the extracellular side corresponding to the base of the teepee. The overall length of the pore is 4.5 nm and its diameter varies along its distance. From inside the cell the pore begins as a water-filled tunnel of 1.8 nm length (inner pore) surrounded by predominantly nonpolar side chains pointing to the pore axis. The diameter of this region is sufficiently wide to allow the passage of fully hydrated cations. This long entry way then opens

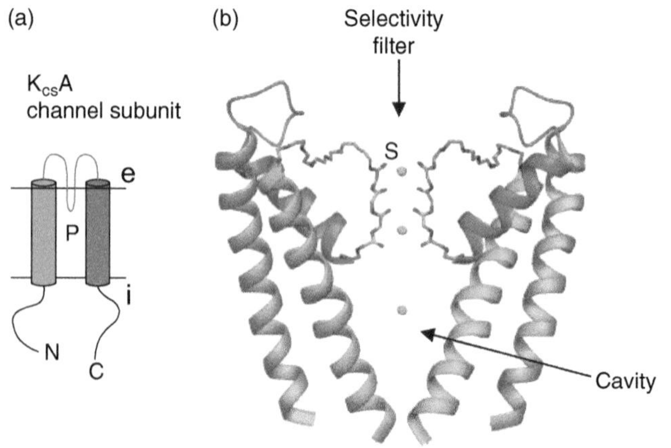

FIGURE A3.2 (a) Membrane topology of the KcsA channel subunit showing the two transmembrane segments and the pore loop (P). (b) Two diametrically opposed subunits of KcsA are depicted to show the cavity in the membrane with 3 ions in the cavity, S sites (ion-binding sites) of the selectivity filter (shown in sticks). Adapted from Noskov SY, Roux B (2006) Ion selectivity in potassium channels. *Biophys. Chem.* **124**, *279–291, with permission.*

to a wider water-filled cavity (1 nm across). Beyond this vestibule is the 1.2 nm long selectivity filter. After this, the pore opens widely to the extracellular side of the membrane.

What are the respective roles of the parts of the pore?

The pore comprises a wide, non-polar aqueous cavity on the intracellular side, leading up, on the extracellular side, to a narrow pore that is 1.2 nm long and lined exclusively by main chain carbonyl oxygens formed by the residues corresponding to the signature sequence TTVGYG common to all K^+ channels.

Electrostatic calculations show that when an ion is moved along a narrow pore through a membrane it must cross an energy barrier that is maximal at the membrane center. A K^+ ion can move throughout the inner pore and cavity and still remain mostly hydrated, owing to the large diameter of these regions. The role of the inner pore and the cavity is to lower the electrostatic barrier. The cavity overcomes the electrostatic destabilization from the low dielectric bilayer by simply surrounding an ion with polarizable water. Another feature that contributes to the stabilization of the cation at the bilayer center are the four pore helices which point directly at the center of the cavity. The amino to carboxyl orientation of these helices imposes a negative electrostatic (cation attractive) potential via the helix dipole effect. These two mechanisms (large aqueous cavity and oriented helices) serve to stabilize a cation in the hydrophobic membrane interior.

The selectivity filter that follows, in contrast, is lined exclusively by polar main-chain atoms. They create a stack of sequential carbonyl oxygen rings which provide multiple closely spaced binding sites (S) for cations separated by 0.3–0.4 nm. This selectivity filter attracts K^+ ions and allows them to move.

Why are cations permeant and not anions?

As might have been anticipated for a cation channel, both the intracellular and extracellular entryways are negatively charged by acidic amino acids that raise the local concentration of cations while lowering the concentration of anions.

Why are K^+ ions at least 1000 times more permeant than Na^+ ions?

The selectivity filter is so narrow that a K^+ ion evidently dehydrates to enter into it and only a single K^+ ion can pass through at one time. To compensate for the energy cost of dehydration, the carbonyl oxygen atoms come in very close contact with the ion and act like surrogate water – they substitute for the hydration waters of K^+. This filter is too large to accommodate an Na^+ ion with its smaller radius (main chain oxygens are spatially inflexible and their relative distances to the center of the pore cannot readily be changed). It is proposed that a K^+ ion fits in the filter so precisely that the energetic costs and gains are well balanced.

What drives K^+ ions to move on?

K^+ ions bind simultaneously at two binding sites 0.75 nm apart near the entry and exit point of the selectivity filter. Binding at adjacent sites may provide the repulsive force for ion flow through the selectivity filter: two K^+ ions at close proximity in the selectivity filter repel each other. The repulsion overcomes the strong interaction between ion and protein and allows rapid conduction in the setting of high selectivity. This leads to a rate of diffusion of around 10^8 ions per second.

APPENDIX 3.3 THE NERNST EQUATION

The material in this appendix is adapted from Katz B (ed.) (1966) *Nerve, Muscle and Synapse* (New York: McGraw-Hill). When $V_m = E_{ion}$, a particular ion has an equal tendency to diffuse in one direction according to its concentration gradient as to move in the reverse direction according to membrane potential. The net flux of this ion is null, so the current carried by this ion is null. $V_m = E_{ion}$ means that:

$$\text{osmotic work}(W_o) = \text{electrical work}(W_e) \qquad \text{(a)}$$

The osmotic work required to move one mole of a particular ion from a compartment where its concentration is low to a compartment when its concentration is high is equal to the electrical work needed to move one mole of this ion against the membrane potential in the opposite direction. Here, active diffusion of ions is considered instead of passive diffusion. The electrical work required to move 1 mole of an ion against a potential difference E_{ion} is:

$$W_e = zFE_{ion}, \qquad \text{(b)}$$

where z is the valence of the transported ion, equal to +1 for monovalent cations such as Na^+ or K^+, to −1 for monovalent anions such as Cl^-, and to +2 for divalent cations such as Ca^{2+}. F is the Faraday constant. F for hydrogen is the charge of one hydrogen atom: $F = Ne$. Here N is the Avogadro number, which is $6.022 \times 10^{23}\ mol^{-1}$ (one mole of hydrogen atoms contains 6×10^{23} protons and the same number of electrons), and e is the elementary charge of a proton, which is 1.602×10^{-19} coulombs (C). So $F = 96\,500\ C\ mol^{-1}$. Therefore zF with $z = 1$ is the charge of 1 mole of protons or 1 mole of monovalent cations (Na^+, K^+). The charge of one mole of monovalent anions (Cl^-) is $-F$ ($z = -1$); the charge of 1 mole of divalent cations (Ca^{2+}) is $2F$ ($z = 2$); etc.

The osmotic work required to move 1 mole of ions from a compartment where its concentration is low to a compartment where the concentration is high can be compared to the work done in compressing 1 g equivalent of an ideal gas. The gas is contained in a cylinder with a movable piston. Mechanical work to move the piston is W, calculated from force times distance of displacement of the piston (δl). The force exerted is equal to the pressure p of the gas multiplied by the surface area S of the piston. So the work δW done to displace the piston is $pS\ \delta l$, which equals $p\ \delta v$. Therefore the work done in compressing a gas from a volume v_1 to a volume v_2 is:

$$W = \int_{v2}^{v1} p\,dv. \qquad \text{(c)}$$

The gas law tells us that $pv = RT$ (hence $p = RT/v$), with R the constant of an ideal gas ($R = 8.314$ V C K^{-1} mol^{-1}) and T is the absolute temperature.

Equation (c) can be changed to:

$$W = RT\int_{v2}^{v1} (1/v)\,dv = RT(\ln v_1 - \ln v_2) \qquad \text{(d)}$$
$$= RT\ln(v_1/v_2).$$

By analogy the osmotic work is:

$$W_o = RT\ln([\text{ion}]_e/[\text{ion}]_i) \qquad \text{(e)}$$

From equation (a), $W_o = -W_e$, so from equations (b) and (e) the Nernst equation is obtained:

$$RT\ln([\text{ion}]_e/[\text{ion}]_i) = zFE_{\text{ion}}$$

$$E_{\text{ion}} = (RT/zF)\ln([\text{ion}]_e/[\text{ion}]_i) \qquad \text{(Nernst)}$$

At 20°C, RT/F is about 25 mV, and moving from Neperian logarithms to decimal ones a factor of 2.3 is needed. Hence:

$$E_{\text{ion}} = (58/z)\log_{10}([\text{ion}]_e/[\text{ion}]_i).$$

Of course, this description of the E_{ion} is entirely based on a physical theory of passive ion movements. Transmembrane flux of ions, however, involves active transport of ions as well. For example, the gradients for Na^+ and, in particular, for Ca^{2+} are regulated by complex mechanisms relying on transporters and intracellular sequestration so that the possibility of predicting the precise reversal potential of responses mediated by rises in Na^+ or Ca^{2+} permeability on the basis of their apparent transmembrane concentrations is limited.

CHAPTER

4

The voltage-gated channels of Na^+ action potentials

Constance Hammond

OUTLINE

The ionic basis for nerve excitation was first elucidated in the squid giant axon by Hodgkin and Huxley (1952) using the voltage clamp technique. They made the key observation that two separate voltage-dependent currents underlie the action potential: an early transient inward Na^+ current which depolarizes the membrane, and a delayed outward K^+ current largely responsible for repolarization. This led to a series of experiments that resulted in a quantitative description of impulse generation and propagation in the squid axon.

Nearly 30 years later, Sakmann and Neher, using the patch clamp technique, recorded the activity of the voltage-gated Na^+ and K^+ channels responsible for action potential initiation and propagation. Taking history backwards, action potentials will be explained from the single channel level to the membrane level.

4.1 PROPERTIES OF ACTION POTENTIALS

4.1.1 The different types of action potentials

The action potential is a sudden and transient depolarization of the membrane. The cells that initiate action potentials are called 'excitable cells'. Action potentials can have different shapes; i.e. different amplitudes and durations. In neuronal somas and axons, action potentials have a large amplitude and a small duration: these are the Na^+-dependent action potentials (**Figures 4.1 and 4.2a**). In other neuronal cell bodies, heart ventricular cells and axon terminals, the action potentials have a longer duration with a plateau following the initial peak: these are the Na^+/Ca^{2+}-dependent action potentials (**Figure 4.2b–d**). Finally, in some neuronal dendrites and some endocrine cells, action potentials have a small amplitude and a long duration: these are the Ca^{2+}-dependent action potentials.

Action potentials have common properties; for example they are all initiated in response to a membrane depolarization. They also have differences; for example in the type of ions involved, their amplitude, duration, etc.

4.1.2 Na^+ and K^+ ions participate in the action potential of axons

The activity of the giant axon of the squid is recorded with an intracellular electrode (in current clamp; see

Cellular and Molecular Neurophysiology. DOI: 10.1016/B978-0-12-397032-9.00004-2

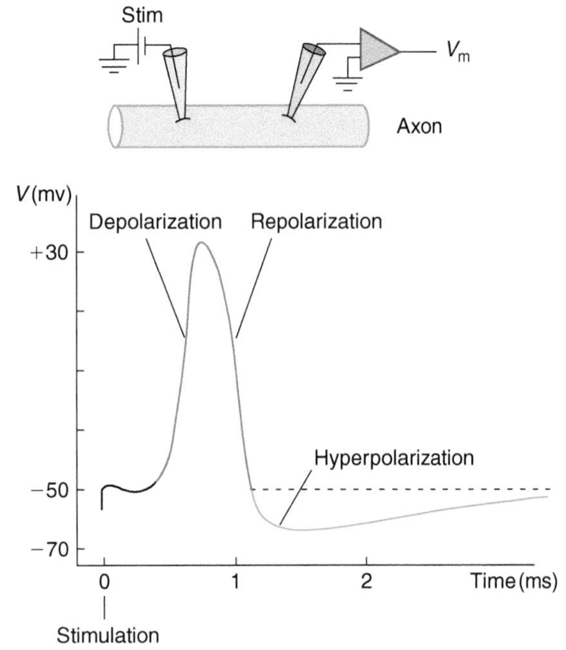

FIGURE 4.1 **Action potential of the giant axon of the squid.** Action potential intracellularly recorded in the giant axon of the squid at resting membrane potential in response to a depolarizing current pulse (the extracellular solution is seawater). The different phases of the action potential are indicated. Adapted from Hodgkin AL, Katz B (1949) The effect of sodium ions on the electrical activity of the giant axon of the squid. *J. Physiol.* **108**, 37–77, with permission.

Appendix 4.1) in the presence of seawater as the external solution.

Na^+ ions participate in the depolarization phase of the action potential

When the extracellular solution is changed from seawater to an Na^+-free solution, the amplitude and rise time of the depolarization phase of the action potential gradually and rapidly decreases, until after 8 s the current pulse can no longer evoke an action potential (**Figure 4.3**). Moreover, in control seawater, tetrodotoxin (TTX), a specific blocker of voltage-gated Na^+ channels, completely blocks action potential initiation (**Figure 4.4a,c**), thus confirming a major role of Na^+ ions.

K^+ ions participate in the repolarization phase of the action potential

Application of tetraethylammonium chloride (TEA), a blocker of K^+ channels, greatly prolongs the duration of the action potential of the squid giant axon without changing the resting membrane potential. The action potential treated with TEA has an initial peak followed by a plateau (**Figure 4.4a,b**) and the prolongation is sometimes 100-fold or more.

4.1.3 Na^+-dependent action potentials are all or none and propagate along the axon with the same amplitude

Depolarizing current pulses are applied through the intracellular recording electrode, at the level of a neuronal soma or axon. We observe that (i) to a certain level of membrane depolarization called the threshold potential, only an ohmic passive response is recorded (**Figure 4.5a,** right); (ii) when the membrane is depolarized just above threshold, an action potential is recorded. Then, increasing the intensity of the stimulating current pulse does not increase the amplitude of the action potential (**Figure 4.5a,** left). The action potential is all or none.

Once initiated, the action potential propagates along the axon with a speed varying from 1 to 100 m s^{-1} according to the type of axon. Intracellular recordings at varying distances from the soma show that the amplitude of the action potential does not attenuate: the action potential propagates without decrement (**Figure 4.5b**).

4.1.4 Questions about the Na^+-dependent action potential

- What are the structural and functional properties of the Na^+ and K^+ channels of the action potential? (Sections 4.2 and 4.3)
- What represents the threshold potential for action potential initiation? (Section 4.4)
- Why is the action potential all or none? (Section 4.4)
- What are the mechanisms of action potential propagation? (Section 4.4)

4.2 THE DEPOLARIZATION PHASE OF Na^+-DEPENDENT ACTION POTENTIALS RESULTS FROM THE TRANSIENT ENTRY OF Na^+ IONS THROUGH VOLTAGE-GATED Na^+ CHANNELS

4.2.1 The Na^+ channel consists of a principal large α-subunit with four internal homologous repeats and auxiliary β-subunits

The primary structures of the *Electrophorus* electroplax Na^+ channel, that of the rat brain, heart and skeletal muscles, have been elucidated by cloning and sequence analysis of the complementary cDNAs. The Na^+ channel in all these structures consists of an α-subunit of approximately 2000 amino acids (260 kDa), composed of four homologous domains (I to IV) separated by surface loops of different lengths. Within each domain there are six segments forming six putative transmembrane α-helices (S1 to S6) and a hairpin-like P-loop also called re-entrant pore loop

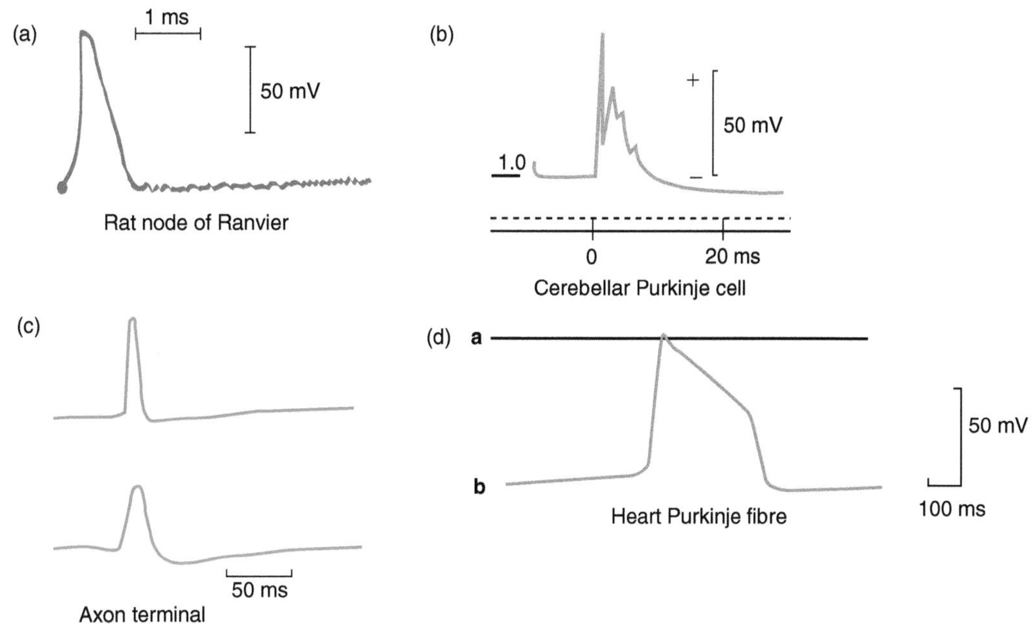

FIGURE 4.2 **Different types of action potentials recorded in excitable cells. (a)** Sodium-dependent action potential intracellularly recorded in a node of Ranvier of a rat nerve fiber. Note the absence of the hyperpolarization phase flowing the action potential. **(b–d)** Sodium–calcium-dependent action potentials. **(b)** Intracellular recording of the complex spike in a cerebellar Purkinje cell in response to climbing fiber stimulation: an initial Na^+-dependent action potential and a later larger slow potential on which are superimposed several small Ca^{2+}-dependent action potentials. The total duration of this complex spike is 5–7 ms. **(c)** Action potential recorded from axon terminals of *Xenopus* hypothalamic neurons (these axon terminals are located in the neurohypophysis) in control conditions (top) and after adding blockers of Na^+ and K^+ channels (TTX and TEA, bottom) in order to unmask the Ca^{2+} component of the spike (this component has a larger duration due to the blockade of some of the K^+ channels). **(d)** Intracellular recording of an action potential from an acutely dissociated dog heart cell (Purkinje fiber). Trace 'a' is recorded when the electrode is outside the cell and represents the trace 0 mV. Trace 'b' is recorded when the electrode is inside the cell. The peak amplitude of the action potential is 75 mV and the total duration 400 ms. All these action potentials are recorded in response to an intracellular depolarizing pulse or to the stimulation of afferents. Note the differences in their durations. Part **(a)** adapted from Brismar T (1980) Potential clamp analysis of membrane currents in rat myelinated nerve fibres. *J. Physiol.* **298**, 171–184, with permission. Parts **(b–d)** adapted from Coraboeuf E, Weidmann S (1949) Potentiel de repos et potentials d'action du muscle cardiaque, mesurés à l'aide d'électrodes internes. *C. R. Soc. Biol.* **143**, 1329–1331; Eccles JC, Llinas R, Sasaki K (1966) The excitatory synaptic action of climbing fibres on the Purkinje cells of the cerebellum. *J. Physiol.* **182**, 268–296; and Obaid AL, Flores R, Salzberg BM (1989) Calcium channels that are required for secretion from intact nerve terminals of vertebrates are sensitive to ω-conotoxin and relatively insensitive to dihydropyridines. *J. Gen. Physiol.* **93**, 715–730, with permission.

between S5 and S6 (**Figure 4.6a**). The four homologous domains probably form a pseudo-tetrameric structure whose central part is the permeation pathway (**Figure 4.6b**). Parts of the α-subunit contributing to pore formation have been identified by site-directed mutagenesis.

Each domain contains a unique segment, the S4 segment, with positively charged residues (arginine or lysine) at every third position with mostly non-polar residues intervening between them (see **Figure 4.15a**). This structure of the S4 segment is strikingly conserved in all the types of Na^+ channels analyzed so far and this led to the suggestion that the S4 segments serve as voltage sensors (see Section 4.2.7).

The α-subunit of mammals is associated with smaller auxiliary subunits named β-subunits. They are small proteins of about 200 amino acid residues (33–36 kDa), with a substantial N-terminal domain, a single putative membrane spanning segment and a C-terminal intracellular domain. The structure of these β-subunits resembles the family of cell adhesion molecules.

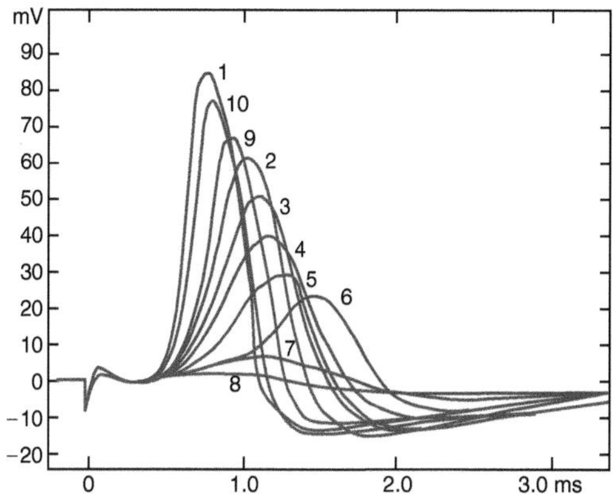

FIGURE 4.3 **The action potential of the squid giant axon is abolished in a Na^+-free external solution. (1)** Control action potential recorded in seawater; **(2–8)** recordings taken at the following times after the application of a dextrose solution (Na-free solution): 2.30, 4.62, 5.86, 6.10, 7.10 and 8.11 s; **(9)** recording taken 9 s after reapplication of seawater; **(10)** recording taken at 90 and 150 s after reapplication of seawater; traces are superimposed. From Hodgkin AL, Katz B (1949) The effect of sodium ions on the electrical activity of the giant axon of the squid. *J. Physiol.* **108**, 37–77, with permission.

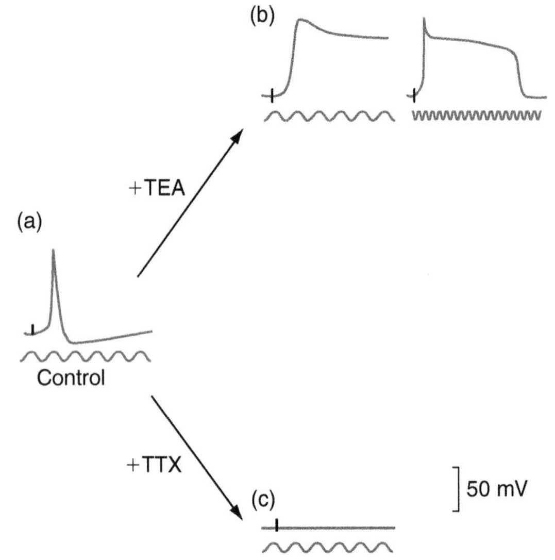

FIGURE 4.4 **Effects of tetrodotoxin (TTX) and tetraethylammonium chloride (TEA) on the action potential of the squid giant axon.** **(a)** Control action potential. **(b)** TEA application lengthens the action potential (left), which then has to be observed on a different timescale (right trace). **(c)** TTX totally abolishes the initiation of the action potential. Adapted from Tasaki I, Hagiwara S (1957) Demonstration of two stable potential states in the squid giant axon under tetraethylammonium chloride. *J. Gen. Physiol.* **40**, 859–885, with permission.

The α-subunit mRNA isolated from rat brain or the α-subunit RNAs transcribed from cloned cDNAs from a rat brain are sufficient to direct the synthesis of functional Na^+ channels when injected into oocytes. These results establish that the protein structures necessary for voltage-gating and ion conductance are contained within the α-subunit itself. However, the properties of these channels are not identical to native Na^+ channels as it has been shown that the auxiliary β-subunits play a role in the targeting and stabilization of the α-subunit in the plasma membrane in specific locations in excitable cells.

The nomenclature of all Na^+ voltage-gated channels is the following: Na_v to indicate the principal permeating ion (Na^+) and the principal physiological regulator (v for voltage), followed by a number that indicates the gene subfamily (currently Na_v1 is the only subfamily). The number following the decimal point identifies the specific channel isoform (e.g. $Na_v1.1$). At present, nine functional isoforms have been identified.

$Na_V1.1$, 1.2, 1.3 and 1.6 are the primary sodium channels in the central nervous system. $Na_V1.7$, 1.8 and 1.9 are

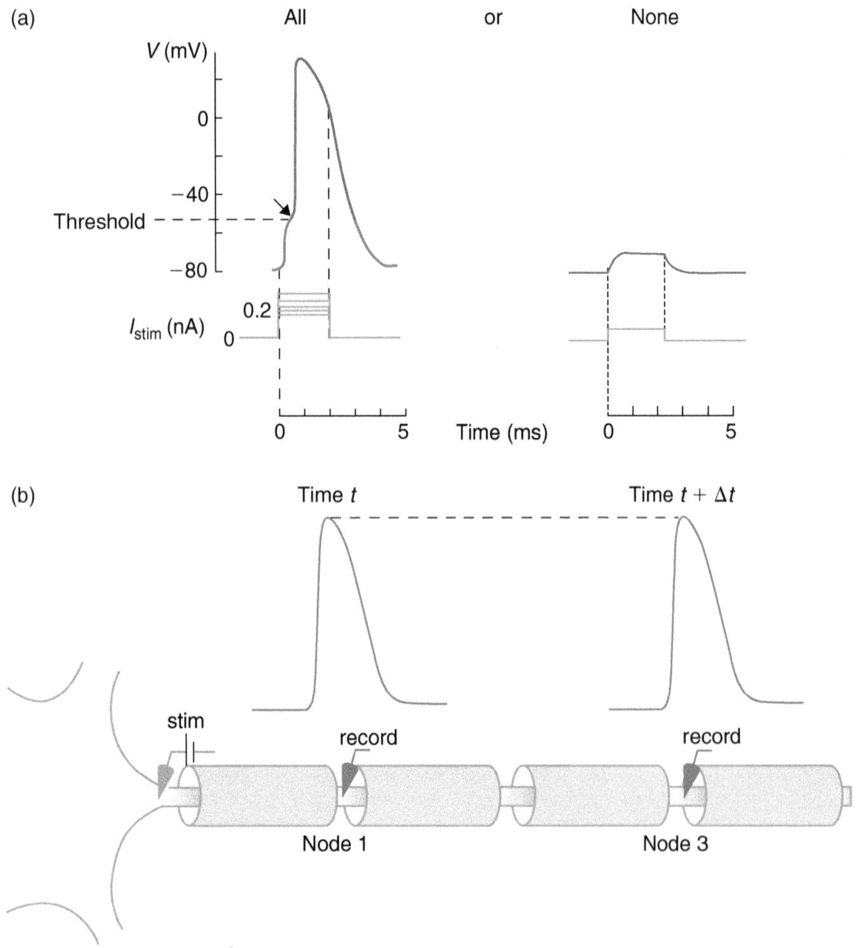

FIGURE 4.5 **Properties of the Na^+-dependent action potential.** **(a)** The response of the membrane to depolarizing current pulses of different amplitudes is recorded with an intracellular electrode. Upper traces are the voltage traces, bottom traces are the current traces. Above 0.2 nA an axon potential is initiated. Increasing the current pulse amplitude does not increase the action potential amplitude (left). With current pulses of smaller amplitudes, no action potential is initiated. **(b)** An action potential is initiated in the soma-initial segment by a depolarizing current pulse (stim). Intracellular recording electrodes inserted along the axon record the action potential at successive nodes at successive times. See text for further explanations.

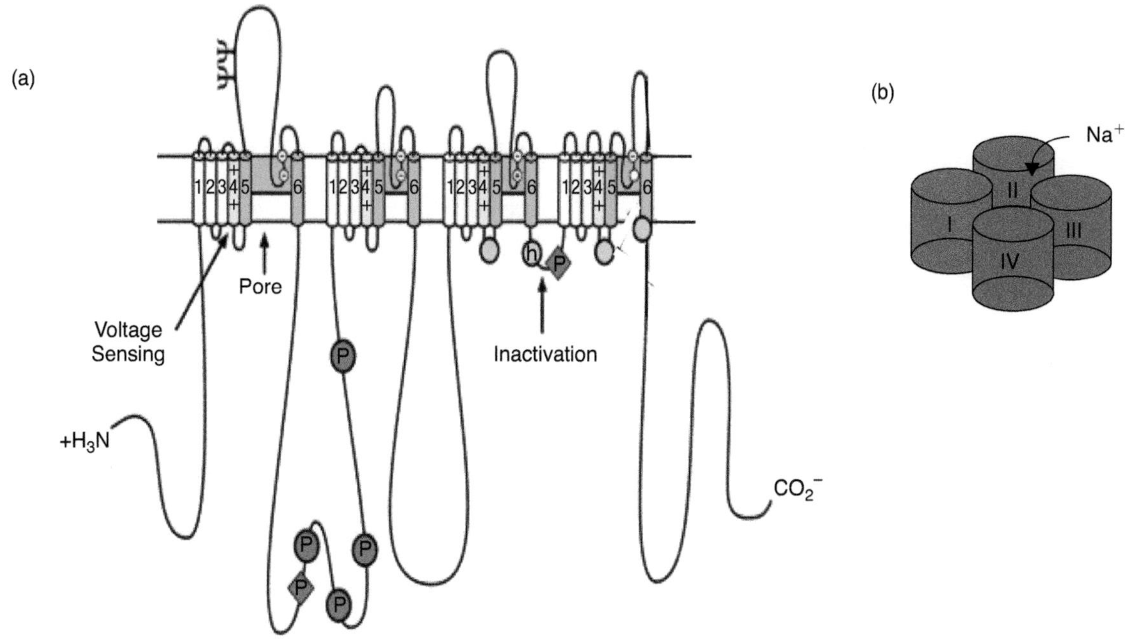

FIGURE 4.6 **The primary structures of the voltage-gated sodium channel α-subunit. (a)** Cylinders represent probable α-helical segments (putative membrane-spanning segments). Bold lines represent the polypeptide chains with length approximately proportional to the number of amino acid residues in the brain sodium channel subtypes. Y, sites of probable N-linked glycosylation; green, pore-lining segments; white circles, the rings of amino residues that form the ion selectivity filter (and the tetrodotoxin binding site); yellow, S4 voltage sensors; the inactivation gate loop contains the inactivation particle (symbolized by h). **(b)** Diagram of the pseudo-tetrameric structure whose central part is the permeation pathway (pore). Adapted from Catterall WA (2012) Voltage-gated sodium channels at 60: structure, function and pathophysiology. *J. Physiol.* **590**, 2577–2589, with permission.

the primary sodium channels in the peripheral nervous system. $Na_V1.4$ is the primary sodium channel in skeletal muscle, whereas $Na_V1.5$ is primary in heart. Most of these sodium channels also have significant levels of expression outside of their primary tissues.

4.2.2 Membrane depolarization favors conformational change of the Na+ channel towards the open state; the Na+ channel then quickly inactivates

The function of the Na^+ channel is to transduce *rapidly* membrane depolarization into an entry of Na^+ ions. The activity of a single Na^+ channel was first recorded by Sigworth and Neher in 1980 from rat muscle cells with the patch clamp technique (cell-attached patch; see Appendix 4.3).

It must be explained that the experimenter does not know before recording it which type of channel(s) is in the patch of membrane isolated under the tip of the pipette. He or she can only increase the chance of recording an Na^+ channel, for example, by studying a membrane where this type of channel is frequently expressed and by pharmacologically blocking the other types of channels that could be activated together with the Na^+ channels (voltage-gated K^+ channels are blocked by TEA). The recorded channel is then identified by its voltage dependence, reversal potential, unitary conductance, ionic permeability, mean open time, etc. Finally, the number of Na^+ channels in the patch of membrane cannot be predicted. Even when pipettes with small tips are used, the probability of recording more than one channel can be high because of the type of membrane patched. For this reason, very few recordings of single native Na^+ channels have been performed. The number of Na^+ channels in a patch is estimated from recordings where the membrane is strongly depolarized in order to increase to its maximum the probability of opening the voltage-gated channels present in the patch.

Voltage-gated Na+ channels of the skeletal muscle fiber

A series of recordings obtained from a single Na^+ channel in response to a 40 mV depolarizing step given every second is shown in **Figure 4.7a** and **c.** The holding potential is around −70 mV (remember that in the cell attached configuration, the membrane potential can only be estimated). A physiological extracellular concentration of Na^+ ions is present in the pipette.

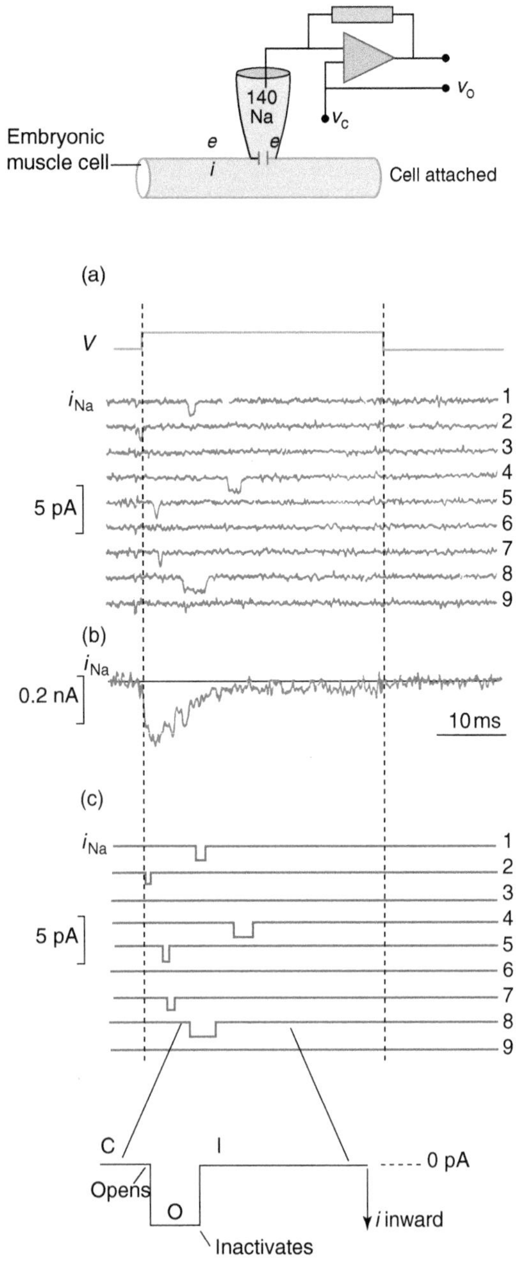

FIGURE 4.7 **Single Na⁺ channel openings in response to adepolarizing step (muscle cell).** The activity of the Na⁺ channel is recorded in patch clamp (cell-attached patch) from an embryonic muscle cell. **(a)** Nine successive recordings of single channel openings (iNa) in response to a 40 mV depolarizing pulse (V trace) given at 1 s intervals from a holding potential 10 mV more hyperpolarized than the resting membrane potential. **(b)** Averaged inward Na⁺ current from 300 elementary Na⁺ currents as in **(a)**. **(c)** The same recordings as in **(a)** are redrawn in order to explain more clearly the different states of the channel. On the bottom line one opening is enlarged. C, closed state; O, open state; I, inactivated state. The solution bathing the extracellular side of the patch or intrapipette solution contains (in mM): 140 NaCl, 1.4 KCl, 2.0 $MgCl_2$, 1 $CaCl_2$ and 20 HEPES at pH 7.4. TEA 5 mM is added to block K⁺ channels and bungarotoxin to block acetylcholine receptors. Adapted from Sigworth FJ, Neher E (1980) Single Na⁺ channel currents observed in rat muscle cells. *Nature* **287**, 447–449, with permission.

At holding potential, no variations in the current traces are recorded. After the onset of the depolarizing step, unitary Na⁺ currents of varying durations but of the same amplitude are recorded (lines 1, 2, 4, 5, 7 and 8) or not recorded (lines 3, 6 and 9). This means that six times out of nine, the Na⁺ channel has opened in response to membrane depolarization. The Na⁺ current has a rectangular shape and is downward. By convention, inward currents of + ions (cations) are represented as downward (inward means that + ions enter the cell; see Section 3.3.3). The histogram of the Na⁺ current amplitude recorded in response to a 40 mV depolarizing step gives a mean amplitude for i_{Na} of around −1.6 pA (see Appendix 4.3).

It is interesting to note that once the channel has opened, there is a low probability that it will reopen during the depolarization period. Moreover, even when the channel does not open at the beginning of the step, the frequency of appearance of Na⁺ currents later in the depolarization is very low; i.e. the Na⁺ channel inactivates.

Rat brain Na⁺ channels

The activity of rat brain Na⁺ channels has been studied in cerebellar Purkinje cells in culture. Each trace of **Figure 4.8a** and **c** shows the unitary Na⁺ currents (i_{Na}) recorded during a 20 ms membrane depolarization to −40 mV (test potential) from a holding potential of −90 mV. Rectangular inward currents occur most frequently at the beginning of the depolarizing step but can also be found at later times (**Figure 4.8a**, line 2). The histogram of the Na⁺ current amplitudes recorded at −40 mV test potential gives a mean amplitude for i_{Na} of around −2 pA (**Figure 4.8d**). Events near −4 pA correspond to double openings, meaning that at least two channels are present in the patch.

The unitary current has a rectangular shape

The rectangular shape of the unitary current means that when the Na⁺ channel opens, the unitary current is nearly immediately maximal. The unitary current then stays constant: the channel stays open for a time which varies; finally, the unitary current goes back to zero though the membrane is still depolarized. The channel may not reopen (**Figures 4.7a,c**) as it is in an inactivated state (I) (**Figure 4.7c**, bottom trace). After being opened by a depolarization, the channel does not go back to the closed state but inactivates. In that state, the pore of the channel is closed (no Na⁺ ions flow through the pore) as in the closed state but the channel cannot reopen immediately (which differs from the closed state). The inactivated channel is refractory to opening unless the membrane repolarizes to allow it to return to the closed (resting) state.

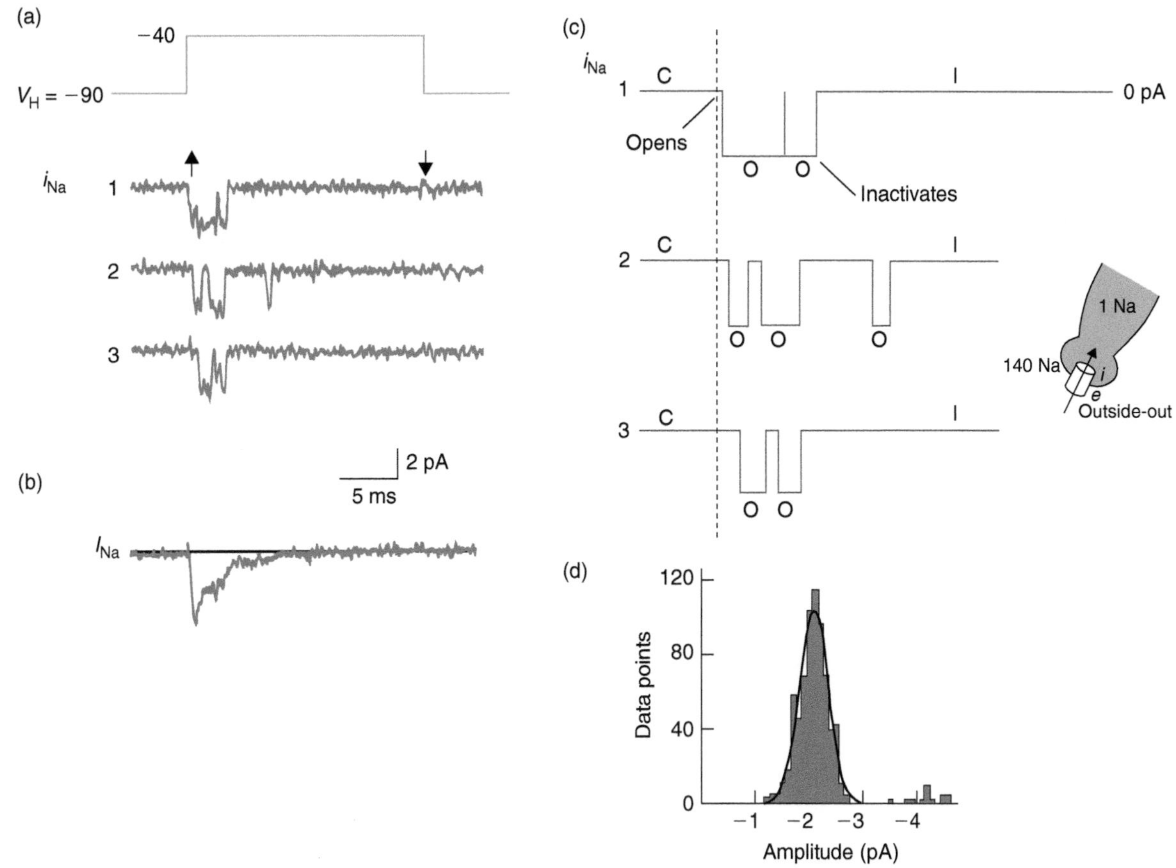

FIGURE 4.8 **Single-channel activity of a voltage-gated Na^+ channel from rat brain neurons.** The activity of an Na^+ channel of a cerebellar Purkinje cell in culture is recorded in patch clamp (outside-out patch) in response to successive depolarizing steps to −40 mV from a holding potential V_H = −90 mV. **(a)** The 20 ms step (upper trace) evokes rectangular inward unitary currents (i_{Na}). **(b)** Average current calculated from all the sweeps which had active Na^+ channels within a set of 25 depolarizations. **(c)** Interpretative drawing on an enlarged scale of the recordings in **(a)**. **(d)** Histogram of elementary amplitudes for recordings as in **(a)**. The continuous line corresponds to the best fit of the data to a single Gaussian distribution. C, closed state; O, open state; I, inactivated state. The solution bathing the outside face of the patch contains (in mM): 140 NaCl, 2.5 KCl, 1 $CaCl_2$, 1 $MgCl_2$, 10 HEPES. The solution bathing the inside of the patch or intrapipette solution contains (in mM): 120 CsF, 10 CsCl, 1 NaCl, 10 EGTA-Cs^+, 10 HEPES-Cs^+. Cs^+ ions are in the pipette instead of K^+ ions in order to block K^+ channels. Adapted from Gähwiler BH, Llano I (1989) Sodium and potassium conductances in somatic membranes of rat Purkinje cells from organotypic cerebellar cultures. *J. Physiol.* **417**, 105–122, with permission.

In other recordings, such as that of **Figure 4.8a** and **c**, the Na^+ channel seems to reopen once or twice before inactivating. This may result from the presence of two (as here) or more channels in the patch so that the unitary currents recorded do not correspond to the same channel. It may also result from a slower inactivation rate of the channel recorded, which in fact opens, closes, reopens and then inactivates.

The unitary current is carried by a few Na^+ ions

How many Na^+ ions enter through a single channel? Knowing that in the preceding example, the unitary Na^+ current has a mean amplitude of −1.6 pA for 1 ms, the number of Na^+ ions flowing through one channel during 1 ms is $1.6 \times 10^{-12}/(1.6 \times 10^{-19} \times 10^3) = 10\,000$ Na^+ ions (since 1 pA = 1 pC s^{-1} and the elementary charge of one electron is 1.6×10^{-19} C). This number, 10^4 ions, is negligible compared with the number of Na^+ ions in the intracellular medium: if $[Na^+]_i$ = 14 mM, knowing that 1 mole represents 6×10^{23} ions, the number of Na^+ ions per liter is $6 \times 10^{23} \times 14 \times 10^{-3} = 10^{22}$ ions l^{-1}. In a neuronal cell body or a section of axon, the volume is of the order of 10^{-12} to 10^{-13} liters. Then the number of Na^+ ions is around 10^9 to 10^{10}.

The Na^+ channel fluctuates between the closed, open and inactivated states

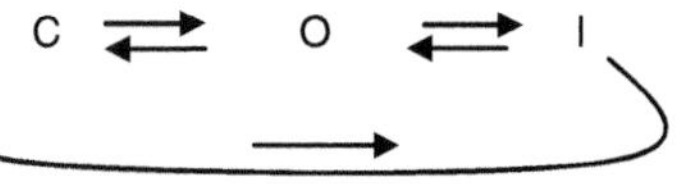

where C is the channel in the closed state, O in the open state and I in the inactivated state. Both C and I states are non-conducting states. The C to O transition is triggered by membrane depolarization. The O to I transition is due to an intrinsic property of the Na^+ channel. The I to C transition occurs when the membrane repolarizes or is already repolarized. In summary, the Na^+ channel opens when the membrane is depolarized, stays open during a mean open time of less than 1 ms, and then usually inactivates.

4.2.3 The time during which the Na^+ channel stays open varies around an average value, τ_o, called the mean open time

In **Figures 4.7a** and **4.8a** we can observe that the periods during which the channel stays open, t_o, are variable. The mean open time of the channel, τ_o, at a given potential is obtained from the frequency histogram of the different t_o at this potential. When this distribution can be fitted by a single exponential, its time constant provides the value of τ_o (see Appendix 4.3). The functional significance of this value is the following: during a time equal to τ_o the channel has a high probability of staying open.

For example, the Na^+ channel of the skeletal muscle fiber stays open during a mean open time of 0.7 ms. For the rat brain Na^+ channel of cerebellar Purkinje cells, the distribution of the durations of the unitary currents recorded at −32 mV can be fitted with a single exponential with a time constant of 0.43 ms ($\tau_o = 0.43$ ms).

4.2.4 The i_{Na}–V relation is linear: the Na^+ channel has a constant unitary conductance γ_{Na}

When the activity of a single Na^+ channel is now recorded at different test potentials, we observe that the amplitude of the inward unitary current diminishes as the membrane is further and further depolarized (see **Figure 4.10a**). In other words, the net entry of Na^+ ions through a single channel diminishes as the membrane depolarizes. The i_{Na}–V relation is obtained by plotting the amplitude of the unitary current (i_{Na}) versus membrane potential (V_m). It is linear between −50 mV and 0 mV (**Figure 4.9a**). For membrane potentials more hyperpolarized than −50 mV, there are no values of i_{Na} since the channel rarely opens or does not open at all. Quantitative data for potentials more depolarized than 0 mV are not available.

The critical point of the current–voltage relation is the membrane potential for which the current is zero; i.e. the reversal potential of the current (E_{rev}). If only Na^+ ions flow through the Na^+ channel, the reversal potential is equal to E_{Na}. From −50 mV to E_{rev}, i_{Na} is inward and its amplitude decreases. This results from the decrease of the Na^+ driving force ($V_m - E_{Na}$) as the membrane approaches the reversal potential for Na^+ ions. For membrane potentials more depolarized than E_{rev}, i_{Na} is now outward (not shown). Above E_{rev}, the amplitude of the outward Na^+ current progressively increases as the driving force for the exit of Na^+ ions increases.

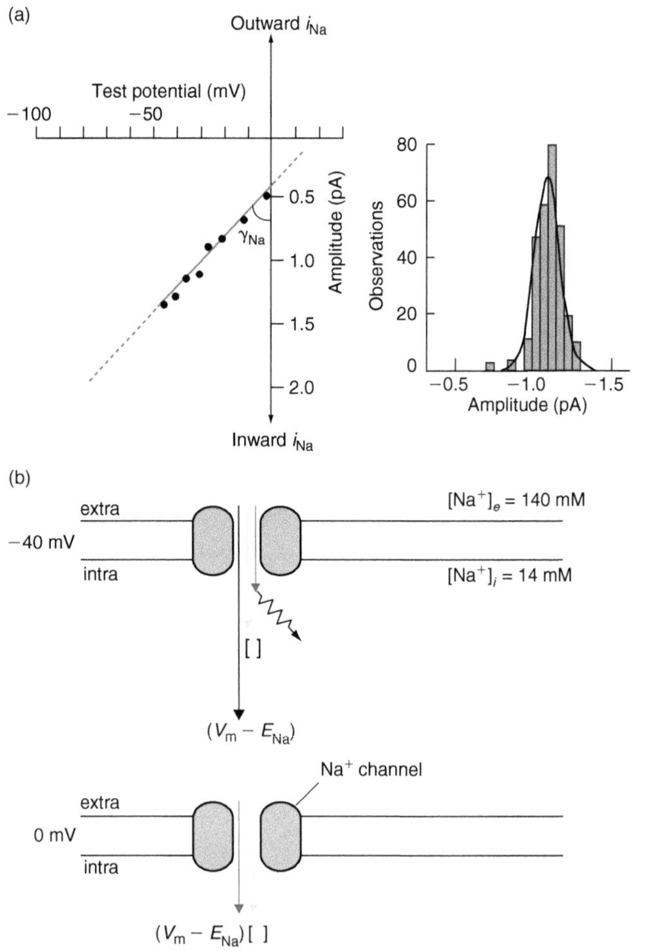

FIGURE 4.9 **The single-channel current/voltage (i_{Na}/V) relation is linear. (a)** The activity of the rat type II Na^+ channel expressed in *Xenopus* oocytes from cDNA is recorded in patch clamp (cell-attached patch). Plot of the unitary current amplitude versus test potential: each point represents the mean of 20–200 unitary current amplitudes measured at one potential (left) as shown at −32 mV (right). The relation is linear between test potentials −50 and 0 mV (holding potential = −90 mV). The slope is γ_{Na} = 19 pS. **(b)** Drawings of an open voltage-gated Na^+ channel to explain the direction and amplitude of the net flux of Na^+ ions at two test potentials (−40 and 0 mV). [ ], force due to the concentration gradient across the membrane; → force due to the electric gradient; $V - E_{Na}$, driving force. The solution bathing the extracellular side of the patch or intrapipette solution contains (in mM): 115 NaCl, 2.5 KCl, 1.8 $CaCl_2$, 10 HEPES. Plot **(a)** adapted from Stühmer W, Methfessel C, Sakmann B *et al.* (1987) Patch clamp characterization of sodium channels expressed from rat brain cDNA. *Eur. Biophys. J.* **14**, 131–138, with permission.

The linear $i_{Na} - V$ relation is described by the equation $i_{Na} = \gamma_{Na}(V_m - E_{Na})$, where V_m is the test potential, E_{Na} is the reversal potential of the Na^+ current, and γ_{Na} is the conductance of a single Na^+ channel (unitary conductance). The value of γ_{Na} is given by the slope of the linear i_{Na}/V curve. It has a constant value at any given membrane potential. This value varies between 5 and 18 pS depending on the preparation.

4.2.5 The probability of the Na^+ channel being in the open state increases with depolarization to a maximal level

An important observation at the single channel level is that the more the membrane is depolarized, the higher is the probability that the Na^+ channel will open. This observation can be made from two types of experiments:

- The activity of a single Na^+ channel is recorded in patch clamp (cell-attached patch). Each depolarizing step is repeated several times and the number of times the Na^+ channel opens is observed (**Figure 4.10a**). With depolarizing steps to −70 mV from a holding potential of −120 mV, the channel very rarely opens and, if it does, the time spent in the open state is very short. In contrast, with depolarizing steps to −40 mV, the Na^+ channels open for each trial.
- The activity of two or three Na^+ channels is recorded in patch clamp (cell-attached patch). In response to depolarizing steps of small amplitude, Na^+ channels do not open or only one Na^+ channel opens at a time. With larger depolarizing steps, the overlapping currents of two or three Na^+ channels can be observed, meaning that this number of Na^+ channels opens with close delays in response to the step (not shown).

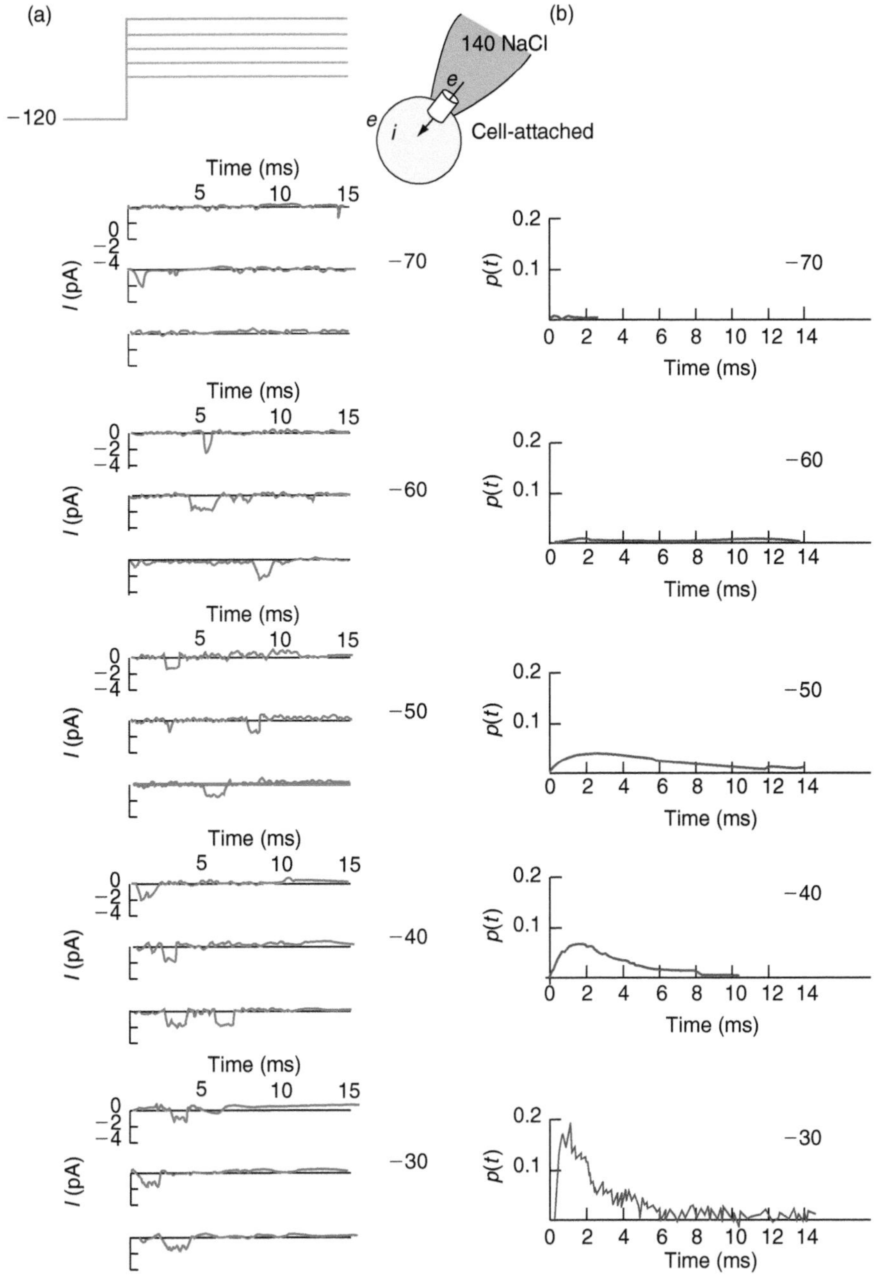

FIGURE 4.10 **The open probability of the voltage-gated Na+ channel is voltage and time dependent.** Single Na^+ channel activity recorded in a mammalian neuroblastoma cell in patch clamp (cell-attached patch). **(a)** In response to a depolarizing step to the indicated potentials from a holding potential of −120 mV, unitary inward currents are recorded. **(b)** Ensemble of averages of single-channel openings at the indicated voltages; 64 to 2000 traces are averaged at each voltage to obtain the time-dependent open probability of a channel ($p_{(t)}$) in response to a depolarization. The open probability at time *t* is calculated according to the equation: $p(t) = I_{Na(t)}/Ni_{Na}$, where $I_{Na(t)}$ is the average current at time *t* at a given voltage, *N* is the number of channels (i.e. the number of averaged recordings of single channel activity) and i_{Na} is the unitary current at a given voltage. At −30 mV the open probability is maximum. The channels inactivate in 4 ms. Adapted from Aldrich RW, Steven CF (1987) Voltage-dependent gating of sodium channels from mammalian neuroblastoma cells. *J. Neurosci.* 7, 418–431, with permission.

From the recordings of **Figure 4.10a**, we can observe that the probability of the Na^+ channel being in the open state varies with the value of the test potential. It also varies with time during the depolarizing step: openings occur more frequently at the beginning of the step. The open probability of Na^+ channels is voltage and time dependent. By averaging a large number of records obtained at each test potential, the open probability (p_t) of the Na^+ channel recorded can be obtained at each time t of the step (**Figure 4.10b**). We observe from these curves that after 4–6 ms the probability of the Na^+ channel being in the open state is very low, even with large depolarizing steps: the Na^+ channel inactivates in 4–6 ms. When we now compare the open probabilities at the different test potentials, we observe that the probability of the Na^+ channel being in the open state at time $t = 2$ ms increases with the amplitude of the depolarizing step.

4.2.6 The macroscopic Na^+ current (I_{Na}) has a steep voltage dependence of activation and inactivates within a few milliseconds

The macroscopic Na^+ current, I_{Na}, is the sum of the unitary currents, i_{Na}, flowing through all the open Na^+ channels of the recorded membrane

At the axon initial segment or at nodes of Ranvier, there are N Na^+ channels that can be activated. We have seen that the unitary Na^+ current flowing through a single Na^+ channel has a rectangular shape. What is the time course of the macroscopic Na^+ current, I_{Na}?

If we assume that the Na^+ channels in one cell are identical and function independently, the sum of many recordings from the same Na^+ channel should show the same properties as the macroscopic Na^+ current measured from thousands of channels with the voltage clamp technique. In **Figure 4.7b**, an average of 300 unitary Na^+ currents elicited by a 40 mV depolarizing pulse is shown. For a given potential, the 'averaged' inward Na^+ current has a fast rising phase and presents a peak at the time $t = 1.5$ ms. The peak corresponds to the time when most of the Na^+ channels are opened at each trial. Then the averaged current decays with time because the Na^+ channel has a low probability of being in the open state later in the step (owing to the inactivation of the Na^+ channel). At each trial, the Na^+ channel does not inactivate exactly at the same time, which explains the progressive decay of the averaged macroscopic Na^+ current. A similar averaged Na^+ current is shown in **Figure 4.8b**. The averaged current does not have a rectangular shape because the Na^+ channel does not open with the same delay and does not inactivate at the same time at each trial.

The *averaged* macroscopic Na^+ current has a time course similar to that of the *recorded* macroscopic Na^+ current from the same type of cell at the same potential. However, the averaged current from 300 Na^+ channels still presents some angles in its time course. In contrast, the macroscopic recorded Na^+ current is smooth. The more numerous are the Na^+ channels opened by the depolarizing step, the smoother is the total Na^+ current. The value of I_{Na} at each time t at a given potential is:

$$I_{Na} = Np_{(t)}\, i_{Na}$$

where N is the number of Na^+ channels in the recorded membrane and $p_{(t)}$ is the open probability at time t of the Na^+ channel; it depends on the membrane potential and on the channel opening and inactivating rate constants. i_{Na} is the unitary Na^+ current and $Np_{(t)}$ is the number of Na^+ channels open at time t.

The I–V relation is bell-shaped though the i–V relation is linear

We have seen that the amplitude of the unitary Na^+ current decreases linearly with depolarization (see **Figure 4.9a**). In contrast, the I_{Na}–V relation is not linear. The macroscopic Na^+ current is recorded from a myelinated rabbit nerve with the double electrode voltage clamp technique. When the amplitude of the peak Na^+ current is plotted against membrane potential, it has a clear bell shape (**Figures 4.11** and **4.12a**).

Analysis of each trace from the smallest depolarizing step to the largest shows that:

- For small steps, the peak current amplitude is small (0.2 nA) and has a slow time to peak (1 ms). At these potentials, the Na^+ driving force is strong but the Na^+ channels have a low probability of opening (**Figure 4.11a**). Therefore, I_{Na} is small since it represents the current through a small number of open Na^+ channels. Moreover, the small number of activated Na^+ channels open with a delay since the depolarization is just *subliminal*. This explains the slow time to peak.
- As the depolarizing steps increase in amplitude (to $-42/-35$ mV), the amplitude of I_{Na} increases to a maximum (-3 nA) and the time to peak decreases to a minimum (0.2 ms). Larger depolarizations increase the probability of the Na^+ channel being in the open state and shorten the delay of opening (**Figure 4.10**). Therefore, though the amplitude of i_{Na} decreases between -63 and -35 mV, the amplitude of I_{Na} increases owing to the large increase of open Na^+ channels.
- After this peak, the amplitude of I_{Na} decreases to zero since the open probability does not increase enough to compensate for the decrease of i_{Na}. The

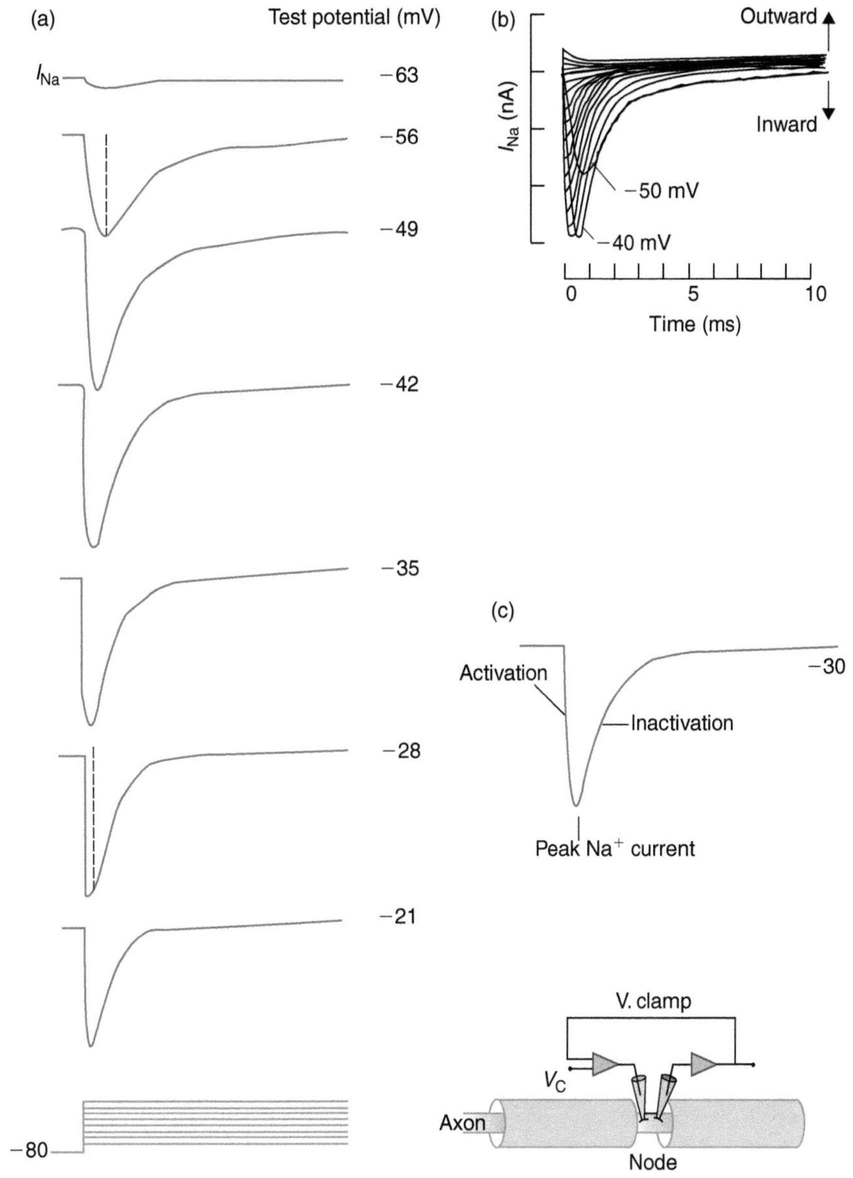

FIGURE 4.11 **Voltage dependence of the macroscopic voltage-gated Na+ current.** The macroscopic voltage-gated Na+ current recorded in a node of a rabbit myelinated nerve in voltage clamp conditions. **(a)** Depolarizing steps from −70 mV to −21 mV from a holding potential of −80 mV evoke macroscopic Na+ currents (I_{Na}) with different time courses and peak amplitudes. The test potential is on the right. Bottom trace is the voltage trace. **(b)** The traces in **(a)** are superimposed and current responses to depolarizing steps from −14 to +55 mV are added. The outward current traces are recorded when the test potential is beyond the reversal potential (+30 mV in this preparation). **(c)** I_{Na} recorded at −30 mV. The rising phase of I_{Na} corresponds to activation of the Na+ channels and the decrease of I_{Na} corresponds to progressive inactivation of the open Na+ channels. The extracellular solution contains (in mM): 154 NaCl, 2.2 $CaCl_2$, 5.6 KCl; pH 7.4. Adapted from Chiu SY, Ritchie JM, Bogart RB, Stagg D (1979) A quantitative description of membrane currents from a rabbit myelinated nerve. *J. Physiol.* **292**, 149–166, with permission.

reversal potential of I_{Na} is the same as that of i_{Na} since it depends only on the extracellular and intracellular concentrations of Na+ ions.
- I_{Na} changes polarity for V_m more depolarized than E_{rev}: it is now an outward current whose amplitude increases with the depolarization **(Figure 4.10b)**.

It is important to note that membrane potentials more depolarized than +20 mV are non-physiological.

Activation and inactivation curves: the threshold potential

Activation rate is the rate at which a macroscopic current turns on in response to a depolarizing voltage step. The Na+ current is recorded in voltage clamp from a node of rabbit nerve. Depolarizing steps from −70 mV to +20 mV are applied from a holding potential of −80 mV. When the ratio of the peak current at each test potential to the maximal peak current (I_{Na}/I_{Namax}) is plotted against test potential, the activation curve of I_{Na} can

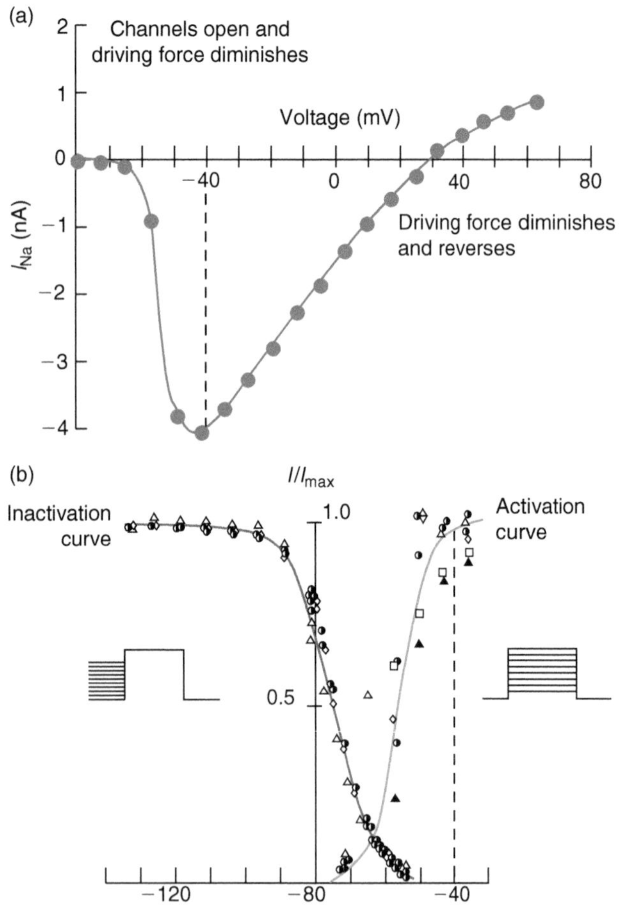

FIGURE 4.12 **Activation–inactivation properties of the macroscopic voltage-gated Na+ current. (a)** The $I_{Na}-V$ relation has a bell shape with a peak at −40 mV and a reversal potential at +30 mV (the average E_{Na} in the rabbit node is +27 mV). **(b)** Activation (right curve) and inactivation (left curve) curves obtained from nine different experiments. The voltage protocols used are shown in insets. In the ordinates, I/I_{max} represents the ratio of the peak Na+ current (I) recorded at the tested potential of the abscissae and the maximal peak Na+ current (I_{max}) recorded in this experiment. It corresponds in the activation curve to the peak current recorded at −40 mV in **Figure 4.11**. From Chiu SY, Ritchie JM, Bogart RB, Stagg D (1979) A quantitative description of membrane currents from a rabbit myelinated nerve. *J. Physiol.* **292**, 149–166, with permission.

be visualized. The distribution is fitted by a sigmoidal curve (**Figure 4.12b**). In this preparation, the threshold of Na+ channel activation is −60 mV. At −40 mV, I_{Na} is already maximal ($I_{Na}/I_{Namax} = 1$). This steepness of activation is a characteristic of the voltage-gated Na+ channels.

Inactivation of a current is the decay of this current during a maintained depolarization. To study inactivation, the membrane is held at varying holding potentials and a depolarizing step to a fixed value is applied where I_{Na} is maximal (0 mV, for example). The amplitude of the peak Na+ current is plotted against the holding potential. I_{Na} begins to inactivate at −90 mV and is fully inactivated at −50 mV. Knowing that the resting membrane potential in this preparation is around −80 mV, some of the Na+ channels are already inactivated at rest.

Ionic selectivity of the Na+ channel

To compare the permeability of the Na+ channel to several monovalent cations, the macroscopic current is recorded at different membrane potentials in the presence of external Na+ ions and when all the external Na+ are replaced by a test cation. Lithium is as permeant as sodium but K+ ions are weakly permeant ($P_K/P_{Na} = 0.048$). Therefore, Na+ channels are highly selective for Na+ ions and only 4% of the current is carried by K+ ions (**Figure 4.13**).

Tetrodotoxin is a selective open Na+ channel blocker

A large number of biological toxins can modify the properties of the Na+ channels. One of these, tetrodotoxin (TTX), which is found in the liver and ovaries of pufferfish (tetrodon), totally abolishes the current through most of

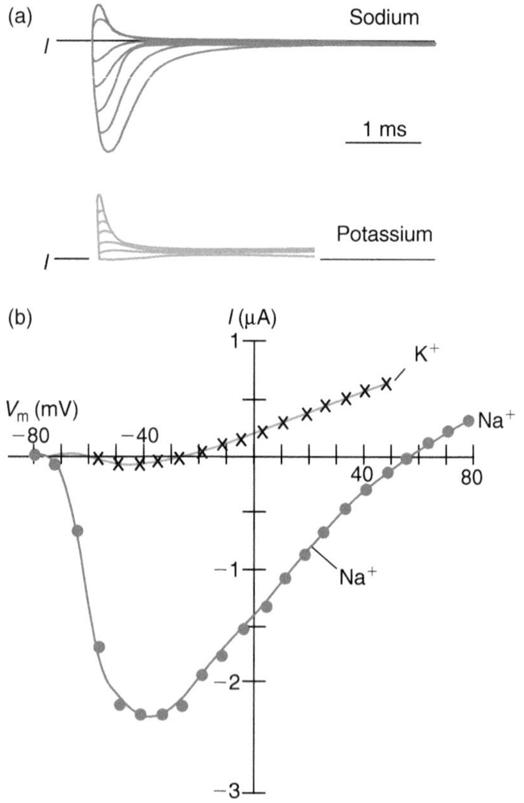

FIGURE 4.13 **Ionic selectivity of the Na+ channel. (a)** The macroscopic Na+ current is recorded with the double-electrode voltage clamp technique in a mammalian skeletal muscle fiber at different test membrane potentials (from −70 to +80 mV) from a holding potential of −80 mV. **(a)** Inward currents in normal Na+–Ringer (sodium) and in a solution where all Na+ ions are replaced by K+ ions (potassium). The other voltage-gated currents are blocked. **(b)** $I-V$ relation of the currents recorded in **(a)**. I is the amplitude of the peak current at each tested potential. Adapted from Pappone PA (1980) Voltage clamp experiments in normal and denervated mammalian skeletal muscle fibers. *J. Physiol.* **306**, 377–410, with permission.

the Na^+ channels (TTX-sensitive Na^+ channels) (see **Figure 4.4**). However, some Na^+ channels are resistant to TTX such as those from the pufferfish. TTX has a binding site supposed to be located near the extracellular mouth of the pore.

A single point mutation of the rat brain Na^+ channel type II, which changes the glutamic acid residue 387 to glutamine (E387Q) in the repeat I, renders the channel insensitive to concentrations of TTX up to tens of micromolars. *Xenopus* oocytes are injected with the wild-type mRNA or the mutant mRNA and the whole cell Na^+ currents are recorded with the double-electrode voltage clamp technique. TTX sensitivity is assessed by perfusing TTX-containing external solutions and by measuring the peak of the whole-cell inward Na^+ current (the peak means the maximal amplitude of the inward Na^+ current measured on the I_{Na}/V relation). The dose–response curves of **Figure 4.14** show that 1 μM of TTX completely abolishes the wild-type Na^+ current, but has no effect on the mutant Na^+ current. The other characteristics of the Na^+ channel are not significantly affected, except for a reduction in the amplitude of the inward current at all potentials tested. All these results suggest that the link between segments S5 and S6 in repeat I of the rat brain Na^+ channel is in close proximity to the channel mouth.

Comparison of the predicted protein sequences of the skeletal muscle sodium channels show that pufferfish Na^+ channels have accumulated several unique substitutions in the otherwise highly conserved pore loop regions of the four domains. Among these substitutions, some are associated with TTX resistance as assessed with patch clamp recordings. What advantages does TTX resistance offer pufferfish? A main advantage is that the high tissue concentrations of TTX act as an effective

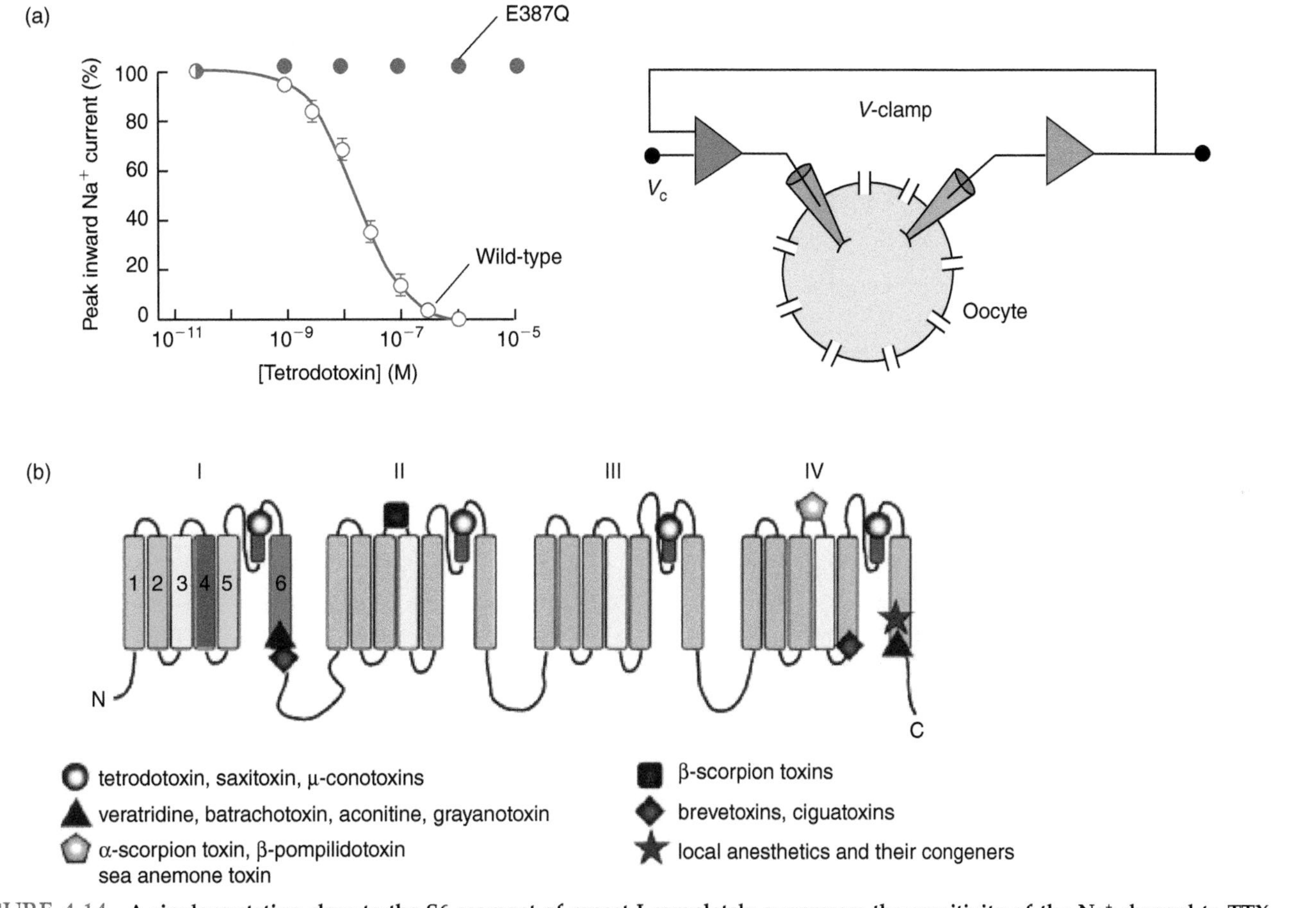

FIGURE 4.14 **A single mutation close to the S6 segment of repeat I completely suppresses the sensitivity of the Na^+ channel to TTX.** **(a)** A mutation of the glutamic acid residue 387 to glutamine (E387Q) is introduced in the rat Na^+ channel type II. *Xenopus* oocytes are injected with either the wild-type mRNA or the mutant mRNA. The macroscopic Na^+ currents are recorded 4–7 days later with the double-electrode voltage clamp technique. Dose–response curves for the wild-type (open circles) and the mutant E387Q (filled circles) to tetrodotoxin (TTX). TTX sensitivity is determined by perfusing TTX-containing external solutions and by measuring the macroscopic peak inward current. The TTX concentration that reduces the wild-type Na^+ current by 50% (IC50) is 18 nM. Data are averaged from 7–8 experiments. **(b)** Topology of drug binding sites on Na^+ channel α-subunit. Each symbol represents the toxin or drug binding site indicated at the bottom of the figure. All these sites have been characterized by site-directed mutagenesis. Part (**a**) from Noda M, Suzuki H, Numa S, Stühmer W (1989) A single point mutation confers tetrodotoxin and saxitoxin insensitivity on the sodium channel II. *FEBS Lett.* **259**, 213–216, with permission. Part (**b**) drawing adapted from Ogata N, Ohishi Y (2002) Molecular diversity of structure and function of the voltage-gated Na^+ channels. *Jpn. J. Pharmacol.* **88**, 365–377, with permission.

chemical defense against predators. Also TTX resistance enables pufferfish to feed on TTX-bearing organisms that are avoided by other fish.

TTX is not the only toxin to target Na^+ channels. Most of these toxins, except for TTX and its congeners that occlude the outer pore of the channel, bind to sites that are related to the activation and inactivation process. They fall into at least five different classes according to their corresponding receptor sites: (i) hydrophilic toxins such as TTX, saxitoxin (STX) and α-conus toxin; (ii) lipid-soluble neurotoxins such as batrachotoxin (BTX), veratridine, aconitine and grayanotoxin; (iii) α-scorpion peptide toxins and sea anemone peptide toxins; (iv) β-scorpion peptide toxins; (v) lipid-soluble brevetoxins and ciguatoxins. They can have opposite effects. For example, toxins (ii) are activators of Na^+ channels and toxins (iii) the α-scorpion toxins block fast inactivation and thereby generate persistent sodium current.

4.2.7 Segment S4, the region between segments S5 and S6, and the region between domains III and IV play a significant role in activation, ion permeation and inactivation, respectively

The major questions about a voltage-gated ionic channel and particularly the Na^+ channel are the following:

- How does the channel open in response to a voltage change?
- How is the permeation pathway designed to define single-channel conductance and ion selectivity?
- How does the channel inactivate?

In order to identify regions of the Na^+ channels involved in these functions, site-directed mutagenesis experiments were performed. The activity of each type of mutated Na^+ channel is analyzed with patch clamp recording techniques.

The membrane spanning segments S5 and S6 and the P-loop are membrane associated and contribute to pore formation

The Na^+ channels are highly selective for Na^+ ions. This selectivity presumably results from negatively charged amino acid residues located in the channel pore. Moreover, these amino acids must be specific to Na^+ channels (i.e. different from the other members of voltage-gated cationic channels such as K^+ and Ca^{2+} channels) to explain their weak permeability to K^+ or Ca^{2+} ions.

Studies using mutagenesis to alter ion channel function have shown that the S5 and S6 segments of each domain, with each S5–S6 pair connected by an intervening loop (P-loop), form the central ion-conducting pore (see **Figure 4.6**). The region of the sodium channel that forms the outer end of the pore was first determined by identifying the amino acid residues that form the binding site of the pore-blocking toxin tetrodotoxin. They are located in the P-loop between S5 and S6. Mutations of the same amino acid residues were shown to control ion selectivity. Moreover, a single amino acid substitution in these regions in repeats III and IV alters the ion selectivity of the Na^+ channel to resemble that of Ca^{2+} channels. These residues would constitute part of the selectivity filter of the channel. Recent studies of the crystal structure of a bacterial Na^+ channel revealed an overall pore architecture with a large external vestibule, a narrow ion selectivity filter containing the amino acid residues shown to determine ion selectivity (in the P-loop), and a large central cavity that is lined by the S6 segments and is water filled. There is a general agreement that the selectivity filter is formed by P-loops; i.e. relatively short polypeptide segments that extend into the aqueous pore from the extracellular side of the membrane. Rather than extending completely across the lipid bilayer, a large portion of the pore loop is near the extracellular face of the channel. Only a short region extends into the membrane to form the selectivity filter. In the case of the voltage-gated Na^+ channel, each of the four homologous domains contributes a loop to the ion-conducting pore.

The S4 segment is the voltage sensor

The S4 segments are positively charged and hydrophobic (**Figure 4.15a**). Moreover, the typical amino acid sequence of S4 (four to eight repeated motifs of a positively charged amino acid residue (usually arginine) followed by two hydrophobic residues) is conserved among the different voltage-gated channels. These observations led to the suggestion that S4 segments have a transmembrane orientation and are voltage sensors. To test this proposed role, positively charged amino acid residues are replaced by neutral or negatively charged residues in the S4 segment of a rat brain Na^+ channel type II. The mutated channels are expressed in *Xenopus* oocytes. When more than three positive residues are mutated in the S4 segments of repeat I or II, no appreciable expression of the mutated channel is obtained. The replacement of only one arginine or lysine residue in segment S4 of repeat I by a glutamine residue shifts the activation curve to more positive potentials (**Figures 4.15b,c**).

It is hypothesized that, at rest, the positive charges in S4 form ion pairs with neighboring negatively charged residues, thereby stabilizing the channel in the non-conducting closed conformation. With a change in the electric field across the membrane, these ion pairs would break as the S4 charges move and new ion pairs would form to stabilize the conducting, open conformation of the channel. The transmembrane position of the S4 segment has been confirmed by mapping the receptor sites for scorpion toxins in detail and showing that these toxins bind to the outer end of the S3–S4 loop of the voltage sensors in both resting and activated states, thereby

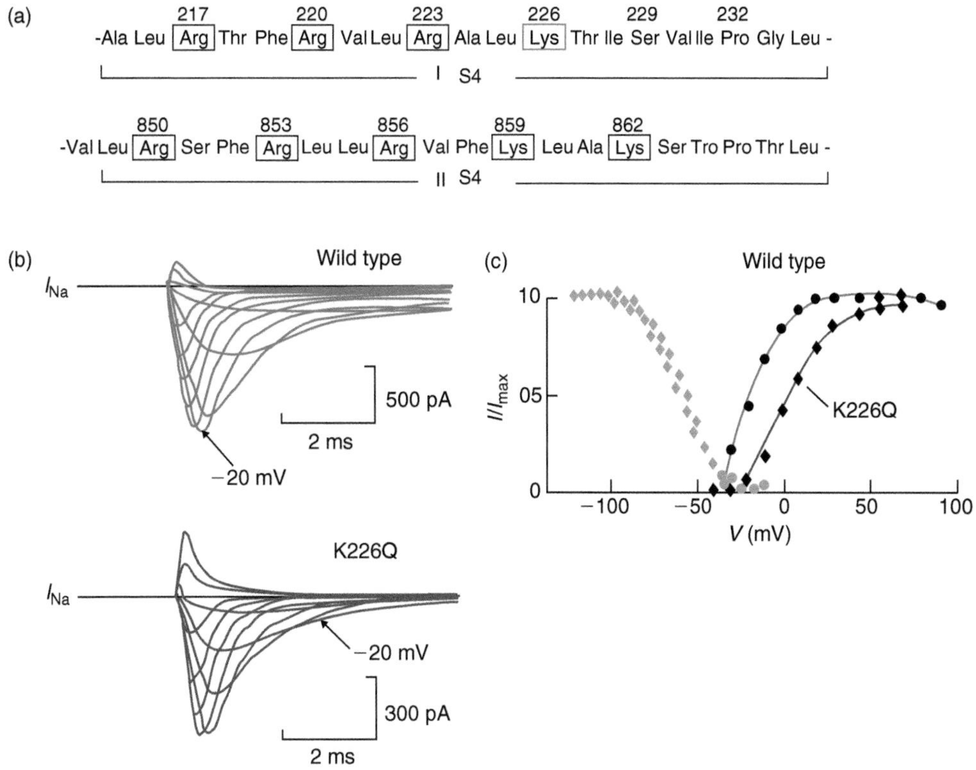

FIGURE 4.15 **Effect of mutations in the S4 segment on Na^+ current activation.** Oocytes are injected with the wild-type rat brain Na^+ channel or with Na^+ channels mutated on the S4 segment. The activity of a population of Na^+ channels is recorded in patch clamp (cell-attached macropatches). **(a)** Aminoacid sequences of segment S4 of the internal repeats I (I S4) and II (II S4) of the wild-type rat Na^+ channel. Positively charged amino acids are boxed with solid lines and the numbers of the relevant residues are given. In the mutated channel studied here the lysine residue in position 226 is replaced by a glutamine residue (K226Q). **(b)** In response to step depolarizations ranging from −60 to +70 mV from a holding potential of −120 mV, a family of macroscopic Na^+ currents is recorded for each type of Na^+ channel. The arrow indicates the response to the test potential −20 mV. Note that at −20 mV the amplitude of the Na^+ current is at its maximum for the wild-type and less than half maximum for the mutated channel. **(c)** Steady-state activation (right) and inactivation (left) curves for the wild-type (circles) and the mutant (diamonds) Na^+ channels. Adapted from Stühmer W, Conti F, Suzuki H *et al.* (1989) Structural parts involved in activation and inactivation of the sodium channel. *Nature* **339**, 597–603, with permission.

establishing that the S4 segment remains in a transmembrane position in both of these states.

The cytoplasmic loop between domains III and IV contains the inactivation particle which, in a voltage-dependent manner, enters the mouth of the Na^+ channel pore and inactivates the channel

The results obtained from three different types of experiments strongly suggest that the short cytoplasmic loop connecting homologous domains III and IV, L_{III-IV} loop (see **Figures 4.6a** and **4.16a**), is involved in inactivation: (i) cytoplasmic application of endopeptidases; (ii) cytoplasmic injection of antibodies directed against a peptide sequence in the region between repeats III and IV; and (iii) cleavage of the region between repeats III and IV (**Figure 4.16a–c**); all strongly reduce or block inactivation. Moreover, in some human pathology where the Na^+ channels poorly inactivate (as shown with single-channel recordings from biopsies), this region is mutated.

Positively charged amino acid residues of this L_{III-IV} loop are not required for inactivation since only the mutation of a hydrophobic sequence, isoleucine–phenylalanine–methionine (IFM), to glutamine completely blocks inactivation. The critical residue of the IFM motif is phenylalanine since its mutation to glutamine slows inactivation 5000-fold. It is proposed that this IFM sequence is directly involved in the conformational change leading to inactivation. It would enter the mouth of the pore, thus occluding it during the process of inactivation. In order to test this hypothesis, the ability of synthetic peptides containing the IFM motif to restore fast inactivation to non-inactivating rat brain Na^+ channels expressed in kidney carcinoma cells is examined. The intrinsic inactivation of Na^+ channels is first made non-functional by a mutation of the IFM motif. When the recording is now performed with a patch pipette containing the synthetic peptide with an IFM motif, the non-inactivating whole cell Na^+ current now inactivates. Since the restored inactivation has the rapid, voltage-dependent time course characteristic of inactivation of the wild-type Na^+ channels, it is proposed that the IFM motif serves as an inactivation particle (**Figure 4.16d**).

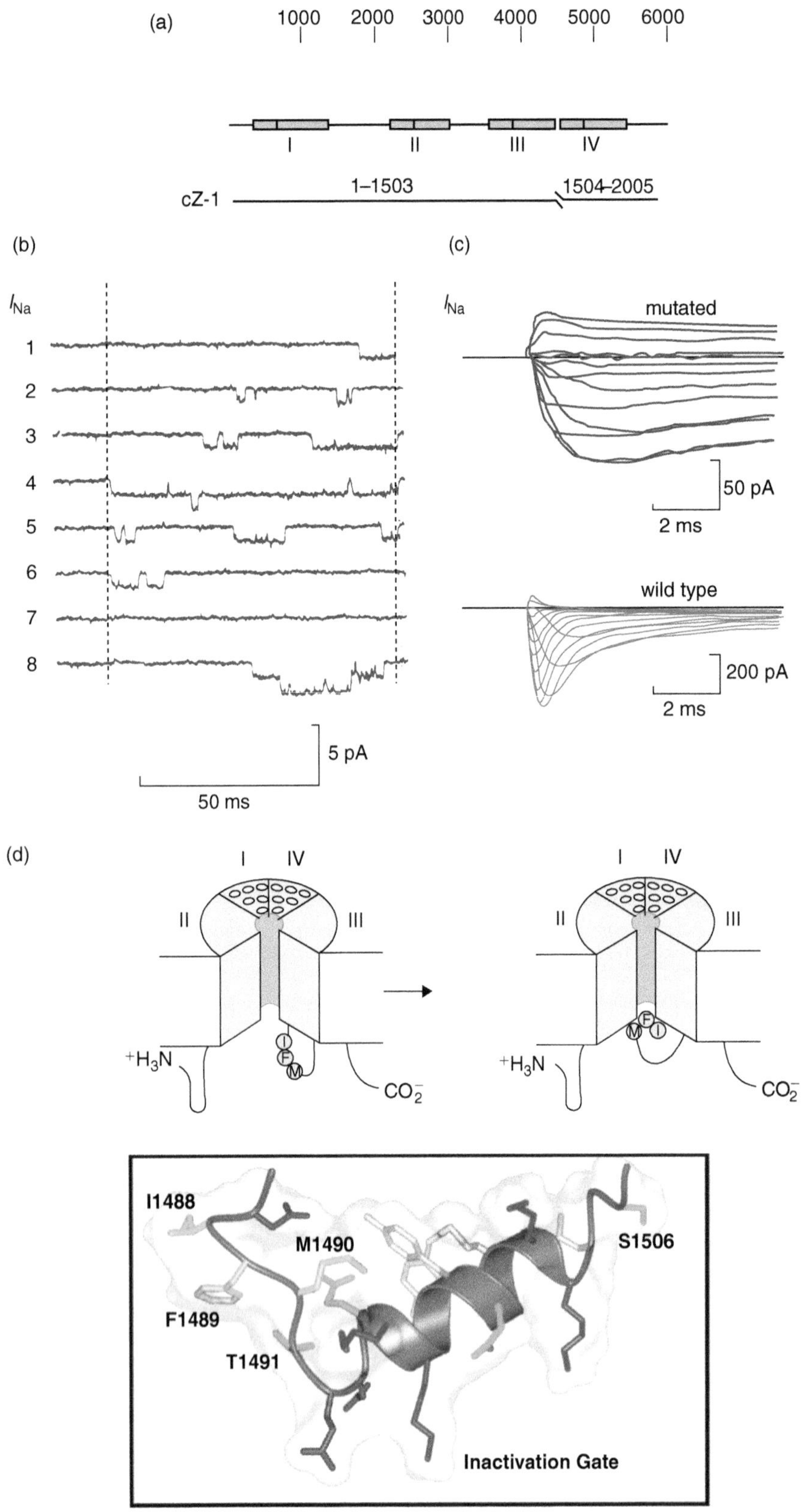

FIGURE 4.16 **Effects of mutations in the region between repeats III and IV on Na$^+$ current inactivation. (a)** Linear representation of the wild-type Na$^+$ channel (upper trace) and the mutated Na$^+$ channel (bottom trace). The mutation consists of a cut with an addition of four to eight residues at each end of the cut. An equimolar mixture of the two mRNAs encoding the adjacent fragments of the Na$^+$ channel protein separated with a cut is injected into oocytes. **(b)** Single-channel recordings of the activity of the mutated Na$^+$ channel in response to a depolarizing step to −20 mV from a holding potential of −100 mV. Note that late single or double openings (line 8) are often recorded. The mean open time τ_o is 5.8 ms and the elementary conductance γ_{Na} is 17.3 pS. **(c)** Macroscopic Na$^+$ currents recorded from the mutated (upper trace) and the wild-type (bottom trace) Na$^+$ channels. **(d)** Model for inactivation of the voltage-gated Na$^+$ channels. The region linking repeats III and IV is depicted as a hinged lid that occludes the transmembrane pore of the Na$^+$ channel during inactivation. Parts (**a**)–(**c**) from Pappone PA (1980) Voltage clamp experiments in normal and denervated mammalian skeletal muscle fibers. *J. Physiol.* **306**, 377–410, with permission. Drawing (**d** top) from West JW, Patton DE, Scheuer T *et al.* (1992) A cluster of hydrophobic amino acid residues required for fast sodium channel inactivation. *Proc. Natl Acad. Sci. USA* **89**, 10910–10914, with permission. Part (**d** bottom) adapted from Catterall WA (2012) Voltage-gated sodium channels at 60: structure, function and pathophysiology. *J. Physiol.* **590**, 2577–2589, with permission.

4.2.8 Conclusion: the consequence of the opening of a population of *N* Na^+ channels is a transient entry of Na^+ ions which depolarizes the membrane above 0 mV

The function of the population of *N* Na^+ channels at the axon initial segment or at nodes of Ranvier is to ensure a *sudden* and *brief* depolarization of the membrane above 0 mV.

Rapid activation of Na^+ channels makes the depolarization phase sudden

In response to a depolarization to the threshold potential, the closed Na^+ channels (**Figure 4.17a**) of the axon initial segment begin to open (**b**). The flux of Na^+ ions through the few open Na^+ channels depolarizes the membrane more and thus triggers the opening of other Na^+ channels (**c**). In consequence, the flux of Na^+ ions increases, depolarizes the membrane more and opens other Na^+ channels until all the *N* Na^+ channels of the segment of membrane are opened (**d**). In (**d**) the depolarization phase is at its peak. Na^+ channels are opened by depolarization and once opened, they contribute to the membrane depolarization and therefore to their activation: it is a self-maintained process.

Rapid inactivation of Na^+ channels makes the depolarization phase brief

Once the Na^+ channels have opened, they begin to inactivate (**e**). Therefore, though the membrane is depolarized, the influx of Na^+ ions diminishes quickly. Therefore, the Na^+-dependent action potential is a spike and does not present a plateau phase. Inactivation is a very important protective mechanism since it prevents potentially toxic persistent depolarization.

4.3 THE REPOLARIZATION PHASE OF THE SODIUM-DEPENDENT ACTION POTENTIAL RESULTS FROM Na^+ CHANNEL INACTIVATION AND PARTLY FROM K^+ CHANNEL ACTIVATION

The participation of a voltage-gated K^+ current in action potential repolarization differs from one preparation to another. For example, in the squid axon, the voltage-gated K^+ current plays an important role in spike repolarization, though in mammalian peripheral nerves this current is almost absent. However, the action potentials of the squid axon and that of mammalian nerves have the same duration. This is because the Na^+ current in mammalian axons inactivates two to three times faster than that of the

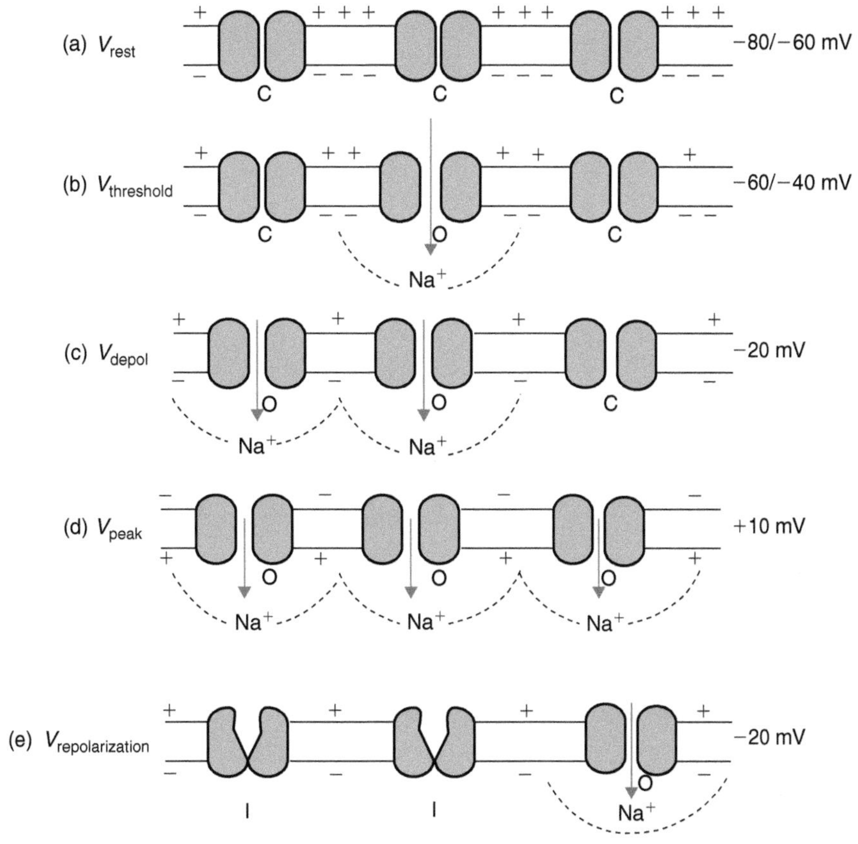

FIGURE 4.17 **Different states of voltage-gated Na^+ channels in relation to the different phases of the Na^+-dependent action potential.** C, closed state; O, open state; I, inactivated state; →, driving force for Na^+ ions.

frog axon. Moreover, the leak K^+ currents are important in mammalian axons (see below).

Voltage-gated K^+ channels can be classified into two major groups based on physiological properties:

- Delayed rectifiers which activate after a delay following membrane depolarization and inactivate slowly;
- A-type channels which are fast activating and fast inactivating.

The first type, the delayed rectifier K^+ channels, plays a role in action potential repolarization. The A-types inactivate too quickly to do so. They play a role in firing patterns and are explained in Chapter 17.

This section will explain the structure and activity of the voltage-gated, delayed rectifier K^+ channels responsible for action potential repolarization in the squid or frog nerves. Then Section 4.4 will explain the other mode of repolarization observed in mammalian nerves in which the delayed rectifier current does not play a significant role.

4.3.1 The delayed rectifier K^+ channel consists of four α-subunits and auxiliary β-subunits

K^+ channels represent an extremely diverse ion channel type. All known K^+ channels are related members of a single protein family. Finding genes responsible for a native K^+ current is not an easy task, because K^+ channels have a great diversity: more than 100 K^+ channel subunits have been identified to date. Among strategies used to identify which genes encode a particular K^+ channel are the single cell reverse transcriptase chain reaction (scRT-PCR) protocol combined with patch clamp recording and the injection of subfamily-specific dominant negative constructs in recorded neurons. Results of such experiments strongly suggested that α-subunits of delayed rectifiers are attributable at least to K_v1, K_v2 and K_v3 subfamilies.

Delayed rectifier K^+ channel α-subunits form homo- or hetero-tetramers in the cell membrane. As for the Na^+ channel, the P-loop linking segments S5 and S6 contributes to the formation of the pore and the auxiliary small β-subunits associated with the α-subunit are considered to be intracellularly located (**Figure 4.18a**).

4.3.2 Membrane depolarization favors the conformational change of the delayed rectifier channel towards the open state

The function of the delayed rectifier channel is to transduce, with a delay, membrane depolarization into an exit of K^+ ions

Single-channel recordings were obtained by Conti and Neher in 1980 from the squid axon. We shall, however, look at recordings obtained from K^+ channels expressed in oocytes or in mammalian cell lines from cDNA encoding a delayed rectifier channel of rat brain. Since the macroscopic currents mediated by these channels have time courses and ionic selectivity resembling those of the classical delayed outward currents described in nerve and muscle, these single-channel recordings are good examples for describing the properties of a delayed rectifier current.

Figure 4.19 shows a current trace obtained from patch clamp recordings (inside-out patch) of a rat brain K^+ channel (RCK1) expressed in a *Xenopus* oocyte. In the presence of physiological extracellular and intracellular K^+ concentrations, a depolarizing voltage step to 0 mV from a holding potential of −60 mV is applied. After the onset of the step, a rectangular pulse of elementary current, upwardly directed, appears. It means that the current is outward; K^+ ions leave the cell. In fact, the driving force for K^+ ions is outward at 0 mV.

It is immediately striking that the gating behavior of the delayed rectifier channel is different from that of the Na^+ channel (compare **Figures 4.7a** or **4.8a** and **4.19**). Here, the rectangular pulse of current lasts the whole depolarizing step with short interruptions during which the current goes back to zero. It indicates that the delayed rectifier channel opens, closes briefly and reopens many times during the depolarizing pulse: the delayed rectifier channel does not inactivate within seconds. Another difference is that the delay of opening of the delayed rectifier is much longer than that of the Na^+ channel, even for large membrane depolarizations (mean delay 4 ms in **Figure 4.20a**).

When the same depolarizing pulse is now applied every 1–2 s, we observe that the delay of channel opening is variable (1–10 ms) but gating properties are the same in all recordings: the channel opens, closes briefly and reopens during the entire depolarizing step (**Figure 4.20a**). Amplitude histograms collected at 0 mV membrane potential from current recordings, such as those shown in **Figure 4.20a**, give a mean amplitude of the unitary currents of +0.8 pA (**Figure 4.20c**). This means that the most frequently occurring main amplitude is +0.8 pA.

Why do delayed rectifier K_v channels open slower than Na_v channels?

Whereas all four voltage-sensitive domains in K_v channels must be activated in order for pore opening to occur, Na_v channel pore opening requires activation of only three voltage-sensitive domains, contained within domains I–III. Movement of the fourth voltage-sensitive domain in Na_v channels is delayed compared with domains I–III, and functionally, it appears to mediate fast inactivation of ionic current through the pore. Thus, part of the difference in activation speed between Na_v and K_v channels may be due to the lesser number of voltage-sensitive domains required to move in Na_v channels.

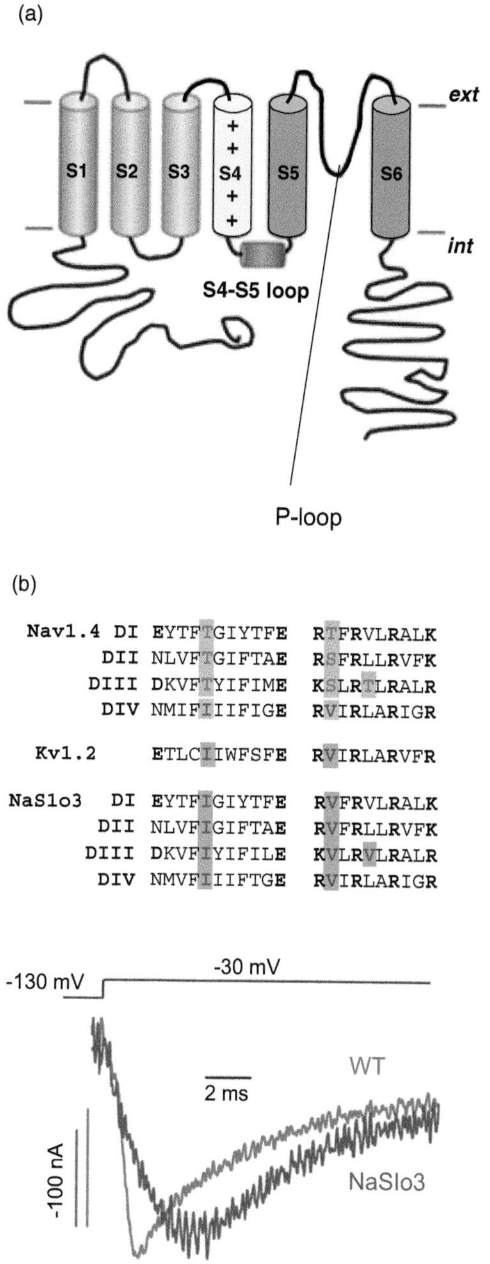

FIGURE 4.18 **Putative transmembrane organization of the α-subunit of the delayed rectifier, voltage-gated K^+ channel. (a)** Diagrammatic representation of the predicted membrane topology of a single $K_v2.1$ α-subunit. S1–S6 represent the transmembrane segments (cylinders represent putative α-helical segments), P-loop represents the amino acid residues that form the bulk of the lining of the channel pore. The voltage-sensing domain is formed by the S1–S4 segments, in which S4 contains a high density of positively charged residues and is the main transmembrane voltage-sensing component. Four of these subunits form a tetrameric structure surrounding a central conduction pathway. **(b)** Portion of the S2 and S4 segments in the $Na_v1.4$ and $K_v1.2$ channels aligned with respect to conserved charged residues (bold black). Rapid $Na_v1.4$ channel contains hydrophilic residues at conserved speed control sites (bold red, T for threonine, S for serine), while slow $K_v1.2$ channel contains hydrophobic residues (bold blue, I for isoleucine, V for valine) at homologous positions. Below are the sequence of the mutated NaSlo3. Normalized ionic current traces for wild-type (WT) $Na_v1.4$ (red) and the NaSlo3 mutant (blue). Part **(a)** from Barros F, Domínguez P, de la Peña P (2012) Cytoplasmic domains and voltage-dependent potassium channel gating. *Front. Pharmacol.* **3**, 49, with permission. Part **(b)** Adapted from Lacroix JJ, Campos FV, Frezza L, Bezanilla F (2013) Molecular bases for the asynchronous activation of sodium and potassium channels required for nerve impulse generation. *Neuron* **79**, 651–657, with permission.

The remaining molecular factors contributing to this difference have been recently defined using combined structure–function analysis (site-directed mutagenesis) with biophysical measurements (voltage clamp techniques to record ionic currents and gating currents) from mammalian $Na_v1.2$ and 1.4 channels and $K_v1.2$ channels expressed in *Xenopus laevis* oocytes. The gating currents reflect the physical rearrangements of the voltage-sensitive domains within each channel type.

Using primary sequence alignments of the voltage sensitive domains (i.e. S1–4 of the four domains I–IV of Na_v channels or S1–4 of the four subunits of K_v channels), highly conserved amino acid differences were identified. In particular, domains I–III of fast activating Na_v channel isoforms contain the key hydrophilic residues threonine (T) or serine (S) in their S2 and S4 segments, whereas slow activating K_v channels contain hydrophobic residues such as isoleucine (I) or valine (V) in the equivalent positions. Noteworthy in the voltage-sensitive domain IV of Na_v channels, which activates slowly, these S2 and S4 positions contain the hydrophobic residues, isoleucine and valine, respectively, and thus resemble the motifs of slowly activating K_v channels (**Figure 4.18b**).

The authors tested whether replacing the two hydrophilic threonine residues in the S2 and S4 segments of domains I, II and III of $Na_v1.4$ channel with hydrophobic residues (I or V) would create a 'slow' activating $Na_v1.4$ channel. They in fact obtained a slow activating

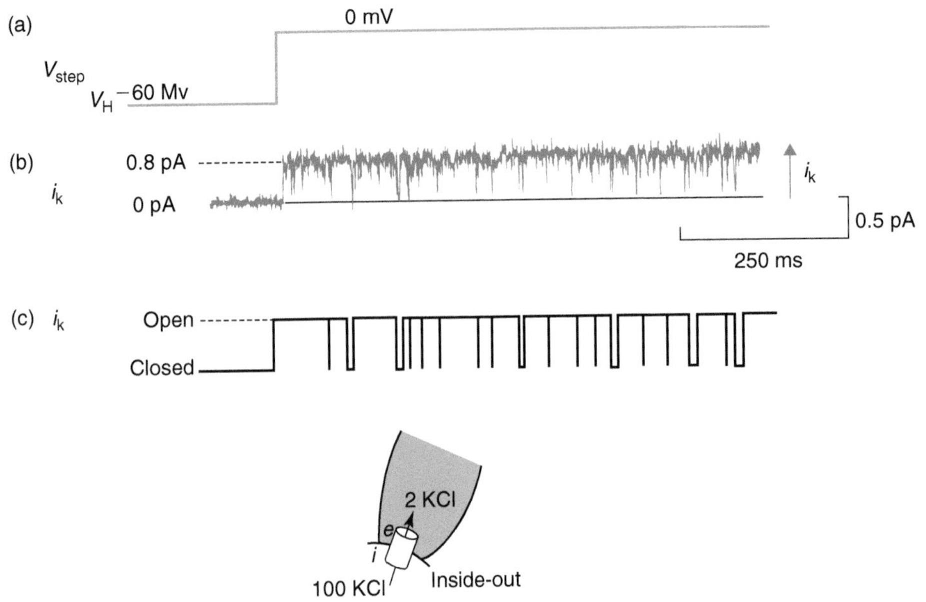

FIGURE 4.19 **Single K⁺ channel openings in response to a depolarizing step.** The activity of a single delayed rectifier channel expressed from rat brain cDNA in a *Xenopus* oocyte is recorded in patch clamp (inside-out patch). A depolarizing step to 0 mV from a holding potential of −60 mV **(a)** evokes the opening of the channel **(b)**. The elementary current is outward. The channel then closes briefly and reopens several times during the depolarization, as shown in the drawing **(c)** that interprets the current trace. Bathing solution or intracellular solution (in mM): 100 KCl, 10 EGTA, 10 HEPES. Pipette solution or extracellular solution (in mM): 115 NaCl, 2 KCl, 1.8 $CaCl_2$, 10 HEPES. Adapted from Stühmer W, Stocker M, Sakmann B *et al.* (1988) Potassium channels expressed from rat brain cDNA have delayed rectifier properties. *FEBS Lett.* **242**, 199–206, with permission.

Na_v channel (NaSlo) (**Figure 4.18b**). Using a variety of amino acid substitutions at the S2 and S4 'speed control' positions, the authors observed that the rates of activation correlated negatively with the hydrophobicity of amino acid side chains present at these locations; that is, the presence of hydrophobic residues slowed down the speed of activation. In summary, K_v channels activate less rapidly than fast activating Na_v channels for

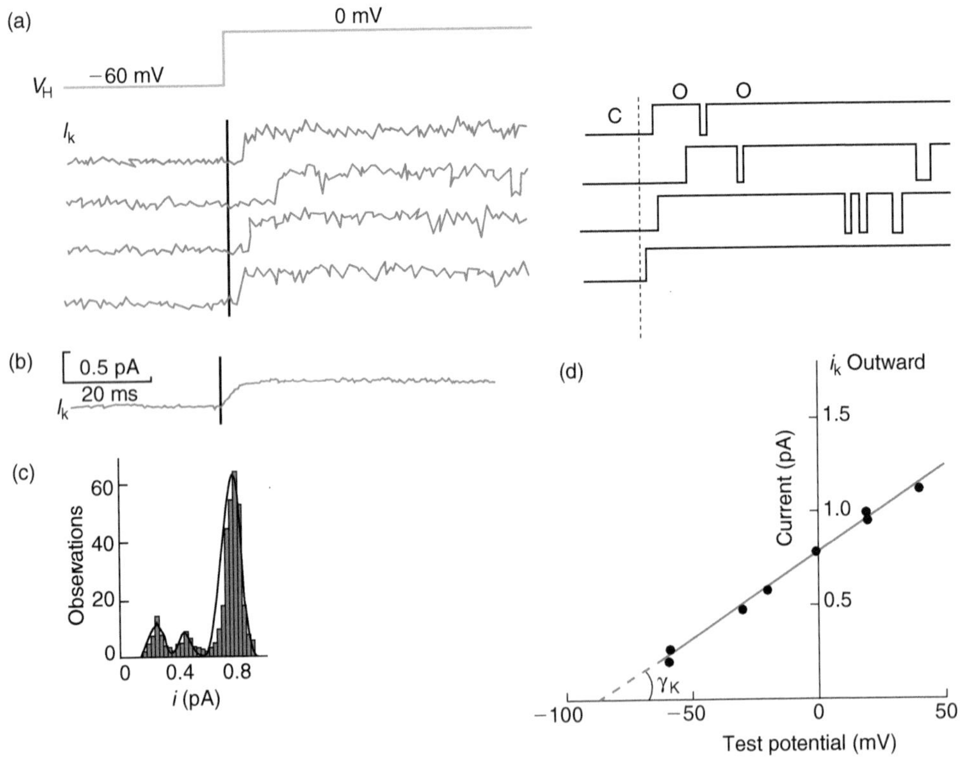

FIGURE 4.20 **Characteristics of the elementary delayed rectifier current.** Same experimental design as in **Figure 4.19**. The patch of membrane contains a single delayed rectifier channel. **(a)** Successive sweeps of outward current responses to depolarizing steps from −60 mV to 0 mV (C for closed state, O for open state of the channel). **(b)** Averaged current from 70 elementary currents as in **(a)**. **(c)** Amplitude histogram of the elementary outward currents recorded at test potential 0 mV. The mean elementary current amplitude observed most frequently is 0.8 pA. **(d)** Single channel current–voltage relation (i_K–V). Each point represents the mean amplitude of at least 20 determinations. The slope is γ_K = 9.3 pS. The reversal potential E_{rev} = −89 mV. From Stühmer W, Stocker M, Sakmann B *et al.* (1988) Potassium channels expressed from rat brain cDNA have delayed rectifier properties. *FEBS Lett.* **242**, 199–206, with permission.

at least two reasons: (1) four K_v subunits must activate vs. three Na_v domains to achieve pore opening and (2) hydrophilic vs. hydrophobic amino acids present at S2 and S4 'speed control' sites in each Na_v domain or K_v subunit have a kinetic influence.

4.3.3 The open probability of the delayed rectifier channel is stable during a depolarization in the range of seconds

The average open time τ_o measured in the patch illustrated in **Figure 4.19** is 4.6 ms. The mean closed time τ_c is 1.5 ms. As seen in **Figures 4.19** and **4.20a**, during a depolarizing pulse to 0 mV the delayed rectifier channel spends much more time in the open state than in the closed state: at 0 mV its average open probability is high ($p_o = 0.76$).

In order to test whether the delayed rectifier channels show some inactivation, long-lasting recordings are performed. Though no significant inactivation is apparent during test pulses in the range of seconds, during long test depolarizations (in the range of minutes) the channel shows steady-state inactivation at positive holding potentials (not shown). Therefore, in the range of seconds, the inactivation of the delayed rectifier channel can be omitted: the channel fluctuates between the closed and open states:

$$C \rightleftharpoons O$$

The transition from the closed (C) state to the open (O) state is triggered by membrane depolarization with a delay. The delayed rectifier channel activates in the range of milliseconds. In comparison, the Na^+ channel activates in the range of submilliseconds. The O to C transitions of the Na^+ channel frequently happen though the membrane is still depolarized. It also happens when the membrane repolarizes.

4.3.4 The K^+ channel has a constant unitary conductance γ_K

In **Figure 4.21a**, unitary currents are shown in response to increasing depolarizing steps from −50 to

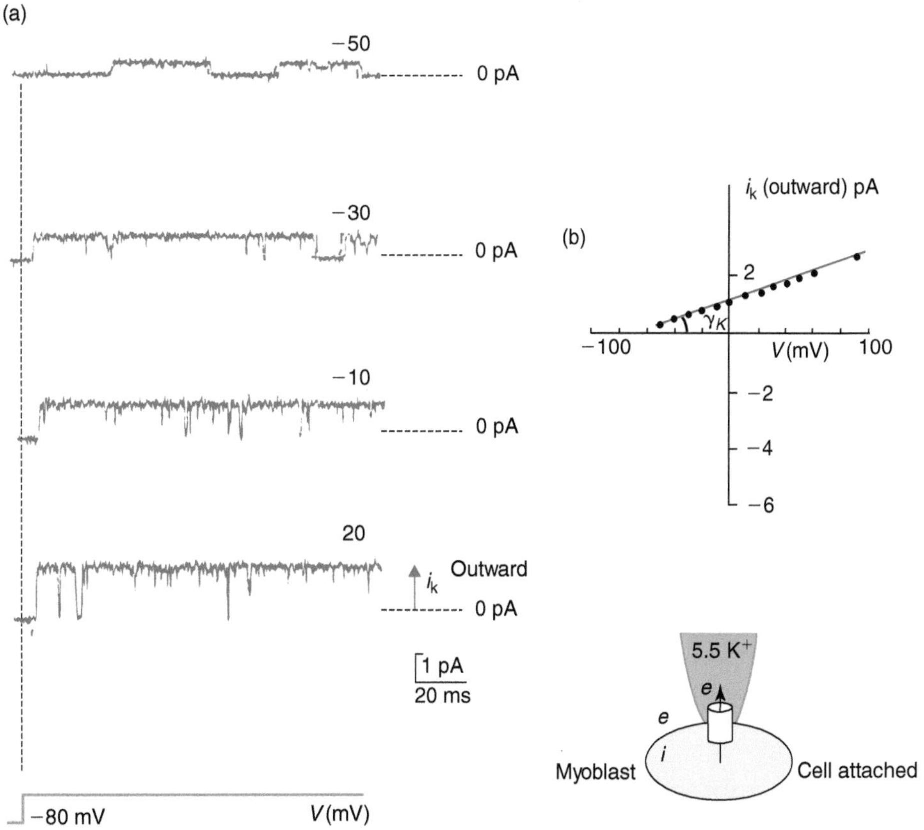

FIGURE 4.21 **The single-channel current/voltage (i_K/V) relation is linear.** Delayed rectifier K^+ channels from rat brain are expressed in a myoblast cell line. **(a)** The activity of a single channel is recorded in patch clamp (cell-attached patch). Unitary currents are recorded at different test potentials (from −50 mV to +20 mV) from a holding potential at −80 mV. Bottom trace is the voltage trace. **(b)** $i_K - V$ relation obtained by plotting the mean amplitude of i_K at the different test potentials tested. i_K reverses at = −75 mV and γ_K = 14 pS. Intrapipette solution (in mM): 145 NaCl, 5.5 KCl, 2 $CaCl_2$, 2 $MgCl_2$, 10 HEPES. Adapted from Koren G, Liman ER, Logothetis DE *et al.* (1990) Gating mechanism of a cloned potassium channel expressed in frog oocytes and mammalian cells. *Neuron* **2**, 39–51, with permission.

+20 mV from a holding potential of −80 mV. We observe that both the amplitude of the unitary current and the time spent by the channel in the open state increase with depolarization.

When the mean amplitude of the unitary K^+ current is plotted versus membrane test potential, a linear i_K/V relation is obtained (**Figures 4.20d** and **4.21b**). This linear i_K/V relation (between −50 and +20 mV) is described by the equation $i_K = \gamma_K(V - E_K)$, where V is the membrane potential, E_K is the reversal potential of the K^+ current, and γ_K is the conductance of the single delayed rectifier K^+ channel, or unitary conductance. Linear back-extrapolation gives a reversal potential value around −90/−80 mV, a value close to E_K calculated from the Nernst equation. This means that from −80 mV to more depolarized potentials, which correspond to the physiological conditions, the K^+ current is outward. For more hyperpolarized potentials, the K^+ current is inward.

The value of γ_K is given by the slope of the linear i_K/V curve. It has a constant value at any given membrane potential. This value varies between 10 and 15 pS depending on the preparation (**Figures 4.20d** and **4.21b**).

4.3.5 The macroscopic delayed rectifier K^+ current (I_K) has a delayed voltage dependence of activation and inactivates within tens of seconds

Whole cell currents in *Xenopus* oocytes expressing delayed rectifier channels start to activate at potentials positive to −30 mV and their amplitude is clearly voltage dependent. When unitary currents recorded from 70 successive depolarizing steps to 0 mV are averaged (**Figure 4.20b**), the macroscopic outward current obtained has a slow time to peak (4 ms) and lasts the entire depolarizing step. It closely resembles the whole cell current recorded with two electrode voltage clamps in the same preparation (rat brain delayed rectifier channels expressed in oocytes; **Figure 4.22a**). The whole cell current amplitude at steady state (once it has reached its maximal amplitude) for a given potential is:

$$I_K = N p_o i_K,$$

where N is the number of delayed rectifier channels in the membrane recorded, p_o the open probability at steady state and i_K the elementary current. The number of open channels Np_o increases with depolarization (to a maximal value) and so does I_K.

The I_K/V relation shows that the whole cell current varies linearly with voltage from a threshold potential which, in this preparation, is around −40 mV (**Figure 4.22b**). When the membrane is more hyperpolarized than the threshold potential, very few channels are open and I_K is equal to zero. For membrane potentials more depolarized than the threshold potential, I_K depends on

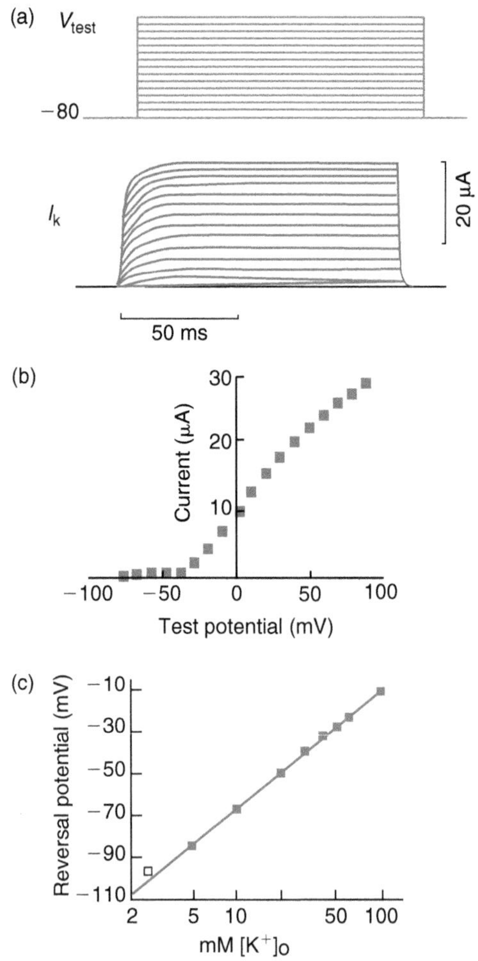

FIGURE 4.22 **Characteristics of the macroscopic delayed rectifier K^+ current.** The activity of N delayed rectifier channels expressed from rat brain cDNA in oocytes recorded in double-electrode voltage clamp. **(a)** In response to depolarizing steps of increasing amplitude (given every 2 s) from a holding potential of −80 mV (upper traces), a non-inactivating outward current of increasing amplitude is recorded (lower traces). **(b)** The amplitude of the current at steady state is plotted against test potential. The potential threshold for its activation is −40 mV. **(c)** The value of the reversal potentials of the macroscopic current is plotted against the extracellular concentration of K^+ ions on a semi-logarithmic scale. The slope is −55 mV. From Stühmer W, Stocker M, Sakmann B *et al.* (1988) Potassium channels expressed from rat brain cDNA have delayed rectifier properties. *FEBS Lett.* **242**, 199–206, with permission.

p_o and the driving force state ($V - E_K$) which augments with depolarization. Once p_o is maximal, I_K augments linearly with depolarization since it depends only on the driving force.

The delayed rectifier channels are selective to K^+ ions

Ion substitution experiments indicate that the reversal potential of I_K depends on the external K^+ ion concentration as expected for a selective K^+ channel. The reversal potential of the whole cell current is measured as in **Figure 4.22b** in the presence of different external concentrations of K^+ ions. These experimental values are plotted against the external K^+ concentration, $[K^+]_o$, on a semi-logarithmic scale. For concentrations ranging from 2.5 (normal frog Ringer) to 100 mM, a linear relation with a slope of 55 mV for a 10-fold change in $[K^+]_o$ is obtained (not shown). These data are well fitted by the Nernst equation. It indicates that the channel has a higher selectivity for K^+ ions over Na^+ and Cl^- ions.

The delayed rectifier channels are blocked by millimolar concentrations of tetraethylammonium (TEA) and by Cs^+ ions. Ammonium ions can pass through most K^+ channels, whereas its quaternary derivative TEA cannot, resulting in the blockade of most of the voltage-gated K^+ channels: TEA is a small open channel blocker. Amino acids in the carboxyl half of the region linking segments S5 and S6 (i.e. adjacent to S6) influence the sensitivity to pore blockers such as TEA.

4.3.6 Conclusion: during an action potential the consequence of the delayed opening of K^+ channels is an exit of K^+ ions, which repolarizes the membrane to resting potential

Owing to their delay of opening, delayed rectifier channels open when the membrane is already depolarized by the entry of Na^+ ions through open voltage-gated Na^+ channels (**Figure 4.23**). Therefore, the exit of K^+ ions does not occur at the same time as the entry of Na^+ ions (see also **Figure 4.24**). This allows the membrane to first depolarize in response to the entry of Na^+ ions and then to repolarize as a consequence of the exit of K^+ ions.

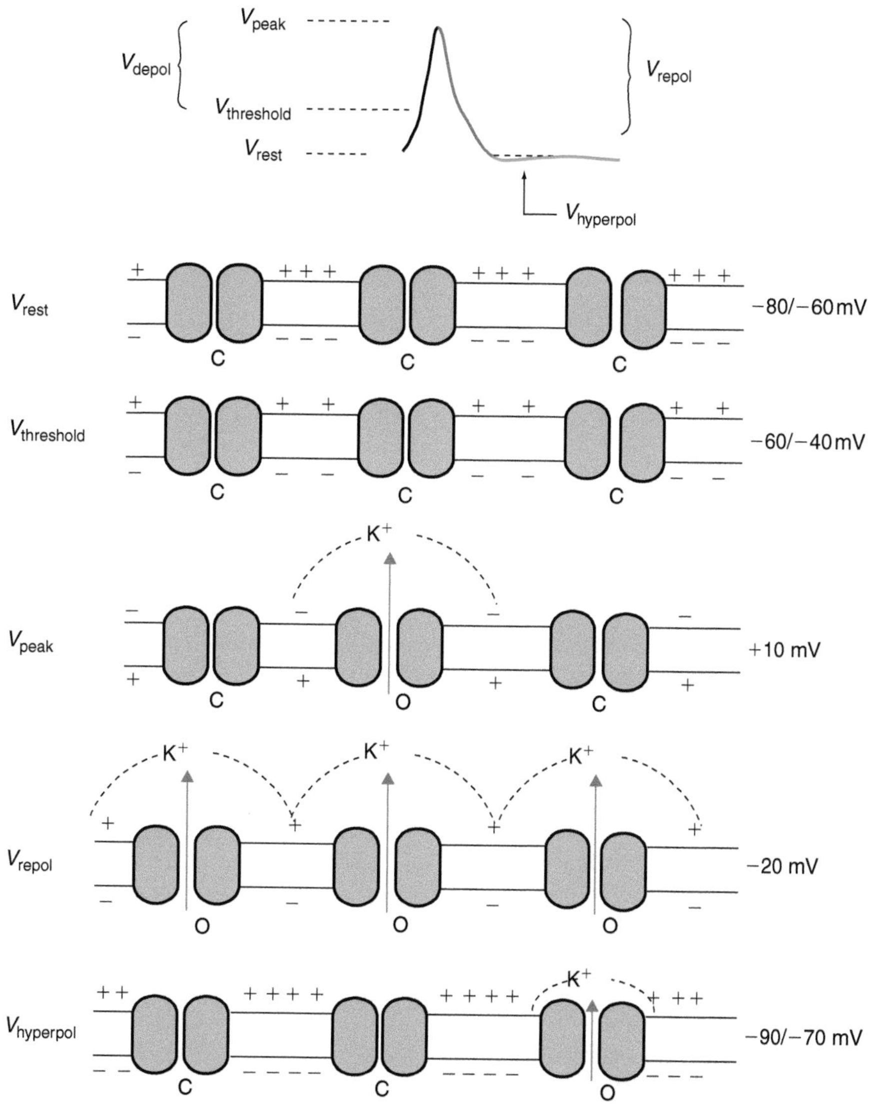

FIGURE 4.23 **States of the delayed rectifier K^+ channels in relation to the different phases of the Na^+-dependent action potential.** C, closed state; O, open state; ↑, driving force for K^+ ions.

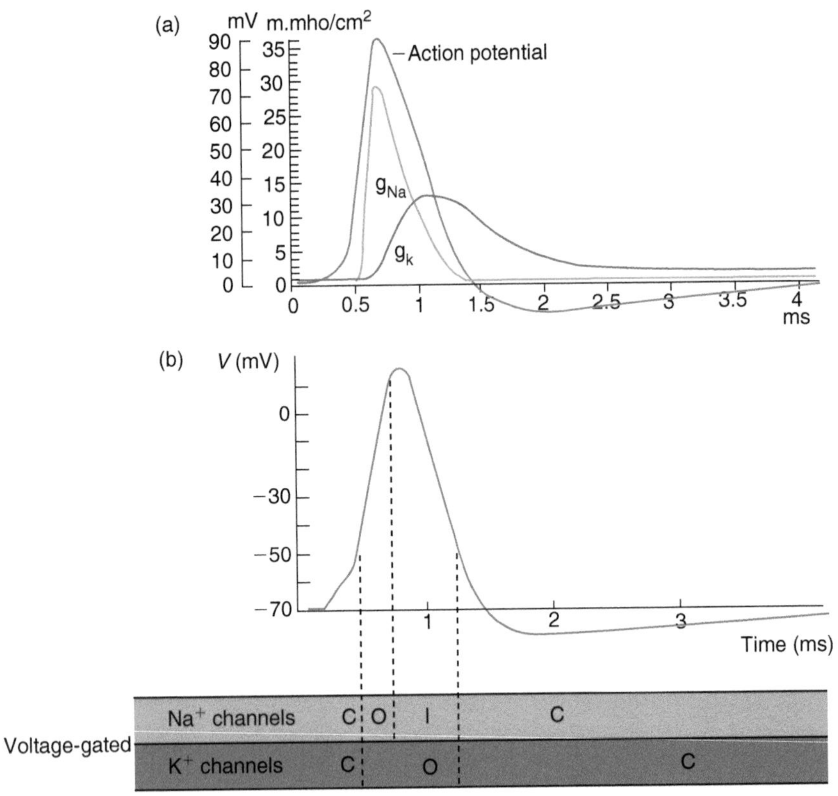

FIGURE 4.24 **Gating of Na^+ and K^+ channels during the Na^+-dependent action potential.** **(a)** Interpretation of the manner in which the conductances to Na^+ and K^+ contribute to the action potential. **(b)** State of the Na^+ and K^+ voltage-gated channels during the course of the action potential. O, channels open; I, channels inactivate; C, channels close or are closed. Trace (**a**) adapted from Hodgkin AL, Huxley AF (1952) A quantitative description of membrane current and its application to conduction and excitation in nerve. *J. Physiol.* **117**, 500–544, with permission.

4.4 SODIUM-DEPENDENT ACTION POTENTIALS ARE INITIATED AT THE AXON INITIAL SEGMENT IN RESPONSE TO A MEMBRANE DEPOLARIZATION AND THEN ACTIVELY PROPAGATE ALONG THE AXON

Na^+-dependent action potentials, because of their short duration (1–5 ms), are also named spikes. Na^+ spikes, for a given cell, have a stable amplitude and duration; they all look alike, and are binary, all-or-none. The pattern of discharge (which is often different from the frequency of discharge) and not individual spikes, carries significant information.

4.4.1 Summary on the Na^+-dependent action potential

The depolarization phase of Na^+ spikes is due to the rapid time to peak inward Na^+ current which flows into the axon initial segment or node. This depolarization is brief because the inward Na^+ current inactivates in milliseconds (**Figure 4.24b**).

In the squid giant axon or frog axon, spike repolarization is associated with an outward K^+ current through delayed rectifier channels (**Figures 4.24** and **4.25**) since TEA application dramatically prolongs the action potential (see **Figure 4.4b**). As pointed out by Hodgkin and Huxley: 'The rapid rise is due almost entirely to Na^+ conductance, but after the peak, the K^+ conductance takes a progressively larger share until, by the beginning of the hyperpolarized phase, the Na^+ conductance has become negligible. The tail of raised conductance that falls away gradually during the positive phase is due solely to K^+ conductance, the small constant leak conductance being of course present throughout.'

In contrast, in rat or rabbit myelinated axons, the action potential is very little affected by the application of TEA. The repolarization phase in these preparations is largely associated with a leak K^+ current. Voltage clamp studies confirm this observation. When the leak current is subtracted, almost no outward current is recorded in rabbit node (**Figure 4.25b**).

However, squid and rabbit nerve action potentials have the same duration (**Figure 4.25a**). In this preparation, the normal resting membrane potential is around −80 mV, which suggests the presence of a large leak K^+ current. Moreover, test depolarizations evoke large outward K^+ currents insensitive to TEA (**Figure 4.26**). How does the action potential repolarize in such preparations? First, the Na^+ currents in the rabbit node inactivate two to three times faster than those in the frog node. Second, the large leak K^+ current present at depolarized membrane potentials repolarizes the membrane. The amplitude of the leak K^+ current augments linearly with depolarization, depending only on the K^+ driving force.

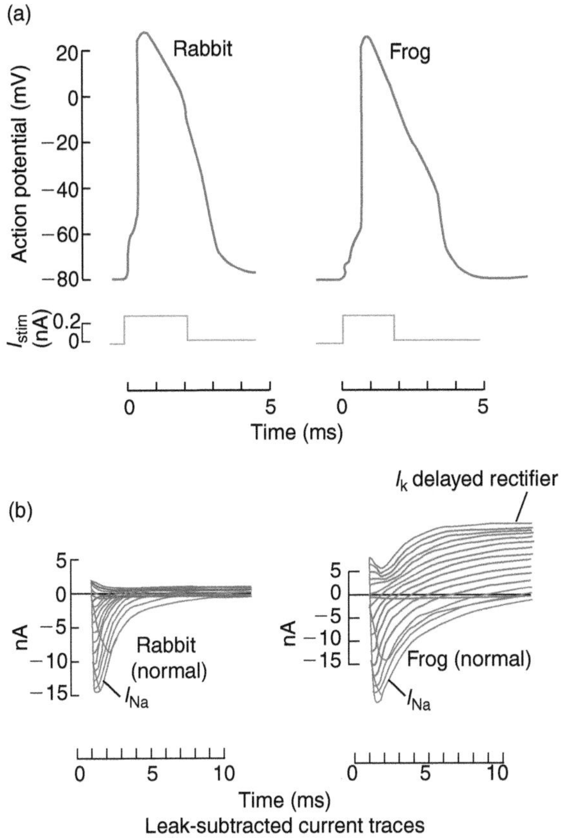

FIGURE 4.25 **The currents underlying the action potentials of the rabbit and frog nerves. (a)** The action potentials are recorded intracellularly at 14°C. Bottom trace is the current of stimulation injected in order to depolarize the membrane to initiate an action potential. **(b)** The currents flowing through the membrane at different voltages recorded in voltage clamp. In the rabbit node, very little outward current is recorded after the large inward Na^+ current. In the frog nerve, a large outward K^+ current is recorded after the large inward Na^+ current. Leak current is subtracted from each trace and does not appear in these recordings. Adapted from Chiu SY, Ritchie JM, Bogart RB, Stagg D (1979) A quantitative description of membrane currents in rabbit myelinated nerve. *J. Physiol.* **292**, 149–166, with permission.

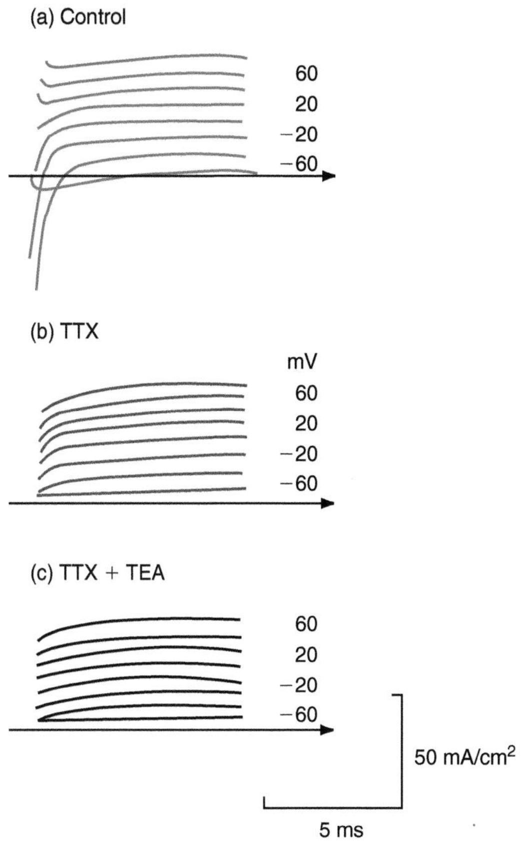

FIGURE 4.26 **TEA-resistant outward current in a mammalian nerve.** The currents evoked by depolarizing steps from −60 to +60 mV from a holding potential of −80 mV are recorded in voltage clamp in a node of Ranvier of an isolated rat nerve fiber. Control inward and outward currents **(a)**, after TTX 25 nM **(b)**, and after TTX 25 nM and TEA 5 mM **(c)** are added to the extracellular solution. The outward current recorded in **(c)** is the leak K^+ current. The delayed outward K^+ current is taken as the difference between the steady-state outward current in **(b)** and the leak current in **(c)**. Adapted from Brismar T (1980) Potential clamp analysis of membrane currents in rat myelinated nerve fibres. *J. Physiol.* **298**, 171–184, with permission.

4.4.2 Depolarization of the membrane to the threshold for voltage-gated Na^+ channel activation has two origins

The inward current which depolarizes the membrane of the initial segment to the threshold potential for voltage-gated Na^+ channel opening is one of the following:

- A depolarizing current resulting from the activity of excitatory afferent synapses (see Chapters 9 and 11) or afferent sensory stimuli (see Chapters 12 and 15). In the first case, the synaptic currents generated at postsynaptic sites in response to synaptic activity summate and, when the resulting current is inward, it can depolarize the membrane to the threshold for spike initiation. In the second case, sensory stimuli are transduced in inward currents that can depolarize the membrane to the threshold for spike initiation.
- An intrinsic regenerative depolarizing current such as, for example, in heart cells or invertebrate neurons.

4.4.3 The site of initiation of Na^+-dependent action potentials is the axon initial segment

The site of initiation was suggested long ago to occur in the axon initial segment since the threshold for spike initiation was the lowest at this level. This has been directly demonstrated with the double patch clamp technique. First, the dendrites and soma belonging to the same Purkinje neuron of the cerebellum are visualized in a rat brain slice. Then the activity is recorded simultaneously at both these sites with two patch electrodes (whole-cell patches). To verify that somatic and dendritic recordings are made from the same cell, the Purkinje cell is filled with two differently colored fluorescent dyes: Cascade blue at the soma and Lucifer yellow at the dendrite. To determine the site of action potential initiation during synaptic activation of Purkinje cells, action potentials are evoked by stimulation of afferent parallel fibers which make synapses on distal dendrites of Purkinje cells (see **Figures 7.8** and **7.9**).

In all Purkinje cells tested, the evoked action potential recorded from the soma has a shorter delay and a greater amplitude than that recorded from a dendrite (**Figure 4.27a**). Moreover, the delay and the difference in amplitude between the somatic spike and the dendritic spike both augment when the distance between the two patch electrodes is increased. This suggests that the site of initiation is proximal to the soma.

Simultaneous whole cell recordings from the soma and the axon initial segment were performed to establish whether action potential initiation is somatic or axonal in origin. Action potentials clearly occur first in the axon initial segment (**Figure 4.27b**). These results suggest that the actual site of Na^+-dependent action potential initiation is in the axon initial segment of Purkinje cells. Experiments carried out by Sakmann *et al.* in other brain regions give the same conclusion for all the neurons tested.

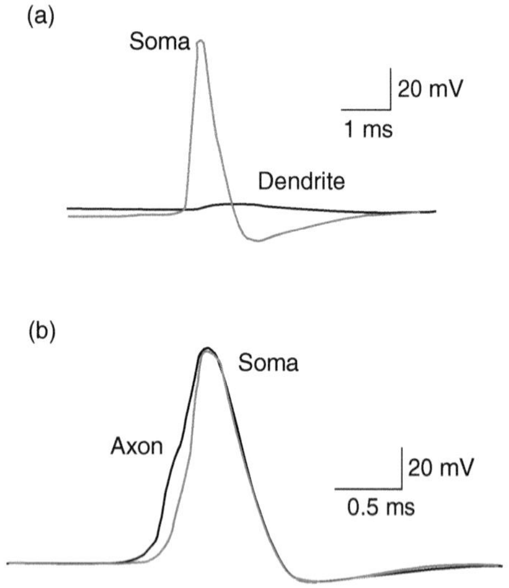

FIGURE 4.27 **The Na^+-dependent action potential is initiated in the axon initial segment in Purkinje cells of the cerebellum.** The activity of a Purkinje cell recorded simultaneously at the level of the soma and **(a)** 117 μm away from the soma at the level of a dendrite, or **(b)** 7 μm away from the soma at the level of the axon initial segment, with the double-patch clamp technique (whole-cell patches). Afferent parallel fibers are stimulated by applying brief voltage pulses to an extracellular patch pipette. In response to the synaptic excitation, an action potential is evoked in the Purkinje cell and recorded at the two different neuronal sites: soma and dendrite **(a)** or soma and axon **(b)**. Adapted from Stuart G, Hauser M (1994) Initiation and spread of sodium action potentials in cerebellar Purkinje cells. *Neuron* **13**, 703–712, with permission.

But why does the spike start in the axon initial segment? Beyond simply referring to the proximal end of the axon, some studies define the AIS by labeling for ankyrin-G, an Na^+-channel clustering molecule. It showed that the length of the axon initial segment varies widely among neuronal types (e.g. 20 μm in rat cerebellar Purkinje cells, 40–70 μm in rat layer 5 cortical pyramidal cells). The subtypes of Na^+ channels and their density are the essential issues in defining the site of spike initiation. The activation voltage ($V_{1/2}$) of the axonal Na^+ channels is more negative than on the soma, suggesting a mechanism for reduced threshold based on channel kinetics. For example, $Na_v1.6$ channels are concentrated in the axon initial segment of pyramidal cells, and these have a more negative $V_{1/2}$ and less inactivation than do $Na_v1.2$ channels found on the soma and the proximal axon. Moreover, patch clamp recordings in outside-out configuration suggest 20-fold differences of Na^+ channel density between the axon initial segment and the soma in pyramidal cells.

The action potential, once initiated, spreads passively back into the soma and dendritic tree. Passively means that it propagates with attenuation since it is not reinitiated in dendrites. Simultaneously, it actively propagates into the axon (see below). In some neurons, for example the pyramidal cells of the neocortex, the action potential actively back-propagates into the dendrites, but this is not a general rule.

4.4.4 The Na^+-dependent action potential actively propagates along the axon to axon terminals

Voltage-gated Na^+ channels are present all along the axon at a sufficient density to allow firing of axon potentials.

The propagation is active

Active means that the action potential is reinitiated at each node of Ranvier for a myelinated axon, or at each point for a non-myelinated axon. The flow of Na^+ ions through the open Na^+ voltage-gated channels of the axon initial segment creates a current that spreads passively along the length of the axon to the first node of Ranvier (**Figure 4.28**). It depolarizes the membrane of the first node to the threshold for action potential initiation. The action potential is now at the level of the first node. The entry of Na^+ ions at this level will depolarize the membrane of the second node and open the closed Na^+ channels. The action potential is now at the level of the second node.

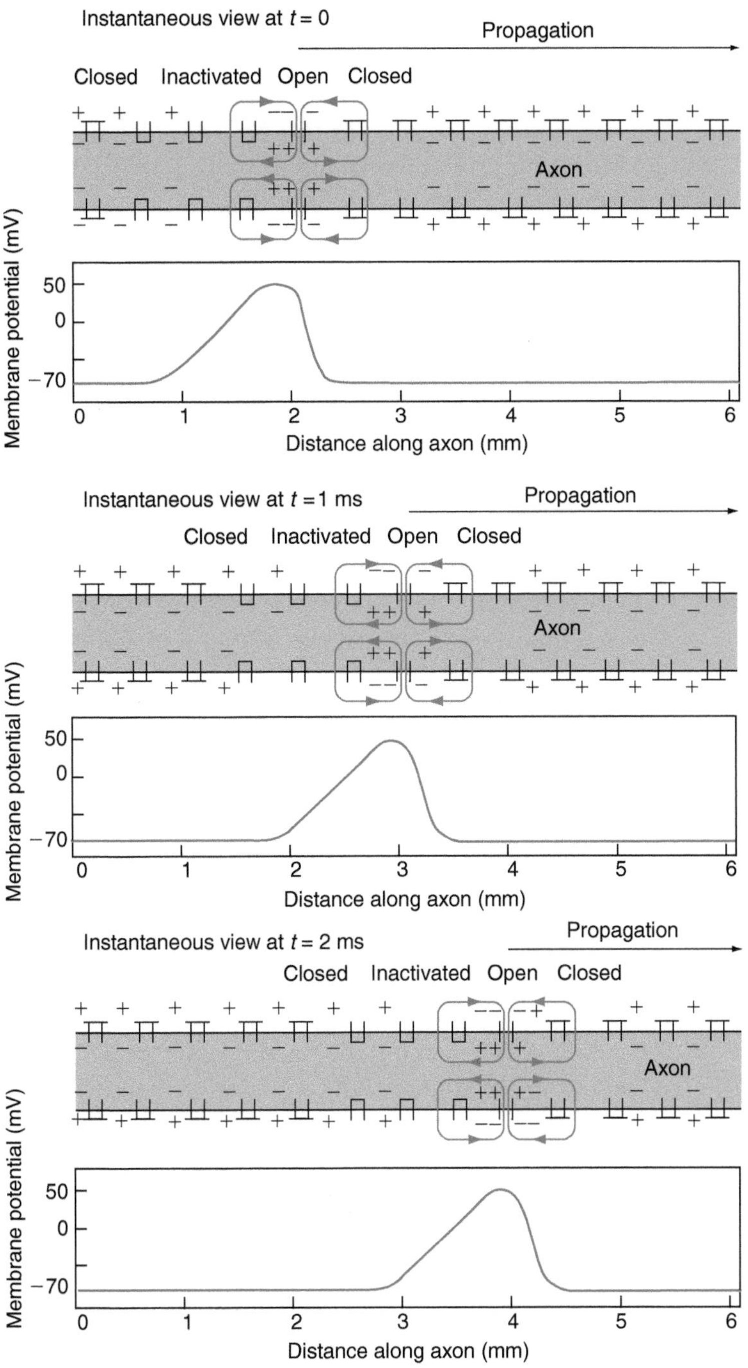

FIGURE 4.28 **Active propagation of the Na^+-dependent action potential in the axon and axon collaterals.** Scheme provided by Alberts B, Bray D, Lewis J *et al.* (1983) *Molecular Biology of the Cell*, New York: Garland Publishing.

The propagation is unidirectional owing to Na^+ channel inactivation

When the axon potential is, for example, at the level of the second node, the voltage-gated Na^+ channels of the first node are in the inactivated state since they have just been activated or are still in the open state (**Figure 4.28**). These Na^+ channels cannot be re-activated. The current lines flowing from the second node will therefore activate only the voltage-gated Na^+ channels of the third node towards axon terminals where the voltage-gated Na^+ channels are in the closed state (**Figure 4.28**). In the axon, under physiological conditions, the action potential cannot back-propagate.

The refractory periods between two action potentials

After one action potential has been initiated, there is a period of time during which a second action potential cannot be initiated or is initiated but has a smaller amplitude (**Figure 4.29**): this period is called the 'refractory

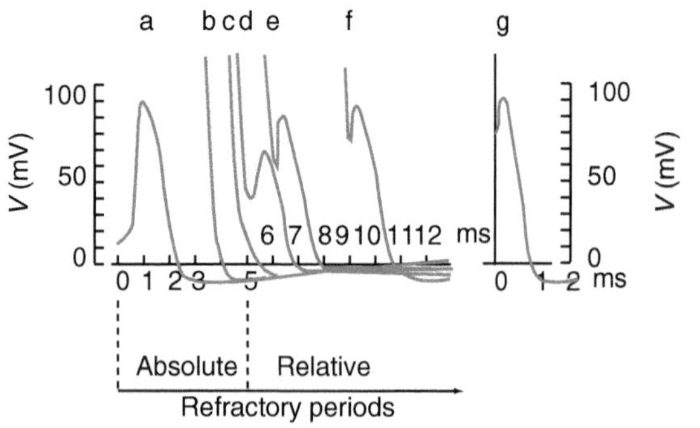

FIGURE 4.29 **The refractory periods.** A first action potential is recorded intracellularly in the squid axon *in vitro* in response to a small depolarizing stimulus **(a)**. Then a second stimulus with an intensity six times greater than that of the first is applied 4, 5, 6 or 9 ms after. The evoked spike is either absent (**b** and **c**; only the stimulation artifact is recorded) or has a smaller amplitude (**d–f**). Finally, when the membrane is back in the resting state, the evoked action potential has the control amplitude **(g)**. Adapted from Hodgkin AL, Huxley AF (1952) A quantitative description of membrane current and its application to conduction and excitation in nerve. *J. Physiol.* **117**, 500–544, with permission.

period' of the membrane. It results from Na^+ channel inactivation. Since the Na^+ channels do not immediately recover from inactivation, they cannot reopen immediately. This means that once the preceding action potential has reached its maximum amplitude, Na^+ channels will not reopen before a certain period of time needed for their de-inactivation (**Figure 4.24b**). This represents the absolute refractory period which lasts in the order of milliseconds.

Then, progressively, the Na^+ channels will recover from inactivation and some will reopen in response to a second depolarization: this is the relative refractory period. This period finishes when all the Na^+ channels at the initial axonal segment or at a node are de-inactivated. This actually protects the membrane from being depolarized all the time and enables the initiation of separate action potentials.

4.4.5 Do the Na^+ and K^+ concentrations change in the extracellular or intracellular media during firing?

Over a short time scale, the external or internal Na^+ or K^+ concentrations do not change during the emission of action potentials. A small number of ions are in fact flowing through the channels during an action potential and the Na–K pump re-establishes continuously the extracellular and intracellular Na^+ and K^+ concentrations at the expense of ATP hydrolysis. Over a longer time scale, during high-frequency trains of action potentials, the K^+ concentration can significantly increase in the external medium. This is due to the very small volume of the extracellular medium surrounding neurons and the limited speed of the Na–K pump. This excess of K^+ ions is buffered by glial cells which are highly permeable to K^+ ions (see Section 2.1.3).

4.4.6 Characteristics of the Na^+-dependent action potential are explained by the properties of the voltage-gated Na^+ channel

The *threshold* for Na^+-dependent action potential initiation results from the fact that voltage-gated Na^+ channels open in response to a depolarization positive to −50/−40 mV.

The Na^+-dependent action potential is *all or none* because voltage-gated Na^+ channels self-activate (see **Figure 4.17**). It propagates *without attenuation* since the density of voltage-gated Na^+ channels is constant along the axon or at nodes of Ranvier. It propagates *unidirectionally* because of the rapid inactivation of voltage-gated Na^+ channels. The instantaneous frequency of Na^+-dependent action potentials is limited by the *refractory periods*, which also results from voltage-gated Na^+ channel inactivation.

4.4.7 The role of the Na^+-dependent action potential is to evoke neurotransmitter release

The role of the Na^+-dependent action potential is to propagate, without attenuation, a strong depolarization to the membrane of the axon terminals. There, this depolarization opens the high-threshold voltage-gated Ca^{2+} channels. The resulting entry of Ca^{2+} ions into axon terminals triggers exocytosis and neurotransmitter release. The probability value of all these phenomena is not 1. This means that the action potential can fail to invade an axon terminal, the Ca^{2+} entry can fail to trigger exocytosis, etc. Neurotransmitter release is explained in Chapter 8.

Further reading

Anderson, P.A.V., Greenberg, R.M., 2001. Phylogeny of ion channels: clues to structure and function. Compar. Biochem. Physiol. B 129, 17–28.

Bender, K.J., Trussell, L.O., 2012. The physiology of the axon initial segment. Annu. Rev. Neurosci. 35, 249–265.

Catterall, W.A., 2000. From ionic currents to molecular mechanisms: the structure and function of voltage gated sodium channels. Neuron 26, 13–25.

Catterall, W.A., 2012. Voltage-gated sodium channels at 60: structure, function and pathophysiology. J. Physiol. 590, 2577–2589.

Eaholtz, G., Scheuer, T., Catterall, W.A., 1994. Restoration of inactivation and block of open sodium channels by an inactivation gate peptide. Neuron 12, 1041–1048.

Goldin, A.L., Barchi, R.L., Caldwell, J.H., et al., 2000. Nomenclature of voltage-gated sodium channels. Neuron 28, 365–368.

Gutman, G.A., Chandy, K.G., Grissmer, S., et al., 2005. International Union of Pharmacology. LIII. Nomenclature and molecular relationships of voltage-gated potassium channels. Pharmacol. Rev. 57, 473–508.

Hamill, O.P., Marty, A., Neher, E., et al., 1981. Improved patch damp technique for high resolution current recording from cells and cell-free membrane patches. Pflügers Arch. 391, 85–100.
Heinemann, S.H., Terlau, H., Stühmer, W., Imoto, K., Numa, S., 1992. Calcium channel characteristics conferred on the sodium channel by single mutations. Nature 356, 441–443.
Hodgkin, A.L., Huxley, A.F., 1952. A quantitative description of membrane current and its application to conduction and excitation in nerve. J. Physiol. (Lond.) 117, 500–544.
Hu, W., Tian, C., Li, T., Yang, M., HouH, Shu, Y., 2009. Distinct contributions of Na(v)1. 6 and Na(v)1. 2 in action potential initiation and backpropagation. Nat. Neurosci. 12, 996–1002.
McKinnon, R., 2003. Potassium channels. FEBBS Lett. 555, 62–65.
Neher, E., Sakmann, B., 1976. Single channel currents recorded from membrane of denervated frog muscle fibres. Nature 260, 779–802.
Noda, M., Ikeda, T., Suzuki, H., et al., 1986. Expression of functional sodium channels from cloned cDNA. Nature 322, 826–828.
Payandeh, J., Scheuer, T., Zheng, N., Catterall, W.A., 2011. The crystal structure of a voltage-gated sodium channel. Nature 475, 353–358.
Qu, Y., Rogers, J.C., Chen, S.F., McCormick, K.A., Scheuer, T., Catterall, W.A., 1999. Functional roles of the extracellular segments of the sodium channel alpha subunit in voltage-dependent gating and modulation by beta1 subunits. J. Biol. Chem. 274, 32647–32654.
Schmidt-Hieber, C., Bischofberger, J., 2010. Fast sodium channel gating supports localized and efficient axonal action potential initiation. J. Neurosci. 30, 10233–10242.
Sokolov, S., Scheuer, T., Catterall, W.A., 2005. Ion permeation through a voltage-sensitive gating pore in brain sodium channels having voltage sensor mutations. Neuron 47, 183–189.
Vassilev, P.M., Scheuer, T., Catterall, W.A., 1988. Identification of an intracellular peptide segment involved in sodium channel inactivation. Science 241, 1658–1661.
Venkatesh, B., Lu, S.Q., Dandona, N., et al., 2005. Genetic basis of tetrodotoxin resistance in pufferfishes. Curr. Biol. 15, 2069–2072.

APPENDIX 4.1 CURRENT CLAMP RECORDING

The current clamp technique, or intracellular recording in current clamp mode, is the traditional method for recording membrane potential: resting membrane potential and membrane potential changes such as action potentials and postsynaptic potentials. Membrane potential changes result from intrinsic or extrinsic currents. Intrinsic currents are synaptic or autorhythmic currents. Extrinsic currents are currents of known amplitude and duration applied by the experimenter through the intracellular recording electrode, in order to mimic currents produced by synaptic inputs.

Current clamp means that the *current applied* through the intracellular electrode is clamped to a constant value by the experimenter. It does not mean that the *current flowing through the membrane* is clamped to a constant value.

How to record membrane potential

The intracellular electrode (or the patch pipette) is connected to a unity-gain amplifier that has an input resistance many orders of magnitude greater than that of the micropipette plus the input resistance of the cell membrane ($R_p + R_m$). The output of the amplifier follows

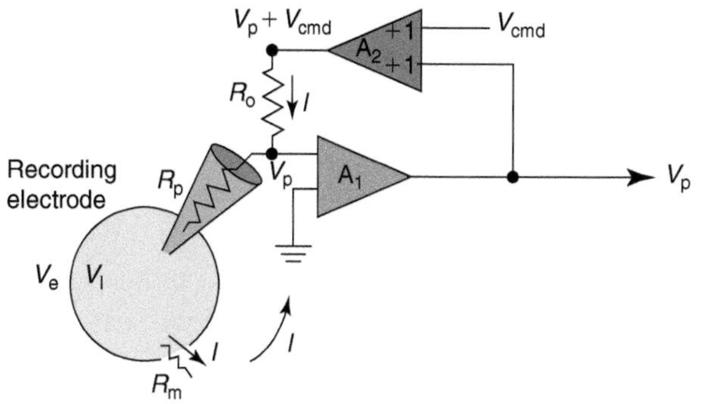

FIGURE A4.1 **A unity gain amplifier A1 and a current source made by adding a second amplifier A2.** The micropipette voltage V_p is measured by A1. The command voltage V_{cmd} and V_p are the inputs of A2 (V_p and V_{cmd} are added). The current I applied by the experimenter in order to induce V_m changes, flows through R_o and is equal to $I = V_p/R_o$ since the voltage across the output resistor R_o is equal to V_{cmd} regardless of V_p. I flows through the micropipette into the cell then out through the cell membrane into the bath grounding electrode. I is here an outward current. Capacitances are ignored. Adapted from *The Axon Guide*, Axon Instruments Inc., 1993.

the voltage at the tip of the intracellular electrode (V_p) (**Figure A4.1**). By definition, membrane potential V_m is equal to $V_i - V_e$ (i for intracellular and e for extracellular). In **Figure A4.1**, $V_i - V_e = V_p - V_{bath} = V_p - V_{ground} = V_p - 0 = V_p$. When a current I is simultaneously passed through the electrode, $V_p = V_m$ as long as the current I is very small in order not to cause a significant voltage drop across R_p (see the last section of this appendix).

How to inject current through the intracellular electrode

In a current injection, a circuit is connected to the input node, the current injected (I) flows down the electrode into the cell (**Figure A4.1**). This current source allows a constant (DC) current to be injected, either outward to depolarize the membrane or inward to hyperpolarize the membrane (**Figure A4.2**). When the recording electrode is filled with KCl, a current that expels K^+ ions into the cell interior depolarizes the membrane (V_m becomes

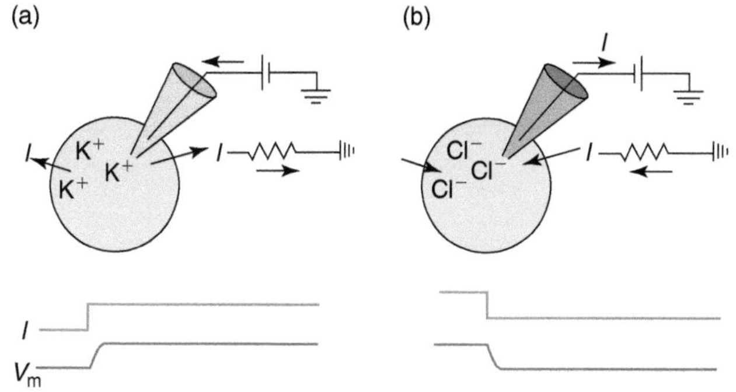

FIGURE A4.2 **(a)** When the recording electrode is filled with KCl, a current expels K^+ ions into the cell interior to depolarize the membrane (V_m becomes less negative). **(b)** A current expels Cl^- ions into the cell interior to hyperpolarize the membrane (V_m becomes more negative).

less negative) (**Figure A4.2a**), whereas a current that expels Cl^- ions into the cell interior hyperpolarizes the membrane (V_m becomes more negative) (**Figure A4.2b**).

Outward means that the current is flowing through the membrane from the inside of the cell to the bath; inward is the opposite

The current source can also be used to inject a short-duration pulse of current: a depolarizing current pulse above threshold to evoke action potential(s) or a low-amplitude depolarizing (**Figure A4.3**) or hyperpolarizing current pulse to measure the input membrane resistance R_m since $\Delta V_m = R_m \times \Delta I$.

How to measure the membrane potential when a current is passed down the electrode

The injected current (I) causes a corresponding voltage drop (IR_p) across the resistance of the pipette (R_p). It is therefore difficult to separate the potential at the tip of the electrode ($V_p = V_m$) from the total potential ($V_p + IR_p$). For example, if $R_p = 50\ M\Omega$ and $I = 0.5$ nA, $IR_p = 25$ mV, a value in the V_m range. A special compensation circuitry can be used to eliminate the micropipette voltage drop IR_p.

APPENDIX 4.2 VOLTAGE CLAMP RECORDING

The voltage clamp technique (or intracellular recording in voltage clamp mode) is a method for recording the current flowing through the cell membrane while the membrane potential is held (clamped) at a constant value by the experimenter. In contrast to the current clamp technique (see Appendix 4.1), voltage clamp does not mimic a process found in nature. However, there are several reasons for performing voltage clamp experiments:

- When studying voltage-gated channels, voltage clamp allows control of a variable (voltage) that determines the opening and closing of these channels.
- By holding the membrane potential constant, the experimenter ensures that the current flowing through the membrane is linearly proportional to the conductance G ($G = 1/R$) being studied. To study, for example, the conductance G_{Na} of the total number (N) of voltage-gated Na^+ channels present in the membrane, K^+ and Ca^{2+} voltage-gated channels are blocked by pharmacological agents, and the current I_{Na} flowing through the membrane, recorded in voltage clamp, is proportional to G_{Na}:

$$I_{Na} = V_m G_{Na} = kG_{Na}, \text{ since } V_m \text{ is constant.}$$

How to clamp the membrane potential at a known and constant value

The aim of the voltage clamp technique is to adjust continuously the membrane potential V_m to the command potential V_{cmd} fixed by the experimenter. To do so, V_m is continuously measured *and* a current I is passed through the cell membrane to keep V_m at the desired value or command potential (V_{cmd}). Two voltage clamp techniques are commonly used. With the two-electrode voltage clamp method, one electrode is used for membrane potential measurement and the other for passing current (**Figure A4.4**). The other method uses just one electrode, in one of the following ways:

- The same electrode is used part time for membrane potential measurement and part time for current injection (also called the discontinuous single-electrode voltage clamp technique, or dSEVC). This is used for cells that are too small to be impaled with two electrodes; it will not be explained here.
- In the patch clamp technique, the same electrode is used full time for simultaneously measuring membrane potential and passing current (see Appendix 4.3).

In the two-electrode voltage clamp technique, the membrane potential is recorded by a unity-gain amplifier A1 connected to the voltage-recording electrode E1. The membrane potential measured, V_m

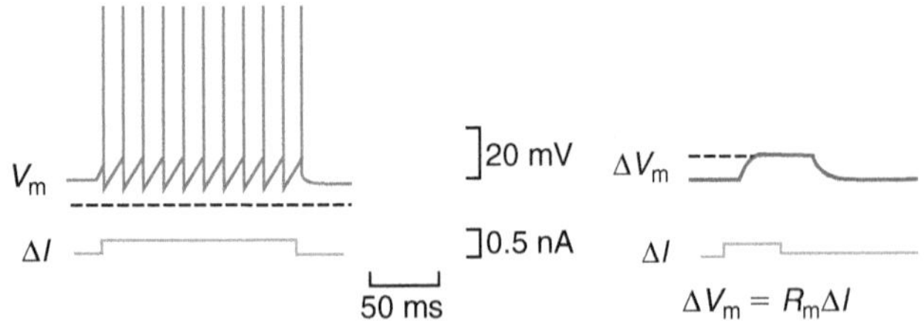

FIGURE A4.3 **Injection of a suprathreshold (left) and subthreshold (right) depolarizing pulse.**

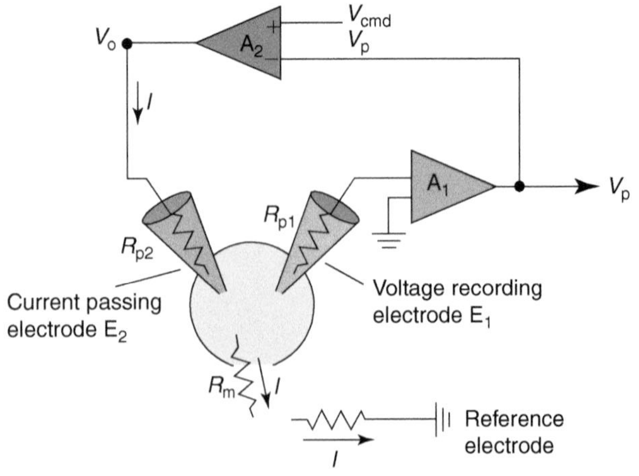

FIGURE A4.4 **Two-electrode voltage clamp.** Adapted from *The Axon Guide*, Axon Instruments Inc., 1993.

(or V_p; see Appendix 4.1) is compared with the command potential V_{cmd} in a high-gain differential amplifier A2. It sends a voltage output V_o proportional to the difference between V_m and V_{cmd}. V_o forces a current I to flow through the current-passing electrode E2 in order to obtain $V_m - V_{cmd} = 0$. The current I represents the total current that flows through the membrane. It is the same at every point of the circuit.

Example of a voltage clamp recording experiment

Two electrodes are placed intracellularly into a neuronal soma (an invertebrate neuron, for example) (**Figure A4.5**). The membrane potential is first held at −80 mV. In this condition, an outward current flows through the membrane in order to maintain the membrane potential at a value more hyperpolarized than V_{rest}. This stable outward current I_{-80} flows through the membrane as long as $V_{cmd} = -80$ mV.

A voltage step to −20 mV is then applied for 100 ms. This depolarizing step opens voltage-gated channels. In the presence of K^+ and Ca^{2+} channel blockers, only a voltage-gated Na^+ current is recorded. To clamp the membrane at the new $V_{cmd} = -20$ mV, a current $I_{(-20)}$ is sent by the amplifier A2. On the rising phase of the step this current is equal to the capacitive current I_c necessary to charge the membrane capacitance to its new value plus the leak current I_L flowing through leak channels (lines 1a and 1b). Since the depolarizing step opens Na^+ voltage-gated channels, an inward current I_{Na} flowing through open Na^+ channels will appear after a small delay (line 1c). Normally, this inward current flowing through the open Na^+ channels, I_{Na}, should depolarize the membrane but, in voltage clamp experiments, it does not: a current constantly equal to I_{Na} but of opposite direction is continuously sent (in the microsecond range) in the circuit to compensate I_{Na} and to clamp the membrane to V_{cmd}. Therefore, once the membrane capacitance is charged, $I_{(-20)} = I_L + I_{Na}$. Usually on recordings I_C is absent owing to the possibility of compensating for it with the voltage clamp amplifier.

Once the membrane capacitance is charged, the total current flowing through the circuit is $I = I_L + I_{Na}$ ($I_c = 0$). Therefore, in all measures of I_{Na}, the leak current I_L must be deduced. To do so, small-amplitude hyperpolarizing or depolarizing steps ($\Delta V_m = \pm 5$ to ± 20 mV) are applied at the beginning and at the end of the experiment. These voltage steps are too small to open voltage-gated channels in order to have $I_{Na} = 0$ and $I = I_L$. If we suppose that I_L is linearly proportional to ΔV_m, then I_L for a ΔV_m of +80 mV (from −80 to 0 mV) is eight times the value of I_L for $\Delta V_m = +10$ mV (see **Figure 3.9d**).

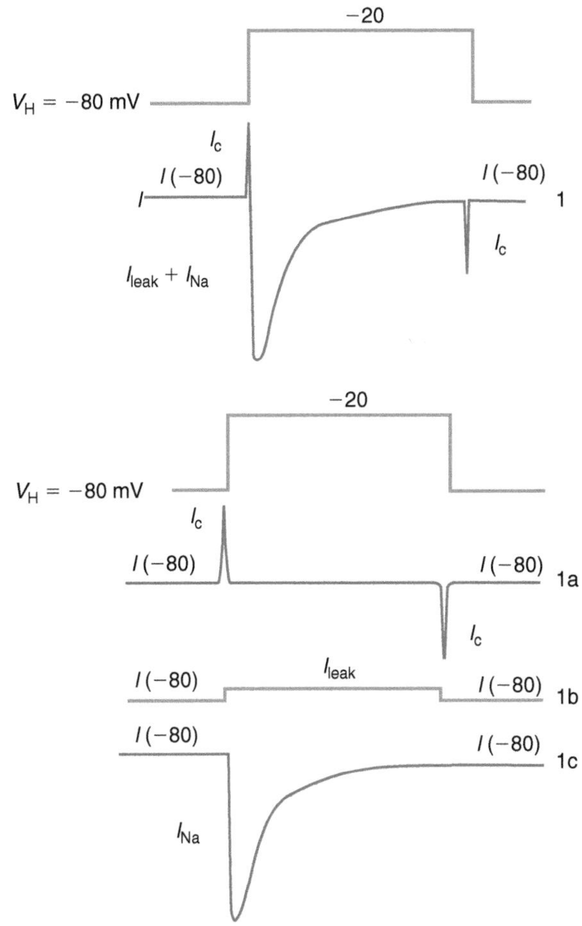

FIGURE A4.5 **Various currents.** (I = 1a + 1b + 1c) evoked by a voltage step to −20 mV (V_H = −80mV) in the presence of K^+ and Ca^{2+} channel blockers.

Is all the membrane surface clamped?

In small and round cells such as pituitary cells, the membrane potential is clamped on all the surface. In contrast, in neurons, because of their geometry, the voltage clamp is not achieved on all the membrane surface: the distal dendritic and axonal membranes are out of control because of their distance from the soma where the intracellular electrodes are usually placed. Such space clamp problems have to be taken into account by the experimenter in the analysis of the results. In the giant axon of the squid, this problem is overcome by inserting two long axial intracellular electrodes into a segment of axon in order to control the membrane potential all along this segment.

APPENDIX 4.3 PATCH CLAMP RECORDING

The patch clamp technique is a variation of the voltage clamp technique. It allows the recording of current flowing through the membrane: either the current flowing through all the channels open in the whole cell membrane or the current flowing through a single channel in a patch of membrane. In this technique, only one electrode is used full time for both voltage recording and passing current (it is a continuous single-electrode voltage clamp technique, or cSEVC). The patch clamp technique was developed by Neher and Sakmann. By applying very low doses of acetylcholine to a patch of muscle membrane, they recorded for the first time, in 1976, the current flowing through a single nicotinic cholinergic receptor channel (nAChR), the unitary nicotinic current.

Some of the advantages of the patch clamp technique are that (i) with all but one configuration (cell-attached configuration) the investigator has access to the intracellular environment (**Figure A4.6**); (ii) it allows the recording of currents from cells too small to be impaled with intracellular microelectrodes; and (iii) it allows the recording of unitary currents (current through a single channel).

A4.3.1 The various patch clamp recording configurations

First, a tight seal between the membrane and the tip of the pipette must be obtained. The tip of a micropipette that has been fire polished to a diameter of about 1 μm is advanced towards a cell until it makes contact with its membrane. Under appropriate conditions, a gentle suction applied to the inside of the pipette causes the formation of a very tight seal between the membrane and the tip of the pipette. This is the cell-attached configuration (**Figure A4.6**). The resistance between the interior of the pipette and the external solution can be very large, of the order of 1 GΩ (10^9 Ω) or more. It means that the interior of the pipette is isolated from the extracellular solution by the seal that is formed.

This very large resistance is necessary for two reasons (**Figure A4.7**):

- It allows the electrical isolation of the membrane patch under the tip of the pipette since practically no current can flow through the seal. This is important because if a fraction of the current passing through the membrane patch leaks

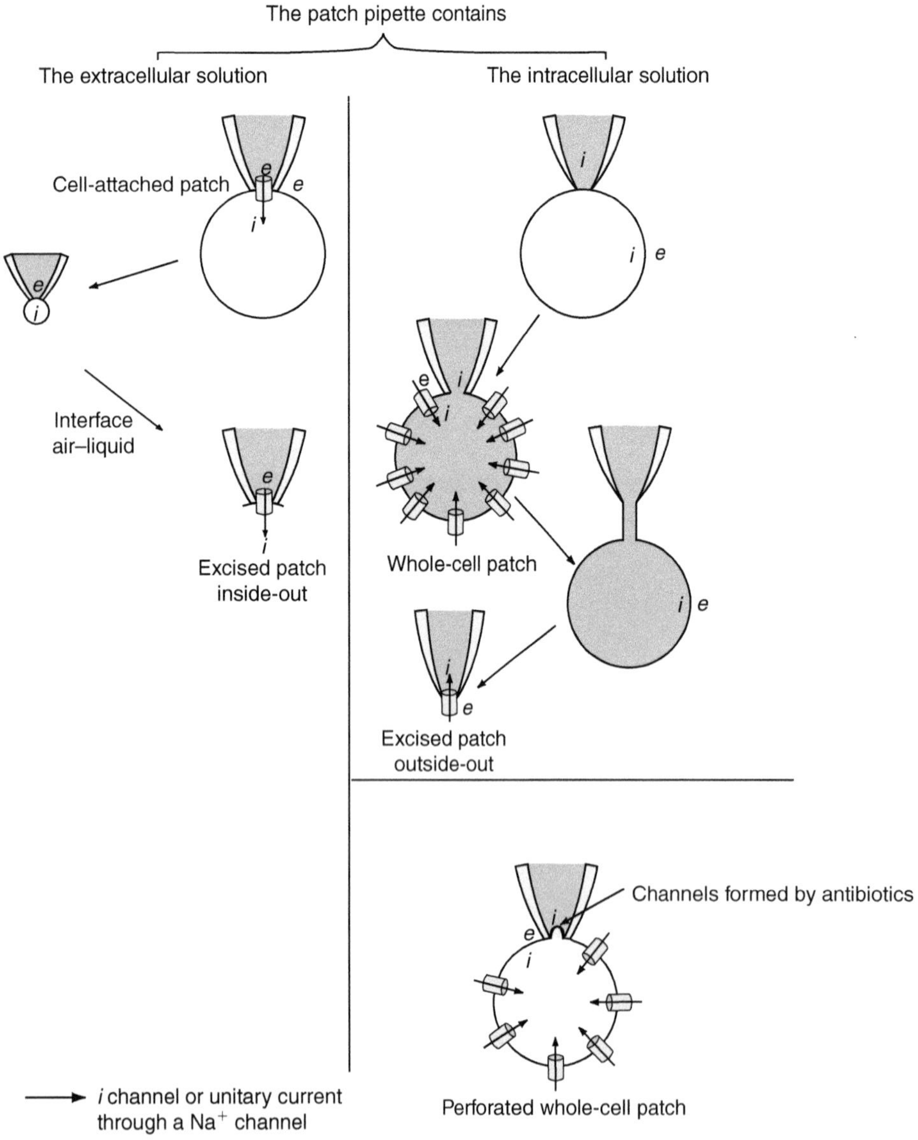

FIGURE A4.6 **Configurations of patch clamp recording.**

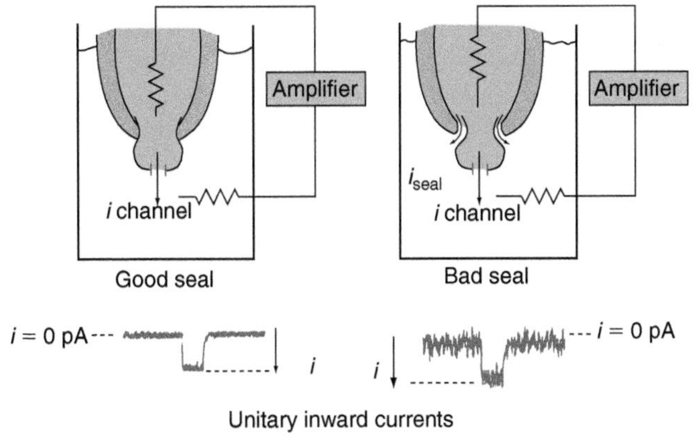

FIGURE A4.7 **Good and bad seals.** From *The Axon Guide*, Axon Instruments Inc., 1993.

out through the seal, it is not measured by the electrode.

- It augments the signal-to-noise ratio since thermal movement of the charges through a bad seal is a source of additional noise in the recording. A good seal thus enables the measurement of the current flowing through one single channel (unitary current) which is of the order of picoamperes.

From the 'cell-attached' configuration (the last to be explained), one can obtain other recording configurations. In total, three of them are used to record unitary currents, and one (whole-cell) to record the current flowing through all the open channels of the whole cell membrane.

Whole-cell configuration

This configuration is obtained from the cell-attached configuration. If a little suction is applied to the interior of the pipette, it may cause the rupture of the membrane patch under the pipette. Consequently, the patch pipette now records the activity of the whole-cell membrane (minus the small ruptured patch of membrane). Rapidly, the intracellular solution equilibrates with that of the pipette, the volume of the latter being many times larger. This is especially true for inorganic ions.

This configuration enables the recording of the current flowing through the N channels open over the entire surface of the cell membrane. Under conditions where all the open channels are of the same type (with the opening of other channels being blocked by pharmacological agents or the voltage conditions), the total current flowing through a population of identical channels can be recorded, such that at steady state:

$$I = Np_o i,$$

where N is the number of identical channels, p_o the probability that these channels are in the open state, Np_o the number of identical channels in the open state, and i the unitary current.

The advantages of this technique over the two-electrode voltage clamp technique are: (i) the recording under voltage clamp from cell bodies too small to be impaled with two electrodes and even one; and (ii) there is a certain control over the composition of the internal environment and a better signal-to-noise ratio. The limitation of this technique is the gradual loss of intracellular components (such as second messengers), which will cause the eventual disappearance of the responses dependent on those components.

Perforated whole-cell configuration

This is a variation of the whole-cell configuration, and also allows the recording of current flowing through the N channels open in the whole membrane but avoids washout of the intracellular solution. This configuration is obtained by introducing into the recording pipette a molecule such as nystatin, amphotericin or gramicidin, which will form channels in the patch of membrane under the tip of the electrode. To record in this configuration, first the cell-attached configuration is obtained and then the experimenter waits for the nystatin channels (or amphotericin or gramicidin channels) to form without applying any suction to the electrode. The channels formed by these molecules are mainly permeable to monovalent ions and thus allow electrical access to the cell's interior. Since these channels are not permeant to molecules as large or larger than glucose, whole-cell recording can be performed without removing the intracellular environment. This is particularly useful when the modulation of ionic channels by second messengers is studied.

In order to evaluate this problem of 'washout', we can calculate the ratio between the cell body volume and the volume of solution at the very end of a pipette. For example, for a cell of 20 μm diameter the volume is: $(4/3)\pi(10 \times 10^{-6})^3 = 4 \times 10^{-15}$ liters. If we consider 1 mm of the tip of the pipette, it contains a volume of the solution approximately equal to 10^{-13} l, which is 100 times larger than the volume of the cell body.

Excised patch configurations

If one wants to record the unitary current i flowing through a single channel and to control simultaneously the composition of the intracellular environment, the so-called excised or cell-free patch configurations have to be used. The *outside-out configuration* is obtained from the whole-cell configuration by gently pulling the pipette away from the cell. This causes the membrane patch to be torn away from the rest of the cell at the same time that its free ends reseal together. In this case, the intracellular environment is that of the pipette, and the extracellular environment is that of the bath. This configuration is used when rapid changes of the extracellular solution are required to test the effects of different ions or pharmacological agents when applied to the extracellular side of the membrane.

The *inside-out configuration* is obtained from the cell-attached configuration by gently pulling the pipette away from the cell, lifting the tip of the pipette from the bath in the air and putting it back into the solution (interface of air–liquid). In this case, the intracellular environment is that of the bath and the extracellular one is that of the pipette (the pipette is filled with a pseudo-extracellular solution). This configuration is used when rapid changes in the composition of the intracellular environment are necessary to test, for example, the effects of different ions, second messengers and pharmacological agents in that environment.

Cell-attached configuration

The intracellular environment is that of the cell itself, and the extracellular environment of the recorded membrane patch is the pipette solution. This configuration enables the recording of current flowing through the channel or channels present in the patch of membrane that is under the pipette and is electrically isolated from the rest of the cell. If one channel opens at a time, then the unitary current *i* flowing through that channel can be recorded. The recordings in cell-attached mode present two limitations: (i) the composition of the intracellular environment is not controlled; and (ii) the value of the membrane potential is not known and can only be estimated.

Let us assume that the voltage in the interior of the patch pipette is maintained at a known value V_p (p = pipette). Since the voltage across the membrane patch is $V_m = V_i - V_e = V_i - V_p$, it will not be known unless V_i, the voltage at the internal side of the membrane, is also known. V_i cannot be measured directly. One way to estimate this value is to measure the resting potential of several identical cells under similar conditions (with intracellular or whole-cell recordings), and to calculate an average V_i from the individual values. Sometimes, however, V_i can be measured when the cell is large enough to allow a two-electrode voltage clamp recording to be made simultaneously with the patch clamp recording (with a *Xenopus* oocyte, for example). Another method consists of replacing the extracellular medium with isotonic K^+ (120–150 mM). The membrane potential under these conditions will be close to 0 mV.

To leave the intracellular composition intact while recording the activity of a single channel is particularly useful for studies of the modulation of an ionic channel by second messengers.

A4.3.2 Principles of the patch clamp recording technique

In the patch clamp technique, as in all voltage clamp techniques, the membrane potential is held constant (i.e. clamped) while the current flowing through a single open channel or many open channels (Np_o) is measured (**Figure A4.8**). In the patch clamp technique, only one micropipette is used full time for both voltage clamping and current recording. How at the same time via the same pipette can the voltage of the membrane be controlled and the current flowing through the membrane be measured?

When an operational amplifier A1 is connected as shown in **Figure A4.8a** with a high megohm resistor R_f (f = feedback), a current-to-voltage converter is obtained. The patch pipette is connected to the negative input and the command voltage (V_{cmd}) to the positive one. The resistor R_f can have two values: $R_f = 1$ GΩ in the whole-cell configuration and 10 GΩ in the excised patch configurations.

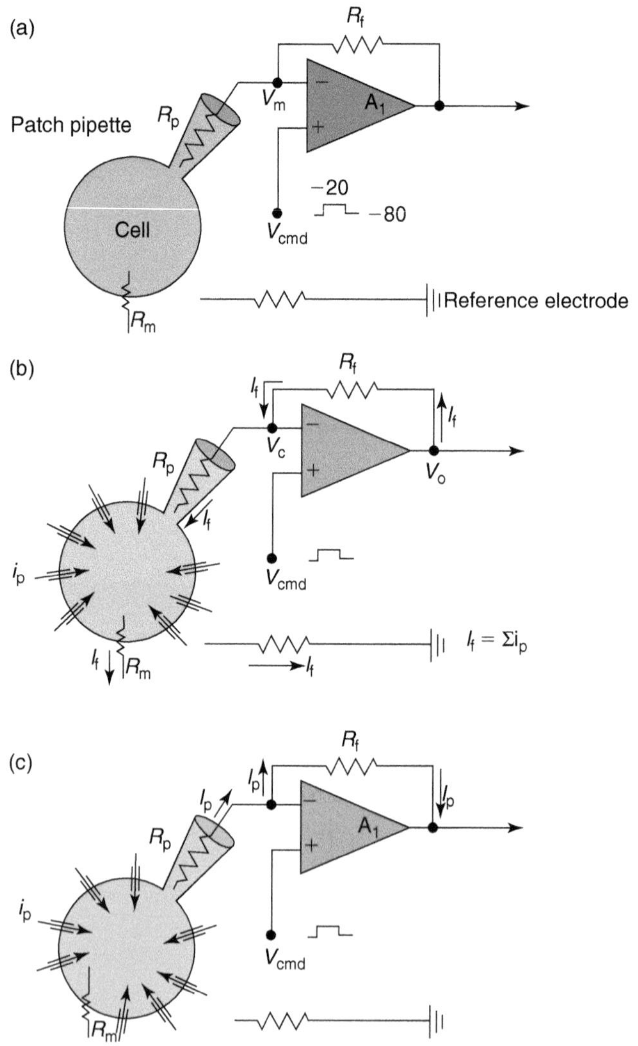

FIGURE A4.8 **Example of a patch clamp recording in the whole-cell configuration. (a)** The amplifier compares V_m to the new $V_{cmd} = -20$ mV. **(b)** The amplifier sends V_o so that $V_m = V_{cmd} = -20$ mV. Owing to the depolarization to −20 mV, the Na^+ channels open and unitary inward currents i_p flow through the *N* open channels ($Ni_p = I_p$). **(c)** The whole-cell current I_p flows through the circuit and is measured as a voltage change.

How the membrane is clamped at a voltage equal to V_{cmd}

R_p represents the electrode resistance and R_m the membrane input resistance (**Figure A4.8a**). Suppose that the membrane potential is first clamped to −80 mV (V_{cmd} = −80 mV), then a voltage step to −20 mV is applied for 100 ms (V_{cmd} = −20 mV for 100 ms). The membrane potential (V_m) has to be clamped quickly to −20 mV ($V_m = V_{cmd}$ = −20 mV) whatever happens to the channels in the membrane (they open or close). The operational amplifier A1 is able to minimize the voltage difference between two inputs to a very small value (0.1 μV or so). A1 compares the value of V_{cmd} (entry +) to that of V_m (entry −). It then sends a voltage output (V_o) in order to obtain $V_m = V_{cmd}$ = −20 mV (**Figure A4.8b**).

What is this value of V_o? Suppose that at the time t of its peak the Na^+ current evoked by the voltage step to −20 mV is I_{Na} = 1 nA. V_o will force a current I = −1 nA to flow through $R_f = 10^9\ \Omega$ in order to clamp the membrane potential: $V_o = R_f I = 10^9 \times 10^{-9} = 1$ V. It is said that V_o = 1 V/nA or 1 mV/pA.

The limits of V_o in patch clamp amplifiers are +15 V and −15 V. This means that V_o cannot be bigger than these values, which is largely compatible with biological experiments where currents through the membrane do not exceed 15 nA.

The amplifier A1 compares V_m with V_{cmd} and sends V_o at a very high speed. This speed has to be very high in order to correct V_m according to V_{cmd} very quickly. The ideal clamp is obtained at the output of the circuit via R_f (black dot V_c on the scheme of **Figure A4.8b**). As in the voltage clamp technique, a capacitive current is present at the beginning and at the end of the voltage step on the current trace and a leak current during the step, but they are not re-explained here.

A4.3.3 The unitary current *i* is a rectangular step of current (see Figure 4.8a,c)

We record, for example, in the outside-out patch clamp configuration the activity of a single voltage-sensitive Na^+ channel. When a positive membrane potential step is applied to depolarize the patch of membrane from −90 mV to −40 mV, an inward current i_{Na} flowing through the open Na^+ channel is recorded (inward current means a current that flows across the membrane from the outside to inside). By convention, inward currents are represented as downward deflections and outward currents as upward deflections.

The membrane depolarization causes activation of the voltage-dependent Na^+ channel, and induces its transition from the closed (C) state (or conformation) to the open (O) state, a transition symbolized by:

$$C \rightleftharpoons O$$

where C is the closed state of the channel (at −90 mV) and O is the open state of the channel (at −40 mV).

While the channel is in the O conformation(at −40 mV), Na^+ ions flow through the channel and an inward current caused by the net influx of Na^+ ions is recorded. This current reaches its maximum value very rapidly. Thus, the maximal net ion flux is established almost instantaneously given the timescale of the recording (of the order of microseconds). The development of the inward current thus appears as a vertical downward deflection.

A delay between the onset of the voltage step and the onset of the current i is observed. This delay has a duration that varies from one depolarizing test pulse to another and also according to the channel under study. This delay is due to the conformational change or changes of the protein. In fact, such changes previous to opening can be multiple:

$$C_1 \rightleftharpoons C_2 \rightleftharpoons C_3 \rightleftharpoons O$$

Notice that the opening delay does not correspond to the intrinsic duration of the process of conformational change, which is extremely short. It corresponds to the statistical nature of the equilibrium between the 2, 3, N closed and open conformations. The opening delay therefore depends on the time spent in each of the different closed states (C_1, C_2, C_3).

The return of the current value to zero corresponds to the closing of the channel. This closure is the result of the transition of the channel protein from the open state (O) to a state in which the channel no longer conducts (state in which the aqueous pore is closed). It can be either a closed state (C), an inactivated state (I) or a desensitized state (D). In the case of the Na^+ channel, the return of the current value to zero is due mainly to the transition of the protein from the open state to the inactivated state (O → I). Before closing for a long time, the channel can also flicker between the open and closed state (C ⇌ O):

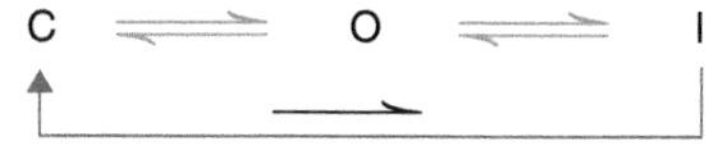

Just as the current reaches its maximum value instantaneously during opening, it also returns instantaneously to its zero value during closing of the pore. Because of this, the unitary current i has a step-like rectangular shape.

A4.3.4 Determination of the conductance of a channel

If we repeat several times the experiment shown in **Figure 4.8a**, we observe that for a given voltage step ΔV, i varies around an average value. The current fluctuations are measured at regular intervals before, during and immediately after the depolarizing voltage pulse. The distribution of the different i values during the voltage pulse describes a Gaussian curve in which the peak corresponds to the average i value (**Figure 4.8d**). There is also a peak around 0 pA (not shown on the figure) which corresponds to the different values of i when the channel is closed. Since the channel is in the closed state most of the time, where i has values around 0 pA, this peak is higher than the one corresponding to $i_{channel}$ (around −2 pA). The width of the peak around 0 pA gives the mean value of the fluctuations resulting from noise. Therefore, the two main reasons for these fluctuations of $i_{channel}$ are: the variations in the noise of the recording system and the changes in the number of ions that cross the channel during a unit of time Δt.

Knowing the average value of i and the reversal potential value of the current (E_{rev}), the average conductance value of the channel under study, γ, can be calculated: $\gamma = i/(V_m - E_{rev})$.

However, there are cases in which the distribution of i for a give membrane potential shows several peaks. Different possibilities should be considered:

- Only one channel is being recorded from but it presents several open conformational states, each one with different conductances. The peaks correspond to the current flowing through these different substrates.
- Two or more channels of the *same* type are present in the patch and their activity recorded. The peaks represent the multiples of i ($2i$, $3i$, etc.).
- Two or more channels of *different* types are present in the patch and their activity is simultaneously recorded. The peaks correspond to the current through different channel types.

A4.3.5 Mean open time of a channel

An ionic channel fluctuates between a closed state (C) and an open state (O):

$$C \underset{\alpha}{\overset{\beta}{\rightleftharpoons}} O$$

where α is the closing rate constant or, more exactly, the number of channel closures per unit of time spent in the open state O. β is the opening rate constant or the number of openings per unit of time spent in the closed state R (α and β are expressed in s^{-1}).

Once activated, the channel remains in the O state for a time t_o, called open time. When the channel opens, the unitary current i is recorded for a certain time t_o. t_o for a given channel studied under identical conditions varies from one recording to another (**Figure A4.9**). t_o is an aleatory variable of an observed duration. When the number of times a value of t_o (in the order of milli- or microseconds) is plotted against the values of t_o, one obtains the open time histogram; i.e. the distribution of the different values of t_o (**Figure A4.10**). This distribution declines and the shorter open times are more frequent than the longer ones.

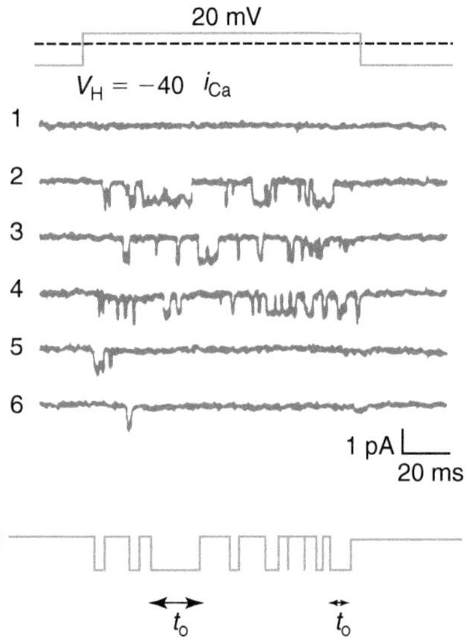

FIGURE A4.9 **Example of the patch clamp recording of a single voltage-dependent Ca^{2+} channel.** In response to a voltage step to +20 mV from a holding potential of −40 mV, the channel opens and closes several times during each of the six trials. Adapted from Fox AP, Nowyky MC, Tsien RW (1987) Single channel recordings of three types of calcium channels in chick sensory neurones. *J. Physiol. (Lond.)* **394**, 173–200, with permission.

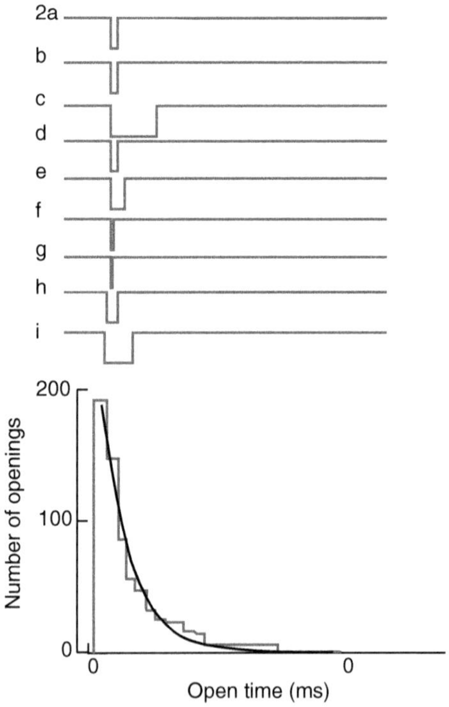

FIGURE A4.10 **Determination of the mean open time of a channel.** Trial 2 of **Figure A4.9** is selected and all the openings are aligned at time 0. τ_o = 1.2 ms.

Why does the distribution of t_o decrease?

The histogram is constructed as follows. At time $t = 0$, all the channels are open (the delay of opening is ignored, all the openings are aligned at time 0; Figure A4.10). As time t increases, the number of channels that remain open can only decrease since channels progressively close. This can also be expressed as follows: the longer the observation time, the lower the probability that the channel is still in the open state. Or, alternatively, the longer the observation time, the closer the probability will be to 1 that the channel will shut (1 is the maximum value used to express a probability). It is not a Gaussian curve because the delay of opening is ignored and all the openings begin at $t = 0$.

Why is the decrementing distribution of t_o exponential?

A channel open at $t = 0$ has a probability of closing at $t + \Delta t$. It has the same probability of closing if it is still open at the beginning of any subsequent observation interval Δt. This type of probability is described mathematically as an exponential function of the observation time. Thus, when the openings of a homogeneous population of channels are studied, the decrease in the number of events is described by a single exponential.

Experimental determination of τ_o, the mean open time of a channel

The mean open time τ_o is the time during which a channel has the highest probability of being in the open state: it corresponds to the sum of all the values that t_o may take, weighted by their corresponding probability values. This value is easy to calculate if the distribution is described by a single exponential. In order to verify that the histogram is actually described by a single exponential, one has to first build the histogram by plotting the number of times a value of t_o is observed as a function of t_o; i.e. number of events = $f(t_o)$.

The exponential that describes the histogram has the form $y = y_0 e^{-t/\tau 0}$, where y is the number of events observed at each time t. This curve will be linear on semilogarithmic coordinates if it is described by a single exponential. The slope can be measured with a regression analysis. It corresponds to the mean open time τ_o of the channel. τ_o is the value of t_o for a number of events equal to $1/e$. It is the 'expected value' of t_o. The expected value of t_o is the sum of all the values of t_o weighted to their corresponding probabilities.

In the case of the conformational changes the value of τ_o provides an estimate of the closure rate constant α, because at steady state $\tau_o = 1/\alpha$. For example, from the open time histogram of the nicotinic receptor channel, we can determine its mean open time τ_o. Knowing that in conditions where the desensitization of the channel is negligible $\tau_o = 1/\alpha$, we can calculate from π_o the closing rate constant of the channel. If $\tau_o = 1.1$ ms, $\alpha = 900\ s^{-1}$. The channel closes 900 times for each second spent in the open state. In other words, there is an average of 900 transitions of the channel to the closed state for each second spend in the open state.

C H A P T E R

5

The voltage-gated channels of Ca^{2+} action potentials: Generalization

Constance Hammond, François Michel

O U T L I N E

Chapter 4 explained the Na^+-dependent action potential propagated by axons. There are two other types of action potentials: (i) the Na^+/Ca^{2+}-dependent action potential present notably in axon terminals or heart muscle cells (see **Figure 4.2d**), where it is responsible for Ca^{2+} entry and an increase of intracellular Ca^{2+} concentration, a necessary prerequisite for neurotransmitter release (secretion) or muscle fiber contraction; and (ii) the Ca^{2+}-dependent action potential (in which Na^+ ions do not participate) in dendrites of Purkinje cells of the cerebellum (see **Figure 17.9**) and in endocrine cells (**Figure 5.1a**). In Purkinje cell dendrites, it depolarizes the membrane and thus modulates neuronal integration; in endocrine cells, it provides a Ca^{2+} entry to trigger hormone secretion.

5.1 PROPERTIES OF Ca^{2+}-DEPENDENT ACTION POTENTIALS

In some neuronal cell bodies, in heart ventricular muscle cells and in axon terminals, the action potentials have a longer duration than Na^+ spikes, with a plateau following the initial peak: these are the Na^+/Ca^{2+}-dependent action potentials (see **Figure 4.2b–d**). In some neuronal dendrites and some endocrine cells, action potentials have a small amplitude and a long duration: these are the Ca^{2+}-dependent action potentials (**Figure 5.1**). All action potentials are initiated in response to a membrane depolarization. Na^+, Na^+/Ca^{2+} and Ca^{2+}-dependent action potentials differ in the type of voltage-gated channels responsible for their depolarization and repolarization phases. We will examine the properties of a Ca^{2+}-dependent action potential.

5.1.1 Ca^{2+} and K^+ ions participate in the action potential of endocrine cells

The activity of pituitary endocrine cells that release growth hormone is recorded in the perforated whole-cell configuration (current clamp mode; see Appendix 4.1). They display a spontaneous activity. When these cells are previously loaded with the Ca^{2+}-sensitive dye Fura-2, changes of intracellular Ca^{2+} concentration can also be quantified (see Appendix 5.1). Simultaneous recording of potential and $[Ca^{2+}]_i$ changes shows that for each action potential there is a corresponding $[Ca^{2+}]_i$ increase (**Figure 5.1a**). This strongly suggests that Ca^{2+} ions are entering the cell during action potentials.

Cellular and Molecular Neurophysiology. DOI: 10.1016/B978-0-12-397032-9.00005-4

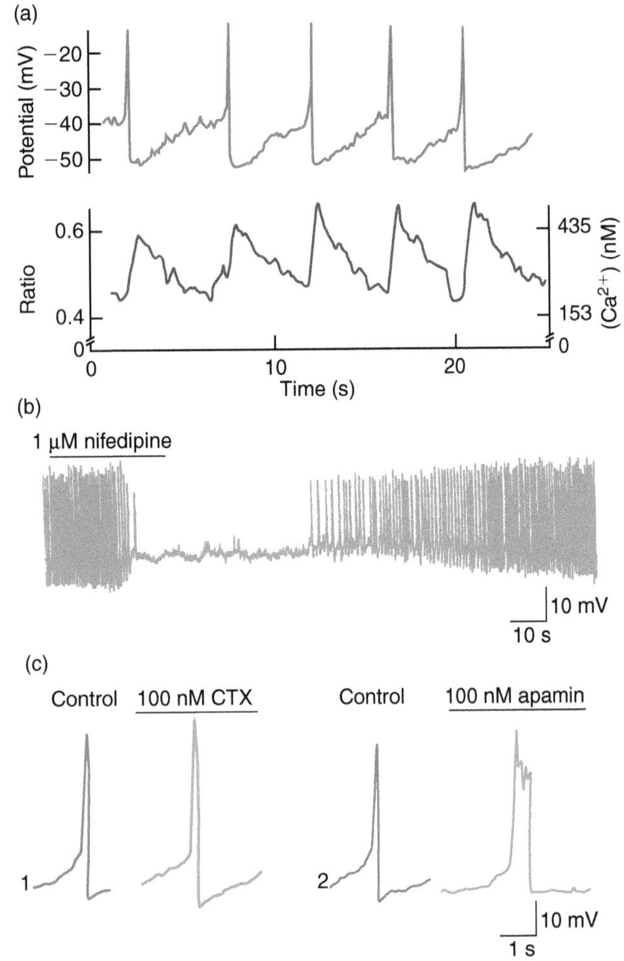

FIGURE 5.1 **The Ca^{2+}-dependent action potential of an endocrine cell.** Growth-hormone secreting cells of the anterior pituitary in culture are loaded with the Ca^{2+}-sensitive dye Fura-2 and their activity is recorded in perforated whole-cell patch configuration (current clamp mode). **(a)** Simultaneous recordings of action potentials (top trace) and cytosolic $[Ca^{2+}]$ oscillations (bottom trace) in control conditions. **(b)** Nifedipine, an L-type Ca^{2+} channel blocker, is applied for 20 s. **(c)** Action potential in the absence and presence of blockers of Ca^{2+}-activated K^+ channels, charybdotoxin (CTX, 1) and apamin (2). Adapted from Kwiecien R, Robert C, Cannon R et al. (1998) Endogenous pacemaker activity of rat tumour somatotrophs. *J. Physiol.* **508**, 883–905, with permission.

Ca^{2+} ions participate in the depolarization phase of the action potential

When the extracellular solution is changed from control extracellular Krebs solution to a Ca^{2+}-free solution, or when nifedipine, an L-type Ca^{2+} channel blocker, is added to the external medium (**Figure 5.1b**), the amplitude and risetime of the depolarization phase of the action potential gradually and rapidly decrease until action potentials are no longer evoked.

K^+ ions participate in the repolarization phase of the action potential

Application of charybdotoxin (CTX) or apamin, blockers of Ca^{2+}-activated K^+ channels, increases the peak amplitude and prolongs the duration of action potentials (**Figure 5.1c**). Note that apamin also blocks the after-spike hyperpolarization (**Figure 5.1c2**).

5.1.2 Questions about the Ca^{2+}-dependent action potential

- What are the structural and functional properties of the Ca^{2+} and K^+ channels involved? (Sections 5.2 and 5.3)
- What represents the threshold potential for Ca^{2+}-dependent action potential initiation? Where are Ca^{2+}-dependent action potentials initiated? (Section 5.4)

5.2 THE TRANSIENT ENTRY OF Ca^{2+} IONS THROUGH VOLTAGE-GATED Ca^{2+} CHANNELS IS RESPONSIBLE FOR THE DEPOLARIZING PHASE OR THE PLATEAU PHASE OF Ca^{2+}-DEPENDENT ACTION POTENTIALS

The voltage-gated Ca^{2+} channels involved in these action potentials are high voltage-activated (HVA) Ca^{2+} channels. There are three main types of such channels: the L-type (L for long lasting), the N-type (N for neuronal or for neither L nor T) and the P-type (P for Purkinje cells where they were first described).

5.2.1 The voltage-gated Ca^{2+} channels are a diverse group of multisubunit proteins

They are composed of a pore-forming α_1-subunit associated with auxiliary subunits. The pore-forming α_1-subunit contains about 2000 amino acid residues (190–250 kDa), and has an amino acid sequence and a predicted transmembrane structure similar to the previously characterized pore-forming α-subunit of Na^+ channels: four repeated domains (I to IV), each of which contains six transmembrane segments (1 to 6) and a pore loop (P) between transmembrane segments S5 and S6 of each domain (**Figure 5.2a**). It incorporates the conduction pore, the voltage sensor, the gating apparatus and the known sites of channel regulation by second messengers, drugs and toxins. The S4 segments of each homologous domain serve as the voltage sensors for activation, moving under the influence of the electric field and initiating a conformational change that opens the pore. The S5 and S6 segments and the membrane-associated pore loop between them form the pore lining of the voltage-gated ion channels. The external end of the pore is lined by the pore loop (P), which contains a pair of glutamate residues in each domain that are required for Ca^{2+} selectivity.

Auxiliary subunits can include a disulfide-linked $\alpha_2\delta$ dimer, an intracellular phosphorylated β-subunit, and a

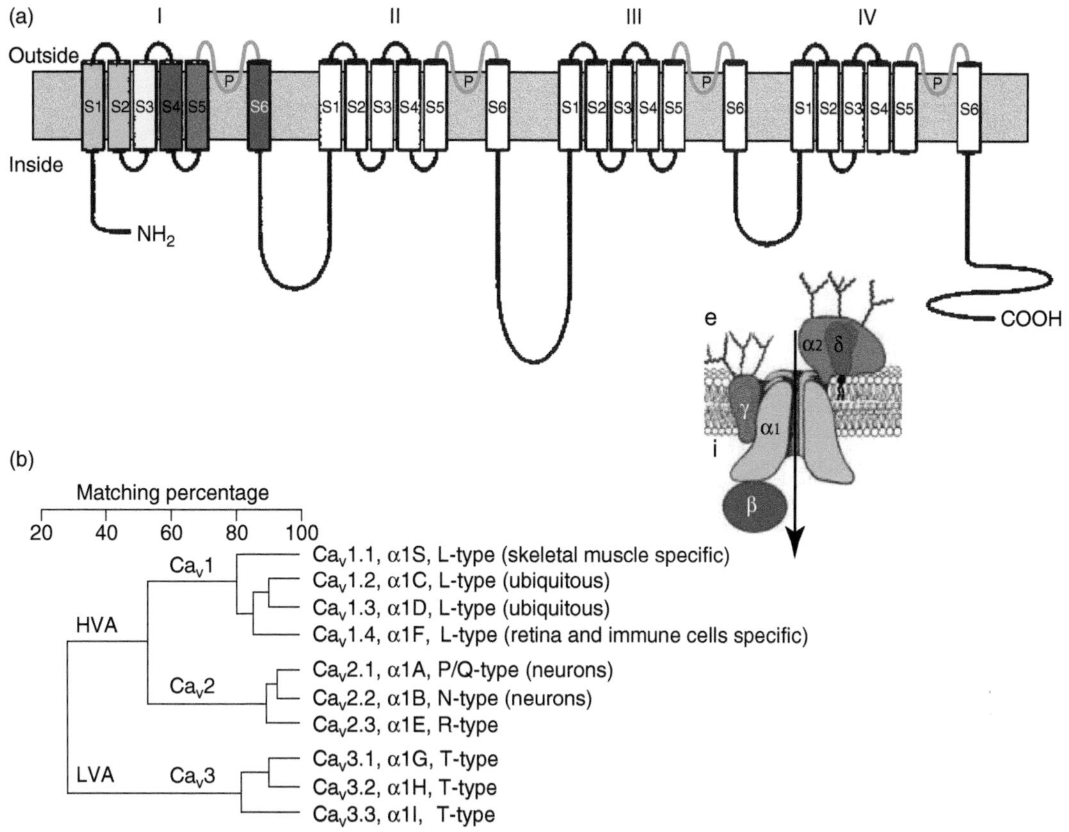

FIGURE 5.2 **Subunits of voltage-gated Ca^{2+} channels.** (**a**) Membrane topology for the α_1-subunit of a cardiac L-type Ca^{2+} channel (P: the P-loops). Inset: The multimeric complexes of subunits formed by an α_1 pore-forming protein associated to three auxiliary subunits (α_2-δ, β, and γ). (**b**) Evolutionary tree of voltage-gated α_1-subunit of Ca^{2+} channels. Low voltage-activated Ca^{2+} channels (LVA) appear to have diverged from an ancestral Ca^{2+} channel before the bifurcation of the high voltage-activated (HVA) channels in Ca_v1 and Ca_v2 subfamilies. Part (**a**) adapted from Sather WA (2003) Permeation and selectivity in calcium channels. *Annu. Rev. Physiol.* **65**, 133–159, with permission. Part (**b**) Adapted from Perez-Reyes E, Cribbs LL, Daud A *et al.* (1998) Molecular characterization of a neuronal low voltage-activated T-type calcium channel. *Nature* **391**, 896–900, with permission.

transmembrane γ-subunit (**Figure 5.2a inset**). They play a role in the expression and gating properties of the Ca^{2+} channels by modulating various properties of the α_1-subunit.

The pharmacological and electrophysiological diversity of Ca^{2+} channels primarily arises from the diversity of α_1-subunits. The primary structure of the different α_1-subunits has been defined by homology screening and their function characterized by expression in mammalian cells or *Xenopus* oocytes. The recent nomenclature divides the Ca^{2+} channels into three structurally and functionally related families (Ca_v1, Ca_v2, Ca_v3) to indicate the principal permeating ion (Ca) and the principal physiological regulator (v for voltage), followed by a number that indicates the gene subfamily (1, 2 or 3). The number following the decimal point identifies the specific channel isoform (e.g. $Ca_v1.1$) (**Figure 5.2b**). High-threshold Ca^{2+} channels comprise L (Ca_v1), P/Q ($Ca_v2.1$), N ($Ca_v2.2$) and R ($Ca_v2.3$)-type Ca^{2+} channels. The Ca_V1 subfamily initiates contraction, secretion, regulation of gene expression, integration of synaptic input in neurons, and synaptic transmission at synapses in specialized sensory cells. The Ca_V2 subfamily is primarily responsible for initiation of synaptic transmission at fast synapses.

How to record the activity of Ca^{2+} channels in isolation

This needs to block the voltage-gated channels that are not permeable to Ca^{2+} ions. Different strategies can be used: in whole-cell or intracellular recordings, tetrodotoxin (TTX) and tetraethylammonium chloride (TEA) are added to the extracellular solution and K^+ ions are replaced by Cs^+ in the intrapipette solution, in order to block voltage-gated Na^+ and K^+ channels. In cell-attached recordings, the patch pipette is filled with a solution containing Ca^{2+} or Ba^{2+} ions as the charge carrier. When Ba^{2+} substitutes for Ca^{2+} in the extracellular solution, the inward currents recorded in response to a depolarizing step are Ba^{2+} currents. Ba^{2+} is often preferred to Ca^{2+} since it carries current twice as effectively as Ca^{2+} and poorly inactivates Ca^{2+} channels (see Section 5.2.3).

As a consequence, unitary Ba^{2+} currents are larger than Ca^{2+} ones and can be studied more easily.

Another challenge is to separate the various types of Ca^{2+} channels in order to record the activity of only one type (since in most of the cells they are co-expressed). These different Ca^{2+} channels are the high voltage-activated L, N and P channels (this chapter) and the low-threshold T channel. T-type Ca^{2+} channels are low threshold-activated channels, also called subliminal Ca^{2+} channels, that can be identified by their low threshold of activation and their rapid inactivation. They are studied with other subliminal channels in Section 14.2.2.

HVA Ca^{2+} channels exhibit overlapping electrophysiological profiles. It is important to separate them in order to study their characteristics and to identify their respective roles in synaptic integration (dendritic Ca^{2+} channels), in transmitter release (Ca^{2+} channels of axon terminals), hormone secretion (Ca^{2+} channels of endocrine cells) and muscle contraction (Ca^{2+} channels of smooth, skeletal or cardiac muscle cells).

HVA Ca^{2+} channels can be separated by using Ca^{2+} channel blockers, which can be subdivided into three general classes: small organic blockers; peptide toxins; and inorganic blockers. Small organic blockers include the dihydropyridines (DHP) that selectively block L-type channels (these channels are also selectively opened by Bay K 8644). Peptide toxins include an ω-conotoxin of the marine snail *Conus geographicus* that selectively blocks N-type channels and a purified polyamine fraction of the funnel-web spider (*Agelenopsis aperta*) venom (FTX) or a peptide component of the same venom, ω-agatoxin IVA (ω-Aga-IVA), that selectively blocks P-type channels. Inorganic blockers include divalent or trivalent metal ions such as cadmium and nickel, but they are not selective and are thus not used to separate the different types of HVA channels.

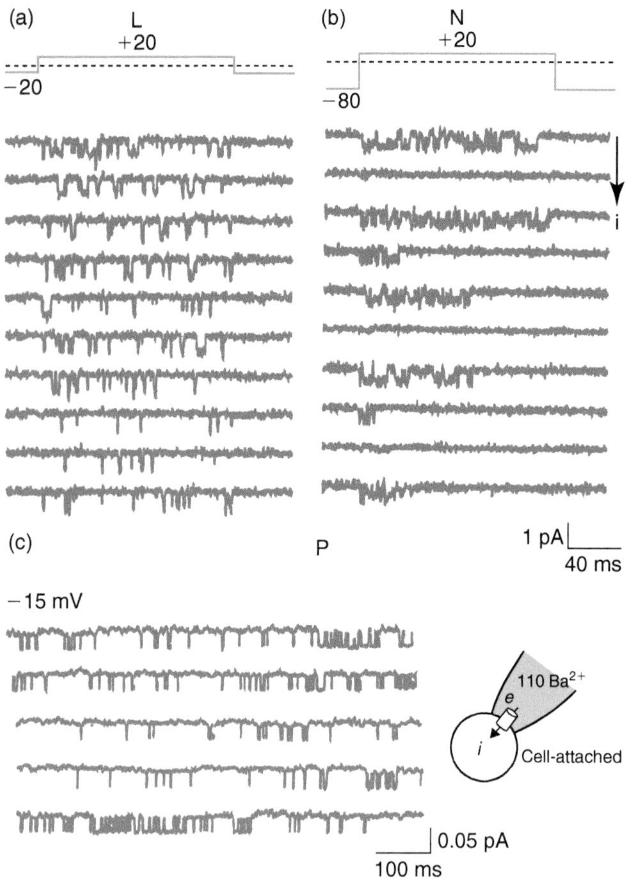

FIGURE 5.3 **Single-channel recordings of the high-threshold Ca^{2+} channels: the L, N and P channels.** The activity of **(a)** single L and **(b)** N Ca^{2+} channels is recorded in patch clamp (cell-attached patches) from dorsal root ganglion cells and that of a single P channel **(c)** is recorded from a lipid bilayer in which a P channel isolated from cerebellum has been incorporated. All recordings are performed with Ba^{2+} (110 or 80 mM) as the charge carrier. In response to a test depolarizing step to +20 mV (**a**,**b**) or at a depolarized holding potential of −15 mV (**c**), unitary inward currents are recorded. Upper traces are voltage and the corresponding unitary current traces are the bottom traces (5–10 trials). V_H = −20 mV in (**a**), −80 mV in (**b**) and −15 mV in (**c**). In (**a**) and (**b**) the intrapipette solution contains: 110 mM $BaCl_2$, 10 mM HEPES and 200 μM TTX. The extracellular solution bathing the membrane outside the patch contains (in mM): 140 K aspartate, 10 K-EGTA, 10 HEPES, 1 $MgCl_2$ in order to zero the cell resting membrane potential. In (**c**) the solution bathing the extracellular side of the bilayer contains (in mM): 80 $BaCl_2$, 10 HEPES. The solution bathing the intracellular side of the bilayer in (**c**) contains (in mM): 120 CsCl, 1 $MgCl_2$, 10 HEPES. Parts (**a**) and (**b**) adapted from Nowycky MC, Fox AP, Tsien RW (1985) Three types of neuronal calcium channel with different calcium agonist sensitivity. *Nature* **316**, 440–443, with permission. Part (**c**) adapted from Llinas R, Sugimori M, Lin JW, Cherksey B (1989) Blocking and isolation of a calcium channel from neurons in mammals and cephalopods utilizing a toxin fraction (FTX) from funnel web spider poison. *Proc. Natl Acad. Sci. USA* **86**, 1689–1693, with permission.

5.2.2 The L-, N- and P-type Ca^{2+} channels open at membrane potentials positive to –20 mV; they are high-threshold Ca^{2+} channels

The L-type Ca^{2+} channel has a large conductance and inactivates very slowly with depolarization

The activity of single L-type Ca^{2+} channels is recorded in sensory neurons of the chick dorsal root ganglion in patch clamp (cell-attached patch with Ba^{2+} as the charge carrier). In response to a test depolarization to +20 mV from a *depolarized* holding potential (−40 to 0 mV), unitary inward Ba^{2+} currents are evoked and recorded throughout the duration of the depolarizing step (**Figure 5.3a**).

The voltage dependence of activation is studied with depolarizations to various test potentials from a holding potential of −40 mV (**Figure 5.4**). With test depolarizations up to +10 mV, openings are rare and of short duration. Activation of the channel becomes significant at +10 mV: openings are more frequent and of longer duration. At all potentials tested, openings are distributed relatively evenly throughout the duration of

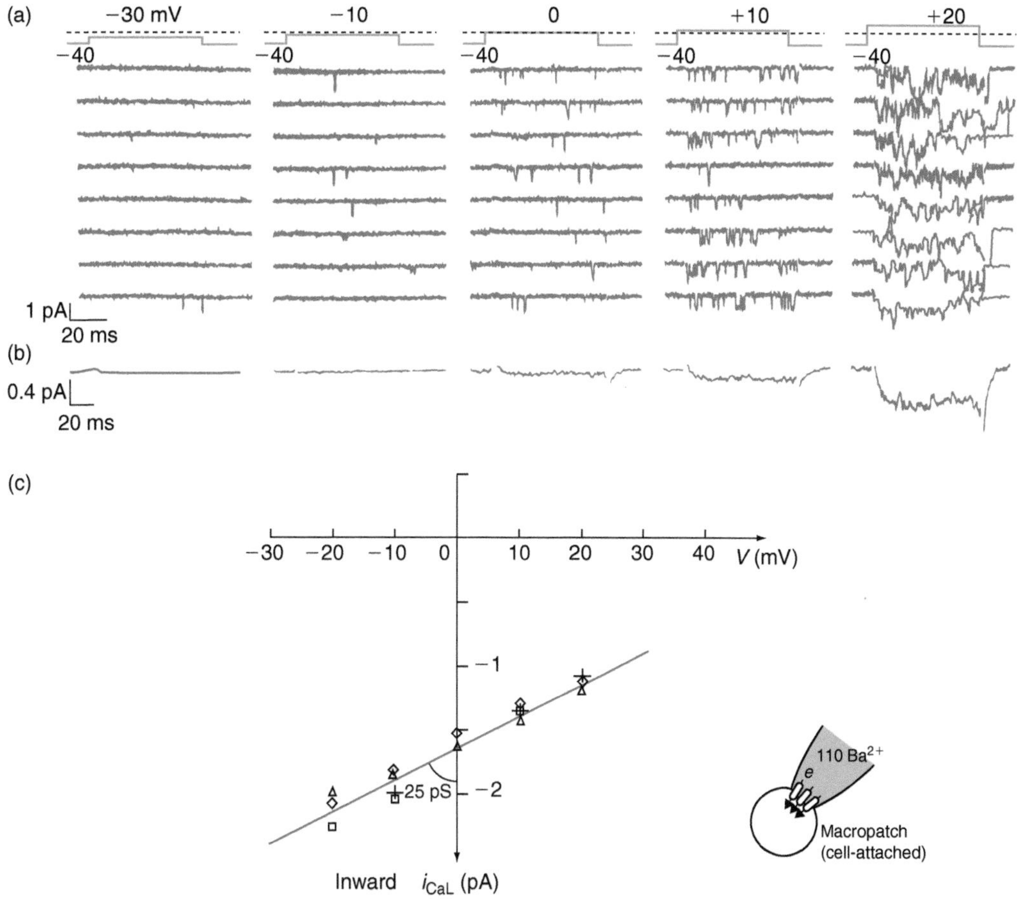

FIGURE 5.4 **Voltage dependence of the unitary L-type Ca^{2+} current. (a)** The activity of L channels (the patch of membrane contains more than one L channel) is recorded in patch clamp (cell-attached patch) in a sensory dorsal root ganglion neuron. The patch is depolarized to −30, −10, 0, +10 and +20 mV from a holding potential of −40 mV. **(b)** Macroscopic current traces obtained by averaging at least 80 corresponding unitary current recordings such as those in **(a)**. The probability of the L channels being in the open state increases with the test depolarization so that at +20 mV, openings of the 4–5 channels present in the patch overlap, leading to a sudden increase in the corresponding macroscopic current. **(c)** The unitary L current amplitude (i_L) is plotted against membrane potential (from −20 to +20 mV) in the absence (+, square) or presence (Δ, lozenge) of Bay K 8644 in the patch pipette. The amplitude of i_L decreases linearly with depolarization between −20 and +20 mV with a slope γ_L = 25 pS. The intrapipette solution contains (in mM): 110 $BaCl_2$, 10 HEPES. The extracellular solution bathing the extracellular side of the membrane outside of the recording pipette contains (in mM): 140 K-aspartate, 10 K-EGTA, 1 $MgCl_2$, 10 HEPES. A symmetric K^+ solution is applied in order to zero the cell resting potential. Adapted from Fox AF, Nowycky MC, Tsien RW (1987) Single-channel recordings of three types of calcium channels in chick sensory neurons. *J. Physiol.* **394**, 173–200, with permission.

the depolarizing step (**Figures 5.3a** and **5.4a**). At −20 mV, the mean single-channel amplitude of the L current (i_L) is around −2 pA. i_L amplitude diminishes linearly with depolarization: the i_L/V relation is linear between −20 and +20 mV. Between these membrane potentials, the unitary conductance, γ_L, is constant and equal to 20–25 pS in 110 mM Ba^{2+}(**Figure 5.4c**).

The main characteristics of L-type channels are (i) their very slow inactivation during a depolarizing step; (ii) their sensitivity to dihydropyridines; and (iii) their loss of activity in excised patches. Bay K 8644 is a dihydropyridine compound that increases dramatically the mean open time of an L-type channel without changing its unitary conductance (**Figure 5.5**). It has no effect on the other Ca^{2+} channel types (see **Figure 5.9**). Bay K 8644 binds to a specific site on the α_1-subunit of L channels and changes the gating mode from brief openings to long-lasting openings even at weakly depolarized potentials (V_{step} = −30 mV). Other dihydropyridine derivatives such as nifedipine, nimodipine and nitrendipine selectively block L channels (see **Figure 5.16**). If all Ca^{2+} channels share the same general structural features, the amino acid residues that confer high affinity for the organic Ca^{2+} antagonists such as verapamil and nitrendipine are present only in the Ca_V1 family of Ca^{2+} channels.

The loss of activity of an L channel in excised patch can be observed in outside-out patches. In response to a test depolarization to +10 mV, the activity of an L channel rapidly disappears (**Figure 5.6**). To determine the nature of the cytoplasmic constituent(s) necessary to restore

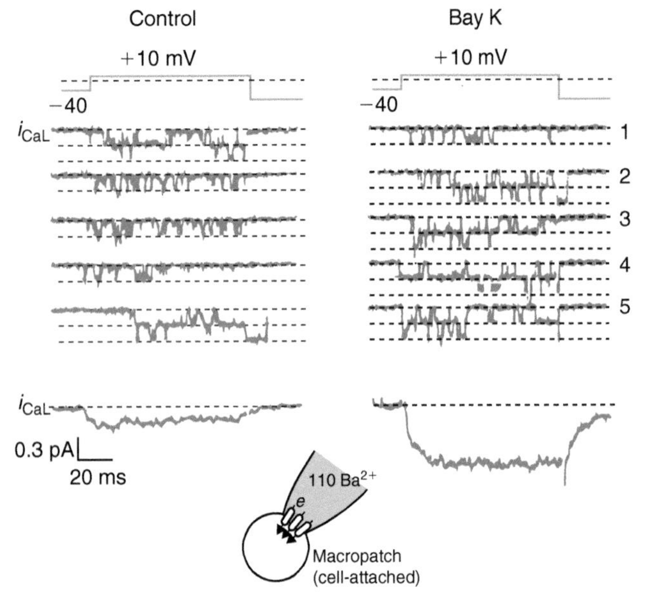

FIGURE 5.5 **Bay K 8644 promotes long-lasting openings of L-type Ca^{2+} channels.** The activity of three L channels is recorded in patch clamp (cell-attached patch). Top traces: a depolarizing step to +10 mV from a holding potential of −40 mV is applied at a low frequency. Middle traces (1 to 5): five consecutive unitary current traces recorded in the absence (left) and presence (right) of 5 μM Bay K 8644 in the bathing solution. Recordings are obtained from the same cell. Dashed line indicates the mean amplitude of the unitary current (−1.28 pA) which is unchanged in the presence of Bay K. Bottom traces: macroscopic current traces obtained by averaging at least 80 corresponding unitary current recordings. Adapted from Fox AP, Nowycky MC, Tsien RW (1987) Single-channel recordings of three types of calcium channels in chick sensory neurones. *J. Physiol.* **394**, 173–200, with permission.

the activity of the L channel, inside-out patches are performed, a configuration that allows a change of the medium bathing the intracellular side of the membrane.

The activity of a single L channel is first recorded in cell-attached configuration in response to a test depolarization to 0 mV (**Figure 5.7**). Then the membrane is pulled out in order to obtain an inside-out patch. The L-type activity rapidly disappears and is not restored by adding ATP–Mg to the intracellular solution. In contrast, when the catalytic subunit of the cAMP-dependent protein kinase (PKA) is added, the L-channel activity reappears (the catalytic subunit of PKA does not need the presence of cAMP to be active). This suggests that PKA directly phosphorylates the L channel thus allowing its activation by the depolarization. It means that, in physiological conditions, the activity of L channels requires the activation of the following cascade: the activation of adenylate cyclase by the α-subunit of the G protein, the formation of cAMP and the subsequent activation of protein kinase A. Other kinases might also play a role. That is how neurotransmitters and a wide variety of hormones modulate L-type Ca^{2+} currents in neurons but also in endocrine cells and in smooth, skeletal and cardiac muscle.

The N-type Ca^{2+} channel inactivates with depolarization in the tens of milliseconds range and has a smaller unitary conductance than the L-type channel

The activity of single N-type channels is recorded in the same preparation in patch clamp (cell-attached

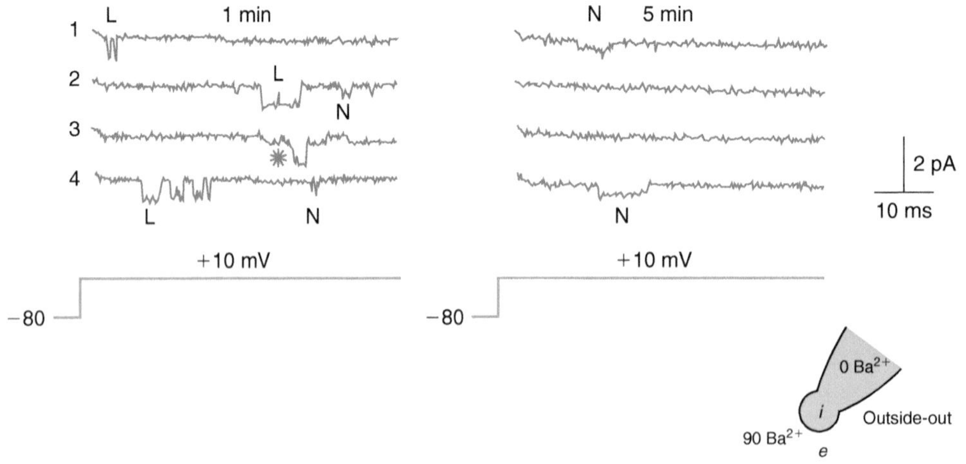

FIGURE 5.6 **In excised patches, the activity of L channels disappears within minutes.** The activity of an L and N channel is recorded in patch clamp (outside-out patches from a pituitary cell line in culture) in response to a depolarizing pulse to +10 mV from a holding potential of −80 mV. Left: One minute after forming the excised patch, the two types of channels open one at a time or their openings overlap (line 3, *). Five minutes after, only the activity of the N-type is still present. The activity of the L-type will not reappear spontaneously. The extracellular solution contains (in mM): 90 $BaCl_2$, 15 TEACl, 2×10^{-3} TTX, 10 HEPES. The intrapipette solution contains (in mM): 120 CsCl, 40 HEPES. Adapted from Armstrong D, Eckert R (1987) Voltage-activated calcium channels that must be phosphorylated to respond to membrane depolarization. *Proc. Natl Acad. Sci. USA* **84**, 2518–2522, with permission.

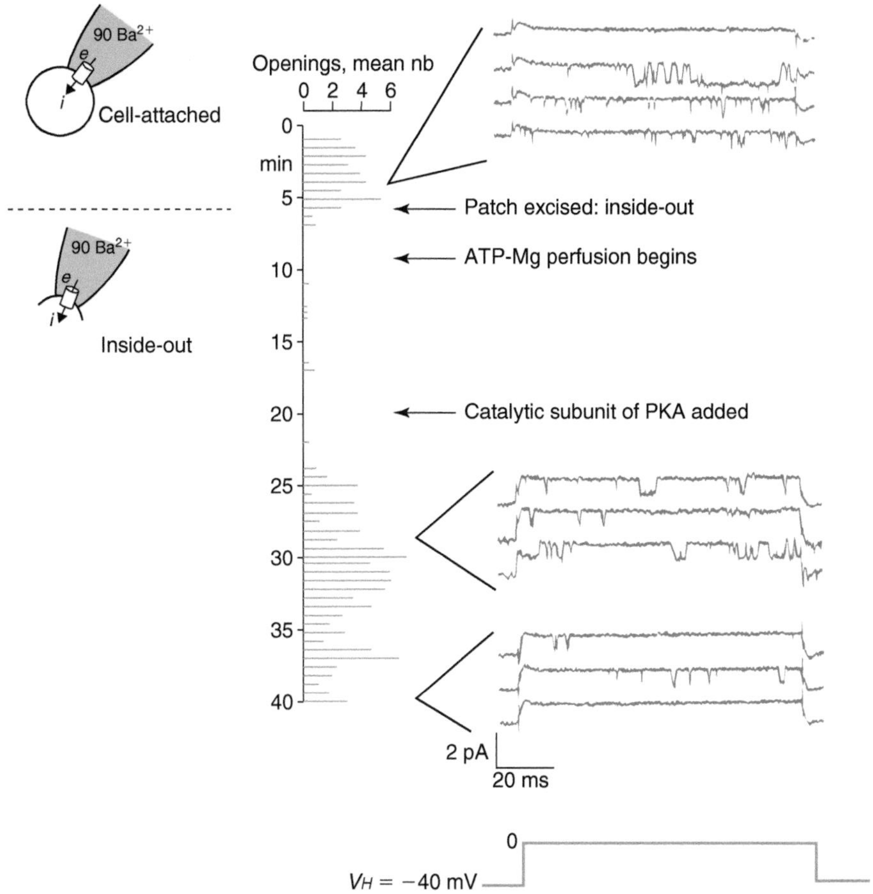

FIGURE 5.7 **Phosphorylation reverses the loss of activity of the L channels in an inside-out patch.** The activity of an L-type channel is recorded in patch clamp (inside-out patch from a pituitary cell line in culture) in response to a depolarizing pulse to 0 mV from a holding potential of −40 mV. The horizontal traces are the unitary current traces and the vertical histogram represents the average number of channel openings per trace, determined over 30 s intervals and plotted versus time of the experiment (0–40 min). After 5 min of recording in the cell-attached configuration, the activity of the channel is recorded in the inside-out configuration. See text for further explanations. The intrapipette solution contains (in mM): 90 $BaCl_2$, 15 TEACl, 2×10^{-3} TTX, 10 HEPES. The solution bathing the intracellular side of the patch contains (in mM): 120 CsCl, 40 HEPES. From Armstrong D, Eckert R (1987) Voltage-activated calcium channels that must be phosphorylated to respond to membrane depolarization. *Proc. Natl Acad. Sci. USA* **84**, 2518–2522, with permission.

patch, with Ba^{2+} as the charge carrier). In contrast to the L channels, N channels inactivate with depolarization. Therefore, their activity has to be recorded in response to a test depolarization from a *hyperpolarized* holding potential (−80 to −60 mV) (**Figure 5.3b**). At holding potentials positive to −40 mV (e.g. −20 mV; **Figure 5.3a**), the N channel(s) is inactivated and its activity is absent in the recordings.

N-channel activity differs from that of the L channel in several aspects:

- N channels often open in bursts and inactivate with time and voltage (see Section 5.2.3);
- Measured at the same test potential, the mean amplitude of the N unitary current is smaller than that of L (e.g. $i_N = -1.22 \pm 0.03$ pA and $i_L = -2.07 \pm 0.09$ at −20 mV; **Figure 5.3a,b**) which makes its mean unitary conductance also smaller (γ_N =13 pS in 110 mM Ba^{2+}; **Figure 5.8b**);
- N channels are insensitive to dihydropyridines but are selectively blocked by ω-conotoxin GVIA;
- N channels do not need to be phosphorylated to open (**Figure 5.6**).

The P-type Ca^{2+} channel differs from the N channel by its pharmacology

The activity of a single P-type channel is recorded from lipid bilayers in which purified P channels from cerebellar Purkinje cells have been incorporated. Ba^{2+} ions are used as the charge carrier. The activity of the P channel is recorded at different steady holding potentials. At −15 mV, the channel opens, closes and reopens during the entire depolarization, showing little time-dependent inactivation (**Figure 5.3c**). The mean unitary conductance, γ_P, is 10–15 pS in 80 mM Ba^{2+}. Recordings performed in dendrites or the soma of cerebellar Purkinje cells with patch clamp techniques (cell-attached patches) gave similar values of the unitary conductance (γ_P = 9–19 pS in 110 mM Ba^{2+}) but, for undetermined reasons, the threshold for activation is at a more depolarized potential (−15 mV) than for isolated P channels inserted in lipid bilayers (−45 mV). When the funnel web toxin fraction (FTX) is added to the recording patch pipette (the intrapipette solution bathes the extracellular side of the patch), only rare high-threshold unitary currents are recorded from Purkinje cell dendrites or soma at all potentials tested (Bay K 8644 or ω-conotoxin have no effect). These results suggest that the P channel is the predominant high-threshold Ca^{2+} channel expressed by Purkinje cells. They also show that the use of selective toxins allows differentiation between P, N and L channels.

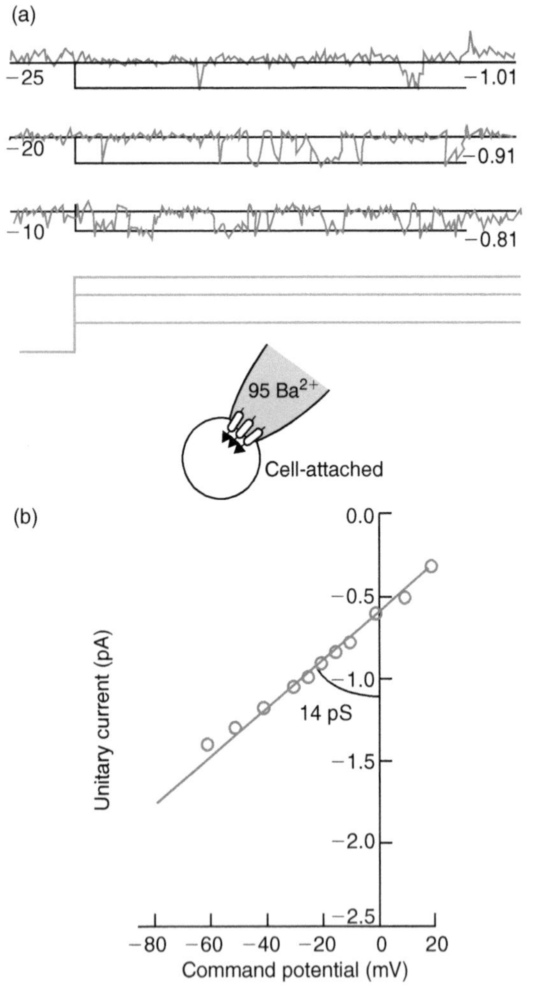

FIGURE 5.8 **Voltage dependence of the unitary N current, i_N. (a)** The activity of an N channel is recorded in patch clamp (cell-attached patch) in a granule cell of the hippocampus. The patch is depolarized to −25, −20 and −10 mV from a holding potential of −80 mV. The amplitude of the unitary current at these voltages is indicated at the end of each recording. **(b)** The unitary N current amplitude (i_N) is plotted against membrane potential (from −60 to +20 mV). The amplitude of i_N decreases linearly with depolarization between −60 and +20 mV with a slope γ_N = 14 pS (n = 14 patches). Adapted from Fisher RE, Gray R, Johnston D (1990) Properties and distribution of single voltage-gated calcium channels in adult hippocampal neurons. *J. Neurophysiol.* **64**, 91–104, with permission.

5.2.3 Macroscopic L-, N- and P-type Ca^{2+} currents activate at a high threshold and inactivate with different time courses

The macroscopic L-, N- and P-type Ca^{2+} currents (I_{Ca}), at time t during a depolarizing voltage step, are equal to: $I_{Ca} = Np_t i_{Ca}$ where N is the number of L, N or P channels in the membrane, p_t is their probability of being open at time t during the depolarizing step, Np_t is the number of open channels at time t during the depolarizing step and i_{Ca} is the unitary L, N or P current. At steady state, $I_{Ca} = Np_o i_{Ca}$, where p_o is the probability of the channel being open at steady state.

The I/V relations for L-, N- and P-type Ca^{2+} currents have a bell shape with a peak amplitude at positive potentials

The I/V relation of the different types of high threshold Ca^{2+} currents is studied in whole-cell recordings in the presence of external Ca^{2+} as the charge carrier. To separate the L, N and P currents, specific blockers are added to the external medium or the membrane potential is clamped at different holding potentials. With this last procedure, the L current can be separated from other Ca^{2+} currents since it can be evoked from depolarized holding potentials. As shown in **Figure 5.9**, the L and N currents averaged from the corresponding unitary currents recorded in 110 mM Ba^{2+} clearly differ in their time course. The averaged N current decays to zero level in 40 ms while the averaged L current remains con-

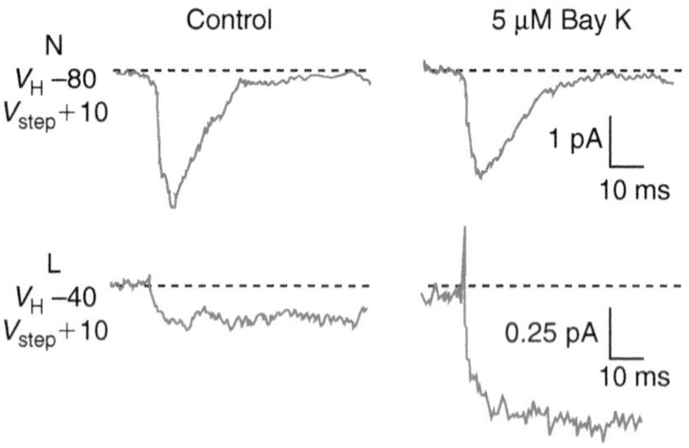

FIGURE 5.9 **Averaged N- and L-type Ca^{2+} currents.** Single-channel N current averages (top traces) and L current averages (bottom traces) from cell-attached recordings of dorsal root ganglion cells with Ba^{2+} as the charge carrier (see also **Figure 5.3a,b**). Currents are averaged before (left) and after (right) exposure to 5 μM Bay K 8644. Voltage steps from −80 to +10 mV (top traces) and from −40 to 110 mV (bottom traces). From Nowycky MC, Fox AP, Tsien RW (1985) Three types of neuronal calcium channel with different calcium agonist sensitivity. *Nature* **316**, 440–443, with permission.

stant during the 120 ms depolarizing step to +10 mV. As already observed (**Figure 5.3a,b**), by holding the membrane at a depolarized potential, the N current inactivates and the L current can be studied in isolation.

The macroscopic N- and L-type Ca^{2+} currents are studied in spinal motoneurons of the chick in patch clamp (whole-cell patch) in the presence of Na^+ and K^+ channel blockers and in the presence of a T-type Ca^{2+} channel blocker. In response to a depolarizing voltage step to +20 mV from a holding potential of −80 mV, a mixed N and L whole-cell current is recorded (**Figure 5.10a**). When the holding potential is depolarized to 0 mV, a voltage step to +20 mV now only evokes the L current (**Figure 5.10b**). The difference current obtained by subtracting the L current from the mixed N and L current gives the N current (**Figure 5.10c**). The I/V relations of these two Ca^{2+} currents have a bell shape with a peak around +20 mV (**Figure 5.10d,e**). For comparison the peak amplitude of the macroscopic Na^+ current is around −40 mV (see **Figure 4.12a**).

The macroscopic P-type Ca^{2+} current is studied in cerebellar Purkinje cells. These neurons express T-, P- and a few L-type Ca^{21} channels. In the presence of Na^+ and K^+ channel blockers and by choosing a holding potential where the low threshold T current is inactivated, the macroscopic P current can be studied. The I_P/V relation has a bell shape. The maximal amplitude is recorded around −10 mV (**Figure 5.11**).

The bell shape of all the I_{Ca}/V relations is explained by the gating properties of the Ca^{2+} channels and the

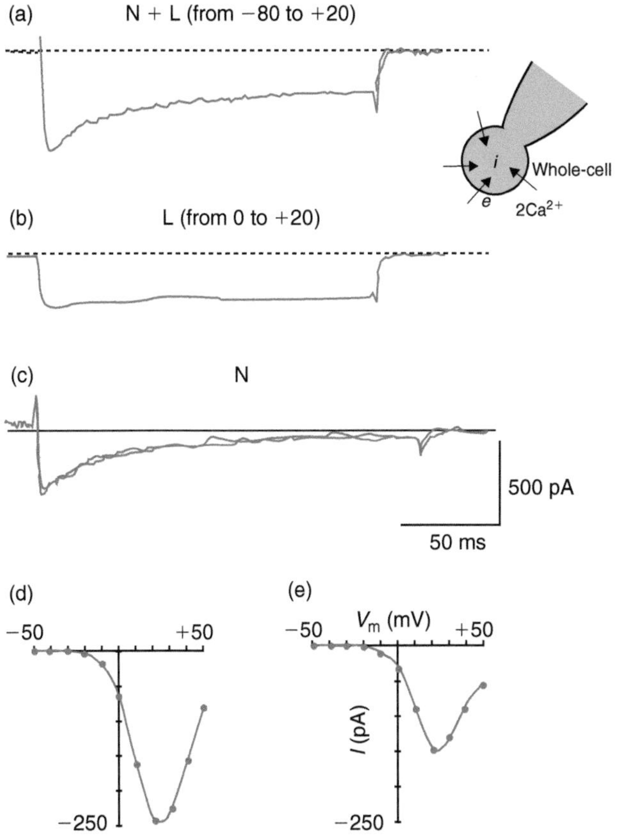

FIGURE 5.10 **N- and L-type macroscopic Ca^{2+} currents. (a)** The mixed N and L macroscopic current is recorded with Ca^{2+} as the charge carrier (whole-cell patch) from chick limb motoneurons in culture in response to a voltage step from −80 to +20 mV. **(b)** The macroscopic L current is recorded in isolation by changing the holding potential to 0 mV. **(c)** The difference current obtained by subtracting the L current (**b**) from the N and L current (**a**) is the N current. **(d)** I/V relation for the L current recorded as in (**b**). **(e)** I/V relation for the N current obtained as the difference current. The intrapipette solution contains (in mM): 140 Cs aspartate, 5 $MgCl_2$, 10 Cs EGTA, 10 HEPES, 0.1 Li_2GTP, 1 MgATP. The bathing solution contains (in mM): 146 NaCl, 2 $CaCl_2$, 5 KCl, 1 $MgCl_2$, 10 HEPES. Adapted from McCobb DP, Best PM, Beam KG (1989) Development alters the expression of calcium currents in chick limb motoneurons. *Neuron* **2**, 1633–1643, with permission.

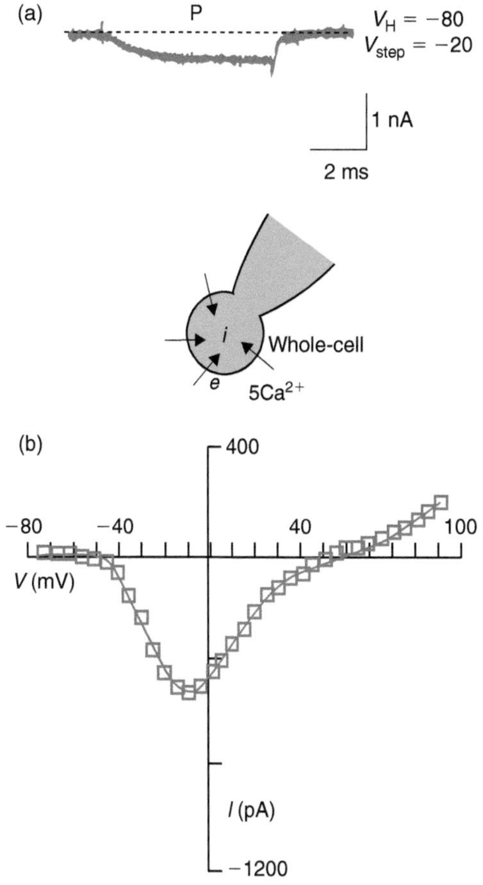

FIGURE 5.11 **P-type macroscopic Ca^{2+} current.** The whole-cell P current recorded from acutely dissociated Purkinje cells (whole-cell patch) with Ca^{2+} as the charge carrier. **(a)** Whole-cell P current recorded in response to a depolarizing pulse to −20 mV from a holding potential of −80 mV. **(b)** I/V relation of the P current. In the recordings the low threshold T-type Ca^{2+} current was either absent, inactivated or subtracted. The intrapipette solution contains (in mM): 120 TEA glutamate, 9 EGTA, 4.5 $MgCl_2$, 9 HEPES. The bathing solution contains (in mM): 5 $CaCl_2$, 154 TEACl, 0.2 $MgCl_2$, 10 glucose, 10 HEPES. Adapted from Reagan LJ (1991) Voltage-dependent calcium currents in Purkinje cells from rat cerebellar vermis. *J. Neurosci.* **7**, 2259–2269, with permission.

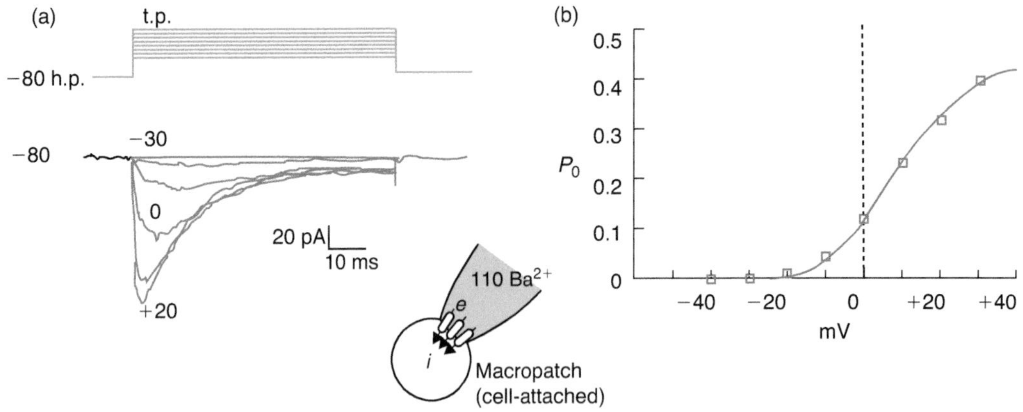

FIGURE 5.12 **The peak opening probability of the N current.** The macroscopic N current is recorded in a dorsal root ganglion neuron from a cell-attached patch containing hundreds of N channels (macropatch). **(a)** Current recordings (bottom traces) in response to test potentials (t.p.) ranging from −30 to +20 mV from a holding potential (h.p.) of −80 mV (upper traces). **(b)** Voltage dependence of the peak opening probability (p_o) from data obtained in (**a**). Values of p_o are obtained by dividing the peak current I by the unitary current i_N obtained at each test potential and by an estimate of the number of channels in the patch (599): $p_o = I/Ni_N$. N was determined by comparison with the single-channel experiment in **Figure 5.3b**, which shows that in response to a depolarization to +20 mV from a holding potential of −80 mV, $p_o = 0.32$ and $i_N = 0.76$ pA. I, the peak current evoked by the same voltage protocol, is 145 pA. $N = I/p_o\, i_N = 145/(0.32 \times 0.76) = 599$ channels. The intrapipette solution contains (in mM): 100 CsCl, 10 Cs-EGTA, 5 $MgCl_2$, 40 HEPES, 2 ATP, 0.25 cAMP; pH = 7.3. The extracellular solution contains (in mM): 10 $CaCl_2$, 135 TEACl, 10 HEPES, 0.2×10^{-3} TTX; pH = 7.3. From Nowycky MC, Fox AP, Tsien RW (1985) Three types of neuronal calcium channel with different calcium agonist sensitivity. *Nature* **316**, 440–443, with permission.

driving force for Ca^{2+} ions. The peak amplitude of I_{Ca} increases from the threshold potential to a maximal amplitude (**Figures 5.10d,e, 5.11b** and **5.12a**) as a result of two opposite factors: the probability of opening which strongly increases with depolarization (**Figure 5.12b**) and the driving force for Ca^{2+} which linearly decreases with depolarization (i_{Ca} linearly diminishes). After a maximum, the peak amplitude of I_{Ca} decreases owing to the progressive decrease of the driving force for Ca^{2+} ions and the increase of the number of inactivated channels. Above +30/+40 mV, the probability of opening (p_o) no longer plays a role since it is maximal (**Figure 5.12b**). I_{Ca} reverses polarity between +50 mV and +100 mV, depending on the preparation studied. This value is well below the theoretical E_{Ca}.

This discrepancy is partly due to the strong asymmetrical concentrations of Ca^{2+} ions. To measure the reversal potential of I_{Ca}, the outward current through Ca^{2+} channels must be measured. This outward current, caused by the extremely small intracellular concentration of Ca^{2+} ions, is carried by Ca^{2+} ions but also by internal K^+ ions, which are around 10^6 times more concentrated than internal Ca^{2+} ions. This permeability of Ca^{2+} channels to K^+ ions 'pulls down' the reversal potential of I_{Ca} towards E_K.

Activation–inactivation properties

Activation properties are analyzed by recording the macroscopic L, N or P currents in response to increasing test depolarizations from a fixed hyperpolarized holding potential (−80 mV, **Figures 5.13b, 5.14b** and **5.15b**). In dorsal ganglion neurons, the L and N currents are half activated around 0 mV (**Figures 5.13c** and **5.14c**) while in Purkinje cells, the P current is half activated around −20 mV (**Figure 5.15c**).

Voltage-gated Ca^{2+} channels show varying degrees of inactivation

Inactivation properties are analyzed by recording the macroscopic L-, N- or P-type Ca^{2+} currents evoked by a voltage step to a fixed potential from various holding potentials (with Ca^{2+} as the charge carrier). The L current is half inactivated around −40 mV (**Figure 5.13a,c**), the N current around −60 mV (**Figure 5.14a,c**) and the P current around −45 mV (**Figure 5.15a,c**).

In summary, L channels generate a large Ca^{2+} current that is activated by large depolarizations to 0/+10 mV and inactivates with a very slow time course during a step. N and P channels generate smaller Ca^{2+} currents that are activated with depolarization to −30/0 mV and inactivate or not during a depolarizing step.

The inactivation process of Ca^{2+} channels can be voltage-dependent, time-dependent *and* calcium-dependent. Voltage-dependent inactivation is observed by changing the holding potential (see **Figures 5.13a, 5.14a** and **5.15a**). Time-dependent inactivation is observed during a long depolarizing step, in the presence of Ba^{2+} as the charge carrier (**Figure 5.16**). Ca^{2+}-dependent inactivation depends on the amount of Ca^{2+} influx through open Ca^{2+} channels. It can be considered as a negative feedback control of Ca^{2+} channels by Ca^{2+} channels.

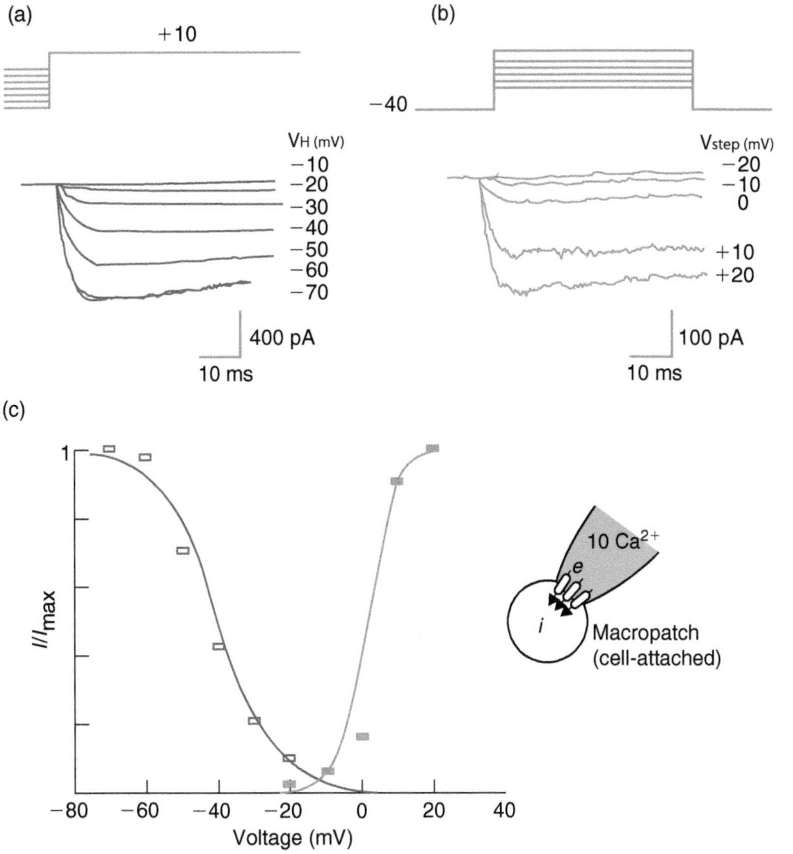

FIGURE 5.13 **Voltage dependence of activation and inactivation of the L-type Ca^{2+} current.** The macroscopic L current is recorded in a cell with very little T or N current. **(a)** Inactivation of the L current with holding potential: a test depolarization to +10 mV is applied from holding potentials (V_H) varying from −70 to −10 mV. **(b)** Activation of the L current with depolarization: test depolarizations (V_{step}) to −30, −20, −10, 0, +10 and +20 mV are applied from a holding potential of −40 mV. **(c)** Activation–inactivation curves obtained from the data in (**b**) and (**a**), respectively. The peak Ca^{2+} current amplitudes (I) are normalized to the maximal current (I_{max} = 1) obtained in each set of experiments and plotted against the holding potential (inactivation curve, blue) or test potential (activation curve, orange). For the activation curve, data are plotted as $I = I_{max}\{1 + \exp[(V_{1/2} - V)/k]\}^{-1}$ and for the inactivation curve as $I = I_{max}\{1 \exp[(V - V_{1/2})/k]\}^{-1}$. $V_{1/2}$ is the voltage at which the current I is half-activated ($I = I_{max}/2$ when $V_{1/2}$ = 2 mV) or half-inactivated ($I = I_{max}/2$ when $V_{1/2}$ = −40 mV). All the recordings are performed in the presence of 10 mM Ca^{2+} in the recording pipette solution which bathes the extracellular side of the channels. Adapted from Fox AP, Nowycky M, Tsien RW (1987) Kinetic and pharmacological properties distinguishing three types of calcium currents in chick sensory neurones. *J. Physiol.* **394**, 149–172, with permission.

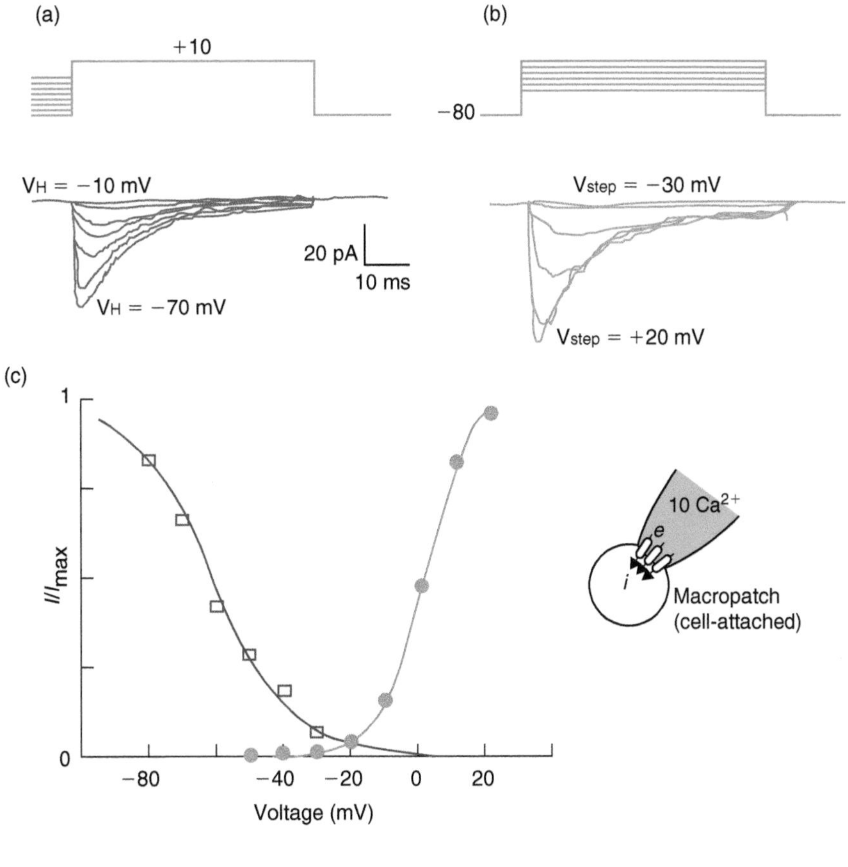

FIGURE 5.14 **Voltage dependence of activation and inactivation of the N-type Ca^{2+} current.** The macroscopic N current is recorded in cell-attached patches containing hundreds of channels (macropatch). **(a)** Inactivation of the N current with holding potential: test depolarization to +10 mV is applied from holding potentials (V_H) varying from −70 to −10 mV. **(b)** Activation of the N current with depolarization: test depolarizations (V_{step}) to −30, −20, −10, 0, +10 and +20 mV are applied from a holding potential of −80 mV. **(c)** Activation–inactivation curves obtained from the data in (**b**) and (**a**), respectively. The peak Ca^{2+} current amplitudes (I) are normalized to the maximal current (I_{max} = 1) obtained in each set of experiments and plotted against the holding (inactivation curve, blue) or test potential (activation curve, orange). For the activation curve, data are plotted as $I = I_{max}\{1 + \exp[(V_{1/2} - V)/k]\}^{-1}$ and for the inactivation curve as $I = I\{1 + \exp[(V - V_{1/2}/k]\}^{-1}$. $V_{1/2}$ is the voltage at which the current I is half-activated ($I = I_{max}/2$ when $V_{1/2}$ = 1.5 mV) or half-inactivated ($I = I_{max}/2$ when $V_{1/2}$ = −61.5 mV). The number of channels is estimated as in **Figure 5.12**. Adapted from Fox AP, Nowycky MC, Tsien RW (1987) Single-channel recordings of three types of calcium channels in chick sensory neurones. *J. Physiol.* **394**, 173–200, with permission.

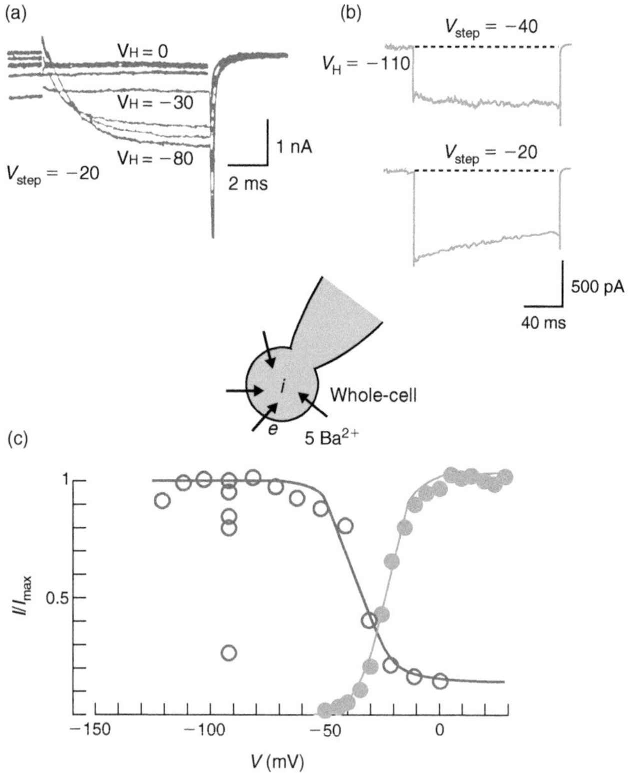

FIGURE 5.15 **Voltage dependence of activation and inactivation of the P-type Ca^{2+} current.** The macroscopic P current is recorded in Purkinje cells (whole-cell patch). The T-type Ca^{2+} current present in these cells is either absent or subtracted. **(a)** Inactivation of the P current with holding potential: a test depolarization (V_{step}) to −20 mV is applied from holding potentials (V_H) of −80, −60, −40, −30, −20, −10 and 0 mV. Holding potential positive to −60 mV elicited steady inward current, which is maximal at −30 mV. **(b)** Activation of the P current with depolarization: test depolarizations (V_{step}) to −40 and −20 mV are applied from a holding potential of −110 mV. **(c)** Activation–inactivation curves obtained from the data obtained in **(b)** and **(a)**, respectively. The peak Ca^{2+} current amplitudes (I) are normalized to the maximal current (I_{max} = 1) obtained in each set of experiments and plotted against the holding (inactivation curve, blue) or test potential (activation curve, orange). For the activation curve, data are plotted as $I = I_{max}\{1 + \exp[(V_{1/2} - V)/k]\}^{-1}$ and for the inactivation curve as $I = I_{max}\{1 + \exp[(V - V_{1/2})/k]\}^{-1}$. $V_{1/2}$ is the voltage at which the current I is half-activated ($I = I_{max}/2$ when $V_{1/2} = -22$ mV) or half-inactivated ($I = I_{max}/2$ when $V_{1/2} = -34$ mV). In all recordings, the extracellular solution contains 5 mM Ba^{2+}. Adapted from Regan L (1991) Voltage-dependent calcium currents in Purkinje cells from rat cerebellar vermis. *J. Neurosci.* **11**, 2259–2269, with permission.

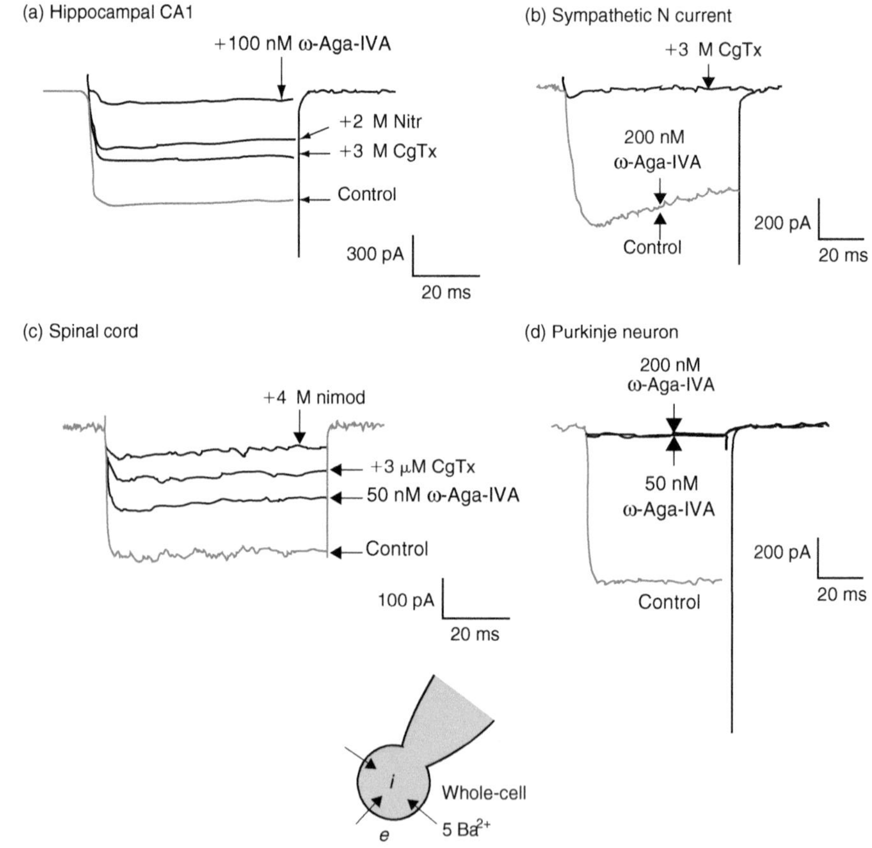

FIGURE 5.16 **Pharmacology of L-, N- and P-type Ca^{2+} channels.** The macroscopic mixed Ca^{2+} currents are recorded in different neurons with Ba^{2+} as the charge carrier (whole-cell patch). High-threshold Ca^{2+} currents are evoked by depolarizations to −30 or −10 mV from a holding potential of −90 or −80 mV. Various blockers or toxins are applied in order to block selectively one type of high-threshold Ca^{2+} current at a time: ω-conotoxin (CgTx, 3 μM) selectively blocks N current, nitrendipine or nimodipine (nitr., nimod., 2–4 μM) selectively blocks L current, and ω-agatoxin (ω-Aga-IVA, 50–200 nM) selectively blocks P current. In hippocampal cells of the CA1 region and in spinal cord interneurons, the high-threshold Ca^{2+} current is a mixed N, L and P current. In sympathetic neurons, it is almost exclusively N and, in Purkinje cells, almost exclusively P. The intrapipette solution contains (in mM): 108 Cs methanesulfonate, 4 $MgCl_2$, 9 EGTA, 9 HEPES, 4 MgATP, 14 creatine phosphate, 1 GTP; pH = 7.4. The extracellular solution contains (in mM): 5 $BaCl_2$, 160 TEACl, 0.1 EGTA, 10 HEPES; pH = 7.4. Adapted from Mintz IM, Adams ME, Bean B (1992) P-type calcium channels in rat central and peripheral neurons. *Neuron* **9**, 85–95, with permission.

Calcium-dependent inactivation

Several lines of evidence point to the existence of a Ca^{2+}-induced inactivation of Ca^{2+} currents:

- The degree of inactivation is proportional to the amplitude and frequency of the Ca^{2+} current;
- Intracellular injection of Ca^{2+} ions into neurons produces inactivation;
- Intracellular injection of Ca^{2+} chelators such as EGTA or BAPTA reduces inactivation (**Figure 5.17**);
- Substitution of Ca^{2+} ions with Sr^{2+} or Ba^{2+} reduces inactivation;
- Very large depolarizations to near E_{Ca}, where the entry of Ca^{2+} ions is small, produce little inactivation.

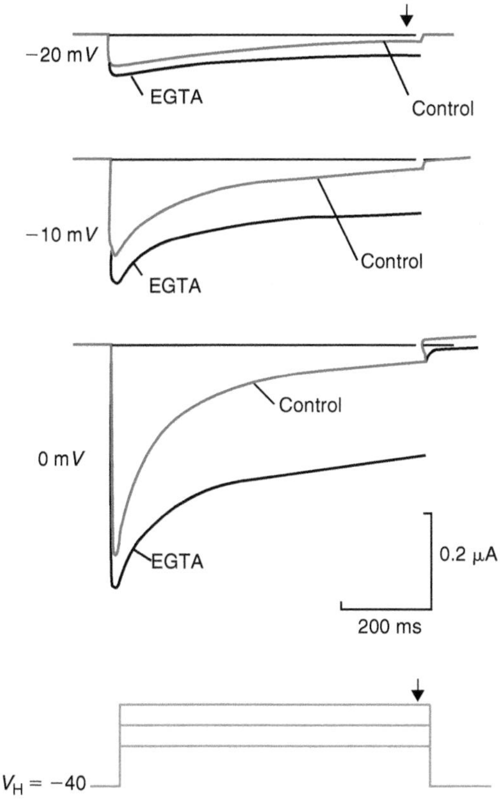

FIGURE 5.17 **Intracellular EGTA slows Ca^{2+}-dependent inactivation of Ca^{2+} channels.** The macroscopic Ca^{2+} current is recorded in axotomized *Aplysia* neurons in double-electrode voltage clamp (axotomy is performed in order to improve space clamp). Control Ca^{2+} currents are recorded in response to step depolarizations to −20, −10 and 0 mV from a holding potential of −40 mV (control traces). Iontophoretic ejection of EGTA (300–500 nA for 4–8 min) increases the peak amplitude of the Ca^{2+} current and slows its inactivation at all potentials tested (EGTA traces). The amplitude of the non-inactivating component of the current is measured at the end of the steps (arrow). Adapted from Chad J, Eckert R, Ewald D (1984) Kinetics of calcium-dependent inactivation of calcium current in voltage-clamped neurones in *Aplysia californica*. *J. Physiol. (Lond.)* **347**, 279–300, with permission.

Recordings of L and N channels in **Figures 5.13** and **5.14** were obtained with Ca^{2+} as the charge carrier and that of P channels in **Figure 5.15** with Ba^{2+} as the charge carrier. Therefore, the inactivation seen in **Figures 5.13** and **5.14** results from three parameters: voltage, time and increase of intracellular Ca^{2+} concentration. In contrast, the inactivation of the P current observed in **Figure 5.15** is a voltage- and time-dependent process.

The macroscopic Ca^{2+} current of *Aplysia* neurons is recorded in voltage clamp. During depolarizing voltage steps, the Ca^{2+} current increases to a peak and then declines to a steady state Ca^{2+} current (a non-inactivating component of current). The buffering of cytoplasmic free Ca^{2+} ions with EGTA increases the amplitude of the peak current and that of the steady-state current (**Figure 5.17**). This shows that the increase of intracellular Ca^{2+} ions resulting from Ca^{2+} entry through Ca^{2+} channels causes Ca^{2+} current inactivation. It also shows that the peak current is probably already decreased in amplitude owing to early development of inactivation.

5.3 THE REPOLARIZATION PHASE OF Ca^{2+}-DEPENDENT ACTION POTENTIALS RESULTS FROM THE ACTIVATION OF K^+ CURRENTS I_K AND I_{KCa}

The K^+ currents involved in calcium spike repolarization are the delayed rectifier (I_K) studied in Chapter 4 and the Ca^{2+}-activated K^+ currents (I_{KCa}). Meech and Strumwasser in 1970 were the first to describe that a microinjection of Ca^{2+} ions into *Aplysia* neurons activates a K^+ conductance and hyperpolarizes the membrane. On the basis of these results, the authors postulated the existence of a Ca^{2+}-activated K^+ conductance. The amount of participation of Ca^{2+}-activated K^+ currents in spike repolarization depends on the cell type.

5.3.1 The Ca^{2+}-activated K^+ currents are classified as big K (BK) channels and small K (SK) channels

Big K channels ($K_{Ca}1.1$) have a high single-channel conductance (100–200 pS) and are sensitive to both voltage *and* Ca^{2+} ions so that their apparent sensitivity to Ca^{2+} ions is increased when the membrane is depolarized. Their activity is blocked by low concentrations of TEA and charybdotoxin, a toxin from scorpion venom. Small K channels ($K_{Ca}2$) have a smaller single-channel conductance (10–20 pS) and are insensitive to TEA and charybdotoxin but sensitive to apamin, a toxin from bee venom which acts as a specific blocker of SK channels. Small K channels are voltage independent and are solely gated by submicromolar concentrations of intracellular Ca^{2+} ions (EC_{50} values of ≈0.3–0.5 μM). Big K and small K channels are very selective for K^+ ions over Na^+ ions and are activated by increases in the concentration of cytoplasmic Ca^{2+} ions.

The channels originally termed 'big' potassium (BK) channels are also called maxi-K channels or SLO family channels, a name derived from the conserved gene that encodes this channel, which was first cloned in *Drosophila melanogaster*. Voltage-clamp recordings of currents in the flight muscles of a *Drosophila* mutant with a severely lethargic phenotype, named *slowpoke*, revealed that the calcium-dependent component of the outward K^+ current was absent, implicating the *slowpoke* (*slo*) gene as the structural locus encoding the channel protein. The mammalian *slo* ortholog *Slo1* was cloned by low-stringency DNA hybridization of a mammalian cDNA library using the *Drosophila slo* cDNA. The conserved protein domains of *Slo1* seem to reflect separate mechanisms for voltage and Ca^{2+} sensing. The primary sequence of the BK α-subunit consists of two distinct regions. The 'core' region and the C-terminus region (see **Figure 5.20a**).

The core region includes seven hydrophobic segments (S0–S6) and resembles a canonical voltage-gated K^+ channel except for the additional S0 segment. It includes the voltage-sensor domain (S1–S4), and the pore domain (S5–S6 with the pore loop). The pore domain of the channel is assigned to the region contained between the S5 and S6 segments, which includes the signature sequence of K^+ channel TVGYG (Thr-Val-Gly-Tyr-Gly). The NH_2-terminus of the protein is placed at the extracellular side of the membrane.

The cytoplasmic C-terminus region comprises about two-thirds of the protein and consists of a pair of RCK domains both with sequence homology to the 'Regulator of K^+ conductance'. The RCK domains from the four BK α-subunits assemble into an octameric gating ring on the intracellular side of the tetrameric channel. Four primary binding sites for Ca^{2+}, called 'calcium bowls', are encoded within the second RCK domain of each monomer. The 'calcium bowl' is a calcium-binding motif that includes a string of conserved aspartate residues (negatively charged). This site is located on the outer perimeter of the gating ring, in a region denominated the assembly interface.

The genes that encode the SK channels belong to the KCNN gene family. SK channels have a similar topology to members of the voltage-gated (K_v) K^+ channel superfamily. They consist of six transmembrane segments (S1–S6), with the pore located between S5 and S6. The S4 segment, which confers voltage sensitivity to the K_v channel, shows in SK channels a reduced number and a disrupted array of positively charged amino acids. The SK channels retain only two of the seven positively charged amino acids that are found in the S4 segment of K_v channels, and only one of these residues corresponds to the four arginine residues that carry the gating charges in K_v channels. These differences in the primary sequence could represent the molecular framework for the observed voltage independence of SK channels. Functional SK channels assemble as tetramers.

SK channels share a common Ca^{2+}-gating mechanism. The pore-forming subunits do not contain an intrinsic Ca^{2+}-binding domain. Rather, Ca^{2+} gating is endowed by a constitutive interaction between the pore-forming subunits and calmodulin. Calmodulin is a small, highly conserved calcium binding protein. It is constitutively bound to each SK subunit, to the intracellular, highly conserved, domain immediately following the sixth transmembrane domain. Each of the four SK subunits harbors a bound calmodulin, and binding of Ca^{2+} ions to calmodulins initiates a conformational rearrangement that results in channel gating.

This was shown by site-directed mutagenesis. If Ca^{2+} ions directly bind to SK channels then negatively charged residues are likely to mediate Ca^{2+} binding. Mutating each of the 21 conserved negatively charged residues within the predicted intracellular domains of the SK channel subunits into a neutral amino acid does not markedly alter Ca^{2+} gating. Only mutations in the C-terminal domain give rise to calcium insensitive channels. Introduction of a stop codon at positions 463, 444 or 421 results in loss of channel function (see **Figure 5.20c**). These results indicate either that Ca^{2+} is bound to this region through an unknown Ca^{2+}-binding motif, or that another protein may interact with the C-terminal domain and mediate Ca^{2+} gating. The affinity of SK channels to Ca^{2+} ions led to the hypothesis that calmodulin, which has a similar affinity, is a good protein candidate for mediating Ca^{2+} sensitivity. Then the interaction between SK channel and calmodulin was confirmed.

5.3.2 Ca^{2+} entering during the depolarization or the plateau phase of Ca^{2+}-dependent action potentials activates K_{Ca} channels

To study Ca^{2+}-activated K^+ channels from rat brain neurons, plasma membrane vesicle preparation is incorporated into planar lipid bilayers. In such conditions, the activity of four distinct types of Ca^{2+}-activated K^+ channels is recorded. We will look at one example of a big K and one example of a small K channel. This preparation allows the recording of single-channel activity (**Figure 5.18**).

The current–voltage relations obtained in the presence of two different extracellular K^+ concentrations show that the current reverses at E_K, the theoretical reversal potential for K^+ ions as expected for a purely K^+-selective channel. The Ca^{2+} dependence is studied by

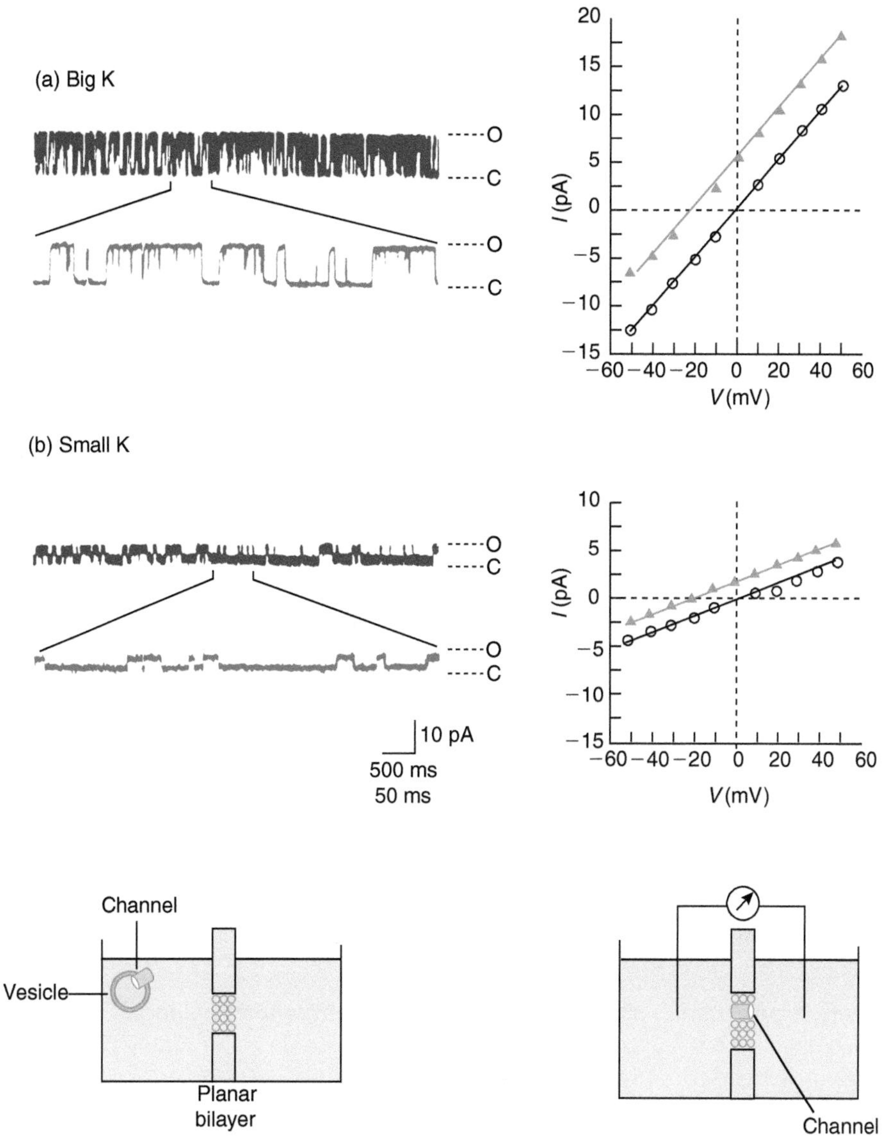

FIGURE 5.18 **Two types of rat brain Ca^{2+}-activated K^+ channels incorporated into lipid bilayers. (a, b)** Left: Single-channel recordings in symmetrical K^+ (the extracellular and intracellular solutions contain 150 mM KCl) at V_H = 40 mV. For all traces channel openings (O) correspond to upward deflections. The recording length of upper traces is 6.4 s and each lower trace is expanded to show a 640 ms recording. Right: I/V relationships for the big K channel (**a**) and the small K channel (**b**) in symmetrical K^+ (150 mM, red circles) or in asymmetrical K^+ (150 mM KCl inside, 50 mM KCl outside) (green triangles). The slope conductance for each of these channels in symmetrical 150 mM KCl is 232 pS (big K channel) and 77 pS (small K channel). All the recordings are performed in the presence of 1.05 mM $CaCl_2$ in the intracellular solution. Adapted from Reinhart PH, Chung S, Levitan IB (1989) A family of calcium-dependent potassium channels from rat brain. *Neuron* **2**, 1031–1041, with permission.

raising the intracellular Ca^{2+} concentration in the range of 0.1–10 μM. Channels are activated by micromolar concentrations of Ca^{2+}. The open probabilities of the big K and small K channels are largely increased when the medium bathing the intracellular side of the membrane contains 0.4 μM Ca^{2+} instead of 0.1 μM (**Figure 5.19a,b**). For comparison, the Ca^{2+}-sensitivity of big K channels from cultured rat skeletal muscle is shown in **Figure 5.19c**. The rat brain big K channels are sensitive to nanomolar concentrations of charybdotoxin (CTX) and millimolar concentrations of extracellular TEA ions (**Figure 5.20**).

The macroscopic Ca^{2+}-activated K^+ currents are recorded from a bullfrog sympathetic neuron in single-electrode voltage clamp mode ($V_H = -28$ mV). The iontophoretic injection of Ca^{2+} ions via the recording electrode triggers an outward current (**Figure 5.21a**). Its amplitude increases when the iontophoretic current is increased; i.e. when the amount of Ca^{2+} ions injected is increased. To study the voltage-dependence and the kinetics of activation of this Ca^{2+}-activated outward current, depolarizing steps from a holding potential of −50 mV are applied in the presence of 2 mM of Ca^{2+} in the extracellular medium (**Figure 5.21b**, 2Ca). Suppression of Ca^{2+} entry by removal of Ca^{2+} ions from the extracellular medium (0 Ca) eliminates an early Ca^{2+}-activated outward current. In the Ca-free medium, only the sigmoidal delayed rectifier K^+ current I_K is recorded. In the presence of external Ca^{2+} ions, both I_K and a $I_{K(Ca)}$ are recorded (**Figure 5.21b**, right). The recorded $I_{K(Ca)}$ corresponds to a big K current also called I_C in some preparations. It has activation kinetics sufficiently rapid to play a role in spike repolarization (**Figure 5.22**).

In nerve terminals at the motor end plate, big K channels are co-localized with voltage-dependent Ca^{2+} channels. They play an important role in repolarizing

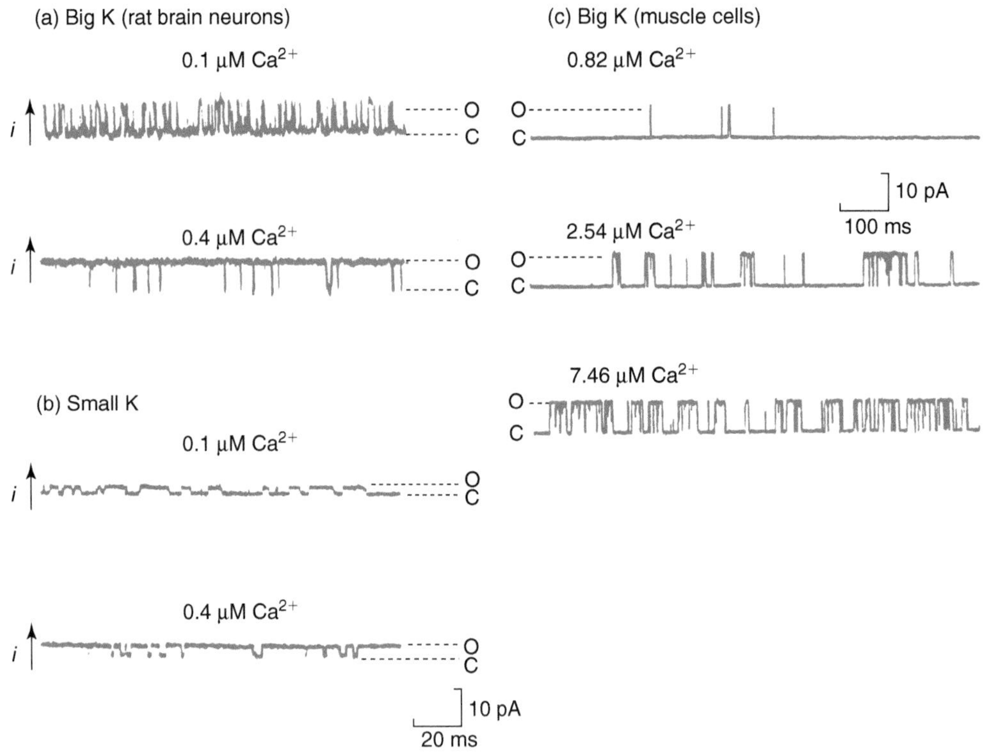

FIGURE 5.19 **Ca^{2+}-dependence of Ca^{2+}-activated K^+ channels. (a, b)** Single-channel activity of Ca^{2+}-activated channels from the rat brain. The activity of the 232 pS big K channel and that of the 77 pS small K channel is recorded in the presence of 0.1 μM Ca^{2+} (upper traces) and 0.4 μM Ca^{2+} (lower traces) in symmetrical 150 mM KCl (V_H = +20 mV). **(c)** Single-channel activity of a big K channel from rat skeletal muscle recorded at three different Ca^{2+} concentrations in symmetrical 140 mM KCl (V_H = +30 mV). O, open state; C, closed state. Part (**a**) from Chad J, Eckert R, Ewald D (1984) Kinetics of calcium-dependent inactivation of calcium current in voltage-clamped neurones in *Aplysia californica*. *J. Physiol. (Lond.)* **347**, 279–300, with permission. Part (**b**) adapted from McManus OB, Magleby KL (1991) Accounting for the calcium-dependent kinetics of single large-conductance Ca^{2+}-activated K^+ channels in rat skeletal muscle. *J. Physiol.* **443**, 739–777, with permission.

the plasma membrane following each action potential. This repolarization resulting from the increased activity of Ca^{2+}-activated K^+ channels closes voltage-dependent Ca^{2+} channels and constitutes an important feedback mechanism for the regulation of voltage-dependent Ca^{2+} entry. K_{Ca} current thereby lowers intracellular Ca^{2+} concentration and dampens neurotransmitter secretion. Conversely, when it is strongly reduced by TEA or apamin, transmitter release is increased. K_{Ca} current is also responsible for the slow after hyperpolarization (AHP) which limits the firing frequency of repetitive Ca^{2+} action potentials.

5.4 CALCIUM-DEPENDENT ACTION POTENTIALS ARE INITIATED IN AXON TERMINALS AND IN DENDRITES

5.4.1 Depolarization of the membrane to the threshold for the activation of L-, N- and P-type Ca^+ channels has two origins

L-, N- and P-type Ca^{2+} channels are high-threshold Ca^{2+} channels. This means that they are activated in response to a relatively large membrane depolarization. In cells (e.g. neurons, heart muscle cells) where the resting membrane potential is around −80/−60 mV, a 40–60 mV depolarization is therefore needed to activate the high-threshold Ca^{2+} channels. Such a membrane depolarization is too large to result directly from the summation of excitatory postsynaptic potentials (EPSPs). It usually results from an Na^+ spike. In heart Purkinje cells, Na^+ entry during the sudden depolarization phase of the action potential depolarizes the membrane to the threshold for L-type Ca^{2+} channel activation: the Na^+-dependent depolarization phase is immediately followed by a Ca^{2+}-dependent plateau (**Figure 4.2d**). In axon terminals, the situation is similar: the Na^+-dependent action potential actively propagates to axon terminals where it depolarizes the membrane to the threshold potential for N- or P-type Ca^{2+} channel activation: an Na^+/Ca^{2+}-dependent action potential is initiated (**Figure 4.2c**).

In cerebellar Purkinje neurons, the situation is somehow different: dendritic P-type Ca^{2+} channels are opened by the large EPSP resulting from climbing fiber EPSP. As a result, Ca^{2+}-dependent action potentials are initiated and actively propagate in dendrites (**Figure 4.2b**; see also Sections 16.2 and 17.3).

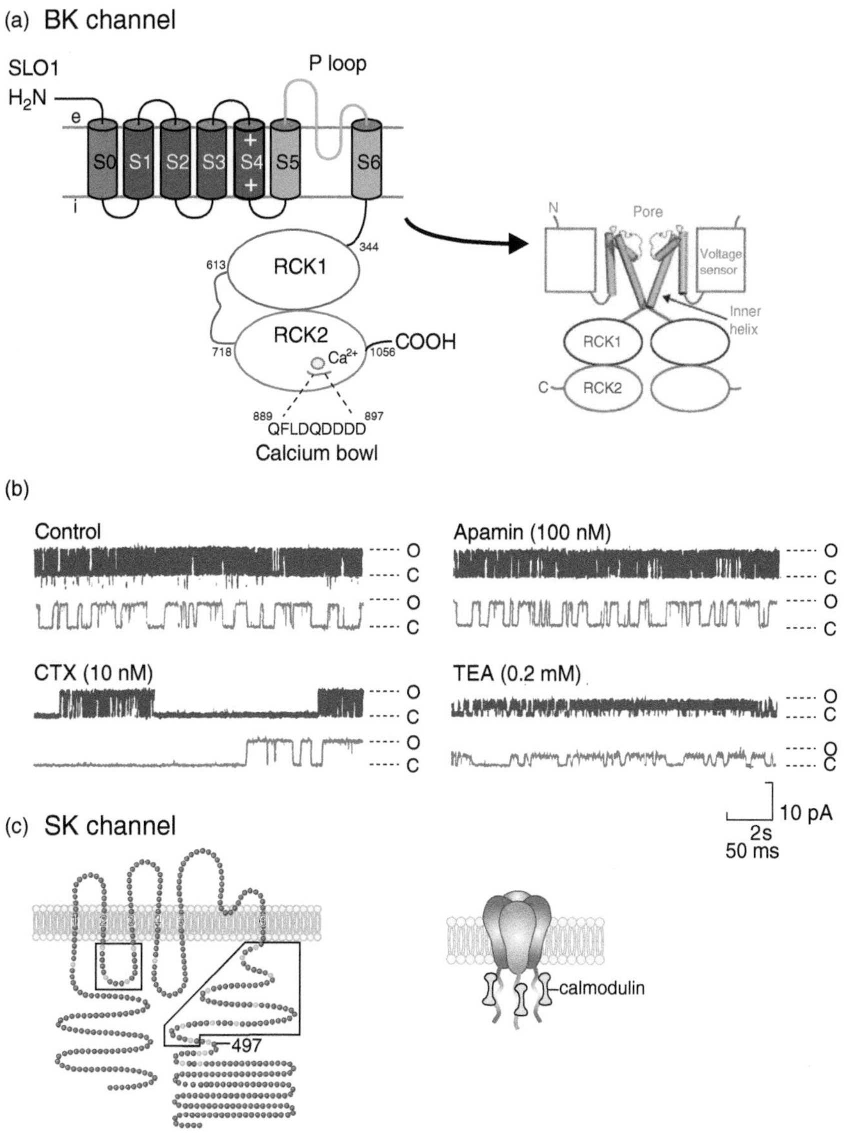

FIGURE 5.20 **The big K channel.** (**a**) Schematic representation of SLO1 α-subunit (BK channel). In the right drawing, only two opposing subunits are shown instead of four, for clarity. (**b**) Single-channel activity of the big K channel (232 pS channel) in symmetrical 150 mM KCl at two different time bases (V_H = +40 mV). From top to bottom and left to right: Control conditions. In the presence in the extracellular solution of, respectively, 10 nM charybdotoxin (CTX), 100 nM apamin, and 0.2 mM tetraethylammonium chloride (TEA). All the recordings are performed in the presence of 1.05 mM Ca^{2+} in the intracellular solution. (**c**) Left, an SK subunit. Each circle represents a single residue, and the positions of the conserved intracellular negatively charged amino acids are in blue. The regions in which multiple charge neutralizations were introduced, that is, the 2–3 loop and the proximal domain of the intracellular C-terminus, are boxed. Right, the core components of SK channels are tetrameric assemblies of the pore-forming subunits, which can be homomeric or heteromeric and have constitutively bound calmodulin (CaM) that mediates Ca^{2+} gating. Part (**a**) from Yuan P, Leonetti MD, Pico AR, Hsiung Y, MacKinnon R (2010) Structure of the human BK channel Ca^{2+}-activation apparatus at 3.0 angström resolution. *Science* **329**, 182–186 and from Yuan P, Leonetti MD, Hsiung Y, MacKinnon R (2012) Open structure of the Ca^{2+} gating ring in the high-conductance Ca^{2+}-activated K^+ channel *Nature* 481, 94–97. Part (**b**) from Chad J, Eckert R, Ewald D (1984) Kinetics of calcium-dependent inactivation of calcium current in voltage-clamped neurones in Aplysia californica. *J. Physiol. (Lond.)* **347**, 279–300, with permission. Part (**c** left) from Xia X-M, Fakler B, Rivard A *et al.* (1998) Mechanism of calcium gating in small-conductance calcium-activated potassium channels. *Nature* **395**, 503–507, with permission.

The cells that do not express voltage-gated Na^+ channels and initiate Ca^{2+}-dependent action potentials (endocrine cells, for example; see **Figure 5.1**) usually present a depolarized resting membrane potential (−50/−40 mV) close to the threshold for L-type Ca^{2+} channel activation. In such cells, the activation of high-threshold Ca^{2+} channels results from a depolarizing current generated by receptor activation or from an intrinsic pacemaker current (for example, activation of the T-type Ca^{2+} current – see Section 14.2.2 – or the turning off of a leak K^+ current).

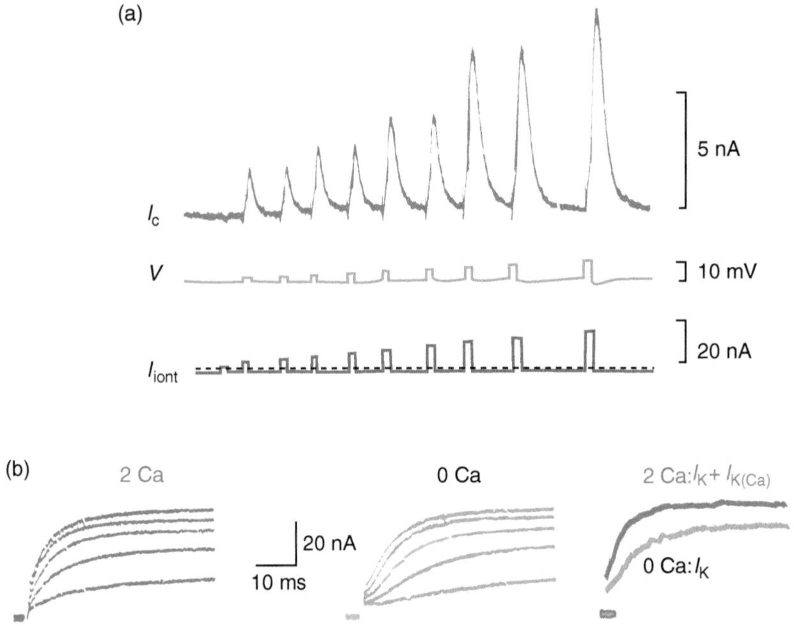

FIGURE 5.21 **The macroscopic Ca^{2+}-activated K^+ current of bullfrog sympathetic neurons.** (**a**) Outward currents (Ic, red) recorded in single-electrode voltage clamp at a holding potential of −28 mV. In response to increasing 0.4 s intracellular iontophoretic injections of Ca^{2+} (I_{iont}, blue) from a microelectrode containing 200 mM $CaCl_2$, increasing outward currents are recorded (red traces). A steady backing current of −2 nA is applied to the iontophoretic pipette (dashed black line = zero iontophoretic current). The cell was bathed in normal Ringer solution. (**b**) Outward currents recorded during voltage steps to −20, −10, 0 and +20 mV from a holding potential of −50 mV in the presence of 2 mM external Ca^{2+} (2 Ca) and a Ca-free external medium (0 Ca). The leak current is subtracted. The two superimposed current traces recorded at the same potential in the presence (2 Ca) or absence (0 Ca) of external Ca^{2+} ions show that an early component of the outward current is present ($I_{K(Ca)}$) in the presence of Ca^{2+} ions. Adapted from Brown DA, Constanti A, Adams PR (1983) Ca^{2+}-activated potassium current in vertebrate sympathetic neurons. *Cell Calcium* **4**, 407–420, with permission.

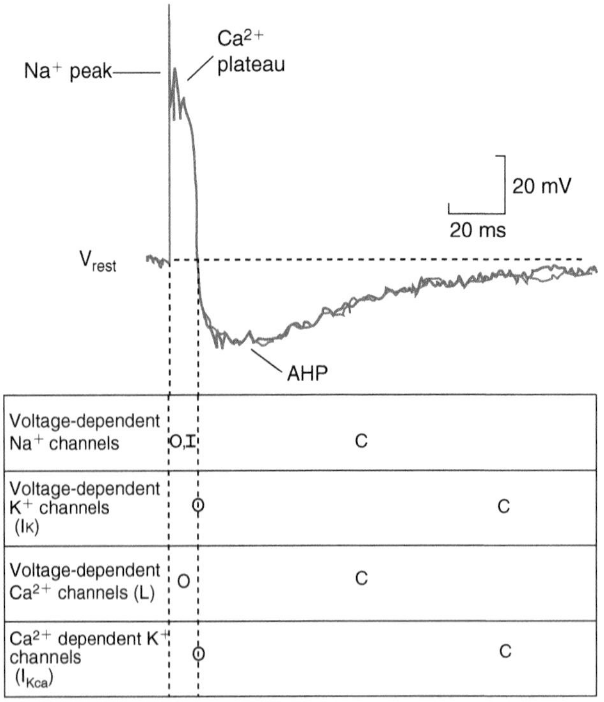

FIGURE 5.22 **States of voltage gates Na^+, Ca^{2+} and K^+ channels.** Different states in relation to the various phases of the Na^+/Ca^{2+}-dependent action potential. Example of the action potential recorded in olivary neurons of the cerebellum. Channels: 0, open; I, inactivate; C, close.

5.4.2 The role of the calcium-dependent action potentials is to provide a local and transient increase of $[Ca^{2+}]_i$ to trigger secretion, contraction and other Ca^{2+}-gated processes

In some neurons, Ca^{2+} entry through high-threshold Ca^{2+} channels participates in the generation of various forms of electrical activity such as dendritic Ca^{2+} spikes (Purkinje cell dendrites) and activation of Ca^{2+}-sensitive channels such as Ca^{2+}-activated K^+ or Cl^- channels. However, the general role of Ca^{2+}-dependent action potentials is to provide a local and transient increase of intracellular Ca^{2+} concentration. Under normal conditions, the intracellular Ca^{2+} concentration is very low, less than 10^{-7} M. The entry of Ca^{2+} ions through Ca^{2+} channels locally and transiently increases the intracellular Ca^{2+} concentration up to 10^{-4} M. This local $[Ca^{2+}]_i$ increase can trigger Ca^{2+}-dependent intracellular events such as exocytosis of synaptic vesicles, granules or sliding of the myofilaments actin and myosin. It thus couples action potentials (excitation) to secretion (neurons and other excitable secretory cells, see Chapter 7) or it couples action potentials to contraction (heart muscle cells). The influx of Ca^{2+} also couples neuronal activity to metabolic processes and induces long-term changes in neuronal and synaptic activity. During development, Ca^{2+} entry regulates outgrowth of axons and dendrites and the retraction of axonal branches during synapse elimination and neuronal cell death.

5.5 A NOTE ON VOLTAGE-GATED CHANNELS AND ACTION POTENTIALS

Voltage-gated Na^+, K^+ and Ca^{2+} channels of action potentials share a similar structure and are all activated by membrane depolarization. The Na^+, Na^+/Ca^{2+} and Ca^{2+} action potentials have a similar pattern: the depolarization phase results from the influx of cations, Na^+ and/or Ca^{2+}, and the repolarization phase results from the inactivation of Na^+ or Ca^{2+} channels together with the efflux of K^+ ions. However, these action potentials have at least one important difference. The Na^+-dependent action potential is all-or-none. In contrast, the Ca^{2+}-dependent action potential is gradual. This reflects different functions. The Na^+-dependent action potential propagates over long distances *without attenuation* in order to transmit information from soma-initial segment to axon terminals where they trigger Ca^{2+}-dependent action potentials. Ca^{2+}-dependent action potentials have the general role of providing a local, *gradual* and transient Ca^{2+} entry.

Further reading

Adelman, J.P., Maylie, J., Sah, P., 2012. Small-conductance Ca^{2+}-activated K^+ channels: form and function. Annu. Rev. Physiol. 74, 245–269.

Berkefeld, H., Fakler, B., Schulte, U., 2010. Ca^{2+}-activated K^+ channels: from protein complexes to function. Physiol. Rev. 90, 1437–1459.

Bertolino, M., Llinas, R.R., 1992. The central role of voltage-activated and receptor-operated calcium channels in neuronal cells. Ann. Rev. Pharmacol. Toxicol. 32, 399–421.

Bourinet, E., Mangoni, M.E., Nargeot, J., 2004. Dissecting the functional role of different isoforms of the L-type Ca^{2+} channel. J. Clin. Invest. 113, 1382–1384.

Catterall, W.A., 2014. Structure and function of voltage-gated sodium channels at atomic resolution. Exp. Physiol. 99, 35–51.

Curtis, B.M., Catterall, W.A., 1984. Purification of the calcium antagonist receptor of the voltage-sensitive calcium channel from skeletal muscle transverse tubules. Biochemistry 23, 2113–2118.

De Leon, M., Wang, Y., Jones, J., et al., 1995. Essential Ca^{2+}-binding motif for Ca^{2+}-sensitive inactivation of L-type Ca^{2+} channels. Science 270, 1502–1506.

Denk, W., Piston, D.W., Webb, W.W., 1995. Two-photon molecular excitation in laser-scanning microscopy. In: Pawley, J.B. (Ed.), Handbook of Biological Microscopy. Plenum Press, New York.

Elkins, T., Ganetzky, B., Wu, C.F., 1986. A Drosophila mutation that eliminates a calcium-dependent potassium current. Proc. Natl Acad. Sci. USA 83, 8415–8419.

Grynkiewicz, G., Poenie, M., Tsien, R.Y., 1985. A new generation of calcium indicators with greatly improved fluorescence properties. J. Biol. Chem. 260, 3440–3448.

Heginbotham, L., Lu, Z., Abramson, T., MacKinnon, R., 1994. Mutations in the K^+ channel signature sequence. Biophys. J. 66, 1061–1067.

Heinemann, S.H., Terlau, H., Stühmer, W., Imoto, K., Numa, S., 1992. Calcium channel characteristics conferred on the sodium channel by single mutations. Nature 356, 441–443.

Hodgkin, A.L., Huxley, A.F., 1952. A quantitative description of membrane current and its application to conduction and excitation in nerve. J. Physiol. (Lond.) 117, 500–544.

Kohler, M., Hirschberg, B., Bond, C.T., et al., 1996. Small-conductance, calcium-activated potassium channels from mammalian brain. Science 273, 1709–1714.

Latorre, R., Morera, F.J., Zaelzer, C., 2010. Allosteric interactions and the modular nature of the voltage- and Ca^{2+}-activated (BK) channel. J. Physiol. 588, 3141–3148.

Meera, P., Wallner, M., Song, M., Toro, L., 1997. Large conductance voltage- and calcium-dependent K+ channel, a distinct member of voltage-dependent ion channels with seven N-terminal transmembrane segments (S0-S6), an extracellular N terminus, and an intracellular (S9-S10) C terminus. Proc. Natl Acad. Sci. USA 94, 14066–14071.

Neher, E., Augustine, G.J., 1992. Calcium gradients and buffers in bovine chromaffin cells. J. Physiol. (Lond.) 450, 273–301.

Schreiber, M., Yuan, A., Salkoff, L., 1999. Transplantable sites confer calcium sensitivity to BK channels. Nat. Neurosci. 2, 416–421.

Stea, A., Soong, T.W., Snuth, T.P., 1995. Determinants of PKC-dependent modulation of a family of neuronal calcium channels. Neuron 15, 929–940.

Stocker, M., 2004. Ca^{2+}-activated K^+ channels: molecular determinants and function of the SK family. Nat. Rev. Neurosci. 5, 758–770.

Stotz, S.C., Zamponi, G.W., 2001. Structural determinants of fast inactivation of high voltage-activated Ca^{21} channels. Trends Neurosci. 24, 176–181.

Takahashi, M., Seagar, M.J., Jones, J.F., Reber, B.F., Catterall, W.A., 1987. Subunit structure of dihydropyridine-sensitive calcium channels from skeletal muscle. Proc. Natl Acad. Sci. USA 84, 5478–5482.

Tan, Y.P., Llano, I., Hopt, A., Wuerrihausen, F., Neher, E., 1999. Fast scanning and efficient photodetection in a simple two-photon microscope. J. Neurosci. Meth. 92, 123–135.
Tsien, R.W., Lipscombe, D., Madison, D.V., Bley, K.R., Fox, A.P., 1988. Multiple types of neuronal calcium channels and their selective modulation. Trends Neurosci. 11, 431–438.
Tsien, R.Y., 1989. Fluorescent probes of cell signalling. Ann. Rev. Neurobiol. 12, 221–253.
Varadi, G., Mori, Y., Mikala, G., Schwartz, A., 1995. Molecular determinants of Ca^{2+} channels function and drug action. Trends Pharmacol. Sci. 16, 43–49.
Yuan, P., Leonetti, M.D., Hsiung, Y., MacKinnon, R., 2011. Open structure of the Ca^{2+} gating ring in the high-conductance Ca2+-activated K+ channel. Nature 481, 94–97.
Zamponi, G.W., Bourinet, E., Nelson, D., Nargeot, J., Snutch, T.P., 1997. Crosstalk between G proteins and protein kinase C mediated by the calcium channel alpha$_1$ subunit. Nature 385, 442–446.
Zhong, H., Li, B., Scheuer, T., Catterall, W.A., 2001. Control of gating mode by a single amino acid residue in transmembrane segment IS3 of the N-type Ca^{2+} channel. Proc. Natl Acad. Sci. USA 98, 4705–4709.

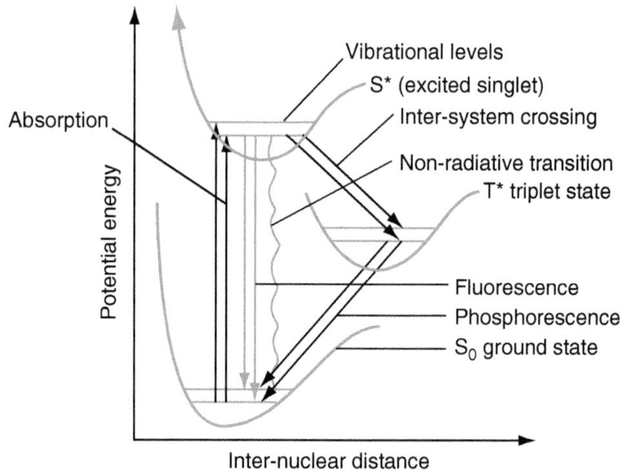

FIGURE A5.1 **Pathways to excitation and de-excitation of an electron.** The rotational levels between the vibrational levels and higher excited states are not shown for the sake of simplicity.

APPENDIX 5.1 FLUORESCENCE MEASUREMENTS OF INTRACELLULAR CA^{2+} CONCENTRATION

A5.1.1 The physical basis of fluorescence

The interaction of light with matter

Light is electromagnetic radiation that oscillates both in space and time, and has electric and magnetic field components that are perpendicular to each other. If for the sake of simplicity one focuses only on the electromagnetic component, it can be seen that the molecule, which is much smaller than the wavelength of light, will be perturbed by light because its electronic charge distribution will be altered by the oscillating electric field component of the light. Without resorting to complicated quantum mechanical calculations, we can say that light will interact with matter via a resonance phenomenon; i.e. the matter will absorb light only if the energy of the incoming photon is exactly equal to the difference between the potential energy of the lowest vibrational level of the ground state and that of one of the vibrational levels of the first excited state (**Figure A5.1**). The absorption of light therefore occurs in discrete amounts termed quanta. The energy E in a quantum of light (a photon) is given by:

$$E = h\nu = hc / \lambda,$$

where h is Planck's constant, ν and λ are the frequency and wavelength of the incoming light, and c is the speed of light in a vacuum. When a quantum of light is absorbed by a molecule, a valence electron will be boosted into a higher energy orbit, called *the excited state.* This phenomenon will take place in 10^{-15} s, resulting in conservation of the molecular coordinates. For the sake of simplicity the rotational energy levels are not taken into account and it is assumed that at room temperature the electrons will be at their lowest vibrational energy level.

The difference of energy between the vibrational levels being typically in the order of 10 kcal mol^{-1}, there is not enough thermal energy to excite a transition to higher vibrational levels at room temperature. One might thus assume that most of the electrons will lie at the lowest vibrational level of the ground state $S_{v=0}$ (S for singlet, as the electrons spins are antiparallel). Because absorptive transitions occur to one of the vibrational levels of the excited state, had there been no interaction with the solvent molecules one could have measured the energy difference between the ground state and each of the vibrational levels of the excited state. This type of spectra can only be obtained for chemical compounds in the gaseous state. The absorption spectra under those circumstances would resemble narrowly separated bands; however, the interaction of the orbital electrons with solvent molecules will broaden those peaks, producing the absorption spectra of the more familiar form.

The return from the excited state

The electrons that have been promoted to one of the vibrational levels of the excited state will lose their vibrational energy through interaction with solvent molecules by a process known as *vibrational relaxation*. This process has a time scale much shorter than the lifetime of the electrons in the excited state (10^{-9} to 10^{-7} s for aromatic molecules). The electrons that have been promoted to the excited state will return to the ground state from the lowest lying excited vibrational state, by one of the following ways.

Fluorescence emission

Some of the electrons in the excited state will return to one of the vibrational levels of the ground state by a radiative transition, whose frequency will be a function of the energy difference separating these levels. If one simply assumes that the energy spacing the vibrational levels of the excited and ground states are similar, one expects the fluorescence emission spectrum to be a mirror image of the absorption spectrum (**Figure A5.2**).

A further expectation will be that the $S_{v=0}$ to $S^*_{v=0}$ absorption will be at the same frequency as the $S^*_{v=0}$ to $S_{v=0}$ emission; however, this is rarely the case, as the absorption process takes place in about 10^{-15} s. The orientation of the solvent molecules with respect to the electronic states will be conserved as well as the quantum coordinates of the molecule; however, as the excited level lifetimes are rather long, the solvent molecules will reorient favorably about the electronic levels, resulting in a difference in the zero–zero frequencies. This difference between $S_{v=0}$ to $S^*_{v=0}$ absorption and $S^*_{v=0}$ to $S_{v=0}$ emission is termed the *Stoke's shift* (**Figure A5.2**).

Non-radiative transition

In this process the excitation energy will be lost mainly by interactions with solvent molecules, resulting in some of the electrons of the excited state returning to the ground state with a non-radiative transition. This process is favored by an increase in temperature, and can explain why increasing the temperature causes a decrease in fluorescence intensities.

Quenching of the excited state

The excitation energy might be lost through interactions, in the form of collisions of quenchers with the electrons in the excited orbital. Typical quenchers such as O_2, I^- and Mn^{2+} ions will quench every time they collide with an excited singlet.

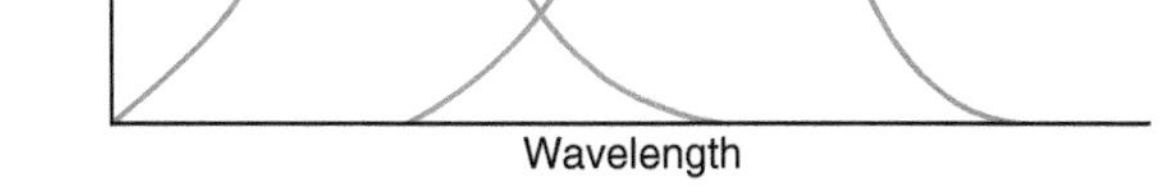

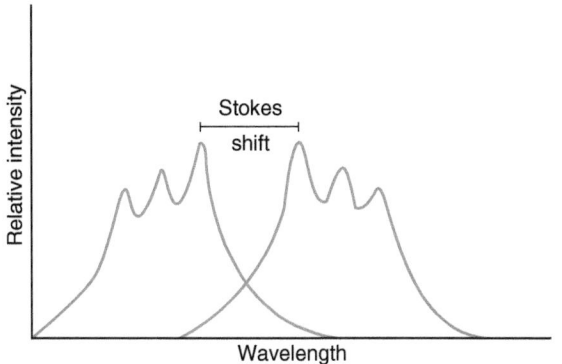

FIGURE A5.2 **Excitation (left) and emission (right) spectra of a hypothetical molecule.** The excitation spectrum has the same peaks as the absorption spectrum; the separation between the individual peaks reflects the potential energy differences between the vibrational levels.

Intersystem crossing

Intersystem crossing is a mechanically forbidden quantum process that occurs by a spin exchange of the electron of the excited singlet state, resulting in an excited triplet state T* (**Figure A5.1**). As this process involves a forbidden transition its probability of occurrence will be extremely low; nevertheless it will occur because the potential energy of the excited triplet is usually lower than that of the excited singlet state. The electron in the excited triplet state can then become de-excited by a non-radiative transition, quenching, or by a radiative transition called *phosphorescence* (the light emitted will be of longer wavelength than fluorescence because of the lower potential energy of the excited triplet). One should note that the return to the ground state necessitates a novel forbidden transition $T^*_{v=0}$ to $S_{v=x}$ (x for any vibrational level of the ground state). The probability of this transition will be extremely low, for the same reasons given above, resulting in a long lifetime of the excited triplet state (seconds to days). This long-lived triplet state will result in a very weak intensity of radiation, will be prone to quenching by collisions with quenchers, and the non-radiative processes will compete well with the phosphorescence. Phosphorescence in solution will rarely be observed. In order to observe phosphorescence at all, one must rigorously remove oxygen from the medium, and should use rigid glasses at very low temperatures, in order to minimize the competing non-radiative processes.

Some of the electrons that have undergone intersystem crossing, and therefore are in the $T^*_{v=0}$ state, may undergo a novel intersystem crossing to the S* level by the thermal energy provided by the solution, provided the energy difference between the T* and S* states is small; the return from the $S^*_{v=0}$ to $S^*_{v=x}$ level by fluorescence emission is called *delayed fluorescence* and has the effect of lengthening the fluorescence lifetime of the molecule beyond what is expected in normal fluorescence emissions.

Photolysis: bleaching and toxicity

The molecules in the excited state undergo certain chemical reactions resulting in the loss of fluorescence; this is called *photobleaching*. It is estimated that a good organic fluorophore can be excited about 10^4 to 10^5 times before it bleaches. Some of the reaction products might be damaging for the cell, resulting in phototoxicity. One of the important ingredients in bleaching is the interaction between the triplet state of the fluorophore (**Figure A5.1**) and molecular oxygen (O_2). The triplet state can transfer its energy to oxygen and bring it to its singlet excited state. Singlet oxygen is a reactive molecule that participates in many kinds of chemical reactions with organic molecules. As a result, the fluorophore loses its ability to fluoresce (it bleaches). In addition, the singlet oxygen can interact with other organic molecules causing phototoxicity for living cells. The minimal intensity of excitation and the minimal exposure time must be used in order to keep photobleaching and phototoxicity to a minimum.

A5.1.2 Fluorescence measurements: general points

Advantages

In the absence of fluorophore, provided there is no background fluorescence, the level of the signal is zero, so that even a very small change of fluorescence of the fluorophore is detected. This might need a large amplification, itself limited by the noise level of the amplifier chain.

Observation of fluorescence emission

The best fluorimeter should maximize collection of the fluorescence emission and minimize collection of excitation light. This is usually achieved by selecting a band of excitation wavelengths located outside the emission spectrum using filters (interference or combination filters), or monochromator on the excitation side and highpass or bandpass filters on the emission side. The emission-side filters pass wavelengths longer than the excitation wavelengths (remember the Stoke's shift).

For the measurement of fluorescence from individual cells, epi-illuminated fluorescence microscopes are used. The epiluminescence technique means that both the excitation and emission light have a common optical path through the objective. The key element of epi-illumination is the dichroic mirror; an interference mirror formed by successive depositions of dielectric layers on a transparent substrate. The dichroic mirror reflects the wavelengths below its cutoff frequency and transmits those that are above the cutoff. This cutoff frequency is chosen so that it reflects all of the excitation wavelengths, and transmits most of the emission wavelengths. The Stoke's shift is an aid in this respect. It is also possible to find polychroic mirrors that allow the simultaneous detection of several chromophores (**Figure A5.3**).

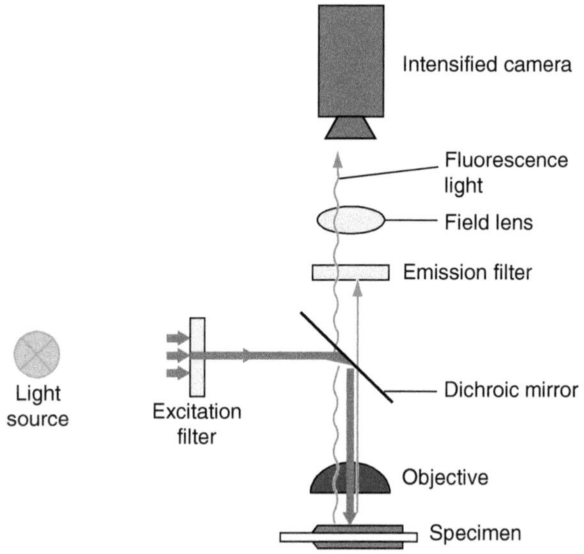

FIGURE A5.3 **Epi-illumination microscopy.**

Confocal and multiphoton microscopy

The principle of confocal imaging was developed by Marvin Minsky in the mid-1950s. In a conventional (i.e. wide-field) fluorescence microscope, the entire specimen is flooded in light from a light source. All parts of the specimen throughout the optical path are excited and the fluorescence is detected by a photodetector or a camera. The confocal microscope uses a laser to provide the excitation light (in order to get very high intensity of excitation and high resolution). The laser beam reflected from the dichroic beam splitter hits two mirrors which are mounted on motors and scan the sample in X and Y directions. Dye in the sample fluoresces and the emitted light captured by the objective lens (epi-fluorescence) is de-scanned by the same moving mirrors. The fluorescence passes through the dichroic and is focused onto the pinhole in an optically conjugate plane. The pinhole in front of the detector eliminates out-of-focus light. Only the fluorescence within the focal plane is detected by the detector; i.e. a photomultiplier tube. In fact, there is not a complete image of the sample at any given instant; only one point of the sample is observed (laser scanning confocal microscopy: LSCM). The detector is connected to a computer which builds up the image, one pixel at a time. In practice, this can be done several times per second (up to 8192 × 8192 pixel image). The time limitation is due to the scanning mirrors but also to the sensitivity of the detectors and the quality of the labeling. With poor labeling one has to scan at a slower speed to improve signal-to-noise ratio. Several features have been developed to improve speed performance (spinning Nipkow disc, multibeam excitation, linned 'pinhole' or resonant scanner), each having good and bad influences on the quality of the information collected. Imaging setups cannot simultaneously acquire images at maximal speed and produce high quality images.

The resolution improvement due to confocal microscopy either in the xy or z dimension and the general use of laser as light source represent a real revolution in imaging biological material. The resolution of the system, i.e. the ability to discriminate two very close points, is given by the quality of the light pathway. The limitation step is the numerical aperture (NA) of the objective. NA is a dimensionless number: $NA = n \sin \theta$ where n is the optical index (1 for dry objective, 1.33 for water immersion and 1.52 for oil immersion) and θ is the half angle of light collected through the front lens of the objective. The larger NA is, the more light and information is collected

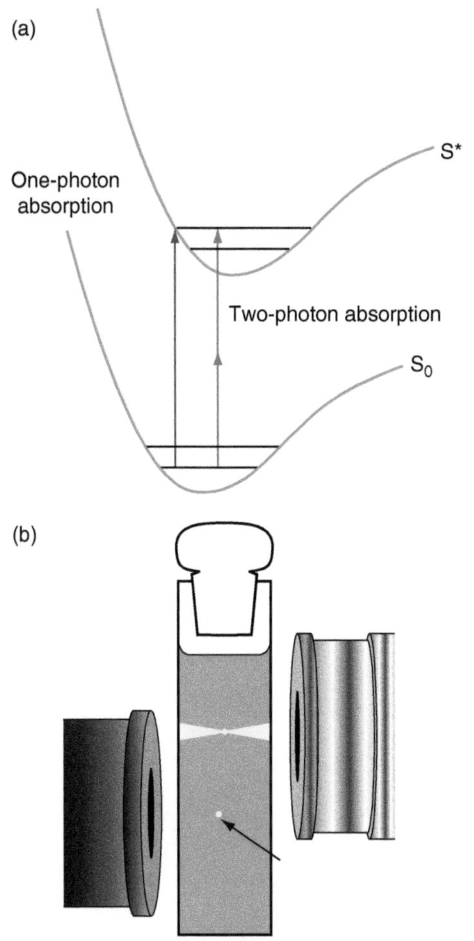

FIGURE A5.4 **Comparison of one-photon and two-photon absorption.** (**a**) (a) Two photons in the red (right) combine their energies to get absorbed as a one blue photon (left). The energies of the photons can be thought as equal to the amplitude of the vectors. The two photons that are absorbed need not have equal energies. (**b**) Fluorescence emission profile produced by one-photon absorption occurs throughout the laser beam focused in a fluorescent solution by the objective on the right. With the two-photon scheme, excitation is limited to the focal point of the objective on the left (shown by the arrow) providing inherent three-dimensional resolution.

and the better the resolution is. Confocal technology increased the resolution by ≈30% compared to wide field microscopy.

Despite these benefits, the confocal approach presents some limitations for specific applications such as deep imaging or long duration experiments in living tissue. Indeed, biological tissues strongly scatter visible light and prevent more than 150 μm depth images by defect of excitation light and scattering of emitted fluorescence. On the other hand, increasing the laser power is not the solution because thermal and photolysis effects will definitively affect the sample. Because the laser beam excites only the fluorophores in its path, scanning the sample induces a large photobleaching effect in the entire light cone (**Figure A5.4b**). One has to deal with the laser power and preservation of the sample.

In order to alleviate some of these limitations Denk and colleagues in 1990 took advantage of an old physical theoretic prediction and new powerful pulsed lasers to develop multiphoton microscopy. In 1931, Maria Goeppert-Mayer (Nobel Prize in Physics 1963) predicted the possibility of simultaneous absorption by a molecule of two photons of long wavelength, combining their energies to cause the transition of the molecule to the excited state (remember that the energy of a photon is inversely proportional to its wavelength: $E = hc/\lambda$). This can be viewed as two IR (near infrared) photons being absorbed simultaneously by a molecule normally excited by UV (**Figure A5.4a**). This technique, however, did not find practical use until the advent of very short pulse-width lasers for the following reasons.

The probability of two-photon absorption is $\approx 10^{31}$ times lower than the probability of one-photon absorption, and therefore does not occur under normal illumination conditions. Typical cross-sections for one-photon absorption are of the order of 10^{-16} cm^2; for the two-photon case they are 10^{-48} cm^4 s^{-1}. The two-photon cross-sections are cited in GM (Goeppert-Mayer) units, with 1 GM being 10^{-50} cm^4 s^{-1}. In order for this absorption to occur, very high density photon fluxes confined to a small volume are needed. This was made possible by the advent of pulsed lasers, which typically generate pulses of 70 to 100 fs (10^{-15} s) width, each at power levels of 500 kW (2–5 W on average) and repetition rates of 80 MHz. For fluorescence measurements of biological samples, 1 mW of laser intensity on the specimen plane is typically used. Under these conditions the excitation is confined to the focal volume only, as the necessary photon flux can only be reached at this plane. This has two very important implications: (i) as the excitation is limited to the focal plane, the emission is also limited to this plane (**Figure A5.4b**), resulting in an intrinsic 'confocal' image. At this time, image reconstruction does not need any pinhole and allows the use of 'non-de-scanned' detectors placed just after the condenser or the objective (transmission or epi-fluorescence). This short-length optical tract is several times more efficient than the confocal one to collect emitted fluorescence and greatly improves recorded signals; (ii) other chromophores in the light cone are not excited, so photodamage and phototoxicity resulting from photolysis of the chromophore are greatly reduced. Light at long wavelengths is less prone to scattering and better penetrates biological tissue, enabling researchers to measure Ca^{2+} dynamics in a non-invasive way and from deeper locations. It is possible to measure Ca^{2+} signals from rat brain at a depth of 800 μm from the surface. Absorption is not limited to two photons only. Triple-photon absorption by UV dyes (DAPI, etc.) or by nucleic acids with cross-sections as large as 10^{-75} cm^6 s^{-2} has been reported. This technique currently has the drawback of requiring rather expensive lasers, but one can expect prices to come down in the future, allowing their routine use.

A5.1.3 Measurement of ion concentration by fluorescence techniques

The main requirement for an indicator to report the concentration of an ion is a change in its optical properties, and at the same time it must be highly specific for the ion in question, at physiological pH values. Furthermore, its binding and release from the ion must be faster than the kinetics of the intracellular ionic changes. Its affinity to that ion should be compatible with physiological conditions and prevent buffering effects. One can therefore envisage the production of probes that will change their absorption, bioluminescence (such as aequorin) or fluorescence properties as a function of ion concentration. Fluorescence is the technique of choice because of its higher sensitivity. In fluorescence measurements, the change in optical property sought to report an ionic concentration might be a change in quantum yield, excitation spectra or emission spectra.

Fluorescence-based Ca^{2+} imaging typically uses small organic dyes. With the increase of genetic engineering tools and systematic optimization of genetically encoded Ca^{2+} indicators (GECIs), a new generation of Ca^{2+} sensing molecules appeared. Several advantages make GECIs the dyes of choice: (i) they are expressed in specific cells and therefore label the targeted cell population only, (ii) they avoid using chemical solvents (such as pluronic acid and DMSO classically used for Ca^{2+} dyes penetration) that can alter cell physiology, and (iii) they are expressed during the animal's lifetime and therefore allow chronic *in vivo* recordings.

Organic indicators for Ca^{2+}

Tsien (1989) developed many probes sensitive to free Ca^{2+} ion concentration. The common property of these probes is that they are all fluorescent derivatives of the calcium chelator BAPTA, which in turn is an aromatic analogue of the commonly used calcium chelator (EGTA, ethyleneglycol bis (13-aminoether) -N,N,N′,N′ tetra-acetic acid) (**Figure A5.5**). The probes form an octahedral complex, with the calcium ion at the center of the plane formed by the COO^- groups of the carboxylic acid. The binding and unbinding of the ion induces a strain or relaxation on the electron cloud of the aromatic groups, which in turn results in changes of the spectral properties of the reporter chromophore.

Once these probes are synthesized, several strategies were developed to make them penetrate the cell, including direct injection of the chromophore salt through the patch-clamp pipette allowing Ca^{2+} measurement combined with electrophysiological monitoring from a single neuron. Another way is to neutralize carboxylic groups with ester residues to form acetoxymethylester variety (AM). This neutral molecule diffuses through the plasma membrane and is de-esterified enzymatically to produce functional chromophore (**Figure A5.5**); this is convenient to load a population of cells either *in vivo* or *in vitro*.

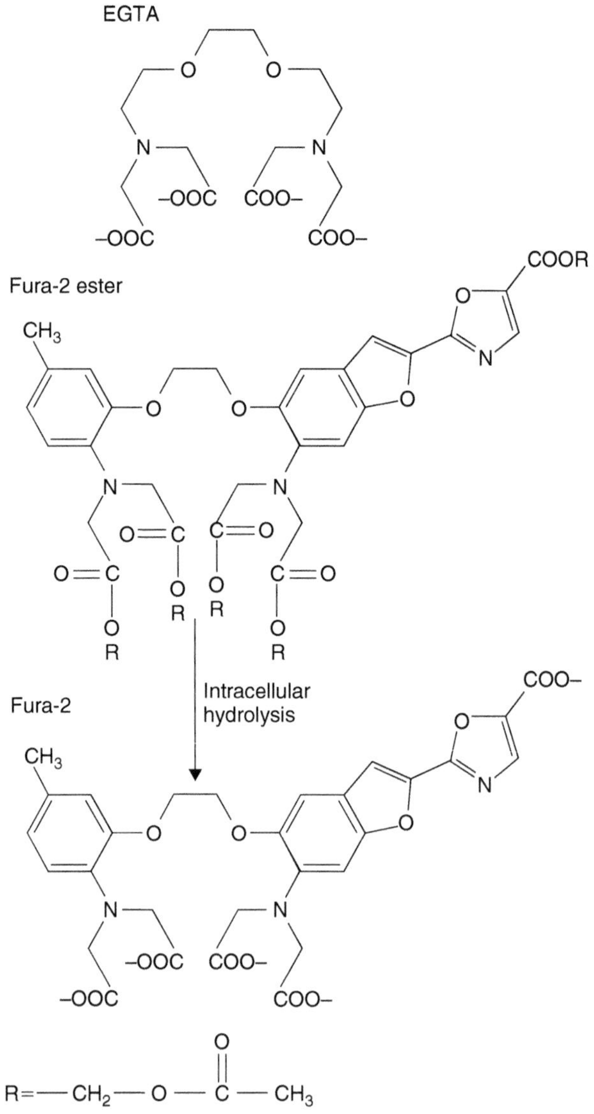

FIGURE A5.5 **Chemical structure of Fura-2.** Note the similarities between Fura-2 and the acetoxymethylester variety Fura-2AM and EGTA. The AM variety is membrane permeant, and is de-esterified by intracellular esterases, liberating Fura-2, formaldehyde and acetate ions.

Three such reporter chromophores have found much use in the measurement of intracellular free Ca^{2+} concentrations, namely INDO, Fura-2 and FLUO-4. Each of these probes has a certain number of advantages over the others, depending on the measurement technique sought. INDO and Fura-2 are ratiometric probes; i.e. the change in spectral properties occur at two different wavelengths, and by measuring the fluorescence intensities at these two wavelengths and taking their ratio, one can calculate the absolute value of the free Ca^{2+} ion concentration within the cytosol, given by the following formula (see Grynkiewicz *et al.*, 1985):

$$[Ca^{2+}] = K_i\{(R - R_{min}) / (R_{max} - R)\},$$

where R_{min} is the ratio at two wavelengths at zero ion concentration, R_{max} is the ratio at 'infinite' ion concentration, R is the ratio of the measurements and K_i is a constant unifying instrumental parameters together with the K_D of the chromophore for calcium. The major advantage of ratiometric probes is the fact that they are insensitive to the intensity of the emitted light, which changes from the center to the periphery of most of the cells. This is because of differences in thickness at the center and towards the edges, so there are more chromophores in the center than at the edges.

INDO's emission properties at 405 nm and 480 nm change upon binding to Ca^{2+} (λ_{exc} = 350 nm). The two emission intensities can easily be measured by using a beam splitter, two interference filters and two photomultipliers; it is fairly difficult to envisage the use of two intensified cameras to form an image unless one uses a specifically split CCD array. Therefore INDO has been applied in processes that require either rapid determination of the free calcium concentration (i.e. cell sorting), or where the kinetics of the free calcium change are fast.

Fura-2, upon binding to calcium, undergoes a change in its absorption spectrum and therefore in its excitation spectrum; namely the emission intensity (collected at λ_{em} = 510 nm and higher) increases at λ_{exc} = 340 nm and decreases at λ_{exc} = 380 nm (**Figure A5.6**). A typical property of all the indicators that undergo either an excitation or emission shift is the presence of an 'isosbestic point', namely the presence of a 'unique point' in the spectrum when the parameter concentration is changed (Ca^{2+} in the case of Fura-2). The isosbestic point is only present when two species are in equilibrium (in our case calcium-bound and free forms of Fura-2 or of INDO). The absence of this point can be taken as an indicator for the contamination by another ion. This point appears at 360 nm for Fura-2 (**Figure A5.6**). As Fura-2 undergoes a change in its absorption properties, alternating the excitation filters at the two chosen wavelengths, mostly at 340 and 380 nm, and collecting the emission above 510 nm with an intensified camera, one can construct the free calcium image, or the time series of the changing free Ca^{2+} concentration in a living cell, by calculating the free Ca^{2+}concentration at each pixel (picture element). One is not limited to these two wavelengths; it might even be advantageous to take the images at longer wavelengths than 340 nm as most

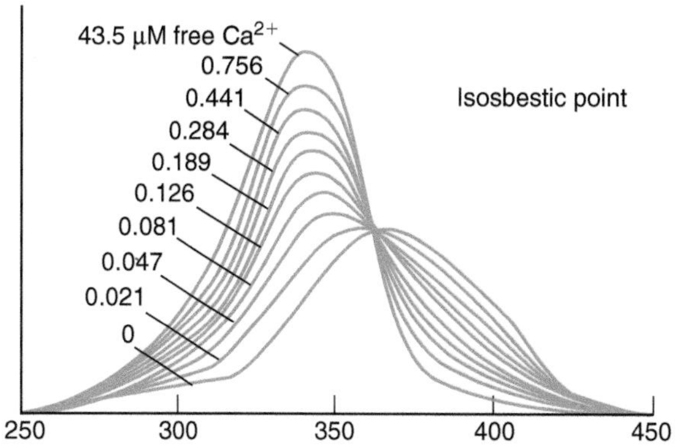

FIGURE A5.6 **Excitation spectral changes of Fura-2 as a function of Ca^{2+} concentration.** Each curve represents the intensity of fluorescence emitted by Fura-2 (at λ = 510 nm) as a function of the wavelength of excitation (from 250 to 450 nm) and for a given Ca^{2+} concentration (from 0 to 43.5 µM). Knowing that Fura-2 + Ca^{2+} $\rightleftharpoons$ Fura-2-Ca, the curve obtained in the presence of the maximal Ca^{2+} concentration (43.5 µM) represents the excitation spectrum of the bound form of Fura (Fura-2-Ca). In contrast, the curve obtained for the minimal Ca^{2+} concentration (0 µM) represents the excitation spectrum of the free form of Fura (Fura-2). It appears clearly that measured at λ = 340/350 nm and at λ = 380 nm, the intensity of fluorescence emitted by Fura-2 varies with the ratio free/bound forms of Fura; i.e., with Ca^{2+} concentration.

of the optical path of old fluorescence microscopes is opaque to this wavelength.

With the advent of confocal microscopy, an indicator with absorption properties in the visible part of the spectrum was needed (confocal microscopes use laser scanning). Ultraviolet (340 nm) lasers have been recently developed but combined with the adaptation of the microscope optics they are still too expensive. Moreover, UV irradiation is damageable to cells. FLUO-4, Calcium Green and Oregon Green were developed to respond to this need. The main disadvantage of these probes is that their quantum efficiency changes at one wavelength only (≈530 nm, when excited at the 488 nm line of the argon laser) upon binding to Ca^{2+} ions. It is therefore not possible to measure absolute values of Ca^{2+} concentrations directly; nevertheless, if the resting level of the free Ca^{2+} concentration in the cell is known, the values obtained before stimulation can be used to calculate the approximate value of the free Ca^{2+} concentration under stimulation.

Experience shows that fluorescence intensity always seems too low, and it is tempting to increase the concentration of the reporter molecule inside the cell to overcome this problem. In the case of Ca^{2+} measurements this will have the adverse effect of buffering free Ca^{2+} ions and to prevent its rise (BAPTA-like backbone). It is necessary to find a compromise between the signal-to-noise level and the buffering of the ion in general, the best approach being the use of the least amount of indicator required for the job.

Proteinic indicators for Ca^{2+} (GECI)

There are numerous reviews on GECIs in excitable or non-excitable cells such as lymphocytes, cardiomyocytes, neurons, etc. It is almost impossible to compare the different GECIs described in the literature because their characterizations were not standardized. Many parameters influence experimental conditions such as cell types, Ca^{2+} sensitivity or potential toxicity. GECI are divided into two families based on their mechanisms of Ca^{2+} detection. Initially, the idea was to take advantage of FRET phenomenon to monitor intracellular Ca^{2+}. For the sake of simplicity we can tell that FRET works if two fluorophores (one donor and one acceptor) are as close as the donor can transfer part of its excitation energy to the acceptor in a non-radiative manner. Two fluorescent proteins (CFP and YFP, for example) attached with a conformational Ca^{2+}-sensitive linker such as calmodulin-binding peptide M13 make the GECI (Cameleon by Miyawaki). Another linker like troponin-C has also been used by Heim and Griesbeck (2004). The group of Tsien proposed the circular permutation and insertion of the Ca^{2+}-sensitive linker in the fluorescent protein. The linker is inserted in an adapted place to allow spectral properties changing reversibly with the amount of Ca^{2+} (PeriCaM by Miyawaki *et al.* (1997) or GCaMP by Nakai and colleagues (2001) are good examples). These GECIs are often used as equivalents to single excitation/emission organic dyes. Recent studies by Chen *et al.* (2013) with GCaMP6 variant demonstrate that it allows detection of single action potential Ca^{2+} signature in somata and also of its propagation in the dendritic tree *in vivo* through a rat cranial window. The selection of the appropriate GECI for a particular application remains challenging because many parameters are in balance.

Indicators for Na^+ and K^+

SBFI and PBFI are designed around a crown ether chelator to which benzofuranyl chromophores are linked, conferring to those molecules the same spectroscopic properties as Fura. Hence the same filter sets can be used as for Fura. The cavity size of the crown ether is the factor that determines the specificity of the molecule for Na^+ or K^+. The specificities of both SBFI and PBFI for their respective ions is much smaller than that of Fura for Ca^{2+}, and the K_D changes as a function of the concentration of the other ion, the ionic strength, pH and temperature.

Sodium Green is designed around a crown ether chelator to which two dichlorofluorescein chromophores are linked, resulting in similar spectroscopic properties as Calcium Green (i.e. excited at 488 nm). The cavity size of the crown ether results in a greater selectivity for Na^+ over K^+ compared with SBFI – 41 versus 18 times, respectively. The spectral properties, however, result in emission changes at one wavelength only, so ratiometric measurements with this reporter molecule are not possible.

All of the cation reporter molecules suffer K_D changes as a result of intracellular interactions as mentioned above, so they need to be calibrated *in situ* using pore-forming antibiotics like gramicidin and loading the cells with known ionic conditions. Another point that must be borne in mind is the fact that protein dye interactions might dampen or completely eliminate the signals.

Indicators for Cl^-

All of the chloride indicators are based on methoxyquinolinium derivatives and report the chloride by the diffusion-limited collisional quenching of the chromophore in the excited state interacting with the halide ion. The quenching is not accompanied by spectral shifts, so ratiometric measurements are not possible. As the quenching depends on collisional encounters of the halide ion, it is very sensitive to intracellular viscosity and temperature. The quenching efficiency is greater for the other halides such as Br^- and I^-.

Genetically encoded chloride sensors have been created, taking advantage of YFP Cl^- sensitivity. A ratiometric dye called clomeleon was produced in 2000 by Kuner and Augustine. One major drawback of this dye is its pH sensitivity, which greatly affects its fluorescence. Recent optimization, with a Cl^- sensor by Markova and colleagues (2008) and with mClY by Zhong and colleagues (2014), overcomes these difficulties. They obtained a chloride sensor with low sensitivity to pH (pKa = 5.9), a rather high chloride affinity (14 mM) and a bleach time constant of 175 seconds (which is 15-fold more than the bleach time constant of YFP).

Organic voltage sensitive dyes (VSDs)

Voltage sensitive dyes (VSDs) enable the measurement of membrane potential in cells or organelles that are too small for microelectrode impalement. Moreover, these probes can map variations of membrane potential with spatial resolution and sampling frequency that are difficult to achieve using microelectrodes, such as cell microdomains or full network studies. Potentiometric probes include many chemical structures (styryl, carbocyanines and rhodamines, oxonols, for example); the existence of numerous dyes analogues reflects the observation that no single dye provides the optimal response under all experimental conditions. Selecting the best voltage sensitive probe might be empirical and the choice of the class of dye is determined by different factors such as accumulation in cells, response mechanisms, toxicity or kinetics of the electrical events observed. VSDs are divided into two categories concerning this last parameter: (i) fast-response probes that are sufficiently fast to detect transient (millisecond) potential changes in excitable cells – however, the magnitude of their potential-dependent fluorescence change is often small; they typically show a 2–10% fluorescence change per 100 mV; (ii) slow-response probes, the magnitude of their optical responses is much larger than that of fast-response probes (typically a 1% fluorescence change per mV). Slow-response probes are suitable for detecting changes in average membrane potentials of non-excitable cells caused by respiratory activity, ion-channel permeability, drug binding and other factors.

Genetically encoded voltage indicators (GEVIs)

None of the former probes were completely satisfying and often suffered from phototoxicity. GEVIs have the advantage of expression specificity. A precursor of GEVIs was produced by Siegel and Isacoff as early as 1997. After several improvement steps and the discovery of genetical optimization of Archaerhodopsin 3 (Arch) from *Halorubrum sodomense*, Cohen and collaborators (Hochbaum *et al.*, 2014) generated the QuasArs sensor series with better photophysical properties than VSDs.

Further reading

Baird, G.S., Zacharias, D.A., Tsien, R.Y., 1999. Circular permutation and receptor insertion within green fluorescent proteins. Proc. Natl Acad. Sci. USA 96, 11241–11246.

Chen, T.W., Wardill, T.J., Sun, Y., Pulver, S.R., Renninger, S.L., Baohan, A., et al., 2013. Ultrasensitive fluorescent proteins for imaging neuronal activity. Nature 499, 295–300.

Denk, W., Strickler, J.H., Webb, W.W., 1990. Two-photon laser scanning fluorescence microscopy. Science 248, 73–76.

Heim, N., Griesbeck, O., 2004. Genetically encoded indicators of cellular calcium dynamics based on troponin C and green fluorescent protein. J. Biol. Chem. 279, 14280–14286.

Hochbaum, D.R., Zhao, Y., Farhi, S.L., Klapoetke, N., Werley, C.A., Kapoor, V., et al., 2014. All-optical electrophysiology in mammalian neurons using engineered microbial rhodopsins. Nat. Methods 11 (8), 825–833.

Kuner, T., Augustine, G.J., 2000. A genetically encoded ratiometric indicator for chloride: capturing chloride transients in cultured hippocampal neurons. Neuron 27, 447–459.

Markova, O., Mukhtarov, M., Real, E., Jacob, Y., Bregestovski, P., 2008. Genetically encoded chloride indicator with improved sensitivity. J. Neurosci. Methods 170, 67–76.

Miyawaki, A., Llopis, J., Heim, R., McCaffery, J.M., Adams, J.A., Ikura, M., Tsien, R.Y., 1997. Fluorescent indicators for Ca2+ based on green fluorescent proteins and calmodulin. Nature 388, 882–887.

Nagai, T., Sawano, A., Park, E.S., Miyawaki, A., 2001. Circularly permuted green fluorescent proteins engineered to sense Ca^{2+}. Proc Natl Acad Sci USA 98, 3197–3202.

Nakai, J., Ohkura, M., Imoto, K., 2001. A high signal-to-noise Ca(2+) probe composed of a single green fluorescent protein. Nat. Biotechnol. 19 (2), 137–141.

Siegel, M.S., Isacoff, E.Y., 1997. A genetically encoded optical probe of membrane voltage. Neuron 19, 735–741.

Tsien, R.Y., 1989. Fluorescent probes of cell signaling. Annu. Rev. Neurosci. 12, 227–253.

Zhong, S., Navaratnam, D., Santos-Sacchi, J., 2014. A genetically-encoded YFP sensor with enhanced chloride sensitivity, photostability and reduced pH interference demonstrates augmented transmembrane chloride movement by gerbil prestin (SLC26a5). PLoS One 9 (6), e99095.

Websites

http://www.molecularexpressions.com/primer/techniques/fluorescence/filters.html for full explanation of the filters.

http://micro.magnet.fsu.edu/primer/techniques/fluorescence/filters.html

http://micro.magnet.fsu.edu/index.html

http://www.lifetechnologies.com/fr/fr/home/references/molecular-probes-the-handbook.html

APPENDIX 5.2 TAIL CURRENTS

Tail currents are observed in voltage or patch clamp experiments. 'Tail' means that the voltage-gated current is observed at the end of a depolarizing voltage step, upon sudden removal of the depolarization of the membrane. Tail currents do not exist in physiological conditions; they are 'experimental artifacts'. However, there are several reasons for studying tail currents: they are tools for determining characteristics of currents such as reversal potential and inactivation rate constants. Tail currents were first described by Hodgkin and Huxley (1952) in the squid giant axon.

Single-channel tail current

In patch clamp recordings of the activity of a single voltage-gated channel, a unitary current of much larger amplitude is occasionally observed at the end of the voltage step (**Figure A5.7**). It corresponds to the current flowing through a channel that is not yet closed at the end of the depolarizing step. Therefore, tail currents are recorded for voltage-gated channels that do not rapidly close or inactivate during a depolarizing step, such as delayed rectifier K^+ or L-type Ca^{2+} channels.

The activity of an L-type Ca^{2+} channel is recorded in patch clamp (cell-attached patch) in the presence of the selective agonist Bay K 8644. On stepping back the membrane to the holding potential, the L-type Ca^{2+} channel opened by the preceding depolarization does not immediately close since the transition $O \rightleftharpoons C$ is not immediate. The inward unitary Ca^{2+} current recorded at this moment is larger (**Figure A5.7**) because of the larger driving force upon removal of depolarization than

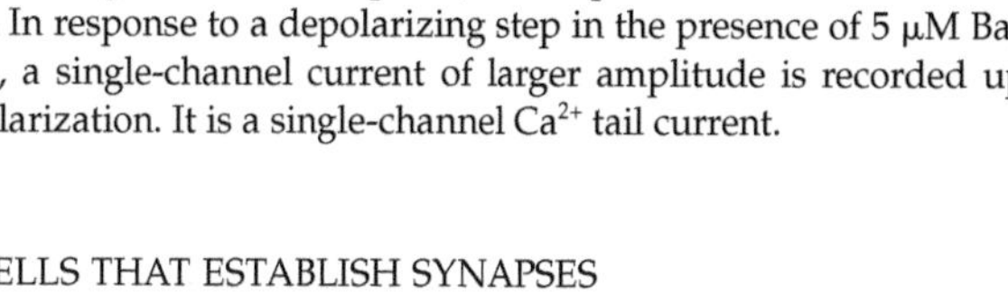

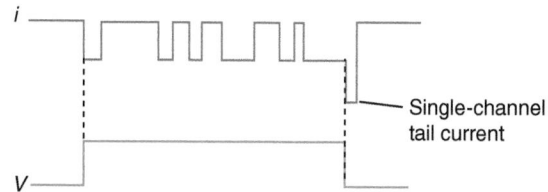

FIGURE A5.7 **Activity of an L-type Ca^{2+} channel.** Recorded in patch clamp (cell-attached patch) in the presence of 110 mM external Ba^{2+}. In response to a depolarizing step in the presence of 5 μM Bay K 8644, a single-channel current of larger amplitude is recorded upon repolarization. It is a single-channel Ca^{2+} tail current.

during the depolarizing step: during the depolarizing step to 0 mV, $i_{Ca} = \gamma_{Ca}(V_m - E_{Ca}) = \gamma_{Ca}(0 - 50) = -50\gamma_{Ca}$; upon removal of depolarization $i_{Ca} = \gamma_{Ca}(V_m - E_{Ca}) = \gamma_{Ca}(-60 - 50) = -110\gamma_{Ca}$.

Then, after a few milliseconds, owing to closing of the channel, the tail current returns to zero (the voltage-gated channel closes in response to the repolarization of the membrane).

Whole-cell tail current

In voltage or whole-cell patch clamp recordings (in the presence of Na^+ and K^+ channel blockers), a voltage step to 0 mV from a holding potential of −40 mV activates a number N of L-type Ca^{2+} current. At the end of the voltage step, a Ca^{2+} current of larger amplitude and small duration is always recorded: the tail Ca^{2+} current (**Figure A5.8**). Then the amplitude of this tail current progressively diminishes. The peak of the whole-cell tail current has a larger amplitude than that of the whole-cell current recorded during the voltage step since the driving force for Ca^{2+} ions is larger upon removal of depolarization than during the depolarization, as explained above.

The tail current diminishes progressively owing to the progressive closure of the N open Ca^{2+} channels: the channels do not all close at the same time once the membrane is repolarized. The whole-cell tail current of **Figure A5.8** represents the summation of hundreds to thousands of recordings of single-channel tail currents.

In **Figures A5.7** and **A5.8**, the tail currents are inward. The direction of a tail current (as for any type of current) depends on the sign of the driving force; i.e. the value of membrane potential upon repolarization (V_H) and that of the reversal potential of the current (E_{rev}) which depends on the ions flowing through the open channels. By varying the voltage at the end of the depolarizing step, the tail current varies in amplitude and direction (inward to outward or the reverse) and it is possible to determine the reversal potential of the tail current under study: when $V_H = E_{rev}$ the tail current is equal to zero (**Figure A5.9**). This value of E_{rev} is the same for the tail current and the current recorded during the voltage step since it concerns the same channels.

The voltage protocol of **Figure A5.9** allows the determination of E_{rev} and consequently identification of the type of ions that carry the current E_{rev} can also be determined directly by changing the voltage-step value. However, for K^+ channels, for example, E_{rev} is near −100 mV, a membrane potential where the open probability of voltage-gated channels is very low. By using tail currents, this problem is overcome.

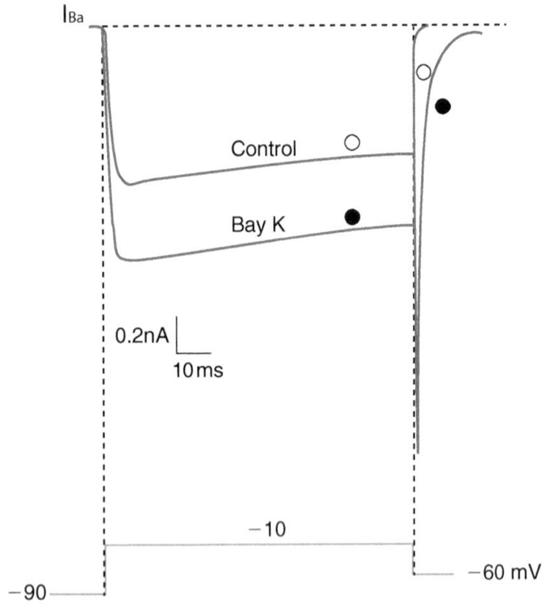

FIGURE A5.8 **Activity of a dorsal root ganglion neuron.** Recorded in single-electrode voltage clamp in the presence of Na^+ and K^+ channel blockers and 2 mM external Ba^{2+}. A depolarization to −10 mV followed by a repolarization to −60 mV is applied to the membrane from a holding potential V_H = −90 mV. The depolarizing step evokes an inward Ba^{2+} current followed by an inward Ba^{2+} tail current (control, open circle). The presence of 1 μm Bay K 8644 increases the amplitude of the Ba^{2+} current during the step. It also prolongs the Ba^{2+} tail current (black circle). Adapted from Carbone E, Formenti A, Pollo A (1990) Multiple actions of Bay K 8644 on high-threshold Ca channels in adult rat sensory neurons. *Neurosci. Lett.* **111**, 315–320, with permission.

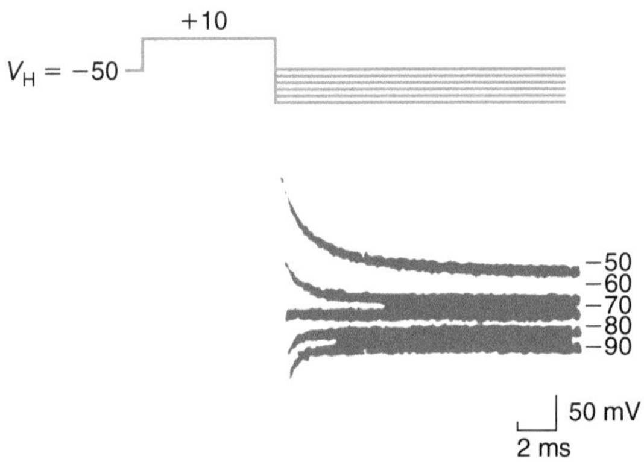

FIGURE A5.9 **Activity of a chick dorsal root ganglion cell.** Recorded in single-electrode voltage clamp, in the presence of 1 μm TTX and 10 mM Co^{2+} to block, respectively, Ca^{2+} and Na^+ channels. Depolarizations to +10 mV from a holding potential of −50 mV followed by successive repolarizations to −50, −60, −70, −80 and −90 mV are applied (V traces). Bottom I traces show the K^+ tail currents at the corresponding membrane potentials (the outward K^+ current during the step is not shown). The reversal potential of the K^+ tail current is −70 mV. It indicates the value of E_K in these cells. To separate the ionic tail current from the capacitive current, the latter was subtracted from the total current by digital summation of the currents elicited with identical depolarizing and hyperpolarizing test pulses. Adapted from Dunlap K, Fischbach GD (1981) Neurotransmitters decrease the calcium conductance activated by depolarization of embryonic chick sensory neurones. *J. Physiol.* **317**, 519–535, with permission.

C H A P T E R

6

The chemical synapses

Constance Hammond, Monique Esclapez

OUTLINE

In 1888, Ramon y Cajal suggested that the contacts between the axon terminals of a neuron and the dendrites or the perikaryon of another neuron are the points at which information flows from one neuron to the other: 'Les articulations ou contacts utiles et efficaces entre neurones ne s'effectuent qu'entre cylindre-axiles, collatérales ou terminales d'un neurone et les prolongements ou le corps cellulaire d'un autre neurone.' The term *synapse* was introduced by Sherrington (1897) to describe these zones of contact between neurons, specialized in the transmission of information.

In fact, the term 'synapse' is not used exclusively to describe connections between neurons (interneuronal connections) but also those between neurons and effector cells such as muscular and glandular cells (neuroeffector synapses) and those between receptive cells and neurons (**Figure 6.1**). These contacts are the points where the information is transmitted from one cell to the other: synaptic transmission.

According to morphological and functional criteria, there are various types of synapses, including chemical, electrical and mixed types.

- *Chemical synapses* are characterized morphologically by the existence of a space between the plasma membranes of the connected cells. These spaces are called *synaptic clefts*. In this case, a molecule – the neurotransmitter – conveys information between the presynaptic cell and the postsynaptic cell. Chemical synapses will be described in this chapter (**Figure 6.2a**). Some of the chemical synapses have particular characteristics:
  - *Reciprocal synapses* are formed by the juxtaposition of two chemical synapses oriented in the reverse direction to each other (**Figure 6.2d**). They are extremely rare.
  - *Glomeruli* are formed by a group of chemical synapses. In some cases, a group of dendrites form chemical synapses with the axon they surround (**Figure 6.2e**). In other cases, numerous axon terminals form synapses with the dendrite they surround.
  - *Electrical synapses* or gap junctions are characterized by the apposition of the plasma membranes of the connected cells. In this case,

Cellular and Molecular Neurophysiology. DOI: 10.1016/B978-0-12-397032-9.00006-6

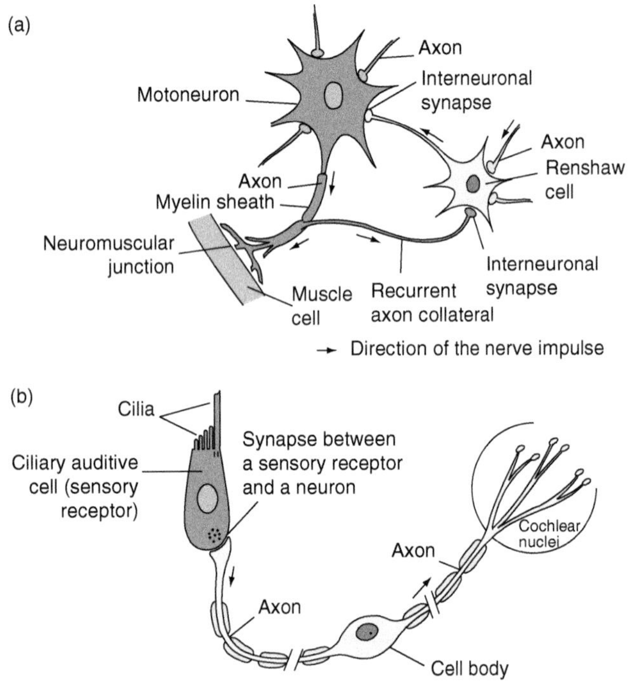

FIGURE 6.1 **Types of cells connected by chemical synapses. (a)** Interneuronal synapses and neuromuscular junction. Example of synapses between a motoneuron (Golgi type I neuron that innervates striated muscle fibers) and a Renshaw cell (a Golgi type II neuron in the spinal cord) and between a motoneuron and a striated muscle cell. **(b)** Synapse between a sensory receptor and a neuron. Example of synapses between an auditory receptive cell (ciliary cell in the cochlea) and a primary sensory neuron whose cell body is located in the spiral ganglion. This neuron is free of dendrites and has a T-shaped axon that drives sensory information from the periphery to the central nervous system. Drawing **(a)** from Eckert R, Randall D, Augustine G (1988) *Anim Physiol.* New York: W. A. Freeman, with permission.

the ions flow directly from one cell to the other without the use of a chemical transmitter. Gap junctions also allow the exchange of small-diameter intracellular molecules such as second messengers and metabolites (**Figure 6.2b**). These synapses are common between glial cells in the mammalian central nervous system.

- *Mixed synapses* are formed by the juxtaposition of a chemical synapse and a gap junction (**Figure 6.2c**). In mammals, between neurons, these synapses are more common than the electrical synapses.

6.1 THE SYNAPTIC COMPLEX'S THREE COMPONENTS: PRESYNAPTIC ELEMENT, SYNAPTIC CLEFT AND POSTSYNAPTIC ELEMENT

This section takes as an example the interneuronal chemical synapses. Under an electron microscope, a section of brain tissue taken from a region of the central nervous system rich in cell bodies and dendrites (gray matter) reveals many synaptic contacts at the surface of a dendritic shaft or of a dendritic spine (**Figure 6.3**, arrows). One of these synaptic contacts represents a synaptic complex (**Figure 6.4a**). The synaptic complex includes three components: the presynaptic element, the synaptic cleft and the postsynaptic element. The synaptic complex is *the non-reducible basic unit* of each chemical synapse since it includes the minimal requirement for efficient chemical synaptic transmission.

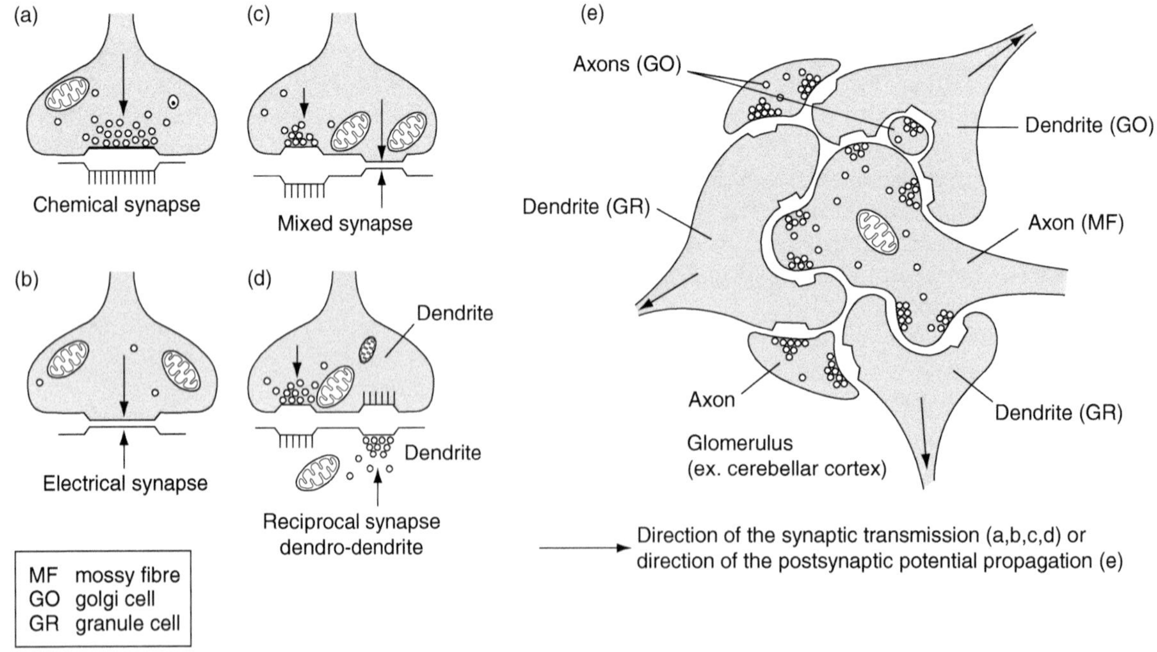

FIGURE 6.2 **Types of synapses.** See text for explanations. MF, mossy fiber; GO, Golgi cell; GR, granule cell. Parts **(a)**–**(d)** from Bodian D (1972) Neuronal junctions: a revolutionary decade. *Anat. Rec.*, **174**, 73–82, with permission. Drawing **(e)** from Steiger U (1967) Uber den Feinbau des Neuropils im Corpus pedunculatum des Waldaneise. *Z. Zelforsch.* **81**, 511–536, with permission.

6.1.1 The pre- and postsynaptic elements are morphologically and functionally specialized

The presynaptic element is characterized by the presence of numerous mitochondria and synaptic vesicles which store the neurotransmitter (**Figures 6.3** and **6.4**). Two types of synaptic vesicles are described: the clear vesicles (40–50 nm in diameter) and the dense-core vesicles or dense granules, which have an electron-dense core (40–60 nm in diameter). Occasionally, under the presynaptic membrane can be seen an electron-dense zone with a geometry more or less distinguishable, the presynaptic grid (see **Figure 6.10**). It corresponds to a particular organization of the cytoskeleton which is related to the exocytotic machinery.

The postsynaptic element in the interneuronal synapses is characterized by a submembranous electron-dense zone, which most probably corresponds to the region where the postsynaptic receptors are anchored. In cases where the postsynaptic element is non-neuronal, we shall see that various other postsynaptic specializations exist.

The synaptic complex displays a particular asymmetric structure, the synaptic vesicles being present only in the presynaptic element. This structural asymmetry suggests a functional asymmetry.

6.1.2 General functional model of the synaptic complex

A general functional model of chemical synaptic transmission is as follows. The newly synthesized neurotransmitter molecules are stored in the synaptic vesicles present in the presynaptic element. In a non-depolarized presynaptic element, the voltage-sensitive Ca^{2+} channels are closed and Ca^{2+} ions cannot enter the intracellular space. The exocytosis of synaptic vesicles is normally triggered by an increase of the intracellular Ca^{2+} concentration. Then, while the presynaptic membrane is at rest (i.e. as long as it is not depolarized by the arrival of an action potential), the probability of exocytosis of a synaptic vesicle and the release of its content into the synaptic cleft is very low: the neurotransmitter molecules are not released in significant quantities into the synaptic cleft. There is no synaptic transmission.

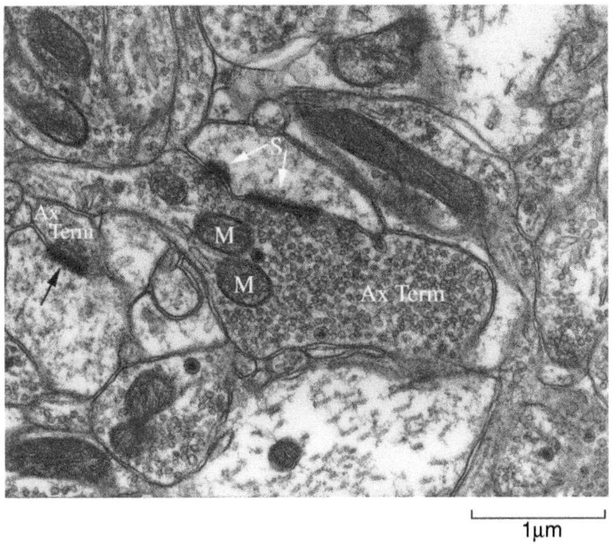

FIGURE 6.3 **Axo-spinous synapses.** One of which (center) is a 'perforated' synapse. Microphotography of a section of the hippocampus (molecular layer of the fascia dentata) observed under the electron microscope. Two synaptic boutons (Ax Term.) filled with synaptic vesicles and forming one or two asymmetric synaptic contacts (arrows) with dendritic spines (S) of pyramidal neurons can be visualized. M: mitochondries, Bar: 1 micrometer (microphotography Alfonso Represa).

Now, when Na^+-dependent action potentials (AP) propagate to axon terminals, they induce a depolarization of the presynaptic membrane (**Figure 6.4a**, 1). This results in opening of the voltage-sensitive Ca^{2+} channels present in the presynaptic membrane (2). Ca^{2+} entry through the opened channels evokes an increase of the intracellular Ca^{2+} concentration ($[Ca^{2+}]_i$), a factor required for triggering exocytosis of synaptic vesicles.

Thus, the probability of exocytosis of synaptic vesicles is strongly increased. This results in fusion of a docked vesicle(s) with the presynaptic plasma membrane (3) and release of the neurotransmitter molecules in the synaptic cleft (extracellular medium; 4). Once released into the synaptic cleft, neurotransmitter molecules bind to an ensemble of receptors: postsynaptic receptors (5a; receptor-channels and G-protein coupled receptors), presynaptic receptors (G-protein coupled receptors, 5b neurotransmitter transporters), glial receptors (5c; neurotransmitter transporters), and in some synapses enzymes (5d) that degrade the neurotransmitter molecules present in the synaptic cleft.

All these receptors are proteins that bear specific receptor sites to the neurotransmitter. By binding to postsynaptic receptor channels (5a) or to postsynaptic receptors coupled to G proteins, the neurotransmitter will induce the movement of ions through postsynaptic channels (6) and a postsynaptic current. At that stage, the synaptic transmission is completed. By binding to transporters present in the neuronal and glial membranes or in the cleft, the neurotransmitter is rapidly eliminated from the synaptic cleft. In the presynaptic element, the neurotransmitter is taken back into vesicles or degraded. The membrane is recycled by an endocytotic process (5e). All the events here called (5) are simultaneous.

To refill the synaptic vesicles, the neurotransmitter has also to be synthesized *de novo*. Neurotransmitters are generally synthesized in axon terminals from a precursor present in the axon terminals or taken up from blood. The enzymes necessary for its synthesis are synthesized in the soma and carried via the anterograde axonal transport to the axon terminals (see Section 1.3.2). Neurotransmitters are actively transported from the cytoplasm to the intravesicular compartment via transporters specific

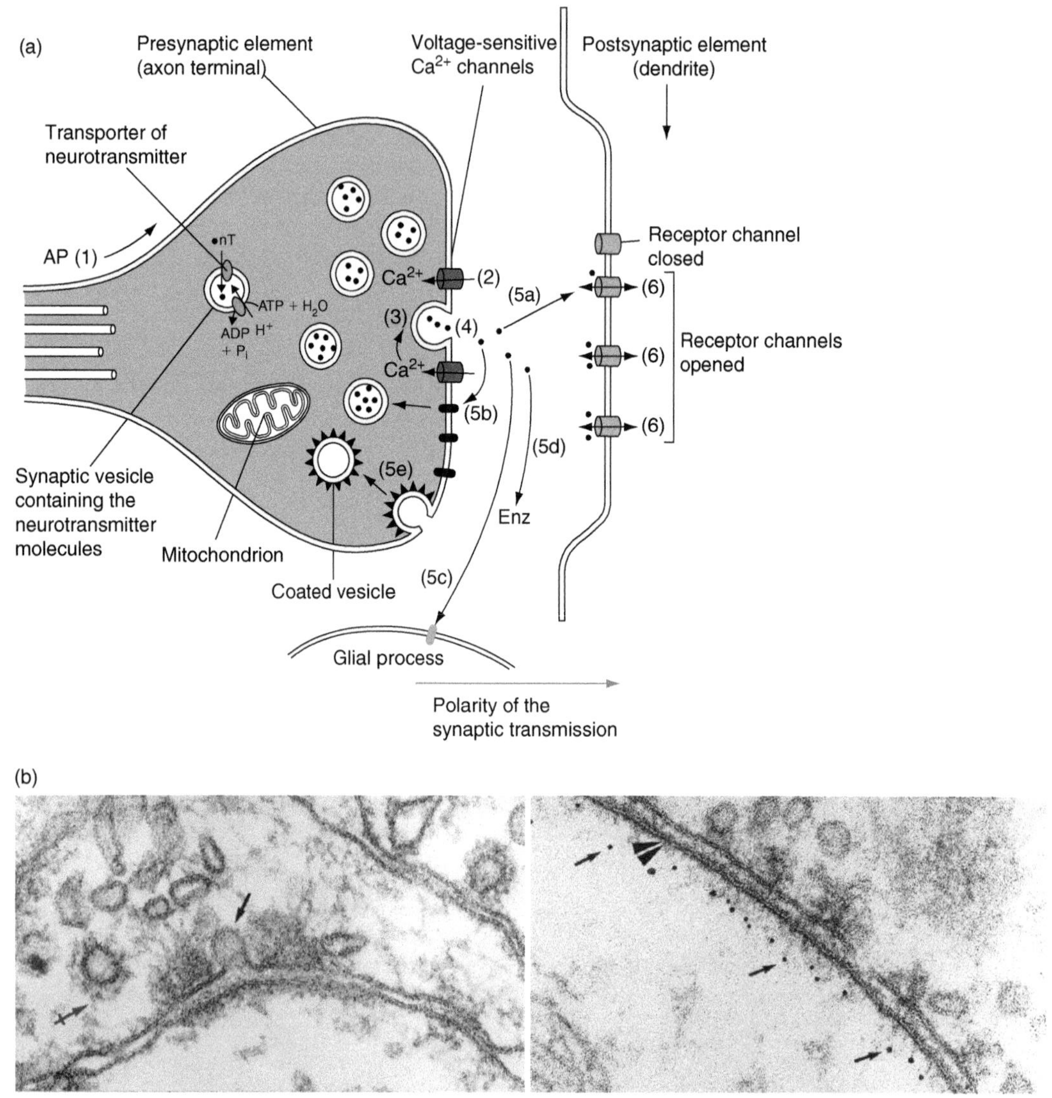

FIGURE 6.4 **Pre- and postsynaptic specializations. (a)** Schematic of the synaptic transmission (AP, action potential; see text for explanation). **(b)** Electron photomicrographs of transverse sections at the level of synaptic complexes. Left: A figure of exocytosis (long arrow) between two dense presynaptic projections. A coated vesicle (crossed arrow) characteristic of the recycling of the membrane is also seen (inhibitory synapse afferent to the Mauthner cell in the fish). Right: Postsynaptic localization of the glycine receptors, visualized with gold particles associated to a specific monoclonal antibody (single arrow). These are lined up at a distance from the membrane; this space originates mainly from the aggregation of the antibodies used to label indirectly the receptors (inhibitory synapse afferent to a motoneuron in the spinal cord of the rat). (**b**, left) From Triller A, Korn H (1985) Activity-dependent deformations of presynaptic grids at central synapses. *J. Neurocytol.* **14**, 177–192, with permission. (**b**, right) From Triller A, Cluzeaud F, Pfeiffer F, Korn H (1986) Distribution and transmembrane organization of glycine receptor at central synapses: an immunocytochemical touch. In Levi-Montalcini R *et al.* (eds) *Molecular Aspects of Neurobiology*, Berlin: Springer Verlag, with permission.

to each neurotransmitter. However, neurotransmitter peptides are synthesized as an inactive precursor form in the neuronal soma, and are carried to the axonal terminals via anterograde axonal transport (see **Figure A6.1**).

This general scheme is of course oversimplified. For example, the presynaptic element can contain more than a single neurotransmitter (as shown with immunocytochemistry with antibodies against neurotransmitters, enzymes of neurotransmitter synthesis, or against transporters of neurotransmitters, see Appendix 6.2); the intracellular concentration of Ca^{2+} ions can be increased also by the release of such ions from intracellular stores; and the role of presynaptic receptors has been ignored, for which there is evidence in the majority of synapses.

As we have seen, the *presynaptic element* contains the machinery for the synthesis, storage, release and inactivation of neurotransmitter(s). The presynaptic active zone is the complex formed by the synaptic vesicles and the region of the presynaptic membrane where exocytosis occurs (**Figure 6.4b**). Various methods can be used to characterize the neurotransmitter(s) present in a presynaptic element: immunohistochemical methods that identify the synthesis enzyme of the neurotransmitter in various parts of the neuron, and the *in situ* hybridization technique that identifies the mRNA coding for the synthesis enzyme of the neurotransmitter (see Appendix 6.2). However, identification of a substance as a neurotransmitter requires experimental proof of a number of other criteria (see Appendix 6.1). If these criteria have not been satisfied, the substance is called a *putative* neurotransmitter.

The *postsynaptic element* is specialized to receive information. Its plasma membrane contains proteins that are receptors for the neurotransmitter: receptor channels (**Figure 6.4c**) and G-protein linked receptors. Various methods can be used to characterize the receptors present in the postsynaptic membrane: radioautographic techniques with monoclonal antibodies, or extremely rarely anti-idiotype antibodies. Postembedding immunogold labeling (see **Figure 6.6d**) and, more recently, freeze fracture replica immunogold labeling (see **Figure 6.6e**) allow visualization and exact localization of postsynaptic receptors with electron microsocopy.

In most cases, synaptic transmission is unidirectional (or polarized): it propagates only from the presynaptic element, which contains the neurotransmitter, to the postsynaptic element at the surface of which are receptors for the neurotransmitter (**Figure 6.4a**). In the case of dendro-dendritic synapses (olfactory bulb of the rat), we recognize two juxtaposed synaptic complexes that work in opposite polarities; these are the reciprocal synapses (**Figure 6.2d**). However, it is worth noting that, here also, the synaptic transmission is polarized in each of the synaptic complexes.

6.1.3 Complementarity between the neurotransmitter stored and released by the presynaptic element and the nature of receptors in the postsynaptic membrane

In all synapses, receptors present in the postsynaptic membrane are those that specifically recognize the neurotransmitter released from the corresponding presynaptic element. For example, in glutamatergic synapses, glutamate receptors are found highly concentrated in the membrane of the corresponding postsynaptic element. Efficient synaptic transmission requires, in fact, specific localization of receptors on the postsynaptic membrane apposed to the transmitter release site. This is the case even in synapses where the pre- and postsynaptic cells have different embryonic origin (as in a nerve–muscle junction, for example). This pre–post complementarity requires, at least, the following steps to be completed: targeting, anchoring and clustering of postsynaptic receptors.

Targeting of receptors to a specific postsynaptic membrane

How do neurons target specific proteins to specialized neuronal subdomains? We shall take the example of metabotropic glutamate receptors (G-protein linked receptors of glutamate). They are a homologous family of differentially targeted receptors. Among mGluRs (see Chapter 12), mGluR1a and mGluR2 are targeted to dendrites and excluded from axons, whereas mGluR7 is targeted to dendrites and axons. In order to study the peptide sequence that could be responsible for this differential targeting, native or chimeric mGluRs are expressed, one at a time, in cultured hippocampal neurons, from viral vectors. The distribution of these expressed mGluRs is then checked by labeling mGluR with a specific antibody coupled to a fluorescent marker (Texas Red) and by labeling axons with a tau antibody (green) or dendrites with an MAP2 antibody (green).

First, the distribution of expressed mGluRs is checked. The selective distribution of endogenous mGluRs is reproduced for expressed mGluRs: axon exclusion of mGluR1a and mGluR2 and axon targeting for mGluR7 (all three are also targeted to dendrites). What mediates the axon exclusion of mGluR1a and mGluR2? The working hypothesis is that, since the C-terminal cytoplasmic domain of these receptors is the most divergent region of the primary sequence of mGluRs, it is involved in targeting. To answer this question, the distribution of chimeric mGluRs, such as mGluR2tail7 and mGluR7tail2 constructs, is studied (**Figure 6.5a**). To be sure that tails are intact in the constructs, antibodies against tail2 or tail7 are tested. As they still recognize the tails, the C-terminal epitope of chimeric mGluR forms correctly from the transfected chimeric cDNA.

Analysis of these chimeric constructs reveals that the C-terminal cytoplasmic domain of mGluRs contains the axon/dendrite targeting information. The mGluR2tail7 – containing the backbone of mGluR2 up to and including the seventh transmembrane domain followed by the C-terminal 65-amino-acid domain of mGluR7 – is targeted to axons (**Figure 6.5b**, left column). The reciprocal chimera mGluR7tail2 is present in dendrites but excluded from axons (**Figure 6.5b**, right column). To more narrowly define the targeting signal only the 30-amino-acid distal part of the C-terminal domain between mGluR2 and mGluR7 is swapped. These constructs are not as efficiently targeted as the first ones, showing that axon exclusion of mGluR7 versus mGluR2 is dependent on the 65-amino-acid C-terminal sequences primarily and not exclusively on the more distal amino acids. The mGluR2 C-terminus is required for axon exclusion and the mGluR7 C-terminus is required for axon targeting of the native proteins. These are 'axon exclusion' and 'axon targeting' signals. The mGluR targeting signals may function at any stage in targeting: sorting into specific vesicles from the *trans*-Golgi network, transport by

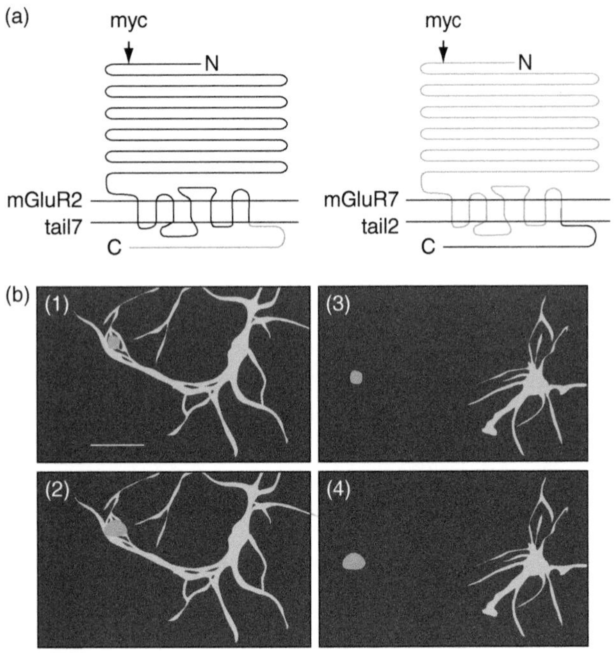

FIGURE 6.5 **Differential targeting of mGluR chimera. (a)** Schematics of mGluR chimera (mGluR2tail7 and mGluR7tail2) primary structures with mGluR2 structure shown in black and mGluR7 structure shown in green. The *myc* ten amino acid epitope tag (arrow) is inserted in the N-terminal extracellular domains, three amino acids past the signal sequence. **(b)** Left: Expressed *myc*-mGluR2tail7 recognized by surface labeling with the *myc* antibody (1) and labeling after permeabilization with an antibody against the C-terminus of mGluR7 (2). Both epitopes give the same distribution pattern, with labeling the full extent of the transfected neuron on the right and the axon of the transfected neuron in contact with a non-transfected neuron on the left. Right: expressed *myc* mGluR7tail2 was recognized by surface labeling with the *myc* antibody (3) and labeling after permeabilization with an antibody against the C-terminus of mGluR2 (4). Again, both epitopes give the same distribution pattern, with labeling of the somato-dendritic domain of the transfected neuron on the right. Scale bar, 50 mm. From Nash Stowell J, Craig AM (1999) Axon/dendrite targeting of metabotropic glutamate receptors by their cytoplasmic carboxy-terminal domains. *Neuron* **22**, 525–536, with permission.

association with specific motors, selection of plasma membrane addition sites, etc. When mGluR1a and mGluR2 are not detected at the surface of the axon, they are also not detected in the axoplasm. Therefore, the mGluR targeting signals such as 'axon exclusion' may act at an early stage, such as sorting out into vesicles directly targeted to dendrites.

Anchoring and clustering of receptors in the postsynaptic membrane

A single neuron may receive input from thousands of synaptic connections on its cell body and dendrites. To integrate these signals rapidly and specifically, the neuron anchors a high concentration of receptors at postsynaptic sites, matching the correct receptor with the neurotransmitter released from the presynaptic terminal. The mechanism of site-specific receptor clustering has been most thoroughly investigated at the neuromuscular junction (a cholinergic synapse; see Section 6.3). Rapsyn is believed to be one of the molecules responsible for nicotinic cholinergic receptors (nAChR) clustering. For glycine receptors (GlyR), postsynaptic clustering is dependent on gephyrin, a 93 kDa channel-associated protein, which is totally unrelated to rapsyn. We shall take here the example of the identification of anchor proteins for excitatory glutamatergic synapses and, in particular, for the ionotropic glutamatergic NMDA (*N*-methyl-D-aspartate) receptor (see Chapter 10).

Excitatory synapses are characterized by a morphological and functional specialization of the postsynaptic membrane called the postsynaptic density (PSD), which is usually located at the tip of the dendritic spine. The generally accepted roles for the PSD are to mediate the apposition of pre- and postsynaptic membranes, to cluster postsynaptic receptors, and to couple the activation of postsynaptic receptors to biochemical signaling events in the postsynaptic neuron.

NMDARs are composed of NR1 and NR2 subunits, which co-translationally assemble in the endoplasmic reticulum to form functional channels. A conspicuous feature of the NR2 subunits is their extended, intracellular C-terminal sequence distal to the last transmembrane region. The working hypothesis is that this region participates in anchoring. In an attempt to identify molecules that can mediate the association between NMDA receptors and cytoskeleton, the yeast two-hybrid system is used to search for such gene products that bind to the intracellular C-terminal tails of NR2 subunits at synapses.

The NR2 subunits are found to interact specifically with a family of membrane-associated synaptic proteins. In mammals, this family includes PSD-95, PSD-93 and PSD-95-synapse-associated-proteins (SAP). In their N-terminal half, this family of proteins is characterized by the presence of three domains with a length of approximately 90 amino acids, termed PDZ domains; they are therefore called PDZ-containing proteins (P for PSD-95, D for *dlg* and Z for ZO-1, the first proteins to be identified with these domains) (**Figure 6.6a**). PDZ repeats are protein-binding sites that recognize a short consensus peptide sequence of NR2 subunits (**Figure 6.6b**). PDZ domains also bind to intracellular proteins (**Figure 6.6b**). Synaptic NMDARs are thus concentrated in postsynaptic densities (PSDs), where they are structurally organized (and spatially restricted) in a large macromolecular signaling complex composed of kinases, phosphatases, and adaptor and scaffolding proteins. Scaffolding proteins serve to structurally organize and localize proteins to the PSD (**Figure 6.6d,e**) and physically link receptors such as NMDARs in close proximity to protein kinases (like CaM kinase II activated by Ca^{2+} influx through NMDA receptors), phosphatases, and other receptors.

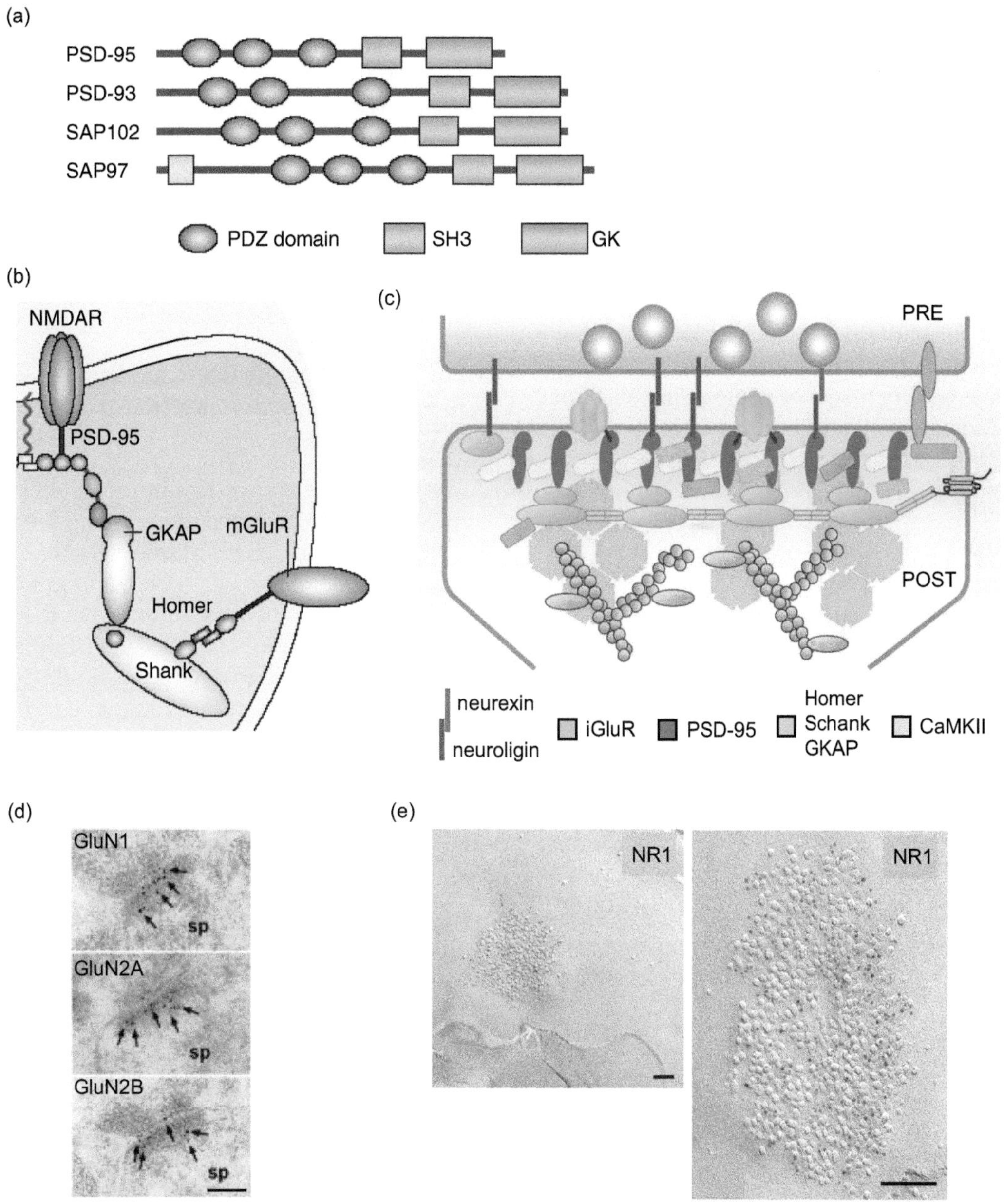

FIGURE 6.6 **PDZ-containing proteins and anchoring of glutamate receptors.** Schematic diagram of PDZ proteins. PDZ domains are often found in scaffold proteins as multiple tandem arrays and/or linked to other kinds of modular protein-interaction domain. PDZ domains are shown as purple ellipses. GK, guanylate-kinase-like domain; SH3, Src homology 3 domain. (**b**) Schematic of possible interactions in the postsynaptic element between the NMDA receptor (NMDAR) and anchor proteins: PSD-95, GKAP guanylate kinase-associated protein, Shank SH3 and ankyrin repeat-containing protein. GKAP scaffolding proteins interact with the carboxy terminal GK-like domain of PSD-95 family proteins. Homer is a family of scaffolding proteins that associate with Shank and group I metabotropic glutamate receptors (mGluR1 and mGluR5), thereby linking Shank with mGluRs. (**c**) Schematic diagram of the major proteins of the postsynaptic density of excitatory synapses. Direct protein interactions are indicated by direct contacts or overlaps between the proteins. The relative numbers of the proteins shown correlate roughly with their relative abundance. (**d**) Electron micrographs of postembedding immunogold labeling for NMDAR subunits (black particles). Serial ultra thin sections show that GluN1, GluN2A and GluN2B subunits are present in the postsynaptic active zone of a glutamatergic synapse. Sp: spine head; Scale bar 100 nm. (**e**) Electron micrographs at two different magnifications of a freeze fracture replica of the postsynaptic membrane immunolabeled with an antibody that recognizes the NR1 subunit of NMDA receptors (see Appendix 6.2). The postsynaptic membrane specializations (clusters of intramembrane particles) show strong immunoreactivity for NR1 in the superficial spinal dorsal horn of rats. NR1 subunits are labeled with 5 nm gold particles. Scale bars: 100 nm. Parts (**a**) and (**b**) The metabotropic glutamate receptor (mGluR) is anchored via Homer. From Kim E, Sheng M (2004) PDZ domain proteins of synapses. *Nat. Rev. Neurosci.* **5**, 771–781. Part (**c**) Sheng M, Kim E (2011) The postsynaptic organization of synapses. *Cold Spring Harb Perspect Biol* **3**:a005678. Part (**d**) Szabadits E, Cserép C, Szönyi A, Fukazawa Y, Shigemoto R, Watanabe M, *et al.* (2011) NMDA receptors in hippocampal GABAergic synapses and their role in nitric oxide signaling. *J. Neurosci.* **31**, 5893–5904. Part (**e**) Antal M, Fukazawa Y, Eordogh M, Muszil D, Molnar E, Itakura M, Takahashi M, Shigemoto R. (2008) Numbers, densities, and colocalization of AMPA- and NMDA-type glutamate receptors at individual synapses in the superficial spinal dorsal horn of rats. *J. Neurosci.* **28**, 9692–9701.

Maintenance of synapses between neurons

Maintenance of synapses between neurons is achieved by cell adhesion proteins such as neuroligins and neurexins. Neurexins were originally discovered in 1992 as receptors for α-latrotoxin, a component of black widow spider venom. A few years later, neuroligins were identified as neurexin binding partners. Neuroligins are strictly localized at the postsynaptic membrane. They act as ligands for neurexins, which are mostly found at the presynaptic membrane. The cytoplasmic tails of neuroligins bind to PSD scaffold proteins including PSD-95, whereas presynaptic neurexins bind to other types of scaffolding proteins. Such sets of protein–protein interactions provide a trans-synaptic link between the PSD and the presynaptic active zone (**Figure 6.6c**).

6.2 THE INTERNEURONAL SYNAPSES

6.2.1 In the CNS, the most common synapses are those where an axon terminal is the presynaptic element

As described in Section 1.1.2, the axon terminals are either *terminal boutons* (**Figure 6.3a**), which are terminals of axonal branches, or *boutons en passant* (**Figure 6.3b**), which appear as swellings located along the non-myelinated axons and at the nodes of Ranvier along myelinated axons. These two types of axon terminals form synaptic contacts with various neuronal postsynaptic elements: a dendrite (axo-dendritic synapse), a soma (axo-somatic synapse) or an axon (axo-axonic synapse) (**Figure 6.7**). More rarely, there are synapses in which the presynaptic element is a dendrite (dendro-dendritic synapse; see **Figure 6.2d**) or a soma (soma-somatic or soma-dendritic synapses).

6.2.2 At low magnification, the axo-dendritic synaptic contacts display features implying various functions

We will consider as an example the cerebellar cortex, a layered structure in which the cells and their afferents are well characterized. The Purkinje cells are the single 'output' cells of the cerebellar cortex (Golgi type I neurons, **Figure 6.8a**) which send their axons to the deep cerebellar nuclei. They have a cell body with a large diameter (20–30 μm) from which emerges a single dendritic trunk that gives rise to numerous spiny dendritic branches which arborize in the molecular layer. The dendritic tree is planar, and the dendritic branches extend mainly in

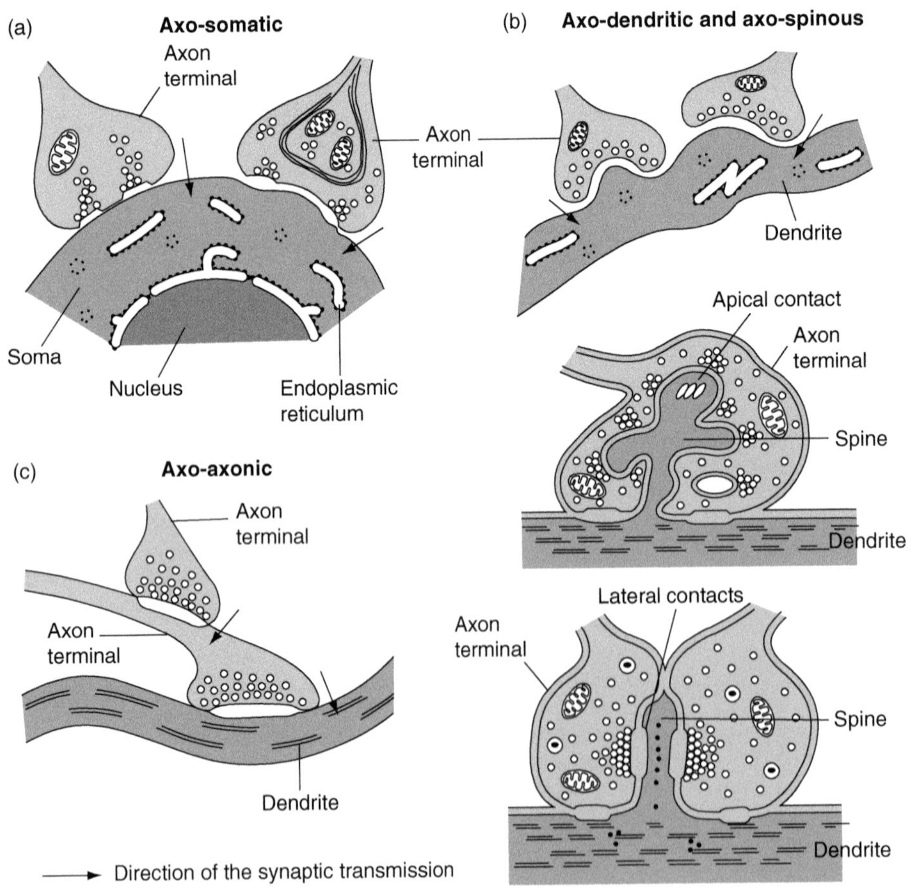

FIGURE 6.7 **Types of interneuronal synapses. (a)** Terminal boutons forming axo-somatic synapses. **(b)** Top to bottom: Synapses between terminal boutons and a smooth dendritic branch (axo-dendritic synapse) and two examples of indented synapses between terminal boutons and a dendritic spine (axo-spinous synapses). **(c)** Synapse between an axon terminal and a terminal axon collateral (axo-axonic synapse, extremely rare). The 'postsynaptic' axon terminal is itself 'presynaptic' to a dendrite. From Hamlyn LH (1972) The fine structure of the mossy fiber endings in the hippocampus of the rabbit. *J. Anat.* **96**, 112–120, with permission.

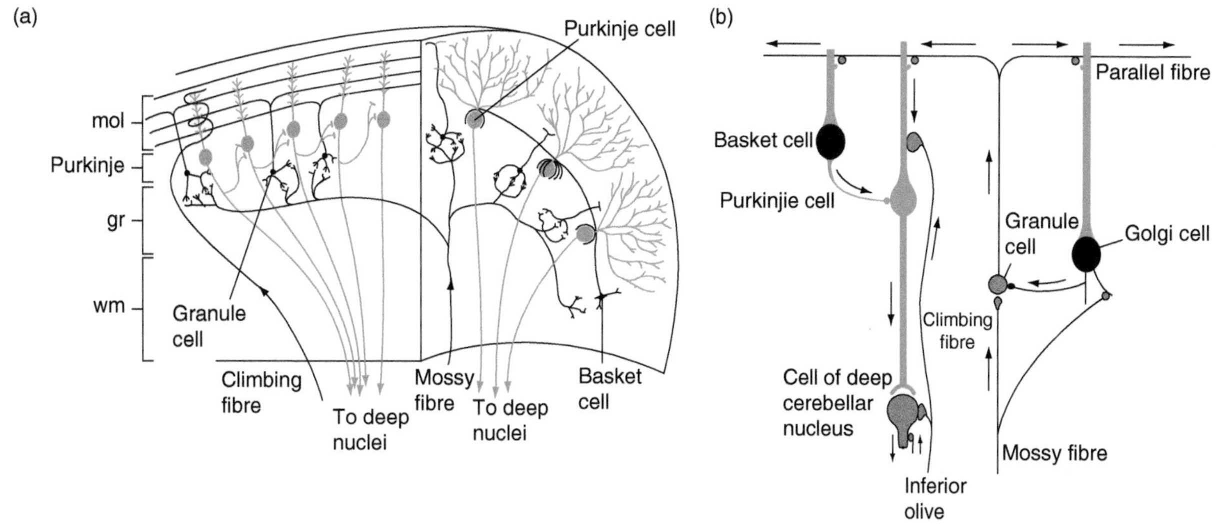

FIGURE 6.8 **Synaptic connections in the cerebellar cortex. (a)** Diagram of cells in a folium of the cerebellum. The layers are depicted: from the surface (pia-mater) to the depth, the molecular layer (mol), the layer of Purkinje cells, the granular layer (gr) and the white matter (wm). By comparing the drawings of the Purkinje cells in the sagittal plane with those in the transverse plane, notice that the dendritic tree of the Purkinje cells is planar. A climbing fiber which arborizes along the dendrites of one of the Purkinje cells is shown. It synapses directly with one Purkinje cell. Mossy fibers synapse with many granule cells. The axons of the granule cells enter the molecular layer and bifurcate in a T to form the parallel fibers running lengthwise in the folium. Granule cell axons synapse with Purkinje cells and basket cells. **(b)** Schematic representation of the principal synaptic connections within the cerebellar cortex. Inhibitory GABAergic Purkinje cells are in green; inhibitory GABAergic interneurons are in black; excitatory neurons are in grey. Drawing **(a)** from Gardner E (1975) *Fundamentals of Neurology*, Philadelphia: W.B. Saunders. Drawing **(b)** from Eccles JC (1973) *J. Physiol. (Lond.)* **299**, 1–3, with permission.

the transverse plane. The neurotransmitter of Purkinje cells is γ-aminobutyric acid (GABA).

Purkinje cells receive two types of excitatory afferents: the climbing fibers (axons of the neurons in the inferior olivary nucleus) and the parallel fibers (axons of the granule cells in the cerebellar cortex). The inhibitory afferents arise mainly from the numerous local circuit neurons in this structure: the basket cells and the stellate cells (**Figure 6.8b**).

A single climbing fiber innervates each single Purkinje cell. The climbing fiber gives rise to numerous axon collaterals that 'fit' the shape of the postsynaptic dendritic tree: the axon collaterals 'climb' along the dendrites (**Figures 6.8a** and **6.9a,b**) forming numerous synaptic contacts with the soma and the dendrites of the Purkinje cell. These contacts are axo-dendritic or axo-spinous, the presynaptic element being a terminal bouton. Such a synaptic organization implies that this excitatory afferent is very efficient: a single action potential along the climbing fiber can in fact induce a response in the Purkinje cell.

The axons of the granule cells form very different synaptic contacts with the Purkinje cells. The axons enter the molecular layer where they bifurcate and extend for 2 mm in a plane perpendicular to the plane of the dendritic tree of the Purkinje cell and form what are called parallel fibers (**Figure 6.8**). Parallel fibers form a few 'en passant' synapses (axo-spinous synapses between axonal varicosities and distal dendritic spines), with the numerous Purkinje cells (about 50). Therefore, each Purkinje cell receives synaptic contacts from about 200 000 parallel fibers. The consequence of such a synaptic organization is as follows. Activation of a parallel fiber cannot induce a Purkinje cell response since the activation of one or a few of these excitatory synapses cannot trigger postsynaptic action potentials; numerous parallel fibers converging on to a single Purkinje cell must be activated to induce a response in this cell.

The basket cells are local circuit neurons (Golgi type II neurons) which inhibit the activity of Purkinje cells. The axons of these neurons project to a large number of Purkinje cells and give rise to numerous axon collaterals which form 'baskets' around the soma of Purkinje cells. The axonal branches extend further and terminate 'en pinceau' around the initial segment of the Purkinje cells' axon (**Figures 6.8** and **6.9a,c**). Such an organization allows inhibition of the Purkinje cells at a strategic point where the sodium action potentials arise. This represents an efficient way of counteracting the excitatory potentials propagating along the dendritic branches to the initial axonal segment.

6.2.3 Interneuronal synapses display ultrastructural characteristics that vary between two extremes: types 1 and 2

A classification of the synapses on the basis of the form of their synaptic complex observed under electron microscopy was proposed by Gray (1959). This author

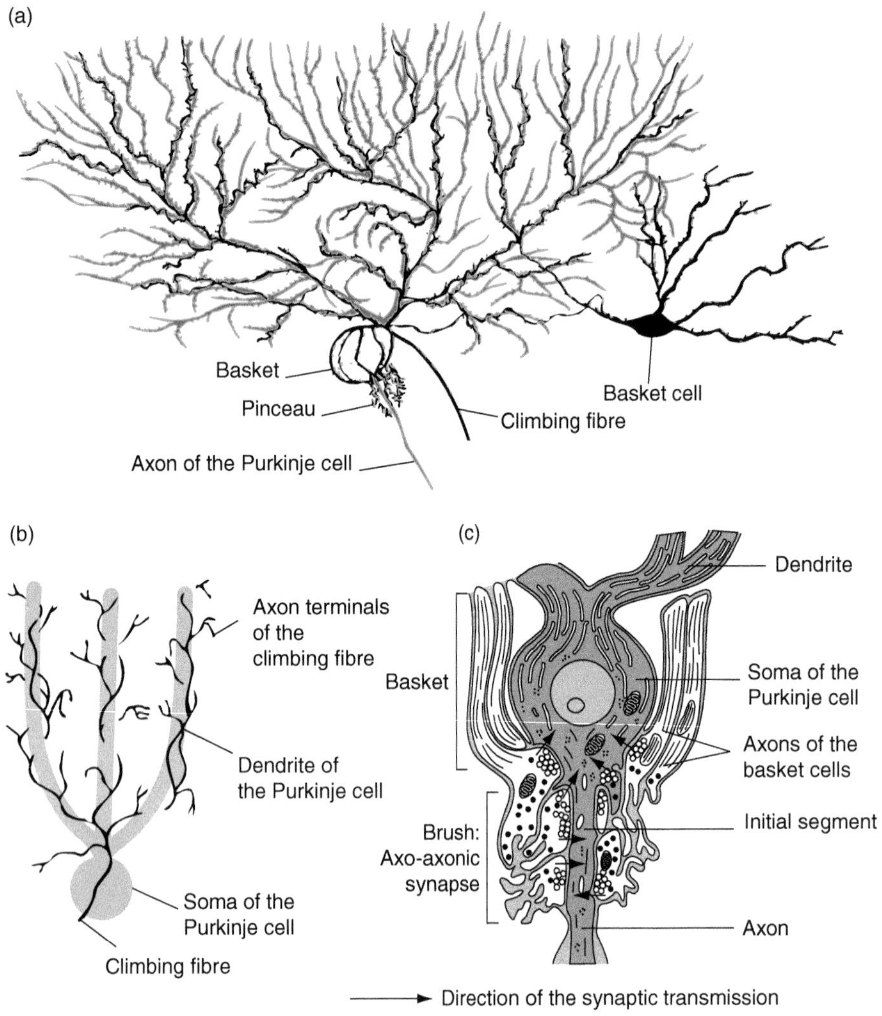

FIGURE 6.9 **Varieties of synaptic arrangements at the level of a Purkinje cell.** Representation on a single drawing **(a)** and in two separated schematic drawings **(b** and **c)** of the synaptic arrangements between a climbing fiber and a Purkinje cell (**a** and **b**) and between a basket cell and a Purkinje cell (**a** and **c**). Part (**a**) from Chan-Palay V, Palay S (1974) *Cerebellar Cortex: Cytology and Organization*, Berlin: Springer Verlag, with permission. Part (**b**) from Scheibel ME, Scheibel AB (1958) *Electroenceph. Clin. Neurophysiol.* **Suppl. 10**, 43–50, with permission. Part (**c**) from Hamori J, Szentagothai J (1965) The Purkinje cell baskets: ultrastructure of an inhibitory synapse. *Acad. Biol. Hung.* **15**, 465–479, with permission.

described two types of synaptic complexes in the cerebral cortex which he named types 1 and 2 (**Figure 6.10**).

- Type 1 synapses are asymmetrical because they have a prominent accumulation of electron-dense material on the postsynaptic side. These synapses are found more often on dendritic spines or distal dendritic branches. The presynaptic element contains round vesicles and the synaptic cleft is about 30 nm wide.
- Type 2 synapses are symmetrical because they have electron-dense zones of the same size in both the pre- and postsynaptic elements. The presynaptic element contains oval-shaped vesicles and the synaptic cleft is narrow. These synapses are more commonly found at the surface of dendritic trunks and soma.

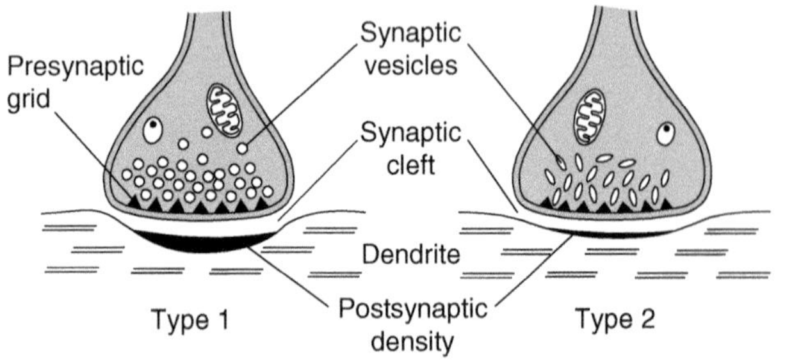

FIGURE 6.10 **Schematic representation of type 1 (asymmetric) and type 2 (symmetric) synapses according to Gray.**

On the basis of correlations between physiological and morphological data obtained in the cerebellar cortex, Gray proposed that type 1 synapses are excitatory whereas type 2 synapses are inhibitory.

In the central nervous system (CNS), types 1 and 2 synapses are the extremes of a morphological continuum since synaptic complexes may have intermediate forms and display features that characterize both types of synapse; e.g. a large synaptic cleft (type 1) and a narrow postsynaptic density (type 2). In addition, it has been shown that the form of the synaptic vesicles is dependent on the fixation technique used.

6.3 THE NEUROMUSCULAR JUNCTION IS THE GROUP OF SYNAPTIC CONTACTS BETWEEN THE TERMINAL ARBORIZATION OF A MOTOR AXON AND A STRIATED MUSCLE FIBER

The motoneurons or motor neurons have their cell body located in motor nuclei of the brainstem or in the ventral horn of the spinal cord. The axons of these neurons are myelinated and form the cranial and spinal nerves that innervate the skeletal striated muscles (see Section 1.4.1). In general, a single striated muscle fiber is innervated by one motoneuron but a single motoneuron can innervate many muscle fibers. The myelin sheath of each axon is interrupted at the zone where the axon arborizes at the surface of the muscle fiber. At this point, the thin non-myelinated axonal branches possess numerous varicosities which are located in the depression at the surface of the muscle fiber: the synaptic gutter. The axon terminals are covered by the non-myelinating Schwann cells (**Figure 6.11**; see also Section 2.3.2).

6.3.1 In the axon terminals, the synaptic vesicles are concentrated at the level of the electron-dense bars; they contain acetylcholine

The neuromuscular junction is formed by the juxtaposition of the terminals of a motor axon and the corresponding subsynaptic domains of a striated muscle fiber, these two elements being separated by a 50–100 nm-wide cleft. In a transverse section of a neuromuscular junction observed under electron microscopy (**Figure 6.12a**), the vesicles in the presynaptic element are small (40–60 nm diameter), clear and contain acetylcholine, the neurotransmitter of all the neuromuscular junctions. Larger vesicles (80–120 nm diameter) that contain an electron-dense material are also present but in a much lower proportion (1% of the total population). The vesicles are aggregated in the presynaptic zones where an electron-dense material is present, the dense bars. These dense bars are functionally homologous to the presynaptic grid of interneuronal synapses. They are 100 nm wide and are located perpendicularly to the largest axis of each axonal branch. The vesicles are aligned along each side of these bars. The complex of dense bars and synaptic vesicles forms a presynaptic active zone (Couteaux, 1960). There are many active zones per varicosity. They are located opposite to the folds of the postsynaptic plasma membrane. Each active zone with the folds of the sarcolemma in front of them forms a synaptic complex. Therefore, the neuromuscular junction contains numerous synaptic complexes.

Synthesis of acetylcholine takes place in the cytoplasm of the presynaptic element from two precursors: choline and acetyl coenzyme A (acetyl CoA). The reaction is

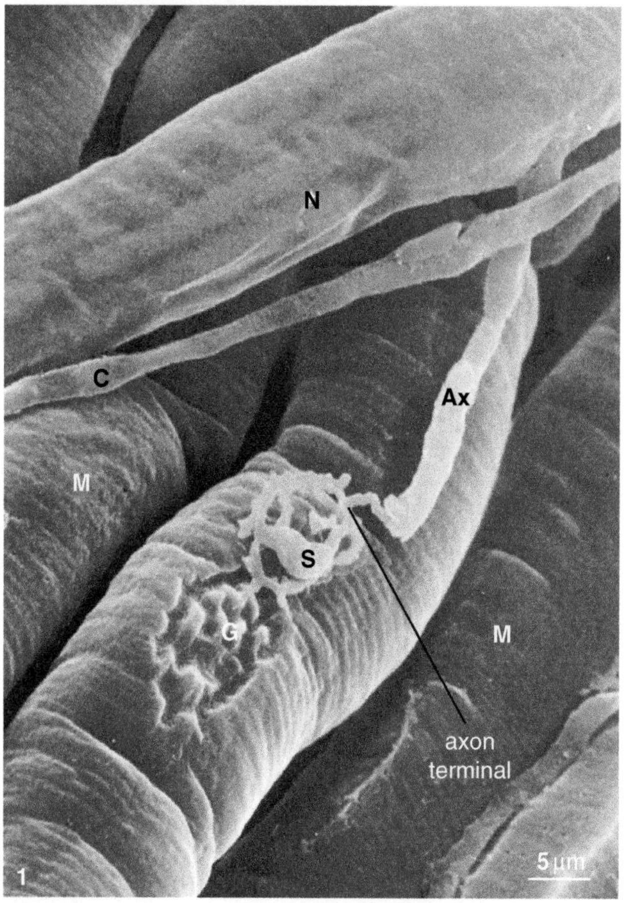

FIGURE 6.11 **The neuromuscular junction. Photograph of a rat neuromuscular junction observed under the scanning electron microscope.** The terminal part of the axon (ax) is detached from the muscle cell (M) in order to show the synaptic gutter (G); c, capillary; N, motor nerve; S, nucleus of a Schwann cell. From Matsuda Y *et al.* (1988) Scanning electron microscopic study of denervated and reinnervated neuromuscular junction. *Muscle Nerve* **11**, 1266–1271, with permission.

catalyzed by choline acetyltransferase (CAT). Acetylcholine is transported actively into synaptic vesicles where it is stored (see **Figure 8.20a**). The protein responsible for this active transport is a transporter which uses the energy of the proton (H^+) gradient. This gradient of protons is established by active transport of H^+ ions from the cytoplasm towards the interior of the vesicles by an H^+-ATPase pump.

6.3.2 The synaptic cleft is narrow and occupied by a basal lamina which contains acetylcholinesterase

The postsynaptic muscular membrane (sarcolemma) is covered, on the extracellular surface, with a layer of electron-dense material, the basal lamina (**Figure 6.12a,c**). This lamina, which follows the folds of the sarcolemma, is a conjunctive tissue secreted by the non-myelinating Schwann cells covering the axon terminals. It contains, *inter alia,* collagen, proteoglycans and laminin.

Acetylcholinesterases are glycoproteins synthesized in the soma and carried to the terminals via anterograde axonal transport. They are inserted into the presynaptic membrane and the basal lamina. They display an important structural polymorphism (**Figure 6.12b**): they have a globular form (G) or an asymmetric form (A). These different forms have distinct localizations. Globular forms (G) are anchored in the pre- or postsynaptic membrane (these are ectoenzymes) and are secreted as a soluble protein into the synaptic cleft. Asymmetric forms (A) are anchored in the basal lamina (**Figure 6.12c**). The molecules of acetylcholine, released in the synaptic cleft when the neuromuscular junction is activated, cross the basal lamina through its loose stitches. But a part of the acetylcholine molecules is also degraded before being fixed to postsynaptic receptors, by the acetylcholinesterase inserted in the basal lamina. The other part is quickly degraded after its fixation. Acetylcholinesterases hydrolyze acetylcholine into acetic acid and choline. Choline is taken up by presynaptic terminals for the synthesis of new molecules of acetylcholine. This degradation system of acetylcholine is a very efficient system for inactivation of a neurotransmitter.

6.3.3 Nicotinic receptors for acetylcholine are abundant in the crests of the folds in the postsynaptic membrane

The plasma membrane of muscle cells, the sarcolemma, presents numerous folds in mammalian neuromuscular junctions. By using a radioactive ligand for a type of acetylcholine nicotinic receptor, α-bungarotoxin labeled with a radioactive isotope or a fluorescent molecule, it has been shown that the radioactive material accumulates predominantly in the crests of the folds in the sarcolemma. Immunocytochemical techniques produce similar results. Other studies have shown that they are anchored to the underlying cytoskeleton (see the following section).

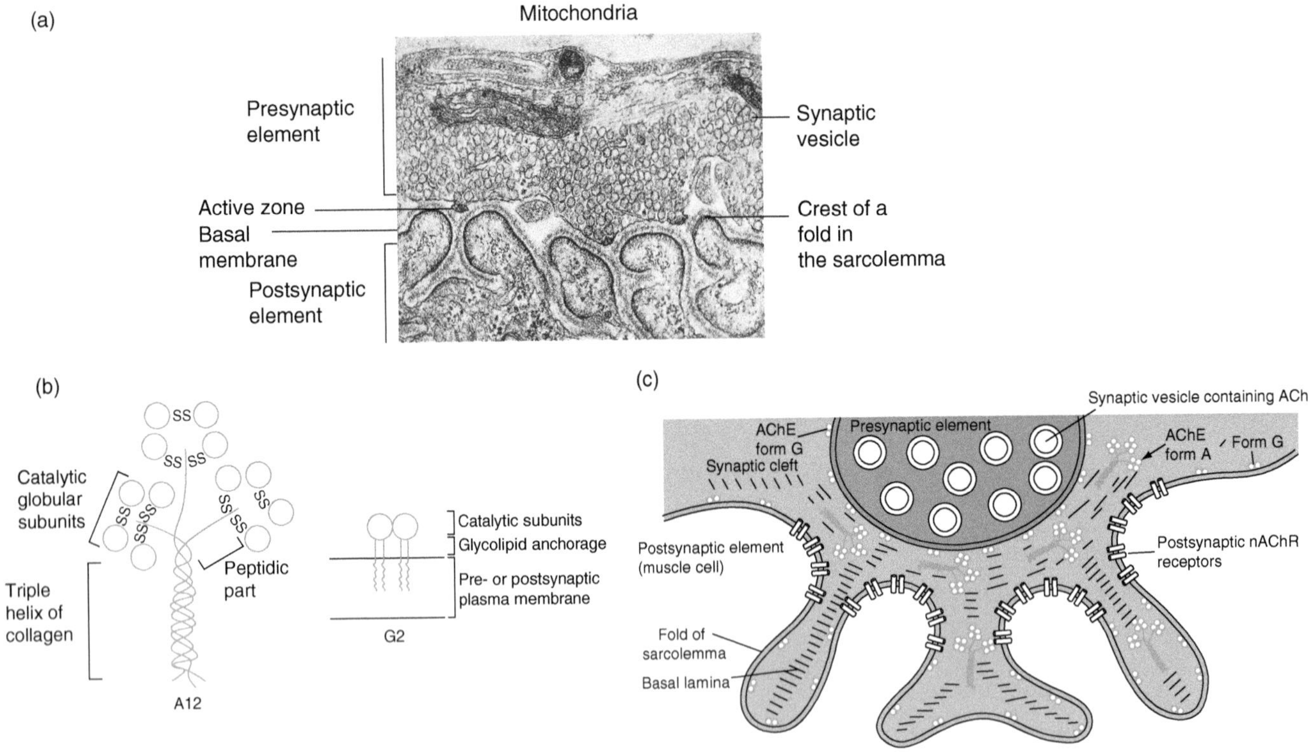

FIGURE 6.12 **Ultrastructure of a neuromuscular junction and the location of acetylcholinesterases. (a)** Microphotography of the neuromuscular junction of a batrachian observed under the electron microscope. In the axon terminal can be seen mitochondria and numerous vesicles. The axonal plasma membrane displays signs of exocytosis (active zones). The basal lamina is located in the synaptic cleft. The postsynaptic muscle cell membrane has numerous folds. **(b)** Schematic of the asymmetric (A12) and globular (G2) forms of acetylcholinesterase (AChE) (top). The index number of A or G indicates the number of catalytic subunits. The asymmetric forms consist of a collagen tail, three peptide parts and catalytic subunits. The globular forms consist of one or more catalytic subunits (hydrophilic domain) and a glycolipid part (hydrophobic domain) which permits their insertion in the lipid bilayer. **(c)** Location of acetylcholinesterase in the neuromuscular junction. The A forms are synthesized in the motoneurons and secreted into the synaptic cleft where they are associated with the basal lamina. The globular forms are synthesized in the motoneurons and inserted into the presynaptic plasma membrane or secreted into the synaptic cleft. Microphotograph **(a)** by Pécot-Dechavassine. Drawings **(b)** and **(c)** from Berkaloff A, Naquet R, Demaille J (eds) (1987) *Biologie 1990: Enjeux et Problématiques*, Paris: CNRS, with permission.

The nicotinic receptor is a transmembrane glycoprotein comprising four homologous subunits assembled into a heterologous $\alpha_2\ \beta\ \gamma\ \delta$ pentamer. It is a receptor channel permeable to cations whose activation results in the net entry of positively charged ions and in depolarization of the postsynaptic membrane. The structure and functional characteristics of the muscular nicotinic receptors are given in Chapter 8.

6.3.4 Mechanisms involved in the accumulation of postsynaptic receptors in the folds of the postsynaptic muscular membrane

The acetylcholine nicotinic receptors are, in the adult neuromuscular junction, present in high density (about 10 000 molecules per μm) in the postsynaptic regions and occur in a much lower density in the non-synaptic membrane (extrajunctional membrane). Under the nerve terminal, the muscle cell is free of the myofilaments actin and myosin. At this level, four to eight cell nuclei are found, the fundamental nuclei (Ranvier, 1875). The myonuclei located outside the postsynaptic region (extrasynaptic) are the sarcoplasmic nuclei. The formation of this well organized subsynaptic domain – which concerns not only the nicotinic receptors but also the Golgi apparatus and the cytoskeleton (it also comprises the organization of the basal lamina and the distribution of the asymmetric form of acetylcholinesterase in the synaptic cleft) – occurs in numerous steps during maturation of the neuromuscular junction (**Figure 6.13a**):

- There is an increase in the number of nicotinic receptors (1 and 2) during fusion of the myoblasts to form myotubes, owing to the neosynthesis of these receptors. They have an even distribution over the membrane surface. This phenomenon is independent

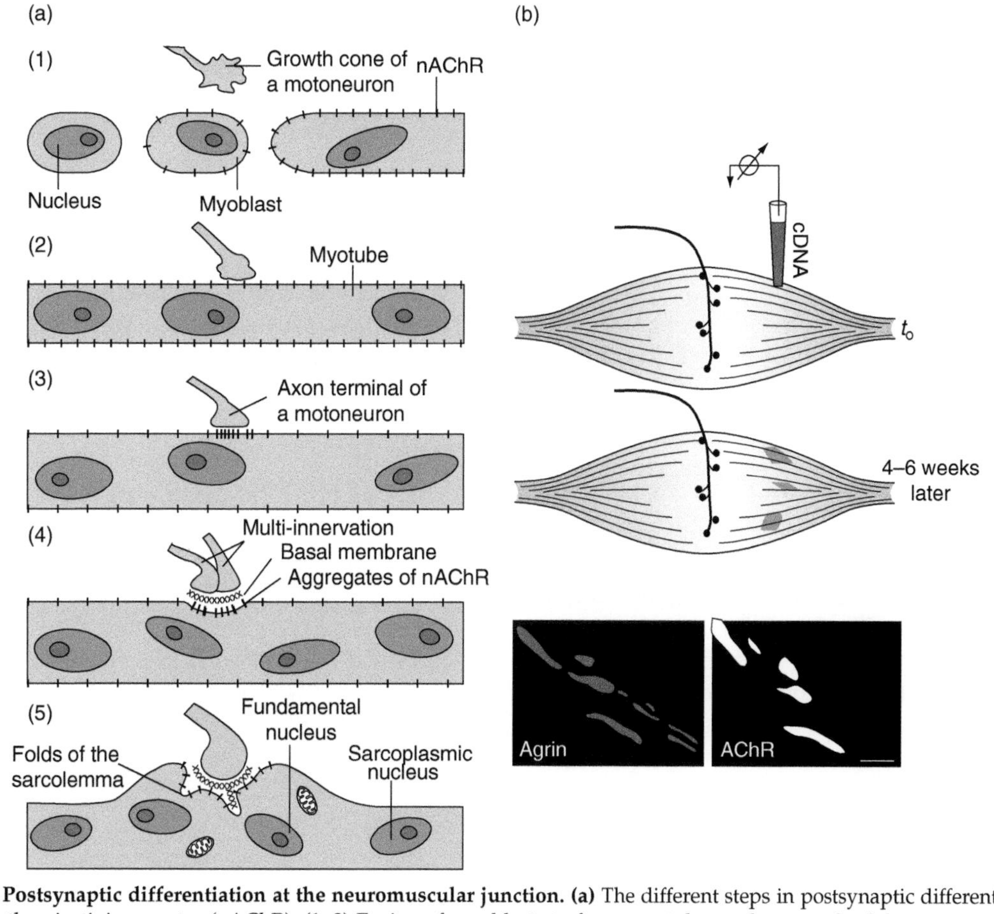

FIGURE 6.13 **Postsynaptic differentiation at the neuromuscular junction. (a)** The different steps in postsynaptic differentiation. The black thin bars indicate the nicotinic receptor (nAChR). (1, 2) Fusion of myoblasts to form myotubes and approach of the axon growth cone; (3) the growth cone forms contact with the myotube and induces the clustering of nicotinic receptors at this level; (4) numerous motor terminals converge towards a single aggregate of nicotinic receptors; but (5) a single terminal stabilizes and the folds of the sarcolemma develop. **(b)** Injection of expression constructs that encode full-length neural agrin into single muscle fibers *in vivo* (t_0). After 4–6 weeks, muscle fibers have a deposit of chick neural agrin in muscle basal lamina at ectopic sites. Staining of the muscle anti-chick agrin antibodies reveals the depositing of neural agrin as visualized in optical longitudinal sections through ectopic sites (Agrin). In this area, ectopic neural agrin induces aggregation of nAChRs (AChR) as visualized by *in situ* hybridization. Scale bar, 15 μm. Drawing **(a)** from Laufer R, Changeux JP (1989) Activity-dependent regulation of gene expression in muscle and neuronal cells. *Mol. Neurobiol.* **3**, 1–53, with permission. Part **(b)** from Ruegg MA, Bixby JL (1998) Agrin orchestrates synaptic differentiation at the vertebrate neuromuscular junction. *Trend. Neurosci.* **21**, 22–27, with permission.

of the neuromuscular activity since it is not affected by the injection *in ovo* of nicotinic antagonists such as curare.

- There is formation of aggregates of nicotinic receptors under the nerve terminal (3–5) and disappearance of extrajunctional receptors (5). Upon innervation, nAChR rapidly accumulates under the nerve endings. *In situ* hybridization experiments with a genomic coding probe (see Appendix 6.2) have shown that in innervated 15-day-old chick muscle, the nAChR α-subunit mRNAs accumulate under the nerve endings. More precisely, accumulation of the mRNAs increases around the subsynaptic (fundamental) nuclei and decreases around the sarcoplasmic nuclei. This can be interpreted as a differential expression of the nAChR α-subunit gene in the fundamental and sarcoplasmic nuclei. The presence of motor nerve and muscle activity are both crucial for the regulation of nAChR mRNA levels in the developing fiber.
- Distribution of the Golgi apparatus, studied by using a monoclonal antibody directed against it, shows a similar evolution. In cultured myotubes, the Golgi apparatus is associated with every nucleus. Conversely, in 15-day-old innervated chick muscle, the Golgi apparatus is now restricted to discrete, highly focused regions that appear to co-distribute with endplates (revealed by fluorescein isothiocyanate conjugated α-bungarotoxin, a labeled ligand of nAChR).
- There is stabilization of nicotinic receptors in the postsynaptic membrane (5).

These observations raise questions about the nature of the signaling pathways that underlie such reorganization. Is there activation of second messengers by anterograde signals from the nerve endings that would lead to positive regulation of the expression of the nicotinic receptor in the junctional regions and negative regulation in the extrajunctional regions? Are there retrograde signals too?

Aggregation of proteins at the nerve–muscle contact depends, in fact, on instructive signals that are released by the motor axon. More than 20 years ago, McMahan and colleagues identified the basal lamina as the carrier of the information necessary to induce pre- and post-synaptic specializations during neuromuscular regeneration. A protein, agrin, was purified from basal lamina extracts of the cholinergic synapse of the electric organ of *Torpedo californica* (see **Figure 8.1**). When added to cultured myotubes, soluble agrin induces the aggregation of acetylcholine receptors. This led McMahan to formulate the following hypothesis: 'Agrin is released from motor neurons, binds to a receptor on the muscle cell surface and induces postsynaptic specializations. Subsequent binding of agrin to synaptic basal lamina will then immobilize agrin.'

The fact that neural agrin is necessary and sufficient for postsynaptic differentiation is confirmed by the following experiment. Agrin is a protein of 225 kDa, consisting of domains found in other basal lamina proteins. The region of agrin necessary and sufficient to bind to the basal lamina (it binds in fact to laminins) maps to the amino-terminus end of the molecule. The most carboxy-terminus is necessary and sufficient for its nAChR-aggregating activity. Agrin mRNA undergoes alternative splicing at several sites, two of which modulate agrin's ability to induce nAChR clustering in cultured muscle cells. In innervated adult rat muscle, injection of expression constructs that encode full-length chick neural agrin is sufficient to induce postsynaptic specializations: after 4–6 weeks, staining with anti-chick agrin fluorescent antibodies reveals the deposit of neural agrin in basal lamina at the ectopic site of injection (**Figure 6.13b**). In this area, aggregation of nAChRs is observed with immunocytochemistry. Therefore, ectopic expression of recombinant agrin in adult muscle *in vivo* induces the formation of postsynaptic-like structures that closely resemble the muscle endplate.

Which molecule is the agrin receptor(s) and what are the second messengers activated by agrin and responsible for postsynaptic differentiation? Agrin initiates a signaling cascade which is still under study. Interestingly, voltage-dependent Na^+ channels are also concentrated at the synapse where they are restricted to the depth of postjunctional folds. This clustering pathway also involves agrin.

6.4 THE SYNAPSE BETWEEN THE VEGETATIVE POSTGANGLIONIC NEURON AND THE SMOOTH MUSCLE CELL

Smooth muscle cells are present in most of the visceral organs (digestive system, uterus, bladder, etc.) but also in the wall of blood vessels and around the hair follicles. They are innervated by postganglionic neurons of the autonomic nervous system (orthosympathetic neurons and parasympathetic neurons) (**Figure 6.14**).

6.4.1 The presynaptic element is a varicosity of the postganglionic axon

The axons of the postganglionic neurons are not myelinated. Before contacting the smooth muscles, the axons divide into numerous thin filaments 0.1–0.5 μm in diameter which travel alone or in fascicles over long distances along the smooth muscle cells (**Figures 6.14**

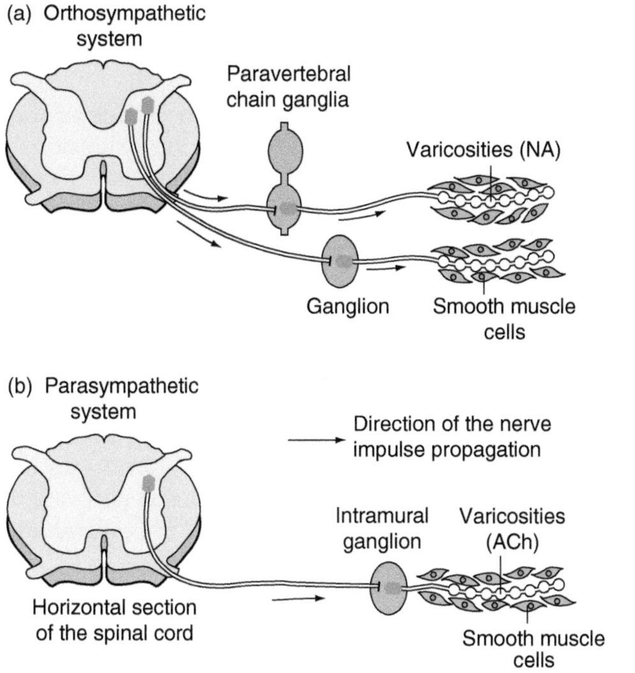

FIGURE 6.14 **The orthosympathetic (a) and parasympathetic systems (b).** The synapses between postganglionic neurons and smooth muscle fibers are shown. The axon terminals of postganglionic neurons are varicosities and the synaptic contacts are of the 'boutons en passant' type. ACh, acetylcholine; NA, noradrenaline; arrows show the direction of action potential propagation.

and **6.15a**). Each of these filaments has swellings or varicosities 0.5–2.0 μm in diameter spaced 3–5 μm apart. The varicosities are the presynaptic elements. The varicosities contain mitochondria and numerous synaptic vesicles whereas the intervaricose segments contain many elements of the cytoskeleton (**Figure 6.15b**). There are no electron-dense regions in the presynaptic membrane, which suggests the absence of a preferential zone for exocytosis (active zone) in these synapses.

The axonal varicosities contain a large number of small granular synaptic vesicles with an electron-dense central core (30–50 nm diameter), but also some large granular vesicles (60–120 nm diameter) and small agranular vesicles. The neurotransmitter of the orthosympathetic postganglionic neurons is noradrenaline (NA) (**Figure 6.14a**). It is stored in small and large granular vesicles. Noradrenaline is a catecholamine (as dopamine and adrenaline) synthesized from the amino acid tyrosine.

The noradrenaline receptors present in the postsynaptic membrane of smooth muscle cells are G-protein-coupled receptors. The inactivation of noradrenaline released from nerve terminals is, to a large extent, achieved by reuptake by the catecholaminergic neurons or the nearby glial cells. It is recycled into the synaptic vesicles or degraded by specific enzymes such as monoamine oxidase (MAO). Some of the catecholamines are degraded in the synaptic cleft by catechol-O-methyl-transferase (COMT).

Acetylcholine (ACh) is the neurotransmitter of the parasympathetic postganglionic neurons (**Figure 6.14b**). The varicosities of these axons contain mainly small agranular vesicles but also large granular vesicles. Acetylcholine is stored in small agranular vesicles. The acetylcholine receptors present in the postsynaptic membrane of smooth muscle cells are cholinergic muscarinic receptors (mAChR); they are protein-G-coupled receptors.

6.4.2 The width of the synaptic cleft is very variable

Where the synaptic cleft is narrowest, in the vas deferens or in the pupil, for example, it measures between 15 and 20 nm. However, in the wall of blood vessels, the closest contacts are spaced 50–100 nm apart.

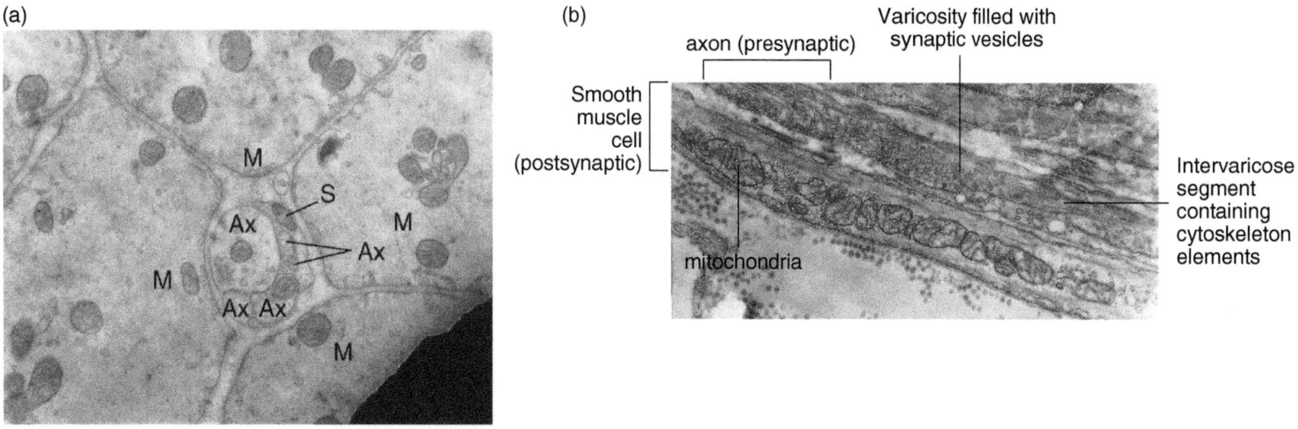

FIGURE 6.15 **A nerve–smooth muscle synapse. (a)** Microphotograph of smooth muscle cells of the intestine (M) and postganglionic axon fascicles (parasympathetic nervous system, Ax) which are half-covered by a Schwann cell (S), sectioned in the transverse plane and observed under the electron microscope. Note the width of the synaptic cleft. **(b)** Longitudinal section of a postganglionic axon showing a varicosity filled with vesicles and an intervaricose region. The postsynaptic smooth muscle cell contains numerous mitochondria. Microphotography by Jacques Taxi.

6.4.3 The autonomous postganglionic synapse is specialized to ensure a widespread effect of the neurotransmitter

The large width of the synaptic cleft results in a widespread effect of the neurotransmitter on the postsynaptic membrane compared with the neuromuscular junction or central synapses where secretion of the neurotransmitter is focused on a small postsynaptic region. Moreover, in some autonomous synapses, there is no distinguishable specialization of the presynaptic membrane, which suggests that the vesicles have no preferential site for exocytosis. Formation of a dense plexus by the postganglionic axons also contributes to the extended diffusion of presynaptic messages. Therefore, activation of a postganglionic neuron results in activation of numerous postsynaptic cells. Finally, the presence of numerous gap junctions, which connect smooth muscle cells, permits spread of the synaptic response to neighboring muscle cells, even to those not innervated.

6.5 EXAMPLE OF A NEUROGLANDULAR SYNAPSE

This section considers the synapse between an orthosympathetic preganglionic neuron and the chromaffin cell of the adrenal medulla (**Figure 6.16**).

The adrenal medulla is the central part of the adrenal gland, the endocrine gland located above each kidney. It is formed by secretory cells, which are called chromaffin cells since they are colored by chromium salts. The adrenal medulla is innervated by orthosympathetic preganglionic neurons which have axons that form the splanchnic nerve. When this nerve is stimulated, the chromaffin cells secrete essentially adrenaline but also noradrenaline and enkephalins (endogenous opioid peptides). These hormones are then transported via the blood to numerous target tissues and mainly the heart and blood vessels.

The *presynaptic element* of this synapse is the axon terminal of the splanchnic orthosympathetic preganglionic neurons. Their cell bodies are located in the intermediate horn of the spinal cord. Most of the axons are non-myelinated and are surrounded by the extensions of the non-myelinating Schwann cells. This glial sheath is present until the axon collaterals penetrate the junctional space. The axon terminals are mainly terminal boutons with a diameter ranging between 1 and 3 μm. They contain clear vesicles (10–60 nm diameter) as well as some dense-core and granular vesicles (25–115 nm diameter). Acetylcholine is the neurotransmitter of this synapse. The terminal boutons form a large variety of synaptic contacts with the chromaffin cells. They are characterized by a narrow synaptic cleft (15–20 nm) and the presence of electron-dense pre- and postsynaptic zones similar to those observed at the level of the central interneuronal synapses.

In the *postsynaptic region*, the cytoplasm of the chromaffin cells is free of chromaffin granules, organelles that store hormones in the adrenal medulla. The acetylcholine receptors present in the postsynaptic membrane are cholinergic nicotinic receptor channels. In the rest of the chromaffin cell cytoplasm there are numerous chromaffin granules. These granules are colored by chromium salts which react with adrenaline to form a yellow–brown precipitate.

6.6 SUMMARY

Chemical synapses are connections between two neurons or between a neuron and a non-neuronal cell (muscle cell, glandular cell, sensory cell). The synaptic complex is the non-reducible basic unit of each chemical synapse as it represents the minimal requirement for an efficient chemical synaptic transmission. It includes three elements: the presynaptic element (such as an axon terminal), a synaptic cleft, and a postsynaptic element (such as a dendritic spine).

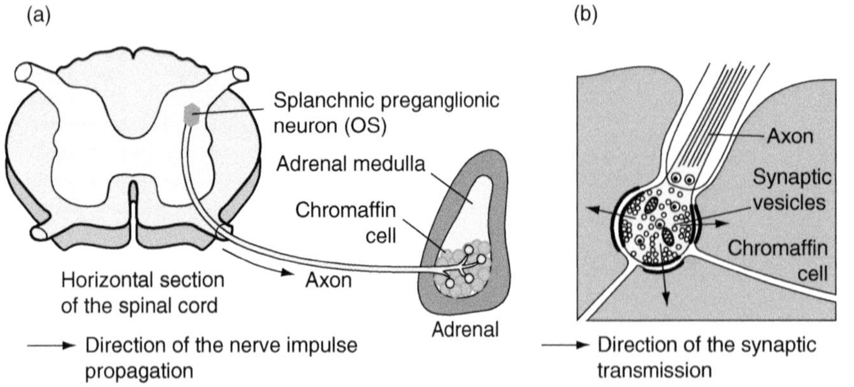

FIGURE 6.16 **Synapses between an orthosympathetic preganglionic neuron and chromaffin cells in the adrenal medulla. (a)** The cell bodies of preganglionic neurons are localized in the spinal cord and their axons form the orthosympathetic splanchnic nerve. These neurons innervate the chromaffin cells. **(b)** Postganglionic axon terminal forming numerous contacts with chromaffin cells. Drawing **(b)** from Coupland RE (1965) Electron microscopic observations on the structure of the rat adrenal medulla: II. Normal innervation. *J. Anat. (Lond.)* **99**, 255–272, with permission.

The *presynaptic element* is characterized by (i) an active zone (i.e. a specialized presynaptic membrane area) where the density of Ca^{2+} channels is high and where occurs the fusion of synaptic vesicles (exocytosis), and (ii) a nearby cytoplasmic region where the synaptic vesicles are found close to the presynaptic membrane, with a particular cytoskeletal arrangement. The regulated release of neurotransmitter occurs at active zones. However, the active zone is not a characteristic of all synapses. Monoaminergic synapses and peptidergic synapses do not always have discernible active zones. The presence of an active zone would be a clue to focal neurotransmitter release.

The *postsynaptic element* is characterized, in interneuronal synapses, by a sub-membranous electron-dense zone (postsynaptic density), which corresponds to the region where the postsynaptic receptors are anchored. There is a strict complementarity between the neurotransmitter released by the presynaptic element and the postsynaptic receptors inserted in the postsynaptic membrane. This includes specific targeting, anchoring and clustering of postsynaptic receptors.

In most cases, the ultrastructure of chemical synapses is asymmetric, synaptic vesicles that contain the neurotransmitter(s) being present only in the presynaptic element. Synaptic transmission is unidirectional – it occurs from the presynaptic element to the postsynaptic one.

Further reading

Betz, H., 1999. Structure and functions of inhibitory and excitatory glycine receptors. Ann. NY Acad. Sci. 868, 667.

Brenman, J.E., Topinka, J.R., Cooper, E.C., et al., 1998. Localization of postsynaptic density-93 to dendritic microtubules and interaction with microtubule-associated protein 1A. J. Neurosci. 18, 8805–8813.

Cohen, I., Rimer, M., Lomo, T., McMahan, U.J., 1997. Agrin-induced postsynaptic-like apparatus in skeletal muscle fibers in vivo. Mol. Cell. Neurosci. 9, 237–253.

Couteaux, R., 1998. Early days in the research to localize skeletal muscle actylcholinesterases. J. Physiol. (Paris) 92, 59–62.

Fujimoto, K., 1995. Freeze-fracture replica electron microscopy combined with SDS digestion for cytochemical labeling of integral membrane proteins. Application to the immunogold labeling of intercellular junctional complexes. J. Cell Sci. 108, 3443–3449.

Ichtchenko, K., Hata, Y., Nguyen, T., et al., 1995. Neuroligin 1: a splice site-specific ligand for beta-neurexins. Cell 81, 435–443.

Irie, M., Hata, Y., Takeuchi, M., et al., 1997. Binding of neuroligins to PSD-95. Science 277, 1511–1515.

Massoulie, J., Anselmet, A., Bon, S., et al., 1998. Acetylcholinesterase: C-terminal domains, molecular forms and functional localization. J. Physiol. (Paris) 92, 183–190.

Niethammer, M., Sheng, M., 1998. Identification of ion channel-associated proteins using the yeast two-hybrid system. Meth. Enzymol. 293, 104–122.

Nitkin, R.M., Smith, M.A., Magill, C., et al., 1987. Identification of agrin, a synaptic organizing protein from Torpedo electric organ. J. Cell Biol. 105, 2471–2478.

Sanes, J.R., 1998. Agrin receptors at the skeletal neuromuscular junction. Ann. NY Acad. Sci. 841, 1–13.

Severs, N.J., 2007. Freeze-fracture electron microscopy. Nat. Protoc. 2, 547–576.

Ushkaryov, Y.A., Petrenko, A.G., Geppert, M., Sudhof, T.C., 1992. Neurexins: synaptic cell surface proteins related to the alpha latrotoxin receptor and laminin. Science 257, 50–56.

Wang, Z.Z., Mathias, A., Gautam, M., Hall, Z.W., 1999. Metabolic stabilization of muscle nicotinic acetylcholine receptor by rapsyn. J. Neurosci. 19, 1998–2007.

APPENDIX 6.1 NEUROTRANSMITTERS, AGONISTS AND ANTAGONISTS

Neurotransmitters are molecules of varied nature: quaternary amines, amino acids, catecholamines or peptides, which are released by neurons at chemical synapses. They transmit a message from a neuron to another neuron, or to an effector cell, or a message from a sensory cell to a neuron.

A6.1.1 Criteria to be satisfied before a molecule can be identified as a neurotransmitter

Identification of a substance as a neurotransmitter requires the experimental proof of several criteria. If these are not satisfied, the term *putative* neurotransmitter is used. The criteria are:

- The putative neurotransmitter must be present in the presynaptic element.
- The precursors and enzymes necessary for *synthesis* of the putative neurotransmitter must be present in the presynaptic neuron.
- The putative neurotransmitter must be released in response to activation of the presynaptic neuron and in a quantity sufficient to produce a postsynaptic response. This release should be dependent on Ca^{2+} ions.
- There should be *binding to specific postsynaptic receptors*: (i) specific receptors of the neurotransmitter are present in the postsynaptic membrane; (ii) application of the substance at the level of the postsynaptic element reproduces the response obtained by stimulation of the presynaptic neuron; and (iii) drugs which specifically block or potentiate the postsynaptic response have the same effects on the response induced by the application of the putative neurotransmitter.
- The elements of the synaptic nervous tissue (pre- or postsynaptic elements, glial cells, basal membrane) must possess one or several mechanisms for *inactivation* of the putative neurotransmitter.

Currently, few molecules have satisfied all these criteria to be firmly identified as a neurotransmitter at a particular synapse. In most cases, there is no more than fragmentary evidence owing to technical limitations.

A6.1.2 Types of neurotransmitter

Acetylcholine: a quaternary amine

In the peripheral nervous system, acetylcholine is the neurotransmitter of all the synapses between motoneurons and striated muscle cells, of all the synapses between preganglionic and postganglionic neurons of the para- and orthosympathetic systems, and of all the synapses between parasympathetic postganglionic neurons and effector cells (see **Figures 6.12, 6.14** and **6.16**). It is also a neurotransmitter in the central nervous system. Choline acetyltransferase (CAT), the enzyme required for acetylcholine synthesis, is a specific marker of cholinergic neurons. Antibodies against vesicular acetylcholine transporter (VAChT) an ≈70 kDa protein that actively transports ACh from cytoplasm into synaptic vesicles allows labeling of cholinergic vesicles and therefore cholinergic terminals. Using immunocytochemical or *in situ* hybridization techniques (see Appendix 6.2), one can visualize cholinergic neuronal pathways by labeling choline acetyltransferase or its mRNA. At the same time, the acetylcholine receptors can be localized.

Amino acids: glutamate, GABA (γ-aminobutyric acid) and glycine

In contrast to other neurotransmitters, glutamate also plays an important role in cellular metabolism (in intermediary metabolism, in the synthesis of proteins and as a precursor of GABA). It is, therefore, present in all neurons and its identification as a neurotransmitter poses several problems.

In fact, evidence for the enzymes of its synthesis or degradation cannot represent a valid criterion for the identification of glutamatergic neurons. These difficulties can be overcome as glutamate is present in much higher concentrations in neurons where it plays a neurotransmitter role. In addition, these neurons have the property of selectively recapturing glutamate with the help of a high-affinity transport system, and localization of this transport system can be used to identify glutamate neurons. Glutamic acid decarboxylase, the enzyme required for GABA synthesis, is a good marker of GABAergic neurons. In the CNS, two isoforms of glutamic acid decarboxylase (GAD67 and GAD65), each encoded by a different gene and highly conserved among vertebrates, are co-expressed in most GABA neurons. Also, antibodies against the vesicular GABA transporter (VGAT) that actively transports GABA from cytoplasm into synaptic vesicles allows labeling of GABAergic vesicles and therefore GABAergic terminals. Thus, GABAergic neurons (cell bodies, axon terminals and fibers) can be visualized by the localization of these two forms of GADs or of VGAT or their mRNAs by immunocytochemistry and *in situ* hybridization, respectively, or by immunohistochemical detection of GABA itself.

Monoamines

These are classified as catecholamines (adrenaline, noradrenaline, dopamine), indolamine (serotonin) and imidazole (histamine). Adrenaline, noradrenaline and dopamine are all catecholamines. Their structure has a common part, the catechol nucleus (a benzene ring with two adjacent substituted hydroxyl groups). They are synthesized from a common precursor, tyrosine. Serotonin or 5-hydroxytryptamine (5HT) is synthesized from tryptophan, a neutral amino acid. Also, antibodies against the vesicular monoamine transporter (VMAT) that actively transports monoamines from cytoplasm into synaptic vesicles allows labeling of monoaminergic vesicles and therefore monoaminergic terminals.

Neuropeptides

Peptides that are present in neurons with a supposed role in synaptic transmission are called neuropeptides. They are, for example, opioid peptides (enkephalins, dynorphin, β-endorphin) or they are peptides that have been first identified in the gastrointestinal tract (substance P, cholecystokinin, vasoactive intestinal peptide (VIP)) or in the hypothalamo-hypophyseal complex (luteinizing hormone releasing hormone (LHRH), somatostatin, adrenocorticotropic hormone (ACTH), vasopressin) or they are also circulating hormones (corticotropin or ACTH, insulin), before they were suggested as neurotransmitters in the central nervous system. It seems reasonable that other neuropeptides await discovery. These peptides were proposed to be neurotransmitters on the basis of their presence and synthesis in the neurons as well as by their release from axonal terminals by a Ca^{2+}-dependent mechanism. For some peptides other criteria have also been demonstrated.

The differences in the chemical nature of neurotransmitters have a fundamental consequence. Non-peptidic neurotransmitters are synthesized in axonal terminals: a precursor (or precursors) synthesized by the neuron or taken up from the extracellular medium is transformed into a neurotransmitter via an enzymatic reaction in axon terminals. The synthesis enzyme(s) is (are) synthesized in the cell body and transported to axon terminals via axonal transport. The newly synthesized neurotransmitter is then actively transported inside the synaptic vesicles (**Figure A6.1a**). Peptidic neurotransmitters are synthesized in the cell body since axon terminals, being deprived of the organelles responsible for protein synthesis, cannot themselves synthesize neuropeptides (**Figure A6.1b**). These are synthesized in cell bodies in the form of larger peptides called *precursors*. These precursors are then transported to the axon terminals by fast axonal transport. Cleavage of the precursors into neuroactive peptides is carried out by vesicular peptidases during anterograde axonal transport. Since these precursors have no biological activity, regulation of peptidase

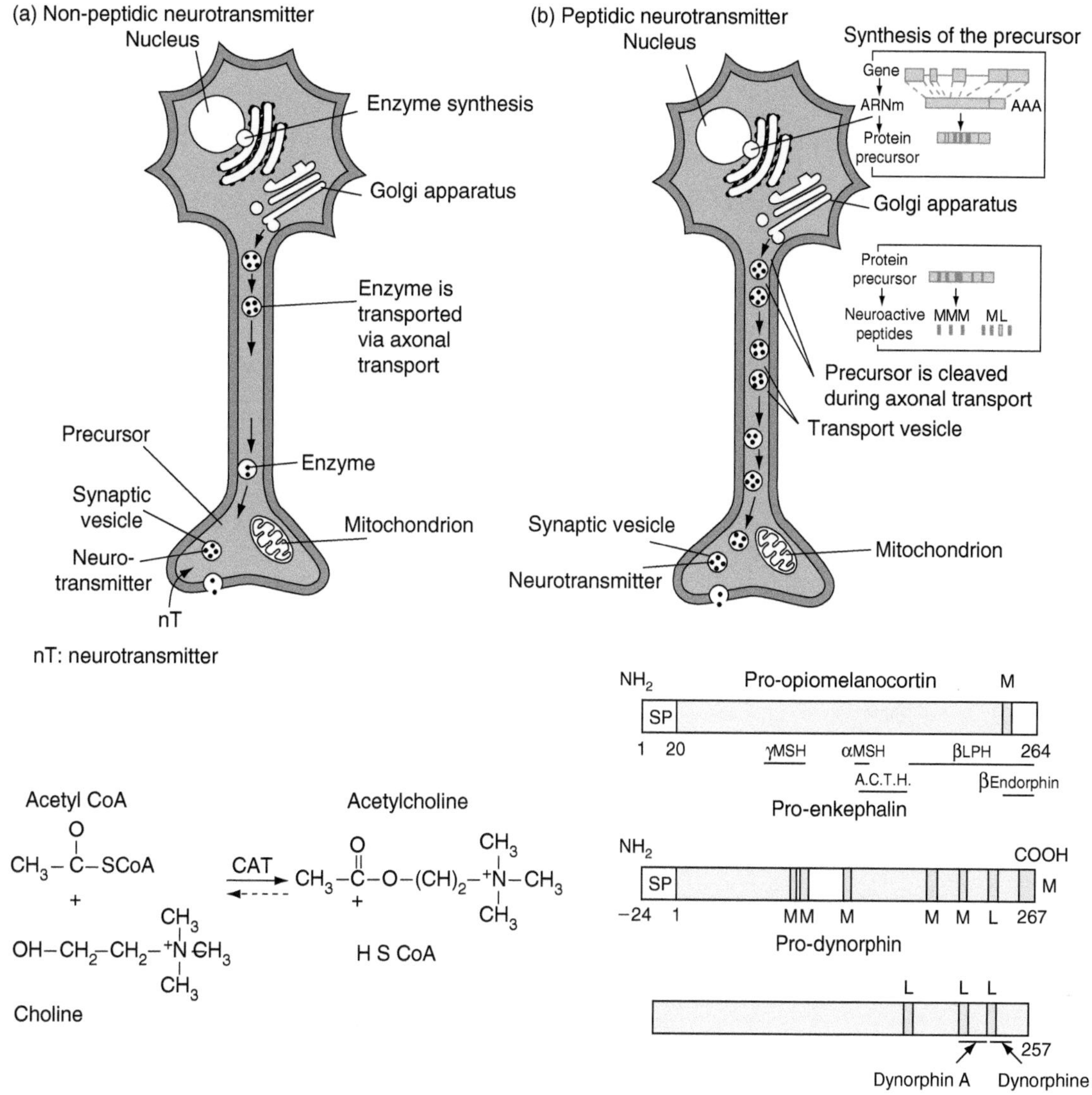

FIGURE A6.1 **Synthesis of non-peptidic and peptidic neurotransmitters. (a)** Non-peptidic type (example, acetylcholine). **(b)** Peptidic type (example, opioid peptides). Bottom left: The synthesis reaction of acetylcholine from acetylcoenzyme A. Bottom right: Precursors of endomorphines are pro-opiomelanocortin, pro-enkephalin A and pro-dynorphin. The peptides they contain are shown. The numbers indicate the position of the peptides along the protein. CAT, choline acetyltransferase; HSCoA, coenzyme A; MSH, melanocyte stimulating hormone; ACTH, corticotropin; enk, enkephalin; L, leu-enkephalin; M, met-enkephalin; LPH, lipotropin.

activity seems to be an important factor in the regulation of the synthesis of peptidic neurotransmitters.

Concerning their mode of inactivation, most of the neurotransmitters are taken up by axon terminals or glial cells via specific transporters. The major exception is acetylcholine, which is degraded in the synaptic cleft by acetylcholinesterases. Since enzymatic degradation is more rapid than a transporter reaction, acetylcholine is much more rapidly inactivated than the other neurotransmitters.

A6.1.3 Agonists and antagonists of a receptor

An *agonist* is a molecule (drug, neurotransmitter, hormone) that binds to a specific receptor, activates the receptor and thus elicits a physiological response:

$$A + R \underset{k_{-1}}{\overset{k_{+1}}{\rightleftharpoons}} AR \rightleftharpoons AR^* - - - \rightarrow \text{physiological response}$$

where A is the agonist, R is the free receptor, AR is the agonist–receptor complex and AR^* is the activated state of the receptor bound to the agonist. k_{+1} and k_{-1} measure the rate at which association and dissociation occur. An agonist (and an antagonist) is defined in relation to a receptor and not to a neurotransmitter. For example, an agonist of nicotinic acetylcholine receptors such as nicotine is not an agonist of muscarinic acetylcholine receptors though both receptors are activated by acetylcholine.

An *antagonist* is a molecule that prevents the effect of the agonist. A *competitive antagonist* (C) is a receptor antagonist that acts by binding reversibly to an agonist receptor site (R). It does not activate the receptor and thus does not elicit a physiological response:

$$B + R \underset{k_{-1B}}{\overset{k_{+1B}}{\rightleftharpoons}} BR - - - \rightarrow \text{no physiological response}$$

The effect of a *reversible competitive antagonist* can be reversed when the agonist concentration is increased since the agonist (A) and the reversible competitive antagonist (B) compete for the same receptor site (R). Competition means that the receptor can bind only one molecule (A or B) at a time. An *irreversible competitive antagonist* is a receptor antagonist that dissociates from the receptor slowly or not at all. For this reason, its effect cannot be reversed when the agonist concentration is increased.

APPENDIX 6.2 IDENTIFICATION AND LOCALIZATION OF NEUROTRANSMITTERS AND THEIR RECEPTORS

A6.2.1 Immunohis(cy)tochemistry

Principle and definitions

Immunohistochemical techniques are based on the high specificity of the antigen–antibody reaction. They consist of the detection of an antigen present in histological or cellular structures by application on tissues or cells of its specific antibody or antiserum. The complex antigen–antibody formed is then visualized under light or electron microscopy by means of various methods of detection described below.

The antigen is an endogenous molecule able to induce the formation of antibodies when injected into a foreign body. The antigen will be recognized specifically by these antibodies. Antigens are endogenous proteins such as synthesizing enzymes or receptors for neurotransmitters. Small neuro-active peptides or amino acid neurotransmitters can become antigenic after being conjugated to a carrier protein or a polysaccharide.

The antibodies are immunoglobulins of type G (IgG) or type M (IgM), Y-shaped proteins that display two binding sites for the antigen. These two binding sites recognize a very short amino acid sequence of the antigen. This sequence is called an antigenic determinant. The term 'hapten' is used to describe an amino acid sequence that binds specifically to the binding site of the antibody but cannot induce on its own an immune response (example: an amino acid neurotransmitter like GABA).

Two families of antibodies are commonly used in immunohistochemistry: polyclonal antibodies (or antiserum) and monoclonal antibodies. A *monoclonal antibody* is the product of a single B lymphocyte clone (**Figure A6.2a**). It is made of a population of identical

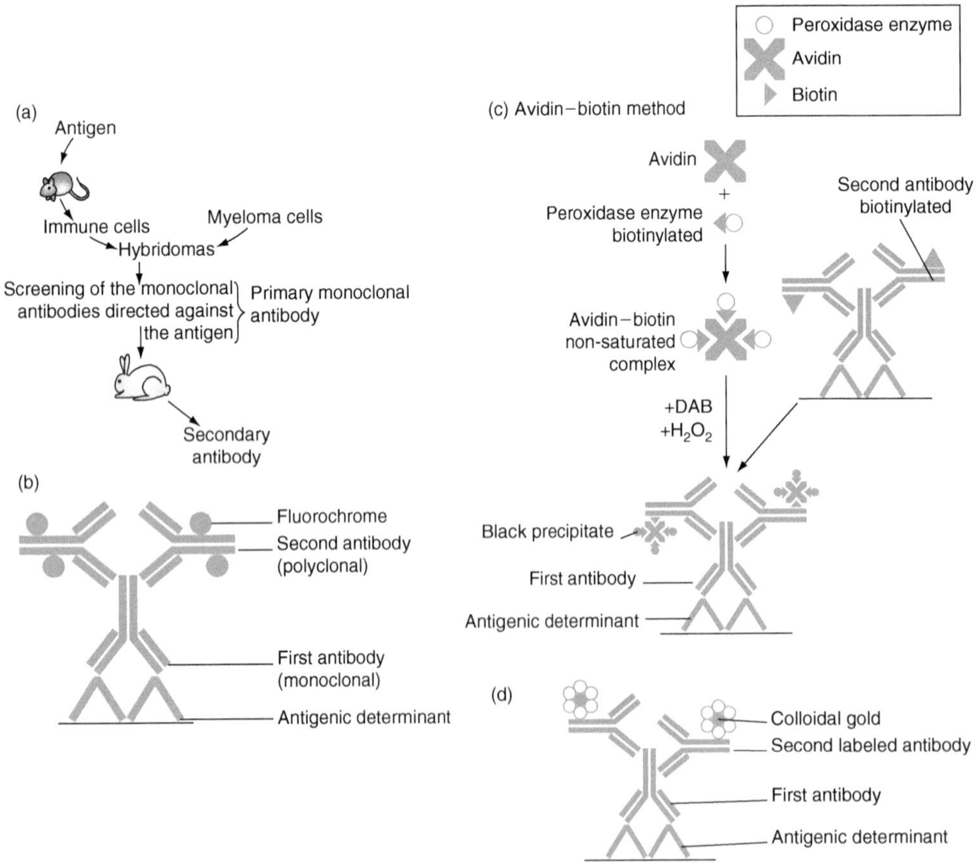

FIGURE A6.2 **Synthesis and labeling of secondary antibodies raised against monoclonal antibodies specific to a neuronal antigen.**

antibody molecules, each of them recognizing the same antigenic determinant (or hapten) on the antigen. A *polyclonal antibody* consists of a heterogeneous family of antibodies that recognize different antigenic determinants on the same antigen. They are generated in a host animal (usually a rabbit) after its immunization by injection of the antigen. The antibody (polyclonal or monoclonal) used to recognize an antigen into the tissues is called the *primary antibody*.

Antibodies are themselves antigenic, so it is possible to produce antibodies that will recognize antigenic determinants on various regions of an antibody. Antibodies directed against 'primary antibodies' are called *secondary antibodies*. They are anti-IgG or anti-IgM antibodies. These secondary antibodies can be labeled and are then used to detect the antigen/primary-antibody complex.

Among the different antigenic determinants of an antibody, those that are associated with the antigen-binding site are called *idiotypes*. Secondary antibodies directed against the specific antigen-binding sites (idiotypes) of a primary antibody are called *anti-idiotype antibodies*. These anti-idiotype antibodies are also useful tools for the localization of receptors.

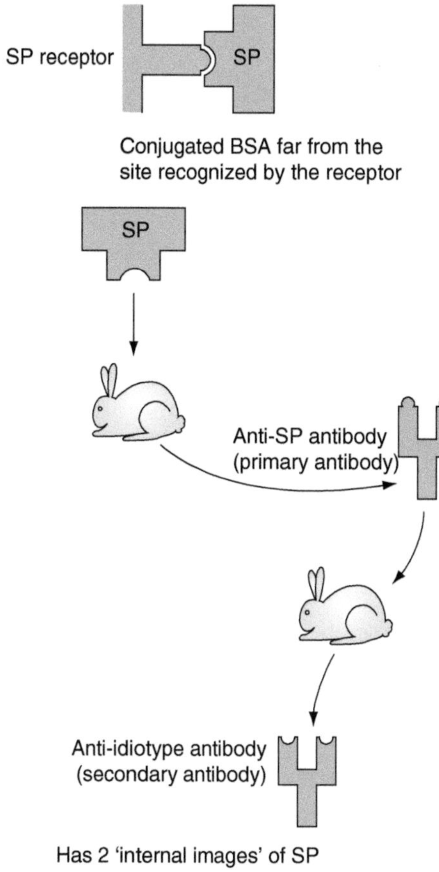

FIGURE A6.3 **Synthesis of anti-idiotype antibodies for substance P (SP).**

Applications

Localization of neurons synthesizing a specific neurotransmitter

If we want to localize, for example, the cholinergic neurons in a section of brain tissue, the approach is to reveal the neurons that contain the synthesizing enzyme for acetylcholine, choline acetyltransferase (ChAT). Sections of brain tissue are incubated with a primary antibody directed against ChAT. The antibodies will bind specifically to the antigen into sections and a stable antigen–antibody complex will be formed only in the neurons that contain ChAT. After washing the sections to remove the antibodies that did not link with the antigen, the antigen–antibody complex is detected according to one of the methods described below (detection).

Localization of receptors of a neurotransmitter

Primary antibodies directed against a purified receptor can be used to localize a specific type of receptor on sections of brain tissue, similarly to the localization of synthesis enzymes of neurons.

A second method for localization of receptors uses the anti-idiotype antibodies (**Figure A6.3**). Anti-idiotype antibodies are generated against primary antibodies specific to the ligand of the receptor. For example, for the localization of substance P receptors, anti-idiotype antibodies are generated against the antibody specific to substance P. Anti-idiotype antibodies are used since their antigen-binding sites have structural similarities with the ligand itself, of which they constitute a sort of 'molecular image'. This property allows them to bind the biological receptor. This method displays the advantage to enable receptor antibodies to be obtained without a pre-purification of the receptor. The receptor can therefore be localized and its stereo-specificity studied.

Detection of the antigen–antibody complex

Most of the detection methods use markers (labels). The marker is bound to the secondary antibody to obtain *labeled secondary antibodies* which, by reacting with the *antigen–primary antibody complex*, will allow visualization of the antigen under light or electron microscopy. The markers used for light microscopy are (i) fluorochromes that can be detected with a microscope equipped with an epifluorescence system, or (ii) enzymes (such as peroxidase) that will induce a chromogen reaction in the presence of its substrate (such as the diaminobenzide (DAB)). For electron microscopy, electron-dense compounds such as colloidal gold particles are used to label secondary antibodies (**Figures A6.2d** and **6.6d**).

Why use labeling of secondary antibodies and not primary ones?

Labeling of antibodies (markers are directly conjugated to the primary antibodies) displays many disadvantages. The labeling of the antibody reduces significantly its capacity to recognize the antigen, thus its specificity. Moreover, there is only one molecule of marker for a single antigen–antibody complex. Therefore this technique is poorly sensitive and cannot be used to localize antigens that are present in small quantities in neurons. In addition, for each antigen studied, it is necessary to label the corresponding antibody. For these reasons, indirect labeling methods have been developed to increase the sensitivity of detection by amplification of the labeling. All of them use as a first step an unlabeled primary antibody that binds to its antigen into the section. In the following step, the primary antibody is recognized by a secondary antibody which is raised in another species. The secondary antibody is a serum anti-heterologous IgG (or IgM) that binds to many antigenic sites on the primary antibody.

Three main methods commonly used

In the first method (**Figure A6.2b,d**), the secondary antibody is labeled with one of the markers previously described. Since many labeled secondary antibodies bind to a single primary antibody molecule, this technique allows an increase in the labeling and consequently better visualization of the antigen.

In the peroxidase–anti-peroxidase (PAP) detection method, the molecules of secondary antibody are unlabeled and applied in excess, allowing one of their antigen binding sites to be left free. A third step consists in incubating the tissue in a solution containing the peroxidase–anti-peroxidase complex formed by several molecules of peroxidase and antibodies directed against those molecules. The antibodies of the PAP complex are raised in the same species as the primary antibody and thus are recognized by the free antigen-binding sites of the secondary antibody.

The avidin (or streptavidin)–biotin method (**Figure A6.2c**) uses the very high affinity of a little hydrosoluble vitamin, biotin, for the protein avidin. The tissue is incubated in the presence of the secondary antibody previously conjugated to biotin molecules (biotinylated secondary antibodies), then in the presence of the preformed complex consisting of molecules of biotin and avidin covalently bound with markers such as peroxidase or fluorochromes.

Both PAP and avidin–biotin methods allow a strong amplification of the labeling and are the most sensitive immunohis(cy)tochemical techniques.

A6.2.2 *In situ* hybridization

Principle

The aim of these techniques is the detection of a specific nucleic acid sequence in cells on histological sections or in cultured cells. *In situ* hybridization techniques are based on the capacity of all nucleic or ribonucleic acids sequences (ADN or ARN) to bind to a complementary sequence. The sequence of nucleic acids to recognize may correspond to chromosomal DNA, called 'hybridization on chromosomes'. But, in general, the term '*in situ* hybridization' relates to the detection of messenger RNA (mRNA). This detection or recognition is made possible by the use of a probe that corresponds to a sequence of nucleic acids complementary to the DNA or RNA that is to be detected. The probe is labeled with a marker. When they are in the presence of each other, the specific labeled probe and the endogenous RNA (or DNA) recognized by the probe hybridize (because of the complementary sequences). The hybrids thus formed are detected by means of the marker linked to the probe.

Application

The *in situ* hybridization technique allows visualization of gene transcripts and, therefore, localization of the potential site of synthesis for the protein or the peptide coded by this gene. In the case of nerve cells, this technique can localize neurons that express a gene coding for a neurotransmitter (if it is a peptide), for a synthesizing or degradative enzyme of a specific neurotransmitter, or for a receptor for a neurotransmitter. For example, it has been used to localize neurons that contain the mRNA coding for the precursors of the enkephalins. Moreover, the role of various factors that regulate the expression of these peptides can be studied because this technique can be used to analyze variations in the level of transcripted mRNA in relation to the activity of the neuron, the presence of hormonal factors, etc.

Probes

The most commonly used types are double-stranded DNA, single-stranded DNA, single-stranded RNA and oligodeoxyribonucleotides. The latter three, which include only complementary strands (antisense) to the targeted sequence (cellular mRNA), provide the highest sensitivity. The labeling of the probe is performed in general during probe synthesis by incorporation of markers into the probe.

Double-stranded DNA probes are cDNAs (complementary DNA to cellular mRNA). They are obtained by reverse transcription of cellular RNAs by means of a reverse transcriptase, an enzyme which is able to make complementary single-stranded DNA chains from RNA templates. This is followed by second-strand synthesis using a DNA polymerase. cDNAs are inserted into plasmid vectors, this cDNA library is then screened to identify and isolate the cDNA of interest. This cDNA is then amplified in bacteria or by the polymerase chain reaction (PCR) using oligoprimers on opposite strands. The double-stranded DNA probe can be labeled by different techniques: the nick-translation method consists of inducing cuts in the double-strand DNA with a DNAse and repairing these cuts by incorporation of labeled and unlabeled desoxynucleotides in the presence of DNA polymerase; the random primed method uses, after denaturation of the two strands of cDNA, the ability of the fragment of DNA polymerase to copy single-stranded DNA templates primed with random hexa-nucleotide mixture. Finally, the probes can be labeled during the PCR reaction in the presence of labeled nucleotides. Random priming and PCR give the highest efficiency of labeling.

Double-stranded DNA probes need to be denatured before use for hybridization. These probes are less sensitive than the single-stranded type because many of the two strands can reappariate during the hybridization reaction instead of hybridizing with the target.

Single-stranded DNA probes are preferentially obtained by PCR-based methods using a specific primer from the complementary strand of the RNA transcript and a mixture of labeled and unlabeled desoxynucleotides. Probes which do not reappariate with themselves are thus more sensitive and produce less background noise than double-stranded DNA probes.

Single-stranded RNA probes are produced by *in vitro* transcription of specific cDNAs sub-cloned into a transcription vector (plasmid containing the appropriate polymerase initiation site), by means of an RNAse polymerase. Before this transcription step, the plasmid is linearized with a restriction enzyme to avoid the transcription of plasmid sequences that will cause high backgrounds. The labeling is performed during the transcription by incorporation of labeled nucleotides. The labeled transcript is in general hydrolyzed to obtain probes of approximately 150–200 nucleotides in length.

Among the different types of probe, RNA probes are the most sensitive. Since RNA–RNA hybrids are more stable than DNA–RNA ones, strong specific staining with low background can be achieved with RNA probes by using post-hybridization treatment with RNAse and high-temperature washes.

Oligonucleotides are obtained through automated chemical synthesis. They are small (typically 20–30 bases in length) single-stranded DNA probes. Labeling of the probes is performed during synthesis or by adding a tail of labeled nucleotides. Their small size gives them good access to the targeted nucleic acid sequence but limits their sensitivity. However, they are useful when target abundance is high, when gene-specific probes cannot be obtained otherwise, or when only protein sequence information is available.

Markers

Two major families of markers are used for probe labeling: radioactive labels which are detected by autoradiography, and non-radioactive labels which are detected by immunocytochemistry. For many years, radioisotopes have been the only markers available to label nucleotides. Among the different radioisotopes used to label probes, ^{35}S is the most commonly used for radioactive *in situ* hybridization. ^{35}S-labeled probes give a resolution of about one cell diameter and relatively rapid results (the time exposure for detection of hybrids is about one week). Despite their high sensitivity, radiolabeled probes have disadvantages such as safety measures required during experimental procedures, limited utilization time, and limited spatial resolution due to scattering of emitted radiation.

More recently, non-radioactive *in situ* hybridization techniques have been introduced with the development of hapten-labeled nucleotides that can be well incorporated during probe synthesis. Biotin- or dioxigenin-labeled probes are the most commonly used for cellular mRNA detection. Fluorescent labeling is successfully used for chromosomal *in situ* hybridization. Non-radioactive labeled probes are stable, give rapid results and display high levels of cellular resolution comparable to that obtained with immunocytochemistry. Furthermore, they open up new opportunities with the possibility of using different labels for simultaneous detection of different sequences in the same tissue.

Detection of the hybrids

For radioactive probes, a low-resolution signal can be obtained by placing the tissue or cells mounted on slides in contact with X-ray film for overnight exposure. This step allows one to control the efficiency of the reaction. If satisfactory, a greater resolution is obtained by dipping the slides in a liquid photographic emulsion, exposed for one or several weeks and developed. In general, a nuclear stain of cells is performed (e.g. with Toluidine Blue) before observation under a microscope with brightfield or darkfield illumination.

Detection of hybrids labeled with non-radioactive probes is performed by immunohistochemistry using specific antibody conjugated with an enzyme such as peroxidase or alkaline phosphatase that will give a color precipitate in the presence of their substrates or by immunofluorescence using antibody conjugated with fluorochromes.

A6.2.3 Freeze fracture electron microscopy

Freeze fracture electron microscopy is unique among electron microscope techniques in revealing the structural organization of the entire membrane components. This technique consists in physically fracturing (breaking apart) blocks of frozen tissue in order to reveal membrane structural details exposed by the fracture plane. These membrane components are visualized by deposition onto their surface of a thin platinum–carbon layer, creating a replica.

Such replication by vapor deposition of platinum and carbon (Pt/C) reveals transmembrane proteins as intramembrane particles. It is necessary to remove all biological remnants from the replica to allow electron beam penetration and image formation in the transmission electron microscope. Gentle washing with SDS detergent removes most biological material, leaving only those macromolecules that are coated with Pt/C strongly adsorbed to the Pt/C film. These can be then immunogold labeled (see **Figure 6.6e**). With primary antibodies now made in many different species, and gold labels made in many different but uniform sizes, freeze fracture replicas now allow simultaneous visualization, identification and mapping of multiple proteins in complex CNS tissues.

CHAPTER

7

Neurotransmitter release

Constance Hammond, Oussama El Far, Michael Seagar

OUTLINE

The neuron is a secretory cell. The secretory product, the neurotransmitter, is released at the level of chemical synapses (see Chapter 6 and Appendix 6.1). Neurotransmitters achieve the transmission of information at the level of chemical synapses between neurons, neurons and muscle cells, neurons and glandular cells, and sensory receptors and neurons.

Neurotransmitters synthesized by the neuron are stored in the presynaptic element, inside the synaptic vesicles. In the absence of presynaptic activity, the probability of a neurotransmitter being released in the synaptic cleft is very low. This probability increases strongly when the presynaptic element is depolarized by an action potential. The vesicle hypothesis of neurotransmitter release, first formulated by Del Castillo and Katz (1954), is the generally accepted theory of neurotransmitter release (see Appendix 7.1). It states that the neurotransmitter molecules released in the synaptic cleft are those stored in synaptic vesicles (see **Figure 6.4a**). Many recent studies have confirmed the existence of vesicular release, such as data obtained with combined capacitance measurements and amperometry or optical analysis of labeled synaptic vesicles.

Many presynaptic elements contain large dense-core vesicles as well as small synaptic vesicles (see Section 6.1.1). Demonstration at the frog neuromuscular junction that exocytosis of small and large dense-core vesicles can be dissociated pharmacologically strongly suggests the existence of differences in the mechanisms that regulate exocytosis of the two types of secretory vesicles. However, the two systems share at least some common mechanisms for final fusion since both are sensitive to botulinum toxins. This chapter focuses on Ca^{2+}-regulated release of neurotransmitter from small synaptic vesicles.

The work of B. Katz and his collaborators led also to the formulation of the hypothesis that neurotransmitter release from presynaptic vesicles is triggered by elevations of the intracellular Ca^{2+} concentration ($[Ca^{2+}]_i$). Since then, this local rise of $[Ca^{2+}]_i$ has been clearly established as one of the prerequisites for neurotransmitter release. The molecular mechanisms responsible for the coupling between Ca^{2+} ion influx and exocytosis are being elucidated. This includes the identification of the proteins involved in exocytosis, the steps regulating exocytosis and their order of appearance in the

Cellular and Molecular Neurophysiology. DOI: 10.1016/B978-0-12-397032-9.00007-8

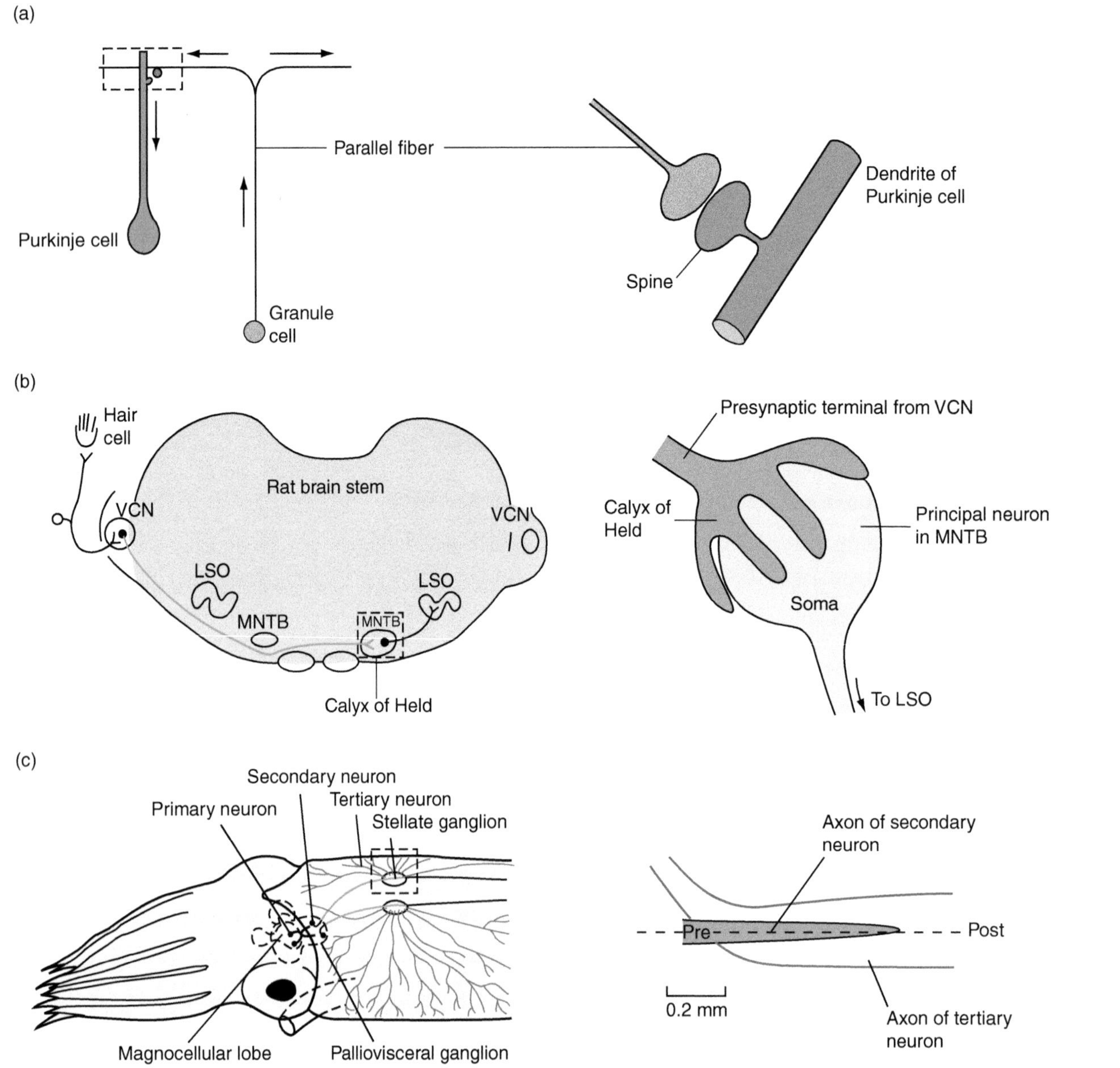

FIGURE 7.1 **Three examples of preparations in which synaptic transmission has been studied. (a)** *The cerebellar cortex.* Left: Schematic showing the connections between a granular cell axon (parallel fiber) and a Purkinje cell. These synapses are axo-spinous (right). **(b)** *The calyx of Held.* Left: Frontal section of the brainstem drawn at the level of the 8th nerve. The axon collaterals of the globular cells of the ventral cochlear nucleus (VCN) project to the neurons of the contralateral medial nucleus of the trapezoid body (MNTB). This synapse is axo-somatic and each MNTB neuron receives only one axon terminal that forms the calyx of Held (right). **(c)** *Squid giant synapse.* Secondary neurons, that receive sensory information from the primary ones (left), establish giant axo-axonic synapses with tertiary neurons (right). The tertiary neurons are responsible for contraction of the mantle muscles thus permitting expulsion of water and propelling the animal out of the danger zone. The dotted square indicates the region enlarged on the right of the figure. LSO, lateral superior olive. Drawing **(b)** adapted from Forsythe ID, Barnes-Davies M, Brew HM (1995) The calyx of Held: a model for transmission at mammalian glutamatergic synapses. In: *Excitatory Aminoacids and Synaptic Transmission*, 2nd edn, New York: Academic Press, with permission. Drawing **(c)** from Llinas R (1982) Calcium in synaptic transmission *Sci. Am.* **247**, 56–65, with permission.

phenomenon. It is noteworthy that even when all the proper conditions come together (a presynaptic spike, opening of Ca^{2+} channels, Ca^{2+}entry), exocytosis of a synaptic vesicle is still not guaranteed (see Appendix 7.2).

The examples in this chapter will be taken mainly from studies on the glutamatergic synapses of the mammalian central nervous system (**Figure 7.1a,b**), and on other synapses that have been examined owing to the large diameter of their presynaptic element, such as the squid giant synapse (**Figure 7.1c**), the neuromuscular junction (see **Figures 6.11** and **6.12a**) and the central synapse, the calyx of Held (**Figure 7.1b**).

7.1 OBSERVATIONS AND QUESTIONS

7.1.1 Quantitative data on synapse morphology and synaptic transmission

The regulated release of neurotransmitter occurs at the active zone of a synapse. Synapses of the mammalian central nervous system generally exhibit one or two active zones (or release sites) per bouton (**Figure 7.2a,b**); but there are exceptions, such as the perforated synapse of the hippocampus (see **Figure 6.3**) or the calyx of Held in the medial trapezoid body of the brainstem (**Figure 7.1b**) that contain more than one active zone (**Figure 7.2c**) (around 4–5 for the former and at least 200 for the latter). Giant synapses such as in the squid (**Figure 7.1c**) comprise around 4400 active zones and the neuromuscular junction (see **Figure 6.12a**) from 300 to 1000 active zones. Synapses of the mammalian central nervous system have small dimensions. In the hippocampus, synaptic terminals are rarely more than 1–5 μm wide, but there are exceptions. Vesicles have diameters of 25–60 nm. Pre- and postsynaptic elements are distant by 10–30 nm (cleft height). Postsynaptic density areas range from 0.01 to 0.5 μm^2. In three dimensions, the surfaces of presynaptic and postsynaptic membranes have the shape of two plates facing one another.

Postsynaptic responses are either a depolarization or a hyperpolarization of the postsynaptic membrane. The former is called the *excitatory postsynaptic potential* (EPSP; **Figure 7.3a**) since it brings the membrane potential closer to the spike threshold, and the latter is called the *inhibitory postsynaptic potential* (IPSP) since it does the opposite. These responses, being voltage changes, are recorded in current clamp. The underlying currents can be recorded in voltage clamp; they are the *excitatory* or *inward* (EPSC; **Figure 7.3b**) and the *inhibitory* or *outward* (IPSC) postsynaptic currents. In this chapter, only excitatory responses are studied.

In response to a single spike in the presynaptic axon, a small-amplitude EPSP (or EPSC) is recorded; it is called 'single-spike EPSP' (or EPSC). It has the following characteristics: the synaptic delay ranges from 500 μs to 1 ms (**Figure 7.3b**); the amplitude of the single-spike EPSP can reach a few mV (**Figure 7.3a**) depending on the number of boutons and active zones activated by the presynaptic spike (**Figure 7.2**). The half-duration of a single-spike EPSP is in the order of milliseconds to tens of milliseconds. Each single presynaptic spike does not necessarily

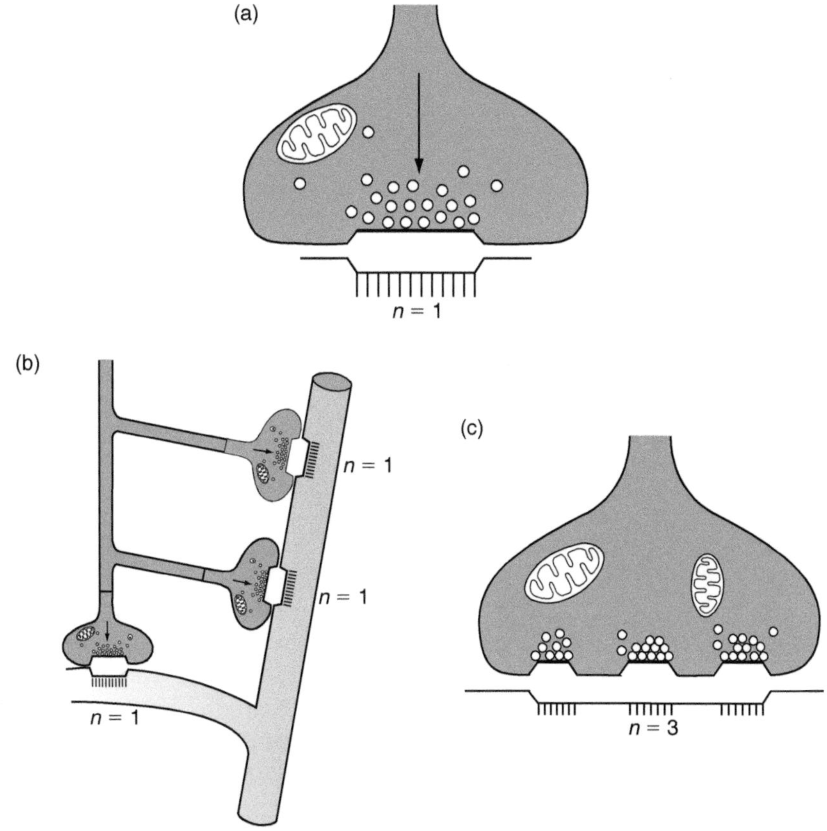

FIGURE 7.2 **Number *n* of active zones per presynaptic terminal.** Drawings of synaptic boutons of the central nervous system with presynaptic active zone(s) and postsynaptic membranes. The active zones are represented as a black bar with adjacent synaptic vesicles. The number of active zones per bouton is $n = 1$ in **(a)** and **(b)** but $n = 3$ in **(c)**. The synaptic connections are such that a single presynaptic spike will activate a maximum of one active zone in (**a**), and a maximum of three in (**b**) and (**c**).

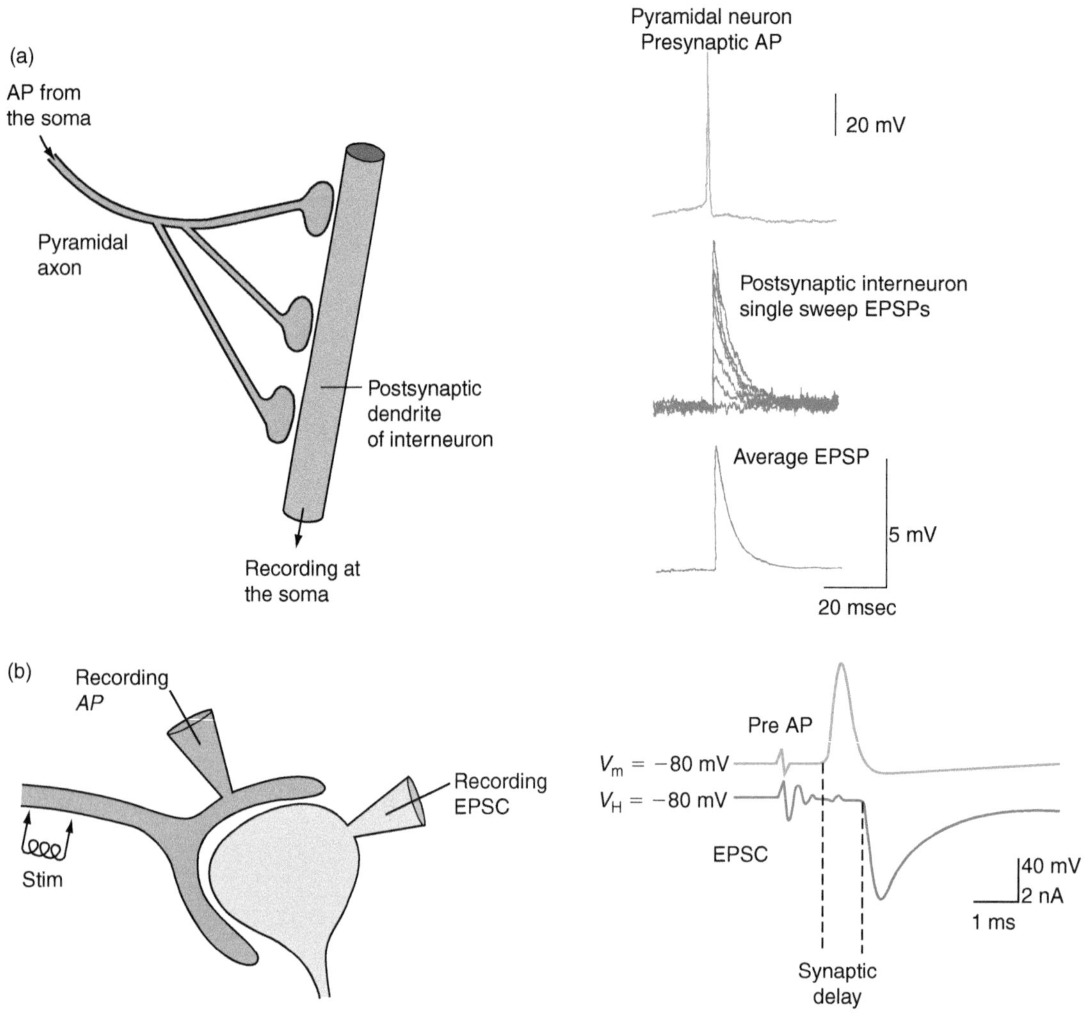

FIGURE 7.3 **Basic properties of synaptic transmission. (a)** In the neocortex, pyramidal neurons are Golgi type I neurons, whose main axons leave the cortex after giving off collaterals. Left: These collaterals establish excitatory synapses with dendrites of local interneurons (Golgi type II neurons). Three synapses are represented but their exact number in the experiment performed at right was not determined. Right: In response to each presynaptic action potential (AP) evoked in the pyramidal neuron (top trace), a single-spike EPSP is recorded in the postsynaptic interneuron: it fluctuates in amplitude from 0 mV (failure) to 5 mV (middle traces); EPSPs recorded in response to four successive presynaptic spikes are superimposed (bottom traces). $V_m = -75$ mV in middle and bottom traces. **(b)** The synaptic delay between a presynaptic action potential (AP) evoked by stimulation of the afferent axon and recorded in the calyx of Held, and the postsynaptic response (excitatory postsynaptic current, EPSC) recorded in the postsynaptic soma varies from 500 μs to 1 ms. Drawing (**a**) from Thomson AM, personal communication. Drawing (**b**) from Borst JGG, Helmchen F, Sakmann B (1995) Pre- and postsynaptic whole-cell recordings in the medial nucleus of the trapezoid body of the rat. *J. Physiol.* **489**, 825–840, with permission.

evoke a postsynaptic potential – there are failures of synaptic transmission (**Figure 7.3a**).

In summary, the synapse is an electrochemical unit specialized to function on a distance scale of micrometers and a timescale of submilliseconds.

7.1.2 Ways of estimating neurotransmitter release in central mammalian synapses

Recording of the postsynaptic response

If one considers the schematic representation of synaptic transmission in **Figure 6.4a**, neurotransmitter release corresponds to steps 2, 3 and 4. One way to estimate neurotransmitter release is to measure the postsynaptic response that it evokes. It is an *indirect measure* since it includes events following release, such as neurotransmitter diffusion from the pre- to the postsynaptic element, binding of neurotransmitter molecules to postsynaptic receptors (step 5a), and induction of the postsynaptic current (step 6). In addition, to be a reliable detector of release events the postsynaptic responses (EPSP or IPSP) should not activate postsynaptic voltage-dependent currents that would amplify or decrease them. Most of the data on neurotransmitter release explained here have been obtained from recordings of postsynaptic responses to a single presynaptic action potential.

Other techniques

Other techniques can be used to monitor transmitter release from peripheral synapses, large synaptosomes or endocrine cells; they are the patch clamp technique and amperometry. The first consists of measuring the membrane capacitance that is directly proportional to the membrane surface area. Upon fusion of a secretory vesicle, membrane area and therefore capacitance increases stepwise by an amount equal to the vesicle or granule membrane area. Amperometry monitors the release of some secretory products by measuring the oxidation of electroactive substances (dopamine, serotonin) with a carbon fiber microelectrode placed near the cell.

7.1.3 Questions

Considering the general functional model of synaptic transmission, which states that exocytosis of synaptic vesicles is triggered by an increase of the intracellular Ca^{2+} concentration in the presynaptic element ($[Ca^{2+}]_i$) (in **Figure 6.4a**), the following questions can be asked:

- Is $[Ca^{2+}]_i$ increase a prerequisite for transmitter release? Which type of Ca^{2+} channels are present in the presynaptic membrane and in response to which signal do they open (step 2)?
- Is the presynaptic $[Ca^{2+}]_i$ increase local and transient in response to a presynaptic spike (step 3)?
- How does $[Ca^{2+}]_i$ increase trigger exocytosis? Why is it so rapid (step 3)?
- How do vesicles fuse with the presynaptic membrane in response to a single presynaptic spike (step 4)?
- How much neurotransmitter is released into the synaptic cleft from a single vesicle (step 4)? What is the neurotransmitter lifetime in the cleft?
- What is the mechanism underlying the clearance of the transmitter from the synaptic cleft (steps 5)?

To answer the above questions, we will study the processes of transmitter release in chronological order, from depolarization of the presynaptic membrane by an action potential to the release of transmitter from synaptic vesicles (presynaptic processes I and II; Sections 7.2 and 7.3). Processes occurring in the synaptic cleft, just after transmitter release, are studied in Section 7.4. Details on the quantal and probabilistic nature of neurotransmitter release are given in Appendices 7.1 and 7.2.

7.2 PRESYNAPTIC PROCESSES I: FROM PRESYNAPTIC SPIKE TO $[Ca^{2+}]_i$ INCREASE

7.2.1 The presynaptic Na^+-dependent spike depolarizes the presynaptic membrane, opens presynaptic Ca^{2+} channels and triggers Ca^{2+} entry

In a resting presynaptic element, Ca^{2+} ions are present at a very low concentration, 10^{-8} to 10^{-7} M. This intracellular Ca^{2+} concentration, $[Ca^{2+}]_i$, is at least 10 000 times lower than the extracellular one (see **Figure 3.2**). It is maintained at this resting level by various Ca^{2+} clearance mechanisms (see Section 7.2.4). In response to a presynaptic action potential $[Ca^{2+}]_i$ suddenly increases in the presynaptic element (see **Figure 6.4a**, step 1, and **Figure 7.6**).

What is the origin of $[Ca^{2+}]_i$ increase in the presynaptic element? Is $[Ca^{2+}]_i$ increase a prerequisite for transmitter release?

In an extracellular medium deprived of Ca^{2+} ions or containing Ca^{2+} channel blockers such as Co^{2+} or Cd^{2+} ions, presynaptic $[Ca^{2+}]_i$ increase and postsynaptic response are no longer observed although presynaptic action potentials are unchanged. This suggests that external Ca^{2+} ions enter the presynaptic element through voltage-gated Ca^{2+} channels (which are blocked by $Co^{2+}$and Cd^{2+}; see Section 5.1). It also shows that presynaptic $[Ca^{2+}]_i$ increase is a prerequisite for synaptic transmission. This has been shown to be valid for all chemical synapses that have been studied.

What triggers the opening of voltage-gated Ca^{2+} channels?

The following hypothesis has been proposed. The brief membrane depolarization that occurs in the ascending phase of each Na^+-dependent presynaptic spike triggers the opening of voltage-dependent Ca^{2+} channels and allows subsequent Ca^{2+} influx into the presynaptic element.

In order to check that a presynaptic depolarization can trigger Ca^{2+}entry, Llinas and co-workers (1966) performed the following experiment in the squid giant synapse. They introduced two microelectrodes, one in the presynaptic element to inject a depolarizing current and one in the postsynaptic element to record its activity, and blocked Na^+-dependent spikes with tetrodotoxin (TTX). In such conditions, direct depolarization of the presynaptic membrane, though it fails to evoke a presynaptic spike (because of TTX), evokes a presynaptic $[Ca^{2+}]_i$ increase and a postsynaptic response. A presynaptic membrane depolarization can thus trigger Ca^{2+} channel opening and neurotransmitter release.

7.2.2 Ca^{2+} enters the presynaptic bouton during the time course of the presynaptic spike through high-voltage-activated Ca^{2+} channels (N and P/Q types)

The calyx of Held is an axo-somatic synapse located in the rat brainstem, in the medial nucleus of the trapezoid body (MNTB), a nucleus that participates in sound localization. It is the largest synapse in the mammalian central nervous system. The presynaptic axon originates in the contralateral cochlear nucleus and each MNTB postsynaptic neuron receives only one calyx (see **Figure 7.1b**). Synaptic transmission is glutamatergic.

When does Ca^{2+} enter the presynaptic element in response to a presynaptic spike?

In order to record the presynaptic Ca^{2+} current, the presynaptic membrane is clamped at $V_H = -80$ mV by means of two whole-cell electrodes. To isolate the presynaptic Ca^{2+} current, voltage-dependent Na^+ and K^+ currents are blocked with TTX and tetraethlyammonium chloride (TEA), respectively. An action potential waveform is injected into the presynaptic terminal through one of the whole-cell electrodes. It evokes a presynaptic Ca^{2+} current that is recorded by the second whole-cell electrode. Recordings show that Ca^{2+} influx is tightly associated with the repolarizing phase of the action potential (**Figure 7.4**): it is essentially a tail current (see

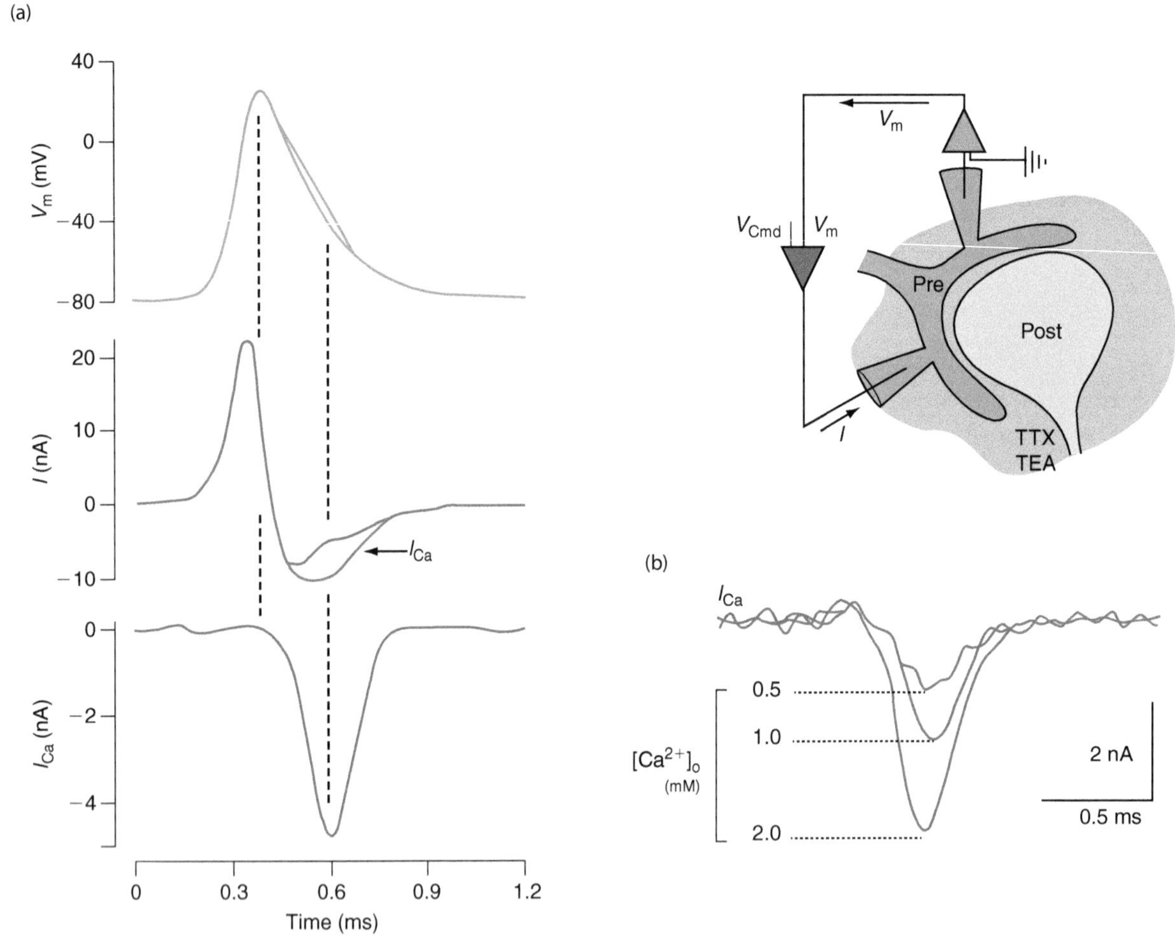

FIGURE 7.4 **Ca^{2+} current flows into presynaptic terminal during the repolarizing phase of the presynaptic spike.** Two whole-cell electrodes are positioned in the calyx of Held; one measures the membrane potential (V) and the other one injects current (I). It is a two-electrode voltage clamp configuration. The preparation is bathed in 2 mM external Ca^{2+} with TTX and TEA to block the voltage-gated Na^+ and K^+ currents. **(a)** The voltage clamp command (V_{Cmd}) is an action potential waveform (V) from a holding potential of −80 mV. A reduced and inverted action potential waveform is also applied, scaled and re-inverted to measure the passive current. Top, recorded voltages. The two action potential waveforms are superimposed. Due to series resistance, the repolarization is somewhat slower during the full action potential than during the scaled action potential. Middle, currents. The current flowing during the full-sized action potential has a larger inward component (labeled I_{Ca}). The two passive transients overlay well. Bottom, calcium current. The calcium current is obtained by subtracting the passive current from the current measured during the full action potential. All traces are the average of 11. Vertical dotted lines denote peak of the action potential waveform and of the calcium current. **(b)** This Ca^{2+} current is reduced in 1 or 0.5 mM external Ca^{2+} ($[Ca^{2+}]_o$). Part **(a)** from Borst JG, Sakmann B (1998) Calcium current during a single action potential in a large presynaptic terminal of the rat brainstem. *J. Physiol.* **506**, 143–157, with permission. Part **(b)** adapted from Borst JGG, Sakmann B (1996) Calcium influx and transmitter release in a fast CNS synapse. *Nature* **383**, 431–434, with permission.

Appendix 5.2) that activates shortly after the peak of the action potential and ends before repolarization is complete. It has a peak amplitude of 2.6 ± 0.2 nA and a half-width of about 350 ms. The delay between the beginning of the action potential and that of Ca^{2+} current is about 500 μs at 23–24°C.

Which types of Ca^{2+} channels are involved in transmitter release?

This was first investigated in the frog neuromuscular junction and then in other different preparations. Antibodies against ω-conotoxin GVIA, which selectively binds to N-type Ca^{2+} channels, were seen to label active zones on the terminals of motoneurons. In central synapses, to examine the Ca^{2+} channels responsible for Ca^{2+} influx and transmitter release, pharmacological agents that selectively block a type of Ca^{2+} channel have been tested on the amplitude of the presynaptic Ca^{2+} increase and the postsynaptic response.

Consider the example of the glutamatergic synapse between the axons of granule cells (parallel fibers) and Purkinje cells in the rat cerebellum (see **Figures 6.8** and **7.1a**). The presynaptic Ca^{2+} concentration is determined with the Ca^{2+}-sensitive dye magfura, a low-affinity Ca^{2+}-sensitive dye that emits light in the presence of free Ca^{2+} with a sensitivity of 10^{-4} M (see Appendix 5.1). The transmitter release is estimated from the postsynaptic excitatory current recorded in voltage clamp (whole-cell configuration) from the Purkinje soma. Parallel fibers are excited by a stimulating electrode placed in the molecular layer (**Figure 7.5a**). Ca^{2+} entry in response to presynaptic stimulation is measured as a fluorescence signal.

A single stimulus produces an abrupt change in fluorescence (a fluorescence transient) which returns to resting levels within a few hundreds of milliseconds. At saturating concentration, ω-conotoxin GVIA (0.5 μM) inhibits by 27.0 ± 1.7% the fluorescence transient elicited by the stimulation (**Figure 7.5b**, top traces). In comparison, ω-agatoxin IVA (200 nM) reduces the amplitude of the transient by 50.1 ± 0.9% (**Figure 7.5c**, top traces) and nimodipine (5 μM), an L-type channel blocker, has no effect. Simultaneous application of the two toxins has an additive effect and inhibits Ca^{2+} influx by 77 ± 3% (**Figure 7.5d**). In conclusion, at this cerebellar synapse, the ω-conotoxin-sensitive N-type and the ω-agatoxin-sensitive P/Q-type Ca^{2+} channels are both present in the presynaptic membrane and allow around 80% of Ca^{2+} entry in response to a presynaptic spike.

For all synapses studied so far, Ca^{2+} enters presynaptic terminals mainly through N- and/or P/Q-type Ca^{2+} channels. The functional properties of these channels impose at least one constraint: since N- and P/Q-type Ca^{2+} channels are high-voltage-activated channels, Ca^{2+} enters presynaptic terminals only in response to a *large* membrane depolarization (such as the depolarizing phase of the Na^+ action potential).

7.2.3 Presynaptic $[Ca^{2+}]$ increase is transient and restricted to micro- or nanodomains close to docked vesicles

How does Ca^{2+} rise in a presynaptic terminal, uniformly or in domains?

This has been studied in the squid giant synapse (**Figure 7.1c**) with Ca^{2+} imaging techniques. A low-affinity Ca^{2+}-sensitive dye (*n*-aequorin-J, a protein that emits light in the presence of free Ca^{2+} ions with a minimum sensitivity of 10^{-4} M; see Appendix 5.1) is injected into the presynaptic terminal in order to visualize only zones where Ca^{2+} concentration is high. Then the presynaptic axon is stimulated at 10 Hz. Multiple fluorescent domains, equally spaced, with a mean size of 0.250 to 0.375 μm^2 (mean 0.313 μm^2) are seen in the presynaptic terminal (**Figure 7.6**). Their number 4500 is quite close to the average number of release sites in this terminal (4400). These microdomains are located at active zones of the presynaptic plasma membrane. This observation is evidence for the fact that voltage-dependent Ca^{2+} channels are clustered at active zones. Since the presynaptic stimulation used in this experiment is not an action potential, it does not indicate the physiological Ca^{2+} concentration in presynaptic microdomains.

Where are presynaptic Ca^{2+} channels located?

The rapidity with which neurotransmitter release can be triggered after Ca^{2+} influx (within 200 μs) makes it likely that Ca^{2+} ions act at a very short distance from the Ca^{2+} channels. This suggested that: (i) Ca^{2+} channels and release sites are located close to each other; and (ii) there exists a stable complex between synaptic vesicles and plasma membrane, preassembled in the resting state, before $[Ca^{2+}]_i$ increases. Ca^{2+} diffusion is too restricted and delay of exocytosis too short to allow for vesicle movement before fusion with the plasma membrane (Ca^{2+} ions diffuse no more than a few vesicle diameters into the cytoplasm). In other words, presumably synaptic vesicles available for rapid transmitter release must be predocked in the vicinity of Ca^{2+} channels.

The localization of Ca^{2+} channels relative to the position of transmitter release sites was first investigated with imaging (**Figure 7.7a**) and immunocytochemical (**Figure 7.7b**) techniques. In the frog neuromuscular junction, the presynaptic nerve terminal is a long structure (several hundreds of micrometers) characterized by the presence of neurotransmitter release sites or active zones spaced at regular 1 μm intervals. Directly across the synaptic cleft just facing active zones are clusters of nicotinic acetylcholine receptors (nAChR) located on the edge of the postjunctional folds (see **Figure 6.12a**). The preparation is double-labeled to disclose both postsynaptic nAChR and presynaptic N-type Ca^{2+} channels. The idea is that the localization of postsynaptic nAChRs indicates exactly the localization of

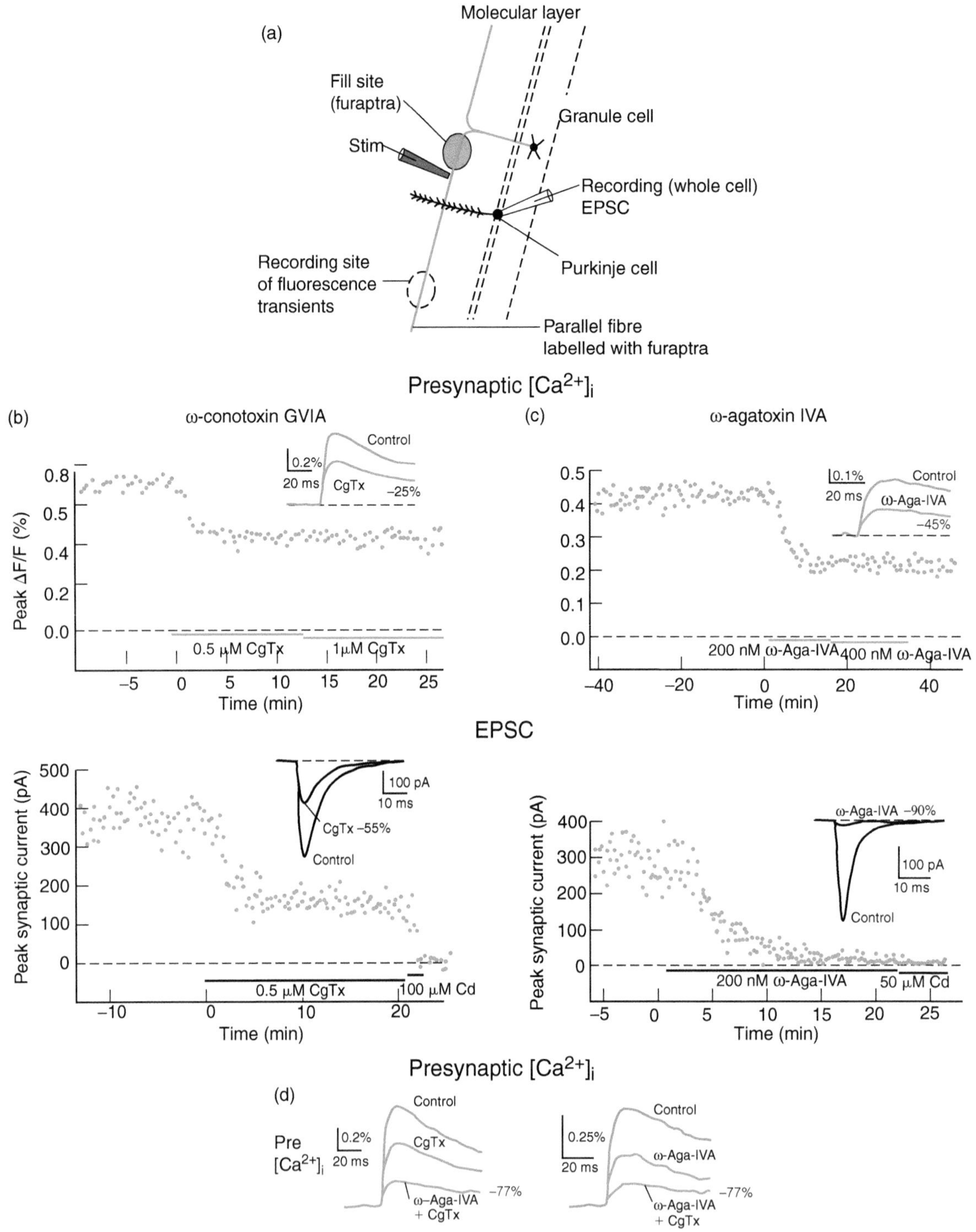

FIGURE 7.5 **N- and P/Q-type Ca^{2+} channel blockers reduce presynaptic Ca^{2+} influx and synaptic transmission at a cerebellar synapse. (a)** Transverse cerebellar slice with the relative locations of labeling with the dye furaptra (fill site), stimulus electrode and recording sites (whole-cell recording of EPSC from a Purkinje soma and recording of presynaptic fluorescence transients in the molecular layer). **(b)** Amplitude of furaptra fluorescence transients ($\Delta F/F$) in the presence of increasing concentrations of ω-conotoxin GVIA (CgTx) to block the N-type Ca^{2+} current (top traces). Concomitant recording of the postsynaptic current is shown in the bottom trace. Each furaptra transient is elicited by a single stimulus of the parallel fiber tract and is a measure of the presynaptic Ca^{2+} influx. The inset shows superimposed fluorescence transients in control conditions and after addition of 0.5 and 1 μM CgTx. **(c)** The same experiment as in **(b)** but in the presence of increasing concentrations of ω-agatoxin IVA (ω-Aga IVA) to block the P/Q-type Ca^{2+} current. The inset shows superimposed fluorescence transients in control conditions and after addition of 200 and 400 nM ω-Aga IVA. Concomitant recording of the postsynaptic current is shown in the bottom trace. **(d)** Additive effects of the sequential application of saturating concentrations of the toxins on the furaptra transients. Adapted from Mintz I, Sabatini BL, Regehr WG (1995) Calcium control of transmitter release at a cerebellar synapse. *Neuron* **15**, 675–688, with permission.

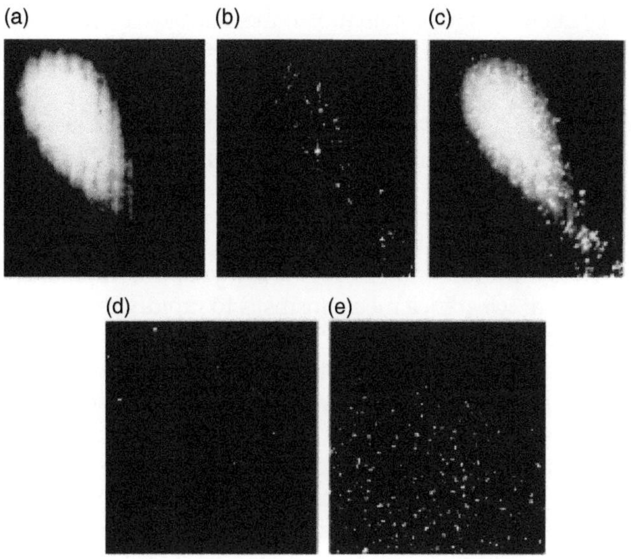

FIGURE 7.6 **Microdomains of Ca^{2+} increase in the presynaptic terminal of the squid giant synapse.** (**a**) Fluorescence image of a presynaptic terminal injected with *n*-aequorin-J. When the presynaptic fiber is fully loaded, it is continuously stimulated at 10 Hz for 10 s. (**b**) The acquisition during these 10 s of tetanic stimulation reveals stable quantum emission domains that appear as white spots. The background fluorescence shown in (**a**) disappears in (**b**) due to subtraction. (**c**) Superposition of the fluorescent images in (**a**) and (**b**) reveals that the distribution of the microdomains of high calcium coincides with the presynaptic terminal. Emission domains in an unstimulated terminal (**d**) and in the same terminal during tetanic stimulation (**e**) are shown at high magnification. Adapted from Llinas R, Sugimori M, Silver RB (1992) Microdomains of high calcium concentration in a presynaptic terminal. *Science* **256**, 677–679, with permission.

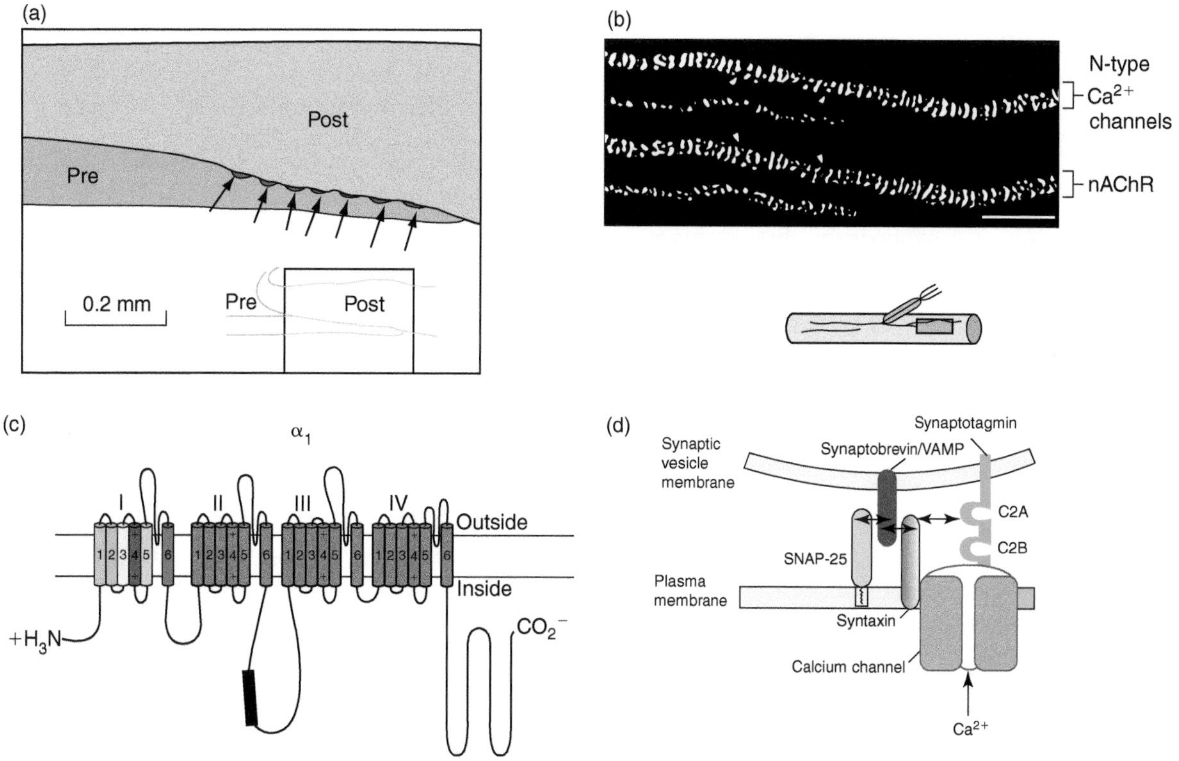

FIGURE 7.7 **Presynaptic N-type Ca^{2+} channels are clustered at active zones. (a)** In the squid giant synapse, presynaptic zones of [Ca^{2+}]$_i$ increase in response to a train of brief presynaptic stimuli (0.5 s at 80 Hz) are visualized with the FURA-2 technique. They are localized at active zones. The diagram illustrates the synapse and the box indicates the region studied. **(b)** In the frog neuromuscular junction, N-type Ca^{2+} channels and nicotinic acetylcholine receptors (nAChR) are labeled with two different selective toxins coupled to different fluorescent dyes. The preparation is viewed with a confocal laser microscope. The diagram illustrates the structure of the neuromuscular junction and the box indicates the region scanned by the microscope. The images showing the distribution of presynaptic Ca^{2+} channels (top) and postsynaptic nAChRs (bottom) are separated for clarity but they are in fact superimposed. **(c)** Predicted topological structure of the α_1-subunit of N-type Ca^{2+} channel. The rectangle indicates the synprint site. **(d)** Theoretical model of interaction of presynaptic $Ca_v2.2$ and 2.1 channels (N- and P/Q-type Ca^{2+} channels) with SNARE proteins (syntaxin, SNAP-25 and synaptotagmin) at the presynaptic plasma membrane. Part (**b**) from Robitaille R, Adler EM, Charlton MP (1990) Strategic location of calcium channels at transmitter release sites of frog neuromuscular synapses. *Neuron* **5**, 773–779, with permission. Part (**c**) from Sheng ZH, Rettig J, Takahashi M, Catterall WA (1994) Identification of a syntaxin-binding site on N-type calcium channels. *Neuron* **13**, 1303–1313, with permission. Part (**d**) from Catterall WA (2000) Structure and regulation of voltage-gated Ca^{2+}channels. *Annu. Rev. Cell Dev. Biol* **16**, 521–555, with permission.

presynaptic active zones. Ca^{2+} channels are labeled with biotinylated ω-conotoxin, a specific and irreversible blocker of N-type Ca^{2+} channels and of synaptic transmission at the frog neuromuscular junction. To reveal ω-conotoxin-sensitive Ca^{2+} channel labeling, preparations are then incubated with streptavidin Texas Red (see Appendix 6.2) which fluoresces red. When nerve terminals are removed by pulling off branches of the motor nerve, the Ca^{2+} channel labeling totally disappears, indicating that ω-conotoxin binding sites are strictly located on presynaptic terminals. Postsynaptic nicotinic receptors nAChR are labeled with α-bungarotoxin coupled to boron dipyrromethanedifluoride which fluoresces green. Under the light microscope, each fluorescent band of presynaptic Ca^{2+} channels is matched by a fluorescent stain of postsynaptic nAChR (**Figure 7.7b**). Bands of labeled Ca^{2+} channels and labeled nAChRs are thus almost perfectly aligned, suggesting that Ca^{2+} channels are clustered in the membrane of presynaptic active zones, opposite the postjunctional folds.

The clustering of presynaptic N- and P/Q-type Ca^{2+} channels ($Ca_v2.2$ and $Ca_v2.1$) at active zones has been confirmed by the discovery of a physical link between these Ca^{2+} channels and syntaxin, an integral protein of the presynaptic plasma membrane involved in vesicle docking at the presynaptic active zones (see Section 7.3.2). This was shown (i) by co-precipitation of N and P/Q channels (labeled with their specific toxins) with syntaxin (labeled with a specific antibody), (ii) by the identification of a syntaxin-binding domain on N- and P/Q-type α_1-subunits, and (iii) by inhibition of synaptic transmission after injection of peptide inhibitors into presynaptic neurons. N- and P/Q-type Ca^{2+} channels interact directly with SNARE proteins through the *synaptic protein interaction* (synprint) site, which resides in the N-terminal half of the L_{II-III} cytoplasmic loop connecting domains II and III of their α_1-subunit (**Figure 7.7c**). The synprint site of $Ca_v2.2$ binds the plasma membrane proteins syntaxin and SNAP-25 as well as the vesicle protein synaptotagmin (**Figures 7.7d** and **7.12**). Injection of peptide inhibitors of this interaction into presynaptic neurons inhibits synaptic transmission, consistent with the conclusion that this interaction is required to position docked synaptic vesicles near Ca^{2+} channels for effective fast exocytosis. These results define a second functional activity of the presynaptic Ca^{2+} channel: targeting docked synaptic vesicles to a source of Ca^{2+} for effective transmitter release (**Figure 7.7d**) and making release as fast as possible.

These results exemplify some general principles of rapid Ca^{2+} signaling in neurotransmitter release:

- Ca^{2+} entry into the presynaptic element occurs in close proximity to the exocytotic apparatus.
- Clustering of Ca^{2+} channels close to release sites ensures that a Ca^{2+} signal is rapidly available (in the hundreds of microseconds timescale) to the nearby Ca^{2+}-sensitive proteins, which initiate transmitter release.

7.2.4 Ca^{2+} clearance makes presynaptic $[Ca^{2+}]_i$ increase transient: it shapes its amplitude and duration

Ca^{2+} clearance is the removal of Ca^{2+} ions (in excess compared to the resting state) from the presynaptic terminal. The aim of Ca^{2+} clearance mechanisms is to rapidly re-establish the resting level of $[Ca^{2+}]_i$. Ca^{2+} clearance is achieved by proteins that extrude Ca^{2+} ions toward the extracellular space or toward organelles such as the endoplasmic reticulum or mitochondria: these are Ca^{2+} pumps (Ca^{2+}-ATPases) and Ca^{2+}-transporters (Na–Ca exchanger) (see **Figure A3.1** and **Figure 7.8**). Ca^{2+} ions that enter the presynaptic terminal will also rapidly bind to cytosolic proteins (Ca^{2+}-buffers, Ca-B). Because such binding confiscates Ca^{2+} ions, it can rapidly diminish freely diffusing Ca^{2+} ions.

But such buffering is not a real clearance since Ca^{2+} ions will unbind from these proteins; it is a temporary clearance. Unbound Ca^{2+} ions will have then to be extruded by pumps and transporters. Therefore, the more numerous the number of Ca^{2+} pumps and transporters, the more efficient and rapid is Ca^{2+} clearance.

Extrusion of Ca^{2+} to the extracellular medium by the plasma membrane Ca-ATPase (PMCA) pump and by the Na–Ca exchanger

The former uses the hydrolysis of ATP as a source of energy and is independent of the extracellular Na^+ concentration. The latter is driven by the Na^+ electrochemical gradient across the plasma membrane and is thus sensitive to extracellular Na^+ concentration. The Ca-ATPase pump is proposed to be a low-capacity high-affinity system (K_D = 0.2–0.3 μM) whereas the Na–Ca exchanger would have a high-capacity low-affinity system (K_D = 0.5–1.0 μM). The Ca-ATPase pump would thus be the most efficient system in the presence of a low presynaptic activity, and the two systems would act in synergy to regulate the intracellular Ca^{2+} concentration after a train of action potentials.

Sequestration of Ca^{2+} ions in smooth endoplasmic reticulum and mitochondria

This is achieved by sarco-endoplasmic Ca-ATPase (SERCA) pumps present in the membrane of these organelles. The smooth endoplasmic reticulum is a Ca^{2+} storage compartment. In the different cell types studied, the smooth endoplasmic reticulum Ca-ATPase pump has a better affinity for Ca^{2+} than that of the mitochondrion. This latter would function in rare situations in cases of massive Ca^{2+} entry. Noteworthy, following an appropriate signal (such as the formation of inositol trisphosphate, IP_3), the Ca^{2+} ions stored in these compartments

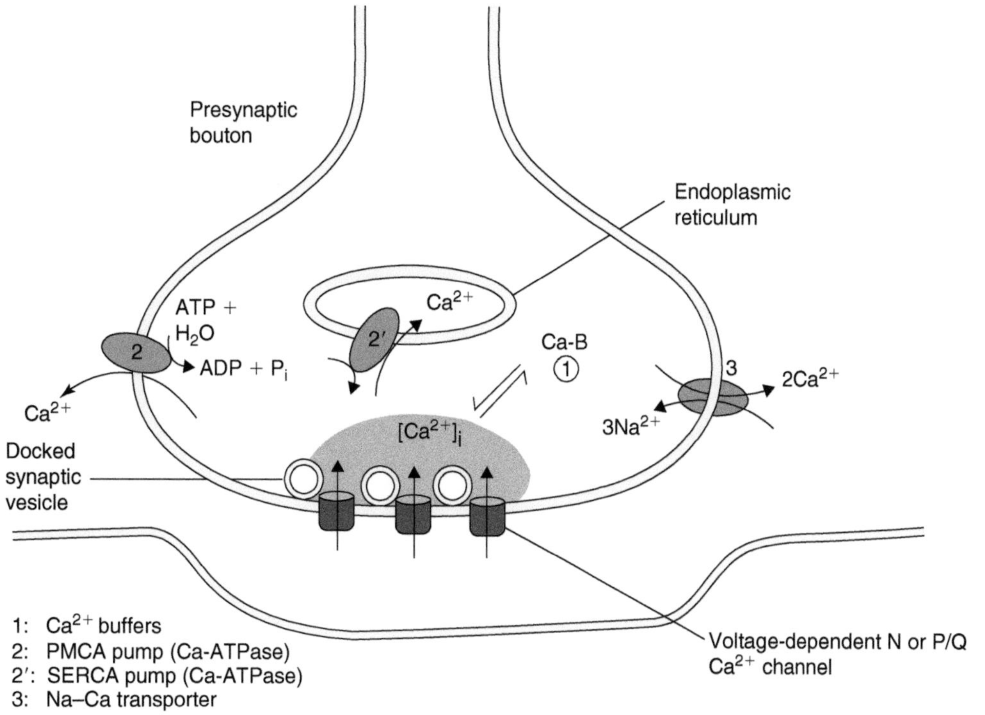

FIGURE 7.8 **Ca^{2+} clearance mechanisms in a presynaptic terminal.** While Ca^{2+} ions enter at the level of the presynaptic active zone through high-voltage-activated Ca^{2+} channels, they are rapidly buffered by cytoplasmic Ca^{2+}-binding proteins (Ca-B) (1). Ca^{2+} ions are also actively cleared from the intracellular medium towards the extracellular medium via Ca-ATPases of the plasma membrane (PMCA pumps) (2) and the Na–Ca exchangers (3). They are also cleared by active transport toward endoplasmic reticulum via another type of Ca-ATPase (SERCA pumps) (2′). This clearing has a time constant of the order of tens of milliseconds to seconds.

can be released in the cytoplasm through Ca^{2+}-permeable channels.

Ca^{2+} buffering by cytosolic proteins

Different cytosolic proteins have the ability to bind Ca^{2+} with a high affinity. These proteins have, in general, a low molecular weight and act primarily as Ca^{2+} buffers (such as parvalbumin and calbindin) or subserve messenger functions (calmodulin). Parvalbumin is strongly expressed in many GABAergic neurons in the mammalian central nervous system (the co-localization of GABA and parvalbumin has been shown immunohistochemically using highly specific antibodies to GABA and parvalbumin), whereas in other neurons the concentration of parvalbumin is much lower. The high concentration of parvalbumin might have a consequence on a neuron's ability to rapidly buffer Ca^{2+}. This is especially important for neurons that have a high tonic activity since trains of action potentials trigger repetitive $[Ca^{2+}]_i$ increases in their synaptic terminals.

Calbindin is a protein that was originally found in the gut, where it binds Ca^{2+} and is vitamin D dependent. Its presence has been shown in neurons of the mammalian central nervous system, notably the Purkinje cells of the cerebellar cortex and the dopaminergic neurons of the substantia nigra. Calmodulin has a high affinity for Ca^{2+} and a role of intracellular messenger. The buffering of free intracellular Ca^{2+} ions by cytoplasmic calcium-binding proteins is a very efficient system responsible for the rapid disappearance of free Ca^{2+} ions.

The relative contribution of the clearance systems: example of Purkinje cells

Cerebellar Purkinje cells are GABAergic neurons that have powerful systems to control $[Ca^{2+}]_i$. Immunocytochemical studies demonstrate considerable amounts of cytosolic Ca^{2+}-binding proteins, particularly calbindin D_{28K} and parvalbumin. There are also numerous Ca^{2+} pumps localized in the endoplasmic reticulum (SERCA pumps) or the plasma membrane (PMCA pumps). In order to understand the respective roles of these clearance systems, Purkinje cells are loaded with the fluorescent Ca^{2+} dye FURA-2 (see Appendix 5.1), $[Ca^{2+}]_i$ transients are evoked by direct membrane depolarization and measured by microfluorometry, and clearance systems are pharmacologically inhibited one at a time. Since all these experiments are achieved with the use of a whole-cell electrode, the duration of the study is limited to 25 minutes in order to avoid the washing out of intracellular constituents that would give an artefactual diminution of $[Ca^{2+}]_i$.

The contribution of SERCA pumps to the amplitude and decay phase of $[Ca^{2+}]_i$ transients is studied by applying cyclopiazonic acid (CPA) or thapsigargin, specific inhibitors of this ATPase. For blocking PMCA pumps, 5,6-succinimidyl

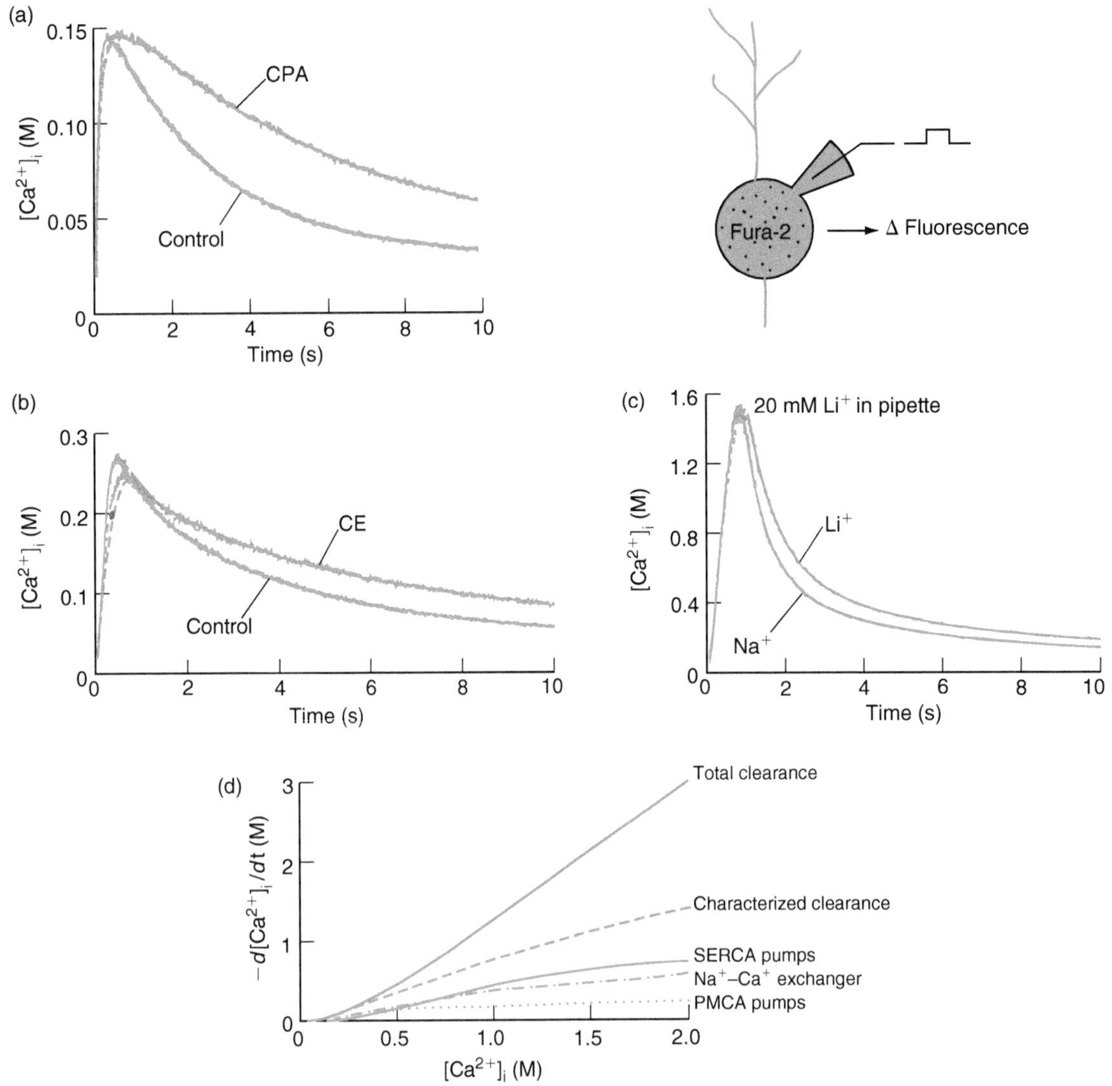

FIGURE 7.9 **Relative contribution of the different mechanisms of Ca^{2+} clearance in Purkinje somata.** $[Ca^{2+}]_i$ transients are evoked in FURA-2-loaded Purkinje cells by a depolarizing current pulse of varying duration (60–250 ms). Effects on $[Ca^{2+}]_i$ transients of **(a)** cyclopiazonic acid (CPA), a blocker of SERCA pumps, **(b)** 5,6-succinimidyl carboxyeosin (CE), a blocker of PMCA pumps, and **(c)** Li^{2+} saline, a blocker of Na–Ca exchanger. **(d)** Total Ca^{2+} clearance rate is presented in comparison with the rate of the different components characterized in the above experiments. Clearance rate is plotted as a function of the $[Ca^{2+}]_i$ in the range between 50 nM and 2 μM. The clearance rate is calculated as follows: (i) The decay phase of each transient is fitted by a single or double exponential function and the derivative function ($d[Ca^{2+}]_i/dt$) is calculated from the fit. (ii) $2d[Ca^{2+}]_i/dt$ is then plotted as a function of the $[Ca^{2+}]_i$ values obtained from the experimental fit. (iii) The plots from transients with equal peak $[Ca^{2+}]_i$ in each condition (control *versus* inhibitor) are pooled and fitted with a polynomial function of fifth to seventh order. Adapted from Fierro L, DiPolo R, Llano I (1998) Intracellular calcium clearance in Purkinje cell somata from rat cerebellar slices. *J. Physiol.* **510**, 499–512, with permission.

carboxyeosin (CE) is applied; and for blocking the Na–Ca exchanger, external Na^+ is replaced with Li^+, choline or N-methyl-D-glucamine, cations that cannot substitute for Na^+ in the exchange reaction. The rate of decay of $[Ca^{2+}]_i$ transients with similar peak values is compared in control and experimental conditions in order to calculate the rate of clearance. None of these inhibitors affects resting $[Ca^{2+}]_I$ levels, indicating that the passive leak of Ca^{2+} into the somata is small. For low-intensity $[Ca^{2+}]_I$ transients (0.5 μM at the peak), the proportion of intracellular Ca^{2+} removed by SERCA pumps, PMCA pumps and the Na–Ca exchanger is balanced (**Figure 7.9**). They equally remove 78% of the intracellular Ca^{2+}.

7.3 PRESYNAPTIC PROCESSES II: FROM $[Ca^{2+}]$ INCREASE TO SYNAPTIC VESICLE FUSION

Nerve terminals have three functionally distinct synaptic vesicle pools: (i) the readily-releasable pool (≈1%) corresponds to a few vesicles docked at the plasma membrane, which only require Ca^{2+} influx to trigger release (exocytosis) of their contents; (ii) the recycling pool (10–15%), at physiological stimulation frequencies, undergoes continuous cycles of docking, fusion, endocytosis and loading; (iii) finally, the reserve pool accounts for 80–90% of the total population and corresponds to vesicles that only respond to intense stimulation. This pool is probably mobilized when the recycling pool is depleted.

7.3.1 Overview of the hypothetical vesicle cycle in presynaptic terminals

From observations and experiments described in the following sections, synaptic vesicle traffic in nerve terminals is considered to involve several hypothetical stages (**Figure 7.10**).

- *Targeting and docking*. After they fill with neurotransmitter by active transport, synaptic vesicles dock at morphologically defined sites of the presynaptic plasma membrane.
- *Priming*. Through a series of molecular processes, docked vesicles acquire Ca^{2+}-sensitivity and become competent for fusion; i.e. they enter the readily-releasable pool.
- *Fusion*. The local increase of $[Ca^{2+}]_i$ triggers exocytosis of docked vesicles; i.e. fusion of synaptic vesicle membrane with the presynaptic plasma membrane and release of the vesicular content through a fusion pore.
- *Retrieval and recycling*. Vesicles are recycled via clathrin-dependent endocytosis.

Comparison of neurotransmitter release with other secretory systems shows that targeting, docking and fusion of vesicles involve common mechanisms. Synaptic transmission makes use of a mechanism that is common to biology.

7.3.2 Docking: a subpopulation of synaptic vesicles is docked to the active zone close to Ca^{2+} channels by means of specific pairing of vesicular and plasma membrane proteins

How do synaptic vesicles recognize the presynaptic plasma membrane for docking?

Selective targeting of a vesicle to its correct destination was proposed by G. Palade in 1970 to result from specific recognition sites between vesicle and plasma membranes. Later, the molecular basis of this specific interaction was studied by J. Rothman and colleagues in a cell-free preparation of Golgi stacks that reconstitutes vesicle-mediated transport between Golgi cisternae, a model of constitutive vesicle fusion with a membrane. The experiments described briefly here led to the discovery and purification of several proteins crucial for exocytosis.

Everything began with the discovery that *N*-ethylmaleimide (NEM) blocks the fusion of Golgi vesicles with Golgi stacks: the vesicles still bud off from cisternae but the released vesicles no longer fuse with the next stack membrane; they accumulate docked to the target membrane. This suggested the involvement of an NEM-sensitive fusion protein (NSF) in the fusion step. That protein was purified according to its ability to restore fusion after NEM inactivation. NSF is a 76 kDa protein, a water-soluble ATPase with two distinct ATP-binding sites. Adaptor proteins required for NSF function were subsequently purified and named soluble NSF-attachment

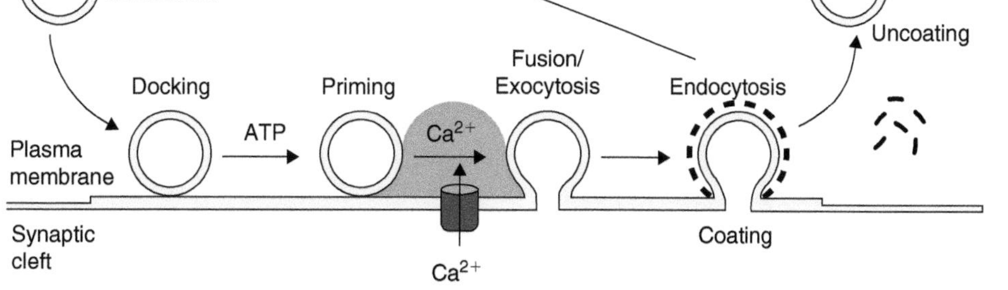

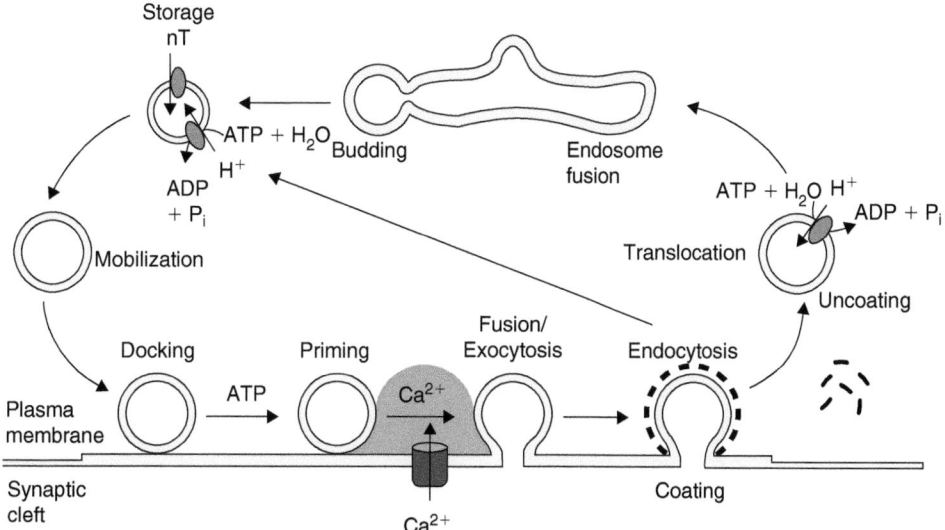

FIGURE 7.10 **Diagram of the hypothetical synaptic vesicle cycle in a presynaptic terminal.** The same synaptic vesicle is shown at different stages. Sites of docking, priming and fusion have been separated for clarity. NT, neurotransmitter. Adapted from Südhof TC (1995) The synaptic vesicle cycle: a cascade of protein–protein interactions. *Nature* **375**, 645–653, with permission.

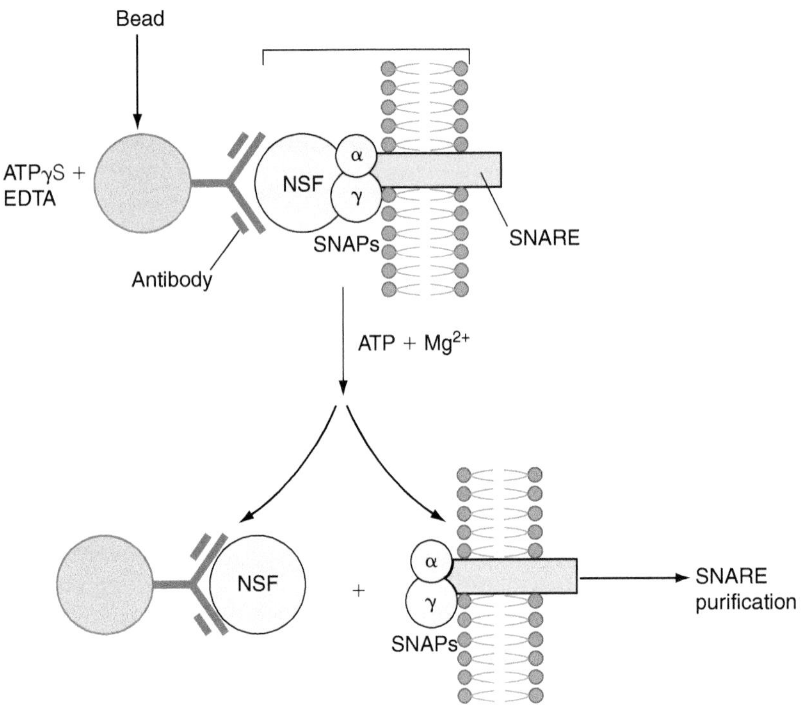

FIGURE 7.11 **The SNARE discovery.** Scheme of the experiment that identified the integral membrane proteins (SNAREs) of the vesicle and presynaptic plasma membranes of brain synapses. A stable complex between the NSF protein, the SNAPs and their membrane receptors can be isolated in the presence of a non-hydrolyzable analog of ATP (ATPγS). Inversely, the membrane-bound form of the NSF protein is released from the membranes to the cytoplasm by ATP hydrolysis (i.e. in the presence of ATP and Mg^{2+}). Solubilized brain membranes and NSF–SNAPs were immobilized on beads via a specific anti-*myc* antibody (NSF is tagged with the marker *myc*), in the presence of the non-hydrolyzable analog of ATP, ATPγS, and in the absence of Mg^{2+}. The stable complex [NSF–SNAP–membrane proteins] was thus captured. It was then dissociated in the presence of ATP and Mg^{2+} in NSF on the one hand and the complex SNAPs-membrane proteins on the other hand. Eluted proteins were collected and SNAREs characterized. For clarity, solubilized membrane proteins are represented by a black rectangle inserted in a lipid bilayer. From Rothman JE (1994) Intracellular membrane fusion. *Adv. Second Messenger Phosphoprotein Res.* **29**, 81–96, with permission.

proteins or SNAPs. There are three forms, α, β, and γ-SNAPs. The existence of such a membrane-bound form of NSF + SNAPs suggested that these proteins recognized specific receptors situated in the membrane.

The Rothman group used immobilized α-SNAP and NSF to isolate the SNAP receptors (SNAREs) (**Figure 7.11**). Since the complex NSF–SNAP is attached to membranes via SNAREs in the absence of hydrolyzable ATP, this property was utilized to purify SNAREs of the vesicle membrane (v-SNARE; v for vesicle) and SNAREs of the target presynaptic plasma membrane (t-SNARE; t for target). In the vesicle membrane, synaptobrevin (VAMP 2) was thus identified as a v-SNARE. In the presynaptic membrane, syntaxin 1 and SNAP-25 (SyNaptosome-Associated Protein 25, a protein of 25 kDa which has no relation to the similarly named SNAPs) were thus identified as t-SNAREs (**Figure 7.12a**). The cytoplasmic domains of the SNARE proteins assemble into a SNARE complex. The associating segments are called SNARE motifs and contain 60–70 residues. The SNARE complex is composed of four α-helices. In neurons, SNAP-25 contributes two helices to the SNARE complex while syntaxin 1 and synaptobrevin each contribute one helix. These three SNAREs, syntaxin 1, SNAP-25 and VAMP 2, form a stable heterotrimeric complex (**Figure 7.12b**) that has been proposed to mediate the docking of synaptic vesicles at the presynaptic plasma membrane.

The crucial function of SNARE proteins in numerous membrane trafficking pathways has been established in various organisms. Each transport vesicle has its own specific v-SNARE that pairs up in a unique match with a cognate t-SNARE found only at the intended target membrane. (NB: SNAREs are often categorized as Q-SNAREs or R-SNAREs based on their molecular properties. Q-SNAREs have a glutamine residue (Q in the single letter code) at the central zero-layer of their SNARE motif, while R-SNAREs have an arginine (R). Syntaxin 1 and SNAP-25 are Q-SNAREs while synaptobrevin is an R-SNARE.) The SNAREs differ from one vesicle transport system to the other, while NSF proteins and soluble NSF attachment proteins (SNAPs) are very general cytoplasmic proteins.

In summary, at the docking stage, the membrane of synaptic vesicles and the target plasma membrane are tied together via the binding of v-SNAREs (synaptobrevin) and t-SNAREs (syntaxin, SNAP-25) that are respectively vesicular and plasma membrane proteins. The cytoplasmic domains of these proteins assemble into a four-helix bundle that pulls the vesicle and target membranes together.

7.3.3 Multiple Ca^{2+} ions must bind to Ca^{2+} receptor(s) to initiate vesicle fusion (exocytosis)

SNAREs constitute the core of a conserved fusion machine, but additional accessory proteins must serve to regulate the fusion reaction, in particular at least one protein that senses Ca^{2+}, since in Ca^{2+}-regulated exocytosis as in neuronal synapses, the docked vesicles do not fuse with a high probability until a significant $[Ca^{2+}]_i$ rise occurs. The delay between $[Ca^{2+}]_i$ rise and the postsynaptic response can be as short as 60 to 200 μs, placing

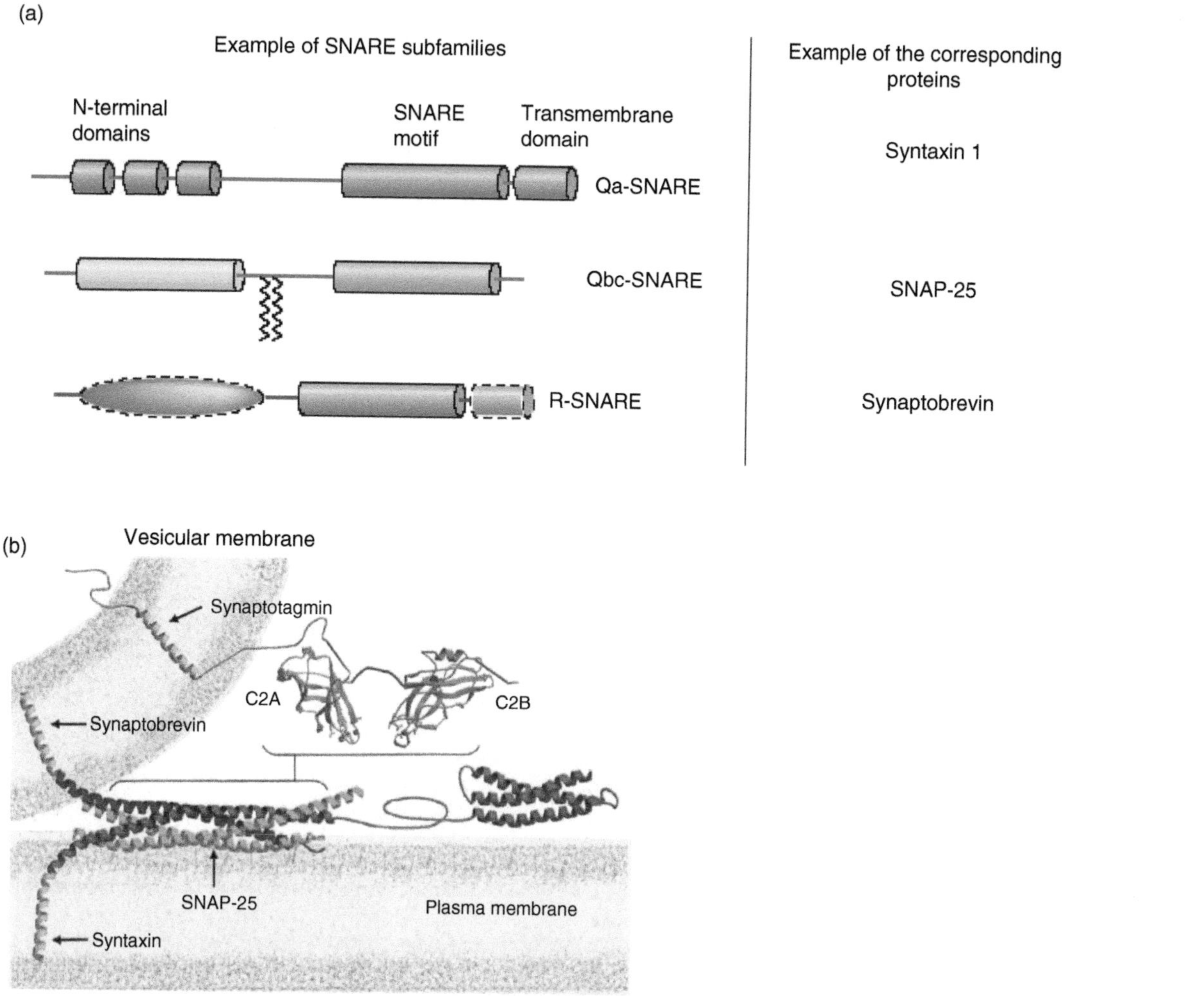

FIGURE 7.12 **The structures of three SNARE subfamilies and model of a trans-SNARE complex.** **(a)** Qa-SNAREs have N-terminal antiparallel three-helix bundles. Qbc-SNAREs represent a small subfamily of SNAREs – the SNAP-25 (25-kDa synaptosome-associated protein) subfamily – that contain two SNARE motifs connected by a linker that is frequently palmitoylated (zig-zag lines), and most of the members of this subfamily function in constitutive or regulated exocytosis. The various N-terminal domains of R-SNAREs are represented by a basic oval shape. Dashed domain borders highlight domains that are missing in some subfamily members. **(b)** Model of a *trans*-SNARE complex and synaptotagmin. The regions that interact are indicated with *brackets*. Part **(a)** adapted from Jahn R, Scheller RH (2006) SNAREs-engines for membrane fusion. *Nat. Rev. Mol. Cell Biol.* 7, 631–643, with permission. Part **(b)** adapted from Littleton JT, Bai J, Vyas B *et al.* (2001) Synaptotagmin mutants revel essential functions for the C2B domain in Ca-triggered fusion and recycling of synaptic vesicles *in vivo*. *J. Neurosci.* **21**, 1421–1433, with permission.

strong kinetic constraints on the transduction pathway that ends up in exocytosis and release of neurotransmitter by the presynaptic terminal.

What is the local Ca^{2+} concentration required to trigger vesicle fusion?

The Ca^{2+} sensor (also called Ca^{2+} receptor) should have a low affinity for Ca^{2+} ions, in the order of tens or hundreds of micromolars depending on the nerve terminal that is under study.

Why is Ca^{2+} entry close to docked vesicles and release sites?

As explained in Section 7.2.3, SNARES are tightly coupled to either N- or P/Q-type Ca^{2+} channels. This ensures Ca^{2+} entry close to docked vesicles and release sites (**Figure 7.7**).

What is the identity of the Ca^{2+} receptors that transduce the $[Ca^{2+}]_i$ rise to trigger exocytosis?

Synaptotagmin 1, a 65 kDa synaptic vesicle protein, is the major Ca^{2+} receptor in presynaptic exocytosis. It has a large cytoplasmic domain composed of tandem Ca^{2+}-binding motifs called C2 domains (C2A and C2B) (**Figures 7.7d** and **7.12b**). It spans the vesicle membrane once, near its N-terminus, and possesses a short intravesicular domain. Synaptotagmin 1 seems to bind a total of five Ca^{2+} ions, three by C2A and two by C2B. Synaptotagmins form a multigene family with more

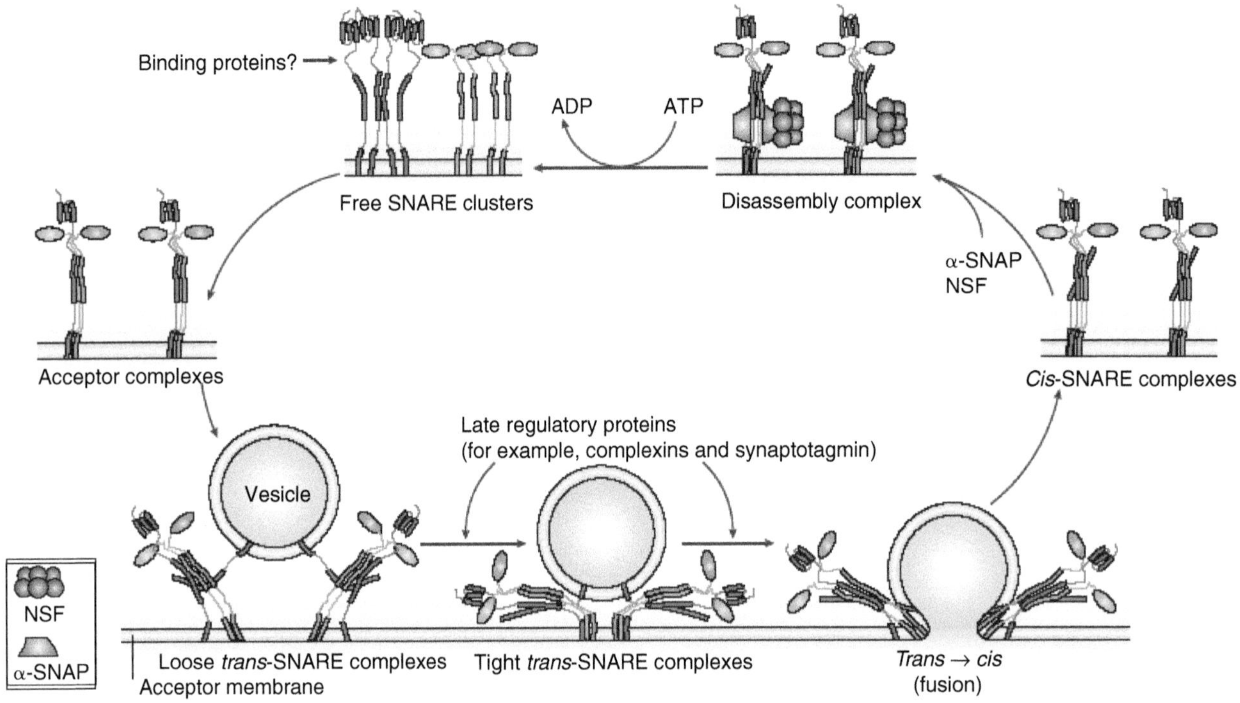

FIGURE 7.13 **Hypothetical SNARE conformational cycle during vesicle docking and fusion.** As an example, we consider three Q-SNAREs (Q-soluble *N*-ethylmaleimide-sensitive factor attachment protein receptors) on an acceptor membrane (red and green) and an R-SNARE on a vesicle (blue). Q-SNAREs, which are organized in clusters (top left), assemble into acceptor complexes. Acceptor complexes interact with the vesicular R-SNAREs through the N-terminal end of the SNARE motifs, and this nucleates the formation of a four-helical *trans*-complex. *Trans*-complexes proceed from a loose state (in which only the N-terminal portion of the SNARE motifs are 'zipped up') to a tight state (in which the zippering process is mostly completed), and this is followed by the opening of the fusion pore. In regulated exocytosis, these transition states are controlled by late regulatory proteins that include complexins (small proteins that bind to the surface of SNARE complexes) and synaptotagmin (which is activated by an influx of calcium). During fusion, the strained *trans*-complex relaxes into a *cis*-configuration. *Cis*-complexes are disassembled by the AAA1 (ATPases associated with various cellular activities) protein NSF (*N*-ethylmaleimide-sensitive factor) together with SNAPs (soluble NSF attachment proteins) that function as co-factors. The R- and Q-SNAREs are then separated by sorting (e.g. by endocytosis). Adapted from Jahn R, Scheller RH (2006) SNAREs – engines for membrane fusion. *Nat. Rev. Mol. Cell Biol.* 7, 631–643, with permission.

than 15 members. Synaptotagmin co-immunoprecipitates with syntaxin from rat brain extracts and binds to isolated t-SNAREs. Binding occurs at the base of t-SNAREs in the region that assembles into SNARE complexes (**Figure 7.12b**). This binding site suggests that synaptotagmin regulates SNARE function. The C2-domains of synaptotagmin also bind lipid bilayers in a Ca^{2+}-dependent manner and can penetrate membranes that contain phosphatidylserine. Phosphatidylserine is mainly present in the inner layer of the plasma membrane.

7.3.4 Fusion: from Ca^{2+} binding to exocytosis

Fusion is the process by which two lipid membranes merge. Fusion between synaptic vesicles and presynaptic plasma membrane leads to release of vesicle contents (neurotransmitters) into the synaptic cleft by exocytosis. Synaptic vesicle fusion is regulated by an increase in $[Ca^{2+}]_I$ (as opposed to constitutive fusion which is Ca^{2+}-independent).

How does binding of Ca^{2+} to its receptor(s) initiate exocytosis?

One model (**Figure 7.13**) proposes that in the absence of Ca^{2+}, v- and t-SNAREs partially preassemble into a ring-like structure. In response to $[Ca^{2+}]_i$ rise, synaptotagmin binds Ca^{2+} ions and, by changing conformation, drives complete assembly of trans-SNARE complexes, bringing the membranes that are destined to fuse into close proximity such as in the center of the ring, so the hydratation barrier at the surface of the membranes is overcome and the membranes touch.

Other interpretations have been proposed. For example, synaptotagmin would bind to t-SNAREs and block SNARE assembly as long as $[Ca^{2+}]_i$ does not rise. Then, upon binding of Ca^{2+} ions, synatotagmin would change conformation and let SNAREs assemble. In this model, Ca^{2+}-free synaptotagmin acts as a fusion clamp. Complexin (also called synaphin), a soluble synaptotagmin and SNARE-complex binding protein is also thought to inhibit full SNARE complex zipping by binding to a groove in partially assembled SNARE complexes (**Figure 7.13**).

Whichever the mechanism, one must keep in mind that even when all the proper conditions come together (a presynaptic spike, opening of Ca^{2+} channels, Ca^{2+} entry, Ca^{2+} binding), the average probability of the exocytosis of a synaptic vesicle still remains below 1 (see Appendix 7.2). In fact, the mean release probability in the CNS is between 0.05 and 0.5 according to the synapse studied. In other words, Ca^{2+}-triggered exocytosis is *inefficient*: only one out of two to 20 action potentials leads to exocytosis.

Exocytosis involves the formation of a fusion pore

The earliest attempts to characterize the fusion pore came from the ultrastructural studies of Heuser and Reese in 1981. When exocytosis has been captured by rapid freezing (in the presence of pharmacological agents to prolong the excitation of the nerve terminal, see Section 7.3.5), images suggested the presence of a narrow fusion pore. Later, the development of patch-clamp capacitance techniques demonstrated the formation of a fusion pore during exocytosis. When chromaffin cells that release adrenalin are excited, small step-like increases of membrane capacitance of about 1 fF are recorded. These capacitance steps correspond to exocytosis of large dense core vesicles and, for each exocytosis, to the formation of a fusion pore: an increase of capacitance corresponds to an increase of membrane surface that occurs if the vesicle fuses, even transiently, with the plasma membrane. After formation, the pore would expand in order to allow rapid transmitter release in the cleft. Theoretical models propose that the fusion pore expands at a rate approaching 100 nm ms^{-1} within a few tens of microseconds to achieve the observed transmitter release time course.

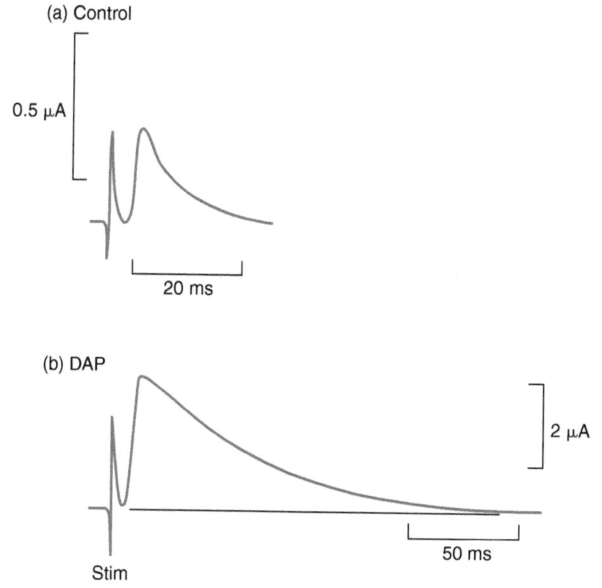

FIGURE 7.14 Diaminopyridine (DAP) increases the duration and amplitude of motor endplate current. The postsynaptic endplate current of a frog sartorius muscle cell is evoked by stimulation (2 μA intensity, 5 ms duration) of the motor nerve (V_H = −90 mV). (**a**) The average amplitude of the evoked postsynaptic current is 0.5 μA in control Ringer solution. (**b**) The postsynaptic current evoked by the same stimulation in the presence of DAP (1 mM) is always greater than 1 μA and can reach 3.3 μA. These inward currents are represented upwardly, which is unusual. Adapted from Katz B, Miledi R (1979) Estimates of quantal content during chemical potentialization of transmitter release. *Proc. R. Soc. Lond. B* **205**, 369–378, with permission.

7.3.5 Pharmacology of neurotransmitter release

Agents blocking K^+ currents potentiate neurotransmitter release

Voltage-gated K^+ channels (see Section 4.3) and Ca^{2+}-activated K^+ channels ($K_{(Ca)}$; Section 5.3) are present at the presynaptic membrane. The outflow of K^+ ions through these channels repolarizes the presynaptic membrane and limits Ca^{2+} entry (in amplitude and duration). At the neuromuscular junction, simultaneous labeling with fluorescent toxins specific for presynaptic $K_{(Ca)}$ and Ca^{2+} channels showed that they are localized close to one another, at presynaptic active zones. This organization ensures a rapid activation of these K^+ channels during Ca^{2+} entry triggered by an action potential.

When these K^+ channels are blocked, Ca^{2+} entry is increased and so is transmitter release. At the neuromuscular junction, potassium channel blockers such as TEA and 3,4-diaminopyridine, when used in conjunction with TTX (to block voltage-dependent Na^+ channels and thus inhibit repetitive nerve impulses and muscle contraction), potentiate the postsynaptic response (**Figure 7.14**). In the presence of K^+ blocking agents, the endplate current has an average time of decline much longer than that of the miniature current. This is due in part to the fact that, in this case, the release of vesicles is asynchronous and there is a temporal spread of quantal release during the entire presynaptic calcium spike. On the other hand, the massive release possibly saturates the enzyme acetylcholinesterase and, as a result, there is a repeated activation of postsynaptic nicotinic receptors.

Example of botulinum toxins

Botulinum toxins block synaptic transmission at peripheral synapses, notably the neuromuscular junction and their effect often persists for several months. Without treatment the patient can die from asphyxia (paralysis of respiratory muscles) and, in cases where the patient survives, he may suffer from muscle atrophy owing to the non-functioning of muscle cells. Botulinum neurotoxins, of which there are eight different types (A to H), are derived from the microorganism *Clostridium botulinum*. Botulinum neurotoxin A is the most toxic protein known to humans. Botulinum toxins are proteases, each of which cleaves a single target at a single site. These proteins have a heavy chain (100 kDa) linked by a single disulfide bond to a light chain (50 kDa). The heavy chain is responsible for the selective binding of the toxin to neuronal cells and for the penetration of the light chain into neurons; the light chain bears the protease activity. It contains a consensus sequence of the catalytic site of metallopeptidases and has a Zn^{2+}-dependent endopeptidase activity.

Botulinum toxins enter nerve terminals and decrease the number of acetylcholine-containing vesicles released into the synaptic cleft, without affecting acetylcholine synthesis or action potential conduction in the motor nerve. The effect of the toxin can be reversed by increasing Ca^{2+} concentration in the extracellular medium or by adding TEA or 3,4-diaminopyridine, agents that potentiate Ca^{2+} influx. Knowing these facts, it was hypothesized that botulinum toxin affected either Ca^{2+} entry or the coupling between intracellular Ca^{2+} concentration increase and exocytosis of synaptic vesicles. To verify the first proposition, Ca^{2+} entry was recorded in terminals 'paralyzed' by botulinum toxins. The results showed that presynaptic Ca^{2+} current is not significantly changed (**Figure 7.15a**). The botulinum toxin therefore acts presynaptically to decrease acetylcholine release, after the entry of Ca^{2+} ions.

In order to identify the intracellular target of botulinum toxins, synaptic vesicles from rat cerebral cortex were purified. Among the many proteins detected in these vesicles, one protein band was altered by

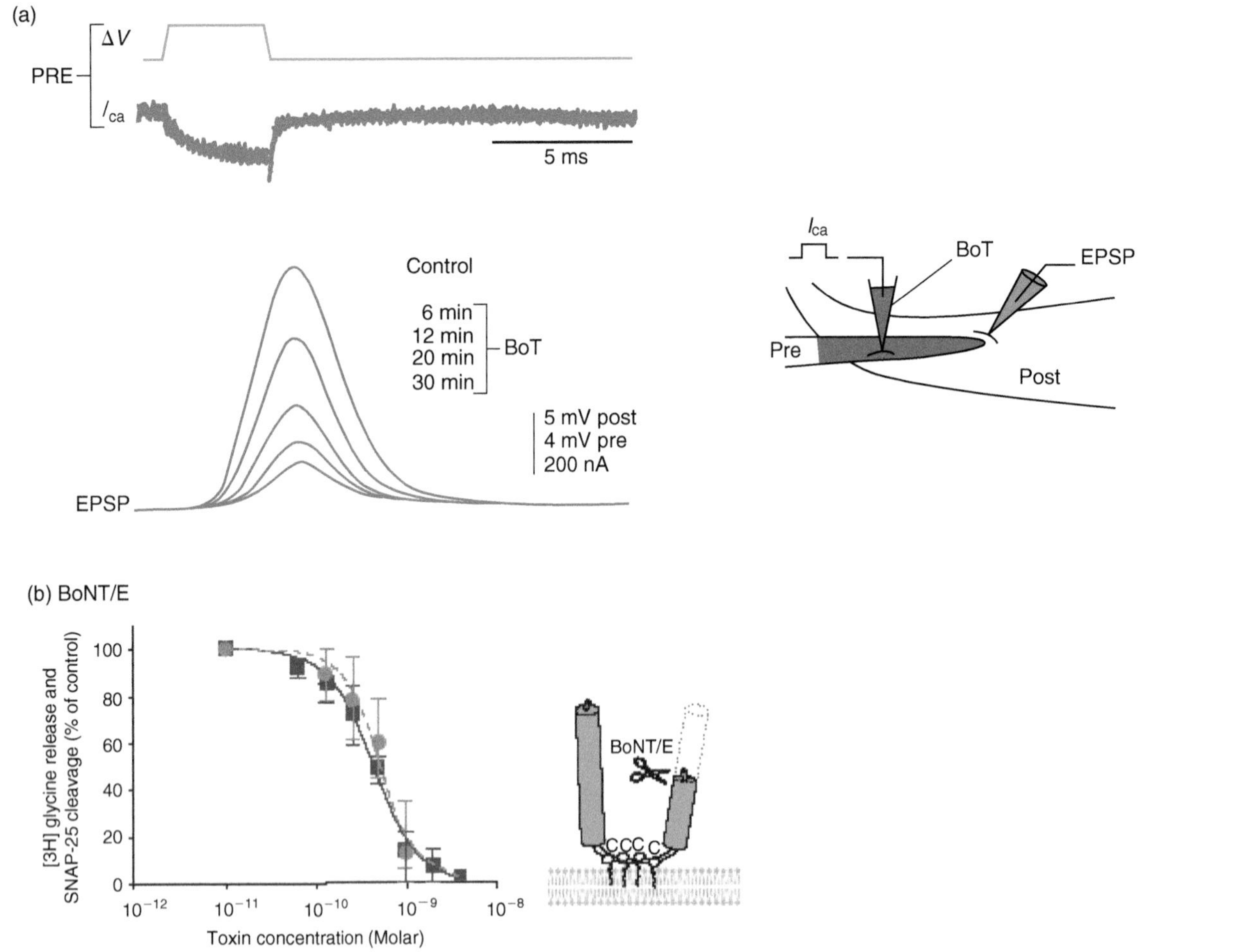

FIGURE 7.15 **Botulinum toxins strongly decrease synaptic transmission without affecting presynaptic Ca^{2+} current. (a)** In the squid giant synapse, at the stellate ganglion, presynaptic and postsynaptic intracellular electrodes are implanted to allow simultaneous recording of the presynaptic Ca^{2+} current (voltage clamp mode) and the postsynaptic response (EPSP, current clamp mode). A presynaptic voltage step (ΔV) evokes a presynaptic Ca^{2+} current (I_{Ca}) and, after a delay, a postsynaptic response (EPSP, control). After injection of botulinum toxin (BoT) through the presynaptic electrode, the EPSP decreases with time. Note that in the same time the presynaptic Ca^{2+} current is unchanged. **(b)** Dose–response curve of the effect of botulinum neurotoxin E (BoNT/E) on SNAP-25 proteolysis (purple squares, continuous line) and glycine release (green circles, broken line) in cultured neurons. The right-hand panel shows a schema of SNAP-25 and the position of BoNT/E cleavage. Part **(a)** adapted from Marsal J, Ruiz-Montasell B, Blasi J *et al.* (1997) Block of transmitter release by botulinum C1 action on syntaxin at the squid giant synapse. *Proc. Natl. Acad. Sci. USA* **94**, 14871–14876, with permission. Part **(b)** adapted from Keller JE *et al.* (2001) Uptake of botulinum neurotoxin into cultured neurons. *Biochemistry* **43**, 526–532, with permission.

incubation with botulinum toxin. The electrophoretic mobility of this band corresponds to that of synaptobrevin, the v-SNARE which plays a role in vesicle fusion. Syntaxin 1 and SNAP-25 are also targets for botulinum toxin. Botulinum toxins B, D, F and G are specific for synaptobrevin, botulinum toxin C cleaves syntaxin 1, and botulinum toxins A and E are specific for SNAP-25 (**Figure 7.15b**).

Botulinum toxins specifically bind to nerve terminals. After endocytotic uptake, they deliver their zinc-endopeptidase light chain inside the cytosol, where it specifically cleaves a SNARE protein at a single site. The cleavage of one of the SNAREs greatly reduces the probability of neurotransmitter release.

7.4 PROCESSES IN THE SYNAPTIC CLEFT: FROM TRANSMITTER RELEASE IN THE CLEFT TO TRANSMITTER CLEARANCE FROM THE CLEFT

The time course of a transmitter in the cleft depends on the balance between the amount of transmitter released per unit of time and the efficacy of clearance mechanisms that clear transmitter molecules from the cleft.

7.4.1 The amount of neurotransmitter released in the synaptic cleft

The amount of transmitter released per vesicle and per unit of time in the synaptic cleft depends on (i) the concentration of the transmitter in the exocytotic vesicle, (ii) the volume of the vesicle, and (iii) the rate of transmitter release through the vesicle fusion pore into the cleft (i.e. the dimension of the fusion pore).

What is the concentration of neurotransmitter in a synaptic vesicle?

Consider the example of synaptic vesicles that contain glutamate as a neurotransmitter. To isolate synaptic vesicles, antibodies against a vesicular protein (such as synaptophysin) are immobilized on the surface of non-porous methacrylate microbeads. Using these immunobeads, synaptic vesicles are isolated. To avoid the loss of glutamate, the vesicular H^+ gradient (responsible for the transport of neurotransmitter molecules into vesicles; see Appendix 3.1) is preserved by adding an ATP-regenerating system. Under these conditions, high levels of glutamate are found in vesicles: 0.8 μmoles of glutamate per milligram of synaptophysin. Knowing that synaptophysin represents 7% of total vesicle protein, it gives 60 nmoles of glutamate per milligram of protein. This gives an intravesicular concentration of 60 mM, assuming an internal volume of 1 μl mg^{-1} of protein. The concentration of glutamate in synaptic vesicles is estimated at 60–210 mM depending on the preparation studied. The concentrations of other transmitters such as GABA and glycine are not known.

Is the vesicle content stable? If yes, what mechanisms regulate the amount of neurotransmitter per vesicle?

Miniature spontaneous postsynaptic responses that correspond to the exocytosis of a very small number of vesicles (say one, two or three; see Appendix 7.1) have quite stable amplitude. This suggests that the vesicle content is relatively constant. Therefore, at a given synapse, synaptic vesicles would contain the same amount of transmitter. Active transport into synaptic vesicles is achieved by proteins that couple the uptake of transmitter to the movement of H^+ in the opposite direction (along its electrochemical gradient). A vesicular H^+-ATPase provides the H^+ electrochemical gradient that drives transmitter uptake. Thus, to achieve a stable vesicular content, the number of synaptic transporters in the vesicle membrane, the activity of the H^+-ATPase and the cytoplasmic concentration of transmitter must be stable.

How does the transmitter diffuse from the vesicle into the cleft?

The detailed nature of the vesicle fusion process remains unclear. In particular, the formation of the fusion pore, its opening rate, as well as how molecules diffuse out from the vesicle are not known. It has been hypothesized that two distinct modes of fusion can occur. The first mode is classical 'full-fusion' exocytosis in which the synaptic vesicle collapses entirely into the plasma membrane releasing its entire contents. The second involves transitory opening of a fusion pore with only partial release. In this mode, called 'kiss-and-run', neurotransmitter release and refilling can occur without complete incorporation of the vesicle into the target membrane. While evidence for the 'kiss-and-run' mode has been obtained in large dense core vesicle (LDCV) fusion, the role of this mode in synaptic vesicle fusion is still controversial.

What is the peak concentration of the neurotransmitter in the cleft?

The time course of neurotransmitter concentration can be evaluated experimentally. The technique utilizes the non-equilibrium displacement of a competitive antagonist following the synaptic release of transmitter. A specific antagonist of the postsynaptic ionotropic receptors that mediate transmission is applied. Attenuation of synaptic response amplitude is measured at one or more concentrations of a rapidly dissociating competitive antagonist and a dose–response curve is constructed for the inhibition of synaptic transmission. Transmitter peak concentration in the cleft would be around 1 mM for glutamate and achieved in around 20 μs.

7.4.2 Transmitter time course in the synaptic cleft is brief and depends mainly on a transmitter binding to target proteins

The speed of transmitter clearance from the synaptic cleft is a fundamental parameter influencing many aspects of synaptic function. The amount of time the neurotransmitter stays in the cleft depends on (i) the amount of transmitter released, (ii) the transmitter diffusion coefficient, the geometry of the cleft and adjacent extra synaptic space, (iii) the distribution and affinity of transmitter binding sites, and (iv) the transporter uptake rate and/or the turnover rate of degradative enzymes. Therefore the transmitter time course varies significantly from synapse to synapse.

Theoretical models predict that within 50 μs the transmitter is evenly distributed throughout the cleft and by 500 μs the cleft is clear of the transmitter. Only transporters, degradative enzymes and diffusion achieve the real removal of neurotransmitter molecules from the cleft. Binding to pre- and postsynaptic receptors is a temporary clearance (buffering) since transmitter molecules will be back in the cleft as soon as they unbind from receptors (as already seen for Ca^{2+}clearance; see Section 7.2.4). The turnover rate for known neurotransmitter transporters is in the range 1–15 s^{-1}: a single transporter requires at least 60 ms to complete its cycle. This is extremely slow compared with the turnover rate for acetylcholinesterase which is in the order of 10^4 s^{-1}. It shows how highly efficient this enzyme is in removing acetylcholine from the cleft at nicotinic synapses (e.g. the neuromuscular junction). Therefore, for glutamate and GABA, which are not enzymatically degraded in the cleft, uptake transporters are too slow to achieve rapid removal of transmitter molecules from the cleft.

In central synapses, rapid buffering of transmitter molecules from the cleft arises from the binding of these molecules to receptor proteins: postsynaptic receptor channels (ionotropic receptors), and pre- and postsynaptic G-protein-linked receptors (metabotropic receptors). These receptors bind the transmitter molecules tightly enough (i.e. with a high affinity) to prevent release from binding sites for tens of milliseconds, sufficient time for transporters to become less saturated. In other words, each transporter stands ready to bind a transmitter molecule as soon as it is released from a receptor. Otherwise, neurotransmitter molecules would be released a second time in the cleft and evoke a second postsynaptic response or amplify the duration of the first one.

7.5 SUMMARY (FIGURES 7.16 AND 7.17)

The answers to the questions raised in Section 7.1.3 are as follows:

- The opening of presynaptic high-voltage-activated Ca^{2+} channels (N- and P/Q-type Ca^{2+} channels) is triggered by membrane depolarization that occurs during the depolarizing phase of presynaptic action potentials.
- Presynaptic $[Ca^{2+}]_i$ increase is local in response to a presynaptic spike, due to the clustering of N- and P/Q-type Ca^{2+} channels close to docked vesicles, at release sites (active zones). The $[Ca^{2+}]_i$ increase is transient since Ca^{2+} channels open and close quickly and Ca^{2+} ions are cleared from the active zones by diffusion, binding to receptor proteins (cytoplasmic Ca^{2+}-binding proteins and transmembrane proteins such as Ca^{2+}-ATPase or Na–Ca exchanger).
- Exocytosis of a docked vesicle is triggered when local $[Ca^{2+}]_i$ increase is around 0.5–40 μM. The first step includes binding of multiple Ca^{2+} ions to Ca^{2+}-binding protein(s) (Ca^{2+}-sensor(s) such as synaptotagmin 1). Although the exact mechanism of vesicle fusion is not yet known, the assembly of trans-SNARE complexes is a critical step. Exocytosis is rapidly triggered since Ca^{2+} entry is close to release sites and each docked and primed vesicle is ready to fuse.
- The transmitter is released in the synaptic cleft through a fusion pore.
- The neurotransmitter molecules present in the cleft bind to specific receptors such as pre- and postsynaptic ionotropic and metabotropic receptors. Binding to postsynaptic ionotropic receptors results in an EPSC or IPSC.
- The clearance of the neurotransmitter molecules (total removal from the cleft) is achieved by their binding to neuronal and glial transporters, and/or to degradative enzymes, and by their diffusion out of the cleft. Theoretical models predict that by 500 μs the cleft is clear of neurotransmitter molecules.

Neurotransmitter release, including steps from Ca^{2+} entry into the presynaptic terminal to vesicle exocytosis, is achieved by a cascade of chemically-gated events (**Table 7.1**). It differs fundamentally from electrical events such as action potentials that are achieved by a cascade of voltage-gated events. Once presynaptic Ca^{2+} channels open, synaptic transmission is determined by the different affinity constants of the reactions between ligands and receptors that underlie synaptic transmission: this includes binding of Ca^{2+} to Ca^{2+} sensor(s), and binding of neurotransmitter molecules to pre- and postsynaptic receptors (receptor channels,

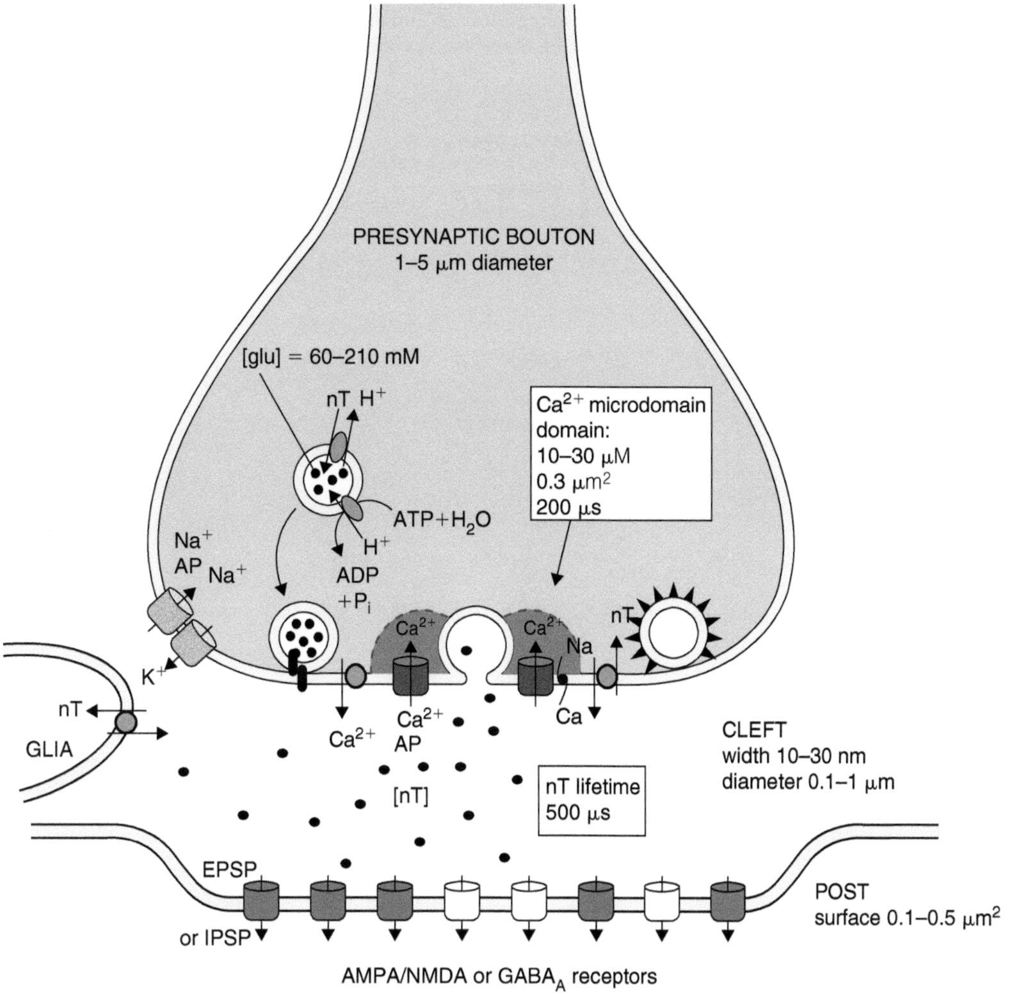

FIGURE 7.16 **Cascade of events leading to neurotransmitter release and its clearance from the synaptic cleft.** Schematic of some steps of synaptic transmission (between the presynaptic action potential and the postsynaptic response, EPSP or IPSP). In the presynaptic and glial plasma membranes, only one example of each channel, pump or transporter, is represented owing to the lack of space. In the postsynaptic membrane, several examples of ionotropic glutamatergic (AMPA and NMDA) or GABAergic ($GABA_A$) channels are represented. nT, neurotransmitter.

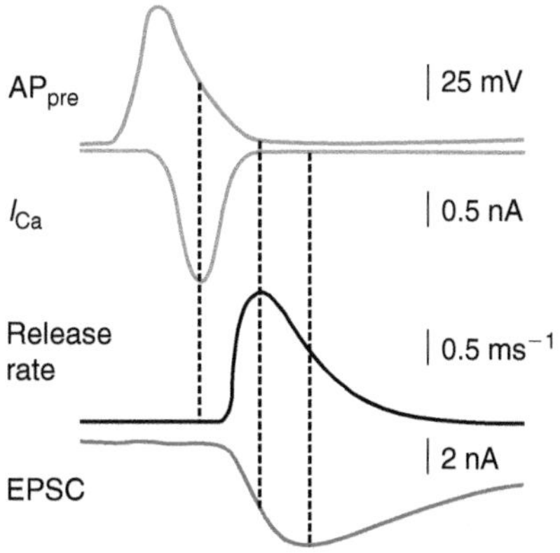

FIGURE 7.17 **Summary (example of the calyx of Held).** Time course of the signaling cascade, showing (top to bottom) the presynaptic action potential (AP) waveform and resulting Ca^{2+} current (I_{Ca}), (inferred) release rate and postsynaptic EPSC. Adapted from Meinrenken CJ, Borst JGG, Sakmann B (2003) Local routes revisited: the space and time-dependence of the Ca^{2+} signal for phasic transmitter release at the rat calyx of Held. *J. Physiol.* **547**, 665–689 with permission.

TABLE 7.1 Some of the reactions (voltage-gated and ligand-gated) from the opening of N- and P/Q-type Ca^{2+} channels to transmitter release into the synaptic cleft and its clearance from the synaptic cleft. The numbers indicate the values of K_D or EC_{50}. All the vesicle steps have been omitted

Location of reactions	Type of reaction	Steps of Figure 6.4a			Reactions	Effect
PRE	Voltage-gated	Step 2	N & P/Q channels closed	action potential $\rightleftharpoons$	N & P/Q channels open	Ca^{2+} entry
		Step 3a	$_{2\text{–}5}Ca^{2+}$ + sensors	10–30 μM $\rightleftharpoons$	$_{2\text{–}4}$Ca-sensor	Exocytosis
	Ligand-gated					
		Step 3b	Ca^{2+} + pumps	0.25–1.50 mM $\rightleftharpoons$	Ca-pumps	Ca^{2+}
			Ca^{2+} + transporters		Ca-transporters clearance	
CLEFT	Ligand-gated	Step 5a/6	2Glu +AMPAR	0.25–1.50 mM $\rightleftharpoons$	Glu_2-AMPAR	EPSP
			2Glu + NMDAR	1 μM $\rightleftharpoons$	Glu_2-NMDAR	
		Step 5a/6	2Gaba + $GABA_A$	8–40 μM $\rightleftharpoons$	$Gaba_2$-$GABA_A$	IPSP
		Step 5b/5c	nT + transporter	high affinity $\rightleftharpoons$	nT + transporter	nT uptake

G-protein-coupled receptors, uptake transporters). Transmitter release from presynaptic elements is a Ca^{2+}-regulated, multiprotein process.

Further reading

Baumert, M., Maycox, P.R., Navone, F., et al., 1989. Synaptobrevin: an integral membrane protein of 18,000 daltons in small synaptic vesicles of rat brain. EMBO J. 8, 379–384.

Bennett, M.K., Calakos, N., Scheller, R.H., 1992. Syntaxin: a synaptic protein implicated in docking of synaptic vesicles at presynaptic active zones. Science 257, 255–259.

Bollmann, J.H., Sakmann, B., Borst, J.G.G., 2000. Calcium sensitivity of glutamate release in a calyx-type terminal. Science 289, 953–957.

Burger, P.M., Mehl, E., Cameron, P.L., et al., 1989. Synaptic vesicles immunoisolated from rat cerebral cortex contain high levels of glutamate. Neuron 3, 715–720.

Calakos, N., Scheller, R.H., 1996. Synaptic vesicles, docking and fusion: a molecular description. Physiol. Rev. 76, 1–29.

Jackson, M.B., Chapman, E.R., 2006. Fusion pores and fusion machines in Ca^{2+}-triggered exocytosis. Annu. Rev. Biophys. Biomol. Struct. 35, 135–160.

Kochubey, O., Lou, X., Schneggenburger, R., 2011. Regulation of transmitter release by Ca(2+) and synaptotagmin: insights from a large CNS synapse. Trends Neurosci. 34, 237–246.

Leveque, C., el Far, O., Martin-Moutot, N., et al., 1994. Purification of the N-type calcium channel associated with syntaxin and synaptotagmin: a complex implicated in synaptic vesicle exocytosis. J. Biol. Chem. 269, 6306–6312.

Littleton, J.T., Barnard, J.O., Titus, S.A., et al., 2001. SNARE-complex disassembly by NSF follows synaptic-vesicle fusion. Proc. Natl Acad. Sci. USA 98, 12233–12238.

Matthew, W.D., Tsavaler, L., Reichardt, L.F., 1981. Identification of a synaptic vesicle-specific membrane protein with a wide distribution in neuronal and neurosecretory tissue. J. Cell. Biol. 91, 257–269.

Miledi, R., 1973. Transmitter release induced by injection of calcium into nerve terminals. Proc. R. Soc. Lond. B. Biol. Sci. 183, 421–425.

Oyler, G.A., Higgins, G.A., Hart, R.A., et al., 1989. The identification of a novel synaptosomal-associated protein, SNAP-25, differentially expressed by neuronal subpopulations. J. Cell Biol. 109, 3039–3052.

Prado, V.F., Martins-Silva, C., de Castro, B.M., et al., 2006. Mice deficient for the vesicular acetylcholine transporter are myasthenic and have deficits in object and social recognition. Neuron 51, 601–612.

Rettig, J., Heinemann, C., Ashery, U., et al., 1997. Alteration of Ca^{2+} dependence of neurotransmitter release by disruption of Ca^{2+} channel/syntaxin interaction. J. Neurosci. 17, 6647–6656.

Rizzoli, S.O., Betz, W.J., 2005. Synaptic vesicle pools. Nat. Rev. Neurosci. 6, 57–69.

Rothman, J.E., 1994. Intracellular membrane fusion. Adv. Second Messenger Phosphoprotein Res. 29, 81–96.

Schneggenburger, R., Han, Y., Kochubey, O., 2012. Ca(2+) channels and transmitter release at the active zone. Cell Calcium 52, 199–207.

Sheng, Z.H., Westenbroek, R.E., Catterall, W.A., 1998. Physical link and functional coupling of presynaptic calcium channels and the synaptic vesicle docking/fusion machinery. J. Bioenerg. Biomembr. 30, 335–345.

Sheng, Z.H., Yokoyama, C.T., Catterall, W.A., 1997. Interaction of the synprint site of N-type Ca^{2+}-channels with the C2B domain of synaptotagmin I. Proc. Natl Acad. Sci. USA 94, 5405–5410.

Söllner, T., Whiteheart, S.W., Brunner, M., et al., 1993. SNAP receptors implicated in vesicle targeting and fusion. Nature 362, 318–324.

Song, H., Ming, G., Fon, E., et al., 1997. Expression of a putative vesicular acetylcholine transporter facilitates quantal transmitter packaging. Neuron 18, 815–826.

Südhof, T.C., Rothman, J.E., 2009. Membrane fusion: grappling with SNARE and SM proteins. Science 323, 474–477.

Weber, T., Zemelman, B.V., McNew, J.A., et al., 1998. SNAREpins: minimal machinery for membrane fusion. Cell 92, 759–772.
Whiteheart, S.W., Rossnagel, K., Buhrow, S.A., et al., 1994. *N*-ethymaleimide-sensitive fusion protein: a trimeric ATPase whose hydrolysis of ATP is required for membrane fusion. J. Cell Biol. 126, 945–954.
Zhang, X., Kim-Miller, M.J., Fukuda, M., et al., 2002. Ca^{2+}-dependent synaptotagmin binding to SNAP-25 is essential for Ca-triggered exocytosis. Neuron 34, 599–611.

APPENDIX 7.1 QUANTAL NATURE OF NEUROTRANSMITTER RELEASE

A7.1.1 Spontaneous release of acetylcholine at the neuromuscular junction evokes miniature endplate potentials: the notion of quanta

At the neuromuscular junction (also called the motor endplate), when an intracellular electrode is implanted in a muscle fiber at the level of the postsynaptic membrane in the presence of TTX and in the absence of extracellular Ca^{2+}, spontaneous postsynaptic potentials of very small amplitude (0.5–1.0 mV on average) are recorded, though presynaptic spikes and synaptic transmission are blocked (**Figure A7.1a**). They occur randomly at a low frequency (about 1 s^{-1}). They have been called 'miniature' endplate potentials (mEPP) by Fatt and Katz (1952). As explained by Katz (1966): 'Except for their spontaneous occurrence and their small size, the miniatures are indistinguishable from the EPSPs evoked by presynaptic nerve stimulation.' For example, curare suppresses them and acetylcholinesterase inhibitors enhance their amplitude and duration with the same doses and to approximately the same extent. Normally, miniature potentials are well below the firing level of the muscle cell and so remain localized and produce no contraction. Since they disappear after the motor nerve has been cut and in the presence of botulinum toxin, they are evoked by presynaptic release of Ach. The interpretation is that motor nerve terminals at rest are in a state of intermittent secretory activity: they liberate small quantities of ACh at random intervals at an average rate of about one per second.

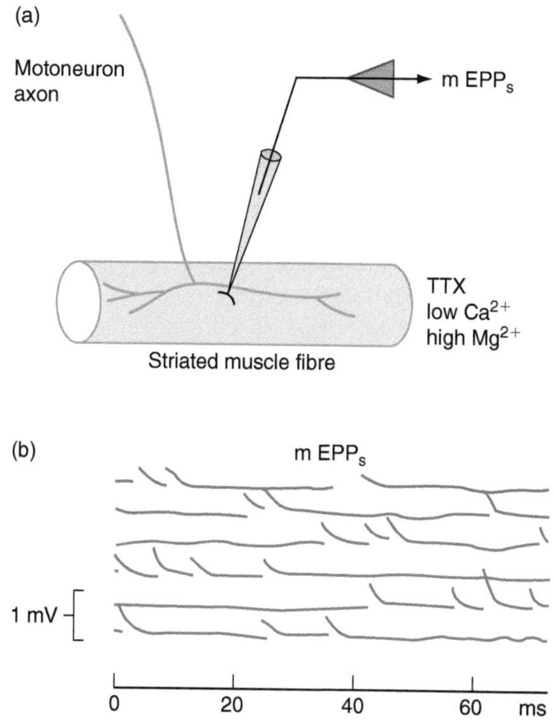

FIGURE A7.1 **Miniature postsynaptic potentials.** Miniature endplate potentials (right mEPPS) recorded at the frog neuromuscular junction as shown **(a)**. Recordings are obtained in the presence of a low external Ca^{2+} concentration **(b)**. Adapted from Fatt P, Katz B (1952) Spontaneous subthreshold activity at motor nerve endings. *J. Physiol. (Lond.)* **117**, 109–128, with permission.

Miniature endplate potentials, being the smallest recorded event with relatively constant amplitude and an all-or-nothing characteristic, were named *quanta* (with reference to quantum physics) by Del Castillo and Katz in 1954. Considering the fast rise time of miniatures, they hypothesized that miniatures arise from the synchronous action of a packet of a large number of ACh molecules at a time: 'At this stage, the characteristic presynaptic vesicles were revealed by electron microscope and the suggestion arose that they could be the subcellular particles in which the transmitter is stored and from which it is released in an all-or-none fashion.' This was confirmed by the observation of exocytosis at active zones of the neuromuscular junction (Couteaux and Pécot-Dechavassine, 1970).

However, direct visualization of exocytosis at a neuromuscular junction was achieved only 20 years later by using electron microscopy combined with new methods to rapidly freeze nerve terminals (Heuser *et al.*, 1979). The tissue is freeze-fractured during synaptic activity at precise times following nerve stimulation. A metal replica of the presynaptic membrane after fracture is observed under the electron microscope. Images of exocytosis are observed only in the presence of 4-aminopyridine (a blocker of K^+ channels which prolongs depolarization of the presynaptic membrane; see Section 7.3.5), so that important ACh release occurs. In the absence of drugs, rearrangement of presynaptic intramembranous particles is the most common ultrastructural observation. This low probability of observation of exocytosis in the absence of 4-aminopyridine is not surprising, considering the low probability of exocytosis at a synaptic complex at a given time (see Appendix 7.2).

Similar spontaneous miniature potentials are recorded from synapses of the central nervous system. They are called 'miniature postsynaptic potentials' (mEPSPs and mIPSPs) or 'miniature currents' (mEPSCs and mIPSCs) (**Figure A7.1b**).

The theory of vesicular release of neurotransmitter, or the quantal nature of chemical transmission, states that one quantum equals one vesicle. The size of a quantum is designated by q.

A7.1.2 The quantal composition of EPSPs and IPSPs

At the neuromuscular junction, a quantum produces a 0.5–1 mV miniature endplate potential (mEPP). In response to a presynaptic action potential and in the presence of control concentrations of external Ca^{2+}, an endplate potential (EPP) is recorded: it has a much larger amplitude (50–70 mV) than miniatures. The quantal theory assumes that EPPs are made up of n quanta released simultaneously (200–300 mEPPs). This evidence is obtained by lowering external Ca^{2+} concentration and thus reducing the amplitude of evoked EPPs. If one lowers the normal Ca^{2+} concentration and adds magnesium to the muscle bath, the amount of acetylcholine delivered by an impulse can be reduced to a very low level and, under these experimental conditions, the quantal composition of the endplate potential becomes immediately apparent (see Appendix 7.2). This result has been confirmed in various synapses of vertebrates and invertebrates.

APPENDIX 7.2 THE PROBABILISTIC NATURE OF NEUROTRANSMITTER RELEASE: THE NEUROMUSCULAR JUNCTION AS A MODEL

Neurotransmitter release is probabilistic. In response to a presynaptic action potential, each docked synaptic vesicle has a probability p to fuse with the presynaptic plasma membrane (i.e. to undergo exocytosis). This probability p varies from 0 to 1 (0, p, 1). This means that even when all the proper conditions are fulfilled, synaptic transmission can still fail (see **Figure 7.3a**). What fails exactly is vesicle exocytosis. Of course, when chemical transmission at a single synapse is achieved by only one active zone (mammalian CNS synapses), failures are much more commonly observed than when it is achieved by a large number of active zones (neuromuscular junction).

The first studies on the probability of neurotransmitter release were performed on the neuromuscular junction, as this preparation offers the possibility of simultaneously recording the activity of pre- and postsynaptic elements and of manipulating the parameters related to neurotransmitter release, here acetylcholine release. At the motor endplate level, the number of active zones is estimated at between 300 and 1000. For this reason, the recorded postsynaptic response is global, representing the summation of evoked responses from many active zones.

In the absence of any nerve stimulation, miniature endplate postsynaptic potentials of 0.5–1.0 mV average amplitude are recorded. They occur randomly. These miniature endplate potentials, which are the smallest recorded events and have relatively constant amplitude, were named 'quanta' by Del Castillo and Katz. They proposed that each quantum corresponds to the content of one synaptic vesicle. The size of a quantum is q.

In contrast, in response to nerve stimulation, the probability of synaptic vesicle exocytosis is very high and the size of the endplate potential is in the order of tens of millivolts. In order to test whether EPSPs are made up of quanta, the size of EPSPs is reduced by immersing the preparation in an extracellular medium containing a low Ca^{2+} concentration. In these conditions, following motor nerve stimulation, one can record postsynaptic depolarizations (motor endplate potentials) with a low and variable amplitude and also numerous failures (absence of postsynaptic response) (**Figure A7.2a**). These amplitude variations are in graduated steps, each one corresponding to a quantum of amplitude q (**Figure A7.2b**).

The postsynaptic response is constantly a multiple of 0, 1, 2, ..., x quanta, x always being a whole number. If one admits that a quantum corresponds to the release of a synaptic vesicle, it appears that 0, 1, 2, ..., x synaptic vesicles are released (x being a natural number as each synaptic vesicle is an entity). This produces fluctuations in the amplitude of the postsynaptic response. The distribution of these fluctuations on a graph shows the presence of regularly spaced peaks, the first three peaks clearly corresponding to amplitudes $1q$ (0.4 mV), $2q$ (0.8 mV) and $3q$ (1.2 mV), but the presence of other peaks ($4q$, ..., xq) is less evident (**Figure A7.2c**). The first two amplitudes ($1q$ and $2q$) are more frequent than others. In other words, the probability of recording postsynaptic potentials of amplitude $1q$ or $2q$ is greater than that of recording potentials of amplitude $3q$ or above in the presence of a low external Ca^{2+} concentration.

The demonstration that postsynaptic potentials are composed of discrete units has necessitated the application of statistical tests to the experimental data. Poisson's law was applied to test that each response is made up of discrete (0, 1, 2, 3, ...) units. As stated earlier, the results shown in **Figure A7.2** were obtained in conditions where p is reduced (reduced extracellular Ca^{2+} concentration). This is a necessary condition for the use of the Poisson distribution. In other words, each time an action potential invades the axon terminals, it causes the release of a few vesicles out of a very large available population. In the Poisson distribution, the probability of observing a postsynaptic potential composed of x miniature potentials $p(x)$ is:

$$p(x) = \frac{e^{-m} m^x}{x!}$$

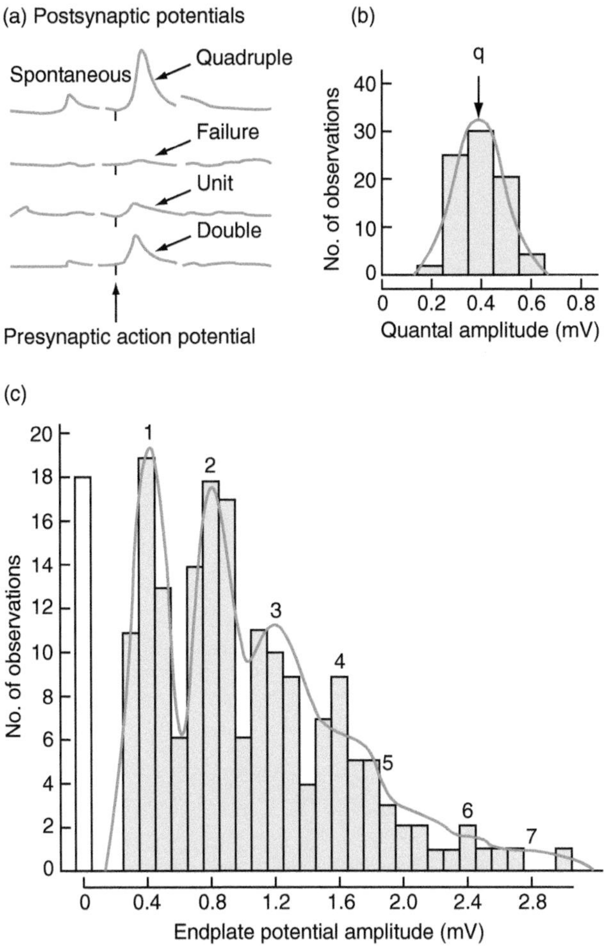

FIGURE A7.2 **Demonstration of the probabilistic nature of acetylcholine release at the neuromuscular junction. (a)** Recordings of spontaneous miniatures and evoked endplate potentials (mEPP and EPP). The nerve–muscle preparation is bathed in a low-Ca^{2+} high-Mg^{2+} concentration medium. In such conditions, spontaneous mEPPs have nearly the same unitary amplitude whereas evoked EPPs have an amplitude that is a multiple of the mEPP amplitude. **(b)** The distribution of mEPP amplitude is unimodal and their average amplitude is $q = 0.4$ mV. **(c)** Histogram of evoked EPPs recorded in response to a presynaptic action potential. The cases where no postsynaptic response is recorded (failures) are represented at 0 mV (18 cases). The theoretical distribution calculated from Poisson's law (represented by the green line) fits the distribution of the amplitude of recorded EPPs (histogram). Numbers on top of the histogram peaks indicate equivalent mEPP charges. Part **(a)** adapted from Liley AW (1956) The quantal component of the mammalian endplate potential. *J. Physiol. (Lond.)* **133,** 571–587, with permission. Parts **(b)** and **(c)** from Boyd IA, Martin AR (1956) The endplate potential in mammalian muscle. *J. Physiol. (Lond.)* **132,** 74–91, with permission.

where m is the mean number of quanta (miniature potentials) that compose the postsynaptic response (i.e. the average number of released vesicles in response to a presynaptic spike). N is the total number of observations (the number of recordings of the postsynaptic potential in response to a presynaptic spike) and $n(x)$ is the number of times that the recorded postsynaptic potentials is composed of x miniature potentials (amplitude of the postsynaptic response = xq). The probability that a postsynaptic potential is composed of x miniature potentials, $p(x)$, is equal to the number of times this event is observed, $n(x)$, over the total number of experiments, N:

$$p(x) = \frac{n(x)}{N} \qquad (1)$$

When N is large enough, $Np(x)$ is close to the observed number of responses which contain x quanta (which are made up of a summation of x miniatures).

The difficulty here is to determine the value of m, the mean number of quanta that compose the postsynaptic response (also called the average quantal content). To determine m, two methods can be used.

First, given that the amplitude of miniature potentials is a unit, one can calculate:

$$m = \frac{\text{average amplitude of evoked responses}}{\text{average amplitude of miniature potentials}}$$

This method is used in Chapter 8.

The second option is the so-called failure method. In conditions where p is artificially reduced (reduced external Ca^{2+} concentration), the number of times the synaptic transmission fails is high: numerous presynaptic action potentials are not followed by vesicle release. In these cases of failure, $x = 0$ (the postsynaptic responses of null amplitude composed of 0 miniature potentials). From equation (1), $p(0)$ is the number of failures over the total number of stimulations N; it is large and equal to:

$$p(0) = e^{-m} = \frac{\text{number of failures}}{\text{number of simulations}} = n(0)/N$$

Therefore, $m = \ln(N/n(0))$. The $n(0)/N$ ratio, determined from experimental results (**Figure A7.2c**), leads to an easy deduction of m.

With m known, p can be calculated for each value of x, and a theoretical curve is drawn, showing the distribution of postsynaptic potentials. **Figure A7.2c** shows the correlation between experimental data and the Poisson distribution. Quantal acetylcholine release at the neuromuscular junction is a valid model. The following hypothesis has been proposed.

Acetylcholine release is a discontinuous quantal phenomenon, each quantum corresponding to the total content of one synaptic vesicle. The probability p that the postsynaptic response will be composed of 1, 2, …, x quanta depends on the experimental conditions (composition of the extracellular medium) and on the frequency of presynaptic activity. Once again, the results in **Figure A7.2** have been obtained in a medium where p is reduced, a condition necessary to the use of the Poisson distribution.

PART II

IONOTROPIC AND METABOTROPIC RECEPTORS IN SYNAPTIC TRANSMISSION

C H A P T E R

8

The ionotropic nicotinic acetylcholine receptors

Constance Hammond

OUTLINE

In the pioneering work on the sensitivity of striated and smooth muscles to the plant alkaloids nicotine and muscarine, John Newport Langley suggested the existence of two classes of chemical receptors: muscarinic and nicotinic. Otto Loewi then showed that the natural neurotransmitter he named vagusstoff (substance of the vagus, later identified as acetylcholine by Henry Dale), mimicked the effects of both muscarine and nicotine. With electrophysiological recordings at the neuromuscular junction, Robert Katz described for the first time the electrical responses of muscle cells evoked by acetylcholine.

The nicotinic acetylcholine receptor (nAChR) is a glycoprotein present at nicotinic cholinergic synapses. The preparation that has been used most extensively to study the nicotinic receptor is the electric organ of the electric ray, *Torpedo* (Torpedo nAChR; **Figure 8.1a**), or of the electric eel, in part because this preparation is extremely rich in nicotinic receptors, and because snake venom α-toxins had been identified as highly selective markers of nAChRs. In mammals, nAChRs have been mostly studied at the neuromuscular junction (muscle nAChR; **Figure 8.1b**) but also in the peripheral nervous system (synapses between pre- and postganglionic neurons of the autonomic nervous system) and, more recently, in the central nervous system where they are also present (neuronal nAChR; see Appendix 8.1).

8.1 OBSERVATIONS

The axon of a motoneuron is stimulated in the presence of a low Ca^{2+} concentration (0.5 mM) and a high Mg^{2+} concentration (6 mM) in the extracellular medium to reduce synaptic transmission. In this condition, a postsynaptic depolarizing potential is recorded with the use of an intracellular electrode implanted in the muscle fiber at the level of the neuromuscular junction (current clamp mode) (**Figure 8.2a**). It is an excitatory postsynaptic potential (EPSP), also called endplate potential (EPP). Its amplitude varies with the intensity of stimulation. In contrast, when the postsynaptic recording electrode is far from the endplate, no response is recorded (**Figure 8.2b**).

The low Ca^{2+} and high Mg^{2+} concentrations in the extracellular medium are a necessary condition to avoid muscle fiber contraction during recording. The number of active zones per neuromuscular junction is in fact so high (around 300 to 1000) that a presynaptic axonal spike always triggers a very large EPP that in turn depolarizes the muscle membrane to the threshold potential for voltage-gated Na^+ channels opening and thus evokes a postsynaptic action potential (not shown) and muscle fiber contraction. The elimination of the muscle fiber action potential by lowering the release of acetylcholine

Cellular and Molecular Neurophysiology. DOI: 10.1016/B978-0-12-397032-9.00008-X

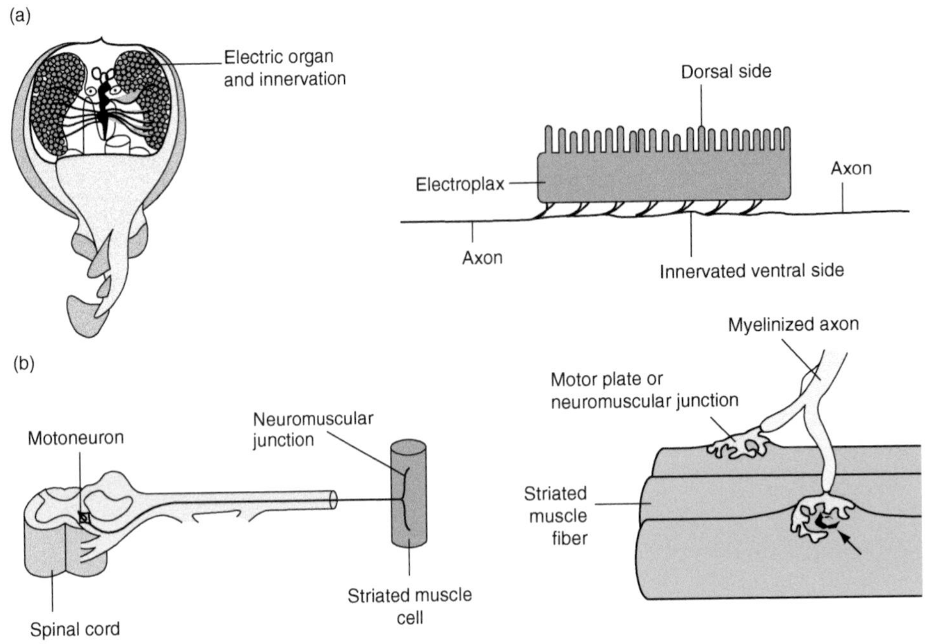

FIGURE 8.1 **Examples of preparations in which nicotinic receptors have been extensively studied.** (**a**) The electric organ of the electric ray. On a dissected *Torpedo* (left) we can see the electric organs and their innervation. These organs constitute electroplax membranes (right) which are modified muscle cells that do not contract. Nicotinic receptors are present at the command neuron's synapse level, on the ventral side of the postsynaptic membrane of the electroplax. The electroplax are simultaneously activated and the summation of their electric discharges can be of the order of 500 V. (**b**) The neuromuscular junction. Striated muscle cells are innervated by motoneurons whose cell bodies are located in the ventral horn of the spinal cord (horizontal section, left). In mammals, each muscle cell is innervated by one nerve fiber. As the axon makes contact with the muscle cell, it loses its myelin sheath and divides into several branches that are covered by unmyelinated Schwann cells. The thick arrow (right) points to one terminal that has been lifted to show the postsynaptic folds where nicotinic receptors are located.

from presynaptic terminals allows the recording of EPP in isolation.

Questions

- When released in the synaptic cleft, to which type(s) of postsynaptic receptor do the molecules of acetylcholine bind?
- How is the binding of acetylcholine transduced into a depolarization of the postsynaptic membrane?
- Why is the postsynaptic depolarization recorded only at the level of the neuromuscular junction?

8.2 THE TORPEDO OR MUSCLE NICOTINIC RECEPTOR OF ACETYLCHOLINE IS A HETEROLOGOUS PENTAMER $\alpha_2\beta\gamma\delta$

The only type of acetylcholine receptor present in the postsynaptic membrane of neuromuscular junctions or at electrical synapses of *Torpedo* is the nicotinic receptor (nAChR). It is named 'nicotinic' after its sensitivity to nicotine (nicotine is an agonist of nAChRs).

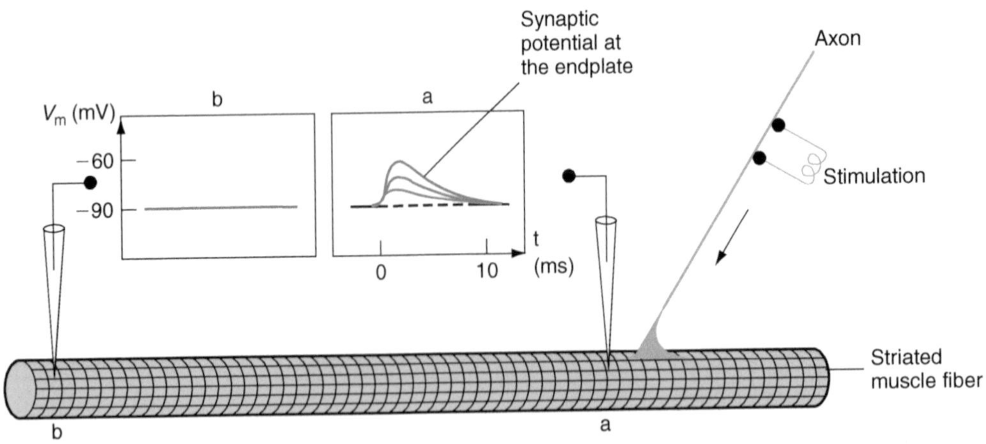

FIGURE 8.2 **The endplate potential.** The axon of a motoneuron is stimulated and the postsynaptic response is recorded in the presence in the bath of a low Ca (0.5 mM) and a high Mg (6 mM) concentration. Two intracellular electrodes are implanted in the muscle fiber, one close to the neuromuscular junction (**a**) and the other more than 5 mm away (**b**). The evoked response is a transient depolarization called an endplate potential.

8.2.1 Nicotinic receptors have a rosette shape with an aqueous pore in the center

Under the electron microscope, the nicotinic receptor of the neuromuscular junction, located in the postsynaptic muscular membrane, has a rosette shape with an 8–9 nm diameter and a central depression of diameter 1.5–2.5 nm. This depression corresponds to the channel portion of the protein (**Figure 8.3**). Each rosette is made up of five regions of high electronic density arranged around an axis perpendicular to the plane of the plasma membrane. In transverse section, the rosette appears as a cylinder 11 nm long, extending beyond each side of the membrane (6 nm towards the synaptic cleft and 1.5 nm towards the cytoplasm).

8.2.2 The four subunits of the nicotinic receptor are assembled as a pentamer $\alpha_2\beta\gamma\delta$

The nicotinic receptor is normally purified from the electric organ of *Torpedo* or the electric eel. A 290–300 kDa glycoprotein is obtained when this purification is performed on an affinity column using an agarose bound cholinergic ligand (**Figure 8.4**). When this glycoprotein is incorporated into a planar lipid bilayer or into lipid vesicles, it presents the same functional characteristics as the native receptor: when acetylcholine is present in the extracellular side at a concentration of 10^{-5}–10^{-4} mol l^{-1}, it induces the passage of cations across the bilayer.

The nicotinic receptor is composed of four glycopolypeptide subunits α, β, γ, δ

In the presence of the detergent SDS (sodium dodecyl sulfate), the 290–300 kDa protein dissociates into four different subunits which migrate on a polyacrylamide gel as molecules with apparent molecular weights of 38 kDa (α), 49 kDa (β), 57 kDa (γ) and 64 kDa (δ) (**Figure 8.5a**). The same experiment carried out with nicotinic receptors obtained from the neuromuscular junction shows very similar results. At the vertebrate neuromuscular junction, receptors are either of the embryonic form, composed of α, β, γ, and δ subunits in a 2:1:1:1 ratio (**Figure 8.5b**), or of the adult form composed of α, β, δ, and ε subunits, also in a 2:1:1:1 ratio (**see Figure A8.3**).

Genes coding for each subunit of the nicotinic receptor of the electric ray and of the mammalian receptor have been cloned. When the corresponding mRNAs are injected into *Xenopus* oocytes, functional nicotinic receptors are synthesized and incorporated into the oocyte membrane. It has, therefore, been confirmed that the subunits αβγ and δ are sufficient to obtain a functional nicotinic receptor containing the acetylcholine receptor sites and the elements that form the ionic channel (**Figure 8.6**). The nicotinic receptors are part of the cys-loop superfamily of ligand-gated ion channels which include receptors for the inhibitory transmitters GABA (see Chapter 9) and glycine and one type of serotonin receptor. The cys-loop receptors are named after a characteristic loop formed by

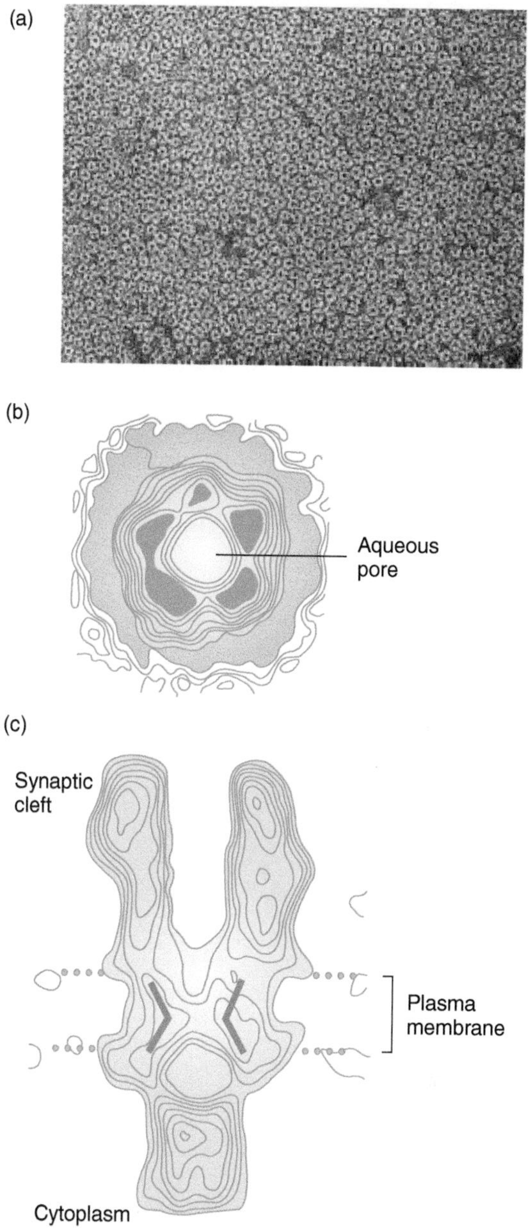

FIGURE 8.3 **The nicotinic receptor has a rosette shape.** (**a**) Membrane surface of *Torpedo* electric cells (electroplax). Each rosette constitutes one nicotinic receptor. (**b**) This computer reconstructed image of a single nicotinic receptor provides a more detailed view (superior view). (**c**) Electron microscopic analysis of tubular crystals of *Torpedo* nAChR viewed from the side. Part (**a**) from Cartaud J, Benedetti EL, Sobel A, Chargeux JP (1978) A morphological study of the cholinergic receptor protein from Torpedo narmorata in its membrane environment and in its detergent-extracted purified form. *J. Cell Sci.* **29**, 313–337, with permission. Part (**b**) from Bon F *et al.* (1982) Orientation relative de deux oligornères constituant la forme lourde du récepteur de l'acétylcholine chez la torpille marbrée. *C. R. Acad. Sci.* **295**, 199, with permission. Part (**c**) from Unwin N (1993) The nicotinic acetylcholine receptor at 9 Å resolution. *J. Mol. Biol.* **229**, 1101–1124, with permission.

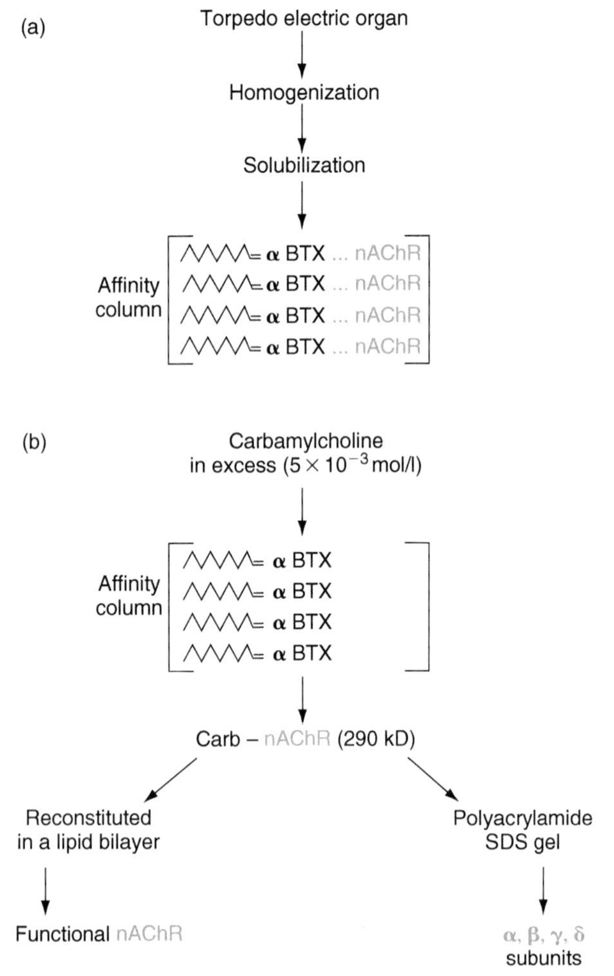

FIGURE 8.4 **Stages of affinity column purification of the nicotinic receptor.** (**a**) The electric organ of the electric ray is homogenized and membrane proteins solubilized. The resulting extract is run through an affinity column, on to whose sepharose (^^^ =) a nicotinic cholinergic ligand α-bungarotoxin (α-BTX) has been covalently bound. Owing to their affinity to α-BTX, the nicotinic receptors bind to it. (**b**) In order to recover the nicotinic receptors, another nicotinic ligand, carbamylcholine (Carb), is run in excess through the column to displace the binding of α-BTX to the receptor. Carbamylcholine-bound nicotinic receptor is obtained at the outflow of the column. Carbamylcholine is eliminated by dialysis and the nicotinic receptor is thus obtained in an isolated form. The nicotinic receptor can then be reincorporated into a lipid bilayer to study its functional characteristics. It may also be treated with a detergent (SDS) to dissociate its subunits.

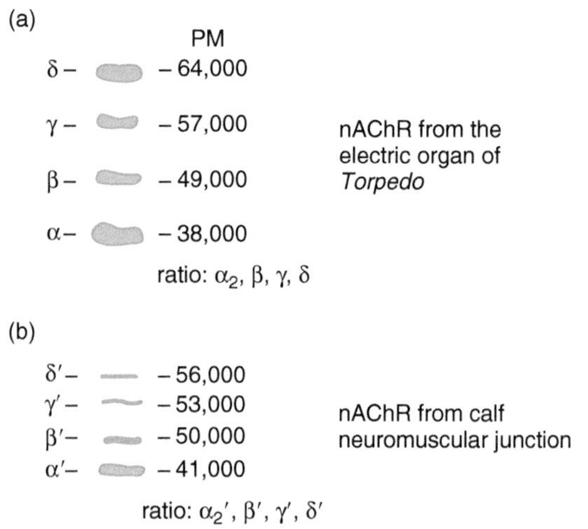

FIGURE 8.5 **Separation of the different nicotinic receptor subunits on a gel.** The subunits of the purified nicotinic receptor have been dissociated with the detergent sodium dodecyl sulfate (SDS) and separated on a polyacrylamide gel: subunits of (**a**) the nicotinic receptor of electric organ, and of (**b**) the calf neuromuscular junction. The four subunits obtained, α, β, γ, δ and α′, β′, γ′, δ′, have similar molecular weights. From Anholt R, Lindstrom J, Montal M (1984) The molecular basis of neurotransmission: structure and function of the nAChR. In Martinosi A (ed.) *The Enzymes of Biological Membranes*, New York: Plenum Press, with permission.

a disulfide bond between two cysteine residues in the N terminal extracellular domain. The nAChR subunit proteins vary in length with a well-conserved transmembrane topology:

- an NH_2-terminal region which forms a large hydrophilic domain of about 200 amino acids; the α subunits (α1–9) possess two adjacent cysteines essential for acetylcholine binding whereas the non-α referred to as βγε or δ do not;
- three hydrophobic sequences (M1, M2 and M3) with short connecting hydrophilic loops;
- a large hydrophilic domain which varies significantly from one subunit to another and contains functional phosphorylation sites;
- a fourth hydrophobic sequence (M4) and a short carboxy-terminal tail.

A model of the transmembrane organization common to all subunits has been proposed (**Figure 8.6b**): the hydrophilic NH_2-terminal domain is located on the extracellular side of the membrane (in the synaptic cleft), the second hydrophilic domain is located on the cytoplasmic side and the four hydrophobic sequences are membrane-spanning segments. Each subunit therefore crosses the membrane four times and both the amino- and carboxy-terminals are oriented towards the synaptic cleft. The cys-loop, after which the nACh receptors are named, is formed from 13 amino acids contained by a disulfide bond. It is situated at the base of the extracellular domains and is important for communication between the neurotransmitter binding sites and the ion channel.

8.2.3 Each α-subunit contains one acetylcholine receptor site located in the hydrophilic NH_2-terminal domain

Before the structure of the nicotinic receptor was known, it had been demonstrated that two acetylcholine molecules had to be bound to the receptor in order to initiate an ionic flux (see Section 8.3). It seemed logical

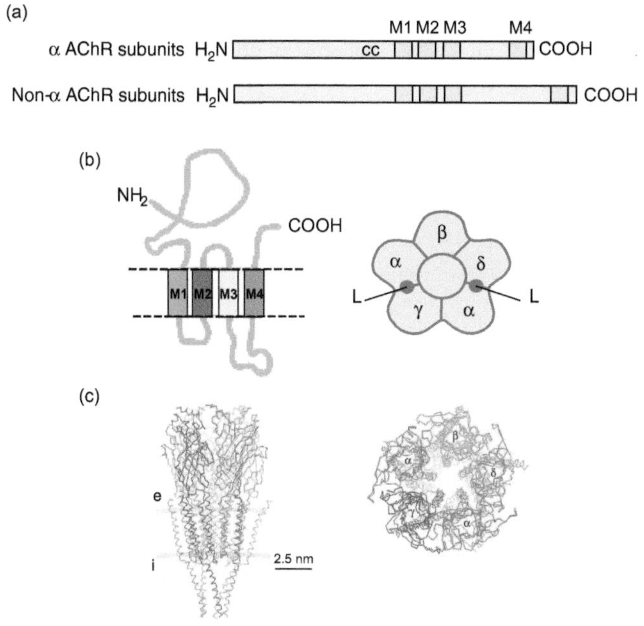

FIGURE 8.6 **Muscle-type nicotinic receptor (nAChR) model.** Schematic representation of the primary sequence of the α- (α_1–α_9) and non-α- (β_1–β_4, γ, ε, and δ) subunits of the nAChR. M1–M4, transmembrane domains; CC, Cys-Cys pair found in the α-subunits from both muscle- and neuronal-type nAChRs. (**b**) Diagram of the tertiary organization of nAChRs. (Left) Schematic representation of the transmembrane organization of a single subunit. (Right) Schematic representation of the oligomeric organization of muscle-type nAChR. The pentameric nAChR is formed by two α-subunits and three non-α-subunits. The two ligand binding sites (L) are located at the interfaces of one α--subunit and one non-α-subunit. (**c**) Architecture of the extracellular and transmembrane domains of the nAChR from electron micrographs at 4 Å (0.4 nm) resolution. (Left) View from the side, parallel with the membrane plane of a single receptor. Individual subunits are in different colours (α, red; β, green; γ, blue; δ, light blue). The extracellular domains are β-sandwiches formed from two anti-parallel β-sheets perpendicular to the membrane topped by an α-helix. The transmembrane domains of each subunit contain four α-helices. (Right) Plan view from the synaptic cleft. The five subunits form a ring. A water filled vestibule runs through the extracellular domains down to the channel and through the membrane. The four M2 domains (in bright colors) of the four subunits line the channel. Parts (**a**,**b**) adapted from Arias HR (2000) Localization of agonist and competitive antagonist binding sites on nicotinic acetylcholine receptors. *Neurochem. Internatl.* **36**, 595–645, with permission. Part (**c**) adapted from Unwin N (2005) Refined structure of the nicotinic acetylcholine receptor at 4 Å resolution. *J. Mol. Biol.* **346**, 967–989, with permission.

that these two sites involve identical subunits; i.e. the two α-subunits. Additionally, based on the organization of the hydrophilic sequences, it was proposed that this site is located in the large hydrophilic NH_2-terminal domain which is exposed to the synaptic cleft (**Figures 8.6b** and **8.7**).

This proposal has been confirmed by covalent binding studies of cholinergic agonists on α-subunits, isolated either from nicotinic receptor-rich membranes, or expressed in frog oocytes from the corresponding mRNA. Of the four subunit types αβγ and δ the α-subunits have

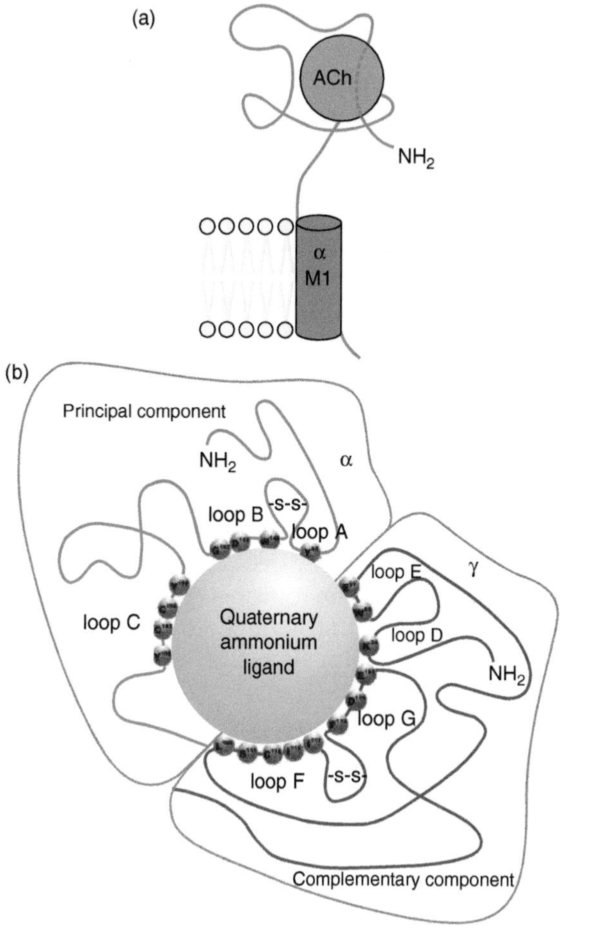

FIGURE 8.7 **Model of the acetylcholine binding site.** (**a**) Lateral and (**b**) top view of a schematic model for the proposed folding of the extracellular domain involved in the binding site of several quaternary-ammonium compounds, including the native neurotransmitter ACh and the competitive antagonist curare. The low-affinity ACh binding site is located in the αγ-interface. Its counterpart, the high-affinity ACh binding site, which is located in the αδ-interface, is not included in this model. The large sphere represents a quaternary ammonium-containing molecule. The principal component of the ligand binding site is located in the α-subunit which contributes with loops A, B, and C. The residues involved in the binding site are represented by small spheres in one letter code. Loop A is mainly formed by residue Y_{93}. Loop B is molded by amino acids W_{149}, Y_{152} and probably G_{153}. Loop C is shaped by residues Y_{190}, C_{192}, C_{193}, and Y_{198}. The disulfide bond indicated in the αγ-subunit as S-S represents the link between C_{128} and C_{142}. The complementary component of the ligand binding site is located on either the γ or the δ-subunit. Part (**b**) adapted from Arias HR (2000) Localization of agonist and competitive antagonist binding sites on nicotinic acetylcholine receptors. *Neurochem. Internatl.* **36**, 595–645, with permission.

been shown to be the main contributors to cholinergic agonist binding.

The next step was to determine which amino acids are part of the acetylcholine receptor site. To this end, labeled cholinergic ligands are used. These ligands are able to bind covalently to the acetylcholine receptor sites. One of the most used is MBTA (4-(*N*-maleimido) benzyl trimethylammonium iodide) which binds covalently to α-subunit receptor sites after reduction of disulfide bridges. Once labeled, the α-chain is sequenced and the labeled regions identified. In this way, a region containing cysteines 192 and 193 was identified and proposed as one of the potential sites of interaction with cholinergic ligands (**Figure 8.7b**).

Other data have provided additional evidence of the participation of cysteine residues 192 and 193 in the acetylcholine receptor site. In the first place, these cysteine residues are present only in the α-subunits. Furthermore, when frog oocytes are injected with mRNA coding for α-subunits that have been mutated at the level of cysteines 192 and 193 (serines replaced for cysteines), the α-subunits obtained are unable to bind cholinergic ligands.

However, all these results have the shortcoming of having been obtained from preparations previously treated with disulfide bond-reducing agents (such as dithiothreitol). This treatment is necessary in order to allow the covalent binding of the cholinergic ligand MBTA to the receptor site but it alters the receptor site selectivity for cholinergic ligands.

In order to obtain a more detailed map of the native protein's acetylcholine receptor site, a labeled photoactivated cholinergic ligand has been used: ^{3}H-DDF (para-*N*,*N*-dimethylamino benzene diazonium fluoroborate). ^{3}H-DDF is a competitive antagonist of acetylcholine which, once photoactivated, binds covalently (irreversibly) to the acetylcholine receptor sites. This reaction is carried out on the whole nicotinic receptor channel and the α-subunits are then isolated, the segments labeled by ^{3}H-DDF are purified and their sequence analyzed. This led to the demonstration that the residues tyr (Y) 93, trp (W) 149, tyr (Y) 190, cys (C) 192 and cys (C) 193, all labeled by ^{3}H-DDF, are part of the acetylcholine receptor site. This labeling is in fact inhibited by other nicotinic agonists and competitive antagonists (**Figure 8.7b**). This result is valid for the nicotinic receptor of the electric organ as well as for that of the neuromuscular junction. Finally, data obtained from crystallization with different ligands of the ACh binding protein (AChBP), a water-soluble protein expressed by mollusks and analogous to the extracellular ligand-binding domain of the cys-loop receptors, or of the extracellular domain of the α7 subunit, have extended the results obtained with the photolabeling approach.

8.2.4 The pore of the ion channel is lined by the M2 transmembrane segments of each of the five subunits

The ion channel can be considered as functionally equivalent to the active site of allosteric enzymes: its states (open, closed, blocked) are determined by the effectors of the receptor (binding of agonists, competitive antagonists and non-competitive antagonists; see Appendix 6.1). Concerning the channel structure, the question is which of the four hydrophobic membrane spanning segments M1 to M4 (**Figure 8.6**) are part of the walls of the ionic channel. On the basis of the hypothesis that non-competitive inhibitors bind to a high-affinity site located inside the open ion channel (channel blockers; see Section 8.6.3), photoactivable non-competitive inhibitors are used to label residues participating in the walls of the ion channel (this is a similar approach to that used for determination of the ACh binding site). Radioactive chlorpromazine activated with ultraviolet light labels serine, leucine and threonine residues from the M2 membrane-spanning segment from all subunits of the *Torpedo* acetylcholine receptor. These results point to a contribution of the M2 membrane-spanning segment to the walls of the ion channel. Electron microscopy images of the open conformation of the channel identified five rods bordering the ion channel, attributed to the M2 segment. The axis of the rods are around 1.8 and 1.15 nm from the axis of the pore, at the upper and lower faces, respectively, in agreement with a funnel-shaped pore with a minimal diameter of around 1 nm located at the cytoplasmic border of M2.

In the M2 segments of nAChR subunits, there are remarkable amino acids highly conserved among the nAChR subunits sequenced to date and which, in the proposed model, form rings: a ring of negatively charged amino acids (Asp/Glu residues) that repel negative ions, three rings of hydrophobic leucine/valine, two rings of polar serine/threonine and, finally, a cytoplasmic ring of negatively charged amino acids (**Figure 8.8**).

Site-directed mutagenesis of some amino acids located in the M2 segment confirmed the contribution of M2 segments to the regulation of ion transport through the nicotinic channel. Chimeric cDNAs are constructed to add or substitute amino acids in the M2 segment. The results obtained will be explained in detail in the section describing the study of ionic selectivity (Section 8.3.2).

8.3 BINDING OF TWO ACETYLCHOLINE MOLECULES FAVORS CONFORMATIONAL CHANGE OF THE PROTEIN TOWARDS THE OPEN STATE OF THE CATIONIC CHANNEL

8.3.1 Demonstration of the binding of two acetylcholine molecules

It has been demonstrated that two acetylcholine molecules must bind to the receptor to trigger the opening of the channel and allow cations to flow through. The proof of this has been obtained from dose–response curves. The response to acetylcholine (i.e. the flux of cations measured at very short intervals after the application of

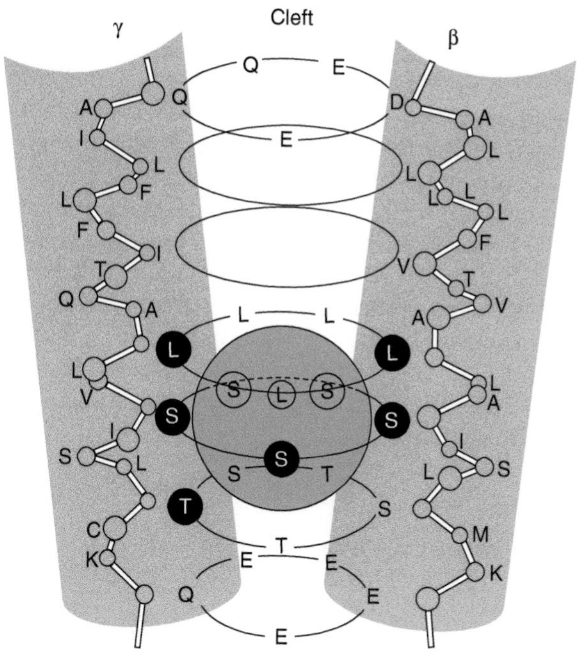

FIGURE 8.8 **Characteristic amino acid residues along the M2 segment of the subunits of the nAChR.** The M2 membrane-spanning segments are symmetrically arranged around the central axis of the molecule (two of them are represented). The relative position of the α-carbons of the amino acids is shown as one-letter code. E, glutamic acid, S, serine, T, threonine, L, leucine, Q, glutamine. The data accumulated to date suggest that the upper part of the channel, the α-helical component, acts as a water pore, whereas the lower loop component contributes to the selectivity filter of the ion channel. Adapted from Revah F, Galzi JL, Giraudat J *et al.* (1990) The noncompetitive blocker [^{3}H]-chlorpromazine labels three amino acids of the acetylcholine receptor γ subunit implications for the α-helical organization of the M2 segments and the structure of the ion channel. *Proc. Natl Acad. Sci. USA* **87**, 4675–4679, with permission.

acetylcholine), or the opening probability of the channel (see below), is proportional to the square of the acetylcholine concentration:

$$\text{Response} = f(\text{ACh})^2$$

However, this demonstration is obscured by the consequences of receptor desensitization, which have to be eliminated from the recordings (see Section 8.4). For this reason, this demonstration will not be presented in detail here. The conformational change of the protein towards the open state is clearly favored when two acetylcholine molecules bind to the receptor. The following model accounts for these observations (however, as we shall see in Section 8.4, this model is in fact much more complex):

$$R \rightleftharpoons AR \rightleftharpoons A_2R \rightleftharpoons A_2R^* \rightarrow \text{cationic current}$$

where R is the nicotinic receptor in its closed configuration; R* is the nicotinic receptor in its open configuration; and A is acetylcholine.

The rate of isomerization between R and R* lies in the microsecond to millisecond timescale. The passage of cations through the open channel is the result of the conformational change ($A_2R \rightleftharpoons A_2R^*$). Electrophysiological techniques (patch clamp recordings of the unitary cationic current flowing through a single channel) can be used to study this flow of cations. Based on the results obtained with electrophysiological techniques, we shall look at the properties of the nicotinic channel and of the protein conformational changes.

8.3.2 The nicotinic channel has a selective permeability to cations: its unitary conductance is constant

When the unitary current crossing a nicotinic channel in the presence of acetylcholine is recorded in patch clamp, all the preparations tested show an inward current at negative holding potentials (and under physiological ionic conditions) (**Figure 8.9**).

The nicotinic current reverses at 0 mV

If the imposed membrane potential (V_m) is varied between −100 mV and +80 mV while recording unitary currents with the patch clamp technique (i_{ACh}), one can trace an i_{ACh}/V curve (**Figure 8.9**). This curve is approximately linear between −80 and +80 mV. The measured current is inward for negative voltages and outward for positive voltages.

The i_{ACh}/V curve crosses the voltage axis at a value where the current is zero. This value is called the *reversal potential of the nicotinic response*, or E_{ACh}. The value of this reversal potential is close to 0 mV in the experimental conditions of **Figure 8.10** but may vary slightly towards negative voltages depending on the preparation.

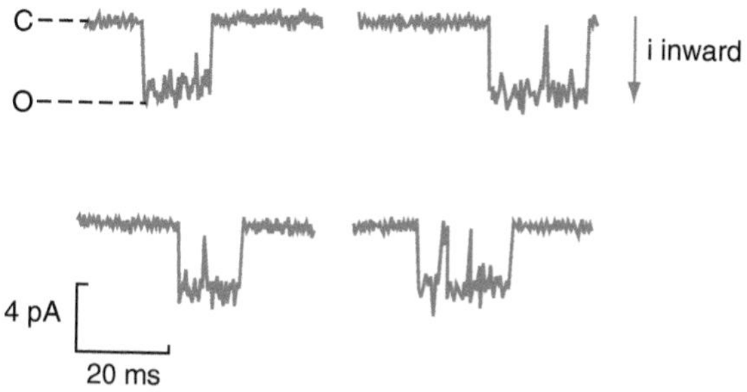

FIGURE 8.9 **Patch clamp recording of nicotinic receptor activity in rat sympathetic neurons.** Channel activity is recorded in the attached-cell configuration. While the membrane is kept at a negative potential an inward current is recorded in the presence of acetylcholine under physiological conditions. C, closed channel; O, open channel. Adapted from Colquhoun D, Ogden DC, Mathie A (1987) Nicotinic acetylcholine receptors of nerve and muscle: functional aspects. *TIPS* **8**, 465–472, with permission.

The unitary conductance is constant

The linear i_{ACh}/V relationship observed in **Figure 8.10b** (between −80 and +80 mV) is described by the equation: $i_{ACh} = \gamma_{ACh}(V_m - E_{ACh})$, where V_m is the membrane potential, E_{ACh} is the reversal potential of the nicotinic response, and γ_{ACh} is the conductance of a single nicotinic channel, or its unitary conductance. The value of γ_{ACh} is given by the slope of the linear i_{ACh}/V curve. It has a constant value at any given membrane potential. This value varies between 35 and 55 pS depending on the preparation and is a fundamental property of a nicotinic channel.

The nicotinic channel is a cationic channel

The reversal potential of the nicotinic response (E_{ACh} = 0 mV) does not correspond to the equilibrium potentials of any of the ions in solution. It is not an Na^+ channel because, in the experimental conditions of **Figure 8.10**, $E_{Na} = 58 \log(160/3) = +100$ mV. It is not a K^+ channel either, since $E_K = 58 \log(3/160) = -100$ mV. And it is not a Cl^- channel because, if the chloride ions are replaced by large anions that cannot cross the channel, such as SO_4^{2-}, no reversal potential change of the nicotinic response is observed.

By performing extracellular ionic substitution experiments, it has been shown that the nicotinic channel is permeable to Na^+, K^+, Ca^{2+} and Mg^{2+} ions. However, Ca^{2+} and Mg^{2+} ions contribute only a small fraction to the nicotinic current, which is essentially due to the flux of Na^+ and K^+ ions through the open channel. If different cations cross the same channel and have similar permeabilities, we define:

$$E_{cations} = 58 \log([cations]_e/[cations]_i),$$

where [cations] = $[Na^+]$ + $[K^+]$. In our case, we obtain $E_{cations}$ = 0 mV. In other words, $E_{cations} = E_{ACh}$ = 0 mV.

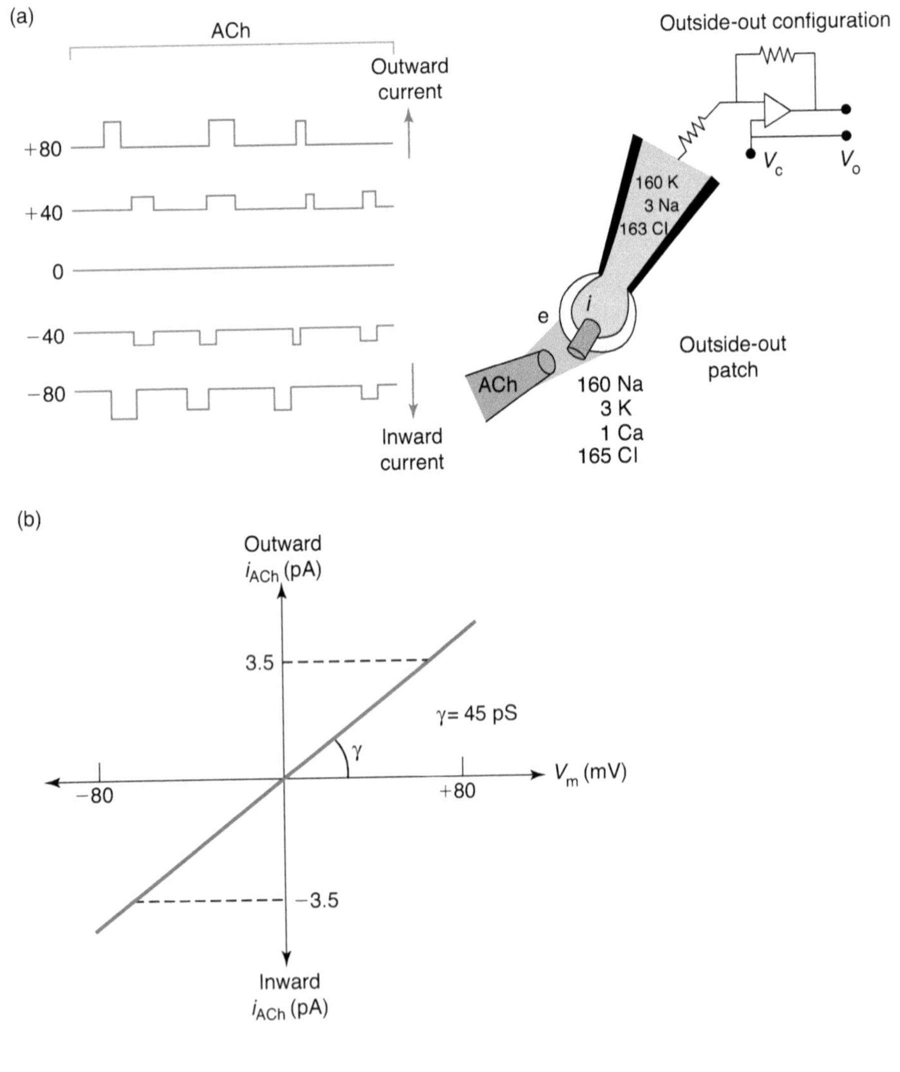

FIGURE 8.10 **Nicotinic unitary current recorded in patch clamp (outside-out configuration) at various membrane potentials (from −80 mV to +80 mV).** (**a**) Nicotinic unitary current recorded at different membrane potentials in response to the application of acetylcholine (ACh). A downward deflection indicates an inward current and an upward deflection indicates an outward current. (**b**) i_{ACh}/V curve obtained from the average values of i_{ACh} at each membrane potential V_m. The curve reverses at 0 mV and the slope corresponds to the unitary conductance γ.

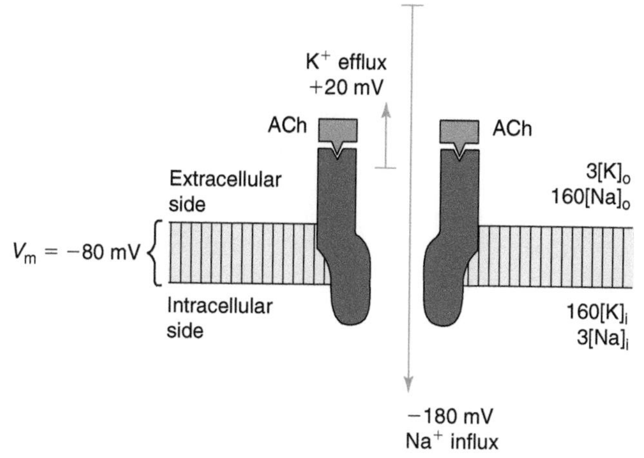

FIGURE 8.11 **Determination and vectorial representation of the Na^+ ion driving force ($V_m - E_{Na}$) and K^+ ion driving force ($V_m - E_K$) for a membrane potential of −80 mV.** Observe that 90% of the current is due to Na^+ ions. The net flux of positive charges is inward. This explains why ACh induces an inward current at V_m = −80 mV.

In which direction do Na^+ and K^+ ions cross the open nicotinic channel at different membrane potentials?

When the membrane is at a voltage of −80 mV, the Na^+ ion driving force is inward and equal to −180 mV, while the K^+ ion driving force is outward and equal to +20 mV (**Figure 8.11**). If channels open, more Na^+ will enter the cell than K^+ ions will leave it. The net flux of positively charged ions is, thus, inward: an inward current is recorded (**Figures 8.9** and **8.10a**).

If the same reasoning is followed for different membrane potential values, the same result is obtained as with the i_{ACh}/V curve: the unitary current is inward for negative membrane potentials (net flux of positive charges is inward), and the unitary current is outward for positive membrane potentials (net flux of positive charges is outward) (**Figures 8.10a** and **8.12**).

Effect of a decrease in $[Na^+]_e$

When extracellular Na^+ ions are partially replaced with a non-permeant substance such as sucrose (without changing the osmotic pressure), the reversal potential of the nicotinic response shifts towards more negative potentials (**Figure 8.13**). This is explained by the fact that, at this point, extracellular Na^+ ions contribute less to the nicotinic current, and the nicotinic reversal potential E_{Ach} shifts towards the K^+ equilibrium potential (E_K).

Substitution of K^+ ions for extracellular Na^+ ions

When almost all extracellular Na^+ ions are replaced by extracellular K^+ ions, the I/V curve obtained superimposes on the control I/V curve (**Figure 8.14**). Thus, extracellular K^+ ions can replace extracellular Na^+ ions; i.e. the nicotinic channel does not distinguish between Na^+ and K^+ ions. In other words, it presents similar permeabilities for both ions. For this reason, the reversal potential of the nicotinic response is independent of the relative concentrations of extracellular Na^+ and K^+ ions. It depends solely on the sum of these concentrations.

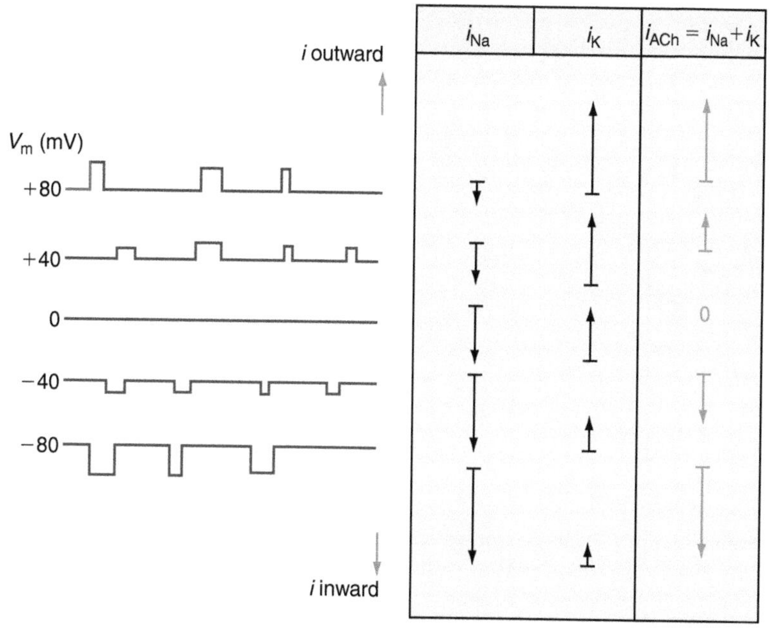

FIGURE 8.12 **Evolution of sodium and potassium currents (i_{Na} and i_K) through the nicotinic channel as a function of membrane potential V_m.** The current induced by acetylcholine i_{ACh} corresponds to the sum of two currents: $i_{ACh} = i_{Na} + i_K$. When $i_{Na} = -i_K$ the current is zero. This occurs at the reversal potential of the nicotinic response ($V_m = E_{ACh}$). A deflection of the traces (left) or an arrow of the chart (right) in the downward direction represents an inward current, and in the upward direction an outward current.

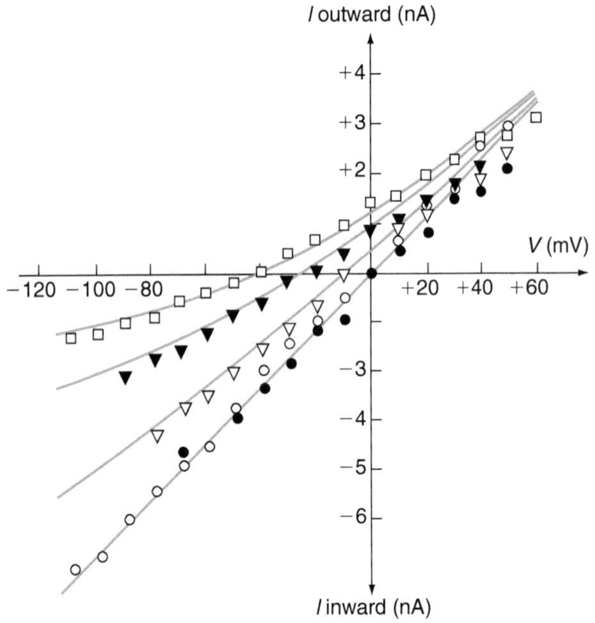

FIGURE 8.13 **Effect of lowering Na^+ concentration.** The composition of the external environment is: 5 mM of K^+; 0.1 mM of Ca^{2+}; and 21 (□), 46 (▼), 96 (▼) or 146 (• and ○) mM of Na^+. Each point represents the average of 25 measurements of I_{ACh}. When $[Na^+]_e$ decreases, the reversal potential of the nicotinic response shifts towards the K^+ equilibrium potential. Control E_{ACh}: (•,○) 58 log(146 + 5)/[cations]$_i$; (▼) 58 log(96 + 5)/[cations]$_i$; (▼) 58 log(46 + 5)/[cations]$_i$; (□) 58 log(21 + 5)/[cations]$_i$. Adapted from Linder TM, Quastel DMJ (1978) A voltage clamp study of the permeability change induced by quanta of transmitter at the mouse endplate. *J. Physiol. (Lond.)* **281**, 535–556, with permission.

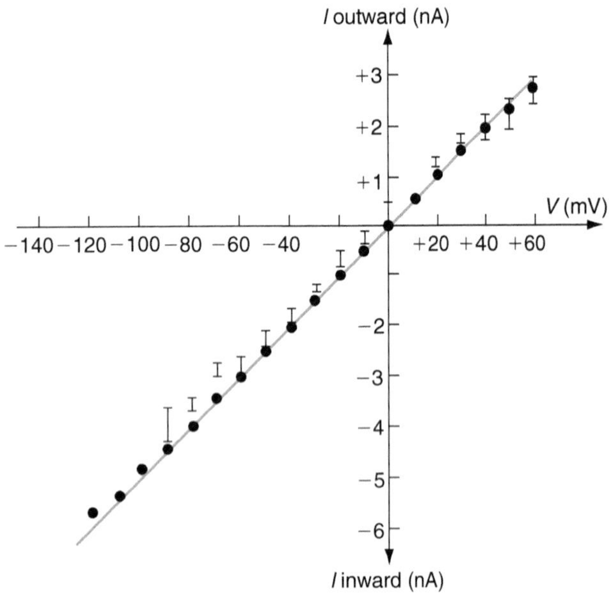

FIGURE 8.14 **Replacing external Na^+ for K^+ ions has no effect.** Control I/V curve (•) 146 mM of Na^+ and 5 mM of K^+, and (I) 2 mM of Na^+ and 149 mM of K^+. Observe that K^+ ions can replace Na^+ ions without affecting the I/V curve. The nicotinic channel does not distinguish between these two cations. Adapted from Linder TM, Quastel DMJ (1978) A voltage clamp study of the permeability change induced by quanta of transmitter at the mouse endplate. *J. Physiol. (Lond.)* **281**, 535–556, with permission.

Mutations in the M2 membrane-spanning segment can convert ion selectivity from cationic to anionic

The question was: do substitutions and/or additions of amino acids within (or near) the M2 segment from a nicotinic α-subunit (here the neuronal α_7-subunit) with homologous amino acids of the glycine receptor suffice to convert α-subunit ion-channel selectivity from cationic to anionic? (The glycine receptor is a receptor channel selectively permeable to anions, Cl^- ions.) The M2 sequences of α-subunits of a cationic channel (nAChR) and anionic channels (GlyR and $GABA_AR$) show similarities at the level of the threonine (244) and leucine (247) rings and differences at the level of rings of negative amino acids, Glu 237 and Glu 258 (**Figure 8.15a**). A chimeric cDNA encoding the α_7-subunit of neuronal nAChR is constructed in which, in the M2 segment, a proline residue is added at position 236 bis, and amino acids at positions 237, 240, 251, 254, 255 and 258 are exchanged with those found in the M2 segment of the glycine receptor α-subunit (nAChRα_7*; **Figure 8.15a**). Interestingly, glutamates (E) 237 and 258, which form negative rings in the nAChR (repelling negative ions), are exchanged with alanine (A) and asparagine (N) residues, respectively. The chimeric cDNA is injected into oocytes that express homomeric mutated nAChR* (formed by five identical mutated α_7-subunits). The current recorded with double-electrode voltage clamp (see Appendix 4.2) is compared with that recorded in oocytes expressing wild-type (non-mutated) homomeric AChR.

Ionic currents recorded in response to ACh application (100 μM, 2 s duration) from oocytes expressing wild-type homomeric α_7nAChR (in the presence of a Ca^{2+} chelator inside the oocyte) reverses around +3 mV. Substitution of 90% of the external chloride ions did not change the value of the reversal potential. These data thus support the conclusion that wild-type homomeric αnAChR, like native nAChR, is selective for cations.

Ionic currents recorded in response to ACh application (100 μM, 2 s duration) from oocytes expressing mutated homomeric α*nAChR reverses around −20 mV. Substitution of 90% external chloride ions by isethionate (an impermeant anion) shifts the reversal potential towards positive voltage (around +30 mV) (**Figure 8.15b**, left). This shift is well described by the Goldman–Hodgkin relationship for chloride-specific channels (**Figure 8.15b**, right). This indicates that ACh-activated currents are almost entirely carried by chloride ions in oocytes expressing mutated homomeric nAChR*. Then, introducing appropriate amino acid residues from the putative channel domain of a chloride-selective GlyR α-subunit into that of a cation selective nAChR α-subunit allows the design of an ACh-gated channel now selective for chloride. This confirms that the M2 segment forms the walls of the channel and strongly suggests that the exchanged residues face the lumen of the channel.

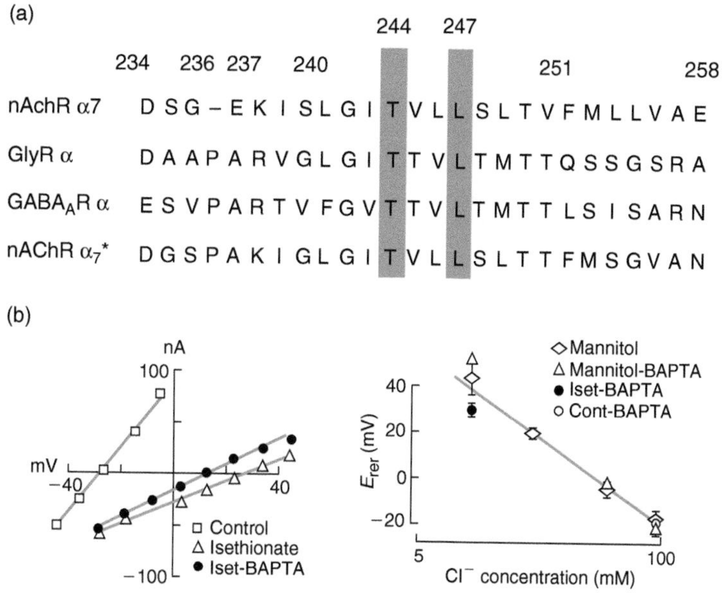

FIGURE 8.15 **Mutations in the M2 segment of an α-subunit of the nAChR convert ion selectivity of the homomeric nAChR from cationic to anionic.** (**a**) Comparison of M2 sequences from subunits of the cation selective nicotinic α_7 receptor subunit with those of the anion-selective glycine α_1, $GABA_A$ α_1 and β_1 and mutated nicotinic α_7* receptor subunits. (**b**, left) *I*/*V* relationship of the α_7* mutant receptor is first determined in control conditions (control) with 2 s ACh (100 μM) applications (outside-out patch clamp recording of a patch of membrane containing a large number of nicotinic receptors called macropatch). Then, 90% of chloride ions of the extracellular medium were replaced by the non-permeant anion isethionate and the *I*/*V* curve determined (isethionate). This last experiment was also performed in the presence of a chelator of Ca^{2+} ions (BAPTA) injected inside the cell (iset-BAPTA) in order to reduce secondary currents that could be triggered by the entry of Ca^{2+} ions through the nicotinic receptors. (**b**, right) Reversal potential values as a function of the logarithm of external chloride concentration (92, 50.5, 19.75 and 9.5 μM external chloride after substitution of NaCl by mannitol or isethionate). The solid line corresponds to the theoretical Nernst relation. From Galzi JL, Devillers-Thiéry A, Hussy N (1992) Mutations in the channel domain of a neuronal nicotinic receptor convert ion selectivity from cationic to anionic. *Nature* **359**, 500–505, with permission.

8.3.3 The time during which the channel stays open varies around an average value τ_o, the mean open time, and is a characteristic of each nicotinic receptor

When recording in patch clamp from myotubes (embryonic muscle cells) or from denervated muscle cells, in the presence of very small doses of acetylcholine, openings of the nicotinic channels separated by periods of silence are observed (**Figure 8.16**). The nicotinic receptor switches between states in which the channel is closed and the unitary current is zero (R, AR, A_2R), and a state in which the channel is open and shows a measurable unitary current (A_2R^*) (**Figure 8.17**). These conformational changes can be modeled as:

$$\underbrace{R \rightleftharpoons AR \rightleftharpoons A_2R}_{\text{closed channel}} \rightleftharpoons \underbrace{A_2R^*}_{\text{open channel}}$$

where A is acetylcholine or any nicotinic agonist; R is the receptor in the closed conformation; and R* is the receptor in the open conformation.

In **Figure 8.16a**, one observes that the periods during which the channel is open, t_o, are variable. To obtain the mean open time of the channel, τ_o, one can build a frequency histogram of the different t_o. The exponential curve obtained provides the value of τ_o (see **Figure 8.16b** and Appendix 4.3). The functional significance of this value is as follows: during a time equal to τ_o the channel has a high probability of being open.

τ_o is a characteristic of the nicotinic receptor channel type

Nicotinic receptors from the electric organ of *Torpedo* and from the calf neuromuscular junction can be studied in patch clamp after the expression of the corresponding mRNA injected into *Xenopus* oocytes (outside-out configuration). Recording of such channels shows that electric organ and neuromuscular junction channels present very similar conductances (40 and 42 pS) but very different mean open times (τ_o = 0.6 and 7.6 ms, respectively).

Another example is given by the study of nicotinic receptors from fetal or adult bovine muscle. A study of the subunit structure of the bovine muscle nAChR showed the presence of the α, β, γ and δ-subunits as in the case for *Torpedo* electroplax nAChR. In addition, a novel subunit termed the ε-subunit has been discovered by cloning and sequencing the DNA complementary to the muscle mRNA encoding it. The ε-subunit shows higher sequence homology with the γ-subunit than with any other subunit. In order to study the properties of the

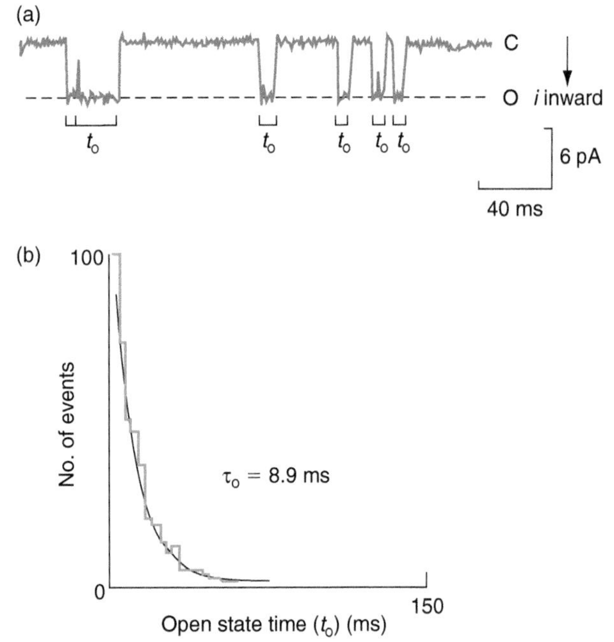

FIGURE 8.16 **Patch clamp recording (attached-cell configuration) of myotube nicotinic receptor channel activity (V_m = −170 mV).** (**a**) Myotubes (embryonic muscle cells) are recorded in the presence of a low concentration of acetylcholine (200 nM). At this concentration, the channels open during periods t_o. This recording does not correspond to a single nAChR because one finds approximately 100 000 nAChR per patch. The repeated openings (downward deflections) correspond, therefore, to the opening of different nAChR. However, all the nAChR being identical, it seems as though the activity of the same nAChR was recorded. (**b**) The mean open time τ_o can thus be calculated. C, closed channel; unitary current is zero. O, open channel; inward unitary current (downward deflections).

γ and ε-subunits, various combinations of the subunit-specific mRNAs are injected in *Xenopus* oocytes and their functional properties are studied in the presence of acetylcholine.

Figure 8.18a shows recordings of ACh-activated single channels from outside-out patches isolated from oocytes injected with the α, β, γ and δ-subunit-specific mRNAs (left) or with the α, β, ε and δ-subunit-specific mRNAs (right). The conductance and mean open time τ_o (**Figure 8.18c,d**) of the channels formed in a given oocyte differ in relation to the mRNA combination with which it was injected. This suggests that a single subunit can change the conductance and gating properties of the nAChR channel.

To compare the two classes of nAChR channels produced in *Xenopus* oocytes with native bovine nAChR channels, the ACh-activated channels of fetal and adult bovine muscle are recorded. **Figure 8.18b** shows ACh-activated single currents from outside-out patches of native fetal (left) and adult (right) bovine muscle. nAChR single-channel current in fetal muscle is similar to that of nAChRγ whereas the nAChR single-channel current in adult muscle is similar to that of nAChRε (compare **Figure 8.18a,b**). This suggests that the nAChR channel in fetal muscle is assembled from α, β, γ and δ-subunits whereas the endplate channel in adult muscle is assembled from the α, β, ε and δ-subunits. To study this developmental change in the contents of the five nAChR-subunit mRNAs in bovine muscle, total RNA is extracted from the diaphragm muscle at various stages of fetal and postnatal development. It is then subjected to blot hybridization analysis using the respective cDNA probes. The results show that the contents of the γ and ε-subunit mRNAs vary markedly during muscle development showing reciprocal changes: the γ-subunit mRNA is abundant at earlier fetal stages (3–5 months' gestation), but is hardly or not detectable after birth; conversely, considerable amounts of the ε-subunit mRNA appear only at postnatal stages and is not detectable at earlier fetal stages (3–4 months' gestation). Therefore, the replacement of the γ-subunit by the ε-subunit in the nAChR complex is responsible for the changes in the properties of the nAChR channel that occur during muscle development. This phenomenon of subunit replacement during development cannot be generalized to all the other mammalian nAChR.

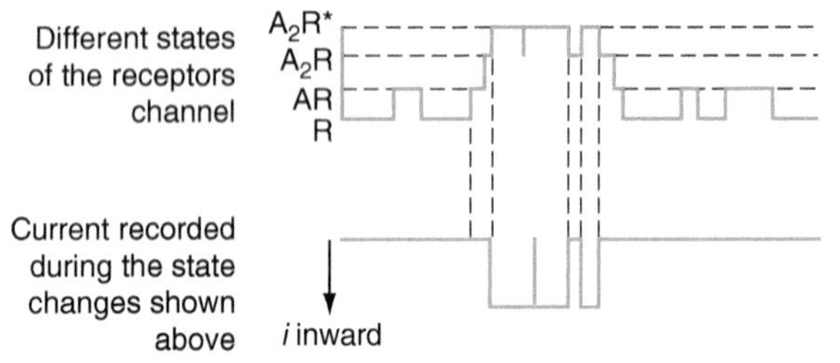

FIGURE 8.17 **Correlation between the nicotinic current and the states of the channel.** The channel opens only (inward current, lower trace) when the protein is in the A_2R^* state. The rapid fluctuations between states A_2R^* and A_2R correspond to short-lived closures. When the receptor channel loses one or two of its acetylcholine molecules, the closures last longer. Adapted from Colquhoun D, Ogden DC, Mathie A (1987) Nicotinic acetylcholine receptors of nerve and muscle: functional aspects. *TIPS* **8**, 465–472, with permission.

8.4 THE NICOTINIC RECEPTOR DESENSITIZES

During the recording of a nicotinic receptor channel with the patch clamp technique (whole-cell configuration) in the presence of a high and constant concentration of acetylcholine, there is a progressive diminution of the total current I_{ACh} (**Figure 8.19a**). This decrease in current corresponds to the progressive desensitization of the nicotinic receptor present in the membrane.

When recording unitary nicotinic currents (outside-out or cell-attached configuration) in the presence of a strong concentration of acetylcholine, there are

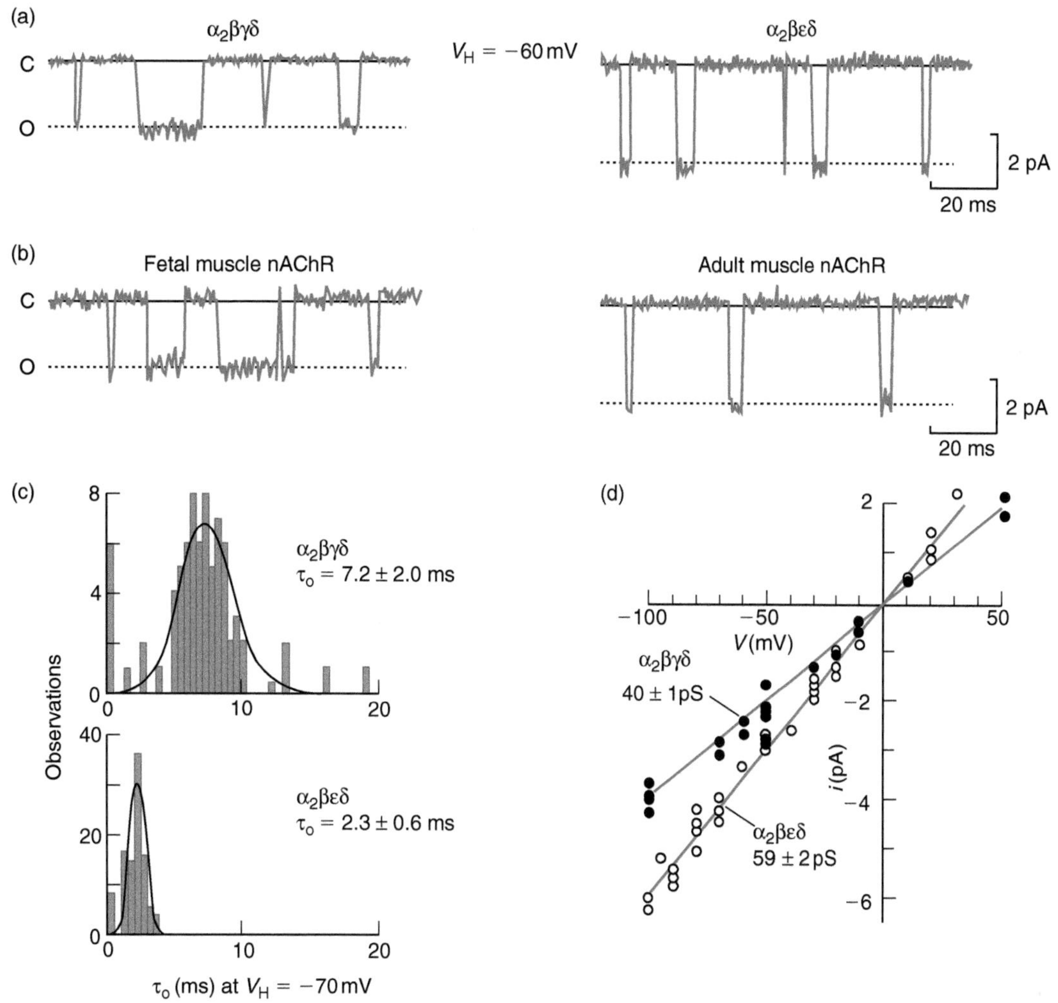

FIGURE 8.18 **The subunit structure participates in determining the nicotinic channel conductance and mean open time.** See text for explanations. C, closed channel; O, open channel. Adapted from Mishina M, Takai T, Imoto K *et al.* (1986) Molecular distinction between fetal and adult forms of muscle acetylcholine receptor. *Nature* **321**, 406–411, with permission.

repeated openings separated by long periods of silence (**Figure 8.19b**). These sequences of openings are known as *unitary current bursts*. Within a burst, the protein rapidly fluctuates between the closed and open states, symbolized as follows:

$$\underbrace{R \rightleftharpoons AR \rightleftharpoons A_2R}_{\text{closed channel states}} \rightleftharpoons \underbrace{A_2R^*}_{\text{open channel state}}$$

The long silent periods correspond to desensitization of the receptor in the presence of acetylcholine. In the desensitized state, the nicotinic receptor is refractory to activation. Consequently, the channel does not open despite the fact that two molecules of ACh are bound to the receptor.

In summary, desensitization is a phenomenon that renders the nicotinic receptor incapable of being activated by its agonists. The desensitized nicotinic receptor presents two main characteristics: (i) a high affinity for ACh and (ii) a closed ionic channel: the unitary current is equal to 0 pA (long silent periods).

$$A_2R \rightleftharpoons A_2R^* \underset{k_{-3}}{\overset{k_3}{\rightleftharpoons}} A_2D_1$$
$$A_2R^* \underset{k_{-4}}{\overset{k_4}{\rightleftharpoons}} A_2D_2$$

These two states D_1 and D_2, which are closed channel states, are distinguished from one another by their affinity constants for ACh (**Table 8.1**) and by the rate constants $k_3 = 0.01\ s^{-1}$ and $k_4 = 1\ s^{-1}$.

The concept of nicotinic receptor desensitization has been proposed by different authors, notably by Katz and Thesleff, from measurements of the time course of the global synaptic response during an iontophoretic application of acetylcholine.

The process of desensitization appears slowly and is slowly reversible. This is an intrinsic property of the

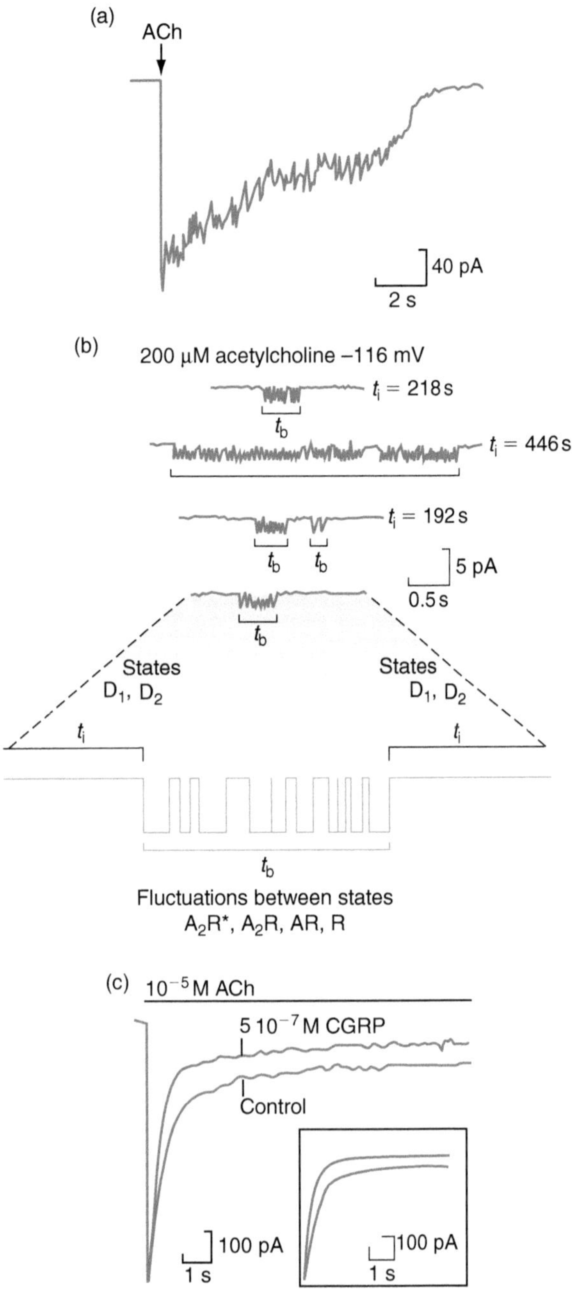

FIGURE 8.19 **Nicotinic receptor desensitization.** (**a**) Patch clamp recording (whole-cell configuration) of the nicotinic current from adrenal chromaffin cells (V_m = −70 mV). In the presence of a high concentration of acetylcholine (20 μM) the inward current reaches a peak of 235 pA, and then decreases despite a constant acetylcholine concentration. This current corresponds to the sum of several unitary currents crossing all the activated nicotinic channels. The decrease in current is due to the desensitization of a large fraction of cellular nicotinic receptors. (**b**) Single-channel recordings (cell-attached or outside-out configurations) illustrating another consequence of desensitization. The membrane patch is exposed to a high concentration of acetylcholine (200 μM) for an extended period. Under these conditions, nicotinic receptors present in the patch desensitize. After a certain time, one of the channels reopens and fluctuates between the states A_2R^*, A_2R, AR and R during a time t_b (duration of the burst of openings) before desensitizing again for a duration t_i (interburst duration). The traces shown correspond to segments of a continuous recording. The duration of the desensitized periods t_i between two successive traces is indicated at the end of each trace (218, 466 and 192 s). (**c**) Cultured muscle cell recording (whole-cell configuration). Nicotinic current evoked by the application of 10 μM acetylcholine recorded in the presence of 500 nM CGRP (calcitonin gene-related peptide) and in the absence of the peptide (control) (V_m = −60 mV). The nicotinic current reaches a peak with a 200 ms delay and then begins to decrease. The sum of two exponentials can describe this decrease in current. CGRP increases the speed of the fast component. Part (**a**) from Clapham DE, Neher E (1984) Trifluoperazine reduces inward ionic currents and secretion by separate mechanisms in bovine chromaffin cells. *J. Physiol. (Lond.)* **353**, 541–564, with permission. Part (**b**) from Colquhoun D, Ogden DC, Mathie A (1987) Nicotinic acetylcholine receptors of nerve and muscle: functional aspects. *TIPS* **8**, 465–472, with permission. Part (**c**) from Mulle C, Benoit P, Pinset C *et al.* (1988) Calcitonin gene-related peptide enhances the rate of desensitization of the nicotinic acetylcholine receptor in cultured mouse muscle cells. *Proc. Natl Acad. Sci. USA* **85**, 5728–5732, with permission.

protein. In order to study in patch clamp the states R and R* of the nicotinic receptor channel, it is necessary to choose conditions under which desensitization is negligible. To this end, researchers work with very low doses of acetylcholine (of the order of the nanomoles per liter) because high doses favor the conformational change of the protein into desensitized states (**Figure 8.16**). An alternative approach is the use of high doses of acetylcholine (of the order of micromoles per liter; **Figure 8.19b**). In this case, the receptors desensitize (silent periods known as interburst periods) and eventually one or several of the channels open and close repetitively (opening bursts) before re-desensitizing. Desensitization can be minimized by excluding the first and the last opening during the bursting periods, and thus the values calculated for τ_o are then related only to the R* state.

The rate of desensitization of the nicotinic receptor seems to be related to its state of phosphorylation. In fact, studies of the ionic flux through nicotinic receptors incorporated into liposomes have shown that an increase in the level of phosphorylation of the receptors by cyclic AMP augments the desensitization rate of these receptors. In the neuromuscular junctions, a peptide present in the motoneurons is released at the same time as acetylcholine. This peptide, CGRP (calcitonin gene-related peptide), is capable of increasing the level of cyclic

TABLE 8.1 Affinity constants for acetylcholine

State	Affinity constant for ACh K_D
R	10 μM to 1 mM
D_1	1 μM
D_2	3–10 nM

AMP in cultured embryonic muscle cells, consequently increasing the number of phosphorylated nicotinic receptors. In patch clamp recordings (whole-cell configuration) of embryonic muscle cells, the simultaneous application of this peptide and acetylcholine accelerates the rapid phase of desensitization of the nicotinic receptors (**Figure 8.19c**). In unitary recordings (cell-attached configuration), CGRP decreases the opening frequency of the nicotinic channels (while at the same time leaving unaffected their mean open time and unitary conductance). These effects are mimicked by the application of substances that augment the intracellular cyclic AMP level (such as forskolin). The following hypothesis has been proposed. CGRP activates a specific membrane receptor. This leads to an increase in the intracellular cyclic AMP concentration and an activation of protein kinase A. Protein kinase A, directly or indirectly, phosphorylates certain subunits of the nicotinic receptor, leading to a rapid desensitization of the receptors.

Generalization

Nicotinic receptors are allosteric receptors as defined by Monod, Wyman and Changeux (MWC model, 1965). The MWC model hypothesizes the following:

- nAChRs are oligomers made up of a finite number of identical subunits that occupy equivalent positions and, as a consequence, possess at least one axis of rotational symmetry.
- nAChRs can exist spontaneously in the four freely interconvertible and discrete conformational states described above (R, R*, D_1, D_2), even in the absence of ligand. The closed states are R, AR and A_2R; the open states are R*, AR* and A_2R^*; the desensitized states are D_1, AD_1, A_2 and D_2, AD_2, A_2D_2.
- The affinity and activity of the stereospecific sites carried by the nAChR differ between these four states.

This gives a tetrahedral model for interactions between the four conformational states instead of the sequential model previously described for simplicity:

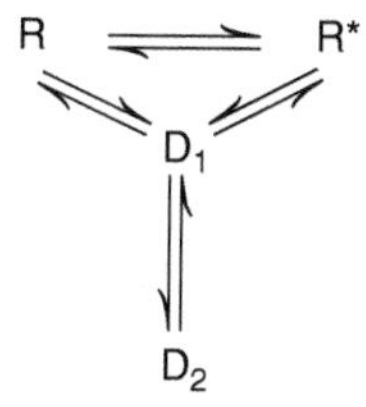

As a result, gating of the nAChR cannot be viewed solely as a ligand-triggered process but as reflecting an intrinsic structural transition of the receptor molecule, which may even occur in the absence of ligand. Moreover, at low agonist concentrations, desensitized states can be stabilized under conditions of negligible channel opening.

8.5 nAChR-MEDIATED SYNAPTIC TRANSMISSION AT THE NEUROMUSCULAR JUNCTION

A nicotinic synaptic current is evoked by a brief augmentation of the concentration of acetylcholine in the synaptic cleft. This increase, caused by the asynchronous release of synaptic vesicles, is brief because (**Figure 8.20a**): (i) the release is brief; and (ii) acetylcholine rapidly disappears from the synaptic cleft. In fact, when acetylcholine is released into the synaptic cleft, it may either bind to acetylcholine-gated channels, diffuse out of the synaptic cleft, or be rapidly degraded by acetylcholinesterase.

During the analysis of synaptic currents induced by the release of endogenous acetylcholine, the desensitized states of the receptor can be neglected because of the rapid elimination of acetylcholine from the synaptic cleft (in the order of microseconds). The model for this is:

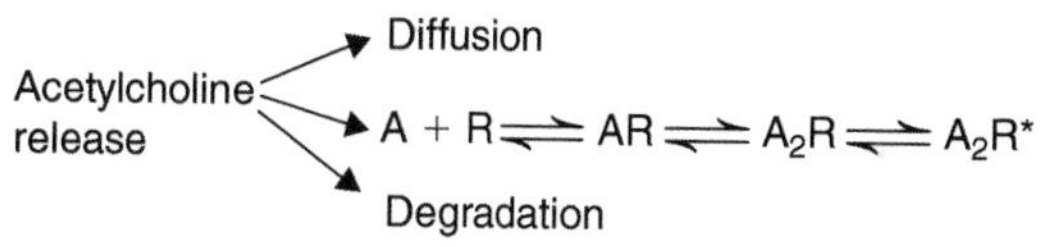

This is very different from what occurs during the recording of the activity of a nicotinic receptor in a patch of membrane in the inside-out configuration, when the acetylcholine is continuously present in the patch pipette, or the recording of the activity of a nicotinic receptor channel in a patch of membrane in the outside-out configuration in response to acetylcholine pressure applied from another pipette. Even the shortest applications in this case are of the order of tens of milliseconds.

8.5.1 Miniature and endplate synaptic currents are recorded at the neuromuscular junction

Two main types of postsynaptic currents can be recorded at any synapse: spontaneous and evoked synaptic currents. The former is recorded in the absence of presynaptic stimulation whereas the latter represents the response to a presynaptic stimulation. Spontaneous synaptic currents can be further divided into those evoked by spontaneous presynaptic spikes in a Ca^{2+}-dependent manner, and miniature currents evoked even

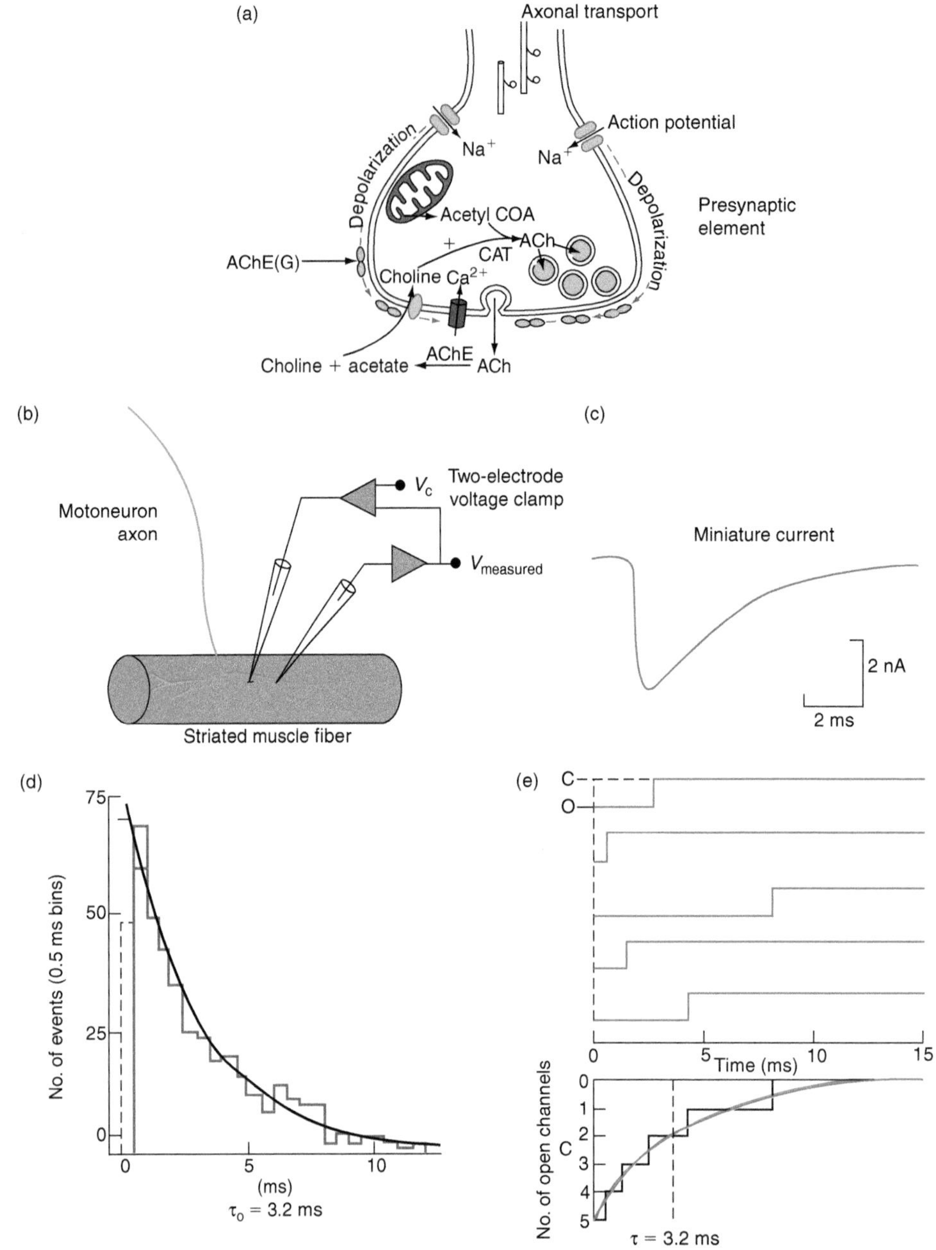

FIGURE 8.20 **A cholinergic presynaptic terminal and miniature nicotinic current.** (**a**) Functional scheme of the presynaptic component of the neuromuscular nicotinic cholinergic synapse. The enzymes choline acetyl transferase (CAT) and acetylcholinesterase (AChE) are synthesized in the cell body of the motoneuron and carried to axon terminals via anterograde axonal transport. Acetylcholine (Ach) is synthesized in axon terminals from choline and acetyl coenzyme A (acetylCOA). About 50% of the released choline is recaptured by the presynaptic terminals. Acetylcholine is actively transported from the cytoplasm into synaptic vesicles via a vesicular Ach carrier using the H^+ gradient as an energy source (antiport). (**b**) Miniature current recorded in two-electrode voltage clamp from a normally innervated muscle fiber in the absence of stimulation and in the presence of TTX. (**c**) The neuromuscular junction miniature current I_{ACh} is the current crossing N nicotinic channels activated by spontaneously released endogenous acetylcholine. The current rising phase is fast owing to the quasi-synchronous activation of the N nicotinic channels. The exponential falling phase of the current is slower ($V_m = -80$ mV). (**d**) Since the opening of the channels is synchronous, their closure appears after a variable time t_0 whose distribution is exponential. The mean open time of the N channels is $\tau_o = 3.2$ ms. (**e**) The same value of τ_o is obtained for the time constant of the falling phase (τ) of the total current I_{ACh}. In the example given in (**d**), $N = 5$, but N is in fact always larger (see text). Parts (**c**) and (**d**) adapted from Colquhoun D (1981) How fast do drugs work? *TIPS* **2**, 212–217, with permission.

in the absence of spontaneous spikes (in the presence of tetrodotoxin (TTX)). Miniature currents have a very small amplitude and correspond to the release of one or a few synaptic vesicles (see Appendix 7.1).

Miniature currents

Miniature currents are the currents recorded at the neuromuscular junction in the total absence of stimulation of the motor nerve and in the presence of TTX (**Figure 8.20b**). These currents are evoked by the spontaneous liberation of acetylcholine from the presynaptic terminal. Thus, if an innervated muscle fiber in the absence of nerve stimulation is recorded from under voltage clamp (**Figure 8.20b**), from time to time a miniature current will be recorded (**Figure 8.20c**). The recorded current is due to the spontaneous release of a synaptic vesicle or quantum of acetylcholine (1 vesicle = 1 quantum; see Chapter 7). This current is inward at −80 mV and has a maximum amplitude of about 4 nA.

How many receptor channels are opened by acetylcholine at the peak of a miniature current? Knowing the amplitude of the unitary current and the amplitude of a miniature current at the same membrane potential, we can calculate the number of nicotinic receptor channels opened by acetylcholine at the peak of the miniature response.

At $V_m = -80$ mV, we have:

$$i = 2.5\text{pA}, \text{and}\, I = 4\text{nA} = 4 \times 103^{\circ}\,\text{pA};$$

i.e. $4.10^3/2.5$ = 1600 nicotinic receptor channels opened by acetylcholine.

Since two molecules of acetylcholine are needed to open one channel, the average number of acetylcholine molecules released is 1600 × 2 = 3200 molecules of ACh per vesicle. In other words, 1 quantum equals about 3000 molecules of ACh.

What is the time course of a miniature current? The time it takes to reach the maximal amplitude of the miniature current is approximately 100 μs, while it takes longer to disappear. The decrease of the current has an exponential time course with a time constant of the order of milliseconds (**Figure 8.20e**). This current decrease depends solely on τ_o (**Figure 8.20d,e**).

Motor endplate current

The endplate current is recorded (in voltage clamp) at the neuromuscular junction while the motor nerve is being stimulated (**Figure 8.21**). At $V_m = -80$ mV the current is inward and has an amplitude of approximately 400 nA.

How many channels are opened at the peak of the motor endplate current? The motor endplate current is composed of 400 nA/4 nA = 100 miniature currents produced by 1600 × 100 = 16 × 10^4 nicotinic receptors opened by released acetylcholine.

What is the time course of the motor endplate current? Approximately 100 vesicles are released in an asynchronous manner by the stimulated presynaptic terminal. This is the reason why the time it takes to reach the maximal or peak amplitude of this current is relatively longer than the time it takes to reach the peak of a miniature current (300 μs instead of 100 μs).

In current clamp recordings, this endplate inward current depolarizes the postsynaptic membrane thus evoking the endplate potential recorded in **Figure 8.2.**

8.5.2 Synaptic currents are the sum of unitary currents appearing with variable delays and durations

There is a variable delay in the appearance of current flow through each one of the postsynaptic receptor channels. The reason for this is that the synaptic vesicles are released in an asynchronous manner. Furthermore, acetylcholine molecules must diffuse for a certain time before they reach a free receptor channel. We have seen that the concentration of acetylcholine in the synaptic cleft decreases so rapidly that a receptor channel has very few chances of being reopened a second time by binding again two acetylcholine molecules.

Determination of the value of the total synaptic current, I_{ACh}, at steady state is:

$$I_{\text{ACh}} = Np_o \text{iACh},$$

where N is the number of nic otinic channels in the membrane, i_{ACh} is the unitary current, and p_o is the open-state probability of the channel and depends on the acetylcholine concentration and on the receptor channel opening (β) and closing (α) rate constants.

In the model below we have the following rate constants:

$$\text{R} \underset{k_{21}}{\overset{k_1}{\rightleftharpoons}} \text{AR} \underset{k_{22}}{\overset{k_2}{\rightleftharpoons}} \text{A}_2\text{R} \underset{\text{a}}{\overset{\text{b}}{\rightleftharpoons}} \text{A}_2\text{R*}$$

where α and β are the closing and opening rate constants of the channel. The channel's probability of being in the open state at steady state (see Appendix 4.3) is then:

$$p_o = \beta'/(\beta' + \alpha),$$

where β′ is the apparent opening rate constant of the channel, which depends on β but also on the rate constants of the preceding stages k_1 and k_2(β′ = f[ACh]). Thus, if α is short (i.e. the mean open time τ_o is long (because $\tau_o = 1/\alpha$)), then p_o is high (it approaches its maximum value), and I_{ACh} is large. The falling phase of the total current is exponential, with a time constant equal to τ_o.

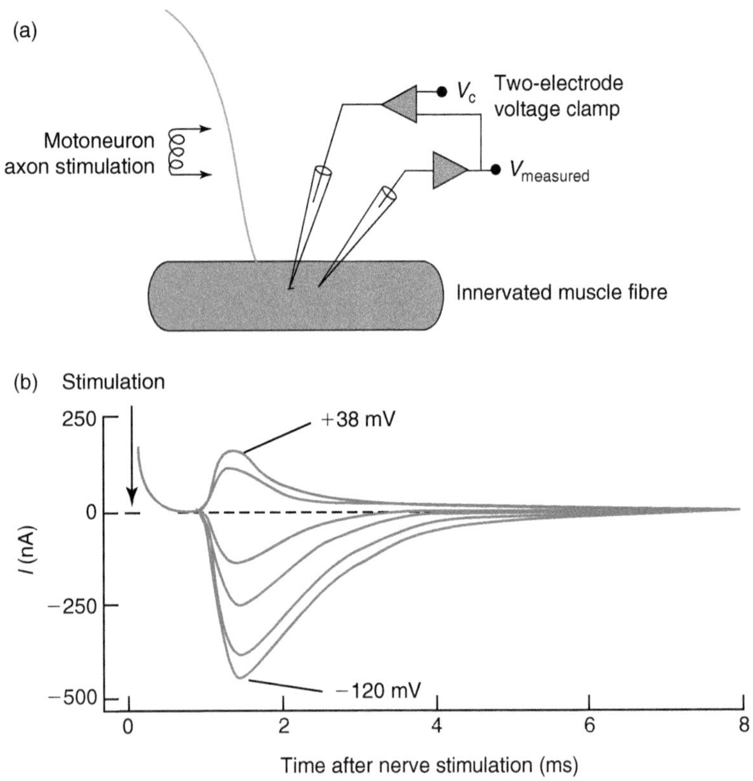

FIGURE 8.21 **Motor endplate current.** (a) Current recorded in response to a stimulation of the nerve fiber under two-electrode voltage clamp. (b) This current is inward for negative membrane potentials and outward for positive voltages. As in the case of the unitary current, the motor endplate current reverses around 0 mV (frog's neuromuscular junction). Muscular action potentials are blocked by voltage clamping the corresponding muscle region. The muscle contraction induced by the inward current can be blocked by the destruction of T tubules (with a hyperosmotic shock). Adapted from Magleby KL, Stevens CF (1972) A quantitative description of end plate currents. *J. Physiol. (Lond.)* **223**, 173–197, with permission.

Let us assume that at a time t a certain number of ionic channels are opened more or less synchronously (this is the case of miniature currents). Because acetylcholine disappears very rapidly from the synaptic cleft, a channel has very few chances of reopening. Each one of the open channels has an opening duration t_o. We have seen that the duration t_o during which each channel remains open can be described by an exponential distribution (see **Figure 8.16**). **Figure 8.20e** shows that the total current crossing N channels decreases exponentially with a time constant equal to τ_o.

In conclusion, the falling phase of the total current is not due to the progressive disappearance of acetylcholine from the synaptic cleft (because it disappears with a time constant of the order of microseconds) but depends only on τ_o, an intrinsic property of the channel.

8.6 NICOTINIC TRANSMISSION PHARMACOLOGY

8.6.1 Nicotinic agonists

Nicotinic receptor agonists (see Appendix 6.1) bind to the same receptor site as acetylcholine and favor the conformational changes of the protein towards the open state. These agonists are, for example, suberyldicholine, carbachol and PTMA (phenyl trimethyl ammonium) (**Figure 8.22a**). The application of one of these on a patch of muscle membrane (outside-out configuration) leads to the onset of a current whose amplitude at each membrane potential tested is equal to the current evoked by acetylcholine. However, the duration of the openings of the channel depends on the agonist used (**Figure 8.22b,c**).

8.6.2 Competitive nicotinic antagonists

Competitive antagonists (see Appendix 6.1) bind to the same receptor site as acetylcholine but *do not favor* its conformational change towards the open state. By binding to the acetylcholine receptor sites, competitive antagonists prevent acetylcholine from binding to its receptor sites and activating the nAChR. They decrease the number of sites available to acetylcholine and, therefore, decrease or completely block (depending on the dose used) the nicotinic cholinergic response. A distinction is made between competitive antagonists whose effect is reversible ((+)tubocurarine) and those whose effect is irreversible (DDF).

The application of (+)tubocurarine on a patch of muscle membrane in the outside-out recording configuration and in the presence of a low dose of acetylcholine causes a drop in the opening frequency of the channels. This occurs because (+)tubocurarine reduces the number of receptor sites available for acetylcholine. It should be noted, however, that the amplitude of the unitary current i_{Ach} evoked in the presence or absence of (+)tubocurarine is identical. The application of (+) tubocurarine on an isolated nerve–muscle preparation

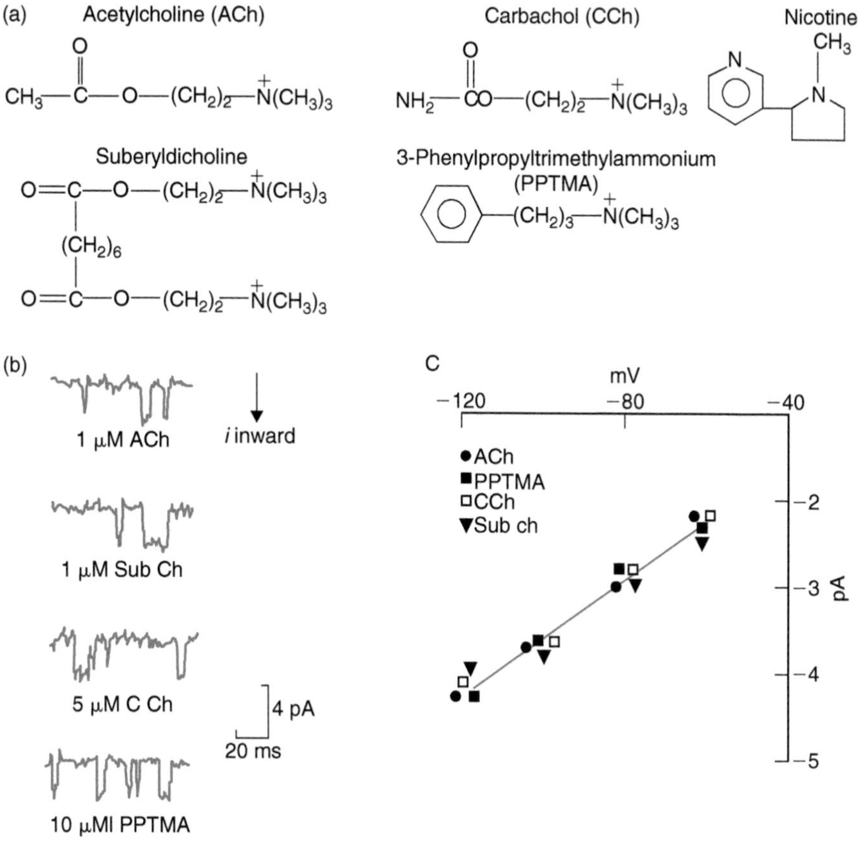

FIGURE 8.22 **Unitary currents evoked by nicotinic agonists.** (**a**) Structure of the nicotinic agonists tested. (**b**) Inward unitary currents evoked by different nicotinic agonists (V_m = −80 mV) recorded in patch clamp (outside-out configuration) from isolated rat myotubes. Solutions (in mM): intracellular 150 KCl, 5 Na_2EGTA, 0.5 $CaCl_2$; extracellular 135 NaCl, 5.4 KCl. (**c**) i/V curves built from results similar to those shown in (**b**), but at different membrane potentials, are completely superimposable. The slope of each curve gives a unitary conductance γ of approximately 34 pS. Adapted from Gardner P, Ogden DC, Colquhoun D (1984) Conductances of single ion channels opened by nicotinic agonists are indistinguishable. *Nature* **289**, 160–163, with permission.

induces a reduction in the amplitude of the spontaneous miniature currents (currents evoked by the endogenous and spontaneous liberation of acetylcholine). The effect of (+)tubocurarine can be reversed by elevating the concentration of acetylcholine applied or released. Since the binding of (+)tubocurarine to the nicotinic receptor is reversible, increasing the acetylcholine concentration will increase the probability that the receptor sites are occupied by acetylcholine.

The binding of α-bungarotoxin (venom from the snake *Bungarus multicinctus*) to the nicotinic receptor of the neuromuscular junction is very stable. For this reason, this toxin is used as a marker of acetylcholine receptor sites in this preparation. This labeling permits localizing and counting of these receptors. Labeled α-bungarotoxin also allows identification of the receptor during its purification process (see **Figure 8.4**).

8.6.3 Channel blockers

Channel blockers are substances that bind to the aqueous pore of the open receptor, preventing the passage of cations through it. Among these substances are procaine and its derivatives (QX 222, lidocaine, benzocaine), and also histrionicotoxin and chlorpromazine.

Application of acetylcholine in the presence of benzocaine to a muscle membrane patch in the outside-out recording configuration evokes the onset of opening bursts. These bursts of unitary currents are due to the numerous fluctuations of the receptor between its open and its closed state (**Figure 8.23a**). The following model describes this process:

$$R \rightleftharpoons R^* \rightleftharpoons R^*B,$$

where R represents the nicotinic receptor in its closed state, R* the nicotinic receptor in its open state, and R*B the nicotinic receptor in the open but blocked state. In the R* state, the cations cross the aqueous pore, while in the states R and R*B the ions cannot cross it.

In the presence of benzocaine, a change in the kinetics of the falling phase of the current (**Figure 8.23b**) is observed during the recording of spontaneous miniature currents of an isolated nerve–muscle preparation (V_m = −100 mV). In the absence of benzocaine, the falling phase is described by a single exponential with a time constant of τ = 3.8 ms. In the presence of benzocaine, the falling phase is described by two exponentials with time constants τ_1 = 1.0 ms and τ_2 = 7.6 ms. The explanation of this effect is that the channels opened by acetylcholine are very rapidly blocked by benzocaine, which quickly blocks the unitary current. Thus, one observes a fast initial decrement of the miniature current (with a shorter time constant than in the absence of benzocaine). The channels then reopen and reblock repeatedly. This

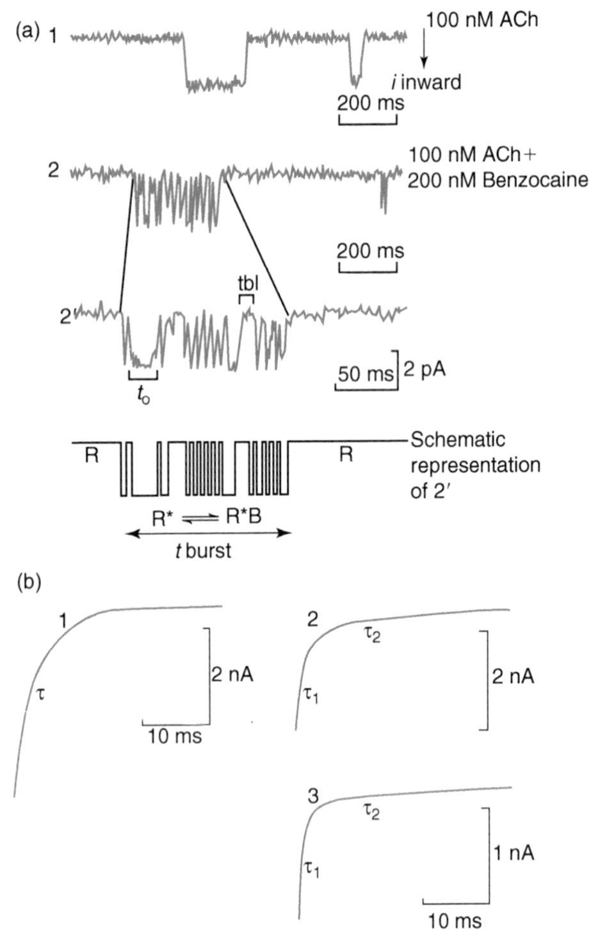

FIGURE 8.23 **Benzocaine effect on the time course of ACh evoked unitary and miniature nicotinic currents.** (**a**) Patch clamp recording (cell-attached configuration) of nicotinic unitary currents evoked by the application of ACh (i_{ACh}). (1) In the presence of 100 nM of ACh, the channels open for a mean duration of $\tau_o = 19$ ms ($V_m = -110$ mV). (2) In the presence of 100 nM of ACh + 200 μM of benzocaine, bursts of openings ($V_m = -130$ mV) are recorded. (2′) The same recording as (2) but with a different timescale and with an added diagram of the openings and closings of the channel. t_o, time during which the channel stays open; t_{bl}, time during which the channel is blocked by benzocaine; t_{burst} duration of a burst of openings. The histograms of t_o and of t_{bl} are described by a single exponential with the following average values: $\tau_o = 2.8$ ms, and $\tau_{bl} = 3.5$ ms (extrajunctional muscle membrane of 4- to 6-week-old muscle cells). (**b**) Two-electrode voltage clamp recording of miniature nicotinic currents. Each curve corresponds to the average of 8 to 14 miniature currents ($V_m = -100$ mV). (1) In the absence of benzocaine, the miniature currents reach their maximum in approximately 1 ms. Their decrement is described by a single exponential with a time constant $\tau = 3.8$ ms. (2) In the presence of extracellular benzocaine (300 μM, 15 min), the peak amplitude of miniature currents decreases, and the falling phase is described by two exponentials with time constants $\tau_1 = 1.0$ ms and $\tau_2 = 7.6$ ms. (3) In the presence of a higher concentration of benzocaine (500 μM, 17 min), the amplitude of the miniature current peak is further diminished and the time constants of the falling phase become $\tau_1 = 0.7$ ms and $\tau_2 = 11.6$ ms (frog cutaneous pectoris muscle). Parts (**a**) and (**b**) adapted from Ogden DC, Siegelbaum SA, Colquhoun D (1981) Block of acetylcholine-activated ion channels by an uncharged local anesthetic. *Nature* **289**, 596–598, with permission.

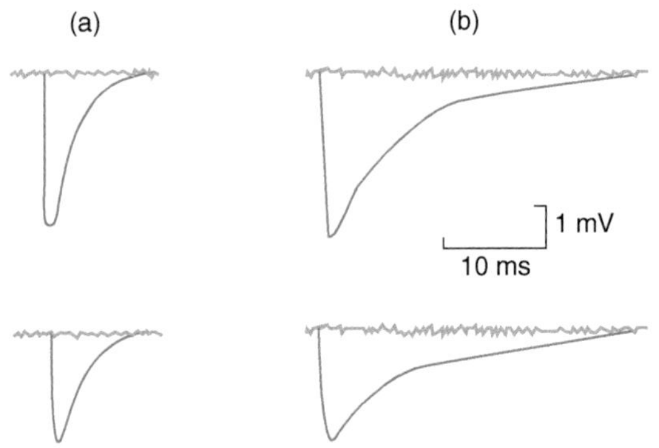

FIGURE 8.24 **Effect of the acetylcholinesterase inhibitor prostigmine on the duration of miniature currents.** Recordings (**a**) in the absence of and (**b**) in the presence of prostigmine (10^{-6} g ml^{-1}), showing that prostigmine augments the duration of the miniature current. Adapted from Katz B, Miledi R (1973) The binding of acetylcholine to receptors and its removal from the synaptic cleft. *J. Physiol. (Lond.)* **231**, 549–574, with permission.

increases the duration of the miniature current and one observes a second slower decrementing phase (with time constant τ_2).

8.6.4 Acetylcholinesterase inhibitors

These inhibitors have a reversible effect, as in the case of prostigmine, or an irreversible effect, as in the case of DFP (difluorophosphate). The application of prostigmine to an isolated nerve–muscle preparation significantly increases the miniature current duration (**Figure 8.24**). In the presence of prostigmine, acetylcholine molecules degrade much more slowly, and thus are able to bind repeatedly and trigger the reopening of nicotinic receptors. This repeated binding considerably increases the duration of the miniature current falling phase. As we have already seen (see Section 8.5), the miniature current time constant in the absence of acetylcholinesterase inhibitors reflects the nicotinic receptor average open time. However, the average open time of the nicotinic receptor is clearly not the same in the presence of prostigmine.

8.7 SUMMARY

Upon release of acetylcholine in the synaptic cleft of the neuromuscular junction, two molecules of acetylcholine bind to each postsynaptic nAChRs, at specific sites. Upon ACh binding, each nAChR undergoes fast activation leading to an open-channel state, and a slow desensitization reaction leading to a closed-channel state refractory to activation. nAChR opening occurs in the millisecond range, fast desensitization in the 0.1 s range and slow desensitization in the minute range.

To which type(s) of postsynaptic receptor do the molecules of acetylcholine bind?

At the neuromuscular junction, ACh binds to the muscle-type nAChR. Muscle nAChRs have a fixed composition $[\alpha 1]_2[\beta 1][\delta][\gamma \text{ or } \varepsilon]$ in vertebrates. Each subunit contains a large N-terminal hydrophilic domain exposed to the synaptic cleft, followed by three transmembrane segments (M1 to M3), a large intracellular loop and a C-terminal transmembrane segment (M4). Acetylcholine binding sites are located at the interface between α and non-α-subunits in the N-terminal regions. The ion channel is lined by the M2 segment from each of the five subunits.

How is the binding of acetylcholine transduced into a depolarization of the postsynaptic membrane?

Upon binding of two molecules of ACh, nAChRs undergo a fast conformational change to a state where the pore is open. The pore is selectively permeable to cations. At −90 to −80 mV, the resting potential of muscle fibers, the electrochemical gradient ($V_m - E_{ion}$) for Na^+ and Ca^{2+} ions is large whereas that for K^+ ions it is small. As a result, there is a net inward flux of cations through open nAChRs measured as a transient inward current. This inward current transiently depolarizes the muscular membrane and thus evokes the endplate potential (EPP). EPP is transient (it lasts several milliseconds) owing to the rapid closing of nAChRs and the rapid elimination of ACh from the synaptic cleft by degradation by acetylcholinesterases. In physiological conditions, desensitization of nAChRs does not play a major role.

Why is the postsynaptic depolarization restricted to the postsynaptic membrane?

nAChRs are restricted to the membrane of the postsynaptic element where they are anchored by cytoskeletal proteins. Therefore, the ACh-evoked postsynaptic inward current is triggered at the level of the postsynaptic element only. It is then passively conducted along the postsynaptic muscular membrane. This conduction is decremental (see **Chapter 15**) such as that at several millimeters from the junction, the EPP amplitude is nearly close to zero.

The function of the nicotinic receptor is to ensure rapid synaptic transmission. This is achieved by converting the binding of two acetylcholine molecules into a rapid and transient increase in cationic permeability. This permeability increase is made possible by conformational changes of the receptor channel: it transiently switches from the state in which the channel is closed into a state in which the channel is open. The nicotinic acetylcholine receptor also presents allosteric binding sites, topographically distinct from the neurotransmitter binding site, to which a variety of pharmacological agents and physiological ligands can bind. In doing so they regulate the transitions between the different states of the nAChR.

Further reading

Abakas, M.H., Kaufmann, C., Archdeacon, P., Karlin, A., 1995. Identification of acetylcholine receptor channel-lining residues in the entire M2 segment of the α subunit. Neuron 13, 919–927.

Changeux, J.P., Kasai, M., Huchet, M., Meunier, J.C., 1970. Extraction from electric tissue of gymnotus of a protein presenting several typical properties characteristic of the physiological receptor of acetylcholine. C R Acad. Sci. 270, 2864–2867.

Changeux, J.P., 2013. 50 years of allosteric interactions: the twists and turns of the models. Nat. Rev. Mol. Cell. Biol. 14, 819–829.

Corringer, P.J., Bertrand, S., Bohler, S., et al., 1998. Critical elements determining diversity in agonist binding and desensitization of neuronal nicotinic acetylcholine receptors. J. Neurosci. 15, 648–657.

Couturier, S., Bertrand, D., Matter, J.M., et al., 1990. A neuronal nicotinic acetylcholine receptor subunit (α_7) is developmentally regulated and forms a homo-oligo-meric channel blocked by α-BTX. Neuron 5, 847–856.

Czajikowski, C., Karlin A, 1995. Structure of the nicotinic acetylcholine-binding site: identification of acidic residues in the δ subunit with 0.9 nm of the α subunit-binding site disulfide. J. Biol. Chem. 270, 3160–3164.

Hurst, R., Rollema, H., Bertrand, D., 2013. Nicotinic acetylcholine receptors: from basic science to therapeutics. Pharmacol. Therapeut. 137, 22–54.

Devillers-Thiery, A., Giraudat, J., Bentaboulet, M., Changeux, J.P., 1983. Complete mRNA coding sequence of the acetylcholine binding alpha-subunit of Torpedo marmorata acetylcholine receptor: a model for the transmembrane organization of the polypeptide chain. Proc. Natl Acad. Sci. USA 80, 2067–2071.

Gay, E.A., Giniatullin, R., Skorinkin, A., Yakel, J.L., 2008. Aromatic residues at position 55 of rat alpha7 nicotinic acetylcholine receptors are critical for maintaining rapid desensitization. J. Physiol. 586, 1105–1115.

Katz, B., Miledi R., 1973. The binding of acetylcholine to receptors and its removal from the synaptic cleft. J. Physiol. (Lond.) 231, 549–574.

Langley, J.N., 1905. On the reaction of cells and nerve-endings to certain poisons, chiefly as regards the reaction of striated muscle to nicotine and to curare. J. Physiol. 33, 374–413.

Leonard, R.J., Labarca, C.G., Charnet, P., et al., 1988. Evidence that the M2 membrane-spanning region lines the ion channel pore of the nicotinic receptor. Science 242, 1578–1581.

Miller, P.S., Smart, T.G., 2010. Binding, activation and modulation of cys-loop receptors. Trends Pharmacol. Sci. 31, 161–174.

Monod, J., Wyman, J., Changeux, J.P., 1965. On the nature of allosteric transitions: a plausible model. J. Mol. Biol. 12, 88–118.

Murray, N., Zheng, Y.C., Mandel, G., et al.,1995. A single site on the ε-subunit is responsible for the change in ACh receptor channel conductance during skeletal muscle development. Neuron 14, 865–870.

Roerig, B., Nelson, D.A., Katz, L.C., 1997. Fast synaptic signalling by nicotinic acetylcholine and serotonin 5-HT3 receptors in developing visual cortex. J. Neurosci. 17, 8353–8362.

Unwin, N., 1995. Acetylcholine receptor channel imaged in the open state. Nature 373, 37–43.

APPENDIX 8.1 THE NEURONAL NICOTINIC RECEPTORS

To identify the presence of cholinergic nicotinic synapses in the central nervous system, researchers recorded responses of central neurons to the local application of nicotinic agonists. **Figure A8.1a** shows a current clamp recording from a neuron in a rat brain slice (interneuron of the CA1 region of the hippocampus, see **Figure 19.5**), illustrating the membrane potential change in response to ACh application. ACh evoked a depolarization sufficient to bring the membrane potential of the neuron to threshold for action potential firing. When this interneuron was voltage clamped to −70 mV, ACh evoked an inward current (**Figure A8.1b**). These data demonstrate that ACh can excite inhibitory interneurons of the hippocampal formation by activation of an inward current at near-resting membrane potentials. This effect of ACh is by direct action on the interneuron (postsynaptic effect) because it was not inhibited by blockers of synaptic transmission, including tetrodotoxin, glutamate receptor antagonists (CNQX and APV), and $GABA_A$ receptor antagonist (bicuculline). The inward current response to ACh and the resulting excitation suggested the involvement of cationic receptor channels such as nAChRs. To further investigate this, ACh was applied to interneurons voltage clamped to different membrane potentials to obtain a current–voltage relationship (**Figure 8.1c**). The membrane potential at which ACh produced zero net current flow was between −10 and 0 mV, a reversal potential consistent with the activation of a nonselective cation conductance by ACh. To investigate the pharmacology of the response to ACh, nicotinic receptor-selective ligands were utilized. Application of the nAChR agonist nicotine mimicked the effect of ACh in activating inward current responses (**Figure 8.1d**). Mecamylamine, an nAChR antagonist caused a marked inhibition of the inward current response to ACh (**Figure 8.1e**). The response to ACh was also inhibited by α-bungarotoxin (**Figure 8.1f**). In contrast, the muscarinic receptor antagonist atropine (10 μM) had no effect on the inward current response to ACh (not shown).

To identify whether central nAChRs are permeable to Ca^{2+} ions, single-channel currents evoked by ACh in pure external Ca^{2+} medium are recorded in outside-out patches from freshly dissociated neurons of the rat central nervous system (habenula nucleus) (**Figure A8.2a**). The presence of a current shows that Ca^{2+} permeates these neuronal nAChRs channels. Similarly, in whole-cell recordings, ACh also evokes an inward current when all external cations are replaced with Ca^{2+} (**Figure A8.2b**). When habenula neurons are loaded with FURA-2 (see Appendix 5.1), an increase of $[Ca^{2+}]_i$ up to the micromolar range is observed upon acetylcholine application. This increase is reversibly abolished when Ca^{2+} is removed from the perfusion medium. To exclude a possible involvement of Ca^{2+} entry through voltage-gated Ca^{2+} channels opened by the depolarization subsequent to the activation of nAChRs, the membrane is clamped at −60 mV. In this condition, application of high K^+ (140 mM) external medium, which would depolarize a poorly clamped neuronal membrane, yields no detectable increase of $[Ca^{2+}]_i$. In contrast, application of nicotine evokes an inward whole-cell current and a concomitant rapid increase of $[Ca^{21}]_i$ up to the micromolar range (**Figure A8.2c**).

nAChRs have now been identified in many presynaptic and postsynaptic membranes of CNS synapses. Neuronal nAChRs form a heterogeneous family of subtypes form by five subunits encoded by nine α- (α_2–α_{10}) and three β- (β_2–β_4) subunit genes. Two main subfamilies of neuronal nAChRs have been identified so far (**Figure A8.3**):

1. Neuronal homomeric nAChRs which bind α-bungarotoxin and consist of $\alpha_7\alpha_8\alpha_9$ or α_{10}-subunits. They can be homopentameric $\alpha_7\alpha_8$ and α_9) or heteropentameric α_7 and α_8 or α_9 and α_{10}). Their relative Ca/Na permeability (pCa/Na) is around 10, showing that these receptors are permeable to Ca^{2+} ions as determined by using current and fluorescence measurements (for memory pCa/Na of muscle nAChRs is around 0.2). For native neuronal nAChRs from α_7 gene product, another property is the very rapid desensitization of the whole-cell current. For example, in neurons of the chick peripheral nervous system (ciliary ganglion), fast perfusion of nicotine evokes a large, quickly desensitizing current, strongly depressed by α-bungarotoxin (**Figure A8.4a,b**). When tryptophane in position 55 (W55) in the extracellular domain of rat α_7 nAChR is mutated to alanine (A), the receptor desensitizes much more slowly (**Figure 8.4c**). Also the replacement of the proline residue at position 180 (P180) of the rat α_7 nAChR by either threonine (α_7-P180T) or serine (α_7-P180S) slows the rate of onset of desensitization dramatically, similar to that of the W55A mutation. All these results indicate that rapid desensitization of the wild-type rat α_7 nAChR is facilitated by the presence of the tryptophane and proline residues at positions 55 and 180, respectively. The ligand-binding sites on the homopentameric receptors are present at the interface formed by opposite sides of the same subunit and it is thought that they have five identical ligand-binding sites per receptor molecule (**Figure A8.3**, right).
2. Neuronal heteropentameric nAChRs which do not bind α-bungarotoxin and consist of α- and β-subunits. Their relative Ca/Na permeability is around 2. The heteropentameric receptors have two binding

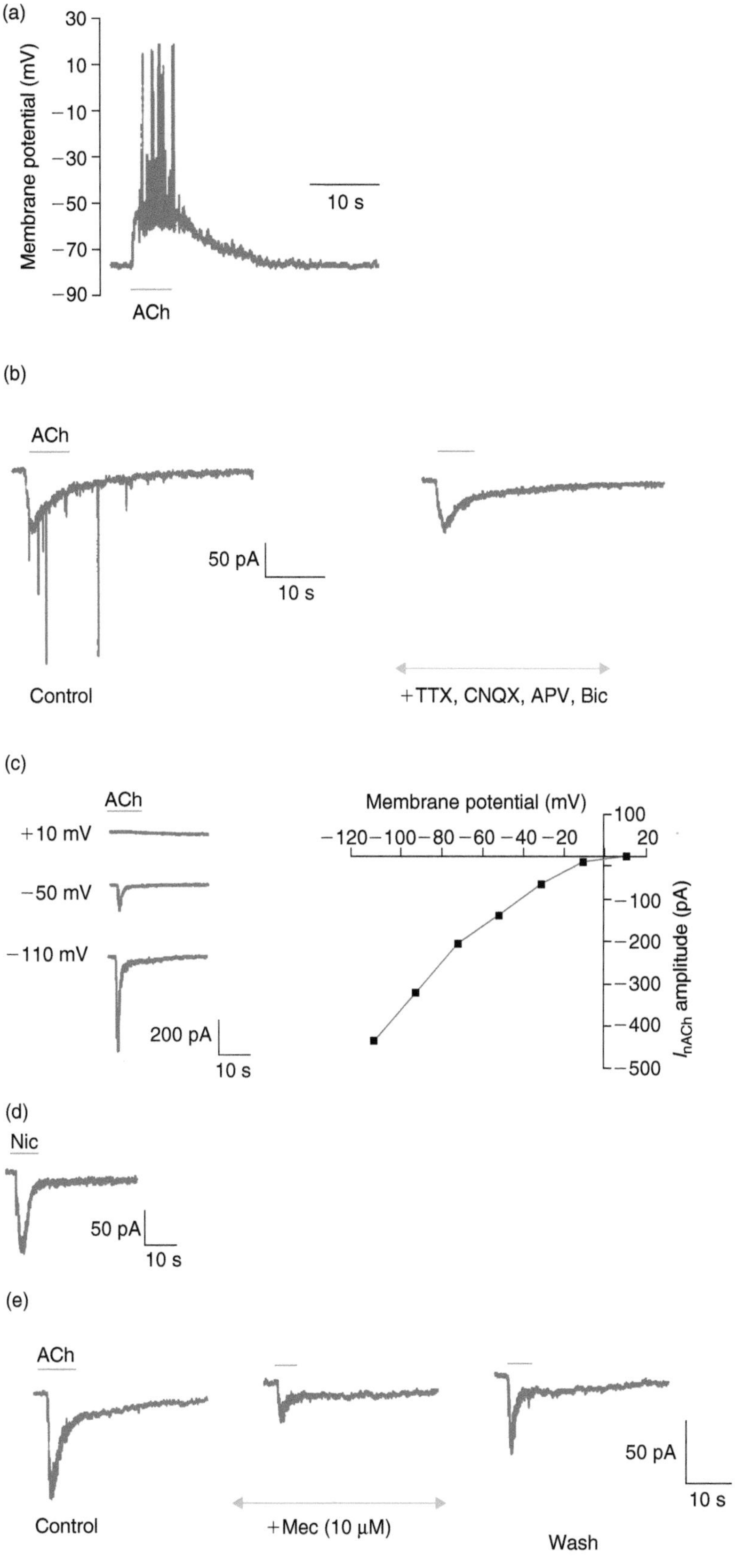

FIGURE A8.1 **Direct excitation of hippocampal interneurones by ACh-activated cation current and pharmacological characterization of the effect of ACh.** (**a**) Current-clamp recording (using potassium gluconate intracellular solution) from a CA1 interneuron, showing the depolarization and action potential firing in response to ACh (100 μM, 5 s, bars). (**b**) Voltage-clamp recording (CA1 interneuron, cell held at −70 mV) demonstrating that the effect of ACh (100 μM, 5 s, bar) persists in the presence of inhibitors of synaptic transmission (TTX, CNQX, APV, Bic). (**c**) Membrane potential was changed from −110 to 10 mV in order to obtain a steady-state current voltage relationship for the effect of ACh (100 μM, 5 s, bar). Example traces at membrane potentials −110, −50 and 10 mV are shown (left). (**d**) Effect of the nAChR agonist nicotine (100 μM, 5 s) on a CA1 interneuron voltage clamped to −70 mV. (**e**) Effect of nAChR antagonists on the responses of interneurons to ACh. Mecamylamine (Mec, 10 μM) reversibly inhibited the response to ACh (100 μM, 5 s, bars) in this voltage-clamped CA1 interneuron by 75%. (**f**) Pre-treatment of slices with α-bungarotoxin (α-BgTx, 100 μM, 10 min minimum) significantly reduced ACh responses of CA1 interneurons; control slices were matched from the same rats. The response to ACh averaged −59 ± 13 pA (n = 4) in control slices and −11 ± 5 pA (n = 4) in α-BgTX-treated slices, corresponding to approximately 80% inhibition, *$P < 0.05$. From Jones S, Yakel JL (1997) Functional nicotinic ACh receptors on interneurons of the rat hippocampus. *J. Physiol.* **504**, 603–610, with permission.

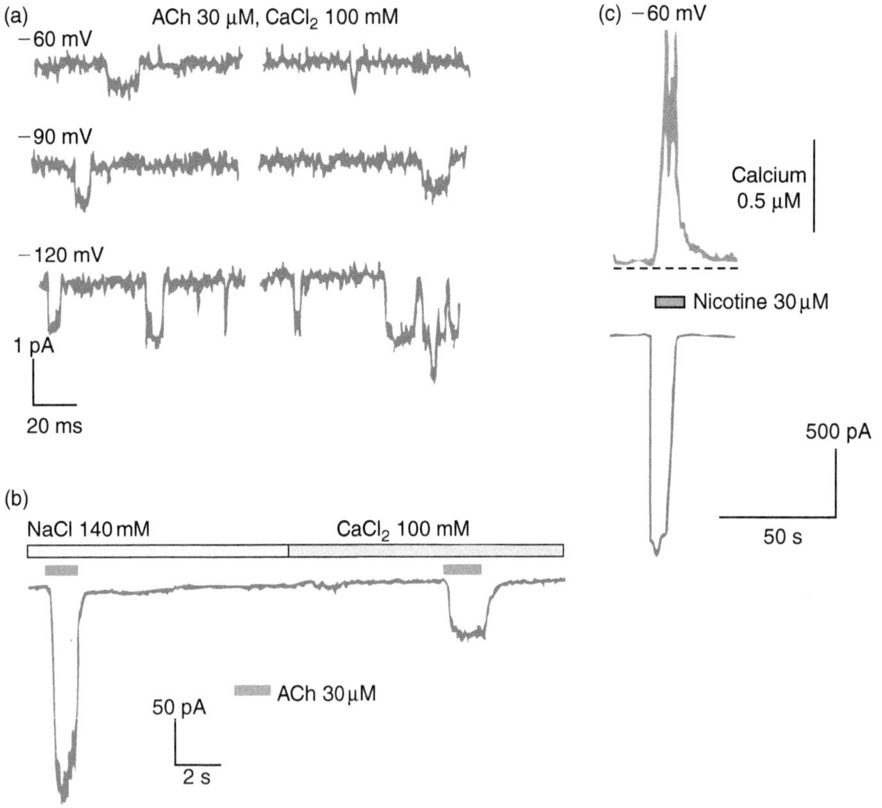

FIGURE A8.2 **Ca2+ permeates neuronal nAChR channels.** (**a**) Single-channel currents evoked by ACh in outside-out patches of habenula neurons at three holding potentials (−60, −90, −120 mV) in the presence of a pure $CaCl_2$ external medium. (**b**) Whole-cell currents evoked by ACh in control (140 mM NaCl, 1 mM $CaCl_2$) and pure $CaCl_2$ medium. (**c**) Increase in $[Ca^{2+}]_i$ (top trace) and whole-cell current (bottom trace) evoked by a 10 s application of nicotine (30 μM) at $V_H = -60$ mV. Adapted from Mulle C, Choquet D, Korn H, Changeux JP (1992) Calcium influx through nicotinic receptor in rat central neurons: its relevance to cellular regulation. *Neuron* **8**, 135–143, with permission.

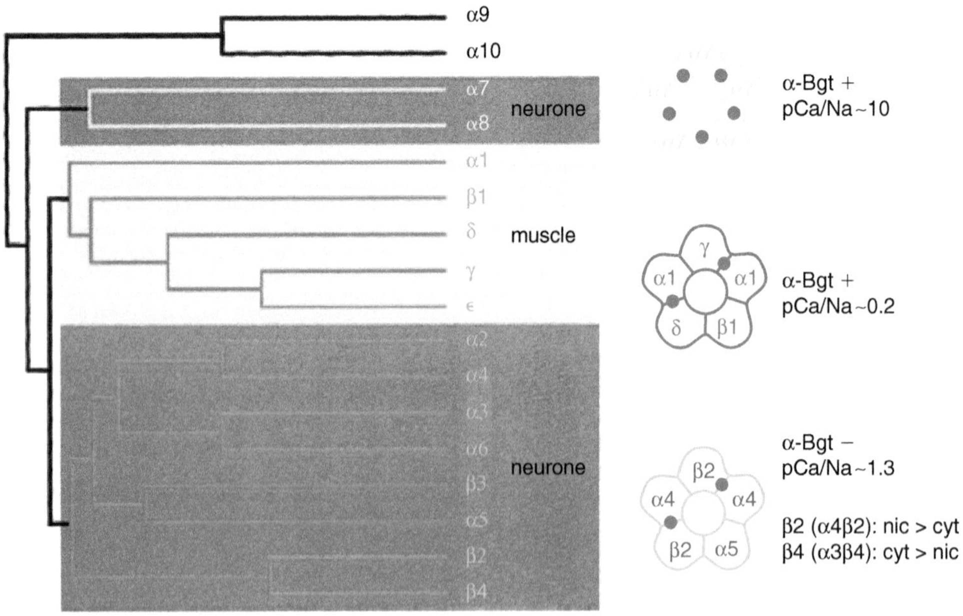

FIGURE A8.3 **Phylogeny of nAChRs and properties of receptors.** Red dots indicate the agonist binding sites. pCa/Na, relative calcium/sodium permeability. From Le Novere N, Corringer PJ, Changeux JP (2002) The diversity of subunit composition in nAChRs: evolutionary origins, physiologic and pharmacologic consequences. *J. Neurobiol.* **53**, 447–456, with permission.

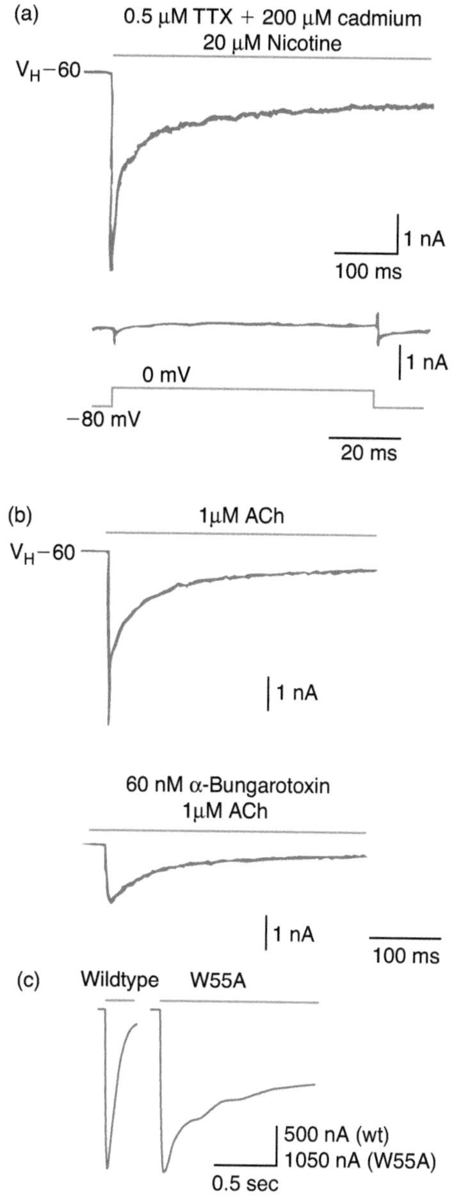

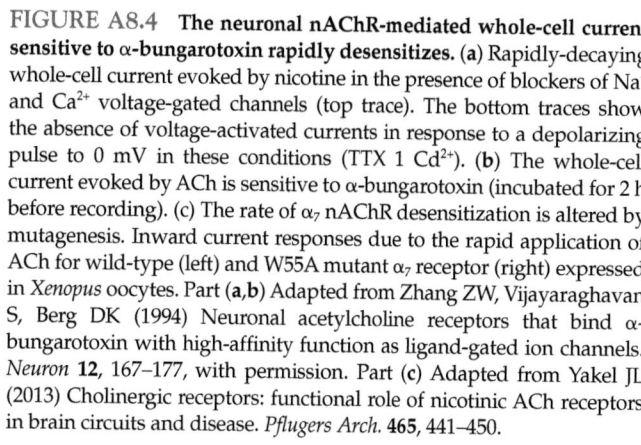

FIGURE A8.4 **The neuronal nAChR-mediated whole-cell current sensitive to α-bungarotoxin rapidly desensitizes.** (**a**) Rapidly-decaying whole-cell current evoked by nicotine in the presence of blockers of Na^+ and Ca^{2+} voltage-gated channels (top trace). The bottom traces show the absence of voltage-activated currents in response to a depolarizing pulse to 0 mV in these conditions (TTX 1 Cd^{2+}). (**b**) The whole-cell current evoked by ACh is sensitive to α-bungarotoxin (incubated for 2 h before recording). (c) The rate of α_7 nAChR desensitization is altered by mutagenesis. Inward current responses due to the rapid application of ACh for wild-type (left) and W55A mutant α_7 receptor (right) expressed in *Xenopus* oocytes. Part (**a,b**) Adapted from Zhang ZW, Vijayaraghavan S, Berg DK (1994) Neuronal acetylcholine receptors that bind α-bungarotoxin with high-affinity function as ligand-gated ion channels. *Neuron* **12**, 167–177, with permission. Part (**c**) Adapted from Yakel JL (2013) Cholinergic receptors: functional role of nicotinic ACh receptors in brain circuits and disease. *Pflugers Arch.* **465**, 441–450.

sites per receptor molecule located at the interface between an α- and β-subunit. These receptors usually comprise: (i) two α-subunits carrying the principal component of the ACh-binding site $\alpha_2\alpha_3\alpha_4$ or α_6; (ii) two non-α-subunits carrying the complementary component of the ACh-binding site β_2 or β_4; and (iii) a fifth subunit that does not participate in ACh binding $\alpha_5\beta_3$ but also β_2 or β_4 (**Figure A8.3**, right).

The identification of the subtype composition of native neuronal nAChRs is currently based on a combination of technical approaches, such as *in situ* hybridization, single-cell PCR, pharmacology and electrophysiology.

Central nicotinic receptors are mostly localized in the membrane of axon terminals (presynaptic nAChRs) where, via the increase of the intracellular concentration of Ca^{2+} ions, they can enhance the release of the transmitter present in the vesicles of the axon terminal. Only rare cases of central synapses with a postsynaptic localization of nAChRs have been reported. In that case, nAChRs participate in fast excitatory synaptic transmission (**Figure A8.1**). In contrast to the NMDA subtype of glutamate receptors that provide a current with an accompanying Ca^{2+} influx only when the membrane is depolarized enough to relieve the Mg^{2+} block (see Chapter 10), the nAChRs do the same but at negative membrane potentials, in a situation of strong driving forces for Na^+ and Ca^{2+} ions. However, the rapid desensitization of α-bungarotoxin-sensitive nAChRs is a way to control ACh-induced current and Ca^{2+} entry.

CHAPTER

9

The ionotropic $GABA_A$ receptor

Constance Hammond

OUTLINE

K. Krnjevic and S. Schwartz observed in 1967 that γ-aminobutyric acid (GABA), applied by microiontophoresis on intracellularly recorded cortical neurons, evokes a hyperpolarization of the neuronal membrane which has properties similar to synaptically evoked hyperpolarization. This led to the hypothesis that GABA mediates inhibitory synaptic transmission in the *adult* vertebrate central nervous system.

GABA released by presynaptic terminals activates two main types of receptors: (i) ionotropic receptors called $GABA_A$ and (ii) metabotropic receptors, i.e. receptors coupled to GTP-binding proteins called $GABA_B$ receptors.

The aim of the present chapter is to study how a $GABA_A$ receptor converts the binding of two GABA molecules to a rapid and transient change of the membrane potential in the *adult* vertebrate central nervous system. $GABA_A$ receptors of the developing brain will be explained in Chapter 20.

9.1 OBSERVATIONS AND QUESTIONS

In the rat hippocampus, in response to a single spike evoked in a presynaptic GABAergic interneuron, a transient hyperpolarization of the membrane mediated by $GABA_A$ receptors is recorded in the postsynaptic pyramidal cell (in the presence of blockers of $GABA_B$ receptors). This hyperpolarization of synaptic origin is called the 'single-spike inhibitory postsynaptic potential' (single-spike IPSP). This $GABA_A$-mediated IPSP has the following characteristics (**Figure 9.1**). It is:

- totally blocked in the presence of bicuculline thus showing it is mediated by $GABA_A$ receptors;
- potentiated by diazepam (a benzodiazepine, an anxiolytic and anticonvulsant drug);
- potentiated by pentobarbitone sodium (a barbiturate, a sedative and anticonvulsant drug).

These observations raise several questions:

- How does a $GABA_A$ receptor mediate the binding of GABA into a transient hyperpolarization of the membrane; i.e. how is the agonist binding signal transmitted down to the channel (coupling), and transduced in opening of the ion channel (gating), and what are the permeant ions, does GABA induce the entry of negatively charged ions or does it induce the exit of positively charged ions?

Cellular and Molecular Neurophysiology. DOI: 10.1016/B978-0-12-397032-9.00009-1

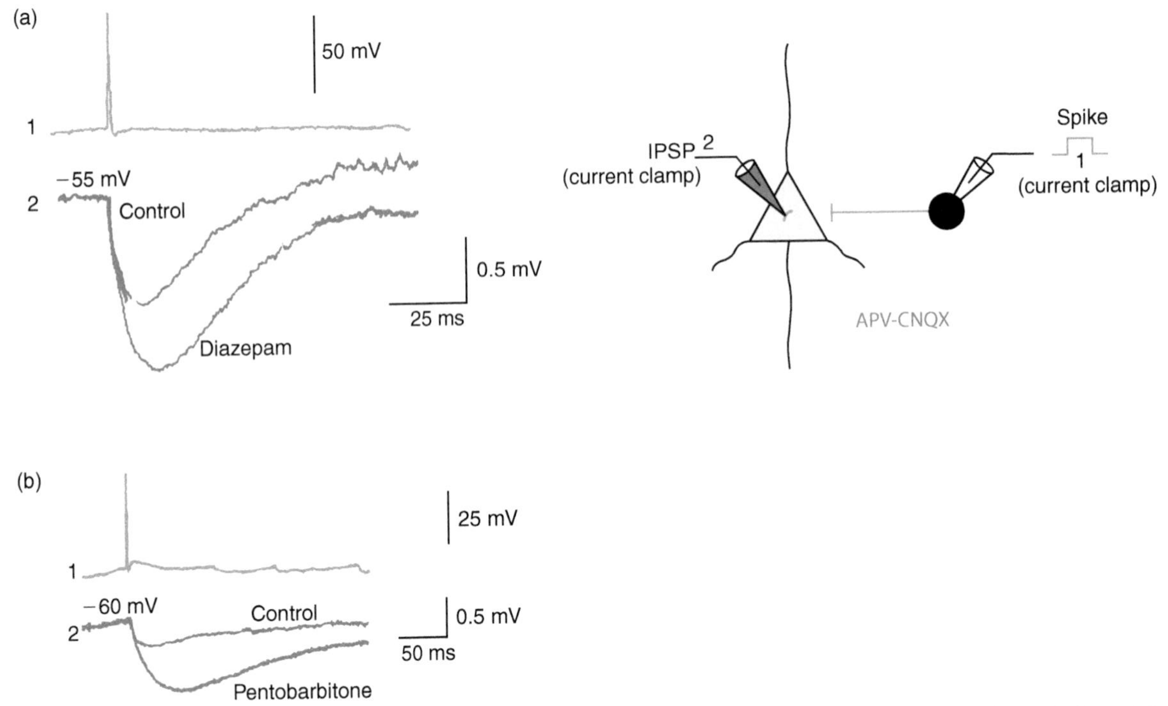

FIGURE 9.1 **GABA$_A$ receptor-mediated IPSPs recorded in pyramidal cells of the hippocampus and two examples of their modulation.** The activity of a pair of connected neurons, a presynaptic GABAergic interneuron and postsynaptic pyramidal cell, is recorded with two intracellular electrodes. **(a)** A single spike is triggered in the presynaptic GABAergic interneuron in response to a current pulse (1). It evokes an IPSP (averaged single-spike IPSP; 2 control) in the postsynaptic pyramidal neuron. This IPSP is increased in amplitude and duration in the presence of a benzodiazepine (2; diazepam 1 μM). **(b)** Similar experiment in a different cell pair with pentobarbitone sodium (250 μM), a barbiturate. Adapted from Pawelzik H, Bannister AP, Deuchars J et al. (1999) Modulation of bistratified cell IPSPs and basket cell IPSP by pentobarbitone sodium, diazepam and Zn^{2+}: dual recordings in slices of adult hippocampus. *Eur. J. Neurosci.* **11**, 3552–3564, with permission.

- Do benzodiazepines and barbiturates act directly on GABA$_A$ receptors? Are there selective and distinct binding sites on the receptor for each of these drugs? How do they potentiate the hyperpolarizing effect of GABA? Are there other modulators of GABA$_A$ receptors?

9.2 GABA$_A$ RECEPTORS ARE HETERO-OLIGOMERIC PROTEINS WITH A STRUCTURAL HETEROGENEITY

9.2.1 The diversity of GABA$_A$ receptor subunits

The first subunits to be identified were α_1 and β_1. A GABA$_A$ receptor purified on affinity columns from calf, pig, rat or chick brain dissociates in the presence of a detergent into two major subunits which migrate on polyacrylamide gels and have apparent molecular masses of 53 kDa (α_1-subunit) and 56 kDa (β_1-subunit). Electrophoretic studies based on receptors purified from different regions of the central nervous system show the presence of multiple bands corresponding to apparent molecular masses of 48–53 kDa and of 55–57 kDa. This strongly suggests the occurrence of several isoforms of α- and β-subunits.

Demonstration of the structural heterogeneity of the GABA$_A$ receptor came from cloning of cDNAs for different α- and β-subunits. The diversity of GABA$_A$ receptor subunits (α, β, γ, δ, ε, π, ρ) and the existence of different isoforms for the different subunits were then revealed. All subunits are similar in size, contain about 450–550 amino acids and are strongly conserved among species (**Figure 9.2a**). A high percentage of sequential identity (70–80%) is found between subunit isoforms (between α and between β isoforms, for example). Sequential identity is also found, but to a lesser extent (30–40%), between subunit families. To date, 19 isoforms of mammalian GABA$_A$ receptor subunits have been cloned: α_{1-6}, β_{1-3}, γ_{1-3}, δ, ε, π, θ, ρ_{1-3}. This multiplicity of subunits provides a daunting number of potential subunit combinations. The common elements of the subunit structure include (**Figure 9.2b**):

- a large N-terminal hydrophilic domain exposed to the synaptic cleft (extracellular);
- three highly conserved hydrophobic transmembrane domains (M1, M2, M3), a large, poorly conserved, hydrophilic domain of variable size and amino acid sequence, with putative phosphorylation sites, separating the M3 and M4 segments and located in the cytoplasm and a fourth transmembrane domain

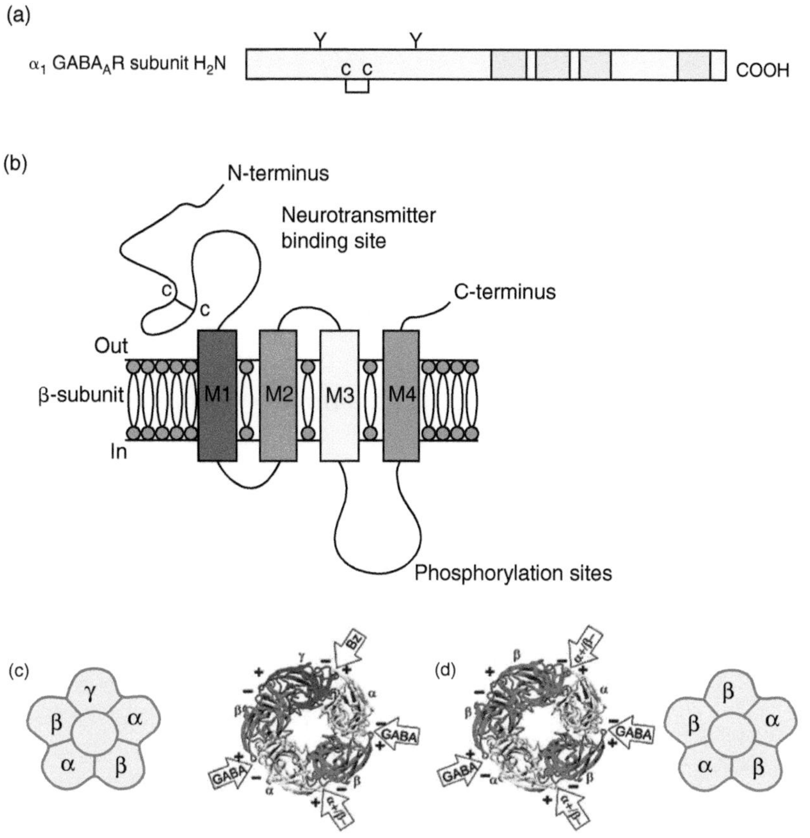

FIGURE 9.2 **The $GABA_A$ receptor subunits. (a)** Schematic illustration of the primary sequence of the α_1-subunit from the $GABA_AR$; C-C, Cys-Cys bridge; Y, oligosaccharide groups. **(b)** Model of transmembrane organization of $GABA_A$ receptor subunits. The large hydrophilic N-terminal domain is exposed to the synaptic cleft and carries the neurotransmitter site and glycosylation sites. The four M segments span the membrane. The hydrophilic domain separating M3 and M4 faces the cytoplasm and contains phosphorylation sites. The C-terminus is extracellular. **(c)** Schematic representation of the oligomeric organization of a $GABA_A$ receptor formed by two α-subunits, two β-subunits and one γ-subunit and **(d)** formed by two α- and three β-subunits. The structures are also shown in ribbon representations viewed from the extracellular side. The plus (+) and the minus (−) side of each subunit is indicated. The two GABA sites are located at the $\beta^+\alpha^-$ interfaces, and the benzodiazepine (BZ) binding site is located at the $\alpha^+\gamma^-$ interface. Part **(a)** adapted from Arias HR (2000) Localization of agonist and competitive antagonist binding sites on nicotinic acetylcholine receptors. *Neurochem. Internatl* **36**, 595–645, with permission. Part **(b)** from Lovinger DM (1997) Alcohols and neurotransmitter gated ion channels: past, present and future. *Naunyn-Schmiederberg's Arch. Pharmacol.* **356**, 267–282, with permission. Parts **(c)** and **(d)** adapted from Sieghart W, Ramerstorfer J, Sarto-Jackson I, Varagic Z, Ernst M (2012) A novel $GABA_A$ receptor pharmacology: drugs interacting with the $\alpha^+\beta^-$ interface. *Br. J. Pharmacol.* **166**, 476–485, with permission.

(M4) with a relatively short and variable extracellular C-terminal.

The segment M2 of each of the subunits composing the $GABA_A$ receptor is thought to line the channel (as for the nAChR) and to contribute to ion selectivity and transport. Apparently, a small number of amino acids within the M2 sequence is responsible for anionic versus cationic permeability (see Section 8.3.2 and **Figure 8.15**). The M3–M4 linker is the intracellular domain that binds the cytoskeleton.

$GABA_A$ receptors (with nAChRs) are part of the cys-loop pentameric ligand-gated ion channel superfamily: their α-subunits have a characteristic cysteine–cysteine pair in the N-terminal extracellular domain (**Figure 9.2a,b**) and the disulfide bond between the two cysteines forms a characteristic loop.

9.2.2 Subunit composition of native $GABA_A$ receptors and their binding characteristics

To form a $GABA_A$ receptor, with the large number of known subunits taken five at a time, thousands of combinations are possible. Two approaches are currently used to elucidate which subunit combinations exist:

- a comparative study of the functional properties of receptors expressed in oocytes or in transfected mammalian cells from known combinations of cloned subunits (with the restriction that *Xenopus* oocyte does not automatically assemble a channel composed of all injected subunits);
- a comparative study of the distribution of the various subunit mRNAs in the brain using the *in situ* hybridization technique (see Appendix 6.2).

Surprisingly, the transient expression in transfected cells of identical α- or β-subunits gives functional homomeric $GABA_A$ receptors; i.e. receptors which induce a current in the presence of GABA. This current is blocked by $GABA_A$ antagonists and potentiated by barbiturates but is unaffected by benzodiazepines. These properties can be attributed to the conserved structural features of all the subunits. However, these channels resulting from expression of single subunits are assembled inefficiently (are rare and slightly detectable) and it is unlikely that native receptors are formed from identical subunits. Expression of a γ-subunit together with an α- and a β-subunit in transfected cells gives rise to the expression of a $GABA_A$ receptor with all the features of the homomeric receptor with in addition the sensitivity to benzodiazepines. This does not necessarily imply that the receptor site for benzodiazepines is situated on the γ-subunit, but that expression of the latter is required for the action of benzodiazepines. The major receptor subtype of the $GABA_A$ receptor in adults consists of α_1-, β_2- and γ_2-subunits and the most likely stoichiometry is two α-subunits, two β-subunits and one γ-subunit (β α β α γ) (**Figure 9.2c**). In summary, functional receptors can consist of α- and β-subunits alone; however, native receptors include an additional 'modulatory' subunit as γ, δ, ε, π, or θ.

In contrast, ρ-subunits are capable of forming functional homo-oligomeric receptors. Such receptor assemblies derived from various isoforms of the ρ-subunit are present in the retina.

9.3 BINDING OF TWO GABA MOLECULES LEADS TO A CONFORMATIONAL CHANGE OF THE $GABA_A$ RECEPTOR INTO AN OPEN STATE; THE $GABA_A$ RECEPTOR DESENSITIZES

9.3.1 GABA binding site

The amino acids identified by site-directed mutagenesis to affect channel activation by GABA are in the β-subunit: glutamate (E) 155 and nearby residues tyrosine (Y) 157 and threonine (T) 160. For example, cysteine (C) substitution of β_2 Glu^{155} alters both channel-gating properties and impairs agonist binding as it results in spontaneously opened $GABA_A$ channels. A model of the $GABA_AR$ agonist-binding site predicts that β_2 Glu^{155} interacts with the positively charged moiety of GABA. In the α_1-subunit, the mutation of phenylalanine (F) 64 to leucine (L) also impairs activation of the GABA channel indicating a role for this α-subunit residue in GABA binding.

Therefore, identified domains of the β-subunit and of the neighboring α-subunit contribute to the GABA-binding site. It is thought that most αβγ receptors are pentameric with a stoichiometry 2α, 2β and 1γ, which is consistent with data indicating that GABA sites are located at the interface between α- and β-subunits and that there are two GABA sites per receptor (**Figure 9.2c,d**). $GABA_A$ receptors have a relatively low affinity for GABA, of the order of 10–20 μM.

9.3.2 Evidence for the binding of two GABA molecules

Analysis of dose–response curves suggests the binding of two GABA molecules prior to opening of the channel. The response studied, the peak amplitude of the total current I_{GABA} evoked by GABA in whole-cell patch clamp recording, is proportional to the square of the dose of GABA (but only at low doses of GABA):

$$I_{GABA} = f[GABA]^2.$$

At very low doses of GABA, when receptor desensitization is negligible, it seems that upon binding of two GABA molecules to the receptor, the conformational change of the receptor channel to an open state is favored. These observations can be accounted for by the following model:

$$2G + R \underset{k_{-1}}{\overset{k_1}{\rightleftharpoons}} G + GR \underset{k_{-2}}{\overset{k_2}{\rightleftharpoons}} G_2R \underset{\alpha}{\overset{\beta}{\rightleftharpoons}} G_2R^*$$

where G is GABA; R is the $GABA_A$ receptor in the closed configuration; GR or G_2R is the mono- or doubly-liganded $GABA_A$ receptor in the closed configuration; and G_2R^* is the doubly-liganded $GABA_A$ receptor in the open configuration.

9.3.3 The $GABA_A$ channel is selectively permeable to Cl^- ions

The reversal potential of the GABA current varies with the Cl^- equilibrium potential, E_{Cl}

The ionic selectivity of the channel is studied in outside-out patch clamp recordings from cultured spinal neurons. This patch clamp configuration allows control of the membrane potential as well as the composition of the intracellular fluid. When the intracellular and extracellular solutions contain the same Cl^- concentration (145 mM), the unitary current evoked by GABA reverses at E_{rev} = 0 mV (**Figures 9.3a–c** and **9.5a**). In this case, E_{Cl} is also equal to 0 mV:

$$E_{Cl} = -58 \log(145/145) = 0\,\text{mV}.$$

If part of the intracellular Cl^- is replaced with non-permeant anions such as isethionate ($HO\text{-}CH_2\text{-}CH_2\text{-}SO_3$), for a 10-fold change in intracellular Cl^- concentration a shift in the reversal potential of approximately 56 mV is observed (**Figure 9.3d**). This value approaches very closely that of 58 mV predicted by the Nernst equation for E_{Cl} at 20°C:

$$E_{Cl} = -58 \log(145/14.5) = -58\,\text{mV}.$$

Finally, changes in extracellular Na^+ or K^+ concentration have very little effect on the reversal potential of the $GABA_A$ response. Taken together, these results demonstrate that the $GABA_A$ channel is selectively permeable to Cl^-.

In physiological extracellular and intracellular solutions, the $GABA_A$ current recorded in isolated spinal neurons reverses at −60 mV

Using the technique of patch clamp recording one can record the unitary currents (i_{GABA}) across the $GABA_A$ channel (spinal neurons in culture, cell-attached configuration). The GABA present in the solution inside the recording pipette (5 μM) evokes outward single-channel currents at −30, 0 and +20 mV (**Figure 9.4a**). The magnitude of the single-channel current increases with depolarization suggesting that the reversal potential for the $GABA_A$ response is negative to −30 mV.

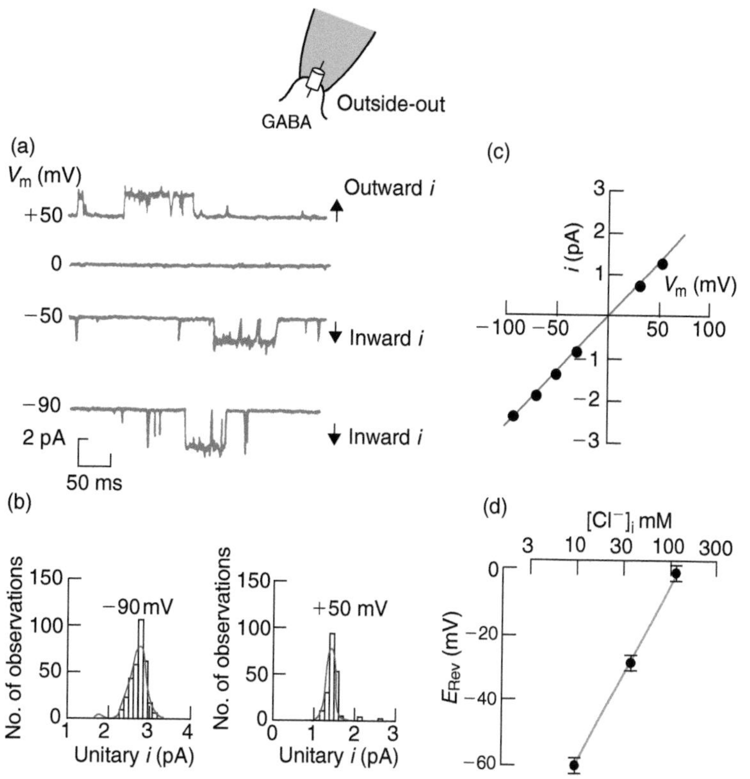

FIGURE 9.3 **Variations of the reversal potential of the $GABA_A$ response as a function of the Cl^- equilibrium potential.** The single-channel current i flowing across the $GABA_A$ channel is recorded in cultured mouse spinal neurons (outside-out patch clamp recording; equal concentrations of Cl^- on both sides of the patch: 145 mM). **(a)** In the presence of GABA (10 mM), the single-channel current i is outward at $V_m = +50$ mV (upward deflection), null at $V_m = 0$ mV and inward at $V_m = -50$ mV or −90 mV (downward deflections). **(b)** The distribution of single channel currents i in different patches of membrane held at $V_m = -90$ mV (left) and +50 mV (right) shows the existence of a single peak of current of −2.70 ± 0.17 pA and 1.48 ± 0.10 pA, respectively. These two values give a single channel conductance γ equal to 30 pS ($\gamma = i/V_m$ as $E_{rev} = 0$ mV). **(c)** i/V curve obtained by averaging the most frequently observed single-channel currents. It is a straight line according to the equation $i = \gamma(V_m - E_{rev})$. The relationship is linear between $V_m = -90$ mV and +50 mV and the slope is $\gamma = 30$ pS. **(d)** Reversal potential of the $GABA_A$ response (in mV) as a function of the intracellular Cl^- concentration $[Cl^-]_i$ (in mM). Each point represents the mean value of E_{rev} from four different cells. Note that, at the three $[Cl^-]_i$ tested, E_{rev} (experimental value) is very close to E_{Cl} (calculated by the Nernst equation):

$[Cl^-]_i$ (mM)	E_{Cl} (mV)	E_{rev} (mV)
14.5	-58	-56
45	-29	-28
145	0	0

Parts (a)–(c) adapted from Borman J, Hamill OP, Sakmann B (1987) Mechanism of anion permeation through channels gated by glycine and γ-aminobutyric acid in mouse cultured spinal neurones, *J. Physiol. (Lond.)* **385**, 246–286, with permission. Part (d) from Sakmann B, Borman J, Hamill OP (1983) Ion transport by single receptor channels, *Cold Spring Harbor Symposia in Quantitative Biology* **XLVIII**, 247–257, with permission.

The i/V curve, obtained by plotting the unitary current i_{GABA} against the membrane potential V, shows in this experiment a reversal potential of the GABA-induced current around −60 mV (**Figure 9.4**). Thus, at a potential close to the resting membrane potential (−60 mV), the current evoked by GABA is not detectable. At potentials more depolarized than rest (e.g. −30 mV) (**Figure 9.5b**), an outward current is recorded whose magnitude increases with depolarization of the postsynaptic membrane.

As E_{Cl} is close to the resting membrane potential (−60 mV) in physiological intracellular and extracellular solutions, the electrochemical gradient for the Cl^- ions ($V - E_{Cl}$) for $V = E_{Cl} = -60$ mV is close to 0 mV. The net flux of Cl^- ions at a potential close to rest is therefore null or very small: no current is recorded even though the $GABA_A$ channels are open. On the other hand, as the membrane potential depolarizes, the net flux of Cl^- ions becomes inward. An inward net flux of negative charges corresponds to an outward current. At potentials more depolarized than V_{rest}, an outward current is recorded (**Figure 9.4**). At potentials more hyperpolarized than −60 mV, i_{GABA} is inward (the net flux of Cl^- ions is outward) but of very small amplitude.

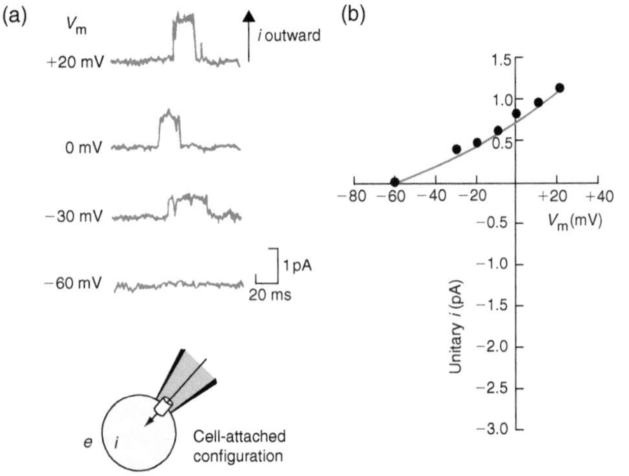

FIGURE 9.4 **Activity of a single GABA$_A$ receptor channel in physiological solutions.** The single-channel GABA$_A$ current recorded in cell-attached configuration in rat spinal neurons. **(a)** At $V_m = -30$, 0 and +20 mV respectively, the GABA present in the patch pipette at a concentration of 10 μM elicits an outward current (upward deflection). This current increases with depolarization of the patch. At $V_m = -60$ mV, no current is recorded. **(b)** i/V curve obtained by plotting the amplitude of the recorded unitary current i (pA) against the membrane potential V_m (mV). The intracellular medium is the physiological cytosol and the extracellular or intrapipette solution contains 144.6 mM Cl^-. The intracellular Cl^- concentration is estimated at 13 mM, which gives a value of −60 mV for the Cl^- reversal potential: $E_{Cl} = -58 \log(144.6/13) = -60$ mV. As all Na^+ ions are replaced by K^+ ions in the extracellular solution, the K^+ concentration is similar in both solutions, which gives a reversal potential for the K^+ current near 0 mV. The membrane potential values indicated in (**a**) and (**b**) are evaluated on the basis that the value 0 mV is the potential at which the K^+ currents across the K^+ channels are zero. From Sakmann B, Bormann J, Hamill OP (1983) Ion transport by single receptor channels, *Cold Spring Harbor Symposia in Quantitative Biology* **XLVIII**, 247–257, with permission.

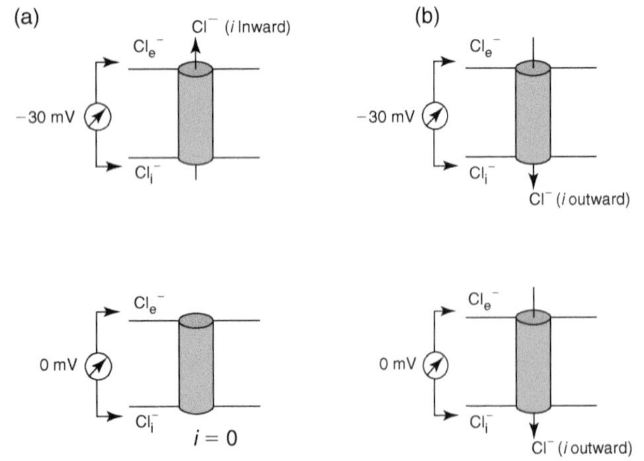

FIGURE 9.5 **Variations of the flux of Cl^- ions as a function of membrane potential and intracellular and extracellular Cl^- concentrations. (a)** Symmetrical media: $[Cl^-]_e = [Cl^-]_i = 145$ mM; $E_{Cl} = 0$ mV. **(b)** Physiological media: $[Cl^-]_e = 145$ mM; $[Cl^-]_i = 14.5$ mM; $E_{Cl} = -58$ mV.

9.3.4 The single-channel conductance of GABA$_A$ channels is constant in symmetrical Cl^- solutions, but varies as a function of potential in asymmetrical solutions

Experiments have been performed on mouse spinal neurons in culture, in conditions of equal intra- and extracellular Cl^- concentration (145 mM) to minimize rectification (variation of the conductance γ as a function of membrane potential). Histograms of single-channel currents evoked by GABA and recorded at $V = +50$ mV (outward i_{GABA}) and $V_m = -90$ mV (inward i_{GABA}) show at each potential a single peak of current equal to +1.48 and −2.7 pA respectively (**Figure 9.3b**). From these values of i_{GABA}, the mean single-channel conductance γ_{GABA} can be calculated, as $i_{GABA} = \gamma_{GABA}(V_m - E_{rev})$ and $E_{rev} = 0$ mV in these conditions. A value of 30 pS is obtained for both experimental conditions. This value of γ_{GABA} is also the slope of the i_{GABA}/V curve obtained by averaging the most frequent single-channel current i_{GABA} recorded at each membrane potential studied. This curve, based on the equation $i = \gamma(V - E_{rev})$ is linear between −90 mV and +50 mV and has a slope of 30 pS (**Figure 9.3c**).

However, in physiological conditions, when Cl^- concentration is approximately 10-fold lower in the intracellular than in the extracellular fluid, the unitary conductance γ_{GABA} varies with the membrane potential. The conductance in fact decreases progressively as the outward Cl^- current decreases (**Figure 9.4b**). This phenomenon is called *rectification*. This rectification (non-symmetrical inward and outward currents) results from the difference in Cl^- concentration on either side of the membrane.

9.3.5 Mean open time of the GABA$_A$ channel

With patch clamp recording of GABA$_A$ channels (in chromaffin cells of the adrenal medulla or cultured hippocampal neurons, in outside-out configuration), two types of openings are observed in the presence of low concentrations of GABA (**Figure 9.6a**): (i) brief openings, and (ii) longer duration openings interrupted by brief periods of closure: such a group of repeated openings and closures is called a burst of openings.

Brief openings (triangles)

Brief openings have a mean duration, τ_o, of 2.5 ms and contribute little to the total current.

Bursts of openings (open circles)

A burst is defined as a sequence of openings each one having a duration t_o, separated by brief closures of duration t_c. Brief durations are defined as less than 5 ms in the example illustrated in **Figure 9.6**. The duration of each burst t_b is $\Sigma t_o + \Sigma t_c$ and its mean duration τ_b is equal to 20–50 ms depending on the preparation used

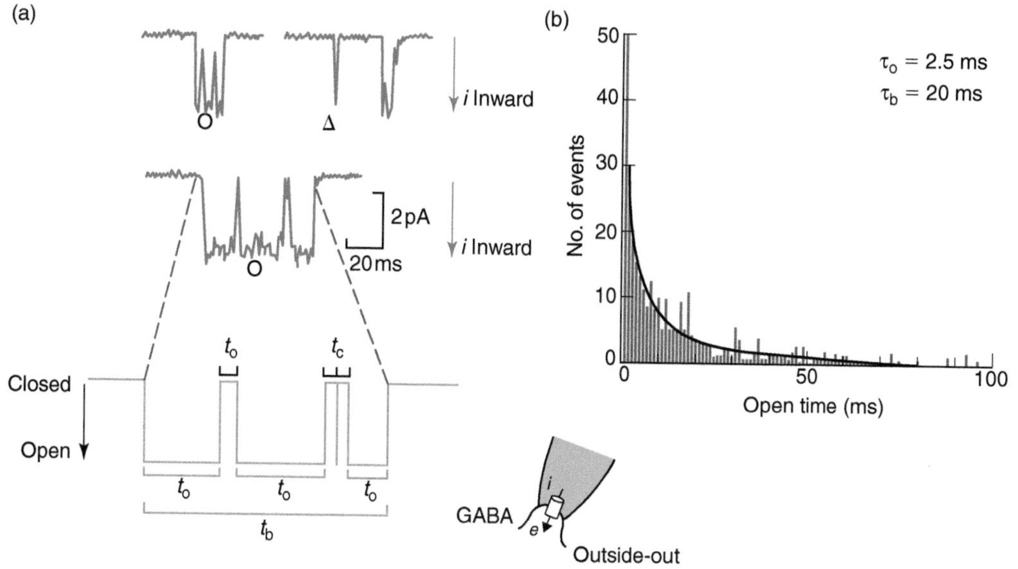

FIGURE 9.6 **Mean open time of $GABA_A$ channels.** Patch-clamp recording of the activity of the $GABA_A$ receptor channels from chromaffin cells of the adrenal medulla (outside-out configuration). The intracellular and extracellular Cl^- concentrations are similar and the membrane potential is maintained at −70 mV. **(a)** Inward unitary currents through a single $GABA_A$ channel evoked by GABA (10 μM). Brief openings (triangle) and bursts of openings (O, long duration openings interrupted by short closures defined in this experiment as less than 5 ms). **(b)** Histogram of open times measured in a homogeneous population of channels (mean value of $i = -2.9$ pA). The open times plotted on the graph represent the duration of short openings (t_o) and the duration of bursts of openings (t_b). The histogram is described by the sum of two exponentials with decay time constants of $\tau_o = 2.5$ ms and $\tau_b = 20$ ms. τ_o corresponds to the mean open time of short openings and τ_b to the mean open time of bursts of openings. From Borman J, Clapham DE (1985) γ-aminobutyric acid receptor channels in adrenal chromaffin cells: a patch clamp study, *Proc. Natl Acad. Sci. USA* **82**, 2168–2172, with permission.

(**Figure 9.6b**). The openings and the brief closures observed within each burst in the presence of GABA are thought to correspond to fluctuations of the receptor between the double-liganded open state and the double-liganded closed state (before the two molecules of GABA leave the receptor site). Thus, upon a single activation by two molecules of GABA, the double-liganded receptor would open and close several times:

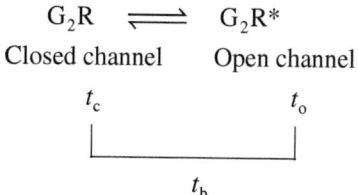

$$G_2R \rightleftharpoons G_2R^*$$

Closed channel | Open channel

t_c t_o

t_b

Silent periods

Silent periods separate single openings or bursts; they are periods during which the channel is closed and the unitary current is zero. In the presence of very low concentrations of GABA (when the receptor has a low probability to desensitize), they correspond to the G_2R, GR and R states of the $GABA_A$ receptor. Recordings in outside-out configuration show R and R* states of the $GABA_A$ receptor. Opening characteristics of the $GABA_A$ receptor are very similar to those of the nicotinic receptor (see Section 8.3.3). However, the short openings observed within bursts are approximately twice as abundant in the case of the $GABA_A$ receptor. This is explained by the fact that opening (β) and closing (α) rate constants have much closer values in the case of the $GABA_A$ receptor (see Appendix 9.1) than in the case of the nicotinic receptor (see Appendix 4.3).

9.3.6 The $GABA_A$ receptor desensitizes

Recordings in outside-out configuration show a rundown of the frequency of opening of the $GABA_A$ channels upon prolonged application of GABA (0.5 μM), whereas neither the intensity of the unitary current nor the mean open time of the channels τ_o appears to be affected (**Figure 9.7a**). Considering that $I = Np_oi$, if p_o (open probability of the channel) decreases as a result of a decrease in the frequency of opening events, the current I_{GABA} decreases even though i_{GABA} remains constant. Similarly, G_{GABA}, the total conductance, decreases (since $G = Np_o\gamma$). As the $GABA_A$ receptors gradually desensitize, p_o becomes progressively smaller with time, I_{GABA} as well as G_{GABA} gradually decrease to practically zero. This is confirmed in recordings of the total current I_{GABA} evoked by a prolonged application of GABA: the amplitude of I_{GABA} decreases with time as well as G_{GABA} (**Figure 9.7b**). This rundown of the $GABA_A$ response, which increases with increasing concentrations of GABA, is largely attributed to the desensitization of the $GABA_A$ receptor.

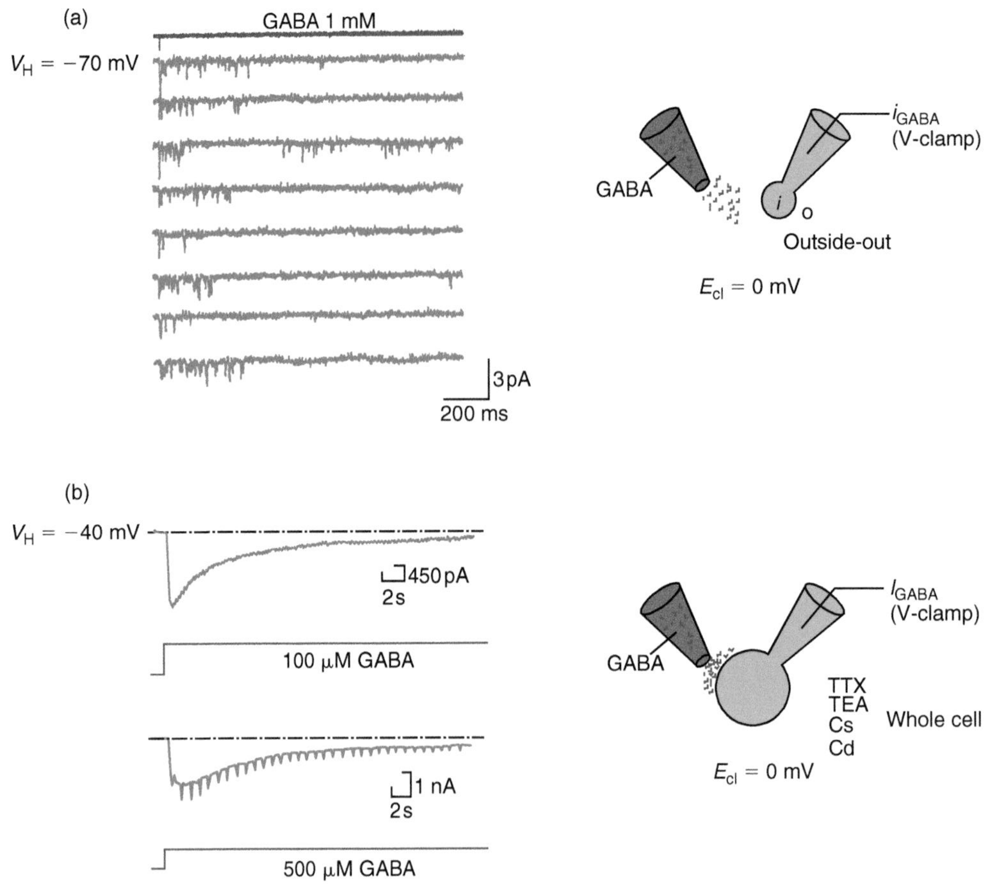

FIGURE 9.7 **Desensitization of the GABA$_A$ receptor. (a)** Outside-out patch excised from a cell transfected with $\alpha_1\beta_2\gamma_2$ cDNAs. A 2 ms pulse of GABA (1 mM) evokes single GABA$_A$ channel activity. **(b, top)** Patch clamp recording (whole-cell configuration, $V_H = -40$ mV) from a chick cerebral neuron. A prolonged application of GABA at high concentration (100 μM) evokes a total current I_{GABA} which decreases with time to almost zero. The total current I_{GABA} corresponds to the sum of the unitary currents i_{GABA}, passing through the open GABA$_A$ channels, while the other currents have been blocked with TTX and TEA as well as with Cs^+ and Cd^{2+} ions. **(b, bottom)** The same experiment in the presence of 500 μM of GABA and with hyperpolarizing voltage steps applied at a constant rate. The decrease in amplitude of the step current during the GABA$_A$ response shows that the decrease of I_{GABA} is associated with a decrease in G_m (as $i_{step} = Gm\ V_{step}$, V_{step} being constant), a decrease of i_{step} implies a decrease of G_m. There are symmetrical Cl^- concentrations in (**a**) and (**b**). Part (**a**) from Zhu WJ, Wang JF, Corsi L, Vicini S (1998) Lanthanum-mediated modification of GABA$_A$ receptor deactivation, desensitization and inhibitory synaptic currents in rat cerebellar neurons, *J. Physiol. (Lond.)* **511**, 647–661, with permission. Part (**b**) from Weiss DS, Barnes EM, Hablitz JJ (1988) Whole-cell and single-channel recordings of GABA-gated currents in cultured chick cerebral neurons, *J. Neurophysiol.* **59**, 495–513, with permission.

Therefore, upon long-lasting activation by GABA, the doubly-liganded receptor goes into at least one desensitization state:

$$\begin{array}{llll} \text{States of doubly liganded receptor:} & G_2R \underset{\alpha}{\overset{\beta}{\rightleftharpoons}} & G_2R^* \underset{k_{-3}}{\overset{k_3}{\rightleftharpoons}} & G_2D \\ \text{State of the channel} & \text{closed} & \text{open} & \text{closed} \end{array}$$

9.4 PHARMACOLOGY OF THE GABA$_A$ RECEPTOR

Benzodiazepines, barbiturates and neurosteroids enhance GABA$_A$ receptor current, whereas bicuculline, picrotoxin and β-carbolines reduce GABA$_A$ current, by binding to specific sites on the GABA$_A$ receptor channels.

9.4.1 Bicuculline and picrotoxin reversibly decrease total GABA$_A$ current; they are respectively competitive and non-competitive antagonists of the GABA$_A$ receptor

Excised outside-out patches are obtained from spinal cord neurons and held at −75 mV to prevent spontaneous openings of voltage-gated channels (**Figure 9.8a**). Recordings are performed in symmetrical chloride solutions. Prior to GABA application, occasional brief spontaneous currents are recorded (1). Following GABA application, bursting inward chloride currents are evoked (2). These GABA-induced bursting currents are reversibly reduced in frequency by the concomitant application of bicuculline (3) or picrotoxin (4) (**Figure 9.8b**). In whole-cell recordings, this effect is recorded as a decrease of the amplitude of the total current I_{GABA}: if p_o decreases, $Np_o\ i_{GABA} = I_{GABA}$ decreases.

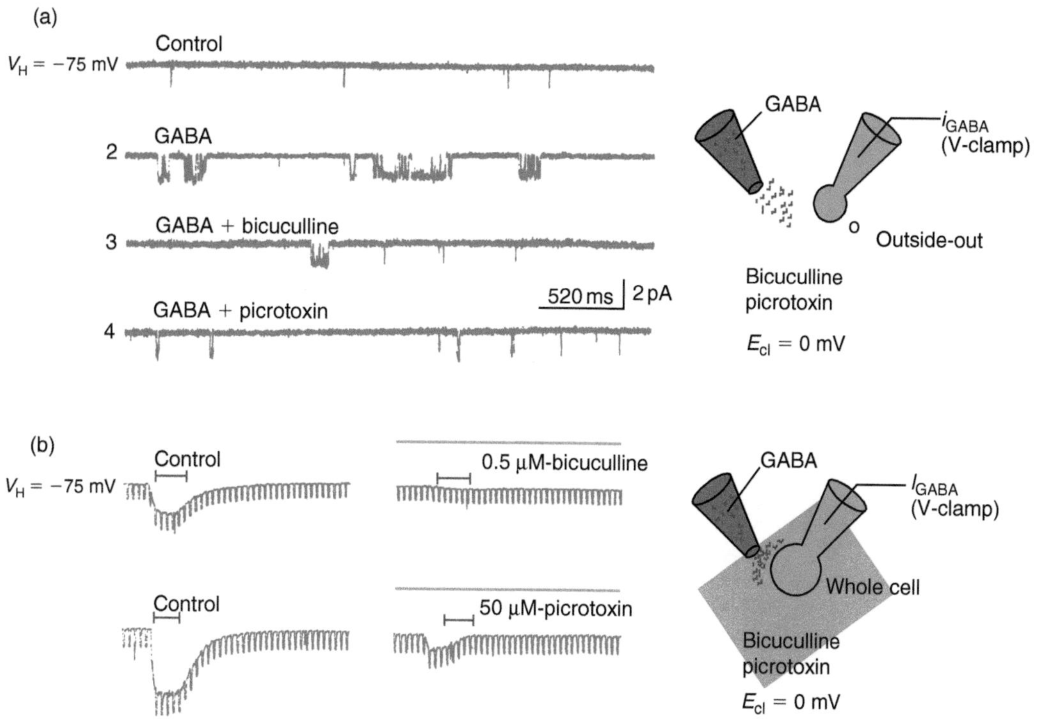

FIGURE 9.8 **Antagonists of the GABA$_A$ receptor. (a)** Activity of an outside-out patch from spinal cord neurons recorded in symmetrical Cl^- solutions. GABA (2 μM) is applied in the presence of bicuculline (0.2 μM) or picrotoxin (10 μM). **(b)** The same experiment in the whole-cell configuration. Steps are applied as in **Figure 9.7** to evaluate membrane conductance G_m during the response. Adapted from MacDonald RL, Rogers CJ, Twyman RE (1989) Kinetic properties of the GABA$_A$ receptor main conductance state of mouse spinal cord neurones in culture. *J. Physiol. (Lond.)* **410**, 479–499, with permission.

When the dose of GABA is increased and the dose of antagonist is kept constant, the inhibition by bicuculline is reduced whereas that by picrotoxin is unchanged (not shown). This shows that bicuculline is a competitive antagonist whereas picrotoxin is a non-competitive antagonist. Bicuculline binds to the same receptor sites as GABA. It is selective for the GABA$_A$ receptor and therefore serves as a good tool to identify GABA$_A$-mediated responses.

Picrotoxin in contrast binds to the ionic channel (it is a channel blocker). Its binding site involves the M2 segment, the region thought to line the chloride ion channel. The exact location of picrotoxin binding to ionophore is still unknown but its sensitivity to mutations in residues 2, 3 and 6 of M2 suggests that the site contains residues 2–6. Mutation of the highly conserved threonine residue at the 6 position in M2 of either α_1-, β_2- or γ_2-subunits of $\alpha_1\beta_2\gamma_3$ GABA receptors abolishes antagonism by picrotoxin at concentrations up to 100 μM. Importantly, while amino acid composition of residue 2 is variable in different ionotropic receptors, the composition of residue 6 is highly conservative, implying that it is crucial for picrotoxin binding to ionophore, and most likely representing the epicenter of its binding pocket. Both bicuculline and picrotoxin are potent convulsants when administered intravenously or intraventricularly.

9.4.2 Benzodiazepines, barbiturates and neurosteroids reversibly potentiate total GABA$_A$ current; they are allosteric agonists at the GABA$_A$ receptor

Benzodiazepines and barbiturates are two classes of clinically active agents (**Figure 9.9a,b**). Barbiturates are hypnotic and anti-epileptic agents and the benzodiazepines are anxiolytic agents, muscle relaxants and anticonvulsants. Various progesterone metabolites that are synthesized in the brain and thus called endogenous neurosteroids act directly on the GABA$_A$ receptor. Examples are allopregnanolone (**Figure 9.10c**) and tetrahydrodeoxycorticosterone.

Benzodiazepines, barbiturates and neurosteroids bind to the GABA$_A$ receptor at specific receptor sites

First, it was shown that co-expression of α- or β-subunits with a γ-subunit is required for the positive modulation of GABA-evoked Cl^- currents by benzodiazepines and for photoaffinity labeling of the benzodiazepine receptor site (see Section 9.2.2). It is now well established that benzodiazepines bind to the $\alpha^+\gamma^-$ extracellular interface **(Figure 9.2c)**.

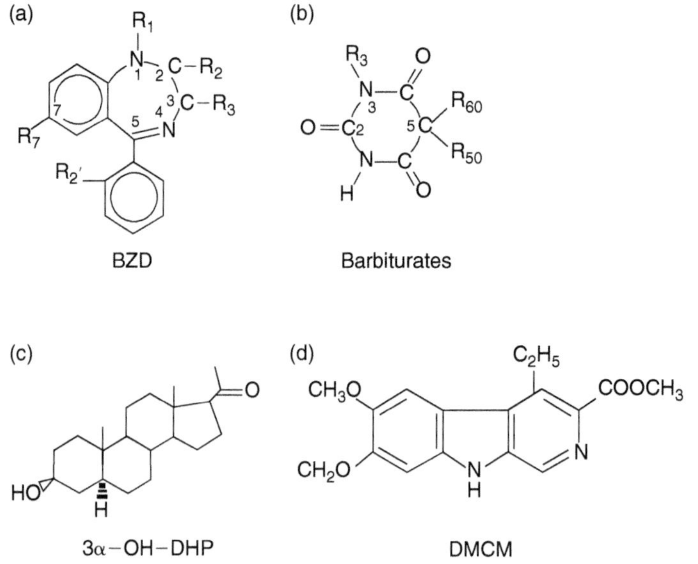

FIGURE 9.9 **Allosteric modulators of the GABA$_A$ receptor. (a)** Structure of benzodiazepines (BZD). For diazepam, the radicals are: R$_1$CH$_3$, R$_2$O, R$_3$H, R$_7$Cl, R$_2$9H. **(b)** Structure of barbituric acid derivatives. For pentobarbital, the radicals are: R$_{5a}$ethyl, R$_{5b}$ H and R$_3$phenyl. **(c)** Structure of a neurosteroid, allopregnenolone (3α-OH-DHP). **(d)** Structure of a β-carboline, methyl 6, 7-dimethoxyl-4-ethyl β-carboline 3 carboxylate (DMCM).

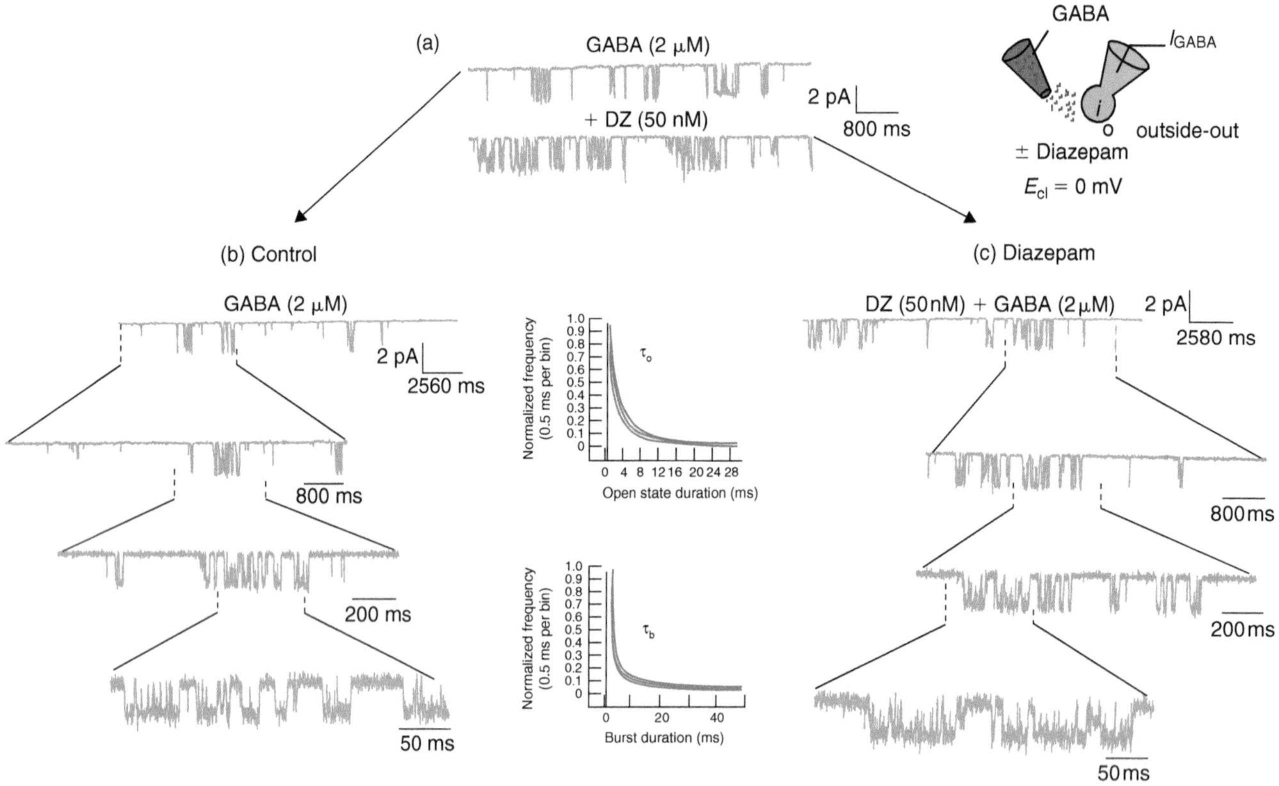

FIGURE 9.10 **GABA$_A$ single-channel current in the presence or absence of diazepam. (a)** Bursting inward currents in outside-out patches from spinal cord neurons evolved by GABA alone or GABA with diazepam (DZ). **(b,c)** The same experiment at increasing time resolution to demonstrate typical features of the unitary current. Open duration/frequency and bursts duration/frequency histograms for GABA are not significantly altered by addition of diazepam from 20 to 1000 nM (middle histograms). From Rogers CJ, Twyman RE, MacDonald RL (1994) Benzodiazepine and β-carboline regulation of single GABA$_A$ receptor channels of mouse spinal neurones in culture. *J. Physiol. (Lond.)* **475**, 69–82, with permission.

Benzodiazepines, barbiturates and neurosteroids potentiate the GABA$_A$ response

To test whether *benzodiazepines* have a direct effect on the GABA$_A$ receptor channel and to identify their effect, the unitary current i_{GABA} is recorded in voltage clamp ($V_H = -75$ mV) in outside-out patches (cultured spinal cord neurons). In the presence of GABA, the benzodiazepine diazepam (DZ, 50 nM) increases the opening frequency of the channel but does not change i_{GABA} amplitude nor the time spent by the channel in the open configuration at each opening (**Figure 9.11**).

Consistent with this finding, diazepam decreases the mean closed time τ_c; i.e. the time spent by the channel in the closed configuration (**Figure 9.12c**). With decreasing or increasing doses, the effect of diazepam was less pronounced (U-shaped concentration dependency). Diazepam (50 nM) also increases the burst frequency without changing the mean burst duration (τ_b) nor the mean number of openings per burst. All the currents evoked by GABA alone or GABA with diazepam are blocked by bicuculline, thus showing that they are mediated by the GABA$_A$ receptor. Since the i_{GABA}/V relationship shows that the unitary conductance is unchanged, it is hypothesized that diazepam alters the gating properties of the GABA$_A$ receptor channel: it increases the probability of the channel being in the open state, p_o.

The equation $Np_o i_{GABA} = I_{GABA}$ tells us that when p_o increases, I_{GABA} increases. Therefore, benzodiazepines should increase the amplitude of the total current I_{GABA} recorded in the whole-cell configuration (**Figure 9.13**). The I/V curves show that the total currents I_{GABA} evoked in the presence or absence of benzodiazepines reverse at the same potential (not shown). This indicates that the potentiation of I_{GABA} by these drugs is not the result of a change in the ion selectivity of the channel. Benzodiazepine agonists bind to a site distinct from that of GABA. It was first hypothesized that benzodiazepines do so by allosterically decreasing the GABA concentration needed to elicit half-maximal channel activity (EC_{50}). Another explanation is that benzodiazepines increase the adoption of a transitional receptor state, the preactivated state. This state occurs after GABA binding, before proceeding to channel activation.

Barbiturates also increase the bicuculline-sensitive current I_{GABA} but via a different mechanism: they do not increase the frequency of GABA$_A$ channel openings, instead they increase the duration of single openings and bursts of native GABA$_A$ receptors (**Figure 9.11a**) or transfected $\alpha_1\ \beta_1\ \gamma_2$ receptors (**Figure 9.11b**, right). An increase in the time spent in the open configuration at each opening results in an increase of the probability of the channel being in the open state and therefore in an increase of I_{GABA} (see **Figure 9.13**).

Neurosteroids at physiological concentrations increase the total bicuculline-sensitive current I_{GABA}. To study the mechanism of action, the effect of neurosteroids on single GABA$_A$ channel activity is recorded. Transformed human embryonic kidney cells 293 are transfected with the GABA$_A$ subunit combination $\alpha_1\ \beta_1\ \gamma_2$. The activity of expressed single GABA$_A$ channels is recorded in the outside-out configuration with symmetrical Cl^- concentrations on both sides of the patch of membrane. Pregnenolone and allopregnenolone (3α-OH-DHP) in combination with GABA increases the GABA$_A$ channel activity: it increases the number of active channels in the patch and the channel open probability (**Figure 9.11b,c**). On native GABA$_A$ receptors of spinal cord neurons, pregnenolone in combination with GABA also increases the duration of single and burst openings. Either one or both these effects on frequency of openings or opening duration result in an increase of p_o and thus an increase of I_{GABA} (see **Figure 9.13**).

In conclusion, benzodiazepines, barbiturates and neurosteroids can be considered as allosteric agonists of the GABA$_A$ receptor: they modulate the efficacy of activation of the receptor by GABA. They act via distinct receptor sites on the GABA$_A$ receptor and via different mechanisms.

9.4.3 β-Carbolines reversibly decrease total GABA$_A$ current; they bind at the benzodiazepine site and are inverse agonists of the GABA$_A$ receptor

β-Carbolines such as methyl-6,7-dimethoxyl-4-ethyl-β-carboline-3-carboxylate (DMCM) are convulsant and anxiogenic drugs. They bind to the benzodiazepine receptor site but have reverse effects: they are called 'benzodiazepine inverse agonists' (Table 9.1).

The activity of outside-out patches of spinal cord neurons in culture is recorded in voltage clamp. When DMCM is applied with GABA, it decreases the number of GABA$_A$ receptor openings compared with what is observed with GABA alone (**Figure 9.12a**). DMCM (20–100 nM) reduces single openings as well as burst frequency. However, the number of openings per burst is unchanged. The time spent by the GABA$_A$ channel in the open or bursting states is also unchanged but the time spent in the closed state is increased, consistent with a decrease in opening frequency (**Figure 9.12b,c**). Like diazepam, DMCM does not alter GABA$_A$ receptor single-channel conductance nor single-channel open or burst properties. DMCM decreases p_o and thus decreases I_{GABA}. Since burst frequency, but not intraburst opening frequency, is altered, it is unlikely that receptor channel opening rates (α and β) are altered by diazepam or DMCM.

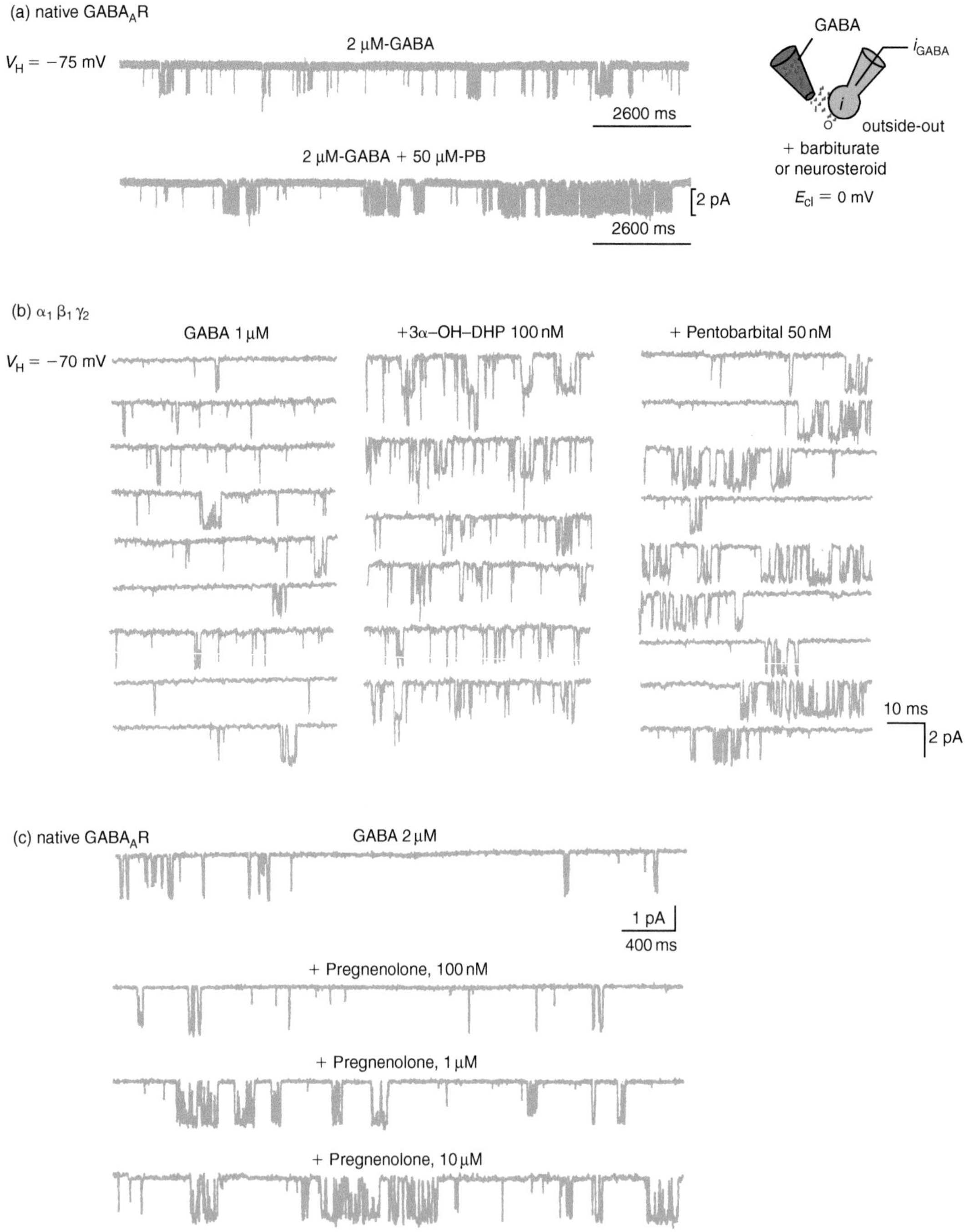

FIGURE 9.11 **GABA$_A$ single-channel current in the presence or absence of barbiturates or neurosteroids. (a)** Unitary GABA$_A$ currents in outside-out patches of spinal cord neurons evoked by GABA alone or GABA with phenobarbitone (PB). **(b)** Unitary GABA$_A$ currents recorded in outside-out patches excised from $\alpha_1\beta_1\gamma_2$ transfected cells evoked by GABA alone (left) or in combination with the neurosteroid 3α-OH-DHP (center) or pentobarbital (right). **(c)** Unitary GABA$_A$ currents in outside-out patches of spinal cord neurons evoked by GABA alone or GABA with pregnenolone at increasing concentration. Part (**a**) from MacDonald RL, Rogers CJ, Twyman RE (1989) Barbiturate regulation of kinetic properties of the GABA$_A$ receptor channel of mouse spinal neurones in culture, *J. Physiol. (Lond.)* **417**, 483–500, with permission. Part (**b**) from Twyman RE, MacDonald RL (1992) Neurosteroid regulation of GABA$_A$ receptor single channel kinetic properties of mouse spinal cord neurons in culture, *J. Physiol. (Lond.)* **456**, 215–245, with permission. Part (**c**) from Puia G *et al.* (1990) Neurosteroids act on recombinant human GABA$_A$ receptors, *Neuron* **4**, 759–765, with permission.

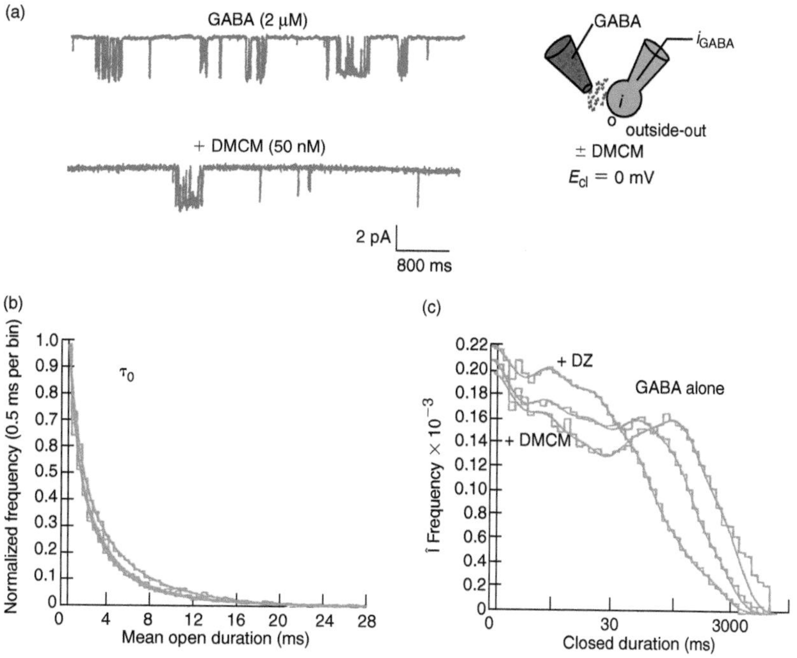

FIGURE 9.12 **$GABA_A$ single-channel current in the presence or absence of a β-carboline. (a)** Unitary $GABA_A$ currents in outside-out patches of spinal cord neurons evoked by GABA alone or GABA with the β-carboline DMCM. **(b)** Histograms for open durations are plotted for GABA (2 μM, uppermost curve, $\tau_o = 4.10 \pm 0.03$ ms) and for GABA and DMCM (20–100 nM, $\tau_o' = 4.40 \pm 0.07$ ms). **(c)** Closed duration-frequency histograms for GABA (2 μM), for GABA with diazepam (DZ, 100 nM) and for GABA with DMCM (100 nM). DZ shifts long closed durations to shorter durations while DMCM shifts long closed durations to longer ones. From Rogers CJ, Twyman RE, MacDonald RL (1994) Benzodiazepine and β-carboline regulation of single $GABA_A$ receptor channels of mouse spinal neurones in culture, *J. Physiol. (Lond.)* **475**, 69–82, with permission.

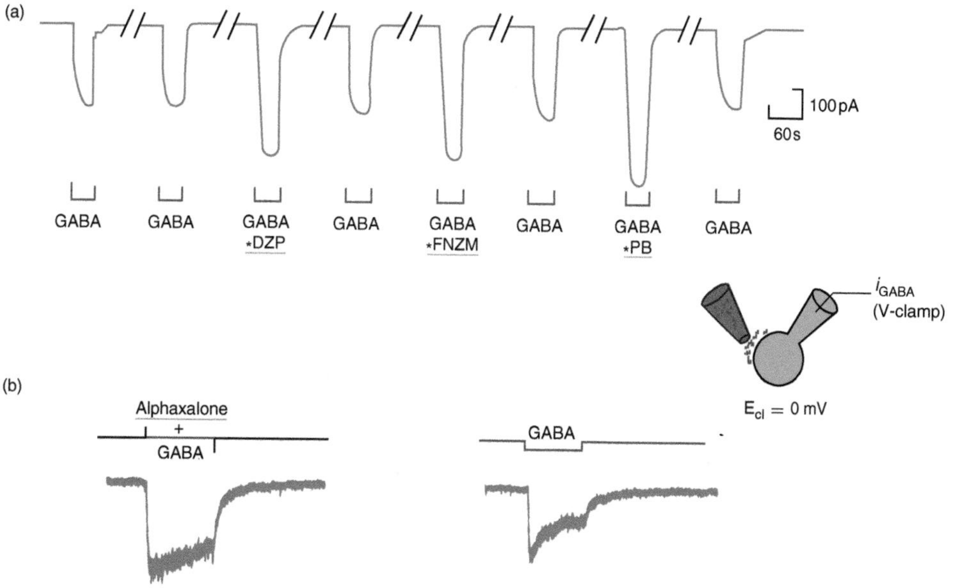

FIGURE 9.13 **Potentiation of the total current I_{GABA} by benzodiazepines, barbiturates and neurosteroids. (a)** $GABA_A$ receptors are expressed in transfected cells with the α, β and γ cDNA subunits. This model is interesting because these cells do not normally express GABA receptors (neither $GABA_A$ nor $GABA_B$); the GABA applied in the bath activates therefore only the number (*N*) of $GABA_A$ receptors expressed. On the other hand, it is a model where the presynaptic release of GABA is excluded because of the absence of synapses. The total current I_{GABA} is recorded using the patch clamp technique (whole-cell configuration). The total current recorded at $V_m = -60$ mV (GABA, 10 μM) is inward and is significantly potentiated by the simultaneous application of the two benzodiazepines, diazepam (DZP, 1 μM) and flunitrazepam (FNZM, 1 μM) and by pentobarbital (PB, 50 μM). **(b)** Whole-cell $GABA_A$ current evoked by 10 μM of GABA in spinal neurons in culture in the absence (right) or in the presence (left) of the neurosteroid alphaxalone (3α-hydroxy-5α-pregnane-11,20-dione, 1 μM). In **(a)** and **(b)**, E_{Cl} is estimated to be near 0 mV. Part **(a)** adapted from Pritchett DB, Southeimer H, Shivers BD, *et al.* (1989) Importance of a novel $GABA_A$ receptor subunit for benzodiazepine pharmacology, *Nature* **338**, 582–585, with permission. Part **(b)** from Barker JL, Harrison NL, Lange GD, Owen DG (1987) Potentiation of γ-aminobutyric-acid-activated chloride conductance by a steroid anaesthetic in cultured rat spinal neurones, *J. Physiol. (Lond.)* **386**, 485–501, with permission.

TABLE 9.1 Pharmacology of GABA$_A$ receptors

	GABA site	Benzodiazepine site
Selective agonists	Muscimol	Flunitrazepam
	Isoguvacine	
Inverse agonists	–	β-Carbolines
Competitive antagonists	Bicuculline	–
	Gabazine	

9.5 GABA$_A$-MEDIATED SYNAPTIC TRANSMISSION

9.5.1 The GABAergic synapse

To identify a synapse as GABAergic, several techniques are used such as immunocytochemistry for the GABA synthetic enzyme, glutamate decarboxylase (GAD). Now, to identify a synaptic response as mediated by GABA$_A$ receptors, the simplest test is the block by bicuculline. Also benzodiazepines significantly potentiate and prolong GABA$_A$-mediated IPSPs (see **Figure 9.1a**). Because of the selective action of benzodiazepines on the GABA$_A$ channel, these substances can be used experimentally together with bicuculline to identify a GABA$_A$ response.

When GABA is released in the synaptic cleft, it can (**Figure 9.14**):

- bind to postsynaptic GABA$_A$ receptors;
- bind to postsynaptic or presynaptic GABA$_B$ receptors;
- bind to glial and neuronal transporters and thus be taken up by presynaptic elements or glial cells;
- diffuse away from the synaptic cleft.

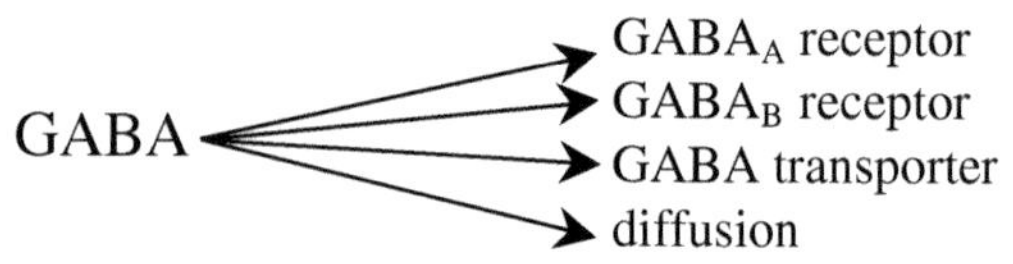

9.5.2 The synaptic GABA$_A$-mediated current is the sum of unitary currents appearing with variable delays and durations

When GABA binds to postsynaptic GABA$_A$ receptors at resting potential, it generally evokes an inhibitory postsynaptic potential (IPSP).

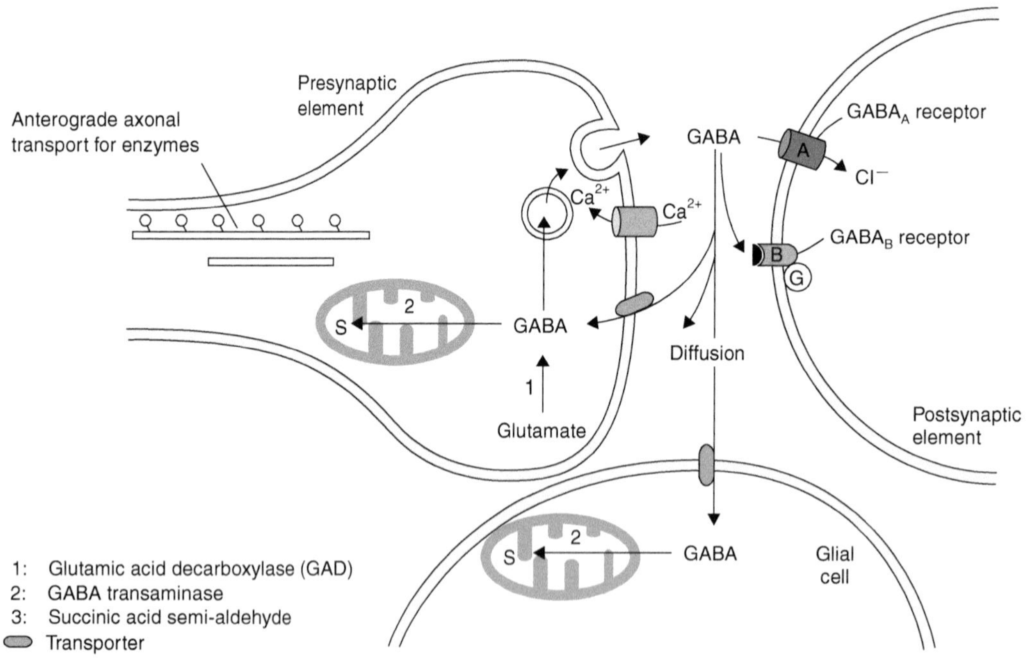

FIGURE 9.14 **The GABAergic synapse.** Functional scheme of a GABAergic synapse where the ionotropic (receptor-channel) GABA$_A$ receptors and the metabotropic (G-protein linked) GABA$_A$ receptors are co-localized. Presynaptic receptors are omitted. In order to study in isolation the GABA$_A$ response, GABA$_A$ receptors are selectively blocked. GABA is formed by the irreversible decarboxylation of glutamate catalyzed by the enzyme glutamic acid decarboxylase (GAD) and is metabolized by the mitochondrial enzyme GABA transaminase (GABA-T) into succinic acid semi-aldehyde (S). Enzymes are synthesized in the soma and carried to axon terminals via fast anterograde axonal transport. GABA is synthesized in the cytoplasm and transported actively into synaptic vesicles by a vesicular carrier. A percentage of the GABA released in the synaptic cleft is uptaken into presynaptic terminals and glial cells by GABA transporters which co-transport Na^+Cl^-. These transports are inhibited by nipecotic acid or β-alanine.

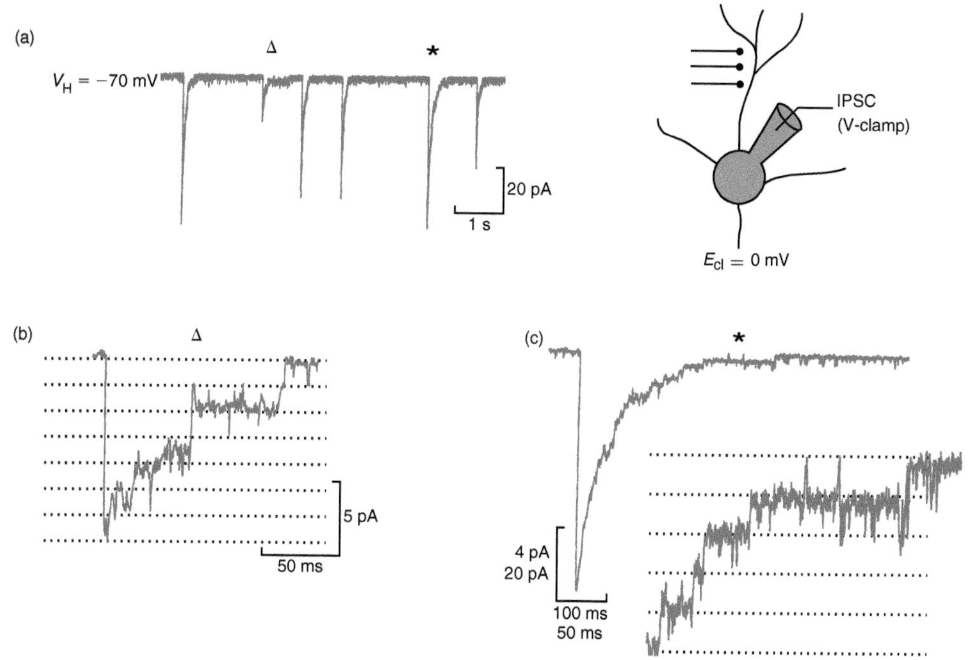

FIGURE 9.15 **Inhibitory postsynaptic currents (IPSCs) result from the summation of unitary GABA$_A$-mediated currents. (a)** Spontaneous IPSCs are recorded from a cerebellar granule cell (whole-cell configuration, voltage clamp mode). IPSCs are inward at −70 mV since E_{Cl} in this experiment is at 0 mV. **(b,c)** Two IPSCs of different amplitude from trace **(a)** are enlarged to show that unitary step currents can be resolved in their decay phase. Ionotropic glutamatergic transmission is pharmacologically blocked. From Brickley SG, Cull-Candy SG, Farrant M (1999) Single-channel properties of synaptic and extrasynaptic GABA$_A$ receptors suggest differential targeting of receptor subtypes. *J. Neurosci.* **19**, 2960–2973, with permission.

From single GABA$_A$ current to IPSC

In a slice of hippocampus, for example, GABAergic afferents are still spontaneously active and evoke spontaneous synaptic GABA$_A$-mediated postsynaptic currents (called sIPSCs) that can be recorded in whole-cell configuration in voltage clamp. Since they are blocked by bicuculline, they are mediated by postsynaptic GABA$_A$ receptors. In symmetrical Cl^- solutions, at a holding potential of −70 mV, GABA$_A$-mediated currents are inward (outward flow of Cl^- ions) (**Figure 9.15a**). When sIPSCs are observed at a higher magnification and a faster time base, unitary current steps can be identified. For example, during a very small sIPSC (**Figure 9.15b**), the peak is an integer of 7 unitary currents. During a larger IPSC, steps cannot be identified at the level of the peak but are clear during the decay phase (**Figure 9.15c**). sIPSCs result from the activation of N postsynaptic GABA$_A$ receptors.

Single-spike IPSC

When simultaneous intracellular recordings of a pair of connected neurons are performed, a presynaptic GABAergic interneuron and a postsynaptic pyramidal cell, each presynaptic action potential can be correlated with the postsynaptic current that it evokes, called *single-spike IPSC*. This postsynaptic current is outward at −30 mV in control intracellular Cl^- concentration. Its amplitude varies at each trial due to the fact that a presynaptic action potential activates, at each trial, a variable number of active zones from the total number of active zones established by the presynaptic axon with the postsynaptic membrane (see Appendix 7.2).

Single-spike IPSP

The same experiment performed in current-clamp mode allows recordings of the postsynaptic variation of potential resulting from the postsynaptic GABA$_A$-mediated current (**Figure 9.16**). In response to a single presynaptic action potential (trace 1), an inhibitory postsynaptic potential called *single-spike IPSP* (traces 2; V_m = −62 to −66 mV) is recorded. It is a transient hyperpolarization of the membrane. The amplitude of this single-spike IPSP varies in amplitude at each trial, from 0.6 to 4.2 mV. This is due to the fact that a presynaptic action potential activates, at each trial, a variable number of active zones from the total number of active zones established by the presynaptic axon. The conductance G_{IPSP} can be calculated knowing I_{IPSP}, V and E_{Cl}. In this experiment, G_{IPSP} = 6.7 nS ± 2.3 nS at the peak of the IPSP. Knowing that the unitary conductance of the GABA$_A$ channel is estimated in these neurons to be γ_{GABA} = 20–30 pS, it can be deduced that approximately 300 GABA$_A$ channels are open at the peak of the single-spike IPSP ($G_{IPSP} = Np_o \gamma_{GABA}$).

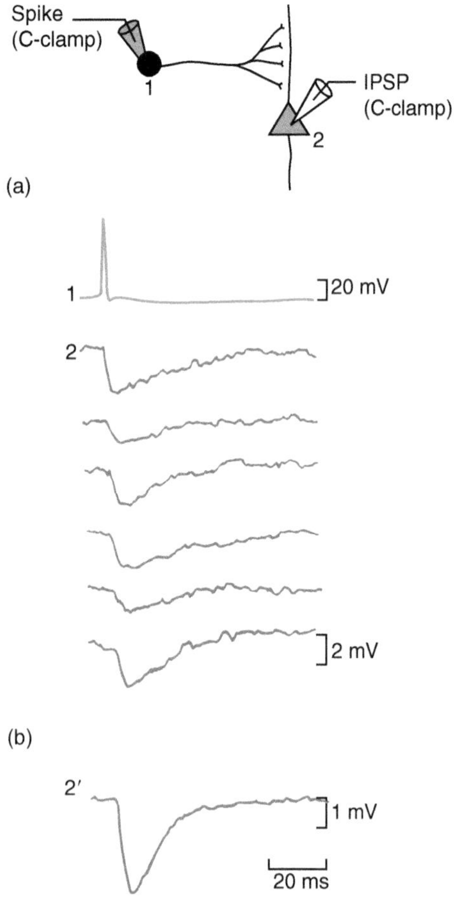

FIGURE 9.16 **Characteristics of a single-spike IPSP mediated by GABA$_A$ receptors. (a)** The activity of a pair of connected presynaptic GABAergic interneuron and postsynaptic pyramidal cell of the hippocampus *in vitro* is recorded with two intracellular electrodes. In response to each action potential evoked in neuron 1 (presynaptic GABAergic interneuron), a single-spike IPSP is recorded in neuron 2 (postsynaptic pyramidal neuron) with a mean latency of 0.7 ± 0.2 ms. Each single-spike IPSP results from the activation of several boutons which are not all activated at each trial as shown by their variable amplitude (mean amplitude = −2.1 ± 0.7 mV). They are blocked by picrotoxin (10^{-4} M, not shown), a GABA$_A$ channel blocker. **(b)** Average of 20 IPSPs obtained by triggering each trace from the peak of the presynaptic action potential. Adapted from Miles R, Wong RKS (1984) Unitary inhibitory synaptic potentials in the guinea-pig hippocampus in vitro. *J. Physiol.* **356**, 97–113, with permission.

9.5.3 Generalization: the consequences of the synaptic activation of GABA$_A$ receptors depend on the relative values of E_{Cl} and V_m (see Appendix 9.2)

It has been explained above that synaptically released GABA evokes a transient hyperpolarization of the postsynaptic membrane, called IPSP, via the activation of GABA$_A$ receptors. However, this is not always the case.

When E_{Cl} is more hyperpolarized than V_{rest}, GABA$_A$ receptor activation leads to a hyperpolarizing postsynaptic potential (IPSP) and inhibition of the postsynaptic activity

In recordings with electrodes filled with potassium acetate (instead of potassium chloride, in order not to change the intracellular concentration of Cl^-), depending on the neuron studied, E_{Cl} can be more hyperpolarized than the postsynaptic resting membrane potential (V_{rest}) and a hyperpolarizing postsynaptic potential (IPSP) is recorded in response to the stimulation of GABAergic afferent fibers (**Figures 9.16b** and **9.17a**).

When E_{Cl} is close to V_{rest}, activation of GABA$_A$ receptor leads to a 'silent inhibition' of postsynaptic activity

When E_{Cl} is close to V_{rest}, the electrochemical gradient for Cl^- is very weak or null and no IPSP is observed even though GABA$_A$ channels are open. However, GABA still has an inhibitory effect on postsynaptic activity, as any EPSP occurring during the effect of GABA is strongly inhibited (see Section 13.2.3). This inhibitory effect of GABA is called a *shunting effect*. It is due to an increase in the membrane conductance (G_{IPSP}) during the silent inhibition owing to the opening of GABA$_A$ channels. If this effect is large, the membrane resistance decreases and any other synaptic current evoked at this time will produce only a small change in membrane potential (according to Ohm's law, when I_{EPSP} is constant but R_m decreases, then V_{EPSP} decreases). The silent GABA$_A$ inhibition reduces the amplitude of postsynaptic depolarizations and consequently prevents the generation of postsynaptic action potentials (**Figure 9.17b**). It must be noted that this shunting effect of GABA is always present, whatever the effect of GABA (hyperpolarization, silence or depolarization), and it attenuates other concomitant depolarizations or hyperpolarization.

When E_{Cl} is more depolarized than V_{rest} but below the threshold for action potential generation, GABA$_A$ receptor activation leads to a depolarizing current and an inhibition of postsynaptic activity

When E_{Cl} is more depolarized than V_{rest}, the activation of GABA$_A$ receptors causes an inward current (outward flow of Cl^- ions) and a depolarization of the membrane. This depolarization does not excite the membrane; i.e. does not trigger action potentials when E_{Cl} is more negative than $V_{threshold}$. As long as E_{Cl} remains below the threshold for activation of voltage-dependent Na^+ channels, postsynaptic activity is inhibited (**Figure 9.17c**). In this case, GABA$_A$ current adds to other depolarizing current flowing through the membrane but once V_m becomes more depolarized than E_{Cl}, the GABA$_A$ current becomes hyperpolarizing.

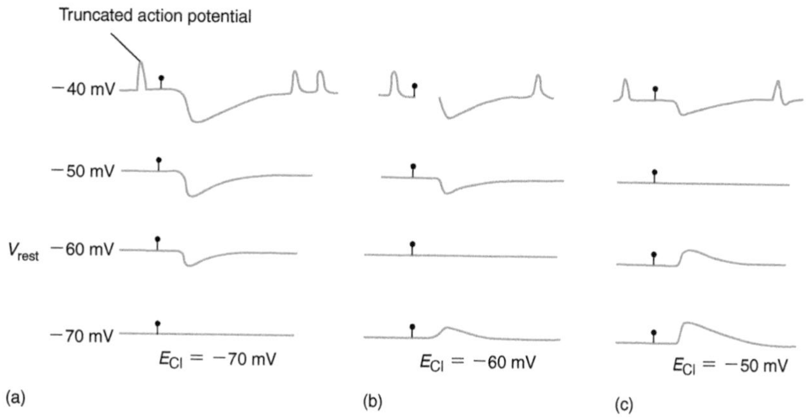

FIGURE 9.17 **Reversal potential of GABA$_A$-mediated IPSPs as a function of membrane potential.** E_{Cl} is **(a)** more negative (−70 mV), **(b)** equal (−60 mV) or **(c)** more positive (−50 mV) than the resting membrane potential of the neuron. A resting membrane potential for the neuron of −60 mV and a threshold for action potential generation at −40 mV are assumed. Action potentials are truncated owing to their large amplitude. In the three cases represented, inhibition of activity is produced during the IPSP (upper trace).

When E_{Cl} is more depolarized than V_{rest} and above the threshold for action potential generation, GABA$_A$ receptor activation leads to a depolarizing current and an excitation of postsynaptic activity

When E_{Cl} is more depolarized than the threshold for activation of voltage-dependent Na^+ channels, GABA$_A$ receptor activation causes a postsynaptic depolarization or EPSP and evokes spikes. Note that such a depolarizing and *excitatory* effect of GABA is present in very young GABAergic synapses (see Chapter 20).

9.5.4 What shapes the decay phase of GABA$_A$-mediated currents?

GABA$_A$ receptor-mediated IPSCs peak rapidly (in 0.5–5 ms) and usually decay with two time constants ranging from a few milliseconds to tens or hundreds of milliseconds. To determine the factors responsible for the duration of IPSC is interesting since IPSC decay determines the time course of the resistive shunt or hyperpolarization that prevents neuronal firing in response to excitatory inputs.

The duration of the synaptic current is determined by the period during which each activated GABA$_A$ receptor remains in the G_2R^* state. Various factors may determine the duration of the G_2R^* state and therefore the time course of the postsynaptic GABA$_A$ response:

Case 1: The GABA$_A$ receptor deactivates due to unbinding of GABA and does not reopen because GABA has disappeared from the cleft:

$$G_2R^* \rightharpoonup G_2R \rightleftharpoons 2G + R$$

Case 2: The GABA$_A$ receptor deactivates due to unbinding of GABA but it reactivates several times before closing since removal of GABA from the synaptic cleft by neuronal/glial uptake or diffusion is very slow:

$$G_2R^* \rightleftharpoons G_2R \rightleftharpoons 2G + R$$

Case 3: The GABA$_A$ receptor desensitizes:

$$G_2R^* \rightleftharpoons G_2D$$

If the decay of the synaptic current is due to rapid closure of the GABA$_A$ channels without any reopenings (case 1), this implies that the concentration of GABA decreases very rapidly in the synaptic cleft after its release (as a consequence, for example, of rapid diffusion or very efficient uptake mechanisms). In this case, the channels have very little chance of being reactivated by the repeated binding of two molecules of GABA: the time constant of the decay of the postsynaptic current is τ_o; i.e. the mean open time of the channels which depends only on the closing time constant of the channels (in the absence of desensitization). GABA$_A$ receptors have a low affinity for GABA (10–20 μM) and display brief openings and bursts (0.2–25 ms).

Assuming a brief increase of GABA concentration in the synaptic cleft, these properties predict an IPSC decay of no more than 10–25 ms, much shorter than is frequently observed. This discrepancy could indicate that GABA stays longer in the cleft (case 2) or synaptically activated GABA$_A$ receptors visit desensitized states (case 3).

If the decay of the synaptic current is due to the slow disappearance of GABA from the vicinity of the receptors (removal due to GABA uptake and/or diffusion), this implies, in contrast, a prolonged presence of GABA in the synaptic cleft. The prolonged presence of GABA may result from a slow mechanism of uptake or from a restricted diffusion of GABA away from the synaptic cleft. Hence, GABA may reactivate GABA$_A$ channels

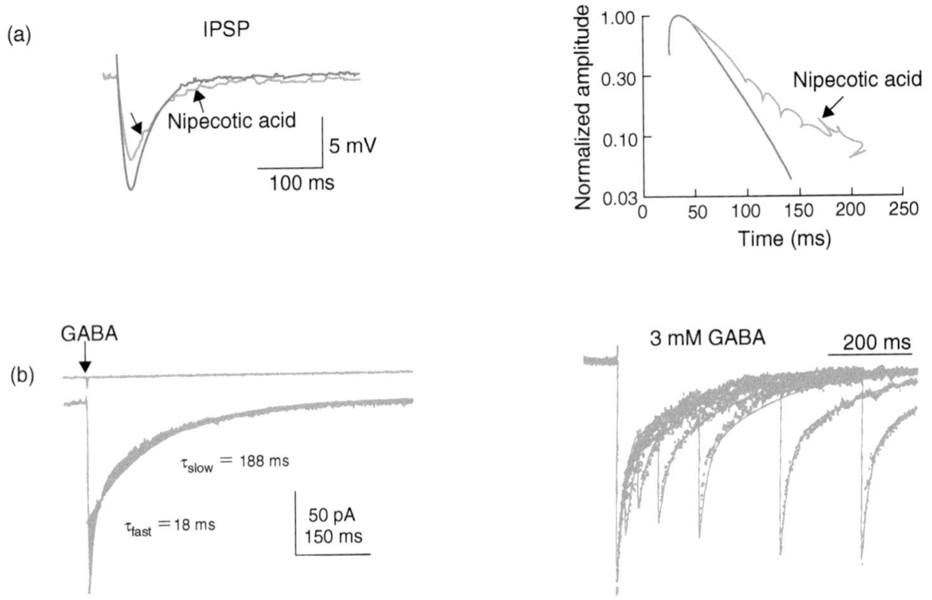

FIGURE 9.18 **Modulation of the decay phase of GABA$_A$-mediated IPSPs and current.**

which have previously opened and the decay time constant will exceed the value of τ_o. However, uptake inhibitors have very little effect on GABA$_A$ synaptic responses.

When applied by micro-iontophoresis in hippocampal slices, nipecotic acid, a neuronal and glial uptake inhibitor, has only a weak effect on the duration of the postsynaptic potential (IPSP) evoked by stimulation of GABAergic afferent fibers: it prolongs the later phase of the IPSP (**Figure 9.18a**). Thus, as mentioned earlier, the uptake process may be too slow to have much influence on the time course of the IPSP.

If the decay of the synaptic current is due to the transition of the channel to a desensitized state, the decay time constant would be related to the kinetics of desensitization. Moreover, if there is a prolonged presence of the GABA in the synaptic cleft, and rapid kinetics of desensitization and recovery from desensitization, the channels may reopen once they have recovered from desensitization. To test this hypothesis, GABA is very briefly applied (for 1–10 ms) to outside-out patches of pyramidal neurons. The ensemble average of patch currents decay with bi-exponential kinetics similar to that of the IPSC (**Figure 9.18b**). When pairs of 1–3 ms GABA pulses are given at variable intervals, the second pulse evokes a smaller peak current than the first pulse, demonstrating that channels do not return to the unbound state immediately after closing but rather enter an agonist-insensitive (desensitized) state. This suggests that after a brief pulse, GABA occupies receptors long enough for many channels to accumulate in desensitized state(s). A rapidly-equilibrating desensitized state(s) A prolongs the GABA$_A$-mediated IPSC and provides a mechanism for low-affinity receptors to support long-lasting currents.

In the case of the nicotinic channel, the removal of acetylcholine is very rapid and desensitization of the nAChR is slow relative to deactivation and does not affect the shape of the endplate current. The kinetic properties of the deactivation of the channel mostly determine the time course of the endplate current (see Section 8.5.2). In the case of the GABA$_A$ receptor, the situation appears to be more complex. For very low-amplitude postsynaptic currents, the mean open time of the GABA$_A$ channel may determine their time course, the released GABA being rapidly removed by diffusion from the cleft. However, for larger postsynaptic currents, evoked by a greater presynaptic release of GABA, the slower removal of GABA from the synaptic cleft together with the desensitization of the channel that recovers over the course of deactivation can prolong synaptic currents.

9.6 SUMMARY

The GABA$_A$ receptor channel belongs to the cys-loop family of ligand-gated channels and is activated by γ-aminobutyric acid, the neurotransmitter at numerous synapses in the mammalian central nervous system. The GABA$_A$ receptor pentamer is formed by a combination of α-, β-, γ-, δ-, ε-, π-, θ-subunits. Native receptors consist of α- and β-subunits and usually include an additional subunit, typically γ, δ, or ε. The GABA$_A$ receptor comprises the GABA receptor sites on its surface, the elements that make the ionic channel selectively permeable to chloride ions, as well as all the elements necessary for interactions between different functional domains. Thus, the GABA receptor sites and the chloride channel are part of the same unique protein.

How does GABA$_A$ receptor mediate the binding of GABA into a transient hyperpolarization of the membrane?

The fixation of GABA to the GABA$_A$ receptor induces a conformational change of the receptor and opening of the GABA$_A$ channel. This channel opens in bursts (it rapidly fluctuates between closed and open bi-liganded states). It is permeable to Cl^- ions. Depending on the membrane potential and the concentration of Cl^- in the extracellular and intracellular media, there is an influx or an efflux of Cl^-. In adult neurons, there is generally an influx of Cl^-; i.e. an outward current which hyperpolarizes the membrane, in particular when the membrane is previously depolarized by an EPSP. This hyperpolarization is the inhibitory postsynaptic potential (IPSP) mediated by GABA$_A$ receptors. Even when the membrane is not hyperpolarized by GABA (when $E_{Cl} = V_m$), the opening of GABA$_A$ channels is inhibitory: opening of GABA$_A$ channels reduces membrane resistance and thus reduces the depolarizing effect of concomitant inward currents ($\Delta V_m = R_m \Delta I_{inward}$; when ΔI_{inward} is constant and R_m diminishes, ΔV_m is reduced).

The precise measurement of E_{Cl} and V_{rest} (see Appendix 9.2) in many different central neurons is important to understand the behavior of the membrane in response to GABA.

Do benzodiazepines and barbiturates act directly on GABA$_A$ receptors? Are there selective and distinct binding sites on the receptor for each of these drugs? How do they potentiate the hyperpolarizing effect of GABA? Are there other modulators of GABA$_A$ receptors?

Aside from the GABA receptor sites, the GABA$_A$ receptor contains a variety of topographically distinct receptor sites capable of recognizing clinically active substances, such as benzodiazepines (anxiolytics and anticonvulsants), barbiturates (sedatives and anticonvulsants) and neurosteroids (see **Table 9.1**). There are selective allosteric binding sites on GABA$_A$ receptors such as (i) the benzodiazepine site that recognizes the allosteric agonists benzodiazepines and the inverse agonists β-carbolines, and (ii) the barbiturate site. Benzodiazepines increase the opening frequency of the GABA$_A$ receptor whereas barbiturates increase the duration of each opening; they both potentiate GABA$_A$ current. In contrast, inverse agonists depress GABA$_A$ current. GABA$_A$ receptors are probably a pentameric assembly derived from a combination of three different types of subunits. This leads not only to structural heterogeneity but also to pharmacological heterogeneity of the GABA$_A$ receptors, especially regarding the sensitivity to benzodiazepines.

Further reading

Campo-Soria, C., Chang, Y., Weiss, D.S., 2006. Mechanism of action of benzodiazepines on GABA$_A$ receptors. Br. J. Pharmacol. 148, 984–990.

Colquhoun, D., Hawkes, A.G., 1977. Relaxations and fluctuations of membrane currents that flow through drug-operated channels. Proc. R. Soc. Lond. B Biol. Sci. 199, 231–262.

Ernst, M., Bruckner, S., Boresch, S., Sieghart, W., 2005. Comparative models of GABA$_A$ receptor extracellular and transmembrane domains: important insights in pharmacology and function. Mol. Pharmacol. 68, 1291–1300.

Gurley, D., Amin, J., Ross, P.C., Weiss, D.S., White, G., 1995. Point mutations in the M2 region of the α, β or γ subunit of the GABA$_A$ channel that abolish block by picrotoxin. Receptors Channels 3, 13–20.

Kash, T.L., Jenkins, A., Kelley, J.C., et al., 2003. Coupling of agonist binding to channel gating in the GABA(A) receptor. Nature 421, 272–275.

Krnjevic, K., Schwartz, S., 1967. The action of γ-aminobutyric acid on cortical neurones. Exp. Brain Res. 3, 320–336.

Mody, I., Pearce, R.A., 2004. Diversity of inhibitory neurotransmission through GABA(A) receptors. Trends Neurosci. 27, 569–575.

Newell, J.G., McDevitt, R.A., Czajkowski, C., 2004. Mutation of glutamate 155 of the GABA$_A$ receptor β_2 subunit produces a spontaneously open channel: a trigger for channel activation. J. Neurosci. 24, 11226–11235.

Polenzani, L., Woodward, R.M., Miledi, R., 1991. Expression of mammalian γ-aminobutyric acid receptors with distinct pharmacology in *Xenopus* oocytes. Proc. Natl Acad. Sci. USA 88, 4318–4322.

Pritchett, D.B., Sontheimer, H., Shivers, D.B., et al., 1989. Importance of a novel GABAA receptor subunit for benzodiazepine pharmacology. Nature 338, 582–585.

Schofield, P.R., Darlison, M.G., Fujita, N., et al., 1987. Sequence identity and functional expression of the GABA-A receptor shows a ligand-gated receptor super-family. Nature 328, 221–227.

Sedelnikova, A., Erkkila, B., Harris, H., Zakharkin, S.O., Weiss, D.S., 2006. Stoichiometry of a pore mutation that abolishes picrotoxin-mediated antagonism of the GABA$_A$ receptor. J. Physiol. 577, 569–577.

Sigel, E., 2002. Mapping of the benzodiazepine recognition site on GABA receptors. Curr. Top. Med. Chem. 2, 833–839.

APPENDIX 9.1 MEAN OPEN TIME AND MEAN BURST DURATION OF THE GABA$_A$ SINGLE-CHANNEL CURRENT

The conformational changes of the GABA$_A$ channel are modeled as follows:

$$2G + R \underset{k_{-1}}{\overset{k_1}{\rightleftharpoons}} GR \underset{k_{-2}}{\overset{k_2}{\rightleftharpoons}} G_2R \underset{\alpha}{\overset{\beta}{\rightleftharpoons}} G_2R^*$$

Upon application of GABA, the conformational change of the GABA$_A$ receptor towards the G_2R^* state is favored (opening of the channel). However, it has been found that the GABA$_A$ receptor closes and opens rapidly several times upon opening: these are bursts of openings (see **Figure 9.7**). The short-duration closures represent the fluctuation of the receptor between the G_2R and G_2R^* states. During these bursts, the receptor returns much less frequently to the GR state, and even less frequently to the R state, before reopening.

The mean open time is $\tau_o = 1/\alpha$. When the receptor is in the G_2R^* state, it can only transfer to the G_2R state with a rate constant α (this is true when desensitization is negligible). τ_o is calculated experimentally from the different open times t_o within each burst.

The mean closed time within bursts is $\tau_c = 1/(\beta + 2k_{-2})$. When the receptor is in the G_2R state, it can either reopen with the rate constant β or transfer to the GR state with a rate constant $2k_{-2}$. The average number of short closures per burst is $nf = \beta/2k_{-2}$ (see Colquhoun and Hawkes, 1977).

Once the values of τ_o, τ_c and nf are experimentally defined, the values of α and β can be deduced: $\alpha = 50\ s^{-1}$; $\beta = 330\ s^{-1}$.

Note that α and β differ by a factor of about 6 (whereas for the acetylcholine nicotinic receptor nAChR, they differ by a factor of 40). This illustrates numerically the fact that fluctuations between the double-liganded open and closed states are more frequent for the GABA$_A$ receptor, the probability that the channel opens in a given time being only six times greater than the probability that the channel closes.

APPENDIX 9.2 NON-INVASIVE MEASUREMENTS OF MEMBRANE POTENTIAL AND OF THE REVERSAL POTENTIAL OF THE GABA$_A$ CURRENT USING CELL-ATTACHED RECORDINGS OF SINGLE CHANNELS

Whether GABA depolarizes or hyperpolarizes neurons depends on the value of the resting membrane potential (V_{rest}) and the reversal potential of the GABA$_A$ receptor-mediated responses ($E_{GABA(A)}$). The difference between $E_{GABA(A)}$ and V_{rest} is called GABA driving force ($DF_{GABA(A)}$). GABA causes neuronal depolarization or hyperpolarization if $DF_{GABA(A)}$ is positive or negative, respectively. When $E_{GABA(A)}$ equals V_{rest}, $DF_{GABA(A)}$ is null and the GABA$_A$ receptor-mediated response is isoelectric and associated with a drop in the membrane resistance without any evident change in the membrane potential.

To determine the action of GABA in a given neuron, one has to measure $E_{GABA(A)}$ and V_{rest}. However, conventional intracellular and whole-cell recording techniques introduce numbers of errors in the measurements of $E_{GABA(A)}$ and V_{rest}. These include alterations in the intracellular ionic composition, liquid junction potentials, space-clamp problems and neuronal depolarization due to the short-circuit effect of the conductance between the

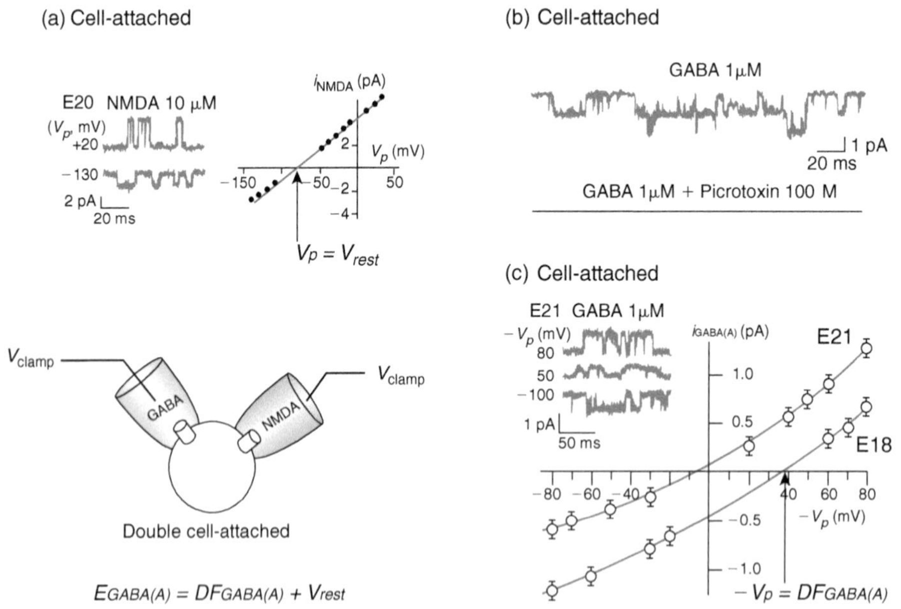

FIGURE A9.1 **(a)** Cell-attached recordings of the activity of single NMDA channels with 10 μM of NMDA in the patch pipette (left). $i/(V_p)$ relationship of the unitary NMDA current through NMDA channels (right). **(b)** Cell-attached recordings of the activity of single GABA$_A$ channels with 1 μM of GABA in the patch pipette (upper trace); the unitary currents are not observed in the presence of the GABA$_A$ antagonist picrotoxin (100 μM) together with GABA in the patch pipette (lower trace). **(c)** $i/(-V_p)$ relationships of the unitary currents through GABA$_A$ channels in two cells, one at embryonic day 21 (E21) and one at E18; their reversal potential correspond to the GABA$_A$ driving forces ($DF_{GABA(A)}$).

electrode and cell membrane. These errors are most important in small cells but they are not negligible in large neurons. To overcome the problems of conventional techniques, cell-attached recordings of ionic channels can be used to measure $E_{GABA(A)}$ and V_{rest} from intact cells in a non-invasive manner. Such measurements were used in the investigation of the action of oxytocin on GABA signaling at birth (**see Chapter 20**).

V_{rest} can be measured by studying the reversal potential of currents with a known reversal potential. For example, it is known that the reversal potential of the current through NMDA channels is close to 0 mV (see Chapter 10). In cell-attached configuration of patch clamp recordings, the patch of membrane under the recording electrode contains a variety of ionic channels. Addition of NMDA to the pipette solution causes activation of currents through NMDA channels. The relationship between i_{NMDA} and the extracellular potential applied to the patch of membrane (V_p) is plotted. This curve [$i_{NMDA} = f(V_p)$] gives the value of V_p when $i_{NMDA} = 0$ pA. At this value of V_p, single-channel NMDA current is null because $V_m = V_p - V_{rest} = 0$ mV. This allows estimation of V_{rest} ($V_{rest} = V_p$).

To estimate $E_{GABA(A)}$, the relationship between the single-channel $GABA_A$ current ($i_{GABA(A)}$) and V_p is plotted. Addition of GABA to the pipette solution causes activation of currents through $GABA_A$ channels which are abolished when picrotoxin is added in the pipette solution (**Figure A9.1b**). The relationship [$i_{GABA(A)} = f(V_p)$] gives the value of V_p when $i_{GABA(A)} = 0$ pA (**Figure A9.1c**). By definition, when $i_{GABA(A)}$ is null, $V_m = E_{GABA(A)}$. Therefore, when $i_{GABA(A)} = 0$ pA, $V_m = V_p - V_{rest} = E_{GABA(A)}$ (i.e. $E_{GABA(A)} - V_{rest} = -V_p$). By definition, $E_{GABA(A)} - V_{rest} = DF_{GABA(A)}$, the driving force of chloride ions through the $GABA_A$ channel. Therefore, when $i_{GABA(A)} = 0$ pA, $DF_{GABA(A)} = -Vp$. Knowing V_{rest} and $DF_{GABA(A)}$, it is easy to calculate $E_{GABA(A)} = DF_{GABA(A)} + V_{rest}$. In addition, the slopes of the $i_{NMDA} - V_p$ and $i_{GABA(A)} - V_p$ relationships provide an estimate of the conductance of the recorded NMDA and $GABA_A$ channels, respectively. Simultaneous recordings of $GABA_A$ and NMDA channels from the same neuron (with double cell-attached recordings) enable determination of $DF_{GABA(A)}$ and V_{rest} in the same neuron. Other channels with a known reversal potential (such as potassium channels) can also be used to measure V_{rest}.

CHAPTER

10

The ionotropic glutamate receptors

Constance Hammond

OUTLINE

From the original observations of Curtis and collaborators (1961), it is known that glutamate has a depolarizing effect on neurons. Glutamate is with GABA the major neurotransmitter in the vertebrate central nervous system. It activates two main types of postsynaptic receptors: (i) ionotropic glutamate receptors (iGluRs) that are ligand-gated channels, and (ii) metabotropic glutamate receptors (mGluRs) that are receptors coupled to GTP-binding proteins. The latter type is covered in Chapter 14.

The tightly regulated glutamate release from one neuron is detected by the glutamate receptors localized in the postsynaptic membrane of the adjacent neuron. Specificity of synaptic signaling by glutamate in space and time is conferred by the precise positioning of synapses and by the neuron-specific expression of a subset of genes encoding glutamate receptors.

10.1 THERE ARE THREE DIFFERENT TYPES OF IONOTROPIC GLUTAMATE RECEPTORS. THEY HAVE A COMMON STRUCTURE AND ALL PARTICIPATE IN FAST GLUTAMATERGIC SYNAPTIC TRANSMISSION

10.1.1 Ionotropic glutamate receptors are named after their selective or preferential agonist. They share a common structure

Based on agonist preference, iGluRs can be pharmacologically categorized into those that form receptors that are activated by the synthetic agonist N-methyl-D-aspartate (NMDA) and those that are not (**Figure 10.1a,b**). Non-NMDA receptors are further categorized by their

Cellular and Molecular Neurophysiology. DOI: 10.1016/B978-0-12-397032-9.00010-8

affinity for the synthetic agonist α-amino-3-hydroxy-5-methyl-4-isoxazole propionate (AMPA) versus the naturally occurring neurotoxin kainate. Agonists and antagonists that selectively act on one or the other have been developed in recent years.

Despite varying degrees of sequence identity, the major structural features are conserved in all known iGluR subunits. NMDA, AMPA and kainate receptors are glycoproteins composed of several subunits. To date, molecular cloning has identified 18 cDNAs, four for AMPA receptor subunits (termed GluA1–4), five for kainate receptor subunits (termed GluK1–5) and seven for NMDA receptor subunits (termed GluN1, GluN2A–D, GluN3A–B) (**Figure 10.1c**). iGluR subunits have in common a large

(a)

iGluRs	Non-NMDA receptors		NMDA receptors	
	AMPA receptors	KA receptors	Glu site	Gly site
Selective agonists	AMPA	Kainate (< 1 μM) ATPA (GluK1)	NMDA	Glycine
Non-selective agonists	Glutamate		Glutamate	–
Selective antagonists	GYKI 53655 GYKI 52466 NBQX (1 μM)	SYM 2081 UBP 310 (GLUK2/K5)	D-AP5	5,7-dichloro-kynurenate
Non-selective antagonists	CNQX DNQX NBQX (>1 μM)		–	CNQX
Channel blockers			MK 801 ketamine	

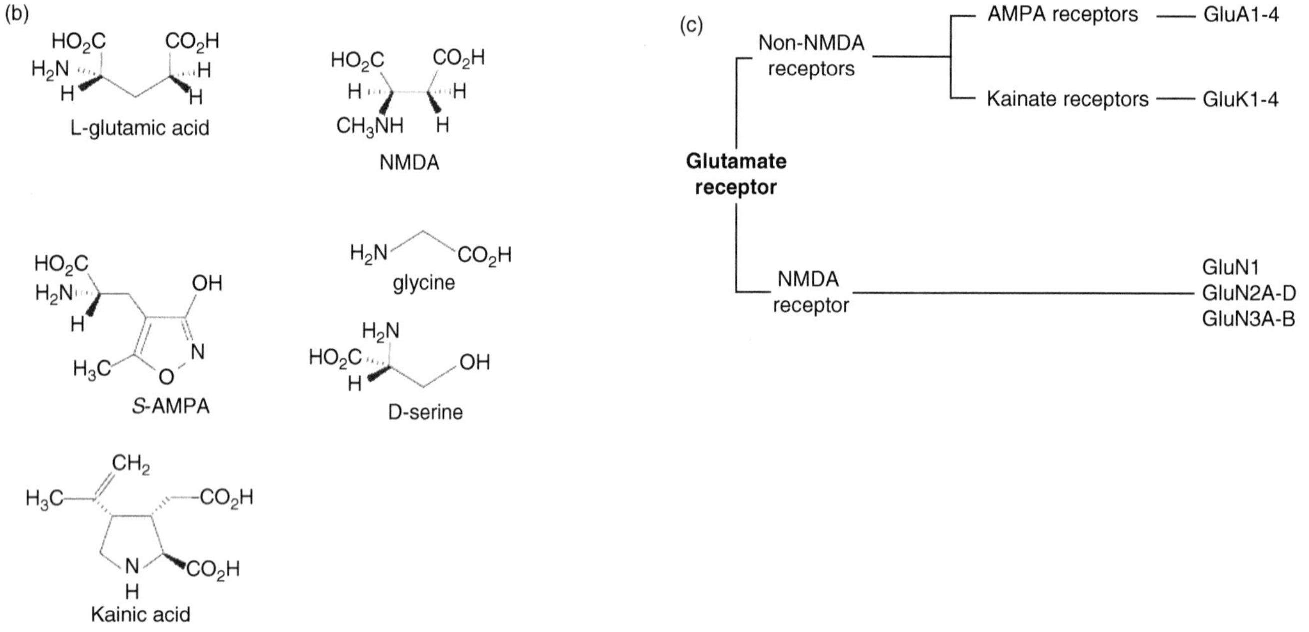

FIGURE 10.1 **Pharmacology of ionotropic glutamate receptors. (a)** AMPA: α-amino-3-hydroxy-5-methyl-4-isoxazolepropionic acid; ATPA : (*RS*)-2-amino-3-(3-hydroxy-5-*tert*-butylisoxazol-4-yl)propanoic acid selective agonist of GluK1-containing receptors; CNQX : 6-cyano-7-nitroquinoxaline-2,3-dione; D-AP5 :DL-2-amino-5-phosphonopentanoic acid; DNQX: 6,7-dinitroquinoxaline-2,3-dione; GYKI 52466: 4-(8-methyl-9*H*-1,3-dioxolo[4,5-*h*][2,3]benzodiazepin-5-yl)-benzenamine dihydrochloride; GYKI 53655: 1-(4-aminophenyl)-3-methylcarbamyl-4-methyl-3,4-dihydro-7,8-methylenedioxy-5*H*-2,3-benzodiazepine hydrochloride; MK 801:(5S,10R)-(+)-5-methyl-10,11-dihydro-5H-dibenzo[a,d]cyclohepten-5,10-imine hydrogen maleate or dizocilpine hydrogen maleate; NBQX: 2,3-dioxo-6-nitro-1,2,3,4-tetrahydrobenzo[*f*]quinoxaline-7-sulfonamide; SYM 2081: (2*S*,4*R*)-4-methylglutamic acid; UBP 310: (S)-1-(2-amino-2-carboxyethyl)-3-(2-carboxy-thiophene-3-yl-methyl)-5-methylpyrimidine-2,4-dione selective antagonist of GluK1/GluK3. **(b)** Agonists of iGluRs. **(c)** Family of ionotropic glutamate receptor subunits. Part **(c)** adapted from Wollmuth LP, Sobolevsky AI (2004) Structure and gating of the glutamate receptor ion channel. *Trend. Neurosci.* **27**, 321–328, with permission.

extracellular N-terminus domain and four hydrophobic segments. Immunocytochemical and biochemical studies have indicated that the C-terminus is intracellularly located. When N-glycosylation consensus sequences were introduced at different sites along the entire length of a GluA1 subunit, to test which part of the protein was extracellularly located (glycosylation at a particular site is taken as a proof of its external location), the hypothesis was put forward that the receptor has only three transmembrane domains, corresponding to M1, M3 and M4. In this model, M2 does not span the membrane but is considered to lie in close proximity to the intracellular surface of the plasma membrane and to have a hairpin structure (P-loop). Furthermore, the entire region between M3 and M4 is extracellular. In summary, a typical iGluR subunit consists of a bilobed amino-terminal domain (ATD) that participates in subtype-specific receptor assembly, trafficking and modulation, a ligand-binding domain (LBD) central to agonist/competitive antagonist binding and to activation gating, a transmembrane domain (TMD) consisting of three membrane-spanning segments (M1, M3, M4) and a re-entrant pore loop (P-loop or M2), that forms the membrane-spanning ion channel, and a cytoplasmic carboxy-terminal domain (CTD) of variable length involved in receptor trafficking and coupling to signaling cascades. This model is applicable to all iGluR subunits (**Figure 10.2**). The total number of amino acids per subunit ranges from 900 to over 1480. The difference in subunit size is almost entirely accounted for by differences in the length of the intracellular carboxyl (C)-terminal domain (CTD).

10.1.2 The binding site

First, researchers identified two extracellular domains called S1 and S2 as responsible for agonist specificity.

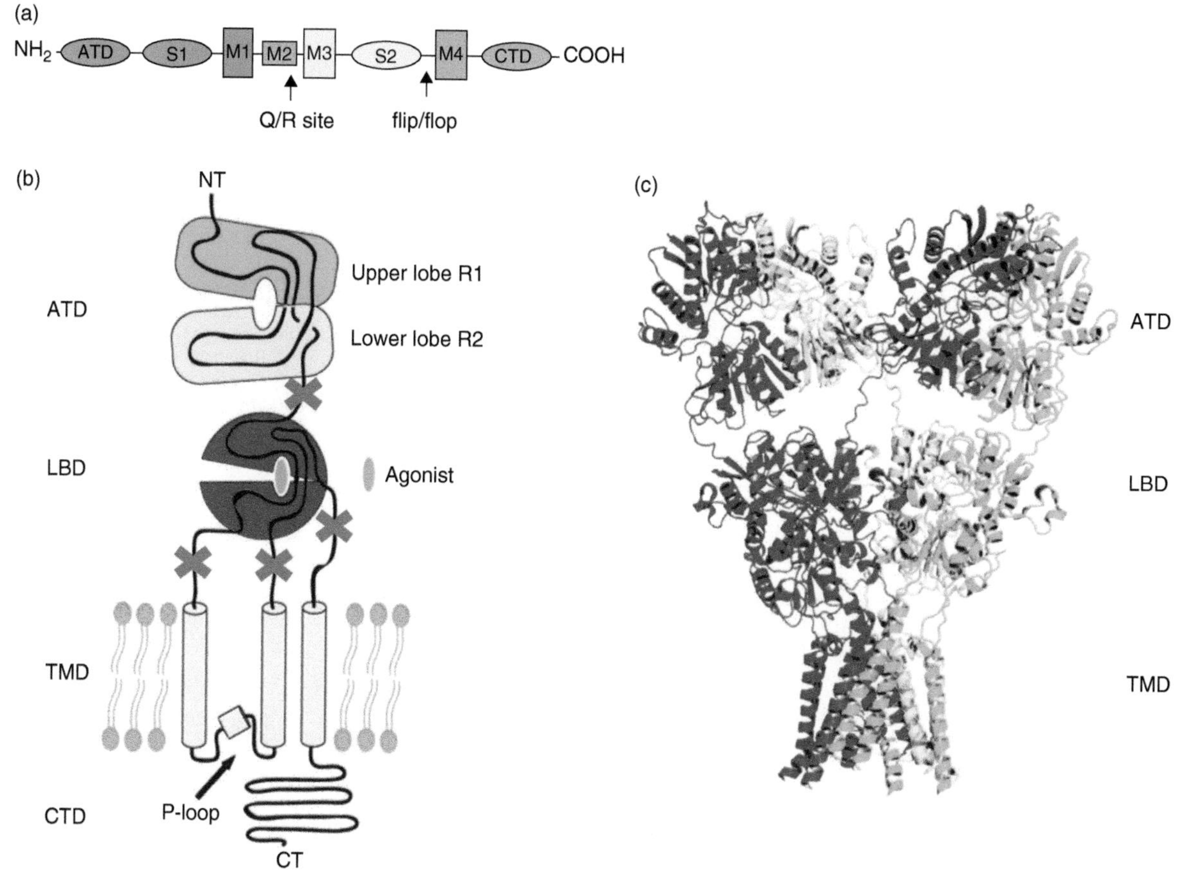

FIGURE 10.2 **Organization of domains and subunits in iGluRs. (a)** See text. **(b)** iGluR subunits are composed of distinct domains including the amino terminal domain (ATD), ligand-binding domain (LBD), transmembrane domain (TMD), and carboxyl terminal domain (CTD). The ion channel is formed by the membrane-embedded domains 1, 2 (P-loop), 3 and 4. The iGluR S1S2 constructs are generated by deleting the ATD, coupling the end of S1 to the beginning of S2 via a Gly-Thr (GT) linker and deleting the final transmembrane segment by ending the polypeptide near the end of S2 (X). For GluA and GluK receptor subunits and for GluN2 subunit, the S1S2 complex forms the glutamate-binding site (agonist), whereas the GluN1 subunit forms the glycine-binding site. **(c)** Crystal structure of the homotetrameric full-length GluA2 receptors showing the pattern of subunit arrangement and domain organization in the tetrameric assembly. The four subunits are colored as blue, yellow, green and magenta. Part **(b)** from Furukawa H (2012) Structure and function of glutamate receptor amino terminal domains. *J. Physiol.* **590**, 63–72, with permission. Part **(c)** from Sobolevsky AI, Rosconi MP, Gouaux E (2009) X-ray structure, symmetry and mechanism of an AMPA-subtype glutamate receptor. *Nature* **462**, 745–756, with permission.

Across iGluR subunits, these regions show a high degree of sequence homology. This ligand-binding domain consists of two regions termed S1 (N-terminal of M1 transmembrane spanning domain) and S2 (between regions M3 and M4). Both these domains participate to LBD (**Figure 10.2**). Then researchers designed a water-soluble mini receptor, able to bind ligands with affinities and selectivities similar to those observed with intact AMPA receptors. It consisted of the agonist-binding core, the S1S2 domains linked together by a short, flexible, peptide linker. They made this S1S2 construct from one AMPA receptor secreted as a soluble protein from either insect or *E. coli* cells. Once this type of construct was produced in sufficiently large quantities in a functionally active state, they could obtain well-ordered crystals and perform structural analysis by X-ray diffraction.

The crystal structure of the GluA2 S1S2–kainate complex showed that the receptor fragment has a 'clamshell'-like structure and that the ligand resides in a cavity formed between both domains, i.e. in the cleft between each shell. Furthermore, it allowed definition of the location of the amino-acid residues that interact with the agonists glutamate, AMPA, kainate and the competitive antagonist DNQX. But the important discovery was that the 'clamshell', in the presence of AMPA and glutamate, was around 20 degrees more closed in comparison to the state in the presence of the antagonist DNQX, whereas in the presence of the partial agonist kainate, the degree of closure was intermediate. This suggested that the fundamental conformational change involved in the activation of the ion channel was the closure of the clamshell. Similar studies of the binding core of NMDA subunits revealed that it too retains a general clamshell-like architecture reminiscent of the other AMPA and kainate subunit-binding pockets.

In summary, the overall fold of the iGluR ligand-binding domains for AMPA, kainate and NMDA receptors is nearly identical. Key amino acid side chains that interact with the ligand α-amino and α-carboxy groups are the same in all four iGluR families. What differs are the amino acids that interact with the glutamate γ-carboxy group or, in the case of GluN1 subunits, prevent the binding of glutamate. In the agonist-bound complex of all iGluRs, the ligand is buried in the interior of the protein but the volume of the ligand-binding cavity varies substantially. Within a tetrameric receptor complex, the ATDs and LBDs assemble as dimers, with the full receptor operating as a dimer-of-dimers.

10.1.3 The three ionotropic receptors participate in fast glutamatergic synaptic transmission

In the cat neocortex, in response to a single spike in a presynaptic pyramidal neuron, a transient depolarization of the membrane mediated by ionotropic glutamate receptors is recorded in the postsynaptic interneuron. This depolarization of synaptic origin is called a *single spike excitatory postsynaptic potential* (single-spike EPSP) (**Figure 10.3a**). Note the small amplitude of this EPSP. It is insensitive to APV (the antagonist of NMDA receptors) and totally abolished by CNQX (an antagonist of both AMPA and kainate receptors) (not shown), thus showing that it results from the activation of postsynaptic non-NMDA receptors.

In the rat hippocampus, a large-amplitude EPSP is recorded from an interneuron in response to the stimulation of the presynaptic glutamatergic pyramidal neuron. This EPSP presents two components: an early one abolished by CNQX and a late one abolished by APV. In the presence of D-APV, there is a clear reduction in the EPSP duration (**Figure 10.3b**). The early component of the EPSP that remains is D-APV-insensitive; i.e. mediated by non-NMDA receptors as it is abolished by CNQX, a non-selective antagonist of AMPA and kainate receptors (not shown).

In the same preparation, the non-NMDA receptor-mediated component recorded in the presence of D-APV can be further separated into AMPA receptor-mediated and kainate receptor-mediated. Thanks to GYKI 53655, a selective antagonist of AMPA receptors, the kainate component of the EPSP is thus revealed (**Figure 10.3c**). It is a very small-amplitude component that is antagonized by LY 293558, a selective antagonist of GluK1-containing kainate receptors.

These observations raise several questions:

- Do all ionotropic glutamate receptors have the same properties, for example the same sensitivity to glutamate, the same ionic permeability?
- Do they have the same conditions of activation, deactivation, and desensitization?
- Do postsynaptic elements contain more than one iGluR type?

10.2 AMPA RECEPTORS ARE AN ENSEMBLE OF CATIONIC RECEPTOR-CHANNELS WITH DIFFERENT PERMEABILITIES TO Ca^{2+} IONS

10.2.1 The diversity of AMPA receptors results from subunit combination, alternative splicing and post-transcriptional nuclear editing

Cloning studies have demonstrated that AMPA selective ionotropic glutamate receptors are built from the four closely related subunits GluA1, GluA2, GluA3 and GluA4. The four predicted polypeptide sequences, each approximately 900 amino acids in length, revealed similarities between 70% (GluA1 and 2) and 73% (GluA2 and

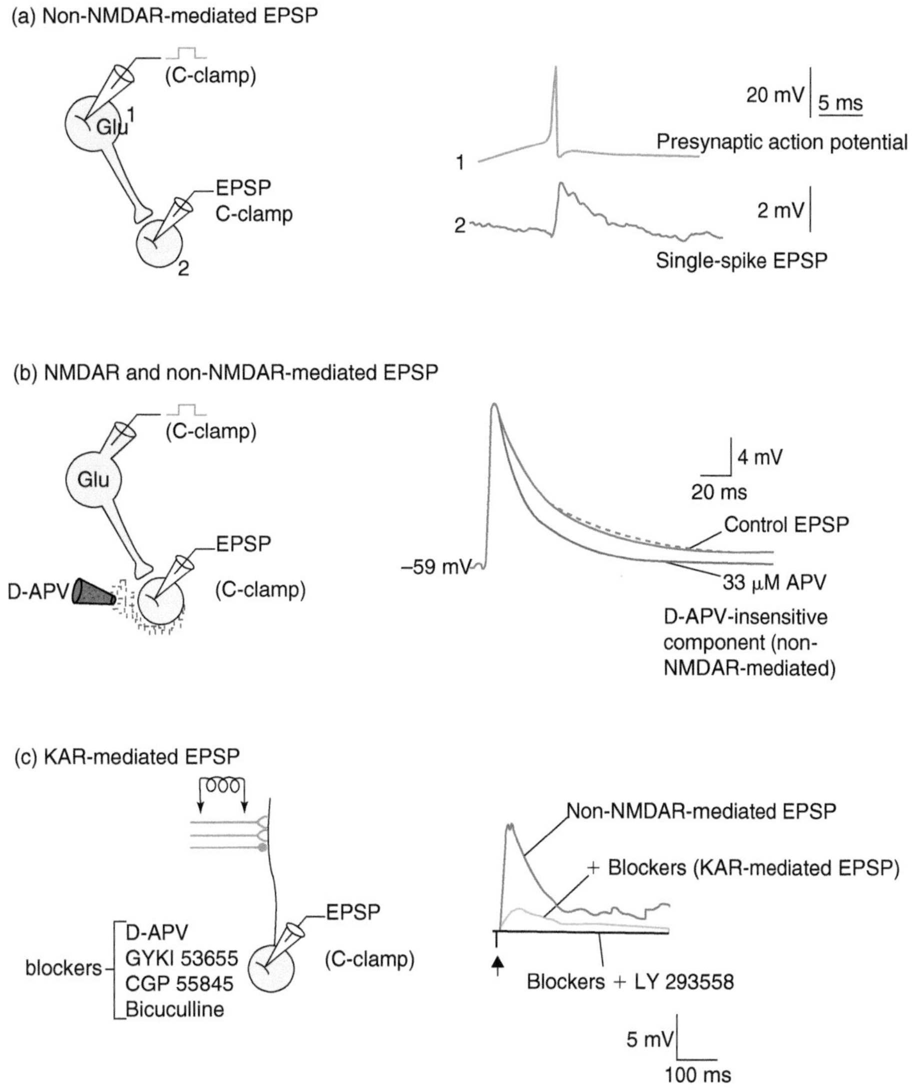

FIGURE 10.3 **iGluR-mediated EPSPs and their components. (a)** Current clamp recordings of a presynaptic glutamatergic pyramidal neuron and a postsynaptic interneuron in the cat neocortex *in vitro*. A spike triggered by a current pulse in the presynaptic neuron evokes a single-spike EPSP in the postsynaptic interneuron (intracellular recordings). **(b)** The same experiment in the hippocampus *in vitro*. In response to a stronger stimulation of the presynaptic pyramidal neuron and in the absence of external Mg^{2+}, a postsynaptic EPSP of larger amplitude (control) with a fast rising phase and a long duration (sometimes up to 500 ms) is recorded. The same experiment in the presence of APV (APV, 33 μM). Picrotoxin is added to the extracellular solution in order to block $GABA_A$ synaptic receptors. **(c)** In response to the stimulation of afferent fibers, a control EPSP is recorded from a hippocampal interneuron. In the presence of blockers of NMDA (APV), AMPA (GYKI 53655), $GABA_B$ (CGP 55845) and $GABA_A$ (bicuculline) receptors, a low-amplitude component is still present. It is mediated by kainate receptors since it is totally blocked by LY 293558. Part (**a**) from Buhl EH, Tamas G, Szilagyi T *et al.* (1997) Effect, number and location of synapses made by single pyramidal cells onto spiny interneurons of cat visual cortex. *J. Physiol.* (*Lond.*) **500**, 689–713, with permission. Part (**b**) from Forsythe ID, Westbrook GL (1988) Slow excitatory postsynaptic currents mediated by *N*-methyl-D-aspartate receptors on cultured mouse central neurons. *J. Physiol.* (*Lond.*) **396**, 515–533, with permission. Part (**c**) from Cossart R, Esclapez M, Hirsch J *et al.* (1998) GluR5 kainate receptor activation in interneurons increases tonic inhibition of pyramidal cells. *Nat. Neurosci.* **1**, 470–478, with permission.

3). These subunits when expressed *in vitro* constitute a high-affinity ^{3}H AMPA and low-affinity kainate receptor type of glutamate-gated ion channel. These different subunits are abundantly and differentially expressed in the brain, as revealed by *in situ* hybridization studies (see Appendix 6.2). Although these different iGluR subunits exhibit some ability to form homomeric channels when expressed by themselves in *Xenopus* oocytes or cultured mammalian cells, it is considered likely that channels are formed *in vivo* by different combinations of subunits. Thus, with four receptor subunits there is already a very large number of potential combinations.

The diversity of iGluRs results not only from subunit combinations but also from two genetic processes: *alternative splicing* and *editing* of the pre-messenger RNA (or primary transcript). Alternative splicing concerns the 38-amino-acid sequence preceding the most C-terminal putative transmembrane domain M4 of each of the four receptor subunits (**Figure 10.2a**). This small segment has been shown to exist in different versions (with different amino acid sequences), encoded by adjacent exons of the receptor genes. As a consequence, each of the four subunits may exist in different molecular forms

Editing is a post-transcriptional change of one or more bases in the pre-mRNA such that the codon(s) encoded by the gene and the codon(s) present in the mRNA differ. It has been established that the sequences necessary for editing lie in the introns. Thus, only primary transcripts can be edited. Therefore, editing is not a regulatory mechanism for mature mRNA, but results from post-transcriptional nuclear editing. In AMPA receptors, editing concerns only the GluA2 subunit. The GluA2 subunit possesses an arginine (R) in the M2 putative membrane spanning segment at position 586; whereas in GluA1, GluA3 and GluA4 subunits, glutamine (Q) lies in the homologous position. This functional critical position is referred to as the Q/R site (see **Figure 10.2a**). The arginine codon (CGG) is not found in the GluA2 gene. It is introduced into the GluA2 mRNA by an adenosine to inosine conversion in the respective glutamine codon (CAG) of the GluA2 transcript by a double-stranded RNA adenosine deaminase. The inosine is subsequently read as a guanosine, resulting in the change of codon identity (CGG). Editing is developmentally regulated such that 99% of the GluA2 subunit in postnatal stages is in the edited (R) form. The consequences of the GluA2 editing at the Q/R site are analyzed in Section 10.2.3.

10.2.2 The native AMPA receptor is permeable to cations and has a unitary conductance of 8 pS

When quisqualate (1–10 μM) is applied in the extracellular milieu of cultured central neurons recorded in the outside-out patch clamp configuration, an inward unitary current, i_q, is observed at $V_m = -60$ mV (**Figure 10.4a**). If the application of quisqualate is repeated on different membrane patches recorded at the same membrane potential, one observes that the recorded inward unitary currents are not a homogeneous population. The unitary conductance values corresponding to the four peaks of **Figure 10.4b** are calculated from the different i_q, recorded with the equation $i_q = \gamma_q(V - E_{rev})$ given that $E_{rev} = 0$ mV. One set of similar unitary currents appears with a higher frequency than the others, the one of $\gamma_q = 8$ pS. Plotting the mean unitary current amplitude of this population as a function of the membrane potential, we obtain an i_q/V relation (**Figure 10.4c**) that follows the equation $i_q = \gamma_q(V - E_{rev})$. Between −80 mV and +80 mV this curve is linear (i.e. voltage independent), has a slope of 8 pS, and shows a quisqualate current reversal potential of approximately 0 mV.

These results show that the majority of the channels activated by quisqualate have a unitary conductance of 8 pS. This conductance is only slightly or not at all sensitive to membrane potential variations. Additionally, the unitary current i_q reverses at a value close to 0 mV if the extracellular and intracellular environments contain similar concentrations of monovalent cations (**Figure 10.4c**). This suggests that the quisqualate-activated channel is permeable to cations.

To test this hypothesis, the reversal potential of the current is recorded at different intracellular and extracellular concentrations of Na^+ and K^+ ions. When the extracellular Na^+ concentration is lowered from 140 to 50 mM by replacing Na^+ ions with choline ions, the reversal potential becomes negative (it shifts from 0 mV to about −20 mV). When Cs^+ ions substitute for intracellular K^+ ions, the reversal potential is not affected. These results suggest that the quisqualate-activated channel is permeable to Na^+, K^+ and Cs^+ ions, and impermeable to choline ions.

The quisqualate-activated channel shows a low permeability to divalent cations, especially to Ca^{2+}. Thus, variations by a factor of 20 of the extracellular Ca^{2+} concentration have no effect on the reversal potential of the quisqualate current recorded from neurons in the whole-cell configuration. Likewise, only small changes of the photometrically recorded intracellular Ca^{2+} concentration (see Appendix 6.1) can be measured during a quisqualate-evoked response at a constant voltage of −60 mV. It should be noted that it is essential to carry out these recordings in cells maintained at membrane potentials lower than the activation threshold of the voltage-sensitive Ca^{2+} channel in order to prevent Ca^{2+} inflow through these channels.

In conclusion, the native quisqualate-activated channel, recorded in spinal neurons in culture, is permeable to monovalent cations: the application of quisqualate at a membrane potential of $V_m = -60$ mV evokes a unitary inward current that results from the inflow of Na^+ ions and an outflow of K^+ ions through the same channel (the Na^+ inflow is stronger than the K^+ outflow). This AMPA receptor is a classic cationic channel receptor: it has a negligible permeability to Ca^{2+} ions ($P_{Ca}/P_{Na} = 0.1$) and its conductance is only weakly voltage dependent.

However, studies performed in other preparations show that some AMPA receptors are permeable to Ca^{2+} ions. The above pioneering electrophysiological experiments which characterized the properties of native AMPA receptor-channels were carried out before molecular cloning of glutamate receptor subunits had been achieved, which showed that the Ca^{2+} permeability of AMPA receptor channels varies with their subunit composition. In the example of **Figure 10.4**, the native AMPA receptors studied probably contained in their structure the edited form of GluA2 (GluA2(R)) as explained below.

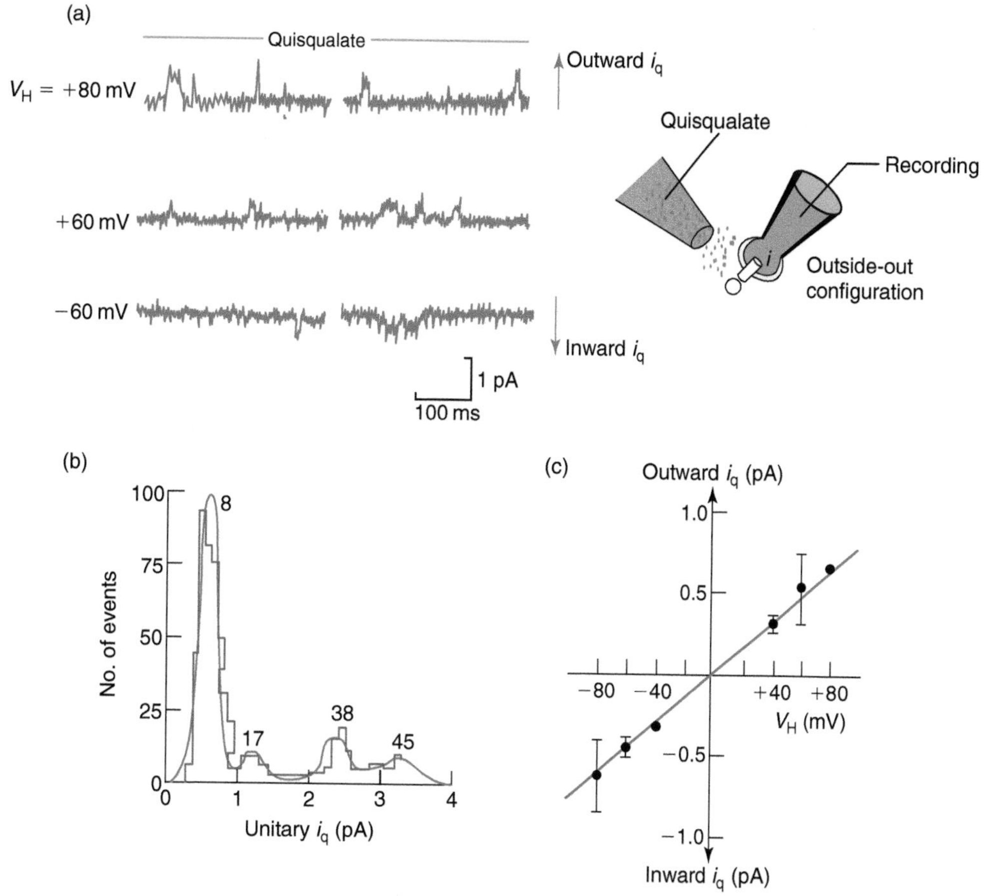

FIGURE 10.4 **Electrophysiological properties of native AMPA receptor-channel. (a)** Patch clamp recording (outside-out configuration) of the activity of a quisqualate-activated channel. When the membrane is held at −60 mV, the unitary current i_q is inward (downward deflection). At +60 mV or +80 mV, i_q is outward (upward deflection). **(b)** Unitary current amplitude histogram (in pA). Currents recorded at the same voltage but from different membrane patches. **(c)** i_q/V curve obtained from the averages of unitary currents recorded from a homogeneous population of channels (8 pS population). Intrapipette solution (in mM): 140 CsCI, 5 K-EGTA, 0.5 $CaCl_2$; extracelluar solution: 140 NaCl, 2.8 KCl, 1 $CaCl_2$. Parts **(a)** and **(c)** from Ascher P, Nowak L (1988) Quisqualate and kainate-activated channels in mouse central neurons in culture. *J. Physiol. (Lond.)* **399**, 227–245, with permission. Part **(b)** from Cull-Candy SG, Usowicz MM (1987) Patch clamp recording from single glutamate-receptor channels. *TIPS* **8**, 218–224, with permission.

10.2.3 AMPA receptors are permeable to Na^+, K^+ and Ca^{2+} ions unless the edited form of GluA2 is present; in the latter case, AMPA receptors are impermeable to Ca^{2+} ions

Comparison of the amino-acid sequence of the GluA2 subunit with that of the GluA1, 3, 4 subunits revealed a marked difference in the P-loop (M2) segment. A positively charged arginine (R) is present at position 586 of the GluA2 subunit, whereas a neutral glutamine (Q) is found at the homologous position in the other subunits. This position is termed the Q/R site. GluA2 is the only AMPAR subunit concerned by RNA editing by adenosine deaminase 2 at the Q/R site, in the pore lining region. This change from glutamine (Q, CAG codon) to arginine (R, CGG codon) (**Figure 10.5a**) has a major consequence on the channel permeability: those channels with a GluA2(R) subunit are Ca^{2+}-impermeable while those with a GluA2(Q) or no GluA2 subunit are Ca^{2+}-permeable.

To show this, the current response to glutamate of homomeric GluAs expressed in transfected cells is studied in the presence of extracellular solutions containing Na^+ or Ca^{2+} as the only cations. In cells expressing the GluA2(R) subunit only, the glutamate-evoked current is present in high Na^+ solution and nearly absent in high Ca^{2+} solution, indicating that this homomeric channel has a low divalent/monovalent permeability ratio. Moreover, heteromeric GluA2(R) + GluA1 subunit association forms Ca^{2+}-impermeable oligomeric channels in oocytes (**Figure 10.5b**). The situation is different in the absence of the GluA2(R) subunit since homomeric GluA1, GluA3 or heteromeric GluA1 + GluA3 channels allow the influx of Ca^{2+}. Moreover, mutational analysis indicates that the Ca^{2+} permeability of channels assembled from GluA2 subunits in which the arginine is replaced by glutamine (GluA2(R586Q)) is high, whereas the Ca^{2+} permeability of channels assembled from GluA4 subunits in which the glutamine is replaced by arginine (GluA4(Q587R) is low.

Therefore, the presence of a positively charged side chain of one amino acid (R) determines the divalent/monovalent permeability ratio. GluA2(R) dominates the properties of ion flow through the heteromeric iGluR channel. The positive charge of arginine (R) at the Q/R disrupts the interactions between Ca^{2+} ions and the channel pore, hindering the flow of the cation through the channel.

10.2.4 AMPA current through GluA2-lacking AMPA receptors displays inward rectification

Ca^{2+}-permeable AMPA receptors display inward rectification (i.e. they exhibit a reduced outward current at depolarizing membrane potentials; **Figure 10.5c**), which arises from fast voltage-dependent channel block by intracellular polyamines: positively charged spermine (and, to a lesser extent, spermidine), as well as polyamine spider toxins such as argiotoxin and Joro spider toxin, selectively block GluA2-lacking receptors. They do so because they are attracted by a negatively charged ring of carbonyl-oxygen groups provided by the glutamine residues of the GluA1, GluA3 or GluA4 subunits but are repelled by the positively charged arginine residue in the GluA2 subunit. In some cases, Ca^{2+}-permeable AMPA receptors display a 'double rectification': inward current at negative potentials, very low conductance between −20 and 30 mV and an outward current at potentials >40 mV. This recovery of ion flow at very positive potentials appears to be due to the fact that polyamines pass right through the channel from the inside to the outside when the driving force is large enough.

The role of polyamines was discovered by comparing *I*/*V* curves in whole-cell and outside-out configurations. When recording the whole-cell glutamate-induced current from HEK 293 cells transfected with GluA(Q) subunits, strong inward rectification is observed (**Figure 10.5c, left**). Inward rectification, however, is eliminated in outside-out membrane patches (**Figure 10.5c, right**) suggesting that a diffusible substance in the cytoplasm confers inward rectification. From studies of rectification of K^+ channels, the authors tried Mg^{2+} (1 mM) or spermine (10 μM) in the pipette solution (intracellular medium). Mg^{2+} had no effect but spermine restored strong inward rectification in a manner qualitatively similar to responses obtained with whole-cell recording (**Figure 10.5c, left**).

Another difference between AMPA receptors concerns the amplitude and kinetics of the synaptic current. AMPA receptors that lack GluA2 subunits exhibit larger single channel conductance than GluA2-containing receptors. Synaptic currents mediated via GluA2-lacking AMPARs typically exhibit rapid rise and decay kinetics. Incorporation of GluA2 subunits into AMPA receptors prolongs the decay time of synaptic currents.

10.2.5 Summary

AMPARs are grouped into two functionally distinct tetrameric assemblies based on the inclusion or exclusion of the GluA2 receptor subunit. GluA2-containing receptors are thought to be the most abundant AMPAR in the CNS, typified by their small unitary events, Ca^{2+} impermeability and insensitivity to polyamine block. In contrast, GluA2-lacking AMPARs exhibit large unitary conductance, marked divalent permeability and nano- to micromolar polyamine affinity.

10.3 KAINATE RECEPTORS ARE AN ENSEMBLE OF CATIONIC RECEPTOR CHANNELS WITH DIFFERENT PERMEABILITIES TO Ca^{2+} IONS

Kainic acid is a powerful neurotoxin which kills neurons by means of overexcitation. It is isolated from a seaweed known for its potency at killing intestinal worms. The word 'kainic' is derived from the Japanese *kaininso* which means the 'ghost of the sea'. At nanomolar concentrations, kainate is a preferential agonist of kainate receptors but, at higher concentrations, it also activates AMPA receptors (see **Figure 10.1a**).

10.3.1 The diversity of kainate receptors

A family of kainate receptors has been cloned and five subunits termed GluK1–3 (ex GluR5–7) and GluK4–5 (ex KA1 and KA2) have been identified (see **Figure 10.1c**). GluK1 to GluK3 represent subunits with a low-affinity kainate-binding site and a dissociation constant in the range of 50–100 nM, whereas GluK4 and GluK5 correspond to subunits with a high-affinity kainate-binding site (K_D of 5–15 nM). GluK1 to GluK3 are of similar size and share 75–80% amino acid sequence identity with each other and around 40% with AMPA receptors. GluK4 and GluK5 share 70% amino acid sequence identity with each other and around 40% with either GluA1–GluA4 or GluK1–GluK3.

Kainate receptors are tetrameric combinations of five subunits. Of these, GluK1–3 can form functional homomeric or heteromeric receptors. In contrast, GluK4 or GluK5 expression does not generate agonist-sensitive

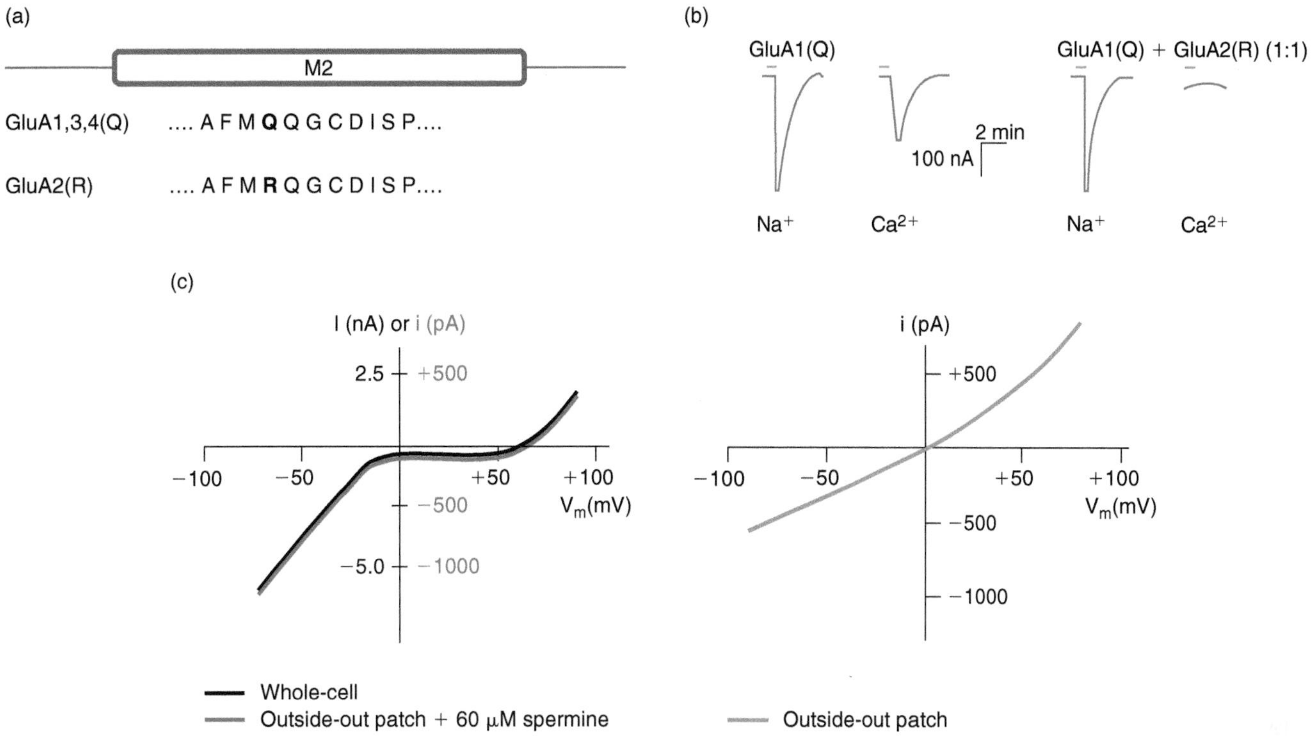

FIGURE 10.5 **Functional properties of AMPA receptors. (a)** Linear representation of the M2 segment (P-loop) of GluA receptors and site of editing (Q/R site) in this segment. **(b)** Comparison of whole-cell currents evoked by pulse application (25 s, bars) of a glutamate agonist to homomeric GluA1(Q) (left) or heteromeric GluA1(Q) + GluA2(R) (right) channels expressed in oocytes and recorded in normal Ringer (Na^+) and Ca^{2+}-Ringer (Ca^{2+}) solutions. Oocytes were injected with a single GluA subunit cRNA (2 ng) or a combination of two types of GluA subunit cRNA (2 ng + 2 ng for 1:1 combination). Intrapipette solution (in mM): 250 CsCl, 250 CsF, 100 EGTA. Na^+-external solution (in mM): 115 NaCl, 2.5 KCl, 1.8 $CaCl_2$, 10 Hepes; Ca^{2+}-external solution (in mM): 10 $CaCl_2$, 10 Hepes. **(c)** *I*/*V* plots for whole-cell and outside-out patch responses to glutamate recorded from GluA2-lacking channels expressed in HEK cells. (Left) Whole-cell responses are recorded on average 100 s after breakthrough, using a polyamine-free internal solution (red trace). Data from outside-out patches (black trace) are recorded using 60 μM spermine in the intrapipette solution; this concentration closely matches the complex *I*/*V* relationship of whole-cell responses over the range −100 to +100 mV. The two *I*/*V* plots are superimposed. (Right) *I*/*V* plot of the glutamate-induced current recorded in outside-out patches. The *I*/*V* plot is roughly linear. Part (b) from Hollmann M, Hartley M, Heinemann S (1991) Ca^{2+} permeability of KA-AMPA-gated glutamate receptor channels depends on subunit composition. *Science* **252**, 851–853, with permission. Part (c) from Koh DS, Burnashev N, Jonas P (1995) Block of native Ca(2+)-permeable AMPA receptors in rat brain by intracellular polyamines generates double rectification. *J. Physiol.* **486**, 305–312 and Bowie D, MayerML (1995) Inward rectification of both AMPA and kainate subtype glutamate receptors generated by polyamine-mediated ion channel block. *Neuron* **15**, 453–462.

channels. They participate in heteromeric receptors, partnering any of the GluK1–3 subunits.

Like those for AMPA receptors, the mRNAs encoding for kainate receptors are subject to editing and/or alternative splicing. GluK1 and GluK2 mRNA can be edited at a Q/R site in the M2 segment (see **Figure 10.2a**), but not GluK3, GluK4 and GluK5. In analogy with the GluA2 subunit of AMPA receptors, the presence of an arginine residue results in receptors that have low permeability to Ca^{2+}. However, in contrast to GluA2, the Q/R site editing is incomplete during development and significant amounts of both edited and non-edited versions of GluK1 and GluK2 coexist in adult brain. Alternative splicing of GluK1, GluK2 and GluK3 further adds to receptor diversity whereas the GluK4 and GluK5 subunits do not undergo any known process of alternative splicing or RNA editing.

10.3.2 Native kainate receptors are permeable to cations

One way to study kainate channels in isolation is to apply kainate (or the agonist SYM 2081) in the presence of an antagonist of AMPA receptors, the 2,3 benzodiazepine GYKI 53655. Granule cells of the rat cerebellar cortex are dissociated and plated in culture dishes. Whole-cell recordings are performed in voltage clamp mode in order to record the total kainate current (I_{kai}). Concanavalin A is added in the extracellular solution to reduce kainate receptor desensitization. Kainate 10 μM in the presence of GYKI 53655 evokes an inward current when the membrane is held at −80 mV (**Figure 10.6a**). The current/voltage curve obtained by varying the holding potential is shown in **Figure 10.6b**. I_{kai} reverses around 0 mV, suggesting that it is carried by cations (in the presence of

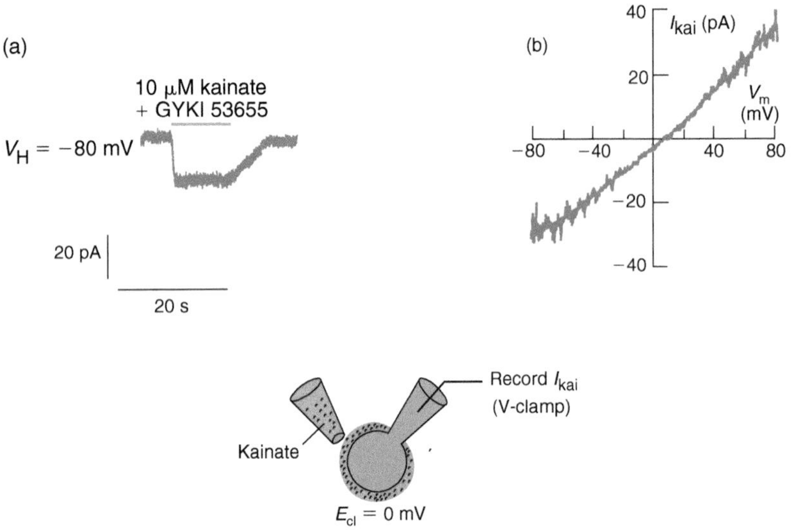

FIGURE 10.6 **Whole-cell kainate current in cerebellar granule cells. (a)** Whole-cell current (I_{kai}) evoked by the application of 10 μm of kainate in the presence of GYKI 53655 (100 μM) in a concanavalin A-treated granule cell. **(b)** I_{kai}/V relationship. I_{kai} is measured during 500 ms voltage ramps from −80 to +80 mV. From Pemberton KE, Belcher SM, Ripellino JA, Howe JR (1998) High affinity kainate-type ion channels in rat cerebellar granule cells. *J. Physiol. (Lond.)* **510**, 401–420, with permission.

control extra- and intracellular solutions). Experiments in high Na^+ or Cs^+ solutions have confirmed the cationic permeability of kainate channels. Editing of either M1 or M2 sites changes the Ca^{2+} permeability of GluK2 subunits (**Figure 10.7a**).

Editing at the Q/R site affects the I_{kai}/V relationship in hippocampal neurons. In cells expressing GluK2(Q), the non-edited form of GluK2 at the M2 site, as shown by single-cell RT-PCR, the current/voltage relationship shows a strong inward rectification (**Figure 10.7b**); whereas in a cell expressing the edited form, GluK2(R), the curve is almost linear (**Figure 10.7c**). This shows a clear relationship between the rectification properties of native kainate receptors and editing of Q/R site in the GluK2 subunit mRNA. One of the possible physiological consequences for the presence of rectification is that in a membrane depolarized to around −20 mV, the GluK2(Q)-mediated current is extremely small.

10.4 NMDA RECEPTORS ARE CATIONIC-RECEPTOR-CHANNELS HIGHLY PERMEABLE TO Ca^{2+} IONS; THEY ARE BLOCKED BY Mg^{2+} IONS AT VOLTAGES CLOSE TO THE RESTING POTENTIAL, WHICH CONFERS STRONG VOLTAGE DEPENDENCE

10.4.1 Molecular biology of NMDA receptors

Molecular cloning has identified to date seven cDNAs encoding GluN1, GluN2A–D and GluN3A–B subunits of the NMDA receptor (see **Figure 10.1c**), the deduced amino acid sequences of which are 18% (GluN1 and GluN2), 55% (GluN2A and GluN2C), 70% (GluN2A and GluN2B) or around 25% (GluN3 and GluN1 or GluN3 and GluN2) identical. When the *Xenopus* oocyte system and transfected mammalian cells are employed to study the functional properties of these subunits, large currents are measured only in oocytes co-expressing GluN1 and GluN2 subunits. Actually, native NMDA receptors are obligate hetero-oligomers and functional NMDA receptors are composed of the constitutive GluN1 subunit and one or more of four different GluN2 subunits. Most NMDARs are tetramers thought to be composed of two GluN1 and two GluN2 subunits. The presence and role of GluN3 in native NMDA receptors are poorly understood.

Recombinant heteromeric NMDA receptors display different properties depending on which of the four GluN2 subunits are assembled with GluN1. Site-directed mutagenesis has revealed that the GluN2 subunit carries the binding site for glutamate whereas the homologous domain of the GluN1 subunit carries the binding site for the co-agonist glycine (see **Figure 10.2b**).

The crystal structures of the ligand-binding core of GluN2A bound to glutamate and GluN1-GluN2A heterodimer bound to glycine and glutamate gave important data. The first one defined the determinants of glutamate and NMDA recognition whereas the second one suggested a mechanism for ligand-induced ion channel opening. Also, the crystal structure of the GluN1 subunit glycine-binding domain (GluN1 S1S2) revealed that it too retains a general clamshell-like architecture reminiscent of the other AMPA and kainate subunit-binding pockets. Why does the binding core of GluN1 recognize glycine and not glutamate? One amino acid in the GluN1 subunit glycine-binding pocket, a Trp side chain at position 731, has a key role in selectivity by occupying space required for the glutamate γ-carboxy group; as a result, glutamate cannot bind to the GluN1 subunit.

GluN1 and GluN2 subunits carry in the M2 segment (which forms a re-entrant P-loop) an asparagine residue in a position homologous to the Q/R site of AMPA receptors and GluN2 subunits carry an asparagine at the

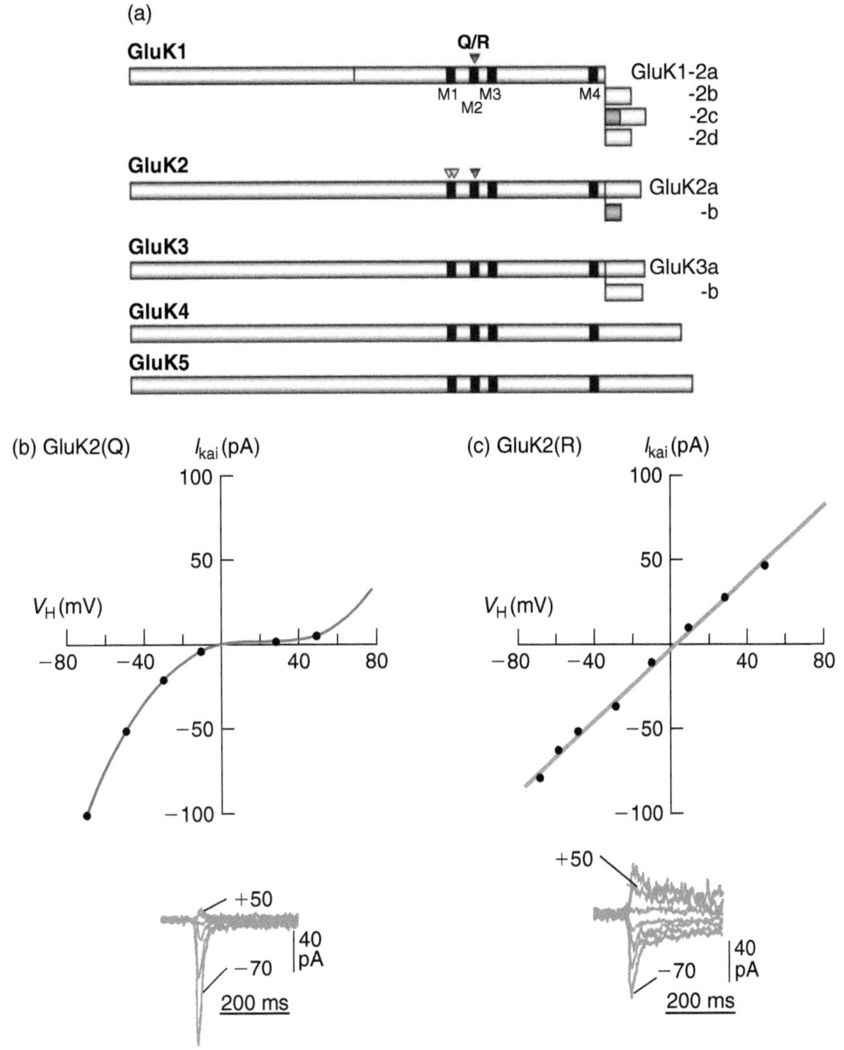

FIGURE 10.7 **Correlation of functional properties of native kainate receptors and RNA editing of the Q/R site of the GluR6. (a)** KAR subunits (GluK1–5) and splice variants. Black boxes represent membrane domains (M1–M4). Triangles depict sites of RNA editing, including the 'Q/R' site within both GluK1 and GluK2, which controls ion permeability of the channel. **(b)** Expressing homomeric GluK2(Q) and **(c)** expressing homomeric GluK2(R). Kainate (300 μM) is rapidly applied while holding the membrane potential at different voltages, from −70 to +50 mV. Insets show the current traces at these different voltages. Part (**a**) from Contractor A, Mulle C, Swanson GT (2011) Kainate receptors coming of age: milestones of two decades of research. *Trends Neurosci.* **34**, 154–163, with permission. Part (**b**) from Ruano D, Lambolez B, Rossier J *et al.* (1995) Kainate receptor subunits expressed in single cultured hippocampal neurons: molecular and functional variants by RNA editing. *Neuron* **14**, 1009–1017, with permission.

N + 1 site (see **Figure 10.11a**). Expression of modified subunits in *Xenopus* oocytes showed that these asparagines are crucial for the particular properties of divalent ion permeation of NMDA channels.

10.4.2 Native NMDA receptors have a high unitary conductance of 50 pS or a lower one of 17–35 pS depending on subunit composition

The experiments related in this section allowed the characterization of the electrophysiological properties of native NMDA channels. They were carried out before molecular cloning of NMDA receptor subunits had been achieved. As has already been pointed out, the NMDA channel is blocked by extracellular Mg^{2+} ions at voltages close to the resting potential of the cell. Mg^{2+} ions block the channel in the open state thus preventing the passage of other ions. The concentrations of Mg^{2+} that produce a significant block are similar to the concentrations of Mg^{2+} normally present in the extracellular milieu. For the sake of clarity, we shall first look at the conductance and permeability properties of the NMDA channel in the absence of extracellular Mg^{2+}. Subsequently, we shall consider the nature of the changes that occur in a medium containing Mg^{2+} ions.

Let us look at patch clamp recordings (outside-out configuration) of cultured central neurons in the absence of Mg^{2+} ions. The application of NMDA (10 μM) in the extracellular milieu at $V = -60$ mV induces an inward unitary current, i_N (**Figure 10.8a**). The i_N/V relation obtained under these conditions is linear (between −80 mV and +60 mV) and is described by the equation $i_N = \gamma_N (V - E_{rev})$ (**Figure 10.8b**). The slope of this curve corresponds to the unitary conductance of the NMDA channel, γ_N. The average value of γ_N is in the range 40–50 pS. The unitary conductance of NMDA channels is only slightly voltage dependent in the absence of Mg^{2+} ions.

We now know that the frequently described 50 pS openings are associated with GluN2A- or GluN2B-containing receptors whereas GluN2C- or GluN2D-containing receptors display low-conductance (17–35 pS) openings.

10.4.3 The NMDA channel is highly permeable to monovalent cations and to Ca^{2+}

The i_N/V relation shows that i_N reverses at a membrane potential value close to 0 mV when the extracellular and intracellular cationic concentrations are similar. This value suggests that the NMDA channel is permeable to cations. Replacing any of the monovalent cations (Na^+, K^+ or Cs^+) by another induces only minor changes in the reversal potential; i.e. the channel discriminates only slightly between the different monovalent cations.

In order to establish whether the NMDA channel is permeable to Ca^{2+} or not, two types of experiments have been performed. The first type of experiment consisted of photometric measurements (see Appendix 5.1) of the variations of intracellular Ca^{2+} in response to NMDA. To carry out these experiments and to prevent the activation of voltage-dependent Ca^{2+} channels, the activity of a spinal neuron is recorded in patch clamp (whole-cell configuration) at a holding potential of −60 mV (**Figure 10.9a**). Under these conditions, the intracellular Ca^{2+} concentration strongly increases during an NMDA-evoked response. This augmentation is selectively blocked by the NMDAR antagonist AP-5 or the NMDA channel blocker MK 801. Since it disappears in Ca^{2+}-free external solution, it clearly results from an influx of Ca^{2+} through NMDA channels and is not due to a release of these ions from intracellular storage pools of Ca^{2+}. Also,

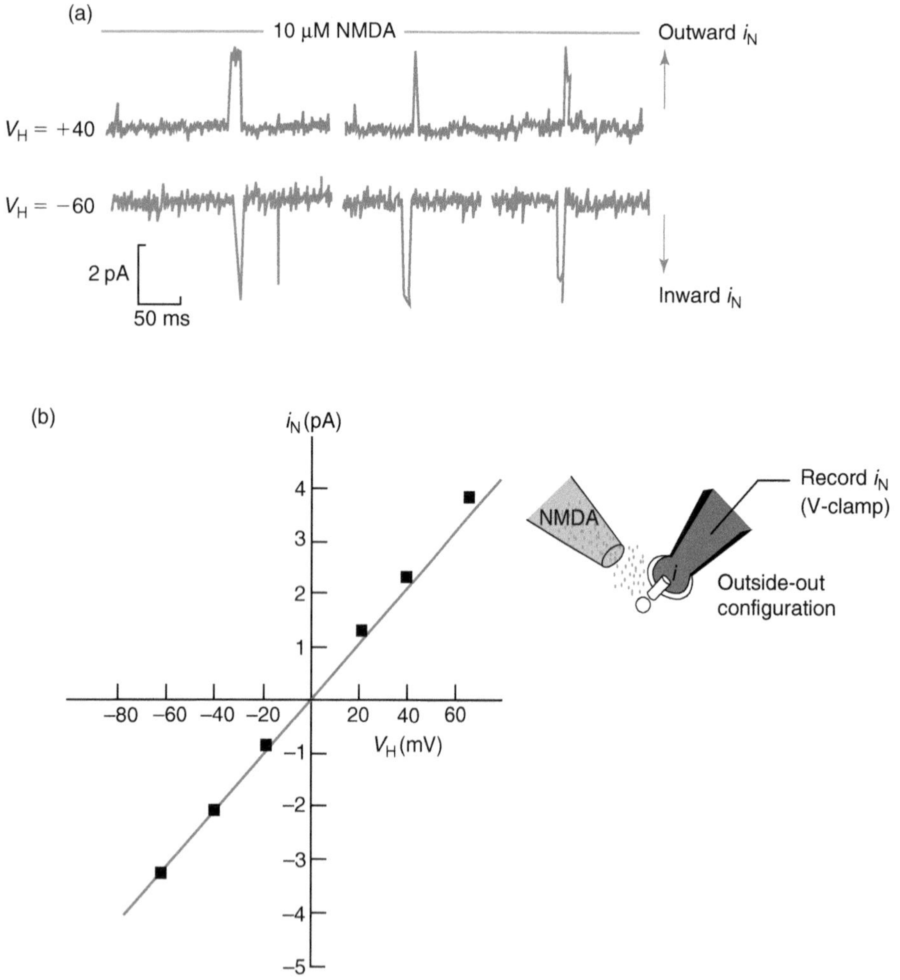

FIGURE 10.8 **Unitary NMDA current in the absence of extracellular Mg^{2+}.** **(a)** Outside-out patch clamp recordings of the activity of an NMDA (10 μM) activated channel at two holding potentials, −60 and +40 mV. **(b)** i_N/V relation obtained from the averages of unitary currents i_N recorded from a homogeneous population of channels (a population that shows a 40–50 pS unitary conductance). Intrapipette solution (in mM): 140 CsCl, 5K-EGTA, 0.5 $CaCl_2$; extracellular solution: 140 NaCl, 2.8 KCl, 1 $CaCl_2$. Part (**a**) from Ascher P, Bergestovski P, Nowak L (1988) *N*-methyl-D-aspartate-activated channels of mouse central neurons in magnesium free solutions. *J. Physiol. (Lond.)* **399**, 207–226, with permission. Part (**b**) from Cull-Candy SG, Usowicz MM (1987) Patch clamp recording from single glutamate-receptor channels. *TIPS* **8**, 218–224, with permission.

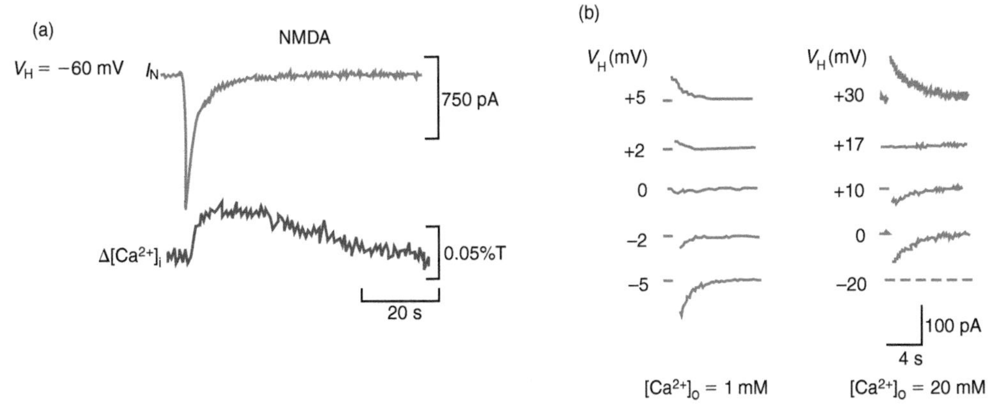

FIGURE 10.9 **Optical measurements of intracellular Ca^{2+} concentration changes during an NMDA-evoked response.** The Ca^{2+}-sensitive dye Arsenazo III is used. The absorption coefficient of this dye varies at certain wavelengths when it complexes Ca^{2+} ions. The activity of cultured spinal neurons is recorded in the whole-cell patch clamp configuration in the absence of external Mg^{2+}. **(a)** Pressure application of 1 μM of NMDA (20 ms) in the presence of 2.5 mM of Ca^{2+} in the extracellular milieu evokes an inward current I_N (top trace). During this response there is an increase of $[Ca^{2+}]_i$ (bottom trace). **(b)** Reversal potential of the whole-cell NMDA current (I_N) as a function of extracellular $[Ca^{2+}]_o$. Currents activated by the application of 1 mM of NMDA are recorded at different membrane potentials in the presence of 1 mM (left) or 20 mM (right) of $[Ca^{2+}]_o$. Part (**a**) from Mayer ML, MacDermott AB, Westbrook GL *et al.* (1987) Agonist- and voltage-gated calcium entry in cultured mouse spinal chord neurons under voltage clamp using Arsenazo III. *J. Neurosci.* **7**, 3230–3244, with permission. Part (**b**) from MacDermott AB, Mayer ML, Westbrook GL *et al.* (1986) NMDA-receptor activation increases cytoplasmic calcium concentration in cultured spinal neurons. *Nature* **321**, 519–522, with permission.

changes in the extracellular Ca^{2+} concentration are accompanied by changes in the reversal potential of the macroscopic NMDA current (**Figure 10.9b**). Furthermore, an inward single channel current is recorded at −60 mV under conditions where Ca^{2+} is the only cation present in the extracellular milieu, i.e. when Ca^{2+} ions are the only ions that can carry this current. All these results indicate that Ca^{2+} ions actually carry part of the NMDA current.

What does molecular biology tell us about Ca^{2+} permeability?

The molecular substrate for Ca^{2+} permeability of NMDA channels is analyzed by exchanging (by site-directed mutagenesis) either glutamine (Q) or arginine (R) for asparagine (N) in the M2 domain of GluN1 or GluN2A subunits (see **Figure 10.11a**). Wild-type and mutant GluN subunits are co-expressed by cells transfected with cDNAs. Whole-cell currents are activated by application of L-glutamate to transfected cells expressing heteromeric wild-type or mutant NMDA receptors and differences in Ca^{2+} permeability are analyzed. Replacing the asparagine (N) by arginine (R) in the GluN1 subunit generates 'mutant GluN1-wild-type GluN2A' channels that do not exhibit a measurable Ca^{2+} permeability (see **Figure 10.11b**). In high Ca^{2+} solution, glutamate evokes a small outward current at V_H = −60 mV, indicating a low Ca^{2+} permeability of the mutant channel. Thus, when the positively charged arginine (R) occupies the critical position in M2 of the GluN1 subunit, Ca^{2+} ions appear to be prevented from entering the channel, suggesting that the size and the charge of the amino acid present at this critical position in the M2 segment are important for Ca^{2+} permeability.

Models propose that the M2 segment (P-loop) forms the narrowest part of the pore toward the intracellular aspect of the channel, and creates the selectivity filter. M1, M3 and M4 residues participate in forming the large extracellular vestibule just external to the selectivity filter.

10.4.4 NMDA channels are blocked by physiological concentrations of extracellular Mg^{2+} ions; this block is voltage-dependent

Single NMDA channel current evoked by NMDA (10 μM) is recorded in the presence of increasing concentrations of extracellular Mg^{2+} ions (in μM: 0, 10, 50, 100) (**Figure 10.10a**). At V_H = −60 mV, NMDA evokes an inward unitary current whose amplitude remains constant at all the Mg^{2+} concentrations tested (i_N = −2.7 pA). However, whereas in the absence of Mg^{2+} the NMDA channel opens for periods of several milliseconds (0), in the presence of Mg^{2+} (10, 50, 100 μM), the recordings show bursts of short openings during which the channel fluctuates between the open state (mean duration t_o) and blocked state (mean duration t_{bl}). At an Mg^{2+} concentration of 100 μM, the unitary current appears to decrease. This is due to the high Mg^{2+} blocking frequency which makes channel openings too brief to be resolved by the recording system. The repeated closures of the channel (which correspond to fluctuations between the open and the blocked states) strongly diminish the average time during which the channel is open. Note that when the recorded unitary current is outward at $V_H = +40$ mV, the presence of Mg^{2+}, even at a concentration of 100 μM, has no effect on the channel open time. The most interesting aspect of this block is its voltage dependence. Mg^{2+} block becomes progressively stronger as voltage is made more and more hyperpolarized while at positive voltages (where the current i_N is outward) block by physiological Mg^{2+}concentrations is hardly visible.

The properties of Mg^{2+} block of unitary currents can be used to predict macroscopic currents (whole-cell configuration). Recall that $I_N = Np_o i_N$. We know that i_N is approximately constant at a given membrane potential irrespective of the Mg^{2+} concentration, and that the mean open time of each channel (and therefore the open state probability p_o of the channel as well) decreases as a function of the Mg^{2+} concentration. From the recordings in **Figure 10.10a** we can predict that at negative potentials, in the presence of Mg^{2+}, the macroscopic current I_N will be small.

As a matter of fact, the I_N/V curve described by the equation $I_N = G_N(V - E_{rev})$ in the presence of extracellular Mg^{2+} ions (500 μM) is not linear. This non-linearity appears at negative voltages; i.e. when the current is inward (**Figure 10.10b**). Furthermore, we observe that for V_H between −35 and −80 mV, the current amplitude diminishes (region of negative conductance) instead of increasing, as would be expected from the increasing electrochemical gradients for Na^+ and Ca^{2+} and the decreasing electrochemical gradient for K^+. This peculiar property of the I_N/V curve in this region of voltage is due to the block of the channel by Mg^{2+}. Since Mg^{2+} ions are normally found in the extracellular milieu at concentrations of approximately 1 mM, at membrane potentials close to the resting potential a majority of NMDA channels are blocked.

Mechanism of action of Mg^{2+} ion block: a hypothesis

Mg^{2+} ions block open NMDA channels, thus preventing the passage of Na^+, Ca^{2+} and K^+ ions. The probability that an Mg^{2+} ion will enter the NMDA channel increases

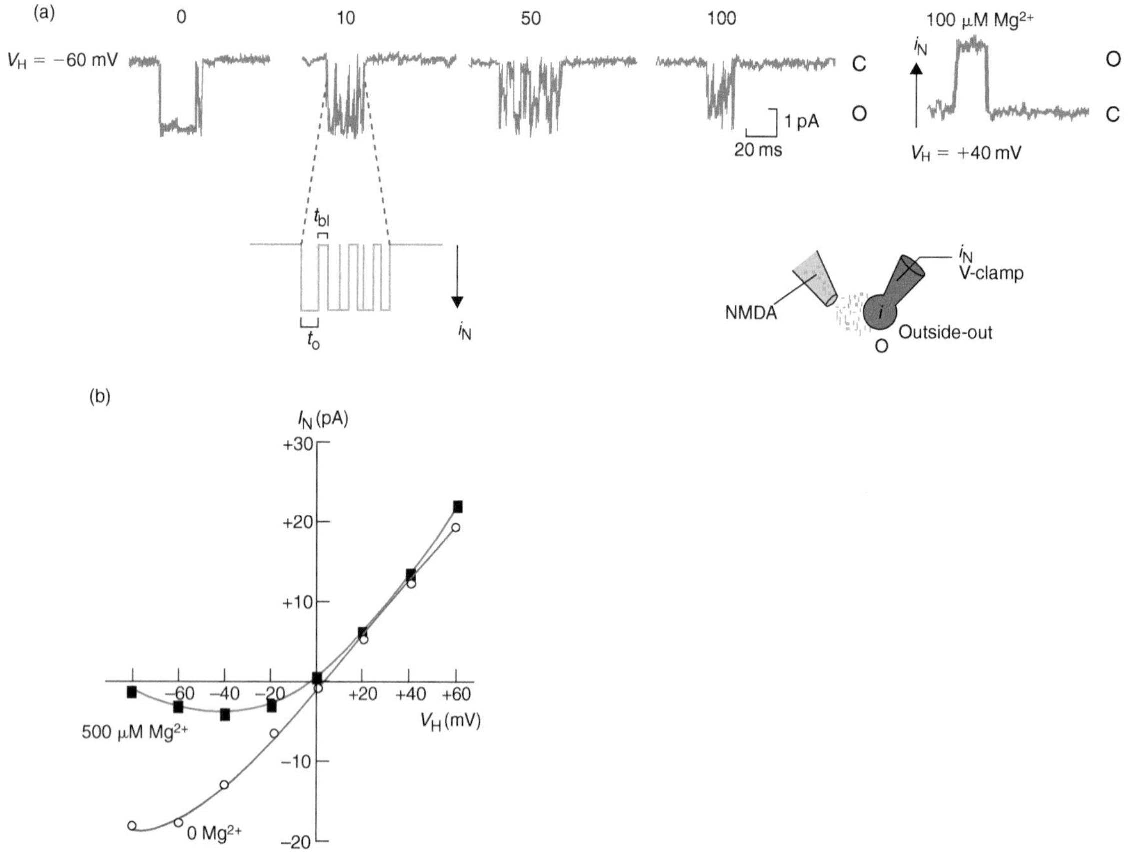

FIGURE 10.10 **NMDA channel block by extracellular Mg^{2+} ions. (a)** Outside-out patch clamp recording of the activity of cultured central neurons. Application of NMDA (10 μM) in the absence of external Mg^{2+} ions (O) and in the presence of Mg^{2+} (10, 50, 100 μM) at $V_H = -60$ mV. At $V_H = +40$ mV, i_N is outward. **(b)** Voltage sensitivity of the NMDA response in the presence of extracellular Mg^{2+} ions. The total current I_N is recorded in the whole-cell patch clamp configuration in the absence (o) and in the presence (■) of 500 μM of Mg^{2+}. Part **(a)** from Ascher P, Nowak L (1988) The role of divalent cations in the *N*-methyl-D-aspartate responses of the mouse central neurons in culture. *J. Physiol. (Lond.)* **399**, 247–266, with permission. Part **(b)** from Nowak L, Bregestovski P, Ascher P *et al.* (1984) Magnesium free glutamate-activated channels in mouse central neurones. *Nature* **307**, 463–465, with permission. C – Closed state; O – Open state.

with the level of membrane hyperpolarization: the greater the electrical gradient, the stronger are the Mg^{2+} ions attracted into the channel. For this reason, the block of NMDA channels by Mg^{2+} is voltage sensitive. This block can be symbolized as follows:

$$R + NMDA \rightleftharpoons NMDA - R* \xrightleftharpoons{+Mg^{2+}} NMDA - R* - Mg$$

where R is the NMDA receptor in the closed state, R* is the NMDA receptor in the open state, and R*–Mg is the open NMDA receptor blocked by Mg^{2+} ions. The reaction $R* + Mg^{2+} \rightleftharpoons R* - Mg$ is strongly favored to the right when $[Mg^{2+}]$ is increased and when V is hyperpolarized.

Why is the NMDA channel permeable to Ca^{2+} ions and blocked by Mg^{2+} ions?

One can separate the effects of cations into two groups:

- those, like Ca^{2+}, which pass through the NMDA channel (e.g. Ba^{2+}, Cd^{2+});
- those which mimic the Mg^{2+} effect, i.e. block the NMDA channel (e.g. Co^{2+}, Ni^{2+}, Mn^{2+}).

The difference between the ions that pass through the channel and those that block it coincides with the difference in the speed with which the water molecules surrounding these ions can exchange with other water molecules of the aqueous solution. In fact, this exchange is a thousand times faster for the group of permeable (Ca^{2+}-like) ions than for the group of blocking (Mg^{2+}-like) ions.

These differences have led to the suggestion that both ions can cross the channel but only in their dehydrated forms. The following model of the channel has been proposed. The channel has a large extracellular entrance and presents a narrow constriction towards the intracellular side through which the ions can cross only in their dehydrated form. The narrow constriction is formed by non-homologous asparagine residues: the N site of the GluN1 subunit and an adjacent one, to the N site in GluN2, the N + 1 asparagine (**Figure 10.11a**). The size of hydrated Mg^{2+}, at least 0.7 nm, is larger than the estimated 0.55 nm pore size of NMDA receptor channels suggesting that Mg^{2+} block at the narrow constriction could arise by steric occlusion. Because of the slow rate of dehydration of the cations from the Mg^{2+} group, these ions are trapped in the interior of the channel, thus blocking it. Another hypothesis for the blockade by Mg^{2+} ions is that a high affinity binding site for Mg^{2+} ions exists inside the channel so that Mg^{2+} ions would cross the channel slowly and thus block it for the time of passage.

What does molecular biology tell us about Mg^{2+} block?

The molecular substrate for the hypothesis of a high affinity site in the NMDA channel for Mg^{2+} ion binding is analyzed by exchanging (by site-directed mutagenesis) either glutamine (Q) or arginine (R) for asparagine (N) in the M2 domain of GluN1 or GluN2A. Wild-type and mutant GluN subunits are co-expressed by cells transfected with cDNAs. Whole-cell currents are activated by application of L-glutamate to transfected cells expressing heteromeric wild-type or mutant NMDA receptors, and differences in Ca^{2+} or Mg^{2+} permeability and channel block by extracellular Mg^{2+} are analyzed. Replacing the asparagine (N) by arginine (R) in the GluN1 subunit generates 'mutant GluN1-wild-type GluN2A' channels that do not exhibit a measurable Ca^{2+} permeability (**Figure 10.11b**), as already pointed out. Whole-cell I_N/V curves in (1) divalent ion-free external solution and (2) after adding 0.5 mM of Mg^{2+} to the external solution superimpose almost completely, showing that the mutant channel is not blocked by external Mg^{2+} ions (**Figure 10.11c**). Conversely, the presence of glutamine (Q) instead of asparagine (N) in the GluN2A subunit generated 'wild-type GluN1-mutant GluN2A' channels with increased Mg^{2+} permeability and thus reduced sensitivity to block by extracellular Mg^{2+}. The whole-cell current evoked by glutamate is recorded in (1) divalent ion-free Ringer and (2) after addition of 0.1 mM of Mg^{2+} as a function of membrane potential. The I_N/V relations show that the wild-type channel consisting of GluN1 and GluN2A subunits (**Figure 10.11d**) is blocked by external Mg^{2+}, while the mutant channel comprising wild-type GluN1 and mutant GluN2A (N595Q) subunits is permeable to Mg^{2+} since it is only slightly blocked by external Mg^{2+} (**Figure 10.11e**). Moreover, substitutions of the two adjacent asparagines in the GluN2A subunit strongly reduce the block. These effects show little dependence on pore size thus suggesting that the block does not arise by steric occlusion. In summary, the asparagines are critical in all native NMDAR subtypes for Ca^{2+} permeability and Mg^{2+} block.

Why do channel properties fundamental to NMDAR function (Mg^{2+} block, Ca^{2+} permeability, and single channel conductance) vary among NMDAR subtypes?

The four GluN2 subunits (GluN2A–D) contribute to four diheteromeric NMDAR subtypes, GluN1/2A, GluN1/2B, GluN1/2C and GluN1/2D, that have different channel properties. GluN1/2A and GluN1/2B receptors exhibit higher Mg^{2+} affinities (4–5-fold) and higher single-channel conductances (≈50 pS) than do GluN1/2C and GluN1/2D receptors (which exhibit lower Mg^{2+} affinities and lower single-channel conductances ≈35 pS). Likewise, GluN1/2A and GluN1/2B receptors have higher Ca^{2+} permeabilities than do GluN1/2C and GluN1/2D receptors. To understand these differences, sites in the M1–M4 regions were targeted for mutagenesis. The authors selected sites containing the same residue in GluN2A and GluN2B subunits, but a different

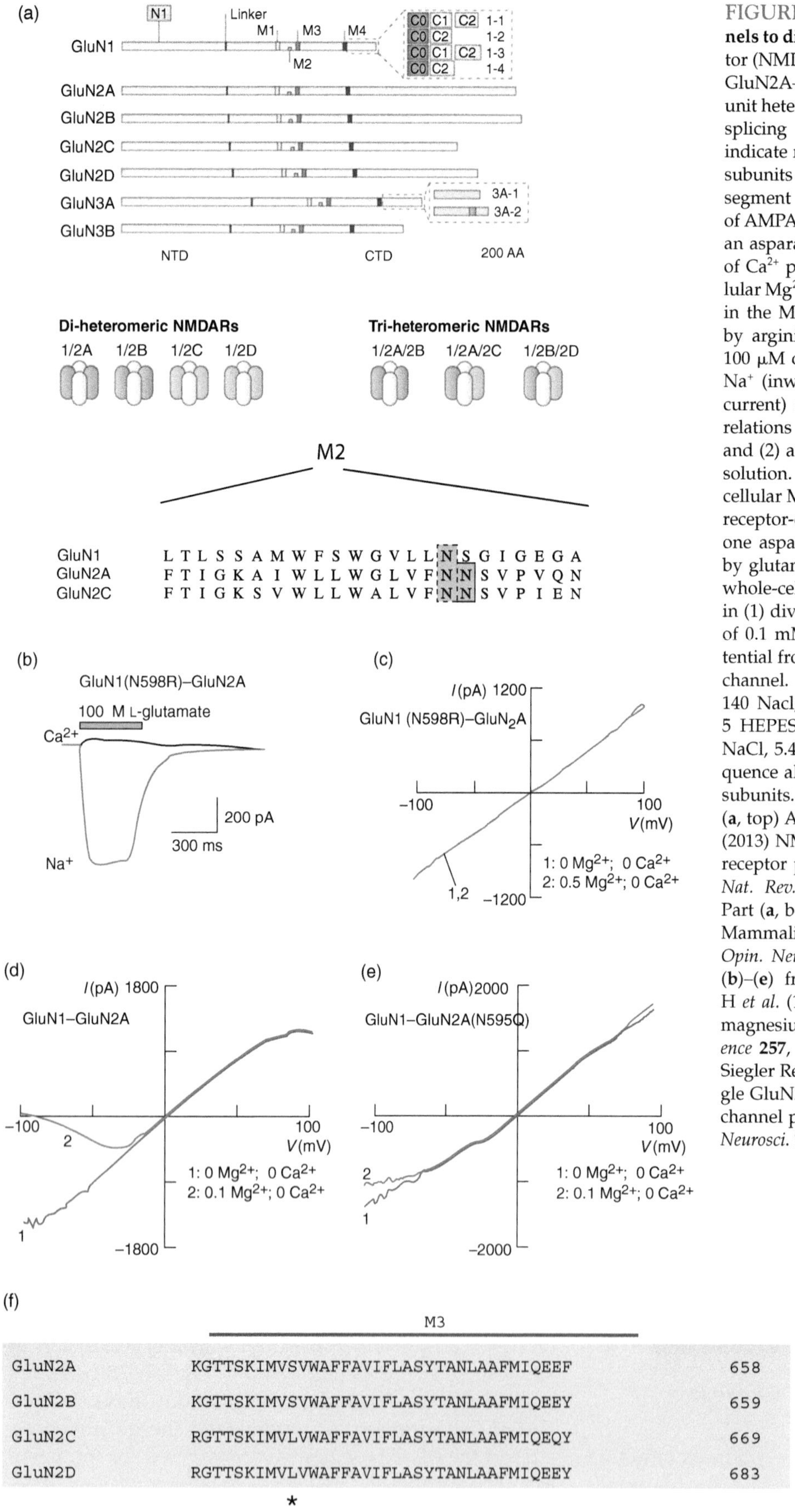

FIGURE 10.11 **Permeability of NMDA channels to divalent cations.** (**a**, top) Seven NMDA receptor (NMDAR) subunits have been identified: GluN1, GluN2A–GluN2D and GluN3A and GluN3B. Subunit heterogeneity is further enhanced by alternative splicing of GluN1 and GluN3A subunits. M1–M4 indicate membrane segments. (**a**, bottom) The GluN subunits carry an asparagine residue (N) in the M2 segment in a position homologous to the Q/R site of AMPAR and KAR. The GluN2 subunits also carry an asparagine residue at site N + 1. (**b,c**) Reduction of Ca^{2+} permeability and channel block by extracellular Mg^{2+} in a mutant channel where asparagine (N) in the M2 segment of the NR1 subunit is replaced by arginine (R). **(b)** Whole-cell current elicited by 100 μM of glutamate (bar) at $V_H = -60$ mV in high Na^+ (inward current) or high Ca^{2+} (small outward current) extracellular solution. **(c)** Whole-cell I_N/V relations in (1) divalent ion-free external solution and (2) after adding 0.5 mM of Mg^{2+} to the external solution. (**d,e**) Difference in channel block by extracellular Mg^{2+} between wild-type and mutant NMDA receptor-channels. In the GluN2A (N595Q) subunit, one asparagine (N) in the M2 segment is replaced by glutamine (Q) by site-directed mutagenesis. The whole-cell current evoked by glutamate is recorded in (1) divalent ion-free Ringer and (2) after addition of 0.1 mM of Mg^{2+} as a function of membrane potential from **(d)** wild-type channels and **(e)** a mutant channel. Extracellular high Na^+ solution (in mM): 140 Nacl, 5 HEPES: high Ca^{2+} solution: 110 $CaCl_2$, 5 HEPES. Divalent ion-free Ringer's solution: 135 NaCl, 5.4 KCl, 5 HEPES. (**f**) Amino acid residue sequence alignment of the M3 segment of GluN2A–D subunits. The asterisk marks the GluN2 S/L site. Part (**a**, top) Adapted from Paoletti P, Bellone C, Zhou Q (2013) NMDA receptor subunit diversity: impact on receptor properties, synaptic plasticity and disease. *Nat. Rev. Neurosci.* **14**, 383–400, with permission. Part (**a**, bottom) from Wisden W, Seeburg PH (1993) Mammalian ionotropic glutamate receptors. *Curr. Opin. Neurobiol.* **3**, 291–298, with permission. Parts (**b**)–(**e**) from Burnashev N, Schoepfer R, Monyer H *et al.* (1992) Control of calcium permeability and magnesium blockade in the NMDA receptor. *Science* **257**, 1415–1419, with permission. Part (**f**) from Siegler Retchless B, Gao W, Johnson JW (2012) A single GluN2 subunit residue controls NMDA receptor channel properties via intersubunit interaction. *Nat. Neurosci.* **15,** 406–413, with permission.

residue in both GluN2C and GluN2D subunits. Then, in GluN2A subunits, they substituted the wild-type residue with the residue found in GluN2C and GluN2D subunits.

They thus identified a strategic serine/leucine (S/L) site in the GluN2 M3 transmembrane segment (**Figure 10.11f**). NMDARS in which the GluN2C and GluN2D leucine (L) was substituted in place of the GluN2A and GluN2B serine (S) at GluN2A(S632) in M3, resulted in Mg^{2+} IC_{50} that were strikingly similar to GluN1/2D receptor Mg^{2+} IC_{50}. The authors performed similar measurements to evaluate the Ca^{2+} permeability of GluN1/2A(S632L) and GluN1/2D(L657S) receptors. The P_{Ca}/P_{Cs} ratio for GluN1/2A(S632L) receptors is lower than the ratio for GluN1/2A receptors, and the P_{Ca}/P_{Cs} ratio for GluN1/2D(L657S) receptors is higher than the ratio for GluN1/2D receptors. Thus, the GluN2 S/L site regulates the NMDAR subtype dependence of selective permeability to Ca^{2+} as well as Mg^{2+} IC_{50}.

Finally, the authors investigated the single channel conductance with the same strategy. The main state conductance of wild-type GluN1/2A receptors is around 55 pS, whereas that of GluN1/2A(S632L) receptors is around 35 pS, significantly different from the GluN1/2A receptor main state conductance, but not significantly different from that of GluN1/2D receptors (around 37 pS). The GluN1/2D(L657S) receptor main state conductance is around 55 pS, significantly different from the GluN1/2D receptor main state conductance, but not significantly different from that of GluN1/2A receptors.

How does the GluN2 S/L site transmit its effects to the pore? Due to its location at the base of the M3 region, the GluN2 S/L site is unlikely to directly interact with permeant or blocking ions. This led to the hypothesis that the GluN2 S/L site exerts its influence over Mg^{2+} block, Ca^{2+} permeability, and single-channel conductance through interactions with amino acid residues that are closer to the pore. Using structural homology models, the authors predicted that the GluN2 S/L site conveys its effects on the pore by interacting with the P-loop of the adjacent GluN1 subunit. Results from mutant cycle experiments support the model's prediction, identifying GluN1(W608) as a P-loop residue coupled to the GluN2 S/L site.

All the above results show that the different Mg^{2+} inhibition, Ca^{2+} permeability and single-channel conductance of NMDAR subtypes all depend on a naturally occurring amino acid residue substitution at the GluN2 S/L site in the M3 segment (**Figure 10.11f**). Nevertheless, the asparagines in the M2 segment (**Figure 10.11a–e**) are critical (in all NMDAR subtypes) for Ca^{2+} permeability and Mg^{2+} block. Mutations at the GluN2 S/L site do not eliminate Mg^{2+} inhibition – the maximal change in IC_{50} observed when the S/L site residue is mutated to tryptophan is ≈10-fold. In contrast, critical asparagine mutations can eliminate Mg^{2+} block (**Figure 10.11c–e**).

10.4.5 Glycine is a co-agonist of NMDA receptors

Glycine is an amino acid that acts as an inhibitory neurotransmitter at certain central nervous system synapses in vertebrates. However, this amino acid also plays a role in modulating the NMDA response. The effect of glycine on the NMDA receptor channel was discovered in cultured central neurons recorded in the whole-cell patch clamp configuration. When NMDA was applied by slow perfusion the response was much larger than when NMDA was rapidly perfused into the bath. The following hypothesis was proposed. The cultured cells (neurons and glia) tonically release a substance that accumulates in the bath due to the slow perfusion and potentiates the NMDA response. In order to characterize the active substance present in the medium, a variety of treatments was applied. It was established that its activity was still present after heating the medium to 90°C, and that its molecular weight is below 700 Da. After testing the most common amino acids, glycine proved to be the most effective in reproducing the effects of the conditioning medium on the NMDA response (**Figure 10.12a**).

Patch clamp outside-out recordings showed that glycine potentiates the NMDA response by augmenting the NMDA receptor channel opening frequency (thus increasing the open state probability of the channel, p_o). The molecular mechanisms of this potentiating effect remain to be determined. Nevertheless, the fact that glycine has an effect on NMDA receptor-channels recorded from excised outside-out patches rules out the mediation of its effect by a diffusible, intracellular second messenger. It was then suggested that glycine would in fact be indispensable for the activation of the NMDA receptor channel by its agonists. When the activity of NMDA receptor channels expressed in oocytes is recorded in the whole-cell patch clamp configuration, a current in response to NMDA application is observed only when glycine is also present in the bath (**Figure 10.12b**).

How can these results be interpreted from a physiological perspective? Glycine is in fact present at relatively high concentrations in the cerebrospinal fluid (several μM). This level is close to the concentration required to produce its maximum effect. However, a high-affinity glycine pump may lower extracellular glycine concentration at the level of glutamatergic synapses. These transporters, via the modulation of extracellular glycine concentration, could play a significant role in determining the NMDA response.

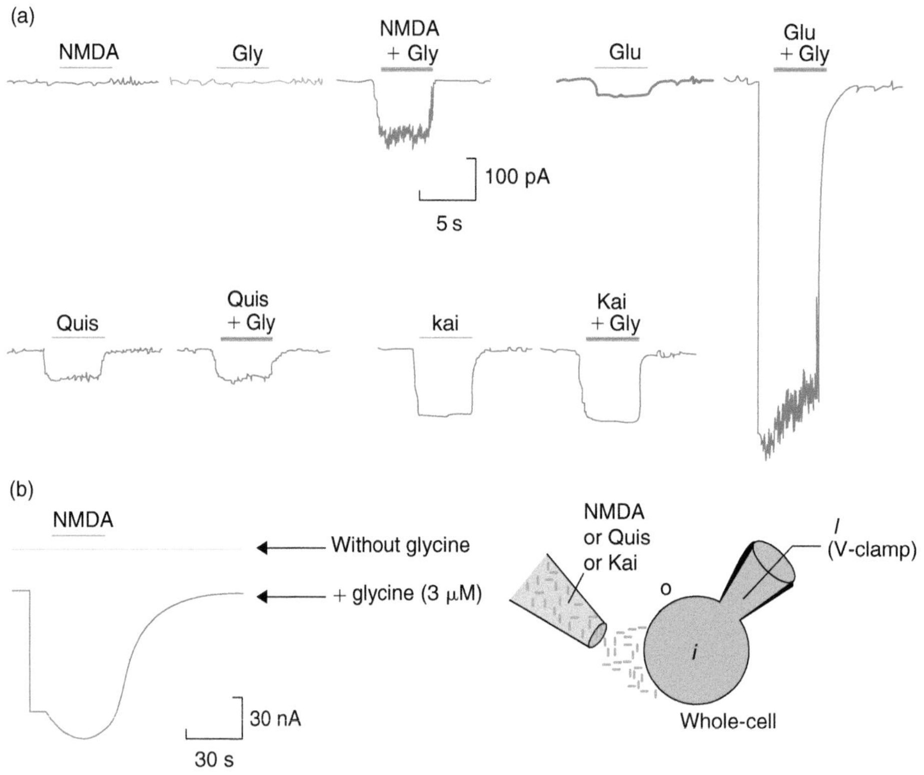

FIGURE 10.12 **Potentiation of the NMDA response by glycine.** (**a**, top) Whole-cell currents evoked in cultured central neurons in response to 10 μM of NMDA or 10 μM of glutamate at $V_H = -50$ mV in the absence or presence of 1 μM of glycine (Gly). (**a**, bottom) The same experiment with quisqualate (Quis) or kainate (Kai) applications. Glycine by itself does not trigger an inward current at any concentration, through either NMDA or non-NMDA channels. (**b**) Whole-cell inward current (66 ± 13 nA) evoked in *Xenopus* oocytes which express NMDA receptors, in response to 300 μM of NMDA at $V_H = -60$ mV in the absence of presence of 3 μM of glycine. In **(a)** and **(b)** the extracellular solution is devoid of Mg^{2+} ions. Part (**a**) from Johnson JW, Ascher P (1987) Glycine potentiates the NMDA response in cultured mouse brain neurons. *Nature* **325**, 529–531, with permission. Part (**b**) from Kleckner N, Dingledine R (1988) Requirements for glycine in activation of NMDA receptors expressed in *Xenopus* oocytes. *Science* **241**, 835–837, with permission.

10.4.6 Conclusions on NMDA receptors

The NMDA receptor-channels are unique among the glutamate receptors since:

- Their ion channel is subject to a voltage-dependent block by Mg^{2+};
- They are highly Ca^{2+}-permeable;
- They display unusually slow kinetics owing to slow glutamate unbinding;
- Their activation requires the presence not only of glutamate but also of a co-agonist (glycine or D-serine); glutamate binds to GluN2A–D subunits while glycine binds to GluN1 subunits.
- They are equipped with an array of modulatory sites conferring an exquisite sensitivity to the extracellular microenvironment.

NMDA receptors require at least two conditions for their activation: the binding of both glutamate and the co-agonist glycine to the receptor *and* the depolarization of the membrane. It should be noted that the voltage sensitivity of the NMDA channels (due to an extrinsic ion, namely Mg^{2+}) differs radically from that of voltage-dependent Na^+ and Ca^{2+} channels. In the two latter cases, the voltage sensitivity is an intrinsic property of the protein which does not require extracellular or intracellular ligands.

Di-heteromeric GluN1-GluN2A or GluN1-GluN2B receptors generate 'high-conductance' channel openings (≈50 pS) with high sensitivity to Mg^{2+} blockade (half-maximal inhibitory concentration (IC_{50}) of ≈15 μM at −70 mV) and high Ca^{2+} permeability (*p*Ca/*p*Cs of ≈7.5) compared to di-heteromeric GluN1-GluN2C or GluN1-GluN2D receptors which generate 'low-conductance' channel openings (17–35 pS) with a lower sensitivity to Mg^{2+} blockade (IC_{50} of ≈80 μM at −70 mV) and a lower Ca^{2+} permeability (*p*Ca/*p*Cs of ≈4.5). These marked differences are all controlled by a single GluN2 residue in the M3 segment.

The voltage sensitivity of the NMDA channel has important physiological implications. Since these channels are blocked by Mg^{2+} ions at voltages close to the resting potential of the cell, does the presence of the neurotransmitter in the synaptic cleft suffice to evoke a postsynaptic NMDA response? Mg^{2+} channel block of NMDARs inhibits current influx through the majority of agonist-bound, open NMDARs at resting membrane potentials, but this block is relieved by depolarization. Thus, substantial current flow through NMDARs requires coincident presynaptic activity (glutamate release) and postsynaptic activity (depolarization to relieve Mg^{2+} channel block), conferring on NMDARs a coincidence detection capability. Knowing that the non-NMDA and the NMDA receptors coexist in the postsynaptic membrane, what is the consequence of non-NMDA receptor-induced depolarization on the activation of NMDA receptors? These questions are analyzed in the following section.

10.5 SYNAPTIC RESPONSES TO GLUTAMATE ARE MEDIATED BY NMDA AND NON-NMDA RECEPTORS

10.5.1 Glutamate receptors are co-localized in the postsynaptic membrane of glutamatergic synapses

The glutamatergic synapses typically exhibit an electron-dense postsynaptic density where ionotropic glutamate receptors are concentrated as shown by immunogold labeling. To date, among iGluRs, only NMDA and AMPA receptors can be labeled for electron microscopy observation. In contrast to iGluRs, metabotropic glutamate receptors (mGluRs) appear to occur at highest concentrations in the perisynaptic annulus; i.e. the narrow zone surrounding the postsynaptic specialization (**Figure 10.13**). The receptors localized in the postsynaptic specialization are directly apposed to the presynaptic active zone. The enrichment of iGluRs at the site of the postsynaptic specialization reflects their roles in mediating fast glutamatergic transmission. Such enrichment depends on anchoring synaptic proteins (see Chapter 8).

Glutamate released into the synaptic cleft diffuses to the postsynaptic membrane and binds to postsynaptic NMDA as well as non-NMDA receptors (**Figure 10.13**) and evokes a synaptic response which is a postsynaptic excitatory current (EPSC). In turn, this EPSC depolarizes the membrane; i.e. it evokes an excitatory postsynaptic potential (EPSP; see **Figure 10.3**). *In vivo*, when the membrane potential is near the resting potential of the cell, a large fraction of the NMDA receptors are blocked by Mg^{2+} ions present in the synaptic cleft. Therefore, glutamate first activates non-NMDA receptors.

- Is the synaptic response to glutamate mediated by non-NMDA receptors only?
- If not, under which conditions are NMDA receptors activated by synaptically released glutamate and how do they contribute to the synaptic response?

To answer these questions it is necessary to differentiate, in the postsynaptic current, between the component due to the activation of non-NMDA and that due to NMDA receptors (**Figure 10.13**, inset).

10.5.2 The glutamatergic postsynaptic current is inward and can have at least two components in the absence of extracellular Mg^{2+} ions

A global postsynaptic inward current (excitatory postsynaptic current; EPSC), evoked by the stimulation of a presynaptic glutamatergic neuron, is recorded in the whole-cell configuration (voltage clamp mode). When the presynaptic neuron is stimulated in the absence of extracellular Mg^{2+} ions and the postsynaptic membrane is held at −46 mV, an inward current showing two components is recorded (**Figure 10.14a**):

- an early component: an initial peak of current of great amplitude and rapid inactivation;
- a late component: a current of smaller amplitude that inactivates slowly.

To identify the NMDA and non-NMDA components of the postsynaptic current we can make use of the different properties of the non-NMDA and the NMDA channels summarized in **Figure 10.1a**. The non-NMDA component is not affected by different concentrations of extracellular Mg^{2+} ions, nor by the presence of APV, but disappears in the presence of CNQX, an antagonist of non-NMDA receptors. On the other hand, the NMDA component is present in a medium devoid of Mg^{2+} ions but disappears in the presence of APV, the competitive antagonist of NMDA receptors.

In the presence of Mg^{2+} ions, the late component is largely attenuated at negative potentials but is present at all positive voltages (**Figure 10.14b**, black square). In the absence of Mg^{2+} ions and the presence of APV in the extracellular environment (**Figure 10.14c**), the late component disappears at all voltages tested. These results strongly suggest that the late component of the synaptic current results from the activation of NMDA receptors (APV-sensitive and blocked by Mg^{2+} at negative voltages). The reverse experiment in the presence of CNQX is not shown.

Which receptor channels contribute to the non-NMDA component of the synaptic current?

The relative participation of AMPA and kainate receptors in the non-NMDA component of the synaptic glutamatergic current required the discovery of selective antagonists as CNQX is not selective of either one of these receptors. Only after the fortuitous discovery that the 2,3-benzodiazepine muscle relaxant GYKI 53655 is a specific AMPA antagonist (see **Figure 10.1a**) could physiologists begin to distinguish

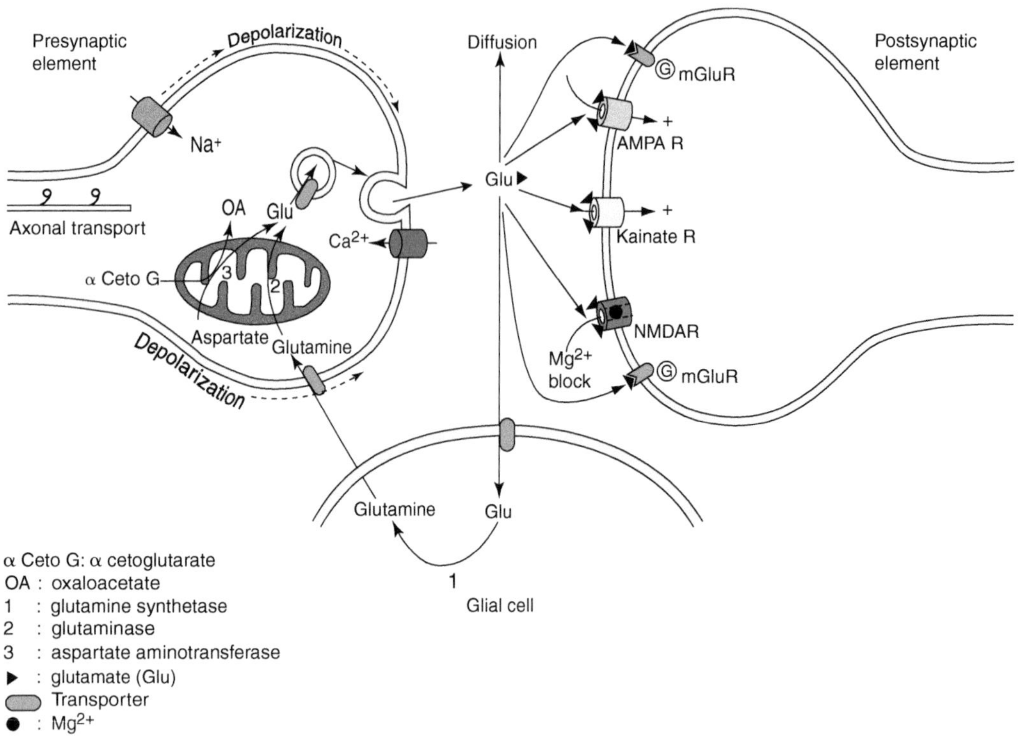

FIGURE 10.13 **The glutamatergic synapse.** Functional scheme of a glutamatergic synapse where ionotropic and metabotropic glutamate receptors are co-localized. Presynaptic receptors are omitted. The enzymes (1 to 3) and mitochondria are carried to axon terminals via anterograde axonal transports. Glutamate synthesized in mitochondria of the presynaptic element is transported actively into synaptic vesicles by a vesicular carrier. A percentage of the glutamate released in the synaptic cleft is uptaken into presynaptic terminals and glial cells by transporters. Inset shows iGluRs antagonists. Inset from Mody I (1998) Interneurons and the ghost of the sea. *Nat. Neurosci.* **1**, 434–436, with permission.

between the separate activation of AMPA and kainate receptors at synapses. In many glutamatergic synapses, the non-NMDA postsynaptic current results solely from AMPA receptors. But there are few places where synaptically activated kainate receptors were identified. One of them is the CA3 region of the hippocampus where pyramidal neurons and interneurons receive excitatory input from glutamatergic mossy fibers (see **Figure 7.3**).

EPSCs evoked by the stimulation of afferents are recorded from interneurons in the whole-cell configuration (voltage clamp mode) in the continuous presence of D-APV to block NMDA receptors. A large, rapidly decaying control EPSC with a small, long-lasting tail is recorded (**Figure 10.15a**). Application of GYKI, the AMPA antagonist, blocks the rapid component but the slow component is mostly unaffected. The subsequent addition of the AMPA/kainate antagonist CNQX blocks the GYKI-resistant slow component. When the synapse between afferent glutamatergic fibers and pyramidal CA1 neurons is now studied, no GYKI-resistant (kainate-mediated) component can be shown, although the EPSC in pyramidal cells is more than twice as large as the EPSC in interneurons (**Figure 10.15b**). This indicates that, at this synapse in CA1, kainate receptors are absent and that the non-NMDA component of the EPSC is mediated only by AMPA receptors.

10.5.3 The glutamatergic postsynaptic depolarization (EPSP) has at least two components in the absence of extracellular Mg^{2+} ions

The postsynaptic response to the stimulation of the presynaptic neurons (same preparation as above) is recorded in whole-cell configuration (current clamp mode to leave the voltage free to vary) in the absence of Mg^{2+} ions. Such stimulation evokes a postsynaptic depolarization (EPSP), which results from the evoked synaptic inward current through postsynaptic ionotropic glutamate receptors. As in the case of the EPSC, the EPSP shows two identifiable components (see **Figure 10.3b**). In the presence of APV, the early component is only slightly affected or not at all. This APV-insensitive early component is a result of the early synaptic inward current; i.e. the inward current through non-NMDA receptors. In these conditions (absence of Mg^{2+} ions, presence of APV), the duration of the EPSP is reduced. The difference between the APV-insensitive component of the EPSP and the total EPSP corresponds to the APV-sensitive component;

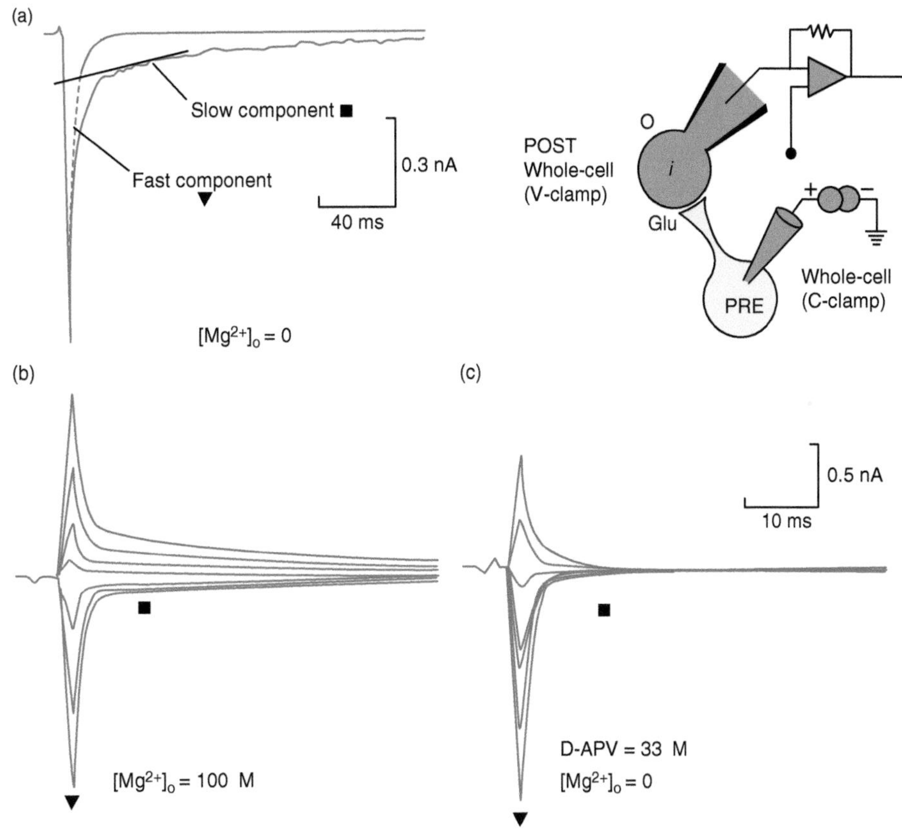

FIGURE 10.14 **Postsynaptic inward current evoked by the stimulation of a glutamatergic presynaptic neuron. (a)** Whole-cell postsynaptic inward current (EPSC) recorded $V_H = -46$ mV in the absence of Mg^{2+}, in response to the activation of a presynaptic glutamatergic neuron. The peak current decays with a time constant $\tau_1 = 4.2$ ms and the slow components decays with a time constant $\tau_2 = 81.8$ ms. **(b)** The EPSC is recorded at different V_H in the presence of Mg^{2+} (100 μM). **(c)** The EPSC is recorded at different V_H in the presence of 33 μM of D-APV and in the absence of extracellular Mg^{2+}. Picrotoxin (10–100 μM) is added to the extracellular solution to block GABAergic inhibitory synaptic activity. From Forsythe ID, Westbrook GL (1988) Slow excitatory postsynaptic currents mediated by N-methyl-D-aspartate receptors on cultured mouse central neurons. *J. Physiol. (Lond.)* **396**, 515–533, with permission.

i.e. the component resulting from the inward current through NMDA receptors.

In summary, the component resulting from the activation of the NMDA receptors has a slower rising phase and lasts longer than the component mediated by the non-NMDA receptors. Thus, when the NMDA receptors are activated, the peak of the EPSP is not always affected but the duration of the EPSP is much longer. As for EPSCs, the non-NMDA component of glutamatergic EPSPs is mediated either by AMPA receptors alone or by both AMPA and kainate receptors (see **Figure 10.3c**).

10.5.4 Synaptic depolarization recorded in physiological conditions: factors controlling NMDA receptor activation

The recordings of **Figure 10.14** have shown the presence of two components in the synaptic current and depolarization when the extracellular medium is Mg^{2+}-free. What is the situation in physiological conditions when the extracellular physiological milieu has an Mg^{2+} concentration of approximately 1 mM? Since at this Mg^{2+} concentration and at membrane potentials close to the resting potential of the cell, most of the NMDA channels are closed, under which conditions will NMDA receptors participate to synaptic transmission?

It seems unlikely that the extracellular Mg^{2+} concentration *in vivo* will vary sufficiently to allow the 'unblocking' of NMDA receptors. However, depolarizations reduce the level of Mg^{2+} block of the NMDA channel. Thus, one can imagine that a depolarization of the membrane is precisely what allows the NMDA channels to become 'unblocked'. A depolarization can be the consequence of the activation of other receptors present in the postsynaptic membrane, such as non-NMDA receptors. It can also result from the activation of a subpopulation of NMDA receptors that are not blocked at the resting potential. This hypothesis can be summarized as follows.

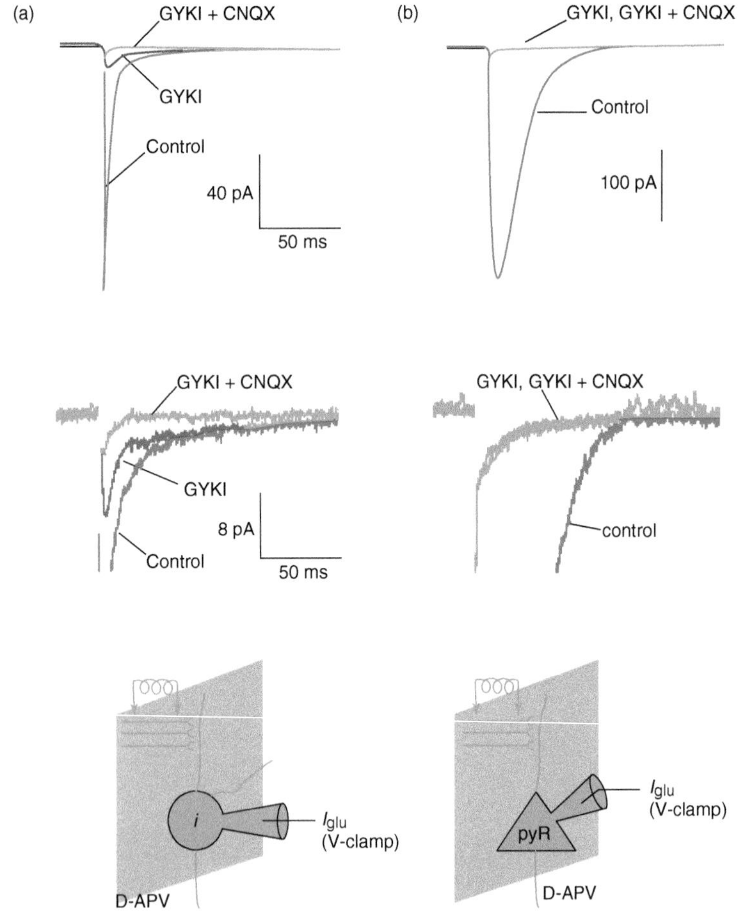

FIGURE 10.15 **The AMPA- and kainate-mediated component of EPSCs.** Experiments performed in slices of the rat hippocampus (CA1 region). **(a)** Averaged EPSCs recorded from an interneuron (I) in control conditions (continuous presence of 100 μM of D-APV), after bath application of 70 μM of GYKI 53655 and after addition of 100 μM of CNQX. Middle traces are the same EPSCs at high gain. **(b)** The same experiment performed in pyramidal cells (pyR). V_H in **(a)** and **(b)** is −80 mV. From Frerking M, Malenka RC, Nicoll RA (1998) Synaptic activation of kainate receptors on hippocampal interneurons. *Nat. Neurosci.* **1**, 479–486, with permission.

When NMDA and non-NMDA receptors coexist in the postsynaptic membrane

When the glutamate concentration is sufficiently high to activate non-NMDA receptor channels, a current is generated through these channels and an APV-insensitive depolarization is recorded.

If this non-NMDA mediated depolarization is not strong enough to allow 'unblocking' of the NMDA receptors, only the early component of the depolarization (non-NMDA component) is recorded (see **Figure 10.3a**).

If this non-NMDA mediated depolarization is sufficiently strong to 'unblock' NMDA receptors, it triggers the activation of an inward NMDA current through these channels and an additional depolarization of the membrane. This depolarization allows the 'unblocking' of additional NMDA receptors which, activated by glutamate, evoke an enhanced depolarization. The more depolarized the membrane, the higher the number of NMDA receptors activated by glutamate. Note that this regenerative phenomenon, due to the voltage sensitivity of the NMDA receptors (associated with the negative-slope region of the I_N/V curve), reminds us of a similar phenomenon observed with the action potential generating Na^+ channels. In the present case, an important postsynaptic depolarization made up of the non-NMDA early component and the NMDA late component is recorded. However, the NMDA component not only prolongs the EPSP but also allows a significant influx of Ca^{2+} ions. These ions have numerous roles: one of them is the activation of channels sensitive to intracellular Ca^{2+} ions. Another role of intracellular Ca^{2+} is as a second messenger. Consequently, it participates in the regulation of a number of intracellular Ca^{2+}-sensitive processes.

When NMDA receptors are the only receptors present in the postsynaptic membrane

In certain preparations, the postsynaptic depolarization recorded in response to the endogenous release of glutamate shows only one component, the NMDA component. This has led to the assumption that not all the NMDA receptors are blocked by Mg^{2+} ions at resting membrane potential. For example, channels containing GluN2A or GluN2B are more sensitive to Mg^{2+} block compared with GluN2C or GluN2D-containing channels. The mechanism of NMDA receptor activation in this case would be the following. When the glutamate concentration in the synaptic cleft is high enough to activate the few NMDA receptors that are not blocked by Mg^{2+} at the resting potential, a small inward current is activated. This current produces a small depolarization of the membrane which allows the 'unblocking' of additional NMDA receptors and, as in the previous example, this triggers a regenerative phenomenon. The resulting Ca^{2+} influx through the open NMDA channels triggers Ca^{2+}-dependent processes.

10.6 SUMMARY

Glutamate receptors comprise the glutamate receptor sites on their surface, the elements that make the ionic channel selectively permeable to cations, as well as all the elements necessary for interactions between different functional domains. Thus, the agonist receptor sites and the cationic channel are part of the same unique protein. The function of postsynaptic iGluRs is to mediate fast excitatory synaptic transmission by converting the binding of glutamate to a rapid and transient increase in cationic permeability and a subsequent membrane depolarization. NMDA, AMPA and kainate receptors are co-expressed in many neurons. Therefore, to study them separately, patch clamp techniques and the use of selective agonists for each receptor type have proven to be particularly useful.

Do all ionotropic glutamate receptors have the same properties; for example, the same sensitivity to glutamate, the same ionic permeability?

NMDA and kainate receptors have a higher affinity (around 1 μM) for glutamate than do AMPA receptors (250–1500 μM). All iGluRs are cationic channels permeable to Na^+ and K^+. Some AMPA receptors and all NMDA receptors are also permeable to Ca^{2+}.

Do all these receptors have co-agonists acting at modulatory sites?

No – to date, only NMDA receptors have been shown to contain a co-agonist binding site (for glycine).

What are the exact conditions of the activation of the different iGluRs?

AMPA receptors show fast gating kinetics, desensitize strongly and are typically poorly permeable to Ca^{2+} ions. Therefore, once glutamate is released in the cleft, AMPA receptors open rapidly and briefly, allowing a transient K^+ efflux + Na^+ influx (recorded as an inward current at physiological membrane potentials) and sometimes a Ca^{2+} influx, through the postsynaptic membrane. This AMPA receptor-mediated current or EPSC generates a fast-rising EPSP (or a fast rising EPSP component) with a time to peak in the order of 1 ms.

When kainate receptors are present in the postsynaptic element, they allow a transient K^+ efflux + Na^+ influx (recorded as an inward current at physiological membrane potentials) and sometimes a Ca^{2+} influx, through the postsynaptic membrane. The kainate receptor-mediated EPSC is smaller and slower (time to peak of the order of 5–10 ms) than the AMPA-mediated one, thus giving a slow-rising, low-amplitude EPSP component.

NMDA receptors are unusual ligand-gated channels because their activation not only requires the binding of two agonists, glutamate and glycine, but also demands the relief of Mg^{2+} block by depolarization. NMDA receptors have a complex role based on three properties: they open slowly (they require more than 2 ms to open) and remain open longer than AMPA receptors (they slowly deactivate and weakly desensitize); this slow time course allows the summation of responses to events tens of milliseconds apart. NMDA receptors function as a coincidence detector: NMDA current occurs only with coincident presynaptic release of glutamate and postsynaptic depolarization; i.e. only when agonist binding and cell depolarization take place simultaneously. Once open, NMDA channels allow a transient K^+ efflux + Na^+ influx + Ca^{2+} influx (recorded as an inward current at physiological membrane potentials) through the postsynaptic membrane. The resulting EPSC triggers a slow-rising, long-duration EPSP (or EPSP component) and the resulting increase of intracellular Ca^{2+} concentration triggers a cascade of molecular events in the postsynaptic cell.

Note that the risetime of EPSPs depends on the agonist binding rate and on the opening rate of postsynaptic receptors. The amplitude of EPSPs depends on the number of open channels in the postsynaptic element; i.e. on the number N of receptors present in the membrane, on the open probability (p_o) of the channel, and on the concentration of neurotransmitter in the synaptic cleft.

Further reading

Armstrong, N., Sun, Y., Chen, G.Q., Gouaux, E., 1998. Structure of a glutamate receptor ligand binding core in complex with kainate. Nature 395, 913–917.

Burnashev, N., Monyer, H., Seeburg, P.H., Sakmann, B., 1992. Divalent ion permeability of AMPA receptor channels is dominated by the edited form of a single subunit. Neuron 8, 189–198.

Furukawa, H., 2012. Structure and function of glutamate receptor amino terminal domains. J. Physiol. 590, 63–72.

Geiger, J.R., Melcher, T., Koh, D.S., et al., 1995. Relative abundance of subunit mRNAs determines gating and Ca^{2+} permeability of AMPA receptors in principal neurons and interneurons in rat CNS. Neuron 15, 193–204.

Gregor, P., Mano, I., Maoz, I., et al., 1989. Molecular structure of the chick cerebellar kainate-binding subunit of a putative glutamate receptor. Nature 342, 689–692.

Hirai, H., Kirsch, J., Laube, B., et al., 1996. The glycine binding site of the *N*-methyl-D-aspartate receptor subunit NR1: identification of novel determinants of co-agonist potentiation in the extracellular M3-M4 loop region. Proc. Natl Acad. Sci. USA 93, 6031–6036.

Hume, R.I., Dingledine, R., Heinemann, S.F., 1991. Identification of a site in glutamate receptor subunits that controls calcium permeability. Science 253, 1028–1031.
Kuusinen, A., Arvola, M., Keinänen, K., 1995. Molecular dissection of the agonist binding site of an AMPA receptor. EMBO J. 14, 6327–6332.
Laube, B., Hirai, H., Sturgess, M., et al., 1997. Molecular determinants of agonist discrimination by NMDA receptor subunits: analysis of the glutamate binding site on the NR2B subunit. Neuron 18, 493–503.
Liu, S.J., Savtchouk, I., 2012. Ca^{2+} permeable AMPA receptors switch allegiances: mechanisms and consequences. J. Physiol. 590, 13–20.
Pinheiro, P.S., Lanore, F., Veran, J., et al., 2013. Selective block of postsynaptic kainate receptors reveals their function at hippocampal mossy fiber synapses. Cereb. Cortex 23, 323–331.
Sobolevsky, A.I., Rosconi, M.P., Gouaux, E., 2009. X-ray structure, symmetry and mechanism of an AMPA-subtype glutamate receptor. Nature 462, 745–756.
Sommer, B., Kohler, M., Sprengel, R., Seeburg, P.H., 1991. RNA editing in brain controls a determinant of ion flow in glutamate-gated channels. Cell 67, 11–19.
Stern-Bach, Y., Bettler, B., Hartley, M., Sheppard, P.O., O'Hara, P.J., Heinemann, S.F., 1994. Agonist selectivity of glutamate receptors is specified by two domains structurally related to bacterial amino-acid binding proteins. Neuron 13, 1345–1357.
Swanson, G.T., Kamboj, S.K., Cull-Candy, S.G., 1997. Single-channel properties of recombinant AMPA receptors depend on RNA editing, splice variation, and subunit composition. J. Neurosci. 17, 58–69.
Wada, K., Dechesne, C.J., Shimasaki, S., et al., 1989. Sequence and expression of a frog complementary DNA encoding a kainate binding protein. Nature 342, 684–689.

CHAPTER

11

The metabotropic $GABA_B$ receptors

David Mott

OUTLINE

Gamma-aminobutyric acid (GABA) is the primary inhibitory neurotransmitter in the mammalian central nervous system. It is found in virtually every area of the brain. It exerts fast and powerful synaptic inhibition by acting on $GABA_A$ receptors. These receptors are directly coupled to an integral chloride channel and produce inhibition by increasing the membrane chloride conductance. This form of synaptic inhibition is critical for maintaining and shaping neuronal communication.

However, like other neurotransmitters that activate fast, ionotropic responses lasting for milliseconds, GABA can also activate a second class of receptors which produce slow synaptic responses capable of lasting for seconds. The receptors producing these slow, metabotropic responses are designated $GABA_B$ receptors. $GABA_B$ receptors play a major role in regulating neurotransmission, which makes them potentially important therapeutic targets in the treatment of a variety of neurological conditions, including epilepsy, spasticity, pain and psychiatric illness. $GABA_B$ receptors are G-protein coupled to a number of cellular effector mechanisms, including adenylyl cyclase, voltage-dependent calcium channels and inwardly rectifying potassium channels. These different effectors enable $GABA_B$ receptors to produce not only inhibition, but a diversity of other effects on neuronal function. Thus, $GABA_B$ receptors enable GABA to modulate neuronal activity in a fashion that is not possible through $GABA_A$ receptors alone.

This chapter will focus on $GABA_B$ receptors and the different effects that these receptors can have on cellular function.

11.1 $GABA_B$ RECEPTORS WERE ORIGINALLY DISCOVERED BECAUSE OF THEIR INSENSITIVITY TO BICUCULLINE AND THEIR SENSITIVITY TO BACLOFEN

The discovery of $GABA_B$ receptors was made possible by the development in the early 1970s of the compound β-parachlorophenyl GABA (baclofen). Baclofen is a GABA analog which can be orally administered and will penetrate the blood–brain barrier. It was hoped that after gaining access to the brain this compound would act on GABA receptors and be an effective anticonvulsant. Indeed, baclofen did mimic many of the actions of GABA and was found to reduce skeletal muscle tone and inhibit spinal reflex activity, making it a successful agent in treating spinal cord spasticity. Yet, despite these similarities with GABA, several important differences between the actions of GABA and baclofen were reported, the most notable of which was the insensitivity of

Cellular and Molecular Neurophysiology. DOI: 10.1016/B978-0-12-397032-9.00011-X the actions of baclofen to the classical GABA antagonist bicuculline.

It was at this time that Norman Bowery and his colleagues found that application of GABA decreased the release of norepinephrine from a preparation of rat isolated atrium. Interestingly, this effect of GABA was insensitive to bicuculline and was not mimicked by classical GABA agonists, such as isoguvacine and THIP. Bowery and his colleagues found similar results when they measured the effect of GABA on the release of norepinephrine in another peripheral preparation, the rat isolated anococcygeus muscle. In both of these preparations the GABA analog baclofen mimicked the action of GABA by depressing the release of norepinephrine in a dose-dependent manner (**Figure 11.1**). Furthermore, neither the effect of GABA nor that of baclofen appeared to be mediated by an increase in chloride conductance, suggesting that a receptor other than the classical GABA receptor was responsible for the presynaptic inhibition of norepinephrine release.

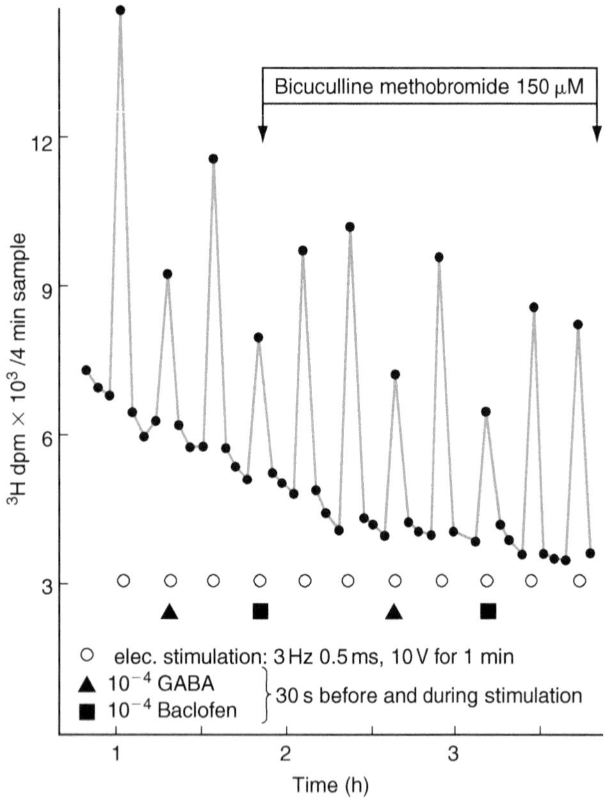

FIGURE 11.1 **GABA and baclofen suppress ^{3}H-norepinephrine release from the rat atrium.** The release of ^{3}H-norepinephrine was assessed by taking samples of the superfusate every 4 minutes and measuring the tritium content (in dpm) by liquid scintillation spectrometry. Electrical stimuli were delivered to the tissue at times indicated by the open circles. These stimuli caused the release of ^{3}H-norepinephrine and so increased the tritium content of the sample. GABA (filled triangles) and baclofen (filled squares) reduced the release of ^{3}H-norepinephrine by the stimulus. The effect of these drugs was insensitive to co-application of bicuculline methobromide. From Bowery NG, Doble A, Hill DR, Hudson AL, Turnbull MJ, Warrington R (1981) Structure/activity studies at a baclofen-sensitive, bicuculline-insensitive GABA receptor. In DeFeudis FV, Mandel P (eds), *Amino Acid Neurotransmitters*. New York: Raven Press, with permission.

To determine whether this bicuculline-insensitive action of GABA was confined to the periphery, Bowery and co-workers tested the effect of GABA and baclofen on potassium-evoked norepinephrine release from brain slices. They found that, as in the periphery, GABA suppressed norepinephrine release by acting on a bicuculline-insensitive receptor that was separate from the classical bicuculline-sensitive GABA receptor. This action of GABA was mimicked by the GABA analog baclofen, but not by other known GABA agonists. Radio-ligand receptor binding in brain using ^{3}H-GABA demonstrated two distinct binding sites for GABA with different distributions. These results led Bowery and his co-workers in 1981 to propose the existence of a new class of GABA receptor, which they termed the $GABA_B$ receptor, while designating the classical GABA receptor as the $GABA_A$ receptor.

$GABA_B$ receptor pharmacology

GABA is the endogenous agonist at both $GABA_A$ and $GABA_B$ receptors. $GABA_B$ receptors are pharmacologically distinguished from $GABA_A$ receptors by their insensitivity to the $GABA_A$ antagonist bicuculline and their selective activation by the prototypic agonist baclofen (**Figure 11.2**). Baclofen activates $GABA_B$ receptors in a stereospecific manner with the (−) isomer being about 100 times more potent than the (+) isomer. In contrast, $GABA_B$ receptors are not sensitive to classical agonists at the $GABA_A$ receptor, such as muscimol and isoguvacine, or to modulators of $GABA_A$ receptors such as benzodiazepines, barbiturates and neurosteroids.

The discovery of selective $GABA_B$ receptor antagonists with increased receptor affinity and improved pharmacokinetic profile has been an important element in establishing the significance and structure of $GABA_B$ receptors. The first $GABA_B$ receptor antagonists, phaclofen and 2-hydroxysaclofen (**Figure 11.2**), represented a major breakthrough in the study of $GABA_B$ receptors even though they possessed relatively low potencies. Subsequently, Froestl and co-workers introduced CGP35348, the first $GABA_B$ receptor antagonist capable of crossing the blood–brain barrier. This was soon followed by CGP36742, the first orally active $GABA_B$ receptor antagonist. Although these compounds displayed rather low potency, Froestl and co-workers found that the substitution of a dichlorobenzene moiety into these antagonist molecules increased their affinities by about 10 000-fold. This breakthrough resulted in the production of a host of compounds, such as CGP52432, CGP55845, CGP64213 and CGP71872 (**Figure 11.2**), which had affinities in the nanomolar and even subnanomolar range. This series of compounds eventually led to the development of the radio-iodinated, high-affinity antagonist [^{125}I]-CGP64213, which was used to clone $GABA_{B1}$.

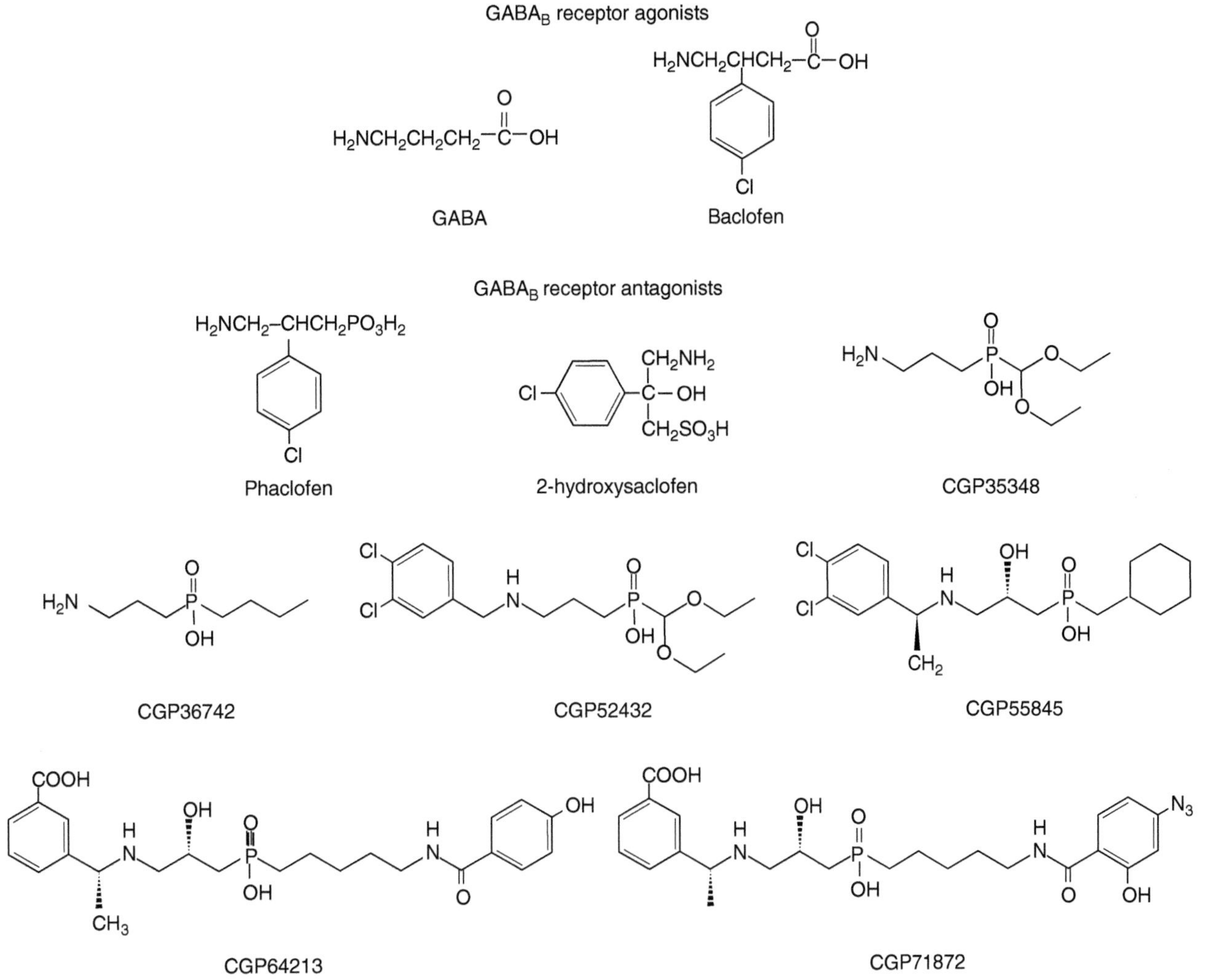

FIGURE 11.2 **Structures of selected GABA$_B$ receptor agonists and antagonists.** Adapted from Mott DD, Lewis DV (1994) The pharmacology and function of central GABA$_B$ receptors. *Int. Rev. Neurobiol.* **39**, 97–223, with permission.

11.2 STRUCTURE OF THE GABA$_B$ RECEPTOR

Using a high-affinity antagonist, the structural properties of the GABA$_B$ receptor were characterized by expression cloning. Expression of a fully functional GABA$_B$ receptor was found to require coupling between two separate and distinct gene products, GABA$_{B1}$ and GABA$_{B2}$. GABA$_B$ receptors are thus the first example of a functional heterodimeric metabotropic receptor.

11.2.1 GABA$_B$ receptors belong to Family 3 G-protein-coupled receptors

Cloning of the GABA$_B$ receptor

In 1997, Bettler and colleagues successfully cloned the first GABA$_B$ subunit, which they named GABA$_{B1}$. The derived sequence of GABA$_{B1}$ indicated that it shares no significant sequence similarity to GABA$_A$ or GABA$_C$ receptors, but is distantly related to Family 3 G-protein-coupled receptors (GPCRs). This family of receptors includes metabotropic glutamate receptors (mGluRs), the Ca^{2+}-sensing receptor, a family of pheromone receptors and certain mammalian taste receptors. Like other Family 3 GPCRs, GABA$_{B1}$ subunits have several characteristic features, including a large extracellular amino-terminus which plays a critical role in ligand binding, followed by seven closely spaced transmembrane domains, indicative of GPCRs. When compared to mGluRs, GABA$_{B1}$ shares only 18–23% sequence homology; however, hydrophobicity profiles indicate clear conservation of structural architecture between these receptors. The N-terminal extracellular domain of both mGluRs and GABA$_{B1}$ shares limited but significant similarity with bacterial periplasmic amino-acid-binding proteins (PBP) such as the leucine-binding protein (LBP). However, the intracellular loops of the GABA$_B$ receptor are not as well conserved as in other Family 3 receptors. In particular, most cysteine residues, which are highly conserved in other Family 3 receptors, are not conserved in $GABA_{B1}$.

11.2.2 $GABA_B$ receptors are heterodimers

$GABA_{B1}$ receptors are non-functional

Whereas $GABA_{B1}$ displays binding and biochemical characteristics similar to those of native $GABA_B$ receptors, several important discrepancies were noted between these cloned receptors and native $GABA_B$ receptors. For example, the affinity of agonists, but not antagonists, was 100–150-fold lower for $GABA_{B1}$ than for native receptors. Most importantly, when expressed in cell lines, $GABA_{B1}$ coupled only weakly to adenylyl cyclase and did not couple to other effector systems, such as calcium or potassium channels. The reason for the failure of $GABA_{B1}$ to produce functional receptors was examined using epitope-tagged versions of $GABA_{B1}$ to study the cellular distribution of the receptor protein. It was found that $GABA_{B1}$ was retained in the endoplasmic reticulum and therefore failed to reach the cell surface. Thus, it appeared as though $GABA_{B1}$ required additional information for functional targeting to the plasma membrane.

Fully functional $GABA_B$ receptors require coupling between $GABA_{B1}$ and $GABA_{B2}$

The failure of $GABA_{B1}$ to produce functional $GABA_B$ receptors inspired an intensive search for other related genes, ultimately resulting in the discovery of a second $GABA_B$ receptor gene, termed $GABA_{B2}$. This receptor subtype was 35% homologous with $GABA_{B1}$ and exhibited many of the structural features of $GABA_{B1}$, including a large molecular weight, an extended extracellular N-terminus and seven transmembrane spanning domains.

Importantly, it was found that this receptor must be co-expressed with $GABA_{B1}$ to form a fully functional $GABA_B$ receptor. Co-expression of $GABA_{B1}$ and $GABA_{B2}$ resulted in efficient surface expression of the receptor and the agonist affinity of these heterodimeric receptors was similar to that of native $GABA_B$ receptors. This finding represented the first evidence for heterodimerization among GPCRs. Recombinant heteromeric $GABA_B$ receptors are fully functional and display robust coupling to all prominent effector systems of native $GABA_B$ receptors (**Figure 11.3**).

The existence of $GABA_B$ heterodimers in neurons was confirmed in immunoprecipitation experiments. In these experiments, antibodies raised against $GABA_{B2}$ efficiently co-precipitated the $GABA_{B1}$ proteins from cortical

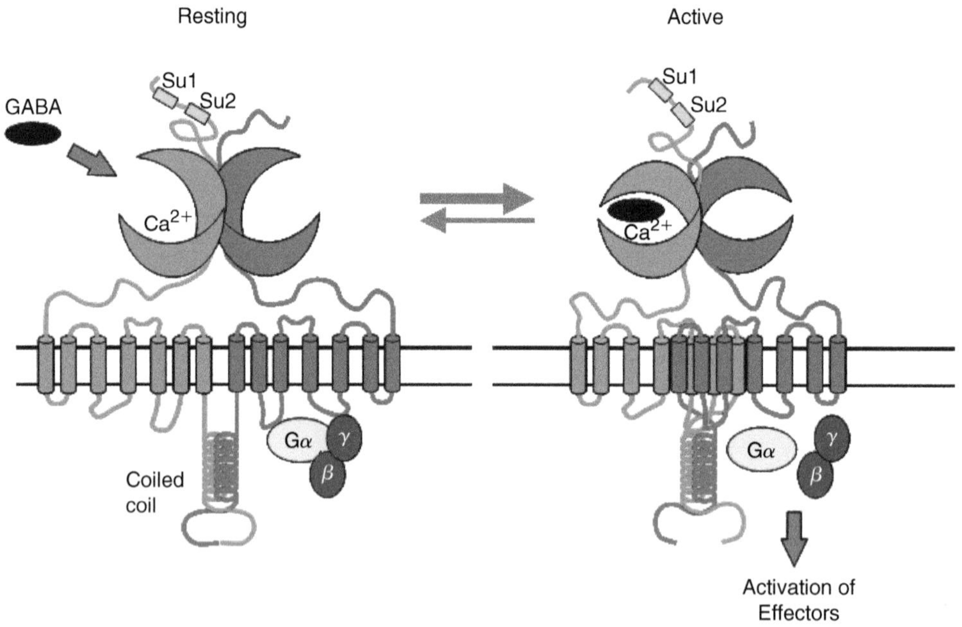

FIGURE 11.3 **$GABA_B$ receptors are functional heterodimers.** Agonist binding activates the $GABA_B$ receptor heterodimer causing a conformational change that results in receptor coupling to effector systems. The $GABA_{B1}$ (green) and $GABA_{B2}$ (red) subunits interact with each other through their carboxy-terminal coiled-coil domains. GABA binds only to the $GABA_{B1}$ subunit whereas G proteins bind to the $GABA_{B2}$ subunit. In the inactive resting state, the ligand-binding pocket in the $GABA_{B1}$ extracellular domain is open and the transmembrane domains of $GABA_{B1}$ and $GABA_{B2}$ are apart. Agonist binding to $GABA_{B1}$ causes the ligand-binding pocket to close and the receptor to become activated. The binding of GABA within the binding pocket and the closed conformation of the receptor are stabilized by calcium. The activated receptor undergoes a conformational change in which the transmembrane domains of $GABA_{B1}$ and $GABA_{B2}$ move closer together. This conformational change is necessary for the initiation of downstream effector signaling. The sushi domains in the amino-terminus of the $GABA_{B1a}$ subunit are indicated (Su1 and Su2). These domains are important for receptor targeting in the cell and are not present in $GABA_{B1b}$. Adapted from Calver AR, Davies CH, Pangalos M (2002) $GABA_B$ receptors: from monogamy to promiscuity. *Neurosignals* **11**, 299–314, copyright S Karger AG, Basel, with permission.

membranes. Conversely, antibodies that recognize $GABA_{B1}$ co-precipitated the $GABA_{B2}$ receptor. Thus, native $GABA_B$ receptors appear to be heterodimers composed of $GABA_{B1}$ and $GABA_{B2}$ which interact in a stoichiometry of 1:1.

Genetic studies have found that mice lacking the $GABA_{B1}$ or $GABA_{B2}$ gene show a loss of all typical $GABA_B$ responses. These findings indicate that $GABA_{B1}$ and $GABA_{B2}$ alone can account for all of the classical $GABA_B$ functions and strongly suggest that the existence of additional obligatory receptor subunits is unlikely. In addition, $GABA_{B1}$ and $GABA_{B2}$ protein are substantially downregulated in $GABA_{B2}$ and $GABA_{B1}$ knockout mice, respectively. The degradation of the partner subunit in each of these knockout mouse lines indicates the importance of heteromeric assembly for stable expression of each subunit. It also suggests that in wild-type mice virtually all $GABA_{B1}$ protein is associated with $GABA_{B2}$. These studies support the conclusion that native $GABA_B$ responses are predominantly mediated by heteromeric receptors derived from $GABA_{B1}$ and $GABA_{B2}$ genes.

11.2.3 Surface expression of $GABA_B$ receptors requires coupling between $GABA_{B1}$ and $GABA_{B2}$

For both recombinant and native $GABA_B$ receptors, the interaction of $GABA_{B1}$ and $GABA_{B2}$ within the cell is critical for the correct assembly of the heterodimer on the cell surface. It is now known that $GABA_{B1}$ is prevented from traveling to the cell surface by an endoplasmic reticulum (ER)-retention signal in its cytoplasmic tail. The ER-retention signal in $GABA_{B1}$ is located in an α-helical coiled-coil domain in the carboxy-terminus of the peptide. A coiled-coil domain is also present in the carboxy-terminus of $GABA_{B2}$. $GABA_{B1}$ and $GABA_{B2}$ form a tightly coupled heterodimer via an interaction of these coiled-coil domains. Formation of this heterodimer masks the ER-retention signal in $GABA_{B1}$ from its ER-anchoring mechanism, allowing the heteromeric receptor to travel to the cell surface. The ER-retention motif therefore ensures that only correctly assembled receptor complexes traffic to the cell surface. ER-retention signals are also observed in other multisubunit proteins, where they serve as a quality-control mechanism.

$GABA_B$ receptors are the first functional heterodimers to be identified within the metabotropic class. Only among the ionotropic receptors have heterodimers been previously recognized (i.e. $GABA_A$ receptors). Other members of the Family 3 GPCRs, including metabotropic glutamate receptors and the Ca^{2+}-sensing receptor, have previously been reported to form homodimers. As opposed to the $GABA_B$ receptor, dimer formation for these receptors has been shown to be caused by the disulfide interaction of cysteine residues in the extracellular N-terminal domain. These cysteine residues are absent in the $GABA_B$ receptor, which dimerizes predominantly through an interaction of coiled-coil domains in the carboxy-terminus (**Figure 11.3**). That such closely related receptors have evolved different mechanisms of dimerization suggests that dimerization is important for this class of receptors.

11.2.4 The GABA-binding site is in the extracellular amino-terminal domain of $GABA_{B1}$

The ligand-binding domain for Family 3 GPCRs is located in the extracellular amino-terminus of the receptor in a region with significant homology to bacterial periplasmic-binding proteins, such as the leucine-binding protein (LBP). Activation of $GABA_B$ receptors occurs exclusively by GABA binding to the $GABA_{B1}$ subunit. The $GABA_{B2}$ subunit does not bind agonist or antagonist. The function of the extracellular domain of $GABA_{B2}$ is not well understood; however, it has been suggested to interact with and enhance the affinity of $GABA_{B1}$ agonist-binding site for GABA.

To understand better the agonist-binding domain of $GABA_{B1}$, Bettler and colleagues constructed chimeric receptors which contain the amino-terminus of $GABA_{B1}$ on the body of the $mGlu_1$ receptor. They found that these chimeric receptors and wild-type $GABA_B$ receptors possessed similar binding affinities for $GABA_B$ receptor ligands. Furthermore, radio-iodinated antagonist binding affinities were also unaltered in $GABA_{B1}$ truncation mutants in which the entire carboxy-terminus after the first transmembrane domain was deleted. Finally, when the amino-terminus of the $GABA_{B1}$ subunit was produced as a soluble miniprotein it bound radiolabeled $GABA_B$ receptor antagonist with a similar affinity to control wild-type receptor. These studies indicate that, like the amino-terminal domain of other Family 3 GPCRs, the amino-terminal domain of $GABA_{B1}$ is both necessary and sufficient for ligand binding.

Mutagenesis studies support the LBP-like domain in the N-terminus of $GABA_{B1}$ as a critical region for ligand binding. Mutation of several key residues in this area markedly alters the affinity of the $GABA_B$ receptor for antagonists, suggesting that the architecture of this region bears structural homology to that of LBP. Three-dimensional modeling of the $GABA_B$-binding domain based on the known crystal structure of LBP supports a Venus flytrap model for receptor activation. According to this model, the ligand-binding site is formed in a groove between two large globular domains in the N-terminus of $GABA_{B1}$. Activation of the receptor results from the closure of these two lobes upon agonist binding, similar to a Venus flytrap. Closure of these lobes produces a conformational change in the protein complex that allows activation of the G-protein-coupled signaling system (**Figure 11.3**). This model is similar to that proposed for other members of Family 3 GPCRs.

Ligand binding to GABA$_B$ receptors requires the presence of divalent cations

Like other members of Family 3 GPCRs, such as the Ca^{2+}-sensing receptor and metabotropic glutamate receptor, GABA$_B$ receptors are sensitive to calcium. This differs from GABA$_A$ receptors which have no such requirement. Calcium binding to the GABA$_{B1}$ subunit of the receptor allosterically potentiates the action of GABA by stabilizing the closed conformational state of the agonist-binding domain (**Figure 11.3**). Interestingly, other divalent cations, including Hg^{2+}, Pb^{2+}, Cd^{2+} and Zn^{2+}, inhibit GABA-receptor binding. The effects of calcium on the GABA$_B$ receptor are agonist dependent in that calcium more strongly potentiates the effect of GABA than baclofen. Mutational analysis has revealed that a specific highly conserved serine residue (S269) in the GABA$_{B1}$ ligand-binding site is critical for the effect of calcium. Possibly, calcium binding to this residue helps to optimally position GABA in the agonist-binding pocket. Because GABA$_B$ receptors have a high sensitivity to calcium, the calcium-binding site on GABA$_{B1}$ is saturated with calcium under normal physiological conditions. Therefore, calcium modulation of GABA binding to the GABA$_B$ receptor would potentially play a role only during times when extracellular calcium concentrations are low, such as during ischemia or epileptic seizures.

11.2.5 GABA$_{B2}$ subunits couple to inhibitory G proteins

Guanyl nucleotide-binding proteins (G proteins) carry signals from activated membrane receptors to effector enzymes and channels. These molecules enable a single receptor to be functionally connected to a variety of different effector mechanisms in a single cell or to different effectors in different cells. Coupling of GABA$_B$ receptors and G proteins was originally deduced from binding studies of ^{3}H-GABA and ^{3}H-baclofen to crude synaptic membranes prepared using whole rat brain. In these experiments, the addition of guanyl nucleotides, such as GTP, did not affect the binding of ^{3}H-GABA to GABA$_A$ receptors, but potently inhibited GABA$_B$-receptor binding (**Figure 11.4**). This effect was concentration dependent and was not mimicked by adenosine 5′-triphosphate (ATP), indicating that it was specific for guanyl nucleotides. The inhibition of ligand binding produced by GTP was caused by a decrease in GABA$_B$ receptor affinity and not a decrease in the number of available GABA$_B$ receptors. It was concluded that the addition of GTP promoted the dissociation of the G protein from the receptor, causing the receptor to revert to its low-affinity conformation. Thus, GABA$_B$ receptors appeared to couple to G proteins.

The identity of the G proteins coupled to GABA$_B$ receptors was established through two different experiments. First, it was observed that inhibition of

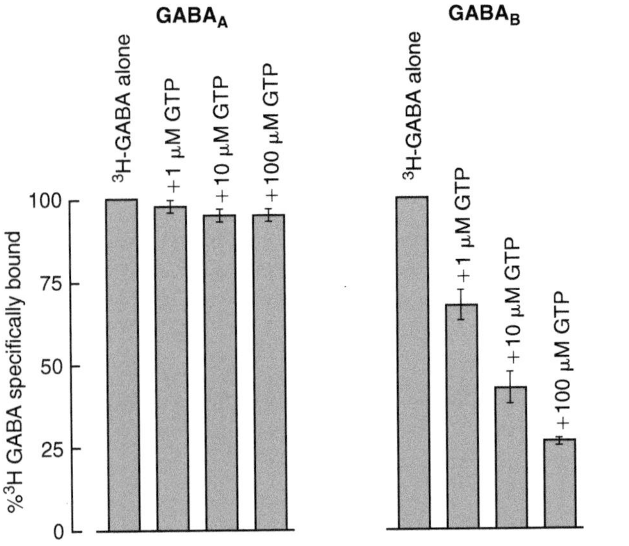

FIGURE 11.4 **The effect of GTP on ^{3}H-GABA binding to GABA$_A$ and GABA$_B$ receptors.** ^{3}H-GABA binding to crude synaptic membranes from whole rat brain was measured in the presence of either isoguvacine or baclofen to saturate GABA$_A$ and GABA$_B$ receptors, respectively. The addition of increasing concentration of GTP had no effect on GABA$_A$ receptor binding but produced a concentration-dependent inhibition of GABA$_B$ receptor binding. From Hill DR, Bowery NG, Hudson AL (1984) Inhibition of GABA$_B$ receptor binding by guanyl nucleotides. *J. Neurochem.* **42**, 652–657, with permission.

GABA$_B$-receptor binding by GTP was blocked by pertussis toxin. This demonstrated that GABA$_B$ receptors are functionally coupled to the inhibitory G proteins, G_i and/or G_o. This finding was further confirmed using cloned heteromeric GABA$_B$ receptors (GABA$_{B1}$/GABA$_{B2}$) expressed with chimeric G_q proteins in human embryonic kidney cells (HEK 293). Wild-type G_q protein activates phospholipase C (PLC). Ordinarily GABA$_B$ receptors do not stimulate PLC activity, indicating that they do not couple to G_q protein. PLC activity produced by GABA$_B$ receptor activation was then measured following the addition of chimeric G_q proteins in which the five carboxy-terminal residues of the G_q α-subunit had been exchanged for those of either $G_i\alpha$, $G_o\alpha$, or $G_z\alpha$ protein. The five carboxy-terminal residues of the Gα-subunit are critical for coupling of G proteins to receptors. Only those chimeric G_q proteins containing the coupling sites of $G_i\alpha$ or $G_o\alpha$ protein were able to activate PLC, indicating that only G_i and G_o proteins interact with the GABA$_B$ receptor.

A large number of studies have examined the molecular determinants of receptor-G protein-coupling selectivity in GABA$_B$ receptor subunits. It is now established that G proteins bind to the heptahelical region of the GABA$_{B2}$ subunit. In contrast, the heptahelical region of GABA$_{B1}$ does not bind G protein but enhances the efficiency of the interaction of G protein with the GABA$_{B2}$ subunit. Thus, the GABA$_B$ receptor is an obligate heterodimer in which the GABA$_{B1}$ subunit is necessary for agonist binding while the GABA$_{B2}$ subunit is required for G-protein signaling (**Figure 11.3**).

In Family 3 GPCRs, the second intracellular loop (i2) plays a critical role in the interaction of the receptor with G proteins. In the GABA$_B$ receptor, the i2 loop in GABA$_{B2}$ is critical for G-protein coupling as well. Sequence comparison between the i2 loops of GABA$_{B1}$ and GABA$_{B2}$ revealed several important differences. Thus, exchanging the i2 loop between GABA$_{B1}$ and GABA$_{B2}$ prevented receptor function. This finding indicated that the i2 loops on the GABA$_{B2}$ subunit need to be correctly positioned relative to other intracellular domains for proper G-protein coupling. Mutational analysis confirmed the importance of the i2 loop in GABA$_{B2}$. Furthermore, these studies found that mutation of a critical lysine residue (K686) in the third intracellular loop of GABA$_{B2}$ also suppressed coupling of G proteins to the GABA$_B$ receptor. This lysine residue is important for functional coupling of G proteins to GABA$_B$ receptors and appears to serve a similar function in other Family 3 GPCRs as well.

11.2.6 Molecular diversity of GABA$_B$ receptors arises from GABA$_{B1}$ isoforms

When it became apparent that only two genes encoded all GABA$_B$ receptor subunits, a search began for subunit isoforms. To date, no isoforms of GABA$_{B2}$ have been identified. In contrast, two predominant isoforms of GABA$_{B1}$, termed GABA$_{B1a}$ and GABA$_{B1b}$, have emerged. Numerous studies in recombinant systems have concluded that GABA$_{B1a}$ and GABA$_{B1b}$ isoforms exhibit no unique functional or pharmacological properties. GABA$_{B1a}$ and GABA$_{B1b}$ isoforms are generated by differential promoter usage within the GABA$_{B1}$ gene. GABA$_{B1a}$ differs from GABA$_{B1b}$ in having a longer amino-terminus that contains a pair of short consensus repeats, also known as sushi repeats (**Figure 11.3**). Sushi repeats were originally identified in complement proteins and are involved in protein–protein interactions. These sushi domains have been proposed to play a role in targeting GABA$_B$ receptors to specific sites within the cell.

11.2.7 GABA$_B$ receptors are located throughout the brain at both presynaptic and postsynaptic sites

GABA$_B$ receptors can be found in most regions of the brain (**Figure 11.5**). In the majority of these areas, the number of GABA$_B$ receptors is either less than or equal to the number of GABA$_A$ receptors. However, there are a few brain regions, such as the brainstem and certain thalamic nuclei, where GABA$_B$ receptors can account for up to 90% of the total GABA binding sites. Autoradiography or antibody labeling of GABA$_{B1}$ and GABA$_{B2}$ suggests that these subunits are similarly distributed; however, there are some brain regions, such as the caudate putamen or olfactory bulb, where GABA$_{B1}$ is in much greater abundance than GABA$_{B2}$. The brain regions possessing the highest density of GABA$_B$ receptors are the thalamic nuclei, the molecular layer of the cerebellum, the cerebral cortex and the interpeduncular nucleus. GABA$_B$ receptors are also found in high density in laminae II and III of the spinal cord.

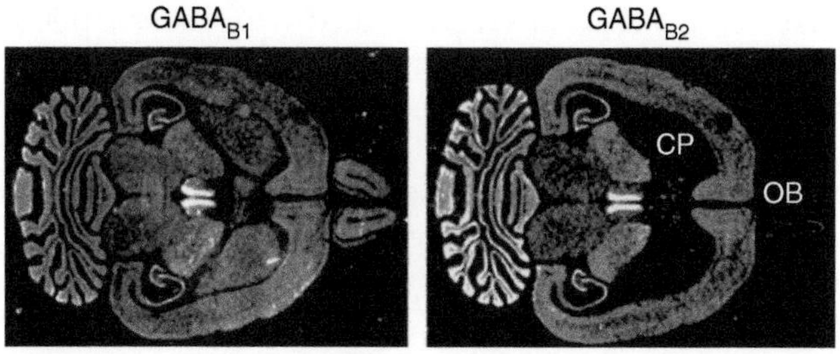

FIGURE 11.5 **Distribution of GABA$_{B1}$ and GABA$_{B2}$ subunit mRNA in the rat brain.** GABA$_{B1}$ and GABA$_{B2}$ transcripts are present at similar levels throughout the brain, except in caudate putamen (CP) and olfactory bulb (OB) where GABA$_{B2}$ is less abundant than GABA$_{B1}$ mRNA. From Bettler B, Kaupmann K, Mosbacher J, Gassmann M (2004) Molecular structure and physiological functions of GABA$_B$ receptors. *Physiol. Rev.* **84**, 835–867, with permission.

Both electrophysiological recordings and immunogold electron microscopic techniques have been used to investigate the subcellular localization of GABA$_B$ receptors. These receptors are present on presynaptic terminals, where they modulate the release of a variety of different neurotransmitters, and on postsynaptic membranes, where they inhibit excitatory neurotransmission. Presynaptically, GABA$_B$ subunits are located in the extrasynaptic membrane and near the active zones in presynaptic glutamatergic and GABAergic terminals, supporting a close link with the transmitter release machinery. Postsynaptically, GABA$_B$ receptors are located on both dendritic shafts as well as the extrasynaptic membrane of dendritic spines. Dendritic spines form the majority of excitatory synapses and the presence of GABA$_B$ receptors on these structures as well as on glutamatergic terminals suggests a close coupling of excitatory and inhibitory systems. The subcellular localization of GABA$_B$ receptors also appears to depend upon the brain region examined. For example, immunogold electron microscopic studies have revealed that, in the cerebellum, GABA$_B$ receptors are enriched in synapses whereas, in thalamic nuclei, GABA$_B$ receptors are found in extrasynaptic membrane, having no enrichment in synapses. Presumably, GABA$_B$ receptors in extrasynaptic membrane or on excitatory terminals would be activated by GABA spilling over from neighboring inhibitory terminals.

In situ hybridization techniques have suggested a differential localization of GABA$_{B1a}$ and GABA$_{B1b}$ to pre- and postsynaptic sites. These studies have suggested that

$GABA_{B1a}$ is more closely associated with presynaptic receptors, whereas $GABA_{B1b}$ may participate in the formation of postsynaptic $GABA_B$ receptors. The mechanism underlying this differential distribution involves the sushi domains on the $GABA_{B1a}$ isoform. These sushi domains interact with cytosolic proteins containing axon-sorting signals, causing $GABA_{B1a}$-containing $GABA_B$ receptors to traffic to the axon terminal. Once in the terminal the sushi domains interact with an extracellular binding partner to hold the receptors close to the release machinery.

Both $GABA_{B1a}$- and $GABA_{B1b}$-containing $GABA_B$ receptors are found in the dendritic compartment. $GABA_{B1b}$-containing $GABA_B$ receptors distribute selectively to dendritic spines where they cluster with G-protein-activated inwardly rectifying potassium (Kir) channels (previously termed GIRK channels). Accordingly, experiments using knockout mice that lack either the $GABA_{B1a}$ or $GABA_{B1b}$ isoform demonstrate that $GABA_{B1b}$-containing $GABA_B$ receptors generate larger synaptic responses than do $GABA_{B1a}$-containing $GABA_B$ receptors. Both $GABA_{B1a}$ and $GABA_{B1b}$ isoforms are found in dendritic shafts. The differential distribution of $GABA_{B1}$ isoforms to axonal and dendritic compartments may allow the strength of pre- and postsynaptic $GABA_B$ responses to be independently regulated. The role of presynaptic and postsynaptic $GABA_B$ receptors will be discussed in greater detail later in this chapter (see Section 11.6).

11.2.8 $GABA_B$ receptor properties are regulated by auxiliary subunits

Recent studies have identified auxiliary proteins that assemble with native $GABA_B$ receptors and alter their kinetic and pharmacological properties. These auxiliary proteins were identified using unbiased proteomics approaches which revealed that $GABA_B$ receptors in the brain were high molecular mass complexes of $GABA_{B1}$, $GABA_{B2}$ and members of a subfamily of the KCTD (potassium channel tetramerization domain-containing) proteins. KCTD proteins 8, 12, 12b and 16 can associate tightly with the carboxy-terminus of $GABA_{B2}$ forming a stable receptor complex. When expressed in heterologous cells these KCTD proteins increase agonist potency and alter the G-protein signaling of the receptors by accelerating onset of potassium currents and promoting their desensitization in a KCTD-subtype-specific manner. Similarly, knockdown of KCTD proteins in neurons markedly prolongs the risetime and reduces desensitization of the $GABA_B$-receptor-mediated potassium current. The ability of each of the KCTD proteins to confer unique properties on the $GABA_B$ receptor markedly increases the diversity of $GABA_B$ receptor types. *In situ* hybridization experiments have shown that the four KCTD proteins have unique, but overlapping expression patterns in the adult brain with most neurons expressing at least one KCTD and some neurons expressing multiple subtypes. These findings suggest that most neurons express $GABA_B$ receptors containing KCTDs. Furthermore, the expression of KCTDs varies during development. Thus, $GABA_B$ receptor signaling would be expected to differ in distinct brain regions or neuronal populations and during development due to the distinct spatial and temporal distribution pattern of KCTDs.

11.3 SUMMARY

Much has been learned about the $GABA_B$ receptor since its initial description more than 30 years ago. $GABA_B$ receptors represent the first example of a heterodimeric GPCR. These receptors require both $GABA_{B1}$ and $GABA_{B2}$ subunits for efficient surface expression and function. Furthermore, the pharmacology and kinetics of the $GABA_B$ receptor are modulated by its association with auxiliary KCTD subunits. This architecture in which the receptor is formed by both principal and auxiliary subunits is well established for ionotropic receptors, but rare among G-protein-coupled receptors. The principal subunits each serve a distinct role in the receptor. The $GABA_{B1}$ subunit contains the GABA-binding domain, but this subunit is trapped in the endoplasmic reticulum by an ER-retention signal in its carboxy-terminal. The $GABA_{B2}$ subunit interacts with $GABA_{B1}$, masking this ER-retention signal and allowing the heterodimeric receptor to be trafficked to the surface. Once at the surface, the $GABA_{B2}$ subunit links to the G protein, allowing for a fully functional receptor. Although $GABA_{B2}$ does not bind GABA, it interacts with the GABA-binding domain on the $GABA_{B1}$ subunit, enhancing its affinity for GABA. Similarly, $GABA_{B1}$ interacts with the $GABA_{B2}$ subunit promoting its association with the G protein. Thus, the two subunits act in concert to link GABA binding to activation of downstream effectors. We will discuss effectors linked to $GABA_B$ receptors in the next section.

11.4 $GABA_B$ RECEPTORS ARE G-PROTEIN COUPLED TO A VARIETY OF DIFFERENT EFFECTOR MECHANISMS

$GABA_B$ receptors have the potential to produce a variety of different neuronal responses because they are coupled through inhibitory G proteins to several intracellular effectors (**Figure 11.6**). These different effectors enable GABA, acting through $GABA_B$ receptors, to have a broader range of effects than it could by acting on $GABA_A$ receptors alone. The primary actions of $GABA_B$-receptor activation include modulation of adenylyl cyclase activity, inhibition of voltage-dependent calcium channels and activation of inwardly rectifying potassium channels.

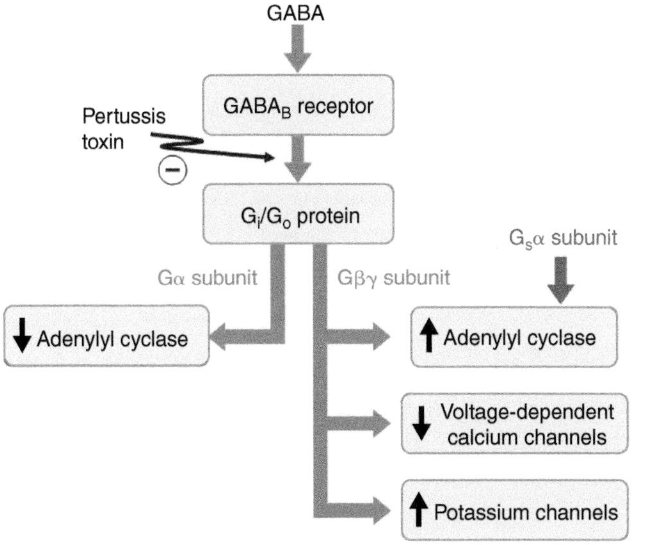

FIGURE 11.6 **A schematic diagram depicting the major effector systems to which GABA$_B$ receptors are coupled.**

11.4.1 GABA$_B$ receptors regulate the activity of adenylyl cyclase

Adenylyl cyclase converts ATP to cyclic AMP. Cyclic AMP, in turn, activates several different target molecules, such as cyclic AMP-dependent protein kinase (protein kinase A or PKA) to regulate cellular functions, including gene transcription, cellular metabolism and synaptic plasticity. Nine isoforms of adenylyl cyclase (types I to IX) have been identified and all are expressed in neurons. The α-subunit of G_i and G_o proteins inhibits several adenylyl cyclase isoforms, including types I, V and VI.

GABA$_B$ receptors are negatively coupled to adenylyl cyclase through inhibitory G proteins

The ability of GABA$_B$ receptors to couple to inhibitory G proteins suggested that GABA$_B$ receptor activation would inhibit adenylyl cyclase activity through activation of $G_i\alpha$ and/or $G_o\alpha$ proteins. To test this hypothesis, the effect of GABA$_B$-receptor activation on adenylyl cyclase activity was measured by the enzymatic conversion of [α-^{32}P]ATP to cyclic [^{32}P]AMP in crude synaptosomal preparations from a variety of regions of the rat brain. Application of baclofen or GABA caused a decrease in cAMP levels, reflecting a reduction in basal adenylyl cyclase activity (**Figure 11.7a**). This effect was blocked by the GABA$_B$-receptor antagonist, CGP 35348, indicating that it was mediated by GABA$_B$ receptors (**Figure 11.7b**). Application of pertussis toxin dramatically reduced the effect of baclofen on adenylyl cyclase. Since pertussis toxin selectively inactivates G_i/G_o proteins, these results demonstrate that GABA$_B$ receptors are negatively coupled to adenylyl cyclase through one or both of these inhibitory G proteins.

Reconstitution experiments have also been used to demonstrate that GABA$_B$ receptors are negatively coupled to adenylyl cyclase through inhibitory G proteins. Purified phospholipids were combined with purified GABA$_B$ receptor, partially purified G_i/G_o protein, partially purified adenylyl cyclase and GTP to form a reconstituted membrane preparation. This preparation was then incubated with forskolin, to activate the adenylyl cyclase, and either baclofen or GABA, to activate the GABA$_B$ receptors. In theory, during this incubation, the baclofen or GABA should bind to the GABA$_B$ receptor, causing a decrease in the formation of cAMP by adenylyl cyclase as compared to the level of cAMP formation in the absence of baclofen or GABA. This was exactly what happened (**Figure 11.7c**). Furthermore, the inhibitory effect of baclofen and GABA on adenylyl cyclase was antagonized by the addition of the GABA$_B$ receptor antagonist, 2-hydroxysaclofen, demonstrating that the inhibition was mediated by GABA$_B$ receptors.

To demonstrate the necessity of each element in the preparation, partially reconstituted membrane preparations were prepared. As predicted, inhibition of cAMP formation by baclofen or GABA was not observed if either the GABA$_B$ receptor or the G_i/G_o protein was omitted from the preparation. Furthermore, the omission of adenylyl cyclase resulted in the almost complete absence of cAMP formation. The inability of GABA$_B$ receptors to inhibit cAMP formation in the absence of G_i/G_o protein further confirms that GABA$_B$ receptors can negatively couple to adenylyl cyclase through either or both of these G proteins.

GABA$_B$ receptors facilitate neurotransmitter-mediated activation of adenylyl cyclase

In contrast to its direct suppression of cAMP levels through the α-subunit of G_i/G_o proteins, GABA$_B$-receptor activation can also have another seemingly opposite effect on cAMP accumulation. When adenylyl cyclase is stimulated to produce cAMP by a G_s-protein-coupled receptor, GABA$_B$ receptor activation will enhance this increase in cAMP accumulation. For example, addition of baclofen enhances by two- to threefold the increase in cAMP accumulation produced by norepinephrine (β-receptors), adenosine (A2 receptors) or vasoactive intestinal peptide (VIP) receptors (**Figure 11.8**). This effect is contrary to the inhibition of adenylyl cyclase discussed above.

The mechanism of this effect lies in the ability of the βγ-subunit from the G_i/G_o protein, liberated by the activation of GABA$_B$ receptors, to synergize the interaction of $G_s\alpha$ with certain isoforms of adenylyl cyclase, specifically types II, IV and VII. The stimulatory action of GABA$_B$ receptors on cAMP levels represents a form of G-protein cross-talk. It depends upon the simultaneous activation of GABA$_B$ receptors and a G_s-coupled GPCR and the expression of appropriate adenylyl cyclase isoforms. Under these conditions, adenylyl cyclase types II,

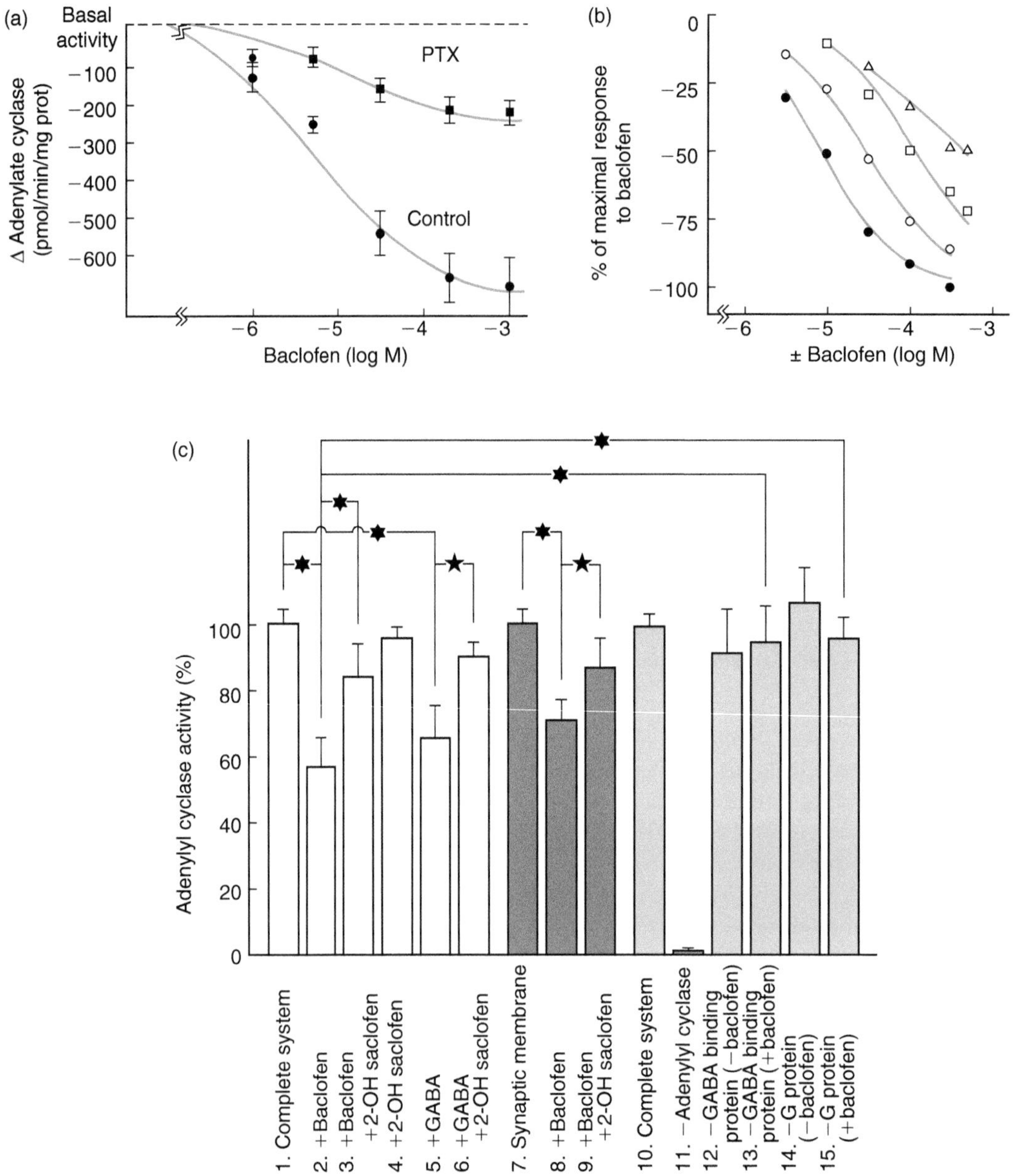

FIGURE 11.7 **GABA$_B$ receptors couple to adenylyl cyclase through inhibitory G proteins. (a)** Adenylyl cyclase activity in membranes of cerebellar granule cells was measured by the conversion of [α-^{32}P]ATP to cyclic [^{32}P]AMP. In control preparations, baclofen decreased the activity of adenylyl cyclase in a concentration-dependent manner (filled circles). Treatment of the membranes with pertussis toxin (PTX) antagonized the effect of baclofen on adenylyl cyclase activity (filled squares). **(b)** The inhibition of adenylyl cyclase activity by baclofen was antagonized in a concentration-specific manner by the GABA$_B$ receptor antagonist CGP 35348. The inhibition produced by increasing concentrations of baclofen alone (filled circles) was compared to that observed in the presence of baclofen plus either 0.6 mM (open circles), 1.5 mM (open squares) or 5 mM (open triangles) CGP 35348. **(c)** The effect of baclofen and GABA on adenylyl cyclase activity measured in reconstituted membranes (open bars), synaptic membranes (dark bars) and partially reconstituted membranes (light bars). Note that the removal of the G protein from the reconstituted system (15) blocked the inhibitory effect of baclofen (2). See text for details. Significant differences are indicated by an asterisk (V) signifying $p < 0.05$ or a star (H) signifying $p < 0.01$. Part (**a**) from Xu J, Wojcik WJ (1986) Gamma aminobutyric acid B receptor-mediated inhibition of adenylate cyclase in cultured cerebellar granule cells: blockade by islet-activating protein. *J. Pharmacol. Exp. Ther.* **239**, 568–573, with permission. Part (**b**) adapted from Holopainen I, Rau C, Wojcik WJ (1992) Proposed antagonists at GABA$_B$ receptors that inhibit adenylyl cyclase in cerebellar granule cell cultures of rat. *Eur. J. Pharmacol. Mol. Pharmacol. Section* **227**, 225–228, with permission. Part (**c**) from Nakayasu H, Nishikawa M, Mizutani H, Kimura H, Kuriyama K (1993) Immunoaffinity purification and characterization of γ-aminobutyric acid (GABA)$_B$ receptor from bovine cerebral cortex. *J. Biol. Chem.* **268**, 8658–8664, with permission.

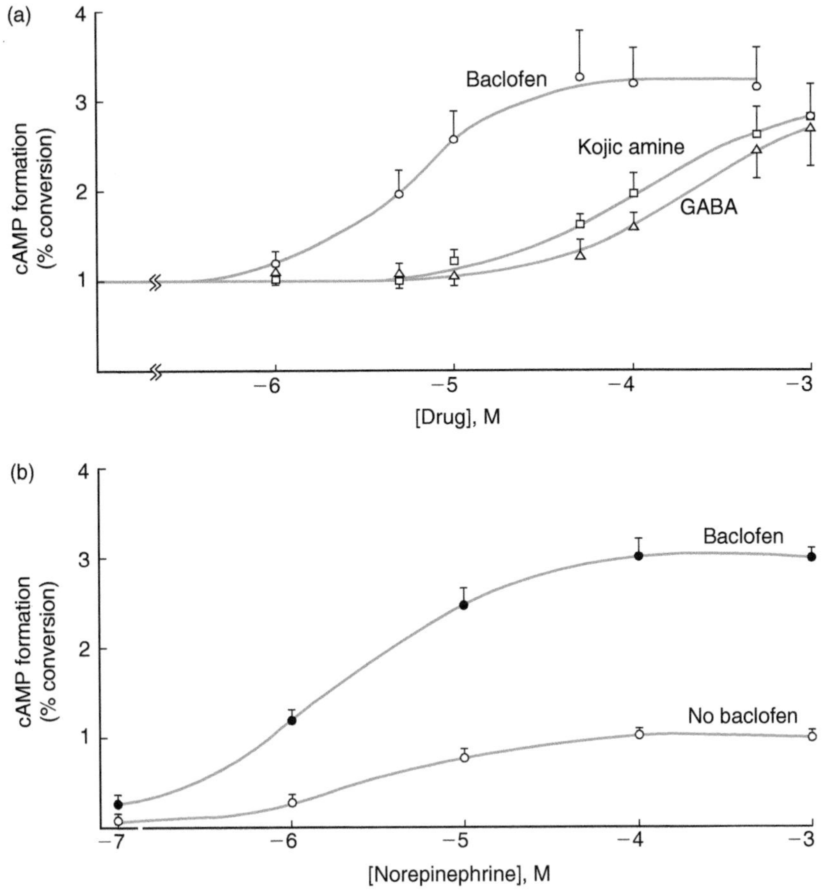

FIGURE 11.8 **The effect of GABA$_B$ receptor activation on norepinephrine stimulated cAMP accumulation. (a)** The effect of different GABA$_B$ receptor agonists on the cAMP accumulation produced by 100 μM norepinephrine in rat brain cerebellar slices is shown. The GABA$_B$ agonists baclofen (open circles), kojic amine (open squares) and GABA (open triangles) were applied at increasing concentrations in the presence of norepinephrine. All of these GABA$_B$ agonists enhanced cAMP formation produced by the norepinephrine. **(b)** Baclofen (100 μM) potentiates the cAMP formation induced by increasing concentrations of norepinephrine. The effect of norepinephrine alone (open circles) and norepinephrine plus baclofen (filled circles) is shown. From Karbon EW, Duman RS, Enna SJ (1984) GABA$_B$ receptors and norepinephrine-stimulated cAMP production in rat brain cortex. *Brain Res.* **306**, 327–332, with permission.

IV and VII can act as molecular 'coincidence detectors'. The adenylyl cyclase responds only minimally to activation by a single signal but synergistically to the coincident arrival of dual signals through separate pathways.

Accordingly, G protein α- and βγ-subunits liberated by the activation of GABA$_B$ receptors could produce opposing effects on cAMP levels. The α_i/α_o-subunits could directly inhibit one isoform of adenylyl cyclase while the βγ-subunits could synergize the G_s-mediated stimulation of a different adenylyl cyclase. Depending upon the overall balance between inhibitory and stimulatory effects, this could result in a net increase or decrease in cAMP accumulation. Through these mechanisms, GABA$_B$ receptors have the potential to regulate a variety of camp-dependent mechanisms in neurons.

11.4.2 GABA$_B$ receptor activation inhibits voltage-dependent calcium channels

GABA$_B$ receptors are negatively coupled to voltage-dependent calcium channels. By inhibiting calcium entry through these channels GABA$_B$ receptors have the potential to modulate a variety of neuronal functions, perhaps the most significant of which is the ability to regulate neurotransmitter release.

Heterodimeric GABA$_B$ receptors directly inhibit calcium currents

Inhibition of calcium currents by GABA$_B$-receptor activation was first observed in electrophysiological recordings made from neurons in the dorsal root ganglion (DRG). In this preparation, baclofen was found to decrease the calcium-dependent plateau phase of the action potential (**Figure 11.9a**). The effect of baclofen was blocked by a GABA$_B$ antagonist, indicating that it was mediated by GABA$_B$ receptors.

To confirm that GABA$_B$-receptor activation directly inhibited calcium channels the effect of baclofen on calcium currents in voltage-clamped DRG neurons was examined. The calcium current was pharmacologically isolated by application of blockers of sodium and potassium currents. Under these conditions, a depolarizing voltage step from a holding potential of −80 mV evoked a sustained inward calcium current. Baclofen reversibly reduced the amplitude of this current, indicating that GABA$_B$-receptor activation depresses voltage-dependent calcium currents in DRG neurons (**Figure 11.9b,c**). Similar studies have subsequently confirmed that GABA$_B$-receptor activation can inhibit voltage-dependent calcium currents in many different types of both peripheral and central neurons.

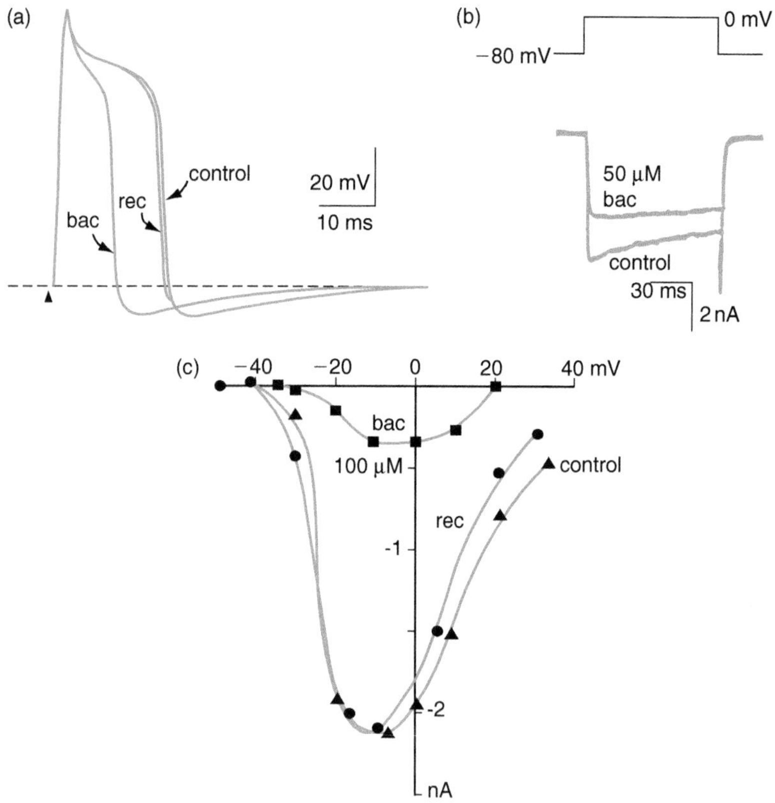

FIGURE 11.9 **Baclofen suppresses voltage-dependent calcium currents in DRG neurons. (a)** The effect of baclofen on the action potential in a DRG neuron. Baclofen (100 μM; bac) reversibly depressed the calcium-dependent plateau phase of the action potential compared to control (con) or wash (rec). **(b)** In the same preparation, 50 μM baclofen depressed the pharmacologically isolated calcium current (bottom). This current was evoked by a depolarizing voltage step from −80 mV to 0 mV (top). See text for details. **(c)** The current–voltage relationship for the voltage-dependent calcium current in DRG neurons is shown in control (con), in 100 μM baclofen (bac) and after 5 minutes of wash (rec). The current–voltage curve represents the amplitude of the calcium current evoked by a voltage step from the holding potential of −80 mV to a variety of test potentials. Baclofen markedly inhibited the calcium current. From Dolphin AC, Huston E, Scott RH (1990) GABA$_B$-mediated inhibition of calcium currents: a possible role in presynaptic inhibition. In Bowery NG, Bittiger H, Olpe H-R (eds) *GABA$_B$ Receptors in Mammalian Function*. Chichester: Wiley, with permission.

A subsequent study addressed the question of whether heterodimerization of GABA$_B$ receptors is required for the coupling of these receptors to calcium channels in neurons. In this study, GABA$_B$-expression constructs were injected into the nuclei of superior cervical ganglion (SCG) neurons, resulting in the expression of GABA$_B$-receptor protein. Baclofen had no effect on calcium currents in uninjected SCG neurons. However, the expression of heterodimeric GABA$_B$ receptors composed of GABA$_{B1}$ plus GABA$_{B2}$ resulted in a marked baclofen-mediated inhibition of calcium channel currents in these cells. The actions of baclofen were blocked by the selective GABA$_B$-receptor antagonist CGP62349, indicating that the effect was mediated by GABA$_B$ receptors. Injection of an antisense construct to block GABA$_{B1}$ expression markedly decreased GABA$_{B1}$ protein levels as well as the inhibitory effects of baclofen on calcium currents. These results suggest that heterodimeric assemblies of GABA$_{B1}$ and GABA$_{B2}$ are necessary for GABA$_B$-receptor-mediated inhibition of calcium-channel currents.

GABA$_B$ receptors inhibit a variety of voltage-gated calcium channels

The voltage-dependent calcium current evoked in a given cell is typically produced by the activation of several different calcium channel types. Thus, partial suppression of this current by GABA$_B$-receptor activation could be produced by a partial inhibition of several different channel types or the complete inhibition of only a single type. Because of the different physiological functions of the various voltage-dependent calcium channels, it is important to determine the type(s) of calcium channel inhibited by GABA$_B$ receptors. This can be accomplished through the use of calcium-channel antagonists that are specific for different calcium channels. The ability of a selective antagonist to occlude inhibition of a calcium current by baclofen indicates that the antagonist and baclofen are acting on the same subset of channels. Alternately, specific calcium channel antagonists can be used to pharmacologically isolate a single type of calcium current and the effect of GABA$_B$-receptor activation assessed. Finally, kinetic analysis of the calcium current inhibited by GABA$_B$ receptors can be used to determine the electrophysiological characteristics of the inhibited current, which can then be compared to the known properties of identified calcium channels.

Using these techniques, GABA$_B$ receptors have been shown to inhibit all types of calcium channels (**Figure 11.10**). Inhibition of N-type and P/Q-type calcium channels by GABA$_B$ receptors is observed in most neurons. In comparison, GABA$_B$-receptor-mediated inhibition of L-type channels is dependent upon the cell type. For example, it is observed in cerebellar granule neurons and hippocampal pyramidal neurons, but not in cerebellar Purkinje neurons, spinal cord neurons or thalamocortical neurons. Similarly, GABA$_B$-receptor-mediated inhibition of T-type calcium channels is also neuron dependent. Baclofen suppresses current through

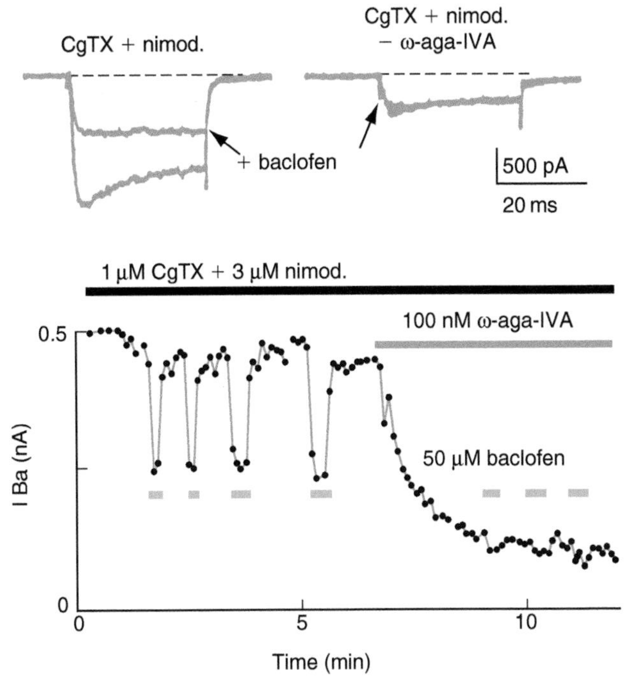

FIGURE 11.10 **Baclofen suppresses the P-type calcium current in cerebellar Purkinje neurons.** In the presence of 1 μM ω-conotoxin (CgTX) and 3 μM nimodipine (nimod.) to block N-type and L-type calcium channels a voltage step from −80 mV to +10 mV elicits an inward calcium current (top left). This current is partially inhibited by 50 μM baclofen. Application of the P-type calcium channel antagonist ω-agatoxin-IVA (ω-aga-IVA; 100 nM) partially blocks the current and occludes any further inhibition by baclofen (top right). The time course of the peak calcium channel current amplitude throughout the experiment is shown below. CgTX and nimod. are applied throughout the experiment (black bar), ω-aga-IVA is applied for the period of time indicated by the green bar. Note that ω-aga-IVA suppressed the calcium current and completely occluded any further inhibition by baclofen, demonstrating that baclofen was acting on the P-type calcium current. In this experiment, barium was exchanged for calcium so the currents that were measured represent barium flux through calcium channels. From Mintz, IM, Bean, BP (1993) GABA$_B$ receptor inhibition of P-type Ca^{2+} channels in central neurons. *Neuron* **10**, 889–898, with permission.

these channels in DRG neurons and interneurons in the stratum lacunosum moleculare of the hippocampus, but not in thalamocortical neurons or pyramidal neurons of the hippocampus. Thus, GABA$_B$-receptor activation has the potential to inhibit a variety of different voltage-dependent calcium channels. The mechanisms that enable cell-type-dependent regulation of these calcium channels by GABA$_B$ receptors are not well understood.

Inhibition of calcium channels is dependent upon G_i/G_o proteins

Several lines of evidence were used to demonstrate the involvement of G proteins in the inhibition of calcium channels by GABA$_B$ receptors (**Figure 11.11**). First, simply omitting GTP from the internal pipette solution during whole-cell recording gradually blocked the effect of baclofen on the calcium current. This occurred because, in the absence of a replacement supply, GTP slowly washed out of the cell during the experiment, thereby inactivating G proteins. Alternately, loading the cell with guanosine 5′-*O*-(2-thiodiphosphate) (GDP-β-S), a GDP analog that inhibits the binding of GTP to G proteins, antagonized the effect of baclofen on calcium currents. Conversely, the effect of baclofen was enhanced when cells were loaded with guanosine 5′-*O*-(γ-thiotriphosphate) (GTP-γ-S), a non-hydrolyzable GTP analog that irreversibly activates G proteins. These findings support a role for G proteins in the coupling of GABA$_B$ receptors to calcium channels. Furthermore, the observation that pertussis toxin blocks the inhibitory effect of baclofen on calcium channels indicated that GABA$_B$ receptors coupled to calcium channels through inhibitory G_i/G_o proteins.

In theory, G_i/G_o proteins could inhibit calcium channels by physically interacting with the channel itself or by resulting in the production of a second messenger molecule which would diffuse to and inhibit the channel. Experimental evidence indicates that G proteins interact directly with N-type and P/Q-type channels. Evidence for this conclusion came from experiments using cell-attached patches in DRG neurons. It was found that baclofen, applied outside the patch pipette, did not affect the amplitude of calcium currents in cell-attached patches. However, in the same cell, baclofen applied inside the patch pipette produced clear inhibition of the calcium current, demonstrating that baclofen was able to inhibit calcium currents in these cells. The inability of baclofen, applied outside the patch pipette, to inhibit calcium channels under the patch indicates that a diffusible second messenger was not involved in the inhibition.

G proteins inhibit calcium currents through a direct interaction of the βγ-subunit with the calcium channel

To identify which G proteins are involved in receptor-mediated inhibition of calcium channels in native systems, a number of studies were performed with blocking antibodies and antisense oligonucleotides complementary to G-protein subunits. These studies suggested that $G_o\alpha$-proteins were primarily responsible for the inhibition. However, other studies suggested a role for both $G_o\alpha$ and $G_i\alpha$. This led to the hypothesis that the species involved was the Gβγ-subunit, which is common to both of these G proteins, rather than any particular Gα-subunit. This idea received experimental support when it was demonstrated that transfection of primary neurons or cell lines with Gβγ, but not Gα-subunits, mimicked the effect of baclofen and produced tonic inhibition of the calcium current. Binding studies using purified Gβγ-dimers and recombinant calcium channel subunits have further demonstrated a direct interaction between Gβγ-subunits and calcium channels. It is now well established that G proteins inhibit N- and P/Q-type calcium currents through a direct interaction of the Gβγ-dimer with the calcium channel.

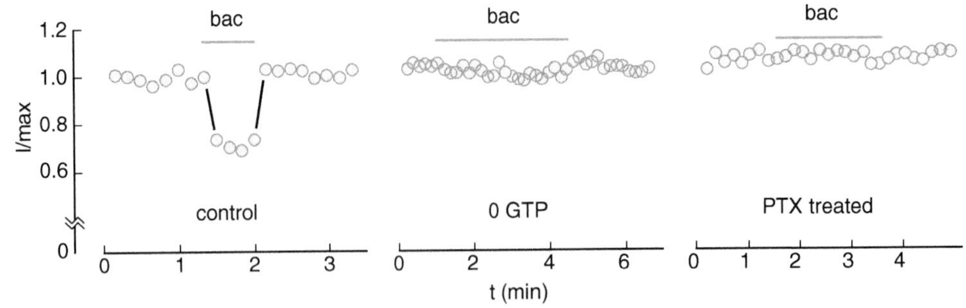

FIGURE 11.11 **GABA$_B$ receptors are coupled to calcium channels through inhibitory G proteins.** Calcium currents in cerebellar granule cells were evoked by stepping from a holding voltage of −80 mV to a test voltage of +10 mV. Current was expressed as a percentage of the maximal current in the cell at the beginning of the experiment. Bath application of baclofen (100 μM; bac) for the time indicated by the bar reduced the size of the calcium current (left). This inhibition was antagonized by removal of GTP from the internal pipette solution (center) and by pretreatment of the neurons 12 to 16 hours earlier with pertussis toxin (PTX; right). These observations indicate that GABA$_B$ receptors mediate inhibition of calcium channels through inhibitory G_i/G_o proteins. From Amico C, Marchetti C, Nobile M, Usai C (1995) Pharmacological types of calcium channels and their modulation by baclofen in cerebellar granules. *J. Neurosci.* **15**, 2839–2848, with permission.

GABA$_B$ receptors inhibit calcium channels by altering their voltage dependence

It was originally proposed that GPCRs, such as GABA$_B$ receptors, that inhibit calcium channels do so by blocking and thereby reducing the number of functional channels. However, subsequent experiments demonstrated that strong depolarization of the neuronal membrane could overcome the transmitter-mediated inhibition of the calcium current. This result cannot be explained by a mechanism which involves a reduction in the number of functional channels. Instead, it appears that receptor activation induces a large shift in the voltage dependence of channel activation. Thus, following exposure to the transmitter, almost all of the channels are still fully functional and can be opened by a strong depolarization. However, a percentage of the channels undergo a shift in voltage dependence so that they are no longer opened by small to moderate depolarizations. In practice, this means that a transmitter, such as GABA, is able to inhibit the calcium current during activation by low to moderate depolarization, but the inhibitory effect of the transmitter is lost during strong depolarizations (**Figure 11.12**). In light of this observation, calcium channels have been proposed to exist in two states termed 'willing' and 'reluctant' to describe their ease of activation. These two states exist in equilibrium according to the following model where $C_{willing}$ is the closed channel in the absence of transmitter, $C_{reluctant}$ is the closed channel in the presence of transmitter, and O is the open channel:

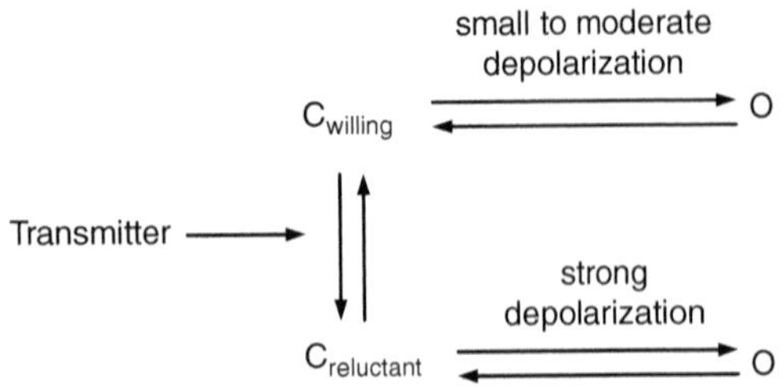

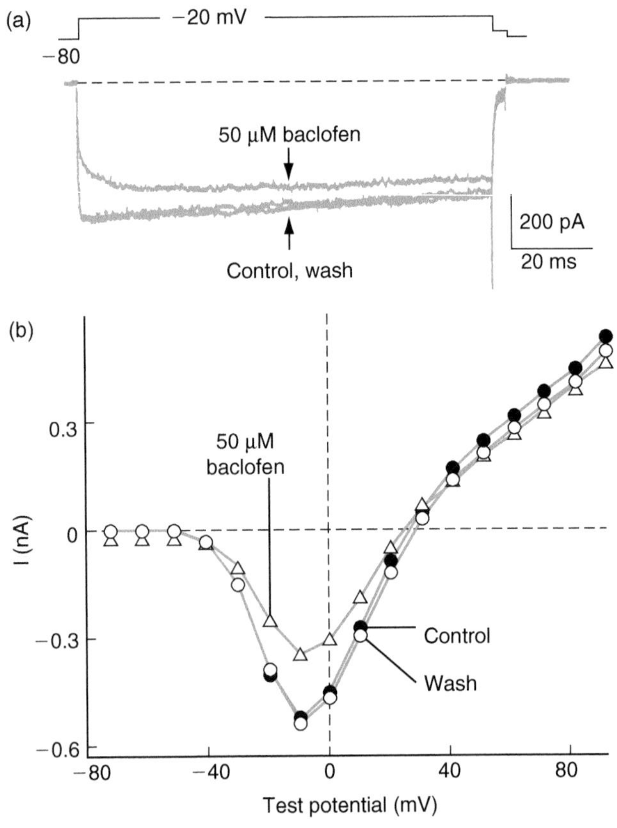

FIGURE 11.12 **Inhibition of the calcium current by baclofen is voltage dependent.** **(a)** In cerebellar Purkinje neurons a voltage step from −80 mV to −20 mV (top) in the presence of 1 μM ω-conotoxin and 3 μM nimodipine evokes a P-type calcium current (bottom). This current is suppressed by the subsequent application of baclofen (50 μM) and recovers following washout of the baclofen. **(b)** Inhibition of P-type calcium currents by baclofen is voltage dependent. This graph shows the amplitude of the calcium current evoked by depolarizing steps to a variety of different test potentials in control (filled circles), 50 μM baclofen (open triangles) and after washout of the baclofen (open circles). The holding potential was −80 mV. Baclofen inhibited the current most effectively when it was evoked with voltage steps to potentials below +20 mV. Strong depolarizations were able to overcome the inhibition produced by baclofen. Adapted from Mintz IM, Bean BP (1993) GABA$_B$ receptor inhibition of P-type Ca^{2+} channels in central neurons. *Neuron* **10**, 889–898, with permission.

Calcium channels predominantly exist in the willing mode in the absence of transmitter and can be activated by small to moderate depolarizations. In contrast, activation of G_i/G_o-coupled receptors, like GABA$_B$ receptors, shifts the balance of equilibrium to favor the 'reluctant' state in which large depolarizations are required to open the channels.

11.4.3 GABA$_B$ receptors activate potassium channels

GABA$_B$ receptors activate potassium currents mediated by G-protein-activated inwardly rectifying potassium (Kir) channels (previously termed GIRK channels). Through these potassium channels, GABA$_B$ receptors play a critical role in regulating neuronal excitability.

GABA$_B$ receptors couple to inwardly rectifying (Kir3) potassium channels

The interaction of GABA$_B$ receptors with potassium channels was initially suggested in experiments showing that baclofen produced a strong outward current and an increase in membrane conductance in voltage-clamped hippocampal pyramidal neurons (**Figure 11.13a,b**). A GABA$_B$-receptor antagonist blocked both the outward current and conductance increase produced by baclofen,

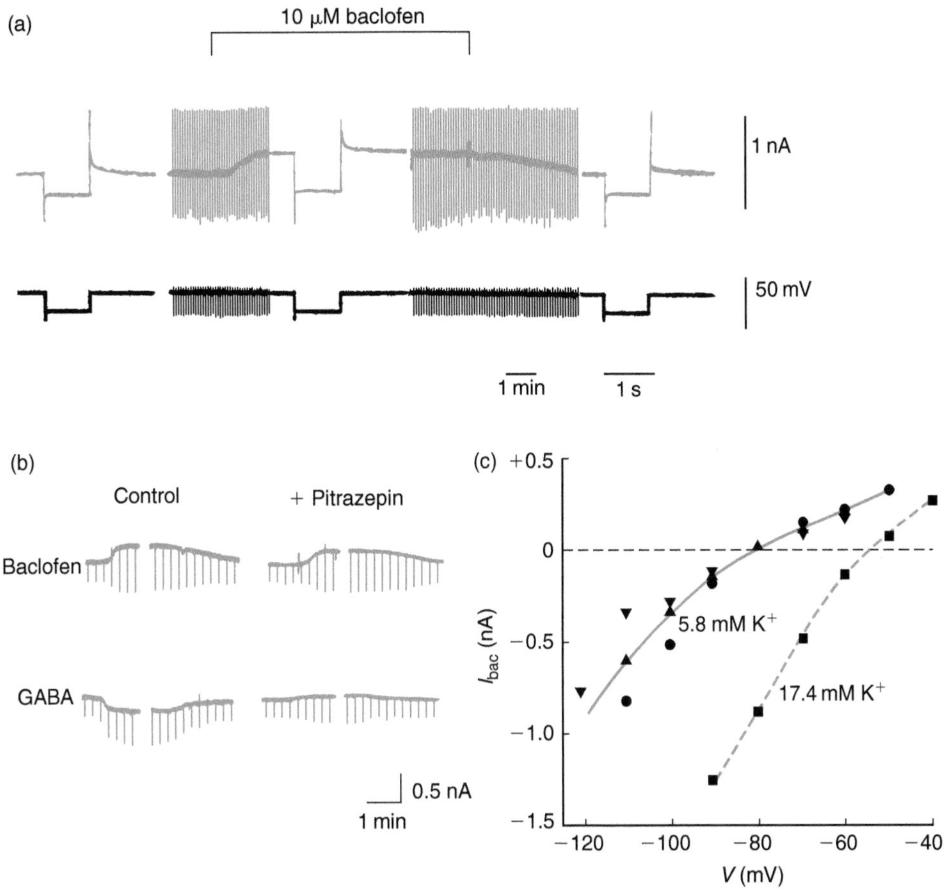

FIGURE 11.13 **GABA$_B$-receptor activation produces an outward current that is mediated by potassium ions. (a)** In a hippocampal pyramidal neuron held in voltage clamp at a membrane potential of −61 mV, baclofen (10 μM) produces an outward current, seen as a gradual upward deflection in the current recording as the baclofen washed into the preparation. Both the current (green) and voltage (black) recordings are shown. Voltage steps lasting 1 s were delivered repetitively during the experiment to assess the membrane conductance. At three points in the experiment the recording was expanded to show better the response to the voltage step. Note that in the presence of baclofen the current response to the voltage step is larger, indicating an increase in membrane conductance. **(b)** Another voltage clamped pyramidal cell shows a similar outward current in response to baclofen (top left). Baclofen increased the membrane conductance of this neuron as indicated by an increase in the amplitude of the current deflection in response to repetitive voltage steps (downward deflections). In this cell, GABA evoked an inward current and conductance increase (bottom left), suggesting that it was primarily acting on GABA$_A$ receptors. Blockade of the GABA$_A$-linked chloride conductance with pitrazepin (10 μM) had no effect on the baclofen-evoked current (top right), but caused the GABA response to become an outward current (bottom right). This occurred because blockade of the GABA$_A$-receptor-mediated inward chloride current enabled the underlying GABA$_B$-receptor-mediated outward current to become visible. **(c)** The current–voltage relationship for the baclofen-evoked current in a pyramidal cell when the extracellular concentration of potassium ions was 5.8 mM (circles, triangles) and 17.4 mM (squares). Altering the extracellular potassium concentration depolarized the reversal potential of the baclofen-evoked current by an amount predicted by the Nernst equation, indicating that this current was mediated by potassium ions. Adapted from Gähwiler BH, Brown, DA (1985) GABA$_B$-receptor-activated K^+ current in voltage-clamped CA3 pyramidal cells in hippocampal cultures. *Proc. Natl Acad. Sci. USA* **82**, 1558–1562, with permission.

confirming that these effects were mediated by $GABA_B$ receptors. The current–voltage curve for the baclofen-mediated current in these pyramidal neurons displayed inward rectification and reversed at a membrane potential of about −80 mV. This reversal potential corresponded well with the calculated equilibrium potential for potassium ions, suggesting that the current was caused by an increase in the potassium conductance of the membrane.

This was further confirmed by measuring the shift in the reversal potential of the $GABA_B$-receptor-mediated current produced by an increase in the extracellular potassium concentration from 5.8 mM to 17.4 mM (**Figure 11.13c**). The reversal potential of the current depolarized at 26 mV, an amount close to that predicted by the Nernst equation (29 mV). This close agreement indicated that the $GABA_B$ current was carried by potassium ions. This conclusion was further confirmed when it was observed that compounds that are known to block potassium channels, such as extracellular barium or intracellular cesium, also blocked the response to baclofen and GABA.

Subsequent studies have supported these early findings and have further demonstrated that $GABA_B$ receptors couple to the Kir3 family of inwardly rectifying potassium channels. Specifically, heteromeric potassium channels composed of Kir3.1/Kir3.2 (GIRK1/GIRK2) or Kir3.1/Kir3.4 (GIRK1/GIRK4) couple with high efficiency to $GABA_B$ receptors in a variety of heterologous systems. Another study, using mice whose Kir3.2 genes were genetically deleted, reported that $GABA_B$-receptor-mediated potassium currents were absent in CA3 hippocampal neurons. Analysis of Kir3 protein levels in these mice revealed a lack of Kir3.2 protein and a substantial reduction in Kir3.1 protein, indicating a critical role for Kir3 proteins in mediating the effect of $GABA_B$-receptor activation. Alternately, examination of the electrophysiological and pharmacological properties of the $GABA_B$-receptor-mediated potassium current in hippocampal CA3 pyramidal neurons revealed that this current shared similar properties to that mediated by Kir3.1/Kir3.2 or Kir3.1/Kir3.4 potassium channels. These studies point to the important role of Kir3 channels in mediating the effects of $GABA_B$ receptors. However, it has also been reported that, in some neurons, baclofen can induce linear or outwardly rectifying potassium conductances, suggesting that channels other than Kir3 can also contribute to the $GABA_B$-receptor-mediated potassium current.

$GABA_B$ receptors are coupled to potassium channels via inhibitory G proteins

Just as they are linked to their other effector systems, $GABA_B$ receptors are coupled to potassium channels through G proteins. This conclusion is based on the observation that GDP-β-S reduced the potassium current produced by baclofen. In contrast, GTP-γ-S mimicked the effect of baclofen. Exposure to pertussis toxin blocked the activation of potassium channels by both baclofen and GABA, indicating that the effect of $GABA_B$ receptors on potassium channels is achieved through either one or both of the inhibitory G proteins, G_i and/or G_o.

The identity of the G protein through which $GABA_B$ receptors couple to potassium channels was further examined using heteromeric $GABA_B$ receptors ($GABA_{B1a}$/$GABA_{B2}$ or $GABA_{B1b}$/$GABA_{B2}$) expressed in HEK293 cells which stably expressed Kir3.1/Kir3.2 potassium channels. In these cells, all endogenous G_i/G_o protein activity was eliminated by pertussis toxin treatment, preventing the $GABA_B$ receptors from activating any potassium current. The introduction of mutant pertussis toxin resistant G_i/G_o proteins in these cells then allowed the determination of those G proteins which would rescue coupling between $GABA_B$ receptors and potassium channels. Interestingly, G-protein coupling by $GABA_{B1a}$- and $GABA_{B1b}$-containing receptors was different. For receptors containing $GABA_{B1a}$, coupling to Kir3 channels was rescued only by the addition of G_{oA} proteins. However, for receptors containing $GABA_{B1b}$, both G_{oA} and G_{i2} proteins rescued Kir3 coupling. Thus, both $GABA_{B1}$ subunits appear able to signal through G_{oA} protein. However, the ability of $GABA_{B1b}$ to also signal through G_{i2} protein suggests differences in receptor–effector coupling.

$GABA_B$ receptors are directly coupled to potassium channels by βγ-subunits of G proteins

Several lines of evidence indicate that G_i/G_o proteins couple to potassium channels via their βγ-subunits. For example, application of purified Gβγ-subunits but not Gα-subunits to the intracellular surface of excised patches of chick embryonic atrial cells activated G-protein-gated potassium channels, suggesting that βγ-subunits carried the functional signal. This suggestion was confirmed by subsequent binding studies which demonstrated a direct interaction between Gβγ-subunits and Kir3.1, Kir3.2 and Kir3.4. Similarly, it was found, using the yeast two hybrid system that the G-protein β-subunit bound directly with the amino-terminus of Kir3.1. Mutational analysis was then used to determine the binding site for Gβγ-subunits on the Kir3 proteins and revealed that βγ-subunits bound to sites on both the amino- and carboxy-terminus of the Kir3 protein.

The ability of Gβγ-dimers to activate Kir3 channels brought into question the molecular determinants of the interaction specificity. For example, in native tissues only G_i/G_o and not G_s proteins activate Kir3 channels and yet all of these G proteins release free βγ-subunits upon receptor stimulation. It is known that receptor specificity does not lie at the level of the βγ-subunit since a variety of different βγ-subunits have been shown to be equally effective at stimulating Kir3 channels. In a mammalian expression system (HEK293 cells), G_i/G_o-coupled receptors but not G_s-coupled receptors activate Kir3.1/Kir3.2 channels, suggesting that the receptor specificity lies at the level of the G-protein α-subunit. This possibility was confirmed by the observation that G_s-coupled receptors could be made to stimulate potassium channels by swapping critical residues on the carboxy-terminus of the $G_s\alpha$-subunit with those of the $G_i\alpha$-subunit. Thus, Gβγ directly controls Kir3 channels; however, Gα determines the specificity of receptor action.

The Gα-subunit determines receptor specificity by facilitating the association between Gβγ and the Kir3 channel. Recent studies using fluorescence resonance energy transfer (FRET) and total internal reflected fluorescence (TIRF) microscopy reported that the Gβγ-complex is closely associated with the Kir3 channel's cytosolic domains at rest and that upon GPCR activation the Gβγ-dimer undergoes a change in its relative position on the channel to promote activation. The Gβγ association with the channel at rest depends on its interaction with Gα. Thus, the specificity of the interaction between Gβγ and the Kir3 channel results from the close association between these proteins and this interaction is facilitated by Gα.

$GABA_B$ receptors and Kir3 channels form a macromolecular signaling complex in lipid rafts

Recent studies indicate that $GABA_B$ receptors and their effector Kir3 channels form tight associations within lipid rafts. Lipid rafts are specialized plasma membrane microdomains enriched in certain lipids that can serve as platforms for signaling molecules. For example, G_i/G_o proteins are enriched in the lipid raft fraction from cerebellar membranes. $GABA_B$ receptors and Kir3 channels are also enriched in lipid rafts. The presence of each of these proteins suggests that lipid rafts serve to cluster $GABA_B$ receptors with their effector and signaling systems. This suggestion received further support from studies using FRET and fluorescently-labeled proteins that report a tight association between $GABA_B$ receptors, Kir3 channels and G proteins in macromolecular signaling complexes. It has been proposed that certain $GABA_{B1}$ isoforms and Kir3 channel compositions preferentially coexist in these signaling complexes, providing a mechanism for functional heterogeneity within the $GABA_B$ system.

$GABA_B$ receptor-activated potassium channels display flickering behavior

Single-channel potassium currents, activated by baclofen or GABA, can be recorded from cell-attached patches of cultured hippocampal neurons. They are blocked by $GABA_B$-, but not $GABA_A$-receptor antagonists, indicating that they are $GABA_B$-receptor dependent. These currents are potassium selective. Thus, alterations in the concentration of potassium ions in the pipette cause a corresponding shift in the reversal potential of the single-channel current. The single-channel current amplitude that occurs with highest probability is about 4 pA (**Figure 11.14**). This corresponds to a conductance of 67 pS.

Kir3 channels that are coupled to $GABA_B$ receptors exhibit complex behavior with a number of different gating modes. Activation of $GABA_B$ receptors alters the gating of these channels such that the channel spends more time in modes with higher open probabilities. When activated, a prominent characteristic of these single-channel currents is a rapid flickering between open and closed states. This flickering appears to show a variety of different subconductance levels (**Figure 11.15a**). These different conductance levels are particularly prominent during wash on and wash out of the baclofen (**Figure 11.15b**). Histograms of the current amplitudes reveal many peaks which appear to occur at multiples

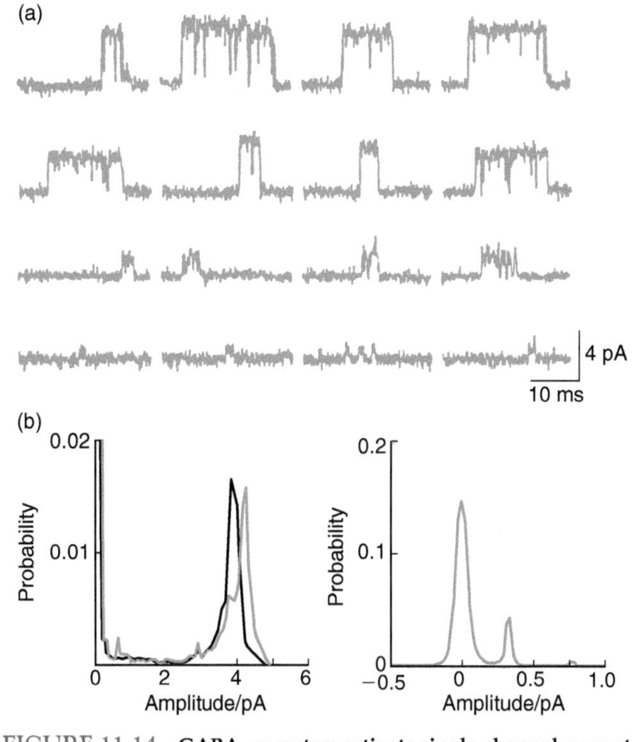

FIGURE 11.14 **$GABA_B$ receptors activate single-channel currents with a mean amplitude of 4 pA. (a)** Examples of single-channel currents evoked by $GABA_B$ receptor activation in cultured hippocampal neurons. Currents were recorded in the cell-attached patch configuration and GABA was applied through the bath to the membrane outside of the patch. Currents in the lower two rows were selected for this figure because of their small amplitude. **(b)** Current amplitude probability histograms of the $GABA_B$-receptor-mediated single-channel current. These histograms were constructed from data collected from the same patch as in (**a**). The graph on the left shows two histograms (black and green lines) representing the current amplitudes taken from two unbroken segments of data in this patch. Note that in both cases a channel with an amplitude of about 4 pA occurred with the greatest probability. The histogram on the right was taken from sections of the data that had the smallest currents. The smaller peak corresponds to elementary channel current with an amplitude of 0.36 pA. From Premkumar LS, Chung S-H, Gage PW (1990) GABA-induced potassium channels in cultured neurons. *Proc. R. Soc. Lond. B* **241**, 153–158, with permission.

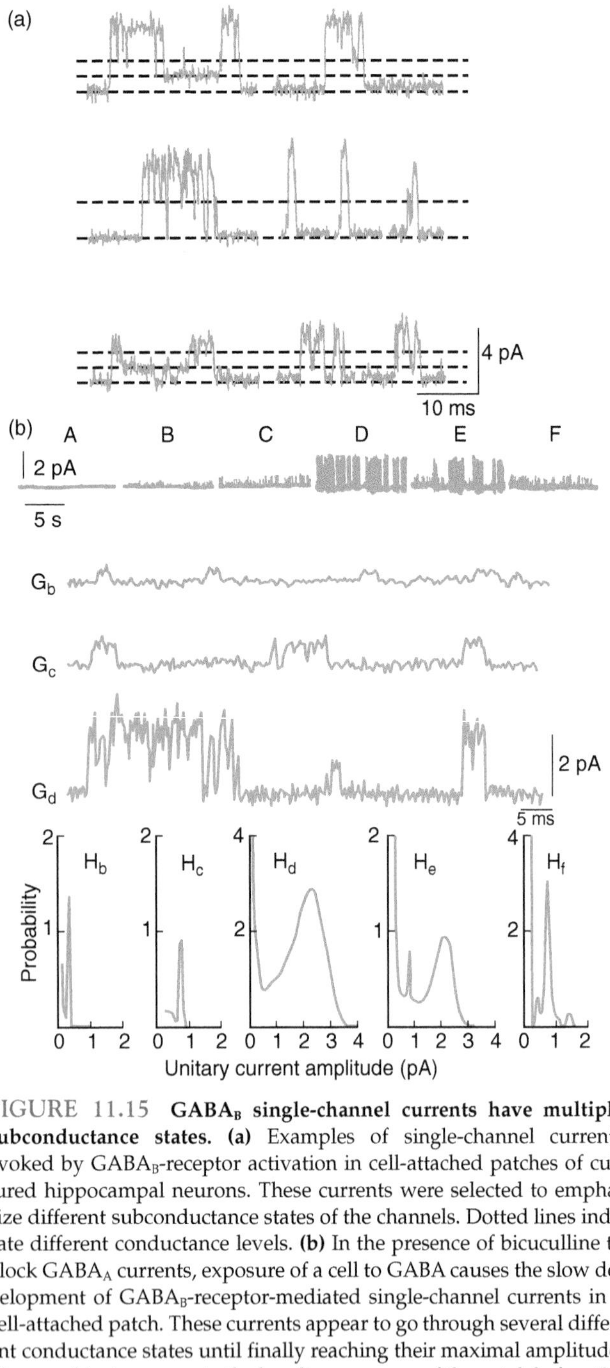

FIGURE 11.15 **GABA$_B$ single-channel currents have multiple subconductance states. (a)** Examples of single-channel currents evoked by GABA$_B$-receptor activation in cell-attached patches of cultured hippocampal neurons. These currents were selected to emphasize different subconductance states of the channels. Dotted lines indicate different conductance levels. **(b)** In the presence of bicuculline to block GABA$_A$ currents, exposure of a cell to GABA causes the slow development of GABA$_B$-receptor-mediated single-channel currents in a cell-attached patch. These currents appear to go through several different conductance states until finally reaching their maximal amplitude. The panel in A represents the baseline response of the patch before the addition of agonist. The panels in B–D show activity in the patch at 25 s (B), 1.5 min (C) and 4 min (D) after the addition of agonist. Panels E and F show patch activity after 5 and 10 minutes of wash, respectively. The records in the middle three rows represent an expansion of a portion of the panels shown in B (G_b), C (G_c), and D (G_d). Finally, the current amplitude probability histograms (H_b–H_f) were produced from data collected at the same time as panels B–F. Note the progressive increase in the GABA$_B$-receptor-mediated current amplitude following the application of GABA. From Premkumar LS, Chung S-H, Gage PW (1990) GABA-induced potassium channels in cultured neurons. *Proc. R. Soc. Lond. B* **241**, 153–158, with permission.

of the smallest peak. This smallest peak represents an elementary current amplitude of 0.36 pA (**Figure 11.14**), corresponding to a conductance of 5–6 pS. Whereas this measurement reflects the conductance coupled to somatic GABA$_B$ receptors, a subsequent study used non-stationary variance analysis to estimate the conductance linked to synaptically activated GABA$_B$ receptors. This study reported a small unitary conductance in the range of 5–12 pS, in agreement with the elementary conductance observed at the single-channel level. This small subconductance state predominates during synaptic GABA$_B$ currents.

In single-channel studies, the kinetics of the current activation and deactivation, even during large events (>4 pA), are extremely rapid. Since the elementary current amplitude is only 0.36 pA, many of these elementary channels would need to open or close synchronously to cause these rapid transitions. However, it is extremely unlikely that these channels would behave independently in such a synchronized fashion. Therefore, it has been suggested that these elementary channels function cooperatively. According to this hypothesis, these elementary co-channels would form oligomers of varying size which would function as a single unit. Activation of the oligomer would cause many of the channels to open simultaneously. The flickering behavior of the current would then represent the transient opening and closing of the elementary channels within the oligomer.

11.5 SUMMARY

GABA$_B$ receptors are coupled through inhibitory G_i/G_o proteins to multiple effector systems. The primary effects of GABA$_B$ receptor activation include inhibition of adenylyl cyclase, inhibition of voltage-dependent calcium channels and activation of inwardly rectifying potassium channels. Future studies may reveal other effector systems to which GABA$_B$ receptors are also coupled. By coupling to these different effector systems, GABA$_B$ receptors enable GABA to have a broader range of effects on neurons than it could by acting only on GABA$_A$ receptors. The discussion so far has focused on the intrinsic properties of the GABA$_B$ receptor and the effector systems to which they are coupled. We will now turn our attention to the role that these receptors play in synaptic activity.

11.6 THE FUNCTIONAL ROLE OF GABA$_B$ RECEPTORS IN SYNAPTIC ACTIVITY

GABAergic synapses in the central nervous system contain both GABA$_A$ and GABA$_B$ receptors capable of responding to the synaptic release of GABA. Once released, the lifetime of GABA in the synaptic cleft is very brief (milliseconds) both because the duration of the release is very short and the GABA that is released quickly diffuses away. In addition, there exists an avid uptake system to actively remove GABA from the synaptic cleft. These systems combine to tightly regulate GABA concentration in the synaptic cleft.

Synaptic activation of $GABA_A$ receptors produces a rapid, synchronous opening of chloride channels, resulting in a fast inhibitory postsynaptic current. In contrast, synaptic activation of $GABA_B$ receptors initiates a second-messenger-mediated process which is considerably slower. Because of the delay inherent in the second-messenger system, GABA has disappeared from the synaptic cleft before the $GABA_B$-receptor-mediated response even begins. Thus, the kinetics of this response are determined not by the binding/unbinding of GABA from the $GABA_B$ receptor but rather by the kinetics of the second-messenger system involved. The effects of $GABA_B$ receptors are exerted by both postsynaptic and presynaptic receptors, which play very different roles in neuronal function. Postsynaptic $GABA_B$ receptors inhibit excitatory transmission primarily by hyperpolarization. In contrast, the primary functional effect of presynaptic receptors is to inhibit the release of neurotransmitter.

11.6.1 Postsynaptic $GABA_B$ receptors produce an inhibitory postsynaptic current (IPSC)

When stimulated by synaptically released GABA, postsynaptic $GABA_B$ receptors increase the potassium conductance of the neuronal membrane. For a neuron near its resting potential, this increase in potassium conductance produces a large hyperpolarization of the membrane which is seen in a whole-cell voltage clamp recording as an outward current. This outward current, termed an inhibitory postsynaptic current (IPSC), is produced by the summation of the elementary current flowing through each of the $GABA_B$-receptor-activated potassium channels (**Figure 11.16a**). The small elementary conductance of the $GABA_B$-coupled potassium channel suggests that a large number of these channels open during an average-sized $GABA_B$ IPSC. In the example shown in **Figure 11.16a**, the conductance of the $GABA_B$ IPSC is 1.25 nS. Therefore, based on an elementary conductance for $GABA_B$-coupled potassium channels of 5–12 pS, it can be calculated that approximately 150 channels opened at the peak of the $GABA_B$ IPSC.

The kinetics of the $GABA_B$ receptor-mediated response are slow

Because it is coupled through a second-messenger system, the $GABA_B$-receptor-mediated hyperpolarization has a time course that is very different from that produced by an ionotropic receptor channel, such as $GABA_A$ (**Figure 11.16b**). Measurements of the time required from stimulation of the presynaptic terminals to the initiation of the postsynaptic hyperpolarization have ranged from 20 to 50 ms. This onset latency is considerably longer than that of the $GABA_A$-receptor-mediated response (<3 ms). The risetime of the $GABA_B$-mediated current is also slow and it does not reach a peak for 130 to 300 ms. This risetime is much slower than the $GABA_A$ response which typically reaches a peak in 1–15 ms. The slower risetime of the $GABA_B$ response is thought to occur because of the asynchronous activation of potassium channels. Finally, the $GABA_B$ response decays back to baseline over the next 400 to 1300 ms. This slow rate of decay may reflect the rate of GTP hydrolysis, suggesting that it is the decline of activated G protein that ultimately terminates the response. In contrast, the $GABA_A$ response decays to baseline much more rapidly (80–220 ms). The prolonged duration of the $GABA_B$ response enables GABA to produce inhibition over a much longer period of time than it could by acting on $GABA_A$ receptors alone.

$GABA_B$ receptors are more sensitive than $GABA_A$ receptors to GABA

Dose–response curves reveal that GABA is much more potent in activating $GABA_B$ than $GABA_A$ receptors. Despite the higher sensitivity of $GABA_B$ receptors to GABA, activation of these receptors requires high intensity or repetitive stimulation of the neuronal network. This situation arises because $GABA_B$ receptors are mostly located extrasynaptically and are not activated until sufficient GABA is released to overcome local uptake systems and spill over onto the receptors. However, the higher sensitivity of $GABA_B$ receptors to GABA enables these receptors to respond to the low concentrations of GABA that are able to reach these extrasynaptic spaces.

The $GABA_B$ IPSC produces inhibition by hyperpolarizing the neuronal membrane

Whole-cell voltage clamp recordings reveal that the maximal peak conductance increase produced by activation of $GABA_A$ receptors is much greater (5- to 10-fold) than that produced by activation of $GABA_B$ receptors. For example, in hippocampal pyramidal neurons, the maximal conductance of the $GABA_A$ IPSC ranges from 90 to 140 nS. This compares to a range of 13 to 19 nS for the maximal conductance of the $GABA_B$ IPSC in these same cells. Similar differences between the maximal conductance values of $GABA_A$- and $GABA_B$-receptor-mediated currents have been reported in other brain regions.

Despite its relatively small conductance, the $GABA_B$ current produces a large hyperpolarization from rest in most neurons. This strong hyperpolarization occurs because activation of $GABA_B$ receptors drives the membrane potential towards the reversal potential for potassium ions. In physiological conditions, the equilibrium

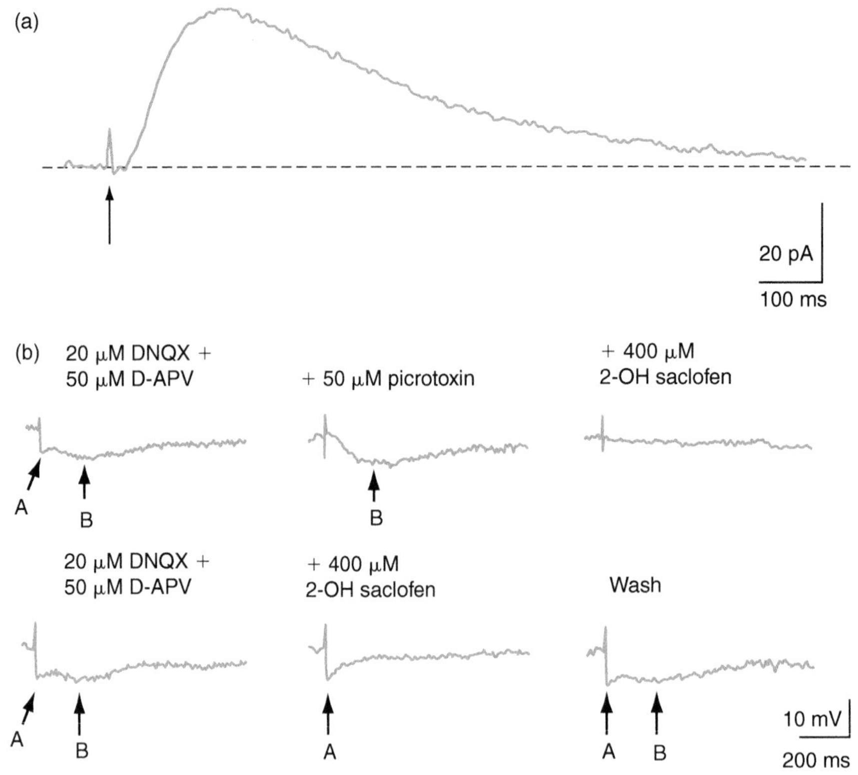

FIGURE 11.16 **Synaptically released GABA activates postsynaptic GABA$_B$ receptors to produce a slow IPSC. (a)** Stimulation of inhibitory fibers evokes a stimulus artefact (arrow) followed by a GABA$_B$-receptor-mediated IPSC in a hippocampal neuron held at a potential of −60 mV in a whole-cell voltage clamp. The GABA$_B$ IPSC was pharmacologically isolated from the excitatory synaptic current using DNQX, which blocks AMPA/kainate receptors, and APV, which blocks NMDA receptors. It was also isolated from the GABA$_A$ inhibitory current using bicuculline which blocks GABA$_A$ receptors. Note the slow onset of the IPSC and its long latency. **(b)** GABA$_A$ and GABA$_B$ inhibitory postsynaptic potentials (IPSPs) were recorded in current clamp from a dentate gyrus granule cell. These hyperpolarizing potentials were evoked by stimulating inhibitory fibers. They were isolated from glutamatergic excitatory potentials by application of DNQX and APV. GABA$_A$ and GABA$_B$ IPSPs are indicated by arrows labeled 'A' and 'B', respectively. In control (top left), a stimulus evoked a stimulus artifact (upward deflection) followed by both a GABA$_A$ and a GABA$_B$ IPSP which can be seen as the fast and slow components of the hyperpolarizing response, respectively. Application of the GABA$_A$ antagonist, picrotoxin blocks the GABA$_A$ IPSP leaving only the slow GABA$_B$ IPSP (top center). The GABA$_B$ antagonist 2-hydroxysaclofen blocks this GABA$_B$ IPSP. Similarly, in another cell, application of 2-hydroxysaclofen to the control response (bottom left) blocks the GABA$_B$ IPSP, leaving an isolated GABA$_A$ IPSP (bottom center). The effect of this antagonist is reversible (bottom right). Note the difference in the time course of the isolated GABA$_B$ IPSP (top center) and the isolated GABA$_A$ IPSP (bottom center). Part (**a**) from Mott DD, Lewis, DV Unpublished observations. Part (**b**) from Mott DD, Lewis DV (1992) GABA$_B$ receptors mediate disinhibition and facilitate long-term potentiation in the dentate gyrus. *Epilepsy Res.* **Suppl. 7**, 119–134, with permission.

potential for potassium ions (−80 to −98 mV) is quite negative relative to the resting membrane potential (−50 to −75 mV) of most cells. Therefore, even though the conductance of the GABA$_B$ IPSC is small, the driving force for potassium can be quite large. In fact, because of this large driving force and the long duration of the GABA$_B$ response, the GABA$_B$ IPSC can move an amount of charge that is close to that carried by the GABA$_A$ IPSC. For example, in granule cells of the dentate gyrus about 8 pC of charge leaves the cell during the GABA$_B$ IPSC. This compares favorably to the 9 to 35 pC that are carried by the GABA$_A$ response in these same cells.

The GABA$_A$ IPSC powerfully inhibits neuronal excitability both by hyperpolarizing the postsynaptic membrane and increasing its conductance. Hyperpolarization moves the postsynaptic membrane away from action potential threshold, whereas the conductance increase produced by the GABA$_A$ IPSC shunts the postsynaptic membrane thereby short-circuiting excitatory responses. This inhibition powerfully suppresses both voltage-dependent and voltage-independent excitatory currents and cannot be overcome by depolarization. In contrast, the GABA$_B$ IPSC produces a large hyperpolarization with a fairly small conductance increase. Thus, it inhibits its neurons primarily through hyperpolarization. This hyperpolarizing inhibition is effective in suppressing voltage-dependent currents, such as NMDA-receptor-mediated responses. However, since it can be overcome by neuronal depolarization, it does not effectively inhibit voltage-independent currents. Inhibition produced by GABA$_B$ receptors has been suggested to be a more modulatory form of inhibition than that produced by GABA$_A$ receptors, enabling a fine-tuning of neuronal function. Thus, GABA$_B$-receptor-mediated inhibition differs in both kinetics and function from inhibition produced by GABA$_A$ receptors.

11.6.2 Presynaptic GABA$_B$ receptors inhibit the release of many different transmitters

GABA$_B$ receptors are located on presynaptic terminals where they inhibit the release of a variety of neurotransmitters, including GABA, glutamate, dopamine, serotonin and norepinephrine. Inhibition of transmitter release by synaptically released GABA is dramatically enhanced following pharmacological blockade of GABA uptake, indicating that released GABA has to overcome uptake in order to reach these presynaptic GABA$_B$ receptors. By activating presynaptic GABA$_B$ receptors, synaptically released GABA can inhibit transmitter release at the inhibitory terminal from which the GABA was originally released (homosynaptic depression) as well as at neighboring inhibitory and/or excitatory terminals (heterosynaptic depression).

Presynaptic GABA$_B$ receptors inhibit the release of GABA

Inhibition of GABA release by presynaptic GABA$_B$ receptors has been especially well examined. It has been conclusively demonstrated that synaptically released GABA can feed back onto presynaptic GABA$_B$ receptors located on the activated GABAergic terminal as well as on other neighboring GABAergic terminals. These presynaptic GABA$_B$ receptors can then suppress the subsequent release of GABA, causing both GABA$_A$ IPSCs and GABA$_B$ IPSCs to be smaller.

This effect can be clearly observed in a cortical neuron using whole-cell voltage clamp to record GABA$_A$ IPSCs (**Figure 11.17**). During the delivery of paired electrical stimuli to GABAergic axons, the first stimulus of the pair evokes the release of GABA, resulting in the production of a GABA$_A$ IPSC. However, this released GABA also activates presynaptic GABA$_B$ receptors on the inhibitory terminals suppressing the release of further GABA. Thus, a second identical stimulus delivered 300 ms later evokes a GABA$_A$ IPSC that is greatly reduced. Application of the GABA$_B$ antagonist, 2-hydroxysaclofen, blocks the presynaptic GABA$_B$ receptors, preventing the reduction in the second IPSC. The ability of released GABA to act on presynaptic GABA$_B$ receptors to suppress subsequent GABA$_A$ IPSCs endows the GABAergic system with a powerful feedback mechanism capable of suppressing GABAergic inhibition in an activity-dependent manner.

The time course of the depression of GABA release is similar to the time course of the postsynaptic GABA$_B$ IPSC

Just like the postsynaptic effect of GABA$_B$ receptors, the time course of the inhibition of GABA release by GABA$_B$ receptors reflects a second-messenger-coupled mechanism. Following the stimulation of an inhibitory pathway, the onset of the presynaptic inhibition is slow, reaching a peak about 200 ms after the initial stimulus (Figure 11.18). The duration of the effect is also quite prolonged and can extend for up to several seconds. Thus, although GABA has a brief lifetime in the synaptic cleft, the activation of presynaptic GABA$_B$ receptors by this GABA enables it to modulate the subsequent release of transmitter for a more prolonged time.

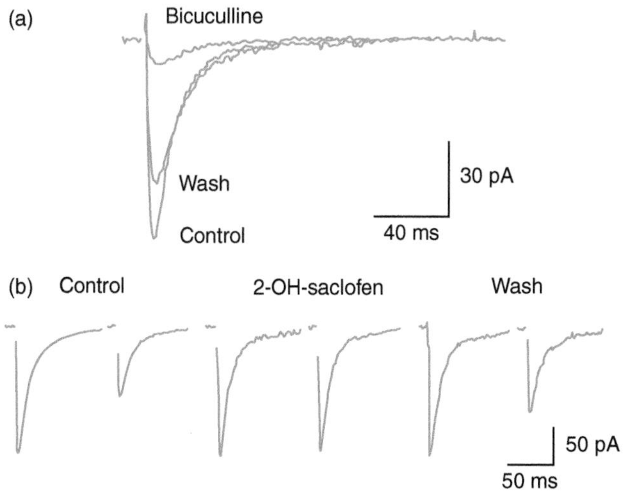

FIGURE 11.17 **Presynaptic GABA$_B$ receptors mediate paired pulse depression of IPSCs. (a)** The pharmacologically isolated GABA$_A$ IPSC in a neuron in the somatosensory cortex is reversibly blocked by bicuculline. This GABA$_A$ IPSC was evoked by electrical stimulation of inhibitory fibers and recorded in whole-cell voltage clamp. It was isolated from the excitatory synaptic current by application of CNQX and APV, antagonists of AMPA/kainate and NMDA receptors, respectively. The neuron was held at a membrane potential of −70 mV, causing the GABA$_A$ IPSC to be an inward current. **(b)** Under these same conditions, if paired stimuli are delivered 300 ms apart, the GABA$_A$ IPSC evoked by the second stimulus of the pair is reduced (left). This reduction of the second IPSC is blocked by the GABA$_B$ antagonist, 2-hydroxysaclofen (center). This effect is reversible after washout of the antagonist (right). The cell recorded in this experiment was from a young (postnatal day 10) rat. In animals of this young age, the postsynaptic GABA$_B$ response is developmentally immature, whereas the presynaptic GABA$_B$ response is fully developed. This difference in development explains why no GABA$_B$ IPSC is evident in these recordings and gives further confirmation that postsynaptic GABA$_B$ receptors are not necessary for paired pulse depression of IPSCs. In older animals, postsynaptic GABA$_B$ IPSCs are fully developed and both GABA$_A$ and GABA$_B$ IPSCs are depressed by presynaptic GABA$_B$ receptors (see Figure 11.18). Adapted from Fukuda A, Mody I, Prince DA (1993) Differential ontogenesis of presynaptic and postsynaptic GABA$_B$ inhibition in rat somatosensory cortex. *J. Neurophysiol.* **70**, 448–452, with permission.

GABA$_B$ receptors suppress transmitter release by directly targeting the release machinery and by inhibiting voltage-dependent calcium channels

GABA$_B$ receptors suppress transmitter release through multiple mechanisms. They have been shown

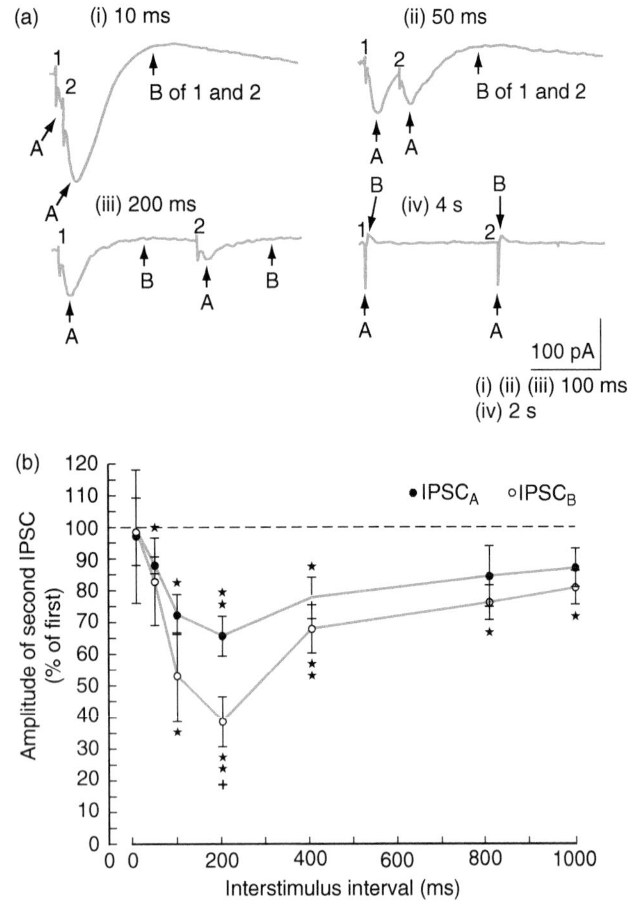

FIGURE 11.18 **Paired pulse depression of IPSCs is maximal when stimuli are delivered 200 ms apart. (a)** Isolated GABA$_A$ and GABA$_B$ IPSCs were recorded in whole-cell voltage clamp from a dentate gyrus granule cell. Since the cell was held at a membrane potential of −80 mV, the GABA$_A$ IPSC is an inward current whereas the GABA$_B$ IPSC is an outward current. Inhibitory fibers were electrically stimulated to evoke IPSCs. Paired stimuli were delivered at increasing intervals to determine the time course of the inhibition of GABA release produced by presynaptic GABA$_B$ receptors. Responses to paired stimuli at four different intervals are shown. In this cell, suppression of the second IPSC was greatest when the stimuli were delivered 200 ms apart. GABA$_A$ and GABA$_B$ IPSCs are indicated by arrows labeled 'A' and 'B', respectively. **(b)** Graph of the averaged data obtained from six cells showing the time course of the suppression of the second IPSC. For both the GABA$_A$ and GABA$_B$ IPSC the second response of the pair was maximally depressed when the stimuli were delivered about 200 ms apart. Stars indicate a significant depression of the IPSC (*$p < 0.05$, **$p < 0.01$). The cross (+) indicates that the GABA$_B$ IPSC was significantly more depressed than the GABA$_A$ IPSC. From Mott DD, Xie CW, Wilson WA, Swartzwelder HS, Lewis DV (1993) GABA$_B$ autoreceptors mediate activity-dependent disinhibition and enhance signal transmission in the dentate gyrus. *J. Neurophysiol.* **69**, 674–691, with permission.

to directly target the release machinery. GABA$_B$-receptor activation releases Gβγ-subunits that directly interact with SNARE (soluble N-ethylmaleimide-sensitive factor attachment protein (SNAP) receptor) proteins to limit vesicle fusion. In addition, they suppress neurotransmitter release through a voltage-dependent inhibition of N- and P/Q-type calcium channels. These calcium channels are expressed presynaptically and have been implicated in the control of neurotransmitter release. It has been reported that activation of GABA$_B$ receptor inhibits up to 50% of the calcium current. However, because of the non-linear relationship between calcium concentration in the presynaptic terminal and transmitter release, this 50% reduction in calcium current is often sufficient to inhibit neurotransmitter release by more than 90%.

In theory, GABA$_B$ receptors could suppress calcium currents by directly inhibiting calcium channels or by activating Kir3 channels that would hyperpolarize the presynaptic terminal and oppose the depolarization necessary for calcium-channel activation. Two lines of evidence argue in support of a direct effect of GABA$_B$ receptors on presynaptic calcium channels. First, at many terminals, GABA$_B$-receptor activation blocks some but not all types of calcium channel. This differential inhibition suggests a direct effect of GABA$_B$ receptors on distinct calcium channel types within the terminal. In contrast, the activation of a potassium conductance by GABA$_B$ receptors would cause a general decrease in all components of the calcium influx. Second, electrophysiological recordings from the giant nerve terminals (calyces of Held) in the medial nucleus of the trapezoid body have demonstrated that baclofen-mediated suppression of synaptic transmission was associated with a reduction in presynaptic calcium current but not with activation of a potassium current in the presynaptic terminal. These observations strongly argue that, at least at these synapses, presynaptic GABA$_B$ receptors suppress transmitter release by directly inhibiting calcium channels.

Several observations suggested that presynaptic GABA$_B$ receptors in these giant terminals were coupled to calcium channels via G_i/G_o proteins. First, loading of the presynaptic terminal with GDP-β-S blocked the effect of baclofen on calcium currents. In contrast, GTP-γ-S suppressed presynaptic calcium currents and occluded the effect of baclofen. Second, inhibition of calcium channels by baclofen was blocked by N-ethylmaleimide, a sulfhydryl alkylating agent which uncouples G_i/G_o proteins from their receptors. These results directly indicate that at this giant synapse GABA$_B$ receptors suppress synaptic transmission by inhibiting presynaptic calcium channels and that GABA$_B$ receptors couple to these calcium channels via G_i/G_o proteins.

As discussed previously (Section 11.4.3), GABA$_B$-receptor-mediated inhibition of calcium channels is voltage dependent. Strong depolarization of presynaptic terminals relieves the inhibition. The extent of depolarization of the presynaptic terminal and the level of GABA$_B$-mediated inhibition of transmitter release are therefore regulated by action potential frequency. During high-frequency activity, depolarization of the presynaptic terminal would relieve GABA$_B$ inhibition of calcium channels and restore neurotransmitter release. At inhibitory terminals, the relief of $GABA_B$-mediated inhibition of GABA release during high-frequency activity may serve as a feedback mechanism to prevent overexcitation.

11.7 SUMMARY

$GABA_B$ receptors enable GABA to produce a variety of effects on neuronal function. These receptors are located both pre- and postsynaptically where they can be activated by synaptically released GABA. Postsynaptic $GABA_B$ receptors generate a slow inhibitory current which is carried by potassium ions. This current produces a hyperpolarizing inhibition which effectively inhibits voltage-dependent conductances, such as the NMDA-receptor-mediated current. Presynaptic $GABA_B$ receptors inhibit the release of a variety of different neurotransmitters, including glutamate and GABA. The ability of $GABA_B$ receptors to regulate GABA release provides an important mechanism for the feedback control of both $GABA_A$ and $GABA_B$ inhibition. Thus, by acting at both pre- and postsynaptic sites, $GABA_B$ receptors have the potential to produce profound changes in neuronal function.

Further reading

Bettler, B., Kaupmann, K., Mosbacher, J., Gassmann, M., 2004. Molecular structure and physiological functions of $GABA_B$ receptors. Physiol. Rev. 84, 835–867.

Bettler, B., Tiao, J.Y., 2006. Molecular diversity, trafficking and subcellular localization of $GABA_B$ receptors. Pharmacol. Ther. 110, 533–543.

Bowery, N.G., Bettler, B., Froestl, W., et al., 2002. International Union of Pharmacology. XXXIII. Mammalian gamma-aminobutyric $acid_B$ receptors: structure and function. Pharmacol. Rev. 54, 247–264.

Couve, A., Filippov, A.K., Connolly, C.N., Bettler, B., Brown, D.A., Moss, S.J., 1998. Intracellular retention of recombinant $GABA_B$ receptors. J. Biol. Chem. 273, 26361–26367.

De Koninck, Y., Mody, I., 1997. Endogenous GABA activates small-conductance K^+ channels underlying slow IPSCs in rat hippocampal neurons. J. Neurophysiol. 77, 2202–2208.

Filippov, A.K., Couve, A., Pangalos, M.N., Walsh, F.S., Brown, D.A., Moss, S.J., 2000. Heteromeric assembly of $GABA_{B1}$ and $GABA_{B2}$ receptor subunits inhibits Ca^{2+} current in sympathetic neurons. J. Neurosci. 20, 2867–2874.

Gassmann, M., Bettler, B, 2012. Regulation of neuronal $GABA_B$ receptor functions by subunit composition. Nat. Rev. Neurosci. 13, 380–394.

Ikeda, S.R., 1996. Voltage-dependent modulation of N-type calcium channels by G-protein beta gamma subunits. Nature 380, 255–258.

Kaupmann, K., Huggel, K., Heid, J., et al., 1997. Expression cloning of $GABA_B$ receptors uncovers similarity to metabotropic glutamate receptors. Nature 386, 239–246.

Kaupmann, K., Malitschek, B., Schuler, V., et al., 1998. $GABA_B$-receptor subtypes assemble into functional heteromeric complexes. Nature 396, 683–687.

Malitschek, B., Schweizer, C., Keir, M., et al., 1999. The N-terminal domain of γ-aminobutyic $acid_B$ receptors is sufficient to specify agonist and antagonist binding. Mol. Pharmacol. 56, 448–454.

Parmentier, M.-L., Prézeau, L., Bockaert, J., Pin, J.-P., 2002. A model for the functioning of Family 3 GPCRs. Trends Pharmacol. Sci. 23, 268–274.

Schuler, V., Lüscher, C., Blanchet, C., et al., 2001. Epilepsy, hyperalgesia, impaired memory, and loss of pre- and postsynaptic $GABA_B$ responses in mice lacking $GABA_{B1}$. Neuron 31, 47–58.

Schwenk, J., Metz, M., Zolles, G., et al., 2010. Native $GABA_B$ receptors are heteromultimers with a family of auxiliary subunits. Nature 465, 231–235.

Takahashi, T., Kajikawa, Y., Tsujimoto, T., 1998. G-protein-coupled modulation of presynaptic calcium currents and transmitter release by a $GABA_B$ receptor. J. Neurosci. 18, 3138–3146.

White, J.H., Wise, A., Main, M.J., et al., 1998. Heterodimerization is required for the formation of a functional $GABA_B$ receptor. Nature 396, 679–682.

Yamada, M., Inanobe, A., Kurachi, Y., 1998. G protein regulation of potassium ion channels. Pharmacol. Rev. 50, 723–757.

C H A P T E R

12

The metabotropic glutamate receptors

Laurent Fagni, Jean-Philippe Pin

OUTLINE

Earlier chapters have described how glutamate is the neurotransmitter of most excitatory synapses in the CNS. Initially, the actions of glutamate in the nervous system were thought to be solely mediated by ligand-gated ion channels, also called ionotropic glutamate receptors (iGluRs). However, in the mid-1980s, several groups were looking for other types of glutamate receptors since glutamate was able to bind on brain membranes at specific sites different from the known iGluRs. At the same time, a new signaling cascade for G-protein-coupled receptors was discovered. It involves a phospholipase C that degrades phospholipids from the plasma membrane to produce two second messengers: inositol triphosphate and diacylglycerol (**Figure 12.1a**). Then two groups found that glutamate was able to activate this transduction cascade by activating a receptor different from the iGluRs (**Figure 12.1b**), both in cultured neurons and in brain tissues. Indeed, this response was neither mimicked by iGluR-selective agonists nor inhibited by iGluR-selective antagonists. Two years later, this receptor was successfully expressed in *Xenopus* oocytes after injection of rat brain mRNA, as illustrated by activation of an oscillatory chloride current in these cells (**Figure 12.1c**), a response typical for PLC-coupled receptors expressed in these cells. These data firmly demonstrated the existence of G-protein-coupled (or 'metabotropic') glutamate receptors (mGluRs). Today, it is well established that these receptors are involved in important brain physiological functions such as learning and memory, and play pivotal roles in neuropathologies such as pain, drug addiction, schizophrenia, anxiety, epilepsy, Alzheimer's and Parkinson's diseases. This chapter discusses the structural features and functions of mGluRs.

12.1 THE IDENTIFICATION OF THE EIGHT METABOTROPIC GLUTAMATE RECEPTOR SUBTYPES

A common way to clone a given gene is to utilize the function of the encoded protein to screen a library of cDNAs expressed in a heterologous system, such as

Cellular and Molecular Neurophysiology. DOI: 10.1016/B978-0-12-397032-9.00012-1

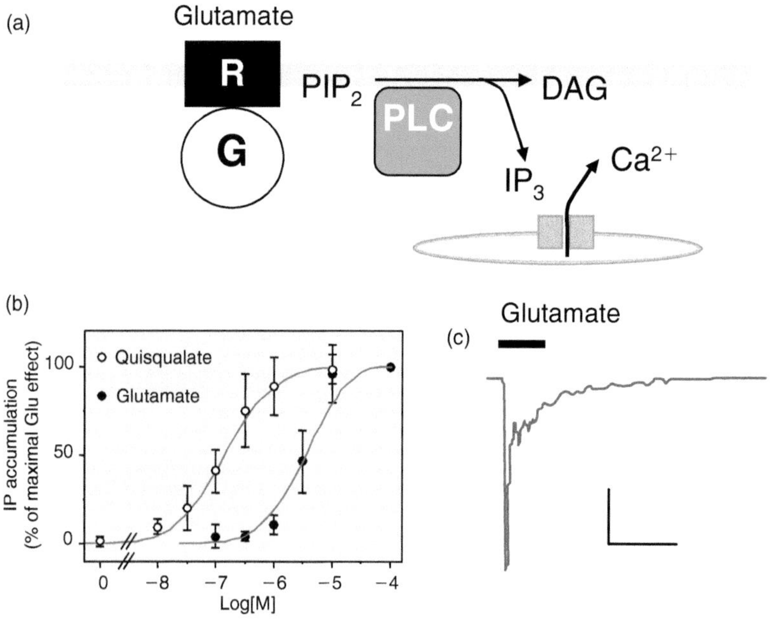

FIGURE 12.1 **Glutamate activates G-protein-coupled receptors producing the second messenger inositol triphosphates (IP_3). (a)** Metabotropic glutamate receptors (mGluRs) activate the enzyme PLC by stimulating specific G proteins. This enzyme degrades the membrane phospholipid phosphatidyl-inositol-biphosphate (PIP_2) to produce IP_3 and di-acylglycerol (DAG). IP_3 then acts on intracellular receptor channels and allows the release of Ca^{2+} from intracellular stores. **(b)** The first evidence for the existence of mGlu receptors came when glutamate was found to stimulate the production of IP_3 in cultured neurons and brain tissues. **(c)** Further evidence for the existence of PLC-coupled mGluRs was the observation that glutamate activates chloride currents through the release of Ca^{2+} from intracellular stores in *Xenopus* oocytes injected with rat brain mRNA (scale bars: vertical: 200 nA; horizontal: 20 s). Part (**b**) adapted from Sladeczek F, Pin J-P, Recasens M *et al.* (1985) Glutamate stimulates inositol phosphate formation in striatal neurons. *Nature* **317**, 717–719, with permission.

Xenopus laevis oocytes or cell lines. The first mGluR was cloned by injecting small pools of cDNA into oocytes and was identified using electrophysiological responses resulting from inositol phosphate production. It was named mGluR1. Hydrophobicity analysis of the deduced amino-acid sequence suggested seven transmembrane domains similar to all G-protein-coupled receptors. However, mGluR1 shared no sequence homology to these receptors, indicating that it belonged to a distinct G-protein-coupled-receptor family. Cloning of other members of the mGluR family was then accomplished using low-stringency hybridization of DNA probes derived from mGluR1, and looking for mGluRs presumably containing similar sequences. This approach allowed the identification of seven additional genes encoding mGluR2 to mGluR8. Alternative splicing of many of these mGluR genes results in further diversity among mGluR proteins with different C-terminal intracellular tails. The sequencing of both human and mouse genomes did not reveal additional homologous proteins aside from the more distant $GABA_B$, calcium-sensing, basic amino acid and sweet and monosodium glutamate taste receptors.

The cloning of eight mGluR subtypes immensely expanded the study of these receptors as clones were expressed in heterologous systems to determine their coupling to second-messenger systems and establish their pharmacological profiles. Based on sequence homology and nature of the coupled G protein, mGluR subtypes could be classified into three different groups (**Figure 12.2a**). The group-I mGluRs include mGluR1 and mGluR5. These receptors share approximately 60% sequence identity. They are most potently activated by quisqualate and are selectively activated by 3,5-dihydroxyphenylglycine (DHPG). The group-II mGluRs include mGluR2 and mGluR3 subtypes. They display about 70% sequence identity with each other, but less than 50% homology with the six other mGluR clones. They are potently and selectively activated by (2S,2′R,3′R)-2-(2′,3′-dicarboxycyclopropyl)glycine (DCG-IV) and LY354740. The group-III mGluRs (mGluR4, mGluR6, mGluR7 and mGluR8) share approximately 70% sequence identity within the group, and less than 50% identity with the other four mGluRs. L-2-amino-4-phosphonobutyric acid (L-AP4) is the most potent and selective agonist of group-III mGluRs.

Several variants generated by alternative splicing have been identified. Among those reported in the literature, 3, 2 and 2 variants for mGluR1, mGluR5 and mGluR7, respectively, have been well characterized and demonstrated *in vivo* (**Figure 12.2b**). All these variants differ at the level of their intracellular tail that is the site of interaction of several intracellular proteins, which regulate the trafficking, function, and desensitization properties of these proteins.

12.2 HOW DO METABOTROPIC GLUTAMATE RECEPTORS CARRY OUT THEIR FUNCTION? STRUCTURE–FUNCTION STUDIES OF METABOTROPIC GLUTAMATE RECEPTORS

The main function of mGluRs is to activate G proteins upon glutamate binding. Hydrophobicity analysis of various mGluRs suggests a large extracellular N-terminus with a signal peptide, followed by seven transmembrane domains, and a cytoplasmic C-terminal tail. While the seven transmembrane domains are a general feature

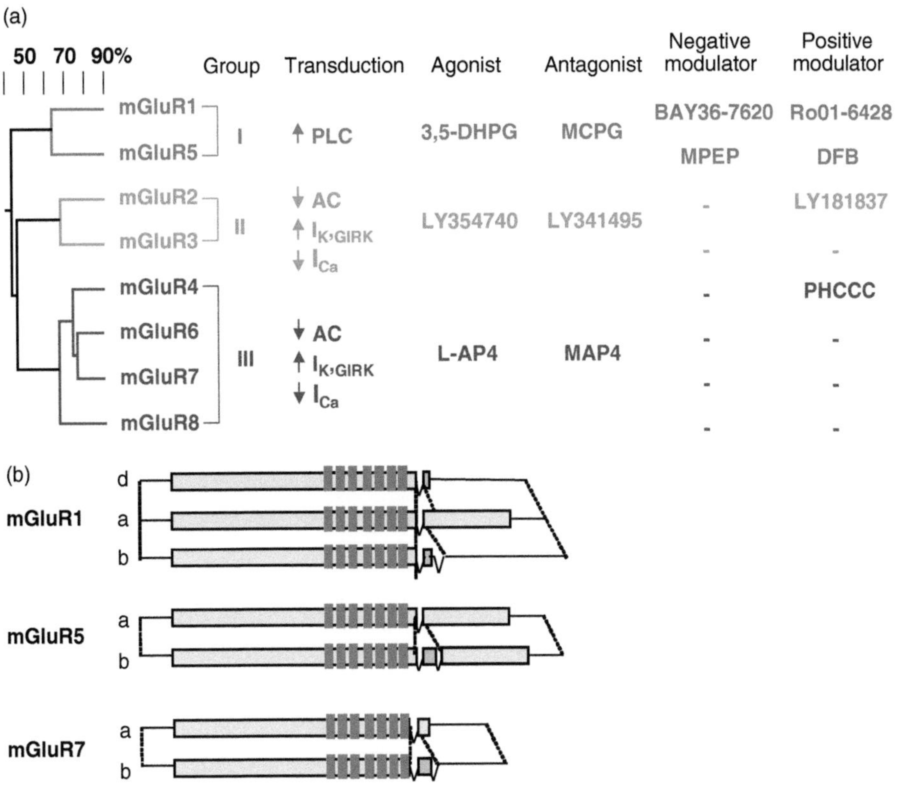

FIGURE 12.2 **Classification, coupling properties and pharmacology of mGluRs.** **(a)** mGluRs can be divided into three groups based on amino-acid sequence similarity (left), signaling properties (center) and pharmacology (right). Agonists and antagonists selective for each group of mGluRs are indicated. The recently identified negative and positive modulators are indicated. Note these modulators are selective for a single mGluR subtype. **(b)** Schematic representation of the well characterized mGluR variants resulting from alternative splicing of the pre-mRNA. The 7TM coding region is indicated with the seven red vertical bars. Identical mRNA sequences are linked by dotted lines.

of G-protein-coupled receptors, the large N-terminal domain of mGluRs is structurally divergent. A simple early hypothesis postulated that the N-terminal domain is responsible for agonist binding, while the transmembrane domains and cytoplasmic loops are responsible for G-protein coupling and signaling. This was confirmed by constructing various chimeric receptors made of domains taken from mGluRs with different properties. For example, an mGluR chimera made of the extracellular domain of mGluR2 (group-II) followed by the remainder of mGluR1 (group-I) amino-acid sequence has the agonist profile of group-II mGluRs (activation by DCG-IV), but the coupling properties of group-I mGluRs (stimulation of PI hydrolysis; see below) (**Figure 12.3a**).

Further support for the idea that the N-terminal domain mediates agonist binding is the finding that this domain shares weak sequence homology with bacterial periplasmic amino-acid binding proteins. Based on the known crystal structure of these bacterial proteins, the N-terminal domain of mGluRs was modeled as a 'Venus flytrap'-like structure, which clamps together upon ligand binding. Consistent with this model, point mutations of residues that are critical for ligand binding dramatically reduced agonist binding. Further evidence for the N-terminal domain being responsible for agonist binding was the demonstration that this domain produced alone as a soluble protein binds agonist with an affinity similar to that measured on the full-length mGluR. Eventually, crystals were obtained from this purified soluble protein and allowed the resolution of the atomic structure of this part of mGluR1 (**Figure 12.4**). This structure was solved both in its empty form and with bound glutamate. These studies confirmed the Venus flytrap bilobate structure with the glutamate binding site located in the cavity between the two lobes, and bring important information on the understanding of the activation process of this protein (see below). As proposed, based on the homology with bacterial amino-acid binding proteins, this protein adopts a closed conformation in the presence of glutamate.

Metabotropic glutamate receptors must somehow convert the binding of glutamate into G-protein activation. Regions of G-protein-coupled receptors that are responsible for G-protein activation have been extensively studied, namely in β-adrenergic receptors and rhodopsin. Specific regions of G-protein-coupled receptors determine the efficacy of transducing ligand binding to G-protein activation, while other regions specify which G protein is activated. In most G-protein-coupled receptors, the second and third intracellular loops play critical roles in coupling efficacy and specificity. Conversely, mGluRs chimera studies have identified the second intracellular loop as the major determinant of coupling specificity, while all of the other intracellular loops are involved in coupling efficacy. For example, a chimera of mGluR3 with the second intracellular loop and cytoplasmic tail of mGluR1 couples to phospholipase C (PLC), thus exhibiting the G-protein specificity of mGluR1 (**Figure 12.3b**).

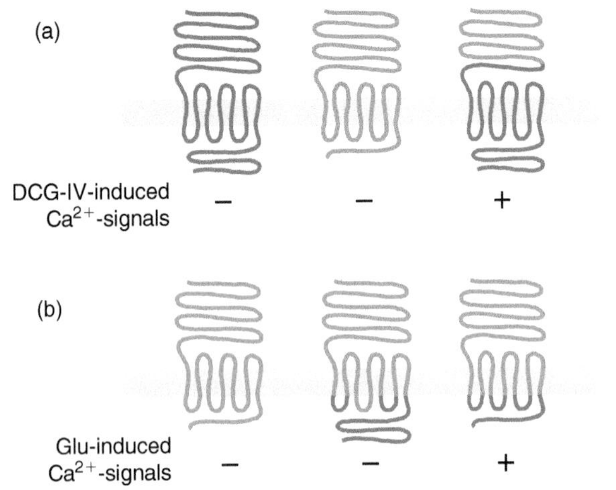

FIGURE 12.3 **Determination of the major functional domains of mGluRs using the chimeric approach. (a)** Swapping the two third N-terminal portions of the extracellular domain of the group-I mGluR1 (red) with those of the group-II mGluR2 (green) generated a chimeric receptor that couples to PLC, like the group-I mGluR1, but is activated by the group-II agonist DCG-IV. This indicates that the N-terminal domain of mGluRs is responsible for agonist recognition. **(b)** Swapping most intracellular parts, except the second intracellular loop, of the adenylyl-cyclase-coupled mGluR3 (green) with those of the PLC-coupled mGluR1 (red) was not sufficient to allow the chimeric receptor to activate PLC and generate intracellular Ca^{2+} signals (central). However, swapping of the second intracellular loop and other intracellular parts resulted in chimeric receptors that activate PLC-like mGluR1 (right). This indicates that the second intracellular loop plays a critical role in specifying PLC activation by mGluR1, whereas the other intracellular parts are required for an efficient coupling to PLC. Part **(a)** adapted from Takahashi K, Tsuchida K, Taneba Y *et al.* (1993) Role of the large extracellular domain of metabotropic glutamate receptors in agonist selectivity determination. *J. Biol. Chem.* **266**, 19341–19345, with permission. Part **(b)** adapted from Gomeza J, Joly C, Kuhn R *et al.* (1996) The second intracellular loop of metabotropic glutamate receptor 1 cooperates with other intracellular domain to control coupling to G proteins. *J. Biol. Chem.* **271**, 2199–2205, with permission.

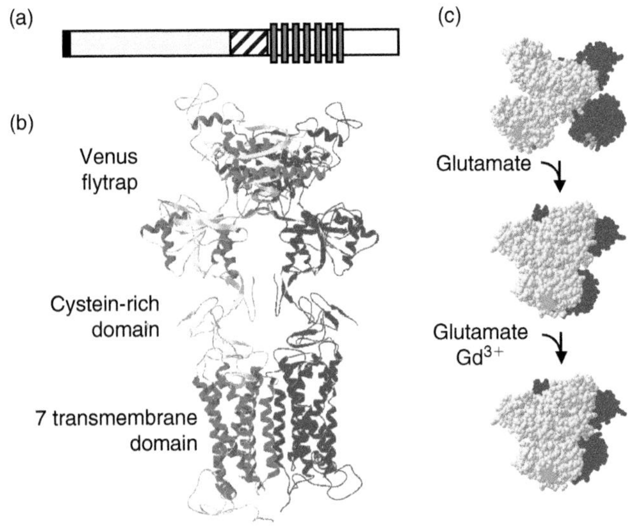

FIGURE 12.4 **Structure and activation mechanism of mGluRs. (a)** Schematic view of the structural domains of mGluRs as determined by sequence homology searches: from the N-terminal end (left) to the C-terminus (right): signal peptide (in black), the Venus flytrap domain (gray), the cysteine-rich domain (hatched), the seven transmembrane domain (with seven vertical bars) and the intracellular C-terminal segment (white). **(b)** Structural view of an mGluR based on the solved structure of the Venus flytrap domain of mGluR1, on the 3D model of the cysteine-rich region, and the solved structure of the prototypical G-protein-coupled receptor rhodopsin. Note the receptor is a constitutive dimer: the subunit in the front is colored according to its secondary structure (helices in red, strands in yellow and loops in gray), while the other subunit in the back is in blue. The dimer of the Venus flytrap domain is that observed in the absence of bound agonist (inactive state), while the dimer of 7TMs is based on the proposed dimerization mode of rhodopsin in its inactive state. **(c)** The three identified conformations of the dimer of Venus flytrap of mGluR1: top, the inactive unliganded form; upon binding of at least one molecule of glutamate, closure of the yellow subunit results in a major change in the relative orientation of the domains, leading to a partial activation of PLC; bottom, upon binding of glutamate in the blue subunit and cation (such as Gd^{3+}) binding at the interface between the two subunits, a third state is obtained in which both domains are closed, and this corresponds to the fully active state of the receptor.

At first glance, the regions responsible for G-protein-coupling specificity, lack of sequence homology between mGluRs and the other G-protein-coupled receptors, suggest a specific structure for the mGluRs. However, the second intracellular loop of mGluRs is predicted to form amphipathic α-helices similar to the third intracellular loop of the other G-protein-coupled receptors. Furthermore, mGluRs, as all other G-protein-coupled receptors, recognize the same amino acid residues on G-protein α-subunits. Therefore, the general structural strategy utilized to activate G proteins is probably shared among all G-protein-coupled receptors, although the specific regions involved in this process may differ.

Accordingly, the general structure of mGluRs consists of a Venus flytrap domain connected, through a cysteine-rich region, to a prototypical seven transmembrane spanning domain (**Figure 12.4**). But how can the closure of the Venus flytrap domain by glutamate activate the transmembrane domain leading to G-protein stimulation? Much information to answer that question came from the solved structure of the extracellular domain of mGluR1. Indeed, the structure revealed a dimer of Venus flytrap domains, their N-terminal lobes contacting each other through a hydrophobic interface. This was consistent with previous biochemical studies indicating that mGluRs are constitutive dimers linked by a disulfide bridge both in transfected cells and in neurons. The structure of the active dimer with bound glutamate revealed a major conformational change resulting from agonist binding. Not only were the Venus flytraps closed, but also an important rotation of one flytrap compared to the other was found (**Figure 12.4**). Such a movement was then proposed to force the seven transmembrane domains to also move one compared to the other. This was confirmed using energy transfer technology that allows estimation of the relative distance between proteins fused to the cyan fluorescent protein (CFP) and yellow fluorescent protein (YFP). More recent studies revealed that the change in conformation is different whether one or two glutamate molecules are bound to the dimer, and whether cations also interact with the complex. Indeed, a single agonist per mGluR1 dimer allows a partial activation of the PLC pathway, but full activation of the adenylyl cyclase. Two agonists per dimer are required for the full activation of PLC. This represents one of the first examples of clear evidence that different conformations of a G-protein-coupled receptor exist, which leads to the activation of different signaling pathways (**Figure 12.4**).

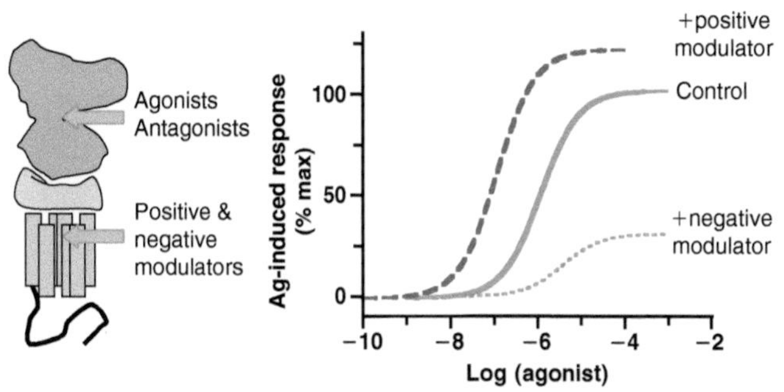

FIGURE 12.5 **Modulation of mGluR function by positive and negative modulators.** On the left, a scheme of the general structure of an mGluR subunit shows that whereas agonists and antagonists bind to the Venus flytrap domain, allosteric modulators (both positive and negative) bind to the 7TM domain. On the right, the effects of the positive and negative modulators on the dose effect of an agonist are illustrated. The positive modulators increase both the potency (the dose–response curve is shifted to the left) and efficacy (increase in the maximal effect of agonist) of agonists. The negative modulators act as non-competitive antagonists, decreasing the maximal effect of agonists.

12.3 HOW TO IDENTIFY SELECTIVE COMPOUNDS ACTING AT THE METABOTROPIC GLUTAMATE RECEPTOR – TOWARDS THE DEVELOPMENT OF NEW THERAPEUTIC DRUGS

After the cloning of mGluRs, a search for selective ligands started with the aim of identifying their physiological roles, and to validate these receptors as new targets for the treatment of psychiatric and neurological diseases. At first, most laboratories synthesized glutamate derivatives and examined their effects on the eight mGluRs. These studies led to the discovery of both agonists and antagonists, some of which displayed a nanomolar affinity such as the group-II agonist LY354740. However, most of these molecules, and especially the antagonists, display a low affinity and a poor selectivity. Indeed, due to the high conservation of the glutamate binding site, most of the identified compounds were only selective for mGluRs from one group, but none could be considered as really selective for a single receptor. These studies, however, bring further information on the mechanism of activation of these receptors since most antagonists were found to possess extrafunctional groups that prevent the closure of the Venus flytrap domain.

Due to the difficulty in identifying potent and selective compounds derived from glutamate, pharmaceutical companies started high-throughput screening campaigns based on functional assays using cell lines expressing these receptors. To do so, most companies used specific fluorescent probes that respond to intracellular Ca^{2+}, as an indication of PLC activation. Such signals could not only be recorded with the PLC-coupled group-I mGluRs, but also with the group-II and group-III receptors co-expressed with modified G-protein α-subunits allowing their coupling to PLC. Such assays were conducted in microplates containing 384 or even 1536 wells, such that millions of molecules could be tested for activity in a few weeks. This approach led to the discovery of very selective compounds with chemical structures totally different from that of glutamate. Not surprisingly, these molecules were found to bind at a site distinct from that of glutamate, not located in the Venus flytrap domain, but in the seven transmembrane domain (**Figure 12.5**). As such these molecules were called allosteric modulators.

Two types of allosteric modulators were identified. The first ones, called negative modulators, act as non-competitive antagonists, decreasing the maximal response generated by glutamate without affecting its potency (**Figure 12.5**). The first compounds identified were MPEP, a highly selective mGluR5 inhibitor, and BAY36-7620, a selective mGluR1 inhibitor (**Figure 12.2**). The second type of modulators facilitates the action of glutamate by increasing both its potency and its efficacy, and has no agonist activity on their own (**Figure 12.5**). Like the negative modulators, these molecules are highly selective for a given mGluR (**Figure 12.2**). Such compounds, called positive modulators, offer a number of advantages over agonists for therapeutic applications. Whereas an agonist activates the receptor in any cells that express it, and in a sustained manner, the positive modulators only facilitate the action of glutamate when and where released in the brain. As such, these positive modulators maintain the normal biological rhythm of the targeted receptor, and do not permanently activate it. Accordingly, and in contrast to what is expected when using agonists, the signaling system (the receptor itself or the signaling network involved) targeted by these positive modulators will less likely desensitize. The advantage of using positive modulators for therapeutic intervention is well illustrated by the potent anxiolytic activity of benzodiazepines, which are positive modulators of the $GABA_A$ receptors, whereas the pure agonists are devoid of such effects.

12.4 WHAT BIOCHEMICAL MEANS DO METABOTROPIC GLUTAMATE RECEPTORS UTILIZE TO ELICIT PHYSIOLOGICAL CHANGES IN THE NERVOUS SYSTEM? SIGNAL TRANSDUCTION STUDIES OF METABOTROPIC GLUTAMATE RECEPTORS

Once activated by glutamate, mGluRs initiate a host of intracellular biochemical cascades, which eventually change the physiological behavior of the cell. As any G-protein-coupled receptor, mGluRs activate heterotrimeric G proteins. These consist of a GTP hydrolyzing α-subunit and a membrane-bound complex of β- and γ-subunits. Once the G-protein-coupled receptors are activated, the G-protein α-subunit exchanges GDP with GTP and dissociates from the βγ-complex. Both the activated GTP-bound α-subunit and the free βγ-complex now initiate various downstream processes. The α-subunit activates membrane-bound enzymes, whereas βγ-subunits can directly modulate the activity of ion channels.

As mentioned earlier, the discovery of mGluRs was initiated by the finding that glutamate can stimulate PI hydrolysis. Phospholipids are major constituents of cell membranes that play important roles in intracellular signaling. Phospholipases (PLCs) hydrolyze phospholipids, including the PI phosphatidyl-inositol diphosphate (PIP2), into inositol triphosphate (IP3) and membrane-bound diacylglycerol (DAG). Thus, a simple way to follow phospholipase C activity is to load cells with radioactively labeled inositol, which is then incorporated into membrane phospholipids, and determine the release of inositol phosphate from the membrane or organic fraction to the cytoplasmic or aqueous fraction. In neurons loaded with ^{3}H-inositol, glutamate increases the aqueous fraction of radiolabeled inositol phosphate, indicating PLC hydrolysis of PIP2.

Based on pharmacological studies of native and recombinant mGluRs, it is now known that group-I mGluRs couple to G_q protein, whose α-subunit activates PLC. This results in the production of IP3, which releases Ca^{2+} from internal stores. The released Ca^{2+}, along with the concomitantly synthesized DAG, activates protein kinase C (PKC). DAG may also be further processed to yield various lipid messengers, including arachidonic acid. Thus, group-I mGluRs can exert their effects through intracellular Ca^{2+}, lipid messengers and PKC activation. The rapid rise in intracellular Ca^{2+} concentration is now widely used to record the activity of PLC-coupled receptors using fluorescent Ca^{2+} binding molecules such as Fluo4.

Activation of PLC is not, however, an exclusive group-I mGluR pathway in native systems. For instance, in cultured glial cells, these receptors activate mitogen-activated protein kinase (MAPK), possibly through phosphoinositide 3-kinase (PI3K). In cerebellar neurons, mGluR1 can trigger a direct functional coupling between intracellular ryanodine-sensitive receptors and plasma membrane L-type Ca^{2+} channels through a G-protein-dependent but unidentified pathway (**Figures 12.6, 12.7**). In CA3 hippocampal neurons, functional pharmacological studies have shown that mGluR1 can mobilize an Src-family tyrosine kinase pathway, independently of G-protein activation.

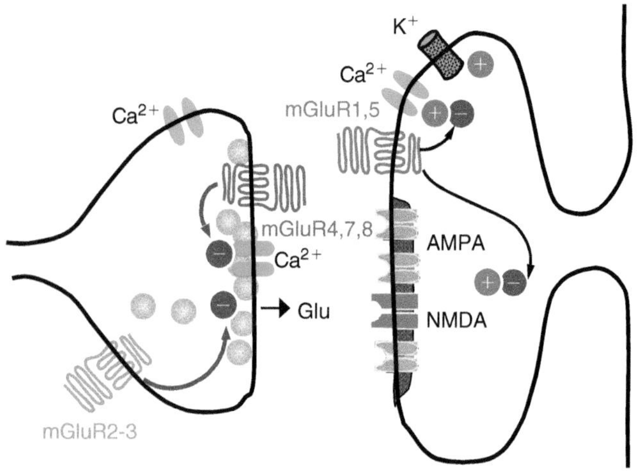

FIGURE 12.6 **Schematic representation of a glutamatergic synapse with the modulatory roles of mGluRs.** Note group-I mGluRs (1 and 5) are mostly located in the postsynaptic spines, on the side of the postsynaptic density, from where they regulate postsynaptic ionotropic glutamate receptors, as well as Ca^{2+} and K^{+} channels. Group-II (mGluR2) and group-III (mGluR4, 7 and 8) are mostly presynaptic, regulating neurotransmitter release by, at least, inhibiting Ca^{2+} channels.

The α-subunits of G_s and G_o/G_i proteins can stimulate and inhibit the adenylyl cyclase (AC) activity, respectively. Group-II and group-III mGluRs couple to G_i/G_o proteins. Similarly to PLC, AC activity was measured by loading living cells with radiolabeled adenine, which is then incorporated into intracellular ATP pools and finally cAMP. Nowadays, several non-radioactive and high-throughput assays are being used to measure cAMP produced in the cells. Sensors have also been developed that allow the visualization of cAMP concentration in living cells in real time. These sensors are made with cAMP binding proteins fused to fluorescent proteins. Activity of both group-II and group-III mGluRs inhibits forskolin-stimulated cAMP accumulation in heterologous expression systems, neuronal cultures and brain slices. A more relevant issue is the effect of group-II and group-III mGluRs on neurotransmitter-induced increases in cAMP accumulation in native systems. Group-II mGluR agonists potentiate cAMP accumulation induced by stimulation of β-adrenergic receptors and other G_s-coupled receptors. In this case, released G-protein βγ-subunits may actually potentiate type-II AC activation by G_s-protein α-subunits. Conversely, group-III mGluRs inhibit neurotransmitter-induced increases in cAMP, through the G_o/G_i-protein

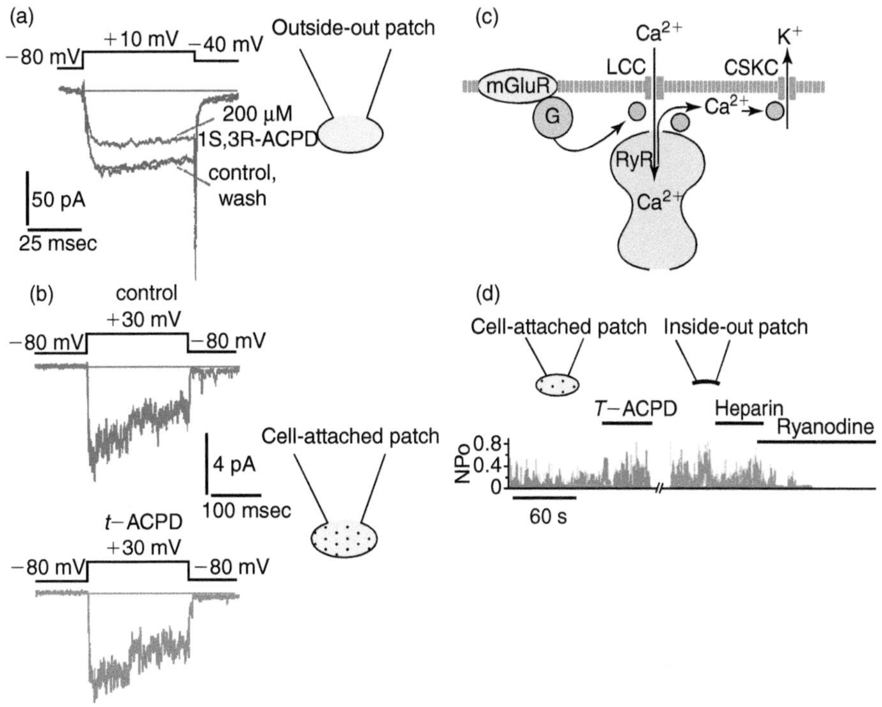

FIGURE 12.7 **mGluR-mediated inhibition and activation of voltage-sensitive Ca^{2+} currents in neurons. (a)** The upper panel shows a recording of macroscopic Ca^{2+} currents from a rat CA3 pyramidal cell outside-out patch. The membrane was held at −80 mV and stepped to +10 mV, resulting in activation of voltage-sensitive Ca^{2+} channels. Application of the group-I mGluR agonist, t-ACPD, to the surface of the patch induced in a reversible reduction of the Ca^{2+} current. **(b)** However, when macroscopic Ca^{2+} currents were recorded in the cell-attached mode and the agonist was applied outside the patch, t-ACPD had no effect. These findings suggest that group-I mGluRs acted through a membrane delimited mechanism rather than readily diffusible second messenger. **(c)** In cerebellar granule cells, mGluR1 triggers a tight coupling between ryanodine-sensitive receptors (RyR) and membrane L-type Ca^{2+} channels (LCC), in a G-protein-dependent manner. The release of Ca^{2+} from intracellular ryanodine-sensitive stores activates a Ca^{2+}-dependent K^+ conductance (CSKC) located in close proximity to LCC and RYRs. **(d)** The opening probability of L-type Ca^{2+} channels (NPo) was monitored in a cell-attached patch and after excision of the patch into the inside-out configuration. The agonist, t-ACPD (100 µM), induced opening of LCC recorded in the cell-attached configuration. LCC remained active in the excised patch and the activity was blocked by ryanodine, but not the IP3 receptor antagonist heparin, when applied to the intracellular side of the recorded patch. This experiment confirmed the model described in (**c**). Part (**b**) adapted from Swartz KJ, Bean BP (1992) Inhibition of Ca^{2+} channels in rat CA3 pyramidal neurons by a metabotropic glutamate receptor. *J. Neurosci.* **12**, 4358–4371, with permission. Part (**d**) modified from Fagni L, Chavis P, Ango F (2000) Complex interactions mGluRs, intracellular Ca^{2+} stores and ion channels in neurons. *Trends Neurosci.* **23**, 80–88, with permission.

α-subunit. Therefore, group-II and group-III mGluRs may exert their effects by altering intracellular cAMP levels. However, repeated attempts to find physiological roles for group-II and group-III mGluRs involving changes in intracellular cAMP in native systems have been unsuccessful, with one exception. In hippocampal glial cells, a form of glial–neuron signaling occurs in which group-II mGluRs potentiate cAMP formation induced by β-adrenergic receptors. Cyclic AMP metabolites are then released by glia and activate adenosine receptors on nearby neurons, thus modulating synaptic transmission.

Alternative avenues are becoming increasingly evident. For instance, the group-III presynaptic receptor mGluR7 mobilizes two separate cAMP-independent pathways and inhibits P/Q-type Ca^{2+} channels in hippocampal and cerebellar neurons. First, mGluR7 can activate PLC, probably through G_i/G_o-protein βγ-subunits, which results in PKC activity. The activated PKC directly or indirectly inhibits the Ca^{2+} channels (**Figure 12.8b**). Second, the Ca^{2+}–calmodulin complex and G-protein βγ-subunits bind to the C-terminus of mGluR7 in a mutually exclusive manner. Thus, when Ca^{2+}–calmodulin interacts with the receptor, the released G-protein βγ-subunits can directly interact with and inhibit the Ca^{2+} channels (**Figure 12.8c**). Group-II and group-III mGluRs can activate GIRK (G-protein-coupled inwardly rectifying K^+) channels expressed in *Xenopus* oocytes through direct interaction of endogenously released G-protein βγ-subunits with the channels. The relative dearth of evidence for group-II and group-III mGluR modulation of cAMP levels leading to physiological effects may indicate that direct inhibition of P/Q-type Ca^{2+} channels or activation of GIRK channels by G-protein βγ-subunits plays an equal or even greater role than modulation of AC activity. Indeed group-II and group-III mGluRs seem to couple to these channels more efficiently than to inhibition of cAMP formation in both expression systems and native neurons.

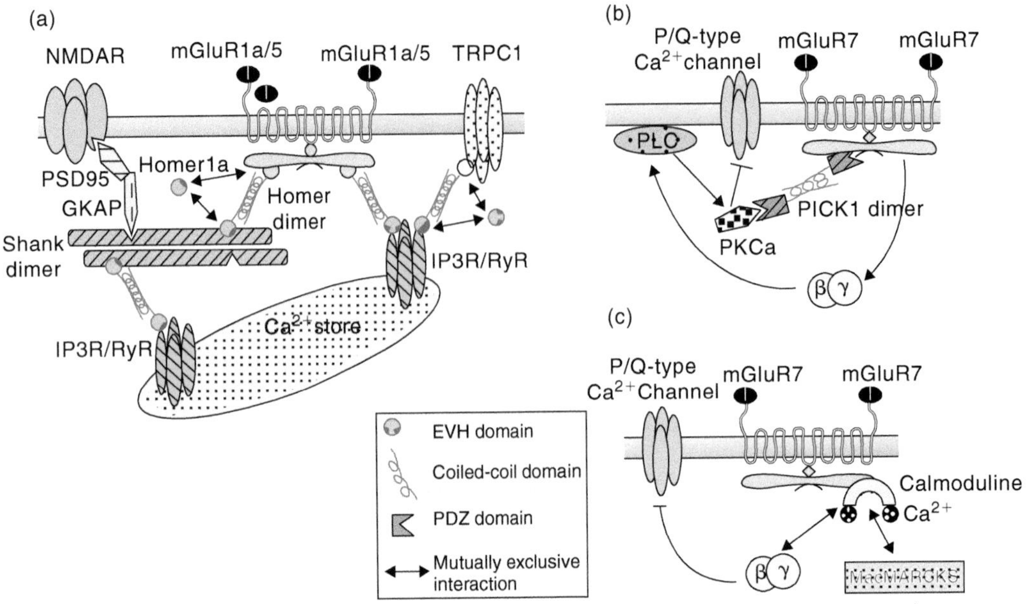

FIGURE 12.8 **Current models for post- and presynaptic multiprotein mGluR complexes. (a)** Homer proteins bind to Ca^{2+}-permeable store-operated TRPC1 channel, postsynaptic mGluR1a and mGluR5, as well as two sites of IP3 and ryanodine receptors (IP3R/RyR). Homer can also interact with the shank protein, which associates with GKAP-PSD95-NMDA-receptor complex through a PDZ interaction. The confinement of these receptors and channels by homer-based protein–protein interactions increases the functionality of the system and optimizes mGluR1a mGluR5 intracellular Ca^{2+} signaling. **(b)** The adaptor protein PICK1 interacts with the C-terminus of mGluR7 and PKCa via its PDZ domain and dimerizes through its coiled-coil domain. Thus, it physically links mGluR7 to PKCa. Stimulation of mGluR7 releases G-protein βγ-subunits, which results in phospholipase C (PLC) activation in neurons. The PLC pathway triggers PKC activity, which then directly or indirectly inhibits P/Q-type Ca^{2+} channels. Proper functioning of this cascade requires the integrity of the mGluR7–PICK1–PKCa complex, presumably because of the necessity for mGluR7 to be in close proximity to its effector, PKCa (see Perrog *et al.* (2002) *EMBO J.* **21**, 2990–2999). **(c)** Ca^{2+}–calmodulin complex and G-protein βγ-subunits undergo competitive binding on the C-terminus of presynaptic mGluR7. The Ca^{2+} influx from voltage-gated channels activates calmodulin, which then binds to mGluR7 and releases pre-bound G-protein βγ-subunits from the mGluR7 C-terminus. Free G-protein βγ-subunits are thus available for direct inhibition of P/Q-type Ca^{2+} channels (see Bertaso *et al.* (2006) *J. Neurochem.* **99**, 288–298). These examples illustrate the importance of mGluR multiprotein complexes in the proper function of these receptors.

12.5 HOW IS THE ACTIVITY OF METABOTROPIC GLUTAMATE RECEPTORS MODULATED? STUDIES OF mGluR DESENSITIZATION

Soon after the discovery that group-I mGluRs underlie glutamate-stimulated PI hydrolysis, it was discovered that pre-incubation of neuronal cultures or brain slices with group-I agonists decreases the PI hydrolysis response to subsequent exposures of agonist. This phenomenon commonly occurs with many G-protein-coupled receptors and is referred to as desensitization. This phenomenon has been well characterized for many G-protein-coupled receptors and involved interaction of the activated receptor with various intracellular proteins. First, the activated receptor is recognized by G-protein-coupled receptor specific kinases (GRK) that phosphorylate the receptor at various places, especially within the C-terminal tail. The phosphorylated receptor gains affinity for a protein called β-arrestin (β-arrestin 1 and β-arrestin 2) that prevents G-protein activation. This allows the recruitment of additional proteins to the receptor and activation of new signaling pathways. Most importantly, the recruitment of β-arrestin permits incorporation of the receptor into specific membrane microdomains from where it will be internalized into endosomes. The internalized receptor will then either be recycled and re-targeted to the cell surface, or sent to lysosomes to be degraded.

Such a desensitization cascade has indeed been reported for group-I mGluRs, although differences were observed. Indeed, activated mGluR1 or mGluR5 were shown to recruit GRK and β-arrestin both in transfected cells and in neurons, resulting in the internalization of the receptor. However, GRK2 (one subtype of the GRKs) can inhibit receptor signaling without phosphorylating the receptor. This effect is independent of β-arrestin recruitment and likely results from a direct competition of GRK2 and the G protein on the receptor. In contrast, GRK4 kinase activity is needed for this GRK to desensitize mGluR1 in Purkinje neurons. Other proteins such as optineurin, a protein that interacts with the protein huntingtin involved in Huntington's disease, have also been shown to uncouple group-I mGluRs from its G protein.

In addition to these direct desensitization and internalization mechanisms resulting from the activation of mGluRs, heterologous desensitization can also occur independently of the receptor activation. The most common mechanism for reduced coupling of the receptor to G protein is phosphorylation of the receptor itself, demonstrated also for many other G-protein-coupled receptors. For example, β-adrenergic receptors desensitize when phosphorylated by cAMP-dependent protein kinase (PKA) activated either by the β-adrenergic receptor itself or by other receptors. Similarly, group-I mGluRs desensitization can be mediated by the direct phosphorylation of the receptor by protein kinase C (PKC) whether activated by the receptor itself or by other receptors. As such, this represents not only a mechanism for negative feedback, but also a mechanism for cross-talk between different neurotransmitter pathways that also utilize PKC-dependent signaling.

Much less is known about desensitization and internalization properties of group-II and group-III mGluRs. A few studies indicated that these receptors may not desensitize in neurons. However, mGluR4 and mGluR7 desensitization and internalization was observed after PKC activation by mGuR7 itself or other receptors. GRK2 was also shown to uncouple mGluR4 from the MAPK pathway without affecting its ability to inhibit adenylyl cyclase.

12.6 METABOTROPIC GLUTAMATE RECEPTORS MODULATE NEURONAL EXCITABILITY

MGluRs modulate two major neuronal functions: excitability and synaptic transmission. Altered excitability is mainly exhibited by changes in the threshold for action potential firing or firing pattern. A powerful mechanism for altering neuronal excitability is to modulate K^+ channels. Potentiation of K^+-channel function leads to reduced excitability, while inhibition of these channels results in enhanced excitability. Spike accommodation is defined as a reduction in spike frequency during a suprathreshold depolarizing stimulus. This phenomenon strongly depends on K^+-channel activity and it is a mechanism by which neurons control their excitability. Fast and slower activating K^+ currents are involved in spike frequency maintenance and accommodation. One of these is $I_{K,AHP}$ (K^+ after-hyperpolarization current), which is activated by rises in intracellular Ca^{2+} occurring during repetitive firing. Inhibition of $I_{K,AHP}$ leads to reduced K^+ efflux and firing fails to accommodate. An early experiment showed that application of glutamate and related group-I mGluR agonists onto hippocampal neurons inhibits $I_{K,AHP}$ and spike accommodation (**Figure 12.9**). As the agonist did not affect the $I_{K,AHP}$ concomitant Ca^{2+} increase, it was concluded that the inhibition of $I_{K,AHP}$ must be mediated through a direct inhibition of the K^+ channels by mGluR5. Several groups have shown that group-I mGluR agonists depolarize neurons by inhibiting voltage-independent K^+ conductances termed $I_{K,leak}$ and $I_{K,M}$. Group-I mGluRs can also excite neurons by activating non-selective cation conductances which have not always been identified, but which include Na^+/Ca^{2+} exchangers and receptor-operated TRPC channels. Thus, it is tempting to conclude that group-I mGluRs rather increase neuronal excitability, whereas group-II and group-III mGluR agonists induce a K^+ conductance-based hyperpolarization. However, the physiological situation may be more complex than that, as the increase in intracellular Ca^{2+} that results from mGluR1 stimulation can also activate Ca^{2+}-sensitive K^+ channels in cerebellar neurons (**Figure 12.7**) and this was shown to result in reduced cell excitability.

Modulation of Ca^{2+} channels by mGluRs also bridges changes in neuronal excitability. Ca^{2+} channels help generating action potential upstroke in certain neurons and also mediate the Ca^{2+} influx required for synaptic transmission (**Figure 12.6**). Furthermore, Ca^{2+} influx at various locations within neurons results in the activation of multiple signal-transduction pathways. In general, group-II and group-III mGluRs inhibit Ca^{2+} currents, and the specific subtype of Ca^{2+} current varies depending on the preparation. In hippocampal CA3 pyramidal neurons, group-I mGluR agonists also inhibit N-type Ca^{2+} currents. This effect would involve a membrane delimited mechanism, such as direct binding of G-protein βγ-subunits to the pore-forming subunit of Ca^{2+} channels. In cerebellar neurons, group-I mGluR agonists can trigger a direct functional coupling between ryanodine receptors and L-type Ca^{2+} channels upon voltage activation of the channels (**Figure 12.7a,b**). This leads to local increase in intracellular Ca^{2+} and opening of a Ca^{2+}-sensitive K^+ channel located in close proximity to the activated Ca^{2+} channels (**Figure 12.7c,d**). These examples point out the complexity of the mechanisms by which mGluRs can control channel activity and neuronal excitability. It clearly appears that the subcellular localization of mGluRs and that of the regulated ion channels specifies the effects of these receptors on cell excitability.

12.7 METABOTROPIC GLUTAMATE RECEPTORS MEDIATE AND MODULATE SYNAPTIC TRANSMISSION

Besides altering neuronal excitability, mGluRs are also able to mediate slow excitatory postsynaptic responses. Repetitive stimulation of glutamatergic afferents in slices induce a slow EPSC in cerebellar Purkinje neurons and hippocampal CA3 pyramidal cells that results from activation of group-I mGluRs (likely mGluR1) and opening of TRPC1 channels (**Figures 12.8** and **12.10**). In the hippocampus, but not cerebellum, this effect is mediated via mobilization of a G-protein-independent tyrosine kinase pathway.

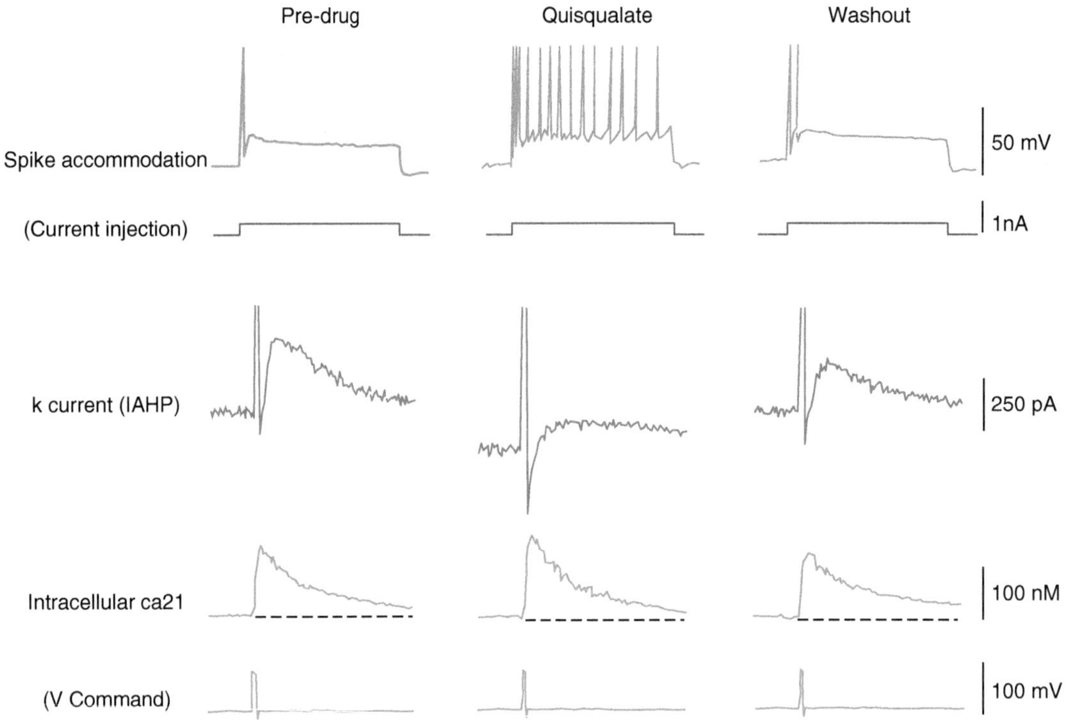

FIGURE 12.9 **Modulation of action potential accommodation by inhibition of a Ca^{2+}-dependent K^+ current, rather than Ca^{2+} influx.** The top row shows the action potential firing pattern elicited by prolonged depolarizing current injection in hippocampal neurons. Only one to two action potentials normally fire even in the presence of a suprathreshold stimulus. Such accommodation is due to depolarization-induced Ca^{2+} influx and activation of $I_{K,AHP}$. The middle row shows the late outward K^+ current $I_{K,AHP}$ under voltage clamp conditions, while the bottom row shows the Ca^{2+} influx as measured by fluorescence-based techniques. The spike accommodation is reversibly blocked by the group-I mGluR agonist quisqualate, as is the $I_{K,AHP}$ current. However, the magnitude of the intracellular Ca^{2+} response is not altered by quisqualate, indicating that group-I mGluRs directly modulate the $I_{K,AHP}$ channel, rather than the Ca^{2+} influx that activates the channel. Adapted from Charpak S, Gahwiler BH, Do KQ, Knopfel T (1990) Potassium conductances in hippocampal neurons blocked by excitatory amino acid transmitters. *Nature* **347**, 765–767, with permission.

mGluRs can also modulate synaptic transmission. In general, mGluR effects on synaptic transmission can be divided simply into presynaptic and postsynaptic (**Figure 12.6**). Presynaptically, group-II and group-III mGluRs typically act as glutamate autoreceptors, decreasing glutamate release from presynaptic nerve terminals, typically by inhibiting presynaptic N-type Ca^{2+} channels. The P/Q-type Ca^{2+} channels are also typically presynaptic and mGluR7 inhibits excitatory synaptic events through specific blockade of these channels. Conversely, in the hippocampus, it appears that mGluR4 reduces frequency with no effect on amplitude of mEPSCs, indicating a reduction in presynaptic Ca^{2+}-independent release probability. Presynaptic mGluRs can also act as heteroreceptors, on GABAergic nerve terminals, where they reduce GABA release.

mGluRs can execute subtype- and location-specific postsynaptic roles and control neurotransmission in a paracrine manner. Activation of mGluR1 in cerebellar Purkinje cells induces the release of endocannabinoids, which then retrogradely act on presynaptic type-1 cannabinoid (CB1) receptors located on both excitatory (climbing fiber) and inhibitory (GABA interneuron) axon terminals, and inhibit neurotransmitter release. In cerebellar Purkinje cells, postsynaptic mGluR1 plays a pivotal role in synaptic plasticity. Purkinje cells receive excitatory inputs from climbing fibers and parallel fibers. When these inputs are conjunctively stimulated, it results in a long-term depression (LTD) of the parallel fiber–Purkinje cell synaptic response. Based on a large body of evidence, it has been established that LTD of the parallel fiber–Purkinje cell synapse results from iGluR and group-I mGluR mobilization of voltage-gated Ca^{2+}-channel activation in the Purkinje cell. In this effect, mGluR contribution is to raise intracellular Ca^{2+} stores, which results in PKC activation, protein kinase G-activation through nitric oxide (NO) production, and guanylyl-cyclase activation. In the hippocampus, group-I mGluRs can also induce LTD, but pharmacological and gene deletion studies raised controversial results regarding involvement of intracellular Ca^{2+} stores. In the nucleus accumbens, the mGluR1-mediated production of endocannabinoid and retrograde inhibition of neurotransmitter release would contribute to LTD. Group-II and group-III mGluRs can also induce LTD in different regions of the brain: in the nucleus accumbens, striatum, hippocampal CA3 region.

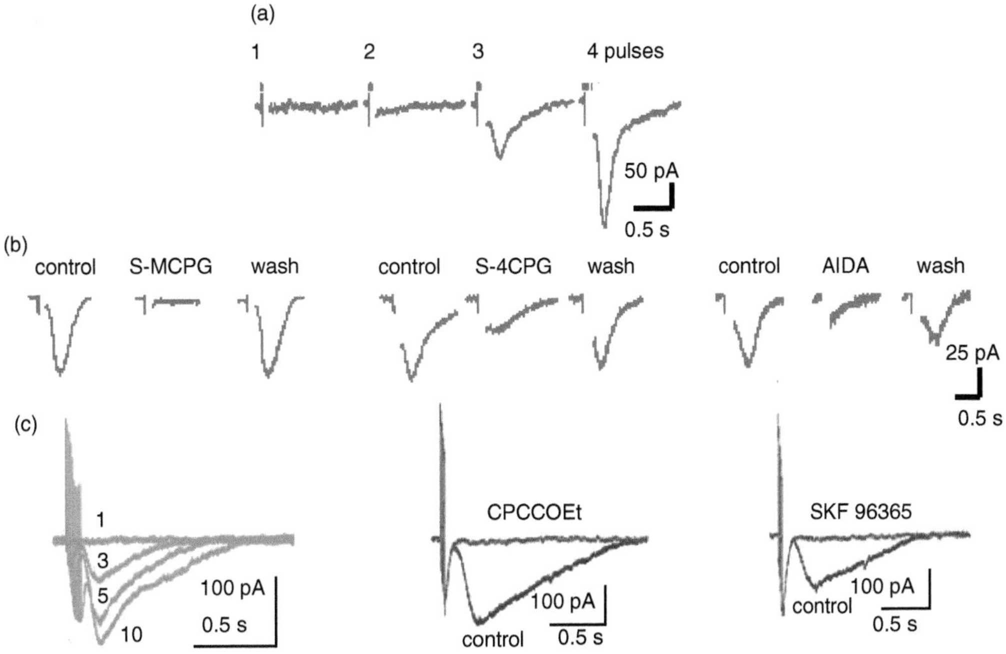

FIGURE 12.10 **mGluR1 mediates slow EPSCs in hippocampal CA3 pyramidal cells and cerebellar Purkinje cells. (a)** Activity-dependent induction of the slow EPSC in hippocampal CA3 pyramidal cells is induced by repetitive stimulation of afferent mossy fibers. **(b)** The evoked EPSC amplitude is reduced by mGluR1 antagonists S-MCPG (500 μM), S-4CPG (100 μM) and AIDA (200 μM). **(c)** (Left) A similar slow EPSC can be elicited in cerebellar Purkinje cells by repetitive (100 Hz) parallel fiber stimulation, in the presence of the AMPA- and GABA-receptor antagonists, CNQX and gabazine, respectively. (Center and right) The slow EPSC evoked by 5 successive pulses was blocked by the mGluR1 antagonist CPCCOEt (100 μM, center) and TRPC1-channel antagonist SKF96365 (30 μM; right). Note that at least three successive stimulation pulses (100 Hz) are required to induce the mGluR1-dependent slow EPSC, in both hippocampal neuron and cerebellar Purkinje cell. The amplitude of the EPSC then increased with the number of pulses. Part (**b**) adapted from Heuss C, Scanziani M, Gahwiller BH, Gerber U (1999) G-protein-independent signalling mediated by metabotropic glutamate receptors. *Nat. Neurosci.* **2**, 1070–1077, with permission. Part (**c**) adapted from Kim SJ, Kim YS, Yuan JP, Petralla RS, Worley PF, Linden DJ (2003) Activation of the TRPC1 cation channel by metabotropic glutamate receptor mGluR1. *Nature* **426**, 285–291, with permission.

Long-term potentiation (LTP) is characterized by a long-lasting increase in excitatory postsynaptic response, which can be induced by repetitive stimulation of afferents. In the CA1 area of the hippocampus, induction of LTP is an NMDA receptor-dependent mechanism. Consistent with the observation that group-I mGluRs potentiate NMDA currents, group-I mGluR agonists potentiate the NMDA component of EPSCs and promote LTP (**Figure 12.8**). Furthermore, mice lacking the *mGluR5* gene are specifically deficient in the NMDA component of LTP, while having normal non-NMDA LTP.

12.8 PRE- AND POSTSYNAPTIC FUNCTIONAL ASSEMBLY OF METABOTROPIC GLUTAMATE RECEPTORS

In order to carry out their functions, mGluRs must properly localize in neurons (**Figure 12.6**). Immunological and functional studies show a predominant postsynaptic localization of group-I mGluRs, presynaptic localization of mGluR7, and pre- and postsynaptic localization of the other mGluR subtypes in the brain. Ultrastructurally, glutamatergic postsynaptic membrane contains an electron-dense zone that has been named postsynaptic density (PSD). Postsynaptic mGluRs lie within the PSD, except group-I mGluRs which form an annulus surrounding the PSD. The presynaptic mGluRs are localized in the symmetrically positioned active zone.

Recent studies have begun to uncover the biochemical nature of the PSD. It consists of a large multimeric protein complex that comprises not only postsynaptic glutamate receptors, but also scaffolding proteins that physically link these receptors to downstream signaling molecules and cytoskeleton. Analogous to the PSD, the presynaptic active zone consists of a protein matrix that assembles receptors and channels to their signaling pathways, including those involved in neurotransmitter release. Homer, calmodulin and PDZ domain proteins have been characterized as mGluRs partners (**Figure 12.8**). For instance, the EVH domain of homer proteins can interact with a PPxxF consensus sequence that is present in the C-terminus of mGluR1a and mGluR5, but also ryanodine receptors, IP3 receptors and TRPC1 channels. The coiled-domain of homer allows multimerization of the protein. In general, PDZ interactions depend on the last three C-terminal amino acids (PDZ motif) of one partner that fits within the PDZ hydrophobic pocket (PDZ domain) of the other partner. For instance, the PDZ domain-containing protein tamalin was found to bind the C-terminus of mGluR1a and mGluR5, whereas the PDZ proteins PICK1 and GRIP bind to mGluR7.

Mutation/deletion and overexpression experiments have shown that these proteins play important roles in the synaptic localization, membrane targeting and function of mGluRs. As such, homer multimers can form a large multiprotein complex with mGluR1a/5, IP3/ryanodine receptors and TRPC1 channels (**Figure 12.8**). The homer1a isoform is induced following epileptic activity, acute cocaine intake or inflammatory pain, and disrupts the complex. This is functionally important as disruption of the complex can alter the intracellular Ca^{2+} signaling and induce agonist-independent/constitutive activity of mGluR1a and mGluR5. Thus, transgenic mice that constitutively express homer1a or mice that have been infused with viral vectors carrying homer1a transcript display reduced LTP and hyposensitivity to epileptogenic agents and cocaine. Conversely, preventing activity-dependent upregulation of homer1a with small interference RNA exacerbates inflammatory pain in mice. Thus, homer1a blunts the physiological actions of mGluR1a/5.

Recently, homer was also found to bind the shank family of PSD proteins, whose members also contain the PPxxF motif. Shank binds to GKAP (guanylate kinase-associated protein) through its PDZ domain, thereby cross-linking the homer-group-I mGluR with the GKAP/PSD-95/NMDA receptor complex (**Figure 12.8**). One can now envision the beginnings of a large PSD complex comprising group-I mGluRs and NMDA receptors, linked by scaffolding and effector proteins. Thus, this shank-based physical coupling may provide the structural correlate for a functional cross-talk between these receptors. Group-I mGluRs can also directly interact with adenosine A1 receptor, thus adding a second-order complexity in the interplay between glutamatergic and other neurotransmitter systems. Finally, mGluR1 can directly interact with the $Ca_v2.1$ subunit of P/Q-type Ca^{2+} channels, suggesting a more direct control of these channels by glutamate.

The notion that proper functioning of receptors and channels requires their association with multiprotein complexes also applies to the presynaptic mGluR7 subtype. The C-terminal PDZ motif of this receptor can interact with the N-terminal PDZ domain of PICK1. PICK1 can dimerize through its C-terminal coiled-coil domain and interact with PKCα (**Figure 12.8b**). The integrity of this complex is required for inhibition of Ca^{2+} channels and synaptic transmission by mGluR7, probably because of the necessity for mGluR7 to be in close vicinity of its effector, PKC, to mediate its effect. MGluR7 can also block Ca^{2+} channels and neurotransmission through a separate pathway that requires interaction of the receptor with Ca^{2+}–calmodulin. Another protein called MacMARCKS can bind mGluR7 and causes the release of Ca^{2+}–calmodulin from the C-terminus of the receptor (**Figure 12.8c**), thus blunting the Ca^{2+}–calmodulin-dependent mGluR7 signaling. Collectively, these examples show the importance of multiprotein complexes in the localization and function of mGluRs.

12.9 PHYSIOLOGICAL ROLES OF METABOTROPIC GLUTAMATE RECEPTOR – A STUDY OF KNOCKOUT MODELS

A way to study the physiological role of a protein is to knock out the corresponding gene and to characterize the phenotype of the genetically modified animal (**Table 12.1**). Such an approach, combined with pharmacological and functional studies, has been used to study the physiological roles of mGluRs. The mGluR1 knockout mouse has no gross anatomical or basic electrophysiological abnormalities, except poly-innervation of cerebellar Purkinje cells by climbing fibers, impaired cerebellar LTD and impaired hippocampal mossy fiber LTP. These functional disorders are accompanied by deficient spatial learning and eye-blink conditioning, as well as severe motor coordination impairment. This phenotype clearly shows an important function of this receptor in cerebellar development and function, as well as in motor learning tasks.

TABLE 12.1 mGluR knockout phenotypes

mGluR subtype	Phenotype knockout
mGluR1	Ataxia, impaired cerebellar LTD, impaired hippocampal mossy fiber LTP
mGluR2	Impaired hippocampal mossy fiber LTD
mGluR4	Increased susceptibility to absence epilepsy, impaired paired-pulse facilitation and post-tetanic potentiation, impaired rotating rod motor learning
mGluR5	Impaired fear conditioning, impaired behavioral effects of cocaine
mGluR6	Altered electroretinogram
mGluR7	Altered theta rhythm and impaired working memory, deficits in taste aversion and fear responses, increased epileptic seizure susceptibility
mGluR8	Increased anxiety, altered photoreceptor function

The mGluR5 knockout mouse displays altered synaptic functions in lateral amygdala and, as expected, this is accompanied by impaired fear conditioning. The mGluR5 deletion suppresses reinforcing properties of cocaine administration, without affecting the dopamine response in the nucleus accumbens. This suggests a potential therapeutic value of mGluR5 in the treatment of cocaine addiction. The mGluR5 knockout phenotype is different from that of the mGluR1 knockout, although mGluR5 is functionally related to mGluR1. This observation is indeed not quite surprising if we consider that these receptors display mirror distribution in the brain. Pharmacological studies have also shown that mGluR5 regulates extrapyramidal motor functions and may be a potential target for the treatment of Parkinson's disease. The mGluR1a/5–homer complex mediates a component of hyperalgesia and might be therapeutically targeted to prevent and treat inflammatory pain. Finally, group-I mGluRs have also been involved in temporal lobe epilepsy.

The mGluR2 knockout mouse shows no histological changes and no alteration in basal neurotransmission. The NMDA-receptor-mediated LTD was, however, almost abolished at the hippocampal mossy fiber pathway. Nevertheless, the transgenic animal performs normally in the water-maze learning task. Thus, mGluR2 is essential for mossy fiber hippocampal LTD and this phenomenon does not seem to be required for spatial learning.

The mGluR4 subtype is highly expressed presynaptically on thalamo-cortical neurons that are implicated in absence epilepsy. The mGluR4-deficient mouse is completely resistant to generalized absence seizures induced by $GABA_A$-receptor antagonists and displays increased evoked glutamate release in structures that are involved in absence epilepsy (ventrobasal thalamus, nucleus reticularis thalami and cerebral cortex laminae IV–VI). This indicates a role of mGluR4 on neurotransmitter release and its possible implication in absence seizures. The mGluR4 knockout mouse also shows altered short-term synaptic plasticity and poorly performs in rotating rod motor learning. Therefore, a function of mGluR4 may be to control synaptic efficacy and to support learning of complex motor tasks.

The mGluR6 subtype is exclusively expressed on retinal ON and OFF bipolar cells. Light inhibits (hyperpolarizes) the photoreceptors (the rods and cones) and shuts down glutamate release onto the secondary retinal bipolar cells. Now, what is the sign-inverting mechanism that translates the inhibitory photoreceptor response into an excitatory signal suitable for transmission of visual information to the visual cortex? Knockout experiments have shown that the sign inversion is mediated by mGluR6. Glutamate released from photoreceptors binds to mGluR6 and a G protein called transducin is activated, which causes phosphodiesterase to break down cyclic GMP. This triggers the closing of cyclic GMP-gated cation-selective channels, resulting in hyperpolarization. Light triggers depolarization of the bipolar cell and transmission of visual information by inhibiting glutamate release and shutdown of this pathway.

The mGluR8 knockout mouse shows increased anxiety-related behavior, suggesting a role of this receptor in response to a novel stressful environment. The mGluR8 subtype is ubiquitously expressed in the brain, but also in the retina where it is located on rod photoreceptors. Its activation leads to a decrease in glutamate release, probably via inhibition of voltage-gated Ca^{2+} channels. Therefore, its function would be to control neurotransmitter release from rod spherules. It is expected that the mGluR8-deficient mouse displays altered visual perception.

MGluR7 deletion in mice increases amplitude and power of the electroencephalographic theta rhythm. This change is accompanied by deficits in working memory, conditioned taste aversion and fear-conditioned response. The mGluR7-deficient mouse also shows increased epileptic seizure susceptibility. The mGluR7 subtype therefore plays a crucial role in neuronal excitability and specific cognitive functions.

12.10 SUMMARY

- 'Metabotropic' glutamate receptors were first hypothesized based on glutamate-stimulated, G-protein-dependent PI hydrolysis in neuronal preparations.
- mGluR1 was first cloned using expression cloning. Subsequently, mGluR2 to mGluR8 were cloned using homology-based techniques.
- mGluRs can be divided into three groups based on sequence homology and signaling:
  - *Group I*, mGluR1 and 5 couple to G_q protein and stimulate IP hydrolysis.
  - *Group II*, mGluR2 and 3, as well as *Group III*, mGluR4, 6, 7 and 8, couple to G_i/G_o proteins, inhibit AC and activate GIRK channels.
- mGluRs are large and complex proteins made of an extracellular ligand-binding domain connected through a cysteine-rich region to a seven transmembrane domain activating G proteins.
- Structure–function studies revealed a dimeric functioning of these proteins, agonist binding resulting in a change in the relative position of the two protomers, switching the transmembrane domains into their active state.
- High-throughput screening approaches identified allosteric modulators of mGluRs, which either inhibit or facilitate receptor activation by acting in the seven transmembrane domains. These compounds are very

selective, and have the advantage of maintaining the normal biological activity of the receptor *in vivo*.

- mGluRs can control neuronal excitability by modulating K^+ and Ca^{2+} conductances via a large variety of intracellular messengers. These include mainly G-protein βγ-subunits, Ca^{2+} ions and protein kinases.
- Group-I mGluR function is downregulated by PKC-mediated heterologous desensitization. This allows for negative feedback as well as for cross-talk with other neurotransmitter systems.
- mGluRs act as presynaptic auto- and hetero-receptors, reducing the release of many neurotransmitters. The major mechanism for presynaptic reduction of transmitter release is inhibition of Ca^{2+} channels and reduced release probability.
- Postsynaptically, mGluRs have a dual action on synaptic transmission. They can induce slow excitatory synaptic responses and/or modulate the efficacy of the fast synaptic transmission mediated by ionotropic glutamate receptors. For example, mGluRs are involved in long-term depression (LTD) and long-term potentiation (LTP) of excitatory synaptic transmission, in cerebellum and hippocampus.
- The synaptic localization and functions of mGluRs depend on multiprotein complexes that include the receptors themselves and various scaffolding, signaling and effector molecules.

Further reading

Aiba, A., Kano, M., Chen, C., et al., 1994. Deficient cerebellar long-term depression and impaired motor learning in mGluR1 mutant mice. Cell 79, 377–388.

Ango, F., Prezeau, L., Muller, T., et al., 2001. Agonist-independent activation of metabotropic glutamate receptors by the intracellular protein Homer. Nature 411, 962–965.

Charpak, S., Gahwiler, B.H., Do, K.Q., Knopfel, T., 1990. Potassium conductances in hippocampal neurons blocked by excitatory amino-acid transmitters. Nature 347, 765–767.

Conn, P.J., Pin, J.-P., 1997. Pharmacology and functions of metabotropic glutamate receptors. Ann. Rev. Pharmacol. Toxicol. 37, 205–237.

Dhami, G.K., Ferguson, S.S., 2006. Regulation of metabotropic glutamate receptor signalling, desensitization and endocytosis. Pharmacol. Ther. 111, 260–271.

El Far, O., Betz, H., 2002. G-protein coupled receptors for neurotransmitter amino acids: C-terminal tails, crowded signalosomes. Biochem. J. (Review) 365, 329–336.

Fagni, L., Worley, P., Ango, F., 2002. Homer as both a scaffold and transduction molecule. STKE (Review) (137), RE8.

Gereau, R.W., Conn, P.J., 1995. Multiple presynaptic metabotropic glutamate receptors modulate excitatory and inhibitory synaptic transmission in hippocampal area CA1. J. Neurosci. 15, 6879–6889.

Gereau, R.W., Heinemann, S.F., 1998. Role of protein kinase C phosphorylation in rapid desensitization of metabotropic glutamate receptor 5. Neuron 20, 143–151.

Gomeza, J., Joly, C., Kuhn, R., et al., 1996. The second intracellular loop of metabotropic glutamate receptor 1 cooperates with other intracellular domains to control coupling to G-proteins. J. Biol. Chem. 271, 2199–2205.

Goudet, C., Binet, V., Prezeau, L., Pin, J.-P., 2004. Allosteric modulators of class-C G-protein coupled receptors open new possibilities for therapeutic application. Drug Discov. Today: Ther. Strat. 1, 125–133.

Goudet, C., Gaven, F., Kniazeff, J., et al., 2004. Heptahelical domain of metabotropic glutamate receptor 5 behaves like rhodopsin-like receptors. Proc. Natl Acad. Sci. USA 101, 378–383.

Jia, Z., Lu, Y., Henderson, J., et al., 1998. Selective abolition of the NMDA component of long-term potentiation in mice lacking mGluR5. Learn. Mem. 5, 331–343.

Kim, E., Sheng, M., 2004. PDZ domain proteins of synapses. Nat. Rev. Mol. Cell. Biol. 4, 833–841.

Kniazeff, J., Bessis, A.S., Maurel, D., Ansanay, H., Prezeau, L., Pin, J.-P., 2004. Closed state of both binding domains of homodimeric mGlu receptors is required for full activity. Nat. Struct. Mol. Biol. 11, 706–713.

Kunishima, N., Shimada, Y., Tsuji, Y., et al., 2000. Structural basis of glutamate recognition by a dimeric metabotropic glutamate receptor. Nature 407, 971–977.

Pagano, A., Rüegg, D., Litschig, S., et al., 2000. The non-competitive antagonists 2-methyl-6-(phenylethynyl)pyridine and 7-hydroxyiminocyclopropan [b]chromen-1a-carboxylic acid ethyl ester interact with overlapping binding pockets in the transmembrane region of group I metabotropic glutamate receptors. J. Biol. Chem. 275, 33750–33758.

Pin, J.-P., Acher, F., 2002. The metabotropic glutamate receptors: structure, activation mechanism and pharmacology. Cur. Drug Targets CNS Neur. Dis. 1, 297–317.

Pin, J.-P., Joly, C., Heinemann, S.F., Bockaert, J., 1994. Domains involved in the specificity of G-protein activation in phospholipase C-coupled metabotropic glutamate receptors. EMBO J. 13, 342–348.

Pin, J.-P., Kniazeff, J., Liu, J., et al., 2005. Allosteric functioning of dimeric class C G-protein coupled receptors. FEBS J. 272, 2947–2955.

Sladeczek, F., Pin, J.-P., Recasens, M., et al., 1985. Glutamate stimulates inositol phosphate formation in striatal neurons. Nature 317, 717–719.

Smitt, P.S., Kinoshita, A., Leeuw, B.D., et al., 2000. Paraneoplastic cerebellar ataxia due to autoantibodies against a glutamate receptor. N. Engl. J. Med. 342, 21–27.

Swartz, K.J., Bean, B.P., 1992. Inhibition of calcium channels in rat CA3 pyramidal neurons by a metabotropic glutamate receptor. J. Neurosci. 12, 4358–4371.

Takahashi, K., Tsuchida, K., Taneba, Y., et al., 1993. Role of the large extracellular domain of metabotropic glutamate receptors in agonist selectivity determination. J. Biol. Chem. 258, 19341–19345.

Tateyama, M., Abe, H., Nakata, H., Saito, O., Kubo, Y., 2004. Ligand-induced rearrangement of the dimeric metabotropic glutamate receptor 1alpha. Nat. Struct. Mol. Biol. 11, 637–642.

Tateyama, M., Kubo, Y., 2006. Dual signalling is differentially activated by different active states of the metabotropic glutamate receptor 1. Proc. Natl Acad. Sci. USA 103, 1124–1128.

Tu, J.C., Xiao, B., Naisbitt, S., et al., 1999. Coupling of mGluR/homer and PSD-95 complexes by the Shank family of postsynaptic density proteins. Neuron 23, 583–592.

Tu, J.C., Xiao, B., Yuan, J.P., et al., 1998. Homer binds a novel proline-rich motif and links group 1 metabotropic glutamate receptors with IP3 receptors. Neuron 21, 717–726.

PART III

SOMATO-DENDRITIC PROCESSING AND PLASTICITY OF POSTSYNAPTIC POTENTIALS

CHAPTER

13

Somato-dendritic processing of postsynaptic potentials I: Passive properties of dendrites

Constance Hammond

OUTLINE

Neurons of the mammalian central nervous system receive many afferents which contact different parts of their somato-dendritic arborization. When these afferents are activated, if their combined effect is depolarizing enough, they trigger the firing of sodium action potentials in the postsynaptic neuron. Classically, it is accepted that these action potentials are generated at a central point in the neuron, at the level of the initial segment of the axon (action potential generating zone; see Section 4.4.3 and **Figure 13.1**). The action potential propagates back to the soma and forward to the axon and axon terminals.

Action potentials are the response of the postsynaptic neuron. This response may be simple, consisting of a single action potential. In this case, it can be described by a single characteristic: its latency. However, the postsynaptic response is generally more complex, consisting of several action potentials. It can then be described by several parameters: the latency of the first action potential, the duration of the response, the frequency of the action potentials that compose the response and the overall form – the pattern or configuration – of the response (see **Figure 14.1**).

The events that lead to a postsynaptic response can be separated into several stages. When the afferent synapses are activated, an excitatory or inhibitory current is generated at the subsynaptic membrane, as a result of activation of receptor channels by the neurotransmitter(s). These postsynaptic currents propagate through the dendrites to the soma and to the initial segment of the postsynaptic neuron. In the course of their propagation, the postsynaptic currents summate. If the sum of the postsynaptic currents is sufficient to depolarize the membrane of the initial segment as far as the threshold potential for activation of the voltage-sensitive sodium channels, a response is triggered in the postsynaptic neuron (**Figure 13.1**).

However, we will see in the following chapters that the presence of a postsynaptic response and the characteristics of this response are not the result only of the integration of different currents of synaptic origin over the somato-dendritic tree. In fact, the response of the postsynaptic neuron is the result of two types of currents: currents across receptor channels in the postsynaptic membrane evoked by neurotransmitters and currents across voltage-sensitive channels present in the non-synaptic membrane. Currents of the first type are generated strictly at the postsynaptic membrane and their presence and duration is determined essentially by the interaction between the transmitter and the receptor channel and the intrinsic properties of the receptor channel. Currents of the second type are generated at the non-synaptic membrane (dendritic, somatic or at the initial segment) by voltage changes resulting from currents of synaptic origin, or from currents generated during the first action potential that is fired. The voltage-gated channels responsible for this second type of current are

Cellular and Molecular Neurophysiology. DOI: 10.1016/B978-0-12-397032-9.00013-3

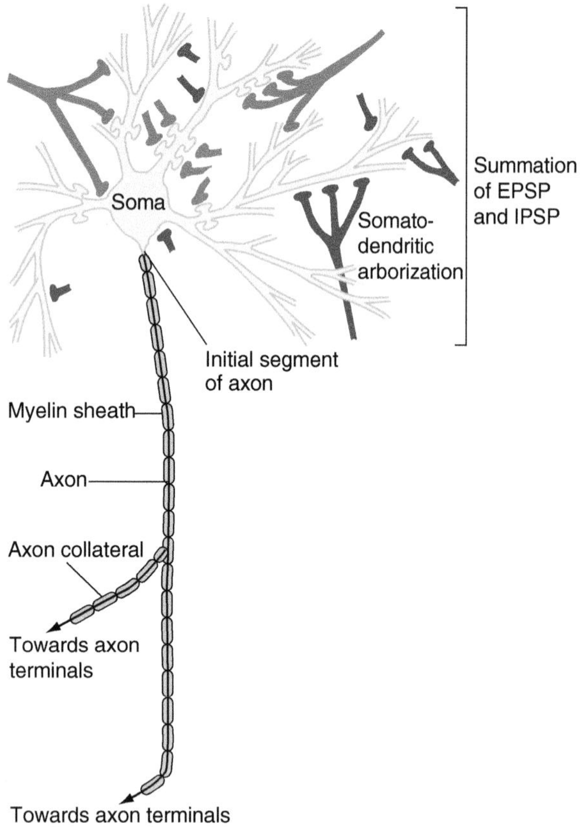

FIGURE 13.1 **Schematic of a neuron and some of its afferents.** Afferent fibers establish synaptic contacts on spines and dendritic branches which are situated at different distances from the soma of the postsynaptic neuron. When these afferents are activated, the depolarizing or hyperpolarizing postsynaptic currents are conducted towards the soma and initial segment of the axon. It is at this level that the response of the postsynaptic neuron is generated. The response is then conducted along the axon and its collateral branches.

different from those of the sodium action potential and are generally activated in the sub-threshold range of membrane potentials. The duration of these subliminal voltage-gated currents is determined by the gating properties of the corresponding channels.

This chapter looks at the conduction and the summation of synaptic currents (first type of currents) over the dendritic tree. The characteristics of the diverse non-synaptic, subliminal currents (second type of currents) together with their role in the pattern of the postsynaptic discharge will be studied in the following chapters.

13.1 PROPAGATION OF EXCITATORY AND INHIBITORY POSTSYNAPTIC POTENTIALS THROUGH THE DENDRITIC ARBORIZATION

Excitatory and inhibitory postsynaptic potentials result, respectively, from depolarizing or hyperpolarizing currents through channels opened by neurotransmitters (receptor channels) in the postsynaptic membrane. These currents are generated over the somato-dendritic tree, at sites more or less distant from the soma (distal dendritic sites or proximal dendritic sites). Once generated, the postsynaptic currents propagate along the length of the dendrites to the soma. For a long time it was thought that postsynaptic currents propagated passively and decrementally along the dendrites: passively because dendrites do not generate action potentials, the propagation of the signal depending only on the cable properties of the dendrite; and decrementally because the signal attenuates as it propagates, owing to the leakage properties of the membrane. From this it would be expected that depolarizations evoked by distal excitatory synapses would be smaller in amplitude at the soma and would have a longer risetime than depolarizations evoked by proximal synapses.

In fact, it seems that propagation is not always passive and not always decremental. There may be at least two types of propagation of postsynaptic currents through dendrites:

- a passive and decremental propagation, which implies an attenuation of distal postsynaptic currents;
- a passive but only slightly decremental propagation, which occurs where the cable properties of the dendrite are very good and involve no attenuation, or a weak attenuation, of distal postsynaptic currents.

These two alternatives are treated in Sections 13.1.2 and 13.1.3.

13.1.1 The complexity of synaptic organization (Figure 13.1)

Presynaptic complexity

A presynaptic afferent axon gives off many axon terminals (terminal boutons or 'en passant' terminals). In this way, it generally establishes several synaptic contacts with the postsynaptic neuron. In addition, the postsynaptic neuron receives synapses coming from many other presynaptic axons. It is thus possible to distinguish several levels of complexity in postsynaptic potentials:

- the postsynaptic potential evoked in the absence of presynaptic action potential (in the presence of tetrodotoxin (TTX)) by the spontaneous fusion of a synaptic vesicle with the presynaptic membrane: miniature postsynaptic potential;
- the postsynaptic potential representing the sum of postsynaptic potentials generated by synaptic boutons coming from the same presynaptic axon: unitary postsynaptic potential;
- the postsynaptic potential representing the sum of all the postsynaptic potentials generated at all the active synaptic boutons: composite postsynaptic potential.

Postsynaptic complexity

Different dendritic postsynaptic regions (spines, branches and main trunks) are not equivalent. The diameter of dendritic trunks is greater than that of branches, particularly distal branches. Thus, different dendritic compartments have different resistances (note that $R = \rho l/s$, ρ being the resistivity, l the length and s the cross-section of the dendrite). This means that spines with a neck, or a very small diameter pedicle, have a high resistance. Consequently, synaptic currents generated at different points do not give the same potential change: for the same inward current I, the amplitude of the resulting postsynaptic depolarization (V_{EPSP}) will be greater for the dendritic regions where the resistance $r_m = 1/g_m$ is large ($V_{EPSP} = I_{EPSP}/g_m$).

Complexities of the propagation of postsynaptic action potentials

Postsynaptic potentials (EPSPs and IPSPs) propagate along the dendrites to the action potential initiation zone, which is generally situated in the initial region of the axon (initial segment). Depending on the cable properties of the dendrites, the postsynaptic potentials can change their characteristics (amplitude, risetime) during their propagation.

13.1.2 Passive decremental propagation of postsynaptic potentials

'Decremental' means that the postsynaptic potentials attenuate as they propagate. This implies that the postsynaptic potentials are not regenerated at each point along the dendrites, as is the action potential as it travels along the axon. This passive propagation depends on the cable properties of the dendrite. In order to estimate quantitatively the modifications of postsynaptic potentials in the course of their conduction, a theoretical model of the passive properties of membrane potential changes was first established by Wilfred Rall from data obtained on the squid giant axon. Thus, a postsynaptic potential conducted with decrement (i) reduces in amplitude, and (ii) has a risetime (rt) which gets longer as it is propagated along the dendrites (**Figure 13.2**).

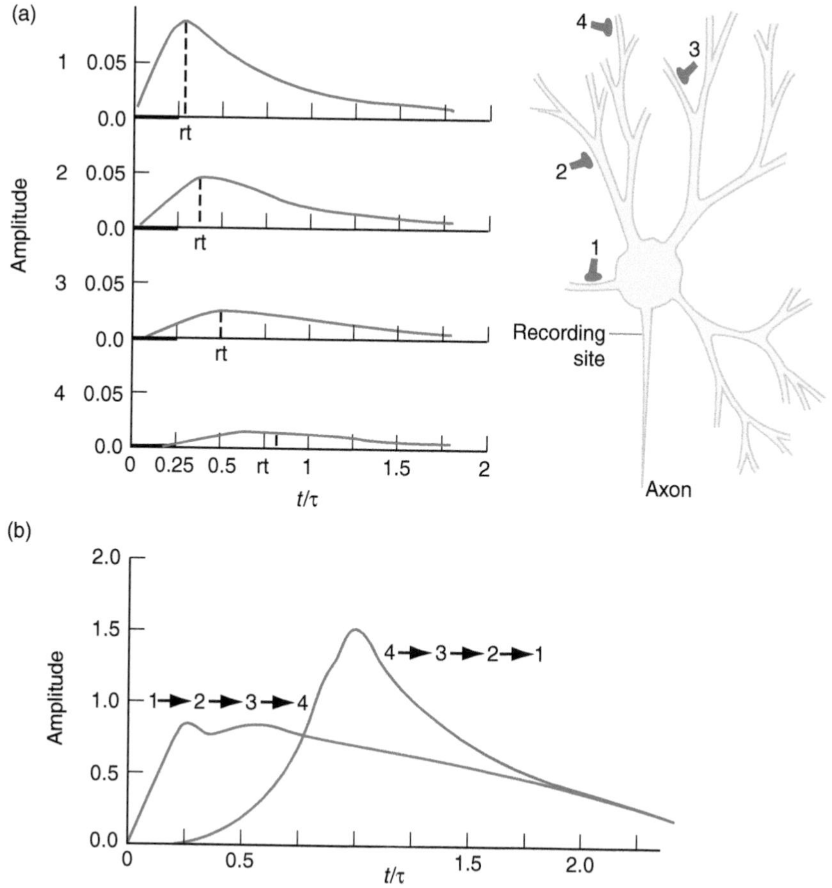

FIGURE 13.2 **Theoretical model of decremental conduction of excitatory postsynaptic potentials (EPSP) along dendrites.** **(a)** Four EPSPs numbered 1 to 4 are generated at the instant t between $t = 0$ and $t = 0.25$ ms (black bar in simulation diagram on left), at different sites within the dendritic tree (schematic drawing on right). At the site of generation, these EPSPs are identical in amplitude and duration. After conduction along the dendrites, their shapes are different (theoretical recordings at the level of the soma, simulation diagram on left). It can be observed that the further away the site of generation of the EPSP (case 4), the smaller is its amplitude and the longer is its risetime (rt) when it arrives at the level of the soma (compare the theoretical recordings 1 to 4). **(b)** Theoretical model of the linear summation of EPSP (see text for explanation). From Rall W (1977) *Handbook of Physiology*, vol. 1, part 1, Bethesda, MA: American Physiological Society, with permission.

The reduction in amplitude of the postsynaptic current as it gets further from the generation site is due to the fact that the current flows not only longitudinally along the dendrite but also transversely across the channels that are open in the dendritic membrane potential. This 'leak' of ions towards the extracellular medium results in a reduction in the postsynaptic current and a consequent reduction in the amplitude of the postsynaptic potential. Thus, the fewer the number of channels open in the dendritic membrane, the higher will be the value of r_m, the better will be the cable properties of the dendrite and the less will be the reduction in amplitude of postsynaptic potentials of distal origin.

The increase in the risetime of the postsynaptic potentials is due to the fact that part of the postsynaptic current serves to charge the capacity of each unit of membrane along the dendrite. The consequence of this is a change in the time course of the postsynaptic current: as it gets further from its point of generation, its risetime becomes longer (it can also be said that the speed of rising becomes slower).

13.1.3 Passive and non-decremental propagation of postsynaptic potentials

This type of propagation means that postsynaptic potentials are conducted passively along the dendrites but, because of the good cable properties of the dendritic arborization, they are almost unattenuated as they propagate. Thus, in the model of the synapse of Ia afferent fibers with spinal motoneurons, it has been shown that the unitary EPSPs evoked by the activity of afferent fibers and recorded in the soma have very similar amplitudes even though their risetimes may be different; i.e. when they are generated at different distances from the soma. This implies that, in this model, there must be local dendritic mechanisms that allow an almost non-attenuating conduction of the distal postsynaptic potentials.

13.2 SUMMATION OF EXCITATORY AND INHIBITORY POSTSYNAPTIC POTENTIALS

13.2.1 Linear and non-linear summation of excitatory postsynaptic potentials

In general, many excitatory synaptic afferents converge on a single neuron. At each excitatory synapse that is activated, there is an inward current of positive charges. When the membrane potential is not held at a fixed value, this inward current of positive charges depolarizes the postsynaptic membrane: this is the postsynaptic potential, or EPSP (see, for example, the current clamp recordings of the synaptic response to glutamate in Chapter 10).

A unitary EPSP (meaning one caused by the activation of a single afferent fiber; Section 13.1.1) cannot trigger action potentials. EPSPs generated in isolation are too small in amplitude to depolarize the membrane of the initial segment to the threshold potential for the opening of voltage-sensitive Na^+ channels. However, if many EPSPs generated at different sites in the dendritic arborization arrive more or less simultaneously at the level of the initial segment, the probability that they will generate action potentials becomes much greater. This is due to the fact that the EPSPs summate.

Linear summation of excitatory postsynaptic potentials

The term 'linear summation' means that the composite EPSP (see Section 13.1.1) resulting from the activity of several excitatory synapses has an amplitude that is equal to the geometric sum of the different EPSPs contributing to it. This is true when the EPSPs are generated at sites that are sufficiently far or isolated from one another to avoid interactions between them (on different dendritic branches, or on different dendritic spines, for instance).

A postsynaptic neuron generally receives many excitatory synapses at different points on its somato-dendritic arborization (see **Figures 13.1 and 13.2a**). These EPSPs summate as they propagate, in a temporo-spatial manner. To grasp this phenomenon, it must be understood that the EPSPs generated at different sites in the dendritic arborization and conducted to the initial segment of the axon can arrive spread out in time. The offset between the EPSPs will depend on the distances between the generation sites and on the respective times at which they were generated. The examples demonstrated here are based on theoretical calculations of the cable properties of dendrites. These data give a qualitative understanding of the phenomenon of summation but do not constitute a real experimental demonstration.

Let us consider the example of four EPSPs of the same amplitude, generated at different sites in the dendritic arborization, at times such that their arrivals at the initial segment are offset in time. **Figure 13.2b** shows the 'composite EPSP' obtained in two cases of arrival sequences. In the case $1 \rightarrow 2 \rightarrow 3 \rightarrow 4$, the four EPSPs are generated at the same time *t*; but since some are generated at more distal sites, their arrivals at the initial segment are staggered, the most proximal arriving first and the most distal arriving last. In the case $4 \rightarrow 3 \rightarrow 2 \rightarrow 1$, the most distal EPSPs are generated well before the proximal EPSPs, so that the distal EPSPs arrive before the proximal EPSPs. The theoretical results show that in the first case, in which the proximal EPSPs occur first and are followed by the more distal EPSPs, the 'composite EPSP' has a short latency, a long duration and a small amplitude; while in the second case, in which the distal EPSPs arrive before the proximal EPSPs, the 'composite EPSP' has a long latency and a large amplitude (**Figure 13.2b**).

The following experiment supports theoretical calculations. Simultaneous whole-cell recordings are performed from three interconnected pyramidal cells. Two of these neurons are presynaptic to the third. The composite EPSP evoked when the two presynaptic cells are stimulated simultaneously is equal to the linear sum of the unitary EPSPs evoked when each cell is stimulated separately. Subsequent morphological reconstruction of the pre- and postsynaptic neurons are performed to confirm that the presynaptic terminals are on different branches of the postsynaptic cell's basal dendrites and hence probably electrotonically distant from each other.

Non-linear summation of excitatory postsynaptic potentials

The term 'non-linear summation' means that the 'composite EPSP' has an amplitude that is not equal to the geometric sum of the different EPSPs contributing to it. This occurs, for instance, when two EPSPs are generated at the same site or at sites that are close.

Let us take the example of two excitatory synapses whose neurotransmitter is glutamate and which are situated close together on the same dendritic segment (**Figure 13.3**), supposing that the membrane potential of the dendritic segment is V_m. When synapse 1 is active alone, $EPSP_1$ is recorded, due to the excitatory postsynaptic current I_1, such that $I_1 = g_{cations} (V_m - E_{cations})$, whose amplitude is $V_{EPSP1} = I_1/g_m$, where g_m is the membrane conductance (**Figure 13.3a**). When synapse 2 is active alone, $EPSP_2$ is recorded at level 2, due to the postsynaptic current I_2, such that $I_2 = g_{cations} (V_m - E_{cations})$, whose amplitude is $V_{EPSP2} = I_2/g_m$ (**Figure 13.3b**). If we suppose that when the two EPSPs are generated separately, $V_{EPSP1} = V_{EPSP2}$, what is the amplitude of the 'composite EPSP' when the two synapses are active at the same time?

When synapse 1 is activated first, $EPSP_1$ is recorded in the postsynaptic element and will be conducted passively to neighboring regions (**Figure 13.3a**). At time $t + \Delta t$, $EPSP_1$ arrives at the postsynaptic element 2. The membrane of the postsynaptic element 2 is then at a potential V_m' which is more positive than V_m (**Figure 13.3c**). If at this moment ($t + \Delta t$) synapse 2 is active, the postsynaptic current I_2' will be smaller than if it had taken place independently from I_1 because the electrochemical gradient of Na^+ and Ca^{2+} ions is reduced. $I_2' = g_{cations}(V_m' - E_{cations})$ and $I_2' < I_2$ because $(V_m' - E_{cations}) < (V_m - E_{cations})$. The 'composite EPSP' will have an amplitude less than the geometric sum $EPSP_1 + EPSP_2$ (**Figure 13.3c**).

This is also the case when a single excitatory synapse is activated repetitively by the arrival of high-frequency presynaptic action potentials. When an excitatory postsynaptic current is generated before the preceding current has ended, it has a smaller amplitude because the postsynaptic membrane is depolarized. Thus, during high-frequency activation, successive excitatory postsynaptic potentials have amplitudes that are smaller and smaller.

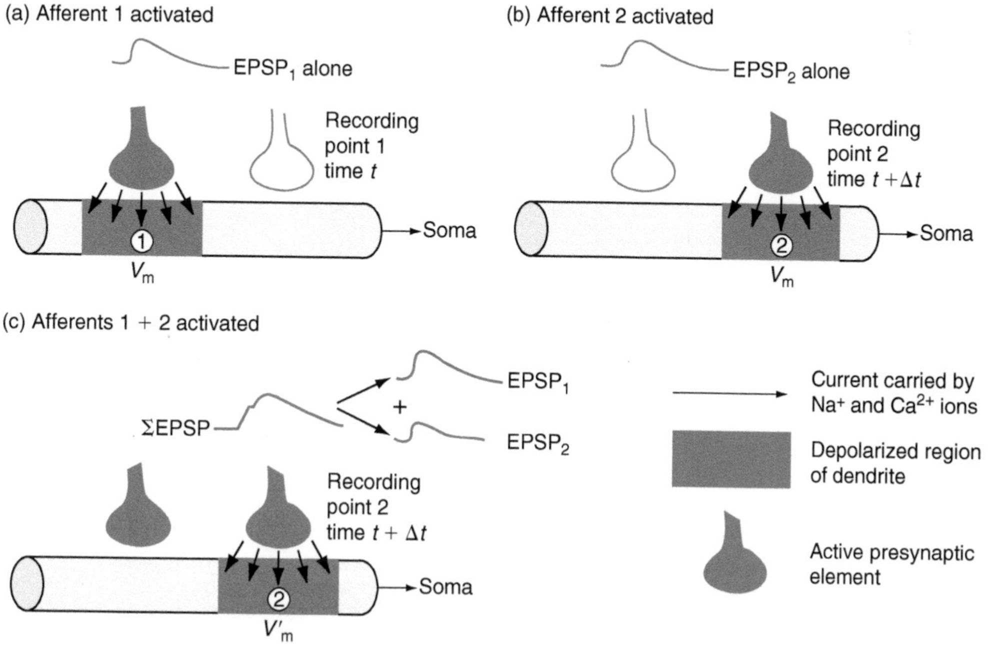

FIGURE 13.3 **Non-linear summation of excitatory postsynaptic potentials.** Suppose that there are two excitatory synapses situated close together on the same dendritic segment. **(a)** When afferent 1 is activated at time t, a depolarization of the postsynaptic membrane 1 is recorded at time t ($EPSP_1$ alone). This depolarization propagates in the two directions away from 1. **(b)** When afferent 2 is activated, at time $t + \Delta t$, a depolarization of the postsynaptic membrane 2 is recorded ($EPSP_2$ alone). **(c)** When the two afferents 1 and 2 are activated as before, but together, one at time t and the other at time $t + \Delta t$, a depolarization of the postsynaptic membrane 2 ($\Sigma EPSP$) is recorded at time $t + \Delta t$, which does not correspond to the geometric sum $EPSP_1$ alone + $EPSP_2$ alone, since $EPSP_2'$ has an amplitude which is smaller than $EPSP_2$ (see text for explanation).

13.2.2 Linear and non-linear summation of inhibitory postsynaptic potentials

When inhibitory synapses are active they cause, in the postsynaptic membrane, an outward postsynaptic current of positive charges (carried by K^+ ions) or an inward current of negative charges (carried by Cl^- ions) which hyperpolarizes the membrane: this is the inhibitory postsynaptic potential, or IPSP.

Linear summation of IPSPs is symmetrically the same as linear summation of EPSPs. Non-linear summation of IPSPs is symmetrically the same as non-linear summation of EPSPs.

13.2.3 The integration of excitatory and inhibitory postsynaptic potentials partly determines the configuration of the postsynaptic discharge

In order for an action potential to be triggered at the initial segment, the membrane of the initial segment must be depolarized to the threshold potential for the opening of voltage-sensitive Na^+ channels. It is also necessary for this depolarization to have a relatively rapid risetime so that the Na^+ channels do not inactivate during the depolarization. The characteristics of depolarization of the initial segment (amplitude, duration, risetime) result partly from the summation of excitatory and inhibitory postsynaptic potentials.

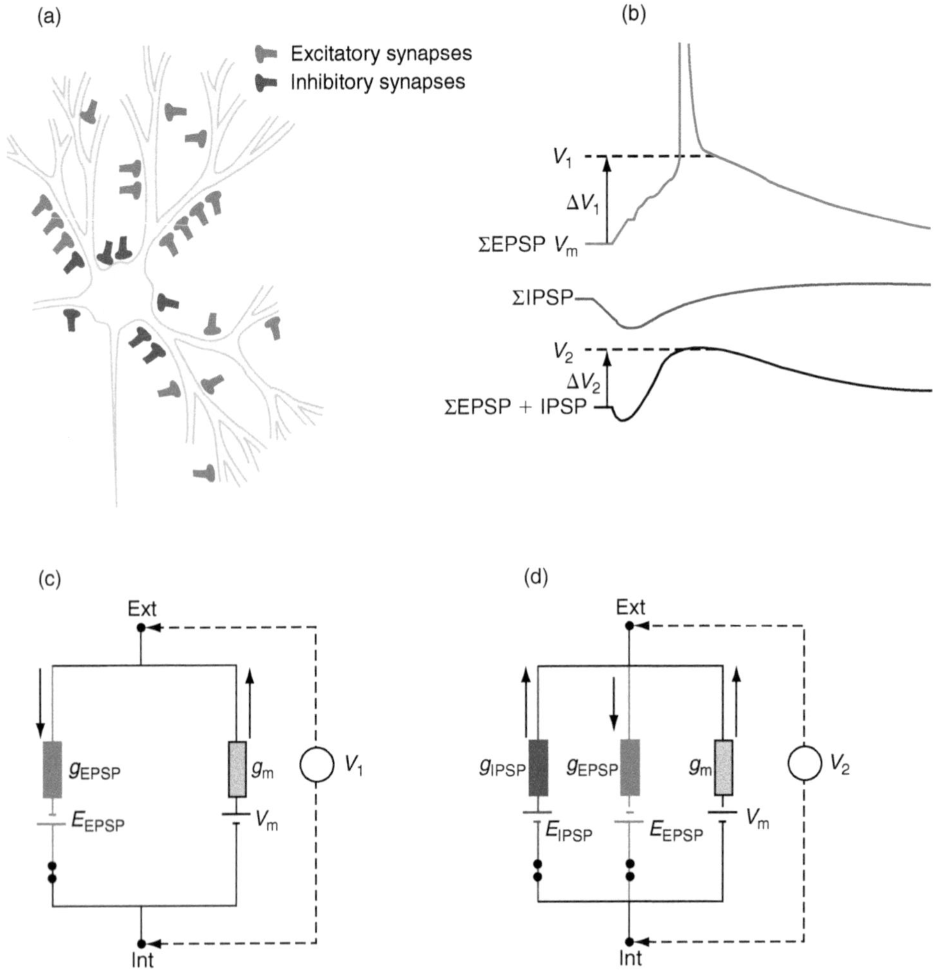

FIGURE 13.4 **Integration of excitatory (EPSP) and inhibitory (IPSP) postsynaptic potentials.** **(a)** Suppose that on a dendritic tree, there are glutaminergic excitatory synapses which are situated distally, and $GABA_B$-type inhibitory synapses which are situated proximally, and that all of these are active at the same instant *t*. **(b)** If only the excitatory synapses are active, a depolarization, a composite EPSP (ΣEPSP) will be recorded at the soma which corresponds to the linear and non-linear summation of all the different EPSPs (top trace). We will suppose that the ΣEPSP has an amplitude that is sufficient to trigger an action potential (upper trace). If only the inhibitory synapses are active, a hyperpolarization, a composite IPSP (ΣIPSP) will be recorded at the soma which corresponds to the linear and non-linear summation of all the different IPSPs (middle trace). When all these different synapses are activated at the same time *t*, a depolarization preceded by a hyperpolarization, a composite PSP, will be recorded at the soma, corresponding to the sum of the different synaptic potentials (ΣEPSP + ΣIPSP) (bottom trace). In this case, the amplitude of the depolarization is no longer sufficient to trigger an action potential. **(c)** Electrical equivalent of the membrane at the level of the initial segment, for an EPSP alone. **(d)** Electrical equivalent for the membrane when an EPSP and an IPSP summate. The currents I_{EPSP} and I_{IPSP} are opposite and subtract from one another. By comparing with **(c)**, it is observed that I_{EPSP} in **(c)** is greater than $I_{EPSP} + I_{IPSP}$ in **(d)**, and $\Delta V_1 > \Delta V_2$.

Integration of depolarizing (excitatory) postsynaptic potential with hyperpolarizing (inhibitory) postsynaptic potential

A hyperpolarizing postsynaptic potential is due to a current whose reversal potential is more negative than the resting membrane potential of the cell. This type of inhibition is generally due to the opening of K^+ channels ($GABA_B$-type inhibition). Since the equilibrium potential of K^+ ions is more negative than the resting membrane potential, the opening of K^+ channels gives rise to an outward current (an exit of positive charges) and to a hyperpolarization of the membrane; i.e. an IPSP. If this IPSP is concomitant with an EPSP, it will reduce the amplitude of the EPSP. This type of summation of EPSP and IPSP is summarized in **Figure 13.4.**

Integration of depolarizing (excitatory) postsynaptic potential and silent (inhibitory) postsynaptic potential

A silent postsynaptic potential is due to a current whose reversal potential is close to the resting potential of the cell. Generally, this is caused by a current of Cl^- ions through $GABA_A$ channels (see Chapter 9). When the equilibrium potential of Cl^- ions is close to the membrane resting potential, the opening of Cl^- channels does not reveal a hyperpolarizing current at the resting potential (from which comes the term 'silent' for this inhibition). However, when the membrane is depolarized by an EPSP, the inhibition is no longer silent, but becomes hyperpolarizing. The result is the reduction or even the complete suppression of the EPSP (**Figure 13.5**).

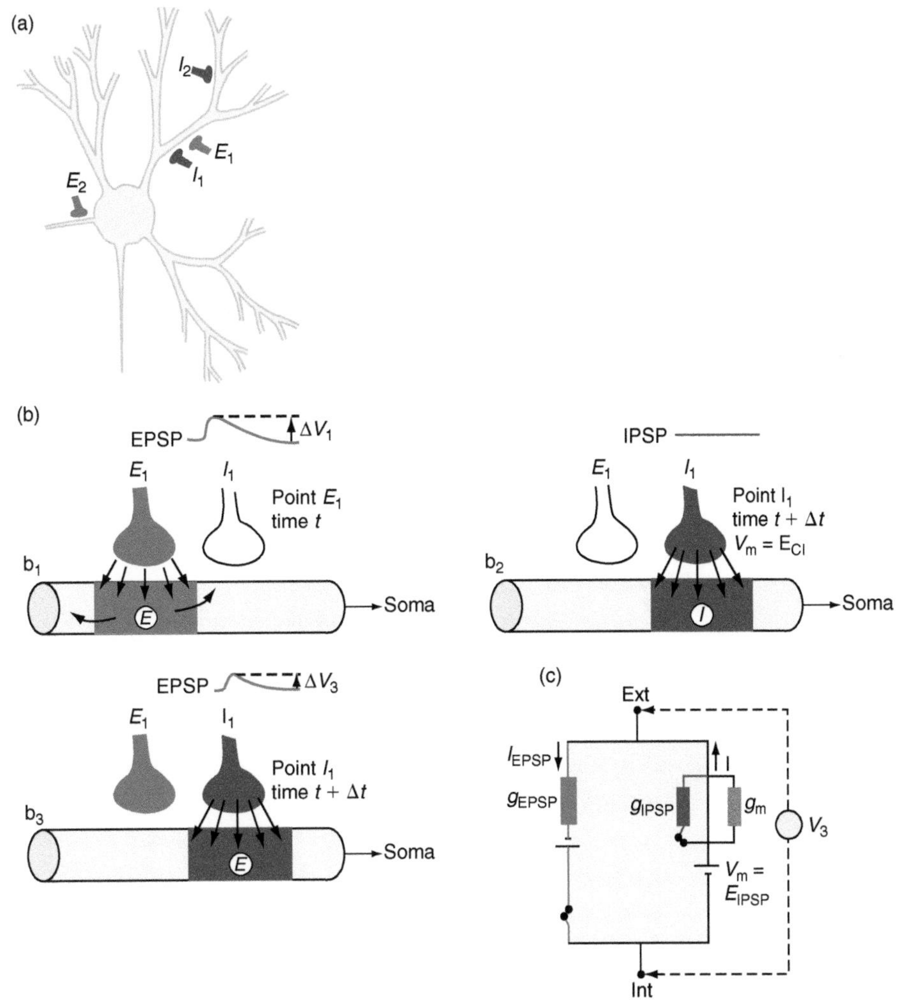

FIGURE 13.5 **Role of silent inhibition. (a)** This diagram shows two synapses, one glutamatergic with postsynaptic AMPA receptors (E_1) and the other GABAergic with $GABA_A$ postsynaptic receptors (I_1), situated close to one another on the same dendritic segment, such that the inhibitory synapse is closer to the soma than the excitatory synapse. **(b)** When the excitatory synapse is excited alone, an EPSP of ΔV_1 in amplitude is recorded (b_1). When the inhibitory synapse is activated alone, no change in potential is recorded because $V_m = E_{Cl}$ (b_2). When both synapses are activated, the EPSP which propagates towards the soma is reduced in amplitude (amplitude ΔV_3), or even cancelled out. This type of inhibition is selective because it only acts on excitatory synapses that are situated distally. **(c)** Electrical equivalent of the membrane at the dendritic segment. If this is compared with **Figure 13.4c**, it can be seen that $\Delta V_3 < \Delta V_1$ because $g_m + g_{IPSP} > g_m$.

Integration of depolarizing (excitatory) postsynaptic potential and depolarizing inhibitory postsynaptic potential

A depolarizing inhibitory postsynaptic potential is due to a synaptic current whose reversal potential is more positive than the resting potential of the membrane but more negative than the threshold for the opening of the Na^+ channels of the action potential. This is generally due to the opening of Cl^- channels in cells in which the reversal potential for Cl^- ions is situated between the resting potential and the threshold potential for the opening of the Na^+ channels of the action potential. Thus, when the membrane is at its resting potential, this Cl^- current causes a slight depolarization of the membrane, but does not trigger action potentials. When the membrane is depolarized (by an EPSP) above the inversion potential of Cl^- ions, this current causes a hyperpolarization of the membrane and an inhibition of the EPSP.

13.3 SUMMARY

Several types of inhibition appear over the length of the somato-dendritic arborization and these limit the effect of excitatory synapses. The opening or non-opening of the Na^+ channels of the action potential and, in consequence, the generation of action potentials which will constitute the response of the postsynaptic neuron, are the result of this summation of excitatory and inhibitory postsynaptic potentials. However, the characteristics of the response of the postsynaptic neuron are determined not only by the amplitude and duration of the depolarization of synaptic origin but also by the characteristics of the membrane of the initial segment, also known as 'input–output' characteristics.

Further reading

Buhl, E.H., Halasy, K., Somogyi, P., 1994. Diverse sources of hippocampal unitary inhibitory postsynaptic potentials and the number of release sites. Nature 368, 823–828.

Cauller, L.J., Connors B.W., 1992. Functions of very distal dendrites. In: McKenna, T.M., Davis, J., Zornetzer, S.E. (Eds.), Single Neuron Computation. Academic Press, New York.

Gulledge, A.T., Kampa, B.M., Stuart, G.J., 2005. Synaptic integration in dendritic trees. J. Neurobiol. 64, 75–90.

Larkum, M.E., Launey, T., Dityatev, A., Lüscher, H.R., 1998. Integration of excitatory postsynaptic potentials in dendrites of motoneurons of rat spinal cord slice cultures. J. Neurophysiol. 80, 924–935.

Miles, R., Toth, K., Gulyas, A.I., Hajos, N., Freund, T.F., 1996. Differences between somatic and dendritic inhibition in the hippocampus. Neuron 16, 815–823.

Rall, W., 1977. Core conductor theory and cable properties of neurons. In: Brookhart, J.M., Mountcastle, V.B., Kandel, E.R., Geiger, S.R. (Eds.), Handbook of Physiology, vol. 1, part 1. American Physiological Society, Bethesda, MD.

Redman, S.J., 1973. The attenuation of passively propagating dendritic potentials in a motoneuron cable model. J. Physiol. 234, 637–664.

Reyes, A., 2001. Influence of dendritic conductances on the input-output properties of neurons. Annu. Rev. Neurosci. 24, 653–675.

Shepherd, G.M., 1994. The significance of real neuroarchitectures for neural network simulations. In: Schwartz, E.L. (Ed.), Computational Neuroscience. Oxford University Press, New York.

Spruston, N., Johnston, D., 1992. Perforated patch clamp analysis of the passive membrane properties of three classes of hippocampal neurons. J. Neurophysiol. 67, 508–528.

Spruston, N., Jaffe, D.B., Johnston, D., 1994. Dendritic attenuation of synaptic potentials and currents: the role of passive membrane properties. Trends Neurosci. 17, 161–166.

CHAPTER

14

Subliminal voltage-gated currents of the somato-dendritic membrane

Constance Hammond, Laurent Aniksztejn

OUTLINE

Not all neurons respond in the same way when they are activated by a depolarizing current pulse or when they are hyperpolarized: it can be said that they do not have the same pattern of firing (**Figure 14.1**). When depolarized, spinal motoneurons may respond with a low-frequency regular discharge, certain pyramidal neurons of the hippocampus with a burst of action potentials followed by a long silence, neurons of the inferior olive nucleus with an irregular sustained activity of bursts of action potentials. Conversely, when hyperpolarized, hippocampal neurons become silent whereas thalamic and subthalamic neurons discharge bursts of action potentials. It should be noted that the term 'firing pattern' is not equivalent to the term 'discharge frequency', except in cases where the response consists of action potentials generated at a regular frequency. In this case only, the mean value of the discharge frequency is sufficient to describe the response of the neuron. In other cases, the mean frequency value has no significance.

Neurons, as pointed out by Llinas (1988), are not interchangeable; i.e. a neuron cannot be functionally replaced by one of another type even if their synaptic connectivity, type of afferent neurotransmitters and receptors to these neurotransmitters are identical. The activity of a neuronal network is related not only to the *excitatory and inhibitory interactions* among neurons but also to their *intrinsic electrical properties*. The 'personality' of a neuron is defined by its input–output characteristics; i.e. its firing pattern (output) in response to a depolarization or a hyperpolarization (input).

Input–output characteristics are the result of a rich repertoire of ionic currents other than those of the action potentials. These currents, inward or outward, are called *subliminal voltage-gated currents* because they are activated at voltages sub-threshold to that of action potentials. They are located either in the dendritic or the somatic membrane or both. In many experiments, recordings are performed at the level of the soma and the question may remain as to where these currents are generated: at the level of the dendrites, the soma, or the initial segment of the axon?

14.1 OBSERVATIONS AND QUESTIONS

Figure 14.1 shows the different responses of central neurons recorded in current clamp mode, in response to a depolarizing current pulse and at different membrane potentials.

- What are the currents activated during the depolarizing current pulse that stop the firing of hippocampal neurons after 100–200 ms? Are these currents activated by action potentials?
- What are the currents activated by the depolarizing current pulse that delay the firing of motoneurons innervating the ink gland of *Aplysia*?
- What are the currents that make a thalamic neuron fire in the bursting mode? Why are they activated at a hyperpolarized membrane potential?

Cellular and Molecular Neurophysiology. DOI: 10.1016/B978-0-12-397032-9.00014-5

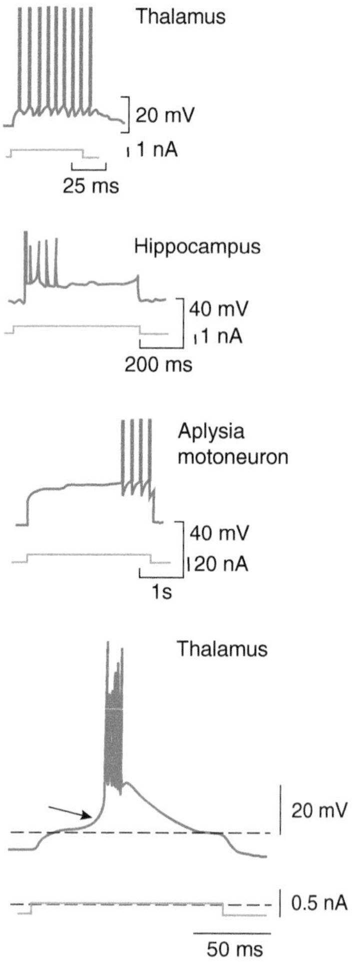

FIGURE 14.1 **Different neuronal responses to a depolarizing current pulse.** From top to bottom: thalamo-cortical neuron, pyramidal neuron of the hippocampus, motoneuron innervating the ink gland of *Aplysia*, and again a thalamo-cortical neuron but from a more hyperpolarized membrane potential.

To answer these questions, the main subliminal currents must be first explained individually (this chapter). Their influence on postsynaptic potentials and firing patterns are explained in Chapters 15–17.

14.2 THE SUBLIMINAL VOLTAGE-GATED CURRENTS THAT DEPOLARIZE THE MEMBRANE

To depolarize and excite a membrane at rest, currents have to be inward and to be turned on at potentials more negative than the threshold for the opening of voltage-gated Na^+ channels of the action potential. Three types of inward subliminal currents will be explained: the persistent Na^+ current (I_{NaP}), the transient Ca^{2+} current (I_{CaT}) and the hyperpolarization-activated cationic current (I_h).

14.2.1 The persistent inward Na^+ current, I_{NaP}

I_{NaP} is a slowly inactivating current. It is thus called 'P' for persistent in comparison with the fast-inactivating Na^+ current of the action potential (I_{Na}) which is transient. I_{NaP} is activated at potentials of about 10 mV negative to I_{Na}; i.e. at sub-threshold potentials. I_{NaP} is present in many vertebrate central neurons and, in particular, in Purkinje cells. I_{NaP} was first described by Prince and coworkers in hippocampal (1979) and neocortical (1982) neurons in slices, as an increase in slope resistance at potentials 10–15 mV positive to resting potential (the I_{NaP}/V relation showed an inward rectification). In the presence of tetrodotoxin (TTX) in the extracellular medium or when external Na^+ is replaced by the impermeant ion choline, this inward rectification disappears. An easy and quick way to isolate the I_{NaP} and reveal its voltage dependence is the use of the slow ramp (20–50 mV/s) protocol since it allows inactivation of I_{Na}.

Structure of the main channel subunit

An important question is whether persistent and fast inactivating Na-currents arise from different sets of sodium channels or whether the persistent Na-current results from different gatings of the same channel type.

To determine whether the I_{NaP} channel is a non-inactivating subtype of the classical fast-inactivating Na^+ channel, or is a different channel, the Na^+ channel α-subunit transcripts expressed in Purkinje cells are studied using single-cell reverse transcription-PCR (RT-PCR). Purkinje cells have been chosen because the two different voltage-gated Na^+ currents are recorded from them: one responsible for the fast depolarization phase of action potentials and a second responsible for the TTX-sensitive prolonged potential plateau. mRNA transcripts for two α-subunits have been found. $Na_v1.6$ (also named cerIII) is suggested to be responsible for I_{NaP}; the other, $Na_v1.1$ (rat brain I), is responsible for I_{Na} (see **Figure 4.6**). This result favors the hypothesis that I_{NaP} of Purkinje cells does not result from an Na^+ channel with multiple gating states but corresponds to a different channel.

However, this question is still open. The molecular nature of the channels carrying persistent Na^+ current is largely unknown. The amino-acid sequence responsible for fast inactivation and the location of the activation gate are known. The part which forms the pore-forming subunit and the part which is responsible for the TTX-sensitivity are also known. However, we do not know which particular amino-acid sequence is responsible for the non-inactivating properties of the persistent channels.

Gating properties and ionic nature

In whole-cell recordings, in the presence of Cd^{2+} (200 μM) and K^+-channel blockers (20 mM tetraethylammonium chloride (TEA), 2 mM Cs^+), a slowly inactivating Na^+ current is recorded in response to incremental depolarizing steps between −60 and −50 mV from a holding potential of −90 mV (**Figure 14.2a**). Note that at more depolarized levels, the fast-inactivating Na^+ current superimposes on the slow one. I_{NaP} has a fast activation time of 2–4 ms, and once it is evoked it is present for several hundreds of milliseconds and is totally and reversibly blocked by TTX (1 μM) (**Figure 14.2b**). This observation, together with the lack of effect of extracellular Co^{2+} on this current, suggests that it is an inward current which uses Na^+ as the charge carrier. The I/V relation gives an estimated reversal potential of +49.1 ± 1.3 mV (**Figure 14.2c**).

To determine the properties of the channels responsible for whole-cell I_{Na} and I_{NaP}, Na^+-channel behavior is examined in stellate cell somata (entorhinal cortex), using the cell-attached recording configuration. Patches contain many Na^+channels (macropatch experiment). A 500 ms depolarizing step elicits a large transient inward current in the first few milliseconds after the start of the step because of the superimposed opening of rapidly activating and inactivating transient Na^+ channels (**Figure 14.2d, top**). When many consecutive traces are averaged (**Figure 14.2d bottom**), the transient channel activity produces a transient ensemble current that closely resembles I_{Na} recorded in whole-cell experiments, in terms of both time course (compare **Figure 14.2a** and **d**) and I–V relationship (**Figure 14.2e, black dots**). The depolarizing step also elicits a lower level of sustained channel activity, characterized by repeated episodes of prolonged openings throughout the 500 ms depolarization (**Figure 14.2d, top**). Ensemble averages of the late openings produce a measurable persistent current (**Figure 14.2d, bottom**). The mean I–V relationship for ensemble persistent currents (**Figure 14.2e, open circles**) is similar to that determined from whole-cell recordings of I_{NaP} (compare **Figures 14.2c and 14.2e, open circles**). In all experiments, both transient and persistent single-channel currents are absent when 1 μM TTX is included in the patch pipette (data not shown), indicating that both types of channel behavior are caused by TTX-sensitive Na^+ channels. Together, these data indicate that the non-inactivating channel activity recorded in macropatches is responsible for whole-cell I_{NaP}. The single-channel conductance determined for late channel openings in eight macropatches is 19.3 ± 2.3 pS.

The voltage dependence of activation is determined by applying different voltage steps from a holding potential V_H = −90 mV. The normalized peak current amplitude is plotted against the test potential to give the activation curve. In suprachiasmatic neurons, I_{NaP} is half-activated at V_m = −43 mV, a potential at least 10 mV more negative than $V_{1/2}$ of I_{Na} (**Figures 14.2e** and **14.2f**). To study the inactivation properties of I_{NaP}, the membrane is clamped at different holding potentials and the current in response to a depolarizing step to −50 mV is recorded. A plot of peak current amplitude against holding potential gives a measure of the voltage dependence of inactivation. In these neurons, I_{NaP} is half-inactivated at $V_{1/2}$ = −68 mV (**Figure 14.3**). De-inactivation of I_{NaP} requires hyperpolarization of the membrane.

Pharmacology

I_{NaP} is sensitive to TTX at nanomolar doses in the extracellular solution and to internal QX-314 (a derivative of lidocaine) injected into the intracellular medium. These toxins also block I_{Na}, which is a problem when studying the role of I_{NaP} on the pattern of discharge of a neuron. There is no selective blocker of I_{NaP}. To study I_{NaP} in isolation, Ca^{2+} and K^+ channels must be blocked by Cd^{2+}, Co^{2+} or Ni^{2+} (200 μM to 1 mM), 4-AP (1 mM), TEA (10 mM), Ba^{2+} (1 mM). In these conditions I_{Na} will be also recorded at some voltage steps but is easily recognized by its high amplitude and fast inactivation (see **Figures 14.2a** and **d**).

Summary

I_{NaP} is a TTX-sensitive Na^+ current that activates at around −60 to −50 mV. I_{NaP} can be distinguished from the transient Na^+ current of action potential, I_{Na}, by its low threshold of activation and its slow inactivation.

14.2.2 The low-threshold transient Ca^{2+} current, I_{CaT}

The ability of neurons to fire low-threshold Ca^{2+} spikes suggested the existence of a low-voltage-activated Ca^{2+} current (I_{CaT}). I_{CaT} is called 'T' for transient since once activated it rapidly inactivates. It was originally described by Carbone and Lux in 1984 but its existence had first been suggested by Eccles and co-workers in 1964 to explain their observation of a period of enhanced excitability following membrane hyperpolarization in some central neurons.

Structure of the main channel subunit

The Ca_v3 family of α_1-subunits (see **Figure 5.2b**) conducts T-type Ca^{2+} currents, which are activated and inactivated more rapidly and at more negative membrane potentials than other Ca^{2+} current types. Three Ca_v3 different α_1-subunits have been described: $Ca_v3.1$ (α_{1G}), $Ca_v3.2$ (α_{1H}) and $Ca_v3.3$ (α_{1I}). Their primary structure shows the presence of four homologous domains, each containing six putative transmembrane segments (S1–S6) and a P-loop between segments 5 and 6. In contrast to the α_1-subunits of high-voltage-activated Ca^{2+} channels (L, N, P), T-channel α_1-subunits contain a large extracellular loop located between S5 and the P-loop. Ca_v3 α_1-subunits are only distantly related to these homologs with less than 25% amino-acid sequence identity.

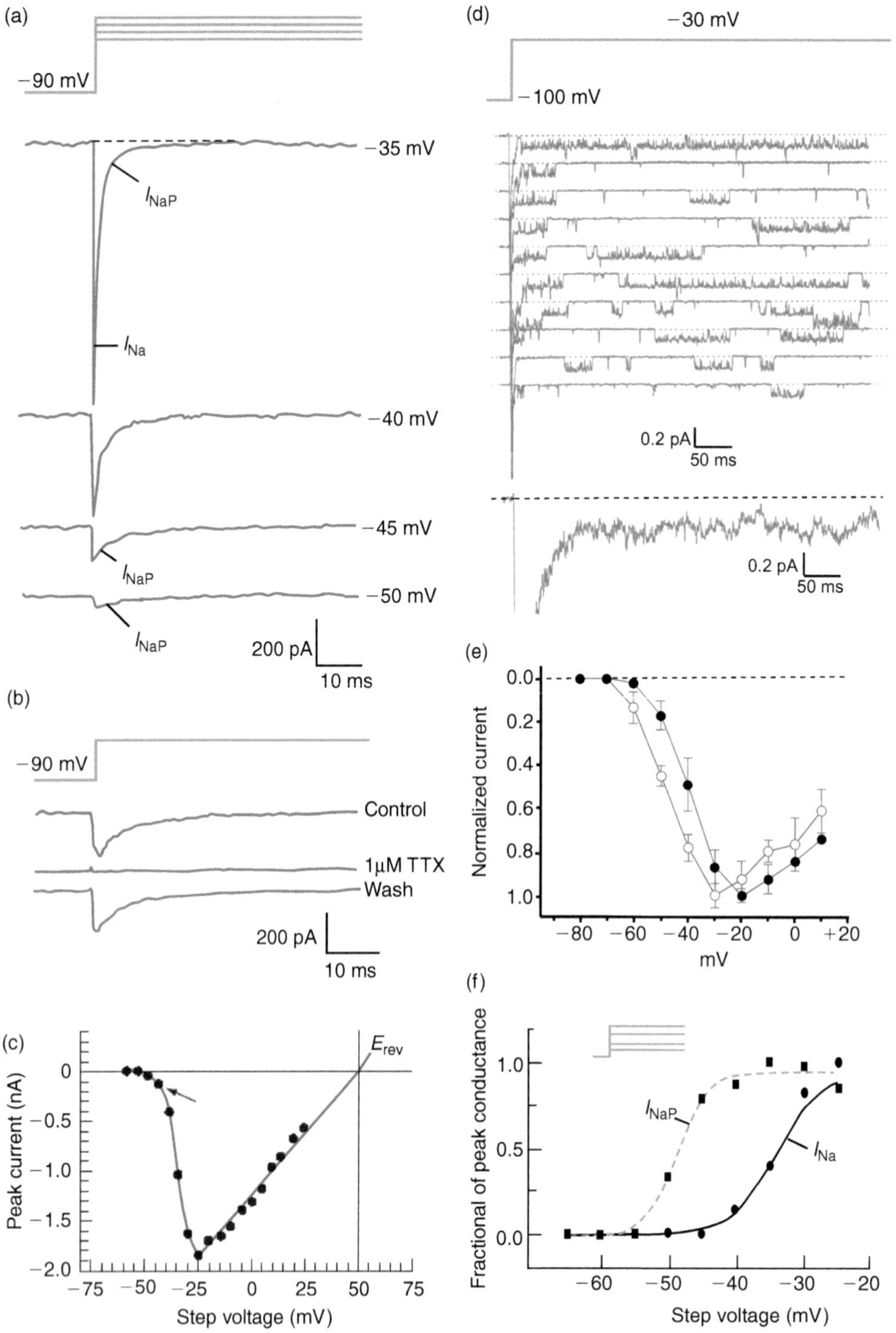

FIGURE 14.2 **Activation properties of the slowly inactivating Na⁺ current, I_{NaP}. (a–c,f)** The activity of neurons of the suprachiasmatic nucleus is recorded in slices (whole-cell configuration, voltage clamp mode). **(a)** Incremental steps from a holding potential of −90 mV. **(b)** Reversible block by TTX. **(c)** *I*/*V* plot for the peak I_{NaP} measured from records in **(a)**. **(d,e)** The activity of acutely dissociated stellate neurons of the entorhinial cortex is recorded in cell-attached configuration, voltage clamp mode. **(d, top)** Consecutive traces elicited by 500 ms depolarizing pulses to −30 mV. **(d, Bottom)** The trace shows the average of a set of 20 consecutive traces. **(e)** Mean current–voltage relationships for the transient (black dots; $n = 7$) and persistent (open circles; $n = 7$) Na^+ currents. **(f)** Activation curve ($V_{1/2} = -43$ mV). Parts (**a–c,f**) adapted from Pennartz CMA, Bierlaagh MA, Geurtsen AMS (1997) Cellular mechanisms underlying spontaneous firing in rat suprachiasmatic nucleus: involvement of a slowly inactivating component of sodium current. *J. Neurophysiol.* **78**, 1811–1825, with permission. Parts (**d,e**) adapted from Magistretti J, Ragsdale DS, Alonso A (1999) High conductance sustained single-channel activity responsible for the low-threshold persistent Na^+ current in entorhinal cortex neurons. *J. Neurosci.* **19**, 7334–7341, with permission.

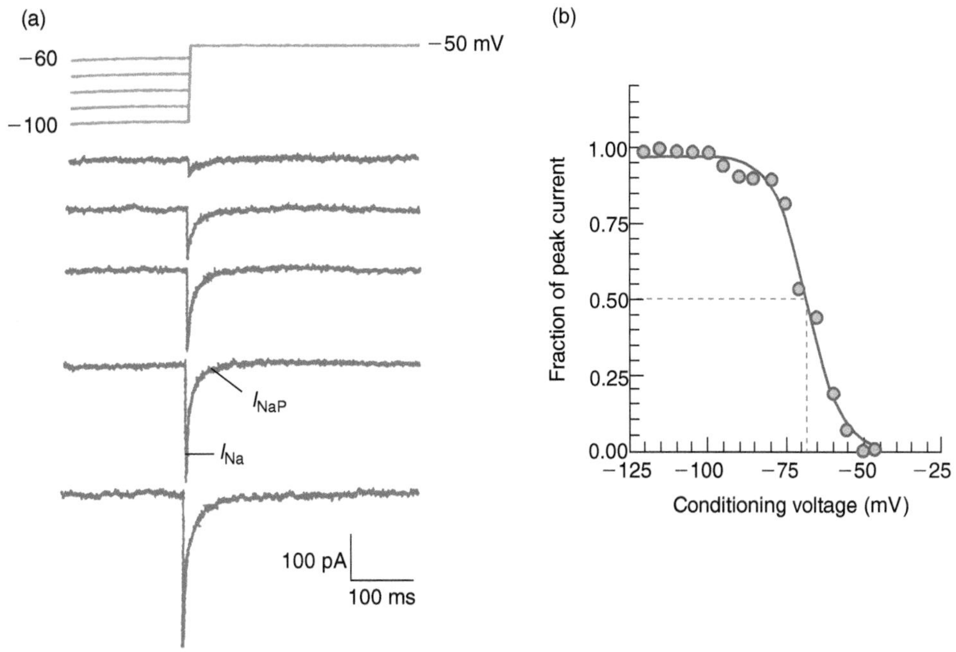

FIGURE 14.3 **Inactivation properties of the slowly inactivating Na^+ current, I_{NaP}.** The activity of neurons of the suprachiasmatic nucleus is recorded in slices (whole-cell configuration, voltage clamp mode). **(a)** Currents recorded in response to a fixed step to −50 mV from incremental holding potentials. **(b)** Inactivation curve ($V_{1/2} = -68$ mV, dotted line). Adapted from Pennartz CMA, Bierlaagh MA, Geurtsen AMS (1997) Cellular mechanisms underlying spontaneous firing in rat suprachiasmatic nucleus: involvement of a slowly inactivating component of sodium current. *J. Neurophysiol.* **78**, 1811–1825, with permission.

Single-channel conductance

The activity of chick dorsal root ganglion cells in culture is recorded in patch clamp (cell-attached patch). The patch pipette contains 110 mM of $BaCl_2$ instead of 2 mM of $CaCl_2$ for the following reasons: Ca^{2+} channels are permeable to Ba^{2+}; Ba^{2+} does not inactivate Ca^{2+} channels; Ba^{2+} does not activate Ca^{2+}-dependent channels such as Ca^{2+}-activated K^+ channels. Ba^{2+} current, at such high concentration of charge carrier, 110 mM, through Ca^{2+} channels has in general a large amplitude and is more easy to study. Moreover, when Ba^{2+} is the only external cation, Na^+ currents are not recorded.

The membrane is held at a hyperpolarized potential (V_H = −80 mV) to keep T channels non-inactivated. When depolarizing voltage steps of small amplitude (30–60 mV) are applied to the patch, unitary inward Ba^{2+} currents are recorded. These unitary inward currents have a small amplitude (**Figure 14.4**) and the unitary conductance γ_T is very low compared with that of high-threshold-activated Ca^{2+} channels: it varies between 5 and 9 pS in 110 mM of Ba^{2+}. At physiological concentrations of external Ca^{2+} (2 mM), the expected conductance would be around 1 pS. Single-channel T currents are present at the beginning of the depolarizing step and then disappear, though the membrane is still depolarized, consistent with a rapid inactivation of T channels. T channels are more resistant to rundown of activity following patch excision or whole-cell recording than are L-type Ca^{2+} channels.

Gating properties and ionic nature

The voltage-dependence of the T current is studied in whole-cell recordings. In order to study the *I*/*V* relation of the T current in isolation, the other Ca^{2+} currents present in the cell, such as the high-threshold L-, N- and P-type Ca^{2+} currents (see Chapter 5), are pharmacologically blocked. However, in some preparations, T channels predominate and can be studied in the absence of blockers. For example, during development, embryonic hippocampal neurons in culture first express T-type Ca^{2+} channels and then, with neurite extension, also express high-threshold Ca^{2+} channels.

To study the activation properties of I_{CaT}, the membrane potential is held at −90 mV. The *I*/*V* relation shows that I_{CaT} activates in response to depolarizations positive to −55 mV and is maximal around −10 mV when recorded in the presence of 10 mM of external Ca^{2+} (**Figure 14.4c**). The voltage dependence of activation is determined by applying different voltage steps from a holding potential V_H = −105 mV (**Figure 14.5b**). The normalized peak current amplitude is plotted against the test potential (**Figure 14.5c**). In chick sensory neurons, it is half-activated at −51 mV (in 10 mM of external Ca^{2+}). Note that the rate of activation is highly voltage dependent.

During a 150 ms depolarizing pulse to −10 mV, I_{CaT} rapidly inactivates (in 50 ms): it is transient (**Figure 14.4c**). To study the inactivation properties of I_{CaT}, the membrane is clamped at different holding potential amplitudes and the current in response to a depolarizing step to −35 mV

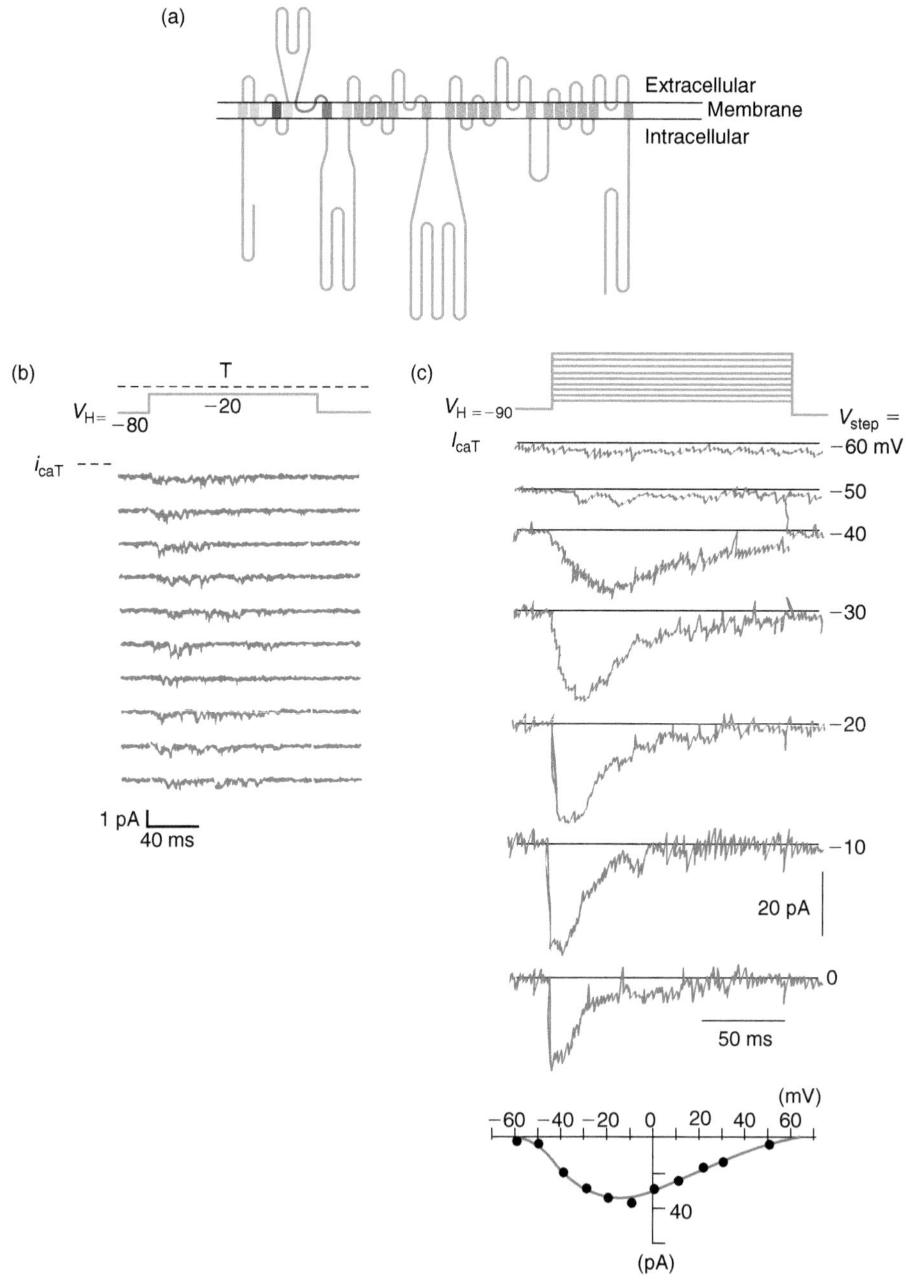

FIGURE 14.4 **T-type Ca^{2+} channel and current, I_T. (a)** Predicted topology of the rat α_{1I}-subunit. **(b)** Activity of a single T channel recorded from dorsal root ganglion cell bodies in patch clamp (cell-attached patch). Mean i_{CaT} amplitude at –20 mV is –0.62 ± 0.03 pA. **(c)** *I/V* relation of the whole-cell T current from a freshly plated (5 h) dissociated rat hippocampal neuron recorded in 10 mM of external Ca^{2+}. Part **(a)** adapted from Lee JH, Daud AN, Cribbs LL *et al.* (1999) Cloning and expression of a novel member of the low-voltage-activated T-type calcium channel family. *J. Neurosci.* **19**, 1912–1921, with permission. Part **(b)** adapted from Nowycky MC, Fox AP, Tsien RW (1985) Three types of neuronal calcium channels with different calcium agonist sensitivity. *Nature* **316**, 440–443, with permission. Part **(c)** adapted from Yaari Y, Hamon B, Lux HD (1987) Development of two types of calcium channels in cultured mammalian hippocampal neurons. *Science* **235**, 680–682, with permission.

is recorded (**Figure 14.5a**). A plot of peak current against holding potential (**Figure 14.5c**) gives a measure of the voltage dependence of inactivation. In chick sensory neurons, I_{CaT} is half-inactivated at −78 mV (in 10 mM of external Ca^{2+}).

The removal of inactivation (de-inactivation) of the T current is time and voltage dependent. Time dependence of inactivation is studied in lateral geniculate cells *in vitro* (in single-electrode voltage clamp) with the following protocol. The membrane is held at −55 mV to completely inactivate the T current. A voltage step to −95 mV of variable duration is then applied. On stepping back to −55 mV, the peak amplitude of the T current is measured (I) and compared with its maximum amplitude (I_{max}). At 35°C, 500–600 ms at −95 mV are needed to totally remove inactivation. The removal of

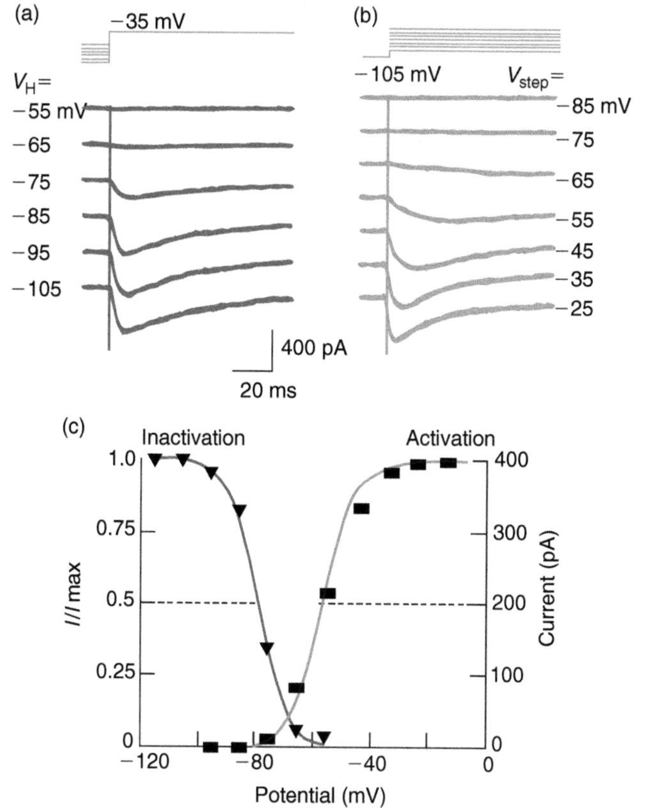

FIGURE 14.5 **Voltage dependence of activation and inactivation of the T current.** Recordings in 5–10 mM of external Ca^{2+}: **(a)** inactivation; **(b)** activation. See text for explanation. **(c, left)** I_{max} is the maximal peak current amplitude obtained at $V_H = -105$ mV. **(c, right)** I_{max} is the maximal peak current amplitude obtained at $V_{step} = -35$ to -20 mV. Adapted from Fox AP, Nowycky MC, Tsien RW (1987) Kinetic and pharmacological properties distinguishing three types of calcium currents in chick sensory neurones. *J. Physiol.* **394**, 149–172, with permission.

inactivation of the T current is also voltage dependent: at potentials close to −55 mV the de-inactivation is less complete than at more hyperpolarized potentials.

Pharmacology

There are no highly specific antagonists or toxins for the T-type Ca^{2+} channels. Low concentrations of the inorganic cation Ni^{2+} (20–50 μM) strongly depress I_{CaT}. Ethosuximide and amiloride have been also reported to reduce I_{CaT} in some preparations. Specific toxins acting at high-threshold-activated (HVA) channels are inefficient in T channels. To study I_{CaT} in isolation, Na^+ and K^+ channels must be blocked by TTX (1 μM), 4-AP (1 mM), TEA (10 mM), Ba^{2+} (1 mM) and HVA Ca^{2+} channels by their specific blockers (nifedipin, ω-conotoxin GVIA, ω-agatoxin).

In neurons, dopamine and other neurotransmitters inhibit T-type Ca^{2+} currents via second messengers. A number of serine/threonine kinases, calmodulin-dependent protein kinases and tyrosine kinases have been found to be involved in the regulation of T-type calcium channels. The site of interaction is in the intracellular loop connecting domains II and III. Since single-channel current amplitude is not affected, inhibition is suggested to result from reduction in the probability of opening (P_o).

Summary

T-type Ca^{2+} current is an inward current carried by Ca^{2+} ions; the activation threshold is around resting membrane potential (−50 to −60 mV), 10–20 mV negative to spike threshold. It is named 'T' for transient (owing to fast voltage-dependent inactivation) and also for tiny (owing to small single-channel conductance). T-type current is a low-threshold-activated (LVA) Ca^{2+} current that can be distinguished from the HVA L-, N- and P-type Ca^{2+} currents by the following criteria: it is activated after small depolarizations of the membrane (low voltage of activation); it is transient; it has a tiny single-channel conductance; and it slowly switches from the inactivated to the closed state upon repolarization of the membrane (generating a slow deactivation tail current). It is totally inactivated at potentials close to the resting potential and is de-inactivated during a transient hyperpolarization of the membrane. Therefore, it is fully activated by a depolarization only when the membrane potential has been previously maintained at a potential more hyperpolarized than resting membrane potential. T-type Ca^{2+} current is therefore well suited for rhythmic firing of action potentials (see Chapter 17) and for generation of large Ca^{2+} transients because of its activation at negative membrane potentials where the driving force for Ca^{2+} entry is large.

14.2.3 The hyperpolarization-activated cationic current, I_h, I_f, I_Q

I_h has an unusual voltage dependence since it is activated upon hyperpolarization of the membrane beyond resting membrane potential. For this reason, it has several names, 'h' for hyperpolarization, 'f' for funny (in the sinoatrial node of the heart) and 'Q' for queer, in some early studies in view of its odd electrophysiological behavior and its undefined functional significance. It was originally observed by Ito and co-workers in 1962 in cat motoneurons as a non-ohmic behavior of the I/V relation in the hyperpolarizing direction.

Structure of the main channel subunit

The I_h channel is a family of channels whose name is HCN: *h*yperpolarization-activated, *c*yclic *n*ucleotide-modulated channels. There are four known HCN subunit isoforms, HCN1–4, which combine to form tetrameric channels. HCN channels belong with the cyclic-nucleotide-gated (CNG) channels to the superfamily of K^+ channels. They contain the conserved motifs of K^+ voltage-gated channels, including the S1–S6 segments, a charged S4 voltage sensor and a pore-lining P-loop (**Figure 14.6a**). In addition, all family members

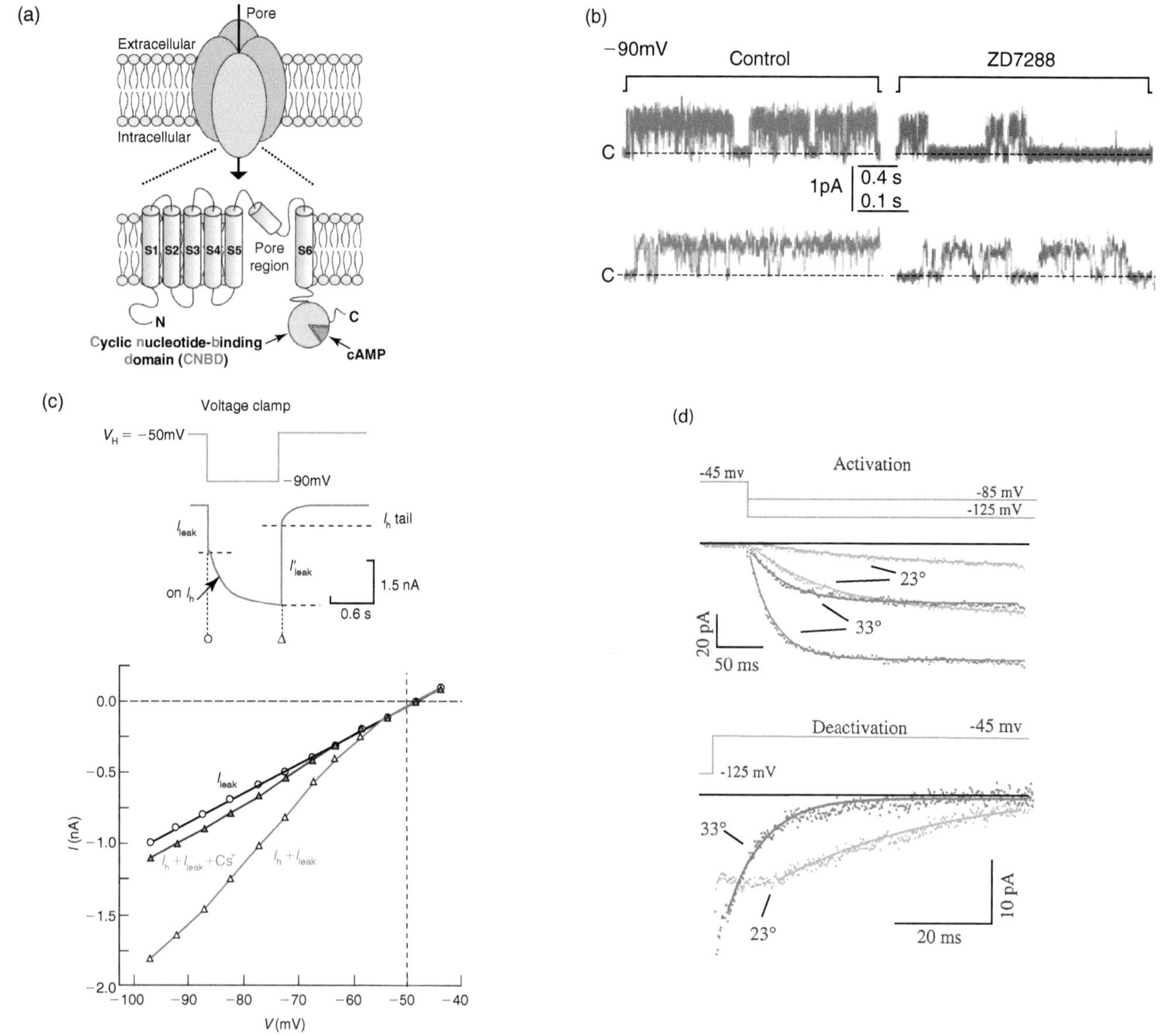

FIGURE 14.6 **The hyperpolarization-activated H channel and current, I_h. (a)** Topology of cyclic nucleotide-modulated ion channels. HCN channels are composed of four subunits (top). One subunit consists of six transmembrane segments (S1–S6) including an amino- and carboxy-terminal region (bottom). The C-terminal region contains a CNBD. The C-linker region of each subunit is composed of six α-helices that directly connect the CNBD to transmembrane segment S6. CNBDs consist of four α-helices. The cAMP molecule is shown as a stick model. **(b)** Representative traces of channel openings elicited by hyperpolarization steps from −30 mV to −90 mV in inside-out patches in control, and 50 μM ZD7288 conditions. Based on inside-out patch configuration, positive (upward) deflections represent inward current. Closed state is indicated by C and dashed line. ZD7288 reduces the probability of opening. **(c)** The activity of a thalamic neuron is recorded in slices (intracellular recording, voltage clamp mode). (Top) Relaxation experiment. The extracellular medium contains 14.5 mM of K^+ and 116 mM of Na^+. (Bottom) I/V relation obtained from relaxation experiment: instantaneous leak current is plotted as circles, steady-state current (I_{leak} + I_h), measured at the end of the voltage steps, is plotted as triangles. Open symbols represent currents under control conditions and filled symbols are currents after application of Cs^+. **(d)** Activation and deactivation kinetics of I_h are highly temperature dependent. (Top) The activation rate of currents evoked by step hyperpolarizations from −45 to −85 and to −125 mV increases nearly fivefold for a 10°C increase in bath temperature. Curves are well fit by single exponential functions (solid lines). (Bottom) The deactivation rate of currents present after repolarization from −125 to −45 mV show a similar temperature dependence. After a short plateau region the current decays are well fit by single exponential functions (solid lines). The currents shown are from a single dendritic recording (260 μm). Part (**a**) adapted from Schünke S, Stoldt M (2013) Structural snapshot of cyclic nucleotide binding domains from cyclic nucleotide-sensitive ion channels. *Biol. Chem.* **394** (11), 1439–1451. Part (**b**) adapted from Simeone TA, Rho JM, Baram TZ (2005) Single channel properties of hyperpolarization-activated cation currents in acutely dissociated rat hippocampal neurones. *J. Physiol.* **568**, 371–380. Part (**c**) adapted from McCormick DA, Pape HC (1990) Properties of a hyperpolarization-activated cation current and its role in rhythmic oscillation in thalamic relay neurones. *J. Physiol.* **431**, 291–318, with permission. Part (**d**) from Magee JC (1998) Dendritic hyperpolarization-activated currents modify the integrative properties of hippocampal CA1 pyramidal neurons. *J. Neurosci.* **18**, 7613–7624.

contain a conserved cyclic nucleotide-binding (CNB) domain in their carboxy-terminus. This domain is homologous to the CNB domain of protein kinases and of CNG channels (see **Table 3.1**), showing that the gating of I_h channels is directly regulated by cyclic nucleotides such as cAMP or cGMP.

Single channel current

CA1 pyramidal neurons are acutely dissociated from rat hippocampal slices. Pyramidal neurons are identified by the presence of a relatively long apical dendrite, pyramidal soma, and basal processes. Inside-out patches are excised from their somata. Unitary H-channel currents are recorded in these excised inside-out patches. To record single H channels in isolation, the pipette solution contains blockers of voltage- and ligand-gated channels, including TTX (1 μM to block Na^+ channels), TEA (10 mM to block K^+ channels), $BaCl_2$ (1 mM to block I_{KIR}), $CdCl_2$ (0.2 mM to block Ca^{2+} channels), picrotoxin (10 μM to block $GABA_A$ receptors), DNQX (10 μM to block AMPA receptors), and APV (20 μM to block NMDA receptors). In response to consecutive steps (2.5 s duration) from −30 mV to −90 mV, single channels currents are recorded (**Figure 14.6b**). Based on inside-out patch configuration, positive (upward) deflections represent inward current. To confirm the identity of the channels, the internal blocker ZD7288 (50 μM) is added to the bath solution (since the internal side of the membrane is external in inside-out patches) (**Figure 14.6b**). These single channel recordings were performed long after the discovery of I_h. This is why the gating properties of I_h are explained below for whole-cell recordings.

Gating properties and ionic nature

The type of voltage clamp experiment that allows one to record I_h is called a *relaxation experiment*. The activity of thalamic neurons in slices is recorded under a two-electrode voltage clamp (**Figure 14.6c**). A 1.5 s hyperpolarizing voltage step to −90 mV while the membrane potential is held at $V_H = -60$ mV evokes an instantaneous inward current (I_{leak}) followed by a slowly developing inward current which shows no inactivation with time (on I_h). I_{leak} reflects the leak current through channels open at $V_H = -50$ mV (mostly K^+ channels). The following slow inward current reflects the slow opening of I_h channels. When the membrane is then repolarized to −60 mV at the end of the voltage step, an instantaneous current (I'_{leak}) is recorded followed by an inward tail current (tail I_h). The instantaneous current reflects the leak current through channels open at −90 mV. The tail current reflects the kinetics of closure (deactivation) of I_h channels. I' is larger than I_{leak} because at −90 mV not only the leak channels are open but also the H (I_h) channels that have been opened by the hyperpolarization.

By varying the amplitude of voltage steps, I_h is shown to activate at between −45 and −60 mV and to be half-activated at around −75 and −85 mV (**Figure 14.6c**). I_h reverses at around −50 to −30 mV depending on the neuronal type. Decreasing the external Na^+ concentration from 153 to 26 mM reduces the amplitude of I_h (**Figure 14.7a**). When Na^+ ions are totally replaced by the non-permeant cation choline, I_h disappears almost completely (since E_K is around −90 mV). Similarly, raising the external K^+ concentration from 2.5 mM to 12.5 mM enhances I_h (not shown). These results indicate that I_h is carried by both Na^+ and K^+ ions, which is consistent with the extrapolated reversal potential. H channels are in fact four times more permeable to K^+ than to Na^+.

The kinetics of activation and deactivation of I_h are voltage dependent (from ≈50 ms at −75 mV to ≈16 ms at −125 mV for activation time constant and 30 ms at −70 mV to ≈7 ms at −30 mV for tail current time constant) and temperature dependent (Q_{10} = ≈4.5) (**Figure 14.6d**). The amplitude of the current is slightly dependent on temperature ($Q_{10} = 1.7$).

Pharmacology

Application of cAMP to the internal surface of an inside-out patch induces a reversible increase in the magnitude of the inward current during a step to −100 mV (**Figure 14.7b**). This effect of cAMP is due to a positive shift in the steady-state activation curve of I_h current by 2 to 10 mV. I_h is completely blocked in the presence of 1–3 mM of Cs^+ in the extracellular solution (**Figures 14.6c and 14.7c**). A bradycardic agent ZM 227189 (10–100 μM; Zeneca) or ZD 7288 (10–20 μM) (**Figure 14.6b**) selectively blocks I_h. In contrast, I_h is insensitive to external Ba^{2+}, TTX, TEA and 4-AP that effectively block Na^+ or K^+ channels. That is why to study I_h in isolation, Ca^{2+}, Na^+ and K^+ channels can be blocked by Cd^{2+}, Co^{2+} or Ni^{2+} (200 μM to 1 mM), TTX (1 μM), 4-AP (1 mM), TEA (10 mM), Ba^{2+} (1 mM). I_h is enhanced by lamotrigin (50 μM), an antiepileptic drug which shifts I_h voltage dependence of activation to more depolarized levels, closer to the neuronal resting potential.

Summary

I_h is carried by Na^+ and K^+ ions and is a voltage- and time-dependent current. It activates with a slow time course upon hyperpolarization, at potentials more negative than −50 mV. I_h exerts a depolarizing action because it is a cationic current. I_h does not inactivate, making a proportion of channels constantly open at resting membrane potential (typically around −70 mV). When present in the soma, I_h frequently functions as a pacemaker current, triggering a depolarizing ramp after hyperpolarizing events, such as inhibitory postsynaptic potentials. I_h is not uniformly distributed between soma and dendrites. In cortical pyramidal cells, the density of H channels increases sevenfold from the soma to the distal

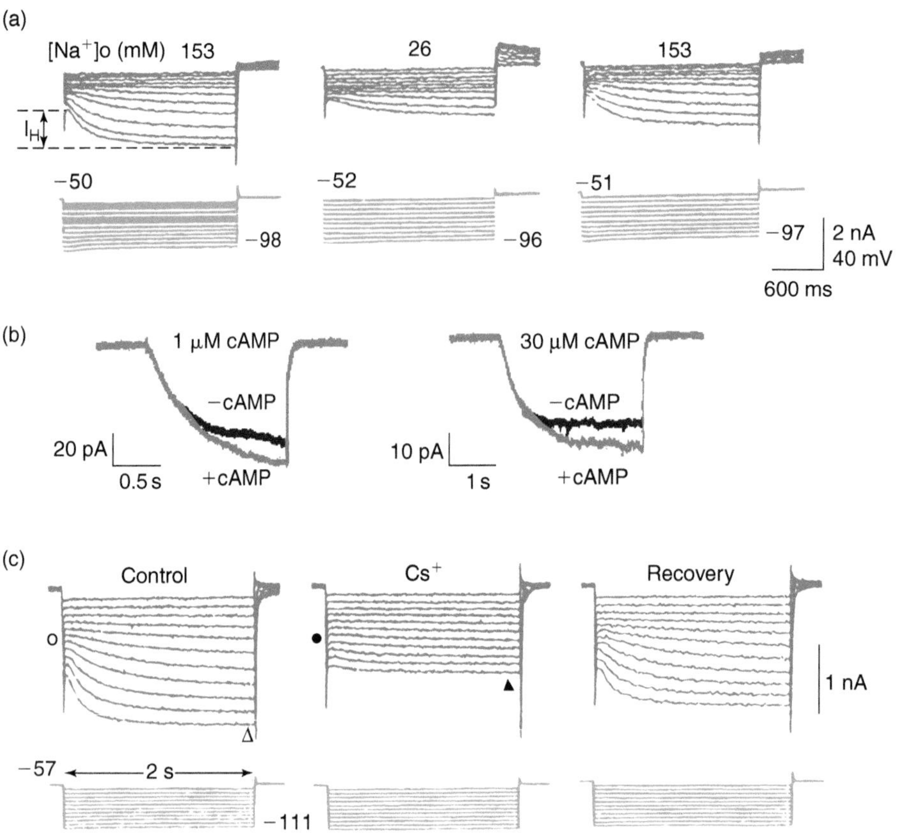

FIGURE 14.7 **Ionic selectivity and pharmacology of I_h.** The activity of thalamic neurons is recorded in slices (intracellular recording, voltage-clamp mode). **(a)** Effect on I_h of changing the extracellular concentration of Na^+ ions. A family of I_h currents (upper traces) are evoked by stepping membrane potential to hyperpolarized potentials from $V_H = -50$ mV (bottom traces), in control conditions (153 mM, left and right) and during reduced Na^+ concentration (26 mM, center). It should be noted that there is no change in the baseline current. **(b)** The mBCNG-1 channel is expressed in *Xenopus* oocytes. The I_h current is evoked in inside-out macropatches by a hyperpolarizing step to −100 mV in the absence or presence of cAMP in the bath. **(c)** Similar experiment as in (**a**), in control conditions and in the presence of Cs^+. Parts (**a**) and (**c**) adapted from McCormick DA, Pape HC (1990) Properties of a hyperpolarization-activated cation current and its role in rhythmic oscillation in thalamic relay neurones. *J. Physiol.* **431**, 291–318, with permission. Part (**b**) adapted from Santoro B, Liu DT, Yao H *et al.* (1998) Identification of a gene encoding a hyperpolarization-activated pacemaker channel of brain. *Cell* **93**, 717–729, with permission.

apical dendrites. This has consequences on dendritic integration (see Section 15.3). I_h is directly modulated by internal cyclic nucleotides such as cAMP or cGMP. It is reversibly decreased by extracellular Cs^+ or ZD 7288 and enhanced by lamotrigin.

14.3 THE SUBLIMINAL VOLTAGE-GATED CURRENTS THAT HYPERPOLARIZE THE MEMBRANE

The common characteristic of these subliminal currents is to be outward, carried by K^+ ions and turned on at potentials more negative than the threshold for the opening of voltage-gated Na^+ channels of the action potential. Four types of outward K^+ currents will be explained: the early K^+ currents (I_A, I_D), the K^+ currents activated by intracellular Ca^{2+} (I_{KCa}), the muscarine-sensitive K^+ current (I_M) and the inward rectifier K^+ current (I_{KIR}). All K channels operate as complexes comprised of pore-forming α-subunits plus a number of associated ancillary subunits. α-Subunits tetramerize to form a central pore.

14.3.1 The rapidly inactivating transient K^+ current: I_A or I_{Af}

I_A is a K^+ current which rapidly activates and inactivates in response to depolarizing steps from holding potentials negative to the resting membrane potential. It was originally described in 1971 by Connor and Stevens, in molluskan neurons, and termed 'A current'.

Structure of the main channel subunit

The I_A channel belongs to the family of *Shal* K^+ channels (or K_v4, they are gated by transmembrane voltage, hence the K_v nomenclature). They have *six putative membrane spanning segments* designated S1–S6, flanked by intracellular domains of variable length and a pore domain P between S5 and S6. The Shal-type family in mammals is comprised of three distinct genes: $K_v4.1$, $K_v4.2$ and $K_v4.3$.

Gating properties and ionic nature

The activity of medullary or hippocampal neurons in culture is recorded in voltage clamp mode in a medium containing TTX (to block Na^+ currents) and TEA (to block the K^+ currents of the delayed rectification), Cd^{2+} (200 μM) to depress Ca^{2+}-dependent currents, carbachol (50 μM) to block I_M (see Section 14.3.4) and Cs^+ (1–3 mM) to block I_h. When the membrane potential is maintained at −70 mV, depolarizing voltage steps (of 10 to 58 mV amplitude) evoke an outward current whose amplitude increases with the amplitude of the depolarization (**Figure 14.8b**). This current activates rapidly (within milliseconds) and then inactivates rapidly and exponentially during the step. This is the 'A current', I_A. Inactivation of I_A is time and voltage dependent. It appears after a few milliseconds, which makes I_A short lasting. In this respect, I_A resembles more the I_{Na} of the action potential than the current of the delayed rectification, I_{KDR}. The current plateau which follows the peak of the I_A current represents the sum of the leak current and of the outward currents which are not blocked by TEA.

When a hyperpolarizing voltage step ($V_{hyperpol}$) of varying amplitude is now applied during the 50 ms before the depolarizing voltage step (V_{step}) to −20 mV (**Figure 14.8a**), it can be seen that I_A is inactivated when the membrane potential is maintained at a value more positive than −50 mV; at these membrane potentials I_A cannot be activated by a depolarization. Also, I_A is de-inactivated when the membrane potential is maintained at a value more negative than −50 mV and can then be activated by depolarizations to membrane potentials positive to −50 mV and its amplitude increases with the difference $V_{hyperpol} - V_{step}$. Activation and inactivation curves (**Figure 14.8c**) have been constructed from the data in traces (**b**) and (**a**), respectively.

Pharmacology

I_A is blocked by application of 4-aminopyridine (4-AP, 1–3 mM) in the extracellular medium. It is insensitive to TEA, Cs^+ and Ba^{2+}. Thus, there are two ways of blocking I_A, either by depolarizing the membrane above −50 mV, or by applying 4-aminopyridine. To study I_A, Na^+ and Ca^{2+} channels must be blocked by TTX (1 μM), Cd^{2+}, Co^{2+} or Ni^{2+} (200 μM to 1 mM).

Summary

I_A is a fast-inactivating (in the order of milliseconds) K^+ current. The threshold potential for its activation is situated at around −60 to −45 mV; i.e. at a value slightly more negative than the threshold potential for the inward Na^+ current of the action potential. I_A can be fully activated by a depolarization only when the membrane potential has been previously maintained at a potential more hyperpolarized than −60 mV (I_A is inactivated at resting membrane potential). It has characteristics which distinguish it from the K^+ currents of the delayed rectification: it is a rapidly activating and inactivating K^+ current and has pharmacological properties different from that of the delayed rectifier currents.

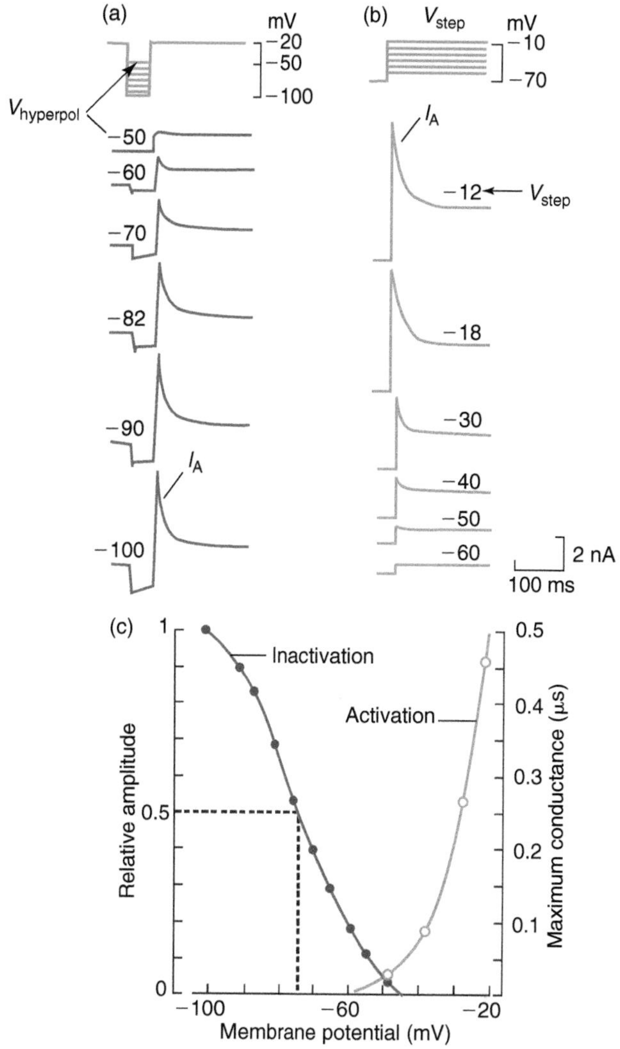

FIGURE 14.8 **Activation–inactivation properties of I_A.** Voltage clamp analysis of I_A in cultured mouse spinal neurons bathed in a medium containing TTX (1 μM) and TEA (25 mM). **(a)** Hyperpolarizing conditioning pulses 50 ms in duration (green steps) remove current inactivation obtained by holding at −20 mV (purple traces). **(b)** Rapid activation of a transient outward current (orange traces) as the membrane potential is stepped from −70 mV to depolarized potentials (green steps). **(c)** Data obtained in **(a)** and **(b)** have been used to construct activation (orange) and inactivation (purple) curves. From Segal M, Rogawski MA, Barker JL (1984) A transient potassium conductance regulates the excitability of cultured hippocampal and spinal neurons. *J. Neurosci.* **4**, 604–609, with permission.

14.3.2 The slowly inactivating transient K^+ current, I_D or I_{As}

In addition to I_A, a K^+ current that also activates rapidly (within milliseconds) but slowly inactivates (over seconds) was first described by Storm (1988) in hippocampal pyramidal cells. It was termed 'D current' because it delays the cell firing.

Gating properties

The activity of the pyramidal cells of the hippocampus is intracellularly recorded in brain slices. Voltage clamp experiments are performed in the presence of external Cd^{2+} (200 μM) to depress Ca^{2+}-dependent currents, carbachol (50 μM) to block I_M (see Section 14.3.4) and Cs^+ (1–3 mM) to block I_h. When, in such conditions, a depolarizing voltage step to −26 mV is applied from a holding potential $V_H = -80$ mV, an outward current consisting of two components is recorded: a fast-inactivating component, sensitive to high doses of 4-aminopyridine (4-AP, 1–3 mM) and a slowly inactivating component sensitive to low doses (40 μM) of 4-AP (**Figure 14.9a**). The first component corresponds to I_A and the second one is termed I_D.

In order to construct the activation–inactivation curves of I_D, the same protocol as that explained for I_A is applied (**Figure 14.9b,c**). They show that I_D is half-inactivated at −88 mV, and is inactivated when the membrane potential is maintained at a value more positive than −50 mV (at these membrane potentials I_D cannot be activated by a depolarization). Further, I_D is de-inactivated when the membrane potential is maintained at a value more negative than −60 mV (it can then be activated by depolarizations to membrane potentials more positive than −70 mV). Therefore, I_D contrasts with I_A in having a threshold for both activation and inactivation 10–20 mV more negative (compare **Figures 14.8c** and **14.9c**).

Pharmacology

I_D is much more sensitive to 4-aminopyridine than I_A, the latter requiring 1–3 mM for a block and the former 30–40 μM. I_D is insensitive to TEA and Cs^+ ions. Thus, there are two ways of blocking I_D, either by depolarizing the membrane above −70 mV or by applying 40 μM of 4-aminopyridine in the extracellular medium. To study I_D, Na^+ and Ca^{2+} channels must be blocked by TTX (1 μM), Cd^{2+}, Co^{2+} or Ni^{2+} (200 μM to 1 mM).

Summary

I_D has characteristics which distinguish it from the other K^+ currents. It inactivates more slowly (in the order of seconds) than the transient current I_A and activates more rapidly and at more negative potentials than the currents of the delayed rectification I_{KDR}. It has also different pharmacological properties.

14.3.3 The K^+ currents activated by intracellular Ca^{2+} ions, I_{KCa}

I_{KCa} currents are outward K^+ currents sensitive to the intracellular concentration of Ca^{2+} ($[Ca^{2+}]_i$). In vertebrate neurons, these currents may be more or less sensitive to voltage, but for all of them an increase of the intracellular Ca^{2+} concentration is a necessary prerequisite to their activation. $[Ca^{2+}]_i$ increase may be the result of Ca^{2+} entry through voltage-dependent Ca^{2+} channels opened by depolarization, or Ca^{2+} entry through cationic receptor channels largely permeable to Ca^{2+} ions such as the

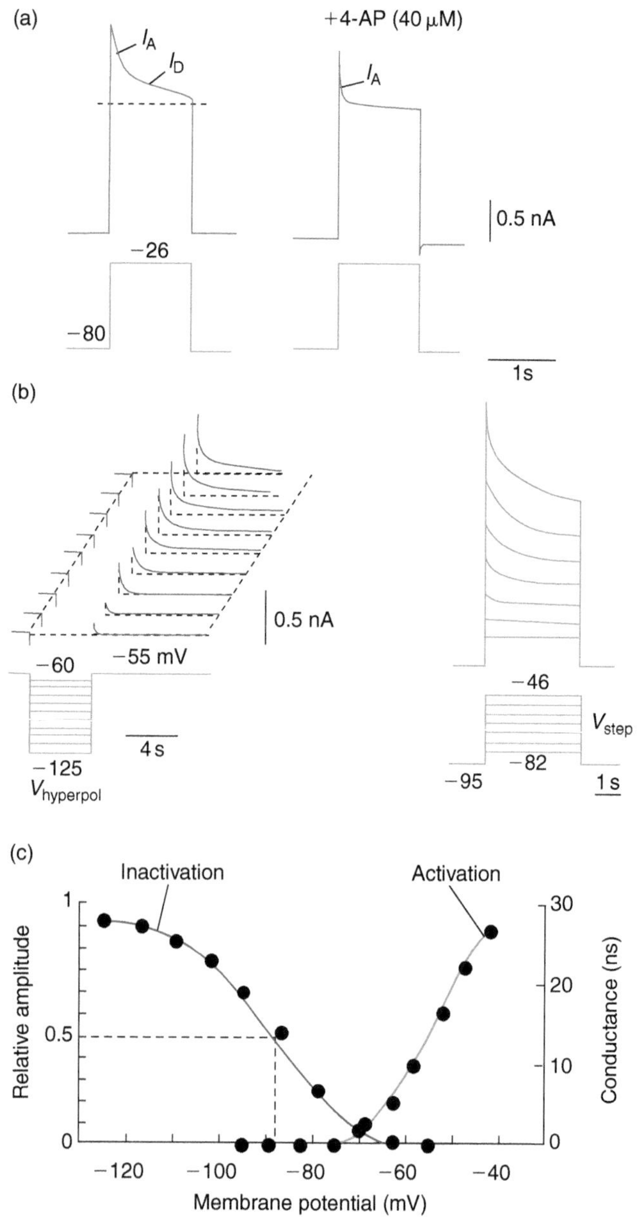

FIGURE 14.9 **Activation–inactivation properties of I_D. (a)** The transient outward K^+ current elicited by a depolarizing step from −80 mV consists of two components, a fast one I_A and a slow one I_D. In the presence of 4-aminopyridine (+4-AP), I_D is blocked. **(b)** Records of I_A and I_D showing voltage dependence of inactivation (purple traces) and activation (orange traces). The amplitude of I_D is measured 50 ms after the step onset, when I_A has inactivated completely. **(c)** Plot of data from **(b)**. I_D is half-inactivated at −88 mV (dotted line). From Storm JF (1988) Temporal integration by a slowly inactivating K^+ current in hippocampal neurons. *Nature* **336**, 379–381, with permission.

NMDA-type glutamate receptors, or Ca^{2+} release from intracellular stores. I_{KCa} have been explained in Chapter 5.

14.3.4 The K^+ current sensitive to muscarine, I_M

I_M is a depolarization-activated K^+ current originally described in frog sympathetic neurons by Brown and Adams in 1980, and studied since in a variety of other vertebrate neurons. I_M was so called because it is inhibited by muscarinic acetylcholine receptor agonists such as the alkaloid muscarine. It is therefore under the control of muscarinic cholinergic receptors.

Structure of M channels

M channels are composed of homomeric or heteromeric assembly of four K_v7 subunits encoded in the brain by *KCNQ2–5* genes. Thus, functional M channels are generated by homomeric association of $K_v7.2$ subunits, $K_v7.3$ subunits (in rodents), $K_v7.4$ or $K_v7.5$ subunits. M channels result also from heteromeric association of two $K_v7.2$ with two 7.3 subunits, an assembly which underlies M current in many neurons and gives rise to a much larger channel current than assemblies of $K_v7.3$ with 7.4 subunits or $K_v7.3$ with $K_v7.5$ subunits. Each subunit consists of six transmembrane segments (S1–S6), forming the voltage sensor (S1–S4) and pore (S5–P–S6) domains. The N- and C-terminal domains are intracellularly located (**Figure 14.10a**). The C-terminus is composed of different modules which are binding sites for several regulatory molecules such as: phosphatidylinositol 4,5-bisphosphate (PIP2), the binding of which is required for the function of the channel; calmodulin, which plays a critical role for heteromeric assembly of subunits, channel trafficking, and control of PIP2 affinity; syntaxin 1A, which binds to the $K_v7.2$ subunit and reduces channel open probability; and ankyrin G, the binding of which is required for the targeting of K_v7 channels to the axonal membrane (initial segment and nodes of Ranvier, see below).

Gating properties and ionic nature

I_M can be recorded in response to two types of voltage clamp protocols (in the presence of TTX). In the first protocol, the membrane potential is stepped from −60 to −30 mV. At −60 mV, most of the M channels are closed. A step to −30 mV reveals, superimposed on the leak current (measured from the response to a symmetrical hyperpolarizing step to −90 mV), a slowly developing outward current due to the slow opening of M channels in response to membrane depolarization (**Figure 14.10b**).

The second type of protocol involves relaxation experiments (**Figure 14.10c**). The membrane is held at $V_H = -30$ mV, a potential at which M channels remain open and contribute a steady outward current. A negative step to −60 mV causes an instantaneous inward current (I_{leak}) followed by an inward current which slowly develops (I_M). When the membrane is repolarized to −30 mV, at the end of the voltage step, an instantaneous inward current (I'_{leak}) is recorded followed by a slow outward tail current (tail I_M).

The explanation of the recordings of **Figure 14.10c** is the following. When the membrane is stepped from 30 to −60 mV, all the M channels open at −30 mV do not close immediately or at the same time. There is an instantaneous diminution of the outward current (recorded as an instantaneous inward current, I_{leak}). It represents the current through channels open at $V_H = -30$ mV; i.e. M channels and 'leak channels'. It therefore depends on the number of channels open at −30 mV. Then there appears an exponential diminution of outward current or slow inward relaxation (I_M) which reflects the kinetic of M channels closure in response to the hyperpolarizing step to −60 mV. When the membrane is stepped back to −30 mV, the M channels do not open instantaneously. There is a first instantaneous outward current (I'_{leak}) which represents the current through channels open at −60 mV ('leak channels' only since M channels are closed). This instantaneous outward current is smaller than the fast inward one recorded from −30 to −60 mV. It clearly indicates that M channels had closed in response to the preceding step from −30 to −60 mV (when the ohmic current is smaller in response to the same ΔV, it means that the membrane conductance is smaller). Then an outward tail current (tail I_M) appears through the M channels which slowly open again in response to the depolarization. The time course of the outward current evoked by the return to the steady holding potential can be distorted by the presence of other K^+ currents activated by this protocol (I_{KDR}, I_A, I_D and I_{KCa}), particularly when the membrane is stepped back to V_H from a potential more negative than −70 mV. The M channels seem to be fully closed at membrane potentials more negative than −60 mV, because in response to hyperpolarizing commands from −60 mV ($V_H = -60$ mV) only ohmic (passive) current is recorded (not shown). With increasing step commands, the inward relaxation reverses in direction at step potentials between −70 and −100 mV. This reversal potential shifts to a more positive value on raising external K^+ concentration. Thus, I_M is largely a K^+ current.

Subcellular localization and function of M channels

M (K_v7) channels are mainly distributed at axon initial segments and nodes of Ranvier where they are co-clustered with Na^+ channels. They are also found in the somato-dendritic compartment but three- to fivefold times less than in the axon. I_M controls different aspects of neuronal excitability according to subcellular localization of M channels. At the axon initial segment, M channels control resting membrane potential and spike

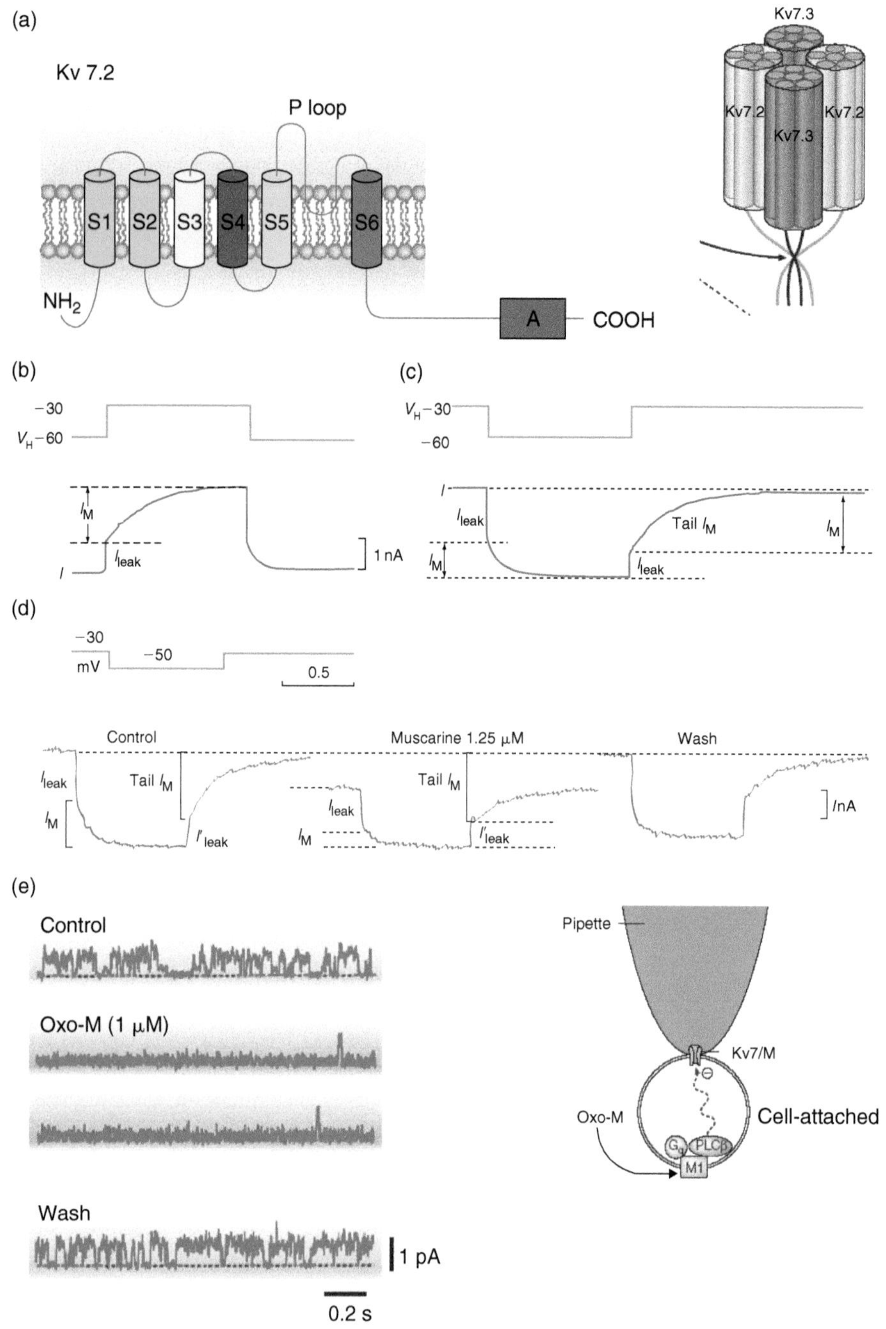

FIGURE 14.10 **Characteristics of the M current, I_M. (a)** Membrane topology of K_v7/KCNQ channel subunits. They have a conventional Shaker-like K^+ channel structure with a long intracellular carboxy-terminal tail. Four such subunits make up a functional K_v7 channel. All five ($K_v7.1$–$K_v7.5$) K_v7 channel subunits can form homomeric channels, whereas the formation of heteromers is restricted to certain combinations. **(b–e)** M currents are recorded from a sympathetic neuron in voltage clamp mode. **(b,c)** See text. **(d)** Effect of muscarine on I_M. The membrane is held at −30 mV and stepped to −50 mV. Muscarine evokes an inward current (the current trace is lower): it reduces the steady outward current through M channels open at −30 mV. In contrast, the baseline level attained at −60 mV, at the end of the command step, remains the same: muscarine does not produce an inward current at voltages where the M channels are normally shut. Inward and outward relaxations are largely depressed. **(e)** shows remote signaling by muscarinic acetylcholine receptors. Bath application of the muscarinic agonist oxotremorine-methiodide (Oxo-M) strongly depresses unitary M currents in sympathetic neurons. Channel activity recovers fully after Oxo-M is washed out. Parts **(a,e)** from Delmas P, Brown DA (2005) Pathways modulating neural KCNQ/M (Kv7) potassium channels. *Nat. Rev. Neurosci.* **6**, 850–862, with permission. Parts **(b,c)** adapted from Brown D, Adams PR (1980) Muscarinic suppression of a novel voltage-sensitive K^+ current in a vertebrate neuron. *Nature* **283**, 673–676; and Adams PR, Brown DA, Constanti A (1982) M currents and other potassium currents in bullfrog sympathetic neurones. *J. Physiol.* **330**, 537–572, with permission. Part **(d)** adapted from Adams PR, Brown DA, Constanti A (1982) Pharmacological inhibition of the M-current. *J. Physiol.* **332**, 223–262, with permission.

threshold and prevent spontaneous firing of the neuron. At nodes of Ranvier, I_M increases the availability of Na^+ channels and therefore the amplitude of the propagating action potential because it favors Na^+ channel de-inactivation by hyperpolarizing membrane potential. At the somato-dendritic level, I_M reduces spike after-depolarization and contributes to the medium and slow after-hyperpolarizations that follow a burst of action potentials. Thus, I_M serves as a brake for neuronal firing.

Pharmacology

K_v7 channels are blocked by XE-991 (10,10-*bis*(4-pyridinylmethyl)-9(10*H*)-anthracenonedihydrochloride) (IC_{50} = 0.6–0.98 μM), linopirdinedihydrochloride (IC_{50} = 4–7 μM), and TEA (IC_{50} = 1–5 mM). I_M is enhanced by retigabine (0.1–10 μM), a compound marketed as an anti-epileptic drug for the treatment of resistant partial onset seizures. It enhances I_M via its interaction with several specific residues located in the S5 and S6 segments. It stabilizes the open state of the channel and produces a hyperpolarizing shift of the current–voltage relationship. Zinc pyrithione (1–10 μM) also interacts with specific residues located in the pore domain and augments all K_v7 channel-mediated currents except for those containing $K_v7.3$. It produces a hyperpolarizing shift of the current–voltage relationship and increases single channel open probability.

I_M is sensitive to different neurotransmitters acting on metabotropic receptors like muscarinic receptors in which activation by specific agonists decrease I_M as shown in **Figure 14.10d**. The consequent loss of the steady outward current under muscarine generates a steady inward current at −30 mV (difference between baseline control and muscarine) and a step to −50 mV now reveals mostly the leak current (I_{leak}) showing that most of the M channels are already closed. In single-channel patch-clamp recordings from sympathetic neurons, stimulating muscarine receptors with the muscarinic agonist oxotremorine-methiodide (Oxo-M) reversibly closes M channels inside the patch. These findings indicate that inhibition of M-channel activity involves a molecule that is capable of diffusing into (or out of) the membrane region circumscribed by the patch electrode (**Figure 14.10e**). There is now evidence that muscarinic-receptor activation blocks M channels via a decrease of PIP2.

Summary

I_M is activated by depolarizations to membrane potentials positive to −60 mV. It activates slowly (within hundreds of milliseconds) and does not inactivate. I_M differs from the delayed rectifier current I_{KDR} involved in spike repolarization by having a 40 mV more negative activation threshold and slower kinetics. It differs from I_A and I_D transient K^+ currents because it does not inactivate. It is turned off by agonists of metabotropic receptors coupled to G_q/11 proteins, for example M1 muscarinic acetylcholine receptors.

14.3.5 The inward rectifier K^+ current, I_{KIR}

I_{KIR} is a K^+ current that was originally described in skeletal muscle fibers by Sir Bernard Katz (1949). It is a Ba^{2+}-sensitive K^+ current with an I/V relation showing rectification when the current is inward (at potentials more hyperpolarized than −90 mV). I_{Kir} is modulated by G proteins.

Structure of the main channel subunit

These channels form a new channel-gene superfamily: inwardly rectifying K^+ subunits possess only two putative transmembrane segments, M1 and M2, and a highly conserved pore loop located between M1 and M2, which correspond to transmembrane regions S5 and S6 together with the P-loop of the voltage-gated K^+ channels with six transmembrane domains (such as K_v channels) (**Figure 14.11a**). Two-thirds of the amino-acid sequence is characterized by large hydrophilic amino- and carboxy-terminal domains that extend from 90 to over 200 amino acids into the cytoplasm, respectively. These intracellular domains are of crucial importance for channel modulation by intracellular regulators, but also for establishing the strong voltage dependence of inward rectification. Each receptor is made up of four Kir subunits

The high-resolution crystal structures of KcsA, a bacterial homolog of potassium channels closely related to eukaryotic Kir channels, confirmed many of the architectural features of the Kir channel family (see Chapter 3).

Gating properties and ionic nature

The activity of the neurons of the accumbens nucleus is recorded intracellularly. The resting potential is very negative, around −85 mV. In a single-electrode voltage clamp, when the current evoked by hyperpolarizing steps is recorded in the presence of varying concentrations of external K^+, a series of I/V curves is obtained (**Figure 14.11b**). These curves increase in steepness between −50 and −120 mV. Thus, the permeability to K^+ is high when the current is inward ($V - E_K$ is negative) and low when the current is outward ($V - E_K$ is positive). In other words, inwardly rectifying channels conduct more efficiently when the membrane is negative to E_K.

Increasing the external K^+ concentration increases the slope of the I/V curve (i.e. the conductance) and shifts the reversal potential to more positive values, thus showing that K^+ ions participate in the current responsible for the inward rectification.

Rectification in native channels is due in part to voltage-dependent block by cytoplasmic Mg^{2+} ions and polyamines. The block by intracellular Mg^{2+} ions was

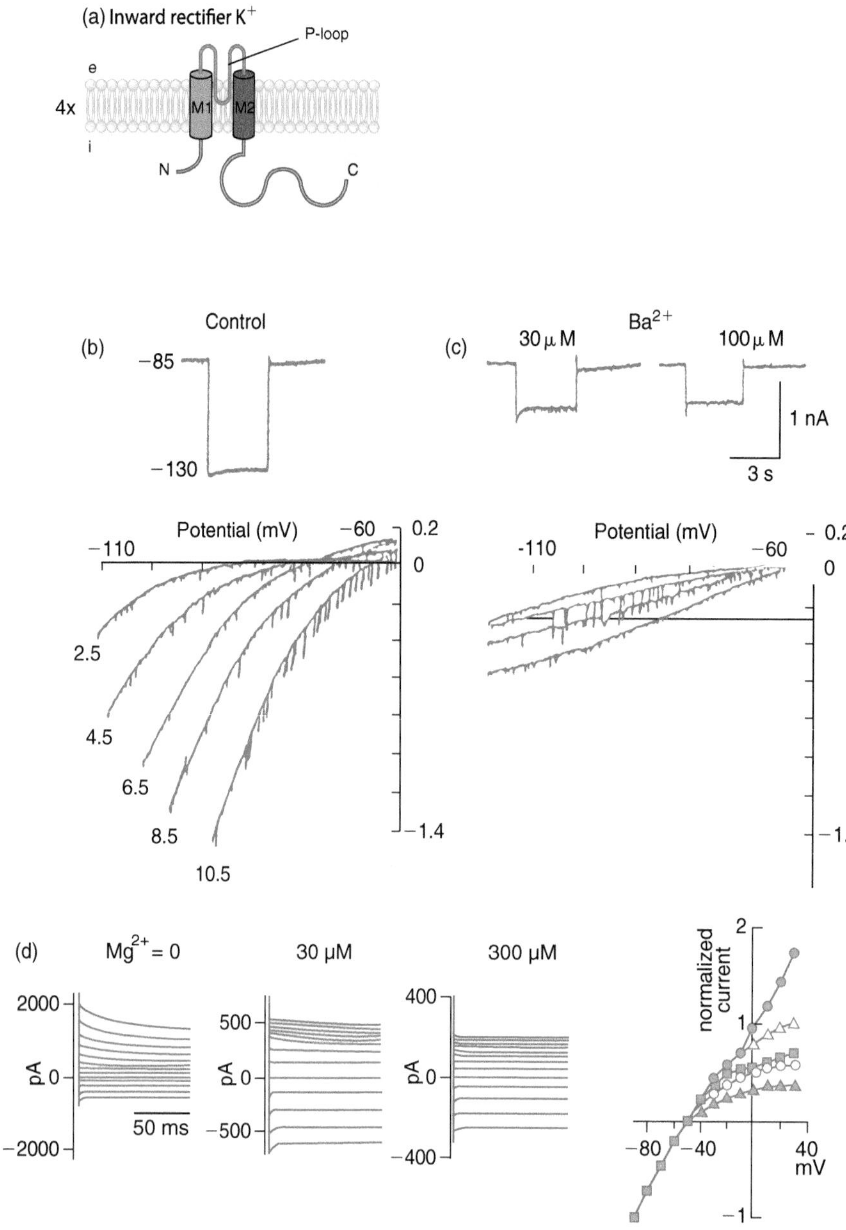

FIGURE 14.11 **Characteristics of the inward rectifier current, I_{Kir}.** Membrane topology of a single subunit of an inward rectifier K$^+$ channel. **(b,c)** Currents are recorded from neurons of the nucleus accumbens, in single-electrode voltage clamp mode. Inward current recorded in response to a hyperpolarizing step to −130 mV from a holding potential of −85 mV in control external K$^+$ concentration (2.5 mM) (**b**, top traces). *I/V* relations in five different external K$^+$ concentrations (**b**, bottom traces). The amplitude of the current has been measured at the end of 3–5 s steps, at steady state. Same current as in (**b**) recorded in the presence of 30 and 100 μM of Ba^{2+} **(c**, top traces**)**. *I/V* relations as in (**b**) but in the presence of 30 μM of Ba^{2+} **(c**, bottom traces**)**. **(d)** Macroscopic currents recorded from an excised inside-out patch of HEK293T cells transfected with Kir2.1. $[K^+]_e$ in the pipette is 20 mM, and $[K^+]_i$ in the bath is 140 mM ($E_K = -49$ mV). The holding potential is −50 mV and step pulses from +30 down to −90 mV are applied by 10 mV decrements. The current amplitudes are measured at 500 ms from the beginning of the step pulses and normalized by the values at −90 mV. The symbols used are: 0 μM MgA, black dot; 3 μM, open triangle; 30 μM, black square; 300 μM, open circle; 3 μM, black triangle. Part (**a**) adapted from Swartz KJ (2004) Towards a structural view of gating in potassium channels. *Nat. Rev. Neurosci.* **5**, 905–916, with permission. Parts (**b,c**) adapted from Uchimura N, Cherubini E, North A (1989) Inward rectification in rat nucleus accumbens. *J. Neurophysiol.* **62**, 1280–1286, with permission. Part (**d**) from Kubo Y, Murata Y (2001) Control of rectification and permeation by two distinct sites after the second transmembrane region in Kir2.1 K$^+$ channel. *J. Physiol.*, **531**, 645–660, with permission.

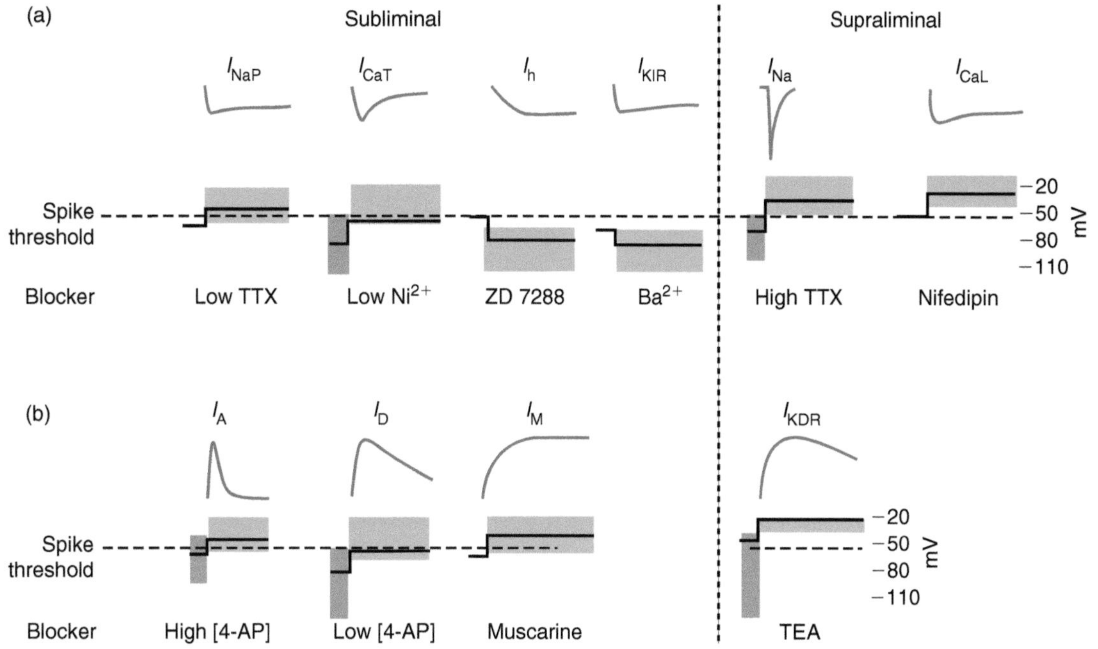

FIGURE 14.12 **Comparison between subliminal and supraliminal voltage-gated currents. (a)** Inward and **(b)** outward voltage-gated currents. Subliminal currents are on the left and high-threshold-activated currents are on the right. Currents (top traces) are shown in response to typical voltage steps (solid black lines). The approximate voltage ranges of activation and inactivation are shown in green. The effective blocking agents are indicated below. Part (**b**) adapted from Storm JF (1988) Temporal integration of a slowly inactivating K^+ current in hippocampal neurons. *Nature* **336**, 379–381, with permission.

discovered by Matsuda in cardiac ventricular cells that express high levels of Kir channels. When in whole-cell recordings the intrapipette contains artificial cerebrospinal fluid with 0 mM Mg^{2+}, the *I/V* curve becomes ohmic with symmetrical outward and inward currents. This was later confirmed with neuronal Kir 2.1 channels (**Figure 14.11d**). In excised patches, a small amount of Mg^{2+} ions (0.1 mM) on the cytoplasmic side diminishes the open channel outward currents which become flickery or noisy but have no effect on the inward currents. This suggests that Mg^{2+} acts as an open channel blocker (as for NMDA receptors).

Mg^{2+} ions are not the only blocking particles. Strong inwardly rectifying potassium channels are blocked by intracellular polyamines in a voltage-dependent manner. The affinity and voltage dependence of block varies with the identity of the blocking polyamine, spermine generally being the most potent and voltage-dependent blocker.

Pharmacology

Ba^{2+} (30–100 μM) causes an inward current at the resting potential (due to the blockade of the outward I_{Kir} present at rest) and decreases I_{Kir} at all potentials tested. Ba^{2+} thus linearizes the *I/V* curves (**Figure 14.11c**).

Summary

I_{Kir} is activated at resting membrane potential. Kir channels conduct more current in the inward (when the membrane is more hyperpolarized than E_K) than in the outward direction (when the membrane is more depolarized than E_K). Hyperpolarization above the K^+ equilibrium potential hardly occurs under physiological conditions at the mammalian plasma membrane. Consequently, despite their name, Kir channels primarily conduct outward potassium currents. This rectification property of the inward rectifier K^+ channel is chiefly due to the block of outward current by cytoplasmic Mg^{2+} and polyamines. Owing to rectification, I_{Kir} has a low amplitude in the outward direction. Kir-mediated outward K^+ fluxes prevent action potential firing by small electrical stimuli. I_{Kir} maintains membrane potential close to E_K.

14.4 CONCLUSIONS

Comparison between inward and outward voltage-gated currents is shown in **Figure 14.12**. Subliminal currents are on the left and supraliminal currents on the right. I_{CaT} has voltage properties similar to those of I_A and I_D but with opposite functions. I_h and I_M have symmetrical voltage properties and opposite functions.

Depending on their location, subliminal currents are activated by different signals. When located in dendrites, they are activated by a depolarization (EPSP) or a hyperpolarization (IPSP) of synaptic origin. When located in the soma–initial segment membrane, they are activated by the first action potential generated or by the hyperpolarization that follows an action potential (after spike hyperpolarization).

Depending on their location, subliminal voltage-gated currents also have different roles. When present in the dendritic membrane they may boost or counteract EPSPs or IPSPs (see Chapter 15), but when present in the soma–initial segment membrane they underlie intrinsic firing patterns, modulate synaptically driven firing patterns or participate in network oscillations (see Chapters 17, 19 and 20). When present in the whole neuronal membrane, subliminal currents that are activated around rest and that do not rapidly inactive (I_h, I_M, I_{KIR}) also determine resting membrane potential (see Chapter 3).

Further reading

Araki, T., Ito, M., Oshima, T., 1962. Potential changes produced by application of current steps in motoneurones. Nature 191, 1104–1105.

Baldwin, T.J., Tsaur, M.L., Lopez, G.A., et al., 1991. Characterization of a mammalian cDNA for an inactivating voltage-sensitive K^+ channel. Neuron 7, 471–483.

Battefeld, A., Tran, B.T., Gavrilis, J., Cooper, E.C., Kole, M.H., 2014. Heteromeric Kv7.2/7.3 channels differentially regulate action potential initiation and conduction in neocortical myelinated axons. J. Neurosci 34, 3719–3732.

Bruce, P., Bean, B.P., 2009. Inhibition by an excitatory conductance: a paradox explained. Nat. Neurosci. 12, 530–532.

Biel, M., Wahl-Schott, C., Michalakis, C., Zong, X., 2009. Hyperpolarization-activated cation channels: from genes to function. Physiol. Rev. 89, 847–885.

Birnbaum, S.G., Varga, A.W., Yuan, L.L., et al., 2004. Structure and function of Kv4-family transient potassium channels. Physiol. Rev. 84, 803–833.

Carbone, E., Lux, H.D., 1984. A low voltage-activated, fully inactivating Ca^{2+} channel in vertebrate sensory neurons. Nature 310, 501–502.

Cheong, E., Shin, H.S., 2013. T-type Ca^{2+} channels in normal and abnormal brain functions. Physiol. Rev. 93, 961–992.

Connor, J.A., Stevens, C.F., 1971. Prediction of repetitive firing behaviour from voltage-clamp data on an isolated neurone soma. J. Physiol. 213, 31–53.

Crill, W.E., 1996. Persistent sodium current in mammalian central neurons. Ann. Rev. Physiol. 58, 349–362.

Delmas, P., Brown, D.A., 2005. Pathways modulating neural KCNQ/M (Kv7) potassium channels. Nat. Rev. Neurosci. 6, 850–862.

DePuy, S.D., Yao, J., Hu, C., et al., 2006. The molecular basis for T-type Ca^{2+}channel inhibition by G protein $\beta 2\gamma 2$ subunits. Proc. Natl Acad. Sci. USA 103, 14590–14595.

Devaux, J.J., Kleopa, K.A., Cooper, E.C., Scherer, S.S., 2004. KCNQ2 is a nodal K^+ channel. J. Neurosci. 24, 1236–1244.

French, C.R., Sah, P., Buckett, K.J., Gage, P.W., 1990. A voltage-dependent persistent sodium current in mammalian hippocampal neurons. Gen. Physiol. 95, 1139–1157.

Hotson, J.R., Prince, D.A., Schwartzkroin, P.A., 1979. Anomalous inward rectification in hippocampal neurons. J. Neurophysiol. 42, 889–895.

Huguenard, J.R., 1996. Low-threshold calcium currents in central nervous system neurons. Ann. Rev. Physiol. 58, 329–348.

Kay, A.R., Sugimori, M., Llinas, R., 1998. Kinetic and stochastic nature of a persistent sodium current in mature guinea pig cerebellar Purkinje cells. J. Neurophysiol. 80, 1167–1179.

Kiss, T., 2008. Persistent Na^+ channels: origin and function. Acta Biol. Hung. 59 (Suppl.), 1–12.

Lopatin, A.N., Makhina, E.N., Nichols, C.G., 1994. Potassium channel block by cytoplasmic polyamines as the mechanism of intrinsic rectification. Nature 372, 366–369.

MacKinnon, R., 2003. Potassium channels. FEBS Lett. 555, 62–65.

Matsuda, H., Saigusa, A., Irisawa, H., 1987. Ohmic conductance through the inwardly rectifying K^+ channel and blocking by internal Mg^{2+}. Nature 325, 156–158.

Miller, A.G., Aldrich, R.W., 1996. Conversion of a delayed rectifier K^+ channel by three amino acids substitution. Neuron 16, 853–858.

Pan, Z., Kao, T., Horvath, Z., et al., 2006. A common ankyrin-G-based mechanism retains KCNQ and Nav channels at electrically active domains of the axon. J. Neurosci. 26, 2599–2613.

Perez-Reyes, E., Cribbs, L.L., Daud, A., et al., 1998. Molecular characterization of a neuronal low-voltage-activated T-type calcium channel. Nature 391, 896–900.

Song, W.J., 2002. Genes responsible for native depolarization-activated K^+ currents in neurons. Neurosci. Res. 42, 7–14.

Shah, M.M., Migliore, M., Valencia, I., Cooper, E.C., Brown, D.A., 2008. Functional significance of axonal Kv7 channels in hippocampal pyramidal neurons. Proc. Natl Acad. Sci. USA 105, 7869–7874.

Standen, N.B., Stanfield, P.R., 1978. A potential- and time-dependent blockade of inward rectification in frog skeletal muscle fibres by barium and strontium ions. J. Physiol. (Lond.) 280, 169–191.

Yue, C., Yaari Y., 2004. KCNQ/M channels control spike after depolarization and burst generation in hippocampal neurons. J. Neurosci. 24, 4614–4624.

CHAPTER

15

Somato-dendritic processing of postsynaptic potentials II. Role of sub-threshold depolarizing voltage-gated currents

Constance Hammond, Dominique Debanne, Jean-Marc Goaillard

OUTLINE

Chapter 13 showed that in a dendritic tree in which currents propagate passively, EPSPs are attenuated in amplitude and slowed in time course as they spread to the soma due to passive properties of the dendrites. In summary, the influence of an EPSP on neuronal output (firing), which relies on its ability to depolarize the axon, depends upon the initial size and shape of the synaptic response, as well as how the cable properties of the dendritic tree filter the response as it spreads from the synapse to the site of action potential generation. As a consequence, excitatory or inhibitory synapses located on distal dendrites should be less efficient in depolarizing or hyperpolarizing the soma–initial segment region. In other words, the combination of the large variation in synaptic distance and the cable-filtering properties of dendrites should, in theory, cause the amplitude and temporal characteristics of functionally similar inputs to be highly variable at the final integration site.

Although theoretical analyses have predicted such a clear location-dependent variability of synaptic input, there is now considerable evidence indicating that the shape of EPSPs may be relatively independent of synapse location (e.g. in pyramidal neurons of the CA1 region of the hippocampus, **Figure 15.1**). The ability to simultaneously record synaptic activity from several locations on the same neuron (distal or proximal dendrites, soma) and the advent of imaging techniques with high spatial and temporal resolution now give the opportunity to understand the real dendritic processing of synaptic information. The hypothesis was then formulated: if some dendrites do not behave as simple cables, is it because their membranes express voltage-gated

Cellular and Molecular Neurophysiology. DOI: 10.1016/B978-0-12-397032-9.00015-7

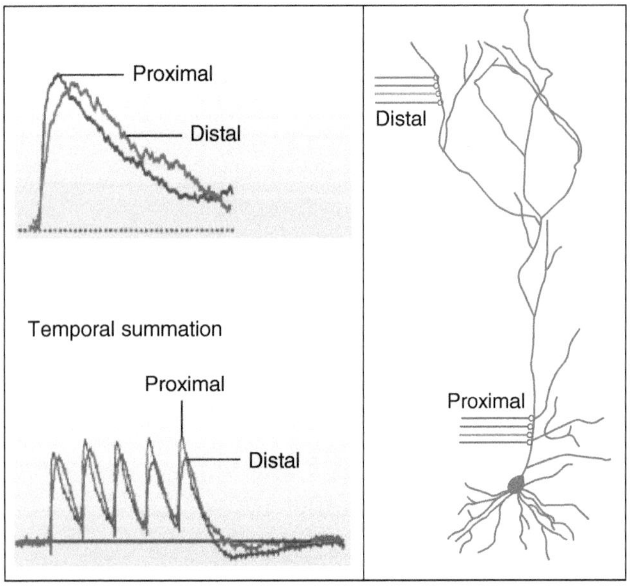

FIGURE 15.1 **Synaptic integration in CA1 pyramidal neurons is independent of location.** Upper traces: The average unitary excitatory postsynaptic potentials (EPSPs) recorded at the soma of a neuron receiving distal and proximal input. Somatic EPSP amplitude is similar in spite of location differences. Lower traces: The amount of temporal summation at the soma is the same for a 50 Hz train of stimuli applied to distal (300 μm) or to proximal Schaffer collateral inputs (50 μm). (**b**) A CA1 pyramidal neuron showing the location of proximal and distal synaptic inputs across the dendritic arbor. Adapted from Magee JC (2000) Dendritic integration of excitatory synaptic input. *Nat. Neurosci. Rev.* **1**, 181–190, with permission.

channels? Experiments were thus designed to answer the following questions:

- Are sub-threshold voltage-gated currents present in dendritic membranes?
- Are they distributed uniformly over somato-dendritic membranes, or are the electrogenic properties in dendrites fundamentally different from those in the soma?
- How do these dendritic or somatic currents shape excitatory postsynaptic potentials and affect input summation?
- Are the dendrites and soma the only cellular compartments involved in the integration of synaptic inputs?

To address these questions, the models used are brain regions organized in layers such as the neocortex or the hippocampus. In such regions, dendritic recordings are much easier than in other structures because the dendritic layer is easily recognizable from the somatic layer. To answer the first two questions, the activity of sub-threshold voltage-gated channels must be recorded in patches of dendritic membrane. Then, the amplitude of sub-threshold currents recorded from similar-sized patches of dendritic and somatic membranes must be compared. Experiments are thus performed in brain slices to allow outside-out, dendrite-attached or soma-attached recordings. For technical reasons, dendritic recording is limited to dendritic branches with a diameter greater than 1 μm. To answer the third question, the EPSPs evoked at different locations of the dendritic tree are simultaneously recorded in whole-cell somatic and dendritic patch clamp configuration combined or not with imaging techniques. To answer the last question, focal application of ion channel blockers is used to determine the distribution of specific ion channels in the different cellular compartments.

15.1 PERSISTENT Na^+ CHANNELS ARE PRESENT IN THE AXO-SOMATIC REGION OF NEOCORTICAL NEURONS; I_{NaP} BOOSTS $EPSP_S$

The persistent Na^+ current I_{NaP} is a tetrodotoxin- (TTX) sensitive Na^+ current that activates below spike threshold and slowly inactivates (see Section 14.2.1). To record Na^+ current in isolation, the solution bathing the extracellular face of the patch contains NaCl. Tetraethylammonium chloride (TEA) is included in the external solution or CsCl in the internal solution to strongly reduce outward K^+ currents. Voltage-gated Ca^{2+} currents (and therefore Ca^{2+}-activated currents) are blocked by substitution of Mn^{2+} for Ca^{2+} in the extracellular medium. Persistent Na^+ current is identified by inward polarity, its low threshold of activation (10–15 mV negative to I_{Na}), and its blockade by TTX. It can be easily differentiated from the Na^+ current of action potentials (I_{Na}) as the latter activates at a higher threshold, has a far greater amplitude and inactivates much faster.

15.1.1 Persistent Na^+ channels are present in the dendrites and soma of pyramidal neurons of the neocortex

Dendritic recordings

Single Na^+-channel activity is recorded in patch clamp from dendrites of acutely isolated pyramidal neurons of the neocortex (dendrite-attached recordings). Depolarizing voltage steps to −60/+10 mV are applied from a holding potential of −100 mV. They evoke Na^+-channel openings in all the recorded patches (**Figure 15.2a**). The most prominent activity in multichannel patches consists of early, short-lived openings clustered within the first few milliseconds that correspond to I_{Na}. In addition, a different Na^+-channel activity consisting of prolonged or late openings is recorded (I_{NaP}). When many consecutive traces are averaged, this persistent channel activity is able to produce sizable net inward current even for 500 ms. The I/V relationship

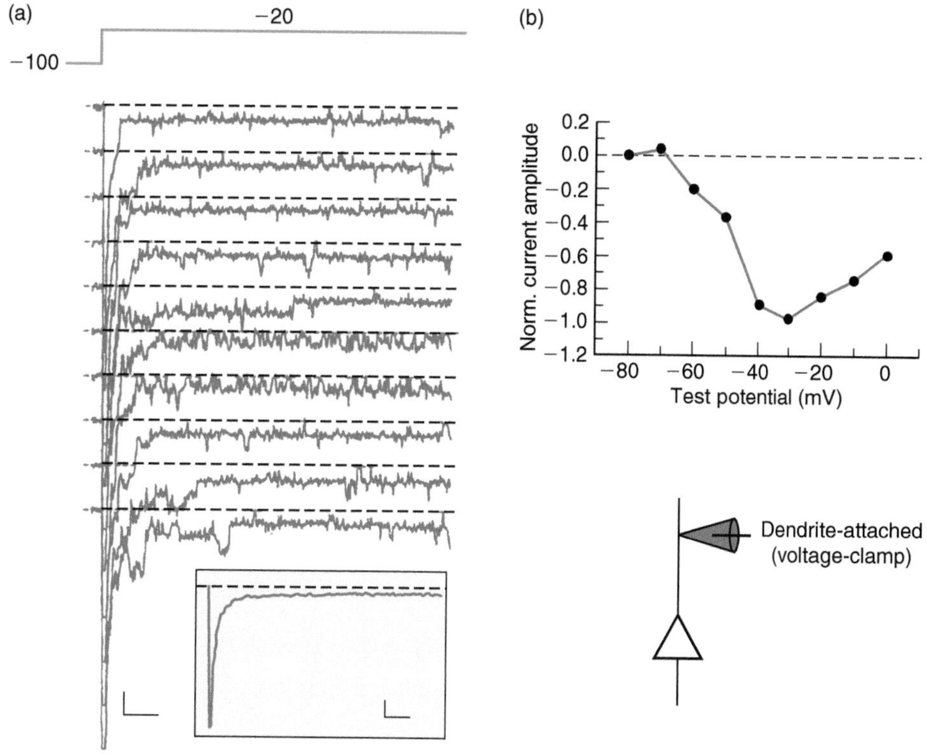

FIGURE 15.2 **Persistent Na^+ channel activity in dendrites of cortical pyramidal neurons. (a)** Na^+-channel currents evoked by a 50 ms depolarizing pulse. The current traces shown are consecutive sweeps (scale bar 2 pA and 5 ms). *Insets*: Ensemble average current obtained from 20 consecutive sweeps (scale bar 2.5 pA and 5 ms). **(b)** Voltage dependence of the persistent component of ensemble average currents obtained as in **(a)**. The plot is normalized to the absolute value of its peak amplitude. Adapted from Magistretti J, Ragsdale DS, Alonso A (1999) Direct demonstration of persistent Na^+ channel activity in dendritic processes of mammalian cortical neurones. *J. Physiol. (Lond.)* **521**, 629–636, with permission.

of the persistent component (**Figure 15.2b**) is not different from that of the macroscopic current recorded in the same cells.

Somatic recordings

Pyramidal neurons of the neocortex are loaded with the Na^+-sensitive, membrane-impermeant fluorescent dye SBFI (sodium benzofuramisophthalate). A slow depolarizing ramp is applied through the somatic intracellular electrode to a final depolarization (around −50 mV) that is known to activate I_{NaP} and is sub-threshold for action potential initiation. **Figure 15.3a** (top trace, arrow) shows the sub-threshold inward current evoked by the depolarizing ramp (middle trace), which totally disappears in the presence of TTX (**Figure 15.3b**, top trace). During activation, a TTX-sensitive increase of intracellular Na^+ concentration is observed in the soma (**Figure 15.3a,b**, bottom traces) as well as in the proximal part of the apical dendrite (not shown). This strongly suggests that I_{NaP} channels are present in the somatic membrane. However, this experiment does not allow one to localize the exact site(s) of I_{NaP} generation since the rise of Na^+ in the proximal dendrite can result from Na^+ diffusion from the soma, and vice versa.

15.1.2 Are persistent Na^+ channels activated by EPSPs? Where does I_{NaP} boost EPSP amplitude, in the dendrites, in the soma or in the axon?

Activation of I_{NaP} by local synaptic inputs is tested by simultaneous whole-cell dendritic and somatic recordings (in current clamp mode) made from the same pyramidal neuron of the neocortex. Dendritic EPSPs can be evoked either by (i) stimulation of afferents or by (ii) intradendritic current injection (simulated EPSP). These EPSPs must be sub-threshold (they must not trigger Na^+ action potentials), in order to evoke only the low-threshold, TTX-sensitive, persistent Na^+ current (I_{NaP}) and not the voltage-gated Na^+ current of action potentials. The role of I_{NaP} on these EPSPs is then deduced by studying the effect of TTX on the amplitude and duration of EPSPs. However, since TTX affects postsynaptic Na^+ channels as well as presynaptic ones, it profoundly alters synaptic transmission. Such a study therefore needs to bypass synaptic transmission by using only simulated EPSPs (protocol (ii)). The voltage change occurring during an EPSP is simulated by dendritic current injection with a time course similar to that of an excitatory postsynaptic current (EPSC). To do so, EPSCs are first recorded and

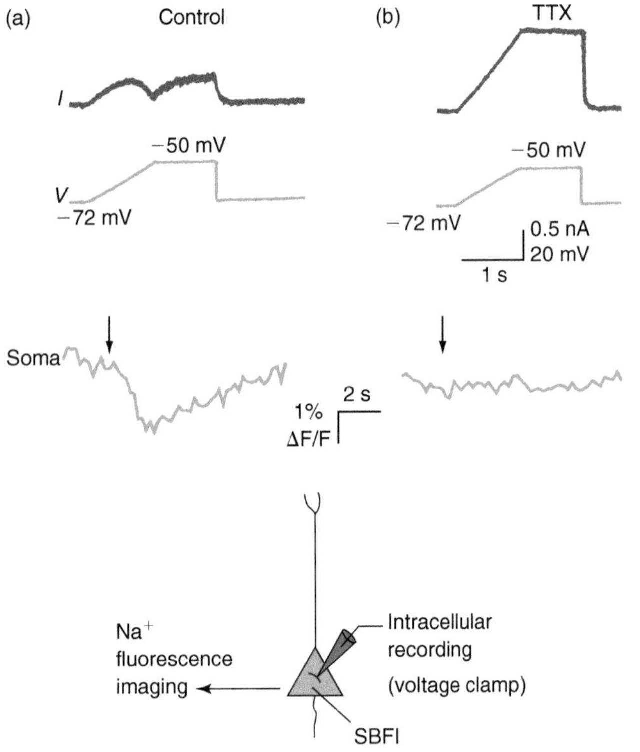

FIGURE 15.3 **I_{NaP} activation and the resultant SBFI fluorescence changes in the soma of a pyramidal neuron of the neocortex. (a)** Intracellular recording (voltage clamp mode) of the current evoked by a depolarizing ramp from −72 mV to a 1 s constant step at −50 mV (top traces). The corresponding decrease of SBFI fluorescence in the soma is shown in the bottom trace. The arrow indicates time of voltage clamp. A decrease of SBFI fluorescence reflects an increase of intracellular Na^+ concentration. **(b)** The same experiment in the presence of 1 μM of TTX in the bath. Adapted from Mittman T, Linton SM, Schwindt P, Crill W (1997) Evidence of persistent Na^+ current in apical dendrites of rat neocortical neurons from imaging of Na^+-sensitive dye. *J. Neurophysiol.* **78**, 1188–1192, with permission.

then simulated. Simulated EPSPs generated by dendritic current injections are recorded both at their site of generation (dendritic site) and in the soma. Bath and local applications of TTX are used to determine whether I_{NaP} is involved in the amplification of simulated EPSPs and to localize where EPSPs are amplified, locally in the dendrites or in the soma region.

When the amplitude of EPSPs recorded at the soma is greater than 5 mV, bath application of TTX causes a substantial reduction in the peak amplitude (≈30%) and a 53 ± 5% reduction of the EPSP area (when V_{rest} = −65 mV). At which site does a synaptic signal experience amplification while it travels from the dendrites to the axon initial segment? The site of EPSP amplification is tested by local application of TTX to either the site of simulated EPSP generation in the dendrites or to the somatic region. Local application of TTX is achieved by pressure ejection of TTX from a patch pipette, the tip of which is placed close to either the dendritic recording site or the soma. To minimize the spread of TTX, a low concentration of TTX (100 nM) is used. That this TTX application does in fact block dendritic Na^+ channels is verified by its ability to reduce the amplitude of backpropagated Na^+ action potentials (see Chapter 16). I_{NaP} seems to be mostly located in the soma: local application of TTX to the dendritic recording site has little or no effect on the simulated EPSP, but when applied to the soma TTX reduces the somatic EPSP peak amplitude and integral.

More recent experiments using even more precise and localized applications of TTX in fact indicate that I_{NaP} in a layer V pyramidal neuron is primarily generated in the proximal axon (**Figure 15.4**), and that the somatic blockade (**Figure 15.4b**) could be due to the spillover of TTX onto the proximal region of the axon. In these experiments, the persistent Na^+ current is evoked by a sub-threshold ramp of voltage applied through the recording electrode in the soma. A low concentration of TTX (100 nM) was applied either on the dendrite, the soma or the axon. No effect on I_{NaP} was observed when the TTX was locally applied on the apical dendrite (**Figure 15.4a**) or the soma (**Figure 15.4b**). However, application on the axon produced a large reduction in the amplitude of I_{NaP} (**Figure 15.4c**). The graph expressing the effect of TTX as a function of the distance of application in the dendrite, soma and axon (**Figure 15.4d**) clearly shows that reduction in the amplitude of I_{NaP} occurs only when TTX is applied on the axon.

Therefore, in the neocortex, EPSPs activate I_{NaP} and, in turn, I_{NaP} boosts EPSPs in amplitude and duration. This amplification mainly occurs in the axo-somatic region. However, assuming a resting membrane potential of −65/−70 mV, to activate I_{NaP}, EPSPs must depolarize the membrane by 5–15 mV. Only summed EPSPs can reach this amplitude.

15.2 T-TYPE Ca^{2+} CHANNELS ARE PRESENT IN THE DENDRITES OF CORTICAL NEURONS; I_{CaT} BOOSTS EPSPs

The T-type Ca^{2+} current is an amiloride- and Ni^{2+}-sensitive Ca^{2+} current which activates below spike threshold (low voltage-activated or LVA), inactivates rapidly with time and is totally inactivated at −40 mV (see Section 14.2.2). In order to record Ca^{2+} current in isolation, the solution bathing the extracellular face of the patch contains Ba^{2+} (110–120 mM) as the charge carrier and TEA and TTX for blocking K^+ and Na^+ currents, respectively. T-type Ca^{2+} current is identified by inward polarity, unitary current amplitude, activation–inactivation characteristics, sensitivity to Ni^{2+} or amiloride, and insensitivity to dihydropyridines (L-type blocker), ω-conotoxins or funnel web toxin (N-type and P-type blockers).

Are the pharmacological tools used in all the above experiments sufficiently selective to allow the conclusion

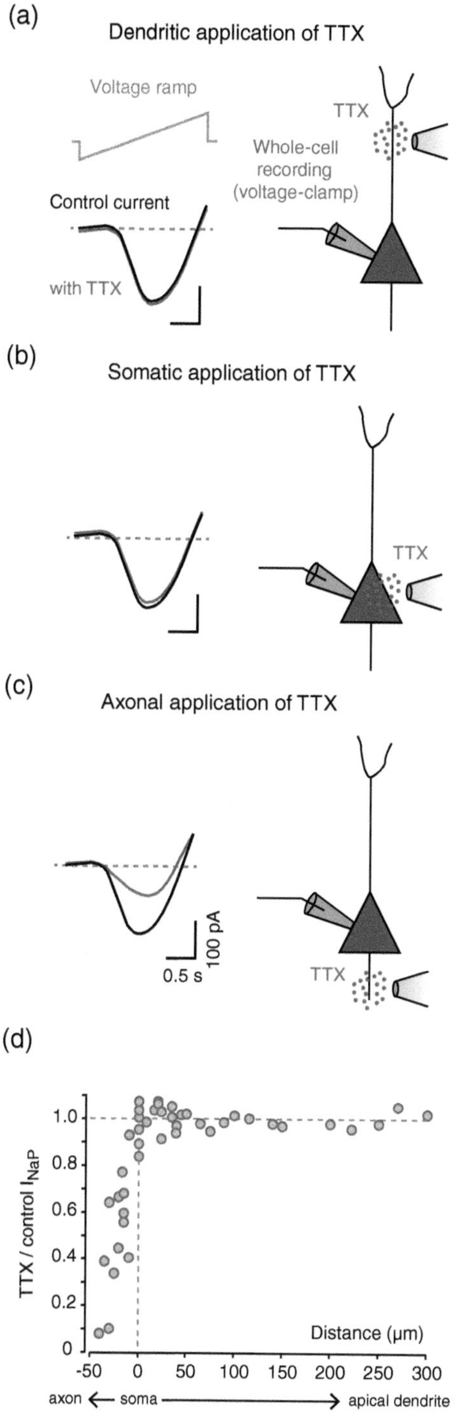

FIGURE 15.4 **I_{NaP} is predominantly expressed in the axons of layer V neocortical pyramidal neurons. (a)** Schematic depicting the effect of local dendritic application of TTX on the amplitude of the persistent sodium current (I_{NaP}) in layer V neocortical pyramidal neurons. I_{NaP} is measured in response to a voltage ramp using whole-cell somatic voltage-clamp recordings. Local application of TTX on the apical dendrite does not modify the amplitude of I_{NaP}. **(b)** Same experiment as in **(a)** showing the lack of effect of local somatic application of TTX on I_{NaP}. **(c)** In contrast to dendritic **(a)** and somatic **(b)** applications, local application of TTX on the axon strongly reduces the amplitude of I_{NaP} (top traces). **(d)** Scatter plot showing the percentage of inhibition of I_{NaP} by TTX as a function of the location of the TTX application. While axonal applications (left, negative distances from the soma) strongly reduce I_{NaP} amplitude, somatic (0 μm) and dendritic (positive distances from the soma) have negligible effects on I_{NaP} amplitude. Adapted from Astman N, Gutnick MJ, Fleidervish IA (2006), Persistent sodium current in layer 5 neocortical neurons is primarily generated in the proximal axon. *J. Neurosci.* **26**, 3465–3473.

that the observed Ca^{2+} current is I_{CaT}? The answer is no. These experiments do not exclude some partial contribution of a dendritic R-type Ca^{2+} current (I_{CaR}) which is also sensitive to Ni^{2+} and amiloride. However, since I_{CaR} is a high-threshold-activated current – it activates at higher depolarized potentials than I_{CaT} – it has been considered in the following experiments that the Ca^{2+} current activated in dendrites by step depolarizations or by EPSPs is I_{CaT}.

15.2.1 T-type Ca^{2+} channels are present in the dendrites of pyramidal neurons of the hippocampus

The activity of a patch of apical dendritic membrane is recorded in the dendrite-attached configuration (voltage clamp mode). In response to depolarizing steps to −15 mV from a hyperpolarized potential (−85 mV, to de-inactivate T channels), channel openings are

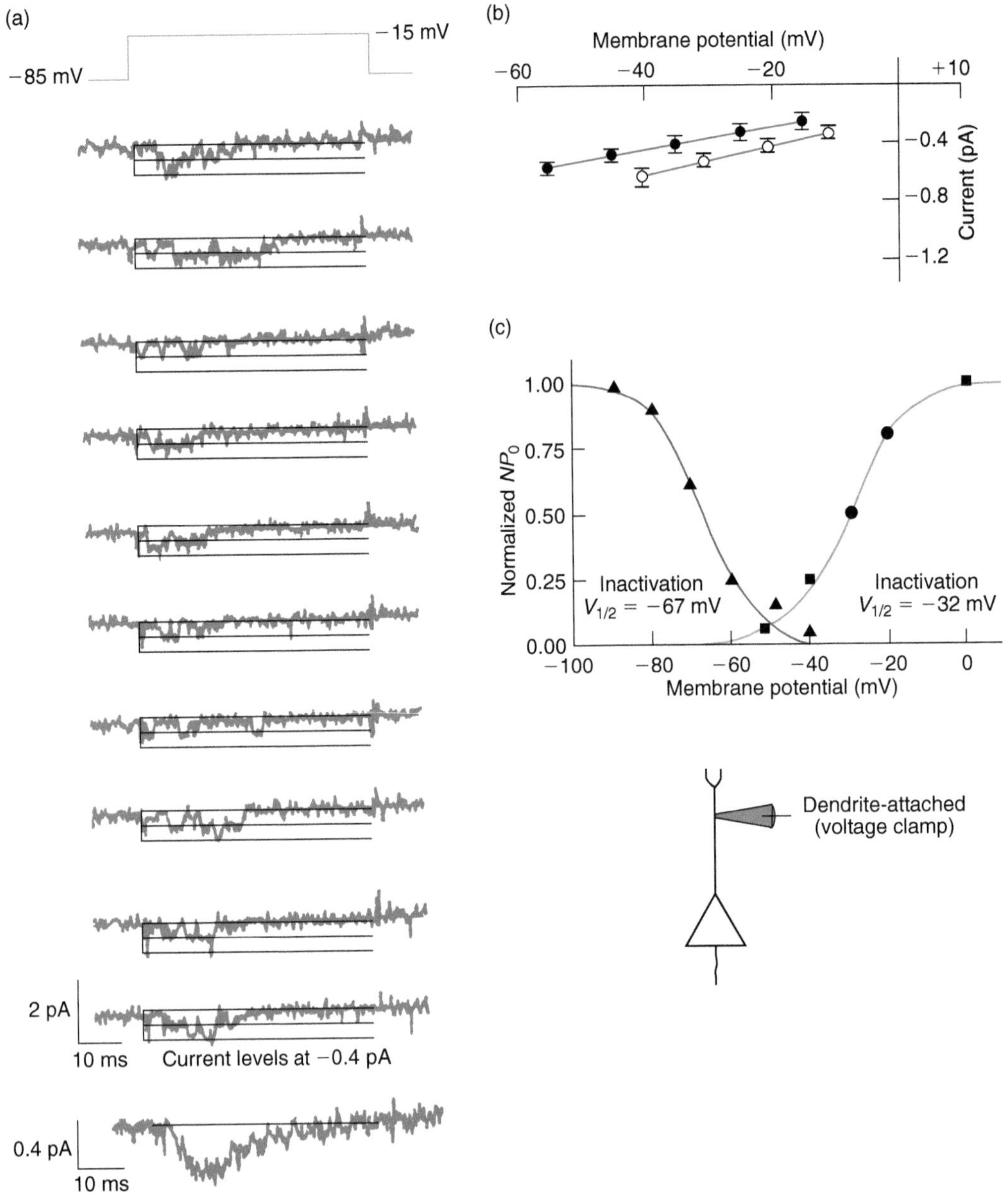

FIGURE 15.5 **Dendritic low-voltage-activated Ca^{2+} channel activity in pyramidal neurons of the hippocampus. (a)** Consecutive sweeps of T-type Ca^{2+}-channel activity recorded from a dendrite-attached patch (voltage clamp mode) in response to 60 ms depolarizing steps to −15 mV (V_H = −85 mV). Bottom trace is the ensemble average (104 sweeps) demonstrating significant inactivation during the 60 ms depolarizing step (110 mM of Ba^{2+} in the recording solution). **(b)** i_T/V plot of T-type Ca^{2+}-channel activity. Unitary current amplitude is plotted as a function of membrane potential for patches recorded with either 20 mM (•) or 110 mM (○) Ba^{2+} as charge carrier. The slope (unitary conductance) γ_T is between 7 pS (20 mM of Ba^{2+}) and 11 pS (110 mM of Ba^{2+}). **(c)** Representative steady-state activation (j) and inactivation (m) plots for dendritic LVA Ca^{2+} channels recorded in 20 mM of Ba^{2+}. Adapted from Magee JC, Johnston D (1995) Characterization of single voltage-gated Na^+ and Ca^{2+} channels in apical dendrites of rat CA1 pyramidal neurons. *J. Physiol. (Lond.)* **487**, 67–90, with permission.

recorded. They occur mostly at the beginning of the depolarizing step, are of small unitary current amplitude (**Figure 15.5a**), and the i_T/V plot gives a unitary conductance γ_T of 7–11 pS (**Figure 15.5b**). They are sensitive to Ni^{2+} and amiloride (not shown). These data, together with the activation–inactivation characteristics (**Figure 15.5c**) reveal that T-type Ca^{2+} channels are present within the apical dendrite of pyramidal neurons. They are similar in basic characteristics to T-type Ca^{2+} channels recorded from many neuronal somata (compare with **Figure 14.5**).

15.2.2 Dendritic T-type Ca^{2+} channels are activated by EPSPs; in turn, I_{CaT} boosts EPSPs amplitude

Activation of dendritic I_{CaT} by local synaptic inputs is tested by simultaneous dendrite-attached and whole-cell somatic recordings from the same pyramidal neuron of the CA1 region. Sub-threshold EPSPs are evoked by Schaffer collateral stimulation. These EPSPs must be sub-threshold (they must not trigger Na^+ action potentials), in order to evoke only the low-threshold, Ni^{2+}-sensitive, Ca^{2+} current (I_{CaT}) and not the high-voltage-activated Ca^{2+} currents such as the L-, N- or P/Q-type currents. EPSPs are recorded from the soma (in current clamp mode) after propagation in the dendritic tree. Single-channel T-type Ca^{2+} currents are recorded from the patch of dendritic membrane (in voltage clamp mode). If channel openings only occur during EPSPs, they are considered to have been triggered by it.

In response to Schaffer collateral stimulation, the activity of single Ca^{2+} channels is recorded (**Figure 15.6a**). These single-channel currents are not recorded when the Ca^{2+} channel blocker $CdCl_2$ (0.5 mM) is present in the pipette (not shown). Single-channel openings are most often observed near the peak or falling phases of the EPSPs. EPSP-activated channel openings display small unitary current amplitude and slope conductance ($\gamma = 9.0 \pm 1.6$ pS) characteristic of T-type dendritic channels (see **Figure 15.5**). EPSPs with a peak amplitude of 10 mV (at the somatic recording site) are necessary for activation of T-type dendritic Ca^{2+} channels. When a 4 s hyperpolarizing prepulse is applied 400 ms before synaptic stimulation, the open probability (p_o) of T-type Ca^{2+} channels is increased in a voltage-dependent manner (**Figure 15.6b**). This suggests that a large proportion of the T-type Ca^{2+}-channel population is inactivated at resting potential. Therefore,

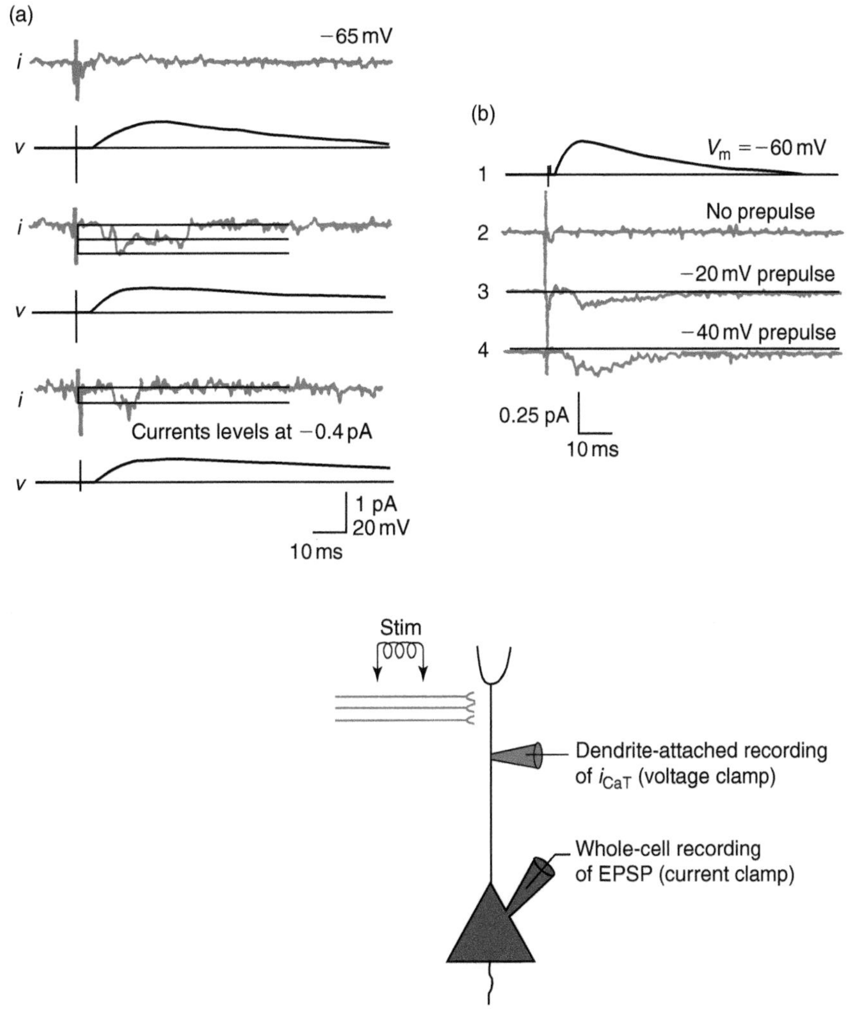

FIGURE 15.6 **Synaptic activation of LVA Ca^{2+} channels in hippocampal CA1 pyramidal neurons.** Sub-threshold EPSPs are evoked by Schaffer collateral stimulation and are recorded from the soma (in current clamp mode) after propagation in the dendritic tree. **(a)** Consecutive sweeps of dendrite-attached recordings (voltage clamp mode) with the patch held at −65 mV showing Ca^{2+}-channel activity recorded at the dendritic site (*i*, top traces) and of sub-threshold EPSPs (*v*, bottom traces) recorded at the somatic site (whole-cell configuration). **(b)** Hyperpolarizing prepulses (not shown) increase the activation of T-type Ca^{2+} channels by an EPSP. Ensemble average of 50 consecutive current traces without prepulse (2), ensemble average of 60 consecutive current traces after a 4 s prepulse of −20 mV (3), and ensemble average of 60 consecutive traces after a 4 s prepulse of −40 mV (4). The patch is returned to a holding potential that is 10 mV depolarized from resting potential 400 ms before synaptic stimulation in order to evoke an EPSP of similar amplitude (1) in all trials. Adapted from Magee JC, Johnston D (1995) Synaptic activation of voltage-gated channels in the dendrites of hippocampal pyramidal neurons. *Science* **268**, 301–304, with permission.

membrane hyperpolarization (as during IPSPs), by allowing channel de-inactivation, is necessary for maximal channel activation by EPSPs. Thus, the contribution of LVA Ca^{2+} channels to EPSP amplitude would be particularly enhanced for EPSPs occurring after hyperpolarizing IPSPs.

Another way to address the question of the activation of T-type Ca^{2+} current by EPSPs is to measure Ni^{2+}-sensitive intradendritic $[Ca^{2+}]_i$ increase during sub-threshold EPSPs. Whole-cell recordings in the soma are performed in conjunction with high-speed fluorescence imaging. To measure changes in intracellular Ca^{2+} concentration, the fluorescent indicator FURA-2 is included in the pipette solution. Detectable increases in Ca^{2+} concentration are observed in response to as few as two consecutive synaptic stimulations (50 Hz) but a short train of five stimuli provides a very reproducible increase above baseline ($2.2 \pm 0.5\%\ \Delta F/F$) (**Figure 15.7a**).

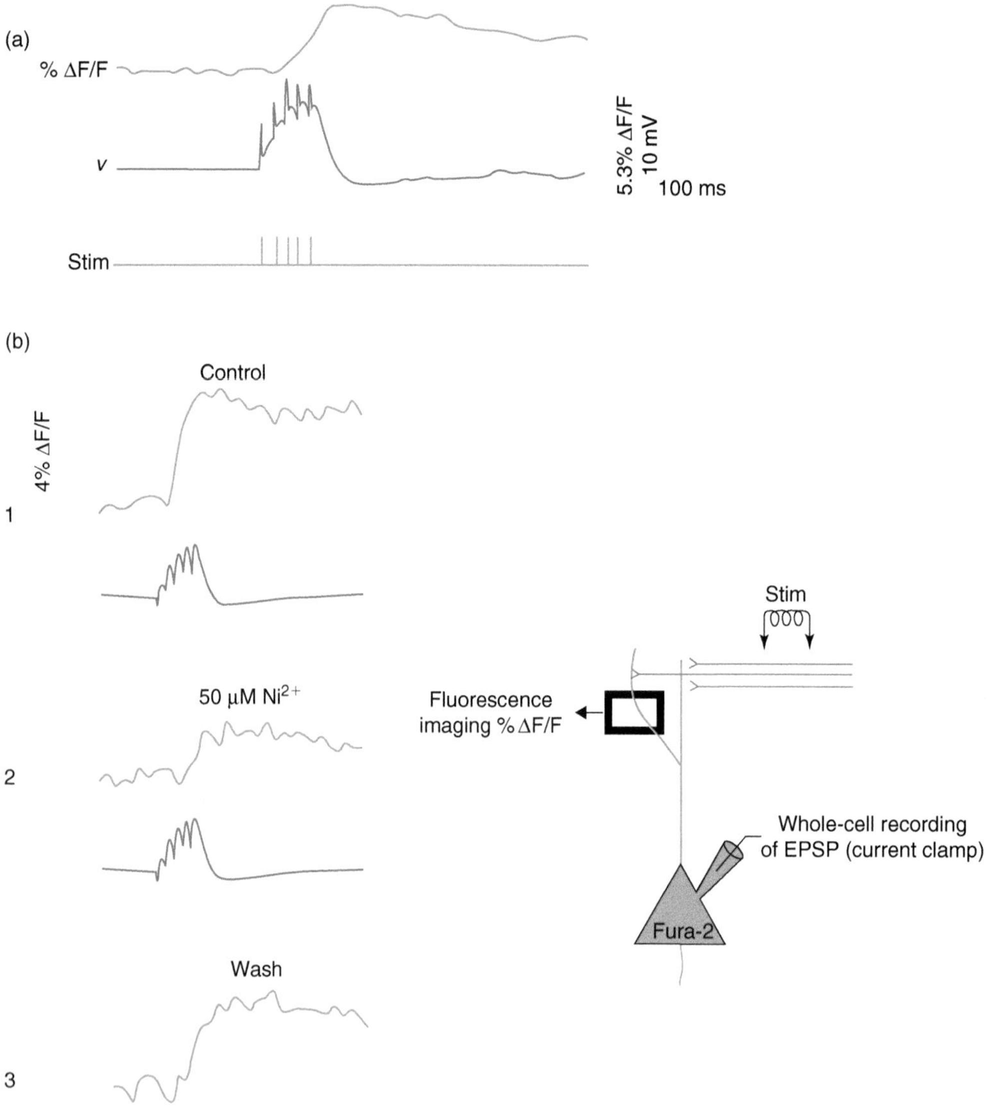

FIGURE 15.7 **Sub-threshold EPSPs cause a localized, Ni^{2+}-sensitive elevation of intradendritic Ca^{2+} concentration.** Sub-threshold EPSPs are evoked by stimulation of afferents close to the dendrite under study. **(a)** Time course of percentage change in FURA-2 fluorescence in a dendrite (%$\Delta F/F$, top trace) evoked by a short train of five EPSPs and somatic voltage recordings (V) of the five EPSPs (whole-cell configuration, current clamp mode, bottom trace). The fluorescence trace is from the region delimited by the small black frame on the schematic representation of the FURA-2 loaded neuron. **(b)** Localized percent change in FURA-2 fluorescence (%$\Delta F/F$, top trace) induced by a short train of five EPSPs and somatic voltage recordings of the five EPSPs (bottom traces) in the absence (1), presence (2) and 20 min after washing (3) of 50 μM of $NiCl_2$. The somatic recording of EPSPs (bottom traces) is unaffected by Ni^{2+} application. All traces in the figure are averages of five consecutive sweeps. Adapted from Magee JC, Christofi G, Miyakawa H *et al.* (1995) Subthreshold synaptic activation of voltage-gated Ca^{2+} channels mediates a localized Ca^{2+} influx into the dendrites of hippocampal pyramidal neurons. *J. Neurophysiol.* **74**, 1335–1342, with permission.

The rise in $[Ca^{21}]_i$ continues throughout the course of the synaptic stimulation and begins to decay back to baseline several milliseconds after the end of the EPSP train. It thus appears that sub-threshold stimulations of sufficient amplitude result in a transient elevation of intradendritic $[Ca^{2+}]_i$. $[Ca^{2+}]_i$ transients are localized primarily to the area of the synaptic input (not shown). The localized nature of these $[Ca^{2+}]_i$ signals implies that the largest changes in $[Ca^{2+}]_i$ occur in the dendrites where the synaptic input is located and that this signal attenuates as it approaches the soma. Through which types of dendritic Ca^{2+} channels are Ca^{2+} ions entering the cell; or does this $[Ca^{2+}]_i$ increase result from the release of intradendritic Ca^{2+} stores? For the first hypothesis, the candidates are Ca^{2+}-permeable receptor channels (NMDA or AMPA receptors; see Sections 10.2 and 10.4) and voltage-gated Ca^{2+} channels. Application of APV (50 μM), a specific antagonist of NMDA channels, has very little effect on the sub-threshold Ca^{2+} signals as long as the EPSP amplitude is maintained constant. Therefore, 50 μM of APV is included in the bath solution for the remainder of the experiment. In contrast, membrane hyperpolarization to around −100 mV during synaptic stimulation prevents the synaptically-induced rise in intradendritic Ca^{2+} concentration, indicating that Ca^{2+} entry is voltage dependent. All these data demonstrate that $[Ca^{2+}]_i$ signals result from Ca^{2+} influx but not through NMDA or AMPA receptors (antagonists of AMPA receptors cannot be tested since they would cancel the EPSP which is an AMPA-mediated EPSP). This influx is then likely to occur through voltage-gated ion channels. The primary candidate is the T-type Ca^{2+} channel. The effect of 50 μM of Ni^{2+} is therefore tested. When bath is applied, such a concentration of Ni^{2+} produces a 54 ± 5% block of the synaptically-induced influx of Ca^{2+} and this block is completely reversible with washout of Ni^{2+} (**Figure 15.7b**).

What are the consequences of this local intradendritic Ca^{2+} increase? Does dendritic I_{CaT} boost EPSPs? To address this question, EPSPs are evoked far out on the apical dendrite and their shape is recorded at the soma with the dendritic I_{CaT} active or partially suppressed by local pressure application of I_{CaT} blockers such as Ni^{2+} or amiloride. EPSPs are evoked by afferent fiber stimulation at a frequency of 0.2 Hz and are recorded at the level of the soma (whole-cell configuration). To visualize the approximate spread of Ni^{2+} (5 μM) or amiloride (50 μM) in the tissue, both drugs are dissolved in 2% food color solution. In control experiments, dendritic pressure application of food color solution alone produces negligible reductions of EPSP amplitude. To study the role of dendritic I_{CaT} with minimum contamination by somatic I_{CaT}, the membrane potential is set to −70 mV at the soma and stimulation amplitude is adjusted to obtain EPSP peak amplitudes at the soma of 7 mV on average. Under these conditions, somatic EPSP amplitude should be too small to activate LVA Ca^{2+} channels. Dendritic amiloride application reduces EPSP amplitude by 27 ± 2% and Ni^{2+} application reduces it by 33 ± 2.9% (**Figure 15.8**, top traces). The effects of both amiloride and Ni^{2+} reverse within 15–20 min of drug washout. Hyperpolarization of the membrane to −90 mV attenuates the effect of both antagonists (**Figure 15.8**, bottom traces). However, any of the observed effects can be due to a presynaptic action of Ni^{2+} or amiloride.

In order to check this, EPSPs are recorded extracellularly near the dendrite, at the level of afferent stimulation in control conditions and in the presence of blockers. Bath application of both drugs, at a concentration 10 times higher than the concentration reached during focal application around the apical dendrite, fails to impair synaptic transmission. Thereby this excludes any presynaptic action of Ni^{2+} or amiloride. Therefore, in CA1 pyramidal neurons, EPSPs activate a T-type Ca^{2+} current that can indeed alter the weight of EPSPs. This amplification occurs in dendritic regions. However, as for persistent Na^+ channels, assuming a resting membrane potential of −65/−70 mV, to activate I_{CaT}, EPSPs must depolarize the membrane by 5–10 mV in order to activate I_{CaT}. Only summed EPSPs can reach this amplitude.

15.3 THE HYPERPOLARIZATION-ACTIVATED CATIONIC CURRENT I_h IS PRESENT IN DENDRITES OF HIPPOCAMPAL PYRAMIDAL NEURONS; I_h ATTENUATES EPSPs

The hyperpolarization-activated cation current I_h is a Cs^+-sensitive current turned on by hyperpolarization. It is inward at potentials more hyperpolarized than its reversal potential (between −50 and −30 mV), and does not inactivate (see Section 14.2.3). Because of its voltage range of activation (typically for voltages more hyperpolarized than −55/−60 mV), I_h is active at resting membrane potential and deactivates during membrane depolarization. To record I_h, the solution bathing the extracellular face of the membrane and the pipette solution contain control concentrations of Na^+, K^+ and Ca^{2+} ions.

15.3.1 H-type cationic channels are expressed in dendrites of pyramidal neurons of the hippocampus

The basic biophysical properties and the subcellular distribution of I_h are investigated in cell-attached configuration and voltage clamp mode. Long duration (1–3 s) hyperpolarizing steps evoke inward currents from cell-attached macropatches obtained from both the soma and apical dendritic regions but with different amplitudes (**Figure 15.9a**). These inward currents are slowly activating, non-inactivating and slowly

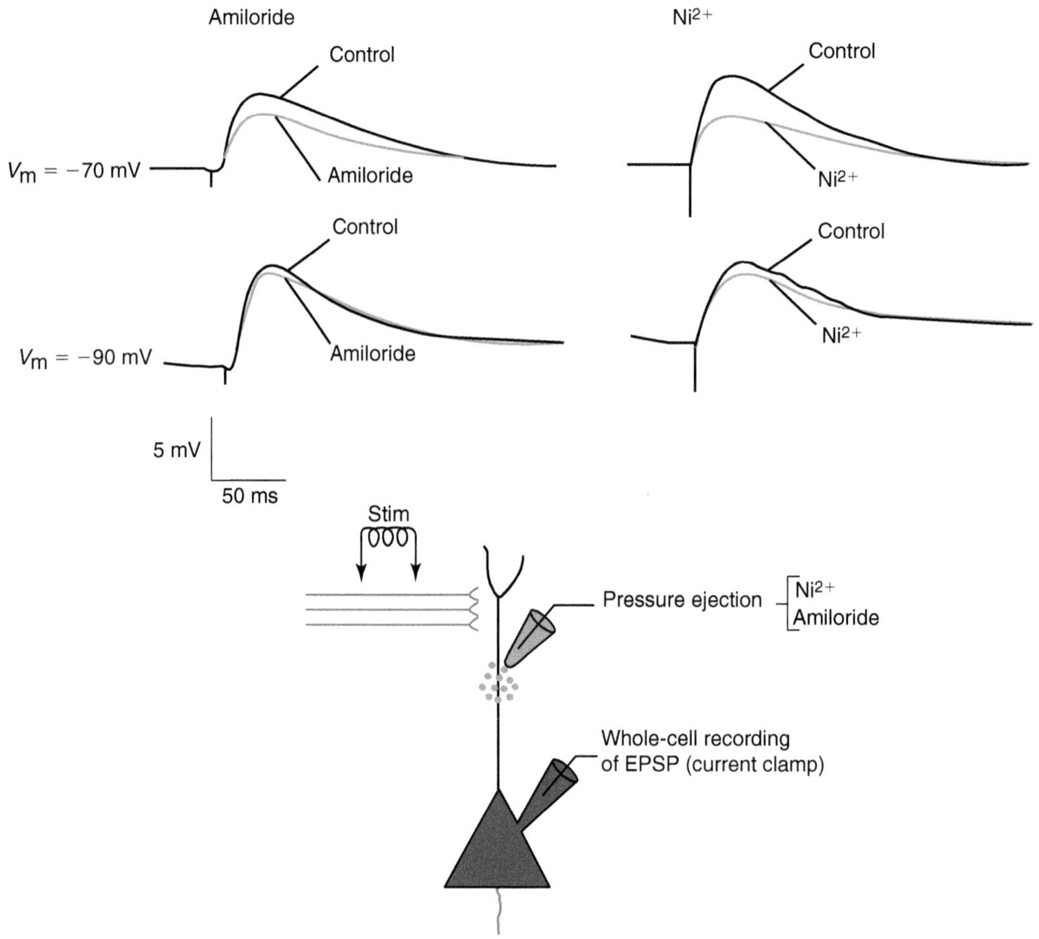

FIGURE 15.8 **EPSPs in hippocampal pyramidal dendrites are amplified by an amiloride- and Ni^{2+}-sensitive Ca^{2+} current.** EPSPs are evoked by stimulation of afferent fibers in the outer stratum radiatum. EPSPs are recorded from the soma (whole-cell configuration, current clamp mode) at two different membrane potentials (−70 and −90 mV) adjusted by current injection through the whole-cell pipette. Superimposed traces of averaged EPSPs ($n = 50$) recorded before and during local dendritic application of amiloride (50 μM, left) or Ni^{2+} (5 μM, right) show that both drugs reduce EPSPs recorded at −70 mV but do not significantly reduce them at −90 mV. Adapted from Gillessen T, Alzheimer C (1997) Amplification of EPSPs by low Ni^{2+} and amiloride-sensitive Ca^{2+} channels in apical dendrites of rat CA1 pyramidal neurons. *J. Neurophysiol.* **77**, 1639–1643, with permission.

deactivating (as seen on tail currents; see the inset of **Figure 15.9d** and Appendix 5.2). Currents begin to activate near −60 mV and steady-state current amplitude increases in an approximately linear manner with membrane hyperpolarization up to −140 mV (**Figure 15.9b**). Inclusion of 5 mM of Cs^+ in the external recording solution totally blocks the current, thus showing that it is an I_h current (**Figure 15.9c**).

The steady-state current amplitude at −130mV progressively increases with distance away from the soma (soma: 8.9 ± 1.6pA, $n + 21$; dendrite 300–350 μm away from the soma: 62.3 ± 8.5pA, $n = 14$) (the mean dendritic length is 500 μm). The mean current can be converted to mean current density (per μm^2) by normalizing to a 5 μm^2 patch area. It is 1.8 ± 0.3 pA μm^{-2} at the soma as compared with a density of 12.5 ± 1.7 pA μm^{-2} recorded from dendrites located 300–350 μm away from the soma. Therefore, the density of I_h increases over sixfold in 350 μm towards distal dendrites. Even with these elevated I_h densities, absolute I_h density is quite small compared with other dendritic channel densities, K^+ channels in particular. In pyramidal cells of the cortex, where recordings have been performed up to 800 μm from the soma, this density increases more than 10 times (**Figure 15.10**).

In conclusion, the ionic selectivity (data not shown), voltage ranges of activation and kinetics of activation (and lack of inactivation), as well as the sensitivity to external Cs^+ all fall within the ranges reported for a wide variety of central and peripheral I_h in neurons, as well as in cardiac cell types. Moreover, a six- to 13-fold increase in I_h density is found across the somato-dendritic axis.

15.3.2 Dendritic H-type cationic channels are activated by IPSPs; in turn, I_h decreases EPSP amplitude

The impact of I_h channels on the shape and propagation of sub-threshold voltage signals is determined by

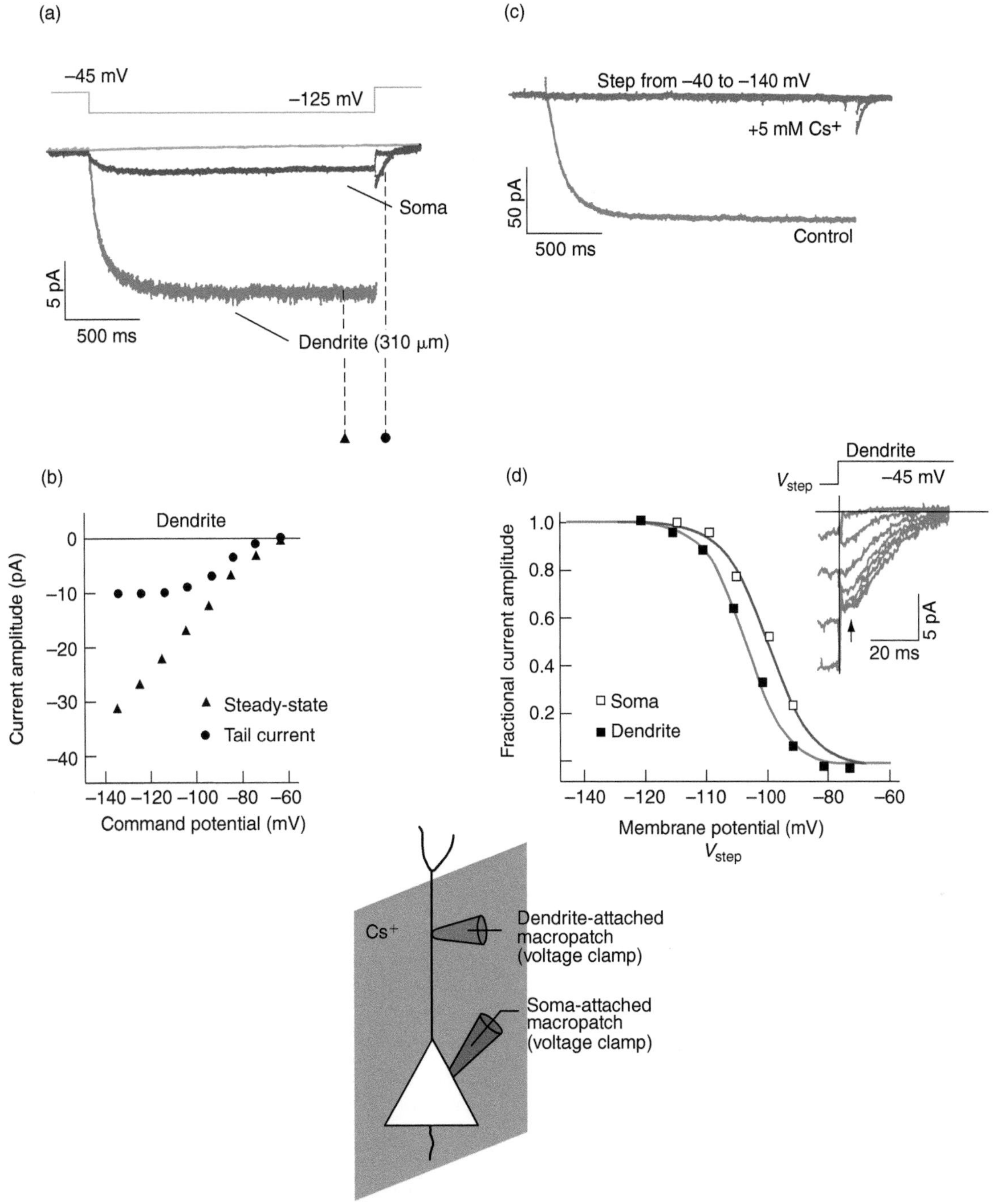

FIGURE 15.9 **Dendritic and somatic hyperpolarization-activated cation current I_h in pyramidal neurons of the hippocampus. (a)** In a dendrite-attached macropatch located in the apical dendrite (310 μm from the soma), hyperpolarizing steps to −125 mV (V_H = −45 mV) evoke inward currents that are larger than those recorded from the soma with similar-sized pipettes. **(b)** *I/V* plots for steady-state inward current measured 900 ms after the start of the step (▲) and for inward tail current measured 5 ms after the end of the step (•). **(c)** Blockade by 5 mM of external Cs^+ of the inward current evoked by a hyperpolarizing step to −140 mV. **(d)** Activation curves generated from the tail currents (inset). The dendritic curve ($V_{1/2}$ = −89 mV) is shifted 6 mV hyperpolarized with respect to the somatic curve ($V_{1/2}$ = −83 mV). Command potentials (V_{step}) are given in 10 mV increments from −65 to −135 mV. Adapted from Magee JC (1998) Dendritic hyperpolarization-activated currents modify the integrative properties of hippocampal CA1 pyramidal neurons. *J. Neurosci.* **18**, 7613–7624, with permission.

using simultaneous whole-cell current clamp recordings from both the soma and dendrites. EPSPs are simulated by dendritic current injection. Under control conditions, current injections in the dendritic compartment result in EPSP-shaped voltage transients, the amplitude and kinetics of which are filtered significantly as they propagate from the dendritic injection site to the recording somatic site (**Figure 15.11a**). When the amplitude of the simulated EPSP is 8.0 ± 0.5 mV at the dendritic recording site, it attenuates to 3.0 ± 0.2 mV at the somatic recording site.

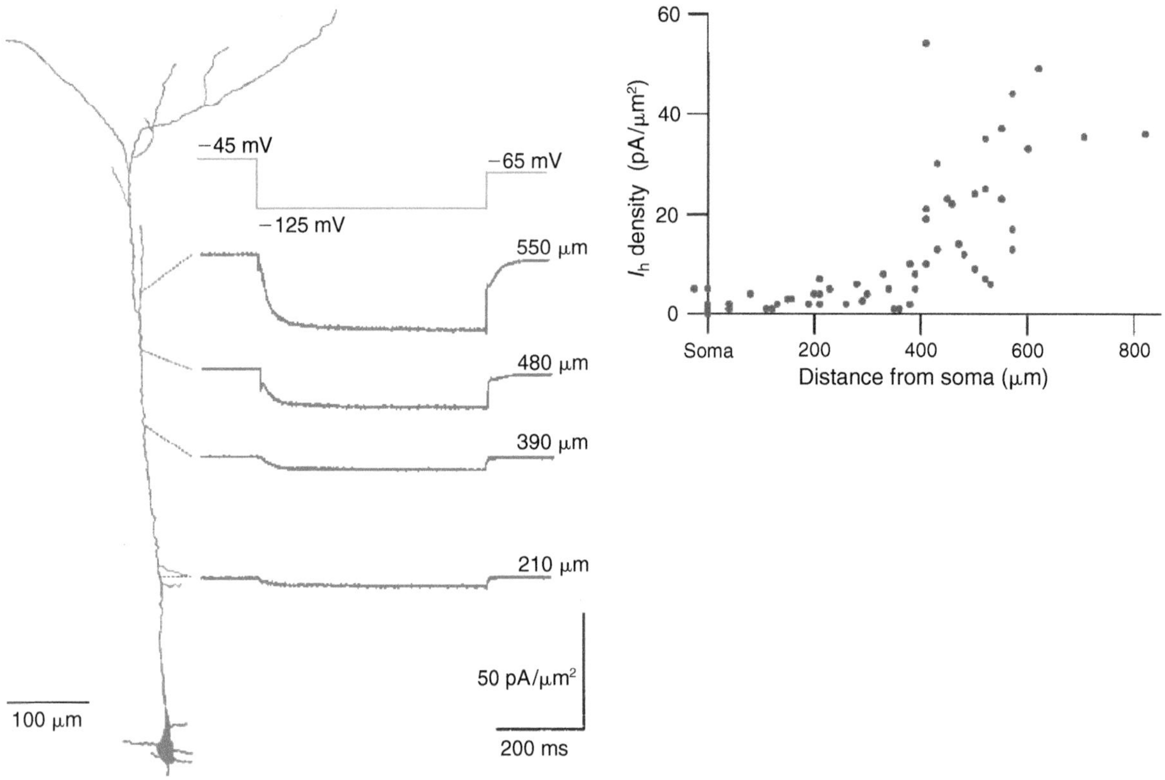

FIGURE 15.10 **I_h is situated on the distal apical dendrite. (a)** Consecutive cell-attached patches along the apical dendrite of a layer V pyramidal neuron at different distances from the soma using a high-K^+ pipette solution. Hyperpolarizing voltage commands to approximately −125 mV resulted in the activation of tiny I_h currents at distances smaller than 400 μm from the soma, while at more distal recording sites a large I_h current flow could be induced. After the cell-attached recordings, the cell was filled with biocytin by going to whole-cell mode and the resting membrane potential measured. **(b)** The I_h current densities from 60 cell-attached recordings were plotted against their distance from the soma. While on the basal dendrites, the soma, and the apical dendrite <400 μm nearly no I_h currents could be found, more distal recordings <820 μm showed a marked non-linear increase in I_h density. Adapted from Berger T, Larkum ME, Luscher HR (2001) High I_h channel density in the distal apical dendrite of layer V pyramidal cells increases bidirectional attenuation of EPSPs. *J. Neurophysiol.* **85**, 855–868, with permission.

When the EPSP duration is 15 ± 0.8 ms at the dendritic site, it extends to 39 ± 2 ms at the somatic site. Repetitive dendritic current injections are also given to mimic repetitive synaptic inputs. These events are filtered similarly by dendritic arborizations (**Figure 15.11b**). When the peak amplitude is 24 ± 3 mV at the dendritic site, it attenuates to 8 ± 1 mV at the somatic site. When the duration is 136 ± 1 ms at the dendritic site, it extends to 154 ± 2 ms at the somatic site. H channel blockade with external Cs^+ increases single EPSP amplitude by 7 ± 4% at the dendritic site and by 10 ± 2% at the somatic site. It also increases EPSP duration by 10 ± 2% at the dendritic site and by 38 ± 9% at the somatic site (**Figure 15.11a**). The increase in amplitude of somatic EPSPs during I_h blockade is mostly the result of the increase of input membrane resistance (due to H channel closure).

For repetitive EPSPs, the presence of external Cs^+ increases the amplitude by 22 ± 5% at the dendritic site and by 42 ± 8% at the somatic site. The associated increase in duration is 3 ± 1% at the dendritic and 7 ± 2% at the somatic sites (**Figure 15.11b**). Therefore, for single EPSPs, the amount of amplitude attenuation occurring between the dendrites and soma is the same in the presence or absence of I_h. In contrast, for repetitive EPSPs, I_h reduces the peak amplitude reached during the train by a factor of 2. How does I_h attenuate EPSPs summation? This is due to the deactivation of I_h during the summation of the dendritic EPSPs because the depolarization produced by the EPSPs reaches a membrane potential where H channels close. The closure of the channel exerts a hyperpolarizing influence and counterbalances the depolarization produced by the EPSPs. This influence will be stronger at distal sites because of the large number of H channels at this level. It is dependent on the frequency of stimulation because of the slow kinetic of H channel deactivation. Thus, the summed EPSPs must have an amplitude and a duration sufficient to deactivate I_h. At very high frequency of stimulation, dendritic I_h has less effect on temporal integration because the summation of the EPSPs overcomes the inhibitory effect of I_h deactivation.

In conclusion, dendritic I_h decreases the amount of current transmitted from the dendrites to the soma in particular for summed EPSPs. Also, since I_h density is six- to 13-fold higher in distal dendrites, the absolute effectiveness

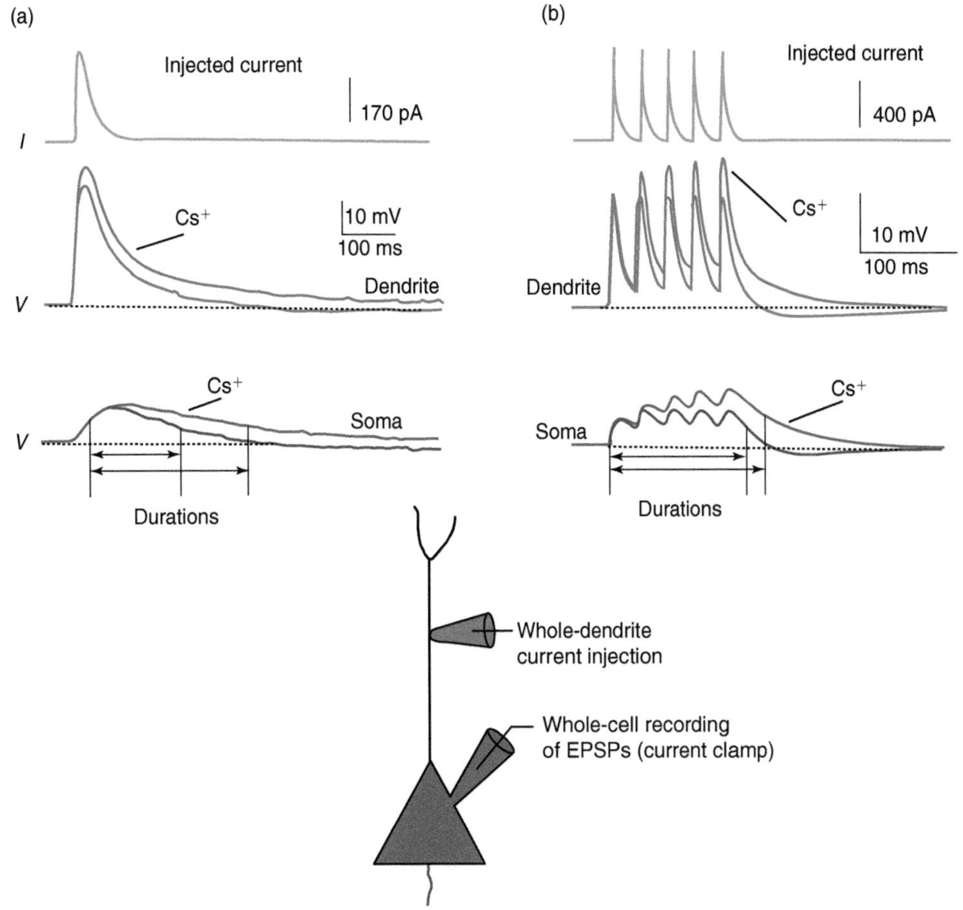

FIGURE 15.11 **EPSP amplitude, duration and summation are all regulated by I_h in hippocampal pyramidal neurons.** EPSPs are generated by injection of an exponentially rising and falling voltage waveform into the current clamp input of the amplifier (dendritic current injections are performed 250 μm away from the soma). Simultaneous whole-cell recordings are performed from the dendrite and the soma of the same pyramidal neuron. **(a)** A single current injection into the dendritic electrode produces an EPSP-shaped transient, the amplitude and duration of which is increased in the presence of 3 mM of external Cs^+. **(b)** Repetitive current injections produce a train of EPSP-shaped voltage transients, the peak amplitude and duration of which are also increased in the presence of 3 mM of external Cs^+. Adapted from Magee JC (1998) Dendritic hyperpolarization-activated currents modify the integrative properties of hippocampal CA1 pyramidal neurons. *J. Neurosci.* **18**, 7613–7624, with permission.

of distal synaptic inputs (i.e. the total charge transferred from synapse to soma) is reduced by the increasingly large I_h conductance. Notably, in pyramidal cortical neurons of layer V, the high I_h channel density in the apical tuft increases the electrotonic distance between this distal dendritic compartment and the somatic compartment in comparison to a passive dendrite. These findings suggest that integrations of synaptic inputs in the apical tuft and in the basal dendrites occur spatially independently.

15.4 A-TYPE K^+ CHANNELS ARE PRESENT IN THE DENDRITES OF HIPPOCAMPAL NEURONS; I_A ATTENUATES EPSPs

The A-type K^+ current is a 4-aminopyridine-sensitive K^+ current that activates below spike threshold. I_A can be recorded after blocking Na^+ channels with TTX and Ca^{2+} channels with Cd^{2+} and Ni^{2+}. The A-type current is a fast activating K^+ current that inactivates rapidly (in a few tens of milliseconds), and therefore belongs to the family of transient K^+ currents.

15.4.1 A-type K^+ channels are expressed in the dendrites of hippocampal pyramidal neurons

The biophysical properties and subcellular distribution of I_A in CA1 pyramidal neurons are investigated in cell-attached configuration, using the voltage clamp mode of the patch-clamp technique. Depolarizing steps of voltage evoke outward currents. Recordings reveal a high density of outward current composed of two distinct components. The first component is a transient component that rapidly activates and rapidly inactivates. The second component is sustained. The density of the transient outward current increases linearly with distance from the soma while the sustained component

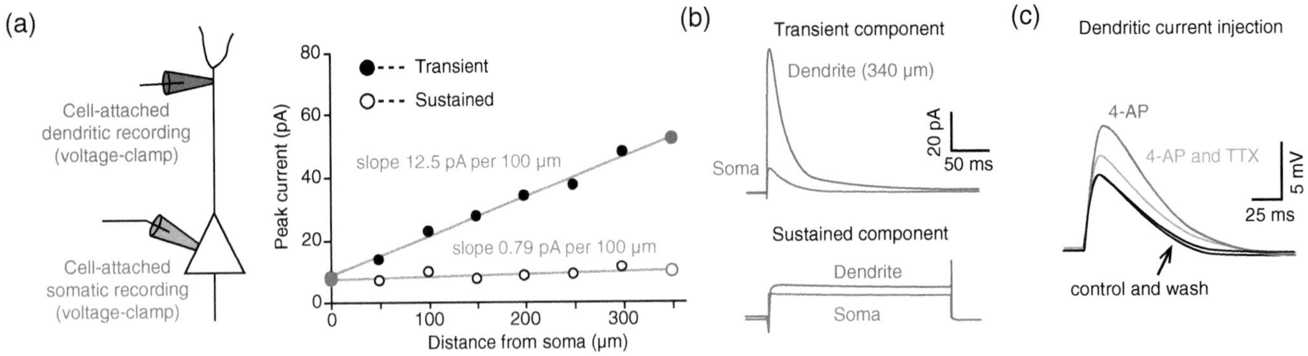

FIGURE 15.12 **I_A is strongly expressed in hippocampal pyramidal dendrites and attenuates EPSPs. (a)** Left, schematic depicting the experiment design: cell-attached voltage-clamp recordings of potassium currents were performed at the somatic (blue electrode) and dendritic (red electrode) levels. Right, scatter plot showing the distribution of amplitudes of the transient and sustained potassium currents in hippocampal pyramidal dendrites as a function of the recording distance from the soma. While the transient component (I_A, filled circles) shows a strong increase in expression between the soma (blue circles) and the distal dendrite (red circles), the sustained component shows a stable amplitude along the entire dendrite. **(b)** Simultaneous dendritic (top traces) and somatic (bottom traces) voltage-clamp recordings illustrating the differences in the distribution of the transient and sustained currents that can be observed in a hippocampal pyramidal neuron. Red traces correspond to distal dendritic recordings and blue traces to somatic recordings (same as in panel **(a)**). **(c)** Effect of the pharmacological blockade of I_A on the integration of simulated dendritic EPSPs. EPSPs recorded in the soma in response to dendritic current injection are amplified when I_A is blocked using 4-AP (red trace). Although this effect involves the TTX-sensitive sodium current, the effect of 4-AP persists in the presence of TTX, demonstrating that I_A blockade on its own has an effect on the dendritic integration of EPSPs. The amplification revealed in 4-AP demonstrates that I_A attenuates EPSPs in control conditions. Adapted from Hoffman DA, Magee JC, Colbert CM, Johnston D (1997) K+ channel regulation of signal propagation in dendrites of hippocampal pyramidal neurons. *Nature* 387, 869–875, with permission.

remains constant along the somato-dendritic axis (**Figure 15.12a**). The peak current density of transient channels increases from an average of 8 pA per patch in the soma to 52 pA in the distal dendrites for a voltage step from −85 to −55 mV. The current density of the sustained component is constant from the soma (8 pA per patch) to distal dendrites (10 pA per patch). Representative traces from the soma and distal dendrite of both the transient and sustained components are shown in **Figure 15.12b**.

15.4.2 A-type K^+ current attenuates sub-threshold EPSPs generated in the dendrites

The high density of dendritic A-type K^+ channels has a substantial effect on sub-threshold synaptic events propagating from the dendrite to the soma. Application of 4-aminopyridine increases the amplitude and duration of EPSP-shaped voltage transients induced by dendritic current injection (**Figure 15.12c**). Application of TTX after A-channel blockade revealed that a large fraction of the 4-aminopyridine-induced increase in EPSP amplitude is due to Na^+-channel activation. Nonetheless, specific blockade of I_A in the presence of TTX demonstrates that dendritic A-type channels are able to attenuate EPSPs in the absence of Na^+-current activation. Thus dendritic A-type channels act to counteract EPSP amplification produced by persistent Na^+ currents.

15.5 FUNCTIONAL CONSEQUENCES

The sub-threshold voltage-gated I_{NaP} and I_{CaT} present in the postsynaptic membrane along the dendro-axonal axis of pyramidal neurons of the neocortex or the hippocampus boost the effects of local EPSPs by acting as voltage amplifiers. This could be one solution to overcome the passive decay of EPSPs en route to the axon. This argument, however, is somewhat somatocentric; i.e. the emphasis is put on how dendrites amplify events so that they are bigger in the soma. An alternative viewpoint is that these channels are more important for dendritic interactions in the immediate vicinity of the synaptic inputs. For example, multiple EPSPs occurring on the same branch and within a narrow time frame should activate voltage-gated channels more strongly than a single EPSP and produce a much bigger response than would occur if EPSPs were on separate branches. However, inhibitory currents such as the A-type K^+ current may partly counteract the EPSP amplification due to Na^+- and Ca^{2+}-channel activation.

Under physiological conditions, an EPSP-evoked $[Ca^{2+}]_i$ transient would occur mainly after the summation of a number of unitary EPSPs and thus would represent the integrated result of dendritic activity at a given moment and at a given location. A number of possible physiological functions for dendritic $[Ca^{2+}]_i$ transients exists. Intracellular Ca^{2+} may activate biochemical pathways, Ca^{2+}-induced Ca^{2+} release as well as Ca^{2+}-activated

K^+ currents present in the dendritic membrane. Such outward current would change the shape of the EPSP. Ca^{2+} is also implicated in postsynaptically induced forms of plasticity such as long-term potentiation or long-term depression (see Chapter 18).

Voltage-gated channels activated by sub-threshold depolarization also influence both spatial and temporal summation of EPSPs. For example, I_h strongly determines summation of EPSPs.

15.6 CONCLUSIONS

We can now answer the four questions asked in the introduction:

- Sub-threshold depolarizing voltage-gated currents are present in the dendritic membrane of some CNS neurons. These are the currents I_h, I_A and I_{CaT}. I_{NaP} is mainly located in the proximal region of the axon.
- They are not distributed uniformly over somato-dendritic membranes. In pyramidal neurons of the neocortex, I_{NaP} seems more efficient at the proximal region of the axon. In contrast, I_{CaT} is primarily located in the dendrites of pyramidal neurons of the hippocampus. Moreover, in pyramidal neurons of the hippocampus and of the layer V of the cortex, density of I_A and I_h increases from the soma to distal dendrites.
- I_{NaP} and I_{CaT} activation boosts EPSP amplitude and duration. Moreover, I_{CaT} activation induces a transient and local increase of intradendritic Ca^{2+} concentration. In contrast, I_A and I_h attenuate EPSP waveform. The case of I_h is peculiar. It is I_h deactivation that generates an outward current which produces a hyperpolarization that shortens the duration of local EPSPs and attenuates their temporal summation.
- Although it has long been thought that most currents involved in synaptic integration are located in the dendrites (such as I_{CaT}, I_h and I_A), the findings on I_{NaP} presented in this section suggest that the proximal region of the axon might also be involved in the integration of synaptic inputs. Therefore, although most of the integration occurs in the dendrites and the soma, all cellular compartments upstream of the axon initial segment (including the proximal region of the axon) seem to play a role in the integration of synaptic inputs.

It seems that although dendritic voltage-gated sub-threshold currents may have a limited impact on the amplitude of unitary synaptic input, they actively shape repetitive synaptic potentials of larger amplitude. Finally, one must keep in mind that the state of voltage-gated channels, closed, open or inactivated, depends on the history of the membrane. If a segment of dendritic membrane has been previously depolarized before synaptic activity, the voltage-gated channels present in this segment of dendritic membrane will be already inactivated and will not play a role. Therefore there is a dynamic aspect in the active properties of dendrites.

Further reading

Astman, N., Gutnick, M.J., Fleidervish, I.A., 2006. Persistent sodium current in layer 5 neocortical neurons is primarily generated in the proximal axon. J. Neurosci. 26, 3465–3473.

Gulledge, A.T., Kampa, B.M., Stuart, G.J., 2005. Synaptic integration in dendritic trees. J. Neurobiol. 64, 75–90.

Hausser, M., Spruston, N., Stuart, G.J., 2000. Diversity and dynamic of dendritic processing. Science 290, 739–744.

Hoffman, D.A., Magee, J.C., Colbert, C.M., Johnston, D., 1997. K+ channel regulation of signal propagation in dendrites of hippocampal pyramidal neurons. Nature 387, 869–875.

Isomura, Y., Fujiwara-Tsukamoto, Y., Imanishi, M., et al., 2002. Distance-dependent Ni^{2+}-sensitivity of synaptic plasticity in apical dendrites of hippocampal CA1 pyramidal cells. J. Neurophysiol. 87, 1169–1174.

Johnston, D., Magee, J.C., Colbert, C.M., Christie, B.R., 1996. Active properties of neuronal dendrites. Annu. Rev. Neurosci. 19, 165–186.

Judkewitz, B., Roth, A., Hausser, M., 2006. Dendritic enlightment: using patterned two-photon uncaging to reveal the secrets of the brain's smallest dendrites. Neuron 50, 180–183.

Lipowsky, R., Gillessen, T., Alzheimer, C., 1996. Dendritic Na^+ channels amplify EPSPs in hippocampal CA1 pyramidal cells. J. Neurophysiol. 76, 2181–2190.

Magee, J.C., 1999. Dendritic I_h normalizes temporal summation in hippocampal CA1 neurons. Nat. Neurosci. 2, 508–514.

Magee, J.C., Cook, E.P., 2000. Somatic EPSP amplitude is independent of synapse location in hippocampal pyramidal neurons. Nat. Neurosci. 3, 895–903.

Markram, H., Sakmann, B., 1994. Calcium transients in dendrites of neocortical neuron evoked by single subthreshold excitatory postsynaptic potentials via low-voltage-activated calcium channels. Proc. Natl Acad. Sci. USA 91, 5207–5211.

Mouginot, D., Bossu, J.L., Gähwiler, B.H., 1997. Low-threshold Ca^{2+} currents in dendritic recordings from Purkinje cells in rat cerebellar slice cultures. J. Neurosci. 17, 160–170.

Oviedo, H., Reyes, A.D., 2002. Boosting of neuronal firing evoked with asynchronous and synchronous inputs to the dendrite. Nat. Neurosci. 5, 261–266.

Schwindt, P.C., Crill, W.E., 1995. Amplification of synaptic current by persistent sodium conductance in apical dendrite of neocortical neurons. J. Neurosci. 74, 2220–2224.

CHAPTER

16

Somato-dendritic processing of postsynaptic potentials III. Role of high-voltage-activated depolarizing currents

Constance Hammond, Jean-Marc Goaillard, Dominique Debanne

OUTLINE

Dendrites of neurons of the mammalian central nervous system (CNS) have long been considered as electrically passive structures which funnel postsynaptic potentials to the soma and axon initial segment, the site of action potential initiation. However, the recording of dendritic action potentials (at first with intracellular electrodes) from dendrites of some neurons of the mammalian central nervous system (**Figure 16.1**) indicated that these dendrites express high-threshold-activated Na^+ or Ca^{2+} channels. This led to the suggestion that, in these neurons, synaptic integration is not solely governed by (passive) cable properties of dendrites. In the previous chapter, we studied the role of subliminal voltage-gated currents in the shaping of postsynaptic potentials. This chapter will examine the roles of high-voltage-activated currents.

Dendritic events have recently come under direct experimental scrutiny with the use of dendritic patch recordings and by the advent of imaging techniques with high spatial and temporal resolution. This permitted the design of experiments to answer the following questions:

- Are high-voltage-activated (HVA) currents present in the dendritic membrane of some CNS neurons?
- Are they distributed uniformly over the soma-dendritic membrane so that electrogenic properties in dendrites are not fundamentally different from that in soma?
- How do the currents affect synaptic potentials and input summation?

This chapter looks at experiments performed on four types of central neurons, on pyramidal neurons of the neocortex or hippocampus, dopaminergic neurons of substantia nigra pars compacta, and Purkinje cells of the cerebellar cortex.

Cellular and Molecular Neurophysiology. DOI: 10.1016/B978-0-12-397032-9.00016-9

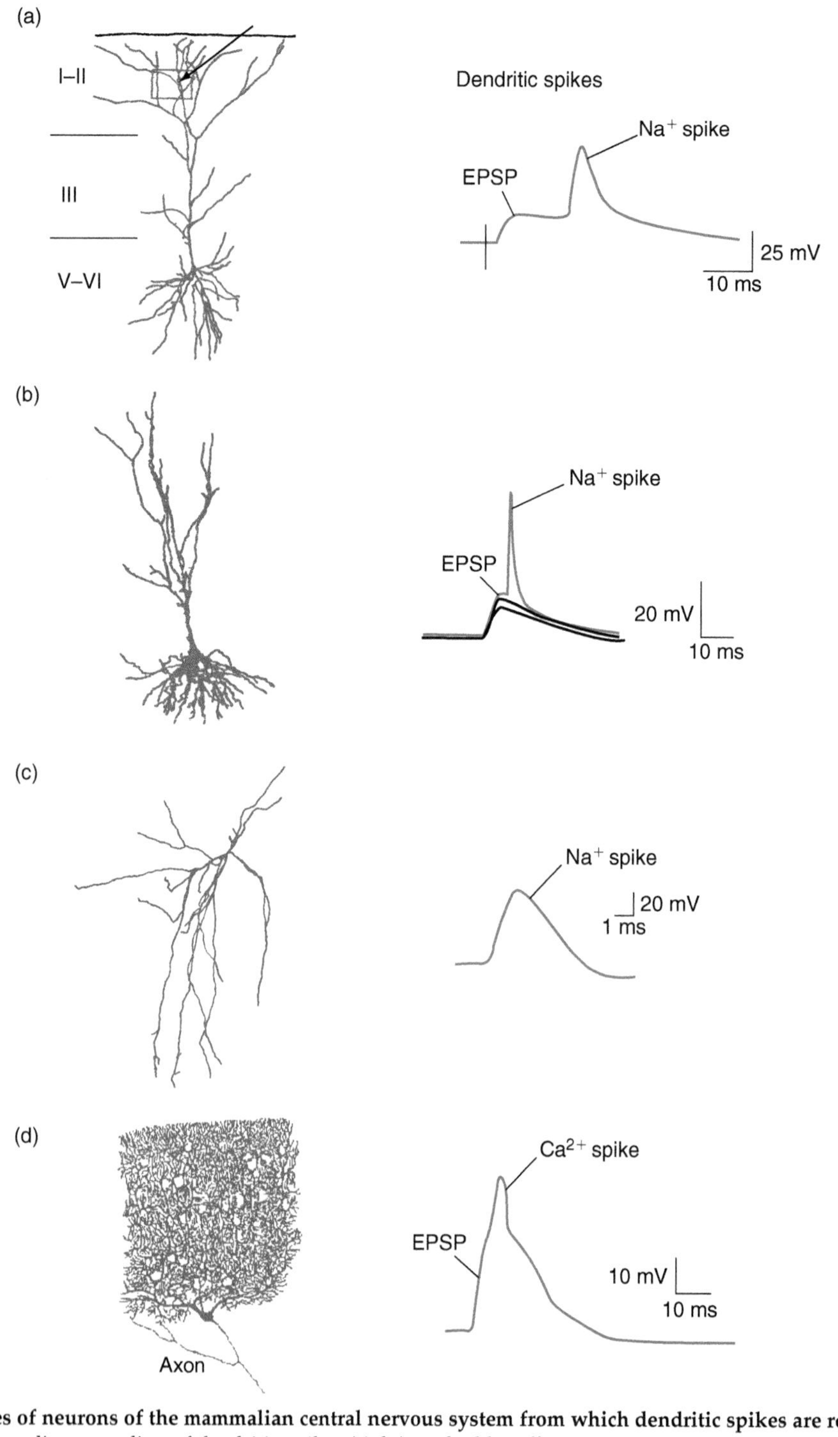

FIGURE 16.1 **Examples of neurons of the mammalian central nervous system from which dendritic spikes are recorded.** Drawing of neurons (left) with the corresponding recording of dendritic spikes (right) evoked by afferent stimulation. **(a)** Pyramidal neuron of the neocortex. **(b)** Pyramidal neuron of the hippocampus. **(c)** Dopaminergic neurons of the substantia nigra. **(d)** Purkinje cell of the cerebellar cortex. Part (**a**) adapted from Seamans JK, Gorelova N, Yang CR (1997) Contribution of voltage-gated Ca^{2+} channels in the proximal versus distal dendrites to synaptic integration in prefrontal cortical neurons. *J. Neurosci.* **17**, 5936–5948, with permission. Part (**b**) drawing by Taras Pankevitch and Roustem Khazipov and adapted from Tsubokawa H, Ross WN (1996) IPSPs modulate spike backpropagation and associated $[Ca^{2+}]_i$ changes in dendrites of hippocampal CA1 pyramidal neurons. *J. Neurophysiol.* **76**, 2896–2906, with permission. Part (**c**) drawing by Jérôme Yelnik and adapted from Häusser M, Stuart G, Racca C, Sakmann B (1995) Axonal initiation and active dendrite propagation of action potentials in substantia nigra neurons. *Neuron* **15**, 637–647, with permission. Part (**d**) adapted from Callaway JC, Lasser-Ross N, Ross WN (1995) IPSPs strongly inhibit climbing fiber-activated $[Ca^{2+}]_i$ increases in the dendrites of cerebellar Purkinje neurons. *J. Neurosci.* **15**, 2777–2787, with permission.

16.1 HIGH-VOLTAGE-ACTIVATED Na^+ AND/OR Ca^{2+} CHANNELS ARE PRESENT IN THE DENDRITIC MEMBRANE OF SOME CNS NEURONS, BUT ARE THEY DISTRIBUTED WITH COMPARABLE DENSITIES IN SOMA AND DENDRITES?

One way to answer the above questions is first to identify the activity of HVA channels in patches of dendritic membrane and then to compare the amplitude of the HVA current recorded from similar-sized patches of dendritic and somatic membranes. Experiments are performed in brain slices and recordings are either in the outside-out or dendrite-attached configuration. For technical reasons, this type of patch recording is limited to dendritic branches with a diameter greater than 1 μm. In order to record Na^+ channels, the solution bathing the extracellular face of the patch contains NaCl; and in order to strongly reduce outward K^+ currents, tetraethylammonium chloride (TEACl) is added in the external solution or CsCl in the internal solution. Na^+ channel activity is identified by inward current polarity, voltage-dependent channel gating, unitary current amplitude, its blockade by tetrodotoxin (TTX) and the lack of effect of Cd^{2+}.

16.1.1 High-voltage-activated Na^+ channels are present in some dendrites

Pyramidal neurons of the hippocampus

In every dendrite-attached patch, Na^+-channel activity is consistently found and more than a single channel is always recorded (**Figure 16.2a**). Na^+ channels are opened by depolarizations of about 15 mV from rest. The i_{Na}/V relationship shows that Na^+ channels have a unitary conductance γ_{Na} of 15 pS and a unitary current that reverses at E_{rev} = +54 mV (**Figure 16.2b**). This value is close to the calculated Nernst equilibrium potential assuming an intracellular Na^+ concentration of 10 mM (extracellular concentration is 110 mM). These data, together with the activation–inactivation characteristics of dendritic I_{Na} (**Figure 16.2c**), reveal that HVA Na^+ channels present in the apical dendrites of pyramidal neurons have basic characteristics similar to HVA Na^+ channels recorded in many soma (compare with **Figure 4.8**) with a difference concerning the inactivation properties (slow inactivation and slow recovery from inactivation for dendritic channels). This explains the decrease of amplitude of repetitive dendritic action potentials.

Pyramidal neurons of the neocortex

Outside-out macropatches of dendritic membrane are excised at different distances from the soma, up to 500 μm. In response to step depolarizations applied through the recording electrode, an inward current that rapidly inactivates and totally disappears in the presence of TTX in the bath is recorded (**Figure 16.3a**). This is observed whether patches are excised from proximal or more distal dendrites. Moreover, this TTX-sensitive Na^+ current has a similar amplitude to patches taken from dendritic or somatic membranes, thus suggesting a similar somatic and dendritic density of Na^+ channels in both membranes.

Dopaminergic neurons of the pars compacta of the substantia nigra

Outside-out patches are excised from somatic or dendritic membranes. In an attempt to maintain constant patch membrane area, all recordings are made with pipettes of similar size. Multichannel TTX-sensitive Na^+ currents are recorded from both patches. The average peak Na^+ current in somatic patches is 5.0 ± 1.3 pA and that in dendritic patches is 3.6 ± 1.6 pA. Again, in these neurons, there is a similar density of Na^+ channels in dendritic and somatic membranes.

Purkinje cells of the cerebellar cortex

The situation in these cells is fundamentally different. As suggested by the absence of large-amplitude Na^+ action potentials in dendrites of Purkinje cells, there is an extremely low-amplitude, TTX-sensitive Na^+ current in outside-out macropatches excised from dendrites (1.9 ± 0.4 pA) compared with that in patches excised from the soma (12.4 ± 1.5 pA) (**Figure 16.3b**). In fact, Na^+-channel density steeply declines in dendrites with distance from the soma. These results were confirmed by experiments in Purkinje cells loaded with the Na^+ indicator SBFI (sodium benzofuram isophthalate). During Na^+ spikes, changes in the intracellular Na^+ concentration were detected only in soma and not in dendrites. In these cells, Na^+ channels are distributed non-uniformly over the somatic and dendritic membrane.

Conclusions

In pyramidal neurons of the neocortex or the hippocampus and in dopaminergic neurons of the substantia nigra, there is a similar density of TTX-sensitive Na^+ channels in somatic and dendritic membranes up to several hundreds of μm from the soma. In contrast, there is a low density of Na^+ channels in the dendritic membrane of Purkinje cells. It must be pointed out that most CNS dendrites contain a low density of HVA Na^+ channels. Pyramidal neurons of the neocortex or the hippocampus and substantia nigra neurons are exceptions.

16.1.2 Dendritic Na^+ channels are opened by backpropagating Na^+ action potentials

The question concerning the role of dendritic Na^+ channels in dendritic Na^+ action potentials is the following: do dendritic Na^+ channels allow the initiation of Na^+ action potentials in dendrites in response to synaptic activity (**Figure 16.4a**), or do dendritic Na^+ channels only

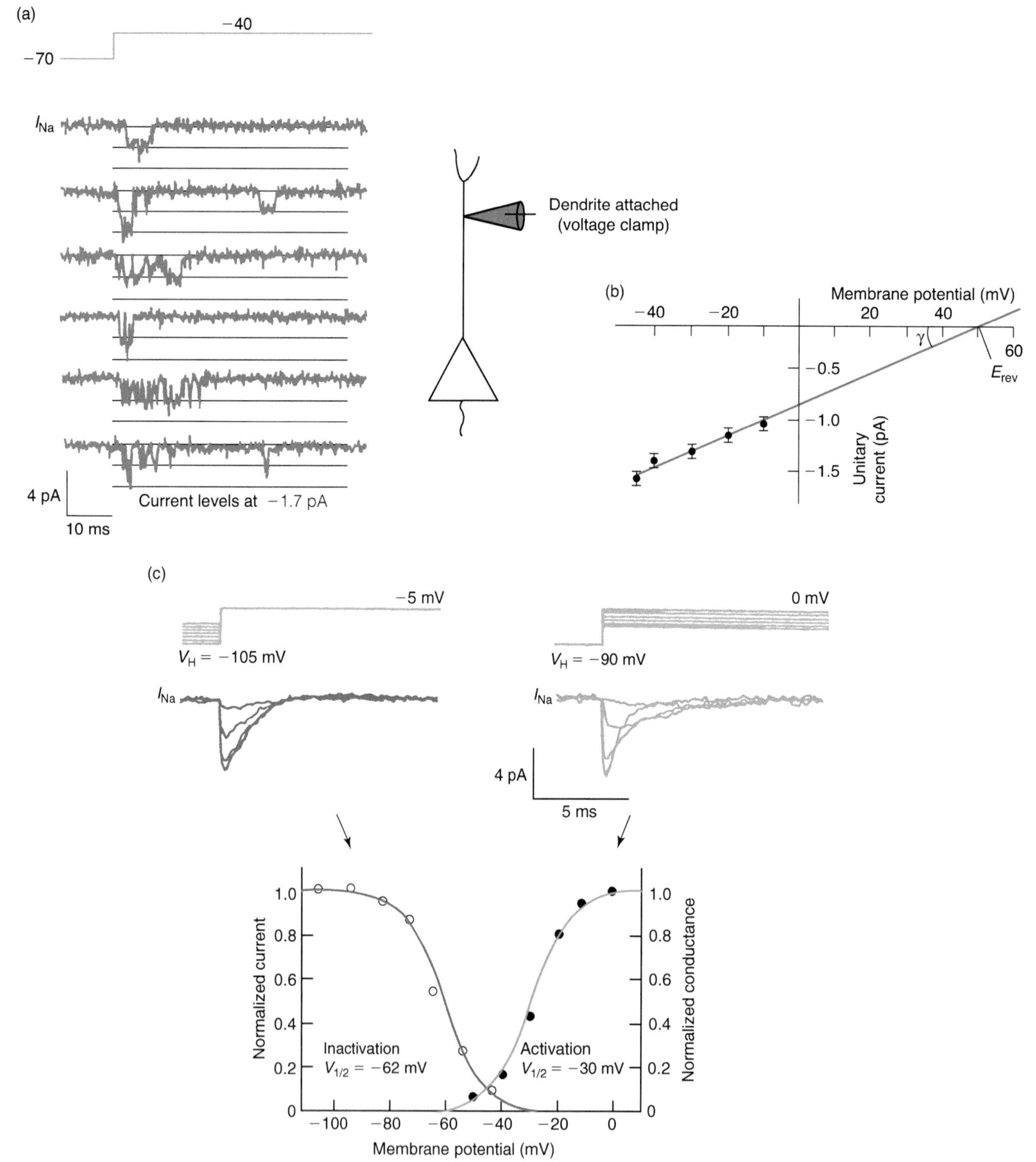

FIGURE 16.2 **Characteristics of dendritic Na^+ channels in a pyramidal neuron of the hippocampus. (a)** Consecutive sweeps showing Na^+ channel openings (dendrite-attached configuration, voltage clamp mode) in response to step depolarizations to −40 mV (V_H = −70 mV). Most of the channel openings occur at the beginning of the step but there are some late reopenings. **(b)** Current–voltage plot of Na^+ channel activity. Unitary current amplitude from a total of 27 patches is plotted as a function of membrane potential. Bars are standard error of the mean (SEM). The slope indicates a unitary conductance γ of 15 pS and the extrapolated reversal potential E_{rev} is +54 mV. **(c)** Dendritic Na^+ channel steady-state activation and inactivation characteristics. Activation is tested by applying depolarizing steps to −65 to 0 mV from V_H = −90 mV. Inactivation is tested by applying a depolarizing step to −5 mV from a V_H varying from −105 to −45 mV. The representative steady-state activation (black circle) and inactivation (open circle) plots for dendritic Na^+ channels indicate that they are half-activated at $V_{1/2}$ = −30 mV and half-inactivated at $V_{1/2}$ = −62 mV. Adapted from Magee JC, Johnston D (1995) Characterization of single voltage-gated Na^+ and Ca^{2+} channels in apical dendrites of rat CA1 pyramidal neurons. *J. Physiol. (Lond.)* **487**, 67–90, with permission.

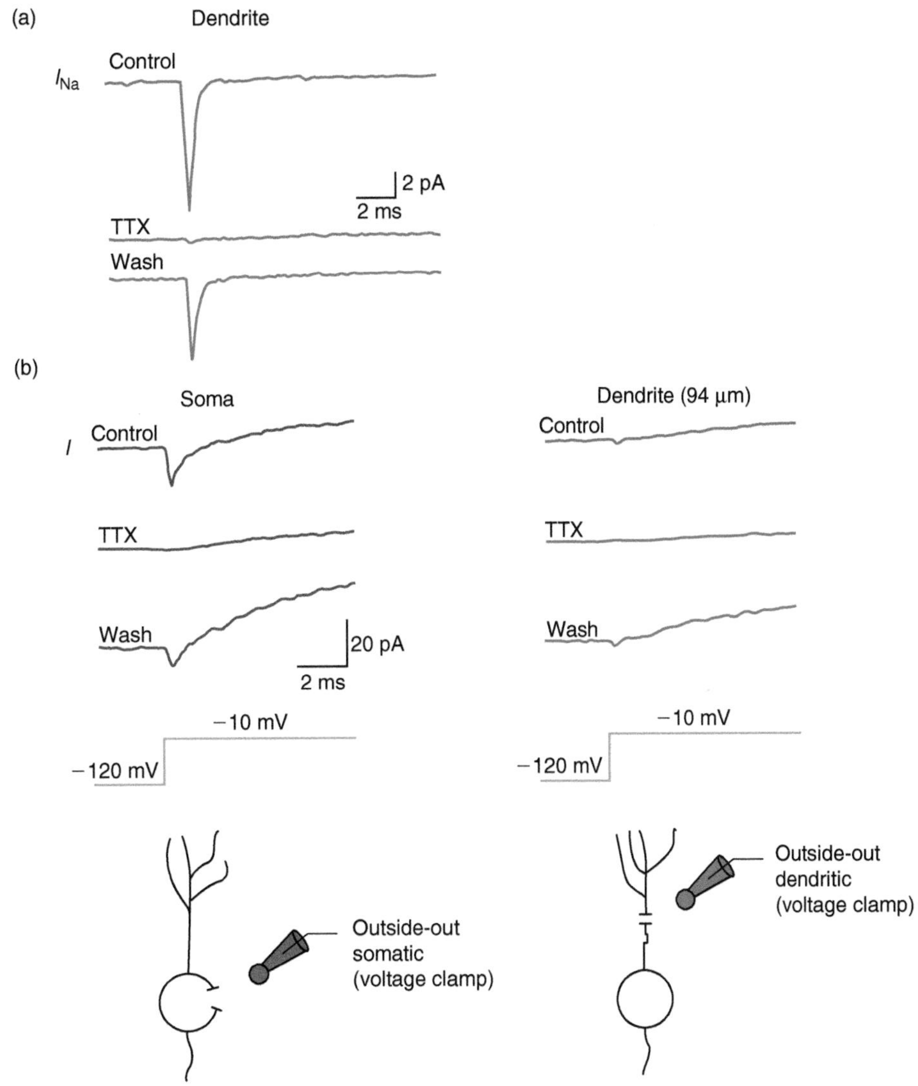

FIGURE 16.3 **TTX-sensitive inward currents in dendrites of pyramidal neurons of the neocortex and Purkinje cells of the cerebellar cortex. (a)** Neocortex. Rapidly inactivating inward current evoked by a depolarizing step to −10 mV (V_H = −90 mV) in an outside-out dendritic macropatch excised from the apical dendrite of a layer V pyramidal neuron (439 μm from the soma) (control). This current is reversibly blocked in the presence of 500 nM of TTX in the external solution. **(b)** Purkinje cells. Voltage-activated currents evoked by a depolarizing step to −10 mV (V_H = −120 mV) in outside-out macropatches excised from either the soma (left) or dendrite (right, 94 μm from the soma) of Purkinje cells using similar-sized patch pipettes. A rapidly inactivating inward current followed by an outward current that is more prominent in the somatic membrane are recorded (control). Rapidly inactivating currents in both somatic and dendritic patches are reversibly blocked by the presence of 500 nM of TTX in the extracellular medium (TTX). Part (**a**) adapted from Stuart G, Sakmann B (1994) Active propagation of somatic action potentials into neocortical pyramidal cell dendrites. *Nature* **367**, 69–72, with permission. Part (**b**) adapted from Stuart GJ, Häusser M (1994) Initiation and spread of sodium action potentials in cerebellar Purkinje cells. *Neuron* **13**, 703–712, with permission.

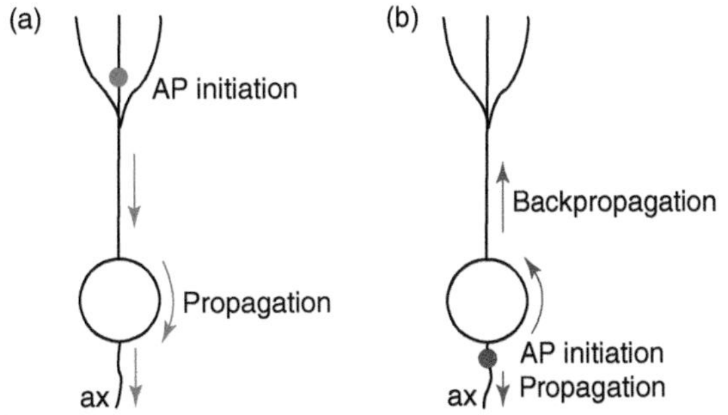

FIGURE 16.4 **Schematic drawings of two hypotheses concerning the site of dendritic Na⁺ spike initiation. (a)** In response to dendritic EPSPs, Na⁺ action potential (AP) is locally initiated in dendrites (black point) and then actively propagates to the soma-initial segment and along the axon. **(b)** In response to dendritic EPSPs, Na⁺ action potential is first initiated at the soma-initial segment (black point) and then actively propagates along the axon and backpropagates (actively or passively) into the dendritic tree.

allow active backpropagation of Na^+ action potentials first initiated in the axon hillock region (**Figure 16.4b**)? To further explain the latter hypothesis it must be assumed that, in general, Na^+ action potentials, once they have been initiated, actively (i.e. in a regenerative manner) propagate along the axon (orthodromic propagation; see Section 4.4) and, at the same time, passively propagate into the soma and dendritic tree (passive backpropagation). Therefore, the question is not 'do Na^+ action potentials backpropagate in the dendritic tree?' but rather 'do they backpropagate *actively* in the dendritic tree of neurons that contain HVA Na^+ channels in their dendrites?'

Pyramidal neurons of the neocortex

Simultaneous whole-cell recordings (current clamp mode) are made from the soma and apical dendrite of the same pyramidal neuron in slices *in vitro*. To confirm that the recorded dendrite and soma belong to the same neuron, the cell is simultaneously filled from the somatic and the dendritic recording sites with different colored fluorescent dyes present in the recording pipettes. In response to suprathreshold synaptic stimulation, action potential initiation occurs first at or near the soma (**Figure 16.5a**). Simultaneous recordings obtained from the soma and axon hillock of the same cell further show that initiation occurs first in the axon, possibly as a result of differences in geometry between soma and axon, as well as possible differences in the density, distribution and properties of voltage-activated Na^+ channels in these structures.

Action potentials have also been observed to be generated in distal dendrites of neocortical pyramidal neurons in response to stimulation of afferents. However, these Na^+ spikes attenuate as they spread to the soma and axon. As a consequence, Na^+ action potentials are always initiated in the axon before the soma even when synaptic activation is intense enough to initiate dendritic regenerative potentials. Na^+ action potentials then actively propagate along the axon. Do they also actively backpropagate in the dendritic tree (**Figure 16.4b**)?

To investigate whether voltage-activated Na^+ channels aid the backpropagation of somatic action potentials into the dendrites, the internal Na^+ channel blocker QX-314 is included in the dendritic patch pipette during simultaneous somatic and dendritic recording. Following establishment of the dendritic patch, dendritic action potentials are observed to decrease progressively in amplitude before any change is observed in the amplitude or time course of the somatic action potential. This suggests that dendritic Na^+ channels boost the amplitude of dendritic action potentials as they backpropagate into the dendritic tree.

To compare the expected attenuation of dendritic action potentials in the presence and absence of Na^+ dendritic channels, TTX is applied in the bath. Since action potentials can no longer be evoked in this condition by current injection in the soma (Na^+ channels are blocked), a voltage command simulating an action potential is applied at the soma. The amount of attenuation is compared with that of action potentials evoked in the soma in the absence of TTX in the extracellular medium (**Figure 16.5b**, 1,2). On average, from these experiments, the amplitude of evoked dendritic action potentials is 70% of that of somatic action potentials, whereas in the presence of TTX it represents only 30% (**Figure 16.5b**, 2,3). These results show unequivocally that there is a regenerative (active) backpropagation of somatic action potentials in the dendrites of layer-V pyramidal neurons via the activation of TTX-sensitive dendritic Na^+ channels.

Dopaminergic neurons of the substantia nigra pars compacta

The axon of substantia nigra dopaminergic neurons emerges from the soma (25%) or a dendrite (75%). Axon-soma separations range from 0 to 250 μm. Axons emerge from the largest diameter dendrite since the diameter of axon-bearing dendrites is approximately twice as thick as that of non-axon-bearing dendrites. To determine the site of axon potential initiation (dendritic or axonal), simultaneous whole-cell recordings (current clamp mode) are made under visual control from the soma and dendrite of nigral dopaminergic neurons in slices. In many nigral dopaminergic cells, action potential is observed to occur first at the dendritic recording site (**Figure 16.6a**) and, in some cases, it is observed to occur first at the soma. To visualize the neuron recorded, the somatic pipette is filled with a biocytin-containing solution. Morphological examination of biocytin-filled neurons shows that in every case where the action potential is observed to occur first in the dendrite, the axon of the neuron is found to emerge from the dendrite from which the recording had been made (**Figure 16.6a**). When the action potential is observed to occur first in the soma, the axon is found to originate either from the soma or from a dendrite other than the one from which the dendritic recording had been made. Finally, in cases where the action potential appears to be simultaneous at the somatic and dendritic recording site, the axon is found to emerge from the dendrite in between the two recording pipettes. These findings indicate that the site of action potential initiation is always the axon hillock.

To determine whether Na^+ dendritic channels support the regenerative backpropagation of Na^+-dependent action potentials, the amplitude of action potentials evoked in control extracellular solution is compared with that of a voltage waveform (simulated action potential) injected in the soma in the presence of TTX in the bath. In all such experiments, the attenuation of a simulated action potential waveform injected in the soma in the presence of TTX is greater than that of the action potential evoked by somatic current injection in the absence of TTX (**Figure 16.6b**). These results suggest that dendritic Na^+ channels support the *active* backpropagation of Na^+

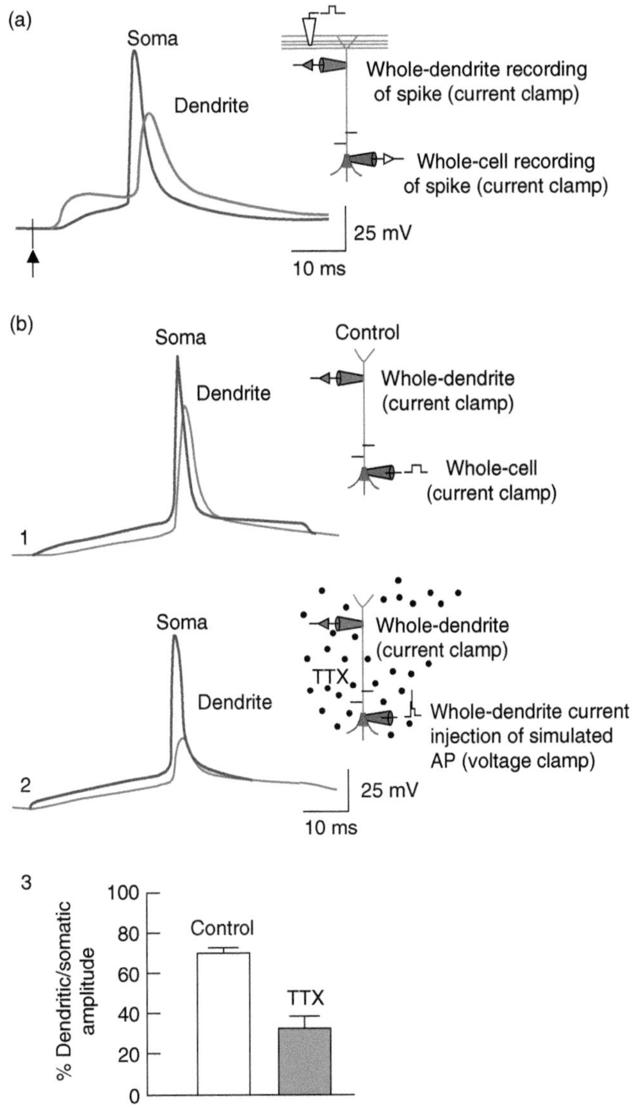

FIGURE 16.5 **Site of initiation of Na^+ action potential and its active backpropagation into the dendritic tree of pyramidal neurons of the neocortex.** **(a)** Na^+ action potential evoked by distal synaptic stimulation in layer I and simultaneously recorded from the soma and a dendrite (dendritic recording is 525 μm from the soma). **(b)** Comparison of active and passive propagation of Na^+ action potential in the apical dendrite studied with simultaneous somatic and dendritic recordings. **(b_1)** An action potential is evoked in the soma by a depolarizing current pulse (200 pA, soma). It propagates in the apical dendrite where it is recorded (dendrite, 310 μm from the soma). **(b_2)** A simulated action potential waveform is injected in the soma in the presence of 1 μM of TTX. The somatic voltage response (soma) propagates passively in the dendrites where it is recorded at the same location as in b_1 but in the presence of TTX (dendrite). The simulated somatic action potential is recorded later at the soma with a second somatic recording pipette (soma). **(b_3)** Histogram of the average amplitude of dendritic action potentials recorded as in b_1 (open column) and of dendritic responses recorded as in b_2 (black column). Data are expressed as a percentage of the response recorded at the soma ± SEM; dendritic recordings 165–470 μm from the soma. Adapted from Stuart G, Sakmann B (1994) Active propagation of somatic action potentials into neocortical pyramidal cell dendrites. *Nature* **367**, 69–72, with permission.

action potentials in the dendritic tree of nigral dopaminergic neurons.

Another type of experiment gave similar results. Glutamate is pressure applied at a dendritic site to mimic synaptic inputs. TTX is applied either at the site of dendritic glutamate application or in the region of the axon hillock and soma (**Figure 16.7**). The activity of nigral dopaminergic neurons is recorded in whole-cell configuration. Application of TTX to the site of dendritic excitation does not alter the frequency, latency or threshold of glutamate-evoked burst firing (**Figure 16.7a**). The application of vehicle alone to the site of dendritic excitation has no effect on glutamate-evoked activity (data not shown). In contrast, application of TTX to the axo-somatic region dramatically reduces burst firing (**Figure 16.7b**). Taken together these data suggest that glutamate-evoked burst firing is initiated in the axon initial segment in reponse to dendritic activation of nigral dopaminergic neurons.

The emergence of the axon from a dendrite rather than from the soma may have interesting consequences. It reverses the normal direction of propagation of the action potential in that the action potential will travel from the

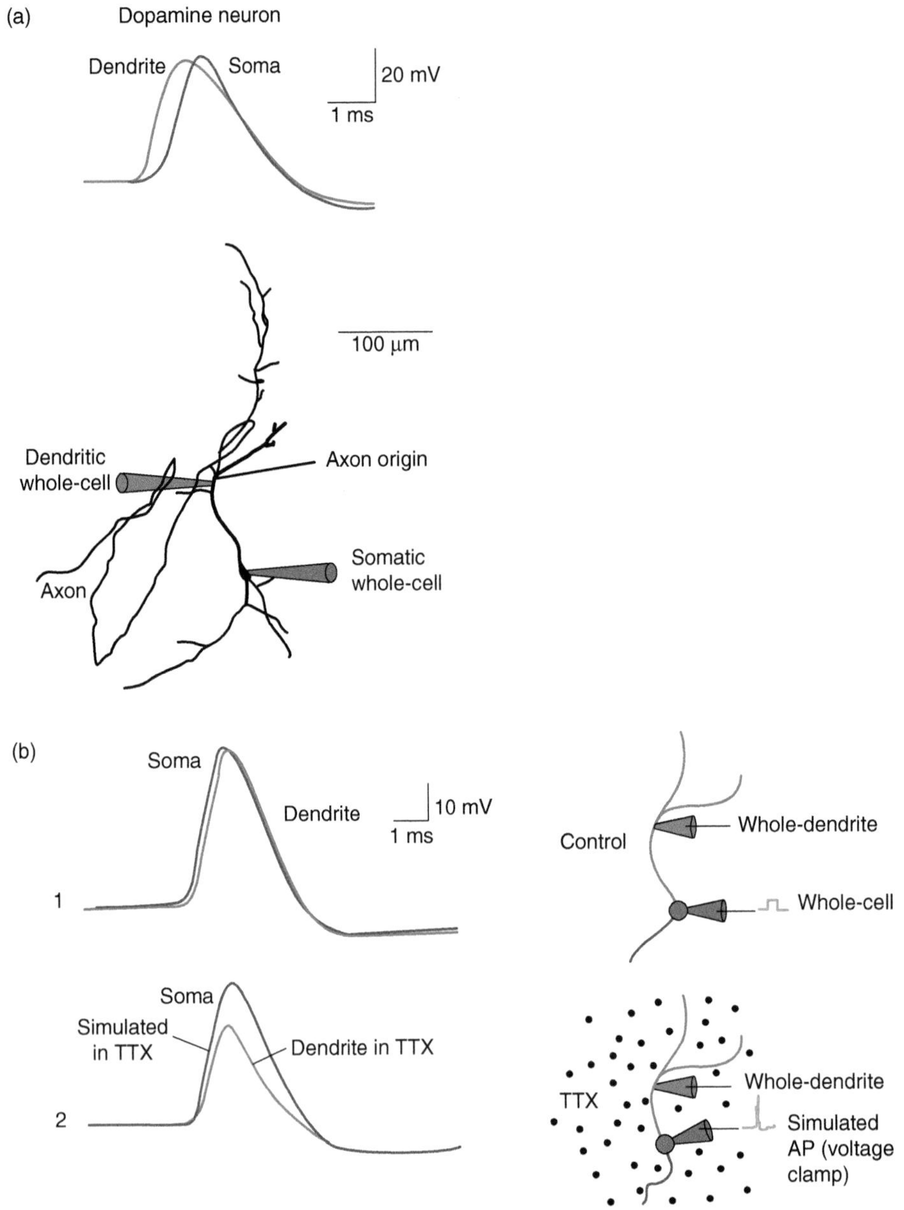

FIGURE 16.6 **Site of initiation of Na$^+$ action potential and its active backpropagation in the dendritic tree of dopaminergic neurons of the substantia nigra. (a)** Spontaneous Na$^+$ action potential recorded simultaneously at the soma and dendrite (top) and the morphological reconstruction of the filled recorded neuron (below) with the location of the somatic and dendritic pipettes. The axon origin is indicated. The action potential is observed to occur first at the dendritic recording site, 195 μm from the soma; the axon of this cell emerges from the dendrite from which the dendritic recording is made (215 μm from the soma). **(b) (b_1)** An action potential is evoked in the soma by a depolarizing current pulse (200 pA, soma). It propagates in the dendrite where it is recorded (dendrite, 100 μm from the soma). **(b_2)** A simulated action potential waveform is injected in the soma in the presence of 1 μM of TTX. The somatic voltage response (soma) propagates passively in the dendrites where it is recorded at the same location as in $\mathbf{b_1}$ but in the presence of TTX (dendrite). The simulated somatic action potential is recorded later at the soma with a second somatic recording pipette (soma). Adapted from Häusser M, Stuart G, Eacca C, Sakmann B (1995) Axonal initiation and active dendritic propagation of action potentials in substantia nigra neurons. *Neuron* **15**, 637–647, with permission.

dendritic tree toward the soma. Consequently, the dendrite bearing the axon will experience the action potential before it spreads into the soma and other dendrites. In these neurons, the final site of integration prior to the axon will not be at the soma but rather in the dendrites at the point where the axon emerges. This suggests that synapses made on the axon-bearing dendrite will be in an electrotonically privileged position (the concept of a 'privileged dendrite').

Dendritic Na_v channels underlie the active backpropagation of action potentials, which leads to the activation of 'high voltage-activated' dendritic Ca^{2+} channels, increase of intradendritic Ca^{2+} concentration and exocytosis of dendritic dopamine-containing vesicles. Dopamine in turn negatively modulates dopaminergic neuron activity via D2 dopamine receptor (GIRK). This is critical for the negative feedback control of firing in

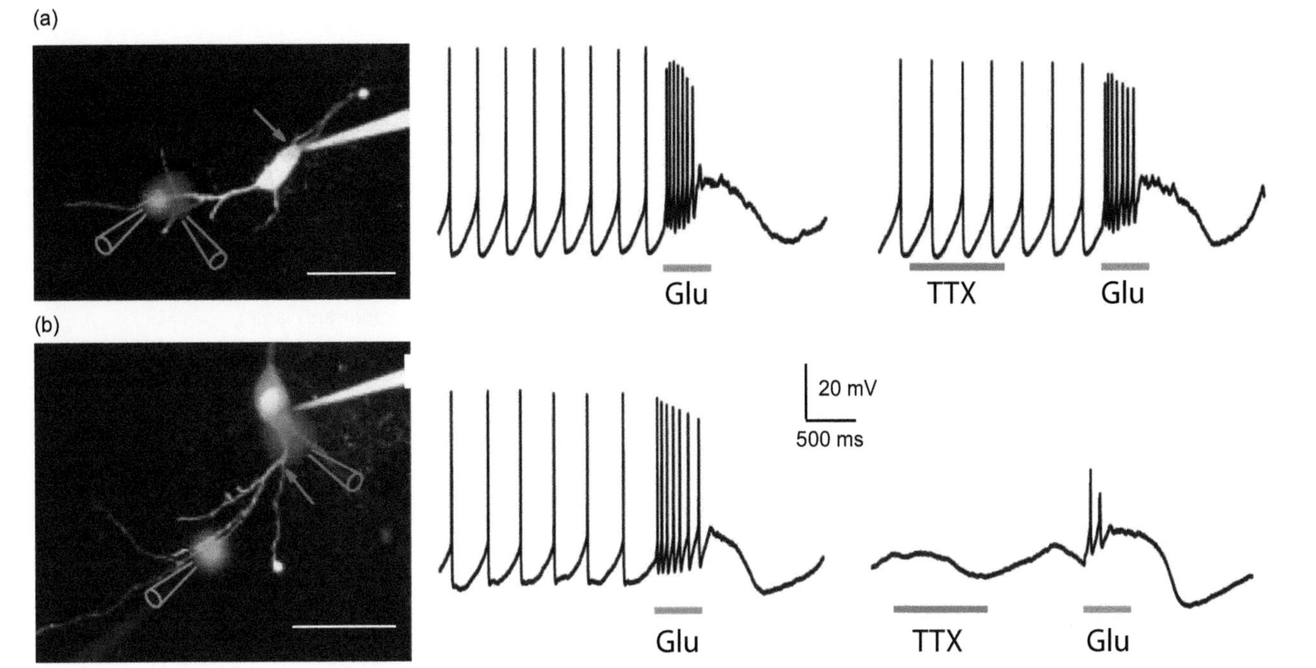

FIGURE 16.7 **Initiation of action potential bursts is impaired by blockade of axosomatic Nav channels but not dendritic Nav channels.** Schematic drawings in **(a)** and **(b)** show the site of pressure application of glutamate at dendritic sites (green pipette) and that of TTX (red pipette) next to the site of dendritic glutamate application in **(a)** or close to the axon hillock (red arrow) in **(b)**. The somatic recording pipette is white. Scale bar: 50 μm. **(a)** Application of TTX (red) 1 s prior to the dendritic pulsed application of glutamate (green) has little effect (right) on evoked activity compared to that evoked under control conditions (left). **(b)** In contrast, axo-somatic application of TTX reduces the intensity of burst firing evoked by dendritic application of glutamate (right) compared to that evoked under control conditions (left). TTX is applied 1 s prior to glutamate application. From Blythe SN, Wokosin D, Atherton JF, Bevan MD (2009) Cellular mechanisms underlying burst firing in substantia nigra dopamine neurons. *J. Neurosci.* **29**, 15531–15541, with permission.

nigral dopaminergic neurons. Clearly, this homeostatic mechanism would be compromised if dopamine were released by nigral dendrites in response to dendritic synaptic excitation instead of axonal activity.

Purkinje cells

In these cells, Na^+ action potentials are initiated in the axon and decrease markedly with increasing distance from the soma, as shown with simultaneous somatic and dendritic recordings. On average, the amplitude of somatic Na^+ action potentials is 78 mV whereas that of dendritic Na^+ action potentials is only a few millivolts at distances greater than 100 μm from the soma (**Figure 16.8a**). This strongly suggests that, in these neurons, Na^+ action potentials spread *passively* into the dendritic tree. In fact, in the presence of TTX, a simulated somatic action potential waveform attenuates in a similar manner as the synaptically evoked action potential (**Figure 16.8b**). This represents a striking contrast with neocortical layer-V pyramidal cells or nigral dopaminergic neurons in which somatic Na^+ action potentials *actively* backpropagate into the dendrites. This marked attenuation of Na^+ action potentials is consistent with the observed low Na^+ current density in the dendrites of Purkinje cells compared with that found in the soma.

Conclusions

When TTX-sensitive voltage-gated Na^+ channels are present in high density in the dendritic membrane, they allow *active* backpropagation of Na^+ action potentials in the dendritic tree. This is, for example, the case with dendrites of pyramidal neurons of the neocortex and hippocampus and of dopaminergic neurons of the substantia nigra. In contrast, the active backpropagation of Na^+ action potentials does not exist in dendrites that have in their membrane a low density of Na^+ channels, like Purkinje cells of the cerebellum. In this latter case, which is in fact the general case, Na^+ action potentials backpropagate *passively* (with decrement) in the dendritic tree.

16.2 HIGH-VOLTAGE-ACTIVATED Ca^{2+} CHANNELS ARE PRESENT IN THE DENDRITIC MEMBRANE OF SOME CNS NEURONS, BUT ARE THEY DISTRIBUTED WITH COMPARABLE DENSITIES IN SOMA AND DENDRITES?

16.2.1 High-voltage-activated Ca^{2+} channels are present in some dendrites

In order to record Ca^{2+} channels in isolation, the solution bathing the extracellular face of the patch contains

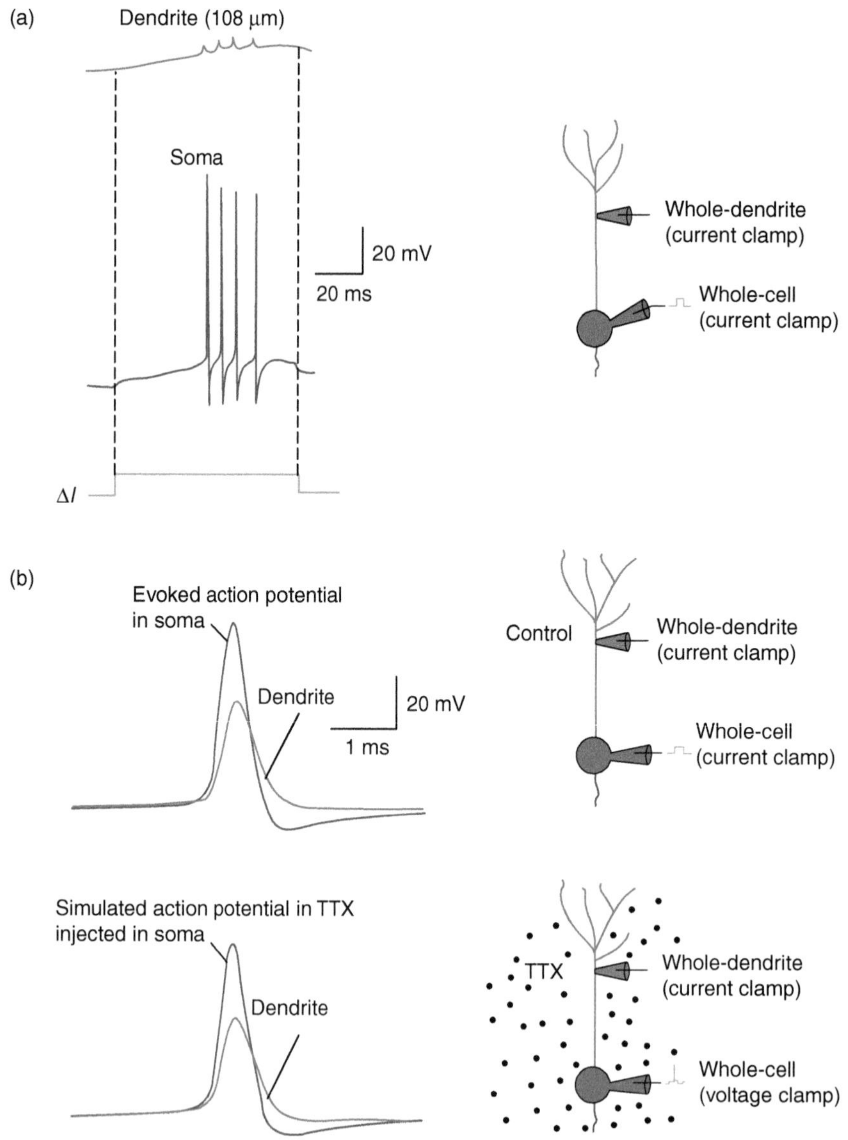

FIGURE 16.8 **Passive propagation of Na^+ action potentials in the dendritic tree of Purkinje cells. (a)** Simultaneous recordings at the soma and dendrite (108 μm from the soma) of a train of Na^+ action potentials evoked by a somatic long depolarizing current pulse (100 pA). **(b) (b_1)** An action potential is evoked in the soma by a depolarizing current pulse (soma). It propagates in the dendrite where it is recorded (dendrite, 47 μm from the soma). **(b_2)** A simulated action potential waveform is injected in the soma in the presence of 1 μM of TTX. The somatic voltage response propagates passively in the dendrites where it is recorded at the same location as in $\mathbf{b_1}$ but in the presence of TTX (dendrite). The simulated somatic action potential is recorded later at the soma with a second somatic recording pipette (soma). Adapted from Stuart GJ, Häusser M (1994) Initiation and spread of sodium action potentials in cerebellar Purkinje cells. *Neuron* **13**, 703–712, with permission.

Ba^{2+} as the charge carrier and TEACl and TTX for blocking K^+ and Na^+ currents, respectively. Ca^{2+}-channel activity is identified by inward current polarity, voltage-dependent channel gating, unitary current amplitude, single-channel behavior and its blockade by Cd^{2+}.

Purkinje cells of the cerebellar cortex

To determine whether the dendrites of Purkinje cells contain HVA Ca^{2+} channels, dendrite-attached patch recordings are performed in slices. Patches always show the activity of several channels, thus suggesting a tight clustering of Ca^{2+} channels in the dendritic membrane. **Figure 16.9a** shows the I/V relationship of a multichannel inward current carried by 10 mM of Ba^{2+} and evoked by a voltage ramp from −80 to +80 mV applied to a dendrite-attached macropatch. This dendritic Ba^{2+} current activates at −35 mV and is maximal around 0 mV. This HVA current is insensitive to the presence in the pipette solution of the specific blocker of N-type Ca^{2+} channels, ω-conotoxin GVIA (ωCgTx) and to the L-type channel opener (Bay K 8644). To test whether it is a P/Q-type Ca^{2+} current, a specific blocker, the funnel web spider

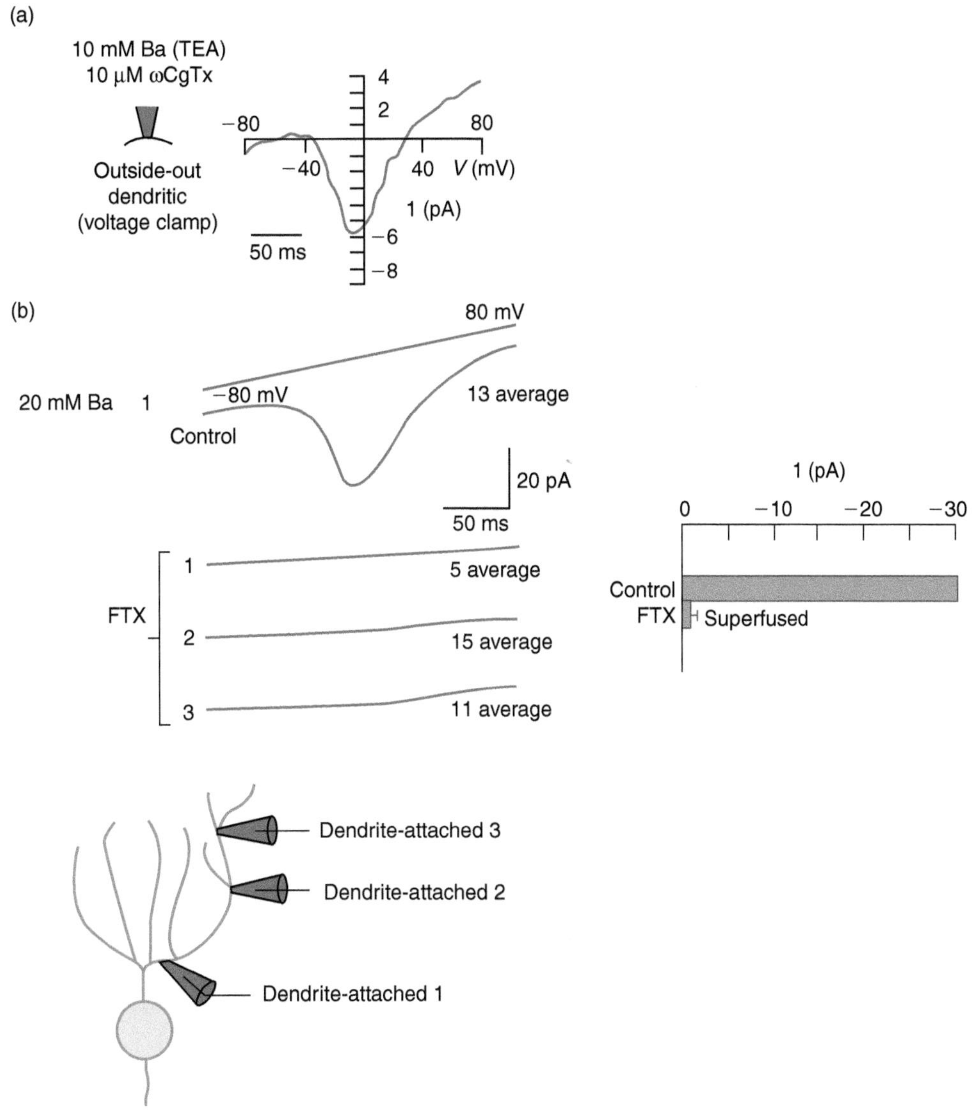

FIGURE 16.9 **P-type Ca^{2+} channel current in dendrites of Purkinje cells. (a)** In the presence of 10 μM of ω-CgTx added to the 10 mM of Ba^{2+} pipette solution, currents are evoked in an outside-out macropatch of dendritic membrane by a depolarizing voltage ramp from −80 to +80 mV. The *I/V* plot shows that the evoked inward current peaks at −9 mV and activates at −44 mV. **(b)** Currents carried by 20 mM of Ba^{2+} evoked by voltage ramps (from −80 to +80 mV) in dendrite-attached macropatches in different conditions (left). Top: Averaged current in control conditions. Lower traces: Funnel web toxin (FTX) is first applied in the extracellular medium, then the patch is performed. Three different dendrite-attached patch recordings are shown (the approximate positions of the recording pipettes are indicated). The averaged currents recorded show the absence of inward Ba^{2+} current (right). Adapted from Usowicz MM, Sugimori M, Cherksey B, Llinas R (1992) P-type calcium channels in the somata and dendrites of adult cerebellar Purkinje cells. *Neuron* **9**, 1185–1199, with permission.

toxin (FTX), is applied. Owing to the patch configuration, drugs must either be included in the pipette or be superfused over the cell before dendrite-attached recording, and a population of dendrite-attached recordings in control conditions is compared with the same number of recordings in the presence of the Ca^{2+}-channel blocker. Funnel web toxin is the only drug that blocks the dendritic Ca^{2+} current (**Figure 16.9b**), thus showing that the dendritic Ba^{2+} current recorded is carried through P/Q-type Ca^{2+} channels. Their characteristics are close to that of P/Q-type channels recorded in Purkinje cell somata (see Section 5.2.2).

Pyramidal neurons of the hippocampus

In contrast to the above findings, in pyramidal neurons of the hippocampus, there is a heterogeneous distribution of different types of Ca^{2+} channels within the soma-proximal dendritic trunks and more distal dendrites. Recordings of single Ca^{2+} channels in dendrite-attached patches show that L-type Ca^{2+} channels (sensitive to dihydropyridines) are observed at fairly high density only in the first 50 μm from the soma and at extremely low density in more distal dendritic patches where mainly Ni^{2+}-sensitive, T-type Ca^{2+} channels are present. Therefore, HVA Ca^{2+} channels in these cells would be confined to the soma and very proximal dendrites.

Conclusions

Dendrites of Purkinje cells contain a high density of HVA Ca^{2+} channels of the P/Q type. In contrast, HVA Ca^{2+} channels are present at low density in dendrites of pyramidal neurons of the neocortex or the hippocampus. It must be pointed out that in most CNS dendrites there is a low density of HVA Ca^{2+} channels. Purkinje cells are exceptions.

16.2.2 High-voltage-activated Ca^{2+} channels of Purkinje cell dendrites are opened by climbing fiber EPSP; this initiates Ca^{2+} action potentials in the dendritic tree of Purkinje cells

The cerebellar Purkinje cells receive two kinds of excitatory inputs, a single powerful climbing fiber (CF) and many thousands of small parallel fibers (PF). The CF synapse arises from an axonal projection from the inferior olive, a brainstem nucleus. The CF synapse is composed of around 300 synaptic contacts located on the largest dendritic branches (thick and smooth dendrites) and on the smaller spiny dendrites (see **Figures 6.8** and **6.9**). The pioneering work of Llinas and Nicholson (1976) with intradendritic recordings showed that activation of this single distributed synapse evokes a large, all-or-none EPSP surmounted with one or two Ca^{2+}-dependent action potentials (complex spike; **Figure 16.10a**). They are Ca^{2+} spikes since they disappear in the presence of Cd^{2+}. They are characterized by a rather slow onset, but a large amplitude at dendritic level which fluctuates between 30 and 60 mV. Their time course is much longer than that of Na^+ spikes. Their threshold is lower at the dendritic level: depolarizations at around 10 mV are sufficient to generate Ca^{2+} dendritic spikes at the dendritic level while 20 mV depolarizations are required at the somatic level to evoke them. This complex spike then evokes bursts of Na^+ action potentials in the axon.

The climbing fiber-evoked EPSP underlying this complex spike can be uncovered in dendritic recordings by evoking a simultaneous IPSP by stimulation of interneurons (**Figure 16.10b**). Climbing fiber EPSP has a lower amplitude and a longer duration than the complex spike. This suggests that activation of a dendritic voltage-gated depolarizing current(s) amplifies the CF EPSP. Many data suggest that this dendritic depolarizing current is a P-type Ca^{2+} current. First, P channels are present in the dendritic membrane (see Section 16.2.1). Second, the complex spike is accompanied by a transient rise in intracellular Ca^{2+} concentration which is most prominent at dendritic locations (**Figure 16.10c**).

Modeling of the complex spike shows the currents underlying the CF-evoked complex spike and their sequence of activation (**Figure 16.11**). These currents are the large CF synaptic inward current (resulting from the summation of around 300 unitary synaptic currents through glutamate AMPA receptors). CF-induced synaptic inward current depolarizes the dendritic membrane and thus activates P-type Ca^{2+} channels over large regions of the dendrites. The resulting Ca^{2+} current is responsible for almost all of the resulting additional depolarization and for dendritic Ca^{2+} spikes. Ca^{2+} spikes are generated at multiple sites along the dendritic tree, which explains why the CF EPSP recorded in a dendrite is sometimes surmounted by more than one Ca^{2+} spike. In turn, Ca^{2+} entry increases intradendritic Ca^{2+} concentration and thus activates the Ca^{2+}-activated K^+ outward current in the whole dendritic tree. This repolarizes the complex spike. In conclusion, activation of the dendritic P-type Ca^{2+} current boosts the amplitude of the CF EPSP and, by activating K^+ current, leads to a faster repolarization of the EPSP (**Figure 16.10b**).

16.2.3 Dendritic high-voltage-activated Ca^{2+} channels are opened by backpropagating Na^+ action potentials

Pyramidal neurons of the hippocampus

To test whether dendritic HVA Ca^{2+} channels are activated by sub-threshold EPSPs or by higher amplitude depolarizations such as backpropagating Na^+ action potentials, simultaneous dendrite-attached (voltage clamp mode) and whole-cell somatic (current clamp mode) recordings are performed in the same neuron (**Figure 16.12a**). Excitatory postsynaptic potentials are evoked by Schaffer collateral stimulation. Na^+ spikes are evoked by intrasomatic injection of a depolarizing current pulse. EPSPs and spikes are recorded from the soma while channel openings are recorded from the dendritic patch of membrane. Ca^{2+}-channel activity present in the dendritic patch is first recorded and Ca^{2+} channels are classified as HVA or LVA (low-voltage-activated, also called 'subliminal') channels. Two types of Ca^{2+} channels are encountered regularly on dendrites greater than 100 μm from the soma, essentially the LVA T-type Ca^{2+} channels and less frequently the HVA L-type Ca^{2+} channels. Only the T-type channels are opened in response to sub-threshold EPSPs. Instead, somatically generated action potentials or trains of suprathreshold synaptic stimulation are required for HVA channel activation (**Figure 16.12a**).

Dendritic HVA channel openings are observed during and after the repolarization phase of somatically generated Na^+ action potentials. This strongly suggests that dendritic HVA Ca^{2+} channels are opened by Na^+ action potentials backpropagating into the dendrites. Openings occur tens of milliseconds after the action potential. This provides an influx of Ca^{2+} throughout an extended

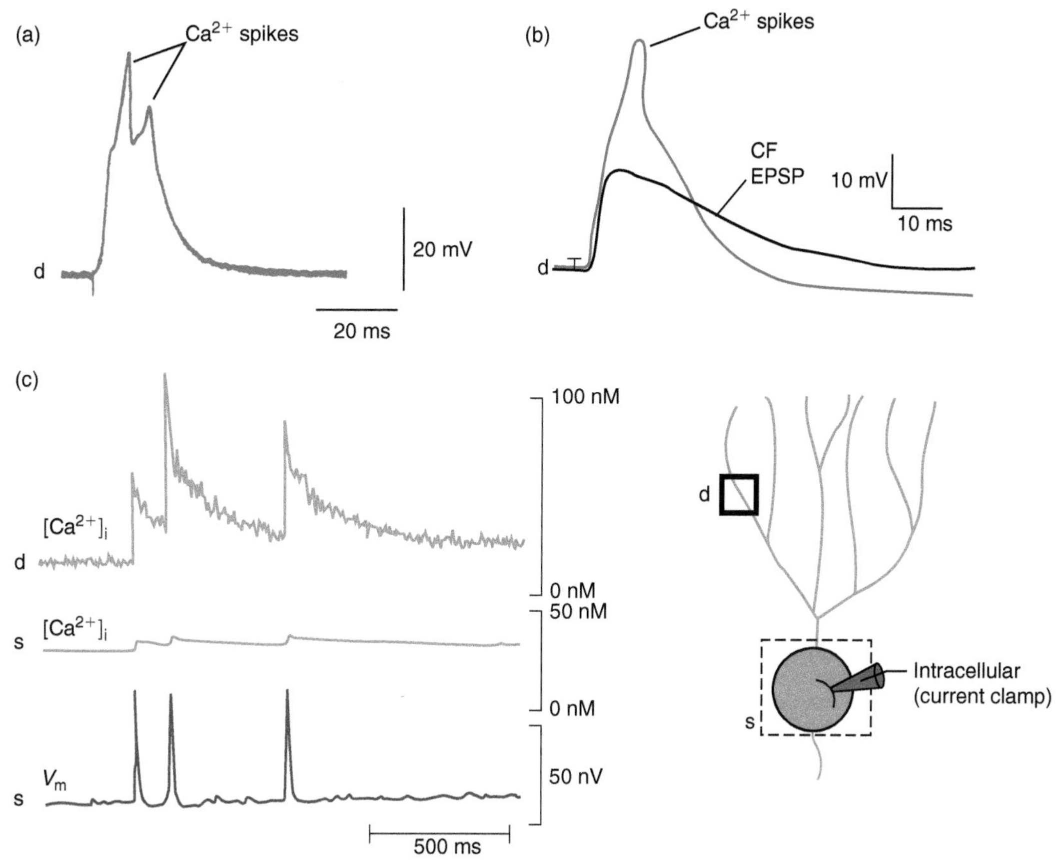

FIGURE 16.10 **P-type Ca^{2+} current activated by climbing fiber EPSP in dendrites of Purkinje cells. (a)** Intradendritic recording of the synaptically evoked climbing fiber response that is surmounted by two Ca^{2+} spikes (intracellular recording in current clamp mode). **(b)** Intradendritic recording at resting potential of the climbing fiber EPSP showing a 2–3 ms-wide Ca^{2+} spike and of the climbing fiber EPSP recorded during a concomitant IPSP (CF EPSP) (intracellular recording in current clamp mode). **(c)** Time course of $[Ca^{2+}]_i$ recorded at a dendritic (d, top trace) and somatic (s, middle trace) site during spontaneous climbing fiber responses (s, bottom trace) recorded with simultaneous microfluorometric measurements of cytosolic free calcium concentration and intracellular (intrasomatic) electrophysiological recordings (current clamp mode). Part (**a**) adapted from Llinas R, Sugimori M (1980) Electrophysiological properties of *in vitro* Purkinje cell dendrites in mammalian cerebellar slices. *J. Physiol. (Lond.)* **305**, 197–213, with permission. Part (**b**) adapted from Callaway JC, Lasser-Ross N, Ross WN (1995) IPSPs strongly inhibit climbing fiber-activated $[Ca^{2+}]_i$ increases in the dendrites of cerebellar Purkinje neurons. *J. Neurosci.* **15**, 2777–2787, with permission. Part (**c**) adapted from Knöpfel T, Vranesic I, Staub C, Gähwiler BH (1991) Climbing fiber response in olivo-cerebellar slice cultures: II. Dynamics of cytosolic calcium in Purkinje cells. *Eur. J. Neurophysiol.* **3**, 343–348, with permission.

portion of the dendritic tree (defined by the extent of action potential propagation). The spatial domain of the effects of these HVA Ca^{2+} channels will therefore be much more extensive compared with a local opening of Ca^{2+} channels by EPSPs. Thus, the HVA Ca^{2+} channels in the CA1 apical dendrites may modify synaptic strength over broad areas of the dendrites (see Chapter 18).

Conclusions

When HVA Ca^{2+} channels are present with a high density in the dendritic membrane, they allow *generation and propagation* of Ca^{2+} action potentials in dendrites. This is, for example, the case with dendrites of Purkinje cells. In contrast, Ca^{2+} action potentials do not exist in dendrites that contain in their membrane a low density of HVA Ca^{2+} channels, such as pyramidal neurons of the hippocampus.

16.3 FUNCTIONAL CONSEQUENCES

16.3.1 Amplification of distal synaptic responses by dendritic HVA currents counteracts their attenuation due to passive propagation to the soma

High-voltage-activated Na^+ and Ca^{2+} channels opened by EPSPs boost the effect of local synaptic inputs by acting as either voltage or current amplifiers. This could be one solution to overcome the passive decay of EPSPs, en route to the soma. This argument is somewhat somatocentric; i.e. the emphasis is on how dendrites might amplify events so that they are bigger in the soma. An alternative viewpoint is that these channels are more important for dendritic interactions in the immediate vicinity of synaptic inputs. For example, multiple EPSPs

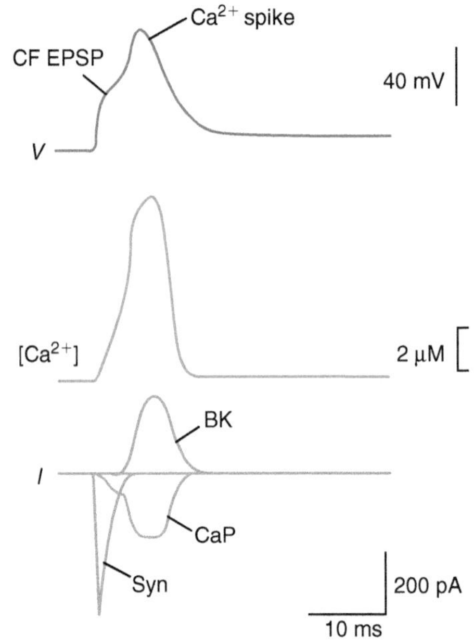

FIGURE 16.11 **Dendritic currents underlying the climbing fiber-evoked EPSP in Purkinje cells. (a)** Modeling of the CF response recorded in current clamp mode in a dendrite (top trace) and the underlying $[Ca^{2+}]_i$ transient (middle trace) and currents (*I*, bottom traces). The underlying currents are the synaptic glutamatergic current (Syn) which generates the climbing fiber EPSP, depolarizes the dendritic membrane and thus activates the dendritic P-type Ca^{2+} current (CaP) which further depolarizes the dendritic membrane, amplifies the EPSP and generates a Ca^{2+} spike (shown in top trace). The resultant $[Ca^{2+}]_i$ increase (middle trace) activates the BK current (BK, bottom trace) which rapidly repolarizes the membrane. Adapted from De Schutter E, Bower JM (1994) An active membrane model of the cerebellar Purkinje cell: II. Simulation of synaptic response. *J. Neurophysiol.* **71**, 401–419, with permission.

occurring on the same branch and within a narrow time should activate voltage-gated channels more strongly than a single EPSP and produce a much bigger response than would occur if EPSPs were on separate branches.

16.3.2 Active backpropagation of Na^+ spikes in the dendritic tree depolarizes the dendritic membrane, with multiple consequences

Most of the consequences of the presence of Na^+ spikes (large amplitude depolarizations) in the dendrites are still hypotheses. The only well-demonstrated one is the opening of dendritic Ca^{2+} channels and the consequent increase in intradendritic Ca^{2+} concentration. Such an increase will have by itself other consequences.

A retrograde signal that activates voltage-sensitive dendritic Ca^{2+} channels

In the hippocampus, Na^+ action potentials open dendritic Ca^{2+} channels, leading to a widespread influx of Ca^{2+} in the dendrites. In order to localize and quantify the increase of intradendritic Ca^{2+} concentration resulting from backpropagated Na^+ action potentials, pyramidal neurons of the hippocampus are loaded with FURA-2 and a train of action potentials is evoked by somatic depolarization through the whole-cell recording electrode. The evoked Ca^{2+} influx is thus visualized in the dendrite under fluorescence observation. Then to identify the type of Ca^{2+} channel involved and its localization along the dendrite, the same experiment is repeated in the presence of specific Ca^{2+}-channel blockers. Finally, a control experiment is performed in the absence of extracellular Ca^{2+}. Results show that $[Ca^{2+}]_i$ transients are largest in the proximal dendrites and smaller changes occur in more distal dendritic regions (**Figure 16.12b**).

One particular role for an intradendritic increase of Ca^{2+} concentration is found in dopaminergic neurons of the substantia nigra. $[Ca^{2+}]_i$ increase triggers transmitter release from dendrites (a *presynaptic* effect). In these cells, clusters of synaptic vesicles containing dopamine are present in dendrites that behave in certain sites as presynaptic elements. Dendritic release of dopamine is Ca^{2+} dependent and TTX sensitive. Backpropagated action potentials may thus provide the stimulus (i.e. intradendritic $[Ca^{2+}]_i$ increase) to trigger dopamine release and evoke synaptic transmission from nigral dendrites to postsynaptic sites.

Apart from this very particular case, intradendritic $[Ca^{2+}]_i$ increase will have a *postsynaptic* effect. Intracellular Ca^{2+} activates biochemical pathways, Ca^{2+}-induced Ca^{2+} release as well as Ca^{2+}-activated K^+ currents present in the dendritic membrane. Such outward current changes the shape of EPSPs. Intracellular Ca^{2+} is also implicated in postsynaptically induced forms of plasticity such as long-term potentiation or depression (see Chapter 18).

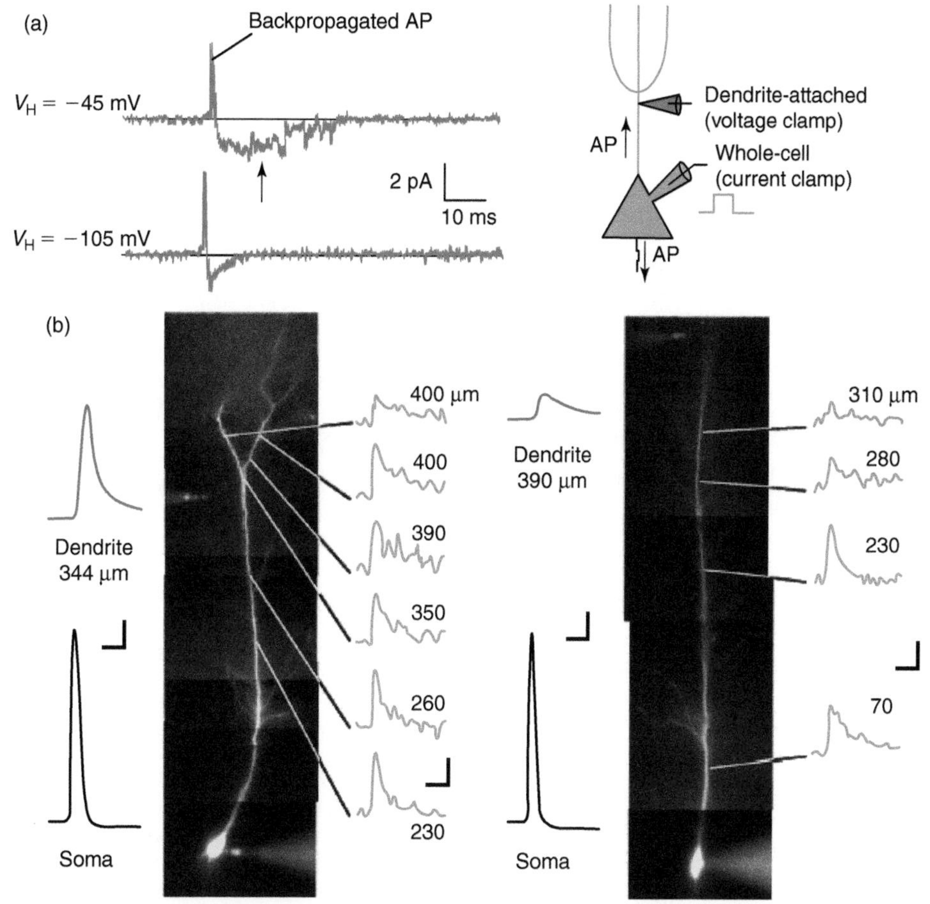

FIGURE 16.12 **Activation of dendritic HVA Ca^{2+} channels by backpropagated Na^+ action potential in hippocampal pyramidal neurons.** Distal dendritic calcium influx is correlated with the efficacy of action potential backpropagation. **(a)** An action potential is evoked in the soma by a depolarizing current pulse. It backpropagates in the dendrites and is recorded at a dendritic site as a capacitative current at two different holding potentials (backpropagated AP). When the dendritic patch is held 20 mV more depolarized than resting potential (−45 mV), numerous openings of channels are observed following the action potential (arrow). In contrast, when the dendritic membrane is held at −105 mV, the backpropagated action potential does not evoke channel openings. **(b)** Spike-induced $[Ca^{2+}]_i$ transients in a FURA-2 loaded neuron. Single action potentials (e.g. *left*) that propagate efficiently to the distal dendrite trigger robust calcium influx (expressed as the relative change in fluorescence, DF/F) in both proximal and distal dendritic compartments. *Right*: A different pyramidal neuron filled with Fura-2 exhibits weak action-potential backpropagation in the distal dendrites. The associated calcium influx shows significant attenuation in distal dendritic regions. Physiology scale bars: 20 mV, 1 ms. Imaging scale bars: 5% DF/F, 300 ms. Part (**a**) adapted from Magee JC, Johnston D (1995) Synaptic activation of voltage-gated channels in the dendrites of hippocampal pyramidal neurons. *Science* **268**, 301–304, with permission. Part (**b**) adapted from Golding NL, Kath WL, Spruston N (2001) Dichotomy of action-potential backpropagation in CA1 pyramidal neuron dendrites. *J. Neurophysiol.* **86**, 2998–3010, with permission.

A retrograde signal that amplifies NMDA-mediated synaptic currents

The transient depolarization due to backpropagated Na^+ spikes may relieve the voltage-dependent Mg^{2+} block of NMDA receptor channels and amplify the signal mediated by these channels (see Section 10.4).

A retrograde signal that shunts ongoing synaptic integration

Backpropagated Na^+ action potentials act as a signal to the dendritic tree that the axon has fired. This transient depolarization, by reducing the electrochemical gradient for cations, will diminish ongoing postsynaptic excitatory currents. Moreover, dendritic action potentials will open voltage-sensitive channels and thus diminish the resistance of the dendritic membrane and shunt ongoing synaptic integration. Finally, the rise in dendritic Ca^{2+} concentration could also transiently shunt out parts of the dendritic tree by opening Ca^{2+}-activated K^+ currents.

16.3.3 Initiation of Ca^{2+} spikes in the dendritic tree of Purkinje cells evokes a widespread intradendritic $[Ca^{2+}]$ increase

As seen above, intradendritic $[Ca^{2+}]$ increase will have a *postsynaptic* effect in Purkinje cells. It activates biochemical pathways, Ca^{2+}-induced Ca^{2+} release as well as Ca^{2+}-activated K^+ currents present in the dendritic membrane. Such outward currents change the shape of EPSPs. Intracellular Ca^{2+} is also implicated in postsynaptically induced forms of plasticity such as long-term depression which has been extensively studied in Purkinje cells (see Chapter 18).

16.4 CONCLUSIONS

High-voltage-activated channels have been shown in the dendritic membrane of some CNS neurons such as pyramidal neurons of the neocortex and hippocampus, dopaminergic neurons of the substantia nigra pars compacta, and Purkinje cells of the cerebellar cortex. To answer the questions asked in the introduction:

- High voltage-gated currents are present in the dendritic membrane of some CNS neurons. These are the depolarizing currents I_{Na}, I_{CaP} and I_{CaL}.
- They are not all distributed equally over somato-dendritic membranes. I_{Na} is present at the same density in the somatic and dendritic membranes in pyramidal neurons of the neocortex or hippocampus and in dopaminergic neurons of the substantia nigra but is nearly absent in dendrites of Purkinje cells. I_{CaP} is present at the same density in the somatic and dendritic membranes in Purkinje cells. In contrast, I_{CaL} is mostly present in the somatic and very proximal dendritic membranes of pyramidal neurons of the hippocampus.
- Since I_{Na}, I_{CaP} and I_{CaL} are activated at high voltage, they can be activated only by already large summed EPSPs or by backpropagating Na^+ action potentials (for I_{CaP} and I_{CaL}).
- I_{Na} supports the active backpropagation of Na^+ action potentials in pyramidal neurons of the neocortex or hippocampus and in dopaminergic neurons of the substantia nigra in response to suprathreshold EPSPs (these Na^+ action potentials are first initiated at the axon initial segment). I_{CaP} boosts climbing fiber EPSP and supports the initiation and active propagation of Ca^{2+} action potentials in dendrites of Purkinje cells. Direct (by EPSPs) or indirect (via dendritic Na^+ action potentials) activation of I_{CaP} and I_{CaL} induces a transient increase of intradendritic Ca^{2+} concentration that is more or less localized depending on the neuron considered.

Further reading

Callaway, J.C., Ross, W.N., 1997. Spatial distribution of synaptically activated sodium concentration changes in cerebellar Purkinje neurons. J. Neurophysiol. 77, 145–152.

Christie, B.R., Eliot, L.S., Ito, K., et al., 1995. Different Ca^{2+} channels in soma and dendrites of hippocampal pyramidal neurons mediate spike-induced Ca^{2+} influx. J. Neurophysiol. 73, 2553–2557.

Colbert, C.M., Magee, J.C., Hoffman, D.A., Johnston D., 1997. Slow recovery from inactivation of Na^+ channels underlies the activity-dependent attenuation of dendritic action potentials in hippocampal CA1 pyramidal neurons. J. Neurosci. 17, 6512–6521.

Hausser, M., Spruston, N., Stuart, G.J., 2000. Diversity and dynamics of dendritic signaling. Science 290, 739–744.

Johnston, D., Magee, J.C., Colbert, C.M., Christie, B.R., 1996. Active properties of neuronal dendrites. Annu. Rev. Neurosci. 19, 165–186.

Kuczewski, N., Porcher, C., Lessmann, V., Medina, I., Gaiarsa, J.L., 2008. Back-propagating action potential: a key contributor in activity-dependent dendritic release of BDNF. Commun. Integr. Biol. 1, 153–155.

Lüscher, H.R., Larkum, M.E., 1998. Modeling action potential initiation and back-propagation in dendrites of cultured rat motoneurons. J. Neurophysiol. 80, 715–729.

Lacey, M.G., Mercuri, N.B., North, R.A., 1987. Dopamine acts on D2 receptors to increase potassium conductance in neurones of the rat substantia nigra zona compacta. J. Physiol. 392, 397–416.

Markram, H., Helm, P.J., Sakmann, B., 1995. Dendritic calcium transients evoked by single backpropagating action potentials in rat neocortical pyramidal neurons. J. Physiol. (Lond.) 485, 1–20.

Miyakawa, H., Lev-Ram, V., Lasser-Ross, N., Ross, W.N., 1992. Calcium transients evoked by climbing fiber and parallel fiber synaptic inputs in guinea pig cerebellar Purkinje neurons. J. Neurophysiol. 68, 1178–1188.

Schiller, J., Schiller, Y., Stuart, G., Sakmann, B., 1997. Calcium action potentials restricted to distal apical dendrites of rat neocortical pyramidal neurons. J. Physiol. (Lond.) 505, 605–616.

Stuart, G., Schiller, J., Sakmann, B., 1997. Action potential initiation and propagation in rat neocortical pyramidal neurons. J. Physiol. (Lond.) 505, 617–632.

Stuart, G., Spruston, N., Sakmann, B., Hausser, M., 1997. Action potential initiation and backpropagation in neurons of the mammalian CNS. Trends Neurosci. 20, 125–131.

Vetter, P.R., Roth, A., Hausser, M., 2000. Propagation of action potentials in dendrites depends on dendritic morphology. J. Neurophysiol. 85, 926–937.

Williams, S.R., Stuart, G.J., 2000. Backpropagation of physiological spike trains in neocortical pyramidal neurons: implications for temporal coding in dendrites. J. Neurosci. 20, 8238–8246.

Yuste, R., Tank, D.W., 1996. Dendritic integration in mammalian neurons, a century after Cajal. Neuron 16, 701–716.

CHAPTER

17

Firing patterns of neurons

Constance Hammond

OUTLINE

The electrical activity of a neuron is related not only to the excitatory and inhibitory synaptic inputs that it receives, but also to its intrinsic electrophysiological membrane properties; i.e. the subliminal voltage-gated channels present in its dendritic, somatic and initial segment membranes and activated in the near-threshold range of membrane potential. As a result, the same postsynaptic depolarizing current will trigger different firing patterns according to the neuronal cell type recorded. In brief, the firing pattern (output) of a neuron results from the integration of synaptic currents (input) and subliminal voltage-gated currents present in the somatic and dendritic membrane. This concept was stated simply by Rodolpho Llinas in 1990: 'Nerve cells are not interchangeable: a neuron of a given kind cannot be functionally replaced by one of another type even if their synaptic connectivity and the type of neurotransmitter outputs are identical.'

In this chapter we shall consider the mammalian central nervous system to demonstrate how these intrinsic electrophysiological properties determine the firing patterns of neurons. We shall study the mechanisms underlying the firing patterns of medium spiny neurons of the striatum, of inferior olivary neurons, of Purkinje cells of the cerebellar cortex, and of thalamic and subthalamic neurons.

17.1 MEDIUM SPINY PROJECTION NEURONS OF THE STRIATUM

The striatum belongs to the basal ganglia, a set of interconnected nuclei that play a role in the memory of learned motor acts. The striatum is the main input stage to the basal ganglia since it receives and processes inputs arising from many cortical areas. The striatum contains several types of neurons among which medium spiny neurons (MSNs) are the most numerous. Medium spiny neurons are projection (Golgi type I) neurons that use GABA as a neurotransmitter. They project to globus pallidus and substantia nigra, the two output nuclei of the basal ganglia. They receive numerous inputs. Excitatory glutamatergic inputs come from neocortical and thalamic neurons. Several thousands of these, from nearly as many different afferent neurons, impinge on medium spiny neurons. Inhibitory GABAergic inputs come from local interneurons (Golgi type II) and from neurons of other basal ganglia nuclei. *In vivo*, medium spiny neurons exhibit different types of spontaneous activity depending on the arousal state of the animal.

17.1.1 Medium spiny neurons have a hyperpolarized resting membrane potential due to the inward rectifier K^+ current

In vitro, in the presence of blockers of synaptic transmission, MSN resting membrane potential is stable at around −80 mV. This is attributable to the presence of a strong and rapidly activating inwardly rectifying potassium-selective current (I_{KIR}; see Section 14.3.5) (**Figure 17.1**). Therefore, the membrane of these cells is 'clamped' near the K^+ equilibrium potential by the K^+-selective inward rectifier current that is open at rest, in the absence of afferent synaptic activity. In brief, the membrane potential is determined by I_{KIR} which

Cellular and Molecular Neurophysiology. DOI: 10.1016/B978-0-12-397032-9.00017-0

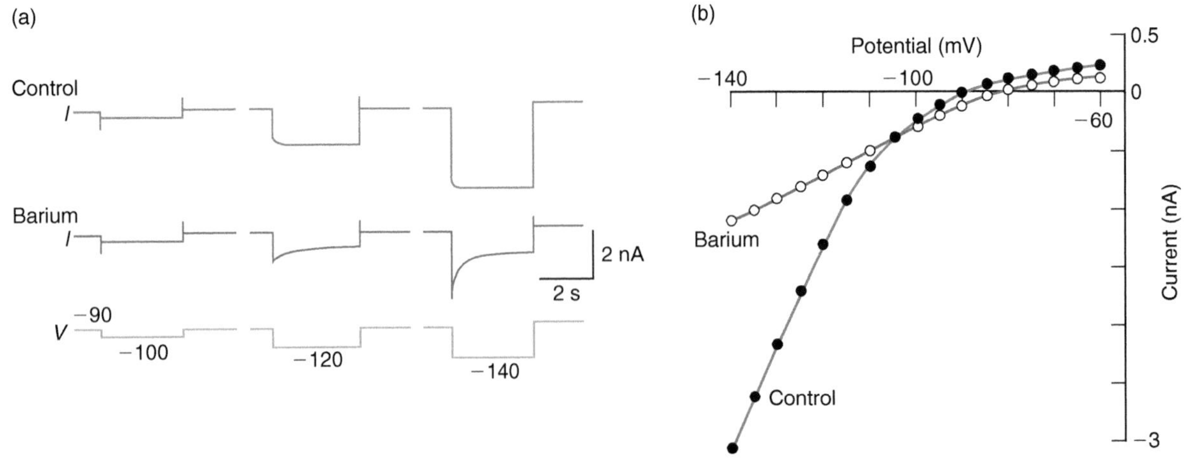

FIGURE 17.1 **The inward rectification K$^+$ current of medium spiny neurons. (a)** Membrane current (I) evoked by hyperpolarizing steps (V) to the indicated potentials (in mV), in control and in the presence of barium (10 μM). **(b)** I/V relations constructed from the steady-state I currents recorded in **(a)**. Note the reversal potential at around −90 mV. Adapted from Uchimura N, Cherubini E, North RA (1989) Inward rectification in rat nucleus accumbens neurons. *J. Neurophysiol.* **62**, 1280–1286, with permission.

dominates the other currents in the absence of synaptic currents. As a result, these neurons are characterized by a low input resistance at resting membrane potential.

17.1.2 When activated by a depolarizing current pulse, the response of medium spiny neurons *in vitro* is a long-latency regular discharge

The long-latency response of medium spiny neurons can be observed *in vitro* in response to an intracellular current pulse that mimics a depolarizing synaptic input (**Figure 17.2**). In the presence of a low dose of 4-aminopyridine (4-AP, 30–100 μM), known to preferentially block the slowly inactivating transient K$^+$ current (called I_{As} or I_D) (see Section 14.3.2), the latency of the first spike in response to a 400 ms depolarizing current pulse is largely reduced.

Voltage clamp recordings have shown that neostriatal medium spiny neurons possess at least three types of depolarization-activated K$^+$ currents. There are the two types of transient A currents; i.e. the fast- (I_{Af} or I_A) and slow- (I_{As} or I_D) inactivating, activated at subthreshold membrane potentials (around −65 mV) and both sensitive to 4-AP (see Sections 14.3.1 and 14.3.2). There is also a non-inactivating current (I_{KDR}) available at more depolarized potentials (−20 to −30 mV) and relatively resistant to 4-AP but blocked by tetraethylammonium chloride (TEA; see Section 4.3). The importance of these voltage-dependent K$^+$ currents in opposing depolarization and firing is also indicated by the large increase in the amplitude of the evoked depolarization after such currents are poisoned by intracellular injection of cesium (cells depolarize to a mean potential of −30 mV instead of −55 mV in control solution). Why does the response consist of a regular discharge with no adaptation? Adaptation (slowing of spike frequency inside a train) results from the progressive summation of the Ca^{2+}-activated K$^+$ current that underlies the slow after-hyperpolarization. In medium spiny neurons, this current is weak or absent.

Why do medium spiny neurons have a unique firing pattern? Rhythmic bursting currents are either suppressed in these neurons by the presence of K$^+$ currents at both hyperpolarized and depolarized potentials, or are absent.

17.1.3 When recorded in freely moving rodents, the activity of medium spiny neurons varies with the arousal state of the animal

Intracellular recordings are performed from MSNs during the different states of vigilance of anesthetic-free rats. These states are identified by their specific patterns of electromyographic (EMG) and electroencephalographic (EEG) activity (top and middle traces in **Figure 17.3a–c**). Wakefulness is identified by a sustained EMG activity and a low-amplitude desynchronized EEG (**Figure 17.3a**). Slow-wave sleep periods are characterized by a mild muscle activity without phasic contractions and high-amplitude, low frequency EEG waves (**Figure 17.3b**). Paradoxical sleep emerges from slow-wave sleep and is characterized by an almost complete loss of nuchal muscle tone and the occurrence of relatively large-amplitude EEG waves at 5–9 Hz (**Figure 17.3c**). Epochs of paradoxical sleep are relatively short and rare in the rat.

Intracellular recordings of MSN activity during these different states are shown in the bottom traces of **Figure 17.3a–c**. The waking state activity of MSNs is characterized by an irregular firing pattern (**Figure 17.3a**). In contrast, during slow-wave sleep (SWS), MSNs display rhythmic firing (**Figure 17.3b**). The paradoxical

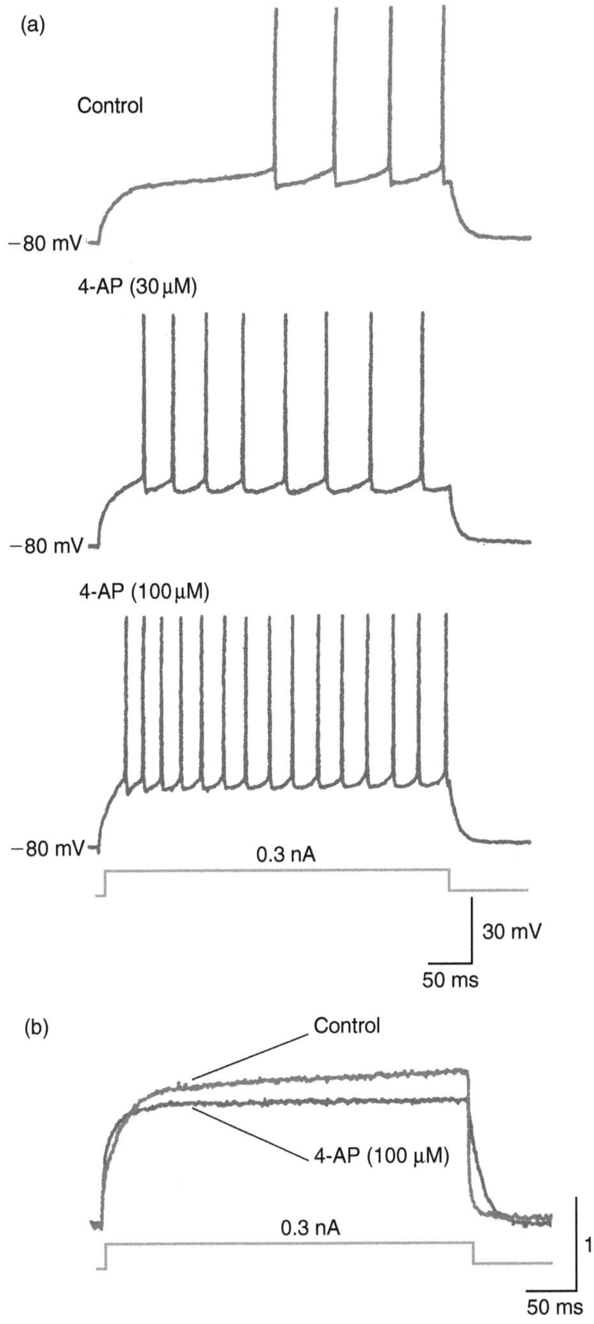

FIGURE 17.2 **The long-latency discharge of medium spiny neurons. (a)** A suprathreshold current pulse is delivered in control conditions and in the presence of 4-AP. Between pulses, the cell membrane is hyperpolarized back to the original resting membrane potential (−80 mV). 4-AP decreases the first spike latency and increases the frequency of discharge. **(b)** Comparison of the voltage deflections produced by a sub-threshold 0.5 nA current pulse (400 ms duration) in the presence of TTX shows that 4-AP reduces the slope of the ramp potential and decreases the apparent time constant of the membrane (average of four responses). Adapted from Nisenbaum ES, Xu ZC, Wilson CJ (1994) Contribution of a slowly inactivating potassium current to the transition to firing of neostriatal spiny projection neurons. *J. Neurophysiol.* **71**, 1174–1189, with permission.

sleep state is characterized by the absence of MSN firing (**Figure 17.3c**).

MSN irregular firing during wakefulness

During the waking state, most of the recorded MSNs fire action potentials with a mean firing rate around 3 Hz. The intracellular activity of MSNs during wakefulness is characterized by the occurrence of depolarizing envelopes, of variable amplitude and duration, on which are superimposed high-frequency depolarizing events leading to a random firing activity. The high-frequency noise-like fluctuations probably originate from the relatively uncorrelated firing of many converging corticostriatal neurons. However, MSN firing is not correlated with cortical activity (EEG).

MSN oscillatory firing during slow-wave sleep

Most of the recorded MSNs display a rhythmic firing pattern during slow-wave sleep due to periodic transitions of their membrane potential between two states: a silent hyperpolarized state (down state) and a suprathreshold depolarized state (up state). Up states last few hundreds of milliseconds and generate single spikes or trains of action potentials. The existence of two states is attested by the bimodal distribution of MSN membrane potential computed from continuous records of hundreds of seconds: the mean potential during up states is around −65 mV, whereas during down state it is around −75 mV (**Figure 17.3b**). In contrast to wakefulness, cross-correlation between intracellular striatal depolarizations and cortical electroencephalographic activity during slow-wave sleep reveals a strong temporal coherence between membrane fluctuations and cortical EEG rhythm.

The ionic mechanisms underlying the fluctuations between a silent down state and a depolarized active up state has been analyzed in anesthetized animals where the same oscillatory activity of MSNs is recorded. The down state does not result from a tonic inhibitory afferent synaptic activity since it is still observed when bicuculline (the $GABA_A$-receptor antagonist) is iontophoretically applied near the recording electrode. It results from the presence of a strong and rapidly activating inwardly rectifying potassium-selective current (I_{KIR}; see above) (and also from the silence of corticostriatal neurons). The up state, in contrast, absolutely requires the integrity of excitatory synaptic inputs to MSNs. When blockers of synaptic transmission are iontophoretically applied near the recording electrode, up-state transitions are abolished. Up-state transitions depend on the synchronous activity of excitatory afferents arising from the cortico- and/or thalamo-striatal pathways. In brief, during the up states the membrane potential results from the interaction between strong depolarizing synaptic currents and intrinsic voltage-dependent subliminal currents.

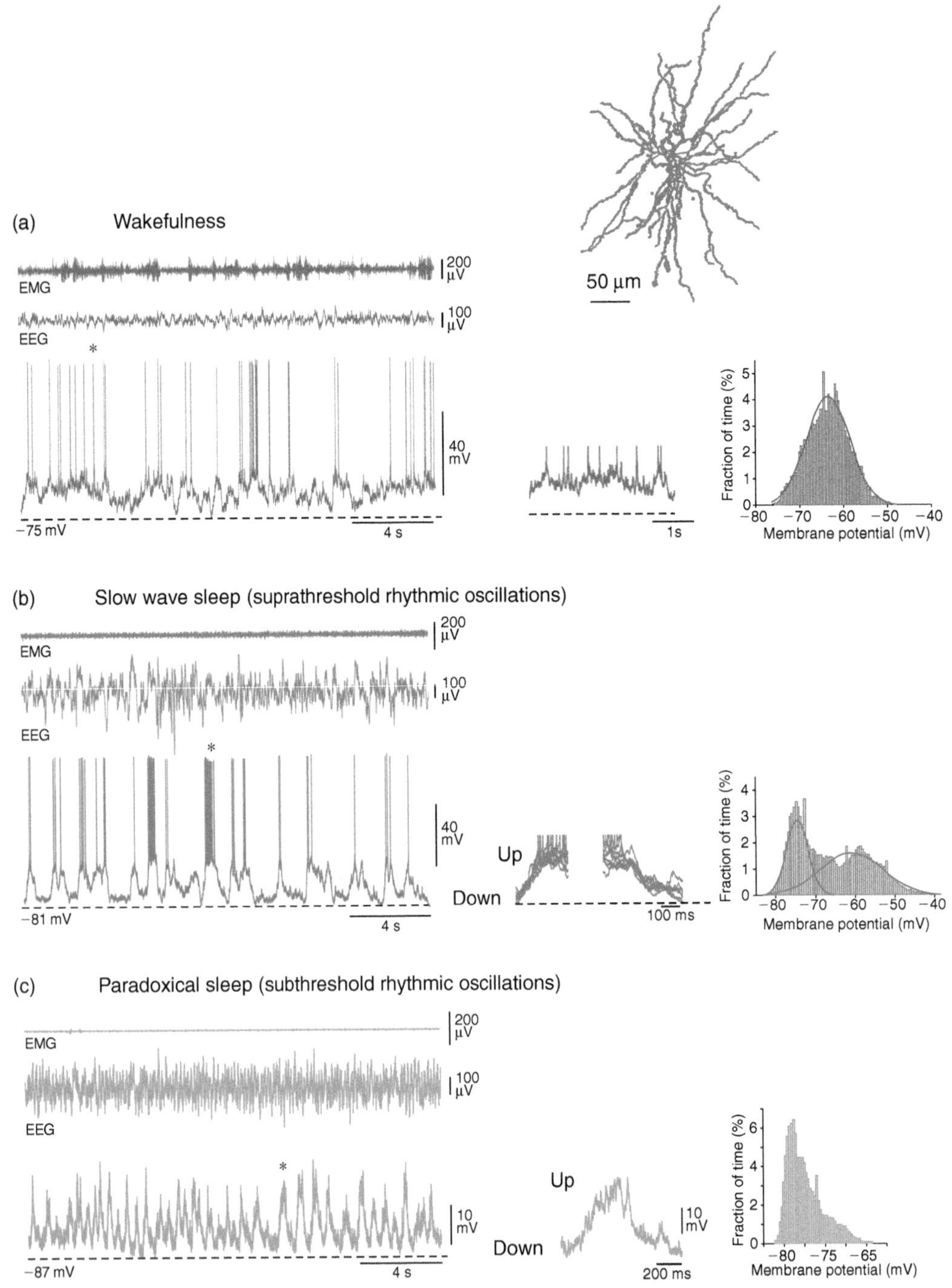

FIGURE 17.3 **Firing patterns of medium spiny neurons of the striatum according to the arousal state of the rat. (Left column a–c)** *In vivo* intracellular recording of the activity of spiny projection neurons of the striatum (bottom traces) together with the corresponding electroencephalogram (EEG) to check arousal state and electromyogram (EMG) to check spontaneous movements, during wakefulness **(a)**, slow-wave sleep **(b)** and paradoxical sleep **(c)** in anesthetic free rats. **(Middle column a–c)** Expansion of the membrane fluctuations as indicated by the asterisk in the left columns **(a)**, **(b)** nine superimposed traces, truncated records and **(c)** truncated record. **(Right column a–c)** Membrane potential distribution (bin size, 0.5 mV) calculated from the recording depicted in the left column. Recording is unimodal during wakefulness **(a)**, is fitted by a double Gaussian ($r^2 = 0.97$) during slow-wave sleep **(b)** and is unimodal but skewed toward more hyperpolarized potentials during paradoxical sleep. Drawing by Jérôme Yelnik. Parts (**a**, **b**, **c**) from Mahon S, Vautrelle N, Pezard L, Slaght SJ, Deniau JM, Chouvet G, Charpier (2006) Distinct patterns of striatal medium spiny neuron activity during the natural sleep–wake cycle. *J. Neurosci.*, **26**, 12587–12595, with permission.

MSN silent membrane oscillations during paradoxical sleep

MSNs recorded during periods of paradoxical sleep are silent but their membrane potential shows large amplitude (10–30 mV) sub-threshold oscillations occurring at 1–2 Hz. However, the distribution of MSN membrane potential shows a unimodal distribution: MSN membrane potential varies around a hyperpolarized state at around −80 mV (**Figure 17.3c**). The rhythmic sub-threshold oscillations probably result from the summation of high-frequency depolarizing afferent synaptic events. The difference in rhythmicity between the striatal membrane oscillations and the cortical EEG remains to be elucidated.

Summary

In the absence of afferent synaptic activity, medium spiny neurons have a hyperpolarized membrane potential and are silent. When depolarized, the triggering or not of action potentials as well as the latency of this discharge results from the interaction between the depolarizing glutamatergic synaptic current (mediated by AMPA receptors) and the intrinsic subliminal voltage-gated K^+ currents that oppose depolarization. MSNs show two different types of spiking according to the arousal state of the animal: irregular spiking during waking state and rhythmic spiking during slow-wave sleep. These two types are determined by the activity of afferent synapses.

17.2 INFERIOR OLIVARY CELLS

The inferior olive (IO) is a brainstem nucleus whose neurons innervate and monosynaptically excite cerebellar Purkinje cells through characteristic axonal terminations known as *climbing fibers* (see **Figures 6.8** and **6.9**). They use an excitatory amino acid as a neurotransmitter, most probably glutamate.

17.2.1 Inferior olivary cells are silent at rest in the absence of afferent activity

In slices *in vitro*, extracellular and intracellular recordings reveal that inferior olivary neurons of the principal olive are generally silent at the resting membrane potential but can spontaneously display sequences of membrane oscillations in response to afferent synaptic activity (**Figure 17.4a**). Non-oscillating inferior olive cells have a resting membrane potential of −55 to −60 mV.

In response to stimulation of afferents or to intracellular current pulses, a typical rhythmic bursting activity is recorded with a frequency varying from 3 to 12 Hz, depending on membrane potential. Intracellular recordings in slices *in vitro* revealed that these cells have the intrinsic properties necessary to oscillate endogenously.

17.2.2 When depolarized, inferior olivary cells oscillate at a low frequency (3–6 Hz) *in vitro*

When inferior olivary cells are slightly depolarized, their response to a depolarizing current pulse is characterized by an initial fast-rising spike (1 ms duration) which is prolonged to 10–15 ms by a plateau (ADP: after-spike depolarization) on which small action potentials are sometimes superimposed (**Figure 17.4**). It is followed by a large-amplitude long-lasting (150–200 ms) after-hyperpolarization (AHP) which silences the spike-generating activity and terminates in a rebound depolarization (arrowhead). The rebound depolarization may evoke another complex action potential: these cells have oscillatory membrane properties. Owing to their difference of threshold potential of initiation, the peak and plateau are called *high-threshold spike* (HTS), whereas the rebound depolarization is called *low-threshold spike* (LTS).

Pioneering *in vitro* studies by Llinas and Yarom in 1981 described the ionic currents that underlie the endogenous oscillatory properties of single inferior olivary neurons. The analysis of the currents responsible for this discharge configuration gives the following description. To record the low- and high-threshold spikes together or the low-threshold spike in isolation, the membrane potential is respectively maintained at a depolarized potential (**Figures 17.4b** and **17.5a**) or a hyperpolarized potential (**Figure 17.5b**).

- Tetrodotoxin (TTX) abolishes the peak of the action potential, showing that it results from the activation of the voltage-dependent I_{Na} (**Figures 17.4b** and **17.5a**).
- Ca^{2+}-channel blockers decrease the after-depolarization (ADP), the small superimposed action potentials, the AHP and the rebound depolarization, but leave intact the early Na^+-dependent spike (**Figures 17.4c** and **17.5b**). This shows that ADP, AHP and rebound depolarization are all Ca^{2+} dependent. The plateau is the result of the activation of a high-threshold Ca^{2+} current since the depolarization required to evoke it in the presence of TTX is high (see current trace I in **Figure 17.4b**, compare control and TTX). This current, localized in the dendrites, is activated by the fast Na^+-dependent action potential.
- The AHP is dependent on the amplitude of the ADP (**Figure 17.5a**, compare a_1 and a_2) and is blocked by external Ba^{2+} ions. It results from the activation of the Ca^{2+}-sensitive K^+ currents (I_{KCa}).
- The rebound depolarization or low-threshold spike is suppressed by Ca^{2+}-channel blockers (**Figure 17.5b**) and is activated at sub-threshold potentials. It

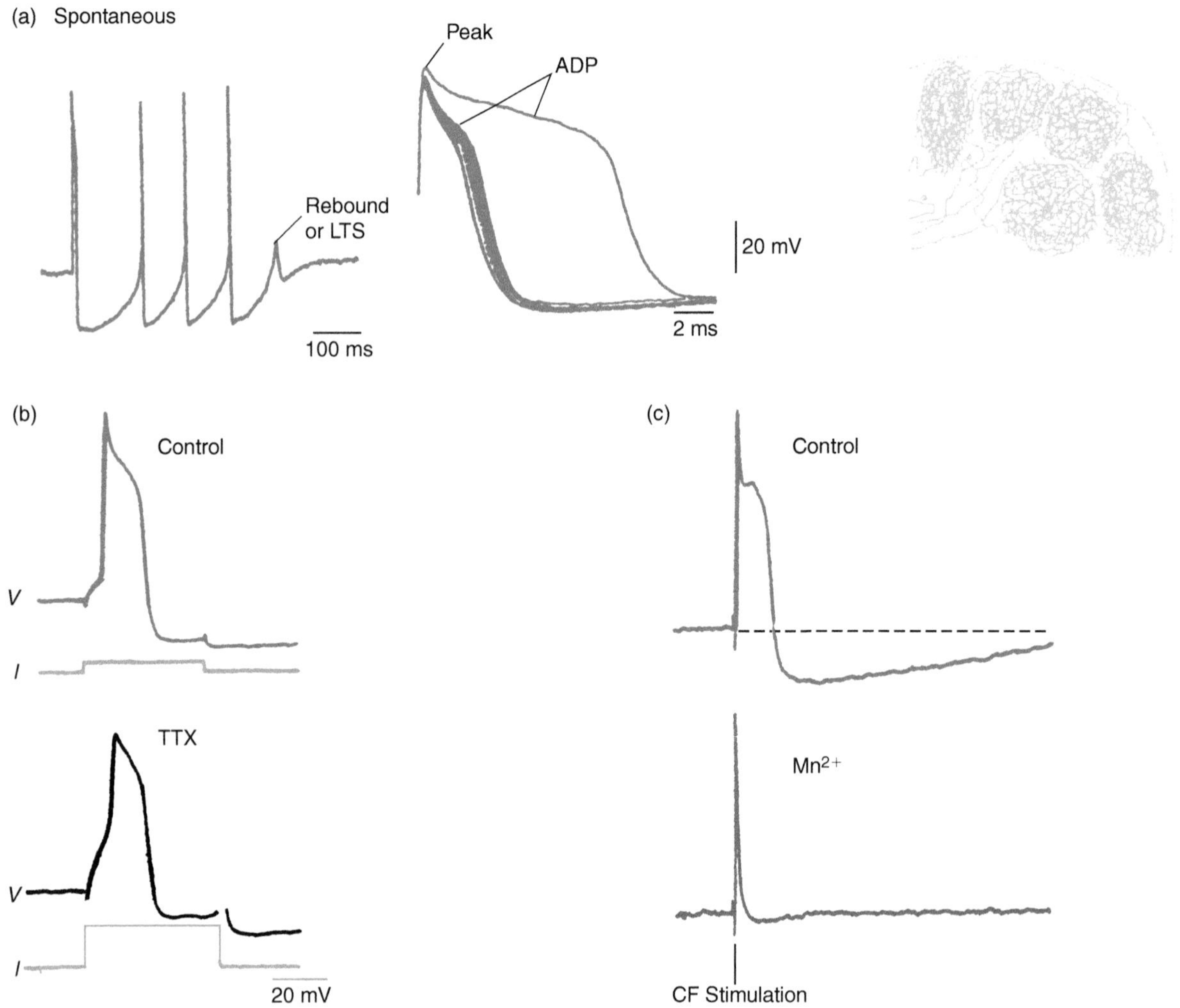

FIGURE 17.4 **Complex spikes of inferior olivary neurons.** The activity of olivary neuron is intracellularly recorded under current clamp in cerebellar slices (inset represents five of these neurons). **(a)** Spontaneous low-frequency train of spikes from an olivary neuron, displayed at two different sweep speeds. The action potentials shown at left are displayed superimposed at right at a faster sweep speed. The first action potential which arises from the resting membrane potential level has a slightly higher amplitude at the peak and a rather prolonged plateau (after-spike depolarization, ADP) which is followed by an after-hyperpolarization. The rest of the spikes in the train become progressively shorter until failure of spike generation occurs (arrow, left) and the train terminates. **(b)** Effect of TTX (left) and Mn^{2+} (right) on the different parts of the complex spike evoked in two olivary neurons either by a depolarizing intracellular current pulse (left) or climbing fiber stimulation (CF, right). Part (**a**) adapted from Llinas R, Yarom Y (1986) Oscillatory properties of guinea-pig inferior olivary neurones and their pharmacological modulation: an *in vitro* study. *J. Physiol. (Lond.)* **376**, 163–182, with permission. Part (**b**) adapted from Llinas R, Yarom Y. (1981) Electrophysiology of mammalian olivary neurones *in vitro*: different types of voltage-dependent ionic conductances. *J. Physiol. (Lond.)* **315**, 549–567, with permission. Drawing by Ramon Y Cajal, 1911.

is due to the activation of a low-threshold Ca^{2+} current (I_{CaT}) localized at the level of the soma. This current is de-inactivated during the period of after-hyperpolarization and activated when the hyperpolarization decreases. In slices from knockout mice that lack the gene for the pore-forming α1G subunit of the T-type calcium channel ($Ca_v3.1^{-/-}$) and their littermate wild-type (WT) mice, the low-threshold calcium spike is absent in IO neurons from $Ca_v3.1^{-/-}$ mice whereas the high threshold calcium spike is not affected. Though I_h is still present the sustained endogenous oscillation following rebound potentials is absent because I_h is not strong enough to depolarize the membrane to the threshold potential of Na^+ spikes.

Summary

Inferior olivary neurons are silent at rest. When depolarized to the threshold potential of the voltage-sensitive Na^+ channels, a sodium action potential is generated in the soma–initial segment region and the dendritic membrane is depolarized up to the level of activation of the high-threshold Ca^{2+} channels. The entry of Ca^{2+} ions through these channels causes a dendritic calcium plateau (ADP) and then the activation of Ca^{2+}-sensitive K^+ channels. The resulting I_{KCa} hyperpolarizes the membrane (AHP). This

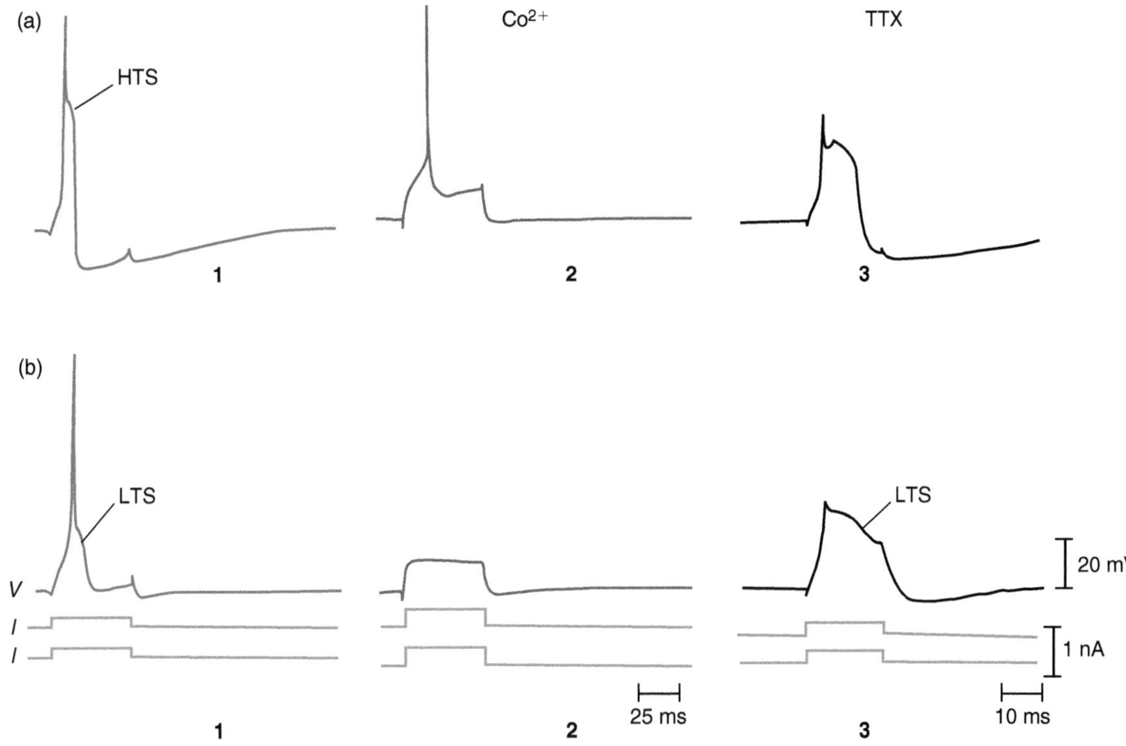

FIGURE 17.5 **The high- and low-threshold Ca^{2+} spikes of inferior olivary neurons.** Effect of membrane potential on excitability. A depolarizing current pulse of constant amplitude evokes **(a, 1)** a high-threshold Ca^{2+} spike (HTS) at resting membrane potential and **(b, 1)** a low-threshold Ca^{2+} spike (LTS) at a more hyperpolarized potential. Note that the ADP and AHP are smaller in **(b)** than in **(a)**. From left to right, effect of Co^{2+} and TTX in the same conditions. Adapted from Llinas R, Yarom Y (1981) Electrophysiology of mammalian olivary neurones *in vitro*: different types of voltage-dependent ionic conductances. *J. Physiol. (Lond.)* **315**, 549–567, with permission.

after-spike hyperpolarization allows the de-inactivation of the somatic T-type Ca^{2+} channels. As the amplitude of the AHP diminishes and the membrane potentials return to baseline, the low-threshold Ca^{2+} current (I_{CaT}) is activated, generates a 'low-threshold' Ca^{2+}-dependent spike, which reinitiates the cycle by activating again the Na^+/Ca^{2+} action potential (sodium spike–ADP sequence). The cycle can thus repeat itself at 3–6 Hz without any external intervention (**Figure 17.6a**).

17.2.3 When hyperpolarized, inferior olivary cells oscillate at a higher frequency (8–10 Hz) *in vitro*

When inferior olivary cells are slightly hyperpolarized, their response to a depolarizing current pulse is characterized by cycles of low-threshold Ca^{2+} spikes activating one or two fast Na^+ spikes and followed by a pronounced after-hyperpolarization, at a frequency of 9–12 Hz (**Figure 17.6b**). The enhancement of rhythmic oscillations with hyperpolarization suggests that a depolarizing current such as I_h may contribute to these oscillations. I_h is activated upon membrane hyperpolarization, is carried by both Na^+ and K^+ ions and has a reversal potential around −30 to −40 mV. Therefore, at hyperpolarized potentials, I_h is inward and depolarizing.

On the basis of pharmacological experiments, the following sequence of events is proposed to explain oscillations at hyperpolarized potentials. Activation of a somatic low-threshold Ca^{2+} spike which generates one or two Na^+ spikes is followed by an AHP, mediated largely by the activation of an apamin-sensitive Ca^{2+}-activated K^+ current. In addition, during the low-threshold Ca^{2+} spike, a portion of I_h is deactivated; this facilitates the generation of the AHP by allowing it to reach more negative potentials. The AHP subsequently results in two important effects: removal of inactivation of I_T and the activation of I_h. Activation of I_h depolarizes the membrane toward the threshold of activation of I_T and subsequently promotes the generation of a low-threshold Ca^{2+} spike and associated Na^+ action potentials and therefore reinitiates the oscillation (**Figure 17.6b**). In hyperpolarized olivary cells, 9–12 Hz oscillations are recorded owing to a decrease of the involvement of the high-threshold Ca^{2+} current, resulting in a shortening of the duration of the AHP.

When the membrane potential is in a region between rhythmic oscillations at hyperpolarized and depolarized membrane potentials, inferior olivary cells are silent.

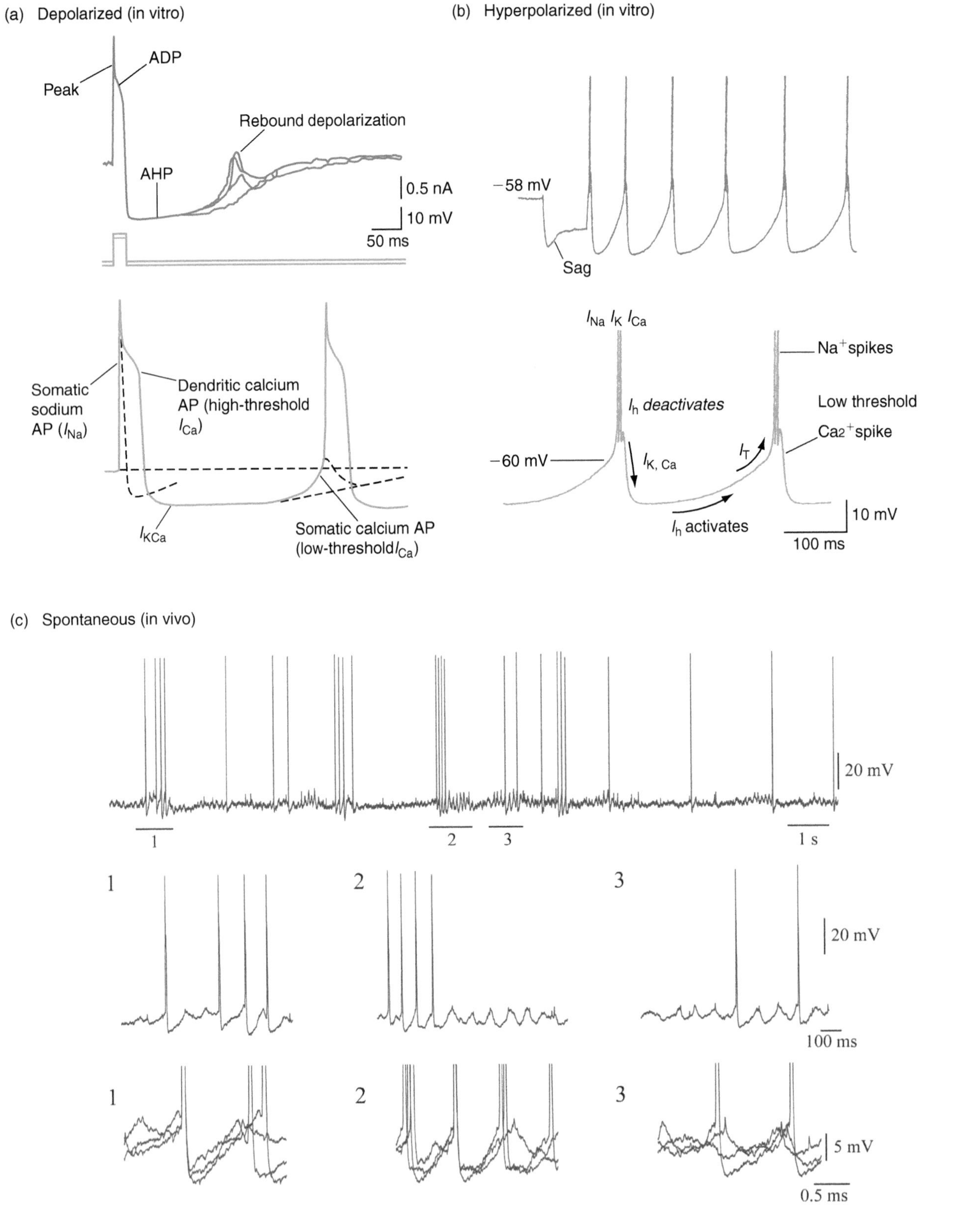

FIGURE 17.6 **Ionic currents underlying the discharge configuration of inferior olivary neurons. (a)** In slightly depolarized cells, direct stimulation of the neuron by injecting a depolarizing current step evokes a sequence consisting of a TTX-sensitive action potential, followed by Ca^{2+}-dependent events, a plateau (ADP), a period of after-hyperpolarization (AHP) and a depolarizing rebound of variable amplitude (four superimposed top traces). Schematic of this discharge configuration and indication of the different currents sequentially activated (see text for explanation) (bottom trace). **(b)** Direct intracellular injection of a hyperpolarizing current pulse is associated with a depolarizing sag and the generation of a rhythmic sequence of low-threshold Ca^{2+} spikes (top trace). Schematic of this discharge configuration and indication of the different currents sequentially activated (see text for explanation) (bottom trace). **(c)** Typical intracellular recording of the activity of an olivary neuron *in vivo* in an anesthetized rat. (Top trace) Sub-threshold oscillatory activity and epochs of suprathreshold activity. (Middle) The three marked areas in **(c) top** are shown at higher magnification. (Bottom) Superposition of the spikes from each epoch reveals the close association between the sub-threshold activity and the spikes. Note that in panel 3, two sub-threshold events are superimposed on the spike by aligning the peaks of sub-threshold events with the deflection point of the spike. Part **(a)** adapted from Llinas R, Yarom Y (1981) Properties and distribution of ionic conductances generating electroresponsiveness of mammalian inferior olive neurons *in vitro*. *J. Physiol. (Lond.)* **315**, 569–584, with permission. Part **(b)** adapted from Bal T, McCormick D (1997) Synchronized oscillations in the inferior olive are controlled by the hyperpolarization-activated cation current I_h. *J. Neurophysiol.* **77**, 3145–3156, with permission. Part **(c)** from Chorev E, Yarom Y, Lampl I (2007) Rhythmic episodes of subthreshold membrane potential oscillations in the rat inferior olive nuclei *in vivo*. *J. Neurosci.* **27**, 5043–5052, with permission.

17.2.4 *In vivo*, in anesthetized rats, the membrane potential of inferior olivary cells spontaneously oscillates

Intracellular recordings from inferior olive neurons *in vivo*, in anesthetized animals, show that 85% of the inferior olivary neurons have spontaneous sub-threshold membrane oscillations with spikes generated on the depolarization phase of the oscillations. Prolonged recordings of several minutes reveal that the spikes are not uniformly distributed, but clustered in distinct groups separated by quiescent periods (**Figure 17.6c**).

All of the spontaneous action potentials recorded are of the high-threshold type (**Figure 17.6a**). This contrasts with *in vitro* findings where low-threshold responses are frequently encountered (**Figure 17.6b**). This is not due to different intrinsic properties, since low-threshold spikes can be induced by hyperpolarizing the membrane *in vivo*. In *in vivo* conditions, cells are usually bombarded with glutamatergic inputs, resulting in significant depolarization of the dendrites. Thus, the threshold of high-threshold action potential is quickly reached.

The prevalence of high-threshold spikes, which generate prolonged somatic depolarization, has significant functional implications. High threshold action potentials triggered by single EPSPs evoked by white matter stimulation are followed by a plateau phase, on which is superimposed one to six small (<10 mV) wavelets at high frequency. The plateau duration and therefore the number of wavelets depends on the amplitude of the oscillation at the moment where the high threshold spike is generated. Thus, a high-threshold action potential triggered by a single synaptic stimulus triggers a burst of 1–6 axonal spikes. The number of axonal spikes codes for the amplitude of membrane oscillations. Axonal spikes backpropagate to the soma where they are seen as wavelets and orthodromically propagate along the climbing fibers to the Purkinje cells of the cerebellar cortex. However, there are failures in the orthodromic propagation and not all spikes of the burst reach axon terminals.

17.3 PURKINJE CELLS ARE PACEMAKER NEURONS THAT RESPOND BY A COMPLEX SPIKE FOLLOWED BY A PERIOD OF SILENCE

Purkinje cells are located in the cerebellar cortex in the so-called Purkinje cell layer (see **Figure 6.8**). The dendritic tree of Purkinje cells in the rat receives about 175 000 excitatory glutamatergic synaptic contacts from parallel fibers of granule cells and around 300 from a single climbing fiber of an inferior olivary neuron. They also receive about 1500 GABAergic inputs from local interneurons. However, even when disconnected from these inputs, Purkinje cells present a tonic, single-spike, spontaneous activity – thus called *intrinsic*.

17.3.1 Purkinje cells present an intrinsic tonic firing that depends on a persistent Na^+ current

Cerebellar Purkinje neurons *in vivo* show high-frequency, regular spontaneous firing that is independent of synaptic activity since it is still recorded in cerebellar slice preparations or cultured Purkinje neurons when synaptic activity is blocked or in isolated Purkinje neurons (**Figure 17.7a**). TTX abolishes this intrinsic firing in all cells tested (**Figure 17.7b**), whereas Ca^{2+}-channel blockers did not suppress it (not shown), suggesting that it consists of Na^+-dependent spikes. How are these spikes generated in the absence of synaptic activity?

Spikes are generated by the spontaneous depolarization that, between consecutive spikes, depolarizes the membrane from the peak of the AHP to the threshold potential of the following spike. This phase of slow depolarization is called *pacemaker potential* or *pacemaker depolarization*, by analogy with pacemaker activity of cardiac cells. To identify the ionic currents that flow during spontaneous activity, previously recorded action potentials are used as voltage commands, and ionic currents during these voltage commands are recorded in voltage clamp (**Figure 17.7c**). This shows that pacemaker depolarization depends mainly on a persistent TTX-sensitive Na^+ current (I_{NaP}) (see Section 14.2.1) present in the cell body of Purkinje cells.

Another key factor allowing spontaneous firing is the lack of active K^+ currents between −70 and −50 mV which allows a high input membrane resistance during the pacemaker depolarization (note that these membrane properties are just the opposite of that of striatal medium spiny neurons; see Section 17.1.1). Thus, initially a small Na^+ current can depolarize the membrane to the threshold potential of Na^+ spikes. Moreover, the cationic I_h present in these cells may also play a role at the beginning of the pacemaker depolarization. The K^+ currents that repolarize the spikes in Purkinje neurons are notable for their very fast deactivation so that the membrane does not hyperpolarize very deeply (there is not a prominent AHP). The rapid deactivation of K^+ current also returns the input resistance to a high value within milliseconds so that the small interspike I_h and I_{NaP} can effectively depolarize the membrane for another action potential. These Na^+ spikes then passively propagate in the dendritic tree (see **Figure 17.9**).

17.3.2 Purkinje cells respond to climbing fiber activation by a complex spike

We will study the response of Purkinje cells to one of its excitatory afferents, the climbing fibers that are the

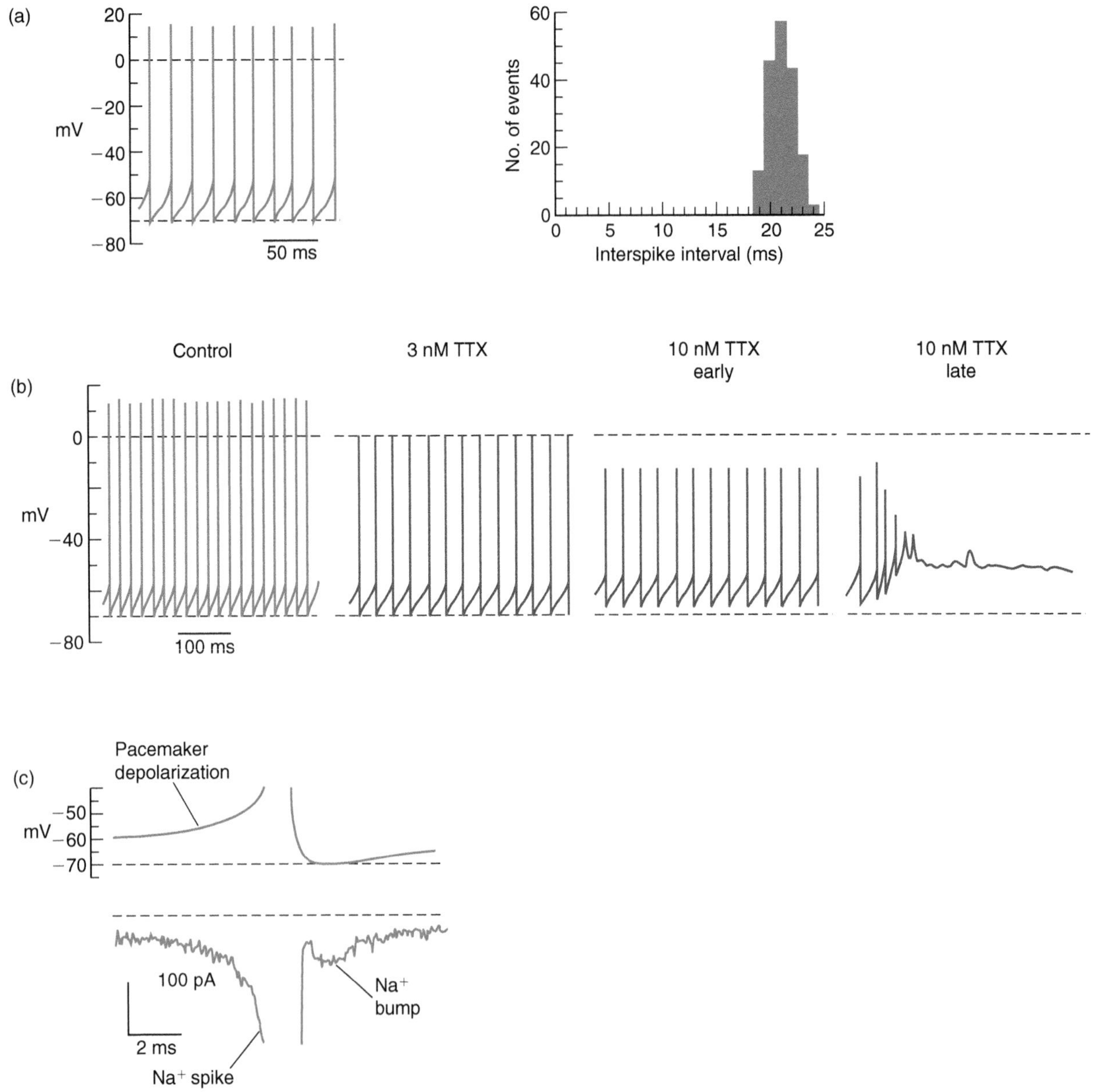

FIGURE 17.7 **The intrinsic tonic firing of isolated Purkinje cells. (a)** Spontaneous action potentials recorded from an isolated Purkinje neuron in control conditions (left) and interspike interval histogram for the same cell (right). Dotted lines indicate −70 and 0 mV. **(b)** Spontaneous firing in control extracellular medium and in the presence of TTX as indicated. This cell continues to fire for some time in 10 nM of TTX (early) before silencing and resting at −51 mV (late). **(c)** Kinetics of Na^+ currents evoked by the spike train protocol. The spike train in **(a)** is used as a command voltage (top trace) and the currents evoked are recorded in voltage clamp (bottom trace). The first 13 ms are shown. The arrow indicates the bump of Na^+ current that occurs when the action potential command reaches its trough. Spike and bump Na^+ currents are sensitive to TTX (not shown). Adapted from Raman IM, Bean BP (1999) Ionic currents underlying spontaneous action potentials in isolated Purkinje neurons. *J. Neurosci.* **19**, 1663–1674, with permission.

axons of inferior olivary neurons. In the adult, a single climbing fiber innervates each Purkinje cell. This innervation has the following particular characteristic. The climbing fiber winds itself around the dendrites making a great number of 'en passant' boutons along its course (see **Figures 6.8** and **6.9**). These synapses are excitatory and the neurotransmitter is the excitatory amino acid glutamate. The activation of a climbing fiber thus causes a massive all-or-none depolarization of the dendritic arborization and an activation of the high-threshold Ca^{2+} channels present at different points along the dendrites (see **Figures 16.10** and **17.9**). Thus, in response to the activation of a climbing fiber, several Ca^{2+} action potentials are generated in the dendrites. These Ca^{2+} action potentials propagate passively along the dendrites, summing together and depolarizing the axon initial segment to the threshold for triggering sodium action potentials (**Figure 17.8a**). Ca^{2+} spikes force the cell to

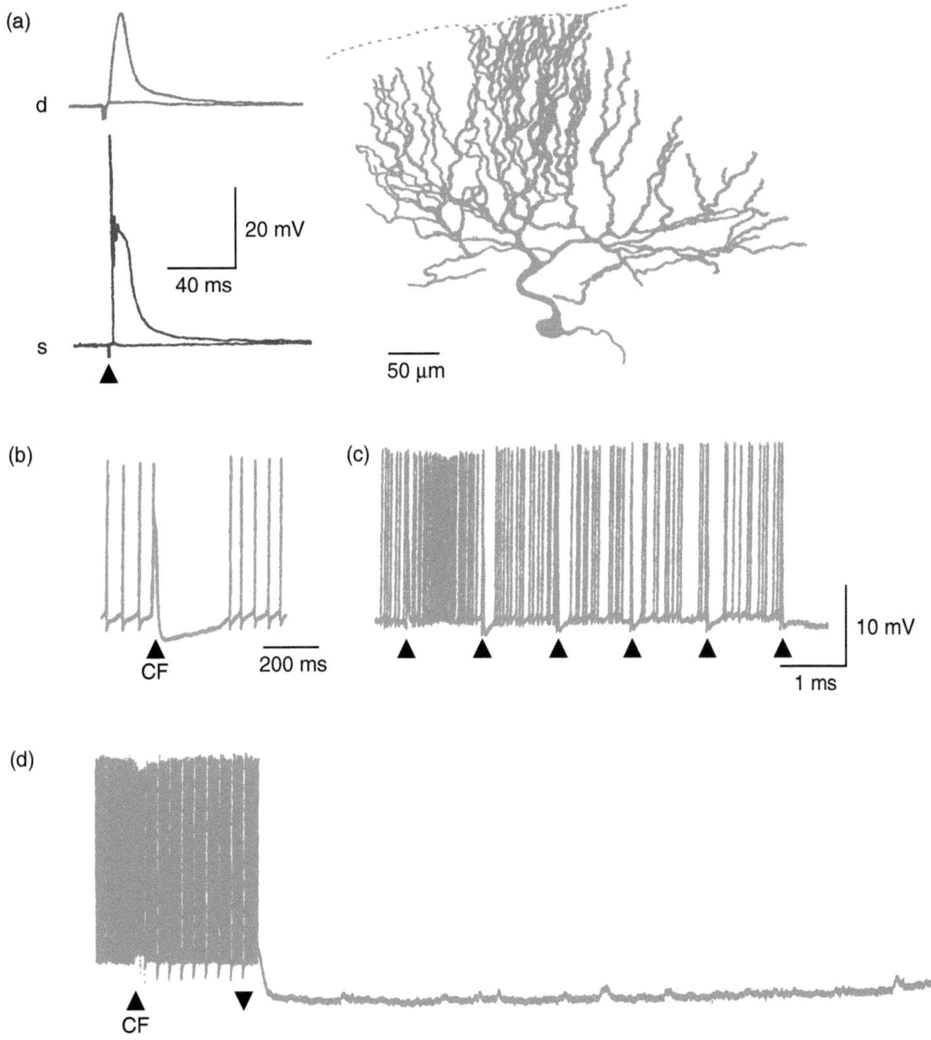

FIGURE 17.8 **Climbing fiber response of Purkinje cells and its after-effect. (a)** All-or-none dendritic (d, top) and somatic (s, bottom) climbing fiber response. The position of the traces relative to the drawing of the recorded Purkinje cell indicates the recording sites. **(b)** Climbing fiber (CF) response followed by a transient inactivation of spontaneous firing. **(c)** Climbing fiber stimulation at 1 Hz (arrowheads). **(d)** At a slower sweep speed, the long-lasting hyperpolarization following a train of climbing fiber stimulation at 1 Hz is shown. Adapted from Hounsgaard J, Mitgaard J (1989) Synaptic control of excitability in turtle cerebellar Purkinje cells. *J. Physiol. (Lond.)* **409**, 157–170, with permission.

respond with a high-frequency burst of Na^+ spikes at the level of the soma and axon. Afferent information coming from the inferior olive is thus amplified.

The climbing fiber response is followed by a long-lasting hyperpolarization (**Figure 17.8b**) that is abolished in the presence of TEA, but unaffected by TTX, thus suggesting that it is mediated by a voltage-sensitive, Ca^{2+}-dependent K^+ current. Repeated activation of the climbing fiber gradually induces an additional hyperpolarization with a much longer time course that is accompanied by a reduction of the frequency of Na^+ spikes (**Figure 17.8c**). Therefore, climbing fiber responses are potent regulators of Purkinje cell excitability. For example, in cells with a high spontaneous firing rate, climbing fiber responses evoked at 10 Hz shift the membrane potential by 10–15 mV to a level well below the threshold for Na^+ spikes (**Figure 17.8d**). The non-linear membrane properties of the soma-dendritic membrane of Purkinje cells are such that only small changes in current are needed to shift the membrane potential in the depolarizing or hyperpolarizing direction.

17.3.3 Purkinje cell activity *in vivo*

In vivo, Purkinje neurons fire action potentials tonically at high rates, despite the presence of high levels of spontaneous inhibitory synaptic activity. Superimposed upon this tonic firing are complex spikes generated by the periodic activation of the climbing fibers originating from the inferior olive as well as increases in firing rate caused by single spikes evoked by activity in granule cells, the origin of the parallel fibers. In freely moving animals, or animals trained for a non-periodic task, complex spikes are randomly evoked and do not show

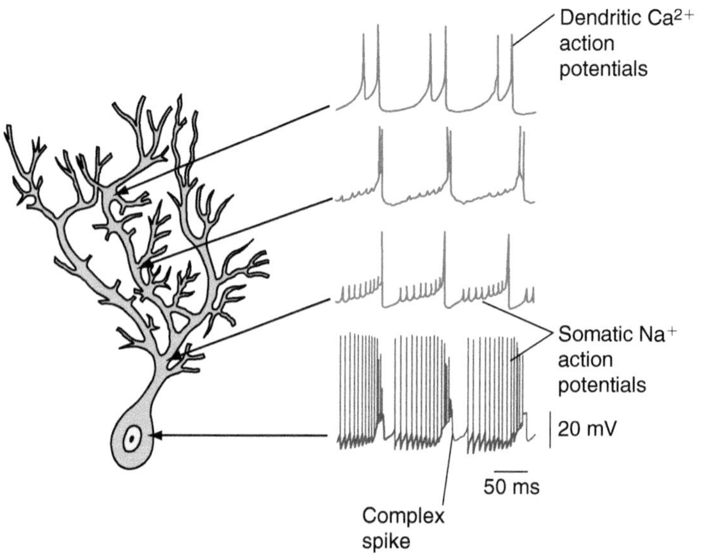

FIGURE 17.9 **Integration of Na$^+$ and Ca^{2+} action potentials in Purkinje cells.** The activity of Purkinje cells is recorded intracellularly in the soma and in three different regions of the dendritic tree (current clamp mode) in cerebellar slices. Na$^+$-dependent action potentials are spontaneously evoked at the soma–axon hillock region. They passively backpropagate in the dendritic tree (note their rapid and strong diminution in amplitude). Ca^{2+}-dependent action potentials evoked at different points of the dendritic tree (in response to climbing fiber activation) propagate passively to the axon hillock region where they evoke the complex response followed by a period of cell silence. Adapted from Llinas R, Sugimori M (1980) Electrophysiological properties of *in vitro* Purkinje cell dendrites in mammalian cerebellar slices. *J. Physiol. (Lond.)* **305**, 197–213, with permission.

rhythmicity, questioning the role of inferior olive periodic oscillations.

Summary

Purkinje cells are not silent at rest, they display a tonic firing mode of Na$^+$-dependent action potentials that depends on the depolarizing drive of an intrinsic persistent Na$^+$ current. These action potentials passively backpropagate in the dendritic tree. Purkinje cells produce two distinct forms of action potential output: simple and complex spikes. Simple spikes are intrinsically generated or are driven by parallel fiber input, while complex spikes are activated by climbing fiber input. In response to climbing fiber activation, an Na$^+$–Ca^{2+} spike is evoked (resulting from the activation of the high-threshold dendritic Ca^{2+} current and of the somatic Na$^+$ current). It is followed by a long-lasting inhibition of intrinsic tonic firing due to the activation of Ca^{2+}-activated K$^+$ currents (**Figure 17.9**).

17.4 THALAMIC AND SUBTHALAMIC NEURONS

The *thalamus* relays and integrates information destined for the cerebral cortex (see **Figure 1.14**). It is formed from many nuclei which are classically separated into two groups: the specific nuclei and the non-specific nuclei, according to whether they project to a localized area of the cerebral cortex or to several functionally different areas. When recorded in brain slices *in vitro*, the thalamocortical and thalamic reticular neurons have complex intrinsic properties that allow them to display two firing patterns, a tonic one and a bursting one, depending on membrane potential (**Figure 17.10a**). Similarly, *in vivo*, during periods of slow-wave sleep, rhythmic burst firing is prevalent, whereas waking activity is dominated by the occurrence of trains of action potentials.

The *subthalamic nucleus* (STN) is part of the basal ganglia. Its name comes from its localization ventral to the thalamus. STN controls the output of basal ganglia to the thalamus. It contains a homogeneous population of Golgi type I neurons that use glutamate as a neurotransmitter and project to both substantia nigra and the two pallidal segments. Like thalamo-cortical neurons, when recorded in brain slices *in vitro*, STN neurons display two firing patterns, a tonic one and a bursting one, depending on the membrane potential (**Figure 17.10b**). In awake resting animals, STN neurons have a tonic mode of discharge; whereas during and after limb and eye movements as well as in a parkinsonian state, STN neurons discharge bursts of high-frequency spikes.

In both neuronal types, tonic firing is recorded at more depolarized potential than burst firing (**Figure 17.10**). Modulation of the membrane potential by the activity of afferents thus plays an important role in the triggering of either one of the discharge configurations.

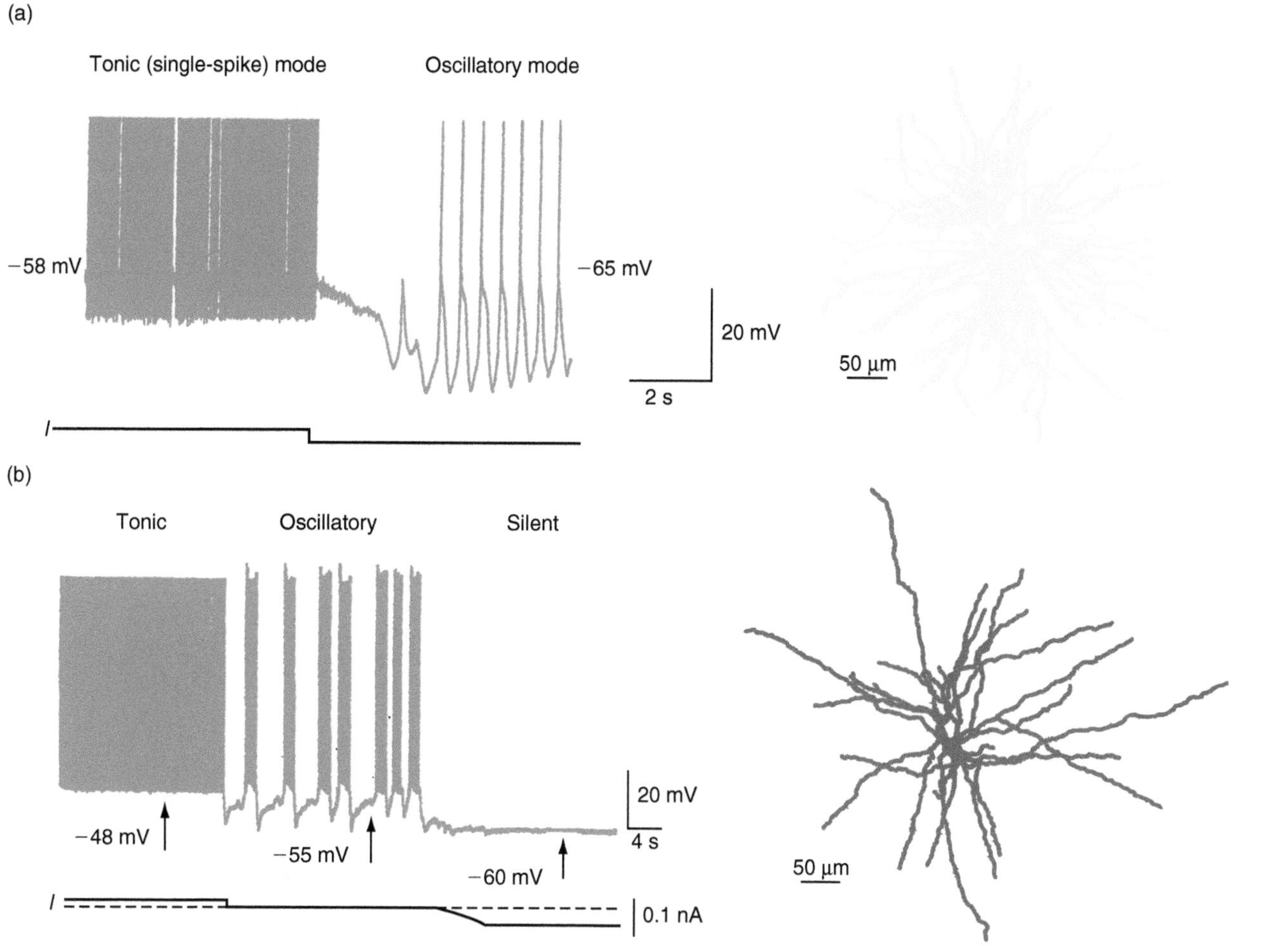

FIGURE 17.10 **The two states of activity of thalamic and subthalamic neurons. (a)** The activity of a thalamocortical neuron (inset) is recorded in current clamp. When depolarized to −58 mV with intracellular injection of current, the neuron displays the tonic firing mode and switches to the oscillatory bursting mode when hyperpolarized. **(b)** The same protocol applied to a subthalamic neuron (inset) allows one to record the two firing modes. When the membrane is further hyperpolarized, the cell becomes silent. Part (**a**) adapted from McCormick DA, Pape HC (1990) Properties of a hyperpolarization-activated cation current and its role in rhythmic oscillation in thalamic relay neurons. *J. Physiol. (Lond.)* **431**, 291–318, with permission. Part (**b**) adapted from Beurrier C, Congar P, Bioulac B, Hammond C (1999) Subthalamic neurons switch from single-spike activity to burst-firing mode. *J. Neurosci.* **19**, 599–609, with permission. Drawings by Jérôme Yelnik.

17.4.1 The intrinsic tonic (single-spike) mode depends on a persistent Na^+ current

Tonic activity of STN neurons recorded in slices *in vitro* is still present when blockers of synaptic transmission are added in the external medium, such as the Ca^{2+}-channel blockers Co^{2+}, Cd^{2+} or Mn^{2+} (**Figure 17.11a**). This shows that the single-spike mode results from a cascade of voltage-gated currents intrinsic to the membrane. As in Purkinje neurons, spikes are generated by the spontaneous depolarization that, between consecutive spikes, depolarizes the membrane from the peak of the AHP to the threshold potential of the following spike. TTX (1 μM) abolishes spontaneous firing, indicating that it consists of Na^+ spikes. Interestingly, at the onset of action of TTX, a few sub-threshold slow depolarizations that normally lead to spike firing are still observed (**Figure 17.11b**). As in Purkinje cells, this phase of slow depolarization, the pacemaker depolarization, depends mainly on the activation of a persistent TTX-sensitive Na^+ current (I_{NaP}) present in these neurons and which presents a voltage dependency that allows it to be activated in the pacemaker range (**Figure 17.11c**).

The same ionic mechanism underlies tonic firing in thalamic neurons (see **Figure 17.14a**). It is important to note that in both cells the other key factor that allows spontaneous firing is the weak presence of active K^+ currents between −70 and −50 mV which allows a high input membrane resistance. Thus, a small Na^+ current can depolarize the membrane to the threshold potential of Na^+ spikes. Moreover, in thalamic and subthalamic neurons, there is a significant contribution of I_h (see Section 14.2.3) to the resting membrane potential as shown by the hyperpolarizing effect of external Cs^+. This hyperpolarization is, in general, large enough to move the cell into the burst mode of action potential generation (see

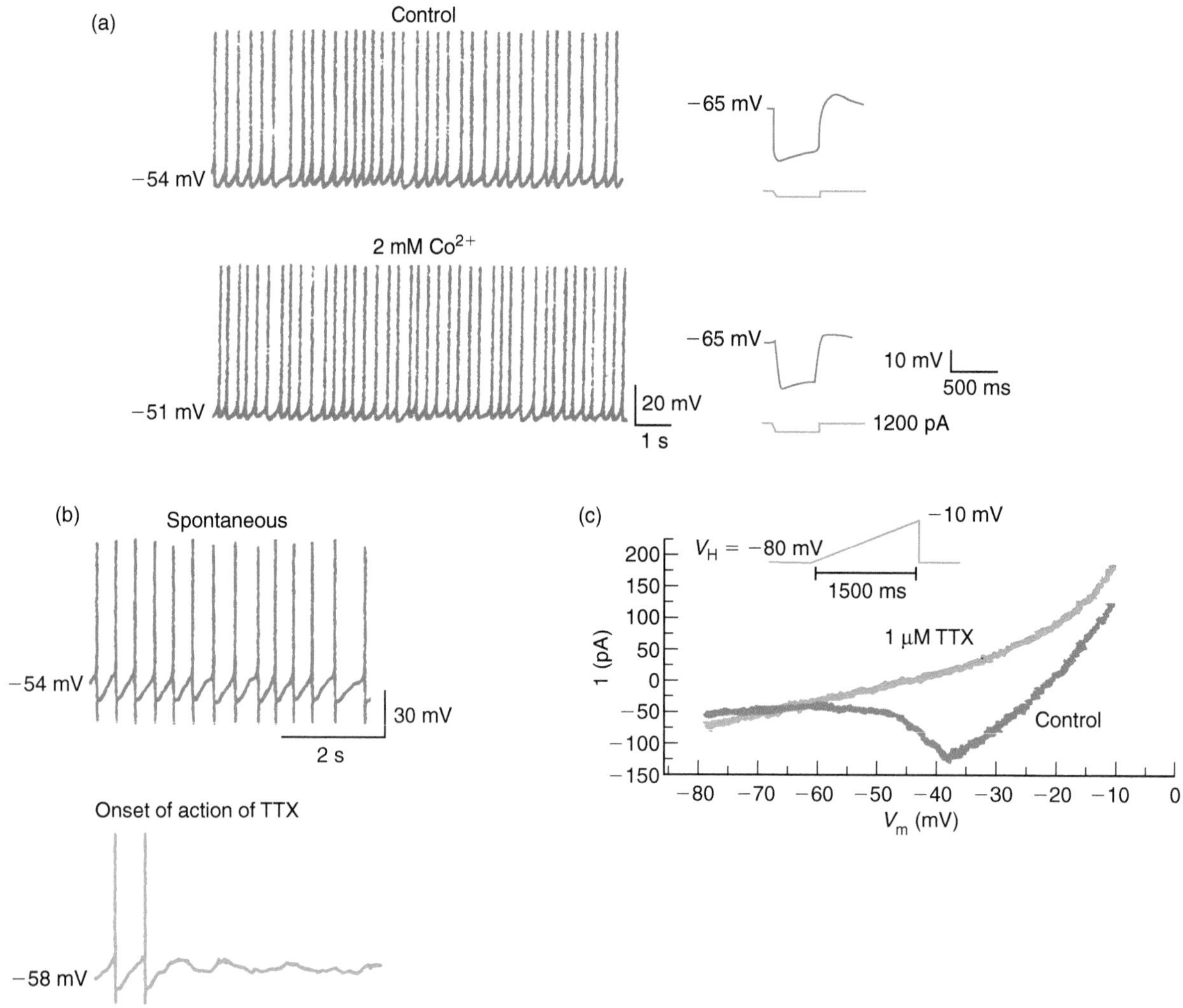

FIGURE 17.11 **Na^+ currents are critical for intrinsic tonic firing mode of subthalamic neurons. (a)** Tonic activity of an STN neuron recorded in control medium and during application of Co^{2+} (left). Right traces show that the low-threshold Ca^{2+} spike evoked at the break of a hyperpolarization pulse is strongly decreased in Co^{2+} to attest that Ca^{2+} channels are effectively blocked in these conditions. **(b)** Tonic activity recorded in control medium and at the onset of TTX (1 μM) application. **(c)** Persistent Na^+ current recorded in whole-cell patch clamp in response to a depolarizing ramp (5 mV s^{-1}) in the absence (control) and presence of TTX. Parts (**a**) and (**c**) adapted from Beurrier C, Bioulac B, Hammond C (2000) Slowly inactivating sodium current (I_{NaP}) underlies single-spike activity in rat subthalamic nucleus. *J. Neurophysiol.* **83**, 1951–1957, with permission. Part (**b**) adapted from Bevan MD, Wilson CJ (1999) Mechanisms underlying spontaneous oscillation and rhythmic firing in rat subthalamic neurons. *J. Neurosci.* **19**, 7617–7628, with permission.

Figure 17.14), suggesting that the fraction of I_h open at rest depolarizes the membrane and maintains it in a stable state where the neuron discharges in the single-spike mode.

17.4.2 The bursting mode depends on a cascade of subliminal inward currents: I_h, I_{CaT}, I_{CAN}

The burst of action potentials which rise from the peak of each slow depolarization or low-threshold spike disappear in the presence of TTX. In contrast, LTS is not affected by TTX but disappears in the presence of Ca^{2+}-channel blockers (**Figure 17.12**). This demonstrates that the fast action potentials are sodium spikes and that LTS results from a Ca^{2+} current. This slow depolarization appears only when the membrane has been previously hyperpolarized for at least 150 ms, suggesting that it results from a low-threshold-activated T-type Ca^{2+} current (I_{CaT}). This current is normally inactivated at resting membrane potential (or at potentials more depolarized than resting potential) and is de-inactivated by a transient hyperpolarization of the membrane. The low-threshold spike leads to the activation of a high-threshold Ca^{2+} current, the entry of Ca^{2+} ions (probably in the dendrites) and the activation of Ca^{2+}-sensitive K^+ currents (I_{KCa}). Each action potential is followed by a phase of after-hyperpolarization (see **Figure 17.14b**).

The hyperpolarization-activated cationic current (I_h) known to be present and activated in the oscillatory range in thalamic neurons also plays a role. For example,

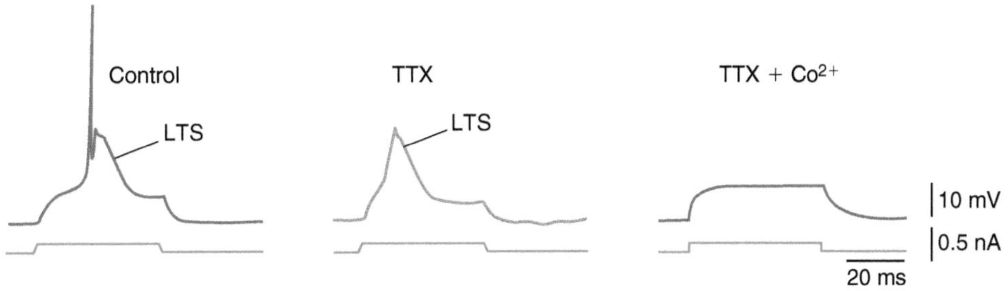

FIGURE 17.12 **Thalamic oscillations depend on a low-threshold Ca^{2+} spike (LTS).** When the membrane of a thalamocortical neuron is hyperpolarized to −65 mV, a depolarizing current pulse evokes an LTS that is insensitive to TTX and abolished by Co^{2+} (1 mM). Note the presence of a TTX-sensitive Na^+ spike in control conditions. Adapted from Llinas R, Jahnsen H (1982) Electrophysiology of thalamic neurons *in vitro*. *Nature* **297**, 406–408, with permission.

application of small amounts of Cs^+ hyperpolarizes the membrane and reduces the AHP. As already said, a fraction of I_h is open at rest and depolarizes the membrane. In addition, during the low-threshold Ca^{2+} spike and the generation of action potentials, a portion of I_h is deactivated (owing to the depolarization). This deactivation of I_h facilitates the generation of the AHP by allowing it to reach more negative potentials (see **Figure 17.14b**). This pronounced AHP subsequently results in two important effects: removal of inactivation of I_{CaT} and the activation of I_h. The latter in turn depolarizes the membrane potential toward the threshold for activation of I_{CaT} and subsequently promotes the generation of a low-threshold Ca^{2+} spike and associated Na^+-dependent action potentials.

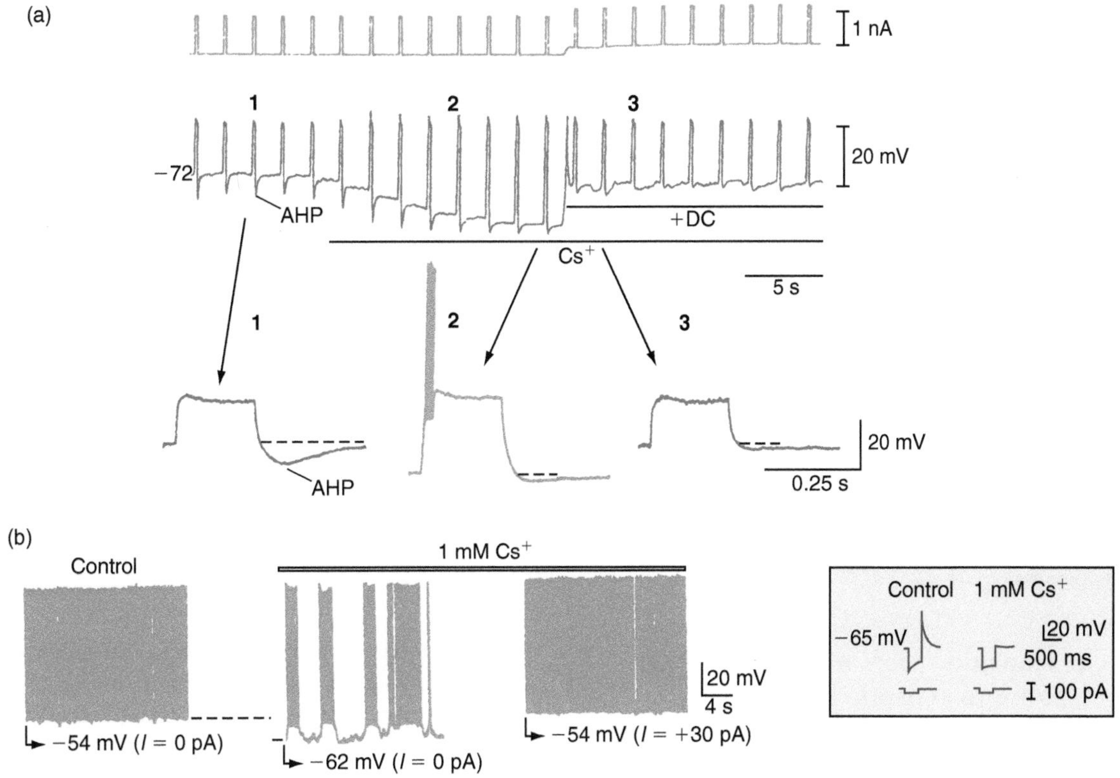

FIGURE 17.13 **Contribution of I_h to resting potential and firing mode. (a)** Thalamocortical neuron. A depolarizing current pulse from resting potential (−72 mV) which does not result in an LTS (1) or the generation of action potential is applied. Cs application results in a substantial hyperpolarization of the membrane that de-inactivates the LTS thereby activating a burst of spikes (2). Compensation for the hyperpolarization with intracellular injection of current (+ DC) reveals that the AHP is nearly abolished during Cs^+ (3). **(b)** Subthalamic (STN) neuron. In control conditions, at rest, an STN neuron discharges in the single-spike mode. Bath application of Cs^+ hyperpolarizes the membrane by 8 mV and shifts STN activity to burst firing mode. Continuous injection of positive current shifts the membrane potential back to the control value and to single-spike activity, though Cs^+ is still present. Concomitantly, the depolarizing sag in response to negative current pulse is strongly decreased as well as the depolarizing rebound seen at the break of pulse, to attest that I_h is strongly reduced (insets). Part **(a)** adapted from McCormick DA, Pape HC (1990) Properties of a hyperpolarization-activated cation current and its role in rhythmic oscillation in thalamic relay neurons. *J. Physiol. (Lond.)* **431**, 291–318, with permission. Part **(b)** adapted from Beurrier C, Bioulac B, Hammond C (2000) Slowly inactivating sodium current (I_{NaP}) underlies single-spike activity in rat subthalamic nucleus. *J. Neurophysiol.* **83**, 1951–1957, with permission.

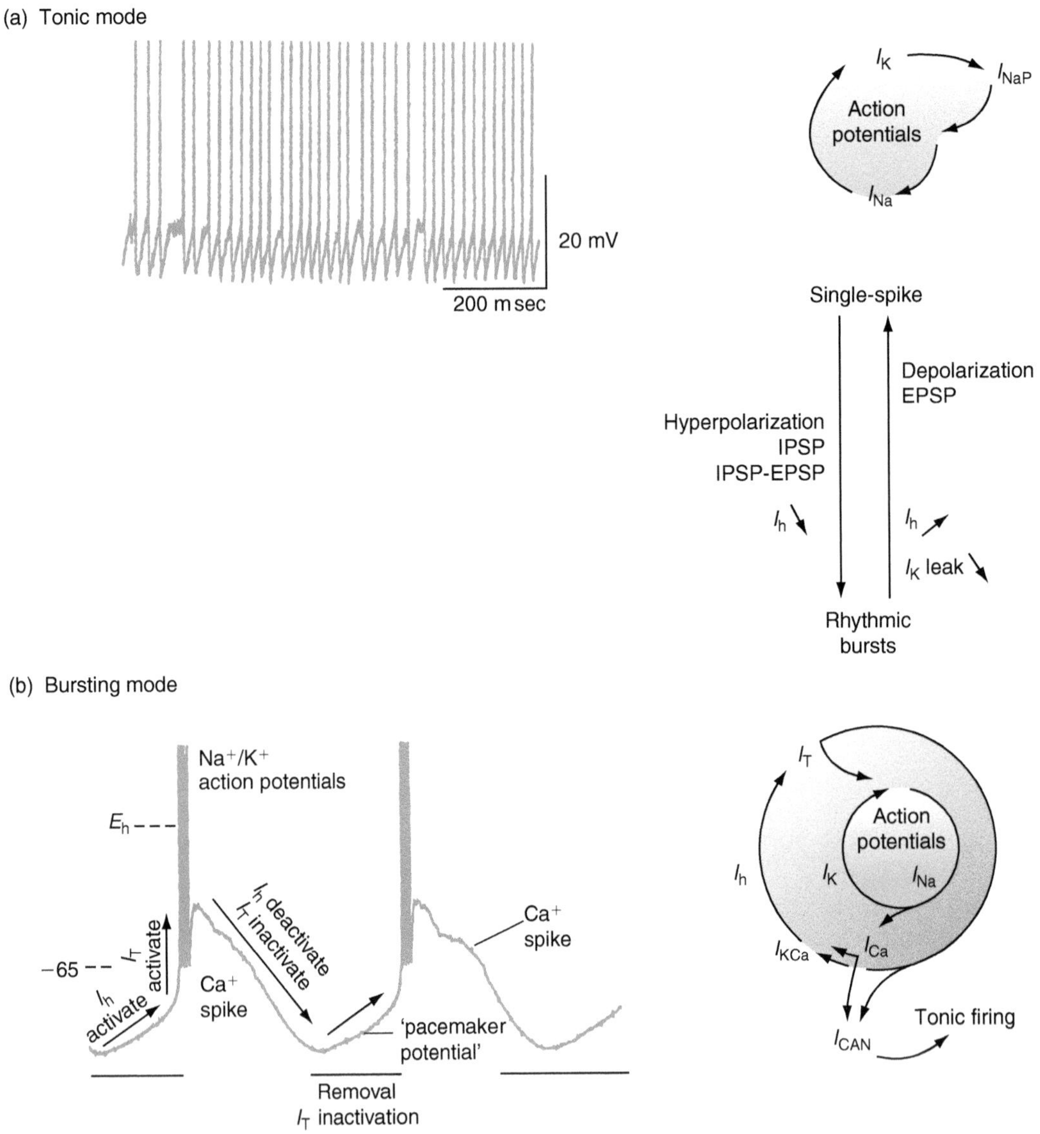

FIGURE 17.14 **Currents underlying the tonic and burst firing modes.** Recordings and scheme of the ionic basis of **(a)** the tonic mode and **(b)** the bursting mode of thalamic neurons. Adapted from McCormick DA, Pape HC (1990) Properties of a hyperpolarization-activated cation current and its role in rhythmic oscillation in thalamic relay neurons. *J. Physiol. (Lond.)* **431**, 291–318; and Bal T, McCormick DA (1993) Ionic mechanisms of rhythmic burst firing and tonic activity in the nucleus reticularis thalami, a mammalian pacemaker. *J. Physiol. (Lond.)* **468**, 669–691, with permission.

17.4.3 The transition from one mode to the other in response to synaptic inputs

When do thalamic and STN neurons discharge in a single spike? At resting membrane potential or at potentials more positive than rest, I_{CaT} (see Section 14.2.2) is inactivated and the regular frequency firing pattern can thus occur. In this state, an EPSP evokes a regular train of discharge.

When do thalamic and STN neurons discharge in bursting mode? Bursting mode requires that the membrane is at a potential more negative than the resting potential, so that I_{CaT} is de-inactivated and thus may be activated. Bursting state is present as long as the membrane is hyperpolarized. For example, at the break of an IPSP or in response to an EPSP evoked during or after an IPSP, a short sequence of bursts is recorded. In this case, the bursting mode is transient; it is not a stable state. Unless hyperpolarized, thalamic and STN neurons discharge in single-spike mode. We see here that IPSP does not always mean inhibition of the postsynaptic neuron: when neurons have the ability to oscillate, an IPSP can evoke a burst of spikes (i.e. an excitation). This is observed, for example, in STN neurons, during and after the execution of a conditioned movement.

In vivo, thalamocortical neurons discharge in the single-spike or bursting mode, depending on the waking state of the animal: a stable bursting mode is observed

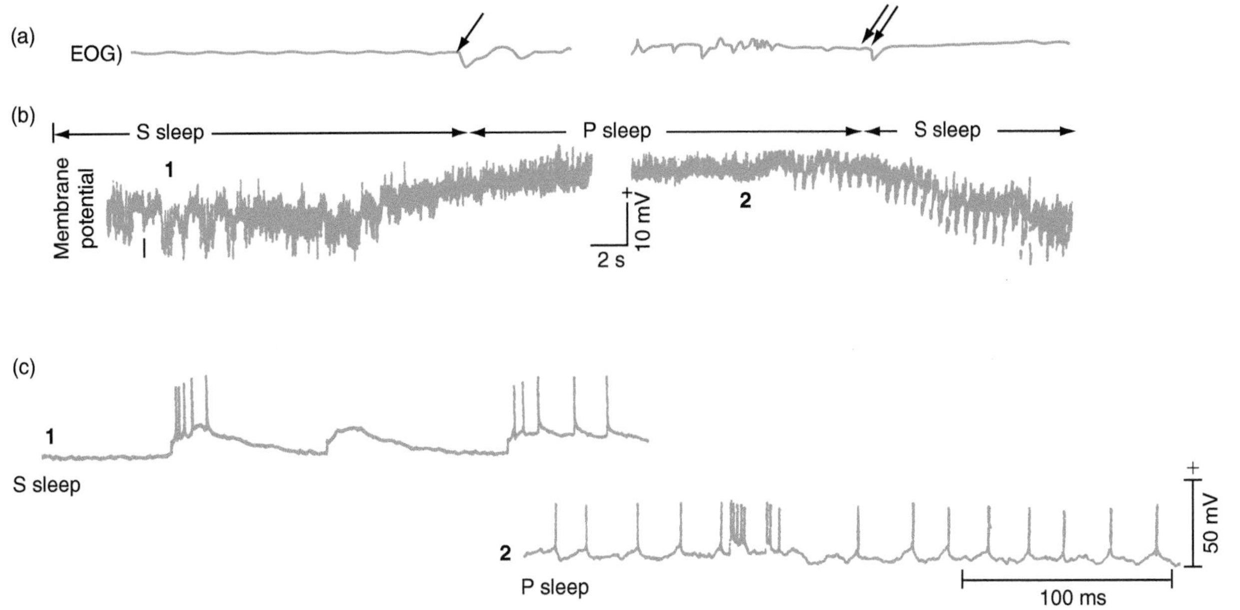

FIGURE 17.15 ***In vivo*, thalamic neurons display the single-spike or bursting mode, in relation to behavioral state.** Simultaneous display of **(a)** eye movements (electro-oculogram, EOG) and **(b)** membrane potential of an intracellularly recorded thalamic neuron during slow-wave sleep (S sleep) and paradoxical sleep (P sleep) in an intact animal. **(b)** The neuron is already depolarized by 8 mV, when the animal enters P sleep (first eye movement, arrow). Depolarization is maintained throughout P sleep. Upon last eye movement (double arrow), membrane potential repolarizes as the animal goes back to S sleep (the trace is filtered at 0–75 Hz). **(c)** Enlarged sequences (labeled 1 and 2 under trace (**b**)) of spontaneous activities: 1, bursting mode during S sleep (hyperpolarized resting potential); 2, single-spike mode during P sleep (depolarized resting potential). Adapted from Hirsch J, Fourment A, Marc ME (1983) Sleep-related variations of membrane potential in the lateral geniculate body relay neurons of the cat. *Brain Res.* **259**, 308–312, with permission.

during slow-wave sleep, a stable single-spike mode during waking or paradoxical sleep (see **Figure 17.15**). The transition from the electroencephalogram (EEG)-synchronized sleep to the waking or rapid-eye-movement (REM)-sleep state (paradoxical sleep) occurs with a progressive depolarization of thalamocortical cells and the abolition of intracellular slow oscillations (LTS) and burst firing, and the appearance of tonic activity. Such changes can be mimicked by the activation of muscarinic or glutamatergic metabotropic receptors that reduce a resting leak K^+ current in thalamocortical neurons (**Figure 17.14**). The modulation of I_h can also play a role as seen in **Figure 17.13**. For example, activation of serotoninergic and β-adrenergic metabotropic receptors shifts the voltage-dependence of I_h to more positive membrane potentials. This reduces the ability of cells to oscillate. Together these results suggest that the release of acetylcholine, glutamate, serotonin and norepinephrine abolishes sleep-related activity in thalamocortical networks and facilitates the single-spike activity typical of the waking state.

Summary

Thalamic and subthalamic neurons can function either as relay systems or as oscillators. During oscillations, afferent informations have a low probability of evoking a response. For example, when thalamic neurons are oscillating during slow-wave sleep, there is a marked diminution of responsiveness of thalamic neurons to activation of their receptive fields, 'presumably owing to the hyperpolarized state of these neurons, the interrupting effects of spontaneous thalamocortical rhythms and the frequency limitations of the burst firing mode'. Oscillations are also recorded in pathological conditions: in the STN of parkinsonian patients and in the thalamocortical networks during absence epileptic seizures. Noteworthy, during these oscillations, motor or sensory processing is unimpaired.

Further reading

Bal, T., Von Krosigk, M., McCormick, D.A., 1994. From cellular to network mechanisms of a thalamic synchronized oscillation. In: Buzski, G. (Ed.), Temporal Coding in the Brain. Springer-Verlag, Berlin.

Byrne, J.H., 1980. Analysis of ionic conductance mechanisms in motor cells mediating inking behavior in *Aplysia californica*. J. Neurophysiol. 43, 630–650.

Choi, S., Yu, E., Kim, D., et al., 2010. Subthreshold membrane potential oscillations in inferior olive neurons are dynamically regulated by P/Q- and T-type calcium channels: a study in mutant mice. J. Physiol. 588, 3031–3043.

Crépel, F., Pénit-Soria, J., 1986. Inward rectification and low threshold calcium conductance in rat cerebellar Purkinje cells: an in vitro study. J. Physiol. (Lond.) 372, 1–23.

Hakimian, S., Norris, S.A., Greger, B., Keating, J.G., Anderson, C.H., Thach, W.T., 2008. Time and frequency characteristics of Purkinje cell complex spikes in the awake monkey performing a nonperiodic task. J. Neurophysiol. 100, 1032–1040.

Llinas, R.R., 1988. The intrinsic electrophysiological properties of mammalian neurons: insights into central nervous system function. Science 242, 1654–1664.

Llinas, R.R., Jahnsen, H., 1982. Electrophysiology of mammalian thalamic neurons. Nature 297, 406–408.

Mathy, A., Ho, S.S., Davie, J.T., Duguid, I.C., Clark, B.A., Häusser, M., 2009. Encoding of oscillations by axonal bursts in inferior olive neurons. Neuron 62, 388–399.

McCormick, D.A., Bal, T., 1997. Sleep and arousal: thalamocortical mechanisms. Ann. Rev. Neurosci. 20, 185–215.

Wilson, C.J., Kawaguchi, Y., 1996. The origins of two state spontaneous membrane fluctuations of neostriatal spiny neurons. J. Neurosci. 16, 2397–2410.

CHAPTER

18

Synaptic plasticity

Constance Hammond, Jean-Marc Goaillard, Dominique Debanne, Jean-Luc Gaiarsa

OUTLINE

Synaptic responses undergo short- and long-term modifications. This chapter examines the mechanisms underlying plasticity in adult synapses. Developmental forms of plasticity are not covered here. The first form of long-term changes of synaptic efficacy is called hebbian plasticity and comprises long-term potentiation (LTP) and long-term depression (LTD). Moreover, there are several forms of LTP and LTD classified by their mechanisms and their mode of induction. We have chosen three examples, the NMDA receptor-dependent LTP, the metabotropic glutamate receptors (mGluR)-dependent LTD in the cerebellum and the spike-timing dependent plasticity (STDP). The second, more recently discovered, form of long-term changes of synaptic efficacy is called homeostatic plasticity and occurs in response to prolonged changes in activity. Homeostatic plasticity can affect both excitatory and inhibitory synapses, but we have chosen the example of synaptic scaling at glutamatergic synapses to illustrate this type of plasticity. Before explaining these different forms of long-term synaptic changes, we shall examine the meaning of 'long term' versus 'short term'.

18.1 SHORT-TERM POTENTIATION (STP) OF A CHOLINERGIC SYNAPTIC RESPONSE AS AN EXAMPLE OF SHORT-TERM PLASTICITY: THE CHOLINERGIC RESPONSE OF MUSCLE CELLS TO MOTONEURON STIMULATION

Repetitive high-frequency (>15 Hz) stimulation of the presynaptic element (motoneuron) leads to a short-term potentiation (STP) of the postsynaptic response of the muscle cell. As shown in **Figure 18.1**, successive stimulations produce in these conditions excitatory postsynaptic currents (EPSC) of greater and greater amplitudes. This phenomenon, first discovered at the neuromuscular junction, is also observed at the squid giant synapse and in mammalian afferent synapses to motoneurons.

Cellular and Molecular Neurophysiology. DOI: 10.1016/B978-0-12-397032-9.00018-2

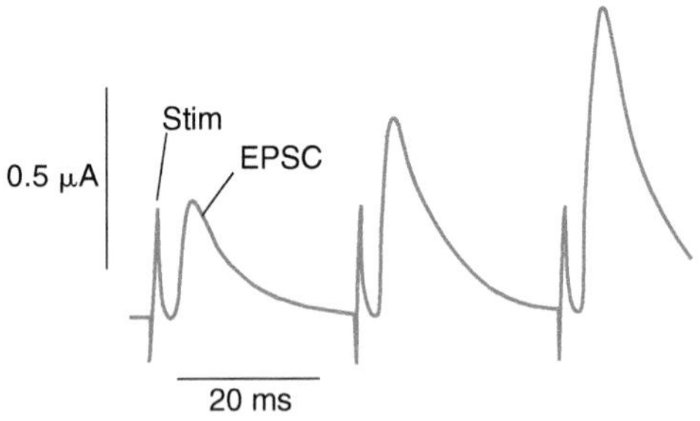

FIGURE 18.1 **Presynaptic facilitation at the frog neuromuscular junction.** The activity of a frog sartorius muscle cell is recorded in normal Ringer solution (V_m = −90 mV). The motor endplate currents are evoked by repetitive stimulations (stim) of the motor nerve (2 μA intensity, 5 ms duration). The average current intensity (EPSC) in response to the first stimulation is 0.5 μA. This amplitude gradually rises following second and third stimulations. The inward currents are represented upwardly, which is unusual. Adapted from Katz B, Miledi R (1979) Estimates of quantal content during chemical potentiation of transmitter release. *Proc. R. Soc. Lond.* **B205**, 369–378, with permission.

In the squid giant synapse, synaptic facilitation has the following characteristics. When the presynaptic element repeatedly fires, an increase of the postsynaptic response amplitude is observed. This increase diminishes with a time constant of the order of tens of milliseconds. Simultaneous recordings of presynaptic action potentials, presynaptic Ca^{2+} current (I_{Ca}), variations of the intracellular Ca^{2+} concentration and postsynaptic depolarization show that the postsynaptic response amplitude increases when:

- the amplitude and duration of presynaptic spikes are unchanged;
- the amplitude of the presynaptic I_{Ca} evoked by each presynaptic depolarizing pulse or action potential is constant;
- the increase of the presynaptic intracellular Ca^{2+} concentration is identical in response to each depolarizing pulse or action potential.

The increase of intracellular Ca^{2+} concentration ($[Ca^{2+}]_i$) in the presynaptic element slowly disappears, in about one second, whereas the Ca^{2+} current and the release of the neurotransmitter both last about 1 ms. Katz and Miledi, in 1965, were the first to propose that STP is due to residual Ca^{2+} ions still present in the presynaptic active zone when the second presynaptic spike occurs. The following hypothesis was proposed. Ca^{2+} ions enter the presynaptic element through voltage-gated Ca^{2+} channels opened by the depolarization. The intracellular Ca^{2+} concentration is very high at active zones at the end of the action potential. These Ca^{2+} ions act rapidly and locally on target molecules to trigger the exocytosis of synaptic vesicles with a probability p. At the same time, the Ca^{2+} ions are also buffered in the cytoplasm and are actively transported to the extracellular medium or into organelles (see Section 7.2.4). But a residual and quite high $[Ca^{2+}]_i$ is still present close to the presynaptic membrane for some time. This $[Ca^{2+}]_i$ value is not high enough to trigger neurotransmitter release but, added to the incoming increase of $[Ca^{2+}]_i$ accompanying the arrival of the second action potential (when the delay between the two action potentials is short), it increases neurotransmitter release probability to the second action potential, and thus causes potentiation of the postsynaptic response.

STP can also be induced by high-frequency stimulation (conditioning tetanus) of the afferent motoneuron (model of the crayfish neuromuscular junction) (**Figure 18.2a**). In that case, the postsynaptic response

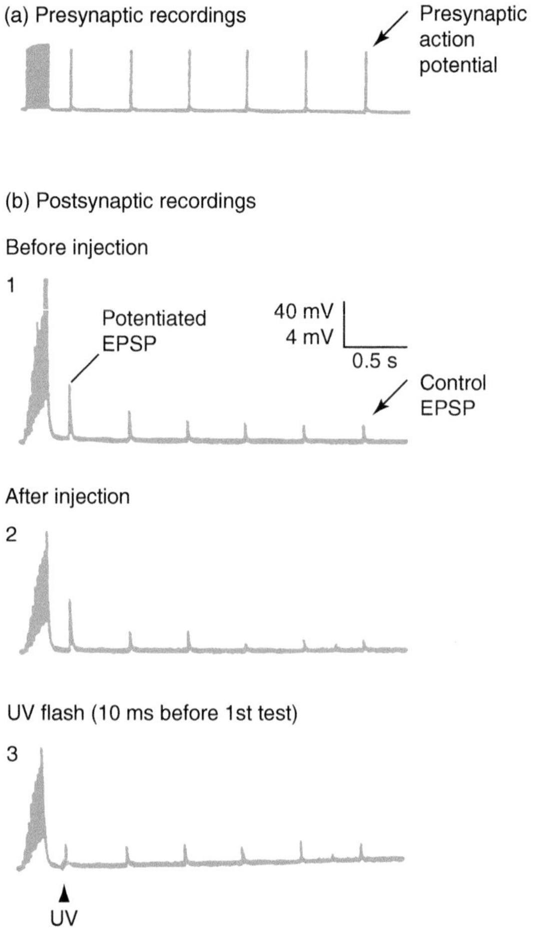

FIGURE 18.2 **Rapid reduction of residual Ca^{2+} ions quickly eliminates STP.** The activity of the crayfish dactyl opener muscle cell is intracellularly recorded in current clamp mode in response to the stimulation of an axonal branch of the presynaptic motoneuron. The electrode positioned inside the presynaptic axon is filled with diazo-2 (50 mM) and fluorescein (10 mM) in KCl (3 M) in order to both stimulate the presynaptic axon and to fill it with the Ca^{2+}chelator. A conditioning tetanus (10 stimuli at 50 Hz) followed by a single stimulus at 2 Hz is applied to the axon. **(a)** Action potentials recorded from the preterminal axon branch. **(b)** The response (EPSP) of the postsynaptic muscle cell is recorded in control conditions (1), after the intracellular injection of diazo-2 (2) and after photolysis of diazo-2 by an ultraviolet flash given after the tetanus (3). Adapted from Kamiya H, Zucker RS (1994) Residual Ca^{2+} and short-term synaptic plasticity. *Nature* **371**, 603–606, with permission.

(EPSP) that is recorded at regular intervals after the tetanus is potentiated, and then decays to control amplitude within 1.5 s (**Figure 18.2b**, 1). In order to test the hypothesis of Katz and Miledi, a photolabile Ca^{2+} chelator, diazo-2, is injected into the presynaptic terminals. The motoneuron is penetrated at the level of an axon branch with a microelectrode containing KCl (to record presynaptic activity), the photolabile Ca^{2+} chelator diazo-2 (to chelate Ca^{2+} ions with an affinity of 150 nM after photolysis) and fluorescein (to monitor the progress of injection). First, the control STP is recorded (**Figure 18.2b**, 1). Then diazo-2 is injected into the presynaptic axon in order to test that before photolysis diazo-2 has little effect on STP since the unphotolyzed chelator has a low affinity for Ca^{2+} ions (**Figure 18.2b**, 2). An ultraviolet flash is given after the tetanus in order to produce a chelator with 150 nM affinity: the STP of the postsynaptic response is prevented (**Figure 18.2b**, 3).

These results show that STP is due to residual-free Ca^{2+} ions following presynaptic activity. What are the molecular targets of Ca^{2+} action in short-term plasticity? Are Ca^{2+}-sensor proteins other than synaptotagmins (the main trigger for Ca^{2+}-dependent vesicle fusion) involved? Many candidates have been identified among vesicular, plasma membrane and cytoplasmic proteins of the presynaptic element (see Section 7.3) and it is likely that multiple molecular mechanisms contribute to the effects of residual Ca^{2+}.

18.2 LONG-TERM POTENTIATION (LTP) OF A GLUTAMATERGIC SYNAPTIC RESPONSE: EXAMPLE OF THE GLUTAMATERGIC SYNAPTIC RESPONSE OF PYRAMIDAL NEURONS OF THE CA1 REGION OF THE HIPPOCAMPUS TO SCHAFFER COLLATERAL ACTIVATION

18.2.1 The Schaffer collaterals are axon collaterals of CA3 pyramidal neurons which form glutamatergic excitatory synapses with dendrites of CA1 pyramidal neurons

The hippocampus is a telencephalic structure with a rostro-caudal extension in the rat. It is composed of two closely interconnected crescent-like regions, Ammon's horn and the dentate gyrus (**Figure 18.3a**). Ammon's horn is formed by a layer of principal neurons, the pyramidal neurons, and is subdivided into three regions called CA1, CA2 and CA3 (CA for *c*ornu *a*mmonis). The dentate gyrus is formed by a layer of principal neurons called *granular cells*. Numerous interneurons are present in each region (see also Chapter 19).

The pyramidal cells of CA3 have branched axons. One branch leaves the hippocampus and projects to other structures. The other branches are recurrent collaterals that form synapses with dendrites of CA1 pyramidal neurons. These collaterals run in bundles and form the Schaffer collateral pathway; and their terminals form asymmetrical synapses with the numerous spines of CA1 dendrites. These synapses are excitatory and the neurotransmitter is glutamate. Owing to the laminar organization of the hippocampal structure, it is possible to stimulate selectively the Schaffer collateral pathway (stim) and to record the evoked excitatory postsynaptic potential (EPSP) in a CA1 pyramidal soma either *in vivo* or *in vitro* (**Figure 18.3a**). EPSPs can be recorded from a single neuron (intracellular or whole-cell somatic recording) or extracellularly in the dendritic layer from a population of neurons (field EPSPs).

We shall restrict our study of long-term potentiation (LTP) to the synaptic response of CA1 pyramidal neurons to Schaffer collateral stimulation in hippocampal slices, recorded in the presence of bicuculline (an antagonist at $GABA_A$ receptors) in order to prevent the participation of GABAergic inhibitory responses resulting from interneuron activation.

18.2.2 Observation of the long-term potentiation of the Schaffer collateral-mediated EPSP

The pioneering observation was made by Bliss and Lomo, in 1973, that high-frequency stimulation of Schaffer collaterals in the rat hippocampus *in vivo* produces an increase of the amplitude of the Shaffer collateral-mediated EPSP recorded from the postsynaptic pyramidal neuron (**Figure 18.3b**). The exact protocol is the following. A single stimulus applied repeatedly at a very low frequency (0.02–0.03 Hz) to Schaffer collaterals evokes stable control EPSPs in the postsynaptic pyramidal neuron. These control responses can be averaged to give a mean control EPSP. Control EPSPs are largely mediated by non-NMDA receptors since they are nearly completely abolished by CNQX (not shown). A tetanic stimulation is then applied to the Schaffer collateral pathway through the same stimulating electrode (one train of 1 s duration, composed of 50–100 stimuli at 100 Hz). After this tetanus, the same single stimulus applied repeatedly at the same very low frequency, through the same stimulating electrode, now evokes 'post-tetanic' EPSPs of larger amplitude: the EPSP is potentiated. Since this potentiation lasts from minutes to hours it is a long-term potentiation (LTP).

Is LTP restricted to the synapses that have been tetanized?

Two EPSPs evoked in one pyramidal neuron in response to the stimulation of two different Schaffer collateral inputs are recorded (**Figure 18.4a**). When only one input (S_1) is tetanized, the response evoked by a single shock at S_1 is potentiated (LTP of $EPSP_1$) whereas the response evoked in the same pyramidal neuron at S_2 ($EPSP_2$) is not potentiated: LTP is synapse specific.

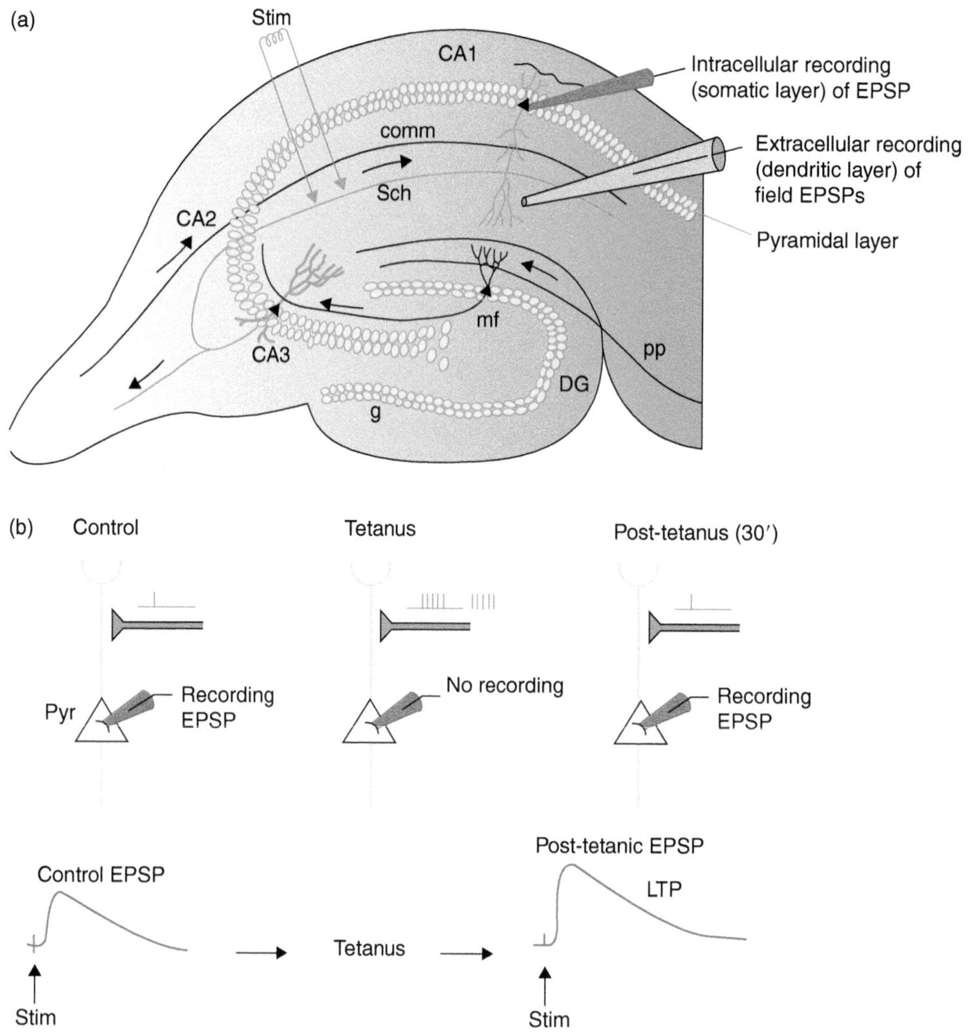

FIGURE 18.3 **Tetanic LTP in the hippocampus is induced by high-frequency stimulation of afferent fibers. (a)** Coronal section of the rat hippocampus showing the major excitatory connections. CA1, CA2, CA3 regions of the hippocampus; DG, dentate gyrus layer composed of granular cells (g) which send their axons (mossy fibers, mf) to CA3 pyramidal dendrites. CA3 pyramidal cells send axon collaterals, called Schaffer collaterals (Sch), to CA1 pyramidal apical dendrites. The tetanic stimulation (stim, e.g. 1–4 trains of 10 stimulations at 100 Hz applied every 1 s) is applied to Schaffer collaterals and the AMPA-mediated postsynaptic response is recorded intracellularly in the soma of a pyramidal cell (EPSP or EPSC) and/or extracellularly in the layer of CA1 pyramidal dendrites. comm, commissural fibers; pp, perforant path. **(b)** A CA1 pyramidal neuron represented upside down compared with its position in the coronal section and the afferent Schaffer collaterals. The AMPA-mediated EPSP evoked by a single stimulation of Schaffer collateral (one vertical bar) is intracellularly recorded in the presence of bicuculline. After a tetanic stimulation of the Schaffer collaterals (shown as high-frequency bars), a potentiation of the glutamatergic EPSP evoked by a single stimulation is recorded. Adapted from Kauer JA, Malenka RC, Nicoll RA (1988) Persistent postsynaptic modification mediates long-term potentiation in the hippocampus. *Neuron* **1**, 911–917, with permission.

In other words, when generated at one set of synapses by repetitive activation, LTP does not normally occur in other synapses on the same cell.

18.2.3 Long-term potentiation (LTP) of the glutamatergic EPSP recorded in CA1 pyramidal neurons results from an increase of synaptic efficacy (or synaptic strength)

LTP of the Schaffer collateral-mediated EPSP can have several origins. We shall study some of the hypotheses one by one.

Does LTP result from a non-specific change of postsynaptic cell excitability?

To test this hypothesis, a pulse of depolarizing current is injected directly into the pyramidal cell. The response of the membrane is the same before and after the tetanus, at all potentials tested. Therefore, LTP does not result from a change of the total resistance of the postsynaptic membrane. It is also not due to a persistent reduction of the inhibitory GABAergic responses since it is still observed in the presence of bicuculline, a $GABA_A$-receptor antagonist.

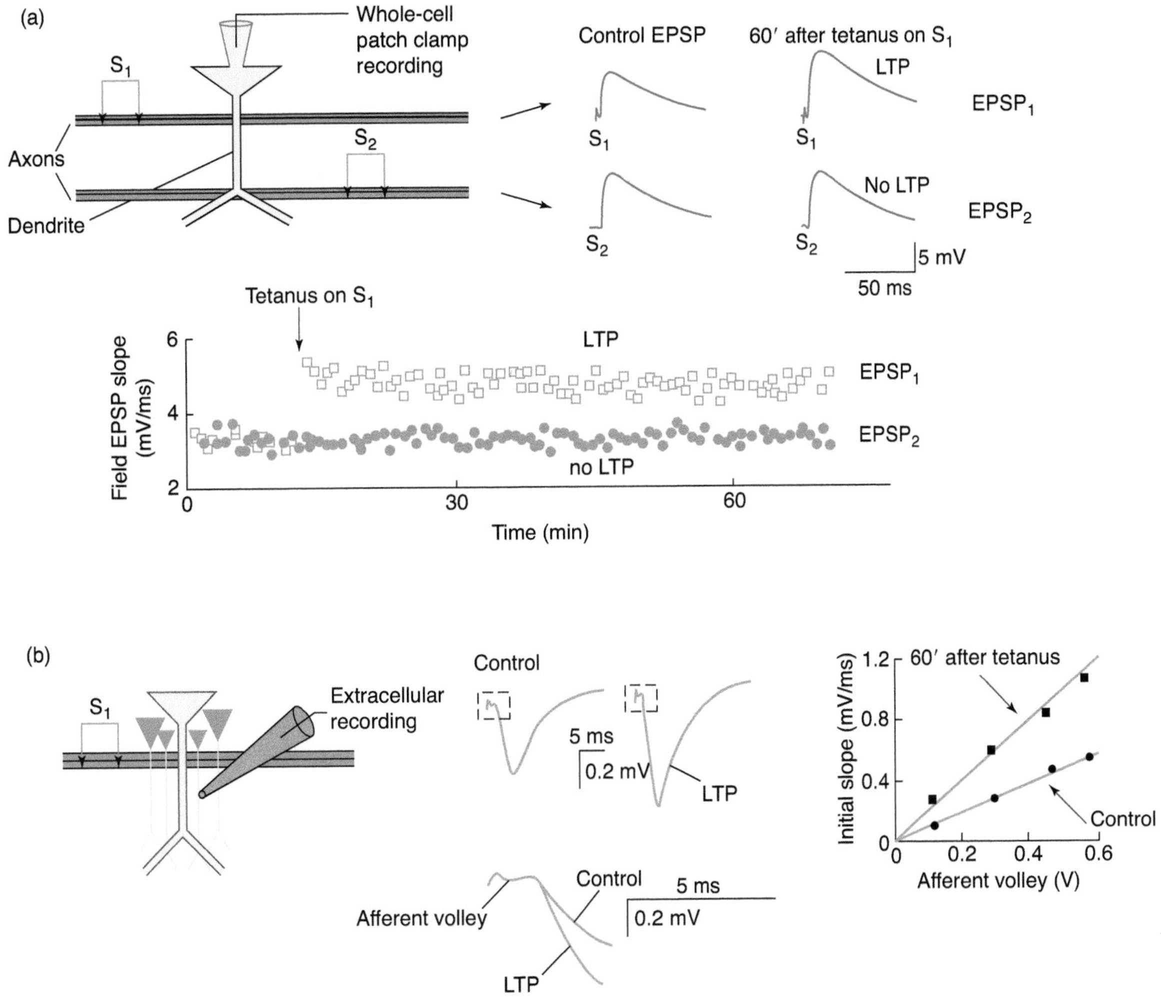

FIGURE 18.4 **Tetanic LTP is synapse specific. (a)** The postsynaptic responses (control $EPSP_1$ and $EPSP_2$) of a single pyramidal neuron are intracellularly recorded in current clamp mode (whole-cell patch) in response to stimulations S_1 and S_2. Then, stimulus S_1 is tetanized but not stimulus S_2. Sixty minutes after the tetanus on S_1, $EPSP_1$ and $EPSP_2$ are recorded again. The diagram illustrates the time course of the initial slope of $EPSP_1$ and $EPSP_2$ before and after the tetanus on S_1. **(b)** Extracellular recording of the response of a population of CA1 pyramidal neurons to stimulation (S_1) of afferent Schaffer collaterals. The stimulation S_1 evokes an afferent volley (the extracellular recording of presynaptic action potentials in all stimulated afferent axons) and a field EPSP (the extracellular recording of the postsynaptic response of pyramidal neurons). Sixty minutes after a tetanus (two trains of 100 Hz, 1 s duration, 30 s interval) applied through the same stimulating electrode, the field EPSP is recorded. Note the increased initial slope 60 min after the tetanus (enlarged dotted squares). The input/output curves depict the amplitude of the afferent volley versus the initial slope of the field EPSP. Part (**a**) from L. Aniksztejn and (**b**) from H. Gozlan, personal communications.

Does LTP result from an increase in the number of stimulated axons?

One way to answer this question is to record simultaneously the presynaptic action potentials (afferent volley) and the postsynaptic response. This is possible with extracellular recordings at the level of apical dendrites of pyramidal cells. A single stimulus applied repeatedly at a low frequency (0.02–0.03 Hz) to the Schaffer collaterals evokes a stable 'control' field EPSP recorded by an extracellular electrode placed in the dendritic field of CA1 pyramidal neurons (**Figures 18.3a** and **18.4b**, left and middle). A field EPSP corresponds to the response of a population of pyramidal neurons situated close to the recording electrode and connected to the stimulated axons. Its slope is proportional to the amplitude of the currents generated in the postsynaptic neurons. After the stimulating artefact, before the field EPSP develops, the afferent volley is recorded (inset).

As a control intracellular EPSP, the control field EPSP is mediated predominantly by non-NMDA receptors since it is nearly completely abolished by the bath application of CNQX (not shown), a selective antagonist of AMPA receptors. A tetanic stimulation (one train of 1 s duration, composed of 50–100 stimuli at 100 Hz) is then applied to the Schaffer collateral pathway through the stimulating electrode. After this tetanus, the same single stimulus, again through the same stimulating electrode, now evokes a 'post-tetanic' field EPSP of larger amplitude and with a steeper initial slope than the control one: the field EPSP is potentiated (LTP). The value of the initial slope of a field EPSP (or of an intracellular EPSP) is an accurate index of the changes of the monosynaptically evoked postsynaptic excitatory response since the field EPSP (as well as the intracellular EPSP) can be composed of monosynaptic as well as polysynaptic unitary EPSPs. This potentiation of EPSP amplitude and slope is persistent: it lasts hours when recorded in the *in vitro* hippocampal slice preparation and days when induced in the freely moving animal. However, the afferent volley (the presynaptic component) is unchanged.

LTP is an increase of synaptic strength

Following a tetanic stimulation, the presynaptic component (the afferent volley) is unchanged whereas the peak amplitude and the initial slope of the postsynaptic one (field EPSPs) are potentiated (by 30% and 200%, respectively). The input/output curve depicting the initial slope of the field EPSPs versus afferent volley amplitude has a different slope before (control) and 60 minutes after the tetanus (**Figure 18.4b**, right). This result shows that potentiation of the postsynaptic response does not result from an increase of the number of stimulated axons but from a genuine increase in synaptic efficacy: the same input evokes an enhanced output.

LTP consists of two phases: induction and maintenance

LTP-generating mechanisms are classically separated into two phases: a brief induction phase (1–20 s) which occurs during tetanus, and a following expression phase; i.e. the mechanisms sustaining the persistent enhancement of synaptic efficacy. Therefore, LTP is triggered rapidly (within seconds) whereas it is maintained for long periods of time (for hours in *in vitro* preparations and days *in vivo*). We will now analyze these two phases.

18.2.4 Induction of LTP results from a transient enhancement of glutamate release and a rise in postsynaptic intracellular Ca^{2+} concentration

Why is tetanic stimulation necessary to induce LTP? What does tetanic stimulation add to a single shock stimulation?

Tetanic stimulation evokes a large release of glutamate from Schaffer collateral terminals compared with a single shock. Glutamate released in synaptic clefts during tetanus binds to non-NMDA- and NMDA-receptor channels but also to the metabotropic glutamate receptors (receptors linked to G proteins) present in the postsynaptic membrane (see **Figure 18.7**). The fact that tetanic stimulation can be replaced by a pairing diagram consisting of low-frequency stimulation of Schaffer collaterals combined with the intracellular depolarization of the postsynaptic neuron suggests that one of the roles of the tetanus is to depolarize the postsynaptic membrane. This is confirmed by the following experiment. When the postsynaptic potential is hyperpolarized during the tetanic stimulation, LTP is not induced. The tetanus-induced depolarization is the large EPSP recorded during the tetanus and which results from the strong activation of AMPA receptors (see **Figures 18.5(2)** and **(4)** and **18.7**).

What does induce postsynaptic depolarization?

Several observations led to the conclusion that in CA1, induction of LTP is not just depolarization dependent but also NMDA-receptor dependent: the application of APV, the selective antagonist of NMDA receptors, during the tetanic simulation prevents the induction of LTP (**Figure 18.5**). In contrast, antagonists of non-NMDA receptors such as CNQX, applied during the tetanus, do not prevent the induction of LTP. Therefore, induction of LTP is voltage and NMDA-receptor dependent. These results confirm that, in the CA1 region of the hippocampus, tetanus induces a postsynaptic *depolarization* generated by the enhancement of glutamate release. This postsynaptic depolarization is a necessary prerequisite for LTP induction, because it allows the activation of NMDA receptors.

What is the role of postsynaptic NMDA-receptor activation?

To explain the APV-sensitivity of LTP, the following hypothesis is proposed. The postsynaptic depolarization evoked by the tetanic stimulation allows the activation of postsynaptic NMDA receptors and the subsequent Ca^{2+} entry into the postsynaptic element. In fact, during tetanic stimulation, an increase of $[Ca^{2+}]_i$ is observed in the dendrites of the postsynaptic pyramidal cells as visualized with a fluorescent calcium-sensitive dye. When this transient elevation of $[Ca^{2+}]_i$ is prevented by the intracellular injection of a Ca^{2+}-chelator agent (BAPTA) in the recorded pyramidal cell before the tetanus or by a strong postsynaptic depolarization which decreases the driving force for Ca^{2+} entry, LTP is not observed or is reduced. A simultaneous extracellular recording of the field EPSP shows that LTP is, however, generated in the other stimulated cells (which were not injected with BAPTA or depolarized). These results indicate that an increase of $[Ca^{2+}]_i$ is essential for the induction of LTP.

For how long must $[Ca^{2+}]_i$ remain increased in the postsynaptic element to trigger LTP?

The duration for which $[Ca^{2+}]_i$ must remain elevated to induce LTP was tested by injecting into the recorded

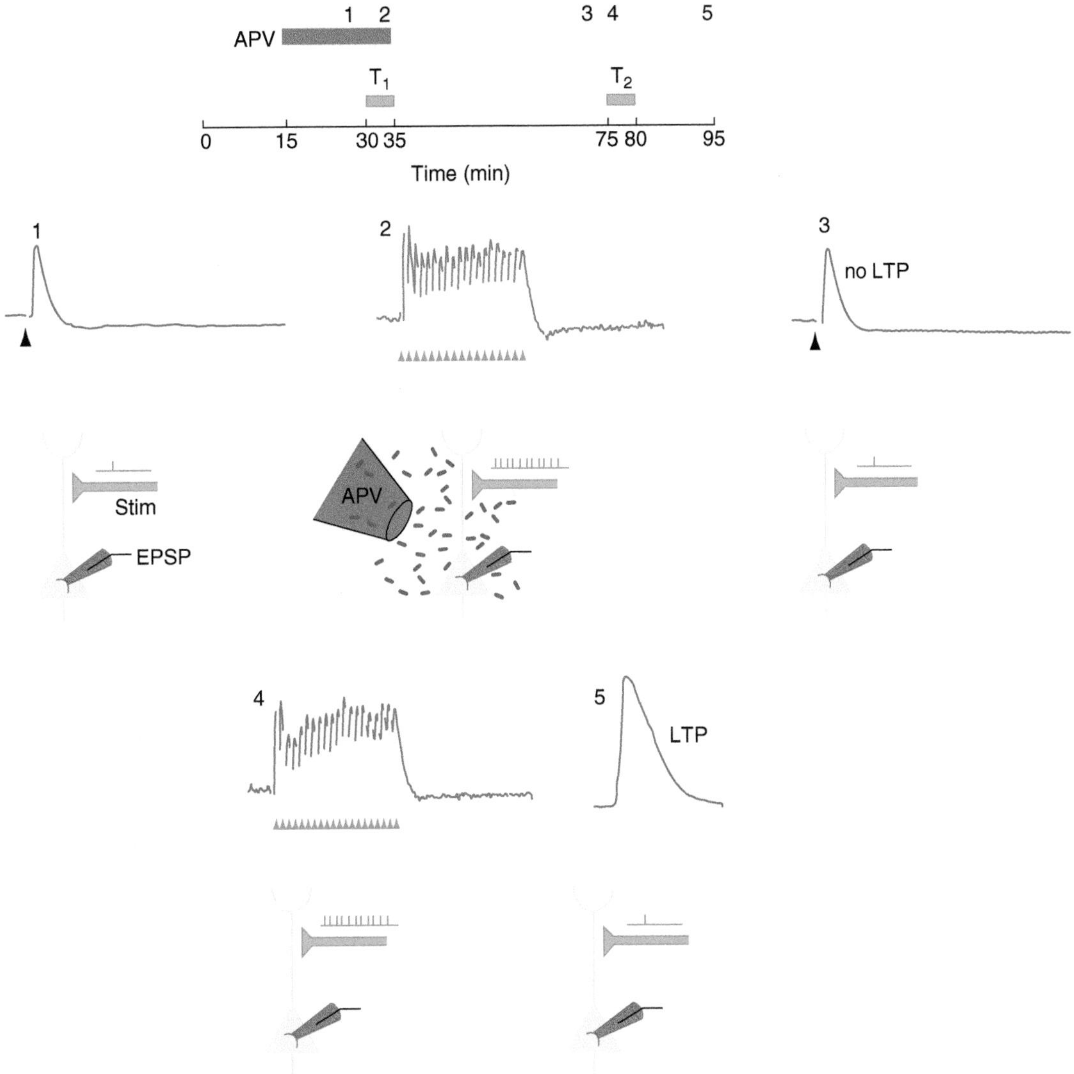

FIGURE 18.5 **NMDA receptor activation is required for LTP induction.** An intracellular glutamatergic EPSP is evoked by Schaffer collateral stimulation (1). D-APV (20 μM, black bar) is applied before and during the tetanus (T_1, 2). LTP is not induced since the EPSP recorded one hour after wash of APV (3) has the same peak amplitude as the control one (1). A second tetanus (T_2, 4) is applied in the absence of D-APV; the EPSP is now potentiated (LTP, 5). Note that APV evokes only a small change of the depolarization of the membrane during the tetanus (compare 2 and 4). T_1 and T_2 are identical periods of tetanic stimulation composed of 10–12 high-frequency trains at 30 s intervals. Each train comprised 20 stimulations at 100 Hz. Adapted from Collingridge GL, Herron CE, Lester RAJ (1988) Frequency-dependent N-methyl-D-aspartate receptor-mediated synaptic transmission in rat hippocampus. *J. Physiol.* **399**, 301–312, with permission.

neuron a photosensitive Ca^{2+} chelator. This compound, diazo-4, has a low affinity for Ca^{2+} (K_D = 89 μM) which can be suddenly (in 100–400 μs) increased (K_D = 0.55 μM) when a UV flash inducing the photolysis of its diazo-acetyl groups is applied to the cell (**Figure 18.6a**). Thus, introduction of diazo-4 into a cell does not affect ambient Ca^{2+} levels before the application of UV light. The manipulation of the delay between the LTP-inducing tetanus and photolysis of diazo-4 allows one to determine the minimum duration of postsynaptic $[Ca^{2+}]_i$ increase necessary to induce LTP. When Ca^{2+} is chelated by diazo-4 photolysis 2.5 s or more after the tetanus, LTP is still induced (**Figure 18.6b**). In contrast, if the UV flash follows the 1 s duration tetanus without delay, the induction of LTP is prevented (**Figure 18.6c**). Therefore, an increase of $[Ca^{2+}]_i$ lasting at most 2.5 s (1 s during the tetanus plus 1.5 s after) is sufficient for LTP induction.

The hypothetical model for LTP induction

The following model is proposed to explain the induction of LTP in the CA1 region of the hippocampus.

BEFORE TETANUS

A single shock evokes the release of glutamate from the stimulated terminal. Glutamate activates postsynaptic non-NMDA and metabotropic glutamate receptors.

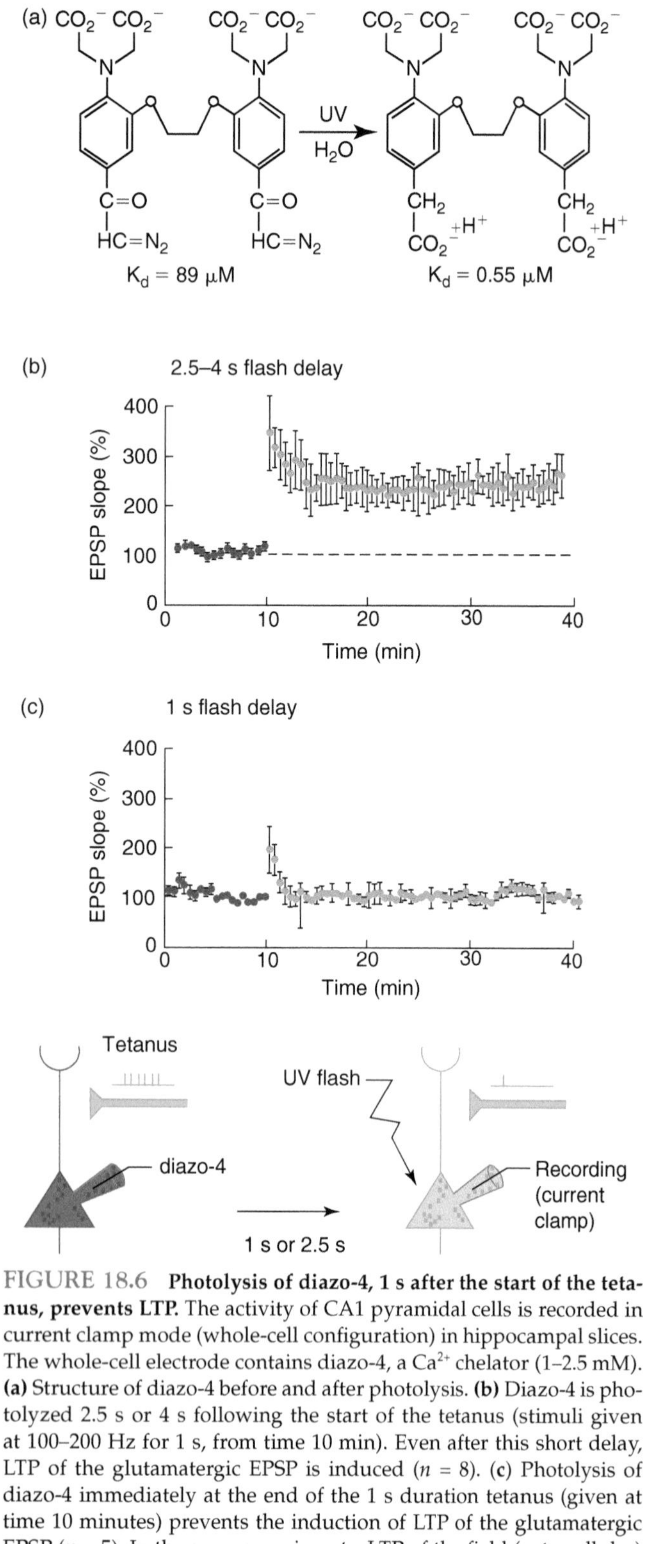

FIGURE 18.6 **Photolysis of diazo-4, 1 s after the start of the tetanus, prevents LTP.** The activity of CA1 pyramidal cells is recorded in current clamp mode (whole-cell configuration) in hippocampal slices. The whole-cell electrode contains diazo-4, a Ca^{2+} chelator (1–2.5 mM). **(a)** Structure of diazo-4 before and after photolysis. **(b)** Diazo-4 is photolyzed 2.5 s or 4 s following the start of the tetanus (stimuli given at 100–200 Hz for 1 s, from time 10 min). Even after this short delay, LTP of the glutamatergic EPSP is induced (n = 8). **(c)** Photolysis of diazo-4 immediately at the end of the 1 s duration tetanus (given at time 10 minutes) prevents the induction of LTP of the glutamatergic EPSP (n = 5). In the same experiments, LTP of the field (extracellular) EPSP is observed (not shown). Adapted from Malenka RC, Lancaster B, Zucker RS (1992) Temporal limits on the rise in postsynaptic calcium required for the induction of long-term potentiation. *Neuron* **9**, 121–128, with permission.

NMDA receptors, owing to Mg^{2+} block, are weakly activated and contribute little to the basal EPSP (**Figure 18.7a**).

DURING TETANUS

The high-frequency stimulation (tetanus) activates a certain number of afferent axons (**Figure 18.7b**). This enhances the release of glutamate from the stimulated terminals, thus evoking a postsynaptic depolarization due to the inward current through postsynaptic non-NMDA (AMPA) receptors and probably also the activation of metabotropic glutamate receptors (mGluR). Activation of AMPA receptors depolarizes the postsynaptic elements (spines) to the point where the Mg^{2+} blockade of the NMDA receptors is removed, thus allowing the influx of Ca^{2+} ions into the spines through NMDA channels and a further depolarization of the membrane. The depolarization of synaptic origin can also bring the postsynaptic membrane to the threshold for dendritic voltage-gated Ca^{2+}-channel activation allowing an additional Ca^{2+} entry. The short-lasting (few seconds) rise in $[Ca^{2+}]_i$ resulting from NMDA- and Ca^{2+}-channel activation provides the necessary trigger for the subsequent events: activation of Ca^{2+}-dependent protein kinases and other Ca^{2+}-dependent processes, which lead to the expression of LTP; i.e. a persistent increase in synaptic efficacy (see maintenance or expression, Section 18.2.5).

Metabotropic glutamate receptors are modulators that regulate the threshold of induction of NMDAR-dependent LTP

In the presence of t-ACPD, a selective agonist of mGluRs, a sub-threshold tetanus (which alone triggers only short-term potentiation), now generates an LTP (**Figure 18.8**). This effect is blocked by APV, a selective antagonist of NMDA receptors and by protein kinase C (PKC) inhibitors. It indicates that activation of mGluRs reduces the threshold of LTP induction, an effect mediated by NMDA receptors and protein kinase C. This effect is specific to mGluRs since application in similar conditions of agonists of iGluRs, AMPA or NMDA, in addition to the sub-threshold tetanus, fails to trigger LTP. In control situations (in the absence of tetanus), a link between mGluRs, protein kinase C and NMDA receptors is suggested by numerous experiments.

In intracellular recordings of pyramidal neurons of the CA1 region of the hippocampus, the mGluR agonist t-ACPD enhances the current generated by NMDA but not by AMPA applications, in the presence of tetrodotoxin (TTX) (to block action potentials and therefore network activity) and K^+ channels blockers.

This effect is blocked by the intracellular injection of a protein kinase C inhibitor. The intracellular injection of protein kinase C enhances the NMDA-receptor-mediated current.

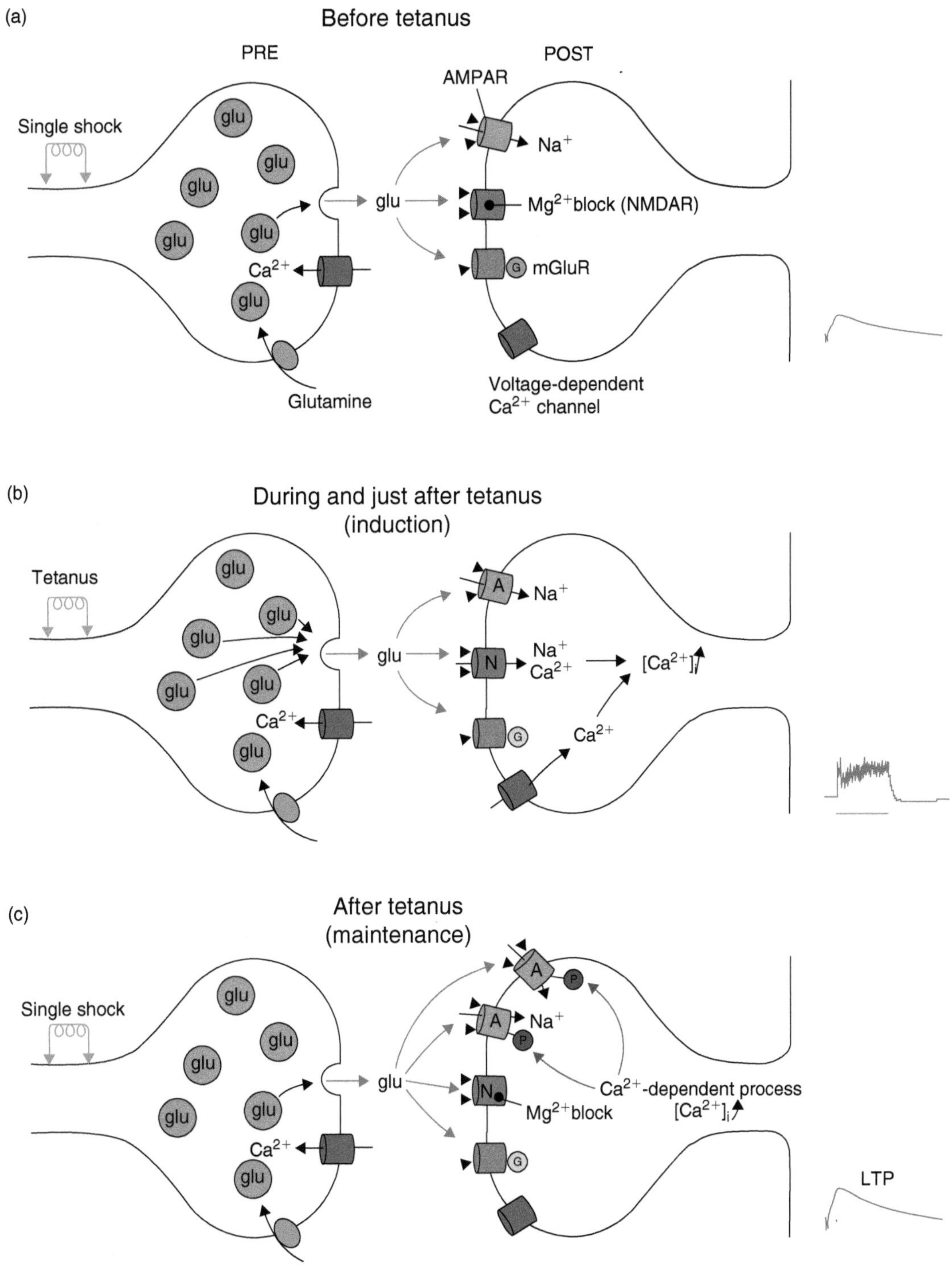

FIGURE 18.7 **Schematic on the role of NMDA receptors and intracellular Ca^{2+} ions in the induction and maintenance of LTP.** See text for explanation.

- Protein kinase C phosphorylation sites are present on NMDA receptors.
- In oocytes transfected with cDNAs coding for mGluRs and NMDA receptors, the mGluR agonist t-ACPD increases NMDA currents, an effect blocked by protein kinase C inhibitors.
- In a wide range of cell types, kinases and phosphatases modulate rapidly and reversibly NMDA-receptor activity.

These results suggest that the activation of postsynaptic mGluRs enhances (via protein kinase C) the

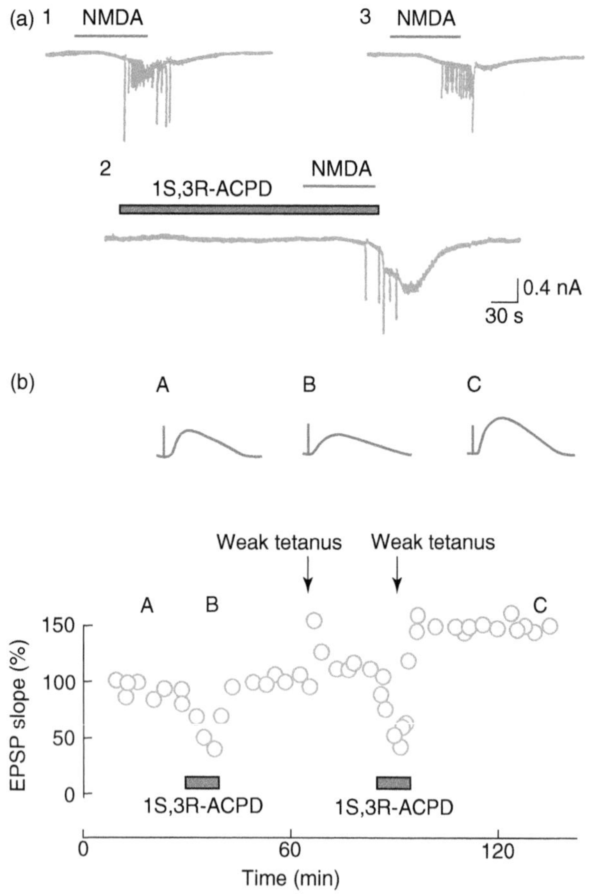

FIGURE 18.8 **mGluRs activation potentiates NMDA-mediated currents and facilitates LTP induction of AMPA-mediated EPSP. (a)** The activity of a CA1 pyramidal neuron is intracellularly recorded in single-electrode voltage clamp mode in slices. The external solution contains TTX to block synaptic activity and K-channel blockers and the intracellular electrode is filled with CsCl. Bath application of NMDA (10 μM, 90 s) evokes an inward current (1) with rapid inward voltage-gated Ca^{2+} currents evoked in unclamped regions of the neuronal membrane. Bath application of 1S,3R-ACPD (50 μM, 4 min), an mGluR agonist, before and during NMDA application (10 μM, 90 s) potentiates the NMDA-evoked current (2). This effect is reversible since 5 minutes after washing NMDA (10 μM, 90 s) evokes an inward current (3) of similar amplitude to the one observed in control (1). **(b)** The activity of a CA1 pyramidal neuron is intracellularly recorded in current clamp mode in slices. The diagram shows the amplitude of the initial slope of the AMPA-mediated EPSP recorded in response to Schaffer collateral stimulation. Bath application of 1S,3R-ACPD (50 μM, 2 min) reversibly depresses the EPSP (compare B with A). A sub-threshold tetanic stimulation of Schaffer collaterals (weak tetanus: stimuli at 50 Hz for 0.5 s) induces a short-term potentiation of the EPSP (trace not shown). The same weak tetanus given during bath application of 1S,3R-ACPD (50 μM, 2 min) now induces a long-term potentiation of the EPSP (C). Part (**a**) adapted from Ben Ari Y, Aniksztejn L (1995) Role of glutamate metabotropic receptors in long-term potentiation in the hippocampus. *Sem. Neurosci.* **7**, 127–135, with permission. Part (**b**) adapted from Aniksztejn L, Otani S, Ben Ari Y (1992) Quisqualate metabotropic receptors modulate NMDA currents and facilitates induction of LTP through protein kinase C. *Eur. J. Neurosci.* **4**, 500–505, with permission.

postsynaptic NMDA-receptor-mediated current (activated by the release of glutamate evoked by a sub-threshold tetanus applied to the afferents). This enhancement of NMDA-receptor-mediated response, together with the activation of AMPA receptors, induces LTP of the glutamatergic AMPA-receptor-mediated response.

18.2.5 Expression of LTP (also called maintenance) involves a persistent enhancement of the AMPA component of the EPSP

Owing to the Mg^{2+} block of NMDA receptors, the control glutamatergic EPSP recorded in CA1 pyramidal neurons (in the presence of physiological concentrations of Mg^{2+}) is mainly mediated by non-NMDA receptors (**Figure 18.7a**) since it is negligibly affected by APV (the selective antagonist at NMDA receptors) and nearly completely blocked by CNQX. The same analysis of the relative contribution of NMDA and non-NMDA receptors was applied to the potentiated EPSP after a tetanus.

The EPSC (postsynaptic excitatory current) evoked in CA1 pyramidal neurons in response to Schaffer collateral stimulation is recorded with patch clamp techniques (whole-cell patch) at two different holding potentials. At $V_H = -80$ mV, the EPSC is mainly mediated by AMPA receptors owing to the Mg^{2+} block of NMDA receptors at this hyperpolarized potential. In contrast, at $V_H = +30$ mV, the control EPSC (which is inverted since the reversal potential of the glutamate response is 0 mV) is mixed and mainly mediated by NMDA receptors as shown by the small effect of CNQX (**Figure 18.9a**). The early rising phase of the EPSC is mainly mediated by AMPA receptors, while the current measured 100 ms after the stimulation is mainly mediated by NMDA receptors.

The recorded cell is subjected to a procedure which induces LTP. After the tetanus, the membrane potential is returned to −80 mV and the test stimulation of Schaffer collaterals is regularly applied to verify that the EPSC is now potentiated (LTP has been induced) (**Figure 18.9b**). This potentiated EPSC is also recorded at 130 mV in order to evaluate the amplitudes of the early AMPA and the late NMDA components. The early component (AMPA-mediated) approximately doubles while the late component (NMDA-mediated) remains rather stable (**Figure 18.9c**). Therefore, LTP of the glutamatergic response, in this experiment, is primarily mediated by an enhancement of the AMPA component of the synaptic current.

This differential effect of the tetanus can be explained by:

- an increase in the density of AMPA receptors in the synaptic cleft (clustering);
- a change in the properties of AMPA receptors (affinity, unitary current amplitude);
- an increase in the effective spread of synaptic current from dendritic spines into dendrites (a change of diameter of the neck of the spines, for example).

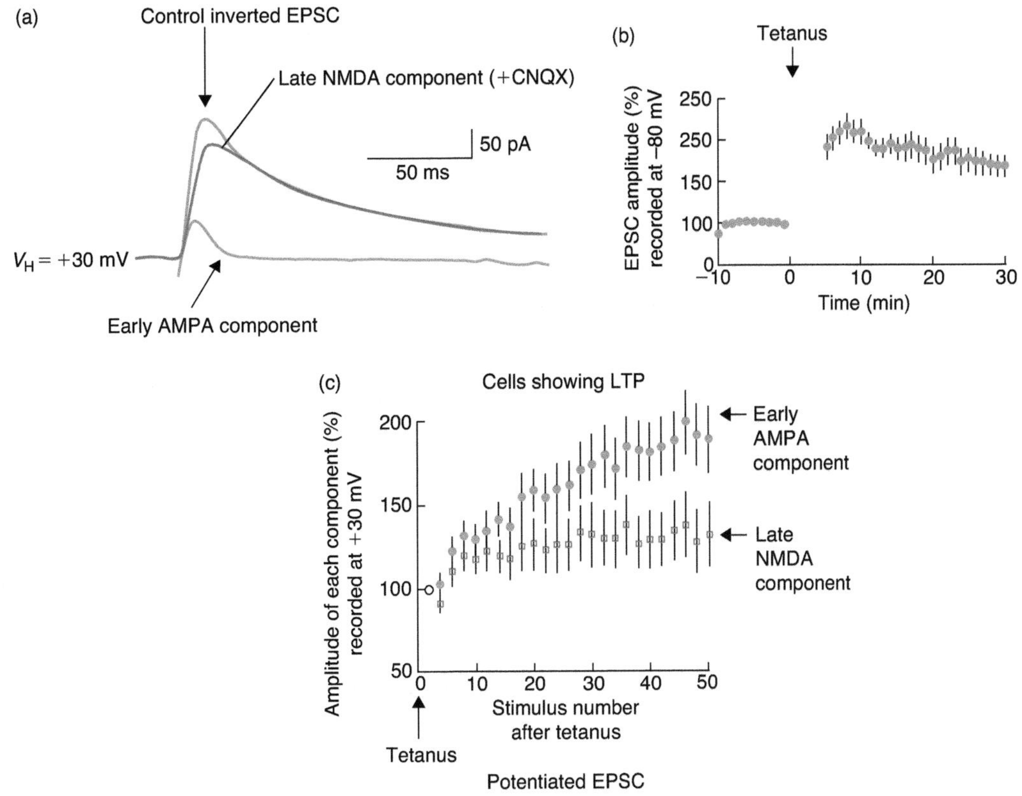

FIGURE 18.9 **Differential enhancement of the non-NMDA and NMDA components in LTP.** The excitatory postsynaptic current (EPSC) evoked by Schaffer collateral stimulation is recorded in a CA1 pyramidal neuron in hippocampal slices with patch clamp techniques (whole-cell patch). **(a)** At V_H = +30 mV, the EPSC is inverted (control). Application of the non-NMDA-selective antagonist (CNQX) selectively reduces the early component of the current, leaving the late component (NMDA-receptor-mediated) unaffected. The subtracted record (sub = control − CNQX insensitive) illustrates the time course of the AMPA-receptor-mediated component (CNQX sensitive). **(b)** The EPSC peak amplitude (expressed as a percentage of the control) is plotted against time, before and after the tetanus applied to evoke LTP (V_H = −80 mV, except during the tetanus). The total EPSC is clearly potentiated by the procedure. **(c)** The EPSC peak amplitude, expressed as a percentage of the control, is measured from just after the induction of LTP to 30 min after (V_H = +30 mV). The early (CNQX sensitive) AMPA-receptor-mediated component is clearly potentiated while the late (CNQX insensitive) NMDA-receptor-mediated component is not significantly potentiated. Recordings in **(b)** and **(c)** are from the same cell; to obtain the curves in **(b)** and **(c)**, the membrane potential is continuously shifted from −80 to +30 mV. Adapted from Perkel DJ, Nicoll RA (1993) Evidence for all or none regulation of neurotransmitter release: implications for long-term potentiation. *J. Physiol.* **471**, 481–500, with permission.

This differential enhancement of the two components of the EPSC favors the hypothesis that *expression* of LTP requires postsynaptic mechanisms and does not result exclusively from a presynaptic mechanism, such as a persistent enhancement of glutamate release. If the expression of LTP resulted only from a presynaptic mechanism, a similar increase of both components of the EPSC should have been observed (assuming that AMPA and NMDA receptors are co-localized in the same postsynaptic membrane). As shown above, LTP *induction* clearly requires postsynaptic events: activation of NMDA receptors, increase of postsynaptic $[Ca^{2+}]_i$. Therefore, if LTP expression were presynaptic, that would imply that some message must be sent from the postsynaptic spines to the presynaptic elements. This retrograde messenger would be generated postsynaptically and would trigger a sustained enhanced release of glutamate by the presynaptic element. The identity of any such messenger remains elusive. It appears now safe to state that the major mechanism for the expression of LTP involves a postsynaptic mechanism (see below).

The Ca^{2+} signal is translated into an increase in synaptic strength by biochemical pathways

What are the biochemical pathways activated by intradendritic Ca^{2+} increase that are key components absolutely required for translating the Ca^{2+} signal into an increase in synaptic strength? Amongst the kinases, the Ca^{2+} calmodulin-dependent protein kinase II (CaMKII) plays a crucial role:

- CaMKII is found in high concentrations in the postsynaptic density in spines, near postsynaptic glutamate receptors.
- Injection of inhibitors of CaMKII in the postsynaptic cell, or genetic deletion of a critical CaMKII subunit, block the ability to generate LTP.

- When autophosphorylated on Thr(286), the activity of CaMKII is no longer dependent on Ca^{2+} calmodulin. This allows its activity to continue long after the Ca^{2+} signal has returned to baseline.
- Replacement of endogenous CaMKII by a form of CaMKII containing a Thr286 point mutation (by the use of genetic techniques) blocks LTP.
- Finally, CaMKII directly phosphorylates the AMPA receptor *in situ*.

Several other protein kinases have been suggested to contribute to LTP, including protein kinase C (PKC) since its selective inhibition by intracellular injection of a PKC inhibitory peptide (PKCI) prevents LTP induction.

Expression of LTP involves the phosphorylation and persistent upregulation of AMPA receptors

A persistent enhancement of the AMPA component of the EPSP could result from a persistent modification in the function *and/or* the number of postsynaptic AMPA receptors (see **Figure 18.7c**). The former hypothesis implies either an increase in unitary current amplitude or an increase in the probability of opening of the AMPA channel. Such a change in receptor function generally involves a phosphorylation by a serine or a threonine kinase. In fact, induction of LTP specifically increases the phosphorylation of Ser831 of the GluR1 subunit, an effect that is blocked by a CaMKII inhibitor (AMPA receptors in CA1 pyramidal cells are heteromers composed primarily of GluR1 and GluR2 subunits). Moreover, genetic deletion of GluR1 subunit prevents the generation of LTP in CA1 pyramidal cells.

In agreement with the second hypothesis is the physiological and anatomical evidence of a rapid and selective upregulation of AMPA receptors after the induction of LTP. For example, when the GluR1 subunit of AMPA receptors is tagged with green fluorescent protein (GFP) and transiently expressed in hippocampal CA1 neurons, its distribution can be observed with a two-photon laser scanning microscope (see Appendix 5.1). After a tetanus, a rapid delivery of tagged receptors to dendritic spines is observed. The increased number of AMPA receptors in the plasma membrane at synapses is achieved via activity-dependent changes in AMPA receptor trafficking. However, the source of the AMPARs that are delivered to synapses during LTP is still unknown as are the detailed mechanisms that deliver and retain AMPARs in the postsynaptic density.

18.2.6 Multiple ways to induce LTP, multiple forms of LTP and multiple ways to block LTP induction

Although synchronous activation of a number of presynaptic fibers by high-frequency stimulation is the most reliable way to evoke LTP (**Figure 18.10a**), LTP can be also evoked *in vitro* by a combination of low-frequency stimulation of presynaptic afferents and postsynaptic injection of a depolarizing current pulse to activate NMDA receptors and voltage-gated Ca^{2+} channels (**Figure 18.10b**); and by a combination of low-frequency stimulation of presynaptic afferents and bath application of a selective agonist at metabotropic glutamate receptors (**Figure 18.10c**). All these forms of LTP induction are blocked by bath application of APV, an antagonist at NMDA receptors, by intracellular injection of a Ca^{2+} chelator (BAPTA), or by intracellular injection of a hyperpolarizing current pulse during tetanic stimulation (**Figure 18.10d**).

Another form of LTP is induced by bath application of K^+-channel blockers, such as tetraethylammonium chloride (TEA), that depolarizes the presynaptic elements and enhances transmitter release (**Figure 18.10e**). This form of LTP also requires a rise of the postsynaptic intracellular Ca^{2+} concentration. This rise is produced by the entry of Ca^{2+} ions through voltage-gated Ca^{2+} channels activated by the depolarization resulting from the closure of K^+ channels by TEA. In contrast to tetanus LTP, TEA-induced LTP is not synapse specific since all the synapses are activated by bath application of TEA. Moreover, TEA-induced LTP is NMDA-receptor independent. It is blocked by bath application of CNQX (an antagonist at AMPA receptors) or of Ca^{2+}-channel blockers, or by intracellular injection of a Ca^{2+} chelator (**Figure 18.10f**).

The observation that a rise of the intracellular Ca^{2+} concentration is a necessary prerequisite for LTP induction raises the possibility that a wide range of physiological or pathological processes known to evoke a rise of $[Ca^{2+}]_i$ would trigger long-lasting changes of synaptic efficacy. Both seizures, which generate synchronized giant paroxysmal activity, and anoxic–ischemic episodes which generate LTP of NMDA-receptor-mediated EPSPs (anoxic LTP), are in fact associated with $[Ca^{2+}]_i$ rises and long-lasting changes of synaptic efficacy. In such cases, LTP of excitatory synaptic transmission may participate in the pathological consequences of these insults.

18.2.7 Summary: principal features of LTP in the Schaffer collateral–pyramidal cell glutamatergic transmission

- LTP is a long-lasting phenomenon, persisting for hours *in vitro* and days or weeks in the intact animal.
- LTP is synapse specific.
- LTP results from an increase in the synaptic response without changes in the number of stimulated presynaptic axons.
- LTP does not result from a persistent change of postsynaptic cell excitability.
- LTP does not result from a persistent reduction of the inhibitory GABAergic responses.

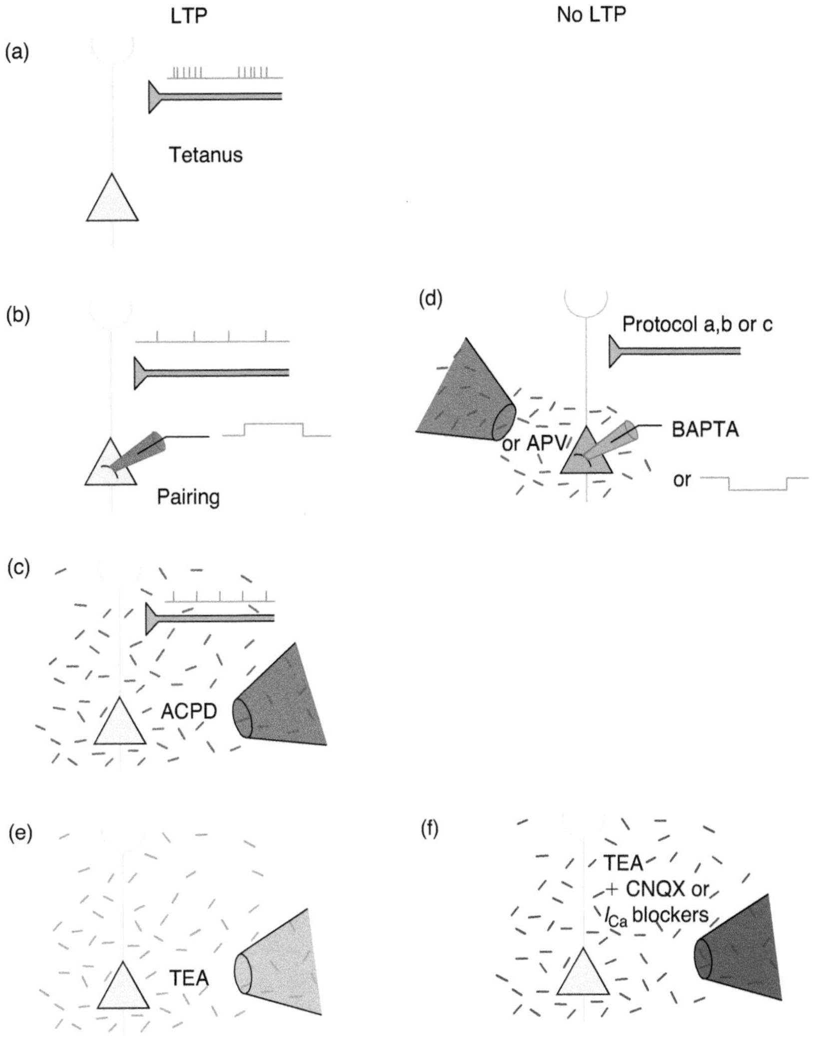

FIGURE 18.10 **Hippocampal LTP. (a–d)** Tetanic LTP and **(e,f)** TEA-induced LTP. The multiple ways of induction are shown in **(a–c)** and **(e)**, and blockade of induction in **(d)** and **(f)**.

- LTP consists of two phases: a brief induction phase (1–20 s) followed by the expression phase; i.e. the mechanisms sustaining the persistent enhancement of synaptic efficacy.
- *Induction* of LTP is voltage and NMDA dependent. It requires depolarization of the postsynaptic membrane (resulting from AMPA-receptor activation) *and* activation of NMDA receptors which leads to Ca^{2+} entry into the postsynaptic spine and postsynaptic $[Ca^{2+}]_i$ increase. The NMDA receptor acts as a detector of coincident activity in the postsynaptic cells as it opens efficiently only when glutamate is released from the presynaptic terminal and the postsynaptic cell is strongly depolarized (to relieve Mg^{2+} block). Channel opening produces a rise in Ca^{2+} that is largely restricted to the dendritic spine onto which the active synapse terminates. This Ca^{2+} elevation is both necessary and sufficient for LTP induction.
- *Maintenance* of LTP, at least during the initial phase, results from the triggering of a Ca^{2+}-dependent cascade of events leading to postsynaptic modifications of AMPA receptor function and density, and to persistent enhancement of the synaptic glutamatergic response (LTP).

18.3 THE LONG-TERM DEPRESSION (LTD) OF A GLUTAMATERGIC RESPONSE: EXAMPLE OF THE RESPONSE OF PURKINJE CELLS OF THE CEREBELLUM TO PARALLEL FIBER STIMULATION

Purkinje cells represent the single output neurons of the cerebellar cortex. Each of them receives two distinct excitatory inputs, one from parallel fibers (axons of granule cells) and the other from a climbing fiber (axons of the contralateral inferior olive cells). These two types of inputs display distinct characteristics. A single climbing fiber terminates on each Purkinje cell. This powerful one-to-one excitatory input makes multiple synapses

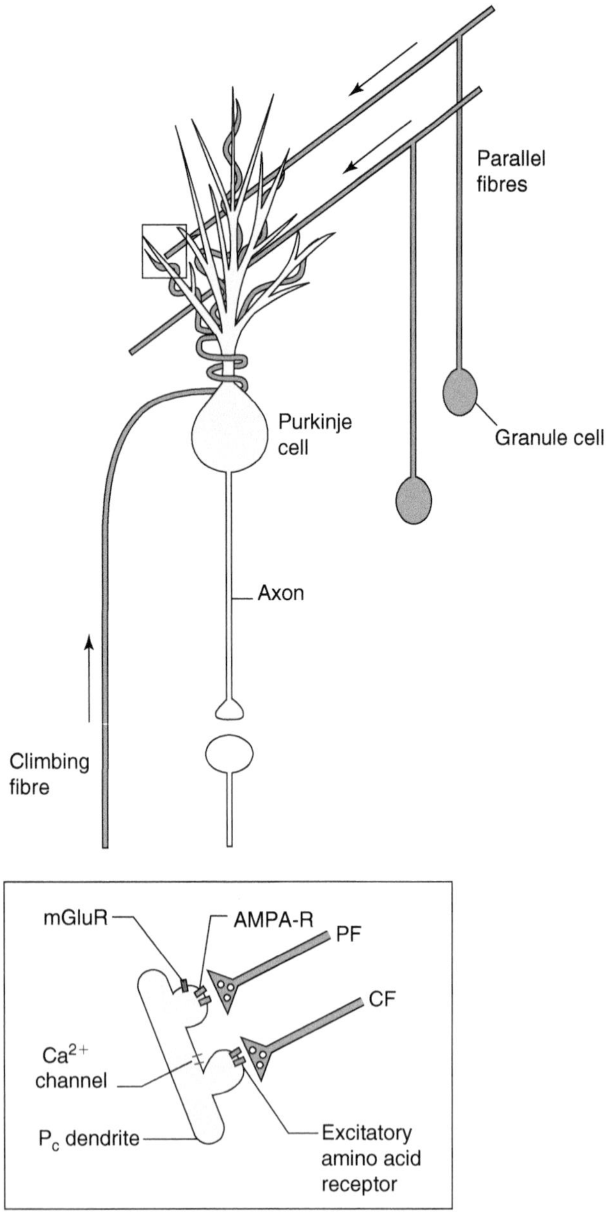

FIGURE 18.11 **Simplified neural circuit in the cerebellar cortex.** Inset shows a more detailed view of the synaptic contacts between a parallel or a climbing fiber terminal and the Purkinje cell dendrite. AMPA-R, AMPA receptor; mGluR, metabotropic glutamate receptors; VDCC, voltage-dependent calcium channels; CF, climbing fiber; PF, parallel fiber; Pc, Purkinje cell. Adapted from Daniel H, Levenes C, Crépel F (1998) Cellular mechanisms of cerebellar LTD. *Trends Neurosci.* **9**, 401–407, with permission.

on the soma and proximal dendrites of the Purkinje cell (**Figure 18.11**; see also **Figures 6.8** and **6.9**). In contrast, many parallel fibers converge on each Purkinje cell but each fiber makes few synapses on each Purkinje cell.

The putative neurotransmitter at parallel-fiber and climbing-fiber synapses is glutamate. Fast excitatory synaptic transmission at these synapses is mediated entirely by non-NMDA ionotropic glutamatergic receptors, since both synapses lack NMDA receptors in the adult – in marked contrast to most other neurons in the brain (**Figure 18.11**, inset). In addition, parallel-fiber/Purkinje-cell synapses also bear mGluR1 receptors known to be coupled to phospholipase C, activation of which leads to production of inositol trisphosphate (IP_3) and diacylglycerol (DAG).

The dual arrangement of the two excitatory synaptic inputs raises the question of the role of the powerful input (climbing fiber) on the weaker input (parallel fibers). The co-activation of climbing-fiber and parallel-fiber inputs induces a persistent decrease in the efficacy of the parallel-fiber/Purkinje-cell synapse. This decrease of efficacy is called *long-term depression* (LTD). With the experimental advantages of *in vitro* brain slices and culture preparations, cerebellar LTD constitutes a simple model to study activity-dependent changes confined to excitatory synapses.

18.3.1 The long-term depression of a postsynaptic response (EPSC or EPSP) is a decrease of synaptic efficacy

Ito and co-workers (1982) were the first to demonstrate in the rabbit cerebellum *in vivo* that conjunctive stimulation of the afferent climbing and parallel fibers leads to an LTD of synaptic transmission at parallel-fiber/Purkinje-cell synapses. In other terms, LTD is the attenuation of the Purkinje cell response to parallel fibers after the conjunctive stimulation of parallel and climbing fibers.

The activity of a Purkinje cell is intracellularly recorded in current clamp mode in rat cerebellar slices. The stimulation of parallel fibers evokes an EPSP resulting from the activation of postsynaptic AMPA receptors by glutamate released from the stimulated terminals since it is totally blocked by CNQX, a selective AMPA-receptor antagonist. After recording this control parallel-fiber-mediated EPSP for several minutes, parallel fibers are then stimulated in conjunction with the climbing fiber at low frequency, 4 Hz for 25 s. After this conjunctive stimulation, the same stimulation of parallel fibers as in the control now evokes a smaller EPSP (**Figure 18.12**). The parallel-fiber-mediated EPSP stays attenuated for the rest of the recording session. It is a long-term depression.

The persistent decrease of the parallel-fiber-mediated synaptic response can be also studied in another *in vitro* preparation, a culture of Purkinje cells, granule cells and an inferior olivary explant (**Figure 18.13a**). The parallel-fiber-mediated postsynaptic excitatory current (EPSC) is first recorded in voltage clamp (**Figure 18.13b**). The repetitive conjunctive stimulation of a single granule cell (whose axon is a parallel fiber) and the inferior olivary explant (which sends an axon, the climbing fiber, to the recorded Purkinje cell) is then applied while the Purkinje cell activity is recorded in current clamp mode. When switching back to voltage clamp mode, the parallel-fiber-mediated EPSC (in response to granule cell

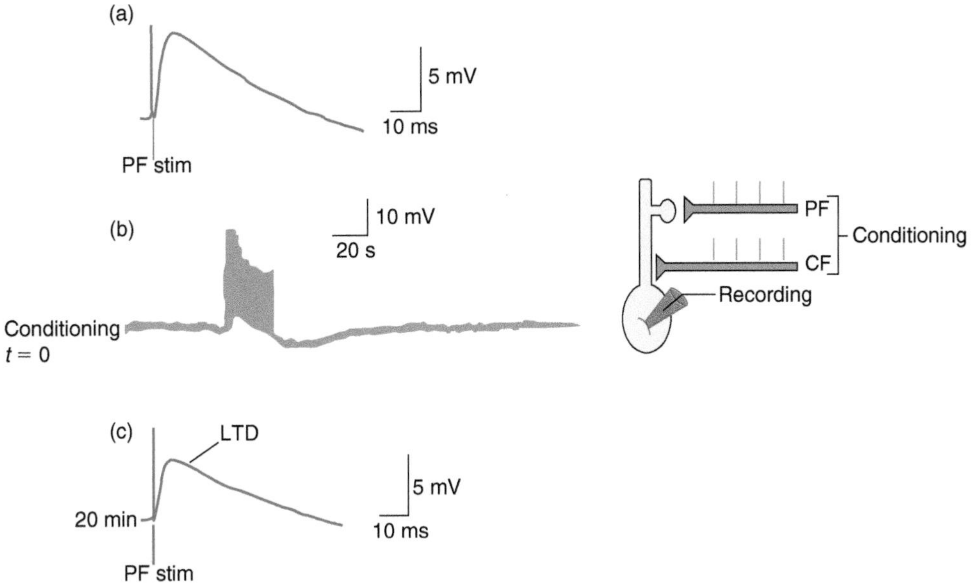

FIGURE 18.12 **LTD of the parallel-fiber-mediated EPSP.** The activity of a Purkinje cell is intracellularly recorded in current clamp mode in a cerebellar slice. **(a)** The EPSP in response to PF stimulation (the stimulating electrode is placed in the superficial molecular layer) is recorded in the presence of picrotoxin (40 μM) to block IPSPs mediated by local interneurons. **(b)** CF and PF are then stimulated conjointly at 4 Hz for 25 s. To stimulate climbing fibers, a second electrode is placed in the white matter. **(c)** Twenty minutes after the end of conditioning stimulation, the EPSP recorded in response to PF stimulation is still depressed in amplitude. Adapted from Sakurai M (1990) Calcium is an intracellular mediator of the climbing fiber in induction of cerebellar long-term depression. *Proc. Natl Acad. Sci. USA* **87**, 3383–3385, with permission.

stimulation) is persistently decreased. This *in vitro* preparation allows the stimulation of a single presynaptic granule cell before and after LTD induction. Therefore, it can be demonstrated that LTD is observed though the number of parallel fibers stimulated before and after the conditioning stimulation is identical (a depressed EPSC or EPSP could in fact result from a decrease in the number of stimulated axons).

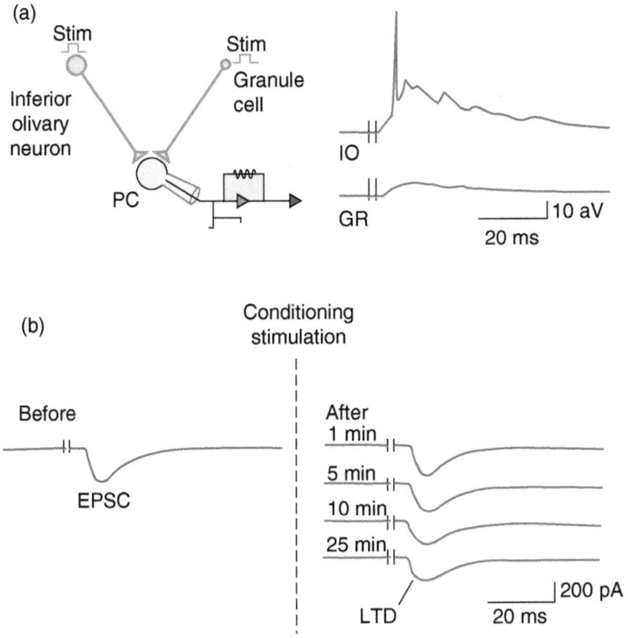

FIGURE 18.13 **Cerebellar LTD is observed when a single parallel fiber is stimulated.** **(a)** The activity of a Purkinje cell is recorded in patch clamp (whole-cell patch) in co-cultures of rat cerebellar Purkinje cells (PC), granule cells and an explant of inferior olivary neurons, to record the evoked postsynaptic current (EPSC). The conditioning stimulation consists of the conjunctive stimulation of a single granule cell (GR) and the inferior olivary explant (IO) at 2 Hz for 20 s while the Purkinje cell membrane is recorded in current clamp mode. **(b)** The Purkinje cell membrane is held at $V_H = -50$ mV for 1 min in voltage clamp mode and the response (EPSC) to the activation of a single granule cell is recorded before and 1, 5, 10 and 25 min after the conditioning stimulation. Adapted from Hirano T (1990) Depression and potentiation of the synaptic transmission between a granule cell and a Purkinje cell in rat LTD 20 ms cerebellar culture. *Neurosci. Lett.* **119**, 141–144, with permission.

18.3.2 Induction of LTD requires a rise in postsynaptic intracellular Ca^{2+} concentration and the activation of postsynaptic AMPA receptors

As already studied in Sections 16.2.2 and 17.3, the response of a Purkinje cell to the activation of its afferent climbing fiber is an all-or-none response composed of an initial depolarization, an overshooting action potential and following depolarizing humps. Since the activation of the afferent climbing fiber potently activates the voltage-gated Ca^{2+} channels present in the membrane of Purkinje dendrites (**Figure 18.11**, inset), it was supposed that the consequent rise in intradendritic Ca^{2+} concentration played a role in LTD. In fact, *in vivo* experiments have shown that hyperpolarization of the membrane by the activation of stellate cells during co-stimulation of parallel and climbing fibers prevents the occurrence of LTD.

$[Ca^{2+}]_i$ rises during co-stimulation

In order to simultaneously record the synaptic responses and the intracellular Ca^{2+} concentration, the activity of a Purkinje cell is recorded in patch clamp (whole-cell patch) in the presence of a fluorescent calcium dye, FURA-2, injected into the cell (**Figure 18.14**; see also Appendix 5.1). First, the control EPSC in response to parallel fiber stimulation is recorded in the Purkinje cell. Then, parallel fibers and climbing fibers are co-stimulated in phase at a low frequency (1–4 Hz;

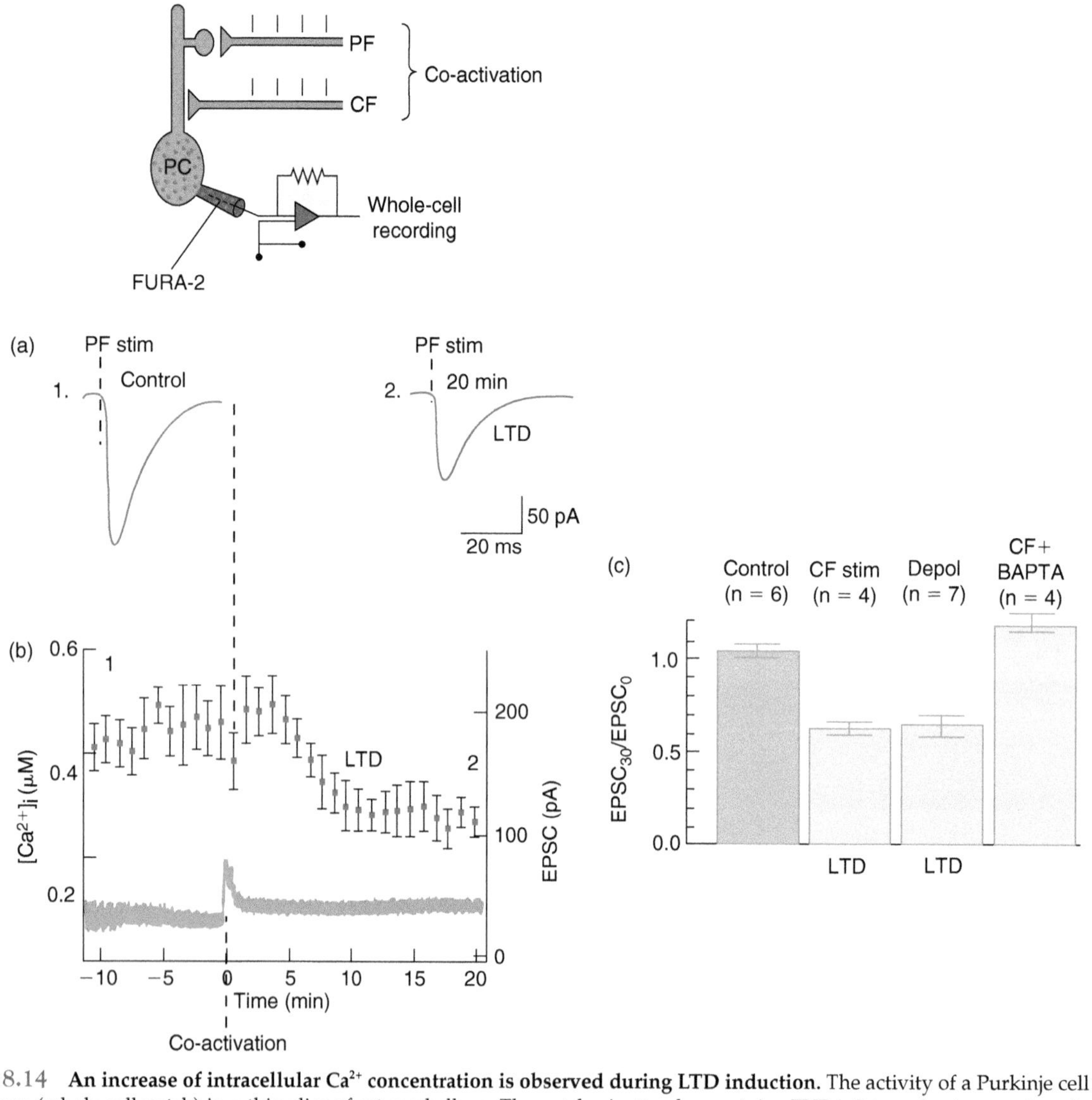

FIGURE 18.14 **An increase of intracellular Ca^{2+} concentration is observed during LTD induction.** The activity of a Purkinje cell is recorded in patch clamp (whole-cell patch) in a thin slice of rat cerebellum. The patch pipette also contains FURA-2 in order to record on-line the intracellular Ca^{2+} concentration. **(a)** The excitatory postsynaptic current (EPSC) recorded in voltage clamp in response to parallel fiber stimulation (PF stim, 1 Hz) is recorded before (control) and 20 min after the conditioning stimulation (co-activation: conjunctive stimulation of parallel and climbing fibers while the Pc membrane is recorded in current clamp mode). **(b)** Time course of changes in parallel-fiber-mediated EPSC amplitude (top curve) and in $[Ca^{2+}]_i$ (bottom curve). The conditioning stimulation (given at time 0) induces an LTD of the EPSC (with a delay) and an immediate transient rise of $[Ca^{2+}]_i$. Note that the stimulation of parallel fibers before the conditioning stimulation does not induce significant changes of $[Ca^{2+}]_i$. **(c)** Average changes in PF-mediated EPSC expressed as the ratio of EPSC amplitude before ($EPSC_0$) and 30 min ($EPSC3_0$) after co-activation, in four different conditions: no co-activation (control), co-activation (CF stim), pairing (depol) and co-activation in the presence of BAPTA (CF + BAPTA) in the patch electrode. From Konnerth A, Dreessen J, Augustine GJ (1992) Brief dendritic calcium signals initiate long-lasting synaptic depression in cerebellar Purkinje cells. *Proc. Natl Acad. Sci. USA* **89**, 7051–7055, with permission.

dotted line). Five minutes after this conditioning stimulation, the response to the same parallel fiber stimulation recorded from the same Purkinje cell begins to decrease and stays attenuated thereafter. Cerebellar LTD is associated with an increase of Ca^{2+} concentration in Purkinje cell dendrites during the conditioning stimulation.

LTD of the response to parallel fiber is not observed when the parallel fibers are stimulated alone at 1–4 Hz; what adds the climbing fiber stimulation?

Voltage-gated Ca^{2+} channels located in the membrane of Purkinje cell dendrites are activated by the climbing-fiber-mediated EPSP. This suggests that the resulting increase of intradendritic Ca^{2+} concentration is a necessary prerequisite for LTD induction. This hypothesis is tested by hyperpolarizing the Purkinje cell membrane during the co-stimulation or by injecting a Ca^{2+} chelator into the Purkinje cell before the co-stimulation (**Figure 18.15**), or by removing the external Ca^{2+} ions, in order to prevent the rise of intradendritic Ca^{2+} concentration: all these procedures block LTD induction. Along the same lines, climbing fiber stimulation can be replaced by direct intracellular depolarization of the Purkinje cell which evokes Ca^{2+} spikes. This is called the 'pairing protocol' (**Figure 18.16a**). In conclusion, LTD of synaptic transmission at parallel/Purkinje-cell synapses is triggered by a rise of intracellular Ca^{2+} concentration resulting from Ca^{2+} entry in Purkinje cell dendrites through voltage-gated Ca^{2+} channels opened by the membrane depolarization during co-stimulation of climbing and parallel fibers.

LTD of the response to parallel fiber is not observed when the climbing fiber is stimulated alone at 1–4 Hz; what adds the parallel fiber stimulation?

The glutamate released from parallel fiber terminals activates the non-NMDA receptors present in the postsynaptic membrane (ionotropic AMPA receptors and metabotropic glutamate receptors, mGluR1; **Figure 18.11**, inset). AMPA receptors mediate the excitatory response (EPSP or EPSC) evoked by parallel fiber stimulation since it is totally blocked by the application of CNQX, a selective antagonist of this class of receptors. In order to test the role of non-NMDA receptors in LTD induction, CNQX is bath applied during or after a pairing protocol (direct Purkinje cell depolarization with

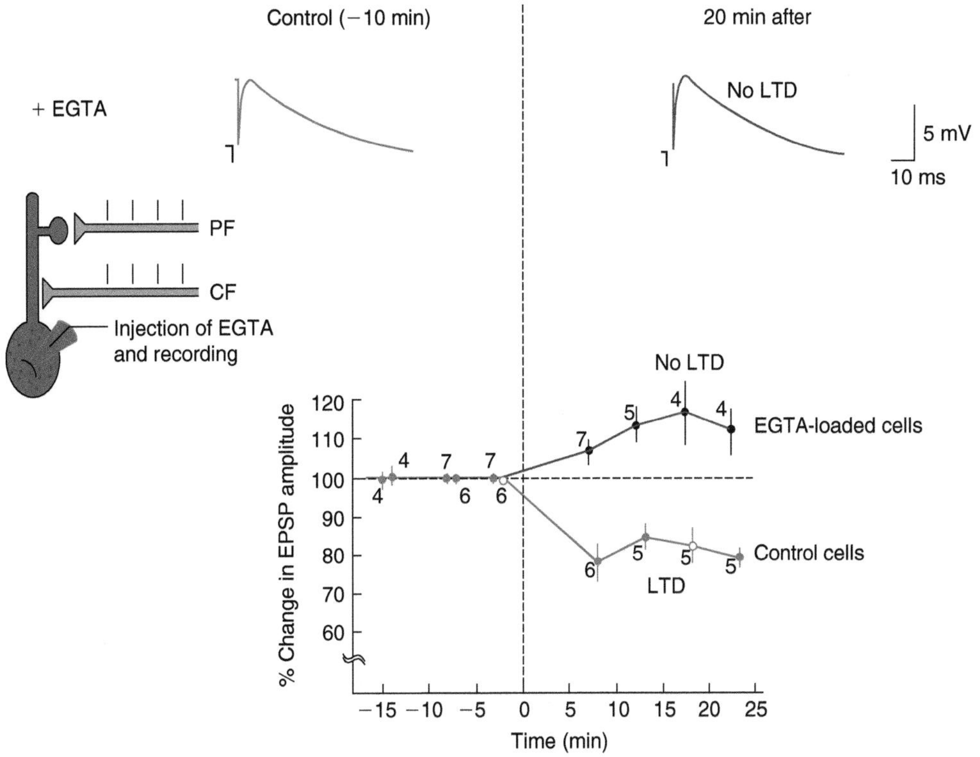

FIGURE 18.15 **The induction of cerebellar LTD requires an increase of intracellular Ca^{2+} concentration.** The activity of a Purkinje cell is intracellularly recorded (current clamp mode) in a guinea pig cerebellar slice. The amplitude of the EPSP recorded in response to parallel fiber stimulation is recorded before and after the conditioning stimulation (conjunctive stimulation of PF and CF at 4 Hz for 25 s) in control cells (white circles). The same experiment is performed after intracellular injection of the Ca^{2+} chelator EGTA into the recorded Purkinje cells (black circles). The respective averaged EPSPs recorded in the presence of EGTA are shown in the insets. The time 0 represents the end of conjunctive stimulation. The values at each plotted point represent the number of cells recorded. Adapted from Sakurai M (1990) Calcium is an intracellular mediator of the climbing fiber in induction of cerebellar long-term depression. *Proc. Natl Acad. Sci. USA* **87**, 3383–3385, with permission.

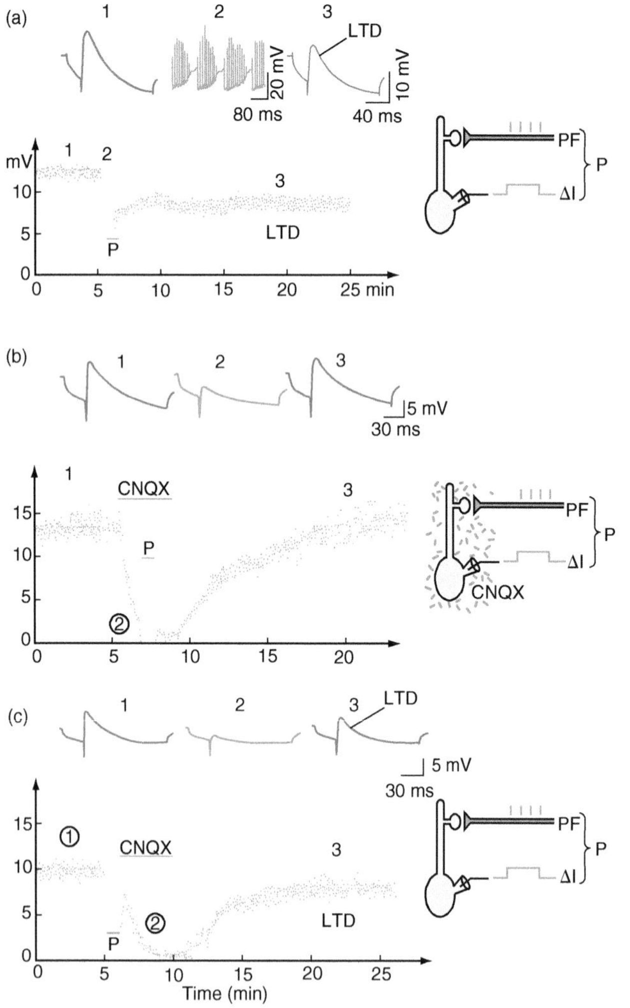

FIGURE 18.16 **Induction of cerebellar LTD requires the activation of postsynaptic AMPA receptors.** The activity of a Purkinje cell is recorded in patch clamp (whole-cell patch, current clamp mode) in cerebellar thin slices. The PF-mediated EPSP is evoked during a hyperpolarizing current pulse before and after the pairing in order to test the variation of membrane resistance during the experiment. **(a, top traces)** A control EPSP is recorded in response to parallel fiber stimulation (1). After the pairing (P) – i.e. intracellular depolarizing pulses to evoke Ca^{2+} spikes in conjunction with parallel fiber stimulation (2) – an LTD of the parallel-fiber-mediated EPSP is observed (3). Note the change in calibrations between 1, 3 and 2. **(a, bottom trace)** Plot of the EPSP amplitude against time. **(b)** The same experiment as in **(a)** but in the presence of CNQX (4 μM) in the bath before, during and after pairing (P). During CNQX application the parallel-fiber-mediated EPSP (2) is completely blocked since it is mediated by AMPA receptors. **(c)** The same experiment as in **(b)** but CNQX is bath applied after pairing (P). Part **(a)** adapted from Hémart N, Daniel H, Jaillard D *et al.* (1995) Receptors and second messengers involved in long-term depression in rat cerebellar slices *in vitro*: a reappraisal. *Eur. J. Neurosci.* 7, 45–53, with permission.

parallel fiber stimulation). The blockade of non-NMDA receptors during the pairing protocol prevents LTD induction while it has no effect after the pairing protocol (once LTD is induced) (**Figure 18.16b,c**).

In order to test the role of the non-NMDA receptors in LTD induction, parallel fiber stimulation can also be replaced by external application of agonists at non-NMDA receptors. Parallel fiber stimulation during the conditioning stimulus can be replaced by the application on the Purkinje cell dendrites of glutamate or quisqualate (agonists on *both* AMPA and metabotropic receptors, or a solution containing *both* AMPA and an agonist of metabotropic receptors) (see **Figure 18.21c,d**). The activation of AMPA receptors alone by AMPA or the application of NMDA is ineffective. This is in keeping with the recent demonstration that antibodies directed against the mGluR1 subunit block LTD induction in cultured Purkinje cells. The final demonstration of the participation of mGluR to LTD induction in acute cerebellar slices has been given recently by showing that LTD of the parallel-fiber-mediated EPSP is markedly impaired in knockout mice lacking mGluR1 (see **Figure 18.21f** and Section 12.4). In conclusion, the activation of the parallel fibers during the conjunctive or pairing stimulation allows the release of glutamate and the activation of both AMPA and metabotropic glutamatergic postsynaptic receptors.

These, with the concomitant rise in intracellular calcium concentration, are possibly the necessary and sufficient processes for LTD induction, since the conditioning stimulation can be replaced by a direct depolarization of the Purkinje cell membrane to activate voltage-dependent Ca^{2+} channels (to mimic climbing fiber stimulation) and the concomitant application of agonists of AMPA and metabotropic receptors (to mimic parallel fiber stimulation) (see **Figure 18.21d**).

18.3.3 The expression of LTD involves a persistent desensitization of postsynaptic AMPA receptors

The fact that co-activation of Purkinje cells by climbing fiber stimulation and iontophoretic application of glutamate on Purkinje cell dendrites induces a long-lasting decrease of the response to this agonist led Masao Ito to postulate that LTD of parallel-fiber-mediated EPSP or EPSC is due to a long-term desensitization of ionotropic glutamate receptors of Purkinje cells (a desensitized state is a state where the probability of the channel opening is very low). This would explain the decrease in synaptic efficacy.

In Purkinje cells in cerebellar slices, a pairing procedure known to induce LTD of the synaptic response induces a long-lasting decrease of the response to iontophoretic application of glutamate (or quisqualate, not shown) but not of aspartate (**Figure 18.17**). This suggests that LTD of synaptic transmission between parallel fibers and Purkinje cells is accompanied by LTD of the responsiveness of Purkinje cells to glutamate or quisqualate, whereas that to aspartate is unaffected. The observed decrease in efficacy of glutamate or

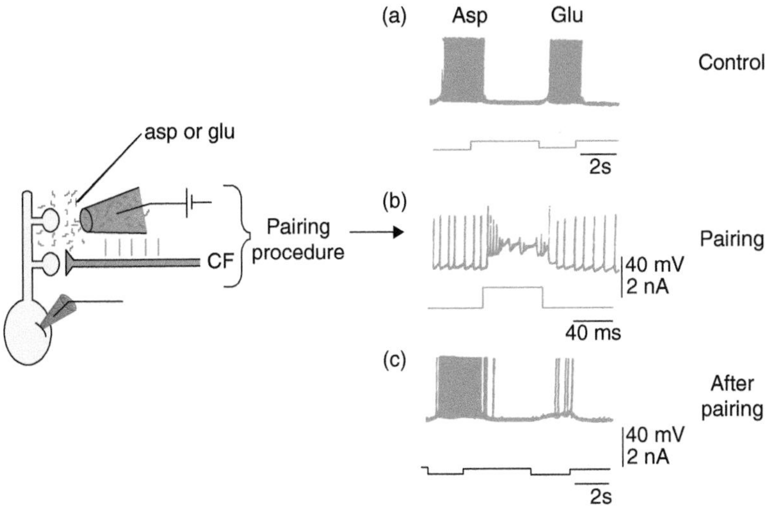

FIGURE 18.17 **The postsynaptic glutamate response is selectively depressed.** The response of a Purkinje cell to iontophoretic application of glutamate (glu) or aspartate (asp) is intracellularly recorded (current clamp mode, $V_m = -65$ mV) in cerebellar slices. **(a)** Glutamate or aspartate are alternatively ejected in the dendritic field of the recorded Purkinje cell. They both evoke a transient membrane depolarization which reaches the firing level. **(b)** The conditioning stimulation used to induce LTD consists of climbing fiber stimulation (2–4 Hz) paired for 1 min with the ejections of glutamate and aspartate at 2 min intervals. **(c)** Twenty minutes after the pairing procedure, the response to glutamate is selectively depressed (the response to aspartate is left unaffected). Adapted from Crépel F, Krupa M (1988) Activation of protein kinase C induces a long-term depression of glutamate sensitivity of cerebellar Purkinje cells: an in vitro study. *Brain Res.* **458**, 397–401, with permission.

quisqualate in activating Purkinje cells could involve a desensitization of AMPA receptors. What are the mediators between Ca^{2+} entry and the long-term changes of AMPA receptors?

18.3.4 Second messengers are required for LTD induction

The metabotropic receptors mGluR1 are abundantly expressed in Purkinje cells. These receptors are coupled to phospholipase C and their activation leads to the formation of inositol trisphosphate and diacylglycerol. Moreover, Ca^{2+}-dependent PKC is also expressed abundantly in Purkinje cells. Therefore, during the conditioning stimulus, the Ca^{2+}-dependent kinases such as protein kinase C can be activated by both the increase of intracellular Ca^{2+} concentration due to climbing fiber activation (see **Figure 18.14b**) and the activation of mGluR1 by glutamate released from parallel fibers. The role of protein kinase C in LTD is tested by injecting into the recorded Purkinje cell a selective inhibitor of protein kinase C (PKC 19-36) before the conditioning stimulus (**Figure 18.18**). In such conditions, LTD is not induced. Moreover, selective expression of a PKC inhibitor in Purkinje cells in transgenic mice leads to a complete blockade of LTD induction, supporting the hypothesis that activation of PKC is necessary for LTD induction.

Antibodies directed against mGluR1 as well as mGluR1 knockout mice were used to demonstrate the role of these metabotropic receptors in LTD induction: in both preparations long-term depression at the parallel-fiber/Purkinje-cell synapse is absent. These preparations also permitted testing of the possible role of internal Ca^{2+} stores: is the combination of direct activation of IP_3-sensitive Ca^{2+} stores in Purkinje cell dendrites and a conventional pairing protocol in mGluR1-deficient mice (mGluR1$^{-/-}$) sufficient to rescue LTD in the cerebellum of these animals by bypassing the disrupted mGluR1 (**Figure 18.19**)? Caged-IP_3 and the fluorescent Ca^{2+}-sensitive dye fluo-3 are present in the whole-cell pipette. The recording session starts 30–45 minutes after whole-cell 'break in' to allow diffusion of the compounds

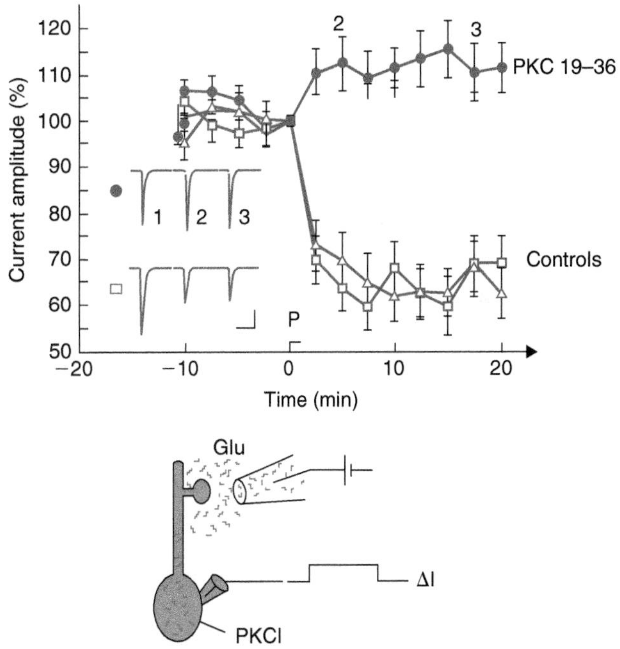

FIGURE 18.18 **Protein kinase C inhibition prevents LTD induction.** The activity of a Purkinje cell is recorded in patch clamp (whole-cell patch) in the presence of the selective PKC inhibitor, PKC 19-36 (circles) or a non-inhibitory control peptide (triangle) or the intracellular solution only (square) in the patch pipette. The control EPSC evoked by parallel fiber stimulation is recorded for 10 min and the conditioning stimulation is applied at $t = 0$. It consists of the conjunctive application of glutamate and intracellular depolarization. LTD of the EPSC is not observed in cells dialyzed with PKC 19-36. Scale bars: 100 pA, 2 s. Adapted from Linden DJ, Connor JA (1991) Participation of postsynaptic PKC in cerebellar long-term depression in culture. *Science* **254**, 1656–1659, with permission.

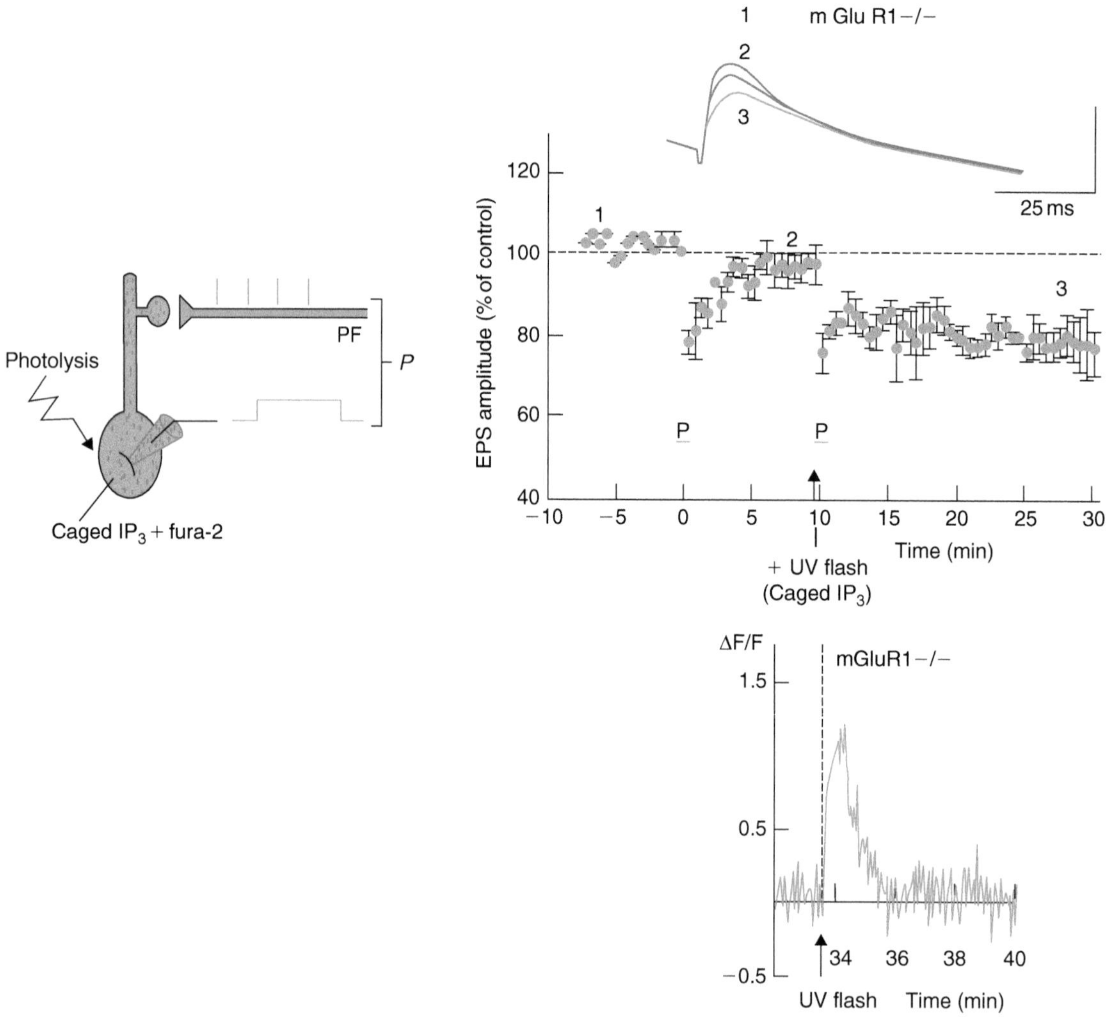

FIGURE 18.19 **LTD in mGluR1-deficient Purkinje cells.** The central plot represents the normalized amplitude of PF-mediated EPSPs against time before and after two successive pairing protocols (P), first at $t = 0$ in the control condition and then at $t = 10$ minutes, combined with photolysis of caged IP3. Each point is the mean ± SEM of separate experiments in four cells. The top inset represents superimposed averaged EPSPs recorded at the indicated times. The bottom inset represents the Ca-induced fluorescence change evoked by photorelease of caged IP3 in a FURA-2-loaded cell. Adapted from Daniel H, Levenes C, Fagni L *et al.* (1999) Inositol-1,4,5-trisphosphate-mediated rescue of cerebellar long-term depression in subtype 1 metabotropic glutamate receptor mutant mouse. *Neuroscience* **91**, 1–6, with permission.

in the dendrites of the recorded Purkinje cell *in vitro* (in cerebellar slices). Control parallel-fiber EPSPs are recorded in control conditions (trace 1). Pairing (simultaneous depolarization of the Purkinje cell and parallel fiber stimulation) is first performed in the absence of photolysis of caged-IP_3. It induces only a transient depression of the EPSP (trace 2). A second pairing is then performed with concomitant photolysis of caged-IP_3. The transient intracellular Ca^{2+} increase in response to a UV flash is visualized by the change of fluorescence of fluo-3 (bottom inset, $\Delta F/F$). After such pairing, parallel-fiber EPSPs are depressed by 76.2 ± 8.2% even 20 minutes after the pairing period (trace 3). The same protocol in the presence of the inhibitory PKC 19-36 peptide fails to induce LTD (not shown). This demonstrates that the impairment of LTD in mGluR1-deficient Purkinje cells is caused by the lack of functional mGluR1 preventing the second-messenger cascade activation. It also suggests that the combination of Ca^{2+} influx through voltage-gated Ca^{2+} channels (in response to Purkinje cell membrane depolarization) and Ca^{2+} release from IP_3-sensitive Ca^{2+} stores is capable of restoring LTD in mGluR1 knockout mice.

The hypothesis is the following. During the conditioning stimulus, the formation of diacylglycerol following activation of mGluR1, together with the cytosolic Ca^{2+} increase due to the activation of voltage-gated channels (and perhaps the release of Ca^{2+} ions from internal stores due to the formation of IP_3), leads to the activation of protein kinase C. This, with other second-messenger cascades, would lead directly or indirectly to phosphorylation of AMPA receptors and activation of their transition to a stable desensitized state and thus to LTD (**Figure 18.20**).

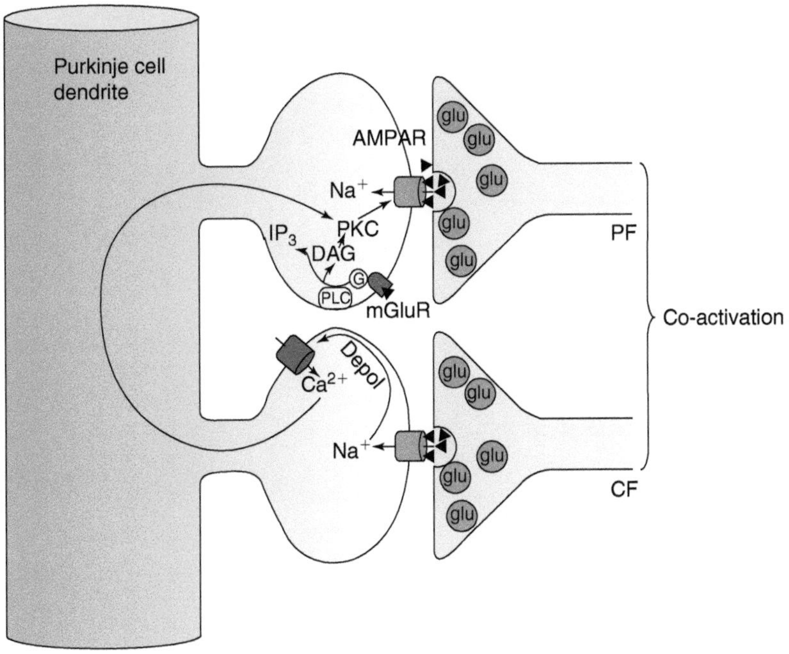

FIGURE 18.20 **Schematic of some of the putative mechanisms of cerebellar LTD induction.**

18.3.5 The different ways to induce or block cerebellar LTD

Long-term depression of the response of a Purkinje cell to parallel fiber activation can be *induced* by (**Figure 18.21a–d**):

- conjunctive stimulation of the afferent parallel and climbing fibers;
- conjunctive stimulation of the parallel fibers and intracellular injection of a depolarizing current (which evokes Ca^{2+} spikes) into the Purkinje cell;
- conjunctive iontophoretic application of glutamate, quisqualate or AMPA + t-ACPD to the Purkinje cell dendrites and stimulation of its afferent climbing fiber;
- conjunctive iontophoretic application of glutamate or quisqualate or AMPA + t-ACPD to the Purkinje cell dendrites and intracellular injection of depolarizing current (which evokes Ca^{2+} spikes) into the Purkinje cell.

Long-term depression of the response of a Purkinje cell to parallel fiber activation can be *blocked* by (**Figure 18.21e–g**):

- intracellular injection of a Ca^{2+} chelator into the Purkinje cell or injection of a hyperpolarizing current into the Purkinje cell during the conjunctive stimulation or the pairing protocol;
- bath application of CNQX or the lack of mGluR1 in the cerebellum and notably in Purkinje cell membrane (mGluR1 gene-deficient mice are obtained by disrupting the mGluR1 gene);
- intracellular injection in the Purkinje cell or bath application of an inhibitor of protein kinase C before the conditioning stimulus.

18.3.6 Summary: principal features of LTD in parallel-fiber/Purkinje cell glutamatergic transmission

- LTD results from a decrease of the parallel-fiber-mediated EPSC or EPSP without changes in the number of afferent axons stimulated: it is a depression of the synaptic efficacy.
- LTD is a very long-lasting phenomenon since it persists for the duration of the experiment, up to several hours.
- LTD is input specific: it is restricted to those parallel-fiber synapses activated at the same time as climbing fibers.
- LTD is associated with a large increase of Ca^{2+} concentration in Purkinje cell dendrites which occurs during the conjunctive stimulation of parallel and climbing fibers. Climbing-fiber synapses are very powerful and their activation leads to a large rise in intracellular Ca^{2+} that is permissive for LTD.
- A key signal that distinguishes active from inactive parallel-fiber synapse, and which is required to trigger LTD, is activation of group 1 mGluRs. Induction of cerebellar LTD requires activation of mGluR1.
- LTD is expressed as a depression of AMPA-mediated current at the parallel-fiber/Purkinje-cell synapses activated at the same time as climbing fibers. It results from the long-term desensitization of AMPA receptors which requires the activation of protein kinase C.

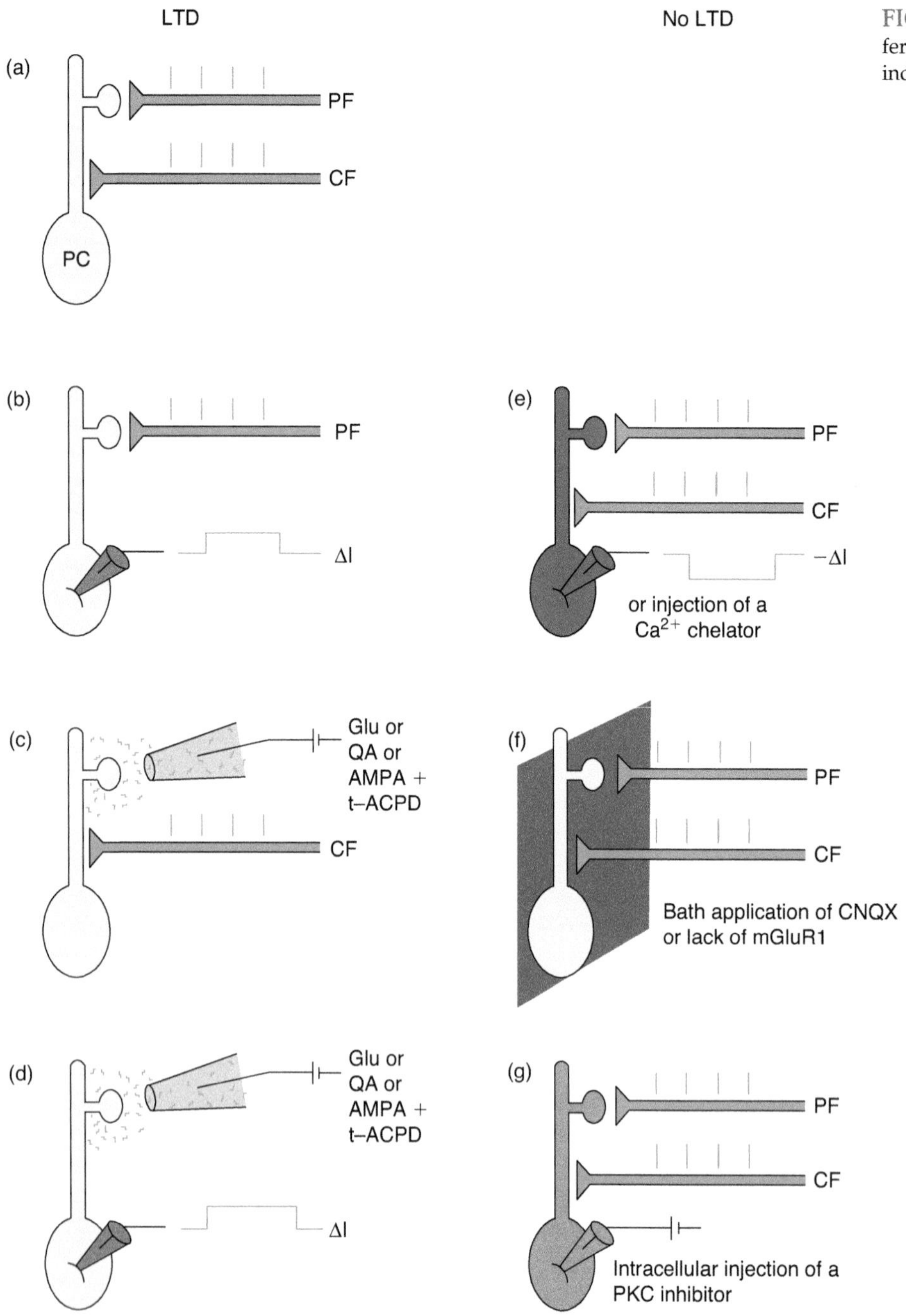

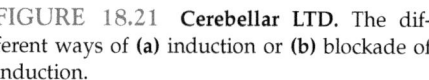

FIGURE 18.21 **Cerebellar LTD.** The different ways of **(a)** induction or **(b)** blockade of induction.

18.4 LONG-TERM SYNAPTIC MODIFICATION INDUCED BY RELATIVE TIMING BETWEEN PRE- AND POSTSYNAPTIC ACTIVITY: SPIKE-TIMING-DEPENDENT PLASTICITY (STDP)

While LTP and LTD are usually induced by coordinated repetitive stimulation of several synaptic inputs, or by stimulation of synaptic inputs paired with postsynaptic depolarization, these induction protocols may not represent physiological patterns of activity. However, long-term synaptic modifications can also be induced at glutamatergic synapses by repetitive association between presynaptic and postsynaptic spiking activity, by eliciting single spikes in the presynaptic and postsynaptic neurons. The polarity of the synaptic change is determined by the relative timing of pre- and postsynaptic spiking activity. This plasticity has therefore been called spike-timing-dependent plasticity or STDP.

18.4.1 Induction of LTP by positive correlation

Following the principle of Hebb and the Pavlovian associative learning, it has been shown that activation of a synapse +5 to +30 ms before postsynaptic firing leads to the reinforcement of this synapse (**Figure 18.22a,b**). This

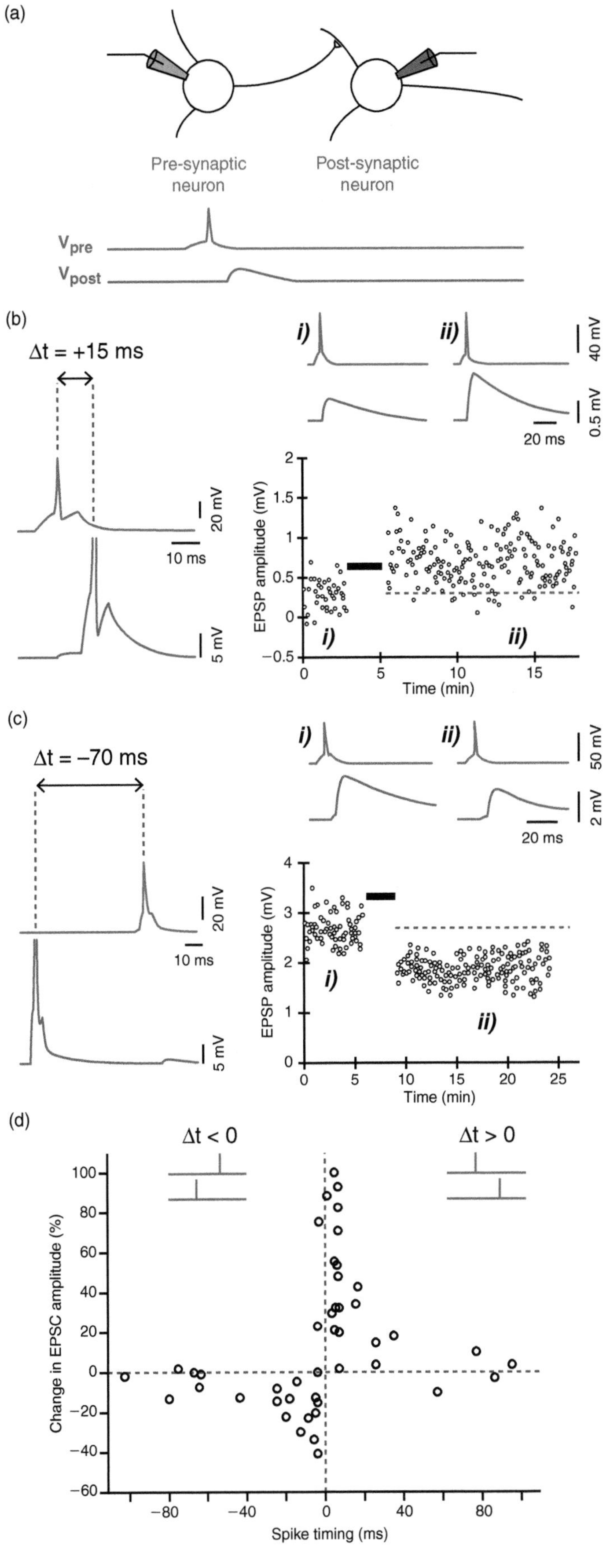

FIGURE 18.22 Spike-timing-dependent plasticity (STDP) at hippocampal synapses. (a) Two neurons connected by a synapse are recorded intracellularly. An action potential in the presynaptic neuron (blue) elicits an excitatory postsynaptic potential in the postsynaptic cell (red). **(b)** Long-term synaptic potentiation induced by positive correlation. Left, induction protocol (pairing indicated by horizontal bar in graph) consisting in single presynaptic action potential elicited 15 ms before a single postsynaptic action potential (repeated 50 times at 0.3 Hz). Right, time-course of the potentiation. Averaged traces from the labeled time points are illustrated above the graph. **(c)** Long-term synaptic depression induced by negative correlation. Left, induction protocol repeated 100 times at 0.3 Hz. Note that here, the postsynaptic spike precedes the presynaptic action potential (70 ms). Right, time-course of the depression. Averaged traces from the labeled time points are illustrated above the graph. Adapted from Debanne *et al.* (1998) *J. Physiol.* **(d)** STDP curve. The percentage of synaptic change is expressed as a function of the spike timing (the presynaptic spike is taken as the time reference). Synaptic potentiation is obtained for positive timing and depression for negative timing. Parts (**b**) and (**c**) adapted from Debanne, D., Gähwiler, B.H., Thompson, S.M., 1998. Long-term synaptic plasticity between pairs of CA3 pyramidal cells in hippocampal slice cultures. *J. Physiol.* **507**, 237–247, with permission. Part (**d**) adapted from Bi and Poo (1998) Synaptic modifications in cultured hippocampal neurons: dependence on spike timing, synaptic strength, and postsynaptic cell type. *J. Neurosci.* 18, 10464–10472.

elementary association of positively correlated pre- and postsynaptic events needs to be repeated a few tens of times to produce a persistent change in synaptic strength. In contrast to conventional LTP, this form of potentiation can be induced during low-frequency stimulation. Like conventional LTP, this potentiation requires the activation of postsynaptic NMDA receptors. The timing-dependent feature of this plasticity can be demonstrated by the fact that no LTP is induced if the delay between the presynaptic activation and postsynaptic spiking is increased (for example, to +100 ms).

18.4.2 Induction of LTD by negative correlation

Long-term depression is induced if the synapse is repeatedly activated a few milliseconds after postsynaptic firing, i.e. by negatively correlated pre- and postsynaptic events (**Figure 18.22c**). Here again, this elementary association needs to be repeated a few tens of times to produce a persistent depression. The timing-dependent feature of this plasticity can be demonstrated by the fact that no LTD is induced if the delay between the presynaptic activation and postsynaptic spiking is increased (for example, +100 ms).

18.4.3 The STDP rule

Synaptic strength at glutamatergic synapses therefore depends on the precise timing of pre- and postsynaptic spikes, thus resulting in an asymmetric plasticity rule, also called the STDP rule (**Figure 18.22d**). This plasticity rule has been confirmed in different brain areas, both *in vitro* and *in vivo*, and is considered today by theoreticians as the most important learning rule to account for functional plasticity in the brain.

18.4.4 Summary: principal features of STDP

- STDP is a form of synaptic plasticity that depends on the relative timing between pre- and postsynaptic spikes.
- LTP is induced by repeated positive correlation whereas LTD is induced by repeated negative correlation.
- LTP requires the activation of postsynaptic NMDA receptors.

18.5 THE HOMEOSTATIC PLASTICITY OF A GLUTAMATERGIC RESPONSE: EXAMPLE OF THE SYNAPTIC SCALING AT NEOCORTICAL GLUTAMATERGIC SYNAPSES

The term homeostasis was introduced by the American physiologist Walter B. Cannon to describe the ensemble of complex mechanisms enabling biological organisms to maintain steady states for their most critical parameters (temperature in warm-blooded animals, blood pressure, plasmatic glucose concentration…) in the face of environmental variations. Homeostatic plasticity therefore refers to a form of plasticity that helps neurons maintain a stable level of excitability in spite of variations in network activity. The best example of homeostatic synaptic plasticity is the synaptic scaling occurring at the glutamatergic synapses received by neocortical pyramidal neurons, described originally by Turrigiano and collaborators (1998). In that form of plasticity, the average strength of glutamatergic synapses is regulated in response to chronic changes in network activity, such that imposed increases in network activity induce a decrease in excitatory synaptic strength while imposed decreases in activity induce an increase in excitatory synaptic strength. Therefore, disturbing network activity triggers compensatory changes in synaptic strength that return average firing rates back to their control value.

18.5.1 Up-scaling of excitatory synaptic strength induced by chronic blockade of network activity

In this experiment, primary cultures of neocortical neurons are submitted to a chronic blockade of activity using either TTX or CNQX. While TTX blocks the sodium channels underlying the action potential and therefore all spiking activity in the network, CNQX, a specific blocker of AMPA receptors, blocks all excitatory synaptic activity. TTX or CNQX is applied during 48 hours in the culture dish, therefore blocking action potentials or excitatory synaptic activity in all the neurons. Excitatory synaptic activity in the network is monitored by measuring the frequency and amplitude of miniature excitatory postsynaptic currents (EPSCs) in several neurons using the patch-clamp technique. Excitatory synaptic activity after chronic blockade (TTX in the example presented in **Figure 18.23**) is compared to cultures kept for the same duration but not pharmacologically treated. After 48 hours of TTX application, the average amplitude of miniature EPSCs is significantly increased (**Figure 18.23a,b**). More interestingly, the distribution of amplitudes of the miniature EPSCs indicates that all EPSCs are affected by the same ratio, i.e. that all synaptic events have been scaled up by the same factor following the blockade of activity (multiplicative change). As a consequence, the curves of EPSC cumulative frequencies (TTX vs control) can be superimposed by simple scaling, therefore the term 'synaptic scaling' (**Figure 18.23c**). While the chronic blockade of AMPA receptors using CNQX induces synaptic scaling, blocking NMDA receptors using APV is unable to trigger synaptic scaling. Therefore, NMDA receptors, which are involved in the induction of hebbian plasticity, do not seem to be involved in the induction of homeostatic plasticity.

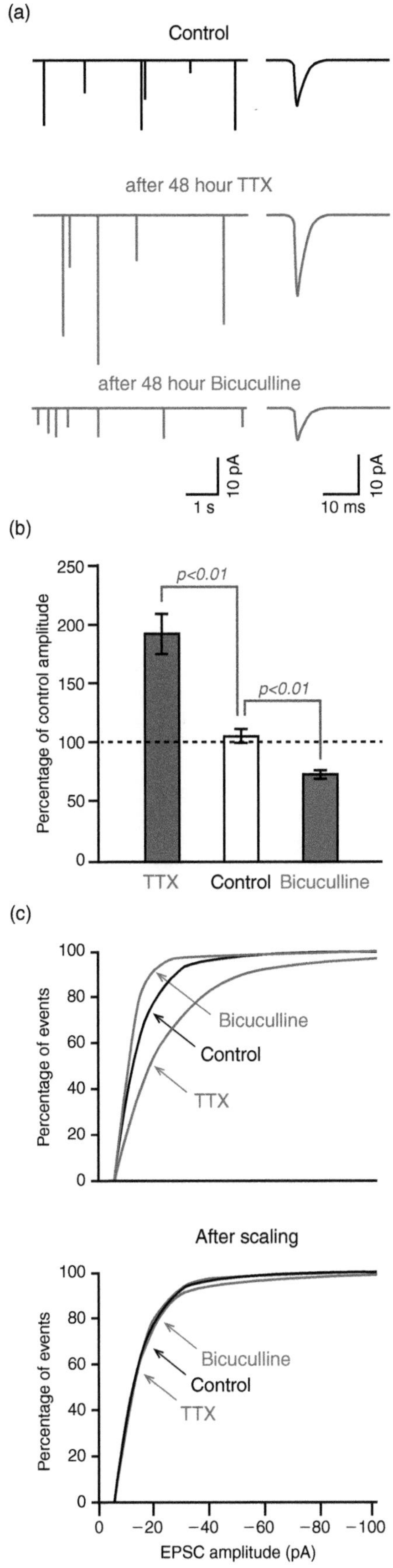

FIGURE 18.23 **Synaptic scaling of excitatory events in neocortical neurons. (a)** Representative recordings of AMPA miniature EPSCs from neocortical neurons grown in control condition (black traces), in the presence of TTX (red traces) or in the presence of bicuculline (blue traces). Averaged miniature EPSC waveforms are shown on the right. **(b)** Average miniature EPSC amplitudes expressed as a percentage of control indicate that TTX and bicuculline amplitudes differ significantly from control (one-way ANOVA). **(c)** Cumulative amplitude histograms for miniature EPSCs recorded under each condition. Bottom curves have been obtained by scaling all values in the TTX or bicuculline conditions by the same factor. The scaling factor was >1 for bicuculline, <1 for TTX. Adapted from Turrigiano GG, Leslie KR, Desai NS, Rutherford LC, Nelson SB (1998) Activity-dependent scaling of quantal amplitude in neocortical neurons. *Nature* **391**, 892-896, with permission.

18.5.2 Down-scaling of excitatory synaptic strength induced by chronic enhancement of network activity

In this experiment, primary cultures of neocortical neurons are submitted to a chronic enhancement of activity using bicuculline, a specific blocker of GABAergic neurotransmission. By suppressing all inhibitory synaptic transmission, bicuculline induces a significant increase in network activity. Bicuculline is applied over 48 hours, the same duration used for chronic blockade of activity (previous section). Excitatory synaptic strength in the network is monitored by measuring the frequency and amplitude of miniature EPSCs using the patch-clamp technique. Excitatory synaptic activity after chronic application of bicuculline is compared to cultures kept for the same duration but not pharmacologically treated. After 48 hours of bicuculline application, the average amplitude of miniature EPSCs is significantly decreased compared to untreated cultures (**Figure 18.23a,b**). Again, the distribution of amplitudes of the miniature EPSCs indicates that all EPSCs are affected by the same ratio, i.e. that all synaptic events have been scaled down by the same factor following activity enhancement (multiplicative change): the curves of EPSC cumulative frequencies can be superimposed by simple scaling (**Figure 18.23c**). As with chronic activity blockade (section 18.5.1), the synaptic scaling in response to bicuculline application affects both AMPA and NMDA receptors.

18.5.3 Synaptic scaling also affects inhibitory synaptic transmission

Synaptic scaling has not only been observed at excitatory synapses, but also at inhibitory neocortical synapses. Using primary cultures, Kilman and collaborators (2002) demonstrated that activity deprivation induced by TTX application induces a scaling-down of inhibitory synaptic strength. Again, this synaptic scaling relies on a multiplicative change in the number of receptors clustered at the synapse. Here, the number of $GABA_A$ receptors is decreased by chronic activity blockade.

18.5.4 Summary: principal features of homeostatic synaptic scaling

- Synaptic scaling is induced by chronic blockade or enhancement of network activity.
- The changes in synaptic strength compensate the imposed changes in network activity: while activity blockade induces an increase in excitatory synaptic strength, activity enhancement induces a decrease in excitatory synaptic strength.
- Synaptic scaling represents an increase or a decrease in the number of postsynaptic receptors, which are proportional to the starting number of receptors.
- Synaptic scaling at excitatory synapses depends on the activation of AMPA receptors, but not NMDA receptors.
- Synaptic scaling at excitatory synapses affects both the number of AMPA receptors and the number of NMDA receptors.
- Synaptic scaling affects both excitatory and inhibitory synapses.

Further reading

Aiba, A., Kano, M., Chen, C., et al., 1994. Deficient cerebellar long-term depression and impaired motor learning in mGluR1 mutant mice. Cell 79, 377–388.

Bi, G.Q., Poo, M.M., 1998. Synaptic modifications in cultured hippocampal neurons: dependence on spike timing, synaptic strength and post-synaptic cell type. J. Neurosci. 18, 10464–10472.

Bredt, D.S., Nicoll, R.A., 2003. AMPA receptor trafficking at excitatory synapses. Neuron 40, 361–379.

Carroll, R.C., Lissin, D.V., Zastrowvon, M., et al., 1999. Rapid redistribution of glutamate receptors contributes to long-term depression in hippocampal cultures. Nat. Neurosci. 2, 454–460.

Conquet, F., Bashir, Z.I., Davies, C.H., et al., 1994. Motor deficit and impairment of synaptic plasticity in mice lacking mGluR1. Nature 372, 237–243.

Crépel, F., Audinat, E., Daniel, H., et al., 1994. Cellular locus of the nitric oxide-synthase involved in cerebellar long-term depression induced by high external potassium concentration. Neuropharmacology 33, 1399–1405.

Debanne, D., Gähwiler, B.H., Thompson, S.M., 1998. Long-term synaptic plasticity between pairs of CA3 pyramidal cells in hippocampal slice cultures. J. Physiol. 507, 237–247.

De Zeeuw, C.I., Hansel, C., Bian, F., et al., 1998. Expression of a protein kinase C inhibitor in Purkinje cells blocks cerebellar LTD and adaptation of the vestibuloocular reflex. Neuron 20, 495–508.

Hartell, N.A., 2001. Receptors, second messengers and protein kinases required for heterosynaptic cerebellar long-term depression. Neuropharmacology 40, 148–161.

Ito, M., Sakurai, M., Tongroach, P., 1982. Climbing fibre induced depression of both mossy fibre responsiveness and glutamate sensitivity of cerebellar Purkinje cells. J. Physiol. 324, 113–134.

Kilman, V., Van Rossum, M.C.W., Turrigiano, G.G., 2002. Activity deprivation reduces miniature IPSC amplitude by decreasing the number of postsynaptic $GABA_A$ receptors clustered at neocortical synapses. J. Neurosci. 22, 1328–1337.

Lev-Ram, V., Makings, L.R., Keitz, P.F., et al., 1995. Long-term depression in cerebellar Purkinje neurons results from coincidence of nitric oxide and depolarization induced Ca^{2+} transients. Neuron 15, 407–415.

Linden, D.J., Dickinson, M.H., Smeyne, M., Connor, J.A., 1991. A long-term depression of AMPA currents in cultured cerebellar Purkinje neurons. Neuron 7, 81–89.

Lisman, J., Schulman, H., Cline, H., 2002. The molecular basis of CaMKII function in synaptic and behavioral memory. Nat. Rev. Neurosci. 3, 175–190.

Malenka, R.C., Bear, M.F., 2004. LTP and LTD: an embarrassment of riches. Neuron 44, 5–21.

Schuman, E.M., Dynes, J.L., Steward, O., 2006. Synaptic regulation of dendritic mRNAs. J. Neurosci. 26, 7143–7146.

Shi, S.H., Hayashi, Y., Petralia, R.S., et al., 1999. Rapid spine delivery and redistribution of AMPA receptors after synaptic NMDA receptors activation. Science 284, 1811–1816.
Shigemoto, R., Abe, T., Nomura, S., et al., 1994. Antibodies inactivating mGluR1 metabotropic glutamate receptor block long-term depression in cultures Purkinje cells. Neuron 12, 1245–1255.
Turrigiano, G.G., Leslie, K.R., Desai, N.S., Rutherford, L.C., Nelson, S.B., 1998. Activity-dependent scaling of quantal amplitude in neocortical neurons. Nature 391, 892–896.
Turrigiano, G.G., Nelson, S.B., 2004. Homeostatic plasticity in the developing nervous system. Nat. Rev. Neurosci. 2, 97–107.

APPENDIX 18.1 DEPOLARIZATION-INDUCED SUPPRESSION OF INHIBITION (DSI): AN EXAMPLE OF SHORT-TERM PLASTICITY AT GABAergic SYNAPSES

Inhibitory synaptic responses also undergo short- and long-term modifications. This appendix examines the mechanisms underlying short-term plasticity at inhibitory synapses.

A18.1.1 Observation of depolarization-induced suppression of inhibition

This form of short-term plasticity at GABAergic synapses was first reported more than a decade ago in the cerebellum. The pioneering observation was made by Alain Marty and collaborators in 1991: the direct depolarization of cerebellar Purkinje cells induces a decrease of the amplitude of spontaneous inhibitory postsynaptic currents (IPSCs). The experimental protocol is the following (**Figure A18.1**). Evoked and spontaneous inhibitory postsynaptic currents (IPSCs) are recorded from Purkinje cells *in vitro* using the whole-cell patch clamp technique in the continuous presence of ionotropic glutamate receptor antagonists. The IPSCs are entirely blocked by bath application of bicuculline, showing that they are mediated by $GABA_A$ receptors. After a stable control period, a train of depolarizing voltage steps (8 pulses from −70 mV to 120 mV, 100 ms duration) are applied through the recording pipette. This train of pulses leads to a transient decrease of the amplitude of evoked and spontaneous GABAergic IPSCs. The inhibition of GABAergic activity is quantified as a function of time by counting the sum of the amplitudes of spontaneous IPSCs during sampling intervals of 2 s. A similar effect is induced by a single depolarizing pulse of 0.1–1 s duration. This phenomenon, termed depolarization-induced suppression of inhibition (DSI), develops within seconds following the voltage pulses and lasts tens of seconds. DSI is not restricted to cerebellar $GABA_A$ synapses since it is also observed for hippocampal $GABA_A$ synapses following the depolarization of CA1 pyramidal neurons. Experiments performed in either preparation contributed to the description of DSI that follows.

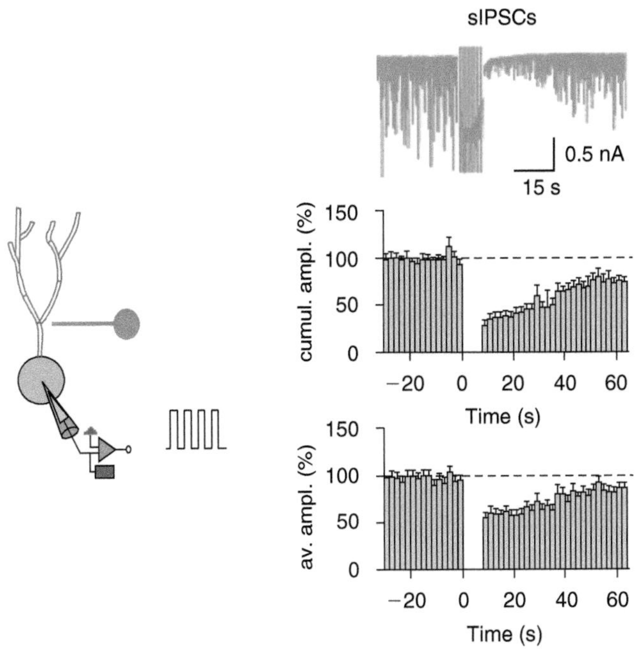

FIGURE A18.1 **Effects of postsynaptic depolarization on the spontaneous synaptic activity in Purkinje cells recorded in the presence of blockers of glutamatergic synaptic currents: evidence for depolarization-induced suppression of inhibition.** Whole-cell recording of cerebellar Purkinje cell *in vitro*. The top trace shows a typical DSI protocol. Spontaneous GABA-IPSCs are recorded in the presence of glutamatergic receptor antagonists. After a control period, the Purkinje cell is depolarized 8 times to 0 mV for 100 ms, at 1 Hz; the trace shows the consequent dramatic inhibition of sIPSCs and, thereafter, their recovery phase over 60 s. In this cell, four such protocols were averaged, yielding 79.6 ± 0.5% inhibition in the sIPSCs cumulative amplitude (cumul. ampl.), calculated over the first 10 s after the end of the pulse train. Decreases in the frequency and in the average amplitude (av. ampl.) contributed in equal measure to this reduction: 53.6 ± 4.3 and 55.2 ± 4.3%, respectively. The two bottom graphs show the time course of the DSI of cumulative and average amplitudes for $n = 14$ experiments performed with this protocol. Adapted from Diana MA, Marty A (2003) Characterization of depolarization-induced suppression of inhibition using paired interneuron–Purkinje cell recordings. *J. Neurosci.* **23**, 5906–5918, with permission.

A18.1.2 DSI requires a postsynaptic rise of intracellular calcium concentration

It was quickly recognized that an increase in intracellular Ca^{2+} concentration, through the activation of voltage-dependent Ca^{2+} channels (VDCCs), is necessary and sufficient to trigger DSI.

- When the postsynaptic neuron is loaded with the fast Ca^{2+} buffer BAPTA, the depolarization fails to trigger DSI. Moreover, release of Ca^{2+} ions via flash photolysis of caged Ca^{2+} is sufficient to induce DSI (**Figure A18.2**). These observations show that an increase in $[Ca^{2+}]_i$ is required to trigger DSI.

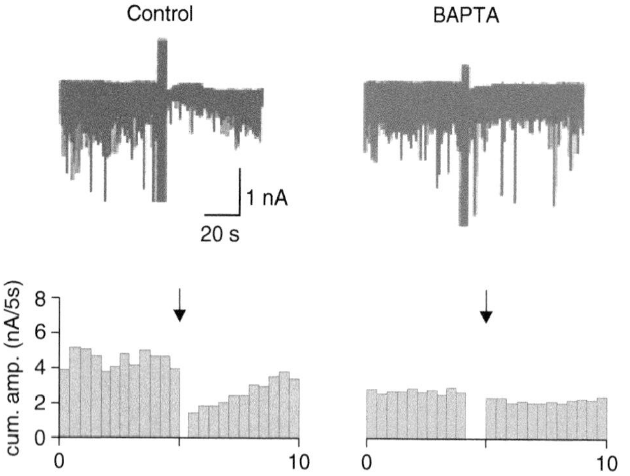

FIGURE A18.2 **DSI requires a postsynaptic rise of intracellular Ca^{2+} concentration.** Whole-cell recording of a cerebellar Purkinje cell. Top traces show the effect of repeated depolarizing voltage pulse on spontaneous IPSCs in control cell (left) and in cells loaded with the fast Ca^{2+} chelator BAPTA (right). Similar depolarization has no effect on spontaneous IPSCs. Adapted from Glitsch M, Parra P, Llano I (2000) The retrograde inhibition of IPSCs in rat cerebellar Purkinje cells is highly sensitive to intracellular Ca^{2+}. *Eur. J. Neurosci.* **12**, 987–993, with permission.

- Fura2 Ca^{2+} measurements combined with whole-cell patch clamp recordings of cerebellar Purkinje cells indicate that the extent of DSI is a function of the peak postsynaptic intracellular Ca^{2+} concentration.
- DSI does not occur when Ca^{2+} ions are removed from the extracellular solution, showing that an influx of Ca^{2+} is necessary to induce DSI.
- DSI is observed in the presence of D-APV, a selective antagonist of NMDA receptors but not in the presence of blockers of voltage-dependent Ca^{2+} channels.

A18.1.3 DSI results from a transient suppression of GABA release

What are the mechanisms underlying the transient decrease of GABAergic synaptic activity? DSI can be accounted for by alterations in the density or properties of postsynaptic $GABA_A$ receptors, or by a presynaptic decrease of transmitter release.

The presynaptic nature of DSI was rapidly established on the basis of the following experiments. The amplitude of miniature GABAergic IPSCs or the postsynaptic response to GABA applications are not affected by the DSI protocol. These observations strongly suggest that the postsynaptic sensitivity to GABA is not affected. In contrast, the widely used indicators of presynaptic modifications are affected. Thus, during DSI the percentage of synaptic failures increases and the frequency of miniature IPSCs decreases, indicating a decrease in release probability.

A18.1.4 Endocannabinoid as a retrograde messenger in DSI

Altogether these observations indicate that the postsynaptic $[Ca^{2+}]_i$ rise triggered by the depolarization of the target (recorded) neuron leads to a transient suppression of GABA release from presynaptic terminals. It was thus hypothesized that a retrograde signal is sent from the postsynaptic neuron to the presynaptic inhibitory terminals. To approach this question, the effect of depolarizing voltage pulses is analyzed in paired recordings of neighboring Purkinje cells separated by 25–150 μm. Simultaneous recordings from two neighboring Purkinje cells revealed that a depolarization applied to one cell results in the transient suppression of inhibition in both recorded cells. Thus, DSI is not restricted to the stimulated neuron but can spread to its neighbors.

The following step was to identify the retrograde messenger of DSI. In some studies it was proposed that glutamate is released by the postsynaptic cell and by acting on metabotropic glutamatergic receptors (mGluRs) inhibits the release of GABA. However, in 2001 Wilson and Nicoll demonstrated that endocannabinoids and the subsequent activation of type 1 cannabinoid receptors (CB1Rs) are the retrograde signal of DSI in the hippocampus. A similar conclusion was reached about the cerebellum.

This conclusion is supported by the following observations:

- CB1 receptor agonist induces a presynaptic inhibition of GABAergic synaptic postsynaptic currents similar to that occurring during DSI. Moreover, this effect is only observed at synapses that have the ability to express DSI. In CB1-insensitive GABAergic synapses, likely lacking CB1 receptors, DSI is not observed. DSI is abolished by the application of CB1 receptor antagonists (**Figure A18.3**) and not observed in CB1 receptor knockout mice. These observations clearly show that the activation of CB1Rs is required to trigger DSI.
- Photolysis of caged Ca^{2+} inside a pyramidal neuron triggers DSI that is completely abolished by CB1R antagonist. This observation demonstrates that a postsynaptic rise of $[Ca^{2+}]_i$ leads to the release of endocannabinoids that activate presynaptic CB1Rs and inhibit GABA release.
- The retrograde messenger, like CB1 receptor, uses a G-protein signaling pathway to induce DSI. In the rat hippocampus, DSI is blocked by pre-treatment with pertussis toxin or with N-ethylmaleimide (NEM), which blocks pertussis-toxin-sensitive G proteins. However, bath applied NEM and pertussis toxin have access to both presynaptic and postsynaptic sites. To determine the locus of the relevant G protein, the postsynaptic cell is loaded with a

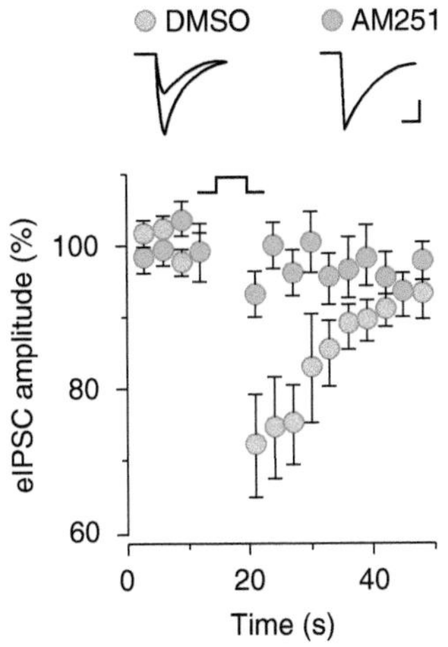

FIGURE A18.3 **Activation of the cannabinoid receptor CB1 is required to trigger the depolarization-induced suppression of inhibition.** Whole-cell recording of hippocampal pyramidal cells. Hippocampal slices are pre-incubated in DMSO alone (d) or in the presence of a CB1 receptor antagonist (AM251) dissolved in DMSO (s). In the presence of the CB1R antagonist, a 5 s depolarizing step results in little or no suppression of evoked IPSCs (eIPSCs), while in control DMSO experiment, the same depolarization leads to a robust reduction. Inserts show average eIPSCs for the 10 s before and the 10 s after the application of the depolarizing step. Scale bars: 200 pA, 20 ms. Adapted from Wilson RI, Nicoll RA (2001) Endogenous cannabinoids mediate retrograde signalling at hippocampal synapses. *Nature* **410**, 588–592, with permission.

non-hydrolyzable GTP analog. In these cells, DSI is not affected by NEM or pertussis toxin, indicating that only the activation of presynaptic G proteins is required for DSI induction.

- Loading the postsynaptic target neuron with botulinum toxin that blocks membrane fusion and vesicular release does not prevent DSI. This observation is consistent with the mode of release of endocannabinoids, anandamide and 2-AG, that exit the cell by diffusion across the membrane or by passive transport but not by vesicular fusion.

A18.1.5 Functional significance of DSI

A train of action potentials in the Purkinje cells or the activation of their afferent glutamatergic climbing fiber induces DSI of afferent $GABA_A$ activity, indicating that this form of short-term plasticity might play a role in controlling neuronal excitability *in vivo*. Similarly, a single Ca^{2+} spike is sufficient to produce DSI in the hippocampus. These observations indicate that DSI modulates the strength of GABAergic synapses *in vivo*.

Further reading

Diana, M., Marty, A., 2004. Endocannabinoid-mediated short-term synaptic plasticity: depolarization-induced suppression of inhibition (DSI) and depolarization-induced suppression of excitation (DSE). Br. J. Pharmacol. 142, 9–19.

Llano, I., Leresche, N., Marty, A., 1991. Calcium entry increases the sensitivity of cerebellar Purkinje cells to applied GABA and decreases inhibitory synaptic currents. Neuron 6, 565–574.

Piomelli, D., 2003. The molecular logic of endocannabinoid signaling. Nat. Rev. Neurosci. 11, 873–874.

Pitler, T.A., Alger, B.E., 1992. Postsynaptic spike firing reduces synaptic GABA responses in hippocampal pyramidal cells. J. Neurosci. 12, 4122–4132.

Vincent, P., Marty, A., 1993. Neighboring cerebellar Purkinje cells communicate via retrograde inhibition of common presynaptic interneurons. Neuron 11, 885–893.

PART IV

THE HIPPOCAMPAL NETWORK

CHAPTER

19

The adult hippocampal network

Constance Hammond

OUTLINE

The hippocampus is part of the limbic system which mediates emotions and aspects of learning and memory. In the rat, it is a rostro-caudal structure (**Figure 19.1**) whereas in primates it is strictly localized in the temporal lobe.

The hippocampus is composed of two interconnected crescent-like regions (**Figure 19.2a,b**): the Ammon's horn (cornu ammonis) and the dentate gyrus (DG, also called fascia dentata). On a coronal section, Ammon's horn of the rodent can be further subdivided into three regions, CA1, CA2 and CA3 (CA for cornu ammonis) (**Figure 19.2b**). In humans, another subdivision exists, CA4, which is located between the two bands of the dentate gyrus. Ammon's horn and dentate gyrus both contain a layer of principal neurons that are projection neurons (Golgi type I): the pyramidal cell layer and the granular cell layer, respectively. In the dentate gyrus, the regions are the stratum moleculare (MOL) and the hilus (**Figure 19.2b**). Numerous local interneurons (Golgi type II) are present in each region. Principal cells use glutamate as a neurotransmitter whereas interneurons use GABA.

19.1 OBSERVATIONS AND QUESTIONS

What is a network?

In nuclei of the central nervous system, various types of neurons are generally present: projection neurons (Golgi type I) whose axons project to neurons located outside the nucleus, and local interneurons (Golgi type II) whose axons project to neurons located inside the nucleus. These neuronal types are connected to each other (intrinsic connections): they form a network. Each network receives afferents from neurons located in other nuclei (extrinsic connections).

Are networks completely different from one nucleus to another or are there some fundamental principles of organization?

In the neocortex and the Ammon horn of the hippocampus, Golgi type I neurons are the pyramidal cells, in the cerebellar cortex they are the Purkinje cells, and in the striatum they are the medium spiny neurons. Pyramidal cells are glutamatergic whereas Purkinje cells and medium spiny neurons are GABAergic. Aside from these principal cells, a large variety of local GABAergic interneurons is present in all these nuclei.

Does the precise knowledge of intrinsic and extrinsic connections as well as the firing patterns of neurons allow us to explain how network oscillations are generated?

In the hippocampus of the freely moving rat, several types of oscillations (network activities) are recorded from populations of neurons *in vivo* (extracellular

Cellular and Molecular Neurophysiology. DOI: 10.1016/B978-0-12-397032-9.00019-4

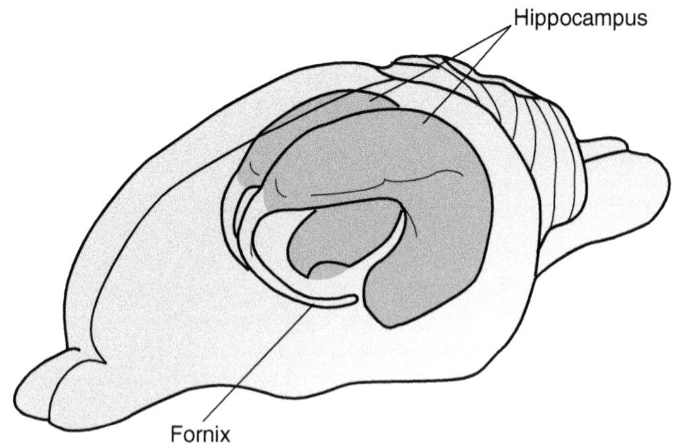

FIGURE 19.1 **Schematic of the localization of the two hippocampi inside a rat brain.**

recordings). For example, a rhythmic slow activity called 'theta' (5–10 Hz) is recorded during exploratory behavior, such as sniffing, rearing and walking, and the paradoxical phase of sleep; once the animal stays still, or during consummatory behaviors or slow-wave sleep, intermittent sharp waves (SPW) are recorded in the dendritic layer of CA1–CA3 (**Figure 19.3** and **Section 19.5**). These oscillations are network oscillations.

Neuronal oscillations have two main origins. They can be *intrinsic* to the neuron when they result from the activation of a cascade of currents intrinsic to the membrane, as described for thalamic and subthalamic neurons in Section 17.4. Intrinsic oscillations propagate in the network via gap junctions as in inferior olive and/or via chemical synapses. Oscillations can be *extrinsic* when they result from the activity of a group of synaptically interconnected neurons, from the activity of their chemical synapses, as is the case for hippocampal neurons.

The aim of the present chapter is to give a description of the adult hippocampal network and to explain how it can generate oscillations.

19.2 THE HIPPOCAMPAL CIRCUITRY

19.2.1 Ammon's horn

Ammon's horn is a curved structure. It has a laminar organization with five layers. Owing to its U-shape, the layers (and pyramidal cells) are upside down in CA1 compared with CA3 (see **Figures 19.2b** and **19.6b**).

Principal cells are called the pyramidal cells; they use glutamate as a neurotransmitter

The principal cells, the pyramidal cells, have their soma aligned in a thin layer called the *pyramidal cell layer* or *stratum pyramidale* (P, **Figure 19.2b**). The name of these cells comes from the clear triangular shape of their cell body (**Figure 19.4a**). The cell body has a diameter of

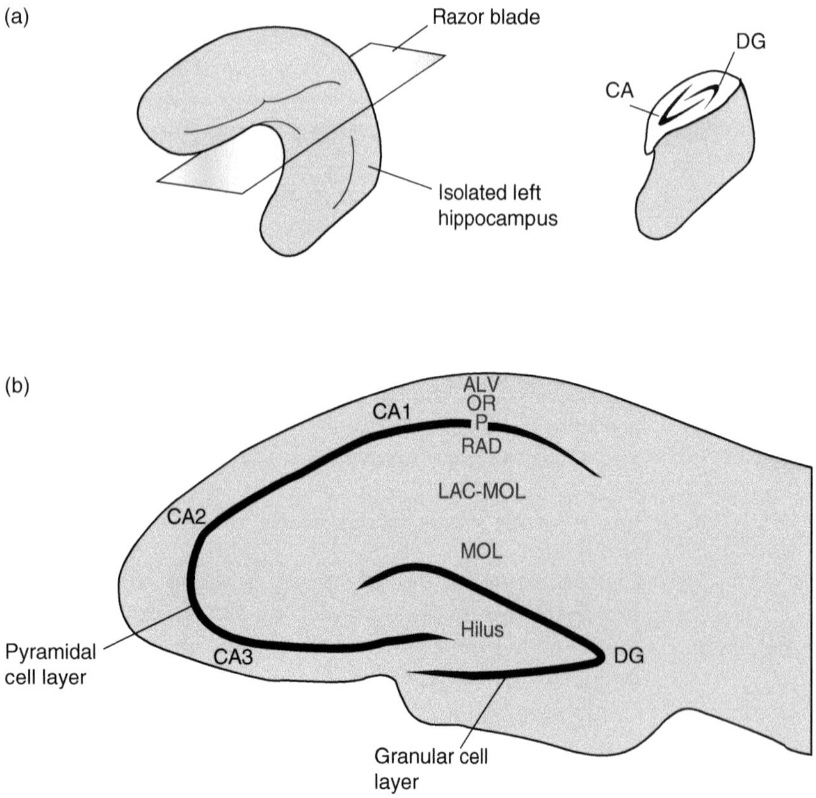

FIGURE 19.2 **Structure of the rat hippocampus. (a)** Schematic of the slice preparation protocol. **(b)** Schematic transverse section of the hippocampus. ALV: stratum alveus, OR: stratum oriens, P: stratum pyramidale, RAD: stratum radiatum, LAC-MOL: stratum lacunosum moleculare, MOL: stratum moleculare.

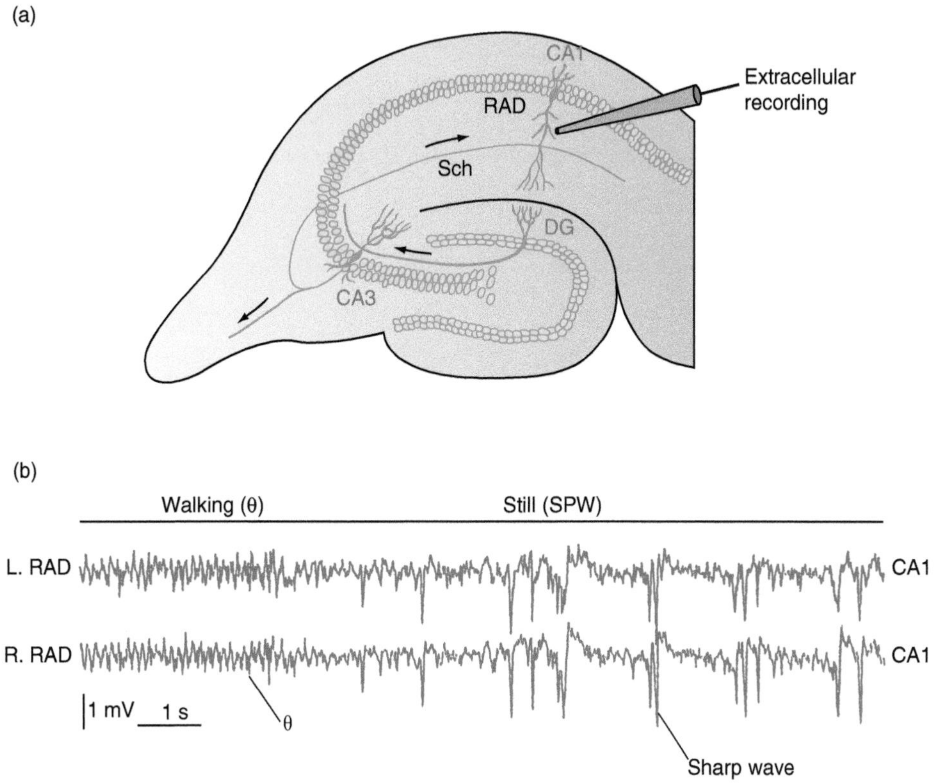

FIGURE 19.3 **Extracellular field recordings of hippocampal oscillations in a freely moving rat during transition from walking to immobility.** An extracellular recording electrode is implanted in the stratum radiatum (RAD) of the CA1 region of each hippocampus, the left (L) and the right (R). During exploratory activity (walking), regular theta waves are recorded (θ); during immobility, large monophasic sharp waves (SPW) are recorded. Note the bilaterally synchronous nature of SPW. Sch: Schaffer collaterals, i.e. axons of CA3 pyramidal cells. Adapted from Buzsáki G (1989) Two-stage model of memory trace formation: a role for 'noisy' brain states. *Neuroscience* **31**, 551–570, with permission.

20 μm. Three main dendritic trunks emerge from the cell body, one apical and two basal. Apical dendrites extend in the stratum radiatum (so called because of the radial organization of apical dendrites from all pyramidal cells) and arborize in *stratum lacunosum moleculare* (LAC-MOL). Basal dendrites ramify in *stratum oriens* (OR). Dendrites have numerous spines. In CA3, the proximal part of apical dendrites of pyramidal cells presents giant spines (thorny excrescences; **Figure 19.4b**) that are the postsynaptic elements of the synapses with granular cells of dentate gyrus. Axons of pyramidal cells run in *stratum alveus* (ALV) where they emit numerous collaterals before leaving the hippocampus.

Several types of inhibitory interneurons innervate pyramidal neurons; they use GABA as a neurotransmitter

The activity of pyramidal cells is modulated not only by glutamatergic afferences but also by ones coming from local inhibitory interneurons. Four main types of inhibitory interneurons have been described in Ammon's horn, all GABAergic. They have their cell bodies in the five layers of the CA regions which are from the external to the internal part of the hippocampus (**Figures 19.2b** and **19.5a**, left): stratum alveus (ALV), stratum oriens (OR), stratum pyramidale (P), stratum radiatum (RAD) and stratum lacunosum moleculare (LAC-MOL). Interneurons can be classified according to their site(s) of termination on pyramidal neurons (**Figure 19.5a**, right), but also from their firing pattern, neurochemical content, birth date or spatial origin:

- Basket cells (BC) innervate the soma and proximal dendrites located in stratum pyramidale and radiatum. Cell bodies of basket cells are mainly located in stratum pyramidale. Some of them are also cholecystokinin positive. They all originate from the medial ganglionic eminence;
- Bistratified cells (BiC) innervate both apical and basal dendrites on their proximal part located in stratum radiatum and oriens. Cell bodies of bistratified cells are located in stratum oriens/radiatum or stratum pyramidale. They are somatostatin positive and originate from the medial ganglionic eminence;
- Oriens-lacunosum moleculare cells (O-LMC) innervate distal apical dendrites located in stratum lacunosum moleculare. Cell bodies of O-LMC are located in

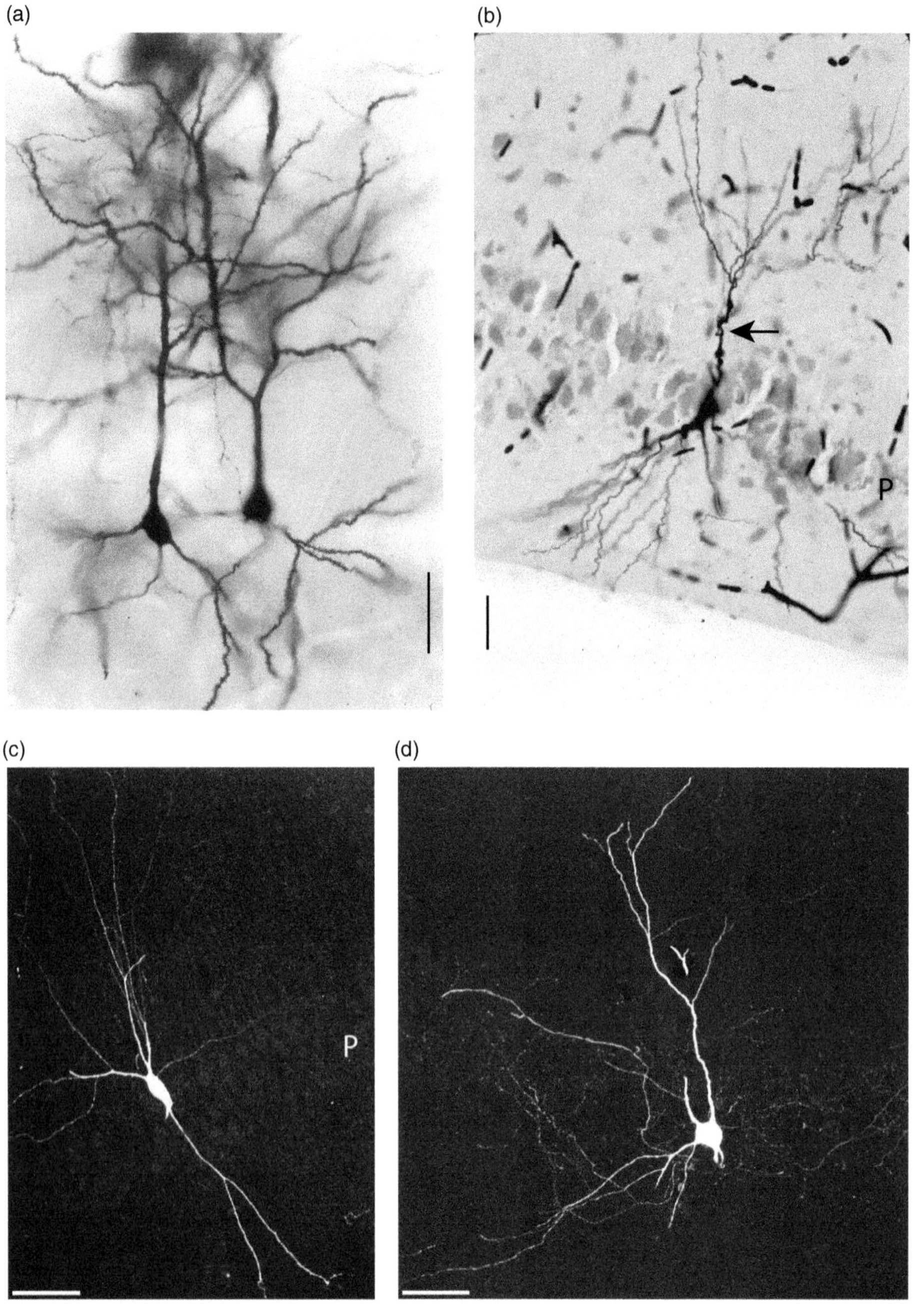

FIGURE 19.4 **Photomicrographs of stained Golgi CA1 and CA3 pyramidal neurons and CA1 interneurons. (a)** CA1 pyramidal neuron. **(b)** CA3 pyramidal neuron. The underlying Nissl coloration shows the density of neuronal cell bodies in the pyramidal layer. The arrow points to the giant dendritic spines. **(c,d)** Examples of CA1 GABAergic interneurons labeled with GFP. Their cell body is located inside or close to the pyramidal cell layer (P). The thin labeling corresponds to axon and axon collaterals. Scales: 50 μm in **a**–**d**. Photographs: (**a**) by Olivier Robain; (**b**) by Jean Luc Gaiarsa; (**c,d**) by Agnès Baude.

stratum oriens. They are somatostatin positive and originate from the medial ganglionic eminence;

- Axo-axonic cells (AAC), also called chandelier cells, innervate exclusively the axon initial segment. Cell bodies of axo-axonic cells are located within or near stratum pyramidale.

When interneurons are activated by extrinsic afferences, they participate in feedforward inhibition; when they are activated by recurrent axon collaterals of pyramidal cells they participate in feedback inhibition (**Figure 19.5b**). Some interneurons like O-LMC are involved only in feedback inhibition since they are activated only by axon collaterals from principal cells.

19.2.2 The dentate gyrus

The dentate gyrus is also a curved structure, with a U-shape and a three-layer organization (see **Figure 19.2b**).

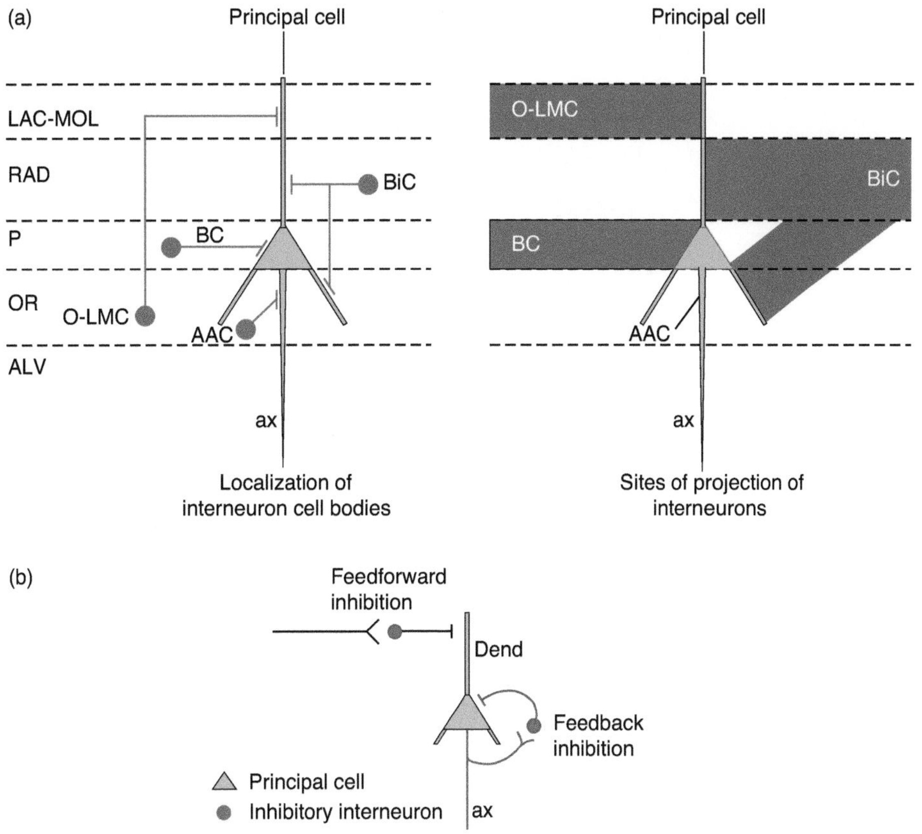

FIGURE 19.5 **Intrinsic connections in CA1 and CA3. (a)** Schematic of a pyramidal cell indicating the localization of the cell bodies of the different interneurons (left) and the segregated postsynaptic domains innervated by the distinct presynaptic interneurons (right). **(b)** Illustration of feedforward and feedback inhibition. See text for abbreviations. Part (a) adapted from Maccaferri G, Roberts DB, Szucs P *et al.* (2000) Cell surface domain specific postsynaptic currents evoked by identified GABAergic neurones in rat hippocampus *in vitro*. *J. Physiol.* **524**, 91–116, with permission.

Principal cells called the granular cells use glutamate as a neurotransmitter

The principal cells, called the granular cells, have their soma densely packed in a thin layer, the granular cell layer. Somas have a small diameter (14–18 μm) and are ovoid. Dendritic trees emerge from the apical pole of somas and form the molecular layer. Axons, called mossy fibers, have a small diameter (0.5 μm) and are not myelinated. They emerge from the basal pole of somas, divide in the hilus in numerous collaterals that contact local interneurons, and cross the hilus to make synapses with CA3 pyramidal cells.

Several types of inhibitory interneurons innervate granular cells; they use GABA as a neurotransmitter and their cell body is located in the three layers of DG

The same four types of interneurons as those found in Ammon's horn have been described in the dentate gyrus. Interneurons located in stratum pyramidale in CA are located in the granular layer in DG. Similarly, interneurons located in stratum oriens of CA are in the hilus of DG and those in stratum radiatum of CA are in stratum moleculare of DG.

19.2.3 Principal cells form a tri-neural excitatory circuit

The main circuit inside the hippocampal formation involves the principal cells: granular cells of DG, pyramidal cells of CA3 and pyramidal cells of CA1. All these cells use an excitatory amino acid as a neurotransmitter. First, granular cells project on to CA3 pyramidal cells (**Figure 19.6a**). Their axon (called mossy fibers) terminates on the proximal portion of CA3 apical dendrites, on to giant spines. This restricted zone of projection forms the stratum lucidum (LUC), a sublayer of the radiatum that exists only in the CA3 region. Synapses between mossy fibers and dendritic spines of CA3 pyramidal cells are giant synapses (see **Figure 6.3**). In turn, CA3 pyramidal cells send axon collaterals, called the Schaffer collaterals, to CA1 pyramidal cells, on the distal part of their apical dendrites, at the level of stratum lacunosum moleculare. In coronal slices *in vitro*, all these circuits are present since they are organized in the transverse plane (**Figure 19.6b**).

In addition, pyramidal cells of the CA3 and CA1 regions emit local axon collaterals that contact local interneurons. Similarly, granular cells emit axon collaterals that locally innervate interneurons. Moreover, in CA3,

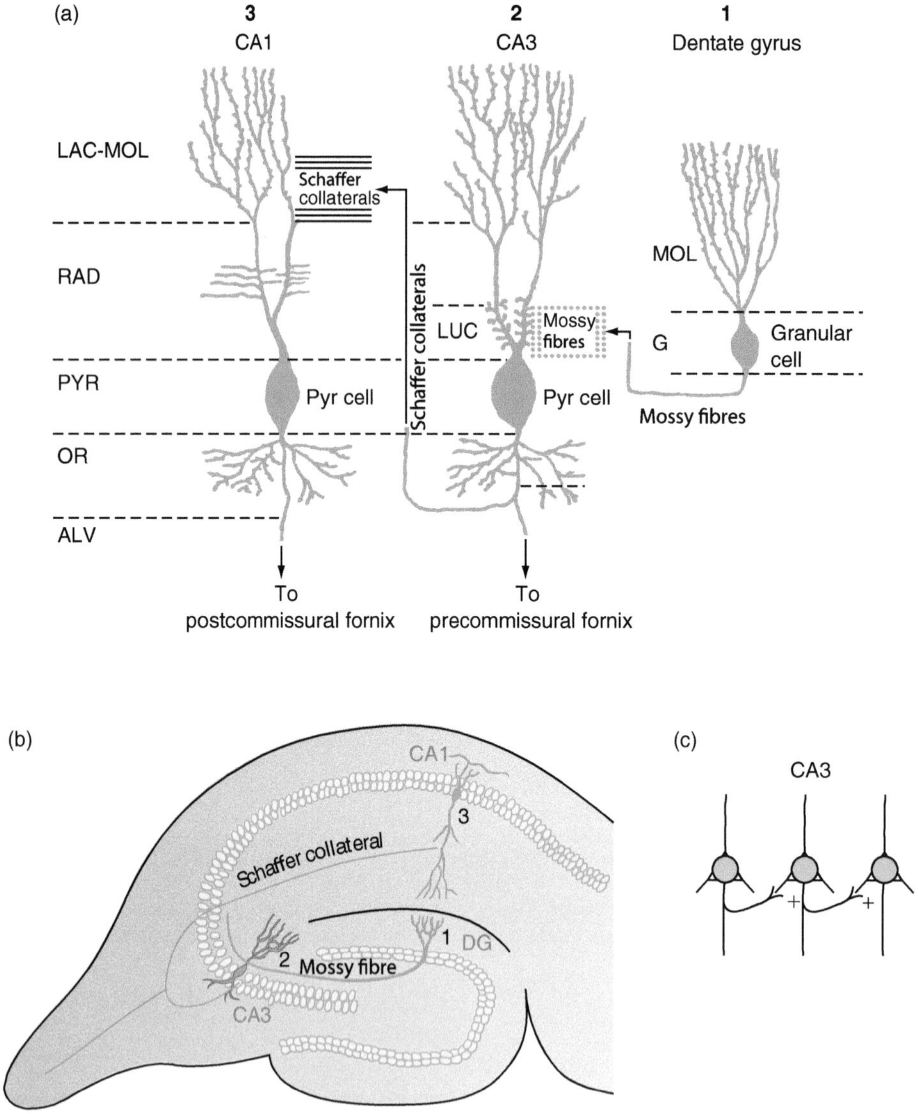

FIGURE 19.6 **The tri-neuronal circuit between principal cells.** **(a)** Sites of termination of axons of principal cells on target principal cells (which are CA1 and CA3 pyramidal cells). Axons of granular cells are called mossy fibers. Axonal collaterals of CA3 pyramidal cells are called Schaffer collaterals. **(b)** The tri-neuronal circuit is organized in the transverse plane. LUC, stratum lucidum of CA3; pyr cell, pyramidal cell. **(c)** Illustration of recurrent excitation. Part (**a**) adapted from Altman J, Brunner RL, Bayer SA (1973) The hippocampus and behavioral maturation. *Behav. Biol.* **8**, 557–596, with permission.

pyramidal cells are connected to each other by excitatory recurrent collaterals (**Figure 19.6c**). Therefore, local circuits superimpose on the main tri-neuronal excitatory circuit (**Figure 19.7**). Local circuits are detailed below in Sections 19.3 and 19.4.

19.2.4 Extrinsic afferences to principal cells and interneurons

The two major pathways that convey afferent glutamatergic inputs to the hippocampus are the *perforant* pathway coming from entorhinal cortex and the *fornix* coming from the medial septum and anterior thalamus. Moreover, the two hippocampi are interconnected by the *commissural* pathway. They also receive serotoninergic afferents from the raphe nuclei. These connections and their functions will not be studied here.

19.3 ACTIVATION OF INTERNEURONS EVOKES INHIBITORY GABAergic RESPONSES IN POSTSYNAPTIC PYRAMIDAL CELLS

To study the response of a postsynaptic pyramidal neuron to a presynaptic interneuron, the activity of these connected neurons is recorded concomitantly. Two neurons that are connected are called a pair.

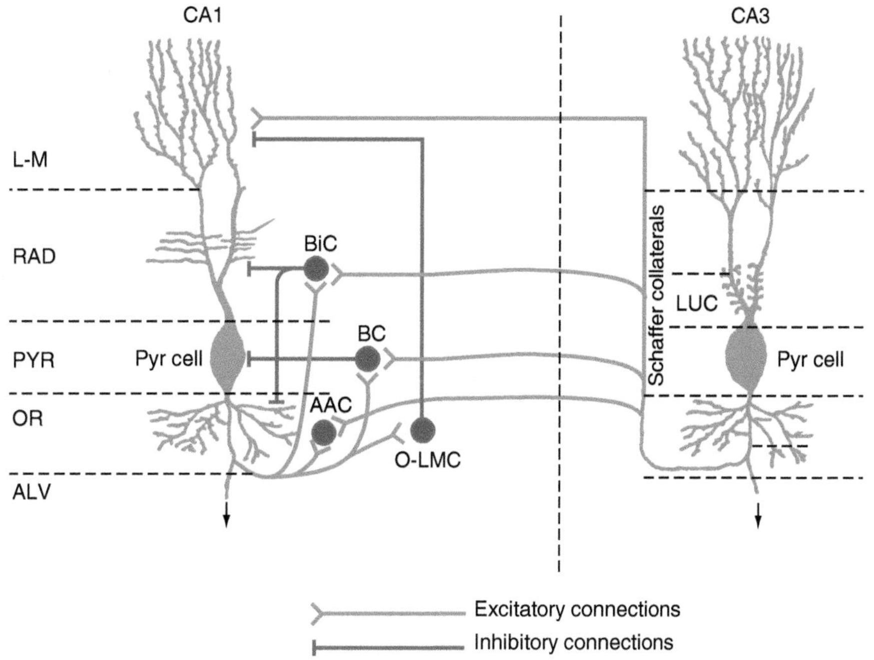

FIGURE 19.7 **Schematic of the synaptic circuitry in the CA1 region of the hippocampus and afferent connections from Schaffer collaterals of CA3 pyramidal cells.** Adapted from Altman J, Brunner RL, Bayer SA (1973) The hippocampus and behavioural maturation. *Behav. Biol.* **8**, 557–596, with permission.

19.3.1 Experimental protocol to study pairs of neurons

The hippocampus is an adequate structure to study pairs of neurons. Thanks to its laminar organization, the localization of the cell bodies of the different neurons is strictly organized. For example, in slices of the rat hippocampus, when an electrode is inserted in the pyramidal layer of the CA1 region, the probability of impaling or patching the cell body of a pyramidal neuron is high and that of an interneuron (a basket cell, for example) around ten times less (**Figure 19.5a**, left). Conversely, when the electrode is inserted in stratum oriens or radiatum, the probability of impaling or patching a pyramidal cell is close to 0 whereas that for an interneuron (O-LMC, AAC, BiC) is very high, close to 1.

To record the activity of an interneuron–pyramidal cell pair, one electrode (electrode 1) is placed in stratum oriens or radiatum or pyramidale to patch or impale an interneuron, and the other electrode (electrode 2) is placed in the stratum pyramidale to patch or impale a pyramidal cell (**Figure 19.8**). The activity of the interneuron is always recorded in current clamp mode and the activity of the pyramidal cell is recorded either in voltage clamp (to record the inhibitory postsynaptic current or IPSC) or current clamp (to record the inhibitory postsynaptic potential or IPSP). Recordings are performed with either intracellular or whole-cell electrodes. Interneurons and pyramidal cells are identified during the recording session. To do so, spontaneous action potentials and evoked responses are recorded in current clamp mode. Interneurons are characterized by the presence of an after-hyperpolarization following their action potentials and by their response to a long-lasting depolarizing current pulse which lacks spike frequency accommodation (**Figure 19.8** – compare recordings in (**a**) and (**b**)).

To check a connection between the interneuron and the pyramidal cell, electrode 1 is used as the stimulatory electrode and electrode 2 as the recording one. Spontaneous firing of the recorded interneuron is prevented by the continuous injection of a hyperpolarizing current. A suprathreshold square current pulse is injected into the interneuron to evoke action potentials (an example is given in **Figure 19.9**). If an IPSC or IPSP is evoked in the pyramidal cell in response to interneuron stimulation, the two neurons are thus identified as a pair of synaptically coupled neurons. If no synaptic response is recorded, electrode 1 is left in place and electrode 2 is changed (or the reverse). Another interneuron or pyramidal cell is patched (or impaled) and stimulated. When a pair is found, the study of the synaptic response can begin.

After the recording session, the type of interneurons recorded is identified on morphological criteria. To do this, electrodes 1 and 2 are filled with biocytin which diffuses into the cell. Slices are then fixed and biocytin-filled cells are visualized by the avidin-biotinylated horseradish peroxidase or fluorescent streptavidin method (see Appendix 6.2). The dendritic tree and axonal arborization of the recorded neurons are drawn by reconstruction from serial 60 μm thick sections under a microscope. This also allows one to check that the two neurons are connected and to count the number of contacts. Then, under electron

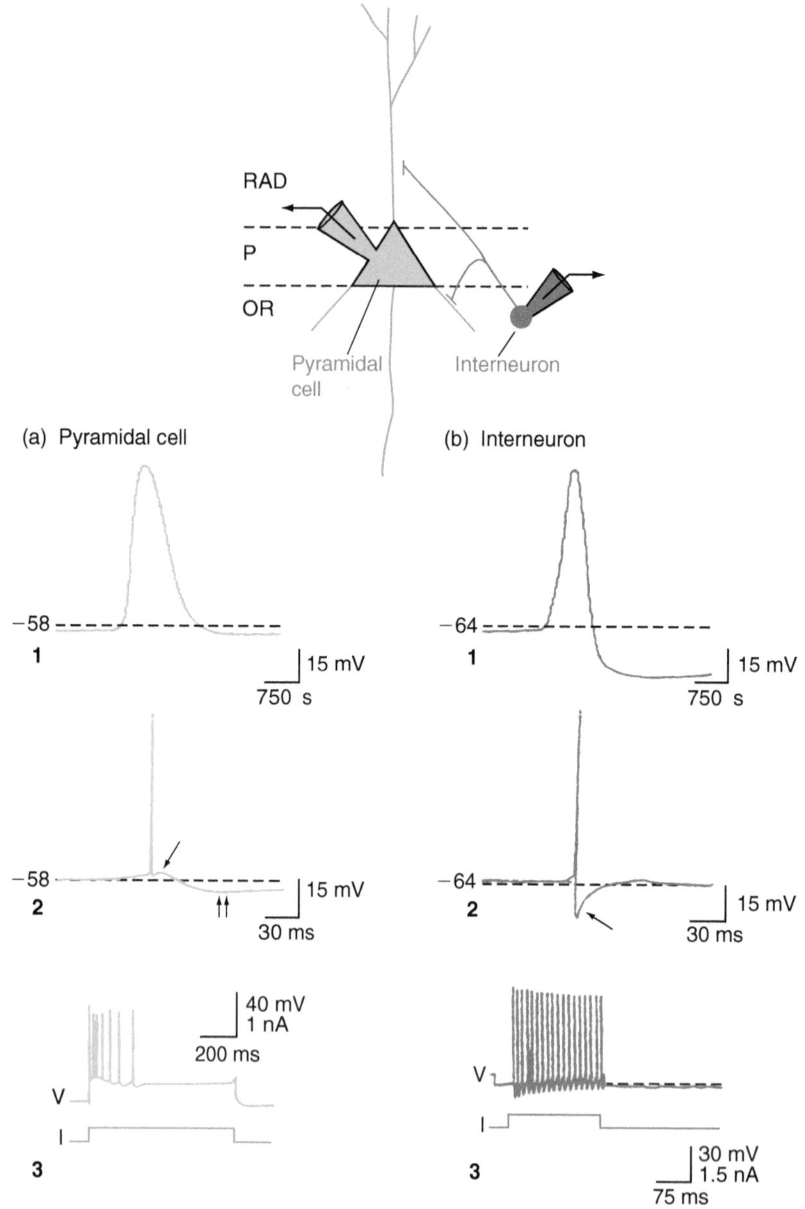

FIGURE 19.8 **Physiological characteristics that differentiate pyramidal neurons from interneurons. (a)** Action potential of a pyramidal neuron at a fast (1) and a slow (2) time base to show the presence of an after-spike depolarization (arrow) followed by a slow after-spike hyperpolarization (double arrow). The bottom trace (3) shows the response of a pyramidal neuron to a depolarizing current pulse. **(b)** Action potential of an interneuron recorded in the stratum oriens (OR) at a fast (1) and a slow (2) time base to show the presence of a fast after-spike-hyperpolarization (arrow). The bottom trace (3) shows the response of a pyramidal neuron **(a)** or an interneuron **(b)** to a depolarizing current pulse applied via the recording whole cell electrode. Adapted from Lacaille JC, Williams S (1990) Membrane properties of interneurons in stratum oriens-alveus of the CA1 region of rat hippocampus *in vitro*. *Neuroscience* **36**, 349–359, with permission.

microscopy, the type of synapses between the two neurons and the number of active zones can be precisely analyzed.

In summary, the basis for the selection of pairs of connected neurons are: (i) the presence of a short latency (monosynaptic) IPSC or IPSP in the pyramidal cell following an action potential in the putative interneuron; (ii) stable recordings from both cells for sufficient time to obtain an averaged IPSC or IPSP; and (iii) recovery of at least part of the biocytin-labeled interneuron to allow its identification.

19.3.2 Unitary inhibitory postsynaptic currents (IPSCs) evoked by different types of interneurons are all $GABA_A$-mediated but have different kinetics when recorded at the level of the soma

To study the synaptic current evoked in a pyramidal neuron in response to a single action potential in the presynaptic interneuron, the activity of both neurons is recorded in whole-cell configuration.

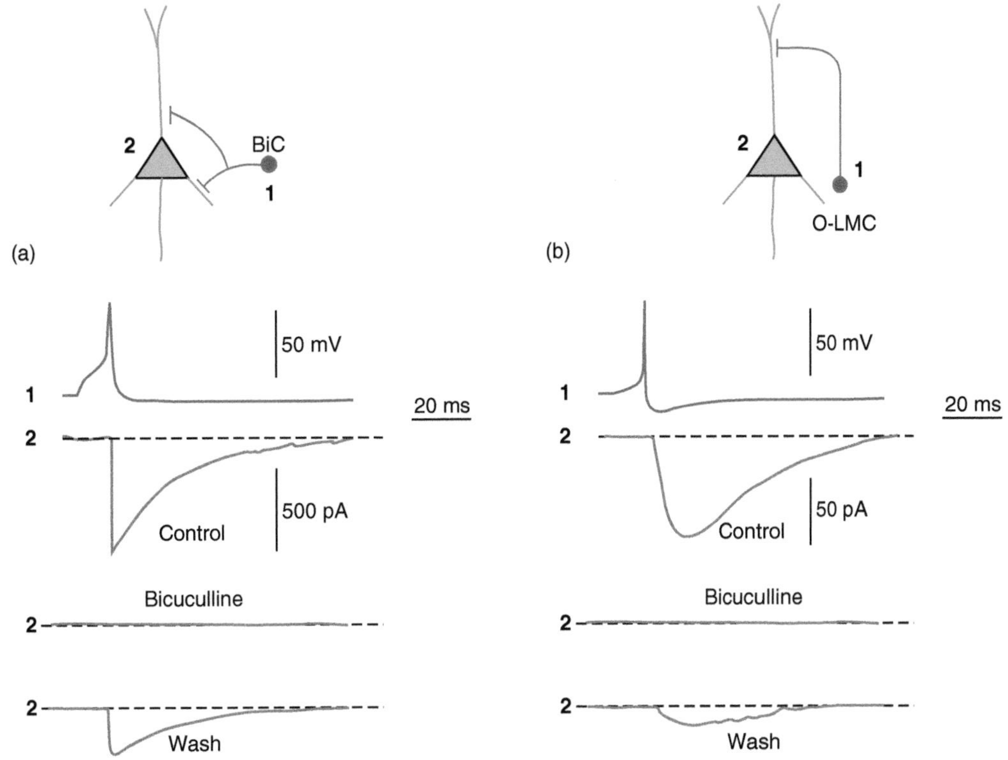

FIGURE 19.9 **Unitary IPSCs (uIPSCs) evoked in pyramidal cells in response to different types of interneurons are all mediated by $GABA_A$ receptors.** Averaged uIPSC evoked in a pyramidal neuron (2) in response to a single spike in a bistratified interneuron (1a, BiC) or an oriens lacunosum moleculare interneuron (1b, O-LMC) in control conditions, in the presence of bicuculline (Bicu, 10 μM) and after partial washout of the drug (Wash). Adapted from Maccaferri G, Roberts DB, Szucs P *et al.* (2000) Cell surface domain specific postsynaptic currents evoked by identified GABAergic neurones in rat hippocampus *in vitro*. *J. Physiol.* **524**, 91–116, with permission.

Whole-cell patch recordings

The intrapipette solution of electrode 1 (interneuron) is designed to allow the recording of action potentials (in mM): 130 K gluconate, $2MgCl_2$, 0.1 EGTA, 2 ATP, 0.3 GTP, 10 Hepes and 0.5% biocytin. Action potentials are generated in the interneuron by injection of a suprathreshold square current pulse at 0.1–1 Hz.

The intrapipette solution of electrode 2 (pyramidal cell) is designed to record $GABA_A$-mediated currents in isolation (in mM): 100 CsCl, $2MgCl_2$, 0.1 EGTA, 2 ATP, 0.3 GTP, 40 Hepes, 5 QX-314 and 0.5% biocytin, at a pH of 7.2. QX-314 and Cs^+ strongly reduce voltage-gated Na^+ and K^+ currents, respectively. Ionotropic glutamate receptors are blocked by DNQX 20 μM and D-AP5 50 μM in the bath. The synaptic current is recorded in voltage clamp mode ($V_H = -70$ mV). Internal Cl^- concentration is 104 mM and the external concentration is 135 mM, which gives a reversal potential for Cl^- ions close to 0 mV (recall that in physiological conditions E_{Cl} is around −70 mV; see Section 9.3.4). Therefore, at $V_H = -70$ mV, when $GABA_A$ receptors open, there is an outflow of Cl^- through channels permeable to Cl^- ions; it is recorded as an inward current (an inward current is by convention an inward movement of positive charges; see Section 3.3.3).

Unitary IPSCs evoked by different types of interneurons are all mediated by $GABA_A$ receptors

When a single action potential is evoked in the presynaptic interneuron, an inward current is recorded in the postsynaptic pyramidal cell (**Figure 19.9**). This inward current is totally blocked by bicuculline, thus showing that it is mediated by $GABA_A$ receptors. This holds true for all the following pairs: BiC–pyr, O-LMC–pyr, BC–pyr and AAC–pyr.

These $GABA_A$-mediated currents are called *inhibitory postsynaptic currents* (IPSCs), though they are inward, because in control Cl^- conditions they would be outward and thus inhibitory. An IPSC which results from a single action potential in the presynaptic neuron is called a *unitary IPSC* (uIPSC).

Proximally and distally generated unitary IPSCs have different kinetic parameters

The unitary IPSCs of **Figure 19.9** are recorded at the level of the soma of pyramidal cells. When evoked in distal dendrites by O-LMC, unitary IPSCs are passively conducted along the apical dendrite before being recorded in the soma. In contrast, unitary IPSCs evoked at the level of the soma by basket cells (BC) or at the axon initial segment by axo-axonic cells (AAC) are

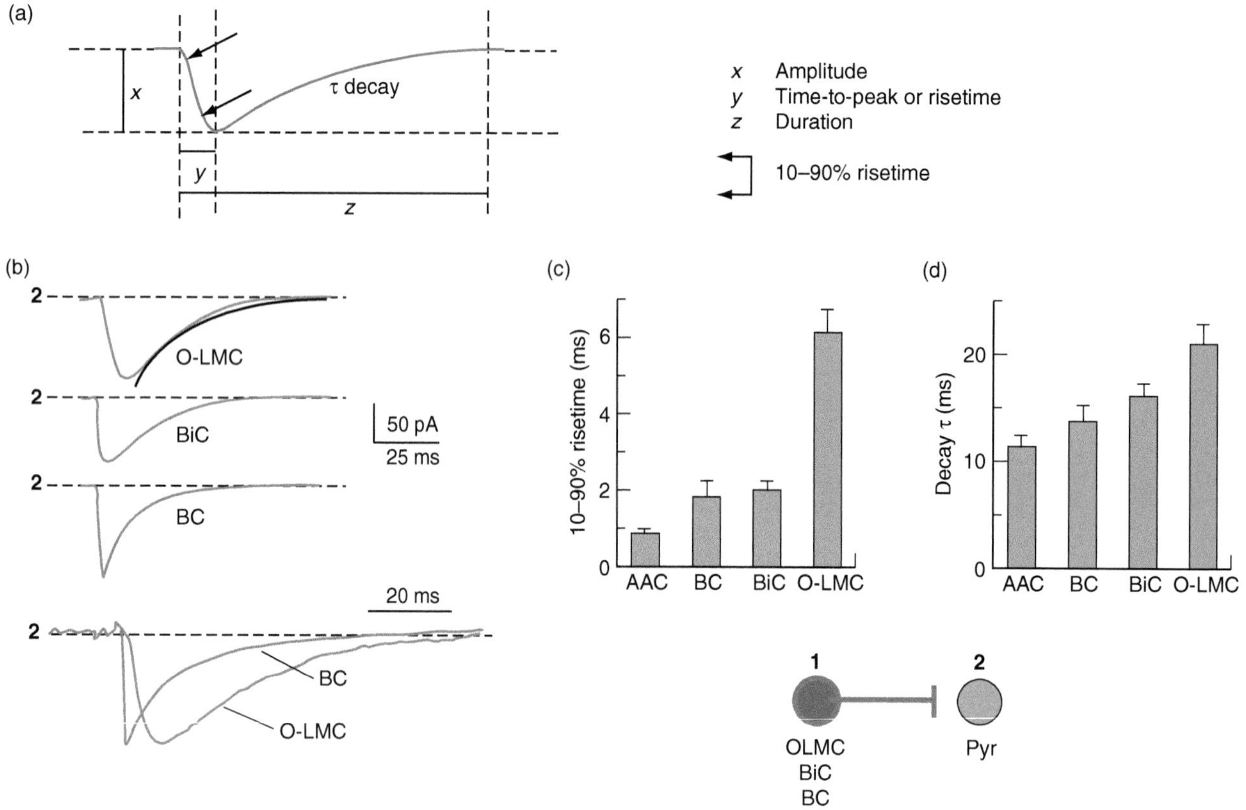

FIGURE 19.10 **Kinetic parameters of unitary IPSCs evoked in a pyramidal cell in response to different types of interneurons. (a)** Definition of the parameters of postsynaptic currents. **(b)** Comparison of the risetime (or time to peak) and decay phase of three different uIPSCs. The decay phase of the uIPSC generated by an O-LM cell is fitted by an exponential (black trace) to obtain the value of τ. Bottom trace shows superimposed averaged uIPSCs generated by a presynaptic basket cell (BC) or oriens lacunosum moleculare cell (O-LMC) and scaled at the same amplitude. Note the much longer time to peak of the IPSC evoked by the O-LM cell in comparison with that of the basket cell. **(c)** Histogram of the risetimes (10–90%) and **(d)** of the decay (τ, time to 63% of decay) of the uIPSCs generated by different classes of presynaptic interneurons. Adapted from Maccaferri G, Roberts DB, Szucs P *et al.* (2000) Cell surface domain specific postsynaptic currents evoked by identified GABAergic neurones in rat hippocampus *in vitro*. *J. Physiol.* **524**, 91–116, with permission.

generated at sites close to the recording electrode. As shown in **Figure 19.10**, distally evoked unitary IPSCs have a slower time to peak (also called the risetime) than those evoked in proximal dendrites and soma. Unitary IPSCs evoked by axo-axonic cells have a fast risetime (0.8 ± 0.1 ms) whereas those evoked by basket cells have a slower risetime; and those evoked in the most distal pyramidal dendrites by oriens lacunosum moleculare cells have a very slow risetime (6.2 ± 0.6 ms). The kinetics of unitary IPSCs recorded in the soma of pyramidal cells reflect the domain of innervation: this can result from electrotonic dendritic filtering (see Section 13.1) and/or the lack of voltage clamp of the more distal locations and/or site-specific subunit composition of $GABA_A$ receptors.

In summary, in Ammon's horn, the synapses established by interneurons on the soma (BC), on proximal or distal dendrites (BiC, O-LMC) or the axon initial segment (AAC) of pyramidal cells are all inhibitory and $GABA_A$-mediated. This means that $GABA_A$ synapses are present all along the somato-dendritic tree and axon initial segment of pyramidal cells. Depending on where they are generated on the dendritic tree, $GABA_A$-mediated IPSCs have different risetimes.

19.3.3 $GABA_A$-mediated IPSCs generate IPSPs in postsynaptic pyramidal cells

IPSCs generate transient hyperpolarizations of the postsynaptic membrane, called *i*nhibitory *p*ostsynaptic *p*otentials (IPSPs). To study the IPSPs evoked in a pyramidal neuron in response to a presynaptic interneuron in physiological conditions, the activity of both neurons is recorded with intracellular electrodes filled with 1.5 M of KCH_3SO_4 (and 2% biocytin) so as not to change the internal Cl^- concentration. In these conditions, the reversal potential for Cl^- ions is around −70 mV, which is close to that *in vivo*.

In **Figure 19.11b**, a single spike in the presynaptic interneuron (bistratified cell) evokes a transient hyperpolarization in the postsynaptic pyramidal neuron; this is called inhibitory (IPSP) because it hyperpolarizes the membrane to a potential far from the threshold of spike initiation. It is unitary (uIPSP) since it is evoked by a

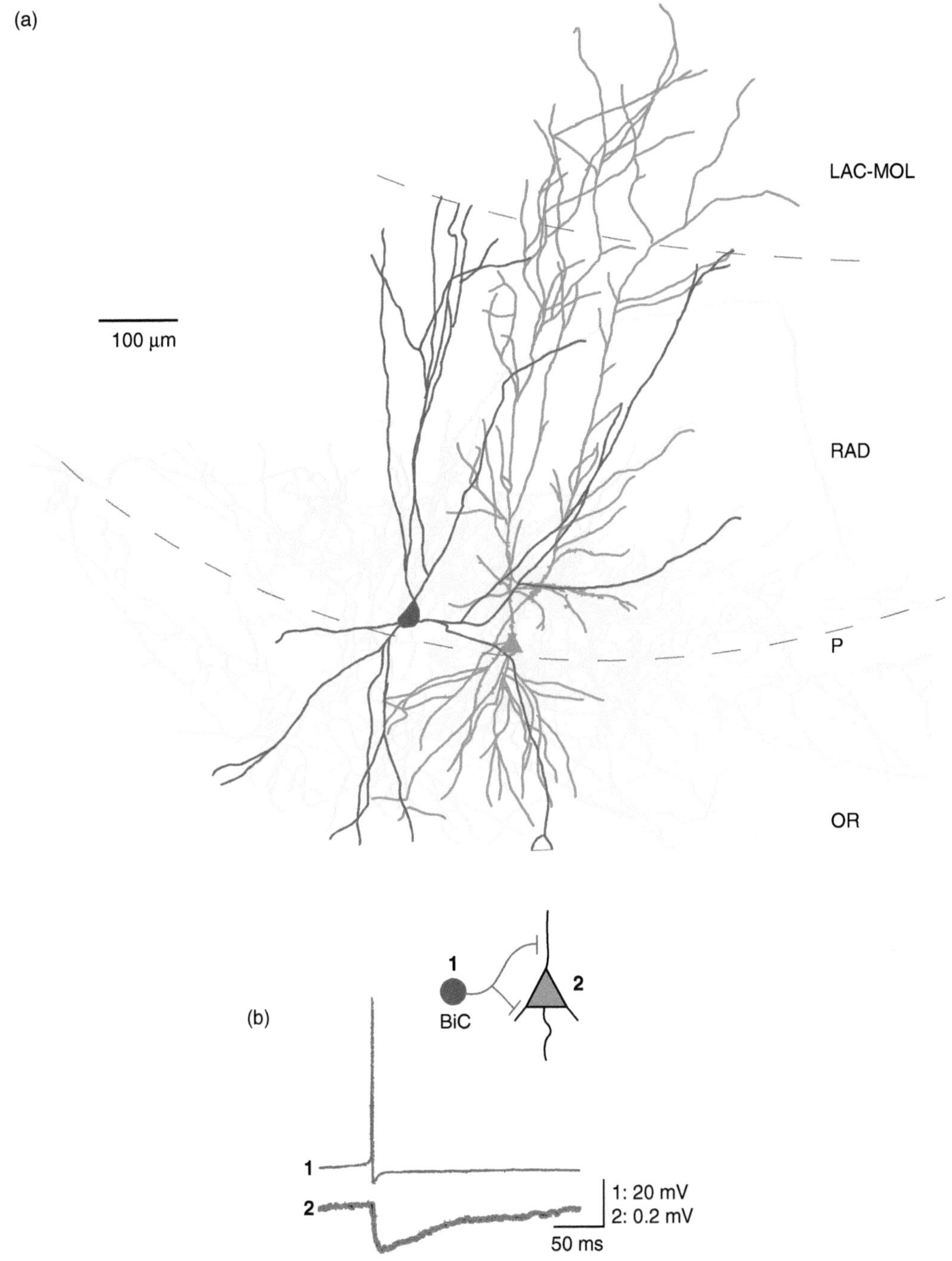

FIGURE 19.11 **Unitary IPSP evoked in a CA1 pyramidal cell in response to a bistratified cell (BiC). (a)** Reconstruction of the biocytin-filled presynaptic interneuron (somato-dendritic tree in blue, axon in yellow) and a postsynaptic pyramidal cell (green). **(b)** An action potential in BiC (1) elicits a small-amplitude, short-latency unitary IPSP in the postsynaptic pyramidal cell (2). (Trace 2 is an averaged unitary IPSP.) Adapted from Buhl EH, Halasy K, Somogyi P (1994) Diverse sources of hippocampal unitary inhibitory postsynaptic potentials and the number of synaptic release sites. *Nature* **368**, 823–828, with permission.

single presynaptic spike. To evoke this uIPSP, the presynaptic action potential first propagates to the numerous synaptic terminals of the interneuron and evokes the release of GABA from all or some of these terminals. Therefore, a unitary IPSC or IPSP can result from the activation of one or more release sites, depending on the number of synapses established by the presynaptic interneuron on the recorded postsynaptic pyramidal neuron and on the number of active zones per synapse (see **Figure 7.2**). A study under electron microscopy then reveals the exact number of synaptic complexes since a single synaptic bouton may establish multiple synaptic complexes (around 60 to 140 terminals from a single basket cell).

In the example of **Figure 19.11a**, there are six synaptic contacts between the axon of the presynaptic basket cell and the postsynaptic pyramidal neuron. Electron microscopy shows that these six synaptic contacts correspond to six synaptic complexes (there is a single active zone per bouton). Therefore, a maximum of six release sites is responsible for this unitary IPSP. This allows one to calculate the average amplitude of an IPSP evoked by the activity of one release site only: it is around 30 μV (220 μV divided by 6) which is a very small hyperpolarization. In **Figure 19.12**, the IPSP evoked by an axo-axonic cell corresponds to the activity of eight synaptic contacts between a presynaptic axo-axonic cell and a pyramidal neuron. This $GABA_A$-mediated IPSP reverses at −78 mV (range −65 to −78 mV), which is close to E_{Cl}. Morphological studies show that inhibitory interneurons establish an average of 5–30 synaptic contacts with a single postsynaptic pyramidal cell.

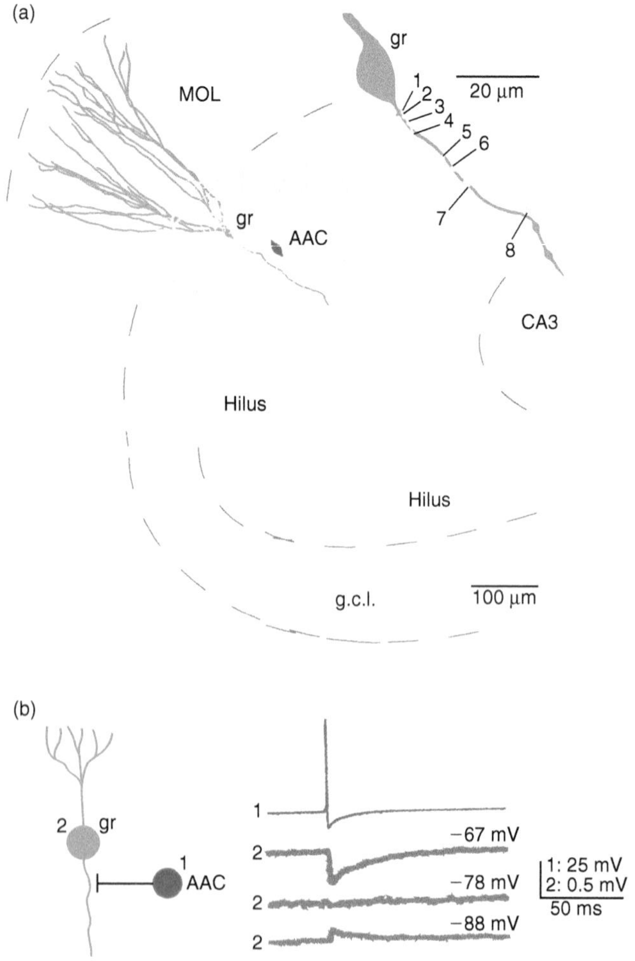

FIGURE 19.12 **Unitary IPSP evoked in a granular cell in response to an axo-axonic cell (AAC) and location of contact sites. (a)** Reconstruction of the biocytin-filled presynaptic interneuron (soma in blue, axon in yellow, dendrites absent) and postsynaptic granular cell (green). The top right shows the location of the eight contact sites between the GABAergic axon (yellow) and the axon initial segment of the postsynaptic granular cell (green). **(b)** An action potential in AAC (1) elicits a small-amplitude, short-latency unitary IPSP in the postsynaptic granular cell (2) that reverses at around −78 mV (traces 2 show averaged EPSPs). Adapted from Buhl EH, Halasy K, Somogyi P (1994) Diverse sources of hippocampal unitary inhibitory postsynaptic potentials and the number of synaptic release sites. *Nature* **368**, 823–828, with permission.

19.3.4 $GABA_B$-mediated IPSPs are also recorded in pyramidal neurons in response to strong interneuron stimulation

The activity of a CA3 pyramidal cell is intracellularly recorded with electrodes filled with 2 M of $KMeSO_4$ so as not to change the intracellular concentration of Cl^-. A stimulating electrode is placed in the hilus (it stimulates local interneurons and excitatory afferents from granular cells of the dentate gyrus). In response to the stimulation, a complex synaptic response is recorded (**Figure 19.13**): a prior EPSP (inset) followed by a biphasic IPSP (early

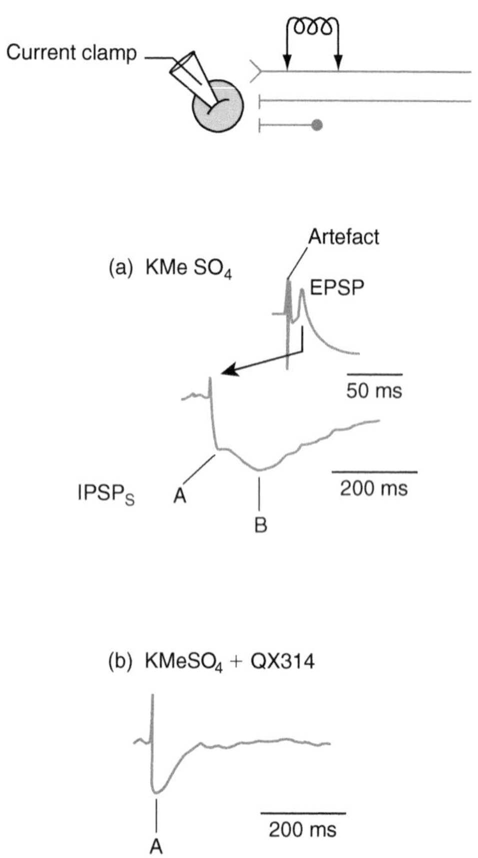

FIGURE 19.13 **$GABA_A$ and $GABA_B$ receptor-mediated IPSPs in pyramidal cells. (a)** CA3 pyramidal cells respond to a hilar stimulation by a biphasic IPSP preceded by an EPSP (inset). The biphasic IPSP consist of an early (A) and a late (B) IPSP. **(b)** In the presence of QX 314 (50 mM) in the pipette solution, stimulation no longer evokes the late IPSP whereas the early one (A) is spared (as well as the EPSP, not shown). The pipette is filled with potassium methyl sulfate ($KMeSO_4$). Adapted from McLean HA, Ben Ari Y, Gaiarsa JL (1995) NMDA-dependent $GABA_A$-mediated polysynaptic potentials in the neonatal rat hippocampal CA3 region. *Eur. J. Neurosci.* **7**, 1442–1448, with permission.

and late). When QX-314, a derivative of lidocaine that blocks voltage-gated Na^+ currents (fast and persistent) and $GABA_B$ receptor-activated K^+ current (see Chapter 11), the late component disappears. In contrast, in the presence of bicuculline in the bath, the early component disappears (not shown). This shows that the early IPSP is mediated by $GABA_A$ receptors whereas the late phase is mediated by $GABA_B$ receptors.

Postsynaptic $GABA_B$ receptor-mediated IPSPs are recorded only in response to a strong activation of interneurons. This effect is absent in the response to a single spike in interneurons, suggesting that a larger release of GABA in the synaptic cleft is necessary to activate postsynaptic $GABA_B$ receptors. This late IPSP prolongs the inhibition of pyramidal cells by GABAergic interneurons.

19.4 ACTIVATION OF PRINCIPAL CELLS EVOKES EXCITATORY GLUTAMATERGIC RESPONSES IN POSTSYNAPTIC INTERNEURONS AND OTHER PRINCIPAL CELLS (SYNCHRONIZATION IN CA3)

To study the physiological response of a postsynaptic neuron to an action potential in the presynaptic pyramidal neuron, the activity of these connected neurons is recorded concomitantly with intracellular or whole-cell electrodes. Electrodes 1 and 2 are filled with 4% biocytin in 0.5 M of potassium acetate. In these conditions, the reversal potential for Cl^- ions is around −70 mV, which is close to that *in vivo*.

19.4.1 Pyramidal neurons evoke AMPA-mediated EPSPs in interneurons

EPSPs elicited in interneurons in response to a single action potential in the presynaptic pyramidal cell have a mean amplitude of 1–4 mV and a time to peak (rise-time) of 1.5–4 ms. They are totally blocked by CNQX, the selective blocker of AMPA receptors (not shown). They fluctuate in amplitude at all synapses examined and sometimes fail (**Figure 19.14a**). This latter observation suggests that there is a low probability of release or the existence of a few release sites. In fact, under light microscopy, in all pairs studied, a single synaptic contact is identified between the filled pyramidal cell and interneuron. Electron microscopy shows that each contact has a single active zone.

Studies are performed in the hippocampus on a large number of cells in order to obtain quantitative data: pyramidal cells are filled *in vivo* with neurobiotin, and parvalbumin-containing interneurons (basket and axo-axonic cells) are revealed by immunocytochemistry. This study confirms that each filled pyramidal axon establishes a single contact with parvalbumine-containing interneurons. By counting the boutons terminating on interneurons, it has been shown that over 1000 excitatory synapses terminate on a single inhibitory cell, suggesting that more than 1000 pyramidal cells converge onto one interneuron. This excitatory drive presumably contributes to the high frequency of the spontaneous firing of hippocampal interneurons.

Interestingly, single pyramidal cell action potentials cause inhibitory cells to fire at resting membrane potential with a probability of 0.4 (**Figure 19.15**). The mean interval between pre- and postsynaptic spikes is 2.9 ± 0.7 ms. Factors contributing to spike-to-spike transmission are the low firing threshold of inhibitory interneurons, their depolarized resting membrane potential and the large EPSCs elicited by pyramidal cells leading to large EPSPs (of the order of the millivolts) owing to the high input membrane resistance of interneurons. Moreover, CA3 pyramidal cells have a tendency to discharge in bursts of several action potentials (with 5–10 ms intervals). This has as a consequence that the reliability of transmission is enhanced and temporal summation occurs when the interval between presynaptic spikes is shorter than the time course of EPSPs (**Figure 19.14b**).

19.4.2 EPSPs in interneurons lead to feedback inhibition of pyramidal neurons

When two pyramidal cells of the CA3 region are recorded simultaneously, firing in one pyramidal cell can evoke an IPSP in the other pyramidal cell (**Figure 19.16a**). The mean IPSP latency is 3.5 ± 0.7 ms and the mean amplitude 1.9 ± 0.6 mV. Bicuculline (a $GABA_A$ receptor antagonist) as well as CNQX (an AMPA/KA receptor antagonist) completely suppresses these evoked IPSPs. This experiment excludes the possibility that pyramidal cells establish monosynaptic inhibitory connections. It shows in contrast that these IPSPs result from a bisynaptic connection between pyramidal cells: the first synapse is glutamatergic and the second one GABAergic. There are two ways to block this IPSP: to block synaptic transmission at the first synapse with CNQX or at the second synapse with bicuculline.

19.4.3 CA3 pyramidal neurons are monosynaptically connected via glutamatergic synapses

Neighboring pyramidal cells of CA3 are monosynaptically connected via axon collaterals (see **Figure 19.6c**). In pairs of pyramidal neurons, presynaptic pyramidal action potentials evoke EPSPs in the postsynaptic pyramidal cell (**Figure 19.17a**) that are sensitive to CNQX. These EPSPs are considered to be monosynaptic on the

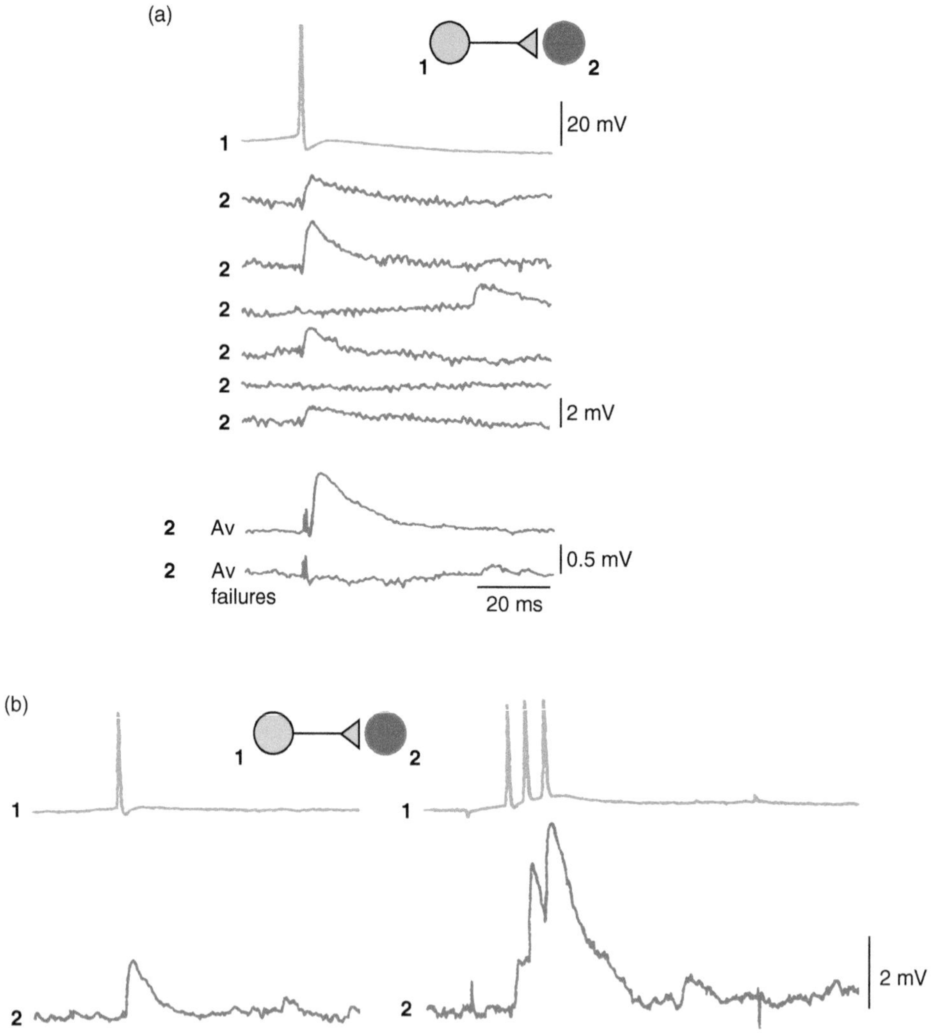

FIGURE 19.14 **Unitary EPSP evoked in an inhibitory interneuron in response to a CA3 pyramidal cell. (a)** A single spike in the presynaptic pyramidal cell (1) evokes a unitary EPSP in the postsynaptic interneuron (2) that fluctuates in amplitude and sometimes fails. An averaged EPSP and an averaged trace of failures ($n = 38$) are shown below. Av: average. **(b)** EPSPs initiated in an inhibitory interneuron in response to a single spike (left) or a train of three spikes (right) in the presynaptic pyramidal cell (1). On the right there is a temporal summation of the EPSPs. Adapted from Miles R (1990) Synaptic excitation of inhibitory cells by single CA3 hippocampal pyramidal cells of the guinea pig *in vitro*. *J. Physiol.* **428**, 61–77, with permission.

basis of their mean latency (range 0.8–1.2 ms) and the proportion of transmission failures (small in monosynaptic connections).

19.4.4 Overview of intrinsic hippocampal circuits

The main circuit is the tri-neuronal circuit between the excitatory principal cells (see **Figure 19.6a,b**). In each region, dentate gyrus, CA1 or CA3, axon collaterals of principal cells excite inhibitory interneurons which, in turn, inhibit other principal cells (feedforward inhibition). Interneurons are also directly activated by extrinsic excitatory afferences (feedforward inhibition). In CA3, principal cells (pyramidal cells) are monosynaptically connected via excitatory recurrent axon collaterals (see **Figure 19.6c**).

19.5 OSCILLATIONS IN THE HIPPOCAMPAL NETWORK: EXAMPLE OF SHARP WAVES (SPW)

Gyorgy Buzsaki discovered sharp waves (SPW) in 1983 when recording from CA1 stratum radiatum and a pyramidal cell layer simultaneously in the freely moving rat. When the rat is exploring, typical theta waves are present. When the animal becomes immobile or goes to sleep (slow-wave stage), large-amplitude intermittent sharp waves of 40–120 ms replace theta oscillations (see **Figure 19.3**). The frequency of the intermittent sharp waves ranges from 0.02 to 3 Hz. Recordings in layers of the CA1 region with extracellular electrodes show that sharp wave amplitude is maximal in stratum radiatum, the layer where apical dendrites of CA1 pyramidal

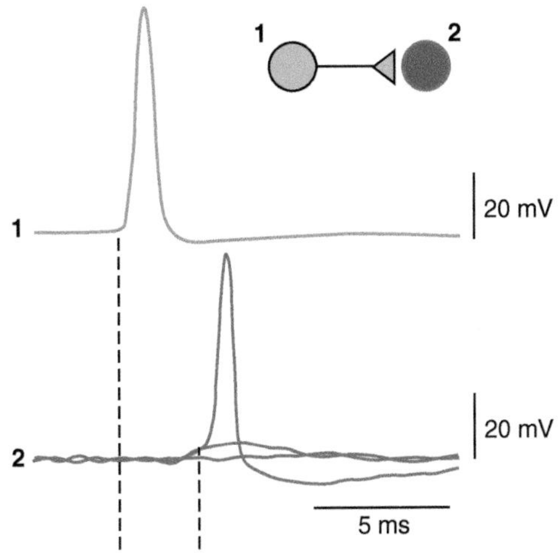

FIGURE 19.15 **Spike to spike transmission at an excitatory synapse between a CA3 pyramidal neuron and a postsynaptic inhibitory interneuron.** In response to successive single spikes in a presynaptic pyramidal cell (1, only one spike is displayed), one transmission failure, one unitary EPSP and one unitary EPSP that causes postsynaptic firing are recorded in the postsynaptic interneuron (2, three superimposed traces). Adapted from Miles R (1990) Synaptic excitation of inhibitory cells by single CA3 hippocampal pyramidal cells of the guinea pig *in vitro*. *J. Physiol.* **428**, 61–77, with permission.

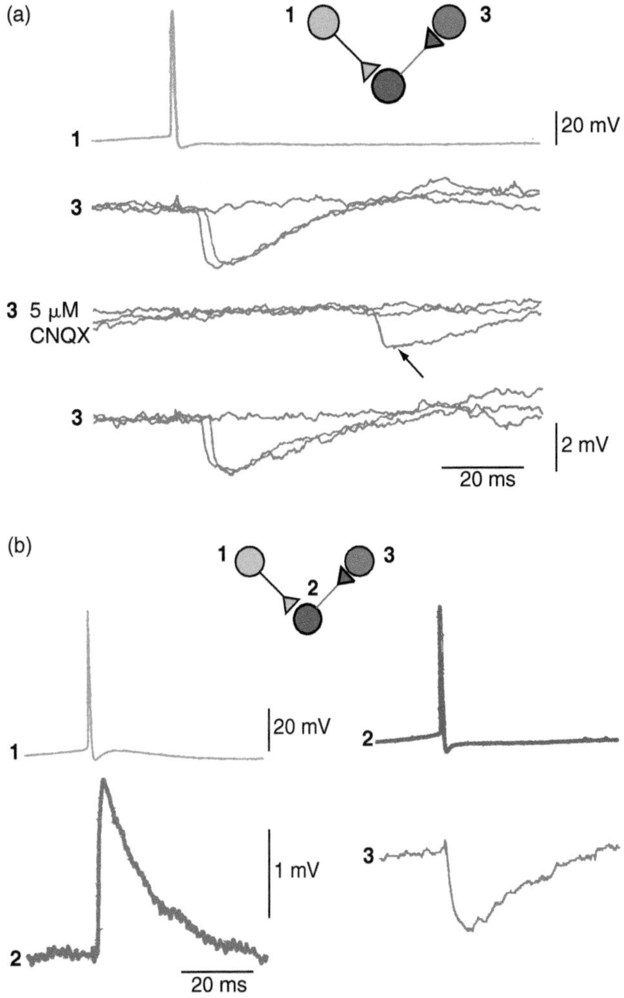

FIGURE 19.16 **Feedforward inhibition between two CA3 pyramidal cells. (a)** In response to a single spike in one pyramidal cell (1), an IPSP is recorded in a neighboring pyramidal cell (3). Upper trace 2 shows three superimposed responses to three presynaptic action potentials (trace 1, only one spike is displayed). There is one transmission failure and two IPSPs. Middle trace 3 shows the effect of CNQX in the bath: the evoked IPSPs are suppressed but spontaneous ones are still present (arrow). Bottom trace 3 shows the return to control solution. **(b)** Sequential recordings of the three connected cells. Pyramidal cell 1 activates interneuron 2 (average of uEPSP, $n = 40$) and interneuron 2 inhibits cell 3 (average of uIPSP, $n = 40$). Adapted from Miles R (1990) Synaptic excitation of inhibitory cells by single CA3 hippocampal pyramidal cells of the guinea pig *in vitro*. *J. Physiol.* **428**, 61–77, with permission.

neurons extend (**Figure 19.18a,b**). The immediate cause of sharp waves in the CA1 region is the synchronous discharge of a large number of CA3 pyramidal neurons. This results in the near-simultaneous depolarization of CA1 pyramidal cells via Schaffer collaterals which terminate dominantly on apical CA1 dendrites (see **Figure 19.6**) (recall that the extracellular recording of EPSPs from a population of neurons, called field EPSP, appears as a downward deflection in extracellular recordings; see **Figure 18.4b**). In brief, sharp waves represent a coherent depolarization of the apical dendrites of CA1 pyramidal neurons.

However, the synchronous discharge of a large number of CA3 pyramidal neurons also directly activates interneurons (see **Figure 19.7**). Therefore, concurrent with sharp waves, pyramidal cells and interneurons are activated synchronously (interneurons are even activated earlier than target pyramidal cells, because of their lower spike threshold). As a result, dendrites of CA1 pyramidal cells respond by summed EPSPs (see **Figure 19.14b**) to bursts of action potentials in Schaffer collaterals and by summed IPSPs in response to bursts of action potentials in axons of interneurons. How do CA1 pyramidal cells finally respond?

Whenever a sharp wave is present in the radiatum, a 140–200 Hz field oscillation is present in the pyramidal cell layer (**Figure 19.18b**). This is best shown by filtering the field potential below 50 Hz (to suppress oscillations at a frequency under 50 Hz). This reveals high-frequency oscillations that form a ripple **(Figure 19.18c)**. These oscillations are most prominent in the pyramidal layer (**Figure 19.18**, trace 5). The fast field oscillation is believed to represent summed fast IPSPs in the somata of pyramidal cells brought about by the activated interneurons.

The explanation is the following. Interneurons, including basket and axo-axonic (chandelier) cells, fire together (perhaps coupled by gap junctions) at around 200 Hz and impose a series of IPSPs on the somata of pyramidal cells. Some pyramidal cells are excited through their dendrites strongly enough so that excitation can overcome somato-axonal inhibition. However, because

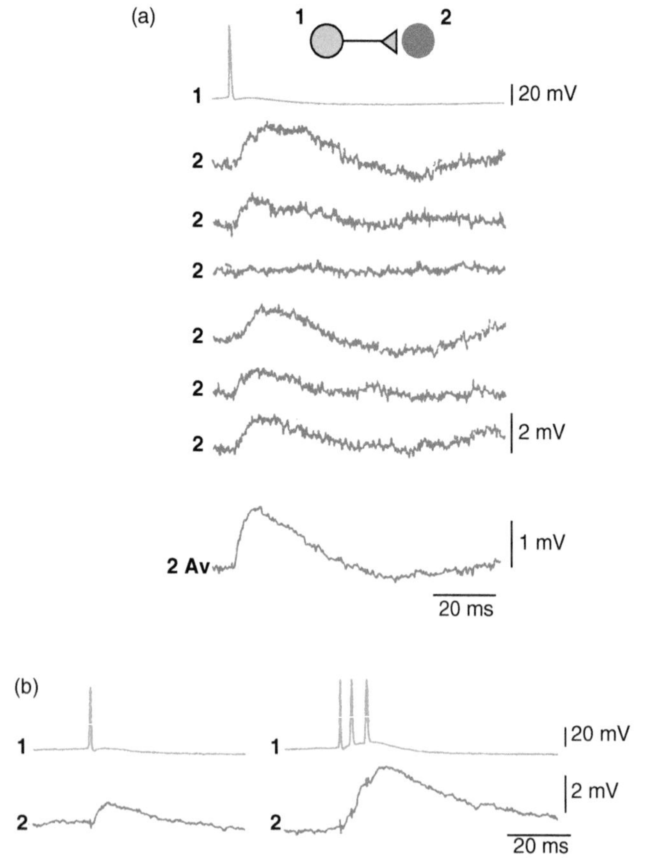

FIGURE 19.17 **Unitary EPSP evoked in a CA3 pyramidal cell in response to a CA3 pyramidal cell. (a)** Monosynaptic unitary EPSP evoked in a postsynaptic pyramidal cell (2) in response to a single spike in the presynaptic pyramidal cell (1). The EPSP fluctuates in amplitude and sometimes fails. Bottom trace shows an averaged (Av) EPSP. **(b)** EPSPs initiated in a postsynaptic pyramidal cell (2) in response to a single spike (left) or a train of three spikes (right) in the presynaptic pyramidal cell (1). On the right there is a temporal summation of the EPSPs. Part (**a**) adapted from Miles R, Wong RKS (1986) Excitatory synaptic interactions between CA3 neurones in the guinea pig hippocampus. *J. Physiol.* **373**, 397–418, with permission. Part (**b**) adapted from Miles R (1990) Synaptic excitation of inhibitory cells by single CA3 hippocampal pyramidal cells of the guinea pig *in vitro*. *J. Physiol.* **428**, 61–77, with permission.

of these series of IPSPs in the soma and axon initial segment membrane, the spike(s) emerge at the periods where inhibition is least (i.e. out of phase with the spikes of the interneurons; **Figure 19.19**). In short, inhibition does not necessarily prevent firing but serves to time the occurrence of spikes.

Sharp waves are envisaged as an endogenous mechanism for consolidating synaptic changes and transferring information from the hippocampus to neocortex during sleep. The strong depolarization of pyramidal cell dendrites during a sharp wave burst enhances the size of fast dendritic spikes (see Section 16.3.2). The large membrane depolarization also triggers Ca^{2+} spikes which, in turn, may alter the weights of the simultaneously active nearby synapses (see Section 16.3.3).

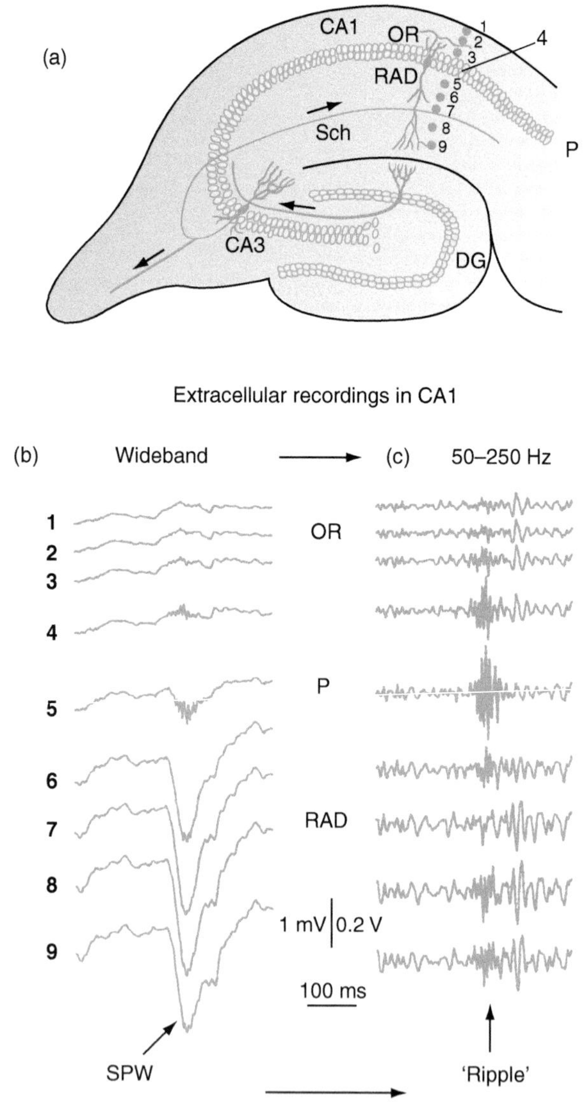

FIGURE 19.18 **Sharp waves (SPW) in CA1 of the awake immobile rat and the high-frequency oscillations (ripples). (a)** The extracellular activity of a population of neurons is recorded with nine extracellular electrodes in the CA1 region. **(b)** Extracellular recordings show that the sharp waves are the most pronounced in the apical dendritic layer (stratum radiatum, RAD). **(c)** When the recordings in (**b**) are filtered in order to leave only the events with a frequency between 50 and 250 Hz, high-frequency oscillations that form a ripple are revealed. Ripples are particularly prominent in the pyramidal layer (p). Note that the amplitude scale is increased 5-fold between (**b**) and (**c**). Adapted from Ylinen A, Bragin A, Nadasdy Z *et al.* (1995) Sharp wave-associated high-frequency oscillation (200 Hz) in the intact hippocampus: network and intracellular mechanisms. *J. Neurosci.* **15**, 30–46, with permission.

19.6 SUMMARY

What is a neuronal network?

A neuronal network is formed by neurons from the same nucleus. It is described by the connections between these neurons, the type of synapse (glutamatergic, GABAergic, etc.), and the arrangement of the synapses on somato-dendritic trees.

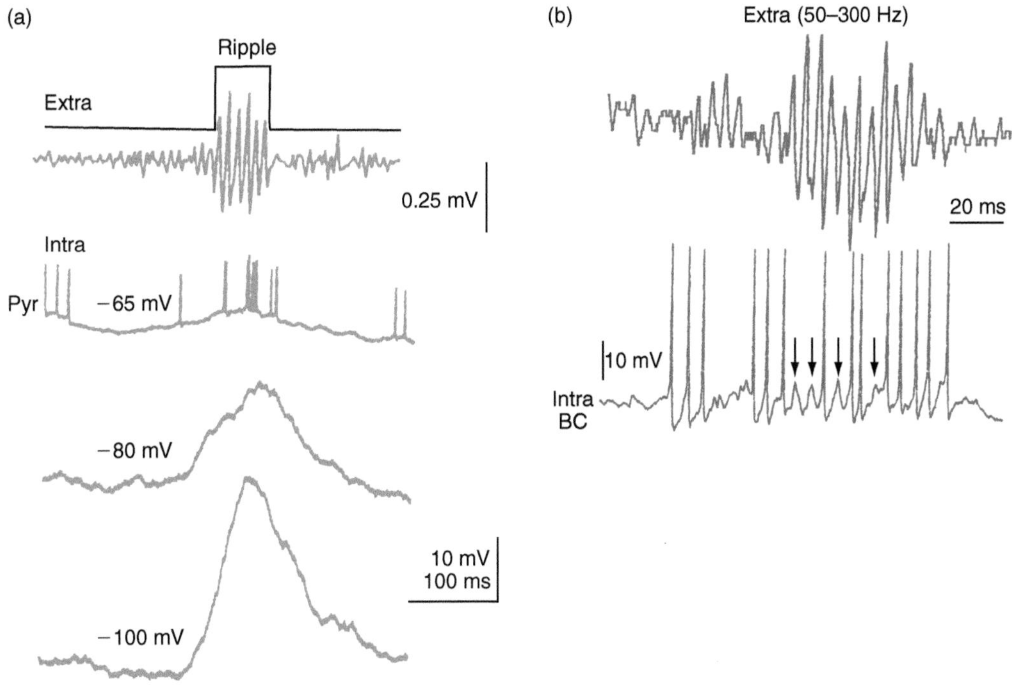

FIGURE 19.19 **Intracellular activity of a pyramidal neuron and an interneuron during high-frequency oscillations (ripples).** Intracellular recording **(a)** from a CA1 pyramidal neuron and **(b)** from a CA1 basket cell (BC) during a single ripple event. In **(a)**, membrane hyperpolarization of the pyramidal cell from −65 to −100 mV reveals a strong depolarization force during the ripple. In **(b)** arrows indicate spike failures. Part **(a)** adapted from Ylinen A, Bragin A, Nadasdy Z *et al.* (1995) Sharp wave-associated high-frequency oscillation (200 Hz) in the intact hippocampus: network and intracellular mechanisms. *J. Neurosci.* **15**, 30–46, with permission. Part **(b)** adapted from Freund TF, Buzsaki G (1996) Interneurons in the hippocampus. *Hippocampus* **6**, 347–470, with permission.

Are networks completely different from one nucleus to another or are there some fundamental principles of organization?

Networks have in common a basic organization. There are always principal cells and most often interneurons (the subthalamic nucleus is, for example, devoid of interneurons whereas striatum contains many different types of these). Interneurons are always connected so as to provide feedforward and feedback inhibitions.

Afferent extrinsic connections impinge on to principal cells and interneurons.

Does the precise knowledge of intrinsic and extrinsic connections as well as the firing patterns of neurons allow us to explain how network oscillations are generated?

Yes, it does, if the pattern of afferent activity in extrinsic afferent axons is known.

Further reading

Buzsáki, G., Moser, E.I., 2013. Memory, navigation and theta rhythm in the hippocampal-entorhinal system. Nat. Neurosci. 16, 130–138.

Buzsáki, G., Silva, F.L., 2012. High frequency oscillations in the intact brain. Prog. Neurobiol. 98, 241–249.

Csicsvari, J., Hirase, H., Czurko, A., et al., 1999. Oscillatory coupling of hippocampal pyramidal cells and interneurons in the behaving rat. J. Neurosci. 19, 274–287.

Kepecs, A., Fishell, G., 2014. Interneuron cell types are fit to function. Nature 505, 318–326.

Klausberger, T., Somogyi, P., 2008. Neuronal diversity and temporal dynamics: the unity of hippocampal circuit operations. Science 321, 53–57.

Maccaferri, G., 2005. Stratum oriens horizontal interneurone diversity and hippocampal network dynamics. J. Physiol. 562, 73–80.

Maccaferri, G., Lacaille, J.C., 2003. Interneuron diversity series: hippocampal interneuron classifications – making things as simple as possible, not simpler. Trends Neurosci. 26, 564–571.

Nadasdy, Z., Hirase, H., Czurko, A., et al., 1999. Replay and time compression of recurring spike sequences in the hippocampus. J. Neurosci. 19, 9497–9507.

Scharfman, H.E., Myers, C.E., 2013. Hilar mossy cells of the dentate gyrus: a historical perspective. Front. Neural Circuits 6, 106.

Takács, V.T., Szönyi, A., Freund, T.F., Nyiri, G., Gulyás, A.I., 2014. Quantitative ultrastructural analysis of basket and axo-axonic cell terminals in the mouse hippocampus. Brain Struct. Funct. Epub ahead of print, DOI: 10.1007/s00429-013-0692-6.

CHAPTER

20

Maturation of the hippocampal network

Yehezkel Ben Ari

OUTLINE

Developing neurons and circuits have several unique features and mechanisms that differ from those in adults. First, several processes and cascades occur in the developing but seldom in the adult brain, including cell migration, differentiation, programmed cell death etc. Also, the subunit composition of ionotropic and metabotropic receptor channels or of voltage-gated ionic channels is often different in developing neurons. Ionic currents follow developmental sequences usually shifting from long lasting decay time constants to shorter ones. This chapter describes some of the sequential events that take place during the construction of the hippocampal network that has been extensively investigated. It concentrates on the maturation of the main neuronal elements of the hippocampus, pyramidal neurons and interneurons, and their transmitters glutamate and GABA which, in the adult, provide most of the excitatory and inhibitory drives, respectively. The properties of electrical activity that result from this maturation are described and compared with what is observed in the adult hippocampus. Although centered primarily on GABAergic signals, the conclusions apply to many other systems that also follow developmental sequences.

20.1 GABAergic NEURONS AND GABAergic SYNAPSES DEVELOP PRIOR TO GLUTAMATERGIC ONES

20.1.1 GABAergic interneurons divide and arborize prior to pyramidal neurons and granular cells

In adults, GABAergic and glutamatergic signals equilibrate in order to prevent seizures. What is the situation during brain maturation? Do interneurons and pyramidal neurons become functional at the same developmental stage or are they sequentially functional? This was investigated using the bromodeoxyuridine (BrdU) technique. When BrdU is injected systemically, it is incorporated in the DNA of cells in the process of division. It is therefore possible to label neuronal ensembles according to their postmitotic age and determine their migration speed. This showed that GABAergic interneurons divide prior to the principal neurons (CA1–CA3 pyramidal neurons and granular cells of the dentate gyrus) (**Figure 20.1**). Thus, in the rat, interneurons divide between E13 (embryonic age, 13 days) and E17, whereas pyramidal neurons divide between E16 and E21 (there

Cellular and Molecular Neurophysiology. DOI: 10.1016/B978-0-12-397032-9.00020-0

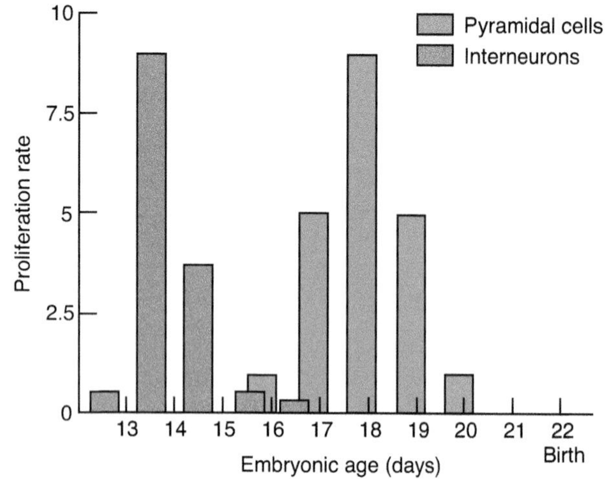

FIGURE 20.1 **Histogram showing the proliferation rate of pyramidal cells (green) and interneurons (gray) according to the embryonic age of the rat.** Adapted from Bayer SA (1980) Development of the hippocampal region in the rat: I. Neurogenesis examined with ^{3}H-thymidine autoradiography. *J. Comp. Neurol.* **190**, 87–114, with permission.

is an additional difference between CA3 and CA1 pyramidal neurons, the former reaching maturity earlier). More recent techniques relying on genetic birth dating of neurons suggest even earlier development of GABAergic interneurons, further reinforcing their interventions early in hippocampal activities. The granule cells of the fascia dentata have a primarily postnatal division; it is estimated that over 85% of the granule cells in the rat will divide in the three-week period following birth. Interneurons are mature at an earlier stage than the bulk of principal cells.

To determine if this is also manifested by a sequential maturation of the axonal and dendritic arbors, it is possible to patch-clamp interneurons and pyramidal neurons and to inject a dye in the intracellular medium. Such studies show that interneurons indeed mature and arborize at an earlier stage than pyramidal neurons. The interneuronal circuit is therefore in a situation to exert an important modulatory role on the growth of pyramidal cells and the formation of the hippocampal network. This observation raises the following question: 'Since GABAergic interneurons are mature before glutamatergic pyramidal neurons, do these interneurons establish synapses before the glutamatergic ones on to target pyramidal neurons?'

20.1.2 GABAergic synapses are established before glutamatergic ones on to pyramidal cells

To determine the formation of GABA and glutamate synapses, the activity of pyramidal neurons is recorded in the whole-cell configuration (voltage clamp mode) in slices at an early stage – say at birth – and the properties of the postsynaptic currents (PSCs) that occur spontaneously or in response to electrical stimulation are determined. This is combined to morphological reconstruction of the recorded neurons (marked by intracellular injection of biocytin). At birth (P0, postnatal day 0), pyramidal neurons in the rat hippocampus are composed of three populations (**Figure 20.2**):

- Eighty percent of the neurons have a soma and an axon but essentially no apical or basal dendrites. These neurons are 'silent' in that no spontaneous or evoked synaptic current is recorded from them even in response to strong electrical stimuli (**Figure 20.2a,** silent neurons). These neurons do, however, express extrasynaptic receptors since bath applications of GABA or glutamate agonists evoke the usual currents (not shown) observed in more adult neurons, confirming (see above) that the expression of receptors precedes that of functional synapses. They are identified as neurons (and not glia) by their ability to generate spikes in response to an intracellular depolarization.
- Ten percent of the pyramidal neurons have a bigger soma, an axon and a small apical dendrite restricted to the initial part of the stratum radiatum and no basal dendrite. In these neurons, $GABA_A$-receptor-mediated PSCs are recorded but glutamate-receptor-mediated EPSCs are absent (the synaptic response is fully abolished in the presence of bicuculline) (**Figure 20.2b,** GABA neurons). There are only GABAergic synapses established on these neurons: they are thus of the 'GABA only' type. In these whole-cell experiments, the reversal potential of Cl^- is 0 mV, which explains why we speak of PSC rather than IPSC.
- Ten percent of the neurons have an extensively arborized apical dendrite that penetrates to the most distal part of the apical dendrite (lacunosum moleculare) and a more developed basal dendrite. In these neurons, $GABA_A$- *and* glutamate-receptor-mediated PSCs are recorded as shown by the sequential use of selective antagonists (**Figure 20.2c,** GABA + Glu neurons). Thus, there are GABA and glutamate synapses established on these neurons: they are of the 'GABA + Glu' type.

Similar recordings in embryonic slices (E19) indicate that, at this age, virtually all pyramidal neurons (over 90%) are 'silent', instead of 80% at P0. All these results show that GABAergic synapses are established prior to glutamatergic ones on pyramidal neurons. Moreover, GABAergic synapses are formed only when the pyramidal neurons have an apical dendrite (glutamatergic synapses are established on the pyramidal neurons when the apical dendrite reaches the stratum lacunosum moleculare). Parallel immunocytochemical data confirm that synaptic markers such as synaptophysin (**Figure 20.3**) or synapsin or markers of GABAergic terminals are first observed at birth at the level of the apical dendrites of

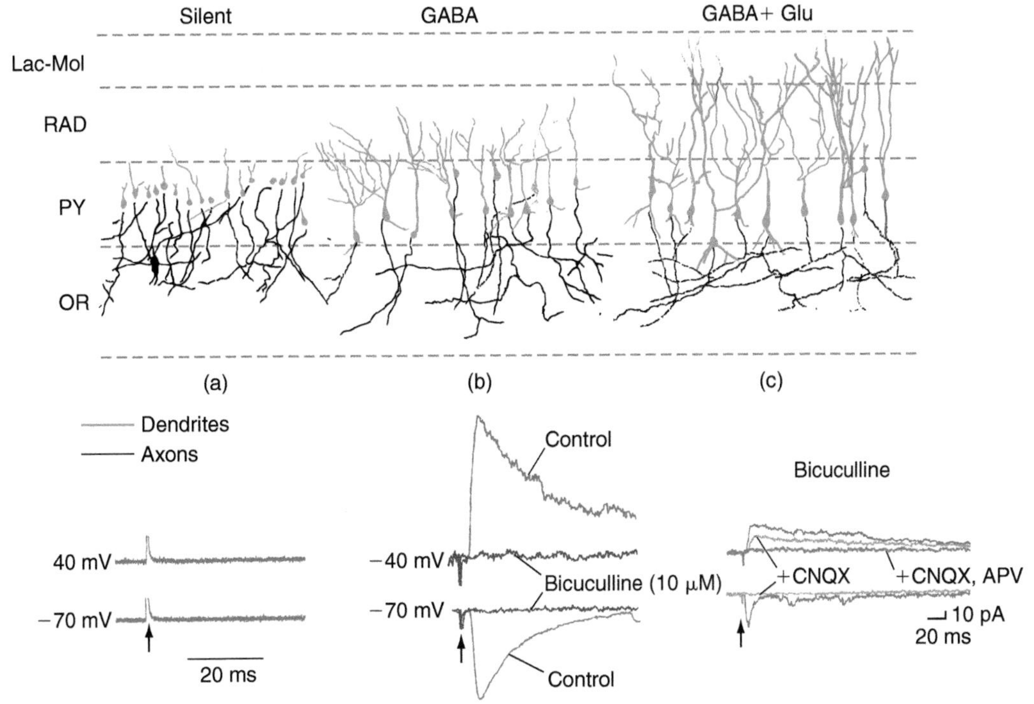

FIGURE 20.2 **Pyramidal cells at P0, grouped accordingly to their synaptic properties.** The activity of pyramidal cells is recorded at P0 in whole-cell configuration (voltage clamp mode) in response to stratum radiatum stimulation. Recorded pyramidal cells are injected with biocytin and reconstructed with a camera lucida. Neurons are grouped according to their synaptic properties. **(a)** Silent cells displaying no synaptic current. **(b)** Cells displaying a bicuculline-sensitive ($GABA_A$-mediated) synaptic current only. **(c)** Cells displaying bicuculline- ($GABA_A$), CNQX- (AMPA) and APV- (NMDA) sensitive synaptic currents. GABA only **(b)** and GABA + Glu **(c)** pyramidal cells differ essentially by the presence of an apical dendrite in the stratum Lac-Mol. Green: somato-dendritic tree; black: axon and axonal arborization. Layers of the hippocampus from top to bottom: Lac-Mol, stratum lacunosum moleculare; RAD, stratum radiatum; PYR, stratum pyramidale; OR, stratum oriens. Arrows indicate the stimulating artifact. Adapted from Tyzio R, Represa A, Jorquera I *et al.* (1999) The establishment of GABAergic and glutamatergic synapses on CA1 pyramidal neurons is sequential and correlates with the development of the apical dendrite. *J. Neurosci.* **19**, 10372–10382, with permission.

pyramidal neurons. They are observed neither in the pyramidal layer nor in the stratum oriens at this early stage.

A similar study performed in immature GABAergic interneurons indicates a similar sequence in which GABAergic synapses are formed before glutamatergic synapses. However, the sequence in interneurons occurs earlier: more than 90% of interneurons are innervated at birth at a time when most pyramidal neurons are silent. Therefore, GABAergic neurons and synapses provide most of the synaptic drive at an initial stage.

20.1.3 Sequential expression of GABA and glutamate synapses is also observed in the hippocampus of subhuman primates *in utero*

Similar studies have been performed in fetal and embryonic macaque rhesus hippocampus during the second part of gestation (birth takes place around E165 in this species) in order to understand whether the GABA–Glu sequence of innervations applies also to other species, and in particular to non-human

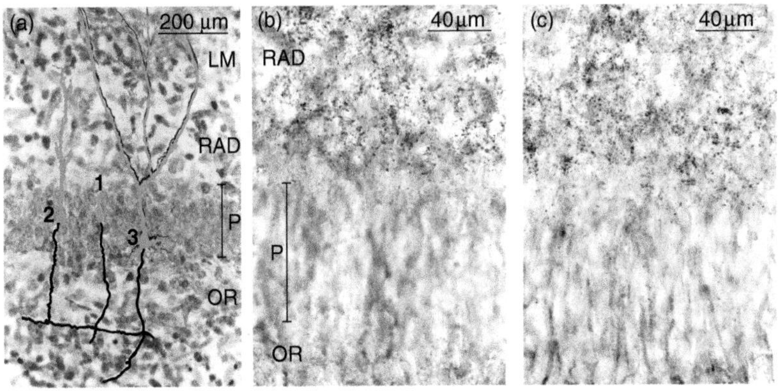

FIGURE 20.3 **Presence of synaptic boutons in stratum radiatum but not in stratum pyramidale at P0. (a)** CA1 hippocampal section stained with cresyl violet (shown here as gray). Three pyramidal cells are shown: silent (1, middle), GABA only (2, left) and GABA + Glu (3, right), to demonstrate the distribution of the dendrites within all the layers. Other sections from the same hippocampus but at a different scale depict immunolabeling with **(b)** synaptophysin and **(c)** synapsin (see also **Figure 20.2**). The labeling is observed in stratum radiatum but not pyramidale. LM, stratum lacunosum moleculare; RAD, stratum radiatum; P, stratum pyramidale; OR, stratum oriens.

primates. The monkey embryos were removed by C-sections (between E85 and E154), the brain dissected and hippocampal slices obtained. The activity of pyramidal cells is recorded in voltage clamp (whole-cell configuration) and neurons are filled with biocytin and reconstructed. As for the postnatal rat, fetal pyramidal neurons of the macaque can be divided into three populations (**Figure 20.4**):

- 'Silent' neurons that have an axon but a dendritic arbor restricted to stratum pyramidale (**Figure 20.4a**). 'Silent' neurons generate sodium action potentials when depolarized by intracellular current injection (they are neurons, not glia). Bath-applied GABA or glutamate agonists evoke currents (not shown), indicating that

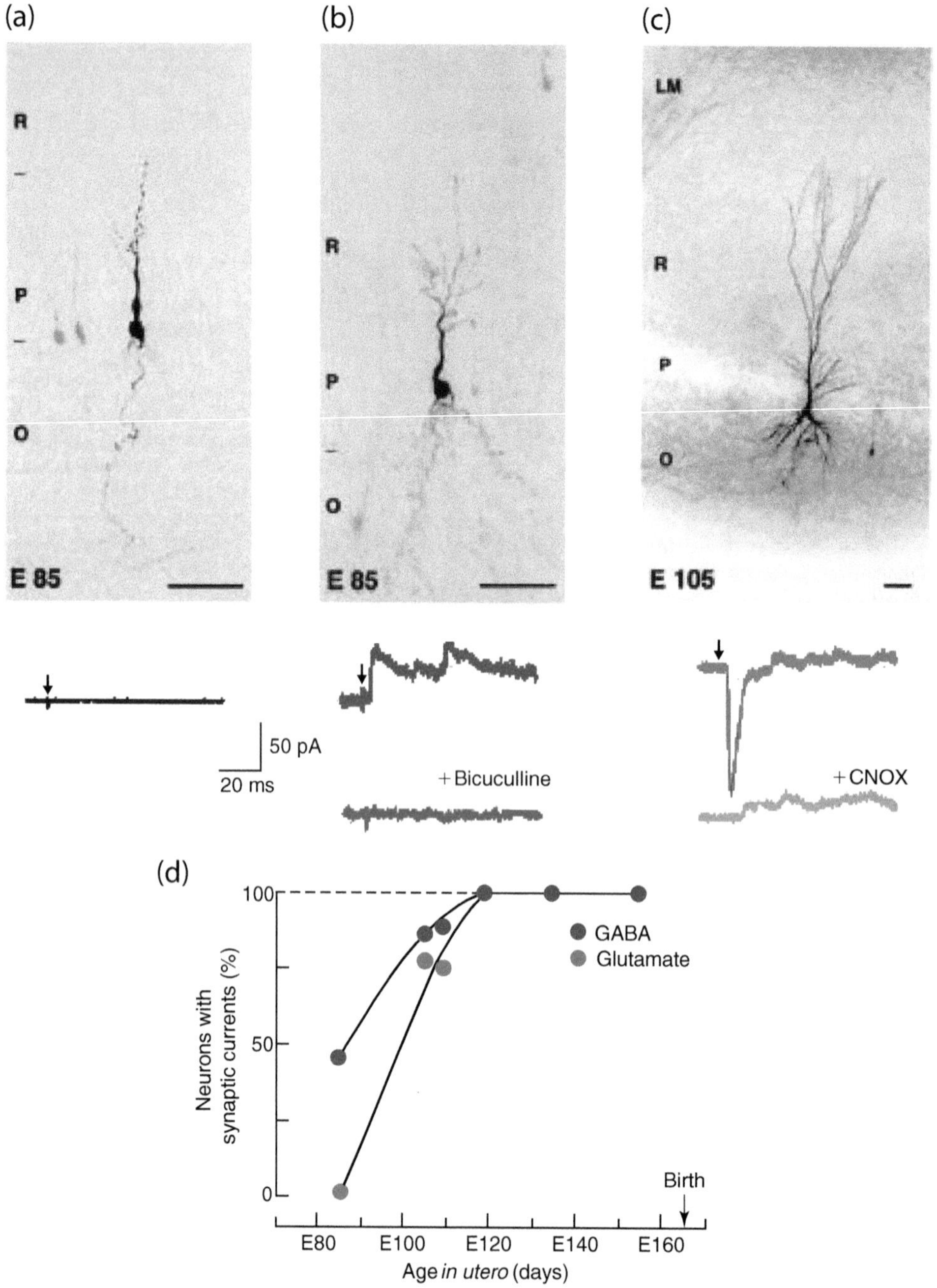

FIGURE 20.4 **Monkey pyramidal cells at embryonic ages.** The activity of pyramidal cells is recorded in whole-cell configuration (voltage clamp mode) in response to stratum radiatum stimulation. Recorded cells are injected with biocylin and reconstructed with a camera lucida. **(a)** Silent cells at E85 displaying no synaptic current, **(b)** cell at E85 displaying bicuculline-sensitive ($GABA_A$-mediated) synaptic currents only, and **(c)** cell at E105 displaying CNQX-sensitive (AMPA-mediated) synaptic currents. Arrows indicate the stimulating artifact. **(d)** Developmental curve to depict the progressive expression at the embryonic stage of GABA and glutamate synapses. LM, stratum lacunosum moleculare; R, stratum radiatum; P, stratum pyramidale; O, stratum oriens. E means embryonic day. Khazipov R, Esclapez M, Caillard O, Bernard C, Khalilov I, Tyzio R, *et al.* (2001) Early development of neuronal activity in the primate hippocampus in utero. *J. Neurosci.* **21**, 9770–9781.

the postsynaptic receptors are functional but not the synapses.

- Neurons that have axons and small apical dendrites express only $GABA_A$-mediated PSCs (**Figure 20.4b**). There are only GABAergic synapses established on these neurons; they are thus of the 'GABA only' type.
- Neurons that have an arborized apical dendrite as well as basal dendrites exhibit both $GABA_A$- and glutamate-mediated PSCs (**Figure 20.4c**). Thus, there are GABA and Glu synapses established on these neurons; they are of the 'GABA + Glu' type.

This sequence is observed in both CA1 and CA3, with the only difference that the latter matures earlier. Quantification of the percentage of neurons expressing GABAergic synapses and glutamatergic synapses indicates that at the beginning of mid-gestation over 50% of CA1 pyramidal neurons have functional GABAergic but not glutamatergic synapses (**Figure 20.4d**). In contrast, a few weeks later, one month before birth, all pyramidal neurons have both GABA and glutamate synapses. This results from the high speed of establishment of glutamatergic synapses during this period. Therefore, the sequence described for the rat hippocampus applies also to that of monkeys and probably of humans too. However, the gestation period during which the sequence is established is different: the first week postnatal in the rat, and the beginning of the second part of the gestation period in the rhesus monkey.

20.1.4 Questions about the sequential maturation of GABA and glutamate synapses

To summarize, GABAergic synapses between interneurons and the dendrites of pyramidal neurons are the first synapses to be established on pyramidal neurons of the hipppocampus (**Figure 20.5**). Glutamatergic synapses are formed at a later stage (such as synapses between two pyramidal cells or between extrinsic glutamatergic

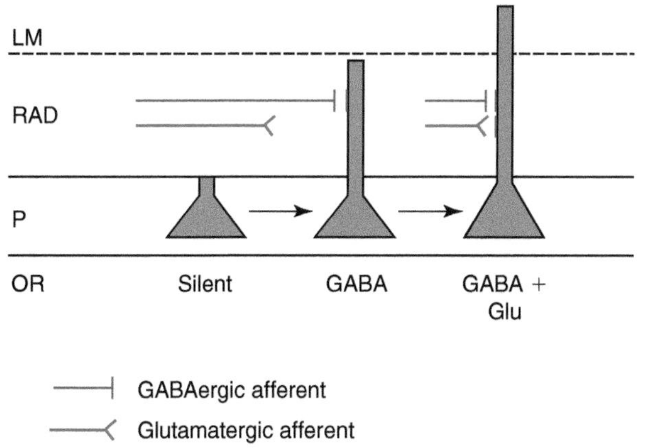

FIGURE 20.5 **Schematic of the different stages of maturation of CA1 pyramidal cells at P0.** See the caption to **Figure 20.2.**

fiber tracts and pyramidal cells). This sequential expression suggests that the developmental stage of the target determines whether or not synapses will be established with presynaptic axons. This also suggests that the rules governing the formation of GABA and glutamate synapses differ, the latter requiring more mature postsynaptic targets.

These observations in turn raise the following questions:

- Does the activation of $GABA_A$ receptors in immature neurons evoke a current and a potential change identical to that in adults, or does the neonatal $GABA_A$ synaptic response have different properties?
- What are the consequences of these developmentally regulated features on the electrical properties of the immature network? How does the immature network discharge?

20.2 $GABA_A$ RECEPTOR-MEDIATED RESPONSES DIFFER IN DEVELOPING AND MATURE BRAINS

20.2.1 Activation of $GABA_A$ receptors is depolarizing and excitatory in immature networks because of a high intracellular concentration of chloride

In the adult hippocampus, activation of $GABA_A$ receptors at rest evokes an inward flow of Cl^- ions across the postsynaptic membrane (i.e. an outward current) that results in membrane hyperpolarization (see **Figures 9.16** and **19.9**), a decrease in membrane resistance and a shunt effect (see **Figure 9.17** and Section 9.5.3). $GABA_A$-mediated inhibition is the key element that provides the basis for the coordinated synchronized neuronal activity. Removal of this inhibition leads in the adult hippocampus to the generation of epileptiform activities. In contrast, in the immature hippocampus, $GABA_A$ receptors have a totally different function – they mediate excitation (EPSPs).

$GABA_A$-receptor activation evokes a depolarization and bursts of action potentials in the immature hippocampus

To determine the properties of the $GABA_A$-mediated response in immature neurons, the activity of pyramidal cells is recorded in embryonic or early neonatal slices. One puzzling observation is that brief applications of the $GABA_A$ receptor antagonist bicuculline, which in the adult generates epileptiform activity, silences in contrast ongoing activity in slices at an early developmental stage (**Figure 20.6a**), suggesting that GABA exerts an excitatory action at this stage. In keeping with this, intracellular and whole-cell recordings (current clamp mode) show that $GABA_A$-receptor activation leads to a

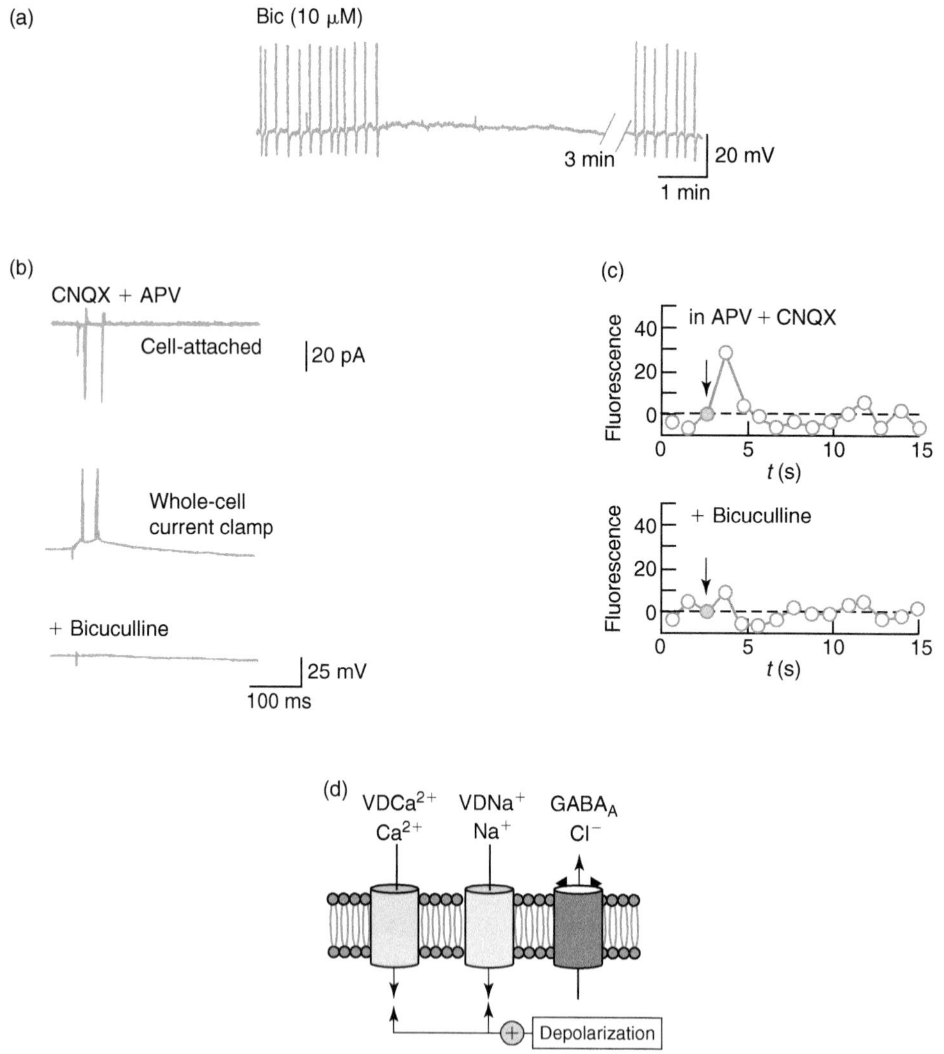

FIGURE 20.6 **Synaptic activation of $GABA_A$ receptors is depolarizing in neonatal rat hippocampus.** The activity of CA3 hippocampal neurons recorded in slices *in vitro.* **(a)** Effect of bicuculline on the spontaneous activity of a pyramidal cell recorded at −70 mV with an intracellular electrode filled with K methyl sulfate. **(b)** Response of an interneuron to stimulation in stratum radiatum in the continuous presence of CNQX (10 μM) and APV (50 μM). The excitatory response is recorded from the same interneuron in the cell-attached and whole-cell configuration (current clamp mode). It is totally abolished by bicuculline (10 μM). **(c)** In the same conditions, electrical stimulation (arrow) evokes an increase of $[Ca^{2+}]_i$ that is abolished by bicuculline, in a pyramidal neuron loaded with fluo-3. **(d)** $GABA_A$ receptor activation leads to membrane depolarizing and thus opening of voltage-dependent (VD) Na^+ and Ca^{2+} channels. Part (**a**) adapted from Gaiarsa JL, Coradetti R, Ben-Ari Y, Cherubini E (1990) GABA mediated synaptic events in neonatal rat CA3 pyramidal neurons *in vitro*: modulation by NMDA and non-NMDA receptors. In: Ben Ari Y (ed.) *Excitatory Amino Acids and Neuronal Plasticity*, New York: Plenum Press. Part (**b**) adapted from Ben-Ari Y, Khazipov R, Leinekugel X *et al.* (1997) $GABA_A$, NMDA and AMPA receptors: a developmentally regulated 'ménage à trois'. *Trends Neurosci.* **20**, 523–529. Part (**c**) adapted from Khazipov R. Leinekugel X, Khalilov I *et al.* (1997) Synchronization of GABAergic interneuronal network in CA3 subfield of neonatal rat hippocampal slices. *J. Physiol. (Lond.)* **498**, 763–772; all with permission.

depolarization of the membrane and the generation of sodium action potential(s). This is also observed in cell-attached recordings in which the intracellular concentration of Cl^- is not altered (**Figure 20.6b**). Therefore, GABA depolarizes immature neurons.

$GABA_A$-receptor-mediated depolarization evokes Ca^{2+} entry through both the voltage-gated Ca^{2+} and NMDA channels

Activation of $GABA_A$ synapses in immature neurons leads to an increase of intracellular Ca^{2+} concentration ($[Ca^{2+}]_i$), as shown by Ca^{2+} imaging techniques (**Figure 20.6c**). In order to understand the underlying mechanism (recall that $GABA_A$ receptor channels are not permeable to Ca^{2+} ions), two main hypotheses can be tested: the $[Ca^{2+}]_i$ increase results either from the activation of voltage-gated Ca^{2+} channels or from the activation of NMDA receptor channels.

To test the first hypothesis, synaptic $GABA_A$ receptors are activated by stimulation of afferents in the presence of CNQX + APV, the blockers of ionotropic glutamate receptors. An increase of intracellular Ca^{2+} concentration is still observed and is blocked by the subsequent application of bicuculline (**Figure 20.6c**) or of antagonists of Ca^{2+} channels such as nifedipin or D-600 (not shown). Therefore, the activation of neonatal GABA$_A$ receptors results in a $[Ca^{2+}]_i$ increase due, at least partly, to Ca^{2+} entry through voltage-gated Ca^{2+} channels (**Figure 20.6d**).

Conversely, in the presence of antagonists of voltage-gated Ca^{2+} channels, synaptic GABA$_A$-receptor activation still induces a small increase of $[Ca^{2+}]_i$ that is abolished by APV, the selective antagonist of NMDA receptor channels (not shown). This suggests that the GABA$_A$-mediated depolarization can remove the voltage-dependent Mg^{2+} block of NMDA channels (see **Figure 20.6d**). This was confirmed by recordings of single NMDA-channel activity and cell-attached recordings of the synaptic responses evoked in the presence of an AMPA receptor antagonist.

There is a synergy between GABA$_A$ and NMDA receptor channels in the immature hippocampus

Therefore, in immature neurons, there is a synergistic action between GABA$_A$ and NMDA receptors. This stands in contrast with the adult situation. In the neonatal hippocampus, GABAergic synapses act much like AMPA-receptor-mediated synapses at a later developmental stage: they provide the excitatory drive required to generate sodium and calcium action potentials as well as to activate NMDA receptors (see **Figure 20.7a,b**).

There are, however, additional factors to take into account to fully comprehend the operative mode of the immature circuit. Even when GABA is depolarizing and excitatory, there is an inhibitory component due to a shunt mechanism suggesting that the net effects of the activation of GABA receptors is dual, depending on many factors including the timing between the arrival of GABA and glutamate signals. Also, at a given age of the rat, owing to the heterogeneity of hippocampal neurons, GABA will exert different actions in different neurons – depolarizing in one and hyperpolarizing in the other – presumably because they are in a different developmental stage. Thus, recording from different neurons in the same slice can reveal a cocktail of effects, including in some a net excitatory action, in others a dual effect (excitatory and inhibitory). It is also important to note that the depolarization produced by GABA can also activate other voltage-gated currents and the convergence of these actions will reach spike threshold.

The developmental curve of the concentration of $[Cl^-]_i$ is exponential

Why does GABA depolarize immature neurons? At least two hypotheses can be proposed: the GABA$_A$ channel is permeable to cations in immature neurons or the GABA$_A$ channel is permeable to Cl^- ions in immature neurons but the Cl^- driving force is reversed, due to an increased intracellular concentration of Cl^- ions.

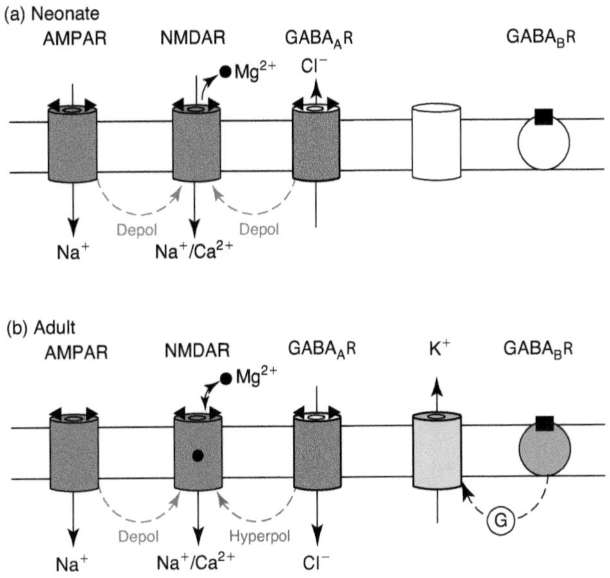

FIGURE 20.7 **Major developmental changes in the GABA–glutamate interactions.** See text for explanations. Adapted from Leinekugel X, Medina I, Khalilov I *et al.* (1997) Ca^{2+} oscillations mediated by the synergistic excitatory actions of GABA$_A$ and NMDA receptors in the neonatal hippocampus. *Neuron* **18**, 243–255, with permission.

To test the second hypothesis, it is important to use a non-invasive technique that does not perturb $[Cl^-]_i$. Indeed, patch clamp or intracellular techniques modify $[Cl^-]_i$ and the polarity of the currents generated by GABA. In contrast, by recording in cell-attached configuration the openings of single GABA$_A$ channels, the concentration of $[Cl^-]_i$ can be determined at different ages. The value of the resting membrane potential is determined by means of recordings of the activity of single NMDA channels: since the NMDA current reverses at 0 mV, it is possible to calculate the genuine resting membrane potential (**Figure 20.8b**). The difference between the two values enables calculation of the driving force for GABA (DF_{GABA}) (see Appendix 9.2). During maturation, neurons from the hippocampus show a progressive reduction of DF_{GABA} with a shift of $[Cl^-]_i$ from 25–30 mM (embryonic) to 7–8 mM at the end of the first postnatal week (**Figure 20.8a**). This corresponds to a shift of the reversal potential of GABA$_A$ current from 40 mV above V_{rest} (embryonic) to a few mV close to V_{rest} (adult). In the former case, GABA will generate spikes, in the latter it will inhibit their generation. This curve differs in various neuronal populations according to their developmental properties but even within the same population it will differ according to the degree of development and differentiation of the neuron. However, it is suggested that this curve is universal, i.e. observed in every animal species and brain structure. This results from the early expression of the chloride importer NKCC1 in immature neurons whereas the main chloride exporter KCC2 has a delayed expression.

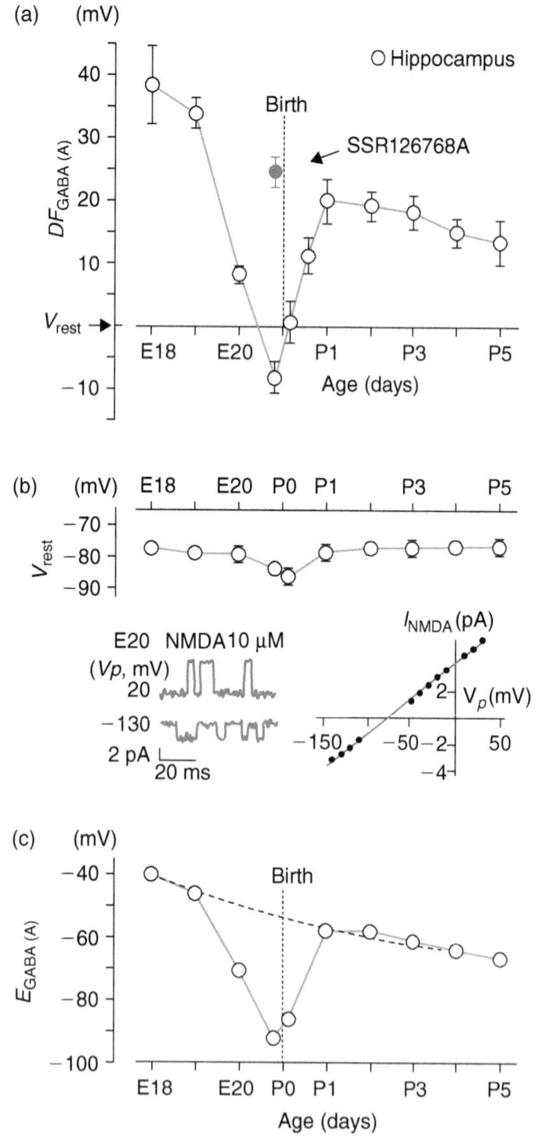

FIGURE 20.8 **Developmental profile of $E_{GABA(A)}$ in the rat. (a)** Summary plot of the age-dependence of $DF_{GABA(A)}$ inferred from the recordings of single-channel $GABA_A$ currents [mean ± SEM; 209 CA3 pyramidal cells (o); 6 to 24 patches for each point]. Red code – pretreatment with SSR126768A, an antagonist of oxytocin ($n = 25$ hippocampal patches). **(b)** Age dependence of the resting membrane potential (V_{rest}) of CA3 pyramidal cells inferred from the reversal of single NMDA channels recorded in cell-attached mode ($n = 84$ cells; 4 to 12 patches for each point). **(c)** Age dependence of the $GABA_A$ reversal potential ($E_{GABA(A)} = V_{rest} + DF_{GABA(A)}$). Note a transient hyperpolarizing shift of $E_{GABA(A)}$ near birth. See Appendix 9.2 for the explanation of the method. Adapted from Tyzio R, Cossart R *et al.* (2006) Maternal oxytocin triggers a transient inhibitory switch in GABA signaling in the fetal brain during delivery. *Science* **314**, 1788–1792.

The developmental curve is interrupted around delivery

When the developmental decline of $[Cl^-]_i$ is investigated in detail, a brief interruption of the curve is observed around delivery time, shortly before and after. This abrupt fall of $[Cl^-]_i$ is associated with a powerful inhibitory action of GABA (**Figure 20.8a**). This period is brief as the curve resumes its exponential decline subsequently with a higher $[Cl^-]_i$ and depolarizing and excitatory actions of GABA. Why is the curve interrupted transiently, $[Cl^-]_i$ dramatically reduced and GABA inhibitory during delivery? As delivery is initiated by an intense release of hormones, the possible role of oxytocin – a hormone released by the mother to trigger uterine contractions and labor – is tested **(Figure 20.8)**. Blocking oxytocin receptors suppresses the transient reduction of $[Cl^-]_i$ and the excitatory to inhibitory shift of the actions of GABA **(Figure 20.8c)**. Therefore, the same maternal agent that triggers labor also induces a shift of the actions of GABA during that critical period. Although the exact mechanisms of this action have not been determined, it involves the co-transporters of chloride since blocking the main importer of chloride in immature neurons NKCC1 (see above) prevents the effects of the hormone. Interestingly, this oxytocin-mediated abrupt and brief reduction of $[Cl^-]_i$ exerts a neuroprotective and analgesic action on the newborn's brain illustrating the importance of $[Cl^-]_i$ levels during delivery.

20.2.2 $GABA_B$-receptor-mediated IPSCs have a delayed expression in immature neurons

Maturation is also associated with alterations in the development of inhibition mediated by $GABA_B$ (as well as adenosine or serotonin) receptors. In adults, activation of these receptors exerts a powerful control at both postsynaptic and presynaptic levels: at the postsynaptic level it generates a large hyperpolarization due to the activation of K^+ channels via a G protein (it reverses at E_K). At the presynaptic level, activation of these receptors reduces the release of GABA and the amplitude of the GABAergic synaptic currents via a reduction of a presynaptic Ca^{2+} current and/or the activation of K^+ channels in axon terminals. During maturation, the postsynaptic $GABA_B$ receptors are not functional at an early stage (at birth and until P5–P6 in pyramidal neurons) (**Figure 20.9**). Binding studies show that the receptors are present, but intracellular injection of GTPγS to activate G proteins fails to activate $GABA_B$-receptor-mediated currents. Therefore the absence of $GABA_B$-mediated currents is not due to a delayed expression of the receptors but more likely to a delayed coupling of the $GABA_B$ receptors to G proteins and K^+ channels. In keeping with this, the other members of this family (metabotropic receptors) are also not operative at birth.

In contrast, the presynaptic mechanisms are operational already before birth in pyramidal neurons; the activation of $GABA_B$ (or adenosine and serotonin receptors) leads to a reduction of the PSCs in pyramidal neurons or interneurons (not shown). Therefore, the developing circuit operates with the two main postsynaptic

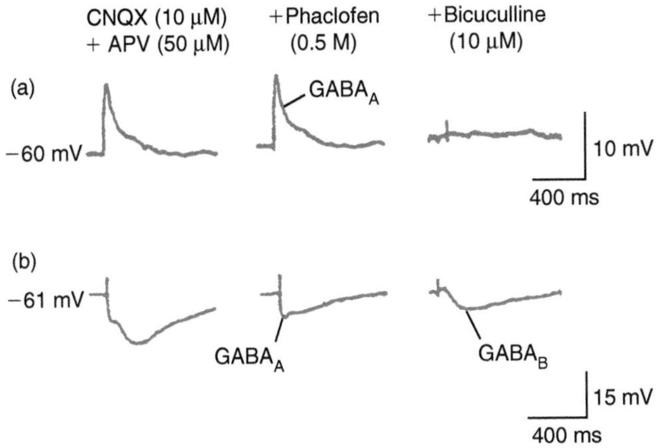

FIGURE 20.9 **$GABA_B$-mediated response is absent in neonate and present in adult pyramidal cells.** The activity of CA3 pyramidal cells recorded with an intracellular electrode filled with K methyl sulfate in the continuous presence of blockers of ionotropic glutamate receptors (CNQX + APV). **(a)** In the neonate: In response to electrical stimulation of stratum radiatum, a depolarizing response is recorded in the neonatal pyramidal cell. This response is mediated by $GABA_A$ receptors since it is unaffected by application of the $GABA_B$ antagonist phaclofen but is totally abolished by the $GABA_A$ antagonist bicuculline. **(b)** In the adult: In contrast, stimulation induces a biphasic hyperpolarization. Phaclofen reduces the late component ($GABA_B$) and leaves intact the early one ($GABA_A$) whereas bicuculline suppresses the early one and leaves intact the late one. Adapted from Gaiarsa JL, Tseeb V, Ben-Ari Y (1995) Postnatal development of pre- and postsynaptic $GABA_B$-mediated inhibitions in the CA3 hippocampal region of the rat. *J. Neurophysiol.* **73**, 246–255, with permission.

receptor-mediated inhibitory mechanisms, $GABA_A$ and $GABA_B$, being poorly developed or acting in a reversed manner. In neonatal hippocampus, the principal mode of operation of transmitter-gated inhibition relies on a presynaptic control of transmitter release.

20.3 MATURATION OF COHERENT NETWORKS ACTIVITIES

20.3.1 Network-driven giant depolarizing potentials (GDPs) provide most of the synaptic activity in the neonatal hippocampus

Electrical activity in neonatal rats (postnatal days P0–P8) is characterized by the presence of spontaneous network-driven *giant depolarizing potentials* (GDPs) that provide most of the synaptic activity. GDPs are large and long-lasting (several hundreds of milliseconds) synaptic potentials giving rise to bursts of spikes that occur repetitively at a frequency varying between 0.05 and 0.2 Hz (**Figure 20.9a**). GDPs are recorded in the vast majority of neurons (both pyramidal neurons and interneurons) in hippocampal slices obtained from the brain of rats aged between birth and 2 weeks postnatal (**Figure 20.9c**). GDPs also prevail *in utero* in the monkey during the second half of gestation (E85–E135). Therefore, GDPs and other similar patterns that provide the first coherent pattern of network activity constitute a universal transient phase and are replaced subsequently by a more diversified panel of patterns that enable adults networks to generate a complex repertoire of diversified behaviors (see below).

20.3.2 Giant depolarizing potentials result from GABAergic and glutamatergic synaptic activity

GDPs are network-driven (in contrast to pacemaker patterns that are generated by the recorded neuron independently of its synaptic inputs), since (i) they are blocked by tetrodotoxin (TTX) (**Figure 20.10b**); (ii) their amplitude but not their frequency is modified by alterations of the resting membrane potential as expected from a synaptic current in contrast to an endogenous pacemaker oscillation; (iii) they are often blocked by the $GABA_A$-receptor antagonist bicuculline (see **Figure 20.6a**) but also by ionotropic glutamate receptor antagonists (CNQX and APV), suggesting that both GABA and glutamate participate in their generation; (iv) they are blocked by an antagonist of the chloride importer NKCC1 that reduces $[Cl^-]_i$ levels indicating that the depolarizing actions of GABA are instrumental in the generation of GDPs; and (v) they disappear roughly around the end of the first postnatal week when the depolarizing/hyperpolarizing shift has taken place (**Figure 20.10c**).

Determination of the currents underlying GDPs with patch clamp recordings (voltage-clamp mode) reveals that $GABA_A$ currents are present either alone or in conjunction with AMPA- and NMDA-receptor-mediated currents. The relative participation of GABA and glutamate most likely depends on the maturational stage of the recorded neuron, reflecting the heterogeneity of the neuronal population.

The mechanism underlying the generation of GDPs therefore includes a network-driven barrage of depolarizing GABA and glutamate postsynaptic currents impinging on to pyramidal neurons and in turn leading to a recurrent GABAergic excitation from the GABAergic interneurons. This is also suggested by paired recordings from a pyramidal neuron and an interneuron that show synchronous GDPs in connected neurons and very few action potentials generated outside the GDPs (**Figure 20.11a**). Therefore, GDPs are triggered by the combined depolarizing effects of GABA- and glutamate-mediated currents and by a basic circuit that includes feedforward excitation of pyramidal neurons and interneurons followed by the recurrent depolarization produced by the recurrent collaterals of GABAergic interneurons (**Figure 20.11b**).

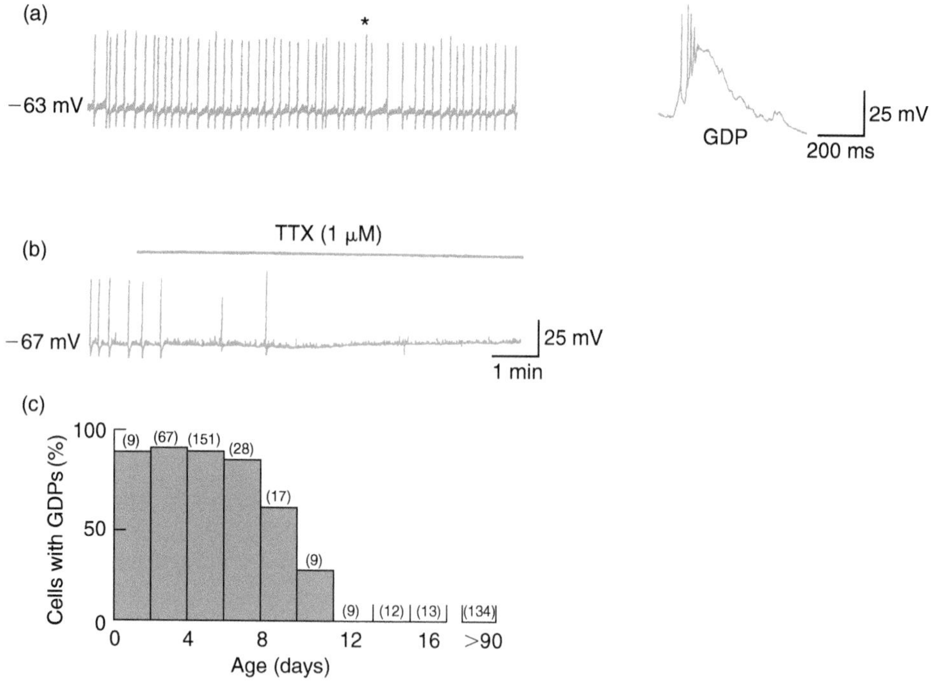

FIGURE 20.10 **Spontaneous giant depolarizing potentials (GDPs) are generated in a polysynaptic circuit in neonatal hippocampus. (a)** Spontaneous GDPs are recorded from a CA3 pyramidal neuron with an intracellular electrode filled with K methyl sulfate. The inset shows a GDP on an extended timescale. **(b)** Spontaneous GDPs are blocked by TTX. **(c)** Histogram of the number of pyramidal cells with GDPs as a function of the age of the rat. The number of recorded cells is in parentheses. Part **(a)** JL Gaiarsa, personal communication. Parts **(b)** and **(c)** adapted from Ben-Ari Y, Cherubini E, Corradetti, Gaiarsa JL (1989) Giant synaptic potentials in immature rat CA3 hippocampal neurones. *J. Physiol. (Lond.)* **416**, 303–325, with permission.

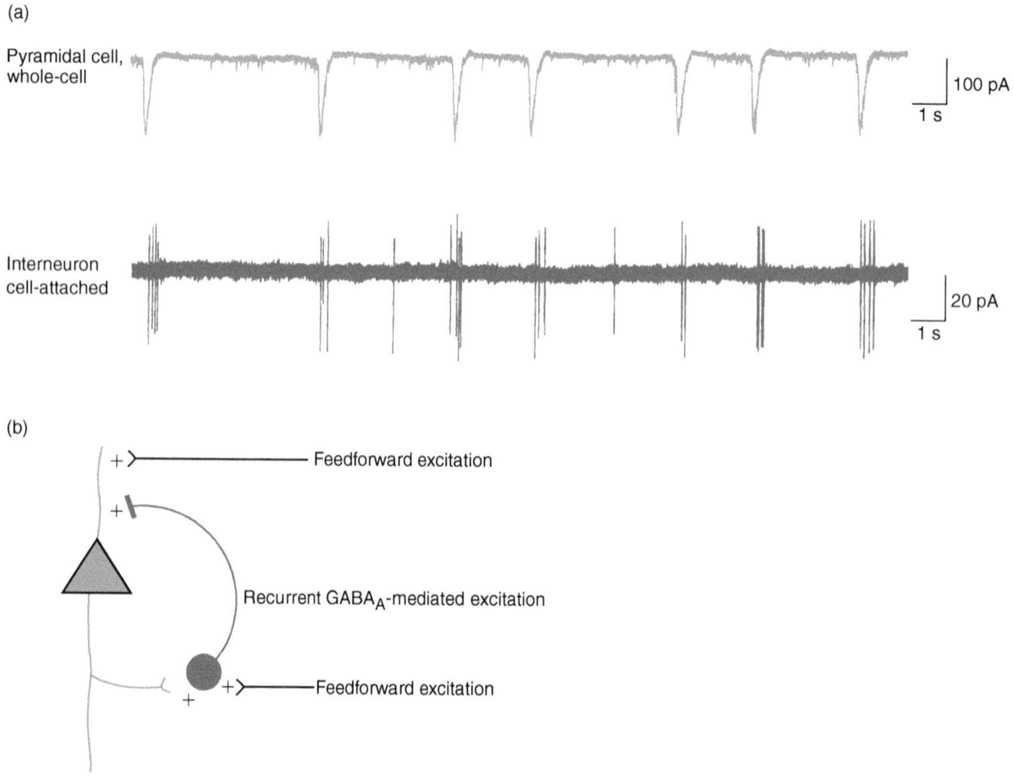

FIGURE 20.11 **GDPs are synchronous in pyramidal cells and interneurons. (a)** Dual recordings of spontaneous currents (voltage clamp mode) in a CA3 pyramidal neuron (whole-cell configuration, upper trace) and a neighboring interneuron (cell-attached configuration, lower trace). Note that bursts of currents of action potentials in the interneuron are synchronous with currents of GDPs in the pyramidal cell. **(b)** Schematic explaining the generation of GDPs. Part **(a)** adapted from Khazipov R, Leinekugel X, Khalilov I *et al.* (1997) Synchronization of GABAergic interneuronal network in CA3 subfield of neonatal rat hippocampal slices. *J. Physiol. (Lond.)* **498**, 763–772, with permission.

20.3.3 Giant depolarizing potentials are generated in the septal pole of the immature hippocampus and then propagate to the entire structure

Paired recordings in slices show that virtually all neurons have GDPs that are fully synchronous in neurons that are not too distant. To study the mechanism of the generation and propagation of GDPs, it is possible to record from an intact hippocampus superfused *in vitro*. In this preparation, the entire hippocampi are dissected and placed in a conventional *in vitro* chamber (**Figure 20.12a**). With multiple whole-cell (and extracellular field) recordings it is shown that *GDPs propagate with a septo-temporal gradient of automaticity*. This was shown by recording field potentials with a multiple electrode array along the rostro-caudal axis. This type of experiment shows that there is a rostro-caudal latency (**Figure 20.12a**). In addition, transection of the hippocampus in two along the longitudinal axis reveals that in both hemisected hippocampi there is a rostro-caudal latency (**Figure 20.12b,c**). Therefore, the rostral pole of the hippocampus, being the most active, paces the rhythm of the entire structure. Since GDPs are present in isolated portions of hippocampus, including mini-slices of CA3 or dentate gyrus subfields, the neuronal elements required for their generation are present in local neuronal circuits. However, the anterior parts of the hippocampus have a higher frequency of GDPs than the caudal ones.

This situation has some similarities with the generation of rhythmic activity in the cardiac muscle in which different parts, including sino-atrial node, atrio-ventricular node, His bundle and Purkinje fibers, have auto-rhythmic potentials that allow them to discharge periodically. However, the sino-atrial node is the normal pacemaker owing to a higher rhythm of activity. The mechanisms underlying the role of the anterior parts are not presently known but are likely to be due to a rostro-caudal gradient of maturation.

Using the two interconnected hippocampi *in vitro*, it is also possible to determine the hemispheric propagation of GDPs. Paired whole-cell recordings from two neurons in each hippocampus reveal that, even during the first few days after birth, GDPs can propagate from one hippocampus to the other and back, suggesting that the commissural connections are mature. This also suggests that the two hippocampi can synchronize each other at an early developmental stage.

In contrast, structures connected to the hippocampus, such as the septum or the entorhinal cortex, do not generate GDPs if they are disconnected from the hippocampus, but they do express GDPs originating in and propagating from the hippocampus. Therefore, the hippocampus must constitute a major source of network-driven synaptic activity that can modulate the electrical activity of brain structures with which it is connected.

20.3.4 Hypotheses on the role of the sequential expression of GABA- and glutamate-mediated currents and of giant depolarizing potentials

The earlier expression of GABAergic current, its depolarizing action and the resulting GDPs are key properties of developing networks. Similar effects of GABA have been described in several brain and peripheral structures, and GDP-like events also predominate in virtually

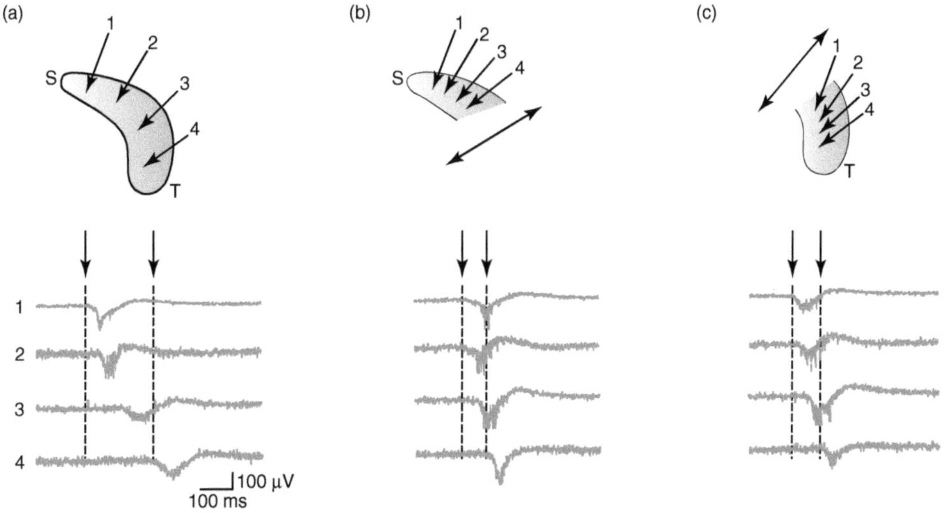

FIGURE 20.12 **GDPs propagate in the rostro-caudal direction in immature hippocampus.** The spontaneous GDPs of four populations of hippocampal neurons are simultaneously recorded with four extracellular electrodes in **(a)** the intact hippocampus, or in **(b)** the isolated rostral or **(c)** caudal halves of the hippocampus of neonatal rats. In extracellular recordings, GDPs are recorded as inward field potentials. Adapted from Leinekugel X, Khalilov I, Ben-Ari Y, Khazipov R (1998) Giant depolarizing potentials: the septal pole of the hippocampus paces the activity of the developing intact septohippocampal complex *in vitro*. *J. Neurosci.* **18**, 6349–6357, with permission.

all the structures studied so far, including spinal cord and neocortex. This raises the question of the role of this sequential expression of synapses, of the depolarizing actions of GABA, and of the GDPs. The following considerations should be taken into account:

- Owing to their long durations, $GABA_A$-mediated PSCs are highly suitable for summation (tens of milliseconds, in contrast to the milliseconds duration of AMPA-mediated PSCs). This is of importance in immature neurons that possess very few synapses initially, so that there are few spontaneous PSCs to summate if the excitatory drive is provided only by the brief AMPA-mediated PSCs. Furthermore, even if more depolarized than the resting potential, the reversal potential for GABAergic currents is closer to rest than that of glutamatergic PSCs. This and the shunting mechanism, which is inherent to the operation of GABAergic currents, prevent the occurrence of too strong depolarizations and of excitotoxic stimuli that occur when glutamatergic receptors are repetitively stimulated.
- The combined actions of GABA and glutamate facilitate the generation of a large increase of $[Ca^{2+}]_i$ as a result of the activation of voltage-gated Ca^{2+} channels and the removal of the Mg^{2+} blockade from NMDA channels (**Figure 20.7**). In keeping with this, studies in neonatal slices using Ca^{2+} imaging techniques indicate that GDPs are associated with important $[Ca^{2+}]_i$ oscillations (**Figure 20.13**). Other observations suggest that a rise of $[Ca^{2+}]_i$ is needed for dendritic growth and synapse formation.
- Network-driven oscillations like the GDPs may participate in the formation of functional neuronal units like the formation of visual columns. This may follow the Hebbian rule 'neurons that fire together wire together'. Network-driven oscillations provide a suitable way of organizing these units as both presynaptic and postsynaptic neurons will be excited in a synchronized manner. It is likely that these immature patterns of oscillations play an important role in electrically interconnecting neurons that will become part of an ensemble of neurons subsequently.

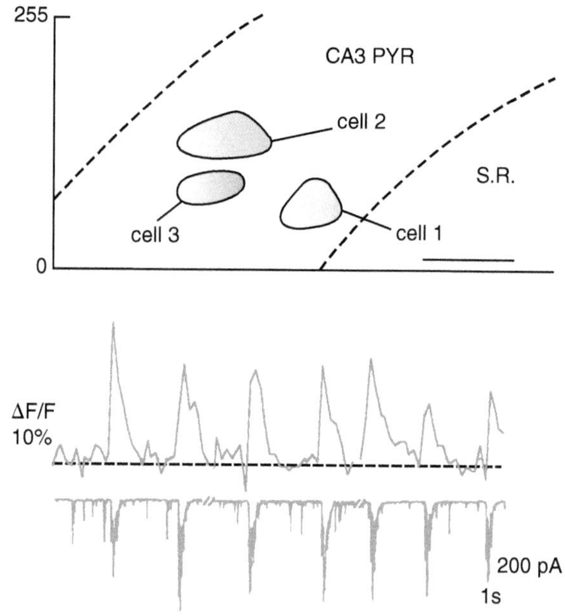

FIGURE 20.13 **Synchronous spontaneous Ca^{2+} oscillations in CA3 pyramidal neurons.** CA3 pyramidal neurons in slices are loaded with fluo-3 and their activity is simultaneously recorded in whole-cell configuration (voltage clamp model). Each spontaneous GDP current is concomitant to a transient increase of intracellular Ca^{2+} concentration. Adapted from Leinekugel X, Medina I, Khalilov I *et al.* (1997) Ca^{2+} oscillations mediated by the synergistic excitatory actions of $GABA_A$ and NMDA receptors in the neonatal hippocampus. *Neuron* **18**, 243–255, with permission.

20.3.5 Non-synaptic intrinsic currents precede the expression of GDPs

GDPs are the first synapse-driven currents recorded in the hippocampus and other brain structures but they are not the first currents observed at earlier developmental stages. Using a dynamic two-photons microscope to record calcium currents in hundreds of neurons, a tri-phasic sequence was unraveled: (i) neurons first generate intrinsic calcium currents that do not propagate to other neurons; then (ii) around delivery, neurons start generating intrinsic non-synapse driven synchronized calcium plateaux in small cell assemblies (SPAs); then (iii) the first synapse driven mediated pattern, i.e. the GDPs, appear. Furthermore, genetic fate and birth dating techniques reveal that early born GABAergic interneurons act as hub neurons orchestrating the generation and occurrence of GDPs (**Figure 20.14**). Therefore, early born GABAergic neurons provide the initial drive for the generation of GDPs. Collectively, these observations illustrate the importance of GABAergic signals in the maturation of hippocampal networks.

20.4 CONCLUSIONS

The first synapses to be established in the hippocampus are the GABAergic synapses between interneurons and pyramidal cells or between two interneurons. In these synapses, transmission is mediated by $GABA_A$ receptors.

Activation of $GABA_A$ receptors evokes a depolarization of the postsynaptic membrane, in contrast to what is observed in mature neurons where $GABA_A$ receptors

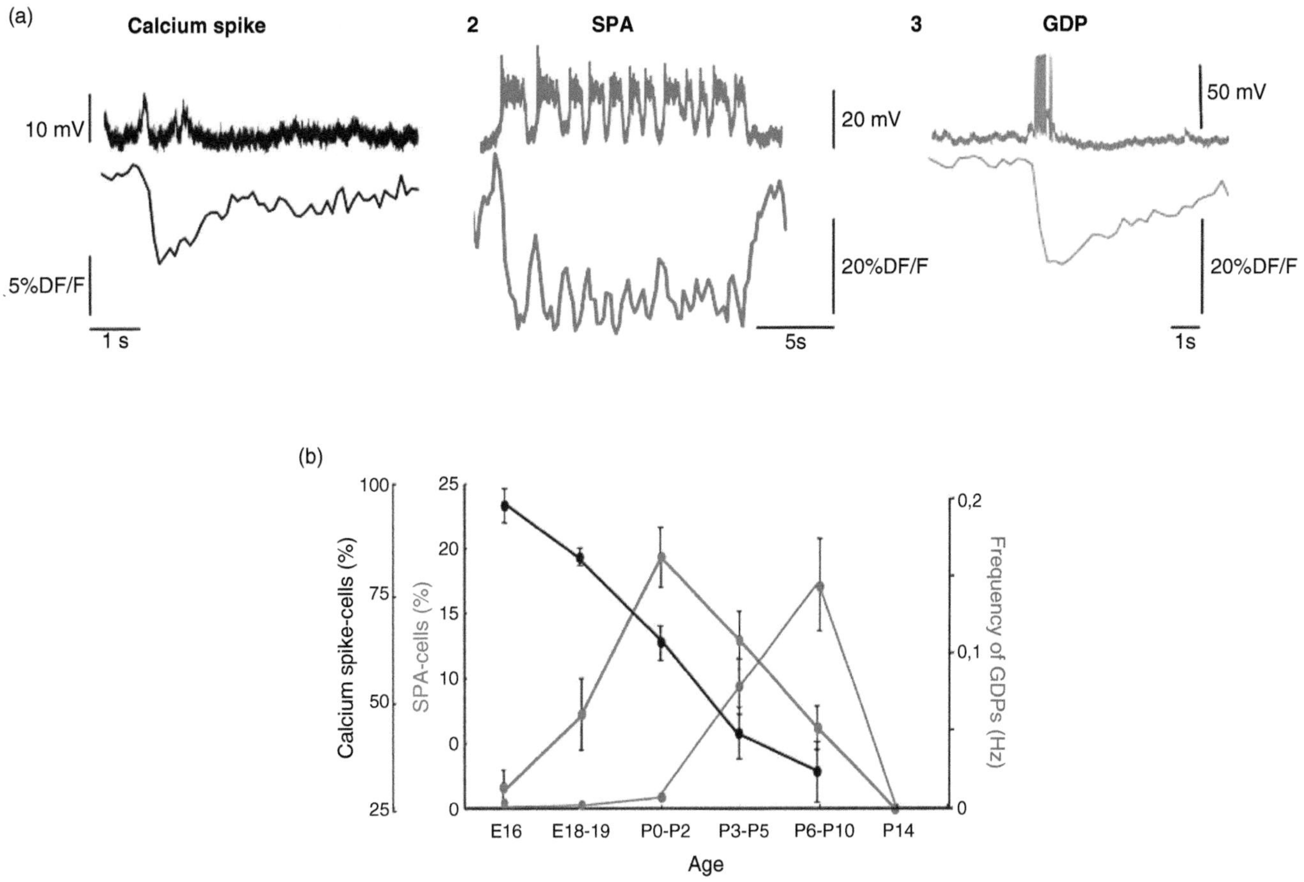

FIGURE 20.14 **Developmental features of the three dominant forms of primary activity in the CA1 region of the developing hippocampus. (a)** Simultaneous current-clamp recording ($V_{rest} = -60$ mV, top) and calcium imaging (bottom trace) of the three types of immature calcium activities in three representative neurons: calcium spikes, SPAs and GDPs. **(b)** Graph indicates the fraction of calcium-spike cells (black), SPA cells (red) relative to the number of active cells, as well as the frequency of GDPs (blue), for six successive age groups. Error bars indicate SEMs. From Crépel V, Aronov D, Jorquera I, Represa A, Ben-Ari Y, Cossart R (2007). A parturition-associated nonsynaptic coherent activity pattern in the developing hippocampus. *Neuron* **54**, 105–120, with permission.

mediate inhibition (i.e. membrane hyperpolarization or silent inhibition). This $GABA_A$-mediated depolarization is strong enough to activate Na^+, Ca^{2+} and NMDA voltage-gated channels and thus to evoke action potentials and Ca^{2+} entry.

$GABA_B$ receptors are not active in the postsynaptic membrane of immature GABAergic synapses owing to uncoupling to G proteins and target K^+ channels. In contrast, $GABA_B$ receptors are functional in the presynaptic membrane where they mediate presynaptic inhibition of GABA release.

As a consequence, $GABA_A$ receptors in the immature hippocampus play the role of AMPA receptors in adult networks, and the only transmitter-mediated synaptic inhibition present in immature hippocampal neurons is a $GABA_B$-mediated presynaptic one. The immature hippocampal network displays spontaneous discharges, owing to $GABA_A$-mediated giant depolarizations, which periodically allow Ca^{2+} entry and transient increases of intracellular Ca^{2+} concentration in hippocampal neurons.

Further reading

Ben-Ari, Y., 2002. Excitatory actions of GABA during development: the nature of the nurture. Nat. Rev. Neurosci. 3, 728–739.

Ben-Ari, Y., Gaiarsa, J.L., Khazipov, R., 2007. A pioneer transmitter that excites immature neurons and generate primitive oscillations. Physiol. Rev. 87, 1215–1284.

Bonifazi, P., Goldin, M., Picardo, M.A., et al., 2009. GABAergic hub neurons orchestrate synchrony in developing hippocampal networks. Science 326, 1419–1424.

Hollrigel, G.S., Ross, S.T., Soltesz, I., 1998. Temporal patterns and depolarizing actions of spontaneous $GABA_A$ receptor activation in granule cells of the early postnatal dentate gyrus. J. Neurophysiol. 80, 2340–2351.

LoTurco, J.J., Owens, D.F., Heath, M.J., et al., 1995. GABA and glutamate depolarize cortical progenitor cells and inhibit DNA synthesis. Neuron 15, 1287–1298.

Menendez de la Prida, L., Bolea, S., Sanchez-Andres, JV., 1998. Origin of the synchronized network activity in the rabbit developing hippocampus. Eur. J. Neurosci. 10, 899–906.

Obrietan, K., van den Pol, A.N., 1998. $GABA_B$ receptor-mediated inhibition of $GABA_A$ receptor calcium elevations in developing hypothalamic neurons. J. Neurophysiol. 79, 1360–1370.

Owens, D.F., Boyce, L.H., Davis, M.B., Kriegstein A.R., 1996. Excitatory GABA responses in embryonic and neonatal cortical slices

demonstrated by gramicidin perforated-patch recordings and calcium imaging. J. Neurosci. 16, 6414–6423.
Owens, D.F., Kriegstein, A.R., 2002. Is there more to GABA than synaptic inhibition? *Nat*. Rev. Neurosci. 3, 715–727.
Rivera, C., Voipio, J., Payne, J.A., et al., 1999. The K^+/Cl^- co-transporter KCC2 renders GABA hyperpolarizing during neuronal maturation. Nature 397, 251–255.
Rohrbough, J., Spitzer, N.C., 1996. Regulation of intracellular Cl^- levels by Na^+-dependent Cl^- cotransport distinguishes depolarizing from hyperpolarizing $GABA_A$ receptor-mediated responses in spinal neurons. J. Neurosci. 16, 82–91.
Tyzio, R., Nardou, R., Ferrari, D.C., et al., 2014. Oxytocin-mediated GABA inhibition during delivery attenuates autism pathogenesis in rodent offspring. Science 343, 675–679.

Contributors

Laurent Aniksztejn (Chapters 14.2.3 and 14.3.4) Directeur de Recherche Inserm, Institut de Neurobiologie de la Méditerranée, Marseille, France

Yehezkel Ben Ari (Chapter 20) Directeur de Recherche Inserm, Institut de Neurobiologie de la Méditerranée, Marseille, France

Elena Avignone (Chapter 2.2) Maitre de Conférence, Institut Interdisciplinaire de Neurosciences, UMR 5297, Université Bordeaux Segalen, Bordeaux, France

Myrian Cayre (Chapter 2.3) Directeur de Recherche CNRS, Institut de Biologie du Développement de Marseille, Marseille, France

Dominique Debanne (Chapters 15, 16, 18) Directeur de Recherche Inserm, Unité de Neurobiologie des canaux ioniques et de la synapse UMR 1072, Faculté de Médecine, Aix-Marseille Université, Marseille, France

Oussama El Far (Chapter 7) Chargé de recherche Inserm, Unité de Neurobiologie des canaux ioniques et de la synapse UMR 1072, Faculté de Médecine, Aix-Marseille Université, Marseille, France

Monique Esclapez, PhD (Appendix 6.2) Chargée de recherche Inserm, Institut de Neurosciences des Systèmes, INSERM U751, Faculté de Medecine Timone, Marseille, France

Laurent Fagni, PhD (Chapter 12) Institut de Génomique Fonctionnelle, Université Montpellier I & II, Montpellier, France

Jean-Luc Gaiarsa (Appendix 18.1) Directeur de Recherche CNRS,Institut de Neurobiologie de la Méditerranée, Marseille, France

Jean-Marc Goaillard (Chapters 15, 16, 18) Directeur de Recherche Inserm, Unité de Neurobiologie des canaux Ioniques et de la Synapse, UMR 1072, Faculté de Médecine, Marseille, France

Constance Hammond (Chapters 1–10, 13–19) Directeur de Recherche Inserm, Institut de Neurobiologie de la Méditerranée, Marseille, France

Roustem Khazipov (Appendix 9.2) Directeur de Recherche Inserm, Institut de Neurobiologie de la Méditerranée, Marseille, France

François Michel (Appendix 5.1) Ingénieur, Université Aix-Marseille II, Directeur de Recherche Inserm, Institut de Neurobiologie de la Méditerranée, Marseille, France

David D. Mott, PhD (Chapter 11) Assistant Professor, Department of Pharmacology, Physiology and Neuroscience, School of Medicine, University of South Carolina, Columbia, SC, USA

Aude Panatier (Chapter 2.1) Chargée de recherche CNRS, Neurocentre Magendie INSERM U862, Physiopathologie de la plasticité neuronale Université Bordeaux 2, Bordeaux, France

Jean-Philippe Pin, PhD (Chapter 12) Directeur de Recherche – CNRS, Institut de Génomique Fonctionnelle, Dépt. de Pharmacologie Moléculaire, UMR 5203 CNRS – U 661 INSERM – Université Montpellier I & II, Montpellier, France

Michael Seagar (Chapter 7) Directeur de recherche Inserm, Unité de Neurobiologie des canaux ioniques et de la synapse UMR 1072, Faculté de Médecine, Aix-Marseille Université, Marseille, France

Index

T

U

V

W

X

Y

CPI Antony Rowe
Chippenham, UK
2015-11-24 21:27

Part III
Using Current Knowledge of the Vitamins

20. Sources of the Vitamins

21. Assessing Vitamin Status

Preface to the Fifth Edition

Understanding the vitamins is key to understanding nutrition. The history of their discovery and the continuing elucidation of their roles in health is the history of the emergence of nutrition as a science from the areas of physiology, biochemistry, medicine, and agriculture.

Capturing the understanding that grew out of that history is both a challenge and a privilege. For us, it involved months of reviewing thousands of publications and looking for clear ways to present complex information without overstating present understanding.

Producing this fifth edition of *The Vitamins* benefitted from the inclusion of a coauthor, which we believe brought a new prospective to the text. James studied the first edition of the *The Vitamins* as a masters student at the University of New Hampshire in 1997. He encountered the second edition of the text as Jerry's student at Cornell University in 2001. We are hopeful that the dynamic relationship we have enjoyed, as student/mentor, colleagues, friends, and now coauthors, has resulted in the most effective edition of this text, as both a reference and a teaching aid.

In writing this fifth edition of *The Vitamins*, we were mindful of comments from users of previous editions, which prompted several changes that we believe enhanced the book. We reorganized several chapters, which reduced their number. We emphasized roles of the gut microbiome in several places of importance. We added sections on biomarkers of vitamin status and modestly expanded the section on biofortification. We added, redrew, and updated several tables and figures. We used extensive footnoting as a means of including explanatory notes as well as for citing primary sources.

We are grateful for the professional assistance from editors, Ms. Jaclyn Truesdell, Ms. Megan Ball, and Ms. Caroline Johnson of Elsevier.

We enjoyed writing this fifth edition of *The Vitamins* together. We hope you will find it useful.

Gerald F. Combs, Jr.
Topsham, Maine
James P. McClung
Westborough, Massachusetts
June 2016

How to Use This Book

TO THE HEALTH PROFESSIONAL

The Vitamins is designed as a one-stop source of comprehensive, current information on the vitamins. In it you will find information on the history of vitamin discovery, the chemical properties of the vitamins and their isomers and metabolites, the utilization and metabolism of vitamins, the consequences of their deficient and excessive intakes, biomarkers of vitamin status, and the health roles if particular vitamins in beyond the traditional deficiencies. You will find examples of classical and current research findings as well as citations to recent key publications in the footnotes. You may find Appendix particularly useful, as it lists the vitamin contents of a most common foods. Please let us know of any ways you see we might enhance *The Vitamins*.

TO STUDENTS AND INSTRUCTORS

The Vitamins is also intended as a teaching text for an upper-level college course within a nutrition or health-related curriculum; however, it will also be useful as a workbook for self-paced study of the vitamins. It has several features that are designed to enhance its usefulness to students as well as instructors. Here is how we suggest using it.

To the student When you use this text, make sure to have by your side a notebook, pencil (not pen—you may want to make changes in the notes you take). Then, before reading each chapter, take a few moments to go over the "Anchoring Concepts and Learning Objectives" on the chapter title page. *Anchoring Concepts* are the ideas fundamental to the subject matter of the chapter, the concepts to which the new ones presented in the chapter will be related. Those in the first several chapters should already be very familiar to you; if not, then it will be necessary for you to do some background reading or discussion until you feel comfortable in your understanding of these basic ideas. You will find that most chapters are designed to build upon the understanding gained through previous chapters; in most cases, the *Anchoring Concepts* of a chapter relate to the *Learning Objectives* of previous chapters. Pay attention to the *Learning Objectives*; they are the key elements of understanding what the chapter is intended to support. Keeping the *Learning Objectives* in mind as you go through each chapter will help you maintain focus on those elements. Next, read through the Vocabulary list and *mark* any terms that are unfamiliar or about which you feel unsure. Then, make a list of *your own questions* about the topic of the chapter.

As you read through the text, look for items related to your questions and for unfamiliar terms. You will be able to find key terms in bold-faced type, and you should be able to get a good feel for their meanings from the contexts of their uses. If this is not sufficient for any particular term, then look it up in a medical dictionary. Do not wait to do this. Cultivate the habit of being bothered by not understanding something—this will help you enormously in years to come.

As you proceed through the text, note what information the layout is designed to convey. First, note that the major sections of each chapter are indicated with a bold heading. This is done to help you *scan* for particular information. Also note that the footnoted information is largely supplementary and not essential to the understanding of the key concepts presented. Therefore, the text may be read at two levels: at the basic level, one should be able to ignore the footnotes and still get the key concepts; at the more detailed level, one should be able to pick up more background, particularly key citations to the primary literature, from the footnotes. Refer back frequently to your own list of questions and "target" vocabulary words; when you find an answer or can make a deduction, make a note. Do not be reluctant to write in the book, particularly to put a concept into your own words, or to note something you find important or do not fully understand. Studies show that to be an effective learning technique.

When you have completed a chapter, take sometime to list what you see as the key points—those that you would cover in a formal presentation. Then, skim back over the chapter.

You will find that Chapters 6–19 each have one or more Case Studies comprised of more clinical case reports abstracted from the medical literature. For each, use the associated questions to focus your thinking on the features that relate to vitamin functions. As you do so, try to ignore the obvious connection with the subject of the chapter; put yourself in the position of the attending physician who was called upon to diagnose the problem without prior knowledge that it involved any particular nutrient, much less a certain vitamin. The Case Study in Chapter 21 is different; it is a fictional but highly plausible scenario that calls for a nonobvious decision. Additional case studies are listed in Appendix B.

Take sometime and go through the Study Questions and Exercises at the end of each chapter. These, too, are designed to direct your thinking back to the key concepts of the respective chapter and to facilitate integration of those concepts with those you already have.

We have made a point in Chapter 1 of using the technique of *concept mapping* do demonstrate the integration of complex subject matter. We have found the *concept map* to be a powerful teaching/learning tool. If you have had no previous experience with this device, then it will be worth your while to consult *Learning How to Learn.*[1]

When you have done all of this for a chapter, then deal with your questions. Discuss them with fellow students or look them up. To assist you in the latter, a short reading list is included at the end of each chapter. With the exception of Chapter 2, which lists papers of landmark significance to the discovery of the vitamins, the reading lists consist of key reviews in prominent scientific journals. These reviews and the papers cited in the footnotes will help you find primary research papers on topics of specific interest.

After you have followed all of these steps, *reread the chapter*. You will find this last step to be extraordinarily useful in gaining a command of the material.

Last, but certainly not least, have *fun* with this fascinating aspect of the field of nutrition!

To the instructor The format of this text reflects the way GFC taught a course called "The Vitamins" for some 29 years at Cornell University. To that end, some experiences in using *The Vitamins* as a text for my course may be of interest to you.

I have found that *every* student comes to the study of the vitamins with *some* background knowledge of the subject, although those backgrounds are generally incomplete and frequently include areas of misinformation. This is true for upper-level nutrition majors and for students from other fields, the difference being largely one of magnitude. This is also true for instructors, most of whom come to the field with specific expertise that relates to only a subset of the subject matter.

You can demonstrate this in the following exercise, best done of the first day of class. Raise your index finger (best done with a bit of dramatic flair) and say "vitamin A." Hold that pose for 10 s and then ask "*What came to mind when I said 'vitamin A'?*" Without fail, someone will say "vision" or "carrots," and then an older graduate student may add "toxic." When it looks safe to chime in, others will add what will build to an array of descriptors that, collectively, are more relevant to vitamin A than any is individually. Most of the answers, by far, will relate to the clinical symptoms of vitamin A deficiency and the sources of vitamin A in diets. Catch each answer by dashing it on to a large sticky note and then stick the note haphazardly to a blackboard or wall. If you hear something complex or a cluster of concepts, make sure to question the contributor until you hear one or more individual concepts, which you can record on individual sticky notes. This approach *never fails* to stimulate further answers, and it is common that a group of 15–20 students will generate a list of twice that number of concepts before the momentum fades. Having used sticky notes, it is easy to move them into clusters and, thus, to use the activity to construct a *concept map* of "Vitamin A" based solely on the knowledge that the students, collectively, brought into the room. This exercise can demonstrate an empowering idea that, having at least *some* background on the subject and being motivated (by any of a number of reasons) to learn more, *every* learner brings to the study of the vitamins a unique perspective which may not be readily apparent.

We are convinced that meaningful learning is served when both instructor and students come to understand each others' various perspectives. This has two benefits in teaching the vitamins. First, it is in the instructor's interest to know the students' ideas and levels of understanding concerning issues of vitamin need, vitamin function, etc., such that these can be built upon and modified as may be appropriate. Second, many upper-level students have interesting experiences (through personal or family histories, their own research, information from other courses, etc.) that can be valuable contributions to classroom discussions. These experiences are assets that can reduce the temptation to fall back on the "instructor knows all" notion, which we all know to be false. To identify student perspectives, it is useful to assign on the first class period, for submission at the second class, a written autobiographical sketch. Distribute your own as a model, and ask each student to write "as much or as little" as he or she cares to, recognizing that you will distribute to the class copies of whatever is submitted. The biographical sketches will range from a few sentences that reveal little of a personal nature to longer ones that provide many good insights about their authors; *everyone* will help you to get to know your students personally and to get a better idea of their understandings of the vitamins and of their expectations of the course. The exercise serves the students in a similar manner, thus promoting a group dynamic that facilitates classroom discussions.

The Vitamins can be used as a typical text from which you can make regular reading assignments as preparation for each class. This will free you of the need for lecturing in favor of an open discussion format. In fact, this approach allows more information to be covered, as even a brilliant lecturer simply cannot cover the vitamins in any real depth

1. Novak, J.D., Gowin, D.B., 1984. Learning How to Learn. Cambridge, University Press, New York, NY, pp. 199

within the limits of traditional class periods. This was the original motivation for putting that information into this text, which has allowed shifting responsibility for learning to the student to glean from assigned reading. This allows class time to be used to facilitate learning through discussions of issues of student interest or concern. Often, this means that certain points were not clear upon reading or that the reading itself stimulated questions not specifically addressed in the text. Usually, these questions are nicely handled by eliciting the views and understandings of other students and by your giving supplementary information.

With this approach, the instructor's class preparation involves the collation of research data that will supplement the discussion in the text, and the identification of questions that can initiate discussions. In developing questions, it may be useful to prepare your own concept maps of the subject matter and to ask rather simple questions about the linkages between concepts, e.g., "*How does the mode of enteric absorption of the tocopherols relate to what we know about its physiochemical properties?*" If you are unfamiliar with concept mapping, then consult "*Learning How to Learn*" and experiment with the technique to determine whether it can assist you in your teaching.

The Study Questions and Exercises or Case Studies can be used to give weekly written assignments to keep students focused on the topic and prevent them from letting the course slide until exam time. More importantly, there is learning associated with the thought that necessarily goes into such written assignments. To support that learning, make a point of going over each assignment briefly at the beginning of the class at which it is due and return it by the *next* class with your written comments. You will find that the *Case Studies* are abstracted from actual clinical reports; students enjoy and do well on these assignments.

The model we used in teaching *The Vitamins* at Cornell was to evaluate student's performance on the basis of class participation, weekly written assignments, a review of a recent research paper, and either one or two examinations. To allow each student to pursue a topic of specific individual interest, students were asked to review a research paper published within the last year, using the style of *Nutrition Reviews*. Students were asked to make a short (10 min) presentation of each in class. Their reviews were evaluated on the basis of critical analysis and on the importance of the paper to the field. This assignment was also well received. Because many students are inexperienced in research and will, thus, feel uncomfortable in criticizing it, it is helpful to conduct in advance a discussion of the general principles of experimental design and statistical inference. Exams were also concept-oriented: students were given brief case descriptions and actual experimental data, and were asked to lay out diagnostic strategies, develop hypotheses, design means of hypothesis testing and interpretation of results, etc. Many students may prefer the more familiar short-answer test; such inertia can be overcome by using examples in class discussions and or homework assignments.

The Vitamins was been of great value in enhancing the teaching of the course by that name at Cornell. Thus, it is our sincere wish that it will assist you similarly in your teaching. Please let us know how it meets your needs and how we might enhance it for that purpose.

Gerald F. Combs, Jr.
James P. McClung

Part I

Perspectives on the Vitamins in Nutrition

Chapter 1

What Is a Vitamin?

Chapter Outline

Anchoring Concepts

1. Certain factors, called *nutrients*, are necessary for normal physiological function of animals, including humans. Some nutrients cannot be synthesized adequately by the host and must therefore be obtained from the external chemical environment; these are referred to as *dietary essential nutrients*.
2. *Diseases* involving physiological dysfunction, often accompanied by morphological changes, can result from insufficient intakes of dietary essential nutrients.

Imagination is more important than knowledge.

A. Einstein

LEARNING OBJECTIVES

1. To understand the classic meaning of the term *vitamin* as it is used in the field of nutrition.
2. To understand that the term *vitamin* describes both a concept of fundamental importance in nutrition as well as any member of a rather heterogeneous array of nutrients, any one of which may not fully satisfy the classic definition.
3. To understand that some compounds are vitamins for one species and not another, and that some are vitamins only under specific dietary or environmental conditions.
4. To understand the concepts *vitamer* and *provitamin*.

VOCABULARY

Vitamer
Vitamin
Provitamin

1. THINKING ABOUT VITAMINS

Among the nutrients required for the many physiologic functions essential to life are the vitamins. Unlike other nutrients, the vitamins do not serve structural functions, nor does their catabolism provide significant energy. Instead, the physiologic functions of vitamins are highly specific, and, for that reason, they are required in only small amounts in the diet. The common food forms of most vitamins require some metabolic activation to their functional forms.

Although the vitamins share these general characteristics, they show few close chemical or functional similarities; their categorization as vitamins is strictly empirical. Consider also that, whereas several vitamins function as enzyme cofactors (vitamins A, K, and C; thiamin; niacin; riboflavin; vitamin B_6; biotin; pantothenic acid; folate; and vitamin B_{12}), not all enzyme cofactors are vitamins.[1] Some vitamins function as biological antioxidants (vitamins E and C), and several function as cofactors in metabolic oxidation–reduction reactions (vitamins E, K, and C; niacin; riboflavin; and pantothenic acid). Two vitamins (vitamins A and D) function as hormones; one of them (vitamin A) also serves as a photoreceptive cofactor in vision.

2. VITAMIN: A REVOLUTIONARY CONCEPT

Everyday Word or Revolutionary Idea?

The term *vitamin*, today a common word in everyday language, was born of a revolution in thinking about the interrelationships of diet and health that occurred at the

1. Other enzyme cofactors are biosynthesized, e.g., heme, coenzyme Q, and lipoic acid.

The Vitamins. http://dx.doi.org/10.1016/B978-0-12-802965-7.00001-0

beginning of the 20th century. That revolution involved the growing realization of two phenomena that are now taken for granted, even by the nonscientist:

1. Diets are sources of many important nutrients.
2. Insufficient intakes of specific nutrients can cause certain diseases.

In today's world each of these concepts may seem self-evident, but in a world still responding to and greatly influenced by the important discoveries in microbiology made in the 19th century, each represented a major departure from contemporaneous thinking in the area of health. Nineteenth-century physiologists perceived foods and diets as sources of only four types of nutrients: protein, fat, carbohydrate, ash,[2] and water. After all, these accounted for very nearly 100% of the mass of most foods. With this view, it is understandable that, at the turn of the century, experimental findings that now can be seen as indicating the presence of hitherto unrecognized nutrients were interpreted instead as substantiating the presence of natural antidotes to unidentified disease-causing microbes.

Important discoveries in science have ways of directing, even entrapping, one's view of the world; resisting this tendency depends on critical and constantly questioning minds. That such minds were involved in early nutrition research is evidenced by the spirited debates and frequent polemics that ensued over discoveries of apparently beneficial new dietary factors. Still, the systematic development of what emerged as nutritional science depended on a new intellectual construct for interpreting such experimental observations.

Vitamin or Vitamine?

The elucidation of the nature of what was later to be called *thiamin* occasioned the proposition of just such a new construct in physiology.[3] Aware of the impact of what was a departure from prevailing thought, its author, the Polish biochemist Casimir Funk, chose to generalize from his findings on the chemical nature of that "vital amine" to suggest the term **vitamine** as a generic descriptor for many such **accessory factors** associated with diets. That the factors soon to be elucidated comprised a somewhat chemically heterogeneous group, not all of which were nitrogenous, does not diminish the importance of the introduction of what was first presented as the *vitamine theory*, later to become a key concept in nutrition: the vitamin.

The term vitamin has been defined in various ways. While the very concept of a vitamin was crucial to progress in understanding human physiology and nutrition, the actual definition of a vitamin has evolved in consequence of that understanding.

2. The residue from combustion, i.e., minerals.
3. This is a clear example of what T.H. Kuhn called a "scientific revolution" (Kuhn, T.H., 1968. *The Structure of Scientific Revolutions*. University of Chicago Press, Chicago, IL.), i.e., the discarding of an old paradigm with the invention of a new one.

3. AN OPERATING DEFINITION OF A VITAMIN

A vitamin is defined as follows (Fig. 1.1). A vitamin

- is an *organic compound* distinct from fats, carbohydrates, and proteins
- is a *natural component of foods* in which it is usually present in minute amounts
- is essential, also usually in minute amounts, for *normal physiological function* (i.e., maintenance, growth, development, and/or production)
- prevents a *specific deficiency syndrome*, which occurs when it is absent or underutilized
- is *not synthesized by the host* in amounts adequate to meet normal physiological needs.

This definition will be useful in the study of vitamins, as it effectively distinguishes this class of nutrients from others (e.g., proteins and amino acids, essential fatty acids, and minerals) and indicates the needs in various normal physiological functions. It also denotes the specificity of deficiency syndromes by which the vitamins were discovered. Further, it places the vitamins in that portion of the external chemical environment on which animals (including humans) must depend for survival, thus distinguishing vitamins from hormones.

Some Caveats

It will quickly become clear, however, that, despite its utility, this operating definition has limitations, notably with respect to the last clause. Many species can, indeed, synthesize at least some of the vitamins, although not always at the levels required to prevent deficiency disorders. Four examples illustrate this point:

Vitamin C: Most animal species have the ability to synthesize ascorbic acid. Only those few that lack the enzyme L-gulonolactone oxidase (e.g., the guinea pig, humans) cannot. For those species, ascorbic acid is properly be called vitamin C.

Vitamin D: Individuals exposed to modest amounts of sunlight can produce cholecalciferol, which functions as a hormone. Only individuals without sufficient exposure to ultraviolet light (e.g., livestock raised in indoor confinement, people spending most of their days indoors) require dietary sources of vitamin D.

Choline: Most animal species have the metabolic capacity to synthesize choline; however, some (e.g., the chick, the rat) may not be able to employ that capacity if they are fed insufficient amounts of methyl donor compounds. In

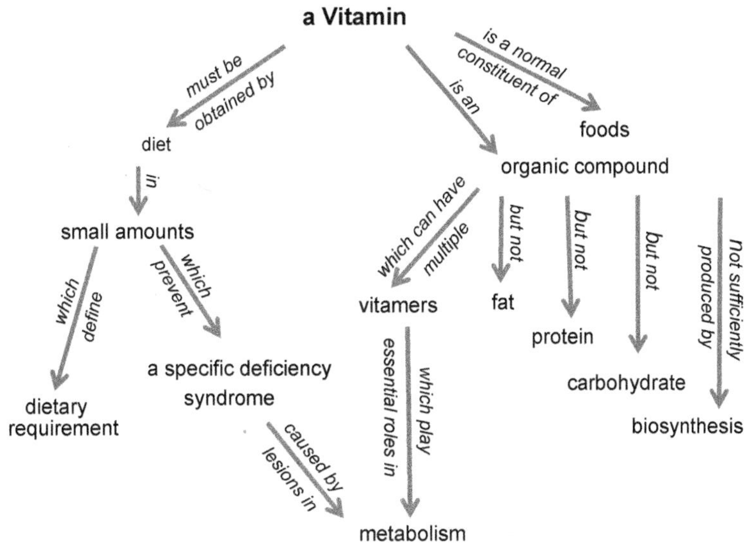

FIGURE 1.1 Concept map of a Vitamin.[4]

addition, some (e.g., the chick) do not develop that capacity completely until they are several weeks of age. Thus, for the young chick and for individuals of other species fed diets providing limited methyl groups, choline is a vitamin.

Niacin: All animal species can synthesize nicotinic acid mononucleotide from the amino acid tryptophan. Only those for which this metabolic conversion is particularly inefficient (e.g., the cat, fishes) and others fed low dietary levels of tryptophan require a dietary source of *niacin*.

With these counterexamples in mind, the definition of a vitamin has specific connotations for animal species, stage of development, diet or nutritional status, and physical environmental conditions.[5]

The "Vitamin Caveat"

- Some compounds are vitamins for one species and not another.
- Some compounds are vitamins only under specific dietary or environmental conditions.

4. THE RECOGNIZED VITAMINS

Thirteen substances or groups of substances are now generally recognized as vitamins (Table 1.1); others have been proposed.[6] In some cases, the familiar name is actually the generic descriptor for a family of chemically related compounds having qualitatively comparable metabolic activities. For example, the term *vitamin E* refers to those analogs of tocol or tocotrienol[7] that are active in preventing such syndromes as fetal resorption in the rat and myopathies in the chick. In these cases, the members of the same vitamin family are called *vitamers*. Some carotenoids can be metabolized to yield the metabolically active form of vitamin A; such a precursor of an actual vitamin is called a **provitamin**.

5. STUDY QUESTIONS AND EXERCISES

1. What are the key features that define a vitamin?
2. What are the fundamental differences between vitamins and other classes of nutrients… between vitamins and hormones?
3. Detail, citing a specific example, a situation in which a vitamin may be nutritionally essential for one species but not another.
4. Using key words and phrases, list briefly what you know about each of the recognized vitamins.

4. The concept map can be a useful device for organizing thought, as its discipline can serve to assist in identifying the nature and extent of concepts related to the one in question. A concept map should be laid out as a hierarchy of related concepts with the superordinate concept at the top and all relationships between concepts identified with a verb phrase. Thus, it can be "read" from top to bottom. One of the authors (GFC) has used concept mapping in graduate-level teaching, both as a group exercise and testing device. For a useful discussion of the educational value of the concept map, the reader is referred to *Learning How to Learn*, 1984, J.D. Novak and D.B. Gowin, Cornell University Press, Ithaca, NY, pp. 199.

5. For this reason, it is correct to refer to vitamin C for the nutrition of humans but ascorbic acid for the nutrition of livestock.

6. These include such factors as inositol, carnitine, bioflavonoids, pangamic acid, and laetrile, for some of which there is evidence of vitamin-like activity (Chapter 19).

7. Tocol is 3,4-dihydro-2-methyl-2-(4,8,12-trimethyltridecyl)-6-chromanol; tocotrienol is the analog with double bonds at the 3, 7, and 11′ positions on the phytol side chain (Chapter 7).

TABLE 1.1 The Vitamins: Their Vitamers, Provitamins, and Functions

Group	Vitamers	Provitamins	Physiological functions
Vitamin A	Retinol Retinal Retinoic acid	β-Carotene Cryptoxanthin	Visual pigments; epithelial cell differentiation
Vitamin D	Cholecalciferol (D_3) Ergocalciferol (D_2)		Calcium homeostasis; bone metabolism; transcription factor
Vitamin E	α-Tocopherol γ-Tocopherol		Membrane antioxidant
Vitamin K	Phylloquinones (K_1) Menaquinones (K_2) Menadione (K_3)		Blood clotting; calcium metabolism
Vitamin C	Ascorbic acid Dehydroascorbic acid		Reductant in hydroxylations in the formation of collagen and carnitine, and in the metabolism of drugs and steroids
Vitamin B_1	Thiamin		Coenzyme for decarboxylations of 2-keto acids (e.g., pyruvate) and transketolations
Vitamin B_2	Riboflavin		Coenzyme in redox reactions of fatty acids and the tricarboxylic acid (TCA) cycle
Niacin	Nicotinic acid Nicotinamide		Coenzyme for several dehydrogenases
Vitamin B_6	Pyridoxol Pyridoxal Pyridoxamine		Coenzyme in amino acid metabolism
Folic acid	Folic acid Polyglutamyl folacins		Coenzyme in single-carbon metabolism
Biotin	Biotin		Coenzyme for carboxylations
Pantothenic acid	Pantothenic acid		Coenzyme in fatty acid metabolism
Vitamin B_{12}	Cobalamin		Coenzyme in the metabolism of propionate, amino acids, and single-carbon units

Chapter 2

Discovery of the Vitamins

Chapter Outline

Anchoring Concepts

1. A scientific theory is a plausible explanation for a set of observed phenomena; because theories cannot be tested directly, their acceptance relies on a preponderance of supporting evidence.
2. A scientific hypothesis is a tentative supposition that is assumed for the purposes of argument or testing and is thus used in the generation of evidence by which theories can be evaluated.
3. An empirical approach to understanding the world involves the generation of theories strictly by observation, whereas an experimental approach involves the undertaking of operations (experiments) to test the truthfulness of hypotheses.
4. Physiology is that branch of biology seeks to elucidate the processes, activities, and phenomena of life and living organisms, while biochemistry seeks to elucidate the molecular bases for such phenomena.
5. The field of nutrition is derived from both of these disciplines; it seeks to elucidate the processes by which animals or plants take in and utilize food substances.

When science is recognized as a framework of evolving concepts and contingent methods for gaining new knowledge, we see the very human character of science, for it is creative individuals operating from the totality of their experiences who enlarge and modify the conceptual framework of science.

J.D. Novak.[1]

1. Joseph D. Novak (b. 1932) is a prominent American educator known for his research on human learning, knowledge creation, and knowledge representation. Prof. Novak, spent most of his career at Cornell University where he and his colleagues developed the technique of Concept Mapping as a means of representing science knowledge.

LEARNING OBJECTIVES

1. To understand the nature of the process of discovery in the field of nutrition.
2. To recognize the major forces in the emergence of **nutrition science**.
3. To understand the impact of the **vitamine theory**, as an intellectual construct, on that process of discovery.
4. To understand that the discoveries of the vitamins proceeded along indirect lines, most often through the seemingly unrelated efforts of many people.
5. To recognize the key events in the discovery of each of the vitamins.
6. To become familiar with the basic terminology of the vitamins and their associated deficiency disorders.

VOCABULARY

Accessory factor
Anemia
Animal model
Animal protein factor
Ascorbic acid
β-Carotene
Beriberi
Biotin
Black tongue disease
Cholecalciferol
Choline
Dermatitis
Ergocalciferol
Fat-soluble A
Filtrate factor
Flavin

The Vitamins. http://dx.doi.org/10.1016/B978-0-12-802965-7.00002-2

Folic acid
Germ theory
Hemorrhage
Lactoflavin
Niacin
Night blindness
Ovoflavin
Pantothenic acid
Pellagra
Polyneuritis
Prothrombin
Provitamin
Purified diet
Pyridoxine
Retinen
Riboflavin
Rickets
Scurvy
Thiamin
Vitamin A
Vitamin B
Vitamin B complex
Vitamin B_{12}
Vitamin B_2
Vitamin B_6
Vitamin C
Vitamin D
Vitamin E
Vitamin K
Vitamine
Vitamine theory
Water-soluble B
Xerophthalmia

1. THE EMERGENCE OF NUTRITION AS A SCIENCE

In the span of only five decades commencing at the very end of the 19th century, the vitamins were discovered. Their discoveries were the result of the activities of hundreds of people that can be viewed retrospectively as having followed discrete branches of intellectual progress. Those branches radiated from ideas originally derived inductively from observations in the natural world, each starting from the recognition of a relationship between diet and health. Subsequently, branches were pruned through repeated analysis and deduction—a process that both produced and proceeded from the fundamental approaches used in experimental nutrition today. Once pruned, the limb of discovery may appear straight to the naive observer. Scientific discovery, however, does not occur that way; rather, it tends to follow a zigzag course, with many participants contributing many branches. In fact, the contemporaneous view of each participant may be that of a thicket of tangled hypotheses and facts. The seemingly straightforward appearance of the emergent limb of discovery is but an illusion achieved by discarding the dead branches of false starts and unsupported hypotheses, each of which can be instructive about the process of scientific discovery.

With the discovery of the vitamins, therefore, nutrition moved from a largely observational activity to one that relied increasingly on hypothesis testing through experimentation; it moved from empiricism to science. Both the process of scientific discovery and the course of the development of nutrition as a scientific discipline are perhaps best illustrated by the history of the discovery of the vitamins.

2. THE PROCESS OF DISCOVERY IN NUTRITIONAL SCIENCE

Empiricism and Experiment

History demonstrates that the process of scientific discovery begins with the synthesis of general ideas about the natural world from observations of particulars within it—i.e., an *empirical phase*. In the discovery of the vitamins, this initial phase was characterized by the recognition of associations between diet and human diseases, namely night blindness, scurvy, beriberi, rickets, and pellagra, each of which was long prevalent in various societies. The next phase in the process of discovery involved the use of these generalizations to form hypotheses that could be tested experimentally—i.e., the *experimental phase*. In the discovery of the vitamins, this phase necessitated the development of two key tools of modern experimental nutrition: the **animal model** and the **purified diet**. The availability of both of these tools proved to be necessary for the discovery of each vitamin; in cases where an animal model was late to be developed (e.g., for pellagra), the elucidation of the identity of the vitamin was substantially delayed.

3. THE EMPIRICAL PHASE OF VITAMIN DISCOVERY

The major barrier to entering the empirical phase of nutritional inquiry proved to be the security provided by prescientific attitudes about foods that persisted through the 19th century. Many societies had observed that human populations in markedly contrasting parts of the world tended to experience similar health standards despite the fact that they subsisted on very different diets. These observations were taken by 19th-century physiologists to indicate that health was not particularly affected by the kinds of foods consumed. Foods were thought important as sources of the only nutrients known at the time: *protein*, *available energy*, and *ash*. While the "chemical revolution," led by the French scientist Antoine Lavoisier,[2] started probing the elemental components and metabolic fates of these nutrients, the widely read ideas of the German chemist Justus von Liebig[3] resulted in protein being recognized as the only essential nutrient, supporting both tissue growth and repair as well as energy production. In the middle part of the century, attention was drawn further from potential relationships of diet and health by the major discoveries of Pasteur,[4] Liebig,[5] Koch,[6] and others in microbiology. For the first time, several diseases, first anthrax and then others, could be understood in terms of a microbial etiology. By the end of the century, germ theory, which proved to be of immense value in medicine, directed hypotheses for the etiologies of most diseases. The impact of this understanding as a barrier to entering the inductive phase of nutritional discovery is illustrated by the case of the Dutch physician Christiaan Eijkman,[7] who found a water-soluble factor from rice bran to prevent a beriberi-like disease in chickens (now known to be the vitamin thiamin) and concluded that he had discovered a "pharmacological antidote" against the beriberi "microbe" presumed to be present in rice.

Diseases Linked to Diet

Nevertheless, while they appeared to have little effect on the prevailing views concerning the etiology of human disease, by the late 1800s empirical associations had been made between diet and the diseases scurvy, rickets, pellagra, and night blindness.

Scurvy has been known that scurvy, the disease involving apathy, weakness, sore gums, painful joints, and multiple hemorrhages, could be prevented by including in the diet green vegetables or fruits. Descriptions of cases in such sources as the Eber papyrus (c.1150 BCE) and writings of Hippocrates (c.420 BCE) are often cited to indicate that scurvy was prevalent in those ancient populations. Indeed, signs of the disease are said to have been found in the skeletal remains of primitive humans. Scurvy was common in northern Europe during the Middle Ages, a time when local agriculture provided few sources of vitamin C that lasted through the winter. In northern Europe, it was treated by eating cresses and spruce leaves. Scurvy was very highly prevalent among seamen, particularly those on ocean voyages to Asia during which they subsisted for months at a time on dried and salted foods. The Portuguese explorer Vasco da Gama reported losing more than 60% of his crew of 160 sailors in his voyage around the Cape of Good Hope in 1498. In 1535–1536, the French explorer Jacques Cartier reported that signs of scurvy were present in all but three of his crew of 103 men (25 of whom died) during his second Newfoundland expedition. In 1595–1597, the first Dutch East Indies fleet lost two-thirds of its seamen due to scurvy. In 1593, the British admiral Richard Hawkins wrote that, during his career, he had seen some 10,000 seamen die of the disease.

The link between scurvy and preserved foods was long evident to seafarers. The first report of a cure for the disease appears to have been Cartier's description of the rapidly successful treatment of his crew with an infusion of the bark of Arborvitae (*Thuja occidentalis*) prepared by the indigenous Hurons of Newfoundland. By 1601, the consumption of berries, vegetables, scurvy-grass (*Cochlearia officinalis*, which contains as much ascorbic acid as orange juice), and citrus fruits or juices was recognized as effective in preventing the disease. In that year, the English privateer Sir James Lancaster introduced regular issues of lemon juice (three spoonfuls each morning) on one of his found ships, finding significantly less scurvy among treated sailors. Nevertheless, the prestigious London College of Physicians viewed scurvy as a "putrid" disease in which affected tissues became alkaline and stated that other acids could be as effective as lemon juice in treating the disease. Accordingly, in the mid-1600s British ship's surgeons were supplied with vitriol (dilute sulfuric acid).

Against this background, in 1747, James Lind, a Scottish physician serving in the British Royal Navy, conducted what has been cited as the first controlled clinical trial to compare various therapies recommended for scurvy in British sailors at sea. Lind's report, published 6 years later, described 12 sailors with scurvy whom he assigned in pairs to 2-week regimens including either lemons and oranges, vitriol, vinegar, or other putative remedies. His results were

2. Antoine-Laurent de Lavoisier (1743–1794) is often considered the "father of modern chemistry", as his work changed that science from a qualitative to a quantitative one. He is best known for his discovery of oxygen and its role in combustion.

3. In his widely read book, *Animal Chemistry, or Organic Chemistry in its Application to Physiology and Pathology*, Liebig argued that the energy needed for the contraction of muscles, in which he was able to find no carbohydrate or fat, must come only from the breakdown of protein. Protein, therefore, was the only true nutrient.

4. Louis Pasteur (1822–1895) was a French pioneering microbiologist. He disproved the doctrine of "spontaneous generation" of microbial life and advanced "germ theory." He discovered the principles of vaccination, fermentation and developed the process of heat-killing of microbes in liquids is now called "pasteurization".

5. Justus von Liebig (1803–1873) was a German chemist who made major contributions to agricultural and biological chemistry, elucidated the importance of nitrogen in plant nutrition, and introduced laboratory experience in teaching chemistry.

6. Robert Koch (1843–1910) was a German physician who identified the causative agents of tuberculosis, cholera and anthrax, and formulated the general principles ("Koch's Postulates") for linking specific microorganisms to specific diseases. In 1905, he received the Nobel Prize for Physiology or Medicine.

7. Christiaan Eijkman (1858–1930) was trained in the Netherlands and served as a medical officer in the Dutch Indies. After contracting malaria in 1885, he returned to Amsterdam where he worked in the laboratories of Forster and, then, Kock (Berlin). In Koch's laboratory he met another Dutch physician C.A. Pekelharing whom he assisted in a second period of service in the Indies investigating beriberi. They proposed establishing a medical laboratory of which Eijkman was named director and Director of the Javanese Medical School, which ultimately became the University of Indonesia.

clear: the pair treated with lemons and oranges recovered almost completely within 6 days; whereas, no other treatment resulted in any improvement. In 1753, he published his now-classic work "A Treatise on Scurvy," which had great impact on the medical thought of the time, as it detailed past work on the subject (most of which was anecdotal) and also presented the results of his experiments. Lind believed that citrus contained "a saponaceous, attenuating and resolving virtue" that helped free skin perspiration that had become clogged by sea air; however, his results were taken as establishing the value of fresh fruits in treating the disease. Still, it was not until the 1790s that the British Navy had made it a regular practice to issue daily rations of lemon juice to all seamen—a measure that gave rise to the term "limey"[8] as a slang expression for a British seaman. In the early part of the 19th century, there remained no doubt of a dietary cause and cure of scurvy; even so, it would be more than a century before its etiology and metabolic basis would be elucidated. Outbreaks of scurvy continued in cases of food shortages: in British prisons, during the California gold rush, among troops in the Crimean War, among prisoners in the American Civil War, among citizens during the Siege of Paris in 1871, and among polar explorers in the early 20th century.

It is said that signs consistent with **beriberi** (e.g., initial weakness and loss of feeling in the legs leading to heart failure, breathlessness, and, in some cases, edema) are described in ancient Chinese herbals (~2600 BCE). Certainly, beriberi was an historic disease prevalent in many Asian populations subsisting on diets in which polished (i.e., "white" or dehulled) rice is the major food. For example, in the 1860s, the Japanese navy experienced the disease affecting 30–40% of its seamen. Interesting clinical experiments conducted in the 1870s with sailors by Dr Kanehiro Takaki, a British trained surgeon who later became Director General of the Japanese Naval Medical Service, first noted an association between beriberi and diet: Japanese sailors were issued lower protein diets than their counterparts in European navies, which had not experienced the disease. Takaki conducted an uncontrolled study at sea in which he modified sailors' rations to increase protein intake by including more meat, condensed milk, bread, and vegetables at the expense of rice. This cut both the incidence and severity of beriberi dramatically, which he interpreted as confirmation of the disease being caused by insufficient dietary protein. The adoption of Takaki's dietary recommendations by the Japanese navy was effective—eliminating the disease as a shipboard problem by 1880—despite the fact that his conclusion, reasonable in the light of contemporaneous knowledge, later proved to be incorrect.

Rickets, the disease of growing bones, presents in children as deformations of the long bones (e.g., bowed legs, knock knees, and curvatures of the upper and/or lower arms), swollen joints, and/or enlarged heads. It is generally associated with the urbanization and industrialization of human societies. Its appearance on a wide scale was more recent and more restricted geographically than that of either scurvy or beriberi. The first written account of the disease is believed to be that of Daniel Whistler,[9] who wrote on the subject in his medical thesis in 1645. A complete description of the disease was published shortly thereafter (in 1650) by the Cambridge professor Francis Glisson,[10] so it is clear that by the middle of the 17th-century rickets had become a public health problem in England. However, rickets appears not to have affected earlier societies, at least not on such a scale. Studies in the late 1800s by the Scottish physician T.A. Palm[11] showed that the mummified remains of Egyptian dead bore no signs of the disease. By the latter part of the century, the incidence of rickets among children in London exceeded one-third; by the turn of the century, estimates of prevalence were as high as 80% and rickets had become known as the "English disease." Noting the absence of rickets in southern Europe, Palm in 1888 was the first to point out that rickets was prevalent only where there is relatively little sunlight (e.g., in the northern latitudes). He suggested that sunlight exposure prevented rickets, but others held that the disease had other causes—e.g., heredity or syphilis. Through the turn of the century, much of the Western medical community remained either unaware or skeptical of a food remedy that had long been popular among the peoples of the Baltic and North Sea coasts, and that had been used to treat adult rickets in the Manchester Infirmary by 1848: cod liver oil. Not until the 1920s would the confusion over the etiology of rickets become clear.

Pellagra, the disease characterized by lesions of the skin and mouth, and by gastrointestinal and mental disturbances, also became prevalent in human societies fairly recently. There appears to have been no record of the disease, even in folk traditions, before the 18th century. Its first documented description, in 1735, was that of the Spanish physician Gaspar Casal. His observations were disseminated by the French physician François Thiery, whom he met some years later after having been appointed as physician to the court of King Philip V. In 1755, Thiery published a brief account of Casal's observations in the *Journal de Vandermonde*; this became the first published report on the

8. That lemons were often called *limes* has been a source of confusion to many writers on this topic.

9. Whistler (1619–1684) was an English physician. His thesis at the Royal College of Physicians was the first printed book on rickets.

10. Francis Glisson (1599–1677) was a British physician and anatomist who wrote a text on pediatric rickets.

11. Theobold A. Palm (1849–?) was a Scottish physician born to missionary parents in Ceylon. After studying medicine at Edinburgh University, he served as a medical missionary in Japan, where he noted the absence of rickets, in marked contrast to the prevalence of that condition he found in Britain on his return in 1884. In 1888, he commented on Britain's "want of light" in a letter to the British Medical Journal in which he went on to recommend "the systematic use of sunbaths" as a rickets therapy.

disease. Casal's own description was included in his book on the epidemic and endemic diseases of northern Spain, *Historia Natural y Medico de el Principado de Asturias*, which was published in 1762, i.e., 3 years after his death. Casal regarded the disease, popularly called *mal de la rosa*, as a peculiar form of leprosy. He associated it with poverty and with the consumption of spoiled corn (maize).

In 1771, a similar dermatological disorder was described by the Italian physician Francesco Frapolli. In his work *Animadversiones in Morbum Volgo Pelagrum*, he reported the disease to be prevalent in northern Italy. In that region corn, recently introduced from America, had become a popular crop, displacing rye as the major grain. The local name for the disease was "pelagra," meaning rough skin. There is some evidence that it had been seen as early as 1740. By 1784 the prevalence of pelagra (now spelled pellagra) in that area was so great that a hospital was established in Legano for its treatment. Success in the treatment of pellagra appears to have been attributed to factors other than diet—e.g., rest, fresh air, water, and sunshine. Nevertheless, the disease continued to be associated with poverty and the consumption of corn-based diets.

Following the finding of pellagra in Italy, the disease was reported in France in 1829 by the French physician Jean-Marie Hameau. It was not until 1845 that another French physician Théophile Roussel associated pellagra with Casal's *mal de la rosa* and proposed that these diseases, including a similar disease called *flemma salada*,[12] were related or identical. To substantiate his hypothesis, Roussel spent 7 months of 1847 in the area where Casal had worked in northern Spain[13] investigating *mal de la rosa* cases; on his return, he presented to the French Academy of Medicine evidence in support of his conclusion. Subsequently, pellagra, as it had come to be called, was reported in Romania by Theodari in 1858, and in Egypt by the British physician Pruner-Bey in 1874. It was a curiosity, not to be explained for years, that pellagra was never endemic in the Yucatán Peninsula, where the cultivation of corn originated. The disease was not reported there until 1896.

It is not known how long pellagra had been endemic in the United States; however, it became common early in the 20th century. In 1912, American physician J.W. Babcock examined the records of the state hospital of South Carolina and concluded that the disease had occurred there as early as 1828. It is generally believed that pellagra also appeared during or after the American Civil War (1861–1865), in association with food shortages in the southern states. It is clear from George Searcy's 1907 report to the American Medical Association that the disease was endemic at least in Alabama.[14] By 1909, it had been identified in more than 20 states, several of which had impaneled Pellagra Commissions, and a national conference on the disease was held in South Carolina.

Since it first appeared, pellagra was associated with poverty and with the dependence on corn as the major staple food. Ideas were proffered that it was caused by a toxin associated with spoiled corn, yet by the turn of the century other hypotheses were also popular. These included the suggestion of an infectious agent with, perhaps, an insect vector.

Night blindness, the inability to see under low levels of light, was one of the first recorded medical conditions. Writings of Ancient Greek, Roman, and Arab physicians show that animal liver was known to be effective in both the prevention and cure of the disease. The Eber papyrus (c.1550 BCE)[15] described its treatment by the squeezing of liquid from a lamb's liver (now known to be a good source of vitamin A in well-nourished animals) directly into the eyes of the affected patient. The use of liver for the prevention of night blindness became a part of the folk cultures of most seafaring communities. In the 1860s, the French physicians, Hubbenet and, later, Bitot, each noted the presence of small, foamy white spots on the outer aspects of the conjunctiva of patients with night blindness—those lesions have become known as "Bitot's spots." Corneal ulceration, now known to be a related condition resulting in permanent blindness, was recognized in the 18th and 19th centuries in association with protein energy malnutrition as well as such diseases as meningitis, tuberculosis, and typhoid fever. In Russia, it occurred during long Lenten fasts. In the 1880s, cod liver oil was found to be effective in curing both night blindness and early corneal lesions; by the end of the century, cod liver oil, meat, and milk were used routinely in Europe to treat both conditions. It was not until the early 1900s, however, that the dietary nature of night blindness, and the corneal lesions that typically ensued, was understood—not until the "active lipid" was investigated, i.e., the factor in cod liver oil that supported growth and prevented night blindness and xerophthalmia in the rat.

Ideas Prevalent by 1900

Thus, by the beginning of the 20th century, four different diseases had been linked with certain types of diet. Further,

12. Literally meaning "salty phlegm," this condition involved gastrointestinal signs, delirium, and a form of dementia. It did not, however, occur in areas where maize was the major staple food; this, and disagreement over the similarities of symptoms, caused Roussel's proposal of a relationship between these diseases to be challenged by his colleague Arnault Costallat. From Costallat's letters describing *flemma salada* in Spain in 1861, it is apparent that he considered it to be a form of acrodynia, then thought to be due to ergot poisoning.

13. Casal practiced in the town of Oviedo in the Asturias of northern Spain.

14. Sercy, a physician at the Mount Vernon Insane Hospital in Mobile, Alabama, reported 88 cases of pellagra at that institution in 1906.

15. The Eber Papyrus, named for the German egyptologist who discovered it, is among the oldest extant medical papyri of ancient Egypt. Written in c.1550 BCE, the 20 m long scroll is thought to be copied from earlier texts. It is housed at the University of Liepzig.

TABLE 2.1 Diet–Disease Relationships Recognized by 1900

Disease	Associated Diet	Recognized Prevention
Scurvy	Salted (preserved) foods	Fresh fruits, vegetables
Beriberi	Polished rice-based	Meats, vegetables
Rickets	Few "good" fats	Eggs, cod liver oil
Pellagra	Corn-based	None
Night blindness	None	Cod liver oil

by 1900, it was apparent that at least two, and possibly three, could be cured by changes in diet (Table 2.1).

Other diseases, in addition to those listed in Table 2.1, had been known since ancient times to respond to what is now called diet therapy. Unfortunately, much of this knowledge was overlooked, and its significance was not fully appreciated by a medical community galvanized by the new germ theory of disease. Alternative theories for the etiologies of these diseases were popular. Thus, as the 20th century began, it was widely held that scurvy, beriberi, and rickets were each caused by a bacterium or bacterial toxin rather than by the simple absence of something required for normal health. Some held that rickets might also be due to hypothyroidism, while others thought it to be brought on by lack of exercise or excessive production of lactic acid. These theories died hard and had lingering deaths. In explanation of the lack of interest in the clues presented by the diet–disease associations outlined above, Harris (1955) mused: "Perhaps the reason is that it seems easier for the human mind to believe that ill is caused by some positive evil agency, rather than by any mere absence of any beneficial property."

Limitations of Empiricism

In actuality, the process of discovery of the vitamins had moved about as far as it could in its empirical phase. Further advances in understanding the etiologies of these diseases would require the rigorous testing of the various hypotheses—i.e., entrance into the deductive phase of nutritional discovery. That movement, however, required tools for productive scientific experimentation—tools that had not been available previously.

4. THE EXPERIMENTAL PHASE OF VITAMIN DISCOVERY

In a world where one cannot examine all possible cases (i.e., use strictly inductive reasoning), natural truths can be learned only by inference from premises already known to be true (i.e., through deduction). Both the inductive and deductive approaches may be linked; that is, probable conclusions derived from observation may be used as hypotheses for testing deductively in the process of scientific experimentation.

Requirements of Nutrition Science

For scientific experimentation to yield informative results, it must be both **repeatable** and **relevant**. The value of the first point, **repeatability**, should be self-evident. Inasmuch as natural truths are held to be constant, nonrepeatable results cannot be construed to reveal them. The value of the second point, **relevance**, becomes increasingly important when it is infeasible to test a hypothesis in its real-world context. In such circumstances, it becomes necessary to employ a representation of the context of ultimate interest—a construct known in science as a **model**. Models are born of practical necessity, but they must be developed carefully to serve as analogs of situations that cannot be studied directly.

Defined Diets Provided Repeatability

Repeatability in nutrition experimentation became possible with the use of **diets of defined composition**. The most useful type of defined diet that emerged in nutrition research was the **purified diet**. Diets of this type were formulated using highly refined ingredients (e.g., isolated proteins, refined sugars and starches, refined fats) for which the chemical composition could be tested and quantified. It was the use of defined diets that facilitated experimental nutrition; such diets could be prepared over and over by the same or other investigators to yield comparable results. Results obtained through the use of defined diets were repeatable and, therefore, predictable.

Appropriate Animal Models Provided Relevance

Relevance in nutrition research became possible with the identification of **animal models**[16] appropriate to diseases of interest in human medicine or to physiological processes of

16. In nutrition and other biomedical research, an animal model consists of the experimental production in a conveniently managed animal species of biochemical and/or clinical changes that are comparable to those occurring in another species of primary interest but that may be infeasible, unethical, or uneconomical to study directly. Animal models are, frequently, easily managed and rapidly growing species with small body weights (e.g., rodents, chicks, rabbits); however, they may also be larger species (e.g., monkeys, sheep), depending on the target problem and species they are selected to represent. In any case, background information on the biology and husbandry should be available. The selection and/or development of an animal model should be based primarily on representation of the biological problem of interest without undue consideration of the practicalities of cost and availability.

interest in human medicine or animal production. The first of these was discovered quite by chance by keen observers studying human disease. Ultimately, the use of animal models would lead to the discovery of each of the vitamins, as well as to the elucidation of the nutritional roles and metabolic functions of each of the approximately 40 nutrients. The careful use of appropriate animal models made possible studies that would otherwise be infeasible or unthinkable in human subjects or in other animal species of interest.

Major Forces in the Emergence of Nutritional Science

- Recognition that certain diseases were related to diet
- Development of appropriate animal models
- Use of defined diets

An Animal Model for Beriberi

The analytical phase of vitamin discovery, indeed modern nutrition research itself, was entered with the finding of an animal model for beriberi in the 1890s. In 1886, Dutch authorities sent a commission led by Cornelius Pekelharing to their East Indian colony (now Indonesia) to find the cause of beriberi, which had become such a problem among Dutch soldiers and sailors as to interrupt military operations in Atjeh, Sumatra. Pekelharing took an army surgeon stationed in Batavia (now Jakarta), Christiaan Eijkman, whom he had met when each was on study leave (Pekelharing from his faculty post at the University of Utrecht, and Eijkman as a medical graduate from the University of Amsterdam) in the laboratory of the great bacteriologist, Robert Koch. The team, unaware of Takaki's work, expected to find a bacterium as the cause, and was therefore disappointed, after 8 months of searching, to uncover no such evidence. They concluded, "Beriberi has been attributed to an insufficient nourishment and to misery: but the destruction of the peripheral nervous system on such a large scale is not caused by hunger or grief. The true cause must be something coming from the outside, but is it a poison or an infection?"

However, looking for a poison, they observed, would be very difficult, whereas they had techniques for looking for a microorganism that had been successful for other diseases. Thus, they tried to culture organisms from blood smears from patients and to create the disease in monkeys, rabbits, and dogs by inoculations of blood, saliva, and tissues from patients and cadavers. When single injections produced no effects, they used multiple injection regimens. Despite the development of abscesses at the point of some injections, it appeared that multiple inoculations could produce some nerve degeneration in rabbits and dogs. Pekelharing concluded that beriberi was indeed an infectious disease, but an unusual one requiring repeated reinfection of the host. Before returning to Holland, Pekelharing persuaded the Dutch military to allow Eijkman to continue working on the beriberi problem.

The facilities used by the Commission at the Military Hospital Batavia became a new Laboratory for Bacteriology and Pathology of the colonial government, and Eijkman was named as director, with one assistant. His efforts in 1888 to infect rabbits and monkeys with Pekelharing's *micrococcus* were altogether unsuccessful, causing him to posit that beriberi must require a long time before the appearance of signs. The following year, he started using chickens as his animal model. Later in the year, he noted that many, regardless of whether they had been inoculated, lost weight, and started walking with a staggering gait. Some developed difficulty standing and died. Eijkman noted on autopsy no abnormalities of the heart, brain, or spinal cord, but microscopic degeneration of the peripheral nerves, particularly in the legs. The latter were signs he had observed in people dying of beriberi. He was unable, though, to culture any consistent type of bacteria from the blood of affected animals. It would have been easy for Eijkman to dismiss the thought that this avian disease, which he called "polyneuritis," might be related to beriberi.

Serendipity or a Keen Eye?

After persisting in his flock for some 5 months, the disease suddenly disappeared. Eijkman reviewed his records and found that in June, shortly before the chickens had started to show paralysis, a change in their diet had been occasioned by failure of a shipment of feed grade brown (unpolished) rice to arrive. His assistant had used, instead, white (polished) rice from the hospital kitchen. It turned out that this extravagance had been discovered a few months earlier by a new hospital superintendent, who had ordered it stopped. When Eijkman again fed the chickens brown rice, he found affected animals recovered completely within days.

With this clue, Eijkman immediately turned to the chicken as the animal model for his studies. He found chicks showed signs of **polyneuritis** within days of being fed polished rice, and that their signs disappeared even more quickly if they were then fed unpolished rice. It was clear that there was something associated with rice polishings that protected chickens from the disease. After discussing these results, Eijkman's colleague Adolphe Verdeman, the physician inspector of prisons in the colony, surveyed the use of polished and unpolished rice and the incidence of beriberi among inmates. His results (Table 2.2), later confirmed by others in similar epidemiological investigations, demonstrated the advantage enjoyed by prisoners eating unpolished rice: they were much less likely to contract beriberi. This information, in conjunction with his experimental findings with chickens, allowed Eijkman to investigate, by means of bioassay, the beriberi-protective factor apparently associated with rice husks.

TABLE 2.2 Beriberi in Javanese Prisons c.1890

Diet	Population	Cases	Prevalence (Cases/10,000 People)
Polished rice	150,266	4200	279.5
Partially polished rice	35,082	85	24.2
Unpolished rice	96,530	86	8.9

Antiberiberi Factor Is Announced

Eijkman used this animal model in a series of investigations in 1890–1897 and found that the antipolyneuritis factor could be extracted from rice hulls with water or alcohol, that it was dialyzable, but that it was rather easily destroyed with moist heat. He concluded that the water-soluble factor was a "pharmacological antidote" to the "beriberi microbe," which, although still not identified, he thought to be present in the rice kernel proper. Apparently, Gerrit Grijns,[17] who continued that work after Eijkman returned to Holland, came to interpret these findings somewhat differently. Grijns went on to show that polyneuritis could be prevented by including mung bean (*Vigna radiata*) in the diet; this led to mung beans being found effective in treating beriberi. In 1901, Grijns suggested, for the first time, that beriberi-producing diets "lacked a certain substance of importance in the metabolism of the central nervous system." Subsequently, Eijkman came to share Grijn's view; in 1906, the two investigators published a now-classic paper in which they wrote, "There is present in rice polishings a substance different from protein, and salts, which is indispensable to health and the lack of which causes nutritional polyneuritis."

5. THE VITAMINE THEORY

Defined Diets Revealed Needs for Accessory Factors

The announcement of the antiberiberi factor constituted the first recognition of the concept of the vitamin, although the term itself was yet to be coined. At the time of Eijkman's studies, but a world removed and wholly separate, others were finding that animals would not survive when fed "synthetic" or "artificial" diets formulated with purified fats, proteins, carbohydrates, and salts—i.e., containing all of the nutrients then known to be constituents of natural foods.

17. Grijns (1865–1944) was a Dutch physician trained at the University of Utrecht. He assisted Eijkman in Batavia and continued that work when Eijkman, having contracted malaria, returned to Holland in 1896.

Such a finding was first reported by the Russian surgeon Nikolai Lunin, in 1888, who found that the addition of milk to a synthetic diet supported the survival of mice. Lunin concluded, "A natural food such as milk must, therefore, contain besides these known principal ingredients small quantities of other and unknown substances essential to life."

Lunnin's finding was soon confirmed by several other investigators. By 1912, Rhömann in Germany, Socin in Switzerland, Pekelharing in The Netherlands, and Hopkins in England had each demonstrated that the addition of milk to purified diets corrected the impairments in growth and survival that were otherwise produced in laboratory rodents. The German physiologist Wilhelm Stepp took another experimental approach. He found it possible to extract, from bread and milk, factors required for animal growth. Although Pekelharing's 1905 observations, published in Dutch, lay unnoticed by many investigators, his conclusions about what Hopkins had called the accessory factor in milk alluded to the modern concept of a vitamin: "If this substance is absent, the organism loses the power properly to assimilate the well known principal parts of food, the appetite is lost and with apparent abundance the animals die of want. Undoubtedly this substance not only occurs in milk but in all sorts of foodstuffs, both of vegetable and animal origin."

Perhaps the most important of the early studies with defined diets were those of the Cambridge biochemist Frederick Gowland Hopkins.[18] His studies demonstrated that the growth-promoting activities of accessory factors were independent of appetite, and that such factors prepared from milk or yeast were biologically active in very small amounts.

Two Lines of Inquiry

Therefore, by 1912, two independently developed lines of inquiry had revealed that foods contained beneficial factor(s) in addition to the nutrients known at the time. That these factor(s) were present and active in minute amounts was apparent from the fact that almost all of the mass of food was composed of the known nutrients.

Two Lines of Inquiry Leading to the Discovery of the Vitamins

- The study of substances that prevent deficiency diseases
- The study of accessory factors required by animals fed purified diets.

18. Sir Frederick Gowland Hopkins (1861–1947), is known for his work at Cambridge University, which involved not only classic work on accessory growth factors (for which he shared, with Christiaan Eijkman, the 1929 Nobel Prize in Medicine or Physiology), but also the discoveries of glutathione and tryptophan.

Comments by Hopkins in 1906 indicate that he saw connections between the accessory factors and the deficiency diseases. On the subject of the accessory growth factors in foods, he wrote, "No animal can live on a mixture of pure protein, fat and carbohydrate, and even when the necessary inorganic material is carefully supplied the animal still cannot flourish. The animal is adjusted to live either on plant tissues or the tissues of other animals, and these contain countless substances other than protein, carbohydrates and fats. In diseases such as rickets, and particularly scurvy, we have had for years knowledge of a dietetic factor; but though we know how to benefit these conditions empirically, the real errors in the diet are to this day quite obscure ... They are, however, certainly of the kind which comprises these minimal qualitative factors that I am considering."

Hopkins demonstrated the presence of a factor(s) in milk that stimulated the growth of animals fed diets containing all of the then-known nutrients (Fig. 2.1).

The Lines Converge

The discovery by Eijkman and Grijns had stimulated efforts by investigators in several countries to isolate the antiberiberi factor in rice husks. Umetaro Suzuki, of Imperial University Agricultural College in Tokyo, succeeded in preparing a concentrated extract from rice bran for the treatment of polyneuritis and beriberi. He called the active fraction "oryzanin" but could not achieve its purification in crystalline form. Casimir Funk,[19] a chemist at the Lister Institute in London, concluded from the various conditions in which it could be extracted and then precipitated that the antipolyneuritis factor in rice husks was an organic base and, therefore, nitrogenous in nature. When he appeared to have isolated the factor, Funk coined a new word for it, with the specific intent of promoting the new concept in nutrition to which Hopkins had alluded. Having evidence that the factor was an organic base, and therefore an *amine*, Funk chose the term **vitamine**[20] because it was clearly *vital*, i.e., pertaining to life.

Funk's Theory

In 1912, Funk published his landmark paper presenting the **vitamine theory**; in it he proposed, in what some have referred to as a leap of faith, four different vitamines. That the concept was not a new one, and that not all of these factors later proved to be amines (hence, the change to **vitamin**[21]) are far less important than the focus the newly coined term gave to the diet–health relationship. Funk was not unaware of the importance of the term itself; he wrote, "I must admit that when I chose the name "vitamine" I was well aware that these substances might later prove not all to be of an amine nature. However, it was necessary for me to use a name that would sound well and serve as a 'catch-word.'"[22]

Funk's Vitamines

- Antiberiberi vitamine
- Antirickets vitamine
- Antiscurvy vitamine
- Antipellagra vitamine

Impact of the New Concept

The vitamine theory opened new possibilities in nutrition research by providing a new intellectual construct for interpreting observations of the natural world. No longer was the elucidation of the etiologies of diseases to be constrained by the germ theory. Thus, Funk's greatest contribution involves not the data generated in his laboratory, but rather the theory produced from his thoughtful review of information already in the medical literature of the time. This fact caused Harris (1955) to observe, "The interpreter may be as useful to science as the discoverer. I refer here to any man[23] who is able to take a broad view of what has already been done by others, to collect evidence and discern through it all some common connecting link."

The real impact of Funk's theory was to provide a new concept for interpreting diet-related phenomena. As the educational psychologist Novak[24] observed more recently, "As our conceptual and emotional frameworks change, we see different things in the same material."

Still, it was not clear by 1912 whether the accessory factors were the same as the vitamines. In fact, until 1915, there was a considerable debate concerning whether the growth factor for the rat was a single or multiple entity (it was already clear that there was more than one vitamine). Some investigators were able to demonstrate it in yeast and not butter; others found it in butter and not yeast. Some showed it to be identical with the antipolyneuritis factor; others showed that it was clearly different.

19. Funk (1884–1957) was born in Poland and studied in Switzerland, Paris and Berlin.

20. Harris (1955) reported that the word *vitamine* was suggested to Funk by his friend, Dr Max Nierenstein, Reader in Biochemistry at the University of Bristol.

21. The dropping of the *e* from *vitamine* is said to have been the suggestion of J.C. Drummond.

22. Funk, C. (1912). The etiology of the deficiency diseases. *J. State Med.* 20, 341–368.

23. Harris's word choice reveals him as a product of his times. Because it is clear that the process of intellectual discovery to which Harris refers does not recognize gender, it is more appropriate to read this word as *person*.

24. Novak, J.D. (1977) "A Theory of Education," Cornell University Press, Ithaca, NY.

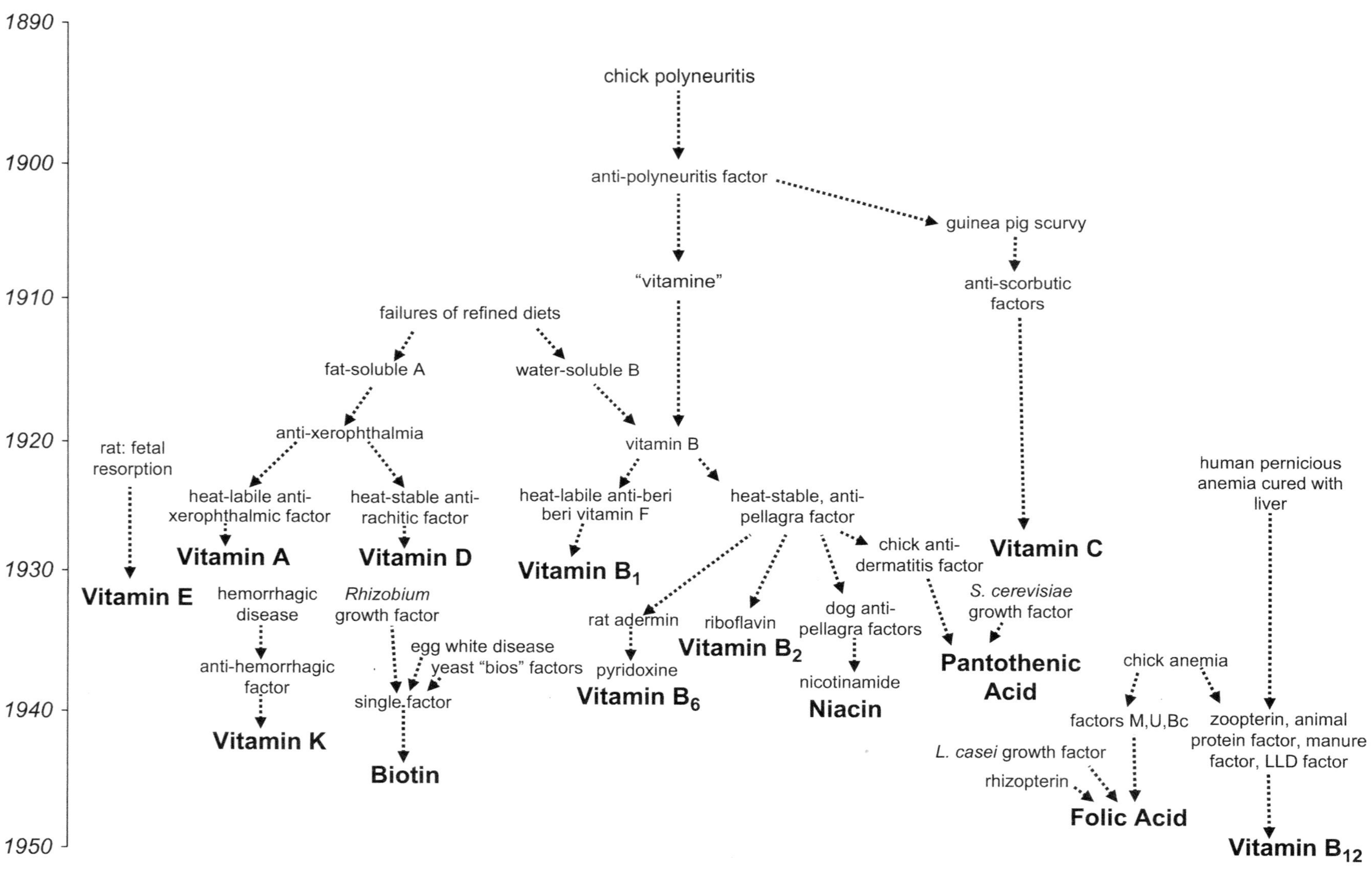

FIGURE 2.1 The cascade of vitamin discovery.

TABLE 2.3 McCollum's Rat Growth Factors

Factor	Found in	Not Found in
Fat-soluble A	Milk fat, egg yolk	Lard, olive oil
Water-soluble B	Wheat, milk, egg yolk	Polished rice

There Is More Than One Accessory Factor

The debate was resolved by the landmark studies of the American biochemist Elmer McCollum[25] and his volunteer assistant Marguerite Davis[26] at the University of Wisconsin in 1913–1915. Using diets based on casein and lactose, they demonstrated that at least two different additional growth factors were required to support normal growth of the rat. One factor could be extracted with ether from egg or butterfat (but not olive or cottonseed oils) but was nonsaponifiable; it appeared to be the same factor shown earlier by Wilhelm Stepp,[27] and by Thomas Osborne[28] and Lafayette Mendel[29] in the same year, to be required to sustain growth of the rat. The second factor was extractable with water and prevented polyneuritis in chickens and pigeons. McCollum called these factors **fat-soluble A** and **water-soluble B**, respectively (Table 2.3).

Accessory Factors Prevent Disease

Subsequent studies conducted by McCollum's group showed that the ocular disorders (i.e., **xerophthalmia**[30]) that developed in rats, dogs, and chicks fed fat-free diets could be prevented by feeding them cod liver oil, butter, or preparations of fat-soluble A, which then became known as the **antixerophthalmic factor**. Shortly, it was found that the so-called water-soluble B material was not only required for normal growth of the rat but also prevented polyneuritis in the chick. Therefore, it was clear that water-soluble B was identical to or at least contained Funk's antiberiberi vitamine; hence, it became known as **vitamine B**.

Accessory Factors Are the Same as Vitamines

With these discoveries, it became apparent that the biological activities of the accessory factors and the vitamines were likely to be due to the same compounds. The concept of a vitamine was thus generalized to include nonnitrogenous compounds, and the antipolyneuritis vitamine became **vitamin B**.

Elucidation of the Vitamines

So it was, through the agencies of several factors, a useful new intellectual construct, the use of defined diets, and the availability of appropriate animal models, that nutrition emerged as a scientific discipline. By 1915, thinking about diet and health had been forever changed, and it was clear that the earlier notions about the required nutrients had been incomplete. Therefore, it should not be surprising to find, by the 1920s, mounting interest in the many questions generated by what had become sound nutritional research. That interest and the further research activity it engendered resulted, over the brief span of only five decades, in the development of a fundamental understanding of the identities and functions of about 40 nutrients, one-third of which are considered vitamins.

Crooked Paths to Discovery

The paths leading to the discovery of the vitamins wandered from Java with the findings of Eijkman in the 1890s, to England with Funk's theory in 1912, to the United States with the recognition of fat-soluble A and water-soluble B in 1915. By that time the paths had already branched, and for the next four decades they would branch again and again as scientists from many laboratories and many nations would pursue many unexplained responses to diet among many types of animal model. Some of these pursuits appeared to fail; however, in aggregate, all laid the groundwork of understanding on which the discoveries of those factors now recognized to be vitamins were based. When viewed in retrospect, the path to that recognition may seem deceptively straight—but it most definitely was not. The way was branched and crooked; in many cases, progress was made by several different investigators traveling in apparently different directions. The following recounts the highlights of the exciting search for the elucidation of the vitamins.

---

25. Elmer Verner McCollum (1879–1967) received his doctorate at Yale and worked on dietary protein quality with Osborne and Mendel there. In 1909, he joined the faculty of the University of Wisconsin, where his work on growth-promoting factors in deproteinized milk led to the recognition of vitamin A. In 1917, he moved to Johns Hopkins University. McCollum opposed Funk's term "vitamine" on the basis that all essential nutrients were vital.

26. Marguerite Davis (1887–1967) was a graduate student with McCollum. When she and McCollum had shown that water-soluble B was not a single compound, she gave the components letter names, thus, starting that tradition of naming vitamins.

27. Wilhelm Stepp (1882–1964) was a professor of medicine at several German Universities (Strassburg, Jena, Breslau and Munich).

28. Thomas Burr Osborne (1859–1929) was a professor of chemistry who spent his career at the Connecticut Agricultural Experiment Station studying protein quality and nutritional requirements. His collaboration with Mendel led to the recognition of the essentiality of amino acids.

29. Lafayette Benedict Mendel (1872–1935) was a professor of physiological chemistry at Yale University who worked with Osborne to determine why rats could not survive on diets of only purified carbohydrates, fats and proteins.

30. Xerophthalmia, from the Greek *xeros* ("dry") and *ophthalmos* ("eye"), involves dryness of the eyeball owing to atrophy of the periocular glands, hyperkeratosis of the conjunctiva, and, ultimately, inflammation and edema of the cornea, which leads to infection, ulceration, and blindness.

6. ELUCIDATION OF THE VITAMINS

New Animal Model Reveals New Vitamin: "C"

Eijkman's report of polyneuritis in the chicken and an animal model for beriberi stimulated researchers Axel Holst and Theodor Frölich at the University of Christiana in Oslo, who were interested in **shipboard beriberi**, a common problem among Norwegian seamen. Working with pigeons, they found a beriberi diet to produce the polyneuritis described by Eijkman; however, they considered that condition very different from the disease of sailors. In 1907, they attempted to produce the disease in another experimental animal species: the common Victorian household pet, the guinea pig. Contrary to their expectations, they failed to produce, by feeding that species a cereal-based diet, anything resembling beriberi; instead, they observed the familiar signs of scurvy. Eijkman's work suggested to them that, like beriberi, scurvy too might be due to a dietary deficiency. Having discovered, quite by chance, one of the few possible animal species in which scurvy could be produced,[31] Holst and Frölich had produced something of tremendous value—an animal model of scurvy[32]—showing that lesions could be prevented by feeding apples, (unboiled) cabbage, potatoes, and lemon juice.

This finding led Henriette Chick and Ruth Skelton of the Lister Institute, in the second decade of the 20th century, to use the guinea pig to develop a bioassay for the determination of the antiscorbutic activity in foods, and S.S. Zilva and colleagues (also at the Lister Institute) to isolate from lemons the crude factor that had come to be known as **vitamin C**. It was soon found that vitamin C could reduce the dye 2,6-dichloroindophenol, but the reducing activity determined with that reagent did not always correlate with the antiscorbutic activity determined by bioassay. Subsequently, it was found that the vitamin was reversibly oxidized, but that both the reduced and oxidized forms had antiscorbutic activity.

In 1932, Albert Szent-Györgi, a Hungarian scientist working in Hopkins' laboratory at Cambridge University, and Glen King at the University of Pittsburgh established that the antiscorbutic factor was identical with the reductant *hexuronic acid*,[33] now called **ascorbic acid**. Szent-Györgi had isolated it in crystalline form from adrenal cortex, while King had isolated it from cabbage and citrus juice.[34] After Szent-Györgi returned to Hungary to take a professorship, he was joined by an American-born Hungarian, J. Svirbely, who had been working in King's laboratory. Szent-Györgi had isolated c.500 grams of crystalline hexuronic acid from peppers, and then 25 g of the vitamin from adrenal glands, making samples available to other laboratories. On April 1, 1932, King and Waugh reported that their crystals protected guinea pigs from scurvy; 2 weeks later, Svirbely and Szent-Györgi reported virtually the same results. The following year, the chemical structure of ascorbic acid was elucidated by the groups of Haworth in Birmingham and Karrer in Zurich, both of which also achieved its synthesis.

Fat-Soluble A: Actually Two Factors

Pursuing the characterization of fat-soluble A, by 1919 McCollum's group[35] and others had found that, in addition to supporting growth for the rat, the factor also prevented xerophthalmia and night blindness in that species. In 1920, Drummond called the active lipid **vitamin A**.[36] This factor was present in cod liver oil, which at the turn of the century had been shown to prevent both xerophthalmia and night blindness—which Bitot, some 40 years earlier, had concluded had the same underlying cause.

Vitamin A Prevents Rickets?

Undoubtedly influenced by the recent recognition of vitamin A, Edward Mellanby, who had worked with Hopkins, undertook to produce a dietary model of rickets. For this he used puppies, which the Scottish physician Findley found developed rickets if kept indoors.[37] Mellanby fed a low-fat diet based on oatmeal with limited milk intake to puppies that he kept indoors; the puppies developed the

31. Their finding was, indeed, fortuitous, as vitamin C is now known to be an essential dietary nutrient only for the guinea pig, primates, fishes, some fruit-eating bats, and some passeriform birds. Had they used the rat, the mouse or the chick in their study, vitamin C might have remained unrecognized for perhaps quite a while.

32. In fact, scorbutic signs had been observed in the guinea pig more than a decade earlier, when a U.S. Department of Agriculture pathologist noted in an annual report: "When guinea pigs are fed with cereals (bran and oats mixed), without any grass, clover or succulent vegetables, such as cabbage, a peculiar disease, chiefly recognizable by subcutaneous extravasation of blood, carries them off in four to eight weeks." That this observation was not published for a wider scientific audience meant that it failed to influence the elucidation of the etiology of scurvy.

33. It is said that when Szent-Györgi first isolated the compound, he was at a loss for a name for it. Knowing it to be a sugar, but otherwise ignorant of its identity, he proposed the name *ignose*, which was disqualified by an editor who did not appreciate the humor of the Hungarian chemist. Ultimately, the names **ascorbic acid** and **vitamin C**, by which several groups had come to refer to the antiscorbutic factor, were adopted.

34. The reports of both groups (King, C.G., Waugh, W.S., 1932. Science 75, 357–358; Svirbely, J.L., Szent-Györgi, A., 1932. Biochem. J. 26, 865–870) appeared within 2 weeks of one another in 1932. In fact, Svirbely had recently joined Szent-Györgi's group, having come from King's laboratory. In 1937, King and Szent-Györgi shared the Nobel Prize for their work in the isolation and identification of vitamin C.

35. In 1917, McCollum moved to the newly established School of Public Health at Johns Hopkins University.

36. In 1920, J.C. Drummond proposed the use of the names *vitamin A* and *vitamin B* for McCollum's factors, and the use of the letters C, D, etc., for any vitamins subsequently to be discovered.

37. Exposing infants to sunlight is a traditional practice in many cultures and had been a folk treatment for rickets in northern Europe.

marked skeletal deformities characteristic of rickets. When he found that these deformities could be prevented by feeding cod liver oil or butterfat without allowing the puppies outdoors, he concluded that rickets, too, was caused by a deficiency of vitamin A, which McCollum had discovered in those materials.

New Vitamin: "D"

McCollum, however, suspected that the antirachitic factor present in cod liver oil was different from vitamin A. Having moved to the Johns Hopkins University in Baltimore, he conducted an experiment in which he subjected cod liver oil to aeration and heating (100°C for 14 h), after which he tested its antixerophthalmic and antirachitic activities with rat and chick bioassays, respectively. He found that heating had destroyed the antixerophthalmic (vitamin A) activity, but that cod liver oil had retained antirachitic activity. McCollum called the heat-stable factor **vitamin D**.

β-Carotene, a Provitamin

At about the same time (1919), Steenbock in Wisconsin pointed out that the vitamin A activities of plant materials seemed to correlate with their contents of yellow pigments. He suggested that the plant pigment **carotene** was responsible for the vitamin A activity of such materials. Yet the vitamin A activity in organic extracts of liver was colorless. Therefore, Steenbock suggested that carotene could not be vitamin A, but that it may be converted metabolically to the actual vitamin. This hypothesis was not substantiated until 1929, when von Euler and Karrer in Stockholm demonstrated growth responses to carotene in rats fed vitamin A-deficient diets. Further, Moore in England demonstrated, in the rat, a dose–response relationship between dietary β-carotene and hepatic vitamin A concentration. This proved that β-carotene is, indeed, a **provitamin**.

Vitamin A Linked to Vision

In the early 1930s, the first indications of the molecular mechanism of the visual process were produced by George Wald, of Harvard University but working in Germany at the time, who isolated the chromophore **retinen** from bleached retinas.[38] A decade later, Morton in Liverpool found that the chromophore was the aldehyde form of vitamin A—**retinaldehyde**. Just after Wald's discovery, Karrer's group in Zurich elucidated the structures of both β-carotene and vitamin A. In 1937, Holmes and Corbett succeeded in crystallizing vitamin A from fish liver. In 1942, Baxter and Robeson crystallized *retinol* and several of its esters; in 1947, they crystallized the 13-*cis*-isomer. Isler's group in Basel achieved the synthesis of retinol in the same year and that of β-carotene 3 years later.

The Nature of Vitamin D

McCollum's discovery of the antirachitic factor he called vitamin D in cod liver oil, which was made possible through the use of animal models, was actually a *rediscovery*, as that material had been long recognized as an effective medicine for rickets in children. Still, the nature of the disease was the subject of considerable debate, particularly after 1919, when Huldschinsky, a physician in Vienna, demonstrated the efficacy of ultraviolet light in healing rickets. This confusion was clarified by the findings in 1923 of Goldblatt and Soames, who demonstrated that when livers from rachitic rats were irradiated with ultraviolet light, they could cure rickets when fed to rachitic, nonirradiated rats. The next year, Steenbock's group demonstrated the prevention of rickets in rats by ultraviolet irradiation of either the animals themselves *or* their food. Further, the light-produced antirachitic factor was associated with the fat-soluble portion of the diet.[39]

Vitamers D

The ability to produce vitamin D (which could be bioassayed using both rat and chick animal models) by irradiating lipids led to the finding that large quantities of the vitamin could be produced by irradiating plant sterols. This led Askew's and Windaus's groups, in the early 1930s, to the isolation and identification of the vitamin produced by irradiation of **ergosterol**. Steenbock's group, however, found that while the rachitic chick responded appropriately to irradiated products of cod liver oil or the animal sterol **cholesterol**, that animal did *not* respond to the vitamin D so produced from ergosterol. On the basis of this apparent lack of equivalence, Wadell suggested in 1934 that the irradiated products of ergosterol and cholesterol were different. Subsequently, Windaus's group synthesized 7-dehydrocholesterol and isolated a vitamin D-active product of its irradiation. In 1936, they reported its structure, showing it to be a side chain isomer of the form of the vitamin produced from plant sterols. Thus, two forms of vitamin D were found: **ergocalciferol** (from ergosterol), which was called vitamin D_2,[40] and **cholecalciferol** (from cholesterol), which

38. For this and other discoveries of the basic chemical and physiological processes in vision, George Wald was awarded, with Haldan K. Hartline (of the United States) and R. Grant (of Sweden), the Nobel Prize in Chemistry in 1967.

39. This discovery, i.e., that by ultraviolet irradiation it was possible to induce vitamin D activity in such foods as milk, bread, meats, and butter, led to the widespread use of this practice, which has resulted in the virtual eradication of rickets as a public health problem.

40. Windaus's group had earlier isolated a form of the vitamin he had called vitamin D_1, which had turned out to be an irradiation-breakdown product, **lumisterol**.

was called vitamin D_3. While it was clear that the vitamers D had important metabolic roles in calcification, insights concerning the molecular mechanisms of the vitamin would not come until the 1960s. Then, it became apparent that neither vitamer was metabolically active per se; each is converted in vivo to metabolites that participate in a system of calcium homeostasis that continues to be of great interest to the biomedical community. With this understanding, it became apparent that vitamin D_3 was actually a steroid hormone.[41]

Multiple Identities of Water-Soluble B

By the 1920s, it was apparent that the antipolyneuritis factor, called water-soluble B and present in such materials as yeasts, was not a single substance. This was demonstrated by the finding that fresh yeast could prevent both beriberi and pellagra. However, the antipolyneuritis factor in yeast was unstable to heat, while such treatment did not alter the efficacy of yeast to prevent **dermatitis** in rodents. This caused Goldberger to suggest that the then-called vitamin B was actually at least *two* vitamins: the antipolyneuritis vitamin and a new antipellagra vitamin.

In 1926, the heat-labile antipolyneuritis/beriberi factor was first crystallized by Jansen and Donath, working in the Eijkman Institute (which replaced Eijkman's simple facilities) in Batavia. They called the factor *aneurin*. Their work was facilitated by the use of the small rice bird (*Munia maja*) as an animal model in which they developed a rapid bioassay for antipolyneuritic activity.[42] Six years later, Windaus's group isolated the factor from yeast, perhaps the richest source of it. In the same year (1932), the chemical structure was determined by R.R. Williams, who named it **thiamin**—i.e., the vitamin containing sulfur (*thios*, in Greek). Noting that deficient subjects showed high blood levels of pyruvate and lactate after exercise, in 1936 Rudolph Peters of Oxford University used, for the first time, the term "biochemical lesion" to describe the effects of the dietary deficiency. Shortly thereafter, methods of synthesis were achieved by several groups, including those of Williams, Andersag and Westphal, and Todd. In 1937, thiamin diphosphate (thiamin pyrophosphate) was isolated by Lohmann and Schuster, who showed it to be identical to the *cocarboxylase* that had been isolated earlier by Auhagen. That many research groups were actively engaged in the research on the antipolyneuritis/beriberi factor is evidence of intense international interest due to the widespread prevalence of beriberi.

The characterization of thiamin clarified the distinction of the antiberiberi factor from the antipellagra activity. The latter was not found in maize (corn), which contained appreciable amounts of thiamin. Goldberger called the two substances the "A-N factor" (antineuritic) and the "P-P factor" (pellagra-preventive). Others called these factors vitamins F (for Funk) and G (for Goldberger), respectively, but these terms did not last.[43] By the mid-1920s the terms **vitamin B_1** and ***vitamin B_2*** had been rather widely adapted for these factors, respectively; this practice was codified in 1927 by the Accessory Food Factors Committee of the British Medical Research Council.

Vitamin B_2: A Complex of Several Factors

That the thermostable second nutritional factor in yeast, which by that time was called vitamin B_2, was not a single substance, and was not immediately recognized, giving rise to considerable confusion and delay in the elucidation of its chemical identity (identities). It should be noted that efforts to fractionate the heat-stable factor were guided almost exclusively by bioassays with experimental animal models. Yet, different species yielded discrepant responses to preparations of the factor. When such variation in responses among species was finally appreciated, it became clear that vitamin B_2 actually included *several* heat-stable factors. Vitamin B_2, as then defined, was indeed a complex.

Components of the Vitamin B_2 Complex

- The P-P factor (preventing pellagra in humans and pellagra-like diseases in dogs, monkeys, and pigs)
- A growth factor for the rat
- A pellagra-preventing factor for the rat
- An antidermatitis factor for the chick

Vitamin B_2 Complex Yields Riboflavin

The first substance in the vitamin B_2 complex to be elucidated was the heat-stable, water-soluble rat growth factor, which was isolated by Kuhn, György, and Wagner-Jauregg at the Kaiser Wilhelm Institute in 1933. Those investigators found that thiamin-free extracts of autoclaved yeast, liver, or rice bran prevented the growth failure of rats fed a thiamin-supplemented diet. Further, they noted that a yellow-green fluorescence in each extract promoted rat growth, and that

41. 1,25-dihydroxycholecalciferol meets the standard definition of a hormone in as much as it is produced and transported through the circulation to exert biological activity in distal organs.

42. The animals, which consumed only 2 grams of feed daily, showed a high (98+%) incidence of polyneuritis within 9–13 days if fed white polished rice. The delay of onset of signs gave them a useful bioassay of antipolyneuritic activity suitable for use with small amounts of test materials. This point is not trivial, inasmuch as there is only about a teaspoon of thiamin in a ton of rice bran. The bioassay of Jansen and Donath was sufficiently responsive for 10 μg of active material to be curative.

43. In fact, the name *vitamin F* was later used, with some debate as to the appropriateness of the term, to describe essential fatty acids. The name *vitamin G* has been dropped completely.

the intensity of fluorescence was proportional to the effect on growth. This observation enabled them to develop a rapid chemical assay that, in conjunction with their bioassay, they exploited to isolate the factor from egg white in 1933. They called it **ovoflavin**. The same group then isolated, by the same procedure, a yellow-green fluorescent growth-promoting compound from whey (which they called **lactoflavin**). This procedure involved the adsorption of the active factor on fuller's earth,[44] from which it could be eluted with base.[45] At the same time, Ellinger and Koschara, at the University of Düsseldorf, isolated similar substances from liver, kidney, muscle, and yeast, and Booher in the United States isolated the factor from whey. These water-soluble growth factors became designated as **flavins**.[46] By 1934, Kuhn's group had determined the structure of the so-called flavins. These substances were thus found to be identical; because each contained a ribose-like (ribotyl) moiety attached to an isoalloxazine nucleus, the term **riboflavin** was adopted. Riboflavin was synthesized by Kuhn's group (then at the University of Heidelberg) and by Karrer's group at Zurich in 1935. As the first component of the vitamin B_2 complex, it is also referred to as vitamin B_2; however, that should not be confused with the earlier designation of the P-P factor.

Vitamin B_2 Complex Yields Niacin

Progress in the identification of the P-P factor was retarded by two factors: the pervasive influence of the germ theory of disease and the lack of an animal model. The former made acceptance of evidence suggesting a nutritional origin of the disease a long and difficult undertaking. The latter precluded the rigorous testing of hypotheses for the etiology of the disease in a timely and highly controlled manner. These challenges were met by Joseph Goldberger, a U.S. Public Health Service bacteriologist who, in 1914, was put in charge of the Service's pellagra program.

Pellagra: An Infectious Disease?

Goldberger's first study[47] is now a classic. He studied a Jackson, Mississippi, orphanage in which pellagra was endemic. He noted that whereas the disease was prevalent among the inmates, it was absent among the staff, including the nurses and physicians who cared for patients; this suggested to him that pellagra was not an infectious disease. Noting that the food available to the professional staff was much different from that served to the inmates (the former included meat and milk not available to the inmates), Goldberger suspected that an unbalanced diet was responsible for the disease. He secured funds to supply meat and milk to inmates for a 2-year period of study. The results were dramatic: pellagra soon disappeared, and no new cases were reported for the duration of the study. However, when funds expired at the end of the study and the institution was forced to return to its former meal program, pellagra reappeared. While the evidence from this uncontrolled experiment galvanized Goldberger's conviction that pellagra was a dietary disease, it was not sufficient to affect a medical community that thought the disease likely to be an infection.

Over the course of two decades, Goldberger worked to elucidate the dietary basis of pellagra. Among his efforts to demonstrate that the disease was not infectious was the exposure, by ingestion and injection, of himself, his wife, and 14 volunteers to urine, feces, and biological fluids from **pellagrins**.[48] He also experimented with 12 male prisoners who volunteered to consume a diet (based on corn and other cereals, but containing no meat or dairy products) that he thought might produce pellagra: within 5 months half of the subjects had developed dermatitis on the scrotum, and some also showed lesions on their hands.[49] The negative results of these radical experiments, plus the finding that therapy with oral supplements of the amino acids cysteine and tryptophan was effective in controlling the disease, led, by the early 1920s, to the establishment of a dietary origin of pellagra. Further progress was hindered by the lack of an appropriate animal model. Although pellagra-like diseases had been identified in several species, most proved not to be useful as biological assays (indeed, most of these later proved to be manifestations of deficiencies of other vitamins of the B_2 complex and to be wholly unrelated to pellagra in humans).

The identification of a useful animal model for pellagra came from Goldberger's discovery in 1922 that maintaining dogs on diets essentially the same as those associated with human pellagra resulted in the animals developing a necrotic degeneration of the tongue called **black tongue disease**. This animal model for the disease led to the final solution of the problem.

44. Floridin, a nonplastic variety of kaolin containing an aluminum magnesium silicate. The material is useful as a decolorizing medium. Its name comes from an ancient process of cleaning or *fulling* wool, in which a slurry of earth or clay was used to remove oil and particulate dirt.

45. By this procedure, the albumen from 10,000 eggs yielded c.30 mg of riboflavin.

46. Initially, the term **flavin** was used with a prefix that indicated the source material; for example, ovoflavin, hepatoflavin, and lactoflavin designated the substances isolated from egg white, liver, and milk, respectively.

47. See the listing of papers of key historical significance, in Recommended Reading at the end of this chapter.

48. People with pellagra.

49. Goldberger conducted this study with the approval of prison authorities. As compensation for participation, volunteers were offered release at the end of the 6 mo. experimental period, which each exercised upon the conclusion of the study without evaluation. For that reason, Goldberger was unable to demonstrate to a doubting medical community that the unbalanced diet had, indeed, produced pellagra.

Impact of an Animal Model for Pellagra

This finding made possible experimentation that would lead rather quickly to an understanding of the etiology to the disease. Goldberger's group soon found that yeast, wheat germ, and liver would prevent canine black tongue and produce dramatic recoveries in pellagra patients. By the early 1930s, it was established that the human pellagra and canine black tongue curative factor was heat-stable and could be separated from the other B_2 complex components by filtration through fuller's earth, which adsorbed only the latter. Thus, the P-P factor became known as the *filtrate factor*. In 1937, Elvehjem isolated *nicotinamide* from liver extracts that had high antiblack tongue activity and showed that nicotinamide and **nicotinic acid** each cured canine black tongue. Both compounds are now called **niacin**. In the same year, several groups went on to show the curative effect of nicotinic acid against human pellagra.

It is ironic that the antipellagra factor was already well known to chemists of the time. Some 70 years earlier, the German chemist Huber had prepared nicotinic acid by the oxidation of nicotine with nitric acid. Funk had isolated the compound from yeast and rice bran in his search for the antiberiberi factor; however, because it had no effect on beriberi, nicotinic acid remained, for two decades, an entity with unappreciated biological importance. This view changed in the mid-1930s, when Warburg and Christiaan isolated nicotinamide from the hydrogen-transporting coenzymes I and II,[50] giving the first clue to its importance in metabolism. Within a year, Elvehjem had discovered its nutritional significance.

B_2 Complex Yields Pyridoxine

During the course of their work leading to the successful isolation of riboflavin, Kuhn and colleagues noticed an anomalous relationship between the growth-promoting and fluorescence activities of their extracts: the correlation of the two activities diminished at high levels of the former. Further, the addition of nonfluorescent extracts was necessary for the growth-promoting activity of riboflavin. They interpreted these findings as evidence for a second component of the heat-stable complex—one that was removed during the purification of riboflavin. These factors were also known to prevent dermatoses in the rat, an activity called **adermin**; however, the lack of a critical assay that could differentiate between the various components of the B_2 complex led to a considerable confusion.

In 1934, György proffered a definition of what he called **vitamin B_6 activity**[51] as the factor that prevented what had formerly been called **acrodynia** or **rat pellagra**, which was a symmetrical florid dermatitis spreading over the limbs and trunk, with redness and swelling of the paws and ears. His definition effectively distinguished these signs from those produced by riboflavin deficiency, which involves lesions on the head and chest, and inflammation of the eyelids and nostrils. The focus provided by György's definition strengthened the use of the rat in the bioassay of vitamin B_6 activity by clarifying its end point. Within 2 years, partial purification of **vitamin B_6** had been achieved by his group; and in 1938 (only 4 years after the recognition of the vitamin), the isolation of vitamin B_6 in crystalline form was achieved by five research groups. The chemical structure of the substance was quickly elucidated as 3-hydroxy-4,5-bis-(hydroxymethyl)-2-methylpyridine. In 1939, Folkers achieved the synthesis of this compound, which György called **pyridoxine**.

B_2 Complex Yields Pantothenic Acid

In the course of studying the growth factor called vitamin B_2, Norris and Ringrose at Cornell described, in 1930, a pellagra-like syndrome of the chick. The lesions could be prevented with aqueous extracts of yeast or liver, then recognized to contain the B_2 complex. In studies of B_2 complex-related growth factors for chicks and rats, Jukes and colleagues at Berkeley found positive responses to a thermostable factor that, unlike pyridoxine, was not adsorbed by fuller's earth from an acid solution. They referred to it as their **filtrate factor**.

At the same time, and quite independently, the University of Texas microbiologist R.J. Williams was pursuing studies of the essential nutrients for *Saccharomyces cerevisiae* and other yeasts. His group found a potent growth factor that they could isolate from a wide variety of plant and animal tissues.[52] They called it **pantothenic acid**, meaning "found everywhere," and also referred to the substance as **vitamin B_3**. Later in the decade, Snell's group found that several lactic and propionic acid bacteria require a growth factor that had the same properties. Jukes recognized that his filtrate factor, Norris's chick antidermatitis factor, and the unknown factors required by yeasts and bacteria were identical. He demonstrated that both his filtrate factor and pantothenic acid obtained from Williams could prevent dermatitis in the chick. Pantothenic acid was isolated and its chemical structure was determined by Williams's group in 1939. The chemical synthesis of the vitamin was achieved by Folkers the following year. The

50. Nicotinamide adenine dinucleotide (NAD) and nicotinamide adenine dinucleotide phosphate (NADP), respectively.

51. György defined vitamin B_6 activity as "that part of the vitamin B-complex responsible for the cure of a specific dermatitis developed by rats on a vitamin-free diet supplemented with vitamin B1, and lactoflavin."

52. The first isolation of pantothenic acid employed 250 kilograms of sheep liver. The autolysate was treated with fuller's earth; the factor was adsorbed to Norite and eluted with ammonia. Brucine salts were formed and were extracted with chloroform–water, after which the brucine salt of pantothenic acid was converted to the calcium salt. The yield was 3 grams of material with c. 40% purity.

TABLE 2.4 Factors Leading to the Discovery of Pantothenic Acid

Factor	Bioassay
Filtrate factor	Chick growth
Chick antidermatitis factor	Prevention of skin lesions and poor feather development in chicks
Pantothenic acid	Growth of *Saccharomyces cerevisiae* and other yeasts

various factors leading to the discovery of pantothenic acid are presented in Table 2.4.

A Fat-Soluble, Antisterility Factor: Vitamin E

Interest in the nutritional properties of lipids was stimulated by the resolution of fat-soluble A into vitamins A and D by the early 1920s. Several groups found that supplementation with the newly discovered vitamins A, C, and D and thiamin markedly improved the performance of animals fed purified diets containing adequate amounts of protein, carbohydrate, and known required minerals. However, H.M. Evans and Katherine Bishop, at the University of California, observed that rats fed such supplemented diets seldom reproduced normally. They found that fertility was abnormally low in both males (which showed testicular degeneration) and females (which showed impaired placental function and failed to carry their fetuses to term).[53] Dystrophy of skeletal and smooth muscles of the uterus was also noted. In 1922, these investigators reported that the addition of small amounts of yeast or fresh lettuce to the purified diet would restore fertility to females and prevent infertility in animals of both sexes. They designated the unknown fertility factor as *factor X*. Using the prevention of **gestation resorption** as the bioassay, Evans and Bishop found factor X activity in such unrelated materials as dried alfalfa, wheat germ, oats, meats, and milk fat, from which it was extractable with organic solvents. They distinguished the new fat-soluble factor from the known fat-soluble vitamins by showing that single droplets of wheat germ oil administered daily completely prevented gestation resorption, whereas cod liver oil, known to be a rich source of vitamins A and D, failed to do so.[54] In 1924, Sure, at the University of Arkansas, confirmed this work, concluding that the fat-soluble factor was a new vitamin, which he called **vitamin E**.

53. The vitamin E-deficient rat carries her fetuses quite well until a fairly late stage of pregnancy, at which time they die and are resorbed. This syndrome is distinctive; it termed **gestation resorption**.

54. In fact, Evans and Bishop found that cod liver oil actually increased the severity of the gestation resorption syndrome, a phenomenon now understood on the basis of the antagonistic actions of high concentrations of the fat-soluble vitamins.

A Classic Touch in Coining Tocopherol

Soon, Evans was able to prepare a potent concentrate of vitamin E from the unsaponifiable lipids of wheat germ oil; others prepared similar vitamin E-active concentrates from lettuce lipids. By the early 1930s, Olcott and Mattill at the University of Iowa had found that such preparations, which prevented the gestation resorption syndrome in rats, also had chemical antioxidant properties that could be assayed in vitro.[55] In 1936, Evans isolated from unsaponifiable wheat germ lipids allophanic acid esters of three alcohols, one of which had very high biological vitamin E activity. Two years later, Fernholz showed that the latter alcohol had a phytyl side chain and a hydroquinone moiety and proposed the chemical structure of the new vitamin. Evans coined the term **tocopherol**, which he derived from the Greek words *tokos* ("childbirth") and *pherein* ("to bear");[56] he used the suffix *-ol* to indicate that the factor is an alcohol. He also named the three alcohols α-, β-, and γ-tocopherol. In 1938, synthesis of the most active vitamer, α-tocopherol, was achieved by the groups of Karrer, Smith, and Bergel. A decade later another vitamer, δ-tocopherol, was isolated from soybean oil; not until 1959 were the **tocotrienols** described.[57]

Antihemorrhagic Factor: Vitamin K

In the 1920s, Henrik Dam, at the University of Copenhagen, undertook studies to determine whether cholesterol was an essential dietary lipid. In 1929, Dam reported that chicks fed diets consisting of food that had been extracted with nonpolar solvents to remove sterols developed subdural, subcutaneous, or intramuscular **hemorrhages**, **anemia**, and abnormally long blood-clotting times. A similar syndrome in chicks fed ether-extracted fish meal was reported by McFarlane's group, which at the time was attempting to determine the chick's requirements for vitamins A and D. They found that nonextracted fish meal completely prevented the clotting defect. Holst and Holbrook found that cabbage prevented the syndrome, which they took as evidence of an involvement of vitamin C. By the mid-1930s, Dam had shown that the clotting

55. Although the potencies of the vitamin preparations in the in vivo (rat gestation resorption) and in vitro (antioxidant) assays were not always well correlated.

56. Evans wrote in 1962 that he was assisted in the coining of the name for vitamin E by George M. Calhoun, Professor of Greek and a colleague at the University of California. It was Calhoun who suggested the Greek roots of this now-familiar name.

57. The tocotrienols differ from the tocopherols only by the presence of three conjugated double bonds in their phytyl side chains.

defect was also prevented by a fat-soluble factor present in green leaves and certain vegetables, and distinct from vitamins A, C, D, and E. He named the fat-soluble factor **vitamin K**.[58]

At that time, Herman Almquist and Robert Stokstad, at the University of California, found that the hemorrhagic disease of chicks fed a diet based on ether-extracted fish meal and brewers' yeast, polished rice, cod liver oil, and essential minerals was prevented by a factor present in ether extracts of alfalfa, and that was also produced during microbial spoilage of fish meal and wheat bran. Dam's colleague, Schønheyder, discovered the reason for prolonged blood-clotting times of vitamin K-deficient animals. He found that the clotting defect did not involve a deficiency of tissue thrombokinase or plasma fibrinogen, or an accumulation of plasma anticoagulants; he also determined that affected chicks showed relatively poor thrombin responses to exogenous thromboplastin. The latter observation suggested inadequate amounts of the clotting factor **prothrombin**, a factor already known to be important in the prevention of hemorrhages.

In 1936, Dam partially purified chick plasma prothrombin and showed its concentration to be depressed in vitamin K-deficient chicks. It would be several decades before this finding was fully understood.[59] Nevertheless, the clotting defect in the chick model served as a useful bioassay tool. When chicks were fed foodstuffs containing the new vitamin, their prothrombin values were normalized; hence, clotting time was returned to normal and the hemorrhagic disease was cured. The productive use of this bioassay led to the elucidation of the vitamin and its functions.

Vitamers K

Vitamin K was first isolated from alfalfa by Dam in collaboration with Paul Karrer at the University of Zurich in 1939. They found that the active substance, which was a yellow oil, was a quinone. The structure of this form of the vitamin (called **vitamin K_1**) was elucidated by Doisy's group at the University of St Louis, and by Karrer's, Almquist's and Feiser's groups in the same year. Soon, Doisy's group isolated a second form of the vitamin from putrified fish meal; this vitamer (called **vitamin K_2**) was crystalline. Subsequent studies demonstrated that this vitamer too differs from vitamin K_1 by having an unsaturated isoprenoid side chain at the 3-position of the naphthoquinone ring; in addition, putrified fish meal was found to contain several vitamin K_2-like substances with polyprenyl groups of differing chain lengths. Syntheses of vitamins K_2 were later achieved by Isler's and Folker's groups. A strictly synthetic analog of vitamers K_1 and K_2, consisting of the methylated head group alone (i.e., 2-methyl-1,4-naphthoquinone), was shown by Ansbacher and Fernholz to have high antihemorrhagic activity in the chick bioassay. It is, therefore, referred to as **vitamin K_3**.

Bios Yields Biotin

During the 1930s, independent studies of a yeast growth factor (called **bios IIb**[60]), a growth- and respiration-promoting factor for *Rhizobium trifolii* (called **coenzyme R**), and a factor that protected the rat against hair loss and skin lesions induced by raw egg white feeding (called **vitamin H**[61]) converged in an unexpected way. Kögl's group isolated the yeast growth factor from egg yolk and named it **biotin**. In 1940, György, du Vigneaud, and colleagues showed that vitamin H prepared from liver was remarkably similar to Kögl's egg yolk biotin.[62] The chemical structure of biotin was elucidated in 1942 by du Vigneaud's group at Cornell Medical College;[63] its complete synthesis was achieved by

58. Dam cited the fact that the next letter of the alphabet that had not previously been used to designate a known or proposed vitamin-like activity was also the first letter in the German or Danish phrase *koagulation facktor*, and was thus a most appropriate designator for the antihemorrhagic vitamin. The phrase was soon shortened to *K factor* and, hence, **vitamin K**.

59. It should be remembered that, at the time of this work, the biochemical mechanisms involved in clotting were incompletely understood. Of the many proteins now known to be involved in the process, only prothrombin and fibrinogen had been definitely characterized. It would not be until the early 1950s that the remainder of the now-classic clotting factors would be clearly demonstrated and that, of these, factors VII, IX, and X would be shown to be dependent on vitamin K. While these early studies effectively established that vitamin K deficiency results in impaired prothrombin activity, that finding would be interpreted as indicative of a vitamin K-dependent activation of the protein to its functional form.

60. **Bios IIb** was one of three essential growth factors for yeasts that had been identified by Wilders at the turn of the century in response to the great controversy that raged between Pasteur and Liebig. In 1860, Pasteur had declared that yeast could be grown in solutions containing only water, sugar, yeast ash (i.e., minerals), and ammonium tartrate; he noted, however, the growth-promoting activities of *albuminoid materials* in such cultures. Liebig challenged the possibility of growing yeast in the absence of such materials. Although Pasteur's position was dominant through the close of the century, Wilders presented evidence that proved that cultivation of yeast actually did require the presence of a little wort, yeast water, peptone, or beef extract. (Wilders showed that an inoculum the size of a bacteriological loopful, which lacked sufficient amounts of these factors, was unsuccessful, whereas an inoculum the size of a pea grew successfully.) Wilders used the term *bios* to describe the new activity required for yeast growth. For three decades, investigators undertook to characterize Wilders's bios factors. By the mid-1920s, three factors had been identified: *bios I*, which was later identified as meso-inositol; **bios IIa**, which was replaced by pantothenic acid in some strains and by β-alanine plus leucine in others; and **bios IIb**, which was identified as biotin.

61. György used the designation *H* after the German word *haut* (skin).

62. For a time, the factors obtained from egg yolk and liver were called α-biotin and β-biotin, respectively. They were reported as having different melting points and optical rotations. Subsequent studies, however, clearly demonstrated that such differences do not exist, nor do preparations from these sources exhibit different activities in microbiological systems.

63. du Vigneaud was to receive a Nobel Prize in Medicine for his work on the metabolism of methionine and methyl groups.

TABLE 2.5 Factors Leading to the Discovery of Biotin

Factor	Bioassay
Bios IIb	Yeast growth
Coenzyme R	*Rhizobium trifolii* growth
Vitamin H	Prevention of hair loss and skin lesions in rats fed raw egg white

Folkers in the following year. A summary of the factors leading to the discovery of biotin is presented in Table 2.5.

Antianemia Factors

The last discoveries that led to the elucidation of new vitamins involved findings of anemias of dietary origin. The first of these was reported in 1931 by Lucy Wills's group as a **tropical macrocytic anemia**[64] observed in women in Bombay, India, which was often a complication of pregnancy. They found that the anemia could be treated effectively by supplementing the women's diet with an extract of autolyzed yeast.[65] Wills and associates found that a macrocytic anemia could be produced in monkeys by feeding them food similar to that consumed by the women in Bombay. Further, the monkey anemia could be cured by oral administration of yeast or liver extract, or by parenteral administration of extract of liver; these treatments also cured human patients. The antianemia activity in these materials thus became known as the **Wills factor**.

Vitamin M?

Elucidation of the Wills factor involved the convergence of several lines of research, some of which appeared to be unrelated. The first of these came in 1935 from the studies of Day and colleagues at the University of Arkansas Medical School, who endeavored to produce riboflavin deficiency in monkeys. They fed their animals a cooked diet consisting of polished rice, wheat, washed casein, cod liver oil, a mixture of salts, and an orange; quite unexpectedly, they found them to develop anemia, leukopenia,[66] ulceration of the gums, diarrhea, and increased susceptibility to bacillary dysentery. They found that the syndrome did not respond to thiamin, riboflavin, or nicotinic acid; however, it could be prevented by daily supplements of either 10 grams of brewers' yeast or 2 grams of a dried hog liver–stomach preparation. Day named the protective factor in brewers' yeast **vitamin M** (for monkey).

Factors U and R, and Vitamin B_c

In the late 1930s, three groups (Robert Stokstad's at the University of California, Leo Norris's at Cornell, and Albert Hogan's at the University of Missouri) reported syndromes characterized by anemia in chicks fed highly purified diets. The anemias were found to respond to dietary supplements of yeast, alfalfa, and wheat bran. Stokstad and Manning called this unknown *factor U*; Baurenfeind and Norris at Cornell called it **factor R**. Shortly thereafter, Hogan and Parrott discovered an antianemic substance in liver extracts; they called it **vitamin B_c**.[67] At the time (1939), it was not clear to what extent these factors may have been related.

Yeast Growth Related to Anemia?

At the same time, the microbiologists Snell and Peterson, who were studying the bios factors required by yeasts, reported the existence of an unidentified water-soluble factor that was necessary for the growth of *Lactobacillus casei*. This factor was present in liver and yeast, from which it could be prepared by adsorption to and then elution from Norit;[68] for a while they called it the *yeast Norit factor*, but it quickly became known as the *L. casei factor*. Hutchings and colleagues at the University of Wisconsin further purified the factor from liver and found it to stimulate chick growth; this suggested a possible identity of the bacterial and chick factors. The factor from liver was found to stimulate the growth of both *Lactobacillus helveticus* and *Streptococcus fecalis* R.,[69] whereas the yeast-derived factor was twice as potent for *L. helveticus* as it was for *S. fecalis*. Thus, it became popular to refer to these as the "liver *L. casei* factor" and the "yeast (or fermentation) *L. casei* factor."

Snell's group found that many green leafy materials were potent sources of something with the microbiological effects of the **Norit eluate factor**—extracts promoted the growth of both *S. fecalis* and *L. casei*. They named the factor, by virtue of its sources, **folic acid**. In 1943, a fermentation product

64. A **macrocytic anemia** is one in which the number of circulating erythrocytes is below normal, but the mean size of those present is greater than normal (normal range, 82–92 μm³). Macrocytic anemias occur in such syndromes as pernicious anemia, sprue, celiac disease, and macrocytic anemia of pregnancy. Wills' studies of the macrocytic anemia in her monkey model revealed megaloblastic arrest (i.e., failure of the large, nucleated, embryonic erythrocyte precursor cell type to mature) in the erythropoietic tissues of the bone marrow, and a marked reticulocytosis (i.e., the presence of young red blood cells in numbers greater than normal [usually <1%], occurring during active blood regeneration); both signs were eliminated coincidentally on the administration of extracts of yeast or liver.

65. Wills's yeast extract was not particularly potent, as they needed to administer 4 g two to four times daily to cure the anemia.

66. Leukopenia refers to any situation in which the total number of leukocytes (i.e., white blood cells) in the circulating blood is less than normal, which is generally c.5000 per mm³.

67. Hogan and Parrott used the subscript *c* to designate this factor as one required by the chick.

68. A carbon-based filtering agent.

69. *Streptococcus fecalis* was then called *Streptococcus lactis* R.

was isolated that stimulated the growth of *S. fecalis* but not *L. casei*; this was called the **SLR factor** and, later, **rhizopterin**.

Who's on First?

It was far from clear in the early 1940s whether any of these factors were at all related, as folic acid appeared to be active for both microorganisms and animals, whereas concentrates of vitamin M, factors R and U, and vitamin B_c appeared to be effective only for animals. Clues to solving the puzzle came from the studies of Mims and associates at the University of Arkansas Medical School, who showed that incubation of vitamin M concentrates in the presence of rat liver enzymes caused a marked increase in the folic acid activity (i.e., assayed using *S. casei* and *Streptococcus lactis* R.) of the preparation. Subsequent work showed such "activation" enzymes to be present in both hog kidney and chick pancreas. Charkey, of the Cornell group, found that incubation of their factor R preparations with rat or chick liver enzymes produced large increases in their folic acid potencies for microorganisms. These studies indicated for the first time that at least some of these various substances may be related.

TABLE 2.6 Factors Leading to the Discovery of Folic Acid

Factor	Bioassay
Wills' factor	Cure of anemia in humans
Vitamin M	Prevention of anemia in monkeys
Vitamin B_c	Prevention of anemia in chicks
Factor R	Prevention of anemia in chicks
Factor U	Prevention of anemia in chicks
Yeast Norit factor	Growth of *Lactobacillus casei*
L. casei factor	Growth of *L. casei*
SLR factor	Growth of *Rhizobium* species
Rhizopterin	Growth of *Rhizobium* species
Folic acid	Growth of *Streptococcus fecalis* and *L. casei*

Derivatives of Pteroylglutamic Acid

The real key to solving what was clearly the most complicated puzzle in the discovery of the vitamins came in 1943 with the isolation of **pteroylglutamic acid** from liver by Stokstad's group at the Lederle Laboratories of American Cyanamid, Inc., and by Piffner's group at Parke-Davis, Inc. Stokstad's group achieved the synthesis of the compound in 1946. Soon it was found that pteroylmonoglutamic acid was indeed the substance that had been variously identified in liver as factor U, vitamin M, vitamin B_c, and the liver *L. casei* factor. The yeast *L. casei* factor was found to be the diglutamyl derivative (pteroyldiglutamic acid) and the liver-derived vitamin B_c was the hexaglutamyl derivative (pteroylhexaglutamic acid). Others of these factors (the *SLR factor*) were subsequently found to be single-carbon metabolites of pteroylglutamic acid. These various compounds thus became known generically as **folic acid**. A summary of the factors leading to the discovery of folic acid is presented in Table 2.6.

Antipernicious Anemia Factor

The second nutritional anemia that was found to involve a vitamin deficiency was the fatal condition of human patients that was first described by J.S. Combe in 1822 and became known as **pernicious anemia**.[70] The first real breakthrough toward understanding the etiology of pernicious anemia did not come until 1926, when Minot and Murphy found that lightly cooked liver, which the prominent hematologist G.H. Whipple had found to accelerate the regeneration of blood in dogs made anemic by exsanguination, was highly effective as therapy for the disease.[51,71,72] This indicated that liver contained a factor necessary for hemoglobin synthesis.

Intrinsic and Extrinsic Factors

Soon, studies of the antipernicious anemia factor in liver revealed that its enteric absorption depended on yet another factor in the gastric juice, which W.B. Castle, in 1928, called the **intrinsic factor**, to distinguish it from the *extrinsic factor* in liver. Biochemists then commenced a long endeavor to isolate the antipernicious anemia factor from liver. The isolation of the factor was necessarily slow and arduous for the reason that the only bioassay available was the hematopoietic response of human pernicious anemia patients, which was frequently not available. No animal model had been found, and a bioassay could not be replaced by a chemical reaction or physical method because, as is now known, this most potent vitamin is active at exceedingly low concentrations. Therefore, it was most important to the elucidation of the antipernicious anemia factor when, in 1947, Mary Shorb of the University of Maryland found that it

70. This condition has also been called **Addison's anemia** after T. Addison, who described it in great detail in 1949, and **Biemer's anemia**, after A. Biemier, who reported the disease in Zurich in 1872 and coined the term **pernicious anemia**.

71. Minot and Murphy treated 45 pernicious anemia patients with 120–240 g of lightly cooked liver per day. The patients' mean erythrocyte count increased from 1.47×10^6/mL before treatment to 3.4×10^6/mL and 4.6×10^6/mL after 1 and 2 mos. of treatment, respectively.

72. Whipple, Minot, and Murphy shared the 1934 Nobel Prize in Medicine for the discovery of whole liver therapy for pernicious anemia.

was also required for the growth of *Lactobacillus lactis* Dorner.[73] With Shorb's microbiological assay, isolation of the factor, by that time named **vitamin B_{12}** by the Merck group, proceeded rapidly.

Animal Protein Factors

At about the same time, animal growth responses to factors associated with animal proteins or manure were reported as American animal nutritionists sought to eliminate expensive and scarce animal by-products from the diets of livestock. Norris's group at Cornell attributed responses of this time to an **animal protein factor**; the factor in liver necessary for rat growth was called **factor X** by Cary and **zoopherin**[74] by Zucker and Zucker. It soon became evident that these factors were probably identical. Stokstad's group found the factor in manure and isolated an organism from poultry manure that would synthesize a factor that was effective both in promoting chick growth and in treating pernicious anemia. That the antipernicious anemia factor was produced microbiologically was important, in that it led to an economical means of industrial production of vitamin B_{12}.

Vitamin B_{12} Isolated

By the late 1940s, Combs[75] and Norris, using chick growth as their bioassay procedure, were fairly close to the isolation of vitamin B_{12}. However, in 1948, Folkers at Merck, using the *L. lactis* Dorner assay, succeeded in first isolating the antipernicious anemia factor in crystalline form. This achievement was accomplished in the same year by Lester Smith's group at the Glaxo Laboratories in England (who found their pink crystals to contain cobalt), assaying their material on pernicious anemia patients in relapse.[76] The elucidation of the complex chemical structure of vitamin B_{12} was finally achieved in 1955 by Dorothy Hodgkin's group at Oxford with the use of X-ray crystallography. In the early 1960s, several groups accomplished the partial synthesis of the vitamin; it was not until 1970 that the de novo synthesis of vitamin B_{12} was finally achieved by Woodward and Eschenmoser. A summary of the factors leading to the discovery of vitamin B_{12} is presented in Table 2.7.

TABLE 2.7 Factors Leading to the Discovery of Vitamin B_{12}

Factor	Bioassay
Extrinsic factor	Cure of anemia in humans
LLD factor	Growth of *L. lactis* Dorner
Vitamin B_{12}	Growth of *L. lactis* Dorner
Animal protein factor	Growth of chicks
Factor X	Growth of rats
Zoopherin	Growth of rats

TABLE 2.8 Timelines for the Discoveries of the Vitamins

Vitamin	Proposed	Isolated	Structure Determined	Synthesis Achieved
Thiamin	1906	1926	1932	1933
Vitamin C	1907	1926	1932	1933
Vitamin A	1915	1937	1942	1947
Vitamin D	1919	1932	1932 (D_2)	1932
			1936 (D_3)	1936
Vitamin E	1922	1936	1938	1938
Niacin	1926	1937	1937	1867[a]
Vitamin B_{12}	1926	1948	1955	1970
Biotin	1926	1939	1942	1943
Vitamin K	1929	1939	1939	1940
Pantothenic acid	1931	1939	1939	1940
Folate	1931	1939	1943	1946
Riboflavin	1933	1933	1934	1935
Vitamin B_6	1934	1936	1938	1939

[a]*Much of the chemistry of nicotinic acid was known before its nutritional roles were recognized.*

Vitamins Discovered in Only Five Decades

Beginning with the concept of a vitamin, which emerged with Eijkman's proposal of an antipolyneuritis factor in 1906, the elucidation of the vitamins continued through the isolation of vitamin B_{12} in potent form in 1948 (Table 2.8). Thus, the identification of the presently recognized vitamins was achieved within a period of only 42 years! For some vitamins (e.g., pyridoxine) for which convenient animal models were available, discoveries came rapidly; for others (e.g., niacin, vitamin B_{12}) for which animal models were late to be found, the pace of scientific progress was much slower (Fig. 2.1). These paths of discovery were marked by nearly a dozen Nobel Prizes (Table 2.9).

73. For a time, this was referred to as the **LLD factor**.
74. The term **zoopherin** carries the connotation: "to carry on an animal species."
75. Characterization of the animal protein factor was the subject of the senior author's father's doctoral thesis in Norris's laboratory at Cornell in the late 1940s.
76. Friedrich (1988) has pointed out that it should be no surprise that the first isolations of vitamin B_{12} were accomplished in industrial laboratories, as the task required industrial-scale facilities to handle the enormous amounts of starting material that were needed. For example, the Merck group used a ton of liver to obtain 20 mg of crystalline material.

TABLE 2.9 Nobel Prizes Awarded for Research on Vitamins

Year of Award	Recipients	Discovery
Prizes in Medicine and Physiology		
1929	Christiaan Eijkman and Frederick G. Hopkins	Discovery of the antineuritic vitamin; discovery of the growth-stimulating vitamins
1934	George H. Whipple, George R. Minot, and William P. Murphy	Discoveries concerning liver therapy against pernicious anemia
1937	Albert von Szent-Györgi and Charles G. King	Discoveries in connection with the biological combustion, with especial reference to vitamin C, and the catalysis of fumaric acid
1943	Henrik Dam and Edward ADoisy	Discovery of vitamin K; discovery of the chemical nature of vitamin K
1953	Fritz A. Lipmann	Discovery of coenzyme A and its importance in intermediary metabolism
1955	Hugo Theorell	Discoveries relating to the nature and mode of action of oxidizing enzymes
1964	Feodor Lynen and Konrad Bloch	Discoveries concerning the mechanism and regulation of cholesterol and fatty acid metabolism
Prizes in Chemistry		
1928	Adolf Windaus	Studies on the constitution of the sterols and their connection with the vitamins
1937	Paul Karrer and Walter N. Haworth	Research on the constitution of carotenoids, flavins, and vitamins A and B; researches into the constitution of carbohydrates and vitamin C
1938	Richard Kuhn	Work on carotenoids and vitamins
1957	Alexander Todd	Work on the structure of nucleotides (including vitamin B_{12})
1964	Dorothy C. Hodgkin	Elucidation of the structure of vitamin B_{12}
1965	Robert B. Woodward	Chemical synthesis of vitamin B_{12}
1967	George Wald, H.K. Hartline, and R. Grant	Discoveries of the basic chemical and physiological processes in vision

7. VITAMIN TERMINOLOGY

The terminology of the vitamins can be as daunting as that of any other scientific field. Many vitamins carry alphabetic or alphanumeric designations, yet the sequence of such designations has an arbitrary appearance by virtue of its many gaps and inconsistent application to all of the vitamins. This situation notwithstanding, the logic underlying the terminology of the vitamins becomes apparent when it is viewed in terms of the history of vitamin discovery. The familiar designations in use today are, in most cases, the surviving terms coined by earlier researchers on the paths to vitamin discovery. Thus, because McCollum and Davis used the letters A and B to distinguish the lipid-soluble antixerophthalmic factor from the water-soluble antineuritic and growth activity that was subsequently found to consist of several vitamins, such chemically and physiologically unrelated substances as thiamin, riboflavin, pyridoxine, and cobalamins (in fact, all water-soluble vitamins except ascorbic acid, which was designated before the vitamin B complex was partitioned) are all called B vitamins. In the case of folic acid, certainly the name survived its competitors by virtue of its relatively attractive sound (e.g., vs **rhizopterin**). Therefore, the accepted designations for the vitamins, in most cases, have relevance only to the history and chronology of their discovery and not to their chemical or metabolic similarities. The discovery of the vitamins left a path littered with designations of "vitamins," "factors," and other terms, most of which have been discarded (see Appendix A for a complete listing).

8. OTHER FACTORS SOMETIMES CALLED VITAMINS

Several other factors have, at various times or under certain conditions, been called vitamins. Many remain today only as historic markers of once incompletely explained phenomena, now better understood. Today, some factors would appear to satisfy, for at least some species, the operating definition of a vitamin; although in practice that term

TABLE 2.10 Vitamin-Like Factors

Substance	Biological Activity
Choline	Component of the primary membrane structural component phosphatidylcholine and the neurotransmitter acetylcholine; contributor to single-C metabolism; essential for normal growth and bone development in young poultry; can spare methionine in many animal species; and thus can be essential in diets that provide limited methyl groups.
Nonprovitamin A Carotenoids	
Flavonoids	Reported to reduce capillary fragility, and inhibit in vitro aldolase reductase (has a role in diabetic cataracts) and *o*-methyltransferase (inactivates epinephrine and norepinephrine)
Carnitine	Essential for transport of fatty acyl CoA from cytoplasm to mitochondria for β oxidation; synthesized by most species except some insects, which require a dietary source for growth
myo- i nositol	Component of phosphatidylinositol; prevents diet-induced lipodystrophies due to impaired lipid transport in gerbils and rats; essential for some microbes, gerbils, and certain fishes
Ubiquinones	Group includes a component of the mitochondrial respiratory chain; are antioxidants and can spare vitamin E in preventing anemia in monkeys, and in maintaining sperm motility in birds
Orotic Acid	
p-Aminobenzoic acid	Essential growth factor for several microbes, in which it functions as a provitamin of folic acid; reported to reverse diet- or hydroquinone-induced achromotrichia in rats and to ameliorate rickettsial infections
Lipoic acid	Cofactor in oxidative decarboxylation of α-keto acids; essential for growth of several microbes but inconsistent effects on animal growth
Pyrroloquinoline quinone	Component of certain bacterial and mammalian metallooxidoreductases; deprivation impairs growth, causes skin lesions in mice

is restricted to those factors required by higher organisms.[77] These vitamin-like factors (Table 2.10) are discussed in Chapter 19. At various times, of course, other factors have been represented as vitamins; however, no solid evidence has been sustained to support such claims (Table 2.11).

9. MODERN HISTORY OF THE VITAMINS

While the first half of the 20th century was an exciting period of vitamin discovery, the subsequent history of this field has been characterized by the generation of the huge amount of additional information needed to use the vitamins to improve human and animal health and to optimize the efficiency of producing food animals.[78] This work has revealed that some vitamins are widely underconsumed, that some may be associated with chronic disease, and that several are produced by the gut microbiome. Many recent studies have strived to elucidate advanced functional roles (such as transcriptional regulation) for the vitamins beyond those associated with their original discovery. Still, there are concerns about how best to assess vitamin intake, vitamin status, and the contents and stabilities of vitamin isomers in foods.

77. Organic growth-promoting substances required only by microorganisms are frequently called **nutrilites**.

78. By the end of 2015, nearly 350,000 scientific papers on vitamins were listed in PubMed.gov (of the U.S. National Library of Medicine).

TABLE 2.11 Inactive Factors *Not* Considered Vitamins

Substance	Purported Biological Activity
Laetrile	A cyanogenic glycoside with unsubstantiated claims of antitumorigenicity
Gerovital	Unsubstantiated antiaging elixir
Orotic acid	Normal metabolic intermediate of pyrimidine biosynthesis with hypocholesterolemic activity
Pangamic acid	Ill-defined substance(s), originally derived from apricot pits, with unsubstantiated claims for a variety of health benefits

Have that all the vitamins have been discovered?[79] Perhaps, if one were to hold to the classical definition of "vitamin." Still, the field has already granted a considerable

79. When the senior author (GFC) was an assistant professor (in the mid-1970s), he discussed with his Dad how unlikely it had become that a graduate student could be asked to undertake thesis research on "unidentified growth factors" (UGFs). He asked his Dad why, when he was a graduate student in the latter 1940s, he had thought it profitable to undertake such research. The senior Dr Combs pointed out that "Every UGF had proven to be an essential nutrient, so we had confidence that the 'animal protein factor' would also be one." He was correct, of course; his work in the Cornell laboratory of Prof. Leo Norris contributed directly to the elucidation of vitamin B_{12}.

TABLE 2.12 Contemporary Vitamin Research Needs

Area	Needs
Analytical and physical chemistry	• Better understandings of chemical and biological potencies and stabilities (to storage, processing, and cooking) of the vitamins, their various vitamers, and chemical derivatives • Better analytical methods for measuring vitamin contents of food
Biochemistry and molecular biology	• More complete understanding of the molecular mechanisms of vitamin action, including roles in gene expression • More complete elucidation of the pathways of vitamin metabolism • More complete understanding of the interactions with other nutrients and/or factors (e.g., disease, oxidative stress, genotype) that affect vitamin functions and needs • Understanding of the contributions of the gut microbiome to vitamin nutritional status and the effects of diet composition
Nutritional surveillance and epidemiology	• Development of informative biomarkers of status for underconsumed vitamins (e.g., vitamin A, riboflavin, folate) • Better understanding of vitamin intakes and status of populations and at-risk subgroups • Better understanding of the relationships of vitamin intake/status and disease risks
Nutrition and dietetics	• More complete understanding of the quantitative requirements for vitamins for individuals at all life stages • More complete understanding of genotypes with particular vitamin needs (e.g., choline) • More complete understanding of the role of the gut microbiome as a source of vitamins (e.g., vitamin K) for the host
Medicine	• More complete understanding of the roles of vitamins in etiology and/or management of chronic (e.g., cancer, heart disease), congenital (e.g., neural tube defects), and infectious diseases • More complete understanding of the needs for vitamins over the life cycle • More complete understanding of risks of supranutritional intakes of certain vitamins (e.g., vitamin A, vitamin D, folate)
Agriculture and international development	• Development of smallholder farming/gardening systems and other food-based approaches that support nutritional requirements with respect to vitamins (e.g., Vitamin A, Vitamin C, folate) and other nutrients • Development of foods with increased contents of underconsumed vitamins and other micronutrients
Food science and technology	• Development of food processing techniques that retain vitamins in food • Development of novel means of vitamin supplementation and fortification to enhance enteric absorption efficiency (e.g., nanoparticles)

latitude in that definition by admitting vitamin D and vitamin C to vitamin status. Why not choline? Or lycopene? In fact, it may not serve the field to be concerned as to whether these and other bioactive factors in foods can be called vitamins. It will be better to remain open to reinterpreting the notion "vitamin" in light of emerging knowledge of the metabolic roles of bioactive factors in foods with documentable roles in metabolism and health. Contemporary understanding of vitamins and their roles in nutrition and health is the subject of the following chapters. Those chapters will show the areas in which that understanding is incomplete. Those areas must be the foci of future research (Table 2.12).

10. STUDY QUESTIONS AND EXERCISES

1. How did the vitamin theory influence the interpretation of findings concerning diet and health associations?
2. For each vitamin, list the key empirical observations that led to its initial recognition.
3. In what general ways were animal models employed in the discovery of the vitamins? What ethical issues must be addressed in this type of research?
4. Which vitamins were discovered as results of efforts to use chemically defined diets for raising animals? How would you go about developing such a diet?
5. Which vitamins were discovered primarily through human experimentation? What ethical issues must be addressed in this type of research?
6. Prepare a concept map illustrating the interrelationships of the various prevalent ideas and the many goals, approaches, and outcomes that resulted in the discovery of the vitamins.

RECOMMENDED READING

General History of the Vitamins

Baron, J.H., 2009. Sailor's scurvy before and after James Lind – a reassessment. Nutr. Rev. 67, 315–332.

Carpenter, K.J., 1986. The History of Scurvy and Vitamin C. Cambridge University Press, Cambridge, MA. 288 pp.

Carpenter, K.J., 2000. Beriberi, White Rice and Vitamin B: A Disease, a Cause, and a Cure. University of California Press, Los Angeles, CA. 282 pp.

Carpenter, K.J., 2003. A short history of nutritional science: Part 1 (1785–1885). J. Nutr. 133, 638–645.

Carpenter, K.J., 2003. A short history of nutritional science: Part 2 (1885–1912). J. Nutr. 133, 975–984.

Carpenter, K.J., 2003. A short history of nutritional science: Part 3 (1912–1944). J. Nutr. 133, 3023–3032.

Carpenter, K.J., 2003. A short history of nutritional science: Part 4 (1945–1985). J. Nutr. 133, 3331–3342.

Funk, C., 1912. The etiology of the deficiency diseases. J. State Med. 20, 341–368.

Goldblith, S.A., Joslyn, M.A. (Eds.), 1964. Milestones in Nutrition. AVI, Westport, CT. 797 pp.

György, P., 1954. Early experiences with riboflavin – a retrospect. Nutr. Rev. 12, 97–104.

Harris, L.J., 1955. Vitamins in Theory and Practice, fourth ed. Cambridge University Press, Cambridge, MA. 366 pp.

Hoffbrand, A.V., Weir, D.G., 2001. Historical review: the history of folic acid. Br. J. Hematol. 113, 579–589.

Lepkovsky, S., 1954. Early experiences with pyridoxine – a retrospect. Nutr. Rev. 12, 257–260.

McKay, C.M., 1973. Notes on the History of Nutrition Research. Hans Huber, Berne. 234 pp.

Northrop-Clewes, C.A., Thurnham, D., 2012. The discovery and characterization of riboflavin. Ann. Nutr. Metab. 61, 224–230.

Olson, J.A., 1994. Vitamins: the tortuous path from needs to fantasies. J. Nutr. 124, 1771S–1776S.

Roe, D.A., 1973. A Plague of Corn: The Social History of Pellagra. Cornell University Press, Ithaca, NY. 217 pp.

Sebrell, W.H., 1981. History of pellagra. Fed. Proc. 40, 1520–1522.

Smith, E.L., 1952. The discovery and identification of vitamin B_{12}. Proc. Nutr. Soc. 6, 295–299.

Sommer, A., 2008. Vitamin A deficiency and clinical disease: an historical overview. J. Nutr. 138, 1835–1839.

Terres, M., 1964. Goldberger on Pellagra. Louisiana State University Press, Baton Rouge. 395 pp.

Wald, G., 1968. Molecular basis of visual excitation. Science. 162, 230–239.

Wolf, G., 2001. The discovery of the visual function of vitamin A. J. Nutr. 131, 1647–1650.

Papers of Key Historical Significance

Vitamin A

McCollum, E.V., Davis, M., 1913. The necessity of certain lipins in the diet during growth. J. Biol. Chem. 15, 167–173.

Osborne, T.B., Mendel, L.B., 1917. The role of vitamins in the diet. J. Biol. Chem. 31, 149–163.

Wald, G., 1933. Vitamin A in the retina. Nature 132, 316–323.

Steenbock, H., 1919. A review of certain researches relating to the occurrence and chemical nature of vitamin A. Yale J. Med. 4, 563–578.

Vitamin D

McCollum, E.V., Simmonds, N., Pitz, W., 1916. The relation of the unidentified dietary factors, the fat-soluble A, and water-soluble B, of the diet to the growth promoting properties of milk. J. Biol. Chem. 27, 33–43.

Vitamin E

Evans, H.M., Bishop, K.S., 1922. On the existence of a hitherto unrecognized dietary factor essential for reproduction. Science 56, 650–651.

Olcott, H.S., Mattill, H.A., 1931. The unsaponifiable lipids of lettuce: II. Fractionation. J. Biol. Chem. 93, 65–70.

Vitamin K

Dam, H., 1929. Cholesterinositoffwechsel in huhnereirn und huhnchen. Biochem. Z. 215, 475–492.

Ascorbic acid

Holst, A., Frölich, T., 1907. Experimental studies relating to ship-beriberi and scurvy. II. On the etiology o scurvy. J. Hyg. Camb. 7, 634–671.

King, C.G., Waugh, W.A., 1932. The chemical nature of vitamin C. Science 75, 357–358.

Svirbely, J.L., Szent-Györgi, A., 1932. Hexuronic acid as the antiscorbutic factor. Nature 129, 576–583.

Thiamin

Eijkman, C., 1889. Over de oorzaak der Beri-Beri. Geneeskd. Tijdschr. Nederl. 29, 76–87.

Eijkman, C., Vorderman, A.G., 1898. Fe bestrijding der beri-beri. Koninklijke Akademie van Wenteschappen, Amsertadm, Verslag. Wis-en Natuurkundige Afd. 6, 6–11.

Riboflavin

Kuhn, R., Györgi, P., Wagner-Juregg, T., 1933. Über eine neue Klass von Naturefarbstoffen (Vorläufige Mitteilung). Ber. Dtsch. Chem. Ges. 66, 567–580.

Niacin

Elvehjem, C., Madden, R., Strong, F., et al., 1938. The isolation and identification of the anti-black tongue factor. J. Biol. Chem. 123, 137–149.

Goldberger, J., 1922. The relation of diet to pellagra. J. Am. Med. Assoc. 78, 1676–1680.

Warburg, O., Christian, W., 1935. Co-ferment problem. Biochem. Z. 275, 464–470.

Biotin

Kögl, F., Toennis, B., 1936. Über das bios-problem: Darstellung von krystallisiertem Biotin aus eigelb. Z. Physiol. Chem. 242, 43–73.

Pantothenic Acid

Norris, L.C., Ringrose, A.T., 1930. The occurrence of a pellagrous-like syndrome in chicks. Science 71, 643.

Williams, R.J., Lyman, C.M., Goodyear, G.H., et al., 1933. "Pantotheenic acid", a growth determinant of universal biological occurrence. J. Am. Chem. Soc. 55, 2912–2927.

Jukes, T.H., 1939. Pantothenic acid and the filtrate (chick anti-dermatitis) factor. J. Am. Chem. Soc. 61, 975–976.

Folate

Wills, L., 1931. Treatment of "pernicious anaemia of pregnancy" and "tropical anaemia" with special reference to yeast extract as a curative agent. Br. Med. J. 1, 1059–1064.

Mitchell, H.K., Snell, E.E., Williams, R.J., 1941. The concentration of "folic acid". J. Am. Chem. Soc. 63, 2284.

Mimms, V., Totter, J.R., Day, P.L., 1944. A method for the determination of substances enzymatically convertible to the factor stimulating Streptococcus lactis R. J. Biol. Chem. 155, 401–405.

Vitamin B_{12}

Castle, W.B., 1929. Observations on the etiologic relationship of achylia gastrica to pernicious anemia. I. The effect of the administration to patients with pernicious anemia of beef muscle after incubation with normal human gastric juice. Am. J. Med. Sci. 178, 748–763.

Minot, G.R., Murphy, W.P., 1926. Treatment of pernicious anemia by a special diet. JAMA 87, 470–476.

Shorb, M.S., 1948. Activity of vitamin B12 for the growth of Lactobacillus lactis. Science 107, 397–398.

Chapter 3

General Properties of Vitamins

Chapter Outline

Anchoring Concepts

1. The chemical composition and structure of a substance determine both its physical properties and chemical reactivity.
2. The physicochemical properties of a substance determine the ways in which it acts and is acted on in biological systems.
3. Substances tend to be partitioned between hydrophilic regions (plasma, cytosol, and mitochondrial matrix space) and hydrophobic regions (membranes, bulk lipid droplets) of biological systems on the basis of their relative solubilities; overcoming such partitioning requires actions of agents (micelles, binding, or transport proteins) that serve to alter their effective solubilities.
4. Isomers and analogs of a given substance may not have equivalent biological activities.

La vie est une fonction chimique.

A.L. Lavoisier[1]

LEARNING OBJECTIVES

1. To understand that the term **vitamin** refers to a family of compounds, i.e., structural analogs, with qualitatively similar biological activities but often quantitatively different potencies.
2. To become familiar with the chemical structures and physical properties of vitamins.
3. To understand the relationship between the physicochemical properties of vitamins and their stabilities, and how these properties affect their means of enteric absorption, transport, and tissue storage.
4. To become familiar with the general nature of vitamin metabolism.

VOCABULARY

Adenosylcobalamin
Ascorbic acid
β-Carotene
Binding proteins
Bioavailability
Biopotency
Biotin
Carotenoid
Cholecalciferol
6-Chromanol nucleus
Chylomicrons
Cobalamin
Corrin nucleus
Cyanocobalamin
Ergocalciferol
FAD
FMN
Folacin
Folic acid
HDL
LDL
Lipoproteins
Menadione
Menaquinone
Methylcobalamin
Micelle

1. Antoine-Laurent de Lavoisier (1743–94), while best known for his discovery of oxygen and its role in combustion, also discovered hydrogen and sulfur, recognized the constancy of mass in reactions that change the form of matter, and made important contributions to chemical nomenclature. A nobleman by birth and administrator of the Ferme Générale (profits from which funded his research) of the Ancien Régime, Lavoisier was sent to the guillotine at the height of the French Revolution.

The Vitamins. http://dx.doi.org/10.1016/B978-0-12-802965-7.00003-4

Microbiota
Microbiome
NAD
Naphthoquinone nucleus
Niacin
Nicotinamide
Nicotinic acid
NMN
Pantothenic acid
Phylloquinone
Portomicron
Pteridine
Pteroylglutamic acid
Pyridoxine
Pyridoxol
Retinal
Retinoic acid
Retinoid
Retinol
Riboflavin
Simon's metabolites
Steroid
Tetrahydrofolic acid
Thiamin
Thiamin pyrophosphate
Tocol
Tocopherol
Tocotrienol
Vitamin A
Vitamin B_2
Vitamin B_6
Vitamin B_{12}
Vitamin C
Vitamin D
Vitamin D_2
Vitamin D_3
Vitamin E
Vitamin K
Vitamin K_3
VLDL

1. VITAMIN NOMENCLATURE

The vitamins are organic, low molecular weight substances that have key roles in metabolism. Few of the vitamins are single substances; almost all are families of chemically related substances, i.e., **vitamers**, sharing qualitative (but not necessarily quantitative) biological activities. Thus, the vitamers comprising a vitamin family may vary in biopotency, and the common vitamin name is actually a generic descriptor for all of the relevant vitamers. Otherwise, vitamin families are chemically heterogeneous.

The nomenclature of the vitamins is in many cases complicated, reflecting both the terminology that evolved nonsystematically during the course of their discovery, as well as more recent efforts to standardize the vocabulary of the field. The standards for vitamin nomenclature policy were established by the International Union of Nutritional Sciences in 1978.[2] This policy distinguishes between generic descriptors used to describe families of compounds having vitamin activity (e.g., **vitamin D**) and to modify such terms as **activity** and **deficiency**, and trivial names used to identify specific compounds (e.g., **ergocalciferol**). These recommendations have been adopted by the Commission on Nomenclature of the International Union of Pure and Applied Chemists, the International Union of Biochemists, and the Committee on Nomenclature of the American Society for Nutrition. The latter organization publishes the policy every few years.[3] According to this accepted nomenclature, the vitamins are described as follows:

Vitamin A is the generic descriptor for compounds with the biological activity of retinol, formally derived from a monocyclic parent compound containing five C—C double bonds and a functional group at the terminus of the acyclic portion. Due to their structural similarities to retinol, they are called **retinoids**; those with vitamin A activity occur naturally in three forms: the alcohol **retinol**, the aldehyde **retinal** (also **retinaldehyde**), and the acid **retinoic acid**. β-Carotene and some other polyisoprenoid plant pigments (called **carotenoids** due to their relation to the carotenes) yield retinoids on metabolism and thus also have vitamin A activity; these are called **provitamin A carotenoids** and include β-carotene, a retinol dimer.

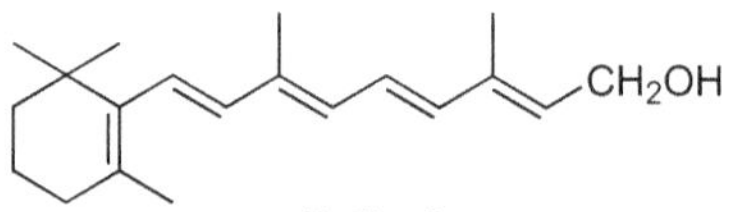

Retinol

Vitamin D is the generic descriptor for all steroids qualitatively exhibiting qualitatively the biological activity of **cholecalciferol**. These compounds are derived in vivo by photolysis of the B ring of 7-dehydrocholesterol but retain the intact A, C, and D steroid rings. Vitamin D-active compounds with a 9-carbon side chain containing a single double bond are derivatives of **ergocalciferol**, also called **vitamin D_2**, which can be produced by photolysis of plant sterols. Vitamin D-active compounds with an 8-carbon side chain and no double bonds are derivatives of **cholecalciferol**, also called **vitamin D_3**, which is produced metabolically through a natural process of photolysis of 7-dehydrocholesterol on the surface of skin exposed to ultraviolet irradiation, e.g., sunlight.

2. Anonymous, 1978. Nutr. Abstr. Rev. 48, 831–835.
3. Anonymous, 1990. J. Nutr. 120, 12–19.

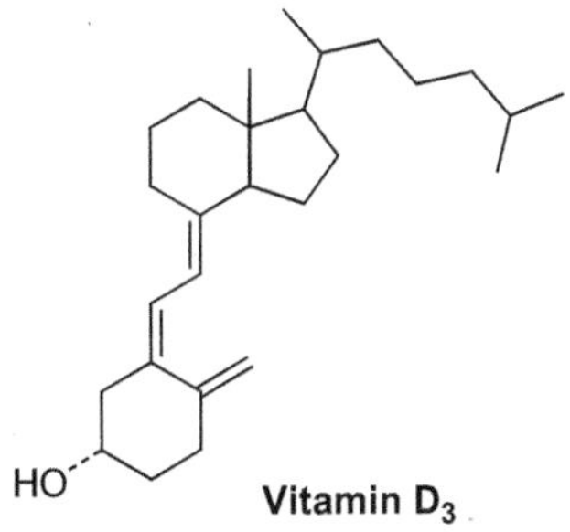

Vitamin D_3

Vitamin E is the generic descriptor for all tocol and tocotrienol derivatives that exhibit qualitatively the biological activity of **α-tocopherol**. These compounds are isoprenoid side chain derivatives of 6-chromanol, **tocols** with side chains consisting of three fully saturated isopentyl units, **tocopherols** comprising the mono-, di-, and trimethyl tocols; and **tocotrienols** being 6-chromanol derivatives with a similar side chain containing three double bonds.

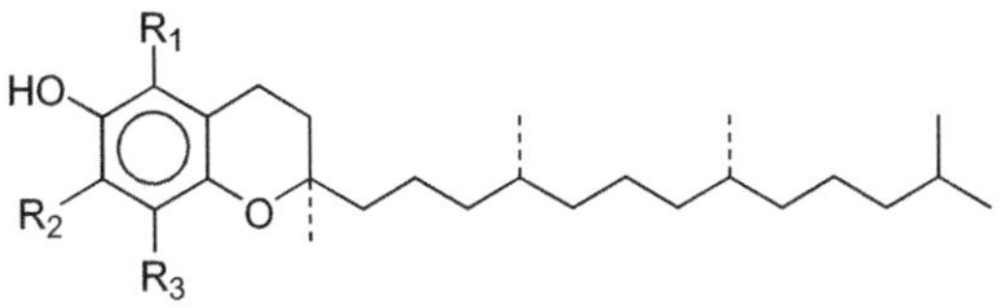

General Structure of the Tocopherols

Vitamin K is the generic descriptor for 2-methyl-1,4-naphthoquinone and its derivatives exhibiting qualitatively the biological activity of phylloquinone. Three groups of vitamers occur, each consisting of variably substituted naphthoquinone ring compounds. The **phylloquinone**[4] group is synthesized by green plants; it includes forms with phytyl and further alkylated side chains having saturated isoprenoid units with a double bond only on the proximal isoprene unit. They are designated as *K-n,* n indicating the number of side chain isoprenoid units. The **menaquinone** group is synthesized by bacteria; it includes vitamers with side chains consisting of variable numbers of isoprenoid units each with a double bond. They are designated *MK*[5]*-n* to indicate side chain length. The synthetic compound 2-methyl-1,4-naphthoquinone (i.e., without a side chain) is called **menadione**.[6] It does not exist naturally but has biological activity by virtue of the fact that human and other animals can alkylate it to produce such metabolites as MK-*4*.

4. Formerly called **phytylmenaquinones**, or **vitamin K1**; the latter term is still encountered.

5. Formerly called **prenylmenaquinones**.

6. Formerly called **vitamin K3**.

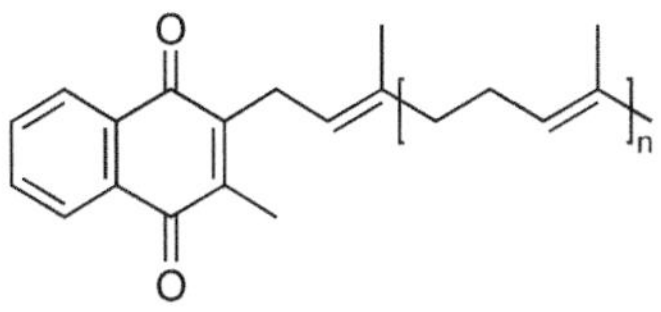

The Phylloquinones

Vitamin C is the generic descriptor for compounds exhibiting qualitatively the biological activity of ascorbic acid, i.e., 2,3-didehydro-l-threo-hexano-1,4-lactone.

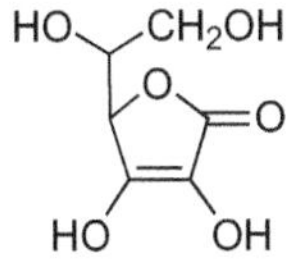

Ascorbic Acid

Thiamin is the trivial designation of the compound 3-[(4-amino-2-methyl-5-pyrimidinyl)methyl]-5-(2-hydroxyethyl)-4-methylthiazolium.

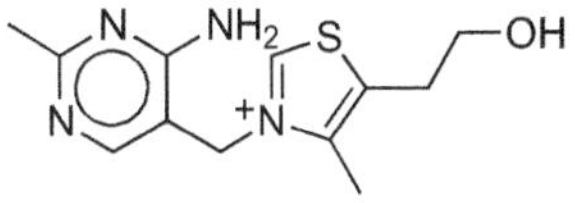

Thiamin

Riboflavin is the trivial designation of the compound 7,8-dimethyl-10-(1′-d-ribityl)isoalloxazine, formerly known as **vitamin B_2**, vitamin G, lactoflavin, or riboflavin.

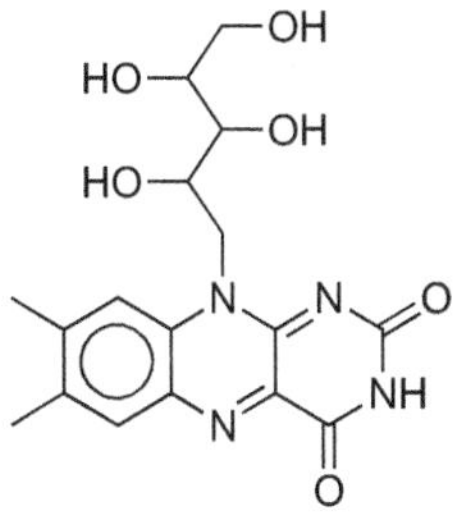

Riboflavin

Niacin is the generic descriptor for pyridine 3-carboxylic acid and derivatives exhibiting qualitatively the biological activity of nicotinamide.[7]

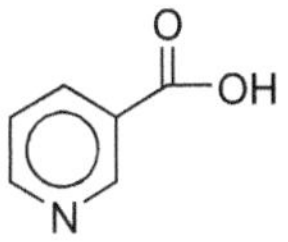

Nicotinic Acid

7. This compound is sometimes called niacinamide.

Vitamin B_6 is the generic descriptor for 3-hydroxy-2-methylpyridine derivatives exhibiting the biological activity of pyridoxine. The term **pyridoxine** is the trivial designation of the single vitamin B_6-active compound, 3-hydroxy-4,5-bis(hydroxymethyl)-2-methylpyridine, formerly called **adermin** or **pyridoxol**.

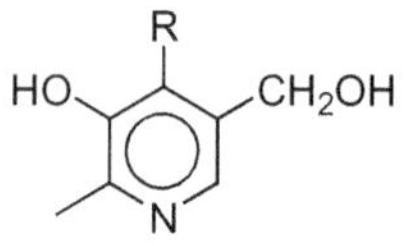

Pyridoxine

Biotin is the trivial designation of the compound *cis*-hexahydro-2-oxo-1*H*-thieno[3,4-*d*]imidazole-4-pentanoic acid, formerly known as vitamin **H** or **coenzyme R**.

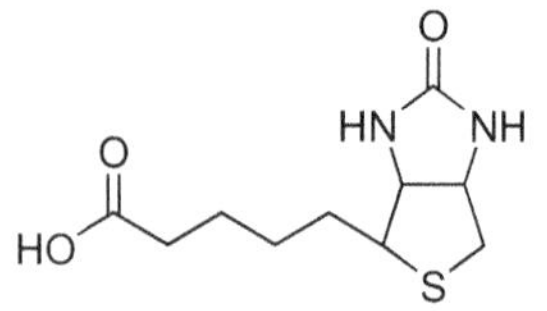

Biotin

Pantothenic acid is the trivial designation for the compound dihydroxy-β,β-dimethylbutyryl-β-alanine, formerly known as pantoyl-β-alanine.

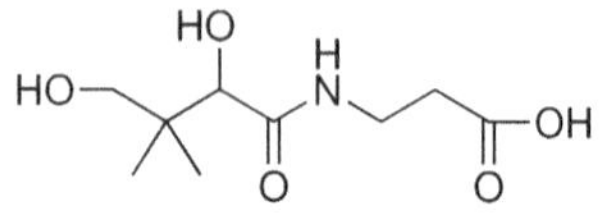

Pantothenic Acid

Folate is the generic descriptor for **folic acid** (pteroylmonoglutamic acid) and related compounds exhibiting the biological activity of folic acid. The terms **folacin, folic acids,** and **folates** are used only as general terms for this group of heterocyclic compounds based on the *N*-[(6-pteridinyl) methyl]-*p*-aminobenzoic acid skeleton conjugated with one or more L-glutamic acid residues. The reduced compound tetrahydropteroylglutamic acid is called **tetrahydrofolic acid**; its single carbon derivatives are named according to the specific carbon moiety bound.

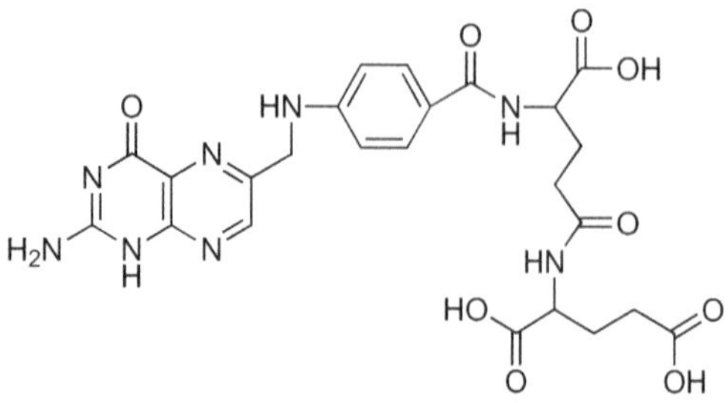

Pteroylglutamic Acid

Vitamin B_{12} is the generic descriptor for cobalamins exhibiting the qualitative biological activity of **cyanocobalamin** (also called **cobalamin**). The cobalamins corrinoids (i.e., compounds containing a cyclic nucleus comprised of four pyrrole rings similar to the porphyrin) with a central cobalt atom that can bind small ligands and nucleotides. The unliganded, reduced form is **cob(I)alamin**. The two functional forms of the vitamin are **methylcobalamin** with a methyl group and a**denosylcobalamin** with a 5′-deoxyadenosyl grouping. Several synthetic vitamers are metabolically active by virtue of their being converted either of the functional forms; these include cyanocobalamin with a cyano group (CN^-), **aquacobalamin**[8] with a bound water molecule, **hydroxocobalamin**[9] with a hydroxo (OH) group, and **nitritocobalamin**[10] with a nitrite group.

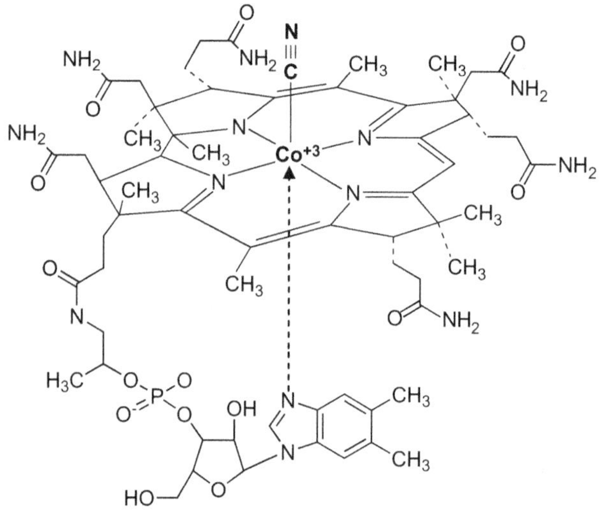

Cyanocobalamin

Few of the vitamins are biologically active without metabolic conversion to another species and/or binding to a specific protein. Thus, any consideration of the vitamins in nutrition involves, for each vitamin group, a number of vitamers and metabolites; some of these are important in the practical sense for food and diet supplementation (Table 3.1); whereas, others are important in the physiological sense as they participate in metabolism.

2. CHEMICAL AND PHYSICAL PROPERTIES OF THE VITAMINS

It has been convenient to classify the vitamins on the basis of their physical properties (Table 3.2), as being either fat

8. Formerly, vitamin B_{12a}.
9. Formerly vitamin B_{12b}.
10. Formerly, vitamin B_{12c}.

TABLE 3.1 Relevant Forms of the Vitamins

Vitamin	Representative	Metabolically Active Forms	Important Dietary Forms
Vitamin A	Retinol	Retinol	Retinyl palmitate, retinyl acetate, provitamins (β-carotene, other carotenoids)
		Retinal	
		Retinoic acid	
Vitamin D	Cholecalciferol	25-OH-cholecalciferol	Cholecalciferol, ergocalciferol
		1,25-$(OH)_2$-cholecalciferol	
Vitamin E	α-Tocopherol	α-, β-, γ-, δ-Tocopherols	*R,R,R*-α-Tocopherol, all-*rac*-α-tocopheryl acetate
Vitamin K	Phylloquinone	Phylloquinones (K_n)	Phylloquinones (K_n), menaquinones (MK_n), menadione sodium bisulfite complex
		Menaquinones (MK_n)	
Vitamin C	Ascorbic acid	Ascorbic acid	L-ascorbic acid, sodium ascorbate
		Dehydroascorbic acid	
Thiamin	Thiamin	Thiamin pyrophosphate	Thiamin, thiamin pyrophosphate, thiamin disulfide, thiamin HCl, thiamin mononitrate
Riboflavin	Riboflavin	Flavin mononucleotide (FMN)	FMN, FAD, flavoproteins, riboflavin
		Flavin adenine dinucleotide (FAD)	
Niacin	Nicotinamide	Nicotinamide adenine dinucleotide (NAD)	NAD, NADP, nicotinamide, nicotinic acid
		Nicotinamide adenine dinucleotide phosphate (NADP)	
Vitamin B_6	Pyridoxine	Pyridoxal 5′-phosphate, pyridoxamine 5′-phosphate	Pyridoxal HCl, pyridoxal- 5′-phosphate, pyridoxamine-5′-phosphate
Biotin	*d*-Biotin	*d*-Biotin	Biocytin, *d*-biotin
Pantothenic acid	Pantothenic acid	Coenzyme A	Calcium pantothenate, coenzyme A, acyl CoAs
Folate	Pteroylglutamic acid	Pteroylpolyglutamates	Pteroyl poly- and monoglutamates
Vitamin B_{12}	Cyanocobalamin	Methylcobalamin5′-deoxyadenosylcobalamin	Cyano-, aqua-, hydroxo-, methyl-, and 5′-deoxyadenosylcobalamins

TABLE 3.2 Physical Properties of the Vitamins

Vitamin	Vitamer	MW	Solubility		Absorption Max	Molar Absorptivity		Fluorescence		Melting Point (°C)	Color/Form
			Organic[a]	H_2O		Absorptivity ε	$A^{1\%}_{1\,cm}$	Excitation (nm)	Emission (nm)		
Vitamin A	All-*trans*-retinol	286.4	+	–	325	52,300	1845	325	470	62–64	Yellow/crystal
	11-*cis*-retinol	286.4	+	–	319	34,900	1220				
	13-*cis*-retinol	286.4	+	–	328	48,300	1189				
	Retinal	284.4	+	–	373		1548			61–64	Orange/crystal
	All-*trans*-retinoic acid	300.4	+	sl[b]	350	45,300	1510			180–182	Yellow/crystal
	13-*cis*-Retinoic acid	300.4	+	sl[b]	354	39,800	1325			180–182	Yellow/crystal
	All-*trans*-retinyl acetate	312.0	+	sl[b]	326		1550			57–58	Yellow/crystal
	All-*trans*-retinyl palmitate	508.0	+	sl[b]	325–328		975			28–29	
Provitamin A	β-Carotene	536.9	+	–	453	2592	139			183	Purple/crystal
	α-Carotene	536.9	+	–	444	2800				187	Purple/crystal
Vitamin D	Vitamin D_2	396.7	+	–	264	18,300	459	No fluorescence		115–118	White/crystal
	Vitamin D_3	384.6	+	–	265	19,400	462	No fluorescence		84–85	White/crystal
	25(OH) vitamin D_3	400.7	+	–	265	18,000	449	No fluorescence			
	1α,25$(OH)_2$vitamin D_3	416.6	+	–	264	19,000	418	No fluorescence			
Vitamin E	α-Tocopherol	430.7	+	–	292	3265	75.8	295	320	2.5	Yellow/oil
	β-Tocopherol	416.7	+	–	296	3725	89.4	297	322		Yellow/oil
	γ-Tocopherol	416.7	+	–	298	3809	91.4	297	322	−2.4	Yellow/oil
	δ-Tocopherol	402.7	+	–	298		91–92	297	322		Yellow/oil
	α-Tocopheryl acetate	472.8	+	–	286	1891–2080	40–44	290	323		Yellow/oil
	α-Tocopheryl succinate	530.8	+	–	286	2044	38.5	285	310		
	α-Tocotrienol	424.7	+	–	292	3652	86.0				
	β-Tocotrienol	410.6	+	–	296	3540	86.2	290	323		Yellow/oil
	γ-Tocotrienol	410.6	+	–	298	3737	91.0	290	324		Yellow/oil
	δ-Tocotrienol	396.6	+	–	298	3403	85.8	292	324		Yellow/oil

Vitamin K	Vitamin K_1	450.7	+	–	242	17,900	396	No fluorescence			Yellow/oil
					248	18,900	419				
					260	17,300	383				
					269	17,400	387				
					325	3100	68				
	Vitamin $K_{2(20)}$	444.7	+	–	248	19,500	439	No fluorescence		35	Yellow/crystal
	Vitamin $K_{2(30)}$	580.0	+	–	243	17,600	304	No fluorescence		50	Yellow/crystal
					248	18,600	320				
					261	16,800	290				
					270	16,900	292				
	Vitamin $K_{2(35)}$	649.2	+	–	243	18,000	278	No fluorescence		54	Yellow/crystal
					248	19,100	195				
					261	17,300	266				
					270	30,300	467				
					325–328	3100	48				
	Vitamin K_3	172.2	+	–						105–107	Yellow/crystal
Vitamin C	Ascorbic acid	176.1	–	+	245	12,200	695	No fluorescence		190–192	White/crystal
	Calcium ascorbate	390.3	–	+							White/crystal
	Sodium ascorbate	198.1	–	xs[e]						218[c]	White/crystal
	Ascorbyl palmitate	414.5	–	+							White/crystal
Thiamin	Thiamin disulfide	562.7	–	sl[b]				No fluorescence		177	Yellow/crystal
	Thiamin HCl	337.3	–	xs[e]						246–250	White/crystal
	Thiamin mononitrate	327.4	–	+						196–200[c]	White/crystal
	Thiamin monophosphate	344.3	–								
	Thiamin pyrophosphate	424.3	–							220–222[c]	
	Thiamin triphosphate	504.3	–							228–232[c]	

Continued

TABLE 3.2 Physical Properties of the Vitamins—cont'd

Vitamin	Vitamer	MW	Solubility		Absorption Max	Molar Absorptivity		Fluorescence		Melting Point (°C)	Color/Form
			Organic[a]	H_2O		Absorptivity ε	$A^{1\%}_{1cm}$	Excitation (nm)	Emission (nm)		
Riboflavin[d]	Riboflavin	376.4	–	+	260	27,700	736	360, 465	521	278[c]	Orange-yellow/ crystal
					375	10,600	282				
					450	12,200	324				
	Riboflavin-5′-phosphate	456.4	–	+	260	27,100	594	440–500	530		Orange-yellow/ crystal
	FAD	785.6	–	+	260	37,000	471	440–500	530		
					375	9300	118				
					450	11,300	144				
Niacin	Nicotinic acid	123.1	–	+	260	2800	227	No fluorescence		237	White/crystal
	Nicotinamide	122.1	–	xs[e]	261	5800	478	No fluorescence		128–131	White/crystal
Vitamin B_6	Pyridoxal HCl	203.6	–	+	390	200	9.8	330[f]	382	165[c]	White/crystal
					318	8128	399	310	365[g]		
	Pyridoxine	169.2			254	3891	23	320	380	160	
					324	7244	428	332	400		
	Pyridoxol HCl	205.6	–	+	253	3700	180			206–208	White/crystal
					290	8400	408				
					292	7720	375				
					325	7100	345				
	Pyridoxamine di-HCl	241.1	–	+	253	4571	190	320	370[g]	226–227	
					328	7763	322	337	400[h]		
	Pyridoxal 5′-phosphate	247.1	–	+	330	2500	101	365	423[h]		
								360	430[g]		
					388	4900	198	330	410[f]		

Biotin	*d*-Biotin	244.3	–	+	204	(Very weak)		No fluorescence		232–233	Colorless/crystal
Pantothenic acid	Pantothenic acid	219.2	–	xs[e]	204	(Very weak)		No fluorescence			Clear/oil
	Calcium pantothenate	467.5	–	–	No chromophore			No fluorescence		195–196[c]	White/crystal
	D-pantothenol	205.3		sl[b]	No chromophore			No fluorescence			Clear/oil
Folate	Folic acid	441.1			282	27,000	612	363	450–460[g]		
					350	7000	159				
	Tetrahyrdofolate	445.4			297	27,000	606	305–310	360[h]		
	10-Formyl FH_4	473.5			288	18,200	384	313	360[f]		
	5-Formyl FH_4	473.5			287	31,500	665	314	365[f]		
	5-Methyl FH_4	459.5			290	32,000	697				
	5-Formimino FH_4	472.5			285	35,400	749	308	360[f]		
	5,10-Methenyl FH_4	456.4			352	25,000	548	370	470[h]		
	5,10-Methylene FH_4	457.5			294	32,000	700				
Vitamin B_{12}	Cyanocobalamin	1355.4	–	xs[e]	278	8700	115	No fluorescence			Dark red/crystal
					261	27,600	204				
					551	8700	64				
	Hydroxylcobalamin (B_{12a})	1346.4	–	+	279	19,000	141	No fluorescence			Dark red/crystal
					325	11,400	85				
					359	20,600	153				
					516	8900	66				
					537	9500	71				
	Aquacobalamin (B_{12b})	1347.0	–	+	274	20,600	153				
					317	6100	45				
					351	26,500	197				
					499	8100	60				
	Nitrocobalamin (B_{12c})	1374.6	–	+	352	21,000	153				Red/crystal

Continued

TABLE 3.2 Physical Properties of the Vitamins—cont'd

Vitamin	Vitamer	MW	Solubility		Absorption Max	Molar Absorptivity		Fluorescence		Melting Point (°C)	Color/Form
			Organic[a]	H_2O		Absorptivity ε	$A^{1\%}_{1cm}$	Excitation (nm)	Emission (nm)		
					357	19,100	139				
					528	8400	60				
					535	8700	63				
	Methylcobalamin	1344.4	–	+	266	19,900	148				Red/crystal
					342	14,400	107				
					522	9400	70				
	Adenosylcobalamin cobamide	1579.6	–	+	288	18,100	115				Yellow-orange/ crystal
					340	12,300	78				
					375	10,900	60				
					522	800	51				

[a] *In organic solvents, fats, and oils.*
[b] *sl, Slightly soluble.*
[c] *Decomposes at this temperature.*
[d] *Fluoresces.*
[e] *xs, Freely soluble.*
[f] *Neutral pH.*
[g] *Alkaline pH.*
[h] *Acidic pH.*

soluble or water soluble.[11] In fact, this way of classifying the vitamins recapitulates the history of their discovery, calling to mind McCollum's "fat-soluble A" and "water-soluble B." The water-soluble vitamins tend to have one or more **polar** or ionizable groups (carboxyl, keto, hydroxyl, amino, or phosphate), whereas the fat-soluble vitamins have predominantly aromatic and aliphatic characters.

Fat-soluble vitamins—soluble in nonpolar solvents:
Vitamins A, D, E, and K
Water-soluble vitamins—soluble in polar solvents:
Vitamin C, thiamin, riboflavin, niacin, pyridoxine, pantothenic acid, biotin, folate, and vitamin B_{12}

The fat-soluble vitamins have some traits in common, in that each is composed either entirely or primarily of five-carbon **isoprenoid** units (i.e., related to *isoprene*, 2-methyl-1,3-butadiene) derived initially from acetyl CoA in those plant and animal species capable of their biosynthesis. In contrast, the water-soluble vitamins have, in general, few similarities of structure. The routes of their biosyntheses in capable species do not share as many common pathways.

Vitamin Stability

For the use of vitamins as food/feed additives, in dietary supplements, and as pharmaceuticals, stability is a prime concern. In general, the fat-soluble vitamins, vitamin C, thiamin, riboflavin, and biotin are poorly stable to oxidation. They must be protected from heat, oxygen, metal ions (particularly Fe^{2+} and Cu^{2+}), polyunsaturated lipids undergoing peroxidation, and ultraviolet light. Antioxidants are frequently used in formulations of these vitamins. For vitamins A and E, the more stable esterified forms are used for these purposes. Because of the instabilities of their naturally occurring vitamers, the amounts of the fat-soluble vitamins in natural foods and feedstuffs are highly variable, being greatly affected by the conditions of food production and processing. Niacin, vitamin B_6, pantothenic acid, folate, and vitamin B_{12} tend to be more stable under most practical conditions (Table 3.3). Some vitamins can undergo degradation by reacting to factors in foods during sample storage and/or preparation: ascorbic acid with plant ascorbic acid oxidase; thiamin with sulfites or with plant or microbial thiaminases; folates with nitrites; and pantothenic acid with microbial pantothenases.

3. PHYSIOLOGICAL UTILIZATION OF THE VITAMINS

Vitamin Absorption

The means by which the vitamins are absorbed are determined by their chemical and associated physical properties. The fat-soluble vitamins (and hydrophobic substances such as carotenoids and cholesterol), which are not soluble in the aqueous environment of the alimentary canal, are associated with and dissolved in other lipid materials. In the upper portion of the gastrointestinal tract, they are dissolved in the bulk lipid phases of the emulsions that are formed of dietary fats[12] by the mechanical actions of mastication and gastric churning. Emulsion oil droplets, however, are generally too large (e.g., 1000 Å) to gain the intimate proximity to the absorptive surfaces of the small intestine that is necessary to facilitate the diffusion of these substances into the hydrophobic environment of the brush border membranes of intestinal mucosal cells. However, **lipase**, which is present in the intestinal lumen following synthesis in and export from the pancreas via the pancreatic duct, binds to the surface of emulsion oil droplets, where it catalyzes the hydrolytic removal of the α and α′ fatty acids from triglycerides, which make up the bulk of the lipid material in these large particles. The products of this process (i.e., free fatty acids and β-monoglycerides) have strong polar regions or charged groups and, thus, will dissolve monomerically in this aqueous environment. However, they also have

---

11. The term solubility refers to the interactions of solutes and solvents. A soluble substance can disperse on a molecular level within a solvent. Solvents such as water, which can either donate or accept electrons, are said to be polar; whereas solvents (e.g., many organic solvents) incapable of such interactions are said to be nonpolar. Polar solvents such as water can either donate or accept electrons; whereas nonpolar solvents (e.g., many organic solvents) are incapable of such interactions. Compounds that are polar or that have charged or ionic character are soluble in polar solvents and are called hydrophilic; they are insoluble in nonpolar organic solvents and are, hence, also called lipophobic. Molecules that do not contain polar or ionizable groups tend to be insoluble in water (are hydrophobic) but soluble in nonpolar organic solvents (are lipophilic). Some large molecules (e.g., phospholipids, fatty acids, and bile salts) that have local areas of charge or ionic bond density, as well as other areas without charged groups, exhibit both polar and nonpolar characteristics. They are called amphipaths; having both hydrophilic and lipophilic internal regions, they tend to align along the interfaces of mixed polar/nonpolar phases. Amphipathic molecules are important in facilitating the dispersion of hydrophobic substances in aqueous environments; they do this by surrounding those substances, forming the submicroscopic structure called the mixed micelle.

12. It follows that fat-soluble vitamins are not well absorbed from low-fat diets; Studies have demonstrated markedly less absorption of carotenoids (Brown, M.J., Ferruzzi, M.G., Nguyen, M.L., Cooper, D.A., Eldridge, A.L., Schwartz, S.J., White, W.S., 2004. Am. J. Clin. Nutr. 80, 396–403) and vitamin D3 (Dawson-Huges, B., Harris, S.S., Lichtenstein, A.H., Dolnikowski, G., Palermo, N.J., Rasmussen, H., 2015. J. Acad. Nutr. Diet. 115, 225–230) from fat-free meals compared to fat-containing controls. However, the minimum amount of fat required is not clear. One study (Roodenburg, A.J., Leenen, R., van het Hof, K.H., Weststrate, J.A., Tilburg, L.B., 2000. Am. J. Clin. Nutr. 71, 1187–1193) found that as little as 3–5 g fat per meal may be sufficient for optimal absorption of provitamin A carotenoids, although the utilization of vitamins from plant foods is likely to require higher levels of fat to render accessible fat-soluble nutrients from the plant structures in which they are present.

TABLE 3.3 Stabilities of the Vitamins

		Unstable to						
Vitamin	**Vitamer**	**UV**	**Heat**[a]	**O_2**	**Acid**	**Alkali**	**Metals**[b]	**To Enhance Stability**
Vitamin A	Retinol	+		+	+		+	Keep in the dark, exclude O_2, use antioxidants
	Retinal			+	+		+	Exclude O_2, use antioxidants
	Retinoic acid							Good stability
	Dehydroretinol			+				Exclude O_2, use antioxidants
	Retinyl esters							Good stability
	β-Carotene	+		+				Keep in the dark, exclude O_2, use antioxidants
Vitamin D	D_2	+	+[c]	+	+		+	Keep cool, in the dark, exclude O_2, use antioxidants
	D_3	+	+[c]	+	+	+	+	Keep cool, in the dark, exclude O_2, use antioxidants
Vitamin E	Tocopherols		+	+	+	+	+	Keep cool, at neutral pH
	Tocopheryl esters				+	+		Good stability
Vitamin K	K	+		+		+	+	Avoid reductants[c], work in subdued light
	MK	+		+		+	+	Avoid reductants[c], work in subdued light
	Menadione	+				+	+	Avoid reductants[c], work in subdued light
Vitamin C	Ascorbic acid			+[b]		+	+	Exclude O_2, at neutral pH
Thiamin	Disulfide form		+	+	+	+	+	Keep at neutral pH[d]
	Hydrochloride[e]		+	+	+	+	+	Exclude O_2, at neutral pH[d]
Riboflavin	Riboflavin	+[f]	+			+	+	Keep in the dark, at pH 1.5–4[d]
Niacin	Nicotinic acid							Good stability
	Nicotinamide							Good stability
Vitamin B_6	Pyridoxal	+	+					Keep cool, work in subdued light
	Pyridoxol HCl			+		+		Good stability
Biotin	Biotin			+		+		Exclude O_2, at pH 4–9, use antioxidants, work in subdued light
Pantothenic acid	Free acid[g]	+		+		+		Cool, neutral pH
	Calcium salt[e]		+					Exclude O_2, at pH 5–7
Folate	FH_4	+	+	+	+[h]		+	Good stability[d]
Vitamin B_{12}	Cyano-B_{12}	+			+[i]		+[j]	Good stability[c] at pH 4–7

[a] *That is, 100°C.*
[b] *In solution with Fe^{3+} and Cu^{2+}.*
[c] *Isomerization to the previtamin form may be unavoidable, but tachysterol can also be formed under acid conditions in samples exposed to light.*
[d] *Unstable to reducing agents.*
[e] *Slightly hygroscopic.*
[f] *Especially in alkaline solution.*
[g] *Very hygroscopic.*
[h] *pH$<$5.*
[i] *pH$<$3.*
[j] *pH$>$9.*

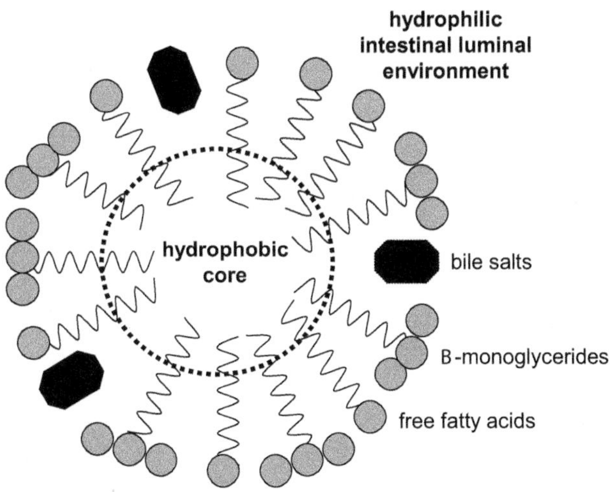

FIGURE 3.1 Mixed micelles form in the intestinal lumen by the spontaneous association of the products of triglyceride digestion, β-monoglycerides and free fatty acids, and bile salts. Their hydrophobic cores provide environments for the fat-soluble vitamins (A, D, E, and K) and other lipophilic dietary components. The absorption across the intestinal microvillar surface is facilitated by their very small size (10–50 Å diameter).

long-chain hydrocarbon nonpolar regions; therefore, when certain concentrations (**critical micellar concentrations**) are achieved, these species and bile salts, which have similar properties, combine spontaneously to form small particles called **micelles** (Fig. 3.1). Mixed micelles, thus, contain free fatty acids, β-monoglycerides, and bile salts in which the nonpolar regions of each are associated interiorly, and the polar or charged regions of each are oriented externally and are associated with the aqueous phase. The core of the mixed micelle is hydrophobic and serves to solubilize the fat-soluble vitamins and other nonpolar lipid substances. Because they are small (10–50 Å in diameter), mixed micelles can gain close proximity to microvillar surfaces of intestinal mucosa, facilitating the diffusion of their contents into and across those membranes. Because the enteric absorption of the fat-soluble vitamins depends on micellar dispersion, it is impaired under conditions of lipid malabsorption or very low dietary fat (<10 g/day).

The water-soluble vitamins, which are soluble in the polar environment of the intestinal lumen, can be taken up by the absorptive surface of the gut more directly. Some (vitamin C, thiamin, niacin, vitamin B_6, biotin, pantothenic acid, folate, and vitamin B_{12}) are absorbed as the result of passive diffusion; others are absorbed via specific carriers as a means of overcoming concentration gradients unfavorable to simple diffusion. Several (vitamin C, vitamin B_{12}, thiamin, niacin, and folate) are absorbed via carrier-dependent mechanisms at low doses[13] and by simple diffusion (albeit at lower efficiency) at high doses (Table 3.4).

13. These processes show apparent Km values in the range of 0.1–300 μM. [The Km parameter is a measure of the affinity of an enzyme for its substrate. Expressed as the substrate concentration at half-maximal velocity of the particular enzyme-catalyzed reaction, the Km is inversely related to the affinity of enzyme–substrate binding.]

The absorption of at least three water-soluble vitamins (vitamin C, riboflavin, and vitamin B_6) appears to be regulated in part by the dietary supply of the vitamin in a feedback manner. Thus, it has been questioned whether high doses of one vitamin/vitamer may antagonize the absorption of related vitamins. There is some evidence for such mutual antagonisms among the fat-soluble vitamins, as well as in the case of α-tocopherol, a high intake of which antagonizes the utilization of the related γ-vitamer.

Vitamin Transport

The mechanisms of postabsorptive transport of the vitamins also vary according to their particular physical and chemical properties (Table 3.5). The degree of solubility in the aqueous environments of the blood plasma and lymph is a major determinant of ways in which the vitamins are transported from the site of absorption (the small intestine) to the liver and peripheral organs. The fat-soluble vitamins, because they are insoluble in plasma and lymph, depend on carriers that are soluble in those aqueous environments. These vitamins, therefore, are associated with the lipid-rich **chylomicrons**[14] that are elaborated in intestinal mucosal cells, largely of reesterified triglycerides from free fatty acids and β-monoglycerides that have just been absorbed. As the lipids in these particles are transferred to other **lipoproteins**[15] in the liver, some of the fat-soluble vitamins (vitamins E and K) are also transferred to those carriers. Others (vitamins A and D) are transported from the liver to peripheral tissues by specific **binding proteins** of hepatic origin (Table 3.6). Some of the water-soluble vitamins are transported by protein carriers in the plasma and therefore are not found free in solution. Some (riboflavin, vitamin B_6) are carried via weak, nonspecific binding to albumin, and may thus be displaced by other substances (e.g., ethanol) that also bind to that protein. Others are tightly associated with certain immunoglobulins (riboflavin) or bind

14. Chylomicrons are the largest (~1 μm diameter) and the lightest of the blood lipids. They consist mainly of triglyceride with smaller amounts of cholesterol, phospholipid, protein, and the fat-soluble vitamins. They are normally synthesized in the intestinal mucosal cells and serve to transport lipids to tissues. In mammals, these particles are secreted into the lymphatic drainage of the small intestine (hence their name). However, in birds, fishes, and reptiles, they are secreted directly into the renal portal circulation; therefore, in these species they are referred to as **portomicrons**. In either case, they are cleared from the plasma by the liver, and their lipid contents are either deposited in hepatic stores (e.g., vitamin A) or released back into the plasma bound to more dense particles called **lipoproteins**.

15. As the name would imply, a lipoprotein is a lipid–protein combination with the solubility characteristics of a protein (i.e., soluble in the aqueous environment of the blood plasma) and hence involved in lipid transport. Four classes of lipoproteins, each defined empirically on the basis of density, are found in the plasma: chylomicrons/portomicrons, high-density lipoproteins (**HDLs**), low-density lipoproteins (**LDLs**), and very low-density lipoproteins (**VLDLs**). The latter three classes are also known by names derived from the method of electrophoretic separation, i.e., α-, β-, and pre-β-lipoproteins, respectively.

TABLE 3.4 Enteric Absorption of the Vitamins[a]

Vitamer	Digestion	Site	Enterocytic Metabolism	Efficiency (%)	Conditions of Potential Malabsorption
Micelle-Dependent Diffusion					
Retinol	—	D, J	Esterification	80–90	Pancreatic insufficiency (pancreatitis, selenium deficiency, cystic fibrosis, cancer), β-carotene cleavage, biliary atresia, obstructive jaundice, celiac disease, very low-fat diet
Retinyl esters	Deesterified	D, J	Reesterification		
		D, J	Esterification	50–60	Pancreatic or biliary insufficiency
Vitamins D	—	D, J	—	~50	Pancreatic or biliary insufficiency
Tocopherols	—	D, J	—	20–80	Pancreatic or biliary insufficiency
Tocopherol esters	Deesterified[b]	D, J	—	20–80	Pancreatic or biliary insufficiency
MKs	—	D, J	—	10–70	Pancreatic or biliary insufficiency
Menadione	—	D, J	—	10–70	Pancreatic or biliary insufficiency
Active Transport					
Phylloquinone	—	D, J	—	~80	Pancreatic or biliary insufficiency
Ascorbic acid	—	I	—	70–80	D-isoascorbic acid
Thiamin	—	D	Phosphorylation		Pyrithiamin, excess ethanol
Thiamin di-P	Dephosphorylation[b]	D	Phosphorylation		Pyrithiamin, excess ethanol
Riboflavin	—	J	Phosphorylation		
FMN, FAD	Hydrolysis[b]	J	Phosphorylation		
Flavoproteins	Hydrolysis[b]	J	Phosphorylation		
Folylmono-glu	—	J	Glutamation		Celiac sprue
Folylpoly-glu	Hydrolysis[b]	J	Glutamation		Celiac sprue
Vitamin B_{12}	Hydrolysis[b]	I	Adenosinylation, methylation	>90	Intrinsic factor deficiency (pernicious anemia)
Facilitated Diffusion[c]					
Nicotinic acid	—	J		>90[d]	
Nicotinamide	—	J		~100[d]	
Niacytin	Hydrolysis[b]	J			
NAD(P)	Hydrolysis[b]	J			

Biotin	—	J			Biotinidase deficiency, consumption of raw egg white (avidin)
Biocytin	Hydrolysis[b]	J			Biotinidase deficiency, consumption of raw egg white (avidin)
Pantothenate	—				
Coenzyme A	Hydrolysis[b]				
Simple Diffusion					
Ascorbic acid[e]	—	D, J, I	—	<50	
Thiamin[e,f]	—	J	Phosphorylation		
Nicotinic acid	—	J	—		
Nicotinamide	—	J	—		
Pyridoxol	—	J	Phosphorylation		
Pyridoxal	—	J	Phosphorylation		
Pyridoxamine	—	J	Phosphorylation		
Biotin	—	D, J	—	>95	Consumption of raw egg white (avidin)
Pantothenate	—		—		
Folylmono-glu[e]	—	J	Glutamation		
Vitamin B_{12}[e]	—	D, J	Adenosinylation, methylation	~1	

[a]D, *Duodenum;* folylmono-glu, *folylmonoglutamate;* folylpoly-glu, *folylpolyglutamate;* I, *ileum;* J, *jejunum;* thiamin di-P, *thiamin diphosphate.*
[b]*Yields vitamin in absorbable form.*
[c]*Na^+-dependent saturable processes.*
[d]*Estimate may include contribution of simple diffusion.*
[e]*Simple diffusion important only at high doses.*
[f]*Symport with Na^+.*

TABLE 3.5 Postabsorptive Transport of Vitamins in the Body

Vehicle	Vitamin	Form Transported
Lipoprotein Bound		
Chylomicrons[a]	Vitamin A	Retinyl esters
	Vitamin A	β-Carotene
	Vitamin D	Vitamin D[b]
	Vitamin E	Tocopherols
	Vitamin K	K, MK, menadione
VLDL[c]/HDL[d]	Vitamin E	Tocopherols
	Vitamin K	Mainly MK-4
Associated Nonspecifically With Proteins		
Albumin	Riboflavin	Free riboflavin, FMN
	Vitamin B_6	Pyridoxal, pyridoxal phosphate
Immunoglobulins[e]	Riboflavin	Free riboflavin
Bound to Specific Binding Proteins		
Retinol BP (RBP)	Vitamin A	All-*trans*-retinol
Transcalciferin (vitamin D BP)	Vitamin D	D_2; D_3; 25-OH-D; 1,25-$(OH)_2$-D; 24,25-$(OH)_2$-D
Thiamin BP	Thiamin	Free thiamin
Riboflavin BP	Riboflavin	Riboflavin
Biotinidase	Biotin	Free biotin
Folate BP	Folate	Folate
Transcobalamin II	Vitamin B_{12}	Methylcobalamin
Transcobalamin III	Vitamin B_{12}	Vitamin B_{12}
Carried in Erythrocytes		
Erythrocyte membranes	Vitamin E	Tocopherols
Erythrocytes	Vitamin B_6	Pyridoxal phosphate
	Pantothenic acid	Coenzyme A
Free in Plasma		
—	Vitamin C	Ascorbic acid
—	Thiamin	Free thiamin, thiamin pyrophosphate
—	Riboflavin	FMN
—	Pantothenic acid	Pantothenic acid
—	Biotin	Free biotin
—	Niacin	Nicotinic acid, nicotinamide
—	Folate	Pteroylmonoglutamates[f]
Bound to Specific Intracellular Binding Proteins		
Cellular RBP (CRBP)	Vitamin A	All-*trans*-retinol
Cellular RBP, type II (CRBPII)	Vitamin A	All-*trans*-retinol
Interstitial RBP (IRBP)	Vitamin A	All-*trans*-retinol
Cellular retinal BP (CRALBP)	Vitamin A	All-*trans*-retinal

TABLE 3.5 Postabsorptive Transport of Vitamins in the Body—cont'd

Vehicle	Vitamin	Form Transported
Cellular retinoic acid BP (CRABP)	Vitamin A	All-*trans*-retinoic acid
Vitamin D receptor	Vitamin D	1,25-$(OH)_2$-D
Vitamin E BP	Vitamin E	Tocopherols
Flavoproteins	Riboflavin	FMN, FAD
Transcobalamin I	Vitamin B_{12}	Vitamin B_{12}

[a]*In mammals, lipids are absorbed into the lymphatic circulation, where they are transported to the liver and other tissues as large lipoprotein particles called* chylomicra *(singular,* chylomicron*); in birds, reptiles, and fishes, lipids are absorbed directly into the hepatic portal circulation and the analogous lipoprotein particle is called a* ***portomicron****.*
[b]*Representation of vitamin D without a subscript is meant to refer to both major forms of the vitamin: ergocalciferol (D_2) and cholecalciferol (D_3).*
[c]*VLDL, very low-density lipoprotein.*
[d]*HDL, high-density lipoprotein.*
[e]*For example, IgG, IgM, and IgA.*
[f]*Especially 5-CH_3-tetrahydrofolic acid.*

TABLE 3.6 Tissue Distribution of the Vitamins

Vitamin	Predominant Storage Form(s)	Depot(s)
Vitamin A	Retinyl esters (e.g., palmitate)	Liver
Vitamin D	D_3; 25-OH-D	Plasma, adipose, muscle
Vitamin E	α-Tocopherol	Adipose, adrenal, testes, platelets, other tissues
Vitamin K	K-4, MK-4	Liver
	MK-4	All tissues
Vitamin C	Ascorbic acid	Adrenals, leukocytes
Thiamin	Thiamin pyrophosphate[a]	Heart, kidney, brain, muscle
Riboflavin	FAD[b]	Liver, kidney, heart
Vitamin B_6	Pyridoxal phosphate[b]	Liver[c], kidney[c], heart[c]
Vitamin B_{12}	Methylcobalamin	Liver[d], kidney[c], heart[c], spleen[c], brain[c]
Niacin	No appreciable storage	—
Biotin	No appreciable storage[b]	—
Pantothenic acid	No appreciable storage	—
Folate	No appreciable storage	—

[a]*The amounts in the body are composed of the enzyme-bound coenzyme.*
[b]*Small amounts of the vitamin are found in these tissues.*
[c]*Predominant depot.*

to specific proteins involved in their transport (riboflavin, vitamins A, D, E, and B_{12}). Several vitamins (e.g., vitamin C, thiamin, niacin, riboflavin, pantothenic acid, biotin, and folate) are transported in free solution in the plasma.

Tissue Distribution of the Vitamins

The retention and distribution of the vitamins among the various tissues also vary according to their general physical and chemical properties (Table 3.6). In general, the fat-soluble vitamins are well retained; they tend to be stored in association with tissue lipids. For that reason, lipid-rich tissues such as adipose and liver frequently have appreciable stores of the fat-soluble vitamins. Storage of these vitamins means that animals may be able to accommodate widely variable intakes without consequence by mobilizing their tissue stores in times of low dietary intakes.

In contrast, the water-soluble vitamins tend to be excreted rapidly in the urine and not retained well. Few of these vitamins are stored to any appreciable extent. The notable exception is vitamin B_{12}, which, under normal circumstances, can accumulate in the liver in amounts adequate to satisfy the nutritional needs of the host for periods of years.

4. METABOLISM OF THE VITAMINS

Some Vitamins Have Limited Biosynthesis

By definition, the vitamins as a group of nutrients are obligate factors in the diet (i.e., the chemical environment) of an organism. Nevertheless, some vitamins do not quite fit that general definition because they may, in fact, be biosynthesized regularly by certain species, and under certain circumstances by other species. This is the "**vitamin caveat**" (Chapter 1). The biosynthesis of such vitamins (Table 3.7) thus depends on the availability, either from dietary or metabolic sources, of appropriate precursors. Examples include the following:

- adequate free tryptophan is required for niacin production in species capable of substantive tryptophan–niacin conversion;
- the presence of 7-dehydrocholesterol in the surface layers of the skin is required for its conversion to vitamin D_3 in individuals exposed to UV light; and
- flux through the gulonic acid pathway is needed to produce ascorbic acid in those (most nonprimate) species in which that pathway is intact.

Some vitamins can be synthesized by the **microbiota** of the hindgut. Until fairly recently, this source of vitamins has not been regarded as having immediate physiological relevance, as it was thought that vitamins synthesized by the gut microbiota were not absorbed in the large intestine. However, recent studies and a greater appreciation of earlier work with animals models[16] have changed that view, and it is now clear that the hindgut microbiota can contribute meaningful amounts of at least six vitamins: vitamin K[17], thiamin, riboflavin, pyridoxine, biotin, and folate. Metagenomic studies have shown that microbiomes relatively rich in *Bacteroides* spp. tend to have enriched capacities of the synthesis of biotin and riboflavin, while those relatively rich in *Prevotella* spp. tend to have enriched capacities for the synthesis of thiamin and folate. The microbial synthesis of these vitamins in the gut is also affected by the nature of an individual's diet; greatest synthesis can be expected when diets are rich in soluble fiber, e.g., diets rich in plant foods and whole grains.

16. In 1914, Cooper (J. Hygiene 14,12-22) showed that feeding a fecal extract could cure pigeons of their polyneuritis.
17. As menaquinones, MKn.

Most Vitamins Require Metabolic Activation

Only a few vitamins are directly metabolically active: vitamin E, some vitamers K (e.g., MK-4), and vitamin C. The others require metabolic activation or linkage to a cofunctional species (e.g., an enzyme) (Table 3.8). The transformation of dietary forms of the vitamins into their respective, metabolically active forms may involve substantive modification of a vitamin's chemical structure and/or its combination with another species. Thus, factors that affect the metabolic (i.e., enzymatic) activation of vitamins to their functional species can have profound influences on their nutritional efficacy.

Vitamin Binding to Proteins

Some vitamins, even some requiring metabolic activation, are biologically active only when bound to a specific protein (Table 3.9). This often occurs when the vitamin serves as the prosthetic group of an enzyme, remaining bound to the enzyme protein during catalysis. Vitamins of this type are properly called **coenzymes**. In other cases, vitamins participate in enzymatic catalysis but are not firmly bound to enzyme protein during the reaction; they are more properly called **cosubstrates**. This distinction, however, does not address the mechanism, but only the tightness, of binding.[18] Therefore, the term coenzyme has come to be used to describe enzyme cofactors of both types. Other vitamins bind to specific nuclear receptors to elicit transcriptional modulation of one or more protein products (Table 3.10).

Vitamin Excretion

In general, the fat-soluble vitamins, which tend to be retained in hydrophobic environments, are excreted with the feces via the enterohepatic circulation[19] (Table 3.11). Exceptions include vitamins A and E, which to some extent have water-soluble metabolites (e.g., short-chain derivatives of retinoic acid; and the so-called ***Simon's metabolites*** [carboxylchromanol metabolites] of vitamin E), and menadione, which can be metabolized to a polar salt; these vitamin metabolites are excreted in the urine. In contrast, the water-soluble vitamins are generally excreted in the urine, both in intact forms (riboflavin, pantothenic acid) and as water-soluble metabolites (vitamin C, thiamin, niacin, riboflavin, pyridoxine, biotin, folate, and vitamin B_{12}).

18. For example, the associations of NAD and NADP with certain oxidoreductases are weaker than those of FMN and FAD with the flavoprotein oxidoreductases.
19. These substances are discharged from the liver with the bile; the amounts that are not subsequently reabsorbed are eliminated with the feces.

TABLE 3.7 Vitamins That Can Be Biosynthesized by the Host

Vitamin	Precursor	Route	Conditions Increasing Dietary Need
Niacin	Tryptophan	Conversion to nicotinamide mononucleotide (NMN) via picolinic acid	Low 3-OH-anthranilic acid oxidase activity
			High picolinic acid carboxylase activity
			Low dietary tryptophan
			High dietary leucine[a]
Vitamin D_3	7-Dehydrocholesterol	UV photolysis	Insufficient sunlight/UV exposure
Vitamin C[b]	Glucose	Gulonic acid pathway	L-gulonolactone oxidase deficiency

[a]*The role of leucine as an effector of the conversion of tryptophan to niacin is controversial (Chapter 12).*
[b]*Humans and other higher primates, guinea pigs, the Indian fruit bat, and a few other species are capable of vitamin C biosynthesis.*

TABLE 3.8 Vitamins That Must Be Activated Metabolically[a]

Vitamin	Active Form(s)	Activation Step	Condition(s) Increasing Need
Vitamin A	Retinol	Retinal reductase; hydrolase	Protein insufficiency
	11-*cis*-Retinol	Retinyl isomerase	
	11-*cis*-Retinal	Alcohol dehydrogenase	Zinc insufficiency
Vitamin D	1,25-$(OH)_2$-D	Vitamin D 25-hydroxylase; 25-OH-D 1-hydroxylase	Hepatic failure
			Renal failure, lead exposure, estrogen deficiency, anticonvulsant drug therapy
Vitamin K	All forms	Dealkylation of Ks, MKs; alkylation of Ks, MKs, menadione	Hepatic failure
Thiamin	Thiamin-diP	Phosphorylation	High carbohydrate intake
Riboflavin	FMN, FAD	Phosphorylation, adenosylation	
Vitamin B_6	Pyridoxal-P	Phosphorylation; oxidation	High protein intake
Niacin	NAD(H)	Amidation (nicotinic acid)	Low tryptophan intake NADP(H)
Pantothenic acid	Coenzyme A	Phosphorylation; decarboxylation;	
		ATP condensation; peptide bonding	
	ACP	Phosphorylation; peptide bonding	
Folate	C_1-FH_4	Reduction; addition of C_1	
Vitamin B_{12}	Methyl-B_{12}	Cobalamin methylation	Folate deficiency
		CH_3 group insufficiency	
	5′-Deoxyadenosyl-B_{12}	Adenosylation	

[a]*ACP, acyl carrier protein; C_1-FH_4, tetrahydrofolic acid;* Thiamin-diP, *thiamin pyrophosphate.*

5. METABOLIC FUNCTIONS OF THE VITAMINS

Vitamins Serve Five Basic Functions

The 13 families of nutritionally important substances called *vitamins* comprise two to three times that number of practically important vitamers, which function in metabolism in five general, and not mutually exclusive ways:

- As **coenzymes**—metabolites that link to enzymes and are required for their catalytic activity.
- As **H^+/e^- donors/acceptors**—factors that can undergo changes in oxidation state in metabolism by being

TABLE 3.9 Vitamins That Must Be Linked to Enzymes and Other Proteins

Vitamin	Form(s) Linked
Biotin	Biotin
Vitamin B_{12}	Methylcobalamin, adenosylcobalamin
Vitamin A	11-*cis*-Retinal
Thiamin	Thiamin pyrophosphate
Riboflavin	FMN, FAD
Niacin	NAD, NADP
Vitamin B_6	Pyridoxal phosphate
Pantothenic acid	Acyl carrier protein
Folate	Tetrahydrofolic acid (FH_4)

TABLE 3.10 Vitamins That Have Nuclear Receptor Proteins

Vitamin	Form(s) linked	Receptor
Vitamin A	All-*trans*-retinoic acid	Retinoic acid receptors (RARs)
	9-*cis*-Retinoic acid	
	9-*cis*-Retinoic acid	Retinoid X receptors (RXRs)
Vitamin D	1,25-$(OH)_2$-Vitamin D_3	Vitamin D receptor (VDR)

oxidized (losing electrons to an acceptor) or reduced (accepting electrons to a donor acceptor) in metabolism.

- As **antioxidants**—factors that inhibit oxidative processes, which produce free radicals that can start chain reactions that damage lipids and proteins and affect cellular function. Antioxidants interrupt such processes by being oxidized themselves and thus removing free radicals.
- As **hormones**—metabolites released by cells or glands in one part of the body that affect cell function in another part of the body.
- As effectors of **gene transcription**—factors that affect the first step in gene expression, the process of "transcription" by which a complementary RNA copy of a DNA sequence is made.

The type of metabolic function of any particular vitamer or vitamin family is dependent on its tissue/cellular distribution and its chemical reactivity, both of which are direct or indirect functions of its chemical structure. For example, the antioxidative function of vitamin E reflects the ability of that vitamin to form semistable radical intermediates; its lipophilicity allows vitamin E to discharge this antioxidant function within the hydrophobic regions of biomembranes, thus protecting polyunsaturated membrane phospholipids. Similarly, the redox function of riboflavin is due to its ability to undergo reversible reduction/oxidation involving a radical anion intermediate. These functions (summarized in Table 3.12), and the fundamental aspects of their significance in nutrition and health, are the subjects of Chapters 6–18.

6. VITAMIN BIOAVAILABILITY

Several factors affect the biological activities of vitamins (Table 3.13). This includes multiple vitamers, which may differ in stability, accessibility from food matrices, and efficiency of enteric absorption and intrinsic metabolic potency. These factors can confound interpretations of the results of vitamin analyses of foods. For this reason, it is useful to distinguish between the intrinsic biological activity of a vitamer, i.e., its *biopotency*, and the actual extent to which that vitamer is absorbed and utilized at the cellular level, i.e., its *bioavailability*.[20] Considerations of relative bioavailability, which can be determined in bioassays that employ reference vitamers of known biopotency, have led to the expression of vitamin contents in foods using standardized unitage based on activities compared to those reference vitamers, e.g., international units, retinol equivalents, α-tocopherol equivalents.

Vitamin Bioavailability

This describes that fraction of ingested vitamin that is absorbed, retained, and metabolized through normal pathways in a form(s) that can be utilized for normal physiologic functions.

7. VITAMIN ANALYSIS

Vitamins are analyzed in foods/feedstuffs and biological specimens for different purposes:

- **Foods/feedstuffs**—for ascertaining potency for food labeling and nutrient database development, for evaluating storage stability, and for estimating vitamin intakes;
- **Biological specimens**—for determining vitamin bioavailability and for vitamin nutritional status.

Various methods are available for the quantitative determination of the vitamins (Table 3.14). Because many vitamers are bound to proteins (e.g., ε-pyridoxyllysine, niacytin) or other factors (e.g., pyridoxine-5′-β-D-glucoside) in foods or can be entrapped in food matrices (e.g., folates), their

20. Some authors have used the term **bioefficacy** with a similar connotation.

TABLE 3.11 Excretory Forms of the Vitamins

Vitamin	Urinary Form(s)	Fecal Form(s)
Vitamin A	Retinoic acid, acidic short-chain forms	Retinoyl glucuronides; intact-chain products
Vitamin D		25,26-$(OH)_2$-D; 25-$(OH)_2$-D-23,26-lactone
Vitamin E	Some carboxylchromanol metabolites	Tocopheryl quinone; tocopheronic acid and its lactone
Vitamin K K's and MK's		Vitamin K-2,3-epoxide; 2-CH_3-3(5′-carboxy-3′-CH_3-2′-pentenyl)-1,4-naphthoquinone; 2-CH_3-3(3′-carboxy-3′-methylpropyl)-1,4-napthoquinone; other unidentified metabolites
Menadione	Menadiol phosphate; menadiol sulfate	Menadiol glucuronide
Vitamin C[a]	Ascorbate-2-sulfate; oxalic acid; 2,3-diketogulonic acid	
Thiamin	Thiamin; thiamin disulfide; thiamin pyrophosphate; thiochrome	
	2-Methyl-4-amino-5-pyrimidine carboxylic acid; 4-methyl-thiazole-5-acetic acid	
	2-Methyl-4-amino-5-hydroxymethyl pyrimidine; 5-(2-hydroxyethyl)-4-methylthiazole	
	3-(2′-Methyl-4-amino-5′-pyrimidinylmethyl)-4-methylthiazole-5-acetic acid	
	2-Methyl-4-amino-5-formylaminomethylpyrimidine; other minor metabolites	
Riboflavin	Riboflavin; 7- and 8-hydroxmethylriboflavins; 8β-sulfonylriboflavin, riboflavinyl peptide ester; 10-hydroxyethylflavin, lumiflavin, 10-formyl-methylflavin; 10-carboxymethylflavin; lumichrome	
Niacin	N^1-methylnicotinamide; nicotinuric acid; nicotinamide-N^1-oxide; N^1-methylnicotinamide-N^1-oxide; N^1-methyl-4-pyridone-3-carboxamide; N^1-methyl-2-pyridone-5-carboxamide	
Vitamin B_6	Pyridoxol, pyridoxal, pyridoxamine, and respective phosphates; 4-pyridoxic acid and its lactone; 5-pyridoxic acid	
Biotin	Biotin; *bis-nor*-biotin; biotin *d*- and *l*-sulfoxide	
Pantothenic acid	Pantothenic acid	
Folate	Pteroylglutamic acid; 5-methyl-pteroylglutamic acid; 10-HCO-FH_4; pteridine acetamidobenzoylglutamic acid	Intact folates
Vitamin B_{12}	Cobalamin	Cobalamin

[a]*Substantial amounts are also oxidized to CO_2 and are excreted across the lungs.*

extraction necessitates disruption of those complexes and separation from interfering substances. This must be done in ways that are both quantitative and accommodate the intrinsic characteristics of each vitamer. Accordingly, conditions of sample extraction must stabilize the vitamin(s) of interest to yield accurate results. Chromatographic separations have proven useful for determining vitamins A, D, E, K, C, thiamin, riboflavin, niacin, and vitamin B_6. They depend on separation by phase partitioning (liquid–liquid[21] or gas–liquid[22]) of vitamers for specificity, ascertained by comparison to authentic standards, and a suitable means of detection (e.g., ultraviolet–visible absorption, fluorescence, electrochemical reactivity, and mass spectrometry)

21. High-performance liquid chromatography, HPLC.
22. Gas-liquid chromatography, GLC.

TABLE 3.12 Metabolic Functions of the Vitamins

Vitamin	Functions
Coenzymes	
Vitamin A	Rhodopsin conformational change following light-induced bleaching
Vitamin K	Vitamin K-dependent peptide-glutamyl carboxylase
Vitamin C	Cytochrome *P*-450-dependent oxidations (drug and cholesterol metabolism, steroid hydroxylations)
Thiamin	Cofactor of α-keto acid decarboxylases and transketolase
Niacin	NAD(H)/NADP(H) used by more than 30 dehydrogenases in the metabolism of carbohydrates (e.g., glucose-6-phosphate dehydrogenase), lipids (e.g., α-glycerol phosphate dehydrogenase), protein (e.g., glutamate dehydrogenase); Krebs cycle, rhodopsin synthesis (alcohol dehydrogenase)
Riboflavin	FMN: L-amino acid oxidase, lactate dehydrogenase, pyridoxine (pyridoxamine); 5′-phosphate oxidase
	FAD: D-amino acid and glucose oxidases, succinic and acetyl CoA dehydrogenases; glutathione, vitamin K, and cytochrome reductases
Vitamin B_6	Metabolism of amino acids (aminotransferases, deaminases, decarboxylases, desulfhydrases), porphyrins (δ-aminolevulinic acid synthase), glycogen (glycogen phosphorylase), and epinephrine (tyrosine decarboxylase)
Biotin	Carboxylations (pyruvate, acetyl CoA, propionyl CoA, 3-methylcrotonyl CoA carboxylases), and transcarboxyl-ations (methylmalonyl CoA carboxymethyltransferase)
Pantothenic acid	Fatty acid synthesis/oxidation
Folate	Single-carbon metabolism (serine–glycine conversion, histidine degradation, purine synthesis, methyl group synthesis)
Vitamin B_{12}	Methylmalonyl CoA mutase, N^5-CH_3-FH_4:homocysteine methyltransferase
H^+/e^- Donors/Acceptors (Cofactors)	
Vitamin K	Converts the epoxide form in the carboxylation of peptide glutamyl residues
Vitamin C	Oxidizes dehydroascorbic acid in hydroxylation reactions
Niacin	Interconverts NAD^+/NAD(H) and $NADP^+$/NADP(H) couples in several dehydrogenase reactions
Riboflavin	Interconverts FMN/FMNH/$FMNH_2$ and FAD/FADH/$FADH_2$ systems in several oxidases
Pantothenic acid	Oxidizes coenzyme A in the synthesis/oxidation of fatty acids
Antioxidants	
Vitamin E	Protects polyunsaturated membrane phospholipids and other substances from oxidative damage via conversion of tocopherol to tocopheroxyl radical and, then, to tocopheryl quinone
Vitamin C	Protects cytosolic substances from oxidative damage
Hormones	
Vitamin A	Signals coordinate metabolic responses of several tissues
Vitamin D	Signals coordinate metabolism important in calcium homeostasis
Effectors of Gene Transcription	
Vitamin A	Bind nuclear retinoid receptors to signal transcription of multiple pathways
Vitamin D	Regulates transcription of some 50 genes associated with many aspects of metabolism

TABLE 3.13 Several Factors Affecting Vitamin Bioavailability

Factor	Examples
Extrinsic Factors	
Differing inherent biopotencies	Cholecalciferol (vitamin D_3) is nearly 10-fold more biopotent than ergocalciferol (vitamin D_2) for humans; the difference is greater for the chick
	All-*rac*-α-tocopherol has 50% of the biopotency of R,R,R-α-tocopherol
	Pyridoxine-5′-β-D-glucoside has half the biopotency of pyridoxine for humans
	α-Carotene, which yields upon central cleavage only a single mole of retinol, has half of the biopotency of β-carotene, which yield 2 mol of retinol
Losses	The vitamin C content of potatoes can drop by one-third within 1 month of storage
	NADH and NADPH appear to be unstable in the acidic conditions of the stomach
	Pyridoxine can bind to proteins during food processing/storage to form ε-pyridoxyllysine, which has half of the biopotency of the parent vitamer
	10-Formyltetrahyrdofolic acid is susceptible to oxidation to 10-formyldihydrofolic acid as well as interconversion to 5,10-methenyltetrahydrofolic acid under acidic conditions
Dietary/food effects	Vitamin A, provitamin A carotenoids, and vitamin D are poorly absorbed from very low-fat diets
	ε-Pyridoxyllysine, which has half of the biopotency of pyridoxine, is present in many plant tissues
	Niacin bound to proteins in some foods (e.g., maize, as niacytin) is not bioavailable unless hydrolyzed during food preparation/cooking to yield niacin
	Some folates in foods can be entrapped in their food matrix and be less biopotent than predicted based on chemical analysis
Intrinsic Factors	
Physiological effects	Provitamin A carotenoids can be less bioavailable than expected based on chemical analysis due to incomplete intestinal cleavage to retinol
	Folic acid (tetrahydrofolic acid) is generally better absorbed than most food forms of (polyglutamated) folate, which requires deconjugation to be absorbed
Health status	Folate absorption is impaired in patients with enteritis, e.g., sprue
	Absorption of vitamin B_{12} is reduced in individuals with digestive disorders, including many older persons who experience diminished gastric parietal cell function

TABLE 3.14 Methods of Vitamin Analysis

Vitamin	Sample Preparation	Instrumental Analysis		Immunological Analysis	Microbiological Assay
		Separation	Detection		
Vitamin A	Direct solvent extraction; alkaline hydrolysis[a], extraction into organic solvents	HPLC[b]	UV absorption, MS[c]	ELISA[d]	
Vitamin D	Alkaline hydrolysis with extraction into organic solvents	HPLC	UV, MS	ELISA	
Vitamin E	Alkaline hydrolysis with extraction into organic solvents	HPLC	Fluorescence, UV, MS	ELISA	
Vitamin K	Direct solvent extraction; supercritical fluid	HPLC	UV	ELISA	
	Extraction[e]; enzymatic hydrolysis				
Vitamin C	Acid hydrolysis	HPLC, IEC[f], MECC[g]	UV, MS		
Thiamin	Acid hydrolysis; enzymatic hydrolysis[h] absorption	IEC, GLC[i], HPLC	FID[k], MS	Fluorescence, UV, ELISA	*Lactobacillus viridescens* (12706)[j]
Riboflavin	Acid hydrolysis	HPLC, MECC	Fluorescence, UV, MS		*Lactobacillus casei* subsp. *rhamnosus* (7469)
					Enterococcus faecalis (10100)
Niacin	Alkaline hydrolysis	IEC, HPLC GLC	UV, FID		*Lactobacillus plantarum* (8014)[l]
		MECC			*Leuconostoc mesenteroides* subsp. *mesenteroides* (9135)
Vitamin B_6	Acid hydrolysis	HPLC, IEC, GLC	Fluorescence, UV	ELISA	*Saccharomyces carlsbergensis* (9080)
		MECC	FID, MS		*Kloeckera apiculata* (8714)
Pantothenic acid	Alkaline hydrolysis; enzymatic hydrolysis	GLC	FID	ELISA	*Lactobacillus plantarum* (8014)[l]

Folate	Enzymatic hydrolysis[o]	IEC	MS	ELISA	*Lactobacillus casei* subsp. *rhamnosus* (7469)[m]
					Enterococcus hirae (8043)[n]
Biotin	Acid hydrolysis; enzymatic hydrolysis[p]		MS	ELISA	*Lactobacillus plantarum* (8014)
Vitamin B_{12}	Direct solvent extraction	MECC	UV absorption	ELISA	*Lactobacilus delbrueckii*, subsp. *lactis* (4797)

[a]*Saponification.*
[b]*High-performance liquid chromatography (HPLC).*
[c]*Mass spectrometry(MS).*
[d]*Enzyme-linked immunosorbent assay (ELISA).*
[e]*Supercritical fluids are gases held above its critical temperature and critical pressure, which confers solvating properties similar to organic solvents with very low viscosities and very high diffusivities.*
[f]*Ion-exchange chromatography (IEC).*
[g]*Micellar electrokinetic capillary electrophoresis or chromatography (MECC).*
[h]*Thiaminase or other phosphatase.*
[i]*Gas–liquid chromatography (GLC).*
[j]*American Type Culture Collection number.*
[k]*Flame ionization detection (FID, used with GLC separation).*
[l]*Responds to nicotinic acid only.*
[m]*Responds to free vitamer only.*
[n]Responds to all vitamers; yields "total" folate activity.
[o]*Folyl conjugase.*
[p]*Papain.*

for sensitivity. Microbiological assays are available for thiamin, riboflavin, niacin, vitamin B_6, pantothenic acid, biotin, folate, and vitamin B_{12}. These methods are based on the absolute requirement of certain microorganisms for particular vitamins for multiplication, which can be measured turbidimetrically or by the evolution of CO_2 from substrate provided in the growth media. Some forms of vitamins A, E, and C can be measured by chemical colorimetric reactions; however, only the dye reduction methods for ascorbic acid have appropriate specificity and reliability to be recommended.[23] Competitive protein-binding assays have been developed for biotin, folate, and vitamin B_{12}.[24] Enzyme-linked immunosorbent assays have been developed for vitamins A, D, E, K, B_6, and B_{12}, as well as pantothenic acid, biotin, and folate. While microbiological and immunological methods are generally economical, they lack the capacity of liquid chromatography–tandem mass spectrometry (LC-MS/MS) for unequivocal identification of multiple specific vitamers based on characteristic patterns of ion daughters produced instrumentally.

8. STUDY QUESTIONS AND EXERCISES

1. Prepare a concept map of the relationships between the chemical structures, the physical properties, and the modes of absorption, transport, and tissue distributions of the vitamins.
2. For each vitamin, identify the key feature(s) of its chemical structure. How is/are this/these feature(s) related to the stability and/or biologic activity of the vitamin?
3. Discuss the general differences between the fat-soluble and water-soluble vitamins, and the implications of those differences in diet formulation and meal preparation.
4. Which vitamins would you suspect might be in shortest supply in the diets of livestock? in your own diet? Explain your answer in terms of the physicochemical properties of the vitamins.
5. Which vitamins would you expect to be stored well in the body? Which would you expect to be unstable in foods or feeds?
6. What factors would you expect to influence the absorption of specific vitamins?

23. Vitamin A: The Carr–Price method, based on the time-sensitive production of a blue complex of retinol and antimony trichloride, is no longer recommended due to its lack of specificity and negative bias. Vitamin E: The Fe^{2+}-dependent reduction of a fat-soluble dye such as bathophenanthroline by vitamin E is not recommended due to its lack of specificity, although many interfering substances can be partitioned into aqueous solvents during sample preparation. **Vitamin C:** The reaction of ascorbic acid with the dye 2,4-dinitrophenolindolphenol remains a useful method due to the fact that most interfering substances can be partitioned into organic solvents during sample preparation.

24. Biotin and avidin; folate and folate-binding protein; vitamin B12 and transcobalamin or R proteins.

RECOMMENDED READING

Vitamin Nomenclature

Anonymous, 1987. Nomenclature policy: generic descriptors and trivial names for vitamins and related compounds. J. Nutr. 120, 12–19.

Vitamin Chemistry and Analysis

Ball, G.F.M., 2005. Vitamins in Foods: Analysis, Bioavailability and Stability. CRC Press, New York. 824 pp.

De Leenheer, A.P., Lambert, W., 2000. Modern Chromatographic Analysis of Vitamins: Revised and Expanded. CRC Press, New York. 632 pp.

Eitenmiller, R.R., Landen, W.O., 2007. Vitamin Analysis for the Health and Food Sciences, second ed. CRC Press, New York. 660 pp.

Zempleni, J., Suttie, J.W., Gregory, J.F., et al., 2014. Handbook of Vitamins, fifth ed. CRC Press, New York. 593 pp.

Vitamin Bioavailability

Gregory, J.F., 2012. Accounting for differences in the bioactivity and bioavailability of vitamers. Food Nutr. Res. 56, 5809–5820.

Chapter 4

Vitamin Deficiency

Chapter Outline

Anchoring Concepts

1. A **disease** is an interruption or perversion of function of any of the organs with characteristic signs and/or symptoms caused by specific biochemical and morphological changes.
2. **Deficient intakes** of essential nutrients can cause disease.

These diseases ... were considered for years either as intoxication by food or as infectious diseases, and twenty years of experimental work were necessary to show that diseases occur which are caused by a deficiency of some essential substance in the food.

C. Funk[1]

LEARNING OBJECTIVES

1. To understand the concept of **vitamin deficiency**.
2. To understand that deficient intakes of vitamins lead to **sequences of lesions** involving changes starting at the biochemical level, progressing to affect cellular and tissue function, and, ultimately, resulting in morphological changes.
3. To appreciate the range of possible morphological changes in organ systems that can be caused by vitamin deficiencies.
4. To get an overview of *specific clinical signs and symptoms* of deficiencies of each vitamin in animals, including humans, as background for further study of the vitamins.
5. To appreciate the relationships of clinical manifestations of vitamin deficiencies and lesions in the biochemical functions of those vitamins.

1. Casimir Funk (1884–1967) was a Polish born chemist who worked at the Lister Institute in London and is credited with formulating the vitamin concept, which he called "vitamines," and later elucidated the chemical structure of thiamin.

VOCABULARY

Achlorhydria
Achromotrichia
Acrodynia
Age pigments
Alopecia
Anemia
Anorexia
Arteriosclerosis
Ataxia
Beriberi
Bradycardia
Brown bowel disease
Brown fat disease
Cage layer fatigue
Capillary fragility
Cardiomyopathy
Cataract
Cervical paralysis
Cheilosis
Chondrodystrophy
Cirrhosis
Clinical signs
Clubbed down
Convulsion
Cornification
Curled toe paralysis
Dermatitis
Desquamation
Dystrophy
Edema
Encephalomalacia
Encephalopathy
Exudative diathesis
Fatty liver and kidney syndrome
Geographical tongue
Glossitis

The Vitamins. http://dx.doi.org/10.1016/B978-0-12-802965-7.00004-6

Hyperkeratosis
Hypovitaminosis
Inflammation
Keratomalacia
Leukopenia
Lipofuscin(osis)
Malabsorption
Mulberry heart disease
Myopathy
Necrosis
Nephritis
Neuropathy
Night blindness
Nyctalopia
Nystagmus
Opisthotonos
Osteomalacia
Osteoporosis
Pellagra
Perosis
Photophobia
Polyneuritis
Retrolental fibroplasia
Rickets
Scurvy
Steatitis
Stomatitis
Symptom
Vitamin deficiency
Wernicke–Korsakoff syndrome
White muscle disease
Xerophthalmia
Xerosis

1. THE CONCEPT OF VITAMIN DEFICIENCY

What Is Meant by the Term *Vitamin Deficiency*?

Because the gross functional and morphological changes caused by deprivation of the vitamins were the source of their discovery as important nutrients, these signs have become the focus of attention for many with interests in human and/or veterinary health. Indeed, freedom from clinical diseases caused by insufficient vitamin nutriture has generally been used as the main criterion by which vitamin requirements have been defined. The expression **vitamin deficiency** therefore simply refers to the basic condition of **hypovitaminosis**. Vitamin deficiency is distinct from but underlies the various biochemical changes, physiological and/or functional impairments, or other overt disease signs by which the need for a vitamin is defined.

A vitamin deficiency is … the shortage of supply of a vitamin relative to its needs by a particular organism.

Vitamin Deficiencies Involve Cascades of Progressive Changes

The diseases associated with low intakes of particular vitamins are clinical manifestations of a progressive sequence of lesions that result from initial biochemical perturbations (e.g., diminished enzyme activity due to lack of a coenzyme or cosubstrate; membrane dysfunction due to lack of a stabilizing factor) that lead first to cellular and subsequently to tissue and organ dysfunction. Thus, the early stages of vitamin deficiency are subclinical and detectable only with biochemical indicators. If uncorrected, these marginal changes lead to characteristic clinical (observable) signs (Fig. 4.1). This cascade can be generalized in four stages.

The Four Stages of Vitamin Deficiency

Subclinical (marginal) deficiency	**Stage I Depletion of vitamin stores**, which leads to…
	Stage II Cellular metabolic changes, which lead to…
Clinical (observable) deficiency	**Stage III Functional defects**, which ultimately produce…
	Stage IV Morphological changes.

Marks[2] illustrated this point with the results of a study of thiamin depletion in human volunteers (Fig. 4.2). When subjects were fed a thiamin-free diet, no changes of any type were detected for 5–10 days, after which the first signs of decreased saturation of erythrocyte transketolase with its essential cofactor, thiamin pyrophosphate (TPP), were noted. Not until nearly 200 days of depletion—i.e., long after tissue thiamin levels and transketolase–TPP saturation had declined—were classic **clinical signs**[3] of thiamin deficiency (**anorexia**, weight loss, malaise, insomnia, hyperirritability) detected.

Marginal deficiencies of vitamins in which the impacts of poor vitamin status are not readily observed without chemical or biochemical testing are often referred to as **subclinical deficiencies** for that reason. Subclinical deficiencies involve depleted reserves or localized abnormalities without the presence of overt functional or morphological defects. The traditional perspective has been that the

2. Marks, J., 1968. The Vitamins in Health and Disease: A Modern Reappraisal. Churchill, London, UK.
3. A symptom is a change, whereas a clinical sign is a change detectable by a trained observer, e.g., a physician.

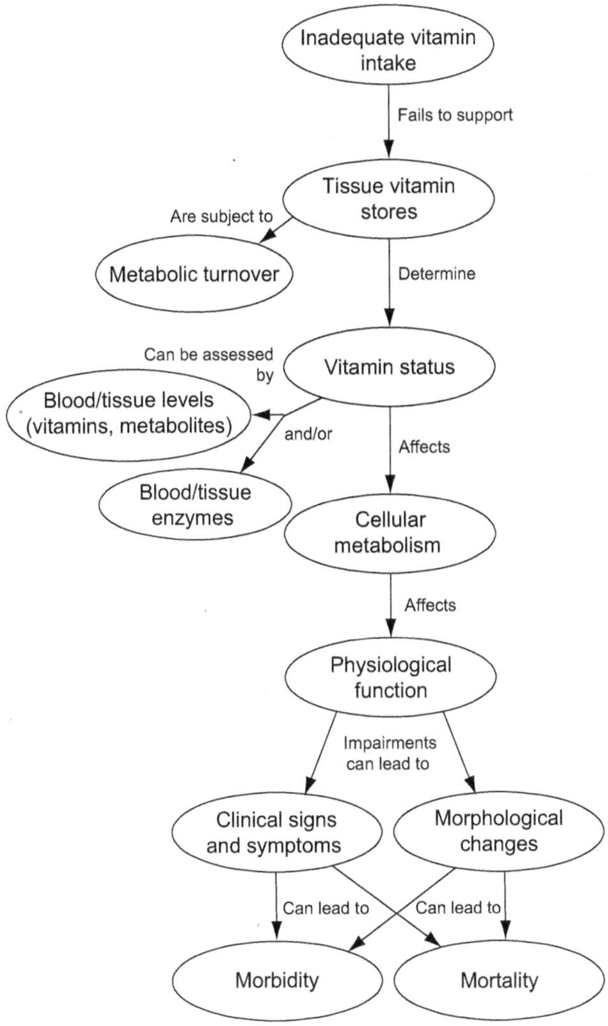

FIGURE 4.1 Concept map of effects of inadequate vitamin intake.

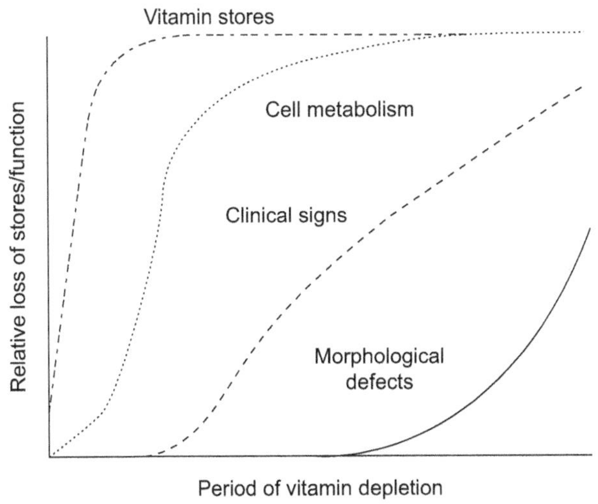

FIGURE 4.2 The four progressive stages of deficiency, starting with tissue depletion and ending with morphological changes. *From Marks, J., 1968. The Vitamins in Health and Disease: A Modern Reappraisal. Churchill, London, UK.*

absence of overt, clinical manifestations of deficiency constitutes good nutrition; this perspective ignores the importance of preventing the early functional impairments that can progress to overt clinical signs. Therefore, the modern view of nutritional adequacy must focus on the maintenance of normal metabolism and, in several cases, body reserves as criteria of adequate vitamin status.[4]

2. CLINICAL MANIFESTATIONS OF VITAMIN DEFICIENCIES

Many Organ Systems Can Be Affected by Vitamin Deficiencies

Every organ system of the body can be the target of a vitamin deficiency. Some vitamin deficiencies affect certain organs preferentially (e.g., vitamin D deficiency chiefly affects calcified tissues); others affect several or many organs in various ways. Because the diagnosis of a vitamin deficiency involves its differentiation from other potential causes of similar clinical signs, it is useful to consider the various morphologic lesions caused by vitamin deficiencies from an organ system perspective. After all, anatomical and/or functional changes in organs are the initial presentations of deficiencies of each of the vitamins. This point is illustrated in Table 4.1, which details the organ systems affected by vitamin deficiencies.

Manifestations of Biochemical Lesions

Relationships Between Biochemical Lesions and Clinical Diseases of Vitamin Deficiencies

The clinical signs and symptoms that characterize the vitamin deficiency diseases are manifestations of underlying impairments (i.e., *lesions*) in biochemical function that result from insufficient vitamin supply (Table 4.2). This is a fundamental concept in understanding the roles of the vitamins in nutrition and health. Hypovitaminosis of sufficient magnitude and duration is causally related to the morphological and/or functional changes associated with the latter stages of vitamin deficiency. While the validity of this concept may be apparent in the abstract, documentary evidence for it in the case of each of the vitamin deficiency diseases, that is, direct cause–effect linkages of specific biochemical lesions and clinical changes, is, for many vitamins, not complete.

Vitamin A offers a case in point of this fact. While the role of vitamin A in preventing nyctalopia (night blindness)

4. By such criteria, marginal vitamin status in the United States appears to be quite prevalent, even though the prevalence of clinically significant vitamin deficiencies appears to be very low. Estimates of marginal status with respect to one or more vitamins have been as high as 15% of adolescents, 12% of persons 65 years of age and older, and 20% of dieters.

TABLE 4.1 Organ Systems Affected by Vitamin Deficiencies in Humans and Other Animals

Organ Systems	Vitamin A	Vitamin D	Vitamin E	Vitamin K	Ascorbic Acid	Thiamin	Riboflavin	Niacin	Pyridoxine	Biotin	Pantothenic Acid	Folate	Vitamin B_{12}
General													
Appetite	+	+	+		+[a]	+	+	+	+	+			
Growth	+	+	+	+	+[a]	+	+	+	+	+	+	+	+
Integument													
Skin	+						+	+					
Hair, nails, feathers	+						+			+			
Musculature													
Skeletal muscles			+										
Heart						+							
Gizzard			+										
Vascular System													
Vessels			+	+	+[a]				+				
Blood cells			+				+					+	+
Clotting system				+									
Gastrointestinal Tract													
Stomach						+		+					
Mouth							+						
Tongue								+					
Small intestine			+	+		+							
Colon							+	+					
Skeletal System													
Bone	+	+			+[a]		+	+					
Teeth		+			+[a]								
Vital organs													
Liver			+			+							

Kidney	+						+		+				
Thymus											+		
Adrenals							+				+		
Pancreas									+				
Adipose			+										
Ocular System													
Eye	+		+			+	+						
Reproductive System													
Vagina	+												
Uterus			+										
Ovary	+						+						
Egg		+											
Testes	+												
Fetus	+						+						
Nervous System													
General	+	+				+	+	+	+		+		
Spinal cord	+					+							
Brain			+										
Peripheral nerves			+										+
Psychological, emotional						+		+					

[a]*Only human, higher primates, the guinea pig, and some birds are affected by ascorbic acid deprivation.*

TABLE 4.2 The Underlying Biological Functions of the Vitamins

Vitamin	Active Form(s)	Deficiency Disorders	Important Biological Functions or Reactions
Vitamin A	Retinol, retinal, retinoic acid	Night blindness, xerophthalmia, keratomalacia, impaired growth	Photosensitive retinal pigment
			Regulation of epithelial cell differentiation
			Regulation of gene transcription
Vitamin D	1,25-$(OH)_2$-D	Rickets, osteomalacia	Promotion of intestinal calcium absorption, mobilization of calcium from bone, stimulation of renal calcium resorption, regulation of PTH secretion, possible function in muscle
Vitamin E	a-Tocopherol	Nerve, muscle degeneration	Antioxidant protector for membranes
Vitamin K	K_n, MK_n	Impaired blood clotting	Cosubstrate for γ-carboxylation of glutamyl residues of several clotting factors and other calcium-binding proteins
Vitamin C	Ascorbic acid, dehydroascorbic acid	Scurvy	Cosubstrate for hydroxylations in collagen synthesis, drug, and steroid metabolism
Thiamin	Thiamin pyrophosphate	Beriberi, polyneuritis, Wernicke–Korsakoff syndrome	Coenzyme for oxidative decarboxylation of 2-keto acids (e.g., pyruvate, 2-keto-glutarate); coenzyme for pyruvate decarboxylase and transketolase
Riboflavin	FMN, FAD	Dermatitis	Coenzymes for numerous flavoproteins that catalyze redox reactions in fatty acid synthesis/degradation, TCA cycle
Niacin	NAD(H), NADP(H)	Pellagra	Cosubstrates for hydrogen transfer catalyzed by many dehydrogenases, e.g., TCA cycle respiratory chain
Pyridoxine	Pyridoxal-5′-phosphate	Signs vary with species	Coenzyme for metabolism of amino acids, e.g., side chain, decarboxylation, transamination, racemization
Folate	Polyglutamyl tetrahydrofolates	Megaloblastic anemia	Coenzyme for transfer of single-carbon units, e.g., formyl and hydroxymethyl groups in purine synthesis
Biotin	1′-*N*-carboxybiotin	Dermatitis	Coenzyme for carboxylations, e.g., acetyl CoA/malonyl CoA conversion
Pantothenic acid	Coenzyme A	Signs vary with species	Cosubstrate for activation/transfer of acyl groups to form esters, amides, citrate, triglycerides, etc.
	Acyl carrier protein		Coenzyme for fatty acid biosynthesis
Vitamin B12	5′-deoxyadenosyl-B12	Megaloblastic anemia	Coenzyme for conversion of methylmalonyl CoA to succinyl CoA
	Methyl-B_{12}	Impaired growth	Methyl group transfer from 5-CH_3-FH_4 to homocysteine in methionine synthesis

5-CH_3-FH_4, 5-methyltetrahydrofolic acid; *CoA*, coenzyme A; *PTH*, parathyroid hormone; *TCA*, tricarboxylic acid cycle.

is clear from presently available knowledge of the essentiality of retinal as the prosthetic group of rhodopsin and several other photosensitive visual receptors in the retina, the amount of vitamin A in the retina, and thus available for visual function, is only about 1% of the total amount of vitamin A in the body. Further, it is clear from the clinical signs of vitamin A deficiency that the vitamin has other essential functions unrelated to vision, especially relating to the integrity and differentiation of epithelial cells. However, although evidence indicates that vitamin A is involved in the metabolism of mucopolysaccharides and other essential intermediates, present knowledge does not adequately explain the mechanism(s) of action of vitamin A in supporting growth and in maintaining epithelia. It has been said that 99% of our information about the mode of action of vitamin A concerns only 1% of the vitamin A in the body.

The ongoing search for a more complete understanding of the mechanisms of vitamin action is therefore largely based on the study of biochemical correlates of changes in physiological function or morphology effected by changes in vitamin status. Most of this knowledge has come from direct experimentation, mostly with animal models. Also edifying in this regard has been information acquired from observations of individuals with different rare, naturally occurring, hereditary anomalies involving vitamin-dependent enzymes and transport proteins. Most of the documented inborn metabolic errors (Table 4.3) involve specific mutations manifest as either a loss or aberration in single factors in vitamin metabolism—a highly targeted situation not readily produced experimentally.[5]

Diagnosing Vitamin Deficiencies

Vitamin deficiencies can be diagnosed based on the organ system affected, the specific clinical signs and/or biochemical lesion, and reference to accepted biomarkers of status (Tables 4.3 and 4.4 can be helpful). This requires a three-step analysis:

1. Identify prospective vitamin deficiencies based on mapping of **signs/symptoms** to those reported in the scientific literature and considering relevant demographic and environmental predictors.
2. Use the appropriate **clinical biochemical markers** to exclude possibilities.
3. Determine the actual deficiency(ies) involved based on **responses to treatment.**

3. CAUSES OF VITAMIN DEFICIENCIES

Primary and Secondary Causes of Vitamin Deficiencies

The balance of vitamin supply and biological need of an individual is called **vitamin status**. Reductions in vitamin status can be produced either by reductions in effective vitamin supply or by increases in effective vitamin need. Vitamin deficiency occurs when vitamin status is reduced to the point of having metabolic impact (i.e., stage II); if not corrected, continued reductions in vitamin status lead inevitably to the observable stages of vitamin deficiency (stages III and IV), at which point serious clinical and morphological changes can manifest. When these changes occur as a result of the failure to ingest a vitamin in sufficient amounts to meet physiological needs, the condition is called a **primary deficiency**. When these changes come about as a result of the failure to absorb or otherwise utilize a vitamin owing to an environmental condition or physiological state, and not to insufficient consumption of the vitamin, the condition is called a **secondary deficiency**.

Causes of Vitamin Deficiencies in Humans

Many of the ways in which vitamin deficiencies can develop are interrelated. For example, poverty and disempowerment are often accompanied by lack of nutrition knowledge and result in a poor diet. People living alone, especially the elderly and others with chronic disease, tend to consume foods that require little preparation and that may not provide adequate nutrition. Despite these potential causes of vitamin deficiency, in most of the technologically developed parts of the world the general level of nutrition is high. In those areas, relative few persons can be expected to show signs of vitamin deficiency; those that do present such signs frequently have a potentiating condition that affects either their consumption of food or their utilization of nutrients. In low-income parts of the world, however, food insecurity is still the largest single cause of general malnutrition today, including deficiencies of multiple nutrients.

High-Risk Groups for Vitamin Deficiencies
- Pregnant women
- Infants and young children
- Elderly people
- Vegetarians
- Food-insecure people
- People with intestinal parasites or infections
- Dieters
- Smokers

Primary deficiencies in humans, therefore, tend to have psychosocial and technological causes:

- Poor food habits
- Poverty (i.e., low food-purchasing power)
- Ignorance (i.e., lack of sound nutrition information)
- Lack of total food (e.g., crop failure)
- Lack of vitamin-rich foods (e.g., consumption of highly refined foods)
- Vitamin destruction (e.g., during storage, processing, and/or cooking)
- Anorexia (e.g., homebound elderly, infirm, dental problems)
- Food taboos and fads (e.g., fasting, avoidance of certain foods)
- Apathy (lack of incentive to prepare adequate meals).

Whereas, secondary deficiencies in humans typically have biological causes:

- Poor digestion (e.g., **achlorhydria**—absence of stomach acid)
- **Malabsorption** (impaired intestinal absorption of nutrients; e.g., as a result of diarrhea, intestinal infection, parasites, and pancreatitis)
- Impaired metabolic utilization (e.g., certain drug therapies)
- Increased metabolic need (e.g., pregnancy, lactation, rapid growth, infection, and nutrient imbalance)
- Increased vitamin excretion (e.g., diuresis, lactation, and excessive sweating).

5. However, it is theoretically possible to produce transgenic animal models with similar metabolic anomalies.

TABLE 4.3 Vitamin-Responsive Inborn Metabolic Lesions

Curative Vitamin	Missing Protein or Metabolic Step Affected	Clinical Condition
Vitamin A	Apolipoprotein B	Abetalipoproteinemia; low tissue levels of retinoids
Vitamin D	Receptor	Unresponsive to 1,25(OH)$_2$-D; osteomalacia
Vitamin E	Apolipoprotein B	Abetalipoproteinemia; low tissue levels of tocopherols
Thiamin	Branched-chain 2-oxoacid dehydrogenase	Maple syrup urine disease
	Pyruvate metabolism	Lactic acidemia; neurological anomalies
Riboflavin	Methemoglobin reductase	Methemoglobinemia
	Electron transfer flavoprotein	Multiple lack of acyl CoA dehydrogenations, excretion of acyl CoA metabolites, i.e., metabolic acidosis
Niacin	Abnormal neurotransmission	Psychiatric disorders, tryptophan malabsorption, abnormal tryptophan metabolism
Pyridoxine	Cystathionine β-synthase	Homocysteinuria
	Cystathionine γ-lyase	Cystathioninuria; neurological disorders
	Kynureninase	Xanthurenic aciduria
Folate	Enteric absorption	Megaloblastic anemia, mental disorder
	Methylene-FH$_4$-reductase	Homocysteinuria, neurological disorders
	Glutamate formiminotransferase	Urinary excretion of FIGLU[a]
	Homocysteine/methionine conversion	Schizophrenia
	Tetrahydrobiopterin–phenylalanine hydrolase	Mental retardation, PKU[b]
	Dihydrobiopteridine reductase	PKU, severe neurological disorders
	Tetrahydrobiopterin formation	PKU, severe neurological disorders
Biotin	Biotinidase	Alopecia, skin rash, cramps, acidemia, developmental disorders, excess urinary biotin, and biocytin
	Propionyl CoA carboxylase	Propionic acidemia
	3-Methylcrotonyl CoA carboxylase	3-Methylcrontonylglycinuria
	Pyruvate carboxylase	*Leigh disease*, accumulation of lactate and pyruvate
	Acetyl CoA carboxylase	Severe brain damage
	Holocarboxylase synthase	Lack of multiple carboxylase activities, urinary excretion of metabolites
Vitamin B$_{12}$	Intrinsic factor/enteric absorption	Juvenile pernicious anemia
	Transcobalamin	Megaloblastic anemia, growth impairment
	Methylmalonyl CoA mutase	Methylmalonic acidemia

[a]FIGLU, *formiminoglutamic acid.*
[b]PKU, *phenylketonuria.*

Causes of Vitamin Deficiencies in Animals

Many of the same primary and secondary causal factors that result in vitamin deficiencies in humans can also produce vitamin deficiencies in animals. In livestock, however, most of the serious cases of vitamin deficiency in animals are due to human errors involving improper or careless animal husbandry.

Primary deficiencies in animals typically have physical causes:

- Improperly formulated diet (i.e., error in vitamin premix formulation)
- Feed mixing error (e.g., omission of vitamin from vitamin premix)

TABLE 4.4 Diagnosis of Vitamin Deficiencies in Humans and Other Animals

Signs and Symptoms by Organ System and Organ			Vitamin Deficiencies Possibly Involved	Humans Affected	Other Species Affected	Vitamin A	Vtiamin D	Vitamin E	Vitamin K	Ascorbic acid	Thiamin	Riboflavin	Niacin	Pyridoxine	Biotin	Pantothenic acid	Folate	Vitamin B_{12}
System	Organ	Signs/Symptoms																
					Criteria for deficient/low status (humans)	Plasma retinol <10 µg/dl (<5 mos., >17 yrs); <20 µg/dl (5 mos.-17 yrs)	Plasma 25(OH)D3 <3 ng/ml	Plasma a-tocopherol <3.5 µg/ml	Clotting time >10 min	Plasma ascorbic acid <2 µg/ml; WBC ascorbic acid <8 µg/ml	RBC transketolase TPP-stimulation >25%; urine thiamin <40 µg/24 hr	RBC GSH reductase FAD-stimulation >40%; urine riboflavin <40 µg/24 hr	Urine N'-methyl-2-pyridone-5-carboxamide <1 µg/24 hr	Plasma pyridoxal phosphate <60 nM	Blood biotin <0.4 ng/ml; urine biotin <10 µg/24 hr	Plasma pantothenic acid <6 µg/dl; urine pantothenic acid <1 mg/24 hrs	Plasma folates <3 ng/ml; RBC folates <140 ng/ml	Plasma Vitamin B_{12} <100 pg/ml
General		General weakness	Vitamin A		Cat													
			Vitamin D	+														
			Ascorbic acid	+														
			Thiamine	+	Rat													
			Riboflavin		Pig, dog, fox													
			Niacin	+	Chick													
			Pyridoxine		Rat, chick													
			Pantothenic acid	+	Chick													
		Reduced appetite	Vitamin A		Rat, chick, mouse, pig, calf													
			Vitamin D		Rat, chick, mouse, pig, calf													
			Vitamin K		Rat, chick, mouse, pig, calf													
			Ascorbic acid		Guinea pig													
			Thiamin[a]	+	Rat, chick, mouse, pig, calf													
			Riboflavin		Rat, chick, mouse, pig, calf													
			Niacin	+	Rat, chick, mouse, pig, calf													
			Pyridoxine		Rat, chick, mouse, pig, calf													
			Biotin		Rat, chick, mouse, pig, calf													
		Growth retardation	Ascorbic acid		Guinea pig													
			Other Vitamins		Rat, mouse, chick, dog, calf													
Integument																		
Skin	Dermis	Scaly dermatitis[b]	Vitamin A		Cattle	+												
			Riboflavin		Pig							+						

Continued

TABLE 4.4 Diagnosis of Vitamin Deficiencies in Humans and Other Animals—cont'd

Signs and Symptoms by Organ System and Organ			Vitamin Deficiencies Possibly Involved	Humans Affected	Other Species Affected	Vitamin A	Vtiamin D	Vitamin E	Vitamin K	Ascorbic acid	Thiamin	Riboflavin	Niacin	Pyridoxine	Biotin	Pantothenic acid	Folate	Vitamin B_{12}
System	Organ	Signs/Symptoms																
			Pyridoxine		Rat (acrodynia[c])									+				
			Biotin		Rat, mouse, hamster, cat, mink, fox										+			
			Pantothenic acid		Rat										+			
		Cracking dermatitis	Niacin	+ (Pellagra)	Chick								+					
			Biotin		Pig, poultry, monkey										+			
			Pantothenic acid		Chick (feet)										+			
		Desquamation[d]	Riboflavin	+	Monkey, rat, chick, dog							+						
			Niacin	+	Chick								+					
			Pyridoxine	+										+				
			Biotin		Rat										+			
		Hyperkeratosis[e]	Riboflavin		Rat							+						
			Niacin	+ (Pellagra)	Chick								+					
			Biotin		Rat, mouse, hamster										+			
		Hyperpigmentation	Niacin	+ (Pellagra)									+					
		Photosensitization	Niacin	+ (Pellagra)									+					
Hair, nails, feathers		Rough	Vitamin A		Cattle, poultry (feathers)	+												
			Biotin		Poultry (feathers)													
		Achromatrichia[f]	Biotin		Rat, rabbit, cat, mink, fox, monkey										+			
			Pantothenic acid		Rat										+			
		Alopecia[g]	Riboflavin		Rat, pig, calf							+						
			Niacin		Rat, pig								+					
			Biotin		Rat, mouse, hamster, rabbit, pig, chick, cat, mink, fox: (spectacle eye)										+			
			Pantothenic acid		Rat										+			

		"Blood"-caked whiskers[h]	Pantothenic acid		Rat							+	
		Impaired growth	Biotin		Poultry							+	
			Folate		Poultry								+
Musculature													
Skeletal muscles	Myopathy[i]		Vitamin E		Rat, guinea pig, pig, rabbit, chick, duck, calf, horse, goat, salmon, mink, catfish (white muscle dis,); lamb (stiff lamb dis.)	+							
			Ascorbic acid	+ (Scurvy)	Guinea pig			+					
			Thamin	+ (Beriberi)	Rat				+				
			Pantothenic acid		Pig							+	
Heart	Rhythm	Bradycardia	Thiamin	+ (Beriberi)					+				
	Muscle	Cardiomyopathy[j]	Thiamin	+ (Beriberi)	Rat				+				
			Vitamin E		Pig (mulberry heart dis.); guinea pig, rabbit, rat, dog, calf, lamb, goat	+							
Gizzard[k]		Myopathy	Vitamin E		Turkey poults, ducklings	+							
Vascular system													
Vessels	General	Arteriosclerosis[l]	Pyridoxine		Monkey						+		
	Capillary	Edema[m]	Vitamin E		Chick (exudative diathesis), pig (visceral edema)	+							
			Thiamin	+ (Beriberi)					+				
		Hemorrhage	Vitamin K		Poultry		+						
			Ascorbic acid	+ (Scurvy)	Guinea pig, monkey			+					
Blood cells	Erythrocyte	Hemolytic anemia[n]	Vitamin E	+	Pig, monkey	+							
		Hemorrhagic anemia[o]	Vitamin K		Rat, chick		+						
		Normocytic hypochromic anemia[p]	Riboflavin		Monkey, baboon					+			

Continued

TABLE 4.4 Diagnosis of Vitamin Deficiencies in Humans and Other Animals—cont'd

Signs and Symptoms by Organ System and Organ			Vitamin Deficiencies Possibly Involved	Humans Affected	Other Species Affected													
System	Organ	Signs/Symptoms				Vitamin A	Vtiamin D	Vitamin E	Vitamin K	Ascorbic acid	Thiamin	Riboflavin	Niacin	Pyridoxine	Biotin	Pantothenic acid	Folate	Vitamin B_{12}
		Megaloblastic anemia[q]	Folate	+	Rat, chick												+	
		Megaloblastic anemia	Vitamin B_{12}	+	Rat, chick													+
		Fragility	Vitamin E	+	Rat, pig, monkey			+										
	Leukocyte	Leukopenia[r]	Riboflavin	+								+						
			Folate		Rat, guinea pig												+	
	Platelet	Thrombocytosis[s]	Vitamin E		Rat			+										
		Excess aggregation	Vitamin E		Rat			+										
Clotting system	Prolonged clotting time		Vitamin K	+	Rat, chick, pig, calf				+									
Gastrointestinal tract																		
Stomach	Epithelium	Achlorhydria[t]	Niacin	+ (Pellagra)									+					
		Gastric distress	Thiamin	+ (Beriberi)							+							
Mouth		Stomatitis[u]	Riboflavin	+	Calf							+						
			Niacin	+ (Pellagra)									+					
			Biotin		Chick										+			
			Pantothenic acid		Chick										+			
		Cheliosis[v]	Riboflavin	+								+						
Tongue		Glossitis[w]	Niacin	+ (Pellagra)									+					
			Riboflavin	+	Rat							+						
			Niacin	+ (Pellagra)									+					
Small intestine	Mucosa	Inflammation	Thiamin		Rat						+							
			Riboflavin		Chick, dog, pig							+						
			Niacin		Chick								+					
		Ulcer	Thiamin		Rat						+							

	Enterocyte	Lipofuscinosis[x]	Vitamin E		Dog (brown bowel disease)			+								
		Hemorrhage	Vitamin K		Poultry				+							
			Thiamin		Rat						+					
			Niacin		Dog								+			
Colon		Diarrhea	Riboflavin		Chick, dog, pig, calf							+				
			Niacin	+ (Pellagra)	Dog, pig, poultry								+			
			Vitamin B_{12}		Young pigs											+
		Constipation	Niacin	+ (Pellagra)									+			
Skeletal system																
Bone	Periosteum	Excessive growth	Vitamin A		Pig, dog, calf, horse, sheep	+										
	Epiphyses	Undermineralization malformations	Vitamin D	+ (Osteomalacia[y]: rickets in children)	Chick, dog, calf (rickets)		+									
			Ascorbic acid	Children						+						
	Cortical bone	Demineralization, increased fractures	Vitamin D	+ (Osteomalacia, osteoporosis[z])	Laying hen (caged layer fatigue)		+									
		Chondrodystrophy[aa]	Niacin		Chick, poult (perosis)								+			
			Biotin		Chick, poult (perosis)										+	
		Congenital deformities	Riboflavin		Rat							+				
			Pyridoxine		Rat									+		
Teeth	Dentin	Caries	Vitamin D	Children			+									
			Pyridoxine	+										+		
Vital organs																
Liver	Hepatocyte	Necrosis	Vitamin E		Pig (hepatosis dietetica); rat, mouse			+								
		Steatosis[bb]	Thiamin		Rat						+					
			Biotin		Chick (fatty liver and kidney syndrome)										+	
			Pantothenic acid		Chick, dog										+	
		Cirrhosis[cc]	Choline		Rat, dog, monkey											

Continued

TABLE 4.4 Diagnosis of Vitamin Deficiencies in Humans and Other Animals—cont'd

Signs and Symptoms by Organ System and Organ			Vitamin Deficiencies Possibly Involved	Humans Affected	Other Species Affected	Vitamin A	Vtiamin D	Vitamin E	Vitamin K	Ascorbic acid	Thiamin	Riboflavin	Niacin	Pyridoxine	Biotin	Pantothenic acid	Folate	Vitamin B_{12}
System	Organ	Signs/Symptoms																
Kidney	Nephron	Nephritis[dd]	Vitamin A	+		+												
		Steatosis	Riboflavin		Pig							+						
			Biotin		Chick										+			
		Hemorrhagic necrosis[ee]	Choline		Rat, mouse, pig, rabbit, calf													
		Calculi	Pyridoxine		Rat									+				
Thymus	Thymocyte	Necrosis	Pantothenic acid		Chick, dog										+			
Adrenals		Hypertrophy	Pantothenic acid		Pig										+			
		Hemorrhage	Riboflavin		Pig							+						
		Lymphoid necrosis	Riboflavin		Baboon							+						
			Pantothenic acid		Rat										+			
Pancreas	Eyelets	Insulin insufficiency	Pyrodoxine		Rat									+				
Adipose	Adipocyte	Lipofuscinosis	Vitamin E		Rat, mouse, hamster, cat, pig, mink (brown fat disease)			+										
Ocular system																		
Eye	Eyeball	Nystagmus[ff]	Thiamin	+ (Wernicke-Korsakoff synd.)							+							
	Retina	Photophobia	Riboflavin	+								+						
		Nyctalopia[gg]	Vitamin A	+	Rat, pig, cat, sheep	+												
		Pigmented retinopathy	Vitamin E	+	Rat, dog, cat, monkey			+										
	Cornea	Xerophthalmia[hh]	Vitamin A	+	Rat, calf	+												
		Keratomalacia[ii]	Vitamin A	+	Rat, calf	+												
	Lens	Cataract[jj]	Vitamin E		Rabbit, turkey embryo			+										

Reproductive system																	
Vagina	Epithelium	Cornification[kk]	Vitamin A	+	Rat	+											
Uterus	Epithelium	Lipofuscinosis	Vitamin E		Rat			+									
Ovary	????	Degeneration	Vitamin A		Poultry	+											
	Estrus	Anestrus[ll]	Riboflavin		Rat						+						
		Low egg production	Vitamin A		Poultry	+											
			Riboflavin		Poultry						+						
			Pyridoxine		Poultry								+				
Egg	Shell	Thinning	Vitamin D		Poultry		+										
Testes	Germinal epithelium	Degeneration	Vitamin A		Rat, bull, cat	+											
			Vitamin E		Rat, rooster, dog, pig, guinea pig, hamster, rabbit, monkey			+									
Fetus		Developmental abnormalities	Riboflavin		Rat, chick (clubbed down)						+						
			Folate		Chick (parrot beak)											+	
		Death	Vitamin A		Poultry	+											
			Vitamin E		Rat			+									
			Riboflavin		Chick						+						
			Folate		Chick											+	
			Vitamin B_{12}		Poultry												+
Nervous system																	
General		Ataxia[mm]	Vitamin A		Chick, pig, calf, sheep	+											
			Vitamin E	+	Chick			+									
			Thiamin	+ (Wernicke-Korsakoff synd.[nn])	Rat, mouse, chick, pig, rabbit, calf, monkey					+							
			Riboflavin		Rat, pig						+						
		Tremors	Niacin	+ (Pellagra)								+					
		Tetany	Vitamin D	+ (Rickets)	Chick, pig		+										
		Abnormal gait	Pantothenic acid		Pig, dog (goose-stepping)									+			

Continued

TABLE 4.4 Diagnosis of Vitamin Deficiencies in Humans and Other Animals—cont'd

Signs and Symptoms by Organ System and Organ			Vitamin Deficiencies Possibly Involved	Humans Affected	Other Species Affected	Vitamin A	Vitamin D	Vitamin E	Vitamin K	Ascorbic acid	Thiamin	Riboflavin	Niacin	Pyridoxine	Biotin	Pantothenic acid	Folate	Vitamin B_{12}
System	Organ	Signs/Symptoms																
		Seizures	Pyridoxine	Infants	Rat									+				
		Paralysis	Riboflavin		Chick (curled toe paralysis); rat							+						
			Pyridoxine		Chick									+				
			Pantothenic acid		Poult (cervical paralysis); chick										+			
		Irritability	Thiamin	+ (Beriberi)							+							
			Niacin	+ (Pellagra)									+					
			Pyridoxine	+	Rat									+				
			Vitamin B_{12}		Pig													+
Spinal cord	Cerebro-spinal fluid	Excess pressure	Vitamin A		Chick, pig, calf	+												
		Opithotonus[oo]	Thiamin		Chick, pigeon (star gazing)						+							
Brain		Encephalopathy[pp]	Vitamin E		Chick (encephalomalacia[qq])			+										
			Thiamin	+ (Wernicke-Korsakoff synd.)							+							
Peripheral nerves		Neuropathy	Vitamin E		Rat, dog, duck, monkey			+										
			Thiamin	+ (Beriberi)	Chick (polyneuritis[rr])						+							
			Riboflavin	+								+						
			Niacin		Rat, pig								+					
			Pyridoxine		Rat									+				
			Pantothenic acid	+ (Burning feet)												+		
			Vitamin B_{12}	+														+

Psychological, emotional						
	Depression	Thiamin	+ (Beriberi)	+		
		Niacin	+ (Pellagra)		+	
	Anxiety	Thiamin	+ (Beriberi)	+		
	Dizziness	Niacin	+ (Pellagra)		+	
	Irritability	Thiamin	+ (Beriberi)	+		
		Niacin	+ (Pellagra)		+	
		Pyridoxine	+			+
	Dementia	Niacin	+ (Pellagra)		+	
	Psychosis	Thiamin	+ (Wernicke-Korsakoff synd.)	+		

[a] *Severe initiation of rapid onset.*
[b] *Inflammation of the skin.*
[c] *Swelling and* **necrosis** *(i.e., tissue and/or organ death) of the paws, tips of the ears and nose, and lips.*
[d] *Shedding of skin.*
[e] *Thickening of the stratum corneum.*
[f] *Loss of normal pigment from hair or feathers.*
[g] *Loss of hair or feathers.*
[h] *Whiskers accumulate porphyrins shed in tears.*
[i] *General term for disease of muscle.*
[j] *General term for disease of the heart muscle.*
[k] *Muscular portion of the forestomach of birds.*
[l] *General term for hardening, i.e., loss of elasticity, of medium or large arteries.*
[m] *Abnormal fluid retention.*
[n] *Abnormally low erythrocyte count due to their fragility, rupture, and clearance.*
[o] *Abnormally low erythrocyte count due to hemorhhage.*
[p] *Abnormally low hemoglobin content in otherwise normal erythrocytes.*
[q] *Abnormally low erythrocyte count due to impaired DNA synthesis, with erythroblast growth without division, i.e., forming macrocytes.*
[r] *Abnormally low white blood cell count.*
[s] *Abnormally high platelet count.*
[t] *Lack of gastric acid production due to dysfunction of gastric parietal cells.*
[u] *Inflammation of the oral mucosa (soft tissues of the mouth).*
[v] *Angular stomatitis, i.e., inflammatory lesions (cracks, fissures) at the labial commissure (corners of the mouth).*
[w] *Inflammation of the tongue.*
[x] *Accummulation of lipid oxidation products.*

Continued

TABLE 4.4 Diagnosis of Vitamin Deficiencies in Humans and Other Animals—cont'd

[y]*Demineralization leading to softening of bones.*
[z]*Progressive demineralization leading to thinning of bones, as in rickets in children.*
[aa]*Disorders of cartilaginous components of growing ends of bones.*
[bb]*Abnormal intracellular retention of lipids.*
[cc]*Chronic liver disease involving replacement of normal tissue with fibrosis and presence of regenerative nodules.*
[dd]*Inflammation of nephrons.*
[ee]*Cell death.*
[ff]*Involuntary eye movement.*
[gg]*Night blindness, i.e., difficulty seeing in low light.*
[hh]*Failure to produce tears due to dysfunction of lacrimal glands, resulting in dryness and thickening of the conjunctiva and cornea and leading to ulceration and blindness.*
[ii]*Drying and clouding of the cornea due to xerophthalmia.*
[jj]*Clouding of the lens.*
[kk]*Formation of an epidermal barrier in stratified squamous epithelial tissue by increased expression of keratin proteins.*
[ll]*Cessation of female ovulatory cycle.*
[mm]*Gross lack of muscular coordination.*
[nn]*Confusion, ataxia, nystagmus, and double vision due to damage in thalamus and hypothalamus (Wernicke's encephalopathy) progressing to loss of memory, confabulation, and hallucination (Korsokoff syndrome).*
[oo]*State of severe hyperextension and spasm of the axial muscles along the spinal column, causing an individual's head, neck, and spinal column to assume an arching position.*
[pp]*Term for global brain disease.*
[qq]*Degenerative disease of brain involving function: blindness, ataxia, circling, and terminal coma.*
[rr]*Neuropathy affecting multiple peripheral nerves.*

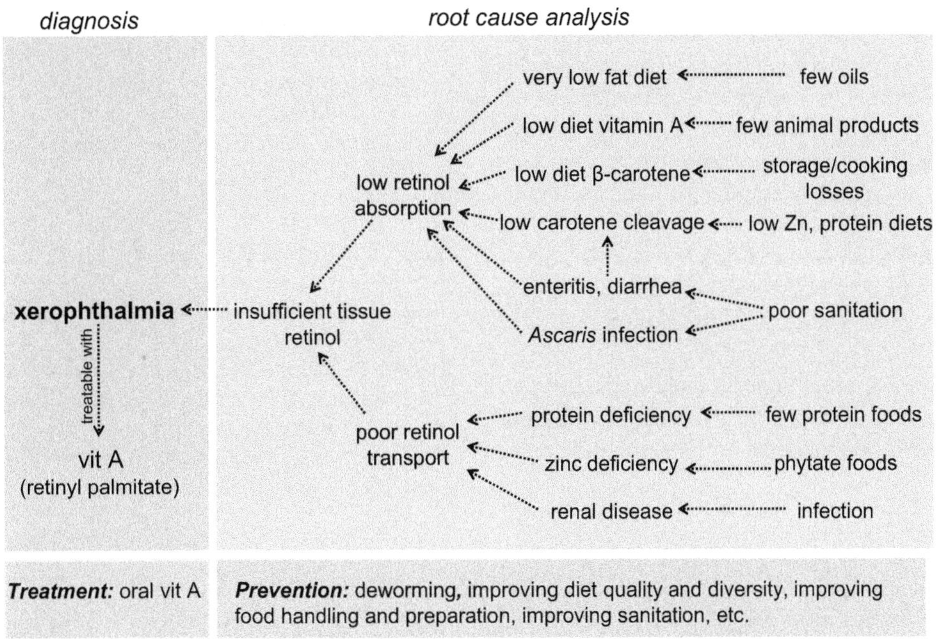

FIGURE 4.3 **Treatment** of nutritional deficiency and **prevention** of its occurrence call for different types of analyses. Cases demand a medical approach in which treatment is based on a diagnosis. Prevention calls for root cause analyses using systems approaches to identify and address the underlying factors.

- Vitamin losses (e.g., during pelleting, extrusion, and/or storage)
- Poor access to feed (e.g., competition for limited feeder space, improper feeder placement, and breakdown of feed delivery system).

Secondary deficiencies in animals tend have biological and social causes:

- Poor feed intake (e.g., inappetence, poor feed palatability, and heat stress)
- Other deficiencies (e.g., deficiencies of protein and/or zinc can impair retinol transport)
- Impaired digestion
- Malabsorption (e.g., diarrhea, parasites, and intestinal infection due to poor hygiene)
- Impaired postabsorptive utilization (e.g., certain drug therapies)
- Increased metabolic demand (e.g., infection, low environmental temperature, egg/milk production, rapid growth, pregnancy, and lactation).

Two Types of Vitamin Deficiencies

- *Primary deficiencies*… involve failures to ingest a vitamin in sufficient amounts to meet physiological needs.
- *Secondary deficiencies*… involve failures to absorb or otherwise utilize a vitamin postabsorptively.

Making Interventions Effective

The management of vitamin deficiencies[6] is no different from that of other diseases—treatment is generally most effective when administered during the early stages of cellular biochemical abnormality, rather than waiting for the manifestation of clinical signs.[7] For this reason, the early detection of insufficient vitamin status using biochemical indicators has been, and will continue to be, a very important activity in the clinical assessment of vitamin status.

Vitamin deficiency disorders are often treated using a medical/pharmacological approach involving a two-step analysis: (1) diagnosis of the deficiency and (2) treatment with an appropriate form of the relevant vitamin. This approach offers the advantages of speed and efficacy, which are often significant in the context of treating subjects in need. However, they typically do not address the multiple, underlying causes of deficiency; these must be addressed for the sustainable prevention of vitamin deficiency, especially in populations. This goal calls for considering the deficiency in the biological, social, demographic, and environmental contexts in which it occurs, to the end of identifying root causes—i.e., the underlying, contributing conditions, one or more of which will likely to be amenable to change. An example of a root cause analysis of xerophthalmia due to vitamin A deficiency is shown in Fig. 4.3.

6. This discussion, of course, is relevant to any class of nutrients.
7. Marks (1968) makes this point clearly with the example of diabetes, which should be treated once hypoglycemia is detected, thus reducing the danger of diabetic arteriosclerosis and retinopathy.

4. STUDY QUESTIONS AND EXERCISES

1. For a major organ system, discuss the means by which vitamin deficiencies may affect its function.
2. List the clinical signs that have special diagnostic value (i.e., are specifically associated with insufficient status with respect to certain vitamins) for specific vitamin deficiencies.
3. For a fat-soluble and a water-soluble vitamin, discuss the relationships between tissue distribution of the vitamin and organ site specificity of the clinical signs of its deficiency.
4. List the animal species and deficiency diseases that, because they show specificity for certain vitamins, might be particularly useful in vitamin metabolism research.
5. Develop a decision tree for determining whether lesions of a particular organ system may be due to insufficient intakes of one or more vitamins.
6. Detail differences between primary and secondary deficiencies and their causes citing specific examples in humans or animals.

RECOMMENDED READING

Marks, J., 1968. The Vitamins in Health and Disease. Churchill, London, 183 pp.

Ross, A.C., Caballero, B., Cousins, R.J., et al. (Eds.), 2012. Modern Nutrition in Health and Disease, eleventh ed. Lippincott Williams and Wilkins, Baltimore, 2069 pp.

Zemplini, J., Suttie, J.W., Gregory, J.F., et al., 2014. Handbook of Vitamins, fifth ed. CRC Press, New York, 592 pp.

Chapter 5

Vitamin Needs and Safety

Chapter Outline

Anchoring Concepts

1. The vitamins have many metabolic role(s) essential to normal physiological function; these roles can be compromised by quantitatively insufficient or temporarily irregular vitamin intakes.
2. Vitamin needs can be determined by monitoring responses of parameters related to the metabolic functions and/or body reserves of the vitamins.
3. Quantitative data are available describing vitamin contents of many common foods and feedstuffs.
4. Vitamins are frequently used in human feeding, in animal diets, and in treating certain clinical conditions at levels in excess of their requirements.
5. Several of the vitamins, most notably vitamins A and D, can produce adverse physiological effects when consumed in excessive amounts.

Nutriment is both food and poison. The dosage makes it either poison or remedy.

Paracelsus[1]

LEARNING OBJECTIVES

1. To understand the concepts of minimum requirement, optimal requirement, and allowance as used with respect to vitamins.
2. To understand the basis for establishing allowances for vitamins in human and animal feeding.
3. To understand the concept of upper safe use level with respect to the vitamins.
4. To understand the margins of safety above their respective requirements for intakes of each of the vitamins.
5. To understand the signs/symptoms of vitamin toxicities in humans and animals.

VOCABULARY

Adequate intake (AI)
Calcinosis
Carotenodermia
Daily values (DVs)
Dietary Guidelines for Americans (DGA)
Dietary reference intake (DRI)
Dietary standards
Estimated average requirement (EAR)
Estimated safe and adequate daily dietary intake (ESADDI)
Food and Agricultural Organization (FAO)
Food and Nutrition Board
Hypervitaminosis
Institute of Medicine (IOM)
Lowest observed adverse effect level (LOAEL)
Margin of safety
Metabolic profiling
Minimum requirement
National Academy of Sciences
National Research Council
No observed adverse effect level (NOAEL)
Nutrient allowances
Nutritional essentiality
Optimal requirement
Oxaluria
Protective nutrient intake
Range of safe intake
Recommended Daily Allowance (RDA)
Recommended Dietary Intake (RDI)

1. Paracelsus (1493–1541) was the Latinized name of the Swiss German philosopher, physician, botanist, and general occultist born Philippus Aureolus Theophrastus Bombastus von Hohenheim, who is generally credited as the founder of the field of Toxicology.

The Vitamins. http://dx.doi.org/10.1016/B978-0-12-802965-7.00005-8

Recommended Nutrient Intake (RNI)
Safety index (SI)
Tolerable upper intake limit (UL)
Toxic threshold upper limit (UL)
World Health Organization (WHO)

1. DIETARY STANDARDS FOR VITAMINS

Purposes of Dietary Standards

The need to formulate healthy diets for both humans and animals has stimulated the translation of current nutrition knowledge into a variety of **dietary standards** for the intakes of specific nutrients. As these are typically developed by committees of experts reviewing the pertinent scientific literature, they are frequently referred to as **dietary recommendations**. Formally, they may be called **Recommended Daily Allowances (RDAs)** or **Recommended Dietary Intakes (RDIs)**.

Regardless of the ways they may be named, dietary standards differ from nutrient requirements, although they are derived from the latter. Dietary standards are relevant to populations; they describe the average amounts of particular nutrients that should satisfy the needs of almost all healthy individuals in defined groups. In contrast, **nutrient requirements** are relevant to individuals; they describe amounts of particular nutrients that satisfy certain criteria related to the metabolic activity of those nutrients or to general physiological function. Because recommended allowances and intakes are designed to satisfy the needs of groups of individuals whose nutrient requirements vary, by definition they exceed the average requirement.

Determining Nutrient Requirements

The nutrient requirement is a theoretical construct that describes the intake of a particular nutrient that supports a body pool of the nutrient and/or its metabolically active forms adequate to maintain normal physiological function. In practice, it is generally used in reference to the lowest intake that supports normal function, that is, the **minimum requirement**. Minimum requirements, while seemingly physiologically relevant, are difficult to define and impossible to measure with any reasonable precision. They can vary according to the criteria by which they are defined. This problem is illustrated by the widely varying estimates of the vitamin A requirement for calves; various estimates may be derived by different criteria (Table 5.1).

TABLE 5.1 Vitamin A Requirements of Calves Based on Different Criteria

Criteria	Estimated Requirement (IU/day)
Prevention of nyctalopia	20
Normal growth	32
Normal serum retinol levels	40
Moderate hepatic retinyl ester reserves	250
Substantial hepatic retinyl ester reserves	1024

From Marks, J., 1968. The Vitamins in Health and Disease: A Modern Reappraisal. Churchill, London, p. 32.

Therefore, to be relevant to the overall health of specific populations, estimates of minimum nutrient requirements must be based on responses of obvious physiological importance. For many nutrients (e.g., the indispensable amino acids), it may be appropriate to define minimum requirements on the basis of a fairly nonspecific parameter such as growth. For vitamins, however, it is appropriate to define minimum requirements on the basis of biomarkers more specifically related to their respective metabolic functions, such as enzyme activities and tissue concentrations, as these can reflect changes at the earlier stages of vitamin deficiencies. The most useful biomarkers of vitamin status are those that respond early to deprivation of the vitamin, as such changes can be used to detect suboptimal vitamin status at the early and most readily corrected stages.

Quantifying the minimum requirement, even with the use of an appropriate biomarker, is not straightforward. It generally requires an experimental approach that may be possible using an animal model, but is seldom feasible with human subjects.[2] In such studies with animals, test subjects are fed a basal diet constructed to be deficient in the nutrient of interest but otherwise adequate with respect to all known nutrients. Various treatment groups receive this diet supplemented with known amounts of the nutrient of interest. This may necessitate the use of uncommon foods/feedstuffs such that the diet bears little similarity to those used in practice; it often means that the test nutrient is provided in free form, which may not resemble its form in practical foods and feedstuffs.[3] Even with these caveats, the level of nutrient intake to be identified as the minimum requirement is not always clear, as the optimal value for that biomarker is usually a matter of judgment.

Most responses of specific nutrient-depleted animals to input of the relevant nutrient appear to be curvilinear (Fig. 5.1, right panel). However, in most nutrient requirement

2. Studies with human volunteers in which diet composition and level of consumption are both well controlled in reference to the caloric needs of individuals generally require fairly long observational periods and call for facilities (metabolic kitchen, clinical assessment, and in-house wards) that are available at relatively few institutions. As a result, they tend to be costly and not widely conducted. Instead, observational epidemiological approaches are often used to impute the levels of nutrient intake of apparently healthy people.

3. e.g., Menadione instead of phylloquinones; tetrahydrofolic acid instead of mixed folates.

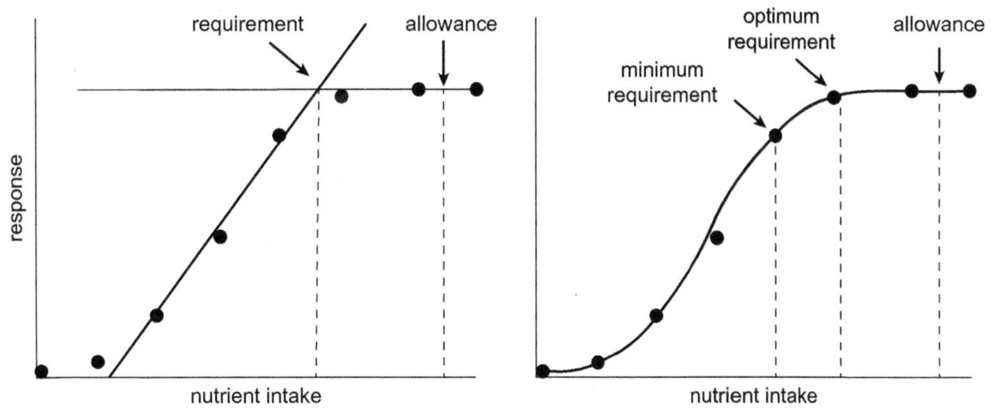

FIGURE 5.1 Requirements and allowances for nutrients are determined from the responses of physiologically meaningful parameters to the level of nutrient intake.

experiments, both rectilinear and curvilinear models usually fit equally well. For this reason, many investigators have used rectilinear models to impute requirements (e.g., the *x* value of the intercept of the two linear regressions of the observed data in broken-line regression analyses[4]; Fig. 5.1, left panel). Others, however, have used curvilinear models, which consider the variations in the experimental population of both the measured response and the nutrient need for maintenance (usually related to body size). From the proposition of curvilinearity, it follows that no value can be properly described as the "requirement" of the test populations. Nevertheless, the approach can be used to determine the risk of not fully satisfying the requirement for given proportions of the experimental population. The level of intake associated with acceptable risk of deficiency (a matter of judgment) is frequently called the "optimum requirement" or "level of optimum intake." In public health, such levels are determined on the basis of assumptions regarding putative health risks and in consideration of interindividual variation. In livestock production, where the cost of feeding accounts for as much as three-quarters of the total cost of production, it is necessary to determine intakes of the more costly nutrients (protein, limiting amino acids, energy) that optimize economic efficiency.

Factors Affecting Vitamin Needs

Many factors can affect nutrient requirements (Tables 5.2 and 5.3), such that those of individuals with the same general characteristics can vary substantially. For most nutrients the requirements of individuals in given populations appear to be normally distributed. For this reason, in the absence of clear information, it is reasonable to assume that the variations in vitamin requirements are similar to those typically observed in biological systems in being normally distributed with a coefficients of variation of 10–15%.

4. This approach offers the advantage of rendering a requirement value that is derived mathematically from the observed data; however, that value tends to be in the region of greatest variation in the input–response curve.

Developing Vitamin Allowances

Because nutrient requirements, even for the best cases, are quantitative estimates based on data of uncertain precision derived from a limited number of subjects, these values have limited practical usefulness. In practice, **nutrient allowances**, or recommended intakes (RIs), are far more useful. They are selected to meet the needs of those individuals with the greatest requirements. That is, an allowance is set at the right-hand tail of the natural distribution of requirements. An allowance exceeds the **estimated average requirement (EAR)** for the population by an increment sometimes referred to as a **margin of safety**. Allowances for vitamins, particularly in livestock feeding, have often been set on the basis of practical experience. Quantitative approaches have been used in establishing allowances for both animals and humans; such parameters are generally described in statistical terms relating to the proportion of the target population's requirements would be met by the recommended level of intake. For example, committees of the U.S. Food and Nutrition Board and WHO/FAO[5] have set allowances at 2 standard deviations (SD) above the EAR (Fig. 5.2), a decision to meet the needs of approximately 97.5% of the population.[6] This method has yielded satisfactory results, likely due in part to the generous estimates of EARs generally made by expert committees.

Allowances, therefore, are derived from estimates of EARs (of typical individuals) made from actual biological data, usually from nutritional experiments. Because they are used as standards for populations, allowances are developed in consideration of risk of nutrient deficiency. Therefore, allowances are relevant to specified populations with their

5. World Health Organization and Food and Agricultural Organization, respectively, of the United Nations.

6. A notable exception is in the setting of allowances for energy; these are typically set at the estimated average requirements of classes of individuals, for the reason that, unlike other nutrients, both intake and expenditure of energy appear to be regulated such that free-living individuals with free access to food maintain (at least very nearly) energy balance.

TABLE 5.2 Factors Affecting Vitamin Needs

Factor	Examples
Physiological determinants	Active growth
	Pregnancy
	Lactation
	Aging
	Intraindividual variation
	Level of physical activity
	Obesity
Hereditary conditions	Inborn vitamin-dependent diseases
	Polymorphisms of vitamin transporters, receptors, vitamin-dependent enzymes, and enzymes of vitamin metabolism
Conditions causing maldigestion/malabsoprtion	Pancreatitis
	Gastrorintestinal surgery
	Endocrine disorders (e.g., diabetes, hypoparathyroidism, congenital or acquired hemolytic, Addison's disease)
	Hepatobiliary disease
	Intestinal resection/bypass
	Pernicious anemia
	Regional ileitis
	Radiation injury
	Kwashiorkor
	Pellagra
	Gluten enteropathy
	Intestinal parasitism (e.g., hookworm, *strongyloides*, *Giardia lamblia*, *Dibothriocephalus latus*)
	Enteritis
	Cystic fibrosis
	Certain drug treatments
Hypermetabolic states	Thyrotoxicosis
	Pyrexial disease
	Various infections
Conditions causing decreased nutrient utilization	Chronic liver disease
	Chronic renal disease
Conditions involving increased cell turnover	Congenital and acquired hemolytic anemias
	Sickle cell disease
Conditions increasing nutrient turnover/loss	Extensive burns
	Bullous dermatoses
	Entheropathy
	Nephrosis
	Surgery
	Hemodialysis
	Smoking

characteristic food habits and inherent variations in nutrient requirements. For example, the RDAs established by the U.S. Food and Nutrition Board are implicitly intended to relate to the US population. These recommendations were originally developed to facilitate the wartime planning of food supplies but have become a key source of information for making food and health policy in the United States and elsewhere (Table 5.4).

TABLE 5.3 Physiologically Significant Drug–Vitamin Interactions

Vitamin	Drugs
Vitamin A	Diuretic: spironolactone
	Bile acid sequestrant: Cholestyramine, colestipol
	Laxative: Phenolphthalein, mineral oil
Vitamin D	Antibacterial: isoniazid
	Anticonvulsant: phenytoin, diphenylhydantoin, primidone
	Bile acid sequestrant: colestipol
	Laxative: phenolphthalein, mineral oil
Vitamin E	Smoking
Vitamin K	Anticoagulant: warfarin
	Anticonvulsant: phenytoin, diphenylhydantoin, primidone
	Bile acid sequestrant: colestipol
	Immunosuppressant: cyclosporins
	Laxative: mineral oil, phenolphthalein
Vitamin C	Antiinflammatory: aspirin
	Oral contraceptives
	Smoking
Thiamin	Alcohol
Riboflavin	Antibacterial: boric acid
	Tranquilizer: chlorpromazine
Niacin	Antibacterial: isoniazid
	Antiinflammatant: phenylbutazone
Vitamin B_6	Analytical reagent: thiosemicarbazide
	Antibacterial: isoniazid
	Anticholinergic, anti-parkinsonian: L-dopa
	Antihypertensive: hydralazine
	Chelating agents, antiarthritic: penicillamine
	Alcohol
	Oral contraceptives
	Smoking
Biotin	None reported
Pantothenic acid	None reported

Continued

TABLE 5.3 Physiologically Significant Drug–Vitamin Interactions—cont'd

Vitamin	Drugs
Folate	Antacid: sodium bicarbonate, aluminum hydroxide
	Antibacterial: sulfasalazine, trimethoprim
	Anticonvulsant: phenytoin
	Antiinflammatant: sulfasalazine, aspirin
	Antimalarial: pyrimethamine
	Antineoplastic: methotrexate
	Bile acid sequestrant: cholestyramine, colestipol
	Diuretic: triamterene
	Alcohol
	Oral contraceptives
Vitamin B_{12}	Antihyperglycemics: biguanides
	Antibacterials: *p*-aminosalicylic acid, neomycin
	Antihistaminic: cimetidine, ranitidine
	Antiinflammatant, gout suppressant: colchicine
	Bile acid sequestrant: cholestyramine, colestipol

A Member of the Food and Nutrition Board, Dr Alfred E. Harper, Observed:

There is not always agreement on the criteria for deciding when a requirement has been met. If the requirement is considered to be the minimal amount that will maintain normal physiological function and reduce the risk of impaired health from nutritional inadequacy to essentially zero, then we are left with questions such as: 'What is normal physiological function?; ' 'What is health?'; and 'What degree of reserve or stores of the nutrient is adequate?' Differences in judgment on such issues are to be expected.

Differences Between Requirements and Allowances

Confusion surrounds the allowances for the vitamins (and other nutrients) that have been developed by various expert committees. Some questions arise, particularly concerning dietary recommendations for livestock, because the rationales for such values are frequently not presented. A fairly common example is the mistaken impression, on the part of formulators of animal feeds, that vitamin allowances are requirements; this mistake can lead to vitamin overfortification of those feed vitamins. Other questions arise over the publication of differing recommendations by different committees of experts, all of whom consider the same basic data in their respective reviews of the pertinent literature. This situation results from the paucity of clear and compelling data on nutrient requirements; differences in environmental conditions and food supplies; and the lack of consensus on such issues as criteria for defining requirements, appropriate margins of safety, and whether standards should be based on intakes of food as consumed or as purchased. These considerations make the variable factor of scientific judgment important in estimating the nature of nutrient requirements. Thus, dietary recommendations are revised periodically[7] as new information becomes available.

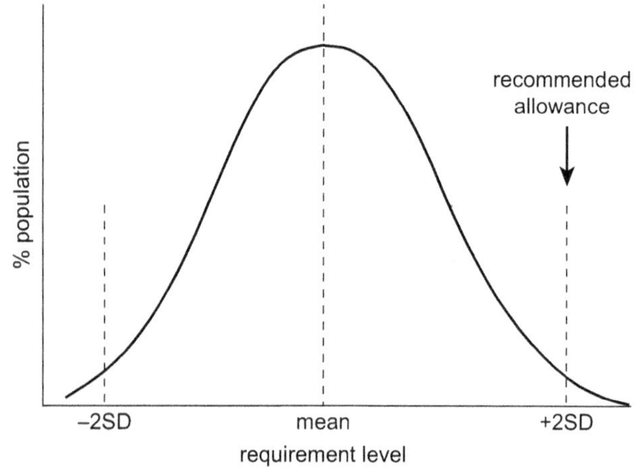

FIGURE 5.2 The "mean plus 2 SD" conversion algorithm for determining recommended daily allowances.

7. As evidence has grown, expert committees typically have reduced the levels of their recommendations for nutrient allowances. This likely reflects the basically conservative nature of the committee system used for these purposes, whereby the paucity of data tends to be handled by generously estimating quantitative needs.

TABLE 5.4 History of the Recommended Daily Allowances (RDAs) for Vitamins[a]

	1941	1948	1957	1968	1976	1980	1989	1997–2001	2010
Vitamin A (mg RE)	1000	1000	1000	1000	1000	1000	1000	900	b
Vitamin D (IU/μg)	—	—	—	400 IU	400 IU	5 μg	5 μg	[10][c]	b
Vitamin E (IU/mg)	—	—	—	30 IU	15 IU	10 IU	10 IU	15 mg	15 mg
Vitamin K (μg)	—	—	—	—	—	—	80	[120][c]	b
Vitamin C (mg)	75	75	70	60	45	60	60	90	b
Thiamin (mg)	2.3	1.5	0.9	1.3	1.4	1.4	1.5	1.2	b
Riboflavin (mg)	3.3	1.8	1.3	1.7	1.6	1.6	1.7	1.3	b
Niacin (mg)	23	15	15	17	18	18	19	16	b
Vitamin B_6 (mg)	—	—	—	2.0	2.0	2.2	2.0	1.7	b
Pantothenic acid (mg)	—	—	—	—	—	—	—	[5][c]	b
Biotin (mg)	—	—	—	—	—	—	—	[30][c]	b
Folate (μg)	—	—	—	400	400	400	200	400	b
Vitamin B_{12} (μg)	—	—	—	3.0	3.0	5.0	2.0	2.4	b

[a]*Values shown are for males, 25–50 years of age or (for vitamin D in 2010) 31–50 years of age.*
[b]*Values set in 1997–2001 remain in use.*
[c]*RDAs not available for these vitamins; values shown are adequate intakes (AIs).*

The RDA Concept

In one sense, the RDA construct is somewhat archaic in that it fails to pertain to biological functions of nutrients that may be nonspecific or nontraditional, in the context of being outside the known functions of nutrients. The conceptual framework upon which the RDA was derived is being replaced by a new, more individualistic view of nutrition that relates more broadly to health. This view is the basis of problems that have become apparent concerning the RDA. To retain the practical utility of the RDA, it will be necessary to reconstruct it; such reconstruction must be based on new paradigms for nutritional science that, informed by the "genomics revolution," explicitly consider individual metabolic characteristics.

The RDA was developed to facilitate food planning for the US population.[8] It is a product of the central concept of the field of nutrition, **nutritional essentiality**, which has been used to describe those factors in the external chemical environment that are specifically required for normal metabolic functions and, accordingly, those exogenous sources on which organisms depend for normal physiologic functions (e.g., growth, reproductive success, survival, and freedom from certain clinical/metabolic disorders). The vitamins are among the more than 40 such factors generally considered to be nutritionally essential, i.e., indispensable in the diets of animals and humans. Deprivation of any one of these is manifested by clinical signs that are usually specific in nature. Nutritional essentiality has been based on empirical findings that nutrients function to prevent ill health in very specific ways. Under this paradigm, nutrient deficiency diseases have contributed to the development of our knowledge of nutrition: their specific prevention has been used both to define nutrient essentiality and to quantify nutrient needs. Indeed, a nutrient has not been considered essential unless a clinical disease has been related specifically to its deprivation. Therefore, as the term has been used, nutritional essentiality clearly connotes the specific prevention of deficiency disease. This connotation is troublesome in dealing with issues of diet and health, as the essentiality paradigm does not pertain to functions of nutrients that are either nonspecific or nontraditional, i.e., outside the known functions of nutrients. This connotation, as expressed in the quantitative estimation of population-based nutrient needs, the RDAs, now serves to limit the essentiality paradigm as a conceptual framework. Modern nutrition is cast in a different context—one in which optimum health is more broadly conceptualized.

Questions about nutrient allowances for humans arise, owing to the application of those values to purposes for

8. The RDAs were originated in 1941 for use in planning U.S. food policy during World War II.

which they were not intended. The RDAs are now used for many other purposes: evaluating nutritional adequacy of diets; evaluating results of dietary surveys; setting standards for food assistance programs; institutional feeding programs, and food and nutrition regulations; developing food and nutrition education programs; formulating new food products and special dietary foods. Many of these uses have revealed limitations of the RDAs: not dealing with associations of diet and chronic and degenerative diseases; not including guidelines for appropriate intakes of fat, cholesterol, and fiber; and for not providing guidance for food selection and prevention of obesity. These limitations stem from that fundamental misunderstanding that, while the RDAs may be used in certain programs to implement sound public health policy, they are not intended to be policy recommendations per se. In fact, RDAs cannot serve as general dietary guidelines; by definition, they are reference standards dealing with nutrients, whereas dietary guidelines deal primarily with foods. The RDAs are, in fact, standards on which sound dietary guidelines, such as the **Dietary Guidelines for Americans (DGAs)**,[9] are to be based.

It should be kept in mind that the RDAs, like other nutrient allowances, are intended to relate to intakes of nutrients as part of normal diets[10] of specified population subgroups. They are intended to be average daily intakes based on periods as short as 3 days (for nutrients with fast turnover rates) to several weeks or months (for nutrients with slower turnover rates).

Questions Concerning RDAs:

- Which level of nutrient need should define a requirement—the level that supports all/some dependent enzymes at 50, 80, or 100% of maximal expression?
- Can a nutrient be *conditionally essential* (e.g., choline for individuals with low methionine intakes; glutamine for surgical patients)?
- How should an individual's varying nutrient requirements be described (e.g., effects of infection, oxidative stress)?
- Can nutrients be considered required for their nonspecific effects (e.g., antioxidants)?
- Can a nonnutrient be considered required (e.g., dietary fiber)?

9. The DGAs are issued every 5 years by the U.S. Departments of Agriculture (USDA) and Health and Human Services (HHS) based on expert consultations reviewing the state of nutrition understanding relative to the health of all Americans age 2 years and over.

10. The RDA subcommittee emphasized that the RDAs can typically be met or closely approximated by diets that are based on the consumption of different foods from diverse food groups that contain adequate energy.

Considering Nonclassical Functions of Nutrients

For some dietary factors, functions influencing the risk of chronic disease have been suggested by epidemiological and experimental animal model studies and, to a lesser extent, clinical trials. Reduced risks of such diseases have been associated with increased intakes and/or status of several vitamins. The metabolic bases of these linkages remain to be elucidated; indeed, these areas are among the most active in contemporary nutritional science:

- **Cancer** and foods containing vitamin A or vitamin C, intakes of riboflavin, and plasma levels or intakes of α-tocopherol, carotenoids, and 25-OH-vitamin D
- **Cardiovascular disease** and intakes of vitamin C, vitamin E, and β-carotene
- **Neural tube defects** and periconceptual folate intake
- **Diabetes and multiple sclerosis** and plasma 25-OH-vitamin D level
- **Psoriasis** response to vitamin D treatment.

The case of the apparent effects of antioxidants illustrates the limitation of the RDA construct. Current thinking is that antioxidant nutrients (vitamins E and C, selenium, and, perhaps, β-carotene) participate in a system of protection against the deleterious metabolic effects of free radicals. Because many diseases are thought to involve enhanced free-radical production, protection from oxidative stress is thought to be critical to normal physiologic function. According to this hypothesis, antioxidants would be expected to suppress radical-induced DNA damage involved in the initiation of carcinogenesis, to inhibit the oxidation of cholesterol in low-density lipoproteins (LDLs) in atherosclerosis, and to inhibit the oxidation of lens proteins in cataracts. Antioxidant nutrients have been shown to enhance immune functions, which may also contribute to reduced risks of cancer as well as infectious disease. Many of these antioxidant effects do not appear to have the specificity connoted by the essentiality paradigm. For example, the complementary natures of the antioxidant functions of vitamins E and C and selenium suggest that anyone may spare needs for the others in protecting against subcellular free-radical damage and LDL oxidation. It is likely that, through such biochemical mechanisms, the antioxidant nutrients may be modifiers of disease risk rather than primary agents in disease etiologies. However nonspecific they may be, such effects raise legitimate questions concerning nutrient need—questions not easily addressed under the essentiality paradigm or translated into RDAs.

New Paradigms for Nutrition

The term "essentiality" has become rather elastic in its application. Nutrients have come to be described as "required" or "essential" for particular functions. Some are called "dispensable" or "indispensable" under specific conditions; several are recognized as "beneficial" at levels greater than those that are considered to be "required." Indeed, the translation of nutritional knowledge into dietary guidance requires such language. However, the emergence of this sort of terminology indicates that the essentiality paradigm is, in fact, being displaced by a new conceptualization of nutrition.

It is likely that new paradigms of nutrition will encompass an individualized view of organisms that recognizes both endogenous and exogenous conditions as determinants of the nature and amounts of factors available from the external chemical environment that must be obtained to support definable health outcomes. Accordingly, such factors will be considered as nutrients *if* and *when* their activities, in the metabolism of the host and/or the associated microflora, are beneficial to those outcomes. This view will recognize different outcomes as being appropriate for various individuals, both within and between a species/population. Thus, freedom from overt physiological dysfunction as well as reduced risk of chronic diseases will be important outcomes in human nutrition; whereas such outcomes as maximal growth rate, optimal efficiency of feed utilization, and minimal susceptibility to infection will be priorities in livestock nutrition.

The old paradigm is being outgrown at an increasing pace with progress in the modern field of molecular biology. It has now become clear that some nutrients function as gene regulators and that predisposition to disease can have genetic bases. The mapping of the human genome and human microbiome has led to the development of powerful tools to study individual metabolic characteristics. As **metabolic profiling** becomes more feasible, it will become possible to address individuals' nutritional needs on the basis of their peculiar metabolic characteristics. Not only clinicians, but also dieticians, will be able to ask such questions as whether an individual has sodium-sensitive hypertension, a cystathionine β-synthase mutation, or the methylenetetrahydrofolate reductase C677T/C genotype. The time is quickly approaching when it will be possible to identify disease predisposition, metabolic characteristics, and specific dietary needs of individuals based on rapid, genomic/metabolomic analyses. As that becomes practicable, the population-based paradigm will lose much of its value.

Reconstructing the RDA

This crisis in conceptualization was manifest in the lively discussion concerning the need for new approaches to the development of dietary recommendations that preceded the development of the **dietary reference intakes (DRIs)** in the 1990s. The challenge was to recreate the RDA as a useful construct under the emerging paradigm that addressed both the prevention of overt nutritional deficiencies as well as the maintenance of health. It became clear that DRIs must accommodate the possibility that a nutrient can have beneficial action at levels above those previously thought to be "required" for normal physiologic function.[11]

2. VITAMIN ALLOWANCES FOR HUMANS

The first nutrient allowances were published 50 years ago by the U.S. National Academy of Sciences. Based on available information, those RDAs have since been revised periodically. Since the first publication of RDAs, similar dietary standards have been produced by several countries and international organizations. For the reasons mentioned previously, the various recommendations tend to be similar but not always identical. For example, most are based on food as consumed; however, some are based on food as purchased, making them appear higher than those of other countries.

Actionable Information Needed for Any Nutrient:

- amount that prevents overt deficiency disease
- amount that can have other health benefits
- amount that can have specific health risks.

RDAs

The RDAs are probably the most widely referenced of the dietary standards, and the set that is most comprehensive with respect to the vitamins. It is worth noting that the RDAs for vitamins are still not complete; that is, quantitative recommendations on some (e.g., vitamin D, vitamin K, biotin, pantothenic acid) have not been made owing to a still-insufficient information base. In 1980, the problem of dealing with nutrients known to be essential for humans but for which insufficient data are available was handled by including provisional recommendations. The 9th and 10th editions of the RDAs included **Estimated Safe and Adequate Daily Dietary Intake (ESADDI)** ranges of daily dietary intakes for

11. The need is best illustrated by the case of the essential trace element selenium (Se). The RDAs for Se (now 55 μg for both women and men) were first established in 1989 based on the amount sufficient for maximal expression of Se-dependent glutathione peroxidase in young men. Since then, several clinical trials and a large number of animal tumor model studies have found antitumorigenic activities of higher intakes of Se, e.g., >200 μg/day (Clark, L.C., Combs Jr., G.F., Turnbull, B., et al., 1996. J. Am. Med. Assoc. 276 (1957)). While the RDA addresses the classical nutritional need for Se (for selenoprotein expression), it is mute of the prospect of cancer risk reduction.

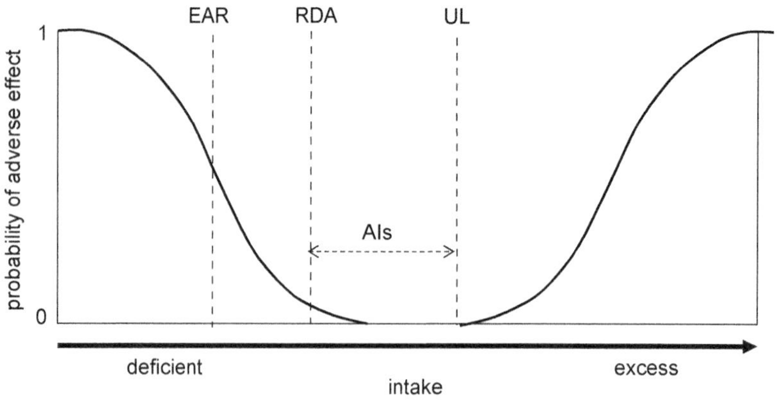

FIGURE 5.3 Conceptual basis for DRIs (dietary reference intakes).

such nutrients. This terminology is no longer used; instead, estimates of **adequate intakes (AIs)** are now used in cases where available data are judged to be insufficient for developing RDAs.

The setting of dietary allowances is an exercise of experts who evaluate published scientific literature. That different expert panels can reach different conclusions from the same body of published data is evidenced by the differences in national dietary allowances. That the growing body of relevant data also changes over time is evidenced by the changes in RDAs over the history of that institution (Table 5.4). For example, only in 1968 were RDAs established for vitamins D, E, C, and B_{12} and folate. In 1989, an RDA for vitamin K was first set; however that value, as well as that for vitamin D, were replaced with AI values in 2000. Only in 2010 were RDAs for vitamin D established.[12]

Accordingly, the setting of dietary allowances is a continuing process. There is a growing appreciation of the need and advantages of harmonizing these processes as carried out in different countries.[13]

Dietary Reference Intakes

The most recent (2010) edition of dietary allowances for vitamin D and calcium builds on the broad, previous (1997–2001) edition of dietary allowances (Table 21-5) produced by the U.S. Food and Nutrition Board. The former were developed in the 1990s by a series of workshops that addressed the conceptual framework upon which that work was based. This resulted in an expansion of the former RDAs with a system of DRIs.[14] This system involved four types of reference values (Fig. 5.3):

- **Estimated average requirements (EARs)**—Intakes to meet the requirements of half of the healthy individuals in each age–sex-specific demographic subgroup of the American population.
- **Recommended Daily Allowances (RDAs)**—Average daily intake level sufficient to meet the requirements of nearly all (97–98%) of the healthy individuals in each age–sex-specific demographic subgroup. The RDA is calculated from the EAR:

$$RDA = EAR + 2SD_{EAR}$$

where SD_{EAR} is the standard deviation of the EAR

While the RDA resembles the historic construct; in fact, it is different in that it assumes the SD_{EAR} to be 10% of the EAR, whereas a value to 15% had been used previously. For this reason, many of the more recent RDAs are lower than earlier ones.

- **Adequate intakes (AIs)**—Observed and/or experimentally determined approximations of nutrient intakes of groups of healthy individuals and extrapolated to each age–sex demographic subgroup. Used when data are

12. These were met with instant dispute by some researchers who thought they had been set too low: Hall, L.M., Kimlin, M.G., Aronov, P.A., et al., 2010. J. Nutr. 140 (542); Heaney, R.P., Holick, M.F., 2011. J. Bone Mineral Res. 26 (455).
13. Fairweather-Tait, S., Gurinović, M., van Ommen, B., et al., 2010. Eur. J. Clin. Nutr. 64 (S26).

14. Questions concerning the means of developing consistent and reliable standards led the 10th RDA Committee to review the scientific basis of the entire RDA table. The Committee recommended a lower RDA for vitamins A (reducing the RDA for men 1000–700 IU, and that for women from 800–600 IUand 600 IU) and vitamin C (reducing the RDA for men from 60 to 40 mg, that for women from 60 to 30 mg, and that for infants from 35 to 25 mg). It was reported that these reductions were resisted by the U.S. Food and Nutrition Board, which had been advised by another subcommittee to increase the intakes of these nutrients based on cancer risk reduction potential. In an unexpected move, the Board elected not to accept the recommendations of the RDA Committee. This move prompted lively discussion in the Nutrition community ultimately resulting in a rethinking of the RDA construct and the development of the DRIs.

judged to be insufficient for the estimation of an EAR and subsequent calculation of an RDA.

- **Tolerable upper intake limits (ULs)**—The highest level of daily intake that is likely to pose no risks of adverse health effects to almost all healthy individuals in a each age–sex-specific demographic subgroup. The use of ULs will facilitate the development of recommendations of nutrient intakes at what might be called "supranutritional" levels when such intakes have been shown to have health benefits. Pertinent to this consideration is emerging understanding of roles of at least several vitamins in reducing chronic disease risk.

This approach to the development of dietary allowances presumes the availability of empirical data for the distribution of individual nutrient requirements, necessary for calculating both the EAR and the SD_{EAR}. However, such data are available for very few nutrients. Thus, the DRI process involved a consensus opinion to assume that the distributions of individual nutrient requirements are each normal with a coefficient of variation (CV) of 10%, with only two exceptions: for vitamin A, CV = 20%; for niacin, CV = 15%.

International Standards

The FAO and WHO have established standards for energy, protein, calcium, iron, and eight of the vitamins (Table 5.6). This system of recommendations is intended for international use and, thus, to be relevant to varied population groups. It includes reference values similar to those used by the Food and Nutrition Board: Requirements similar to EARs, **Recommended Nutrient Intakes (RNIs),** similar to RDA; and Upper Tolerable Nutrient Intake Levels similar to the ULs. In addition, the FAO/WHO system also provides a value applicable for nutrients that may be protective against a specified nutritional or health risk of publish health relevance, **Protective nutrient intakes** (Table 5.5).

3. VITAMIN ALLOWANCES FOR ANIMALS

Public Versus Private Information

The development of livestock production for the economical production of human food and fiber has superimposed practical needs on the formulation of animal feeds that do not exist in human nutrition. Notably, this involves needs for accurate data for both nutrient requirements of animals and nutrient contents of feedstuffs. The availability of such data enables commercial animal nutritionists to formulate nutritionally balanced feeds using computer-based linear programming techniques. Often that capacity enables livestock enterprises to be competitive in a context where the cost of feeding can be the largest single cost of production.[15] Thus, research in food and animal nutrition has, over the past few decades, moved progressively out of the public sector and into the research divisions of agribusinesses with immediate interests in generating such data.[16] The result is that a diminishing proportion of practical animal nutrition data (particularly in the area of amino acid nutrition) remains in the public sector and is, thus, available to the scrutiny of experts. As a consequence, two types of dietary standards are in use. The first is the standard developed by review of open data available in the scientific literature; the second is the standard developed through in-house testing and/or practical experience by animal producers. Whereas the former data are in the public domain, the latter usually are not.

Public information on nutrient allowances is reviewed by expert committees in the United States, the United Kingdom, and several other countries under programs charged with the responsibility of establishing nutrient recommendations on the basis of the best available data. Perhaps the most widely used source of such recommendations is the Committee on Animal Nutrition of the U.S. National Research Council (Table 5.7). Through expert subcommittees, each dedicated to a particular species, the NRC maintains the periodic review of nutrient standards, many of which serve as the bases of recommendations for animal feed formulation throughout the world. Currently, many gaps remain in our knowledge of vitamin requirements. This is particularly true for ruminant species, for which the substantial ruminal destruction of vitamins appears to be compensated by adequate microbial synthesis, and for several nonruminant species that are not widely used for commercial purposes. Therefore, many of the standards for vitamins and other nutrients are imputed from available data on related species; in part for this reason, the requirements for some nutrients (e.g., selenium) appear to be very similar among many species.

4. USES OF VITAMINS ABOVE REQUIRED LEVELS

Typical Uses Exceed Requirements

Most normal diets that include varieties of foods can be expected to provide supplies of vitamins that meet those levels required to prevent clinical signs of deficiencies. In addition, most intentional uses of vitamins are designed to exceed those requirements for most individuals. Indeed, that is the principle by which vitamin allowances are set. Thus, the formulation of diets, the planning of meals, the vitamin fortification of foods, and the designing of vitamin supplements are all designed to provide vitamins at levels contributing to total intakes that exceed the requirements of

15. For example, feed costs for broiler chickens can comprise 60–70% of the total cost of producing poultry meat.

16. This is markedly different from human nutrition research, which continues to be seen as a public good such that the expansion of knowledge has come largely from public-sponsored research.

TABLE 5.5 Food and Nutrition Board Recommended Daily Allowances (RDAs) for Vitamins

Age–Sex Group	Vitamin A (μg)[a]	Vitamin D (μg)	Vitamin E (mg)[b]	Vitamin K (μg)	Vitamin C (mg)	Thiamin (mg)	Riboflavin (mg)	Niacin (mg)[c]	Vitamin B_6 (μg)	Pantothenic Acid (μg)	Biotin (μg)	Folate (μg)[d]	Vitamin B_{12} (μg)
Infants													
0–6 months	[400][e]	10	4	[2.0]	[40]	[0.2]	[0.3]	[2[b]]	[0.1]	[1.7]	[5]	[65]	[0.4]
7–11 months	[500]	10	5	[2.5]	[50]	[0.3]	[0.4]	[4]	[0.3]	[1.8]	[6]	[80]	[0.5]
Children													
1–3 years	300	15	6	[30]	15	0.5	0.5	6	0.5	2	[8]	[150]	0.9
4–8 years	400	15	7	[55]	25	0.6	0.6	8	0.6	3	[12]	[200]	1.2
Males													
9–13 years	600	15	11	[60]	45	0.9	0.9	12	1.0	4	[20]	[300]	1.8
14–18 years	900	15	15	[75]	75	1.2	1.3	16	1.3	5	[25]	[400]	2.4
19–30 years	900	15	15	[120]	90	1.2	1.3	16	1.3	5	[30]	[400]	2.4
31–50 years	900	15	15	[120]	90	1.2	1.3	16	1.3	5	[30]	[400]	2.4
51–70 years	900	15	15	[120]	90	1.2	1.3	16	1.7	5	[30]	[400]	2.4
>70 years	900	20	15	[120]	90	1.2	1.3	16	1.7	5	[30]	[400]	2.4
Females													
9–13 years	600	15	11	[60]	45	0.9	0.9	12	1.0	4	[20]	[300]	1.8
14–18 years	700	15	15	[75]	65	1.0	1.0	14	1.2	5	[25]	[400]	2.4
19–30 years	700	15	15	[90]	75	1.1	1.1	14	1.3	5	[30]	[400]	2.4
31–50 years	700	15	15	[90]	75	1.1	1.1	14	1.3	5	[30]	[400]	2.4
51–70 years	700	15	15	[90]	75	1.1	1.1	14	1.5	5	[30]	[400]	2.4
>70 years	700	20	15	[90]	75	1.1	1.1	14	1.5	5	[30]	[400]	2.4

Pregnancy														
≤18 years	750	15	15	[75]	80	1.4	1.4	18	1.9	6		[30]	[600]	2.6
19–30 years	770	15	15	[90]	85	1.4	1.4	18	1.9	6		[30]	[600]	2.6
31–50 years	770	15	15	[90]	85	1.4	1.4	18	1.9	6		[30]	[600]	2.6
Lactation														
≤18 years	1200	15	19	[75]	115	1.4	1.6	17	2.0	7		[35]	[550]	2.8
19–30 years	1300	15	19	[90]	120	1.4	1.6	17	2.0	7		[35]	[550]	2.8
31–50 years	1300	15	19	[90]	120	1.4	1.6	17	2.0	7		[35]	[550]	2.8

[a] *Retinol equivalents.*
[b] *α-Tocopherol.*
[c] *Niacin equivalents.*
[d] *Folate equivalents.*
[e] *Brackets indicate cases for which RDAs have not been set; AIs are given instead.*
From Food and Nutrition Board, 1997. Dietary Reference Intakes for Calcium, Phosphorus, Magnesium, Vitamin D and Fluoride. National Academy Press, Washington, DC, 432 pp.; Food and Nutrition Board, 2000. Dietary Reference Intakes for Thiamin, Riboflavin, Niacin, Vitamin B_6, Folate, Vitamin B_{12}, Pantothenic Acid, Biotin and Choline. National Academy Press, Washington, DC, 564 pp.; Food and Nutrition Board, 2000. Dietary Reference Intakes for Vitamin C, Vitamin E, Selenium and Carotenoids. National Academy Press, Washington, DC, 506 pp.; Food and Nutrition Board, 2001. Dietary Reference Intakes for Vitamin A, Vitamin K, Arsenic, Boron, Chromium, Copper, Iodine, Iron, Manganese, Molybdenum, Nickel, Silicon, Vanadium and Zinc. National Academy Press, Washington, DC, 773 pp.; Food and Nutrition Board, 2010. Dietary Reference Intakes: Calcium, Vitamin D. National Academy Press, Washington, DC, 1115 pp.

TABLE 5.6 FAO/WHO Recommended Nutrient Intakes (RNIs) for Vitamins[a]

Age-Sex Group	Vitamin A (μg)[b]	Vitamin D (μg)	Vitamin E (mg)	Vitamin K (μg)	Vitamin C (mg)	Thiamin (mg)	Riboflavin (mg)	Niacin (mg)[c]	Vitamin B_6 (μg)	Pantothenic Acid (μg)	Biotin (μg)	Folate (μg)	Vitamin B_{12} (μg)
Infants													
0–6 months	375	5	2.7	5	25	0.2	0.3	2[d]	0.1	1.7	5	80	0.4
7–11 months	400	5	2.7	10	30	0.3	0.4	4	0.3	1.8	6	80	0.5
Children													
1–3 years	400	5	5	15	30	0.5	0.5	6	0.5	2	8	160	0.9
4–6 years	450	5	5	20	30	0.6	0.6	8	0.6	3	12	200	1.2
7–9 years	500	5	7	25	35	0.9	0.9	12	1.0	4	20	300	1.8
Adolescents, 10–18 years													
Males	600	5	10	35–65	40	1.2	1.3	16	1.3	5	25	400	2.4
Females	600	5	7.5	35–55	40	1.1	1.0	16	1.2	5	25	400	2.4
Adults													
Males: 19–50 years	600	5	10	65	45	1.2	1.3	16	1.3	5	30	400	2.4
>50 years	600	10	10	65	45	1.2	1.3	16	1.7	5	30	400	2.4
Females: 19–50 years	500	5	7.5	55	45	1.1	1.1	14	1.3	5	30	400	2.4
51–65 years	500	10	7.5	55	45	1.1	1.1	14	1.5	5	30	400	2.4
Older Adults, >65 years													
Men[c]	600	15	10	65	45	1.2	1.3	16	1.7	5	–	400	2.4
Women[c]	600	15	7.5	55	45	1.1	1.1	14	1.5	5	–	400	2.4
Pregnancy	800	5	–	55	55	1.4	1.4	18	1.9	6	30	600	2.6
Lactation	850	5	–	55	70	1.5	1.6	17	2.0	7	35	500	2.8

[a]*Joint WHO/FAO Expert Consultation, 2001. Human Vitamin and Mineral Requirements. Food and Agricultural Org., Rome, pp. 286.*
[b]*Retinol equivalents.*
[c]*Niacin equivalents.*
[d]*Preformed niacin.*

TABLE 5.7 Estimated Vitamin Requirements (units/kg Diet) of Domestic and Laboratory Animals

Species	Vitamin A (IU)	Vitamin D (IU)	Vitamin E (mg)[a]	Vitamin K (μg)[b]	Vitamin C (mg)	Thiamin (mg)	Riboflavin (mg)	Niacin (mg)	Vitamin B_6 (mg)	Folate (mg)	Pantothenate (mg)	Biotin (μg)	Vitamin B_{12} (μg)	Choline (g)
Birds														
Chickens														
Growing chicks	1500	200	10	0.5		1.8	3.6	27	2.5–3	0.55	10	0.1–0.15	3–9	0.5–1.3
Laying hens	4000	500	5	0.5		0.8	2.2	10	3	0.25	2.2	0.1	4	
Breeding hens	4000	500	10	0.5		0.8	3.8	10	4.5	0.25	10	0.15	4	
Ducks														
Growing	4000	220		0.4			4	55	2.6		11			
Breeding	4000	500		0.4			4	40	3		11			
Geese														
Growing	1500	200				2.5–4		35–55			15			
Breeding	4000	200				4	20							
Pheasants							3.5	40–60			10			1–1.5
Quail														
Growing bobwhite							3.8	30			13			1.5
Breeding bobwhite							4	20			15			1.0
Growing coturnix	5000	1200	12	1		2	4	40	3	1	10	0.3	3	2.0
Breeding coturnix	5000	1200	25	1		2	4	20	3	1	15	0.15	3	1.5
Turkeys														
Growing poults	4000	900	12	0.8–1		2	3.6	40–70	3–4.5	0.7–1	9–11	0.1–0.2	3	0.8–1.9
Breeding hens	4000	900	25	1		2	4	30	4	1	16	0.15	3	1.0
Cats	10,000	1000	80			5	5	45	4	1	10	0.5	20	2.0

Continued

TABLE 5.7 Estimated Vitamin Requirements (units/kg Diet) of Domestic and Laboratory Animals—cont'd

Species	Vitamin A (IU)	Vitamin D (IU)	Vitamin E (mg)[a]	Vitamin K (μg)[b]	Vitamin C (mg)	Thiamin (mg)	Riboflavin (mg)	Niacin (mg)	Vitamin B_6 (mg)	Folate (mg)	Pantothenate (mg)	Biotin (μg)	Vitamin B_{12} (μg)	Choline (g)
Cattle														
Dry heifers	2200	300												
Dairy bulls	2200	300												
Lactating cows	3200	300												
Beef cattle	2200	300												
Dogs	5000	275	50			1	2.2	11.4	1	0.18	10	0.1	22	1.25
Fishes														
Bream									5–6		30–50	1		4.0
Carp	10,000		300				7	28	5–6		10–20			
Catfish	2000	1000	30		60	1	9	14	3	5	40	1	20	3.0
Coldwater spp.	2500	2400	30	10	100	10	20	150	10					
Foxes	2440					1	5.5	9.6	1.8	0.2	7.4			
Goats	60[c]	12.9[c]												
Guinea pigs	23,333	1000	50	5	200	2	3	10	3	4	20	0.3	10	1.0
Hamsters	3636	2484	3	4		20	15	90	6	2	40	0.6	20	2.0
Horses														
Ponies	25[c]													
Pregnant mares	50[c]													
Lactating mares	55–65[c]													
Yearling	40[c]													
2-year olds	30[c]													
Mice	500	150	20	3		5	7	10	1	0.5	10	0.2	10	0.6

Mink	5930		27			1.3	1.6	20	1.6	0.5	8	0.12	32.6	
Primates[d]	15,000	2000	50		0.1		5	50	2.5	0.2	10	0.1		
Rabbits														
Growing	580		40					180	39					1.2
Pregnant	>1160		40	0.2										
Lactating			40											
Rats	4000	1000	30	0.5		4	3	20	6	1	8		50	1.0
Sheep														
Ewes														
Early pregnancy	26[c]	5.6[c]												
Late pregnancy/ lactating	35[c]	5.6[c]												
Rams	43[c]	5.6[c]												
Lambs														
Early weaned	35[c]	6.6[c]												
Finishing	26[c]	5.5[c]												
Shrimp					10		120			120	120			0.6
Swine														
Growing	2200	200	11	2		1.3	2.2–3	10–22	1.5	0.6	11–13	0.1	22	0.4–1.1
Bred gilt/ sow	4000	200	10	2			3	10	1	0.6	12	0.1	15	1.25
Lactating gilt/sow	2000	200	10	2			3	10	1	0.6	12	0.1	15	1.25
Boars	4000	200	10	2			3	10	1	0.6	12	0.1	15	1.25

[a]*α-Tocopherol.*
[b]*Menadione.*
[c]*Unlike almost all of the other values in this table, this requirement is expressed in international units (IU) per kilogram body weight.*
[d]*Nonhuman species.*

From National Research Council, 2015. Nutrient Requirements of Poultry. National Academy Press, Washington, DC; National Research Council, 2006. Nutrient Requirements of Dogs and Cats. National Academy Press, Washington, DC; National Research Council, 2008. Nutrient Requirements of Dairy Cattle, seventh ed. (rev.). National Academy Press, Washington, DC; National Research Council, 2000. Nutrient Requirements of Beef Cattle, seventh ed. (rev.). National Academy Press, Washington, DC; National Research Council, 2011. Nutrient Requirements of Fish and Shrimp. National Academy Press, Washington, DC; National Research Council, 1982. Nutrient Requirements of Mink and Foxes, second ed. (rev.). National Academy Press, Washington, DC; National Research Council, 2006. Nutrient Requirements of Small Ruminants: Sheep, Goats, Cervids and New World Camelids. National Academy Press, Washington, DC; National Research Council, 1995. Nutrient Requirements of Laboratory Animals, fourth ed. (rev.). National Academy Press, Washington, DC; National Research Council, 2007. Nutrient Requirements of Horses, sixth ed. (rev.). National Academy Press, Washington, DC; National Research Council, 2003. Nutrient Requirements of Nonhuman Primates, second ed. (rev.). National Academy Press, Washington, DC; National Research Council, 1977. Nutrient Requirements of Rabbits, second ed. (rev.). National Academy Press, Washington, DC; National Research Council, 2012. Nutrient Requirements of Swine, eleventh ed. (rev.). National Academy Press, Washington, DC.

most individuals by some **margin of safety**. This approach minimizes the probability of producing vitamin deficiencies in populations.

Some clinical conditions require the use of vitamin supplements at levels greater than those normally used to accommodate the usual margins of safety. These include specific vitamin deficiency disorders (e.g., xerophthalmia, rickets, and polyneuritis, encephalopathy related to alcohol abuse) and certain rare inherited metabolic defects (e.g., vitamin B_6-responsive cystathionase deficiency, vitamin B_{12}-responsive transcobalamin II deficiency, biotin-responsive biotinidase deficiency).[17] In such cases, vitamins are prescribed at doses that far exceed requirement levels.

Elevated doses of vitamins are also frequently prescribed by physicians or are taken as over-the-counter supplements by affected individuals in the treatment of certain other pathological states including neurological pains, psychosis, alopecia, anemia, asthenia, premenstrual tension, carpal tunnel syndrome, and prevention of the common cold. Although the efficacies of vitamin supplementation in most of these conditions remain untested in randomized, controlled trials, vitamin prophylaxis, and/or therapy for at least some conditions is perceived as effective by many in the medical community as well as in the general public. This view supports the widespread use of oral vitamin supplements at dosages greater than 50–100 times the RDAs.[18]

5. HYPERVITAMINOSES

Factors Affecting Vitamin Toxicity

Several factors can affect the toxicity of any vitamin. These include the route of exposure, the dose regimen (number of doses and intervals between doses), the general health of the subject, and potential effects of food and drugs. For example, parenteral routes of vitamin administration may increase the toxic potential of high vitamin doses, as the normal routes of controlled absorption and hepatic first-pass metabolism may be circumvented. Large single doses of the water-soluble vitamins are rarely toxic, as they are generally rapidly excreted, thus minimally affecting tissue reserves. However, repeated multiple doses of these compounds can produce adverse effects. In contrast, single large doses of the fat-soluble vitamins can produce large tissue stores that can steadily release toxic amounts of the vitamin thereafter. Some disease states, such as those involving malabsorption, can reduce the potential for vitamin toxicity; however, most increase that potential by compromising the subject's ability to metabolize and excrete the vitamin,[19] or by rendering the subject particularly susceptible to **hypervi taminosis**.[20] Foods and some drugs can reduce the absorption of certain vitamins, thus reducing their toxicities.

The signs of intoxication for each vitamin vary with the species affected and the timecourse of overexposure (Tables 5.8 and 5.9). Nevertheless, certain signs or syndromes are characteristic for each vitamin:

> **Hypervitaminosis A**—The potential for vitamin A intoxication is greater than those for other hypervitaminoses, as its range of safe intakes is relatively small. For humans, acute exposures as low as 25 times the RDA are thought to be potentially intoxicating, although actual cases of hypervitaminosis A have been very rare[21] at chronic doses less than about 9000 µg of retinol equivalents per day. Hypervitaminosis A occurs when plasma retinol levels exceed 3 µmol/L (caused by increases in retinyl esters), which in humans can occur in response to single large doses (>660,000 IU for adults, >330,000 IU for children), or after doses >100,000 IU/day have been taken for several months.

Acute toxicity. Children with hypervitaminosis A develop transient (1–2 days) signs: nausea, vomiting, signs due to increased cerebrospinal fluid pressure (headache, vertigo, blurred, or double vision), and muscular incoordination. Studies have found that 3–9% of children given high, single therapeutic doses (200,000 IU) show transient nausea, vomiting, headache, and general irritability; a similar percentage of younger children may show fontanelle bulging, which subsides in 48–96 h.

Chronic toxicity. Chronic hypervitaminosis A occurs with recurrent exposures exceeding 12,500 IU (infants)–33,000 IU (adult). The early sign is commonly dry lips (cheilitis), which is often followed by dryness and fragility of the nasal mucosa, dry eyes, and conjunctivitis. Skin lesions include dryness, pruritis, erythema, scaling, peeling of the palms and soles, hair loss (alopecia), and nail fragility. Headache, nausea, and vomiting (signs of increased intracranial pressure) can also occur. Infants and young children can show painful periostitis. In animals, adverse effects have been reported at intakes

17. Other examples are given in Chapter 4.

18. For example, several studies have shown that athletes and their coaches generally believe that athletes require higher levels of vitamins than nonathletes. This attitude appears to affect their behavior, as athletes use vitamin (and mineral) supplements with greater frequencies than the general public. One study found that 84% of international Olympic competitors used vitamin supplements. Despite this widespread belief, it remains unclear whether any of the vitamins at levels of intake greater than RDAs can affect athletic performance.

19. For example, individuals with liver damage (e.g., alcoholic cirrhosis, viral hepatitis) have increased plasma levels of free (unbound) retinol and a higher incidence of adverse reactions to large doses of vitamin A.

20. For example, patients with nephrocalcinosis are particularly susceptible to hypervitaminosis D.

21. According to Bendich (1989. Am. J. Clin. Nutr. 49 (358)), fewer than 10 cases per year were reported in 1976–1987. Several of those occurred in individuals with concurrent hepatic damage due to drug exposure, viral hepatitis, or protein-energy malnutrition.

TABLE 5.8 Signs and Symptoms of Vitamin Toxicities in Humans

Vitamin	Children	Adults
Vitamin A	*Acute*: Anorexia, bulging fontanelles, lethargy, high intracranial fluid pressure, irritability, nausea, vomiting	*Acute*: Abdominal pain, anorexia, blurred vision, lethargy, headache, hypercalcemia, irritability, muscular weakness, nausea, vomiting, peripheral neuritis, skin desquamation
	Chronic: Alopecia, anorexia, bone pain, bulging fontanelles, cheilitis, craniotabes, hepatomegaly, hyperostosis, photophobia, premature epiphyseal closure, pruritus, skin desquamation, erythema	*Chronic*: Alopecia, anorexia, ataxia, bone pain, cheilitis, conjunctivitis, diarrhea, diplopia, dry mucous membranes, dysuria, edema, high CSF pressure, fever, headache, hepatomegaly, hyperostosis, insomnia, irritability, lethargy, menstrual abnormalities, muscular pain and weakness, nausea, vomiting, polydypsia, pruritus, skin desquamation, erythema, splenomegaly, weight loss
Vitamin D	Anorexia, diarrhea, hypercalcemia, irritability, lassitude, muscular weakness, neurological abnormalities, pain, polydypsia, polyuria, poor weight gain, renal impairment	Anorexia, bone demineralization, constipation, hypercalcemia, muscular weakness and pain, nausea, vomiting, polyuria, renal calculi
Vitamin E	No adverse effects reported	Mild gastrointestinal distress, some nausea, coagulopathies in patients receiving anticonvulsants
Vitamin K[a]	No adverse effects reported	No adverse effects reported
Vitamin C	No adverse effects reported	Gastrointestinal distress, diarrhea, oxaluria
Thiamin[b]	No adverse effects reported	Headache, muscular weakness, paralysis, cardiac arrhythmia, convulsions, allergic reactions
Riboflavin	No adverse effects reported	No adverse effects reported
Niacin	No adverse effects reported	Vessel dilation, itching, headache, anorexia, liver damage, jaundice, cardiac arrhythmia
Vitamin B_6	No adverse effects reported	Neuropathy, skin lesions
Pantothenic acid	No adverse effects reported	Diarrhea[c]
Biotin	No adverse effects reported	No adverse effects reported
Folate	No adverse effects reported	Allergic reactions[c]
Vitamin B_{12}	No adverse effects reported	Allergic reactions[c]

[a]*Adverse effects observed only for menadione; phylloquinone, and the menaquinones appear to have negligible toxicities.*
[b]*Adverse effects have been observed only when the vitamin was administered parenterally; none when it has been given orally.*
[c]*This sign has been observed in only a few cases.*

as low as 10 times the RDA; but intoxication typically follows chronic intakes of 100- to 1000-fold RDA levels. The most frequently observed signs are loss of appetite, loss of weight or reduced growth, skeletal malformations, spontaneous fractures, and internal hemorrhages. Most signs can be reversed by discontinuing excessive exposure to the vitamin. Ruminants appear to tolerate high intakes of vitamin A better than nonruminants, apparently due to destruction of the vitamin by the rumen microflora. That retinoids can be embryotoxic raises concerns about the safety of high-level vitamin A supplementation for pregnant animals and humans. High doses of retinol, all-*trans*-retinoic acid, or 13-*cis*-retinoic acid can disrupt cephalic neural crest cell activity, producing craniofacial, central nervous system, and cardiovascular and thymus malformations. Fetal malformations have been reported in cases of oral use of 20,000–25,000 IU/day all-*trans*-retinoic acid in treating *acne vulgaris*. Regular intakes exceeding 10,000 IU/day (preformed vitamin A) has been associated with increased risk of birth defects in a small cohort of women with very high vitamin A intakes (mean > 21,000 IU/day). Rare cases of premature closure of lower limb epiphyses have been reported in animals, e.g., "hyena disease" in calves.

The toxicities of carotenoids appear to be low. Regular intakes as great as 30 mg β-carotene per day are without side effects other than carotenodermia.

Hypervitaminosis D—Vitamin D_3 has been found safe for pregnant and lactating women and their children at oral doses of 100,000 IU/day; however, intakes as low as 50 times the RDA have been reported to be toxic to

TABLE 5.9 Signs of Vitamin Toxicities in Animals

Vitamin	Sign	Species
Vitamin A	Alopecia	Rat, mouse
	Anorexia	Cat, cattle, chicken, turkey
	Cartilage abnormalities	Rabbit
	Convulsions	Monkey
	Elevated heart rate	Cattle
	Fetal malformations	Hamster, monkey, mouse, rat
	Hepatomegaly	Rat
	Gingivitis	Cat
	Irritability	Cat
	Lethargy	Cat, monkey
	Reduced CSF pressure	Cattle, goat, pig
	Poor growth	Chicken, pig, turkey
	Skeletal abnormalities	Cat, cattle, chicken, dog, duck, mouse, pig, rabbit, rat, turkey, horse
Vitamin D[a]	Anorexia	Cattle, chicken, fox, pig, rat
	Bone abnormalities	Pig, sheep
	Cardiovascular calcinosis	Cattle, dog, fox, horse, monkey, mouse, pig, rat, sheep, rabbit
	Renal calcinosis	Cattle, chicken, dog, fox, horse, monkey, mouse, pig, rat, sheep, turkey
	Cardiac dysfunction	Cattle, pig
	Hypercalcemia	Cattle, chicken, dog, fox, horse, monkey, mouse, pig, rat, sheep, trout
	Hyperphosphatemia	Horse, pig
	Hypertension	Dog
	Myopathy	Fox, pig
	Poor growth, weight loss	Catfish, chicken, horse, mouse, pig, rat
	Lethality	Cattle
Vitamin E	Atherosclerotic lesions	Rabbit
	Bone demineralization	Chicken, rat
	Cardiomegaly	Rat
	Hepatomegaly	Chicken
	Hyperalbuminemia	Rat
	Hypertriglyceridemia	Rat
	Hypocholesterolemia	Rat
	Impaired muscular function	Chicken
	Increased hepatic vitamin A	Chicken, rat
	Increased prothrombin time	Chicken
	Reduced adrenal weight	Rat
	Poor growth	Chicken
	Increased hematocrit	Rat
	Reticulocytosis	Chicken
	Splenomegaly	Rat

TABLE 5.9 Signs of Vitamin Toxicities in Animals—cont'd

Vitamin	Sign	Species
Vitamin K[b,c]	Anemia[c]	Dog
	Renal failure[c]	Horse
	Lethality	Chicken,[c,d] mouse,[c,d] rat[c]
Vitamin C	Anemia	Mink
	Bone demineralization	Guinea pig
	Decreased circulating thyroid hormone	Rat
	Liver congestion	Guinea pig
	Oxaluria	Rat
Thiamin	Respiratory distress (i.p. dose)	Rat
	Cyanosis (i.p. dose)	Rat
	Epileptiform convulsions (i.p. dose)	Rat
Riboflavin	No adverse effects reported for oral doses lethality (parental dose)	Rat
Niacin	Impaired growth	Chicken (embryo)
	Developmental abnormalities	Chicken (embryo), mouse (fetus)
	Liver damage	Mouse
	Mucocutaneous lesions	Chicken
	Myocardial damage	Mouse
	Decreased weight gain	Chicken
	Lethality (i.p. dose)	Chicken (embryo), mouse[d]
Vitamin B_6	Anorexia	Dog
	Ataxia	Dog, rat
	Convulsions	Rat
	Lassitude	Dog
	Muscular weakness	Dog
	Neurologic impairment	Dog
	Vomiting	Dog
	Lethality	Mouse, rat
Pantothenic acid	No adverse effects reported for oral doses lethality (i.p. dose)	Rat
Biotin	No adverse effects reported for oral doses Irregular estrus (i.p. dose)	Rat
	Fetal resorption (i.p. dose)	Rat
Folate	No adverse effects reported for oral doses	
	Epileptiform convulsions (i.p. dose)	Rat
	Renal hypertrophy (i.p. dose)	Rat
Vitamin B_{12}	No adverse effects reported for oral doses Irregular estrus (i.p. dose)	Rat
	Fetal resorption (i.p. dose)	Rat
	Reduced fetal weights (i.p. dose)	Rat

[a]*Vitamin D_3 is much more toxic than vitamin D_2.*
[b]*Only menadione produces adverse effects; phylloquinone and the menaquinones have negligible toxicities.*
[c]*These effects observed after parenteral administration of the vitamin.*
[d]*Nicotinamide is more toxic than nicotinic acid.*

humans, particularly in children. Affected individuals show anorexia, vomiting, headache, drowsiness, diarrhea, and polyuria. There have been no documented cases of hypervitaminosis D due to excessive sunlight exposure. Excessive intakes of vitamin D increase circulating levels of 25-OH-D_3, which at high levels appears to bind VDR, thus, bypassing the regulation of the 25-OH-D_3-1-hydroxylase to induce transcriptional responses normally signaled only by 1,25-$(OH)_2$-D_3. Hypervitaminosis D is characterized by increases in both the enteric absorption and bone resorption of calcium. This produced hypercalcemia and, ultimately, calcinosis, i.e., deposition of calcium and phosphate in soft tissues (heart, kidney, and vascular and respiratory systems). Thus, the risk of hypervitaminosis D depends on concomitant intakes of calcium and phosphorus and is in conditions, such as chronic inflammation, in which the normal feedback regulation of the renal 25-OH-D_3-1-hydroxylase is compromised. Studies with animals indicate that vitamin D_3 is 10–20 times more toxic than vitamin D_2,[22] apparently because it is more readily metabolized than the latter to the 25-hydroxy metabolites.

Hypervitaminosis E—Vitamin E is one of the *least toxic* of the vitamins. Both animals and humans appear to be able to tolerate high levels of exposure. For humans, daily doses as high as 400 IU have been be considered harmless, and large oral doses as great as 3200 IU have not been found to have consistent ill effects. There have been isolated reports of headache, fatigue, nausea, double vision, muscular weakness, mild creatinuria, and gastrointestinal distress in humans consuming as much as 1000 IU per day. For animals, doses at least two orders of magnitude above nutritional requirements, e.g., to 1000–2000 IU/kg, are without untoward effects. Studies with animals indicate that excessive dosages of tocopherols exert most, if not all, of their adverse effects by antagonizing the utilization of the other fat-soluble vitamins: reducing hepatic vitamin A storage, impairing bone mineralization and producing coagulopathies. In each case, these signs could be corrected with supplements of the appropriate vitamin (A, D, and K, respectively). These effects appear to involve impaired absorption, and inhibition of retinyl ester hydrolase and vitamin K-dependent carboxylations.

Hypervitaminosis K—The toxic potentials of the naturally occurring forms of vitamin K are negligible. Phylloquinone exhibits no adverse effects when administered to animals in massive doses by any route. The menaquinones are similarly thought to have little, if any, toxicity. The synthetic vitamer menadione, when administered parenterally, can at high doses produce fatal anemia, hyperbilirubinemia, and severe jaundice. However, its toxic threshold appears to be at least three orders of magnitude greater than nutritionally required levels. At such doses, menadione appears to cause oxidative stress by reduction to the semiquinone radical, which, in the presence of O_2, is reoxidized to the quinone, resulting in the formation of the superoxide radical anion. Menadione can also react with free sulfhydryl groups; thus, high levels may deplete reduced glutathione (GSH) levels. The horse appears to be particularly vulnerable to menadione toxicity. Parenteral doses of 2–8 mg/kg have been found to be lethal in that species, whereas the parenteral LD_{50}[23] values for most other species are an order of magnitude greater than that.

Hypervitaminosis C—The only adverse effects of large doses of vitamin C that have been consistently observed in humans are gastrointestinal disturbances and diarrhea occurring at levels of intake nearly 20–80 times the RDAs. Concern has also been expressed that excess ascorbic acid may be prooxidative, may competitively inhibit the renal reabsorption of uric acid, may enhance the enteric destruction of vitamin B_{12}, may enhance the enteric absorption of nonheme iron (thus leading to iron overload), may produce mutagenic effects, and may increase ascorbate catabolism that would persist after returning to lower intakes of the vitamin. Present knowledge indicates that most, if not all, of these concerns are not warranted. That ascorbic acid can enhance the enteric absorption of dietary iron has led to concern that megadoses may lead to progressive iron accumulation in iron-replete individuals (iron storage disease). This hypothesis has not been supported by results of studies with animal models. Nevertheless, patients with hemochromatosis or other forms of excess iron accumulation should avoid taking vitamin C supplements with their meals.

Perhaps the greatest concern associated with high intakes of vitamin C concerns increased oxalate production. In humans, unlike other animals, oxalate is a major metabolite of ascorbic acid, accounting for 35–50% of the 35–40 mg of oxalate excreted in the urine each day.[24] The health concern is that high vitamin C intake may lead to increased oxalate production and, thus, to increased risk of urinary calculi.[25] Metabolic studies have indicated that

22. That is, vitamin D3 can produce effects comparable to those of vitamin D2 at doses representing only 5–10% of the latter.

23. The LD50 value is a useful parameter indicative of the degree of toxicity of a compound. It is defined as the lethal dose for 50% of a reference population and is calculated from experimental dose–survival data using the profit analysis.

24. The balance of urinary oxalate comes mainly from the degradation of glycine (about 40% of the total); but some also can come from the diet (5–10%).

25. There is some question as to whether oxalate may have been produced as an artifact of the analytical procedure.

the turnover of ascorbic acid is limited for which reason high intakes of vitamin C would not be expected to greatly affect oxalate production. Clinical studies have revealed slight oxaluria in patients given daily multiple-gram doses of vitamin C. It is not clear whether this effect is clinically significant, as its magnitude is low and within normal variation.[26] Nevertheless, prudence dictates the avoidance of doses greater than 1000 mg of vitamin C for individuals with a history of forming renal stones. Little information is available on vitamin C toxicity in animals, although acute LD_{50} (50% lethal dose) values for most species and routes of administration appear to be at least several grams per kilogram of body weight. Dietary vitamin C intakes 100–1000 times the allowance levels appear safe for most species.

Thiamin hypervitaminosis—The toxic potential of thiamin appears to be low, particularly when administered orally. Parenteral doses of the vitamin at 100–200 times the RDAs have been reported to cause intoxication in humans, characterized by headache, convulsions, muscular weakness, paralysis, cardiac arrhythmia, and allergic reactions. Most of the available information pertinent to its toxic potential is for thiamin hydrochloride. At very high doses (1000-fold levels required to prevent deficiency signs) that form can be fatal by suppressing the respiratory center. Such doses of the vitamin to animals produce curare-like signs, suggestive of blocked nerve transmission: restlessness, epileptiform convulsions, cyanosis, and dyspnea. Lower levels, ≤300 mg/day, are used therapeutically in humans without adverse reactions.

Riboflavin hypervitaminosis—High oral doses of riboflavin are very safe, probably owing to the relatively poor absorption of the vitamin at high levels. Oral riboflavin doses as great as 2–10 g/kg body weight produce no adverse effects in dogs and rats. The vitamin is somewhat more toxic when administered parenterally. The LD_{50} (50% lethal dose) values for the rat given riboflavin by the intraperitoneal, subcutaneous, and oral routes have been estimated to be 0.6, 5, and >10 g/kg, respectively. No adverse effects in humans have been reported.

Niacin hypervitaminosis—*acute toxicity*. In humans, small doses (10 mg) of nicotinic acid can cause flushing, although this effect is not associated with other seriously adverse reactions. At high dosages (two to four g/day), nicotinic acid can cause vasodilation, itching, nausea, vomiting, headaches, and, less frequently, skin lesions. These responses appears to be mediated by the niacin receptor, which is expressed by macrophages and bone marrow-derived cells of the skin. They can be minimized by using a slow-release formulation of nicotinic acid or by using a cyclooxygenase inhibitor (e.g., aspirin, indomethacin). High nicotinic acid doses have been reported to cause itching urticaria (hives), gastrointestinal discomfort (heartburn, nausea, vomiting, rarely diarrhea) in humans. Nicotinamide only rarely produces these reactions. Many patients have taken daily oral doses of 200–1000 mg for periods of years with only occasional side effects (skin rashes, hyperpigmentation, reduced glucose tolerance in diabetics, some liver dysfunction) at the higher dosages. Doses 50–100 times the RDAs are considered safe.

Chronic toxicity. The longer-term effects of high nicotinic acid doses include cases of insulin resistance, which may involve a rebound in lipolysis that results in increased free fatty acid levels. Transient hepatic dysfunction has also been reported. Chronic, high intakes of nicotinamide may deplete methyl groups (to excrete the vitamin), which would be exacerbated by low intakes of methyl donors, methionine and choline, and suboptimal status with respect to folate and/or vitamin B_{12}. Available information on the niacin tolerances of animals is scant but suggests that toxicity requires daily doses greater than 350–500 mg of nicotinic acid equivalents per kilogram body weight.

Hypervitaminosis B_6—The toxicity of vitamin B_6 appears to be relatively low, with intakes as great as 100 times the RDAs having been used safely by many people. Very high doses of the vitamin (several grams per day) have been shown to induce reversible sensory neuropathies marked by changes in gait and peripheral sensation. The primary target appears to be the peripheral nervous system; although massive doses of the vitamin have produced convulsions in rats, central nervous abnormalities have not been reported frequently in humans. Reports of individuals taking massive doses of the vitamin (>2 g/day) indicate that the earliest detectable signs are ataxia and loss of small motor control. Doses up to 750 mg/day for extended periods of time (years) have been found safe. The vitamin can increase the conversion of L-dopa to dopamine to interfere with the former drug in the management of Parkinson's disease in those not taking a decarboxylase inhibitor. Substantial information concerning the safety of large doses of vitamin B_6 in animals is available only for the dog and the rat. Doses less than 1000 times the allowance levels are safe for those species and, by inference, for other animal species.

Biotin hypervitaminosis—Biotin is generally regarded as nontoxic. Adverse effects of large doses of biotin have not been reported in humans or animals given the vitamin at doses as high as 200 mg orally or 20 mg intravenously. Limited data suggest that biotin is safe for most people at doses as great as 500 times the RDAs and for animals at probably more than 1000 times allowance levels. Animal

26. Forty percent of subjects given 2 g of ascorbic acid daily showed increases in urinary oxalate excretion by more than 10% (Chai, W., Liebman, M., Kynast-Gales, S., et al., 2004. Am. J. Kidney Dis. 44, 1060–1066).

studies have revealed few, if any, indications of toxicity, and it is probable that animals can tolerate the vitamin at doses at least an order of magnitude greater than nutritional levels.

Pantothenic acid hypervitaminosis—Pantothenic acid is generally regarded as nontoxic. A few reports indicate diarrhea occurring in humans consuming 10–20 g of the vitamin per day. Thus, pantothenic acid is thought to be safe for humans at doses at least 100 times the RDAs. No adverse reactions have been reported in any animal species following the ingestion of large doses of the vitamin. It has been estimated that animals can tolerate doses of pantothenic acid as great as at least 1000 times their respective nutritional requirements for the vitamin. Parenteral administration of very large amounts (e.g., 1 g per kg body weight) of the calcium salt has been shown to be lethal to rats.

Folate hypervitaminosis—Folate is generally regarded as nontoxic. Other than a few cases of apparent allergic reactions, the only purported adverse effect in humans (interference with the enteric absorption of zinc) is not supported with adequate data. Intakes of 400 mg of folate per day for several months have been tolerated without side effects in humans, indicating that levels at least as great as 2000 times the RDAs are safe. No adverse effects of high oral doses of folate have been reported in animals, although parenteral administration of pharmacologic amounts (e.g., 250 mg/kg, which is about 1000 times the dietary requirement) has produced epileptic responses and renal hypertrophy in rats. High-folate treatment has been found to exacerbate teratogenic effects of nutritional Zn deficiency.

Hypervitaminosis B_{12}—Vitamin B_{12} has no appreciable toxicity. No adverse reactions have been reported for humans or animals given high levels of the vitamin. Upper safe limits of vitamin B_{12} use are, therefore, highly speculative; it appears that doses at least as great as 1000 times the RDAs/allowances are safe for humans and animals.

6. SAFE INTAKES OF VITAMINS

The risks of adverse effects (toxicity) of the vitamins, like those of any other potentially toxic compounds, are functions of dose level. In general, the risk–dosage function is curvilinear, indicating a **hazard threshold** for vitamin dosage at some level greater than the requirement for that vitamin. Thus, a dosage increment exists between the level required to prevent deficiency and that sufficient to produce adverse effects. That increment, the **range of safe intake**, is bounded on the low-dosage side by the allowance, and on the high-dosage side by the upper safe limit, each of which is set on the basis of similar considerations of risk of adverse effects within the population (Fig. 5.4).

Quantifying Safe Intakes

There is no standard algorithm for quantifying the ranges of safe intakes of vitamins, but an approach developed for environmental substances that cause systemic toxicities has recently been employed for this purpose with nutritionally essential inorganic elements. This approach involves the imputation of an acceptable daily intake (ADI)[27] based on the application of a safety factor (SF)[28] to an experimentally determined highest **no observed adverse effect level (NOAEL)** of exposure to the substance. In the absence of sufficient data to ascertain an NOAEL, an experimentally determined **lowest observed adverse effect level (LOAEL)** is used:

$$ADI = LOAEL \div SF$$

An extension of this approach is to express the comparative safety of nutrients using a **safety index (SI)**. This index is analogous to the therapeutic index (TI) used for drugs; it is the ratio of the minimum toxic dose and the RI derived from the RDA:

$$SI = LOAEL \div RI$$

This approach was used by Hathcock[29] to express quantitatively the safety limits of several vitamins for humans (Table 5.10).

The DRIs of the U.S. Food and Nutrition Board (1997–2001, 2010) addressed the safety of high doses of essential nutrients with the UL, defined as the highest level of daily intake that is likely to pose *no* risks of adverse health effects to almost all healthy individuals in each age–sex-specific demographic subgroup. In this context "adverse effect" is defined as any significant alteration in structure or function. It should be noted that the Food and Nutrition Board chose to use the term "tolerable intake" to avoid implying possible beneficial effects of intakes greater than the RDA.[30] The ULs are based on chronic intakes. They are derived through a multistep process:

1. Hazard identification—involving the systematic evaluation of all information pertaining to adverse effects of the nutrient;

27. The U.S. Environmental Protection Agency has replaced the ADI with the reference dose (RfD), a name the agency considers to be more value neutral, i.e., avoiding any implication that the exposure is completely safe or acceptable.

28. SF values are selected according to the quality and generalizability of the reported data in the case selected as the reference standard. Higher values, e.g., 100, may be used if animal data are extrapolated to humans; whereas lower values, e.g., 1 or 3, may be used if solid clinical data are available.

29. Hathcock, J., 1993. Nutr. Rev. 51, 278–285.

30. *See* Food and Nutrition Board, 1998. Dietary Reference Intakes: A Risk Assessment Model for Establishing Upper Intake Levels for Nutrients. National Academy Press, Washington, DC, 71 pp.

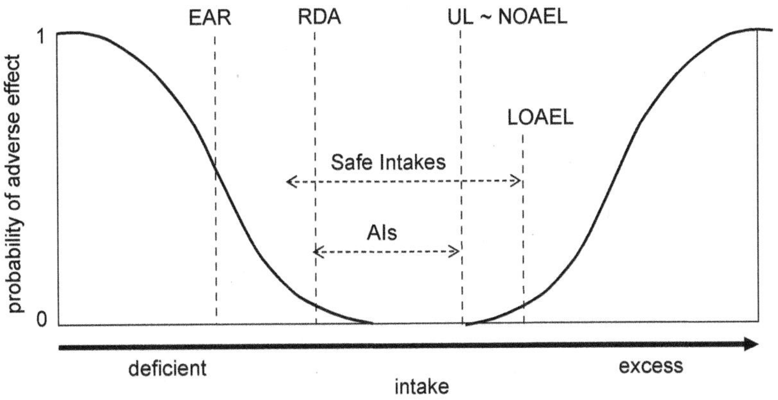

FIGURE 5.4 Vitamin safety follows a biphasic dose–response curve: just as very low intakes of vitamins can produce deficiency disorders, very high intakes can also have potential to produce adverse effects. The inflection points are the RDA and UL (which, in principle, should be comparable to the "no observed adverse effect level," NOAEL) and the "low observed adverse effect level", LOAEL, i.e., the upper end of the range of safe intakes.

TABLE 5.10 Use of a Safety Index to Quantitate the Toxic Potentials of Selected Vitamins for Humans

Parameter	Vitamin A	Vitamin D	Vitamin C	Niacin	Vitamin B_6
RDI (Recommended dietary intake)[a]	3300 IU	20 µg	60 mg	20 mg	2 mg
LOAEL (lowest observed adverse effect level)	25,000 IU	250 µg	2000 mg	500 mg	50 mg
Safety index (SI)	7.6	12.5	33	25	25

[a]*The greatest RDA for persons ≥4 years of age, excluding pregnant and lactating women.*
From Hathcock, J.N., 1993. J. Nutr. Rev. 51 (278); Hathcock, J.N., Shao, A., Vieth, R., et al., 2007. Am. J. Clin. Nutr. 85 (6).

2. Dose–response assessment—involving the determination of the relationship between level of nutrient intake and incidence/severity of adverse effects;
3. Intake assessment—involving the evaluation of the distribution of nutrient intakes in the general population; and
4. Risk characterization—involving the expression of conclusions from the previous steps in terms of the fraction of the exposed population having nutrients in excess of the estimated UL.

In practice, ULs are set at less than the respective LOAELs and no greater than the NOAELs (Fig. 5.4) from which they are derived, subject to uncertainty factors (UFs) used to characterize the level of uncertainty associated with extrapolating from observed data to the general population.[31] The ULs for the vitamins are presented in Table 5.9. Table 5.10 presents the authors' recommended upper safe vitamin intakes for animals.

Ranges of Safe Vitamin Intakes

The vitamins fall into four categories of relative toxicity at levels of exposure above typical allowances (Tables 5.11 and 5.12):

- **Greatest toxic potential**—vitamin A, vitamin D
- **Moderate toxic potential**—niacin
- **Low toxic potential**—vitamin E, vitamin C, thiamin, riboflavin, vitamin B_6
- **Negligible toxic potential**—vitamin K, pantothenic acid, biotin, folate, vitamin B_{12}.

Under circumstances of vitamin use at levels appreciably greater than the standard allowances (RDAs for humans or recommended use levels for animals), prudence dictates giving special consideration to those vitamins with greatest potentials for toxicity (those in the first two or three categories). In practice, it may only be necessary to consider the most potentially toxic vitamins of the first category (vitamins A and D).

31. Small UFs (close to one) are used in cases where little population variability is expected for the adverse effects, where extrapolation from primary data is not believed to under-predict the average human response, and where a LOAEL is available. Larger UFs (as high as 10) are used in cases where the expected variability is great, where extrapolation is necessary from primary animal data, and where a LOAEL is not available and a NOAEL value must be used.

TABLE 5.11 Food and Nutrition Board Tolerable Upper Intake Limits (ULs) for Vitamins

Age–Sex Group	Vitamin A (μg)[a]	Vitamin D (μg)	Vitamin E (mg)[b]	Vitamin K (μg)	Vitamin C (mg)	Thiamin (mg)	Riboflavin (mg)	Niacin (mg)[c]	Vitamin B_6 (μg)	Pantothenic Acid (μg)	Biotin (μg)	Folate (μg)[d]	Vitamin B_{12} (μg)
Infants													
0–11 months	600	25–38	–[e]	–[e]	–[e]	–[e]	–[e]		–[e]	–[e]	–[e]	–[e]	–[e]
Children													
1–3 years	600	63	200	–[e]	400	–[e]	–[e]	10	–[e]	–[e]	–[e]	300	–[e]
4–8 years	600	75	300	–[e]	650	–[e]	–[e]	15	–[e]	–[e]	–[e]	400	–[e]
Males													
9–13 years	1700	100	600	–[e]	1200	–[e]	–[e]	20	–[e]	–[e]	–[e]	600	–[e]
14–18 years	2800	100	800	–[e]	1800	–[e]	–[e]	30	–[e]	–[e]	–[e]	800	–[e]
19+years	3000	100	1000	–[e]	2000	–[e]	–[e]	35	–[e]	–[e]	–[e]	1000	–[e]
Females													
9–13 years	1700	100	600	–[e]	1200	–[e]	–[e]	20	–[e]	–[e]	–[e]	600	–[e]
14–18 years	2800	100	800	–[e]	1800	–[e]	–[e]	30	–[e]	–[e]	–[e]	800	–[e]
>18 years	3000	100	1000	–[e]	2000	–[e]	–[e]	35	–[e]	–[e]	–[e]	1000	–[e]
Pregnancy													
≤18 years	2800	100	800	–[e]	1800	–[e]	–[e]	30	–[e]	–[e]	–[e]	800	–[e]
>18+years	2800	100	1000	–[e]	2000	–[e]	–[e]	35	–[e]	–[e]	–[e]	1000	–[e]
Lactation													
≤18 years	2800	100	800	–[e]	1800	–[e]	–[e]	30	–[e]	–[e]	–[e]	800	–[e]
>18+years	3000	100	1000	–[e]	2000	–[e]	–[e]	35	–[e]	–[e]	–[e]	1000	–[e]

[a] *Retinol equivalents.*
[b] *α-Tocopherol.*
[c] *Niacin equivalents.*
[d] *Folate equivalents.*
[e] *UL not established.*

From Food and Nutrition Board, 1997. Dietary Reference Intakes for Calcium, Phosphorus, Magnesium, Vitamin D and Fluoride. National Academy Press, Washington, DC, 432 pp.; Food and Nutrition Board, 2000. Dietary Reference Intakes for Thiamin, Riboflavin, Niacin, Vitamin B_6, Folate, Vitamin B_{12}, Pantothenic Acid, Biotin and Choline. National Academy Press, Washington, DC, 564 pp.; Food and Nutrition Board, 2000. Dietary Reference Intakes for Vitamin C, Vitamin E, Selenium and Carotenoids. National Academy Press, Washington, DC, 506 pp.; Food and Nutrition Board, 2001. Dietary Reference Intakes for Vitamin A, Vitamin K, Arsenic, Boron, Chromium, Copper, Iodine, Iron, Manganese, Molybdenum, Nickel, Silicon, Vanadium and Zinc. National Academy Press, Washington, DC, 773 pp.; Food and Nutrition Board, 2010. Dietary Reference Intakes: Calcium, Vitamin D. National Academy Press, Washington, DC, 1115 pp.

TABLE 5.12 Recommended Upper Safe Intakes of the Vitamins for Animals

Vitamin	Safe Intake (Multiple of Allowance Level[a])
High Toxic Potential	
Vitamin A	10[b]–30[c]
Vitamin D	10–20[d]
Moderate Toxic Potential	
Niacin[e]	50–100
Low Toxic Potential	
Vitamin E	100
Vitamin C	100–1000
Thiamin	500
Riboflavin	100–500
Vitamin B_6	100–1000
Negligible Toxic Potential	
Vitamin K[f]	1000
Pantothenic acid	1000
Biotin	1000
Folate	1000
Vitamin B_{12}	1000

[a]*From Committee on Animal Nutrition, 1987. Vitamin Tolerance in Animals. National Academy Press, Washington, DC.*
[b]*For nonruminant species.*
[c]*For ruminant species.*
[d]*Vitamin D_3 is more toxic than vitamin D_2.*
[e]*Nicotinamide is more toxic than nicotinic acid.*
[f]*Only menadione has significant (low) toxicity.*

7. STUDY QUESTIONS AND EXERCISES

1. Prepare a concept map illustrating the relationships of the concepts of minimal and optimal nutrient requirements and nutrient allowances to the concepts of physiological function and health.
2. What issues relate to the consideration of nutritional status in such areas as immune function or chronic and degenerative diseases in the development of dietary standards?
3. Which vitamins are most likely to present potential for hazards for humans?
4. Use specific examples to discuss the relationship of the toxic potential of vitamins to their absorption and metabolic disposition.
5. Prepare a concept map illustrating the relationships of the concepts of minimal and optimal nutrient requirements and nutrient allowances to the concepts of physiological function and health.
6. What considerations are necessary in applying the DRIs to individuals?

RECOMMENDED READING

Beaton, G.H., 2005. When is an individual an individual vs. a member of a group? An issue in application of the Dietary Reference Intakes. Nutr. Rev. 64, 211–225.

Chan, L.N., 2006. Drug-nutrient interactions. In: Shils, M.E., Shike, M., Caballero, et al. (Eds.), Modern Nutrition in Health and Disease, tenth ed. Lippincott, New York, pp. 1539–1553 (Chapter 97).

Dwyer, J.T., 2012. Dietary standards and guidelines: similarities and differences among countries. In: Erdman Jr., J.W., Macdonald, I.A., Zeisel, S.H. (Eds.), Present Knowledge in Nutrition, tenth ed. ILSI, Washington, DC, pp. 1110–1134 (Chapter 65).

Food and Nutrition Board, 1997. Dietary Reference Intakes for Calcium, Phosphorus, Magnesium, Vitamin D and Fluoride. National Academy Press, Washington, DC. 432 pp.

Food and Nutrition Board, 1998. Dietary Reference Intakes: A Risk Assessment Model for Establishing Upper Intake Levels for Nutrients. National Academy Press, Washington, DC. 71 pp.

Food and Nutrition Board, 2000a. Dietary Reference Intakes for Thiamin, Riboflavin, Niacin, Vitamin B_6, Folate, Vitamin B_{12}, Pantothenic Acid, Biotin and Choline. National Academy Press, Washington, DC. 564 pp.

Food and Nutrition Board, 2000b. Dietary Reference Intakes for Vitamin C, Vitamin E, Selenium and Carotenoids. National Academy Press, Washington, DC. 506 pp.

Food and Nutrition Board, 2001. Dietary Reference Intakes for Vitamin A, Vitamin K, Arsenic, Boron, Chromium, Copper, Iodine, Iron, Manganese, Molybdenum, Nickel, Silicon, Vanadium and Zinc. National Academy Press, Washington, DC. 773 pp.

Food and Nutrition Board, 2003a. Dietary Reference Intakes: Applications in Dietary Planning. National Academy Press, Washington, DC. 237 pp.

Food and Nutrition Board, 2003b. Dietary Reference Intakes: Guiding Principles for Nutrition Labeling and Fortification. National Academy Press, Washington, DC. 205 pp.

Food and Nutrition Board, 2010. Dietary Reference Intakes, Calcium, Vitamin D. National Academy Press, Washington, DC. 1115 pp.

Hathcock, J.N., 1997. Vitamins and minerals: efficacy and safety. Am. J. Clin. Nutr. 66, 427–437.

Joint WHO/FAO Consultation, 2002. Diet, Nutrition and the Prevention of Chronic Diseases. World Health Org., Geneva. 149 pp.

Joint WHO/FAO Expert Consultation, 2001. Human Vitamin and Mineral Requirements. Food and Agricultural Org., Rome. 286 pp.

Joost, H.G., Gibney, M.J., Cashman, K.D., et al., 2007. Personalized nutrition: status and perspectives. Br. J. Nutr. 98, 26–31.

King, J.C., 2007. An evidence-based approach for establishing dietary guidelines. J. Nutr. 137, 480–483.

Mason, P., 2007. One is okay, more is better? Pharmacological aspects and safe limits of nutritional supplements. Proc. Nutr. Soc. 66, 493–507.

Mattys, C., Bucchini, L., Busstra, M.C., et al., 2006. Dietary standards in the United States. In: Bowman, B.A., Russell, R.M. (Eds.), Present Knowledge in Nutrition, vol. II. ninth ed. ILSI, Washington, DC, pp. 859–875 (Chapter 63).

Murphy, S., 2006. The recommended dietary allowance (RDA) should not be abandoned: an individual is *both* an individual and a member of a group. Nutr. Rev. 64, 313–318.

Rosenberg, I.H., 2007. Challenges and opportunities in the translation of the science of vitamins. Am. J. Clin. Nutr. 85, 325S–327S.

Russel, R.M., 2008. Current framework for DRI development: what are the pros and cons? Nutr. Rev. 66, 455–458.

Subcommittee on Vitamin Tolerance, National Research Council, 1987. Vitamin Tolerance in Animals. National Academy Press, Washington, DC.
Trumbo, P.R., 2013. Dietary reference intakes: cases of appropriate and inappropriate uses. Nutr. Rev. 71, 657–664.
Walter, P., Hornig, D.H., Moser, U. (Eds.), 2001. Functions of Vitamins Beyond Recommended Daily Allowances. Karger, Basel. 214 pp.
Weisell, R., Albert, J., 2012. The role of United Nations Agencies in establishing international dietary standards. In: Erdman Jr., J.W., Macdonald, I.A., Zeisel, S.H. (Eds.), Present Knowledge in Nutrition, tenth ed. ILSI, Washington, DC, pp. 1135–1150 (Chapter 66).
Yates, A.A., 2006. Dietary reference intakes: rationale and applications. In: Shils, M.E., Shike, M., Caballero, B., et al. (Eds.), Modern Nutrition in Health and Disease, tenth ed. Lippincott, New York, pp. 1655–1672 (Chapter 104).

Part II

Considering the Individual Vitamins

Chapter 6

Vitamin A

Chapter Outline

Anchoring Concepts

1. Vitamin A is the generic descriptor for compounds with the qualitative biological activity of **retinol**, i.e., retinoids and some (provitamin A) carotenoids.
2. Vitamin A-active substances are hydrophobic and, thus, are insoluble in aqueous environments (intestinal lumen, plasma, interstitial fluid, and cytosol). Accordingly, vitamin A-active substances are absorbed by micelle-dependent diffusion.
3. Vitamin A was discovered by its ability to prevent **xerophthalmia**.

Nobody was willing to accept that two cents worth of vitamin A was going to reduce childhood mortality by a third or half... A lot of people had spent their lives studying the complex amalgam of elements leading to childhood deaths, and here we were suggesting that we can cut right through this complex, causal web and give two cents worth of vitamin A and prevent those deaths. It didn't sit well.

Al Sommer[1]

1. Alfred Sommer (b. 1942) is professor and Dean Emeritus of the Bloomberg School of Public Health, Johns Hopkins University. In the 1970s, he conducted studies of the impacts of vitamin A deficiency on children in Indonesia. In reanalyzing those results some time later, he noted that the survival of children with vitamin A-deficiency blindness was much lower than those without blindness. He went on to demonstrate that vitamin A treatment, which was known to prevent blindness, also prevented deaths. His remarkable discovery shifted the paradigm for the role of vitamin A in nutrition and health.

LEARNING OBJECTIVES

1. To understand the nature of the various sources of vitamin A in foods.
2. To understand the means of vitamin A absorption from the small intestine.
3. To become familiar with the carriers involved in the extra- and intracellular transport of vitamin A.
4. To understand the metabolic conversions involved in the activation and degradation of vitamin A in its absorption, transport and storage, cellular function, and excretion.
5. To understand current knowledge of the biochemical mechanisms of action of vitamin A and their relationships to vitamin A deficiency diseases.
6. To understand the physiologic implications of high doses of vitamin A.

VOCABULARY

Abetalipoproteinemia
Acyl-CoA:retinol acyltransferase (ARAT)
Aldehyde dehydrogenase
Bleaching
α-Carotene
β-Carotene
β-Carotene 15,15′-oxygenase
γ-Carotene
Canthaxanthin
Carotenodermia
Carotenoid
cGMP phosphodiesterase

The Vitamins. http://dx.doi.org/10.1016/B978-0-12-802965-7.00006-X

Chylomicron
β-Ionone nucleus
Conjunctival impression cytology
CRABP (cellular retinoic acid-binding protein)
CRABP(II) (cellular retinoic acid-binding protein type II)
CRALBP (cellular retinal-binding protein)
CRBP (cellular retinol-binding protein)
CRBP(II) (cellular retinol-binding protein type II)
β-Cryptoxanthin
3,4-Didehydroretinol
Glycoproteins
High-density lipoproteins (HDLs)
Holo-RBP4
Hyperkeratosis
International unit (IU)
Iodopsins
IRBP (interphotoreceptor retinol-binding protein)
Keratomalacia
Low-density lipoproteins (LDLs)
Lecithin–retinol acyltransferase (LRAT)
Lycopene
Measles blindness
Melanopsin
Metarhodopsin II
Modified relative dose–response (MRDR) test
Night blindness
Nyctalopia
Opsins
Pancreatic nonspecific lipase
Peroxisome-proliferator activation receptor (PPAR)
Protein–calorie malnutrition
Provitamin A
Relative dose–response (RDR) test
all-*trans*-Retinal
Retinal isomerase
Retinal oxidase
Retinal reductase
9-*cis*-Retinoic acid
11-*cis*-Retinoic acid
13-*cis*-Retinoic acid
all-*trans*-Retinoic acid
all-*trans*-Retinyl phosphate
apo-RBP4
Retinoic acid receptors (RARs)
Retinoic acid response elements (RAREs)
Retinoids
Retinoid X receptors (RXRs)
all-*trans*-Retinol
13-*cis*-Retinol
Retinol dehydrogenases
Retinol equivalents (RE)
Retinol phosphorylase
Retinyl ester hydrolase
Retinyl β-glucuronide
Retinyl acetate
Retinyl palmitate
Retinyl phosphate
Retinyl stearate
Rhodopsin
STRA6
Thyroid hormone (T_3)
Transducin
Transgenic
Transthyretin
Very low-density lipoproteins (VLDLs)
Xerophthalmia
Xerosis
Zeaxanthin

1. SIGNIFICANCE OF VITAMIN A

Vitamin A is a nutrient of global importance. More than 254 million people are estimated to have deficient serum retinol levels, i.e., <0.7 μM and/or related eye disease. This includes 69% of people in southeast Asia, 49% of people in Africa, and a more out of every four to five living in other regions. Vitamin A deficiency remains the single most important cause of childhood blindness in developing countries. In the mid-1990s, nearly 14 million preschool children (three-quarters from south Asia) were estimated to have clinical eye disease due to vitamin A deficiency. Within a decade that prevalence had declined; yet, more than 5 million children remained affected. If untreated, two-thirds of such children die within months of going blind, due to increased susceptibility to infectious diseases (e.g., measles, severe diarrhea, dysentery, and respiratory diseases) caused by the deficiency. Yet, vitamin A supplementation has been consistently effective[2] in reducing mortality in at-risk children; so much so that vitamin A intervention has been removed from the realm of research and into that of programming.[3]

Despite these gains, the prevalence of subclinical deficiency (serum retinol levels <0.7 μM) has increased. One-third of the world's preschool children appear to be growing up with insufficient vitamin A (Table 6.1). More than 19 million pregnant women in developing countries are vitamin A-deficient; one-third is affected by night blindness.

Subclinical vitamin A deficiency is also associated with increased child mortality in at least 122 countries in Africa, southern and Southeast Asia, and some parts of Latin America and the western Pacific. High rates of morbidity and mortality have long been associated with vitamin A deficiency; recent intervention trials have indicated that

2. 200,000 IU as retinyl palmitate every 6 months.
3. A meta-analysis of 17 clinical trials showed vitamin A supplementation to reduce mortality by an average of 24% (Imdad, A., Herzer, K., Mayo-Wilson, E., et al., 2010. Cochrane Database Syst. Rev. CD008524).

TABLE 6.1 Global Prevalence of Vitamin A Deficiency Among Preschool Children and Pregnant Women

	Children (0–5 years)		Pregnant Women	
Region	**Night Blindness % (millions)**	**Low-Serum Retinol[a] % (millions)**	**Night Blindness (millions)**	**Low Serum Retinol[a] % (millions)**
Africa	2.0 (2.55)	44.4 (56.4)	9.8 (3.02)	13.5 (4.18)
Americas	0.6 (0.36)	15.6 (8.68)	4.4 (0.50)	2.0 (0.23)
Southeast Asia	0.5 (1.01)	49.9 (91.5)	9.9 (3.84)	17.3 (6.69)
Europe	0.8 (0.24)	19.7 (5.81)	3.5 (0.22)	11.6 (0.72)
Eastern Mediterranean	1.2 (0.77)	20.4 (13.2)	7.2 (1.09)	16.1 (2.42)
Western Pacific (including China)	0.2 (0.26)	12.9 (14.3)	4.8 (1.09)	21.5 (4.90)
Global	0.9 (5.17)	33.3 (190)	7.8 (9.75)	15.3 (19.1)

[a]*Serum retinol <0.7 μM.*
After WHO Global Database on Vitamin A Deficiency, 2009. Global Prevalence of Vitamin A Deficiency in Populations at Risk 1995–2005, WHO, Geneva, 55 pp.

providing vitamin A can reduce child mortality by about 25% and birth-related, maternal mortality by 40%. Vitamin A deficiency in these areas does not necessarily imply insufficient national or regional supplies of food vitamin A, as vitamin A deficiency can also be caused by insufficient dietary intakes of protein, fats, and oils. Still, most studies indicate that children with histories of xerophthalmia consume fewer dark green leafy vegetables than their counterparts without such histories.

2. PROPERTIES OF VITAMIN A

Vitamin A is the generic descriptor for compounds with the qualitative biological activity of retinol. These compounds are formally derived from a monocyclic parent with five carbon–carbon double bonds and a functional group at the terminus of the acyclic portion. Due to their structural similarities to retinol, they are called **retinoids**. Those with vitamin A activity have the following features of chemical structure:

- a substituted β-ionone nucleus [4-(2,6,6-trimethyl-2-cyclohexen-1-yl)-3-buten-2-one]; a side chain composed of three isoprenoid units joined head to tail at the 6-position of the β-ionone nucleus;
- a conjugated double-bond system among the side chain and 5,6-nucleus carbon atoms.

All three basic forms (retinol, retinal, and retinoic acid) can occur as two variants: with the β-ionone nucleus (vitamin A_1) or with the dehydrogenated β-ionone nucleus (vitamin A_2). Because the former is both quantitatively and qualitatively more important as a source of vitamin A activity, the term vitamin A typically refers to vitamin A_1.

Chemical structures of the vitamin A group:

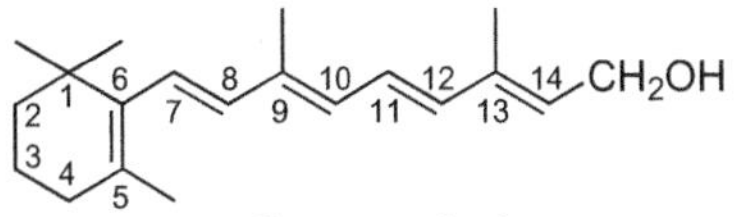

all-*trans*-retinol

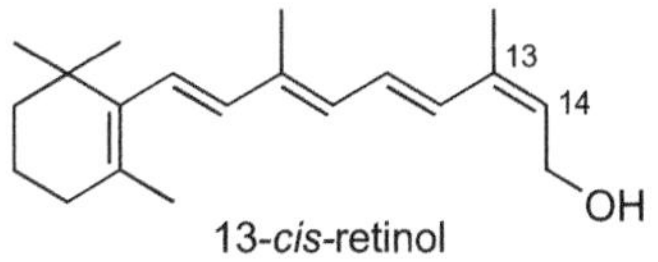

13-*cis*-retinol

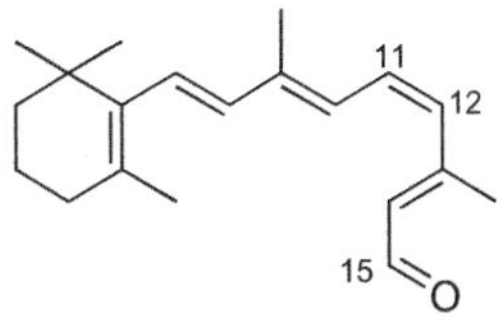

11-*cis*-retinal

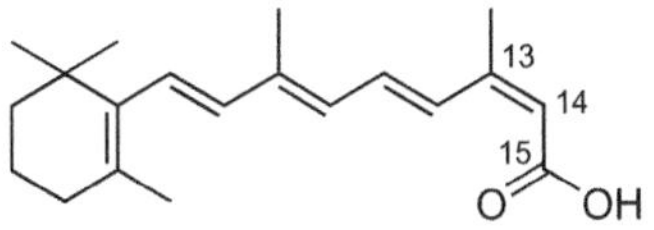

13-*cis*-retinoic acid

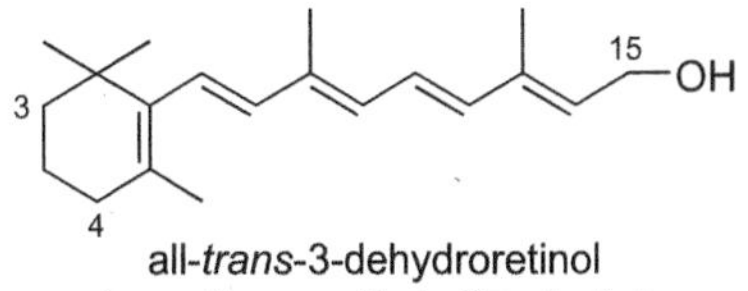

all-*trans*-3-dehydroretinol
(sometimes called **vitamin A_2**)

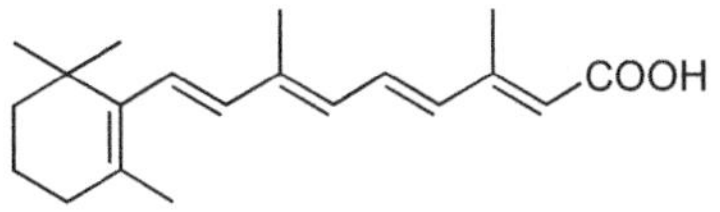

all-*trans*-retinoic acid

Vitamin A-active retinoids occur in three forms:
the alcohol... **retinol**
the aldehyde... **retinal** (also *retinaldehyde*)
the acid... **retinoic acid**.

Some compounds of the class of polyisoprenoid plant pigments called **carotenoids**, owing to their relation to the carotenes, yield retinoids metabolically and, thus, also have vitamin A activity. These **provitamin A** carotenoids include β-carotene, a tail-conjoined retinoid dimer.

Chemical structures of provitamins A:

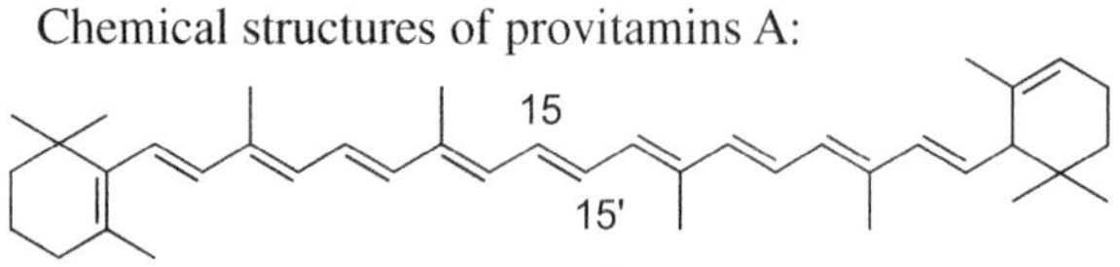

α-carotene

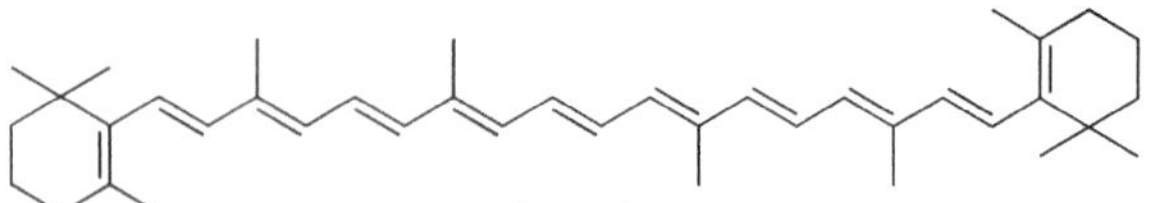
β-carotene

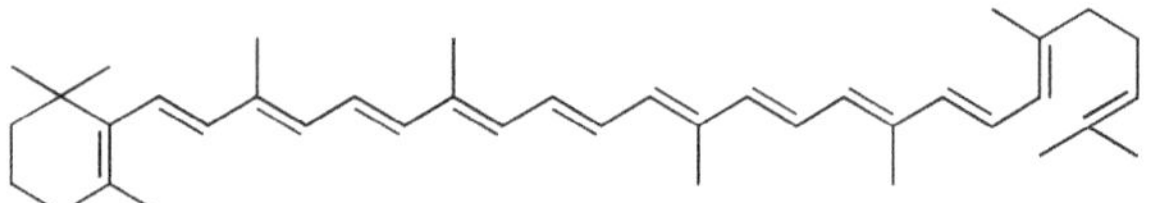
γ-carotene

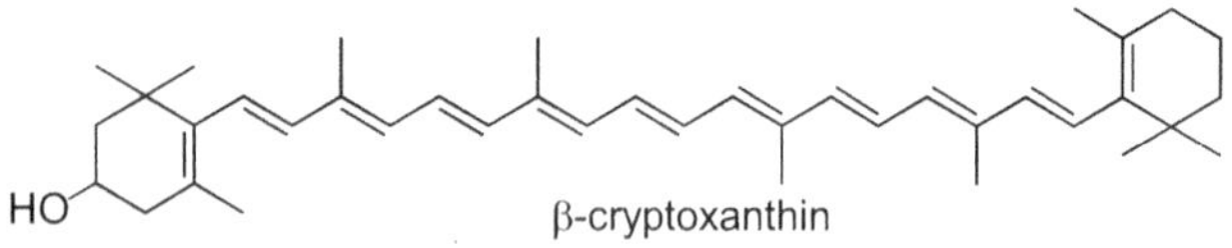

β-cryptoxanthin

Chemical Properties of Vitamin A

Of the 16 stereoisomers of vitamin A made possible by the four side chain double bonds, most of the potential *cis* isomers are sterically hindered; thus, only a few isomers are known. In solution, retinoids and carotenoids can undergo slow conversion by light, heat, and iodine through *cis–trans* isomerism of the side chain double bonds (e.g., in aqueous solution, all-*trans*-retinol spontaneously isomerizes to an equilibrium mixture containing one-third *cis* forms).

Contrary to what might be expected by their larger number of double bonds, carotenoids in both plants and animals occur almost exclusively in the all-*trans* form. These conjugated polyene systems absorb light, and, in the case of the carotenoids, appear to quench free radicals weakly. For the retinoids, the functional group at position 15 determines specific chemical reactivity. Thus, retinol can be oxidized to retinal and retinoic acid or esterified with organic acids; retinal can be oxidized to retinoic acid or reduced to retinol; and retinoic acid can be esterified with organic alcohols. Retinol and retinal each undergo color reactions with such reagents as antimony trichloride, trifluoroacetic acid, and trichloroacetic acids, which were formerly used as the basis of their chemical analyses by the Carr–Price reaction.[4]

The vitamers A are insoluble in water, but soluble in ethanol, and freely soluble in organic solvents including fats and oils. Most are crystallizable but have low melting points (e.g., retinol, 62–64°C; retinal, 65°C). Both retinoids and carotenoids have strong absorption spectra. Vitamin A and the provitamin A carotenoids are very sensitive to oxygen in air, especially in the presence of light and heat; therefore, isolation of these compounds requires the exclusion of air (e.g., sparging with an inert gas) and the presence of a protective antioxidant (e.g., α-tocopherol). The esterified retinoids and carotenoids in native plant matrices are fairly stable.

3. SOURCES OF VITAMIN A

Dietary Sources of Vitamin A

Vitamin A exists in natural products in many different forms. It exists as preformed **retinoids**, which are stored in animal tissues, and as provitamin A **carotenoids**, which are synthesized as pigments by many plants and are found in green, orange, and yellow plant tissues. In milk, meat, and eggs, vitamin A exists in several forms, mainly as long-chain fatty acid esters of **retinol**, the predominant one being **retinyl palmitate**.

Foods. Provitamin A carotenoids are present in both plant and animal food products; in animal products their occurrence results from dietary exposure. Carotenoid pigments are widespread among diverse animal species, with more than 500 different compounds estimated. About 60 of these have provitamin A activity, i.e., those that can be cleaved by animals to yield at least one molecule of retinol.[5] In practice, however, only five or six of these provitamins A are commonly encountered in foods.

Of the some 600 carotenoids in nature, only about 50 have provitamin A activity—those that can be cleaved metabolically to yield at least one molecule of retinol. A half-dozen of these are common in foods.

Therefore, actual vitamin A intakes depend on the patterns of consumption of vitamin A-bearing animal food products and provitamin A-bearing fruits and vegetables

4. Reaction of antimony trichloride with vitamin A in chloroform, which yields a quantifiable blue color.
5. These are synthesized by plants from isopentyl diphosphate and its isomer dimethylallyl diphosphate. The condensation of those precursors (in a 3:1 ratio, respectively) yields geranylgeranyl pyrophosphate two molecules of which are condensed to form phytoene, a colorless 40-carbon tetraterpenoid. Phytoene undergoes a series of desaturation and isomerization reactions to yield all-*trans*-lycopene, which is cyclized to generate β-carotene the hydroxylation which yields α-carotene and β-cryptoxanthin.

TABLE 6.2 Sources of Vitamin A in Foods

	Percentage Distribution of Vitamin A Activity		
Food	**Retinol**	**β-Carotene**	**Non-β-carotenoids**
Animal Foods			
Red meats	90	10	
Poultry meat	90	10	
Fish and shellfish	90	10	
Eggs	90	10	
Milk, milk products	70	30	
Fats and oils	90	10	
Plant Foods			
Maize, yellow		40	60
Legumes and seeds		50	50
Green vegetables		75	25
Yellow vegetables[a]		85	15
Pale sweet potatoes		50	50
Yellow fruits[b]		85	15
Other fruits		75	25
Red palm oil		65	35
Other vegetable oils		50	50

[a]*e.g., Carrots and deep-orange sweet potatoes.*
[b]*e.g., Apricots.*
After Leung, W., Flores, M., 1980. Food Composition Table for Use in Latin America. Institute of Nutrition of Central America and Panama, Guatemala City, Guatemala; and Interdepartmental Committee on Nutrition for National Defense, Washington, DC.

(Table 6.2), the relative contributions of which are influenced by food availability and personal food habits. Nursing infants consume both preformed vitamin A and provitamin A, especially if their mothers have adequate vitamin A intakes. Two-thirds of the vitamin A consumed by American omnivores come from carrots, organ meats, fortified breakfast cereals, cheese, margarine, tomatoes, and eggs. Sixty percent of the vitamin A consumed in the Netherlands comes from meats, fats, and oils. Half of the vitamin A consumed by Inuits comes from the livers of fish, seals, and caribou. Most of the vitamin A in the diets of vegetarians and of individuals in low-income countries comes from plant foods (red palm oil, dark/medium green leaves, yellow/orange fruits, and yellow maize).

Fortified foods. Certain foods are fortified with retinyl esters in many countries: milk, margarine, formula foods, and in some cases wheat flour. In the United States, this practice is regulated by the Food and Drug Administration. The use of agricultural technologies to enhance the micronutrient contents of staple foods—an effort referred to as **biofortification** (*see* Chapter 20, Sources of the Vitamins. V. Biofortification). This has involved the use of molecular biological techniques to produce a rice variety ("Golden Rice") containing appreciable amounts (>35 µg/g) of β-carotene, and traditional plant breeding to make substantial improvements in the β-carotene contents of several crops: high β-carotene carrot, orange-fleshed sweet potato, yellow cassava, and high-β-carotene maize.

Breast milk. Breast milk is the key source of vitamin A for the nursing infant. Retinoid and carotenoid contents of milk depend on the stage of lactation and the vitamin A status of the mother, the patterns of carotenoids in colostrum tend to reflect those in maternal low-density lipoproteins (LDLs), while patterns in mature milk (19 days) reflect those in maternal high-density lipoproteins (HDLs).[6] Breast milk from well nourished, vitamin A-adequate mothers typically drops from c.5–7 µM in colostrum, to c.3–5 µM in transitional milk, to 1.4–2.6 µM in mature milk. These levels are enough to meet the infant's immediate metabolic needs while also supporting the development of adequate

6. Schwiegert, F.J., Bathe, K., Chen, F., et al., 2004. Eur. J. Nutr. 43, 39–44.

vitamin A stores.[7] Such an infant will consume over the first 6 months of life, nearly 60 times as much vitamin A from breast milk (c.300 µmoles) than it accumulated throughout gestation. Vitamin A-deficient mothers, however, produce breast milk that is low in the vitamin; in vitamin A-deficient areas of the world, levels average c.1 µM (levels <1.05 µM/L are considered indicative of maternal vitamin A deficiency[8]). This level appears to be sufficient to meet an infant's immediate metabolic requirements, but higher levels (at least 1.75 µM) are required to support adequate vitamin A stores to protect against the development of xerophthalmia during weaning.

Microbiome. There is no evidence for biosynthesis of vitamin A by the gut microbiome; however, a recent study the possibility of producing a vitamin A-producing probiotic. Wassaf et al.[9] inserted plant genes coding for four key enzymes in the β-carotene biosynthetic pathway into an intestine-adapted mutant strain of *E. coli*. When fed to mice lacking the capacity to cleave β-carotene, the altered bacteria was increased the β-carotene contents of host plasma and liver.

Bioavailability

Although 1 mol of β-carotene can, in theory, be converted (by cleavage of the C–15=C–15′ bond) to 2 mol of retinol, the physiological efficiency of this process appears to be much less. Until recently, the efficiency of bioconversion of β-carotene to retinol was assumed to be about 50%, and that of intestinal absorption was assumed to be about 33%; thus, β-carotene was regarded as having one-sixth the vitamin A value of retinol. Accordingly, carotenoids that yield only 1 mol of retinol metabolically were regarded as having one-twelfth the vitamin A value of retinol. This logic was the basis for older dietary recommendations, which used equivalency ratios of 6:1 and 12:1 in setting retinol equivalency values for provitamin A carotenoids in supplements and foods, respectively.

Subsequent research has shown that the bioconversion of food carotenoids to vitamin A can vary considerably (10–90%) (Table 6.3). Isotope dilution studies have shown purified β-carotene in oils and nutritional supplements to be utilized at about half the efficiency of retinol, while β-carotene from plant foods is utilized with much lower efficiencies than previously thought. Retinol equivalency ratios from 3.8:1 to 28:1 have been reported for humans. Low bioconversion appears to be particularly true in resource-poor countries in which children rely almost entirely on the conversion of β-carotene from fruits and vegetables for their vitamin A.[10] In addition to vitamin A status, which inversely affects carotenoid conversion, several other factors can reduce conversion efficiency: low fat intakes, intestinal roundworms, recurrent diarrhea, tropical enteropathy, and other factors that affect the absorptive function of the intestinal epithelium and intestinal transit time. Accordingly, West and colleagues have suggested the use of ratios of 21:1 for mixed diets (12:1 for fruits and 26:1 for vegetables) in such contexts.[11] While the supporting data are sparse, in 2001 the IOM revised its estimates of the vitamin A biopotency of carotenoids to the figures presented in Table 6.4.

7. The normal weight (c.3.2 kg) infant of a well-nourished, vitamin A-adequate, mother is born with hepatic vitamin A stores of c.5 µmoles.
8. Stoltzfus, R.J., Underwood, B.A., 1995. Bull. WHO 73, 703–711.
9. Wassaf, L., Wirawan, R., Chikindas, M., et al., 2014. J. Nutr. 144, 608–613.

Expressing Vitamin A Activities

Because vitamin A exists in foods and supplements in many different forms of differing biopotencies, the reporting of vitamin A activity in foods requires some means of standardization. Three systems are used for this purpose: **retinol equivalents (RE)**[12] for food applications and **international units (IU)** for pharmaceutical applications.

For reporting food vitamin A activity—retinol equivalents (or retinol activity equivalents, RAE)[13]

1 RE = 1 µg all-*trans*-retinol.
= 2 µg all-*trans*-β-carotene in dietary supplements.
= 12 µg all-*trans*-β-carotene in foods.
= 24 (12–26) µg other provitamin A carotenoids (α-carotene, β-cryptoxanthin) in foods[3]

For pharmaceutical applications—international units.

1 IU[14] = 0.3 µg all-*trans*-retinol.
= 0.344 µg all-*trans*-retinyl acetate.
= 0.55 µg all-*trans*-retinyl palmitate.

10. de Pee, S., West, C.E., Muhilal, X., et al., 1995. Lancet 346, 75–81.
11. The data of van Lieshout, M., West, C.E., van Breeman, R.B., 2001. Am. J. Clin. Nutr. 77, 12–28 suggest a ratio of 2.6 based on stable isotope (or circulating retinol) dilution studies in more than a hundred Indonesian children. Other isotope dilution studies show the conversion of β-carotene to retinol by humans to be quite variable: from individual equivalency ratios of 2:1 to 12:1. (Wang, Z., Yin, S., Zhao, X. et al., 2004. Br. J. Nutr. 91:121–131; Haskell, M.J., Jamil, K.M., Hassan, F., et al., 2004. Am. J. Clin. Nutr. 80:705–714), and from 6:1 to 13:1. De Pee (de Pee, S, West, K.P., Permaesih, D., et al., 1998. Am. J. Clin. Nutr. 68, 1058–1067) suggested that in developing countries this conversion ratio may be as great as 21:1. This implies limits to the contributions of horticultural approaches (e.g., "home-gardening" programs) to solving problems of vitamin A deficiency.
12. These equivalencies were established by the Food and Nutrition Board in 2001. USDA National Nutrient Database for Standard Reference lists both IU and RE; FAO tables list µg of retinol and β-carotene. INCAP (Instituto de Nutrición de Centroamérica y Panamá) tables list vitamin A as µg retinol; those values are not the same as RE values, as a factor of 0.5 was used to convert β-carotene to RE.
13. Proposed in 2001 by the Institute of Medicine, US National Academy of Sciences.
14. Sometimes called USP Unit, as it was adopted by the United States Pharmacopeia.

TABLE 6.3 Apparent Uptake Without Conversion to Vitamin A of Major Food Provitamin A Carotenoids in Humans

	α-Carotene	β-Carotene	β-Cryptoxanthin
Estimated dietary intake (μM/day)	1.18±0.12	6.79±0.36	0.45±0.04
Plasma concentration (μM)	0.10±0.01	0.40±0.04	0.16±0.02
Relative utilization as vitamin A (vs. β-carotene)	0.60	1.00	0.14

After Pooled analysis of several studies; Burri, B.J., Chang, J.S.T., Neidlinger, T.R., 2011. Br. J. Nutr. 105, 212–219.

TABLE 6.4 Relative Biopotencies of Vitamin A and Related Compounds

Compound	Relative Biopotency[a]
All-*trans*-Retinol	100
All-*trans*-Retinal	100
cis-Retinol isomers	23–75
Retinyl esters	10–100
3-Dehydrovitamin A	30
β-Carotene	50
α-Carotene	26
γ-Carotene	21
β–Cryptoxanthin	28
Zeaxanthin	0

[a]*Most relative biopotencies were determined by liver storage bioassays with chicks and/or rats. In the case of 3-dehydrovitamin A, biopotency was assessed using liver storage by fish. In each case, the responses were standardized to that of all-trans-retinol.*

Foods Rich in Vitamin A

Several foods contain vitamin A activity (Table 6.5). It is estimated that carotene from vegetables contributes two-thirds of dietary vitamin A worldwide and more than 80% in developing countries. Other than green and yellow vegetables, few other foods are rich sources of vitamin A, those being liver, oily fishes, and vitamin A-fortified products such as margarine. It should be noted that, for vitamin A and other vitamins that are susceptible to breakdown during storage and cooking, values given in food composition tables are probably high estimates of amounts actually encountered in practical circumstances.

4. ABSORPTION OF VITAMIN A

Absorption of Retinoids

Most of the preformed vitamin A in the diet is in the form of **retinyl esters**, but only free retinoids appear to be taken up by the enterocyte. The absorption occurs in three steps:

1. **Hydrolysis of esters**. Retinyl esters are hydrolyzed in the stomach and lumen of the small intestine to yield retinol; this step is catalyzed by hydrolases produced by the gastric lining, pancreatic lipases situated on the mucosal brush border,[15] and esterases intrinsic to the mucosal brush border.
2. **Micellar solubilization**. The retinoids, being hydrophobic, depend on micellar solubilization for their dispersion in the aqueous environment of the small intestinal lumen. For this reason, they likely to be poorly utilized from low-fat diets. The absorption of retinol esters appears to be fairly high (75–100%); the process is appreciably less efficient at very high vitamin A doses.
3. **Mucosal uptake**
   a. **Lymphatic uptake**. Retinol is taken up by mucosal cells by a saturable process thought to involve the multidomain transmembrane protein **STRA6**,[16] and/or **retinoid-binding protein 2 (RBP2)**.
   b. **Nonlymphatic uptake**. Studies have shown that retinoids can also be absorbed via nonlymphatic pathways. Rats with ligated thoracic ducts retain the ability to accumulate vitamin A in their livers. That such animals fed retinyl esters show greater concentrations of retinol in their portal blood than in their aortic blood suggests that, in mammals, the portal system may be an important alternative route of vitamin A absorption when the normal lymphatic pathway is blocked. This phenomenon corresponds to the route of vitamin A absorption in birds, fishes, and reptiles, which, lacking lymphatic drainage of the intestine, rely strictly on portal absorption.

15. One of these activities appears to be the same enzyme that catalyzes the intralumenal hydrolysis of cholesteryl esters; it is a relatively nonspecific carboxylic ester hydrolase. It has been given various names in the literature, the most common being **pancreatic nonspecific lipase** and **cholesteryl esterase**.
16. This protein was named because its expression is stimulated by retinoic acid.

TABLE 6.5 Vitamin A Contents of Foods

Food	Vitamin A (IUg/100 g)
Grains	
Cornmeal	214
Oats	0
Rice	0
Wheat flour	9
Wheat bran	9
Vegetables	
Asparagus	1006
Beans, green	633
Broccoli	623
Cabbage	98
Carrots	16,700
Cauliflower	0
Kale	13,600
Peas, green	765
Potatoes	0
Tomatoes	830
Fruits	
Apples	54
Apricots	1925
Bananas	64
Grapes	100
Oranges	250
Pears	25
Pineapples	38
Meats	
Beef	0–37
Chicken	80–200
Duck	80–210
Pork	0–37
Trout	50
Salmon	55–195
Liver	
Beef liver	16,900
Pork liver	18,000
Dairy Products and Eggs	
Cheese	8–1240
Milk	160–200
Eggs	540
Other	
Human milk	210

After USDA National Nutrient Database for Standard Reference, Release 28 (http://www.ars.usda.gov/ba/bhnrc/ndl).

Absorption of Provitamin A Carotenoids

The major sources of vitamin A activity for most populations are the provitamin A carotenoids. Their utilization involves three steps:

1. **Release from food matrices**. A major factor limiting the utilization of carotenoids from food sources is their release from physical food matrices. Carotenoids can occur in cytosolic crystalline complexes or in chromoplasts and chloroplasts, where they are associated with proteins, polysaccharides, fibers, and phenolic compounds. Their release from chromoplasts appears to occur more readily than from chloroplasts and is facilitated by the presence of lipid. Many carotenoid complexes are resistant to digestion without heat treatment; therefore, cooking tends to improve the bioavailability of provitamin A compounds in plant food.
2. **Micellar solubilization**. The enteric absorption of carotenoids depends on their solubilization in mixed lipid micelles the formation of which requires the consumption and digestion of lipids. Absorption, particularly of the less polar carotenoids, can be impaired by the presence of undigested lipids or sucrose polyesters in the intestinal lumen. Gastric acidity may also be an affector, as patients with pharmaceutically obliterated gastric acid production showed reduced blood responses to test doses of β-carotene.[17] Carotenoids (and likely retinoids) are not well absorbed from low-fat meals.[18] One study found that as little as 3–5 g fat per meal may be sufficient for optimal absorption of provitamin A carotenoids,[19] although higher amounts of fat are likely to be necessary to render vitamin A accessible from plant matrices, e.g., at least 10 g/day.
3. **Mucosal uptake**. Uptake of carotenoids from micelles has been thought to involve the diffusion directly through the plasma membranes of the enterocytes; however, the process may actually be carrier-mediated. Reboul has pointed out that careful studies have demonstrated saturable uptake, and that passive diffusion cannot explain either the high interindividual variability in carotenoid absorption observed in humans (5–65%), or the antagonism of carotenoid absorption by tocopherols and other carotenoids.[20] Mucosal uptake appears to be impaired by soluble dietary fiber and, likely, other factors that interfere with the contact of the micelle with the mucosal brush

17. This finding has implications for millions of people, as atrophic gastritis and hypochlorhydria are common, particularly among older people.
18. In fact, Brown and colleagues found the use of fat-free dressing to completely block the absorption of β-carotene from fresh vegetable salad. (Brown, M.J., Ferruzi, M.G., Nguyen, M.L., et al., 2004. Am. J. Clin. Nutr. 80, 396–403).
19. Roodenburg, A.J., Leenen, R., van het Hof, K.H., et al., 2000. Am. J. Clin. Nutr. 71, 1187–1193.
20. Raboul, E., 2013. Nutrients 5, 3563–3581.

border. It is likely that lipid transporters may facilitate carotenoid uptake.[21]

Provitamin A Carotenoid Metabolism Linked to Absorption

While β-carotene can cross the mucosal epithelial cell intact, most is metabolized within the cell. Carotenoid absorption typically results in the accumulation in enterocytes of more all-*trans*-β-carotene than 9-*cis*-β-carotene. This suggests enterocytic capacity for *cis–trans* isomerization, which is also indicated by the fact that humans given 9-*cis*-β-carotene show detectable levels of 9-*trans*-retinol in their plasma. The capacity for isomerization would serve to limit the distribution of 9-*cis*-retinoids to tissues and render both isomers of β-carotene capable of being metabolized to **retinal**, thus serving as effective provitamins A.

Fewer than 10% of naturally occurring carotenoids are provitamins A, those capable of yielding retinal upon hydrolysis. This metabolism is catalyzed by **β-carotene oxygenases (BCOs)**[22] (Fig. 6.1). Most of this bioconversion occurs via the central cleavage of the polyene moiety by a predominantly cytosolic enzyme, **β-carotene 15,15′-oxygenase (BCO1)**, found in the intestinal mucosa, liver, and corpus luteum. BCO1, sometimes also called **carotene cleavage enzyme**, cleaves β-carotene into two molecules of retinal. Several variants of BCO1 have been identified; these vary in specific activity by as much as 100%. It contains ferrous iron (Fe^{++}) linked to a histidinyl residue at the axis of a seven-bladed, β-propeller chain fold covered by a dome structure formed by six large loops in the protein. Upon binding within that structure, the three consecutive *trans* double bonds of the carotenoid are isomerized to a *cis-trans-cis* conformation, leading to the oxygen cleavage of the central *trans* bond. Accordingly, the activity of BCO1 can be affected by intakes of iron and factors affecting iron utilization (e.g., copper, fructose). Expression of BCO1 is repressed by the intestinal transcription factor ISX, which is induced by retinoic acid. This factor also appears to repress the expression of a receptor (scavenger receptor B type 1, SR-B1) thought to facilitate the intestinal absorption of lipids including β-carotene. By this mechanism, both the absorption of β-carotene, as well as its cleavage to produce retinal, are reduced under conditions of vitamin A adequacy.

However, the BCOs are not highly specific for β-carotene. Then can cleave other carotenoids; those have provitamin A activities to the extent that they can also yield retinal. Apocarotenals yield retinal; epoxy carotenoids are

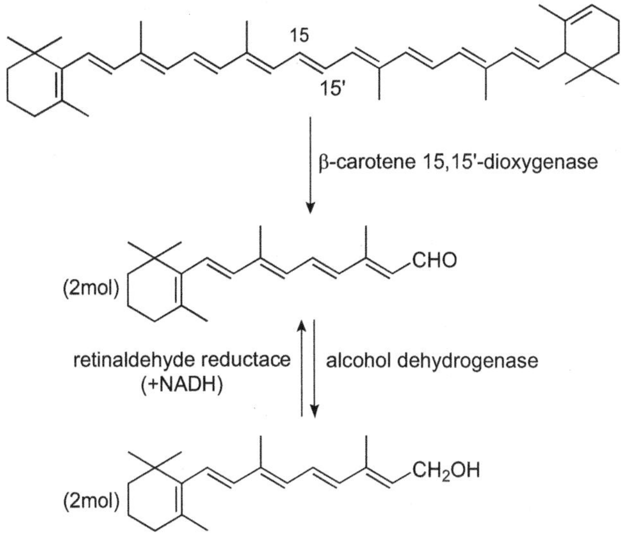

FIGURE 6.1 Bioconversion of provitamins A to retinal.

not metabolized. The reaction requires molecular oxygen, which reacts with the two central carbons (C-15 and C-15′), followed by cleavage of the C—C bond. It is inhibited by sulfhydryl group inhibitors and by chelators of ferrous iron (Fe^{++}). The enzyme has been found in a wide variety of animal species;[23] enzyme activities were found to be greatest in herbivores (e.g., guinea pig, rabbit), intermediate in omnivores (e.g., chicken, tortoise, fish), and absent in the only carnivore studied (cat). The enzyme activity is enhanced by the consumption of triglycerides,[24] suggesting that its regulation involves fatty acids. It is diminished by high intakes of β-carotene and protein deprivation, is induced by vitamin A deficiency, and can be inhibited by quercetin and other flavonols. The symmetric, central cleavage of β-carotene is highly variable between individuals (35–90%). In the bovine corpus luteum, which also contains a high amount of β-carotene, BCO1 activity has been shown to vary with the estrous cycle, showing a maximum on the day of ovulation. Studies with the rat indicate that the activity is stimulated by vitamin A deprivation and reduced by dietary protein restriction.

Low BCO1 activities are associated with the absorption of intact carotenoids; this phenomenon is responsible for the yellow-colored adipose tissue, caused by the deposition of absorbed carotenoids, in cattle. Thus, at low doses β-carotene is essentially quantitatively converted to vitamin A by rodents, pigs, and chicks; cats, in contrast, cannot perform the conversion, and therefore β-carotene cannot support their vitamin A needs.

The asymmetric cleavage of carotenoids also occurs by a second intestinal mucosal enzyme, β-carotene oxygenase

21. Candidates include two class B scavenger receptors, SR-BI and cluster determinant 36 (CD36); a cholesterol transporter, Neimann–Pick like C1 protein (NPV1L1); and a gut-specific transcription factor ISX, which may repress SR-BI expression in that organ.

22. More than 100 enzymes in this group are known. Two occur in animals: β-carotene-15,15'-oxygenase and β-carotene-9′,10′-oxygenase.

23. The β-carotene 15,15′-dioxygenase has also been identified in *Halobacterium halobium* and related halobacteria, which use retinal, coupled with an opsin-like protein, to form bacteriorhodopsin, an energy-generating, light-dependent proton pump.

24. Which also increase CRBP(II) levels.

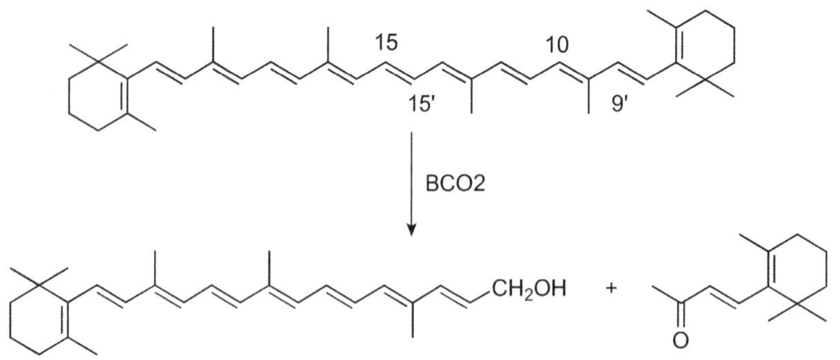

FIGURE 6.2 Asymmetric (or eccentric) cleavage of β-carotene by the β-carotene oxygenase 2 (BCO2) yields apo-10′-β-carotenal and β-ionone.

2 (BCO2), although this appears to be a quantitatively minor pathway. This enzyme cleaves the carotene-9′, 10′-bond to form apo-10′-β-carotenal (Fig 6.2), which can be chain-shortened directly to yield retinal or, first, oxidized to the corresponding apocarotenoic acids and, then, chain-shortened to yield **retinoic acid**.[25] It has been suggested that BCO2 functions in the metabolism of acylic, nonprovitamin A carotenoids such as lycopene, which accumulates when the enzyme is not expressed. Intestinal enzymes can cleave 9-*cis*-β-carotene (which comprises 8–20% of the β-carotene in fruits and vegetables but seems less well utilized than the all-*trans* isomer) to 9-*cis*-retinal which, in turn, appears to be oxidized to 9-*cis*-retinoic acid.

The turnover of carotenoids in the body occurs via first-order mechanisms that differ for individual carotenoids. For example, in humans the biological half-life of β-carotene has been determined to be 37 days, whereas those of other carotenoids vary from 26 days (lycopene) to 76 days (lutein).

Mucosal Metabolism of Retinol

Retinol formed either from the hydrolysis of dietary retinyl esters or from the reduction of retinal cleaved from β-carotene[26] is absorbed by facilitated diffusion via a specific transporter.[27] Retinal produced by the central cleavage of β-carotene is reduced in the intestinal mucosa to retinol (Fig. 6.1) by **retinaldehyde reductase**, which is also found in the liver and eye. The reduction requires a reduced pyridine nucleotide (NADH/NADPH) as a cofactor and has an apparent K_m of 20 mM. It can also be catalyzed by a **short-chain alcohol dehydrogenase/aldehyde reductase**, and there is some debate concerning whether the two activities reside on the same enzyme.

Retinol is quickly reesterified with long-chain fatty acids in the intestinal mucosa whereupon retinyl esters are transported to the liver (i.e., 80–90% of a retinol dose[28]). The composition of lymph retinyl esters is remarkably independent of the fatty acid composition of the most recent meal. **Retinyl palmitate** typically comprises about half of the total esters, with **retinyl stearate** comprising about a quarter and retinyl oleate/linoleate present in small amounts. Two pathways for the enzymatic reesterification of retinol have been identified in the microsomal fraction of the intestinal mucosa (Fig. 6.3): a low-affinity route involving uncomplexed retinol and catalyzed by **acyl-CoA:retinol acyltransferase (ARAT)**; a high-affinity route involving retinol complexed with a specific binding protein, **cellular retinol-binding protein type II [CRBP(II)[29]]** and catalyzed by **lecithin–retinol acyltransferase** (**LRAT**). The expression of LRAT mRNA is induced by retinoic acid and downregulated by vitamin A depletion. It has been suggested that LRAT serves to esterify low doses of retinol, whereas ARAT serves to esterify excess retinol, when CRBP(II) becomes saturated. The identification of a retinoic acid-responsive element in the promoter region of the CRBP(II) gene suggests that the transcription of that gene may be positively regulated by retinoic acid, leading to increased CRBP(II) levels at high vitamin A doses. Experiments have shown that CRBP(II) expression is enhanced under conditions of stimulated absorption of fats, especially unsaturated fatty acids.

5. TRANSPORT OF VITAMIN A

Retinyl Esters Conveyed by Chylomicra in Lymph

Retinyl esters are secreted from the intestinal mucosal cells in the hydrophobic cores of chylomicron particles, by which

25. Studies of these processes are complicated by the inherent instability of carotenoids under aerobic conditions; many of the products thought to be produced enzymatically can also be produced by autoxidation.

26. It has estimated that humans convert 35–71% of absorbed β-carotene to retinyl esters.

27. This protein transports both all-*trans*-retinol and 3-dehydroretinol. Other retinoids appear to be taken up by enterocytes by passive diffusion.

28. Humans fed radiolabeled β-carotene absorbed some unchanged directly into the lymph, with only 60–70% of the label appearing in the retinyl ester fraction.

29. CRBP(II) is a 15.6 kDa protein that constitutes about 1% of the total soluble protein of the rat enterocyte.

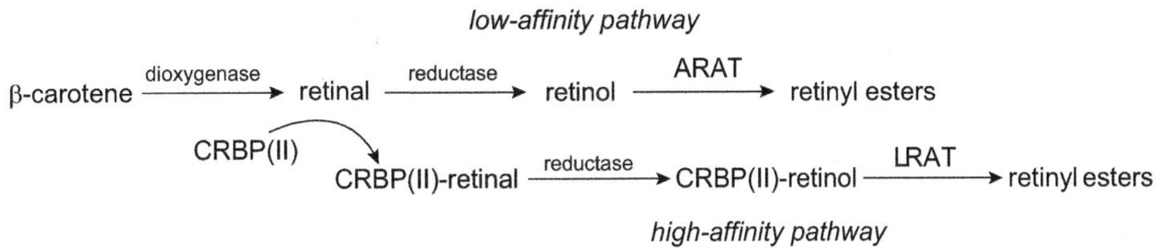

FIGURE 6.3 Intestinal metabolism of vitamin A.

TABLE 6.6 Distribution of Carotenoids in Human Lipoproteins

Carotenoid	VLDL (Very Low-Density Lipoprotein)	LDL (Low-Density Lipoprotein)	HDL (High-Density Lipoprotein)
Zeaxanthin/lutein (%)	16	31	53
Cryptoxanthin (%)	19	42	39
Lycopene (%)	10	73	17
α-Carotene (%)	16	58	26
β-Carotene (%)	11	67	22

After Reddy, P.P., Clevidance, B.A., Berlin, E., Taylor, P.R., Bieri, J.G., Smith, J.C., 1989. FASEB J. 3, A955.

absorbed vitamin A is transported to the liver through the lymphatic circulation, ultimately entering the plasma[30] compartment through the thoracic duct. Carnivorous species in general, and the dog in particular, typically show high plasma levels. Retinyl esters are almost quantitatively retained in the extrahepatic processing of chylomicra to their remnants; therefore, chylomicron remnants are richer in vitamin A than are chylomicra. Retinyl and cholesteryl esters can undergo exchange reactions between lipoproteins including chylomicra in rabbit and human plasma by virtue of a **cholesteryl ester transfer protein** peculiar to those species.[31] Although this kind of lipid transfer is probably physiologically important in those species, the demonstrable transfer involving chylomicra is unlikely to be a normal physiological process.

Transport of Provitamin A carotenoids. Carotenoids appear to be transported across the intestinal mucosa by a facilitated process similar to that of cholesterol. They are not metabolized in the epithelium but are transported from that organ by chylomicra via the lymphatic circulation to the liver, where they are transferred to lipoproteins. It is thought that strongly nonpolar species such as β-carotene and lycopene are dissolved in the chylomicron core; whereas species with polar functional groups may exist at least partially at the surface of the particle. Such differences in spatial distribution would be expected to affect transfer to lipoproteins during circulation and tissue uptake. Indeed, the distribution of carotenoids among the lipoprotein classes reflects their various physical characteristics, with the hydrocarbon carotenoids being transported primarily in LDLs and the more polar carotenoids being transported in a more evenly distributed manner among LDLs and HDLs (Table 6.6). It is thought that small amounts of the nonpolar carotenoids are transferred from chylomicron cores to HDLs during the lipolysis of the triglycerides carried by the former particles; however, because HDL transports only a small fraction of plasma β-carotene, the carrying capacity of the latter particles for hydrocarbon carotenoids would appear to be small. Therefore, it is thought that β-carotene is retained by the chylomicron remnants to be internalized by the liver for subsequent secretion in very low-density lipoproteins (VLDLs).

Abetalipoproteinemia. Absorption of vitamin A and other fat-soluble vitamins is a particular problem in patients with **abetalipoproteinemia**, a rare autosomal recessive disorder characterized by general lipid malabsorption, acanthosis (diffuse epidermal hyperplasia), and hypocholesterolemia. These patients lack apo B and consequently

30. On entering the plasma, chylomicra acquire apolipoproteins C and E from high-density lipoproteins (HDLs). Acquisition of one of these (apo C-II) activates lipoprotein lipase at the surface of extrahepatic capillary endothelia; that lipase hydrolyses the core triglycerides, causing them to shrink and transfer surface components (e.g., apo A-I, apo A-II, some phospholipid) to HDLs and fatty acids to serum albumin, and to lose apo A-IV and fatty acids to the plasma and other tissues. These processes leave a smaller particle, a **chylomicron remnant**, which is depleted of triglyceride but relatively enriched in cholesteryl esters, phospholipids, and proteins (including apo B and apo E). Chylomicron remnants are removed from the circulation almost entirely by the liver by a rapid, high-affinity receptor-mediated process stimulated by apo E.

31. This protein has not been found in several other mammalian species examined.

cannot synthesize any of the apo B-containing lipoproteins (i.e., LDLs, VLDLs, and chylomicra). Having no chylomicra, they show hypolipidemia and low plasma vitamin A levels. However, when given oral vitamin A supplementation, their plasma levels are normal. Although the basis of this response is not clear, it has been suggested that these patients can transport retinol from the absorptive cells via their remaining lipoprotein (HDLs), possibly by the portal circulation.

Vitamin: An Uptake by the Liver

Most of the recently absorbed vitamin A is taken up by the liver from chylomicron remnants, which hepatic parenchymal cells[32] remove from the circulation via a high-affinity receptor-mediated process stimulated by apo E.[33] Because this process is rapid and nearly quantitative, vitamin A (mostly as retinyl esters with smaller amounts of β-carotene) circulates in chylomicra only for a short time.[34] After being taken up by the liver retinyl esters are hydrolyzed to yield retinol in parenchymal cells from which it is transferred by a retinol-binding protein to stellate cells,[35] which also contain appreciable amounts of triglycerides, phospholipids, free fatty acids, and cholesterol. There it is re-esterified and stored in droplets (some that are membrane-bound and appear to be derived from lysosomes and other, larger ones not associated with membranes). It is likely that the reesterification of retinol proceeds by a reaction similar to that of the intestinal microsomal acyl CoA:retinol acyltransferase (ARAT). The liver thus serves as the primary storage depot for vitamin A, normally containing 50–80% of the total amount of the vitamin in the body.[36] Most of this (80–90%) is stored in stellate cells, which account for only about 2% of total liver volume. The balance stored in parenchymal cells. These are the only types of hepatocytes that contain retinyl ester hydrolase activities. Almost all (about 95%) of hepatic vitamin A occurs as long-chain retinyl esters, the predominant one being retinyl palmitate. Kinetic studies of vitamin A turnover indicate the presence, in both liver and extrahepatic tissues, of two effective pools (i.e., fast- and slow-turnover pools) of the vitamin. Of rat liver retinoids, 98% were in the slow-turnover pool (retinyl esters of stellate cells), with the balance corresponding to the retinyl esters of parenchymal cells.

In addition to retinol ester hydrolyases and ARAT, stellate cells contain two other retinoid-related proteins: **cellular retinoid-binding protein (CRBP)** and **cellular retinoic acid-binding protein (CRABP)**. The storage or retinyl esters appears not to depend on the expression of CRBP, as **transgenic** mice that overexpressed CRBP in several organs have not shown elevated vitamin A stores in those organs. The metabolism of vitamin A by hepatic cytosolic retinal dehydrogenase increases with increasing hepatic retinyl ester stores.

Mobilization from the liver. Vitamin A is mobilized as retinol from the liver by hydrolysis of hepatic retinyl esters.[37] This mobilization accounts for about 55% of the retinol discharged to the plasma, the balance coming from recycling from extrahepatic tissues. The retinyl ester hydrolase involved in this process remains poorly characterized; it shows extreme variation between individuals.[38] The activity of this enzyme is known to be low in protein-deficient animals and has been found to be inhibited, at least in vitro, by vitamins E and K.[39]

Extracellular transport. The transport of mobilized retinol from the liver to peripheral tissues is thought to depend on a specific carrier, **retinol-binding protein (RBP4)**.[40] Human RBP4 consists of a single polypeptide chain of 182 amino acid residues, with a molecular mass of 21 kDa. Like other RBPs,[41] it is classified as a member of the lipocalin family of lipid-binding proteins. These are composed of an eight-stranded, antiparallel β-sheet that is folded inward to form a hydrogen-bonded, β-barrel that comprises the ligand-binding domain the entrance of which is flanked by a single loop scaffold. Within this domain, a single molecule of all-*trans*-retinol is completely encapsulated, being stabilized by hydrophobic interactions of the β-ionone ring and the isoprenoid chain with the amino acids lining the interior of the barrel structure. This structure protects the vitamin from oxidation during transport. RBP4 is synthesized as a 24-kDa *pre-RBP4* by parenchymal cells, which also convert it to RBP4 by the cotranslational removal

32. The parenchymal cell is the predominant cell type of the liver, comprising more than 90% of organ volume.

33. Chylomicron remnants are recognized by high-affinity receptors for their apo E moiety.

34. Their remnants are degraded in hepatic parenchymal lysosomes.

35. Are also called **pericytes, fat-storing cells, interstitial cells, lipocytes, Ito cells,** or **vitamin A-storing cells**.

36. Mean hepatic stores have been reported in the range of 171–723 μg/g in children and 0–320 μg/g in adults (Panel on Micronutrients, Food and Nutrition Board [2002] Dietary Reference Intakes for Vitamin A, Vitamin K, Arsenic, Boron, Chromium, Copper, Iodine, Iron, Manganese, Molybdenum, Nickel, Silicon, Vanadium and Zinc. Washington, DC: National Academy Press, p. 95.).

37. Retinol oxidation also produces some retinoic acid most of which in the plasma is bound to albumen.

38. In the rat, hepatic retinyl ester hydrolase activities can vary by 50-fold among individuals and by 60-fold among different sections of the same liver.

39. Each vitamin has been shown to act as a competitive inhibitor of the hydrolase. This effect may explain the apparently impaired hepatic vitamin A mobilization by animals fed very high levels of vitamin E.

40. That oral α-retinol, a structural isomer of retinol not bound by RBP4, could support the deposition of that isomer in the liver and milk in lactating sows. (Dever, J.T., Surles, R.L. and Davis, C.R., 2010. J. Nutr. 141, 42–47) suggests RBP4-independent transport of the vitamin.

41. Cellular retinal- and retinoic acid-binding proteins.

of a 3.5-kDa polypeptide.[42] This protein product (apo-RBP4) is secreted in a 1:1 complex with all-*trans*-retinol (holo-RBP4). Stellate cells also contain low amounts of RBP4; however, it is not clear whether they synthesize it or their apo-RBP4 derives from parenchymal cells to mobilize retinol to the circulation. According to the latter view, stellate cells may be important in the control of retinol storage and mobilization, a complex process that is thought to involve retinoid-regulated expression of CRBPs.

The secretion of holo-RBP4 from the liver is regulated in women by estrogen level,[43] and in all individuals by vitamin A status (i.e., liver vitamin A stores), protein, and zinc status; deficiencies of each markedly reduce RBP4 secretion and, thus, reduce circulating levels of retinol (Table 6.7). In cases of protein–energy malnutrition, RBP4 levels (and, thus, serum retinol levels) can be decreased by as much as 50%. Except in the postprandial state, virtually all plasma vitamin A is bound to RBP4. In the plasma, almost all RBP4 forms a 1:1 complex with **transthyretin**[44] (TTR, a tetrameric, 55-kDa protein that strongly binds four thyroxine molecules). The formation of the relatively large RBP4–TTR complex reduces the loss of RBP4 by glomerular filtration.[45] This effect may also involve **megalin**,[46] which has been shown to bind the RBP4–TTR complex. The kidney appears to be the only site of RBP4 catabolism, which normally turns over rapidly, in 11–16h in humans.[47]

TABLE 6.7 Percentage Distribution of Vitamin A in Sera of Fasted Humans

Fraction	Retinol (%)	Retinyl Palmitate (%)
RBP4 (retinol-binding protein 4)	77	–
VLDL (very low-density lipoprotein)	6	71
LDL (low-density lipoprotein)	8	29
HDL (high-density lipoprotein)	9	–
Total	100	100[a]

[a]*This represents only about 5% of the total circulating vitamin A.*

Computer modeling studies indicate that more than half of hepatically released holo-RBP4 comes from apo-RBP4 recycled from RBP4–TTR complexes. Apo-RBP4 is not secreted from the liver. Vitamin A-deficient animals continue to synthesize apo-RBP4, but the absence of retinol inhibits its secretion (a small amount of denatured apo-RBP4 is always found in the plasma). Owing to this hepatic accumulation of apo-RBP4, vitamin A-deficient individuals may show a transient overshooting of normal plasma RBP4 levels on vitamin A realimentation.

Other factors can alter the synthesis of RBP4 to reduce the amount of the carrier available for binding retinol and secretion into the plasma. Dietary deficiencies of protein (e.g., **protein–energy malnutrition**) and/or zinc can reduce the hepatic synthesis of apo-RBP4. Because, RBP4 is a negative acute phase reactant, subclinical infections, or inflammation can also decrease circulating retinol levels. Thus, low-serum retinol levels in malnourished individuals may not be strictly indicative of a dietary vitamin A deficiency. Also, because vitamin A deprivation leads to reductions in plasma RBP4 only after the depletion of hepatic retinyl ester stores (i.e., reduced retinol availability), the use of plasma RBP4/retinol as a parameter of nutritional vitamin A status can yield false-negative results in cases of vitamin A deprivation of short duration.

A two-point mutation in the human RBP4 gene has been shown to result in markedly impaired circulating retinol levels; surprisingly, this is associated with no signs other than night blindness.[48] This suggests that other pathways are also important in supplying cells with retinol, presumably via retinyl esters and/or β-carotene, and/or with retinoic acid. Indeed, the genetic ablation of RBP4 did not impair the ability of β-carotene to prevent signs of vitamin A deficiency in the mouse, indicating that the provitamin, which is transported independently of RBP4, can serve as a tissue source of retinol.[49] That the systemic functions of vitamin A can be discharged by retinoic acid, which is ineffective in supporting vision, indicates that the metabolic role of RBP4 is to deliver retinol to the pigment epithelium as a direct requirement of visual function and to other cells as a precursor to retinoic acid. Retinoic acid is not transported by RBP4, but it is normally present in the plasma, albeit at very low concentrations (1–3ng/mL), tightly bound to albumin.

42. Retinol-binding proteins isolated from several species (rat, chick, dog, rabbit, cow, monkey, human) have similar sizes and binding properties.

43. Seasonally breeding animals show threefold higher plasma RBP4 levels in the estrous compared with the anestrous phase. Women using oral contraceptive steroids frequently show elevated plasma RBP4 levels.

44. Formerly, *prealbumin*.

45. The half-life of *holo*-RBP4–transthyretin complex in human adult males has been found to be 11–16 hr; whereas, that of free RBP4 was only 3.5h. Half-lives of both increase (i.e., turnover decreases) under conditions of severe protein–calorie malnutrition.

46. Megalin is a 600kDa transmembrane protein in the LDL-receptor family.

47. For this reason, patients with chronic renal disease show markedly elevated plasma levels of both RBP4 (which show a half-life 10- to 15-fold normal) and retinol, although transthyretin levels remain normal. Turnover studies have indicated that some retinol recirculates to the liver, perhaps via transfer to lipoproteins.

48. Biesalski, H.K., Frank, J., Beck, S.C., et al., 1999. Am. J. Clin. Nutr. 69, 931–936.

49. Kim, Y., Wassef, S., Chang, S., et al., 2011. FASEB J. 25, 1641–1652.

It is presumed that the cellular uptake of retinoic acid from serum albumin is very efficient.

RPB4 is also expressed in adipose and other tissues, although liver is the predominant source of the protein in circulation. Appreciable amounts of vitamin A are stored in adipocytes (15–20% of total body store), more than half as retinyl esters. That visceral fat expresses more RBP4 than subcutaneous fat makes serum apo-RBP4 a candidate biomarker for visceral adiposity. Unlike other tissues, which take up retinol from RBP4, adipocytes take up retinyl esters from chylomicra. Studies with the rat have shown that the mobilization of vitamin A from adipocytes also differs from that process in other cells. A cAMP-sensitive, hormone-dependent lipase converts adipocyte retinyl esters to retinal in a manner analogous to the liberation of free fatty acids from adipocyte triglyceride depots.

RBP4 secreted from adipocytes has been found to be elevated in overweight/obese individuals, although much of this appears not to be bound to retinol.[50] That elevated apo-RBP4 levels may be related to the development of type 2 diabetes was suggested by the finding on increased gluconeogenic capacity on mice treated with recombinant RPB4. Studies with humans have found plasma RPB4 level to be positively correlated with body mass index (BMI, kg/m^2) (Fig 6.4), the degree of insulin resistance and impaired glucose tolerance in subjects with obesity or family histories of type 2 diabetes.[51] In contrast, plasma RBP4 levels have been found to be reduced by treatments that reduce insulin resistance, e.g., exercise training, gastric bypass surgery, and positively correlated with the expression of p85[52] in adipose tissue. These observations reflect increased apo-RBP4 secretion under conditions in which adipocytes downregulate the expression of GLUT4, the insulin-responsive transporter required for cellular uptake of glucose. This evidence suggests that adipocyte-derived RBP4 may acting as an adipokine.[53]

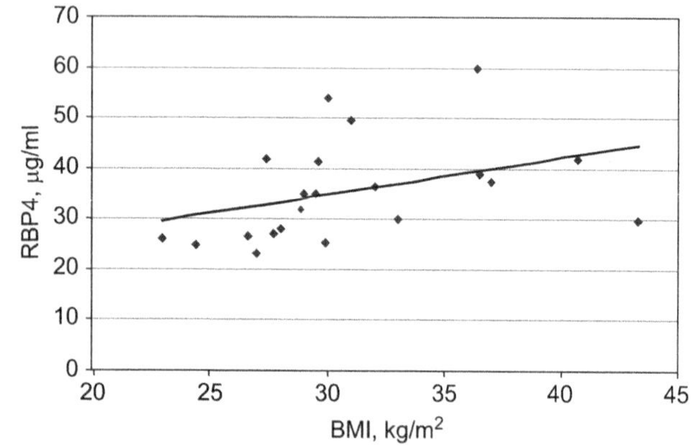

FIGURE 6.4 Positive association of plasma RBP4 level and body mass index (BMI). *Graham, T.E., Yang, Q., Blüher, M., et al., 2006. N. Eng. J. Med. 354, 2552–2563.*

Cellular Uptake of Retinol

Due to their hydrophobic character, the plasma membranes do not present a barrier to retinol uptake and, thus, retinol can enter target cells by nonspecific partitioning into the plasma membrane from holo-RBP4.[54] Nevertheless, most of the vitamin appears to enter cells through specific holo-RBP4-TTR–receptor-mediated mechanisms. The retinol ligand of holo-RBP4-TTR is taken into cells via binding of the complex to a specific cell surface receptor, a multidomain transmembrane protein **STRA6.**[55] STRA6 is expressed in all tissues *except* liver. It is upregulated by retinoic acid.[56] In extrahepatic tissues, STRA6 transfers vitamin A bidirectionally between extracellular and intracellular RBPs, net cellular accumulation being accomplished by trapping the vitamin within the cell through its esterification. The extracellular region of STRA6 binds tightly to the holo-RPB4; however, the presence of TTR inhibits that binding. Thus, it is thought that reductions in circulating TTR, as occur in the acute phase response, release holo-RBP4 for binding to STRA6 to facilitate cellular uptake of retinol. Transfer of retinol into the cell also depends on access to CRBP(I); in cells capable of esterifying retinol,[57] esterification constitutes a sink that drives cellular uptake of the vitamin. In extrahepatic tissues, this process appears to be regulated by all-*trans*-retinoic acid. STRA6 also has a signaling function; when bound to holo-RBP, it activates a Janus kinase (JAK[58]) by catalyzing the phosphorylation of a C-terminal tyrosine residue in that factor.

50. Yang, Q., Graham, T.E., Mody, N., et al., 2005. Nature 436, 356–362.

51. Graham, T.E., Yang, Q., Blűher, M., et al., 2006. New Eng. J. Med. 354, 2552; Broch, M., Vendrell, J., Ricart, W., et al., 2007. Diabetes Care 30, 1802; Chavez, A.O., Coletta, D.K., Kamath, S., et al., 2008. Am. J. Physiol. Endorcrinol. Metab. 296, E768; Chavez, A.O., Coletta, D.K., Kamath, S., et al., 2009. Am. J. Physiol. Endorcrinol. Metab. 296, E768; Kelly, K.R., Kashyap, S.R., O'Leary, V.B., et al., 2009. Obesity 18, 663.

52. p85 is a regulatory subunit of a protein involved in insulin signaling, phosphoinositol 3-kinase (PI3K). PI3K is activated by phosphorylation by insulin receptor substrate-1.

53. Adipokines are cell-signaling proteins (cytokines) secreted by adipocytes. They include several that may be involved in insulin resistance: C-reactive protein (CRP), leptin, adiponectin, resistin, and tumor necrosis factor α (TNFα).

54. Indeed, retinol has been shown to move spontaneously between the layers of artificial phospholipid bilayers.

55. Kelly, M., von Lintig, J., 2015. HepatoBiliary Surg. Nutr. 4, 229–242.

56. The STRA6 gene transcriptionally upregulated by the RARγ/RXRα heterodimer and repressed by RARα. Mutations in the gene produce diffferent developmental defects.

57. i.e., That express lecithin–retinol acyltransferase (LRAT).

58. Originally named "just another kinase," these tyrosine kinases transduce cytokine-mediated signals via the JAK/STAT pathway.

The resulting signaling cascade induces expression of STAT[59] target genes including PPARγ (peroxisome proliferator-activated receptor), which promotes lipid accumulation, and SOCS3,[60] which suppresses insulin signaling.

Another RBP receptor, RBPR2, has been identified. This receptor would appear to play an important role in retinol uptake in most tissues, as STRA6-null mice show reduced levels of vitamin A only in the eye. RBPR2 is a double-chain protein in humans, but a single chain in the mouse. Its primary amino acid structure shows only 18% homology with STRA6. It is thought that this receptor must be active in hepatic uptake of vitamin A. Expressed in liver, RBPR2 is thought to function in the hepatic uptake of vitamin A from chylomicrons, which is quantitatively greater than that of extrahepatic tissues with the exceptions of the mammary gland and bone marrow.

Release of retinol to the target cell increases in the negative charge of the resulting apo-RBP4; this reduces its affinity for TTR, which is subsequently lost. The residual apo-RBP4 can then be filtered by the kidney, where it is degraded. Thus, plasma apo-RBP4 levels are normally low but can be elevated (by about 50%) under conditions of acute renal failure. Studies have shown that apo-RBP4 can be recycled to the holo form; injections of apo-RBP4 into rats produced marked (70–164%) elevations in serum retinol levels. It is thought, therefore, that circulating apo-RBP4 may be a positive feedback signal from peripheral tissues for the hepatic release of retinol, the extent of which response is dependent on the size of hepatic vitamin A stores.

Intracellular Retinoid-Binding Proteins

On entry into the target cell, retinol combines with other binding proteins:

- cellular retinol-binding proteins, CRBP(I) and CRBP(II)
- cellular retinoic acid-binding proteins, CRABP(I) and CRABP(II)
- cellular retinal-binding protein, CRALBP
- interphotoreceptor retinol-binding protein, IRBP

The CRBPs and CRABPs have the same general tertiary structure as the **lipocalins**, a class of low-molecular weight proteins with multistranded β-sheets folded into a deep hydrophobic pocket suitable for binding hydrophobic ligands.[61] Each consists of some 135 amino acid residues and shows with pairwise sequence homologies of 40–74%. Each has been highly conserved (91–96% sequence homology among the human, rat, mouse, pig, and chick proteins). Their genes each contain four exons and three introns, the latter being positioned identically. The proximity of the genes for CRBPs (I) and (II) (only 3 centimorgans) suggests that this pair resulted from the duplication of the same ancestral gene. Their gene products show different tissue distributions: CRBP(I) is among the most abundant cytosolic proteins; it is expressed in most fetal and adult tissues, particularly those of the liver, kidney, lung, choroids plexus. and pigment epithelium; CRBP(II) is expressed only in mature enterocytes in the villi of the mucosal epithelium (especially jejunum) and in the fetal and neonatal liver.

CRALBP and IRBP are in the group of intracellular lipid-binding proteins that include the fatty acid-binding proteins. Like the lipocalins, they also have antiparallel, β-barrel structure that binds lipophilic ligands; however, they bind vitamin A in the reverse orientation from the CRBPs and CRABPs, i.e., with its polar group buried internally and the β-ionone ring close to the surface. The IRBPs can bind three retinol molecules (all other RBPs bind only a single ligand molecule) as well as two long-chain fatty acid molecules.

Tissue levels of the mRNAs for the CRBPs are influenced by vitamin A status. Both CRBP(I) protein and mRNA are reduced by deprivation of the vitamin; however, CRBP(II) protein and mRNA levels are increased by vitamin A deficiency. The CRBP gene appears to be inducible by retinoic acid and **retinoic acid response elements (RAREs)** have been identified in both the CRBP(I) and CRBP(II) promotors. Whether other transcriptional factors also bind to those elements remains to be learned, yet it appears that these genes are responsive to other hormones, including glucocorticoids and 1,25-dihydroxyvitamin D_3, which have been shown to have negative effects on CRBP(I) and CRBP(II), respectively.

Because CRBP(I) is present at high levels in cells that synthesize and secrete RBP4, it has been suggested that it may interact at specific sites to effect the transfer of retinol to RBP4 for release to the general circulation. The synthesis and/or the retinol-binding affinity of CRBP(I) may be affected by thyroid and growth hormones, both of which promote the cellular uptake of retinol.

The intracellular vitamin A-binding proteins appear to be important in the cellular uptake and the intracellular and transcellular transport of vitamin A metabolites. Both the CRBPs and CRABPs serve as carriers of their respective ligands from the cytoplasm into the nucleus, where they are transferred to the chromatin with release of the binding proteins possibly to return to the cytoplasm. The CRBPs appear to have more specialized transport functions in certain tissues. In the liver, their concentrations increase with increasing retinyl ester contents, suggesting that they may function in the transport of retinol from parenchymal cells into the stellate cells, which store retinyl esters. CRBP localization has been found in endothelial cells of the brain microvasculature, in cuboidal cells of the choroid plexus, in

59. Signal transducers and activators of transcription.
60. This gene encodes the protein suppressor of cytokine signaling 3.
61. Other lipocalins bind fatty acids, cholesterol, and biliverdin.

the Sertoli cells of the testis, and in the pigment epithelium of the retina. Because these tight-junctioned cells also have surface receptors for the plasma holo-RBP4–TTR complex, it is thought that their abundant CRBP(I) concentrations are involved in the transport of retinal across the blood–brain, blood–testis, and retinal blood–pigment epithelial barriers. Studies with mice have shown that CRBP(I) is necessary for the hepatic uptake of retinol: CRBP(I)-null individuals exhausted their hepatic retinyl ester stores even when they were fed vitamin A.

CRBP(II) appears to be restricted largely to the enterocytes of the small intestine (particularly, the jejunum). Its abundance in mature enterocytes, where it comprises 1% of the total soluble protein, as well as the absence of CRBP(I) in these cells, suggest that CRBP(II) is involved in enteric absorption of vitamin A, presumably by transporting it across the cell. Both CRBP(II), as well as a high-capacity esterase that esterifies CRBP(II)-bound retinol, have been identified in hepatic parenchymal cells of fetal and newborn rats. After birth, CRBP(II) appears to be replaced by CRBP(I), such that mature animals show none of the former binding protein in that organ. The presence of CRBP(II) in fetal liver corresponds to the increased concentration of retinyl palmitate in that organ at birth.

Some extracellular RBPs appear to serve similar transport functions. These include two low-molecular weight retinoic acid-binding proteins generally related to RBP4 that are secreted into the lumen of the epididymis where they are thought to participate in the delivery of all-*trans*-retinoic acid to sperm, which are rich in CRBP(I).[62] Other retinol-binding proteins are synthesized in the uterine endometrium and secreted into the uterus; these show some sequence homology with RBP4 but are slightly larger. They are thought to be involved in the transport of retinol to the fetus.

CRALBP is a 36 kDa protein that binds 11-*cis*-retinal and appears to be expressed only in pigment epithelial and Müller glial cells of the retina where it is thought to facilitate the intracellular transport of 11-cis-retinal. That it is essential for cone function has been demonstrated in mice: deletion of CRALBP results in severe (20-fold) reduction in cone dark adaptation.[63] This effect was appeared to be due to impaired chromophore recycling by Müller cells, as adenovirus-mediated restoration of CRALBP specifically to Müller cells restored visual sensitivity. Mutations in the CRALBP gene are known to cause forms of retinitis pigmentosa.

IRBP is thought to function in the transport of retinol between the pigment epithelium and photoreceptor cells where it is synthesized and binds to the extracellular matrix of cone outer segments. IRBP is a large (140-kDa) glycoprotein; it can bind 6 mol of long-chain fatty acid in addition to 2 mol of retinol. It has been suggested that its relatively low affinity for retinol, in comparison with the other retinol-binding proteins, facilitates rapid, high-volume transport of that ligand along a series of IRBPs. That IRBP is involved in the visual process, perhaps by delivering the chromophore, is indicated by the fact that its binding specificity shifts from mainly 11-*cis*-retinol to mostly all-*trans*-retinol as eyes become light adapted.[64] The two IRBP binding sites for retinoids have been found to be quite different. One is a strongly hydrophobic binding pocket; the other is a surface site that interacts with retinol via its polar head group. The protein shows higher affinities for all-*trans*-retinol and 11-*cis*-retinal than for other retinoids. Studies have shown that docosahexanoic acid (DHA) induces a rapid and specific release of 11-*cis*-retinal from the IRBP hydrophobic site; whereas, palmitic acid is without effect. This suggests that DHA may function in the targeting of 11-*cis*-retinal to photoreceptor cells, the DHA concentrations of which are much greater than those of pigment epithelial cells.

Retinol Recycling and Homeostasis

The majority of retinol that leaves the plasma appears to be recycled, as plasma turnover rates have been found to exceed utilization rates by more than an order of magnitude. Kinetic studies indicate that in adult rats a retinol molecule recycles via RBP4 12–13 times before its irreversible utilization;[65] in humans, number appears to be much smaller (e.g., 3). In the rat, newly released retinol has been found to circulate in the plasma for 1–3.5 h before leaving that compartment; it recycles in the plasma for a week or more. It is estimated that some 50% of plasma turnover in the rat is to the kidneys, 20% to the liver, and 30% to other tissues. It has been suggested that retinol leaves the plasma bound to RBP4. These parameters are quite different in the neonatal rat, indicating much faster turnover (turnover number 144) and much shorter transit time (0.4 h in plasma.) Although the source of RBP4 for retinol recycling is not established, it is worth noting that mRNA for RBP4 has been identified in many extrahepatic tissues, including kidney.

In healthy individuals, plasma retinol is maintained within a narrow range (40–50 mg/mL in adults; typically about half that in newborn infants) in spite of widely varying intakes of vitamin/provitamin A (Fig. 6.5).[66] This control appears to be effected by several factors: CRBP(II) expression in stellate cells, the esterification of retinol

62. The initial segment of the epididymis contains the greatest concentration of CRBP found in any tissue.

63. Saari, J.C., Nawrot, M., Kennedity, B.N., et al., 2001. Neuron 29, 739–748; Xue, Y., Shen, S., Jui, J., et al., 2015. J. Clin. Invest. 125, 727–738.

64. IRBP is also found in another photosensitive organ, the pineal gland.

65. Lin, L., Green, M.H., Ross, A.C., 2014. J. Nutr. 145, 403–410.

66. In contrast, plasma levels of all-*trans*-retinoic acid and 4-oxo-retinoic acid respond to the level of ingested vitamin A.

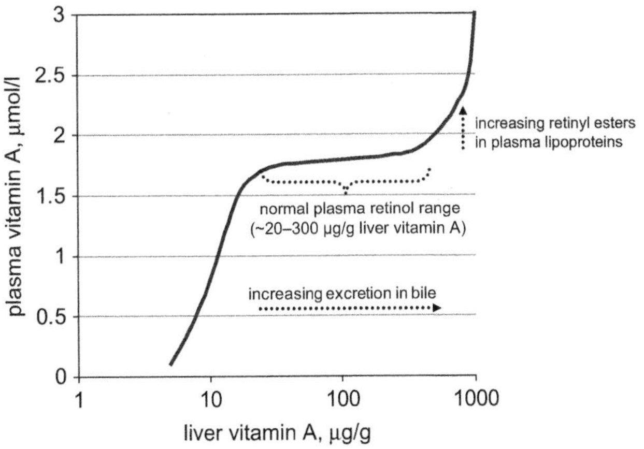

FIGURE 6.5 Regulation of plasma vitamin A levels. *After Olson, J.A., 1984. J. Nat. Cancer Inst. 73, 1439–1446.*

and hydrolyze retinyl esters, and retinol release to and/or removal from the plasma. Liver and kidneys play important roles in these various processes. Renal dysfunction has been shown to increase plasma retinol levels; this may involve a regulatory signal to the liver that alters the secretion of RBP4–retinol. Serum retinol levels can also be affected by nutritional status with respect to zinc, which is required for the hepatic synthesis of RBP4.

Plasma levels of carotenoids, in contrast, do not appear to be regulated but reflect intake of carotenoid-rich foods. Careful studies have revealed, however, cyclic changes of up to nearly 30% in the plasma β-carotene concentrations during the menstrual cycles of women. Whether these fluctuations are physiologically meaningful or whether they relate to fluctuations in plasma lipids is not clear.

Vitamin A in the Eye

Retinol is taken up by the retinal pigment epithelium in preference to uptake by nonocular tissues. However, this process, as in other tissues, involves transfer of retinol from RBP4 by the cellular receptor STRA6, which depends on the functional coupling of STAR6 and LRAT via CRBP. While pigment epithelial and Müller cells store the vitamin in esterified form in lipid droplets,[67] their accumulation appears to be less sensitive than other tissues to high doses of the vitamin. Epithelial reserves are mobilizable, and retinol is transported to rod cells by IRBP for discharge of the visual function of the vitamin. RBP4 is also expressed in the lacrimal glands; retinol appears to reach the cornea via holo-RBP4 secreted in tears. Both retinal pigment epithelial and Müller cells express CRALBP, which is thought to facilitate the intracellular transport of 11-*cis*-retinal and has been demonstrated to be essential for cone function. Retinal pigment epithelial cells also accumulate, by phagocytosis of photoreceptor outer segments, complex mixtures of bis-retinoids as lipofuscin. This process appears to involve non-enzymatic reactions of retinal and is accelerated in some retinal disorders.[68]

Milk Retinol

Retinol is transferred from mother to infant through milk. Studies in animal models suggest that this retinol is drawn preferentially from recently consumed vitamin A, rather than from hepatic stores. The retinoid and carotenoid contents of milk depend on the stage of lactation and the vitamin A status of the mother, the patterns of carotenoids in colostrum tending to reflect those in maternal LDLs, while patterns in mature milk (19 days) reflecting those in maternal HDLs.[69] Retinol concentration of at least 1.75 µM is required to support adequate vitamin A stores to protect against the development of xerophthalmia during weaning. For this reason, vitamin A supplementation of mothers in vitamin A-deficient areas is regarded as a prudent public health strategy. A meta-analysis of randomized controlled trials showed that such measures have reduced infant mortality by 23% in children under 5 years of age in populations at risk to vitamin A deficiency.[70] The success of postpartum maternal supplementation depends on the prevailing breast feeding practices, with the simultaneous promotion of optimal practices (including the feeding of colostrum) being highly effective in improving infant vitamin A status. For example, the administration of 60,000 RE to a low-vitamin A mother can produce a 29% increase in the retinol contents of her breast milk over 6 months.

6. METABOLISM OF VITAMIN A

Metabolic Fates of Retinol

The metabolism of vitamin A (Fig. 6.6) centers around the transport form, retinol, and the various routes of conversion available to it: *esterification*, *conjugation*, *oxidation*, and *isomerization*.

Esterification. Retinol is esterified in the cells of the intestine and most other tissues via enzymes of the endoplasmic reticulum, which use acyl groups from either phosphatidylcholine (LRAT) or acylated coenzyme A (ARAT). These systems show marked specificities for saturated fatty acids, in particular, palmitic acid; thus, the most abundant product is retinyl palmitate.

67. Some investigators referred to these structures as "*retinosomes*".

68. e.g., Stargardt disease (also, fundus flavimaculatus), a recessive juvenile form of macular degeneration.

69. Schwiegert, F.J., Bathe, K., Chen, F., et al., 2004. Eur. J. Nutr. 43, 39–44.

70. Beaton, G.H., Mortorell, R., Aronson, K.J., et al., 1993. Nutrition Policy Discussion Paper No. 13, UN Administrative Committee of Coordination –Subcommittee on Nutrition, New York, 120 pp.

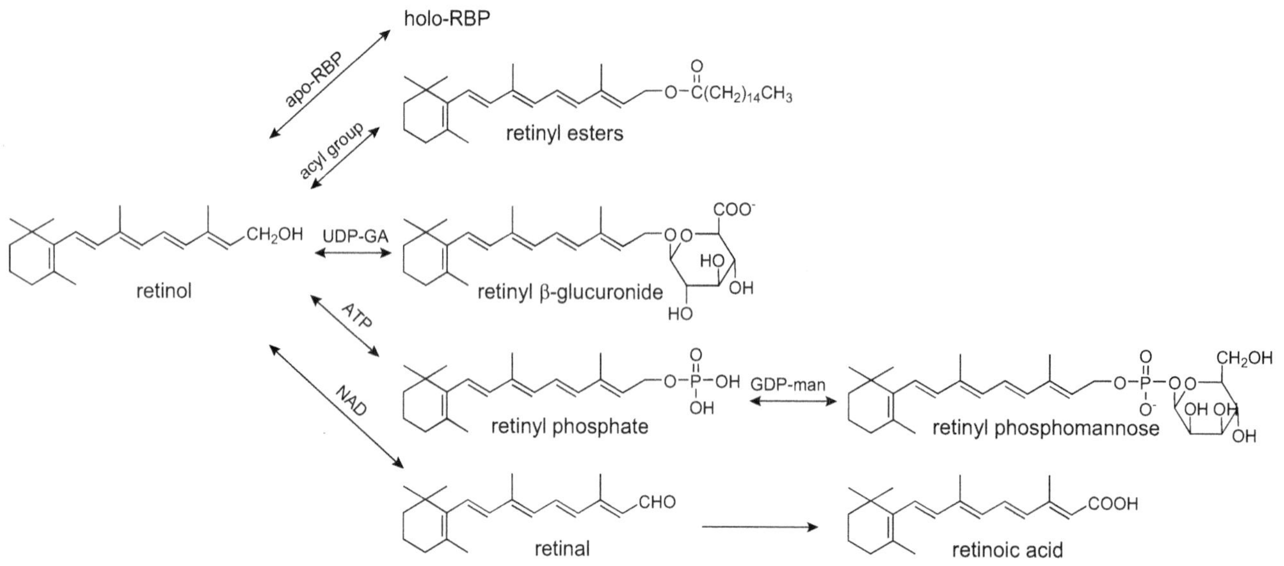

FIGURE 6.6 Metabolic fates of retinol.

Conjugation. Retinol may also be conjugated in either of two ways. The first entails the reaction catalyzed by retinol–UDP–glucuronidase, present in the liver and probably other tissues, which yields retinyl β-glucuronide, a metabolite that is excreted in the bile.[71] The second path of conjugation involves ATP-dependent phosphorylation to yield retinyl phosphate catalyzed by retinol phosphorylase. That product, in the presence of guanosine diphosphomannose (GDP-man), can be converted to the glycoside retinyl phosphomannose, which can transfer its sugar moiety to glycoprotein receptors. However, because only a small amount of retinol appears to undergo phosphorylation in vivo, the physiological significance of this pathway is not clear.

Oxidation. Retinol can also be reversibly oxidized to retinal by multiple NADH-/NADPH- and zinc-dependent retinol dehydrogenases. These cytosolic and microsomal activities are found in many tissues, the greatest being in the testis.[72] A short-chain aldehyde dehydrogenase has been described that can oxidize 9-*cis*- and 11-*cis*-retinol to the corresponding aldehyde. This activity has been identified in several tissues including the retinal pigment epithelium, liver, mammary gland, and kidney. That 9-*cis*-retinol can be converted to 9-cis-retinoic acid is evidenced by the finding of a 9-*cis*-retinol dehydrogenase. The enzyme in both humans and mice is inhibited by 13-*cis*-retinoic acid at levels similar to those found in human plasma, suggesting that 13-*cis*-retinoic acid may affect the regulation of retinoid metabolism.

Retinal can be irreversibly oxidized to retinoic acid by multiple retinal dehydrogenases (RALDHs), which are also aldehyde dehydrogenases. One of these, RALDH2, is critical for embryonic development; it is the first RALDH expressed in the mouse embryo and its genetic ablation produces malformations and death at midgestation.[73] It has been suggested that cytochrome *P*-450 monoxygenases may also produce retinoic acid from retinol through retinal. Because retinoic acid is the active ligand for the nuclear retinoid receptors, it is very likely that this metabolism is tightly regulated.[74] The rate of that reaction is several fold greater than retinol dehydrogenase that, plus the fact that the rate of reduction of retinal back to retinol is also relatively great, results in retinal being present at very low concentrations in tissues. Several retinoic acid isomers have been identified in the plasma of various species;[75] the number having physiological significance is presently unclear.

Isomerization. Interconversion of the most common all-*trans* and various *cis* vitamers occurs in the eye and is a key aspect of the visual function of vitamin A. That process involves light-induced isomerization of 11-*cis*-retinal to all-*trans*-retinal (Fig. 6.8), which alters the affinity of the visual pigment protein opsin for this ligand, leading

71. The former view of retinyl β-glucuronide as an excretory form has changed. About 30% of the amount excreted in the bile is reabsorbed and recycled in an enterohepatic circulation back to the liver, and retinyl β-glucuronide has been found to be produced in many extrahepatic tissues where it can support growth and tissue differentiation.
72. Male rats fed retinoic acid instead of retinol become aspermatogenic and experience testicular atrophy. It has been proposed that retinoic acid is required for spermatogenesis but cannot cross the blood–testis barrier; this is supported by the fact that the rat testis is also rich in CRABP.
73. Niederreither, K., Subbarayan, V., Dolle, P., et al., 1999. Nat. Genet. 21, 444–448.
74. Two microsomal proteins that catalyze this reaction have been isolated from rat liver; one cross-links with holo (but not apo)-CRBP in the presence of NADP.
75. In addition to all-*trans*-, 13-*cis*-, and 9-*cis*-retinoic acid, this number includes the 9,13-*dicis*-, 4-hydroxy-, 4-oxo-, 18-hydroxy-, 3,4-dihydroxy-, and 5,6-epoxy isomers, as well as such derivatives as retinotaurine.

to dissociation linked to nervous signaling and conversion back to the 11-*cis* vitamer by retinal isomerase (RPE65). That enzyme also catalyzes the isomerization of 11-*cis*-retinol to the all-*trans* form. The conversion of all-trans-retinoic acid to 9-cis-retinoic acid has also been demonstrated.

All-*trans*-retinoic acid is produced and catabolized by *unidirectional* processes. While retinol, retinyl esters, and retinal can be oxidized to retinoic acid, retinoic acid *cannot* be converted metabolically to any of the reduced vitamers.

Fates of Retinoic Acid

All-*trans*-retinoic acid is converted to forms that can be readily excreted (Fig. 6.7). It may be directly conjugated by glucuronidation in the intestine, liver, and possibly other tissues to retinyl β-glucuronide. Alternatively, it can be catabolized by several further oxidized products including the 4-hydroxy-, 4-oxo-, 4-dihydroxy-, 18-hyrodroxy-, 3-hyrdoxy-, and 5,6-epoxy metabolites. Several cytochrome *P*-450 enzymes have been implicated in this metabolism. These include different families: CYP2, CYP3, CYP4, and CYP26. Of these, three in the CYP26 family appear to be most important. These are monooxygenases that convert retinoic acid to the 4-hydroxy- (CYP26A), 4-oxo- (CYP26B), and 18-hydroxy- (CYP26C) metabolites. They are expressed early in development and their pattern of expression varies according to tissue and stage of development. The expression of CYP26A is known to upregulated by dietary vitamin A and retinoic acid and downregulated by vitamin A depletion; mice devoid of this isoform have neural tube defects and die shortly after birth.

The oxidative chain-cleavage metabolites are conjugated with glucuronic acid, taurine, or other polar molecules; these, being more polar, can readily be excreted. Glucuronides and retinotaurine comprise a significant portion of the retinoids excreted in the bile. That retinoyl-β-glucuronide has been found to show some vitamin A activity suggests that there may be some hydrolysis to yield retinoic acid.

Role of Retinoid-Binding Proteins in Vitamin A Metabolism

In addition to serving as reserves of retinoids, the CRBPs also serve to modulate vitamin A metabolism by holding the retinoids in ways that render them inaccessible to the oxidizing environment of the cell and by channeling the retinoids via protein–protein interactions among its enzymes (Fig. 6.8, Tables 6.8 and 6.9). Both CRBPs function in directing the metabolism to their bound retinoid ligands by shielding them from some enzymes that would use the free retinoid substrate and by making them accessible to other enzymes important in metabolism. For example, the esterification of retinol by LRAT occurs while the substrate is bound to CRBP(I) or CRBP(II). The abundance of CRBP(I) in the liver and its high affinity for retinol suggest that its presence directs the esterification of the retinoid ligand to the reaction catalyzed by LRAT, rather than that catalyzed by ARAT, which can use only free retinol. This also appears to be the case for CRBP(II), which can bind both retinol and retinal; only when retinal is bound to it can the reducing enzyme **retinal reductase** use that substrate. In addition, the binding of retinol to CRBP(II) greatly reduces the reverse reaction (oxidation to retinal). Thus, by facilitating retinal formation and inhibiting its loss, CRBP(II) seems to direct the retinoid to the appropriate enzymes, which sequentially convert

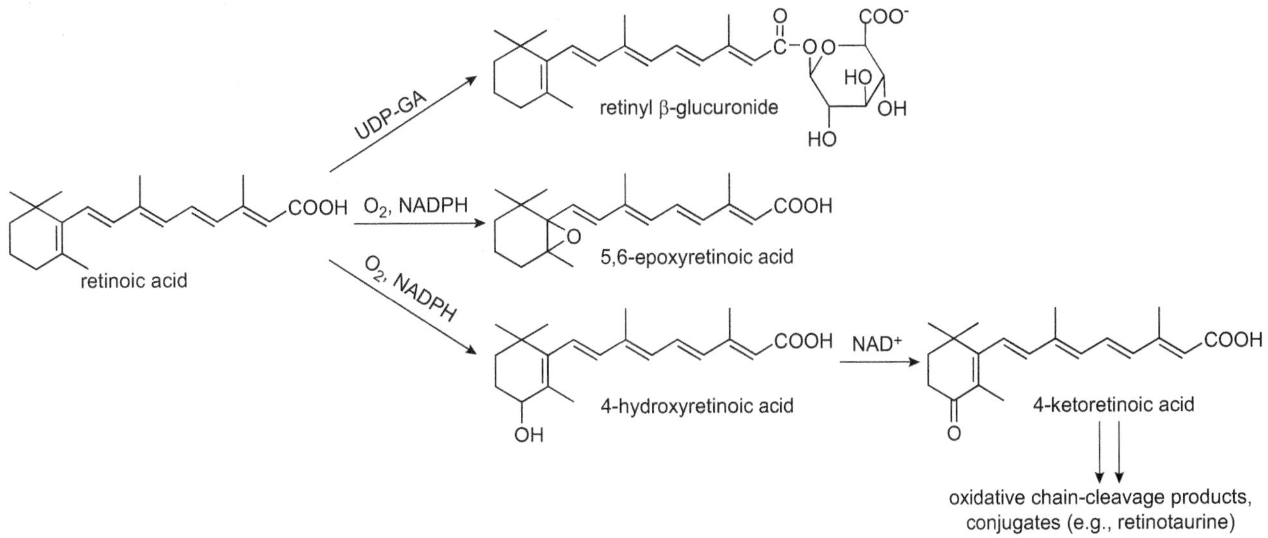

FIGURE 6.7 Catabolism of retinoic acid.

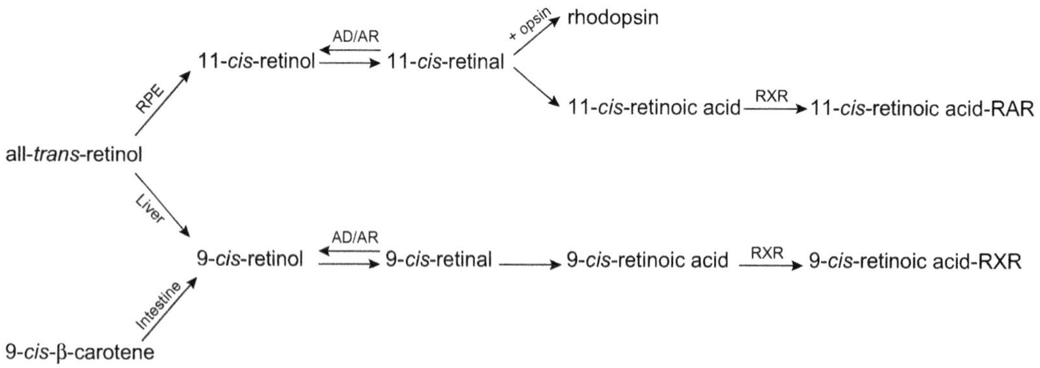

FIGURE 6.8 Roles of binding proteins in vitamin A metabolism.

TABLE 6.8 Vitamin A-Binding Proteins

Binding Protein	Abbreviation	MW (kDa)	Ligand	K_d (nM)	Tissues
Extracellular					
Retinol-BP	RBP4	21.2	All-*trans*-retinol	20	Plasma
Albumin	–	67	All-*trans*-retinoic acid		Plasma
Interphotoreceptor retinol-BP	IRBP	136	All-*trans*-/11-cis-retinol	50–100	Interphotoreceptor space
Intracellular					
Cellular retinol-BP, type I	CRBP(I)	15.7	All-*trans*-retinol	<10	Most *except* heart, adrenal, ileum
			All-*trans*-retinal	50	
Cellular retinol-BP, type II	CRBP(II)	15.6	All-*trans*-retinol(al)	90	Enterocytes, fetal and neonatal liver
Cellular retinal-BP	CRALBP	36	11-*cis*-retinol(al)	15	Retinal pigment epithelial and Müller cells
Cellular retinoic acid-BP, type I	CRABP(I)	15.5	All-*trans*-retinoic acid	0.1	Most *except* liver, jejunum, ileum
Cellular retinoic acid-BP, type II	CRABP(II)	15	All-*trans*-retinoic acid	0.1	Cytosol of embryonic limb bud
Epididymal retinoic acid-BP	–	18.5	All-*trans*-retinoic acid		lumen of epididymis
Uterine retinol-BP	–	22	All-*trans*-retinol		Uterus (sow)
Retinol pigment epithelium protein	RPE65	65	All-trans-retinyl esters		Retinal pigment epithelium
Cellular RBP4 receptor	STRA6	74	All-*trans*-retinol		Extraheptic tissues
Cellular RBP4 receptor 2	RBPR2		All-*trans*-retinol		Liver and other tissues
Nucleus					
Nuclear retinoic acid receptor-α	RARα	50	All-*trans*-retinoic acid		Most *except* adult liver
Nuclear retinoic acid receptor-β	RARβ	50	All-*trans*-retinoic acid		Most *except* adult liver
Nuclear retinoic acid receptor-γ	RARγ	50	All-*trans*-retinoic acid		Most *except* adult liver
Nuclear retinoid X receptor-γ	RXR	50	All-*trans*-retinoic acid		Most *except* adult liver
Rhodopsin	—	41	11-*cis*-retinal		Retina
Melanopsin	—		11-*cis*-retinal		Retina, brain, skin (*Xenopus* sp.)

TABLE 6.9 Apparent Metabolic Functions of the Intracellular Retinoid-Binding Proteins

Binding Protein	Function
CRBP(I) (cellular retinol-binding protein)	Directs retinol to LRAT (lecithin–retinol acyltransferase) and oxidative enzymes; regulates retinyl ester hydrolase
CRBP(II)	Directs retinol to LRAT and oxidative enzymes
CRABPs (I) and (II) (cellular retinoic acid-binding protein)	Regulates intracellular retinoic acid concentrations; direct retinoic acid to catabolizing enzymes
CRALBP (cellular retinal-binding protein)	Regulates enzymatic reactions of the visual cycle

retinal to the esterified form in which it is exported from the enterocytes. The preferential binding of 11-*cis*-retinal by the CRALBPs relative to CRBP(I) appears to be another example of direction of the ligand to its appropriate enzyme, i.e., a microsomal NAD-dependent retinal reductase in the pigment epithelium of the retina that uses only the carrier-bound substrate. In each of these cases, it is likely that the RBP–ligand complex interacts directly with the respective retinoid-metabolizing enzyme.

The CRABPs are noteworthy for their antagonistic effects on the trafficking of retinoic acid, which suggest their roles in regulating intracellular levels of the vitamer. Studies indicate that CRABP(I) facilitates its degradation by CYP26, while CRABP(II) facilitates its transfer to nuclear receptors. Mice in which CRABP genes had been deleted show normal phenotypes and normal susceptibility to the teratogenic effects of high doses of retinoic acid; yet, transgenic animals that overexpress CRABPs show significant pathology (cataracts and pancreatic endocrine tumors).

Excretion of Vitamin A

Vitamin A is excreted in various forms in both the urine and feces. Under normal physiological conditions, the efficiency of enteric absorption of vitamin A is high (80–95%), with 30–60% of the absorbed amount being deposited in esterified form in the liver. The balance of absorbed vitamin A is catabolized (mainly at C-4 of the ring and at C-15 at the end of the side chain[76]) and released in the bile or plasma, where it is removed by the kidney and excreted in the urine (i.e., short-chain, oxidized, conjugated products). About 30% of the biliary metabolites (i.e., retinoyl β-glucuronides) are reabsorbed from the intestine into the enterohepatic circulation back to the liver, but most are excreted in the feces with unabsorbed dietary vitamin A. In general, vitamin A metabolites with intact carbon chains are excreted in the feces, whereas the chain-shortened, acidic metabolites are excreted in the urine. The relative amounts of vitamin A metabolites in the urine and feces, thus, vary with vitamin A intake (i.e., at high intakes fecal excretion may be twice that of the urine) and the hepatic vitamin A reserve (i.e., when reserves are above the low normal level of 20 μg/g, both urinary and fecal excretion vary with the amount of vitamin A in the liver).

76. The chain-terminal carbon atoms (C-14, C-15) can be oxidized to CO_2; retinoic acid is oxidized to CO_2 to a somewhat greater extent than retinol.

7. METABOLIC FUNCTIONS OF VITAMIN A

Feeding provitamin A carotenoids retinyl esters, retinol, and retinal can support the maintenance of healthy epithelial cell differentiation, normal reproductive performance, and visual function. Each of these forms can be metabolized to retinol, retinal, or retinoic acid. But unlike retinol and retinal, retinoic acid cannot be reduced to retinal or retinol. Feeding retinoic acid can support only the systemic functions of vitamin A, e.g., epithelial cell differentiation. These observations and knowledge of retinoid metabolism led to the conclusion that whereas retinal discharges the visual functions, retinoic acid (and, specifically, **all-*trans*-retinoic acid**) must support the systemic functions of the vitamin.

Functional Forms of Vitamin A

Retinol	Transport, reproduction (mammals)
Retinyl esters	Storage
Retinal	Vision
Retinoic acid	Epithelial differentiation, immune function, gene transcription, reproduction

Vitamin A in Vision

The best elucidated function of vitamin A is in the visual process where, as 11-*cis*-retinal, it serves as the photosensitive chromophore of the visual pigments of the ciliated photoreceptor cells of the retina, the rods and cones.[77] Rod cells contain the pigment **rhodopsin**, which has an absorption maximum at 498 nm. Cone cells contain one of three possible **iodopsins**: 64% are "red cones" (absorption maxima at 560–580 nm), 32% are "green cones" (absorption maxima at 530–540 nm), and 2% are "blue cones" (absorption maxima at 420–440 nm). In each case, photoreception is effected by the rapid, light-induced isomerization of 11-*cis*-retinal to the all-*trans* form. That product, present as a protonated Schiff base[78] of a specific lysyl residue of the protein (Fig. 6.9), produces a highly strained conformation, which results in the dissociation of the retinoid from the opsin complex. This process (**bleaching**) is a complex series of reactions, involving progression of the pigment through a series of unstable intermediates of differing conformations[79] and, ultimately, to *N*-retinylidene opsin, which dissociates to all-*trans*-retinal and opsin (Fig. 6.10).

The dissociation of all-*trans*-retinal and opsin is coupled to nervous stimulation of the vision centers of the brain. The bleaching of rhodopsin causes the closing of Na^+ channels in the rod outer segment, thus leading to hyperpolarization of the membrane. This change in membrane potential is transmitted as a nervous impulse along the optic neurons. This response appears to be stimulated by the reaction of an unstable "activated" form of rhodopsin, **metarhodopsin II**, which reacts with **transducin**, a membrane-bound

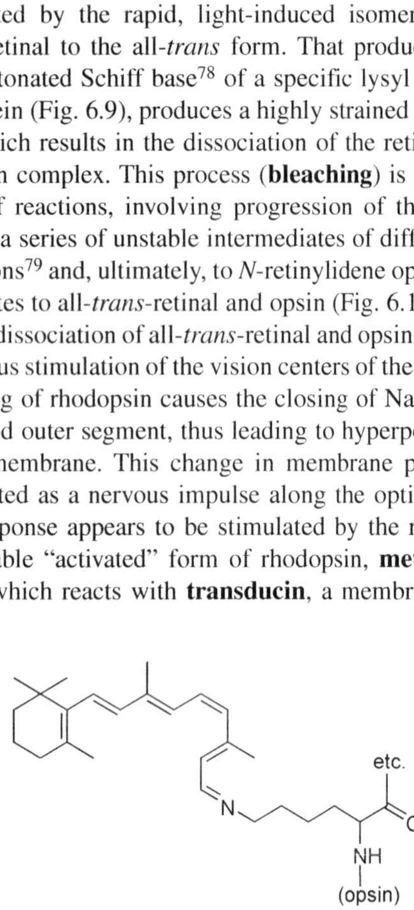

FIGURE 6.9 11-*cis*-Retinal binds to photopigment proteins via a lysyl linkage.

G protein of the rod outer segment disks. This results in the binding of the transducin α-subunit with **cGMP phosphodiesterase**, which activates the latter to catalyze the hydrolysis of cGMP to GMP. Because cGMP maintains Na^+ channels of the rod plasma membrane in the open state, the resulting decrease in its concentration causes a marked reduction in Na^+ influx. This results in hyperpolarization of the membrane and the generation of a nerve impulse through the synaptic terminal of the rod cell.

The visual process is a cyclic one, in that its constituents are regenerated. All-*trans*-retinal can be converted enzymatically in the dark back to the 11-*cis* form. After bleaching, all-*trans*-retinal is rapidly reduced to all-*trans*-retinol, in the rod outer segment. The latter is then transferred (presumably via IRBP) into the retinal pigment epithelial cells, where it is esterified (again, predominantly with palmitic acid) and stored in the bulk lipid of those cells. The regeneration of rhodopsin, which occurs in the dark-adapted eye, involves the simultaneous hydrolysis and isomerization of retinyl esters to yield 11-*cis*-retinol and then 11-*cis*-retinal, which is transferred into the rod outer segment via IRBP. Studies have revealed that the conversion of all-*trans*-retinyl esters to 11-*cis*-retinal in the retinal pigment epithelium is catalyzed by the microsomal membrane protein RPE65.[80] RPE65 is involved in the regeneration of rhodopsin in rods and of the opsins in cones. Its activity appears to be regulated through the addition/release of palmitic acid: palmitoylation of the protein converts the protein from a form soluble in the cytosol to a membrane-bound form, thus controlling its capacity to present its retinoid ligand to cytosolic isomerohydrolase. Nervous recovery is effected by the GTPase activity of the transducin α-subunit, which, by hydrolyzing GTP to GDP, causes the reassociation of transducin subunits and, hence, the loss of its activating effect on cGMP phosphodiesterase. Metarhodopsin II is also removed by phosphorylation to a form incapable of activating transducin and by dissociation to yield opsin and all-*trans*-retinal.

The visual cycle of cones differs from that of rods. Cones have 100-fold lower light sensitivity but 10-fold faster recovery rates compared to rods, suggesting that they may have access to a source of visual chromophore not available to rods. In cones, the oxidation of 11-*cis*-retinol to 11-*cis*-retinal is NADP-dependent, and the isomerization of all-*trans* to 11-*cis*-retinal occurs in a two-step process localized in Müller glial cells. In contrast, the oxidation step in rods is NAD-dependent and the isomerization is a one-step process localized in the retinal pigment epithelium.

Nonvisual opsins. Humans and other mammals have a subset of retinal ganglion cells, as well as cells in the

77. Rods are the most numerous (c.120 million) and are the very sensitive to light intensity, although not to color. Color vision is the provence of the 6–7 million cones, which are be concentrated in the **macula**, particularly in a small (0.3 mm diameter), rod-free area the **fovea centralis**, except for the "blue cones", which are distributed mostly outside of the fovea. While cones comprise only 5% of all photoreceptors in the human eye, they are more numerous in other species, e.g., comprising 60% of the photoreceptors in the chicken eye.

78. A Schiff base is a type of imine with the structure $R_2C=NR'$ often used in coordination complexes. It was described by the German chemist Hugo Schiff (1834–1915) who founded the Chemical Institute of the University of Florence.

79. The conformation of rhodopsin is changed to yield a transient photopigment, **bathorhodopsin**, which, in turn, is converted sequentially to **lumirhodopsin**, **metarhodopsin I**, and (by deprotonation) **metarhodopsin II**.

80. The protein was named for its apparent molecular mass under denaturing conditions, 65 kDa; further studies revealed that its true mass is 61 kDa. Mutations in RPE65 have been found to cause Leber's congenital amaurosis and other forms or autosomal recessive retinitis pigmentosa.

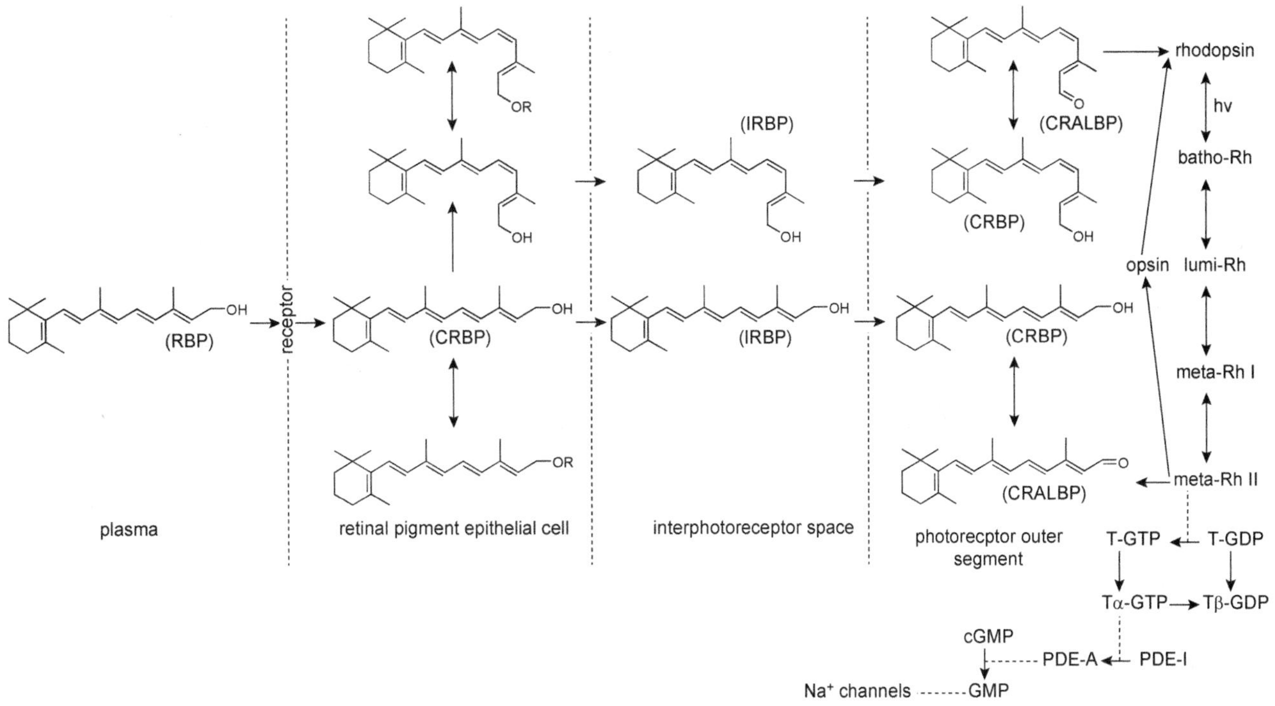

FIGURE 6.10 Vitamin A in the visual cycle. *PDE-A*, phosphodiesterase (active); *PDE-I*, phosphodiesterase (inactive); *Rh*, rhodopsin; *T*, transducin.

suprachiasmatic nucleus of the brain, that express the photopigment **melanopsin**. The light-induced isomerization of its ligand, 11-*trans*-retinal to the 11-cis-form suggests that melanopsin may function in circadian regulation. Other species express various nonvisual opsins (at least 17 have been described) in such tissues as the iris, skin, pineal, and deep brain, where they are thought to function in nonimage-forming photoreceptors that may be involved in photoisomerizations, detection of environmental light, and modulation of arousal states.

Systemic Functions of Vitamin A

Vitamin A supports systemic functions (e.g., corneal integrity, immunity, growth) in which vitamin A acts like a hormone. These functions appear to rely on retinoic acid; because the oxidation of retinal to retinoic acid is irreversible, retinoic acid can support only these systemic functions. Animals fed diets containing retinoic acid as the sole source of vitamin A grow normally and appear healthy in everyway except that they go blind.

Vitamin A-deficient animals die but not from lack of visual pigments. The extraretinal functions of vitamin A are of greater physiological importance than the visual function.

Epithelial differentiation. Chief among the systemic functions of vitamin A is its role in epithelial cell differentiation. Vitamin A-deficient individuals experience replacement of normal mucus-secreting cells by cells that produce keratin, particularly in the conjunctiva and cornea of the eye, the trachea, the skin, and other ectodermal tissues. Less severe effects occur in tissues of mesodermal or endodermal origin. The actions of retinoids in cell differentiation are analogous to those of the steroid hormones, i.e., they bind to the nuclear chromatin to signal transcriptional processes. In fact, studies have revealed that the differentiation of cultured cells can be stimulated by exposure to retinoids, and that abnormal mRNA species are produced by cells cultured in vitamin A-deficient media.[81] Further, retinoic acid has been found to stimulate, synergistically with **thyroid hormone** (T_3)[82], the production of growth hormone in cultured pituitary cells. Some studies with culture cell models have found all-*trans*-retinoic acid treatment to upregulate the expression of certain micro-RNAs,[83] suggesting epigenetic bases for some vitamin A actions.

Immune function. That vitamin A has an important role in immune function is indicated by the fact that vitamin

81. Epidermal keratinocytes cultured in a vitamin A-deficient medium made keratins of higher molecular weight than those made by vitamin A-treated controls. Different mRNA species encoded the different proteins produced under each condition.
82. Triiodothyronine.
83. miRNAs are a class of small (18–23 nucleotides), noncoding RNAs involved in posttranslational regulation of gene expression.

A-deficient animals and humans are typically more susceptible to infection than are individuals of adequate vitamin A nutriture.[84] They show changes in lymphoid organ mass, cell distribution, histology, and lymphocyte characteristics. Vitamin A deficiency leads to histopathological changes that provide environments conducive to secondary infection. This is supported by findings of excess bacteriuria among xerophthalmic compared to nonxerophthalmic, malnourished children in Bangladesh; and negative correlation of plasma retinol level and bacteria adherent to nasopharyngeal cells of children in India. Such outcomes involve impaired epithelial integrity (including mucosal immunity), antibody responses, and lymphocyte differentiation. Underlying these effects are roles of vitamin A in inducing heightened primary immune responses, enhancing memory responses, and accelerating the expansion of the mature B-lymphocyte pool. These functions appear to be discharged by retinoic acid acting at the nuclear level, i.e., by ligating RAR (retinoic acid receptor)–RXR (Retinoid X receptor) to alter the expression of genes affecting immune function. Retinoic acid has, therefore, been called a "*fourth signal*" of the antibody response.[85] In this way, retinoic acid induces changes that imprint antibody-secreting cells by permanently committing them. Studies have shown this function to be particularly important in maintenance of mucosal immunity, which depends on the retinoic acid-dependent modification of B cells by gut-associated dendritic cells within the lymphoid tissues of the small intestine. Both RAR and RXR receptors are constitutively expressed by both B and T cells. Vitamin A, which has long been known to prevent atrophy of the thymus, functions as retinoic acid in at least some subtypes of T cell, including T-helper lymphocytes, the antigen-presenting cells that stimulate B cells to produce immunoglobulin A (IgA). It is known that retinoic acid is a modulator of the differentiation of T cells from interleukin-17-secreting T-helper cells and toward Foxp3$^+$ T regulatory cells, and it may also play a role in determining the development of T cells toward CD^{+4} versus CD^{+8} status.

Reproduction. Studies with animals have demonstrated that vitamin A is required for normal reproduction. For example, rats maintained with retinoic acid grow well and appear healthy but lose reproductive ability, i.e., males show impaired spermatogenesis and females abort and resorb their fetuses. Injection of retinol into the testis restores spermatogenesis, indicating that vitamin A has a direct role in that organ. The chicken has been shown to require retinoic acid for spermatogenesis; all mammalian species examined to date require retinol or retinal. It has been proposed that these effects are secondary to lesions in cellular differentiation and/or hormonal sensitivity. Several researchers have found that vitamin A-deficient dairy cows show reduced corpus luteal production of progesterone and increased intervals between luteinizing hormone peak and ovulation. Some evidence indicates that hormonal parameters respond to oral treatment with β-carotene but *not* preformed vitamin A, suggesting the importance of retinol/retinal production in situ.

Retinoids play fundamental roles as differentiating agents in morphogenesis. Deprivation of vitamin A results in the Japanese quail results in the loss of normal specification of heart left–right asymmetry. That this effect is associated with the decreased expression of RARβ$_2$ in the presumptive cardiogenic mesoderm suggests that retinoids may direct the differentiation of mesoderm into the heart lineage. Studies of the regenerating amphibian limb have revealed profound effects or retinoids in providing positional information to enable cells to differentiate into the pattern of structures relevant to their appropriate spatial locations. On the basis of such observations the morphogenic role of vitamin A was proposed to involve concentration gradients of RARs/RXRs due to differential induction of the receptor by the retinoid, which established positional identity. However, it now appears that embryos have multiple areas with different responsiveness to retinoic acid caused by local differences in the production and binding of, and sensitivity to, retinoic acid. These differences vary among tissues during development, with retinoic acid acting primarily in a paracrine manner in pluripotent cells. In early development retinoic acid signals the posterior neuroectoderm, trunk mesoderm, and foregut endoderm in the organization of the trunk; in later development, it signals development of the other organs including the eye (Table 6.10).

The expression of many proteins results from retinoic acid-induced cell differentiation. Several are induced by activation through RXR/RAR binding: growth hormone[86] (in cultured pituitary cells), the protein laminin (in mouse embryo cells), the respiratory chain-uncoupling protein of brown adipose tissue (suggesting a role in heat production and energy balance), the vitamin K-dependent matrix Gla protein, and the RARs. This indicates autoregulation, i.e., retinoic acid induces its own receptor. In fact, the induction of RARs appears to be differentially selective among various tissues; retinoic acid has been found to induce mainly RARα in hemopoietic cells but RARβ in other tissues.

Bone metabolism. Vitamin A has an essential role in the normal metabolism of bone. Animal studies indicate that both low- and high-vitamin A intakes reduce bone mineral

84. In practice, it can be difficult to ascribe such effects simply to the lack of vitamin A, as deficient individuals generally also have protein–calorie malnutrition which, itself, leads to impaired immune function.

85. The others being signal 1—receptor-binding of antigen; signal 2—receptor-binding costimulatory/accessory factors; signal 3—binding of "*danger signals*", e.g., lipopolysaccharide, to Toll-like receptors.

86. That vitamin A may be required for the expression of growth hormone in humans is suggested by the correlation of plasma retinol and nocturnal growth hormone concentrations in short children.

TABLE 6.10 Genes Regulated by Retinoic Acid in Embryonic Development

Aspect of Development	Expression Induced	Expression Repressed
Hindbrain a.-p.[a] patterning	*Hoxa1, Hoxa3, Hoxb1, Hoxd4, vHnf1*	
Spinal cord motor neuron differentiation	*Pax6, Olig2*	
Early somite formation	*Cdx1*	*Fgf8*
Heart a.-p.[a] patterning		*Fgf8*
Forelimb differentiation		*Fgf8?*
Pancreas differentiation	*Pdx1*	
Lung differentiation	*Hoxa5*	*TGF-β1*
Anterior eye formation	*Pitx2*	
Kidney formation	*Ret*	
Meiosis induction	*Stra8*	

[a]*Anteroposterior.*
After Duester, G., 2008. Cell 134, 921–931.

TABLE 6.11 Prevalence of Vitamin A Deficiency Related to Hemoglobin Status in Indonesian Preschool Children

Hemoglobin (g/dL)	Prevalence of Vitamin A Deficiency (%)[a]
<11.0	54.2
11.1–12.0	43.3
>12.0	34.3

[a]*Based on conjunctival impression cytological assessment.*
After Lloyd-Puryear, M.A., Mahoney, J., Humphrey, F., et al., 1991. Nutr. Rev. 11, 1101–1110.

density.[87] Observational studies in humans, however, have yielded inconsistent results in this regard. Retinoids are thought to be involved in regulating the phenotypic expression of bone-mobilizing cells, osteoclasts, which are reduced in vitamin A deficiency. The consequently unchecked function of bone-forming cells, osteoblasts, would appear to result in excessive deposition of periosteal bone and a reduction in the degradation of glycosaminoglycans. It also appears that 9-*cis*-retinoic acid can serve as an affector of the vitamin D-induced renal calcification involving the vitamin K-dependent matrix γ-carboxyglutamic acid (GLA) protein.

Hematopoiesis. Chronic deprivation of vitamin A leads to anemia. Cross-sectional studies have shown low hemoglobin levels to be associated with the prevalence of xerophthalmia in children (Tables 6.11 and 6.12), and children with mild-to-moderate vitamin A deficiency or mild xerophthalmia have lower circulating hemoglobin levels than nondeficient children. In fact, serum retinol level explains 4–10% of the variation in hemoglobin level among preadolescent children in developing countries. The hematological response to vitamin A deficiency is biphasic, involving an initial fall in both hemoglobin and erythrocyte count due to apparently impaired hemoglobin synthesis, followed by the rise in both variables late in deficiency due to hemoconcentration resulting from dehydration secondary to reduced water intake and/or diarrhea.

The metabolic basis of the role of vitamin A in hematopoiesis appears to involve the mobilization and transport of iron from body stores as well as the enhancement of nonheme–iron bioavailability. The results of cross-sectional studies in developing countries have found serum iron to be positively correlated with serum retinol levels, and animal studies have shown vitamin A deprivation to cause decreases in both hematocrit and hemoglobin levels, which precede other disturbances in iron storage and absorption. Vitamin A deficiency reduces the activity of ceruloplasmin, a copper-dependent protein with ferroxidase activity, which is important in the enteric absorption of iron. This effect appears to occur as the results of a posttranscriptive disruption in the activity. In addition, the results of in vitro studies demonstrate that all-*trans*-retinol induces the differentiation and proliferation of pluripotent hemopoietic cells, which suggest that the anemia of vitamin A deficiency is initiated by impairments in erythropoiesis and accelerated by subsequent impairments in iron metabolism. The presence of vitamin A or β-carotene has been found to increase the enteric absorption of iron from both inorganic and plant sources; this has been explained on the basis of the

87. See review: Henning, P., Conaway, H.H., Lerner, U.H., 2015. Front. Endocrinol. 6, 1–13.

TABLE 6.12 Efficacy of Vitamin A Supplementation on Increasing Hemoglobin Levels in Anemic Subjects

Country	Subject Age	Vitamin A Dosage (μg RE/d)	Follow-up	n	Hemoglobin (g/dL)	
					Baseline	Follow-up
Indonesia[a]	<6 years	0	5 months	240	11.4±1.6	11.2±1.5
		240	5 months	205	11.3±1.6	12.3±1.6*[b]
Guatemala[c]	1–8 years	0	2 months	20	10.4±0.7	10.7±0.6
		2400	2 months	25	10.3±0.8	11.2±0.8*
Indonesia[d]	17–35 years	0	2 months	62	10.4±0.7	10.7±0.6
		3000	2 months	63	10.3±0.8	11.2±0.8*

[a]*Mahilal, P.D., Idjradinata, Y.R., Muheerdiyantiningsih, K.D., 1988. Am. J. Clin. Nutr. 48, 1271–1276.*
[b]*Significantly different from baseline level,* $p<.05$.
[c]*Mejia, L.A., Chew, F., 1988. Am. J. Clin. Nutr. 48, 595–600.*
[d]*Suharno, D., West, C.E., Muhilal, K.D., et al., 1993. Lancet 342, 1325–1328.*

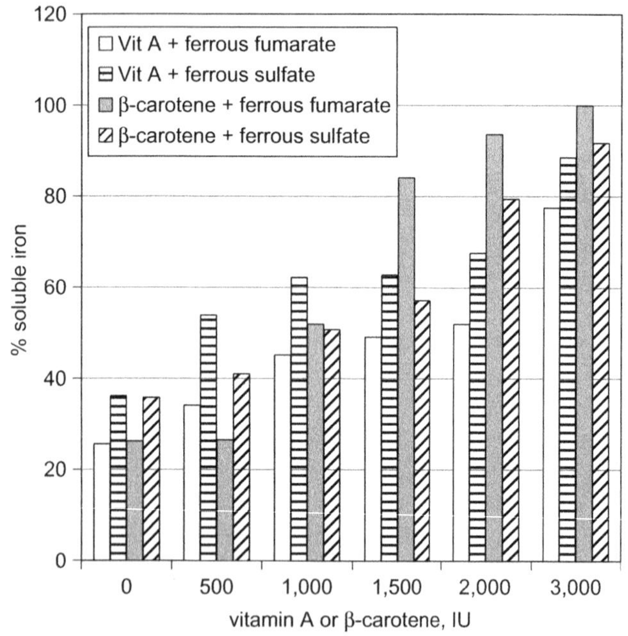

FIGURE 6.11 Effect of vitamin A and β-carotene on iron solubility (at pH 6). *After Garcia-Casal, M.N., Layrisse, Solano, L., et al., 1998. J. Nutr. 128, 646–650.*

TABLE 6.13 Effects of Vitamin A and β-Carotene on the Bioavailability of Nonheme Plant Iron for Humans

Iron Source	Vitamin A (μmol)	β-Carotene (μmol)	Iron Absorption (%)
Rice	0	0	2.1
	1.51	0	4.6[a]
	0	0.58	6.4[a]
	0	0.95	8.8[a]
Corn	0	0	3.0
	0.61	0	6.6[a]
	0	0.67	8.5[a]
	1	1.53	6.3[a]
Wheat	0	0	3.0
	0.66	0	5.5[a]
	0	0.85	8.3[a]
	0	2.06	8.4[a]

[a]$p<.05$, $n=11$–20 *subjects.*
From Garcia-Casal, M.N., Layrisse, M., Solano, L., et al., 1998. J. Nutr. 128, 646–650.

formation of complexes with iron that are soluble in the intestinal lumen, thus, blocking the inhibitory effects of iron absorption of such antagonists as phytates and polyphenols (Fig. 6.11; Table 6.13). Clinical trials have shown intervention with both iron and vitamin A to be more effective on correcting anemia than intervention with iron alone.

Retinoids are involved in the differentiation of myeloid cells into neutrophils, which occurs in the bone marrow; this function appears to involve all-*trans*-retinoic acid, as RARα is the predominant retinoid receptor type found in hematopoietic cells. Vitamin A-deficient animals have been found to sequester retinol in their bone marrow;[88] this response may ensure adequate vitamin A for the growth and differentiation of myeloid cells, which would explain the observation that vitamin A-deficient individuals do not necessarily show neutropenia.[89]

88. Twining, S.S., Schulte, D.P., Wilson, P.M., et al., 1996. J. Nutr. 126, 1618–1626 found vitamin A deprivation to reduce the retinol contents of rat bone marrow by 75%.
89. Abnormally low circulating levels of neutrophils.

Skin health. Vitamin A has a role in the normal health of the skin. Its vitamers, as well as carotenoids, are typically found in greater concentrations in the subcutis than in the plasma (significant amounts are also found in the dermis and epidermis), indicating the uptake of retinol from plasma RBP4. Epithelial cell phenotypes are regulated by hormonal cycles and vitamin A intake; vitamin A deficiency impairs the terminal differentiation of human keratinocytes and causes the skin to be thick, dry, and scaly. It also results in obstruction and enlargement of the hair follicles.[90]

Vitamin A Regulation of Gene Transcription

Vitamin A discharges its systemic functions through the abilities of all-*trans*-retinoic acid and 9-*cis*-retinoic acid to regulate the expression of some 300 genes at specific target sites in the body. The hormone-like regulation of transcription by retinoids is receptor-mediated. Retinoic acid binds to two members of a highly conserved superfamily of proteins that act as nuclear receptors for steroid hormones including 1,25-$(OH)_2$-vitamin D_3 and thyroid hormone (T_3).[91] These nuclear receptors have similar ligand-binding and DNA-binding domains as well as substantial sequence homology. Retinoic acid is thought to interact with them in ways similar to their other ligands, with each receptor binding to regulatory elements upstream from the gene and acting as a ligand-activated transcription factor (Table 6.14). The **RAR**s function by attracting low-molecular weight coactivators, releasing corepressors, and forming obligate heterodimeric complexes with another retinoid receptor, the **RXR**. The RXRs form transcriptionally inactive homotetramers when not bound to ligand; these dissociate to form active homodimers upon retinoid binding. Because the tetramers can bind two DNA recognition sequences simultaneously, the retinoid-induced shift to the liganded dimer results in a change in DNA geometry. The involvement of RXRs with multiple binding partners, including PPARs, the vitamin D receptor (VDR), and farnesoid X receptors (FXR), makes them the master regulators of multiple pathways signaling transcription in response to lipophilic nutrients and hormones (Fig. 6.12). The regulation of gene expression also involves coactivators and corepressors that function through the modification of chromatin.

All isomers of retinoic acid bind the RARs; greatest affinities in vitro are shown by all-*trans*-retinoic acid (K_d = 1–5 nM). Three RAR subtypes have been identified (α, β, and γ). That only all-*trans*-retinoic acid functions as the endogenous ligand for RAR is suggested by the demonstration that growth arrest in mice lacking retinaldehyde dehydrogenase (incapable of producing all-*trans*-retinoic acid) can be rescued by the all-*trans* but not the 9-*cis* vitamer. The ligand-binding domains of the RARs are highly conserved (showing 75% identity in terms of amino acid residues). $RAR\alpha_1$ and $RAR\alpha_2$ share 7 of their 11 exons, suggesting that they arose from a common ancestral RAR gene. Different promoters, however, direct the expression of each in an unusual organization involving the 5′-untranslated regions of the genes divided among different axons. In the case of $RAR\alpha_1$, the 5′ region is encoded in three axons: two contain most of the untranslated region and the third encodes the remainder of that region plus the first 61 amino acids peculiar to $RAR\alpha_1$.

The RXRs show generally weak homology with the RARs, the highest degree of homology (61%) being in their DNA-binding domains. On the basis of homologies with an insect locus, it is thought that the RXRs may have evolved as the original retinoid-signaling system. Three RXR subtypes have been identified α, β, and γ); RXRα responds to somewhat higher retinoic acid levels than other RXR isoforms. Retainaldehyde has also been found to bind weakly RXR (and PPARγ) to inhibit activation (linked to inhibition of adipogenesis). Both RARs and RXRs are found in most tissues; greatest concentrations occur in adrenals, hippocampus, cerebellum, hypothalamus, and testis. Both

TABLE 6.14 Nuclear Retinoic Acid Receptors

Receptor	Isoforms[a]	Ligands
RARα (retinoic acid receptor)	$RAR\alpha_1$, $RAR\alpha_2$	All-*trans*-RA[b], 9-*cis*-RA[c], 13-*cis*-RA[d]
RARβ	$RAR\beta_1$, $RAR\beta_2$, RARβ[e]	All-*trans*-RA, 9-*cis*-RA[c], 13-*cis*-RA[d]
RARγ	$RAR\gamma_1$, $RAR\gamma_2$[c]	All-*trans*-RA, 9-*cis*-RA[c], 13-*cis*-RA[d]
RXRα (retinoid X receptor)		9-*cis*-RA
RXRβ		9-*cis*-RA
RXRγ		9-*cis*-RA
PPARβ/δ (peroxisome proliferator-activated receptor)		All-*trans*-RA[c]
PPARα		All-*trans*-RA[c]
PPARγ		All-*trans*-RA[c]

[a]*Isoforms differ only in their N-terminal regions.*
[b]*Retinoic acid.*
[c]*Binding shown only in vitro.*
[d]*Binding (weak) shown only in vitro.*
[e]*Identified in* Xenopus laevis.

90. That is, follicular hyperkeratosis, which can also be caused by deficiencies of niacin and vitamin A.
91. Triiodothyronine.

the RARs and RXRs can bind **9-*cis*-retinoic acid**[92] in vitro with high affinity (K_d = 10 nM); however, the physiological relevance of this vitamer is unclear as it has not been identified in vivo. RARs are abundant in the brain and pituitary gland (RARα distributed throughout; RXRα and RXRδ in striatal regions with dopaminergic neurons) where they function in regulating expression of the dopamine receptor.

All-*trans*-retinol can also bind **PPAR**s, PPARβ/δ having the greatest affinity. The shuttling of the retinoid to PPARβ/δ is accomplished by the fatty acid-binding protein 5 (FABP5), in contrast to CRABP (II), which delivers retinoids to other receptors. Therefore, the partitioning of retinoid between PPARβ/δ and the RARs and RXRs is a function of the relative amounts of FABP5 and CRABP (II) in cells. Cells in which CRABP (II) predominates express primarily through the RARs; cells with relatively high FABP5 levels express primarily through PPARβ/δ. In the presence of the coactivator SRC-1, the retinoid activation of PPARβ/δ led to the upregulation of expression of 3-phosphoinositide-dependent kinase 1 (PDK1), an activator of the antiapoptotic factor Akt1.[62] Activation of PPARδ, which is downregulated in adipose tissue in obese individuals, induces expression of genes affecting lipid and glucose homeostasis, including the insulin-signaling gene *PDK1*, resulting in improved insulin action. This mechanism is thought to underlie retinoic acid-induced weight loss in obesity-prone mice.[93]

Two types of high-affinity **RAREs** have been identified in the promoter regions of target genes near the transcription start: those that recognize the RXR homodimer and those that recognize the RXR–RAR heterodimer. The RXR–RAR heterodimer binds to RAREs, which consist of direct repeats of the consensus half-site sequence AGGTCA usually separated by five nucleosides. The RXR homodimer binds to cognate retinoid X response elements most of which are direct repeats of AGGTCA with only one nucleoside spacing. Gene expression is effected by activating each response element present in the promoter regions of responsive genes.[94] The RXRs can also form homotetramers as well as dimers with other members of the steroid/thyroid/retinoic acid family; heterodimerization in this system has usually been found to increase the efficiency of interactions with DNA and, thus, transcriptional activation. Further regulation is effected in this signaling system as the RXR–RAR heterodimer appears to repress the transcription-activating function of RXR–RXR.

In the absence of retinoic acid, the aporeceptor pair (RXR–RAR/RXR) binds to the RAREs of target genes, and RAR recruits corepressors that mediate negative transcriptional effects by recruiting histone deacetylase complexes, which modify histone proteins to induce changes in chromatin structure that reduce the accessibility of DNA to transcriptional factors. This process is reversed upon retinoic acid binding: a conformational change in the ligand-binding domain results in the release of the corepressor and the recruitment of coactivators of the AF-2 region of the receptor. Some cofactors interact directly to enhance transcriptional activation, while others can affect the acetylation histone proteins causing the conformational opening of chromatin and the activation of transcription of the target gene. Impairments in the process can lead to carcinogenesis.

The general picture is one of RXRs forming heterodimers with RARs that are activated by retinoids and with other receptors of the same superfamily, i.e., thyroid hormone receptors, the vitamin D_3 receptor, PPAR, and, probably, others yet unidentified (Fig. 6.12). The metabolite 9-*cis*-retinoic acid, which targets RXR, causes the formation of RXR homodimers that recognize certain RAREs. However, the same ligand inhibits the formation of heterodimers of RXR and the thyroid hormone receptor (TR), which reduces the expression of thyroid hormone-responsive genes.[95] In contrast, RXR-specific ligands do not appear to affect the formation of RAR-containing heterodimers. CRABP(II) facilitates interactions of RARα–RXRα heterodimers in forming a gene-bound receptor complex, delivering retinoic acid to its nuclear receptors and acting as a coactivator of the expression of retinoic acid-responsive genes. Retinoid responses appear to be restricted to a subset of retinoid-responsive genes through the action of the orphan COUP (chicken ovalbumin upstream promoter) receptors, which form homodimers that avidly bind several RAREs and repress both RAR–RXR and RXR–RXR activities. The RARs comprise a two-component signaling system for activating transcriptions in which retinoic acid can act in either a paracrine or autocrine manner, as both all-*trans*-retinoic acid and 9-*cis*-retinoic acid can be synthesized within the target cell (from retinol via retinal) or delivered to that cell from the circulation. In either case the ligand is transported intracellularly to the nucleus (via CRABP) and then to the appropriate receptor (RAR, RXR), which then can bind to its cognate RARE to regulate transcription. The RARs are part of a larger system of nuclear receptors (RARs, RXRs, PPARs, VDR, FXR, etc.) that integrates a range of metabolic signaling from retinoids, carotenoid metabolites, and fatty acids in

92. Yet to be identified in vivo, 9-*cis*-retinoic acid is known to be produced from 9-*cis*-retinol by 9-cis-retinol dehydrogenase with subsequent oxidation, by cleavage of 9-*cis*-β-carotene, and from the isomerization of all-*trans*-retinol in the lung.

93. Berry, D.C., Noy, N., 2009. Mol. Cell. Biol. 29, 3286.

94. The RXR–RAR response elements consist of polymorphic arrangements of the nucleotide sequence motif 5′-RG(G/T)TCA-3′. These gene elements are also responsive to thyroxine, suggesting that retinoic acid and thyroid hormone may control overlapping networks of genes.

95. This has been shown for the expression of uridine-5′-diphosphate-glucuronyl transferase, which is involved in the phase II metabolism of xenobiotic and endogenous substrates.

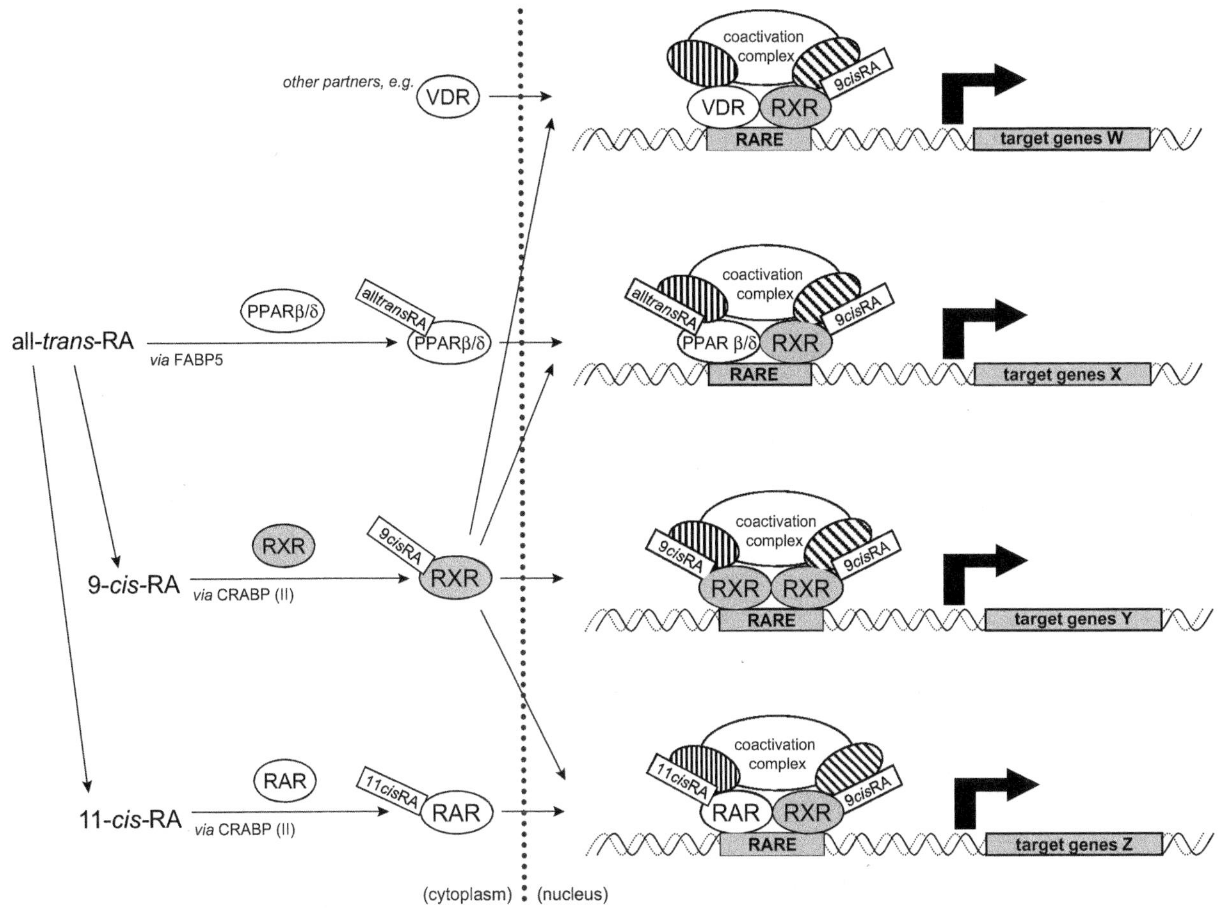

FIGURE 6.12 Participation of retinoid receptors and other nuclear receptors in integrating metabolic signaling in controlling expression of multiple genes. Retinoid activation of receptors that participate in complexes that bind to cognate DNA sequences to cause gene expression (*bold arrows*). The involvement of retinoid receptors (RARs and RXRs) with multiple binding partners allows retinoids to signal transcription of multiple pathways (*see text* for abbreviations.).

controlling the expression of multiple genes and affecting many tissues.

In this system, vitamin A and T_3 appear to play compensatory signaling roles. Studies with rats have shown that deprivation of either factor impairs thyroid signaling in the brain through reduced expression of RAR and TR, as well as a neuronal protein neurogranin.[96] Similarly, regulation of thyroid-stimulating hormone (TSH) has been found to be dependent on both the binding of both TR and RXR (which are activated by T_3 and 9-*cis*-retinoic acid, respectively) to the TSH gene.

Coenzyme Role for Vitamin A?

A coenzyme-like role has been proposed for vitamin A. This hypothesis holds that vitamin A can act as a sugar carrier in the synthesis of **glycoproteins**, which function on the surfaces of cells to effect intercellular adhesion, aggregation, and recognition. Indeed, retinol can be phosphorylated to yield retinyl phosphate, which can accept mannose from GDP-mannose (to form retinyl phosphomannose), which it then donates to a membrane-resident acceptor in the production of glycoproteins. Further, vitamin A-deficient animals appear to synthesize fewer glycoproteins and more abnormal ones (particularly in plasma, intestinal goblet cells, and corneal and trachea epithelial cells) than vitamin A-adequate animals. Still, the physiologic significance of this putative sugar carrier role of vitamin A is not clear. Retinoic acid, the vitamer that supports the systemic functions of the vitamin, cannot serve as a sugar carrier because it cannot be reduced to retinol. Neither can retinyl phosphate accept mannose. It has been proposed that retinoic acid may need to be hydroxylated to a phosphorylatable derivative that could serve as a sugar carrier.

96. Neurogranin is a calmodulin-binding protein expressed in dendritic spines of the brain and participating in protein kinase C signaling.

8. BIOMARKERS OF VITAMIN A STATUS

Adequately nourished individuals have appreciable stores of vitamin A, which tend to mitigate against the effects

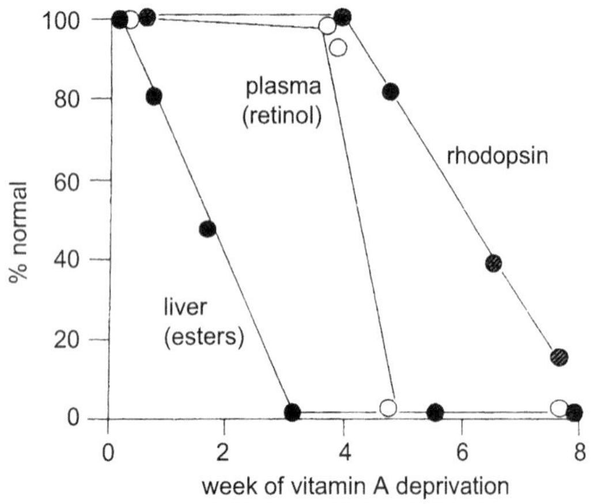

FIGURE 6.13 Hepatic vitamin A stores must be depleted before changes in retinol concentrations in circulation or photopigments occur.

of periods of low intakes of the vitamin. These stores are comprised of fast and slowly turning over pools (Fig. 6.13). As long as hepatic stores are sufficient to provide retinol, the plasma retinol level is minimally affected by vitamin A deprivation. Cellular functions of vitamin A can be expected to change only after plasma retinol–RBP4 concentrations have dropped below about 20 μg/g (0.07 μM). This phenomenon presents challenges in assessing vitamin A status, particularly at the margins of the normal range of plasma retinol concentrations.

Biochemical Assessment

The vitamers A can be accurately quantified in biological specimens using high-performance liquid–liquid partition chromatography (HPLC). Serum retinol is, therefore, a convenient parameter of vitamin A status. The homeostatic control of circulating retinol levels makes this parameter useful only in identifying subjects with chronically low-vitamin A intakes sufficient to exhaust their hepatic stores that support the synthesis of RBP4. Vitamin A-deficient subjects with still-appreciable liver stores will have serum retinol levels in the normal range. Under normal circumstances, 85% of plasma RBP4 exists in the holo-complex in with equimolar amounts of retinol; plasma levels of 0.70–1.05 μM indicate marginal vitamin A status and levels <0.70 μM (20 μg/dL) are considered indicative of vitamin A deficiency.

However, other factors can influence apo-RBP4 levels, affecting its value as a proxy for plasma retinol. Deficiencies of protein and zinc, hepatic steatosis, and injury can impair RBP synthesis in the liver. Hepatic release of apo-RBP to the circulation can be impaired under conditions of inflammation. Febrile illnesses and inflammation can reduce TTR binding to RBP4, allowing it to be lost by renal glomerular filtration. In contrast, obesity, in which RBP4 is released from adipocytes, and renal disease, in which RBP turnover can be less than one-10th of normal, can elevate plasma levels of RBP4, thus biasing assessments toward vitamin A adequacy. Such confounding effects may explain reports that one-fifth of night blind children have serum retinol greater than 0.70 μM and can show xerophthalmia and or conjunctival metaplasia, and that healthy adults depleted of vitamin A can show impaired dark adaptation at serum retinol levels of 20–30 μg/dL. Therefore, assessment of vitamin A status by RBP4 or plasma retinol should consider subject status with respect to general nutritional, inflammation, and general health; it is prudent to include such measures as BMI and, in plasma samples, albumin, C-reactive protein, and α_1-acid glycoprotein (AGP), which can suppress plasma retinol concentrations by as much as 24%.[97]

The concentrations of vitamin A in breast milk (i.e., primarily retinyl palmitate in milk fat) drop in vitamin A deficiency and can, therefore, be used to detect the deficiency in mothers.[98]

Dose–Response Tests

Because of the uncertainties of interpreting serum retinol values, tests have been devised to assess the mobilizable vitamin A capacity of the liver based on serum responses to oral doses of a retinoid.

- Relative **dose–response (RDR) test**. This test assesses hepatic capacity to mobilize vitamin A based on serum responses to oral doses of retinyl ester (retinyl acetate or retinyl palmitate). The test requires two samples of blood: a fasting sample drawn immediately before the administration of the test dose; another sample drawn 5 h later. Retinol is determined in each sample, and the percentage response over baseline is calculated.
- **Modified relative dose–response (MRDR) test**. This test assesses hepatic capacity to mobilize vitamin A based on serum responses to oral doses of **3,4-didehyd roretinol**.[99] It requires only a single blood sample taken 4–6 h after oral administration of the retinoid; both retinol and 3,4-didehydroretinol are determined in the sample, and the response is taken as the molar ratio of the two analytes.

Both tests assume that the appearance of retinoid in the plasma is a function of the amount also entering from endogenous hepatic stores. Therefore, an RDR value of

97. Thurnham, D.J., Northrop-Clewes, C.A., Knowles, J., 2015. J. Nutr. 145, 1137S–1143S.
98. Breast milk retinyl palmitate concentrations typically fall in the range of 1.75–2.45 μM; levels ≤1.05 μM (i.e., ≤8 μg/g milk fat) indicate vitamin A deficiency.
99. Also called **dehydroretinol** and **vitamin A_2**.

≥20% or an MRDR value of 20–30% is taken as indicative of inadequate hepatic storage (<0.07 μmol/g) of the vitamin; children with such low stores are almost certain to be vitamin A deficient. Determination of serum retinoids is accomplished using HPLC; smaller blood samples can be employed by using RPB4 as the endpoint, as that protein determined in very small volumes by enzyme-linked immunoassay. There is, however, a caveat: the use of RBP4 can yield high rates of false positives, particularly in subjects with high BMIs.[100] Such observations are presumably due to the presence in serum of adipose-derived apo-RPB4, which reduces the retinol:RBP4 ratio. Nevertheless, the RDR and MRDR tests remain the instruments of choice as categorical indicators of vitamin A status. While neither give quantitative measures of hepatic vitamin A stores, each can indicate changes in vitamin A status from low to adequate. The MRDR test is particularly useful in assessing impacts of interventions in which serum retinol is not expected to change, e.g., in subjects of marginal status. It has shown that preterm infants generally have low vitamin A stores.

Palmar Scanning

Noninvasive assessment of carotenoid status has been accomplished by scanning the palmar surface of the hand using resonance Raman spectroscopy. This procedure has been shown to be useful in monitoring subject responses to changes in the intake of carotenoid-rich foods;[101] however, as currently developed, it is not specific for provitamin A carotenoids (or vitamin A status).

9. VITAMIN A DEFICIENCY

Like most nutritional deficiencies in human populations, vitamin A deficiency is an outcome of a bio-eco-social system that fails to provide sources of the vitamin in ways that are at once accessible and utilizable. The complexity of this system has been captured in the WHO conceptual model (Fig. 6.14). In humans, vitamin A deficiency is typically associated with malnutrition, particularly protein–energy malnutrition. These conditions have common origins in grossly unbalanced diets and poor hygiene. Accordingly, malnourished children are likely to be deficient in vitamin A and other essential nutrients. Protein deficiency also impairs the synthesis of apo-RBP4, CRBP, and other retinol binding proteins, affecting vitamin A transport and cellular utilization. The storage of vitamin A mitigates the effects of periodic low dietary intakes of the vitamin. Recommended daily allowances for vitamin A (Table 6.15) are set to insure that body stores are sufficient to support adequate plasma retinol–RBP4 concentrations. When they are not, clinical signs of vitamin A deficiency are manifest.

Primary vitamin A deficiency can occur among children and adults who consume diets composed of few

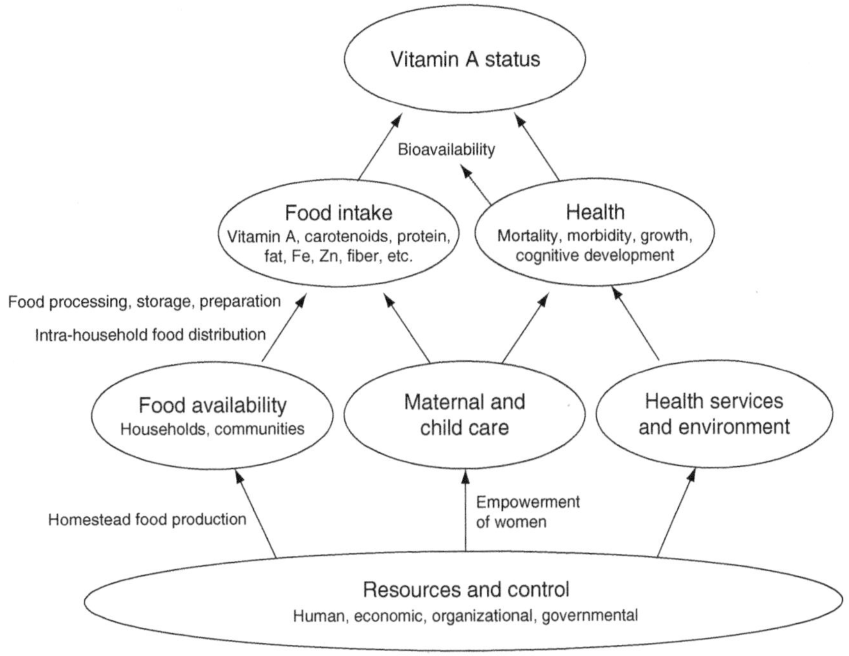

FIGURE 6.14 WHO conceptual framework of causes of vitamin A deficiency in human populations. *After UN Food Nutr. Bull. vol. 19, 1998.*

100. Fujita, M., Brindle, E., Rocha, A., et al., 2009. Am. J. Clin. Nutr. 90, 217–224.

101. Jahns, L., Johnson, L.K., Mayne, S.T., et al., 2014. Am. J. Clin. Nutr. 100, 930–937.

TABLE 6.15 Recommended Daily Allowances for Vitamin A

Age–Sex	US RDA[a] (μg RE/d)	Age–Sex	FAO/WHO RNI[b] (μg RE/d)
0–6 months	[400][c]	0–6 months	375
7–11 months	[500][c]	7–11 months	400
1–3 years	300	1–3 years	400
4–8 years	400	4–6 years	450
7–13 years	600	4–9 years	500
14–>70 years, females	700	10–18 years	600
Males	900	19–65 years, females	500
Pregnancy, ≤18 years	750	Males	600
≥19 years	770	>65 years	600
Lactation, ≤18 years	1200	Pregnancy	800
≥19 years	1300	Lactation	850

[a]*Recommended Daily Allowances; Food and Nutrition Board (2001). Dietary Reference Intakes for Vitamin A, Vitamin K, Arsenic, Boron, Chromium, Copper, Iodine, Iron, Manganese, Molybdenum, Nickel, Silicon, Vanadium and Zinc, National Academy Press, Washington, DC, 773 pp.*
[b]*Recommended Nutrient Intakes; Joint WHO/FAO Expert Consultation, 2001. Human Vitamin and Mineral Requirements. Food and Agricultural Org., Rome, 286 pp.*
[c]*RDA has not been set; average intake (AI) is listed.*

servings of yellow and green vegetables and fruits and liver. For infants and young children, early weaning can increase the risk of primary deficiency. For livestock, it can occur with unsupplemented diets containing low amounts of yellow maize (corn) and corn gluten meal.

Secondary vitamin A deficiency can occur in several ways. One involves chronically impaired enteric absorption of lipids, such as in diseases affecting the exocrine pancreas (e.g., pancreatitis, cystic fibrosis, selenium deficiency) or bile production and release (e.g., biliary atresia, some mycotoxicoses in livestock), or due to the consumption of diets containing very low amounts of fat.[102] Chronic exposure to oxidants can also induce vitamin A depletion; an example is benzo(α)pyrene in cigarette smoke. Nutritional deficiencies of zinc can also impair the absorption, transport, and metabolism of vitamin A, as zinc is essential for the hepatic synthesis or RBP4 and the oxidation of retinol to retinal, which is catalyzed by a zinc-dependent retinol dehydrogenase. Malnourished populations, which typically have low intakes of several essential nutrients including vitamin A and zinc, are at risk to vitamin A deficiency. A prevalence of 25% or more of individuals with plasma retinol levels <0.70 μM (<20 μg/dL) is indicative of population-wide inadequacy with respect to the vitamin.

102. There are few data upon which to base estimates of the minimum amount of dietary fat needed to support the absorption of vitamin A and the other fat-soluble vitamins; in the absence of empirical data, the estimate of 5 g/day is frequently used.

Who is at risk of vitamin A deficiency?

- Infants and children whose mothers were vitamin A deficient during lactation
- Preterm infants whose in utero vitamin A uptake was interrupted
- Children and adults with poor diets and/or malabsorption conditions
- Animals fed unsupplemented grain-based diets free of provitamin A containing feedstuffs.

General Deficiency Signs

Insufficient intakes of vitamin A lead to a sequence of physiological events (Fig. 6.15) that, ultimately, are manifest in several clinical signs (Tables 6.16 and 6.17) and are classified accordingly (Table 6.18) for the purposes of evaluating risk and treatment efficacy (e.g., Table 6.19). Only two signs are unequivocal indicators of vitamin A deficiency; both are ocular lesions:

- **Nyctalopia** is impaired dark adaptation of the retina (night blindness); it can take a year to develop after the initiation of a vitamin A-deficient diet. In individuals of marginal vitamin A status, it can be brought on in a few days by febrile illness (e.g., measles). In either case, it quickly responds to vitamin A treatment.
- **Xerophthalmia** involves permanent morphological changes of the anterior segment of the eye that are not correctable without scarring. These epithelial changes involve dysfunction in the maintenance of normal

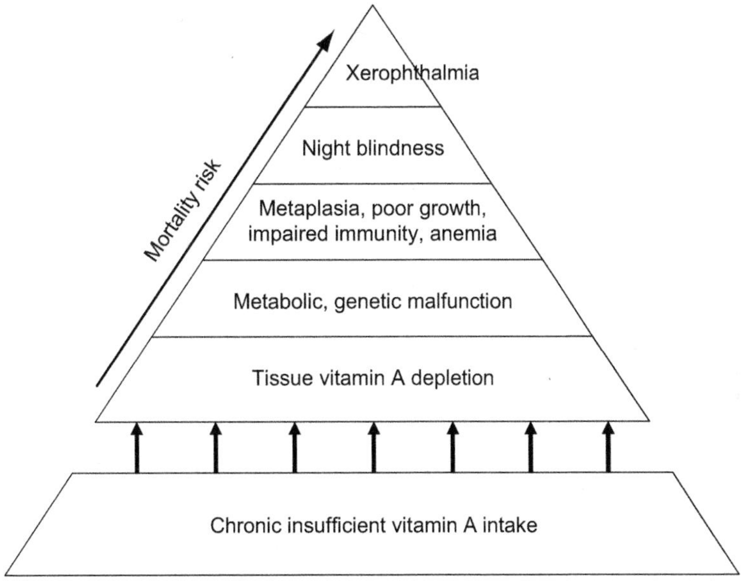

FIGURE 6.15 Progression of vitamin A deficiency. *After West, K.P., 2002. J. Nutr. 132, 2857S–2866S.*

TABLE 6.16 Signs of Vitamin A Deficiency

Organ System	Sign
General	Loss of appetite, retarded growth, drying and keratinization of membranes, infection, death
Dermatologic	Rough scaly skin, rough hair/feathers
Muscular	Weakness
Skeletal	Periosteal overgrowth, restriction of cranial cavity and spinal cord, narrowed foramina
Vital organs	Nephritis
Nervous system	Increased cerebrospinal fluid pressure, ataxia, constricted optic nerve at foramina
Reproductive	Aspermatogenesis, vaginal cornification, fetal death and resorption
Ocular	Nyctalopia, xerophthalmia, keratomalacia, constriction of optic nerve

TABLE 6.17 Stages of Xerophthalmia

Stage	Signs
1. Xerosis	Dryness of conjunctiva Bitot's spots on the conjunctiva near the cornea; ultimately extending to the cornea
2. Keratomalacia	Softening of cornea Ultimate involvement of iris/lens Secondary infection

TABLE 6.18 Clinical Classification of Eye Lesions Caused by Vitamin A Deficiency in Humans

Site Affected	Clinical Sign	Designation
Retina	Night blindness *also* nyctalopia	XN
	Fundus specs	XF
Conjunctiva	Xerosis	X1A
	Bitot's spots	X1B
Cornea	Xerosis	X2
	Ulceration/keratomalacia <1/3 of surface	X3A
	Ulceration/keratomalacia = 1/3 of surface	X3B
	Scar	XS

From WHO.

TABLE 6.19 Relationship of Serum Retinol Level and Clinical Vitamin A Deficiency in Indonesian Preschool Children

Clinical Status		n	Case Frequency, by Serum Retinol Level		
			Deficient	Low	Adequate
XN	X1B		<10 μg/dL	10–20 μg/dL	>20 μg/dL
+	−	174	27%	55%	18%
−	+	51	31%	57%	12%
+	+	79	38%	53%	9%
−	−	252	8%	37%	55%

After Sommer, A., Hussaini, G., Muhilal, L., et al., 1980. Am. J. Clin. Nutr. 33, 887–891.

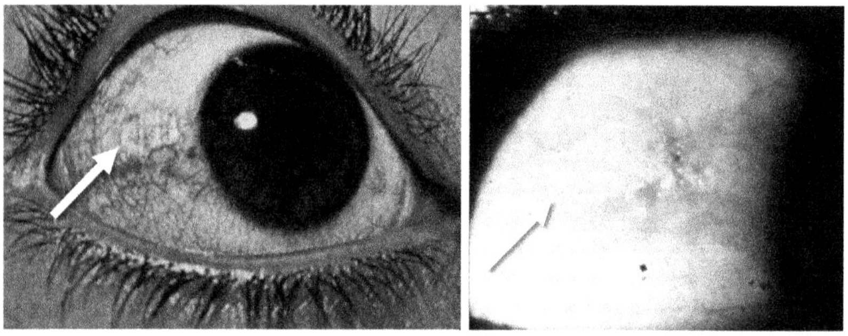

FIGURE 6.16 Bitot's spots on corneas of vitamin A-deficient children. *Courtesy of McLaren, D.S., American University of Beirut.*

epithelial architecture, particularly of the mucous membranes, resulting in keratinizing metaplasia. In the conjunctiva, goblet cells are lost and keratinized material accumulates on the surface over a thickened layer of flattened cells and a prominent granular cell layer. These changes ultimately lead to frank corneal necrosis with dissolution of stromal collagen and loss of keratocytes, i.e., the condition called **keratomalacia**.[103] While xerophthalmia can occur at any age, greatest prevalence is seen in pregnant women and children,[104] particularly those born to vitamin A-deficient mothers thus having low vitamin A stores. Early intervention is necessary to interrupt these progressive lesions in early stages before permanent blindness occurs (Table 6.10).

Deficiency Signs in Humans

Nyctalopia. This is the first functional sign of vitamin A deficiency that can be detected in humans. Dark adaptation is measured by the minimal intensity at which a subject can detect flashes of a white light while the subject fixes on a red light in a darkened environment. Dark adaptation is characterized by an initial decline in the luminance threshold due to cone adaptation, followed at the "rod-cone break" by a second luminance threshold decline. Vitamin A-deficient subjects show longer times to the "rod-cone break," which can occur before symptomatic night blindness. Dark adaptation can also be assessed by the pupillary response to a graduated light stimulus or by electroretinography. Examination by slit lamp can reveal fundus[105] specs, disrupted rod outer segments (with the possible involvement of similarly disrupted cones), and signs suggestive of visual field alterations. The technical expertise and subject burden has meant that these methods have not been widely used.

Ocular lesions. These include conjunctival xerosis with/without **Bitot's spots**[106] (Fig. 6.16), corneal xerosis, ulceration, and/or keratomalacia (Fig. 6.17), and corneal scars can be diagnosed by direct examination of the eye. Morphological changes in epithelial cells blotted from the conjunctival surface can be detected by histologic examination, a procedure called **conjunctival impression**

103. Keratomalacia can also occur in viral infections (e.g., measles) with general malnutrition.
104. Children face heavy demands for vitamin A: rapid growth; morbidity to gastroenteritis, measles, chickpox, pertussis, respiratory infections.
105. The fundus of the eye is the interior surface opposite the lens; it includes the retina, optic disc, macula and fovea, and posterior pole.
106. Described in the 1860's the eponymous Bitot's spots are patches of xerotic conjunctiva with keratin debris and *bacillus* growth.

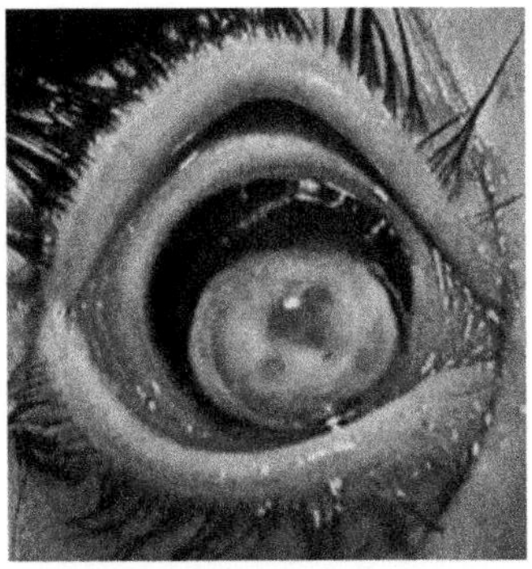

FIGURE 6.17 Keratomalacia in a vitamin A-deficient child. *Courtesy of McLaren, D.S., American University of Beirut.*

cytology.[107] The presence of enlarged, flattened epithelial cells, and few or no goblet cells is indicative of vitamin A deficiency. With progressing vitamin A deficiency, the conjunctival surface takes on a dry, corrugated, irregular surface, ultimately developing an overlay of white, foamy, or "cheesy" material consisting of desquamated keratin and a heavy bacterial growth.[108] Bitot's spots, the sine *qua non* of conjunctival xerosis, are almost always bilateral oval or triangular structures first appearing temporal to the limbus and comprising a thickened, superficial layer of flattened cells usually with a keratinizing surface, a prominent granular cell layer, acanthotic thickening with a disorganized basal cell layer, but with no goblet cells.

The earliest corneal signs of vitamin A deficiency are fine, fluorescein-positive, superficial punctuate keratopathy that usually begin in the inferior aspect of the cornea, particularly infranasally. These lesions can be seen using the slit lamp microscope. Studies have revealed punctuate keratopathy in 60–75% of patients with nyctalopia or vitamin A-responsive conjunctival xerosis. With progressive vitamin A deficiency, the lesions become more numerous and concentrated, involving larger portions of the corneal surface. By the time most of the corneal surface is involved, the lesions are generally apparent by hand-light examination as a haziness and diminished wettability on the corneal surface. At that point the condition is called **corneal xerosis**. In addition to punctuate keratopathy, corneas affected at this level also show stromal edema, again mostly in the inferior aspect. If untreated, the condition progresses to the point of corneal ulceration, frequently characterized by the presence of a single (in a minority of cases, by two to three) sharply defined ulcer varying in depth, usually one-fourth to one-half of corneal thickness, but sometimes deep enough to effect stromal loss. This can lead to deep stromal necrosis characterized by gray-yellow, edematous and cystic lesions varying in size from 2 mm in diameter to covering most of the cornea.

Impaired disease resistance. Vitamin A-deficient individuals are typically more susceptible to infection than are individuals of adequate vitamin A nutriture.[109] They show changes in lymphoid organ mass, cell distribution, histology, and lymphocyte characteristics. Accordingly, low-vitamin A status is frequently associated with increased morbidity and mortality.[110] Such findings were reported in a year-long study of school-age children in Colombia, which found that every 10 μg/dL decrement in plasma retinol was associated with 18% more days of diarrhea and vomiting, 10% more days of coughing and fever, and 6% more doctor visits.[111] A longitudinal study of preschoolers in Indonesia revealed that the overall mortality rate in children with xerophthalmia was 4–5 times that of children with no ocular lesions. Many studies have found positive associations between mild xerophthalmia and risks of diarrhea, respiratory infection, and measles among children. Vitamin A deficiency can reduce child survival by 30% or more,[112] and child mortality increases with increasing severity of the eye disease such that affected children die at nine times the rate of normal children.

Subclinical vitamin A deficiency can also affect resistance to infection. It has been suggested that foci of vitamin A-deficient cells in otherwise normal tissue may provide penetrable sites for bacteria and viruses, thus promoting infection. This hypothesis is consistent with the results of studies showing hypovitaminosis A coupled to secondary coliform infection in cattle, and more frequent bacteriuria among xerophthalmic compared to nonxerophthalmic, malnourished children. That vitamin A-deficient mice can survive enteric coliform infection without eliminating

107. Impression cytology is simple, noninvasive, and useful means of studying the conjunctiva. It involves filter-paper blotting of the anesthetized conjunctiva, followed by fixing, staining, and microscopically examining the lifted cells. Normal conjunctival cells appear in sheets of small, uniform, nonkeratinized epithelial cells with abundant mucin-secreting goblet cells. The procedure facilitates the correct identification of 82–93% of cases of xerophthalmia and 70–90% of unaffected; sensitivity declines and specificity increases when serum retinol or retinol relative dose–response cutoffs are also used in the definition of a case. The use of impression cytology yields estimates of vitamin A deficiency that are 5–10 times the rates of diagnosed xerophthalmia by direct visual examination. Somer suggested that vitamin A deficiency should be considered a public health problem when the prevalence of abnormal impression cytology reaches 20% in either women or children.

108. Most commonly, the "xerosis *Bacillus*," *Corynebacterium xerosis*, aerobic, Gram- and catalase-positive, nonspore-forming rod-shaped bacterium.

109. In practice, it can be difficult to ascribe such effects simply to the lack of vitamin A, as deficient individuals generally also have protein–calorie malnutrition, which, itself, leads to impaired immune function.

110. Although it was not until 1983 that this was noticed. (Sommer, A., Tarwojito, I., Hussaini, G., 1983. Lancet 2, 585–588).

111. Thornton, K., Mora-Plazas, M., Marin, C., et al., 2024. J. Nutr. 144, 496–503.

112. See meta-analyses: Fawzi, W.W., Chalmers, T.C., Herrara, M.G., et al., 1993. Med. Asso. 269, 898–903; Glasziou, P., Mackerras, D., 1993. Br. Med. J. 306, 366–370.

the infection suggests that hypovitaminosis A may create asymptomatic reservoirs for such enteric infections.[113]

Thus, vitamin A deficiency can induce and exacerbate inflammation and lead to histopathological changes that provide environments conducive to secondary infection in loci obstructed by keratinizing debris.[114] Such outcomes involve impairments in epithelial integrity (including mucosal immunity), antibody responses, and lymphocyte differentiation. Vitamin A deficiency impairs adaptive immunity and the development of both B cells and T-helper (Th) cells. Antibody responses, particularly those directed by Th2 cells, are reduced; at the same time chronic inflammatory responses, directed by Th1 cells, are increased, including increased production of proinflammatory cytokines. Children with xerophthalmia were found to have low $CD4^+$:$CD8^+$ ratios as well as other abnormalities in T cell subsets (Table 6.20); these signs reversed upon vitamin A supplementation.

Measles. Measles infection[115] can potentiate subclinical vitamin A deficiency to cause xerophthalmia. "**Measles blindness**" is the leading cause of blindness among children in low-income countries, with as many as 60,000 cases each year. Systematic reviews of clinical intervention trials have confirmed that vitamin A supplementation can reduce mortality in children with measles.[116] The WHO recommends high-dose vitamin A treatment on two consecutive days for measles prophylaxis.

Malaria. Many subjects with malaria[117] are suboptimally nourished with respect to vitamin A and other micronutrients. Low-serum retinol levels are commonly found in patients with malaria. Such findings do not imply causality, as malaria is known to be associated with other factors capable of impairing vitamin A status (reduced food intake, helminth infections). The most direct evidence of a causal linkage comes from studies in Papua New Guinea, Burkina Faso, and Ghana demonstrated vitamin A treatment of apparently nondeficient children to produce 30% reductions in the prevalences of malaria-specific mortality or malarial fever.[118] Such results are supported by findings that the in vivo growth of *Plasmodium falciparum* can be inhibited by retinol, and that retinoic acid treatment of human monocytes can increase phagocytosis of *Plasmodium*-parasitized erythrocytes. It is possible that vitamin A deficiency and infections such as malaria are mutually potentiating, the deficiency compromising resistance to the parasite, and the parasitic infection impairing the utilization of the vitamin.

TABLE 6.20 T Cell Abnormalities in Children

Measure	Without Xerophthalmia	With Xerophthalmia
CD4/CD8	1.11±0.04	0.99±0.05
% CD4/CD45RA (naive)	34.9±1.7	29.9±2.1[a]
% CD4/CD45RO (memory)	18.0±1.1	17.4±1.2
% CD8/CD45RA	37.3±1.7	41.6±2.1
% CD8/CD45RO	7.6±0.6	10.2±0.9[a]
Plasma retinol (μM)	0.84±0.06	0.57±0.04[a]

[a] $p < .05$.
After Semba, R.D., Muhilal, X., Ward, B.J., et al., 1993. Lancet 341, 5–8.

Deficiency Signs in Animals

Animals manifest signs of vitamin A deficiency similar to those of humans, the most notable being nyctalopia and xerophthalmia. Other signs are also of consequence in livestock species, i.e., those involving other systems:

- eyes—excessive lacrimation in cattle and horses
- integument—rough hair coat in cattle; ruffled feathers in poultry; shortened, weakened wool fibers in sheep
- nerves and muscles—seizures in cattle and sheep; unsteady gait in poultry and swine; weakness in horses
- bone—periosteal overgrowth in cattle
- reproduction—reduced conception rates and increased fetal deaths in cattle, swine and poultry; depressed egg production in hens
- disease resistance—increased risk for pinkeye in calves and mastitis in dairy cows
- growth—swine, poultry
- others—and impaired heat tolerance in cattle.

Treatment of Vitamin A Deficiency

Because vitamin A is stored in appreciable amounts in the liver, it can be administered in relatively large, infrequent doses with efficacy. In cases of clear or suspected

113. McDaniel, K., Restori, K.H., Dods, J.W., et al., 2015. Infec. Immun. 83, 2984–2991.
114. Such an example is *Bitot's spots*, which are patches of xerotic conjunctiva with keratin debris and *bacillus* growth. In Indonesia, the presence of Bitot's spots is associated with intestinal worms, such that they are referred to as "*worm feces*."
115. Measles affects an estimated 30 million children each year, causing at least a million deaths.
116. D'Souza, R.M., D'Souza, R., 2004. Cochrane Database Syst. Rev. CD001479; Sudfeld, C.R., Navar, A.M., Halsey, N.A., 2010. Int. J. Epidemiol. 39, 1148S–1155S.
117. Malaria affects populations that are often malnourished, particularly children. In Africa alone, children experience more than 100 million malarial episodes each year, killing some 80,000.
118. Shankar, A.H., Genton, B., Semba, R.D., et al., 1999. Lancet 354, 203–209; Zeba, A.N., Sorgho, H., Rouamba, N., et al., 2008. Nutr. J. 7, 7–14; Owusu-Agyei, S., Newton, S., Mahama, E., et al., 2013. Nutr. J. 12, 131–140.

TABLE 6.21 WHO Recommended Treatment Schedule for Vitamin A Deficiency

		Vitamin A Dose, RE[a], by Age			
Case/Situation	**Time**	**Any Age**	**<6 months**	**6–12 months**	**1 year**
Subjects with xerophthalmia	Day 1		15,000	30,000	60,000
	Day 2		15,000	30,000	60,000
	2–4 weeks		15,000	30,000	60,000
Women reproductive age, night blind or Bitot's spots	Weekly, 3 months	10,000 or 25,000			
Women reproductive age with corneal lesions	Day 1	200,000			
	Day 2	200,000			
	Day 14	200,000			
Subjects with measles	Day 1		15,000	30,000	60,000
	Day 2		15,000	30,000	60,000
Subjects with severe protein–energy malnutrition[b]	Day 1		15,000	30,000	60,000
	Every 4–6 months[c]	60,000			
Asymptomatic high-risk subjects	Once		15,000	30,000	60,000[d]
	Every 4–6 months	60,000			

[a]*Oil solution administered orally.*
[b]*<-3 Z scores for weight:height.*
[c]*Until signs of protein–energy malnutrition subside.*
[d]*Including postpartum mothers.*
After WHO, 1997. Vitamin A supplementation: a guide to their use in the treatment and prevention of vitamin A deficiency and xerophthalmia. second ed. WHO, Geneva.

xerophthalmia, particularly in communities in which the deficiency is prevalent, vitamin A is administered orally in large doses, followed by an additional dose the next day and one-third a few weeks later (Table 6.20). Oral administration of water miscible or oil solutions of the vitamin is as effective as water-miscible preparations administered parenterally. Water-miscible preparations are much more effective than oil solutions when administered parentally, i.e., by intramuscular injection. Topical administration on the skin is ineffective.

High-dose treatment. Vitamin A can be administered safely in relatively high, infrequent oral doses to treat deficient individuals. The WHO dosing schedule (Table 6.21) has been demonstrated safe and effective, the only side effects being transient headache, nausea or vomiting, and diarrhea. Still, high-dose treatment is not recommended for pregnant women due to the risk of teratogenic effects on the fetus (see, *Vitamin A Toxicity*, later in this chapter). Periodic high doses of vitamin A are widely used to prevent deficiency in children living in high risk areas; in those circumstances, its efficacy in preventing xerophthalmia appears very high, i.e., 90%.[119]

Low-dose treatment. Retinyl palmitate in daily doses equal to the RDA, or weekly doses 7 times RDA, are very well absorbed and effective in supporting adequate serum retinol levels. These low-dose regimens greatly reduce the risk of side effects and appear to be particularly effective in reducing the incidence and duration of acute respiratory infections.

Responses to treatment. Night blindness due to vitamin A deficiency responds within hours to days upon the administration of vitamin A (Fig. 6.16), although full recovery of visual function may take weeks and the fading of retinal lesions may take up to 3 months. Active Bitot's spots and the accompanying xerosis responds rapidly to vitamin A treatment; in most cases lesions regress in days and disappear in 2–3 weeks. Punctate keratopathy of the cornea also responds rapidly to vitamin A, improving within a week in response to a large oral dose (Fig. 6.18).

Correction of vitamin A deficiency in children reduces morbidity rates, particularly from diarrhea and measles (Table 6.22). Meta-analyses of community-based, vitamin A intervention studies indicate an average 23% (range: 6–52%)[120] reduction in preschool mortality (Table 6.23).

119. Solon, F.S., Klemm, R.D., Sanchez, L., et al., 2000. Am. J. Clin. Nutr. 72, 738–744.

120. Beaton, G.H., Martorell, R., L'Abbe, et al., 1992. Report to CIDA, University of Toronto; Sommer, A., West, Jr., K.P., Olson, J.A., Ross, C.A., 1996. Vitamin A Deficiency: Health, Survival, and Vision. Oxford University Press, New York, p. 33.

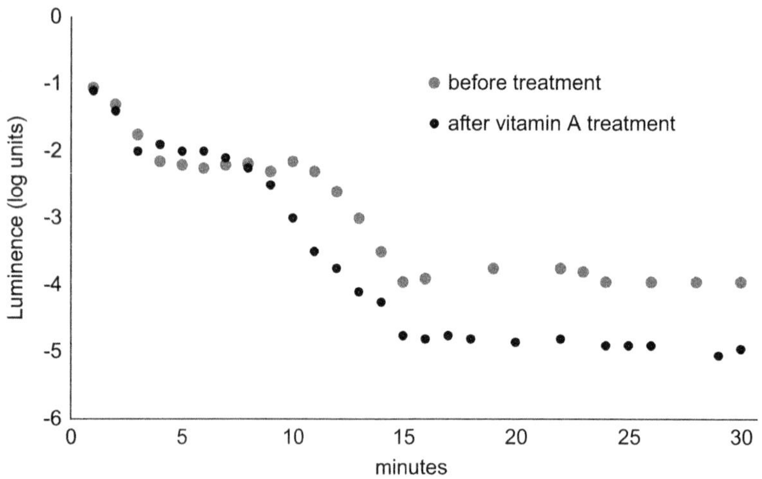

FIGURE 6.18 Dark adaptation in a vitamin A-deficient individual before and after vitamin A treatment. *After Russell, R.M., Multack, R., Smith, V., et al., 1973. Lancet 2, 1161.*

TABLE 6.22 Effects of Vitamin A on Morbidity of Children With Measles in South Africa

	Hospital Morbidity			6-Month Morbidity	
Outcome	**Placebo**	**Vitamin A**	**Outcome**	**Placebo**	**Vitamin A**
Clinical pneumonia (days)	5.7±0.8	3.8±0.4[b]	Weight gain (kg)	2.37±0.24	2.89±0.23[b]
Diarrhea duration (days)	4.5±0.4	3.2±0.7	Diarrheal episodes	6	3
Fever duration (days)	4.2±0.5	3.5±0.3	Respiratory infections	8	3[a]
Clinical recovery (<8 days) (%)	65	96	Pneumonia episodes	3	0[a]
Integrated morbidity score	1.37±0.40	0.24±0.15[b]	Integrated morbidity score	4.12±1.13	0.60±0.22[b]

[a]*$p<.05$.*
[b]*Based on incidence/severity of diarrhea, upper respiratory infections, pneumonia, and laryngotracheobronchitis.*
After Coutsoudis, A., Broughton, M., Coovadia, H.M., 1991. Am. J. Clin. Nutr. 54, 890–895.

TABLE 6.23 Effects of Vitamin A Supplementation on Child Mortality

		Deaths/Total		
Trial	**Months**	**Control**	**Vitamin A**	**Odds Ratio (95% CL)**
Sarlahi, Nepal	12	210/14,143	152/14,487	0.70 (0.57–0.87)
Northern Sudan	18	117/14,294	123/14,446	1.04 (0.81–1.34)
Tamil Nadu, India	12	80/7655	37/7764	0.45 (0.31–0.67)
Aceh, Indonesia	12	130/12,209	101/12,991	0.73 (0.56–0.95)
Hyderabad, India	12	41/8084	39/7691	1.00 (0.64–1.55)
Jumla, Nepal	5	167/3,4111	138/3786	0.73 (0.58–0.93)
Java, Indonesia	12	250/5445	186/5775	0.69 (0.57–0.84)
Bombay, India	48	32/1644	7/1784	0.20 (0.09–0.45)

After Fawzi, W.W., Chalmers, T.C., Merrara, M.G., et al., 1993. J. Am. Med. Assoc. 269, 898–903.

Vitamin A supplementation of children with active, severe, complicated measles has been shown to reduce in-hospital mortality by at least 50%. In other populations, vitamin A treatment has been shown to reduce the symptoms of diarrhea nearly as much (Table 6.24), as well as the symptoms of pneumonia and other infections substantially. Although attendant reductions in morbidity would be expected, studies have shown those effects to be variable. That not all interventions with vitamin A have reduced mortality rates of vitamin A-deficient children is not surprising, as other factors (e.g., time of initiation of breast feeding, poverty, poor sanitation, inadequate diets) clearly contribute to the diminished survival of vitamin A-deficient children. Because enteric pathogens induce unique immune responses not all of which may be comparably affected by the differential action of vitamin A on innate and adaptive immune responses, the efficacy of vitamin A supplementation is likely to depend on the dominant pathogens present in particular communities.

Of the newborn vitamin A supplementation trials conducted to date, vitamin A has been found to reduce mortality in trials with subjects of low-vitamin A status and/or late initiation of breast feeding; whereas, benefits have not been observed in trials with subjects of relatively good vitamin A status and/or early initiation of breast feeding.[121] It is likely that excess mortality occurs not only among xerophthalmic preschoolers but also among those who are mildly to marginally deficient in vitamin A but have not developed corneal lesions. In fact, a large portion of the deaths averted by vitamin A supplementation may be in this low-vitamin A group. Sommer and colleagues have estimated that the improvement of serum retinol levels in mildly deficient, asymptomatic children (with serum retinol levels of 18–20 μg/dL) to serum levels of 30 μg/dL would be expected to reduce mortality by 30–50%. A recent meta-analysis showed a 62% reduction in the risk of measles in response to a two-dose regimen of vitamin A administered with measles vaccination; however, other recent negative findings for vitamin A administered to children at the time of vaccination[122] raise the question of whether vitamin A may benefit mostly children who are not vaccinated, which is often the case in poor countries.

Public health programs. The successes of vitamin A supplementation trials have led to the widespread use of vitamin A capsules: UNICEF estimates that each year more than a half-billion capsules are distributed to some 200 million children in 100 countries. The efficacy of the large-dose, medical approach for reducing child morbidity has been questioned.[123] A meta-analysis of nine randomized controlled trials found that vitamin A supplementation had no consistent effect on the incidence of diarrhea, and may slightly increase the risk of respiratory tract infection,[124] and a very large trial in Ghana found vitamin A supplementation of women without effect on pregnancy-related mortality.[125] West and colleagues[126] noted that, in fact, maternal vitamin A supplementation has reduced maternal mortality in areas of prevalent gestational night blindness and the risk of maternal mortality is very high. Maternal interventions also offer the potential to produce health benefits, such as improved respiratory function, that become apparent in pre-adolescent children.

TABLE 6.24 Effects of Vitamin A Supplementation on Child Cause-Specific Mortality

Study Country	Relative Risk[a] of Death, by Disease		
	Measles	Diarrhea	Respiratory Disease
Indonesia[b]	0.58	0.48	0.67
Nepal[c]	0.24	0.61	1.00
Nepal[d]	0.67	0.65	0.95
Ghana[e]	0.82	0.66	1.00

[a]*Ratio of deaths occurring in the vitamin A-treated group to those occurring in the untreated control group.*
[b]*Rahmathullah, L., Underwood, B.A., Thulasiraj, R.D., et al., 1990. N. Eng. J. Med. 323, 929–935.*
[c]*Reanalysis of data of West, Jr., K.P., Pokhrel, R.P., Katz, J., et al., 1991. Lancet 338, 67–71 cited in Sommer, A., West, Jr., K.P., Olson, J.A., et al., 1996. Vitamin A Deficiency: Health, Survival, and Vision. Oxford University Press, New York, p. 41.*
[d]*Daulaire, N.M.P, Starbuck, E.S., Houston, R.M., et al., 1992. Br. Med. J. 304, 207–210.*
[e]*Ghana VAST Team, 1993. Lancet 342, 7–12.*

10. VITAMIN A IN HEALTH AND DISEASE

Infections

Active infection appears to alter the utilization or, at least, the distribution of vitamin A among tissues. Plasma retinol concentrations drop during malarial attacks, chickenpox, diarrhea, measles, and respiratory disease. Ocular signs of xerophthalmia following measles outbreaks are associated

121. Thurman, D., 2010. Newborn vitamin A dosing and neonatal mortality. Sight Life 1, 19–26.
122. Sudfeld, C.R., Navar, A.M., Halsey, N.A., 2010. Int. J. Epidemiol. 39, 148–155; Benn, C.S., Aaby, P., Nielsen, J., et al., 2009. Am. J. Clin. Nutr. 90, 626–639; Benn, C.S., Rodrigues, A., Yazdanbakhsh, M., et al., 2009. Vaccine 27, 2891–2898.
123. Latham, M., 2010. J. World Pub. Health Nutr. Assoc. 1, 12–24.
124. Grotto, I., Mimouni, M., Gdalevich, M., et al., 2003. J. Pediatrics 142, 297–304.
125. In 1999, West and colleagues. (West, K.P., Katz, J., Khatry, S.K., et al., 1999. Br. Med. J. 318, 570–575) reported vitamin A supplementation to reduce pregnancy-related mortality by 44%, although the apparent effects seemed unrelated to infection/immunity; however, in 2010 Kirkwood and colleagues (Kirkwood, B., Humphrey, J., Moulton, L., et al., 2010. Lancet 376, 1643–1644) found no such protection in an intervention trial with more than 200,000 women in Ghana.
126. West, K.P, Christian, P, Katz, J., et al., 2010. Lancet 376, 873–874.

with declines in plasma retinol levels, depending on the severity and duration of infection, and can be as great as 50%. Episodes of acute infection have been found to be associated with substantive (e.g., eightfold) increases in the urinary excretion of retinol and RBP4. That such insults to vitamin A status can be of clinical significance is indicated by the fact that vitamin A treatment can greatly reduce morbidity and mortality rates in measles and respiratory diseases. Stimulation of immunity and resistance to infection are thought to underlie the observed effects of vitamin A supplements in reducing risks of mortality and morbidity from some forms of diarrhea, measles, HIV infection, and malaria in children. Night blind women have a fivefold increased risk of dying from infections, and low doses of vitamin A have been found to reduce peri- and postpartum mortality in women (Table 6.25), presumably due to reduction in the severity of infections. Indeed, vitamin A supplementation has been found to reduce the incidence of uncomplicated malaria by more than 30%.

HIV. Low-serum retinol levels have been found to be more common among HIV+ than in noninfected individuals and to be highly predictive of vaginal HIV-1 DNA shedding. Whether this is cause or effect is not clear, as serum retinol is known to decrease in the acute-phase response to infection. Intervention studies have found vitamin A supplements effective in improving maternal vitamin A status. Clinical trials with HIV+ children have found vitamin A supplementation to reduce mortality; however, those with HIV+ mothers have not shown benefits in reducing the progress of disease, mother-to-child transmission of the virus, or the prevalence of infection in infants (Table 6.26). That vitamin A may affect other viral infections has been suggested by clinical findings of reduced viremia in some vitamin A-treated subjects with hepatitis C.

TABLE 6.25 Efficacy of Low-Dose Vitamin A Supplements in Reducing Mortality Related to Pregnancy in Nepal

Parameter	Placebo	Vitamin A	β-Carotene
Serum Levels, Midpregnancy[a] (μmol/L)			
Retinol	1.02±0.35	1.30±0.33	1.14±0.39
β-Carotene	0.14±0.12	0.15±0.14	0.20±0.17
Mortality, deaths/100,000 pregnancies (RR, 95% C.L.)			
During pregnancy	235 (1.0)	142 (0.60, 0.26–1.38)	111 (0.47, 0.18–1.20)
0–6 weeks postpartum	359 (1.0)	232 (0.65, 0.34–1.25)	222 (0.62, 0.31–1.23)
7–12 weeks postpartum	110 (1.0)	52 (0.47, 0.13–1.76)	28 (0.25, 0.04–1.42)

[a]*The 3.5 yr. trial involved some 44,646 women who had more than 22,000 pregnancies, 7200–7700 in each treatment group. Vitamin A (retinol) or β-carotene were given in weekly dosages; posthoc analyses revealed that only half of subjects took 80% of the intended doses (7000 IU), suggesting that the study underestimated the potential impact of vitamin A supplementation.*
After West, Jr., K.P., Katz, J., Kharty, S.K., et al., 1999. Br. Med. J. 318, 570–574.

Skin Health

Owing to their similarities to changes observed in vitamin A-deficient animals, certain dermatologic disorders of keratinization (e.g., ichthyosis,[127] Darier's disease,[128] pityriasis rubra pilaris[129]) have been treated with large doses of retinol. Clinical success of such treatments generally has been variable; high doses of the vitamin may be required for efficacy. Greater therapeutic efficacy has been achieved with all-*trans*-retinoic acid, 13-*cis*-retinoic acid, and an ethyl ester of all-*trans*-retinoic acid. The most successful of these has been 13-*cis*-retinoic acid, known generically as **tretinoin**, which is used for the treatment of **acne vulgaris**.[130] The vitamer decreases sebum production, inhibits the development of blackheads, reduces bacterial numbers in both the ducts and surface, and reduces inflammation by inhibiting the chemotactic responses of monocytes and neutrophils. It is also used to treat cystic acne, rosacea, gram-negative folliculitis, pyoderma faciale, hidradenitis suppurativa, and skin cancers.

Retinoids have also been found to produce rapid reductions in the incidence of new nonmelanoma skin cancers in high-risk patients. Therefore, it has been suggested that they may produce regressions of prediagnostic malignant and/or premalignant lesions. Indeed, regressions of cutaneous metastases of malignant melanoma and cutaneous T cell lymphoma have been reported in response to retinoid therapy. Topical treatment with all-*trans*-retinoic acid has been found to protect against photoaging signs by stimulating collagen synthesis and accumulation[131] in the upper papillary dermis, and downregulating the induction by UV light of metalloproteinase expression, thereby increasing collagen replacement. The action of retinoids in psoriasis

127. The term refers to a group of skin disorders, many with genetic components, characterized by varying degrees of dry, thickened, scaling/flaking skin.
128. An autosomally dominant dermatosis characterized by greasy hyperkeratotic papules in seborrheic regions, mucous, and nail changes.
129. Also, "*Devergie's disease*," a chronic disorder of unknown cause characterized by red-orange, scaling plaques, and keratotic follicular papules with itching and severe flaking.
130. Commonly called "*acne*," this chronic skin disease involves the blockade of hair follicles with dead skin cells and skin oils, often involving proliferation of the common skin bacterium *Propionibacterium acnes*. It is most common among teenagers and is thought to have a genetic component.
131. In particular collagen I, which comprises some 85% of dermal collagen.

TABLE 6.26 Vitamin A Supplementation Has Failed to Reduce Mother-To-Child HIV Transmission

		RR[a] to Being HIV+, by Age			
Trial	**Intervention Agent**	**Birth**	**6 weeks**	**12 weeks**	**2 years**
Malawi[b]	Retinol	–	0.96	–	0.84
Tanzania[c]	Retinol + β-carotene	1.60	1.22	–	1.38[d]
South Africa[e]	Retinol + β-carotene	0.85	–	0.91	–

[a]*Ratio of % HIV=children in treatment group to % HIV+ in control group.*
[b]*Kunwenda, N., Motti, P.G., Thaha, T.E., et al., 2002. Clin. Infect. Dis. 35, 618–624.*
[c]*Fawzi, W.W., Msamanga, G.I., Hunter, D., et al., 2002. AIDS 16, 1935–1944.*
[d]*$p<.05$.*
[e]*Coutsoudis, A., Pillay, K., Spooner, E., et al., 1999. AIDS 13, 1517–1524.*

appears to involve thinning of the stratum corneum, reduced keratinocyte proliferation, and reduced inflammation.

The therapeutic value of retinoids is limited by their dose-limiting side effects. Some retinoids, e.g., 13-*cis*-retinoic acid, can be teratogenic, limiting their use especially for women of child-bearing age. Therefore, alternative approaches have been of interest: use of synthetic, mono- and poly-aromatic retinoids; use of inhibitors (imidazoles and triazoles) of the cytochrome *P*-450-dependent 4-hydroxylation of all-*trans*-retinoic acid to sustain its intracellular concentrations. Clinical evaluation of the retinoic acid metabolism inhibitor liarozole,[132] has shown it to be comparably effective as 13-*cis*-retinoic acid in treating psoriasis and ichthyosis; when used topically the azole compound was more effective.

Obesity–Diabetes

Overweight/obese individuals show elevated plasma apo-RBP4 levels (Fig. 6.4).[133] These observations reflect increased apo-RBP4 secretion under conditions in which adipocytes downregulate the expression of GLUT4, the insulin-responsive transporter required for cellular uptake of glucose. This was confirmed by the finding that genetic ablation of adipose-specific GLUT4 increased RBP4 expression in mice. Because downregulation of GLUT4 occurs under conditions of food deprivation, obesity-stimulated secretion of apo-RPB4 may be a signal for restricting glucose uptake by peripheral tissues. Studies in mice have shown that overexpression of RPB4 or treatment with recombinant RPB4 increased expression of the gluconeogenic enzymes in liver and impaired insulin signaling in muscle. This is consistent with findings that obese humans have been found to have elevated plasma levels of RBP4, but normal levels of retinol (Table 6.27), and that prolonged elevation of RBP is associated with increased risk of type 2 diabetes (Table 6.28). Thus, adipocyte-derived apo-RBP4 may act as an adipokine to signal reductions in insulin sensitivity, thus promoting the development of type 2 diabetes. That signaling appears to depend on STRA6; when bound to holo-RBP, activates a JAK/STAT signaling cascade that induces expression of SOCS3, which suppresses insulin signaling. STRA6-null mice do not experience RBP-induced suppression of insulin signaling.[134] It has been suggested that lowering RBP4 should be considered in the prevention/treatment of type 2 diabetes. To that end, fenretinide,[135] a synthetic retinoid that increases urinary excretion of RBP4, has been found to normalize serum RBP4 levels and improve insulin sensitivity and glucose tolerance in obese mice.

Adipose tissue is responsive to vitamin A. Adipose tissue can store vitamin A as retinyl esters, and adipocyte differentiation is regulated by RARs, RXRs, and PPARs. Under conditions of vitamin A deficiency, adipocytes increase their storage of triglycerides and express high levels of RADH1 and mobilize those stores to make retinal and retinoic acid. Vitamin A has been found to be protective against the obesogenic effects of high-fat diets in the mouse. Signaling through RARγ, retinoic acid promotes the maintenance of the preadipocyte phenotype; in mature adipocytes, it upregulates lipid oxidation and energy expenditure. It is not clear, however, whether adipose vitamin A levels are related to obesity, as studies with normal animal models have yielded mixed results. Cattle show declines in plasma retinol at the fattening stage of their growth; but vitamin A deprivation has been found effective in increasing fattening, i.e., muscle marbling, only in animals with a novel single nucleotide polymorphism in aldehyde dehydrogenase I.[136] Still, it has been

132. 6-[(3-chlorophenyl)-imidazol-1-ylmethyl]-1*H*-benzimidazole.
133. Yang, Q., Graham, T.E., Mody, N., et al., 2005. Nature 436, 356–362.
134. Berry, D.C., Jacobs, H., Marwarha, G., et al., 2013. J. Biol. Chem. 34, 24528–24539.
135. 4-Hydroxy(phenyl)retinamide
136. Ward, A.K., McKinnon, J.J., Hendrick, S., et al., 2015. J. Anim. Sci. 90, 2476–2483.

TABLE 6.27 Parameters of Vitamin A Status in Obese and Nonobese Adults

Parameter	Obese	Nonobese
N	76	41
BMI (kg/m²)	34.3±4.0[a]	24.8±3.2
Plasma retinol (μM)	2.23±0.57	2.23±0.58
Plasma RBP (μM)	3.13±0.88[a]	2.67±1.02

[a] $p < .05$.
After Mills, J.P., Furr, H.C., Tanumihardjo, S.A., 2008. Exp. Biol. Med. 233, 1255–1261.

TABLE 6.28 Association of Plasma RBP4 Level and Risk of Type 2 Diabetes in Adults

RBP4 (μg/mL)	Cases/Total	Relative Risk
(Quartile)	N	
<31.3	101/522	1.00
31.3–<38.1	116/525	1.16 (0.86, 1.58)
38.1–<46.1	125/522	1.18 (0.87, 1.61)
≥46.1	165/523	1.75 (1.30, 2.37)
P, trend		<0.001

After Sun, L., Qi, Q., Zong, G., et al., 2014. J. Nutr. 144, 722–728.

suggested that alterations in retinoid metabolism may affect the regulatory activities of PPARγ. For example, studies have found high-fat diets to be more obesogenic for mice lacking BCO1, CRBP(I), or CRBP(III), compared to wild-type mice.

Adipose tissue is also a major storage site of carotenoids, which partition into the bulk lipid droplets of adipocytes. Carotenoid concentrations tend to be inversely related to percentage body fat, due to the fact that the caloric excesses that drive adiposity tend to be unrelated to the intake of carotenoid-rich foods (fruits, vegetables). Therefore, body fat would appear to be a determinant of the tissue distribution of carotenoids including those with provitamin A potential.

Drug Metabolism

That the level of vitamin A intake has been found to affect negatively the genotoxic effect of several chemical carcinogens suggests that the vitamin may play a role in the cytochrome *P*-450-related enzyme system. Indeed, several studies have shown that vitamin A deficiency can reduce hepatic cytochrome *P*-450 contents and related enzyme activities, and vitamin A supplementation has been shown to increase the activities of cytochrome *P*-450 isozymes.[137]

Antioxidant Protection

It has been suggested that actions of vitamin A in supporting the health of the skin and immune systems may involve effects on systems that provide protection against the adverse effects of prooxidants.[138] Yet, it is unlikely that vitamin A, itself, is physiologically significant in this regard, as retinol and retinal cannot quench singlet oxygen ($^1O_2$) and have only weak capacities to scavenge free radicals. It can, however, affect tissue levels of other antioxidants; animal studies have shown that deprivation of vitamin A leads to marked increases in the concentrations of α-tocopherol in the liver and plasma, whereas high intakes of retinyl esters can enhance the bioavailability of selenium, an essential constituent of several glutathione-dependent peroxidases.

Several carotenoids, on the other hand, have been shown to have direct antioxidant activities. These include β-carotene, lycopene, and some oxycarotenoids (zeaxanthin, lutein), which can quench $^1O_2$ or free radicals in the lipid membranes into which they partition (Table 6.29). These antioxidant activities are due to their extended systems of conjugated double bonds, which are thought to delocalize the unpaired electron of a free-radical reactant.[139] At low (physiologic) partial pressures of oxygen, carotenoids can also participate in the reduction of free radicals; xanthophyll carotenoids (lutein, lycopene, and **β-cryptoxanthin**) are more effective than β-carotene and more efficient than α-tocopherol in vitro. Despite these differences, the carotenoids tend to be less plentiful in tissues, for which reason

137. These include CYP3A in rats, rabbits, and guinea pigs, and CYP2A in hamsters.

138. Because aerobic systems rely on O_2 as the terminal electron acceptor for respiration, they must also protect themselves against the deleterious effects of highly reactive O_2 metabolites that can be formed either metabolically or through the action of such physical agents as UV light or ionizing radiation. These reactive oxygen species (ROS) include singlet oxygen ($^1O_2$), superoxide (O_2^-), hydroxyl radical (OH•), and nitric oxide (NO). ROS can react directly or indirectly with polyunsaturated membrane phospholipids (to form scission products), protein thiol groups (to form disulfide bridges), nonprotein thiols (to form disulfides), and DNA (to cause base changes) to alter cellular function. They can also react with polyunsaturated fatty acid components of circulating lipoprotein complexes; such oxidative changes in low-density lipoproteins (LDLs) appear to be important in the development of atherosclerotic lesions. The systems that protect against these oxidative reactions include several reductants (e.g., tocopherols and carotenoids in membranes and lipoprotein complexes; glutathione, ascorbic acid, urate, and bilirubin in the soluble phases of cells) and antioxidant enzymes (e.g., superoxide dismutases, glutathione peroxidases, catalase).

139. This mechanism differs from that of the tocopherols, which donate a hydrogen atom to the lipid-free radical to produce a semistable lipid peroxide; the tocopherols in turn become semiquinone radicals. (The antioxidant function of the tocopherols is discussed in Chapter 8.)

TABLE 6.29 Antioxidant Abilities of Carotenoid and Other Antioxidants

Compound	ROO·Reduction[a]	$^1O_2$ Quenching[a]
Lycopene	–	9×10^9
β-Carotene	1.5×10^9	5×10^9
α-Tocopherol	5×10^8	8×10^7
L-Ascorbate	2×10^8	1×10^7

[a]*Bimolecular rate constants* ($M^{-1}s^{-1}$).
After Sies, H., Stahl, W., 1995. Am. J. Clin. Nutr. 62 (Suppl.), 1315S–1321S.

their contributions to physiologic antioxidant protection are likely to be less important than those of the tocopherols except, perhaps, in cases of high carotenoid intake. Cooperative antioxidant interactions between α-tocopherol and β-carotene have been observed in model systems and it is likely that in vivo carotenoids may serve to protect tocopherols.

The interactions of carotenoids with radicals result in the production of oxidation products of the former and in the bleaching of these pigments;[140] the decomposed product cannot be regenerated metabolically, thus, destroying its provitamin A potential. In ultraviolet (UV)-irradiated skin, lycopene is more susceptible to bleaching than is β-carotene, suggesting that it may be more important in antioxidant protection of dermal tissues. Supplementation with β-carotene has been found to improve antioxidant status in vivo. These results include the following: reduced pentane[141] breath output in smokers; reduced plasma concentrations of malonyldialdehyde[142] in cystic fibrosis; reduced lipid peroxidation products (TBARS) in mice; reduced lethality to cultured cells of prooxidant drugs; and reduced acetaminophen toxicity[143] in mice. These and other nonprovitamin A properties of carotenoids are discussed in greater detail in Chapter 19.

140. This is seen in the loss of pigmentation from the shanks of poultry, which poultry keepers used historically to indicate reproductive status of hens. Immature pullets deposit carotenoids in dermis; whereas, actively laying hens deposit the pigments in the lipids of the developing oocyte. The bleaching of the skin, therefore, is a positive sign of good laying condition.

141. *n*-Pentane (C_5H_{12}) is a scission product of the peroxidative degradation of ω-6 fatty acids.

142. Malonyldialdehyde is also a peroxidative scission product of polyunsaturated fatty acids. It can be detected by reaction with 3-thiobarbituric acid and is the predominant, but not only, reactant in biological specimens. Due to the lack of specificity, results of this analysis are typically expressed as thiobarbituric-reactive substances (TBARS).

143. The microsomal metabolism of acetaminophen, like that of other prooxidant drugs metabolized by cytochrome *P*-450-related enzymes, produces $O_2^{\cdot-}$. Antioxidant status has been shown to affect its acute toxicity in animal models.

Cardiovascular Health

Epidemiologic investigations have repeatedly found inverse relationships between the level of consumption of provitamin A-containing fruits and vegetables and risks of cardiovascular disease. Indeed, plasma retinol levels have been found to be related inversely to the risk of ischemic stroke, and low-plasma β-carotene concentrations are associated with increased risk of myocardial infarction.[144] Such findings have provided the bases for hypothetical actions of vitamin A or, more often, provitamin A carotenoids in chronic disease prevention. Unfortunately, many of these hypotheses have not withstood experimental challenge. Well-designed, randomized, double-blind, clinical intervention trials have found supplements of β-carotene[145] or a combination of β-carotene and/or α-tocopherol[146] to be ineffective in reducing risk of either cardiovascular disease or angina pectoris. In fact, one found an increase in the incidence of angina associated with β-carotene use.

Anticartumorigenesis

Because vitamin A deficiency characteristically results in a failure of differentiation of epithelial cells without impairing proliferation (i.e., the keratinizing of epithelia), it has been reasonable to question the possible role of vitamin A in the etiology of epithelial cell tumors, i.e., carcinomas. The squamous metaplastic changes seen in vitamin A deficiency are morphologically similar to precancerous lesions induced experimentally, and both show downregulation of CRBP(I) expression. Indeed, patients with oral leukoplakia, a precancerous condition of the buccal mucosa, have been found to have lower serum retinol levels than healthy controls, and treatment with retinol has been found to reduce the development of new lesions and to cause remissions in the lesions of some patients. It has been proposed that retinoic acid, which in high doses can inhibit the conversion of papillomas (benign lesions) to carcinomas, can upregulate RARs, which can, in turn, complex with protooncogenes such as *c*-fos to prevent malignant transformation.

Studies with animal tumor models have found vitamin A deficiency to enhance susceptibility to chemical carcinogenesis and large doses of vitamin A (i.e., supranutritional but not toxic) to inhibit carcinogenesis in some models

144. The Physicians Health Study, a prospective study of 22,071 male American physicians (Hak, A.E., Stampfer, M.J., Campos, H., et al., 2003. Circulation 108, 802–807).

145. The Dartmouth Skin Cancer Study involved 1188 male and 532 female Americans treated with 50 mg/day β-carotene for more than 4 years. (Greenberg, E.R., Baron, J.A., Karagas, M.R., et al., 1996. JAMA 275, 699–703.).

146. The Alpha Tocopherol and Beta Carotene (ATBC) Cancer Prevention Trial involved 29,133 Finish male smokers treated with β-carotene and α-tocopherol for nearly 5 years. (Törnwall, M.E., Virtamo, J., Korhonen, P.A., et al., 2004. Eur. Heart J. 25, 1171–1178).

TABLE 6.30 Inhibition by β-Carotene of Chemical Carcinogenesis in Rats: Reduced Hepatic γ-Glutamyltranspeptidase-Positive Foci

Treatment[a]	Foci (Number/cm^2)	Focal Area (% Total Area)
Control	37.1 ± 9.7	1.267 ± 1.121
Retinyl acetate (10 mg/kg/2 days)	34.8 ± 9.6	0.911 ± 0.901
β-Carotene (70 mg/kg/2 days)	20.1 ± 12.5[b]	0.308 ± 0.208

[a]*Rats were also treated with diethylnitrosamine/2-acetylaminofluorene.*
[b]*$p < .05$.*
After Moreno, F.S., Wu, T.S., Penteado, M.V.C., et al., 1995. Int. J. Vit. Nutr. Res. 65, 87–94.

(Table 6.30). Retinoids appear to suppress carcinogenesis and tumor growth by increasing expression of tumor suppressors such as the retinoid signaling protein TRIM16,[147] and inducing apoptosis (programmed cell death) and/or terminal differentiation. The proapoptotic effects appear to be mediated by RARs the target genes of which include players in the intrinsic (caspase cascade initiated by cell stress, DNA damage, or deprivation of growth factor) and the extrinsic (triggered by activation of death receptor-associated caspases) pathways of apoptosis.

Studies have demonstrated the efficacy of retinoic acid in inhibiting the growth of several types of cancer cells and tumors that do not express the RARβ gene even in the presence of physiological levels of vitamin A. The mechanism of silencing of RARβ gene expression is thought to involve hypermethylation of the gene due to loss of heterozygosity of chromosome 3p24, the locus of RARβ, and/or impaired expression of other factors involved in RARβ expression.[148] Loss of RAR function under vitamin A-adequate conditions is associated with different cancers, the best studied of which is acute promyelocytic leukemia (APL[149]). This cancer has been found to result from a nonrandom chromosomal translocation or deletion that leads to the production of a fusion of RARα gene on chromosome 17 to the promyelocytic (PML) gene on chromosome 15. When expressed, the fusion product represses translation and initiates leukemogenesis. This appears to occur through that action of the PML–RAR protein, which is a transcriptional activator of retinoic acid target genes. Studies with one specific target, the tumor suppressor gene RARβ$_2$, have revealed that its promoter contains a high-affinity RARE near the transcription start site, but is inactivated by methylation.[150] It appears that the PML–RAR fusion protein can form a complex with histone deacetylase, which, in turn, becomes oncogenic by recruiting DNA methyltransferases to the promoters of RARβ$_2$ locking them in a stably silenced chromatin state by hypermethylation. Most (80%) APL patients, however, respond to treatment with very high doses of all-*trans*-retinoic acid, resulting in complete remission in more than half of cases. This effect appears to involve retinoic acid causing the dissociation of the PML–RAR protein–histone deacetylase complex, which converts the fusion protein into a transcriptional activator resulting in leukemia cell differentiation. Studies with breast cancer cells[151] suggest that retinoic acid can also cause histone acetylation (due to the release of histone deactylase) of the RARβ$_2$ gene resulting in the inhibition of cell growth. Thus, in sensitive cells retinoic acid can reactivate RARβ$_2$ gene expression through epigenetic means; such reactivation has been shown to suppress malignancy in lung cancer cells.

While the use of retinoic acid, which is rapidly metabolized and eliminated from the body, avoids the problem of chronic hypervitaminosis, its substantial toxicity makes it unsuitable for regular clinical use. Therefore, more than 1500 retinoids have been synthesized and tested for potential anticarcinogenicity.[152] Several[153] have been found to effectively inhibit experimentally induced tumors in several organs of animals[154] and have yielded hopeful results in clinical trials. The consensus, however, is that although retinoids currently available can delay tumorigenesis, they cannot do so at doses that are not, themselves, hazardous.

Epidemiological investigations of vitamin A intake and human cancer have produced mostly negative results, depending on study design. A recent meta-analysis of 15 prospective studies found prostate cancer risk to be positively associated with plasma retinol level.[155] Findings of significant, inverse associations of intakes/plasma level and cancer risk have come mostly from retrospective

147. Tripartite motif protein 16.
148. Possibilities include the orphan receptors nurr77 and COUP-TF, both of which are overexpressed in retinoic acid-resistant cells.
149. APL is characterized by a blockade in myeloid differentiation, resulting in the accumulation in the bone marrow of abnormal promyelocytes and in a coagulopathy involving disseminated intravascular coagulation and fibrinolysis.
150. DiCroce, L, Raker V.A., Corsaro, M., et al., 2002. Science 295, 1079–1082.
151. Sirchia, S.M., Ren, M., Pili, et al., 2002. Cancer Res. 62, 2455–2461.
152. These compounds are formal derivatives of retinal with differing modifications of the isoprenoid side chain (including modification of the polar end and cyclization of the polyene structure), or the cyclic head group (including replacement with other ring systems).
153. For example, 13-*cis*-retinoic acid, *N*-ethylretinamide, *N*-(2-hydroxyethyl)-retinamide, *N*-(4-hydroxyphenyl)-retinamide, etretinate, *N*-(pivaloyloxyphenyl)-retinamide, *N*-(2,3-dihydroxypropyl)-retinamide.
154. Several studies have shown that retinoids can inhibit the initiation and promotion of mammary tumorigenesis induced in rodents by dimethylbenz(α)anthracene or *N*-methyl-*N*-nitrosourea, as well as the induction of **ornithine decarboxylase**, an enzyme the induction of which appears to be essential in the development of neoplasia.
155. Key, T.J., Appleby, P.N., Travis, R.C., et al., 2015. Am. J. Clin. Nutr. 102, 1142–1157.

TABLE 6.31 Results of a Meta-Analysis of Results of 57 Epidemiological Studies of β-Carotene Status and Human Cancer Risk

Design	Pooled Estimate of Risk
Cohort	1.013 (0.884, 1.16)[a]
Nested case–control	0.977 (0.864, 1.105)
Case–control	0.729 (0.640, 0.831)

[a]*Mean (95% confidence limits).*
After Musa-Velosa, K., Card, J.W., Wong, A.W., et al., 2009. Nutr. Rev. 67, 527–545.

case–control studies (Table 6.31). More than 60% of these have indicated reductions in the prevalence of cancers of the lung, colon/rectum, skin, and prostate; whereas, such protective effects have been found in less than 20% of prospective cohort or nested case–control studies. Nevertheless, the results of such surveys have fostered the hypothesis that β-carotene may have some beneficial effect unrelated to its role as a precursor of vitamin A.

That β-carotene may have antitumorigenic effects has been suggested by findings such as that from a case–control study nested within the Nurses' Health Study that found plasma β-carotene level to be inversely related to breast cancer risk in American women.[156] The cancer-chemopreventive potential of supplemental β-carotene/retinoids has been tested in at least five well-designed, placebo-controlled, double-blind clinical intervention trials none of which found significant reduction in cancer in high-risk subjects. In fact, the results of two studies found β-carotene treatment harmful. The first was the Carotene and Retinol Efficacy Trial (CARET), a 12-year study involving more than 18,314 Americans, found no protection against lung or prostate cancer in men, but an increase in lung cancer in women (mostly among former smokers).[157] The second was the Alpha-Tocopherol and Beta-Carotene trial (ATBC),[158] conducted in Finland with more than 29,000 men with histories of smoking, tracked the health impacts of modest supplements of β-carotene (20 mg/day) and/or α-tocopherol (50 mg/day). Within 5–8 years of follow-up, results showed significantly greater total mortality (8%) and lung cancer incidence (18%) among men taking β-carotene in comparison with those not taking that supplement (Table 6.32). The interpretation of these findings is not straightforward, as ample evidence also indicates that β-carotene at these levels is generally safe. It is possible that β-carotene, at these levels of intake, can redox cycle to act as a prooxidant and co-carcinogen in the lungs of smokers. It is also possible that resulting retinoid metabolites may activate RAR to produce antiapoptotic effects (PPARβ/γ target genes include those that activate survival pathways) and/or increase cell proliferation.

TABLE 6.32 Increased Mortality Among Male Smokers Taking β-Carotene

	Mortality Rate (per 10,000 Person-Years)	
Causes of Death	**No β-Carotene**	**β-Carotene**
Lung cancer	30.8	35.6
Other cancers	32.0	33.1
Ischemic heart disease	68.9	77.1
Hemorrhagic stroke	6.0	7.0
Ischemic stroke	6.5	8.0
Other cardiovascular disease	14.8	14.8
Injuries and accidents	19.3	20.3
Other causes	23.5	22.5

After Alpha-Tocopherol, Beta-Carotene Cancer Prevention Study Group, 1994. N. Engl. J. Med. 330, 1029–1035.

APL has been found to respond to all-*trans*-retinol. This hematological malignancy is associated with chromosomal translocations affecting RARα that impair retinoid signaling. Treatment with all-*trans*-retinol produced complete remission in most APL patients.[159] A multicenter trial demonstrated that combined treatment with all-*trans*-retinol and arsenic trioxide produces complete remission in virtually all patients.[160]

11. VITAMIN A TOXICITY

Hypervitaminosis A

The hepatic storage capacity for vitamin A tends to mitigate the development of intoxication due to intakes in excess of physiological needs. However, persistent large overdoses (more than 1000 times the nutritionally required

156. Eliasson, A.H., Liao, X., Rosner, B., et al., 2015. Am. J. Clin. Nutr. 101, 1197–1205.
157. The Beta-Carotene and Retinol Efficacy Trial (CARET) involved 18,314 American male and female current and ex-smokers and asbestos-exposed men supplemented with both β-carotene and retinyl palmitate for 4 years of treatment. (Omenn, G.S., Goodman, G., Thornquist, M., et al., 1996. IARC Sci. Publ. 136, 67–85).
158. Albanes, D., Hainonen, O.P., Huttenen, J.K., et al., 1995. Am. J. Clin. Nutr. 62, 1427S–1430S.
159. Avvisati, G., Tallman, M.S., 2003. Best Pract. Res. Clin. Haematol. 16, 419–427.
160. Lo-Coco, F., Avvisati, G., Vignetti, M., et al., 2013. N. Eng. J. Med. 369, 111–121. This and other studies noted the low prevalences of side effects including fever, dyspnea, weight gain, and hypotension, which was referred to as "*retinoic acid syndrome*." It now appears that these signs depend on the presence of malignant promyelocytes; hence, it is now called "*differentiation syndrome*."

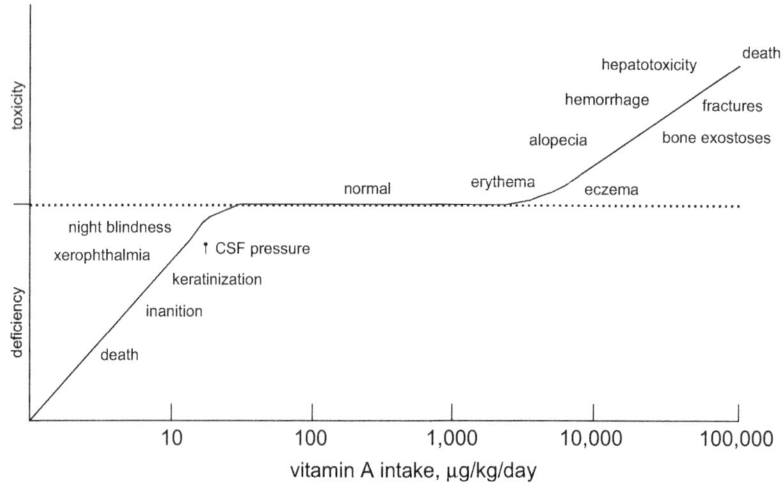

FIGURE 6.19 Both low and high habitual intakes of vitamin A can cause clinical signs.

amount) can exceed the capacity of the liver to store and catabolize and will, thus, produce intoxication[161] (Figs. 6.17 and 6.19). This is marked by the appearance in the plasma of high levels of retinyl esters that, because they are associated with lipoproteins rather than RBP4, are outside the normal strict control of vitamin A transport to extrahepatic tissues.

Five aspects of vitamin A metabolism tend to protect against hypervitaminosis:

1. relatively inefficient conversion of the provitamins A in the gut
2. the unidirectional oxidation of the vitamin to a form (retinoic acid) that is rapidly catabolized and excreted
3. excess capacity of CRBP(II) to bind retinol
4. hepatic storage of vitamin A
5. accelerated catabolism of retinoic acid.

Hypervitaminosis A, therefore, requires high exposures; signs (Table 6.29) are usually reversed on cessation of exposure to the vitamin. Signs of hypervitaminosis A are associated with plasma retinol levels >3 μmol/L and increases in serum retinyl ester levels without substantial increases in circulating holo-RBP4.

Who may be at risk of hypervitaminosis A?

- Children and adults inappropriately using of dietary supplements containing the preformed vitamin, particularly those consuming multiple sources of vitamins.
- Livestock fed incorrect diets in which mixing errors concerning vitamin A have been made.

TABLE 6.33 Signs of Hypervitaminosis A

Organ Affected	Signs
General	Muscle and joint pains, headache
Skin	Erythema, desquamation, alopecia
Mucous membranes	Cheilitis, stomatitis, conjunctivitis
Liver	Dysfunction
Skeletal	Thinning and fracture of long bones

Signs of Toxicity

Acute hypervitaminosis. In humans, signs can be manifest after single large doses (>660,000 IU for adults, >330,000 IU for children), or after doses >100,000 IU/day have been taken for several months (Table 6.33). Children experiencing hypervitaminosis A develop transient (1–2 days) signs: nausea, vomiting, symptoms associated with increased cerebrospinal fluid pressure (headache, vertigo, blurred or double vision), and muscular incoordination. Most field studies have found that 3–9% of children given high single doses (c.200,000 IU) for prophylaxis show transient nausea, vomiting, headache, and general irritability; a similar percentage of younger children may show bulging fontanelles[162] subsiding within 48–96 h.

Those reacting to extremely large doses of the vitamin show drowsiness, loss of appetite, and malaise followed by skin exfoliation, itching (circumocular), and recurrent vomiting. Several studies have found that about 30% of acne patients treated with 13-*cis*-retinoic acid show increased

161. It has been suggested that the hepatic damage following hepatitis B infection may be due to the toxic effects of retinoids, which accumulate in the cholestatic liver.

162. The convex displacement of the infant's "*soft spot*" (the membranous covering of the cranial sutures) caused by fluid accumulation in the skull cavity or increased intracranial pressure.

LDL/HDL ratios which, if persistent, would suggest increased risk of ischemic heart disease. Some 15% of such patients report arthralgia and myalgia.

Chronic hypervitaminosis. Recurrent exposures exceeding 12,500 IU (infants) to 33,000 IU (adult) typically produce many changes; the earliest are likely to involve the skin and mucous membranes (Table 6.31). Dry lips (cheilitis) are a common early sign in humans, often followed by dryness and fragility of the nasal mucosa, leading to dry eyes and conjunctivitis. Skin lesions include dryness, pruritis, erythema, scaling, peeling of the palms and soles, hair loss (alopecia), and nail fragility. Headache, nausea, and vomiting (signs of increased intracranial pressure) have also been reported.

Hypervitaminosis A can also reduce bone mineral density and increase fracture risk,[163] perhaps due to direct membranolytic activities of retinoids, or to their activation of abnormal gene expression. Infants and young children can show painful periostitis[164] and, rarely, premature closure of lower limb epiphyses (manifest as "*hyena disease*" in calves, characterized by shortened hind limbs). Chronic, high intakes of vitamin A may contribute to the pathogenesis of osteoporosis in adults. One study found reduced bone mineral density in association with intakes greater than 0.6 mg RE (2000 IU)/day,[165] a level consumed by at least half of American adults. An analysis of results from more than 72,000 postmenopausal women in the Nurses' Health Study found that individuals reporting intakes of at least 2000 IU/d had twice the risk of hip fracture due to mild or moderate trauma than those reporting intakes less than 500 IU/d.[166] Hypervitaminotic A animals frequently show bone abnormalities that apparently result from changes in impaired osteoclastic activities and enhanced osteoblastic activities resulting in overgrowth of periosteal bone in a nonvitamin D-dependent manner. This, in turn, can lead to impairments in visual function by restricting the optic foramina and pinching the optic nerve, and in motor function by increasing intracranial pressure. Intracranial hypertension is a well-known side effect of therapeutic doses of 13-*cis*-retinoic acid (for acne) and all-*trans*-retinoic acid (acute promyelitic leukemia). That this condition resolves more rapidly if patients undergo weight loss suggests that RBP4 of adipose origin may be involved in triggering these signs.

The accumulation of retinal condensation products is a significant initiating factor in retinal photodamage characterized by progressive retinal cell death. Impaired clearing of all-*trans*-retinal by photoreceptor cells can result in the accumulation of fluorescent bisretinoids by retinal pigment epithelial cells through phagocytosis of photoreceptor outer segments. This process may result in retinal dystrophy and degenerative diseases including Stargardt disease and age-related macular degeneration.

Therapeutic doses of all-*trans*-retinoic acid are generally well tolerated, although some patients experience headache, nausea, and visual changes. In rare cases, muscular stiffness and epileptiform seizures have been reported. A few cases of myocarditis have been reported patients with acute promyelitic leukemia given this vitamer.

Embryotoxic Potential of High Levels of Vitamin A

Retinoids can be toxic to maternally exposed embryos, which limits their therapeutic uses and raises concerns about the safety of high-level vitamin A supplementation for pregnant animals and humans. This is especially true for 13-*cis*-retinoic acid, which is very effective in the treatment of acne but can cause severe disruption of cephalic neural crest cell activity that results in birth defects characterized by craniofacial, central nervous system, cardiovascular, and thymus malformations. Similar effects have been induced in animals by high doses of retinol, all-*trans*-retinoic acid, or 13-*cis*-retinoic acid.

Teratogenicity. Animal model studies suggest that the teratogenic effects of excess vitamin A are due to the embryonic exposure to all-*trans*-retinoic acid (Table 6.34), although such effects can be induced without substantially increasing maternal plasma concentrations of that metabolite. It has been proposed that the mechanism of teratogenic action of retinoids involves elevated production of mRNAs for particular RARs that lead to an imbalance in heterodimers among the various RARs, RXRs, and other hormone receptors, consequently affecting the expression of genes not typically expressed in normal metabolism. Indeed, teratogenic doses of all-*trans*-retinoic acid have been found to produce prolonged increases in RARα_2 mRNA levels.

The critical period for fetal exposure to maternally derived retinoids is during organogenesis, i.e., before many women suspect they are pregnant. This is also before the fetus develops retinoid receptors and binding proteins, which serve to restrict maternal–fetal transfer of retinoids.

Fetal malformations of cranial–neural crest origin have been reported in cases of oral use of all-*trans*-retinoic acid

163. Ribaya-Mercado, J.D., Blumberg, J.B., 2007. Nutr. Rev. 65, 425–428. That these types of bone lesions have been identified in the fossilized skeletal remains of ancient humans suggests that excessive vitamin A is not a new phenomenon.

164. Inflammation of the periosteum, the membranous tissue surrounding bone.

165. The Rancho Bernardo Study of 570 women and 388 men found a U-shaped relationship of bone mineral density and vitamin A intake, with optimal bone mineral density occurring at 2000–2800 IU (0.6–0.9 mg RE) per day. (Promislow, J.H.E., Goodman-Gruen, D., Slymen, D.J., et al., 2002. J. Nutr. 129, 2246–2250).

166. Feskanich, D., Singh, V., Willet, W.C., et al., 2002. JAMA 287, 47–54.

TABLE 6.34 Teratogenicity of Vitamin A in Rodent Models

	Retinyl Palmitate[a]	All-*trans*-Retinoic Acid[a]	
Species	Highest Nonteratogenic	Lowest Teratogenic	Teratogenic
Rat[b]	30	90	6
Mouse[b]	15	50	3
Rabbit[c]	2	5	6
Hamster	–	–	7

[a]*Dosage level (mg/kg/day).*
[b]*Exposed on gestational days 6–15.*
[c]*Exposed on gestational days 6–18.*
After Kamm, J.J., 1982. J. Am. Acad. Dermatol. 64, 552–559.

in treating acne vulgaris and of regular prenatal vitamin A supplements in humans. The latter have generally been linked to daily exposures at or above 20,000–25,000 IU. A retrospective epidemiologic study imputed increased risk of birth defects associated with exposures of about ≥10,000 IU of preformed vitamin A per day;[167] however, the actual observed increase was in a small group of women whose vitamin A intakes exceeded 21,000 IU/day (Table 6.35).

Provitamin A Toxicity. In general, carotenoids have low toxicities. A small intervention study showed that a daily intake of 30 mg of β-carotene from carotene-rich foods produced accumulation of the carotenoid in the skin with consequent yellowing (**carotenodermia**) within 25–42 days of exposure;[168] the effect persisted for at least 14–42 days after cessation of carotene exposure. It is possible that, under highly oxidative conditions, asymmetric cleavage of β-carotene to yield apocarotenals and apocarotenoic acids that can diminish retinoic acid signaling by interfering with the binding of retinoic acid to RAR. This effect has been proposed as the basis for the finding that a regular daily dose of β-carotene increased lung cancer risk among smokers.

Recommended Upper Limits of Exposure

Both the US and the European Union have estimated upper tolerable limits of intakes of preformed vitamin A (Table 6.36).

167. Rothman and colleagues. (Rothman, K.J., Moore, L.L., Singer, M.R., et al., 1995. N. Engl. J. Med. 33, 1369–1373) estimated that threshold using regression techniques. That level of exposure to preformed vitamin A was associated with a birth defect risk of 1 in 57. This report has been criticized for suspected misclassification of malformations. Note: the current RDA for pregnant women of 800 μg of RE (2700 IU) per day.

168. Carotenodermia was diagnosed only after plasma total carotenoid concentrations exceeded 4.0 mg/L.

12. CASE STUDIES

Instructions

Review each of the following case reports, paying special attention to the diagnostic indicators on which the respective treatments were based. Then answer the questions that follow.

Case 1

The physical examination of a 5-month-old boy with severe marasmus[169] showed extreme wasting, apathy, and ocular changes: in the left eye, Bitot's spots, and conjunctival and corneal xerosis; in the right eye, corneal liquefaction and keratomalacia with subsequent prolapse of the iris, extrusion of lens, and loss of vitreous humor. The child was 65 cm tall and weighed 4.5 kg. His malnutrition had begun at cessation of breast feeding at 4 months, after which he experienced weight loss and diarrhea.

Laboratory Results

Parameter	Patient	Normal Range
Hb (hemoglobin)	10.7 g/dL	12–16 g/dL
HCT (hematocrit)	36 ml/dL	35–47 mg/dL
WBC (white blood cells)	15,000/μL	5000–9000/μL
Serum protein	5.6 g/dL	6–8 g/dL
Serum albumin	2.49 g/dL	3.5–5.5 g/dL
Plasma sodium	139 mEq/L	136–145 mEq/L
Plasma potassium	3.5 mEq/L	3.5–5.0 mEq/L
Blood glucose	70 mg/dL	60–100 mg/dL
Total bilirubin	1.1 mg/dL	<1 mg/dL
Serum retinol	5.5 μg/dL	30–60 μg/dL
Serum β-carotene	10.7 μg/dL	50–250 μg/dL
Serum vitamin E	220 μg/dL	500–1500 μg/dL

The child had an infection, showing **otitis media**[170] and *Salmonella* septicemia,[171] which responded to antibiotic treatment in the first week. The patient was given by nasogastric tube an aqueous dispersion of retinyl palmitate (with a nonionic detergent) at the rate of 3000 μg/kg per day for 4 days. This increased his plasma retinol concentration from 5 to 35 μg/dL by the second day, at which level it was

169. Marasmus is characterized by extreme emaciation or general atrophy. It occurs mostly in young children and is caused by extreme undernutrition, primarily lack of energy and protein.

170. Inflammation of the middle ear.

171. Presence in the blood of pathogenic, gram-negative, rod-shaped bacteria of the genus *Salmonella*.

TABLE 6.35 Teratogenic Risk of High Prenatal Exposures to Preformed Vitamin A

Retinol Intake (IU/day)	Pregnancies	Cranial–Neural Crest Defects	Total Defects
0–5000	6410	33 (0.51%)	86 (1.3%)
5001–10,000	12,688	59 (0.47%)	196 (1.5%)
10,001–15,000	3150	20 (0.63%)	42 (1.3%)
>15,000	500	9 (1.80%)	15 (3.0%)

After Rothman, K.J., Moore, L.L., Singer, M.R., et al., 1995. N. Engl. J. Med. 333, 1369–1373.

maintained for the next 12 days. The child responded to general nutritional rehabilitation with a high-protein, high-energy formula that was followed by whole milk supplemented with solid foods. He recovered but was permanently blind in the right eye and was left with a mild corneal opacity in the left eye. He returned to his family after 10 weeks of hospitalization.

Case 2

An obese 15-year-old girl, 152 cm tall and weighing 100 kg, was admitted to the hospital for partial jejunoileal bypass surgery for morbid obesity. She had a past history of obsessive eating that had not been correctable by diet. Except for massive obesity, her physical examination was negative.

Initial Laboratory Results

Parameter	Patient	Normal Range
Hb (hemoglobin)	10.5 g/dL	12–15 g/dL
RBC (red blood cell)	4.5 × 10⁶/μL	4–5 × 10⁶/μL
WBC (white blood cell)	8000/μL	5000–9000/μL
Serum retinol	38 μg/dL	30–60 μg/dL
Serum β-carotene	12 μg/dL	50–300 μg/dL
Serum vitamin E	580 μg/dL	500–1500 μg/dL
Serum 25-OH-D_3	11 ng/dL	8–40 ng/dL

The following test results were within normal ranges: serum electrolytes, calcium, phosphorus, triglycerides, cholesterol, total protein, albumin, total bilirubin, copper, zinc, folic acid, thiamin, and vitamin B_{12}. The patient encountered few postoperative complications except for mild bouts of diarrhea and some fatigue. Over the next year, she lost 45 kg of body weight while ingesting a liberal diet. She reported having three to four stools daily but denied having any objectionable diarrhea or changes in stool appearance. Two years after surgery, she noted the onset of inflammatory horny lesions above her knees and elbows, and she experienced some difficulty in seeing at dusk. The skin lesions failed to respond to topical corticosteroids and oral antihistamine therapy. Because of intensification of these signs, she sought medical help; however, the cause was not determined.

She was readmitted to the hospital, complaining of her skin disorder and night blindness. At that time, she showed evidence of mild liver dysfunction and her serum concentrations of retinol and β-carotene were 16 and 14 μg/dL, respectively. Her fecal fat was 70 g/day (normal, <7 g/day). Biopsies of the skin of her left thigh and right upper arm each showed

TABLE 6.36 Recommended Upper Tolerable Intakes (ULs) of Preformed Vitamin A

US[a]		EU[b]	
Ages (years)	UL (μg RE/day)	Ages (years)	UL (μg RE/day)
0.1	600	1–3	800
1–3	600		
4–8	900	4–6	1100
9–13	1700	7–10	1500
14–18	2800	11–14	2000
		15–17	2600
Adults[c]	3000	Adults	3000

[a]*Food and Nutrition Board, 2001. Dietary Reference Intakes for Vitamin A, Vitamin K, Arsenic, Boron, Chromium, Copper, Iodine, Iron, Manganese, Molybdenum, Nickel, Silicon, Vanadium and Zinc. National Academy Press, Washington, DC, 773 pp.*
[b]*Recommended Nutrient Intakes; Joint WHO/FAO Expert Consultation, 2001. Human Vitamin and Mineral Requirements. Food and Agricultural Org., Rome, 286 pp.*
[c]*Including pregnant and lactating women.*

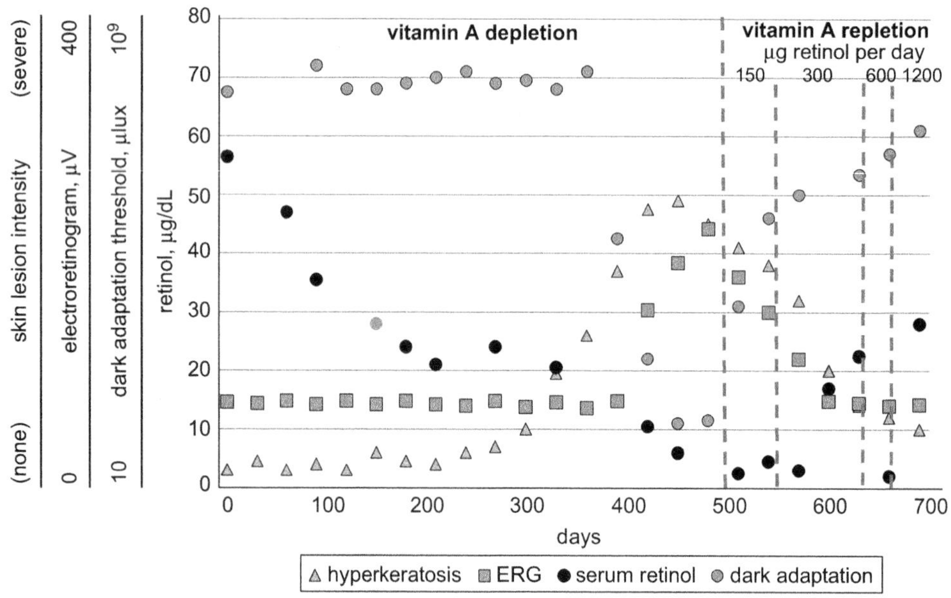

FIGURE 6.20 The volunteer's responses to chronic vitamin A deprivation followed by stepwise repletion with oral supplements of retinol. *After Russel. R.M., Multrak, R., Smityh, V., et al., 1973. Lancet 2, 161–1164.*

hyperkeratosis and horny plugging of dilated follicles. She was treated with 15,000 μg of retinyl palmitate given orally three times daily for 6 months. By 1 month, the follicular hyperkeratosis had cleared and healed with residual pigmentation. By 2 months, the night blindness had subsided. At that time, her serum retinol concentration was 54 μg/dL, β-carotene was 7 μg/dL, α-tocopherol was 1.6 μg/ml, and urinary $[^{57}Co]B_{12}$ was 6.7% (normal, 7–8%). She has been well on a daily oral supplement of 1500 μg of retinyl palmitate.

Case 3

A 41-year-old man was housed in a metabolic ward for 2 years during a clinical investigation of vitamin A deficiency. He weighed 77.3 kg and was healthy by standard criteria (history, physical examination, and laboratory studies). For 505 days, he was fed a casein-based formula diet that contained <10 μg of vitamin A per day. His initial plasma retinol concentration was 58 μg/dL, and his body vitamin A pool, determined by isotope dilution, was 766 mg (10 mg/kg). At the end of 1 year, his plasma retinol had declined to 25 μg/dL and he began to show follicular hyperkeratosis (Fig. 6.20). On day 300, his plasma retinol was 20 μg/dL, and he showed a mild anemia (Hb 12.6 mg/dL). Two months later, by which time his plasma retinol had dropped to 10 μg/dL, he developed night blindness as evidenced by changes in dark adaptation and electroretinogram. When his plasma retinol reached 3 μg/dL, his body vitamin A pool was 377 mg and repletion with vitamin A was begun with increasing doses starting with 150 μg and increasing to 1200 μg of retinol per day over a 145-day period. After receiving 150 μg of retinol per day for 82 days, his night blindness was partially repaired, but his skin keratinization remained and his plasma retinol level was only 8 μg/dL. Then, after receiving 300 μg of retinol per day for 42 days, his follicular hyperkeratosis resolved and his plasma retinol level was 20 μg/dL. At the 600 μg of retinol per day level, his plasma retinol was in the normal range and all signs of vitamin A deficiency disappeared.

Case Questions

1. For each case, what signs/symptoms indicated vitamin A deficiency?
2. Propose hypotheses to explain why the patients of cases 1 and 2 each responded to oral vitamin A treatment even though they had very different medical conditions. Outline tests of those hypotheses.
3. Comment on the value of serum retinol concentration for the diagnosis of nutritional vitamin A status.

13. STUDY QUESTIONS AND EXERCISES

1. Discuss how the absorption, transport, tissue distribution, and intracellular activities of vitamin A relate to the concept of solubility.
2. Construct a flow diagram showing vitamin A, in its various forms, as it passes from ingested food, through the body where it functions in its various physiologic roles, and ultimately to its routes of elimination.
3. Construct a decision tree for the diagnosis of vitamin A deficiency in a human or animal.

4. Night blindness is particularly prevalent among alcoholics. Propose a hypothesis for the metabolic basis of this phenomenon and outline an experimental approach to test it.
5. Construct a figure detailing the mechanism by which vitamin A functions as a transcription factor.
6. Note three physiological mechanisms which may protect against hypervitaminosis A.
7. Discuss the points of control, and intervention possibilities for each, in the WHO conceptual framework for vitamin A deficiency.

RECOMMENDED READING

Berry, D.C., Noy, N., 2012. Signaling by vitamin A and retinol-binding protein in regulation of insulin responses and lipid homeostasis. Biochim. Biophys. Acta 182, 168–176.

Bhat, P.V., Manaolescu, D., 2014. Serum retinol-binding protein, obesity and insulin resistance. In: Dakshinamurti, K., Dakshinamurti, S. (Eds.), Vitamin Binding Proteins: Functional Consequences. CRC Press, New York, pp. 31–48.

Food and Nutrition Board, 2001. Dietary Reference Intakes for Vitamin A, Vitamin K, Arsenic, Boron, Chromium, Copper, Iodine, Iron, Manganese, Molybdenum, Nickel, Silicon, Vanadium and Zinc. National Academy Press, Washington, DC. 773 pp.

Carvalho, J.E., Schubert, M., 2014. Retinoic acid: metabolism, developmental functions and evolution. In: Dakshinamurti, K., Dakshinamurti, S. (Eds.), Vitamin Binding Proteins: Functional Consequences. CRC Press, New York, pp. 1–30.

Chroni, E., Monastrili, A., Tsambo, D., 2010. Neuromuscular adverse effects associated with systemic retinoid dermatotherapy: monitoring and treatment algorithm for clinicians. Drug Saf. 33, 25–34.

Dawson, M.I., Xia, Z., 2012. The retinoid X receptors and their ligands. Biochem. Biophys. Acta 1821, 21–56.

Duester, G., 2008. Retinoic acid synthesis and signaling during early organogenesis. Cell 134, 921–931.

Duriancik, D.M., Lackey, D.E., Hoag, K.A., 2010. Vitamin A as a regulator of antigen presenting-molecules. J. Nutr. 140, 1395–1399.

Frey, S., Vogel, S., 2011. Vitamin A metabolism and adipose tissue biology. Nutrients 3, 27–39.

Graham, T.E., Yang, Q., Blűher, M., et al., 2010. β-Carotene is an important vitamin A source for humans. J. Nutr. 140, 2268S–2285S.

Kiser, P.D., Golczak, M., Palczewski, K., 2014. Chemistry of the retinoid (visual) cycle. Chem. Rev. 114, 194–232.

Kiser, P.D., Palczewski, K., 2010. Membrane-binding and enzymatic properties of RPE65. Progr. Retin. Eye Res. 29, 428–442.

Latham, M., 2010. The great vitamin A fiasco. J. World Pub. Health Nutr. Assoc. 1, 12–24.

Leitz, G., Lange, J., Rimbach, G., 2010. Molecular and dietary regulation of β,β-carotene 15,15′-monooxygenase 1 (BCMO1). Arch. Biochem. Biophys. 502, 8–16.

Lobo, G.P., Hessel, S., Eichinger, A., et al., 2010. ISX is a retinoic acid-sensitive gatekeeper that controls intestinal β-carotene absorption and vitamin A production. FASEB J. 24, 1656–1666.

Di Masi, A., Leboffe, L., De Marinis, E., et al., 2015. Retinoic acid receptors: from molecular mechanisms to cancer therapy. Mol. Asp. Med. 41, 1–115.

Miyabe, Y., Miyabe, C., Nanki, T., 2014. Retinoic acid and immunity. In: Dakshinamurti, K., Dakshinamurti, S. (Eds.), Vitamin Binding Proteins: Functional Consequences. CRC Press, New York, pp. 49–56.

Mora, J.R., von Andrian, U.H., 2009. Role of retinoic acid in the imprinting of gut-homing IgA-secreting cells. Semin. Immunol. 21, 28–35.

Noy, N., 2013. Vitamin A. In: Stipanuk, M.H., Caudill, M.A. (Eds.), Biochemical, Physiological, and Molecular Aspects of Human Nutrition. Elsevier, New York, pp. 683–702.

Noy, N., 2010. Between death and survival: retinoic acid in regulation of apotosis. Ann. Rev. Nutr. 30, 201–217.

Penniston, K.L., Tanumihardjo, S.A., 2006. The acute and chronic toxic effects of vitamin A. Am. J. Clin. Nutr. 83, 191–201.

Pino-Lago, K., Benson, M.J., Noelle, R.J., 2008. Retinoic acid in the immune system. Ann. N.Y. Acad. Sci. 1143, 170–187.

Van Poppel, G., Goldbohm, R.A., 1995. Epidemiologic evidence for β-carotene and cancer prevention. Am. J. Clin. Nutr. 62, 1393S–1402S.

Ramakrishnan, U., Darnton-Hill, I., 2002. Assessment and control of vitamin A deficiency disorders. J. Nutr. 132, 2947S–2953S.

Reboul, F., 2013. Absorption of vitamin A and carotenoids by the enterocyte: focus on transport proteins. Nutrients 5, 3563–3581.

Rhinn, M., Dollé, P., 2012. Retinoic acid signaling during development. Devel 139, 843–858.

Ribaya-Mercado, J.D., Blumberg, J.B., 2007. Vitamin A: is it a risk factor for osteoporosis and bone fracture? Nutr. Rev. 65, 425–438.

Ross, C., Harrison, E.H., 2014. Vitamin A: nutritional aspects of retinoids and carotenoids. In: Zempleni, J., Suttie, J.W., Gregory, J.F., Stover, P.J. (Eds.), Handbook of Vitamins, fifth ed. CRC Press, New York, pp. 1–49.

Saari, J.C., 2012. Vitamin A metabolism in rod and cone visual cycles. Ann. Rev. Nutr. 32, 125–145.

Shumaskaya, M., Wurtzel, E.T., 2013. The carotenoid biosynthetic pathway: thinking in all dimensions. Plant Sci. 208, 58–63.

Solomons, N.W., 2012. In: Erdman Jr., J.W., Macdonald, I.A., Zeisel, S.H. (Eds.), Vitamin a, Chapter 11 in Present Knowledge in Nutrition, tenth ed. ILSI Press, Washington, pp. 149–184.

Sommer, A., 2008. Vitamin A deficiency and clinical disease: an historical overview. J. Nutr. 138, 1835–1839.

Sommerberg, O., Seims, W., Kraemer, K. (Eds.), 2013. Carotenoids and Vitamin a in Translational Medicine. CRC Press, New York. 404 pp.

Soprano, D.R., Soprano, K.J., 1995. Retinoids as teratogens. Annu. Rev. Nutr. 15, 111–132.

Soprano, D.R., Qin, P., Soprano, K.J., 2004. Retinoic acid receptors and cancers. Annu. Rev. Nutr. 24, 201–221.

Tang, G., 2010. Bioconversion of dietary provitamin A carotenoids to vitamin A in humans. Am. J. Nutr. 91, 1468S–1473S.

Tanumihaardjo, S.A., 2011. Biomaerkers of nutrition for development. Am. J. Clin. Nutr. 94, 658S–665S.

Tanumihardjo, S.A., Russell, R.M., Stephensen, C.B., et al., 2016. Biomarkers of nutrition for development – vitamin A review. J. Nutr. 146, 1816–1848.

Wolf, G., 2007. Serum retinol-binding protein: a link between obesity, insulin resistance, and type 2 diabetes. Nutr. Rev. 65, 251–256.

World Health Organization, 2014. WHO Global Database on Vitamin a Deficiency. http://www.who.int/vmnis/database/vitamina/en/.

Chapter 7

Vitamin D

Chapter Outline

Anchoring Concepts

1. Vitamin D is the generic descriptor for **steroids** exhibiting qualitatively the biological activity of cholecalciferol (i.e., vitamin D_3).
2. Most vitamers D are hydrophobic and, thus, are insoluble in aqueous environments (e.g., plasma, interstitial fluids, cytosol).
3. Vitamin D is not required in the diets of animals or humans adequately exposed to sources of **ultraviolet light** (e.g., sunlight).
4. Deficiencies of vitamin D lead to structural lesions of *bone*.

By following the reasoning that vitamin D is not required in the diet under conditions of adequate ultraviolet irradiation of skin and that it is the precursor of a hormone, it is likely that the vitamin is not truly a vitamin but must be regarded as a pro-hormone. These arguments, however, are only semantic; the fact remains that vitamin D is taken in the diet and is an extremely potent substance which prevents a deficiency disease.

Hector DeLuca.[1]

1. Hector DeLuca is a prominent nutritional biochemist whose discoveries were seminal in elucidating the metabolic functions of vitamin D. These included identification of the active forms of the vitamin, 25-hydroxyvitamin D_3 and 1,25-dihydroxyvitamin D_3, and the characterization of the enzymes involved in those pathways. He is a professor Emeritus at the University of Wisconsin, where he has had a long and productive career.

LEARNING OBJECTIVES

1. To understand the nature of the various sources of vitamin D.
2. To understand the means of endogenous production of vitamin D.
3. To understand the means of enteric absorption of vitamin D.
4. To understand the transport and metabolism involved in the activation of vitamin D to its functional forms.
5. To understand the role of vitamin D and other endocrine factors in calcium homeostasis.
6. To understand the roles of vitamin D in noncalcified tissues.
7. To understand the genomic bases of vitamin D action.
8. To understand the physiologic implications of high doses of vitamin D.

VOCABULARY

Apolipoprotein E (apoE)
Cage layer fatigue
Calbindins
Calcidiol
Calcinosis
Calcipotriol
Calcitonin
Calcitriol
Calcitroic acid
Calcium (Ca)

The Vitamins. http://dx.doi.org/10.1016/B978-0-12-802965-7.00007-1

Calcium-binding protein (CaBP)
Calcium-sensing receptor (CaR)
Calmodulin
Cathelicidin
Caveolae
Cholecalciferol
7-Dehydrocholesterol (7-DHC)
1,25-Dihydroxyvitamin D (1,25-$[OH]_2$-vitamin D)
24,25-Dihydroxyvitamin D (24,25-$[OH]_2$-vitamin D)
Diabetes
Diuresis
Epiphyseal plate
Ergocalciferol
Ergosterol
Genu varum
25-Hydroxyvitamin D (25-OH-vitamin D)
Hypercalcemia
Hyperphosphatemia
Hypersensitivity
Hypocalcemia
Hypoparathyroidism
Hypophosphatemia
Lead (Pb)
Lumisterol
Median erythemal dose (MED)
Melanin
Milk fever
25-OH-Vitamin D 1-hydroxylase
Osteoblast
Osteochondrosis
Osteoclast
Osteomalacia
Osteon
Osteopenia
Osteoporosis
Parathyroid gland
Parathyroid hormone (PTH)
Previtamin D
Privational rickets
Privational osteomalacia
Prolactin
Provitamin D
Pseudofracture
Pseudohypoparathyroidism
Psoriasis
Rickets
Sarcopenia
Steroid
Tachysterol
Tibial dyschondroplasia
Transcaltachia
Transcalciferin
24,25,26-Trihydroxyvitamin D (24,25,26-$[OH]_3$-vitamin D)
Varus deformity
Vitamin D-binding protein (DBP)
Vitamin D-dependent rickets types I and II
Vitamin D receptors (VDRs)
Vitamin D resistance
Vitamin D-responsive elements (VDREs)
Vitamin D_2
Vitamin D_3
Vitamin D 25-hydroxylase
Vitamin D-dependent rickets
Vitamin D-resistant rickets
Zinc fingers

1. SIGNIFICANCE OF VITAMIN D

Vitamin D,[2] the "sunshine vitamin," is actually a hormone produced from sterols in the body by the photolytic action of ultraviolet light on the skin; individuals who receive modest exposures to sunlight are able to produce their own vitamin D. However, this is not the case for most people: those living in northern latitudes; those spending most of their days indoors and/or having darker skin; animals being managed in controlled environments. Such individuals must obtain the nutrient from their diets; for them vitamin D is a vitamin in the traditional sense.

Vitamin D plays an important role, along with the essential minerals calcium (Ca), phosphorus (P), and magnesium (Mg), in the maintenance of healthy bones and teeth. Problems in those organs appear to have existed in different past populations throughout the world.[3] Today, nearly 40% of the world's adult population is of insufficient vitamin D status.[4] More than three-quarters of adults in Bangladesh, Scotland, Norway, and Estonia, and at least half of adults in several other countries (Denmark, Korea, New Zealand, Brazil), have been found to be of low-vitamin D status.[5] Huge numbers of infants in many countries are also known to be of low-vitamin D status, including most infants in Iran, Pakistan, India, and Turkey; but for many parts of the world, particularly, Africa and South America, little information is available. Vitamin D deficiency remains a global public health problem in all age groups, the most visible impacts being the increased risk of the metabolic bone diseases, rickets and osteoporosis.

2. The convention used in this chapter is that the terms "vitamin D" or "D" (without subscripts) are used in contexts in which the major vitamers (D_3, or **cholecalciferol**; D_2, or **ergocalciferol**) are equivalent.

3. Evidence of low bone mass has been found in many ancient populations. That such cases have not always been associated with osteoporotic fractures suggests that Ca intakes of earlier peoples may have been greater than those today (Nelson, D.A., Sauer, N.J., Agarwal, S.C., 2004. Evolutionary Aspects of Bone Health. In: Holick, M.F., Dawson-Hughes, B. (Eds.), Nutrition and Bone Health. Humana Press, Totowa, NJ, pp. 3–18).

4. Hilger, J., Friedel, A., Herr, R., et al., 2014. Br. J. Nutr. 111, 23–45; Palacios, C., Gonzalez, L., 2014. J. Steroid Biochem. Mol. Biol. 144, 138–145.

5. Have plasma 25-hydroxycholecalciferol concentrations <50 nM.

Rickets, the deforming and debilitating disease involving delayed or failed endochondral ossification (mineralization at the growth plates) of the long bones, remains a problem in many countries, having been reported at prevalences as great as 10% among infants exclusively fed breast milk and children with little sun exposure. Shockingly high prevalences have been reported in Yemen (27%), Ethiopia (42%), Tibet (66%), and Mongolia (70%).[6] This disability can significantly impair ambulation and work productivity; in some societies, it can be socially stigmatizing.

The global prevalence of **osteoporosis** (loss of bone leading to increased bone fragility) in adults is not well documented. The disease is estimated to affect 10 million Americans ≥50 years of age; the number with low bone mass (**osteopenia**) is estimated to exceed 61 million by 2020. Fracture incidence varies globally from about 50/100,000 for men and women in central Africa, to nearly 400 (men) and >900 (women) per 100,000 in northern Europe.[7] Two million osteoporosis-related fractures[8] occurred in the United States in 2005; by 2025, that number is expected to reach 3 million with an associated medical cost of $16–25 billion. Each year 250,000 Americans (mostly women) are hospitalized for hip fractures; three-fourths do not have full recoveries, half will be dependent on a cane or walker, 40% will require assisted care, and 20% will die of complications within a year. Women are increasingly susceptible to osteoporosis after the onset of menopause when their rate of bone loss can increase as much as 10-fold.[9] It is estimated that, among Americans over 50 years, women experience 75–80% of all hip fractures.

Vitamin D status affects more than bone. It also functions in the regulation of cellular development and differentiation of most cells, in the regulation of the parathyroid gland and the immune system function, in the skin, in cancer prevention, and in the metabolism of foreign compounds.

2. PROPERTIES OF VITAMIN D

Vitamin D is the generic descriptor for all steroids qualitatively exhibiting the biological activity of **cholecalciferol**. These compounds contain the intact A, C, and D steroid rings, ultimately derived in vivo by photolysis of the B ring of 7-dehydrocholesterol (7-DHC). That process frees the A ring from the rigid structure of the C and D rings, yielding conformational mobility in which the A ring undergoes rapid interconversion between two chair configurations. Vitamin D-active compounds have either of two types of isoprenoid side chains attached to the steroid nucleus at C-17 of the D ring:

- A nine-carbon, monounsaturated (i.e., containing one double bond) side chain. Vitamin D-active compounds with this structure are derivatives of **ergocalciferol**, is also called **vitamin D$_2$**. This vitamer can be produced synthetically by the photolysis of plant sterols.
- An eight-carbon, saturated (i.e., containing no double bond) side-chain. Vitamin D-active compounds with this structure are derivatives of **cholecalciferol**, also called **vitamin D$_3$**, which is produced metabolically through a natural process of photolysis of 7-DHC on the surface of skin exposed to ultraviolet irradiation, e.g., sunlight. The metabolically active vitamers are side chain-substituted, open-ring **steroids** with a *cis*-triene structure with hydroxylated carbons at ring position 1 and side chain position 25.

Chemical structures of the vitamin D group:

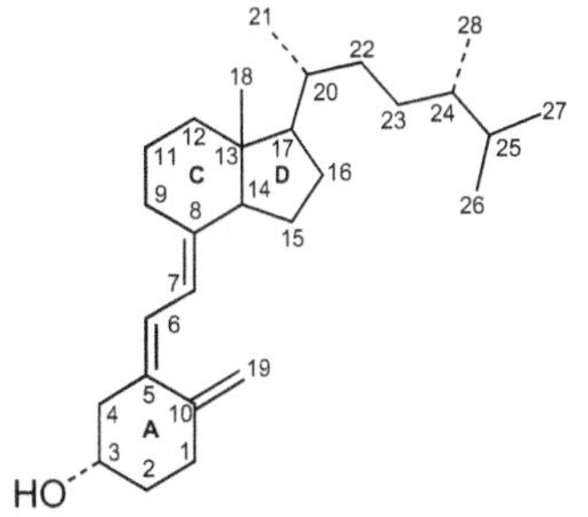

Vitamin D$_3$ (cholecalciferol)

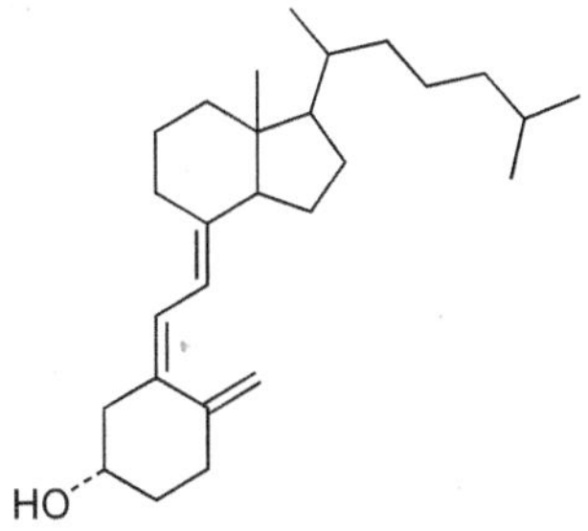

Vitamin D$_2$ (ergocalciferol)

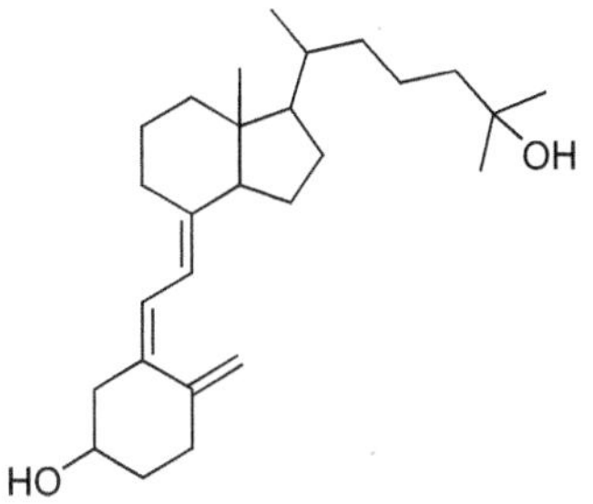

25-OH-Vitamin D$_3$

6. Prentice, A., 2008. Nutr. Rev. 66, S153–S164.
7. Dhanwal, D.K., Dennison, E.M., Harvey, N.C., et al., 2011. Indian J. Orthop. 45, 15–22.
8. Fractures of the vertebrae, hip, forearm, leg, and ankle, in that order, are the most common, although often asymptomatic. Canadian health statistics show the risk of radial fractures for men and women to be about 25/100,000 until the fifth decade of life, when it increases in women to more than 200/100,000 by the seventh and eighth decades.
9. Women can lose as much as one-fifth of their total bone mass within the first 5–7 years of menopause. By their eighth and ninth decades, many have lost 30–50% of their bone mass. Aging men also lose bone, but at a lower rate and starting from a generally larger bone mass. By their eighth and ninth decades, men typically have lost 20–30% of their peak bone mass.

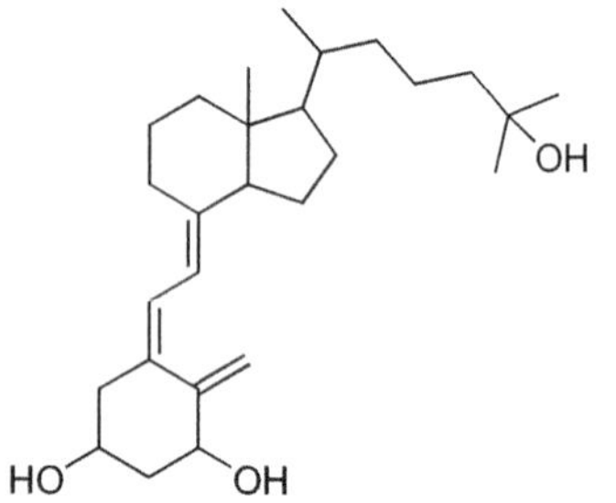

1,25-$(OH)_2$-Vitamin D_3

Chemical Properties of Vitamin D

Unlike the ring-intact steroids, vitamin D-active compounds tend to exist in extended conformations due to the 180° rotation of the A ring about the 6,7 single bond (in solution, the stretched and closed conformations are thought to come to an equilibrium favoring the former). The hydroxyl group on C-3 is, thus, in the β-position (i.e., above the plane of the A ring) in the closed forms, and in the α-position (i.e., below the plane of the A ring) in the stretched forms. Rotation about the 5,6 double bond can also occur by the action of light or iodine to interconvert the biologically active 5,6-*cis* compounds to 5,6-*trans* compounds, which show little or no vitamin D activity.

Vitamers D_2 and D_3 are white-yellow powders that are insoluble in water, moderately soluble in fats, oils, and ethanol; and freely soluble in acetone, ether, and petroleum ether. Each shows strong ultraviolet (UV) absorption, with a maximum at 264 nm. Each is sensitive to O_2, light, and iodine. Heat or mild acidity can convert each to the respective 5,6-*trans* and other inactive forms. While the vitamin is stable in dry form, in organic solvents, and in most plant oils (owing to the presence of α-tocopherol, which serves as a protective antioxidant), its thermal and photo lability can result in losses during such preparatory procedures as saponification with refluxing. Therefore, it is often necessary to use inert gas environments, light-tight sealed containers, and protective antioxidants when isolating the vitamin. In solution, both D_2 and D_3 undergo reversible, temperature-dependent isomerization to previtamin D, each forming an equilibrium mixture of both vitamers.[10]

3. SOURCES OF VITAMIN D

Biosynthesis of Vitamin D_3

Cholecalciferol (D_3) can be biosynthesized by a two-step process involving ultraviolet light (UVB, 290–310 nm) acting on **7-DHC** in the skin (Fig. 7.1):

1. **Photoactivation**—The activation reaction involves light-induced electrolytic ring opening of 7-dehydrocholesterol due to absorption of light by the 5,7-diene of the B ring of the sterol nucleus[11] to produce s-*trans*, s-*cis*-previtamin D_3. This process is optimal with light in the UVB range (295–300 nm); however, shorter wavelength irradiation can also cause this photoisomerization.[12]
2. **Thermal isomerization**—Once formed, previtamin D_3 can photochemically convert to **lumisterol** or **tachysterol**, but its lowest energy prospect is to undergo a thermal rearrangement involving a C19 to C9 hydrogen shift to form **cholecalciferol** (vitamin D_3). The reversible nature of this rearrangement means that previtamin D_3 and vitamin D_3 are always in dynamic equilibrium; under physiological conditions, the equilibrium favors vitamin D_3.

This process in vivo appears to convert only 5–15% of the available 7-DHC to vitamin D_3.[13] Yield is affected by the physical properties of the skin and of the environment; thus, it differs between individuals and species and shows great variation according to time of day, season, and latitude. It has been estimated that adults can biosynthesize as much as 15 µg (0.6 IU) D_3 per day, i.e., 10–25% of their total vitamin D intakes at the summer peak.[14]

The provitamin D sterol, 7-DHC, is both a precursor to and product of cholesterol (via different pathways). It is synthesized in the sebaceous glands of the skin and secreted onto the surface, where it is reabsorbed into the various layers of the epidermis and dermis to be localized mostly in their membranes.[15] The skin contains high concentrations of the sterol (200 times that of liver); in humans the epidermis and dermis in humans have similar 7-DHC contents, although most previtamin D_3 is found in the epidermis. In humans, epidermal 7-DHC concentrations are greatest

10. For example, within 30 days a 93% D, 7% pre-D mixture is established at 20°C; at 100°C a 72% D, 28% pre-D is established in only 30 min.

11. For this reason, the vitamers D are called **secosteroids**, denoting the "broken" ring by the Latin prefix *seco* (to cut).

12. UVC light (200–280 nm) was shown to be effective in curing rickets in the rat (Knudson, A., Benford, F., 1938. J. Biol. Chem. 124, 287–299). It does not penetrate the atmosphere but is used for its germicidal properties.

13. Excess irradiation does little to increase the efficacy of this conversion. Instead, it increases the production of biologically inactive forms (e.g., lumisterol-3, tachysterol-3, and 5,6-*trans*-vitamin D_3).

14. Heaney, R.P., Armas, L.A.G., French, C., 2013. J. Nutr. 143, 571–575.

15. The skin is composed of three layers: the **epidermis**, the **dermis**, and a layer of insulating **subcutaneous fat**. The epidermis is comprised mostly of keratinocytes, which form its tough outer layer, the **stratum corneum** (horny layer) and are continually replaced from a basal layer of cells, the **Malpighian layer** (named for Marcello Malpighi, 1628–94, the "father of microscopic anatomy and histology"). The epidermis also contains a smaller numbers of melanocytes that produce melanin, which filters out UV irradiation from solar exposure and is the major contributor to skin color, and dendritic (antigen-presenting) Langerhans cells, named for the German anatomist Paul Langerhans, 1847–88, who first described them). The thick underlying dermis is comprised of fibrous and elastic tissue (collagen, elastin, and fibrillin); it also contains sensory nerve endings, sweat glands, sebaceous glands, hair follicles, and blood vessels.

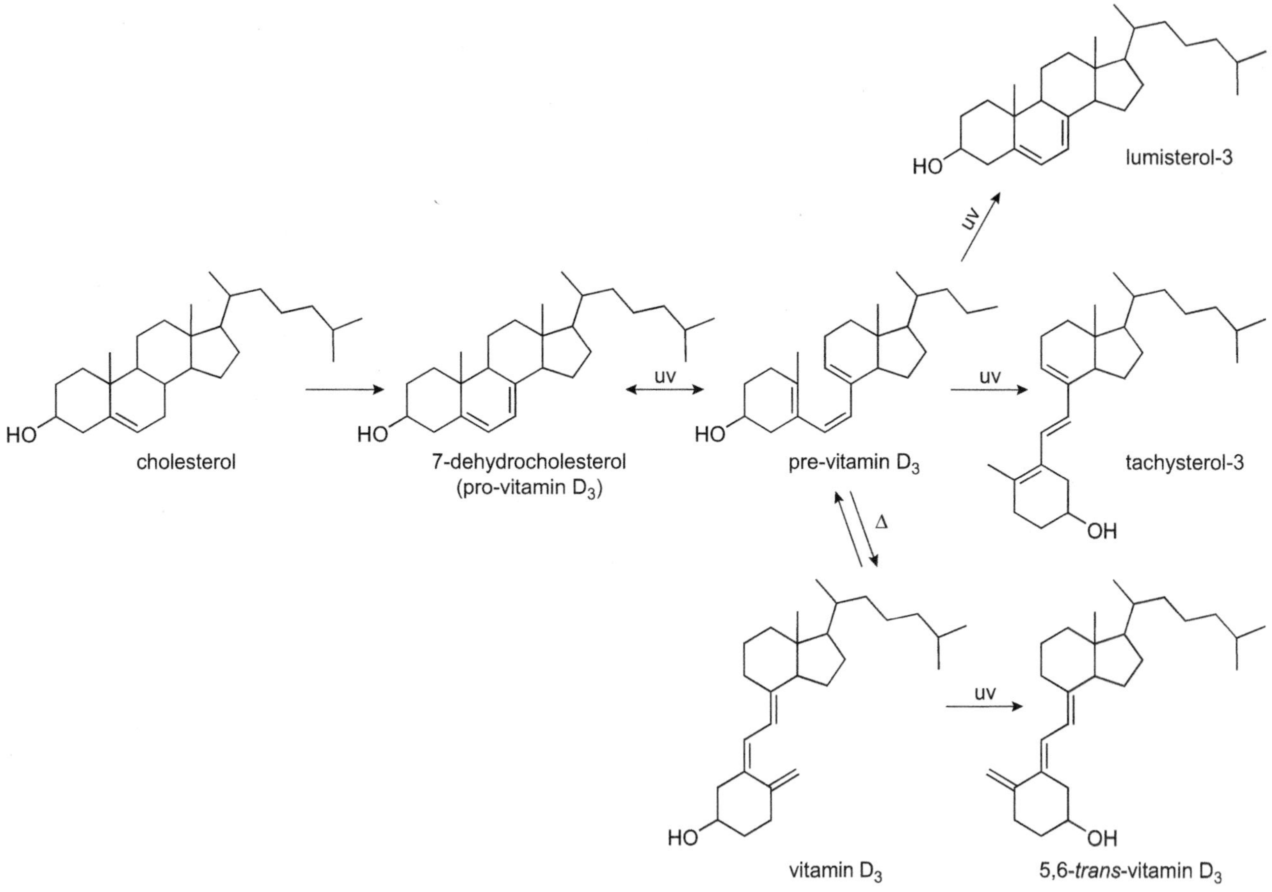

FIGURE 7.1 Biosynthesis of vitamin D_3 and its photolytic by-products.

in the deeper **Malpighian layer**; whereas, in the rat it is distributed more superficially in the **stratum corneum**. The associated differences in UVB penetrance mean that the D_3-biogenic effect of UVB is less efficient in humans than in the rat. Both the thickness and 7-DHC content of skin decline with age, reducing the D_3-biogenic capacity by twofold or more (Fig. 7.2).

Vitamin D biosynthesis is, thus, determined by environmental exposure to UV light, which also can increase risk to skin cancers in individuals experiencing episodes of severe burning.[16] The amount of sunlight required to support adequate vitamin D status is substantially less than that which increases skin cancer risk. Holick has estimated that exposure of only 6–10% of body surface to a **median erythemal dose (MED)**[17] of sunlight can be equivalent to consuming 600–1000 IU (15–25 mg) of vitamin D. He recommends exposures of one-quarter of that amount of

16. UV exposure is a risk factor for skin cancers, 70–85% of which are basal cell carcinomas. Melanomas, while far less prevalent, account for most skin cancer deaths.

17. MED is a measure of skin sensitivity to light, 1 MED I the amount of light that causes the skin to turn slightly pink (**erythema**) due to infiltration of erythrocytes into the exposed skin—the sign of a mild sunburn.

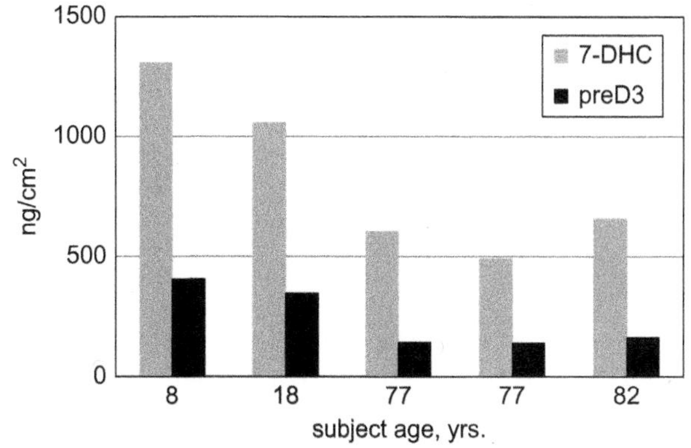

FIGURE 7.2 Epidermal contents of 7-dehydrocholesterol (7-DHC) and previtamin D_3 (pre-D3) of five subjects of varying ages. *After MacLaughlin, J., Holick, M.F., 1985. J. Clin. Invest. 76, 1536–1538.*

sunlight two to three times weekly to support the synthesis of physiologically relevant amounts, c.15,000 IU (375 mg) per week, of the vitamin.[18] Because sunlight exposure varies with latitude and season, D_3 biosynthesis also varies

18. Holick, M.F., 2001. Lancet 357, 4–6; Holick, M.F., Jenkins, M., 2003. The UV Advantage. iBooks, Inc., New York, p. 93.

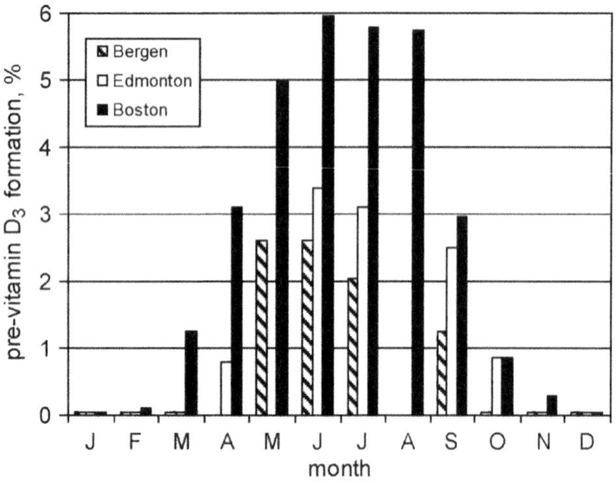

FIGURE 7.3 Seasonality of vitamin D biosynthesis (measures as conversion of 7-dehydrocholesterol in response to 1 h exposure to sunlight) at three different latitudes (Bergen: 60°N; Edmonton: 52°N; Boston: 42°N). *After Holick, M.F., 2008. Nutr. Rev. 66, S182.*

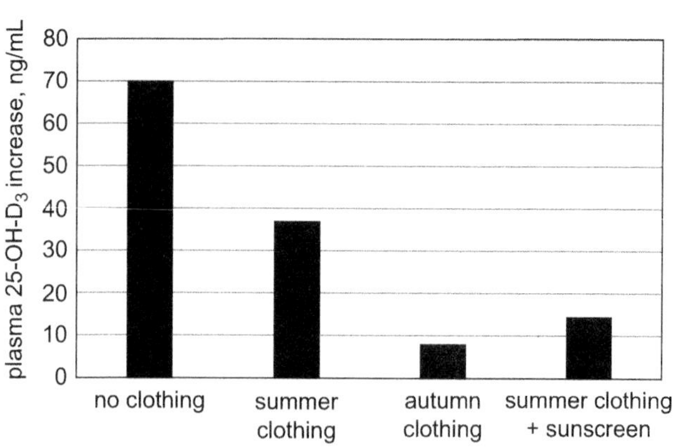

FIGURE 7.4 Effects of clothing and sunscreen on plasma 25-OH-D_3 responses of humans to 1 MED (median erythemal dose) of UVB. *After Matsuoka, L.Y., Wortsman, J., Dannenbaerg, M.J., et al., 1992. J. Clin. Endocrinol. Med. 75, 1099–1103.*

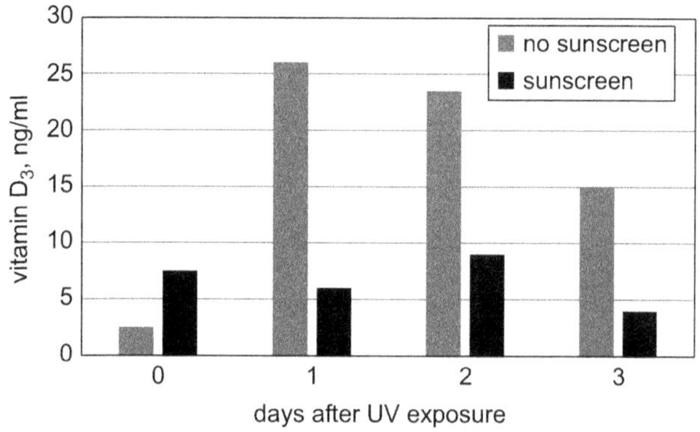

FIGURE 7.5 Effects of sunscreen (SPF8) application on circulating vitamin D_3 responses of humans to 1 MED (median erythemal dose) of UVB radiation. *After Matsuoka, L.Y., Wortsman, J., Hanifan, N., et al., 1987. J. Clin. Endocrinol. Med. 64, 1165–1168.*

according to those factors. Vitamin D-producing UV irradiation depends on the zenith angle of the sun, being greatest at noon (60% occurs between 10 a.m. and 2 p.m.), reaching an annual peak at midsummer (Fig. 7.3), and declining with the distance from the earth's equator. In winter, individuals living above 40° N/S[19] have virtually no vitamin D_3 biogenesis.

Holick's Rule

Sun exposure of one-quarter MED over one-quarter of the body is equivalent to 1000 μg (40 IU) oral vitamin D_3.

The vitamin D_3-biogenic capacity of skin is diminished by factors that block UV penetrance. The epidermal pigment **melanin** efficiently absorbs UVB;[20] therefore, dark-skinned individuals can have lower circulating levels of 25-OH-D_3 and require greater UV doses for comparable vitamin D_3 biosynthesis compared to light-skinned individuals. Compared to a person with light, type 1 skin (easily sunburned; never tan), an individual with dark, type 5 or 6 skin (seldom or never burn; always tan) can require 5–10 times as much solar exposure to produce the same amount of vitamin D_3.[21] This difference would appear to have evolutionary basis, light skin being an adaptation of early humans dispersing into the higher latitudes where risks of suboptimal vitamin D status were significant.

Determinants of Vitamin D_3 Biogenesis

- *Exogenous factors* affecting sunlight exposure: season, latitude, time outdoors, sunscreen use
- *Endogenous factors* affecting UVB responsiveness: skin thickness, pigmentation.

Physical factors that reduce the exposure of the skin to UV light also reduce the biosynthesis of vitamin D_3 (Figs. 7.4 and 7.5). These include factors associated with lifestyle of humans (e.g., clothing, indoor living [glass and plexiglass absorb UV light], sunscreen use) and practical management of livestock (e.g., confined indoor housing). Properly applied topical sunscreens with sun

19. That is, the latitudes of Denver, Philadelphia, Toledo (Spain), Ankara and Beijing in the northern hemisphere, and San Martin de los Andes (Argentina) in the southern hemisphere. *Note*: the entire African and Australian continents lie north of 40°S.

20. Melanin absorbs radiation over a broad range, 290–700 nm. It is thought to be an evolutionary adaptation to protect against hypervitaminosis D induced by solar exposure in tropical latitudes—a trait that was lost in populations migrating to areas distant from the equator.

21. Bauer, J.M., Freyberg, R.H., 1946. J. Am. Med. Assoc. 130, 1208–1215.

TABLE 7.1 Contributions of Sun and Dietary Vitamin D to the Vitamin D Status of Older Women

	Low Sunlight Exposure		High Sunlight Exposure	
	Low Vitamin D	High Vitamin D	Low Vitamin D	High Vitamin D
Summer				
Vitamin D intake (mg/day)	3.58 ± 0.53	16.05 ± 1.38[a]	4.08 ± 0.65	14.53 ± 1.15[a]
Plasma 25-OH-D_3 (n*M*)	44 ± 6	80 ± 8[a]	57 ± 5	74 ± 6[a]
Winter				
Vitamin D intake (mg/day)	3.48 ± 0.45	16.33 ± 1.33[a]	4.40 ± 0.78	14.63 ± 1.43[a]
Plasma 25-OH-D_3 (n*M*)	35 ± 3	81 ± 7[a]	42 ± 4	64 ± 4[a]

[a]$p < 0.05$.
Adapted from *Salamone, L.M., Dallal, G.E., Zantos, D., et al., 1993. Am. J. Clin. Nutr. 58, 80–86.*

protection factors of 8 and greater have been shown to reduce cutaneous vitamin D_3 production by >95%.[22]

People living in northern or southern latitudes typically show seasonal changes in their serum 25-OH-D_3 levels. Greatest concentrations occur in the autumn, i.e., after a summer of relatively great solar exposure. This indicates that sunlight is generally more important than diet as a source of this critical nutrient (Table 7.1). However, many people do not spend sufficient time outside to meet their needs for vitamin D, particularly those with darker skin types and older people.

For individuals with adequate exposure to sunlight, vitamin D_3 cannot be considered a vitamin at all—it is a prohormone produced in the skin. However, environmental and lifestyle factors and, for livestock, management systems, can limit vitamin D_3 biogenesis, rendering them in need of exogenous (dietary) vitamin D. Under such conditions, in which endogenous synthesis is not sufficient to meet needs, vitamin D becomes a vitamin in the traditional sense.

Vitamin D Seasonality for free-living people.

- Plasma levels of 25-OH-D_3 and 24,25-$(OH)_2$-D_3 vary with the season: highest in late summer and lowest in late winter.
- Plasma levels of 1,25-$(OH)_2$-D_3 are fairly constant year-round.

Dietary Sources of Vitamin D

Vitamin D, as either **ergocalciferol (vitamin D_2)** or cholecalciferol (vitamin D_3), is not widely present in nature; however, its provitamins are common is both plants and animals. Ergocalciferol and its precursor **ergosterol** are found in plants, fungi,[23] molds, lichens, and some invertebrates (e.g., snails and worms). In fact, some microorganisms are quite rich in ergosterol, in which it may comprise as much as 10% of the total dry matter.[24] Ergosterol occurs in higher vertebrates only to the extent that they consume it and, then, only in low amounts. The actual distribution of ergocalciferol in nature is much more limited and variable than that of ergosterol (e.g., grass hays and alfalfa contain vitamin D only after they have been cut and left to dry in the sun). Cholecalciferol is widely distributed in animals but has an extremely limited distribution in plants. In animals, tissue cholecalciferol concentrations are dependent on the vitamin D_3 content of the diet and/or the exposure to sunlight. Few foods, however, are rich in the vitamin.

Animal Tissues

The richest natural sources are fish liver and oils[25] in which vitamin D occurs in free form and as long-chain fatty acid esters. Fatty fishes, which are high in food chains in which lower trophic level organisms feed on ergosterol-containing plants, can provide significant amounts of vitamin D.[26] In contrast, the amounts of vitamin D in farm-raised fish will depend on the levels of the vitamin added as a supplement

22. Holick pointed out that because few people apply sunscreens in more than half of recommended amounts, their use does little to limit cutaneous vitamin D production (Holick, M.F., 2004. Am. J. Clin. Nutr. 79, 362–371).

23. Ergosterol was named for the fungus, **ergot** (*Claviceps* spp.) that grows parasitically on rye and related grasses.

24. Provitamin D_3 accounts for virtually all of the sterols in *Aspergillus niger* and 80% of those in *Saccharomyces* cerevisiae (brewers' yeast).

25. Many fish oils have vitamin D_3 concentrations of c.50 μg/g; oils of cod, tuna, and mackerel can contain 20 times that amount. Marine mammals that consume large quantities of these cold water fish accumulate the vitamin in their livers, which is an important dietary source of vitamin D for people that consume seals and whales, e.g., Inuits, Faroe Islanders.

26. A meta-analysis found that the consumption of recommended amounts of fatty fishes increased plasma 25-OH-D_3 levels by some 6.8 nM; whereas, comparable intakes of lean fish had <30% of that effect (Lehmann, U., Gjessing, H.R., Hirche, F., et al., 2015. Am. J. Clin. Nutr. 102, 837–847).

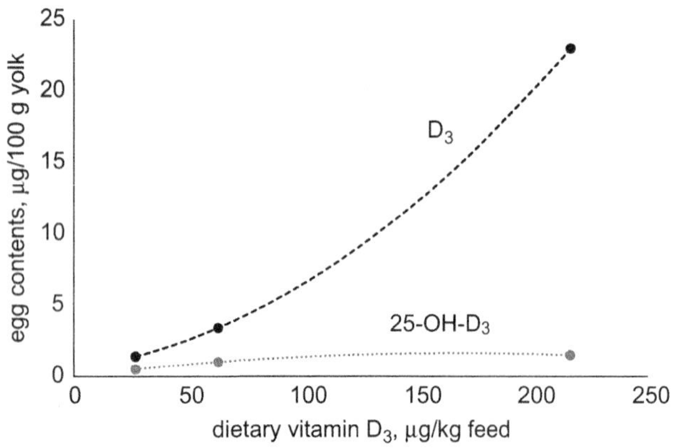

FIGURE 7.6 Effects dietary vitamin D level (X) on vitamin D contents of eggs (D_3,Y); y = 0.0003x^2 + 0.0359x. *Estimated from data presented by Mattila, P., Lehikoinen, K., Kiiskinen, T., et al., 1999. J. Agric. Food Chem. 47, 4089–4092.*

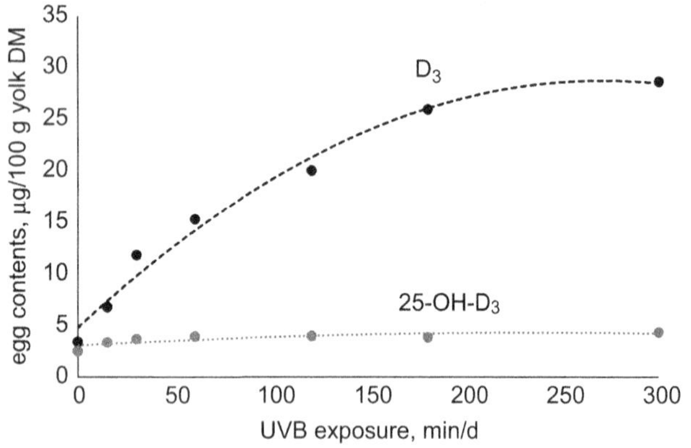

FIGURE 7.7 Effects of UVB exposure (4 weeks, X) on vitamin D contents of eggs (Y); for D_3, Y = −0.0003x^2 + 0.1765x + 4.7809 *(Estimated from Kühn, J., Schutkowski, A., Hirche, F., et al., 2015. J. Steroid Biochem. Mol. Biol. 148, 7–13.).* This effect was seen in the three- to fourfold greater vitamin D_3 contents of eggs from free-range hens compared to those from confined hens but only during the late summer months. Adapted from *Kühn, J., Schutkowski, A., Kluge, H., et al., 2014. Nutrition 30, 481–484.*

to their formulated feeds. This is also true for other livestock. The vitamin D contents of eggs varies according to amounts of the vitamin added to the feed (Fig. 7.6) (addition of 25-OH-D_3 has only a small effect[27]), as well as the amount of solar exposure of the laying hens (Fig. 7.7). Variations in such management practices are likely to explain the differences in egg vitamin D_3 contents observed between individual farms and regions, and over time.[28]

27. Mattila, P., Valkonen, E., Valaja, J., 2011. J. Agric. Food Chem. 59, 8298–8303.

28. Eggs produced commercially in 12 US states showed >12-fold variation in vitamin D_3 content. Between 2000–2001 and 2010, average vitamin D_3 content of US eggs increased by 60% (Exler, J., Phillips, K.M., Patterson, K.Y., et al., 2013. J. Agric. Food Chem. Anal. 29, 110–116). Hens in confined, environmentally managed conditions have virtually no exposure to sunlight; whereas, those managed in courtyard or free-range situations can have significant sun exposure.

Meats tend to have relatively low and variable vitamin D contents (chicken, 0–14 µg/kg; beef, 0–9 µg/kg; pork, 1–23 µg/kg; lamb, 1–61 µg/kg; veal, 0–50 µg/kg) as does milk (0.3–1 µg/kg) if not fortified.[29]

Plant Tissues

With a few notable exceptions, vitamin D_3 is not found in plants. Those exceptions include the species in the families Solanaceae (*Solanum glaucophyllum, Solanum malacoxylon, Solanum torvum, Solanum verbascifolium, Cestrum diurnum*, and *Nierembergia veitchii*) in which the vitamin occurs as water-soluble β-glycosides of vitamin D_3, 25-hydroxyvitamin D_3 (25-OH-D_3), and 1,25-dihydroxyvitamin D_3 (1,25-[OH]$_2$-D_3).[30] Vitamin D_3 has been identified species of the families Gramineae (*Trisetum flavescens* and *Dactylis glomerata*) and Leguminoseae (*Medicago sativa*).

Irradiated Mushrooms

Mushrooms are rich in ergosterol, which makes them a potential source of vitamin D_2 if exposed to sunlight or UVB.[31] That conversion is fairly rapid, requiring short exposures (60–95 min).[32] Accordingly, many mushroom producers use postharvest UVB exposure to produce edible mushrooms that contain significant amounts of D_2. Commercial species have been found to contain 0.03–63 µg (0.001–2.5 IU) vitamin D_2 per g fresh weight; the greatest amounts have been found in irradiated maitake (*Grifola frondosa*) and portabella (*Agaricus bisporus*).[33]

Breast Milk

There are relatively few published data on the vitamin D contents of breast milk, despite the fact that breastfeeding without adequate sunlight exposure or vitamin D supplementation appears to be a risk factor for vitamin

29. Schmid, A., Walther, B., 2013. Adv. Nutr. 4, 453–462.

30. The search for the factor causing calcinosis in ruminants grazing on these Solanaceous species led to the discovery that plants could contain vitamin D. In addition, both D-25-hydroxylase and 25-OH-D-1-hydroxylase were found in *Solanum malacoxylon*, and the grass *Trisetum flavescens*, which produces vitamin D_3 when exposed to UVB. Thus, it appears that these species produce and metabolize vitamin D in ways similar to animals.

31. Mushrooms also contain 22,23-dihydroergosterol; hence, irradiated mushrooms also contain 22-dihydroergocalciferol, a metabolite of unknown physiological significance (Phillips, K.M., Horst, R.L., Koszewski, N.J., et al., 2012. PLoS One 7, 1–10).

32. Studies with the oyster mushroom (*Pleurotus ostreatus*) showed that optimal vitamin D_2 yield is realized within 60 min using 310–320 nm light at 11.5 W/m^2 at 20°C. Beyond that point, concentrations of lumisterol-2, tachysterol-2, and previtamin D_2 increased to the point of accounting for one-sixth of the total steroid products (Krings, U., Berger, R.G., 2014. Food Chem. 149, 10–14).

33. Phillips, K.M., Ruggio, D.M., Horst, R.L., et al., 2011. J. Agric. Food Chem. 59, 7841–7853.

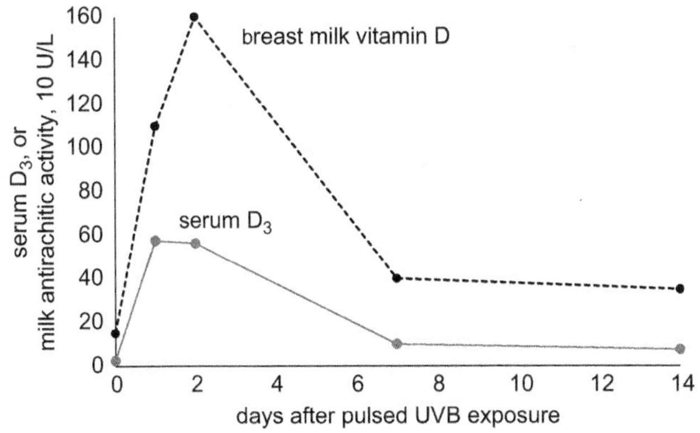

FIGURE 7.8 Related responses of vitamin D in breast milk and maternal serum to a pulsed exposure (1.5 ME.D) to UVB. *(After Specker, B.L., Tsang, R.C., Hollis, B.W., 1985. Am. J. Dis. Child. 139, 1134–1137.).* Note: Antirachitic activity was calculated from analyzed vitamers: 1 U = 25 ng D_2 or D_3, or 5 pg 25-OH-D_2 and 25-OH-D_3.

D-deficiency rickets in children. Available reports have shown breast milk vitamin D concentrations to be highly variable between studies; this is likely due to the inclusion of samples from mothers with differing vitamin D intakes and with different UVB exposures. For example, breast milk from Finish mothers was ninefold greater in the summer compared to winter.[34] Specker and colleagues demonstrated that total breast milk vitamin D concentrations correlated with vitamin D intake are responsive to UVB exposure (Fig. 7.8) and can be lower for African-American women than for white American women (Table 7.2).

25-OH-D in Foods

Foods derived from animals also contain 25-OH-D, which is not measured in conventional food analyses. An expert panel of the NIH Office of Dietary Supplements found that including 25-OH-D in estimates of vitamin D intake of American adults would increase those estimates by 1.7–2.9 μg/day.[35]

Fortified Foods and Dietary Supplements

Even when consumed in appreciable amounts, the low amounts of vitamin D in foods (Table 7.2) make it very difficult to achieve adequate vitamin D nutriture from those foods. Thus, it has become the practice in many countries to fortify certain frequently consumed foods. Some countries mandate the addition of vitamin D (and vitamin A) to certain foods (e.g., margarine) and regulate the voluntary addition of vitamin A to others (e.g., baked goods, breakfast cereals, milk,[36] yogurt, cheeses, orange juice, and infant foods). Both D_2 and D_3 are used in such fortification. Some other foods may be enriched indirectly as the result of the supplementation of animal feeds with the vitamin. A meta-analysis of 16 clinical interventions with vitamin D_3 from fortified foods found significant heterogeneity in the plasma 25-OH-D_3 response due to such factors as dose, latitude, and baseline plasma 25-OH-D_3 concentration.[37] Overall, they found that every 1 μg vitamin D_3 ingested from fortified foods was associated with an average increase in plasma 25-OH-D_3 concentration of 1.2 nM.

Vitamin D is also added in many formulated nutritional supplements. Multivitamin supplements typically contain 400 IU (10 mg) vitamin D, and pharmaceutical preparations can contain as much as 50,000 IU (1250 mg) vitamin D_2 per capsule/tablet. These can provide important vitamin D nutrition; in many countries their use has increased in recent years. However, it can be difficult to identify the actual use of supplements and fortified foods; this can lead to food consumption surveys underestimating total intakes of vitamin D.

Vitamin D Analogues

Many nucleus or side chain derivatives of D_2 or D_3 have been developed for treating vitamin D refractory rickets, hypocalcemia and osteodystrophy secondary to chronic renal disease, and some types of cancer (Table 7.3). The goal has been to develop drugs active in vitamin D-dependent regulation of cell proliferation and growth while having low calcemic potential, thus avoiding the calcinosis caused by high levels of the vitamins.

Vitamin D Bioavailability

The vitamers D vary in bioavailability due to differences in biopotency, i.e., not concerning absorption or transport (Table 7.4). Vitamers requiring metabolic activation (cholecalciferol and ergocalciferol) are less biopotent than those proximal to the points of metabolic function (e.g., 25-OH-vitamin D). Humans and many other species have been thought not to discriminate between these vitamers. Indeed, many studies have found D_3 and D_2 in foods or pure form to produce equivalent responses in increasing plasma levels of their respective 25-hydroxylated intermediate, although a recent meta-analysis found bolus doses

34. Ala-Houhala, M., Koskinen, T., Parvainen, M.T., et al., 1988. Am. J. Clin. Nutr. 48, 1057–1060.
35. Taylor, C.L., Roseland, J.M., Coates, P.M., et al., 2016. J. Nutr. 146, 855–856.

36. In the late 1950s, the American Medical Association recommended that fluid milk be fortified with 400 IU (10 μg) per quart. The US Food and Drug Administration has specified that milk contains 400–600 IU/qt. However, a survey in the 1980s found that most fortified milk products failed to contain the specified amounts of vitamin D (Tanner, J.T., Smith, J., Defibaugh, P., et al., 1988. J. Assoc. Off. Anal. Chem. 71, 601–610). Most European countries to not fortify milk with vitamin D.
37. Black, L.J., Seamans, K.M., Cashman, K.D. et al., 2012. J. Nutr. 142, 1102–1108.

TABLE 7.2 Breast Milk Vitamin D Concentrations in American Women

Donors	D_3	25-OH-D_3	D_2	25-OH-D_2
African-American (n = 10)	36 (22–58)[a,b]	87 (75–101)[b]	54 (22–134)[b]	66 (52–85)[b]
White American (n = 14)	268 (127–567)	124 (97–159)	290 (164–512)	82 (64–105)

[a]*Mean (95% C.L.).*
[b]*Significantly different from White American, p < .05.*
Specker, B.L., Tsang, R.C., Hollis, B.W., 1985. Am. J. Dis. Child. 139, 1134–1137.

TABLE 7.3 Bioactive Vitamin D Analogues

1-OH-analogues	1-OH-D_3
Vitamin D_2 derivatives	1-OH-D_2; 1,25,28-$(OH)_3$-D_2; 1,24*S*-$(OH)_2$-D_2
Side-chain derivatives	Cyclopropane ring (side chain carbons 25, 26, and 27) derivatives, e.g., calcipitriol 20-epi and 20-methyl side chain derivatives analogues with one or more added carbons in the side chain or on the branching methyl groups unsaturated side chain derivatives, e.g., $C_{16}{=}C_{17}$, $C_{22}{=}C_{23}$ oxa-containing (oxygen-for-carbon substitution), e.g., 22-oxa-calcitriol fluorinated derivatives, e.g., F_6-1,25-$(OH)_2$-D_3
Other derivatives	Dehydrotachysterol-2

of D_3 produced greater 25-OH-D responses than D_2 in humans.[38] Some species (avians) are known to distinguish between D_2 and D_3, greatly in favor of the latter; these species require D_3.

Expressing Vitamin D Activities

Because vitamin D exists in foods and supplements in different forms of differing biopotencies, the reporting of vitamin D activity in foods requires some means of standardization. For this purpose, an international unit (**IU**) has been defined:

1 IU = 0.025 mg of vitamin D2 or vitamin D3.

Foods Rich in Vitamin D

The richest food sources of vitamin D are oily fishes and their products, dairy products, irradiated mushrooms, and fortified foods oils (Table 7.5).

38. Tripkovic, L., Lambert, H., Hart, K., et al., 2012. Am. J. Clin. Nutr. 95, 1357–1364.

TABLE 7.4 Relative Biopotencies of Vitamin D-Active Compounds

Compound	Relative Biopotency, %[a]
Vitamin D_3 (cholecalciferol)	100
Vitamin D_2 (ergocalciferol)	100 (mammals)[b]
	10 (birds)[c]
Dihydrotachysterol[d]	5–10
25-OH-Cholecalciferol[e]	200–500
1,25-$(OH)_2$-Cholecalciferol[e]	500–1000
1α-OH-Cholecalciferol[f]	500–1000

[a]*Results of bioassays of rickets prevention in chicks and/or rats.*
[b]*Biopotencies of vitamins D_2 and D_3 are equivalent for mammalian species.*
[c]*Biopotency of vitamin D_2 is very low for chicks, which cannot use this vitamer effectively.*
[d]*A sterol generated by the irradiation of ergosterol.*
[e]*Normal metabolite of vitamin D_3; the analogous metabolite of vitamin D_2 is also formed and is comparably active in nonavian species.*
[f]*A synthetic analog.*

4. ENTERIC ABSORPTION OF VITAMIN D

Micelle-Dependent Passive Diffusion

Vitamin D is absorbed from the small intestine by nonsaturable passive diffusion that is dependent on micellar solubilization and, hence, the presence of fat[39] and bile salts. The fastest absorption appears to be in the upper portions of the small intestine (duodenum and ileum); however, the greatest amount of vitamin D is probably absorbed in the distal region where food has a longer transit time. Like other hydrophobic substances absorbed by micelle-dependent diffusion in mammals, vitamin D enters the lymphatic circulation[40] predominantly (90% of

39. The apparent absorption of D_3 by healthy adults was one-third less from a fat-free meal than from meals containing fat (Dawson-Hughes, B., Harris, S., Lichtenstein, A.H., et al., 2015. J. Acad. Nutr. Diet. 115, 225–230.

40. In birds, reptiles, and fishes, vitamin D, like other lipids, is absorbed into the portal circulation via portomicra.

TABLE 7.5 Vitamin D Activities in Foods

Food	Vitamin D, IU/100 g
Human milk	3
Dairy Products	
Milk	0–51[a]
Butter	0
Cheese	0–24
Cream	44
Eggs	82
Fish Products	
Cod	46
Cod liver oil	100,000
Herring	214
Mackerel	292
Salmon	670–685
Sardines	193
Shrimp	4–5
Beef liver	49
Meats	
Beef	1–16
Pork	1–104
Chicken	3–15
Chicken skin	24
Other	
Mushrooms	0–25,000[b]

[a]US regulations specify that milk be fortified with 400 IU of vitamin D_3 per quart (about 37 IU/100 ml).
[b]If UV-irradiated; upper limit from published literature.
USDA National Nutrient Database for Standard Reference, Release 28 (http://www.ars.usda.gov/ba/bhnrc/ndl).

the absorbed amount) in association with chylomicra (or portomicra), with most of the balance being associated with α-globulins.[41] The efficiency of enteric absorption of dietary vitamin D appears to be about 50%. Newly absorbed vitamin D is released by enterocytes into the lymphatic circulation in chylomicra. The polar metabolites (25-OH-D; 1,25-[OH]$_2$-D) appear to be transported by DBP (vitamin D-binding protein) in the portal blood; only small amounts (13% of 25-OH-D; 1% of 1,25-[OH]$_2$-D) appear in the lymph in chylomicra.

41. probably the **vitamin D-binding protein (DBP)**.

5. TRANSPORT OF VITAMIN D

Vitamin D_3 formed in the skin is translocated into the dermal capillary bed, then into the general circulation, by a process that appears to be selective for D_3, leaving previtamin D_3 in the epidermis. A major player in this translocation is the specific **DBP** in plasma to which all of the vitamin of biogenic origin is bound (Table 7.6). In contrast, enterically absorbed vitamin D is present in the circulation bound to both DBP and lipoproteins in comparable amounts, apparently being transferred during the process of chylomicron degradation in the liver. Studies have demonstrated that circulating levels of the hepatic metabolite, 25-OH-D_3, are affected by **apolipoprotein E (apoE)** genotype (Fig. 7.9); humans with the *APOEε4*[42] allele have also been found to have greater serum 25-OH-D_3 concentrations than noncarriers.

Vitamin D-Binding Protein

Like other sterols, vitamin D is transported in the plasma largely in association with protein. While some birds and mammals transport vitamin D in association with albumin, and fishes with cartilaginous skeletons (e.g., sharks and rays) transport it in association with plasma lipoproteins, most species[43] use a protein that has been called **transcalciferin** or, more commonly, DBP (Table 7.6). DBP is in the same gene family as albumin and α-fetoprotein. In mammals, it is a glycosylated, cysteine-rich, α-globulin of 458 amino acids with a molecular weight of 55–58 kD, depending on its glycosylation state. It has three internally homologous α-helical domains and exists as multiple isoforms[44] due to differences in both the primary structure of the protein (involving the presence/absence of *N*-acetylneuraminic acid on a threonine residue at position 420) and the carbohydrate moiety that is added posttranslationally. Three alleles are common. The chicken has two distinct DBPs (54 kDa and 60 kDa) each of which preferentially binds vitamin D_3 and its metabolites versus the respective vitamin D_2 analogues.

DBP binds vitamins D and metabolites stoichiometrically, with ligand binding dependent on the *cis*-triene structure and C_3-hydroxyl grouping. In adequately nourished individuals, DBP binds some 88% of the 25-OH-D_3 in serum with an affinity and order of magnitude greater than those of 1,25-(OH)$_2$-D. The concentration of DBP in the plasma, typically 4–8 μM,

42. APOE is a key regulator of lipid metabolism. The APOE ε4 allele occurs in 3–40% of populations; in Western populations, it is a risk factor for age-related morbidity and mortality due to cardiovascular disease and late-onset Alzheimer's disease.
43. More than 140 species in five classes have been examined.
44. Humans DBP is identical to group-specific component ($G_c$protein), a genetic marker useful in epidemiologic and forensic studies.

TABLE 7.6 Distribution of Vitamin D Metabolites in Human Plasma

Metabolite	% Distribution			Normal Concentration
	DBP (Vitamin D-Binding Protein)	Lipoproteins	Albumin	
Vitamin D_3	60	40	0	2–4 ng/mL
25-OH-D_3	98	2	0	15–38 ng/mL
24,25-$(OH)_2$-D_3	98	2	0	—
1,25-$(OH)_2$-D_3	62	15	23	20–40 pg/mL

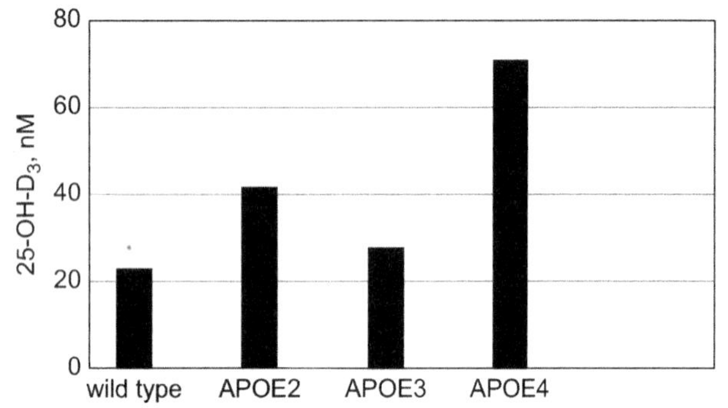

FIGURE 7.9 Relationship of apolipoprotein E genotype and serum 25-OH-D level in the mouse. *After Huebbe, P., Nebel, A., Siegert, S., et al., 2011. FASEB J. 25, 3262–3270.*

greatly exceeds that of 25-OH-D_3 (c.50 nM) and is remarkably constant, as it is unaffected by sex, age, or vitamin D status. The excess binding capacity means that only 5% of DBP actually carries the vitamin. That DBP can also bind the plasma actin monomer and fatty acids, which suggests that the vitamin D transport protein may also have other functions in metabolism.[45] The turnover of DBP in the plasma is much shorter than that of 25-OH-D_3 (1–3 days[46] versus 45 days[47]), indicating that the ligand is recycled. The extended half-life of 25-OH-D appears to be due to its megalin-dependent uptake and subsequent release by myocytes, such that muscle comprises a large extravascular pool protecting 25-OH-D from degradation.[48] DBP can be taken up endocytotically by the renal proximal tubule, being internalized from the glomerular filtrate via binding to the multiligand binding receptors megalin and cubilin.[49] This system results in delivery of 25-OH-D_3 to the cell with catabolism of DBP. Being synthesized by the liver, DBP is depressed in patients with hepatic disease. Its expression is increased by trauma, estrogen therapy, and pregnancy. It does not appear to cross the placenta; fetal DBP is immunologically distinct from the maternal protein.

In addition to facilitating the peripheral distribution of vitamin D obtained from the diet, DBP functions to mobilize the vitamin produced endogenously in the skin. Indeed, vitamin D_3 found in the skin is bound to DBP.[50] It has been suggested that the efficiency of endogenously produced vitamin D_3 is greater than that given orally for the reason that the former enters the circulation strictly via DBP, whereas the latter enters as complexes of DBP as well as chylomicra. This would indicate that oral vitamin D remains longer in the liver and is, thus, more quickly catabolized to excretory forms. In support of this hypothesis, it has been noted that high oral doses of vitamin D can lead to very high levels of 25-OH-D_3 (>400 ng/ml) associated with intoxication; whereas intensive UV irradiation can rarely produce plasma 25-OH-D_3 concentrations greater than one-fifth that level, and hypervitaminosis D has never been reported from excessive irradiation. The DBP protein has also been found on the surfaces of lymphocytes and macrophages, although the functional significance of such binding is not clear.

While DBP binds both D_2 and D_3 and their metabolites, genetic variants differ only in their binding of D_3.[51] The

45. DBP has a high affinity for the actin monomer. The DBP–actin complex (which can interfere with the assay of 25-OH-D-1-hydroxylase in kidney homogenates unless large amounts of 25-OH-D_3 are used) is cleared from the circulation at three times the rate of nonliganded DBP, suggesting that DBP may function in extracellular scavenging.

46. Haddad, J.G., Fraser, D.R., Lawson, D.E.M., 1981. J. Clin. Invest. 67, 1550–1560.

47. Clements, M.R., Davies, M., Fraser, D.R., 1987. J. Clin. Endocrinol. Metab. 73, 659–674.

48. Aboud, M., Gordaon-Thomson, C., Hoy, A.J., 2014. J. Steroid Biochem. Mol. Biol. 14, 232–236.

49. These multiligand receptors are expressed in the apical parts epithelial cells of renal proximal tubules, colocalized in endocytic compartments (coated pits and endosomes). Megalin is a 600 kDa transmembrane protein of the LDL-receptor family; cubilin is a 460 kDa peripheral membrane protein identical to the intrinsic factor–vitamin B_{12} receptor of in the small intestinal epithelium. Both proteins function in the renal reabsorption of proteins including VDB, RBP, transcobalamin (vitamin B_{12} transporter), and transferrin (iron transporter). In the thyroid, megalin appears to function in thyroid hormone transport; in the parathyroid, it is thought to be a Ca^{2+} sensor.

50. DBP has a very low affinity for lumisterols or tachysterols, thus not mobilizing from the skin these forms produced under conditions of excessive irradiation.

51. Nimitphong, H., Saetun, S., Chanprasertyotin, S., et al., 2013. Nutr. J. 12, 39–46.

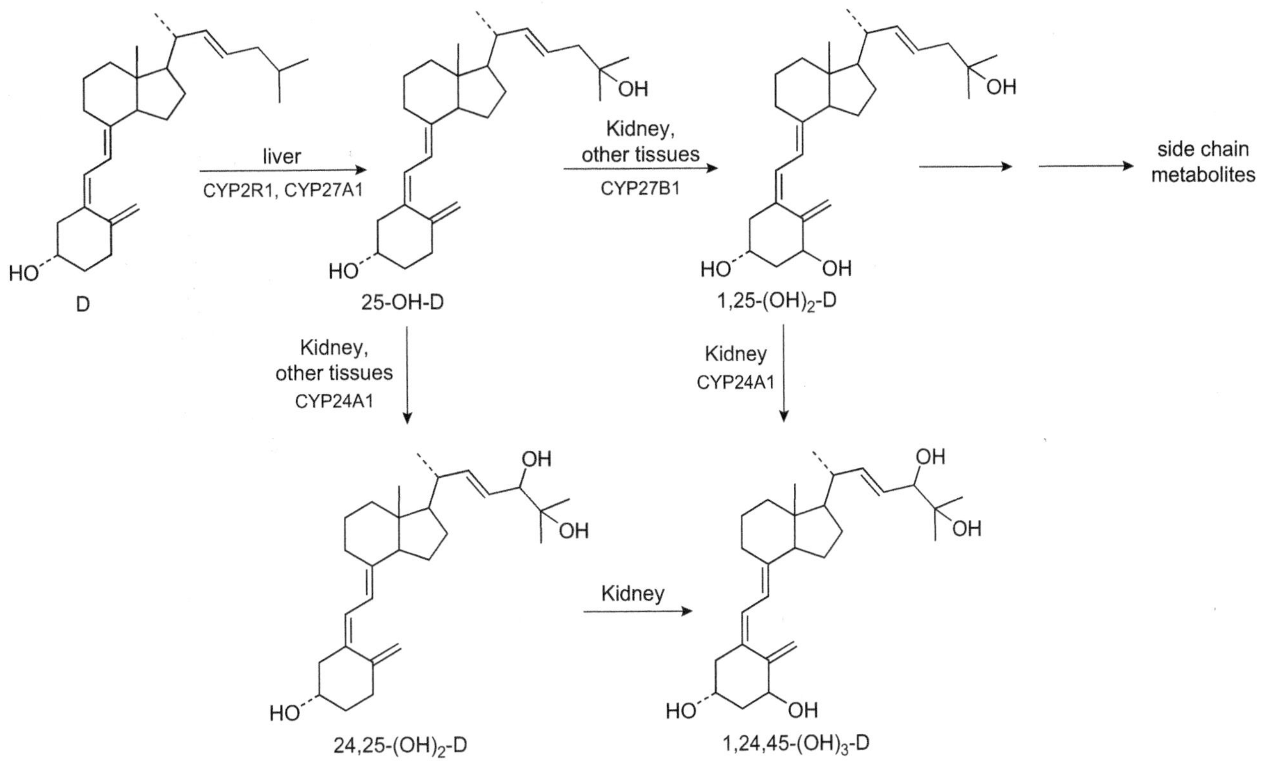

FIGURE 7.10 Vitamin D metabolism.

frequency of rare alleles of DBP polymorphisms has been found to be inversely related to circulating 25-OH-D_3 level, with effects comparable in magnitude to those of vitamin D intake.[52] Different patterns of DBP polymorphisms appear to explain 80% of the difference in plasma DBP levels and 10% of the difference in plasma 25-OH-D_3 levels between black and white subjects.

Tissue Distribution

Unlike the other fat-soluble vitamins, vitamin D is *not* stored by the mammalian liver except in some fishes. It reaches the liver within a few hours after being absorbed across the gut or synthesized in the skin, but from the liver it is distributed relatively evenly among the various tissues, where it resides in hydrophobic compartments. Therefore, fatty tissues such as adipose show slightly greater concentrations. However, in that tissue the vitamin is found in the bulk lipid phase, from which it is only slowly mobilized. About half of the total vitamin D in the tissues occurs as the parent vitamin D_3 species, with the next most abundant form, 25-OH-D_3, representing 20% of the total. In the plasma, however, the latter metabolite predominates by several fold.[53] Tissues including those of the kidneys, liver, lungs, aorta, and heart also tend to accumulate 25-OH-D_3.[54] It is thought that the uneven tissue distribution of vitamin D, in its various forms, relates to differences in both tissue lipid content and tissue-associated vitamin D-binding proteins, the latter fraction being the smaller of the two intracellular pools of the vitamin.

The concentrations of both 25-OH-D_3 and 1,25-$(OH)_2$-D_3 are lower in the cord sera of fetuses and newborn infants than in the sera of their mothers. That fetal 25-OH-D_3 levels correlate with maternal levels (and show the same seasonal variations) suggest that the metabolite crosses the placenta. Such a correlation is not apparent for 1,25-$(OH)_2$-D_3. The extent of transplacental movement of the latter metabolite is not known; however, the placenta appears to be able to produce it from maternally derived 25-OH-D_3.

6. METABOLISM OF VITAMIN D

Metabolic Activation

The metabolism of vitamin D involves its conversions to different hydroxylated products, each of which is more polar than its parent (Fig. 7.10).[55] The production of these

52. Sinotte, M., Diorio, C., Berube, S., et al., 2009. Am. J. Clin. Nutr. 89, 634–640.

53. The next most abundant is 24,25-$(OH)_2$-D_3.

54. These organs are susceptible to calcification under conditions of hypervitaminosis D.

55. It was the observation in the late 1960s of radioactive peaks migrating ahead of vitamin D in gel filtration of plasma from animals given radiolabeled cholecalciferol that first evidenced the conversion of vitamin D3 to other species some of which were ultimately found to be metabolically active.

metabolites, some of which are metabolically active forms of the vitamin, explains the lag time that is commonly observed between the administration of the vitamin and the earliest biological response.

25-Hydroxylation

Most of the vitamin D taken up by the liver from either DBP or lipoproteins is converted by hydroxylation of side chain carbon C-25 to yield 25-OH-D_3 (also called **calcidiol**), the major circulating form of the vitamin. This metabolism occurs in the liver in mammals but in both liver and kidney in birds. That activity, vitamin D 25-hydroxylase, involves at least six cytochrome *P*-450-dependent mixed-function oxygenases[56] of two general types: five low-affinity, high-capacity enzymes associated with the endoplasmic reticulum[57] (CYP2R1, CYP2J2/3, CYP3A4, CYP2D25, and CYP2C11) and one high-affinity, low-capacity enzyme located in the mitochondria (CYP27A1). CYP27A1 also occurs in kidney and bone suggesting extrahepatic 25-hydroxylation of vitamin D_3. The presence of two different mechanisms of 25-hydroxylation would appear to facilitate the maintenance of adequate vitamin D status under both deficient and excessive conditions of vitamin D intake/production. That CYP27A1 is not necessary for vitamin D function is indicated by the fact that, in the mouse, its genetic deletion does not impair bone health. The most physiologically important 25-hydroxylase, at least in humans, appears to be CYP2R1. An individual presenting with vitamin D-dependent rickets was found to have a transition mutation that abolished the 25-hydroxylase activity, and genome-wide studies that have found significant associations of *cyp2r1* polymorphisms and circulating 25-OH-D_3 levels.

25-Hydroxyvitamin D_3 is not retained within the cell but is released to the plasma where it accumulates by binding with DBP. At normal plasma concentrations of this metabolite, only small amounts of 25-OH-D_3 from this pool enter tissues. Therefore, the circulating level of 25-OH-D_3, normally 10–40 ng/mL (25–125 nM), is a good indicator of general vitamin D status.

1-Hydroxylation

25-OH-D_3 is further hydroxylated at the C-1 position of the A ring to yield 1,25-$(OH)_2$-D_3 (also called **calcitriol**). Being produced at a site, the kidney, distant from its target tissue to which it is transported in the blood, 1,25-$(OH)_2$-D_3 has been considered a hormone. The 1-hydroxylase uses $NADPH_2$ as the electron donor and has three constituent proteins: ferridoxin, ferridoxin reductase, and a cytochrome *P*-450 isoform CYP27B1. The complex is located primarily in renal cortical mitochondria but also in mitochondrial and microsomal fractions of at least some extrarenal tissues (e.g., bone cells, keratinocytes,[58] liver, and placenta[59]) where it is thought to provide capacity for local production of the active metabolite. Despite its key role in discharging the actions of vitamin D, the tightly regulated production and relatively fast turnover (4–6 h in serum) of 1,25-$(OH)_2$-D_3 render it not useful as a biomarker of vitamin D status. Circulating levels tend to be about 40 pg/mL (100 nM). An autosomal recessive defects in the *cyp27b1* gene produce **vitamin D-dependent rickets type I**. Affected individuals have normal plasma levels of 25-OH-D_3 but low levels of 1,25-$(OH)_2$-D_3.

Epimerization

25-Hydroxyvitamin D can undergo epimerization[60] at the C3 position[61] to 3-epi-25-(OH)-D_3 in which the C3-hydroxyl moiety extends above the plane of the steroid "A" ring instead of below that plane, as in the normal metabolite. 3-Epi-25-(OH)-D_3 accounts for some 25% of total circulating 25-(OH)-D_3,[62] leading to overestimation of vitamin D status by radioimmunoassay and mass spectrometric methods incapable of discriminating between the epimers. The mechanism of epimerization has not been characterized, but studies have shown that 3-epi-25-(OH)-D_3 is metabolized by the renal 1-hydroxylase to produce 3-epi-1,25-$(OH)_2$-D_3, which binds VDR (vitamin D receptor).

Catabolism

24-Hydroxylation

1,25-$(OH)_2$-D_3 is short-lived; it upregulates its rapid metabolism by the 24-hydroxylase, CYP24A1, which attacks its side chain to produce 1,24,25-$(OH)_3$-D_3 and 1,23, 24,25-$(OH)_4$-D_3. CYP24A1 can also hydroxylate 25-OH-D_3 to produce 24,25-$(OH)_2$-D_3. The 24-hydroxylase has a 10-fold greater affinity for 1,25-$(OH)_2$-D_3 than for 25-OH-D_3, but the 1000-fold excess of the latter in the plasma suggests that the primary physiological significance of the

56. A mixed-function oxygenase uses molecular oxygen (O_2) but incorporates only one oxygen atom into the substrate.
57. In the rat, the microsomal hydroxylase is fivefold more active in males than females; in males it may involve cytochrome P4502C11, which is not expressed by females.
58. When exposed to sunlight/UVB, keratinocytes produce vitamin D_3, from which they can make 1,25-$(OH)_2$-D_3. Their contribution to the circulating pool of 1,25-$(OH)_2$-D_3 is substantially less than kidney but can be significant in patients with renal disease in whom reduced plasma 1,25-$(OH)_2$-D_3 levels induce increased expression of the skin 1-hydroxylase.
59. Maternal levels of 25-OH-D_3 increase during the third trimester of gestation. This presumably assists the mother in providing calcium for the mineralization of the fetal skeleton.
60. i.e., Changing configuration about a single asymmetric center.
61. The 3-epimer of 25-(OH)-D_3 has the C3-hydroxyl moiety extending above the plane of the steroid "A" ring; whereas, in the normal metabolite that grouping extends below the plane.
62. Karras, S.N., Shah, I., Petroczi, A., et al., 2013. Nutr. J. 12, 15–77.

hydroxylase may be in clearing excess 25-OH-D_3. Studies have shown that 24,25-$(OH)_2$-D_3 inhibits the 25-OH-D_3-mediated signaling of Ca and phosphate transport and promotes bone formation and mineralization. The greatest activity of CYP24A1 is found in renal mitochondria; its expression is upregulated in chronic renal disease and 1,25-$(OH)_2$-D_3. Both 1,25-$(OH)_2$-D_3 and 24,25-$(OH)_2$-D_3 are produced under conditions of vitamin D adequacy and normal Ca homeostasis. Calcitriol is a major biliary metabolite of the vitamin. That 24,25-$(OH)_2$-D_3 inhibits the stimulatory effect of parathyroid hormone (PTH)[63] on bone resorption by **osteoclasts** and suggests it may participate in local osteotropic control in bone.

Other Hydroxylations

More than 40 other hydroxylation metabolites of vitamin D have been identified. One, 24R,25-$(OH)_2$-D, appears to have a specific role in embryonic survival and fracture healing.[64] Most others appear to be physiologically inactive and are converted to excretable forms. Some 95% of vitamin D excretion occurs via the bile, with side chain hydroxylation products accounting for nearly one-fifth. Most do not bind DBP; they and their chain-shortened products are cleared from the circulation and converted to excretory forms including fatty esters and glucuronides. Two that are bound by DPB are 25,26-$(OH)_2$-D_3, detectable in the plasma after D_3, and the 26, 23-lactone of 25-OH-D_3 (apparently produced from 25-OH-D_3 by extrahepatic CYP24A1), which accumulates in hypervitaminotic D animals and is the major excretory metabolite in some species (guinea pig, opossum).

Regulation of Vitamin D Metabolism

Vitamin D has been found to have a relatively long physiological half-life, about 2 months. This likely reflects its partial sequestration in adipose, as the half-lives of its circulating (25-OH-D_3) and metabolically active (1,25-$[OH]_2$-D_3) forms are shorter, about 15–45 days and 15 h, respectively.[65] Circulating levels of 1,25-$(OH)_2$-D_3 are regulated at about 40 pg/mL (100 nM).

The 1-hydroxylase is the rate-limiting step in vitamin D metabolism. It is upregulated by the following:

- Low plasma Ca^{2+}—via Ca^{2+}-receptor-mediated stimulation of PTH release by the parathyroid[66]
- Low plasma P—via a pituitary signaling (superstimulation occurs if plasma Ca^{2+} is also low).

It is downregulated by the following:

- Its product, 1,25-(OH)2-D3;
- High plasma P—causes bone cells to secrete fibroblast growth factor 23 (FGF23),[67] which acts as a downregulator.

Other factors affect these responses. PTH and CT both stimulate the renal 1-hydroxylase; however, PTH has a rapid effect mediated by cAMP, while CT has a relatively slow effect apparently mediated by transcription.[68] Estrogen appears to have a role, as ovariectomy reduces the synthesis of 1,25-$(OH)_2$-D_3 by rat kidney. That effect appears to be mediated via PTH, as parathyroidectomy blocks the effect of estrogen in stimulating 1,25-$(OH)_2$-D_3 production. In contrast, regulation of the 1-hydroxylase in extrarenal tissues (e.g., macrophages) appears to be insensitive to PTH but to be stimulated by cytokines such as interferon-gamma and lipopolysaccharide.

The 25-hydroxylase is poorly regulated. It shows little or no feedback inhibition by 25-OH-D_3. Its expression increases with increasing hepatic vitamin D_3 level and exposure to inducers of cytochrome *P*-450 (phenobarbital, diphenylhydantoin)[69] but is inhibited by isoniazid.[70]

Catabolism of 1,25-$(OH)_2$-D_3 is tightly regulated to prevent hypercalcemia and hyperphosphatemia. This is accomplished by PTH, serum P, and factors affecting the principle catabolizing enzyme, the hepatic 24-hydroxylase (CYP24A1). The 1-hydroxylase is inhibited by strontium, the DPB–actin complex, and is feedback inhibited by 1,25-$(OH)_2$-D_3. Thus, when circulating levels of 1,25-$(OH)_2$-D_3 are high, its renal synthesis is low. The tight regulation of the 1-hydroxylase activity results in the maintenance of nearly constant plasma concentrations of 1,25-$(OH)_2$-D_3, which activates its own breakdown by stimulating the transcription of the 24-hydroxylase gene. That stimulation is suppressed in conditions of low serum P.

Epigenetic Regulation

Vitamin D metabolism appears to be regulated, in part, through epigenetic mechanisms. Methylation of the promoter of CYP2R1 (D_3 25-hydroxylase) was observed in leukocytes from vitamin D-deficient individuals; that effect was reversed upon return to vitamin D adequacy. Methylation of

63. PTH is a small (9.6 kDa) protein with high sequence homology across species.
64. St. Arnaud, R., 2010. J. Steroid Biochem. Mol. Biol. 121, 254–256.
65. Jones, G., 2008. Am. J. Clin. Nutr. 88, S582–S586.
66. It has been suggested that some elderly people who cannot adapt to low a Ca diet by increasing enteric Ca absorption may suffer impaired PTH-dependent 1,25-$(OH)_2$-D_3 upregulation.
67. Mutations in the *fgf23* gene are thought to be the cause of autosomal dominant hypophosphatemic rickets.
68. It has been suggested that the function of the CT-sensitive 1-hydroxylase, which is elevated in the fetus, may be to accommodate situations of increased need for 1,25-$(OH)_2$-vitamin D.
69. Antiepileptic agents such as these reduce the biological half-life of vitamin D apparently by enhancing its conversion to 25-OH-D and other hydroxylated products.
70. Patients on long-term isoniazid therapy are at risk to developing bone disease.

the *cyp27b1* gene appears to be the basis of downregulation of 25-OH-D_3 1-hydroxylase expression observed in many cancers. The 24-hydroxylase activity appears to involve tissue-dependent methylation of the *cyp24a1* gene. These effects appear to involve the liganded VDR transactivating or transrepressing genes through interactions with other nuclear factors (e.g., VDR-interacting repressor, histone deacetylase 2) to induce DNA methylation and histone acetylation.

Role of Vitamin D-Binding Protein

DBP is critical in the regulation of vitamin D metabolism, controlling the tissue distribution of vitamin D metabolites. Due to the relative excess of DBP (4–8 μM compared to 50 nM 25-OH-D_3), nearly 90% of circulating vitamin D metabolites are bound to DBP in vitamin D-adequate individuals, while occupying <5% of available binding sites. Further, because of its avid binding of 25-OH-D_3 in the plasma rather than other tissues, concentrations of the free metabolite are maintained at very low levels (10^{-13} M). Both 25-OH-D_3 and 1,25-(OH)2-D_3 can also bind to albumin, albeit with much lower affinities than DBP (Table 7.7). Accordingly, albumin binding is much less effective than DBP in protecting these metabolites against losses by renal filtration as the liganded DBP binds to megalin, which prevents its glomerular filtration. Thus, DBP functions to modulate the availability of these metabolites to the tissues, which appear to receive them from the small pools that are either not protein-bound, or are bound with low affinity to albumin and other non-DBP proteins. It is estimated that while <0.1% of the 25-OH-D_3 in plasma is not bound to proteins, the amount available to extrarenal tissues for in situ synthesis of 1,25-$(OH)_2$-D_3 (i.e., the non-DBP pools) comprises about 10% of total plasma total. Some authors have referred to this as the "bioavailable" pool and have pointed out that this pool appears to be unaffected by differences in circulating DBP and total 25-OH-D_3 levels (Table 7.8).

TABLE 7.7 Relative Affinities of DBP (Vitamin D-Binding Protein) and Albumin for 25-OH-D_3 and 1,25$(OH)_2$-D_3

Protein	Binding Constants, K_a	
	25-OH-D_3	1,25-$(OH)_2$-D_3
DBP	$7\times10^8 M^{-1}$	$4\times10^7 M$
Albumin	$6\times10^5 M^{-1}$	$5.4\times10^4 M^{-1}$

TABLE 7.8 Plasma DBP (Vitamin D-Binding Protein) and 25-OH-D_3 Levels in Black and White Americans

Subjects	DBP, μg/mL	25-OH-D_3, ng/mL	"Bioavailable" 25-OH-D3, ng/mL[a]
Black (n = 1181)	168 ± 3	15.6 ± 0.2	2.9 ± 0.1
White (n = 904)	337 ± 5[b]	25.8 ± 0.4[b]	3.1 ± 0.1

[a]Not bound to DBP.
[b]Mean ± SEM; significantly different from Black mean, $p < .05$.
Powe, C.E., Evans, M.K., Wenger, J., et al., 2013. N. Engl. J. Med. 369, 1991–2000.

Differential Metabolism of Vitamins D_2 and D_3

Although a minor dietary form, D_2 is metabolized analogously to D_3. The enzymes involved in the 1-, 24-, and 25-hydroxylations do not discriminate between these vitamers, and daily doses of modest amounts (1000 IU) of either produce comparable increases in serum 25-OH-D.[71] However, large single doses (50,000 IU) of 25-OH-D_2 are cleared more rapidly than 25-OH-D_3.[72] At high levels of exposure, D_2 is metabolized to a number of mono- (24-), di- (1,24-; 24,26-), and tri- (1,25,28-) hydroxylated metabolites that are not produced from D_3. One study found the half-life of 25-OH-D_2 to be significantly shorter than that of 25-OH-D_3 for subjects in Gambia but not for subjects in the U.K. Such discrepant results may be related to different distributions of genetic variants in DBP, which have been shown to affect responsiveness to D_3 but not to D_2 (Fig. 7.11). For humans and probably most other species, differences in biopotency between D_2 and D_3 are significant only at bolus doses.

Some species, including birds and some New World monkeys, strongly discriminate between D_2 and D_3. For them, D_2 is much less biopotent that D_3; they clear the mono- and dihydroxylated metabolites of D_2 faster than those of D_3. For example, in the chick, the plasma turnover rates of vitamin D_2, 25-OH-D_2, and 1,25-$(OH)_2$-D_2 are 1.5-, 11-, and 33-fold faster than those of the respective vitamin D_3 analogues. These differences in turnover rates are greater than those of the binding affinities to DBP (5-, 3.6-, and 3-fold, respectively), which are greater for D_3 and its metabolites than for D_2 and its metabolites.

7. METABOLIC FUNCTIONS OF VITAMIN D

Vitamin D_3 as a Steroid Hormone

At least some, if not all, of the mechanisms of action of vitamin D fit the classic model of a steroid hormone. That is, it has specific cells in target organs with specific receptor proteins; and the receptor–ligand complex moves to the

71. Holick, M.F., Biancuzzo, R.M., Chen, T.C. et al., 2008. J. Clin. Endocrinol. Metab. 93, 677–681.
72. Armas, L.A., Hollis, B.W., Heaney, R.P., 2004. J. Clin. Endocrinol. Metab. 89, 5387–5391.

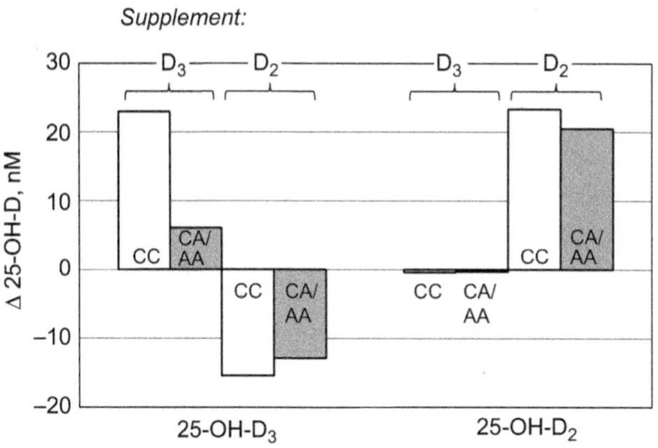

FIGURE 7.11 Changes in plasma 25-OH-D in response to vitamins D_3 and D_2 by humans with different DBP genotypes. *After Nimitphong, H., Saetung, S., Chanprasertyoptin, S., et al., 2013. Nutr. J. 12, 39–49.*

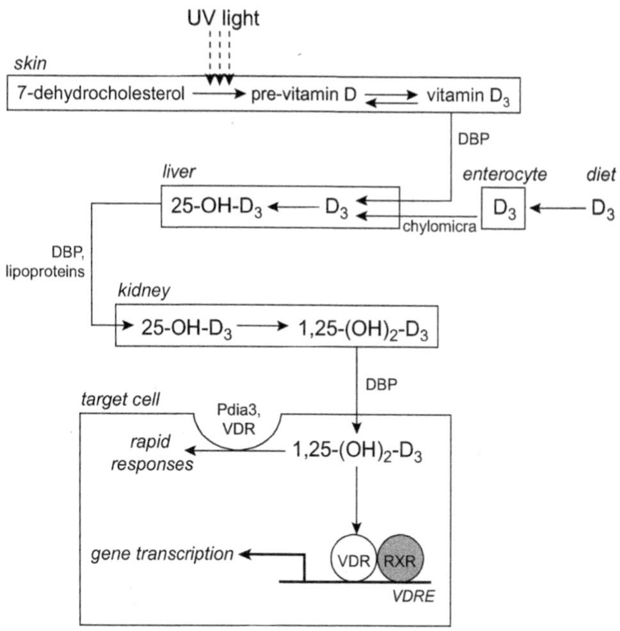

FIGURE 7.12 Metabolic activation of vitamin D (see text for abbreviations).

nucleus, where it binds to the chromatin at specific DNA sequences and stimulates the transcription of certain downstream genes to produce specific mRNAs that encode the synthesis of specific proteins (Fig. 7.12). Vitamin D, itself, is not metabolically active; it must be converted to the active metabolite, 1,25-$(OH)_2$-D_3, which functions in two general ways: by regulating gene expression and by nongenomic signaling of key physiological processes.

Genomic Pathways of Vitamin D Function

Vitamin D Receptor

Target tissues for vitamin D contain a specific nuclear receptor, VDR, which binds 1,25-$(OH)_2$-D_3 with high affinity and 25-OH-D_3 with lower affinity.[73] Autoradiographic studies have shown that 1,25-$(OH)_2$-D_3 is localized in the nuclei of many cell types that contain the VDR. VDRs have been identified in more than 30 different cell types including cells involved in Ca homeostasis (bone, kidney, intestine), immune function, endocrine function, hematopoiesis, skin, and tumors (Table 7.9). Such findings indicate a wide breadth of genomic functions.[74] A rare autosomal recessive defect in VDR manifests as vitamin D-resistant rickets type II characterized by high plasma 1,25-$(OH)_2$-D_3 (differentiating the condition from vitamin D-resistant rickets type I) and, in some families, alopecia. Allelic variation in the VDR gene is associated with reduced bone mineral density and increased fracture risk.[75] In neonates, VDR is absent until weaning.[76] In at least some organs, VDRs appear to be inducible by estrogen. In the guinea pig (and presumably in humans, who also require vitamin C), the number of VDRs can be reduced by deprivation of ascorbic acid.[77]

The VDR is a 51 kDa protein classified in the family of steroid, thyroid hormone, and retinoic acid receptor genes on the basis of its similar primary amino acid structure. VDRs of different species vary in size (e.g., human, 48 kDa; avian, 60 kDa), but their DNA-binding domains show >95% homology between species. Each is a sulfhydryl protein with an N-terminal recognition domain containing a cysteine-rich cluster comprising two "zinc finger" structures.[78] Their ligand-binding domains show less homology between species; each consists of 12 α-helices configured to create a three-layered sandwich structure that encompasses 1,25-$(OH)_2$-D_3 with high affinity in a hydrophobic core. Binding of 1,25-$(OH)_2$-D_3 causes a conformational change in the receptor that allows it to interact with other players in transcription. Multiple polymorphic variations have been identified in the human VDR gene; by affecting mRNA expression, stability, and patterns, variants can have different VDR protein concentrations.

73. The ability of VDR to bind 25-OH-D_3 is likely to be the basis for the ability of anephric patients to increase calcium absorption in response to high doses of vitamin D.

74. That VDR also binds lithocholic acid has caused some to suggest that it may also function as a bile acid sensor.

75. An 18-month study of elderly subjects found VDR *BB* homozygotes more likely to lose bone mineral than *bb* homozygotes (78% versus 31% lost at rates >0.48%/year) (Ferrari, S., Rizzoli, R., Chevalley, T., et al., 1995. Lancet 345, 423–424).

76. VDR abundance varies among various tissues in the range of 200–25,000 copies/cell; VDR contents of intestinal cells are surprisingly low, <2000 copies/cell.

77. Guinea pig intestinal 1,25-$(OH)_2$-D_3 receptors (occupied and unoccupied) were reduced by vitamin C deprivation. This may relate to the development of rickets-like bone changes in vitamin C deficiency.

78. "Zinc fingers" are finger-like structures folded around a zinc atom tetrahedronally coordinated through the sulfhydryls of cysteine residues in the primary structure of the protein. This motif is similar to DNA-binding motifs found in other transcriptional regulating proteins.

TABLE 7.9 Distribution of Known Nuclear Vitamin D Receptors

Organ System	Cell Type
Bone	Osteoblasts
Connective tissue	Cartilage chrondrocytes, fibroblasts, stroma
Alimentary tract	Enterocytes, colonocytes
Liver	Hepatocytes
Kidney	Epithelium (proximal and distal); glomerular podocytes
Heart	Atrial myoendocrine cells, cardiomyocytes
Lung	Bronchial epithelium
Skeletal, smooth muscle	Myocytes
Thymus	Epithelium
Hematolymphopoietic	Activated T and B cells, macrophages, monocytes, spleenocytes, thymus reticular cells, lymphocytes, lymph nodes, tonsillary dendritic cells
Reproductive	Amnion, chorioallantoic membrane, epididymus, ovary, oviduct, placenta, testis, Sertoli and Leydig cells, uterus, yolk sac
Skin	Epidermis, fibroblasts, hair follicles, keratinocytes, melanocytes, sebaceous glands
Nervous	Hippocampus, cerebellar Purkinje and granule cells, bed nucleus, stria terminalis, amygdala central nucleus, sensory ganglia, spinal cord
Immune	Thymus, bone marrow, B cells, T cells
Other endocrine	Adrenal medulla and cortex, pancreatic β cells, pituitary epithelium, thyroid follicles and C cells, parathyroid epithelium, parotid epithelium
Testes	Germ cells
Prostate	Epithelium
Mammary gland	Mammary alveolar and ductal cells
Adipose tissue	Adipocytes
Other	Bladder epithelium, choroid plexus epithelium

These include a *FokI* polymorphism in exon II, *BsmI* and *ApaI* allelic variants in the intron between exons VIII and IX, a *TaqI* restriction fragment polymorphism in exon IX, and a repeat mononucleotide polymorphism in the 3′ untranslated region. Studies have demonstrated contributions of these polymorphisms, as well as epigenetic silencing of VDR expression, to interindividual variations in circulating levels of 25-OH-D_3; to differences in enteric Ca^{2+} absorption and bone mineral density; and to risks of rheumatoid arthritis, osteoarthritis, type 2 diabetes, autoimmune disease, Parkinson's disease, fracture, and cancer.[79]

The liganded VDR is thought to be translocated from the cytosol into the nucleus via interactions with microtubules. This mechanism is thought to involve a sequence followed by other sterols: stimulation of protein kinase C (PKC), PKC-activation of guanylate cyclase, phosphorylation of microtubule-associated proteins, and association of VDR with importins. Like other steroid hormone receptors, the VDR is a *trans*activator of the 1,25-$(OH)_2$-D_3-dependent transcription of mRNAs for various proteins involved in Ca transport, the bone matrix, and cell cycle regulation. The VDR of intestinal epithelial cells can also bind bile acids, ultimately leading to the detoxification of those inducers through upregulation of drug-metabolizing CYP3A enzymes.

Vitamin D-Responsive Elements

Specific DNA promotor sequences act as VDREs. These are similar to the responsive elements mediating gene expression responses of thyroid hormone or retinoic acid; each consists of imperfect direct repeats of six-base pair half-elements.[80] Binding of the 1,25-$(OH)_2$-D_3–VDR complex to VDREs involves one of the retinoid X receptors (RXRs). The preferred active species appears to be a VDR–RXR heterodimer.[81] Most cells contain both VDRs and RXRs; therefore, the availability of 9-*cis*-retinoic acid may determine which set of genes is regulated by 1,25-$(OH)_2$-D_3. Transcriptional regulation of gene expression by vitamin D acting through this system is thought to involve a conformational change in VDR effected by the phosphorylation of a specific serinyl residue upon the binding of 1,25-$(OH)_2$-D_3. This facilitates the recruitment of coactivator proteins that induces chromatin remodeling and exposes domains of the protein capable of interacting with VDREs to influence RNA polymerase-mediated transcription.

Genes Regulated by Vitamin D

Some 50 genes have been identified as being regulated by vitamin D (Table 7.7). These include genes associated with many aspects of metabolism to include vitamin D metabolism as well as cell differentiation and proliferation, energy

79. Utterlinden, 2005. In: Feldman, D., Pike, J.W., Glorieux, F.H. (Eds.), Vitamin D, second ed. Elsevier, New York, NY, pp. 1121–1157.

80. Owing to their direct repeats, these lack the dyad symmetry of the classic steroid hormone-responsive elements.

81. Interaction of these receptor proteins is thought to involve C-terminal dimerization interfaces in both.

metabolism, hormonal signaling, mineral homeostasis, oncogenes, and chromosomal proteins. For most of these, the regulation appears to involve 1,25-$(OH)_2$-D_3-dependent modulation of mRNA levels (i.e., regulation of transcription and/or message stability). To date, 1,25-$(OH)_2$-D_3-regulated transcription has been established for less than a dozen of these genes, and VDREs have been reported for only a few (e.g., calbindin$_{9K}$, integrin$_{\alpha\gamma\beta 3}$, osteocalcin, and the plasma membrane Ca^{2+} pump). Evidence for posttranscriptional regulation of calbindin$_{9K}$ has been presented. From this emerging picture, it is clear that changes in vitamin D status have potential for pleiotropic actions.

The first gene product to be recognized as inducible by 1,25-$(OH)_2$-D_3 was for many years called **Ca-binding protein (CaBP)**. Two forms of CaBP have subsequently been described; these are now called **calbindins**.[82] They are widespread in animal tissues, with greatest concentrations found in avian and mammalian duodenal mucosa, where they can comprise 1–3% of the total soluble protein of the cell. Calbindins function in the enteric absorption of Ca by facilitating the movement of Ca through the enterocytic cytosol while keeping the intracellular concentration of the free Ca^{2+} ion below hazardous levels. Calbindin-D_{9k} can bind two Ca^{2+} atoms, while calbindin-D_{28k} can bind four Ca^{2+} atoms. Calbindin-D_{9k} occurs primarily in mammalian intestinal mucosa but also in kidney, uterus, and placenta; calbindin-D_{28k} occurring in mammalian kidney (distal convoluted tubules), pancreas (β cells) and brain, and avian intestine and kidney. Calbindins are not expressed in vitamin D deficiency but are expressed in response to $1,25(OH)_2$-D_3. That such treatment increases the expression of the protein without affecting its message indicates that vitamin D regulation of calbindin occurs at the translational level.

VDR also downregulates the expression of some genes. These include genes encoding PTH (Table 7.10). This appears to involve binding of VDR homodimers or VDR–RXR heterodimers to a negative response element (nVDRE), which is transcriptionally active in the absence of $1,25(OH)_2$-D_3. It is thought that downregulation occurs by VDR directly binding an activator of the nVDRE.

Vitamin D also appears to have extratranscriptional effects on gene expression. Studies have revealed that 1,25-$(OH)_2$-D can affect DNA methylation, histone acetylation, and microRNA generation, thus, affecting physiological functions via epigenetic effects. The active metabolite has also been found to affect pre-mRNA constitutive and alternative gene splicing by recruiting nuclear receptor

82. Calbindins are members of a large family of Ca^{2+}-binding proteins each with a distinctive helix–loop–helix sequence, the "EF hand." They have been identified in many species. The mammalian form (calbindin-D_{9k}) is a 10kDa protein with two high-affinity Ca^{2+}-binding sites. Avian calbindin-D_{28K} is larger, 30kDa; it is also expressed in the shell gland (uterus) of laying hens.

TABLE 7.10 Genes Known to Be Regulated by Vitamin D

Gene	Tissue in Which Regulation Has Been Demonstrated
Upregulated	
Aldolase subunit B	Chick kidney
Alkaline phosphatase	Chick and rat intestine
ATP synthase	Chick and rat intestine
Calbindin-$D_{28\ kDa}$	Chick brain, kidney, uterus, intestine; mouse kidney; rat kidney, brain
Calbindin-$D_{9\ kDa}$	Chick kidney, skin, bone; rat intestine, skin, bone
Carbonic anhydrase	Marrow; myelomonocytes
CCAT enhancer-binding protein β	Mouse intestine, osteoblasts
Cytochrome *c* oxidase, subunits I, II, and III	Chick and rat intestine
Cytochrome P450 isoform CYP3A	Mouse intestine
Fibronectin	MG-63, TE-85, HL-60 cells
c-Fms	HL-60 cells
c-Fos	HL-60 cells
Glyceraldehyde-3-phosphate dehydrogenase	BT-20 cells
Heat shock protein 70	Peripheral blood monocytes
Integrin$_{\alpha\gamma\beta 3}$	Chick osteoclasts
Interleukin 1	U937 cells
Interleukin 6	U937 cells
Interleukin 3 receptor	MC3T3 cells
c-Ki-Ros	BALB-3T3 cells
Matrix Gla protein	UMR106-01, ROS cells
Metallothionein	Rat keratinocytes
NADH dehydrogenase, subunits II and III	Chick intestine
Nerve growth factor	L-929 cells
Neutrophil-activating polypeptide	HL-60 cells
c-Myc	MG-63 cells
Osteocalcin	ROS cells
Osteopontin	ROS cells
1-OH-D_3 24-hydroxylase	Rat kidney

Continued

TABLE 7.10 Genes Known to Be Regulated by Vitamin D—cont'd

1,25-$(OH)_2$-D_3 receptor	Mouse fibroblasts
Plasma membrane Ca^{2+} pump	Chick intestine
Prolactin	Rat pituitary cells
Protein kinase C	HL-60 cells
Tumor necrosis factor α	U-937, HL-60 cells
Vascular endothelial growth factor (VEGF)	Mouse fibroblasts
Vitamin D receptor	Rat intestine, pituitary
Downregulated	
ATP synthase	Chick kidney
Calcitonin	Rat thyroid gland
CD-23	Peripheral blood monocytes
Collagen, type I	Rat fetal calvaria
Cytochrome *b*	Chick kidney
Cytochrome *c* oxidase. Subunits I, II, and III	Chick kidney
Cytochrome P450 isoform CYP2B1	Mouse liver
Fatty acid-binding protein	Chick intestine
Ferridoxin	Chick kidney
Granulocyte–macrophage colony-stimulating factor	Human T lymphocytes
Histone H_4	HL-60 cells
Interleukin 2	Human T lymphocytes
Interferon γ	Human T lymphocytes
c-Myb	HL-60 cells
c-Myc	HL-60, U937 cells
NADH dehydrogenase subunit I	Chick kidney
25-OH-D_3 1-hydroxylase	Rat kidney
Prepro-PTH (parathyroid hormone)	Rat, bovine parathyroid
Protein kinase inhibitor	Chick kidney
PTH	Rat parathyroid
PTH-related protein	T lymphocytes
Transferrin receptor	PBMCs (peripheral blood mononuclear cells)
α-Tubulin	Chick intestine

TABLE 7.11 Rapid Responses to 1,25-$(OH)_2$-D_3

Response	Organ/Cell
Ca^{2+} transport	Intestinal mucosa, CaCo-2 cells
Protein kinase C (PKC) activation	Intestinal mucosa, chondrocytes, liver, muscle
MAPK activation	Intestinal mucosa, liver, leukemia cells
Ca^{2+} signaling	Adipocytes, pancreatic β cells
Ca^{2+} channel opening	Osteoblasts
G protein activation	Intestinal mucosa
Cell differentiation	Promyelocytic NB4 cells
Src activation	Keratinocytes
Raf activation	Keratinocytes
Insulin secretion	Pancreatic β cells
Increased contraction/relaxation	Cardiomyocytes

Mizwicki, M.T., Norman, W.N., 2009. Sci. Signal. 2, 1–14.

coregulators (e.g., heterogeneous nuclear ribonucleoprotein C, hnRNPC).[83]

Nongenomic Pathways of Vitamin D Function

Some responses to vitamin D occur within seconds to minutes of 1,25-$(OH)_2$-D_3 exposure and are membrane-mediated. This suggests signaling independent of genomic responses, which typically take hours to days. The first such response to be recognized is **transcaltachia**, the rapid transport of Ca^{2+} across the intestinal mucosa. Several other transcription-independent responses have since been demonstrated (Table 7.11). That these rapid responses depend on VDR is indicated by their absence in VDR-null mice. However, evidence indicates the involvement of another receptor, a membrane-associated rapid response steroid-binding protein. That appears to be a protein localized in plasma membrane **caveolae**,[84] protein disulfide isomerase 3 (Pdia3). Convincing evidence shows that binding of 1,25-$(OH)_2$-D_3 to Pdia3 triggers a protein phosphorylation cascade that begins with interaction with phospholipase A_2-activating protein

83. Zhou, R., Chun, R.F., Lisse, T.S., et al., 2015. J. Steroid Biochem. Mol. Biol. 148, 310–317.

84. Caveolae (Latin, "little caves") are small (50–100nm) invaginations of the plasma membrane in endothelial cells, adipocytes, and other cell type. They are comprised of a type of "lipid raft," which are organized microdomains within plasma membranes comprised of glycosphingolipids and protein receptors, and serving as sites for assembling signaling molecules, influencing membrane protein trafficking, etc.

in membrane caveolae, and leads to activation of Ca^{2+}/calmodulin-dependent kinase II (CaMKII) and, ultimately, phospholipase A_2 (PLA_2).[85] The resulting rapid release of arachidonic acid activates PKC either directly or indirectly via its metabolite prostaglandin E_2 (PGE_2). In this view, the binding of 1,25-$(OH)_2$-D_3 to VDR, also in caveolae, stimulates its interaction with and rapid activation of the tyrosine kinase Src.

Calcium and Phosphorus Metabolism

The most clearly elucidated function of vitamin D is in the homeostasis of Ca^{2+} and phosphate. This is effected by a multihormonal system involving the controlled production of 1,25-$(OH)_2$-D_3, which functions in concert with PTH and calcitonin (CT). Regulation of this system occurs at the points of intestinal absorption, bone accretion and mobilization, and renal excretion.

Intestinal Absorption of Ca^{2+}

Calcium is absorbed in the small intestine[86] by both transcellular and paracellular mechanisms. The former is an active, saturable process, occurring in mammals primarily in the duodenum and upper jejunum and constitutes the most important means of absorbing Ca under conditions of low intake of the mineral; the latter is a nonsaturable process occurring throughout the intestine and is the most important means of absorbing Ca when Ca intake is high. The active metabolite, 1,25-$(OH)_2$-D_3, stimulates the enteric absorption of Ca through roles in both mechanisms, although its mechanism in the paracellular process is unclear. The availability of Ca for both processes is affected by both exogenous (e.g., inhibition by food phytates or P) and endogenous (e.g., gastric acid secretion) factors.

The transcellular absorption of Ca^{2+} progresses in three steps, each dependent on 1,25-$(OH)_2$-D_3:

1. **Uptake of Ca^{2+} from the intestinal lumen**. At the microvillus brush border, Ca^{2+} diffuses through a channel or integral membrane transporter (CaT1) gated by the intercellular Ca^{2+} concentration, which affords controlled movement of Ca^{2+} down a steep electrochemical gradient.[87] Vitamin D treatment of cells in culture increases Ca^{2+} uptake, with increased expression of CaT1- and Ca^{2+}-binding proteins (calbindins), which appear to control the microvillar Ca^{2+} channel. Vitamin D also shifts synthesis from phosphatidylethanolamine to phosphatidylcholine, affecting membrane fluidity and increasing the association of the Ca^{2+}-binding protein calmodulin[88] with the brush border.
2. **Translocation of Ca^{2+} across the mucosal cell**. Movement of Ca^{2+} across the enterocyte appears to be facilitated by both calbindins and vesicular transport.
3. **Extrusion of Ca^{2+} into the circulation**. Ca^{2+} moves across the basolateral membrane against a substantial thermodynamic gradient (a 50,000-fold differential in Ca^{2+} concentration and a positive electrical potential)[89] facilitated by a Ca^{2+}-ATPase.[90] This Ca^{2+} pump is stimulated by calmodulin and calbindin. Ca^{+2} is also extruded by a membrane Na+/Ca^{2+} exchanger.[91]

The paracellular absorption of Ca^{+2} is less well understood. Evidence suggests that it is nonsaturable but stimulated by 1,25-$(OH)_2$-D_3 and regulated by the proteins comprising the tight junction complex at the apical region. Diffusion of Ca^{+2} through this pathway occurs when intralumenal Ca^{+2} concentrations exceed 2–6 n*M*. Some researchers have found vitamin D to promote Ca uptake by this pathway, perhaps by 1,25-$(OH)_2$-D_3-mediated activation of PKC, which is known to increase paracellular permeability.

Intestinal Phosphate Absorption

Healthy individuals absorb dietary phosphate (P_i)[92] with efficiencies of 60–65%, most absorption occurring in the duodenum and jejunum. Phosphate is absorbed by two mechanisms: a nonsaturable paracellular pathway and a saturable, energy-requiring, Na^+-dependent process on the mucosal surface that is driven by a Na^+ gradient maintained by the Na^+, K^+-ATPase on the basolateral membrane. The latter process appears to be rate limiting to P_i absorption. Vitamin D, as 1,25-$(OH)_2$-D_3, increases net P_i

85. Doroudi, M., Schwart, Z., Boyan, B.D., 2015. J. Steroid Biochem. Mol. Biol. 147, 81–87.
86. Some 70–80% of enteric calcium absorption occurs in the ileum. The colon may be responsible for 3–8% of Ca absorption.
87. Luminal concentrations of Ca^{2+} can be in the mM range; whereas, intracellular concentrations of the free ion are in the range of 50–100 nM.
88. Calmodulin is a 17 kDa acidic protein (also of the "EF hand" family) expressed in many cell types and subcellular compartments. It can bind as many as four Ca^{2+} ions, which causes conformational change and posttranslational modifications that allow it also to bind to more than 100 target proteins. In this way, calmodulin serves as a major transducer of Ca^{2+} signals in the control of cellular metabolism.
89. It is estimated that the movement of 1 mole of Ca^{2+} against this gradient requires about 9.3 kcal.
90. The CaATPase spans the membrane with a Ca^{2+}-binding domain on the cytoplasmic side. It appears that phosphorylation-induced conformational changes in the protein allow it to form a channel-like opening through which Ca^{2+} is expelled, thus serving as a Ca "pump."
91. The efflux of 1 mole of Ca^{2+} is linked to the influx of 3 moles of Na^+, thus generating negative cytosolic electropotential.
92. P_i is an essential constituent of bone and teeth, which account for some 85% of total body P_i. It is also important in regulating the genes, *PHEX* and *FGF23*. P_i comprises 1% of adult body weight.

uptake by increasing Na^+/P_i cotransport across the mucosal brush border, apparently through upregulation of the transporter.

Renal Resorption of Calcium and Phosphate

Vitamin D, as 1,25-$(OH)_2$-D_3, stimulates the resorption of both P_i and Ca^{2+} in the renal tubule.[93] The quantitative significance of this effect is greater for P_i some 60% of which is reabsorbed in the proximal tubule by a Na^+/P_i cotransport mechanism analogous to those in the intestinal epithelium and in bone. The Na^+/P_i cotransporter, Npt2a, is expressed in proximal tubular cells where the major portion of P_i is reabsorbed; its expression is also upregulated by both PTH and P_i supply. In contrast, Ca^{2+} is reabsorbed mostly in the distal tubule, primarily (80%) by passive, vitamin D-independent, paracellular routes in the proximal tubules and ascending loop of Henle. Some 8000 mg of Ca are filtered at the glomerulus daily,[94] 98% of which is reabsorbed in the tubules. The transcellular process resembles that of the intestine in having a Ca^{2+} channel component, cytosolic Ca^{2+}-binding proteins (calbindins-D_{9k} and -D_{28k}), and a plasma membrane Ca^{2+} ATPase, all of which are located in the distal portions of 1,25-$(OH)_2$-D_3-responsive nephrons.

Calcium and Phosphate Homeostasis

Homeostatic control of Ca and P is dependent on functions of the parathyroid and thyroid glands (Fig. 7.13). Each gland senses serum Ca^{2+} level[95] by a cell surface G protein-like, **Ca^{2+}-sensing receptor (CaR)** in chief cells of the parathyroid and parafollicular cells (C cells) of the thyroid.[96] The CaR appears sensitive to fluctuations in plasma Ca^{2+} of a few percent, responding by regulating the synthesis of two calcitropic hormones: PTH by the parathyroid and CT by the thyroid, which elevate and lower plasma Ca^{2+}, respectively. When Ca^{2+} levels drop the parathyroid loses VDR, which reduces its sensitivity to 1,25-$(OH)_2$-D_3. It also secretes PTH into the circulation, which acts on target tissues to restore plasma Ca^{2+} by stimulating renal tubular Ca^{2+} reabsorption and renal 1-hydroxylation of 25-OH-D_3.[97] The resulting increase in circulating 1,25-$(OH)_2$-D_3 stimulates enteric Ca^{2+}

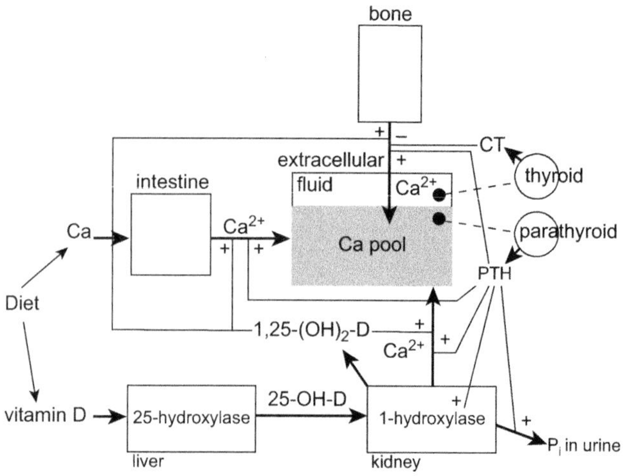

FIGURE 7.13 Calcium homeostasis.

absorption and osteoclastic activity. Demineralization of bone serves to mobilize Ca^{2+} and P_i from that reserve, thus maintaining the homeostasis of those minerals in the plasma.[98] Increasing plasma Ca^{2+} levels evoke signaling by CaR in thyroid C cells to increase the expression of CT, which inhibits bone resorption and, at high doses, increases urinary Ca^{2+} excretion. This system feeds back to regulate the synthesis of PTH through inhibition by 1,25-$(OH)_2$-D_3 binding a negative VDRE near the promotor of the PTH gene. A similar mechanism has been proposed for the downregulation of CT by 1,25-$(OH)_2$-D_3. The interplay of these hormones with Ca^{2+} and 1,25-$(OH)_2$-D_3 produces fine control of circulating Ca^{2+} levels at 4–5.6 mg/dL.[99]

Under **hypercalcemic** conditions, CT is secreted by the thyroid. The hormone suppresses bone mobilization and is also thought to increase the renal excretion of both Ca^{2+} and P_i. In that situation, the 25-OH-vitamin D 1-hydroxylase may be feedback inhibited by 1,25-$(OH)_2$-D_3, which may actually convert to the catalysis of the 24-hydroxylation of 25-OH-D_3. In the case of egg-laying birds, which show relatively high circulating concentrations of 1,25-$(OH)_2$-D_3, the 1-hydroxylase activity remains stimulated by the hormone prolactin.

Secondary Hyperparathyroidism

It is characterized by elevated serum PTH concentrations, is common among elderly people. The condition can reflect some degree of renal insufficiency, with associated reduction in renal 25-OH-vitamin D_3 1-hydroxylase activity. Serum concentrations of PTH can also increase owing to

93. Each day the human kidney filters some 8 g calcium at the glomerulus, 98% of which is reabsorbed.

94. The concentration of ultrafilterable Ca in plasma, c. 1.35 mM (c.55% of total plasma Ca), is similar to that of the glomerular fluid.

95. Although tightly maintained, serum Ca^{2+} levels are normally 10,000 times greater than intracellular ones. Perturbation of this ion gradient by transient increases in intracellular Ca^{2+} concentrations can signal different cellular responses (e.g., transcriptional control, neurotransmitter release, muscular contraction) through the actions of binding proteins.

96. CaR is also expressed in renal tubules.

97. Parathyroidectomized animals cannot mount this 1-hydroxylase response unless treated with PTH.

98. The normal ranges of these parameters in human adults are Ca, 8.5–10.6 mg/dL; P, 2.5–4.5 mg/dL.

99. This corresponds to 8–10 mg total Ca per deciliter, about half of which is present in ionized form (Ca^{2+}).

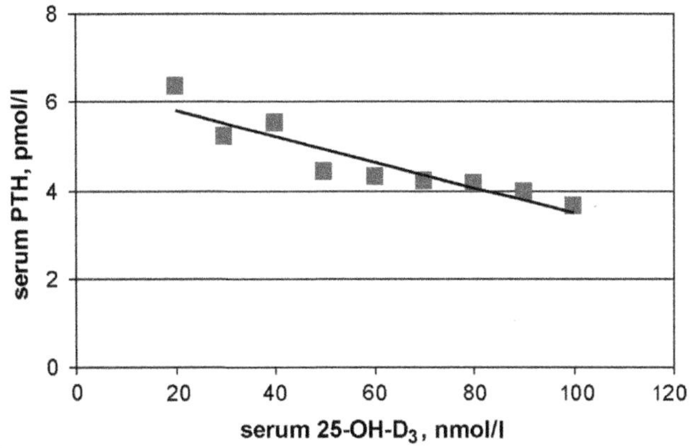

FIGURE 7.14 Relationship of serum PTH and vitamin D status. *Need, A.G., Horowitz, M., Morris, H.A., Nordin, B.C., 2000. Am. J. Clin. Nutr. 71, 1577–1581.*

privational vitamin D deficiency, in which case it is manifested by low circulating levels of 25-OH-D_3 (Fig. 7.14). Accordingly, the PTH levels of people living in northern latitudes are highest during the winter for subjects not taking supplemental vitamin D.

Bone Mineral Turnover

Bone is the predominant target organ for vitamin D, accumulating more than one-quarter of a single dose of the vitamin within a few hours of its administration. That lesions in bone mineralization (rickets, osteomalacia) occur in vitamin D deficiency has long indicated its vital function in the metabolism of this organ.[100] The pattern of vitamin D metabolites in bone differs from that in intestine; whereas the latter contains mainly 1,25-$(OH)_2$-D_3, bone contains mainly 25-OH-D_3 (accounting for >50% of the vitamin D metabolites present, with 1,25-$(OH)_2$-D_3 comprising less than 35%). As in plasma, the level of 24,25-$(OH)_2$-D_3 in bone is fairly constant relative to that of 25-OH-D_3.

Vitamin D contributes to both the formation (mineralization) and the mobilization of bone mineral (demineralization). Bone mineralization is epitaxial, occurring by codeposition of Ca^{2+}, P_i (i.e., $PO_4{}^{3-}$), and hydroxyl (OH^-) ions at multiple sites on the surfaces of preexisting crystals or on topographically similar protein/lipid surfaces. The resulting structure is one comprised of small (<200 Å) crystals with an average chemistry resembling that of hydroxyapatite, $Ca_{10}(PO_4)_6(OH)_2$. Bone mineral may also contain magnesium, sodium, potassium, fluoride, strontium, phosphate, and citrate. The amount of bone mineral is, therefore, a function of the balance of the laying down of mineral[101] by bone-forming cells called **osteoblasts**, and the dissolution of bone crystal[102] by bone-resorbing cells called **osteoclasts**. Calcium deposition in the skeleton involves the intracellular synthesis of collagen and fibrils by osteoblasts, which extrude these fibrils to form the extracellular matrix of bone, portions of which can be mineralized. Bone demineralization is directed by multinucleated osteoclasts that release proteins and lysosomal enzymes that dissolve bone mineral and lyse its organic matrix. The accretion/mobilization of bone Ca^{2+}, therefore, involves the relative activities of osteoblasts and osteoclasts with the bone surface serving, in effect, as a Ca buffer. Bone growth results from the dominance of osteoblastic activity, which in the long bones is organized by the arraying of chondrocytes to affect periosteal apposition along epiphyseal growth plates (Fig. 7.15). Balanced demineralization–mineralization affects the coordinated growth ("modeling") of skeletal bone and the "remodeling" (primarily in the endosteal area) of mature bone to replace damage and prevent senescence.

While the ultimate effects on bone involve both osteoblasts and osteoclasts, vitamin D targets only osteoblasts and osteoprogenitor cells, both of which have VDRs that are stimulated by glucocorticoids. 1,25-Dihydroxyvitamin D_3 affects the expression of several osteoblast genes (Table 7.12). It is generally thought that 24,25-$(OH)_2$-D_3, which is concentrated in the epiphyseal cartilage, may also be involved in this process. While the mechanism remains unclear, it appears to involve PTH[103] and, because PTH stimulates adenylate cyclase activity, perhaps act via cAMP. Vitamin D affects osteoclasts only by indirect means, as osteoclasts are activated by osteoblasts through the induction of the membrane-associated receptor activator of nuclear factor κB (NFκB) ligand (RANKL). Vitamin D also has a role in the differentiation of macrophages to osteoclasts; the number of these giant multinucleated bone-degrading cells is very low in bone from vitamin D-deficient animals.

In the absence of adequate levels of 1,25-$(OH)_2$-D_3, the failure of mineralization and/or net excess of osteoclastic demineralization have structural and functional consequences to bone, as bone density is a primary determinant of bone strength. This ultimately results in the well-known clinical signs (see section Signs of Vitamin D deficiency). Inadequate vitamin D status may also be associated with fracture risk, particularly in older individuals

100. Thus, the involvement of vitamin D in the metabolism of Ca and P was clear, as structural bone contains 99% of total body Ca and 85% of total body P, i.e., 1200 g Ca and 770 g P in a 70 kg man.

101. Mineralization is preceded by secretion of the bone matrix, unmineralized osteoid, by chondrocytes.

102. Dissolution of bone mineral is followed by depolymerization of glycosaminoglycans and digestion of collagen and other bone matrix proteins.

103. Because PTH is secreted in response to hypophosphatemia, deprivation of P_i can also lead to bone demineralization with increases in the activity of 25-OH-vitamin D 1-hydroxylase and accumulation of 1,25-$(OH)_2$-D_3 in target tissues.

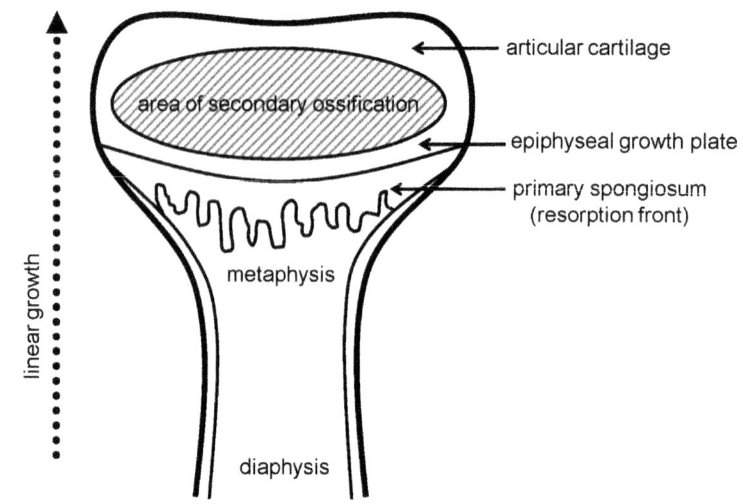

FIGURE 7.15 Linear growth of long bones occurs by deposition of bone along the epiphyseal growth plates comprised of arrays of chondrocytes apical of a highly vascularized zone of spongiform bone.

TABLE 7.12 Vitamin D-Responsive Osteoblast Genes

Collagen
Alkaline phosphatase
Osteocalcin
Osteopontin
Bone sialoprotein-transforming growth factor-β (TGF-β)
Vascular endothelial growth factor (VEGF)
Matrix metalloproteinase-9 (MMP-9)
β3-integrin
Receptor activity of NFκB (RANKL)
Osteopetegrin

TABLE 7.13 Abnormalities in Vitamin D Status of Older Hip Fracture Patients

	Controls[a]	Patients[b]
Parameter	n = 78	n = 120
Dietary Intakes		
Ca (mg/day)	696 ± 273	671 ± 406
Vitamin D (IU/day)	114 ± 44	116 ± 63
Serum Analytes		
Ca (m*M*)	2.35 ± 0.12	2.13 ± 0.16
P_i (m*M*)	1.09 ± 0.15	1.11 ± 0.26
Alkaline phosphatase (units)	2.1 ± 0.5	2.0 ± 0.7
Albumin (g/liter)	41.9 ± 2.8	32.5 ± 4.8[b]
Vitamin D-binding protein (DBP) (mg/liter)	371 ± 44	315 ± 60[b]
25-OH-D_3 (n*M*)	32.9 ± 13.6	18.5 ± 10.6[b]
24,25-$(OH)_2$-D_3 (n*M*)	1.8	0.5[b]
1,25-$(OH)_2$-D_3 (p*M*)	105 ± 3 1	79 ± 46[b]
Parathyroid hormone (PTH) (μg Eq/liter)	0.12 ± 0.05	0.11 ± 0.05

[a]Subjects in both groups were in their eighth decade and had similar histories of sun exposure (one-third with high exposure).
[b]$p < .05$.
Lips, P., van Ginkel, F.C., Jongen, M.H.M., et al., 1987. Am. J. Clin. Nutr. 46, 1005–1010.

(Tables 7.13 and 7.14). A meta-analysis of 12 randomized, controlled trials revealed that vitamin D doses of 700–800 IU (17.5–20 mg) per day reduced the relative risk of hip fracture by 26% and of any nonvertebral fracture by 23%.[104] Risk reductions were not observed for trials that used a lower vitamin dose (400 IU [10 mg]/day).

African-Americans exhibit lower circulating levels of 25-OH-D_3 than white Americans (Fig. 7.16) suggesting increased risk of fracture relative to whites. However, the reverse is true: the prevalences of both hip fracture and osteoporosis of African-Americans are about half that of White Americans.[105] This has been explained on the basis of multiple characteristics of African-Americans that affects fracture risk.[106] Compared to whites, African-Americans tend to have greater peak bone mass, greater

104. Bischoff-Ferrari, H.A., Willett, W.C., Wong, J.B., et al., 2005. J. Am. Med. Assoc. 293, 2257–2264.

105. Barrett-Conner, E., Siris, E.S., Wehren, L.E., et al., 2005. J. Bone Miner. Res. 20, 185–194.

106. Cosman, F., Nieves, J., Dempster, D., et al., 2007. J. Bone Miner. Res. 22, V34–V38; Aloia, J.F., 2008. Am. J. Clin. Nutr. 88, 545S–550S.

TABLE 7.14 Reduction of Fracture Risk in Women by 1,25-$(OH)_2$-Vitamin D_3

Treatment	Year	Women in Study	Women With New Fractures	Number of New Fractures
1,25-$(OH)_2$-D_3	1	262	14	23
	2	236	14	22
	3	213	12	21
Ca	1	253	17	26
	2	240	30[a]	60[b]
	3	219	44[b]	69[b]

[a]*p < .01.*
[b]*p < .001.*
Tilyard, M.W., Spears, G.F.S., Thomson, J., Dovey, S., 1992. N. Engl. J. Med. 326, 357–362.

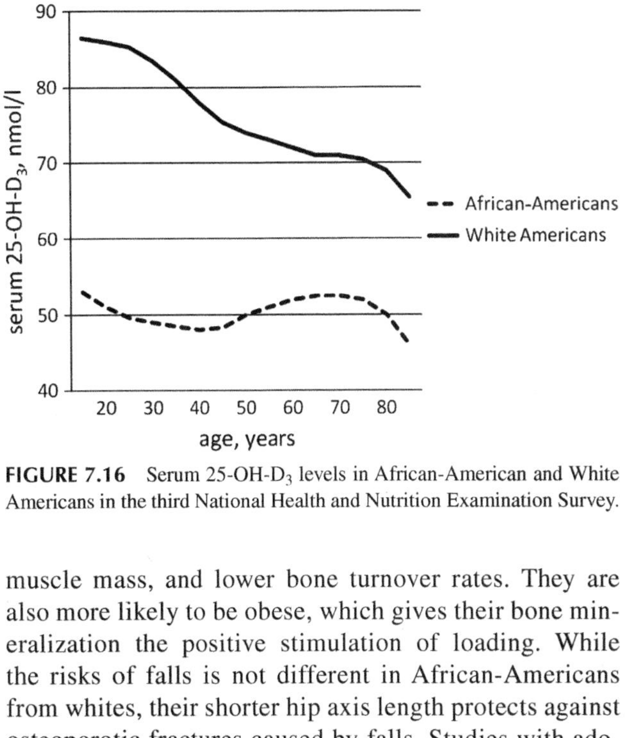

FIGURE 7.16 Serum 25-OH-D_3 levels in African-American and White Americans in the third National Health and Nutrition Examination Survey.

muscle mass, and lower bone turnover rates. They are also more likely to be obese, which gives their bone mineralization the positive stimulation of loading. While the risks of falls is not different in African-Americans from whites, their shorter hip axis length protects against osteoporotic fractures caused by falls. Studies with adolescent girls have found that African-Americans have better enteric Ca absorption and renal Ca conservation that whites. Further, African-American adults have higher levels of PTH without associated bone loss, indicating skeletal resistance to PTH. That these advantages diminish with age are evidenced by the fact that, like other groups, elderly African-Americans with elevated PTH experience bone loss.

VDR genotype appears to contribute significantly to the variation observed in bone mineral density in populations, 80% of which is thought to be due to genetic factors. Two VDR polymorphisms, Fok1 and Bsm1, have been identified as independent risk factors for stress fracture. Individuals with the B- or f-containing genotypes with respect to the each polymorphism were more likely to develop fractures than individuals without those alleles.[107]

Roles of Other Minerals

Vitamin D function can be affected by several other mineral elements:

- **Zinc.** Deprivation of zinc reduces the 1,25-(OH)2-D_3 response to low Ca intake.[108] It has been suggested that zinc may indirectly affect the renal 25-OH-D_3 1-hydroxylase.
- **Iron.** Iron deficiency is associated with low-serum 24,25-(OH)2-D3 levels and reduced 25-OH-D3 responses to supplemental D3. It has been suggested that iron deficiency, which is known to impair the enteric absorption of fat and vitamin A, may also impair the absorption of vitamin D.
- **Lead.** Exposure to lead appears also to impair the 1-hydroxylation of 25-OH-D3. That effect increases the enteric absorption of lead, which can be bound by the vitamin D-responsive protein, calbindin. In children, blood levels of 1,25-(OH)2-D3 and Ca^{2+} have been found to be inversely related to blood lead concentration.[109] 1,25-(OH)2-D3-mediated lead absorption is a chief contributor to the elevated body lead burden observed in Ca deficiency (Table 7.15), although chronic ingestion of lead by Ca-deficient animals reduces 1,25-(OH)2-D3 production. This combination of effects allows the hormone system impaired by chronic lead ingestion to contribute to susceptibility to lead toxicity.[110]

107. Chatzipapas, C., Boikos, S., Drosos, G.I., et al., 2009. Horm. Metab. Res. 41, 635–640; Diogenes, M.E.L., Bezerra, F.F., Cabello, G.M.K., et al., 2010. Eur. J. Appl. Physiol. 108, 31–38.
108. This finding may have clinical significance, as in many parts of the world children are undernourished with respect to both Zn and Ca.
109. The prevalence of lead toxicity among children is seasonal, i.e., greatest in the summer months.
110. Lead poisoning is a serious environmental health issue. In the United States alone, high-blood lead levels are estimated in as many as five million school children and 400,000 pregnant women.

TABLE 7.15 Vitamin D-Stimulated Uptake and Retention of Lead by the Chick

Diet		Kidney		Tibia Ash	
Vitamin D	Pb, %	Ca, ppm	Pb, ppm	Ca, ppm	Pb, ppm
—	0	69.0±8.8	0	33.2±0.4	23.4±14.3
1,25-$(OH)_2$-D_3	0	64.1±1.5	0	36.0±0.2	10.9±6.7
—	0.2	56.6±3.8	4.8±0.5	33.5±0.3	133.1±24.1
1,25-$(OH)_2$-D_3	0.2	80.2±13.4	13.7±2.4	35.5±0.2	335.1±15.8
—	0.8	62.4±2.4	9.2±0.9	32.8±0.2	299.8±4.8
1,25-$(OH)_2$-D_3	0.8	90.1±7.6	32.4±7.6	34.7±0.4	1008.8±71.2

Fullmer, C.S., 1990. Proc. Soc. Exp. Biol. Med. 194, 258–264. Means with like superscripts are not significantly different (p < .05.).

TABLE 7.16 Experimental Evidence for Vitamin D Functions in Noncalcified Tissues

Putative Role	Observations
Cell differentiation	Promotion by 1,25-$(OH)_2$-D_3 of myeloid leukemic precursor cells to differentiate into cells resembling macrophages
Membrane structure	Alteration of the fatty acid composition of enterocytes, reducing their membrane fluidity
Mitochondrial metabolism	Decrease in isocitrate lyase and malate synthase (shown in rachitic chicks)
Muscular function	Stimulation of Ca^{2+} transport into the sarcoplasmic reticulum of cultured myeloblasts by 1,25-$(OH)_2$-D_3 Easing, on treatment with vitamin D, of electrophysiological abnormalities in muscle contraction and relaxation in vitamin D-deficient humans Reduction, on treatment with vitamin D, of muscular weakness in humans
Pancreatic function	Stimulation of insulin production by pancreatic β cells in rats by 1,25-$(OH)_2$-D_3 Impairment of insulin secretion, unrelated to the level of circulating Ca, shown in vitamin D-deficient humans
Immunity	Stimulation of immune cell functions by 1,25-$(OH)_2$-D_3 Control of inflammation by 1,25-$(OH)_2$-D_3-dependent regulation of cytokine production
Neural function	Region-specific enhancement of choline acetyltransferase in rat brain by 1,25-$(OH)_2$-D_3
Skin	Inhibition of DNA synthesis in mouse epidermal cells by 1,25-$(OH)_2$-D_3
Parathyroid function	Inhibition of transcription of PTH gene via interaction of 1,25-$(OH)_2$-D_3 and DNA in parathyroid cells

Vitamin D Functions in Noncalcified Tissues

That 1,25-$(OH)_2$-D_3 and nuclear VDRs occur in tissues not directly involved in Ca^{+2} homeostasis (e.g., pancreatic β cells,[111] skin Malpighian layer cells, specific brain cells, pituitary, muscle, mammary gland, endocrine cells of the stomach, the chorioallantoic membrane surrounding chick embryos) suggests that the vitamin functions in the regulation and differentiation of many cell types (Table 7.16). These functions occur via VDR: at least 100 proteins (including several oncogenes) are known to be regulated by 1,25-$(OH)_2$-D_3. Responses to 1,25-$(OH)_2$-D_3 are observed at concentrations two to three orders of magnitude greater than circulating levels; hence, it is possible that under normal circumstances they may be limited to specific sites of local production of the active metabolite.

111. It is of interest to note that circulating insulin levels are reduced in vitamin D deficiency and respond quickly to treatment with 1,25-$(OH)_2$-D_3.

Muscular Function

That vitamin D plays an important role in muscle is evidenced by the muscle weakness typically experienced by vitamin D-deficient subjects, the presence of VDR in myocytes, and the lack of muscle development in VDR-knockout mice. 1,25-$(OH)_2$-D_3 is essential for the homeostatic control of intracellular Ca^{+2}, affecting both muscle contractility and myogenesis. The former effect involves the transcriptional regulation of various Ca^{+2}-binding proteins, including calbindin-D_{9K}, as well as enzymes related to the synthesis of phosphatidylcholine. Myocytes treated with 1,25-$(OH)_2$-D_3 respond by activating PKC and transporting Ca^{2+} in the sarcoplasmic reticulum, an effect necessary for muscle contraction. They also show upregulation of tyrosine phosphorylation of the MAP kinase cascade,[112] stimulating proliferation and growth. Rapid responses of skeletal muscle have been observed for 1,25-$(OH)_2$-D_3: Ca^{+2} uptake through voltage-dependent channels and Ca^{2+} release by intracellular stores. Thus, it is not surprising that muscular weakness has been noted in subjects with low plasma 25-OH-D_3 levels, and plasma 25-OH-D_3 predicts the rate of strength recovery after exercise-induced muscular injury.[113] This effect may involve activation of mTOR signaling[114] in muscle, which has been shown in 25-OH-D_3-stimulated muscle growth in the chicken.[115] Patients with rickets experience weakness, hypotonia[116], and atrophy. Emerging evidence suggests that, in vitamin D-adequate individuals, skeletal muscle may serve as a storage site for 25-OH-D_3. By supporting muscular function, vitamin D reduces risks of falling and, consequently, fractures.[117]

Immune Function

Studies have shown that 1,25-$(OH)_2$-D_3 supports normal function of both the innate and adaptive components of the immune system as well as the inflammatory cascade.[118] VDRs have been identified in most immune cells, including most antigen-presenting cells (e.g., macrophages, dendritic cells) and $CD4^+$, $CD8^+$ T-lymphocytes. Many of these cells also express the 25-OH-D_3 1- and -24-hydroxylases and are, thus, capable of producing and catabolizing 1,25-$(OH)_2$-D_3. Dendritic cells also express the vitamin D_3-25-hydroxylase. The emerging picture is one of 1,25-$(OH)_2$-D_3, obtained either from local immune cell metabolism or by transport from the kidney, modulating the phenotype and function of dendritic cells by downregulating the expression of costimulatory molecules (CD40, CD80, CD86) and the cytokine IL-12, while upregulating the expression of IL-10. This limits Th1 cell development, promotes the activity of $CD4^+$ suppressor T-cells, and promotes recruitment of regulatory T-cells (Tregs). VDR and 1,25-$(OH)_2$-D_3 also modulate the differentiation of B cell precursors and limit the proinflammatory action of the nuclear factor kappa B (NFκB) pathway. Inflammation is also affected by a 1,25-$(OH)_2$-D_3-induced protein,[119] which forms inflammasome complexes[120] that regulate IL-1β processing. Human trials have found inverse associations of inflammatory markers and plasma 25-OH-D_3 levels.[121] Vitamin D supplementation has been found to upregulate genes in immune response and inflammation pathways in the colon in ways that are mitigated by supplemental dietary calcium.[122]

Macrophages, dendritic cells, and T cells can regulate the production and turnover of 1,25-$(OH)_2$-D_3, which is required for the production of antimicrobial peptides, the **cathelicidins**. They comprise a family of peptides of 12–80 amino acid residues; most have 23–37 residues that form amphoteric α-helices enabling them to enter and disrupt bacterial and fungal membranes and viral envelopes. They also affect the permeability of cell membranes and contribute to the endothelial barrier to pathogens by increasing cell stiffness at sites of infection. They have been shown to reduce susceptibility to several pathogens[123] in both animals and humans. Cathelicidin induction involves initial activation of macrophage Toll-like receptors,[124] which upregulates expression of both VDR and the 25-OH-D_3 1-hydroxylase (CYP27B1), leading to induction of downstream targets of VDR, including cathelicidin which has a VDRE in its promotor. Cathelicidins can also synergize the proinflammatory effects of endogenous mediators. Promotion of cathelicidin secretion in the lungs appears to a means whereby locally generated 1,25-$(OH)_2$-D_3 protects from respiratory infection.

Vitamin D affects the regulation of autoimmunity. This may involve VDR and 1,25-$(OH)_2$-D_3-suppressing

112. This chain of proteins transduces signals from cell surface receptors to DNA in the nucleus. Originally called ERK (extracellular signal-related kinases), it is now call MAP for a prominent constituents, MAPK (mitogen-activated protein kinases).
113. Barker, T., Henricksen, V.T., Martins, T.R., et al., 2013. Nutrients **5**, 1253–1275.
114. mTOR (mammalian target of rapamycin) is a serine/threonine protein kinase that serves as a master regulator of cell growth and proliferation.
115. Vignale, K., Greene, E.S., Caldas, J.V., et al., 2015. J. Nutr. 145, 855–863.
116. Low muscle tone.
117. Falls cause >90% of hip fractures, which increase with age from 30% in subjects >65 years to 50% in subjects >80 years.
118. See reviews: Cantorna, M.T., Snydeer, L., Lin, Y.D. et al., 2015. Nutrients 7, 3011–3021; Baeke, F., Takiishi, T., Korf, H., et al., 2010. Curr. Opin. Pharmacol. 10, 482–496.
119. vitamin D-upregulated protein 1 (VDUP1), a multifunctional, stress-induced protein of the arrestin family of proteins.
120. i.e., Multiprotein oligomers in myeloid cells that promote maturation of inflammatory cytokines.
121. Zanetti, M., Harris, S.S., Dawson-Hughes, B., 2014. Nutr. Rev. 72, 95–98.
122. Protiva, P., Pendyala, S., Nelson, C., et al., 2016. Am. J. Clin. Nutr. 103, 1224–1231.
123. e.g., *Mycobacterium tuberculosis*, Influenza A, *Pseudomonas aeruginosa*, *Bordetella bronchiseptica*, *Aggregatibacter actinomycetemcomitans*, *Candida albicans*.
124. These are membrane-spanning receptors expressed in macrophages and dendritic cells. Their name comes from the toll gene, and the 1985 comment of Christiane Nüsslein-Volhard, "Das ist ja toll!" (That's amazing!) who shared a 1995 Nobel Prize for that discovery.

inflammation and promoting the development NKT and CD8αα T cells, as well as direct effects of cathelicidins.[125]

- **Asthma**. Several, but not all, studies have found low 25-OH-D_3 levels to be associated with increased risk of asthma.[126] It has been suggested that locally produced 1,25-$(OH)_2$-D_3 may regulate sensitivity of lymphocytes and monocytes to glucocorticoids by influencing the expression of the immune response protein FOXP3 in Tregs. This is supported by in vitro studies that have shown D_3 treatment to reduce proliferation of airway smooth muscle cells, a common feature in asthma. That VDR-knockout mice do not show inflammatory responses to experimentally induced asthma suggests that 1,25-$(OH)_2$-D_3 may be immunosuppressive, inhibiting signs/symptoms of T-helper 1 (Th1) cell-driven autoimmune disease.
- **Rheumatoid arthritis (RA)**. One study observed an inverse relationship of vitamin D intake, particularly from supplements, and risk of developing RA.[127] D_3 treatment has been found to prevent experimentally induced RA in animal models, and VDR is known to be expressed by articular chondrocytes in osteoarthritic cartilage.
- **Multiple sclerosis (MS)**. It has been long recognized that MS, which is characterized by immune attacks on the myelin sheaths of nerves, is more prevalent in northern latitudes than in the tropics, being inversely correlated to the numbers of hours of annual or winter sunlight. MS patients tend to have lower plasma 25-OH-D_3 levels than controls. Two studies found D_3 supplements to reduce MS risk by as much as 40%.[128] One study with MS subjects found D_3 supplementation to increase circulating levels of the antiinflammatory cytokine transforming growth factor 1 (TGF-1), suggesting potential for alleviating symptoms.
- **Inflammatory bowel disease**. Patients with Crohn's disease or ulcerative colitis have been found to have lower plasma 25-OH-D_3 levels than unaffected controls. Deficiency of VDR or vitamin D deprivation has been found to exacerbate the development of diarrhea and cachexia in experimental IBD with IL-10 knockout mice; VDR signaling is needed to prevent the proliferation of pathogenic CD8+ T cells.[129]
- **Insulin-dependent diabetes (type 1 diabetes, T1D)**. Vitamin D may play a role in reducing risk of developing T1D, which results from the T-cell dependent destruction of insulin-producing pancreatic β cells by cytokines and free radicals from inflammatory infiltrates. This hypothesis follows from the findings that 1,25-$(OH)_2$-D_3 inhibits production of IL-12, thus suppressing the activity of IL-12-dependent Th1 cells in activating cytotoxic $CD8^+$ lymphocytes and macrophages. T1D prevalence is positively associated with latitude and negatively associated with hours of sunlight. High doses of 1,25-$(OH)_2$-D_3 have been shown to arrest diabetes in the nonobese diabetic mouse.[130] One prospective trial found D_3 doses of 50 µg (2 IU)/day to reduce T1D risk;[131] another detected no benefits using lower doses (<10 µg [0.4 IU]/day).[132] A large, multicenter case–control study[133] and a cohort study[134] both found vitamin D supplementation in infancy to reduce T1D development later in life.
- **Atopic dermatitis**. Low-plasma 25-OH-D_3 levels are associated with increased sensitivity to common food allergins (especially milk and wheat) in infants with atopic dermatitis.

Antioxidant Regulation

Vitamin D appears to function in the regulation of cellular antioxidant balance. Its genomic functions support antioxidant protection. Liganded VDR signals expression of different factors active in scavenging reactive oxygen species (ROS) produced intracellularly;[135] it also inhibits expression of ROS-generating factors.[136] Vitamin D can also have prooxidative effects through both genomic action in signaling expression of NADPH oxidase, and through nongenomic promotion of Ca^{2+} influx into cells, which stimulates the production of reactive nitrogen species. These actions can increase drug sensitivity of cancer cells but also may promote vascular calcification and fat accumulation in the adipocyte (major sources of ROS).

125. Marques, C.D.L., Danta, A.T., Fragoso, T.S., et al., 2010. Braz. J. Rhematol. 50, 67–80.
126. Kerley, C.P., Elnazir, B., Faul, J., et al., 2015. Pulm. Pharmacol. Ther. 32, 75–92.
127. Merlino, L.A., Curtis, J., Mikuls, T.R., et al., 2004. Arthritis Rheumatol. 50, 72–77.
128. Munger, K.L., Zhang, S.M., O'Reilly, E., et al., 2004. Neurology 62, 60–65; Munger, K.L., Levin, L.I., Hollis, B.W., et al., 2006. J. Am. Med. Assoc. 296, 2832–2838.
129. Chen, J., Bruce, D., Cantorna, M.T., 2014. BMC Immunol. 15, 6–17.
130. A 28% reduction in the autoimmune NOD mouse was reported (Mathieu, C., Waer, M., Laureys, J., et al., 1994. Diabetologia 37, 552–558), but no effect was observed in the BB rat model (Mathieu, 1997. In: Feldman, D., Glorieux, F., Pike, J. (Eds.), Vitamin D and Diabetes. Academic Press, San Diego, CA, pp. 1183–1196).
131. Hyppönen, E., Läärä, E., Reunanen, A., et al., 2001. Lancet 358, 1500–1503.
132. Stene, L.C., Ulriksen, J., Magnus, P., Joner, G., 2003. Am. J. Clin. Nutr. 78, 1128–1134.
133. EURODIAB substudy 2 study group, 1999. Diabetologia 42, 51–54.
134. Hyppönen, E., Läärä, E., Reunanen, A., et al., 2001. Lancet 358, 1500–1503.
135. Superoxide dismutase, glutathione peroxidase, catalase, thioredoxin reductase, nuclear regulatory factor 2.
136. e.g., Inducible nitric oxide synthase (iNOS), nitric oxides (NOX).

Cardiovascular Health

Vitamin D may contribute to cardiovascular health. Epidemiological studies have shown that low-circulating 25-OH-D_3 levels, as well as factors known to affect vitamin D status (latitude, altitude, season), are associated with the prevalence of coronary risk factors, with long-term risk of stroke, and with cardiovascular disease (CVD) mortality. Severe vitamin D deficiency is associated with in-hospital cardiovascular mortality in patients with acute coronary disease. These effects may be the result of vitamin D function in the proliferation of vascular smooth muscle, Ca^{2+} homeostasis, regulation of inflammation, and regulation of the renin–angiotensin system.

Neurologic Function

A role of vitamin D in brain development and function was first indicated by the finding of 25-OH-D_3 in cerebrospinal fluid and of expression of both VDR and 25-OH-D_3-1-hydroxylase in neurons and glial cells.[137] Maternal deprivation of vitamin D can alter in brain morphology, stem cell proliferation, and expression of neurotrophic factors in rat pups. Low-vitamin D status in infants has been linked to abnormalities manifest in adulthood, including sensitivity to psychosis-inducing drugs, schizophrenia, and Parkinson's disease; and adequate vitamin D status has been associated with reduced risk of cognitive decline and dementia in older adults.[138] These effects are thought to reflect needs for vitamin D in maintaining dopaminergic neurotransmitter systems, regulating intracellular Ca^{2+} homeostasis, and prevention of oxidative damage.[139] It has been suggested that vitamin D may act as a neurosteroid to protect the brain from secondary insults.

Skin Health

Vitamin D has a paracrine function in the skin. Keratinocytes express 25-(OH)2-D3-1-hydroxylase. Therefore, they cannot only produce D_3 with solar exposure, they can also metabolize it to 1,25-$(OH)_2$-D_3, which has been shown to confer photoprotection by minimizing UV-induced DNA damage, inflammation, and cell death in that organ. Studies have shown that 1,25-$(OH)_2$-D_3 can inhibit proliferation and induce terminal differentiation of cultured keratinocytes. These effects appear to be mediated by VDRs, which are expressed throughout the epidermis as well as in hair follicles and skin immune cells.[140] Among the gene products induced by VDR activation in the skin is cathelicidin,[141] which functions both in the direct killing of pathogens as well as a host response involving cytokine release, inflammation, and cellular immune response. Cathelicidin abnormalities result in skin diseases including atopic dermatitis, rosacea, and psoriasis. Decreased levels of cathelicidin have been observed in atopic dermatitis, and abnormal processing of the cathelicidin peptide has been found to be involved in the inflammatory and vascular responses in rosacea. Patients with hereditary vitamin D-resistant rickets have VDR mutations and show alopecia; however, the basis of that symptom is unclear, as alopecia is not seen in subjects with simple vitamin D deficiency or loss of CYP27B1.

Gut Microbiome

Recent studies have demonstrated a role of vitamin D in regulating the gut microbiome. That vitamin D may protect against enteric infection was indicated by the findings of higher circulating 25-OH-D_3 levels associated with reduced risk to *Clostridium difficile* infection in patients with inflammatory bowel syndromes, and more moderate symptoms in patients with Crohn's disease. Genetic deletion of CYP27B1 (hence, the ability to produce 1,25-$(OH)_2$-D_3) or VDR in the mouse has been shown to increase numbers of *Bacteroidetes* and *Proteobacteria* species at the expense of beneficial members of the *Firmicutes* and *Deferribacteria* phyla. This shift in microbial composition appears to amplify the host response to injury, as was manifest as increased risk to colitis induced by dextran sodium sulfate.[142] The basis of the action of vitamin D in supporting a healthy gut microbiome may be the gut epithelial signaling of VDR, which has been shown to protect the mucosal barrier by inhibiting inflammation-induced epithelial cell apoptosis and resisting the effects of an inflammation-induced VDR-targeting microRNA[143] With emerging evidence of signaling between the microbiome and the gut, the body's largest immune organ, it is possible that effects

137. Interestingly, a few cells appear to express the 1-hydroxylase but not VDR: macrocellular cells in the nucleus basalis of Meynert, and Purkinje cells of the cerebellum.

138. Eyles, D.W., Burne, T.H.J., McGrath, J.J., 2013. Front. Neuroendocrinol. 34, 47–64.

139. Cui, X., Gooch, H., Groves, N.J., et al., 2015. J. Steroid Biochem. Mol. Biol. 148, 305–309; Wrzosek, M., Kakaszkiewicz, J., Wrzosek, M., et al., 2013. Pharmacol. Rep. 65, 271–278.

140. VDRs are expressed by the majority of Langerhans cells, macrophages, and T lymphocytes in skin.

141. The cathelicidins are among 20 or more antimicrobial proteins, including the β-definsins, calprotectin and adrenomedullin. Only the cathelicidins are known to be modulated by vitamin D (*see* Schauber, J., Gallow, R.L., 2008. Exp. Dermatol. 17, 633–639).

142. Ooi, J.H., Li, Y., Rogers, C.J., et al., 2014. J. Nutr. 143, 1679–1686.

143. Li, Y.C., Chen, Y., Du, J., 2015. J. Steroid Biochem. Mol. Biol. 148, 179–183.

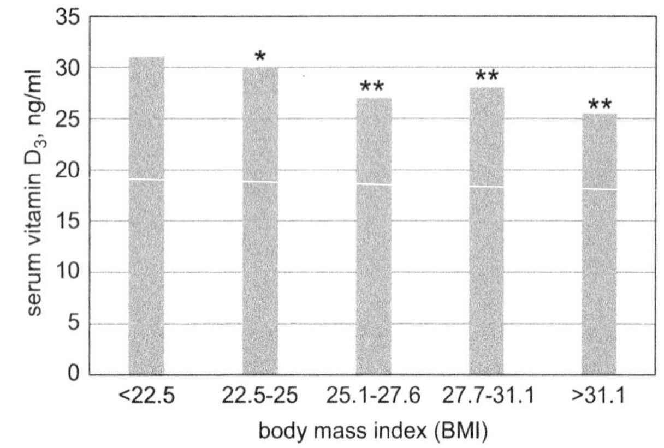

FIGURE 7.17 Inverse relationship of vitamin D status and body mass index apparent in National Health and Nutrition Examination Survey (NHANES) III subjects. Comparisons with lowest quintile: $*p < .05$, $**p < 0.01$. *After Black, P.N., Scrugg, R., 2005. Chest 128, 3792–3798.*

of vitamin D on gut microbial homeostasis may affect immune function of the host.

Adipose Function

Vitamin partitions into bulk lipid depots in adipose tissue where 1,25-$(OH)_2$-D_3 can be produced and functions in the regulation of cytokine production. Sequestration of 25-OH-D_3 is dependent on the degree of adiposity, which is manifest as a reduction of circulating 25-OH-D_3 level (Fig. 7.17). It has been suggested that 25-OH-D_3 evolved as a UVB indicator for species with low-dietary vitamin D intakes; for them, shortening autumnal day length would be detected by declining plasma 25-OH-D_3 levels, which would stimulate accumulation of fat and induce a winter metabolism, which in contemporary health would be the metabolic syndrome. Therefore, it has been proposed that vitamin D deficiency may contribute to obesity[144] and that, by stabilizing 25-OH-D_3 levels, supplemental D_3 prevent obesity. This is supported by findings that 1,25-$(OH)_2$-D_3 can induce apoptosis in adipocytes; that effect appears to involve VDR competing with RXR for binding PPARγ (peroxisome-proliferator activation receptor) and suppressing fat deposition by promoting expression of the steroid-metabolizing enzyme 11β-hydroxysteroid hydroxylase. Studies have shown that serum 25-OH-D_3 levels are inversely correlated with BMI[145], body fat mass and insulin resistance, and directly associated with weight loss from caloric restriction.[146] That vitamin D functions in supporting insulin sensitivity is indicated by the findings that 1,25-$(OH)_2$-D_3 upregulates the glucose transporter 4 (GLUT4) translocation and insulin signaling in adipocytes,[147] and that VDR polymorphisms have been associated with increased visceral adipose tisssue mass, increased plasma adipokine concentrations, high fasting glucose levels and elevated risk to noninsulin-dependent diabetes (type 2 diabetes, T2D).[148] While a few clinical trials have found weight loss to improve 25-OH-D_3 levels and D_3 supplements to improve insulin sensitivity, such effects have not been observed in most trials. As such, the efficacy of vitamin D in preventing T2D may depend on an individual's degree of adiposity as well as chronic inflammatory status.

Pregnancy

Circulating levels of 1,25-$(OH)_2$-D_3 increase by two- to three-fold in the first trimester of pregnancy, apparently due to the expression of the 1-hydroxylase and suppression of the 24-hydroxylase in placental and decidual cells, which also express VDR. It has been suggested that locally produced 1,25-$(OH)_2$-D_3 functions in the placenta to suppress Th1-dependent immunity to facilitate immune tolerance to implantation, affecting successful fetal maintenance; it also appears to support antimicrobial and antiinflammatory functions. These functions are thought to underlie the observation of lower levels of HIV transmission by vitamin D-adequate mothers to their fetuses. That vitamin D functions in normal fetal development is indicated by meta-analyses that found low-maternal serum 25-OH-D_3 levels to increase risks of preeclampsia, gestational diabetes, small-for-gestational age births, and bacterial vaginosis;[149] and studies that have found maternal vitamin D status to predict growth and later bone mass of children.

8. BIOMARKERS OF VITAMIN D STATUS

The most informative indicator of vitamin D status is the concentration of 25-OH-D_3 in serum/plasma.[150] Those levels are normally in the range of 10–40 ng/mL (25–125 nM). Serum PTH levels can also be a useful biomarker of vitamin D status. Maximal PTH levels are seen at plasma 25-OH-D_3 levels of c.100 nM; PTH expression is downregulated at increasing 25-OH-D_3 levels. PTH levels less than 45–65 nM are

144. Foss, Y.J., 2009. Med. Hyp. 72, 314–321.

145. Body mass index; BMI = weight (kg)/height (m^2), or = [weight (lb)/height (in^2)] × 705.

146. Arunabh, S., Pollack, S., Yeh, J., Aloia, J.F., 2003. J. Clin. Endocrinol. Metab. 88, 157–161; Parikh, J., Edelman, G.I., Uwaifo, R.J., et al., 2004. J. Clin. Endocrinol. Metab. 89, 1196–1199.

147. Manna, P., Jain, S.K., 2012. J. Biol. Chem. 287, 42324–42332.

148. Hitman, G.A., Mannan, N., McDermott, M.F., et al., 1998. Diabetes 47, 688–690; Oh, J.-Y., Barrett-Connor, E., 2002. Metabolism 51, 356–359; Ortlepp, J.R., Metrikat, J., Albrecht, M., et al., 2003. Diabet. Med. 20, 451–454; Khan, R.J., Riestra, P., Gebreab, S.Y., 2016. J. Nutr. 146, 1476–1482.

149. Harvey, N.C., Holroyd, C., Ntani, G., et al., 2014. Health Technol. Assess. 18, 1–190; Aghajafari, F., Nagulesapillai, T., Ronksley, P.E., et al., 2013. Br. Med. J. 346, 1169–1183.

150. Neither 1,25-$(OH)_2$-D_3 or DBP are informative as biomarkers of vitamin D status. Circulating levels of 1,25-$(OH)_2$-D_3 are tightly regulated at about 40 pg/mL (100 nM), reflecting vitamin D status only under conditions of severe deficiency and not reflecting the local production of 1,25-$(OH)_2$-D_3 by many extrarenal tissues. Most (>95%) of DBP in the plasma is not bound to vitamers D; it is present in a 100-fold excess compared to 25-OH-D_3.

TABLE 7.17 Criteria of Vitamin D Status

25-OH-D_3 Level		Status
nM	ng/ml	
<50	<20	Deficient
50–75	20–30	Insufficient
76–250	31–100	Adequate
>250	>100	Excess (risk of hypervitaminosis)

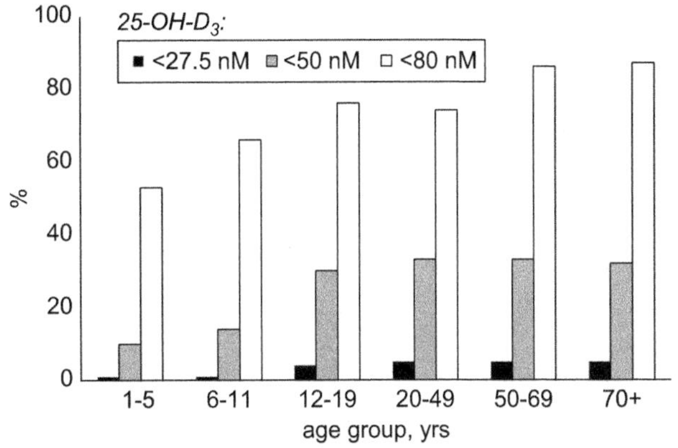

FIGURE 7.18 Prevalence of low-serum 25-OH-D3 levels in Americans, from National Health and Nutrition Examination Survey (NHANES) 2000–2004. *Yetley, E.A., 2008. Am. J. Clin. Nutr. 88, 558S–564S.*

associated with secondary hyperparathyroidism. A network analysis of 12 VDR target genes and 12 clinical and biochemical parameters related to vitamin D status showed PTH level to correlate with most of the other endpoints, and all of the target genes to correlate with serum 25-OH-D_3 level.

The IOM considered 50 nM adequate for bone health in adults and children,[151] although an expert consensus considered the level of 70–80 nM as optimal (Table 7.17, Fig. 7.18).[152] An analysis of multiple endpoints related to bone and dental health, lower extremity function, risks of falls, fractures, and cancer found the optimal serum 25-OH-D_3 level to be 90–100 nM (Fig. 7.19). Such levels require regular daily vitamin D intakes of greater than 1000 IU (25 mg).[153] The maintenance of such serum 25-OH-D_3

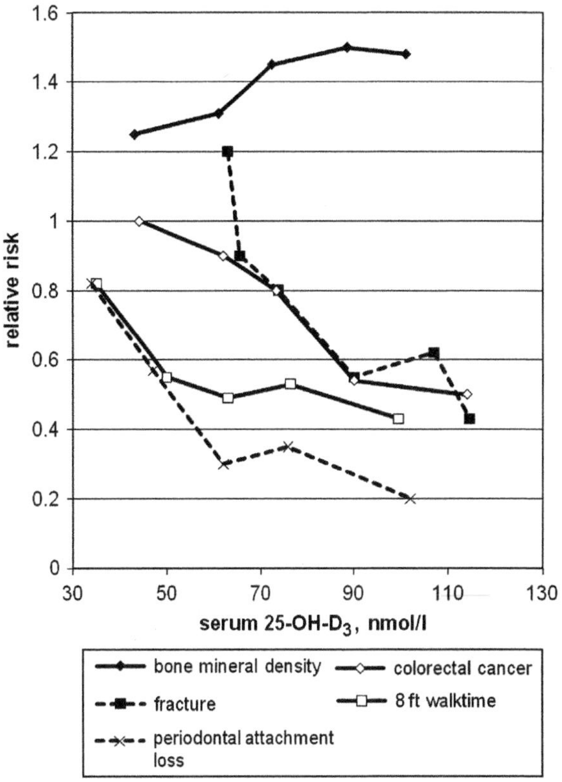

FIGURE 7.19 Relationship of vitamin D status and multiple health risks. *After Bischoff-Ferrari, H.A., Giovannuci, E., Willett, W.C., et al., 2006. Am. J. Clin. Nutr. 84, 18–28.*

levels requires the use of solar radiation, vitamin D supplements, and/or D_2-fortified foods.

Plasma 25-OH-D_3 measurement can be affected by factors in addition to those related to biogenesis and intake of vitamin D. These include polymorphisms of 7-dehydrocholesterol reductase, DBP, and CYP2R1; in some cases their effects can be great as that of deprivation of the vitamin.[154] Pregnancy appears to suppress serum 25-OH-D_3 levels; a study in Boston found that 76% of new mothers (and 81% of their newborns) had serum 25-OH-D_3 levels <20 ng/ml.[155] Plasma 25-OH-D_3 levels are also negatively associated with BMI and percentage body fat, low magnesium status, and recent inflammatory insult.

Recent studies have suggested decline in the vitamin D status of Americans in recent decades. Comparisons of serum 25-OH-D_3 levels measured in sera of participants in the National Health and Examination Surveys (NHANES) conducted in 1988–1994 and 2000–2004 showed an apparent 7.1 nM (c.9%) decline among men. It is likely that methodological changes may have contributed to that difference, as the

151. Institute of Medicine, 2011. Dietary Reference Intakes: Calcium, Vitamin D. National Academy Press, Washington, DC, 1115 pp.

152. Dawson-Hughes, B., Heaney, R.P., Holick, M.F., et al., 2005. Osteoporosis Int. 16, 713–716.

153. An analysis indicated that vitamin D intakes of at least 1000 μg/day are required to bring 50% of healthy American adults above the serum 25-OH-D_3 level of 75 nM (Bischoff-Ferrari, H. A., Giovannucci, E., Willett, W.C., et al., 2006. Am. J. Clin. Nutr. 84, 18–28). Current recommended intakes for vitamin D are 200 and 600 IU/day in younger and older adults, respectively.

154. Sinotte, M., Diorio, C., Berube, S., et al., 2009. Am. J. Clin. Nutr. 89, 634–640; Wang, T.J., Zhang, F., Richards, J.B., et al., 2010. Lancet 376, 180–188.

155. Lee, J.M., Smith, J.R., Philipp, B.L., et al., 2007. Clin. Pediatr. 46, 42–44.

TABLE 7.18 Recommended Vitamin D Intakes

Age–Sex	US RDA[a], µg/day	Age–Sex	FAO/WHO RNI[b], µg/day
0–11 mos.	10	0–11 mos.	5
1–70 years	15	1–18 years	5
>70 years	20	19->65 years, females	7.5
		Males	10
Pregnancy	15	Pregnancy	–
Lactation	15	Lactation	–

[a]*Recommended Dietary Intakes; Food and Nutrition Board, 2010. Dietary Reference Intakes: Calcium, Vitamin D. National Academy Press, Washington, DC, 1115 pp.*
[b]*Recommended Nutrient Intakes, Joint WHO/FAO Expert Consultation, 2001. Human Vitamin and Mineral Requirements. Food and Agricultural Org., Rome, 286 pp.*

manufacturer's reformulation of the assay (an enzyme-linked immunosorbent assay, ELISA) used in those surveys produced 25-OH-D_3 results that 1% lower than the previous method.[156] That the ELISA overestimates 25-OH-D_3 and underestimated 25-OH-D_2 can lead to misclassification of vitamin D status in subjects using supplements containing D_2.[157]

9. VITAMIN D DEFICIENCY

Causes of Vitamin D Deficiency

Vitamin D deficiency can result from inadequate irradiation of the skin, from insufficient intake from the diet, or from impairments in the metabolic activation of the vitamin. Although sunlight can provide the means of biosynthesis of D_3, it is well documented that many people, particularly those in extreme latitudes during the winter months, do not receive sufficient solar irradiation to support adequate vitamin D status. Even people in sunnier climates may not produce adequate D_3 if their lifestyles or health status keep them indoors, or if factors such as air pollution or clothing reduce their exposure to sunlight. Most people, therefore, show strong seasonal fluctuations in plasma 25-OH-D_3 concentration; for some, this can be associated with considerable periods of suboptimal vitamin D status if not corrected by dietary sources of the vitamin. Until the practice of vitamin D fortification of foods became widespread, at least in technologically developed countries, it was difficult to obtain adequate vitamin D from the diet (Table 7.18), as most foods contain only minuscule amounts.[158] Therefore, vitamin D deficiency can have primary (privational) and secondary (nonprivational) causes:

- **Primary causes** involve inadequate vitamin D supply
  - inadequate exposure to sunlight
  - insufficient consumption of foods containing vitamin D
- **Secondary causes** relate to impaired absorption, metabolism, or nuclear binding of the vitamin
  - gastrointestinal diseases (e.g., small bowel disease, gastrectomy, pancreatitis)—involving malabsorption of the vitamin
  - hepatic diseases (biliary cirrhosis, hepatitis)—that reduce activity of the 25-hydroxylase
  - renal diseases—that reduce the activity of the 1-hydroxylase (nephritis, renal failure) or cause loss of 25-OH-D_3 into the urine (nephrotic syndrome)[159]
  - exposure to drugs (e.g., the anticonvulsants phenobarbital, diphenylhydantoin)—that induce the catabolism of both 25-OH-D_3 and 1,25-$(OH)_2$-D_3[160]
  - **hypoparathyroidism**—impairs the ability to respond to hypocalcemia[161] by increasing activity of the 1-hydroxylase
  - *cyp27b1* mutations—resulting in loss of 25-OH-D_3 1-hydroxylase activity in **vitamin D-dependent rickets type I**[162]
  - VDR mutations—impair transcription of vitamin D-regulated genes in **vitamin D-dependent rickets type II**[163]
  - PTH resistance—results in **pseudohypoparathyroidism**, i.e., hypocalcemia without compensating renal retention or bone mobilization of Ca, despite normal PTH secretion[164]
  - **vitamin D resistance**[165]—impairments in both enteric absorption and renal tubular reabsorption of phosphate, hypersensitivity to PTH, and reduced 1-hydroxylation of 25-OH-D_3[166]

156. Looker, A.C., Pfeiffer, C.M., Lacher, D.A., et al., 2008. Am. J. Clin. Nutr. 88, 1519–1527.
157. Nguyen, V.T.Q., Li, X., Castellanos, K.J., et al., 2014. J. AOAC Int. 97, 1048–1055.
158. Eggs are the notable exception. Even cows' milk and human milk contain only very small amounts of vitamin D.
159. Renal tubular degeneration characterized by edema, albuminuria, hypoalbuminemia, and, usually, hypercholesterolemia.
160. Vitamin D supplements (up to 4000 IU/day) are recommended to prevent rickets in children undergoing long-term anticonvulsant therapy.
161. In affected individuals, this leads to hyperphosphatemia. These conditions typically respond to treatment with 1,25-$(OH)_2$-D_3 or high levels of vitamin D_3.
162. This condition can be managed using low doses of 1,25-$(OH)_2$-D_3 or 1-α-OH-D_3.
163. This condition can be managed using high doses of 1,25-$(OH)_2$-D_3 or 1-α-OH-D_3.
164. This condition can be managed using low doses (0.25–3 mg/day) of 1,25-$(OH)_2$-D_3 or 1-α-OH-D_3.
165. Also called **hypophosphatemic rickets/osteomalacia** or **phosphate diabetes**.
166. The condition can be managed using phosphate plus D_3 (25,000–50,000 IU/day) or low doses of 1,25-$(OH)_2$-D_3 or 1-α-OH-D_3.

TABLE 7.19 Signs of Vitamin D Deficiency

Organ System	Rickets	Osteomalacia	Osteoporosis
General	Occurs in young children; loss of apetite, retarded growth	Occurs in older children and adults	Most common in postmenopausal women and older men
Bone/teeth	Failed mineralization, deformation, swollen joints, paint, tenderness, delayed tooth eruption	Demineralized formed bone (e.g., spine, shoulder, ribs, pelvis); fractures (wrist, pelvis); bone pain, tenderness	Loss of trabecular bone with retained structure, high fracture incidence
Skin	Not affected	Not affected	Not affected
Muscle	Weakness, myotonia, pain	Weakness, sarcopenia, pain	Not affected
Vital organs	Not affected	Not affected	Not affected
Nervous	Tetany, ataxia	Not affected	Not affected
Reproductive	Birds: thin eggshell	Low sperm motility and number	Not affected
Ocular	Not affected	Not affected	

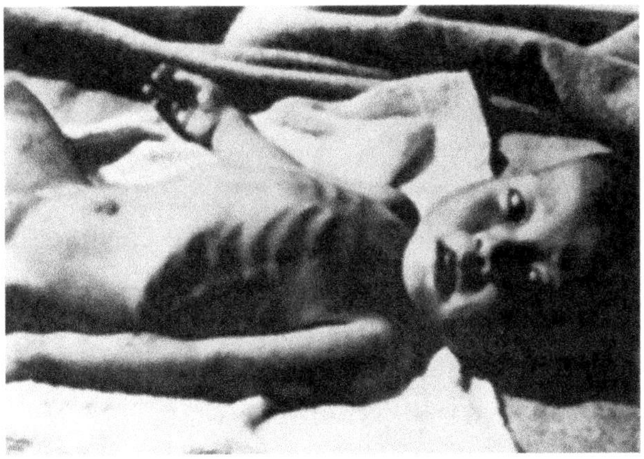

FIGURE 7.20 Rachitic child with beaded ribs.

Signs of Vitamin D Deficiency

Vitamin D deficiency affects several systems, most prominently skeletal and neuromuscular (Table 7.19).

Vitamin D Deficiency in Humans

The most obvious signs of vitamin D deficiency are failures in mineralization of bone; but vitamin D deficiency also increases risks to CVD, respiratory disease, and infection.

Rickets

Rickets first appears in 6--24-month- old children, but can manifest at any time until the closure of the bones' epiphyseal growth plates. It is characterized by impaired mineralization of the growing bones with accompanying bone pain, muscular tenderness, and hypocalcemic tetany. Tooth eruption may be delayed, the fontanelle may close late, and knees and wrists may appear swollen. Affected children develop deformations of their softened, weight-bearing bones, particularly those of the rib cage and legs, characteristic signs being "beaded ribs" (Fig. 7.20), **bow-leg (Genu varum)**, **knock knee**, and **sabre tibia** (forward curvature of the tibia) (Figs. 7.21 and 7.22), which occur in nearly half of cases. Radiography reveals enlarged epiphyseal growth plates resulting from their failure to mineralize and continue growth. Rickets is most frequently associated with low dietary intakes of Ca, as in the lack of access to or avoidance of milk products.[167]

Osteomalacia

Osteomalacia occurs in older children and adults with formed bones whose epiphyseal closure has rendered that region of the bone unaffected by vitamin D deficiency. The signs and symptoms of osteomalacia are more generalized than those of rickets, e.g., muscular weakness and bone tenderness and pain, particularly in the spine, shoulder, ribs, or pelvis. Lesions involve the failure to mineralize bone matrix, which continues to be synthesized by functional osteoblasts; therefore, the condition is characterized by an increase in the ratio of nonmineralized bone to mineralized bone. Radiographic examination reveals abnormally low

167. Despite the use of vitamin D-fortified foods, rickets has reemerged as a public health problem, being reported in at least 22 countries. Cases in Africa and South Asia appear to be caused primarily by deficiencies of calcium, which some have suggested may increase the catabolism of vitamin D. Other cases, however, appear to be due to insufficient vitamin D. These include cases in the United States, 83% of which were described as African-American or black, and 96% of which were breastfed, with only 5% vitamin D supplementation during breastfeeding.

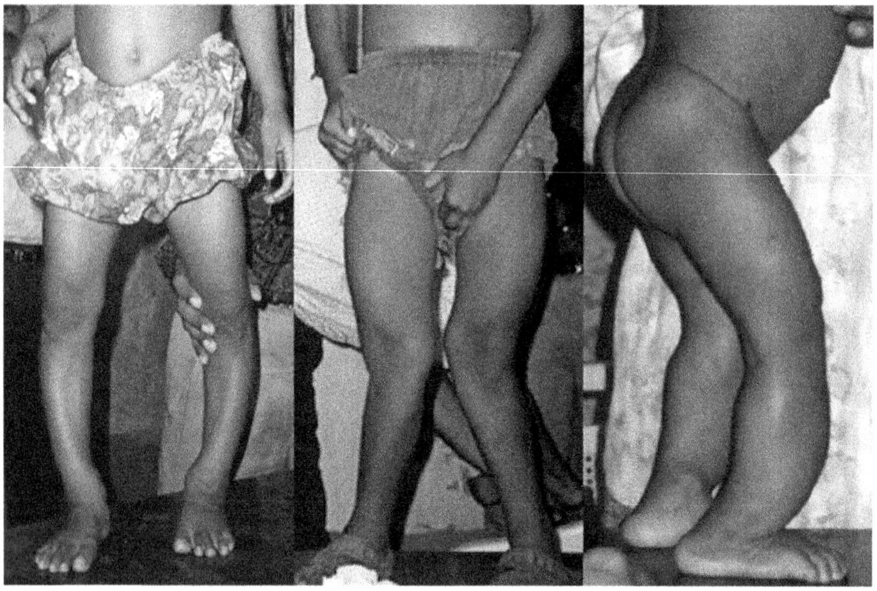

FIGURE 7.21 Rachitic children with genu varum (left), genu valgum (center) and sabre tibia (right).

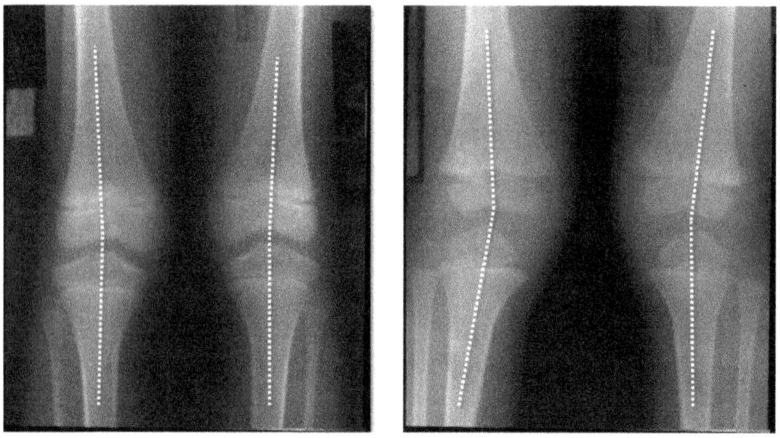

FIGURE 7.22 Knee radiographs of normal (left) and rachitic (right) children. Note the lateral displacement of the lower legs (165° in the child with genu valgum versus 171° in the unaffected child).

bone density (osteopenia) and the presence of pseudofractures, especially in the spine, femur, and humerus. Patients with osteomalacia are at increased risk of fractures of all types, but particularly those of the wrist and pelvis, which are typically caused by falls. Patients with osteomalacia frequently show myopathy primarily of type II muscle fibers (the first to be recruited to avoid falling) resembling the sarcopenia[168] of aging, which is associated with reduction in muscle VDR.

Osteoporosis

Although it is sometime confused with osteomalacia, osteoporosis is a very different disease, being characterized by decreased bone mass with retention of normal histological appearance. Its etiology (loss of trabecular bone with retention of bone structure) is not fully understood; it is a multifactorial disease associated with aging and involving impaired vitamin D metabolism and/or function associated with low or decreasing estrogen levels. It is the most common bone disease of postmenopausal women and also occurs in older men[169] (e.g., nonambulatory geriatrics, postmenopausal women) and in people receiving chronic steroid therapy. In women, bone loss generally begins in the third and fourth decades and accelerates after menopause;

168. Loss of muscle.

169. Osteoporosis affects one-third of women 60–70 years of age, and two-thirds over 80 years, for a total of some 25 million Americans, costing the US economy some $13–18 B per year.

in men, bone loss begins about a decade later. These groups show high incidences of fractures, particularly of the vertebrae, hip, distal radius, and proximal femur. Osteoporosis has two types:

- **Type I osteoporosis**—characterized by distal radial and vertebral fractures; occurring primarily in women of 50–65 years and probably related to postmenopausal decreases in the amount of calcified bone at the fracture site.
- **Type II osteoporosis**—characterized by fractures of the hip, proximal humerus, and pelvis, i.e., where there has been loss of both cortical and trabecular bone; occurring primarily among individuals over 70 years. In women, osteoporosis is characterized by rapid loss of bone (e.g., 0.5–1.5% per year) in the first 5–7 years after menopause.[170] The increased skeletal fragility observed in osteoporosis appears not to be due solely to reductions in bone mass but also involves changes in skeletal architecture and bone remodeling (e.g., losses of trabecular connectivity as well as inefficient and incomplete microdamage repair). Affected individuals show abnormally low circulating levels of 1,25-$(OH)_2$-D_3, suggesting that estrogen loss may impair the renal 1-hydroxylation step, i.e., that the disease may involve a bihormonal deficiency. Studies of the use of various vitamers D in the treatment of osteoporotic patients, most of which have involved low numbers of subjects, have produced inconsistent results. Results of the Nurses' Health Study showed that adequate vitamin D intake (12.5 μg/day) was associated with a 37% lower risk of osteoporotic hip fracture compared to lower intakes of the vitamin.[171]

Musculoskeletal Pain

Deep pain is common among rickets and osteoporosis patients. Some reports have indicated persistent, nonspecific musculoskeletal pain among asymptomatic adults with low circulating levels of 25-OH-D_3; however, a systematic review of published data found no convincing evidence of either low-vitamin D status or latitude being associated with chronic pain prevalence in noncases.[172] Similarly, well-controlled intervention trials have largely yielded negative findings.

170. Therefore, the primary determinant of fracture risk from postmenopausal or senile osteoporosis in older people is the mass of bone each had accumulated during growth and early adulthood. This includes cortical bone, which continues to be accreted after closure of the epiphyses until about the middle of the fourth decade.

171. Feskanich, D., Willett, W.C., Colditz, G.A., et al., 2003. Am. J. Clin. Nutr. 77, 504–511.

172. Straube, Andrew Moore, R., Derry, S., et al., 2009. Pain 141, 10–13.

Vitamin D Deficiency in Renal Patients

Low circulating levels of 25-OH-D_3 are frequently observed in patients with chronic renal disease and those with nephrotic syndrome and normal renal function. Some studies have found treatment with vitamin D analogues to reduce proteinuria in patients with chronic renal disease. Chronic renal disease can lead to bone disease, **renal osteopathy**, a frequent complication in renal patients in which the severity is related to the degree of loss of glomerular filtration rate. The condition is more common in children than adults, probably due to their greater need of growing bone for metabolically active vitamin D, along with phosphate and PTH circulating levels of which also decline in the disease. A meta-analysis of 16 clinical trials concluded that such treatments are effective in increasing serum Ca^{2+} and decreasing serum PTH but did ineffective in reducing either the need for dialysis or survival.[173]

Nonskeletal Effects of Hypovitaminosis D

Studies have shown risks to several nonskeletal diseases to be inversely related to vitamin D status in subjects with plasma 25-OH-D_3 levels <20–50 ng/mL regardless of sun exposure (Table 7.20). Meta-analyses of prospective cohort studies found a nonlinear decrease in all-cause mortality risk with increasing concentrations of plasma 25-OH-D_3 to an apparent optimum in the range of 75–87.5 nM.[174]

Low plasma levels of 25-OH-D_3 are inversely associated with risk factors for falling: postural balance and strength (e.g., leg extension power, quadriceps strength, arm muscle strength, handgrip strength, ability to climb stairs, physical activity).[175] A meta-analysis of 28 observational and cross-sectional studies found elderly fallers to have lower plasma 25-OH-D_3 levels than nonfallers.[176] Other meta-analyses of the eight published randomized trials found supplementation with 700–1000 IU D_3/day to reduce risks of falling by older subjects by 14–19%,[177] although individual studies have reported reductions by nearly 50%. The greatest benefits were observed when vitamin D was given with

173. Palmer, S.C., McGregor, D.O., Craig, J.C., et al., 2009. Cochrane Database Syst. Rev. CD008175.

174. Zittermann, A., Iodice, S., Pilz, S., et al., 2012. Am. J. Clin. Nutr. 95, 91–100; Sempos, C.T., Durazo-Arvizu, R.A., Dawson-Hughes, B., et al., 2013. J. Clin. Endocrinol. Metab. 98, 3001–3009. The latter results suggested a reversal of risk reduction at levels >99 nM.

175. Mowé, M., Haug, E., Bohmer, T., 1999. J. Am. Geriatr. Soc. 47, 220–226; Bischoff, H.A., Stahelin, H.B., Urscheler, N., et al., 1999. Arch. Phys. Med. Rehabil. 80, 54–58; Annweiler, C., Schott-Petelaz, A.M., Berrut, G., et al., 2009. J. Am. Geriatr. Soc. 57, 368–369; Annweiler, C., Montero-Odasso, M., Schot, A.M., et al., 2009. J. Nutr. Health Aging 13, 90–95.

176. Annweiler, C., Beauchet, O., 2014. J. Intern. Med. 277, 16–44.

177. Bischoff-Ferrari, H.A., Dawson-Hughes, B., Staehelin, H.B., et al., 2009. Br. Med. J. 339, 3692–3603; Kalyani, R.R., Stein, B., Valiyil, R., et al., 2010. J. Am. Geriatr. Soc. 58, 1299–1310.

TABLE 7.20 Relationship of vitamin D Status and Mortality

	Serum 25-OH-D Category, nM					P-linear Trend
Cause	<30	30–<50	50–<70	70–<90	90	
All causes	1	0.90 (0.79, 1.03)	0.78 (0.68, 0.90)	0.80 (0.68, 0.94)	0.73 (0.59, 0.90)	<0.0001
Cardiovascular	1	0.95 (0.74, 1.21)	0.84 (0.65, 1.09)	0.82 (0.61, 1.11)	0.73 (0.49, 1.09)	<0.03
Cancer	1	1.11 (0.89, 1.40)	0.94 (0.74, 1.19)	0.96 (0.73, 1.25)	0.90 (0.65, 1.26)	NS
Respiratory diseases	1	0.48 (0.33, 0.70)	0.36 (0.24, 0.55)	0.42 (0.26, 0.69)	0.24 (0.11, 0.54)	<0.0001

Khaw, K.T., Luben, R., Wareham, N., 2014. Am. J. Clin. Nutr. 100, 1361–1370.

supplemental Ca (Fig. 7.27), and when plasma 25-OH-D_3 levels were raised to ≥60 nM.

Vitamin D deficiency appears to increase risk to noninsulin-dependent diabetes (type 2 diabetes, T2D). Subclinical, chronic inflammation has been associated with insulin resistance, which has been found to be inversely related to serum 25-OH-D_3 concentrations over a wide range,[178] and serum 25-OH-D_3 levels have been found to be inversely correlated with insulin resistance, body fat mass, and BMI. The Third National Health and Nutrition Examination Survey (NHANES III) showed serum of 25-OH-D_3 level to be inversely associated with diabetes risk in a multiethnic sample of over 6000 adults.[179] Swedish researchers have found T2D incidence to be highest during the winter months when circulating 25-OH-D_3 levels are lowest.[180] Two VDR polymorphisms, *BsmI* and *ApaI*, as well as low-plasma VDB levels have been associated with high fasting glucose levels, hyperinsulinemia, and elevated T2D risk.[181] The relationship of vitamin D status and T2D risk appears to be greatest among overweight/obese individuals, as prediabetics with low-plasma 25-OH-D_3 levels shows the greatest insulin resistance.[182] It has been suggested that the exacerbation of the T2D risk of low-vitamin D status by adiposity may be due to the sequestration of vitamin by partitioning into bulk lipid depots in adipose tissue (Fig. 7.30). Indeed, the 25-OH-D_3 of subcutaneous white adipose tissue correlates with serum plasma 25-OH-D_3 levels in obese adults. This sequestration has been proposed as suppressing the plasma 25-OH-D_3 response to dietary or biogenic vitamin D, reducing the apparent bioavailability of dietary sources of the vitamin.

178. Chiu, K.C., Chu, A., Go, V., Saad, M.F., 2004. Am. J. Clin. Nutr. 79, 820–825.
179. Scragg, R., Sowers, M., Bell, C., 2004. Diabetes Care 27, 2813–2818.
180. Berger, B., Stenstrom, G., Sundkist, G., 1999. Diabetes Care 22, 773–777.
181. Hitman, G.A., Mannan, N., McDermott, M.F., et al., 1998. Diabetes 47, 688–690; Oh, J.-Y., Barrett-Connor, E., 2002. Metabolism 51, 356–359; Ortlepp, J.R., Metrikat, J., Albrecht, M., et al., 2003. Diabet. Med. 20, 451–454.
182. Abbasi, F., Blasey, C., Feldman, D., et al., 2015. J. Nutr. 145, 71–719.

Low-vitamin D status has also been associated with increased susceptibility to infections,[183] as well as increased risks of autoimmune and CVD, depressive symptoms, schizophrenia and Parkinson's disease, cognitive decline and dementia, periodontal disease, and ocular diseases.[184] Hospital mortality has been found greater for patients with low 25-OH-D3 levels (<20 ng/mL) than of vitamin D-adequate patients.

Serum 25-OH-D_3 levels are more frequently low in pregnant women than in nonpregnant women. A study in Boston found that 76% of new mothers (and 81% of their newborns) had serum 25-OH-D_3 levels <20 ng/mL.[185] Low-vitamin D status is generally more common among black than white pregnant women; black infants are four times as likely as white infants to be exposed in utero to vitamin D deficiency.[186] Systematic reviews and meta-analyses of observational studies have found low maternal serum 25-OH-D_3 levels to be associated with increased risks of preeclampsia, gestational diabetes, small-for-gestational age births, and bacterial vaginosis, although there is clear heterogeneity in many of these responses.[187] Some studies have found maternal vitamin D status to predict growth and later bone mass of children.

Vitamin D Deficiency in Animals

Rickets

Vitamin D-deficient, growing animals show rickets (Figs. 7.23 and 7.24). Species at greatest risk are those that experience rapid early growth, such as the chick. Rachitic signs

183. e.g., Pneumonia, and infections of *Mycobacterium tuberculosis*, *Leishmania major*, hepatitis C, Varicella-Zoster virus, and HIV.
184. e.g., Myopia, age-related macular degeneration, diabetic retinopathy, and uveitis.
185. Lee, J.M., Smith, J.R., Philipp, B.L., et al., 2007. Clin. Pediatr. 46, 42–44.
186. Bodnar, L.M., Simhan, H.N., Powers, R.W., et al., 2007. J. Nutr. 137, 447–452.
187. Harvey, N.C., Holroyd, C., Ntani, G., et al., 2014. Health Technol. Assess. 18, 1–190; Aghajafari, F., Nagulesapillai, T., Ronksley, P.E., et al., 2013. Br. Med. J. 346, f1169–f1183.

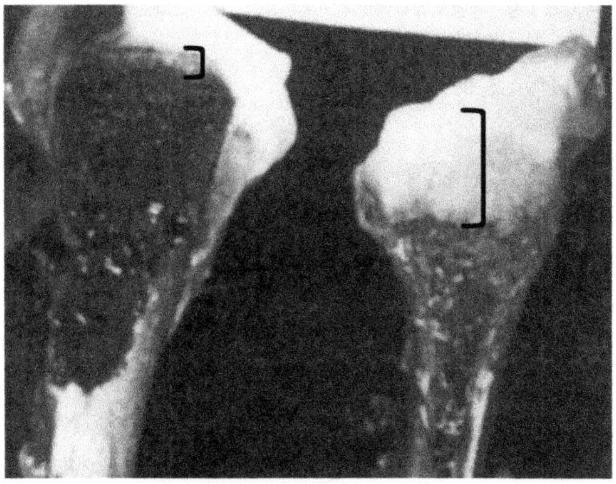

FIGURE 7.23 Tibiae of normal (*left*) and rachitic (*right*) chicks dissected to show epiphyseal growth plates (brackets). Note proliferation cartilage replacing vascularized spongiform bone in the affected tibia.

FIGURE 7.24 Rachitic puppy.

are similar in all affected species: impaired mineralization of the growing bones with structural deformation in weight-bearing bones.

Osteoporosis

Older vitamin D-deficient animals show the undermineralization of bones that characterizes osteoporosis. This can be a practical problem in the high-producing laying hen, in which it is called **cage layer fatigue.** The condition is associated with reductions in egg production, feed intake, and efficiency of feed utilization and survival.[188]

188. In well-managed flocks, it is common for a hen to lay >300 eggs in a year, with 40 of these laid during the first 40days after commencing egg laying. As each eggshell contains about 2g Ca, and the hen is able to absorb only 1.8–1.9g Ca from the diet each day, she experiences a Ca debt of 0.1–0.2g/day during that period. She accommodates this by mobilizing medullary bone; but, as her total skeleton contains only about 35g Ca, chronic demineralization at that rate without either decreasing the rate of egg production or increasing the efficiency of Ca absorption leads to osteoporosis characterized by fractures of the ribs and long bones.

Tibial Dyschondroplasia

There appear to be other situations of impaired renal 1-hydroxylation of 25-OH-D_3, thus limiting the physiological function of the vitamin. One is the failure of bone mineralization seen in rapidly growing, heavy-bodied chickens, and turkeys called tibial dyschondroplasia. The disorder is similar to the condition called osteochondrosis in rapidly growing pigs and horses; it is characterized by the failure of vascularization of the proximal metaphyses of the tibiotarsus and tarsalmetatarsus. It occurs spontaneously, but can be produced in animals made acidotic, that condition reducing the conversion of 1-hydroxylation of 25-OH-D_3. Both the incidence and severity of tibial dyschondroplasia can be reduced by treatment with 1,25-$(OH)_2$-D_3 or 1α-OH-D_3 but not by higher levels of vitamin D_3 alone. That lesions in genetically susceptible poultry lines cannot be completely prevented by treatment with vitamin D metabolites suggests that tibial dyschondroplasia may involve a functional impairment in VDRs.

Milk Fever

High-producing dairy cows can become hypocalcemic at the onset of lactation when they have been fed Ca-rich diets before calving. The condition, called milk fever, occurs when plasma Ca levels decrease to less than about 5.0mg/dL; it is characterized by tetany and coma, which can be fatal. Milk fever results from the inability of the postparturient cow to withstand massive lactational Ca losses by absorbing dietary Ca and mobilizing bone at rates sufficient to support plasma Ca at normal levels. It can be prevented by preparing the pregnant cow for upregulated bone mobilization and enteric Ca absorption. Infield practice, this is done by feeding her a relatively low-Ca diet (100g/day); parenteral treatment with 1,25-$(OH)_2$-D_3 is also effective.

Responses to Treatment

Supplementation and food fortification has proven to be safe and effective in normalizing circulating levels of 25-OH-D_3 and correcting associated physiological defects. Vitamin D is routinely included in the vitamin–mineral supplements of formulated foods for livestock and pets. A meta-analysis of 15 studies found serum 25-OH-D responses of humans to vitamin D-fortified foods varied according to dose level, latitude, and baseline status; but that, on average, a 1 μg/day (40 IU/day) of vitamin D from fortified foods increased serum 25-OH-D level by an average of 1.2nM.[189]

Correction of vitamin D deficiency may be impossible to achieve without the use of large-dose supplements. A meta-analysis of randomized trials noted the inconsistency

189. Black, L., Seamans, K.M., Cashman, K.C., et al., 2012. J. Nutr. 142, 1102–1108.

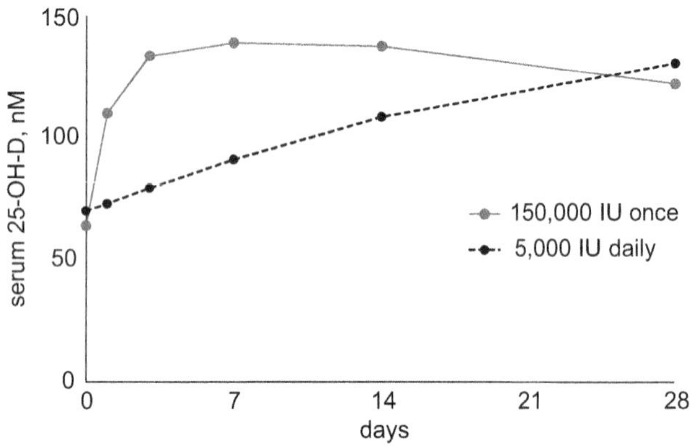

FIGURE 7.25 Comparison of daily versus monthly administration of vitamin D_3 to healthy, nonpregnant, nonlactating, women. *After Meekins, M.E., Oberhelman, S.S., Lee, B.R., et al., 2014. Eur. J. Clin. Nutr. 68, 632–634.*

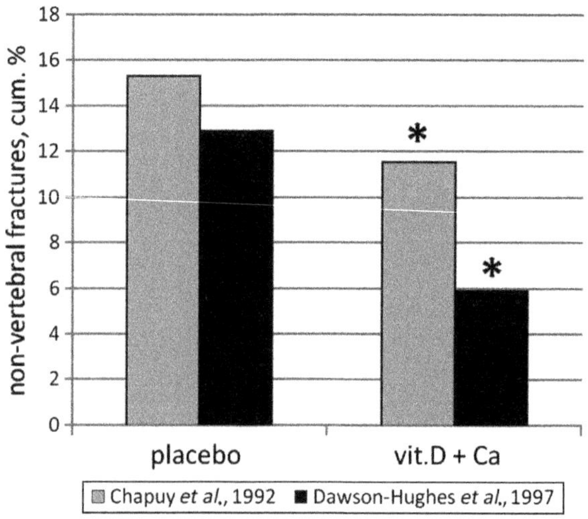

FIGURE 7.26 Prevention of fractures by vitamin D supplementation in two cohorts of older adults *(Chapuy, M.C., Arlo, M.E., Duboeuf, F., et al., 1992. N. Engl. J. Med. 327, 1637–1642.)*: 2790 women, 84 ± 6 years, receiving 1200 mg Ca and 800 IU D_3 daily; *(Dawson-Hughes, B., Harris, S.S., Krall, E.A., et al., 1997. N. Engl. J. Med. 337, 670–676.)*: 389 men and 213 women, 72 ± 5 years, receiving 500 mg Ca and 700 IU D_3 daily). *Significantly different from respective placebo level, $p<.05$.

of effects of vitamin D supplementation on infectious disease; it has been suggested that immunomodulatory efficacy may require higher serum 25-OH-D_3 levels than presently considered adequate. A study with T2D subjects in Finland showed that one-third or more did not respond to low-level (40 or 80 µg/day) D_3 supplements determined by expression of VDR-related genes.[190] Singh and Bonham[191] used responses of some 1300 subjects to develop an algorithm for vitamin D replacement dosing:

D_3 dose, IU/day = [(8.52–desired change in 25-OH-D_3 level, ng/mL) + (0.074 x age, year) − (0.20 x BMI) + (1.74 x albumin level, g/dL) − (0.62 x baseline 25-OH-D_3 level, ng/mL)]/−0.002

Such estimates indicate that the DRI levels of vitamin D intake are grossly inadequate for correcting low-vitamin D status; such conditions require does >2000 IU/day. Doses of 1000 IU have been found effective in reducing bone loss at the hip in postmenopausal women;[192] a meta-analysis of 23 randomized intervention trials in older subjects vitamin D-adequate subjects showed no effects of lower doses on bone mineral content at most sites.[193] A single dose of 150,000 IU D_3 and a lower daily dose of 5000 IU D_3 showed comparable plasma 25-OH-D_3 levels by 28 days, but the high bolus dose supported greater plasma 25-OH-D_3 levels for most of that period (Fig. 7.25). The synthetic vitamin D analogue, 1α,25-(OH)2-2β-(3-hydroxypropyloxyl)-D_3 (eldecalcitol), has been shown to be effective in normalizing plasma 25-OH-D_3 and reducing the incidence of new vertebral fractures in older subjects.[194]

Vitamin D_3 supplementation to correct low/deficient vitamin D status has been found to reduce risks to several outcomes: falling and fractures, including those unrelated to falls (Fig. 7.26), peritonitis and improve survival in renal patients on peritoneal dialysis, and to improve antibacterial immunity in HIV^+ patients, to reduce disease activity in subjects with Crohn's disease, and to improve winter-related atopic dermatitis in children. Correction of low-vitamin D status has been found to improve seizure control in epilepsy[195] and to stabilize Parkinson's disease in individuals with the *Fok*I TT or CT genotypes of VDR.[196]

10. VITAMIN D IN HEALTH AND DISEASE

Immune Dysfunction

Clinical studies have shown that oral and topical applications of either 1,25-$(OH)_2$-D_3 or the synthetic analogue 1α,25-$(OH)_2$-D_3[197] can be effective in managing psoriasis,[198] a Th1-mediated, hyperproliferative disease in which cathelicidin converts otherwise inert self-DNA and -RNA into autoimmune stimuli. Vitamin D treatment

190. Saka, N., Neme, A., Ryynänen, J., et al., 2015. J. Steroid Biochem. Mol. Biol. 148, 275–282.
191. Sing, G., Bonham, A.J., 2014. J. Am. Board Fam. Med. 27, 495–509.
192. MacDonald, H.M., Wood, A.D., Aucott, L.S., et al., 2013. J. Bone Miner. Res. 28, 2202–2213.
193. Reid, I.R., Bolland, M.J., Greay, A., 2014. Lancet 383, 146–155.
194. Noguchi, Y., Kawate, H., Nomura, M., et al., 2013. Clin. Interv. Aging 8, 1313–1321.
195. Holló, A., Clemens, Z., Kamondi, A., et al., 2012. Epilepys Behav. 24, 131–133.
196. Suzuki, M., Yoshioka, M., Hashimoto, M., et al., 2013. Am. J. Clin. Nutr. 97, 1004–1013.
197. Also called **calcipotriol**.
198. Results of a clinical series showed that topical application of 1,25-$(OH)_2$-vitamin D_3 caused complete clearing of lesions in 60% of patients, with an additional 30% of patients showing significant decreases in scale, plaque thickness, and erythema.

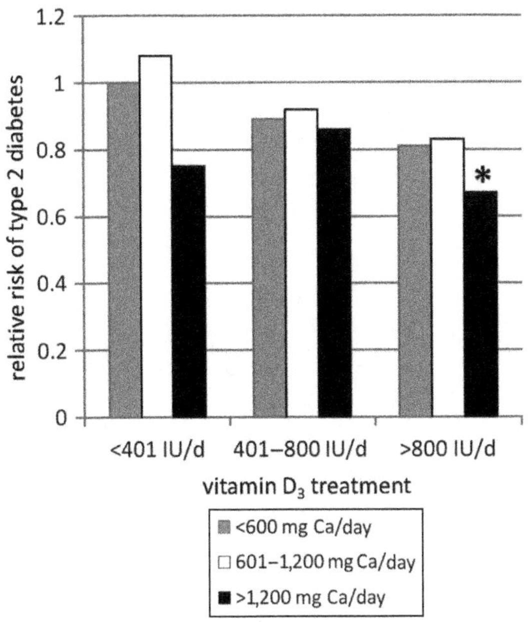

FIGURE 7.27 Reduction of type 2 diabetes risk associated with vitamin D and Ca supplement use by in 83,779 women in the Nurses' Health Study. * indicates significant difference ($p < .05$) from the low vitamin D, low Ca group. *After Pittas, A.G., Dawson-Hughes, B., Lit, T., et al., 2006. Diabetes Care 29, 650–656.*

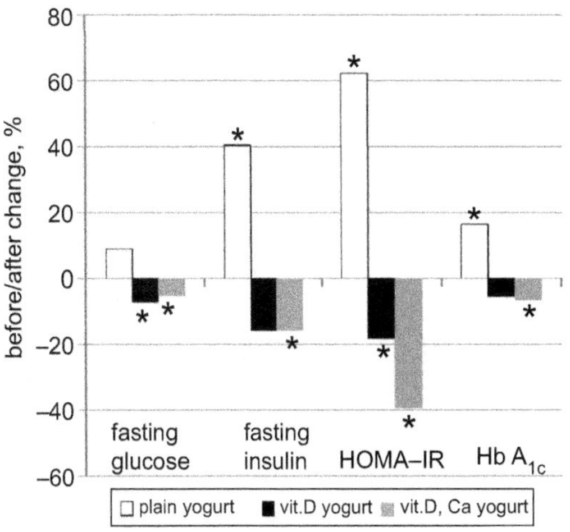

FIGURE 7.28 Effects of 12-week, yogurt-based supplementation with vitamin D (500 IU/d) and Ca (150 mg/day) on parameters of glycemic control in type 2 diabetes patients (fasting serum glucose >126 mg/dL), adults (n = 30 per treatment group); * indicates significant difference ($p < .05$) from the respective baseline value. *After Nikooyeh, B., Neyestani, T.R., Farvid, M., et al., 2011. Am. J. Clin. Nutr. 93, 764–771.*

has also been found to reduce rejection of allografts of heart, kidney, liver, pancreatic islets, small intestine, and skin in animals given high, nonhypercalcemic doses in both the presence and absence of the immunosuppressive drug cyclosporin A. These effects involve the suppression of antibody-presenting cells (particularly dendritic cells and T cells that play major roles in immunological graft rejection) and enhancing numbers of suppressive T ($CD4^+$, $CD25^+$) cells.

Obesity and Type 2 Diabetes

The efficacy of vitamin D in preventing T2D may depend on an individual's degree of adiposity, as the relationship between vitamin D status and T2D risk is strongest for overweight/obese individuals.[199] While a few clinical trials have found weight loss to improve 25-OH-D_3 levels and D_3 supplements to improve insulin sensitivity, such effects have not been observed in most trials, nor has vitamin D supplementation been found to affect body fat accumulation or weight loss. A 20-year follow-up of the Nurses' Health Study cohort found T2D risk to be one-third less for women reporting the use of vitamin D and Ca supplements (Fig. 7.27). An intervention with D_3 in Iran, a country with high reported prevalences of low-vitamin D status, metabolic syndrome, and T2D, found improvement in glycemic control in T2D patients, as indicated by reductions in fasting glucose level and insulin resistance (Fig. 7.29).[200]

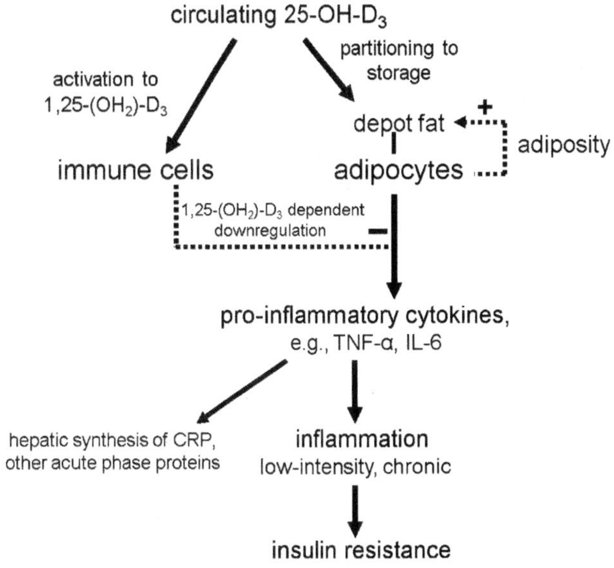

FIGURE 7.29 Hypothetical relationships of vitamin D status and adiposity in affecting insulin resistance. CRP, C-reactive protein; *IL-6*, interleukin-6; *TNF-*, tumor necrosis factor.

Cardiovascular Health

Epidemiological studies have shown that low-circulating 25-OH-D_3 levels as well as factors known to affect vitamin

199. Isaia, G., Giorgino, R., Adami, S., et al., 2001. Diabetes 24, 1496–1503.

200. Insulin resistance was measured by HOMA-IR, the homeostasis model assessment of insulin resistance, i.e., the products of the fasting serum concentrations of insulin and glucose.

D status (latitude, altitude, season) are associated with the prevalence of coronary risk factors and CVD mortality. The 2001–2006 NHANES data showed Americans of low-vitamin D status (serum 25-OH-D_3 <21 ng/mL) had serum homocysteine levels that were inversely related to their 25-OH-D_3 levels.[201] Vitamin D is known to suppress several mechanisms of cardiovascular pathogenesis: proliferation of vascular smooth muscle, vascular calcification, and production of proinflammatory cytokines. Intervention trials have not yielded conclusive results. Most have found vitamin D supplements to have little, if any, effects on blood pressure or other risk factors.

Anticarcinogenesis

Vitamin D was proposed to protect against colon cancer nearly 75 years ago, based on epidemiologic associations with sunlight exposure. Many subsequent studies in Europe and the United States have found positive associations between latitude; sun exposure; and risks of cancers of the prostate, colon, and breast. Residents of the northern United States have nearly a twofold higher risk of total cancer mortality than those of the southern States. The Nurses' Health Study showed a significant inverse linear association between plasma 25-OH-D_3 and colon cancer risk.[202] Risk factors for cancers of the prostate, colon, and breast are related to vitamin D status: dark skin color, northern latitude residence, and increasing age. Highest risks are seen for densely pigmented individuals who also tend to have lower circulating levels of 25-OH-D_3. A meta-analysis found carriers of the VDR *BsmI B* allele to have 6–7% less total cancer.[203] Randomized trials and prospective follow-up studies have found risk reductions of 17–60% per 25 nM increase in 25-OH-D_3 level.[204] A 4-year clinical trial found that a high dose (1000 IU [25 mg]/day) of vitamin D_3 significantly increased the protective effect of supplemental Ca against all-cause cancer (Fig. 6.30).

Anticarcinogenic effects of 1,25-$(OH)_2$-D_3 have been demonstrated in more than a dozen animal models (Table 7.21). These have shown vitamin D to inhibit cancer cell growth, angiogenesis, and metastasis. These effects appear to involve VDRs, which have been identified in many tumors and malignant cell types. VDR activation is known to signal the expression of many genes, including genes involved in differentiation; apoptosis; and inhibiting proliferation, invasiveness, angiogenesis, and metastatic potential. Ablation of VDR in mice increases tissue proliferation and apoptosis, enhances oxidative DNA damage, increases inflammation, and increases activation of oncogenes and loss of tumor-suppressor genes.

201. Amer, M., Qayyum, R., 2014. J. Endocrinol. Metab. 99, 633–638.
202. Feskanich, D., Ma, J., Fuchs, C.S., et al., 2004. Cancer Epidemiol. Biomarkers 13, 1502–1508.
203. Raimondi, S., Johansson, H., Maisonneuve, P., et al., 2009. Carcinogenesis 30, 1170–1180.
204. Pilz, S., Tomaschitz, A., Obermayer-Pietsch, B., et al., 2009. Anticancer Res. 29, 3699–3704.

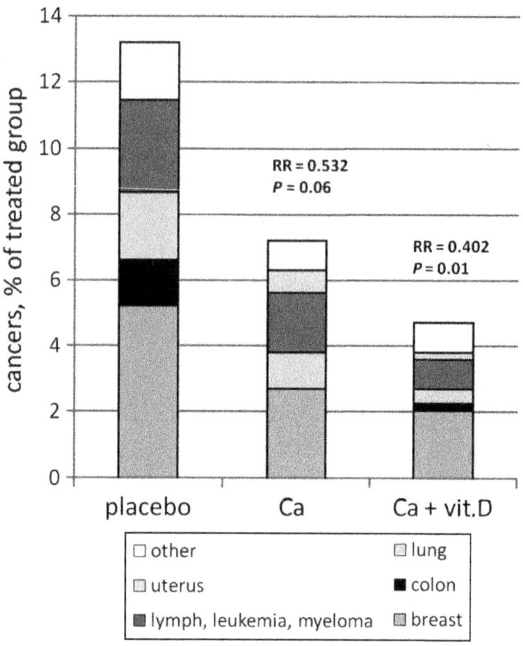

FIGURE 7.30 Four-year cancer incidence in 1179 postmenopausal women randomized to treatment with a placebo, Ca (1400–1500 mg/day) alone or with vitamin D_3 (1100 IU [27.5 mg]/day). *After Lappe, J.M., Travers-Gustafson, D., Davies, K.M., et al., 2007. Am. J. Clin. Nutr. 85, 1586–1591.*

Colorectal Cancer

Mortality to colorectal cancer is significantly higher in the northern and northeasters United States than in the southwest, Hawaii, and Florida. Observational studies have been inconsistent in linking this phenomenon to vitamin D status; however, a meta-analysis has shown significant reduction in colorectal risk associated with serum 25-0H-D_3 levels >20 ng/mL or intakes of at least 1000–2000 IU/day.[205] *In vitro* studies have demonstrated that 1,25-$(OH)_2$-D_3 can both attenuate the growth of rapidly dividing colonic tumor cells and reverse colonocytes from a malignant to a normal phenotype. These effects appear to depend on the activation of VDR, which represses signaling through β-catenin (a mediator of the Wnt pathway) and induces apoptosis. VDR expression appears to decline during the progression of colon cancer; this has been linked to upregulation of transcriptional repressors, which bind to the VDR promotor, and to a shift toward catabolism of both 25-OH-D_3 and 1,25-$(OH)_2$-D_3. These effects appear to involve the suppression of inflammation by the vitamin D system;

205. Gorham, E.D., Garland, C.F., Garland, F.C., et al., 2007. Am. J. Prev. Med. 32, 210.

TABLE 7.21 Animal Tumor Models in Which Anticarinogenic Activity of 1,25-$(OH)_2$-D_3 Has Been Demonstrated

Species	Site	Mode of Induction
Mouse	Skin	Oral treatment: Dimethybenzanthracine, phorbol ester
Mouse (athymic)	Adenocarcinoma	Implantation: CAC-8 cells
	Kaposi sarcoma	Implantation: KS YH-1 cells
	Melanoma	Implantation: human melanoma cells
	Osteosarcoma	Implantaton: LNCaP cells
	Retinoblastoma	Implantation: malignant cells
Mouse, APCmin	Colon	(Spontaneous)
Rat	Mammary	Oral treatment: N-methylnitrosourea, dimethylhydrazine
	Colon	Implantation: human colon cancer cells
		Oral treatment: dimethylhydrazine
	Leydig	Implantation: Leydig tumor cells
	Prostate	Implantation: Dunning LyLu cells
	Walker carcinoma	Implantation: Walker carcinoma cells

supplementation of mice with D_3 and 25-OH-D_3 reduced the development of colon tumors induced by inflammatory stimuli by 50%.[206] High VDR expression has been associated with a more favorable prognosis for colorectal cancer patients.[207] VDR can also bind the secondary bile acid, lithocholic acid, a potent enteric carcinogen. Single nucleotide polymorphisms in the VDR gene have been associated with differences in colon cancer risk. Carriers of *Cdx-2* AA of *FokI* TT showed twice the risk of colorectal cancer compared to other genotypes; those with the *Cdx-2-FokI* A-T, *FokI TaqI* T-G, or *Cdx-2-FokI-TaqI* A-T-G haplotypes showed two- to three fold risks of colon cancer compared to other haplotypes.[208]

206. Murillo, G., Nagpal, N., Tiwari, N., et al., 2010. J. Steroid Biochem. Mol. Biol. 121, 403–407.
207. Evans, S.R.T., Nolla, J., Hanfelt, J., et al., 1998. Clin. Cancer Res. 4, 1591–1595.
208. Ochs-Baalcom, H.M., Cicek, M.S., Thompson, C.L., 2008. Carcinogenesis 24, 1788–1793.

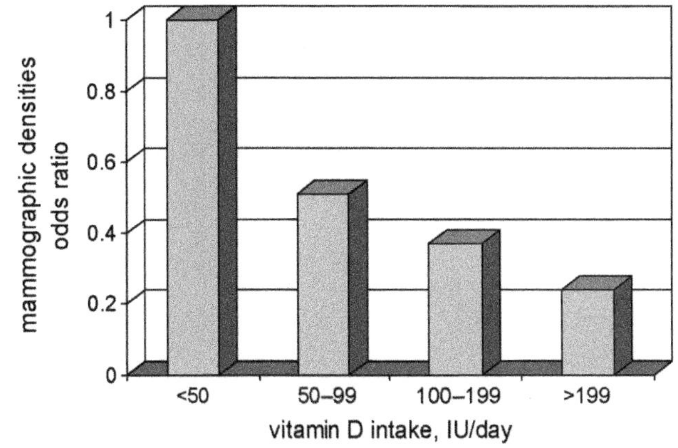

FIGURE 7.31 Inverse relationship of frequency of mammographic breast density and vitamin D status in a cohort of 543 women aged 40–60 years undergoing screening mammographies. *After Berube, S., Diorio, C., Verhoek-Oftedahl, W., et al., 2004. Cancer Epidemiol. Biomarkers Prev. 13, 1466–1472.*

Relatively lower risks of colon (and breast and prostate) cancer have been observed among groups with high consumption of soy foods. It is possible that this may relate to the effects of phytoestrogenic soy isoflavones in upregulating colonocyte production of 1,25-$(OH)_2$-D_3, as genistein has been found to increase expression of the colonic 1-hydroxylase (CYP27B1) while markedly decreasing expression of the 24-hydroxylase (CYP24A1).

Breast Cancer

Studies have shown low-serum 1,25-$(OH)_2$-D_3 levels to be associated with increased risk of breast cancer (Fig. 7.31). Most breast cancer cells express VDR and are growth impaired by physiological concentrations of 25-OH-D_3 or 1,25-$(OH)_2$-D_3. Studies with cultured breast cancer cells have shown that liganded VDR functions in arresting growth, inducing differentiation and apoptosis, and inducing the expression of factors involved in the regulation of cell proliferation, thus, blocking mitogenic signaling including estrogen-driven proliferation.[209] A pooled analysis of two observational studies showed a negative association of vitamin D status and breast cancer risk, with individuals with serum 25-OH-D_3 levels averaging 52 ng/mL having half the risk of those with serum 25-OH-D_3 levels <13 ng/mL and about 60% of the risk of those with serum levels around 20 ng/mL.[210] A meta-analysis found breast cancer risk is less for individuals of the VDR *FokI FF* genotype than for those of the ff genotype.[211] Several studies shown breast cancer incidence and mortality to be inversely correlated

209. e.g., Cyclins C and D1, Kip1, WAF1, c-Fos, C-Myc, c-JUN, and members of the TNF-β family.
210. Garland, C.F., Gorham, E.D., Mohr, S.B., et al., 2007. J. Steroid Biochem. Mol. Biol. 103, 708.
211. Tang, C., Chen, N., Wu, M., et al., 2009. Breast Cancer Res. Treat. 30, 1170.

with serum 25-OH-D_3 level. It has been estimated that, in the United States and Canada, increasing year-round serum 25-OH-D_3 levels to 40–60 ng/mL (100–150 nM) would prevent some 58,000 new cases of breast cancer each year and cut breast cancer mortality in half.[212]

Prostate Cancer

Vitamin D metabolites promote differentiation and inhibit proliferation, invasiveness, and metastasis in prostate cells. These effects are thought to result from VDR activation. High VDR expression has been associated with reduced risk of initiation and progression of prostate cancer in younger men.[213] Studies with rat and human prostate cells have shown 1,25-$(OH)_2$-D_3 to act via VDR synergistically with androgen signaling to upregulate the expression of some 250 genes that are not regulated by either factor alone. Among these genomic changes are differentiation of stem cells to androgen receptor-positive luminal epithelial cells and augmentation of tumor-suppressive micro-RNAs (miRNAs). A meta-analysis found prostate cancer risk is less for carriers of the VDR *BsmI B* allele than those without that allele.[214] Several studies shown the incidences of benign prostatic hyperplasia and prostate cancer incidence and mortality to be inversely correlated with serum 25-OH-D_3 level. An open-label trial in South Carolina found daily supplementation with 4000 IU D_3 for a year to reduce Gleason scores of half of men diagnosed with early-stage, low-risk prostate cancer.[215] It has been estimated that, in the United States and Canada, increasing year-round serum 25-OH-D_3 levels to 40–60 ng/mL (100–150 nM) would prevent some 49,000 new cases of breast cancer each year and cut breast cancer mortality in half.

Skin Cancer

Vitamin D functions in maintaining ordered proliferation of keratinocytes by inhibiting β-catenin signaling and also by protecting them from UV-induced DNA damage by upregulating p53, inhibiting stress-activated kinases, and suppressing nitric oxide production. These effects depend on VDR, as its ablation in the mouse results in increased sensitivity of skin cells to tumorigenesis induced chemically or by UV light. That it is the unbound VDR that is involved in antitumorigenesis is indicated by finding that ablation of *CYP27B1* does enhance skin carcinogenesis. Some, but not all, epidemiological studies have found inverse correlations of 25-OH-D_3 level and risk to melanoma but no clear relationship with nonmelanoma skin cancer.[216]

Other Cancers

That vitamin D may have a role in protecting against leukemia is suggested by the finding that 1,25-$(OH)_2$-D_3 treatment can suppress cell division and induce differentiation in human leukemic cells. This effect involves downregulating the expression of the protooncogene c-*myc*. Case–control studies have provide strong evidence of relationships of vitamin D status and risk to non-Hodgkin lymphoma. A meta-analysis found individuals of the VDR *FokI* TT genotype to have lower risk of skin cancer (including malignant melanoma) than those of the other genotypes.[217] Vitamin D intake was found to be inversely related to lung cancer risk in never-smoking, premenopausal women in the Women's Health Initiative. Evidence is inconsistent for a protective role of vitamin D in ovarian cancer; a long-term, case–control studies have patients with higher prediagnostic serum 25-OH-D levels to have greater risks of surviving ovarian cancer compared to patients of lower vitamin D status.

11. VITAMIN D TOXICITY

Excessive intakes of vitamin D are associated with increases in circulating levels of 25-OH-D_3, the critical metabolite in vitamin D intoxication. This is especially true for D_3, high intakes of which produces greater serum 25-OH-D_3 levels than comparable intakes of D_2, making D_3 10–20 times more toxic than D_2. The basis of toxicity appears to be the ability of 25-OH-D_3 at high levels to compete successfully for VDR binding, bypassing the regulation of the 1-hydroxylase to induce transcriptional responses normally signaled only by 1,25-$(OH)_2$-D_3. Therefore, risk to hypervitaminosis D is increased under conditions in which the normal feedback regulation of the renal 25-OH-D_3-1-hydroxylase is compromised, e.g., chronic inflammation.[218] Recommended upper limits of intake are shown (Table 7.22).

Hypervitaminosis D is characterized by **hypercalcemia** resulting from increased enteric absorption and bone resorption of Ca, with attendant decreases in serum PTH and glomerular filtration. Bone mobilization also results in increased serum levels of zinc from that reserve. Vitamin D-intoxicated individuals show different signs including anorexia, vomiting, headache, drowsiness, diarrhea, and polyuria (Table 7.23). Chronically elevated serum Ca and phosphorus levels ultimately result in **calcinosis**, i.e., the deposition of Ca and P in soft tissues, especially heart and kidney but also the vascular and respiratory systems and

212. Garland, C.F., Gorham, E.G., Mohr, S.B., et al., 2009. Ann. Epidemiol. 19, 468–483.
213. Ahonen, M.H., Tenkanen, L., Teppo, L., et al., 2000. Cancer Causes Control 11, 847–852.
214. Berndt, S.I., Dodson, J.L., Huang, W.Y., et al., 2006. J. Urol. 175, 1613.
215. Hollis, B.W., Marshall, D.T., Savage, S.J., et al., 2013. J. Steroid Biochem. Mol. Biol 136, 233–237.
216. e.g., Actinic keratoses, basal cell carcinomas.
217. Mocellin, S., Nitti, D., 2008. Cancer 113, 2398–2405.
218. Other conditions include tuberculosis and sarcoidosis.

TABLE 7.22 Recommended Upper Tolerable Vitamin D Intakes (UL)

Age–Sex	UL[a], μg/day
0–11 months	25–38
1–3 years	63
4–8 years	75
9–70 + years	100
Pregnancy	100
Lactation	100

[a]*Food and Nutrition Board, 2010. Dietary Reference Intakes: Calcium, Vitamin D. National Academy Press, Washington, DC, 1115 pp.*

TABLE 7.23 Signs and Symptoms of Hypervitaminosis D

Anorexia
Gastrointestinal distress, nausea, vomiting
Headache
Muscular weakness, lameness
Polyuria, polydypsia
Nervousness
Hypercalcemia
Calcinosis

practically all other tissues.[219] Therefore, risk of hypervitaminosis D is not only dependent on exposure to vitamin D but also on concomitant intakes of Ca and phosphorus. That serum 25-OH-D_3 level may have a biphasic relationship with adverse outcomes is suggested by the findings of a prospective study of 6000 older women that showed both frailty and mortality to be lower for subjects with baseline 25-OH-D_3 levels of 20–29 ng/mL (50–72.5 nM) compared to subjects with baseline levels below or above that range.[220]

The IOM[221] suggested that serum 25-OH-D_3 levels greater than 100 nM may be associated with increased risk to all-cause mortality, but a careful examination of the primary data does not support that conclusion. A systematic review of clinical trial results found no evidence for adverse effects for vitamin D_3 doses as high as 10,000 IU/day, and no consistent and reproducible effects, including hypercalcemia, at doses five times that amount.[222] A few cases of hypervitaminosis D have been documented among consumers of milk that, through processing errors, was sporadically fortified with very high levels of the vitamin.[223] Vitamin D_3 has been found safe for pregnant and lactating women and their children at oral doses of 100,000 IU/day.[224] There are no documented cases of hypervitaminosis D due to excessive sunlight exposure.

219. The condition idiopathic infantile hypercalcemia, formerly thought to be due to hypervitaminosis D, appears to be a multifactorial disease with genetic as well as dietary components.

220. Ensrud, K.E., Ewing, S.K., Fredman, L., et al., 2010. J. Clin. Endocrinol. Metab. 95, 5266–5273.

221. Institute of Medicine, 2011. Dietary Reference Intakes: Calcium, Vitamin D. National Academy Press, Washington, DC, 1115 pp.

Hypercalcemia can occur with combined supplementation with vitamin D and Ca. A 7-year study of some 36,000 postmenopausal women found a 17% increase (compared to controls) in renal stones in subjects given a daily supplement of 1000 mg Ca plus 400 IU (10 mg) vitamin D_3.[225] That effect is more likely related to total Ca intake, estimated at 2000 mg/day, as that vitamin D dose would be expected to increase serum 25-OH-D_3 levels by only c.7 nM, i.e., not to levels approaching those (>600 nM) found to produce hypercalcemia. Other clinical intervention trials have used similar vitamin D_3 doses without adverse effects; the few that have observed calciuria have used high D_3 doses (e.g., 2000 IU [50 mg]) in combination with Ca (≥500 mg). A naturally occurring calcinosis in grazing livestock was traced to the consumption of water-soluble glycosides of 1,25-$(OH)_2$-D_3 present in some plants.[226] These compounds appear to be deglycosylated metabolically to yield 1,25-$(OH)_2$-D3, which is 100 times more toxic than the dominant circulating metabolite 25-OH-D_3.

Vitamin D hypersensitivity has been proposed as the basis for Williams–Beuren syndrome, a rare condition of hypercalcemia and Ca hyperabsorption in humans. The syndrome is manifest in infancy; it is characterized by failure to thrive with mental handicap and long-term morbidity. Patients have been found to have normal circulating levels of 25-OH-D_3, but they appear to have exaggerated responses to oral doses of vitamin D_3 and one report presented elevated serum levels of 1,25-$(OH)_2$-D_3 in patients.

222. Hathcock, J. N., Shao, A., Vieth, R., et al., 2007. Am. J. Clin. Nutr. 85, 6–18.

223. Eight cases were described; each consumed a local dairy's milk, which varied in vitamin D content (some contained as much as 245,840–IU of D_3 per liter) (Jacobus, C.H., Holick, M.F., Shao, Q., et al., 1992. N. Engl. J. Med. 326, 1173–1177). US regulations stipulate that milk is to contain 400–IU/qt "within limits of good manufacturing practice"; however, a survey found milk and infant formula preparations rarely contained the amounts of vitamin D stated on the label, owing to both under- and overfortification.

224. Goodenday, L.S., Gordon, G.S., 1971. Ann. Intern. Med. 75, 807–808.

225. Jackson, M.D., Lacroix, A.Z., Gass, M., et al., 2006. N. Engl. J. Med. 354, 669–683.

226. D_3 glycosides have been identified in several species in the families Solanaceae (*Solanum glaucophyllum*, *Solanum malacoxylon*, *Solanum torvum*, *Solanum verbascifolium*, *Cestrum diurnum*, and *Nierembergia veitchii*).

The availability of synthetic 1α-OH-D_3 in recent years has meant that it can be used at very low doses to treat vitamin D-dependent or -resistant osteopathies.[227] This has reduced the risks of hypervitaminosis that attend the use of the massive doses of D_3 that are needed to provide effective therapy in such cases.

12. CASE STUDIES

Instructions

Review each of the following case reports, paying special attention to the diagnostic indicators on which the respective treatments were based. Then answer the questions that follow.

Case 1

When the patient was first evaluated at the National Institutes of Health, he was a thin, short, bowlegged, 20-year-old male. His height at that time was 159 cm (below the first percentile) and he weighed 52 kg. In addition to his dwarfism, he showed a varus deformity of both knees, and he walked with a waddling gait. Radiographs showed diffusely decreased bone density, subperiosteal resorption, and a **pseudofracture**[228] of the left ischiopubic ramus.[229]

Parameter	Patient	Normal Range
Serum Ca	8.0 mg/dL	8.5–10.5 mg/dL
Serum phosphorus	2.2 mg/dL	3.5–4.5 mg/dL
Serum alkaline phosphatase	152 U/mL	<77 U/mL
Urine chromatography	Generalized aminoaciduria	

The patient's history revealed that he had been a normal, full-term infant weighing 3.2 kg. He had been breast-fed and had been given supplementary vitamin D. At 20 months, however, he failed to walk unsupported and was diagnosed as having active rickets, as revealed by genu varum, irregular cupped metaphyses, and widened growth plates,[230] with reductions of both Ca and phosphorus in his blood. The rickets did not respond to oral doses of ergocalciferol (normally effective in treating nutritional rickets), but healing was observed radiographically after intramuscular administration of 1,500,000 IU (37.5 mg) of vitamin D_2 weekly for 5 months. The patient continued to receive vitamin D in the form of cod liver oil, 5000–20,000 IU (125–500 mg) per day. At 4 years of age, corrective surgery was performed for deformities of the tibias and femurs. At age 14, the patient's height was in the 15th percentile. Additional surgery was performed, after which vitamin D therapy was stopped and, over the next 2 years, weakness and severe bone pain became evident. At age 19, bilateral femoral osteotomies[231] were performed again. As an outpatient at the NIH Clinical Center, the patient received oral ergocalciferol, 50,000 IU daily, for the next 6 years and experienced remission of pain and weakness and normalization of serum Ca and phosphorus levels. His height reached 161 cm (63.3 in.), i.e., still below the first percentile. At 27 years of age, his radiographs showed improved density of the skeletal cortices and healing of the pseudofractures, but the patient still showed the clinical stigmata[232] of rickets.

Parameter	Patient	Normal Range
Serum PTH (parathyroid hormone)	0.31 ng/mL	<0.22 ng/mL
Urine cAMP	6 nmol/dL	2.3 ± 1.2 nmol/dL
^{47}Ca absorption	19%	33–43%
Plasma 25-OH-D_3	25 ng/mL	10–40 ng/mL
Plasma 1,25-$(OH)_2$-D_3	213 pg/mL	20–60 pg/mL
Plasma 24,25-$(OH)_2$-D_3	1.0 ng/mL	0.8–3 ng/mL

Two hundred micrograms of 25-OH-D_3 was then given orally daily for 2 weeks. Calcium retention improved, urinary cAMP fell, and plasma phosphorus and Ca rose, each to the normal level. Vitamin D_3 maintenance doses (about 40,000 IU, i.e., 1 mg/day) were given periodically to prevent recurrent osteomalacia.

Case 2

This patient was a sister of the patient described in Case 1. She was first evaluated at the NIH when she was 18 years old. She was a thin female dwarf (147 cm tall, below the first percentile) weighing 44.8 kg. She walked with a waddling gait and had mild bilateral varus deformities of the knees. **Chvostek's sign**[233] was present bilaterally. Analyses of her serum showed 7.0 mg of Ca and 3.0 mg of phosphorus per deciliter and alkaline phosphatase at 110 U/ml. Skeletal radiographs showed delayed ossification of several

227. These diseases include hypoparathyroidism, genetic or acquired hypophosphatemic osteomalacias, renal osteodystrophy, vitamin D-dependent rickets, and osteomalacia associated with liver disease and enteric malabsorption.
228. i.e., New bone detected radiographically as thickening of the periosteum at the site of an injury to the bone.
229. A narrow process of the pelvis.
230. Failure of mineralization of the growing ends of long bones.
231. Surgical correction of bone shape.
232. Residual abnormalities.
233. Facial spasm induced by a slight tap over the facial nerve.

epiphyses and a pseudofracture in the left tibia. Her plasma 25-OH-D was 44 ng/mL, 1,25-$(OH)_2$-D_3 was 280 pg/ml, and 24,25-$(OH)_2$-D_3 was 2.5 ng/mL.

Her history showed that she had been a normal, full-term infant who weighed 3.8 kg at birth. At 5 months of age, she showed radiographic features of rickets. During infancy and childhood, she received vitamin D as cod liver oil, in doses of 2000–10,000 IU/day. She began to walk at 9 months and developed slight varus deformity of both legs. Her rate of growth was at the fifth percentile until the vitamin D was discontinued when she was 11 years old. Within 3 years, her height fell below the first percentile. From ages 15 to 16, the bowing of her legs progressed moderately. When she was 18 years old, at the time of her first admission to the NIH Clinical Center, she was treated with 200 μg of 25-OH-D_3 per day for 2 weeks. During this time, her Ca retention improved, and her serum Ca and P increased. Studies showed that 500 μg 20 (IU) of D_3 per day was required to maintain her plasma Ca in the normal range. At this dose, her 25-OH-D_3 was 141 ng/ml, 1,25-$(OH)_2$-D_3 was 640 pg/mL, and 24,25-$(OH)_2$-D_3 was 3.6 ng/mL (above normal). When she was 24 years old, i.e., 6 years after her first admission to the center, she was readmitted for studies of the effectiveness of oral 1,25-$(OH)_2$-D_3 with a supplement of 800 mg of Ca per day. Serum Ca remained below normal on doses of 2–10 μg of 1,25-$(OH)_2$-D_3 per day. Only when the dose was increased to 14–17 μg of 1,25-$(OH)_2$-D_3 per day did her plasma Ca reach the normal range. PTH remained elevated at 0.40 ng/mL. At these high doses of 1,25-$(OH)_2$-D_3, her plasma 25-OH-D_3 was 26 ng/mL, and her 1,25-$(OH)_2$-D_3 was 400 pg/mL. While on 1,25-$(OH)_2$-D_3, her osteomalacia improved, and serum Ca and P entered normal ranges.

Case Questions

1. What are the common clinical features (physical and biochemical observations, response to treatment, etc.) of these two cases?
2. What can you infer about the nature of vitamin D metabolism in these siblings?
3. Propose a hypothesis to explain these cases of vitamin D-resistant rickets. How might you test this hypothesis?

13. STUDY QUESTIONS AND EXERCISES

1. Construct a flow diagram showing the metabolism of vitamin D to its physiologically active and excretory forms.
2. Construct a "decision tree" for the diagnosis of vitamin D deficiency in a human or animal. How can deficiencies of vitamin D and Ca be distinguished?
3. How does the concept of solubility relate to vitamin D utilization? What features of the chemical structure of vitamin D relate to its utilization?
4. Relate the concept of organ function to the concept of vitamin D utilization/status.
5. Discuss the concept of homeostasis, using vitamin D as an example.
6. Using an evidence-based approach, discuss whether physiological conditions in addition to those caused by vitamin D deficiency should be considered in establishing the RDA for vitamin D.

RECOMMENDED READING

Aloia, J.F., 2008. African Americans, 25-hydroxyvitamin D, and osteoporosis: a paradox. Am. J. Clin. Nutr. 88, 545S–550S.

Annweiler, C., Montero-Odasso, M., Schott, A.M., et al., 2010. Fall prevention and vitamin D in the elderly: an overview of the key role of the non-bone effects. J. Neuroeng. Rehabil. 7, 50–63.

Baeke, F., Takiishi, T., Korf, H., et al., 2010. Vitamin D: modulator of the immune system. Curr. Opin. Pharmacol. 10, 482–496.

Berry, D., Hyppönen, E., 2014. Vitamin D, vitamin D binding protein, and cardiovascular disease (Chapter 7). In: Dakshinamurti, K., Dakshinamurti, S. (Eds.), Vitamin Binding Proteins: Functional Consequences. CRC Press, New York, pp. 107–126.

Bikke, D.D., 2010. Vitamin D: newly discovered actiona require consideration of physiologic requirements. Trends Endocrinol. Metab. 21, 375–384.

Bikke, D.D., 2010. Vitamin D and the skin. J. Bone Min. Metab. 28, 117–130.

Bischoff-Ferrari, H.A., Giovannucci, E., Willett, W.C., et al., 2006. Estimation of optimal serum concentrations of 25-hydroxyvitamin D for multiple outcomes. Am. J. Clin. Nutr. 84, 18–28.

Bonner, F., 2003. Mechanisms of intestinal calcium absorption. J. Cell. Biochem. 88, 387–393.

Borradale, D., Kimlin, M., 2009. Vitamin D in health and disease: an innsight into traditional functions and new roles for the 'sunshine vitamin'. Nutr. Res. Rev. 22, 118–136.

Campbell, F.C., Xu, H., El-Tanani, M., et al., 2010. The yin and yang of vitamin D recpetor (VDR) signaling in neoplastic progression: operational networks and tissue-specific growth control. Biochem. Pharmacol. 79, 1–9.

Chesney, R.W., 2014. The role of vitamin D in infectious processes (Chapter 6). In: Dakshinamurti, K., Dakshinamurti, S. (Eds.), Vitamin Binding Proteins: Functional Consequences. CRC Press, New York, pp. 89–107.

Christakos, S., Barletta, F., Huening, M., et al., 2003. Vitamin D target proteins; function and regulation. J. Cell. Biochem. 88, 238–244.

Clinton, S.K., 2013. Vitamin D (Chapter 31). In: Stipanuk, M.H., Caudill, M.A. (Eds.), Biochemical, Physiological, and Molecular Aspects of Human Nutrition. Elsevier, New York, pp. 703–717.

Crannery, A., Weiler, H.A., O'Donnel, S., et al., 2008. Summary of evidence-based review of vitamin D efficicy and safety in relation to bone health. Am. J. Clin. Nutr. 88, 513S–519S.

Dawson-Hughes, 2004. Calcium and vitamin D for bone health in adults (Chapter 12). In: Holick, M.F., Dawson-Hughes, B. (Eds.), Nutrition and Bone Health. Humana Press, Totowa, NJ, pp. 197–210.

DeLuca, H.F., 2008. Evolution of our understanding of vitamin D. Nutr. Rev. 66, S73–S87.

De Paula, F.J.A., Rosen, C.J., 2010. Vitamin D safety and requirements. Arch. Biochem. Biophys. 523, 64–72.

Dombrowski, Y., Peric, M., Koglin, S., et al., 2010. Control of cutaneous antimicrobial peptides by vitamin D3. Arch. Dermatol. Res. 302, 410–418.

Dror, D.K., Allen, L.H., 2010. Vitamin D inadequacy in pregnancy: biology, outcomes, and interventions. Nutr. Rev. 68, 465–477.

Eyles, D.W., Feron, F., Cui, X., et al., 2009. Psychoneuroendocrinol. 345, S247–S257.

Feldman, D., Pike, J.W., Glorieux, F.H. (Eds.), 2005. Vitamin D, vols. I and II. second ed.Elsevier, New York.

Fetahu, I.S., Höbaus, J., Kállay, E., 2014. Vitamin D and the epigenome. Front. Physiol. 5, 1–12.

Flores, M., 2005. A role of vitamin D in low-intensity chronic inflammation and insulin resistance in type 2 diabetes? Nutr. Res. Rev. 18, 175–182.

Garland, C.F., Gorham, E.G., Mohr, S.B., et al., 2009. Vitamin D for cancer prevention: global perspective. Ann. Epidemiol. 19, 468–483.

Gombart, A.F. (Ed.), 2013. Vitamin D: Oxidative Stress, Immunity, and Aging. CRC Press, New York, p. 440.

Hamilton, B., 2010. Vitamin D and human skeletal muscle. Scand. J. Med. Sci. Sports 20, 182–190.

Hathcock, J.N., Shao, A., Vieth, R., et al., 2007. Risk assessment for vitamin D. Am. J. Clin. Nutr. 85, 6–18.

Hayes, D.P., 2010. Vitamin D and aging. Biogerontology 11, 1–16.

Hewson, M., 2012. Vitamin D and immune function: an overview. Proc. Nutr. Soc. 71, 50–61.

Holick, M.F., 2008. Vitamin D: a D-lightful health perspective. Nutr. Rev. 66, S182–S194.

Holick, M.F. (Ed.), 2010. Vitamin D: Physiology, Molecular Biology, and Clinical Applications, second ed. Humana Press, New York, p. 1155.

Institute of Medicine, 2011. Dietary Reference Intakes: Calcium, Vitamin D. Nat. Acad. Press, Washington, DC, p. 1115.

Jones, G., 2008. Pharmacokinetics of vitamin D toxicity. Am. J. Clin. Nutr. 88, 582S–586S.

Kamen, D.L., Tangpricha, V., 2010. Vitamin D and molecular actions on the immune system: modulation of innate and autoimmunity. J. Mol. Med. 88, 441–450.

Lee, J.H., O'Keefe, J.H., Bell, D., et al., 2008. Vitamin D deficiency: an important, common, and easily treatable cardiovascular risk factor? J. Am. Coll. Cardiol. 52, 1949–1956.

Levine, A., Li, Y.C., 2005. Vitamin D and its analogues: do they protect against cardiovascular disease in patients with kidney disease? Kidney Int. 68, 1973–1981.

Mathieu, C., Gysemans, C., Giulietti, A., et al., 2005. Vitamin D and diabetes. Diabetologia 48, 1247–1257.

Maxwell, C.S., Wood, R.J., 2011. Update on vitamin D and type 2 diabetes. Nutr. Rev. 69, 291–295.

Mizwicki, M.T., Norman, A.W., 2009. The vitamin D sterol-vitamin D receptor ensemble model offers unique insights into both genomic and rapid-response vitamin D signaling. Sci. Signal. 2, 1–14.

Nagpal, S., Na, S., Rathnachalam, R., 2005. Noncalcemic actions of vitamin D receptor ligands. Endocrinol. Rev. 26, 662–687.

Norman, A.W., 2008. A vitamin D nutritional cornucopia: new insights concerning serum 25-hydroxyvitamin D status of the US population. Am. J. Clin. Nutr. 88, 1455–1456.

Norman, A.W., Bouillon, R., 2010. Vitamin D nutritional policy needs a vision for the future. Exp. Biol. Med. 235, 1034–1045.

Norman, A.W., Henry, H.L., 2012. Vitamin D (Chapter 13). In: Erdman Jr., J.W., Macdonald, I.A., Zeisel, S.H. (Eds.), Present Knowledge in Nutrition, tenth ed. ILSI, Washington, DC, pp. 199–213.

Palacios, C., Gonzalez, L., 2014. Is vitamin D deficiency a major global public health problem? J. Steroid Biochem. Mol. Biol. 144, 138–145.

Peterlik, M., Boonen, S., Cross, H.S., et al., 2009. Vitamin D and calcium insufficiency-related chronic diseases: an emerging world-wide public health problem. Int. J. Environ. Res. Public Health 6, 2585–2607.

Prentice, A., 2008. Vitamin D deficiency: a global perspective. Nutr. Rev. 66, S153–S164.

Samuel, S., Sitrin, M.D., 2008. Vitamin D's role in cell proliferatioin and differentiation. Nutr. Rev. 66, S116–S124.

Rojas-Rivera, J., De La Piedra, C., Ramos, A., et al., 2010. The expanding spectrum of biological actions of vitamin D. Nephrol. Dial. Transpl. 25, 2850–2865.

Schuster, I., 2011. Cytochromes P450 are essential palyers in the vitamin D signaling system. Biochem. Biophys. Acta 1814, 186–199.

Schwartz, G.G., 2005. Vitamin D and the epidemiology of prostate cancer. Semin. Dial. 18, 276–289.

Singh, P.K., Campbell, M.J., 2015. Epigenetic regulation of cellular responses toward vitamin D (Chapter 8). In: Ho, E., Domann, F. (Eds.), Nutrition and Epigenetics. CRC Press, Boca Raton, FL, pp. 219–251.

Solomon, A.J., Whitham, R.H., 2010. Multiple sclerosis and vitamin D: a review and recommendations. Curr. Neurol. Neurosci. Rep. 10, 389–396.

Sterling, T.M., Khanal, R.C., Meng, Y., et al., 2014. Rapid pre-genomic responses to Vitamin D (Chapter 5). In: Dakshinamurti, K., Dakshinamurti, S. (Eds.), Vitamin Binding Proteins: Functional Consequences. CRC Press, New York, pp. 71–88.

Teegarden, D., Donkin, S.S., 2009. Vitamin D: emerging new roles in insulin sensitivity. Nutr. Res. Rev. 22, 82–92.

van Etten, E., Stoffels, K., Gysemans, C., et al., 2008. Regulation of vitamin D homeostasis: implications for the immune system. Nutr. Rev. 66, S125–S134.

Wagner, C.L., Greer, F.R., 2008. Prevention of rickets and vitamin D deficiency in infants, children, and adolescents. Am. Acad. Ped. 122, 1142–1152.

Wang, S., 2009. Epidemiology of vitamin D in health and disease. Nutr. Res. Rev. 22, 188–203.

Welsh, J., 2012. Cellular and molecular effects of vitamin D on carcinogenesis. Arch. Biochem. Biophys. 523, 107–114.

Wharton, B., Bishop, N., 2003. Rickets. Lancet 362, 1389–1400.

Willett, A.M., 2005. Vitamin E status and its relationship with parathyroid hormone and bone mineral status in older adults. Proc. Nutr. Soc. 64, 193–203.

Zemel, M.B., Sun, X., 2008. Calcitriol and energy metabolism. Nutr. Rev. 66, S139–S146.

Zhang, R., Naughton, D.P., 2010. Vitamin D in health and disease: current perspectives. Nutr. J. 9, 65–78.

Zheng, W., Teegarden, D., 2014. Vitamin A (Chapter 2). In: Zempleni, J., Suttie, J.W., Gregory, J.F., Stover, P.J. (Eds.), Handbook of Vitamins, fifth ed. CRC Press, New York, pp. 51–88.

Zittermann, A., Schleithoff, S.S., Koerfer, R., 2005. Putting cardiovascular disease and vitamin D insufficiency into perspective. Br. J. Nutr. 94, 483–492.

Chapter 8

Vitamin E

Chapter Outline

Anchoring Concepts

1. Vitamin E is the generic descriptor for all tocopherol and tocotrienol derivatives exhibiting qualitatively the biological activity of α-tocopherol.
2. The E vitamers are hydrophobic and, thus, are insoluble in aqueous environments (e.g., plasma, interstitial fluids, cytosol).
3. By virtue of the phenolic hydrogen on the C-6 ring hydroxyl group, the E vitamers have antioxidant activities.
4. Deficiencies of vitamin E have a wide variety of clinical manifestations in different species.

Vitamin E is a focal point for two broad topics, namely, biological antioxidants and lipid peroxidation damage. Vitamin E is related by its reactions to other biological antioxidants and reducing compounds that stabilize polyunsaturated lipids and minimize lipid peroxidation damage. In vivo lipid peroxidation has been identified as a basic deteriorative reaction in cellular mechanisms of aging processes, in some phases of atherosclerosis, in chlorinated hydrocarbon hepatotoxicity, in ethanol-induced liver injury, and in oxygen toxicity. These processes may be indicative of a universal disease, the chemical-deteriorative effects of which might be slowed by the use of increased amounts of antioxidants.

A. L. Tappel[1]

LEARNING OBJECTIVES

1. To understand the various sources of vitamin E
2. To understand the means of enteric absorption, transport, and cellular uptake of vitamin E
3. To understand the metabolic functions of E vitamers
4. To understand the interrelationships of vitamin E and other nutrients
5. To understand the physiological implications of high doses of vitamin E

VOCABULARY

Abetalipoproteinemia
Antioxidant
Apolipoprotein E (apoE)
Ataxia
Ataxia with vitamin E deficiency
α-Carboxyethylhydroxychroman (α-CEHC)
5′-α-Carboxymethylbutylhydroxychroman (5′-α-CMBHC)
Catalase
Conjugated diene
Cysteine
Cytochrome *P*-450
Encephalomalacia
Ethane
Exudative diathesis
Familial isolated vitamin E deficiency
Foam cells
Free radicals
Free-radical theory of aging
Glutathione (GSH)
Glutathione peroxidases

---

1. Aloys (Al) Tappel is a leader in the field of oxidative damage and antioxidant and was the first to suggest a synergistic relationship of vitamins E and C. His productive career was spent on the faculty of the Department of Food Science and Technology of the University of California at Davis.

The Vitamins. http://dx.doi.org/10.1016/B978-0-12-802965-7.00008-3

Glutathione reductase
Hemolysis
Hemolytic anemia
High-density lipoproteins (HDLs)
Hydroperoxide
Hydroxyl radical (HO•)
Hydrogen peroxide (H_2O_2)
Intraventricular hemorrhage
Ischemia–reperfusion injury
Lipid peroxidation
Lipofuscin
Lipoprotein lipase
Lipoproteins
Liver necrosis
Low-density lipoproteins (LDLs)
Malonyldialdehyde (MDA)
Mitochondrial hormesis
Mulberry heart disease
Myopathy
5-Nitro-tocopherol
Oxidative stress
Oxidized LDLs
Pentane
Peroxide
Peroxyl radical (ROO•)
Phospholipid transfer protein (PLTP)
Polyunsaturated fatty acids (PUFAs)
Prooxidant
Reactive oxygen species (ROS)
Redox tone
Resorption–gestation syndrome
Respiratory burst
Scavenger receptors
Selenium
Simon metabolites
Steatorrhea
Superoxide dismutases
Superoxide radical, $O_2\cdot^-$
Thiobarbituric acid (TBA)
Tocol
Tocopherol-associated protein (TAP)
Tocopherol ω-hydrolases
Tocopherols
α-Tocopherol
α-Tocopherol transfer protein (α-TTP)
β-Tocopherol
γ-Tocopherol
δ-Tocopherol
α-Tocopheronic acid
α-Tocopheronolactone
α-Tocopheroxyl radical
α-Tocopheryl hydroquinone
α-Tocopheryl phosphate
α-Tocopheryl polyethylene glycol-succinate
α-Tocopheryl quinone
α-Tocopheryl succinate
Tocotrienols
Very low density lipoprotein (VLDL)
White muscle disease

1. SIGNIFICANCE OF VITAMIN E

Vitamin E has a fundamental role in the normal metabolism of all cells. Therefore, its deficiency can affect several different organ systems. Its function is related to those of several other nutrients and endogenous factors that, collectively, comprise a multicomponent system that provides protection against the potentially damaging effects of reactive species of oxygen formed during metabolism or that are encountered in the environment. Both the need for vitamin E and the manifestations of its deficiency can be affected by antioxidant nutrients such as **selenium** and vitamin C and by exposure to **prooxidant** factors such as **polyunsaturated fatty acids** (**PUFAs**), air pollution, and ultraviolet (UV) light. Recent evidence indicates that vitamin E may also have nonantioxidant functions in regulating gene expression and cell signaling.

Unlike other vitamins, vitamin E is not only essentially nontoxic but also appears to be beneficial at dose levels appreciably greater than those required to prevent clinical signs of deficiency. Most notably, supranutritional levels of the vitamin have been useful in reducing the oxidation of **low-density lipoproteins** (**LDLs**) and, thus, reducing the risk of atherosclerosis. Although vitamin E is present in most plants, only plant oils are rich sources, and most people consume less than recommended levels.[2] Its low regular intake, ubiquitous and complex nature of its biological function, its demonstrated safety, and its apparent usefulness in combating a variety of **oxidative stress** disorders have generated enormous interest in this vitamin among the basic and clinical science communities and the lay public.

2. PROPERTIES OF VITAMIN E

Vitamin E Structure

The term vitamin E describes **tocopherols** and **tocotrienols**, both of which are isoprenoid side chain derivatives of 6-chromanol that show the biological activity of α-tocopherol. The tocopherols have side chains comprising three fully saturated isopentyl units; the most important of these is **α-tocopherol**. The tocotrienols have side chains containing three double bonds. For these compounds to have

2. Maras, J.E., Bermudez, O.I., Qiao, N., et al., 2004. J. Am. Diet. Assoc. 104, 567–575.

TABLE 8.1 Chromanol Head Group Substituents of Major E Vitamers

Vitamer	R_1	R_2	R_3
α-Tocopherol/α-tocotrienol	CH_3	CH_3	CH_3
β-Tocopherol/β-tocotrienol	CH_3	H	CH_3
γ-Tocopherol/γ-tocotrienol	H	CH_3	CH_3
δ-Tocopherol/δ-tocotrienol	H	H	CH_3
Tocol/tocotrienol	H	H	H

biological vitamin E activity, the obligate structural feature is a free hydroxyl or ester linkage on C-6 of the chromanol nucleus. Hence, the E vitamers are named according to the position and number of methyl groups on their chromanol nuclei (Table 8.1).

Chemical structures of the vitamin E group.

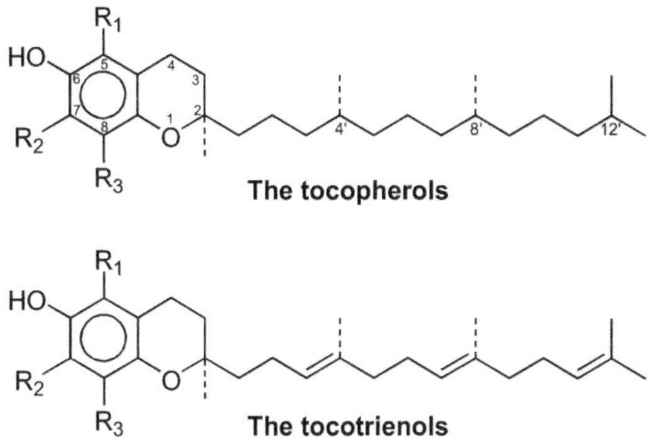

Because the tocopherol side chain contains two chiral centers (C-4′, C-8′) in addition to the one at the point of its attachment to the nucleus (C-2), eight stereoisomers are possible. However, only one stereoisomer occurs naturally: the *R,R,R*-form. The chemical synthesis of vitamin E produces mixtures of other stereoisomers, depending on the starting materials.[3] Although the use of phytol from natural sources (with the *R*-configuration at both the four- and 8-carbons) yields a product racemic at only the 2-position, commercial synthesis of vitamin E now uses primarily a fully synthetic side chain, which yields a mixtures of all eight possible stereoisomers, i.e., 2*RS*, 4′*RS*, and 8′*RS* compounds. This mixture is designated with the prefix all-*rac*- (e.g., all-*rac*-α-tocopherol, all-*rac*-α-tocopheryl acetate). The acetate esters of vitamin E are used in human nutritional supplements and animal feeding, whereas the unesterified (i.e., free alcohol) forms are used as antioxidants in foods and pharmaceuticals. Other forms (e.g., α-tocopheryl hydrogensuccinate, α-tocopheryl polyethylene glycol-succinate) are also used in multivitamin preparations.

3. Through the early 1970s, the commercial synthesis of vitamin E used, as the source of the side chain, isophytol isolated from natural sources (with the *R*-configuration at both the 4- and 8-carbons). Tocopherols so produced were racemic at only the C-2 position. The mixture, 2*RS*-α-tocopherol, was called D,L-α-tocopherol; its acetate ester was the form of commerce and was adopted as the international standard on which the biological activities of other forms of the vitamin are still based.

Vitamin E Chemistry

The tocopherols are light yellow oils at room temperature. They are insoluble in water, but are readily soluble in nonpolar solvents. Tocopherols and their acetates have absorption maxima in the range of 280–300 nm (α-tocopherol, 292 nm); however, their extinction coefficients are not great (70–91). Their fluorescence is significant (excitation, 294 nm; emission, 330 nm), particularly in polar solvents (e.g., diethyl ether or alcohols); this property has analytical utility. Being monoethers of a hydroquinone with a phenolic hydrogen (in the hydroxyl group at position C-6 in the chromanol nucleus), with the ability to accommodate an unpaired electron within the resonance structure of the ring (undergoing transition to a semistable chromanoxyl radical before being converted to tocopheryl quinone), they are good quenchers of free radicals and thus serve as antioxidants.

Vitamin E Stability

The properties that make tocopherols effective antioxidants also render them unstable under aerobic conditions. They are easily oxidized and can be destroyed by peroxides, ozone, and permanganate in a process catalyzed by light and accelerated by PUFAs and metal salts. They are resistant to acids; under anaerobic conditions, they are stable to bases. Tocopheryl esters, by virtue of the blocking of the C-6 hydroxyl group, are very stable in air; therefore, they are the forms of choice as food/feed supplements. Because tocopherol is liberated by the saponification of its esters, extraction and isolation of vitamin E requires the use of protective antioxidants (e.g., propyl gallate, ascorbic acid), metal chelators, inert gas environments, and subdued light.

Vitamin E Biopotency

The E vitamers vary in biopotency, depending on the positions and numbers of their nucleus methyl groups, which determine antioxidant activity, and the conformation of their side chains, which determine distribution and retention in tissues. The vitamer with greatest biopotency is the trimethylated (α) tocopherol (Table 8.2).

Expressing Vitamin E Activity

Vitamin E activity is shown to different degrees by several side chain isomers and methylated analogs of tocopherol and tocotrienol (Table 8.2), the epimeric configuration at the

TABLE 8.2 Relative Biopotencies of Vitamin E–Active Compounds

Trivial Designation	Systematic Name	Biopotency (IU/mg)[a]
R,R,R-α-tocopherol[b]	2*R*,4′*R*,8′*R*-5,7,8-Trimethyltocol	1.49
R,R,R-α-tocopheryl acetate	2*R*,4′*R*,8′*R*-5,7,8-Trimethyltocyl acetate	1.36
All-*rac*-α-tocopherol[c]	2*RS*,4′*RS*,8′*RS*-5,7,8-Trimethyltocol	1.1
All-*rac*-α-tocopheryl acetate	2*RS*,4′*RS*,8′*RS*-5,7,8-Trimethyltocylacetate	1.0
R,R,R-β-tocopherol	2*R*,4′*R*,8′*R*-5,8-Dimethyltocol	0.12
R,R,R-γ-tocopherol	2*R*,4′*R*,8′*R*-5,7-Dimethyltocol	0.05
R-α-tocotrienol	*trans*-2*R*-5,7,8-Trimethyltocotrienol	0.32
R-β-tocotrienol	*trans*-2*R*-5,8-Dimethyltocotrienol	0.05
R-γ-tocotrienol	*trans*-2*R*-5,7-Dimethyltocotrienol	—

[a]*Based chiefly on rat gestation–resorption bioassay data.*
[b]*Formerly called* D-*α-tocopherol.*
[c]*Formerly called* DL-*α-tocopherol, this form remains the international standard despite the fact that it has not been produced commercially for many years.*

TABLE 8.3 Standards for Potency of Major Vitamers E

Vitamer	mg/IU	α-Tocopherol Equivalents (mg)
All-*rac*-α-tocopherol	0.91	0.74
All-*rac*-α-tocopheryl acetate	1	0.67
R,R,R-α-tocopherol	0.67	1
R,R,R-α-tocopheryl acetate	0.74	0.91

2-position being important in determining biological activity. Therefore, the use of an international standard facilitated the referencing of these various sources of vitamin E activity, reflecting differences in their absorption, transport, retention and metabolism, and their intrinsic biopotency. The original preparation "D,L-α-tocopheryl acetate"[4] that served as the international standard has not existed for more than 30 years; R,R,R-α-tocopherol is now used as the international standard (Tables 8.3 and 8.4).[5]

Some of the E vitamers commonly found in foods (**β-** and **γ-tocopherol**, the **tocotrienols**) have little biological activity. The most biopotent vitamer, i.e., the vitamer of greatest interest in nutrition, is α-tocopherol, which occurs naturally as the *RRR* stereoisomer [(*RRR*)-α-tocopherol].

3. SOURCES OF VITAMIN E

Distribution in Foods

Vitamin E is synthesized only by photosynthetic organisms – plants, algae, and some cyanobacteria – where it is thought to function as a protective antioxidant in germination and cold adaptation. All higher plants appear to contain **α-tocopherol** in their leaves and other green parts. Because α-tocopherol is contained mainly in the chloroplasts of plant cells (whereas, the β-, γ-, and δ-vitamers are usually found outside of these particles), green plants tend to contain more vitamin E than yellow plants. The richest food sources are plant oils: wheat germ, sunflower, and safflower oils are rich sources of α-(*RRR*)-tocopherol; corn and soybean oils contain mostly γ-(*RRR*)-tocopherol. Some plant tissues, notably bran and germ fractions[6], can also contain tocotrienols, often in esterified form, unlike the tocopherols which exist only as free alcohols. Animal tissues tend to contain low amounts of α-tocopherol, the highest levels occurring in fatty tissues. These levels vary according to the dietary intake of the vitamin.[7] Because vitamin E occurs naturally in fats and oils, reductions in fat intake can be expected also to reduce vitamin E intake. An amphipathic

4. Through the 1970s, the standard was called D,L-α-tocopheryl acetate; now it would be called (2*RS*)-α-tocopheryl acetate. Because of uncertainty about the proportions of the two diastereoisomers in that mixture, once the supply was exhausted it was impossible to replace it.
5. This system distinguishes only the methylated analogs and not the particular diastereoisomers possible for each.

6. Palm oil and rice bran have high concentrations of tocotrienols; other natural sources include coconut oil, cocoa butter, soybeans, barley, and wheat germ.
7. Muscle from beef fed high levels of vitamin E (e.g., 1300 IU/day) before slaughter can yield vitamin E in excess of 16 nmol/g; this level is effective in reducing postmortem oxidation reactions, thus delaying the onset of meat discoloration because of hemoglobin oxidation and the development of oxidative rancidity.

TABLE 8.4 Relative Biopotencies (%) of Tocopherols and Tocotrienols by Different Bioassays

Vitamer	Prevention of Fetal Resorption (Rat)	Prevention of Hemolysis (Rat)	Prevention of Myopathy (Chick)	Therapy for Myopathy (Rat)
α-Tocopherol	100	100	100	100
β-Tocopherol	25–40	15–27	12	—
γ-Tocopherol	1–11	3–20	5	11
δ-Tocopherol	1	0.3–2	—	—
α-Tocotrienol	28	17–25	—	28
β-Tocotrienol	5	1–5	—	—

metabolite, **α-tocopheryl phosphate**[8], has also been identified at trace levels in foods and animal tissues.

The important sources of vitamin E in human diets and animal feeds are vegetable oils and, to lesser extents, seeds and cereal grains (Table 8.5). The dominant dietary form (70% of tocopherols in American diets) is γ-tocopherol (Tables 8.6 and 8.7). Wheat germ oil is the richest natural source, containing 0.9–1.3 mg of α-tocopherol per gram, i.e., about 60% of its total tocopherols. The seeds and grains from which these oils are derived also contain appreciable amounts of vitamin E. Plants also synthesize tocotrienols. The richest food sources are rice bran oil, in which tocotrienols comprise most of the E vitamers; and palm oil, in which tocotrienols comprise 70% of total E vitamers. Cereals contain small amounts of tocotrienols. Accordingly, cereals in general and wheat germ in particular are good sources of the vitamin. Foods that are formulated with vegetable oils (e.g., margarine, baked products) tend to vary greatly in vitamin E content because of differences in the types of oils used and the thermal stabilities of the E vitamers present.[9] α-Tocopherol is used in dietary supplements.[10] Regardless of the form consumed, α-tocopherol is the main form found in tissues.

The processing of foods and feedstuffs can remove substantial amounts of vitamin E. Vitamin E losses can occur as a result of exposure to peroxidizing lipids formed during the development of oxidative rancidity of fats and to other oxidizing conditions such as drying in the presence of sunlight and air, the addition of organic acids,[11] irradiation, and canning. Milling and refining can reduce the vitamin E content by removal of tocopherol-rich bran and germ fractions and because of the use of bleaching agents (e.g., hypochlorous acid) to improve the baking characteristics of the flour. Some foods (e.g., milk and milk products) also show marked seasonal fluctuations in vitamin E content related to variations in vitamin E intake of the host (e.g., vitamin E intake is greatest when fresh forage is consumed). The many potential sources of vitamin E loss mean that the vitamin E contents of foods and feedstuffs vary considerably.[12]

High–Vitamin E Animal Foods

High-level vitamin E supplementation (e.g., α-tocopheryl acetate fed at levels 10- to 50-fold standard practice) of the diets of poultry, swine, and beef have been found to be effective in increasing the α-tocopherol contents of many tissues. High levels of the vitamin in edible tissues serve to inhibit postmortem oxidative production of off-flavors (oxidative rancidity of lipids) and color (hemoglobin oxidation), which reduce drip loss and increase the effective shelf life of the retail cuts of meat. Studies show that the incorporation of α-tocopherol into muscle is linearly related to α-tocopherol intake to about 220 IU/day for pork and 600 IU/day for beef at which each approached an apparent asymptote of c. 5 μg/g tissue, which corresponded to peroxidizability minima.[13] α-Tocopheryl acetate is often used to protect hens from aflatoxin contaminants of feedstuffs; the α-tocopherol content of eggs has also been increased using high-level vitamin E supplements to laying hen diets. The addition of c. 270 IU α-tocopheryl acetate per kilogram of feed increased egg α-tocopherol threefold, to about 18 mg per egg.[14]

8. Water-soluble and resistant to both acid and alkaline hydrolysis, this metabolite has been missed by traditional methods of vitamin E analysis.

9. Tocotrienols tend to be less stable to high temperatures than tocopherols; baking tends to destroy them selectively.

10. Water-dispersible formulations have been developed for treating lipid-malabsorbing patients.

11. The addition of 1% propionic acid (as an antifungal agent) to fresh grain can destroy up to 90% of its vitamin E.

12. For example, refining losses in edible plant oils are typically 10–40%, but can sometimes be much greater.

13. Sales, J., Koukolová, V., 2011. J. Anim. Sci. 89, 2836–2848.

14. Sujatha, T., Narahari, D., 2011. J. Food Sci. Technol. 48, 494–497.

TABLE 8.5 Significant Food Sources of Vitamin E (≥2 mg/100 g)

Food	Vitamin E (mg/100 g)
Fats and Oils	
Wheat germ oil	149.4
Sunflower oil	41.1
Rice bran oil	32.3
Canola oil	17.5
Palm oil	15.9
Peanut oil	15.7
Olive oil	14.4
Corn oil	14.3
Mayonnaise	11.8
Soybean oil	8.2
Magarine, hard type	3.1
Chicken fat, beef tallow	2.7
Egg yolk	2.3
Butter	2.3
Nuts and Seeds	
Sunflower kernels, dried	35.2
Almonds, dry roasted, unblanched	23.9
Filberts (hazelnuts), dry roasted	15.3
Peanuts, dry roasted	4.9
Pistachio nuts, dry roasted	2.2
Walnuts, dried	2.1
Fish	
Swordfish, dry heat cooked	2.4
Shrimp, moist heat cooked	2.2
Vegetables and Fruits	
Dandelion greens, raw	3.4
Turnip greens, boiled	2.1
Avocados, raw	2.1
Spinach, raw	2.0
Other	
Chilli powder	38.1
Curry powder	25.2
Wheat germ	16.0

From USDA National Nutrient Database for Standard Reference, Release 28. http://www.ars.usda.gov/ba/bhnrc/ndl.

4. ABSORPTION OF VITAMIN E

The primary site of absorption appears to be the medial small intestine. Esterified forms of the vitamin E are hydrolyzed, probably by a mucosal esterase; the predominant forms absorbed are free alcohols. Most studies have shown no appreciable differences in the efficiency of absorption of the acetate ester and free alcohol forms, nor differences in the absorption of the various tocopherol and tocotrienol vitamers. It is clear, however, that regardless of the form absorbed, higher intakes lead to higher amounts of absorption but lower absorption efficiencies (i.e., fractional absorption). At nutritionally important intakes, variable (generally, 20–70%[15]) absorption efficiencies have been reported, with large portions of ingested vitamin E appearing in the feces.

Micelle-Dependent Diffusion

Similar to other hydrophobic substances, vitamin E appears to be absorbed by nonsaturable passive diffusion dependent on micellar solubilization and, hence, the presence of bile salts and pancreatic juice. It is clear that the enteric absorption of vitamin E is dependent on the adequate absorption of lipids, the process requires the presence of fat in the lumen of the gut, and the secretion of pancreatic esterases for the release of free fatty acids from dietary triglycerides, bile acids for the formation of mixed micelles, and esterases for the hydrolytic cleavage of tocopheryl esters when those forms are consumed. Individuals unable to produce pancreatic juice or bile (e.g., patients with biliary obstruction, cholestatic liver disease, pancreatitis, cystic fibrosis, short bowel syndrome) show impaired absorption of vitamin E, and other fat-soluble substances dependent on micelle-facilitated diffusion for their uptake. The micelle-dependent absorption of vitamin E would imply a need for dietary fat to facilitate the process; need for lipid would explain reports of vitamin E in dietary supplements not being well absorbed unless taken with a meal.[16] Studies with radiolabeled α-tocopherol have shown its enteric absorption in humans to be impaired by dietary fat levels less than c. 10% (i.e., 21% of total calories) (Fig. 8.1);[17] however, absorption of that vitamer by the rat was not impaired by feeding a diet containing fat as only 1.6% of total calories. It has been suggested that children can adequately absorb the fat-soluble vitamins with fat intakes as low as 5 g per day. Furthermore, tocopherols can interact with

15. The enteric absorption of γ-tocopherol appears to be only 85% of that of α-tocopherol.

16. Leonard, S.W., Good, C.G., Gugger, T.E., et al., 2004. Am. J. Clin. Nutr. 79, 86–92.

17. Bruno, R.S., Leonard, S.W., Park, S.I., et al., 2006. Am. J. Clin. Nutr. 83, 299–304.

TABLE 8.6 E Vitamers in Fats and Oils

	Tocopherols (%)			Tocotrienols (%)			
Item	α	γ	δ	α	β	γ	δ
Animal Fats							
Lard	>90	<5		<5			
Butter	>90	<10					
Tallow	>90	<10					
Plant Oils							
Soybean	4–18	58–69					
Cotton	51–67	33–49					
Maize	11–24	76–89					
Coconut	14–67		<17	<14	<3	<53	<17
Peanut	48–61	39–52					
Palm	28–50		<9	16–19	4	34–39	<9
Safflower	80–94	6–20					
Olive	65–85				15–35		

From Chow, C.K., 1985. World Rev. Nutr. Diet. 45, 133–166.

TABLE 8.7 E Vitamers in Grains and Oil Seeds

	Tocopherols (%)				Tocotrienols (%)	
Item	α	β	γ	δ	α	γ
Grains						
Maize	6–15		29–55		5–10	34–77
Oats	4–8	<1			10–22	
Milo	4–7		14–17		<1	
Barley	8–10	1–2	3–4		23–28	3
Wheat	8–12	4–6			2–3	
Oil Seeds						
Soybean	1–3		3–33	2–6		trace
Cotton seed	1–18		5–18			1–2

From Cort, W.M., Vicente, T.S., Waysek, E.H., et al., 1983. J. Agric. Food Chem. 31, 1330–1333.

PUFAs in the intestinal lumen; this can result in absorption being stimulated by medium-chain triglycerides and inhibited by linoleic acid.

Role of Mucosal Receptors

Evidence has been presented for roles of cholesterol and lipid transporters in the uptake of α-tocopherol by enterocytes. Several have been suggested: the scavenger receptor class B type I (SR-BI),[18] CD36,[19] NPC1L1,[20]

18. Mice lacking SR-BI show marked reductions in the amounts of α-tocopherol in plasma (particularly, in the HDL fraction) and tissues (Mardones, P., Strobel, P., Miranda, S., et al., 2002. J. Nutr. 132, 443–449).
19. Cluster determinant 36, also called fatty acid translocase.
20. Niemann-Pick C1–like 1; mutations cause a lysosomal disorder, Niemann-Pick type C disease characterized by massive intracellular accumulation of cholesterol and other lipids.

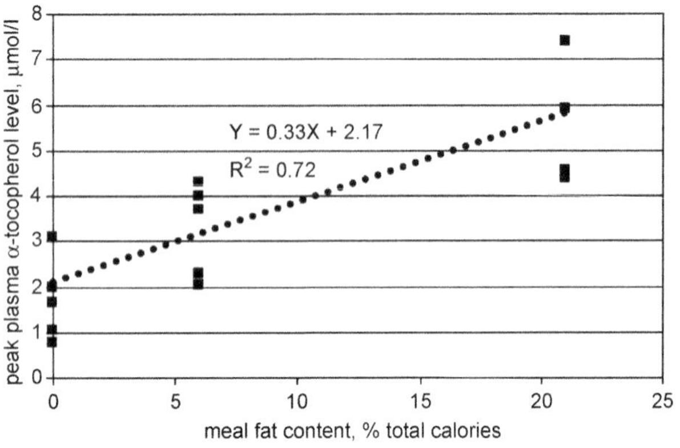

FIGURE 8.1 Effect of the fat level of a meal on the absorption of deuterium-labeled α-tocopherol from that meal by healthy adults. *After Bruno, R.S., Leonard, S.W. Park, S.I., et al., 2006. Am. J. Clin. Nutr. 83, 299–304.*

and ABCA1.[21] ABCA1 has been found to be involved in the export of tocopherols from enterocytes into the lymphatic circulation.[22] Its interaction with α-tocopherol transfer protein (α-TTP) promotes preferential trafficking of α-tocopherol over non–α-vitamers. The inhibition of α-tocopherol enteric absorption by carotenoids, green tea catechins, and γ-tocopherol may involve competitive binding to these receptors.

Uptake Into Lymphatic Circulation

Absorbed vitamin E, similar to other hydrophobic substances, enters the lymphatic circulation[23] in association with nascent triglyceride-rich chylomicra. Studies with radiolabeled compounds have showed preferential lymphatic uptake of α-tocotrienol over α-tocopherol and γ- and δ-tocotrienols. Within the enterocytes, vitamin E combines with other lipids and apolipoproteins to form chylomicra, which are released into the lymphatics in mammals or the portal circulation in birds and reptiles. Thus, the kinetics of vitamin E absorption are biphasic, reflecting the initial uptake of the vitamin by existing chylomicra followed by a lag phase because of the assembly of new chylomicra.

5. TRANSPORT OF VITAMIN E

Being virtually insoluble in aqueous environments, vitamin E is dependent on carriers for its transport to the tissues. Unlike vitamins A and D, vitamin E does not have a specific carrier protein in the plasma. Instead, it is rapidly transferred from chylomicra to plasma lipoproteins to which it binds nonspecifically.

Role of Chylomicra

Vitamin E is transported to the liver by triglyceride-rich chylomicra. The vitamin appears to partition directly into the plasma membranes of parenchymal cells. It may also be taken up with circulating lipoproteins to which the vitamin can transfer as chylomicra are metabolized (Fig. 8.2).

Roles of Lipoproteins

Vitamin E is transported from the liver to peripheral tissues by **VLDLs** synthesized by parenchymal cells. The concentration of vitamin E in plasma is linearly related to the intake of the vitamin up to about 200 mg/day (Fig. 8.3). The diminishing response above that level reflects uptake by the liver for the selective incorporation of vitamin E into nascent VLDLs. Although the majority of the triglyceride-rich VLDL remnants are returned to the liver, some are converted by lipoprotein lipase to **LDLs**. It appears that, during this process, vitamin E also transfers spontaneously to apolipoprotein B (apoB)–containing lipoproteins including the VLDLs, LDLs, and **high-density lipoproteins (HDLs)**. Therefore, plasma tocopherols are distributed among these three lipoprotein classes, with the more abundant LDL and HDL classes comprising the major carriers of vitamin E. As each class of lipoproteins derives its tocopherols ultimately from chylomicra, differences in their α-tocopherol transport comprise the major source of interindividual variation in response to ingested vitamin E. These kinetics are altered by hypercholesterolemia and hypertriglyceridemia in which tocopherol uptake into and turnover in the plasma is reduced. Tocopherol metabolism is also related to **apolipoprotein E (apoE)**, which affects the hepatic binding and catabolism of several classes of lipoproteins. ApoE genotype in the mouse has been found to affect genes encoding for proteins involved in α-tocopherol transport and catabolism; the apoE4 genotype was associated with lower tissue retention of α-tocopherol apparently because of increased retention of the vitamer by LDLs.[24] These differences are accompanied by 15–60% reductions in the levels of mRNA for factors involved in vitamin E binding and transport (SR-BI, LDL receptor, LDL receptor–related protein, ABCA1, the multidrug-resistant transporter) and a more than doubling of the mRNA for CYP3A4, which is involved in tocopherol side chain metabolism.[25] In humans, the apoE4 genotype[26] has been

21. ATP-binding cassette A1.
22. Oram, J.F., Vaughan, A.M., Stocker, R., 2001. J. Biol. Chem. 276, 39,898–39,902.
23. That is, the portal circulation in birds, fishes and reptiles.
24. Huebbe, P., Lodge, J.K., G. Rimbach, 2010. Mol. Nutr. Food Res. 54, 623–630.
25. Huebbe, P., Jofre-Monseny, L, Rimbach, G., 2009. IUBMB Life 61, 453–456.
26. The apoE4 genotype is an important genetic risk factor for age-dependent chronic diseases, including cardiovascular disease and Alzheimer's disease.

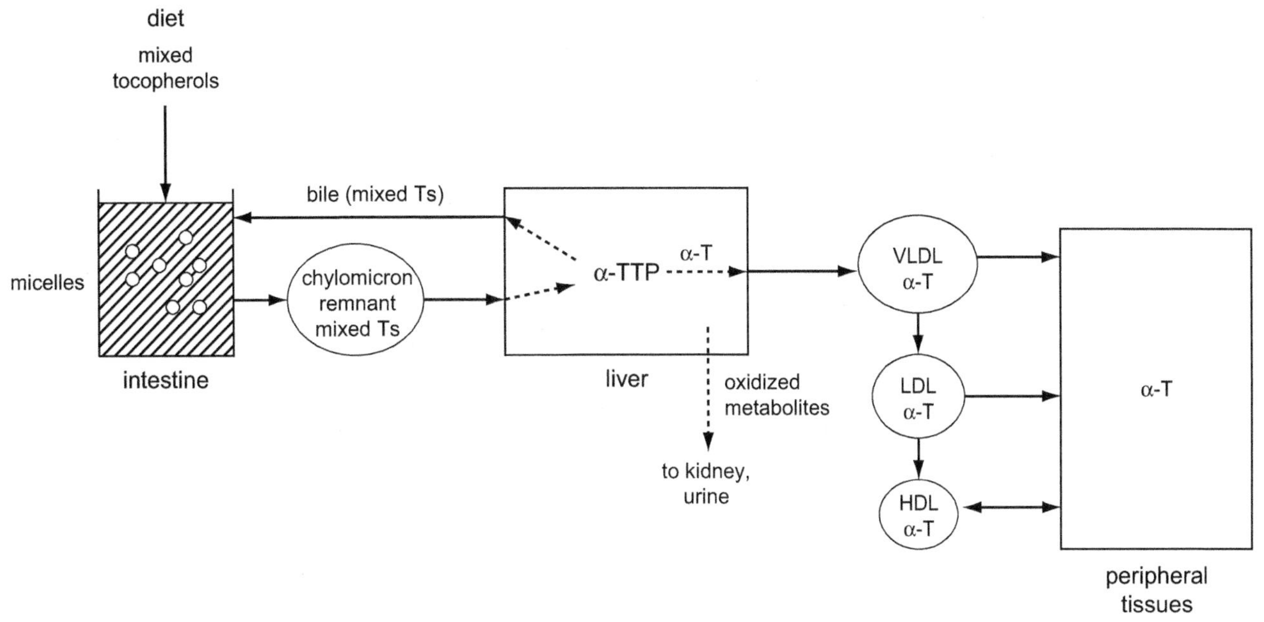

FIGURE 8.2 Absorption and transport of vitamin E. Abbreviations: α-T, α-tocopherol: mixed Ts, mixed tocopherols; **α**-TTP, **α**-tocopherol transfer protein.

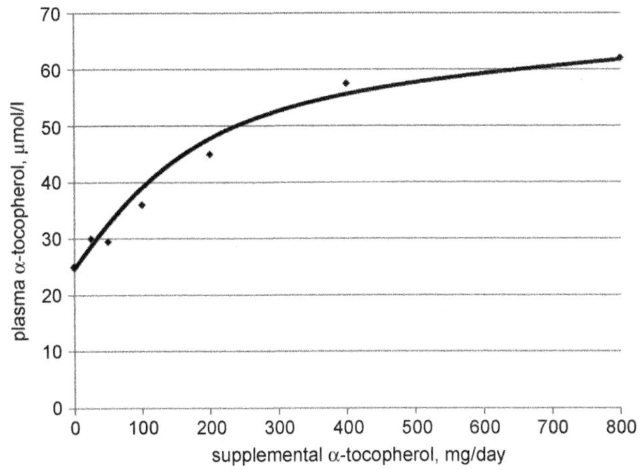

FIGURE 8.3 Plasma response to supplemental α-tocopherol. *After Princen, H.M., van Duyvenvoorde, W. Buytenhek, R., et al., 1995. Arterioscler. Throm. Vasc. Biol. 15, 325–333.*

TABLE 8.8 Relationship of apo E Genotype and Plasma Tocopherol Levels in Children[a]

Genotype	n	α-Tocopherol (μM)[b]	γ-Tocopherol (μM)[b]
E2/2	6	26.5[a] (23.8–29.2)	3.10[a] (2.27–4.22)
E3/2	89	20.8[b] (20.1–21.5)	1.90[b] (1.75–2.07)
E3/3	660	21.3[b] (21.1–21.6)	2.06[a,b] (2.00–2.12)
E4/3	150	21.4[b] (20.9–21.9)	2.05[a,b] (1.92–2.18)
E4/2	8	21.7[a,b] (19.4–23.9)	1.81[a,b] (1.39–2.36)
E4/4	13	19.0[b] (17.2–20.8)	1.84[a,b] (1.49–2.27)

p<.05.
[a]*Ortega, H., Casellia, P., Gomez-Coronado, D., et al., 2005. Am. J. Clin. Nutr. 81, 624–632.*
[b]*Means (95% CI); means with common superscripts are not significantly different,* p>*.05.*

associated with lower plasma vitamin E concentrations (Table 8.8). Tocopherol exchanges rapidly between the lipoproteins, mediated by the **phospholipid transfer protein (PLTP)**[27] and between lipoproteins and erythrocytes, which contain 15–25% of the vitamin E in plasma. Thus, the concentration of vitamin E is correlated with the number of erythrocytes in blood.[28] A fourth of total erythrocyte vitamin E turns over every hour. Postprandial levels of tocopherols exceed those of tocotrienols; this reflects the more rapid metabolic degradation of the latter.

27. Rats, horses, and chicks transport 70–80% of plasma α-tocopherol with HDLs, 18–22% with LDLs, and <8% with VLDLs. Human females, too, transport α-tocopherol preferentially with HDLs; but males transfer most (65%) with LDLs, only 24% with HDLs and 8% with VLDLs.

28. Patients with abetalipoproteinemia are notable exceptions; they may show normal erythrocyte tocopherol concentrations even though their serum tocopherols levels are undetectable.

Cellular Uptake

The cellular uptake of vitamin E appears to occur in the same ways that other lipids are transferred between lipoproteins and cells:

- **Lipase-mediated lipid transfer**. Uptake of α-tocopherol from the amphipathic lipoprotein outer layer is mediated in a directional way by PLTP[29] and lipoprotein lipase-mediated exchange from chylomicra. This route is thought to be important in the uptake of vitamin E by cells that express lipase (adipose, muscle, brain) and particularly important in the transport of α-tocopherol across the blood–brain barrier into the central nervous system.[30]
- **Receptor-mediated endocytosis of lipoproteins**. Evidence suggests that the binding of lipoproteins to specific cell surface receptors must occur to allow the vitamin E to enter cells either by diffusion or with the bulk entrance of lipoprotein-bound lipids. That LDL receptor–deficient cells cannot take up LDL-bound vitamin E at normal rates suggest the involvement of those receptors. Such deficiencies do not necessarily reduce tissue tocopherol levels; but studies in animal models show apoE genotype to be a determinant of both circulating and tissue tocopherol levels.[31] There is also evidence for receptor-mediated uptake of α-tocopherol without uptake of the apolipoprotein, as described for the cellular uptake of cholesterol from HDLs. This appears to involve the scavenger receptors SR-BI[32] and CD36. Polymorphisms of CD36 have been associated with differences in plasma α-tocopherol concentration.[33]
- **Membrane lipid transporter–mediated uptake**. ABCA1, a transporter of cholesterol and other lipids, has been shown to be involved in the export of tocopherols from cells. Its interaction with α-TTP promotes preferential trafficking of α-tocopherol over non–α-vitamers.

Role of the α-Tocopherol Transfer Protein

Although all E vitamers are taken up by the liver, only α-tocopherol is released into the circulation. This is because of the function of a specific tocopherol-binding protein, the **α-tocopherol transfer protein (α-TTP)**. Originally described in rat liver cytosol where it was found to facilitate the transfer of α-tocopherol between microsomes and mitochondria, α-TTP has been identified in liver, brain, spleen, lung, kidney, uterus, and placenta. It is highly conserved: the rat and human liver proteins show 94% sequence homology, and homologies to the interphotoreceptor retinol-binding protein (IRBP), cellular retinal-binding protein (CRALBP), and a PLTP.

This 32-kDa protein binds α-tocopherol with high affinity. It binds *RRR*-α-tocopherol ninefold more avidly than it does the SRR stereoisomer.[34] It can also bind tocotrienols, although with only 12% of the affinity it shows for α-tocopherol, apparently because of the difficulty of the unsaturated phytyl side chains to fit within the ligand-binding pocket. That pocket comprises an N-terminal helical domain and a C-terminal domain; the latter contains a fold that readily accommodates α-tocopherol and four water molecules, two of which are hydrogen bonded to the hydroxyl group on the chroman ring. Binding affinity for tocopherols is also determined by the degree of methylation of the chroman ring, which determines the extent of van der Waals contacts with the pocket. The relative affinities are as follows: α-tocopherol (100) > β-tocopherol (38) > γ-tocopherol (19) > α-tocotrienol (8.5) > δ-tocopherol (5).

The liganded α-TTP acts as a chaperone for α-tocopherol, taking up the vitamer from endocytic vesicles in which it binds to phospholipid membranes and moving it through the cytoplasm to transport vesicles that travel to the plasma membrane such that the vitamer is ultimately secreted complexed to lipoprotein particles in the circulation.[35] The uptake phase of this process is thought to involve the SR-BI; the discharge phase is thought to involve ABCA1, which interacts with α-TTP to release the ligand from its binding pocket and transfer it to apoA1 and HDL. The selectivity of α-TTP for α-tocopherol contributes to the differences in tissue retention and biopotency of these vitamers,[36] explaining the fact that, whereas γ-tocopherol is the dominant dietary form of vitamin E, α-tocopherol constitutes 90% of body vitamin E. Animal models that do not express α-TTP absorb α-tocopherol normally, showing normal levels in chylomicra, but fail to release the vitamer from the liver,

29. Mice lacking PLTP show high plasma levels of α-tocopherol in apoB-containing lipoproteins (Jiang, X.C., Tall, A.R., Qin, S., et al., 2002. J. Biol. Chem. 277, 31,850–31,856).
30. Mice lacking lipoprotein lipase show low brain α-tocopherol levels (although no associated pathologies have been reported; Goti. D., Balazs, Z., Panzenboeck, U., et al., 2002. J. Biol. Chem. 277:28,537–28,544).
31. apoE4 mice show lower tissue α-tocopherol levels than apoE3 mice (Huebbe, P., Lodge, J.K., Ribach, G., 2010. Mol. Nutr. Food Res. 54, 623–630).
32. Mice lacking SR-BI show marked reductions in the amounts of α-tocopherol in plasma (particularly, in the HDL fraction) and tissues (Mardones, P., Stobel, P., Miranda, S., et al., 2002. J. Nutr. 132, 443–449).
33. Lecompte, S., de Edelenyi, F.S., Goumide, L., et al., Am. J. Clin. Nutr. 93, 644–651.
34. The preferential incorporation of the *RRR*-α-isomer into milk by the lactating sow (Lauridson, C., Engel, H., Jensen, S.K., et al., 2002. J. Nutr. 132, 1258–1264) suggests the presence of α-TTP in the mammary gland.
35. Qian, J., Morley, S., Wilson, K., et al., 2005. J. Lipid Res. 46, 2072–2082; Qian, J. Altkinson, J., Manor, D., 2006. Biochem. 45, 8236–8242; Negris, Y., Meydani, M., Zingg, J.M., et al., 2007. Biochim. Biophys. Res. Commun. 359, 348–353.
36. Neither LDL receptor nor lipoprotein lipase mechanisms of vitamin E uptake by cells discriminate between these stereoisomers; yet, α-tocopherol predominates in plasma because of its preferential incorporation into nascent VLDLs, whereas the form often predominating in foods, γ-tocopherol, is left behind only to be more rapidly excreted.

accumulating hepatic stores at the expense of α-tocopherol in peripheral tissues.[37]

The expression of α-TTP occurs predominantly in the liver in adults; but studies in the Zebrafish show that it plays an essential role in embryogenesis. Studies with immortalized human hepatocytes have shown that the α-TTP messenger RNA is increased in response to oxidative stress, hypoxia, agonists of the nuclear receptors retinoid X receptor and peroxisome proliferator-activated receptor alpha, and increased cAMP levels mediated by the cAMP response element-binding transcription factor.[38] Allelic variants in the human α-TTP gene have been associated with differences in circulating α-tocopherol levels.[39] More serious outcomes have been identified in subjects with other α-TTP gene defects:

- **Familial isolated vitamin E** deficiencyhas been identified in a group of Americans with sporadic vitamin E deficiency, with poor incorporation of *RRR*-α-tocopherol into their VLDLs and an inability to discriminate between the *RRR* and *SRR* vitamers. These patients have exceedingly low circulating tocopherol concentrations unless maintained on high-level vitamin E supplements (e.g., 1 g/day). If untreated, they experience progressive peripheral neuropathy (characterized by pathology of the large-caliber axons of sensory neurons) and **ataxia**.
- **Deletion of the terminal 10% of the α-TTP peptide chain** has been identified in several highly consanguineous Tunisian families whose members show low serum tocopherol levels and ataxia both responsive to high-level vitamin E supplements.[40]
- A missense mutation that inserts histidine in place of glutamine at position 101 of the **α-TTP peptide chain has been identified in** Japanese subjects, whose α-TTP has only 11% of the transfer activity of the wild-type protein. Heterozygous individuals show no clinical signs, but have circulating tocopherol levels 25% lower than those of normal subjects.

Other Tocopherol-Binding Proteins. Tocopherols bind other hydrophobic ligand-binding proteins that share the *cis*-retinal binding motif[41] of α-TTP. These proteins include the cellular retinoic acid–binding protein **CRALBP** and the interphotoreceptor retinol–binding protein **IRBP** (*see* Chapter 6), apparently in the same site as retinol, which readily displaces the tocopherol. A related, tocopherol-associated protein (TAP) has been identified;[42] it appears to have the same amino acid sequence as the previously described supernatant protein factor and may be the same protein. TAP has been found in most tissues, with greatest concentrations in liver, brain, and prostate. That TAP may be a transcription factor is suggested by the finding that liganded TAP translocates from the cytosol to the nucleus and activated gene transcription.[43] TAP appears to have a role in regulating cell growth: its knock-down enhanced prostate cell growth, whereas its overexpression suppressed growth of prostate cancer cells.[44] Evidence also suggests that TAP can serve as a tumor suppressor.[45] TAP mRNA has been found to be negatively associated with tumor stage in breast cancer and, hence, a prospective biomarker of the less aggressive breast carcinoma.

Tissue E Vitamers

Tocopherols. The preferential uptake of α-tocopherol results in that vitamer predominating in tissues. Those contents vary among tissues and tend to be related to vitamin E intake, showing no deposition or saturation thresholds (Table 8.9). In fact, increased intake of α-tocopherol displaces non–α-vitamers in tissues (Fig. 8.4). Neural tissues exhibit very efficient retention, i.e., very low apparent turnover rates, of the vitamin.[46] Kinetic studies indicate that tissues have two pools of the vitamin: a *labile*, rapidly turning over pool; and a *fixed*, slowly turning over pool. The labile pools predominate in such tissues as plasma and liver, as the tocopherol contents of those tissues are depleted rapidly under conditions of vitamin E deprivation. Non–*RRR*-α-tocopherols are quickly removed from the plasma. In humans, *RRR*-α-tocopherol remains in plasma nearly four times longer than *SRR*-α-tocopherol (apparent half-lives: 13 h versus 48 h), and three times longer than *RRR*-γ-tocopherol.[47]

Tocotrienols. Tocotrienols can occur in tissues, but in much lower amounts than tocopherols. How they are taken up is not clear. Although tocotrienols can be bound by α-TTP, that binding is much weaker than that of the tocopherols and evidence indicates that they are taken up by other means. Genetic deletion of α-TTP in the mouse has been found to produce classical signs of vitamin E deficiency (midgestational fetal deaths) in α-tocopherol–fed dams,

37. Leonard, S.W., Terasawa, Y., Farese, Jr., R.V., et al., 2002. Am. J. Clin. Nutr. 75, 555–560.
38. Ulatowski, L., Dreussi, C., Noy, N., 2012. Free Radic. Biol. Med. 53, 2318–2326.
39. Wright, M.E., Peters, U., Gunter, M.J., et al., 2011. Cancer Res. 69, 1429–1438.
40. Previously called **Friedreich ataxia,** this condition is now called **ataxia with vitamin E deficiency**.
41. CRAL_TRIO.
42. Stocker, A., Zimmer, S., Spycher, S.E., et al., 1999. IUBMB Life 48, 49–55.
43. Yamauchi, J., Iwamoto, T., Kida, S., 2001. Biochem. Biophys. Res. Commun. 285, 295–299.
44. Ni, J., Wen, X., Yao, J., et al., 2005. Cancer Res. 65, 9807–9816.
45. Wang, X., Ring, B.Z., Seitz, R.S., 2015. BMC Clin. Pathol. 15, 21–31.
46. For example, weanling rats from vitamin E–adequate dams do not show neurologic signs of vitamin E deficiency for as long as 7 weeks when fed a vitamin E–free diet.
47. Traber. M.G., Ramakrishnan, R., Kayden, H.J., 1994. Proc. Natl. Acad. Sci. U.S.A. 91, 10,005–10,008; Leonard, S.W., Paterson, E., Atkinson, J.E., et al., 2005. Free Radic. Biol. Med. 38, 857–866.

TABLE 8.9 Concentrations of α-Tocopherol in Human Tissues

	α-Tocopherol	
Tissue	**Tissue (μg/g)**	**Lipid (μg/g)**
Plasma	9.5	1.4
Erythrocytes	2.3	0.5
Platelets	30	1.3
Adipose	150	0.2
Kidney	7	0.3
Liver	13	0.3
Muscle	19	0.4
Ovary	11	0.6
Uterus	9	0.7
Testis	40	1.0
Heart	20	0.7
Adrenal	132	0.7
Hypophysis	40	1.2

From Machlin, L. J., 1984. Handbook of Vitamins. Machlin, L., (Ed.). Marcel Dekker, New York, p. 99.

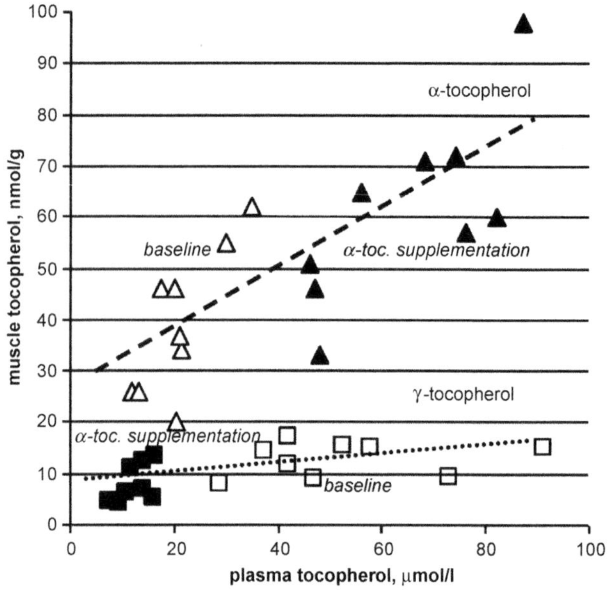

FIGURE 8.4 Correlation of α-tocopherol and γ-tocopherol contents of muscle (*m.* gastrocnemius) and plasma, respectively. Note opposite effects of supplemental α-tocopherol (800 IU/day for 30 days): increased α-tocopherol and reduced γ-tocopherol. *After: Meydani, M., Fielding, R.A., Cannon, J.G., et al., 1997. Nutr. Biochem. 8, 74–81.*

which are prevented by feeding α-tocotrienol.[48] Such findings show that tocotrienols can be transported to peripheral tissues by α-TTP–independent means.

Membrane Vitamin E. In most nonadipose cells, vitamin E (mostly α-tocopherol) is localized almost exclusively in membranes. The highest concentrations are found in the Golgi membranes and lysosomes, where the ratio of vitamin E:phospholipids is approximately 1:65. Other subcellular membranes contain an order of magnitude less vitamin E. It is thought that the vitamins may reside in intimate contact with PUFAs by virtue of their complementary three-dimensional structures (Fig. 8.5). Fluorescence techniques have revealed that vitamin E partitions into membranes where its weak surface-active properties orient it at the interface between the aqueous phase and hydrophobic domain with its phenoxy group being hydrogen bonded to the carbonyl group of the fatty acid ester in the phospholipid bilayer. Dynamic spectroscopic studies have shown that α-tocopherol (and α-tocotrienol) so oriented can rotate about its long axis perpendicular to the plane of the membrane and can diffuse laterally among leaflets of the phospholipid bilayer.

Although vitamin E is the major antioxidant in membranes, research has focused on how the vitamin functions

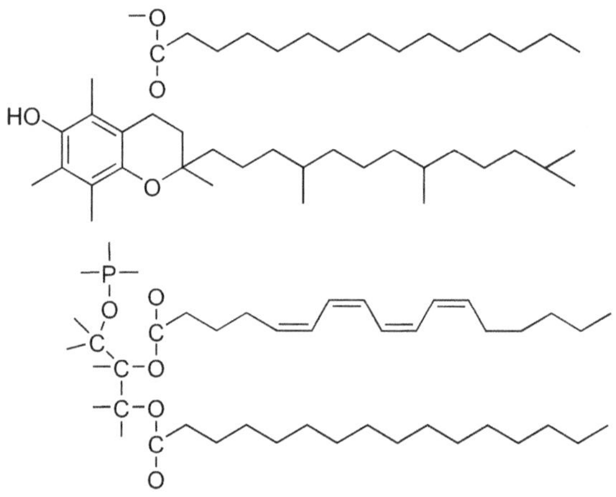

FIGURE 8.5 Proposed interdigitation of tocopherols and polyunsaturated fatty acids in biological membranes.

effectively given the relatively enormous quantities of polyunsaturated lipids also present in membranes. This may be because of the vitamin clustering in membrane locations of greatest need, which α-tocopherol has been found to do by forming complexes with membrane lysolipids, particularly choline lysophosphatides.[49] Such complex formation results in a nonrandom distribution of vitamin E in the phospholipid membrane bilayer, instead of associations with structures analogous to "lipid rafts," i.e., highly disordered,

48. Jishage, K., Arita, M., Igarishi, K., et al., 2001. J. Biol. Chem. 276, 1669–1672.

49. A lysophosphatide results from the partial hydrolysis of a phospholipid (e.g., phosphatidylcholine), which removes one of the fatty acid moieties. Such hydrolysis is catalyzed by phospholipase A_2.

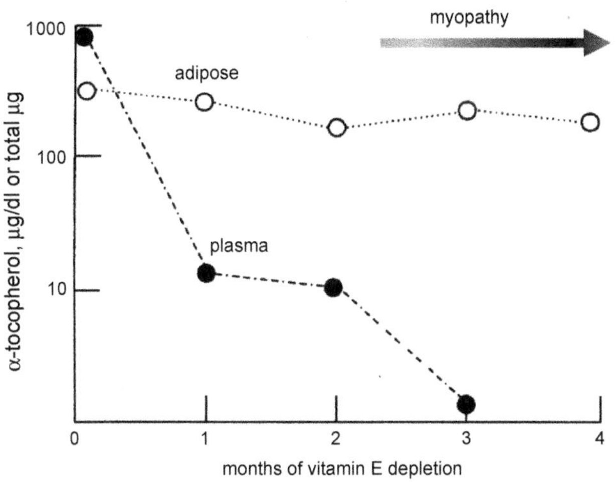

FIGURE 8.6 Retention of α-tocopherol in guinea pig adipose tissue during vitamin E depletion. *Machlin, L.J., Keating, J., Nelson, J., et al., 1979. J. Nutr. 109, 105–109.*

PUFA-rich microdomains depleted of cholesterol and sphingomyelin. Because hydrolytic products are known to have destabilizing effects on membranes, the formation of such complexes of α-tocopherol (but not other isomers) has been shown to stabilize membranes. It has been proposed that, by affecting protein–lipid and/or protein–protein interactions, vitamin E can affect embedded signal transduction pathways.

Adipose tissue. Some 90% of vitamin E in the body is contained in adipose tissue where it resides mostly in the bulk lipids. This constitutes a fixed pool from which the vitamin is slowly mobilized, thus, having long-term physiological significance (Fig. 8.6). After a change in α-tocopherol intake, adipose tocopherols may not reach a new steady state for two or more years; adipose vitamin E levels can be nearly normal even in animals showing clinical signs of vitamin E deficiency. That adipose comprises a sink for vitamin E is indicated by the fact that circulating tocopherols are inversely related to body fat mass and that people on weight-loss programs do not lose vitamin E from their adipose tissues. However, circulating tocopherol levels have been found to increase significantly (10–20%) during intensive exercise, and it has been suggested that vitamin E may be mobilized from its fixed pools by way of the lipolysis induced under such conditions.

6. METABOLISM OF VITAMIN E

Most α-tocopherol is transported to the tissues without metabolic transformation; subsequent metabolism involves head-group and side chain oxidation.[50] The selective accumulation of RRR-α-tocopherol in tissues is the result of discrimination among the various tocopherols consumed. This is effected in two ways: selective retention of α-tocopherol via α-TTP binding and selective metabolism of non–α-vitamers.

Oxidation of the chroman ring. The chromanol hydroxyl group renders tocopherols and tocotrienols capable of undergoing both one- and two-electron oxidations. α-Tocopherol is, thus, converted to α-tocopheryl quinone and (5,6- or 2,3-)-epoxy-α-tocopherylquinones, respectively (Fig. 8.7), thus, enabling it to scavenge free radicals such as peroxynitrate and lipid peroxyl radicals.

Redox Cycling. Oxidation of the chromanol ring is the basis of the in vivo antioxidant function of the vitamin. It involves oxidation primarily to tocopherylquinone, which proceeds through the semistable tocopheroxyl radical intermediate. A significant portion of vitamin E may be recycled in vivo by reduction of tocopheroxyl radical back to tocopherol. This hypothesis is supported by several findings: the very low turnover of α-tocopherol, the slow rate of its depletion in vitamin E–deprived animals, and the relatively low molar ratio of vitamin E to PUFA (about 1:850) in most biological membranes. Several mechanisms have been proposed for the in vivo reduction of tocopheroxyl by various intracellular reductants. *In vitro* studies have demonstrated that this can occur in liposomes by ascorbic acid, in microsomal suspensions by NAD(P)H, and in mitochondrial suspensions by NADH and succinate, with the latter two systems showing synergism with reduced glutathione (GSH) or ubiquinones. Indeed, a membrane-bound tocopheroxyl reductase activity has been suggested. To constitute a physiologically significant pathway in vivo, such a multicomponent system may be expected to link the major reactants, which are compartmentalized within the cell (e.g., ascorbic acid in the cytosol and tocopheroxyl in the membrane). Thus, it is possible that the recycling of tocopherol may be coupled to the shuttle of electrons between one or more donors in the soluble phase of the cell and the radical intermediate in the membrane, resulting in the reduction of the latter.

According to this model, tocopherols and tocotrienols are retained through recycling until the reducing systems in both aqueous and membrane domains become rate limiting, whereupon lipid peroxidation and protein oxidation would increase. Although the monovalent oxidation of tocopherol to the tocopheroxyl radical is reversible (at least in vitro), further oxidation of the radical intermediate is unidirectional. Because tocopherylquinone lacks vitamin E activity, its production represents loss of the vitamin from the system. It can be reduced to α-tocopherylhydroquinone, which can be conjugated with glucuronic acid and secreted in the bile, thus making excretion with the feces the major route of elimination of the vitamin. Under conditions of intakes of nutritional

50. Excretion of the nonmetabolized α-tocopherol occurs only at high doses (e.g., >50 mg), which apparently exceeds the binding capacity of α-TTP.

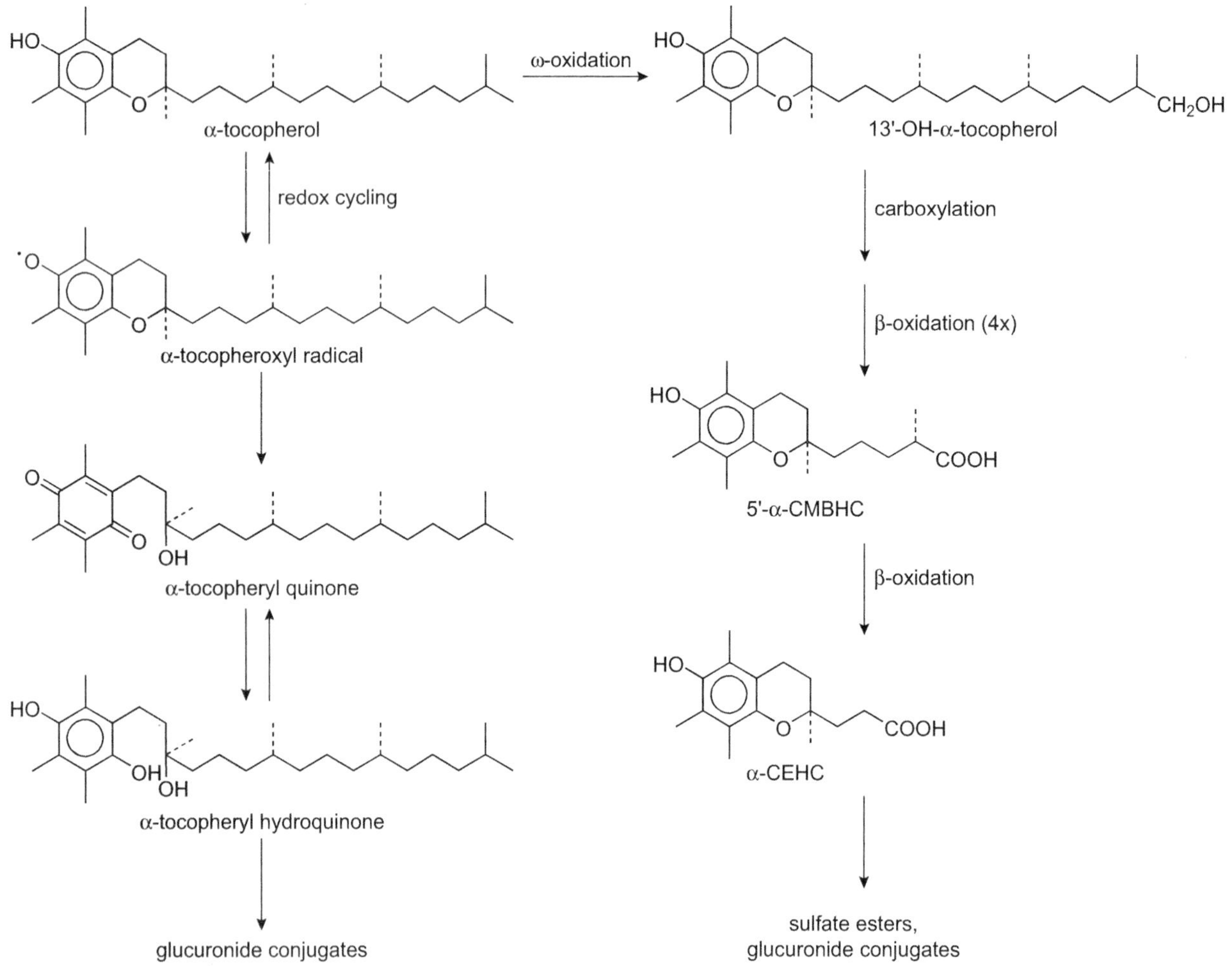

FIGURE 8.7 Vitamin E metabolism (shown with α-tocopherol). *5′-α-CMBHC*, 5′-α-carboxymethylbutylhydroxychroman; *α-CEHC*, α-carboxyethylhydroxychroman.

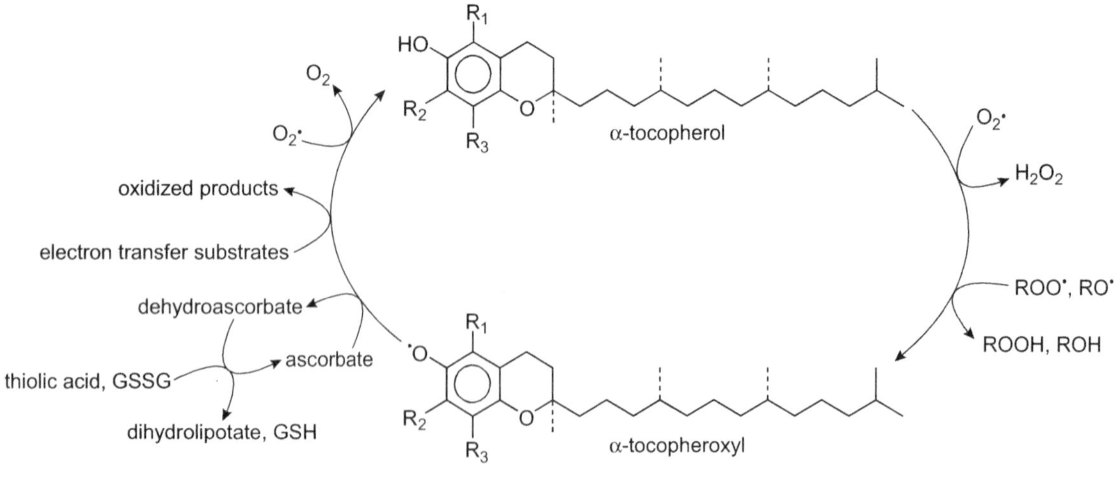

FIGURE 8.8 The vitamin E redox cycle.

levels of vitamin E, less than 1% of the absorbed vitamin is excreted with the urine.

ω-Oxidation of the phytyl side chain. Vitamin E is catabolized to water-soluble metabolites by a cytochrome *P*-450–mediated process initiated by a hydroxylation of a terminal methyl group of the phytyl side chain.[51] This hepatic ω-hydroxylation step, catalyzed by a microsomal cytochrome *P*-450 isoform (CYP4F2 in humans and CYP4F14 in the mouse) also involved in leucotriene ω-hydroxylation, is followed by dehydrogenation to the 13′-chromanol and subsequent truncation of the phytyl side chain through the removal of two- and three-carbon fragments (Fig. 8.7). That other ω-hydrolases exist in extrahepatic tissues is indicated by the fact that genetic deletion of CYP4F14 in the mouse reduced α-tocopherol metabolism by only 70–90%.[52] The products of the side chain oxidation include are excreted in the urine, often as glucuronyl conjugates; they include the folloiwng:[53]

5′(6-hydroxy-2,5,7,8-tetramethylchroman-2-yl)-2-methylpentanoic acid (α-CMBHC)
3-(6-hydroxy-2,5,7,8-tetramethylchroman-2-yl)propanoic acid (α-CEHC)

It has been suggested that these and perhaps other long-chain chromanol metabolites may be more than excretory products; i.e., that they may be metabolic effectors. Support for that hypothesis comes from findings that such long-chain hydroxy- and carboxy-chromanols reduced the uptake of oxidized LDLs by macrophages and induced CD36, the major scavenger receptor for oxidized LDLs in human macrophages.[54]

This pathway catabolizes non–α-tocopherols more extensively than it does α-tocopherol, resulting in much faster turnover of those vitamers.[55] It does, however, appear to be upregulated by high doses of α-tocopherol, suggesting that it is also important in clearing that vitamer. Accordingly, high maternal vitamin E intakes during pregnancy have been found to increase α-CEHC levels in the fetal circulation. At high intakes, α-tocopherol is also excreted in the feces.

51. Sontag, T.J., Parker, R.S., 2007. J. Lipid Res. 48, 1090–1098.
52. Bardowell, S.A., Duan, F., Manor, D., et al., 2012. J. Biol. Chem. 287, 260,077–26,086.
53. **Tocopheronic acid** and **tocopheronolactone**, referred to as **Simon metabolites** after E. J. Simon who described them in the urine of rabbits and humans, were thought to be urinary metabolites of vitamin E. It now appears that they were artifacts resulting from β-oxidation of the vitamin during isolation.
54. Wallert, M., Mosig, S., Rennert, K., et al., 2014. Free Radic. Biol. Med. 68, 42–51.
55. For this reason, the ω-oxidation pathway would appear to contribute, along with the binding specificity of α-TTP, to what has been called the "α-tocopherol phenotype", i.e., the dominance of α-tocopherol in tissues. That *Drosophila*, which appear to lack α-TTP (Parker, R.S., McCormick, C.C., 2005. Biochem. Biophys. Res. Commun. 338, 1537–1541), also show the "α-tocopherol phenotype" suggests the primacy of the ω-oxidation pathway in this regard.

TABLE 8.10 Plasma Tocopherols in Smokers and Nonsmokers

Metabolite	Nonsmokers (n = 19)	Smokers (n = 15)
α-Tocopherol (μM)	16.0 ± 4.0	15.9 ± 5.0
γ-Tocopherol (μM)	1.76 ± 0.98	1.70 ± 0.69
5-Nitro-γ-tocopherol (nM)	4.03 ± 3.10	8.02 ± 3.33*

*$p < .05$.
From Leonard, S.W., Bruno, R.S., Paterson, E., et al., 2003. Free Radic. Biol. Med. 12, 1560–1567.

Other metabolism. The detection of small amounts of **α-tocopheryl phosphate** in tissues of vitamin E–fed animals suggests that the vitamin can be phosphorylated. The metabolic significance of this metabolite is unclear and a kinase has not been identified. Although the metabolic role of α-tocopheryl phosphate is not clear, evidence suggests that it may serve as an active lipid mediator of signal transduction and gene expression.[56] That vitamin E can also be nitrated in vivo is indicated by the occurrence of **5-nitro-γ-tocopherol** in the plasma of cigarette smokers (Table 8.10), presumably because of the high amounts of reactive nitrogen species and the stimulatory effects of cigarette smoke on inflammatory responses. This reaction may be the basis for the enhanced turnover of tocopherols and reduced production of carboxyethylchromanyl metabolites in smokers.

7. METABOLIC FUNCTIONS OF VITAMIN E

Vitamin E as a Biological Antioxidant

The primary nutritional role of vitamin E is as a biological antioxidant. An **antioxidant** is an agent that inhibits oxidation and, thus, prevents such oxidation reactions as the conversion of PUFAs to fatty hydroperoxides, the conversion of free or protein-bound sulfhydryls to disulfides, etc. In reducing **free radicals**, it protects against the potentially deleterious reactions of such highly reactive oxidizing species, thus having functional importance in maintaining membrane integrity in all cells of the body.

Production of Free Radicals and Reactive Oxygen Species (ROS). Free radicals (X•) are produced in cells either by homolytic cleavage of a covalent bond, as in the formation of a C-centered free radical of a PUFA or by a univalent electron transfer reaction. It has been estimated that as much as 5% of inhaled molecular oxygen (O_2) is

56. Zingg, J.M., Meydani, M., Azzi, A., 2010. Mol. Nutr. Food Res. 54, 679–692.

metabolized to yield the so-called reactive oxygen species, i.e., the one- and two-electron reduction products superoxide radical $O_2\cdot^-$ and H_2O_2, respectively. There appear to be three sources of ROS:

- **Normal oxidative metabolism**. The mitochondrial electron transport chain, which involves a flow of electrons from NADH and succinate through a series of electron carriers to cytochrome oxidase reducing oxygen to water, leaks a small amount of electrons that reduce O_2 to $O_2\cdot^-$.
- **Microsomal cytochrome *P*-450 activity**. Several xenobiotic agents are metabolized by the microsomal electron transport chain to radical species (e.g., the herbicide paraquat is converted to an N-centered radical anion) that can react with O_2 to produce $O_2\cdot^-$.
- **Respiratory burst of stimulated phagocytes**. Macrophages produce $O_2\cdot^-$ and H_2O_2 during phagocytosis.

Physiological roles of ROS. ROS play key roles in immune function. On encountering or ingesting a bacterium or other foreign particle, activated[57] neutrophils and macrophages produce large amounts of $O_2\cdot$and H_2O_2 in a process referred to as the "respiratory burst." This involves myeloperoxidase, which catalyzes the H_2O_2-dependent oxidation of halide ions yielding such powerful oxidizing agents as hypochlorous acid and xanthine oxidase, which catalyzes the reaction of xanthine or hypoxanthine with O_2 to generate uric acid. These reactions are important in killing pathogens, but they can also be deleterious to immune cells themselves. If not controlled, they can contribute to the pathogenesis of disease.

Metabolically produced ROS appear to have essential metabolic functions as signaling molecules for the adaptation of skeletal muscle to accommodate the stresses presented by exercise training or periods of disuse. This signaling involves redox-sensitive kinases, phosphatases, and nuclear factor κB (NF-κB), which affect the rate of mitochondrial biogenesis, and the induction of genes related to insulin sensitivity [peroxisome proliferator-activated receptor coactivator (PGC)-1α/β, PPARγ] and ROS defense [superoxide dismutase 1/2 (SOD1/2), glutathione peroxidase (GPX1)]. This system of adaptive responses to oxidative stress facilitates the ultimate development of long-term resistance to that stress (Fig. 8.9). That this system, which has been called **mitochondrial hormesis**,[58] can be impaired by high-level antioxidant treatment was demonstrated by the finding that supplements of vitamins E and C (400 IU α-tocopheryl acetate + two 500-mg doses of ascorbic acid

57. Various cytokines, such as tumor necrosis factor and interferon-γ, can activate phagocytic cells to increase their $O_2\cdot$ generation.

58. Hormesis is the term for a generally favorable biological response to low exposures to stressors/toxins.

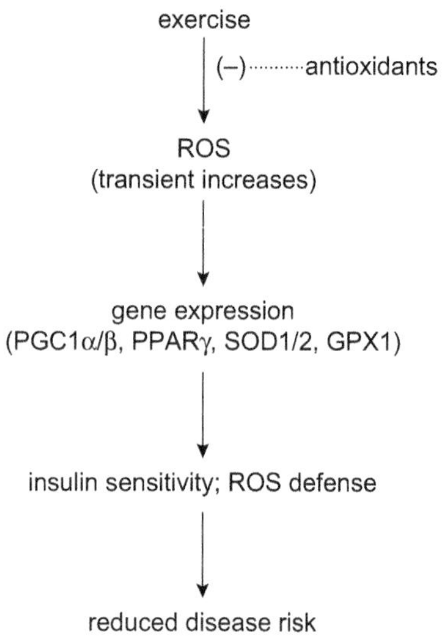

FIGURE 8.9 Mitochondrial hormesis: ROS signaling of insulin sensitivity. *After Ristow, M., Zarse, K., Oberbach, A., et al., 2009. Proc. Natl. Acad. Sci. USA 106, 8665–8670.*

per day) blocked the upregulation of muscle glucose uptake otherwise induced by exercise.[59] This finding raises several questions: Is this effect because of vitamin E, vitamin C, or the combination? What antioxidant dose is required for such effects? What level of redox tone will be beneficial?

Adverse effects of ROS. In the presence of transition metal ions (particularly, Fe^{2+} or Cu^+), $O_2\cdot^-$ and H_2O_2 can react to yield a very highly reactive free-radical species[60], **hydroxyl radical (HO·)** ($O_2\cdot^- + \mathbf{H_2O_2} + \mathbf{Fe^{2+}} \rightarrow \mathbf{O_2} + \mathbf{HO\cdot} + \mathbf{HO^-} + \mathbf{Fe^{3+}}$). These divalent metals can also catalyze the decomposition of H_2O_2 or fatty acyl **hydroperoxide** (ROOH) produced by lipid peroxidation to yield the oxygen-centered radical, RO· or HO·, respectively (**ROOH/HOOH + Fe^{2+} → RO·/HO· + HO^- + Fe^{3+}**). Molecular targets of HO· include:

- **DNA** – to cause oxidative base damage[61]
- **Proteins** – to cause production of carbonyls and other amino acid oxidation products[62]
- **Lipids** – to cause oxidation of PUFAs of membrane phospholipids, the formation of lipid peroxidation products (e.g., MDA, isoprostanes, pentane, ethane), and resulting in membrane dysfunction

Of these, PUFA-containing membrane lipids are particularly susceptible to attack by virtue of their 1,4-pentadiene

59. Ristow, M., Zarse, K., Oberbach, A., et al., 2009. Proc. Natl. Acad. Sci. U.S.A. 106, 8665–8870.

60. Otherwise, neither $O_2\cdot^-$ nor H_2O_2 is highly reactive, and $O_2\cdot^-$ is cleared rapidly (its half-life is c. 1 s).

61. Base damage products such as 8-hydroxydeoxyguanosine (8OHdG), resulting from DNA repair processes, are excreted in the urine. Smokers typically show elevated 8OHdG excretion.

62. For example, methionine sulfoxide, 2-ketohistidine, hydroxylation of tyrosine to DOPA, formylkynurenine, *o*-tyrosine, and protein peroxides.

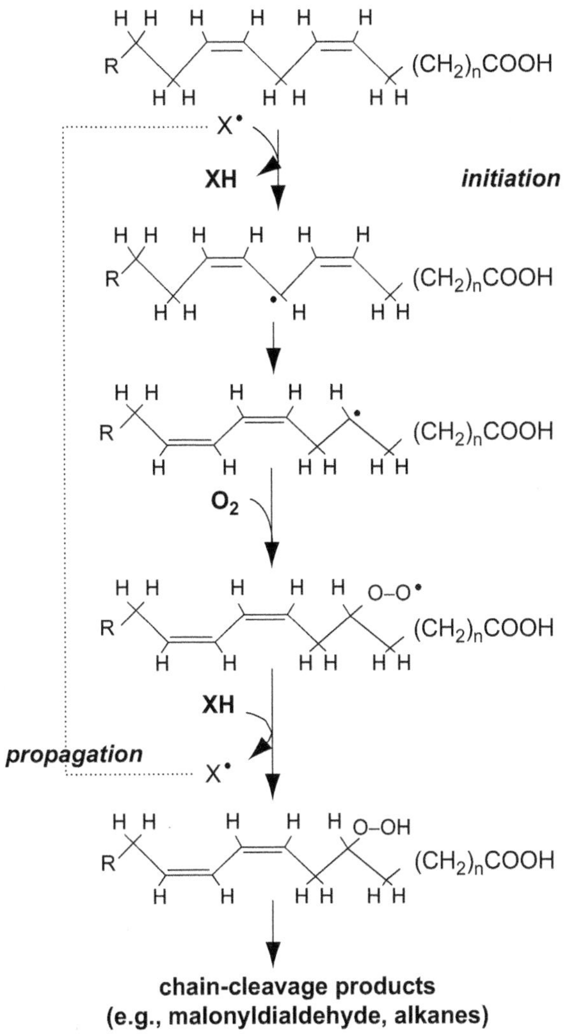

FIGURE 8.10 The self-propagating nature of lipid peroxidation.

systems, which allow abstraction of a complete hydrogen atom (i.e., with its electron) from one of the —CH_2— groups in the carbon chain with the consequent generation of a C-centered free radical (—C—) (Fig. 8.10). This initiation of lipid peroxidation can be accomplished by HO• and, possibly, HOO• (but *not* by H_2O_2 or O_2•⁻). The C-centered radical, being unstable, undergoes molecular rearrangement to form a conjugated diene, which is susceptible to attack by O_2 to yield a peroxyl radical (ROO•). Peroxyl radicals are capable of abstracting a hydrogen atom from other PUFAs and, thus, propagating a chain reaction that can continue until the membrane PUFAs are completely oxidized to hydroperoxides (ROOH).

Fatty acyl hydroperoxides are degraded in the presence of transition metals (Cu^{2+}, Fe^{2+}) and heme and heme proteins (cytochromes, hemoglobin, myoglobin) to release radicals that can continue the chain reaction of lipid peroxidation,[63] also yielding other chain-cleavage products: **MDA**,[64] **pentane**, and **ethane**.[65] This pattern of oxidative degradation of membrane phospholipid PUFAs is believed to disrupt membrane function. Cellular oxidant injury can also occur without significant lipid peroxidation, by oxidative damage to critical macromolecules (DNA and proteins) and decompartmentalization of Ca^{2+}.[66]

In vivo lipid peroxidation has been surprisingly difficult to demonstrate. Little or no evidence of lipid peroxides or their decomposition products have been found in tissues. Although expired breath contains volatile alkanes likely from the decomposition of fatty acyl hydroperoxides, it is difficult to exclude their possible production by gut or skin microbes.

Scavenging Free Radicals. Because of the reactivity of the phenolic hydrogen on its C-6 hydroxyl group and the ability of the chromanol ring system to stabilize an unpaired electron, vitamin E can terminate chain reactions among PUFAs in membranes. This free-radical scavenging property involves donation of the phenolic hydrogen to a fatty acyl free radical (or $O_2\cdot^-$) to prevent the attack of that species on other PUFAs. Tocopherols have great reactivities toward peroxyl and phenoxyl radicals, but can also quench mutagenic electrophiles such as reactive nitrogen oxide species (NO_x). The antioxidant activities of the E vitamers relate to the leaving ability of the phenolic hydrogen; when assessed in vitro, α-tocopherol has the greatest antioxidant activity,[67] followed by the β- and γ-vitamers, which are greater than δ-vitamer.[68] The tocotrienols can also scavenge peroxyl radical. In contrast, NO_x is trapped more effectively by γ-tocopherol than by the α-vitamer.

63. Therefore, a single radical can initiate a chain reaction that may self-propagate repeatedly.

64. Although MDA is a minor product of lipid peroxidation, it has received a great deal of attention because of the ease of measuring it colorimetrically using 3-thiobarbituric acid (TBA), which has been widely used to assess lipid peroxidation. The TBA test faces limitations: much of the MDA it detects may not have been present in the original sample, as lipid peroxides can decompose to MDA during the heating stage of the test; the reaction can also be affected by the presence of iron salts.

65. Pentane and ethane are produced from the oxidative breakdown of ω-6 and ω-3 fatty acids, respectively. Both are excreted across the lungs and can be detected in the breath of vitamin E–deficient subjects.

66. For example, pulmonary injury by the bipyridylium herbicide paraquat involves lipid peroxidation only as a late-stage event.

67. The chemical antioxidant activity of α-tocopherol is about 200-fold that of the commonly used food antioxidant butylated hydroxytoluene.

68. The biological activities of the E vitamers are functions of both their intrinsic chemical antioxidant activities and their efficiencies of absorption and retention. Thus, γ-tocopherol has only 6–16% of the biological activity of the α-vitamer. An exception to this relationship occurs in the case of sesame seed lignans, which potentiate the biopotency of γ-tocopherol. Therefore, sesame oil, which contains only the γ-vitamer, has a biopotency equivalent to that of α-tocopherol. The potentiating factor is believed to be a lignan phenol, sesamolin, to which antiaging properties have been attributed.

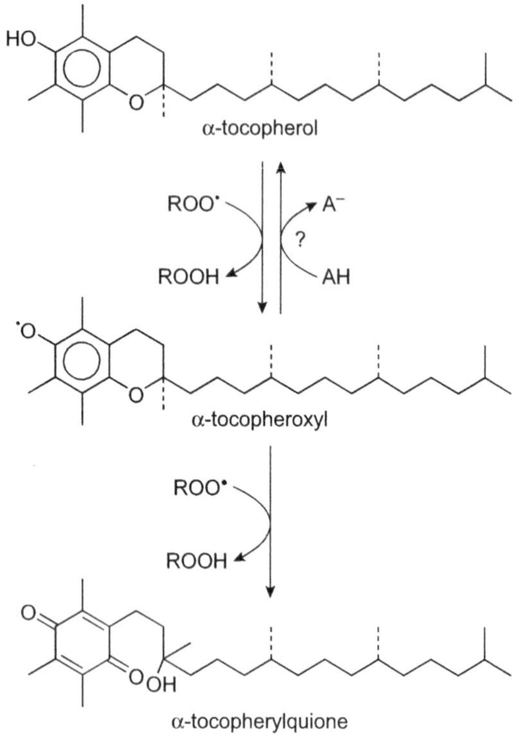

FIGURE 8.11 Oxidation of tocopherols by reaction with peroxyl radicals.

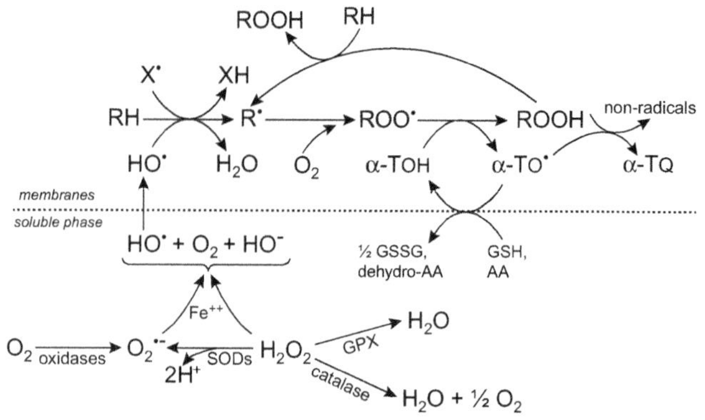

FIGURE 8.12 The cellular antioxidant defense system. *Note:* (a) removal of H_2O_2 to prevent production of ROS; (b) quenching of free radicals by oxidation of α-tocopherol; and (c) regeneration of α-tocopherol by soluble reductants such as reduced glutatione or ascorbic acid. *GSH*, reduced glutathione; *GSSG*, oxidized glutathione; *AA*, ascorbic acid; *α-TOH*, α-tocopherol; *α-TO*, α-tocopheryl radical; *α-TQ*, α-tocopherylquinone; *GPX*, glutathione peroxidase; *SOD*, superoxide dismutase.

In scavenging free radicals, tocopherols and tocotrienols undergo oxidation of their respective alcohol forms to semistable radical intermediates, **tocopheroxyl** (or chromanoxyl) **radicals** (Figs. 8.7 and 8.8). Unlike free radicals formed from PUFAs, the tocopheroxyl radical is relatively unreactive, thus stopping the destructive propagative cycle of lipid peroxidation. In fact, tocopheroxyl is sufficiently stable to react with a second peroxyl radical to form inactive, nonradical products including **tocopherylquinone**[69]. It is, therefore, referred to as a **chain-breaking antioxidant**. Because α-tocopherol can compete for peroxyl radicals much faster than PUFAs, small amounts of the vitamin are able to effect the antioxidant protection of relatively large amounts of PUFAs (Fig. 8.11). **Tocotrienols** are thought to have more potent antioxidant protective potential than tocopherols, as their unsaturated side chains facilitate their more efficient penetration into tissues containing saturated fatty layers, e.g., brain, liver. Factors that increase the production of ROS (e.g., xenobiotic metabolism, ionizing radiation, exposure to prooxidants such as O_3 and NO_2) can increase the metabolic demand for antioxidant protection.

The Antioxidant Defense System. Because it is distributed in membranes, vitamin E serves as a lipid-soluble biological antioxidant with high specificity for loci of potential lipid peroxidation. Its co-transport with polyunsaturated lipids ensures protection of the latter from free-radical attack; circulating tocopherol levels tend to correlate with those of total lipids and cholesterol.[70] However, vitamin E is one of several factors in an antioxidant defense system that protects cells from the damaging effects of oxidative stress (Fig. 8.12). That system includes:

- **Membrane antioxidants** – mostly tocopherols, but also ubiquinones and carotenoids.
- **Soluble antioxidants** – NADPH, NADH, ascorbic acid, GSH and other thiols, uric acid, thioredoxin, bilirubin, polyphenols, and several metal-binding proteins (copper: ceruloplasmin, metallothionein, and albumin; iron: transferrin, ferritin, and myoglobin).
- **Antioxidant enzymes** – SODs,[71] GPXs,[72] thioredoxin reductase,[73] and catalase.[74]

69. Evidence indicates that **tocopherylquinones** can induce apoptosis in cancer cells.

70. Therefore, high plasma vitamin E levels occur in hyperlipidemic conditions (hypothyroidism, diabetes, and hypercholesterolemia), whereas low plasma vitamin E levels occur in conditions involving low plasma lipids (abetalipoproteinemia, protein malnutrition, and cystic fibrosis).

71. The SODs are metalloenzymes. The mitochondrial SOD contains manganese at its active center, whereas the cytosolic SOD contains both copper and zinc as essential cofactors. Although not found in animals, an iron-centered SOD has been identified in blue-green algae.

72. The GPXs contain selenium (Se) at their active centers and depend on adequate Se status for their synthesis. There are four isoforms; each uses reducing equivalents from GSH to reduce H_2O_2 to water, or fatty acyl hydroperoxides to the corresponding fatty alcohols. One isoform is found in membranes and has specificity for esterified hydroperoxides; the others are soluble and have specificities for nonesterified hydroperoxide substrates including H_2O_2. The activities of these enzymes depend on the flavoenzyme **glutathione reductase** to regenerate GSH from its oxidized form (GSSG).

73. Thioredoxin reductase is also a selenoprotein; there are three isoforms.

74. Catalase has an iron redox center. Because its distribution is almost exclusively limited to the peroxisomes/lysosomes, it is not considered of prime importance in antioxidant protection in the cytosol.

In this multicomponent system, vitamin E scavenges radicals within the membrane, where it blocks the initiation and interrupts the propagation of lipid peroxidation. The group of metalloenzymes collectively blocks the initiation of peroxidation from within the soluble phase of the cell: SODs convert $O_2 \cdot^-$ to H_2O_2; catalase and GPXs each further reduce H_2O_2. The aggregate effect of this enzymatic system is to clear $O_2 \cdot^-$ by reducing it fully to H_2O, thus preventing the generation of other, more highly ROS [e.g., HO· and singlet oxygen ($^1O_2$)]. The GPXs can also reduce fatty acyl hydroperoxides to the corresponding fatty alcohols, thus serving to interrupt the propagation of lipid peroxidation.

Some components of this system are endogenous (e.g., NADPH, NADH, and, for most species, ascorbic acid), whereas other components must be obtained, at least in part, from the external chemical environment. The diversity of this system implies the ability to benefit from various antioxidants and other key factors obtained from dietary sources in variable amounts. That the components of the defense system function cooperatively is evidenced by the nutritional "sparing" observed particularly for vitamin E and Se in the etiologies of several deficiency diseases (e.g., **exudative diathesis** in chicks, **liver necrosis** in rats, and **white muscle disease** in lambs and calves). In those species, nutritional deprivation of either vitamin E or Se alone is usually asymptomatic; deficiencies of both nutrients are required to produce disease.

The components of this system respond to changes in cellular redox state, increasing antioxidant protection during cell differentiation. Several lines of evidence indicate that ROS gradients and cellular redox state influence gene expression. Oxidizing conditions have been found to affect cellular ion distribution, expression of chromatin-controlling proteins, and the cytoskeleton and nuclear matrix, which, in turn, affect chromatic configuration and pre-mRNA processing. Thus, it appears that the well-functioning antioxidant defense system serves to maintain low, optimal levels of ROS in cells such that the beneficial effects of prooxidizing conditions are realized and their deleterious effects are minimized. This concept has been called healthy "**redox tone.**"

Because most ROS are produced endogenously by mitochondria, which process 99% of the oxygen utilized by the cell, exercise-related increases in oxidative metabolism[75] are thought to increase needs for vitamin E. Indeed, exercise has been found to increase ROS production, which contributes to fatigue, and lipid peroxidation; both are reduced by vitamin E supplementation. However, low levels of ROS in skeletal muscle are necessary for normal force production,[76] and myocytes respond to oxidative stress by upregulating a variety of proteins involved in the maintenance of cellular integrity. Hence, regular exercise increases the activities of enzymes involved in antioxidant protection (GPXs, mangano-superoxide dismutase, γ-glutamyl synthase, catalase), DNA repair, and cytoprotection (e.g., heat-shock factor-1). This adaptation relies on ROS signaling protein phosphorylation and the binding of redox-regulated transcription factors, NF-κB, and AP-1. ROS also signal mitochondrial biogenesis by stimulating the expression of several proteins including PGC-1α, nuclear factor (erythroid-derived) (NRF)-1, and mitochondrial transcription factor A (mtTFA). This signaling can be blunted by antioxidants. Therefore, trained athletes do not show increases in oxidative stress after accustomed vigorous exercise activities; for them, supplemental antioxidants can prevent metabolic adaptation to training.[77] Studies with humans and animal models have yielded inconsistent results concerning the effects of vitamin E supplementation on exercise performance.

75. O_2 utilization increases 10- to 15-fold during exercise.

76. Reid, M.B., 2001. J. Appl. Physiol. 90, 724–731.

Prooxidant Potential of Vitamin E

α-Tocopherol can also promote lipid peroxidation in LDLs in the absence of other antioxidants (e.g., ascorbic acid, coenzyme Q_{10}, urate). Under such conditions, the single-electron oxidation of tocopherol converts it to the tocopheroxyl radical, which moves into the particle's core where it can abstract hydrogen from a cholesteryl-PUFA ester to yield a peroxyl radical. The presence of secondary antioxidants is needed to prevent LDL oxidation. Otherwise, vitamin E becomes a **chain-transfer agent** to propagate lipid peroxidation in the lipid core. Accordingly, a very high dose (1050 mg/day) of α-tocopherol has been found to increase susceptibility to peroxidation.[78]

Nonantioxidant Functions of Vitamin E

The recognition in the early 1990's that vitamin E could inhibit cell proliferation and protein kinase C (PKC) activity suggested that vitamin E may function in vivo in ways that are unrelated to its function as a biological antioxidant. Subsequent research has demonstrated antiproliferative, proapoptotic, anti-inflammatory, an antiangiogenic effects of tocopherols and tocotrienols that do not appear to involve their antioxidant functions.

Enzyme Regulation. For several enzymes, α-tocopherol appears to participate in complex membrane-based recruitment processes affecting function: inhibition of PKC, NADPH oxidase, phospholipase A_2, protein kinase B/Akt, 5-lipoxygenase, cyclooxygenase A_2, and 3-hydroxymethyl-3-glutaryl-coenzyme A (HMG-CoA) reductase; activation

77. Venditte, P., Napolitano, G., Barone, D., et al., 2014. Free Radic. Res. 48, 1179–1189.

78. Brown, K.M., Morrice, P.C., Duthie, G.G., 1977. Am. J. Clin. Nutr. 65, 496–502.

TABLE 8.11 Vitamin E Target Genes

Function	Gene Product
Tocopherol uptake, metabolism	α-TTP, CYP3A, CYP4F2, HMG-CoA reductase, CRABP-II
Lipid uptake	Srbl, CD36, SR-AI/II, LDL-R, PPARγ
Cholesterol/steroid synthesis	HMG-CoA-r, HMG-CoS, 7DHC, IPδ1, FPPS, 5αR1
Antioxidant defence	γGCS
Cell adhesion	E-selectin, L-selectin, ICAM-1, VCAM-1, integrins, MAC1
Cell growth	Connective tissue growth factor
Extracellular matrix	Tmp1, collagen α1(1), MMP1, MMP9, connective tissue growth factor, glycoprotein IIb
Inflammation	Il-2, Il-4, IL-1-β, TGF-β
Clotting	Christmas factor
Cytoarchitecture	Tmp2, Myh1, Tnni2, Acta1, Krt15
Cell cycle regulation	Cyclin D1, cyclin E1, Bcl12-L1, p27, CD95
Apoptosis	CD95 L, Bcl2-L1
Other functions	Leptin, β-secretase

of protein phosphatase 2A, diacylglycerol kinase, and HMG-CoA reductase. Tocotrienols have been found to impart a variety of effects on cell functions: inhibition of NF-κB, transforming growth factor β (TGF-β), tumor necrosis factor α, IL1β, and P38 signaling; activation of caspases; and downregulation of Bcl2, cyclin D, c-Src, and the Raf/Erk pathway.[79] The effects of tocotrienols would appear to be related to the fact that, because they are inefficiently removed from the liver, they can stimulate stress responses including induction of detoxification and antioxidant genes. These effects appear to be greatest for the undermethylated forms (γ- and δ-) and in hypoxic cells, e.g., tumor cells.

Gene Expression. Vitamin E also appears to participate in transcriptional regulation (Table 8.11). The transcriptional effects of α-tocopherol imply the existence of nuclear receptors for tocopherol and corresponding DNA-responsive elements. One group of nuclear receptors, the pregname X receptor has been found to bind vitamin E; however, the metabolic significance of such binding is presently unclear. Rats deprived of vitamin E show altered patterns of gene expression. Studies have revealed a large number of genes being downregulated, although a number of transport-related genes that were upregulated in liver were downregulated in the cerebral cortex. α -Tocopherol also appears to affect gene expression posttranscriptionally by affecting the expression of small, noncoding RNA (miRNA);[80] this has been demonstrated for miRNA-122 and miRNA-125b.[81] Therefore, the emerging picture is one of α-tocopherol involvement in the regulation of a variety of cellular processes, while inducing its own transport and catabolism, both of which are significantly affected by oxidant/antioxidant status and apoE genotype.

Physiological Functions

Inflammation and Immunity. Vitamin E is essential for optimal function of the immune system. Studies with animals have found that optimization of certain immune parameters requires intakes of at least an order of magnitude greater than those required to prevent clinical signs of deficiency. In humans, supranutritional doses (up to 800 mg α-tocopherol per day) have been found to restore responses to DTH, increase induction of IL-2, and reduce lymphocyte levels of the proinflammatory lipid mediator prostaglandin E_2 (PGE_2) (Table 8.12). Doses as low as 50 mg/day have been associated with reduced incidence of the common cold.[82] Studies with animal models have found vitamin E supplementation to reduce joint swelling, and randomized controlled trials with patients having rheumatoid arthritis (RA) have shown high-level supplementation with the vitamin (100–600 IU/day) to relieve pain and inflammation. Studies in animal models have found tocotrienols to be effective in increasing macrophage production of IL-6, IL-10, IL-1β, and PGE_2, particularly in older animals.[83]

Vitamin E can affect the pathogenesis of several viral infections in which oxidative stress has been implicated. Deprivation of the vitamin increases susceptibility of the mouse to cardiophilic RNA viruses, particularly when animals consume diets containing high amounts of PUFAs, e.g., fish oil (Fig. 8.13). Protection appears to involve suppression of oxidative stress, which provides an environment in which the virus can mutate to more highly virulent forms.[84] This phenomenon has been shown for several RNA viruses including hepatitis, influenza, and AIDS.

Neurologic Function. Vitamin E is essential for neurologic function. It is conserved by neural tissues, to which it is redistributed from other tissues under conditions of deficiency. Neurons are susceptible to deleterious effects of oxidative stress. They contain large amounts of PUFAs and

79. Ahsan, H., Ahad, A., Iqbal, J., et al., 2014. Nutr. Metab. 11, 52–74.

80. Micro-RNAs bind at the mRNA 3′-untranslated region to inhibit translation. Several miRNAs have been identified; each is believed to be capable of binding a 100 different target mRNAs, allowing posttranscriptional silencing of many different genes.

81. Rimbach, G., Moehring, J., Huebbe, P., et al., 2010. Molecules 15, 1746–1761.

82. Hemila, H, Kaprio, J., Albanes, D., et al., 2002. Epidemiol. 13, 32–37.

83. Ren, Z., Pae, M., Dao, M.C., et al., 2010. J. Nutr. 140, 1335–1341.

84. Beck, M.A., Levander, O.A., 1998. Ann. Rev. Nutr. 18, 93–116.

TABLE 8.12 Enhancement of Immune Responses by Vitamin E Supplementation of Healthy Adults

Treatment	Days of Treatment	Vit E in PMNs (nmol)[a]	DTH index (mm)[b]	PMN Proliferation ($\times 10^3$ cpm)[c]	IL-2 Production (kU/L)[d]
Placebo	0	0.14±0.04	16.5±2.2	24.48±2.73	31.8±8.3
	30	0.19±0.03	16.9±2.1	21.95±2.90	37.5±12.5
Vitamin E[e]	0	0.12±0.02	14.2±2.9	20.55±1.93	35.6±9.1
	30	0.39±0.05[f]	18.9±3.5[f]	23.77±2.99[f]	49.6±12.6[f]

[a]*Polymorphonucleocyte α-tocopherol content.*
[b]*Delayed-type hypersensitivity skin test.*
[c]*Concanavalin A–induced proliferation of PMNs.*
[d]*Concanavalin A–induced production of interleukin 2 by PMNs.*
[e]*A total of 800 IU all-rac-α-tocopheryl acetate per day.*
[f]*$p<.05$.*
From Meydani, S.N., Barklund, M.P., Kiu, S., et al., 1990. Am. J. Clin. Nutr. 52, 557–563.

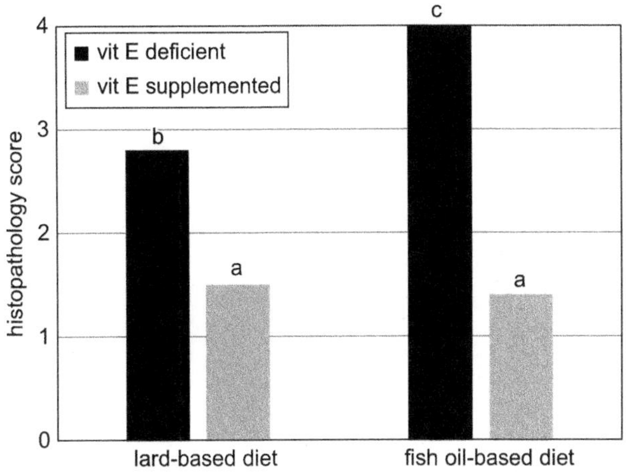

FIGURE 8.13 Cardioprotective effect of α-tocopherol in the mouse: reduction of cardiac damage from Coxsackievirus B3. Means with same superscripts are not significantly different ($p>.05$). *After Beck, M.A., Kolbeck, P.C., Rohr, L.H., et al., 1994. J. Nutr. 124, 345–358.*

iron, but lack extensive antioxidant defense systems. They are terminally differentiated and cannot replicate when damaged. They generate ROS when exposed to redox-cycling drugs (which can cause Parkinson-like neural damage in animal models) or metabolizing dopamine.[85] Exposure to hyperbaric oxygen can cause seizures. Epidemiological studies have found high vitamin E intake to be associated with reduced risks to Alzheimer disease[86] and Parkinson disease.[87]

Cardiovascular Health. Vitamin E serves as an antioxidant in LDLs, which may protect against atherosclerosis.[88] Being rich in both cholesterol and PUFA (Table 8.13), LDLs are susceptible to peroxidation by ROS. Oxidized LDLs stimulate the recruitment, in the subendothelial space of the vessel wall, of monocyte–macrophages that can take up the oxidized particles via scavenger receptors[89] to form the lipid-containing foam cells found in the early stages of atherogenesis. Evidence indicates that vitamin E can also reduce the adherence and aggregation of platelets, to retard the progression of a fatty streak and cell proliferation to advanced lesions (Fig. 8.14).

8. BIOMARKERS OF VITAMIN E STATUS

Plasma α-tocopherol concentration is the most useful biomarker of vitamin E status, particularly at limiting levels. It is directly related to α-tocopherol intake up to about 200 mg/day, but plasma concentrations are inconsistently related at greater vitamin E intakes. As vitamin E is membrane protective, plasma tocopherol levels are inversely related to susceptibility of erythrocytes to oxidative hemolysis. This relationship makes the plasma α-tocopherol level

85. By monoamine oxidase B.
86. Alzheimer disease is the world's most prevalent neurodegenerative disease, affecting an estimated 20–30 million, including almost half of people over the age of 85 years. It is characterized by memory dysfunction, loss of lexical access, temporal and spatial disorientation, and impaired judgment.
87. Parkinson disease is characterized by progressive loss of postural stability, with slowness of movement and tremor.

88. Atherosclerosis is the progressive, focal accumulation of acellular, lipid-containing plaques in the intima of arteries. It is a specific type of arteriosclerosis, although the terms are often used interchangeably. Arteriosclerosis refers to the thickening and loss of elasticity of arteries because of infiltration of the intima by fats and calcific plaques, reducing blood flow to the organs served by affected vessels and leading to such symptoms as angina, cerebrovascular insufficiency, and intermittent claudication.
89. Monocyte–macrophages have few LDL receptors, which are downregulated. Therefore, when incubated with nonoxidized LDLs, they do not form foam cells, as the accumulation of cholesterol further reduces LDL receptor activity. On the other hand, these cells have "**scavenger receptors**" specific for modified LDLs. It is thought that LDL lipid peroxidation products may react with amino acid side chains of apoB to form epitopes that have affinities for the scavenger receptor.

TABLE 8.13 Lipid and Antioxidant Contents of Human Low-Density Lipoproteins

Component	Moles per Mole LDL
Total phospholipids	700 ± 122
Fatty Acids	
Free	26
Total	2700
Triglycerides	170 ± 78
Cholesterol	
Free	600 ± 44
Esters	1600 ± 119
Total	2200
Antioxidants	
α-Tocopherol	6.52
γ-Tocopherol	1.43
Ubiquinonol-10	0.33
β-Carotene	0.27
Lycopene	0.21
Cryptoxanthin	0.13
α-Carotene	0.11

From Keaney, J.F., Frei, B., 1994. Natural Antioxidants in Human Health and Disease. Frei, B. (Ed.). Academic Press, San Diego, p. 306–307.

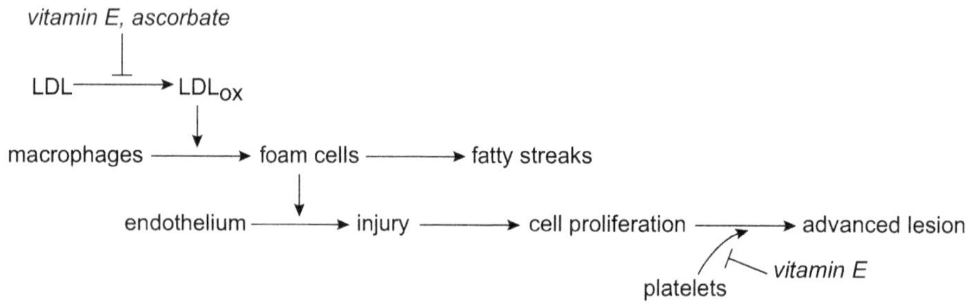

FIGURE 8.14 Model for prevention of atherogenesis by vitamin E.

useful as a parameter of vitamin E status. In individuals, values ≥0.5 mg/dL (≥12 μM) are associated with protection against hemolysis in vitro and are taken to indicate nutritional adequacy The NHANES 1999–2000 data showed plasma α-tocopherol concentrations to average 30 μM (1.3 mg/dL), with some 2% in the deficient range.[90] Plasma concentrations of γ-tocopherol tend to be 10–20% of the level of α-tocopherol. Maternal tocopherol levels increase during pregnancy, but fetal levels remain low, suggesting a barrier to transplacental movement of the vitamin. Infants' serum tocopherol levels are approximately 25% of those of their mothers. They increase to adequate levels within a few weeks after birth, except in infants with impaired abilities to utilize lipids (e.g., premature infants, infants with biliary atresia); they show very low circulating levels of vitamin E (Table 8.14). An analysis of the NHANES 2003–2006 data found plasma tocopherol to be unaffected by inflammation status, but to increase 3–4% during short-term fasting and to be some 20% greater for patients with renal dysfunction compared to matched controls.[91]

90. Ford, E.S., Schleicher, R.L., Mokdad, A.H., et al., 2006. Am. J. Clin. Nutr. 84, 375–383.

91. Haynes, B.M.H., Pfeiffer, C.M., Sternberg, M.R., et al., 2013. J. Nutr. 143, 1001S–1010S.

TABLE 8.14 Serum α-Tocopherol Concentrations in Humans

Group	α-Tocopherol (mg/dL)[a]
Healthy adults	0.85 ± 0.03
Postpartum mothers	1.33 ± 0.40
Infants	
Full term, at delivery	0.22 ± 0.10
Premature, at delivery	0.23 ± 0.10
Premature, at 1 month	0.13 ± 0.05
2 months, breast-fed	0.71 ± 0.25
2 months, bottle-fed	0.33 ± 0.15
5 months	0.42 ± 0.20
2 years	0.58 ± 0.20
Children, 2–12 years	0.72 ± 0.02
Cystic fibrotics, 1–19 years	0.15 ± 0.15
Biliary atresia, 3–15 months	0.10 ± 0.10

[a]*Mean ± SD.*
From Gordon, H.H., Nitowsky, H.M., Tildon, J.T., 1958. Pediatrics 21, 673–681.

Urinary metabolites of α-tocopherol offer potential as biomarkers of vitamin E status. These include α-CMBHC, α-CEHC, and, perhaps, α-tocopheronolactone,[92] and their glucuronyl conjugates.

A novel approach to developing new biomarkers of vitamin E status was taken by West et al.[93] They used a proteomic approach to identify an "α-tocopherome", i.e., a group of proteins expressed in relation to plasma α-tocopherol concentration. Of nearly a 1000 proteins quantifiable in plasma from Nepali children, they identified 6 that explained 71% of the variability in plasma α-tocopherol concentration.[94]

Plasma concentrations of F_2-isoprostanes,[95] which are formed to nonenzymic oxidation of PUFAs, have utility as biomarkers of oxidative stress. Although they are elevated in individuals of low vitamin E status, particularly those also overweight [body mass index (BMI) 25–29.9] or obese (BMI ≥ 30), they are not informative regarding vitamin E

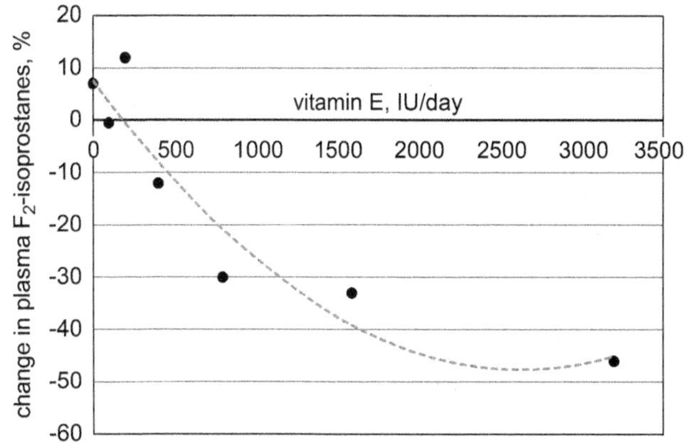

FIGURE 8.15 Reduction in plasma F_2-isoprostane concentration in response to supplemental α-tocopherol. *After Roberts, L.J., Oates, J.A., Linton, M.F., et al., 2007. Free Radic. Biol. Med. 43, 1388–1393.*

status per se. In humans, plasma F_2-isoprostane levels respond to high-dose vitamin E supplementation over a period of weeks (Fig. 8.15). With daily α-tocopherol doses of 800 IU/day, responses were seen only in subjects with relatively high baseline plasma F_2-isoprostane concentrations (>50 μg/mL).[96]

9. VITAMIN E DEFICIENCY

Vitamin E deficiency can have primary (privational) and secondary (nonprivational) causes.

- **primary causes** involve inadequate vitamin E supply
  - dietary patterns that fail to provide vitamin E in adequate amounts
- **secondary causes** relate to impaired absorption, metabolism, or metabolic function of the vitamin
  - **Lipid malabsorption** including those resulting in loss of pancreatic exocrine function (e.g., pancreatitis, pancreatic tumor, nutritional pancreatic atrophy in severe selenium deficiency), those involving a lumenal deficiency of bile (e.g., biliary stasis because of mycotoxicosis, biliary atresia), those caused by defects in lipoprotein metabolism (e.g., **abetalipoproteinemia**[97]), and those typical in prematurity (Table 8.16).
  - **High PUFA intake** increases need for vitamin E. Animals fed high-PUFA diets require more vitamin E than those fed low-PUFA diets. It has been estimated that vitamin E needs increase by 0.18–0.60 mg

92. Although it has been identified in urine samples, it is not clear whether α-tocopheronolactone may be an artifact formed from α-CEHC in sample preparation.

93. West, Jr., K.P., Cole, R.N., Shrestha, S., et al., 2015. J. Nutr. 145, 2645–2656.

94. apoC-III, apoB, pyruvate kinase (muscle), forkhead box 04, unc5 homolog C, regulator of G-protein signaling 8.

95. F_2-isoprostanes are prostaglandin-like compounds formed in vivo from the free radical–catalyzed peroxidation of essential fatty acids, primarily arachidonic acid.

96. Block, G., Jensen, C.D., Morrow, J.D., et al., 2008. Free Radic. Biol. Med. 45, 377–384.

97. Humans with this rare hereditary disorder are unable to produce apoB, an essential component of chylomicra, VLDLs, and LDLs. The absence of these particles from the serum prevents the absorption of vitamin E because of the inability to transport it into the lymphatics. These patients show generalized lipid malabsorption with **steatorrhea** (i.e., excess fat in feces) and have undetectable serum vitamin E levels.

TABLE 8.15 Recommended Vitamin E Intakes

US		FAO/WHO		
Age-Sex	RDA[a] (mg/day)	Age-Sex		RNI[b] (μg/day)
0–6 months	4	0–6 months		2.7
7–11 months	5	7–11 months		2.7
1–3 years	6	1–3 years		5
4–8 years	7	7–9 years		7
9–13 years	11	10–18 years	Females	7.5
14–70+ years	15		Males	10
		19–65+	Females	7.5
			Males	10
Pregnancy	15	Pregnancy		–
Lactation	19	Lactation		–

[a]*Recommended Dietary Intakes; Food and Nutrition Board, 2000. Dietary Reference Intakes for Vitamin C, Vitamin E, Selenium and Carotenoids. National Academy Press, Washington, DC, 506 pp.*
[b]*Recommended Nutrient Intakes; Joint WHO/FAO Expert Consultation, 2001. Human Vitamin and Mineral Requirements. Food and Agricultural Org., Rome, 286 pp.*

α-tocopherol per gram of PUFA consumed; the upper end of that range is frequently cited as a guideline for estimating vitamin E needs.[98]

- **Deficiency of Se**, which spares the need for vitamin E in antioxidant defense. Animals fed low-Se diets generally require more vitamin E than those fed the same diets supplemented with an available source of Se.

Vitamin E needs can be affected by status with respect to other nutrients involved in the cellular antioxidant defense system (sulfur-containing amino acids;[99] copper, zinc, and/or manganese;[100] and riboflavin[101]), and by intake of synthetic antioxidants[102] (e.g., butylated hydroxytoluene,[103] BHA,[104] and DPPD[105]) and, possibly, by vitamin C.[106]

98. There is no consensus among experts in the field as to the quantitation of this obviously important relationship.
99. **Cysteine**, which can be synthesized via transsulfuration from methionine, is needed for the synthesis of glutathione, the substrate for the Se-dependent GPX.
100. These are essential cofactors of the superoxide dismutases.
101. Riboflavin is required for the synthesis of FAD, the coenzyme for glutathione reductase, which is required for regeneration of GSH.
102. Although vitamin E can be replaced by a variety of antioxidants, it should be noted that the effective levels of other antioxidants are considerably greater (two orders of magnitude) than those of α-tocopherol.
103. Butylated hydroxytoluene.
104. Butylated hydroxyanisole.
105. *N,N'*-Diphenyl-*p*-phenylenediamine.
106. The sparing effect of vitamin C is thought to involve its functioning in the reductive recycling of tocopherol.

Recommended intakes of vitamin E have been established (Table 8.15).

Groups at Risk of Vitamin E Deficiency.
Individuals with
very low fat intakes
low intakes of plant oils and nuts
lipid malabsorption syndromes and dyslipidemias

Vitamin E Deficiency Signs in Humans

Vitamin E deficiency is not common in adults. It manifests clinically as hemolytic anemia, i.e., normochromic anemia with reticulocytosis after long periods of depletion of body stores. This was demonstrated in the now-classical vitamin E depletion study by Max Horwitt at Elgin State Hospital.[107] This demonstrated that significant erythrocyte fragility, as evidenced by increased oxidative (H_2O_2-inducible) hemolosis, did not occur when plasma α-tocopherol concentrations exceeded ~0.5 mg/dL and that depletion to such levels required nearly a year of consumption of a low–vitamin E diet (Fig. 8.16).[108]

107. Horwitt (1908–2000) was for many years the Director of the L.B. Mendel Research Laboratory at Elgin State Hospital, Elgin, IL, where he conducted a series of now-classic studies with volunteers. These included some of the first studies of the nutritional aspects of aging, and studies of requirements for thiamin, riboflavin, tryptophan-niacin, and vitamin E.
108. Horwitt, M.K., Harvey, C.C., Duncan, G.D., et al., 1956. Am. J. Clin. Nutr. 4, 408–419.

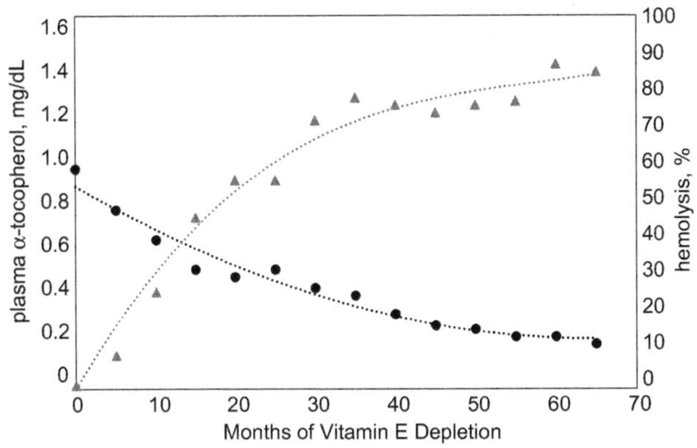

FIGURE 8.16 Relationship of vitamin E depletion and H_2O_2-induced hemolysis in Elgin Project volunteers. *After Horwitt, M.K., 1993. Max K. Horwitt: His Life and Science, M.K. Horwitt, St. Louis, p. 89.*

Chronic deficiency of vitamin E has been described in individuals with variant α-TTPs. These vary from moderate reductions in circulating tocopherol levels to untreatable progressive ataxia, depending on the degree of α-TTPs functionality.[109] Chronic deficiency can produce cerebellar and spinal cord damage manifest as ataxia, impaired reflexes, and impaired proprioreception. Subclinical deficiency can be detected as increased susceptibility to oxidative hemolysis in vitro. In pregnant women, this increases the risk for miscarriage,[110] apparently because of cellular anoxia resulting from primary lesions of the vascular system. In premature infants, vitamin E deficiency is manifest as membrane intraventricular hemorrhage and edema. Recommended intakes of vitamin E have been established (Table 8.15).

Vitamin E Deficiency Signs in Animals

The clinical manifestations of vitamin E deficiency vary considerably between species. In general, however, the targets are the neuromuscular,[111] vascular, and reproductive systems. The various signs of vitamin E deficiency are believed to be manifestations of membrane dysfunction resulting from the oxidative degradation of polyunsaturated membrane phospholipids and/or the disruption of other critical cellular processes.[112] Some signs (e.g., **encephalomalacia** in the chick) (Fig. 8.17) appear to involve local cellular anoxia resulting from primary lesions of the vascular system. Others, such as the nutritional myopathies (Fig. 8.18) appear to involve the lack to protection from oxidative stress. It has also been proposed that some effects (e.g., impaired immune cell functions) may involve loss of control of the oxidative metabolism of arachidonic acid in its conversion to leukotrienes; vitamin E is known to inhibit the 5′-lipoxygenase in that pathway.

Vitamin E deficiency has been shown in experimental animals (and children) to compromise both humoral and cell-mediated immunity. Deficient individuals have PMNs with impaired phagocytic abilities, suppressed oxidative burst and bactericidal activities, and decreased chemotactic responses. They can also show generally suppressed lymphocyte production, impaired T-cell functions, and decreased antibody production. Vitamin E deprivation has been found in animals to increase susceptibility to viral infections and to enhance the virulence of cardiophilic viruses passed through antioxidant-deficient hosts.[113] Vitamin E supplements are used to reduce the risk of mastitis in dairy cows and protect against aflatoxins in poultry diets. These effects are thought to involve loss of redox tone and appear to involve impaired cellular membrane fluidity and enhance PGE_2 production by the host, creating an environment in which viral mutation rates increase.

10. VITAMIN E IN HEALTH AND DISEASE

Oxidative stress plays a role in several conditions that have been associated with relatively low vitamin E. For such conditions, protection would be expected from increasing vitamin E intake. Supranutritional intakes[114] of vitamin E have been found beneficial under some of these conditions.

Cardiovascular Disease

Observational studies have consistently demonstrated benefits of vitamin E on cardiovascular disease risk.[115] Seven of the nine major cohort studies conducted to date (and involving nearly a quarter-million subjects) found inverse associations of vitamin E intake and cardiovascular disease incidence or associated mortality. Two large cohort studies found beneficial effects of high vitamin E intakes achieved through the use of dietary supplements for at least 2 years' duration (Table 8.17).[116] The results

109. Gotoda, T., Arita, M., Arai, H., et al., 1995. N. Engl. J. Med. 333, 1313–1318.

110. Shamin, A.A., Schulze, K., Merrill, R.D., et al., 2015. Am. J. Clin. Nutr. 101, 294–301.

111. Skeletal myopathies of vitamin E–deficient animals entail lesions predominantly involving type I fibers.

112. It is interesting to note a situation in which vitamin E deficiency would appear advantageous: the efficacy of the antimalarial drug derived from Chinese traditional medicine, *qinghaosu* (artemisinin), is enhanced by deprivation of vitamin E. The drug, an endoperoxide, is thought to act against the plasmodial parasite by generating free radicals in vivo. Thus, depriving the patient of vitamin E appears to limit the parasite's access to the protective antioxidant.

113. This involved an increase in viral mutation rate (Levander, O.A., Beck, M.A., 1997. Biol. Trace Elem. Res. 56, 5–21) and was also prevented by selenium.

114. That is, intakes substantially greater than required to prevent signs of nutritional deficiency.

115. *See* review: Cordero, Z., Drogen, D., Weikert, C., et al., 2010. Crit. Rev. Food Sci. Nutr. 50, 420–440.

116. Stampfer, M., Hennekens, C.H., Manson, J.E., et al., 1993. N. Engl. J. Med. 328, 1444–1450; Rimm, E.B., Stampfer, M.J., Ascherio, A., et al., 1993. N. Engl. J. Med. 328, 1450–1456.

TABLE 8.16 Signs of Vitamin E Deficiency

		Responds to		
Organ System	**Sign**	**Vitamin E**	**Selenium**	**Antioxidants**
General	Loss of appetite	+	+	+
	Reduced growth	+	+	+
Dermatologic	None			
Muscular	Myopathies			
	Striated muscles[a]	+	+	
	Cardiac muscle[b]	+	+	
	Smooth muscle[c]	+		
Skeletal	None			
Vital organs	Liver necrosis[d]	+	+	
	Renal degeneration[d]	+		+
Nervous system	Encephalomalacia[e]	+		+
	Areflexia, ataxia[f]	+		
Reproduction	Fetal death[g]	+	+	+
	Testicular degeneration[h]	+	+	
Ocular	Cataract[i]	+		
	Retinopathy[j]	+?		
Vascular	Anemia[j,k]	+		
	RBC hemolysis[l]	+		
	Exudative diathesis[e]	+	+	
	Intraventricular hemorrhage[j]	+		

[a]*Nutritional muscular dystrophies (white muscle diseases) of chicks, rats, guinea pigs, rabbits, dogs, monkeys, minks, sheep, goats, and calves.*
[b]***Mulberry heart disease** (congested heart failure) of pigs.*
[c]*Gizzard **myopathy** of turkeys and ducks.*
[d]*In rats, mice, and pigs.*
[e]*In chicks.*
[f]*In humans with abetalipoproteinemia.*
[g]*In rats, cattle, and sheep.*
[h]*In chickens, rats, rabbits, hamsters, dogs, pigs, and monkeys.*
[i]*Reported only in rats.*
[j]*Low–vitamin E status is suspected in this condition in premature human infants.*
[k]*In monkeys, pigs, and humans.*
[l]*In chicks, rats, rabbits, and humans.*

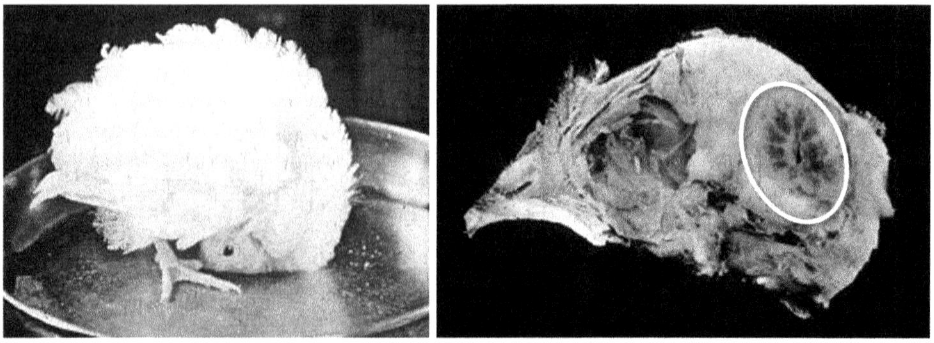

FIGURE 8.17 Encephalomalacia in a vitamin E–deficient chick. Chick is unable to maintain normal posture, tends to fall over, and cannot eat; postmortem examination shows cerebellar hemorrhage.

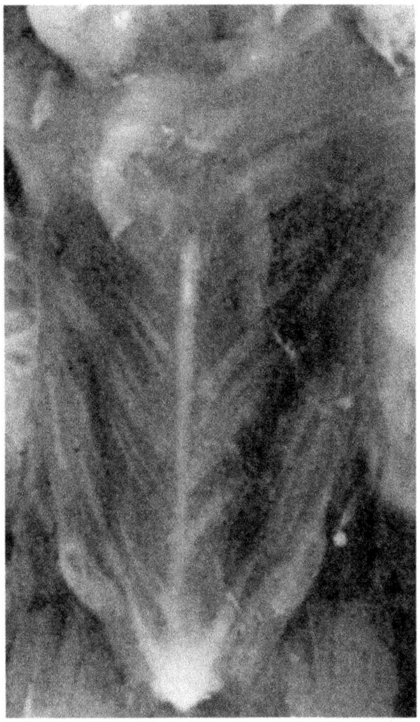

FIGURE 8.18 Nutritional muscular dystrophy in a chick deficient in vitamin E, selenium, and cysteine. Note: white striated breast muscle (*m. pectorales*) because of Zenker-type hyaline degeneration of muscle fibers. This condition, frequently called "white muscle disease", also occurs in vitamin E– and selenium-deficient lambs and calves.

of case–control and cohort studies have, however, been mixed. This is not surprising, given the many sources of variation in such studies: inherent errors in estimating vitamin E intake, variability in cardiovascular risk factors, variability in vitamin E utilization and baseline status, oxidative degradation of tocopherols during sample handling and storage, etc.

That vitamin E may protect against cardiovascular disease would appear likely. It can protect LDLs from oxidative damage, which is thought to be a factor in the etiology of atherosclerosis. Enrichment of LDLs with vitamin E, the predominant antioxidant occurring naturally in those particles, increases the lag phase of their oxidation in vitro, indicating increased resistance to oxidation (Table 8.18). Tocotrienols have been found to produce significant (28%) reductions in circulating triglyceride levels by reducing the upstream regulators of lipid homeostasis genes (Fig. 8.19).[117] When given intravenously, they have been found to inhibit acute platelet-mediated thrombus formation and collagen and ADP-induced platelet aggregation.[118]

The majority of randomized clinical trials have not found vitamin E supplementation to reduce cardiovascular risk. A meta-analysis of randomized trials found vitamin E supplementation to be associated with a 10% reduction in ischemic stroke, but a 22% increase in hemorrhagic stroke.[119] As the intervention agent in most studies has been α-tocopheryl acetate, some have asked whether non–α-vitamers, particularly γ-tocopherol and the tocotrienols, may be effective. It is more likely that beneficial effects may have been missed by not considering disease subtypes or sensitive subgroups, e.g., individuals with polymorphisms with specific genes involved in cellular antioxidant protection. A recent randomized trial found that adults with type 2 diabetes (T2D) showed reduced cardiovascular disease risk in response to supplemental vitamin E *only* if they also had a particular haptoglobin polymorphism, i.e., the *Hp2-2* genotype (Table 8.19).[120,121] Genetic polymorphisms affecting circulating α-tocopherol levels may also affect responsiveness to vitamin E, i.e., those involved in vitamin E transport/retention (α-TTP, TAPs, apoE and A, SR-BI and CD36 scavenger receptors, LDL receptor, PLTP, microsomal triglyceride transfer protein, lipoprotein lipase, and ATP-binding cassette transporter); vitamin E–mediated gene expression (pregnane X receptor); vitamin E metabolism (CYP3A and CYP4F2); and antioxidant metabolism (dehydroascorbate reductase, sodium-coupled ascorbic acid transporters).

Neurodegenerative Disorders. Epidemiological studies have found vitamin E intake to be inversely associated with risks for cognitive impairment[122] and to Alzheimer disease and Parkinson disease, the etiologies of which are thought to involve oxidative stress. Most randomized clinical trials have found vitamin E supplementation to yield inconsistent results with respect to those diseases and amyotrophic lateral sclerosis (ALS),[123] which is thought to

117. Zaiden, N., Yap, W.N., Ong, S., et al., 2010. J. Atheroscler. Thromb. 17, 1019–1032.

118. Qureshi, A.A., Karpen, C.W., Qureshi, N., et al., 2011. Lipids Health Dis. 10, 58–71.

119. For example, Shűrks, M., Glynn, R.J., Rist, P.M., et al., 2010. Br. J. Med. 341, c5702) found supplemental vitamin E not to affect stroke risk, but to reduce risk of ischemic stroke by 10% and increase risk of hemorrhagic stroke by 22%.

120. Milman, U., Blum, S., Shapira, C., et al., 2008. Arterioscler. Thromb. Vasc. Biol. 28, 341–348.

121. Haptoglobin (Hp) complexes with hemoglobin (Hb) to shield its iron center, thus reducing its prooxidative effect. The complex is recognized by the CD163 scavenger receptor and is, thus, cleared from the plasma endocytotically into the reticuloendothelial system. There are two common alleles at the Hp locus, 1 and 2, and functional differences between the Hp1 and Hp2 proteins. Individuals homozygous for Hp2 (*Hp2-2*) have the polymorphism that forms cyclic polymers instead of the linear ones formed by the *Hp1-1* and *Hp1-2* genotypes. *Hp2-2* polymorphisms form complex Hb with 10-fold greater affinity than the linear Hp polymorphisms, bind less avidly CD163, and are, therefore, less efficiently cleared from the plasma. Individuals with the *Hp2-2* phenotype would be expected to have greater needs for the antioxidant effects of vitamin E. Individuals with the *Hp2-2* phenotype and T2D are at increased risk to cardiovascular disease.

122. Mangialasche, F., Solomon, A., Kåreholt, I., et al., 2013. Exp. Gerontol. 48, 148–1435.

123. ALS ("Lou Gherig's Disease") is characterized by profound muscular weakness because of the selective death of upper and lower motor neurons; the disease is ultimately fatal, mostly because of respiratory failure.

TABLE 8.17 High Vitamin E Intakes Associated With Reduced Coronary Heart Disease Risks in Two Cohorts of Americans

	Quintile Group for Vitamin E Intake					
Parameter	1	2	3	4	5	*p* Value for Trend
Women[a] – Total Vitamin E (Diet + Supplements)						
Median intake (IU/day)	2.8	4.2	5.9	17	208	
Relative risk[c]	1.0	1.00	1.15	0.74	0.66	<0.001
Men[b] – Total Vitamin E (Diet + Supplements)						
Median intake (IU/day)	6.4	8.5	11.2	25.2	419	
Relative risk[c]	1.0	0.88	0.77	0.74	0.59	0.001
All – Vitamin E From the Diet						
Range of intakes (IU/day)	1.6–6.9	7.0–9.8	8.2–9.3	9.4–11.0	11.1	
Relative risk[c]	1.0	1.10	1.17	0.97	0.79	0.11
All – Vitamin E From Supplements						
Range of intakes (IU/day)[b]	0	<25	25–99	100–249	≥250	
Relative risk[c]	1.0	0.85	0.78	0.54	0.70	0.22

[a]A total of 39,910 health professionals (139,883 per-years follow-up); Rimm, E.B., Stampfer, M.J., Ascherio, A., et al., 1993. N. Engl. J. Med. 328, 1450–1456.
[b]A total of 87,245 nurses (679,485 per-years follow-up); Stampfer, M., Hennekens, C.H., Manson, J.E., et al., 1993. N. Engl. J. Med. 328, 1444–1450.
[c]Ratio of events in each quintile to those in the lowest quintile; adjusted for age and smoking.

TABLE 8.18 Reduced Low-Density Lipoprotein Susceptibility to Lipid Peroxidation by Oral Vitamin E in Humans

			LDL Oxidation[a]	
Treatment	Time of Sampling (weeks)	LDL α-Tocopherol (μmol/g) Protein	Lag Phase (h)	Rate (μmol/g) Protein (h)
Placebo	0	14.3 ± 5.0	2.1 ± 0.9	396 ± 116
	8	15.8 ± 6.1	2.0 ± 0.8	423 ± 93
Vitamin E[b]	0	13.8 ± 4.1	1.9 ± 0.6	373 ± 96
	8	32.6 ± 11.5[c]	2.9 ± 0.8[b]	367 ± 105

[a]Lipid peroxide formation.
[b]Total of 1200 IU/day as RRR-α-tocopherol.
[c]Significantly different (p < .05) from corresponding placebo value.
From Fuller, C.J., Chandalia, M., Garg, A., et al., 1996. Am. J. Clin. Nutr. 63, 753–759.

involve mitochondrial stress. However, a multicenter trial conducted on patients with Alzheimer disease in the United States found a daily supplement of 2000 IU α-tocopherol to be effective in retarding the functional decline of mild and moderate Alzheimer disease.[124]

124. Dysken, M.W., Sano, M., Asthana, S., et al., 2014. JAMA 311, 33–44.

Inflammatory Diseases

Despite the clear anti-inflammatory effects that have been demonstrated for these vitamers, clinical trials have generally not found α-tocopherol supplements effective in reducing risks to inflammatory diseases (e.g., RA, asthma, and hepatitis) or diseases in which chronic inflammation

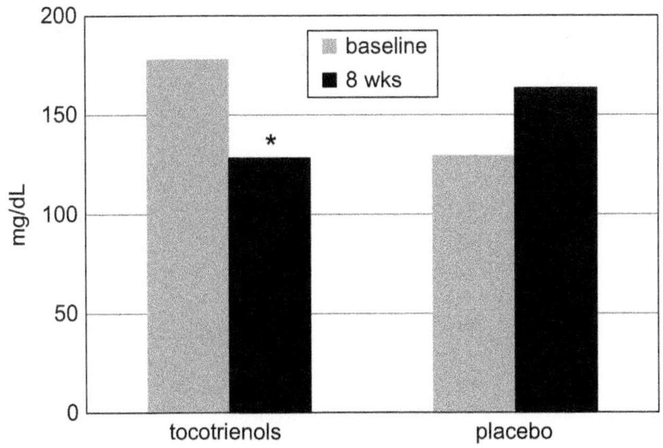

FIGURE 8.19 Reduction of serum triglycerides by mixed tocotrienols in hypercholesterolemic subjects. No effects on serum cholesterol of lipoproteins were noted. *From Zaiden, N., Yap, W.N., Ong, S., et al., 2010. J. Atherocler. Thromb. 17, 1019–1032.*

contributes (e.g., cardiovascular disease, cancer neurodegenerative diseases). The apparently different anti-inflammatory activities of e vitamers have made some to suggest that mixed tocopherols may have potential value in this regard. Studies with animal models of RA have found vitamin E supplementation to reduce joint swelling and γ-tocotrienol to reduce the pathogenesis of collagen-induced arthritis.[125] Randomized controlled trials have shown high-level supplementation with the vitamin (100–600 IU/day) to relieve pain and be anti-inflammatory for patients with RA, but not in reducing the risk of the disease.[126]

Air Pollution and Smoking. Vitamin E can be expected to protect the lungs against the continuous exposure to relatively high concentrations of O_2 and environmental oxidants and irritants. The respiratory epithelium contains vitamin E and relatively high concentrations of vitamin C, urate, GSH, extracellular superoxide dismutase, catalase, and GPX. Individuals living in smog-filled urban areas can be exposed to relatively high levels of ozone (O_3)[127] and NO_2, strong oxidants that provoke inflammatory responses of the airway, i.e., activating neutrophils and evoking macrophage respiratory bursts the overproduction of ROS, which causes peroxidative damage. Because vitamin E deprivation increases the susceptibility of experimental animals to the pathological effects of O_3 and NO_2,[128] it has been suggested that supplements of the vitamin may protect humans against chronic exposure to smog.

Supranutritional doses of vitamin E (up to 560 mg of α-tocopherol per day) have been found to reduce the peroxidation potential of erythrocyte lipids from smokers (Table 8.20). Smoking increases the oxidative burden on the lungs and other tissues because of the sustained exposure to free radicals from tars and gases.[129] This is characterized by increased levels of peroxidation products in the circulation (e.g., MDA) and breath (e.g., ethane, pentane), with decreased levels of ascorbic acid in plasma and leukocytes and of vitamin E in plasma and erythrocytes. High amounts of RNS in cigarette smoke cause nitration of vitamin E; 5-nitro-γ-tocopherol occurs at higher levels in smokers than nonsmokers (Table 8.15). This may be the basis for the enhanced turnover of tocopherols and reduced production of CECH by smokers.

Cataracts.[130] Vitamin E has been shown in animal models to protect against cataracts induced by galactose or aminotriazol and to reduce the photoperoxidation of lens lipids. These effects are thought to involve direct action of the vitamin as an antioxidant, including the maintenance of lens glutathione in the reduced state. Much of cataract lens opacification involves oxidations resulting in loss of sulfhydryls, formation of disulfide and nondisulfide covalent linkages, and oxidation of tryptophan residues. More than a dozen cohort studies have found plasma α-tocopherol level and/or vitamin E intakes to be inversely related to cataract risk.[131] These suggest that 5–10 years of high–vitamin E intake may be required for protection against nuclear cataracts. However, the only randomized controlled trial to have been conducted to date did not find α-tocopherol supplementation (500 IU/day) to reduce incidence of progression of cataract.[132]

Skin. Vitamin E is found in the skin mostly in the lower levels of the stratum corneum, where it is released by sebum. In animal models, the tocopherol content of dermal tissues decreases with UV irradiation, presumably a result of oxidative stress. Topical treatment with vitamin E may increase the hydration of the stratum corneum and confers protection against UV-induced skin damage, as measured by reduced erythemal responses and delayed

125. Radhakrishnan, A., Tudawe, D., Chakravarthi, S., et al., 2014. Exp. Therap. Med. 7, 1408–1414.

126. Karlson, E.W., Shadick, N.A., Cook, N.R., et al., 2008. Arthr. Rheum. 59, 1589–1596.

127. Ambient O_3 level on the top of Mt. Everest or in Los Angeles can be 70–80 ppb, whereas in Grand Forks, ND, the level is <50 ppb.

128. O_3 is produced in photochemical smog from nitrogen dioxide (NO_2) from internal combustion engines, O_2, and gasoline vapors. Both O_3 and NO_2 can generate unstable free radicals that damage lungs through oxidative attack on polyunsaturated membrane phospholipids.

129. Cigarette smoke contains a number of compounds that produce free radicals; it also increases the number of free radical–producing inflammatory cells in the lungs.

130. Cataracts result from the accumulation in the lens of damaged proteins that aggregate and precipitate, resulting in opacification of the lens. Cataract is a significant public health problem in the United States, where a million cataract extractions are performed annually at a cost of some $5 billion. The prevalence of cataracts among Americans increases from about 5% at age 65 years to about 40% at age 75 years. These rates are considerably (up to fivefold) greater in less developed countries.

131. Vishwanathan, R., Johnson, E.J., 2012. In: Erdman, J.W., Macdonald, I.A., Zeisel, S.H., (Eds.) Present Knowledge in Nutrition, tenth ed. Wiley-Blackwell, Ames, IA, p. 946–948.

132. McNeil, J.J., Robman, L., Tikellis, G., et al., 2004. Ophthalmology 111, 75–84.

TABLE 8.19 Relationship of Haptoglobin Genotype and Cardioprotection by Vitamin E

Treatment	Hp Genotype	n Events	Hazard Ratio (95% CI)		p Value
None	Hp2-1	1248	25	1.0	
	Hp1-1	285	6	1.0 (0.4–2.5)	0.92
Placebo	Hp2-2	708	33	2.3 (1.4–3.9)	0.001
Vitamin E	Hp2-2	726	16	1.1 (0.6–2.00)	0.81

From Milman, U., Blum, S., Shapira, C., et al., 2008. Arterioscler. Thromb. Vasc. Biol. 28, 341–348.

TABLE 8.20 Comparison of Effect of Vitamin E on Erythrocyte Lipid Peroxidation in Smokers and Nonsmokers

	Weeks of Vitamin E Administration[a]	Vitamin E in Erythrocytes (μmol/g Hb)	In vitro Lipid Peroxidation (nmol MDA[b]/g Hb)
Nonsmokers	0	20.0 ± 4.5	141 ± 54
	20	36.1 ± 8.2	86 ± 51
Smokers	0	18.0 ± 4.2	291 ± 102
	20	32.8 ± 8.2	108 ± 53

[a]*A total of 70 IU/day.*
[b]*Malonyldialdehyde.*
From Brown, K.M., Morrice, P.E., Duthie, C.G., 1977. Am. J. Clin. Nutr. 65, 496–502.

onset of tumorigenesis. One study found that regular topical application of vitamin E reduced wrinkle amplitude and skin roughness in about half of cases. For these reasons, α-tocopherol and α-tocopheryl acetate are widely used in skin creams and cosmetics.

Type 2 Diabetes (T2D). Vitamin E appears to play protective roles against T2D. α-Tocopherol appears to be less well absorbed and to turnover slower in subjects with metabolic syndrome[133] compared to controls.[134] Studies have found patients with diabetes to have low plasma tocopherols, higher urinary levels of tocopherol metabolites, and greater erythrocyte lipid peroxidation than healthy controls (Table 8.16). A randomized trial reported protection by vitamin E against the development of T2D among subjects with impaired glucose tolerance,[135] and high-level vitamin E supplements (e.g., 900 mg α-tocopherol per day) have been found to improve insulin responsiveness in both normal patients and patients with T2D . Oxidative stress is known to alter cellular serine/threonine kinase activities, reducing insulin signaling by suppressing phosphorylation of insulin receptor substrate-1. Supranutritional levels of α- and γ-tocopherols have been shown to upregulate an endogenous ligand involved in activating PPARγ and upregulate adiponectin, an adipokine that increases insulin sensitivity.[136] Vitamin E has been considered as a factor in protecting against diabetic complications (e.g., retinopathy, cardiac dysfunction) the etiologies of which are thought to involve oxidative stress. Clinical trials have found α-tocopherol supplementation to improve parameters of redox balance and endothelial function in nonobese (BMI < 30) diabetics (Table 8.21).[137]

Tocotrienols have been found effective in normalizing serum glucose and glycated hemoglobin (HbA_{1C}) in a diabetic rat model, and in reducing serum lipids, cholesterol, and LDL-cholesterol in patients with T2D with hyperlipidemia.[138] γ-Tocotrienol has been found to reduce insulin resistance associated with obesity in the rat; these effects appear to be because of reductions in adipose output of proinflammatory cytokines.[139]

133. Metabolic syndrome is associated with risks of developing T2D and cardiovascular disease. It is diagnosed on the basis of an individual having three or more of the following conditions: abdominal obesity, elevated blood pressure, high fasting serum glucose, high serum triglycerides, and low HDL levels.
134. Mah, E., Sapper, T.N., Chitchumroonchokchai, C., et al., 2015. Am. J. Clin. Nutr. 102, 1070–1080.
135. Mayer-Davis, E.J., Costacou, T., King, I., et al., 2002. Diabetes Care 25, 2172–2177.
136. Landrier, J.F., Gouranton, E., El Yazidi, C., et al., 2009. Endocrinol. 150, 5318–5328.
137. Montero, D., Walther, G., Stehouwwer, C.D.A., et al., 2014. Obes. Rev. 15, 107–116.
138. Baliarsingh, S., Bg, Z.H., Ahmad, J., 2005, Atherosclerosis 182, 367–374.
139. Zhao, L., Kang, I., Fang, X. et al., 2015. Int. J. Obesity 39, 438–446.

TABLE 8.21 Diabetes and Tocopherol Status

	Diabetics	Controls
Quilliot, D., Walters, E., Bonte, J.P., et al., 2005. Am. J. Clin. Nutr. 81, 1117–1125.		
Plasma α-tocopherol (μg/mL)	12.9±2.9[a]	17.4±3.7
Plasma LDL (mg/g)	5.5±3.8[a]	6.4±1.3
Sharma, G., Muller, D.P., O'Riordan, S.M., et al., 2013. Free Radic. Biol. Med. 55, 54–62.		
Urine α-CEHC glucuronide (nmol/mmol creatinine)	126±16[a]	73±19
Urine α-CEHC sulfate (nmol/mmol creatinine)	138±33[a]	57±12
Urine α-tocopheronolactore glucuronide (nmol/mmol creatinine)	1098±279[a]	76±13

[a]*Significantly different from control,* $p<.05$.

Preeclampsia. That vitamin E plays a key role in pregnancy is suggested by the fact that circulating α-tocopherol levels correlate positively with fetal growth rate, particularly during the last trimester when O_2 utilization is increased. Increases in lipid peroxides of placental origin in the maternal circulation correlate with the severity of preeclampsia.[140] A systematic review of nine intervention trials found no evidence for vitamin E supplementation during pregnancy affecting risk of preeclampsia.[141]

Altitude. Vitamin E protects against the prooxidative effects of O_3. One study found that a daily supplement of 400 IU vitamin E prevented decreases in anaerobic thresholds of high-altitude mountain climbers, suggesting that exercise at altitude increases the need for vitamin E (Table 8.22).

Ischemia–Reperfusion Injury. Vitamin E and other free-radical scavengers have been found to affect the functions of mitochondria and sarcoplasmic reticula in animal models of myocardial injury induced by ischemia–reperfusion. The injury, occurring in tissues reperfused after a period of ischemia, appears to be because of the oxidative stress of reoxygenation, involving production of ROS. The phenomenon has been demonstrated for several tissues (heart, brain, skin, intestine, and pancreas) and has relevance for the preservation of organs for transplantation. ROS are thought to contribute to milder forms of tissue injury at the time of reperfusion (e.g., myocardial stunning, reperfusion arrhythmias). However, the extent to which free radicals are responsible for the acute tissue damage seen under those circumstances is not clear. Intervention trials with antioxidants have yielded conflicting results, but it is possible that preoperative treatment with vitamin E may be useful in reducing at least some of this type of injury. A randomized clinical trial demonstrated that vitamin E supplements (given by intramuscular injection) can be very effective in protecting premature infants against intraventricular hemorrhage,[142] which involves natural ischemia–reperfusion injury.

TABLE 8.22 Antioxidant Protection by Vitamin E in High-Altitude Climbers

	Change in Pentane Exhalation[a] (%)		
Treatment Group	Median	Lower Quartile	Upper Quartile
Placebo	104	26	122
Vitamin E, 400 IU/day	−3	−7	3

[a]*After 4 weeks at high altitude.*
From Simon-Schnass, I., Pabst, H., 1988. J. Vit. Nutr. Res. 58, 49–56.

Antitumorigenesis

It would seem reasonable to expect that vitamin E may have a role in cancer prevention in as much as at least for some types of cancer are thought to involve ROS. Chemical carcinogenesis typically involves the electrophilic attack of free radicals with DNA. The generation of ROS, mutagenic to mammalian cells in vitro, correlates with the initiation, promotion, and progression of tumors in experimental animal models. MDA has also been found to increase tumor production in animals. Studies with two-stage, UV-, or chemically induced mammary, colon, oral, or skin tumor models have generally shown α-tocopherol to inhibit tumor promotion. These include studies showing topically applied vitamin E to reduce UV-induced skin cancers by as much as 58%.

Evidence indicates that vitamin E, particularly the tocotrienols, can selectively stimulate apoptosis in neoplastic cells.[143] Tocotrienols have been shown to be capable of targeting multiple cell signaling pathways. They

140. That is, pregnancy-induced hypertension associated with proteinuria. Preeclampsia is thought to involve endothelial dysfunction of maternal blood vessels induced by factors released from the placenta. It can develop in the last trimester of pregnancy and has the highest rates of morbidity and mortality of all complications of pregnancy.

141. Conde-Agudelo, A., Romero, R., Kusanovic, J.P., et al., 2011, Am. J. Obstet. Gynecol. 204 (503) e1–e12.

142. Hemorrhage in and around the lateral ventricles of the brain occurs in about 40% of infants born before 33 weeks of gestation.

143. McIntyre, B.S., Briski, K.P., Tirmenstein, M.S., et al., 2000. Lipids 35, 171–180.

have been shown to cause receptor-induced caspase-8 and -3 activation (a typical response to oxidative stress), leading to apoptosis in some cancer cells, and caspase-9 activation by mitochondrial stress in others.[144] They have been found to inhibit cytokine-stimulated activation of NF-κB by inducing its inhibitor, the anti-inflammatory protein A20;[145] to inhibit TGF-β, P38, and β-catenin/Tcf signaling; to downregulate Bcl2, cyclin D, and the Raf/Erk pathway; to inhibit HMGR; to induce DNA fragmentation, to inhibit angiogenesis; and to promote cell cycle arrest.[146] A polymorphism of the scavenger receptor class B member 1 associated with increased circulating levels of vitamin E was also found to be associated with reduced prostate cancer risk.[147] The α-TAP and apparent transcription factor TAP appears to have a role in regulating tumorigenesis. Liganded TAP is thought to serve as a tumor suppressor.[148] Its mRNA is inversely associated with tumor stage in breast cancer. Overexpression of TAPs has been found to suppress growth of prostate cancer cells,[149] apparently by sensitizing them to antiproliferative effects of α-tocopherol and allowing them to accumulate the vitamin.[150]

The results of most studies to date are inconsistent with respect to relationships of vitamin E and cancer incidence. Inverse epidemiological associations have been made between the consumption of vitamin E–rich foods and cancer risks, and longitudinal studies have found circulating tocopherol levels to be only slightly lower (~3%) in patients with cancer than in healthy controls. A pooled analysis of 15 prospective studies found serum/plasma α-tocopherol concentration to be positively associated with prostate cancer risk.[151] A difference in this magnitude was observed between patients with cancer and controls in a large trial in Finland,[152] which also found individuals with relatively low serum α-tocopherol levels to have a 1.5-fold greater risk of cancer than those with higher serum vitamin E levels.

Only a few clinical trials have evaluated the antitumorigenic potential of supplemental vitamin E. One study found the combination of vitamin E, selenium and β-carotene to reduce lung cancer risk by an apparent 45%.[153] Some small trials have suggested that vitamin E may reduce the risk of breast cancer among women with mammary dysplasia; however, those results have not been confirmed in larger trials of benign breast disease. A study with more than 29,000 male smokers found vitamin E treatment (50 mg α-tocopherol per day) to be associated with a 34% reduction in prostate cancer incidence;[154] however, large subsequent trials have not found supplemental vitamin E to affect prostate cancer incidence.[155]

Other Conditions. Studies with animal models have demonstrated that vitamin E can protect against radiation-induced chromosome damage, because of direct ionization of DNA and other cellular targets and indirect effects of ROS that are produced. This effect may be the basis of epidemiological associations between consumption of antioxidant-rich foods and low cancer risk. Vitamin E has frequently proven effective as a therapeutic measure in hemolytic anemia of prematurity, intermittent claudication,[156] and chronic hemolysis in patients with glucose-6-phosphate dehydrogenase deficiency. In veterinary practice, vitamin E (most frequently administered with selenium) has been reported efficacious for treating "tying up"[157] in horses and postpartum placental retention in dairy cows. Formerly, vitamin E was thought to protect against retinopathy of prematurity;[158] however, a controlled clinical trial failed to confirm that. Low vitamin E status has been associated with osteoporosis and fracture risk,[159] supporting the hypothesis that oxidative stress may impair osteoblast function. A systematic review of the relevant scientific literature concluded that tocotrienols may have value in preventing/treating osteoporosis. Of the 11 studies meeting the inclusionary criteria, three

144. *See* review: Sylvester, P.W., 2007. Vit. Horm. 76, 329–356.
145. Wang, Y., Park, N.Y., Jang, Y., et al., 2015. J. Immunol. 195, 126–133.
146. Kannapan, R., Gupta, S.C., Kim, J.H., et al., 2012. Genes Nutr. 7, 43–52; Ling, M.T., Luk, S.U., Al-Ejeh, F., et al., 2012. Carcinogen. 33, 233–249; Xu, W., Du, M., Zhao, Y., et al., 2012. J. Nutr. Biochem. 23, 800–807; Ahsan, H., Ahad, A., Iqbal, J., et al., 2014. Nutr. Metab. 11, 52–74; Shibata, A., Nakagawa, K., Tsuduki, T., 2015. J. Nut. Biochem. 26, 832–840.
147. Major, J.M., Yu, K., Weinstein, S.J., et al., 2014. J. Nutr. 144, 729–733.
148. Wang, X., Ring, B.Z., Seitz, R.S., 2015. BMC Clin. Pathol. 15, 21–31.
149. Ni, J., Wen, X., Yao, J., et al., 2005. Cancer Res. 65, 9807–9816.
150. Morley, S., Thjakur, V., Denielpour, D., et al., 2010. J. Biol. Chem. 285, 35,578–35,589.
151. Key, T.J., Appleby, P.N., Travis, R.C., et al., 2015. Am. J. Clin. Nutr. 145, 1142–1157.
152. That is, the Finnish Mobile Clinic Health Survey, which involved more than 36,000 subjects; Kneckt, P., et al., 1991. J. Clin. Nutr. 53, 283S–289S.
153. Blot, W.J., Li, J.Y., Taylor, P.R., et al., 1993. J. Natl. Cancer Inst. 85, 1483–1492.
154. Albanes, D., Heinonen, O.P., Taylor, P.R., et al., 1996. J. Natl. Cancer Inst. 88, 1560–1570.
155. The HOPE-TOO Trial (Lonn, E., Bosch, J., Yusuf, S., et al., 2005. JAMA 293, 1338–1347); The Physicians' Health Study II (Gaziano, J.M., Glynn, R.J., Christen, W.G., et al., 2009. JAMA 301, 52–62); SELECT (Lippman, S.M., Klein, E.A., Goodman, P.J., et al., 2009. JAMA 301, 39–51); Wang, L., Sesso, H.D., Glynn, R.J., et al., 2014. Am. J. Clin. Nutr. 100, 915–923.
156. Nocturnal leg cramps.
157. That is, rhabdomyolysis involves the breakdown of striated muscle after exercise, characterized by soreness in the gluteal muscles and painful/stiff gait. It most frequently results from having had limited exercise or having been put in a stressful environment.
158. This disorder was formerly called **retrolental fibroplasia**. Its pathogenesis involves exposure to a hyperoxic environment during neonatal oxygen therapy. It can affect as many as 11% of infants with birth weights below 1500 g, resulting in blindness of about one-quarter of them.
159. Michaëlson, K., Wolk, A., Byberg, L., et al., 2014. Am. J. Clin. Nutr. 99, 107–114.

epidemiological studies and eight animal studies reported positive effects of tocotrienols.[160]

Healthy Aging. Studies in animal models have shown that vitamin E status reduces the accumulation of "age pigments", i.e., lipid-soluble, brown-yellow, autofluorescent pigments[161] collectively called lipofuscin, in several tissues (retinal pigment epithelium, heart muscle, brain, skin). That these changes are causally related to aging is suggested by interspecies observations that mammalian life span potentials tend to correlate inversely with metabolic rate and directly with tissue concentrations of tocopherols and other antioxidants (carotenoids, urate, ascorbic acid, and superoxide dismutase activity).[162] It has been proposed that cumulative damage by ROS and associated increases in redox tone lead to gradual decreases in repair capacity likely because of changes in gene expression, diminished immune function, enhanced inflammation, and enhanced programmed cell death.[163] Studies in rats have shown that the age-related decline in the major glucose transporter in neurons (Glut3) is exacerbated by deprivation of vitamin E.

11. VITAMIN E TOXICITY

Vitamin E has been viewed as one of the *least toxic* of the vitamins. Both animals and humans appear to be able to tolerate high levels.[164] For animals, intakes at least two orders of magnitude greater than nutritional requirements, e.g., to 1000–2000 IU/kg, are without untoward effects. For humans, daily doses as high as 400 IU have been be considered harmless, and large oral doses as great as 3200 IU have not been found to have consistent ill effects. These views were challenged by a meta-analysis (of 19 trials) suggesting that vitamin E supplements (≥400 IU/day) may increase all-cause mortality. A more recent meta-analysis, which included a larger number (57) of published trial results, concluded that supplemental vitamin E does *not* affect all-cause mortality at doses up to 5500 IU/day.[165]

It is known that at very high doses of vitamin E can antagonize the functions of other fat-soluble vitamins. Hypervitaminotic E animals have been found to show impaired bone mineralization, reduced hepatic storage of vitamin A, and coagulopathies. In each case, these signs could be corrected with increased dietary supplements of the appropriate vitamin (i.e., vitamins D, A, and K, respectively) and the antagonism seemed to be based at the level of absorption. Isolated reports of negative effects in human subjects consuming up to 1000 IU of vitamin E per day included headache, fatigue, nausea, double vision, muscular weakness, mild creatinuria, and gastrointestinal distress.

Potentially deleterious metabolic effects of high-level vitamin E status include inhibitions of retinyl ester hydrolase and vitamin K–dependent carboxylations. The former effect has been demonstrated in animals, where it results in impaired ability to mobilize vitamin A from hepatic stores. Supranutritional vitamin E treatment of rats has been shown to increase bleeding tendency; however, there is little evidence for comparable effects in humans. Patients given high doses of α-tocopherol (1200 IU/day) have shown increased blood clotting times because of hypoprothrombinemia, but that effect may be of concern only for patients on anticoagulant therapy. Although the metabolic basis of the effect is not clear, it likely involves inhibition of vitamins E and K competing for the CYP4F2-dependent ω-hydroxylation of their structurally similar phytyl side chains, inhibiting the conversion of vitamin K_1 to MK-4 and, thus, the vitamin K–dependent carboxylase (*see* Chapter 9).

Any concerns about the safety of high intakes of vitamin E must be about the use of nutritional supplements, as high intakes are virtually impossible to achieve from foods. Accordingly, the IOM has recommended safe upper limits on vitamin E intakes from supplements (Table 8.23)

TABLE 8.23 Upper Tolerable Limits (UL) for Vitamin E From Supplements

Age Group	UL (mg/day)
Infants	–[a]
Children, 1–3 years	200
4–8	300
9–13	600
14–18	800
Adults, 19+	1000
Pregnancy, ≤18	800
>18	1000
Lactation, ≤18	800
>18	1000

[a]*No UL was set for infants; however, an UL was recommended for premature infants (birth weight ≤1.5 kg) at 21 mg/day.*
From Food and Nutrition Board, 2000. Dietary Reference Intakes for Vitamin C, Vitamin E, Selenium and Carotenoids. National Academy Press, Washington, DC, p. 186–283.

160. Muhammad, N., Borhanuddin, B., Shuid, A.N., et al., 2012. Evid. Based Complem. Altern. Med. ID 250584, 14 pp.

161. Lipofuscins are thought to be condensation products of proteins and lipid. There is some evidence that the pigments isolated from the retinal pigment epithelium contain, at least in part, derivatives of vitamin A (e.g., *N*-retinylidene-*N*-retinylethanolamine).

162. Caloric restriction can increase longevity in animals, perhaps by reducing the metabolic rate, hence, endogenous production of ROS.

163. That is, **apoptosis**.

164. Hathcock, J.N., Azzi, A., Blumberg, J., et al., 2005. Am. J. Nutr. 81, 736–745; Cook-Mills, J.M., May, C.A., 2010. Endocr. Metab. Immune Disord. Drug Targets 10, 348–366.

165. Abner, E.L. Schmitt, F.A., Mendiondo, M.S., et al., 2011. Curr. Aging Sci. 4, 158–170.

12. CASE STUDIES

Instructions

Review the following case reports, paying special attention to the diagnostic indicators on which the treatments were based. Then answer the questions that follow.

Case 1

At birth, a male infant with acidosis[166] and **hemolytic anemia**[167] was diagnosed as having glutathione (GSH) synthetase[168] deficiency associated with 5-keto-prolinuria[169]; he was treated symptomatically. During his second year, he experienced six episodes of bacterial otitis media.[170] His white cell counts fell to 3000–4000 cells/μL during two of these infections, with notable losses of PMNs[171]. Between infections, the child had normal white and differential cell counts, and PMNs were obtained for study.

Functional studies of PMNs showed the following results:

Parameter	Finding
GSH synthetase activity	10% of normal
Phagocytosis of *Staphylococcus aureus*	Less than normal
Iodination of phagocytized zymosan particles	Much less than normal
H_2O_2 production during phagocytosis	Well above normal

The child was then treated daily with 30 IU of all-*rac*-α-tocopheryl acetate per kilogram body weight (about 400 IU/day). His plasma vitamin E concentration rose from 0.34 mg/dL (normal for infants) to 1.03 mg/dL. After 3 months of treatment, the same studies of his PMNs were performed. Although there were no changes in the activity of GSH peroxidase[172] or the concentration of GSH (which remained near 25% of normal during this study) in his plasma, the production of H_2O_2 by his PMNs had declined to normal levels, the iodination of proteins during phagocytosis had increased, and the bactericidal activity toward *Staphylococcus aureus* had increased to the control level. Before his vitamin E therapy, electron microscopy of his neutrophils had revealed defective cytoskeletal structure, with more than the usual number of *microtubules*[173] seen at rest and a disappearance of microtubules seen during phagocytosis. This ultrastructural defect was corrected after vitamin E treatment.

Case 2

A 23-year-old woman with a 10-year history of neurologic disease was admitted complaining of severe ataxia,[174] titubation of the head,[175] and loss of proprioceptive sense in her extremities.[176] Her history revealed that she had experienced difficulty in walking and was unsteady at age 10 years; there was no family history of ataxia, malabsorption, or neurologic disease. At 18 years of age, she had been hospitalized for her neurologic complaints; at that time, she had been below the fifth percentile for both height and weight. Her examination had revealed normal higher intellectual function, speech, and cranial nerve function; but her limbs had been found to be hypotonic[177] with preservation of strength and moderately severe ataxia. Her deep tendon reflexes were absent, plantar responses were abnormal, vibrational sense was absent below the wrists and iliac crests, and joint position sense was defective at the fingers and toes. Laboratory findings at that time had been negative, i.e., she showed no indications of hepatic or renal dysfunction. No etiologic diagnosis was made. Two years later, when she was 20 years old, the patient was reevaluated. By that time, her gait had deteriorated and her proprioceptive loss had become more severe.

Over the next 3 years, her symptoms worsened and, by age 23 years, she had trouble walking unassisted. Still, she showed no sensory, visual, bladder, respiratory, or cardiac signs and ate a normal and nutritious diet. Her only gastrointestinal complaint was of constipation, with bowel movements only once per week. Nerve conduction tests revealed that the action potentials of both her sensory and motor nerves, recorded from the median and ulnar nerves, were normal. Electromyography of the biceps, vastas medialis, and tibialis anterior muscles was normal. However, her cervical and cortical somatosensory-evoked responses to median nerve stimulation were abnormal; there was no peripheral delay and the nature of the response was abnormal. Furthermore, no consistent cortical responses could be recorded after stimulation of the tibial nerve at the ankle. These findings were interpreted as indicating

166. The condition of reduced alkali reserve.
167. Reduced number of erythrocytes per unit blood volume, resulting from their destruction.
168. This is the rate-limiting enzyme in the pathway of the biosynthesis of glutathione (GSH), a tripeptide of glycine, cysteine, and glutamic acid, and the most abundant cellular thiol compound. Oxidized glutathione (GSSG) is a dimer joined by a disulfide bridge between the cysteinyl residues.
169. This is the condition of abnormally high urinary concentrations of 5-ketoproline, the intermediate in the pathway of GSH biosynthesis (the γ-glutamyl cycle).
170. Inflammation of the middle ear.
171. The PMN is a type of white blood cell important in disease resistance, which functions by phagocytizing bacteria and other foreign particles.
172. An enzyme that catalyzes the reduction of hydroperoxides (including H_2O_2) with the concomitant oxidation of glutathione (2 mol of GSH converted to 1 mol of GSSG).
173. A subcellular organelle.
174. Loss of muscular coordination.
175. Unsteadiness.
176. Senses of position, etc., originating from the arms and legs.
177. Having abnormally low tension.

spinocerebellar disease characterized by delayed sensory conduction in the posterior columns.

Routine screening tests failed to detect α-tocopherol in her plasma, although she showed elevated circulating levels of cholesterol (448 mg/dL vs. normal: 150–240 mg/dL) and triglycerides (184 mg/dL vs. normal: 50–150 mg/dL). Her plasma concentrations of 25-hydroxyvitamin D_3 [25-(OH)-D_3], retinol, and vitamin K–dependent clotting factors were in the normal range. Tests of lipid malabsorption showed no abnormality. Her glucose tolerance and pancreatic function (assessed after injections of cholecystokinin and secretin) were also normal.

The patient was given 2 g of α-tocopheryl acetate with an ordinary meal; her plasma α-tocopherol level, which had been nondetectable before the dose, was in the subnormal range 2 h later and she showed a relatively flat absorption curve.[178] She was given the same large dose of the vitamin daily for 2 weeks, at which time her plasma α-tocopherol concentration was 24 µg/mL. When her daily dose was reduced to 800 mg of α-tocopheryl acetate per day for 10 weeks, her plasma level was 1.2 mg/dL, i.e., in the normal range. During this time, she showed marked clinical improvement.

Case Questions

1. What inborn metabolic error(s) was(were) apparent in the first patient?
2. What sign/symptom indicated a vitamin E–related disorder in each case?
3. Why are PMNs useful for studying protection from oxidative stress, as in the first case?
4. What inborn metabolic error might you suspect that led to vitamin E deficiency in the second patient?

13. STUDY QUESTIONS AND EXERCISES

1. Construct a concept map illustrating the nutritional interrelationships of vitamin E and other nutrients.
2. Construct a decision tree for the diagnosis of vitamin E deficiency in a human or animal.
3. What features of the chemical structure of vitamin E relate to its nutritional activity?
4. How might vitamin E utilization be affected by a diet high in polyunsaturated fat? Of a fat-free diet? Of a selenium-deficient diet?
5. Detail the role of vitamin E in maintaining oxidant (peroxide?) tone.
6. What kinds of prooxidants might you expect people or animals to encounter daily?
7. How can nutritional deficiencies of vitamin E and selenium be distinguished?

178. That is, the plot of plasma α-tocopherol concentration versus time.

RECOMMENDED READING

Ahsan, H., Ahad, A., Iqbal, J., et al., 2014. Pharmacological potential of tocotrienols: a review. Nutr. Metab. 11, 52–74.

Atkinson, J., Harroun, T., Wassall, S.R., et al., 2010. The location and behavior of α-tocopherol in membranes. Mol. Nutr. Food Res. 54, 641–651.

Banks, R., Speakman, J.R., Selman, C., 2010. Vitamin E supplementation and mammalian lifespan. Mol. Nutr. Food Res. 54, 719–725.

Birringer, M., 2010. Analysis of vitamin E metabolites in biological specimen. Mol. Nutr. Food Res. 54, 588–598.

Constantinou, C., Papas, A., Constantinou, A.I., 2008. Vitamin E and cancer: an insight into the anticancer activities of vitamin E isomers and analogs. Int. J. Cancer 123, 739–752.

Cutler, R.G., 2005. Oxidative stress profiling: Part I. Its potential importance in the optimization of human health. Ann. N.Y. Acad. Sci. 1055, 93–135.

Di Donato, I., Bianchi, S., Federico, A., 2010. Ataxia with vitamin E deficiency: update of molecular diagnosis. Neurol. Sci. 31, 511–515.

Edrey, Y.H., Salmon, A.B., 2014. Revisiting an age-old question regarding oxidative stress. Free Rad. Biol. Med. 71, 368–371.

Fukuzawa, K., 2008. Dynamics of lipid peroxidation and antioxidation of α-tocopherol in membranes. J. Nutr. Sci. Vitaminol. 54, 273–285.

Gille, L., Staniek, K., Rosenau, T., et al., 2010. Tocopheryl quinones and mitochondria. Mol. Nutr. Food Res. 54, 601–615.

Gohil, K., Vasu, V.T., Cross, C.E., 2010. Dietary α-tocopherol and neuromuscular health: search for optimal dose and molecular mechanisms continues!. Mol. Nutr. Food Res. 54, 693–709.

Gomez-Cabrera, M.C., Salvador-Pascual, A., Cabo, H., et al., 2015. Redox modulation of mitochondriogenesis in exercise. Does antioxidant supplementation blunt the benefits of exercise training? Free Rad. Biol. Med. 86, 37–46.

Gray, B., Swick, J., Ronnenberg, A.G., 2011. Vitamin E and adiponectin: proposed mechanism for vitamin E-induced improvement in insulin sensitivity. Nutr. Rev. 69, 155–161.

Huebbe, P., Lodge, J.K., Rimbach, G., 2010. Implications of apolipoprotein E genotype on inflammation and vitamin E status. Mol. Nutr. Food Res. 54, 623–630.

Ju, J., Picinich, S.C., Yang, Z., et al., 2010. Cancer-preventive activities of tocopherols and tocotrienols. Carcinogenesis 31, 533–542.

Kannappan, R., Gupta, S.C., Kim, J.H., et al., 2012. Tocotrienols fight cancer by targeting multiple cell signaling pathways. Genes Nutr. 7, 43–52.

Lemaire-Ewing, S., Desrumaux, C., Néel, D., et al., 2010. Vitamin E transport, membrane incorporation and cell metabolism: is α-tocopherol in lipid rafts an oar in the lifeboat? Mol. Nutr. Food Res. 54, 631–640.

Mène-Saffrané, L., DellaPenna, D., 2010. Biosynthesis, regulation and functions of tocochromanols in plants. Plant Physiol. Biochem. 48, 301–309.

Mocchegiani, E., Costarelli, L., Giacconi, R., et al., 2014. Vitamin E-gene interactions in aging and inflammatory age-related diseases: implications for treatment. A systematic review. Aging Res. Rev. 14, 81–101.

Muller, D.P.R., 2010. Vitamin E and neurological function. Mol. Nutr. Food Res. 54, 710–718.

Ohnmacht, S., Nava, P., West, R., et al., 2008. Inhibition of oxidative metabolism of tocopherols with ω-*N*-heterocyclic derivatives of vitamin E. Bioorg. Med. Chem. 16, 7631–7638.

Parker, R.S., 2013. Vitamin E. In: Stipanuk, M.H., Caudill, M.A. (Eds.), Biochemical, Physiological and Molecular Aspects of Human Nutrition, third ed. Elsevier, New York, pp. 670–682 (Chapter 29).

Pazdro, R., Burgess, J.R., 2010. The role of vitamin E and oxidative stress in diabetes complications. Mech. Ageing Dev. 131, 276–286.

Rimbach, G., Moehring, J., Huebbe, P., et al., 2010. Gene-regulatory activity of α-tocopherol. Molecules 15, 1746–1761.

Ristow, M., Zarse, K., 2010. How increased oxidative stress promotes longevity and metabolic health: the concept of mitochondrial hormesis (mitohormesis). Exp. Gerontol. 45, 410–418.

Schaffer, S., Müller, W.E., Eckert, G.P., 2005. Tocotrienols: constitutional effects in aging and disease. J. Nutr. 135, 151–154.

Sosa, V., Moliné, T., Somoza, R., et al., 2013. Oxidative stress and cancer: an overview. Ageing Res. Rev. 12, 376–390.

Takada, T., Suzuki, H., 2010. Molecular mechanisms of membrane transport of vitamin E. Mol. Nutr. Food Res. 54, 616–622.

Traber, M.G., 2014. Vitamin E. In: Zemplini, J., Suttie, J.W., Gregory, J.F., et al. (Eds.), Handbook of Vitamins, fifth ed. CRC Press, New York, pp. 125–147 (Chapter 4).

Traber, M.G., Stevens, J.F., 2011. Vitamins C and E: beneficial effects from a mechanistic perspective. Free Rad. Biol. Med. 51, 1000–1013.

Ulatowski, L.M., Manor, D., 2015. Vitamin E and neurodegeneration. Neurobiol. Dis. 84, 78–83.

Zingg, J.M., Meydani, M., Azzi, A., 2014. Vitamins E and C: effects on matrix components in the vascular system. In: Dakshinamurti, K., Dakshinamurti, S. (Eds.), Vitamin-Binding Proteins: Functional Consequences. CRC Press, New York, pp. 127–156 (Chapter 8).

Chapter 9

Vitamin K

Chapter Outline

Anchoring Concepts

1. *Vitamin K* is the generic descriptor for 2-methyl-1,4-naphthoquinone and all its derivatives exhibiting qualitatively the antihemorrhagic activity of phylloquinone.
2. The K vitamers are side chain homologs; each is hydrophobic and, thus, insoluble in such aqueous environments as plasma, interstitial fluids, and cytoplasm.
3. The 1,4-naphthoquinone ring system of vitamin K renders it susceptible to metabolic reduction.
4. Deficiencies of vitamin K have a narrow clinical spectrum: hemorrhagic disorders.

...Then Almquist showed
A substance, phthiocol, from dread T.B.,
Would cure the chicks...
And so the microbes of tuberculosis,
That killed the poet Keats by hemorrhage,
Has yielded forth the clue to save the lives
Of infants bleeding shortly after birth.

T.H. Jukes[1]

1. Thomas H. Jukes (1906–1999) was a British born nutritional biochemist noted for his early work in elucidating the B-vitamins and later pioneering work in the field of molecular evolution. After holding a research position at the Lederle Laboratories of American Cyanamid Co., he spent a long and productive career on the faculty of the University of California at Berkeley where he was a respected and outspoken crusader for sound science in the field of nutrition. He wrote these lines on the occasion of Henrik Dam being awarded the Nobel Prize for elucidating vitamin K; obviously, Jukes, thought that his colleague H.J. Almquist should have shared the prize.

LEARNING OBJECTIVES

1. To understand the nature of the various sources of vitamin K
2. To understand the means of absorption and transport of the K vitamers
3. To understand the metabolic functions of vitamin K
4. To understand the physiological implications of impaired vitamin K status and/or function

VOCABULARY

Atherocalcin
γ-Carboxyglutamate
Chloro-K
Clotting time
Coagulopathy
Collagen
Coprophagy
Coumarins
Dicumarol
Dihydrovitamin K
Dysprothrombinemia
Enterotype
Extrinsic clotting system
Factor II
Factor VII
Factor IX
Factor X

The Vitamins. http://dx.doi.org/10.1016/B978-0-12-802965-7.00009-5

Fibrin
Fibrinogen
Gas6
Gla
Hemorrhage
Hemorrhagic disease of the newborn
Heparin
Hydroxyvitamin K
Hypoprothrombinemia
Intrinsic clotting system
Matrix Gla protein (MGP)
Menadione
Menadione sodium bisulfite complex
Menadione pyridinol bisulfite (MPB)
Menaquinones (MKs)
2-Methyl-1,4-naphthoquinone
Microbiome
Naphthoquinone
Osteocalcin
Periostin
Phylloquinone
PIVKA (protein induced by vitamin K absence)
Protein C
Protein M
Protein S
Protein Z
Serine protease
Stuart factor
Sulfaquinoxaline
Thrombin (factor IIa)
Thromboplastin
Vitamin K deficiency bleeding (VKDB)
Vitamin K–dependent γ-glutamyl carboxylase (VKγGC)
Vitamin K epoxide
Vitamin K epoxide reductase (VKER)
Vitamin K hydroquinone
Vitamin K oxide
Vitamin K quinone
Vitamin K quinone reductase (VKQR)
Warfarin
Zymogen

1. THE SIGNIFICANCE OF VITAMIN K

Vitamin K is synthesized by plants and bacteria, which use it for electron transport and energy production. Animals, however, cannot synthesize the vitamin; still, they require it for blood clotting, bone formation, and other essential functions. These needs are critical to good health; yet enteric microbial synthesis of the menaquinones (MK), including that occurring in the hindgut of humans and other animals (many of whom have **coprophagous**[2] eating habits), renders frank deficiencies of this vitamin rare. Asymptomatic low vitamin K status has been observed in as many as a third of Americans, and premature infants, born with low reserves of the vitamin, face risk of hemorrhagic disease. Vitamin K deficiency can also occur in poultry and other monogastric animals when they are raised on wire or slatted floors and treated with antibiotics that reduce their hindgut microbial synthesis of the vitamin.

The function of vitamin K in blood clotting was the first to be recognized and is widely exploited to reduce risks of thrombosis in cardiac and surgical patients. **Coumarin**-based drugs (e.g., **warfarin**, **dicumarol**)[3] and other inhibitors of the vitamin K oxidation/carboxylation/reduction cycle are valuable for this purpose. In addition, vitamin K has clear roles in the metabolism of both calcified and noncalcified tissues. Emerging evidence indicates that vitamin K functions with other vitamins in the regulation of intracellular Ca^{2+} metabolism, in signal transduction, and in cell proliferation, functions that have profound effects on health status.

2. PROPERTIES OF VITAMIN K

The term vitamin K describes **2-methyl-1,4-naphthoquinone** and its derivatives exhibiting the antihemorrhagic activity of phylloquinone. Naturally occurring forms of the vitamin consist of a side chain–substituted **naphthoquinone nucleus** (at C-3) and are characterized by the type and number of unsaturated isoprene units (not carbon atoms) that form the side chain. For these groups of vitamers, a numeric system is used to indicate side chain length – e.g., the abbreviations K-*n* for the phylloquinones and MK-*n* for the MKs, to indicate the number of isoprenoid units comprising the side chains (Table 9.1). There are three groups of K vitamers:

- **Phylloquinones (K-*n*)** – forms with phytyl and further alkylated side chains consisting of several saturated isoprenoid units. Phylloquinones have only one double bond in their side chains, i.e., on the proximal isoprene unit. These vitamers are synthesized by green plants as a normal component of chloroplasts.

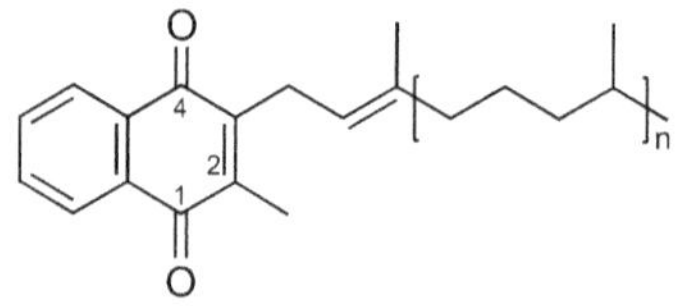

2. The term **coprophagy** describes the ingestion of excrement. This behavior is common in many species and exposes them to nutrients such as vitamin K produced by the microbial flora of their lower guts. Coprophagy can be easily prevented in some species (e.g., chicks) by housing with raised wire floors; it is very difficult to prevent in others (e.g., rats) without the use of such devices as tail cups.

3. These 4-hydroxycoumarins have anticoagulant activities by inhibiting VKER.

TABLE 9.1 Systems of Vitamin K Nomenclature

Chemical Name	Preferred System IUPAC[a]	Other Systems IUNS[b]	Other Systems Traditional
2-Methyl-3-phytyl-1,4-naphthoquinone (*n*)	Phylloquinone-*n* (K-*n*)	Phytylmenaquinone-*n* (PMQ-*n*)	$K_1(n)$
2-Methyl-3-multiprenyl-1,4-naphthoquinone (*n*)	Menaquinone-*n* (MK-*n*)	Prenylmenaquinone-*n* (MQ-*n*)	$K_2(n)$
2-Methyl-1,4-naphthoquinone	Menadione	MK	K_3

[a]*International Union of Pure and Applied Chemists.*
[b]*International Union of Nutritional Sciences.*

- Menaquinones (**MK-*n***) – forms with side chains consisting of variable numbers of isoprenoid units, each containing a double bond. MKs are synthesized only by bacteria (including those of the intestinal microbiota) and some spore-forming *Actinomyces* spp. The predominant vitamers are MK-*6* to MK-*10*; MKs with as many as 13 side chain isoprene units have been identified.

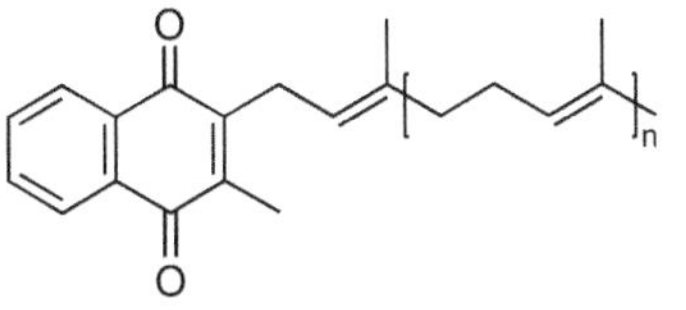

- **Menadione** – does not occur naturally, but is a common synthetic form, 2-methyl-1,4-naphthoquinone. It forms a water-soluble sodium bisulfite addition product, **menadione sodium bisulfite**, the practical utility of which is limited by its instability in complex matrices such as feeds. However, in the presence of excess sodium bisulfite, it crystallizes as a complex with an additional mole of sodium bisulfite (i.e., **menadione sodium bisulfite complex**), which has greater stability and, therefore, is used as a supplement to poultry and swine feeds. A third water-soluble compound is **menadione pyridinol bisulfite** (**MPB**), a salt formed by the addition of dimethylpyridinol.

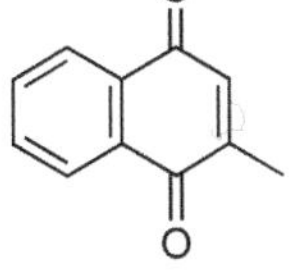

Vitamin K Chemistry

Phylloquinone (K_1) is a yellow oil at room temperature, but the other K vitamers are yellow crystals. The K and MK vitamers and most forms of menadione are insoluble in water, slightly soluble in ethanol, and readily soluble in ether, chloroform, fats, and oils. The K vitamers are sensitive to light and alkali, but are relatively stable to heat and oxidizing environments. Their oxidation proceeds to produce the 2,3-epoxide form. Their reduction (e.g., with sodium hydrogen sulfite) produces the corresponding naphthohydroquinones, which can be reoxidized with mild oxidizing agents. The K vitamers show the characteristic UV spectra of the naphthoquinones, i.e., their oxidized forms having four strong absorption bands in the 240- to 270-nm range. The reduced (hydroquinone) forms show losses of the band near 270 nm and increases of the band around 245 nm. Extinction decreases with increasing side chain length.

Vitamin K Biopotency

The biopotencies of the K vitamers (Table 9.2) depend on both the nature and length of their isoprenoid side chains. In general, the MK-*n* tend to have greater biopotencies than the corresponding phylloquinone (MK-*n*) analogs, and members of each series with four or five isoprenoid side chains are the most biopotent. The reported biopotencies of the menadiones tend to be variable; this may be partly because of varying stabilities of the preparations tested, as well as to whether the vitamin K antagonist **sulfaquinoxaline** was used in the assay diet.

3. SOURCES OF VITAMIN K

Biosynthesis

Synthesis of MK by the gut **microbiome** contributes to vitamin K nutrition in most humans and other animals. Many bacteria in the large intestine can synthesize long-chain MKs.[4] These include the obligate anaerobes of the

4. MK confer essential roles in procaryotes: as electron carriers in respiratory electron transport chains; as antioxidants protecting cellular membranes from peroxidative degradation; and as facilitators of transmembrane movement of molecules.

TABLE 9.2 Relative Biopotencies of Vitamin K–Active Compounds

Compound	Biopotency[a] (%)
Phylloquinones (K-*n*)	
K-*1*[b]	5
K-*2*	10
K-*3*	30
K-*4*	100
K-*5*	80
K-*6*	50
Menaquinones (MK-*n*)	
MK-*2*[b]	15
MK-*3*	40
MK-*4*	100
MK-*5*	120
MK-*6*	100
MK-*7*	70
Forms of Menadione	
Menadione	40–150
Menadione sodium bisulfite complex	50–150
Menadione dimethylpyrimidinol bisulfite	100–160

[a]Relative biopotency based on chick prothrombin/clotting time bioassays using K-4 as the standard.
[b]Indicates the number of side chain isoprenoid units (each containing five carbons).

groups *Bacteroides*,[5] *Eubacterium*, *Propionibacterium*, and *Arachnia* and the facultative anaerobe *Escherichia coli*, which produces mostly MK-*10* and MK-*11* (Table 9.3). The contributions of the gut microbiome to the vitamin K nutriture of the host likely vary according to **enterotype**, i.e., the dominant taxonomic groups comprising that microbiome.[6] Studies of human fecal microbiota found that MK composition was determined mainly by the species composition of *Prevotella* spp. and *Bacteroides* spp.[7] Studies with the rat have shown that enteric production of long-chain MKs can be suppressed by high intakes of phylloquinone or MK-*4*.[8] A quantitative study of colonoscopy patients found total gut MK content to average 1.8 mg (range 0.3–5.1 mg, predominantly MK-*9* and MK-*10*),[9] an amount exceeding nutritional requirements by an order of magnitude. That MKs are absorbed from the large intestine is suggested by the fact that the liver normally contains significant amounts of those vitamers, that germ-free animals have greater dietary requirements for the vitamin than do animals with normal gut microbiomes (Table 9.4), and that prevention of coprophagy is necessary to produce vitamin K deficiency in animals. Humans seldom show signs of vitamin K deficiency; hypoprothrombinemia is rare *except* among patients given antibiotics. Species with short gastrointestinal tracts and very short intestinal transit times (e.g., about 8 h in the chick), less than the generation times of many bacteria, do not have well-colonized guts. Being thus unable to harbor vitamin K–producing bacteria, they depend on their diet as the source of vitamin K.

Dietary Sources

Data for the specific vitamin K contents of foods have been limited by available analytical methods, which have only recently been improved with the development of tandem liquid chromatography–mass spectrometry. Still, it is clear that many foods contribute to meeting vitamin K needs (Table 9.5). Phylloquinones comprise most of the vitamin K in diets, the richest sources being green leafy vegetables (e.g., spinach, kale, broccoli, Brussels sprouts), vegetable oils, and margarine.[10] MKs can comprise as much as a quarter of the vitamin K in many diets; those vitamers largely come from bacterially fermented foods (e.g., cheese, sauerkraut, natto[11]) (Table 9.6), which contain long-chain MKs (particularly MK-*8* and MK-*9*),[12] and poultry and pork products.[13] It is difficult to formulate a normal diet that does not provide some 100 µg of the vitamin per day.

5. The most quantitatively important groups are *Bacteroides* and *Bifidobacteria*, which collectively comprise more than half of the intestinal microfloral mass. *Bifidobacteria* do not produce MKs.
6. Three major human enterotypes have been described based on the species composition of the microbiome (Arumugam, M., Raes, J., Pelletier, E., et al., 2011. Nature 473, 174–180).
7. Karl, J.P., Fu, X., Wang, X., et al., 2015. Am. J. Clin. Nutr. 102, 84–93.
8. Koivu-Tikkanen, T.J., Schurgers, L.J., Thijssen, H.H.W., et al., 2000. Br. J. Nutr. 83, 185–190.
9. Conly, J.M., Stein, K., 1992. Am. J. Gastroenterol. 87, 311–316.
10. Hydrogenation of vegetable oils converts produces 2′,3′-dihydrophylloquinone from phylloquinone, which is less bioactive than the parent, being less well absorbed and not appearing to be metabolized to MK-*4*.
11. A traditional Japanese food made with soy beans fermented with *Bacillus subtilis* var. natto.
12. MK contents can vary because of differences in the MK-biosynthetic capacities of the strains of bacteria used. Various strains of lactic acid bacteria (*Lactococcus lactis* sp., *Leuconostoc lactis*) were found to vary by more than threefold in their capacity to produce long-chain MKs (Morishita, T., Tamura, N., Makino, T., et al., 1999. J. Dairy Sci. 82, 1897–1903).
13. These meats can contain MK-*4* to the extent that they are fed formula feeds in which menadione is used as a feed supplement.

TABLE 9.3 MKs Produced by Dominant Species of Enteric Bacteria

Organism	MK-6	MK-7	MK-8	MK-9	MK-10	MK-11	MK-12
Bacteroides fragilis		+	+	+	++	++	++
Bacteroides vulgatus		+	+	+	++	++	++
Eubacterium lentum	++						
Enterobacter sp.			++				
Enterococcus sp.	+	+	+	++			
Lactococcus lactis		+	++	++			
Leuconostoc lactis		+	++	++			
Veillonella sp.	+	++					

From Mathers, J.C., Fernandez, F., Hill, J.E., et al., 1990. Br. J. Nutr. 63, 639–652.

TABLE 9.4 Impaired Clotting in Germ-Free Rats

Treatment	Prothrombin Time (s)	Hepatic MK-4 (ng/g)
Germ-Free		
Vitamin K deficient	∞	8.3±2.3
+MK-4	5.8±1.4[a]	66.5±25.9
+Menadione	11.1±2.5[a]	12.4±2.0
Conventional		
Vitamin K deficient	12.7±1.6	103.5±44.9
+MK-4	12.5±2.0	207.6±91.3
+Mendione	12.6±0.9	216.2±86.5

[a] $p<.05$.
From Komai, M., Shirakawa, H., Kimura, S., 1987. Int. J. Vit. Nutr. Res. 58, 55–59.

Breast Milk

Breast milk is typically low in vitamin K, providing insufficient amounts of the vitamin to meet the vitamin K needs of infants (Table 9.7).[14] The extent to which such low levels may reflect low maternal intakes of vitamin K is not clear. Vitamin K levels in breast milk can be increased by vitamin K supplementation of the mother; a clinical study found a daily supplement of 5 mg phylloquinone to increase breast

14. In addition, vitamin K appears to be poorly transferred across the placenta. Accordingly, the American Academy of Pediatrics recommended in the early 1960s the administration of vitamin K intramuscularly (1 mg) or orally (1 mg weekly for 12 weeks.) at the time of birth to prevent hemorrhagic disease. This practice is now required by law in the United States and Canada. All commercial infant formulas are supplemented with vitamin K at levels in the range of 50–125 ng/mL.

TABLE 9.5 Vitamin K Contents of Foods

Food	Vitamin K (μg/100 g)
Vegetables	
Asparagus	51
Beans snap	48
Beets	0.2
Broccoli	141
Cabbage	109
Carrots	14
Cauliflower	14
Corn	0.4
Cucumbers	16
Kale	817
Tomatoes	2
Lettuce	103
Peas	30
Potatoes	0.3
Spinach	494
Sweet potatoes	2
Oils	
Canola	71
Corn	2
Olive	60
Peanut	0.7
Soybean	184

Continued

TABLE 9.5 Vitamin K Contents of Foods—cont'd

Food	Vitamin K (μg/100 g)
Fruits	
Apples	2[a]
Bananas	0.05
Cranberries	5
Oranges	0
Peaches	3
Strawberries	2
Meats	
Beef	1–3
Pork	0.1–0.2
Beef liver	3
Dairy Products and Eggs	
Milk, cow's	0.1–0.6
Eggs	0.3
Egg yolk	0.7
Grains	
Rice	0–0.2
Wheat bran	2
Wheat flour	2

[a]~90% in the skin (peel).
From USDA National Nutrient Database for Standard Reference, Release 28. http://www.nal.usda.gov/fnic/foodcomp/search/.

milk phylloquinone concentrations from 27 ± 12 ng/mL at 2 weeks of lactation to 59 ± 25 ng/mL at 26 weeks.[15]

Bioavailability

The bioavailability of vitamin K in most foods has not been determined. Studies have shown that >17% of the phylloquinone in boiled spinach or kale is absorbed by humans.[16] This may relate to its association with the thylakoid membrane in chloroplasts, as the free vitamer is well absorbed (~80%). This suggests that vitamin K may be poorly bioavailable from the foods thought to be the most important sources of the vitamin in most diets, green leafy vegetables. The relative biopotencies of K vitamers appear to depend on the nature of the side chain. Studies have found that phylloquinone concentrations in plasma peak at more than twice the levels of comparable oral doses of MK-*4* and MK-*9*, indicating better short-term bioavailability.[17] Studies using restoration of normal clotting in the vitamin K–deficient chick have shown that when administered orally, phylloquinone, MK-*3*, MK-*4*, and MK-*5* had greater activities than those with longer side chains. This effect reflected the relatively poor absorption of the long-chain vitamers. In contrast, long-chain MKs (especially MK-*9*) had the greatest activities

15. Greer, F.R., Marshall, S.P., Foley, A.L., et al., 1997. Pediatrics 99, 88–92.
16. Gijsbers, B.L.M.G., et al., 1996. Br. J. Nutr. 76, 223; Novotny, J.A., Kurilich, A.C., Britz, S.J., et al., 2010. Br. J. Nutr. 104, 858–862.
17. Schurgers, L.J., Vermeer, C., 2002. Biochim. Biophys. Acta 1570, 27–32.

TABLE 9.6 MKs Produced by Bacteria Used in Food Production

Organism	Food Application	MK-*5*	MK-6	MK-*7*	MK-*8*	MK-9	MK-*10*
Arthrobacter nicotianae	Cheese			+	++	+	
Bacillus subtilis var. *natto*	Natto			++	+		
Brevibacterium linens	Cheese				+		
Hafnia alvei	Cheese						
Lactococcus lactis ssp. *lactis*	Cheeses, sour cream, buttermilk	+		+	+	++	
Lactococcus lactis ssp. *cremoris*	Cheeses, sour cream, buttermilk			+	+	++	
Leuconostoc lactis	Cheese			+	+	++	+
Proprionobacterium shermanii	Cheese					+	
Staphylococcus xylosus	Sausage				+		
Staphylococcus equorum	Meat		+	++	+		

From Morishita, T., Tamura, N., Makino, T., et al., 1999. J. Dairy Sci. 82, 1897–1903; Walther, B., Karl, J.P., Booth, S.L., et al., 2013. Adv. Nutr. 4, 463–473.

TABLE 9.7 Vitamin K Contents of Human Milk

	Vitamin K (nM)
Colostrum (30–81 h)	7.52 ± 5.90[a]
Mature Milk	
1 month	6.98 ± 6.36
3 months	5.14 ± 4.52
6 months	5.76 ± 4.48

[a]*Mean* ± *SD.*
From Canfield, L.M., Hopkinson, J.M., Lima, A.F., et al., 1991. Am. J. Clin. Nutr. 53, 730–735.

TABLE 9.8 Vitamin K Transport in Humans

Fraction	Phylloquinones % Serum Total
Triglyceride-rich fraction	51.4 ± 17.0[a]
LDLs	25.2 ± 7.6
HDLs	23.3 ± 10.9

[a]*Mean* ± *SD.*
From Kohlmeier, M., Solomon, A., Saaupe, J., et al., 1996. J. Nutr. 126, 1192S–1196S.

when administered intracardially to the vitamin K–deficient rat. Of the synthetic forms, some studies have found MPB to be somewhat more effective in chick diets; however, each is generally regarded as of comparable biopotency to MK-*4*.

4. ABSORPTION OF VITAMIN K

Micellar Solubilization

The K vitamers are absorbed from the intestine into the lymphatic (in mammals) or portal (in birds, fishes, and reptiles) circulation by processes that first require that these hydrophobic substances be dispersed in the aqueous lumen of the gut via the formation of mixed micelles, in which they are dissolved. Vitamin K absorption, therefore, depends on normal pancreatic and biliary function, and the presence of some dietary fat. Therefore, conditions resulting in impaired luminal micelle formation (e.g., dietary mineral oil, pancreatic exocrine dysfunction, bile stasis) impair the enteric absorption of the vitamin. The mixtures of phylloquinones and MKs that diets typically contain appear to be absorbed with a wide range of efficiencies, e.g., 5–70%. K vitamers are absorbed across the brush border via noncarrier-mediated passive diffusion, the rates of which are affected by the micellar contents of lipids and bile salts. This occurs in both the distal part of the small intestine and the colon. Thus, noncoprophagous animals, including humans, appear to benefit from the bacterial synthesis of vitamin K in their lower guts by being able to absorb the vitamin at that location. The magnitude of this benefit remains the subject of debate.

5. TRANSPORT OF VITAMIN K

Lipoprotein Carriers

No specific carriers have been identified for any of the K vitamers. Instead, upon absorption into the enterocyte, vitamin K associates with nascent chylomicra, which transport the vitamin in the lymph ultimately to the liver. Vitamin K is rapidly taken up by the liver via an apolipoprotein E (apoE) receptor, which interacts with the apoE on the chylomicron surface. Phylloquinones and MK-*4* have relatively short half-lives in the plasma (c. 17 h); whereas, the longer chain MKs circulate in the plasma for much longer periods (up to 48 h) Ultimately, these vitamers are transferred to triglyceride-rich lipoproteins and, as they lose triglycerides, to low-density lipoproteins (LDLs) and high-density lipoproteins (HDLs) (Table 9.8). Studies have shown these transfers to occur at different rates for different K vitamers. In humans, radiolabeled MK-*4* and MK-*9* were both transferred from the triglyceride-rich fraction to HDLs, but MK-*9* first went to LDLs, which increased its residence time in the plasma.

Cellular Uptake

Uptake of vitamin K is thought to occur in the way that other lipids are taken up into hepatocytes, osteocytes, and other cells. This process involves vitamin K–bearing chylomicron remnants being bound to cell surface receptors of the LDL receptor family by a process referred to as secretion capture. This involves apoE,[18] which is acquired by chylomicron remnants from triglyceride-rich lipoproteins and HDLs, thus allowing it to act as a ligand to facilitate binding of that particle to high-affinity cell surface receptors. This binding leads to cellular internalization, which is facilitated by heparin sulfate proteoglycans. The rapid appearance of radiolabeled metabolites in the urine after an oral dose of labeled phylloquinone shows this to be a rapid process.

18. It has been suggested that apoE genotype may affect the cellular uptake of vitamin K. Experimental tests of that hypothesis have not yielded consistent results; some studies have found carriers of the *APOE4* allele to have the greatest (Yan, I., Shou, B., Nigidikar, S., et al., 2005. Br. J. Nutr. 94, 956–961) or lowest (Saupe, J., Shearer, M.J., Kohlmeier, M., 1993. Am. J. Clin. Nutr. 58, 204–208) circulating levels of phylloquinone.

TABLE 9.9 K Vitamers in Livers of Several Species

Vitamer	Human	Cow	Horse	Dog	Pig
Phylloquinone	+++	+	+++		+
MK-*4*					+
MK-*5*	+				
MK-*6*	+			+	
MK-*7*	+++			+	
MK-*8*	+			+++	+++
MK-*9*	+			+++	+++
MK-*10*	++	+++		+++	+++
MK-*11*	++	+++		+	
MK-*12*		+++		+	
MK-*13*				+	

Tissue Distribution

The liver, the site of synthesis of the vitamin K–dependent coagulation proteins, is the primary storage organ for vitamin K. It rapidly takes up both phylloquinones and MKs, but not menadione, which is instead distributed in other tissues along with other vitamers (Table 9.9). Some 90% of the vitamin K in liver is comprised by MK-*10* and MK-*11*, with only a small portion (10%) as phylloquinones. Extrahepatic tissues of most animals ingesting plant materials contain phylloquinones and MKs with *6–13* isoprenoid units in their side chains. The vitamin is concentrated in several tissues such as the heart, pancreas, and adipose tissue.[19] In brain and kidney, MK-*4* levels typically exceed those of phylloquinone. Vitamin K is localized in cellular membranes (endoplasmic reticulum and mitochondria) where it is depleted only slowly under conditions of vitamin K deprivation. Transplacental movement of vitamin K is poor; it is frequently not detectable in the cord blood from mothers with normal plasma vitamin K levels. Accordingly, newborns are susceptible to hemorrhage.[20]

Animals fed phylloquinone show MK-*4* widely distributed in their tissues. Because MK-*4* is not a bacterial product, its presence in tissues evidences its production from phylloquinone, i.e., interconversion of the phytyl side chain to a geranylgeranyl side chain. Tissues can also contain longer–side chain MKs, even when the sole dietary form is MK-*4*. This raises the possibility of much of the vitamin in tissues being of enteric bacterial origin.

19. Adults undergoing bariatric surgery were found to have phylloquinone levels of 148±72nmol/kg and 175±112nmol/kg in subcutaneous and visceral adipose tissue, respectively (Shea, M.K., Booth, S.L., Gundberg, C.M. et al., 2010. J. Nutr. 140, 1029).

20. Further, because human milk contains less vitamin K than cow's milk, infants who receive only their mother's milk are more susceptible to hemorrhage than are those who drink cow's milk.

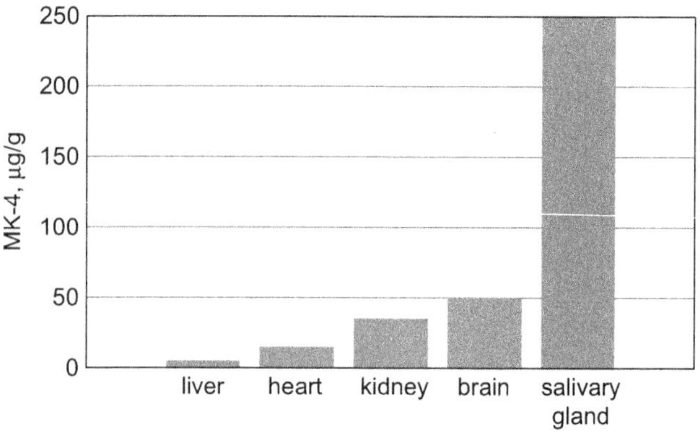

FIGURE 9.1 Detection of MK-*4* in tissues of gnotobiotic, vitamin K–deficient rats treated intraperitoneally with phylloquinone evidences conversion to MK-*4* in vivo. *After Davidson, R.T., Foley, A.L., Engelke, J.A., et al., 1998. J. Nutr. 128, 220–223.*

6. METABOLISM OF VITAMIN K

Side Chain Modification

Dealkylation–Alkylation. That tissues contain MK-*4* when the dietary source of vitamin K is phylloquinone evidences the dealkylation of the phylloquinone side chain. Studies have shown this to occur by first removing the side chain to produce menadione, which is re-alkylated to produce MK-*4*. The conversion appears greatest when phylloquinone is taken orally,[21] suggesting that it may involve the gut microbiome. However, this metabolism does not depend on the microbiome, as evidenced by the finding of MK-*4* in tissues of gnotobiotic animals given phylloquinone intraperitoneally (Fig. 9.1). The conversion has been demonstrated in several tissues of the rat in which MK-*4* levels exceed those of phylloquinone (e.g., brain, pancreas, salivary gland); it appears to be catalyzed by a homolog of a bacterial UbiA-prenyltransferase (UBIAD1), as small interfering RNA against the *UBIAD1* gene inhibited the conversion of phylloquinone to MK-*4* in human cells.[22] UBAID1 is also thought to catalyze the conversion of long-chain MKs to MK-*4*.

Menadione (either from dietary supplements or from microbial degradation of phylloquinone) can be alkylated by a process that uses geranyl pyrophosphate, farnesyl pyrophosphate, or geranylgeranyl pyrophosphate as the

21. Thijssen, H.H., Verrot, L.M., Schurgers, L.J., et al., 2006. Br. J. Nutr. 95, 260–266.

22. Nakagawa, K., Hirota, Y., Sawadaa, N., et al., 2010. Nature 468, 117–121.

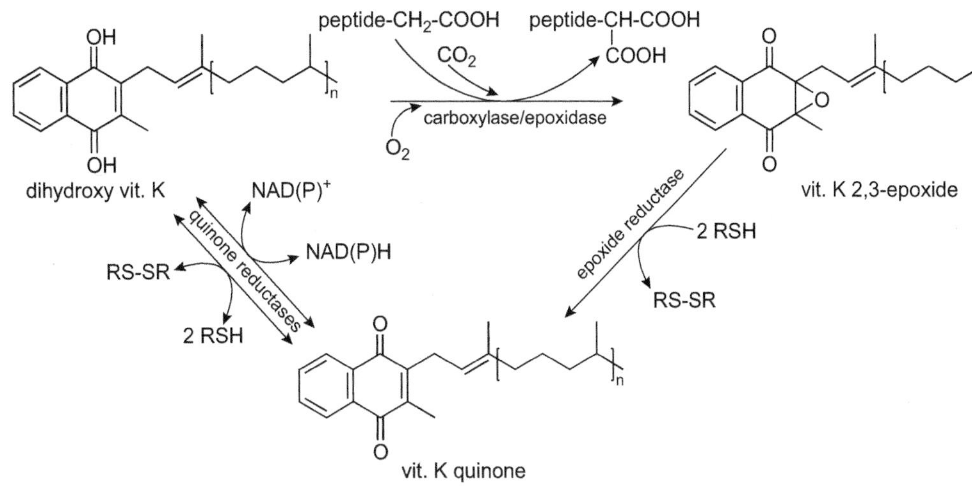

FIGURE 9.2 The vitamin K cycle.

alkyl donor and that can be inhibited by O_2 or warfarin. The main product of the alkylation of menadione is MK-*4*.

Redox Cycling

Vitamin K undergoes cyclic oxidation and reduction coupled to the carboxylation of peptidyl glutamyl residues to produce various functional γ-carboxylated proteins (Fig. 9.2). This redox cycling occurs in three steps:

1. **Vitamin K γ-glutamyl carboxylase (VKγGC)** catalyzes the oxidation of **dihydroxyvitamin K** (including phylloquinone, MK-*4*, and longer–side chain MKs) to the respective K 2,3-epoxide.[23] The enzyme is a 94-kDa protein located in the endoplasmic reticulum and Golgi apparatus. Vitamin K-2,3-epoxide comprises about 10% of the total vitamin K in the normal liver and can be the predominant form in the livers of rats treated with warfarin or other coumarin anticoagulants. Studies of the human, bovine, and rat VKγGC show high (88–94%) sequence homology, each with three γ-carboxyglutyl residues (Gla) per mole of enzyme.[24] Polymorphisms have been related to interindividual variation in sensitivity to warfarin anticoagulation and differences in the γ-carboxylation of osteocalcin.[25]
2. **Vitamin K epoxide reductase (VKER)** catalyzes the reduction of vitamin K-2,3-epoxides to their respective quinones. This is the rate-limiting step in the vitamin K cycle. VKER is an 18-kDa, dithiol-dependent, microsomal enzyme inhibited by coumarins. Genetic variation in the VKER subunit 1 has been shown to account for the variability observed in patient responses to warfarin therapy, to affect circulating levels of phylloquinones and undercarboxylated osteocalcin (ucOC),[26] and to be associated with cardiovascular risk.[27] Subjects with the VKER CG/GG (rs8050894) genotype show increased risk to progressive coronary artery calcification and poorer survival to those effects.[28] The chick has been noted for having relatively low hepatic VKER activity compared to the rat (Table 9.10); that this condition reduces its ability to recycle the vitamin is consistent with its greater hepatic levels of vitamin K-2,3-epoxide, and its greater dietary requirement for vitamin K than the rat.

TABLE 9.10 Species Differences in Vitamin K Metabolism

		V_{max} (μmoles/min/mg)	
Enzyme	**Substrate**	**Chick**	**Rat**
VKγGC	Phylloquinone	14±2	26±1
	MK-*4*	41±3	40±2
VKER	Phylloquinone	26±2	280±2
	MK-*4*	55±7	430±10

From Will, B.H., Usui, Y., Suttie, J.W., 1992. J. Nutr. 122, 2354–2360.

23. Also **vitamin K epoxide** or **vitamin K oxide**.
24. Berkner, K.L., Pudota, B.N., 1998. Proc. Natl. Acad. Sci. USA 95, 466–471.
25. Kinoshita, H., Nakagawa, K., Narusawa, K., 2007. Bone 40, 451–456.
26. Individuals with the minor allele of rs8050894 (G) had significantly higher plasma phylloquinone levels than those with the C allele; GG homozygotes have slightly lower levels of ucOC than other genotypes (Crosier, M.D., Peter, I., Booth, S.L., et al., 2009. J. Nutr. Sci. Vitaminol. 55, 112–119).
27. Individuals with the rs8050894TT genotype had 60% fewer atherothrombotic events than those with other genotypes (Suh, J.W., Baek, S.H., Park, J.S., et al., 2009. Am. Heart Assoc. J. 157, 908–912).
28. Holden, R.M., Booth, S.L., Tuttle, A., et al., 2014. Arterioscler. Thromb. Vasc. Biol. 34, 1591–1596.

3. **Reduction of vitamin K quinone** – This final reductive step leading to the production of the active hydroquinone, dihydroxyvitamin K, can be catalyzed in two ways:
   a. by **vitamin K quinone reductase (VKQR)**, a dithiol-dependent microsomal enzyme inhibited by the coumarin-type anticoagulants
   b. by **DT diaphorase**[29], a microsomal flavoprotein that uses NAD(P)H as a source of reducing equivalents to catalyze the 2-electron reduction of vitamin K quinone and other quinones. The enzyme also catalyzes the reduction of phylloquinones and MKs, but has highest affinity for menadione. In doing so, it protects against menadione cytotoxicity by competing with cytochrome P_{450}–catalyzed single-electron reductions that produce the semiquinone, which can redox cycle and produce ROS. It is relatively insensitive to coumarins, such that reduction of vitamin K quinone persists in anticoagulant-treated individuals.

Catabolism

Catabolism of K vitamers involves metabolism of the isophytyl side; there is no evidence of catabolism of the naphthoquinone ring. The total body pool of phylloquinone, c. 100 mg in an adult, turns over in about 1.5 days. It is thought to be catabolized by side chain removal by the same pathways used to degrade tocopherols: ω-hydroxylation followed by progressive chain shortening by way of β-oxidation. The initial ω-hydroxylation has been found to be catalyzed by a cytochrome *P*-450 isoform, CYP4F2, polymorphisms of which have been found to affect the efficacy of warfarin therapy.[30] The most abundant phylloquinone metabolite is its 2,3-epoxide, formed by the vitamin K–dependent γ-carboxylation of proteins. This metabolite and other phylloquinones and MKs undergo oxidative shortening of the side chain to 5- or 7-carbon carboxylic acids and a variety of other, more extensively degraded metabolites. A fifth of phylloquinone is ultimately excreted in the urine; however, the primary route of excretion of these metabolites is the feces, which contain mostly 7-C and 5-C aglycones and glucuronic acid conjugates excreted via the bile. Warfarin treatment greatly increases the excretion of phylloquinone metabolites in the urine while decreasing the amounts of metabolites in the feces. Little is known about MK metabolism, but it is likely that it also undergoes side chain degradation in the same manner as phylloquinone. MK catabolism, particularly that of the long-chain MKs, appears to be much slower than that of menadione, which is rapidly metabolized and excreted primarily in the urine (e.g., 70% of a physiological dose within 24 h) as the phosphate, sulfate, or glucuronide of menadiol and also in the bile as the glucuronide conjugate. A ring-altered metabolite has been identified: 3-hydroxy-2,3-dihydrophylloquinone, also called **hydroxyvitamin K** (Fig. 9.3).

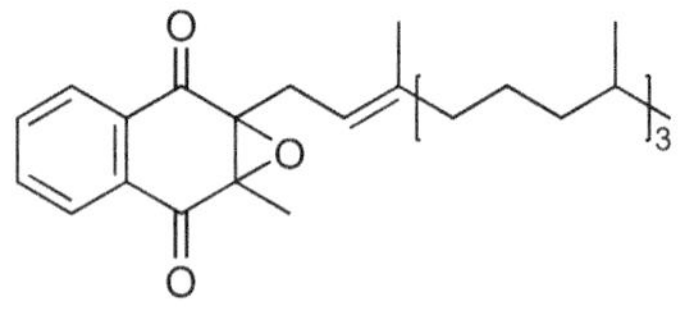

phylloquinone-2,3-epoxide

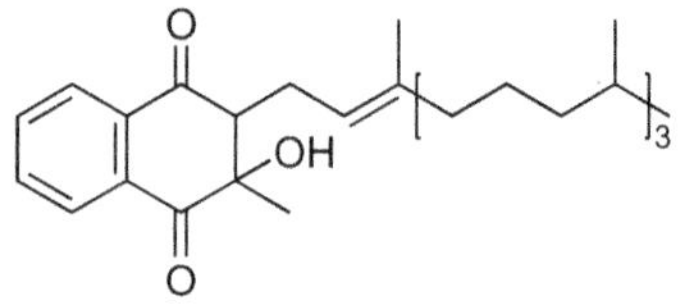

hydroxyphylloquinone

FIGURE 9.3 Ring-altered vitamin K metabolites.

Vitamin K Antagonists

The coumarin anticoagulants were developed after 3,3′-methylbis-(4-hydroxycoumarin) was identified as the active principle in spoiled sweet clover responsible for the hemorrhages and prolonged clotting times of animals consuming that feedstuff. Compounds in this family block the thiol-dependent, redox recycling of the vitamin by inhibiting VKQR. This reduces carboxylation of the Gla protein precursors, including those involved in clotting.[31] Several substituted 4-hydroxycoumarins have been widely used in anticoagulant therapy in clinical medicine, and rodenticides (effective by causing fatal hemorrhaging). The most widely used has been warfarin (3-[a-acetonylbenzyl]-4-hydroxycoumarin)[32], an analog of the naturally occurring hemorrhagic factor dicumarol, and its sodium salt[33].

29. **NAD(P)H:(vitamin K quinone) oxidoreductase; also called** menadione reductase, phylloquinone reductase, and quinone reductase.

30. Carriers of the V433M allele (rs2108622) require higher warfarin doses. This appears to be related to their reduced hepatic capacity to oxidize phylloquinone, leaving them with relatively high hepatic concentrations of the vitamin (McDonald, M.G., Reider, M.J., Nakano, M., et al., 2009. Mol. Pharmacol. 75, 1337–1346).

31. This effect is different from that of another anticoagulant, heparin, a polysaccharide that complexes with thrombin in the plasma to enhance its inactivation.

32. This analog of the naturally occurring vitamin K antagonist, dicumarol, warfarin (4-hydroxy-3-[3-oxo-1-phenylbutyl]-2*H*-1-benzopyran-2-one) was synthesized by Link's group at the University of Wisconsin and named for the Wisconsin Alumni Research Foundation.

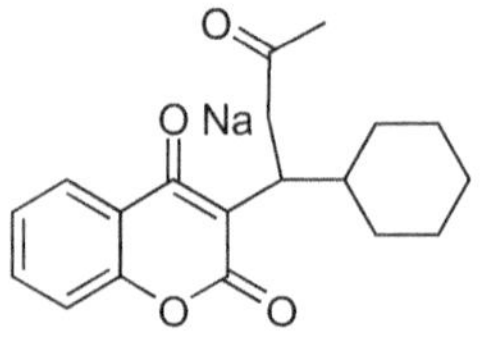

Warfarin

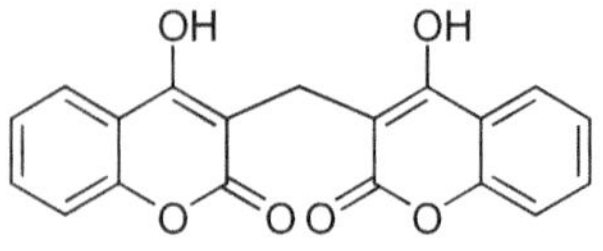

dicumarol

33. Others include ethyl biscoumacetate (3,3′-carboxymethylenebis-[4-hyroxycoumarin] ethyl ester and phenprocoumon (3-[1-phenylproyl]-4-hydroxycoumarin).

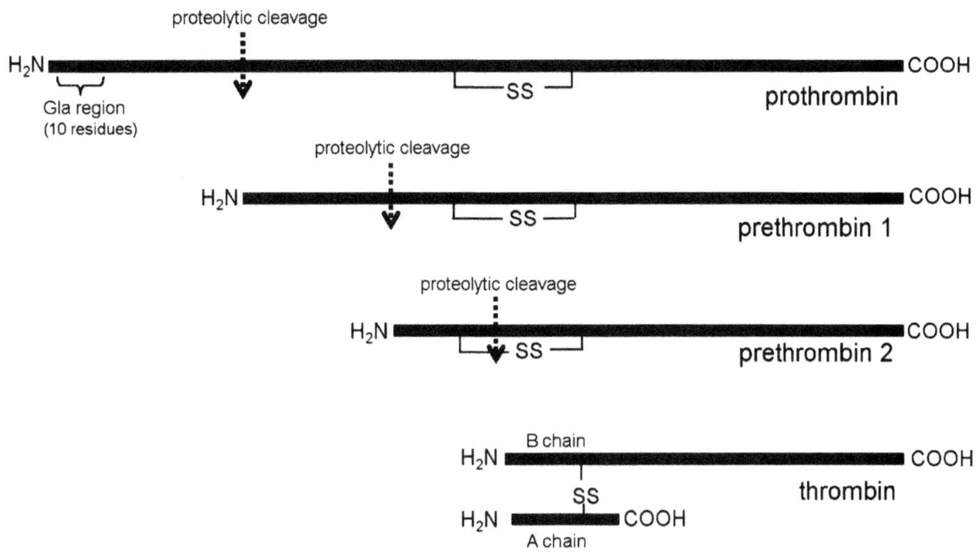

FIGURE 9.4 Thrombin is formed by the successive proteolytic removal of sequences from prothrombin.

Warfarin therapy has been important in preventing strokes; the drug is prescribed for a million patients each year in the United States alone. However, its effectiveness can vary among individuals by as much as 20-fold, and it is estimated that 12% of treated patients continue to experience major bleeding episodes.[34] Warfarin resistance is thought to involve polymorphisms in VKQR and the cytochrome *P*-450 isoform (CYP2C9) that metabolizes the drug. Warfarin-resistant rats have been found to respond to competitive inhibitors of one or more steps in the vitamin K redox cycle: 2-chloro-3-phytyl-1,4-naphthoquinone (chloro-K); 2,3,5,6-tetrachloro-pyridinol and 2-phylloquinone derivatives;[35] and other coumarins.[36]

7. METABOLIC FUNCTIONS OF VITAMIN K

Vitamin K–Dependent γ-Carboxylations

Vitamin K is the cofactor of a specific microsomal carboxylase that uses the energy of oxygenation of vitamin K hydroquinone to drive the γ-carboxylation of peptide-bound glutamic acid residues (Fig. 9.4). This **vitamin K–dependent γ-glutamyl carboxylase (VKγGC)** is a unique, integral membrane protein with no apparent homology with other proteins. It is found in the endoplasmic reticulum predominantly not only in the liver but also in several other organs.[37] In their reduced forms, K vitamers provide reducing equivalents for the reaction, undergoing oxidation to the 2,3-epoxide form. This is coupled to the cleavage of a C–H bond and formation of a carbanion at the γ-position of a peptide-bound glutamyl residue and followed by carboxylation. The VKγGC requires reduced forms of the vitamin (the hydroquinones), CO_2 as the carboxyl precursor, and molecular oxygen (O_2).[38] It is frequently referred to as vitamin K carboxylase/epoxidase to indicate the coupling of the γ-carboxylation step with the conversion of vitamin K to the 2,3-epoxide. Normally, this coupling is tight; full carboxylation of candidate proteins occurs through a progressive mechanism in which all their Glu sites are converted to Gla. However, under conditions of low CO_2 levels or in the absence of peptidyl-Glu, the epoxidation of the vitamin proceeds without concomitant carboxylation, yielding undercarboxylated proteins.

The vitamin K–dependent γ-carboxylation of specific glutamyl residues on the **zymogen**[39] precursors of each blood clotting factor occurs posttranslationally at the N-terminus of the nascent polypeptide. In the case of prothrombin, all 10 glutamyl residues in positions 7–33 (but none of the remaining 33 glutamyl residues) are γ-carboxylated. Carboxylation confers Ca^{2+}-binding capacity, facilitating the formation of Ca^{2+} bridges between the clotting factors

34. Anticoagulant efficacy, typically measured by prothrombin time, i.e., the time for a blood sample to clot after the addition of thromboplastin. As the latter reagents can vary in their sensitivity to clotting factors, such results are normalized to an international average.

35. For example, desmethylphylloquinone, 2-ethylphylloquinone, 2-fluromethylphylloquinone, 2-hydroxymethylphylloquinone, 2-methoxymethylphylloquinone, and 2-trifluromethylphylloquinone.

36. For example, difenacoum (3-[3-*p*-diphenyl-1,2,3,4-tetrahydronaphth-1-yl]-4-hydroxycoumarin); bromodifenacoum (3-[3-{4′-bormodiphenyl-4-yl}-1,2,3,4-tetrahydronaphth-1-yl]-4-hydroxycoumarin).

37. For example, lung, spleen, kidney, testes, bone, placenta, blood vessel wall, and skin.

38. The in vitro activity of the carboxylase is stimulated almost fourfold by pyridoxal phosphate when the substrate is a pentapeptide. It is doubtful whether that cofactor is important in vivo, as no stimulation was observed in the carboxylation of endogenous microsomal proteins.

39. The term "**zymogen**" was coined in 1875 by the German physiologist Rudolf Heidenhain to mean "producing ferment." It is used to indicate an inactive form of an enzyme, which must be converted biochemically to show catalytic activity, i.e., a proenzyme.

and phospholipids on membrane surfaces of blood platelets and endothelial and vascular cells, and between Gla residues (i.e., glutamyl residues that have been carboxylated) to form internal Gla–Gla linkages. Polymorphisms of the VKγGC have been associated with variations in circulating levels of ucOC.[22]

Vitamin K–Dependent Gla Proteins

Vitamin K functions in the posttranslational modification of at least 20 proteins via γ-carboxylation of 10–12 specific glutamate residues to produce **Gla** residues (Table 9.11).[40,41] The Gla proteins function by binding negatively charged phospholipids via Ca^{2+} held by their Gla residues. Each mammalian Gla protein contains a short, carboxylase recognition sequence (cleaved after carboxylation) that binds covalently to glutamate-containing propeptides to enhance catalysis. The VKγGC binds these propeptides with different affinities: compared to the most tightly bound factor X, most other propeptides are bound 2–10 times less tightly, whereas protein C and prothrombin are bound 100 times less tightly.[42] This suggests that competition for the VKγGC underlies variations in the amounts of Gla proteins produced, as has been observed in warfarin-treated cells. In the absence of vitamin K, these proteins can be secreted into the circulation in non- and undercarboxylated forms. These biologically inactive forms continue to be referred to by the name given when they were first recognized, **proteins induced by vitamin K deficiency**.

TABLE 9.11 Vitamin K–Dependent Gla Proteins

Clotting regulatory proteins	Prothrombin (factor II)
	Factor VII
	Factor IX
	Factor X
	Protein C
	Protein M
	Protein S
	Protein Z
Bone proteins	Osteocalcin
	Matrix Gla protein (MGP, periostin)
	Protein S
	Periostin
Transmembrane proteins	Proline-rich Gla protein 1 (PRGP1)
	Proline-rich Gla protein 2 (PRGP2)
	Transmembrane Gla protein 3
	Transmembrane Gla protein 4
Other Gla proteins	Gas6
	Gla-rich protein
	Transthyretin
	Atherocalcin
	PRGPs
	Renal Gla protein

Physiological Functions of the Gla Proteins

Blood clotting. Blood clotting is produced by a complex system of proteins that function to prevent hemorrhage and lead to thrombus formation. The system confers coagulation at the site of injury and curtails the process upon the formation of a clot. The process is initiated by injury to tissues through the release of collagen fibers and a cell surface protein (tissue factor), which interact with vitamin K–dependent Gla proteins in the plasma. These signals are amplified via the clotting pathway, ultimately, to form the clot.

The eight vitamin K–dependent plasma proteins comprising this system (Table 9.12) have homologous amino acid sequences in their first 40 positions, each containing 9–13 Gla residues in their amino-terminal domain.[43] All require Ca^{2+} for activity. Most circulate as a zymogen, i.e., an inactive precursor of the respective functional form, which is a serine protease. Each participates in a cascade of proteolytic activation of a series of factors ultimately leading to the conversion of a soluble protein, fibrinogen, to insoluble fibrin, which cross-links with platelets to form the clot (Figs. 9.4 and 9.5). The activation of proteases in this cascade involves the Ca^{2+}-mediated association of the active protein, its substrate, and another protein factor with a phospholipid surface.

40. This rare amino acid was discovered in studies of the molecular basis of abnormal clotting in vitamin K deficiency. Although it had been known that vitamin K deficiency and 4-hydroxycoumarin anticoagulant treatment each caused hypoprothrombinemia, studies in the early 1970s revealed the presence in each condition of a protein that was antigenically similar to prothrombin but did not bind Ca^{2+} and, therefore, was not functional. Studies of the prothrombin Ca^{2+}-binding sites revealed them to have Gla residues, which were replaced by glutamate residues in the abnormal prothrombin. Subsequently, Gla residues were found in each of the other vitamin K–dependent clotting factors and in several other Ca^{2+}-binding proteins in other tissues.

41. Gla-rich proteins have also been found in the venomous cone snail, some poisonous snakes, and some urochordates. These Gla-proteins serve as paralyzing neurotoxins used to subdue prey.

42. Huber, P., Schmitz, T., Griffen, J., et al., 1990. J. Biol. Chem. 265, 12,467–12,473.

43. Undercarboxylation of the clotting factors is rare, except in cases of patients on anticoagulant therapy.

TABLE 9.12 Characteristics of the Vitamin K–Dependent Plasma Proteins

Parameter	II[a]	VII[b]	IX[c]	X[d]	C	M	S	Z
Concentration (μg/mL)	100	1	3	20	10	<1	1	<1
Molecular mass (kDa)	72	46	55	55	57	50	69	55
Carbohydrate (%)	8	13	26	13	8	+	+	+
Number of chains	1	1	1	2	2	1	1	1
Number of Gla residues	10	10	12	12	11	+	10	13

[a]*Prothrombin.*
[b]*Proconvertin*
[c]*Plasma thromboplastin component, also: Christmas factor, antihemophilic factor B, platelet cofactor II, antiprothrombin II.*
[d]*Stuart factor.*

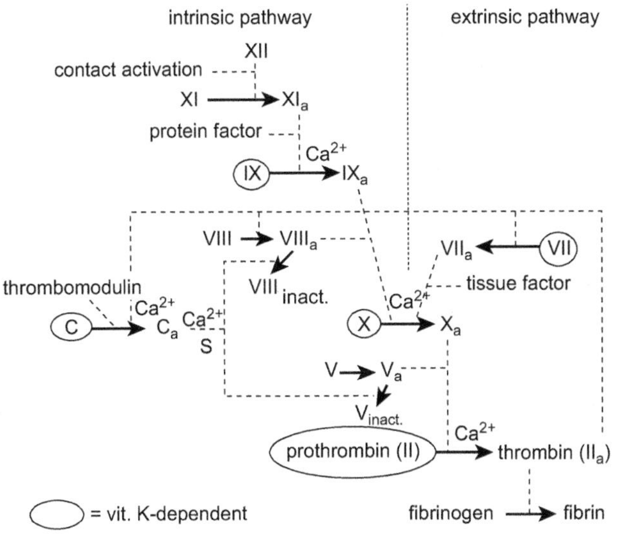

FIGURE 9.5 Roles of vitamin K–dependent proteins (circled factors) in blood clotting.

The key step in this system the activation of **factor X** (also, Stuart factor) by the proteolytic removal of a short polypeptide from the zymogen. This step is effected by two mechanisms:

- **intrinsic clotting system – factor IX** (also, Christmas factor or plasma thromboplastin component) is activated by **plasma thromboplastin** upon contact with a foreign surface.
- **extrinsic clotting system – factor VII**[44] is activated by **tissue thromboplastin** released as the result of tissue injury.

Once activated, factor X[45], after binding Ca^{2+} and phospholipid, can catalyze the activation of other coagulation factors:

- **prothrombin** (factor II) to its active form, **thrombin** (**factor IIa**), which catalyzes the proteolytic change in fibrinogen that renders it insoluble (as fibrin) for clot formation.
- **factor V** to its active form, factor Va; and
- **factor VIII** to its active form factor VIIIa.

Control of clotting is accomplished by the downregulation of thrombin production via thrombin binding to thrombomodulin, which complex activates protein C,[46] which, in turn, inactivates factors Va and VIIIa.

Two components of this system (Proteins S and Z) are not serine proteases. Protein S is found in the plasma both in free form and as a bimolecular complex with a regulatory component (C4b-binding protein) of the complement system. Individuals with inherited protein S deficiency have been reported to have recurrent thromboses. Protein Z is a cofactor for inhibition of activated factor X. Protein Z deficiency has been associated with a bleeding tendency in patients with factor V Leiden mutation.[47]

Bone Health. The presence of vitamin K–dependent Gla proteins in bone suggests a role of the vitamin in bone health. Several studies have shown individuals with

44. Also called **proconvertin**, factor VII is also activated by a high-fat meal and has been associated with increased risk to ischemic heart disease in some studies (e.g., Junker, R., Heinrich, J., Schulte, H., et al., 1997. Arterioscler. Thromb. Vasc. Biol. 1, 1539–1544).

45. Polymorphisms of factor X have been identified; however, these do not appear to affect circulating factor X levels.

46. Protein C also has anti-inflammatory and antiapoptotic activities involving activation of a protease-activated receptor (PAR1) and inhibition of the NF-κB pathway. Individuals with inherited protein C deficiency have elevated risk of thrombosis.

47. This condition involves a point mutation in the gene encoding coagulation factor V; it causes expression of a form of the factor resistant to activated protein C and results in a hypercoagulable state. The mutation occurs in 4–6% of the US population and is associated with increased risk to venous thromboembolism.

low circulating vitamin K levels or low vitamin K intakes to be at elevated risk to osteoporosis or fracture.[48] The Nurses' Health Study, a 10-year prospective study of more than 72,000 women, found the age-adjusted risk of hip fracture to be 30% less in women with vitamin K intakes >109 μg/day compared to those consuming greater amounts.[49] A similar relationship was observed in the Framingham Heart Study: subjects in the highest quartile of vitamin K intake (254 μg/day) had a significant reductions in hip fracture risk compared to those in the lowest quartile (56 μg/day).[50] Low vitamin K status has also been associated with articular cartilage and meniscus damage in men and women.[51] At least a dozen randomized clinical trials have been conducted to determine the efficacy of vitamin K supplementation in reducing bone mineral loss and/or fracture risk. Those have found MK-*4* to reduce fracture risk[52] and improve bone mineral density (Table 9.13).[53]

Such effects appear to be mediated by three vitamin K–dependent Gla proteins in bone:

TABLE 9.13 Efficacy of Vitamin K Supplementation in Reducing Fracture Risk and Bone Mineral Loss in Older Adults

Study	Treatment	n	Fractures (%)	Bone Mineral Density Change (%)
Shiraki et al.[a]	MK-4[d]	86	14.3	−0.5
	Control	94	30.3	−3.3
Iwamoto et al.[b]	MK-4[d]	22	8.7	−0.1
	Control	20	25.0	−1.7
Ishida et al.[c]	MK-4[d]	63	14.3	−1.9
	Control	60	28.3	−3.3

[a]*Shiraki, M., Shiraki, Y., Aoki, C., et al., 2000. J. Bone Min. Res. 15, 515–521.*
[b]*Iwamoto, J. Takeda, T., Ichimura, S., 2001. J. Orthop. Sci. 6, 487–492.*
[c]*Ishida, Y., Kawai, S., 2004. Am. J. Med. 117, 549–555.*
[d]*45 mg/day.*

- **Osteocalcin**[54] is the best characterized vitamin K–dependent protein of calcified tissues, although its function remains unclear. It shows no homology with the vitamin K–dependent plasma proteins, but has been strongly conserved between various species, consisting of a 5.7-kDa protein with three Gla residues in a 49–50 amino acid segment. It binds Ca^{2+} weakly and hydroxyapatite strongly; this serves to maintain its secondary structure while allowing it to bind mineralized bone matrix. It is the second most abundant protein[55] in the bone matrix, comprising ~2% of total bone protein and 10–20% of noncollagen protein. Osteocalcin is expressed relatively late in development, i.e., with the onset of bone mineralization. Its carboxylation is inhibited by warfarin treatment and is stimulated by MK supplementation and (in vitro) 1,25-$(OH)_2$-vitamin D_3. An estimated 20% of osteocalcin is not bound to bone and is free to enter the plasma.
- In most species, osteocalcin is fully carboxylated at each of its Gla sites. This is not the case in humans; analysis of human bone revealed each site to be 67%, 88%, and 93% carboxylated.[56] Similarly, much of the osteocalcin in human plasma is undercarboxylated – estimates approaching 50% have been made. Osteocalcin undercarboxylation may reflect low vitamin K intake; however, high intakes of vitamin K (10-fold the levels thought to be nutritionally required) are needed to support full carboxylation. Plasma levels of ucOC are subject to genetic variability associated with polymorphisms in both the vitamin K–dependent carboxylase and epoxide reductase. Greatest circulating ucOC levels are found in young children and in patients with Paget disease[57] and other disorders involving increased bone resorption/mineralization (Table 9.14). The degree of osteocalcin undercarboxylation has been found to be positively correlated with muscle strength and measures of bone health in older women, suggesting that carboxylated osteocalcin plays a role in muscular skeletal health.[58]
- It has been suggested that osteocalcin may function in the regulation of calcification, perhaps by acting as an attractant for osteoclast progenitor cells. Support for this hypothesis includes that association of plasma ucOC level with increased risk of hip fracture (Table 9.15) and

48. *See* review: Weber, P., 2001. Nutr. 17, 880–887.
49. Feskanich, D., Weber, P., Willett, W.C., et al., 1999. Am. J. Clin. Nutr. 69, 74–79.
50. Booth, S.L., Tuckcer, K.L., Chen, H., et al., 2000. Am. J. Clin. Nutr. 71, 1201–1208.
51. She, M.K., Kritchevsky, S.B., Hsu, F.C., et al., 2014. Osteoarth. Cartilage 23, 370–378.
52. A meta-analysis found seven studies to show an average risk reduction of 60% (Cockayne, S., Adamson, J., Lanham-New, S., et al., 2006. Arch. Intern. Med. 166, 1256–1261).
53. Iwamoto, J., Takeda, T., Sato, Y., 2006. Nutr. Rev. 64, 509–517.
54. Sometimes, "bone GLA protein."
55. Collagen is the most abundant protein in bone.
56. Dowd, T.L., Rosen, J.F., Li, L., et al., 2003. Biochemistry 42, 7769–7779.
57. Paget disease, also called osteitis, affects 3% of adults over 40 years of age. It involves dysfunctional bone remodeling, with bone continually breaking down and rebuilding at rates faster than normal. This results in bone being replaced with soft, porous, highly vascularized bone that can be weak and easily bent, leading to shortening of the affected part of the body, or with excess bone that can be painful and easily fractured. The disease most commonly affects the spine, pelvis, skull, femur, and tibiae.
58. Levinger, I., Scott, D., Nicholson, G.C., et al., 2014. Bone 64, 8–12.

TABLE 9.14 Plasma Osteocalcin Concentrations in Humans

	Osteocalcin (ng/mL)
Children	10–40[a]
Adult men and women <60 years	4–8
Women 60–69 years	7
Women 80–89 years	8
Patients with Paget disease	39
Patients with secondary hyperparathyroidism	47
Patients with osteopenia	9

[a]*Higher levels at 10–15 years.*
From Power, M.J., Fottrell, P.F., 1991. Crit. Rev. Clin. Lab. Sci. 28, 287–335.

TABLE 9.15 Total and Undercarboxylated Osteocalcin in Fracture and Nonfracture Patients[a]

Parameter	Nonfracture	Fracture
n	153	30
Age (years)	82.5±5.9[b]	85.8±6.5[b,c]
Body weight (kg)	56.6±11.5	49.4±10.7
Plasma Osteocalcin (ng/mL)		
Total	6.18±3.34	7.90±4.34[c]
Undercarboxylated	0.89±0.89 (14)[b]	1.47±1.65[c] (19)[b]
Carboxylated	5.29±2.69 (86)[b]	6.43±2.94[c] (81)[b]
25-OH-D_3 (ng/mL)	17.4±14.1	15.9±10.8
Parathyroid hormone (pg/mL)	47.0±23.8	60.0±40.9
Alkaline phosphatase (IU/L)	78±37	92±40[a]

[a]*Szulc, P., Chapuy, M.C., Meunier, P.J., et al., 1996. Bone 18, 487–488.*
[b]*Mean±SD.*
[c]*p<.05.*

from studies in vitro that showed osteocalcin to inhibit the deposition of hydroxyapatite crystals reminiscent of the smaller crystals observed in bones of osteocalcin-deficient mice.[59] Studies have found phylloquinone supplementation to reduce plasma ucOC level, but effects on bone mineral density have been inconsistent. Relatively high intakes of vitamin K (up to 5 mg/day) appear to be required to support full carboxylation. Although osteocalcin undercarboxylation does not impair bone mineralization in animal models, osteocalcin deficiency (because of warfarin treatment or genetic ablation) increases bone mineralization in the rat. Therefore, it has been suggested that osteocalcin may function as a negative regulator of bone formation.[60]

- **Matrix Gla Protein (MGP)**[61] is a small (9.6 kDa), insoluble polypeptide structurally related to osteocalcin, with Ca-binding activity conferred by five Gla residues among its 79–amino acid sequence. It is expressed in many soft tissues, including vascular and smooth muscle cells, but accumulates only in calcified tissues. It has clear affinities for demineralized bone matrix and nonmineralized cartilage, where it is thought to function as an inhibitor in the regulation of calcification. MGP is posttranslationally modified by VKγCG to have five Gla residues and by a casein kinase which phosphorylates three of its serine residues; these groups are thought to participate in Ca-binding. Normally, MGP is thought to be fully carboxylated; however, it has been found incompletely carboxylated (and, thus, inactive) in patients undergoing hemodialysis, presumably because of their loss of vitamin K. Genetic ablation of MGP in the mouse led to early arterial calcification and fatal hemorrhages.[62] The mechanistic basis of this effect remains unclear.
- **Protein S** is synthesized by osteoblasts. It contains a thrombin-sensitive region, an epidermal growth factor–like domain, and a steroid hormone–binding domain. It has been shown to bind tyrosine kinase receptors.[63] A role in bone function was suggested by the finding of severe osteopenia, low bone mineral density, and vertebral compression fractures in two pediatric cases with very low protein S levels.
- **Periostin** is produced and secreted by bone-derived mesenchymal stromal cells and is abundant in mineralized bone nodules.[64] It is believed to function in the formation of the extracellular bone matrix.

Cardiovascular Health. Supplementation with phylloquinone has been found to improve the functional characteristics of the carotid artery,[65] reduce risk of coronary artery calcification,[66] and reduce risk of cardiovascular

59. Boskey, A.L., Gadaleta, S., Gundberg, C., et al., 1998. Bone 23, 187–196.
60. Price, P.A., 1988. Ann. Rev. Nutr. 8, 565–583; Ducy, P., Desbois, C., Boyce, B., et al., 1996. Nature 382, 448–452.
61. Sometimes referred to as **periostin**.
62. Luo, G., Ducy, P., McKee, M.D., et al., 1997. Nature 386, 78–81.
63. Nakamura, Y.S., Hakeda, Y., Takakura, N., et al., 1998. Stem Cells 16, 229–238.
64. Coutu, D.L., Wu, J.H., Monetter, A., et al., 2008. J. Biol. Chem. 283, 17,991–18,001.
65. Braam, L.A., Knapen, M.H., Geusens, P., et al., 2003. Calcif. Tissue Int. 73, 21–26.
66. Shea, M.K., Gundberg, C.M., Meigs, J.B., et al., 2009. Am. J. Clin. Nutr. 90, 1230–1235.

death.[67] Such effects may be mitigated by vitamin K–dependent Gla proteins that impair vascular intimal calcification:

- **MGP** is also expressed in vascular smooth muscle and appears to play a dominant role in maintaining the rate of arterial calcification as low as possible. Its impaired carboxylation by warfarin treatment has been shown to lead to arterial calcification in animal models. Its genetic deletion in the mouse led to the fragmentation and calcification of vascular smooth muscle, the loss of contractility, and death. Mutations in MGP are associated with calcification of cartilage in Keutel syndrome.[68] It is thought that both the Gla and phosphoserine groups of MGP participate in Ca-binding at internal nucleation sites within collagen and elastin fibrils and extracellular matrix components.
- **Atherocalcin** was discovered in calcified atherosclerotic tissue. It has been suggested that atherocalcin may inhibit VKγGC, which is found in the walls of arteries but not veins, and may be involved in the development of atherosclerosis.
- **Osteocalcin**, normally expressed only in bone, is upregulated in arterial calcification.
- **Gas6**[69] is a 75-kDa protein with 11–16 Gla residues. Similar to protein S, it contains an epidermal growth factor–like domain. Gas6 functions as a ligand for the reception tyrosine kinases Ax1 and Sky/Rse and protects cells from apoptosis by activating Ark phosphorylation and inducing MAP kinase[70] activity.
- **Protein S** has a 44% sequence homology with Gas6; but it also has thrombin-sensitive motifs. It has also been suggested as protecting arterial intimal cells from apoptosis.

Other Gla proteins have been identified, although their functions remain unknown:

- **Gas6** is widely distributed in nervous tissue where is thought to protect cells from apoptosis. It has also been shown to have neurotrophic activity toward hippocampal neurons and to promote growth and survival of several types of neural cells. That brain microsomes lack γ-carboxylase activity means that the posttranslational glutamation of Gas6 must occur in other tissues.
- **A Gla-rich protein** has been identified in cartilage.[71] It has more Gla residues (15–16) than other known Gla proteins. Its function remains unknown.
- **Proline-rich Gla proteins** are small (17–23 kDa), single-pass, transmembrane proteins expressed in a variety of extrahepatic tissues.
- **Transthyretin** binds retinol and thyroxine.
- **Renal Gla protein** is thought to have a role in the renal transport/excretion of Ca^{2+}.
- **Thrombin**, factor VII, and **factor Xa** have been shown to interact with tissue factor to activate protease-activated receptors (PARs) to promote platelet aggregation, activate protein C, and promote anti-inflammatory and antiapoptotic responses.
- **Other Gla proteins** have been reported in sperm, urine, hepatic mitochondria, and snake venom.

Prospective functions. Vitamin K may have a role in neurologic function. Warfarin has been shown to reduce brain sphingolipid contents and MK-*4*, which comprises virtually all (98%) the vitamin K in neural tissue, and is strongly correlated with the contents of sphingolipids, sulfatides, sphingomyelin, and gangliosides in neural tissues.[72] That vitamin K can have anti-inflammatory effect is suggested by the observations of phylloquinone reducing the inflammatory response to lipopolysaccharide and of MK-*4* reducing the in vitro production of prostaglandins and IL-6.

8. BIOMARKERS OF VITAMIN K STATUS

Useful biomarkers of vitamin K status include:

- **Plasma phylloquinone** concentration reflects recent dietary intake of that vitamer (Fig. 9.6). Levels are correlated with those of triglycerides and α-tocopherol and have been found to be related to polymorphisms in genes involved in lipoprotein and phylloquinone metabolism.[73] In healthy humans, circulating phylloquinone concentrations are in the range of 0.1–0.7 ng/mL.
- **Prothrombin time** is informative only to detect advanced, subclinical vitamin K deficiency. A 50% loss of plasma prothrombin level is required to affect prothrombin time.
- **Plasma ucOC**, because it is synthesized only by osteoblasts, has been used as a marker of bone formation. High circulating levels predict low bone mineral density and fracture risk and are frequently elevated among postmenopausal women.[74] Percentage undercarboxylation, but not total osteocalcin, can also indicate vitamin K status, as that parameter responds to increasing intakes of the vitamin over the nutritional range (Fig. 9.7).

67. Juanola-Falgarona, M.J., Salas-Salvadó, J., Martínez-González, M.A., et al., 2014. J. Nutr. 144, 743–750.
68. A rare autosomal recessive disorder characterized by cartilage calcification, peripheral pulmonary stenosis, nasal bridge depression, hearing loss, mild mental retardation, and shortened distal phalanges.
69. Named for its gene, Growth Arrest Specific gene 6.
70. Mitogen-activated protein kinase.
71. Viegas, C.S.B., Simes, D.C., Laize, M.K., et al., 2008. J. Biol. Chem. 283, 36,655–36,664.
72. Carrié, I., Portoukalian, J., Vicaretti, R., et al., 2004. J. Nutr. 134, 167–172.
73. Dashti, H.S., Shen, M.K., Smith, C.E., et al., 2014. Am. J. Clin. Nutr. 100, 1462–1469.
74. Affected individuals respond to vitamin K supplements with increased bone formation and decreased bone resorption.

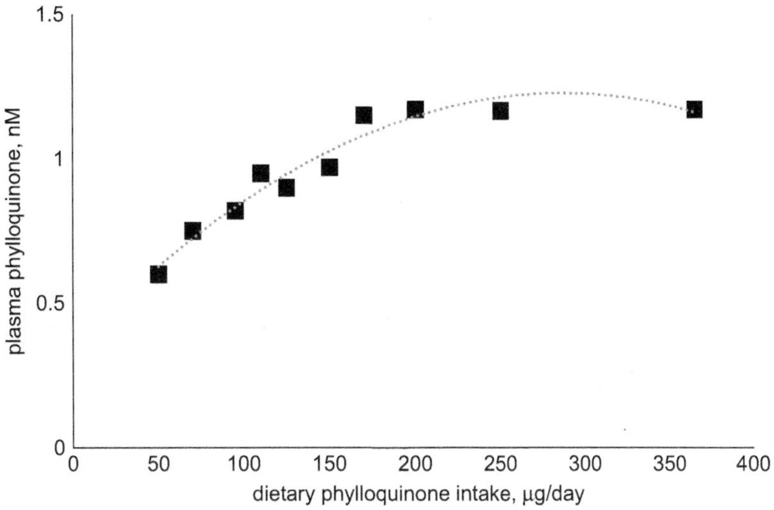

FIGURE 9.6 Relationship of plasma phylloquinone concentration to estimated dietary phylloquinone intake in healthy adults. *After McKeown, N.M., Jacques, P.F., Gundberg, C.M., et al., 2002. J. Nutr. 132, 1329–1334.*

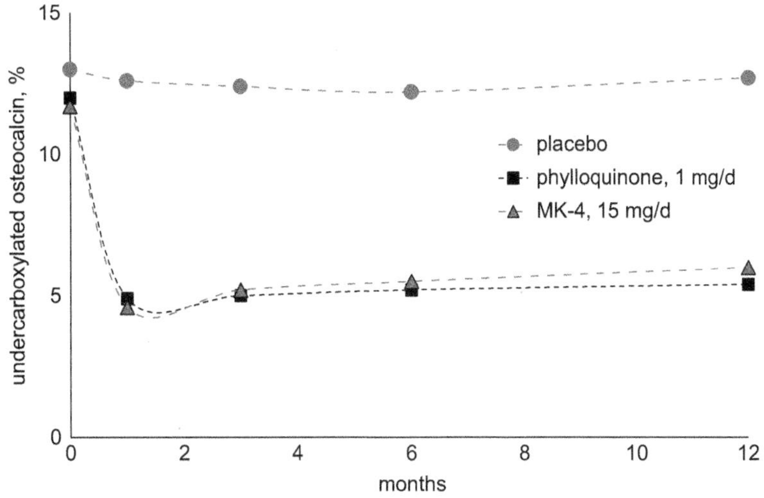

FIGURE 9.7 Effect of phylloquinone supplementation on undercarboxylation of plasma osteocalcin in humans. *After Binkley, N., Harke, J., Krueger, D., et al., 2008. J. Bone Miner. Res. 24, 983–991.*

9. VITAMIN K DEFICIENCY

Vitamin E deficiency can have primary (privational) and secondary (nonprivational) causes.

- **Primary causes** involve inadequate vitamin K supply
  - **Dietary patterns** that fail to provide vitamin K in adequate amounts, e.g., low amounts of green leafy vegetables, cheese, and other fermented foods.
- **Secondary causes** relate to impaired absorption, metabolism, or metabolic function of the vitamin
  - **Very low fat diets** do not support the development of intestinal luminal micelles upon which enteric vitamin K absorption is dependent.
  - **Lipid malabsorption** because of loss of pancreatic exocrine function (e.g., pancreatitis, pancreatic tumor, nutritional pancreatic atrophy in severe Se deficiency, *Ascarid* infection), lumenal deficiencies of bile (e.g., biliary stasis because of mycotoxicosis, biliary atresia), defects in lipoprotein metabolism (e.g., abetalipoproteinemia).
  - **Anticoagulant therapy** including treatment with warfarin, other 4-hydroxycoumarin anticoagulants, or large doses of salicylates, which inhibit the redox cycling of the vitamin.
  - **Insufficient intestinal microbiome** providing little or no vitamin K occurs in neonates and can be produced as a result of antibiotic therapy (with sulfonamides and broad-spectrum antibiotic drugs).
  - **Insufficient placental transport** of vitamin K giving neonates limited vitamin K reserves.

TABLE 9.16 General Signs of Vitamin K Deficiency

Organ System	Sign
General	Decreased growth
Dermatologic	Hemorrhage
Muscular	Hemorrhage
Gastrointestinal	Hemorrhage
Vascular	
Erythrocytes	Anemia
Platelets	Prolonged clotting time

Groups at Risk to Vitamin E Deficiency

Neonates.

Individuals with:
- very low fat intakes
- low intakes of plant oils and nuts
- lipid malabsorption syndromes
- dyslipidemias

Signs of Vitamin K Deficiency

The predominant clinical sign of vitamin K deficiency is **coagulopathy**, presenting as widespread subcutaneous and cerebral hemorrhage (Table 9.16), which can lead to a fatal anemia. The blood shows prolonged clotting time and **hypoprothrombinemia**. Because a 50% loss of plasma prothrombin level is required to affect prothrombin time, prolongation of the latter is a useful biomarker for advanced subclinical vitamin D deficiency.

Vitamin K deficiency can be prevented by consuming the recommended intakes of vitamin K established by expert bodies (Table 9.17).

TABLE 9.17 Recommended Vitamin K Intakes

US		FAO/WHO	
Age, Sex	AI[a] (μg/day)	Age, Sex	RNI[b] (μg/day)
0–6 months	2	0–6 months	5
7–11 months	2.5	7–11 months	10
1–3 years	30	1–3 years	15
4–8 years	55	4–6	20
9–13 years	60	7–9 years	25
14–18 years	75	10–18 years, Female	35–55
>18 years, female	90	Male	35–65
Male	120	>18 years, female	55
Pregnancy, ≤18 years	75	Male	65
>18 years	90	Pregnancy	55
Lactation, ≤18 years	75	Lactation	55
>18 years	90		

[a]Adequate Intakes values are given, as Recommended Dietary Allowances (RDAs) have not been established; Food and Nutrition Board, 2001. Dietary Reference Intakes for Vitamin A, Vitamin K, Arsenic, Boron, Chromium, Copper, Iodine, Iron, Manganese, Molybdenum, Nickel, Silicon, Vanadium and Zinc. National Academy Press, Washington, DC, 773 pp.
[b]Recommended Nutrient Intakes; Joint WHO/FAO Expert Consultation, 2001. Human Vitamin and Mineral Requirements. Food and Agricultural Org., Rome, 286 pp.

Signs of Vitamin K Deficiency in Humans

Vitamin K deficiency presents as **hypoprothrombinemia** and prolonged **clotting time**. Coagulopathies now associated with vitamin K deficiency are not new; they were documented in the 19th century and are likely to have been historic problems. Low vitamin K status has also been associated with elevated risks to osteoporosis and fracture, damage to articular cartilage, and osteoarthritis.[75] These risks in adults are typically greater for older individuals, including those with chronic kidney disease.

Newborns are at increased risk of vitamin K deficiency. The frequency of vitamin K–responsive hemorrhagic disease in 1-month-old infants is 1/4000 overall, but 1/1700 among breast-fed infants. Several factors contribute to their risk:

- **limited vitamin K reserves** because of poor placental transport of the vitamin. Their serum levels are typically about half those of their mothers.
- **no enteric microbial synthesis** of the vitamin, as their **intestines are sterile** for the first few days of life.
- **limited hepatic synthesis of the clotting factors** (Fig. 9.8), e.g., their **p**lasma prothrombin levels are typically a quarter those of their mothers.
- Breast milk is typically an inadequate source of vitamin K.

Some exclusively breast-fed infants not given vitamin K prophylaxis will develop **vitamin K deficiency bleeding (VKDB)**, also called **hemorrhagic disease of the**

75. Misra, D., Booth, S.L., Tolstykh, I., et al., 2013. Am. J. Med. 126, 243–248.

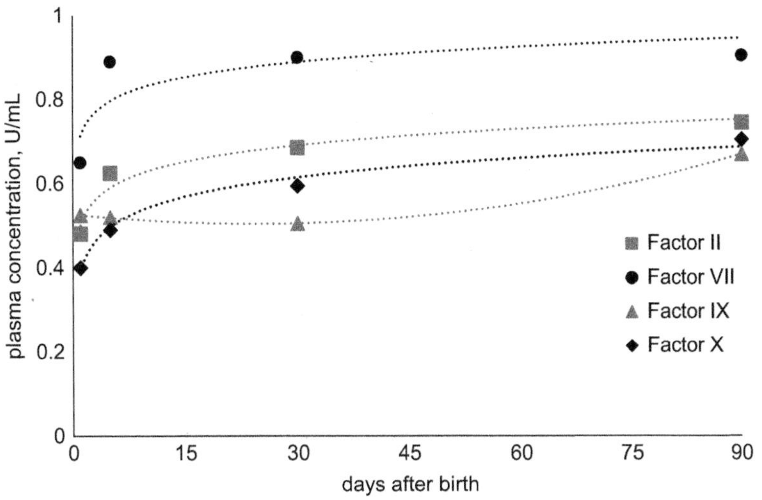

FIGURE 9.8 Developmental expression of vitamin K–dependent coagulation factors in infants. *After Zipursky, A., 1999. Br. J. Haematol. 104, 430–437.*

newborn.[76] This can present in different ways, depending on infant age:[77]

- newborns (first 24 h): cephalohematoma;[78] intracranial, intrathoracic, or intra-abdominal bleeding.
- newborns (first week): generalized ecchymoses;[79] bleeding from the gastrointestinal tract, umbilical cord stump, or circumcision site.
- infants (1–12 weeks): intracranial, skin, or gastrointestinal bleeding.

Infants with disorders involving lipid malabsorption (cystic fibrosis, biliary atresia, α_1-antitrypsin deficiency) can show potentially fatal signs within several weeks: intracranial hemorrhage with liver disease, bilirubinemia, and central nervous system damage. Hemorrhagic disease has also been reported for newborns of mothers on anticonvulsant therapy. It has become a common practice in many countries to treat all infants at birth with phylloquinone administered intramuscularly (0.5–1 mg) or orally (1 mg at birth followed by 50 μg/day for three months). This practice has greatly reduced the incidence of hemorrhagic disease of the newborn. Infants fed formula diets are at lower risk, as the amounts of vitamin K in infant formulas typically exceed those in human milk by as much as 50-fold.

Several congenital disorders of vitamin K–dependent proteins have been identified. Individuals with the VKER CG/GG genotype face increased risk to progressive coronary artery calcification and poorer survivals. Those with mutations in VKQR and VKγGC face **dysprothrombinemia** because of combined deficiencies of the coagulation factors. Genetic variants have also been identified for factor VII, and a congenital deficiency of protein C has been described. These conditions present as a range of spontaneous bleeding symptoms. In some cases, high-level vitamin K supplementation may provide effective management.

Signs of Vitamin K Deficiency in Animals

Monogastric species show hypothrombinemia when deprived of vitamin K. The clinical signs include prolonged clotting times, hemorrhages. Poultry are more likely than other species to show signs of vitamin K. This may be in part because of their hindgut microbial synthesis of MKs (because of their short gut and short transit time) and their susceptibility to intestinal coccidiosis for which sulfaquinoxaline is used. It is also likely because of their relatively low VKQR activities, which results in their inefficient recycling of the vitamin and gives them relatively high needs for dietary vitamin K. The use of sulfa drugs and antibiotics has also been associated clinical signs in young pigs, which are therefore typically given supplemental vitamin K.

Ruminants appear to obtain all their vitamin K needs from their rumen microbiota, which synthesize large amounts of the vitamin. Hypoprothrombinemia and spontaneous bleeding caused by vitamin K deficiency are seen only when they have been exposed to an antagonist such as dicumarol from molded clover in the condition called "spoiled sweet clover disease."

76. Without vitamin K prophylaxis, the risk of hemorrhage for healthy, nontraumatized infants in the first two weeks of life has been estimated to be 1–2/1000, and for older infants a third of that level.
77. VKDB differs from hemophilia by its earlier presentation (within a couple of days after birth) and absence of family history.
78. That is, subperiosteal bleeding.
79. Sheet hemorrhages of the skin, **ecchymoses**, differ from the smaller **petechiae** only in size.

10. VITAMIN K HEALTH AND DISEASE

Antibiotic Therapy

Hypoprothrombinemia has been associated with the use of antibiotics. The prevalence of hypoprothrombinemia increased in the 1980s with the introduction of the β-lactam antibiotics.[80] Although these drugs are administered intravenously, it is possible that they may affect enteric bacterial metabolism via biliary release. They do not alter fecal MKs in all patients, but they increase circulating vitamin K-2,3-epoxide levels in patients treated with vitamin K. The cephalosporin-type antibiotics have been found to inhibit the VKγGC to produce coumarin-like depressions of vitamin K–dependent clotting factors. Unlike the coumarins, the β-lactam antibiotics are weak anticoagulants; their effects are observed only in patients of low vitamin K status.

Anticoagulation Control

Low vitamin K status appears to contribute to unstable anticoagulation control in the use of warfarin in the management of thrombotic disorders. This affects as many as half of the patients and can be reversed by reducing warfarin dose and treating with phylloquinone.[81]

Anticarcinogenesis

That vitamin K status can play an anticarcinogenic role was suggested some six decades ago when MK-*4* treatment was found to increase the survival of patients with inoperable bronchial carcinoma. Since then it has been observed that patients with hepatocellular carcinoma typically have abnormally high circulating levels of under-γ-carboxylated prothrombin. Recently, a large prospective study found that cancer mortality was significantly less in individuals with the highest intakes of phylloquinone, although no effect were observed for MKs;[82] whereas, an earlier study found inverse association of cancer incidence and mortality for intake of MKs, but not phylloquinone.[83] In an 8-year randomized clinical trial, MK (45 mg/day) reduced the risk of hepatocellular carcinoma in 43 women with viral cirrhosis of the liver by 87% compared to controls.[84] Studies with animal models have shown all K vitamers capable of inhibiting tumor cell growth through several mechanisms:

- **Oxidative stress in malignant cells** is thought to be increased by menadione redox cycling.
- **Modulation of transcription factors** by phylloquinone and MKs. In cell culture, these vitamers have been shown to induce proto-oncogenes, increasing the levels of c-myc, c-jun, and c-fos; delaying the cell cycle; and enhancing apoptosis. Menadione can also induce protein tyrosine kinase activation and directly inhibit extracellular signal-regulated kinase protein tyrosine phosphatases. These effects are associated with reduced proliferation.
- **Cell cycle arrest** has been shown to be caused by menadione, which can inhibit cyclin-dependent kinases by binding to sulfhydryls at their active sites. This effect is associated with inhibition of malignant cell proliferation at the G1/S and S/G2 phases of the cell cycle. MKs have been found to affect cyclin function and also manifest as cell cycle inhibition.

Obesity Diabetes

Obesity has been associated with low vitamin K status. Adipose tissue stores vitamin K at relatively high levels.[85] Parameters of glucose metabolism have been inversely associated with serum concentrations of total, but not undercarboxylated, osteocalcin.[86] Phylloquinone levels have been found to vary inversely with percentage body fat in women (Fig. 9.9); higher levels were associated with greater insulin sensitivity and glycemic control as indicated by measures from oral glucose tolerance tests in men and women.[87] A 3-year intervention study found that phylloquinone supplementation reduced insulin resistance in men.[88] However, no significant benefits were found in women in that or another trial.[89]

11. VITAMIN K TOXICITY

No upper tolerable limits have been established for vitamin K. Phylloquinone exhibits no adverse effects when administered to animals in massive doses by any route, although it has been associated with increased risk of chronic kidney disease in humans.[90] The MKs are also thought to have negligible toxicity.

Menadione, however, can be toxic. At high doses - at least three orders of magnitude greater than those levels required for normal physiological function - it can produce

80. This group includes penicillin derivatives (penams), cephalosporins (cephams), monobactams, and carbapenems.
81. Baker, P., Gleghorn, A., Tripp, T., 2006. Br. J. Haematol. 133, 331–336.
82. Juanola-Falgarona, M.J., Salas-Salvadó, J., Martínez-González, M.A., et al., 2014. J. Nutr. 144, 743–750.
83. The European Prospective Investigation into Cancer and Nutrition (Nimptsch, K., Rohrmann, S., Linseisen, J. 2008. Am. J. Clin. Nutr. 87, 985–992).
84. Habu, D., Shiomi, S., Tamori, A., et al., 2004. JAMA 292, 358–361.
85. Adults undergoing bariatric surgery were found to have phylloquinone levels of 148±72 nmol/kg and 175±112 nmol/kg in subcutaneous and visceral adipose tissue, respectively (Shea, M.K., Booth, S.L., Gundberg, C.M., et al., 2010. J. Nutr. 140, 1029–1034).
86. Booth, S.L., Centi, A., Smith, S.R., et al., 2013. Nat. Rev. Endocrinol. 9, 43–55.
87. Yoshida, M., Booth, S.L., Meigs, J.B., et al., 2008. Am. J. Clin. Nutr. 88, 210–215.
88. Yoshida, M., Jacques, P.F., Meigs, J.B., et al., 2008. Diabetes Care 31, 2092–2096.
89. Kumar, R., Binkley, N., Vella, A., 2010. Am. J. Clin. Nutr. 92, 1528–1532.
90. O'Seaghdha, C.M., Hwang, S.J., Holden, R., et al., 2012. Am. J. Nephrol. 36, 68–77.

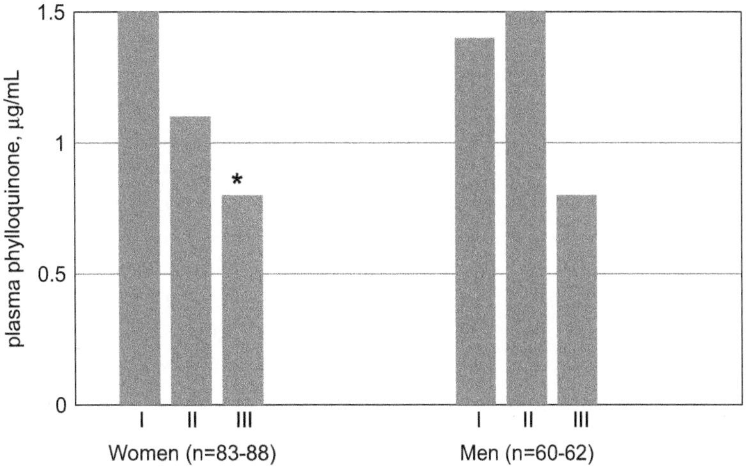

FIGURE 9.9 Relationship of vitamin K status and adiposity (by tertile of % body fat) in older adults (* $p<.05$). *After Shea, M.K., Booth, S.R., Gundberg, C.M., et al., 2020. J. Nutr. 140, 1029–1034.*

oxidative stress and manifests as hemolytic anemia, hyperbilirubinemia, and severe jaundice. This occurs as a result of its redox cycling to produce the superoxide radical anion. At high levels, it can also react with free sulfhydryl groups to deplete reduced glutathione (GSH) and reduce cellular antioxidant capacity.

A review of the US Food and Drug Administration database revealed 2236 adverse reactions reported for 1019 patients receiving intravenous vitamin K from 1968 to 1997.[91] Of those cases, 192 were anaphylactoid reactions and 24 were fatalities; those numbers were only 21 and 4, respectively, for patients given vitamin K doses <5 mg. Persistent, localized eczematous plaque has been reported at the site of injection for some patients given phylloquinone intramuscularly or subcutaneously[92].

12. CASE STUDIES

Instructions

Review the following case report, paying special attention to the diagnostic indicators on which the treatments were based. Then answer the questions that follow.

Case 1

A 60-year-old woman involved in an automobile accident sustained injuries to the head and compound fractures of both legs. She was admitted to the hospital, where she was treated for acute trauma. Her recovery was slow, and for the next 4 months she was drowsy and reluctant to eat. Her diet consisted mainly of orange and glucose drinks with a multivitamin supplement that contained no vitamin K. Her compound fractures became infected and she was treated with a combination of antibiotics (penicillin, gentamicin, tetracycline, and cotrimoxazole). She then developed intermittent diarrhea, which was treated with codeine phosphate. After a month, all antibiotics were stopped, and 46 days later (6 months after the injury), she experienced bleeding from her urethra. At that time other signs were also noted: bruising of the limbs, bleeding gums, and generalized *purpura*.[93] The clinical diagnosis was scurvy until it was learned that the patient was taking 25 mg of ascorbic acid per day via her daily vitamin supplement.

Laboratory Results

Parameter	Patient	Normal Range
Hb	9.0 g/dL	12–16 g/dL
Mean RBC volume	79 fL	80–100 fL
White cells	6.3×10^9/L	$5–10\times10^9$/L
Platelets	320×10^9/L	$150–300\times10^9$/L
Plasma iron	22 µg/dL	72–180 µg/dL
Total iron-binding capacity	123 µg/dL	246–375 µg/dL
Calcium	7.6 mg/dL	8.4–10.4 mg/dL
Inorganic phosphate	2.4 mg/dL	2.4–4.3 mg/dL
Folate	1.6 ng/mL	3–20 ng/mL
Vitamin B_{12}	110 ng/L	150–1000 ng/L
Prothrombin time	273 s	13 s (control)
Thrombin time	10 s	10 s (control)

When her abnormal prothrombin time was noted, specific coagulation assays were performed. These showed that the activity of each of the vitamin K–dependent factors (factors II, VII, IX, and X) was <1% of the normal level and that the activity of factor V was 76% of normal.

91. Fiore, L.D., Sccola, M.A., Cantillon, C.E., et al., 2001. J. Thromb. Thromboysis 11, 175–183.

92. Wilkins, K., DeKoven, J., Assaad, D., 2000. J. Cutan. Med. Surg. 4, 164–168.

93. Subcutaneous hemorrhages.

A xylose tolerance test (to measure small bowel absorption), performed with a single oral dose of 5 g of xylose, showed tolerance within normal limits. A stool culture showed normal fecal flora. The patient was then given phylloquinone (10 mg daily, administered intravenously, for 3 days) and showed a complete recovery of all coagulation factor activities to normal. She was given a high-protein/high-energy diet supplemented with $FeSO_4$ and, for a week, daily oral doses of 10 mg of phylloquinone. Her diarrhea subsided, her wounds healed, and she returned to normal health.

Case 2

A 55-year-old man with arteriosclerotic heart disease and type IV hyperlipoproteinemia was admitted to the hospital with a hemorrhagic syndrome. Six months earlier, he had suffered a myocardial infarction[94] complicated by pulmonary embolism[95] for which he was treated with heparin[96] followed by warfarin. Two months earlier, he had been admitted for a cardiac arrhythmia, at which time his physical examination was normal and chest radiograph showed no abnormalities, but his electrocardiogram showed first-degree atrioventricular block[97] with frequent premature ventricular contractions. At that time, he was taking 5 mg of warfarin per day.

Laboratory Findings 2 Months Before Third Admission[a]

Parameter	Patient	Normal
Prothrombin time	16.6 s	12.7 s
Plasma triglycerides	801 mg/dL	20–150 mg/dL
Serum cholesterol	324 mg/dL	150–250 mg/dL

[a]Blood count, blood urea nitrogen, blood bilirubin, and urinalysis were all normal.

He was treated with warfarin (5 mg/day), digoxin,[98] dip henylhydantoin,[99] furosemide,[100] potassium chloride,[101] and clofibrate.[102] Within a month, quinidine gluconate[103] was substituted for diphenylhydantoin because the patient showed persistent premature ventricular beats, but that drug was discontinued because of diarrhea and procainamide[34] was used instead. At that time, his prothrombin time was 31.5 s, and his warfarin dose was reduced first to half the original dose and then to one-quarter of that level.

At the time of the third admission, the patient appeared well nourished, but had ecchymoses on his arms, abdomen, and pubic area. He had been constipated with hematuria[104] for the preceding 2 days. His physical examination was unremarkable except for occasional premature beats, and his laboratory findings were similar to those observed on his previous admission, with the exception that his prothrombin time had increased to 36.6 s. In questioning the patient, it was learned that he had been taking orally as much as 1200 mg of all-*rac*-α-tocopheryl acetate each day for the preceding 2 months.

Both his warfarin and vitamin E treatments were discontinued, and 2 days later his prothrombin time had dropped to 24.9 s and his ecchymoses began to clear. The patient consented to participate in a clinical trial of vitamin E (800 mg of all-*rac*-α-tocopheryl acetate per day) in addition to the standard regimen of warfarin and clofibrate. The results were as follows:

Effect of Vitamin E on the Activities of the Patient's Coagulation Factors[a]

Activity	Initial Value	+Vit E, 6 weeks	−Vit E, 1 week	Normal Range
Factor II (prothrombin)	11	7	21	60–150
Factor VII[a]	27	16	23	50–150
Factor X[a]	15	10	—	50–150
Prothrombin time (sec)	20.7	29.2	22.3	11.0–12.5

[a]% mean of normal.

Case Questions

1. What signs indicated vitamin K–related problems in each case?
2. What factors probably contributed to the vitamin K deficiency of the patient in case one? Why was phylloquinone, rather than menadione, chosen for treatment of that patient?
3. What factors may have contributed to the coagulopathy of the patient in case two? What might be the basis of the effect of high levels of vitamin E seen in that case?

13. STUDY QUESTIONS AND EXERCISES

1. Construct a concept map to illustrate the ways in which vitamin K affects blood coagulation.
2. Construct a decision tree for the diagnosis of vitamin K deficiency in a human or animal.
3. What features of the chemical structure of vitamin K relate to its metabolic function?

94. Dysfunction because of necrotic changes resulting from an obstruction of a coronary artery.
95. Obstruction or occlusion of a blood vessel by a transported clot.
96. A highly sulfated mucopolysaccharide with specific anticoagulant properties.
97. Impairment of normal conduction between the atria and ventricles.
98. A cardiotonic.
99. A cardiac depressant (and anticonvulsant).
100. A diuretic.
101. That is, to correct for the loss of K^+ induced by the diuretic.
102. An antihyperlipoproteinemic.
103. A cardiac depressant (antiarrhythmic).
104. The presence of blood in the urine.

4. Discuss factors affecting the vitamin K requirement of humans, including infants.
5. What relevance to their vitamin K nutrition would you expect of the rearing of experimental animals in a germ-free environment or fed a fat-free diet?
6. How does the concept of a coenzyme relate to vitamin K?

RECOMMENDED READING

Berkner, K.L., 2008. Vitamin K-dependent carboxylation. Vit. Horm. 78, 131–156.

Beulen, J.W.J., Booth, S.L., van den Heuval, E.G., et al., 2013. The role of menaquinones (vitamin K_2) in human health. Br. J. Nutr. 110, 1357–1368.

Booth, S.L., 2009. Roles for vitamin K beyond coagulation. Ann. Rev. Nutr. 29, 89–110.

Bügel, S., 2008. Vitamin K and bone health in adult humans. Vit. Horm. 78, 393–416.

Dahlbäck, B., Villoutreix, B.O., 2005. The anticoagulation protein C pathway. FEBS Letts 579, 3310–3316.

Danziger, J., 2008. Vitamin K-dependent proteins, warfarin, and vascular calcification. Clin. J. Am. Soc. Neprhol. 3, 1504–1510.

Denisova, N.A., Booth, S.A., 2005. Vitamin K and sphingolipid metabolism: evidence to date. Nutr. Rev. 63, 111–121.

Ferland, G., 2012. Vitamin K. In: Erdman, J.W., Macdonald, I.A., Zeisel, S.H. (Eds.), Present Knowledge in Nutrition, tenth ed. ILSI Press, Washington, DC, pp. 230–247 (Chapter 15).

Greer, F.R., 2010. Vitamin K the basics – what's new? Early Hum. Devel. 86, S43–S47.

Iwamoto, J., 2006. Vitamin K_2 therapy for postmenopausal osteoporosis. Nutrients 6, 1971–1980.

Kaneki, M., Hosoi, T., Ouchi, Y., et al., 2006. Pleiotropic actions of vitamin K: protector of bone health and beyond? Nutr. 22, 845–852.

Mizuta, T., Ozaki, I., 2008. Hepatocellular carcinoma and vitamin K. Vit. Horm. 78, 435–442.

Napolitano, M., Mariani, G., Lapecorella, M., 2010. Hereditary combined deficiency of the vitamin K-dependent clotting factors. J. Rare Dis. 5, 21–29.

Nelsestuen, G.L., Shah, A.M., Harvey, S.B., 2000. Vitamin K-dependent proteins. Vit. Horm. 58, 355–389.

Oldenburg, J., Morinova, M., Műller-Reible, C., et al., 2008. The vitamin K cycle. Vit. Horm. 78, 35–62.

Palaniswamy, C., Aronow, A., Khanagavi, J., et al., 2014. Vitamin K and vascular calcification. In: Dakshinamurti, K., Dakshinamurti, S. (Eds.), Vitamin-Binding Proteins: Functional Consequences. CRC Press, New York, pp. 157–168 (Chapter 9).

Schurgers, L.J., Cranenburg, E.C.M., Vermeer, C., 2008. Matrix Gla-protein: the calcification inhibitor in need of vitamin K. Thromb. Haemost. 100, 593–603.

Schurgers, L.J., Utto, J., Reutelingsperger, C.P., 2013. Vitamin K-dependent carboxylation of matrix Gla-protein: a crucial switch to control ectopic mineralization. Trends Molec. Med. 19, 217–226.

Shearer, M.J., Fu, X., Booth, S., 2012. Vitamin K nutrition, metabolism, and requirements: current concepts and future research. Adv. Nutr. 3, 182–195.

Stafford, D.W., 2005. The vitamin K cycle. Thromb. Haemost. 3, 1873–1878.

Suttie, J.W., 2014. Vitamin K. In: Zemplini, J., Suttie, J.W., Gregory, J.F., Stover, P.J. (Eds.), Handbook of Vitamins, fifth ed. CRC Press, New York, pp. 89–123 (Chapter 3).

Suttie, J.W., 2009. Vitamin K in Health and Disease. CRC Press, New York. 224 pp.

Vermeer, C., Shearer, M.J., Zitterman, A., et al., 2004. Beyond deficiency: potential benefits of increased intakes of vitamin K for bone and vascular health. Eur. J. Nutr. 43, 325–335.

Wallin, R., 2013. Vitamin K. In: Stipanuk, M.H., Caudill, M.A. (Eds.), Biochemical, Physiological and Molecular Aspects of Human Nutrition, third ed. Elsevier, New York, pp. 655–669 (Chapter 28).

Walther, B., Karl, J.P., Booth, S.L., et al., 2013. Menaquinones, bacteria, and the food supply: the relevance of dairy and fermented food products to vitamin K requirements. Adv. Nutr. 4, 463–473.

Chapter 10

Vitamin C

Chapter Outline

Anchoring Concepts

1. Vitamin C is the generic descriptor for all compounds exhibiting qualitatively the biological activity of ascorbic acid.
2. Vitamin C-active compounds are hydrophilic and have an oxidizable/reducible 2,3-enediol grouping.
3. Deficiencies of vitamin C are manifest as connective tissue lesions (e.g., capillary fragility, hemorrhage, muscular weakness).

I still had a gram or so of hexuronic acid. I gave it to [Svirbely] to test for vitaminic activity. I told him that I expected he would find it identical with vitamin C. I always had a strong hunch that this was so but never had tested it. I was not acquainted with animal tests in this field and the whole problem was, for me, too glamorous, and vitamins were, to my mind, theoretically uninteresting. "Vitamin" means that one has to eat it. What one has to eat is the first concern of the chef, not the scientist. Anyway, Svirbely tested hexuronic acid...after one month the result was evident: hexuronic acid was vitamin C.

Albert Szent-Györgyi[1]

LEARNING OBJECTIVES

1. To understand the nature of the various sources of vitamin C.
2. To understand the means of vitamin C synthesis by most species.
3. To understand the means of enteric absorption and transport of vitamin C.
4. To understand the functions of vitamin C in connective tissue metabolism, in drug and steroid metabolism, and in mineral utilization.
5. To understand the physiologic implications of low and high intakes of vitamin C.

VOCABULARY

Antioxidant
Ascorbate
Ascorbate–cytochrome b_5 reductase
Ascorbate phosphate
Ascorbate sulfate
Ascorbic acid
Ascorbyl free radical
Carnitine
Cholesterol 7α-hydroxylase
Collagen
Dehydroascorbic acid
Dehydroascorbic acid reductase
DNA oxidation
Dopamine β-monooxygenase
Ecchymoses
2,3-Enediol
Elastin
Erythorbic acid
Glucose transporters (GLUTs)
Glucuronic acid pathway
Guinea pig
L-gulonolactone oxidase

1. Albert Szent-Györgyi de Nagyrápolt (1893–1986) was a Hungarian physiologist credited with the discovery of vitamin C and elucidating the citric acid cycle in metabolism for which he received the 1937 Nobel Prize in physiology or medicine.

The Vitamins. http://dx.doi.org/10.1016/B978-0-12-802965-7.00010-1

Histamine
Homogentisate 1,2-dioxygenase
Hydroxylysine
4-Hydroxyphenylpyruvate
Hydroxyproline
Hypoascorbemia
Indian fruit bat
Insulin
Iron
Ischemia–reperfusion injury
Lipid peroxidation
Lordosis
Lysyl hydroxylase
Moeller–Barlow disease
Monodehydroascorbate
Monodehydroascorbate reductase
Nitric oxide
Oxalic acid
Oxaluria
Peptidylglycine α-amidating monooxygenase
Petechiae
Prolyl hydroxylases
Prooxidant
Protein oxidation
Rebound scurvy
Red-vented bulbul
L-saccharoascorbic acid
Semiascorbic acid
Scoliosis
Scurvy
Sodium-dependent vitamin C transporters (SVCTs)
Systemic conditioning
ε-N-trimethyllysine hydroxylase and γ-butyrobetaine hydroxylase
Tropoelastin
Tyrosine
Vitamin C

1. THE SIGNIFICANCE OF VITAMIN C

Vitamin C is a dietary essential for only a few species, which, by virtue of a single enzyme deficiency, cannot synthesize it. For most species, **ascorbic acid** is a normal metabolite of glucose, not an essential dietary constituent. Whether it is obtained from exogenous sources biosynthesized by the host, ascorbic acid is important for several physiological functions. Many, if not all, of these functions involve its redox characteristics, such that ascorbic acid, the major water-soluble antioxidant in plasma and tissues, functions with tocopherols, reduced glutathione, and other factors in the antioxidant protection of cells and is thought to support the redox recycling of α-tocopherol and promote the utilization of dietary nonheme iron. It also supports the maintenance of enzyme-bound metals in oxidation states appropriate for their enzymatic functions in the biosynthesis of collagen, carnitine, and nonepinephrine. Compromise of these functions underlies the pathophysiology of vitamin C deficiency. Hypovitaminosis C affects some 5–15% of people worldwide; estimated prevalence in some industrialized countries has been higher.[2] Other beneficial health effects of ascorbic acid have been reported: reductions in hypertension, atherogenesis, diabetic complications, colds and other infections, and carcinogenesis. Although some of these claims have become widely accepted, the empirical evidence remains incomplete for many.

2. PROPERTIES OF VITAMIN C

The term **vitamin C** describes all compounds exhibiting the biological activity of **ascorbic acid** (2,3-didehydro-l-threo-hexano-1,4-lactone; also **L-ascorbic acid**).[3] The vitamin also occurs in the oxidized form, **L-dehydroascorbic acid** or **dehydroascorbic acid**. Biological activity depends on this 6-carbon lactone having a ***2,3*-enediol** structure.

Chemical structure of vitamin C:

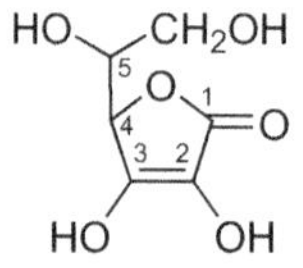

Ascorbic acid

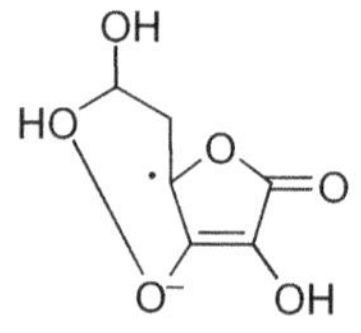

Semidehydroascorbic acid

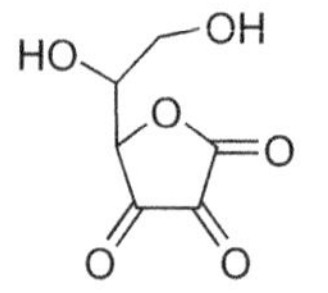

Dehydroascorbic acid

Vitamin C Chemistry

Ascorbic acid is a dibasic acid (pK_a values,[4] 4.1 and 11.8) because both enolic hydroxyl groups can dissociate. It

2. Hampl, J.S., Taylor, C.A., Johnston, C.S., 2004. Am. J. Pub. Health 94, 870–875; Mosdol, A., Erens, B., Brunener, E.J., 2008. J. Pub. Health 30, 456–460; Cahill, L., Corey, P.N., El-Sohemy, A., 2009. Am. J. Epidemiol. 170, 464–471.

3. Formerly, **hexuronic acid**.

4. The quantitative strength of an acid in solution is expressed in terms of its dissociation constant, K_a. The dissociation behavior of an acid is described: $HA+H_2O \leftrightarrow A^-+H_3O^+$, and K_a,=($[A^-]$ $[H_3O^+]$)/($[HA]$ $[H_2O]$), or its log value, pK_a, $-\log 10\ K_a$.

forms salts, the most important of which are the sodium and calcium salts, the aqueous solutions of which are strongly acidic. A strong reducing agent, ascorbic acid is oxidized under mild conditions to dehydroascorbic acid via the radical intermediate **semidehydroascorbic acid** (also, **monodehydroascorbic acid**). The semiquinoid ascorbic acid radical is a strong acid (pK_a −0.45); after the loss of a proton, it becomes a radical anion that, owing to resonance stabilization, is relatively inert but disproportionates to ascorbic acid and dehydroascorbic acid. Thus, the three forms (ascorbic acid, semidehydroascorbic acid, and dehydroascorbic acid) compose a reversible redox system. Thus, it is an effective quencher of free radicals such as singlet oxygen ($^1O_2$). It reduces ferric (Fe^{3+}) to ferrous (Fe^{2+}) iron (and other metals analogously), and the superoxide radical (O_2^-) to H_2O_2 and is oxidized to monodehydroascorbic acid in the process. Ascorbic acid complexes with disulfides (e.g., oxidized glutathione, cystine) but does not reduce those disulfide bonds. At physiological pH, ascorbic acid exists primarily as the ascorbate monoanion, while its reduced form, dehydroascorbic acid, is not ionized.

Dehydroascorbic acid is not ionized in environments of weakly acidic or neutral pH; therefore, it is relatively hydrophobic and is better able to penetrate membranes than is ascorbic acid. In aqueous solution, dehydroascorbic acid is unstable and is degraded by hydrolytic ring opening to yield 2,3-dioxo-L-gulonic acid. Dehydroascorbic acid reacts with several amino acids to form brown colored products, a reaction contributing to the spoilage of food.

TABLE 10.1 Relative Biopotencies of Vitamin C-Active Substances

Compound	Relative Biopotency (%)
Ascorbic acid	100
Ascorbyl-5,6-diacetate	100
Ascorbyl-6-palmitate	100
6-Deoxy-6-chloro-L-ascorbic acid	70–98
Dehydroascorbic acid	80
6-Deoxyascorbic acid	33
Ascorbic acid 2-sulfate	±[a]
Isoascorbic acid	5
L-glucoascorbic acid	3

[a]*This form is active in fishes, which have an intestinal sulfohydrase that liberates ascorbic acid; it is inactive in guinea pigs, rhesus monkeys, and humans, which lack that enzyme.*

Vitamin C Biopotency

Several synthetic analogues of ascorbic acid have been made. Some (e.g., 6-deoxy-L-ascorbic acid) have biological activity, whereas others (e.g., D-isoascorbic acid and L-glucoascorbic acid) have little or no activity. Several esters of ascorbic acid are converted to the vitamin in vivo and thus have good biological activity (e.g., ascorbyl-5,6-diacetate, ascorbyl-6-palmitate, 6-deoxy-6-chloro-L-ascorbic acid; see Table 10.1). Esters of the C-2 position show variable vitamin C activity among different species.

3. SOURCES OF VITAMIN C

Biosynthesis of Ascorbic Acid

Most higher animals (and probably all green plants) can synthesize vitamin C. They make it from glucose via the **glucuronic acid pathway** (Fig. 10.1, Table 10.2). The enzymes of this pathway are localized in the kidneys of amphibians, reptiles, egg-laying mammals, and the more primitive orders of birds; in both the kidneys and livers of many marsupials; but only in the livers of passerine birds and other mammals. The transfer of ascorbic acid synthesis from the kidney to the larger liver has been interpreted as an evolutionary

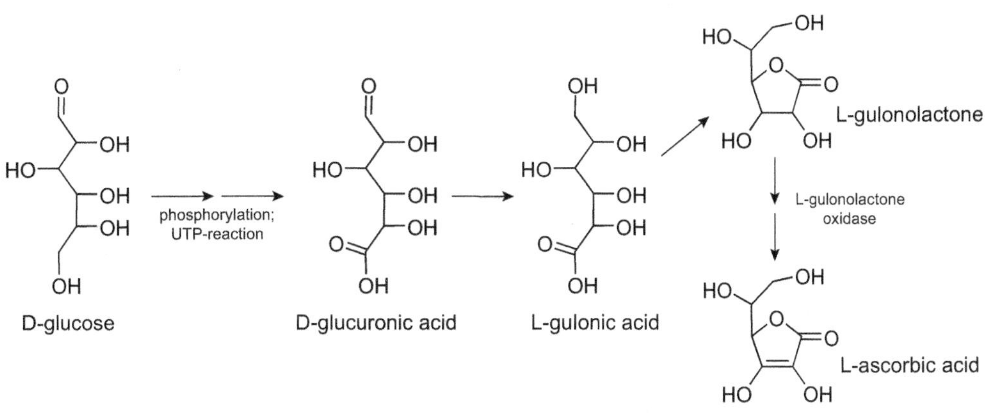

FIGURE 10.1 Biosynthesis of ascorbic acid.

TABLE 10.2 Estimated Rates of Ascorbic Acid Biosynthesis in Several Species

Species	Synthetic Rate, mg/kg BW	$T_{1/2}$,[a] days	Turnover, %/day
Mouse	125	1.4	50
Golden hamster	20	2.7	26
Rat	25	2.6	26
Rabbit	5	3.9	18
Guinea pig	0	3.8	18
Human	0	10–20	3

[a]*half-life in the body.*

adaptation that provided increased synthetic capacity to meet the increased needs associated with homeothermy. The biosynthesis of ascorbic acid is coupled to glycogenolysis. It can be stimulated by xenobiotic compounds including drugs (e.g., barbiturates, aminopyrine, antipyrine, chlorobutanol) and carcinogens (e.g., 3-methylcholanthrene, benzo-α-pyrene) due to induction of the glucuronic acid pathway, which is needed for xenobiotic detoxification by conjugation.[5] It can be inhibited by deficiencies of vitamins A or E, or biotin. Some species may not express this key enzyme early in development; the fetal rat, for example, is incapable of ascorbic acid biosynthesis until the 16th day of gestation. There is no evidence that ascorbic acid can be synthesized by the gut microbiome of any species.

Evolutionary loss of ascorbic acid biosynthetic capacity appears to have occurred in invertebrates, teleost fishes,[6] several species of birds (e.g., **red-vented bulbul**[7]), and some mammals (humans, other primates,[8] **guinea pigs**, most bats,[9] and a few mutant strains of rats[10]). These species do not express the last enzyme in the biosynthetic pathway, **L-gulonolactone oxidase.**[11] That microsomal flavoenzyme catalyzes the oxidation of L-gulonolactone[12] to L-2-ketogulonolactone, which, in turn, yields L-ascorbic acid by spontaneous isomerization. While all species studied have the gene, in some it is so highly mutated that it yields no gene product.[13] The loss of this single enzyme renders ascorbic acid, an otherwise normal metabolite, a vitamin. Therefore, **scurvy** can correctly be considered a congenital metabolic disease, **hypoascorbemia**.

Distribution in Foods

Vitamin C is widely distributed in both plants and animals, occurring mostly (80–90%) as **ascorbic acid** but also as **dehydroascorbic acid**. The proportions of both species tends to vary with food storage time, due to the time-dependent oxidation of ascorbic acid. Fruits, vegetables,[14] and organ meats (e.g., liver and kidney) are generally the best sources; only small amounts are found in muscle meats (Table 10.3). Plants synthesize L-ascorbic acid from carbohydrates; most seeds do not contain ascorbic acid but start to synthesize it on sprouting. Some plants accumulate high levels of the vitamin (e.g., fresh tea leaves, some berries, guava, rose hips). For practical reasons, citrus and other fruits are good daily sources of vitamin C, as they are generally eaten raw and are, therefore, not subjected to cooking procedures that can destroy vitamin C. Ascorbic acid is frequently added at low levels to processed foods to enhance shelf-life or preserve flavor. The analogue, **erythorbic acid,**[15] is also used as a food preservative. While it has no vitamin C activity, it can yield false positives in some analyses for plasma ascorbic acid.[16]

Stability in Foods

The vitamin C contents of most foods decrease dramatically during storage owing to the aggregate effects of several processes by which the vitamin can be destroyed (Table 10.4). Ascorbic acid is susceptible to oxidation to dehydroascorbic acid, which itself can be rapidly and irreversibly degraded at neutral pH by irreversible hydrolytic opening of the lactone ring to yield 2,3-diketogulonic acid. These reactions occur in the presence of O_2, even traces of metal ions, and

5. Because ascorbic acid synthesis and excretion are increased by exposure to xenobiotic inducers of hepatic, cytochrome *P*450-dependent, mixed-function oxidases (MFOs), it has been suggested that the urinary ascorbic acid concentration may be useful as a noninvasive screening parameter of MFO status.
6. Although most fish appear to be able to synthesize ascorbic acid, only carp (*Cyprinidae*) and Australian lungfish (*Neoceratodus forsteri*) appear to be able to do so at rates sufficient to meet their physiologic needs.
7. The bulbuls (*Pycnonotidae*) comprise 13 genera and 109 species distributed in Africa, Madagascar, and southern Asia. While the red-vented bulbul is often cited as being unable to biosynthesize ascorbic acid, it is not known how widely distributed in this family and the class *Aves* is the dietary need for vitamin C.
8. Gorilla, orangutan, gibbon, macaque, marmoset, owl monkey.
9. Including the **Indian fruit bat**, *Pteropus giganteus*, also known as the "flying fox."
10. A gulonolactone null mouse has been developed; and the osteogenic disorder Shionogi rat (ODS-*od*/*od*) derived from the Wistar strain has a dysfunctional form of that enzyme.
11. Whether loss of this enzyme may underlie inabilities of other species to synthesize ascorbic acid is still speculative.
12. Replacement (by injection) of this substrate prevents scurvy in guinea pigs.
13. This may be due to the presence of retrovirus-like sequences, identified in the human gene, that may have caused its activation. It has been suggested that mutations in this gene may have been driven by disadvantageous effects of H_2O_2 generated during the oxidation of gulono-1,4-lactone.
14. Historically, the potato was the best source of vitamin C in North America and Europe.
15. Also referred to as D-isoascorbic acid or D-araboascorbic acid.
16. This is not a problem for blood sampled after an overnight fast, as erythorbic acid is cleared from the blood within 12 h.

TABLE 10.3 Vitamin C Contents of Foods

Food	Vitamin C, mg/100 g
Fruits	
Apple	5
Banana	9
Cherry	7–10
Grapefruit	34
Guava	228
Lemon	53
Melons	8–37
Orange	59
Peach	7
Raspberry	26
Rose hips	426
Strawberry	59
Tangerine	27
Vegetables	
Asparagus	6
Broccoli	89
Cabbage	37
Carrot	3
Cauliflower	48
Celery	3
Collards	35
Corn	7
Kale	120
Leek	12
Potato	11
Onion	7
Pea	40
Parsley	133
Pepper	80–128
Cereals	(none)
Animal Products	
Beef	0
Milk cow	0–1
Milk, human	5

Adapted from Uncooked; USDA National Nutrient Database for Standard Reference, Release 28 http://www.ars.usda.gov/ba/bhnrc/ndl.

TABLE 10.4 Two-Day Storage Losses of Vitamin C

	% Lost	
Food	**4°C**	**20°C**
Beans	33	53
Cauliflower	8	26
Lettuce	36	42
Parsley	13	70
Peas	10	36
Spinach	32	80
Spinach (winter)	7	22

are enhanced by heat and conditions of neutral to alkaline pH. The vitamin is also reduced by exposure to oxidases in plant tissues. Therefore, substantial losses of vitamin C can occur during storage and are enhanced greatly during cooking. For example, stored potatoes lose 50% of their vitamin C within 5 mos. and 65% within 8 mos. of harvest. Apples and cabbage stored for winter can lose 50% and 40%, respectively, of their original vitamin C contents. Losses in cooking are usually greater with such methods as boiling, as the stability of ascorbic acid is much less in aqueous solution. For example, potatoes can lose 40% of their vitamin C content by boiling. Alternatively, quick heating methods can protect food vitamin C by inactivating oxidases, and acidic conditions stabilize dehydroascorbic acid.

Vitamin C Bioavailability

Vitamin C in most foods appears to have biological activities comparable to that of purified L-ascorbic acid at doses in the nutritional range (15–200 mg). At higher doses, bioavailability declines due to declining absorption efficiency. In humans, doses up to 200 mg are nearly completely absorbed, but doses of 1000 mg are utilized with only 50% efficiency. Because dehydroascorbic acid can be reduced metabolically to yield ascorbic acid (after enteric absorption and subsequent cellular uptake), both forms present in foods have vitamin activity. Several synthetic ascorbic acid derivatives also have vitamin C activity and offer advantages of superior chemical stability. Forms, such as ascorbate 2-sulfate, ascorbate 2-monophosphate, ascorbate 2-diphosphate, and ascorbate 2-triphosphate (mixtures of the latter three are referred to as **ascorbate polyphosphate**), are useful as vitamin C supplements for fish diets where the intrinsic instability of ascorbic acid in aqueous environments is a problem. The more highly biopotent of these vitamers appears to be effectively hydrolyzed in the digestive tract and tissues to yield ascorbic acid (Table 10.5).

TABLE 10.5 Vitamin C-Active Derivatives of Ascorbic Acid

Strongly Biopotent[a]	Weakly Biopotent[b]
Ascorbic acid 2-*O*-α-glucoside	L-ascorbyl palmitate
6-Bromo-6-deoxy-L-ascorbic acid	L-ascorbyl-2-sulfate
L-ascorbate 2-phosphate	L-ascorbate-*O*-methyl ether
L-ascorbate 2-triphosphate	

[a]>50% antiscorbutic activity of ascorbic acid.
[b]<50% antiscorbutic activity of ascorbic acid.

4. ABSORPTION OF VITAMIN C

Species that can synthesize ascorbic acid do not have active transport mechanisms for its enteric absorption. They absorb the vitamin across the mucosal brush border strictly by passive diffusion.

Species unable to synthesize ascorbic acid (e.g., humans, guinea pigs) absorb the vitamin by both passive and active means. Passive diffusion is important at high doses. At low doses, the most important means of absorbing the vitamin involves saturable, carrier-mediated active transport mechanisms. Thus, the efficiency of absorption of physiological doses (e.g., ≤180 mg/day for a human adult) of vitamin C is high, 80–90%, and declines markedly at vitamin C doses greater than about 1 g.[17] The reduced and oxidized forms of the vitamin are absorbed by different mechanisms of active transport, which occur throughout the small intestine:

- **Na^+-dependent vitamin C transporters (SVCTs)**[18] move ascorbate by an electrogenic process involving two Na^+ ions per ascorbic acid molecule. This family of surface glycoproteins comprises multiple isoforms; SVCT1 is the predominant form expressed the intestinal mucosa where it is localized on the brush border. It is inhibited by aspirin.[19] Its genetic deletion does not block ascorbic acid uptake,[20] suggesting other means of absorbing the vitamin are available. A genetic variant has been associated with susceptibility to Crohn's disease.[21]
- **Glucose transporter 1 (GLUT1)** facilitates vitamin C uptake by mucosal cells. The uptake of dehydroascorbic acid is 10- to 20-fold faster than that of ascorbic acid.[22] Upon entry into the cell, dehydroascorbic acid is quickly reduced to ascorbate.

Efflux of ascorbate across the basolateral side of the mucosal epithelial cell into the portal circulation can occur in several ways:

- by **SVCT2**-facilitated transport;
- by volume- or Ca^{+2}-sensitive **anion channels** that form pores in the plasma membrane;
- by **glutamate–ascorbate exchange**; and
- by **exocytosis** of ascorbate-containing vesicles and gap junction hemichannels.[23]

5. TRANSPORT OF VITAMIN C

Transport in Reduced Form

Vitamin C is transported in the plasma predominantly (80–90%) as ascorbate. Also present are small amounts of dehydroascorbic acid formed by the oxidation of ascorbate by diffusible oxidants of cellular origin (Fig. 10.2). Plasma ascorbate shows a sigmoid relationship with the level of vitamin C intake, saturation in humans is achieved at daily doses of 1000 mg or more (Fig. 10.3).[24] Levels in healthy humans are typically 30–70 μM and appear to be inversely related to adiposity, although that affect may simply reflect the fact that lower-energy diets tend to be richer in vitamin C-rich fruits and vegetables.[25]

Cellular Uptake

The uptake of vitamin C into cells occurs by the same mechanisms as those responsible for its enteric absorption. Uptake by simple diffusion is negligible due to the charge of ascorbate (which is ionized under physiological conditions) and the oil:water partitioning characteristics of dehydroascorbic acid, which excludes it from lipid membranes.

17. The efficiency of vitamin C absorption declines from about 75% of a 1-g dose, to about 40% of a 3-g dose, and about 24% of a 5-g dose; net absorption plateaus at 1–1.2 g at doses of at least 3 g.
18. These are members of the SLC23 human gene family.
19. For example, in humans, a 900-mg dose of aspirin blocks the expected rises in plasma, leukocyte, and urinary levels of ascorbic acid owing to a simultaneous dose of 500 mg of vitamin C.
20. Corpe, C.P., Tu, H., Eck, P., et al., 2010. J. Clin. Invest. 120, 1069–1083.
21. This variant (rs10063949-G) compromises the absorption and cellular uptake of ascorbate, which is thought to be needed to address the oxidative stress induced by excess ROS produced from aberrant immune responses in individuals genetically susceptible to inflammatory bowel disease (Shaghaghi, M.A., Bernstein, C.N., León, A.S., et al., 2014. Am. J. Clin. Nutr. 99, 378–383.).
22. Studies with cultured cells have shown that D-isoascorbic acid has only 20–30% of the activity of L-ascorbic acid in stimulating collagen production. The basis of this difference involves the much slower cellular uptake of the D-form, as, once inside the cell, both vitamers behaved almost identically.
23. See review: Corti, A., Casini, A.F., Pompella, A., 2010. Arch. Biochem. Biophys. 500, 107–115.
24. Levine, M., Conry-Cantilena, Wang, Y., et al., 1996. Proc. Natl. Acad. Sci. U.S.A. 93, 3704–3079) found that 200-mg/day doses produced only 80% saturation and that RDA-level doses supported plasma ascorbic acid concentrations on the lower third of the response curve.
25. Plasma ascorbic acid levels were inversely related to waist-to-hip ratio and to waist and hip circumferences but not to body mass index in a large European cohort (Canoy, D., Wareham, N., Welch, A., et al., 2005. Am. J. Clin. Nutr. 82, 1203–1209).

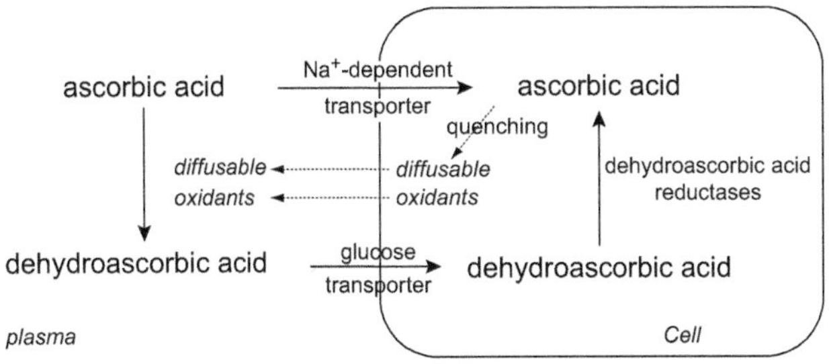

FIGURE 10.2 Redox cycling of ascorbic acid.

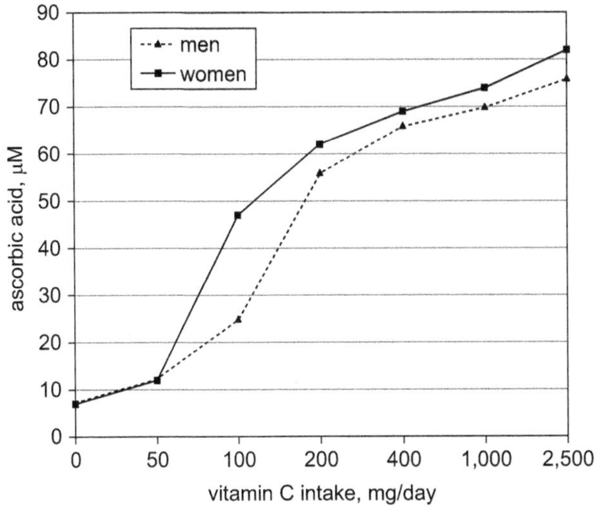

FIGURE 10.3 Relationship of plasma ascorbic acid (steady state) level and vitamin C intake. *From Levine, M., Wang, Y., Padayatty, S.J., et al., 2001. Proc. Natl. Acad. Sci. U.S.A. 98, 9842–9846.*

Nevertheless, cells accumulate ascorbic acid to levels 5- to 100-fold those of plasma; human cells become saturated at vitamin C intakes of about 100 mg/day. This is accomplished by facilitated uptake:

- **Ascorbate is taken up by SVCT1 and SVCT2.**[26] Each transporter is highly specific for ascorbate. SVCT1 is a high-affinity, high-capacity transporter expressed in epithelial tissues including the intestine, liver, and kidney; it is responsible for most ascorbate transport. SVCT2 is a high-affinity, low-capacity transporter expressed in brain, lung, heart, eye, placenta, neuroendocrine and exocrine tissues, and endothelial tissues. It appears to be responsible for the accumulation of ascorbate in tissues[27] and is upregulated under conditions of oxidative stress.[28] Genetic deletion of SVCT2 results in low-placental vitamin C transport and fetal death.[29] Both isoforms contain multiple potential N-glycosylation and protein kinase C (PKC) phosphorylation sites, suggesting regulation via glycosylation and/or PKC pathways. In the absence of ascorbate, the SVCTs can facilitate the unitransport of Na^+, allowing that ion to leak from cells. The SVCTs are noncompetitively inhibited by flavonoids and can be affected by cytokines and steroids. Their expression varies inversely with intracellular ascorbate levels.[30] SVCT1 appears to be principally involved in the maintenance of whole body vitamin C homeostasis by affecting enteric absorption and renal reabsorption;[31] while SVCT2 appears to be of principal importance in the protection of metabolically active tissues from oxidative stress.
- **Dehydroascorbic acid is taken up by GLUTs.** These transmembrane proteins are widely expressed (GLUT1 ubiquitously;[32] GLUT3 predominantly in brain and nerve cells; and GLUT4 predominantly in adipose, and cardiac and skeletal muscle).[33] They have similar affinities for ascorbate and glucose. Therefore, by competing for uptake by the transporter, hyperglycemia inhibits dehydroascorbic acid uptake. Accordingly, diabetics can have abnormally high plasma levels of dehydroascorbic acid.[34] Because the dehydroascorbic acid content of plasma is typically low, GLUT-facilitated uptake is thought to be a means of rapidly scavenging the oxidized form of the vitamin [e.g., as a result of reactive oxygen species (ROS) released by phagocytic cells] from the circulation so that it may be recycled to ascorbate intracellularly. This is best demonstrated by human erythrocytes, which have GLUT but no SVCTs.

Tissue Distribution

Nearly all tissues accumulate vitamin C, including some that lack ascorbic acid-dependent enzymes (Table 10.6).

26. These members of the solute carrier family 2 are designated SLC23A1 and SLC23A2, respectively.
27. Its genetic deletion rendered the mouse mortally depleted of ascorbate in every tissue (Sotirious, S., Gispert, S., Cheng, J. et al., 2002. Nature Med. 8, 514–517).
28. May, J.M., Li, L., Qu, Z., 2010. Mol. Cell. Biochem. 343, 217–222.
29. Harrison, F.E., Dawes, S.M., Meredith, M.E., et al., 2010. Free Rad. Biol. Med. 49, 821–829.
30. MacDonald, L., 2002. Br. J. Nutr. 87, 97–100.
31. Genetic deletion of SVCT1 resulted in massive losses of ascorbate from the plasma into the urine (Corpe, C.P., Tu, H., Eck, P., et al., 2010. J. Clin. Invest. 120, 1069–1083).
32. Congenital deficiency of GLUT1, a rare condition, is manifest in infancy as seizures and delayed development, presumably due to insufficient supply of glucose to the brain.
33. These are also members of the solute carrier family 2, designated SLC2A1, SLC2A3, and SCL2A4, respectively. GLUT2, previously considered only as a high-affinity transporter of glucosamine with low affinities for glucose and fructose, has been shown to transport dehydroascorbate with low affinity in hepatocytes (Mardones, L. Ormazabal, V., Romo, X., et al., 2011. Biochem. Biophys. Res. Commun. 410, 7–12).
34. Impaired cellular uptake of vitamin C, due to competition with glucose, may contribute to pathology in diabetes.

TABLE 10.6 Ascorbic Acid Concentrations of Human Tissues

Tissue	Ascorbic Acid, mg/100 g
Adrenals	30–40
Pituitary	40–50
Liver	10–16
Thymus	10–15
Lungs	7
Kidneys	5–15
Heart	5–15
Muscle	3–4
Brain	3–15
Pancreas	10–15
Lens	25–31
Plasma	0.4–1

Certain cell types (e.g., peripheral mononuclear leukocytes) can accumulate millimolar concentrations. Tissue levels are decreased by virtually all forms of stress, which also stimulate the biosynthesis of the vitamin in those animals able to do so.[35] The concentration of ascorbate in the adrenals is particularly high (72–168 mg/100 g in the cow); one-third of the vitamin is concentrated at the site of catecholamine formation, from which it is released with newly synthesized corticosteroids in response to stress.[36] The ascorbate concentration of brain tissue also tends to be high (5–28 mg/100 g), particularly in regions also rich in catecholamines.[37] Brain ascorbate levels are among the last to be affected by dietary deprivation of vitamin C in nonsynthesizing species. A relatively large amount of ascorbate is also found in the eye where it is thought to protect critical protein sulfhydryl groups from oxidation.[38] The ascorbate levels of white blood cells reach plateaus at vitamin C doses of 2 g/day, with lymphocytes, platelets, monocytes, and neutrophils showing decreasing plateau levels in that order[39]. Leukocyte ascorbate concentrations correlate with tissue levels of the vitamin.[40] There is no stable reserve of vitamin C; excesses are quickly excreted. At saturation, the total body pool of the human has been estimated to be 1.5–5 g,[41] the major fractions being found in the liver and muscles by virtue of their relatively large masses.

35. The ascorbic acid content of brown adipose tissue of rats can increase 60% during periods of cold stress.

36. That is, in response to the release of adrenocorticotropic hormone (ACTH).

37. Phenylalanine and tyrosine metabolites with hormonal functions, including epinephrine (adrenaline), norepinephrine (noradrenaline), and dopamine.

38. Lenses of cataract patients have lower lens ascorbic acid concentrations (e.g., 0–5.5 mg/100 g) than those of healthy patients (e.g., 30 mg/100 g).

39. Levine, M., Wang, Y., Padayatty, S.J., et al., 2001. Proc. Nat. Acad. Sci. U.S.A. 98, 9842–9846.

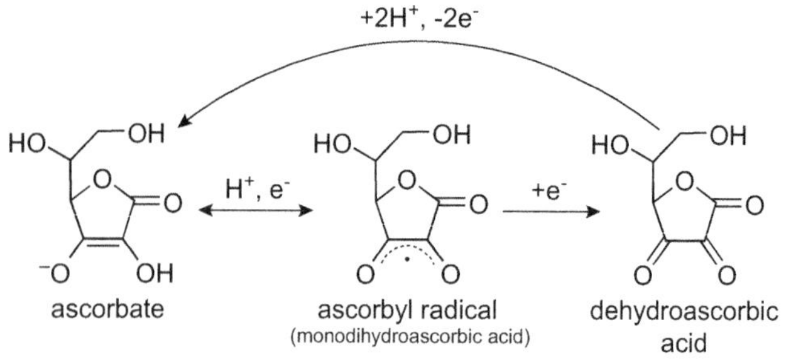

FIGURE 10.4 Oxidation–reduction reactions of vitamin C.

6. METABOLISM OF VITAMIN C

Oxidation

Ascorbate can be oxidized in vivo by two successive losses of single electrons. The first monovalent oxidation results in the formation of the **ascorbyl free radical**.[42] That partially reduced form can establish a reversible electrochemical couple with ascorbate, or it can be further oxidized to dehydroascorbic acid (Fig. 10.4).

The partially oxidized form, ascorbyl radical, can be reduced to dehydroascorbic acid by the cytosolic selenoenzyme thioredoxin reductase to from dehydroascorbic acid. It can also undergo nonenzymatic dismutation of 2 mol of ascorbyl radical to form equimolar amounts of dehydroascorbic acid and ascorbate. The completely oxidized form, dehydroascorbic acid, is unstable at physiological pH. If not reduced back to the ascorbate state, it undergoes irreversible ring opening 2,3-diketo-L-gulonic acid, which can undergo decarboxylation to CO_2 and five-carbon fragments (xylose, xylonic acid, lyxonic acid), or be oxidized to form oxalic acid and 4-C fragments (e.g., threonic acid). In addition, the formation of L-ascorbic acid 2-sulfate from ascorbic acid occurs in humans, fishes, and rats; and the oxidation of the ascorbate 6-carbon to form L-saccharoascorbic acid has been demonstrated in monkeys. Ascorbate may also undergo oxidation by reaction with tocopheroxyl or urate radicals (Fig. 10.5).

40. Leukocyte ascorbic acid concentrations are usually greater in women than men, and decrease with age and in some diseases.

41. The first signs of scurvy are not seen until this reserve is depleted to 300–400 mg.

42. The ascorbyl radical is also called monodehydroascorbic acid; it is relatively stable with a rate constant for its decay of about $10^5 M^{-1} s^{-1}$.

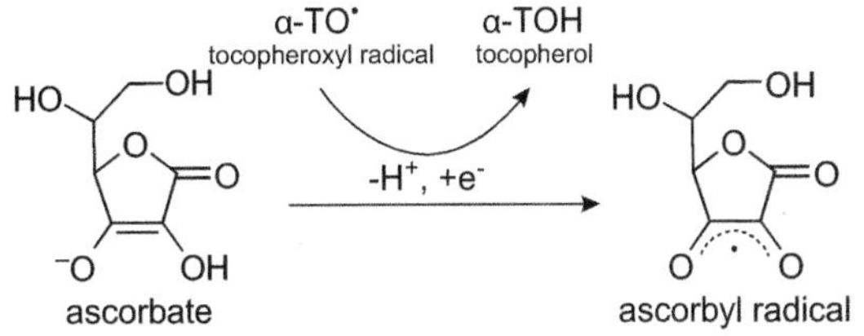

FIGURE 10.5 Coupling of ascorbate oxidation to reduction of α-tocopheroxyl radical.

Ascorbate Regeneration

Ascorbate can be regenerated from dehyroascorbic acid. Multiple reductase activities in mitochondria, endoplasmic reticulum, and erythrocyte plasma membranes promote a favorable ascorbate redox potential, indirectly preserving other antioxidants such as tocopherol. With the presence of an effective scavenging system specific for the oxidized form, a cycle is effectively established to maintain intracellular levels of the reduced vitamin (Fig. 10.2).

Although dehydroascorbic acid appears to have no metabolic function per se, its recycling to ascorbate renders it physiologically important in protecting cells from ROS. This appears to be the basis of ascorbate-stimulating osteoid formation by osteoblasts in response to ROS released by osteoclasts and of vitamin C protection of intestinal mucosa and mitochondria against ROS generated there. Impairments in this recycling can occur in uncontrolled diabetes due to excessive plasma glucose, which competes with dehydroascorbic acid for cellular uptake by GLUTs and leads to reduced intracellular ascorbate levels that weaken antioxidant defenses.

Excretion

Ascorbate is thought to pass unchanged through the glomeruli and to be actively reabsorbed in the tubules by SCVT1. Little, if any, ascorbic acid is excreted in the urine of humans consuming less than 100mg/day and only one-fourth of the dose is excreted at twice that intake. At doses greater than about 500mg/day (i.e., when blood ascorbic acid concentrations exceed 1.2–1.8mg/dL), virtually all ascorbic acid above that level is excreted unchanged in the urine, thus producing no further increases in body ascorbate stores. The fractional excretion of a parenteral dose of ascorbic acid approaches 100% at doses >2g.

The epithelial cells of the renal tubules reabsorb dehydroascorbic acid after it has been filtered from the plasma. Species vary in their routes of disposition of the vitamin. Guinea pigs and rats degrade it almost quantitatively to CO_2,[43] which is lost across the lungs. Humans, however, normally degrade only a very small amount via that route,[44] excreting mostly ascorbic acid, dehydroascorbic acid, and 2,3-diketogulonic acid, with relatively small amounts of oxalate and ascorbate 2-sulfate. Excretion of oxalate is relevant to risk of renal stone formation. It has been thought that healthy individuals convert no more than 1.5% of ingested ascorbic acid to oxalic acid within 24h; however a careful study showed that subjects given ascorbic acid intravenously excreted <0.5% as oxalate.[45] It is estimated that, of the oxalate excreted daily by humans consuming nutritional amounts of vitamin C (e.g., 30–40mg), 35–50% comes from ascorbic acid degradation, the balance coming from glycine and glyoxylate. Not all individuals show increased urinary oxalate excretion in response to ascorbic acid supplementation; only 40% of adults consuming very high doses (1000mg/day) of vitamin C increased their urinary oxalate levels more than 16% (Table 10.7).

TABLE 10.7 Effect of High-level Ascorbic Acid Supplementation on Urinary Oxalate Excretion

Subject Group (n)	Treatment[a]	Oxalate, μmoles
Responders (19)	Control	513 ± 97
	Ascorbic acid, 1000 mg/day	707 ± 165[b]
Nonresponders (29)	Control	560 ± 110
	Ascorbic acid, 1000 mg/day	551 ± 129[b]

[a]*Each subject experienced alternating 6-day control and ascorbic acid treatments.*
[b]*Significantly different (p<.05) from control treatment within subject group.*
Adapted from Massey, L.K., Liebman, M., Kynast-Gales, S.A., 2005. J. Nutr. 135, 1673–1637.

Ascorbic acid is also excreted in the gastric juice. In healthy adults, that concentration is typically three times that of plasma, although it is low in patients with atrophic gastritis or *Heliobacter pylori* infection.[46]

7. METABOLIC FUNCTIONS OF VITAMIN C

The metabolic functions of vitamin C can be categorized as those of its properties as a biochemical antioxidants and those of its properties as an enzyme cosubstrate.

Antioxidant Functions

Ascorbic acid loses electrons easily and, because of its reversible monovalent oxidation to the ascorbyl radical,

43. The C-1 carbon of ascorbic acid is the main source of CO_2 derived from the vitamin, whereas C-1 and C-2 are the precursors of oxalic acid.

44. Degradation by this path is increased in some diseases and can then account for nearly half of ascorbic acid loss.

45. Robitaille, L., Mamer, O.A., Miller, Jr., W.H., et al., 2009. Metab. Clin. Exp. 58, 263–269.

46. Sobala, G.M., Schorah, C.J., Shires, S., et al., 1993. Gut 34, 1038–1041.

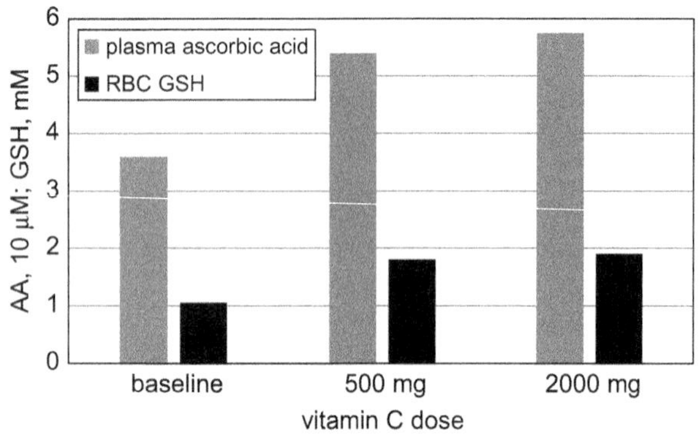

FIGURE 10.6 Enhancement of reduced glutathione (GSH) by vitamin C in men. *After Johnston, S.C., Meyer, C.G., Srilakshmi, J.C., 1993. Am. J. Clin. Nutr. 58, 103–105.*

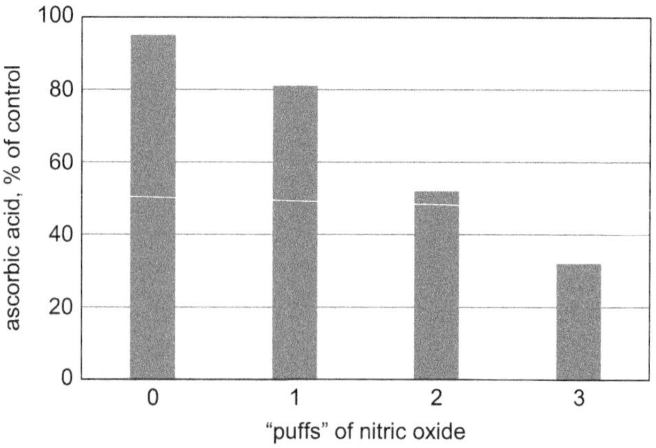

FIGURE 10.7 *In vitro* oxidation of ascorbic acid simulating the pro-oxidative effects of atmospheric nitric oxide. *After Eiserich, J.P., Cross, C.E., van der Vliet, A., 1997. In: Packer, L., Fuchs, J., (Eds.). Vitamin C in Health and Disease. Marcel Dekker, New York, pp. 399–412.*

it can serve as a biochemical redox system. The redox potential of the dehydroascorbic acid–ascorbate couple is in the range of 0.06–0.1V. That of the ascorbyl radical–ascorbate couple is −0.17V. These redox potentials mean that ascorbate can act as an **antioxidant** by reacting with free radicals and undergoing a single-electron oxidation to yield a relatively poorly reactive intermediate, the ascorbyl radical, which disproportionates to ascorbate and dehydroascorbic acid. In this way, ascorbate can reduce toxic ROS ($O_2^{\cdot -}$, $OH^{\cdot}$, $RO_2^{\cdot}$) and RNS ($NO_2^{\cdot}$). Those reactions are of fundamental importance in all aerobic cells, which must defend against the toxicity of the very element depended on as the terminal electron acceptor for energy production via the respiratory chain enzymes. One such reaction is important in extending the antioxidant protection to the hydrophobic regions of cells: ascorbate reduction of the semistable chromanoxyl radical, thus, regenerating the metabolically active form of the lipid antioxidant vitamin E.[47] Such quenching of oxidants protects glutathione in its reduced form (Fig. 10.6).

The antioxidant efficiency of ascorbate is significant at physiological concentrations of the vitamin (20–90 μM). Under those conditions, the predominant reaction is a radical chain-terminating one of ascorbate (AH^-) with a peroxyl radical to yield a hydroperoxide (ROOH) and ascorbyl radical (A^-), which proceeds to reduce a second peroxyl radical and yield the vitamin in its oxidized form, dehydroascorbic acid (A). At low concentrations of the vitamin, 2 moles of peroxyl radical are reduced for every mole of ascorbate consumed:

$$AH^- + RO^{\cdot} \rightarrow A^{\cdot -} + ROO^{\cdot} \ (+H^+ \rightarrow ROOH)$$

$$A^{\cdot -} + R'OO^{\cdot} \rightarrow A + R'OO^- \ (+H^+ \rightarrow R'OOH)$$

47. Evidence for such an effect comes from demonstrations in vitro of the reduction by ascorbic acid of the tocopheroxyl radical to tocopherol, as well as from findings in animals that supplemental vitamin C can increase tissue tocopherol concentrations and spare dietary vitamin E.

Dehydroascorbic acid is inherently unstable, with a half-life of only minutes in physiological conditions and undergoing ring opening to yield 3,4-diketogulonic acid. Therefore, exposure to free radicals can lead to the consumption of vitamin C (Fig. 10.7).[48] This type of direct effect appears to be moderated by the presence of other antioxidants (e.g., reduced glutathione) but is exacerbated by inflammatory oxidants such as $O_2\cdot$, H_2O_2, and hypochlorous acid (HOCl) produced by activated phagocytes. For this reason, smokers, who expose themselves to various highly reactive free radicals in tobacco smoke,[49,50] show a 40% greater turnover of ascorbate than do nonsmokers with similar vitamin C intakes.[51] Even nonsmokers exposed passively to tobacco smoke have been found to have lower circulating ascorbate levels than nonexposed persons.

At relatively high vitamin C concentrations, the slower radical chain-propagating reaction of ascorbyl radical and O_2 become significant. It yields dehydroascorbic acid and superoxide radical, which, in turn, can oxidize ascorbate to return ascorbyl radical:

$$A^{\cdot -} + O_2 \rightarrow A + O_2^{\cdot -}$$

$$O_2^{\cdot -} + AH^- \rightarrow HOO^- + A^{\cdot -}$$

It is thought that, at high-vitamin C concentrations, this two-reaction sequence can develop into a radical chain

48. Eiserich, J.P., Cross, C.E., van der Vliet, A., 1997. In: Packer, L., Fuchs, J., (Eds.), Vitamin C in Health and Disease. Marcel Dekker, New York. pp. 399–412.

49. For example, nitric oxide (NO·); nitrogen dioxide (·NO_2); and alkyl, alkoxyl, and peroxyl radicals.

50. Free radical-mediated processes are thought to be involved in the pathobiology of chronic and degenerative diseases associated with cigarette smoking, e.g., chronic bronchitis, emphysema, cancer, cardiovascular disease.

51. Smith, J.L., Hodges, R.E., 1987. Ann. N.Y. Acad. Sci. 498, 144–152.

autoxidation process that consumes ascorbate, thus, wasting the vitamin. Hence, in aerobic systems, the efficiency of radical quenching of ascorbate is inversely related to the concentration of the vitamin. At physiological concentrations, ascorbate serves as one of the strongest reductants and radical scavengers, reducing oxy, nitro, and ethyl radicals.

Cellular Antioxidant Functions

As the most effective aqueous antioxidant in plasma, interstitial fluids, and soluble phases of cells, ascorbate appears to be the first line of defense against ROS arising in those compartments. Those species include superoxide and hydrogen peroxide arising from activated polymorphonuclear leukocytes or other cells and from gas-phase cigarette smoke,[52] which can promote the oxidation of critical cellular components.

- **Lipid peroxidation.** The LDL (low-density lipoprotein)-protective action of vitamin E appears to be dependent on the presence of ascorbate, which, by reducing the tocopheroxyl radical, prevents the latter from acting prooxidatively, i.e., from abstracting hydrogen from a cholesteryl–polyunsaturated fatty esters to yield peroxyl radicals.
- **Protein oxidation.** At least in vitro, ROS species can oxidize proteins to produce carbonyl derivatives and other oxidative changes associated with loss of function. Whether ascorbate provides such protection in vivo has been suggested as the basis of effects reported for vitamin C in reducing risks to cataracts and other illnesses.
- **DNA oxidation.** Ascorbate contributes to the prevention of oxidative damage to DNA, which is elevated in cells at sites of chronic inflammation and in many preneoplastic lesions. In fact, the continuous attack of DNA by unquenched ROS is believed to contribute to cancer, as elevated steady-state levels of oxidized DNA bases are estimated to cause mutational events.[53] The levels of one base damage product, 8-hydroxy-2'-deoxyguanosine, have been found to be elevated in scorbutic individuals[54] and to be reduced by supplementation with vitamins C and E.[55]
- **NO oxidation.** Ascorbic acid protects **nitric oxide** (NO) from oxidation, supporting the favorable effects of the latter on vascular epithelial function, and lowering blood pressure. This may also involve ascorbate participating in the reductive recycling of tetrahydrobiopterin, an essential cofactor of endothelial nitric oxide synthase. Ascorbate also reacts with nitrites and nitrates formed from NO and commonly found in vegetables and cured foods. In this way, ascorbate prevents the formation of carcinogenic *N*-nitroso compounds.

52. Indeed, genetically scorbutic rats have been found to have elevated levels of LDL lipid peroxidation products (i.e., thiobarbituric-reactive substances, TBARS), which respond to vitamin C supplementation.
53. ~1 per 10^5 bases (Halliwell, B., 2000. Am. J. Clin. Nutr. 72, 1082–1087).
54. Rehman, A., Collis, C.S., Yang, M., et al., 1998. Biochim. Biophys. Res. Commun. 246, 293–298.
55. Moller, P., Viscovich, M., Lykkesfeldt, J., et al., 2004. Eur. J. Nutr. 43, 267–274.

Improving iron utilization. Ascorbic acid can reduce ferric iron (Fe^{3+}) to the ferrous form (Fe^{2+}) and form a stable chelate with the latter. This allows the vitamin to convert the dominant form of iron in the acidic environment of the stomach to a form that is soluble in the alkaline environment of the small intestine. These effects result in increased enteric absorption[56] of both nonheme and heme iron. In these ways, vitamin C increases the bioavailability of iron in foods. Studies with iron-deficient rats, which have upregulated enteric iron absorption, have shown vitamin C to promote the mucosal uptake of iron but not its mucosal transfer. This effect depends on the presence of both ascorbic acid and iron in the gut at the same time, e.g., the consumption of a vitamin C-containing food with the meal. Thus, the low bioavailability of nonheme iron and the iron-antagonistic effects of such food factors as polyphenols and phytates, or of calcium phosphate, can be overcome by the simultaneous consumption of vitamin C (Fig. 10.8).[57] Similarly, ascorbic acid administered parenterally has been found useful as an adjuvant therapy to erythropoietin in hemodialysis patients.

Ascorbic acid also promotes the utilization of heme iron, which appears to involve enhanced incorporation of iron into its intracellular storage form, ferritin.[58] This effect involves facilitation of ferritin synthesis; ascorbate enhances the iron-stimulated translation of ferritin mRNA by maintaining the iron-responsive element-binding protein[59] in its enzymatically active form. Studies with cultured cells have shown that ascorbic acid also enhances the stability of ferritin by blocking its degradation through reduced lysosomal autophagy of the protein. Thus, the decline in ferritin and accumulation of **hemosiderin**[60] in scorbutic animals is reversed by ascorbic acid treatment.[61]

56. This effect can be 200–600%.
57. Anemia, much of it due to iron deficiency, is an enormous global problem, affecting more than 40% of all women. Yet, iron is the fourth most abundant element in the earth's crust and few diets do not contain the element in nonheme form. The problem of iron-deficiency anemia is associated with inadequate iron bioavailability and, thus, vitamin C inadequacy may be a contributing factor.
58. A soluble, iron–protein complex found mainly in the liver, spleen, bone marrow, and reticuloendothelial cells. With 23% iron, it is the main storage form of iron in the body; when that capacity is exceeded, iron accumulates as the insoluble hemosiderin.
59. This is a dual-function protein that also has aconitase activity.
60. A dark yellow, insoluble, granular, iron–storage complex found mainly in the liver, spleen, and bone marrow.
61. The reverse relationship is apparently not significant, i.e., iron loading has been found to have no effect on ascorbic acid catabolism in guinea pigs.

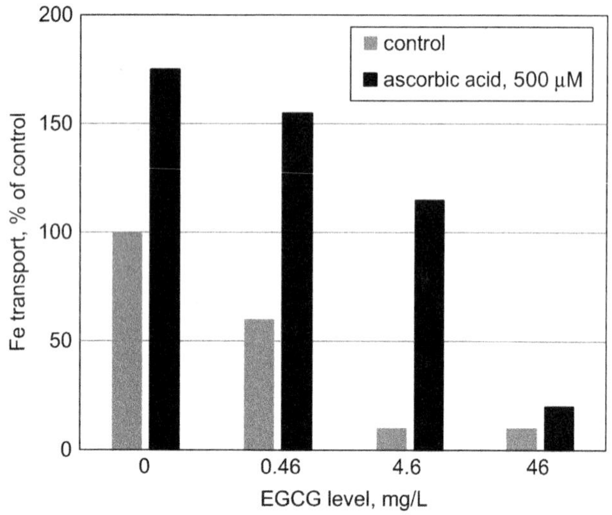

FIGURE 10.8 Effect of vitamin C in promoting iron transport and countering the inhibitory effect of epigallocatecin (EGCG) in the CaCo-2 cell model. *After Kim, E.Y., Ham, S.K., Bradke, D., et al., 2011. J. Nutr. 141, 828–834.*

TABLE 10.8 Relationship of Vitamin C Intake and Cataract Risk

Quintile of Vitamin C Intake, mg/day	Odds Ratio[a]
≤102	1
>102–135	0.88 (0.56–1.40)[b]
>135–164	0.66 (0.41–1.07)
>164–212	0.60 (0.37–1.07)
>212	0.70 (0.44–1.13)

[a]*Ratio of cataracts incidence in each quintile group to that of the lowest (reference) quintile group; P value for trend, 0.04.*
[b]*95% Confidence interval.*
Adapted from Valero, M.P., Fletcher, A.E., De Stavola, B.L., et al., 2002. J. Nutr. 132, 1299–1306.

Interactions with other mineral elements. Ascorbic acid can also interact with several essential trace elements. It can reduce the toxicities of high levels of selenium, copper, nickel, lead, vanadium, and cadmium—elements whose reduced forms are poorly absorbed or more rapidly excreted. In the case of copper, ascorbic acid enhances the postabsorptive utilization of copper for the synthesis of cuproproteins,[62] perhaps by increasing the balance of reduced *versus* oxidized forms of glutathione. Ascorbic acid can also enhance the utilization of low doses of selenium, and increase tissue levels of manganese.

Support of pulmonary function. Its redox properties give ascorbic acid an important role in the antioxidant protection of the lung,[63] which is consistently exposed to high concentrations of oxygen and inhaled toxic gases.[64] Lung parenchymal cells also generate ROS via cytochrome *P*450-dependent metabolism and inflammatory cell invasion. Accordingly, patients with asthma or acute respiratory distress syndrome typically show lower than normal concentrations of ascorbic acid in both plasma and leukocytes.

Support of neurologic function. The brain and spinal cord are among the richest tissues in ascorbic acid contents, with concentrations of 100–500 μM. An estimated 2% of the ascorbic acid in the brain turns over each hour. Plasma ascorbic acid concentrations have been positively associated with cognitive performance in older subjects and with memory in patients with dementia. These relationships may involve protection from inflammation, as such inflammatory mediators as cytokines and free radicals are important in the pathogenesis of neurodegenerative disease. Controlled intervention trials have not been conducted to evaluate the effects of vitamin C on cognitive function, but oral vitamin C was found to improve psychiatric rating in schizophrenics.

Prevention of cataracts. Cataracts, involving opacification of the ocular lens, are thought to result from the cumulative photooxidative effects of ultraviolet light from which the lens is protected by three antioxidants: ascorbic acid, tocopherol, and reduced glutathione. The lens typically contains relatively high concentrations of ascorbic acid (e.g., as much as 30-fold those of plasma), which are lower in aged and cataractous lens.[65] At least 10 cohort studies have found cataract risk to be inversely related to vitamin C intake (Table 10.8).[66] Some, but not all, case–control studies have found inverse associations of cataract risk and ascorbic acid intake and serum ascorbic acid level.[67] A large intervention trial found that high doses of vitamin C (increasing serum ascorbate levels >49 μM) reduced the incidence of cataract by 64%.[68] Scorbutic guinea pigs have been found to develop early cataracts; and ascorbate has been shown to protect against ultraviolet light-induced oxidation of lens proteins.

Diabetes prevention. Diabetic patients typically show lower serum concentrations of ascorbic acid than nondiabetic, healthy controls. Accordingly, reduced serum antioxidant activity has been implicated in the pathogenesis of the

62. Including two that use ascorbate as a cosubstrate (dopamine β-monooxygenase, peptidylglycine α-amidating monooxygenase).
63. See review by Brown, L.A.S., Jones, D.P., 1997. In: Packer, L., Fuchs, J., (Eds.), Vitamin C in Health and Disease. Marcel Dekker, New York, pp. 265–278.
64. For example, ozone, nitric oxide, nitrogen dioxide, cigarette smoke.
65. The ascorbic acid content of the oldest portion of the lens (the nucleus), where most senile cataracts originate, is typically only one-quarter the concentration in the lens cortex.
66. Vishwanathan, R., Johnson, E.J., 2012. In: Erdman, J.W., Macdonald, I.A., Zeisel, S.H., (Eds.), Present Knowledge in Nutrition, tenth ed. Wiley-Blackwell, Ames, IA, pp. 942–946.
67. Simon, J.A., Hudes, E.S., 1999. J. Clin. Epidemiol. 52, 1207–1211.
68. Agte, V., Tarwadi, K., 2010. Ophthalmic. Res. 44, 166–172.

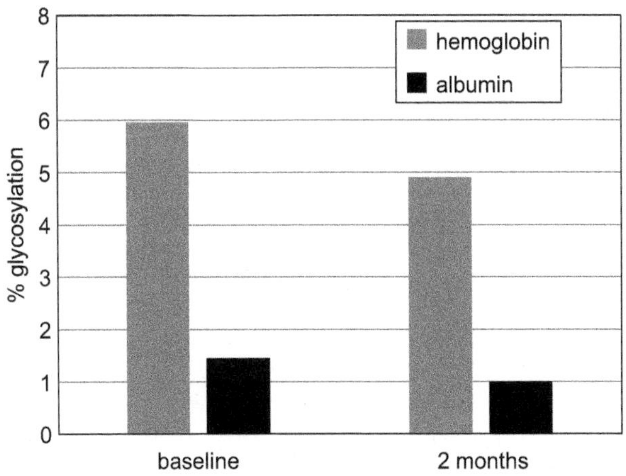

FIGURE 10.9 Protein glycosylation reduced by supplemental vitamin C (1 g/day). *After Davie, S.J., Gould, B.J., Yudkin, J.S., 1992. Diabetes 41, 167–173.*

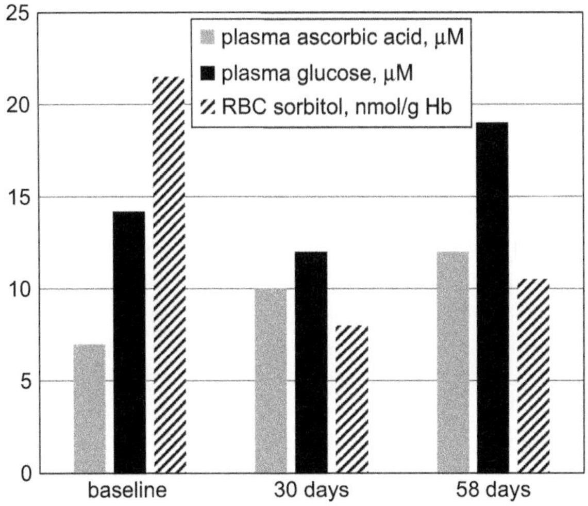

FIGURE 10.10 Responses of diabetics to supplemental vitamin C. *After Cunningham, J.J., Mearkle, P.L., Brown, R.G., 1994. J. Am. Coll. Nutr. 13, 344–350.*

disease. That vitamin C supplementation can reduce glycosylation of plasma proteins (Fig. 10.9) suggests a role of the vitamin in preventing diabetic complications. Intervention trials have shown that vitamin C supplementation can be effective in reducing erythrocyte sorbitol accumulation[69] (Fig. 10.10) and urinary albumin excretion[70] in noninsulin-dependent diabetics, although one found no effect on microvascular reactivity[71]. Treatment with vitamin C has also been shown to prevent arterial hemodynamic changes induced by hyperglycemia.

69. Because the sorbitol or its metabolites is thought to underlie the pathologic complications of diabetes, reduction of tissue sorbitol accumulation is a strategy for managing diabetes.

70. Gaede, P., Poulsen, H.E., Parving, H.H., et al., 2001. Diabetic Med. 18, 756–760.

71. Lu, Q., Bjorkhem, I., Wretlind, B., et al., 2005. Clin. Sci. 108, 507–513.

Prooxidant Potential

In the presence of oxidized metal ions (e.g., Fe^{3+}, Cu^{2+}), high concentrations of ascorbic acid can have prooxidant functions at least in vitro. It does so by donating an electron to reduce such ions to forms that, in turn, can react with O_2 to form oxy radicals (the metal ions being reoxidized in the process):

$$AH^- + Fe^{3+} \rightarrow A^{\cdot -} + Fe^{2+}$$

$$Fe^{2+} + H_2O_2 \rightarrow Fe^{3+} + OH^{\cdot} + OH^-$$

In this way, ascorbate can react with copper or iron salts in vitro and lead to the formation of H_2O_2, $O_2^{\cdot -}$, and OH·, which can damage nucleic acids, proteins, and polyunsaturated fatty acids (PUFAs). Accordingly, iron–ascorbate mixtures are often used to stimulate lipid peroxidation in vitro, and such prooxidative reactions with transition metals are likely to be the basis of the cytotoxic and mutagenic effects of ascorbate observed in isolated cells in vitro. It has been suggested that high serum ascorbate may reduce Fe^{3+} in ferritin[72] to the catalytically active form Fe^{2+}.

The physiological relevance of these prooxidative reactions is unclear. Under physiological conditions, tissue concentrations of ascorbate acid greatly exceed those of dehydroascorbic acid, which greatly exceed those of ascorbyl radical. Thus, the redox potentials of the ascorbyl radical/ascorbate and dehydroascorbic acid/ascorbyl radical couples are sufficient to reduce most oxidizing compounds.

Enzyme Cosubstrate Functions

Ascorbate functions as a cosubstrate for at least 10 enzymes that function in electron transport reactions involved in the synthesis of collagen, norepinephrine,[73] peptide hormones, and carnitine; and the metabolism of tyrosine, xenobiotics, steroids, and fatty acids (Table 10.9). Two of these are monooxygenases, which incorporate a single atom of oxygen into a substrate and require at their active sites copper atoms that are reduced by ascorbate. The others are dioxygenases, which incorporate both atoms of molecular oxygen in different ways;[74] many of these also use α-ketoglutarate as a cosubstrate. In each, the electron acceptor, ascorbyl radical, is subsequently reduced by microsomal **monodehydroascorbate reductase** or **ascorbate–cytochrome b_5 reductase** to regenerate

72. This would require that ascorbate enter the pores of the ferritin protein shell to react with iron on the inner surface.

73. Also, noradrenaline.

74. In most cases, those oxygen atoms are incorporated into different acceptor substrates; in the single case of 4-hydroxyphenylpyruvate dioxygenase, the oxygens are incorporated in different locations in the same acceptor substrate (4-hydroxyphenylpyruvate).

TABLE 10.9 Enzymes That Require Ascorbic Acid as a Cosubstrate

Metabolic Role	Enzyme
Collagen synthesis	Prolyl 4-hydroxylase
	Prolyl 3-hydroxylase
	Lysine hydroxylase
Tyrosine metabolism	4-Hydroxyphenylpyruvate dioxygenase
Catecholamine synthesis	Dopamine β-monooxygenase
Peptide hormone synthesis	Peptidylglycine α-amidating monooxygenase
Carnitine synthesis	γ-Butyrobetaine 2-oxoglutarate 4-dioxygenase
	Trimethyllysine 2-oxoglutarate dioxygenase
Transcriptional responses to hypoxia	HIF-1α prolyl 4-hydroxylases
	HIF-1α asparaginyl hydroxylase
Drug and steroid metabolism	Cholesterol-7α-hydroxylase

ascorbic acid. In many of these functions, ascorbate is not required per se, i.e., it can be replaced by other reductants in vitro.[75] However, these enzymes show high affinities for ascorbate, suggesting that their activities are diminished only by chronic deprivation of the vitamin.

Connective tissue health. Vitamin C is required for wound healing. The vitamin is accumulated at wound sites where it is rapidly utilized.[76] This reflects the function of ascorbic acid in the synthesis of collagen proteins,[77] specifically, in the hydroxylation of specific prolyl and lysyl residues of the unfolded (nonhelical) procollagen chain catalyzed by **prolyl 4-hydroxylase**, **prolyl 3-hydroxylase**, and **lysyl hydroxylase**. These dioxygenases[78] require O_2, Fe^{2+}, and ascorbate, which are stoichiometrically linked to the oxidative decarboxylation of α-ketoglutarate. Each has a specific binding site for ascorbate near its iron center. The role of ascorbate in each reaction is to maintain iron in the reduced state (Fe^{2+}), which dissociates from a critical region (an SH group) of the active site to reactivate the enzyme after catalysis. The posttranslational hydroxylation of these procollagen amino acid residues is necessary for folding into the triple helical structure that can be secreted by fibroblasts. Hydroxyproline residues contribute to the stiffness of the collagen triple helix and hydroxylysine residues bind (via their hydroxyl groups) carbohydrates to form intramolecular cross-links that give structural integrity to the collagen mass. Vitamin C deprivation results in underhydroxylation of procollagen, which accumulates[79] and is degraded; this appears to be the basis of the poor wound healing characteristic of scurvy. Vitamin C-deficient subjects usually show reduced urinary excretion of hydroxyproline. The function of ascorbic acid in collagen synthesis makes the vitamin important in the synthesis of surfactant apoproteins, which have collagen-like domains that require ascorbic acid-dependent hydroxylation for proper folding and stability. Wound repair typically decreases with aging; this had been viewed as indicative of increasing needs for vitamin C by older individuals.

Studies have indicated modest effects of vitamin C deprivation on the hydroxylation of proline in the conversion of the soluble **tropoelastin** to the soluble **elastin**.[80] A component of complement, C1q, resembles collagen in that it contains **hydroxyproline** and **hydroxylysine**. Curiously, vitamin C deprivation reduces overall complement activity but does not affect the synthesis of C1q.

Catecholamine synthesis. Ascorbate serves as an electron donor for **dopamine β-monooxygenase**,[81] a copper enzyme located in the chromaffin vesicles[82] of the adrenal medulla and in adrenergic synapses. The enzyme exists in both membrane-bound and soluble forms; both use O_2 and ascorbate to hydroxylate dopamine to form the neurotransmitter norepinephrine (Fig. 10.11).

Peptide hormone processing. Ascorbate is a cosubstrate copper enzyme, **peptidylglycine α-amidating monooxygenase**, which catalyzes the posttranslational processing of peptides by α-amidation. This process consists of adding an amide group to the C-terminals of physiologically active

75. For example, reduced glutathione, cysteine, tetrahydrofolate, dithiothreitol, 2-mercaptoethanol.

76. Studies with apparently vitamin C-adequate burn patients have shown their plasma ascorbic acid levels to drop to nearly zero after the trauma; this is presumed to reflect the movement of the vitamin to the sites of wound repair.

77. **Collagens**, secreted by fibroblasts and chondrocytes, are the major components of skin, tendons, ligaments, cartilage, the organic substances of bones and teeth, the cornea, and the ground substance between cells. Some 19 types of collagen have been characterized; collectively, they comprise the most abundant type of animal protein, accounting for 25–30% of total body protein.

78. One-half of the O_2 molecule is incorporated into the peptidyl prolyl (or peptidyl lysyl) residue; the other half is incorporated into succinate.

79. Accumulated procollagen also inhibits its own synthesis and mRNA translation.

80. About 1% of the prolyl residues in elastin are hydroxylated. This amount can apparently be increased by vitamin C, suggesting that normal elastin may be underhydroxylated.

81. The specific activity of this enzyme has been found to be abnormally low in schizophrenics with anatomical changes in the brain, suggesting impaired norepinephrine and dopamine neurotransmission in those patients.

82. These vesicles accumulate and store catecholamines in the adrenal medulla; they also contain very high concentrations of ascorbic acid, e.g., 20 mM.

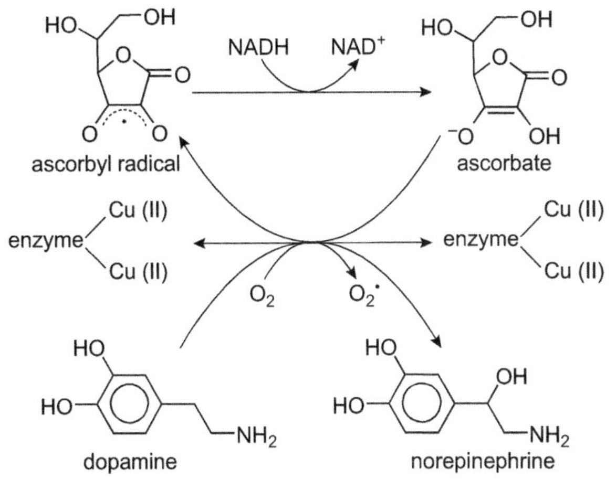

FIGURE 10.11 Role of vitamin C in the conversion of dopamine to norepinephrine by dopamine-β-monooxygenase.

peptides.[83] The enzyme is bifunctional, having two domains that catalyze both of the two steps in the peptide amidation process:

- **peptidylglycine α-hydroxylating monooxygenase**[84] catalyzes the hydroxylation of C-terminal glycine; it is rate limiting to the overall process; that this activity is inhibited by catalase in vitro, which suggests that H_2O_2 is an intermediate in the reaction.
- **peptidyl-α-hydroxyglycine α-amidating lyase** catalyzes the cleavage of the peptidylhydroxyglycine into glyoxylate and the amidated peptide.

Carnitine[85] synthesis. Ascorbic acid is a cosubstrate of two Fe^{2+}-containing hydroxylases involved in the synthesis of carnitine (Fig. 10.12): **ε-N-trimethyllysine hydroxylase** and **γ-butyrobetaine hydroxylase**. While ascorbate supports the greatest activities of these enzymes, is it not essential per se, as each can be driven by excess substrate loads.[86] Scorbutic guinea pigs show low carnitine levels in muscle and heart. Impaired carnitine synthesis is the likely basis of the fatigue, lassitude, and hypertriglyceridemia, which are early signs of vitamin C deficiency.

Tyrosine metabolism. Ascorbate is a cosubstrate of two mixed-function oxidases involved in the oxidative degradation of tyrosine:

- **4-hydroxyphenylpyruvate dioxygenase** catalyzes the oxidation and decarboxylation of the intermediate of tyrosine degradation, 4-hydroxyphenylpyruvic acid to homogentisic acid.
- **homogentisate 1,2-dioxygenase** catalyzes the next step in tyrosine degradation.

By impairing both reactions, vitamin C deficiency can result in tyrosinemia[87] and the excretion of tyrosine metabolites in the urine; both conditions respond to vitamin C supplements.

Transcriptional responses to hypoxia. A group of ascorbate-dependent prolyl 4-hydroxylases and an asparaginyl hydroxylase are involved in the regulation of transcriptional responses to hypoxic stress. Under normal cellular O_2 tensions, the hypoxia-inducible factor 1α (HIG-1α) is targeted for degradation by hydroxylation of proline and asparagine residues. However, under hypoxic conditions, it is not degraded but, instead, dimerizes with HIF-1β to induce hypoxia-responsive genes related to glycolysis, erythropoiesis, and angiogenesis.[88] Therefore, it is thought that, by this mechanism, vitamin C deficiency evokes the hypoxic response.

Vascular endothelial function. The functions of ascorbic acid in collagen synthesis, endothelial growth and survival, and radical scavenging make it important in supporting the vascular bed. Thus, it is thought to contribute to the prevention of endothelial dysfunction leading to inflammatory vascular disease, e.g., atherosclerosis.

Drug and steroid metabolism. Ascorbic acid is thought to be involved in microsomal hydroxylation reactions of drug and steroid metabolism, i.e., those coupled to the microsomal electron transport chain. In these roles, it is likely that ascorbic acid functions as a reducing agent to promote catalytic activity of iron-centered enzymes. The enzymes affected include **cholesterol 7α-hydroxylase**, the hepatic microsomal enzyme involved in the biosynthesis of bile acids; its activity is diminished in the chronically vitamin C-deficient guinea pig and is corrected by feeding vitamin C.[89] Epidemiologic studies have noted significant positive correlations of circulating levels of ascorbic acid and HDL–cholesterol in free-living humans.[90] Vitamin C deprivation reduces hepatic cytochrome *P*450-dependent drug metabolism in the guinea pig, increasing the half-lives of phenobarbital, acetanilide, aniline, and antipyrine. The activities of adrenal mitochondrial and microsomal

83. e.g., bombesin (human gastrin-releasing peptide), calcitonin, cholecystokinin, corticotropin-releasing factor, gastrin, growth hormone-releasing factor, melanotropins, metorphamide, neuropeptide Y, oxytocin, vasoactive intestinal peptide, and vasopressin.

84. This enzyme shares significant sequence homology with dopamine β-hydroxylase.

85. 3-Hydroxy-4-(trimethylazaniumyl)butanoate, which is required for the transport of fatty acids into mitochondria for oxidation to provide energy for the cell.

86. Rebouche, C.J., 1995. Metabolism 44, 1639–1643.

87. Transient tyrosinemia (serum levels >4 mg/dL) occurs frequently in premature infants and involves reduced 4-hydroxyphenylpyruvate dioxygenase activity. Low doses of ascorbic acid usually normalize the condition.

88. Bardos, J.I., Ashcroft, M., 2005. Biochim. Biophys. Acta 1755, 107–120.

89. Guinea pigs fed a high-vitamin C diet (500 mg/kg) show substantial reductions in plasma (~40%) and liver (~15%) cholesterol concentrations. Results of human studies addressing this point have been inconsistent.

90. A study of a healthy, elderly Japanese population found serum ascorbic acid to account for about 5% and 11% of the variation in serum HDL–cholesterol concentrations in men and women, respectively.

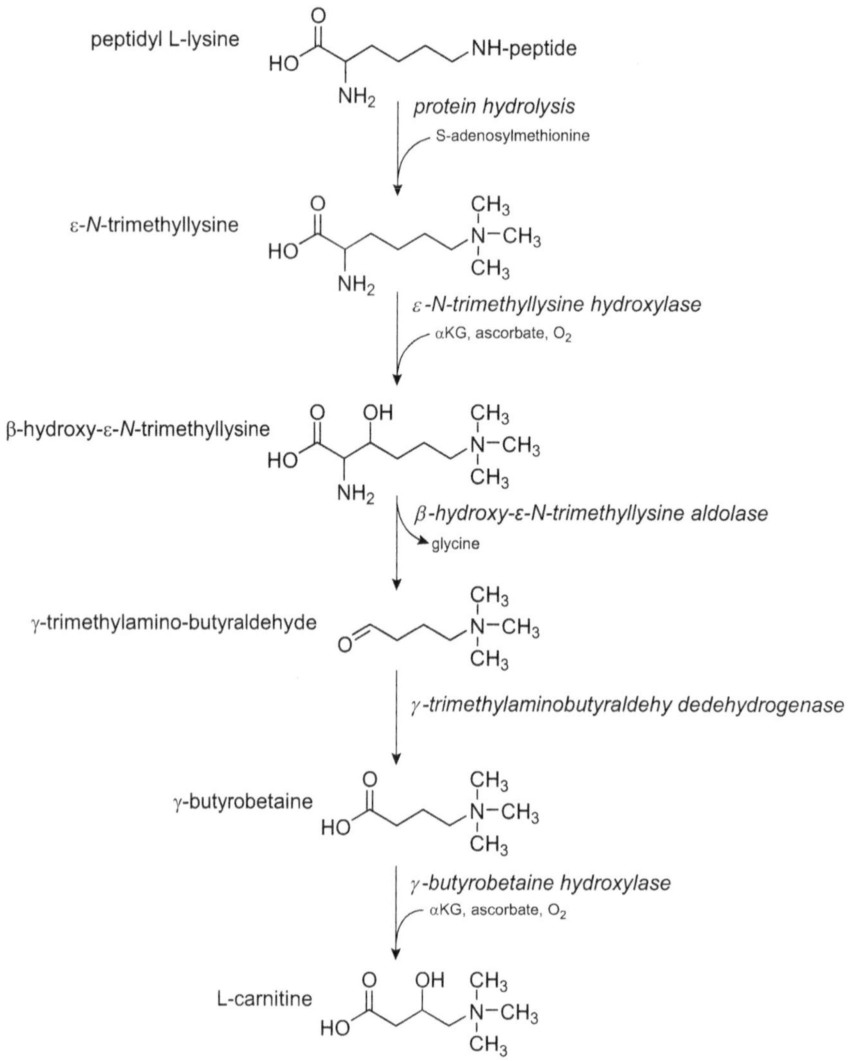

FIGURE 10.12 Ascorbate participates in two steps in the biosynthesis of carnitine from lysine: ε-N-trimethyllysine hydroxylase and γ-butyrobetaine hydroxylase.

steroid hydroxylases are impaired in scorbutic animals and respond to vitamin C therapy.

Immunity and Inflammation

Studies with animal and cell culture models have shown vitamin C to affect immune function in several ways:

- modulation of T cell expression of genes involved in signaling, carbohydrate metabolism, apoptosis, transcription, and immune function;[91]
- support of natural killer cell activity and production of interferons, the proteins that protect cells against viral attack;
- support of positive chemotactic and proliferative responses of neutrophils;
- protection against free radical-mediated protein inactivation associated with the oxidative burst[92] of phagocytic cells;
- support of the synthesis of humoral thymus factor and antibodies of the IgG and IgM classes;
- support of delayed-type hypersensitivity responses.

These functions appear to be affected by vitamin C status: compromised by deprivation of the vitamin; in at least some cases, stimulated by supranutritional doses of vitamin.

91. Grant, M.M., Mistry, N., Lunec, J., et al., 2007. Br. J. Nutr. 97, 19–26.

92. Neutrophils, when stimulated, take up O_2 and generate ROS, which, along with other reactive molecules, kill bacterial pathogens. This **oxidative burst** can be observed in vitro as a rapid consumption of O_2; it also involves the enzymatic generation of bactericidal halogenated molecules via myeloperoxidase. These killing processes are usually localized in intracellular vacuoles containing the phagocytized bacteria.

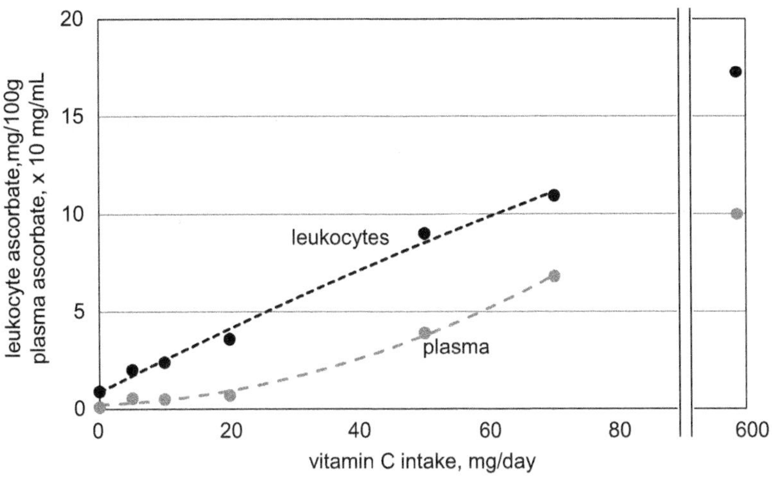

FIGURE 10.13 Relationships of leukocyte and plasma ascorbate concentrations with vitamin C intake. *Note:* leukocyte ascorbate concentrations are an order of magnitude greater than those of plasma. *After Bartley, W., Krebs, H.A., O'Brien, J.R.P., 1953. Special Rep. Series Med. Res. Council 280, 1–179.*

8. BIOMARKERS OF VITAMIN C STATUS

Vitamin C status can be assessed in two ways:

- **Leukocyte ascorbate**—Blood cells contain a substantial fraction of the ascorbic acid in the blood; of these, leukocytes (white cells) have the greatest diagnostic value, as their ascorbic acid concentrations are indicative of dietary intake and tissue levels of the vitamin (Fig. 10.13).[93] White cell ascorbic acid levels increase directly with increasing vitamin C intake, plateauing at vitamin C doses of about 2 g/day. The ascorbic acid contents of other blood cells also increase with vitamin C dose, with lymphocytes, platelets, monocytes, and neutrophils showing lower plateau in that order. Therefore, the use of this biomarker calls for careful preparation to ensure that the analysis is conducted using a uniform cell population.
- **Plasma/serum ascorbate**—These specimens offer the advantage of ease of preparation. Serum ascorbate concentrations of ~60 μM indicate tissue saturation, which requires regular vitamin C intakes of ~100 mg/day (Fig. 10.13). Higher intakes indicate elevated concentrations of the vitamin in extracellular fluids. The common interpretations of plasma ascorbate concentrations are shown in Table 10.10.

Dietary assessment is not recommended for assessment of vitamin C status, particularly for individuals. This approach suffers from the errors inherent in the methodologies for food intake/frequency recalls as well as losses of vitamin C from foods in storage, processing, and cooking.

93. Leukocyte ascorbate concentrations are usually greater in women than men and decrease in both sexes with age.

TABLE 10.10 Physiologic Significance of Plasma Ascorbate Concentration

Vitamin C Status	Plasma Ascorbate, mg/dL (μM)
Tissue ascorbate saturation	0.71–1.5 (40–85)
Adequacy	0.41–0.7 (23–40)
Hypovitaminosis	0.2–0.4 (11–22)
Clinical deficiency	<0.2 (<11)

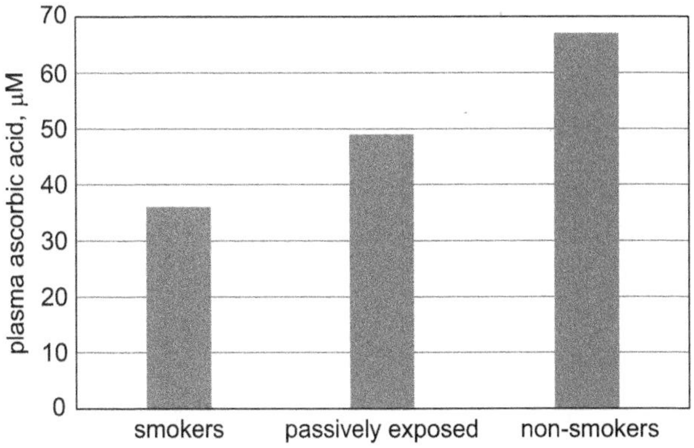

FIGURE 10.14 Effect of smoking on plasma ascorbic acid level. *After Tribble, D.L., Giuliano, L.J., Fortmann, S.P., 1993. Am. J. Clin. Nutr. 58, 886–890.*

Assessment of vitamin C status can be affected by factors unrelated to intake of the vitamin. These include smoking; even passive exposure to tobacco smoke can reduce plasma ascorbic acid concentrations (Fig. 10.14; Table 10.11). Glutathione-S-transferase (GST) genotype has also been found to affect plasma ascorbate level: individuals with the

TABLE 10.11 Effect of Passive Exposure to Tobacco Smoke on Vitamin C Status of Children

Age, Years	Plasma Ascorbic Acid, μM[a]	
	Unexposed	Exposed[b]
2–4	53.0 (50.2–55.8)	47.9 (44.4–51.5)
5–8	53.6 (51.4–55.8)	51.0 (48.5–53.5)
9–12	49.7 (47.4–52.0)	47.7 (45.5–49.9)

[a]*Mean (95% C.I.).*
[b]*A multifactorial ANOVA, with plasma ascorbic acid level adjusted for dietary vitamin C intake, showed the effect of passive exposure to be significant across all age groups, P=.002.*
Adapted from Preston, A.M., Rodriguez, C., Rivera, C.E., et al., 2003. Am. J. Clin. Nutr. 77, 167–172.

GST null genotype are more likely to be clinically deficient if their vitamin C intakes did not meet the Recommended Daily Allowance (RDA).[94]

9. VITAMIN C DEFICIENCY

Determinants of Risk

Vitamin C deficiency can be caused by low dietary intakes, as well as by conditions in which the metabolic demands for ascorbic acid may exceed the rate of its endogenous biosynthesis, thus, increasing the turnover of the vitamin in the body. Such conditions include smoking,[95] environmental/physical stress, chronic disease, and diabetes.

General Signs of Deficiency

In individuals unable to synthesize the vitamin, acute dietary C deficiency is manifested as various signs in the syndrome called scurvy (Table 10.12). The dominant clinical sign is prolonged wound healing time due to diminished rates of collagen synthesis as well as their increased susceptibility to infections. Dietary guidelines for nutritionally adequate intakes of vitamin C have been established (Table 10.13).

TABLE 10.12 General Signs of Vitamin C Deficiency

Organ System	Signs
General	Poor appetite, growth
	Impaired wound healing
	Painful, swollen gums with subsequent tooth loss
	Impaired immunity
	Ill-defined pain in extremities
Muscular	Skeletal muscle atrophy
Vascular	Increased capillary fragility, cutaneous hemorrhages
Nervous	Tenderness; impaired heat resistance

TABLE 10.13 Recommended Vitamin C Intakes

The United States		FAO/WHO	
Age–Sex	RDA[a], mg/day	Age–Sex	RNI[b], mg/day
0–6 months	25	0–6 months	–[c]
7–11 months	30	7–11 months	–[c]
1–6 years	30	1–6 years	30
7–9 years	35	7–9 years	35
10–18 years	40	10–18 years	40
>18 years	45	>18 years	45
Pregnancy	55	Pregnancy	55
Lactation	70	Lactation	70

[a]*Recommended Dietary Allowances; Food and Nutrition Board, 2000. Dietary Reference Intakes for Vitamin C, Vitamin E, Selenium and Carotenoids. National Academy Press, Washington, DC, 506 pp.*
[b]*Recommended Nutrient Intakes; Joint WHO/FAO Expert Consultation, 2001. Human Vitamin and Mineral Requirements. WHO, Rome, 286 pp.*
[c]*No value established.*

Deficiency Signs in Humans

Classic scurvy is manifest in adults after 60–90 days of stopping vitamin C consumption, although other signs can be seen by 30 days.[96] Scurvy presents when the total body pool of vitamin C is reduced to <300 mg from its normal level of ~1500 mg. At that low level, patients show plasma vitamin C levels of 0.13–0.25 mg/dL (normal levels are 0.8–1.4 mg/dL). Signs of the disease occur primarily in mesenchymal tissues. Defects in collagen formation are manifested as hemorrhage (due to deficient formation of intercellular substance) in the skin, mucous membranes, internal organs, and muscles; edema; impaired wound healing; and weakening of collagenous structures in bone, cartilage, teeth,

94. Cahill, L.E., Fontaine-Bisson, B., El-Sohemy, A., 2009. Am. J. Clin. Nutr. 90, 1411–1417.
95. Ascorbic acid concentrations of serum and urine of smokers tend to be about 0.2 mg/dL less than those of nonsmokers; these effects have been observed even after correcting for vitamin C intake, which was found to be about 53 mg/day less in smokers. Further, smokers have higher rates of vitamin C turnover (~100 mg/day) than nonsmokers (~60 mg/day). It is estimated that smokers require 52–68 mg *more* vitamin C per day than nonsmokers to attain comparable plasma ascorbic acid levels.

96. Levine, M., Wang, Y., Padayatty, S.J., et al., 2001. Proc. Nat. Acad. Sci. U.S.A. 98, 9842–9846.

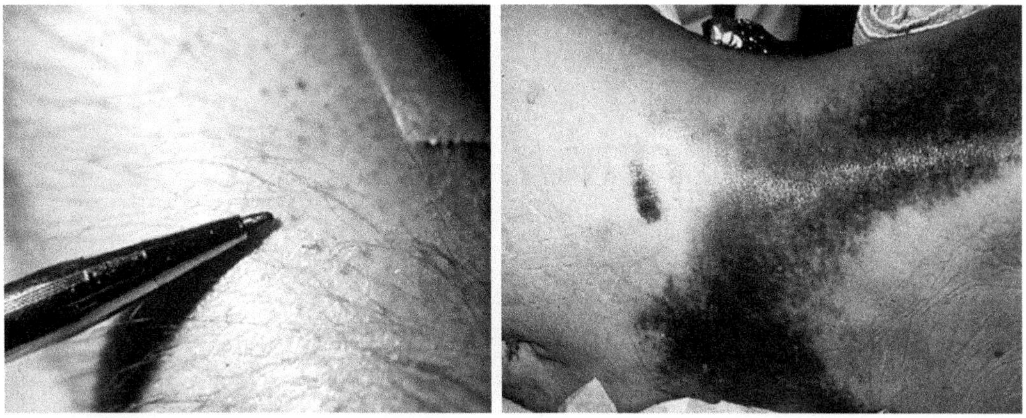

FIGURE 10.15 Hemorrhages in scorbutic patients: petechiae (left), ecchymoses (right). *Courtesy G.F. Combs, Sr.*

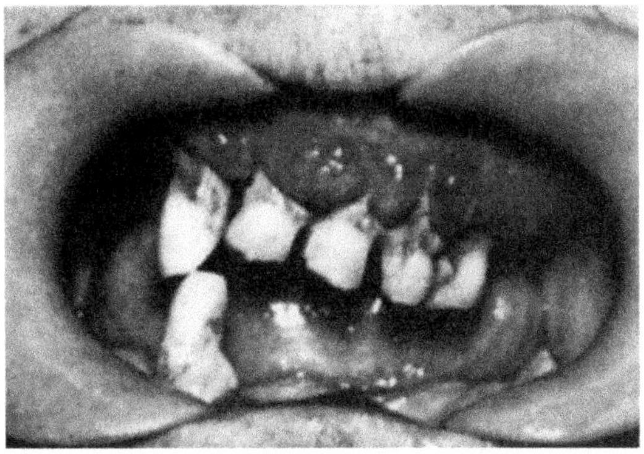

FIGURE 10.16 Scurvy in a middle-aged man. Note: swollen, bleeding gums, and tooth loss. *Courtesy J. Marks, Cambridge University.*

and connective tissues. Scorbutic adults may present with swollen, bleeding gums with tooth loss; that condition may signify accompanying periodontal disease. They also show lethargy, fatigue, rheumatic pains in the legs, muscular atrophy, skin lesions, and hemorrhages in many organs (e.g., skin, gums Fig. 10.16, intestines, subperiosteal tissues, eyes). Cutaneous hemorrhages start as pinpoint, perifollicular **petechiae**, which may coalesce to form large **ecchymoses** (Fig. 10.15). These features are frequently accompanied by psychological changes: hysteria, hypochondria, and depression. Experimental vitamin C deprivation studies conducted in the 1950s and 1960s indicated that the minimum daily dose of vitamin C that would prevent these signs was ≤10 mg.[97] In practice, scurvy is typically observed in subjects who are generally malnourished and show subclinical/clinical signs of thiamin, riboflavin, niacin, and/or pyridoxine hypovitaminoses.

Clinical scurvy was described in an Aboriginal infant who developed the signs at 7 months of age after having been breast-fed by her malnourished mother whose plasma and milk contained only 0.19 mg and 2 mg ascorbic acid per 100 mL, respectively.[98] Pediatric scurvy has also been reported as **Moller–Barlow disease** in nonbreast-fed infants usually at about 6 months of age (when maternally derived stores of vitamin C have been exhausted).[99] That disease is characterized by widening of bone–cartilage boundaries, particularly of the rib cage, stressed epiphyseal cartilage of the extremities, by severe joint pain and, frequently, by anemia and fever. Scorbutic children may present with limp or inability to walk, tenderness of the lower limbs, bleeding of the gums, and petechial hemorrhages.

Hypovitaminosis C. Subclinical deficiency is characterized by plasma ascorbate level <0.75 mg/dL and a total body vitamin C pool <600 mg. It often occurs in the elderly as the product of diminished enteric absorption and increased turnover. This can result in several nonspecific, prescorbutic signs and symptoms: lassitude, fatigue, anorexia, muscular weakness, and increased susceptibility to infection. Epidemiologic data indicate significant associations of low-plasma ascorbic acid concentration with increased risk of ischemic heart disease or hypertension.[100] Low-vitamin C status has been shown to be associated with increased risks of gestational diabetes and premature delivery due to premature rupture of chorioamniotic membranes. This responded to vitamin C supplementation.[101]

97. Bartley, W., Krebs, H.A., O'Brien, J.R.P., 1953. Special Rept. Series, Medical Res. Council. 280, 1–179; Hodges, R.E., Baker, E.M., Hood, J., et al., 1969. Am. J. Clin. Nutr. 22, 535–548; Baker, E.M., Hodges, R.E., Hood, J. et al., 1969. Am. J. Clin. Nutr. 22, 549–556.

98. Kamien, M., Nobile, S., Cameron, P., et al., 1974. Aust. N.Z. J. Med. 4, 126–133.

99. A retrospective study of the 28 cases diagnosed in at the Queen Sirikit National Institute of Child Health from 1995 to 2002, 93% were 1–4 years. of age (Ratanachu-Ek, S., Sukswai, P., Jeerathanyasakun, Y., et al., 2003. J. Med. Assoc. Thai. 86, S734–S740).

100. Patients with each disease are typically of relatively low status with respect to vitamin E and/or copper. In the case of hypertension, a placebo-controlled, double-blind study showed that vitamin C supplements (1 g/day for 3 mos.) significantly reduced systolic and diastolic pressures in borderline hypertensive subjects with normal serum ascorbic acid levels.

101. Casaneuva, E., Ripoll, C., Tolentino, M., et al., 2005. Am. J. Clin. Nutr. 81, 859–863; Borna, S., Borna, H., Daneshbodie, B., 2005. Int. J. Gynecol. Obstet. 90, 16–20.

Hypovitaminosis C appears to compromise the activity of cholesterol 7-α-hydroxylase, the rate-limiting step in the conversion of cholesterol to bile acids. In the guinea pig, this can result in the overproduction of the glycoprotein mucin, apparently as a result of oxidative stress, and the formation of gallstones. Analyses of NHANES data have associated relatively low-serum ascorbic acid concentrations with increased risk of forming gallstones in women (but, curiously, not men). That relationship was U-shaped with the highest prevalence of self-reported gallbladder disease for women with serum ascorbic acid levels in the range of 0.7–1.5 mg/dL.[102] In contrast, an analysis of data from NHANES III by the same group showed a consistent inverse relationship of serum ascorbic acid and gallstone incidence, with vitamin C supplement use being independently associated with a 34% lower prevalence of disease.[103]

Responses to Vitamin C Treatment

Response to vitamin C is dramatic; clinical improvements are seen within a week of supranutritional vitamin C therapy. A repletion study with hypovitaminotic C adults found their serum ascorbate levels to return from nondetectable to tissue saturation levels (~60 μM) within 3 days when given a daily supplement of 1 g of the vitamin; this was after 3 days of a 300 mg/day supplement, which had failed to register a plasma response.[104] Topical application of ascorbic acid has been found useful in treating photodamaged skin, as well as inflammatory conditions of the skin such as acne and eczema.

Deficiency Signs in Animals

In guinea pigs, ascorbic acid deficiency is characterized by intermittent reductions in growth, hematomas (especially of the hind limbs), extremely brittle bones, and abnormalities of epiphyseal bone growth with calcification of bone–cartilage boundaries. Guinea pigs that are deprived of vitamin C also show reduced feed intake and growth, anemia, hemorrhages, altered dentin, and gingivitis. Continued deficiency results in disrupted protein folding and apoptosis in the liver. If not corrected, death usually occurs within 25–30 days. Ascorbic acid deficiency in at least some species of fishes (salmonids and carp) results in spinal curvature (scoliosis[105] and lordosis[106] Fig. 10.17), reduced survival, reduced growth rate, anemia, and hemorrhaging, especially in the fins, tail, muscles, and eyes. Similar signs have been reported in vitamin C-deficient shrimp and eels.

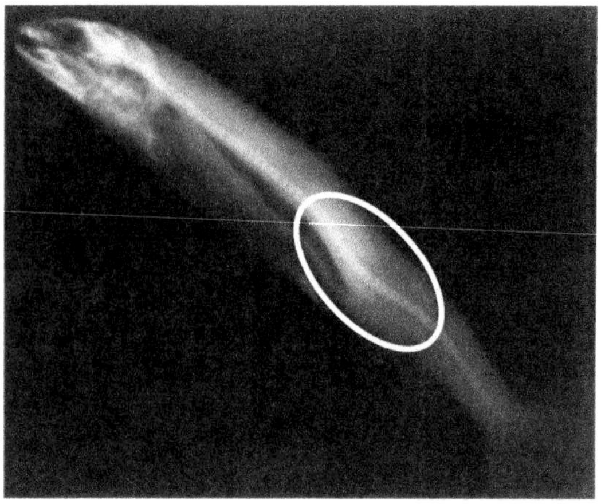

FIGURE 10.17 Radiograph of a vitamin C-deficient trout. Note: lordosis. *Courtesy G.L. Rumsey, Tunison Laboratory of Fish Nutrition, USDI.*

Hypovitaminosis C in animals. Subclinical deficiency in the guinea pig can result in lassitude, anorexia, muscular weakness, loss of reduced glutathione in lymphocytes, loss of α-tocopherol with accumulation of lipid peroxidation products in retinal tissues, hypertriglyceridemia, hypercholesterolemia, and decreased vitamin E concentrations in liver and lungs. These signs respond to vitamin C supplementation.

10. VITAMIN C IN HEALTH AND DISEASE

Vitamin C intakes greater than 100–200 mg/day result in elevated concentrations of the vitamin in extracellular fluids (plasma, connective tissue fluid, humors of the eye). Under such conditions, pharmacologic actions of this antioxidant vitamin may be possible. Clinical studies have found such supranutritional doses of vitamin C to be of some benefits; however, many such studies have compared treated subjects with controls who did not have tissue saturation with the vitamin. Thus, while there appear to be benefits associated with increasing vitamin C intakes to levels that support tissue saturation, evidence supporting benefits of vitamin C intakes above that level is less clear.

Antioxidant Effects

High doses of vitamin C can reduce markers of oxidative stress, which has been implicated as a central mechanism in the development of obesity-related diseases (i.e., cardiovascular disease and type 2 diabetes), and a cause of chronic obstructive pulmonary disease and Alzheimer's disease. Clinical interventions with supranutritional doses

102. Simon, J.A., Hudes, E.S., 1998. Am. J. Pub. Health 88, 1208.
103. Simon, J.A., Hudes, E.S., 2000. Arch. Int. Med. 160, 931.
104. Nobile, S., Woodhill, J.M., 1981. Vitamin C: The mysterious redox-system, a trigger of life?. MTD Press, Lancaster, p. 78.
105. Lateral curvature of the spine.
106. Anteroposterior curvature of the spine.

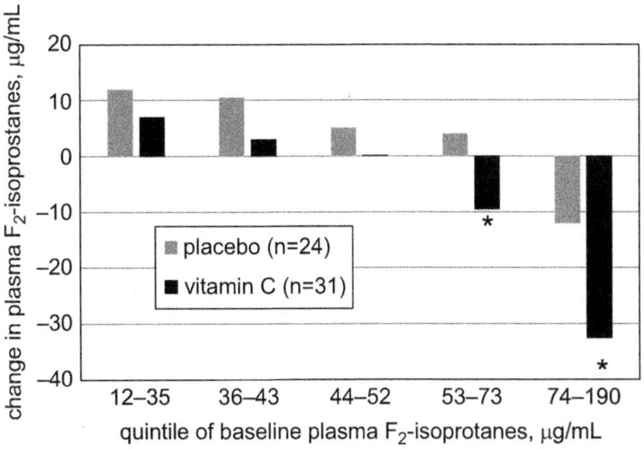

FIGURE 10.18 High-dose vitamin C (1000 mg/day) reduced of plasma F2-isoprostane concentrations in adults showing the highest baseline values. *After Block, G., Jensen, C.D., Morrow, J.D., et al., 2008. Free Rad. Biol. Med. 45, 377–384.*

of vitamin C have been found to reduce plasma concentrations of F_2-isoprostanes, prostaglandin-like products of nonenzymatic free radical attack on PUFAs. These effects are greatest for individuals with relatively high indicators of oxidative stress (Fig. 10.18) and have been found to be affected by factors that contribute to that oxidative stress, e.g., high body mass index (BMI) and smoking.[107]

Antihistamine Effects

High doses of vitamin C can reduce circulating histamine concentrations. On this basis, vitamin C has been used to protect against histamine-induced anaphylactic shock. Ascorbic acid inhibits histamine release and enhances its degradation. It does so by undergoing oxidation to dehydroascorbic acid with the concomitant rupture of the histamine imidazole ring. In cultured cells, this reduces endogenous histamine levels as well as histidine decarboxylase activities, a measure of histamine synthetic capacity. Ascorbic acid can also enhance the synthesis of the prostaglandin E series, members of which mediate histamine sensitivity. Blood histamine concentrations are elevated in several complications of pregnancy associated with marginal ascorbic acid status: preeclampsia,[108] abruption,[109] and prematurity. A large, cross-sectional study indicated that plasma ascorbic acid was inversely associated with biomarkers of inflammation (C-reactive protein) and endothelial dysfunction (tissue plasminogen activator), suggesting that the vitamin has antiinflammatory effects associated with reduced levels of endothelial dysfunction (Table 10.14). Other studies have not observed antiinflammatory effects of vitamin C given to patients.

Common cold. The most widely publicized uses of "megadoses" of vitamin C are in prophylaxis and treatment of the common cold. Large doses (≥1 g) of vitamin C have been advocated for prophylaxis and treatment of the common cold, a use that was first proposed some 25 years ago by Irwin Stone and the Nobel laureate Dr Linus Pauling.[110] Since that time, many controlled clinical studies have been conducted to test that hypothesis (Table 10.15). Whereas many of these have yielded positive results, until recently few have been appropriately designed with respect to blinding, controls, treatment randomization, and statistical power, to make such conclusions unequivocal. In general, most results of well-controlled trials have indicated only small positive effects in reducing the incidence, shortening the duration, and ameliorating the symptoms of the common cold.[111] A meta-analysis of six large clinical trials (including more than 5000 episodes) showed no detectable effects of gram doses of vitamin C on cold incidence, but some evidence of small, protective effects in some subgroups of subjects.[112] A later meta-analysis of 29 randomized, controlled trials noted a consistent benefit of vitamin C supplementation (≥200 mg/day) in reductions of cold duration by 8% in adults and 13.5% in children.[113]

Other Infections

The results of clinical intervention trials with vitamin C on other infections have been inconsistent. Some studies with scorbutic guinea pigs, fishes, and rhesus monkeys have shown vitamin C deficiency to decrease resistance to infections,[114] but several studies have yielded negative results. Studies of ascorbic acid supplementation of species that do not require the vitamin (rodents, birds) have generally shown improved resistance to infection as indicated by increased survival of infected animals, depressed parasitemia, enhanced bacterial clearance, and reduced duration of infection. Several randomized trials

107. Dietrich, M., Block, G., Hudes, M., et al., 2002. Cancer Epidemiol. Biomar. Prev. 11, 7–13; Dietrich, M., Block, G., Benowitz, N.L., et al., 2003. Nutr. Cancer 45, 176–184; Block, G., Jensen, C.D., Morrow, J.D., et al., 2008. Free Rad. Biol. Med. 45, 377–384.
108. The nonconvulsive stage of an acute hypertensive disease of pregnant and puerperal (after childbirth) women.
109. Premature detachment of the placenta.
110. Pauling received two Nobel Prizes: Chemistry, 1954; Peace, 1962.
111. Chalmers, T.C., 1975. Am. J. Med. 58, 532–536.
112. Hemilä, H., 1997. Br. J. Nutr. 77, 59–72.
113. Douglas, R.M. Hemila, H., Chalker, E., et al., 2007 Cochrane Database Syst. Rev. CD000980.
114. e.g., *Mycobacterium tuberculosis*, *Rickettsiae* spp., *Entamoeba histolytica*, and other bacteria, as well as *Candida albicans*.

TABLE 10.14 Relationship of Vitamin C Status and Biomarkers of Inflammation and Endothelial Function

	Quartile[a] of Plasma Ascorbic Acid, μM				
Biomarker	<14.44	14.44–27.11	27.11–40.25	>40.25	*p* Value
CRP[b], mg/L	1.88 (1.73–2.03)[c]	1.73 (1.60–1.80)	1.52 (1.40–1.63)	1.34 (1.23–1.44)	<0.001
Fibrinogen, g/L	3.30 (3.26–3.36)	3.29 (3.24–3.34)	3.18 (3.13–3.23)	3.12 (3.07–3.17)	<0.001
t-PA[d], ng/mL	10.92 (10.63–11.21)	10.66 (10.38–10.93)	10.70 (10.42–10.99)	10.31 (10.03–10.60)	0.01
Blood viscosity, mPa	3.41 (3.39–3.44)	3.41 (3.39–3.43)	3.40 (3.38–3.43)	3.35 (3.33–3.37)	<0.001

[a]*3019 Subjects.*
[b]*C-reactive protein.*
[c]*Mean (95% confidence interval).*
[d]*Tissue plasminogen activator.*
Adapted from Wannamethee, S.G., Lowe, G.D., Rumley, A., et al., 2006. Am. J. Clin. Nutr. 83, 567–574.

TABLE 10.15 Large-Scale, Placebo-Controlled Clinical Trials Do Not Show Vitamin C Protection From Colds

	Vitamin C, g/day		Vitamin C Group		Placebo Group		
Study		months	n	Colds/p/year	N	Colds/p/year	RR (95% CI)
1	1	3	407	5.5	411	5.9	0.93 (0.83–1.04)
2	1	3	339	6.7	349	7.2	0.93 (0.84–1.04)
3	3	9	101	1.7	89	1.8	0.93 (0.73–1.20)
4	2	2	331	11.8	343	11.8	1.00 (0.90–1.12)
5	1	3–6	265	1.2	263	1.2	1.03 (0.80–1.32)
6	1	3	304	8.6	311	8.0	1.08 (0.97–1.21)

Adapted from Hemilä, H., 1997. Br. J. Nutr. 77, 59–72.

have found vitamin C administered to nondeficient individuals to reduce incidence[115] and/or severity[116] of infections other than colds (Table 10.16):

- **Heliobacter pylori**. Randomized trials have shown that vitamin C supplementation can reduce seropositivity for *H. pylori*[117] and protect against the progression of gastric atrophy in seropositive patients.[118] This may be associated with reduced gastric cancer risk for which *H. pylori* is a risk factor.
- **Herpes**. Topical application of ascorbic acid reduced the duration of lesions as well as viral shedding in patients with *Herpes simplex* infections.[119]

115. e.g., Posttransfusion hepatitis, pneumonia, tuberculosis, pharyngitis, laryngitis, tonsillitis, secondary bacterial infections after a common cold episode, and rheumatic fever.
116. e.g., *Herpes labialis*, bronchitis, tonsillitis, rubella, and tuberculosis.
117. Simon, J.A., Hudes, E.S., Perez-Perez, G., et al., 2003. J. Am. Coll. Nutr. 22, 283–289.
118. Sasazuki, S., Sasaki, S., Tsubono, Y., et al., 2003. Cancer Sci. 94, 378–382.
119. Hamuy, R., Berman, B., 1998. Eur. J. Dermatol. 8, 310–319.

Cardiovascular Health

The antioxidant characteristics of ascorbic acid allow it to have an antiatherogenic function in reducing the oxidation of LDLs, a key early event leading to atherosclerosis.[120] Being rich in both cholesterol and PUFAs, LDLs are susceptible to lipid peroxidation by the oxidative attack of ROS. Research has shown that oxidized LDLs stimulate the recruitment, in the subendothelial space of the vessel wall, of monocyte–macrophages that can take up the oxidized particles via scavenger receptors[121] to form the lipid-containing foam

120. **Atherosclerosis** is the focal accumulation of acellular, lipid plaques in the intima of the arteries. The subsequent infiltration by fatty substances (**arteriosclerosis**) and calcific plaques and the consequent reduction in the vessel's lumenal cross-sectional area reduce blood flow to the organs served by the affected vessel, causing such symptoms as angina, cerebrovascular insufficiency, and intermittent claudication.
121. Monocyte–macrophages have few LDL receptors, which are downregulated. When incubated with nonoxidized LDLs they do not form foam cells, as the accumulation of cholesterol further reduces LDL receptor activity. On the other hand, these cells have a specific **scavenger receptor** for modified LDLs. LDL lipid peroxidation products may react with amino acid side chains of apoB to form epitopes with affinities for the scavenger receptor.

TABLE 10.16 Results of Placebo-Controlled, Double-Blinded Studies of Vitamin C and Noncold Infections

Study	Infection	Vitamin C, g/day	Cases/Total Subjects		
			Vitamin C	Placebo	OR (95% CI)
Studies of Infection Incidence					
1[a]	Hepatitis	3.2	6/90	8/85	0.69 (0.26–1.80)
2[b]	Pneumonia	2	1/331	7/343	0.15 (0.01–0.74)
3[c]	Bronchitis	1	8/139	13/140	0.60 (0.27–1.30)
4[c]	Pharyngitis laryngitis and tonsillitis	1	7/139	14/140	0.48 (0.21–1.10)
				Value (n)	
Study	**Infection**	**Vitamin C, g/day**	**Outcome**	**Vitamin C**	**Placebo**
Studies of Infection Severity					
5[d]	*Herpes labialis*	0.6	Days healing	4.21 ± 7[f] (19)	9.72 ± 8 (10)
		1.0	Days healing	4.4 ± 3.9 (19)	9.72 ± 8 (10)
6[e]	Bronchitis	0.2	Decreased score	3.4 ± 1.8 (28)	2.3 ± 2.5 (29)
7[a]	Hepatitis	3.2	SGOT[g] units	474 ± 386 (6)	759 ± 907 (8)

[a]*Kodell, R.G., et al., 1981. Am. J. Clin. Nutr. 34, 20–23.*
[b]*Pitt, H.A., Costrini, A.M., 1996. J. Am. Med. Assoc. 241, 908–911.*
[c]*Ritzel, G., 1961. Helv. Med. Acta 28, 63–68.*
[d]*Terezhalmy, T., et al., 1978. Oral Surg. 45, 56–62.*
[e]*Hunt, C. et al., 1994. Int. J. Vit. Nutr. Res. 64, 212–219.*
[f]*Significantly different from control value,* $p<.05$.
[g]*Serum glutamic–oxaloacetic transaminase; now referred to as aspartate aminotransferase (AST).*

cells found in the early stages of atherogenesis. According to this view, atherogenesis can be reduced by protecting LDLs from free-radical attack, which would appear also to involve both vitamin E in quenching radicals produced in the hydrophobic interior of the LDL particle and ascorbate in recycling vitamin E.

Epidemiologic studies have found cardiovascular disease risk to be inversely associated with vitamin C status. Plasma ascorbic acid concentrations were negatively correlated with plasma malonyldialdehyde concentration and several cardiovascular risk factors: blood pressure (Fig. 10.19), total serum cholesterol, and LDL–cholesterol.[122] An intervention trial found vitamin C supplementation to reduce blood pressure and improve arterial stiffness in patients with noninsulin-dependent diabetes. It has been suggested that ascorbic acid may serve to protect cell membrane pumps from oxidative damage in such ways as to promote ion flux and enhance the vasoactive characteristics of blood vessels.

An analysis of the cohort examined in NHANES II showed that individuals with serum ascorbic acid levels associated with tissue saturation (≥1 mg/dL or ≥60 μM) had risk reductions of 21–27% for coronary heart disease incidence, stroke incidence, and cardiovascular deaths compared to individuals with lower levels of the vitamin. Serum ascorbic acid levels were associated with a 52% reduction in risk to angina without reductions in myocardial infarction or stroke.[123] Two large prospective studies in Japan and Finland found

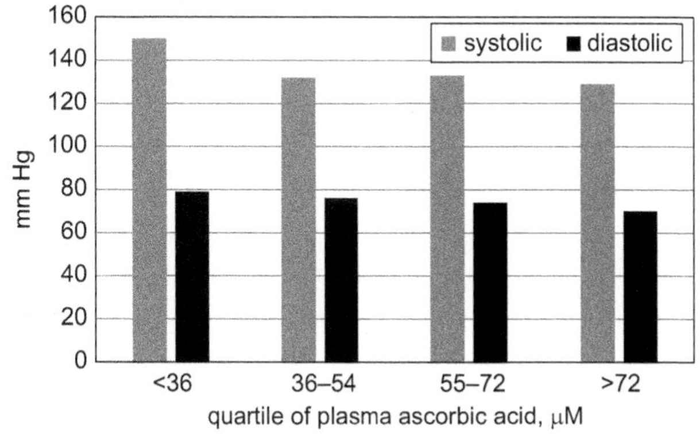

FIGURE 10.19 Relationship between blood pressure and plasma ascorbic acid. *After Choi, E.S.K., Jacques, P.F., Dallai, G.E., et al., 1991. Nutr. Res. 11, 1377–1385.*

122. Toohey, L., Harris, M.A., Allen, K.G., et al., 1996. J. Nutr. 126, 121–128.

123. Simon, J.A., Hudes, E.S., Browner, W.S., 1998. Epidemiology 9, 316–321; Simon, J.A., Hudes, E.S., Tice, J.A., 2001. J. Am. Coll. Nutr. 3, 255–263.

TABLE 10.17 Relationship of Vitamin C Intake and Relative Risk of Coronary Heart Disease: Pooled Results of Nine Studies

Vitamin C Intake, mg/day	Relative Risk[a]
0	1.00
<100	0.97 (0.80–1.18)[b]
100–399	0.82 (0.57–1.18)
400–699	0.94 (0.68–1.32)
≥700	0.75 (0.58–0.98)

[a]*Incidence and mortality.*
[b]*95% C.I.*
Adapted from Knekt, P., Ritz, J., Pereira, M.A., et al., 2004. Am. J. Clin. Nutr. 80, 1508–1520.

serum ascorbic acid levels >0.8 mg/dL to be associated with 30–50% reductions in risks to cerebral infarction and hemorrhagic stroke; these effects were not associated with supplement use, suggesting a role of vitamin C-containing foods, perhaps unrelated to the vitamin per se.[124] Still, protective effects of relatively high vitamin C intakes and, particularly, vitamin C supplement use, are apparent. Individuals in the NHANES I cohort with the highest vitamin C intakes showed 34% less fatal cardiovascular disease compared to subjects with lower estimated vitamin C intakes.[125] A pooled analysis of nine cohorts found fatal coronary heart disease to be reduced by 25% among individuals with vitamin C intakes ≥700 mg/day (Table 10.17).[126] A large, prospective study found that Japanese adults with serum ascorbate concentrations >45 μM had 30–50% less risk of stroke compared to subjects with lower serum levels.[127]

Pulmonary Function

Whether supplemental vitamin C can benefit pulmonary function is not clear. Population studies have found high vitamin C intake to be associated with improved pulmonary function.[128,129] The results of five trials have suggested that vitamin C intake may be inversely related to susceptibility to pneumonia. Three controlled trials have found vitamin C supplements effective in preventing pneumonia; two found that treatment effective in reducing the symptoms of that condition.[130] Half of the dozen clinical intervention trials of vitamin C to date have found improvements in parameters of respiratory function of asthma patients; but these studies have been very small (fewer than 160 patients total). A meta-analysis of randomized, controlled trials revealed no evidence to support the use of vitamin C in the management of asthma.[131]

Anticarcinogenesis

Ascorbic acid has been observed to reduce the binding of polycyclic aromatic carcinogens to DNA and to reduce/delay tumor formation in several animal models. This effect is thought to involve quenching of radical intermediates of carcinogen metabolism. Ascorbic acid is also a potent inhibitor of nitrosamine-induced carcinogenesis, functioning as a nitrite scavenger. This action results from ascorbate reduction of nitrate (the actual nitrosylating agent of free amines) to NO, blocking the formation of nitrosamines.[132] Evidence indicates that ascorbic acid, normally secreted in relatively high concentrations in gastric juice,[133] is a limiting factor in nitrosation reactions, particularly in individuals with gastric pathologies affecting secretion.

Some malignant cells appear to be particularly sensitive to ascorbic acid, namely, those that overexpress GLUT1. Studies have demonstrated this phenomenon for cancer cells with KRAS or BRAF[134] mutations, rapidly take up dehydroascorbic acid.[135] That form can be oxidized to ascorbate to increase ROS and deplete GSH and NADPH. It can also react directly with homocysteine thiolactone, overproduced by cancer cells, converting the latter to mercaptopropionaldehyde, which is lethal to cancer cells.[136]

Protective effects of dietary vitamin C have been detected in two-thirds of the epidemiologic studies in which a dietary vitamin C index has been calculated. In several cases, high vitamin C intake was associated with half the cancer risk associated with low intake. Protective effects have also been detected in a similarly high proportion of studies in which the intake of fruit, but not vitamin C, was assessed. Prediagnostic plasma ascorbate has been found to be inversely associated with risk of gastric adenocarcinoma in a Chinese cohort.[137] Clinical reports have indicated that high doses (10–60 g) can be useful in raising plasma ascorbic acid to levels (*c*.20 mM) capable

124. Yokoyama, T., Date, C., Kokubo, Y., et al., 2000. Stroke 31, 2287–2294; Kurl, S., Tuomainen, T.P., Laukkanen, J.A., et al., 2002. Stroke 33, 1568–1573.
125. Enstrom, J.E., Kanim, L.E., Klein, M.A., 1992. Epidemiology 5, 194–202.
126. Knekt, P., Ritz, J., Pereira, M.A., et al., 2004. Am. J. Clin. Nutr. 80, 1508–1520.
127. Yokoyama, T., Date, C., Kokubo, Y., et al., 2000. Stroke 31, 2287–2294.
128. i.e., Improved forced expiratory volume and improved forced vital capacity.
129. Hu, G., Cassano, P.A., 2000. Am. J. Epidemiol. 151, 975–981.
130. Hemila, H., Louhiala, P., 2007. J. Roy. Soc. Med. 100, 495–498.
131. Ram, F.S., Rowe, B.H., Kaur, B., 2004. Cochrane Database Syst. Rev. 3, CD00993.
132. Vitamin E also has this effect.
133. The concentration of ascorbic acid in gastric juice has often been found to exceed those in the plasma.
134. The *KRAS* and *BRaf* genes encode proteins that function in signaling and directing cell growth.
135. Yun, J., Mullarky, E., Lu, C., et al., 2015. Science 350, 1391–1396.
136. Toohey, J.I., 2008. Cancer Lett. 263, 164–169.
137. Lam, T.K., Freedman, N.D., Fan, J.H., et al., 2013. Am. J. Clin. Nutr. 98, 1289–1297.

TABLE 10.18 Adrenal Depletion of Ascorbic Acid in Laying Hens Under Simulated Adrenal Stress

Treatment	Renal Ascorbate µg/g	Adrenal Ascorbate µg/g	Adrenal Cholesterol mg/g	Adrenal Corticosterone µg/g	Serum Corticosterone µg/L
Control	1.41 ± 0.10	1.02 ± 0.05	6.93 ± 0.25	18 ± 2	4.8 ± 0.4
ACTH[a]	1.14 ± 0.14*[c]	0.77 ± 0.04*	2.57 ± 0.36*	32 ± 3*	4.4 ± 0.4
Dex[b]	1.30 ± 0.06*	0.82 ± 0.08*	8.02 ± 0.83*	17 ± 2	2.4 ± 0.9*

[a]*ACTH, adrenocorticotropic hormone, 2.5 IU/day.*
[b]*Dex, dexamethasone (a suppressor of adrenal corticosterone production), 50 µg/day.*
[c]*Significantly different from control, $p < .05$.*
Adapted from Rumsey, G.L., 1969. Studies of the Effects of Simulated Stress and Ascorbic Acid upon Avian Adrenocortical Function and Egg Shell Metabolism. (Ph.D. thesis). Cornell University, Ithaca, New York.

of killing cancer cells in vitro and in improving outcomes of three small series of cancer patients.[138] A meta-analysis of 10 studies with a total of almost 18,000 subjects found that postdiagnostic supplementation with vitamin C increased survival of breast cancer patients by 15%. For a 100 mg per day increase in vitamin C intake, all-cause mortality was reduced by 27%, and breast cancer mortality was reduced by 22%.[139]

Oxidative Stress

Ischemia–reperfusion injury. Tissues sustain injury upon reperfusion after a period of ischemia, a phenomenon with particular relevance to the preservation of organs for transplantation. ROS are thought to contribute to milder forms of tissue injury at the time of reperfusion (e.g., myocardial stunning, reperfusion arrhythmias); however, it is not clear the extent to which free radicals may also be responsible for the acute tissue damage seen under those circumstances. Vitamin C has been shown to be protective against ischemia–reperfusion injury in animal models,[140] and a randomized trial with men with peripheral artery disease found that intra-arterial administration of ascorbic acid (24 mg/min) completely prevented experimental ischemia–reperfusion-induced endothelial dysfunction.[141]

Exercise. Vigorous physical activity increases ventilation rates and produces oxidative stress,[142] which is thought to affect endothelial function. Studies have shown that antioxidant supplementation can alleviate muscle damage and protein oxidation induced by exercise. Vitamin C treatment prevented acute endothelial dysfunction induced by exercise in patients with intermittent claudication[143] (calf pain during walking). This effect is likely due to the protection of nitric oxide (NO), which mediates endothelium-dependent vasodilation.

Metabolically produced ROS also appear to have essential metabolic functions as signaling molecules for the adaptation of skeletal muscle to accommodate the stresses presented by exercise training or periods of disuse. This signaling involves redox-sensitive kinases, phosphatases, and NF-κB, which affect the rate of mitochondrial biogenesis, as well as the induction of genes related to insulin sensitivity and ROS defense. This system of adaptive responses to oxidative stress facilitates the ultimate development of long-term resistance to that stress. This system, which has been called mitochondrial **hormesis**,[144] can be impaired by high-level antioxidant treatment. Combined supplements of vitamin C (500 mg/day ascorbic acid) and E (400 IU/day α-tocopheryl acetate) blocked the upregulation of muscle glucose uptake otherwise induced by exercise in untrained subjects.[145] This finding raises questions as to what level of "redox tone" may be beneficial.

Environmental stress. Vitamin C deficiency can have benefits under conditions of environmental stress in which the metabolic demands for ascorbic acid may exceed the rate of its endogenous biosynthesis. This is the case in commercial poultry production. Although the chicken does not require the vitamin in the classic sense, under practical conditions of poultry management, the species frequently benefits from ascorbic acid supplements[146] under stressful environmental conditions (e.g., extreme temperature, prevalent disease, crowding, inadequate ventilation) that stimulate adrenal depletion of ascorbic acid (Table 10.18). Controlled experiments with laying hens have shown that supplemental ascorbic acid can improve egg production and eggshell characteristics in laying hens subjected to heat stress.

138. Cameron, E., Pauling, L., 1978. Proc. Natl. Acad. Sci. U.S.A. 75, 4538–4342; Drisko, J., Chapman, J., Hunter, V., 2003. J. Am. Coll. Nutr. 22, 118–123; Padayatty, S., Riordan, H. Hewitt, S., et al., 2006. CMAJ 174, 937–942.
139. Harris, H.R., Orsini, N., Wolk, A., 2014. Eur. J. Cancer 50, 1223–1231.
140. Bailey, D.M., Raman, S., McEneny, J., et al., 2006. Free Rad. Biol. Med. 40, 591–600.
141. Pleiner, J., Schaller, G., Mittermayer, F., et al., 2008. Atherosclerosis 197, 383–391.
142. O_2 utilization increases 10- to 15-fold during exercise.
143. Silvestro, A., Scopacasa, F., Oliva, G., et al., 2002. Atherosclerosis 165, 277–283.
144. Hormesis is the term for a generally favorable biological response to low exposures to stressors/toxins.
145. Ristow, M., Zarse, K., Oberbach, A., et al., 2009. Antioxidants prevent health-promoting effects of physical exercise in humans. Proc. Nat. Acad. Sci. U.S.A. 106, 8665–8670.
146. e.g., 150 mg/kg diet.

TABLE 10.19 Recommended Upper Tolerable Intakes (ULs) of Preformed Vitamin C

Ages, Years	UL, mg/day
<1	–[a]
1–3	400
4–8	650
9–13	1200
14–18	1800
>18 years	2000
Pregnancy	
≤18 years	1800
>18 years	2000
Lactation	
≤18 years	1800
>18 years	2000

[a]*No value established.*
Adapted from Food and Nutrition Board, 2000. Dietary Reference Intakes for Vitamin C, Vitamin E, Selenium and Carotenoids. National Academy Press, Washington, DC, 506 pp.

11. VITAMIN C TOXICITY

No significant adverse effects of ascorbic acid, its various salts and esters, have been identified.[147] High doses of the vitamin have been used both orally and intravenously without incident. Some subjects taking megadoses of the vitamin have reported gastrointestinal disturbances and diarrhea. Upper tolerable intakes for vitamin C have been established (Table 10.19).

Little information is available on vitamin C toxicity in animals, although acute LD_{50} (50% lethal dose) values for most species and routes of administration appear to be at least several grams per kilogram of body weight. A single study showed mink to be very sensitive to hypervitaminosis C, with daily intakes of 100–200 mg of ascorbic acid, pregnant females developed anemia and had reduced litter sizes.

Questions have been raised as to whether megadoses of vitamin C may have risks. Evidence has shown that most, if not all, of these concerns are not warranted.

- **Urinary oxalates**. In humans, unlike other animals, oxalate is a major metabolite of ascorbic acid, which accounts for 35–50% of the 35–40 mg of oxalate excreted in the urine each day.[148] Metabolic studies have demonstrated that healthy individuals convert <1.5% of ingested ascorbic acid to oxalic acid each day. Even when the vitamin was administered intravenously, <0.5% was excreted as <0.5% as oxalate.[149] Still, a case of urolithiasis in a 9-year-old boy was reported. The subject showed extremely high-urinary oxalate concentrations after having been given high doses of vitamin C since age 3. Prohibition of vitamin C supplements reduced his oxalate excretion to normal levels with no recurrence of symptoms for the duration of follow-up (3 years).[150] Some clinical studies have found multiple gram doses of vitamin C to produce oxaluria of low magnitude and within normal variation.[151] While such effects are unlikely to increase risk of forming urinary calculi, prudence dictates avoiding vitamin C doses >1000 mg for individuals with histories of renal stones.
- **Uricosuria**. Ascorbic acid and uric acid are both reabsorbed by the renal tubules, perhaps by the same SCVT. If so, then high levels of ascorbic acid might competitively inhibit uric acid reabsorption. Evidence does not support that hypothesis. A randomized trial with healthy subjects showed that vitamin C (500 mg/day) significantly reduced serum uric acid concentrations and increased the glomerular filtration rate.[152]
- **Vitamin B_{12} stability**. In the 1970s, it was claimed that high levels of ascorbic acid added to test meals resulted in the destruction of food vitamin B_{12}. Studies have not supported that prospect. In fact, the only cobalamin that is sensitive to reduction by ascorbic acid is aquocobalamin, which is not a major form of the vitamin in foods. Several clinical investigations have found high doses of vitamin C not to affect vitamin B_{12} status.
- **Iron overload?** That ascorbic acid can enhance the enteric absorption of dietary iron has led to questions as to whether megadoses of the vitamin C might promote iron accumulation in iron-replete individuals. Such an effect is not expected, as optimal iron absorption is effected with much lower doses of vitamin (25–50 mg per meal). Studies in mice have found ascorbic acid not to enhance prooxidative effects induced by dietary iron; nor did high parenteral doses of ascorbic acid increase prooxidative biomarkers in human subjects.[153] Nevertheless, patients with

147. Hathcock, J.N., Azzi, A., Blumberg, J., et al., 2005. Am. J. Clin. Nutr. 81, 736–745; Elmore, A.R., 2005. Int. J. Toxicol. 24, 51–111.
148. The balance of urinary oxalate comes mainly from the degradation of glycine (about 40% of the total); but some also can come from the diet (5–10%).
149. Robitaille, L., Mamer, O.A., Miller, Jr., W.H., et al., 2009. Metab. Clin. Exp. 58, 263–269.
150. Chen, X., Shen, L., Gu, X., et al., 2014. Urology 84, 922–924.
151. Forty percentage of subjects given 2 g of ascorbic acid daily showed increases in urinary oxalate excretion by more than 10% (Chai, W., Liebman, M., Kynast-Gales, S., et al., 2004. Am. J. Kidney Dis. 44, 1060–1069).
152. Huang, H.Y., Appel, L.J., Choi, M.J., et al., 2005. Arthritis Rheum. 52, 1843–1847. This finding suggests that vitamin C may be beneficial in the management of gout.
153. Up to 7500 mg (Mühlhöfer, A., Mrosek, S., Schlegel, B., et al., 2004. Eur. J. Clin. Nutr. 58, 1151–1158).

hemochromatosis or other forms of excess iron accumulation are recommended to avoid taking vitamin C supplements with their meals.

- **Systemic conditioning**. That chronic intakes of large amounts of vitamin C might lead to persistent upregulation of ascorbic acid catabolism was once proposed. The concern was that such "systemic conditioning" might precipitate "rebound scurvy" in individuals returning to nutritional intakes of the vitamin. That hypothesis was based on uncontrolled observations of a few individuals and what is now widely regarded as erroneous interpretations of experimental results.[154] A well-controlled study with guinea pigs found transient declines in plasma ascorbic acid levels in some animals removed from high-level vitamin C treatments. Controlled studies have not consistently demonstrated such effects. Studies in humans indicate that high doses of ascorbic acid are mostly degraded to CO_2 by the gut microbiome, with major portions also excreted intact in the urine.
- **Mutagenesis**. Vitamin C is not intrinsically mutagenic; however, mutagenesis can be demonstrated in vitro by treating cells with ascorbate and Cu^{2+}, which produces ROS. No evidence of mutagenic effects in vivo has been produced; doses as great as 5000mg have not induced mutations in mice.[155]

12. CASE STUDIES

Instructions

Review each of the following case reports, paying special attention to the diagnostic indicators on which the treatments were based. Then answer the questions that follow.

Case 1

A 26-year-old man volunteered for a 258-day experiment of ascorbic acid metabolism. He was 184cm tall and weighed 84.1kg. His medical history, physical examination, vital signs, and past diet history revealed a healthy individual with no irregularities. During the experiment, his temperature, pulse, and respiration rates were recorded four times daily, and his blood pressure was measured twice daily. He was examined by an internist daily; periodically, he was examined by an ophthalmologist and had chest radiograms and electrocardiograms made. Twenty-four-hour collections of urine and feces were made daily to determine urinary and fecal nitrogen, and for the radioactive assay of ascorbic acid. Samples of expired air were collected for the measurement of radioactivity.

The subject was fed a control diet consisting of soy-based products. The diet provided 2.5mg of ascorbic acid per day, which was supplemented by a daily capsule containing an additional 75mg of ascorbic acid. The subject's body vitamin C pool was labeled with L-[1–^{14}C]ascorbate 1 week before initiating vitamin C depletion; it was calculated to be 1500mg. Beginning on day 14, the diet was changed to a liquid formula containing no vitamin C, as ascertained by actual analysis. This diet, based on vitamin-free casein, provided 3300kcal and supplied protein, fat, and carbohydrate as 15%, 40%, and 45% of total calories, respectively. It was fed from day 14 to day 104, during which time the subject developed signs of scurvy. Ascorbic acid was not detectable in his urine after 30days of depletion. He showed petechiae on day 45, when his vitamin C pool was 150mg and his plasma level was 0.19mg/dL. Spontaneous ecchymoses occurred over days 36–103; these were followed by coiled hairs, gum changes, *hyperkeratosis*,[156] congested follicles and the Sjögren sicca syndrome,[157] dry mouth, and enlarged parotid salivary glands. The subject developed joint pains on day 68 and joint effusions[158] shortly thereafter, when his vitamin C pool was 100mg and his plasma ascorbic acid level was less than 0.16mg/dL. He also had the unusual complication of a bilateral femoral neuropathy, which began on day 71, when his vitamin C pool was 80mg and his plasma ascorbic acid level was 0.15mg/dL. This, accompanied by the joint effusions, was attributed to hemorrhage into the sheaths of both femoral nerves. On day 80, he experienced a rapid increase in weight, from 81 to 84kg, in combination with dyspnea[159] on exertion and swelling of the legs. At this time, his vitamin C pool was 40mg and his plasma level was 0.15mg/dL.

Beginning on day 105, the subject was put on a vitamin C-repletion regimen involving daily doses of 4mg of ascorbic acid. Immediately following this treatment, the edema worsened, urinary output dropped to 340ml/day, and weight increased to 86.6kg on day 109. There was no evidence of pulmonary congestion or cardiac failure. The ascorbic acid-repletion dose was increased to 6.5mg/day on day 111. His edema persisted for 4days, at which time he had a profound diuresis with complete disappearance of the edema by day 133at which time his weight was 77.2kg (he lost 9.4kg of extracellular fluid). From days 101–133, his body ascorbic acid pool increased from 33 to 128mg. The subject was given 6.5mg of ascorbic acid per day from day 133 to day 227. During this time, all his scorbutic manifestations disappeared,

154. For example, enhanced $^{14}CO_2$ excretion from guinea pigs with larger body pools of ascorbic acid was taken as evidence of greater catabolism.
155. Vojdani, A., Bazargan, M., Vojdani, E., et al., 2000. Cancer Detect. Prev. 24, 508–523.

156. A disease of the mouth characterized with variously sized and shaped, grayish white, flat, adherent patches; having diffuse borders, and a smooth surface with no papillary projections, fissures, erosions, or ulcerations.
157. Dry eyes due to reduction in tears, i.e., keratoconjunctivitis.
158. The escape of fluid from the blood vessels or lymphatics into the joint capsule.
159. Subjective difficulty or distress in breathing, frequently rapid breathing.

and his plasma ascorbic acid fluctuated between 0.10 and 0.25 mg/dL. His body pool was restored slowly to an excess of 300 mg. Beginning on day 228, he received 600 mg of ascorbic acid per day, which rapidly repleted his body pool. At the end of the study, his weight was 81 kg and he was discharged from the metabolic ward in excellent health.

Case 2

A 72-year-old man was admitted to the hospital with symptoms of increasing anorexia, epigastric discomfort unrelated to meals, and nonradiating precordial[160] pain. During the year before admission, he had become increasingly weak and easily fatigued, and had lost nearly 13 kg in weight. Six weeks before admission, he began to have sudden attacks of severe substernal pain followed by cough and dyspnea, and 1 month before admission he had a small *hematemesis*[161] and had noted bright red blood in his stools. He had been living alone and his diet during the past year had consisted chiefly of bread and milk with various soups. For a considerable period, he had noted easy bruising of his skin. His past health had been good except for occasional seizures; these began 2 years before admission and involved loss of consciousness, spasmodic twitching of the limbs, and incontinence preceded by abdominal discomfort.

Physical examination on admission revealed a thin, depressed, lethargic man with a rather gray complexion, and numerous petechiae over the arms, legs, and trunk. His blood pressure was 140/80, his pulse was 68, his respiration was 19, and his temperature was 98.8°F. Examination of his head and neck showed an edentulous[162] mouth, foul breath, ulcerated palate, and retracted gums without hemorrhage. He had a large ecchymosis (15 cm in diameter) on his right thigh. Neurological examination was negative.

Laboratory Findings

Parameter	Patient	Normal Range
Hb	13.2 g/dL	15–18 g/dL
WBC	8000/μL	5000–9000/μL
Platelets	140,000/μL	150,000–300,000/μL
Clotting time	5.75 min	5–15 min
Blood urea	48 mg/dL	10–20 mg/dL
Serum protein	7 g/dL	6–8 g/dL
Serum albumin	3.9 g/dL	3.5–5.5 g/dL
Serum ascorbic acid	<0.1 mg/dL	0.4–1.0 mg/dL

160. Relating to the diaphragm and anterior surface of the lower part of the thorax.
161. Vomiting of blood.
162. Toothless.

His heart was not enlarged and there were no heart murmurs; however, his electrocardiogram showed changes typical of an old myocardial infarction.[163] His chest radiograms showed emphysematous[164] and atheromatous[165] changes. His urine contained occasional pus cells with moderate growth of *Escherichia coli*; no abnormal bacilli were seen in the sputum. Sigmoidoscopy revealed no lesions in the distal 25 cm of the bowel. Because of his anorexia, epigastric discomfort, weight loss, and hematemesis, further investigation of the gastrointestinal tract was made using a barium bolus; this revealed a mass and ulcer crater in the prepyloric area of the stomach, suggesting a gastric neoplasm. A laparotomy[166] was planned. The tentative diagnoses were anterior myocardial infarction, suspected cancer of the stomach, epilepsy, and hemorrhagic diathesis[167] (probably scurvy). Accordingly, the patient was given a high-protein diet and ascorbic acid (1 g/day for 2 weeks, then 150 mg/day for a month).

The patient showed marked improvement following ascorbic acid treatment. He no longer showed an air of lassitude; he gained weight and began to relish his meals. His skin hemorrhages rapidly decreased and no new ones appeared. Three weeks after admission, blood disappeared from his feces. At that time, his epilepsy was satisfactorily controlled using phenobarbital, and his liver function tests and blood chemistry were normal.

Laboratory Findings after Vitamin C Treatment

Parameter	Patient
Blood urea	28 mg/dL
Serum protein	6.3 g/dL
Serum albumin	3.9 g/dL
Serum ascorbic acid	1.0 mg/dL

A second radiological examination, conducted 1 month after ascorbic acid treatment, indicated a normal pylorus; this was confirmed by gastroscopy. A biopsy of the previously involved area showed only a natural glandular pattern, with hemorrhage of the superficial layer of the gastric mucosa. The patient was discharged after 8 weeks of hospitalization and was well when seen later in the outpatient

163. Necrotic changes resulting from obstruction of an end artery.
164. Emphysema involves dilation of the pulmonary air vesicles, usually due to atrophy of the septa between the alveoli.
165. Atheroma refers to the focal deposit or degenerative accumulation of soft, pasty, acellular, lipid-containing material frequently found in intimal and subintimal plaques in arteriosclerosis.
166. A surgical procedure involving incision through the abdominal wall.
167. Any of several syndromes showing a tendency to spontaneous hemorrhage, resulting from weakness of the blood vessels and/or a clotting defect.

clinic. The gastric lesion did not recur. It was concluded that what had appeared to be a prepyloric tumor and ulcer had actually been a bleeding site with a hematoma.[168]

Case Study Questions

1. What thresholds are suggested by the results of the first case study for total body ascorbic acid pool size and plasma ascorbic acid concentration associated with freedom from signs of scurvy?
2. Compute the rate of reduction in ascorbic acid body pool size from the observations on the subject of the first case. Was it linear throughout the study?
3. What signs/symptoms did the patient in the second case show that indicated a problem related to vitamin C status?

13. STUDY QUESTIONS AND EXERCISES

1. Construct a concept map illustrating the relationship of the chemical properties and physiological functions of vitamin C.
2. Construct a decision tree for the diagnosis of vitamin C deficiency in humans.
3. What health complications might you expect to be shown by scorbutic individuals?
4. Compare and contrast the antioxidant properties of vitamins C and E.

168. A localized mass of extravasated blood, usually clotted.

RECOMMENDED READING

Aguirre, R., May, J.N., 2008. Inflammation in the vascular bed: importance of vitamin C. Pharmacol. Ther. 119, 96–103.

Corti, A., Casini, A.F., Pompella, A., 2010. Cellular pathways for transport and efflux of ascorbate and dehydroascorbate. Arch. Biochem. Biophys. 500, 107–115.

Drouin, G., Godin, J.R., Pagé, B., 2011. The genetics of vitamin C loss in vertebrates. Curr. Genomics 12, 371–378.

Hathcock, J.N., Azzi, A., Blumberg, J., et al., 2005. Vitamins E and C are safe across a broad range of intakes. Am. J. Clin. Nutr. 81, 736–745.

Johnston, C.S., Steinberg, F.M., Rucker, R.B., 2014. Ascorbic acid. In: Zempleni, J., Suttie, J.W., Gregory, J.F., Stover, P.J. (Eds.), Handbook of Vitamins", fifth ed. CRC Press, New York, pp. 515–549 (Chapter 14).

Lane, D.J.R., Lawen, A., 2009. Ascorbate and plasma membrane electron transport – enzymes vs. efflux. Free Rad. Biol. Med. 47, 485–495.

Lindblad, M., Tvenden-Nyborg, P., Lykksefeldt, J., 2013. Regulation of vitamin C homeostasis during deficiency. Nutrients 5, 2860–2879.

Lykkesfeldt, J., Poulsen, H.E., 2010. Is vitamin C supplementation beneficial? Lessons learned from randomized controlled trials. Br. J. Nutr. 103, 1251–1259.

May, J.M., Harrison, F.E., 2013. Role of vitamin C in the function of the vascular epithelium. Antioxid. Redox Signal. 19, 2068–2083.

Michels, A., Frie, B., 2013. Vitamin C. In: Stipanuk, M.H., Caudill, M.A. (Eds.), Biochemical, Physiological and Molecular Aspects of Human Nutrition, third ed. Elsevier, New York, pp. 626–654 (Chapter 27).

Pauling, L., 1970. Vitamin C and the Common Cold. W.H. Freeman and Company, San Francisco.

Rivas, C.I., Zuniga, F.A., Salas-Burgos, A., Mardones, L., et al., 2008. Vitamin C transporters. J. Physiol. Biochem. 64, 357–375.

Said, H.M., 2011. Intestinal absorption of water-soluble vitamins in health and disease. Biochem. J. 437, 357–372.

Thakar, N.Y., Wolvetoan, E.J., 2015. Ascorbate as a modulator of the epigenome, chapter 7. In: Ho, E., Domann, F. (Eds.), Nutrition and Epigenetics. CRC Press, Boca Raton, FL, pp. 199–218.

Traber, M.G., Stevens, J.F., 2011. Vitamins C and E: beneficial effects from a mechanistic perspective. Free Rad. Biol. Med. 51, 1000–1013.

Chapter 11

Thiamin

Chapter Outline

Anchoring Concepts

1. Thiamin is the trivial designation of a specific compound, 3-(4-amino-2-methylpyrimidin-5-ylmethyl)-5-(2-hydroxyethyl)-4-methylthiazolium, which is sometimes also called **vitamin** $\mathbf{B_1}$.
2. Thiamin is hydrophilic and its protonated form has a quaternary nitrogen center in the thiazole ring.
3. Deficiencies of thiamin are manifest chiefly as neuromuscular disorders.

There is present in rice polishing a substance different from protein and salts, which is indispensable to health and the lack of which causes nutritional polyneuritis.

C. Eijkman and G. Grijns[1]

LEARNING OBJECTIVES

1. To understand the chief natural sources of thiamin
2. To understand the means of absorption and transport of thiamin
3. To understand the biochemical function of thiamin as a coenzyme and the relationship of that function to the physiological activities of the vitamin
4. To understand the physiological implications of low thiamin status.

1. Christiaan Eijkman (1858–1930) was a Dutch physician whose demonstration that beriberi in Java, Dutch East Indies (Indonesia) had a dietary etiology. His assistant at the time, Dutch physician Gerrit Grijns (1865–1944), is thought to have interpreted their results as indicating a nutrient deficiency, presaging the "vitamin theory." For seminal work leading to the recognition of the vitamins, Eijkman shared with Frederick Gowland Hopkins (who had demonstrated "accessory food factors") the 1929 Nobel Prize for Physiology or Medicine.

VOCABULARY

Acute pernicious beriberi
Alcohol
Aleurone
γ-Aminobutyric acid (GABA)
Amprolium
Anorexia
Ataxia
ATPase
Beriberi
Bradycardia
Cardiac beriberi
Cardiac hypertrophy
Chastek paralysis
Cocarboxylase
Confabulation
Dry beriberi
Dyspnea
Encephalopathy
Fescue toxicity
Hexose monophosphate shunt
Infantile beriberi
α-Ketoglutarate dehydrogenase
Maple syrup urine disease
Neuropathy
Nystagmus
Ophthalmoplegia
Opisthotonos
Oxythiamin
Pentose phosphate pathway
Perseveration
Phosphorylase
Polioencephalomalacia

The Vitamins. http://dx.doi.org/10.1016/B978-0-12-802965-7.00011-3

Polyneuritis
Pyrimidine ring
Pyrithiamin
Pyruvate decarboxylase
Pyruvate dehydrogenase
Rapsyn
Shoshin beriberi
Stargazing
Sulfate
Sulfite
Tachycardia
Thiamin-binding protein (TBP)
Thiamin disulfide
Thiaminases
Thiamin diphosphate phosphotransferase
Thiamin monophosphate (TMP)
Thiamin monophosphatase
Thiamin pyrophosphate (TPP)
Thiamin pyrophosphatase
Thiamin pyrophosphokinase
Thiamin-responsive megaloblastic anemia (TRMA)
Thiamin transporters (ThTr1, ThTr2)
Thiamin triphosphate (TTP)
Thiazole ring
Thiochrome
Transketolase
TPP–ATP phosphoryltransferase
Vitamin B_1
Wernicke–Korsakoff syndrome
Wet (edematous) beriberi
Wilson disease

1. THE SIGNIFICANCE OF THIAMIN

Thiamin is essential in carbohydrate metabolism and neural function; it is required for the production of ATP, ribose, NAD,[2] and DNA. Severe thiamin deficiency results in the nerve and heart disease **beriberi**. Less severe deficiency results in nonspecific signs: malaise, loss of weight, irritability, and confusion. Thiamin-deficient animals show inanition and poor general performance and, in severe cases, **polyneuritis**, making thiamin status economically important in livestock production.

Historically, thiamin deficiency has been prevalent among peoples dependent on polished rice as a staple food. Demographic trends indicate that for many people dependence on rice is likely to increase in the future. Rice and rice/wheat crop rotations are now the basis of the food systems currently supporting one-fifth of the world's people, i.e., those in East, South, and Southeast Asia where populations are expected to more than double within the next four decades. The irony is that whole-grain rice and other cereals are not particularly deficient in thiamin; however, the removal of their thiamin-containing **aleurone** cells[3] renders the polished grains, which consist of little more than the carbohydrate-rich endosperm, nearly devoid of thiamin and other vitamins and essential elements. In fact, thiamin-containing rice polishings are often used to fuel the parboiling of the thiamin-deficient grain. Thus, storage technologies that reduce the need to polish rice such that increased reliance on the new, high-yielding cultivars of rice will not lead to expansions of thiamin deficiency among the poor of south Asia.

2. PROPERTIES OF THIAMIN

The term **thiamin** designates the compound 3-[(4-amino-2-methyl-5-pyrimidinyl)methyl]-5-(2-hydroxyethyl)-4-methylthiazolium, formerly known as **vitamin B_1**, **aneurine**, and **thiamine**. The structural features that are essential for its biological activity include an aromatic **pyrimidine ring** with an amino group on C-4 joined to an aromatic **thiazole ring** containing a quaternary N, an open C-2, and a phosphorylatable hydroxyethyl group on C-5.

Chemical Structures of Thiamin Vitamers

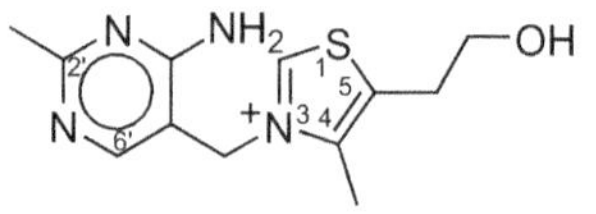

Thiamin (free base)

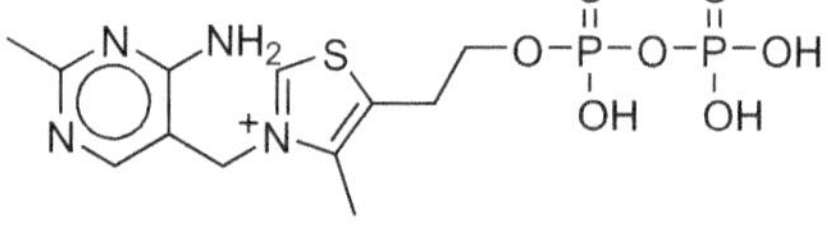

Thiamin pyrophosphate

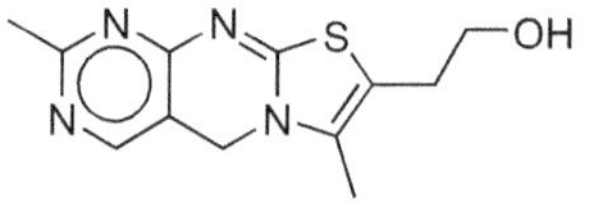

Thiochrome

Thiamin Chemistry

Free thiamin is a white powder with a characteristic "thiazome" odor and bitter taste. It is stable at moderate

2. Nicotinamide adenine dinucleotide, a metabolically active form of the vitamin niacin (see Chapter 13).

3. The aleurone is the outermost layer of the endosperm, surrounding the starchy interior, of grass seeds. In rice and other cultivated cereals, it contains protein-storing vacuoles, and large amounts of oils and minerals, which function in seed development and dormancy. Thus, it is the most nutritious component of most brans. However, rice oil is highly polyunsaturated and, thus, prone to peroxidative rancidity particularly when stored at tropical temperatures. For this reason, milled rice is "polished" by gentle mechanical abrasion to remove the aleurone layer, which greatly increases its useful storage life.

temperatures under slightly acidic conditions but is heat labile at neutral pH and highly unstable under alkaline conditions in which the thiazolium C-2 is vulnerable to ring-opening hydroxyl attack. It is freely soluble in water but practically insoluble in organic solvents. Therefore, the hydrochloride and mononitrate forms are used in commerce. Thiamin hydrochloride (actually, thiamin chloride hydrochloride) is a colorless crystal that is very soluble in water (1 g/mL, thus making it a very suitable form for parenteral administration), soluble in methanol and glycerol, but practically insoluble in acetone, ether, chloroform, and benzene. The protonated salt has two positive charges: one associated with the pyrimidine ring and one associated with the thiazole ring. The mononitrate form is more stable than the hydrochloride form, but it is less soluble in water (27 mg/mL). It is used in food/feed supplementation and in dry pharmaceutical preparations.

Free thiamin is easily oxidized to thiamin disulfide and other derivatives including thiochrome, a yellow biologically inactive product with strong blue fluorescence that can be used for the quantitative determination of thiamin. The thiazole hydroxyethyl group can be phosphorylated in vivo to form thiamin mono-, di-, and triphosphates. Thiamin diphosphate, also called **thiamin pyrophosphate (TPP)**, is the metabolically active form sometimes referred to as **cocarboxylase**. Thiamin antagonists of experimental significance include **pyrithiamin** (the analog consisting of a pyridine moiety replacing the thiazole ring) and **oxythiamin** (the analog consisting of a hydroxyl group replacing the C-4 amino group on the pyrimidine ring).

3. SOURCES OF THIAMIN

Hindgut Microbial Synthesis

Thiamin is produced by bacteria and fungi, including those in the hindgut microbiome. A genomic analysis of 256 representative organisms of the human gut microbiota found more than half capable of de novo synthesis of the vitamin.[4] Those findings suggested that hindgut microbial synthesis may produce some 2.3% of the daily human need for thiamin. It is not clear whether the colon is capable of carrier-mediated absorption of the vitamin, although a high-affinity thiamin transporter (THTR2) has been identified in the colonic mucosa. Therefore, it is likely that thiamin produced in the hindgut may be useful to noncoprophagous animals but that has not been demonstrated. It is clear that rumen microbiome is an important source of thiamin for ruminants. However, because rumen microbes can reduce sulfate to sulfite, high dietary levels of sulfate can have anti-thiamin antagonistic effects for ruminants.

4. Magnúsdóttir, S., Ravchee, D., de Crécy-Lagard, V., et al., 2015. Front. Genet. 6, 148–166.

Distribution in Foods

Thiamin is widely distributed in foods but most contain only low concentrations of the vitamin (Table 11.1). The richest sources are yeasts (e.g., dried brewer's and baker's yeasts) and liver (especially pork liver); however, cereal grains comprise the most important sources of the vitamin in most human diets. Thiamin in foods is considered to be readily available to healthy subjects except in cases of exposure to certain antagonists.

TABLE 11.1 Thiamin Contents of Foods

Food	Thiamin (mg/100 g)
Grains	
Cornmeal	0.39
Oats	0.76
Rice	
Brown	0.18
White, cooked	0.02–0.20
Rye flour	0.29
Wheat	
Whole grain	0.52
White	0
Vegetables	
Asparagus	0.16
Beans, green	0.07
Broccoli	0.07
Cabbage	0.06
Carrots	0.07
Cauliflower	0.05
Kale	0.05
Peas, green	0.27
Potatoes	0.11
Tomatoes	0.04
Fruits	
Apples	0.02
Apricots	0.03
Bananas	0.03
Grapes	0.09
Oranges	0.07
Pears	0.01
Pineapples	0.08

Continued

TABLE 11.1 Thiamin Contents of Foods—cont'd

Food	Thiamin (mg/100 g)
Meats	
Beef	0.02–0.10
Duck	0.26
Pork	0.41–0.92
Cured ham	0.82
Trout	0.15
Salmon	0.02–0.16
Liver beef	0.19
Pork	0.28
Dairy Products and Eggs	
Cheese	0.01–0.15
Milk	0.04–0.05
Eggs	0.04
Other	
Baker's yeast	1.89
Human milk	0.01

From USDA National Nutrient Database for Standard Reference, Release 28 (http://www.ars.usda.gov/ba/bhnrc/ndl).

Whole grains are typically rich in thiamin; however, the vitamin is distributed unevenly in grain tissues. The greatest concentrations of thiamin in grains are typically found in the scutellum (the thin layer between the germ and the endosperm) and the germ. The endosperm (the starchy interior) is quite low in the vitamin. Therefore, milling to degerminate grain, which, because it removes the highly unsaturated oils associated with the germ, yields a product that will not rancidify and, thus, has a longer storage life, also has very low thiamin content. It is estimated that more than one-third of thiamin in the US food supply is provided by grains and grain products, with meats providing about a quarter. In foods derived from plants thiamin occurs predominantly as free thiamin and thiamin monophosphate (TMP); in animal tissues it is found almost entirely (95–98%) in phosphorylated forms, predominantly (80–85%) TPP. TTP comprises only a small portion of total tissue thiamin in most foods, with the notable exceptions of chicken breast muscle and pork skeletal muscle, in which >70% of the vitamin is present at TPP.[5]

Stability in Foods

The stability of food thiamin depends largely on the form and pH. Thiamin is stable in dry preparations and at room temperatures under slightly acidic conditions. However, under conditions of neutral pH, it is susceptible to destruction by heating,[6] and at high temperatures it can undergo Maillard reactions.[7] Accordingly, thiamin is partially lost in cooking. Thiamin is very sensitive to sulfites, which cleave its two ring systems;[8] this reaction can result in the loss of thiamin from sulfite-treated foods, even at low storage temperatures. Thiamin is also sensitive to oxidation by ROS generated by UV light and ionizing radiation (Table 11.2). Protein-bound thiamin, found in animal tissues, is more stable to such losses. Thiamin is stable during frozen storage; substantial losses occur during thawing, however, mainly due to removal via drip fluid.

TABLE 11.2 Thiamin Losses in Food Processing

Procedure	Food	Loss (%)
Convection cooking	Meats	25–85
Baking	Bread	5–35
Heating with water	Vegetables	0–60
Pasteurization	Milk	9–20
Spray drying	Milk	10
Canning	Milk	40
Room temperature storage	Fruits, vegetables	0–20

Thiamin Antagonists

Thiamin in foods can be destroyed or antagonized by several compounds that may occur naturally (Table 11.3). Cases of thiamin deficiency have been found to be related to the ingestion of food containing such antagonists (Table 11.4).[9] They include the following:

- **Thiaminases**—Bacterial thiaminases are exoenzymes, i.e., they are bound to the cell surface; their activities depend on their release from the cell surface, which can occur under acidotic conditions in the rumen. Thiaminases are heat labile and can be rendered ineffective by heat treatment.

5. These tissues also have appreciable amounts of adenylate kinase, which can produce it from TPP using ADP as the phosphate donor.

6. Therefore, the practice of adding sodium bicarbonate to peas or beans for retention of their color in cooking or canning results in large losses of thiamin.

7. Reactions of amino acids and reducing sugars in food matrices; a form of nonenzymatic browning.

8. To yield (6-amino-2-methylprimid-5-yl)methanesulfonic acid and 5-β-hydroxyethyl-4-methylthiazole.

9. Perhaps the best known of these is the condition "**Chastek paralysis**," a neurological disorder described in commercial foxes fed a diet containing raw carp. The syndrome, named for the fox producer, was found to be a manifestation of thiamin deficiency brought on by a microbial thiaminase present in fish gut tissue. Cooking the fish before feeding them to Mr Chastek's foxes prevented the syndrome, apparently by heat denaturing the thiaminase.

TABLE 11.3 Types of Thiaminases

Type	Present in	Mechanism
I	Fresh fish, shellfish, ferns, some bacteria	Displaces pyrimidine methylene group with a nitrogenous base or SH compound to eliminate the thiazole ring
II	Certain bacteria	Hydrolytic cleavage of methylene–thiazole–nitrogen bond to yield the pyrimidine and thiazole moieties

TABLE 11.4 Thiaminase Activities in Seafoods

Seafood	Thiamin Destroyed (mg/100g/h)
Yellowfin tuna	265
Red snapper	265
Skipjack tuna	1000
Mahi mahi	120
Clam	2640

From Hilker, D.M., Peter, G.F., 1966. J. Nutr. 89, 419–426.

- **Thiamin antagonists**
  - *o*- and *p*-Hydroxypolyphenols (e.g., caffeic acid, chlorogenic acid, tannic acid)[10] in ferns, tea, and betel nut react with thiamin to oxidize the thiazole ring yielding the nonabsorbable form **thiamin disulfide**;
  - Some plant flavonoids (e.g., quercetin, rutin) have been reported to antagonize thiamin;
  - Hemin[11] in animal tissues is thought to bind the vitamin.
- **Thiamin analogs**—Several analogs are effective thiamin antagonists, each involving a substitution on either the pyrimidine or thiazole ring[12]. These include the following:
  - **Oxythiamin**—lacks the pyrimidine 4′-amino group essential for the release of aldehyde adducts from the C-2 of the thiazole ring; does not cross the blood–brain barrier; and, therefore, does not affect thiamin-dependent enzymes in the central nervous system.
  - **2-Methylthiamin**—has a methyl group at the 2-position of the thiazole ring; forms an enzymatically inactive complex with TPP enzymes.
  - **Pyrithiamin**—has a pyridine ring structure in place of the thiazole ring; is taken into cells by the thiamin transporter, and competitively inhibits the conversion of thiamin to TPP, increasing the urinary excretion of thiamin.
  - **Amprolium**—has a thiamin-like pyrimidine ring combined through a methylene bridge to a quaternary nitrogen of the pyridine ring. The absence of a hydroxyethyl side chain on that ring prevents the analog from forming diphosphate derivatives, making it a weak thiamin antagonist in animals. An efficient inhibitor of thiamin uptake by bacteria, it is used as an anticoccidial drug for poultry.[13]

4. ABSORPTION OF THIAMIN

Free thiamin must be released from phosphorylated dietary forms to be absorbed across the gut. That dephosphorylation is accomplished by phosphatases and pyrophosphatases in the upper region of the small intestine. The free (nonesterified) vitamin is absorbed by two mechanisms:

- **Active transport** at low luminal concentrations (<2 μM). This saturable, carrier-dependent mechanism is located in the apical brush border of the mucosal epithelium, with greatest activity in the duodenum. Because adrenalectomized rats absorb thiamin poorly,

10. These and related compounds are found in blueberries, red currents, red beets, Brussels sprouts, red cabbage, betel nuts, coffee, and tea.
11. Ferriprotoporphyrin, the nonprotein, Fe^{3+}-containing portion of hemoglobin.
12. The structures of these thiamin antagonists:

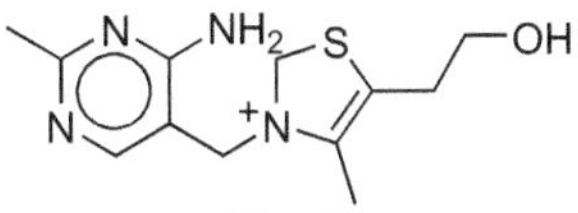

thiamin

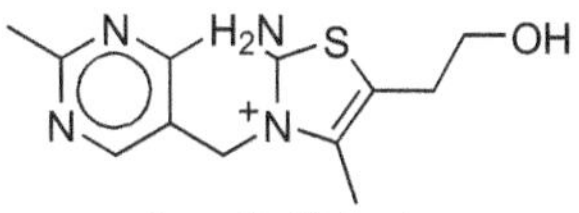

2-methylthiamin

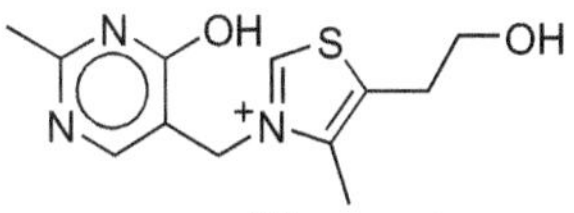

oxythiamin

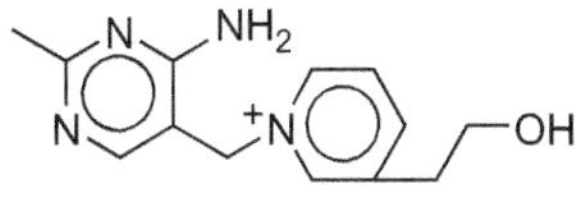

pyrithiamin

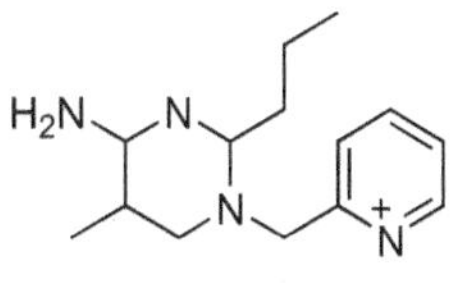

amprolium

13. These compounds are valuable in protecting young poultry from coccidia (i.e., protozoans that can cause intestinal diseases in humans and many animal species). At low doses, they inhibit thiamin transport by the parasite; at higher doses, they can also inhibit thiamin absorption by the host to produce clinical thiamin deficiency.

it is thought that enteric absorption of the vitamin may also be subjected to control by corticosteroid hormones. These transporters can be inhibited by alcohol and pyrithiamine. Involved in this process are two cation carriers, THTR1 and THTR2, the gene products of *slc19a2* and *slc19a3*, respectively.[14,15] Both are thiamin-specific members of the SLC19 (solute carrier proteins) superfamily of transport proteins.[16] Both facilitate the exchange of the thiamin cation for protons, driven by an extracellularly directed proton gradient. These transporters have different functions in the enterocyte.[17] The principle uptake of thiamin across the brush boarder is facilitated by THTR2, which has a relatively high capacity, saturable at thiamin concentrations in the μM range. THTR2 is expressed along the length of the gut with highest expression in the duodenum and jejunum. That it is expressed in colonocytes raises the possibility that humans and other monogastrics may be able to absorb thiamin produced by the hindgut microbiome, although that has not been demonstrated. Located in the basolateral membrane is THTR1, which has a lower transport capacity, being saturated at thiamin concentrations in the nM range.
- **Passive diffusion**—at higher concentrations (e.g., a 2.5 mg dose for a human).

Movement of thiamin across the enterocyte basolateral membrane into the blood is Na^+-dependent, being coupled to the hydrolysis of ATP by a Na^+/K^+ ATPase. While most of the thiamin present in the intestinal mucosa is in phosphorylated form, thiamin arriving on the serosal side of the intestine is largely in the free (nonphosphorylated) monovalent cation. Therefore, the movement of thiamin through the mucosal cell is coupled to its phosphorylation/dephosphorylation.

5. TRANSPORT OF THIAMIN

Thiamin Bound to Serum Proteins

Most of the thiamin in serum is bound nonspecifically to protein, chiefly albumin. About 90% of the total thiamin in blood (typically, 5–12 μg/dL) is contained in erythrocytes.[18] A specific binding protein, **thiamin-binding protein (TBP)**, has been identified in rat serum.[19] With a molecular mass of 38 kDa, TBP binds free thiamin and forms a complex with the riboflavin-binding protein. Like the latter, TBP appears to be regulated by estrogens, i.e., it is inducible in male or ovariectomized rats by parenterally administered estrogen.

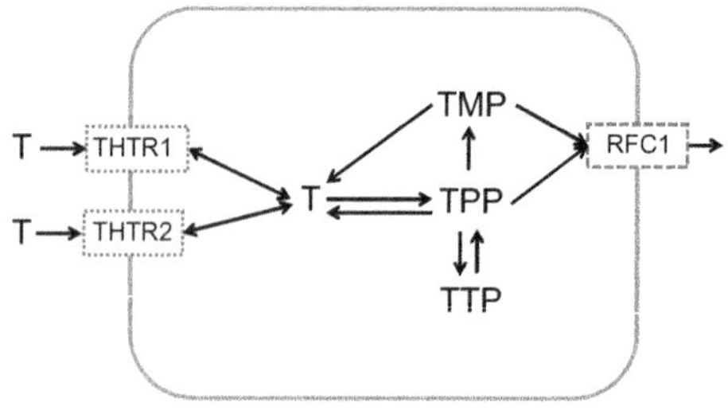

FIGURE 11.1 Membrane transporters involved in thiamin entry and export from cells (see text for abbreviations).

Cellular Uptake

Being hydrophilic and positively charged in the plasma, thiamin does not readily cross the plasma membrane. Cellular entry and export of the positively charged thiamin is facilitated by THTR1 and THTR2 (Fig. 11.1). That THTR1 is particularly important in this regard is indicated by the fact that individuals born with genetic defects in that transporter show **thiamin-responsive megaloblastic anemia (TRMA)**, also characterized by deafness due to nerve damage.[20] A variant of THTR2 was identified in siblings with Wernicke–Korsakoff syndrome and persistent seizures, which signs responded to high doses of thiamin.[21] Another SLC protein, the reduced folate carrier (RFC1, the gene product of *scl25a19*) is involved in the export of negatively charged phosphorylated forms of thiamin. Thiamin is rapidly phosphorylated upon entry into the cell. Hence, TPP comprises most (70–90%) of intracellular thiamin, some 90% of which is bound to proteins. Almost one-third of intracellular TPP enters mitochondria via high-affinity transporters thought to be ThTr1/2 and SLC25A19,[22] which is thought to facilitate entry of TPP by exchange with ATP. Most mitochondrial TPP quickly becomes bound to two proteins (α-ketoglutarate dehydrogenase (α-KGDH) and

14. We are using the convention of citing the gene in lower case italics, and the gene product in all caps except in cases in which the product was named before the gene was identified, e.g., ThTr1.
15. Mutations in the *slc19a2* gene are associated with thiamin-responsive megaloblastic anemia, insulin-dependent diabetes, and sensory-neural hearing loss (TRMA, or Roger's disease). Studies have shown that thiamin deprivation results in the upregulation of *slc19a3* in the mouse. Curiously, a genetic defect in the *slc19a3* gene has been found to cause biotin-responsive basal ganglia disease (Zhang, W.G., Al-Yamani, E., Arcierno, Jr., J.S., et al., 2005. Am. J. Hum. Genet. 77, 16–26).
16. This family also includes the folate transporter, SLC19A1.
17. This is evidenced by the fact that patients with thiamin-responsive megaloblastic anemia (TRMA) due to a mutation in the *slc19a2* gene show normal plasma thiamin levels (Neufeld, E.J., Fleming, J.C., Tartaglini, E., et al., 2001. Blood Cells Mol. Dis. 27, 135–138).
18. Several children who died of SIDS (sudden infant death syndrome) have been found to have very high plasma thiamin concentrations, e.g., fivefold those of infants who died of other diseases. The physiological basis of this effect is unknown, although thiamin deficiency is not thought to be a cause of death in SIDS.
19. TBP has also been identified in rat liver and hens' eggs (yolk and albumen).
20. Some patients respond to high doses (50 mg/day) of thiamin.
21. Kono, S., Miyajima, H., Yoshida, K., et al., 2009. N. Eng. J. Med. 360, 1792–1794.
22. A rare mutation in *slc25a19* causes a fatal microencephaly.

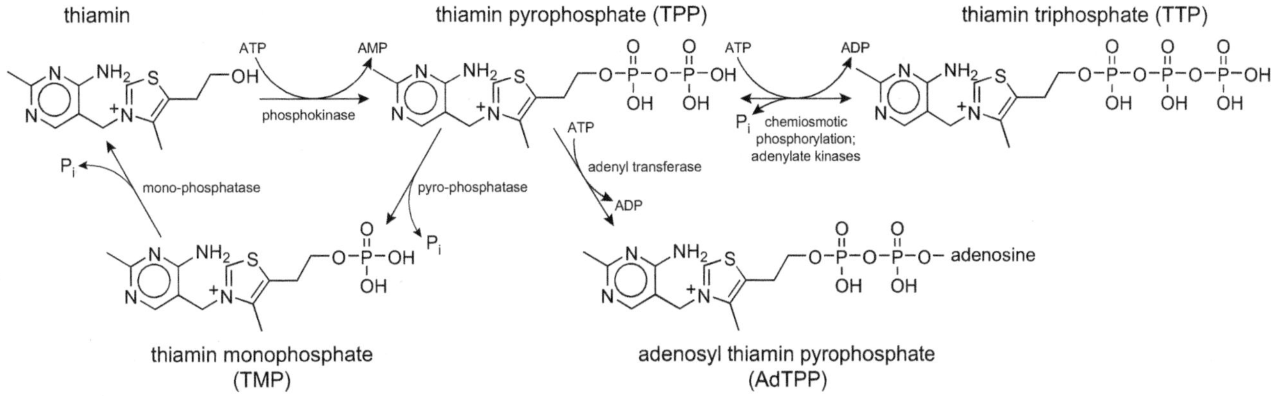

FIGURE 11.2 Metabolic activation of thiamin.

pyruvate dehydrogenase).[23] It can also be dephosphorylated to form TMP, which can move to the cytosol where it can be converted back to thiamin.

Tissue Distribution

That adult human stores only 30–50 mg thiamin, most of which is in skeletal muscle, heart, brain, liver, and kidneys as TPP. Plasma, milk, and cerebrospinal fluid and probably all extracellular fluids contain free thiamin and TMP which, unlike the more highly phosphorylated forms (TPP, TTP), appear capable of crossing cell membranes. Tissue levels of thiamin vary within and between species, with no appreciable storage in any tissue.[24] In infants, blood thiamin levels decline after birth, owing initially to a decrease in free thiamin followed by a decrease in phosphorylated forms. In thiamin-deficient chickens fed the vitamin, heart tissues take up thiamin at much greater rates than liver or brain. In general, the thiamin contents of human tissue tend to be less than those of analogous tissues in other species, particularly the pig, which has relatively high tissue thiamin stores.

While the brain is vulnerable to thiamin deprivation, it does not store substantial amounts of the vitamin. Studies have shown that, during periods of low thiamin intake, the brain is depleted of thiamin more slowly than other tissues; and when given therapeutic amounts of thiamin, it is not repleted as rapidly as other tissues such as liver. Such observations suggest a relatively limited transport of the vitamin across the blood–brain barrier compared to other tissues; unlike the liver, in which thiamine uptake is driven by a Na^+ gradient, crossing the blood–brain barrier appears to be a nonenergy-dependent process driven by intracellular phosphorylation in neural tissues.

23. It is estimated that of the approximate 30 μM thiamin in mitochondria, only 2 μM is not enzyme bound (Bettendorf, L., Mastrogiacomo, J., LaMarch, J., et al., 1996. Mov. Disord. 11, 437–439.

24. Thiamin concentrations are generally greatest in the heart (0.28–0.79 mg/100 g), kidneys (0.24–0.58 mg/100 g), liver (0.20–0.76 mg/100 g), and brain (0.14–0.44 mg/100 g), and are retained longest in the brain.

6. METABOLISM OF THIAMIN

Phosphorylation–Dephosphorylation

Thiamin is phosphorylated in peripheral tissues to form three products (Fig. 11.2):

- **TMP** is produced from thiamin by non-specific phosphatases using ATP.
- **Thiamin diphosphate (TPP)** is formed by **thiamin diphosphokinase (TPK)**[25] using ATP. TPK is a soluble, Mg^{2+}–dependent, cytosolic enzyme with a high affinity for thiamin (K_m 0.1–1 μM) and 10-fold lower affinity for ATP; it functions as a 46–56 kDa homodimer each subunit of which binds a thiamin molecule. The equilibrium of the phosphorylation reaction is driven in favor of TPP by the binding of that product by TK and/or its transport into the mitochondria.[26] TPK is present in liver and brain (particularly, cerebellum and pons) and is inhibitable by pyrithiamine.
- **TTP can be** formed from TPP in two ways:
  - By **adenylate kinases**, particularly in the cytosol of skeletal muscle; this may provide significant amounts of TTP only when tissue concentrations of TTP and AMP are high.
  - By a **chemiosmostic mechanism**[27] similar to that of ATP synthesis in the process of mitochondrial oxidative phosphorylation driven by the protomotive force.[28]
- **Adenosylthiamin pyrophosphate (AdTPP)** and **adenosylthiamin triphosphate (AdTTP)** are thought to be produced by a Mg^{2+} (of Mn^{2+})-dependent adenyltransferase using ATP and ADP.

25. Also, thiamin pyrophosphokinase

26. Accordingly, free cytosolic TPP comprises <10% of total thiamin in the rat brain.

27. Gangolf, M., Wins, P., Thiry, B., et al., 2010. J. Biol. Chem. 285, 583–594.

28. The potential energy associated with proton and voltage gradients across the mitochondrial inner membrane.

The phosphate esters of thiamin can be hydrolyzed by nonspecific phosphatases, including alkaline phosphatase and acid phosphatase, as well as several specific phosphatases:

- **TTP—triphosphatases**. Two specific TTP-phosphatases (membrane-associated and soluble) which hydrolyze TTP to yield TPP have been identified. The soluble enzyme, a 25 kDa monomer, appears to be found in most mammalian tissues with highest expression in neurons and glial cells. Mutations prevent it from being active as a TTPase in the pig, and it has not been found in birds.
- **TPP—diphosphatases**. Several TPPases that can hydrolyze TPP to yield TMP have been identified in association with the endoplasmic reticulum. Two microsomal nucleoside diphosphatases, one activated by ATP, another inhibited by ATP and thought to be a Golgi uridine diphosphatase.
- **TMP hydrolysis**. It is to yield free thiamin appears to be accomplished by nonspecific phosphorhydrolases found in many tissues. No specific TMPase has been reported.

The net result of these phosphorylation/dephosphorylation processes is a large body TPP pool (80% of total) with relatively small pools of TMP and TTP (apparently only in mitochondria). Adenylated forms of TPP and TTP are found at an order of magnitude less than other nucleotides (AMP, ATP, NAD). Thiazole ring analogs with hydroxyethyl groupings (e.g., oxythiamin, pyrithiamin, 2-methylthiamin) compete with the vitamin for phosphorylation.

Catabolism

The turnover of thiamin varies between tissues but is generally high.[29] Thiamin in excess of that which binds in tissues is rapidly excreted. With an estimated half-life of 10–20 days in humans, thiamin deficiency states can deplete tissue stores within a couple of weeks. Studies with fasting and undernourished soldiers have shown that food restriction increases the rate of thiamin excretion.[30] Declines in tissue thiamin levels are thought to involve enhanced degradation of TPP-dependent enzymes in the absence of the vitamin. Numerous metabolites of thiamin have been identified (Table 11.5).

Excretion

Thiamin is excreted in the urine, chiefly as free thiamin and TMP, but also in smaller amounts as TPP, the oxidation product **thiochrome**,[31] more than 20 other metabolites and a 25 kDa thiamin-containing peptide. Metabolites retaining the pyrimidine–thiazole ring linkage account for increasing proportions of total thiamin excretion as thiamin status declines. Urinary losses of thiamin metabolites vary with plasma thiamin levels but increase markedly when renal tubular reabsorption is saturated, which occurs in healthy adults at intakes of 0.3–0.4 mg thiamin per 1000 kcal.[32] Above that threshold, excretion of the vitamin exceeds 100 μg/day, whereas urinary excretion in deficient individuals is < 25 μg/day. Small amounts of the vitamin have also been reported to be lost in sweat.[33]

TABLE 11.5 Urinary Metabolites of Thiamin

Free thiamin
Thiamin disulfide thiamin monophosphate (TMP)
Thiamin diphosphate (TPP)
Thiochrome
Thiamin acetic acid
2-Methyl-4-amino-5-pyrimidine carboxylic acid
4-Methylthiazole-5-acetic acid
2-Methyl-4-aminopyrimidine-5-carboxylic acid
2-Methyl-4-amino-5-hydroxymethylpyrimidine
5-(2-Hydroxyethyl)-4-methylthiazole
3-(2′-Methyl-4-amino-5′-pyrimidinylmethyl)-4-methylthiazole-5-acetic acid
2-Methyl-4-amino-5-formylaminomethylpyrimidine

7. METABOLIC FUNCTIONS OF THIAMIN

Cosubstrate Functions of Thiamin Phosphate Esters

TTP can serve as a phosphate donor and does so for the phosphorylation of certain proteins including the neuromuscular synapse protein, **rapsyn**.[34] It is not clear whether TMP or adenylated derivatives have direct metabolic functions.

Coenzyme Functions of Thiamin Diphosphate

TPP (cocarboxylase) is an essential cofactor for five enzyme complexes (Fig. 11.3). Transketolase catalyzes the

29. Thiamin turnover in rat brain was 0.16–0.055 μg/g/h, depending on the region (Rindi, G., Patrini, C., Comincioli, V., et al., 1980. Brain Res. 181, 369–376).
30. Consolazio, C.F., Johnson, H.L, Krzywicki, J., et al., 1971. Am. J. Clin. Nutr. 24, 1060–1066.
31. The strong fluorescence of thiochrome has been used in the determination of thiamine, which can be oxidized to thiochrome using potassium ferricyanide or cyanogen bromide (Fujiwara, M., Matsui, K., 1953. Anal. Chem. 25, 810–816).
32. Interdepartmental Committee on Nutrition for National Defense, 1963. Manual for Nutrition Surveys, second ed. US Government Printing Office, Washington, DC.
33. Pearson, W.H., 1967. Am. J. Clin. Nutr. 20, 514–521; Sauberlich, H.E., Herman, Y.F., Stevens, C.O., et al., 1979. Am. J. Clin. Nutr. 32, 2237–2244.
34. This receptor-associated protein of the synapse is a 43 kDa protein believed to play a role in anchoring the acetylcholine receptor at sites in synapses.

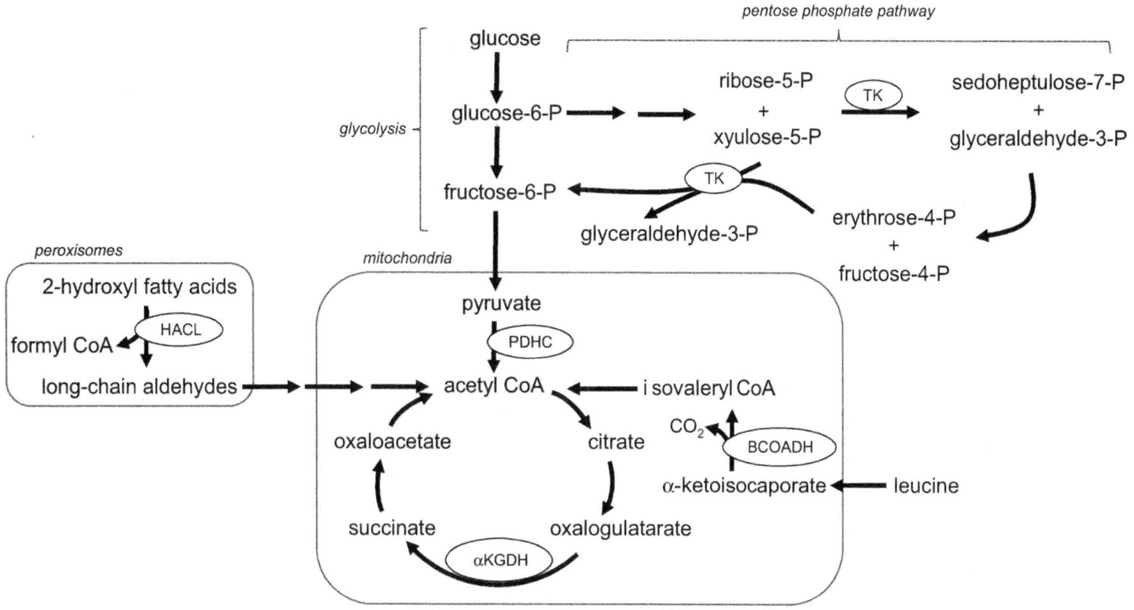

FIGURE 11.3 Roles of TTP-dependent enzymes (ovals) in metabolism. *HACL*, 2-hydroxyacyl CoA lyase; *PDHC*, pyruvate dehydrogenase complex; *TK*, transketolase; *α-KGDH*, α-ketoglutarate dehydrogenase.

transfer of a glycoaldehyde moiety between sugars. Three α-keto acid dehydrogenases catalyze oxidative decarboxylation reactions. A lyase catalyzes the cleavage of 2-hydroxy straight chain fatty acids and 3-methyl branched chain fatty acids after the α-oxidation of the latter. In each of these reactions, TPP serves as a classic coenzyme, binding covalently to the respective holoenzyme, which recognizes both its substituted pyrimidyl and thiazole moieties. Binding is facilitated by Mg^{2+} or some other divalent cation, which is required for enzyme activity. TPP functions as an energy-rich phosphoanhydride with high potential for phosphate group transfer. It is deprotonated to form a carbanion at C-2 of the thiazole ring, which reacts with the polarized 2-carbonyl group of the substrate (an α-keto acid or α-keto sugar) and labializes certain C–C bonds to release CO_2. The remaining adduct reacts by the following:

- protonation to give an active aldehyde addition product (e.g., decarboxylases);
- direct oxidation with suitable electron acceptors—to yield a high-energy, 2-acyl product;
- reacting with oxidized lipoic acid—to yield an acyldihydrolipoate product (e.g., oxidases or dehydrogenases); or
- addition to an aldehyde carbonyl to yield a new ketol.

In higher animals, the decarboxylation is oxidative, producing a carboxylic acid. This involves transfer of the aldehyde from TPP to lipoic acid (forming a 6-*S*-acylated dihydrolipoic acid and free TPP) and then to coenzyme A.[35]

35. Coenzyme A is the metabolically active form of the vitamin **pantothenic acid**.

Transketolase (TK). TK functions at two points in the pentose pathway, which generates pentoses and NADPH from the oxidation of glucose.[36] Both points involve the reversible transfer of 2-C "active glycoaldehyde" fragment from a ketose donor (xylulose-5-phosphate) to an aldose acceptor (ribose-5-phosphate or erythrose-5-phosphate). This is done by the TPP thiazole ring complexing with ketose substrate, releasing glyceraldehyde-3-phosphate and forming a complex with glycoaldehyde, which is transferred to the aldose. By these reactions, thiamin directs glucose either to glycolysis or to the pentose pathway, which generates ribose (required for DNA) while reducing the production of glucose metabolites, a feature of potential import in mitigating diabetic pathology.[37] TK is found in the cytosol of most tissues. It is present in remarkably high amounts in the cornea, where it has been reported to comprise some 10% of total soluble protein. It is present in high amounts in adipose, mammary gland, adrenal cortex, and erythrocytes, all of which rely on carbohydrate metabolism.[38] TK activity depends on its binding to TPP; therefore, responses to thiamin may involve activation of the apoenzyme, That thiamin may also have a direct effect on the genetic expression of the enzyme is suggested by the finding that thiamin deprivation increases the expression of TK mRNA. In thiamin-adequate subjects, TPP binding is at least 85% of

36. This pathway, also called the **hexose monophosphate shunt**, is an important alternate to the glycolysis–Krebs cycle pathway, especially for the production of pentoses for RNA and DNA synthesis and NADPH for the biosynthesis of fatty acids, etc.

37. Thiamin has been found useful in mitigating diabetic nephropathy (Rabbani, N., Alam, S., Riaz, S., et al., 2009. Diabetology 52, 208–211).

38. TK is also highly expressed in pancreatic tumors.

saturation, whereas in thiamin deficiency the percentage of TK bound to TPP is much less. TK isolated from patients with Wernicke–Korsakoff syndrome[39] has been found to have an abnormally low binding affinity for TPP.

α-Keto acid dehydrogenases. TPP is an essential cofactor in multienzyme complexes that catalyze the oxidative decarboxylation of α-keto acids. Each complex is composed of a decarboxylase that binds TPP, a core enzyme that binds lipoic acid, a flavoprotein dihydrolipoamide dehydrogenase that regenerates lipoamide, and one or more regulatory components. There are three classes of this type of TPP-dependent enzyme:

- **Pyruvate dehydrogenase complex (PDHC)**. It converts pyruvate produced from glycolysis to acetyl CoA, a key intermediate in the synthesis of fatty acids and steroids and an acyl donor for numerous acetylation reactions. Its TPP-dependent component is pyruvate dehydrogenase (E1). It also has two non-TPP enzymes, dihydrolipoyl acetyltransferase and dihydrolipoyl dehydrogenase,[40] a kinase and a phosphatase. The latter components regulate enzymatic activity by interconverting the dehydrogenase between active (nonphosphorylated) and inactive (phosphorylated) forms involving three specific serine residues that participate in TPP binding. Regulation also occurs via end product inhibition (by acetyl CoA and NADH). Thiamin status regulates E1 expression; deprivation of thiamin increases the mRNA for the β-subunit of the enzyme. A genetic defect in the E1-α polypeptide produces chronic neurological dysfunction with central nervous system degeneration and, generally, lactic acidosis; most signs respond to high doses of thiamin.
- **α-Ketoglutaric dehydrogenase complex (α-KGDH)**. It converts α-ketoglutarate to succinyl CoA. Its TPP-dependent component is α-ketoglutaric dehydrogenase, which appears to be regulated through stimulation by Ca^{+2} and inhibition by ATP, GTP, NADH, and succinyl CoA.
- **Branched chain α-keto acid dehydrogenase complex (BCKDH)**. It converts branched chain α-keto acids (produced by the transaminations of valine, leucine, and isoleucine) to the corresponding acyl CoAs (isobutyryl-, isovaleryl- and α-methylbutyryl-, respectively), which are subsequently oxidized to yield acetyl and propionyl CoAs. The dehydrogenase is regulated by phosphorylation–dephosphorylation involving a single serine residue. Deprivation of thiamin increases the proportion of dephosphorylated (active) enzyme, thus, serving to mitigate against the metabolic consequences of thiamin deficiency. Genetic defects in subunits of this enzyme complex result in the condition called **maple syrup urine disease (MSUD)**. Signs are manifest in infancy: lethargy, seizures, and, ultimately, mental retardation and a maple syrup odor of the urine due to the presence of the keto acid leucine. Five types of MSUD have been identified. One involves the enzyme's loss of affinity for TPP, reducing its BCKDH activities by 30–40%; this type of MSUD responds to high doses (10–200mg/day) of thiamin.

2-Hydroxyacyl CoA lyase (HACL). TPP is required by this component of the peroxisomal enzyme complex involved in fatty acid catabolism.[41] This enzyme catalyzes the TPP-dependent cleavage of 2-hydroxy fatty acids (e.g., 2-hydroxyoctadecanoic acid[42]) to yield formate and a 1C-shortened aldehyde.

Neurologic Function

Thiamin has a vital role in nerve function, as the signs of thiamin deficiency are mainly neurologic. Thiamin is found in the brain, synaptosomal membranes, and cholinergic nerves. Nervous stimulation by either electrical or chemical means results in the release of thiamin (free thiamin, TMP) associated with the dephosphorylation of its higher phosphate esters. The antagonist pyrithiamin can displace thiamin from nervous tissue and change the electrical activity of the tissue. Irradiation with ultraviolet light at wavelengths absorbed by thiamin destroys the electrical potential of nerve fibers in a manner corrected by thiamin treatment. TTP, which appears to serve as a phosphate donor for phosphorylating synaptic proteins, has been shown to stimulate chloride transport.[43] Brain TTP concentrations tend to be resistant to changes with thiamin deprivation or parenteral thiamin administration, suggesting some degree of homeostatic control in that organ. Thiamin deprivation has been shown to cause oxidative stress, alter neurotransmitter metabolism, and cause dysfunction of the blood–brain barrier in experimental animals.

Thiamin has an essential role in the metabolism of glucose, on which the brain depends as its energy source.[44] TPP is required for the oxidative decarboxylations of pyruvate and α-ketoglutarate, essential steps in energy production via the tricarboxylic acid cycle. However, several

39. Thiamin-responsive encephalopathy characterized by neurological symptoms including ophthalmoplegia (weakness in muscles responsible for eye movements), ataxia, and confusion. Also called Korsakoff's psychosis and alcoholic encephalopathy.

40. The complex in eukaryotes consist of a multiunit structure containing 20–30 heterotetramers of PDHC, each with two α and two β subunits, associated with multiple units of the acyltransferase, a dihydrolipoyl dehydrogenase-binding protein, and homodimers of the dihydrolipoyl dehydrogenase.

41. The complex also catalyzes the chain shortening of these fatty acids to facilitate their β-oxidation.

42. This fatty acid is found in cerebrosides and sulfatides in brain.

43. Bettendorf, L., Kolb, H.A., Schoffeniels, E., 1993. J. Membr. Biol. 136, 281–288.

44. Unlike other tissues, the brain cannot use fatty acids as a source of energy.

TABLE 11.6 Effects of Thiamin Deficiency on Brain Metabolism in the Rat

Parameter	Thiamin Fed	Thiamin Deficient
Body weight (g)	135±4	96±5
Erythrocyte Transketolase (nmol/min/mg)		
Basal activity	6.4±0.6	2.9±0.4[a]
+TPP	6.9±1.4	4.7±0.8
Activation coefficient	1.08±0.08	1.63±0.12[a]
Liver thiamin (nmol/g)	132.0±8.2	6.0±0.7[a]
Brain Analytes		
Thiamin (nmol/g)	12.6±0.6	6.2±0.3[a]
ATP (μmol/g)	2.8±0.1	2.9±0.1
Glutamate (μmol/g)	13.8±0.3	11.3±0.2[a]
α-Ketoglutarate (μmol/g)	0.14±0.01	0.09±0.0[a]
GABA (μmol/g)	1.67±0.05	1.58±0.03[a]

[a]$P<.05$.
From Page, M.G., Ankoma-Say, V., Coulson, W.F., et al., 1989. Br. J. Nutr. 62, 245–253.

experiments have indicated that the depressions of pyruvate and α-KGDH activities that occur in thiamin-deficient animals (Table 11.6) are not of sufficient magnitude to produce the neurological dysfunction associated with the deficiency. That brain ATP levels are unaffected by thiamin deprivation suggests that the metabolic flux though the alternative pathway, the **γ-aminobutyric acid (GABA)** shunt,[45] may be considerably increased in the brains of thiamin-deficient individuals. This suggests that, in addition to its role in the synthesis of that neurotransmitter, the GABA shunt may also yield energy under conditions of thiamin deprivation (Fig. 11.4). Such a phenomenon may explain the anorexia characteristic of thiamin deficiency, as increased GABA flux through the hypothalamus has been shown to inhibit feeding in animals.

It has been suggested that thiamin may be involved the synthesis of myelin. However, the turnover of myelin is much slower ($T_{½}$ 4–5 days) than the response of thiamin therapy (full recovery within 24 h). Therefore, thiamin/TPP would appear to have other functions related to nerve transmission, e.g., roles in regulating Na^+ permeability and maintaining the fixed negative charge on the inner surface of the plasma membrane.

45. *GABA* is synthesized by decarboxylation of glutamate, which is produced by transamination of α-ketoglutarate. GABA can be transaminated to form succinic semialdehyde, which is oxidized to succinate and enters the TCA cycle.

Three conditions indicate key roles of the vitamin in neurologic function:

- **Wernicke–Korsakoff syndrome**. This syndrome consists of Wernicke's encephalopathy and Korsakoff psychosis with signs ranging from mild confusion to coma. Pathology is limited to the central nervous system, with lesions limited to the submedial thalamic nucleus and parts of the cerebellum, particularly the superior cerebellar vermis. Patients frequently have a TK isoform with abnormally low binding affinity for TPP; in some cases this can be overcome with high doses of thiamin. A quarter of patients can be cured by thiamin treatment; it is thought that the balance may have another aberrant TK (or other TPP-dependent enzyme) incapable of binding TPP.
- **Alzheimer's disease**. That thiamin has a role in protecting against Alzheimer's disease is suggested by observations that patients have lower brain activities of TPP-dependent enzymes: 55% less α-KGDH, 70% less PDH, and markedly reduced TK.[46] Patients with frontal lobe degeneration of the non-Alzheimer's type have been found to have only half the cortical TPP levels of unaffected individuals. In an Alzheimer's disease mouse model, thiamin deprivation increased the accumulation of plaques.[47] In another mouse model, the synthetic thiamin precursor benfotiamine[48] was found to both halt the progression of amyloid plaques and promote their regression.[49]
- **Parkinson's disease (PD)**. That thiamin may have a role in protecting against PD is suggested by findings that patients have relatively low cerebrospinal fluid levels of free thiamin (although, normal levels of TPP and TMP) and reduced activities of α-KGDH,[50] which is known to be inhibited by the dopamine oxidation products that are elevated in the disease.[51] PD patients given L-dopa show increased cerebrospinal fluid concentrations of thiamin.[52] Mice given TPP or TTP intrastriatally showed dopamine release.[53] A PD-associated polymorphism in α-KGDH has been suggested.

46. Gibson, G.E., Zhang, H., Sheu, K.F., et al., 1998. Ann. Neurol. 44, 676–681; Gibson, G.E., Pulsinelli, W., Blass, J.P., et al., 1981. Am. J. Med. 70, 1247–1254; Butterworth, R.F., Besnard, A.M., 1990. Metab. Brain Dis. 5, 179–184.
47. Karuppagoundeer, S.S., Xu, H., Shi, Q., et al., 2009. Neurobiol. Aging 30, 1587–1600.
48. *S*-benzoylthiamin O-monophosphate.
49. Pan, X., Gong, N., Zhao, J., et al., 2010. Brain 133, 1342–1351.
50. Mizuno, Y., 1995. Biochim. Biophys. Acta 1271, 265.
51. Cohen, G., Farooqui, R., Kesler, N., 1997. Proc. Nat. Acad. Sci. U.S.A. 94, 4890–4894.
52. Jimenéz- Jimenéz, F.J., Mlina, J.A., Hermánz, A., et al., 1999. Neurosci. Lett. 271, 33–36.
53. Yamashita, H., Zhang, Y.X., Nakamura, S., 1993. Neurosci. Lett. 158, 229–231.

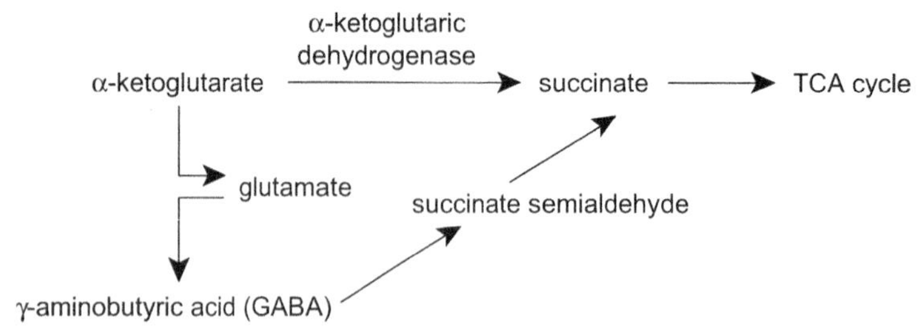

FIGURE 11.4 Flux through the GABA shunt is increased to maintain brain ATP levels in thiamin deficiency.

Vascular Function

Diabetic vascular complications appear to involve insufficiencies of thiamin/TK. TTP has an antidiabetic type role by virtue of its function in TK, which diverts cellular excesses of fructose-6-phosphate and glyceraldehyde-3-phosphate from glycolysis to the hexose monophosphate shunt. This serves to downregulate intracellular glucose levels, thus, avoiding cellular damage. Diabetic subjects have been found to have lower circulating thiamin levels and lower erythrocyte TK activities than healthy controls. Both thiamin and the lipophilic thiamin-derivative benfotiamine[54] have been shown to reduce the accumulation of glycation products and prevent apoptosis in vascular cells cultured under hyperglycemic conditions. Supplementation of either of these compounds has been shown to prevent diabetic cardiomyopathy and neuropathy in the streptozotocin-induced diabetic rat model with moderate insulin treatment.[55] Both have been found effective in preventing vascular dysfunction, oxidative stress, and proteinuria in subjects with type 2 diabetes.[56] Therefore, it has been suggested that diabetes may appropriately be considered a thiamin-deficient state.

Antioxidant Function

The ability to transfer protons from its pyrimidine amino group to its thiazole ring allows thiamin to serve in redox control of cellular pH and to quench prooxidant species. This has been demonstrated in vitro by thiamin prevention of peroxidation in oleic acid or microsomal lipids, with thiamin being oxidized to thiochrome and thiamin disulfide.[57] The physiological relevance of this antioxidant effect is indicated by the fact that thiamin can prevent hepatocyte cytotoxicity and formation of ROS induced by mitochondrial respiratory inhibitors.[58] Oxidative stress of thiamin deficiency is thought to affect immune cell function and to contribute to thiamin-responsive neurologic disorders.

54. *S*-benzoylthiamine O-monophoshate, a synthetic S-acyl derivative of thiamin that is dephosphorylated to yield the lipid-soluble S-benzoylthiamin.

55. Thornally, P.J., Jahan. I., Ng, R., 2001. J. Biochem. 129, 543; Kohda, Y., Shirakawa, H., Yamane, K., et al., 2008. J. Toxicol. Sci. 33, 459–472.

56. Stirban, A., Negrean, M., Mueller-Roesel, M., et al., 2006. Diabetes Care 29, 2064–2071; Arora, S. Lidor, A., Abularrage, C.J., et al., 2006. Ann. Vasc. Surg. 20, 653–658; Riaz, S., Skinner, V., Srai, S.K., 2011. J. Pharmaceut. Biomed. Anal. 54, 817–825.

57. Lukienko, P.I., Mel'nichenko, N.G., Zverinski, I.V., et al., 2000. Bull. Exp. Biol. Med. 130, 874–876.

8. BIOMARKERS OF THIAMIN STATUS

Thiamin status has been assessed in two ways:

- **Degree of TPP saturation of thiamin-dependent enzymes**. This is the most useful means of assessing thiamin status. It takes advantage of the in vitro binding of TPP by erythrocyte transketolase (eTK) from hemolysates. Because thiamin-adequate subjects typically have >85% of eTK bound to TPP, the addition of exogenous TPP should stimulate their eTK activities by no more than 15%. Any stimulation above that level indicates that less eTK was bound to TPP. Therefore, the eTK response to added TPP measures eTK saturation and, thus, thiamin status. This is measured as an activity coefficient:

  eTK activity coefficient = baseline eTK activity/eTK activity with added TPP

  Subjects with eTK activity coefficients <1.15 are considered to be at low risk of thiamin deficiency; those with activity coefficients of 1.15–1.25 or >1.25 are considered to be at moderate and high risks, respectively. Symptoms of beriberi are associated with eTK activity coefficients >1.4.
- **Blood and urinary metabolites**. Thiamin deficiency increases concentrations of both pyruvate and α-ketoglutarate in whole blood[59] or plasma and of methylglyoxal[60] in the urine and cerebrospinal fluid.

58. Mehta, R., Shangari, N., O'Brien, P.J., 2008. Mol. Nutr. Food Res. 52, 379–385.

59. Normal ranges (fasting): pyruvate, 260–450 μg/dL; α-ketoglutarate, 75–170 mg/dL.

60. This appears to result from decreases in reduced glutathione, a cofactor of glyoxylase.

TABLE 11.7 Recommended Thiamin Intakes

US		FAO/WHO	
Age–Sex	RDA[a] (mg/day)	Age–Sex	RNI[b] (μg/day)
0–6 months	[0.2][c]	0–6 months	0.2
7–11 months	[0.3][c]	7–11 months	0.3
1–3 years	0.5	1–3 years	0.5
4–8 years	0.6		
9–13 year females	0.9	4–6 years	0.6
14–18 year females	1.0	7–9 years	0.9
>18 year females	1.1	10–>65 year females	1.1
Males	1.2	Males	1.2
Pregnancy	1.4		1.4
Lactation	1.4		1.5

[a]*Recommended Dietary Requirements; Food and Nutrition Board, 2000. Dietary Reference Intakes for Thiamin, Riboflavin, Niacin, Vitamin B_6, Folate, Vitamin B_{12}, Pantothenic Acid, Biotin and Choline. National Academy Press, Washington, DC, 564 pp.*
[b]*Recommended Nutrient Intakes; Joint WHO/FAO Expert Consultation, 2001. Human Vitamin and Mineral Requirements. Food and Agricultural Org., Rome, 286 pp.*
[c]*RDAs have not been set; AIs are given instead.*

9. THIAMIN DEFICIENCY

Thiamin deficiency can have primary (privational) and secondary (nonprivational) causes. Because thiamin is widely distributed in foods, the latter causes are more common. Thiamin needs (Table 11.7) are directly related to the level of carbohydrate intake.[61] Probably, because overweight/obese individuals tend to have diets high in simple sugars, such individuals tend to have lower circulating thiamin levels than nonobese individuals. A prominent secondary cause of thiamin deficiency is chronically high alcohol consumption, which impairs thiamin absorption and is frequently associated with insufficient dietary intake.

Groups at Risk to Thiamin Deficiency
Older adults
Alcoholics
HIV+/AIDS patients
Subjects with malaria
Pregnant women with prolonged hyperemesis gravidarum.

61. Elmadfa, I., Majchrzak, D., Rust, P., et al., 2001. Int. J. Vitam. Nutr. Res. 71, 217–224.

TABLE 11.8 General Signs of Thiamin Deficiency

Organ system	Signs
General appetite	Severe decrease
Growth	Decrease
Dermatologic	Edema
Muscular	Cardiomyopathy, bradycardia, heart failure, weakness
Gastrointestinal	Inflammation, ulcer
Vital organs	Hepatic steatosis
Nervous	Peripheral neuropathy, opisthotonos

Three ways chronic, excess alcohol consumption can lead to thiamin deficiency:

- **by reducing thiamin intake**—the displacement of thiamin-containing foods
- **by impairing thiamin utilization**—inhibition of thiamin absorption
- **by increasing thiamin need**—high associated carbohydrate intake.

General Signs

Thiamin deficiency in humans and animals is characterized by a predictable range of signs/symptoms including the loss of appetite (**anorexia**), cardiac and neurologic signs (Table 11.8). Many of these signs, particularly in the early phases of deficiency, are nonspecific, common, and frequently overlooked: mental fatigue, emotional lability, paresthesias, generalized weakness, myalgias, back pain, nausea, and reduced physical work capacity. Observational studies have indicated suboptimal thiamin status in nearly one-third of elderly, hospitalized patients, 16–29% of preoperative bariatric surgery patients[62] in the United States, and 45% of HIV+ patients in Switzerland.[63]

Underlying these symptoms are a number of metabolic effects including increased plasma concentrations of pyruvate, lactate, and to a lesser extent α-ketoglutarate (especially after a glucose meal), as well as decreased activities of eTK. These effects result from diminished activities of TPP-dependent enzymes. Increased production of ROS and RNS[64] has been reported in the brains of thiamin-deficient animals. This and the finding that antioxidants

62. Kerns, J.S., Arundel, C., Chawla, L.S., 2015. Adv. Nutr. 6, 147–153.
63. Müri, R.M., von Overbeck, J., Furrer, J., et al., 1999. Clin. Nutr. 18, 375–378.
64. Reactive oxygen species and reactive nitrogen species, respectively.

can attenuate the neurologic effects of thiamin deficiency[65] suggests that oxidative stress plays contributes to the clinical manifestations of the nutritional deficiency. The presentation of thiamin deficiency is variable, affected by such factors as age, caloric (especially carbohydrate) intake, and presence/absence of other micronutrient deficiencies.[66]

Deficiency Signs in Humans

Beriberi. The classic syndrome resulting from thiamin deficiency in humans is beriberi. This disease is prevalent in Southeast Asia, where polished rice is the dietary staple. It appears to be associated with the consumption of diets high in highly digestible carbohydrates but marginal or low in micronutrients. The general symptoms of beriberi are anorexia, cardiac enlargement, lassitude, muscular weakness (with resulting ataxia), paresthesia,[67] loss of knee and ankle jerk responses (with subsequent foot and wrist droop), and dyspnea on exertion. Beriberi occurs in three clinical types:

- **Dry** (neuritic) **beriberi** (Fig. 11.5) occurs primarily in adults; it is characterized by peripheral neuropathy consisting of symmetrical impairment of sensory and motor nerve conduction affecting the distal (more than proximal) parts of the arms and legs. It usually does not have cardiac involvement.
- **Wet** (edematous) **beriberi** (also called **cardiac beriberi**) involves as its prominent signs edema, tachycardia,[68] cardiomegaly, and congestive heart failure; in severe cases, heart failure is the outcome. The onset of this form of beriberi can vary from chronic to acute, in which case it is called **shoshin beriberi** (also called **acute pernicious beriberi)** and is characterized by greatly elevated circulating lactic acid levels.
- **Infantile** (or acute) **beriberi** occurs in breast-fed infants of thiamin-deficient mothers, most frequently at 2–6 months of age. It has a rapid onset and may have both neurologic and cardiac signs with death due to heart failure usually within a few hours. Affected infants are anorectic and regurgitate ingested milk; they may experience vomiting, diarrhea, cyanosis, tachycardia, and convulsions. Their mothers typically show no signs of thiamin deficiency.

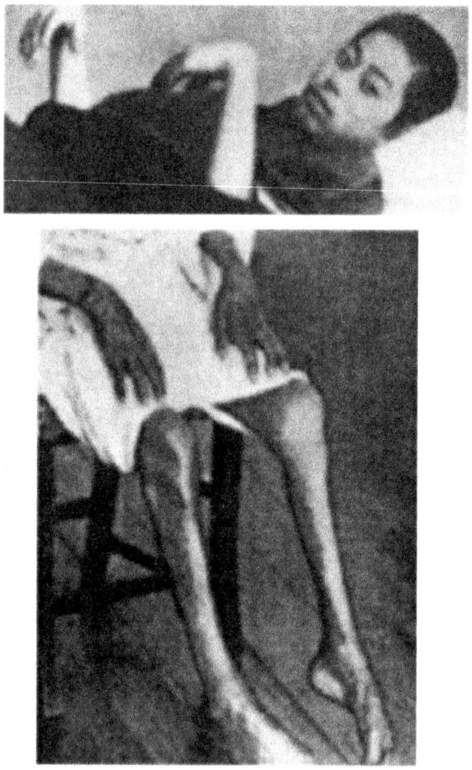

FIGURE 11.5 Neurologic signs of beriberi. *Courtesy Cambridge University Press.*

Wernicke–Korsakoff syndrome.[69] This syndrome condition can be triggered by excessive alcohol consumption in subjects of marginal to deficient thiamin status. It consists of Wernicke's encephalopathy, which includes ophthalmoplegia[70] with lateral or vertical involuntary eye movements (nystagmus[71]) and cerebellar ataxia, and Korsakoff psychosis, which includes severely impaired retentive memory and cognitive function, apathy and confabulation.[72] Pathology is seen in the submedial thalamic nucleus and parts of the cerebellum, particularly the superior cerebellar vermis. The syndrome appears to have a genetic basis, but the affected gene(s) has not been determined. The thiamin-responsive syndrome has been diagnosed in nonalcoholic patients with hyperemesis gravidarum[73] or undergoing dialysis. Patients with the syndrome frequently have a form of TK with low binding affinity for TPP, which can be cured in a quarter of patients by treatment with high oral or intramuscular doses of thiamin. It has been suggested that patients who

65. Pannunzio, P., Hazell, A.S., Pannunzio, M., et al., 2000. J. Neurosci. Res. 62, 286–292.
66. For example, deprivation of magnesium was shown to aggravate the signs of thiamin deficiency in the rat (Dyckner, T., Ek, B., Nyhlin, H., et al., 1985. Acta Med. Scand. 218, 129–131).
67. An abnormal spontaneous sensation, such as burning, pricking, numbness, etc.
68. In contrast to thiamin-deficient animals, which show **bradycardia** (slow heart beat), beriberi patients show **tachycardia** (rapid heart rate, >100 beats/min).
69. This condition has been underdiagnosed by as much as 80% (Harper, C.G., 1979. J. Neurol. Neurosurg. Psychiatry 42, 226–231).
70. Paralysis of one or more of the motor nerves of the eye.
71. Rhythmical oscillation of the eyeballs, either horizontally, rotary, or vertically.
72. Readiness to answer any question fluently with no regard whatever to facts.
73. **Hyperemesis gravidarum** is a severe and intractable form of nausea and vomiting in pregnancy.

fail to respond may have another aberrant TK (or other TPP-dependent enzyme) incapable of binding TPP.

Thiamin-responsive megaloblastic anemia (TRMA). TRMA with diabetes and deafness is a rare, autosomal recessive disorder reported in fewer than three-dozen families. The disorder presents early in childhood with any of the above signs, plus optic atrophy, cardiomyopathy, and stroke-like episodes. The anemia responds to high doses of thiamin[74]. Defects in thiamin transport were reported in TRMA patients.[75] These have been found to be due to mutations of the *slc19a2* gene encoding the high-affinity thiamin transporter.

Other conditions. In women with gestational diabetes, maternal thiamin deficiency correlates with macrosomia (abnormally high infant body weight).[76] Thiamin deficiency has been implicated in cases of sleep apnea and sudden infant death syndrome (SIDS). While this relationship has not been elucidated, it would appear reasonable to expect thiamin to have a role in maintaining the brainstem function governing automatic respiration. In children admitted to a pediatric intensive care unit in Brazil, low blood thiamin level was associated with inflammation.[77] Widespread thiamin depletion in Cuba was reported in 1992–93 during an epidemic of optic and peripheral neuropathy that affected some 50,000 people in a population of 11 million.[78] A large portion (30–70%) of both the cases and the apparently unaffected population showed signs of low thiamin status. The incidence of new cases subsided with the institution of multivitamin supplementation. Still, it is not clear that thiamin deficiency, while widespread in that population, was the cause of the **epidemic neuropathy**.

Thiamin dependency. A few cases of apparent thiamin dependency have been reported. These involved cases of with intermittent episodes of cerebral ataxia, pyruvate dehydrogenase deficiency, branched chain ketoaciduria, and low TK activity coefficients—all of which responded to high doses of thiamin.[79]

Deficiency Signs in Animals

Polyneuritis. The most remarkable sign of thiamin deficiency in most species is anorexia, which is so severe and more specific than any associated with other nutrient deficiencies (apart from that of sodium) that it is a useful diagnostic indicator for thiamin deficiency. Other signs include the secondary effects of reduced total feed intake: weight loss, impaired efficiency of feed utilization, weakness, and hypothermia. The appearance of anorexia correlates with the loss of transketolase activity and precedes changes in pyruvate or α-KGDH activities.

Animals also show neurologic dysfunction due to thiamin deficiency; birds, in particular, show a tetanic retraction of the head called **opisthotonos**, also **stargazing** (Figs. 11.6 and 11.7).[80] Other species generally show **ataxia** and incoordination, which progresses to convulsions and death.

FIGURE 11.6 Opisthotonus in a thiamin-deficient pigeon before (*top*) and after (*bottom*) thiamin treatment. *Courtesy Cambridge University Press.*

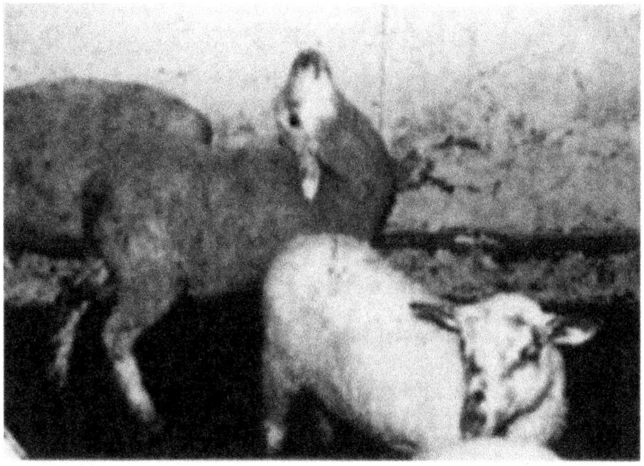

FIGURE 11.7 Opisthotonos in a thiamin-deficient sheep (unaffected sheep in foreground). *Courtesy of M. Hidiroglou, Agricultural Canada, Ottawa, Ontario, Canada.*

74. 20–60 times higher than RDA levels.

75. Poggi, V., Longo, G., DeVizia, et al., 1984. J. Inherit. Metab. Dis. 7, 153–154.

76. Baker, H., Hockstein, S., DeAngelis, B., et al., 2000. Int. J. Vit. Nutr. Res. 70, 317–320.

77. Lima, L.F.P., Leite, H.P., Taddei, J.A., 2011. Am. J. Clin. Nutr. 93, 57–61.

78. Macias-Matos, C., Rodriguez-Ojea, A., Chi, N., et al., 1996. Am. J. Clin. Nutr. 64, 347–353.

79. Lonsdale, D., 2006. Evid. Based Complement. Alternat. Med. 3, 49–59.

80. This sign occurs in young mammals, but it is not usual.

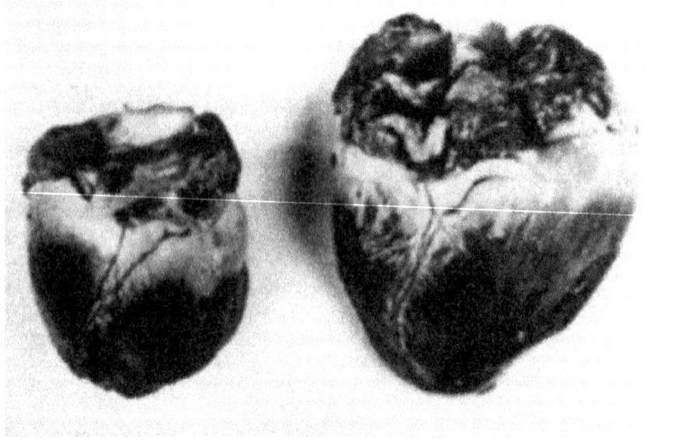

FIGURE 11.8 Hearts from normal (*left*) and thiamin-deficient (*right*) pigs. *Courtesy of T. Cunha, University of Florida.*

These conditions are generally referred to as **polyneuritis**. Most species, but especially dogs and pigs, show **cardiac hypertrophy**[81] (Fig. 11.8), with slowing of the heart rate (**bradycardia**) and signs of congestive heart failure, including labored breathing and edema. Some species also show diarrhea and achlorhydria (rodents), gastrointestinal hemorrhage (pigs), infertility (chickens[82]), high neonatal mortality (pigs), and impaired learning (cats).

The clinical manifestation of thiamin deficiency in young ruminants is the neurologic syndrome called **polioencephalo malacia**,[83] a potentially fatal condition involving inflammation of brain gray matter and presenting as opisthotonos; it readily responds to thiamin treatment.[84] Thiamin deficiency can occur if microbial production of the vitamin is impaired. Such cases have occurred due to the following:

- **Depressed ruminal thiamin synthesis**—as a result of a change in diet that disturbs rumen fermentation;
- **Increased thiamin degradation**—due to an alteration in the microbial population that increases total thiaminase activity. The most important organisms in this regard are those *Bacteroides thiaminolyticus*, *Clostridium sporogenes*, *Megasphaera elsdenii*, *Streptococcus bovis*, other *Clostridium* spp., *Bacillus* spp., and gram-negative cocci, which have thiaminases bound to their cell surfaces as exoenzymes.[85] These can cause significant thiamin losses when they are released into the rumen fluid, which can happen under conditions of sharply declining rumen pH. Accordingly, signs of thiamin deficiency have been observed in animals fed high-concentrate diets that tend to acidify the rumen.
- **Thiamin antagonists**—excess **sulfate** (which rumen microbes reduce to **sulfite**), excess amprolium, or factors contained in bracken fern or endophyte-infected fescue. Thiamin treatment has been found to reduce the signs of summer **fescue toxicity** (reduced performance, elevated body temperature, and rough hair coat) in grazing beef cattle.

81. Enlargement of the heart.
82. Thiamin deficiency impairs the fertility of both roosters (via testicular degeneration) and hens (via impaired oviductal atrophy).
83. Cerebrocortical necrosis.
84. Polioencephalomalacia is thought to be a disease of thiamin deficiency induced by thiaminases synthesized by rumen microbiota or present in certain plants. Affected animals are listless and have uncoordinated movements; they develop progressive blindness and convulsions. The disease is ultimately fatal but responds dramatically to thiamin.
85. Of these, *Bacteroides thiaminolyticus* appears to be of greatest pathogenic importance, as it appears to occur routinely in the ruminal contents and feces of all cases of polioencephalomalacia.

Response to Treatment

Thiamin-deficient humans are typically treated with several days of relatively high doses of the vitamin. Several days of doses are 10–200 mg/day. Cases of cardiac ("wet") beriberi have been treated with parenteral doses of 100 mg for several days.

10. ROLE OF THIAMIN IN HEALTH AND DISEASE

Two thiamin-responsive disorders have been reported:

- **DIDMOAD**. Genetic defects in the calcium channel protein wolframin (encoded by the *wfs1* gene) produce symptoms similar to TRMA syndrome: diabetes insipidus, insulin-dependent diabetes, bilateral progressive optic atrophy and deafness (DIDMOAD), depression, and psychosis. Thiamin supplementation can correct the glucose abnormalities and hematological symptoms.
- **Wilson disease**, due to a mutation of the *atp7b* gene that encodes an ATPase-Cu^{2+} transport protein, leads to toxic accumulation of Cu, which ultimately inhibits two TPP enzymes, α-KGDH and PDH. Supplements of thiamin and lipoate can correct this condition.

Dietary thiamin may also have a role in managing some neurologic diseases:

- **Alzheimer's disease**. The lower brain activities of α-KGDH, PDH, and TK in Alzheimer's disease patients compared to controls may reflect genetic variations in genes encoding portions of those enzymes. Animal studies have demonstrated the potential of thiamin supplementation to arrest and, perhaps, regress the formation of amyloid plaques. Only a few small trials have tested the therapeutic value of thiamin for treating Alzheimer's disease; those have yielded inconsistent results.
- **Parkinson's disease (PD)**. That PD patients given L-dopa showed increased cerebrospinal fluid concentrations of thiamin[86] suggests that supplemental thiamin may have value in managing PD. That hypothesis has not been tested.

86. Jimenéz- Jimenéz, F.J., Mlina, J.A., Hermánz, A., et al., 1999. Neurosci. Lett. 271, 33–36.

11. THIAMIN TOXICITY

Thiamin is generally well tolerated. Therapeutic doses as great as 300 mg/day are used therapeutically (e.g., to treat beriberi, Wernicke–Korsakoff syndrome, etc.) in humans without adverse reactions; however, greater doses have produced allergic reactions, headache, convulsions, weakness, paralysis, and cardiac arrhythmia. In the dog, very high doses (e.g., 1000-fold required levels) of thiamin hydrochloride have been found to be fatal, suppressing respiration and producing curare[87]-like signs suggestive of blocked nerve transmission: restlessness, epileptiform convulsions, cyanosis, and dyspnea.[88] Upper tolerable limits of exposure have not been set for thiamin.

12. CASE STUDIES

Instructions

Review each of the following case reports, paying special attention to the diagnostic indicators on which the treatments were based. Then answer the questions that follow.

Case 1

A 35-year-old man with a history of high alcohol intake for 18 years was admitted to the hospital complaining of massive swelling and shortness of breath on exertion. For several months, he had subsisted almost entirely on beer and whiskey, taking no solid food. He was grossly edematous, slightly jaundiced, and showed transient cyanosis[89] of the lips and nail beds. His heart showed gallop rhythm.[90] His left pleural cavity contained fluid. His liver was enlarged with notable ascites.[91] He had a coarse tremor of the hands and reduced tendon reflexes. His electrocardiogram showed sinus **tachycardia**.[92] His radiogram showed pulmonary edema and cardiac enlargement. He was evaluated by cardiac catheterization.

87. Curare is an extract of various plants (e.g., *Strychnos toxifera, Strychnos castelraei, Strychnos crevauxii, Chondrodendron tomentosum*). Practically inert when administered orally, it is a powerful muscle relaxant when administered intravenously or intramuscularly, exerting its effect by blocking nerve impulses at the myoneural junction. Curare is used experimentally and clinically to produce muscular relaxation during surgery. It was used originally as an arrow poison by indigenous hunters of South America to kill prey by inducing paralysis of the respiratory muscles.

88. Davis, R.E., Icke, G.C., 1983. Adv. Clin. Chem. 23, 93–140.

89. Dark bluish discoloration of the skin resulting from deficient oxygenation of the blood in the lungs or abnormally reduced flow of blood through the capillaries.

90. Triple cadence to the heart sounds at rates of >100 beats/min, indicative of serious myocardial disease.

91. Accumulation of serous fluid.

92. Rapid beating of the heart (>100 beats/min), originating in the sinus node.

Results

Parameter	Patient	Normal Value
Systemic arterial pressure (mm Hg)	100/55	120/80
Systemic venous pressure (cm H_2O)	300	<140
Pulmonary artery pressure (mm Hg)	64/36	<30/<13
Right ventricular pressure (mm Hg)	65/17	<30/<5
O_2 consumption (mL/min)	259	200–250
Peripheral blood O_2 (mL/L)	148	170–210
Pulmonary arterial blood O_2 (mL/L)	126	100–160
Cardiac output (L/min)	11.8	5–7
Blood hemoglobin (g/dL)	11.0	14–19
Cyanide circulation time (s)	12	20
Femoral arterial pyruvate (mg/dL)	1.5	0.8
Femoral arterial lactate (mg/dL)	14.1	4.7
Femoral arterial glucose (mg/dL)	86	74
Femoral arterial lactate (mg/dL)	14.1	4.7

The patient was given thiamin intravenously (10 mg every 6 h) for several days. Improvement was evident by 48 h and continued for 2 weeks. Thirty days later, he or she was free of edema, dyspnea, and cardiomegaly. Cardiac catheterization at that time showed that his or her blood, systemic venous, and all intracardiac pressures, as well as cardiac output, had all returned to normal.

Case 2

Fibroblasts were cultured from skin biopsies from four patients with Wernicke–Korsakoff syndrome and from four healthy control subjects. The properties of transketolase were studied (Table 11.9). The first patient was a 50-year-old woman with a history of chronic alcoholism. She had been admitted to the hospital with disorientation, nystagmus, sixth nerve weakness,[93] ataxia, and malnutrition. Treatment with intravenous thiamin and large oral

TABLE 11.9 Characteristics of Transketolase From Wernicke–Korsakoff Patients

Parameter	Patients	Controls
V_{max} (nmol/min/mg protein)	27±3	17±1
K_m (μM) TPP	195±3	116±2

93. The sixth cranial nerve is the *nervus abducens*, the small motor nerve to the lateral rectus muscle of the eye.

doses of multivitamins had improved her neurologic signs over a few months, but her mental state had deteriorated. She was readmitted with disorientation in both place and time, impaired short-term memory, nystagmus, ataxia, and signs of peripheral neuropathy. She was treated with parenteral thiamin and enteral B vitamins with thiamin; this had improved her general health but had not affected her mental status. The second patient, a 48-year-old man with a 20-year history of chronic alcoholism, was admitted in a severe confusional state. He was disoriented and had severe impairment of recent memory, confabulation, **perseveration**,[94] delusions, nystagmus, and ataxia. Treatment with thiamin and B vitamins had improved his behavior, without affecting his memory.

These results show that the affinity of transketolase for its coenzyme (TPP) in Wernicke–Korsakoff patients was less, by an order of magnitude, than that of controls. Further, this biochemical abnormality persisted in fibroblasts cultured for >20 generations in medium containing excess thiamin and no ethanol. The characteristics of pyruvate and α-KGDHs were similar in fibroblasts from patients and controls.

Case Questions

1. What factors would appear to have contributed to the thiamin deficiencies of these patients?
2. What defect in cardiac energy metabolism would appear to be the basis of the high-output cardiac failure observed in the first case?
3. What evidence suggests that the transketolase abnormality of these patients was hereditary? Would you expect such patients to be more or less susceptible to thiamin deprivation? Explain.

13. STUDY QUESTIONS AND EXERCISES

1. Construct a schematic map of intermediary metabolism showing the enzymatic steps in which TPP is known to function as a coenzyme.
2. Construct a decision tree for the diagnosis of thiamin deficiency in humans or animals.
3. How does the chemical structure of thiamin relate to its biochemical function?
4. What parameters might you measure to assess the thiamin status of a human or animal?
5. Construct a concept map illustrating the possible interrelationships of excessive alcohol intake and thiamin status.

RECOMMENDED READING

Alexander-Kaufman, K., Harper, C., 2009. Transketolase: observations in alcohol-related brain damage research. Int. J. Biochem. Cell Biol. 41, 717–720.

Balakumar, P., Rohilla, A., Krishan, P., et al., 2010. The multifactied therapeutic potential of benfotiamine. Pharmacol. Res. 61, 482–488.

Beltramo, E., Berrone, E., Tarallo, S., et al., 2008. Effects of thiamine and benoftiamine on intracellular glucose metabolism and relevance in the prevention of diabetic complications. Acta Diabetol. 45, 131–141.

Bettendorff, L., 2014. Thiamin, Chapter 7. In: Zemplini, J., Suttie, J.W., Gregory, J.E., et al. (Eds.), Handbook of Vitamins, fifth ed. CRC Press, New York, pp. 267–323.

Manzetti, S., Zhang, J., van der Spoel, D., 2014. Thiamin function, metabolism, uptake and transport. Biochem. 53, 821–835.

Nardone, R., Höller, Y., Storti, M., et al., 2013. Sci. World J. 2013, 8:309143.

Shannon, B., Chipman, D.M., 2009. Reaction mechanisms of thiamin diphosphate enzymes: new insights into the role of a conserved glutamate residue. FEBS J. 276, 2447–2453.

Lu'o'ng, K.V., Nguyễn, L.T.H., 2012. Thiamin and Parkinson's disease. J. Neurol. Sci. 316, 1–8.

94. The constant repetition of a meaningless word or phrase.

Chapter 12

Riboflavin

Chapter Outline

Anchoring Concepts

1. Riboflavin is the trivial designation of a specific compound, 7,8-dimethyl-10-(1′-D-ribityl)-isoalloxazine, sometimes also called **vitamin B_2**.
2. Riboflavin is a yellow, hydrophilic, and tricyclic molecule that is usually phosphorylated (to FMN and FAD) in biological systems.
3. Deficiencies of riboflavin are manifested chiefly as dermal and neural disorders.

In retrospect—the discovery of riboflavin may be considered a scientific windfall. It opened the way to the unraveling of the truly complex vitamin B_2 complex. Perhaps even more significantly, it bridged the gap between an essential constituent and cell enzymes and cellular metabolism. Today, with the general acceptance of this idea, it is not considered surprising that water-soluble vitamins represent essential parts of enzyme systems.

P. György[1]

LEARNING OBJECTIVES

1. To understand the chief natural sources of riboflavin
2. To understand the means of enteric absorption and transport of riboflavin
3. To understand the biochemical function of riboflavin as a component of key redox coenzymes and the relationship of that function to the physiological activities of the vitamin
4. To understand the physiologic implications of low riboflavin status

VOCABULARY

Acyl-CoA dehydrogenase
Adrenodoxin reductase
Alkaline phosphatase
Amino acid oxidases
Cheilosis
Curled toe paralysis
Dehydrogenase
Electron transfer flavoprotein (ETF)
Erythrocyte glutathione reductase
FAD
FAD pyrophosphatase
FAD synthase
Flavin
Flavin adenine dinucleotide (FAD)
Flavin exchange protein (FLX1)
Flavin mononucleotide (FMN)
Flavoenzyme
Flavokinase
Flavoprotein
Flavoproteome
FMN
FMN phosphatase
Geographical tongue
Glossitis
L-Gulonolactone oxidase

1. Paul Gyorgy (1893–1976) was a Hungarian-born American pediatrician and nutritionist known for his discoveries of riboflavin, vitamin B_6, and biotin. In 1975, he was awarded the U.S. National Medal of Science (Biology).

The Vitamins. http://dx.doi.org/10.1016/B978-0-12-802965-7.00012-5

Hypoplastic anemia
Isoalloxazine nucleus
Leukopenia
Lumichrome
Lumiflavin
Monoamine oxidase
Multiple acyl-CoA dehydrogenase deficiency (MADD)
NADH–cytochrome *P*450 reductase
NADH dehydrogenase
Normocytic hypochromic anemia
Ovoflavin
Oxidase
Reticulocytopenia
Riboflavin-binding proteins (RfBPs)
Riboflavin adenine diphosphate
Riboflavin monophosphate
Riboflavin-5′-phosphate
Riboflavin transporters (RFTs)
Riboflavinuria
Riboflavinyl radical
Ribotyl side chain
Stomatitis
Subclinical riboflavin deficiency
Succinate dehydrogenase
Thrombocytopenia
Thyroxine
Ubiquinone reductase
Vitamin B_2

1. THE SIGNIFICANCE OF RIBOFLAVIN

Riboflavin is essential for the intermediary metabolism of carbohydrates, amino acids, and lipids and also supports cellular antioxidant protection. The vitamin discharges these functions in the form of coenzymes that undergo reduction through two sequential single-electron transfer steps. This allows the reactions catalyzed by **flavoproteins** (i.e., **flavoenzymes**) to involve single- as well as dual-electron transfers. This versatility means that flavoproteins serve as switching sites between obligate two electron donors such as the pyridine nucleotides and various obligate one electron acceptors. Because of these fundamental roles of riboflavin in metabolism, a deficiency of the vitamin first manifests itself in tissues with rapid cellular turnover, such as skin and epithelium.

2. PROPERTIES OF RIBOFLAVIN

Riboflavin is among the substituted isoalloxazines synthesized by bacteria, yeasts, and plants. The term riboflavin refers to the compound 7,8-dimethyl-10-(1′-d-ribityl)isoalloxazine[2], which consists of a substituted **isoalloxazine nucleus** with a d-ribityl side chain and reducible nitrogen atoms in nucleus. It is metabolically functional as **flavin mononucleotide (FMN)** and **flavin adenine dinucleotide (FAD)**.[3]

Chemical structures of riboflavin and its metabolically functional forms are as follows:

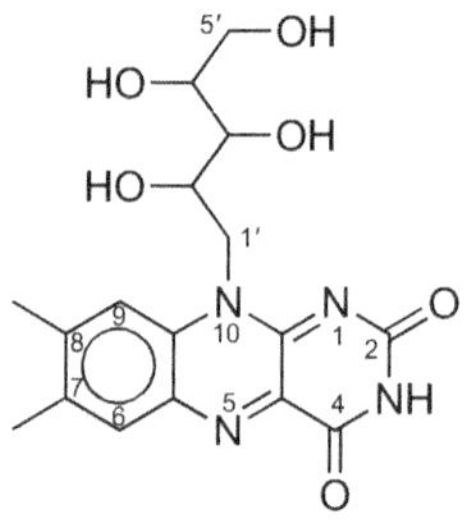

Riboflavin

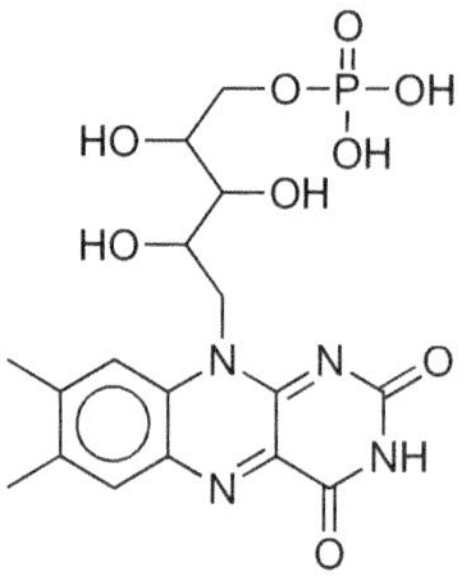

Flavin mononucleotide (FMN)

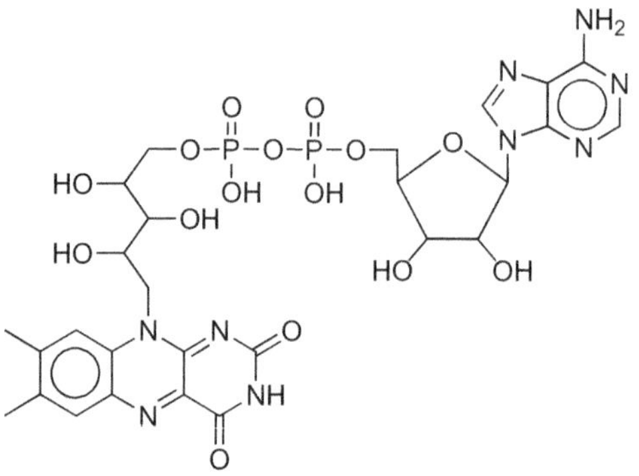

Flavin adenine dinucleotide (FAD)

Riboflavin Chemistry

Riboflavin is a yellow tricyclic molecule that is usually phosphorylated (to FMN and FAD) in biological systems. In FAD, the isoalloxazine and adenine nuclear systems are arranged one above the other and are nearly coplanar. The flavins are light-sensitive, undergoing photochemical degradation of the ribityl side chain, which results in the formation of such breakdown products as **lumiflavin** and lumichrome.[4] Therefore, the handling of riboflavin must be done in the dark or under subdued red light.

2. Formerly known as vitamin B_2, vitamin G, lactoflavin or riboflavin.

3. In fact, these compounds are not nucleotides; they are more properly referred to as **riboflavin monophosphate** and **riboflavin adenine diphosphate**, respectively.

4. 7,8-Dimethylalloxazine, an irradiation product of riboflavin believed also to be produced by intestinal microbiota.

Riboflavin is moderately soluble in water (10–13 mg/dL) and ethanol but insoluble in ether, chloroform, and acetone. It is soluble but unstable under alkaline conditions. Because riboflavin cannot be extracted with the usual organic solvents, it is extracted with chloroform as lumiflavin after photochemical cleavage of the ribityl side chain. Flavins show two absorption bands, at <370 nm and <450 nm, with fluorescence emitting at 520 nm.

The catalytic functions of riboflavin are carried out primarily at positions N-1, N-5, and C-4 of the isoalloxazine nucleus. In addition, the methyl group at C-8 participates in covalent bonding with enzyme proteins. The flavin coenzymes are highly versatile redox cofactors because they can participate in either one- or two-electron redox reactions, thus serving as switching sites between obligate two electron donors (e.g., NAD(H), succinate) and obligate one electron acceptors (e.g., iron–sulfur proteins, heme proteins). They serve this function by undergoing reduction through a two-step sequence involving a radical anion intermediate. Because the latter can also react with molecular oxygen, flavins can also serve as cofactors in the two-electron reduction of O_2 to H_2O and in the reductive four-electron activation and cleavage of O_2 in the monooxygenase reactions. In these redox reactions, riboflavin undergoes changes in its molecular shape, i.e., from a planar oxidized form to a folded reduced form. Differences in the affinities of the associated apoprotein for each shape affect the redox potential of the bound flavin.

Riboflavin antagonists include analogs of the isoalloxazine ring (e.g., diethylriboflavin, dichlororiboflavin, phenothiazine drugs) and the ribityl side chain (e.g., d-araboflavin, D-galactoflavin, 7-ethylriboflavin), bind flavin coenzymes (e.g., adriamycin, tetracycline), or bind riboflavin directly (e.g., boric acid).

3. SOURCES OF RIBOFLAVIN

Hindgut Microbial Synthesis

Riboflavin can be produced by the microbiome of the colon. A genomic analysis of 256 representative organisms of the human gut microbiota found more than half (56%) capable of de novo synthesis of the vitamin.[5] Those findings suggested that hindgut microbial synthesis may produce nearly 3% of the daily human need for riboflavin. However, direct evidence is lacking for the absorption of riboflavin across the colon. It is likely that noncoprophagous animals derive no benefit from this source of the vitamin.

Distribution in Foods

Riboflavin is widely distributed in foods (Table 12.1), where it is present almost exclusively bound to proteins, mainly in

5. Magnúsdóttir, S., Ravchee, D., de Crécy-Lagard, V., et al., 2015. Front. Genet. 6, 148–166.

TABLE 12.1 Riboflavin Contents of Foods

Food	Riboflavin (mg/100 g)
Dairy Products	
Milk	0.17–0.19
Yogurt	0.28
Cheese	
Cheddar	0.43
Cottage	0.17
Meats	
Liver, beef	2.76
Beef	0.09–0.24
Chicken	0.12–0.22
Pork	0.19–0.39
Ham, cured	0.26
Cereals	
Wheat, bran	0.58
Rye flour	0.11
Oats	0.14
Rice	0.01–0.07
Cornmeal	0.20
Vegetables	
Asparagus	0.14
Broccoli	0.12
Cabbage	0.04
Carrots	0.06
Cauliflower	0.06
Lima beans	0.10
Potatoes	0.02
Spinach	0.19
Tomatoes	0.02
Fruits	
Apples	0.03
Bananas	0.07
Oranges	0.05
Peaches	0.03
Strawberries	0.02
Other	
Eggs	0.46

From USDA National Nutrient Database for Standard Reference, Release 28 (http://www.ars.usda.gov/ba/bhnrc/ndl).

the form of FMN and FAD.[6,7] Rapidly growing, green, leafy vegetables are rich in the vitamin; however, meats and dairy products are the most important sources. Animal tissues have been found to contain small amounts of riboflavin-5′-a-D-glucoside, which appears to be as well utilized as free riboflavin. It is estimated that milk and milk products contribute about 50% of the riboflavin in the American diet, with meats, eggs, and legumes contributing a total of about 25%, and fruits and vegetables each contributing about 10%.

Stability

Riboflavin is stable to heat; therefore, most means of heat sterilization, canning, and cooking do not affect the riboflavin contents of foods. However, exposure to light (e.g., sun drying, sunlight exposure of milk in glass bottles, cooking in an open pot) can result in substantial losses, as the vitamin is very sensitive to destruction by light. Thus, exposure of milk in glass bottles to sunlight can result in the destruction of more than half of its riboflavin within a day. Irradiation of food results in the production of reactive oxygen species (ROS, e.g., superoxide, hydroxyl radical) that react with riboflavin to destroy it. The short exposure of meat to sterilizing quantities of γ–irradiation destroys 10–15% of its riboflavin content. Riboflavin photodegradation can be exacerbated by sodium bicarbonate, which is used to preserve vegetable colors. Also, because riboflavin is water soluble, it leaches into water used in cooking and into the drippings of meats. As riboflavin in cereal grains is located primarily in the germ and bran, the milling of such materials,[8] which removes those tissues, results in considerable losses in their contents of the vitamin. For example, about half of the riboflavin in whole grain rice, and more than one-third of riboflavin in whole wheat, is lost when these grains are milled. Parboiled ("converted") rice contains most of the riboflavin of the parent grain, as the steam processing of whole brown rice before milling this product drives vitamins originally present in the germ and aleurone layers into the endosperm, where they are retained.

Bioavailability

The apparent bioavailability of riboflavin can depend on the method of assessment. Because the vitamin is largely removed from the liver after entering the circulation, plasma responses to oral dosing underestimate bioavailability. Better estimates are obtained from assays based on the prevention of clinical signs in animals (e.g., curled toe paralysis in the chick) or on monitoring urinary output. Such assays show that the noncovalently bound forms of riboflavin in foods, FMN, FAD, and free riboflavin appear to be well absorbed. In contrast, covalently bound flavin complexes tend to be more stable to digestion and, thus, less bioavailable; some 10–15% of flavins from plant sources have been found not to be utilized by humans.[9] In general, riboflavin in animal products tends to have a greater bioavailability than that in plant products. While foods contain relatively little free riboflavin, that form is widely consumed in multivitamin supplements and vitamin-fortified cereals.

6. Notable exceptions are milk and eggs, which contain appreciable amounts of free riboflavin.

7. It should be noted that, strictly speaking, FMN is not a nucleotide, nor is FAD a dinucleotide, because each is a D-ribotyl derivative; nevertheless, these names have been accepted.

8. It is the practice in many countries to enrich refined wheat products with several vitamins, including riboflavin, which result in their actually containing *more* riboflavin than the parent grains (e.g., 0.20 mg/100 g *versus* 0.11 mg/100 g). However, rice is usually *not* enriched with riboflavin to avoid coloring the product yellow by this intensely colored vitamin.

4. ABSORPTION OF RIBOFLAVIN

Hydrolysis of Coenzyme Forms

The enteric absorption of riboflavin depends on its being in the free, nonphosphorylated form. This occurs by the actions of nonspecific proteolytic activities of the intestinal lumen, which release riboflavin coenzymes from their protein complexes, and the subsequent hydrolytic activities of several nonspecific brush border pyrophosphatases that liberate riboflavin in free form. Principal contributors among the latter enzymes are the relatively nonspecific alkaline phosphatase,[10] as well as FAD pyrophosphatase (which converts FAD to FMN) and FMN phosphatase (which converts FMN to free riboflavin).

Enteric Absorption of Free Riboflavin

Riboflavin is absorbed across the enterocyte in the free form by highly specific **riboflavin transporters (RFTs)**. These are located on the enterocyte brush border (RFT-1[11]) and basolateral surface (RFT-2), and in intracellular vesicles (RFT-3) and are expressed in several tissues including the proximal small intestine and colon. A genetic defect in RFT-1 was found to manifest as elevated plasma acylcarnitine levels and organic aciduria, suggestive of **multiple acyl-CoA dehydrogenase deficiency (MADD)**.[12] A defect in RFT-2 has been associated with Brown–Vialetto–Van Laere syndrome, which is manifest as a biochemical profile similar to MADD with progressive neurological dysfunction.[13]

9. Decker, K.F., 1993. Ann. Rev. Nutr. 13, 17–41.

10. Alkaline phosphatase appears to have the greatest hydrolytic capacity of the brush border phosphatases.

11. RFT-1 has been identified as G-protein-coupled receptor 172B (GPR172B).

12. Ho, G., Yonezawa, A., Masuda, S., et al., 2011. Human?

13. Bosch, A.M., Abeling, N.G., Ijist, L., et al., 2011. J. Inherit. Metab. Dis. 34, 159–164.

The upper limit of intestinal riboflavin absorption has been estimated to be about 27 mg[14]—more than an order of magnitude greater than the dietary requirement.[15] Riboflavin absorption is enhanced by riboflavin deficiency, by bile salts[16], and by factors that stimulate intestinal motility, e.g., the presence of food. Psyllium fiber reduces riboflavin absorption; but wheat bran has no effect. Alcohol impairs both the digestion of food flavins and the absorption of the free vitamin. Riboflavin absorption is downregulated by high doses of the vitamin, apparently through reduced activity of the riboflavin carrier induced by increased intracellular concentrations of cyclic AMP.

Much of the riboflavin transported into the enterocyte is quickly phosphorylated to FMN. This is accomplished by an ATP-dependent **flavokinase**. Thus, riboflavin enters the portal circulation as both the free vitamin and FMN.

5. TRANSPORT OF RIBOFLAVIN

Protein Carriers

Nonspecific carriers. Riboflavin is transported in the plasma as both free riboflavin and FMN, both of which are mostly bound to plasma proteins. This includes albumin and several immunoglobulins (IgA, IgG, and IgM); collectively, these bind about half of the free riboflavin and 80% of FMN in plasma. This involves hydrogen bonding, the strength of which varies among these proteins as well as among the various species of the vitamin,[17] thus, influencing the distribution of those species to the tissues as well as their clearance across the kidney. The vitamin can be displaced readily from these nonspecific carriers by boric acid,[18] several drugs,[19] lumiflavin, and lumichrome to inhibit its transport to the tissues.

Specific riboflavin-binding proteins (RfBPs). RfBPs have been identified in the plasma of the laying hen and pregnant cows, mice, rats, monkeys, and humans. Of these, the avian RfBP is the best characterized. It is not found in the immature female but is synthesized in the liver under the stimulus of estrogen with the onset of sexual maturity or with induction by estrogen treatment. The avian plasma RfBP is a 37-kDa phosphoglycoprotein with a single binding site for riboflavin. It appears to be one of three products of a single gene, which are variously modified posttranslationally. RfBP synthesized by the liver is N-linked to oligosaccharides and phosphorylated at six serinyl residues prior to being exported to the plasma. This RfBP is antigenically similar in the laying hen and pregnant mice, rats, cows, and humans. The RfBP in avian eggs (albumen and yolk) is synthesized by the oviduct.[20] Male mice produce RfBP in their testicular Leydig cells; it is secreted in response to luteinizing hormone via G-protein-induced production of cyclic AMP.[21] RfBP localizes in the sperm head. Pregnant women show low-circulating plasma levels of RfBP until about 4 months of gestation at which time those levels increase. Placental trophoblasts can also synthesize RfBP. These increases corresponding to two- to threefold increases in the RfBP contents of amniotic fluid. Thus, in each species, RfBP has vital functions in the transplacental/transovarian movement of riboflavin[22] and in the uptake of riboflavin by spermatozoa, as immunoneutralization of the protein terminates pregnancy in females and reduced sperm fertility in males.[23]

Cellular Uptake

Riboflavin uptake into cells occurs by the transfer of the vitamin by RfBP in plasma to RFTs on the plasma membrane. This process is mediated by a Ca^{2+}-dependent RfBP receptor located in clathrin[24]-coated pits on the plasma membrane, which facilitates endocytosis and release of the vitamin, with recycling of the receptor and RfBP being catabolized within endosomes. Receptor-mediated endocytosis has also been implicated in the transport of riboflavin across the placental barrier.[25] RFTs are expressed in the small intestine, colon, placenta, embryonic kidney, testes, prostate, and brain. They show varying sequence homologies and different functionalities. RFT-1 is not sensitive to pH; but RFT-2 has a pH optimum around 6.0. Their apparent K_m values vary in a way suggestive of intracellular riboflavin homeostasis being controlled by

14. Zempleni, J., Galloway, J.R., McCormick, D.B., 1996. Am. J. Clin. Nutr. 63, 54–66.
15. Greater amounts of the vitamin, e.g., from massive doses pass to the hindgut where they are degraded by the microbiota and excreted.
16. Children with biliary atresia (a congenital condition involving the absence or pathological closure of the bile duct) show reduced riboflavin absorption.
17. The binding affinities have been estimated: albumin–riboflavin (one binding site), $K_d = 3.8–10.4$ mM; Ig-riboflavin (two binding sites), $K_d = 2.43$ and 0.068 nM; Ig-FAD (two binding sites), $K_d = 1.73$ and 0.078 nM.
18. Boric acid can produce riboflavinuria and precipitate riboflavin deficiency; some effects of boric acid toxicity can be overcome by feeding riboflavin.
19. For example, ouabain, theophylline, penicillin.
20. It is cotransported into the yolk with the glycoprotein vitellogenin.
21. Subramanian, S., Adiga, P.R., 1996. Mol. Cell Endocrinol. 120, 41–50.
22. The astute observation that a particular hen that produced eggs lacking the normal faint yellow tinge of its otherwise clear albumen led to the discovery of RfBP (which that hen failed to express) as being essential for the transfer of riboflavin to the egg (Winter, W.P., Buss, E.G., Claget, C.O., et al., 1967. Comp. Biochem. Physiol. 22, 889–896).
23. Plasma RfBP or fragments has been suggested as having potential as a vaccine to regulate fertility in both sexes (Adiga, P.R., 1997. Human Reprod. Update 3, 325–332).
24. Named for the Latin "*clatratus*," meaning "like a lattice", this protein plays a major role in the formation of coated vesicles that facilitate endocytosis and exocytosis to allow cells to communicate.
25. Foraker, A.M., Knantwell, C.M., Swan, P.W., 2002. Adv. Drug Deliv. Rev. 55, 1467.

its trafficking into acidic vesicular compartments (RFT-1, 1.38 μM; RFT-2, 0.98 μM; RFT-3, 0.33 μM). A genetic defect in RFT-1 was found to manifest as elevated plasma acylcarnitine levels and organic aciduria, suggestive of MADD.[26] A defect in RFT-2 has been associated with Brown–Vialetto–Van Laere syndrome, which is manifest as a biochemical profile similar to MADD with progressive neurological dysfunction.[27]

Tissue Distribution

Riboflavin is transported into cells in its free form. However, in the tissues, riboflavin is converted to the coenzyme form, predominantly as FMN (60–95% of total flavins) but also as FAD (5–22% of total flavins in most tissues but about 37% in kidney), both of which are found almost exclusively bound to specific flavoproteins. The greatest concentrations of the vitamin are found in the liver, kidney, and heart. In most tissues, free riboflavin comprises <2% of the total flavins. Significant amounts of free riboflavin are found only in retina, urine, and cow's milk,[28] where it is weakly bound to casein. Although the riboflavin content of the brain is not great, the turnover of the vitamin in that tissue is high and the concentration of the vitamin is relatively resistant to gross changes in riboflavin nutriture. These findings suggest a homeostatic mechanism for regulating the riboflavin content of the brain; such a mechanism has been proposed for the choroid plexus,[29] in which riboflavin transport has been found to be inhibited by several of its catabolic products and analogues. It has been estimated that the total body reserve of riboflavin in the adult human is equivalent to the metabolic demands for 2–6 weeks. Riboflavin is found in much lower concentrations in maternal plasma than in cord plasma (in humans, this ratio has been found to be 1:4.7), suggesting the presence of a transplacental transport mechanism.

Tissue RfBPs have been identified in the liver, egg albumen,[30] and egg yolk of the laying hen. Each is similar to the plasma RfBP[31] in that species, differing only in the nature of their carbohydrate[32] contents.[33] A hereditary abnormality in the chicken results in the production of defective RfBPs (in plasma as well as liver and egg). Affected hens show **riboflavinuria** and produce eggs with about half the normal amount of riboflavin and embryos that fail to develop.[34]

6. METABOLISM OF RIBOFLAVIN

Conversion to Coenzyme Forms

After it is taken up by the cell, free riboflavin is converted to its coenzyme forms (Fig. 12.1) in two steps, both of which appear to be regulated by thyroid hormones:

1. **Conversion to FMN** occurs by ATP-dependent phosphorylation to yield **riboflavin-5′-phosphate**, i.e., FMN, in the cytoplasm of most cells. This step is catalyzed by flavokinase[35] (also called riboflavin kinase), which uses ATP or dATP as the phosphate donor and Zn^{2+} as an activator. Flavokinase expression is stimulated by thyroxine. It is downregulated under conditions of dietary riboflavin deficiency and inflammation. The latter effect may be related to overexpression of tumor necrosis factor α (TNFα). Flavokinase is known to bind a TNFα-receptor 1 binding protein that binds a subunit of NADPH oxidase; this multicomponent linking may be important for the activation of NADPH oxidase. Flavokinase has catalytically important sulfhydryl domains, as reducing agents (e.g., reduced glutathione) protect it from inhibition. FMN can be complexed with specific FMN-requiring apoproteins to form functional flavoproteins (Table 12.2).
2. **Conversion to FAD**[36] occurs by the further metabolism of the major portion of FMN to the other coenzyme form, **FAD,** by a second ATP-dependent enzyme, **FAD synthase.**[37] There are two forms of this enzyme, a mitochondrial form (FAD synthase 1) and a cytosolic form (FAD synthase 2). This multicompartment distribution

26. Ho, G., Yonezawa, A., Masuda, S., et al., 2011. Hum. Mutat. 31, E1976–E1984.
27. Bosch, A.M., Abeling, N.G., Ijlst, L., et al., 2011. J. Inherit. Metab. Dis. 34, 159–164.
28. It should be noted that cow's milk differs from human milk in both the amount and form of riboflavin. Cow's milk typically contains 1160–2020 μg of riboflavin per liter, which (like the milk of most other mammals studied) is present mostly as the free vitamin. In contrast, human milk typically contains 120–485 μg of riboflavin per liter (depending on the riboflavin intakes of the mother), which is present mainly as FAD and FMN.
29. The anatomical site of the blood–cerebrospinal fluid barrier.
30. This is the flavoprotein formerly called **ovoflavin**. Comprising nearly 1% of the total protein in egg white, it is the most abundant of any vitamin-binding protein. Unlike the plasma RfBP, which is normally saturated with its ligand, the egg white RfBP is normally less than half-saturated with riboflavin, even when hens are fed diets high in the vitamin. Its bound riboflavin is responsible for the faint yellow tinge of egg albumen.
31. It appears that plasma RfBP, produced and secreted by the liver in response to estrogens, is the precursor to these other binding proteins found in tissues.
32. Primarily in their contents of sialic acid, which occurs in many polysaccharides.
33. It is interesting to note that egg white RfBP forms a 1:1 complex with the thiamin-binding protein (TBP) from the same source.
34. Embryos from hens homozygous for the mutant *rd* allele die of riboflavin deficiency on day 13–14 of incubation but can be rescued by injecting riboflavin or FMN into the eggs.
35. Therefore, hypothyroidism is associated with low flavokinase activity and, accordingly, low tissue levels of FMN and FAD. Hyperthyroidism, in contrast, results in increased flavokinase activity, although tissue levels of FMN and FAD, which appear to be regulated via degradation, are unaffected.
36. Small amounts of FAD derivatives have also been identified: 8α-S-cysteinyl-FAD, 8α-N^1-histidinyl-FAD, 8α-N^3-histidinyl-FAD.
37. This activity is also increased by thyroxine.

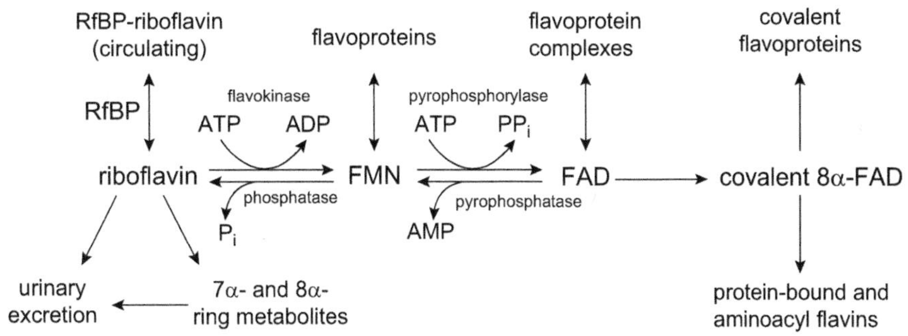

FIGURE 12.1 Riboflavin metabolism.

TABLE 12.2 Covalent Flavoproteins in Animals

Linkage	Enzyme
Histidinyl(N^3)-8α-FAD	Succinate dehydrogenase Dimethylglycine dehyrogenase Sarcosine dehydrogenase
Histidinyl(N^1)-8α-FAD	L-gulonolactone oxidase
Cysteinyl(s)-8α-FAD	Monoamine oxidase

FAD, flavin adenine dinucleotide.

allows FAD to complex, mostly through noncovalent associations with dinucleotide-binding domains, with different dehydrogenases and oxidases within the mitochondria. Noncomplexed FAD is exported from mitochondria into the cytosol by a **flavin exchange protein, FLX1**. FAD synthesis is feedback inhibited by FAD, which is complexed in tissues apparently facilitated by a protein factor. This includes associations via hydrogen bonding with purines,[38] phenols, and indoles (e.g., to peptidyl tryptophan in RfBPs). Less than 10% of FAD is covalently attached to apoenzymes. Linkages of this type involve the riboflavin 8-methyl group, which can form a methylene bridge to the peptide histidyl imidazole function (e.g., in succinic dehydrogenase and sarcosine oxidase) or to the thioether function of a former cysteinyl residue (e.g., in monoamine oxidase.[39]

Glycosylation

The capacity to glycosylate riboflavin has been demonstrated in rat liver. Riboflavin 5′-α-D-glucoside appears to be a metabolically significant metabolite. It is found in the urine of riboflavin-fed rats and has been shown to be comparable to riboflavin as a cellular source of the vitamin.

Catabolism

Flavins that are bound to proteins are resistant to degradation. However, unbound forms are subject to catabolism. Both FAD and FMN are catabolized by intracellular enzymes with different isoforms in mitochondria and cytosol. Thus, FAD is converted to FMN by FAD pyrophosphatase (releasing AMP), and FMN is degraded to free riboflavin by FMN phosphohydrolase. Both FAD and FMN are split to yield free riboflavin by alkaline phosphatase, which is recognized by RFT-2 for export from the cell. Unlike their synthetic analogs, neither FAD pyrophosphatase or FMN phosphohydrolase is not affected by thyroid hormones or riboflavin status.

The degradation of riboflavin per se involves initially its hydroxylation at the 7α- and 8α-positions of the isoalloxazine ring by hepatic microsomal cytochrome *P*450-dependent processes. It is thought that catabolism proceeds by the oxidation and then removal of the methyl groups. The liver, in at least some species, has the ability to form riboflavin α-glycosides. As a result of this metabolism, human blood plasma contains FAD and FMN as the major riboflavin metabolites, as well as small amounts of 7α-hydroxyriboflavin.[40] Side chain oxidation has been observed in bacterial systems but not in higher animals.

Excretion

Riboflavin is rapidly excreted, primarily in the urine. Therefore, dietary needs for the vitamin are determined by its rate of excretion, not metabolism. In a riboflavin-adequate human adult nearly all of a large oral dose of the vitamin will be excreted, with peak concentrations showing in the urine within a couple of hours. Studies in the rat have shown riboflavin to be turned over with a half-life of about 16 days in adequately nourished animals and much longer

38. In FAD, the riboflavin and adenine moieties are predominantly (85%) hydrogen bonded in an intramolecular complex.

39. Another type of linkage involving the 8-methyl group, i.e., a thiohemiacetal linkage, is found in a microbial FAD-containing cytochrome.

40. This compound is also called 7-hydroxymethylriboflavin.

in riboflavin-deficient animals. In normal human adults, the urinary excretion of riboflavin is about 200 μg/24 h; whereas, riboflavin-deficient individuals may excrete only 40–70 μg/24 h. Studies with a diabetic rat model[41] have shown riboflavin excretion to be significantly greater in diabetic individuals than in controls. Riboflavin excretion responds to the level of riboflavin intake; excretion of <27 μg/mg creatinine is generally considered to indicate riboflavin deficiency in adults; however, this parameter tends to reflect current intake of the vitamin rather than total flavin stores.

The vitamin is excreted mainly (60–70%) as the free riboflavin, with smaller amounts of 7α- and 8α-hydroxyr iboflavin,[42] 8α-sulfonylriboflavin, 5′-riboflavinylpeptide, 10-hydroxyethylflavin, riboflavin 5′-α-D-glucoside, lumichrome, and 10-formylmethylflavin. Small amounts of riboflavin degradation products are found in the feces (<5% of an oral dose). As only about 1% of an oral dose of the vitamin is excreted in the bile by humans, most fecal metabolites are thought to be mostly of gut microbial origin. Little, if any, riboflavin is oxidized to CO_2.[43] Ingestion of boric acid, which binds to the riboflavin side chain, increases the urinary excretion of the vitamin.

Riboflavin is secreted into milk mostly as free riboflavin and FAD and the antagonistic metabolite 10-(2′-hydroxyethyl)flavin; amounts depend on the riboflavin intake of the mother. Milk also contains small amounts of other metabolites including 7- and 8-hydroxymethylriboflavins, 10-formylmethylflavin, and lumichrome.

7. METABOLIC FUNCTIONS OF RIBOFLAVIN

Riboflavin functions metabolically as the essential component of the coenzymes FMN and FAD, which act as intermediaries in transfers of electrons in biological oxidation–reduction reactions. More than 100 enzymes are known to bind FAD or FMN in animal and microbial systems. These are encoded by 90 genes.[44] Ten flavoproteins occur in multiple isoforms; allelic variants in two-thirds have been associated with clinical disorders.

Coenzyme Functions

FAD is the cofactor for some 84% of the flavoenzymes; FMN is the cofactor for 16% of them. In most cases, the flavinyl cofactor is bound tightly but noncovalently; a few flavoenzymes[45] bind FAD covalently via histidinyl or cysteinyl linkages to the 8α-position of the isoalloxazine ring. These flavoenzymes include oxidases, which function aerobically, and dehydrogenases, which function anaerobically. Some involve one electron transfers, whereas others involve two electron transfers. This versatility allows flavoproteins to serve as switching sites between obligate two electron donors (e.g., NADH, succinate) and obligate one electron acceptors (e.g., iron–sulfur proteins, heme proteins). Flavoproteins serve this function by undergoing reduction through two single-electron transfer steps (Fig. 12.2) involving a riboflavinyl radical or semiquinone intermediate (with the unpaired electron localized at N-5). Because the radical intermediate can react with molecular oxygen, flavoproteins can also serve as cofactors in the two-electron reduction of O_2 to H_2O, and in the four-electron activation and cleavage of O_2 in monooxygenase reactions.

Collectively, the flavoproteins show great versatility in accepting and transferring one or two electrons with a range of potentials. This feature owes to the variation in the angle between the two planes of the isoalloxazine ring system (intersecting at N-5 and N-10), which is modified by specific protein binding. The flavin-containing dehydrogenases or reductases (their reduced forms) react slowly with molecular oxygen, in contrast to the fast reactions of the flavin-containing oxidases and monooxygenases. In the former reactions, hydroperoxide derivatives of the flavoprotein are cleaved to yield superoxide anion (O_2^-), but in the latter a heterolytic cleavage of the hydroperoxide group occurs to yield the peroxide ion (OOH^-). Many flavoproteins contain a metal (e.g., iron, molybdenum, zinc), and the combination of flavin and metal ion is often involved in the adjustments of these enzymes in transfers between single- and double-electron donors. In some flavoproteins, the means for multiple electron transfers is provided by the presence of multiple flavins as well as metals.

Metabolic Roles

The flavoproteome comprises a large group of enzymes that have central roles in the metabolism of carbohydrates, amino acids, lipids, and the activation of pyridoxine and folate to their functional forms (Table 12.3).[46,47] Others (particularly, glutathione reductase) participate in antioxidant protection by maintaining the glutathione redox cycle and providing the reducing equivalents for neutralizing ROS. Intracellular

41. Streptozotocin-induced diabetes.
42. This compound is also called 8-hydroxymethylriboflavin.
43. Rats have been found to oxidize less than 1% of an oral dose of the vitamin.
44. Including six enzymes involved in riboflavin utilization and metabolism.
45. For example, succinate dehydrogenase, monoamine oxidase, monomethylglycine dehydrogenase.
46. See review: Lienhart, W.D., Gudipati, V., Macheroux, P., 2013. Arch. Biochem. Biophys. 535, 150–162.
47. Riboflavin has also been found to play a role in the regulation of gene expression in bacteria by forming mRNA structures called "riboswitches" that repress conformation to cause premature termination of transcription or inhibit initiation of translation. Analogous function in higher animals has not been reported.

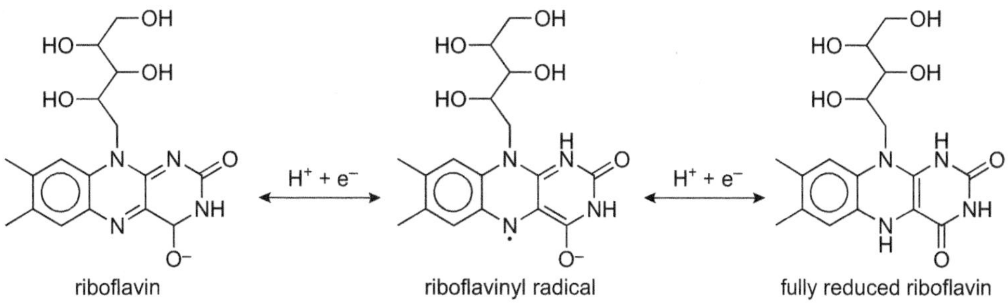

FIGURE 12.2 Two-step, single-electron, redox reactions of riboflavin.

redox balance[48] affects the oxidation state of protein sulfhydryls; complete oxidation (to the sulfenic and sulfinic acid states) is irreversible and can adversely affect protein folding. Such abnormalities are prevented by maintaining GSH:GSSG at adequate levels, which allows protein thiol groups to be protected by S-glutathionylation.

8. BIOMARKERS OF RIBOFLAVIN STATUS

Several clinical biochemical endpoints have been used to assess riboflavin status.[49] Of those, four methods have proven to be the most informative:

- **Degree of FAD/FMN saturation of flavoenzymes** is the most informative biomarker of riboflavin status. The enzyme of choice is typically erythrocyte glutathione reductase (eGR), which responds to riboflavin deprivation. The assay takes advantage of the in vitro binding of FAD by that enzyme in hemolysates. Because riboflavin-adequate subjects typically have 80–90% of eGR bound to FAD, the addition of exogenous FAD should stimulate their eGR activities by no more than XX%; stimulation above that level indicates that less eGR was bound to FAD. Therefore, the eGR response to added FAD measures eGR saturation and, thus, riboflavin status. This is measured as an activity coefficient:

  eGR activity coefficient = baseline eGR activity/eGR activity with added FAD

Subjects with eGR activity coefficients <1.15 are considered to be at low risk of riboflavin deficiency; those with activity coefficients of 1.15–1.25 or >1.25 are considered to be at moderate and high risks, respectively. This biomarker, however, is *not* useful for subjects with the common genetic condition of glucose-6-phosphate dehydrogenase deficiency whose GRs remain saturated with FAD regardless of riboflavin status. **Pyridoxine 5′-phosphate oxidase activity**, which also reflects riboflavin status, has been used for such subjects.

- **Erythrocyte riboflavin content** less than 0.15 mg/L indicates low/deficient riboflavin status.
- **Urinary riboflavin excretion** less than 10% of the amount of the ingested vitamin indicates low/deficient riboflavin status.

9. RIBOFLAVIN DEFICIENCY

Many tissues are affected by riboflavin deficiency (Table 12.4). Therefore, deprivation of the vitamin causes in animals such general signs as loss of appetite, impaired growth, and reduced efficiency of feed utilization, all of which constitute significant costs in animal agriculture. In addition, both animals and humans experiencing riboflavin deficiency show specific epithelial lesions and nervous disorders. These manifestations are accompanied by abnormally low activities of various flavoenzymes. The most rapid and dramatic loss of activity involves eGR. Substantial losses also occur in the activities of flavokinase and FAD synthetase; thus, the biosynthesis of flavoproteins is lost under conditions of riboflavin deprivation. In summary, then, riboflavin deficiency results in impairments in the metabolism of energy, amino acids, and lipids. These metabolic impairments are manifested morphologically as arrays of both general and specific signs/symptoms.

Riboflavin deficiency produces in the small intestine: a hyperproliferative response of the mucosa, characterized by reductions in number of villi, increases in villus length, and increases in the transit rates of enterocytes along the villi. These morphological effects are associated with reduced enteric absorption of dietary iron, resulting in secondary impairments in nutritional iron status in riboflavin-deprived individuals.

Risk Factors for Riboflavin Deficiency

Several factors can contribute to riboflavin deficiency:

- **Inadequate diet** is the most likely cause of riboflavin deficiency. Typically, this involves the low consumption of milk.[50] In industrialized countries, riboflavin deficiency occurs most frequently among alcoholics, whose

48. Intracellular redox potential is maintained by keeping GSH:GSSG 30–100:1 through the use of reducing equivalents from NADPH and cysteine.
49. These have been systematically reviewed (Hoey, L., McNulty, H., Strain, J.J., 2009. Am. J. Clin. Nutr. 89, 1960S–1980S).
50. Children consuming less than a cup of milk per week are likely to be deficient in riboflavin.

TABLE 12.3 Major Components of the Flavoproteome of Humans and Animals

Flavoprotein	Flavin	Metabolic Function
One Electron Transfers		
Mitochondrial electron transfer flavoprotein (ETF)	FAD (flavin adenine dinucleotide)	e^- acceptor for acyl-CoA, branched-chain acyl-CoA, glutaryl-CoA, and sarcosine and dimethylglycine dehydrogenases; links flavoprotein dehydrogenases with respiratory chain *via* ETF–ubiquinone reductase
NADH–ubiquinone reductase	FAD	Transfers reducing equivalent from ETF and ubiquinone in respiratory chain
NADH–cytochrome *P*450 reductase[a]	FMN (flavin mononucleotide)	Transfers reducing equivalent from FMN to cytochrome *P*450
Two Electron Transfers		
Pyridine-Linked Dehydrogenases		
NADP–cytochrome *P*450 reductase[a]	FAD	A key regulatory protein; transfers reducing equivalents from NADP to FAD, then to several acceptor proteins
Adrenodoxin reductase	FAD	Transfers reducing equivalents from NADP to adrenodoxin[b] in Steroid hydroxylation in adrenal cortex
NADP dehydrogenase	FMN	Transfers reducing equivalents from NADP to FMN, then to ubiquinone[c]
NADP-dependent methemoglobin reductase	FAD	Transfers reducing equivalents from NADP to FAD to reduce methemoglobin
NADH–cytochrome b_5 reductase	FAD	Transfers reducing equivalents from NADP to FAD to dehydrogenate stearoyl-CoA to oleoyl-CoA
3-β-Hydroxysterol Δ-24-reductase	FAD	Transfers reducing equivalents from NADPH in the latter steps of cholesterol synthesis
7-Dehydrocholesterol reductase	FAD	Transfers reducing equivalents from NADPH in the latter steps of cholesterol synthesis
Nonpyridine Nucleotide-Dependent Dehydrogenases		
Succinate dehydrogenase	FAD	Transfers reducing equivalents from succinate to ubiquinone yielding fumarate
Acyl-CoA dehydrogenases	FAD	Transfers reducing equivalents from substrate to flavin in the initial step in β-oxidation of fatty acids
Pyridine Nucleotide Oxidoreductases		
Glutathione reductase	FAD	Reduces GSSG to GSH using NADPH
Lipoamide dehydrogenase[d]	FAD	Oxidizes dihydrolipoamide to lipoamide using NAD^+
Reactions of Reduced Flavoproteins With Oxygen		
D-amino acid oxidase	FAD	Dehydrogenates D-amino acid substrates to imino acids, which are hydrolyzed to α-keto acids
L-amino acid oxidase	FMN	Dehydrogenates L-amino acid substrates to imino acids, which are hydrolyzed to α-keto acids
Monoamine oxidase	FAD	Dehydrates biogenic amines[e] to corresponding imines with hydrogen transfer to O_2, forming H_2O_2
Xanthine oxidase	FAD	Oxidizes hypoxanthine and xanthine to uric acid with formation of H_2O_2
L-gulonolactone oxidase	FAD	Oxidizes L-gulonolactone to ascorbic acid
3-Ketosphinganine reductase	FAD	Reduces 3-ketosphinganine to sphinganine in sphingosine biosynthesis
Dihydroceramide desaturase	FAD	Desaturates dihydroceramide to form ceramide in sphingosine biosynthesis
Flavoprotein Monooxygenases		
Microsomal flavoprotein monooxygenase	FAD	Oxidizes N, S, Se, and I centers of various substrates in drug metabolism
Squalene monooxygenase	FAD	Accepts reducing equivalents from ETF to oxidize squalene in the rate-limiting step in cholesterol biosynthesis

[a]*A component of microsomal cytochrome P450, it contains one molecule each of FAD and FMN.*
[b]*An iron–sulfur protein.*
[c]*Also has NADH–ubiquinone reductase activity, reductively releasing iron from ferritin.*
[d]*A component of the pyruvate dehydrogenase and α-ketoglutarate dehydrogenase complexes.*
[e]*For example, serotonin, noradrenaline, benzylamine.*

TABLE 12.4 General Signs of Riboflavin Deficiency

Organ System	Signs
General	
Appetite	Decrease
Growth	Decrease
Dermatologic	Cheilosis, stomatitis
Muscular	Weakness
Gastrointestinal	Inflammation, ulcer
Skeletal	Deformities
Vital organs	Hepatic steatosis
Vascular	
Erythrocytes	Anemia
Nervous	Ataxia, paralysis
Reproductive	
Male	Sterility
Female	Decreased egg production
Fetal	Malformations, death
Ocular	
Retinal	Photophobia
Corneal	Decreased vascularization

TABLE 12.5 Recommended Riboflavin Intakes

The United States		FAO/WHO	
Age–Sex	RDA[a] (mg/day)	Age–Sex	RNI[b] (mg/day)
0–6 months	(0.3)[c]	0–6 months	0.3
7–11 months	(0.4)[c]	7–11 months	0.4
1–3 years	0.5	1–3 years	0.5
4–8 years	0.6	4–6 years	0.6
9–13 years	0.9	7–9 years	0.9
14–18 year females	1.0	10–18 year females	1.0
14–18 year males	1.3	10–18 year males	1.3
>18 year females	1.1	19+ year females	1.1
>18 year males	1.3	19+ year males	1.3
Pregnancy	1.4	Pregnancy	1.4
Lactation	1.6	Lactation	1.6

[a]*Food and Nutrition Board, 2000. Dietary Reference Intakes for Thiamin, Riboflavin, Niacin, Vitamin B_6, Folate, Vitamin B_{12}, Pantothenic Acid, Biotin and Choline. National Academy Press, Washington, DC, 564 pp.*
[b]*Recommended Nutrient Intakes, Joint WHO/FAO Expert Consultation, 2001. Human Vitamin and Mineral Requirements. Food and Agricultural Org., Rome, 286 pp.*
[c]*RDAs have not been set; AdIs are given instead.*

dietary practices are often faulty, leading to this and other deficiencies.

- **Enhanced catabolism** associated with illness or vigorous physical exercise and involving nitrogen loss can increase riboflavin losses.
- **Alcohol** at high intakes appears to antagonize the utilization of FAD from foods.
- **Phototherapy** of infants with hyperbilirubinemia can lead to riboflavin deficiency by photodestruction of the vitamin[51] if the vitamin is not also administered.[52]
- **Diuretics or hemodialysis** can enhance loss of riboflavin (as well as other water-soluble vitamins).

Clinical signs of riboflavin deficiency are rarely seen in the industrialized world. **Subclinical riboflavin deficiency,** i.e., conditions in which intake is in sufficient to keep flavoproteins saturated with their respective flavin cofactors, however, is not uncommon. In fact, it has been estimated that as much as 27% of urban American teenagers of low-socioeconomic status had subclinical riboflavin deficiency. These changes are prevented by adequate regular intakes of the vitamin (Table 12.5).

Deficiency Signs in Humans

Clinical signs of frank riboflavin deficiency are not common; they are manifest only after 3–4 months of deprivation of the vitamin:

- **Dermal lesions** are the first to present **cheilosis,**[53] **angular stomatitis,**[54] **glossitis** (Fig. 12.3),[55] hyperemia,[56] and

51. Phototherapy can be an effective treatment for infants with mild hyperbilirubinemia; however, the mechanism by which it leads to the degradation of bilirubin (to soluble substances that can be excreted) necessarily leads also to the destruction of riboflavin. It is the photoactivation of riboflavin in the patient's plasma that generates singlet oxygen, which reacts with bilirubin. Thus, plasma riboflavin levels of such patients have been found to drop as the result of phototherapy. Riboflavin supplementation prevents such a drop and has been shown to enhance bilirubin destruction.
52. For example, 0.5 mg of riboflavin sodium phosphate per kg body weight per day.

53. Lesions of the lips.
54. Lesions at the corners of the mouth, beginning as white, thickened foci and then developing a macerated appearance; they may ulcerate and then crusted. Healing may leave linear scars.
55. Inflammation of the tongue. This can involve disappearance of filiform papillae and enlargement of fungiform papillae, with the tongue color changing to a deep red. Subjects with this condition, called **geographical tongue,** have soreness of the tongue and loss of taste sensation.
56. Increased amount of blood present.

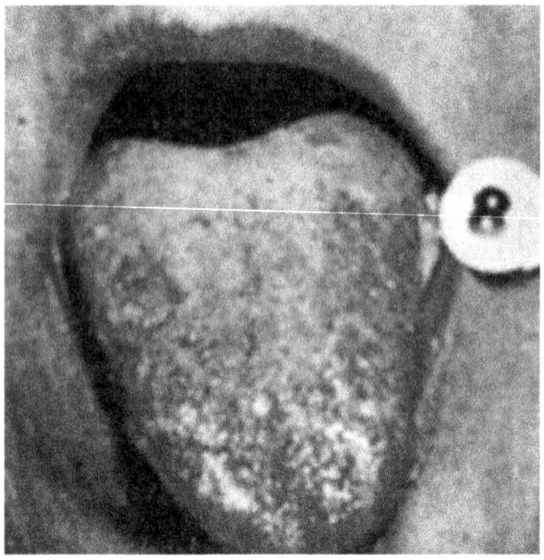

FIGURE 12.3 Geographical tongue in riboflavin deficiency. *Courtesy Cambridge University Press.*

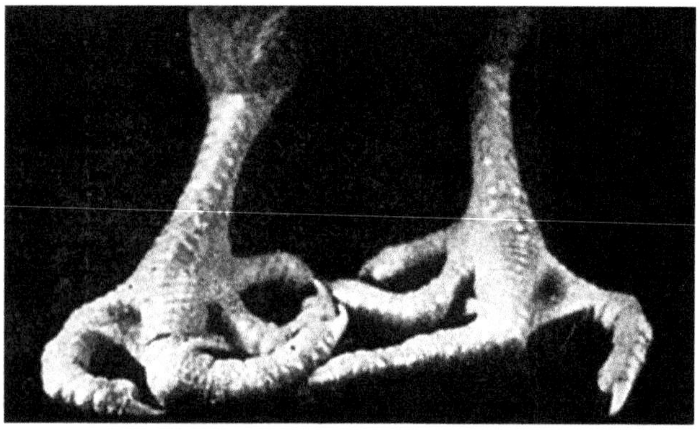

FIGURE 12.4 Curled toe paralysis in a riboflavin-deficient chick. *Courtesy, G.F. Combs, Sr.*

edema[57] of the oral mucosa, seborrheic dermatitis around the nose and mouth and scrotum/vulva. These signs are associated with impaired collagen maturation.

- **Anemia** is later to present **normocytic, hypochromic anemia** with **reticulocytopenia**,[58] **leukopenia,**[59] and **thrombocytopenia**.[60] These signs reflect impaired erythropoiesis, particularly loss of NAD(P)H oxidoreductase, which facilitates cellular uptake of Fe^{2+} by keeping it reduced.
- **Neurological signs** are the last to present demyelinating peripheral neuropathy of the extremities characterized by hyperesthesia,[61] coldness, and pain, as well as decreased sensitivity to touch, temperature, vibration, and position.

Deficiency Signs in Animals

Riboflavin deficiency in animals is potentially fatal. In addition to the general signs already mentioned, animals show other signs that vary with the species. Riboflavin-deficient rodents show dermatologic signs (alopecia, seborrheic inflammation,[62] moderate epidermal hyperkeratosis[63] with atrophy of sebaceous glands) and a generally ragged appearance. Red, swollen lips and abnormal papillae of the tongue are seen. Ocular signs may also be seen (blepharitis,[64] conjunctivitis,[65] and corneal opacity). Feeding a high-fat diet can increase the severity of deficiency signs; high-fat-fed rats showed anestrus, multiple fetal skeletal abnormalities (shortening of the mandible, fusion of ribs, cleft palate, and deformed digits and limbs), paralysis of the hind limbs (degeneration of the myelin sheaths of the sciatic nerves[66]), hydrocephalus,[67] ocular lesions, cardiac malformations, and hydronephrosis.[68]

The riboflavin-deficient chick also experiences myelin degeneration of nerves, affecting the sciatic nerve in particular. This results in an inability to extend the digits, a syndrome called **curled toe paralysis** (Fig. 12.4). In hens, the deficiency involves reductions in both egg production and embryonic survival (decreased hatchability of fertile eggs). Riboflavin-deficient turkeys show severe dermatitis. The deficiency is rapidly fatal in ducks.

Riboflavin-deficient dogs are weak and ataxic. They show dermatitis (chest, abdomen, inner thighs, axillae, and scrotum) and **hypoplastic anemia**[69] with fatty infiltration of the bone marrow. They can have bradycardia and sinus arrhythmia[70] with respiratory failure. Corneal opacity has been reported. The deficiency can be fatal, with collapse and coma. Swine fed a riboflavin-deficient diet grow slowly and develop a scaly dermatitis with alopecia. They can show corneal opacity, cataracts, adrenal hemorrhages, fatty degeneration of the kidney, inflammation of the mucous membranes of the gastrointestinal tract, and nerve

57. Accumulation of excessive fluid in the tissue.
58. Abnormally low number of immature red blood cells in the circulating blood.
59. Abnormally low number of white blood cells in the circulating blood (<5000/ml).
60. Abnormally low number of platelets in the circulating blood.
61. Excessive sensibility to touch, pain, etc.
62. Involving excess oiliness due to excess activity of the sebaceous glands.
63. Hypertrophy of the horny layer of the epidermis.
64. Inflammation of the eyelids.
65. Inflammation of the mucous membrane covering the anterior surface of the eyeball.
66. The nerve situated in the thigh.
67. A condition involving the excessive accumulation of fluid in the cerebral ventricles, dilating these cavities and, in severe cases, thinning the brain and causing a separation of the cranial bones.
68. Dilation of one or both kidneys owing to obstructed urine flow.
69. Progressive nonregenerative anemia resulting from depressed, inadequate functioning of the bone marrow.
70. Irregular heartbeat, with the heart under control of its normal pacemaker, the sinoatrial (S-A) node.

degeneration. In severe cases, deficient individuals can collapse and die.

Riboflavin deficiency in the newborn calf[71] is manifested as redness of the buccal mucosa,[72] angular **stomatitis**,[73] alopecia, diarrhea, excessive tearing and salivation, and inanition. Signs of riboflavin deficiency appear to develop rather slowly in rhesus monkeys. The first signs seen are weight loss (6–8 weeks), followed by dermatologic changes in the mouth, face, legs, and hands and a normocytic hypochromic anemia[74] (2–6 months) and, ultimately, collapse and death with fatty degeneration of the liver. Similar signs have been produced in baboons made riboflavin deficient for experimental purposes.

10. RIBOFLAVIN IN HEALTH AND DISEASE

Vascular Disease

Dietary riboflavin intake has been found to be inversely correlated with serum homocysteine levels.[75] Homocysteinemia has been associated with increased risks to occlusive vascular disease, total and cardiovascular disease-related mortality, stroke, dementia, Alzheimer's disease, fracture, and chronic heart failure.[76] The Framingham Offspring Study found elevated plasma homocysteine levels in subjects with relatively low plasma riboflavin levels (Table 12.6). Such findings are consistent with the role of riboflavin as the essential cofactor (FAD) for methyltetrahydrofolate reductase (MTHFR), which is required for the formation of *N*-5-methyltetrahydrofolate, which, in turn, is required to convert homocysteine to methionine (*see* Chapter 18, Folic Acid). Riboflavin is particularly important in individuals with the 677TT polymorphism of MTHFR, which causes it to lose FAD thus reducing its activity.[77] Individuals with the *TT* genotype[78] have elevated circulating levels of homocysteine, particularly, if they are also low in folate. They are at increased risk for hypertension (Fig. 12.5) and vascular disease in general.[79] Randomized trials have demonstrated that riboflavin supplementation can reduce serum homocysteine levels[80] and blood pressure[81] only in individuals with the 677*TT* genotype.

TABLE 12.6 Relationship of Plasma Riboflavin and Homocysteine Levels Among Subjects in the Framingham Offspring Cohort Study

	Plasma Riboflavin Tertile (nM)		
	<6.89	6.89–10.99	≥11.0
Plasma homocysteine—mean (95% C.I.)	10.3 (9.8–10.8)	9.5 (9.1–10.0)*[a]	9.5 (9.1–10.0)*

[a]$p < .05$, n = 147–152 in each group.
From Jacques, P.F., Boston, A.G., Williams, R.R., et al., 2002. J. Nutr. 132, 283–290.

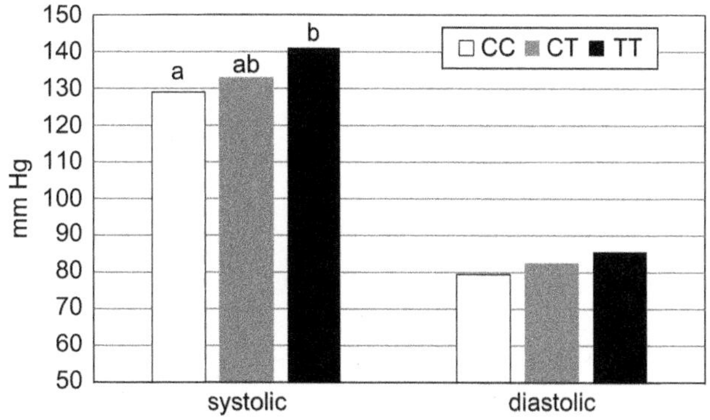

FIGURE 12.5 Relationship of MTHFR (methyltetrahydrofolate reductase) 677T→C genotype and blood pressure. *After Wilson, C.P., Ward, M., McNulty, H. et al., 2012. Am. J. Clin. Nutr. 95, 766–772.*

Anticarcinogenesis

Observational studies have found inverse relationships of riboflavin status and risks of cancers of the esophagus and colon rectum.[82] Riboflavin deprivation has been found to increase aflatoxin B_1-induced DNA damage in the rat, suggesting that suboptimal intakes of the vitamin may enhance carcinogenesis[83]. Plasma levels of RfBP may have utility

71. Ruminants do not normally require a dietary source of riboflavin, as their rumen microbiota synthesize the vitamin in adequate amounts. However, newborn calves and lambs, whose rumen microbiome is not established, require a dietary source of the vitamin, which is supplied by their mothers' milk or by supplements in milk-replacer formula diets.
72. The mucosa of the cheek.
73. Lesions in the corners of the mouth.
74. Anemia involving erythrocytes of normal size but low hemoglobin content.
75. Ganji, G., Kafai, M.R., 2004. Am. J. Clin. Nutr. 80, 1500–1507.
76. Selhub, J., 2006. J. Nutr. 136, 1726S–1730S.
77. McNulty, H., McKinley, M.C., Wilson, B., et al., 2002. Am. J. Clin. Nutr. 76, 436–442.
78. This group has been estimated to comprise 3–32% of various populations.
79. Hustad, S., Ueland, P.M., Volset, S.E., et al., 2000. Clin. Chem. 46, 1065–1072.
80. By as much as 22% (McNulty, H.M., Dowey, L.R.C., et al., 2006. Circulation 113, 74–80).
81. Wilson, C.P., Ward, M., McNulty, H., et al., 2012. Am. J. Clin. Nutr. 95, 766–772.
82. Esophageal squamous cell cancer (He, Y., Shan, B., Song, G., et al., 2009. Asian Pacific J. Cancer Prev. 10, 619–625); colorectal cancer (Figueiredo, J.C., Levine, A.J., Grau, M.V., et al., 2008. Cancer Epidemiol. Biomarkers Prev. 17, 2137–2144; de Vogel, S., Dindore, V., van Engeland, M., et al., 2008. J. Nutr. 138, 2372–2378).
83. Webster, R.P., Gawde, M.D., Bhattacharya, R.K., 1996. Cancer Lett. 98, 129–135.

as a biomarker for some cancers. It has been found to be overexpressed in patients with estrogen-responsive breast cancer, hepatocellular cancer, and prostate cancer.

Malaria

Riboflavin deficiency confers some protection against malaria, decreasing both parasitemia and signs of infection[84] due to the plasmodia autolyzing before they can mature. This protection is thought to involve significant differentials in needs/susceptibilities of the parasite and erythrocytes, the parasite having greater needs for riboflavin,[85] and being more susceptible to oxidative stress.[86] These hypotheses are consistent with genetic defects in certain flavoenzymes being associated with malaria resistance.[87] Flavin analogues[88] that antagonize riboflavin and inhibit glutathione reductase have been shown to have antimalarial activities.[89]

Occular Health

Tryptophan deficiency cataract can be reduced by riboflavin deprivation in the rat. The cataractogenic effect of a low-tryptophan diet appears to the formation of a riboflavinyl tryptophan adduct that accelerates the photooxidation of the amino acid to a prooxidative form.

Mineral Utilization

Riboflavin can form complexes with divalent cations. That correction of riboflavin deficiency can improve the enteric absorption of iron and zinc in the mouse model[90] suggests that riboflavin can be a determinant of the bioavailability of those minerals.

Medical Uses

Riboflavin is photoactive. In its excited state, it can react with amino acids, proteins, or lipids in low-O_2 environments to generate radicals that can, in turn, generate ROS.[91] This property is the basis of riboflavin loss due to photoirradiation treatment of neonatal jaundice.[92] It has also been exploited for medical purposes. These include inducing DNA damage in blood-borne pathogens and catalyzing **corneal collagen cross-linking (CXL)** in the treatment of progressive keratectasia[93] in humans.[94]

High doses of riboflavin in combination with other treatments have been reported to be effective in preventing migraine headaches,[95] although controlled studies have not confirmed that effect.[96]

11. RIBOFLAVIN TOXICITY

The toxicity of riboflavin is *very low* and, thus, problems of hypervitaminosis are not expected. Probably, because it is not well absorbed as high oral doses, riboflavin is essentially nontoxic. Oral riboflavin doses as great as 2–10 g/kg body weight produce no adverse effects in dogs and rats. The vitamin is somewhat more toxic when administered parenterally. The LD_{50} (50% lethal dose) values for the rat given riboflavin by the intraperitoneal, subcutaneous, and oral routes have been estimated to be 0.6, 5, and >10 g/kg, respectively. No tolerable upper limits of exposure have been established.

12. CASE STUDY

Instructions

Review the following summary of a research report, paying special attention to the diagnostic indicators on which the treatments were based. Then answer the questions that follow.

Case

An experiment was conducted to determine the basis of protection by riboflavin deficiency against malarial infection. An animal model, which previously showed such protection against *Plasmodium berghei*, was used. It involved depleting 3-week-old male rats of riboflavin by feeding them a sucrose-based purified diet containing <1 mg of riboflavin per kilogram. A control group

84. Da, B.S., et al., 1988. Eur. J. Clin. Nutr. 42, 227–234.
85. Dutta, P., 1991. J. Protozool. 38, 479–486.
86. Dutta, P., 1993. J. Soc. Pharm. Chem. 23, 11–48.
87. Glutathione reductase, pyridoxine phosphate oxidase, glucose-6-phosphate dehydrogenase (López, C., Saravia, C., Gomez, A., et al., 2010. FASEB J. 467, 1–12).
88. e.g., Galactoflavin, 10-(4′-chlorophenyl)-3-methylflavin and some isoalloxazine derivatives.
89. It has also been proposed that high-dose riboflavin may also be useful in treating malaria. This would involve doses sufficiently large to stimulate NADPH–cytochrome b_5/cytochrome b_5 reductase in the parasite. Because plasmodia typically accumulate large amounts of methemoglobin, upregulation of this flavoenzyme, reduces Fe^{3+} in methemoglobin to Fe^{2+} in hemoglobin, could lead to a futile cycle of hemoglobin oxidation and reduction (Akompong, T., Ghori, N., Haldar, K., 2000. Antimicrob. Agents Chemother. 44, 88–96).
90. Agte, V.V., Paknikar. K.M., Chiplonkar, S.A., et al., 1998. Biol. Trace Elem. Res. 65, 109–116.
91. Marin, C.B., Tso, M., Hadan, C.M., 2002. J. Am. Chem. Soc. 124, 7226–7234.
92. Riboflavin and bilirubin have similar absorption maxima; loss of riboflavin causes loss of eGR, which by compromises ROS protection, causes erythrocyte lysis.
93. A group of noninflammatory disorders involving thinning of the cornea; some forms can sometimes occur after corneal surgery.
94. Dahl, B.J., Spotts, E., Truong, J.Q., 2012. Optometry 83, 33–42.
95. Zencirci, B., 2010. J. Pain. Res. 3, 125–130.
96. MacLennan, S.C., Wade, F.M., Forrest, K.M., et al., 2008. J. Child. Neurol. 23, 1300–1304.

was pair-fed[97] the same basal diet supplemented with 8.5 mg of riboflavin per kilogram.[98] At 6 weeks of age, several biochemical characteristics of erythrocytes (RBCs) were measured, i.e., reduced glutathione levels, activities of antioxidant enzymes, stabilities of erythrocytes to hemolysis (measured by incubating 0.5% suspensions of RBCs with prooxidants [500 μM H_2O_2 or 2.5 μM ferriprotoporphyrin IX] or in a hypotonic medium [151 mOsm] for 1 h at 37°C). Oxidative damage was assessed by measuring H_2O_2-induced production of malonyldialdehyde (MDA). Other studies with this and similar animal models have shown that the riboflavin-deficient group, when infected with the parasite *grows better* and shows *reduced parasitemia* than pair-fed controls.

Results of Biochemical Studies of Erythrocytes

Parameter	Riboflavin Defiscient	Control	*p* Value
Reticulocytes (% total RBCs)	1.50 ± 0.29	1.26 ± 0.37	NS[a]
Hemoglobin (g/dL blood)	14.7 ± 0.6	14.9 ± 0.3	NS
GSH (mmol/g Hb)	7.97 ± 2.89	6.19 ± 2.52	<0.001
Glutathione reductase (mU[b]/mg protein)	42 ± 6	124 ± 16	<0.001
Glutathione reductase activity coefficient	2.37 ± 0.19	1.20 ± 0.08	<0.01
Glutathione peroxidase (mU[b]/g Hb)	918 ± 70	944 ± 62	NS
In Vitro Hemolysis (%)			
H_2O_2-induced	32 ± 9	55 ± 9	<0.05
Hypotonicity	69 ± 4	53 ± 7	<0.01
Ferriprotoporphyrin IX	42 ± 3	29 ± 4	<0.001
Malonyldialdehyde (MDA) (nmol/g Hb)			
Before incubation	25.5 ± 3.8	25.9 ± 3.4	NS
Incubated with H_2O_2	34.8 ± 1.2	42.7 ± 1.8	<0.01

[a]*NS, not significant (p > .05).*
[b]*1 mU = 1 nmol NADPH per min.*

97. **Pair-feeding** is a method of controlling for the effects of reduced food intake that may be secondary to the independent experimental variable (e.g., a nutrient deficiency). It involves the matching of one animal from the experimental treatment group with one of similar body weight from the control group, and the feeding of the latter individual a measured amount of feed equivalent to the amount of feed consumed by the former individual on the previous day. In experiments of more than a few days' duration, this approach normalizes the feed intake of both the experimental and control groups.
98. This level is about three times the amount normally required by the rat.

Case Questions

1. What dependent variables did the investigators measure to confirm that riboflavin deficiency had been produced in their experimental animals?
2. Propose a hypothesis to explain the apparently discrepant results regarding the effects of riboflavin deficiency on erythrocyte stability.
3. Propose a hypothesis for the protective effect of riboflavin deficiency against malarial infection. What other nutrients might you expect to influence susceptibility to this erythrocyte-attacking parasite?

13. STUDY QUESTIONS AND EXERCISES

1. Diagram the general roles of FAD- and FMN-dependent enzymes in various areas of metabolism.
2. Construct a decision tree for the diagnosis of riboflavin deficiency in humans or an animal species.
3. What key feature of the chemistry of riboflavin relates to its biochemical functions in flavoproteins?
4. What diet and lifestyle factors would you expect to affect dietary riboflavin needs? Justify your answer.

RECOMMENDED READING

Lienhart, W.D., Gudipati, V., Macheroux, P., 2013. The human flavoproteome. Arch. Biochem. Biophys. 535, 150–162.
McCormick, D.B., 2012. Riboflavin, Chapter 28. In: Erdman, J.W., Macdonald, I.A., Zeisel, S.H. (Eds.), Present Knowledge in Nutrition, tenth ed. Wiley, New York, pp. 280–306.
Pinto, J.T., Rivlin, R.S., 2014. Riboflavin, Chapter 6. In: Zemplini, J., Suttie, J.W., Gregory, J.F., et al. (Eds.), Handbook of Vitamins, fifth ed. CRC Press, New York, pp. 191–265.
Said, H.M., 2004. Recent advances in carrier-mediated intestinal absorption of water-soluble vitamins. Ann. Rev. Physiol. 66, 419–446.
Stuehr, D.J., Tejero, J., Haque, M.M., 2009. Structural and mechanistic aspects of flavoproteins: electron transfer through the nitric oxide synthase flavoprotein domain. FEBS J. 276, 3959–3974.
Walsh, C.T., Wencewicz, 2013. Flavoenzymes: versatile catalysts in biosynthetic pathways. Nat. Prod. Rep. 30, 175–200.

Chapter 13

Niacin

Chapter Outline

Anchoring Concepts

1. Niacin is the generic descriptor for pyridine 3-carboxylic acid and derivatives exhibiting qualitatively the biological activity of nicotinamide.
2. The two major forms of niacin, nicotinic acid and nicotinamide, are active metabolically as the pyridine nucleotide coenzymes NAD(H) and NADP(H).
3. Deficiencies of niacin are manifest as dermatologic, gastrointestinal, and neurologic changes and can be fatal.

So far as they have been studied, the foodstuffs that appear to be good sources of the black tongue preventive also appear to be good sources of the pellagra preventive...Considering the available evidence as a whole, it would seem highly probable, if not certain, that experimental black tongue and pellagra are essentially identical conditions and, thus, that the preventive of black tongue is identical with the pellagra preventive, or factor P–P.

Joseph Goldberger[1]

LEARNING OBJECTIVES

1. To understand the chief natural sources of niacin
2. To understand the means of enteric absorption and transport of niacin
3. To understand the biochemical function of niacin as a component of coenzymes of different metabolically important redox reactions and the relationship of that function to the physiological activities of the vitamin
4. To understand the factors that can affect low niacin status and the physiological implications of that condition.

VOCABULARY

Acetyl-CoA
ADP-ribosylation
ADP-ribotransferases (ARTs)
α-Amino-β-carboxymuconic-ε-semialdehyde (ACS)
α-Amino-β-carboxymuconic-ε-semialdehyde decarboxylase (ACSD)
Anthranilic acid
Black tongue disease
Casal's collar
Flushing
Formylase
N-Formylkynurenine
"Four Ds" of niacin deficiency
Glucose tolerance factor Hartnup disease
3-Hydroxyanthranilic acid
3-Hydroxyanthranilic acid oxygenase
3-Hydroxykynurenine
Kynurenic acid
Kynurenine
Kynurenine 3-hydroxylase
Leucine
1-Methylnicotinamide
1-Methylnicotinic acid
1-Methyl-2-pyridone 3-carboxamide

1. Joseph Goldberger (1874–1929) was a Hungarian-born American physician who, as member of the U.S. Public Health Service in 1914, demonstrated that pellagra was associated with corn-based diets. Although his work had little immediate public health impact, it ultimately led to the discovery of niacin and the role of the essential amino acid tryptophan.

The Vitamins. http://dx.doi.org/10.1016/B978-0-12-802965-7.00013-7

1-Methyl-4-pyridone 3-carboxamide
1-Methyl-6-pyridone 3-carboxamide
Mono-ADP-ribosyltransferases (ARTs)
NAD^+ kinase
NAD^+ synthetase
$NAD(P)^+$ glycohydrolase
Niacin number
Niacin receptor
Nicotinate phosphoribosyl transferase
Nicotinamide
Nicotinamide adenine dinucleotide (NADH)
Nicotinamide adenine dinucleotide phosphate (NADPH)
Nicotinamide methylase
Nicotinamide *N*-methyltransferase
Nicotinamide mononucleotide (NMN)
Nicotinamide riboside (NR)
Nicotinic acid (NA)
Nixtamalization
Nyacitin
Pellagra
Perosis
Phosphodiesterase
Picolinic acid
Poly(ADP-ribose) polymerases (PARPs)
Pyridine nucleotide
Pyridoxal phosphate
Quinolinate phosphoribosyltransferase
Quinolinic acid
Schizophrenia
Sirtuin
Transaminase
Transhydrogenase
Trigonelline
Tryptophan
Tryptophan pyrrolase
Xanthurenic acid

1. THE SIGNIFICANCE OF NIACIN

Niacin is required for the biosynthesis of the **pyridine nucleotides**, **NAD(H)** and **NADP(H)**, through which the vitamin has key roles in virtually all aspects of metabolism. Historically, niacin deficiency was prevalent among people who relied on maize (corn) as their major food staple; before the availability of inexpensive supplements, the deficiency was also a frequent problem of livestock fed maize-based diets.

Great irony characterizes niacin deficiency. Unlike thiamin deficiency, which also involves an unbalanced cereal-based diet, niacin deficiency more frequently results from poor bioavailability rather than scarcity per se. Hence, paradoxical questions have been asked:

- Why does niacin deficiency occur among individuals who can biosynthesize the vitamin?
- Why did pellagra occur among people eating maize (corn), whereas the disease was unknown in the Americas where maize was a historically important part of the diet?
- Why do maize-based diets produce pellagra, although maize contains an appreciable amount of niacin?
- Why does milk, which contains little niacin, prevent pellagra?
- Why does rice, which contains less niacin than maize, not produce pellagra?

Niacin is also of interest for its health value as a pharmacologic, i.e., multigram, supranutritional dose levels. Understanding the bases of both the physiologic and pharmacologic activities of niacin calls for an appreciation of its complexities, which are manifest differently in various species.

2. PROPERTIES OF NIACIN

The term **niacin** describes pyridine 3-carboxylic acid and its derivatives that exhibit the biological activity of nicotinamide (NAm). That biological activity depends on the following structural features:

- **a pyridine nucleus** substituted with a β-carboxylic acid or a corresponding amine;
- the pyridine nitrogen being able to undergo reversible oxidation/reduction (i.e., quaternary pyridinium ion to/from tertiary amine); and
- open pyridine carbons adjacent to the nuclear nitrogen atom

The coenzyme forms of niacin are the pyridine nucleotides and their reduced forms: nicotinamide adenine dinucleotide (**NAD[H]**) and nicotinamide dinucleotide phosphate (**NADP[H]**).

Chemical structures of niacin:

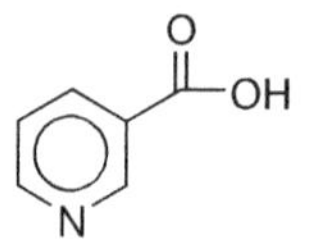

Nicotinic acid

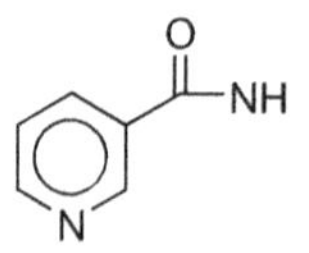

Nicotinamide

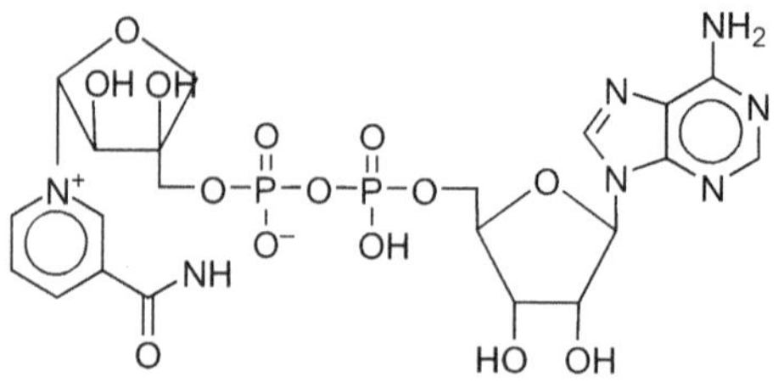

Nicotinamide adenine dinucleotide (NADH)

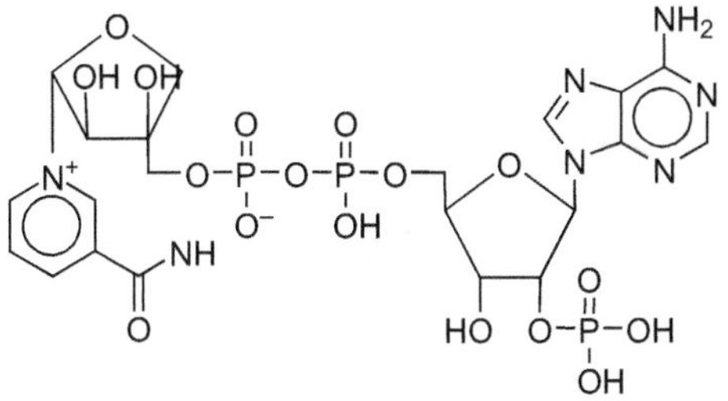

Nicotinamide adenine dinucleotide phosphate (NADPH)

Niacin Chemistry

Nicotinic acid (NA) and **NAm** are colorless and crystalline. Each is insoluble or only sparingly soluble in organic solvents. NA is slightly soluble in water and ethanol; NAm is very soluble in water and moderately soluble in ethanol. The two compounds have similar absorption spectra in water, with an absorption maximum at <262 nm. NA is amphoteric and forms salts with acids as well as bases. Its carboxyl group can form esters and anhydrides, and can be reduced. Both NA and NAm are very stable in dry form, but in solution NAm is hydrolyzed by acids and bases to yield NA.

In each of the coenzyme forms (**NAD[H]** and **NADP[H]**), the electron-withdrawing effect of the *N*-1 atom and the amide group of the oxidized pyridine nucleus enables the pyridine C-4 atom to react with many nucleophilic agents (e.g., sulfite, cyanide, and hydride ions). It is the reaction with hydride ions (H^-) that is the basis of the enzymatic hydrogen transfer by the pyridine nucleotides; the reaction involves the transfer of two electrons in a single step. The hydride transfer of nonenzymatic reactions of the pyridine nucleotides, plus those catalyzed by the pyridine nucleotide-dependent dehydrogenases, is stereospecific with respect to both coenzyme and substrate. At least for reactions of the former type, this stereospecificity results from a specific intramolecular association between the adenine residue and the pyridine nucleus.

Several substituted pyridines are antagonists of niacin in biological systems: pyridine-3-sulfonic acid, 3-acetylpyridine, isonicotinic acid hydrazine, and 6-aminonicotinamide.

3. SOURCES OF NIACIN

Hindgut Microbial Synthesis

Niacin can be produced by the microbiome of the colon. A genomic analysis of 256 representative organisms of the human gut microbiota found most (63%) capable of de novo synthesis of NAD.[2] In addition, some taxa showed the capacity to salvage NAD. Those findings suggested that hindgut microbial synthesis may produce as much as 37% of the daily human need for niacin. However, direct evidence is lacking for the absorption of niacin across the colon. It is likely that a large portion of the microbially produced vitamin is taken up by nonsynthesizing microbes such that noncoprophagous and nonruminant animals derive no benefit from this source of the vitamin.

Distribution in Foods

Niacin occurs in greatest quantities in brewers' yeasts and meats, but significant amounts are also found in many other foods (Table 13.1). The vitamin is distributed unevenly in grains, present mostly in the bran fractions. Niacin occurs predominantly in bound forms, e.g., in plants mostly as protein-bound NA and in animal tissues mostly as NAm in nicotinamide adenine dinucleotide (NAD) and nicotinamide adenine dinucleotide phosphate (NADP). In cow's milk, ~40% of the vitamin is the NAD precursor, nicotinamide riboside (NR). Niacin is added by law to wheat flour and other grain products in the United States.[3]

Stability

Niacin in foods is stable to storage and to most means of food preparation and cooking (e.g., moist heat).

Bioavailability

Niacin is found in many types of foods in forms from which it is not released on digestion, thus rendering it unavailable for absorption. In grains, niacin is present in covalently bound complexes with small peptides and carbohydrates, collectively referred to as **niacytin**.[4] The esterified niacin in these complexes is not normally available; however, its bioavailability can be improved substantially by treatment with base to effect the alkaline hydrolysis of those esters. The tradition in Central American cuisine of soaking and cooking maize in lime[5] water effectively renders available the niacin in that grain.[6] This practice appears to be responsible for effective protection against pellagra in that part of the world. In other foods, niacin is present as a methylated derivative (1-methylnicotinic acid, also called **trigonelline**) that functions as a plant hormone but is also not biologically

2. Magnúsdóttir, S., Ravchee, D., de Crécy-Lagard, V., et al., 2015. Front. Genet. 6, 148–166.

3. Fortification is mandated for niacin, thiamin, riboflavin, folate, and iron.
4. A polysaccharide extracted from wheat bran has been found to contain more than 1% NA bound via an ester linkage to glucose in a complex also containing arabinose, galactose, and xylose. Although NAD^+ and $NADP^+$, both of which are biologically available to humans and animals, are present in early-stage corn, those levels decline as the grain matures and are replaced by NAm and NA as well as forms of very low bioavailability such as bound niacin and trigonelline.
5. Calcium hydroxide.
6. This process, called **nixtamalization**, renders maize (corn) more easily ground, improves its flavor, and reduces mycotoxin content. It is used in making tortillas, hominy, and corn chips. The term itself derives from the Nahuatl (an Aztec dialect) words *nextli* (ashes) and *tamalii* (corn dough).

TABLE 13.1 Niacin Contents of Foods

Food	Niacin (mg/100 g)
Dairy Products	
Milk	0.09–0.1
Yogurt	0.2
Cheeses	0.02–1
Meats	
Beef	2.4–8.5
Chicken	5–11.1
Pork	4–10
Turkey	6.4–9.6
Fish	
Herring	3.3–4.4
Cod	2.5–7.5
Haddock	4.1
Tuna	10.1–10.5
Cereals	
Barley	2.1
Buckwheat	0.9
Cornmeal	3.6
Rice, brown	2.6
Rice, white	0.4
Rye flour	1.7
Wheat germ	6.8
Wheat bran	13.6
Vegetables	
Asparagus	1.08
Beans	0.6–1.3
Broccoli	0.6
Cabbage	0.2
Carrots	1.0
Cauliflower	0.5
Celery	0.3
Corn	1.7
Kale	0.5
Onions	0.1
Peas	0.6–2
Peppers	0.5–1.2
Potatoes	1.4
Soy beans	1.3
Spinach	0.7
Tomatoes	0.6
Fruits	
Apples	0.9
Bananas	0.7
Grapefruit	0.3
Oranges	0.4
Peaches	0.8
Strawberries	0.4
Nuts	
Most nuts	0.3–4.4
Peanuts	12
Other	
Eggs	0.1
Mushrooms	0.2–14
Baker's yeast	12.3

USDA National Nutrient Database for Standard Reference, Release 28. http://www.ars.usda.gov/ba/bhnrc/ndl.

available to animals. This form, however, is heat labile and can be converted to NA by heating.[7]

Importance of Dietary Tryptophan

A substantial amount of niacin can be synthesized from the indispensable amino acid **tryptophan**. Therefore, the niacin adequacy of diets involves both the level of the preformed vitamin and that of its potential precursor tryptophan. This is expressed as niacin equivalence, which includes available niacin as well as tryptophan (Table 13.2).

4. ABSORPTION OF NIACIN

Digestion of NAD(P)

The predominant forms of niacin in most animal-derived foods are the coenzymes NAD(H) and NADP(H). These are digested to release NAm, in which form the vitamin is absorbed (Fig. 13.1). Both coenzyme forms can be

7. Thus, the roasting of coffee beans effectively removes the methyl group from trigonelline, increasing the NA content from 20 to 500 mg/kg. This practice, too, appears to have contributed to the rarity of pellagra in the maize-eating cultures of South and Central America.

TABLE 13.2 The Niacin-Equivalent Contents of Several Foods

Food	Preformed Niacin (mg/1000 kcal)	Tryptophan (mg/1000 kcal)	Niacin Equivalents[a] (mg/1000 kcal)
Cow's milk	1.21	673	12.4
Human milk	2.46	443	9.84
Beef	2.47	1280	23.80
Eggs (whole)	0.60	1150	19.80
Pork	1.15	61	2.17
Wheat flour	2.48	297	7.43
Corn meal	4.97	106	6.74
Corn grits	1.83	70	3.00
Rice	4.52	290	9.35

[a]*Based on a conversion efficiency of 60:1 for humans.*

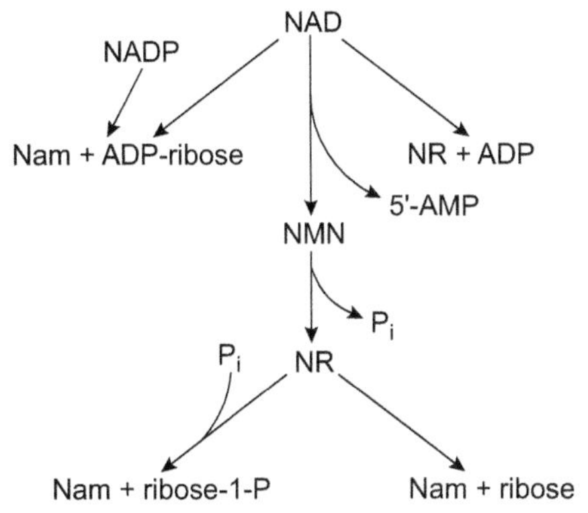

FIGURE 13.1 Metabolic disposition of absorbed niacin.

degraded by the intestinal mucosal enzyme **NAD(P)$^+$ glycohydrolase**, which cleaves the pyridine nucleotides into NAm and ADP-ribose. NAm can also be cleaved at the pyrophosphate bond to yield **nicotinamide mononucleotide (NMN)** and 5′-AMP, or by a **phosphodiesterase** to yield **NR** and ADP. The dephosphorylation of NMN also yields NR, which can be converted to NAm either by hydrolysis (yielding ribose) or phosphorylation (yielding ribose 1-phosphate). The cleavage of NAm to free NA appears to be accomplished by intestinal microorganisms and is believed to be of quantitative importance in niacin absorption.

Facilitated Absorption

Niacin is absorbed in the stomach and small intestine. Studies using everted intestinal sacs prepared from rats have demonstrated that both NA and NAm are absorbed at low concentrations via Na$^+$-dependent, carrier-mediated facilitated diffusion, by the organic ion transporter-10.[8] At high doses, the vitamers appear to be absorbed via passive diffusion, the rate of diffusion of NA being half that of NAm; although the Na$^+$-coupled monocarboxylate transporter SLC5A8 may also contribute. The result is that pharmacologic amounts of the vitamin are absorbed nearly completely.[9] The presence or absence of food in the gut appears to have no effect on niacin absorption. Because NR is not found in plasma, it appears not to be absorbed per se, but first converted to NAm.

5. TRANSPORT OF NIACIN

Free in Plasma

Niacin is transported in the portal circulation as both NA and NAm in unbound forms. Because the NA is converted to NAD(H) and subsequently to NAm in the intestine and liver, circulating levels of NAm tend to exceed those of NA.

Cellular Uptake

Both NA and NAm are taken up by erythrocytes, liver, and most peripheral tissues by facilitated diffusion whereupon they are converted to nucleotides. Tissues vary with respect to their uptake systems: erythrocytes use the anion transport system; renal tubules use a Na$^+$-dependent system; brain uses an energy-dependent system; the choroid plexus appears to have separate systems for accumulating and releasing NA and NAm. A high-affinity, G protein-coupled receptor for NA has been identified in adipose tissue.[10] This "**niacin receptor**" binds

8. Said, H.M., 2011. Biochem. J. **437**, 357–372.
9. In humans at steady state, consuming 3 g of NA per day, 85% of the vitamin is excreted in the urine.
10. Lorensen, A., 2001. Mol. Pharmacol. **59**, 349.

NA only at nonphysiologically high levels;[11] its natural ligand may actually be β-hydroxybutyrate.[12] It is also expressed in spleen and immune cells; in the latter, it is regulated by various cytokines. The receptor appears to play roles in responses (flushing, antihyperlipidemic) to high doses of NA.[13]

Tissue Storage

Niacin is retained in the liver and other tissues, which take it up as NA and/or NAm and convert it to the pyridine nucleotides NAD(H) and NADP(H) (Table 13.3). By far the greater amount is found as NAD(H), most of which, in contrast to NADP(H), is found in the oxidized form (NAD^+).

TABLE 13.3 Pyridine Nucleotide Contents (mg/kg) of Various Organs of Rats

Organ	NAD^+	NADH	$NADP^+$	NADPH
Liver	370	204	6	205
Heart	299	184	4	33
Kidney	223	212	3	54
Brain	133	88	<2	8
Thymus	116	35	<2	12
Lung	108	52	9	18
Pancreas	80	78	<2	12
Testes	80	71	<2	6
Blood	55	36	5	3

Offermanns, K., et al., 1984. Kirk-Othmer Encycl. Chem. Technol. 24, 59–66.

6. METABOLISM OF NIACIN

Niacin Biosynthesis

Tryptophan–niacin conversion. All animal species including humans are capable, to varying degrees, of the de novo synthesis of the metabolically active forms of niacin, NAD(H) and NADP(H). This multistep process (Fig. 13.2) involves as follows:

1. oxidative cleavage of the tryptophan pyrrole ring by **tryptophan pyrrolase** to yield ***N*-formylkynurenine**;
2. removal of the formyl group by **formylase** to yield **kynurenine**;
3. ring hydroxylation of kynurenine by the FAD-dependent **kynurenine 3-hydroxylase** to yield **3-hydroxykynurenine (3-OH-Ky)**;
4. deamination of 3-OH-Ky by a Zn-activated, pyridoxal phosphate-dependent transaminase, to yield **xanthurenic acid**, which can either be excreted in the urine or further metabolized;
5. removal of an alanine residue from the xanthurenic acid side chain by the pyridoxal phosphate-dependent enzyme **kynureninase** to yield **3-hydroxyanthranilic acid (3-OH-AA)**;[14]
6. oxidative ring opening of 3-OH-AA by an Fe^{2+}-dependent dioxygenase, **3-hydroxyanthranilic acid oxygenase (3-HAAO)** to yield the semistable **α-amino-β-carboxymuconic-ε-semialdehyde (ACS)**—a branch point in this metabolism:
   a. It can be converted by **α-amino-β-carboxymuconic-ε-semialdehyde decarboxylase (ACSD)**[15] to **α-aminomuconic-ε-semialdehyde**, which is reduced and further decarboxylated ultimately to yield **acetyl CoA**.
   b. If ACS accumulates, some can spontaneously cyclize in two ways:
      i. with **dehydration** yielding **quinolinic acid**, which is decarboxylated and phosphoribosylated by **quinolinate phosphoribosyl transferase** to yield NMN, which is phosphoadenylated by the ATP-dependent **NAD^+ synthetase** to yield NAD^+;
      ii. with **decarboxylation** yielding **picolinic acid**.

Determinants of tryptophan–niacin conversion. The conversion of tryptophan to NAD is an inefficient process. Balance studies in humans demonstrated that this conversion varies among individuals in the range of 34–86 mg tryptophan per niacin equivalent,[16] on which basis Horwitt recommended the use of a 60:1 ratio for practical purposes.[17] Henceforth, it is normally taken that humans normally require ~60 mg of tryptophan to produce 1 mg of niacin metabolically.[18] This ratio is similarly wide for the chick (45:1) and the rat (50:1), and extremely wide for the duck (175:1). Conversion is reduced under conditions of nutritional iron deficiency and enhanced by niacin deprivation. Niacin-deficient humans are estimated to use nearly 3% of dietary tryptophan for niacin biosynthesis and, thus, are able to satisfy two-thirds of their requirement for the vitamin from the metabolism of

11. Wise, A., Foord, N.J., Fraser, N.J., 2003. J. Biol. Chem. 278, 9869–9876.
12. β-hydroxybutyrate is one of three ketone bodies produced by the liver when glucose cannot be used as an energy substrate. Unlike the others (acetone, acetoacetate), β-hydroxybutyrate has lipolytic activity.
13. It is referred to as HM74A in humans, GPR109A in the rat, and PUMA-G in the mouse.
14. Kynureninas*e* can also convert kynurenine to another urinary metabolite **anthranilic acid**.
15. Previously called picolinic acid carboxylase.
16. Goldsmith, G.A., Rosenthal, H.L., Gibbens, J., et al., 1955. J. Nutr. 56, 371–386; Horwitt, M.K., Harvey, C.C., Rothwell, W.S., et al., 1956. J. Nutr. 60, 1–43S.
17. Horwitt, M.K., 1955. Am. J. Clin. Nutr. 3, 244–245.
18. Hence, food niacin value is defined in terms of niacin equivalents, one unit of which is defined as 1 mg niacin +1/60 mg tryptophan.

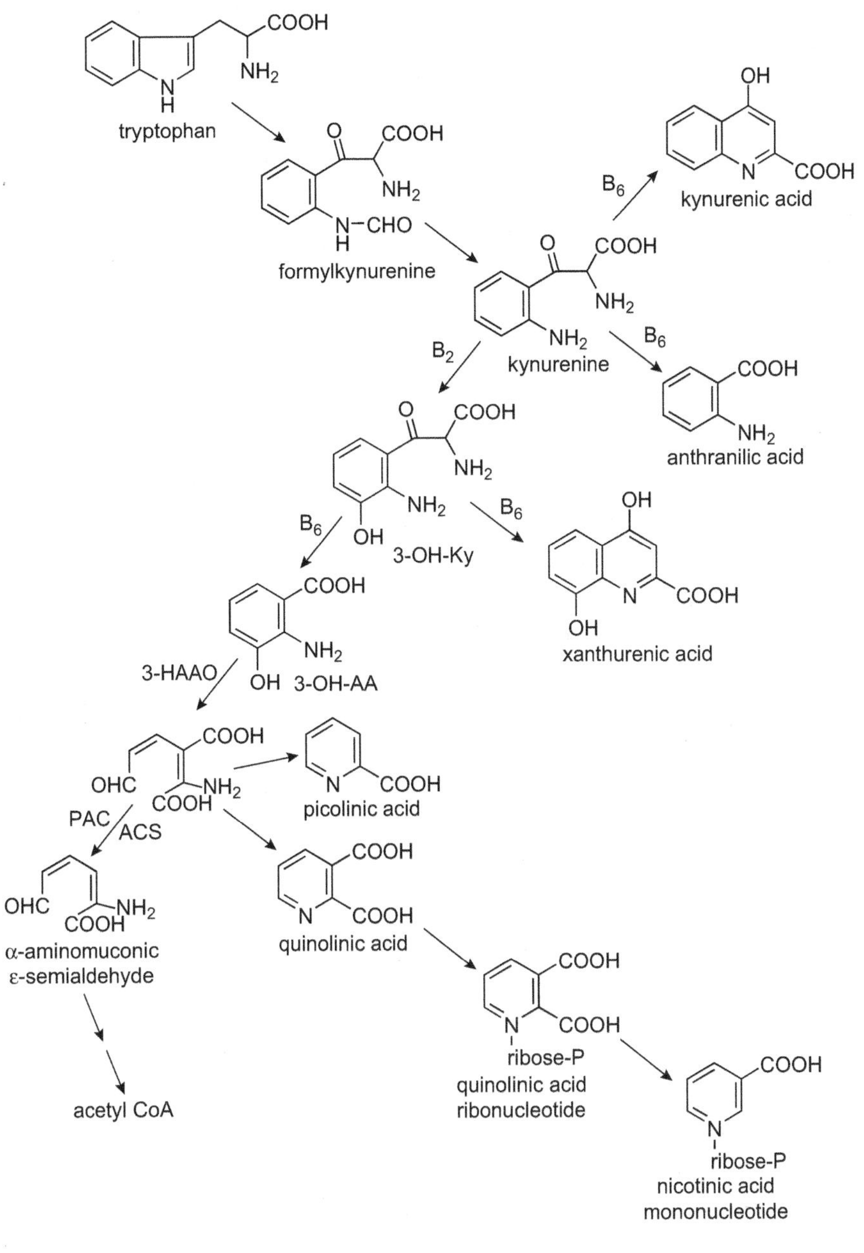

FIGURE 13.2 Metabolic interconversion of tryptophan to niacin. Note: the rate-controlling enzymes, 3-hydroxyanthranilic acid oxidase (3-HAAO) and aminomuconic semialdehyde dehydrogenase (ACSD); the roles of riboflavin (flavin adenine dinucleotide, FAD) and vitamin B_6 (pyridoxyl phosphate, PALP); anthranilic and xanthurenic acids are excreted in the urine.

this indispensable amino acid. Higher niacin-biosynthetic efficiencies are associated with *high* activities of 3-HAAO (enhancing production of ACS, the branch point intermediate in the pathway) and *low* activities of ACSD (which removes the first committed intermediate). Both hepatic activity of ACSD and the ratio of the hepatic activities of 3-HAAO and ACSD vary greatly between species and are inversely correlated with their dietary requirements for preformed niacin (Tables 13.4 and 13.5).

It would appear that protein turnover may preempt niacin synthesis under conditions of limiting tryptophan. In such circumstances the amount of tryptophan available for niacin synthesis would be expected to be low, rendering the calculation of niacin equivalents inaccurate.

Tryptophan–niacin conversion involves pyridoxal phosphate-dependent enzymes at four steps: two transaminases (which catalyze the conversions of kynurenine to kynurenic acid and of 3-hydroxykynurenine to xanthurenic acid) and kynureninase (which catalyzes the conversion of kynurenine to anthranilic acid as well as that of 3-hydroxykynurenine to 3-hydroxyanthranilic acid). While each uses pyridoxal phosphate, only kynureninase is impaired by pyridoxine deprivation. Its affinity for pyridoxal phosphate (K_m,[19] 10^{-3}M) is five orders of magnitude less than those of the transaminases (K_m's about 10^{-8}M); this renders it stripped of its cofactor

19. Michaelis constant; in this case, the concentration of pyridoxal phosphate necessary to support half-maximal enzyme activity.

TABLE 13.4 Relationship Between the 3-HAAO:ACSD (3-Hydroxyanthranilic Acid Oxygenase: α-Amino-β-Carboxymuconic-ε-Semialdehyde Decarboxylase) Ratio and Dietary Niacin Requirement

Animal	3-HAAO:ACSD	Niacin Requirement[a] (mg/kg Diet)
Rat	273	0
Chick:		
Low-niacin requirement strain	48	5
High-niacin requirement strain	27	15
Duck	5.3	40
Cat	5	45
Brook trout, lake trout	2.5	88
Turkey	1.6	70
Rainbow trout, Atlantic salmon	1.3	88
Coho salmon	3.4	175

[a]*Animals fed tryptophan.*
Poston, H.A., Combs, Jr., G.F., 1980. Proc. Soc. Exp. Biol. Med. 163, 452–459.

TABLE 13.5 Variation in Hepatic α-Amino-β-Carboxymuconic-ε-Semialdehyde Decarboxylase (ACSD)[a] Activities in Animals

Animal	ACSD Activity (IU/g)
Cat	50,000
Lizard	29,640
Duck	17,330
Frog	13,730
Turkey	9230
Cow	8300
Pig	7120
Pigeon	6950
Chicken:	
High-niacin requirement strain	5380
Low-niacin requirement strain	3200
Rabbit	4270
Mouse	4200
Guinea pig	3940
Human	3180
Hamster	3140
Rat	1570

[a]*i.e., Picolinic acid carboxylase (PAC).*
DiLorenzo, R. N., 1972. (Ph.D. thesis). Cornell University, Ithaca, New York.

under conditions of pyridoxine deprivation that are not severe enough to reduce cofactor access of the transaminases. Thus, pyridoxine deficiency impairs the overall conversion of tryptophan to niacin by blocking the production of 3-hydroxyanthranilic acid.[20] It does not, however, block the excretion of the urinary metabolites kynurenic acid and xanthurenic acid. This phenomenon has been exploited for the assessment of pyridoxine status by monitoring the urinary excretion of xanthurenic acid after a tryptophan load.

The conversion of tryptophan to niacin is also reduced by high-fat diets or diets containing excess leucine.[21] These effects appear to be due to ketosis, which has been noted as a common feature of diets of individuals with pellagra. NAD synthesis is increased by such factors as caloric restriction and hypoxia and in response to increased **sirtuin**[22] signaling, suggesting that NAD^+ levels may serve as indicators of physiological stress.

20. It has also been suggested that the deficiency of zinc, an essential cofactor of pyridoxal kinase (*see* Chapter 14), may also impair tryptophan–niacin conversion by reducing the production of pyridoxal phosphate.
21. Shastri, N.V., Nayudu, S.G., Nath, M.C., 1968. J. Vitaminol. 14, 198–205; Bender, D.A., 1983. Br. J. Nutr. 50, 25–32.
22. Named for the yeast gene "*silent mating-type information regulation 2*," sirtuins are protein deacetylases or ribosyltransferases that function in the regulation of transcription and apoptosis.

Three sources of pyridine nucleotides. The pyridine nucleotides NAD(H) and NADP(H) are produced from three precursors: NA, NAm, and tryptophan (Fig. 13.3). Whereas NA and NAm are formal intermediates in the biosynthesis of NAD^+ from tryptophan, that step (quinolinate phosphoribosyl transferase) actually leads directly to NAD^+ via NMN. Both NA and NAm are converted to NAD^+ by the same pathway after the latter is deamidated to yield NA. As the nicotinamide deamidase activities of animal tissues are low, this step is thought to be carried out by the intestinal microflora. The resulting NA is then phosphoribosylated (by nicotinate phosphoribosyl transferase), adenylated (by deamido-NAD^+ pyrophosphorylase), and amidated (by NAD synthetase) ultimately to yield NAD^+, which can be phosphorylated by an ATP-dependent NAD^+ kinase to yield $NADP^+$.

Although various tissues of the body are capable of pyridine nucleotide synthesis, there is clearly an exchange between the tissues. This involves primarily NAm, which is rapidly transported between tissues. In the rat, NA appears to be the most important precursor of these coenzymes in the liver, kidneys, brain, and erythrocytes; but in the testes and ovaries NAm appears to be a better precursor. Studies

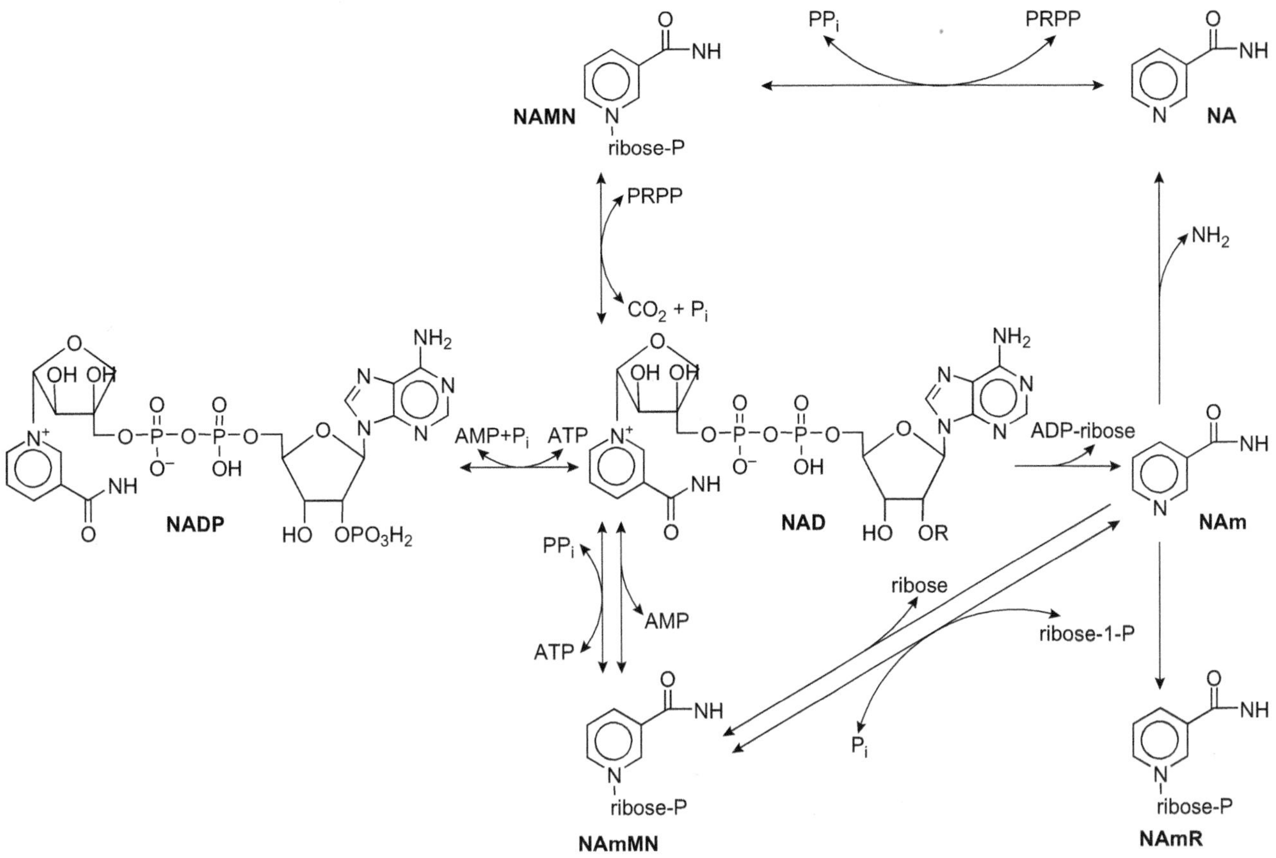

FIGURE 13.3 Niacin metabolism. *NAMN*, nicotinic acid mononucleotide; *NA*, nicotinic acid; *NAm*, nicotinamide; *NAmMN*, nicotinamide mononucleotide; *NAmR*, nicotinamide riboside; *NAD*, nicotinamide dinucleotide; *NADP*, nicotinamide dinucleotide phosphate; *PRPP*, phosphoribosyl pyrophosphate.

with chickens have shown that NAm can be a better dietary source of niacin activity than NA.[23]

Catabolism

The pyridine nucleotides are catabolized by hydrolytic cleavage of their two β-glycosidic bonds, primarily the one at the NAm moiety, by NAD(P)$^+$ glycohydrolase. NAm so released can be deamidated to form NA, in which form it can be reconverted to NAD$^+$. Alternatively, it can be methylated (mainly in the liver) by NAm ***N*-methyltransferase** to yield **1-methylnicotinamide,**[24,25] which can be oxidized to different products that are excreted in the urine.

23. Ohuho, M., Baker, D., 1993. J. Nutr. 123, 201–208, showed NAm to be utilized some 24% better than NA by broiler chickens.

24. NAm methylase activity is very low in fetal rat liver, increasing only upon maturity stimulation of hepatocyte proliferation (e.g., after partial hepatectomy or treatment with thioacetamide). Such increases in enzyme activity are accompanied by drops in tissue NAD$^+$ concentrations, as 1-methylnicotinamide reduces NAD$^+$ synthesis either by inhibiting NAD$^+$ synthetase and/or stimulating NAD(P)$^+$ glycohydrolase. Thus, it is thought that nicotinamide methylase and its product may be involved in the control of hepatocyte proliferation.

25. NA appears not to be methylated by animals. Trigonelline (1-methylnicotinic acid) does appear, however, in the urine of coffee drinkers, owing to its presence in that beverage.

Excretion

Niacin is excreted in appreciable amounts under conditions of supranutritional intake, as both vitamers are actively reabsorbed by the renal glomerulus. Excretion involves different water-soluble metabolites in the urine. At typical levels of intake of the vitamin, the major urinary metabolites are 1-methylnicotinamide[26] and its oxidation product **1-methyl-6-pyridone-3-carboxamide**. Under such conditions, intact NA and NAm, as well as other oxidation products, are also excreted, but in much smaller amounts. Most mammals excrete several metabolites: nicotinamide 1-oxide, 1-methyl-4-pyridone-3-carboxamide, 1-methyl-6-pyridone-3-carboxamide, 6-hydroxynicotinamide, and 6-hydroxynicotinic acid; some species also excrete NA/NAm conjugates of ornithine (2,5-dinicotinyl ornithine by birds only) or glycine (nicotinuric acid by rabbits, guinea pigs, sheep, goats, and calves).

The major urinary metabolite in the rat is **1-methyl-4-pyridone-3-carboxamide**. This metabolite is also found in human urine, but at levels substantially less than 1-methylnicotinamide and **1-methyl-2-pyridone-5-carboxamide**. The urinary metabolite profile can be changed by dietary deprivation of protein and/or amino acids, and it has been

26. Humans normally excrete up to 30 mg of total niacin metabolites daily, of which 7–10 mg is 1-methylnicotinamide.

suggested that the ratio of the pyridone metabolites to 1-methylnicotinamide may have utility as a biomarker for adequate amino acid intake.

At high rates of niacin intake, the vitamin is excreted predominantly (65–85% of total) in unchanged form. At all rates of intake, however, NAm tends to be excreted as its metabolites more extensively than is NA. Further, the biological turnover of each vitamer is determined primarily by its rate of excretion; thus, at high intakes, the half-life of NAm is shorter than that of NA.

7. METABOLIC FUNCTIONS OF NIACIN

Coenzyme Functions

Niacin functions metabolically as the essential component of the enzyme cosubstrates NAD(H)[27] and NADP(H).[28] The most central electron transport carriers of cells, each acts as an intermediate in most of the hydrogen transfers in metabolism, including some 500 reactions in the metabolism of carbohydrates, fatty acids, and amino acids. Each proceeds according to the following general reaction:

$$\mathbf{substrate + NAD(P)^{+} \rightarrow product + NAD\,(P)\,H + H^{+}}$$

The hydrogen transport by the pyridine nucleotides is accomplished by two-electron transfers in which the hydride ion (H^-) serves as a carrier for both electrons. The transfer is stereospecific, involving C-4 of the pyridine ring. The two hydrogen atoms at C-4 of NAD(H) and NADP(H) are not equivalent; each is stereospecifically transferred by the enzymes to the corresponding substrates.[29] In general, stereospecificity is independent of the nature of the substrate and the source of the enzyme, and few regularities are apparent except that dehydrogenases with phosphorylated and nonphosphorylated substrates tend to show opposite ste reospecificities.[30] The reactions catalyzed by the pyridine nucleotide-dependent dehydrogenases occur by the abstraction of the proton from the alcoholic hydroxyl group of the donor substrate, and the transfer of hydride ion from the same carbon atom to the C-4 of NAm. In many cases, this reaction is coupled to a further reaction, such as phosphorylation or decarboxylation. Despite similarities in mechanism and structure,[31] NAD(H) and NADP(H) have quite different metabolic roles and most dehydrogenases have specificity for one or the other.[32]

NAD in redox reactions. The oxidized form NAD^+ serves as a hydrogen acceptor at the C-4 position of the pyridine ring, forming NAD(H) which, in turn, functions as a hydrogen donor to the mitochondrial respiratory chain (TCA cycle) for ATP production (Table 13.6). These reactions include the following:

- glycolytic reactions
- oxidative decarboxylations of pyruvate
- oxidation of acetate in the TCA cycle
- oxidation of ethanol
- β-oxidation of fatty acids
- other cellular oxidations.

NADP(H) in reduction reactions. The phosphorylation of NAD^+ facilitates the separation of oxidation and reduction pathways of niacin cofactors by allowing NADP(H) to serve as a codehydrogenase in the oxidation of physiological fuels.[33] Thus, NADP(H) is maintained in the reduced state, NADPH[34] by the pentose phosphate pathway such that reduction reactions are favored. Many of these also involve flavoproteins.[35] These reactions involve reductive biosyntheses, such as those of fatty acids and steroids (Table 13.6). In addition, NADPH also serves as a codehydrogenase for the oxidation of glucose 6-phosphate in the pentose phosphate pathway.

NAD as a Substrate

ADP-ribosylation. NAD^+ functions in the addition of ADP-ribose to various nucleophilic acceptors. This occurs by its first forming a glycosidic linkage with ADP-ribose, which appears to be released by mitochondria under oxidative stress. The energy released upon breaking that bond is then used to drive mono- and poly-(ADP-ribosyl)ation reactions (Fig. 13.4).

- **Mono(ADP-ribosyl)ation**. NAD^+ serves as the donor of an ADP-ribose moiety to an amino acid residue on an acceptor protein. Originally recognized as properties of bacterial toxins,[36] two groups of **mono-ADP-ribotransferases (ARTs)** have been identified in mammalian cells. **Ecto-ARTs** are secreted or expressed on the outer surfaces of cells in skeletal and cardiac muscles, lung, testes, and lymphatic tissues. They ADP-ribophoshorylate

27. Previously, coenzyme I or diphosphopyridine nucleotide (DPN).
28. Previously, coenzyme II or triphosphopyridine nucleotide (TPN).
29. Because of this phenomenon, the pyridine nucleotide-dependent enzymes are classified according to the side of the dihydropyridine ring to which each transfers hydrogen, i.e., class A and class B.
30. Dehydrogenases with phosphorylated substrates tend to be B-stereospecific, whereas those with small (i.e., no more than three carbon atoms), nonphosphorylated substrates tend to be A-stereospecific.
31. For example, each contains adenosine, which appears to serve as a hydrophobic anchor.
32. A small number of dehydrogenases can use either NAD(H) or NADP(H).
33. For example, glyceraldehyde-3-phosphate, lactate, alcohol, 3-hydroxybutyrate, and pyruvate and α-ketoglutarate dehydrogenases.
34. The $NADP^+$/NADPH couple is largely reduced in animal cells, owing to the **transhydrogenase** activity that catalyzes the energy-dependent exchange of hydride between the pyridine nucleotides coupled to proton transport across the mitochondrial membrane in which it resides (the so-called *redox-driven proton pump*).
35. The first step in most biological redox reactions is the reduction of a flavoprotein by NADPH.
36. Cholera, diphtheria, pertussis, and *pseudomonas* toxins use NAD^+ to catalyze the ADP-ribosylation of host G-proteins and disrupt host cell function.

TABLE 13.6 Some Important Pyridine Nucleotide-Dependent Enzymes of Animals

	Enzyme	
Role	**NAD(H) Dependent**	**NADP(H) Dependent**
Carbohydrate metabolism	3-Phosphoglyceraldehyde dehydrogenase Lactate dehydrogenase Alcohol dehydrogenase	Glucose-6-phosphate dehydrogenase 6-Phosphogluconate dehydrogenase
Lipid metabolism	α-Glycerophosphate dehydrogenase β-Hydroxyacyl-CoA dehydrogenase 3-Hydroxy-3-methylglutaryl-CoA reductase	3-Ketoacyl ACP[a] reductase enoyl-ACP reductase
Amino acid metabolism	Glutamate dehydrogenase	Glutamate dehydrogenase
Other	NADH dehydrogenase/NADH-ubiquinone Dihydrofolate reductase Poly(ADP-ribose) polymerase 4-Hydroxybenzoate hydroxylase NADPH-cytochrome *P*-450 reductase Mono-ADP-ribotransferases	Glutathione reductase Thioredoxin-NADP reductase

[a]*ACP*, acyl carrier protein.

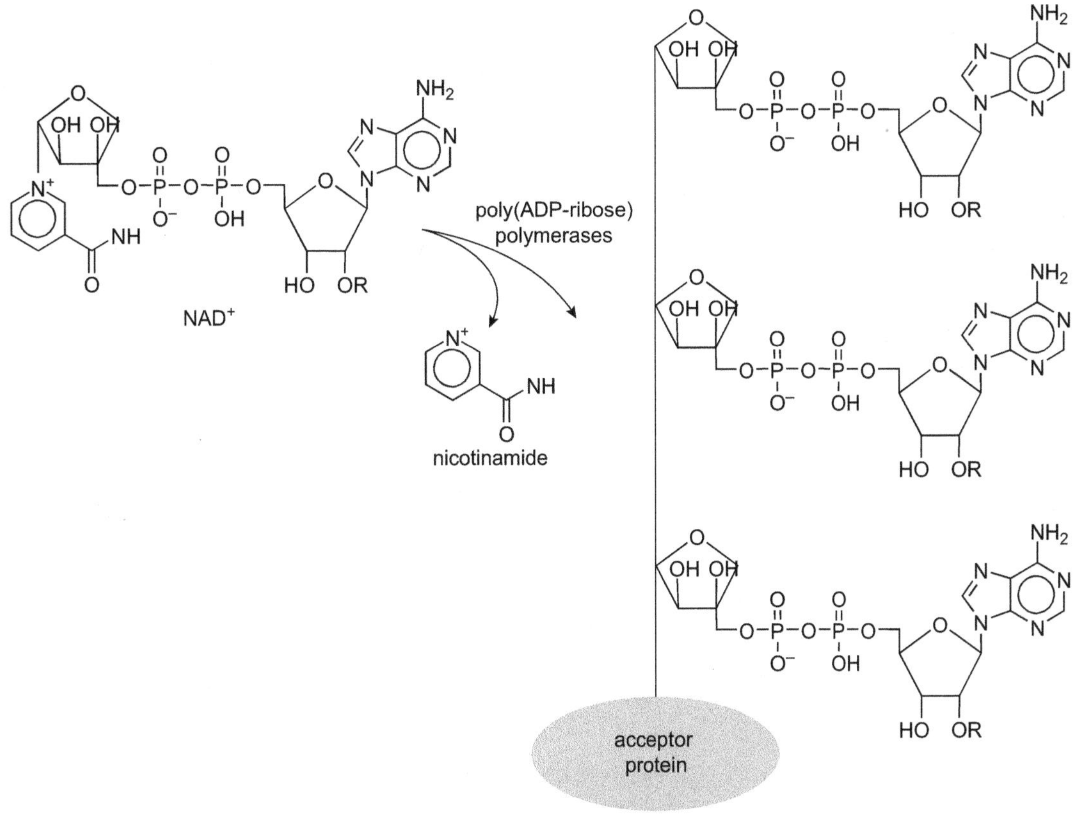

FIGURE 13.4 Poly(ADP-ribosyl)ation of proteins. Branching can occur at ribose links.

several proteins: integrins[37] in the control of myogenesis, defensins[38] in signaling macrophages, and an ATP-gated ion channel[39] to induce apoptosis. Because extracellular NAD concentrations are typically low, it has been suggested that the activities of the ecto-ARTs must depend on NAD being released by damaged cells, which would make NAD a signaling molecule for nearby cell death. **Endo-ARTs** are present in the cytosol

37. Integrins are extracellular receptors that mediate cell–cell attachment and control cell signaling.

38. Defensins are small, cysteine-rich peptides in neutrophils and most epithelial cells that participate in killing phagocytized bacteria by binding to the bacterial cell and forming pore-like structures that facilitate loss of essential ions.

39. P_2X_7.

or inner membranes of cells. They show little homology with the ecto-ARTs. One endo-ART inactivates G-protein β-subunits to serve in cell signaling; others inactivate the protein-folding chaperone, the 78kDa glucose-regulated protein (GRP78), which reduces protein secretion under conditions of cellular stress. In both cases, the inactivation appears to be reversible through the activity of hydrolases, which recycle the acceptor protein by removing its ADP-ribosyl moiety.

- **Poly(ADP-ribosyl)ation**. NAD(H) functions in the formation of ADP-ribose polymers by **poly(ADP-ribose) polymerases (PARPs)**,[40] which are activated by DNA single-strand breaks to catalyze both auto-ADP-ribosylation (to a glutamatyl or aspartyl residue) and, to a lesser extent, ADP-ribosylation of some 30 other acceptor proteins. In this way, PARPs form chains of ADP-ribose sequences on protein monomers, creating branch points at 40–50 unit intervals with new sites for subsequent elongation and increasingly negative charge. The poly(ADP-ribosyl)ation process involves extensive turnover of NAD^+ with concomitant production of NAm for reutilization.

Genomic Stability

Niacin status can affect genomic stability in several ways. First, poly(ADP-ribose) polymers bind to high-affinity binding sites on histones and other nuclear proteins. This has been shown to draw histones away from the DNA, thus, facilitating interactions of exposed DNA with other DNA-binding proteins (polymerases, ligases, helicases, and topoisomerases) involved in replication and repair. Evidence indicates that this process is involved in preventing nonhomologous recombination between two sites of damage. Niacin deprivation appears to impair the activities of all PARPs, resulting in genomic instability.[41] Second, the NAD-dependent deacylation activities of sirtuins 1–6 function in the maintenance of compact chromatin structure and gene silencing by deacylating histones, and in the regulation of many nonhistone proteins including the tumor suppressor p53 and the transcription factors PGC-1α and FOXO1.[42]

Glucose Tolerance Factor

Niacin has been identified as part of the chromium-containing **glucose tolerance factor** of yeast, which enhances the response to insulin. Its role, if any, in that factor is not clear, as free niacin is without effect. It is possible that this activity involves a metal-chelating capacity of NA such as has been reported for zinc and iron.[43]

Affector of Neurotransmission

NAD has been shown to bind to cell surface receptors of colonic enterocytes (P2Y1) and lymphoid cells (P2Y11). Binding has been proposed to inhibit neurotransmission to inhibit colonic contraction.[44]

Niacin-Responsive Genetic Disorders

Polymorphisms affecting the cofactor binding of NAD(P)-dependent proteins can lead to dysfunction if niacin is not plentiful; they can be treated and prevented with high doses of niacin. Polymorphisms affecting several proteins and enzymes have been identified: aldehyde dehydrogenase, glucose-6-phosphate-1-dehydrogenase, and the neutral amino acid transporter. Mutations in the latter result in **Hartnup disease**,[45] a rare familial disorder involving malabsorption of tryptophan (and other amino acids) and characterized by hyperaminoaciduria,[46] a pellagra-like skin rash (precipitated by stress, sunlight, or fever), ataxia and psychiatric disorders ranging from emotional instability to delirium. Nonreabsorbed tryptophan is degraded by gut microbiota to pyruvate and indole, which is reabsorbed and is neurotoxic.

8. BIOMARKERS OF NIACIN STATUS

Niacin Status Can be Assessed in Two Ways

- **Whole blood NAD** levels are sensitive to niacin deprivation and, therefore, can be used as an indicator of niacin status. Because NADP levels remain fairly stable under deficient conditions, the NAD:NADP ratio is frequently used to assess niacin status. This has been called the "**niacin number**."
- **Urinary niacin metabolites**, N^1-methylnicotinamide and N^1-methyl-2-pyridone-5-carboxamide, can indicate niacin status. While the amount of the former has been associated with dermatitis,[47] the ratio of the two metabolites is affected by protein status.

Dietary assessment based on both the preformed vitamin and tryptophan is subject to the limitations of dietary

40. At least 18 PARP genes have been identified.
41. Oei, S.L., Kel, C., Ziegler, M., 2005. Biochem. Cell Biol. **83**, 263–270.
42. Spronck, J.C., Nickerson, J.L., Kirkland, J.B., 2007. Nutr. Cancer **57**, 88–99.
43. Agte, W., Paknikar, K.M., Chiplonkar, S.A., 1997. Biometals **10**, 271–276.
44. Klein, C., Grahnert, A., Abdelrahman, A., et al., 2009. Cell Calcium 46, 263–272.
45. The disease was named for the first case, described in 1951, involving a boy thought to have pellagra. Since that time some 50 proved cases involving 28 families have been described.
46. The presence of abnormally high concentrations of amino acids in the urine.
47. Dilon, J.C., Malfait, P., Demaux, G., et al., 1992. Am. J. Med. 93, 102–104.

recall methods and the uncertainties of niacin bioavailability in foods.

9. NIACIN DEFICIENCY

Determinants of niacin status. Because a substantial amount of niacin can be synthesized from tryptophan, niacin status depends not only on the level of intake of the preformed vitamin but also that of its potential amino acid precursor. Accordingly, niacin deficiency typically occurs in individuals consuming diets low in both of these essential nutrients and, frequently, pyridoxine. Thus, the occurrence of pellagra, as well as niacin-deficiency diseases in animals, is properly viewed as the result of a multifactorial dietary deficiency rather than that of insufficient intake of niacin per se.

In addition to dietary tryptophan and pyridoxine supplies being important determinants of niacin status, it has been suggested that excess intakes of the branched-chain amino acid leucine may antagonize niacin synthesis and/or utilization and, thus, also may be a precipitating factor in the etiology of pellagra. Excess leucine has been shown to inhibit the production of quinolinic acid from tryptophan by isolated rat hepatocytes; however, the magnitude of this effect is small in comparison with the K_m of quinolinate phosphoribosyl transferase for quinolinate, indicating that excess leucine (and/or its metabolites) is unlikely to affect the rate of NAD^+ biosynthesis by the liver. Some studies with intact animals (rats) have produced results supporting the view that excess leucine can impair the synthesis of NAD^+ from tryptophan (either by inhibiting the enzymatic conversion itself or the cellular uptake of the amino acid); however, others have yielded negative results in this regard. Therefore, the relative contribution of high leucine intake to the etiology of pellagra is not clear at present.

Zinc appears to play a role in the pyridoxine-dependent conversion of tryptophan to niacin.[48] Pellagra patients have been found to have low plasma Zn levels, and Zn supplementation increases their urinary excretion of 1-methylnicotinamide and 1-methyl-2-pyridone-5-carboxamide. Studies with rats have shown that treatment of niacin-deficient animals with the metabolic intermediate picolinic acid increases circulating Zn levels.

General signs. Niacin deficiency leads to tissue depletion of NAD(P) that varies with cell turnover. This results in different species-specific signs usually accompanied by loss of appetite and poor growth (Table 13.7).

TABLE 13.7 Signs of Niacin Deficiency

Organ System	Signs
General	Decreased appetite and growth
Dermatologic	Dermatitis, photosensitization
Gastrointestinal	Inflammation, diarrhea, glossitis
Skeletal	Perosis
Vascular	Anemia
Nervous	Ataxia, dementia

Deficiency Signs in Humans

Niacin deficiency in humans results in changes in the skin, gastrointestinal tract, and nervous system. Individuals with this syndrome, **pellagra**, typically complain of a sore tongue, which involves markedly reduced epithelial thickness and diarrhea. The dermatologic changes are most pronounced in the parts of the skin that are exposed to sunlight (face, neck,[49] backs of the hands and fore) (Figs. 13.5 and 13.6). In some patients, lesions resemble early sunburn; in chronic cases the symmetric lesions feature cracking, desquamation,[50] hyperkeratosis, and hyperpigmentation. Lesions of the gastrointestinal tract include angular stomatitis, cheilosis, and glossitis as well as alterations of the buccal mucosa, tongue, esophagus, stomach (resulting in achlor[51]), and intestine (resulting in diarrhea).[52] Pellagra almost always involves anemia.[53] Early neurological symptoms associated with pellagra include anxiety, depression, and fatigue[54]; later symptoms include depression, apathy, headache, dizziness, irritability, and tremors.

Deficiency Signs in Animals

Most niacin-deficient animals show poor growth and reduced efficiency of feed utilized. Pigs and ducks are particularly sensitive to niacin deficiency. Pigs show diarrhea, anemia, and degenerative changes in the intestinal mucosa and nervous tissue[55]; ducks show severely bowed and weakened legs and diarrhea. Niacin-deficient dogs show necrotic

48. Zinc, which is required by pyridoxal phosphokinase, is also related to the function of pyridoxine in this system. Alcoholics, who are typically of low Zn status, can excrete high levels of the niacin metabolites 1-methyl-6-pyridone-3-carboxamide and 1-methylnicotinamide. This excretion can be increased by Zn supplementation, presumably owing to increased pyridoxal phosphate activity and the consequent activation of pyridoxine to the form (pyridoxal phosphate) that facilitates tryptophan–niacin conversion. Zinc deficiency can also reduce the availability of tryptophan for niacin biosynthesis by enhancing its oxidation.

49. This is referred to as **Casal's collar**.

50. The shedding of the epidermis in scales.

51. The absence of hydrochloric acid from the gastric juice, usually due to gastric parietal cell dysfunction.

52. Many of these gastrointestinal changes also occur in schizophrenia.

53. The anemia associated with pellagra is of the macro- or normocytic, hypochromic types.

54. Many of these symptoms also occur in schizophrenia.

55. The syndrome is called "pig pellagra."

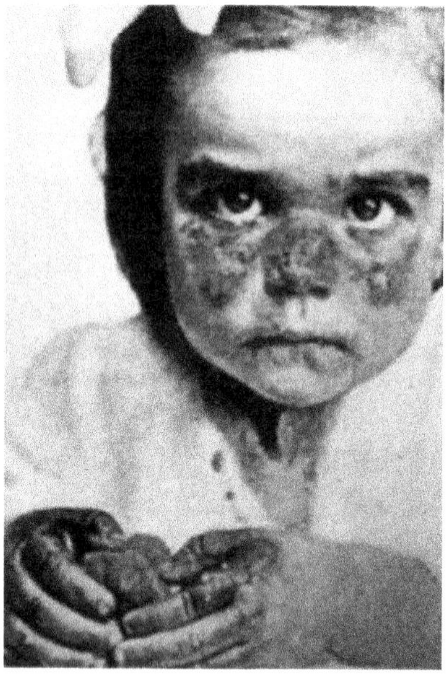

FIGURE 13.5 Pellagra: affected child with facial *"butterfly wing."* *Courtesy Cambridge University Press.*

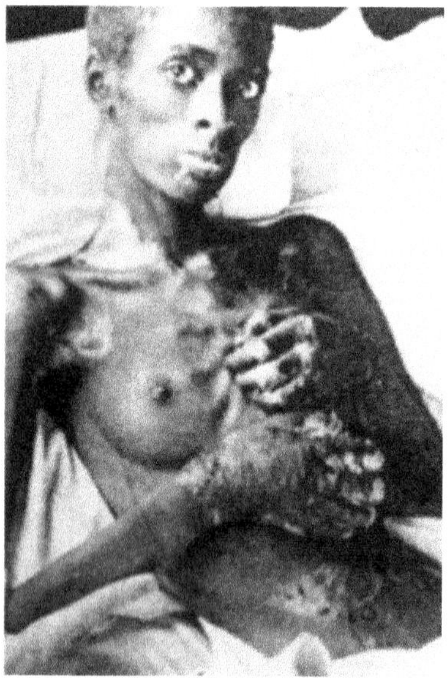

FIGURE 13.6 Pellagra: affected woman with *"pellagra glove."* *Courtesy Cambridge University Press.*

degeneration of the tongue with changes of the buccal mucosa and severe diarrhea.[56] Rodents show alopecia and nerve cell histopathology. Chickens show inflammation of the upper gastrointestinal tract, dermatitis of the legs, reduced feather growth, and **perosis** (Figs. 13.7 and 13.8).[57]

It has been thought that ruminants are not susceptible to niacin deficiency, owing to the synthesis of the vitamin

FIGURE 13.7 Perosis (left leg) in niacin-deficient chick. *Courtesy, M.L. Scott, Cornell University.*

by their rumen microflora. Although that appears to be true for most ruminant species, evidence indicates that fattening beef cattle and some high-producing dairy cows can benefit from niacin supplements under some circumstances. Studies have shown niacin treatment of lactating cows to depress circulating levels of ketones, apparently by reducing lipolysis in adipocytes by a process involving increased cyclic 3′,5′-adenosine monophosphate (cAMP) and, consequently, the concentrations of nonesterified fatty acids in the plasma. That ruminal synthesis of the vitamin may not meet the nutritional needs of the host would appear most likely in circumstances wherein rumen fermentation is altered to enhance energy utilization, with associated reductions in rumen microbial growth.

10. NIACIN IN HEALTH AND DISEASE

Niacin has been associated with a number of health effects unrelated to the signs of niacin deficiency.

Cardiovascular Health

High doses of NA have proven to be among the most useful treatments for hyperlipidemias and hypercholesterolemia. A retrospective evaluation of results from the U.S. Coronary Drug Project showed NA doses of 1–2 g/day to have reduced lethal coronary events, resulting in highly significant reduction of mortality from all causes by 11% (*versus* a placebo). A meta-analysis of 11 clinical trials involving nearly 10,000 subjects found niacin use to be associated with significant reductions in cardiovascular end points, major coronary heart disease events, and stroke incidence.[58] However, in 2011 a large trial of combined NA-statin treatment was stopped after 32 months on the basis of its not showing reductions in cardiovascular events, despite showing the expected anti-hyperlipidemic effects.

High-dose NA treatment reduces all major lipids and apoB-containing lipoproteins (VLDL, LDL), while

56. **Black tongue disease.**
57. Inflammation and misalignment of the tibiotarsal joint (*hock*), in severe cases involving slippage of the Achilles tendon from its condyles, which causes crippling due to an inability to extend the lower leg.

58. Lavigne, P.M., Karras, R.H., 2013. J. Am. Coll. Cardiol. 61, 440–446.

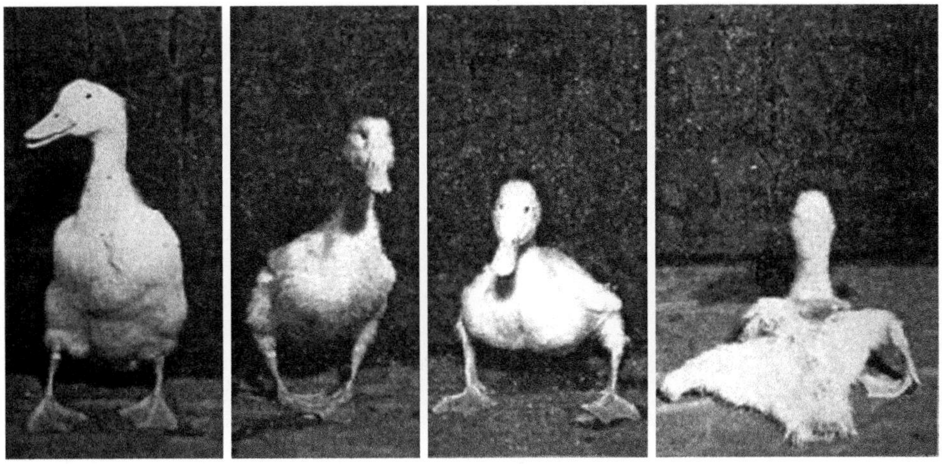

FIGURE 13.8 Leg weakness in ducks fed diets containing adequate (left) or progressively deficient amounts of niacin. *Courtesy M. L. Scott, Cornell University.*

increasing apoA1-containing lipoproteins (HDL) (Tables 13.8 and 13.9). Thus, NA has been considered as an adjunct to statin therapy; limited clinical data indicate that the addition of NA improves lipid profiles in patients with coronary artery disease but does not affect endothelial or microvascular function.[59] The antihyperlipidemic effects of niacin appear to be unrelated to NAD(P) or NAm but instead involve three metabolic phenomena (Fig. 13.9):

- **Reducing hepatic triglyceride synthesis** by NA inhibiting hepatic microsomal diacylglyceride transferase-2 has been demonstrated. That enzyme catalyzes the final reaction in triglyceride synthesis; its inhibition limits the amounts of triglycerides available for the assembly of VLDL.[60] This results in increased degradation of apolipoprotein B and the consequent reduction in both VLDL and its catabolic product, LDL.
- **Reducing the removal of HDL-apoA1** has been demonstrated in cultured cells, with NA inhibiting the catabolism of HDL-apoA1 without affecting apoA-I synthesis. This increases HDL and HDL cholesterol. It has been suggested that the response may involve the putative "HDL catabolism receptor," which may be a β-chain ATP synthase. These increases are thought to reflect reduced exchange of triglycerides and cholesterol esters[61] and the retarded degradation of apoA1.
- **Reducing adipocyte lipolysis** by NA binding to the niacin receptor, GPR109A, which is linked to a G-protein that inhibits adenylate cyclase. That inhibition leads to a decline in cAMP levels, which inhibits the hormone-sensitive lipase and consequently reduces the mobilization of fatty acids from triglycerides in adipose tissue. Reduced release of fatty acids is responsible for at least part of the reduction of hepatic synthesis and secretion of VLDLs and the subsequent decline in circulating LDL levels. Decreased circulating levels of VLDLs are associated with decreased levels of triglycerides and cholesterol. Also contributing to reduced cholesterol levels is a decrease in cholesterol biosynthesis due to NA inhibition of 3-hydroxy-3-methylglutaryl CoA reductase.[62]

TABLE 13.8 Summary of Plasma Lipid Responses to Nicotinic Acid Treatment (>1.5 g/day) of Dyslipidemias

Parameter	Reduction (%)	Increase (%)
Triglycerides	21–44	–
VLDL	25–40	–
LDL cholesterol	2–22	–
HDL cholesterol	–	18–35
Total cholesterol	4–16	–
Lipoprotein Lp(a)	16–36	–

Gille, A., Bodor, E.T., Ahmed, K., et al., 2008. Ann. Rev. Pharmacol. Toxicol. 48, 79–86.

Skin Health

Niacin deficiency increases skin sensitivity to sunlight. This occurs in patients with pellagra and has been produced experimentally in the rainbow trout.[63] This effect appears to be due to inhibition of PARPs and sirtuins, which depend

59. Philpott, A.C., Hubacek, J., Sun, Y.C., et al., 2013. Atherosclerosis 226, 453–458.
60. *See review*: Kamanna, V.S., Kashyap, M.L., 2008. Am. J. Cardiol. 101, 20–26B.
61. This process is mediated by the cholesterol ester transfer protein.
62. DiPalma, J.R., Thayer, W.S., 2001. Ann. Rev. Nutr. 11, 169–176.
63. Poston, H.A., Wolfe, M.J., 1985. J. Fish Dis. 8, 451–460.

TABLE 13.9 Results of Clinical Trials of Nicotinic Acid (≥1 g/day) in Patients on Statins

Study	Subjects (Duration)	Parameter	Placebo	Nicotinic Acid
1[b]	8341 (5 years)	Myocardial infarction	12.2%	8.9%[a]
		Mortality	20.9%	21.2%
1[c]	8441 (15 years follow-up)	Mortality	58.2%	52.0%[a]
2[d]	555 (5 years)	Mortality	29.7%	21.8%[a]
3[e]	146 (2.5 years)	Cardiovascular events	19.2%	4.2%[a]
6[f]	167 (1 year)	Carotid intima-media thickness	0.044 mm	0.014 mm[a]

[a]$p < .05$.
[b]*Coronary Drug Project Research Group, 1975. JAMA 231, 360–366.*
[c]*Canner, P.L., Berge, K.G., Wenger, N.K., 1986. J. Am. Coll. Cardiol. 8, 1245–1252.*
[d]*Carlson, L.A., Rosenhamer, G., 1988. Acta Med. Scan. 223, 405–41.*
[e]*Brown, G., Albers, J.J., Fisher, L.D., 1990. N. Engl. J. Med. 323, 1289–1296.*
[f]*Taylor, A.J., Sullenberger, L.E., Lee, H.J., et al., 2004. Circulation 110, 3512–3519.*

on NAD+ to protect against DNA damage. At high doses, NA has vasodilatory activity; this is due to its binding to hydroxycarboxylic acid receptors, which increase prostaglandin production, increasing microvascular blood flow. Niacin has been used for topical treatment of acne vulgaris and rosacea.[64] Most forms of the vitamin are water soluble and, therefore, not absorbed across the skin; fatty esters of niacin can be absorbed dermally.

Lung Health

Studies in animal models have demonstrated that niacin status can be affected by pulmonary oxidant injury in different ways. Acute oxidative stress, caused by treatment with DNA-damaging agents (lipopolysaccharide, cyclophosphamide, and bleomycin) produces NAD(P) depletion, which can be prevented by supplemental niacin.[65] Nonacute oxidative stress, caused by hyperoxia, increases NAD levels in the lungs and induces poly(ADP-ribose) synthesis even of niacin-deprived animals. These effects increase inflammatory and apoptotic responses.[66]

64. Kirkland, J.B., Millman, C.G., Jacobson, J., 2011. J. Evid. Based Complement. Med. 16, 91–101.
65. Giri, S.N., Blaisdell, R., Rucker, R., et al., 1994. Environ. Health Perspect. 102 S10, 137–147.
66. Kirkland, J.B., 2010. Exp. Biol. Med. 235, 561–568.

Anticarcinogenesis

Metabolic considerations would predict that niacin deprivation might reduce tumorigenesis. Tumor cells are known to be far more dependent on glycolysis than oxidative phosphorylation for generating ATP,[67] and their demands for and turnover of NAD are substantially greater than nontumor cells. However, experimental evidence does not indicate that tumor risk can be reduced by niacin deprivation. In fact, epidemiological evidence has associated marginal niacin intakes and/or the reliance on maize-based diets with increase risks of cancers of the esophagus.[68] Further, studies in animal models have found NA supplementation to reduce yields of esophageal tumors in the *N*-nitrosomethylbenzylamine-treated rat model.[69] Supranutritional doses of niacin have been shown to reduce dramatically the yield of skin tumors in UV-treated mice in a dose-dependent manner that correlated with skin NAD levels.[70]

Neurocognitive Health

NAm has been shown to enhance the effect of tryptophan in supporting brain serotonin levels. It does so by reducing the urinary excretion of tryptophan metabolites and reducing the conversion of tryptophan to niacin. This increases the availability of tryptophan for the synthesis of serotonin, the general effect of which is antidepressive. At the same time, neurons and glial cells have metabolic needs for NAD(P), as they depend on glycolysis for generating ATP. Niacin status has been associated with several neurological disorders:

- **Schizophrenia** is associated with NAD deficiency in critical areas of the brain. Affected individuals oxidize NAm more readily than unaffected people: they excrete greater amounts of 1-methyl-6-pyridone-3-carboxamide. As the excretion of this methylated product is increased by treatment with methylated hallucinogens (e.g., methylated indoles) and is decreased by treatment with tranquilizers, it has been suggested that schizophrenics suffer a depletion of NAm (via its methylation and excretion), which limits NAD+ synthesis. Patients with first episodes show diminished flushing responses to niacin; that response is mediated by vasodilators

67. Ganapathy, V., Thangaraju, M., Prasad, P.D., 2009. Glycolysis in tumor cells is 30 times more active than nontumor cells. Pharmcol. Ther. 121, 29–40.
68. Van Rensburg, S.J., Bradshaw, E.S., Bradshaw, D., et al., 1985. Br. J. Cancer 51, 399–406; Wharendorf, J., Chang-Claude, Q.S., Lian, Y.G., et al., 1989. Lancet 2, 1239–1246; Franceschi, S., Bidoli, E., Baron, A.E., et al., 1990. J. Nat. Cancer. Inst. 82, 1407–1412; Marshall, J.R., Graham, S., Haughery, B.P., 1992. J. Cancer Oral Ocol. 28B, 9–15.
69. Van Rensburg, S.J., Hall, J.M., Gathercole, P.S., 1986. Nutr. Cancer 8, 163–170.
70. Gensler, H.L., Williams, T., Huang, A.C., et al., 1999. Nutr. Cancer 34, 36–43.

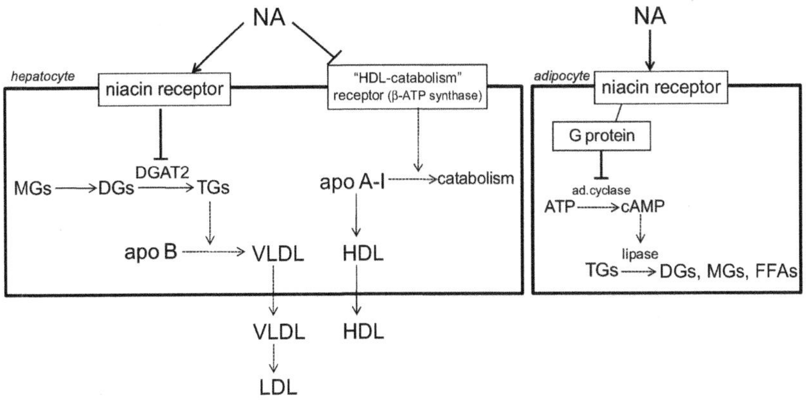

FIGURE 13.9 Schematic representation of apparent metabolic bases for the antihyperlipidemic effects of nicotinic acid. *Dashed arrows* indicate steps reduced by NA. *NA*, nicotinic acid; *MGs*, monoglycerides; *DG*, diglycerides; *TGs*, triacylglycerides; *apo B*, apolipoprotein B; *apo A-I*, apolipoprotein A-I; *DGAT2*, diacylglycerol acyltransferase 2; *ATP*, adenosine triphosphate; *cAMP*, cyclic adenosine monophosphate; *ad. cyclase*, adenylate cyclase; *lipase*, hormone-sensitive lipase.

derived from arachidonic acid, which is typically low in such patients. High doses of NA (e.g., 1 g/day) given with ascorbic acid have been found effective in eliminating psychotic symptoms and preventing relapses of acute cases.

- **Alzheimer's disease** incidence has been associated inversely with niacin intake. A cohort study found that subjects with the greatest intakes of dietary niacin (22 mg/day) experienced cognitive declines over 6 years that averaged 44% less than those of subjects with lower niacin intakes.[71] It is possible that this protection may be due to niacin increasing expression of apoA-1, which is associated with decreased Alzheimer's disease risk, or to sirtuins, which have been shown to reduce amyloid precursor protein in animal models.
- **Parkinson's disease** is characterized by high urinary excretion of methylnicotinamide, suggesting aberrant niacin metabolism.[72]
- **Depression** can be associated with diminished flushing responses to niacin. Studies have shown that 5% of depressed subjects do not show the niacin-induced flushing. Nonresponders are likely to be severely ill, with depressed mood, anxiety, and physical symptoms.[73]
- **Progeny effects of maternal alcohol** (anxiety, neural damage) have been found in animal models to be reduced by maternal NAm treatment.[74]

Type 2 Diabetes (T2D)

T2D is associated with a reduced state of the pyridine nucleotides in the cytosol and mitochondria due to the increased levels of glucose, free fatty acids, lactate, and branched-chain amino acids. This affects gene expression via the NADH-activated transcriptional corepressor, C-terminal binding protein (CtBP), and the NADH enzyme—glyceraldehyde-3-phosphate dehydrogenase. NAm has been found to delay or prevent the development of diabetic signs in the nonobese diabetic mouse model,[75] to decrease the severity of diabetic signs associated with β-cell proliferation induced by partial pancreatectomy,[76] and to protect against diabetes induced by agents[77] that cause DNA strand breakage in β-cells. This appears to be due to preventing extreme NAD(P) depletion, which would be expected to enhance poly(ADP-ribose)-induced signaling leading to inflammation and apoptosis. In clinical trials, NAm has been found to protect high-risk children from developing clinically apparent insulin-dependent diabetes,[78] to improve small artery vasodilatory function in statin-treated T2D patients,[79] and to increase blood glucose levels but not T2D.[80] However, a large randomized trial found NAm ineffective in reducing diabetes risk.[81]

Other Effects

High doses of NA and NAm have been shown to

- reduce the hyperphosphatemia in chronic renal disease and renal dialysis, by inhibiting the Na^+-dependent cotransport of phosphate in both the renal tubule and intestine;[82]

71. Morris, M.C. Evans, D.A., Bienas, J.L., et al., 2004. J. Neurol. Neurosurg. Psychiatr. 75, 1093–1099.
72. Ying, W., 2007. Front. Biosci. 12, 1863–1888.
73. Smensy, S., Baur, K., Rudolph, N., et al., 2010. J. Affect. Disord. 124, 335.
74. Feng, Y., Paul, I.A., LeBlanc, M.H., 2006. Brain Res. Bull. 69, 117–123; Ieraci, A., Herrara, D.G., 2006. PLoS Med. 3, e101.
75. Reddy, S. Bibby, N.J., Elliott, R.B., 1990. Diabetes Res. 15, 95.
76. Yonemura, Y., Takashima, T., Miwa, K., et al., 1984. Diabetes 33, 401.
77. Alloxan, streptozotocin.
78. Elliott, R.B., Chase, H.P., 1991. Diabetologia 34, 362; Manna, R., Milgore, A., Martin, L.S., et al., 1992. Br. J. Clin. Pract. 46, 177–184.
79. Hamilton, S.J., Chew, G.T., Davis, M.E., et al., 2010. Diabetes Vasc. Dis. Res. 7, 296–302.
80. Phan, B.A.P., Muñoz, L., Shadzi, P., et al., 2013. Am. J. Cardiol. 111, 352–355.
81. Gale, E.A., Binley, P.J., Emmett, C.L., et al., 2004. The European-Canadian Nicotinamide Diabetes Intervention Trial. Lancet 363, 925–931.
82. Lenglet, A., Liabeuf, S., Guffroy, P., et al., 2013. Drugs R.D. 13, 165–173.

- reduce hypertensions, probably due to vasodilation; however, results of clinical trials have been inconsistent;[83]
- increase circulating levels of adiponectin; these effects were not accompanied by changes in insulin sensitivity or endothelial function;[84]
- suppress postprandial trigyceridemia, a predictor of cardiovascular events.[85]

11. NIACIN TOXICITY

In general, the toxicity of niacin is low. Nonruminant animals can tolerate oral exposures of at least 10- to 20-fold their normal requirements for the vitamin. The toxic potential of NAm appears to be greater than that of NA, probably by a factor of four. Side effects of high doses appear to result from metabolic disturbances due to the depletion of methyl groups as the result of the metabolism of the vitamin.

Nicotinic Acid

The most common side effect of high-dose NA is **skin flushing** caused by cutaneous vasodilation. This response is transient (30–90min) and accompanied by erythema, tingling, itching, and elevated skin temperature. It affects some 70% of subjects in which it is seen at the beginning of NA therapy and subsides over time with the development of tolerance. Still, for some it can be disagreeable to the point of discontinuing treatment.[86] The flushing response can be evoked by either oral or topical exposure to NA, which triggers COX-1[87]-derived release of prostaglandin D_2 from platelets and dendritic cells. These effects involve the niacin receptor, which is expressed by macrophages and bone marrow-derived cells of the skin. The response can be minimized by using a slow-release formulation of NA or by using a cyclooxygenase inhibitor (e.g., aspirin, indomethacin) prior to taking NA.[88] High doses of NA have also been reported to cause itching urticaria (hives) and gastrointestinal discomfort (heartburn, nausea, vomiting, and rarely diarrhea) in humans. Animal studies have shown that high levels of NAm can raise circulating homocysteine levels, particularly on a high-methionine diet.

The longer-term effects of high NA doses include insulin resistance, which may involve a rebound in lipolysis that results in increased free fatty acid levels. A few cases of transient elevations in the plasma activities of liver enzymes without associated hepatic dysfunction have been reported and chronic doses of NA have been reported to cause hepatic damage.[89]

Nicotinamide

While acute adverse effects of NAm have not been reported for doses used to treat insulin-dependent diabetes (c. 3g/day), larger doses (10g/day) have been found to cause hepatic damage. It is possible that chronic, high intakes of NAm may deplete methyl groups due to the increased demand for methylation to excrete the vitamin. Such effects would be exacerbated by low intakes of methyl donors, methionine and choline, and suboptimal status with respect to folate and/or vitamin B_{12}. NAm can inhibit uricase, depressing intestinal microbial uricolysis, which could lead to uricemia.

12. CASE STUDY

Instructions

Review the following report, paying special attention to the responses to the experimental treatments. Then, answer the questions that follow.

Case

Fourteen patients with alcoholic pellagra and 7 healthy controls, all ranging in age from 21 to 45 years, were studied in the metabolic unit of a hospital. None had severe hepatic dysfunction on the basis of medical history, clinical examination, and routine laboratory tests. The nutritional status of each subject was evaluated at the beginning of the study by clinical examination, anthropometric measurements [body mass index (BMI, weight divided by the square of the height), triceps skinfold thickness, arm and muscle circumference], biochemical tests [24-hr urinary creatinine, serum albumin, total iron-binding capacity (TIBC)], and 24-hr recalls of food consumption. Results indicated that, before admission, the patients with alcoholic pellagra consumed a daily average of 270g of ethanol. Each showed signs of protein–calorie malnutrition (reduced BMI, skinfold thickness, arm and muscle circumference, serum albumin, and TIBC). In addition, their plasma zinc concentrations were significantly lower than those of controls, although their urinary zinc concentrations were not different from the control group.

The pellagra patients were assigned to one of two experimental treatment groups and the healthy controls to another (three treatments, each with n = 7). During the 7-day study, each group received enteral diets prepared from 10% crystalline amino acids (adequate amounts of each, except

83. Bays, H.E., Rader, D.J., 2009. Int. J. Clin. Pract. 63, 1–11.
84. Westphal, S., Borucki, K., Taneva, E., et al., 2007. Atherosclerosis 193, 361–368.
85. Usman, M.H., Qamar, A., Gadi, R., et al., 2012. Am. J. Med. 125, 1026–1035.
86. Dropout for this reason has been seen in 5–20% of patients.
87. Cyclooxygenase-1.
88. 81 mg.
89. Rader, J.I., Calvert, R.J., Hathcock, J.N., 1992. Am. J. Med. 92, 77–83.

for tryptophan) and 85% sucrose, which supplied daily amounts of 0.8 g of protein per kilogram of body weight and 200 kcal/g N. In addition, each patient was given weekly by vein 500 mL of an essential fatty acid emulsion as well as a vitamin–mineral supplement. The diets were administered by intubation directly to the midportion of the duodenum. The control diet was supplemented with tryptophan and the vitamin–mineral supplement contained both niacin and zinc. The diets provided to each group of pellagra patients contained no tryptophan; neither did their vitamin supplement contain niacin. One group of pellagra patients received supplemental zinc (220 mg of $ZnSO_4$) whereas the other did not. Several biochemical measurements were made at the beginning of the experiment and, again, after 4 days. Each of the biochemical measurements was repeated after 7 days of treatment. In most cases, the results showed the same effects but of greater magnitudes.

Results

Subject Group	Parameter	Initial Value	Day 4 Value
Healthy controls	Plasma Zn (mmol/L)	14.2 ± 1.5	16.0 ± 2.2
	Plasma tryptophan (mmol/L)	50.8 ± 12.5	74.3 ± 18.5
	Urine Zn (mmol/day)	7.34 ± 1.38	9.18 ± 2.91
	Urine 6-pyridone[a] (mmol/day)	70 ± 22	640 ± 235
	Urine CH_3-NAm[b] (mmol/day)	78 ± 32	143 ± 48
Pellagra patients	Plasma Zn (mmol/L)	9.9 ± 1.1	9.6 ± 2.0
	Plasma tryptophan (mmol/L)	33.3 ± 15.3	29.5 ± 6.1
	Urine Zn (mmol/day)	9.79 ± 3.06	11.93 ± 10.55
	Urine 6-pyridone[a] (mmol/day)	16 ± 10	19 ± 12
	Urine CH_3-NAm[b] (mmol/day)	6 ± 3	9 ± 6
Pellagra patients fed Zn	Plasma Zn (mmol/L)	9.8 ± 1.0	15.8 ± 3.2
	Plasma tryptophan (mmol/L)	37.3 ± 17.8	23.7 ± 7.6
	Urine Zn (mmol/day)	9.80 ± 3.10	24.02 ± 8.11
	Urine 6-pyridone[a] (mmol/day)	16 ± 11	55 ± 18
	Urine CH_3-NAm[b] (mmol/day)	6 ± 3	33 ± 20

[a]1-Methyl-6-pyridone-3-carboxamide.
[b]1-Methylnicotinamide.

Case Questions

1. What signs support the diagnosis of protein–calorie malnutrition in these alcoholic patients with pellagra?
2. Propose a hypothesis for the mechanism of action of zinc in producing the responses that were observed in these patients with alcoholic pellagra. Outline an experiment (using either pellagra patients or a suitable animal model) to test that hypothesis.
3. List the probable contributing factors to the pellagra observed in these patients.

13. STUDY QUESTIONS AND EXERCISES

1. Diagram the several general areas of metabolism in which NAD(H)- and NADP(H)-dependent enzymes are involved.
2. In general, how do the pyridine nucleotides interact with the flavoproteins in metabolism? What is the fundamental metabolic significance of this interrelationship?
3. Construct a decision tree for the diagnosis of niacin deficiency in humans or an animal species.
4. What key feature of the chemistry of NAm relates to its biochemical functions as an enzyme cosubstrate?
5. What parameters might you use to assess niacin status of a human or animal?
6. Does niacin meet the definition of a vitamin? Why or why not?

RECOMMENDED READING

Al-Mohaissen, M.A., Pun, S.C., Frohlish, J.J., 2010. Niacin: from mechanisms of action to therapeutic uses. Mini Rev. Med. Chem. 10, 204–207.

Alsheikh-Ali, A.A., Karas, R.H., 2008. The safety of niacin in the US Food and Drug Administration adverse effect reporting database. Am. J. Cardiol. 101S, 9B–13B.

Brooks, E.L., Kuvin, J.T., Karas, R.H., 2010. Niacin's role in the statin era. Expert Opin. Pharmacother. 11, 2291–2300.

Bruckert, E., Labreuche, J., Amarenco, P., 2010. Meta-analysis of the effect of nicotinic acid alone or in combination on cardiovascular events and atherosclerosis. Atherosclerosis 210, 353–361.

Farmer, J.S., 2009. Nicotinic acid: a new look at an old drug. Curr. Atheroscler. Rep. 11, 87–92.

Fu, L., Doreswarmy, V., Prakash, R., 2014. The biochemical pathways of central nervous system neural degeneration in niacin deficiency. Neural Regen. Res. 9, 1509–1513.

Julius, U., Fischer, S., 2013. Nicotinic acid as a lipid-modifying drug – a review. Atheroscler. Suppl. 14, 7–13.

Kamanna, V.S., Ganji, S.H., Kashyap, M.L., 2009. The mechanism and mitigation of niacin-induced flushing. Int. J. Clin. Pract. 63, 1369–1377.

Kirkland, J.B., 2014. Niacin. In: Zemplini, J., Suttie, J.W., Gregory, J.F., et al. (Eds.), Handbook of Vitamins, fifth ed. CRC Press, New York, pp. 149–190 (Chapter 5).

Lavinge, P.M., Karas, R.H., 2013. The current state of niacin in cardiovascular disease prevention: a systematic review and meta-regression. J. Am. Coll. Cardiol. 61, 440–446.

Maiese, K., Chong, Z.Z., Hou, J., et al., 2009. The vitamin nicotinamide: translating nutrition into clinical care. Molecules 14, 3446–3485.

Markel, A., 2011. The resurgence of niacin: from nicotinic acid to niaspan/laropiprant. Ind. Med. Assoc. J. 13, 368–374.

Messamore, E., Hoffman, W.F., et al., 2010. Niacin sensitivity and the arachidonic acid pathway in schizophrenia. Schizophr. Res. 122, 248–256.

Niehoff, I.D., Hüther, L., Lebzien, P., 2009. Niacin for dairy cattle: a review. Br. J. Nutr. 101, 5–19.

Penberthy, W.T., Kirkland, J.B., 2012. Niacin. In: (Erdman Jr., J.W., Macdonald, I.A., Zeisel, S.H. (Eds.), Present Knowledge in Nutrition, tenth ed. Wiley-Blackwell, New York, pp. 293–306 (Chapter 19).

Song, W.L., FitzGerald, G.A., 2013. Niacin, an old drug with a new twist. J. Lipid Res. 54, 2586–2594.

Vosper, H., 2009. Niacin: a re-emerging pharmaceutical for the treatment of dyslipidemia. Br. J. Pharmacol. 158, 429–441.

Chapter 14

Vitamin B_6

Chapter Outline

Anchoring Concepts

1. Vitamin B_6 is the generic descriptor for all 3-hydroxy-2-methylpyridine derivatives exhibiting the biological activity of Pn [3-hydroxy-4,5-*bis*(hydroxymethyl)-2-methylpyridine].
2. The metabolically active form of vitamin B_6 is pyridoxal phosphate, which functions as a coenzyme for reactions involving amino acids.
3. Deficiencies of vitamin B_6 are manifested as dermatologic, circulatory, and neurologic changes.

Had we been able to afford Monel metal or stainless steel cages, we would have missed xanthurenic acid.

Samuel Lepkovsky[1]

LEARNING OBJECTIVES

1. To understand the chief natural sources of vitamin B_6.
2. To understand the means of absorption and transport of vitamin B_6.
3. To understand the biochemical function of vitamin B_6 as a coenzyme of different reactions in the metabolism of amino acids, and the relationship of that function to the physiological activities of the vitamin.
4. To understand the physiological implications of low-vitamin B_6 status.

VOCABULARY

Acrodynia
Aldehyde dehydrogenase (NAD+)
Aldehyde oxidase
Alkaline phosphatase
γ-Aminobutyric acid (GABA)
Anthranilic acid
Cheilosis
C-reactive protein
Cystathionine β-synthase
Cystathionine γ-lyase
Cystathioninuria
Decarboxylases
Deoxypyridoxine
Epinephrine
Erythrocyte aspartate aminotransferase (EAAT)
Glossitis
Glycine decarboxylase
Glycogen phosphorylase
Hemoglobin
Homocystinuria
Histamine
Hydrogen sulfide
Hyperoxaluria
Isonicotinic acid hydrazide (INH)
Kynureninase
Methionine load
Norepinephrine

1. Samuel Lepkovsky (1899–1984) was a Polish-born American nutritionist who spent his professional career on the faculty of the University of California at Berkeley. He conducted pioneering work in nutrition, biochemistry, and physiology and is best known for winning the "race for the filtrate factor" by isolating and crystallizing pyridoxine. His observation of a green pigment below the cages of his deficient test animals led to his discovering xanthurenic acid (which reacted with ferric iron on the rusted cage bottoms) in the urine of deficient animals.

The Vitamins. http://dx.doi.org/10.1016/B978-0-12-802965-7.00014-9

Phosphorylases
Premenstrual syndrome
Pyridoxal (Pal)
Pyridoxal dehydrogenase
Pyridoxal kinase
Pyridoxal oxidase
Pyridoxal phosphate (PalP)
Pyridoxal phosphate synthase
Pyridoxamine (Pm)
Pyridoxamine phosphate (PmP)
Pyridoxamine phosphate oxidase
4-Pyridoxic acid
Pyridoxine (Pn)
Pyridoxine glycosides
Pyridoxine hydrochloride
Pyridoxol
Racemases
Schiff base
Schizophrenia
Selenocysteine β-lyase
Selenocysteine γ-lyase
Serine hydroxymethyl transferase
Serine palmitoyltransferase
Serotonin
Sickle cell anemia
Sideroblastic anemia
Steroid hormone–receptor complex
Stomatitis
Transaminases
Tryptophan load test
Vitamin B_6-responsive seizures Xanthurenic acid

1. THE SIGNIFICANCE OF VITAMIN B_6

The biological functions of the three naturally occurring forms of vitamin B_6, **pyridoxine (Pn)**, **pyridoxal (Pal)**, and **pyridoxamine (Pm)**, depend on the metabolism of each to a common coenzyme form, **pyridoxal phosphate (PalP)**. That coenzyme plays critical roles in several aspects of metabolism, giving the vitamin importance in such diverse areas as growth, cognitive development, depression, immune function, fatigue, and steroid hormone activity. Vitamin B_6 is fairly widespread in foods of both plant and animal origin; therefore, problems of primary deficiency are not prevalent. Still, vitamin B_6 status can be antagonized by alcohol and other factors that displace the coenzyme from its various enzymes to increase the rate of its metabolic degradation.

2. PROPERTIES OF VITAMIN B_6

The term **vitamin B_6** describes all 3-hydroxy-2-methylpyridine derivatives exhibiting the biological activity of pyridoxine in the rat. The term **pyridoxine (Pn)** is used to designate one of the vitamin B_6-active compounds, 3-hydroxy-4,5-bis(hydroxymethyl)-2-methylpyridine. Derivatives of 3-hydroxy-2-methyl-5-hydroxypyridine can be biologically active if they have a phosphorylatable 5-hydroxymethyl group, and a substituent on the ring-carbon *para* to the pyridine nitrogen that can be converted to an aldehyde. Two such analogs are biologically active: the aldehyde **pyridoxal (Pal)** and the amine **pyridoxamine (Pm)**.

Structures of vitamin B_6

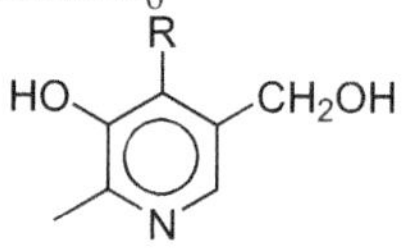

Pyridoxine (Pn): R=CH_2OH
Pyridoxal (Pal): R=CHO
Pyridoxoic acid: R=COOH
Pyridoxamine (Pm): R=CH_2NH_2

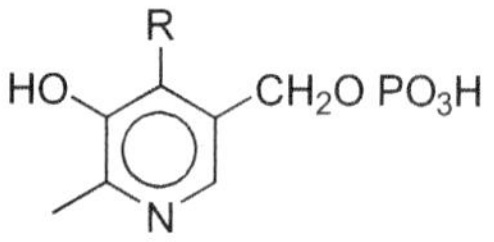

Pyridoxal phosphate (PalP): R=CHO
Pyridoxamine phosphate (PmP): R=CH_2NH_2

Vitamin B_6 Chemistry

Vitamers B_6 are colorless crystals at room temperature. Each is very soluble in water, weakly soluble in ethanol, and either insoluble or sparingly soluble in chloroform. Each is fairly stable in dry form and in solution. Pyridoxine is oxidized in vivo and under mild oxidizing conditions in vitro to yield pyridoxal. The prominent feature of the chemical reactivity of pyridoxal is the ability of its aldehyde group to react with primary amino groups (e.g., of amino acids) to form Schiff bases. The electron-withdrawing effect of the resulting Schiff base labilizes the other bonds on the bound carbon, thus serving as the basis of the catalytic roles of pyridoxal and pyridoxamine.

3. SOURCES OF VITAMIN B_6

Hindgut Microbial Synthesis

Vitamin B_6 can be produced by the microbiome of the colon. A genomic analysis of 256 representative organisms of the human gut microbiota found half capable of de novo synthesis of the vitamin.[2] In addition, the majority of genomes (all phyla except Fusobacterial) also showed the capacity to salvage PalP and its precursors. Those findings suggested that hindgut microbial synthesis may produce as much as

2. Magnúsdóttir, S., Ravchee, D., de Crécy-Lagard, V., et al., 2015. Front. Genet. 6, 148–166.

86% of the daily human need for vitamin B_6, a finding consistent with the fact that primary deficiency of the vitamin is rarely observed. Colonocyte uptake of the vitamin has been demonstrated;[3] however, it is likely that a large portion of the microbially produced vitamin is taken up by nonsynthesizing microbes such that noncoprophagous animals derive little benefit from this source of the vitamin. In contrast, ruminants benefit from their rumen microflora, which produces vitamin B_6 in adequate amounts to meet their needs proximal to where it is absorbed.

Distribution in Foods

Vitamin B_6 is widely distributed in foods, occurring in greatest concentrations in meats, whole grain products (especially wheat), vegetables, and nuts (Table 14.1). In the cereal grains, vitamin B_6 is concentrated primarily in the germ and aleuronic layer. Thus, the refining of grains in the production of flours, which removes much of these fractions, results in substantial reductions in vitamin B_6 content. White bread, therefore, is a poor source of vitamin B_6 unless it is fortified.

The chemical forms of vitamin B_6 tend to vary among foods of plant and animal origin; plant tissues contain mostly Pn (the free alcohol form), whereas animal tissues contain mostly Pal and Pm. A large portion of the vitamin B_6 in many foods is phosphorylated or bound to proteins via the ε-amino groups of lysyl residues or the sulfhydryl groups of cysteinyl residues. The vitamin is also found in glycosylated forms such as 5′-*O*-(β-D-glucopyranosyl) Pn. Vitamin B_6 glycosides are found in varying amounts in different foods but little, if at all, in animal products.

Stability

Vitamin B_6 in foods is stable under acidic conditions but unstable under neutral and alkaline conditions, particularly when exposed to heat or light. Of the several vitamers, Pn is far more stable than either Pal or Pm. Therefore, the cooking and thermal processing losses of vitamin B_6 tend to be highly variable (0–70%), with plant-derived foods (which contain mostly Pn) losing little, if any, of the vitamin and animal products (which contain mostly Pal and Pm) losing substantial amounts. Milk, for example, can lose 30–70% of its inherent vitamin B_6 on drying. The storage losses of naturally occurring vitamin B_6 from many foods and feedstuffs, although they occur at slower rates, can also be substantial (25–50% within a year). Because it is particularly stable, **pyridoxine hydrochloride** is used for food fortification and in multivitamin supplements.

3. Said, Z.M., Subramanian, V.S., Vaziri, N.D., et al., 2008. Am. J. Physiol. Cell Physiol. 294, C1192–C1197.

Bioavailability

The bioavailability of vitamin B_6 in most commonly consumed foods appears to be in the range of 70–80%. However, appreciable amounts of the vitamin in some foods are not biologically available. The determinants of the bioavailability of vitamin B_6 in a food include as follows:

- **Pn glycoside content**—The pyridoxal-5′-β-D-glycosides are poorly digested, being taken up intact, and converted to Pn by a hydrolase in the cytosol. Compared to free Pn, the bioavailabilities of the glycosides have been estimated to be 20–30% in the rat and about 60% in humans. In addition, the presence of Pn glycosides has been found to reduce the utilization of coingested free Pn.
- **Peptide adducts**—Vitamin B_6 can condense with peptide lysyl and/or cysteinyl residues during food processing, cooking, or digestion; such products are less well utilized than the free vitamin. The reductive binding of Pal and Pal 5′-phosphate to e-amino groups of lysyl residues in proteins or peptides produces adducts that are not only biologically unavailable but that also have vitamin B_6-antagonist activity.[4] For example, linatine, the dipeptide of glutamic acid and 1-amino-D-proline in flaxseed, impairs homocysteine metabolism in moderately vitamin B_6-deficient rats.[5] Wheat bran also contains vitamin B_6 in largely unavailable form(s), the presence of which reduces the bioavailability of the vitamin from other foods consumed at the same time.[6] Because plants generally contain complexed forms of Pn, bioavailability of the vitamin of plant foods tends to be greater than that of foods derived from animals.

4. ABSORPTION OF VITAMIN B_6

Digestion of Food Forms

The enteric absorption of vitamin B_6 depends on ingested forms being converted largely to Pm, Pn, and Pal. This involves digestion of the binding proteins in foods followed by metabolism by brush border enzymes:

- **dephosphorylation** of protein-bound PalP and PmP (the major species in animal products) by **alkaline phosphatase** and other phosphatases;
- **deglycosylation** by **lactase–phlorizin hydrolase**.

4. Gregory, J.F., 1980. J. Nutr. 110, 995–1005.
5. Mayengbam, S., Raposo, S., Aliani, M., et al., 2015. J. Nutr. 146, 14–20.
6. Vitamin B_6 is poorly available from the bran fraction of the grain; therefore, the bioavailability of the vitamin from whole wheat bread is less than that of Pn-fortified white bread.

TABLE 14.1 Vitamin B_6 Contents of Foods

Food	Total Vitamin B_6, mg/100g[a]	Vitamin B_6 Vitamers[b]			
		% Glycosylated	% Pn	% Pal	% Pm
Dairy Products					
Milk	0.046–0.05		3	76	21
Yogurt	0.06				
Cheeses	0.01–0.42	4	8	88	
Meats					
Beef	0.13–0.81		16	53	31
Chicken	0.25–0.52	7	74	19	
Lamb	0.13				
Pork	0.28–0.74		8	8	84
Fish					
Haddock	0.33				
Herring	0.17–0.41				
Oysters	0.05–0.10				
Salmon	0.28–0.83		2	9	89
Shrimp	0.10–0.30				
Tuna	0.53		19	69	12
Cereals					
Barley, pearled	0.12		52	42	6
Corn meal	0.30		11	51	38
Oats	0.12		12	49	39
Rice, white	0.09	20	64	19	17
Rice, brown	0.12	23	78	12	10
Wheat, flour, whole	1.30	28	71	16	13
Wheat, white flour	0.04		55	24	21
Vegetables					
Asparagus	0.08				
Beans	0.06–0.20	15–57	62	20	18
Broccoli	0.18	66			
Cabbage	0.12	46	61	31	8
Carrots	0.14	51–86	75	19	6
Cauliflower	0.18	66	16	79	5
Celery	0.07				
Corn	0.09		6	68	26
Onions	0.12				
Peas	0.17	15	47	47	6
Potatoes	0.30	32	68	18	14
Spinach	0.20	50	36	49	15

Continued

TABLE 14.1 Vitamin B_6 Contents of Foods—cont'd

Food	Total Vitamin B_6, mg/100g[a]	Vitamin B_6 Vitamers[b]			
		% Glycosylated	% Pn	% Pal	% Pm
Fruits					
Apples	0.04		61	31	8
Grapefruit	0.04				
Oranges	0.08	47	59	26	15
Peaches	0.03	22	61	30	9
Strawberries	0.05				
Tomatoes	0.08	46	38	29	15
Nuts					
Almonds	0.14				
Pecans	0.19		71	12	17
Walnuts	0.58	7	31	65	4
Other					
Eggs	0.17		0	85	15
Human milk	0.01				

[a]*USDA National Nutrient Database for Standard Reference, Release 28 (http://www.ars.usda.gov/ba/bhnrc/ndl).*
[b]*Leklem, J.E., 1996. In: Ziegler, E.E., Filer, L.J. (Eds.), Present Understanding in Nutrition, seventh ed. ILSI Press, Washington, DC, p. 75; Orr, M.L., 1969. In: Foods: Home Economics Res. Rep. 36. USDA, Washington, DC, 52 pp.*

Diffusion Linked to Phosphorylation

The vitamers Pn, Pal, Pm as well as some Pn glycosides can be absorbed by passive diffusion throughout the gut. For Pn and Pal, the process is driven by the intracellular trapping of the vitamin via the formation of 5′-phosphates through that action of a cytosolic ATP-dependent **Pal kinase**. Intact Pn glycosides taken up by diffusion are later converted to Pn by cytosolic β-glucosidase and then oxidized to PalP.

Uptake by Facilitated Transport

There is also evidence for carrier-mediated absorption of the vitamin.[7]

5. TRANSPORT OF VITAMIN B_6

Plasma Vitamin B_6

Vitamin B_6 is transported primarily in the plasma. Most (>90%) of the circulating vitamin is PalP derived from the hepatic turnover of flavoenzymes. Plasma PalP, typically <1 mmol, comprises a small portion (<0.1%) of total body vitamin B_6. The circulating vitamin is tightly bound to albumin and other plasma proteins via Schiff base linkages.[8]

The vitamin is present in erythrocytes at more than six times the levels in plasma. In erythrocytes, it forms a Schiff base with hemoglobin by binding to the amino group of the N-terminal valine residue of the hemoglobin α-chain. This binding, twice as strong as that to albumin, drives uptake of the vitamin by erythrocytes. Erythrocyte vitamin B_6 levels are particularly high in infants but decline to adult levels by about 5 years of age. PalP content of erythrocytes is often used as a parameter of vitamin B_6 status. In humans and other animals, plasma PalP concentrations decline during pregnancy, as a result of a shift in the distribution of the vitamer in favor of erythrocytes over plasma, as neither the absorption, excretion, or hepatic uptake of the vitamin is affected. Renal failure has been found to reduce the plasma PalP level;[9] whereas, submaximal exercise has been shown to increase it.

7. Said, Z.M., Subramanian, V.S., Vaziri, N.D., et al., 2008. Am. J. Physiol. Cell Physiol. 294, C1192–C1197.

8. Schiff bases are condensation products of aldehydes and ketones with primary amines; they are stable if there is at least one aryl group on either the N or the C that is linked. Vitamin B_6 forms Schiff base linkages with proteins by the bonding of the keto-C of PalP to a peptidyl amino (-NH_2) group. The vitamin also forms a Schiff base with the amino acid substrates of the enzymes for which it functions as a coenzyme; this occurs by the bonding of the amino nitrogen of PalP and the α-C of the substrate.

9. One study showed this depression to be >40% in rats.

TABLE 14.2 Concentrations (nM) of Vitamers B_6 in the Plasma of Several Species

Species	Pal	PalP	Pol	Pm	PmP	Pyridoxic Acid
Pig	29	139	167	—	—	139
Human	62	13	33	6	<3	40
Calf	308	96	50	—	9	91
Sheep	626	57	43	—	466	318
Dog	417	268	66	—	65	109
Cat	2443	139	93	44	271	17

Cellular Uptake

Pal crosses cell membranes more readily than PalP. Its preferential uptake by tissues suggests roles of phosphatases in the cellular retention of the vitamin. After being taken into the cell, the vitamin is phosphorylated by Pal kinase to yield the predominant tissue form, PalP. Small quantities of vitamin B_6 are stored, mainly as PalP, but also as **Pm phosphate**.

Tissue Distribution

The total body pool of vitamin B_6 in the human adult is estimated to be 40–150 mg, constituting a supply sufficient to satisfy normal needs for 20–75 days. This amount is composed of two pools: one with a rapid turnover rate (0.5 day) and a second with a longer turnover rate (25–33 days).[10] Muscle contains most (70–80%) of body's vitamin B_6 in the form of Pal 5′-phosphate bound to **glycogen phosphorylase**. Other tissues (liver, brain, kidney, and spleen) (Table 14.2) contain the vitamin bound to other proteins (e.g., glycogen phosphorylase binding accounts for only 10% of the vitamin in liver) with which it has coenzyme functions. Protein binding is thought to protect PalP from hydrolysis while providing storage of the vitamin.

Moderate exercise has been found to increase plasma PalP concentrations substantially, e.g., by >20% within 20 min. This appears to be related to the increased need for gluconeogenesis, which results in the release of PalP from glycogen phosphorylase. The rapidity of this response suggests either that the vitamer rapidly undergoes hydrolysis, discharge from the muscle, and then rephosphorylation in the liver, or that it is released intact through interstitial fluid.[11]

10. Shane, B., 1978. Human Vitamin B_6 Requirements. National Academies Press, Washington, DC, pp. 111–128.
11. Crozier, P.G., Coredain, L., Sampson, D.A., 1994. Am. J. Clin. Nutr. 40, 552–558.

6. METABOLISM OF VITAMIN B_6

Interconversion of Vitamers

The vitamers B_6 are readily interconverted metabolically by phosphorylation/dephosphorylation, oxidation/reduction, and amination/deamination (Fig. 14.1). Because the non-phosphorylated vitamers cross membranes more readily than their phosphorylated analogues, phosphorylation appears to be an important means of retaining the vitamin intracellularly. Several enzymes are involved in this metabolism:

- **Pyridoxal kinase**—This hepatic enzyme catalyzes the phosphorylation of Pn, Pal, and Pm, yielding the corresponding phosphates. It requires a Zn–ATP complex, the formation of which is facilitated by Zn–metallothionein (MT), and is stimulated by K^+. The role of MT in Pal kinase activity suggests that Zn status may be important in the regulation of vitamin B_6 metabolism. Erythrocyte Pal kinase activity in African-Americans has been reported to be about half that of white Americans, although lymphocytes, granulocytes, and fibroblasts show no such differences. This may indicate reduced erythrocyte retention of vitamin B_6, which depends on the phosphorylation. Pyridoxal kinase also binds the antianxiety drug benzodiazepine,[12] suggesting that the mode of drug action may involve enhancement of neuronal γ-aminobutyrate levels.
- **Alkaline phosphatases**—Phosphorylated forms of the vitamin can be dephosphorylated by membrane-bound alkaline phosphatases in many tissues (e.g., liver, brain, and intestine).
- **Pyridoxamine phosphate oxidase**—This enzyme catalyzes the limiting step in vitamin B_6 metabolism. It requires flavin mononucleotide (FMN); therefore,

12. Hanna, M.C., Turner, A.J., Kirkness, E.F., 1997. J. Biol. Chem. 272, 10756–10760.

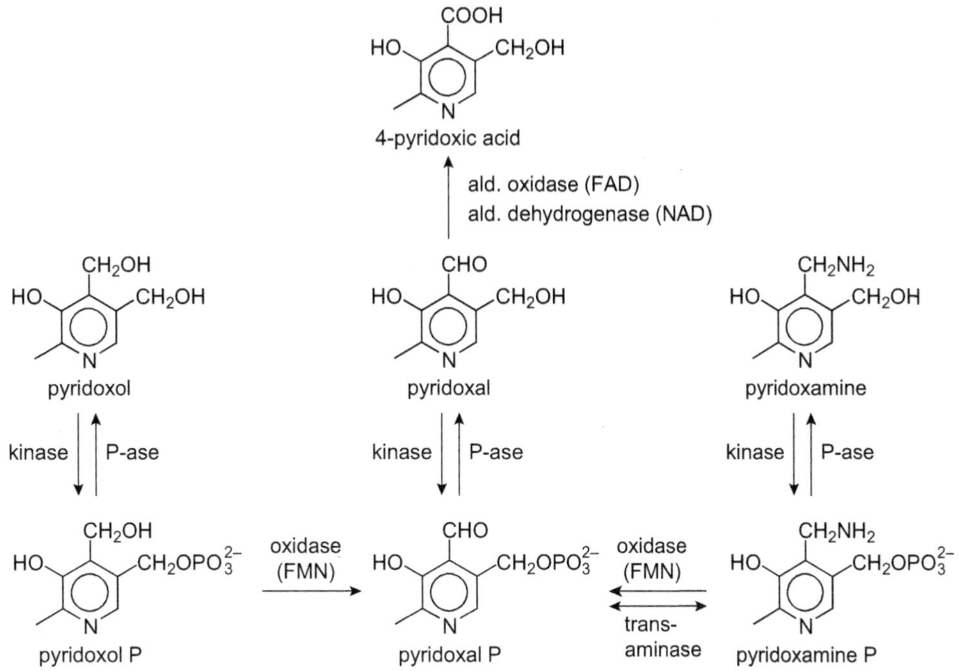

FIGURE 14.1 Metabolism of vitamin B_6.

deprivation of riboflavin may reduce the conversion of Pn and Pm to the active coenzyme PalP.

- **Pyridoxal-5′-phosphate synthase**[13]—This enzyme catalyzes the oxidation of PnP and PalP to PalP.

The liver is the central organ for vitamin B_6 metabolism, containing all of the enzymes involved in its interconversions. The major forms of the vitamin in that organ are PalP and PmP, which are maintained at fairly constant intracellular concentrations in endogenous pools that are not readily accessible to newly formed molecules of those species. The latter, instead, comprise a second pool that is readily mobilized for metabolic conversion (mostly to PalP, Pal, and pyridoxic acid) and release to the blood.

Catabolism

Pyridoxal phosphate is dephosphorylated and oxidized primarily in the liver by the FAD-dependent **aldehyde oxidase** as well as the NAD-dependent **aldehyde dehydrogenase** to yield **4-pyridoxic acid**, the major excretory metabolite. At high intakes, 5-pyridoxic acid is also produced and excreted. Catabolism of the vitamin increases under conditions of systemic inflammation.[14] This is manifest as reductions in PalP levels in several tissues, apparently a consequence of movement of the coenzyme to sites of inflammation where it can function in PalP-dependent enzymes.[15]

Excretion

It has been estimated that humans oxidize 40–60% of ingested vitamin B_6 to 4-pyridoxic acid. In the rat, urinary excretion of 4-pyridoxic acid increases with age in parallel with increases in the hepatic activities of Pal oxidase and Pal dehydrogenase. Small amounts of Pal, Pm, and Pn and their phosphates, as well as the lactone of pyridoxic acid and a ureido–pyridoxyl complex,[16] are also excreted when high doses of the vitamin have been given.[17] Urinary levels of 4-pyridoxic acid are inversely related to protein intake (Table 14.3). This effect appears to be greater for women than for men. However, 4-pyridoxic acid is not detectable in the urine of vitamin B_6-deficient subjects, making it useful in the clinical assessment of vitamin B_6 status.[18,19]

13. i.e., Pyridoxal/pyridoxamine phosphate oxidase.
14. Ulvik, A., Midttun, Ø., Pedersen, E.R., et al., 2014. Am. J. Clin. Nutr. 100, 250–255.
15. Paul, L., Ueland, P.B., Selhub, J., 2014. Nutr. Rev. 71, 239–244.
16. This is formed by the reaction of an amino group of urea with a hydroxyl group of the hemiacetal form of the aldehyde at position 4 of Pal.
17. For example, humans given 100 mg of Pal excrete about 60 mg 4-pyridoxic acid and 2 mg Pal over the next 24 h.
18. In humans, excretion of less than 0.5 mg/day (men) or 0.4 mg/day (women) is considered indicative of inadequate intake of the vitamin. Typical excretion of total vitamin B_6 by adequately nourished humans is 1.2–2.4 mg/day. Of that amount, 0.5–1.2 mg (men) or 0.4–1.1 mg (women) is in the form of 4-pyridoxic acid.
19. Although no explanation has been offered for the correlation, it is of interest that excretion of relatively low amounts (<0.81 mg/24 h) of 4-pyridoxic acid is associated with increased risk of relapse after mastectomy.

TABLE 14.3 Effect of Protein Intake on Vitamin B_6 Status

Dietary Treatment	Intakes		
Protein intake (g/kg)	0.5	1.0	2.0
Vitamin B_6 intake (mg/g protein)	0.04	0.02	0.01
Parameter (adequate value)	% Subjects with low values		
Urinary 4-pyridoxic acid (>3 mmol/day)	11	22	78
Urinary total vitamin B_6 (>0.5 mmol/day)	56	56	67
Plasma PalP (>30 nmol/L)	33	67	78
Urinary xanthurenic acid (<65 mmol/day)	11	11	44

Adapted from Hansen, C.M., Leklem, J.E., Miller, L.T., 1996. J. Nutr. 126, 1891–1901.

Effects of Alcohol and Other Drugs

Several drugs can antagonize vitamin B_6. Among these is alcohol; its degradation product, acetaldehyde, displaces PalP from proteins, resulting in enhanced catabolism of the coenzyme. Acetaldehyde also stimulates the activity of alkaline phosphatase, enhancing the dephosphorylation of PalP. The antituberculosis drug **isonicotinic acid hydrazide**[20] **(INH)** also antagonizes vitamin B_6; it does so by binding the vitamin directly. For this reason, vitamin B_6 must be given to patients treated with INH. Pal kinase binds the antianxiety drug benzodiazepine and can be inhibited by the antiasthmatic drug theophylline. Short-term theophylline therapy induces biochemical signs of vitamin B_6 deficiency due to this effect.

7. METABOLIC FUNCTIONS OF VITAMIN B_6

The metabolically active form of vitamin B_6 is PalP. That form serves as a coenzyme of more than 140 enzymes, most of which are involved in the metabolism of amino acids (Fig. 14.2), functioning in the decarboxylations; transamination, racemization, elimination, and replacement reactions; and β-group interconversions. This group of enzymes shows considerable variation in affinity for PalP, some being more sensitive to deprivation of vitamin B_6 than others.

20. Isoniazid.

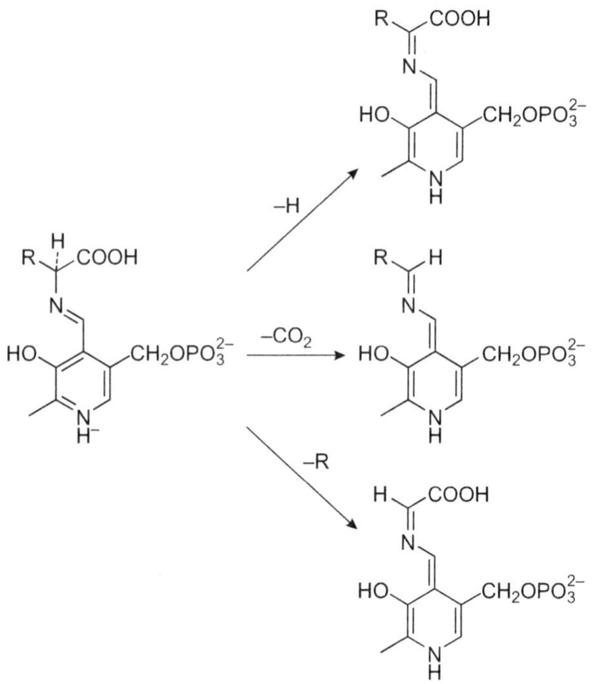

FIGURE 14.2 General reactions of PalP-dependent enzymes in amino acid metabolism.

Mechanisms of Action

Vitamin B_6-dependent enzymes have structural similarities in their coenzyme-binding regions at which PalP or Pm phosphate is bound through the formation of a Schiff base. Accordingly, the mechanisms of the reactions catalyzed by the vitamin B_6-dependent enzymes are also similar. Each involves the binding of an α-carbon of an α-amino acid substrate to the pyridine nitrogen of PalP. The delocalization of the electrons from the α-carbon by the action of the protonated pyridine nitrogen as an electron sink results in the conversion of the former to a carbanion (C^-) at the α-carbon and the labilization of its bonds. This results in the heterolytic cleavage of one of the three bonds to the α-carbon (Table 14.4, Fig. 14.2). The particular bond to be cleaved is determined by the particular PalP-dependent enzyme; each involves the loss of the cationic ligand of an amino acid.

Amino Acid Metabolism

PalP is involved in practically all reactions involved in amino acid metabolism being involved in both their biosynthesis as well as their catabolism:

- **Transaminations**—PalP-dependent transaminases catabolize most amino acids[21]. The response of

21. The only amino acids that are not substrates for PalP-dependent transaminases are threonine, lysine, proline, and hydroxyproline.

TABLE 14.4 Important PalP-Dependent Enzymes of Animals

Type of Reaction	Enzyme
Decarboxylation	Aspartate 1-decarboxylase
	Glutamate decarboxylase
	Ornithine decarboxylase
	Aromatic amino acid decarboxylase
	Histidine decarboxylase
R-group interconversion	Serine hydroxymethyltransferase
	δ-Aminolevulinic acid synthase
Transamination	Aspartate aminotransferase
	Alanine aminotransferase
	γ-Aminobutyrate aminotransferase
	Cysteine aminotransferase
	Tyrosine aminotransferase
	Leucine aminotransferase
	Ornithine aminotransferase
	Glutamine aminotransferase
	Branched-chain amino acid aminotransferase
	Serine–pyruvate aminotransferase
	Aromatic amino acid transferase
	Histidine aminotransferase
Racemization	Cystathionine β-synthase
α,β-Elimination	Serine dehydratase
γ-Elimination	Cystathionine γ-lyase
	Kynureninase

erythrocyte aspartate aminotransferase (EAAT) to in vitro additions of PalP has been used as a biochemical maker of vitamin B_6 status.[22]

- **Transsulfuration**—PalP-dependent enzymes **cystathionine β-synthase** and **cystathionine γ-lyase** catalyze the transsulfuration of methionine to cysteine. Vitamin B_6 deprivation, therefore, reduces the activities of these enzymes, affected individuals show **homocystinuria** (due to impaired conversion to cystathionine) and **cystathioninuria** (due to impaired cleavage of cystathionine to cysteine and α-ketobutyrate). These conditions can be exacerbated for diagnostic purposes by the use of an oral methionine load. Plasma homocysteine concentrations, however, usually do not change in vitamin B_6 deficiency and are therefore not suitable for assessment of vitamin B_6 status.
- **Selenoaminoacid metabolism**—Vitamin B_6 is essential for the utilization of selenium (Se) from the major dietary form, selenomethionine, after that Se is transferred to selenohomocysteine. PalP is a cofactor for two enzymes, **selenocysteine ß-lyase** and **selenocysteine γ-lyase**, which catalyze the elimination of the Se from selenohomocysteine to yield hydrogen selenide (H_2Se). Selenide is the obligate precursor for the incorporation of Se into selenoproteins in the form of selenocysteinyl residues produced during translation.[23]

Single-Carbon Metabolism

PalP-dependent enzymes function in single-carbon metabolism (Fig. 14.3):

- **Serine hydroxymethyl transferase (SHMT)**, in both the cytoplasm and mitochondria, catalyzes the interconversion of glycine and serine, providing single-carbon units to the folate single-carbon pool;
- **Glycine decarboxylase** catalyzes the transfer of single-carbon units to/from tetrahydrofolate, working with SHMT feed the folate single-carbon pool in the form of 5,10-methylenetetrahydrofolate;
- **Cystathionine β-synthetase** catalyzes this first step in the transsulfuration pathway, the addition of serine and homocysteine to yield cystathionine; it is stimulated by S-adenosylmethionine and is minimally affected by vitamin B_6 deprivation;
- **Cystathionine γ-lyase** catalyzes the second step in the transsulfuration pathway, the cleavage of cystathionine to yield cysteine.

Niacin Synthesis

Vitamin B_6 is an essential cofactor for two key enzymes in the synthesis of the vitamin niacin from the indispensable amino acid tryptophan (Fig. 14.4):

- **Kynureninase**—This enzyme catalyzes the removal of an alanyl residue from 3-hydroxykynurenine in the metabolism of tryptophan to the branch point intermediate α-amino-β-carboxymuconic-ε-semialdehyde in the

22. However, EAAT activity coefficients can be affected by factors unrelated to vitamin B_6 status (e.g., intake of protein and alcohol, differences in body protein turnover, certain drugs, genetic polymorphism of the enzyme), which can compromise its use without careful controls.

23. These include the Se-dependent glutathione peroxidases and thioredoxin reductases, which have antioxidant functions; the iodothyronine 5′-deiodinases, which are involved in thyroid hormone metabolism; selenophosphate synthase, which is involved in selenoprotein synthesis; selenoproteins P and W, which are major selenoproteins in plasma and muscle, respectively; and at least a dozen other proteins.

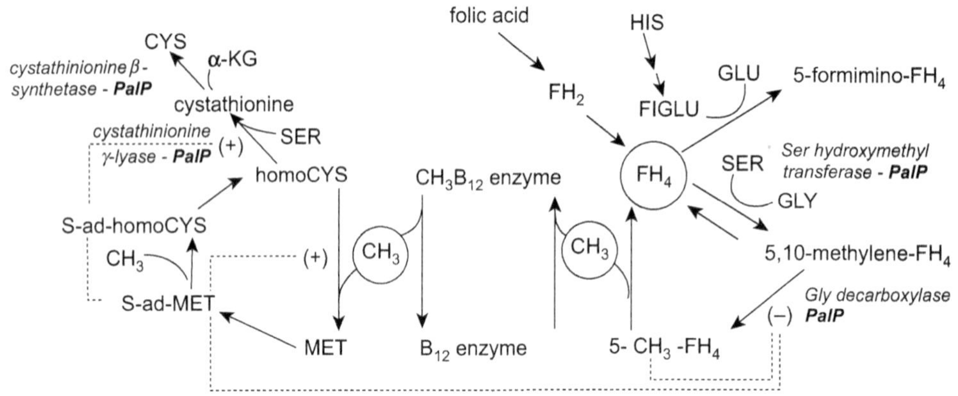

FIGURE 14.3 PalP-dependent enzymes (labeled) in single-carbon metabolism.

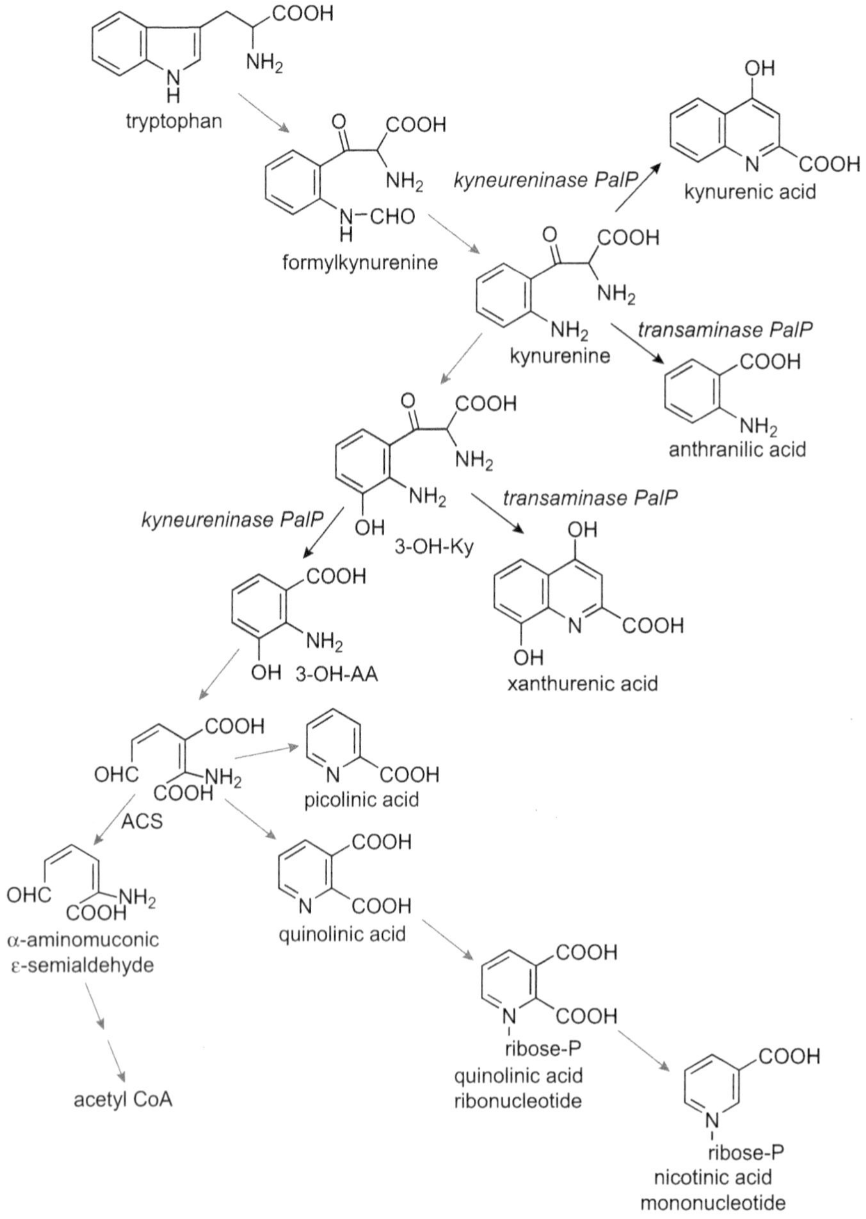

FIGURE 14.4 PalP-dependent enzymes (labeled) in the conversion of tryptophan to NAD.

tryptophan–niacin conversion pathway. Kynureninase also catalyzes the analogous reaction (removal of alanine) using nonhydroxylated kynurenine as substrate and yielding the nonhydroxylated analog of 3-hydroxykynurenine, anthranilic acid.

- **Transaminases**—Vitamin B_6-dependent transaminases metabolize kynurenine and 3-hydroxykynurenine, yielding kynurenic and xanthurenic acids, respectively. The transaminases have much greater binding affinities for PalP than kynureninase[24] and are, therefore, affected preferentially by vitamin B_6 deprivation. This results in blockage in the tryptophan–niacin pathway, with an accumulation of 3-hydroxykynurenine that gets diverted by transamination to yield xanthurenic acid, which appears in the urine.[25] This phenomenon is exploited in the assessment of vitamin B_6 status: deficiency is indicated by urinary excretion of xanthurenic acid after a tryptophan load.

Gluconeogenesis

Vitamin B_6 has two roles in gluconeogenesis:

- **Transaminations**—Amino acid catabolism depends on PalP is a cofactor for transaminases (see "amino acid metabolism" above).
- **Glycogen utilization**—PalP is required as a coenzyme of **glycogen phosphorylase**. Unlike its role in other PalP enzymes, it is the phosphate group of the coenzyme that is catalytically important. It participates in the transfer of inorganic phosphate to the glucose units of glycogen to produce glucose-1-phosphate, which is released. The shift of the enzyme from its inactive to active forms involves an increase in the binding (2–4 moles per mole of enzyme) of PalP. This accounts for more than half of the vitamin B_6 in the body, owing to the abundance of both muscle and glycogen phosphorylase (5% of soluble muscle protein).

24. The Michaelis constants (K_m's) for the transaminases are on the order of 10^{-8} M; whereas the K_m for kynureninase is on the order of 10^{-3} M.

25. Xanthurenic acid was discovered unexpectedly by Lepkovsky (University of California), who during the Great Depression sought to elucidate the nature of rat **adermin**. He wrote of his surprise in finding that the urine voided by his adermin-deficient rats was green, whereas that of his controls were the normal yellow color. In pursuing this observation, he found that urine from deficient animals was normally colored when voided but turned green only on exposure to the rusty dropping pans their limited budget had forced them to use. Thus, he recognized that adermin-deficient rats excreted a metabolite that reacted with Fe^{3+} to form a green derivative. This small event, which might have been missed by someone "too busy" to observe the experimental animals, resulted in Lepkovsky's identifying the metabolite as xanthurenic acid and discovering the role of vitamin B_6 in the tryptophan–niacin conversion pathway. His message: "The investigator has to do more than sit at his desk, outline experiments and examine data."

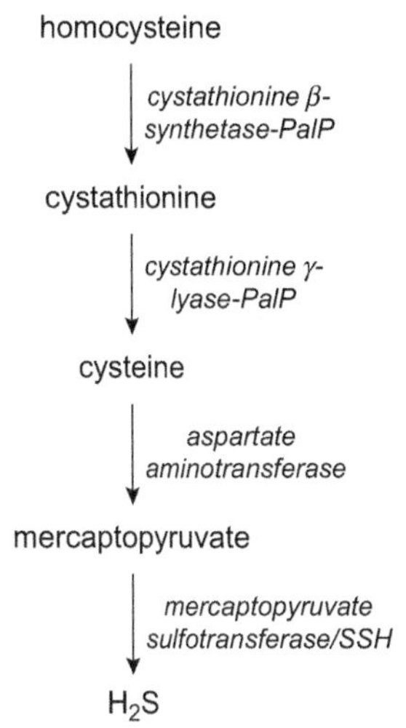

FIGURE 14.5 Role of PalP in the biogenesis of H_2S.

Hydrogen Sulfide Biogenesis

The endogenous mediator, hydrogen sulfide (H_2S), is produced in the heart, kidney, lungs, and the reproductive and central nervous systems. This is accomplished primarily by the transsulfuration pathway, i.e., the PalP enzymes, cystathionine β-synthetase and cystathionine γ-lyase,[26] plus a non-PalP enzyme, 3-mercaptopyruvate sulfotransferase, which can catalyze the desulfuration of cysteine or homocysteine to generate sulfane-S, subsequently reduced to H_2S (Fig. 14.5).[27] Through a process still poorly understood, H_2S is maintained at low, steady-state levels. It functions in paracrine intracellular signaling in the regulation of vascular tone, apoptosis, inflammation, and cellular stress responses. Patients with coronary heart disease have been found to produce significantly lower levels of H_2S than healthy controls, suggesting a role in vascular disease. It appears to exert these effects by regulating Ca^{2+} channels.[28]

Neurologic Function

Vitamin B_6 has a key role in the synthesis of the neurotransmitters: dopamine, norepinephrine, serotonin, and **γ-aminobutyric acid (GABA)**, as well as sphingolipids and polyamines. PalP-dependent enzymes function in the biosynthesis of neurotransmitters: tryptophan decarboxylase and aromatic L-amino acid decarboxylase in the synthesis of serotonin; tyrosine carboxylase in the synthesis of **epinephrine** and **norepinephrine**; glutamate decarboxylase serves in regulating turnover of a major source

26. The PalP-dependent aspartate aminotransferase, which can have cysteine aminotransferase activity, may also be involved.

27. Kabil, O., Banerjee, R., 2014. Antioxid. Redox. Signal. 20, 770–782.

28. Tan, H.H., Wong, P.T.H., Bian, J.S., 2010. Neurochem. Int. 56, 3–10.

of energy for the brain, GABA. The apoenzymes involved in these various steps have widely different affinities for PalP. Therefore, vitamin B_6 deprivation affects preferentially those decarboxylases with low affinities for the coenzyme. Accordingly, moderate deficiency of vitamin B_6 reduces brain serotonin levels without affecting other neurotransmitters.

Animal studies of long-term potentiation, a synaptic model of learning and memory, have revealed that maternal deprivation of the vitamin during gestation and lactation specifically reduces the development of the *N*-methyl-D-aspartate receptor subtype in the young. Although the metabolic basis is not understood, these effects appear to be related to the loss of dendritic arborization in vitamin B_6 deficiency. These lesions are thought to underlie reported effects of impaired learning on the part of the progeny of vitamin B_6-deficient animals and humans.

Histamine Synthesis

PalP functions in the metabolism of the vasodilator and gastric secretagogue histamine as a cofactor for histidine decarboxylase.

Hemoglobin Synthesis and Function

PalP functions in the synthesis of heme from porphyrin precursors as a cofactor for δ-aminolevulinic acid synthase. The vitamin also binds to hemoglobin at two sites on the β chains (the N-terminal valine and Lys-82 residues) and the N-terminal valine residues of the α chains. Binding of Pal or PalP at these sites enhances the O_2-binding capacity of the protein and inhibits the physical deformation of sickle cell hemoglobin.

Lipid Metabolism

Vitamin B_6 is required for the biosynthesis of sphingolipids via the PalP-dependent **serine palmitoyltransferase** and other enzymes in phospholipid synthesis. Diminution in the activities of these enzymes is thought to account for the changes observed in phospholipid contents of linoleic and arachidonic acids in vitamin B_6-deficient animals.

Antioxidant Function

Studies with cell systems have demonstrated antioxidant properties of B_6 vitamers. Pn, Pm, and PalP have been shown to reduce the production of superoxide (O_2^-) and lipid peroxides in response to prooxidative conditions.[29] These effects are thought to be due to the high reactivity of the vitamin with hydroxyl radicals, which preferentially abstract hydrogen atom from either of the vitamin's methanolic carbons (C8 or C9). These reactions can include additions and cyclizations such that the vitamin can be capable of high antioxidant activity by scavenging up to eight hydroxyl radicals.[30]

Cardiovascular Function

Vascular function. That vitamin B_6 plays a role in normal vascular function is evidenced by the fact that low-vitamin B_6 status has been associated with increased risk to coronary artery disease.[31] This relationship has been linked to altered platelet aggregation due to reduced Ca^{+2} influx caused by impaired adenosine-5′-diphosphate receptors and to increased chronic inflammation marked by elevated plasma levels of C-reactive protein.[32] Studies in a rat model found Pm to prevent age-related aortic stiffening and vascular resistance by reducing the formation of collagen cross-linking induced by advanced glycation end products.[33] Such effects would not appear to involve vitamin B_6-dependent enzymes; but, instead, the antioxidant capacity of Pm is to scavenge carbonyls and prevent glycation reactions.[34]

Hypertension. Deprivation of vitamin B_6 has been shown to produce moderate hypertension in the rat. These effects were associated with elevations in plasma levels of epinephrine and norepinephrine, and reduced levels of serotonin in the brain and 5-hydroxytryptophan in nerves.[35] Those observations, and the rapid reversibility of hypertension by vitamin B_6 supplementation, suggest that the condition results from impaired neurotransmitter regulation.

Homocysteinemia. Low-vitamin B_6 status can also cause homocysteinemia as a result of diminished conversion to cystathionine due to impaired activities of the PalP-dependent enzyme cystathionine β-synthase. Homocysteinemia has been associated with increased risks to occlusive vascular disease, total and cardiovascular disease-related mortality, stroke, and chronic heart failure.[36] Low-plasma PalP levels have also been associated with increased risk to vascular disease independent of plasma

29. Jain, S.K., Lim, G., 2001. Free Radic. Biol. Med. 30, 232–237; Mahfouz M.M., Zhou, S.Q., Kummerow, F.A., 2009. Int. J. Vitam. Nutr. Res. 79, 218–229.

30. Matxain, J.M., Padro, D., Ristilä, M., et al., 2009. J. Phys. Chem. B. 113, 9629–9632.

31. Robinson, K., Arheart, D., Refsum, H., 1998. Circulation 97, 437–443.

32. Morris, M.S., Sakakeeny, L., Jacques, P.F., et al., 2010. J. Nutr. 140, 103–110.

33. Wu, E.T., Liang, J.T., Wu, M.S., et al., 2011. Exp. Gerontol. 46, 482–488.

34. Voziyan, P.A., Hodson, B.G., 2005. Cell. Mol. Life Sci. 62, 1671–1681.

35. Viswanathan, M., Paulose, C.S., Lal, K.J., et al., 1990. Neurosci. Lett. 111, 201–205.

36. Selhub, J., 2006. J. Nutr. 136, 1726S–1730S.

TABLE 14.5 Effects of Vitamin B_6 Status on Mitogenic Responses and Interleukin 2 Production by Peripheral Blood Mononucleocytes of Elderly Humans

Parameter	Baseline	B_6-Deprived	B_6-Supplemented
Mitogenic Response to			
Concanavalin A	120	70	190
Phytohemagglutinin	100	70	100
Staphylococcus aureus	115	60	200
IL-2 production (kU/liter)	105	40	145

Adapted from Meydani, S.N., Ribaya-Mercado, J.D., Russell, R.M., et al., 1991. Am. J. Clin. Nutr. 53, 1275–1280.

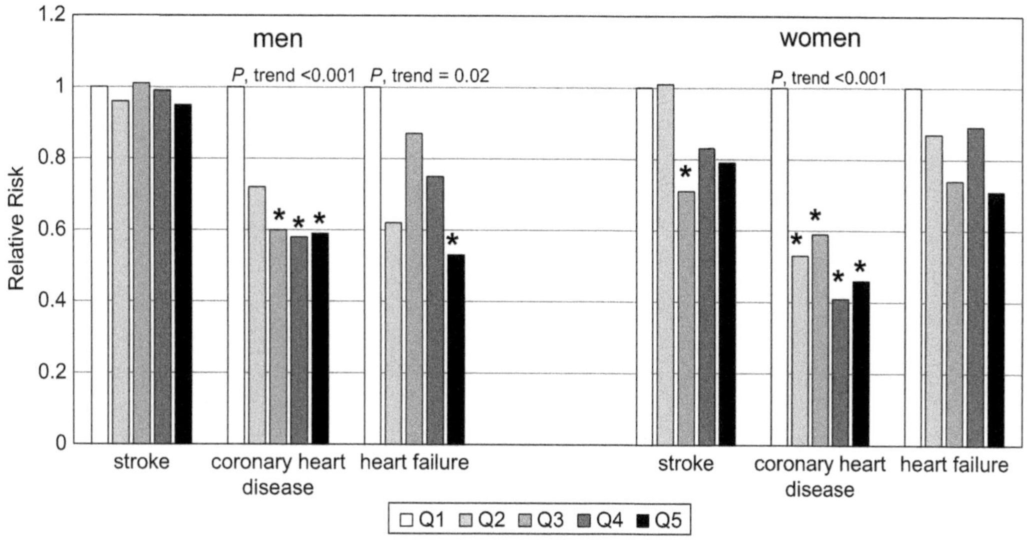

FIGURE 14.6 Cardiovascular Disease Risk by quintile of plasma PalP concentration. *Adapted from Cui, R., Iso, H., Date, C., et al., 2010. Stroke 41, 1285–1289.*

homocysteine level.[37] High doses of vitamin B_6 and folic acid have been found effective in reducing both plasma homocysteine and the incidence of abnormal exercise electrocardiography tests, suggesting reductions in risk of atherosclerotic disease.[38] However, a meta-analysis of 12 randomized trials of vitamin (folate, vitamin B_6, vitamin B_{12}) supplements to lower homocysteine levels in free-living people found no evidence for vitamin B_6 supplements being effective.[39]

Cardiovascular disease. Dietary intakes of vitamin B_6 and plasma levels of PalP have been found to be inversely associated with risk of stroke, coronary heart disease, and heart failure (Table 14.5). A review of six clinical trials with a total of some 8500 chronic kidney disease patients found interventions with vitamin B_6 combined with folate and vitamin B_{12} not to affect risk of cardiovascular mortality (Fig. 14.6).[40]

Immune Function

Vitamin B_6 has a role in the support of immune competence that has not been elucidated. Animal and human studies have demonstrated effects of vitamin B_6 deprivation on both humoral (diminished antibody production) and cell-mediated immune responses (increased lymphocyte proliferation, reduced delayed-type hypersensitivity responses, reduced

37. Robinson, K., Arheart, K., Refsum, H., et al., 1998. Circulation 97, 437–443.
38. Vermuelen, E.G.J., Stehouwer, C.D., Twisk, J.W., et al., 2000 Lancet 355, 517–522.
39. Folic acid produced an average 25% reduction, and vitamin B_{12} produced an average 7% reduction (Clarke, R., Armitage, J., 2000. Semin. Thromb. Hemost. 26, 341–348.).
40. Nursalim, A., Siregar, P., Widyahening, I.S., 2013. Acta Med. Indones. 45, 150–156.

TABLE 14.6 Congenital Disorders of Vitamin B_6-Dependent Metabolism

Disorder	Enzyme Deficiency	Clinical Manifestations
Homocysteinuria[a]	Cystathionine β-synthase	Dislocation of lenses, thromboses, malformation of skeletal and connective tissue, mental retardation
Cystathioninuria	Cystathionine γ-lyase	Mental retardation
GABA deficiency	Glutamate decarboxylase	Seizures
Sideroblastic anemia	δ-Aminolevulinic acid synthase	Anemia, cystathioninuria, xanthurenic aciduria

[a]*Another form is caused by impaired vitamin B_{12}-dependent methionine synthesis.*

T cell-mediated cytotoxicity, reduced cytokine production). Suboptimal status of the vitamin has been linked to declining immunologic changes among the elderly (Table 14.6), persons with human immunodeficiency virus (HIV), and patients with uremia or rheumatoid arthritis. These effects appear to be due to reduced activities of such PalP enzymes as serine transhydroxymethylase and thymidylate synthase, resulting in impaired single-carbon metabolism and reduced DNA synthesis. The vitamin has also been shown to bind to the α-glycoprotein surface receptor (CD4) on a T-helper cells by which it affects photoimmunosuppression.[41] It has also been shown to noncompetitively inhibit HIV-1 reverse transcriptase.[42]

Gene Expression

PalP has been shown to modulate gene expression. Elevated intracellular levels of the vitamin are associated with decreased transcription in responses to glucocorticoid hormones (progesterone, androgens, estrogens). Such diminished responses include hydrocortisone induction of rat liver cytosolic aspartate aminotransferase. Inhibition is caused by the formation of Schiff base linkages of the vitamin to the DNA-binding site of the receptor–steroid complex. This inhibits the ligand binding to the glucocorticoid-responsive element in the regulatory region of the gene. Vitamin B_6 deficiency increases the expression of albumin mRNA sevenfold. The effect appears due to the action of PalP inactivating tissue-specific transcription factors by directly interacting with DNA ligand-binding sites.[43] PalP appears to modulate glycoprotein IIb gene expression by interacting directly to with tissue-specific transcription factors.[44] This results in inhibition of platelet aggregation due to impaired binding of fibrinogen or other adhesion proteins to glycoprotein complexes. PalP has also been shown to suppress mRNA levels for glycogen phosphorylase, apolipoprotein A-1, phenylalanine hydroxylase, glyceraldehyde-3-phosphate dehydrogenase, and β-actin but to decrease mRNA levels for RNA polymerases I and II in the rat model.[45] The effect on glycogen phosphorylase appears to be tissue-specific, as deprivation of the vitamin was found to reduce phosphorylase mRNA levels in muscle but to increase them in liver.[46]

Congenital Disorders of Vitamin B_6 Metabolism

Several rare familial disorders have been identified, each thought to be caused by the expression of deficient amounts or dysfunctional forms of PalP-dependent enzyme (Table 14.7).

- **Homocystinuria** occurs due to a rare (three cases per million), hereditary deficiency of cystathionine β-synthase. The impaired homocysteine catabolism is manifest as elevations in plasma levels of homocysteine, methionine, and cysteine with dislocation of the optic lens,[47] osteoporosis and abnormalities of long bone growth, mental retardation, and thromboembolism. The condition is treated with a low-methionine diet. Half of cases respond to high doses (250–500mg/day) of Pn.[48] Of more than a hundred alleles that have been studied, mutations of the cystathionine β-synthase associated with disease phenotypes have been found in almost one-third.[49] Some mutations, including some of the most frequent ones in the

41. Salhany, J.M., Schopfer, L.M., 1993. J. Biol. Chem. 268, 7643–7645.
42. Mitchell, L.L.W., Cooperman, B.S., 1992. Biochemistry 31, 7707–7713.
43. *See review*: Oka, T., 2001. Nutr. Res. Rev. 14, 257–266.
44. Chang, S.J., Chuang, H.J., Chen, H.H., 1999. J. Nutr. Sci. Vitaminol. 45, 471–479.
45. Oka, T., Komori, N., Kuwahata, M., et al., 1993. FEBS Lett. 331, 162–164.
46. Oka, T., Komori, N., Kuwahata, M., et al., 1994. Experientia 50, 127–129.
47. *Ectopia lentis*
48. Berber, G., Spaeth, G., 1969. J. Pediatr. 75, 463–478.
49. Kraus, J.P., Janosik, M., Kozich, V., et al., 1999. Hum. Mutat. 13, 362–375.

TABLE 14.7 Effects of Nonsteroidal Antiinflammatory Drugs (NSAIDs) on Biomarkers of Vitamin B_6 Status in Rheumatoid Arthritis Patients

Biomarker	Control	NSAIDs, ≤ 6 Months	NSAIDS, > 6 Months
Plasma PalP, nM	42.3 (39.6, 73.8)[a]	35.1 (34.5, 64.9)[a]	29.1 (33.9, 44.9)[a,b]
Plasma Pal, nM	15.3 (12.3, 19.8)	17.1 (16.3, 33.7)	15.9 (15.9, 26.1)
Plasma homocysteine, μM	6.67 (6.50, 8.74)	6.45 (6.156, 8.19)	6.48 (6.47, 7.62)

[a]*Mean (95% C.L.).*
[b]*Significantly different from control (p<.05).*
Adapted from Chang, H.Y., Tang, F.Y., Chen, D.Y., et al., 2013. Am. J. Clin. Nutr. 98, 1440–1449.

human populations studied to date,[50] have been shown to correlate with Pn responsiveness; these would appear to involve the expression of a mutant enzyme with low affinity for PalP.

- **Vitamin B_6-responsive seizures**[51] have been reported.[52] These are manifest as intractable seizures appearing within hours after birth; they are resistant to antiepileptic drugs but stop immediately upon intravenous administration of high levels (100–500 mg) of PalP and are controlled with daily oral doses of PalP (10–30 mg/kg body weight).[53] If untreated, progressive cerebral atrophy ensues. Two types of recessive traits appear to underlie vitamin B_6-responsive seizures. One group involves mutations in the *ALDH7A1* gene, which encodes for antiquitin, an enzyme found in both the cytosol and mitochondria that functions as an aldehyde dehydrogenase in the catabolism of lysine. Deficient expression of antiquitin is characterized by marked elevations in intracellular concentrations of α-amino adipic semialdehyde, piperidine-6-carboxylate, and pipecolic acid, which inhibit the uptake of GABA.[54] The other group involves mutations in Pn/Pm oxidase;[55] these respond only to PalP.
- **Hyperoxaluaria** (Type I) is due to a variant hepatic alanine glyoxylate transferase with abnormally low PalP-binding capacity. High oral doses of vitamin B_6 (e.g., 400 mg/day) reduce hyperoxaluria in some 30% of patients, reducing the risk of formation of oxalate stones and renal injury.[56]

50. Such studies have been conducted only in Europe; no information is available for other populations.
51. Also referred to as pyridoxine-dependent epilepsy.
52. Vitamin B_6 has also been recommended (0.1–1 g/day alone or in combination with tryptophan or magnesium) for reducing seizures in alcoholics and for the treatment of schizophrenia.
53. Gupta, V., Mishra, D., Mathur, I., et al., 2001. J. Pediatr. Child Health 37, 592–596; Gospe, Jr., S.M., 2002. Pediatr. Neurol. 26, 181–185.
54. Jagadeesh, S., Surech, B., Murugan, V., et al., 2013. Paediatr. Int. Child Health 33, 113–115.
55. Gospe, Jr., S.M., 2006. Curr. Opin. Neurol. 19, 148–153.
56. Bhasin, B., Ürekli, H.M., Atta, M.G., 2015. World Rev. Nephrol. 4, 235–244.

TABLE 14.8 Signs of Vitamin B_6 Deficiency

Organ System	Signs
General	Reduced appetite, growth
Dermatologic	Acrodynia, cheilosis, stomatitis, glossitis, seborrheic and scaling dermatitis (around nose)
Muscular	Weakness
Skeletal	Dental caries
Vital organs	Hepatic steatosis
Vascular	Arteriosclerosis, anemia
Nervous	Paralysis, convulsions, peripheral neuropathy
Reproductive	Decreased egg production; fetal malformations, death

8. BIOMARKERS OF VITAMIN B_6 STATUS

Vitamin B_6 status can be assessed in several ways:[57]

- **Blood metabolites**—The most common means of assessing long-term vitamin B_6 status is the measurement of PalP in plasma. This can be accomplished by direct analysis using HPLC or was formerly done using a tyrosine decarboxylase assay. Plasma PalP concentrations ≥30 nM are considered adequate; those of 20–30 nM are generally considered as marginal, i.e., carrying risk of metabolic dysfunction. These measurements are sensitive to the effects of pregnancy, sex, exercise, age, NSAID (nonsteroidal antiinflammatory drug)[58] use (Table 14.8), smoking, and alcohol consumption. Plasma total vitamin B_6 and erythrocyte PalP concentrations have also been used.

57. A microbiological assay employing *Lactobacillus plantarum* is commonly used for the analysis of pantothenic acid. This requires enzymatic pretreatment of specimens to liberate free pantothenic acid.
58. i.e., Nonsteroidal antiinflammatory drugs, such as celecoxib and naproxen.

- **Urinary metabolites**—The urinary excretion of total vitamin B_6, or 4-pyridoxic acid, which comprises half of total vitamin B_6 intake, has been used. The latter is affected by short-term intake as well as tissue stores of the vitamin.
- **Load tests**—Vitamin B_6 function can be assessed by measuring concentrations of downstream metabolites in pathways safely perturbed using large oral doses of key upstream metabolites:
  - **Urinary xanthurenic acid after a tryptophan load**—Because the affinity of kynureninase for PalP is much lower than those of the PalP-dependent transaminases, vitamin B_6 deficiency results in increased production and urinary excretion of xanthurenic acid, which effect is amplified after an oral bolus (2 g) of tryptophan. These results can be affected by gender, estrogens, glucocorticoids, and pregnancy and protein intake, which upregulate tryptophan-2,3-dioxygenase, the rate-limiting enzyme in tryptophan metabolism.
  - **Plasma homocysteine after a methionine load**—Because cystathionine γ-lyase is very sensitive to inadequate PalP supply, plasma homocysteine concentration after an oral bolus (3 g) of methionine, which suppresses homocysteine remethylation, can indicate vitamin B_6 status.
- **Degree of PalP saturation of PalP-dependent enzymes**—Vitamin B_6 status can be assessed by taking advantage of the in vitro binding of PalP by alanine aminotransferase or aspartic aminotransferase from hemolysates. Stimulation of either of these activities upon the addition of exogenous PalP indicates inadequate vitamin B_6 status.

TABLE 14.9 Recommended Vitamin B_6 Intakes

The United States		FAO/WHO	
Age/Sex	RDA[a], µg/day	Age/Sex	RNI[b], µg/day
0–6 months	[0.1][c]	0–6 months	0.1
7–11 months	[0.3][c]	7–11 months	0.3
1–3 year	0.5	1–3 year	0.5
4–8 year	0.6	4–6 year	0.6
9–13 year	1.0	7–9 year	1.0
14–18 year		10–18 year	
females	1.2	females	1.2
Males	1.3	Males	1.3
19–59 year		19–50 year	1.3
females	1.1		
Males	1.3		
>50 year		>50 year	
females	1.5	females	1.5
Males	1.7	Males	1.7
Pregnancy	1.9	Pregnancy	1.9
Lactation	2.0	Lactation	2.0

[a]*Food and Nutrition Board, 2000. Dietary Reference Intakes for Thiamin, Riboflavin, Niacin, Vitamin B_6, Folate, Vitamin B_{12}, Pantothenic Acid, Biotin and Choline. National Academy Press, Washington, DC, 564 pp.*
[b]*Recommended Nutrient Intakes, Joint WHO/FAO Expert Consultation, 2001. Human Vitamin and Mineral Requirements, WHO, Rome, 286 pp.*
[c]*RDA has not been set; Adequate Intake (AI) is given.*

9. VITAMIN B_6 DEFICIENCY

Severe deficiency of vitamin B_6 results in dermatologic and neurologic changes in most species (Table 14.9). Less obvious are the metabolic lesions associated with insufficient activities of the coenzyme PalP. The latter include impaired tryptophan–niacin conversion and impaired transsulfuration of methionine to cysteine. Deprivation of vitamin B_6 also impairs glucose tolerance, due to reduced activities of PalP-dependent transaminases and glycogen phosphorylase, although it may not affect fasting glucose levels. Recommended intakes of vitamin B_6 have been established for humans (Table 14.10).

Deficiency Signs in Humans

Clinical deficiency. Clinical vitamin B_6 deficiency is not common. Vitamin B_6-deficient humans exhibit symptoms that can be quickly corrected by administration of the vitamin: weakness, sleeplessness, nervous disorders (peripheral neuropathies), **cheilosis**,[59] **glossitis**,[60] **stomatitis**, and impaired cell-mediated immunity.[61] Behavioral differences have been associated with low-vitamin B_6 status: a study in Egypt found that mothers of marginal (subclinical) vitamin B_6 status were less responsive to their infants' vocalizations, showed less effective response to infant distress, and were more likely to use older siblings as caregivers than were mothers of better vitamin B_6 status. In addition, studies with volunteers fed a vitamin B_6-free diet or a vitamin B_6 antagonist[62] have shown elevated urinary xanthurenic acid concentrations[63] and increased susceptibility to infection. Because plasma concentrations of PalP decrease with

59. The lesion is morphologically indistinguishable from that produced by riboflavin deficiency.
60. The lesion is morphologically indistinguishable from that produced by niacin deficiency.
61. These can be produced by the antagonist deoxypyridoxine.
62. For example, 4′-deoxyPn.
63. After tryptophan loading, vitamin B_6-deficient subjects also had elevated urinary concentrations of kynurenine, 3-hydroxykynurenine, kynurenic acid, acetylkynurenine, and quinolinic acid.

TABLE 14.10 Relationship of Vitamin B_6 Status and Inflammatory Status in Humans

	Tertile of Plasma PalP, nM			
Characteristic	35 (34, 36)[a]	69 (67, 71)	177 (173, 181)	*p*, Trend
Vitamin B_6 intake, g/day	2.7 (0.9, 4.5)[a]	5.2 (3.4, 6.9)	18.6 (16.8, 20.3)	<.001
C-reactive protein, mg/L	3.1 (2.9, 3.4)[a]	2.1 (1.9, 2.3)	1.8 (1.6, 1.9)	<.001

[a]*Mean (95% C.L.).*
Adapted from Sakakeeny, L., Roubenoff, R., Oben, M., et al., 2012. J. Nutr. 142, 1280–1285.

age, it is expected that elderly people may be at greater risk of vitamin B_6 deficiency than younger people.

Subclinical deficiency. Low-vitamin B_6 status is associated with inflammation as indicated by increased circulating levels of C-reactive protein (Table 14.10)[64] and is observed in chronic inflammatory diseases such as rheumatoid arthritis and inflammatory bowel disease. Low-vitamin B_6 status has also been associated with an increase in the ratio of n-6:n-3 polyunsaturated fatty acids in plasma, a feature associated with increased risk of inflammation.[65] It has been suggested that the reduction in circulating PalP reflects the mobilization of that vitamer to the sites of inflammation where is functions in PalP-dependent enzymes.[66] High-vitamin B_6 intakes are associated with protection against inflammation.[67] Low circulating PalP levels have been reported in patients with asthma, and one small study found vitamin B_6 treatment (100 mg/day) to reduce the severity and frequency of attacks.[68] These effects may be secondary to those of theophylline, which inhibits Pal kinase.

Deficiency Signs in Animals

Vitamin B_6 deficiency in animals is generally manifest as symmetrical scaling dermatitis. In rodents, the condition is called **acrodynia** and is characterized by hyperkeratotic[69] and acanthotic[70] lesions on the tail, paws, face, and upper thorax, as well as by muscular weakness, hyperirritability, anemia, hepatic steatosis,[71] increased urinary oxalate excretion, insulin insufficiency,[72] hypertension, and poor growth. Neurological signs include convulsive seizures (epileptic form) that can be fatal.[73] Reproductive disorders include infertility, fetal malformations,[74] and reduced fetal survival. Some reports indicate effects on blood cholesterol levels and immunity. That tissue carnitine levels are depressed in vitamin B_6-deficient animals has been cited as evidence of a role of the vitamin in carnitine synthesis.

Similar changes are observed in vitamin B_6-deficient individuals of other species. Chickens and turkeys show reduced appetite and poor growth, dermatitis, marked anemia, convulsions, reduced egg production, and low fertility. Pigs show paralysis of the hind limbs, dermatitis, reduced feed intake, and poor growth. Monkeys show an increased incidence of dental caries and altered cholesterol metabolism with arteriosclerotic lesions. Vitamin B_6 deficiency has been reported to cause hyperirritability, hyperactivity, abnormal behavior, and performance deficits in several species. These signs accompany an underlying neuropathology that reduces axonal diameter and dendritic arborization and, thus, impairs nerve conduction velocity.

Ruminants are rarely affected by vitamin B_6 deficiency, as their rumen microflora appears to satisfy their needs for the vitamin. Exceptions are lambs and calves, which, before their rumen microflorae are established, are susceptible to dietary deprivation of vitamin B_6, showing many dermatologic and neurologic changes observed in nonruminant species.

10. VITAMIN B_6 IN HEALTH AND DISEASE

Anticarcinogenesis

Epidemiological studies have yielded inconsistent findings regarding the association of vitamin B_6 intake and cancer risk; however, they have consistently found individuals with higher plasma PalP levels to be 30–50% less likely to develop colorectal cancer risk than individuals with lower plasma PalP levels.[75] Studies in animal models

64. Morris, M.A., Picciano, M.F., Jacques, P.F., et al., 2008. Am. J. Clin. Nutr. 87, 1446–1454; Sakakeeny, L., Roubenoff, R., Oben, M., et al., 2012. J. Nutr. 142, 1280–1285.
65. Zhao, M., Lamers, Y., Ralat, M.A., et al., 2012. J. Nutr. 142, 1791–1797.
66. Lotto, V., Choi, S.W., Friso, S., 2011. Br. J. Nutr. 106, 183–195.
67. Morris, M.S., Sakakeeny, L., Jacques, P.F., et al., 2010. J. Nutr. 140, 103–110.
68. Simon, R.A., Reynolds, R.D., 1988. In: Leklem, J.E., Reynods, R.D. (Eds.), Clinical and Physiological Applications of Vitamin B_6. Alan R. Liss, New York, pp. 307–315.
69. Involving hypertrophy of the horny layer of the epidermis.
70. Involving an increase in the prickle cell layer of the epidermis.
71. This can be precipitated by feeding a vitamin B_6-deficient diet rich in protein.
72. This is believed to be due to reduced pancreatic synthesis of the hormone.
73. Nervous dysfunction is believed to be due to nerve tissue deficiencies of GABA due to decreased activities of the PalP-dependent **glutamate decarboxylase**. The seizures can be controlled by administering either the vitamin or GABA.
74. For example, omphalocele (protrusion of the omentum or intestine through the umbilicus), exencephaly (defective skull formation with the brain partially outside of the cranial cavity), cleft palate, micrognathia (impaired growth of the jaw), splenic hypoplasia.
75. Zhang, X.H., Ma, J., Smith-Warner, S.A., et al., 2007. World Rv. Gastroenterol. 19, 1005–1010.

have found that supranutritional doses of the vitamin can reduce tumorigenesis through effects on cell proliferation and production of reactive oxygen and nitrogen species and angiogenesis.[76] The two, small, randomized intervention trials that have been conducted with human subjects have not found vitamin B_6 supplementation to reduce colorectal cancer risk.[77]

Effects of High-Vitamin B_6 Doses

Vitamin B_6 at relatively high doses has been reported to produce positive effects in a number of conditions affecting individuals who were not apparently deficient in the vitamin:

- **Sideroblastic anemia**. Dosage as great as 200 mg/day (usually as Pn HCl) has been found to stimulate δ-aminolevulinic acid synthase activity and, thus, enhance hematopoiesis in patients.
- **Sickle cell anemia**. A small study found patients to have lower plasma PalP levels than controls, which responded to oral supplementation (100 mg/day) of Pn within 2 months. Both Pn and PalP have been found to protect sickle cells in vitro,[78] but it is not clear whether supplementation with the vitamin may benefit sickle cell anemia patients.
- **Iron storage disease**. Complexes of Pal, which chelate iron (e.g., the isonicotinyl and benzoyl hydrazones), have been found effective in stimulating the excretion of iron in patients with iron-storage disease.
- **Suppression of lactation**. A few studies have reported vitamin B_6 as effective in suppressing lactation probably through the stimulation of dopaminergic activity in the hypothalamus.
- **Adverse drug effects**. Vitamin B_6 is used at doses of 3–5 mg/kg body weight to counteract adverse effects of several types of drugs. The antituberculin drug INH produces a peripheral neuropathy similar to that of vitamin B_6 deficiency by inhibiting the activities of PalP-dependent glutamate decarboxylase and γ-aminobutyrate aminotransferase, which produce and degrade GABA, respectively, in nerve tissue. Two antibiotics antagonize vitamin B_6 by reacting with PalP to form inactive products. Cycloserine reacts with the coenzyme to produce an oxime. Penicillamine produces thiazolidine. L-3,4-dihydroxyphenylalanine antagonizes vitamin B_6 by reacting with PalP to form tetrahydroquinolines. Ethanol increases PalP catabolism. Synthetic estrogens can alter tryptophan–niacin conversion by increasing the synthesis of PalP-dependent enzymes in that pathway, thus, increasing the need for vitamin B_6.
- **Carpal tunnel syndrome**. This disorder, involving pain and paresthesia of the hand, is caused by irritation and compression of the medial nerve by the transverse ligaments of the wrist in ways that are exacerbated by redundant motions. The condition has been associated with low circulating levels of PalP and low-erythrocyte glutamic–oxaloacetic transaminase activities. It has been suggested that such deficiencies lead to edematous changes to and proliferation of the synovia, causing compression of the nerve in the carpal tunnel. Some investigators have reported high doses (50–300 mg/day for 12 weeks) of Pn to be effective as treatment;[79] however, there is no evidence from randomized clinical trials supporting such use of the vitamin.
- **Diabetes**. Several studies have found vitamin B_6 supplementation to improve glucose tolerance. It has been suggested that this may involve the reactivation of kynureninase, which leads to inactivation of insulin through complexation with xanthurenic acid. Studies have also shown that vitamin B_6 supplements (100 mg/day) useful in preventing complications of diabetes mellitus caused by the nonenzymatic glycation of critical proteins.[80] Pyridoxamine has been shown to be a potent inhibitor of the formation of advance glycation products from glycated proteins by scavenging reactive carbonyls; this would appear to be the basis of its protection against the development of renal disease in the diabetic rat model.[81]
- "Chinese restaurant" **syndrome**. The syndrome, which involves headache, sensation of heat, altered heartbeat, nausea, and tightness of the neck induced by oral intake of monosodium glutamate has been reported to respond to Pn (50 mg/day).
- **Premenstrual syndrome**. This syndrome affects some 40% of women 2–3 days before their menstrual flow. It involves tension of the breasts, pain in the lumbar region, thirst, headache, nervous irritability, pelvic congestion, peripheral edema, and usually, nausea and vomiting. Premenstrual syndrome has been reported to respond to vitamin B_6, presumably by affecting levels of the neurotransmitters, serotonin, and GABA that control depression, pain perception, and anxiety. Women experiencing premenstrual symptoms appear to have circulating PalP levels comparable to unaffected women;

76. Komatsu, S., Yanaka, N., Matsubara, K., et al., 2003. Biochem. Biophys. Acta 1647, 127–130.
77. Bønaa, K.H., Njølstad, I., Ueland, P.M., et al., 2006. N. Engl. J. Med. 354, 1578–1588; Ebbing, M., Bleie, Ø., Ueland, P.M., et al., 2008. JAMA 300, 795–804.
78. Kark, J.A., Tarassoff, P.G., Bongiovanni, R., 1983. J. Clin. Invest. 71, 1224–1229.
79. Aufiero, E., Stitik, T.P., Foye, P.M., et al., 2004. Nutr. Rev. 62, 96–104; Goodyear-Smith, F., Arroll, B., 2004. Ann. Fam. Med. 2, 267–273; Ellis, J.M., Pamplin, J., 1999. Vitamin B6 Therapy. Avery Publishing Group, pp. 47–56.
80. Solomon, L.R., Cohen, K., 1989 Diabetes 38, 881–886.
81. Metz, T.O., Alderson, N.L., Thorpe, S. R., et al., 2003. Arch. Biochem. Biophys. 419, 41–49.

nevertheless, high doses of the vitamin have been found to alleviate at least some symptoms in many cases. A review of randomized, clinical trials concluded that Pn doses of up to 100mg/day are likely to be of benefit in treating these symptoms.[82]

- **Nausea and vomiting of pregnancy**. Plasma concentrations of PalP normally decline in the third trimester of pregnancy at which time fetal stores increase, indicating fetal sequestration of the vitamin.[83] While there is no evidence that this contributes to the nausea and vomiting that affects 85% of all pregnancies, randomized clinical trials have shown that the use of Pn used alone (25mg every 8h for 3 days) or in combination with doxylamine[84] significantly reduced nausea and vomiting of pregnancy.[85]

11. VITAMIN B_6 TOXICITY

The toxicity of vitamin B_6 appears to be relatively low, although high doses of the vitamin (several grams per day) have been shown to induce sensory neuropathy marked by changes in gait and peripheral sensation. The primary target, thus, appears to be the peripheral nervous system; although massive doses of the vitamin have produced convulsions in rats, central nervous abnormalities have not been reported frequently in humans. The potential for toxicity resulting from the therapeutic or pharmacologic uses of the vitamin for human disorders (which rarely exceed 50mg/day) is small. Reports of individuals taking massive doses of the vitamin (>2g/day) indicate that the earliest detectable signs were ataxia and loss of small motor control. Many of the signs of vitamin B_6 toxicity resemble those of vitamin B_6 deficiency; it has been proposed that the metabolic basis of each condition involves the tissue-level depletion of PalP. Doses up to at least 500mg/day for extended periods of time (several years) have been found safe.[86] In doses of 10–25mg, vitamin B_6 increases the conversion of L-dopa to dopamine[87] which, unlike its precursor, cannot cross the blood–brain barrier. The vitamin can, thus, interfere with L-dopa in the management of Parkinson's disease; it should not be administered to individuals taking L-dopa without the concomitant administration of a decarboxylase inhibitor. Upper tolerable intakes have not been established for vitamin B_6.

12. CASE STUDIES

Instructions

Review the following case reports, paying special attention to the diagnostic indicators on which the treatments were based. Then, answer the questions that follow.

Case 1

A 16-year-old boy was admitted with *dislocated lenses* and *mental retardation*. 4years earlier, an ophthalmologist had found dislocation of the lenses. On the present occasion, he was thin and blond-headed, with ectopia lentis,[88] an anterior thoracic deformity (*pectus excavatum*[89]), and normal vital signs. His palate was narrow with crowding of his teeth. He had mild scoliosis[90] and genu valgum, which caused him to walk with a toe-in, *Chaplin-like* gait. His neurological examination was within normal limits. On radiography, his spine appeared osteoporotic. His performance on the Stanford–Binet Intelligence Scale gave him a development quotient of 60. His hematology, blood glucose, and blood urea nitrogen values were all within normal limits. His plasma homocysteine level (undetectable in normal patients) was 4.5mg/dL, and his blood *methionine* level was 10-fold normal; the levels of all other amino acids in his blood were within normal limits. Both homocysteine and methionine were increased in his urine, which also contained traces of *S*-adenosylhomocysteine.

The patient was given oral Pn HCl in an ascending dose regimen. Doses up to 150mg/day were without effect but, after the dose had been increased to 325mg/day for 200days, his plasma and urinary homocysteine and methionine levels decreased to normal. These changes were accompanied by a striking change in his hair pigmentation: dark hair grew out from the scalp (the cystine content of the dark hair was nearly double that of the blond hair, 1.5 versus 0.8mEq/mg). On maintenance doses of Pn, he attained relatively normal function, although the connective tissue changes were irreversible.

Case 2

A 27-year-old woman had experienced increasing difficulty in walking. Some 2years earlier, she had been told that

82. Wyatt, K.M., Dimmock, P.W., Jones, P.W., et al., 1999. Br. Med. J. 318, 1375–1381.
83. Contractor, S.F., Shane, B., 1970. Am. J. Obstet. Gynecol. 107, 635–640.
84. A first-generation antihistamine.
85. Sahakian, V., Rouse, D., Sipes, S., et al., 1991. Obstet. Gynecol. 78, 33–36; Madjunkova, S., Maltepe, C., Koren G., 2014. Paediatr. Drugs 16, 199–211.
86. Bendich, A., Cohen, M., 1990. Ann. N.Y. Acad. Sci. 585, 320–330; Mpofu, C., Alani, S.M., Whitehouse, C., et al., 1991. Arch. Dis. Child. 66, 1081–1082.
87. It has been claimed that, via its effect on dopamine, vitamin B_6 can inhibit the release of prolactin, thus inhibiting lactation in nursing mothers. Although this proposal is still highly disputed, there is no evidence that daily doses of less than about 10mg of the vitamin (in multivitamin preparations) has any such effect on lactation.
88. Dislocated lenses.
89. Funnel chest.
90. Lateral curvature of the spine.

vitamin B_6 prevented premenstrual edema and she began taking 500mg/day of *Pn HCl*. After a year, she had increased her intake of the vitamin to 5g/day. During the period of this increased vitamin B_6 intake, she noticed that flexing her neck produced a tingling sensation down her neck and to her legs and soles of her feet.[91] During the 4months immediately before this examination, she had become progressively unsteady when walking, particularly in the dark. Finally, she had become unable to walk without the assistance of a cane. She had also noticed difficulty in handling small objects and changes in the feeling of her lips and tongue; although she reported no other positive sensory symptoms and was not aware of any weaknesses. Her gait was broad-based and stamping, and she was not able to walk at all with her eyes closed. Her muscle strength was normal, but all of her limb reflexes were absent. Her sensations of touch, temperature, pinprick, vibration, and joint position were severely impaired in both the upper and lower limbs. She showed a mild subjective alteration of touch-pressure and pinprick sensation over her cheeks and lips, but not over her forehead. Laboratory findings showed the spinal fluid and other clinical tests to be normal. Electrophysiologic studies revealed that no sensory nerve action potentials could be elicited in her arms and legs, but that motor nerve conduction was normal.

The patient was suspected of having vitamin B_6 intoxication and was asked to stop taking that vitamin. 2months after withdrawal, she reported some improvement and a gain in sensation. By 7months, she could walk steadily without a cane and could stand with her eyes closed. Neurologic examination at that time revealed that, although her strength was normal, her tendon reflexes were absent. Her feet still had severe loss of vibration sensation, despite definite improvements in the senses of joint position, touch, temperature, and pinprick. Electrophysiologic examination revealed that her sensory nerve responses were still absent.

Case Questions

1. Propose a hypothesis consistent with the findings in Case 1 for the congenital metabolic lesion experienced by that patient.
2. Would you expect supplements of methionine and/or cystine to have been effective in treating the patient in Case 1? Defend your answer.
3. If the toxicity of Pn involves its competition, at high levels, with PalP for enzyme-binding sites, which enzymes would you propose as potentially being affected in the condition described in Case 2? Provide a rationale for each of the candidate enzymes on your list.

91. Lhermitte's sign.

13. STUDY QUESTIONS AND EXERCISES

1. Diagram schematically the several steps in amino acid metabolism in which PalP-dependent enzymes are involved.
2. Construct a decision tree for the diagnosis of vitamin B_6 deficiency in humans or an animal species.
3. What key feature of the chemistry of vitamin B_6 relates to its biochemical functions as a coenzyme?
4. What parameters might you measure to assess vitamin B_6 status of a human or animal?
5. What factors might be expected to affect the dietary need for vitamin B_6?

RECOMMENDED READING

Allen, G.F.G., Land, J.M., Heales, S.J.R., 2009. A new perspective on the treatment of aromatic L-amino acid decarboxylase deficiency. Mol. Genet. Metab. 97, 6–14.

Dakshinamurti, S., Dakshimanurti, K., 2014. Vitamin B_6 (Chapter 9). In: Zempleni, J., Suttie, J.W., Gregory, J.F., et al. (Eds.), Handbook of Vitamins, fifth ed. CRC Press, Dekker, New York, pp. 351–395.

Da Silva, V.R., Russell, K.A., Gregory, J.F., 2012. Vitamin B_6 (Chapter 20). In: Erdman, J.W., Macdonald, I.A., Zeisel, S.H. (Eds.), Present Knowledge in Nutrition, tenth ed. Wiley, New York, pp. 307–320.

Gregory, J.F., 1997. Bioavailability of vitamin B-6. Eur. J. Clin. Nutr. 51, S43–S48.

Sánchez-Moreno, C., Jiménez-Excrig, A., Martín, A., 2009. Stroke: roles of B vitamins, homocysteine and antioxidants. Nutr. Res. Rev. 22, 49–67.

Chapter 15

Biotin

Chapter Outline

Anchoring Concepts

1. Biotin is the trivial designation of the compound hexahydro-2-oxo-1*H*-thieno[3,4-*d*]imidazole-4-pentanoic acid.
2. Biotin functions metabolically as a coenzyme for carboxylases, to which it is bound by the carbon at position 2 (C-2) of its thiophene ring via an amide bond to the ε-amino group of a peptidyl lysine residue.
3. Deficiencies of biotin are manifested predominantly as dermatologic lesions.

We started with a bushel of corn, and at the end of the purification process, when the solution was evaporated in a small beaker, nothing could be seen, yet this solution of nothing greatly stimulated growth (of propionic acid bacteria). We now know that the factor was biotin, which is one of the most effective of all vitamins.

H.G. Wood[1]

LEARNING OBJECTIVES

1. To understand the chief natural sources of biotin.
2. To understand the means of absorption and transport of biotin.
3. To understand the biochemical function of biotin as a component of coenzymes of metabolically important carboxylation reactions.
4. To understand the metabolic bases of biotin-responsive disorders, including those related to dietary deprivation of the vitamin and those involving inherited metabolic lesions.

VOCABULARY

Acetyl-CoA carboxylase 1
Acetyl-CoA carboxylase 2
Achromotrichia
Alopecia
Apocarboxylase
Avidin
Biocytin
Biotin-binding proteins
Biotin sulfoxide
Biotinidase
Biotinylation
Biotinyl 5′-adenylate
Bisnorbiotin
Egg white injury
Fatty liver and kidney syndrome (FLKS)
Footpad dermatitis glucokinase
Histone
Holocarboxylase synthetase (HCS)
3-Hydroxyisovalerate
3-Hydroxyisovaleryl carnitine
Kangaroo gait
β-Methylcrotonoyl-CoA carboxylase
Monocarboxylate transporter (MCT1)

1. Harland G. Wood (1907–91) was an American biochemist who spent most of his career at Case Western Reserve University. He is best known for discovering heterotrophic CO_2 fixation and for establishing the mechanism of transcarboxylase—both facilitated by his pioneering use of radiotracers in biochemistry. In 1989, he received the U.S. National Medal of Science.

The Vitamins. http://dx.doi.org/10.1016/B978-0-12-802965-7.00015-0

Multiple carboxylase deficiencies
Ornithine transcarbamylase
Phosphoenolpyruvate carboxykinase (PEPCK)
Propionyl-CoA carboxylase
Pyruvate carboxylase
Sodium-dependent vitamin transporter (SMVT).
Spectacle eye
Streptavidin
Sudden infant death syndrome (SIDS)
Thiophene ring
Transcarboxylase
Ureido nucleus

1. THE SIGNIFICANCE OF BIOTIN

Biotin was discovered in the search for the nutritional factor that prevents **egg white injury** in experimental animals. The biotin antagonist in egg white, **avidin**, produces experimental biotin deficiency in animal models. Practical cases of biotin deficiency in humans were encountered with the advent of total parenteral nutrition (TPN) before the vitamin was routinely added to tube-feeding solutions, and biotin-responsive cases of footpad dermatitis remain a problem in commercial poultry production. Biotin deficiency manifests itself differently in different species, most often with dermatologic lesions. A key feature of biotin metabolism is that it is recycled by proteolytic cleavage from the biotin-dependent carboxylases by the enzyme **biotinidase**. This recycling and the prevalent hindgut microbial synthesis of the vitamin allow quantitative dietary requirements for biotin to be relatively small. Inborn errors of biotin absorption and metabolism have been identified; some respond to large doses of the vitamin.

2. PROPERTIES OF BIOTIN

Biotin is the trivial designation of the compound *cis*-hexahydro-2-oxo-1*H*-thieno[3,4-*d*]imidazole-4-pentanoic acid.[2] It consists of conjoined ureido and tetrahydrothiophene nuclei in which the ureido 3′-nitrogen is sterically hindered, preventing substitution and the ureido 1′-nitrogen is weakly nucleophilic. Biotin functions in metabolism bound to certain enzymes by covalent linkages of the carboxyl group of the biotin side chain with ε-amino groups of peptidyl lysyl residues. This bound form is called **biocytin**.

Chemical Structure of Biotin

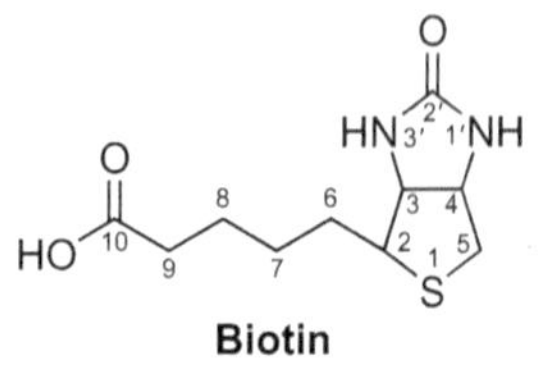

Biotin

2. Formerly, **vitamin H** or **coenzyme R**.

Biotin Chemistry

Biotin is a white crystalline substance that, in dry form, is fairly stable to air, heat, and light. In solution, however, it is sensitive to degradation under strongly acidic or basic conditions. Its structure consists of a planar **ureido nucleus** and a folded **tetrahydrothiophene (thiophane) nucleus**, which results in a boat configuration with a plane of symmetry passing through the S-1, C-2′, and O positions in such a way as to elevate the sulfur atom above the plane of the four carbons. The molecule has three asymmetric centers; however, of the eight possible stereoisomers, only the (+)-isomer (called *d*-biotin) has biological activity. Biotin is covalently bound to its enzymes by an amide bond to the ε-amino group of a lysine residue and C-2 of the thiophane nucleus. This bond is flexible, allowing the coenzyme to move between the active centers of some enzymes. The biotin molecule is activated by polarization of the O and N-1′ atoms of the ureido nucleus. This leads to increased nucleophilicity at N-1′, which promotes the formation of a covalent bond between the electrophilic carbonyl phosphate formed from bicarbonate and ATP and allows biotin to serve as a transport agent for CO_2.

3. SOURCES OF BIOTIN

Hindgut Microbial Synthesis

The microbiota of the monogastric hindgut synthesize significant amounts of biotin. In both rats and humans, fecal excretion of biotin typically exceeds the amount consumed. A genomic analysis of 256 representative organisms of the human gut microbiota found 40% capable of de novo synthesis of the vitamin.[3] In addition, the majority of genomes (all phyla except Fusobacterial) also showed the capacity to salvage PalP and its precursors. Those findings suggested that hindgut microbial synthesis may produce nearly 5% of the daily human need for biotin. Evidence indicates that biotin can be absorbed from the hindgut. Studies with the rat have shown the biotin transport capacity of the colon to be as great as 25% that of the jejunum[4]; and a Na^+-dependent biotin carrier has been identified in human-derived colonic epithelial cells. Accordingly, the dietary biotin requirement of the rat has been determined only under gnotobiotic[5] conditions or with avidin feeding. Still, questions remain as to the nutritional significance of biotin of microbial origin for humans and other non-coprophagous animals, as intracecal treatment with antibiotics to inhibit microbial growth or with lactulose to

3. Magnúsdóttir, S., Ravchee, D., de Crécy-Lagard, V., et al., 2015. Front. Genet. 6, 148–166.
4. Bowman, B.B., Rosenberg, I.H., 1987. J. Nutr. 117, 2121–2128.
5. Germ-free.

stimulate microbial growth failed to affect plasma biotin levels in the pig.[6]

Distribution in Foods

Biotin is widely distributed in foods and feedstuffs, mostly in very low concentrations (Table 15.1). Only a couple of foods (royal jelly,[7] brewers' yeast) contain biotin in large amounts. Milk, liver, egg (egg yolk), and a few vegetables are the most important natural sources of the vitamin in human nutrition; the oilseed meals, alfalfa meal, and dried yeasts are the most important natural sources of the vitamin for the feeding of nonruminant animals. The biotin contents of foods and feedstuffs can be highly variable[8]; for the cereal grains at least, it is influenced by such factors as plant variety, season, and yield (endosperm-to-pericarp ratio). Most foods contain the vitamin as free biotin and as biocytin bound to food proteins.

However, in milk, biotin occurs almost exclusively as the free vitamin in the skim fraction.[9]

Stability

Biotin is unstable to oxidizing conditions and, therefore, is destroyed by heat, especially under conditions that support simultaneous lipid peroxidation.[10] Therefore, such processing techniques as canning, heat curing, and solvent extraction can result in substantial losses of biotin. These losses can be reduced by the use of a food grade antioxidant (e.g., vitamin C, vitamin E, butylated hydroxytoluene, butylated hydroxyanisole).

Bioavailability

Studies of the bioavailability of food biotin to humans have not been conducted. However, biotin bioavailability has been determined experimentally using two types of bioassay: healing of skin lesions in avidin-fed rats, support of growth and maintenance of pyruvate carboxylase activity in chicks. Such assays have shown that the nutritional availability of biotin can be low and highly variable among different foods and feedstuffs (Table 15.2).

TABLE 15.1 Biotin Contents of Foods

Food	Biotin, μg/100 g
Dairy Products	
Milk	2
Cheeses	3–5
Meats	
Beef	3
Chicken	11
Pork	5
Calf kidney	100
Cereals	
Barley	14
Cornmeal	7.9
Oats	24.6
Rye	8.5
Sorghum	28.8
Wheat	10.1
Wheat bran	36
Oilseed Meals	
Rapeseed meal	98.4
Soybean meal	27
Vegetables	
Asparagus	2
Brussels sprouts	0.4
Cabbage	2
Carrots	3
Cauliflower	17
Corn	6
Kale	0.5
Lentils	13
Onions	4
Peas	9
Potatoes	0.1
Soybeans	60
Spinach	7
Tomatoes	4
Fruits	
Apples	1
Bananas	4
Grapefruit	3

Continued

6. Kopinski, J.S., Leibholz, J., Love, R.J., 1989. Br. J. Nutr. 62, 781–788.
7. Royal jelly is secreted by the labial glands of worker honeybees and is rich in biotin (>400 μg/100 g). The few female larvae that are fed royal jelly develop reproductively as queens; whereas, most larvae are fed a mixture of honey and pollen and fail to develop reproductively ability and become workers. The active factor in royal jelly may be a lipid, 10-hydroxy-Δ2-decenoic acid; the role of biotin in royal jelly is unclear.
8. In one study, the biotin contents of multiple samples of corn and meat meal were 56–115 μg/kg (n=59) and 17–323 μg/kg (n=62), respectively.
9. These levels are 20- to 50-fold greater than those found in maternal plasma. Human milk also contains biotinidase, which is presumed to be important in facilitating infant biotin utilization.
10. About 96% of the pure vitamin added to a feed was destroyed within 24 h after the addition of partially peroxidized linolenic acid.

TABLE 15.1 Biotin Contents of Foods—cont'd

Food	Biotin, μg/100 g
Grapes	2
Oranges	1
Peaches	2
Pears	0.1
Strawberries	1.1
Watermelons	4
Nuts	
Peanuts	34
Walnuts	37
Other	
Eggs	20
Brewers' yeast	80
Alfalfa meal	54
Molasses	108

USDA National Nutrient Database for Standard Reference, Release 18.

In general, less than one-half of the biotin present in feedstuffs is biologically available. Although all of the biotin in corn is available, only 20–30% of that in most other grains and none in wheat is available. The bioavailability of biotin in meat products also tends to be very low.

Differences in biotin bioavailability appear to be due to differential susceptibilities to digestion of the various biotin–protein linkages in which the vitamin occurs in foods and feedstuffs. Those linkages involve the formation of covalent bonds between the carboxyl group of the biotin side chain with ε-amino groups of peptidyl lysyl residues, constituting the means by which biotin binds to the enzymes for which it serves as an essential prosthetic group. The utilization of such biocytin biotin thus depends on the hydrolytic digestion of the proteins and/or the hydrolysis of those amide bonds. Biotins in purified preparations, such as those used in dietary supplements, are highly bioavailable.

4. ABSORPTION OF BIOTIN

Digestion of Protein-Bound Biotin

In the digestion of food proteins, protein-bound biotin is released by the hydrolytic action of the intestinal proteases to yield the e-N[1] biotinyl lysine adduct, biocytin, from which free biotin is liberated by the action of an intestinal biotin amide aminohydrolase, **biotinidase**.

Facilitated Transport

At low concentrations, free biotin is absorbed across the enterocyte by two carrier-mediated processes. Its uptake is facilitated by a **Na$^+$-dependent multivitamin transporter** (**SMVT**) bound to the apical membrane (brush border).[11] It can also be inhibited by certain anticonvulsant drugs[12] and ethanol or its major metabolite acetaldehyde (Table 15.3). The intracellular trafficking of SMVT involves distinct trafficking vesicles, the microtubular network, and the microtubule motor protein dynein.[13] The process is not specific for the vitamin, as SMVT also functions in the cellular uptake of pantothenic acid and lipoic acid, which it binds with similar affinities and can inhibit biotin uptake. At the basolateral membrane, another Na$^+$-dependent transporter is involved in translocating biotin to the plasma. The SMVT is regulated by protein kinase C, which can phosphorylate the transporter. Suboptimal SMVT expression is thought to underlie the low biotin absorption observed in alcoholics, pregnant women and patients with inflammatory bowel disease, seborrheic dermatitis or on anticonvulsants, or long-term parenteral nutrition. Four SMVT splicing variants have been identified in the rat.[14]

Passive Diffusion

Both free biotin and nonhydrolyzed biocytin can be absorbed by diffusion, mainly in the jejunum. This becomes physiologically significant only at luminal concentrations >5 μM; even then, biocytin is less well absorbed than the free vitamin.

5. TRANSPORT OF BIOTIN

Transport in Plasma

Biotin is present in low amounts in plasma, most of which is soluble. Less than half is free biotin, the balance being composed of **bisnorbiotin**, **biotin sulfoxide**, and other unidentified metabolites. Some 7% is weakly bound nonspecifically to albumin, α- and β-globulins, and other proteins. An estimated 12% is bound covalently to proteins, predominantly biotinidase, which has both a high-affinity and a low-affinity biotin-binding sites and is thought to function as a biotin transporter. Biotinidase also occurs in human milk and at particularly high levels in colostrum, where it is thought to function in the transport of biotin by the mammary gland to the nursing infant. Biotin is also covalently bound to a plasma glycoprotein present in the serum of the pregnant rat.

11. Encoded by the *SLC5A6* gene.
12. Carbamazepine, primidone.
13. Subramanian, V.S., Marchant, J.S., Said, H.M., 2007. Gastroenterology 132, A583.
14. Said, H.M., 2004. Ann. Rev. Physiol. 66, 419–446.

TABLE 15.2 Biotin Availability in Several Feedstuffs

Feedstuff	Total Biotin[a] (μg/100g)	Available Biotin[b] (μg/100g)	Bioavailability[c] (%)
Barley	10.9	1.2	11
Corn	5.0	6.5	133
Wheat	8.4	0.4	5
Rapeseed meal	93.0	57.4	62
Sunflower seed meal	119.0	41.5	35
Soybean meal	25.8	27.8	108

[a]*Determined by microbiological assay.*
[b]*Determined by chick growth assay.*
[c]*Compared to d-biotin.*
Whitehead, C.C., Armstrong, J.A., Waddington, D., 1982. Br. J. Nutr. 48, 81–88.

TABLE 15.3 Inhibition of Enteric Biotin Transport by Ethanol

Ethanol, %, v/v	Biotin Transport, pmol/g tissue/15min
0	16.89±0.80[a]
0.5	15.16±1.02
1	12.56±1.03[b]
2	11.59±1.16[b]
5	6.61±0.42[a]

[a]*Mean±SD.*
[b]*p>.05.*
Said, H.M., Sharifian, A., Bergherzadeh, A., et al., 1990. Am. J. Clin. Nutr. 52, 1083–1086.

Cellular Uptake

The uptake of biotin into cells is facilitated an SMVT, which also functions in the cellular uptake of pantothenic acid and lipoic acid. Biotin uptake by peripheral blood mononuclear cells, and perhaps other lymphoid cells, appears to be facilitated by and occurs by a **monocarboxylate transporter (MCT1)**, which shows a K_m three orders of magnitude less than SMVT-mediated transport and is not competitively inhibited by either pantothenic or lipoic acids. The intracellular distribution of biotin closely parallels that of its carboxylases. It is found mostly in the cytoplasm (the primary location of acetyl-CoA carboxylase) and mitochondria (in which MCT1 has been detected). A small amount (<1%) is found in the nucleus,[15] which is thought to reflect binding to histones, as it increases (to c.1% of total cellular biotin) with cell proliferation.

Biotin appears to sense and regulate its own intracellular levels. This involves a novel function of **holocarboxylase synthetase (HCS)**, which moves into the nucleus when biotin is available and silences the biotin transporter SMVT by biotinylating histone 4 (H4) at the promoter. Gene silencing results from changes in chromatin structure affecting specific loci.[16] HCS translocation is thought to be regulated by tyrosine kinases and Zn finger proteins that direct the protein to specific chromatin regions.

Tissue Distribution

Appreciable storage of the vitamin appears to occur in the liver, where concentrations of 800–3000ng/g have been found in various species.[17] Most of this appears to be in mitochondrial acetyl-CoA carboxylase. Hepatic stores, however, appear to be poorly mobilized during biotin deprivation and, thus, do not show the reductions measurable in plasma under such conditions. Biotin is transported to the fetus by specific carriers including SMVT. Concentrations of the vitamin in fetal plasma are 3- to 17-fold greater than maternal levels. That milk biotin levels exceed those of maternal plasma by 10- to 100-fold evidences a mammary transport system. **Biotin-binding proteins** have been identified in the egg yolks of many species of birds where they are believed to function in transporting the vitamin into the oocyte, as their binding is weak enough to be reversible.[18] The yolk biotin-binding protein also occurs in the plasma of the laying hen.

15. Stanley, J.S., Griffen, J.B., Zempleni, J., 2001. Eur. J. Biochem. 268, 5424.

16. Gralla, M., Camporeale, G., Zempleni, J., 2008. J. Nutr. Biochem. 19, 400; Bao, B., Pestinger, V., Hassan, Y.I., et al., 2011. J. Nutr. Biochem. 22, 470.

17. These levels contrast with those of plasma/serum, typically c.300ng/L (humans, rats).

18. The yolk biotin-binding protein is a 74.3kDa glycoprotein with an homologous tetrameric structure, each subunit of which binds a biotin molecule. This protein is not to be confused with avidin, the biotin-binding protein of egg white, which irreversibly binds biotin with an affinity three orders of magnitude greater.

6. METABOLISM OF BIOTIN

Linkage to Apoenzymes

Free biotin is attached to its apoenzymes via the formation of an amide linkage to the ε-amino group of a specific lysyl residue. In each of the biotin-dependent enzymes, this binding occurs in a region containing the same amino acid sequence, -Ala-Met-biotinyl-Lys-Met-. The linkage is catalyzed by HCS.[19]

Biotin HCS. HCS is a small, monomeric enzyme that catalyzes the formation of the covalent linkage of the biotin prosthetic group to an ε-amino group of a lysyl residue on the apoenzyme (Fig. 15.1). The process is driven thermodynamically by the hydrolysis of ATP, which occurs in two steps:

1. activation of the vitamin as **biotinyl 5′-adenylate**; and
2. attachment to the **apocarboxylase** by an amide bond with the lysyl residue, with release of AMP.

HCS expression is dependent on adequate biotin status and appears to involve several variants in the range of 62–86 kDa. A high percentage of cellular HCS is present in the nucleus where it is thought to function also in the biotinylation of histones.

Recycling the Vitamin

The normal turnover of the biotin-containing holocarboxylases involves their degradation to yield biocytin, as the biotinyl lysine bond is not hydrolyzed by cellular proteases. Instead, biocytin is cleaved by biotinidase to yield free biotin. Biotinidase is the major biotin-binding protein in plasma; it is also present in breast milk, in which its activity is particularly high in colostrum. The proteolytic liberation of biotin from its bound forms is essential for the reutilization of the vitamin, which is accomplished by its reincorporation into another holoenzyme. Biotinidase may also catalyze the debiotinylation of histones. Mitochondrial biotinyl acetyl-CoA carboxylase may serve as a reservoir to maintain hepatic acetyl-CoA at appropriate levels in the cytosol. This would also provide biotin indirectly to support other biotinyl mitochondrial enzymes under biotin-limiting conditions.

Catabolism

Relatively, little catabolism of biotin is apparent in mammals. A small fraction is oxidized to biotin D- and L-sulfoxides; however, the ureido ring system is not otherwise degraded. The side chain of a larger portion is metabolized via mitochondrial β-oxidation to yield bisnorbiotin and its degradation

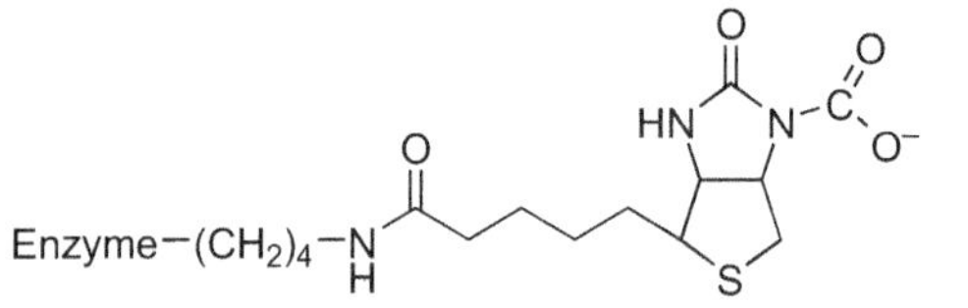

FIGURE 15.1 Biotin is bound covalently to its carboxylases.

products. Evidence suggests that biotin catabolism is greater in smokers than nonsmokers.[20]

Excretion

Biotin is rapidly excreted in the urine (Fig. 15.2). Studies have shown the rat to excrete about 95% of a single oral dose (5 mg/kg) of the vitamin within 24 h. Half of urinary biotin occurs as free biotin, the balance being composed of bisnorbiotin, bisnorbiotin methyl ketone, biotin sulfone, tetranorbiotin-L-sulfoxide, and various side chain products. Although unabsorbed biotin appears in the feces, much fecal biotin is of gut microbial origin and benefits the host by way of hindgut absorption. Thus, at low dietary levels of the vitamin, urinary excretion of biotin can exceed intake. Only a small amount (<2% of an intravenous dose) of biotin is excreted in the bile. The urinary excretion of patients with **achlorhydria** is very low; this may reflect impaired release of bound biotin for subsequent absorption.

7. METABOLIC FUNCTIONS OF BIOTIN

Biotin functions as essential cofactors for five carboxylases. It also functions as a regulator of gene expression, as a substrate for the posttranslational modification of proteins by biotinylation. Evidence also suggests additional biotin-containing proteins.

Carboxylations

Biotin functions in fatty acid metabolism, gluconeogenesis, and amino acid metabolism in key steps involving the transfer of covalently bound, single-carbon units in the most oxidized form, CO_2. This function is implemented by five biotin-dependent carboxylases and transcarboxylases[21] (Table 15.4) to which the biotin prosthetic group is attached by a single enzyme, biotin HCS.

The biotin-dependent carboxylases. Humans and other animals have five biotin-dependent carboxylases (Table 15.4). One (acetyl-CoA carboxylase 1) is cytosolic,

19. This is sometime called "**biotin holoenzyme synthetase**" and, in microorganisms, "**biotin protein ligase**."

20. Sealey, W.M., Teague, A.M., Stratton, S.L., et al., 2004. Am. J. Clin. Nutr. 80, 932.

21. **Transcarboxylase** has been called the "*Mickey Mouse enzyme*," as the electron micrograph image of the bacterial enzyme, with its large single subunit and two smaller flanking subunits, resembles the head of the famous rodent.

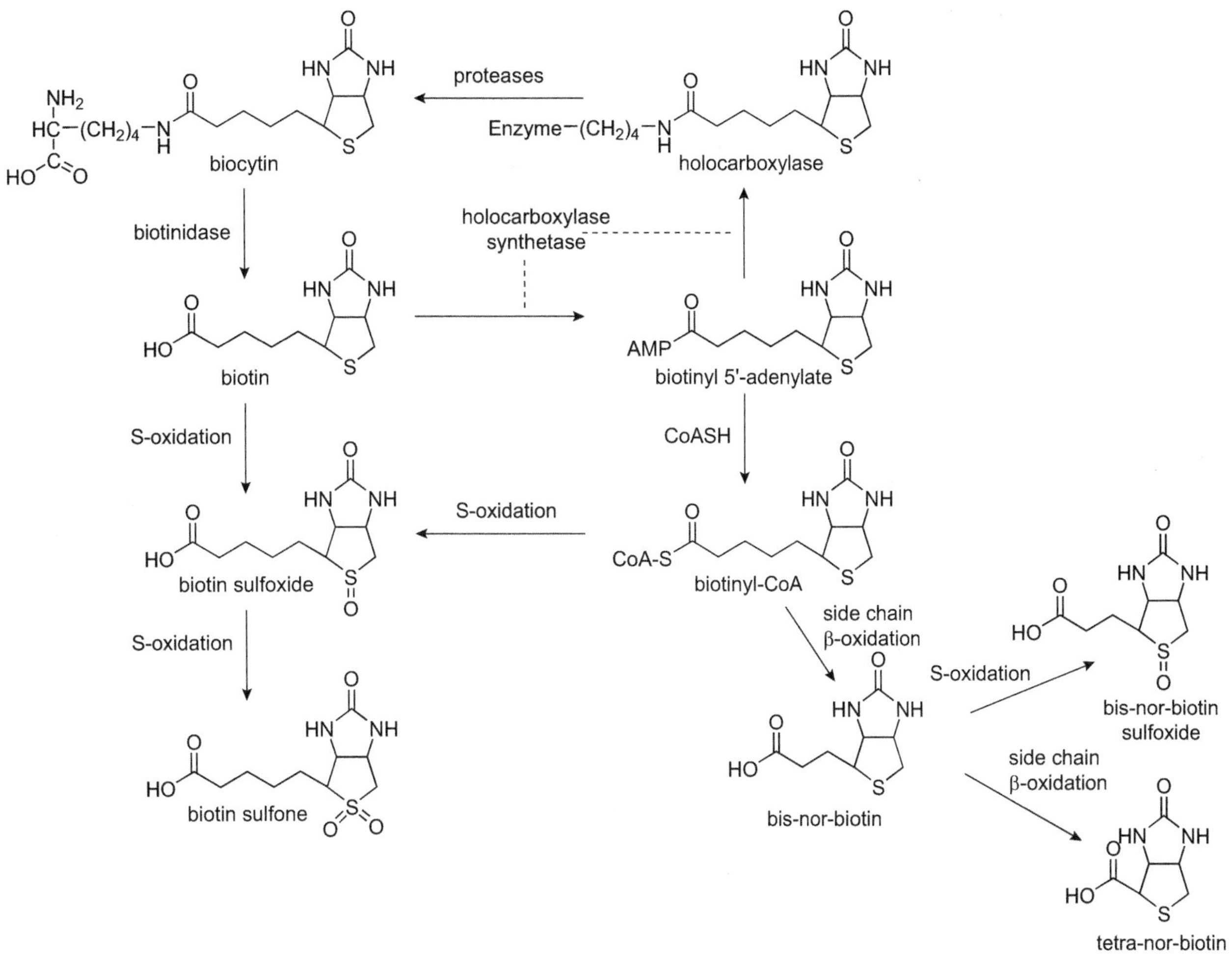

FIGURE 15.2 Biotin metabolism and recycling.

the others are mitochondrial. The catalytic action of each proceeds by a nonclassic, two-site, ping-pong mechanism, with partial reactions being performed by dissimilar subunits:

1. **Carboxylation** (at the carboxylase subsite)—involving the addition of a carboxyl moiety to the biotin ureido-N opposite the side chain, using the bicarbonate/ATP system as the carboxyl donor;
2. **Carboxyl transfer** (at the carboxyl transferase subsite)—involving transfer of the carboxyl group from carboxybiotin to the acceptor substrate.

In each enzyme, the carboxylase and carboxyl transfer two subsites appear to be spatially separated. The transfer of biotinyl CO_2 between those subsites is thought to be facilitated by movement of the prosthetic group back and forth through rotation of one or more of the 10 single bonds on the valeryl lysyl side chain.

Genetic disorders of biotin metabolism. Genetic defects in all of the biotin enzymes have been identified in humans (Table 15.5). These are rare, affecting infants and children, usually with serious consequences. Mutations in genes encoding the three proteins involved in biotin homeostasis (SMVT, biotinidase, HCS) can cause multiple carboxylase deficiencies. Biotinidase deficiency occurs at a frequency of 1 in 60,000 live births, with affected individuals having <30% of normal serum biotinidase activity, compromising their ability to release biotin from dietary proteins and recycle it from endogenous biotinylated proteins. They show the neurological and dermatological symptoms of biotin deficiency, with onset in the first year of life (sometimes within weeks). They also experience additional signs including hearing loss and optic atrophy, which has been interpreted as suggestive of other, still-unidentified functions of biotinidase. This can lead to irreversible neurologic damage but can be prevented by treatment with lifelong, high doses (5–20 mg/day) of biotin. Defects involving the absence of an apocarboxylase do not respond to supplements of the vitamin. They are treated by restricting dietary protein

TABLE 15.4 The Biotin-Dependent Carboxylases of Animals

Enzyme	Metabolic Function
Cytosol	
Acetyl-CoA carboxylase 1 (ACC1)	*Formation of malonyl-CoA from acetyl-CoA for carboxylase fatty acid synthesis in the cytoplasm*—ACC1 requires citrate; it catalyzes the incorporation of bicarbonate into acetyl-CoA to yield malonyl-CoA in the cytosol in the first committed step in the synthesis and elongation of fatty acids. Reduced ACC1 activity due to biotin deprivation impairs lipid synthesis
Mitochondria	
Pyruvate carboxylase (PC)	*Formation of oxaloacetate from pyruvate*—PC is a 130 kDa homologous tetramer. It requires acetyl-CoA and catalyzes the incorporation of bicarbonate into pyruvate to form oxaloacetate, a key step in gluconeogenesis. It also serves to replenish the mitochondrial supply of oxaloacetate to support the TCA (tricarboxylic acid) cycle and the formation of citrate for transport to the cytosol for lipogenesis. Reduced PC activity due to biotin deprivation can lead to fasting hypoglycemia, lactic acidosis, and ketosis
Acetyl-CoA carboxylase 2 (ACC2)	*Formation of malonyl-CoA from acetyl-CoA for carboxylase fatty acid synthesis in the mitochondria*—ACC2 requires citrate; it controls mitochondrial fatty acid oxidation by inhibitory effects of malonyl CoA, which it produces
Propionyl-CoA carboxylase (PCC)	Formation of methylmalonyl-CoA from propionyl-CoA produced by catabolism of some amino acids (e.g., isoleucine) and odd-chain fatty acids—PCC is a dimer with nonidentical subunits. It catalyzes the incorporation of bicarbonate into propionyl-CoA to form methylmalonyl CoA, which isomerizes to succinyl CoA and enters the TCA cycle for energy and glucose production
β-Methylcrotonyl-CoA carboxylase (βMCC)	*Part of the leucine degradation pathway*—βMCC is a dimer with nonidentical subunits. It catalyzes the degradation of leucine. Reduced activity due to biotin deprivation results in shunting the leucine degradation product β-methylcrotonyl-CoA through an alternate catabolic pathway to **3-hydroxyisovaleric acid**, which is excreted in the urine[22]

to limit the production of metabolites upstream of the metabolic lesion.

Gene Expression

Biotin has been shown to be necessary for the normal progression of cells through the cell cycle, with biotin-deficient cells arresting in the G1 phase. Proliferating lymphocytes increases their uptake of biotin, as well as their activities of β-methylcrotonyl-CoA carboxylase and propionyl-CoA carboxylase. These effects would appear to involve effects on gene expression, as biotin has been shown to affect the expression of genes encoding enzymes involved in glucose metabolism (glucokinase, phosphoenolpyruvate carboxykinase, ornithine transcarbamylase), cytokines (IL-2, IL-2 receptor γ, IL-1β, interferon-γ, IL-4[23]), amino acid metabolism (βMCC), and regulation of biotin status (SMVT, HCS). Evidence indicates that biotin status affects the regulation of gene expression by affecting cell signaling and histones.

Signaling by biotin has been demonstrated as follows:

- Biotin deprivation increases the nuclear translocation, binding, and transcriptional activity of **NK-κB**. This increases nuclear levels of p50 and p65, and activities of IκB, which activates genes involved in suppression of apoptosis.[24]
- Biotin increases intracellular production of the key second messenger **nitrous oxide (NO)**, which activates a soluble guanylate cyclase to increase cellular levels of cGMP. Those increases stimulate protein kinase G and lead to phosphorylation and activation of proteins that increase transcription of genes encoding HCS and some carboxylases.[25] It also suppresses endo/sarcoplasmic reticular Ca^{2+}-ATPase 3 (SERCA3), which functions in the transport of Ca^{2+} from the cytosol into the endoplasmic reticulum, to increase cytosolic concentrations of Ca^{+2} and stimulate protein unfolding.[26] That this effect may be important in immune cell function was indicated by the results of a human trial showing biotin supplementation to markedly reduce the expression of SERCA3 in lymphocytes of healthy adults.[27]
- Biotin appears necessary for the expression of the transcription factors **Sp1** and **Sp3**, which are associated

22. Urinary 3-hydroxyisovalerate has, therefore, been proposed as a biomarker of biotin deficiency.
23. Unlike the other proteins listed, which depend on biotin for expression, IL-4 and SMVT are downregulated by biotin.

24. Rodriguez-Melendez, R., Schwab, L.D., Zempleni, J., 2004. Int. J. Vitaminol. Nutr. Res. 74, 209.
25. Solorzano-Vargas, R.S., Pacheco-Alvarez, D., Leon-Del-Rio, A., 2002. Proc. Nat. Acad. Sci. U.S.A. 99, 5325.
26. Griffen, J.B., Rodriguez-Melendez, R., Dode, L., 2006. J. Nutr. Biochem. 17, 272.
27. Weidmann, S., Rodriguez-Melendez, R., Ortega-Cuellar, D., et al., 2004. J. Nutr. Biochem. 15, 433–439.

TABLE 15.5 Genetic Disorders of Biotin Metabolism

Defect	Metabolic Basis	Physiological Effect	Treatment
SMVT (sodium-dependent vitamin transporter) deficiency	Recessive *SLC19A3* mutation	Biotin-responsive basal ganglia disease.[a] *Symptom*: childhood onset of subacute encephalopathy, ataxia, seizures, dystonia, quadriparesis, hyperreflexia	High-dose biotin plus thiamin
Propionyl-CoA carboxylase deficiency	Autosomal recessive lack of enzyme[b]	Propionate accumulation: acidemia, ketoacidosis, hyperammonemia; high urine citrate, 3-OH-propionate, propionyl glycine. *Symptoms*[c]: vomiting, lethargy, hypotonia, mental retardation, cramps	Restrict protein
Pyruvate carboxylase deficiency	Autosomal recessive lack of enzyme[d]	Changes in energy production, gluconeogenesis, and other pathways. *Symptoms*: metabolic acidosis (lactate), hypotonia, mental retardation	None
β-Methylcrotonyl-CoA carboxylase deficiency	Defective enzyme (basis unknown[e])	High urine β-CH_3-crotonylglycine and 3-hydroxyisovaleric acid. *Symptoms*: cramps	Restrict protein
Acetyl-CoA carboxylase deficiency	Lack of enzyme (basis unknown[f])	Aciduria. *Symptoms*: myopathy, neurologic changes	None
Multiple carboxylase deficiency Neonatal type	Autosomal recessive lack of HCS (holocarboxylase synthetase)[g]	Deficiencies of all biotin-containing holocarboxylases; acidosis and aciduria. *Symptoms*: vomiting, lethargy, hypotonia	High-dose biotin; varied results[h]
Juvenile type	Autosomal recessive lack of biotinidase[i]	Deficiencies of all biotin-containing holocarboxylases; acidosis and aciduria. *Symptoms*: skin rash, alopecia, conjunctivitis, ataxia, developmental anomalies, neurological signs	Massive doses of biotin

[a] *20 cases have been reported.*
[b] *Incidence: 1 in 350,000.*
[c] *There is a wide variation in the clinical expression.*
[d] *Fewer than two dozen patients have been described.*
[e] *Three confirmed cases have been reported.*
[f] *One case has been described.*
[g] *Involves failure to link biotin to the apocarboxylases; some 30 HCS mutations have been reported.*
[h] *Response is related to residual HCS activity. Homozygous severe HCS deficiency is fatal unless diagnosed and treated early.*
[i] *Involves failure to release biotin from its bound forms in holocarboxylases; this reduces use of biotin in foods and blocks endogenous recycling of the vitamin.*

with increased expression of cytochrome *P*450 1B1 and reduced expression of SERCA3.

- Biotin activates **jun/fos**[28] signaling which, in turn, activates AP1-dependent pathways.[29] It has been proposed that jun/fos signaling may increase the expression of SMVT, which gene has an AP1-like site.
- Biotin deficiency activates signaling by **tyrosine kinases**, which may contribute to increasing SMVT-mediated biotin uptake.
- Biotin status has been shown to affect the abundance of small, noncoding RNAs in at least some cells: miRNA-539 in cultured human fibroblasts and miRNA-153 in human kidney carcinoma cells.[30]

Histone biotinylation by HCS has been reported to silence several genes. Three histones (H3 and H4 primary targets; H1 to H2A to lesser extents) can be covalently modified by biotinylation at a dozen-specific lysyl residues to affect chromatin structure and stability of repeat regions and transposable elements.[31] Both biotinidase and HCS catalyze the biotinylation of proteins; however, only HCS has been localized in the nucleus, suggesting that it is the physiologically relevant factor in histone biotinylation. While debiotinylation of histones occurs, the mechanism is not clear; it may be catalyzed by an isoform of biotinidase. In humans, <0.1% of histones appear to be biotinylated[32]; nevertheless, the epigenetic impact is significant. One-third of H4 has been found to be biotinylated at lysyl residue

28. Protooncogenes that combine to form the AP-1 transcription factor.
29. Rodriguez-Melendez, R., Griffen, J.B., Zempleni, J., 2006. J. Nutr. 135, 1659.
30. Bao, B., Rodriguez-Melendez, R., Wijeratne, S.S., et al., 2010. J. Nutr. 140, 1546–1551.
31. Reduction of histone biotinylation by biotin-deprivation reduced life span and heat tolerance in *Drosophila melanogaster* (Camporeale, G., Giordano, E., Rendina, R., et al., 2006. J. Nutr. 136, 2735).
32. Bailey, L.M., Ivanov, R.A., Wallace, J.C., et al., 2008. Anal. Biochem. 373, 71.

12 in telomeric repeats,[33] and enrichments in H3 (at lysyl residues 9 or 18) and H4 (at lysyl residue 8) have also been found. While these interactions contribute to the condensation of nuclear chromatin, it has been argued that histone biotinylation may be secondary to HCS interactions with chromatin proteins,[34] placing it in physical proximity to histones, and that histone biotinylation may have little effect on gene repression.[35]

Other Biotin-Containing Proteins

It is possible that there may be unidentified biotin-dependent proteins and enzymes. A mass spectrometric screening of human embryonic kidney cells detected more than a hundred biotinylated proteins.[36] While that number is likely to include proteins biotinylated nonspecifically by HCS, some proteins appeared to have been overrepresented. Those include heat shock proteins and enzymes involved in glycolysis and protein synthesis.

8. BIOMARKERS OF BIOTIN STATUS

Biotin status can be assessed in two ways:

- **Blood and urinary metabolites**—One of the first indicators of biotin deficiency is increased by circulating concentrations of **3-hydroxyisovaleryl carnitine**, and its derivative **carnityl-3-hydoxyisovaleric acid**,[37] which changes as a result of the alternative metabolism of β-methylcrotonyl-CoA by enoyl-CoA hydratase with declining biotin-dependent β-methylcrotonyl-CoA carboxylase activity.[38] Plasma/serum concentrations of biotin and its metabolites are less informative, as they tend to remain stable under conditions of moderate deficiency.
- **Degree of biotin saturation of biotin-dependent enzymes**—This is a useful means of assessing biotin status. It takes advantage of the in vitro binding of biotin by lymphocyte propionyl-CoA carboxylase (PCC). Because biotin-adequate subjects typically have most PCC bound to biotin, the stimulation of PCC activity by added biotin indicates nonsaturation due to suboptimal biotin status. This can be expressed as an activity coefficient:
  - **PCC activity coefficient** = baseline PCC activity/PCC activity with added biotin.

33. Wijeratne, S.S., Camporeale, G., Zempleni, J., 2010. J. Nutr. Biochem. 21, 310.
34. e.g., DNA methyltransferase I; methyl-CpG-binding protein 2; eukaryotic histone-lysine methyltransferase 1.
35. Zemplini, J., Liu, D., Camara, D.T., et al., 2014. Nutr. Rev. 72, 369–376.
36. Xue, J., Zhou, J., Zempleni, J., 2013. Am. J. Physiol. Cell Physiol. 305, C1240–C1245.
37. Mock, D.M., Henrich-Shell, C.L., Carnell, N., et al., 2004. J. Nutr. 134, 317–320.
38. Stratton, S.L., Horvath, T.D., Bogusiewicz, A., et al., 2010. Am. J. Clin. Nutr. 92, 1399–1406.

9. BIOTIN DEFICIENCY

Because biotin is widely distributed among foods and feedstuffs and is synthesized by the gut microbiome, simple deficiencies in animals or humans are rare (Table 15.6). In fact, recommended intakes for biotin have not been established in the United States (Table 15.7). Biotin deficiency can be induced by antagonists, the most prominent is **avidin**, a biotin-binding protein found in egg whites.

Egg White Injury

In the mid-1930s, it was found that biotin supplements prevented the dermatitis and alopecia produced in experimental animals by feeding uncooked egg white. Subsequently, the damaging factor was isolated and named avidin. It is a water-soluble, basic glycoprotein with a molecular mass of

TABLE 15.6 General Signs of Biotin Deficiency

Organ System	Change/Signs
General	Decreased appetite, growth
Dermatologic	Dermatitis, alopecia, achromotrichia
Skeletal	Perosis
Vital organs	Hepatic steatosis FLKS (fatty liver and kidney syndrome in poultry)

TABLE 15.7 Recommended Biotin Intakes

The United States		FAO/WHO	
Age-Sex	AI[a], μg/day	Age-Sex	RNI[b], μg/day
0–6 months	5	0–6 months	5
7–11 months	7	7–11 months	6
1–3 years	8	1–3 years	8
4–8 years	12	4–6 years	12
9–13 years	20	7–9 years	20
14–18 years	25	10–18 years	25
>18 years	30	19+ year	30
Pregnancy	30	Pregnancy	30
Lactation	35	Lactation	35

[a]Adequate Intakes are given, as Recommended Dietary Allowances (RDAs) have not been established; Food and Nutrition Board, 2000. Dietary Reference Intakes for Thiamin, Riboflavin, Niacin, Vitamin B_6, Folate, Vitamin B_{12}, Pantothenic Acid, Biotin and Choline. National Academy Press, Washington, DC, 564 pp.
[b]Recommended Nutrient Intakes; Joint WHO/FAO Expert Consultation, 2001. Human Vitamin and Mineral Requirements. Food Agriculture Org., Rome, 286 pp.

67 kDa. It is a homologous tetramer, each 128-amino acid subunit of which binds a molecule of biotin by linking to two to four tryptophan residues and an adjacent lysine in the subunit binding site. The binding of biotin to avidin is the strongest known noncovalent bond in nature[39]. Avidin is secreted by the oviductal cells of birds, reptiles, and amphibians and, thus, found in the whites of their eggs in which is thought to function as a natural antibiotic, as it is resistant to a broad range of bacterial proteases. It antagonizes biotin by forming with the vitamin a noncovalent complex[40] that is also resistant to pancreatic proteases, thus preventing the absorption of biotin.[41] The avidin–biotin complex is unstable to heat; heating to at least 100°C denatures the protein and releases biotin available for absorption. Therefore, although raw egg white is antagonistic to the utilization of biotin, the cooked product is without effect. The consumption of raw or undercooked whole eggs is probably of little consequence to biotin nutrition, as the biotin-binding capacity of avidin in the egg white is roughly comparable to the biotin content of the egg yolk. However, as a tool to produce experimental biotin deficiency, avidin in the form of dried egg white has been useful.[42]

Deficiency Signs in Humans

Clinical deficiency. Few cases of clinical manifestations of biotin deficiency have been reported in humans. Those have occurred in patients supported by TPN without biotin supplementation, in nursing infants whose mothers' milk contained inadequate supplies of the vitamin,[43] in infants born with congenital biotinidase deficiency, and in adults eating egg whites. Signs included periorificial dermatitis, ketolactic acidosis, conjunctivitis, alopecia, hypotonia, ataxia, seizures, developmental delays, and increased risk to skin infections. One case involved a child fed raw eggs for 6 years. The signs and symptoms included dermatitis, glossitis, anorexia, nausea, depression, hepatic steatosis, and hypercholesterolemia. The impairments of lipid metabolism responded to biotin therapy (Table 15.8).

TABLE 15.8 Effects of Biotin Treatment on Abnormalities in Serum Fatty Acid Concentrations in a Biotin-Deficient Human

Fatty Acid	Normal Values	Biotin-Deficient Patient Values: Before Biotin	Biotin-Deficient Patient Values: After Biotin
18:2ω6	21.56 ± 6.65	9.85[a]	5.36[a]
18:3ω6	0.21 ± 0.27	0.45	0.40
20:3ω6	3.67 ± 1.39	8.66[a]	10.62[a]
20:4ω6	12.49 ± 3.79	9.26	11.72
22:4ω6	1.87 ± 1.01	0.52[a]	0.71[a]
20:3ω9	1.30 ± 1.25	1.05	1.67
18:3ω3	0.21 ± 0.19	0.33	0.18
Total ω6 acids	41.08 ± 5.86	29.42[a]	29.61[a]
Total ω3 acids	5.23 ± 2.16	5.24	4.97
Total ω9 acids	13.14 ± 3.98	17.59[a]	16.4

[a]$p > .05$.
Mock, D.M., Johnson, S.B., Holman, R.T., 1988. J. Nutr. 118, 342–348.

The frequency of **marginal biotin status** (deficiency without clinical manifestation) is not known. Studies with validated biomarkers of biotin status indicate that subclinical biotin deficiency may occur in as many as one-third of pregnancies. That pregnant women experience increased catabolism of the vitamin is indicated by their increased urinary excretion of bisnorbiotin, biotin sulfoxide, and other biotin metabolites. Increased urinary excretion of 3-hydroxyisovaleric acid in late pregnancy has been found to respond to biotin supplementation.[44] Relatively low levels of biotin (versus healthy controls) have been reported in the plasma or urine of patients with partial gastrectomy or other causes of achlorhydria, burn patients, epileptics,[45] elderly individuals, alcoholics,[46] and athletes. It has been suggested that vegetarians may be at risk for deficiency; however, studies have failed to support that hypothesis. In fact, both plasma and urinary biotin levels of strict

39. $K_a = 1015\,M$.
40. Two similar biotin-binding proteins have been identified, both of which show considerable sequence homology with avidin at the biotin-binding site: **streptavidin** from *Streptomyces avidinii* and an epidermal growth factor homolog in the purple sea urchin *Strongylocentrotus purpuratus*.
41. Some cultured mammalian cells (e.g., fibroblasts and HeLa cells) are able to absorb the biotin–avidin complex, using it as a source of the vitamin.
42. Other structural analogs of biotin are also antagonistic to its function: α-dehydrobiotin, 5-(2-thienyl)valeric acid, acidomycin, α-methylbiotin, and α-methyldethiobiotin; several of these are antibiotics.
43. The biotin content of human milk, particularly early in lactation, is often insufficient to meet the demands of infants for which reason it is recommended that nursing mothers take a biotin supplement. That practice substantially increases the biotin contents of their breast milk, e.g., a 3 mg/day supplement increases milk biotin from 1.2–1.5 μg/dL to >33 μg/dL.

44. Mock, D.M., Stadler, D.D., Stratton, S.L., et al., 1997. J. Nutr. 127, 710–716.
45. This may be due to anticonvulsant drug therapy, known side effects of which are dermatitis and ataxia. Some anticonvulsants (e.g., carbamazepine, primidone) are competitive inhibitors of biotin transport across the intestinal brush border.
46. About 15% of alcoholics have plasma biotin concentrations <140 pM, a level shown by only 1% of nonalcoholics.

vegetarians (vegans) and lactoovovegetarians[47] have been found to exceed those of persons eating mixed diets. That hemodialysis can deplete patients of biotin is suggested by the finding that biotin supplementation reduced the severity of muscle cramps in hemodialysis patients.[48]

Deficiency Signs in Animals

Avidin-induced biotin deficiency causes the syndrome originally referred to as **egg white injury**. The major lesions appear to involve impairments in lipid metabolism and energy production. In rats and mice, this is characterized by seborrheic dermatitis and **alopecia**, a hind limb paralysis that results in **kangaroo gait**. In mice and hamsters, it involves teratogenic effects indicated by congenital malformations: cleft palate, micrognathia,[49] micromelia.[50] Fur-bearing animals (mink and fox) show general dermatitis with hyperkeratosis, circumocular alopecia ("spectacle eye"), **achromotrichia** of the underfur, and unsteady gait. Pigs and kittens show weight loss, digestive dysfunction, dermatitis, alopecia, and brittle claws. Guinea pigs and rabbits show weight loss, alopecia, and achromotrichia. Monkeys show severe dermatitis of the face, hands, and feet; alopecia; and watery eyes with encrusted lids. The dermatologic lesions of biotin deficiency relate to impairments of lipid metabolism; affected animals show reductions in skin levels of several long-chain fatty acids (16:0,[51] 16:1, 18:0, 18:1 and 18:2) with concomitant increased in certain others (in particular, 24:1 and 26:1). All species show depressed activities of the biotin-dependent carboxylases, which respond rapidly to biotin therapy.

Biotin deficiency can be produced in chicks by dietary deprivation and seems to occur sporadically in practical poultry production, particularly in northern Europe.[52] This results in impaired growth and reduced efficiency of feed utilization, and is characterized by circumocular alopecia (Fig. 15.3), dermatitis mainly at the corners of the beak, but also of the footpad (Fig. 15.4).[53] In some instances, death occurs suddenly without gross lesions; this condition usually involves hepatic and renal steatosis with hypoglycemia, lethargy, paralysis, and hepatomegaly, and is thus referred to as **fatty liver and kidney syndrome (FLKS)**. The etiology of FLKS appears to be complex, involving such other factors as choline, but seems to involve a marginal deficiency of biotin that impairs gluconeogenesis by limiting the activity of pyruvate carboxylase, especially under circumstances of glycogen depletion brought on by stress.

47. Individuals eating plant-based diets that include dairy products and eggs.
48. Oguma, S., Ando, I., Hirose, T., et al., 2012. Tohoku J. Exp. Med. 227, 217–223.
49. Underdevelopment of the (usually lower) jaw.
50. Undergrowth the limbs.
51. i.e., A 16-carbon fatty acid with no double bonds.
52. In that part of the world, barley and wheat, each of which has little biologically available biotin, are frequently used as major ingredients in poultry diets.
53. **Footpad dermatitis** caused by biotin deficiency is often confused with the dermatologic lesions of the foot caused by pantothenic acid deficiency. Unlike the latter, biotin deficiency lesions are limited to the footpad and do not involve the toes and superior aspect of the foot.

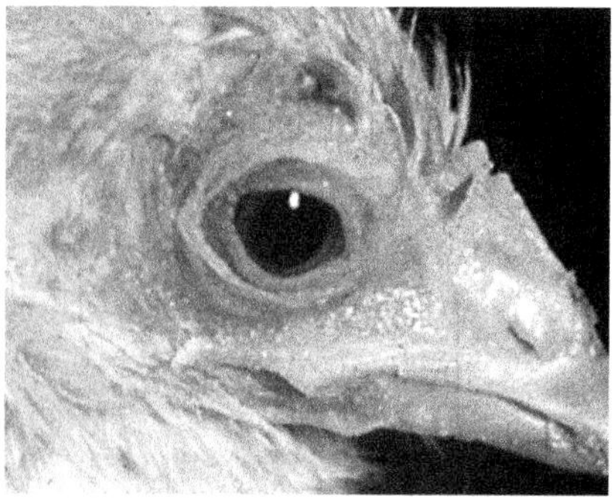

FIGURE 15.3 Loss of circumocular alopecia ("spectacle eye") in the biotin-deficient chick. *Courtesy, G.F. Combs, Sr.*

10. BIOTIN IN HEALTH AND DISEASE

Birth Defects

It has been suggested that marginal biotin status may be teratogenic. Fetal malformations have been produced in mice (Table 15.9) and poultry by feeding maternal diets containing marginal biotin levels, i.e., amounts that did not produce clinical signs of deficiency in the dams. Because humans appear to have relatively poor transport of biotin across the placenta,[54] it has been suggested that human fetuses may be predisposed to biotin deficiency when maternal intakes of the vitamin are marginal. Support for this hypothesis comes from observations that the production of arachidonic acid and prostaglandins, which depend on acetyl-CoA carboxylase and PCC activities, is required for normal palatal plate growth, elevation and fusion in mice, and skeletal development in chicks. That marginal biotin status may be prevalent was suggested by the finding that apparently healthy pregnant women had increased rates of biotin excretion and abnormally low activities of PCC.[55]

Sudden Infant Death Syndrome

It has been suggested that marginal biotin status may play a role in the etiology of sudden infant death syndrome (SIDS), which occurs in human infants at 2–4 months of age. In many ways, SIDS resembles FLKS in the chick, which is caused by biotin deprivation. Studies have shown

54. Schenker, S., Hu, Z., Johnson, R.F., et al., 1993. Alcohol Clin. Exp. Res. 17, 566.
55. Mock, D.M., Stadler, D.D., 1997. J. Am. Coll. Nutr. 16, 252.

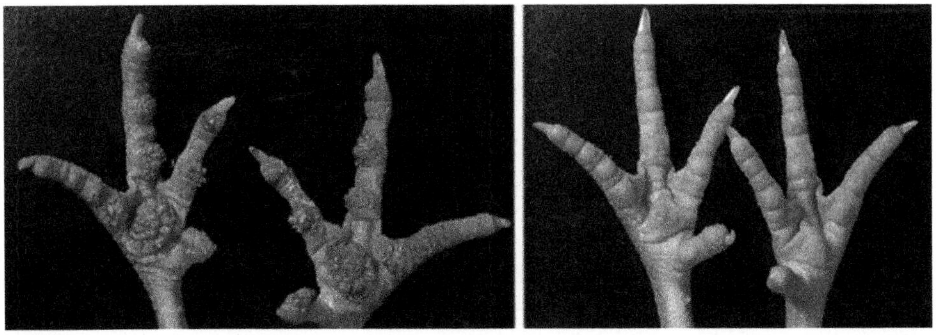

FIGURE 15.4 Footpad dermatitis in the chick (affected, left; biotin-adequate control, right). *Courtesy, G.F. Combs, Sr.*

TABLE 15.9 Effect of Maternal Egg White Feeding on Fetal Malformations in Mice

	Dietary Egg White, %						
Malformation	0	1	2	3	5	10	25
Cleft palate	0.10±0.13	0.25±0.25	2±2	4±2	10±1	11±1	12±0.4
Micrognathia	0	0	0.2±0.2	3±2	9±1	11±1	12±1
Microglossia	0	0	0	0.5±0.5	2±1	6±2	9±3
Hydrocephaly	0	0	0	0.3±0.2	2±1	3±1	3±2
Open eye	0	0	0	0	0.8±0.5	4±1	5±2
Forelimb hypoplasia	0	0	0	7±2	9±2	11±1	12±0.4
Hind limb hypoplasia	0	0	0	5±2	9±2	10±0.8	12±0.4
Pelvic girdle hypoplasia	0	0	0	5±2	9±2	10±1	12±0.5

Mock, D.M., Mock, N.I., Stewart, C.W., et al., 2003. J. Nutr. 133, 2519–2526.

that infants who died of SIDS had significantly lower hepatic concentrations of biotin than did those who died of unrelated causes.[56]

11. BIOTIN TOXICITY

The toxicity of biotin appears to be very low. No cases have been reported of adverse reactions by humans to high levels (doses as high as 200 mg orally or 20 mg intravenously) of the vitamin, as are used in treating seborrheic dermatitis in infants, egg white injury, or inborn errors of metabolism. Animal studies have revealed few, if any, indications of toxicity, and it is probable that animals, including humans, can tolerate the vitamin at doses at least an order of magnitude greater than their respective nutritional requirements. Upper tolerable intakes have not been established for biotin.

56. Johnson, A.R., Hood, R.L., Emery, J.L., 1980. Nature 285, 159.

12. CASE STUDY

Instructions

Review the following case report, paying special attention to the diagnostic indicators on which the treatments were based. Then, answer the questions that follow.

Case

A 12-month-old girl had experienced malrotation[57] and midgut volvulus,[58] resulting in extensive infarction[59] of the small and large bowel at 4 months of age. Her bowel was resected, after which her clinical course was complicated by failure of the anastomosis[60] to heal, peritoneal

57. Failure of normal rotation of the intestinal tract.
58. Twisting of the intestine, causing obstruction.
59. Necrotic changes resulting from obstruction of an end artery.
60. An operative union of two hollow or tubular structures, in this case the divided ends of the intestine.

infection, and intestinal obstruction. After several subsequent surgeries, she was left with only 30 cm of jejunum, 0.5 cm of ileum, and approximately 50% of colon. By 5 months of age, she had lost 1.5 kg in weight and TPN[61] was initiated (providing 125 kcal/kg/day). By the third month of TPN, she had gained 2.9 kg; thereafter, her energy intake was reduced to 60 kcal/kg/day, which sustained her growth within the normal range. Soybean oil emulsion[62] was administered parenterally at least twice weekly in amounts that provided 3.9% of total calories as linoleic acid. Repeated attempts at feeding her orally failed because of vomiting and rapid intestinal transit; therefore, her only source of nutrients was TPN. She had repeated episodes of sepsis and wound infection; broad-spectrum antibiotics were administered virtually continuously from 4 to 11 months of age. Multiple enteroenteric and enterocutaneous fistulas[63] were formed; over 8 months, they provided daily fluid losses >500 mL.

During the third month of TPN, an erythematous[64] rash was noted on the patient's lower eyelids adjacent to the outer canthi.[65] Over the next 3 months, the rash spread became more exfoliative and exuded clear fluid. New lesions appeared in the angles of the mouth, around the nostrils, and in the perineal region.[66] This condition did not respond to topical application of various antibiotics, cortisone, and safflower oil.

During the fifth and sixth months of TPN, the patient lost all body hair developed a waxy pallor, irritability, lethargy, and mild hypotonia.[67] That she was not deficient in essential fatty acids was indicated by the finding that her plasma fatty acid triene-to-tetraene ratio was normal (0.11). During the period from the third to the sixth month, the patient was given parenteral zinc supplements at 7, 30, and 250 times the normal requirement (0.2 mg/day). Her serum zinc concentration increased from 35 to 150 μg/dL (normal, 50–150 μg/dL) and, finally, to greater than 2000 μg/dL without any beneficial effect. Intravenous zinc supplementation was then reduced to 0.4 mg/day. Biotin was determined by a bioassay using *Ochromonas danica*; urinary organic acids were determined by HPLC[68] and GC/MS[69]:

61. Feeding by means other than through the alimentary canal, referring particularly to the introduction of nutrients into veins.
62. e.g., Intralipid.
63. Passages created between one part of the intestine and another (an enteroenteric fistula) or between the intestine and the skin of the abdomen (an enterocutaneous fistula).
64. Marked by redness of the skin owing to inflammation.
65. Corners of the eye.
66. The area between the thighs extending from the coccyx to the pubis.
67. A condition of reduced tension of any muscle, leading to damage by overstretching.
68. High-performance liquid–liquid partition chromatography.
69. Gas–liquid partition chromatography with mass spectrometric detection.

Laboratory Results

Parameter	Patient	Normal Range
Plasma biotin	135 pg/mL	215–750 pg/mL
Urinary biotin excretion	<1 μg/24 h	6–50 μg/24 h
Urinary Organic Acid Excretion		
Methylcitrate	0.1 μmol/mg creatinine	<0.01 μmol/mg creatinine
3-Methylcrotonylglycine	0.7 μmol/mg creatinine	<0.2 μmol/mg creatinine
3-Hydroxyisovalerate	0.35 μmol/mg creatinine	<0.2 μmol/mg creatinine

Treatment with biotin (10 mg/day) was initiated and, after 1 week, the plasma biotin concentration increased to 11,500 pg/mL and organic acid excretion dropped to <0.01 μmol/mg creatinine. After 7 days of biotin supplementation, the rash had improved strikingly and the irritability had resolved. After 2 weeks of supplementation, new hair growth was noted, waxy pallor of the skin was less pronounced, and hypotonia improved. During the next 9 months of biotin therapy, no symptoms and signs of deficiency recurred. The patient's rapid transit time and vomiting did not improve.

Case Questions

1. What signs were first to indicate a problem related to biotin utilization by the patient?
2. What is the relevance of aciduria to considerations of biotin status?
3. How were problems involving essential fatty acids and zinc ruled out in the diagnosis of this condition as biotin deficiency?

13. STUDY QUESTIONS AND EXERCISES

1. Diagram the areas of metabolism in which biotin-dependent carboxylases are involved.
2. Construct a decision tree for the diagnosis of biotin deficiency in humans or an animal species.
3. What key feature of the chemistry of biotin relates to its biochemical function as a carrier of active CO_2?
4. What parameters might you measure to assess biotin status of a human or animal?

RECOMMENDED READING

Beckett, D., 2009. Biotin sensing at the molecular level. J. Nutr. 139, 167–170.

Hassam, Y.I., Zemplini, Y., 2008. A novel, enigmatic histone modification: biotinylation of histones by holocarboxylase synthetase. Nutr. Rev. 66, 721–725.

Marin-Valencia, I., Roe, C.R., et al., 2010. Pyruvate carboxylase deficiency: mechanisms, mimics and anaplerosis. Mol. Gen. Metab. 101, 9–17.

Mock, D.M., 2014. Biotin. In: Zemplini, J., Suttie, J.W., Gregory, J.F., et al. (Eds.), Handbook of Vitamins, fifth ed. CRC Press, New York, pp. 397–419. (Chapter 10).

Said, H.M., 2011. Intestinal absorption of water-soluble vitamins in health and disease. Biochem. Jr. 437, 357–372.

Tang, L., 2013. Structure and function of biotin-dependent carboxylases. Cell Mol. Life Sci. 70, 863–891.

Waldrop, G.L., Holden, H.M., St. Maurice, M., 2012. The enzymes of biotin dependent CO_2 metabolism: what structures reveal about their reaction mechanisms. Protein Sci. 21, 1597–1619.

Wolf, B., 2010. Clinical issues and frequent questions about biotinidase deficiency. Mol. Gen. Metab. 100, 6–13.

Zempleni, J., 2012. Biotin. In: Erdman, J.W., Macdonald, I.A., Zeilsel, S.H. (Eds.), Present Knowledge in Nutrition, tenth ed. Wiley-Blackwell, New York, pp. 359–374. (Chapter 23).

Zempleni, J., Liu, D., Teixeira, G.C., et al., 2014. Mechanisms of gene transcriptional regulation through biotin and biotin-binding proteins in mammals. In: Dakshinamurti, K., Dakshinamurti, S. (Eds.), Vitamin-binding Proteins, pp. 219–228. (Chapter 13).

Zemplini, J., Liu, D., Camara, D.T., et al., 2014. Novel roles of holocarboxylase synthetase in gene regulation and intermediary metabolism. Nutr. Rev. 72, 369–376.

Chapter 16

Pantothenic Acid

Chapter Outline

Anchoring Concepts

1. Pantothenic acid is the trivial designation for the compound dihydroxy-β,β-dimethylbutyryl-β-alanine.
2. Pantothenic acid is metabolically active as the prosthetic group of coenzyme A (CoA) and the acyl carrier protein.
3. Deficiencies of pantothenic acid are manifested as dermal, hepatic, thymic, and neurologic changes.

A pellagrous-like syndrome in chicks has recently been obtained...in an experiment that was originally designed to throw added light upon an unusual type of leg problem occurring in chicks fed semi-synthetic rations....The data obtained in this experiment demonstrate the requirement in another species of the vitamin or vitamins present in autoclaved yeast, occasionally called vitamin B_2, vitamin G or the P–P factor, and indicate that the chick may be a more suitable animal than the white rat for delineating the quantities of this vitamin present in feedstuffs.

L.C. Norris and A.T. Ringrose[1]

LEARNING OBJECTIVES

1. To understand the chief natural sources of pantothenic acid.
2. To understand the means of absorption and transport of pantothenic acid.
3. To understand the biochemical functions of pantothenic acid as components of coenzyme A and the acyl carrier protein.
4. To understand the physiological implications of low-pantothenic acid status.

VOCABULARY

Acetyl CoA
Acyl carrier protein (ACP)
Acyl-CoA synthetase (ACS)
Burning feet syndrome
CoA synthetase
Coenzyme A (CoA)
Dephospho-CoA kinase
Dexpanthenol
Fatty acid synthetase
Malonyl CoA
ω-Methylpantothenic acid
Pantothenate kinase (PanK)
Pantothenic acid
Pantothenol
Pantetheine
Pantetheinase
Pantetheinase
Phosphopantetheine adenylytransferase
Phosphopantetheine-apo-ACP transferase
Phosphopantothenylcysteine decarboxylase
Phosphopantothenylcysteine synthase
4′-Phosphopantetheine

1. Leo Chandler Norris (1891–1986) was a pioneering American nutritionist and a founder of the nutrition programs at Cornell University. He did seminal work on "vitamin G" (later, riboflavin) and the "animal protein factor" (later, vitamin B_{12}), and discovered the essentiality of manganese. One of Norris's students, Arthur T. Ringrose (1908-?), went on to a career as a poultry nutritionist at the University of Kentucky. His students also included Milton L. Scott and Gerald F. Combs, the senior author's major professor and father, respectively.

The Vitamins. http://dx.doi.org/10.1016/B978-0-12-802965-7.00016-2

4′-Phosphopantothenic acid
Propionyl CoA
Sodium-dependent multivitamin transporter (SMVT)
Succinyl CoA

1. THE SIGNIFICANCE OF PANTOTHENIC ACID

Pantothenic acid is widely distributed in many foods. Clinical deficiencies of the vitamin are rare. Pantothenic acid functions as the essential precursor of **coenzyme A (CoA),** which is used by many cellular enzymes, and **acyl carrier protein** (**ACP**). In these forms, pantothenic acid plays essential roles in the metabolism of fatty acids, amino acids, and carbohydrates, and has important roles in the acylation of proteins. Although pantothenic acid is required to produce CoA, rates of CoA synthesis are not affected by deprivation of the vitamin. From such observations, it can be inferred that the vitamin is recycled metabolically; however, definitive understanding of the mechanisms involved remains incomplete.

2. PROPERTIES OF PANTOTHENIC ACID

Pantothenic acid is the trivial designation for the compound dihydroxy-β,β-dimethylbutyryl-β-alanine.[2] It consists of β-alanine joined to 2,4-dihydroxy-3,3-dimethylbutyric acid by an amide linkage, and it is optically active. The vitamin has two metabolically active forms:

- **Coenzyme A (CoA)**[3] in which pantothenic acid has a phosphodiester linkage with adenosine-3′5′-diphosphate
- **ACP** in which pantothenic acid has a phosphodiester linkage with a serinyl residue of the protein.

Chemical structures of pantothenic acid:

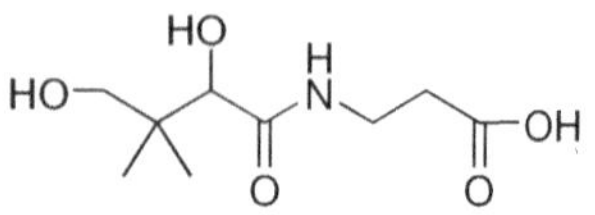

Pantothenic acid

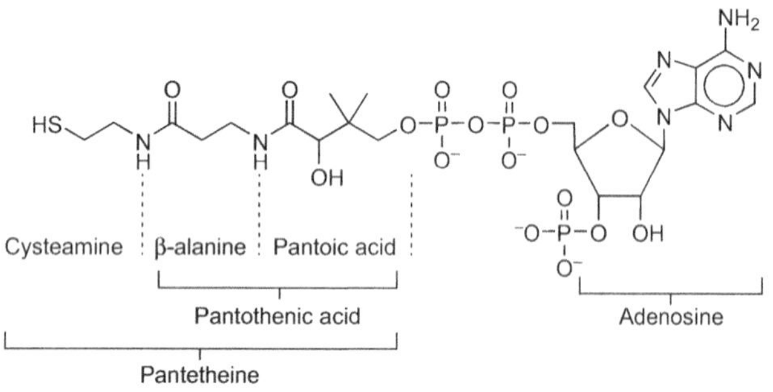

Coenzyme A (showing its constituent parts)

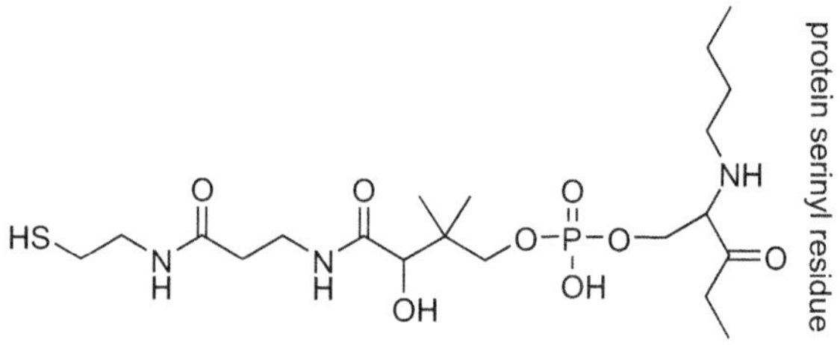

Acyl-carrier protein

Pantothenic Acid Chemistry

Pantothenic acid has an asymmetric center; only the *R*-enantiomer, usually called **d-(+)pantothenic acid**, is biologically active and occurs naturally. Pantothenic acid is a yellow, viscous oil. Its calcium and other salts, however, are colorless and crystalline; calcium pantothenate is the main product of commerce. Neither form is soluble in organic solvents, but each is soluble in water and ethanol. Aqueous solutions of pantothenic acid are unstable to heating under acidic or alkaline conditions, resulting in the hydrolytic cleavage of the molecule (to yield β-alanine and 2,4-dihydroxy-3,3-dimethylbutyrate). The analog panthenol (in which the carboxyl group is replaced by a hydroxymethyl group) is fairly stable in solution. In dry form, the salts are stable to air and light; but they (particularly sodium pantothenate) are hygroscopic.

3. SOURCES OF PANTOTHENIC ACID

Hindgut Microbial Synthesis

Pantothenic acid can be produced by the microbiome of the colon. A genomic analysis of 256 representative organisms of the human gut microbiota found more than half capable of de novo synthesis of the vitamin.[4] However, the total synthetic capacity appeared to be low, e.g., <0.1% of the daily human need. While direct evidence is lacking for the absorption of pantothenic acid across the colon, one study found it necessary to use an antibiotic to produce signs of pantothenic acid deficiency in the mouse.[5] Pantothenic acid has been shown to be produced by some rumen microorganisms.[6] That pantothenic acid deficiency has not been reported in ruminants is consistent with their microbiome being a nutritionally significant source of the vitamin.

Distribution in Foods

As its name implies, pantothenic acid is widely distributed in nature (Table 16.1). It occurs mainly in bound forms

2. Formerly known as *pantoyl*-β-alanine.
3. Studies with liver slices in vitro have demonstrated a correlation between hepatic CoA content and lipid biosynthetic capacity, suggesting that CoA may be a limiting factor in lipogenesis.

4. Magnúsdóttir, S., Ravchee, D., de Crécy-Lagard, V., et al., 2015. Front. Genet. 6, 148–166.
5. Stein, E.D., Diamond, J.M., 1989. J. Nutr. 119, 1973–1983.
6. *Escherichia coli* and *Streptococcus bovis* (Ford, J.E., Perry, K.D., Briggs, C.A.E., 1958. J. Gen. Microbiol. 18, 273–284; Porter, W.G., 1961. Vitamin synthesis in the rumen. In: Lewis, D. (Ed.), Digestive Physiology and Nutrition of the Ruminant. Butterworths, London, p. 226–233).

TABLE 16.1 Pantothenic Acid Contents of Foods

Food	Pantothenic Acid, mg/100g
Dairy Products	
Milk	0.34–0.37
Cheeses	0.08–1.73
Meats	
Beef	0.31–0.67
Pork	0.40–1.71
Chicken giblets	2.97
Cereals	
Cornmeal	0.43
Rice, brown	0.38
Oats	1.35
Wheat flour	0.44
Wheat bran	2.18
Barley	0.14
Vegetables	
Asparagus	0.23
Broccoli	0.57
Cabbage	0.21
Carrots	0.27
Cauliflower	0.67
Lentils	2.14
Potatoes	0.56
Soybeans	0.15
Tomatoes	0.09
Fruits	
Apples	0.06
Bananas	0.33
Grapefruits	0.28
Oranges	0.26
Strawberries	0.13
Nuts	
Cashews	1.27
Peanuts	1.01
Walnuts	1.66
Other	
Eggs	1.53
Mushrooms	0.41–21.9
Bakers' yeast	13.5

From USDA National Nutrient Database for Standard Reference, Release 28 (http://www.ars.usda.gov/ba/bhnrc/ndl).

(CoA, CoA esters, ACP). A glycoside has been identified in tomatoes. Therefore, it must be determined in foods and feedstuffs after enzymatic hydrolysis to liberate the vitamin from CoA. This is done in a two-step procedure using alkaline phosphatase followed by avian hepatic **pantetheinase**, yielding "total" pantothenic acid.

The most important food sources of pantothenic acid are meats (liver and heart are particularly rich). Mushrooms, avocados, broccoli, and some yeasts are also rich in the vitamin; however, cooking, canning, and freezing can produce losses of 35–80%. Whole grains are also good sources; however, the vitamin is localized in the outer layers, thus, it is largely (up to 50%) removed by milling. The most important sources of pantothenic acid for animal feeding are rice and wheat brans, alfalfa, peanut meal, molasses, yeasts, and condensed fish solubles. The richest sources of the vitamin in nature are cold water fish ovaries (>2.3 mg/g)[7] and royal jelly (>0.5 mg/g).[8]

Stability

Pantothenic acid in foods and feedstuffs is fairly stable to ordinary means of cooking and storage. It can, however, be unstable to heat and either alkaline ($pH>7$) or acid ($pH<5$) conditions.[9] Reports indicate losses of 15–50% from cooking meat and of 37–78% from heat-processing vegetables. The alcohol derivative, pantothenol, is more stable; for this reason, it is used as a source of the vitamin in multivitamin supplements.

Bioavailability

The biologic availability of pantothenic acid from foods and feedstuffs is a function of the efficiency of the enteric hydrolysis of its food forms and the absorption of those products. This area has not been well investigated. One study indicated "average" bioavailability of the vitamin in the American diet to be in the range of 40–60%;[10] similar results were obtained for maize meals in another study.[11]

4. ABSORPTION OF PANTOTHENIC ACID

Hydrolysis of Coenzyme Forms

Because pantothenic acid occurs in most foods and feedstuffs as CoA and ACP, the utilization of the vitamin in

7. Tuna, cod.
8. Royal jelly, the food responsible for the diet-induced reproductive development of the queen honeybee, is also the richest natural source of biotin.
9. Pasteurization of milk, because it occurs at neutral pH, does not affect its content of pantothenic acid.
10. Tarr, J.B., Tamura, T., Stokstad, E.L., 1981. Am. J. Clin. Nutr. 34, 1328–1337.
11. Yu, B.H., Kies, C., 1993. Plant Food Hum. Nutr. 43, 87–95.

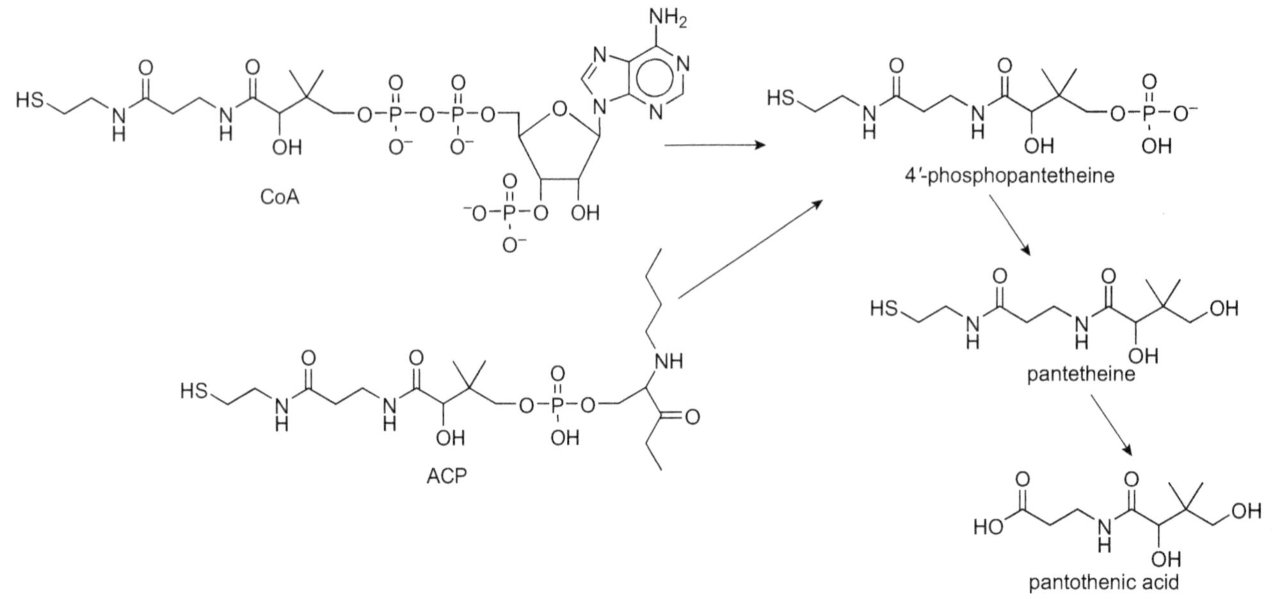

FIGURE 16.1 Liberation of pantothenic acid from coenzyme forms in foods.

foods depends on the hydrolytic digestion of these protein complexes to release the free vitamin. Both CoA and ACP are degraded in the intestinal lumen by hydrolases (pyrophosphatase, phosphatase) to release the vitamin as **4′-phosphopantetheine** (Fig. 16.1). That form is dephosphorylated to yield **pantetheine**, which is absorbed or converted to **pantothenic acid** by another intestinal hydrolase, **pantetheinase.**

Free pantothenic acid is absorbed by two mechanisms:

- **Active transport** at low luminal concentrations. Pantothenic acid[12] is absorbed by a saturable mechanism facilitated by **Na$^+$-dependent multivitamin transporter** (**SMVT**) located on the apical membrane of epithelial brush boarder, particularly villus, cells. This transporter has an apparent K_m of 10–20 μM[13] for pantothenic acid. It also facilitates the uptake of biotin and lipoic acid and can be inhibited by alcohol. The SMVT facilitates the exchange of the pantothenic acid for protons, driven by an extracellularly directed proton gradient. Pantothenic acid appears to be able to cross the enterocyte either in free solution or bound to SMVT via trafficking vesicles and the microtubule network and the motor protein dynein. It is thought that translocation of the vitamin across the basolateral membrane to the portal circulation involves another SMVT.
- **Passive diffusion** at higher luminal concentrations. The alcohol form, **panthenol,** which is oxidized to pantothenic acid in vivo, appears to be diffuse somewhat faster than the acid form.

12. Concentrations after a typical meal have been estimated at 1–2 μM.
13. Said, H.M., 2011. Biochem. J. 437, 357–372.

5. TRANSPORT OF PANTOTHENIC ACID

Plasma and Erythrocytes

Plasma contains the vitamin only in the free acid form. Erythrocytes carry 20–55% of the vitamin in the blood.[14]

Cellular Uptake

Pantothenic acid is taken into cells in its free acid form. In most tissues, this is mediated by the membrane SMVT; however, uptake by erythrocytes and the brain occurs by diffusion.[15] The active uptake of pantothenic acid results in its cellular concentrations being much greater (liver, 10–15 μM; heart ~100 μM) than those of plasma (1–5 μM). Upon cellular uptake, most of the vitamins combine with cysteamine,[16] adenine, and ribose-3′-phosphate, converting it to CoA, the predominant intracellular form, 70–90% of which is in the mitochondria. Erythrocytes metabolize the vitamin to 4′-phosphopantothenic acid, which, lacking the enzymes to produce CoA, they accumulate.

Tissue Distribution

The greatest concentrations of CoA are found in the liver, adrenals, kidneys, brain, heart, and testes.[17] Much of this (70% in liver, 95% in heart) is located in the mitochondria. Tissue CoA concentrations are not affected by deprivation of

14. For example, in the human adult, whole blood contains 1120–1960 ng/mL of total pantothenic acid; of that, plasma contains 211–1096 ng/mL. Blood pantothenic acid levels are generally lower in elderly individuals, e.g., 500–700 ng/mL.
15. Spector, R., 1986. Am. J. Physiol. 250, R292–R297.
16. i.e., Mercaptoethylamine.
17. The human liver typically contains ca. 28 mg of total pantothenic acid (ca.15 μM); that of heart is about 150 μM.

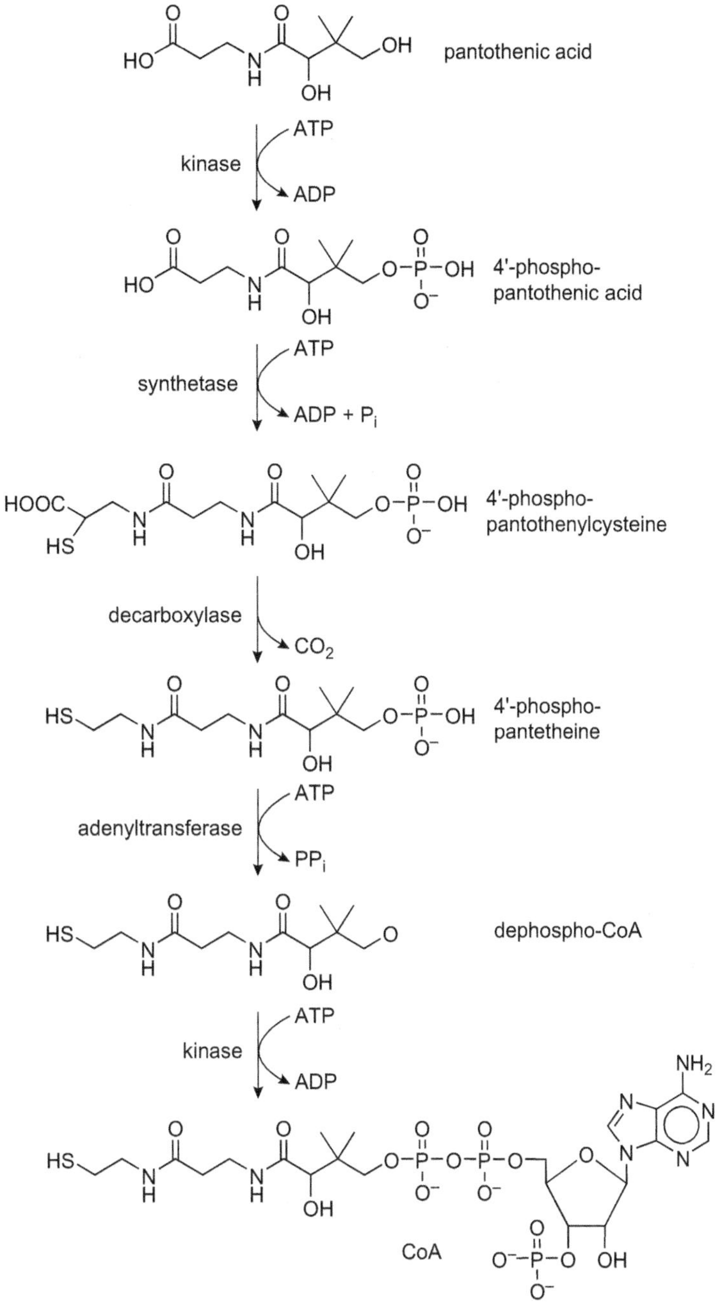

FIGURE 16.2 Biosynthesis of coenzyme A.

the vitamin. This surprising finding has been interpreted as indicating a mechanism for conserving the vitamin by recycling it from the degradation of pantothenate-containing molecules. Pantothenic acid is taken up in the choroid plexus by a specific transport process, which, at low concentrations of the vitamin, involves the partial phosphorylation of the vitamin. The cerebrospinal fluid, because it is constantly renewed in the central nervous system, requires a constant supply of pantothenic acid, which, as CoA, is involved in the synthesis of the neurotransmitter acetylcholine in brain tissue.

6. METABOLISM OF PANTOTHENIC ACID

CoA Synthesis

All tissues have the ability to synthesize CoA from pantothenic acid. At least in rat liver, all of the enzymes in the CoA biosynthetic pathway are found in the cytosol. Four moles of ATP are required for the biosynthesis of a mole of CoA from a single mole of pantothenic acid. The process (Fig. 16.2) is initiated in the cytosol and completed in the mitochondria.

In the cytosol,

1. **Pantothenate kinase (PanK)** catalyzes the ATP-dependent phosphorylation of pantothenic acid to yield **4′-phosphopantothenic acid.** This is the rate-limiting step in CoA synthesis; under normal conditions, it functions far below capacity. Four isoforms have been identified:[18]
   a. **PanK1** expressed primarily in the heart, liver, and kidney.[19]
   b. **PanK2** expressed in all tissues.[20]
   c. **PanK3** expressed (to a high degree) only in liver.
   d. **PanK4** expressed in most tissues, with highest concentrations in muscle.
   The PanKs are inducible[21] and feedback inhibited by 4′-phosphopantothenic acid, CoA esters and, more weakly, CoA and long-chain acyl CoAs all of which are allosteric effectors. Inhibition by CoA esters can be reversed by carnitine. The ethanol metabolite acetaldehyde inhibits the conversion of pantothenic acid to CoA.[22]
2. **Phosphopantothenylcysteine synthetase** catalyzes the ATP-dependent condensation of 4′-phosphopantothenic acid with cysteine to yield **4′-phosphopantothenylcysteine.**
3. **Phosphopantothenylcysteine decarboxylase** catalyzes the decarboxylation of 4′-phosphopantothenylcysteine to yield **4′-phosphopantetheine**, which is transported into the mitochondria.
   In mitochondria, two steps are catalyzed by a single bifunctional enzyme, **CoA synthetase**,[23] located in the inner membrane:
4. **Phosphopantetheine adenyltransferase** catalyzes the ATP-dependent adenylation of 4′-phosphopantetheine to yield **dephospho-CoA.** This reaction is reversible; therefore, at low ATP levels, dephospho-CoA can be degraded to yield ATP.
5. **Dephospho-CoA kinase** catalyzes the ATP-dependent phosphorylation of dephospho-CoA to yield CoA.

The mitochondrial concentration of nonacylated CoA determines the rate of oxidation-dependent energy production. CoA can also enter mitochondria by nonspecific membrane binding as well as by an energy-dependent membrane transporter.[24]

Acyl-CoA Synthesis

CoA serves to activate long-chain fatty acids for various key metabolic roles. These roles include regulation of enzymes and signaling pathways, oxidation to provide cellular energy, and incorporation into acylated proteins and complex lipids. The addition of long-chain fatty acids to CoA is catalyzed by a large family of **acyl-CoA synthetases (ACSs).** This energy-dependent esterification occurs in two steps:

1. Fatty acid + ATP → acyl-AMP + PPi
2. Acyl-AMP + CoASH → acyl CoA + AMP

More than two dozen isoforms of ACS have been identified in mammalian tissues. Most cells have multiple forms; hence, it has been suggested that each form may direct its fatty acid substrates along a specific metabolic route.

Acyl Carrier Protein Synthesis

In higher animals, ACP is associated with a large **fatty acid synthetase** complex composed of two 250 kDa subunits containing several functional domains.[25] The ACP domain is synthesized as an inactive apoprotein but is modified posttranslationally by the addition of the 4′-phosphopantetheine prosthetic group via a phosphoester linkage at a serinyl residue. This modification is catalyzed by **4′-phosphopantetheine-apo-ACP transferase** using CoA as the donor (Fig. 16.3). It is, therefore, likely that ACP synthesis serves as a regulator of intracellular CoA levels.

Pantothenic Acid Recycling

The pantothenic acid components of both CoA and ACP are released metabolically for reutilization. ACP is degraded by an ACP hydrolase that releases 4′-phosphopantetheine yielding apo-ACP. CoA can be catabolized by a nonspecific, phosphate-sensitive, lysosomal phosphatase to dephospho-CoA,

18. Of three distinct types of kinases that occur in various species, those of eukaryotes are grouped in the PanK2 class.
19. The mouse shows two (α, β) variants of PanK1, produced by alternate splicing of the same gene.
20. In humans, PanK2 is located in mitochondria, but its strong inhibition by acetyl CoA ($IC_{50} < 1\ \mu M$) would suggest that it functions at physiological concentrations of acetyl CoA, which exceed that level (Leonardi, R., Rock, C.O., Jackowski, S., et al., 2007. Proc. Natl. Acad. Sci. U.S.A. 104, 1494–1499). However, that PanK2 does, in fact, play a functional role is indicated by that fact that an autosomal recessive PanK2 mutation occurs in Hallervorden–Spatz syndrome, a neurodegenerative disorder presenting as dystonia and optic atrophy or retinopathy (Delgado, R.F., Sanchez, P.R., Speckter, H., et al., 2012. J. Magn. Reson. Imaging 35, 788–794).
21. PanK is induced by the antilipidemic drug clofibrate. Treatment with clofibrate increases hepatic concentrations of CoA, apparently owing to increased synthesis.
22. Alcoholics have been reported to excrete in their urine large percentages of the pantothenic acid they ingest, a condition corrected on ethanol withdrawal.
23. In plants and prokaryotes, these steps are catalyzed by separate enzymes.
24. Tahiliani, A.G., Neely, J.R., 1987. J. Mol. Cell. Cardiol. 19, 1161–1167.
25. Fatty acid synthase complex has several catalytic sites: acetyl transferase, malonyl transferase, 2-oxoacyl synthase, oxoacyl reductase, 3-hydroxyacyl dehydratase, enoyl reductase, and thioester hydrolase.

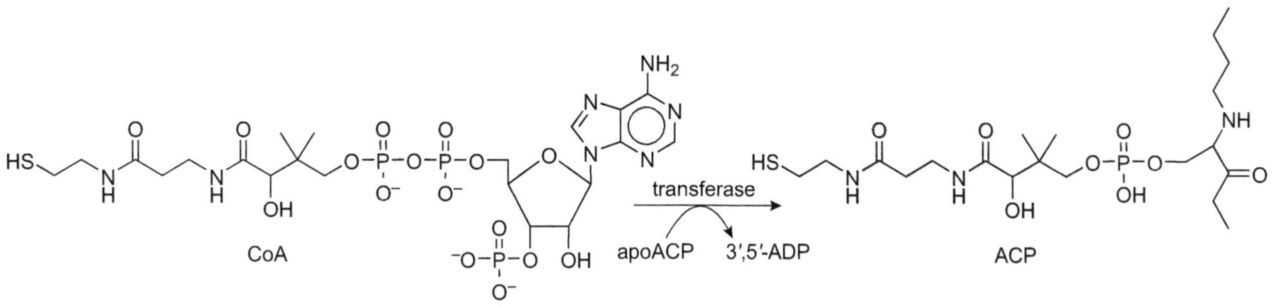

FIGURE 16.3 Coenzyme A provides 4-phosphopantetheine in the biosynthesis of the acyl carrier protein.

which is degraded to 4′-phosphopantetheine by a plasma membrane pyrophosphatase. 4′-Phosphopantetheine from either source is dephosphorylated by microsomal and lysosomal phosphatases to pantetheine, from which pantothenic acid is liberated by membrane **pantetheinases.**[26]

Excretion

Pantothenic acid is excreted mainly in the urine as free pantothenic acid and some 4′-phosphopantothenate; no catabolic products are known. The renal tubular secretion of pantothenic acid, probably by a mechanism common to weak organic acids, results in urinary excretion of the vitamin correlating with dietary intake. An appreciable amount (~15% of daily intake) is oxidized completely and is excreted across the lungs as CO_2. Humans typically excrete in the urine 0.8–8.4 mg of pantothenic acid per day. There appear to be two renal mechanisms for regulating the excretion of pantothenic acid:

- **Active transport** at physiological concentrations of the vitamin in the plasma, with pantothenic acid being reabsorbed by active transport;
- **Tubular secretion** at higher concentrations, tubular reabsorption appears to be the only mechanism for conserving free pantothenic acid in the plasma.

Disorders of Pantothenic Acid Metabolism.

- A polymorphism of PanK2 has been identified as the metabolic basis of an autosomal recessive neurodegenerative disorder, Hallervorden–Spatz syndrome. Affected subjects show dystonia and optic atrophy or retinopathy with the deposition of iron in basal ganglia.[27]

26. These are products of the so-called Vanin (i.e., *VNN1*) gene; they are widely expressed in all tissues, with greatest amounts in kidney, liver, intestine, and lymphoid cells and have significant sequence homology with, but not the activity of, biotinidases. Pantetheinases also appear to have roles in inflammatory responses via their product cysteamine. Studies with animal models in which pantetheinases were genetically deleted have shown reduced gut inflammatory reactions to drug and parasitic stimuli and failed cytokine responses to stress.

27. Gordon, N., 2002. Eur. J. Paediatr. Neurol. 6, 243–247.

7. METABOLIC FUNCTIONS OF PANTOTHENIC ACID

General Functions

Both CoA and 4′-phosphopantetheine in ACP function metabolically as carriers of acyl groups and activators of carbonyl groups in a large number of vital metabolic transformations, including the tricarboxylic acid cycle and the metabolism of fatty acids. In each case, the linkage with the transported acyl group involves the reactive sulfhydryl of the 4′-phosphopantetheinyl prosthetic group.

Different metabolic roles of active forms of pantothenic acid:
- **CoA**—in a broad array of acyl transfer reactions in oxidative energy metabolism and catabolism
- **ACP**—in synthetic reactions.

Acyl CoAs

Scope of functions. Acyl CoAs serve as essential cofactors for some 4% of known enzymes, including at least 100 enzymes involved in intermediary metabolism. CoA functions widely in metabolism in reactions involving either the carboxyl group (e.g., formation of acetylcholine, acetylated amino sugars, acetylated sulfonamides[28]) or the methyl group (e.g., condensation with oxaloacetate to yield citrate) of an acyl CoA. In these reactions, CoA forms high-energy thioester bonds with carboxylic acids, the most important of which is acetic acid, which can come from the metabolism of fatty acids, amino acids, or carbohydrates (Fig. 16.4).

Acetyl CoA, the "*active acetate*" group has many metabolic functions:

- **Acetylations** of alcohols, amines, and amino acids (e.g., choline, sulfonamides, *p*-aminobenzoate).

28. Coenzyme A was discovered as an essential factor for the acetylation of sulfonamide by the liver and for the acetylation of choline in the brain; hence, **coenzyme A** stands for **coenzyme for acetylations**.

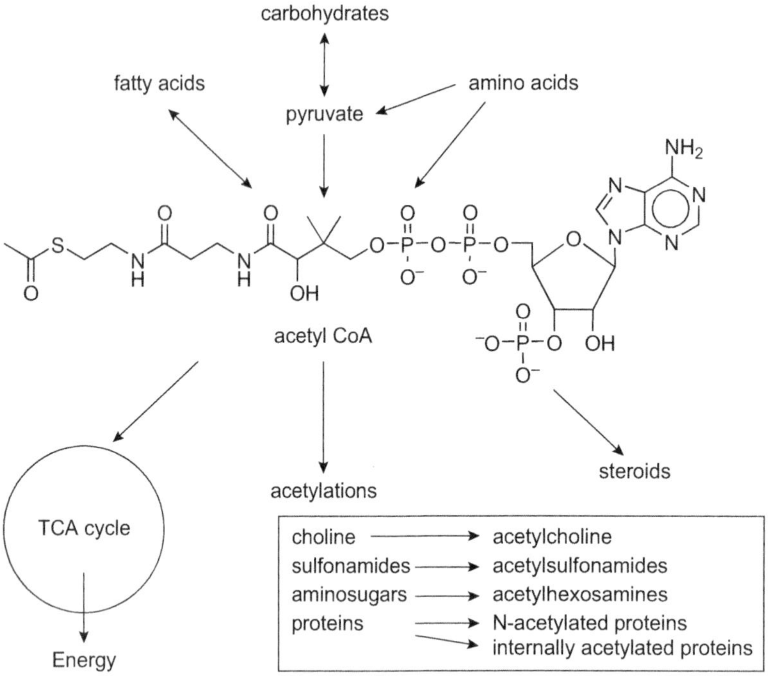

FIGURE 16.4 The central role of acetyl CoA in metabolism.

- **Activation of fatty acids** for incorporation into triglycerides, cholesterol, steroid hormones, prostaglandins, leukotrienes, membrane phospholipids, and regulatory sphingolipids.
- **Synthesis of fatty acids** by addition of 2-carbon fragments from acetyl CoA.
- **Transacylation to carnitine** to form energy-equivalent acylcarnitines capable of being transported into the mitochondria where β-oxidation occurs.[29]
- **Posttranslational long-chain acetylation of proteins,** an estimated half of proteins are acylated[30] at N-terminal residues (most frequently at terminal serinyl or alanyl residues) or internal lysyl residues. This includes the cotranslational processing of peptide hormones from their precursors (e.g., ACTH (adrenocorticotropic hormone) to α-melanocyte-stimulating hormone; β-lipotropin to β-endorphin). In most cases, acylation is without functional significance and can be reversed by NAD-dependent deacylases. In other cases, acylation is required for protein function: acylation of steroid hormone receptors and other regulatory proteins; acylation of α-tubulin, which stabilize microtubules;[31] acylation/deacylation of histones,[32] which affect chromatin packing and, thus, gene expression;[33] and S-acylation of Ras proteins,[34] conferring control of their subcellular trafficking.[35] Most acylations involve palmitic acid[36] (e.g., GTP-binding proteins, protein kinases, membrane receptors, cytoskeletal proteins, mitochondrial proteins) in reversible ester bonds; others involve myristic acid[37] in irreversible amide linkages.
- **Transcriptional regulation** via PPARs (peroxisome proliferator-activated receptors) α, β, and γ.
- **Production of the** "ketone body" acetoacetate derived from fat metabolism when glucose is limiting.

Regulation of acetyl CoA. The abundance of acetyl CoA reflects the energy state of the cell and can affect the activities of many enzymes. The amounts of nonacylated CoA reflect the activities of the various intracellular acyl-CoA thioesterases and acyltransferases in mediating cellular lipid metabolism. They also determine the rate of

29. This is the only means by which long-chain fatty acids can enter the mitochondria for energy production.
30. Stadtman, E.R., 1990. Biochemistry 29, 6323–6331.
31. Acetylation occurs in the a-tubulin after it has been incorporated into the microtubule. It can be induced by such agents as taxol. Acetylated microtubules are more stable to depolymerizing agents such as colchicines.
32. Acetylated histones are enriched in genes that are being actively transcribed.
33. Yasui, K., Matsuyama, T., Ito, T., 2005. Seikagaku 77, 498–504.
34. Ras proteins are located near cell membranes and participate in the regulation of cell division; abnormalities in these proteins can lead to uncontrolled cell division and tumorigenesis.
35. Smotrys, J,E., Linder, M.E., 2004. Annu. Rev. Biochem. 73, 559–587; Rowinsky, E.K., Windle, J.J., Von Hoff, D.D., 1999. J. Clin. Oncol. 17, 3631–3652.
36. *n*-Hexadecanoic acid, C16:0.
37. *n*-Tetradecanoic acid, C14:0.

oxidation-dependent energy production by mitochondria. Steady-state concentrations of CoA (20–150 μM in cytosol; ~2 mM in mitochondria) have been shown to respond to deprivation of food, glucose feeding, and insulin or glucagon treatment. These effects can be countered through the transfer of acyl CoA between subcellular compartments and the reduction of 4′-phosphopantetheine adenyltransferase and dephospho-CoA kinase through a catabolic activity of CoA synthetase.

Acyl Carrier Protein

ACP is a component of the multienzyme complex **fatty acid synthetase**.[38] In ACP, the cofactor functions in two domains, acetyl transferase and malonyl transferase, which transfer the respective acyl groups between 4′-phosphopantetheine at different active sites with successive cycles of condensations and reductions.[39] The nature of the fatty acid synthase complex varies considerably among different species. However, in each, 4′-phosphopantetheine is the prosthetic group for the binding and transfer of the acyl units to a 8.7 kD subunit during catalysis. The sulfhydryl group of the cofactor serves as the point of temporary covalent attachment of the growing fatty acid via a thiol linkage each time an acyl group is added by transfer to the cofactor. In this way, the cofactor appears to function as a swinging arm, allowing the growing fatty acid to reach the various catalytic sites of the enzyme.

8. BIOMARKERS OF PANTOTHENIC ACID STATUS

Pantothenic acid status can be assessed in two ways:[40]

- **Urinary metabolites**—The urinary excretion of pantothenic acid is considered the most reliable indicator of pantothenic acid status.
- **Blood metabolites**—Whole blood or plasma pantothenic acid levels reflect the level of intake of the vitamin. Healthy adults typically show whole blood levels in the range of 1.6–2.7 μM; values < 1 μM indicate suboptimal status.

38. Fatty acid synthase is the name used to identify the multienzyme complex on which the several reactions of fatty acid synthesis (condensations and reductions) occur. In higher animals, the complex is composed of two large (250 kDa) subunits.

39. The seven functional activities of the fatty acid synthase complex are acetyltransferase, malonyltransferase, 3-ketoacyl synthase, 3-ketoacyl reductase, 3-hydroxyacyl dehydratase, enoyl reductase, and thioester hydrolase.

40. A microbiological assay employing *Lactobacillus plantarum* is commonly used for the analysis of pantothenic acid. This requires enzymatic pretreatment of specimens to liberate free pantothenic acid.

TABLE 16.2 General Signs of Pantothenic Acid Deficiency

Organ System	Signs
General	Depressed appetite, growth
Vital organs	Hepatic steatosis, thymic necrosis, adrenal hypertrophy
Dermatologic	Dermatitis, achromotrichia, alopecia
Muscular	Weakness
Gastrointestinal	Ulcers
Nervous	Ataxia, paralysis

9. PANTOTHENIC ACID DEFICIENCY

Deficiencies Rare

Deprivation of pantothenic acid results in metabolic impairments including reduced lipid synthesis and energy production. Signs and symptoms of pantothenic acid deficiency vary among different species; most frequently, they involve the skin, liver, adrenals, and nervous system. Owing to the wide distribution of the vitamin in nature, dietary deficiencies of pantothenic acid are rare; they are more common in circumstances of inadequate intake of basic foods and vitamins and are often associated with (and mistakenly diagnosed as) deficiencies of other vitamins. Understanding of the presentation of pantothenic acid deficiency comes mostly from studies with experimental animals. These have shown a pattern of general deficiency signs (Table 16.2). Recommended intakes for pantothenic acid have been established (Table 16.3).

Antagonists

Pantothenic acid deficiency has been produced experimentally using purified diets free of the vitamin or by administering an antagonist. One antagonist is the analogue **ω-methylpantothenic acid**, which has a methyl group in place of the hydroxymethyl group of the vitamin; this change prevents it from being phosphorylated and inhibits the action of pantothenic acid kinase. Other antagonists include desthio-CoA, in which the terminal sulfhydryl of the active metabolite is replaced with a hydroxyl group, and hopantenate, in which the three-carbon β-alanine moiety of the vitamin is replaced with the four-carbon γ-aminobutyric acid.

Deficiency Signs in Humans

Pantothenic acid deficiency in humans has been observed only in severely malnourished patients and in subjects treated with the antagonist ω-methylpantothenic acid. In cases of the

TABLE 16.3 Recommended Pantothenic Acid Intakes

The United States		FAO/WHO	
Age Status	RDA[a], µg/day	Age Status	RNI[b], µg/day
0–6 months	[1.7][c]	0–6 months	1.7
7–11 months	[1.8][c]	7–11 months	1.8
1–3 years	2	1–3 years	2
4–8 years	3	4–6 years	3
9–13 years	4	7–9 years	4
>13 years	5	>9	5
Pregnancy	6	Pregnancy	6
Lactation	7	Lactation	7

[a]*Recommended Dietary Allowances; Food and Nutrition Board, 2000. Dietary Reference Intakes for Thiamin, Riboflavin, Niacin, Vitamin B6, Folate, Vitamin B12, Pantothenic Acid, Biotin and Choline. National Academy Press, Washington, DC, 564 pp.*
[b]*Recommended Nutrient Intakes; Joint WHO/FAO Expert Consultation, 2001. Human Vitamin and Mineral Requirements. WHO, Rome, 286 pp.*
[c]*RDA has not been established; adequate intake is presented.*

former type, neurologic signs (paresthesia in the toes and sole of the feet) have been reported.[41] Subjects made deficient in pantothenic acid through the use of ω-methylpantothenic acid also developed burning sensations of the feet. In addition, they showed depression, fatigue, insomnia, vomiting, muscular weakness, and sleep and gastrointestinal disturbances. Changes in glucose tolerance, increased sensitivity to insulin, and decreased antibody production have also been reported.

There is some evidence of subclinical pantothenic acid deficiency. Urinary pantothenic acid excretion has been found to be low for pregnant women, adolescents, and the elderly compared with the general population.

Deficiency Signs in Animals

Pantothenic acid deficiency in most species results in reduced growth and reduced efficiency of feed utilization. In rodents, the deficiency presents as scaly dermatitis, achromotrichia, alopecia, and adrenal necrosis.[42] Congenital malformations of offspring of pantothenic acid-deficient dams have been reported. Excess amounts of porphyrins[43] are excreted in the tears of pantothenic acid-deficient rats in a condition called **blood-caked whiskers**. In the chick, deficiency presents as lesions at the corners of the mouth, swollen and encrusted eyelids, dermatitis of the entire foot (with hemorrhagic cracking),[44] poor feathering, fatty liver degeneration, thymic necrosis, and myelin degeneration of the spinal column with paralysis and lethargy.[45] Chicks produced from deficient hens show high rates of embryonic and posthatching mortality. In the dog, deficiency presents as hepatic steatosis, irritability, cramps, ataxia, convulsions, alopecia, and death.[46] In the pig, deficiency presents as dermatitis, acute encephalopathy, hypoglycemia, hyperammonemia, excessive lachrymation, colitis, spastic gait, hypertrophy, and steatosis of multiple organs (e.g., adrenals, liver, heart); and ovarian atrophy with impaired uterine development.[47]

Pantothenic acid deficiency would not be expected in ruminants if ruminal microbial synthesis of the vitamin is significant. In fact, deficiency has not been reported in ruminants, although one study found plasma pantothenic acid concentrations of cows to respond to dietary supplementation with the vitamin in a dose-dependent way.[48]

Marginal deficiency of pantothenic acid in the rat has been found to produce elevated serum levels of triglycerides and free fatty acids. The metabolic basis of this effect is not clear; however, it is possible that it involves a somewhat targeted reduction in cellular CoA concentrations, affecting the deposition of fatty acids in adipocytes (via impaired acyl-CoA synthetase) but not the hepatic production of triglycerides.

10. PANTOTHENIC ACID IN HEALTH AND DISEASE

Benefits have been reported for the use of supplements of pantothenic acid and/or metabolites.

Reduced Serum Cholesterol Level

High doses (500–1200 mg/day) of pantotheine, the dimmer of pantetheine, have been shown to reduce serum concentrations of total and LDL cholesterol and triglycerides, with increases in HDL cholesterol[49]. While the underlying mechanism is unclear, it is thought to involve roles of pantetheine

41. **Burning feet syndrome** was described during World War II in American prisoners held in Japan and the Philippines, who were generally malnourished. That large oral doses of calcium pantothenate provided some improvement suggested that the syndrome involved, at least in part, deficiency of pantothenic acid.
42. Pietrzik, K., Hesse, C.H., Zur, W., et al., 1975. Int. J. Vitam. Nutr. Res. 45, 153–162.
43. e.g., Protoporphyrin IX.
44. These lesions are often confused with the footpad dermatitis caused by biotin deficiency. Unlike the latter, in which lesions are limited to the footpad (i.e., plantar surface), the lesions produced by pantothenic acid deficiency also involve the toes and superior aspect of the foot. Prevention of footpad dermatitis is economically important in poultry production, the US–European market for chicken and duck feet has been estimated to exceed $300 M.
45. Gries, C.L., Scott, M.L., 1972. J. Nutr. 102, 1269–1285.
46. Noda, S., Haratake, J., Sasaki, A., et al., 1991. J. Neurol. Neurosurg. Psychiatr. 51, 582–585.
47. Nelson, R.A., 1968. Am. J. Clin. Nutr. 21, 495–501.
48. Bonomi, A., 2000. Rivista. Sci. dell'Aliment. 29, 321–338.
49. Binaghi, P., Cellina, G., Lo Cicero, G., et al., 1990. Minerva Med. 81, 475–479.

as a cofactor in shunting acetyl groups away from steroid synthesis to oxidative metabolism and/or in reducing triglyceride synthesis through inhibition of hydroxymethylglutaryl-CoA reductase.

Rheumatoid Arthritis

Patients with RA have been found to exhibit lower blood pantothenic acid levels than healthy controls. Nearly 50 years ago, an unblinded trial found relief of symptoms in 20 patients treated with pantothenic acid.[50] A subsequent randomized, controlled trial showed that high doses (up to 2 g/day) of calcium pantothenate reduced the duration of morning stiffness, the degree of disability, and the severity of pain for rheumatoid arthritis patients.[51]

Athletic Performance

While pantothenic acid deficiency is known to reduce exercise endurance in animal models, results of the few studies conducted in humans have been inconsistent. Some results showed improved efficiency of oxygen utilization and reduced lactate acid accumulation in athletes;[52] others showing no benefits.[53]

Wound Healing

Studies in animal models have found pantothenic acid, given orally or topically as pantothenol, to promote the closure of wounds of the skin by facilitating the recruitment of fibroblasts to the injured area.[54] Studies with humans given high, combined doses of pantothenic acid and ascorbic acid have shown no benefits,[55] although a derivative **dexpanthenol** has been found useful in reducing skin dehydration and irritation.[56]

Other Outcomes

It has been suggested that pantothenic acid may have value in treating the systemic autoimmune disease lupus erythematosus, the theoretical argument based on the observation that lupus can be caused by drugs that impair pantothenic acid metabolism. No relevant clinical data have been reported. It has also been proposed that pantothenic acid may have value in the prevention of graying hair.[57] That, too, is without substantiating evidence. There has been interest in the prospect of developing pantothenic acid antagonists for the treatment of malaria, as *Plasmodium falciparum* has been shown to require pantothenic acid, which it typically obtains from erythrocytes.

11. PANTOTHENIC ACID TOXICITY

The toxicity of pantothenic acid is negligible. No adverse reactions have been reported in any species following the ingestion of large doses of the vitamin. Massive doses (e.g., 10 g/day) administered to humans have not produced reactions more severe than mild gastrointestinal distress and diarrhea. Similarly, no deleterious effects have been identified when the vitamin was administered parenterally or topically. It has been estimated that animals can tolerate doses of pantothenic acid as great as at least 100 times their respective nutritional requirements for the vitamin. No upper tolerable intakes for pantothenic acid have been established.

12. CASE STUDY

Review the following experiment, paying special attention to the independent and dependent variables in the design. Then, answer the questions that follow.

Experiment

To evaluate the possible role of pantothenic acid and ascorbic acid in wound healing, a study was conducted assessing the effects of these vitamins on the growth of fibroblasts. Human fibroblasts were obtained from neonatal foreskin; they were cultured in a standard medium supplemented with 10% fetal calf serum and antibiotics.[58] The medium contained no ascorbic acid but contained 4 mg of pantothenic acid per liter. Cells were used between the third and ninth passages. Twenty-four hours before each experiment, the basal medium was replaced by medium supplemented with pantothenic acid (40 mg/L) or pantothenic acid (40 mg/L) plus ascorbic acid (60 mg/L). Cells (1.5×10^5) were plated in 3 mL of culture medium in 28-cm^2 plastic dishes. After incubation, they were collected by adding trypsin and then scraping; they were counted in a hemocytometer. The synthesis of DNA and protein was estimated by measuring the rates of incorporation of radiolabel from [^{3}H]thymidine and [^{14}C]proline, respectively. Total protein was measured in cells (lysed by sonication and solubilized in 0.5 *N* NaOH) and in the culture medium.

50. Subjects were given calcium pantothenate i.m. (Barton-Wright, E.C., Elliot, W.A., 1963. Lancet 2, 862–869).
51. U.S. Practitioner Research Group, 1980. Practitioner 224, 208–2015.
52. Litoff, D., 1985. Med. Sci. Sports Exerc. 17S, 287–294.
53. Nice, C., Reeves, A.G., Brinck-Johnson, T., et al., 1984. J. Sports Med. Phys. Fitness 24, 26–29.
54. Weimann, B.I., Hermann, D., 1999. Int. J. Vitam. Nutr. Res. 69, 113–119.
55. Vaxman, F., Olender, S., Lambert, A., et al., 1995. Eur. Surg. Res. 27, 158–166.
56. Biro, K., Thaci, D., Ochsendorf, F.R., et al., 2003. Contact Dermat. 49, 80–84.
57. That pantothenic acid deficiency can cause achromotrichia in rodents does not imply that graying of human hair is caused by low-pantothenic acid status, nor that supplemental pantothenic acid could have any effect. Indeed, no such effects have been demonstrated.
58. Gentamicin and amphotericin B (Fungizone).

Results After 5 Days of Culture

	Cells	^3H	^{14}C	Cell Protein	Protein in Medium
Treatment	(×10^5)	(10^3 cpm)	(10^3 cpm)	(mg/dish)	(mg/mL)
Control	2.90±0.16	11.6±0.4	1.7±1.0	10.0±1.0	1.93±0.01
+ Pantothenic acid	3.83±0.14[a]	18.7±0.5[a]	2.9±0.1[a]	14.5±0.9[a]	1.93±0.02
+ Pantothenic acid and ascorbic acid	3.74±0.19[a]	18.1±0.8[a]	2.8±0.1[a]	8.1±0.9	2.11±0.01[a]

[a]Significantly different from control value, $p<.05$.

Case Questions

1. Why were thymidine and proline selected as carriers of the radiolabels in this experiment?
2. Why were fibroblasts selected (rather than some other cell type) for use in this study?
3. Assuming that the protein released into the culture medium is largely soluble procollagen, what can be concluded about the effects of pantothenic acid and/or ascorbic acid on collagen synthesis in this system?
4. What implications do these results have regarding wound healing?

13. STUDY QUESTIONS AND EXERCISES

1. Diagram the areas of metabolism in which CoA and ACP (via fatty acid synthase) are involved.
2. Construct a decision tree for the diagnosis of pantothenic acid deficiency in humans or an animal species.
3. What key feature of the chemistry of pantothenic acid relates to its biochemical functions as a carrier of acyl groups?
4. What parameters might you measure to assess pantothenic acid status of a human or animal?

RECOMMENDED READING

Grevengoed, T.J., Klett, E.L., Coleman, R.A., 2014. Acyl-CoA metabolism and partitioning. Annu. Rev. Nutr. 34, 1030.

Hayflick, S.J., 2014. Defective pantothenate metabolism and neurodegeneration. Biochem. Soc. Trans. 42, 1063–1068.

Hunt, M.C., Siponen, M.I., Alexson, S.E.H., 2012. The emerging role of acyl-CoA thioesterases and acyltransferases in regulating peroxisomal lipid metabolism. Biochem. Biophys. Acta 1822, 1397–1410.

Kirby, B., Roman, N., Kobe, B., et al., 2010. Functional and structural properties of mammalian acyl-coenzyme thioesterases. Prog. Lipid Res. 49, 366–377.

Martinez, D.L., Tschiya, Y., Gout, I., 2014. Coenzyme A biosynthetic machinery in mammalian cells. Biochem. Soc. Trans. 42, 1112–1117.

Miller, J.W., Rucker, R.B., 2012. Pantothenic acid (Chapter 24). In: Erdman, J.W., Macdonald, I.A., Zeisel, S.H. (Eds.), Present Knowledge in Nutrition, tenth ed. Wiley-Blackwell, New York, pp. 375–390.

Rucker, R.B., Bauerly, K., 2014. Pantothenic acid (Chapter 8). In: Zempleni, J., Suttie, J.W., Gregory, J.F., et al. (Eds.), Handbook of Vitamins, fifth ed. CRC Press, New York, pp. 325–349.

Chapter 17

Folate

Chapter Outline

Anchoring Concepts

1. Folate is the generic descriptor for folic acid (pteroylmonoglutamic acid) and related compounds exhibiting qualitatively the biological activity of folic acid. The term folates refers generally to the compounds in this group, including mono- and polyglutamates.
2. Folates are active as coenzymes in single-carbon metabolism.
3. Deficiencies of folate are manifested as anemia and dermatologic lesions.

Using Streptococcus lactis *R as a test organism, we have obtained in a highly concentrated and probably nearly pure form an acid nutrilite with interesting physiological properties. Four tons of spinach have been extracted and carried through the first stages of concentration…. This acid, or one with similar chemical and physiological properties, occurs in a number of animal tissues of which liver and kidney are the best sources…. It is especially abundant in green leaves of many kinds, including grass. Because of this fact, we suggest the name "folic acid" (Latin, folium—leaf). Many commercially canned greens are nearly lacking in the substance.*

Mitchell et al.[1]

1. Herschel K. Mitchell (1914–2000) was a biochemist at Stanford and Caltech Universities who did pioneering work in the field of molecular genetics. Esmond E. Snell (1914–2003) was a biochemist at the Universities of Wisconsin, Texas, and California–Berkeley whose work on the nutritional requirements of lactic acid bacteria led to his discoveries of pyridoxal and pyridoxamine; elucidation of the mechanisms of vitamin B_6-dependent enzymes; and to the development of microbiological assays for several vitamins, antivitamins, and growth factors. Roger J. Williams (1893–1988) was an Indian-born American biochemist at the Universities of Oregon, Oregon State and Texas. He is best known for discovering pantothenic acid and naming folic acid.

LEARNING OBJECTIVES

1. To understand the chief natural sources of folates.
2. To understand the means of absorption and transport of the folates.
3. To understand the biochemical functions of the folates as coenzymes in single-carbon metabolism and the relationship of that function to the physiological activities of the vitamin.
4. To understand the metabolic interrelationship of folate and vitamin B_{12} and its physiological implications.

VOCABULARY

p-Acetaminobenzoylglutamate
S-adenosylhomocysteine (SAH)
S-adenosylmethionine (SAM)
p-Aminobenzoylglutamate
Aminopterin
Antifolates
Arsenicosis
Betaine
Carboxypeptidase
Cerebral folate deficiency syndrome
Cervical paralysis
7,8-Dihydrofolate reductase
Dihydrofolic acid (FH_2)
Folacin
Folate
Folate-binding proteins (FBPs)
Folate export pump
Folate receptor (FRs)

The Vitamins. http://dx.doi.org/10.1016/B978-0-12-802965-7.00017-4

Folic acid
Folyl conjugase (folyl γ-glutamyl carboxypeptidase)
Folylpolyglutamates
Folylpolyglutamate synthase
5-Formimino-FH_4
Formiminoglutamate (FIGLU)
5-Formyl-FH_4
10-Formyl-FH_4
γ-Glutamyl hydrolase
Hereditary folate malabsorption (HFM)
Homocysteine (Hcy)
Homocysteinemia
Leukopenia
Macrocytic anemia
Megaloblasts
5,10-Methenyl-FH_4
Methionine synthase
Methionine synthase reductase
Methotrexate
5-Methyl FH_4
Methylation index
5,10-Methylene-FH_4
5,10-Methylene-FH_4 dehydrogenase
5,10-Methylene-FH_4 reductase (MTHFR)
Methyl folate trap
Plasmodium
Methylmalonic acid (MMA)
Neural tube defects (NTDs)
Organic anion transporter (OAT)
Pernicious anemia
Proton-coupled folate transporter (PCFT)
Pteridine
Pterin ring
Pteroylglutamic acid
Pteroylmonoglutamic acid
Purines
Pyrazine nucleus
Reduced folate carrier (RFC)
Serine hydroxymethyltransferase
Single-carbon pool
Sulfa drugs
Tetrahydrofolate reductase
Tetrahydrofolic acid (FH_4)
Tetrahydropteroylglutamic acid
Thymidylate
Thymidylate synthase
Vitamin B_{12}

1. THE SIGNIFICANCE OF FOLATE

Folate is a vitamin that has only recently been appreciated for its importance beyond its essential role in normal metabolism, especially for its relevance to the etiologies of chronic diseases and birth defects. Widely distributed among foods, particularly those of plant foliar origin, this abundant vitamin is underconsumed by people whose food habits do not emphasize plant foods. Intimately related in function with vitamins B_{12} and B_6, its status at the level of subclinical deficiency can be difficult to assess and the full extent of its interrelationships with these vitamins and with amino acids remains incompletely elucidated. Folate deficiency is an important problem in many parts of the world, particularly where there is poverty and malnutrition. It is an important cause of anemia, second only to nutritional iron deficiency.

Evidence shows that marginal folate intakes can support apparently normal circulating folate levels while still limiting single-carbon metabolism. Thus, folate emerged as having an important role in the etiology of homocysteinemia, which was identified as a risk factor for occlusive vascular disease, cancer, and birth defects, particularly **neural tube defects (NTDs)**. In 1998, the U.S. Food and Drug Administration mandated that folic acid be added to all "enriched" cereal grain products (breads, pastas, wheat flours, breakfast cereals, rice) to reduce the prevalence of NTDs. The food system-wide measure increased the folate intakes of Americans, more than doubled circulating levels of the vitamin and was expected to reduce both NTDs and coronary artery disease deaths, while also driving folate supplementation efforts in other countries.

More than two decades of population-based folate, supplementation has seen the prevalence of NTDs decline, indicating that this strategy has been successful. Still, concerns remain about potential risks of treating individuals who are not in need. The first of these was the prospective masking the **macrocytic anemia** of vitamin B_{12} deficiency, which will lead to neuropathy if not corrected. Additional concerns have been added in recent years with growing doubt about the role of **homocysteinemia** in the etiology of cardiovascular disease, with reports of enhanced cognitive impairment and colorectal cancer risk as a consequence of folate supplementation. For these reasons, it is important to understand the role of folate in nutrition and health.

2. PROPERTIES OF FOLATE

Folate Nomenclature

The term **folate** is the generic descriptor for **folic acid** (**pteroylmonoglutamic acid** or **pteroylglutamic acid**) and related compounds exhibiting the biological activity of folic acid. This group, often collectively referred to as **folacin**, **folic acids**, and **folates**, is comprised of large number of pteridine derivatives varying in degree of hydrogenation of pteridine nucleus and capable of binding single-carbon units to nitrogens at position 5 and/or 10. They also have one or more glutamyl residues linked via peptide bonds and are named for the number of glutamyl residues (*n*), using

TABLE 17.1 Key Members of the Folate Family

Vitamer	Abbreviation	R′ (at N-5)	R (at N-10)
Tetrahydrofolic acid	FH_4	H	H
5-Methyltetrahydrofolic acid	$5\text{-}CH_3\text{-}FH_4$	CH_3	H
5,10-Methenyltetrahydrofolic acid	$5,10\text{-}CH^+\text{-}FH_4$	$\text{-}CH^+$-(bridge)	
5,10-Methylenetetrahydrofolic acid	$5,10\text{-}CH_2{=}FH_4$	$\text{-}CH_2{=}$(bridge)	
5-Formyltetrahydrofolic acid	5-HCO-FH_4	HCO	H
10-Formyltetrahydrofolic acid	10-HCO-FH_4	H	HCO
5-Formimintetrahydrofolic acid	5-HCNH-FH_4	HCNH	H

such notations as **PteGlu*n***. The fully reduced compound **tetrahydropteroylglutamic acid** is called **tetrahydrofolic acid**; its single-carbon derivatives are named according to the specific carbon moiety bound. Key members of the folate family are listed in Table 17.1.

Chemical Structures of the Folate Group

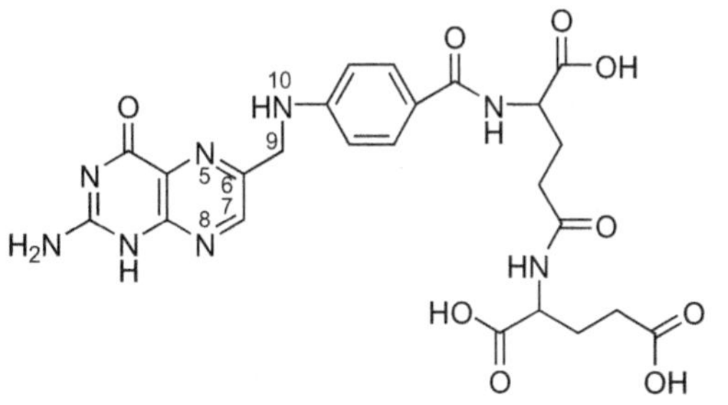

Pteroylglutamic acid

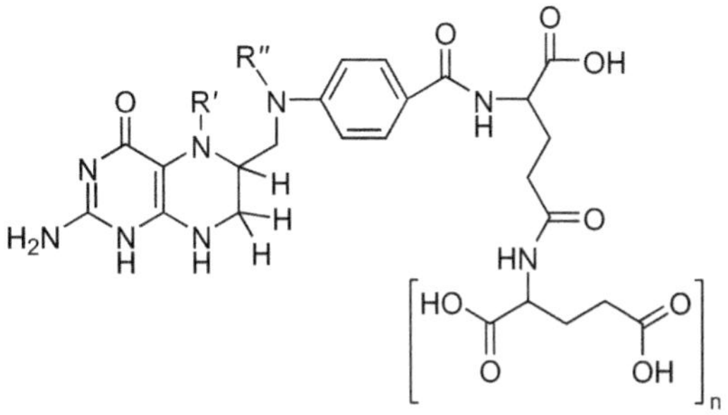

Tetrahydrofolic acid and its derivatives

Folate Chemistry

The folates include a large number of chemically related species. With three reduction states of the **pyrazine nucleus**, six different single-carbon substituents on N-5 and/or N-10, and as many as eight glutamyl residues on the benzene ring, more than 170 different folates are theoretically possible. Not all of these occur in nature, but it has been estimated that as many as 100 different forms are found in animals. The compound called folic acid (pteroylmonoglutamic acid) is probably not present in living cells, rather an artifact of isolation of the vitamin. The folates from most natural sources usually have a single-carbon unit at N-5 and/or N-10; these forms participate in the metabolism of the **single-carbon pool**. The single-carbon units that may be transported and stored by folates can vary in oxidation state from the methyl (e.g., $5\text{-}CH_3\text{-}FH_4$) to the formyl (e.g., 5-HCO-FH_4, 10-HCO-FH_4). Intracellular folates contain poly-γ-glutamyl chains usually of 2–8 glutamyl residues, sometimes extending to 12 in bacteria. Tissues contain enzymes called **conjugases** that hydrolytically remove glutamyl residues to release the monoglutamyl form, folic acid. The folylpolyglutamates are thought to be the active intracellular coenzyme forms. The monoglutamates, which can pass through membranes, appear to be transport forms.

Folates have an asymmetric center at C-6, which provides stereospecificity in the orientation of hydrogen atoms on reduction of the pteridine system; that is, they add to carbons 6 and 7 in positions below the plane of the pyrazine ring. The UV absorption spectra of the folates are characterized by the independent contributions of the pterin and 4-aminobenzoyl moieties; most have absorption maxima in the region of 280–300 nm.

Folic acid (pteroylmonoglutamic acid) is an orange-yellow crystalline substance that is soluble in water but insoluble in ethanol or less polar organic solvents. It is unstable to light, to acidic or alkaline conditions, to reducing agents, and, except in dry form, to heat. It is reduced in vivo enzymatically (or in vitro with a reductant such as dithionite) first to 7,8-dihydrofolic acid (FH_2) and then to FH_4; both of these compounds are unstable in aerobic environments and must be protected by the presence of an antioxidant (e.g., ascorbic acid, 2-mercaptoethanol). Two derivatives of folic acid, each having an amino group in the place of the hydroxyl at C-4, are folate antagonists of biomedical use: **aminopterin** (a rodenticide, 4-aminofolic acid) and **methotrexate** (an antineoplastic agent, 4-amino-N^{10}-methylfolic acid).[2]

3. SOURCES OF FOLATE

Synthesis by the Gut Microbiome

The microflora of the hindgut, particularly *Bacteroides* spp., can synthesize folates in amounts that can contribute significantly to meeting daily needs.[3] A genomic analysis of 256 representative organisms of the human gut microbiota found 43% capable of de novo synthesis of the vitamin, with a total synthetic capacity equivalent to at least 37% of the daily human need.[4] Folate biosynthesis is affected by dietary factors that affect the gut microbiome, e.g., dietary fiber, oligosaccharides, probiotics. Folates can be absorbed across the human colon,[5] with an efficiency of some 46%.[6] Whether other species show similar hindgut absorptive capacities is not clear; pigs fed a prebiotic preferentially used by *Bacteroides* spp. markedly increased their colonic microbial biosynthesis of folate without affecting their circulating levels of the vitamin.[7]

Distribution in Foods

Folates (**folylpolyglutamates**) occur in a wide variety of foods of both plant and animal origin (Table 17.2). Liver, mushrooms, and green leafy vegetables are rich sources of folate in human diets; while oilseed meals (e.g., soybean meal) and animal by-products are important sources of folate in animal feeds. The folates in foods and feedstuffs are almost exclusively in reduced form as polyglutamyl derivatives of **tetrahydrofolic acid (FH_4)**. Very little free folate (folyl monoglutamate) is found in foods or feedstuffs.

Analyses of foods have revealed a wide distribution of polyglutamyl folate derivatives, the predominant forms being **5-methyl-FH_4** and **10-formyl-FH_4**. The folates found in organ meats (e.g., liver and kidney) are about 40% methyl derivatives, whereas that in milk (and erythrocytes) is exclusively the methyl form. Some plant tissues also contain mainly 5-methyl-FH_4 (e.g., lettuce, cabbage, orange juice); others (e.g., soybean) contain relatively little of that form (~15%), the rest occurring as the 5- and 10-formyl derivatives. Most of the folates in cabbage are hexa- and heptaglutamates; whereas, half of those in soybean are monoglutamates. More than one-third of the folates in orange juice are present as monoglutamates and nearly half are present as pentaglutamates. Liver and kidney contain mainly pentaglutamates, and ~60% of the folates in milk are monoglutamates (with only 4–8% each of di- to hepta-glutamates).

Folate Fortification and Supplementation

Since 1998, American law has mandated that all "enriched" cereal grain products (wheat flour, breads, pastas and breakfast cereals, and rice) be fortified with folic acid (140 μg/100 g); it has also permitted addition of folic acid to infant formulae, medical and special dietary foods, meal replacement products, and energy bars and drinks. More than 50 other countries have similar policies for folate fortification of grain products. The US folate fortification program increased folate intakes to medians of some 288 μg/day for adults,[8] and 489–656 μg/day for children 1–13 years[9] and more than doubled circulating levels of the vitamin (Table 17.3).

Human milk. The predominant form of folate in human milk is 5-methyltetrahydrofolate bound to **folate-binding proteins (FBPs)**, which stimulate the enteric absorption of the vitamin. Prenatal supplementation was found to increase folic acid as a percentage of breast milk total folates without increasing that total.[10]

Stability

Most folates in foods and feedstuffs (that is, folates other than **folic acid**[11] and **5-formyl-FH_4**) are easily oxidized and,

2.

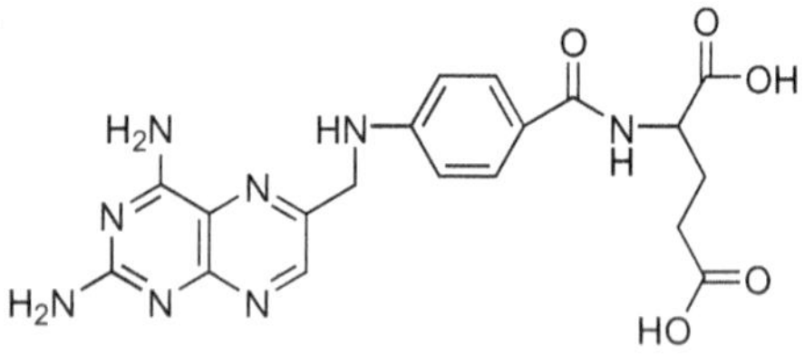

aminopterin

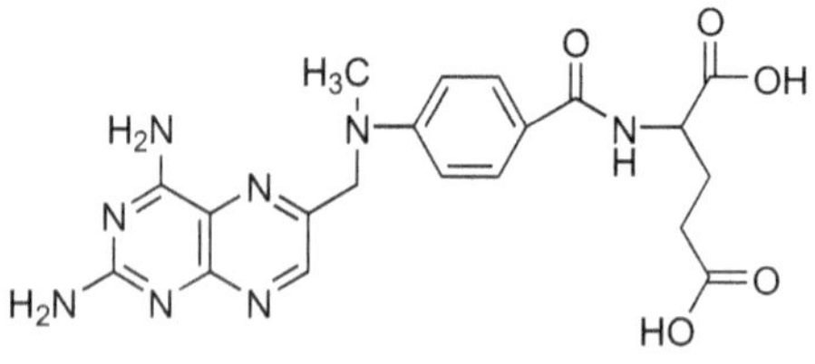

methotrexate

3. Rossi, M., Amaretti, A., Raimondi, S., 2011. Nutrients 3, 118–134.
4. Magnúsdóttir, S., Ravchee, D., de Crécy-Lagard, V., et al., 2015. Front. Genet. 6, 148–166.
5. Aufreiter, S., Gregory, J.F., Pfeiffer, C.M., et al., 2009. Am. J. Clin. Nutr. 90,116–123; Lakoff, A., Fazili, F., Aufreiter, S., et al., 2014. Am. J. Clin. Nutr. 100, 1278–1286.
6. Lakoff, A., Fazili, Z., Aufreiter, S., et al., 2014. Am. J. Clin. Nutr. 100, 1278–1286.
7. Aufreiter, S., Kim, J.H., O'Connor, D.L., 2011. J. Nutr. 141, 366–372.
8. Yang, Q.H., Cogswell, M.E., Hammer, H.C., et al., 2010. Am. J. Clin. Nutr. 91, 64–72.
9. Greater folate intakes of children reflect their greater consumption of fortified breakfast cereals; Bailey, R.L., Dodd, K., Gashche, J.J., et al., 2010. Am. J. Clin. Nutr. 91, 231–237.
10. West, A.A., Yan, J., Perry, C.A., et al., 2012. Am. J. Clin. Nutr. 96, 789–800.
11. Throughout this text, the term **folic acid** is used as the specific trivial name for the compound **pteroylglutamic acid**.

TABLE 17.2 Folate Contents of Foods

Food	Folate, µg/100 g
Dairy Products	
Milk	5
Cheese	5–65
Meats	
Beef	7–10
Chicken	3–9
Pork	0–12
Turkey	6–9
Beef liver	140–1070
Chicken liver	1810
Tuna	15
Cereals	
Barley	14
Cornmeal	25
Rice, white	3
Rice, brown	9
Wheat flour	44
Wheat bran	79
Vegetables	
Asparagus	149
Beans	33–106
Broccoli	63
Cabbage	43
Cauliflower	57
Peas	42
Soybeans	211
Spinach	194
Tomatoes	15
Fruits	
Apples	3
Bananas	20
Oranges	34
Others	
Eggs	47
Bakers' yeast	785

From USDA National Nutrient Database for Standard Reference, Release 28 (http://www.ars.usda.gov/ba/bhnrc/ndl).

TABLE 17.3 Changes in Plasma Folate and Homocysteine Levels in the United States Since Implementation of Folate Fortification of Cereals

Year	Plasma Folate		Plasma Hcy	
	ng/mL	% ≤6.8 nM	µM	% ≤13 µM
1988–94	12.1 ± 0.3[a]	18.4	8.7 ± 0.1	13.2
1999–2000	30.2 ± 0.7[b]	0.8	7.0 ± 0.8	4.5
2001–02	27.8 ± 0.5[b]	0.2	7.3 ± 0.1	4.7

[a]*Mean ± S.E.*
[b]*$p < .05$.*
Ganji, V., Kafai, M.R., 2006. J. Nutr. 136, 153–158.

therefore, are unstable to oxidation under aerobic conditions of storage and processing. Under such conditions (especially in the added presence of heat, UVB light, and/or metal ions), FH_4 derivatives can be oxidized to the corresponding derivatives of **dihydrofolic acid (FH_2)** (partially oxidized) or folic acid (fully oxidized), some of which can react further to yield physiologically inactive compounds. For example, the two predominant folates in fresh foods, 5-methyl-FH_4 and 10-formyl-FH_4, are converted to 5-methyl-5,6-FH_2 and 10-formylfolic acid, respectively. For this reason, 5-methyl-5,6-FH_2 has been found to account for about half of the folate in most prepared foods. Although it can be reduced to the FH_4 form (e.g., by ascorbic acid), in the acidity of normal gastric juice, it isomerizes to yield 5-methyl-5,8-FH_2 which is completely inactive. It is of interest to note that, owing to their gastric acidosis, this isomerization does not occur in pernicious anemia patients, who are, thus, able to utilize the partially oxidized form by absorbing and subsequently activating it to 5-methyl-FH_4. Because some folate derivatives of the latter type can support the growth responses of test microorganisms used to measure folates,[12] some information in the available literature may overestimate the biologically useful folate contents of foods and/or feedstuffs. Substantial losses in the folate contents of food can occur as the result of leaching in cooking water when boiling (losses of total folates of 22% for asparagus and 84% for cauliflower have been observed), as well as oxidation, as described earlier. Due to such losses, green leafy vegetables can lose their value as sources of folates despite their relatively high natural contents of the vitamin. Photodegradation of folates in blood has been observed in patients with psoriasis given phototherapy with high cumulative doses of narrowband UVB

12. *Lactobacillus casei*, *Streptococcus faecium* (formerly, *Streptococcus lactis* R. and *Streptococcus faecalis*, respectively), and *Pediococcus cerevisiae* (formerly, *Leuconostoc citrovorum*) have been used. Of these, *Lactobacillus casei* responds to the widest range of folates.

irradiation.[13] Such degradation is accelerated by endogenous photosensitizers such as flavins and porphyrins but is suppressed by bilirubin and protein binding.

Bioavailability

The biological availability of folates in foods has been difficult to assess. Additional sources of error come from factors affecting the utilization of food folates:

- **antifolates** that bind to the food matrices or inhibit the intestinal brush border folyl conjugase.
- **inherent differences in folyl glutamates.**
- **nutritional status of the host**, e.g., deficiencies of iron and vitamin C can impair the utilization of dietary folate.[14] Vitamin C enhances the utilization of 5-methyl-FH_4 by preventing its oxidative degradation to 5-methyl-FH_2, which does not enter the folate metabolic pool.

Interactions of these factors complicate the task of predicting the bioavailability of dietary folates. This problem is exacerbated by the methodological difficulties in evaluating folate utilization, which has been approached with bioassays with animal models,[15] and with studies in humans using erythrocyte folate response, dilution of stable isotope labeled folate or urinary folate excretion. Each has limitations and sources of error.

The result is that estimates of the bioavailability of food folates (Table 17.4) show high interindividual variation but generally indicate that

- folic acid is virtually completely bioavailable;
- folic acid is similarly highly bioavailable from fortified foods;
- bioavailabilities of food folates appear to range from 10 to 98% of folic acid, although those of most are about 50%;[16] and
- mixed diets have aggregate bioavailability of dietary folates of 50–80%.[17]

13. El-Saie, L.T., Rabie, A.R., Kamel, M.I., et al., 2011. Lasers Med. Sci. 26, 481–485.

14. Some anemic patients respond optimally to oral folate therapy only when they are also given iron. Patients with scurvy often have megaloblastic anemia, apparently owing to impaired utilization of folate. In some scorbutic patients, vitamin C has an antianemic effect; others require folate to correct the anemia.

15. As with any application of information from studies with animal models, the validity of extrapolation is an issue important in assessing folate bioavailability. For example, the rat and many other species have little or no brush border conjugase activity, these species relying on pancreatic conjugase for folate deconjugation. This contrasts with the pig and human, which deconjugate folates primarily by brush border activity.

16. Brouwer, I.A., van Dusseldorp, M., West, C.E., et al., 2001. Nutr. Res. Rev. 14, 267–293; McNulty, H., Pentieva, K., 2004. Proc. Nutr. Soc. 63, 529–536; Mönch, S., Netzel, M., Netzel, G., et al., 2014. R. Soc. Chem. 6, 242–248.

17. Sauberlich, H.E., Kretsch, M.J., Skala, J.H., et al., 1987. Am. J. Clin. Nutr. 46, 1016–1028; Winkels, R.M., Brouwer, I.A., Sieblink E., et al., 2007. Am. J. Clin. Nutr. 85, 465–473.

TABLE 17.4 Individual Variability in Reported Bioavailability Values of Folates in Foods

Food/Feedstuff	Bioavailability, %[a]
Bananas	0–148
Cabbage	0–127
Eggs	35–137
Lima beans	0–181
Liver (goat)	9–135
Orange juice	29–40
Spinach	26–99
Tomatoes	24–71
Wheat germ	0–64
Brewers' yeast	10–100
Soybean meal	0–83

[a]*Results expressed relative to folic acid.*
Baker, H., Jaslow, F.P., Frank, D., 1978. J. Am. Geriatr. Soc. 26, 218–221; Babu, H., Srikantia, S.G., 1976. Am. J. Clin. Nutr. 29, 376–382; Tamura, T., Stokstad, E.R.L., 1973. Br. J. Haematol. 25, 513–532.

4. ABSORPTION OF FOLATE

The efficiency of absorption of dietary folates is about 50% but can vary considerably (10–90%). Folate absorption is a multistep process:

1. **Deconjugation of polyglutamyl folates.** Because the majority of food folates occur as reduced polyglutamates, their absorption depends on their cleavage to mono- or diglutamate forms. This is accomplished by mucosal folyl γ-glutamyl carboxypeptidases, commonly called **folyl conjugases** in the small intestine:[18]
   a. A 700 kDa **brush border exocarboxypeptidase** with an optimum of pH 6.5–7.0. Although present in relatively low amounts, it is most important for the hydrolysis of folylpolyglutamates. A genetic variant has been associated with low-serum folate concentrations and homocysteinemia.
   b. A 75 kDa **intracellular** (lysosomal) **carboxypeptidase** with a pH optimum of pH 4.5–50.

Loss of conjugase activity impairs folate absorption. This can be produced by zinc deficiency or by exposure to naturally occurring conjugase inhibitors in foods such as cabbage, oranges, yeast, beans (red kidney, pinto, lima, navy, soy), lentils, and black-eyed peas (Table 17.5).[19] This

18. Conjugase activities have also been identified in bile, pancreatic juice, kidney, liver, placenta, bone marrow, leukocytes, and plasma, although the physiological importance of these activities is unclear. In the uterus, conjugase activity is induced by estrogen.

19. The conjugase inhibitors in beans and peas reside in the seed coats and are heat-labile.

TABLE 17.5 Inhibition of Jejunal Folyl Conjugase Activities In Vitro by Components of Selected Foods

	% Inhibition, by Conjugase Source	
Food	Pig	Human
Red kidney beans	35.5	15.9
Pinto beans	35.1	33.2
Lima beans	35.6	35.2
Black-eyed peas	25.9	19.3
Yellow cornmeal	35.3	28.3
Wheat bran	−2.0	0
Tomato	8.1	14.2
Banana	45.9	46.0
Cauliflower	25.2	15.3
Spinach	21.1	13.9
Orange juice	80.0	73.4
Egg	11.5	5.3
Milk	13.7	—
Cabbage	12.1	—
Whole wheat flour	0.3	—
Medium rye flour	2.2	—

Bhandari, S.D., Gregory, J.F., 1990. Am. J. Clin. Nutr. 51, 87–94.

is the basis for the low bioavailability of folate in orange juice. Folate absorption can also be reduced by certain drugs including cholestyramine (which binds folates), sal icylazosulfapyridine,[20] diphenylhydantoin,[21] aspirin and other salicylates, and chronic ethanol ingestion.[22]

2. **Active uptake by the enterocyte**. Dietary folates are absorbed in deconjugated form, i.e., as folic acid, 5-methyl-FH_4, and 5-formyl-FH_4.[23] These vitamers are actively transported across the brush border by processes facilitated by two transporters (Fig. 17.1):
   a. **Reduced folate carrier (RFC** or **SLC19A1)** is a member of the solute carrier 19 (SLC19) family of transporters.[24] It is found on the enterocyte apical membrane as a homodimer with independently acting monomers and 12 transmembrane domains. RFC in the enterocyte binds both reduced and oxidized forms of the vitamin with comparable affinities. Its bidirectional transport function is driven by a transmembrane pH gradient; activity is optimal at pH 7.4 and is stimulated by glucose. RFC binds antifolates with affinities at least an order of magnitude greater than those of folate binding. Its expression is upregulated by folate deficiency.
   b. **Proton-coupled folate transporter (PCFT or SLC46A1)** is a Na^+-dependent, high-affinity transporter also of the SLC19 family.[25] It has high affinities for folic acid, 5-methyl-FH_4 and 5-formyl-FH_4, and methotrexate. PCFT-facilitated transport is driven by a transmembrane pH gradient, but functions in the absence of such a gradient, being based on membrane potential and sensitive to folate gradient. Its activity is greatest under acidic conditions (pH optima 5–6) and is, thus, thought to play a major role in facilitating folate uptake in the acidic microenvironment of the jejunum. Loss-of-function mutations in PCFT results in **hereditary folate malabsorption (HFM)**,[26] which is readily corrected by high oral doses of 5-formyl-FH_4 or folic acid.
   c. **Multidrug resistance-associated protein 3 (MRP3)**[27] has been implicated in the enteric absorption of oxidized folates. Its genetic deletion reduced the transport of 5-formyl-FH_4 and 5-methyl-FH_4 across everted duodenal sacs.[28] This suggests that MRP3 may participate in moving folates across the enterocyte basolateral surface where the protein is located.
3. **Passive diffusion into the enterocyte** can account for 20–30% of folate absorption at high folate intakes. Folate diffusion is greatest under acidic conditions in which its molecular charge is reduced. This may be the basis of increased folate absorption observed in individuals with pancreatic exocrine insufficiency; their reduced excretion of bicarbonate and the resulting loss of buffering capacity render the luminal milieu more acidic. Under more basic conditions (pH > 6.0), folate absorption falls off rapidly.

20. Also called azulfidine and sulfasalazine, used to treat inflammatory bowel disorder.
21. Also called dilantin, an anticonvulsant.
22. Other factors may contribute to this phenomenon: enterocytes are known to be sensitive to ethanol toxicity; many chronic alcoholics can have inadequate folate intakes.
23. The dog appears to absorb folylpolyglutamates.
24. This family also includes two thiamin transporters, SLC19A2 and SLC19A3. RFC also transports the folate antagonist methotrexate.
25. PCFT was originally described as a low-affinity heme carrier protein (HCP1), although that is no longer regarded as its primary function. Hence, it is often referred to as PCFT/HCP1.
26. HFM has been reported in some 30 patients, presenting at 2–6 months of age as megaloblastic anemia, mucositis, diarrhea, failure to thrive, recurrent infections, and seizures.
27. The MRPs are members of the ATP-binding cassette (ABC) transporters, several of which (MRPs1–5 and ABCG2) have low-affinity, high capacity for binding folates.
28. Kitamura, Y., Kusuhara, H., Sugiyama, Y., 2010. Pharm. Res. 27, 665–672.

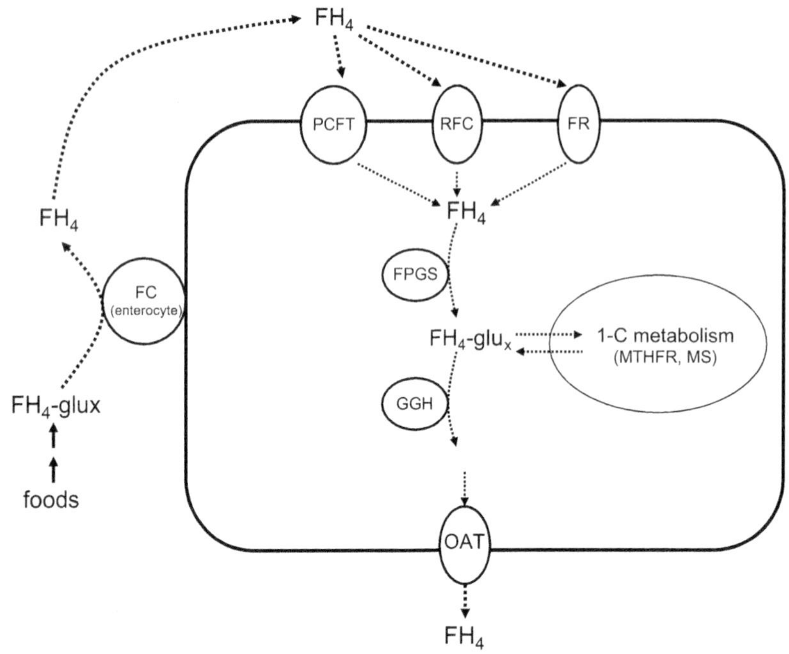

FIGURE 17.1 Facilitated Cellular Uptake and Utilization of Folate. *FC*, folyl conjugase; *FPGS*, folylpolyglutamate synthase; *FR*, folate receptor; *GGH*, γ-glutamyl hydrolase; *MS*, methionine synthase; *MTHFR*, methylenetetrahydrofolate reductase; *OAT*, organic anion transporter; *PCFT*, proton-coupled folate transporter; *RFC*, reduced folate carrier. *After Devos, L., Chanson, A., Liu, Z., et al., 2008. Am. J. Clin. Nutr. 88, 1149–1158.*

4. **Reconjugation**. Absorbed monoglutamyl folates are converted to higher-glutamyl forms by **folylpolyglutamate synthase**, preventing the vitamin being pumped out of the cell.
5. **Exportation**. Absorbed FH_4 can be exported without further metabolism to the portal circulation or after being first alkylated (e.g., by methylation to 5-methyl-FH_4). Both forms are exported across the basolateral cell membrane by a **folate export pump**, which is likely to be a member of the family of multispecific **organic anion transporters (OATs)**, one of which has been shown to transport the folate antagonist methotrexate. Because folate pumps do not transport polyglutamates, those forms are retained within the cell until being deconjugated.

Folates can be absorbed across the colon. This was demonstrated in humans using a $^{13}C_5$-formyl-FH_4 administered in caplets that did not dissolve before reaching the colon. $^{13}C_5$-formyl-FH_4 appeared in the plasma at the rate of 5.8 ± 1.2 nmol/h with a mean absorption of 46%.[29] It is likely that absorption across the colon occurs by the same process as in the small intestine. PCFT expression has been demonstrated in the colon.[30]

29. Lakoff, A., Fazili, Z., Aufreiter, S., et al., 2014. Am. J. Clin. Nutr. 100, 1278–1286.
30. Aufreiter, S., Gregory, J.F., Pfeiffer, C.M., et al., 2009. Am. J. Clin. Nutr. 90,116–123.

5. TRANSPORT OF FOLATE

Erythrocytes

Erythrocytes comprise the largest compartment of circulating folates, containing 670–1800 nM, depending on level of intake of the vitamin. The dominant share of this compartment is 5-methyl-FH_4, which comprises more than 80% of the total. Also present are the metabolite 4α-hydroxy-5-methyl-FH_4 and smaller amounts of FH_4, 5-formyl-FH_4, and 5,10-methenyl-FH_4.[31]

Free in Plasma

Plasma folate concentrations in most species are in the range of 10–30 nM most of which is in free solution. The predominant form in portal plasma is the reduced form, FH_4. This is taken up by the liver, which releases it to the peripheral plasma after converting it primarily to 5-methyl-FH_4 but also to 10-formyl-FH_4. The concentration of 10-formyl-FH_4 is tightly regulated;[32] whereas that of 5-methyl-FH_4 is not. Thus, the latter varies in response to recent folate intake. Therefore, most folate is transported to peripheral tissues in the form of monoglutamyl derivatives. The notable exception is the pig in which FH_4 is the predominant circulating

31. Hartman, B.A., Fazili, Z., Pfeiffer, C.M., et al., 2014. J. Nutr. 144, 1364–1369.
32. In humans, the plasma level is held at ~80 ng/dL.

form.[33] Erythrocytes contain relatively great concentrations of folate, typically 50–100 nM. These stores are accumulated during erythropoiesis; the mature erythrocyte does not take up folate.

Folate transport is reduced by cigarette smoking and chronic alcohol ingestion. Smokers show plasma folate levels that can be nearly half those of nonsmokers.[34] While serum folates have been found to be normal among consumers of moderate amounts of ethanol, more than 80% of chronic alcoholics show abnormally low-serum levels and some 40% show low-erythrocyte levels. This corresponds to a similar incidence (34–42%) of megaloblastosis of the bone marrow in alcoholic patients.[35]

Protein-Bound in Plasma

Folate is also bound to three proteins in the plasma: albumin and an FBP. The high-affinity FBP in plasma is thought to be a solubilized form of tissue FPBs. It is present in low concentrations (binding <10 ng folic acid per deciliter) but is elevated in folate deficiency and pregnancy. Most of FBP-bound plasma folate is likely to be folic acid, which FBP binds avidly and preferentially. Unmetabolized folic acid is detected in nearly all subjects, particularly in those given large (>~260 μg) or consecutive small oral doses (~100 μg) of folic acid.[36] It has been suggested that the presence of unmetabolized folic acid in the circulation reflects a mismatch of supply and cellular demand for the vitamin.

Cellular Uptake

Circulating monoglutamyl folates are taken up by same process by which they were taken up by enterocytes. This involves the folate transporters and receptors (Fig. 17.1):

- **RFCs** are the major transporters of folates into tissue in which they are ubiquitously expressed. In extraintestinal tissues, they bind reduced folates preferentially to nonreduced forms. In renal proximal tubules, RFCs expressed at the apical brush border membrane have antiporter function that favors the transport of folates into cells. In epidermal and dermal cells, their expression is upregulated by exposure to UV irradiation.[37]
- **PCFT** is highly expressed in liver, kidney, colon, brain, retina, and placenta. It functions to facilitate **folate receptor (FR)**-mediated endocytotic uptake of the vitamin.
- High-affinity **FRs** (also called **FBPs**) bind folic acid, 5-methyl-FH_4, and some antifolate drugs stoichiometrically and with high affinities, facilitating their uptake by endocytosis. They occur in three isoforms, some of which function in mediating folate uptake via endocytosis: FR-α (highly expressed in placental epithelia, renal tubules, retinal pigment epithelia, and choroid plexus) and FR-β (highly expressed by spleen, thymus, and CD34+ monocytes).[38] Regulation of FR expression is not well understood, but it is clear that extracellular folate concentration serves as an inverse stimulus to FR expression. That FRs and the folate transporters are localized on opposite aspects of polarized cells facilitate movement of folate across the apical membrane into the cell. In the renal tubule, FR-α functions in reabsorption, efficiently extracts folates from the glomerular filtrate and becoming saturated only at high plasma concentrations of the vitamin.[39] A neuropsychiatric disorder, **cerebral folate deficiency syndrome**, has been described; it involves blockade of folate transport across the blood–brain barrier due to high-affinity autoantibodies against membrane-bound FBPs on the choroid plexus.[40]
- **MRP2 and MRP3** appear to function in the efflux of folates from hepatocytes across their basolateral and canalicular membranes into hepatic sinusoids and bile ducts, respectively.[41]

Within cells, folates are stabilized and protected from oxidative degradation by binding to intracellular proteins. Folate-binding enzymes are typically present in micromolar amounts that exceed folate concentrations by several fold. They bind folates with high affinities, i.e., with dissociation constants in the nanomolar range. Therefore, intracellular concentrations of free folates are exceedingly low.

Tissue Distribution

In humans, the total body content of folate is 5–10 mg, half of which resides in the liver as tetra-, penta-, hexa- and heptaglutamates of 5-methyl-FH_4 and 10-formyl-FH_4.[42] The relative amounts of these single-carbon derivatives

33. The metabolic basis for this anomaly is not clear.
34. These findings probably relate to the inactivation of cobalamins by factors (cyanides, hydrogen sulfide, nitrous oxide) in cigarette smoke.
35. It is likely that effects of such magnitudes have several causes: inhibition of intestinal folyl conjugase activity; decreased urinary folate recovery; displacement of folate-containing foods by alcoholic beverages.
36. Pfeiffer, C.M., Sternberg, M.R., Fazili, Z., et al., 2015. J. Nutr. 145, 520–531.
37. Knott, A., Meilke, H., Koop, U., et al., 2007. Soc. Invest. Dermatol. 127, 2463–2466.
38. Mutations of FR-α are embryonically lethal, but deletion of FR-β produces no phenotype in the mouse. A third isoform, FR-γ, is found on natural killer and TGF-β-induced regulatory T cells.
39. Birn, H., Spiegelstein, O., Christensen, E.I., et al., 2005. J. Am. Soc. Nephrol. 16, 608–615.
40. Ramaekers, V.T., Rothenberg, S.P., Sequeira, J.M., et al., 2005. N. Engl. J. Med. 52, 1985–1991.
41. Kitamura, Y., Hirouchi, H., Kusuhara, H., et al., 2008. J. Pharmacol. Exp. Ther. 327, 465–473; Masud, M., I'isuka, Y., Yamazaki, M., et al., 1997. Cancer Res. 57, 3506–3510.
42. Hepatic reserves of folate should be sufficient to support normal plasma concentrations of the vitamin (>400 ng/dL) for at least 4 weeks. (Signs of megaloblastic anemia are usually not observed within 2–3 months of folate deprivation.) However, some evidence suggests that the release of folate from the liver is independent of nutritional folate status, resulting instead from the death of hepatocytes.

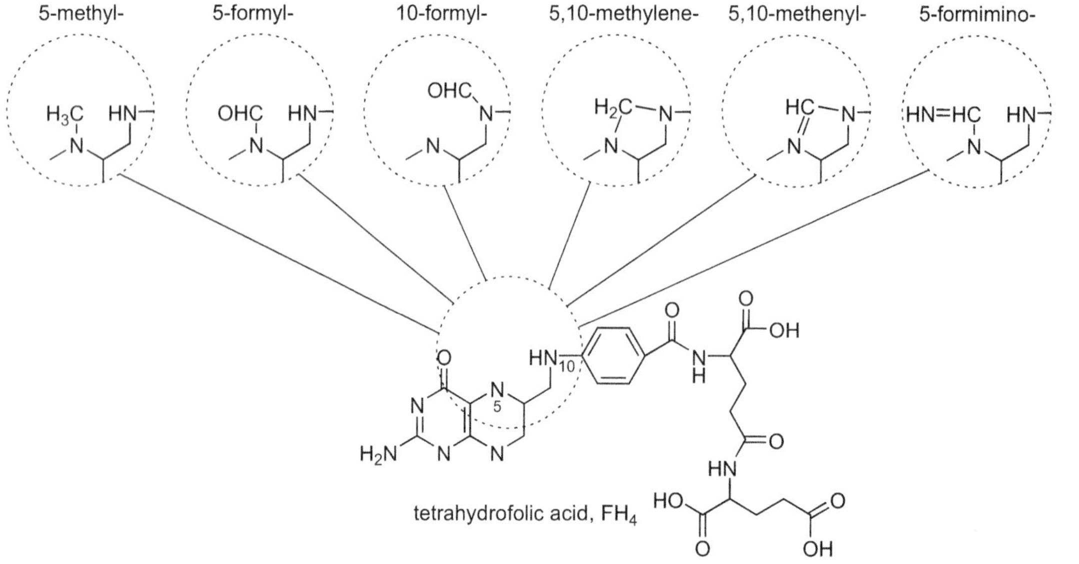

FIGURE 17.2 Single-carbon units carried by folate.

vary among tissues, depending largely on the rate of cell division. In tissues with rapid cell division (e.g., intestinal mucosa, regenerating liver, carcinoma), relatively low concentrations of 5-methyl-FH_4 are found, usually with concomitant elevations in 10-formyl-FH_4. In tissues with low rates of cell division (e.g., normal liver), 5-methyl-FH_4 predominates. Brain folate (mostly 5-methyl-FH_4) levels tend to be very low, with a subcellular distribution (penta- and hexaglutamates mostly in the cytosol and polyglutamates mostly in the mitochondria) the opposite of that found in liver. Folate-deficient animals show relatively low hepatic concentrations of shorter chain length folylpolyglutamates compared with longer chain length folates, suggesting that longer chain length metabolites are better retained within cells. In the rat, uterine concentrations of folates show cyclic variations according to the menstrual cycle, with maxima coincident with peak estrogen levels just before ovulation.[43]

6. METABOLISM OF FOLATE

There are three aspects of folate metabolism:

1. **Reduction of the pteridine ring system** from the two nonreduced states, folic acid and dihydrofolic acid (FH_2), to the fully reduced form tetrahydrofolic acid (FH_4). This form is capable of accepting a single-carbon unit by the action of the cytosolic enzyme **7,8-dihydrofolate reductase** (Fig. 17.2).[44] That activity is found in high amounts in liver and kidney and in rapidly dividing cells (e.g., tumors). It is inhibited by several drugs including the cancer chemotherapeutic drug **methotrexate**.[45]
2. **Reactions of the polyglutamyl side chain** by side chain hydrolysis to the corresponding monoglutamate. These conversions are catalyzed by two enzymes:
   - The ATP-dependent **folylpolyglutamate synthase**[46] catalyzes the conversion of 5-methyl-FH_4 to folylpolyglutamates by linking glutamyl residues to the vitamin by peptide bonds involving the γ-carboxyl groups.[47] The enzyme requires prior reduction of folate to FH_4 or demethylation of the circulating 5-methyl-FH_4 (by vitamin B_{12}-dependent methionine synthase). It is widely distributed at low concentrations in many tissues and is critical in converting the monoglutamyl transport forms of the vitamin to the metabolically active polyglutamyl forms. Mutational loss of the enzyme results in lethal folate deficiency. In most tissues, the activity is rate limiting for folate retention. Cells lacking the enzyme are unable to accumulate the vitamin.[48] Those lacking the mitochondrial enzyme cannot accumulate

43. On the basis of this type of observation, it has been suggested that estrogen enhancement of folate turnover in hormone-dependent tissues may be the basis of the effects of pregnancy and oral contraceptive steroids in potentiating low-folate status.

44. Also called **tetrahydrofolate dehydrogenase**, this 65-kDa NADPH-dependent enzyme can reduce folic acid to FH_2 and, of greater importance, FH_2 to FH_4. The enzyme is potently inhibited by the drug methotrexate.

45. Other inhibitors include the antimalarial drug, pyrimethamine and the antibacterial drug trimethoprim.

46. The mitochondrial and cytosolic forms of the enzyme are encoded by a single gene the transcription of which has alternate start sites and the mRNA of which has alternative translation sites.

47. This enzyme also catalyzes the polyglutamation of the anticancer folate antagonist methotrexate, which enhances its cellular retention. Tumor cells, which have the greatest capacities to perform this side chain elongation reaction, are particularly sensitive to the cytotoxic effects of the antagonist.

48. Because polyglutamation is also necessary for the cellular accumulation and cytotoxic efficacy of antifolates such as methotrexate, decreased folylpolyglutamate synthase activity is associated with clinical resistance to those drugs.

TABLE 17.6 Enzymes Involved in the Acquisition of Single-Carbon Units by Folates

Single-Carbon Unit	Folate Derivative	Enzymes
Methyl group ($-CH_3$)	5-Methyl-FH_4	5,10-Methylene-FH_4 reductase methionine synthase
Methylene group ($=CH_2$)	5,10-Methylene-FH_4	Serine hydroxymethyltransferase 5,10-Methylene-FH_4 dehydrogenase
Methenyl group ($=CH$)	5,10-Methenyl-FH_4	5,10-Methylene-FH_4 dehydrogenase
		5,10-Methenyl-FH_4 cyclohydrolase
		5-Formimino-FH_4 cyclohydrolase
		5-Formyl-FH_4 isomerase
Formimino group ($-CH=NH$)	5-Formimino-FH_4	FH_4 formiminotransferase
Formyl group ($-CH=O$)	5-Formyl-FH_4 10-Formyl-FH_4	FH_4:glutamate transformylase
		5,10-Methenyl-FH_4 cyclohydrolase
		10-Formyl-FH_4 synthase

the vitamin in that subcellular compartment; consequently, they show deficient mitochondrial single-carbon metabolism.

- Folylpolyglutamates are turned over in cells by hydrolysis to shorter chain derivatives by the action of soluble, lysosomal[49] γ-glutamyl carboxypeptidases, also referred to as **folyl conjugases**. Some are zinc metalloenzymes.

3. **Acquisition of single-carbon moieties** at the oxidation levels of formate, formaldehyde, or methanol[50] substituted at the N-5 and/or N-10 positions of the pteridine ring system (Fig. 17.2). The main source of single-C fragments is **serine hydroxymethyltransferase (SHMT)** (Table 17.6), which uses the dispensable amino acid serine[51] as the source of single-C. Each folyl derivative is a donor of its single-C unit in metabolism[52]; thus, by cycling through the acquisition/loss of single-C units, folyl derivatives deliver these species for different metabolic uses. Most single-C folate derivatives in cells are bound to enzymes or FBPs.

49. Lysosomes also contain a folate transporter, which is thought to be active in bringing folypolyglutamates into that vesicle.

50. It should be noted that single carbons at the oxidation level of CO_2 cannot be transported by folates; fully oxidized carbon is transported by biotin and thiamin pyrophosphate.

51. Serine is biosynthesized from glucose in nonlimiting amounts in most cells.

52. Although the route of its biosynthesis is unknown, eukaryotic cells contain significant amounts of 5-formyl-FH_4. That folyl derivative, also called **leucovorin**, **folinic acid** and **citrovorum factor**, is used widely to reverse the toxicity of methotrexate and, more recently, to potentiate the cytotoxic effects of 5-fluorouracil.

Methyl Folate Trap

The major cycle of single-C flux in mammalian tissues is the serine hydroxymethyl transferase/5,10-methylene-FH_4 reductase/methionine synthase cycle, in which the latter reaction is rate limiting (Fig. 17.3). The committed step (5,10-methylene-FH_4 reductase) is feedback-inhibited by **S-adenosylmethionine (SAM)** and product-inhibited by 5-methyl-FH_4. Methionine synthase depends on the transfer of labile methyl groups from 5-methyl-FH_4 to vitamin B_{12}, which, as methyl-B_{12}, serves as the immediate methyl donor for converting homocysteine (Hcy) to methionine. Without adequate vitamin B_{12} to accept methyl groups from 5-methyl-FH_4, that metabolite accumulates at the expense of the other metabolically active folate pools. This is known as the "**methyl folate trap**." The loss of FH_4 that results from this blockage in folate recycling blocks transfer of the histidine–formimino group to folate (as 5-formimino-FH_4) during the catabolism of that amino acid. This results in the accumulation of the intermediate formiminoglutamic acid (FIGLU). Thus, elevated urinary FIGLU levels after an oral histidine load are diagnostic of vitamin B_{12} deficiency.

Catabolism

Tissue folates turn over by the cleavage of the polyglutamates at the C-9 and N-10 bonds to liberate the pteridine and *p*-aminobenzoyl polyglutamate moieties. This results from chemical oxidation of the cofactor both in the intestinal lumen (dietary and enterohepatically recycled folates) as well as in the tissues. Once formed, *p*-aminobenzoylpolyglutamate is degraded, presumably by the action of folyl conjugase, and is acetylated to yield *p*-acetaminobenzoylglutamate and

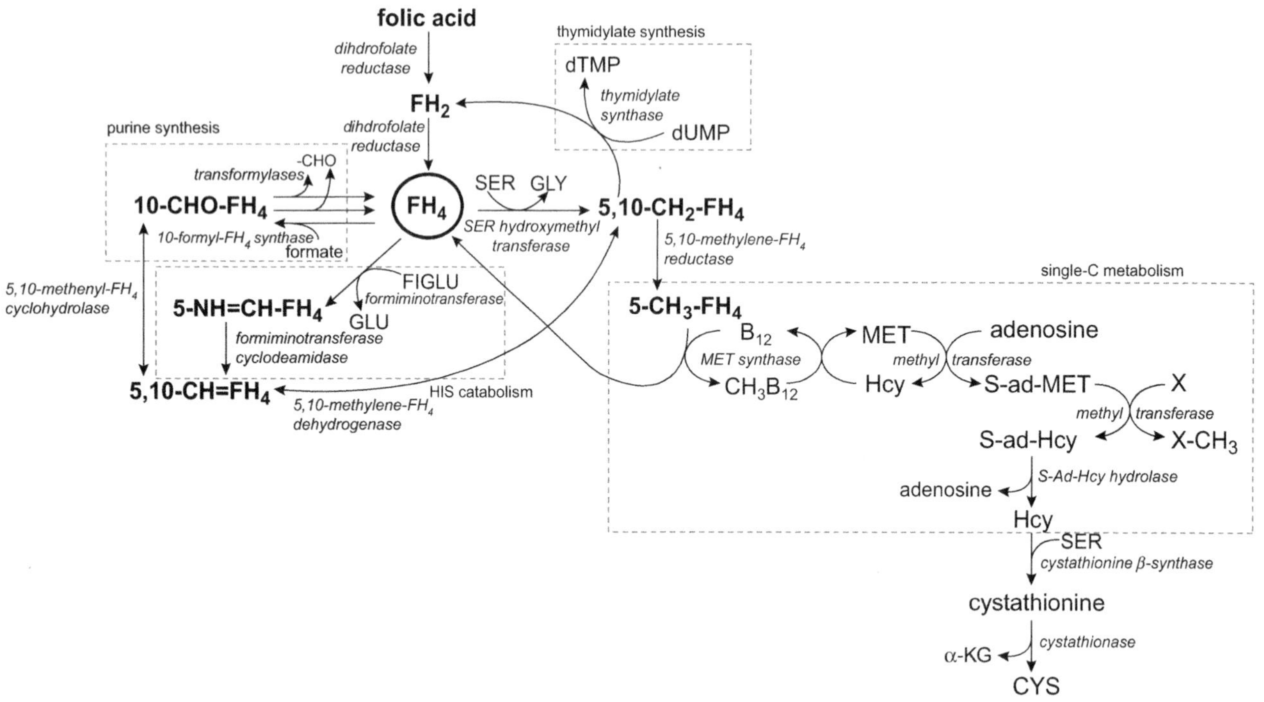

FIGURE 17.3 **Folate roles in metabolism**. Relationship with vitamin B_{12} and the basis of the "**methyl folate trap**" whereby folate accumulates as $5\text{-}CH_3\text{-}FH_4$ when vitamin B_{12} is not available. *CYS*, cysteine; *dTMP*, deoxythymidine monophosphate; *dUMP*, deoxyuridine monophosphate; *FIGLU*, formiminolgutatmic acid; *GLU*, glutamic acid; *Hcy*, homocysteine; *HIS*, histidine; *MET*, methionine; *S-ad-Hcy*, S-adenosylhomocysteine; *S-ad-MET*, S-adenosylmethionine; *SER*, serine; *X*, methyl acceptor; *α-KG*, a-ketoglutarate.

p-acetamiobenzoate. The rate of folate catabolism is related to the rate of intracellular folate utilization, with urinary levels of *p*-acetamiobenzoate correlated with the total body folate pool. Studies of the kinetics of folate turnover using isotopic labels in nonpregnant women have indicated both fast- and slow-turnover body folate pools of which turn over faster with increasing folate intake (Table 17.7). At low-to-moderate intakes, 0.5–1% of total body folate appears to be catabolized and excreted per day. Folate breakdown is greatest under hyperplastic conditions (e.g., rapid growth, pregnancy[53]).

Excretion

Intact folates and the water-soluble side chain metabolites, *p*-acetaminobenzoylglutamate and *p*-acetamiobenzoate, are excreted in the urine and bile. The total urinary excretion of folates and metabolites is small (e.g., ≤1% of total body stores per day). Folate conservation is effected by the reabsorption of 5-methyl-FH_4 by the renal proximal tubule; renal reabsorption appears to be mainly a nonspecific process, although an FBP-mediated process has also been demonstrated. Fecal concentrations of folates are usually rather high—comparable to that excreted in the urine. They are mostly of intestinal microfloral origin, as enterohepatically circulated folates appear to be absorbed quantitatively.

Polymorphisms of Enzymes in Folate Metabolism

Genetic polymorphisms have been identified in most of the major enzymes involved in folate utilization (Table 17.8):[54]

- **Folyl conjugase**. Polymorphism involving a T→C substitution of base pair 1561 has been described.[55] Individuals with the T allele have higher circulating folate levels than those with the C allele.[56]
- **RFC**. Polymorphism involving a G→A substitution of base pair 80 has been described. Individuals homozygous for the G allele tend to have elevated circulating Hcy levels and reduced erythrocyte folate levels.[57]

53. This effect may be an important contributor to the folate-responsive megaloblastic anemia common in pregnancies in parts of Asia, Africa, and Central and South America (where rates as high as 24% have been reported) as well as in the industrialized world (where rates of 2.5–5% have been reported).

54. Polymorphisms of other enzymes involved in folate and single-carbon metabolism have also been described, although none has been shown to affect folate utilization. These include folylpolyglutamate synthase C303T and A2006G, methionine synthase G2756A, and folyl γ-glutamyl hydrolase upstream T→C and C→T polymorphisms.

55. The enzyme is encoded by the glutamate carboxypeptidase II gene (*GCPII*).

56. Afman, L.A., Trijbels, F.J.M., Blom, H.J., 2003. J. Nutr. 133, 75–77.

57. Chango, A., Emery-Fillon, N., de Courcy, G.P., et al., 2000. Mol. Genet. Med. 70, 310–315.

TABLE 17.7 Effect of Dietary Folate on Folate Turnover in Nonpregnant Women

Folate Intake n moles (µg)/day	Total Body Folate Pool µmoles	Catabolic Rate % Body pool/day	Residence Time Days
454 (200)	64.5±2.3	0.47±0.02[a]	212±8[a]
680 (300)	71.5±3.6	0.61±0.04[b]	169±12[b]
907 (400)	73.0±2.4	0.82±0.05[c]	124±7[c]

$p<.05$, n=5–6 per group; means with like superscripts are significantly different.
Gregory, J.F., Williamson, J., Liao, J.F., et al., 1998. J. Nutr. 128, 1896–1906.

TABLE 17.8 Polymorphisms of Proteins Related to Folate Absorption and Metabolism

Enzyme	Polymorphism		Genotype: Frequency	
MTHFR (5,10-methylene-FH_4 reductase)	T6557C	CC: 41%	TT: 18%	CT: 41%
	C1289A	AA: 53%	CC: 9%	AC: 37%
Folyl conjugase	T1561C	CC: 92%	TT: 0[a]	CT: 8%
Reduced folate transporter	G80A	AA: 35%	GG: 18%	AG: 47%
Methionine synthase	G2756A	AA: 59%	GG: 3%	AG: 38%
Methionine synthase reductase	G66A	AA: 28%	GG: 23%	AG: 49%

[a]1 in 625 reported.
Molloy, A.M., 2002. J. Vitam. Nutr. Res. 72, 46–52.

- **MTHFR** (methylenetetrahydrofolate reductase). Three polymorphisms have been identified in the human MTHFR: a C→T substitution of base pair 677, an A→C substitution of base pair 1298, and G1793A.
  - **C677T**. Some 89% of Americans have the C allele. Some 20% of Mexican Americans, 12% of non-Hispanic whites, but only 1% of non-Hispanic blacks are homozygous for the T variant (TT);[58] they have a form of MTHFR with 70% lower enzyme activity, lower affinity for the flavin cofactor, and lower thermal stability[59] than the C/C form of the enzyme (Fig. 17.4). They also show slightly lower plasma folate concentrations, elevated plasma 5,10-methylene-FH_4 concentrations, mild homocysteinemia,[60] and lower global DNA methylation than other genotypes (Table 17.9)—effects that appear to be exacerbated by low folate intakes. Compared to individuals with the CC genotype, those with the TT genotype have greater rates of synthesis of adenine[61] and Hcy,[62] as well as greater risks to colorectal cancer, unipolar depression,

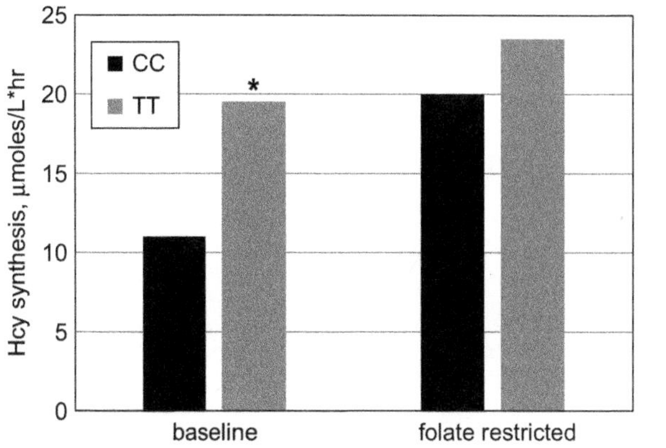

FIGURE 17.4 Effect of MTHFR (5,10-methylene-FH_4 reductase) C677T genotype on homocysteine synthesis by healthy women before and after folate restriction (115 µg/day for 7 weeks). *Difference between genotypes, $p<.05$. *After Davis, S.R., Quinlivan, E.P., Shelnutt, K.P., et al., 2005. J. Nutr. 135, 1045–1050.*

58. Yang, Q.H., Botto, L.D., Gallagher, M., et al., 2008. Am. J. Clin. Nutr. 88, 232–246.
59. For this reason the variant is frequently referred to as the "thermolabile enzyme."
60. Kauwell, G.P.A., Wilsky, C.E., Cerda, J.J., et al., 2000. Metabolism 49, 1440–1443; Hazra, A., Kraft, P., Lazarus, R., et al., 2009. Hum. Mol. Genet. 18, 4677–4687.
61. Quinlivan, E.P., Davis, S.R., Shelnutt, K.P., et al., 2005. J. Nutr. 135, 389–396.
62. Davis, S.R., Quinlivan, E.P., Shelnutt, K.P., et al., 2005. J. Nutr. 135, 1045–1050.

TABLE 17.9 Effect of MTHFR (5,10-methylene-FH_4 Reductase) C677T Genotype on Folate, Homocysteine, and Vitamin B_{12} Status in Humans

	MTHFR C677T Genotype		
Metabolite	CC	TC	TT
Plasma folate, nM	12.8±6.5	12.8±6.7	9.5±3.1
Erythrocyte folate, nM	541±188	517±182	643±186
Plasma Hcy, μM	13.4±3.4	13.2±3.1	17.1±11.5
Plasma vitamin B_{12}, pM	246±130	271±121	233±94

van der Put, N.M., Gabreels, F., Stevens, E.M., et al., 1995. Lancet 346, 1070–1071.

adult acute lymphoblastic leukemia, and ischemic stroke.[63] They also show the greater Hcy-lowering responses to folate supplements.

- **A1298C**. More than 90% of Americans have the A allele.[39] Having the C allele appears to be without significant physiological consequence unless combined with the MTHFR C677T polymorphism; doubly heterozygous individuals have been found to have MTHFR-specific activities of two-thirds those of doubly homozygous individuals, with lower circulating levels of folate and increased circulating levels of Hcy.[64]
- **G1793A**. This polymorphism is less frequent, about 4% of American women had the GG genotype. The functional significance of this polymorphism is unclear.

- **10-Formyl-FH_4 dehydrogenase**. Two intronic single-nucleotide variants have been identified; one has been associated with increased risk of breast cancer in postmenopausal women.[65] The enzyme is known to be epigenetically silenced in cancers.[66]
- **Dihydrofolate reductase**. Three variants have been identified: A 19 base pair deletion in the first intron,[60] a 5′ upstream 9 base pair repeat, and a polymorphism in the 3′-untranslated region. Individuals homozygous for the deletion have impaired tissue folate stores;[67] in mothers the trait is associated with a twofold increase in risk of spina bifida in their children.[68] Individuals with both the deletion and the repeat have lower plasma Hcy concentrations, but their circulating folate levels are unaffected.
- **Methionine synthase**. A polymorphism has been described involving A→G substitution of base pair 2756, which affects the domain involved in methylation of the vitamin B_{12} cofactor. The GG genotype has been associated with reduced plasma Hcy concentrations and increased risks of systemic lupus erythematosus, bipolar disorder, schizophrenia, congenital malformations of the face and spine, and Down syndrome.[69]
- **Methionine synthase reductase**.[70] A polymorphism has been described involving A→G substitution of base pair 66. It shows a prevalence of ~30%. The variant appears to be without physiologic consequence unless present with MTHFR 677TT, in which case serum Hcy concentrations are some 26% lower than those of other genotypes.

Effects of Drugs

Several drugs can impair folate metabolism. These have proven to have clinical applications ranging from the treatment of autoimmune diseases to cancer and malaria—all conditions in which cell proliferation can be suppressed through the inhibition of a folate-dependent step in single-C metabolism.

- **Methotrexate** (amethopterin), a folate analog differing from the vitamin by the presence of an amino group in lieu of the 4-hydroxyl group on the pteridine ring and a methyl group at the N-10 position, has greater affinity than the natural substrate for dihydrofolate reductase, resulting in its inhibition. Accordingly, the drug produces an effective folate deficiency, with reductions in thymidine synthesis and purine levels. This antiproliferative effect is the basis of its use in treating cancer, rheumatoid arthritis, psoriasis, asthma, and inflammatory bowel disease. Because its side effects include those of folate deficiency, methotrexate is usually used with accompanying and carefully monitored folate supplementation to reduce the incidence of side effects. Individuals with the MTHFR 677TT genotype have a higher risk of such adverse effects. In contrast, individuals with the MTHFR 1298CC genotype are at lower risk of side effects.

63. Carr, D.F., Whitely, G., Alfirevic, A., et al., 2009. Pharmacogenomics J. 9, 291–305.
64. Chango, A., Boisson, F., Barbe, F., et al., 2000. Br. J. Nutr. 83, 593–596.
65. Stevens, V.L., McCullough, M.L., Pavluk, A.L., et al., 2007. Cancer Epidemiol. Biomarkers Prev. 16, 1140–1147.
66. Oleinik, N.V., Krupenko, N.I., Krupenko, S.A., 2011. Cancer 2, 130–139.
67. Kalmbach, R.D., Choumenkovitch, S.F., Troen, A.P., et al., 2008. J. Nutr. 138, 2323–2327.
68. Carr, D.F., Whitely, G., Alfirevic, A., et al., 2009. Pharmacogenomics J. 9, 291–305.
69. Stover, P.J., 2011. J. Nutrigenet. Nutrigenomics 4, 293–305.
70. This enzyme that catalyzes the conversion of methionine synthase from its inactive to its active form by regenerating its methylcobalamin.

- **Other drugs** can impair folate status/metabolism. These include anticonvulsants (diphenylhydantoin, phenobarbital), antiinflammatory drugs (sulfasalazine), glycemic control drugs (metformin), and alcohol.

7. METABOLIC FUNCTIONS OF FOLATE

Folates function as enzyme cosubstrates in a network of reactions in the metabolism of amino acids and nucleotides, as well as the formation of the primary methyl donor for biological methylations, SAM. In each of these functions, the fully reduced form, tetrahydrofolic acid (FH_4), serves as an acceptor or donor of a single-carbon unit and is regenerated with the transfer such that continuous cycling can occur (Table 17.10).

Amino Acid Metabolism

Serine–glycine interconversion. FH_4 participates in the reversible interconversion of the dispensable amino acids, serine and glycine. That process is catalyzed by the pyridoxal phosphate-dependent enzyme, SHMT, which exists in three isoforms: SMHT1 and SMHT2α in the cytosol and nucleus, and SMHT2 in the mitochondria. FH_4 accepts a single-C unit from the serine C-3, binding it at the oxidation state of formaldehyde for form 5,10-methylene-FH_4 in the cytosol, and at the oxidation state of formate to form 10-formyl FH_4 in the mitochondria, subsequently entering the cytosol.

Histidine catabolism. Cytosolic formiminotransferase catalyzes the final reaction in the catabolism of histidine by transferring the formimino group from **formiminoglutamate (FIGLU)** to FH_4 to form 5-formimino-FH_4.

Methionine–cysteine transsulfuration. CYS can be produced by transfer of a methyl group from MET to form Hcy, and condensation of the latter with SER to form cystathionine from which α-ketoglutarate is cleaved to yield CYS. The rate-limiting step in this process of transsulfuration of MET to CYS is catalyzed by **MTHFR**, which is allosterically inhibited by SAM.[71]

Methionine regeneration. 5-Methyl-FH_4 is the ultimate methyl donor in the methylation of Hcy to form MET.[72] This occurs throughout the body by the action of the vitamin B_{12}-dependent enzyme MET synthase, which is transiently methylated at its corrin center before transferring the methyl group to Hcy. This process is blocked by deficiency of vitamin B_{12}, which causes secondary folate deficiency due to the accumulation of 5-methyl-FH_4 by the "methyl folate trap." This is the basis by which deficiency of either folate or vitamin B_{12} can cause macrocytic anemia.

Single-Carbon Metabolism

Formation of SAM occurs by moving single-C units from the nonessential amino acids SER and GLY, and the choline metabolites, sarcosine and dimethylglycine. Each is converted to formate, which in the cytoplasm condenses with FH_4 to form 10-formyl-FH_4. The utilization of those single-C units depends on 10-formyl-FH_4 being converted to 5-methyl-FH_4. That conversion is facilitated by low activities of 10-formyl-FH_4 dehydrogenase, which otherwise would deplete the single-C pool by catalyzing the conversion of 10-formyl-FH_4 to CO_2 and FH_4.[73] Ultimately, single-C units are needed in the form of 5-methyl-FH_4, which can pass them through MET to be activated as SAM, the donor of "labile" methyl groups[74] for more than 100 methyltransferases.

SAM-dependent methylations. SAM is the methyl donor for more than 100 methyltransferases, including the following:

- **Methylation of Hcy to regenerate Met** (see previous discussion). This process regenerates the methyl donor SAM from **S-adenosylhomocysteine (SAH)**, which inhibits methyltransferases through tight binding. Thus, the intracellular ratio of SAM:SAH, known as the "methylation index," determines methylation potential, affecting the availability of labile methyl groups to methyltransferases.
- **Methylation of chromatin** is a primary means of regulating gene expression and maintaining genomic integrity. It includes the methylation of DNA and histones catalyzed by methyltransferases that are particularly sensitive to cellular SAM:SAH ratio.[75] Hypomethylation of these factors is believed to alter chromatin structure in ways that affect transcription and can increase the rate of C→T transition mutation.[76]

---

71. It is also an allosteric inhibitor of 5,10-methylene-FH_4 reductase and an activator of the pyridoxal phosphate-dependent enzyme, cystathionine β-synthase, which catalyzes the condensation of Hcy and SER to form cystathionine.

72. Hcy can also be converted to MET in the liver and kidney by a folate-independent process, i.e., by betaine–homocysteine methyltransferase (BHMT), which using betaine (a product of choline oxidation) as the methyl donor.

73. Anguera, M.C., Field, M.S., Perry, C., et al., 2006. J. Biol. Chem. 281, 18335–18342.

74. Most SAM-dependent methylation reactions proceed via SN2 displacement mechanisms; however, some proceed via radical mechanisms that utilize a specialized iron–sulfur cluster to catalyze reductive cleavage yielding the highly reactive 5-adenosyl radical (Zhang, Q., Van der Donk, W.A., Liu, W., 2012. Acc. Chem. Res. 4, 555–564).

75. Waterland, R.A., Jirtle, R.L., 2004. Nutrition 20, 63–68.

76. Methylated CpG sites appear to be at particularly high risk for C→T changes, the most common type of mutational change, which are common in the p53 tumor suppressor gene.

TABLE 17.10 Metabolic Roles of Folate Coenzymes

Folate Coenzyme	Enzyme	Metabolic Role
5,10-Methylene-FH_4	Serine hydroxymethyltransferase	Receipt of a formaldehyde unit in SER catabolism (mitochondrial enzyme important in GLY synthesis)
	Thymidylate synthase	Transfers formaldehyde to C-5 of dUMP to form dTMP in pyrimidines
10-Formyl-FH_4	10-Formyl-FH_4 synthase	Accepts formate from TRY catabolism
	Glycinamide ribonucleotide transformylase	Donates formate in purine synthesis
	5-Amino-4-imidazolecarboxamide transformylase	Donates formate in purine synthesis
	10-Formyl-FH_4 dehydrogenase	Transfers formate for oxidation to CO_2 in HIS catabolism
5-Methyl-FH_4	Methionine synthase	Provides methyl group to convert Hcy to MET
	Glycine *N*-methyltransferase	Transfers methyl group from SAM to GLY in the formation of Hcy
5-Formimino-FH_4	Formiminotransferase	Accepts formimino group from HIS catabolism

- **DNA methylation** consists of the addition of a methyl group from SAM to cytosine bases within DNA CpG islands.[77] DNA hypermethylation is typically associated with gene silencing. Folate deprivation produces chromosomal breaks in megaloblastic bone marrow, reflecting DNA strand breaks and hypomethylation. Studies have found the MTHFR 677TT genotype, which reduces that activity by some 70%, to have reduced genomic DNA methylation.[78] One study found folate deprivation to produce hypomethylation of the *p53* gene and increased genome-wide DNA strand breakage. Such results suggest that folate deprivation may create fragile sites in the genome potentially inducing protooncogene expression. Another study found low-folate subjects with the MTHFR 677TT genotype to have hypomethylation of whole blood DNA, compared to those with the CC genotype; those differences were not apparent among folate-adequate subjects of both genotypes.[79] That folate status affects DNA integrity is supported by findings that circulating folate level is directly related to the length of peripheral lymphocyte telomeres, i.e., the capping chromosomal segments characterized by tandem repeats of DNA and associated proteins, dysfunction of which is associated with age-related disease.[80]
- **Histone methylation** consists of the addition of a methyl group from SAM to lysyl and arginyl residues at the N-termini (i.e., the "histone tails") of histone proteins H3 and H4. This affects the binding of proteins, comprising a means of regulating gene expression and genomic stability. It can be associated with either active transcription or gene silencing. In this aspect of epigenetic control, folate serves both in the donation of the single-C unit as 5-methyl-FH_4, and as the acceptor as FH_4 for formaldehyde generated during oxidative demethylation of histone tails.
- **Transcription factor methylation** has been demonstrated as affecting transcriptional output.[81]
- **RNA methylation** is thought to protect from degradation by RNAses, extending the effective half-lives of RNAs. Evidence suggests that noncoding may also be subject to methylation.[82]
- **Methylation of phosphatidylethanolamine to produce phosphatidylcholine**, the predominantly phospholipid in membranes and lipoproteins. This methylation is dependent on SAM, consuming some 40% of the methyl donor in the liver. Thus, by reducing the flux of methyl groups from 5-methyl-FH_4, folate deficiency can also cause a secondary hepatic deficiency of choline.

77. These are regions of DNA in which the nucleotide composition has a high frequency of linear sequences of cytosine–guanine separated by a phosphate (5′C-p-3′G). In mammals, 70–80% of CpG sites are methylated. CpG islands typically consist of at least >300 base pairs more than half of which are CpGs. These are frequently located near the transcription starts of gene. Methylation of CpG sites within promoters can silence those genes.
78. Castro, R., Rivera, I., Ravasco, P., et al., 2004. J. Med. Genet. 41, 454–458.
79. Friso, S., Choi, S.W., Girelli, D., et al., 2002. Proc. Natl. Acad. Sci. U.S.A. 99, 5606–5611.
80. Paul, L., Catterneo, M., D'Angelo, A., et al., 2009. J. Nutr. 139, 1273–1278.
81. Stark, G.R., Wang, Y., Lu, T., 2011. Cell Res. 21, 375–380.
82. Huang, Y., Ji, L., Huang, Q., et al., 2009. Nature 461, 823–827.

Nucleotide Metabolism

Folate is required for the production of new cells through its functions in the synthesis of purines and thymidylate required for DNA synthesis. Folate deficiency results in the arrest of erythropoiesis prior to the latter stages of differentiation, resulting in apoptotic reduction of cells surviving to postmitotic, terminal stages in the condition called megaloblastic anemia. Anemia can have folate-responsive components in subjects with apparently normal plasma folate levels; therefore, addition of folate to iron supplements can improve the treatment of anemia in pregnancy as well as in undernourished individuals.

Folates are required for the synthesis of purines. While they are not required for the de novo synthesis of pyrimidines, they are required for the synthesis of thymidylate. Both roles are necessary for the de novo synthesis of DNA and, thus, for DNA replication and cell division. Disruption of these functions impairs cell division and results in the macrocytic anemia of folate deficiency.

- **Purine synthesis** (i.e., the synthesis of adenine and guanine) depends on the transfer of formate from 10-formyl-FH_4 to provide the C-2 and C-8 positions of the purine ring (Fig. 17.5). These reactions are catalyzed by aminoimidazolecarboxamide ribonucleotide transformylase and glycinamide ribonucleotide transformylase, respectively.
- **Thymidylate synthesis** depends on the transfer of the single-C unit from 5,10-methylene-FH_4 to deoxyuridine monophosphate, converting it to deoxythymidine monophosphate (dTMP). This step, catalyzed by **thymidylate synthase (TS)**,[83] is rate limiting to DNA replication and, thus, to the normal progression of the cell cycle. TS is expressed only in replicating tissues. During the S phase of the cell cycle TS, FH_2 reductase and SHMT enter the nucleus to effect the folate-dependent synthesis of thymidylate. TS activity limited by deprivation of folate, which results in uracil accumulation in DNA.

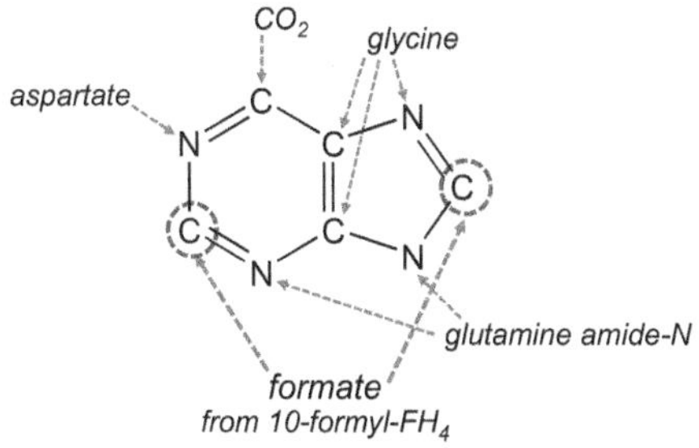

FIGURE 17.5 Sources of purine ring atoms.

83. The antifolate compound 5-fluorodeoxyuridylate complexes with the enzyme and its folate cosubstrate.

Regulation. The regulation of single-C metabolism is effected by the interconversion of oxidation states of the folate intermediates. In mammalian tissues, the β-C of SER is the major source of single-C units for these aspects of metabolism. That C-fragment is accepted by FH_4 to form 5,10-methylene-FH_4 (by **SHMT**), which has a central role in single-C metabolism. It can be used directly for the synthesis of thymidylate (by thymidylate synthase);[84] it can be oxidized to **5,10-methenyl-FH_4** (by **5,10-methylene-FH_4 dehydrogenase**) for the de novo synthesis of purines; or it can be reduced to 5-methyl-FH_4 (by MTHFR) for use in the synthesis of MET. The result is the channeling of single-C units in several directions: to MET, to thymidylate (for DNA synthesis), or to purine synthesis. Because folylpolyglutamates have been found to inhibit a number of the enzymes of single-C metabolism, it has been suggested that variation in their polyglutamate chain lengths (observed under different physiological conditions) may play a regulatory role.

Physiological Functions

Fetal development. Folate is required for normal embryonic growth and development. It plays an important role in promoting closure of the neural tube, defects of which results in malformations of the embryonic brain and/or spinal cord referred to as **NTDs**.[85] The role of folate in promoting neural tube closure likely involves its role in single-C metabolism, as exencephaly can be produced by knocking out SHMT, which is key to the synthesis of thymidylate and, thus, DNA[86] Folate may also function in preventing the misexpression of microRNAs,[87] which confer key roles in development by regulating the expression of certain target mRNAs. microRNAs show distinct expression patterns in the developing brain and are highly

84. This is the sole de novo path of thymidylate synthesis. It is also the only folate-dependent reaction in which the cofactor serves both as a single-carbon donor and as a reducing agent. Thymidylate synthase is the target of the anticancer drug 5-fluorouracil (5-FU); the enzyme converts the drug to 5-fluorodeoxyuridylate, which is incorporated into RNA and is a suicide inhibitor of the synthase, resulting in cellular accumulation of deoxyuridine triphosphate (dUTP) and incorporation of deoxyuridine (dU) into DNA. DNA with this abnormal base is enzymatically cleaved at sites containing it, leading to enhanced DNA breakage.

85. NTDs comprise the most common forms of congenital malformations, with an annual global incidence estimated at >300,000 new cases, >41,000 deaths, and the loss of 2.3 million disability-adjusted life years. These involve developmental failures of the neural structures (brain, spinal cord, cranial bones, vertebral arches, meninges, and overlying skin) formed from the embryonic neural tube in humans 20–28 days after fertilization. The most prominent NTDs are anencephaly and spina bifida. More than 95% of NTD pregnancies occur in families with no history of such defects; but women with one affected pregnancy or with spina bifida themselves face a risk of 3–4% of an NTD in a subsequent pregnancy.

86. Beaudin, A.E., Abarinov, E.V., Noden, D.M., et al., 2011. Am. J. Clin. Nutr. 93, 789–998.

87. miRNAs are small (ca. 22 nucleotides), noncoding transcripts that repress the expression of target mRNAs.

regulated by genomic methylation, deficiencies of which have also been proposed to contribute to NTDs. Folate deficiency does not produce neural tube changes in animal models; however, genetic deletion of methionine synthase reductase resulted in epigenetic changes manifest as congenital malformations in the mouse. Hcy treatment of chick and mouse embryos increases the frequencies of a wide range of congenital malformations including damage to cells of the neural crest.[88]

Homocysteinemia. Folate is required to maintain normal Hcy concentrations, elevations of which can occur through its overproduction from MET and, to a lesser extent, through its impaired disposal by transsulfuration to cystathionine. Both can have congenital causes[89] and can be affected by lifestyle factors[90] and nutritional status with respect to folate, vitamin B_6, and vitamin B_{12}.[91] Hcy can be converted to a thiolactone by methionyl tRNA synthase, in an error-editing reaction that prevents its incorporation into the primary structure of proteins; at high levels the thiolactone can react with protein lysyl residues. Protection against damage to high-density lipoproteins that results from homocysteinylation is effected by a Ca^{+2}-dependent Hcy-thiolactonase associated with those particles.[92] Homocysteinemia also causes displacement of protein-bound cysteine, which changes redox thiol status, probably via thiol–disulfide exchange and redox reactions. It has been suggested that pathogenesis may also involve SAH, which is a potent inhibitor of methyltransferases. SAH is reversibly converted with Hcy, the equilibrium favoring SAH. Therefore, homocysteinemia would be expected to lead to elevated levels of SAH, which appears to be present in the normal circulation at only low amounts (<0.2% of Hcy levels). Dietary intakes of polyunsaturated fatty acids (PUFAs) have been found to be a significant covariate with MTHFR C677T and A1298T genotype in affecting plasma Hcy level.[93] Individuals with 1298AA showed homocysteinemia only with consuming a high-PUFA diet; whereas, individuals with 677TT and the 1298C allele had low Hcy levels even on the high-PUFA diet. Circulating Hcy levels >13 μM[94] have been associated with dysfunction of several types:

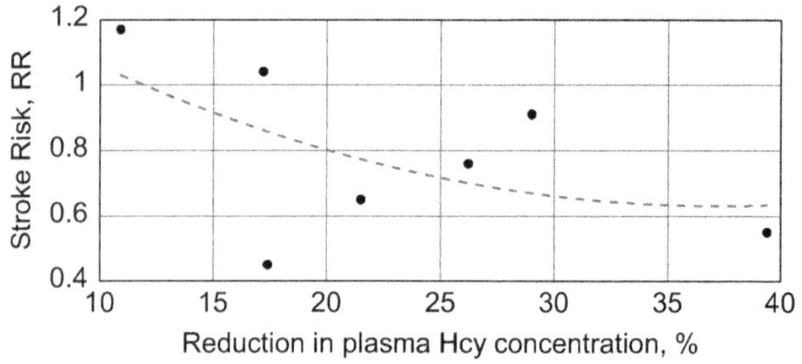

FIGURE 17.6 Results of seven randomized folate intervention trials showing relationship of homocysteine lowering and stroke risk. *After Wang, X., Qin, X., Demirtas, H., et al., 2007. Lancet 369, 1876–1882.*

- **Cardiovascular health**. Epidemiological associations have been made between moderately elevated plasma Hcy concentrations and risks of coronary, peripheral and carotid arterial atherosclerosis, venous thrombosis, carotid thickening, hypertension, and stroke (Fig. 17.6).[95] A meta-analysis of results of 27 cross-sectional and case–control studies[96] attributed 10% of total coronary artery disease to homocysteinemia.[97] That analysis suggested that a 5 μM increase in plasma Hcy level was associated with an increase in the risk of coronary artery disease comparable to a 0.5 mM (20 mg/dL) increase in plasma total cholesterol. A prospective, community-based study found plasma Hcy to be strongly inversely associated with plasma folate level and only weakly associated with plasma levels of vitamin B_{12} and pyridoxal phosphate (Table 17.11).[98] Other meta-analyses have concluded that folate supplementation can reduce risks for progression of carotid intimal thickening and stroke[99] but had no effect on risk of coronary revascularization.[100] Individuals with the MTHFR 677TT genotype are at risk for carotid intima-media thickening, itself a risk factor to vascular disease. They typically have lower levels of folate and higher levels of Hcy than other genotypes (Table 17.12).

88. Van Nil, N.H., Oosterbaan, A.M., Steegers-Theunisswen, R.P.M., 2010. Reprod. Toxicol. 30, 520–531.

89. Inherited deficiencies of cysteine β-synthase and 5,10-methylene-FH_4 reductase have been identified in humans. Genetic determinant of the Hcy response to MET has also been identified (Wernimont, S.M., Clark, A.G., Stover, P.J., et al., 2011. BMC Med. Genet. 12, 150–160; Lievers, K.J.A., Kluijtmans, L.A.J., Blom, J., et al., 2006. Eur. J. Hum. Genet. 14, 1125–1129.)

90. Hcy levels are increased by smoking and chronic alcohol use, and reduced by regular physical activity.

91. Chronic alcoholics have been found to have mean serum Hcy levels, about twice those of nonalcoholics. Chronic ethanol intake appears to interfere with single-C metabolism, and alcoholics are at risk of folate deficiency.

92. Jakubowski, H., 2000. J. Nutr. 130, 377S–381S.

93. Huang, T., Tucker, K.L., Lee, Y.C., et al., 2011. J. Nutr. 141, 654–659.

94. In the NHANES 2003–2006 cohort, such levels were found in 8% of adults ≥20 years and 19% of adults ≥60 years (Center for Disease Control and Prevention. Second National Report on Biochemical Indicators of Diet and Nutrition in the U.S. Population [2012]. CDC, Atlanta).

95. The low prevalence of coronary heart disease among South African blacks has been associated with their typically lower plasma Hcy levels and their demonstrably more effective Hcy clearance after methionine loading.

96. Boushey, C.J., Beresford, S.A., Omenn, G.S., et al., 1995. J. Am. Med. Assoc. 274, 1049–1057.

97. Defined as a plasma Hcy concentration >14 μM.

98. Selhub, J., Jacaques, P.F., Bostom, A.G., et al., 1996. J. Nutr. 126, 1258S–1265S.

99. Yang, H.T., Lee, M., Hong., K.S., et al., 2012. Eur. J. Intern. Med. 23, q745–q754; Qin, X., Xu, M., Zhang, Y., et al., 2012. Atherosclerosis 222, 307–313; Huo, Y., Qin, X., Wang, J., et al., 2012. Int. J. Clin. Pract. 66, 544–551; Wang, X., Qin, X., Demirtas, H., et al., 2007. Lancet 369, 1876–1882.

100. Qin, X., Fan, F., Cui, Y., et al., 2014. Clin. Nutr. 33, 603–612.

TABLE 17.11 Plasma Levels of Homocysteine and Vitamin in Elderly Subjects

Subject Age/Years	Hcy μM	% Elevated	Folate nM	Vitamin B_{12} pM	Pyridoxal Phosphate nM
Men					
67–74	11.8	25.3	9.3	265	52.6
75–79	11.9	26.7	9.3	260	49.6
80+	14.1	48.3	10.0	255	47.6
Trend, *P*	<0.001	<0.001	NS[a]	NS	NS
Women					
67–74	10.7	19.5	10.4	302	59.9
75–79	11.9	28.9	10.2	289	52.2
80+	13.2	41.1	9.7	290	52.1
Trend, *P*	<0.001	<0.001	NS	NS	NS

[a]*NS*, not significant.
From Selhub, J., Jacques, P.F., Wilson, P.W., et al., 1993. J. Am. Med. Assoc. 270, 2693–2698.

TABLE 17.12 Effects of Polymorphisms on Biomarkers of Folate Status

Polymorphism	Plasma Folate, nM	Erythrocyte Folate, nM	Plasma Hcy, μM
MTHFR 677C>T			
CC	12.4 (11.6, 13.2)	889 (851,929)	8.9 (8.6, 9.2)
CT	11.3 (10.6, 11.9)	821 (791, 852)	9.2 (9.0, 9.5)
TT	10.0 (9.2, 11.9)	652 (611, 695)	11.2 (10.5, 11.9)
SLC19A1 80G>A			
GG	11.9 (11.0, 12.9)	861 (815, 910)	9.6 (9.3, 10.0)
GA	11.4 (10.9, 12.0)	804 (775, 833)	9.4 (9.1, 9.7)
AA	10.9 (10.0, 11.9)	776 (733, 822)	9.2 (8.8, 9.6)

Means with 95% CI; means with like superscripts are not significantly different, $p < .05$.
n=120–394; Bueno, O., Molloy, A.M., Fernandez-Ballart, J.D., et al., 2016. J. Nutr. 146, 1–8.

Individuals of that genotype who are also of low riboflavin status are at risk to high blood pressure, presumably due to the lower affinity of the TT enzyme for its flavin cofactor.[101] A study of the NHANES III (1991–94) cohort found individuals with the TT genotype to have a 31% lower risk to cardiovascular disease mortality;[102] however, a meta-analysis of 53 studies showed the TT genotype to be associated with a 20% greater risk of venous thrombosis compared to the CC genotype.[103] Individuals with the 1298CC genotype have a relatively higher risk to cardiovascular disease than those carrying the 1298A allele.

- **Immune function**. Studies with human monocytes demonstrated that in vitro treatment with Hcy activated those cells to express inflammatory cytokines.[104] This suggests that homocysteinemia, by activating monocytes, may contribute to chronic inflammation involved in endothelial cell damage.

101. Yamada, K., Chen, Z., Rozen, R., et al., 2001. Proc. Natl. Acad. Sci. U.S.A. 98, 14853–14858.
102. Yang, Q., Bailey, L., Clarke, R., et al., 2012. Am. J. Clin. Nutr. 95, 1245–1253.
103. Den Heijer, M., Lewington, S., Clarke, R., 2005. J. Thromb. Haemostasis 3, 292–299.
104. Su, S.J., Huang, L.W., Pai, L.S., et al., 2005. Nutrition 21, 994–1002.

- **Bone health**. Homocysteinemia can affect the developing skeleton, producing knockvknees (*genu valgum*) and unusually high arches of the foot (*pes cavus*) in children, with subsequent development of marfanoid features (long limbs) and osteoporosis. Homocysteinemia induced in animals by feeding large amounts of MET or Hcy is accompanied with severe trabecular bone loss with attendant changes in microarchitecture and strength. An analysis of the NHANES 1999–2004 data found homocysteinemia (Hcy >13 μM) to be associated with a twofold increase is risk of lumbar spine osteoporosis.[105] These changes appear to ensue from ROS-dependent activation of osteoclastic bone resorption. Randomized trials using supplements of folate and other B-vitamins (B_6, B_{12}) have found no effects on biomarkers of bone turnover.[106] Some studies have found individuals with the 677TT genotype to have relatively high risks of fracture and relatively lower bone mineral density independent of plasma Hcy level.[107] However, other factors are likely to be involved; a study that found 677TT women to show reduced bone mineral density only if they also had relatively low intakes of folate, vitamin B_{12}, and riboflavin.[108]
- **Age-related decline in physical function**. Subjects in the highest quartile of plasma Hcy concentration had more than four times the risk of being in the worst quartile of decline in physical function.[109]
- **Neurologic function**. Hcy can be neurotoxic. This can occur in several ways: by SAH-inhibition of methyltransferases involved in catecholamine methylation, by Hcy oxidation products acting as agonists of the N-methyl-D-aspartate receptor to cause excitotoxicity, by ROS produced by Hcy oxidation, etc. Folate may also be directly involved in the regulation of neurotransmitter metabolism, as neuropsychiatric subjects with low-erythrocyte folate levels and homocysteinemia show low-cerebral spinal fluid levels of the serotonin metabolite 5-hydroxyindole acetic acid and reduced turnover of dopamine and noradrenaline.[110] It has been suggested that folate may act as a structural analog of tetrahydro biopterin,[111] an essential cofactor in the metabolism of monoamine neurotransmitters. MTHFR and dihydrofolate reductase are thought to function in tetrahydrobiopterin metabolism; the MTHFR 677TT genotype is associated with increased risk to neurological disorders. Thus, homocysteinemia can lead to several types of neurocognitive effects:
  - **Cognitive function**. Homocysteinemia is associated with age-related cognitive decline and risk of developing dementia; however, analyses of the NHANES data suggest that these outcomes are associated with low-vitamin B_{12} status and *not* low-folate status.[112] In fact, they suggest that high-folate status may exacerbate the neuropsychiatric effects of low-vitamin B_{12} status. Cross-sectional studies have indicated low-folate status to be a risk factor for low reading cognition in children[113] and for cognitive decline in aging;[114] and randomized clinical trials have found folic acid supplementation of older adults to improve domains of cognitive function that typically decline with age.[115] Meta-analyses of randomized clinical trials concluded that folic acid yielded no beneficial effects on measures of cognition within 3 years of supplementation.[116] However, supplementation with folic acid and vitamins B_6 and B_{12} reduced brain atrophy progression (a proxy for neuronal injury that would result in cognitive decline) particularly in subjects with higher plasma concentrations of ω-3 fatty acids.
  - **Alzheimer's disease**. Plasma Hcy level is directly associated with plasma level of amyloid β40, a protein associated with aging but not necessarily with Alzheimer's disease (AD). Relatively high intakes of folate have been associated with lower risk to AD[117]
  - **Depression**. Mood changes and other symptoms of depression have frequently been observed in folate deficiency. These symptoms are associated with homocysteinemia; it has been suggested that they reflect Hcy-induced cerebral vascular disease and

105. Bailey, R.L., Looker, A.C., Lu, Z., et al., 2015. Am. J. Clin. Nutr. 102, 687–694.
106. e.g., Green, T.J., McMahon, J.S., Skeaff, C.M., et al., 2007. Am. J. Clin. Nutr. 85, 460–464; van Wijgaarden, J.P., Swart, K.M.A., Enneman, A.W., et al., 2014. Am. J. Clin. Nutr. 100, 1578–1586.
107. Villadsen, M.M., Bunger, M.H., Carstens, M., et al., 2005. Osteoporos. Int. 16, 411–416; Abrahamsen, B., Jorgensen, H.L., Nielsen, T.L., et al., 2006. Bone 38, 215–219; Hong, X., Hsu, Y.U., Terwedow, H., et al., 2007. Bone 40, 737–742.
108. Abrahamsen, B., Madsen, J.S., Tofteng, C.L., et al., 2005. Bone 36, 577–583.
109. Kado, D.M., Bucur, A., Selhub, J. et al. (2002) Am. J. Med. 113, 537–542.
110. Many patients are likely to be exposed to anticonvulsants that antagonize folate metabolism, e.g., phenytoin, carbamazepine, primidone, phenobarbital, valproic acid products, pamotrigine.
111. Both contain a pterin moiety.
112. Selhub, J., Morris, M.S., Jacques, P.F., et al., 2009. Am. J. Clin. Nutr. 89, S702–S706.
113. Nguyen, C.T., Gracely, E.J., Lee, B.K., 2013. J. Nutr. 143, 500–504.
114. Kado, D.M., Karlamangla, A.S., Huang, M.H., et al., 2005. Am. J. Med. 118, 161–167.
115. Elias, M.F., Sullivan, L.M., D'Agostino, R.B., et al., 2005. Am. J. Epidemiol. 162, 644–653; Durga, J., van Boxtel, M.P.J., Schouten, E.G., et al., 2007. Lancet 369, 208–216.
116. Wald, D.S., Kasturiratne, A., Simmonds, M., 2010. Am. J. Med. 123, 522–527; Clarke, R., Bennett, D., Parish, S., et al., 2014. Am. J. Clin. Nutr. 100, 657–666.
117. Luchsinger, J.A., Tang, M.X., Miller, J., et al., 2007. Arch. Neurol. 64, 86–92.

neurotransmitter deficiency. Studies have shown that individuals of the MTHFR 677TT genotype have increased risks of depression.[118] A systematic review of randomized clinical trials suggested that folic acid may benefit the treatment of depression.[119] Victor Herbert[120] experienced depressive mood, irritability, insomnia, fatigue, and forgetfulness after consuming a folate-deficient diet for several months.[121] When he took a folate supplement, those symptoms resolved within 48 h.

- **Schizophrenia.**[122] Both depressed and schizophrenic patients responded to daily doses of 15 mg of 5-methyl-FH_4 with improved clinical and social outcomes compared with placebo controls.[123] Studies have shown that individuals of the MTHFR 677TT genotype have increased risks of schizophrenia.

8. BIOMARKERS OF FOLATE STATUS

Folate status can be assessed by analyses of blood (Fig. 17.7):[124]

- **Serum folate concentration**. Serum folate level is comprised mostly (>80%) of 5-methyl-FH_4. It is responsive to short-term changes in folate intake and is more responsive to intake of folic acid than to intakes of food folates. It also is affected by MTHFR 677 and SLC19A1 genotypes (Table 17.12); MTHFR 677CC individuals have been found to have 13% greater levels than those with the 677 TT genotype.[125] Levels <3 ng/mL are indicative of folate deficiency.

118. Almedia, O.P., McCaul, K., Hankey, G.J., et al., 2008. Arch. Gen. Psychiatr. 65, 1286–1294.
119. Taylor, M.J., Carney, S.M., Goodwin, G.M., et al., 2004. J. Psychopharmacol. 18, 251–256.
120. Victor Herbert, M.D. (1927–2002) was a prominent American nutrition scientist. He was the first to recognize that the anemia common among pregnant women was caused by folate deficiency; that alcoholics were frequently folate deficient; and that vitamin B_{12} deficiency produced the "methyl folate trap." Without missing a beat in his scientific career, he obtained a law degree and took on fraudulent healthy claims about putative nutrients (e.g., "vitamin B_{15}"). He served as President of the American Society for Clinical Nutrition and was the Chief of Hematology and Nutrition Laboratory of the Bronx Veterans Administration Hospital.
121. Herbert, V., Zalusky, R., 1962. J. Clin. Invest. 41, 1263–1276.
122. Nearly two-thirds of patients with megaloblastic anemia due to vitamin B_{12}/folate deficiency show neuropsychiatric complications.
123. Godfrey, P.S., Toone, B.K., Carney, M.W., et al., 1990. Lancet 336, 392–395.
124. Analyses of folates have been complicated by methodologic idiosyncrasies associated with the radioimmunoassay and a microbiological assay that have been used. The latter produced higher values, which are considered to be more accurate. This necessitates adjustment for that difference (Pfeiffer, C.M., Hughes, J.P., Lacher, D.A., et al., 2012. J. Nutr. 142, 886–893).
125. Tsang, B.L., Devine, O.J., Cordero, A.M., et al., 2015. Am. J. Clin. Nutr. 101, 1286–1294.

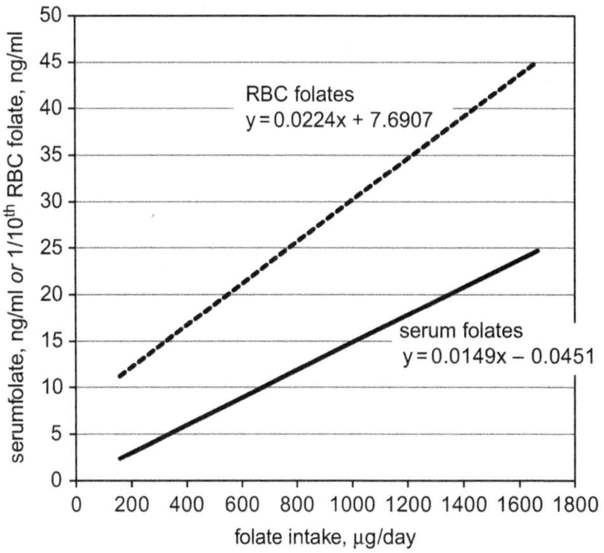

FIGURE 17.7 Relationships of folates in serum and erythrocytes (RBC) to dietary folate intake. Population data from National Health and Nutrition Examination Surveys (NHANES) III and annual NHANES from 1999 to 2004. Adapted from *Quinlivan, E.P., Gregory, J.F., 2007. Am. J. Clin. Nutr. 86, 1773–1779.*

- **Erythrocyte folate concentration**. Erythrocyte folate level indicates long-term folate status, reflecting folate status during erythropoiesis and, thus, particularly the preceding 120 days.[126] Responses to changes in folate intake tend to be greater for women that for men; half of this difference is explained on the basis of differences in body size. Erythrocyte folate levels correlate with hepatic folate levels; thus this biomarker is taken as an indicator of tissue folate stores.[127] Erythrocyte folate level is more responsive to intake of folic acid than to intakes of food folates. It also appears to be affected by MTHFR 677 genotype, as CC individuals have been found to have 16% greater levels than TT individuals. Erythrocyte folate concentrations <140 ng/mL (~320 nM) are indicative of folate deficiency.

Other biomarkers have been used to assess various aspects of folate status or function:

- **Plasma Hcy concentration**. This is not a specific indicator of folate status. While folate deficiency can produce elevated plasma Hcy concentration, that outcome can also be produced by deficiencies of vitamin B_{12} and/or MET, renal insufficiency, and some drugs. These factors must be considered in interpreting plasma Hcy level. In many populations exposed to food folate fortification, elevated plasma Hcy may be more likely to indicate suboptimal vitamin B_{12} status.

126. i.e., The half-life of erythrocytes.
127. Wu, A., Chanarin, I., Slavin, G., et al., 1975. Br. J. Haematol. 29, 469–478.

Homocysteinemia is typically marked by plasma Hcy concentrations >13 μM.

- **Serum unmetabolized folic acid concentration.** Circulating levels of unmetabolized folic acid are highly variable and reflect concentrations of soluble FBP to which that vitamer is preferentially bound. This biomarker typically correlates with serum total folates, particularly in individuals exposed to folic acid fortification/supplementation. It may be most valuable as an indicator of exposure to folic acid-fortified foods.
- **Urinary folates excretion.** Twenty-four hour urinary folate excretion can provide information about average daily folate status. This biomarker is subject to significant interindividual variation part of which has to do with little folate excretion at less than RDA levels of intake.
- **Urinary *p*-aminobenzoylglutamate and *p*-acetamidobenzoylglutamate concentrations.**[128] These oxidative metabolites of folate indicate folate turnover. Their concentrations in 24-h urine samples correlate with the concentrations of folates in serum and erythrocytes[129] but are less sensitive to changes in folate intake than those biomarkers.

9. FOLATE DEFICIENCY

Folate deficiency can have primary (privational) and secondary (nonprivational) causes (Fig. 17.11). Primary causes involve inadequate folate supply, i.e., dietary patterns that fail to provide folate in adequate amounts. Secondary causes relate to impaired absorption, metabolism, or metabolic function of the vitamin (Table 17.13):[130]

Malabsorption

- **Inflammatory bowel diseases** that cause persistent mucosal damage (Crohn disease, ulcerative colitis, tropical sprue, celiac disease).
- **Zinc (Zn) deficiency** reduces the absorption of folylpolyglutamates (but *not* monoglutamates). This is thought to indicate a need for Zn by folate-metabolizing enzymes.
- **Several drugs**, the antiproliferative methotrexate; the anticonvulsants (diphenylhydantoin and phenobarbital); the antiinflammatory sulfasalazine; the diuretic triamterene; the glycemic control drug metformin.
- Chronic alcoholism.

128. Both metabolites can also be determined in plasma/serum.

129. Wolfe, J.M., Baailey, L.B., Herrlinger-Garcia, K., et al., 2003. J. Nutr. 77, 919–923.

130. Examples include *tropical sprue* (inflammation of the mucous membranes of the alimentary tract) and other types of enteritis that involve malabsorption and, usually, diarrhea.

TABLE 17.13 General Signs of Folate Deficiency

Organ System	Signs
General	Reduced appetite, growth
Dermatologic	Alopecia, achromotrichia, dermatitis
Muscular	Weakness
Gastrointestinal	Inflammation
Erythrocytes	Macrocytic anemia
Nervous	Depression, neuropathy, paralysis

Metabolic Impairments

- **Vitamin B_{12} and MET deficiencies**, patients with pernicious anemia[131] generally have low circulating folate levels due to accumulation of the vitamin as 5-methyl-FH_4 in the "methyl folate trap." These low-plasma folate levels cannot be corrected by supplements of MET, although that amino acid, via SAM, inhibits 5,10-methylene-FH_4 reductase to reduce the synthesis of 5-methyl-FH_4.
- **Chronic alcoholism** impairs hepatic MET metabolism by inhibiting the vitamin B_{12}-dependent transmethylation of Hcy.
- Dihydrofolate reductase inhibitors (e.g., methotrexate).
- **Genetic factors**, MTHFR deficiency; glutamate formiminotransferase deficiency; MTHFR polymorphisms.
- **Increased requirements**, hemodialysis, prematurity, pregnancy, lactation.

General Signs of Folate Deficiency

Deficiencies of folate result in impaired biosynthesis of DNA and RNA, and thus in reduced cell division, which is manifested clinically as anemia, dermatologic lesions, and poor growth in most species (Table 17.13). The anemia of folate deficiency is characterized by the presence of large, nucleated erythrocyte-precursor cells called macrocytes and of hypersegmented polymorphonuclear neutrophils,[132] reflecting decreased DNA synthesis and delayed maturation of bone marrow, i.e., megaloblastic erythropoiesis.[133] While clinical deficiency is usually detected as anemia, megaloblastic changes occur in other cells, reflecting growth arrest in the G2 phase of the cell cycle just prior to mitosis.

131. Pernicious anemia is caused by vitamin B_{12} deficiency, resulting from the lack of the intrinsic factor required for the enteric absorption of that vitamin (see Chapter 18).

132. Neutrophil hypersegmentation is defined as >5% five-lobed or any six-lobed cells per 100 granulocytes.

133. This distinguishes the macrocytic anemia of folate or vitamin B_{12} deficiencies from those macrocytic anemias caused by such factors as alcohol abuse, hypothyroidism, heavy smoking, chronic hemolytic anemia, which have normoblastic erythropoiesis.

TABLE 17.14 Recommended Folate Intakes

The United States		FAO/WHO	
Age/Sex	AI[a], μg/day	Age/Sex	RNI[b], μg/day
0–6 months	65	0–11 months	80
7–11 months	80		
1–3 years	150	1–3 years	160
4–8 years	200	4–6 years	200
9–13 years	300	7–9 years	300
>13 years	400	>9 years	400
Pregnancy	600	Pregnancy	600
Lactation	550	Lactation	500

[a]*Adequate Intakes are given, as RDAs have not been established. Food and Nutrition Board, 2000. Dietary Reference Intakes for Thiamin, Riboflavin, Niacin, Vitamin B6, Folate, Vitamin B12, Pantothenic Acid, Biotin and Choline. National Academy Press, Washington, DC, 564 pp.*
[b]*Recommended Nutrient Intakes, Joint WHO/FAO Expert Consultation, 2001. Human Vitamin and Mineral Requirements. WHO, Rome, 286 pp.*

Affected cells have increased DNA content and DNA strand breakage, which is thought to result from the misincorporation of uracil into DNA in place of thymidylate—a potentially mutagenic situation. This sign, which is because it is a manifestation of impaired DNA synthesis, can also be caused by deficiency of vitamin B_{12}.

Severely anemic individuals show weakness, fatigue, difficulty in concentrating, irritability, headache, palpitations, and shortness of breath. Folate deficiency also affects the intestinal epithelium, where impaired DNA synthesis causes megaloblastosis of enterocytes. This is manifested clinically as malabsorption and diarrhea and is a contributor to the clinical picture of tropical sprue. Anemia can have folate-responsive components in subjects with apparently normal plasma folate levels; therefore, addition of folate to iron supplements can improve the treatment of anemia in pregnancy as well as in undernourished individuals.

Deficiency Signs in Humans

Adequate intakes for folate have been established (Table 17.14). Chronic intakes below those levels produce folate deficiency characterized by a sequence of signs, starting with nuclear hypersegmentation of circulating polymorphonuclear leukocytes[134] within about 2 months of deprivation of the vitamin. This is followed by megaloblastic anemia and, then, general weakness, depression, and polyneuropathy. In pregnant women, the deficiency can lead to birth defects or spontaneous abortion. Elderly humans tend to have lower circulating levels of folate, indicating that they may be at increased risk of folate deficiency. This finding appears to involve changes in food habits, which affect intake of the vitamin. Supplements of folate and iron[135] are recommended by the World Health Organization for use in many areas with endemic anemia among women of childbearing age.

Low-Folate Status

That folate-responsive homocysteinemia can be demonstrated in apparently healthy free-living populations suggests the prevalence of undiagnosed suboptimal vitamin status. For example, twice-weekly treatments of elderly subjects with folate (1.1 mg) in combination with vitamin B_{12} (1 mg) and vitamin B_6 (5 mg) have been shown to reduce plasma concentrations of Hcy by as much as half and also to reduce methylmalonic acid (MMA), 2-methylcitric acid, and cystathionine despite the fact that pretreatment plasma levels of those vitamins were not low.

Homocysteinemia. Folate supplementation has been shown to reduce homocysteinemia (Table 17.15), independently reverse endothelial dysfunction, reduce arterial pressure, and increase coronary dilation. It is likely that these effects involve the stimulation of nitric oxide production by 5-methyl-FH_4 and, perhaps, inhibition of lipoprotein oxidation by folic acid. Reduction of serum Hcy is linear up to daily folate intakes of about 0.4 mg, particularly for individuals with relatively high-serum Hcy levels (Fig. 17.8). Greater efficacy may be realized when folate is given in combination with vitamin B_{12}. A meta-analysis of 25 randomized controlled trials showed that daily intakes of 0.8 mg folic acid are required to realize maximal reductions in plasma Hcy levels.

NTDs. Low-maternal folate status is linked with increased risks of NTDs in infants, which affects 1 of every 33 births in the United States and 6% of all births worldwide.[136] NTDs are regarded as multifactorial disorders, but several clinical intervention trials have demonstrated that periconceptional supplemental folate can reduce NTD risk. One of these, a large, well-designed, multicentered trial conducted by the British Medical Research Council found that a daily oral dose of 4 mg of folic acid reduced significantly the incidence of confirmed NTDs among the pregnancies of women at high risk for such disorders.[137] Several subsequent

134. These cytological changes do not become manifest until well after circulating folate levels drop (by 6–8 weeks).

135. 60 mg Fe (as ferrous sulfate) and 2.8 mg folate.

136. CDC, 2015. Morb. Mortal. Wkly. Rep. 64, 1–5

137. The double-blind, randomized clinical trial involved 1817 women, each with a previous affected pregnancy, who were followed in 33 clinics in seven countries. Each subject was randomly assigned to a placebo or a supplement (vitamins A, D, C, B_6; thiamin; riboflavin; and nicotinamide) and/or a placebo or folic acid (4 mg/day) in a complete factorial design and the outcomes of their pregnancies were confirmed. Of 1195 completed pregnancies, 27 had confirmed NTDs; these included 21 cases in both groups not receiving folate, but only 6 cases in both folate groups (relative risk, 0.28; 95% CI, 0.12–0.71). The multivitamin treatment did not significantly affect the incidence of NTDs (MRC Vitamin Research Group, 1991. Lancet 338, 131–137).

TABLE 17.15 Effect of Folate Supplementation on Plasma Homocysteine Levels in Healthy Adults

Treatment	Homocysteine, μM Week 0	Homocysteine, μM Week 13	Change, %	p	% of Subjects With >25% Decrease
Placebo	14.6 (12.7, 16.8)[a]	15.3 (13.2, 17.7)[a]	4.8	.23	10.5
5-Methyl-FH$_4$	13.9 (12.1, 15.9)	11.2 (9.8, 12.8)	−19.4	.001	46.2
Folic acid	13.7 (12.5, 15.1)	10.7 (9.6, 11.9)	−21.9	<.0001	48.7

[a]Mean (95% C.L.).
Zappacosta, B., Mastroiacovo, P., Persichilli, S., et al., 2013. Nutrients 5, 1531–1543.

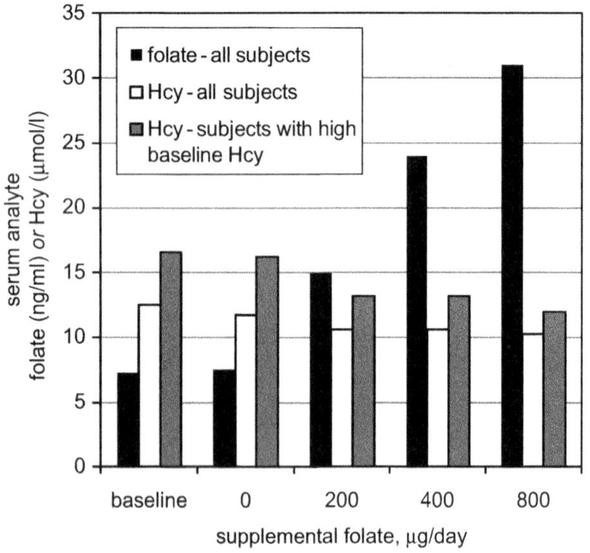

FIGURE 17.8 Effects of folate supplementation on serum levels of folate and homocysteine: results of a 26-week interventions. *After Tighe, P., Ward, M., McNulty, H., et al., 2011. Am. J. Clin. Nutr. 93,11–18.*

studies (Table 17.16) have supported those findings, showing folate supplements to reduce NTD risk by 50–70%. These have included trials conducted in the United States, which found folate supplements (400–4000 μg) effective in preventing NTDs in women with prior NTD pregnancies.[138] While folate supplements do not appear to affect NTD case fatality rates, reductions in NTD incidence are associated with reductions in neonatal deaths. A meta-analysis of eight observational studies indicated that folate supplementation was associated with a 46% reduction in NTD risk, which was associated with a 13% reduction in neonatal deaths.[139] Greatest benefits of folate supplementation have been found in countries with relatively high NTD rates (Fig. 17.9). A trial conducted in China with some 250,000 subjects showed that a daily supplement containing 400 μg folic acid consumed with ≥80% compliance during the periconceptional period was associated with reductions in NTD risk of 85% in a high NTD prevalence area and of 40% in a low prevalence area.[140] A systematic review of 14 folate intervention trials pointed out that not all NTD cases can be prevented by folate, i.e., 8–10 cases per 10,000 live births or abortions appear to be unaffected by increasing folate intake.[141]

Low folate status per se appears to be insufficient to cause NTDs. Instead, maternal low folate status appears to interact with fetal genes and other dietary factors affecting single-C metabolism. In animal models, this involves epigenetic marks on promoters essential for neurogenesis[142] and normal activity of SHMT.[143] In humans, this includes MTHFR 677C > T genotype: 677TT homozygosity has been associated with a fivefold increase in NTD risk for mothers not using multivitamin supplements.[144] A systematic review of clinical trials and observational studies found that erythrocyte folate concentrations in women are inversely associated with the presence of the 677T allele, i.e., CC > CT > TT.[145] Studies have also found DTD risk to be affected by maternal polymorphisms in the RFC (SLC19A1), methylene-FH$_4$ dehydrogenase, MET synthase, and Met synthase reductase.[146] Evidence suggests that NTDs in humans can result from tissue-specific

138. Centers for Disease Control and Prevention, 2000. Morb. Mortal. Wkly. Rep. 49, 1–4; Stevenson, R.E., Allen, R.E., Pai, G.S., et al., 2000. Pediatrics 106, 677–683.
139. Blencowe, H., Cousens, S., Modell, B., et al., 2010. Int. J. Epidemiol. 39:110–120.
140. Berry, R.J., Li., Z., Erickson, J.D., et al., 1999. N. Engl. J. Med. 341, 1485–1490.
141. Heseker, H.B., Mason, J.B., Selhub, J., et al., 2009. Br. J. Nutr. 102, 173–180.
142. i.e., *Hes*1 and *Neurog*2; Ichi, S., Costa, F.F., Bishof, J.M., et al., 2010. J. Biol. Chem. 285, 36922–36932.
143. Beaudin, A.E., Abarinov, E.V., Noden, D.M., et al., 2011. Am. J. Clin. Nutr. 93, 789–798; Beaudin, A.E., Abarinov, E.V., Malysheva, O., et al., 2012. Am. J. Clin. Nutr. 95, 109–114.
144. Botto, L.D., Moore, C.A., Khoury, M.J., et al., 1999. N. Engl. J. Med. 341, 1509–1519.
145. Tsang, B.L., Devine, O.J., Cordero, A.M., et al., 2015. Am. J. Clin. Nutr. 101, 1286–1294; Bueno, O., Molloy, A.M., Fernandez-Ballart, J.D., et al., 2016. J. Nutr. 146, 1–8.
146. Imbard, A., Benoist, J.F., Blom, H.J., 2013. Int. J. Environ. Res. Public Health 10, 4352–4389.

TABLE 17.16 Results of Placebo-Controlled, Clinical Intervention Trials of Folate Supplements in the Prevention of Neural Tube Defects (NTDs)

		NTD Rates, Cases/Total Pregnancies		
Trial	Folate Treatment	Placebo	Treatment	RR (95% CI)
1[a]	4 mg	4/51	2/60	0.42 (0.04–2.97)
2[b]	4 mg ± multivitamins	21/602	6/593	0.34 (0.10–0.74)
3[c]	0.8 mg + multivitamins	2/2104	0/2052	0.00 (0.00–0.85)

[a]*Lawrence, K.M., James, N., Miller, M., Campbell, H., 1981. Br. Med. J. 281, 1542–1511 (women with NTD histories).*
[b]*Milunsky, A., Jick, H., Jick, S.S., et al., 1989. J. Am. Med. Assoc. 262, 2847–2852 (women with NTD histories).*
[c]*Czeizel, A.E., Dudás, I., 1992. J. Am. Med. Assoc. 327, 1832–1835 (women without previous NTD births).*

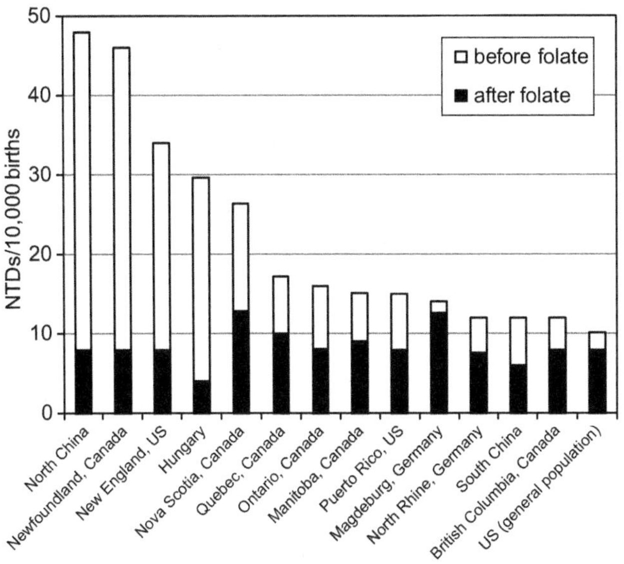

FIGURE 17.9 Reductions in NTD (neural tube defect) risk achieved in folate intervention trials and several countries. *After Heseker, H.B., Mason, J.B., Selhub, J., et al., 2009. Br. J. Nutr. 102, 173–180.*

differential hypermethylation of the fetal genes encoding the folate transporters FBP1 and RFC.[147]

Other effects on fetal growth. Evidence is inconsistent for associations of low-folate status and risks of reduced fetal growth or other congenital defects. Evidence suggests that MTHFR 677TT genotype elevates risk of Down syndrome in individuals also carrying a mutation in methionine synthase reductase.[148] Mothers and fetuses with the heterozygous 677CT genotype appear to have the best chances for viable pregnancies and live births.[149] Women with the MTHFR 1298CC genotype have lower chances of producing healthy infants than women with the 1298AA genotype.[150] The prenatal use of iron-folate supplements has reduced low birth weight in India and Nepal,[151] and the periconceptual use of folate has reduced preterm births in China.[152] A study found that high serum levels of folates and vitamin B_{12} were associated with greater success of assisted reproductive technologies;[153,154] however, a study in Sweden found no benefits of supplemental folate.

Carcinogenesis. Low-folate status has been associated with increased risk of cancers.[155] Women positive for human papilloma virus have a fivefold increase in risk of cervical dysplasia when they also have low serum folates.[156] Two large epidemiological studies have indicated that folate adequacy may reduce the effect of alcohol consumption in elevating breast cancer risk.[157] MTHFR polymorphisms have been related to risks to esophageal cancer, lymphocytic leukemia, and malignant lymphoma.[158] The 1298CC genotype has been associated with moderate reductions in colorectal cancer risk. The 677TT genotype does not appear to affect risk to colorectal adenoma unless folate status is low, in which case it is associated with an increase.[159] Studies in animal models have shown folate deprivation to promote colon carcinogenesis. Meta-analyses of cohort studies have

147. Farkas, S.A., Böttiger, A.K., Isaksson, H.S., et al., 2013. Epigentics 8 (3), 303–316.
148. Hobbs, C.A., Sherman, S.L., Yi, P., et al., 2000. Am. J. Hum. Genet. 67, 623–630.
149. Laanpere, M., Altmäe, S., Straveus-Evers, A., et al., 2009. Nutr. Rev. 68, 99–113.
150. Haggarty, P., McCallum, H., McBain, H., et al., 2006. Lancet 367, 1513–1519.
151. Balarajaan, Y., Subramanian, S.V., Fawzi, W., 2013. J. Nutr. 143, 1309–1315; Nisar, Y.B., Dibley, M.J., Mebrahtu, S., et al., 2015. J. Nutr. 145, 1873–1883.
152. Li, Z., Zhang, L., Ki, H., et al., 2014. Int. J. Epidemiol. 43, 1132–1139.
153. i.e., In vitro fertilization and intracytoplasmic sperm injection.
154. Gaskins, A.J., Chiu, Y.H., Williams, P.L., et al., 2015. Am. J. Clin. Nutr. 102, 943–950.
155. Chen, J., Xu, X., Liu, A., et al., 2010. In: Bailey, L. (Ed.), Folate in Health and Disease, second ed. CRC Press, Boca Raton, pp. 205–234.
156. Butterworth, C.E.J., Haatsh, K.D., Macaluso, M., et al., 1992. J. Am. Med. Assoc. 267, 528–533; Liu, T., Soong, S.J., Wilson, N.P., et al., 1993. Cancer Epidemiol. Biomarkers Prev. 2, 525–530.
157. Zhang, S., Hunter, D.J., Hankinson, S.E., et al., 1999. J. Am. Med. Assoc. 281, 1632–1637; Rohan, T.E., Jain, M.G., Howe, G.R., et al., 2000. J. Natl. Cancer Inst. 92, 266–269.
158. Skibola, C.F., Smith, M.T., Kane, E., et al., 1999. Proc. Natl. Acad. Sci. U.S.A. 96, 12810–12815.
159. Kono, S., Chen, K., 2005. Cancer Sci. 96, 535–542.

found food–folate intakes to be associated with reductions in colorectal cancer risk.[160] However, meta-analyses of several randomized clinical trials concluded that folate supplementation is ineffective in reducing site-specific cancer risks, with the single exception of melanoma.[161]

Folate Supplementation and Fortification

Supplement use. The NHANES data have indicated that some 23% of American adults use folate-containing dietary supplements, whereas the majority (77%) of pregnant women do so.[162] These are typically in the form of a multivitamin/mineral supplement, which provides pregnant women more than 800 μg of the vitamin per day.

Food system-based fortification. In 1992 the U.S. Public Health Service issued a recommendation that all women of childbearing age consume 0.4 mg folic acid daily to reduce their risks of an NTD pregnancy. In the following year, the U.S. Food and Drug Administration (FDA) ruled that all cereal grain products be fortified with 140 μg folic acid per 100 g, and that additions of folic acid be allowed for breakfast cereals, infant formulae, medical and special dietary foods, and meal replacement products. Other countries developed similar policies; those that increased the folate in their food systems experienced significant reductions in the incidence of NTDs: the United States, by 19–31%; Costa Rica, by 63–87%; and Canada, by 47–54%.[163] The US folate-fortification program increased folate intakes and more than doubled circulating levels of the vitamin, reduced plasma Hcy levels, and reduced the incidence of NTDs (Fig. 17.10). Countries using folate fortification in 1999–2002 experienced significantly greater reductions in stroke incidence than those countries without such programs.[164]

Questions about high-folate intakes and cancer risk. That folate functions as a cofactor in nucleotide synthesis suggests the possibility that plentiful supplies of the vitamin may facilitate proliferation in dysplastic and malignant cells. Results of two studies appear to support the prospect that supranutritional folate intake may promote cancer. A prospective study involving 25,400 American women 55–74 years of age found the incidence of breast cancer to be 20% greater for subjects reporting folate intakes ≥400 μg/day compared to those with lower intakes.[165] A placebo-controlled, randomized trial involving 1000 subjects with histories of colorectal adenomas found 2 years of supplementation with 1 mg folic acid to increase the risk of having a recurrent adenoma by 67% and to double the risk of having at least three adenomas.[166] That the institution of nationwide folate fortification in the United States and Canada corresponded with increasing colorectal cancer rates in those countries has been cited as evidence that increased folate intakes may be affecting cancer risk.[167] An analysis of colorectal cancer rates in the United States concluded that the increases observed in the 1990s were unlikely due to folate acid fortification,[168] and a meta-analyses of published data from randomized trials concluded that folic acid, at intakes greater than ~500 μg/day, does not increase cancer risk.[169] Still, those studies may not have included sufficient numbers of the subjects who might be at greatest risk to cancer promotion, i.e., those with prevalent malignancies.[170] This question is informed by a recent clinical study of postpolypectomy patients in which a high-level folate supplement was found to promote changes (increased folate content, reduced global DNA hypomethylation, and reduced DNA uracil misincorporation) in colonocytes adjacent to the polyp site (i.e., in the field that had produced

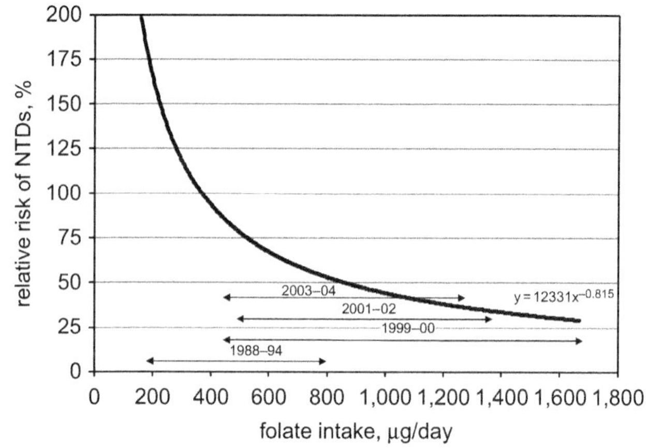

FIGURE 17.10 Relationships of estimated folate intake and risk of neural tube defects showing estimated ranges (10–90% of population) of folate intakes estimated in National Health and Nutrition Examination Surveys (NHANES III, 1988–94 and annual NHANES, 1999–2000, 2001–02, and 2003–04). *Based on combined population data from those surveys, Quinlivan, E.P., Gregory, J.F., 2007. Am. J. Clin. Nutr. 86, 1773–1779.*

160. Sanjoaquin, M.A., Allen, N., Couto, E., et al., 2005. Int. J. Cancer 113, 825; Kim, D.H., Smith-Warner, S.A., Spiegelman, D., 2010. Cancer Causes Control 21, 1919.
161. Ibrahim, E.M., Zekri, J.M., 2010. Med. Oncol. 27, 915–918; Vollset, S.E., Clarke, R., Lewington, S., et al., 2013. Lancet 381, 1029–1036; Qin, X., Cui, Y., Shen, L., et al., 2013. Int. J. Cancer 133, 1033–1042.
162. Vanderwall, C.M., Tangney, C.C., Kwasny, M.J., et al., 2012. J. Acad. Nutr. Diet. 112, 285–290; Branum, A.M., Bailey, R., Singer. B.J., 2013. J. Nutr. 143, 486–492.
163. Yetley, E.A., Rader, J.I., 2004. Nutr. Rev. 62, S50–S59; Chen, L.T., Rivera, M.A., 2004. Nutr. Rev. 62, S40–S43; Mills, J.L., Signore, C., 2004. Birth Defects Res. A 70, 844–845.
164. Yang, Q., Botto, L.D., Erickson, D., et al., 2006. Circulation 113, 1335–1343.
165. Stolzenberg-Solomon, R.Z., Chang, S.C., Leitzmann, M.F., et al., 2006. Am. J. Clin. Nutr. 83, 895–904.
166. Cole, B.F., Baron, J.A., Sandler, R.S., et al., 2007. JAMA 297, 2351–2359.
167. Kim, Y.I., 2007 Am. J. Clin. Nutr. 80, 1123–1128; Mason, J.B., Dickstein, A., Jacques, P.E., et al., 2007. Cancer Epidemiol. Biomarkers Prev. 16, 1325–1329.
168. Keum, N., Giovannucci, E.L., 2014. Am. J. Prev. Med. 46, S65–S72.
169. Vollset, S.E., Clarke, R., Lewington, S., et al., 2013. Lancet 381, 1029–1036.
170. Mason, J.B., 2011. Am. J. Clin. Nutr. 94, 965–966.

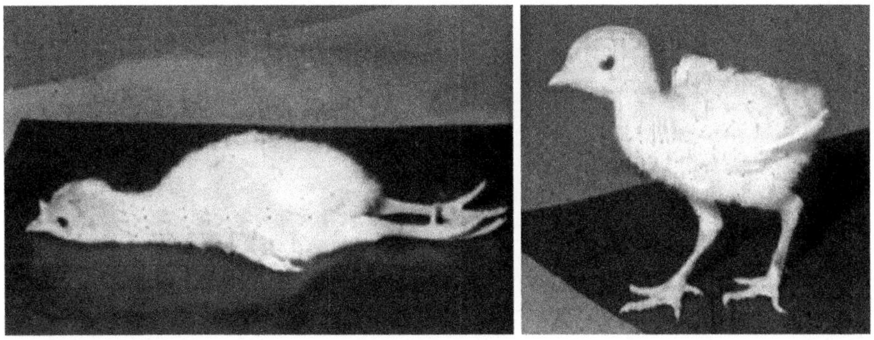

FIGURE 17.11 Cervical paralysis in a folate-deficient turkey poult (left); same poult 15 min after being treated (by injection) with folic acid (right). *Courtesy, G.F., Combs, Sr.*

an adenomatous poly) suggestive of reduced likelihood of mutagenesis and polyp formation.[171]

Deficiency Syndromes in Animals

Folate deficiency in animals is generally associated with poor growth, anemia, and dermatologic lesions involving skin and hair/feathers. In chicks, severe anemia is one of the earliest signs of the deficiency. The anemia is of the macrocytic (megaloblastic) type, involving abnormally large erythrocyte size (the normal range in humans is 82–92 μM^3) due to the presence of **megaloblasts**, which are also seen among the hyperplastic erythroid cells in the bone marrow. Anemia in folate deficiency is followed by **leukopenia** (abnormally low numbers of white blood cells), poor growth, very poor feathering, perosis, lethargy, and reduced feed intake.

Poultry with normally pigmented plumage[172] show achromotrichia due to the deficiency. Folate-deficient turkey poults show a spastic type of **cervical paralysis** in which the neck is held rigid (Fig. 17.11).[173] Folate-deficient guinea pigs show leukopenia and depressed growth; pigs and monkeys show alopecia, dermatitis, leukopenia, anemia, and diarrhea; mink show ulcerative hemorrhagic gastritis, diarrhea, anorexia, and leukopenia. The deficiency is not easily produced in rodents unless a **sulfa drug**[174] or folate antagonist is fed, in which case leukopenia is the main sign.[175] Folate-responsive signs (reduced weight gain, macrocytic anemia) can be produced in catfish by feeding them succinylsulfathiazole. Folate deficiency in the rat has been shown to reduce exocrine function of the pancreas, in which single-carbon metabolism is important.[176]

Folate deficiency is not expected in ruminants with functioning microflora, which produces the vitamin in amounts that are apparently adequate to meet the needs of the host. In fact, nearly all supplemental folate to a dairy diet acid appears to be degraded; high doses of the vitamin (e.g., 0.5 mg/kg of host body weight) are needed to increase serum and milk folate levels.

High-level folate supplementation of the diets of laying hens (e.g., 16 mg/kg of diet) has been an effective in producing eggs enriched in the vitamin for marketing purposes.

10. FOLATE IN HEALTH AND DISEASE

Pernicious anemia. High doses of folate (e.g., 400 μg/day intramuscular; 5 mg/day oral) have been shown to correct the **megaloblastic anemia** of pernicious anemia patients, which are deficient in vitamin B_{12}. This phenomenon renders megaloblastic anemia not useful for diagnosing either vitamin deficiency without accompanying metabolic measurements: FIGLU—elevations indicate folate deficiency, **MMA**—elevations indicate vitamin B_{12} deficiency.

Supplemental folate does not mask the irreversible progression of neurological dysfunction and cognitive decline of vitamin B_{12} deficiency; however, those signs develop over a longer period of time than the anemia produced by the same deficiency. In fact, folate supplementation has been shown to exacerbate the cognitive symptoms of vitamin B_{12} deficiency.[177] Because vitamin B_{12} deficiency affects an estimated 10–15% of the American population over 60 years of age, the amount of folate for the fortification of wheat flour (140 μg/100 g flour) was chosen to provide an amount of added folate (100 μg/person/day) sufficient for only a small proportion of the general population receiving a level (>1 mg/day) capable of masking vitamin B_{12} deficiency.

171. O'Reilly, S., McGlynn, A.P., McNulty, H., et al., 2016. J. Nutr. 146, 933–939.
172. Such breeds include the barred Plymouth Rock, the Rhode Island Red, and the Black Leghorn.
173. Poults with cervical paralysis may not show anemia; the condition is fatal within a couple of days of onset but responds dramatically to parenteral administration of the vitamin.
174. For example, sulfanilamide.
175. Although leukopenia was manifested relatively soon after experimental folate depletion, rats kept alive with small doses of folate eventually also develop macrocytic anemia.
176. Experimental pancreatitis can be produced in that species by treatment with ethionine, an inhibitor of cellular methylation reactions, or by feeding a diet deficient in choline.
177. Morris, M.S., Jacques, P.F., Rosenberg, I.H., 2007. Am. J. Clin. Nutr. 85, 193–200.

Cardiovascular Disease

The lowering of circulating Hcy levels effected by folate supplementation has not been found consistently to reduce risk to cardiovascular disease. A meta-analysis of eight trials involving 37,485 subjects randomized to folate and/or other B-vitamins were used as the intervention agents showed that a 25% reduction in circulating Hcy level for 5 years was not associated with any reductions in cardiovascular events (or death from any cause).[178] In one trial, cardiovascular disease patients who received a combined supplement of folate, vitamin B_6, and vitamin B_{12} showed increased risk for subsequent myocardial infarction.[179] Meta-analyses of trials that used folate as the single intervention agent yielded inconsistent results: one found the vitamin to improve flow-mediated dilatation, suggestive of enhanced vascular function;[180] the other found folate supplementation to be of no benefit in reducing stroke risk.[181] A placebo-controlled, randomized trial demonstrated that a combined supplement containing folate, vitamin B_{12}, and vitamin B_6 significantly reduced the progression of early stage subclinical atherosclerosis (carotid artery intima medial thickening) without homoc ysteinemia.[182]

Immune Function

That folate status may affect immune cell function is suggested by findings that folate deprivation of lymphocytes in vitro causes depletion of interleukin-2 and stimulates p53-independent apoptosis. Similar effects have not been observed in vivo. It has been noted that folate supplementation can stimulate natural killer (NK) cell cytotoxicity among subjects of low-folate status. Curiously, the opposite effect was observed among subjects consuming a high-folate diet, in whom NK cytotoxicity was inversely related to the plasma concentration of the nonmetabolized form of the vitamin, folic acid.[183] The MTHFR 677T allele has been associated with reduced risk to hepatitis B virus (HBV) infection in an HBV endemic area.[184]

178. Clarke, R., Halsey, J., Bennett, D., 2011. J. Inherit. Metab. Dis. 34, 83–91.
179. In the Norwegian Vitamin Trial, 3749 men and women with histories of heart attack were randomized to four combinations of folate, and vitamin B_6 and B_{12} (Bønaa, K.H., Njølstadn I., Ueland, P.M., et al., 2006. N. Engl. J. Med. 354, 1578–1588).
180. De Bree, A., van Mierol, L.A., Draijer, R., 2007. Am. J. Clin. Nutr. 86, 610–617.
181. Lee, M., Hong, K.S., Chang, S.C., et al., 2010. Stroke 41, 1205–1212.
182. Hodis, H.N., Mack, W.J., Dustin, L., et al., 2009. Stroke 40, 730–736.
183. Troen, A.M., Mitchell, B., Sorenson, B., et al., 2006. J. Nutr. 136, 189–194.
184. Bronowicki, J.P., Abdelmouttaleb, I., Peyrin-Biroulet, L., 2008. J. Hepatol. 48, 532–539.

Malaria

The malarial parasite, *Plasmodium falciparum*, requires folate for its own metabolism, including the DNA synthesis required for growth and proliferation.[185] The parasite is capable of synthesizing folate from *p*-aminobenzoic acid and L-glutamate; but it also uses exogenous supplies of folate, such as it finds within the host erythrocyte it invades, to continue its life cycle. Accordingly, antifolates[186] are first-line drugs in the treatment of malaria.[187] The host–parasite competition for folate contributes to the anemia observed in malarial patients, and malaria-induced hemolysis appears to increase the host's need for folate. While prenatal supplements of folate and iron have been found effective in reducing neonatal mortality in malaria-endemic regions, it is likely that those benefits may be limited to situations in which anemia is prevalent and the actual malaria prevalence is low. A robust study in a population with a high prevalence of malaria found supplementation with folate and iron to increase risk of severe illness and death in children who were not iron deficient.[188]

Arsenicosis

Studies in animals have shown folate status to be a determinant of the metabolism and tissue distribution of arsenic (As), which must be methylated to be excreted. Accordingly, urinary concentrations of dimethylarsenate of As-exposed subjects in Bangladesh were found to correlate positively with plasma folate level and negatively with plasma Hcy level; serum As levels were reduced by supplementation with folate.[189] An estimated 140 million people are exposed to As in drinking and irrigation waters in south Asia; many are also inadequately nourished and can be expected to be of low status with respect to folate, vitamin B_{12}, and MET.

Macular Degeneration

High-folate intake has been associated with reduction in risk to developing geographical atrophy, an advanced form of age-related macular degeneration.[190]

185. Globally, malaria causes an estimated 200 million morbid episodes and 2–3 million deaths each year. During pregnancy, the disease also contributes to low birth weight and intrauterine growth retardation.
186. e.g., Pyrimethamine, sulfadoxine.
187. Unfortunately, strains of *P. falciparum* have developed resistance through mutations in their dihydrofolate reductase.
188. Sazawal, S., Black, R.E., Ramsan, M., et al., 2006. Lancet 367, 133–14.
189. Gamble, M.V., Liu, X., Ashan, H., et al., 2005. Environ. Health Perspect. 113, 1683–1688.
190. Merle, B.M.J., Silver, R.E., Rosner, B., et al., 2016. Am. J. Clin. Nutr. 103, 1135–1144.

11. FOLATE TOXICITY

The toxicity of folic acid is negligible. No adverse effects of high oral doses of folate have been reported in humans or animals, although parenteral administration of pharmacologic amounts (e.g., 250 mg/kg, i.e., about 1000 times the dietary requirement) has been shown to produce epileptic responses and renal hypertrophy in rats. Inconsistent results have been reported concerning the effects of high-folate levels (1–10 mg doses) on human epileptics. Some have indicated increases in the frequency or severity of seizures and reduced anticonvulsant effectiveness;[191] whereas, others have shown no effects. Upper tolerable limits have been established for most age groups (Table 17.17).

Folate does have the potential for adverse effects, as it can exacerbate the consequences of vitamin B_{12} deficiency. By circumventing the methyl folate trap, high-folate intakes can provide folic acid directly for DNA synthesis, thus, correcting megaloblastic anemia caused by vitamin B_{12} deficiency. The loss of anemia as a sign of that deficiency can increase the likelihood of it progressing to the point of irreversible neurological damage. Folate supplementation in pregnant women with inadequate vitamin B_{12} intakes increased the risk of their having small-for-gestational-age infants.[192] High-folate intakes have also been found to increase circulating Hcy levels and impair the activities of MET synthase and methylmalonyl-CoA mutase.[193] Mice fed high levels of folic acid (10-fold recommended levels) in low-vitamin B_{12} diets showed reduced MTHFR expression and reduce fetal growth.[194]

TABLE 17.17 Recommended Upper Tolerable Intakes (ULs) of Folate

Ages, Years	UL, µg/day
1–3	300
4–8	400
9–13	600
14–18 females	1000
Males	800
>18 years	1000
Pregnancy	800
Lactation	1000

Food and Nutrition Board of the Institute of Medicine, National Academy of Sciences.

12. CASE STUDY

Instructions

Review the following case report, paying special attention to the diagnostic indicators on which the treatments were based. Then, answer the questions that follow.

Case

A 15-year-old girl was admitted to the hospital because of progressive withdrawal, hallucinations, anorexia, and tremor. Her early growth and development were normal, and she had done average schoolwork until she was 11 years old, when her family moved to a new area. The next year, she experienced considerable difficulty in concentrating and was found to have an IQ of 60. She was placed in a special education program, where she began to fight with other children and have temper tantrums; when punished, she became withdrawn and stopped eating. A year earlier, she had experienced an episode of severe abdominal pain for which no cause could be found, and she was referred to a mental health clinic. Her psychologic examination at that time had revealed inappropriate giggling, poor reality testing, and loss of contact with her surroundings. Her verbal and performance IQs were then 46 and 50, respectively. She was treated with thioridazine[195] and, within 2 weeks, she ate and slept better and was helpful around the house. However, over the succeeding months, while she continued taking thioridazine, her functioning fluctuated and the diagnosis of catatonic schizophrenia was confirmed. Three months before the present admission, she had become progressively withdrawn and drowsy, and needed to be fed, bathed, and dressed. She also experienced visual hallucinations, feelings of persecution, and night terrors. On having a seizure, she was taken to the hospital. Her physical examination on admission revealed a tall, thin girl with fixed stare and catatonic posturing but no neurologic abnormalities. She was mute and withdrawn, incontinent, and appeared to have visual and auditory hallucinations. Her muscle tone varied from normal to diffusely rigid.

On the assumption that her homocystinuria was due to cystathionase deficiency, she was treated with pyridoxine HCl (300 mg/day, orally) for 10 days. Her homocystinuria did not respond; however, her mental status improved and, within 4 days, she was able to conduct some conversation and her hallucinations seemed to decrease. She developed new neurological signs: foot and wrist droop and gradual

191. High doses of folate appear to interfere with diphenylhydantoin absorption.

192. Dwarkanath, P., Barzilay, J.R., Thomas, T., et al., 2013. Am. J. Clin. Nutr. 98, 1450–1458.

193. Selhub, J., Morris, M.S., Jacques, P.F., et al., 2009. Am. J. Clin. Nutr. 89, 702S–706S.

194. Mikael, L.G., Deng, L., Paul, L., et al., 2013. Birth Defects Res. A 97, 47–52; Christensen, K.E., Mikael, L.G., Leung, K.Y., et al., 2015. Am. J. Clin. Nutr. 101, 646–658.

195. An antischizophrenic drug.

loss of reflexes. She was then given folate (20 mg/day orally) for 14 days because of her low-serum folate level. This resulted in a marked decrease in her urinary homocysteine and a progressive improvement in intellectual function over the next 3 months. She remained severely handicapped by her peripheral neuropathy, but she showed no psychotic symptoms. After 5 months of folate and pyridoxine treatment, she was tranquil and retarded but showed no psychotic behavior; she left the hospital against medical advice and without medication.

Laboratory Findings

Parameter	Patient	Normal Range
Electroencephalogram	Diffusely slow	
Spinal Fluid		
Protein (mg/dL)	42	15–45
Cells	None	None
Urine		
Homocysteine	Elevated	
Methionine	Normal	
Serum		
Homocysteine	Elevated	
Methionine	Normal	
Folate (ng/mL)	3	5–21
Vitamin B_{12} (pg/mL)	800	150–900
Hematology		
Hemoglobin (g/dL)	12.1	11.5–14.5
Hematocrit (%)	39.5	37–45
Reticulocytes (%)	1	~1
Bone marrow	No megaloblastosis	

The girl was readmitted to the hospital 7 months later (a year after her first admission) with a 2-month history of progressive withdrawal, hallucinations, delusions, and refusal to eat. The general examination was the same as her first admission, with the exceptions that she had developed hyperreflexia and her peripheral neuropathy had improved slightly. Her mental functioning was at the 2-year-old level. She was incontinent, virtually mute, and had visual and auditory hallucinations. She was diagnosed as having simple schizophrenia of the childhood type. Folate and pyridoxine therapy was started again; it resulted in her decreased Hcy excretion and gradual improvement in mental performance. After 2 months of therapy in the hospital, she was socializing, free of hallucinations, and able to feed herself and recognize her family. At that time, the activities of several enzymes involved in methionine metabolism were measured in her fibroblasts and liver tissue (obtained by biopsy).

Enzyme Activities

Enzyme	Specimen	Enzyme Activity[a]	
		Patient	Normal
Methionine adenyltransferase	Liver	20.6	4.3–14.5
Cystathionine-β-synthase	Fibroblasts	25.9	3.7–65.0
Betaine:Hcy methyltransferase	Liver	26.7	1.2–16.0
5-Methyl FH_4:Hcy methyltransferase	Fibroblasts	3.5	2.9–7.3
5,10-Methylene-FH_4 reductase	Fibroblasts	0.5	1.0–4.6

[a]*Enzyme units.*

Thereafter, she was maintained on oral folate (10 mg/day). She has been free of homocystinuria and psychotic manifestations for several years.

Case Questions

1. On admission of this patient to the hospital, which of her symptoms were consistent with an impairment in a folate-dependent aspect of metabolism?
2. What finding appeared to counterindicate an impairment in folate metabolism in this case?
3. Propose a hypothesis for the metabolic basis of the observed efficacy of oral folate treatment in this case.

13. STUDY QUESTIONS AND EXERCISES

1. Diagram the metabolic conversions involving folates in single-carbon metabolism.
2. Construct a decision tree for the diagnosis of folate deficiency in humans or an animal species. In particular, outline a way to distinguish folate and vitamin B_{12} deficiencies in patients with macrocytic anemia.
3. What key feature of the chemistry of folate relates to its biochemical function as a carrier of single-carbon units?
4. What parameters might you measure to assess folate status of a human or animal?
5. Detail the impact (positive and negative) of food fortification programs increasing the folate intake of populations.

RECOMMENDED READING

Bailey, L.B., da Silva, V., West, A.A., et al., 2012. Folic acid. In: Zempleni, J., Suttie, J.W., Gregory, J.F., et al. (Eds.), Handbook of Vitamins, fifth ed. CRC Press, New York, pp. 421–446 (Chapter 11).

Bailey, L.B., Stover, P.J., McNulty, H., et al., 2015. Biomarkers of nutrition for development – folate review. J. Nutr. 145, S1636–S1680.

Baru, S., Kuizon, S., Junaid, M., 2014. Folic acid supplementation in pregnancy and implications in health and disease. J. Biomed. Sci. 21, 77–86.

Blom, H.J., Smulders, Y., 2011. Overview of homocysteine and folate metabolism, with special references to cardiovascular disease and neural tube defects. J. Inherit. Metab. Dis. 34, 75–81.

Choi, J.H., Yates, Z., Veysey, M., et al., 2014. Contemporary issues surrounding folic acid fortification initiatives. Prev. Nutr. Food Sci. 17, 247–260.

Crider, K.S., Yang, T.P., Berry, R.J., et al., 2012. Folate and DNA methylation: a review of molecular mechanisms and the evidence for folate's role. Adv. Nutr. 3, 21–38.

Da Silva, R.P., Keily, K.B., Al Rajabi, A., 2014. Novel insights on interactions between folate and lipid metabolism. Biofactors 3, 277–283.

Duthie, S.J., 2011. Folate and cancer: how DNA damage, repair and methylation impact colon carcinogenesis. J. Inherit. Metab. Dis. 34, 101–109.

French, M., 2012. Folate (vitamin B_9) and vitamin B_{12} and their function in the maintenance of nuclear and mitochondrial genome integrity. Mutat. Res. 733, 21–33.

Heseker, H.B., Mason, J.B., Selub, J., et al., 2009. Not all cases of neural-tube defect can be prevented by increasing the intake of folic acid. Br. J. Nutr. 102, 173–180.

Imbard, A., Benoist, J.F., Blom, H.J., 2013. Neural tube defects, folic acid and methylation. Int. J. Environ. Res. Public Health 10, 4352–4389.

Kim, S.E., Mashi, S., Lim, Y.I., 2015. Folate, DNA methylation, and colorectal cancer. In: Ho, E., Domann, F. (Eds.), Nutrition and Epigenetics. CRC Press, Boca Raton, FL, pp. 113–161 (Chapter 4).

Manolescu, B.N., Oprea, E., Farcasanu, I.C., et al., 2010. Homocysteine and vitamin therapy in stroke prevention and treatment: a review. Acta Biochim. Pol. 57, 467–477.

McCully, K.S., 2007. Homocysteine, vitamins, and vascular disease prevention. Am. J. Clin. Nutr. 86, 1563S–1568S.

Morris, M.S., 2012. The role of B vitamins in preventing and treating cognitive impairment and decline. Adv. Nutr. 3, 801–812.

Nazki, F.H., Sameer, A.S., Ganaie, B.A., 2014. Folate: metabolism, genes, polymorphisms and the associated diseases. Gene 533, 11–20.

Ohrvik, V.E., Witthoft, C.M., 2011. Human folate bioavailability. Nutrients 3, 475–490.

Osterhues, A., Ali, N.S., Michels, K.B., 2013. The role of folic acid fortification in neural tube defects: a review. Crit. Rev. Food Sci. Nutr. 53, 1180–1190.

Peake, J.N., Copp, A.J., Shawe, J., 2013. Knowledge and periconceptional use of folic acid for prevention of neural tube defects in ethnic communities in the United Kingdom: systematic review and meta-analysis. Birth Defects Res. A 97, 444–451.

Safi, J., Joyeux, L., Chalouhi, G.E., 2012. Periconceptual foliate deficiency and implications in neural tube defects. J. Pregnancy 295083.

Said, H.M., 2011. Intestinal absorption of water-soluble vitamins in health and disease. Biochem. J. 437, 357–372.

Salbaum, J.M., Kappen, C., 2012. Genetic and epigenomic footprints of folate. Prog. Mol. Biol. Transl. Sci. 108, 129–158.

Smulders, Y.M., Blom, H.J., 2011. The homocysteine controversy. J. Inherit. Metab. Dis. 34, 93–99.

Stover, P.J., 2011. Polymorphisms in 1-carbon metabolism, epigenetics and folate-related pathologies. J. Nutrigenet. Nutrgenomics 4, 293–305.

Stover, P.J., Field, M.S., 2011. Trafficking of intracellular folates. Adv. Nutr. 2, 325–331.

Xia, W., Low, P.S., 2010. Folate-targeted therapies for cancer. J. Med. Chem. 53, 6811–6824.

Zhao, R., Diop-Bove, N., Visentin, M., et al., 2011. Mechanisms of membrane transport of folates into cells and across epithelia. Annu. Rev. Nutr. 31, 177–201.

Chapter 18

Vitamin B_{12}

Chapter Outline

Anchoring Concepts

1. *Vitamin B_{12}* is the generic descriptor for all corrinoids (compounds containing the cobalt-centered corrin nucleus) exhibiting the biological activity of cyanocobalamin.
2. Deficiencies of vitamin B_{12} are manifested as anemia and neurologic changes, and can be fatal.
3. The function of vitamin B_{12} in single-carbon metabolism is interrelated with that of folate.

Patients with Addisonian pernicious anemia have…a "conditioned" defect of nutrition. The nutritional defect in such patients is apparently caused by a failure of a reaction that occurs in the normal individual between a substance in the food (extrinsic factor) and a substance in the normal gastric secretion (intrinsic factor).

W. B. Castle and T. H. Ham[1]

LEARNING OBJECTIVES

1. To know the chief natural sources of vitamin B_{12}
2. To understand the means of enteric absorption and transport of vitamin B_{12}
3. To understand the biochemical functions of vitamin B_{12} as a coenzyme in the metabolism of propionate and the biosynthesis of methionine
4. To understand the metabolic interrelationship between vitamin B_{12} and folate
5. To understand the factors that can cause low vitamin B_{12} status, and the physiological implications of that condition

VOCABULARY

Achlorhydria
Adenosylcobalamin
Aquocobalamin
Cobalamins
Cobalt
Corrinoid ring
Cubulin
Cyanocobalamin
Deoxyadenosylcobalamin
Gastric parietal cell
Haptocorrin
Helicobacter pylori
Holotranscobalamin (holoTC)
Homocysteinemia
Homocystinuria
Hydroxycobalamin
Hypochlorhydria
Intrinsic factor (IF)
IF receptor
IF–vitamin B_{12} complex
Imerslund–Gräsbeck syndrome
Lipotrope
Macrocyte
Megaloblastosis
Methionine synthase
Methionine synthase reductase

1. William B. Castle (1897–1990) was a physician and physiologist at Harvard University. He is known for transforming hematology into a dynamic interdisciplinary field with his early discovery of the gastric intrinsic factor, which ultimately led to the identification of vitamin B_{12}. Among his collaborators was a young physician, Thomas Hale Ham (1905–1987), who went on to join the faculty of Western Reserve University where he revolutionized the medical education curriculum with a model that was adopted nationwide.

The Vitamins. http://dx.doi.org/10.1016/B978-0-12-802965-7.00018-6

Methylcobalamin
Methylmalonic aciduria
Methylmalonyl CoA mutase
Methylfolate trap
Methyl-FH_4 methyltransferase
Methylmalonic acid (MMA)
Methylmalonic acidemia
Methylmalonic aciduria
Methylmalonyl CoA mutase
Nitritocobalamin
Ovolactovegetarian
Pepsin
Peripheral neuropathy
Pernicious anemia
Perosis
Pseudovitamin B_{12}
R proteins
Schilling test
Transcobalamin (TC)
Transcobalamin receptor
Vegan
Vitamin B_{12} coenzyme synthetase.

1. SIGNIFICANCE OF VITAMIN B_{12}

Vitamin B_{12} is synthesized by prokaryotic organisms. Animals require the vitamin for critical functions in cellular division and growth. Some animal tissues can store the vitamin in appreciable amounts that are sufficient to meet the needs of the organism for long periods (years) of deprivation. The vitamin is seldom found in foods derived from plants; therefore, non-coprophagous animals and humans that consume strict vegetarian diets are likely to have inadequate intakes of vitamin B_{12}. If prolonged, those will lead to anemia and peripheral neuropathy. Few humans are strict **vegans** (who exclude all foods of animal origin); most consume foods and/or supplements containing vitamin B_{12}. For this reason, frank vitamin B_{12} deficiency is not common. Nevertheless, low vitamin B_{12} status occurs, particularly in individuals with hereditary deficiencies in proteins involved in vitamin B_{12} transport and/or metabolism, or with compromised gastric parietal cell function. Low vitamin B_{12} status limits DNA synthesis, impairs the metabolic utilization of folate, and contributes to homocysteinemia, a risk factor for occlusive vascular disease.

2. PROPERTIES OF VITAMIN B_{12}

Vitamin B_{12} Nomenclature

The term **vitamin B_{12}** is the generic descriptor for all corrinoids (compounds containing the **corrin nucleus**) exhibiting the biological activity of **cyanocobalamin** (also **cobalamin**). The vitamers B_{12} are octahedral cobalt (Co) complexes consisting of a porphyrin-like, cobalt-centered macroring (the corrin nucleus), a nucleotide, and a second Co-bound group (e.g., CH_3, H_2O, CN^-). The corrin nucleus consists of four reduced pyrrole nuclei linked by three methylene bridges and one direct bond. The triply ionized cobalt atom (Co^{3+}) is essential for biological activity; it can form up to six coordinate bonds and is tightly bound to the four pyrrole nitrogen atoms. The central cobalt atom can also bind a small ligand above (α-position) and a nucleotide below (β-position) the plane of the ring system. For example, its α-position ligands include cyano (CN^-) (**cyanocobalamin**), methyl (**methylcobalamin**), 5′-deoxyadenosyl (**adenosylcobalamin**), or hydroxo (OH) (**hydroxocobalamin**[2]) groups. Those, and the unliganded form with a reduced cobalt center (**cob[I]alamin**), are found intracellularly. Other synthetic analogs with vitamin B_{12} activity include forms with aqua- (H_2O) (**aquacobalamin**[3]) or nitrite (**nitritocobalamin**[4]) ligands.

Chemical structures of vitamin B_{12}:

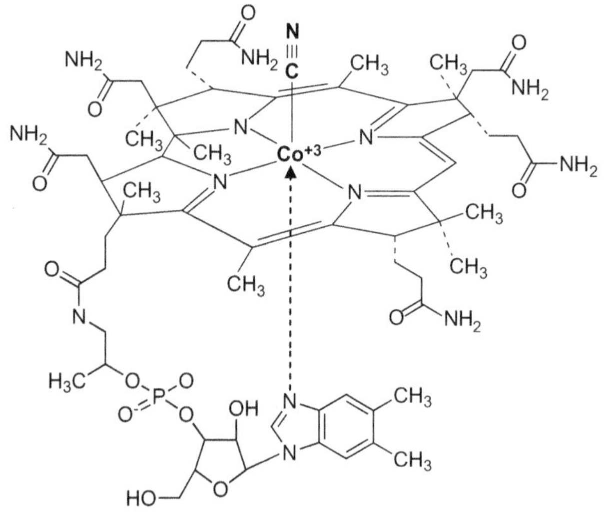

Cyanocobalamin

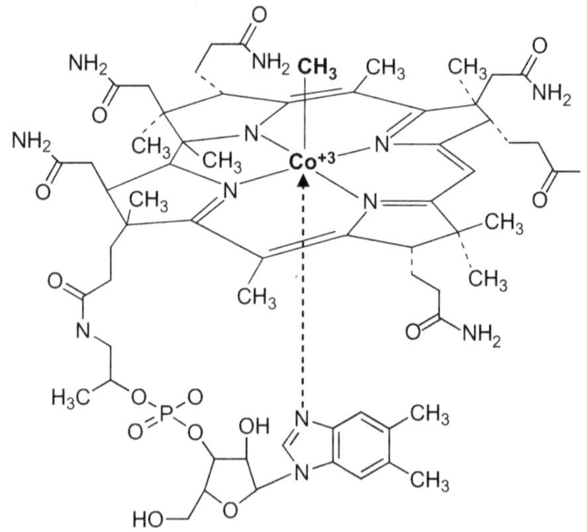

Methylcobalamin

2. formerly, vitamin B_{12b}.
3. formerly vitamin B_{12a}.
4. formerly vitamin B_{12c}.

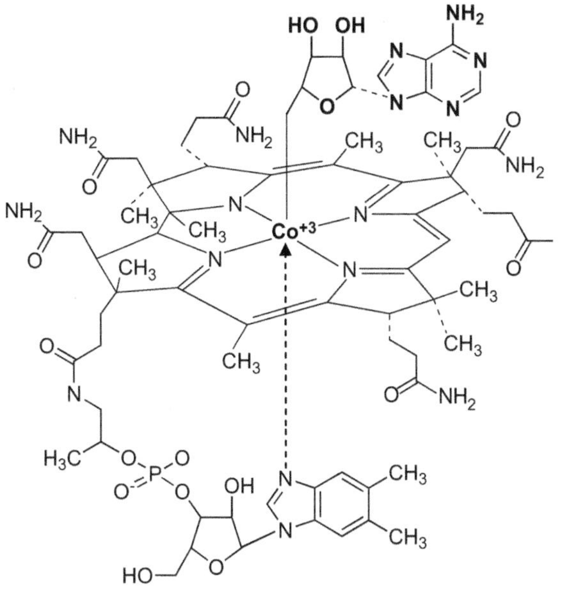

5′-Deoxyadenosylcobalamin

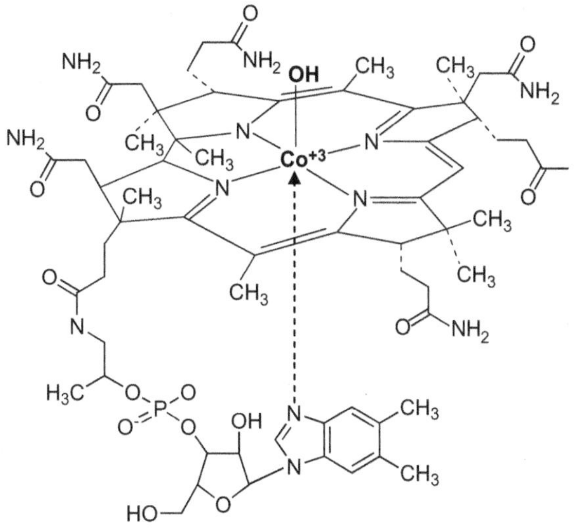

Hydroxocobalamin

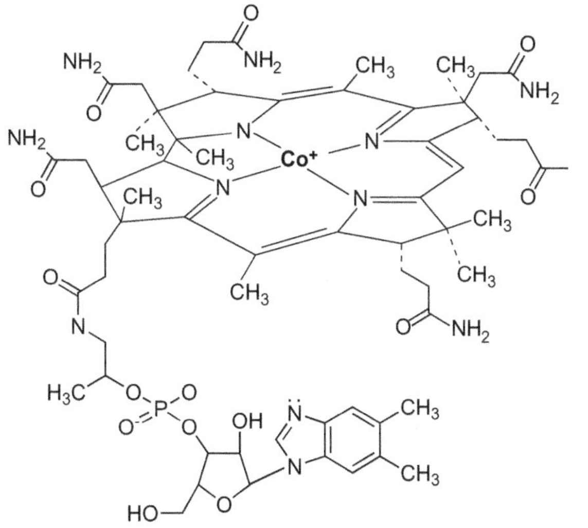

Cob(I)alamin

Vitamin B_{12} Chemistry

The corrinoids are red, red–orange, or yellow crystalline substances that show intense absorption spectra above 300nm owing to the π–π transitions of the corrin nucleus. They are soluble in water and are fairly stable to heat but decompose at temperatures above <210°C without melting. Vitamin B_{12} reacts with ascorbic acid, resulting in the reduction and subsequent degradation of the former, which releases its cobalt atom as the free ion. Cobalamins with relatively strongly bound ligands (e.g., cyano-, methyl-, and adenosylcobalamin) are less reactive and are therefore more stable in the presence of ascorbic acid. The cobalamins are unstable to light. Cyanocobalamin undergoes a photoreplacement of the CN^- ligand with water; the organocobalamins (methyl- and adenosylcobalamin) undergo photoreduction of the cobalt-carbon bond, resulting in the loss of the ligand and the reduction of the corrin cobalt. The vitamin can bind to the vitamin B_{12} enzymes through an imidazole nitrogen of a histidyl residue on the protein, which serves as the ligand to the lower axial position of the cobalt atom instead of the dimethylbenzimidazole grouping.

3. SOURCES OF VITAMIN B_{12}

Synthesis by the Gut Microbiome

Vitamin B_{12} is synthesized by some anaerobic microorganisms, particularly propionic acid bacteria.[5] That synthesis is dependent on an adequate supply of Co. A genomic analysis of 256 representative organisms of the human gut microbiota found 43% capable of de novo synthesis of the vitamin, with a total synthetic capacity equivalent to at least 31% of the daily human need.[6] However, it is not clear that vitamin B_{12} can be absorbed across the human colon, as only 19% of the corrinoids in human stool were found to be absorbable.[7] With sufficient Co from the diet,[8] the rumen microbial synthesis of vitamin B_{12} is substantial.[9] Not only do ruminant species have little, if any, need for preformed vitamin B_{12} in the diet but also their tissues tend to contain

5. *Pseudomonas denitrificans* is widely used for the commercial preparation of vitamin B_{12}. The vitamin is also produced in large amounts by *Propionibacterium freudenreichii* and *Propionibacterium shermanii.* Some lactic acid bacteria have been found to synthesize corrinoids; but they do not appear to release it and, thus, do not appear to contribute to hindgut microbial synthesis of the vitamin.

6. Magnúsdóttir, S., Ravchee, D., de Crécy-Lagard, V., et al., 2015. Front. Genet. 6, 148–166.

7. Albert, R.H., Stabler, S.P., 2008. Am. J. Clin. Nutr. 87, 1324–1335.

8. Beef cattle require 125–160μg/kg dietary dry matter (typically added in a mineral premix or salt lick) to minimize their circulating levels of homocysteine and methylmalonic acid, but higher levels (235–255μg/kg DM) to maximize vitamin B_{12} concentrations in plasma and liver (Stangl, G.I., Schwarz, F.J., Müller, H., et al., 2000. Br. J. Nutr. 84, 645–653).

9. In fact, ruminal infusion with a chelated form of Co has been shown to affect the saturation of cows' milk lipids, reducing the level of fatty acid desaturation. (Leskinen, H., Viitala, S., Mutikaninen, M., et al., 2016. J. Nutr. 146, 976–985.

substantial amounts of the vitamin, making them important potential dietary sources of vitamin B_{12} for meat-eating, nonruminant species.

Distribution in Foods

Because the synthesis of vitamin B_{12} is limited almost exclusively to anaerobic bacteria, the vitamin is found only in foods that have been bacterially fermented or derived from the tissues of animals that have obtained it from their ruminal or intestinal microflora, or ingested it either with their diet or coprophagously. Animal tissues that accumulate vitamin B_{12} are, therefore, excellent food sources of the vitamin (Table 18.1). The richest sources are liver[10] and kidney; other rich sources are dairy products, meats, eggs, fish, and shellfish. The amounts of the vitamin in the tissues of livestock species depend on the feeding management and cut of meat. The principal vitamers in foods are **methylcobalamin**, **deoxyadenosylcobalamin,** and **hydroxycobabalmin**. The richest sources of vitamin B_{12} for animal feedstuffs are animal by-products such as meat and bone meal, fish meal, and whey.

Only a few plant foods contain appreciable amounts of vitamin B_{12}. However, substantial amounts of the vitamin are found in some types of edible algae,[11] particularly green laver (*Enteromorpha* sp.), and purple laver (*Porphyra* sp., i.e., nori), which can contain 12–64 µg/100 g. The microalgae *Chlorella* sp. can vary in vitamin B_{12} content (0–200 + µg/100 g). The fruiting bodies of some mushrooms (*Craterellus cornucopioides*, *Cantharellus cibarius*, *Lentinula edodes*)[12] can contain >5 µg/100 g.[13] While soybeans contain little, if any, vitamin B_{12}, bacterially fermented soy products (e.g., tempeh, natto) can contain significant amounts (~0.75 µg/100 g). Tea leaves can contain vitamin B_{12} (0.1–1.2 µg/100 g). Trace amounts of the vitamin (e.g., 0.14 µg/100 g) have been found in spinach, broccoli, asparagus, and mung bean sprouts, apparently a result of uptake from vitamin B_{12}-containing organic fertilizers. Vitamin B_{12} has been found in cyanobacteria (*Spirulina*, *Aphanizomenon*, *Nostoc*) and a mushroom (*Hericium erinaceus*[14]); but these can also contain **pseudovitamin B_{12}** (7-adeninyl cyanocobamide),[15] which is biologically inactive and may antagonize the utilization of vitamin B_{12}.[16] Compounds with vitamin B_{12}-like activity have been found in bamboo cabbage, spinach, celery, lily bulb, bamboo shoots, and taro.

TABLE 18.1 Food Sources of Vitamin B_{12}

Foods	Vitamin B_{12} (µg/100 g)
Meats	
Beef	1.38–3.17
Beef brain	10.10
Beef kidney	24.9
Beef liver	83.13
Chicken	0.27–0.32
Chicken giblets	9.48
Ham	0.65–1.06
Pork	0.43–1.11
Turkey	0.36–1.65
Dairy Products	
Milk	0.38–0.5
Cheeses	0.29–2.28
Yogurt	0.75
Fish, Sea Food	
Herring	13.14
Salmon	3.26–4.48
Trout	6.3
Tuna	2.55
Clam	40.27
Oysters	16–19.13
Lobster	1.43
Shrimp	1.21–1.87
Vegetables, grains, fruits	None
Other	
Eggs, whole	0.89
Egg whites	0.09
Egg yolk	1.95
Tempeh	0.08

USDA National Nutrient Database for Standard Reference, Release 28 (http://www.ars.usda.gov/ba/bhnrc/ndl).

Breast Milk

The vitamin B_{12} concentration of human milk varies widely (330–320 pg/mL) and is particularly great in colostrum, which contains 10 times as much as mature milk. Most

10. Vitamin B_{12} was discovered as the antipernicious anemia factor in liver.
11. Studies have indicated that vitamin B_{12} is an essential metabolite for half of all algal species.
12. i.e., Black trumpet, golden chanterelle, and shiitake, respectively. Note: significant amounts of vitamin B_{12} were not found in porcini, parasol, oyster, or black morel mushrooms.
13. Watanabe, F., Yabuta, Y., Bito, T., et al., 2014, Nutrients 6, 1861–1873.
14. Lion's mane.
15. Pseudovitamin B_{12} differs from the vitamin by having an adenine moiety replacing the dimethylbenzimidazole.
16. Herbert, V., 1988. Am. J. Clin. Nutr. 48, 852–858.

of the vitamin (mainly methylcobalamin) is bound to R proteins.[17] Initial levels, 260–300pM, decline to by half after the first 12 weeks of lactation. Breast milk vitamin B_{12} levels reflect the level of intake of the vitamin. Milk from strict vegetarian women contains reduced levels as compared to milk from women consuming mixed diets; levels tend to be inversely correlated with the length of time on the vegetarian diet. Oral supplementation with vitamin B_{12} can significantly increase the vitamin B_{12} contents of breast milk and, hence, the vitamin B_{12} intake of nursing infants.[18]

Stability

Vitamin B_{12} is very stable in crystalline form and aqueous solution. High levels of ascorbic acid have been shown to catalyze the oxidation of vitamin B_{12} in the presence of iron to forms that are poorly utilized.

Bioavailability

Vitamin B_{12} is bound to two proteins (enzymes and carriers) in foods. Therefore, its utilization depends on the nature of the food/meal matrix as well as the host's ability to release the vitamin and bind it to proteins that facilitate its enteric absorption. In practice, the bioavailability of vitamin B_{12} in foods is difficult to determine. Bioassays in animal models fed vitamin B_{12}-deficient diets leave questions about applicability to humans, and studies in nondeficient humans require the use of the vitamin labeled with an intrinsic tracer. Further, the microbiological assay commonly used to measure vitamin B_{12} in foods (i.e., *Lactobacillus delbrueckii* growth) appears to yield overestimates by ~30%, due to responses to nonvitamin corrinoids. With those caveats, the bioavailability of vitamin B_{12} from most foods appears to be moderate. Studies have found that about half of the vitamin in most foods is absorbed by individuals with normal gastrointestinal function (Table 18.2). Bioavailability declines at intakes (1.5–2 µg/day) that saturate the active transport of the vitamin across the gut; greater amounts depend on absorption by passive diffusion, a process with only 1% efficiency. Accordingly, about 1% of the vitamin is absorbed from vitamin B_{12} supplements.

TABLE 18.2 Bioavailability of Vitamin B_{12} in Common Foods

Food	Bioavailable (%)
Eggs	4–9
Fish meat	42
Chicken meat	61–66
Lamb meat	56–89
Milk	55–65

Watanabe, F., 2007. Exp. Biol. Med. 232, 1266.

4. ABSORPTION OF VITAMIN B_{12}

Digestion

The naturally occurring vitamin B_{12} in foods is bound in coenzyme form to proteins. The vitamin is released from such complexes on heating, gastric acidification and/or proteolysis (especially by the action of pepsin). Thus, impaired gastric parietal cell function, as in **achlorhydria** or with chronic use of proton pump inhibitors, impairs vitamin B_{12} utilization.

Protein Binding

Free vitamin B_{12} is bound to proteins secreted by the gastric mucosa:

- **R proteins**[19] are glycoproteins that bind vitamin B_{12} to these glycoproteins adventitiously. They are found in plasma, saliva, gastric juice, intestinal contents, tears, cerebrospinal fluid, amniotic fluid, breast milk, leukocytes and erythrocytes in humans, and probably only a few other species. R proteins are members of a family of proteins called **haptocorrins**. While the salivary R protein is the first to bind vitamin B_{12} released from food, it is normally digested by pancreatic proteases in the small intestine to release the vitamin. Patients with pancreatic exocrine insufficiency can have high concentrations of R proteins that render the vitamin poorly absorbed.
- **Intrinsic factor (IF)**[20] is a glycoprotein secreted by gastric parietal cells in the fundus and body of the stomach[21] in response to histamine, gastrin, pentagastrin, and the presence of food. IF is a relatively small protein with a molecular weight of about 50kDa.[22] It is glycosylated (by fucose addition) posttranslationally. It binds the four cobalamins (methyl-, adenosyl-, cyano-, and aquocobalamin) with comparable, high affinities; but it does not bind cobamamides or cobinamides, which remain bound to R proteins and are not absorbed. The binding

17. This contrasts with cow's milk, which, containing no R proteins, typically shows lower concentrations of the vitamin, which is present mainly as adenosylcobalamin.

18. Duggan C., Srinivasan, K., Thomas, T., et al., 2014. J. Nutr. 144, 758–764.

19. These vitamin B_{12}-binding glycoproteins were named for their high electrophoretic mobilities: ***rapid***.

20. IF was identified in the gastric mucosa that was necessary for the utilization of an "extrinsic factor" later identified as vitamin B_{12}.

21. i.e., The same cells that produce gastric acid.

22. e.g., Human 44–63kDa, pig 50–59kDa, according to the carbohydrate moiety isolated with the preparation.

of vitamin B_{12} by IF produces a complex with a smaller molecular radius than that of IF alone; this protects the vitamin from hydrolytic attack by pepsin and chymotrypsin, as well as from side chain modification of the corrin ring by intestinal bacteria.

Mechanisms of Absorption

Carrier-mediated active transport of vitamin B_{12} is efficient (>50%) and quantitatively important at low doses (1–2 μg). Such doses appear in the blood within 3–4 h of consumption. The active transport of vitamin B_{12} depends on the IF–vitamin B_{12} complex binding to a specific brush border receptor in the terminal portion of the ileum, a site it reaches after traveling the length of the small intestine. That receptor[23] consists of two components: the multiligand apical membrane protein cubilin,[24] which binds the IF–vitamin B_{12} complex; and a chaperone, amnionless (AMN), which contributes structure necessary for membrane anchorage in clathrin-coated pits,[25] trafficking to the plasma membrane, and signaling of endocytosis and receptor recycling.[26] After moving into the cell, the IF–vitamin B_{12} complex is thought to be degraded within lysosomes in which free vitamin B_{12} is released.

Deficiency of IF causes pernicious anemia. Patients have a severely limited ability to absorb vitamin B_{12}, excreting 80–100% of oral doses in the feces (vs. 30–60% fecal excretion rates in individuals with adequate IF). Individuals with loss of gastric parietal cell function may be unable to utilize dietary vitamin B_{12}, as these cells produce both IF and acid, both of which are required for the enteric absorption of the vitamin.[27] For this reason, geriatric patients, many of whom are hypoacidic, may be at risk of low vitamin B_{12} status. Mutations in the IF gene can result in failure of its expression or in expression of a defective protein incapable of binding vitamin B_{12}. Affected individuals show macrocytic anemia[28] within the first 3 years of life, which responds to large doses of vitamin B_{12} administered orally or by intramuscular injection. IF secretion can be affected by mutations in the gene encoding fucosyltransferase (FUT2) that catalyzes its posttranslational fucosylation.[29]

Passive diffusion of vitamin B_{12} occurs with low efficiency (~1%) throughout the small intestine and becomes significant only at higher doses. Such doses appear in the blood within minutes of consumption. This passive mechanism is utilized in therapy for pernicious anemia, in which patients are given high doses (>500 μg/day) of vitamin B_{12} per os. For such therapy, the vitamin must be given an hour before or after a meal to avoid competitive binding of the vitamin in food.

Enterohepatic Circulation of Vitamin B_{12}

A significant amount of vitamin B_{12} is released in the bile. In humans, this can be 0.5–5 μg each day,[30] depending on vitamin B_{12} status. Much of this is reabsorbed by the above mechanisms. This capacity to recycle the vitamin reduces dietary need.

5. TRANSPORT OF VITAMIN B_{12}

Transport Proteins

On absorption from the intestine, vitamin B_{12} is initially transported in the plasma as adenosylcobalamin and methylcobalamin bound to two proteins:

- **Plasma haptocorrin,**[31] a 60-kDa R protein, binds most (70–80%) of the vitamin B_{12} in plasma. Plasma haptocorrin is typically 80–90% saturated with its ligand, which turns over slowly (half-life, 9–10 days), becoming available for cellular uptake only over fairly long time frames. A minor variant of this protein, differing only in carbohydrate content, can also be found in plasma. Haptocorrin binds methylcobalamin preferentially, which, therefore, is the predominant circulating form of the vitamin. As most other species lack R proteins, their dominant circulating form is adenosylcobalamin.

Congenital defects in plasma haptocorrin are asymptomatic, suggesting that this form of the vitamin is not physiologically important. Affected individuals show normal absorption and distribution of vitamin B_{12} to their tissues;

23. Genetic defects in these proteins occur in Imerslund–Gräsbeck's syndrome, characterized by vitamin B_{12} malabsorption leading to megaloblastic anemia.
24. Cubilin is a large (460 kDa) membrane protein with no apparent transmembrane segment. It is expressed at high levels in the kidney where it appears to function in the reabsorption of several specific nutrient carriers including albumin, vitamin D-binding protein, transferrin, and apolipoprotein A.
25. Produced by lattices of three clathrin heavy chains and three clathrin light chains, these membrane vesicles facilitate transport of the IF–B_{12} complex across the plasma membrane into the epithelial cell.
26. Fyfe, J.C., Madsen, M., Højrup, P., et al., 2004. Blood 103, 1573–1579.
27. Individuals lacking IF are unable to absorb vitamin B_{12} by active transport. They can be given the vitamin by intramuscular injection (1 μg/day) or in high oral doses (25–2000 μg) to prevent deficiency. Randomized trials have shown that an oral dose regimen of 1000 μg daily for a week, followed by the same dose weekly and, then, monthly can be as effective as intramuscular administration of the vitamin for controlling short-term hematological and neurological responses in deficient patients Butler, C.C., Vidal-Alabell, J., Cannings-John, R., et al., 2006. Fam. Pract. 23, 279–285.
28. Anemia characterized by relatively low cell count with the presence of enlarged erythrocytes produced due to impaired cell division during hematopoiesis.
29. Chery, C., Hehn, A., Mrabet, N., et al., 2013. Biochimie 95, 995–1001.
30. El Kholty, S., Gueant, J.L., Bressler, L., et al., 1991. Gastroenterol. 101, 1399–1408.
31. Formerly, transcobalamin I.

however, they show low circulating levels of the vitamin and can be wrongly diagnosed as vitamin B_{12} deficient if other parameters [MMA, Hcy, FIGLU (formiminoglutatmic acid)] are not considered. The prevalence of plasma haptocorrin defects may be relatively high; one study noted that 15% of apparently healthy subjects had low plasma vitamin B_{12} levels.

Transcobalamin (TC)[32] binds most of the nonhaptocorrin-bound vitamin B_{12} in plasma, i.e., 10–20% of the total. TC is a 38–43 kDa β-globulin protein synthesized in the liver, intestinal mucosa, seminal vesicles, fibroblasts, bone marrow, and macrophages. It is filtered by the kidney and reabsorbed by the proximal tubules. It binds the vitamin stoichiometrically; within 3–4 h of ingestion of the vitamin, TC reaches is typically level of 10–20% saturation with the ligand. Movement of vitamin B_{12} from the intestinal mucosal cell into the plasma depends on the formation of the TC–vitamin B_{12} complex, i.e., **holotranscobalamin (holoTC)**, which turns over rapidly in the enterocyte (half-life c. 6 min). In the plasma, holoTC also turns over fairly rapidly (half-life, 60–90 min), rendering it the primary functional source of vitamin B_{12} for cellular uptake. Within hours, much of the vitamin originally associated with TC becomes bound to plasma haptocorrin[33] and, in humans, to other plasma R proteins.[34] Therefore, holoTC level can be a useful parameter of early-stage vitamin B_{12} deficiency.

Predominant transport forms of vitamin B_{12} differ among species, varying widely in concentration from only hundreds (humans) to thousands (rabbits) of pM. The major circulating vitamer in human plasma is methylcobalamin (60–80% of the total),[35] owing to the fact that haptocorrin and R proteins preferentially bind that vitamer (Table 18.3). However, the major circulating vitamer in other species is adenosylcobalamin, which is bound by TC with comparable affinity to methylcobalamin.

TABLE 18.3 Cobalamins in Normal Human Plasma

	Range (pM)
Total cobalamins	173–545
Methylcobalamin	135–427
Adenosylcobalamin	2–77
Cyanocobalamin	2–48
Aquocobalamin	5–67

Holotranscobalamin Receptor

Membrane-bound receptor proteins for holoTC occur in all cells. The **TC receptor**[36] is a 50-kDa glycoprotein in the low-density lipoprotein receptor family. It has a single holoTC binding site. It is thought that TC receptors mediate the endocytic uptake of holoTC (Fig. 18.1). A soluble form has been identified in human serum.

Intracellular Protein Binding

Upon cellular internalization, holoTC is degraded proteolytically in lysosomes and vitamin B_{12} is released for conversion to methylcobalamin in the cytosol. Virtually, all of the vitamin in the cell is ultimately bound to two vitamin B_{12}-dependent enzymes:

- **methionine synthetase** (also called **methyl-FH_4 methyltransferase**) in the cytosol
- **methylmalonyl CoA mutase** in mitochondria.

Congenital Disorders of Vitamin B_{12} Absorption and Transport

Congenital deficiencies in proteins involved in the absorption and transport of vitamin B_{12} have been described (Table 18.4). These result in tissue-level vitamin B_{12} deficiencies the effects of which are manifest within weeks to years after birth. Most can be managed with high, frequent doses of the vitamin administered intramuscularly or orally. IF gene mutations can result in either IF not being expressed, or in the expression of an IF protein that is functionally inactive or unstable. Affected individuals develop megaloblastic anemia within 1–3 years, i.e., when their maternal stores of vitamin B_{12} are exhausted. Dysfunction of the ileal IF receptor caused by defects in either cubulin or AMT occur in Imerslund–Gräsbeck syndrome,[37,38] a common cause of vitamin B_{12}-associated megaloblastic anemia. A single-nucleotide polymorphism in TC (776C→G) has also been identified. The G allele is most prevalent in Asians (56%) compared to whites (45%) and blacks (36%).[39] Individuals with GG genotype develop severe megaloblastic anemia within the first 5 years of life. They have low circulating levels of both apo- and holoTC, but because most circulating vitamin B_{12} is bound to haptocorrin, their plasma levels of the vitamin are typically normal such that this deficiency can easily

32. Formerly, transcobalamin II.
33. Only by this means does haptocorrin obtain vitamin B_{12}.
34. Due to their affinity for R proteins, the TCs are grouped in a heterogeneous class of proteins called *R binders*.
35. In pernicious anemia patients, methylcobalamin is lost in favor to others forms of the vitamins.
36. Also called CD320.
37. Also called autosomal recessive megaloblastic anemia.
38. Mutant cubilin has been found in Finnish patients; whereas mutant AMN has been found in Norwegian patients.
39. Bowen, R.A., Wong, B.Y., Cole, D.E., 2004. Clin. Biochem. 37, 128–133.

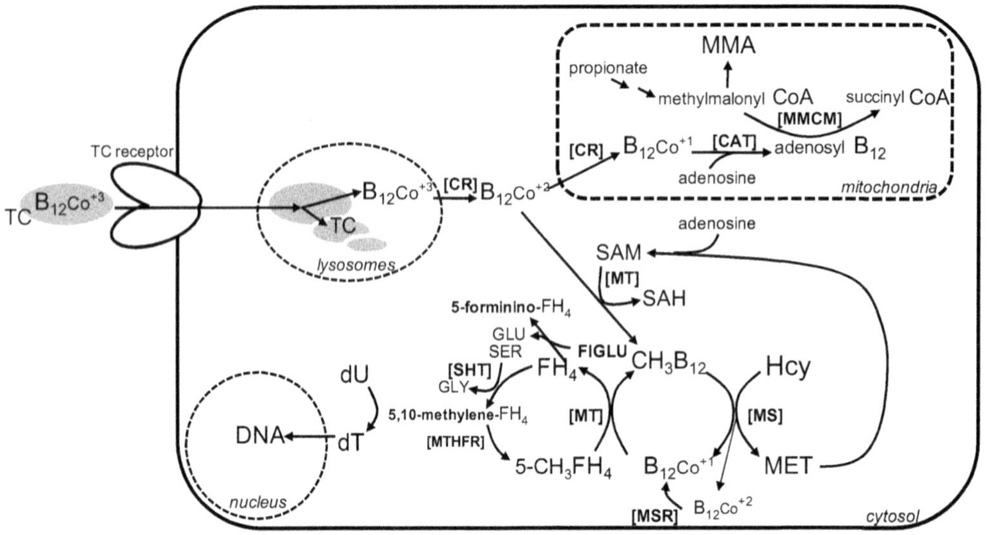

FIGURE 18.1 Uptake and metabolism of vitamin B_{12}, and its relationship with folate in single-carbon metabolism. *TC*, transcobalamin; *MMA*, methylmalonic acid; *SAM*, *S*-adenosylmethionine; *SAH*, *S*-adenosylhomocysteine; *Hcy*, homocysteine; *MET*, methionine; *FH4*, tetrahydrofolic acid; *CH3B12*, methylcobalamin; *5-CH3-FH4*, methyltetrahydrofolic acid; *FIGLU*, formiminoglutamic acid; *GLU*, glutamic acid; *SER*, serine; *dU*, deoxyuridylate; *dT*, deoxythimidylate; *CR*, cobalamin reductases; *CAT*, cobalamin adenosyl transferase; *MMCM*, methylmalonyl CoA mutase; *MT*, methyltransferases; *MS*, methionine synthase; *MSR*, methionine synthase reductase; *SHT*, serine hydroxymethyltransferase; *MTHFR*, methylenetetrahydrofolate reductase.

TABLE 18.4 Congenital Disorders of Vitamin B_{12} Absorption and Transport

Condition	Missing/Deficient Factor	Signs/Symptoms
Lack of intrinsic factor	IF	Megaloblastic anemia presenting at 1–3 years
Imerslund–Gräsbeck syndrome	IF receptor	Specific malabsorption of vitamin B_{12} presenting by 5 years
Lack of transcobalamin	TC	Severe (fatal) megaloblastic anemia presenting early in life
Lack of haptocorrin	Haptocorrin	None

be missed.[40] Patients respond to vitamin B_{12} administered in large doses by intramuscular injection, e.g., 1 mg three times per week. Congenital deficiencies in haptocorrin occur in some 15% of individuals. It is characterized by low circulating levels of vitamin B_{12}; however, that condition is without consequence as levels of the physiological transporter, TC-B_{12}, are unaffected. Still, it can lead to an erroneous impression of vitamin B_{12} deficiency.

Distribution in Tissues

Vitamin B_{12} is the best stored of the vitamins. Under conditions of adequate intake, the vitamin accumulates to appreciable amounts in the body, mainly in the liver (~60% of the total body stores) and muscles (~30% of the total). Body stores vary with the intake of the vitamin but tend to be greater in older subjects. Hepatic concentrations approaching 2 µg/g have been reported in humans; however, a total hepatic reserve of about 1.5 mg is typical.[41] Mean total body stores of vitamin B_{12} in humans are in the range of 2–5 mg. The greatest concentrations of vitamin B_{12} occur in the pituitary gland; kidneys, heart, spleen, and brain also contain substantial amounts; in humans, these organs each contain 20–30 µg of vitamin B_{12}. The great storage and long biological half-life (350–400 days in humans) of the vitamin provide substantial protection against periods of deprivation. The low reserve of the human infant (~25 µg) is sufficient to meet physiological needs for about a year.

40. The 776C>G polymorphism has also been linked to risks of spontaneous abortion and fetal developmental defects (Martinelli, M., Scapoli, L., Palmieri, A., et al., 2006. Hum. Mutat. 27, 294–301), as well as to the onset of Alzheimer's disease McCaddon, A., Blennow, K., Hudson, P., et al., 2004. Dement. Geriatr. Cogn. Disord. 17, 215–221.

41. i.e., c. 1 µg/g.

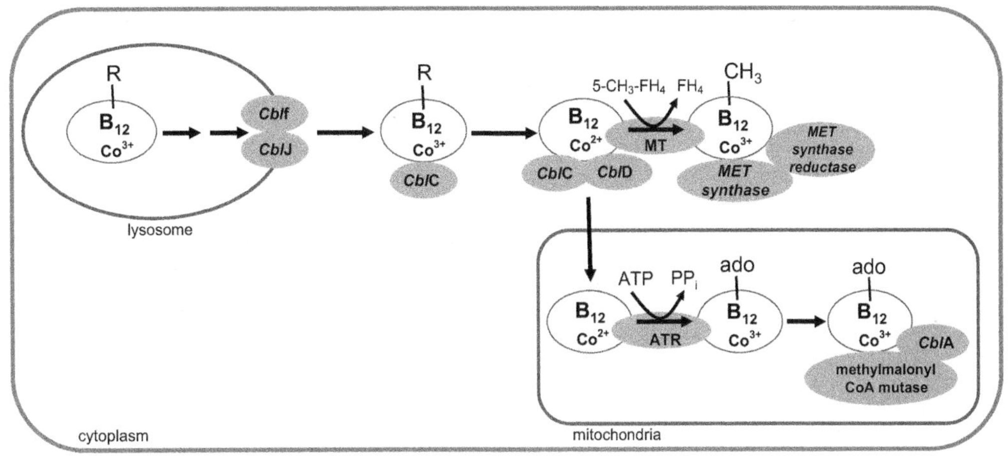

FIGURE 18.2 Intracellular trafficking of vitamin B_{12} by protein chaperones. Exit of the vitamin from the lysosome requires two membrane proteins, *Cbl*F and *Cbl*J. Upon entry into the cytoplasm, the β-axial liganded vitamin is thought to be bound to *Cbl*C [also called cobalamin reductase and MMACHC (for methylmalonic acid type C and homocystinuria)], which forms a complex with *Cbl*D [also called MMADHC (for methylmalonic acid type D and homocystinuria)]. *Cbl*D does not bind the cobalamin but is thought to assist its delivery by *Cbl*C to 5-methyl-FH_4:homocysteine methyltransferase (MT), which produces methylcobalamin that is bound by methionine synthase, which complexes independently with methionine synthase reductase. The mechanism is not clear whereby cobalamin enters the mitochondrion where it is adenosylated by the ATP-dependent cob(I)alamin adenosyltansferase (ATR, also CblB) and then transferred to methylmalonyl CoA mutase. Escape of 5′-adenosine from the mutase active site during catalysis is prevented by a G-protein chaperone *Cbl*A (also called MMAA, for methymalonic aciduria type A) using the binding energy of GTP. *After Gerashim, C., Hannibal, L., Rajabopalan, D., et al., 2013. Biochimie 95, 1023–1032; Gerashim, C., Lofgren, M., Banerjee, R., 2013. J. Biol. Chem. 288, 13186–13193.*

6. METABOLISM OF VITAMIN B_{12}

Intracellular Trafficking

Vitamin B_{12} is delivered to cells in the oxidized from, hydroxycob(III)alamin where it is reduced by thiol- and reduced flavin-dependent reduction of the cobalt center of the vitamin (to Co^{+}) to form cob(I)amin.[42] However, the vitamin is active in metabolism *only* as methyl or 5-deoxyadenosyl derivatives that have either respective group attached covalently to the cobalt atom. Therefore, vitamin B_{12} released from holoTC in lysosomes must enter the cytoplasm to be incorporated as methylcobalamin into methionine synthase, and traverse the cytoplasm to be incorporated as adenosylcobalamin into methylmalonyl CoA mutase. Several protein chaperones are essential to this trafficking (Fig. 18.2).[43]

Activation to Coenzyme Forms

The conversion to these coenzyme forms involves two different enzymatic steps:

- **Methylcobalamin**—Methylation of the vitamin is catalyzed by the cytosolic enzyme **5-methyl-FH_4:homocysteine methyltransferase**. This renders the vitamin, as methylcobalamin, a carrier for the single-C unit used in the regeneration of **methionine (MET)** from **homocysteine (Hcy)**. Methylcobalamin is also produced by recharging the reduced vitamin (Co^{+1}) with a methyl group transferred from 5-methyl FH_4. This cycling risks the occasional oxidation of cobalamin–cobalt (to Co^{+2}), in which case it is reduced back to Co^{+1} by the enzyme **methionine synthase reductase**.
- **Adenosylcobalamin**—Adenosylation of the vitamin occurs in the mitochondrial due to the action of **vitamin B_{12} coenzyme synthetase**, which catalyzes the reaction of cob(II)amin with a deoxyadenosyl moiety derived from ATP. This step depends on the entry of hydroxycobalamin into the mitochondria and its subsequent reduction in sequential, one electron steps involving NADH- and NADPH-linked aquacobalamin reductases[44] to yield cob(II)alamin.

Catabolism

Little, if any, metabolism of the corrinoid ring system is apparent in animals, and vitamin B_{12} is excreted as the intact cobalamin. Apparently, only the free cobalamins (not the methylated or adenosylated forms) in the plasma are available for excretion.

Excretion

Vitamin B_{12} is excreted via both renal and biliary routes at the daily rate of about 0.1–0.2% of total body reserves (in humans this is 2–5 μg/day, thus constituting the daily

42. Also called vitamin B_{12s}.

43. Gerashim, C., Hannibal, L., Rajagopalan, D., et al., 2013. Biochimie 95, 1023–1032.

44. These activities are derived from a cytochrome b_5/cytochrome b_5 reductase complex, and from a cytochrome *P*-450 reductase complex and an associated flavoprotein.

TABLE 18.5 Categories of Congenital Disorders of Vitamin B_{12} Metabolism

Defect	Missing/Deficient Factor (Gene)	Signs/Symptoms
Mitochondrial		
B_{12}-Co^{+3} reduction to B_{12}-Co^{+2} Adenosyl-B_{12} production Production of ado-/methyl B_{12}	Mitochondrial cobalamin reductase Adenosyl transferase Cobalamin reductase	Methylmalonic aciduria Methylmalonic aciduria Homocysteinuria, Methylmalonic aciduria
B_{12} entry into mitochondria Isomerization of methylmalonyl CoA	B_{12} chaperone Methylmalonyl CoA mutase	Homocysteinuria, methylmalonic aciduria methylmalanic aciduria
Cytosolic		
Methionine synthase	Methyl transferase activity	Homocysteinemia, hypomethioninemia, megaloblastic anemia, developmental delay
B_{12}-Co^{+3} reduction to B_{12}-Co^{+2}	Methionine synthase reductase	Homocysteinemia, hypomethioninemia, megaloblastic anemia, developmental delay
Lysosomal		
Lysosome to cytosol B_{12} export	Lysosomal membrane protein	Developmental delay, homocysteinuria, methylmalonic aciduria

requirement for the vitamin).[45] Although it is found in the urine, glomerular filtration of the vitamin is minimal (<0.25 μg/day in humans), and it is thought that urinary cobalamin is derived from the tubular epithelial cells and lymph. Urinary excretion of the vitamin after a small oral dose can be used to assess vitamin B_{12} status; this is called the **Schilling test**. The biliary excretion of the vitamin is substantial, accounting in humans for the secretion into the intestine of 0.5–5 μg g/day. Most (65–75%) of this amount is reabsorbed in the ileum by IF-mediated active transport. This enterohepatic circulation constitutes a highly efficient means of conservation, with biliary vitamin B_{12} contributing only a small amount to the feces.

Congenital Disorders of Vitamin B_{12} Metabolism

Several congenital deficiencies in proteins involved in vitamin B_{12} metabolism, each an autosomal recessive trait, have been reported in humans (Table 18.5). Most present in early childhood as homocysteinemia and/or methylmalonic aciduria, frequently with neurological and psychiatric symptoms. Most can be managed with high, frequent doses of the vitamin administered intramuscularly or orally. Hereditary defects in the vitamin's intracellular protein chaperones can produce homocystinuria (*Cbl*C, *Cbl*D, *Cbl*E, *Cbl*G), methylmalonic aciduria (*Cbl*A, *Cbl*B, *Cbl*C) or combined homocystinuria, and methylmalonic aciduria (*Cbl*F, *Cbl*J, *Cbl*C) (see Fig. 18.2). Of these the most common defect is in cobalamin reductase activity (*Cbl*C); more than 40 mutations have been identified in that gene.[46]

45. Doets, E.L., in't Veld, P.H., Szczerińsk, A., et al., 2012. Ann. Nutr. Metab. 62, 311–322.

7. METABOLIC FUNCTIONS OF VITAMIN B_{12}

Coenzyme Functions

Vitamin B_{12} functions in metabolism in two coenzyme forms: adenosylcobalamin and methylcobalamin. While several vitamin B_{12}-dependent metabolic reactions have been identified in microorganisms,[47] only these two have been discovered in animals. These play key roles in the metabolism of propionate, amino acids, and single carbon.

- **Adenosylcobalamin** is the coenzyme of **methylmalonyl CoA mutase**, which catalyzes the conversion of methylmalonyl CoA to succinyl CoA in the degradation of propionate formed from odd-chain fatty acids, which are particularly important as sources of energy source for ruminants. This reaction involves splitting a carbon–carbon bond of the coenzyme with the formation of a free radical on the coenzyme that can be transferred through an amino acid residue to the substrate. That the

46. More than 400 patients have been described. Early onset (<1 year) patients present with severe neurological, hematological, renal, gastrointestinal, cardiac, and pulmonary symptoms. Late-onset patients present with slowly progressive neurological symptoms.

47. The following microbial enzymes require adenosylcobalamin: glutamate mutase, 2-methylene-glutarate mutase, L-β-lysine mutase, D-α-lysine mutase, D-α-ornithine mutase, leucine mutase, 1,2-dioldehydratase, glyceroldehydratase, ethanolamine deaminase, and ribonucleotide reductase; methylcobalamin is also required for the bacterial formation of methane and acetate.

propionic acid pathway is important in nerve tissue is suggested by the delayed onset of the neurological signs of vitamin B_{12} deficiency effected in animals by dietary supplements of direct (valine, isoleucine) or indirect (methionine) precursors of propionate.[48]

Methylmalonyl CoA mutase is a mitochondrial matrix enzyme, which forms a dimer that binds two adenosylcobalamin molecules. In humans, it is the first vitamin B_{12}-dependent enzyme to be affected by deprivation of vitamin B_{12}. With loss of this activity, vitamin B_{12}-deficient subjects show **methylmalonic aciduria**, especially after being fed odd-chain fatty acids. The accumulation of **MMA** can disrupt normal glucose and glutamic acid metabolism, apparently by inhibiting the tricarboxylic acid (TCA) cycle. Vitamin B_{12} deficiency can cause a reversal of propionyl CoA carboxylase activity, leading to the incorporation of the 3-C propionyl CoA in place of the 2-C acetyl CoA, and resulting in the production of small amounts of odd-chain fatty acids. Increased levels of methylmalonyl CoA can also lead to its incorporation in place of malonyl CoA, resulting in the synthesis of small amounts of methyl-branched chain fatty acids. It has been suggested that the neurological signs of vitamin B_{12} deficiency may result, at least in part, from the production of these abnormal fatty acids in neural tissues.

Several inborn metabolic errors result in decreased in methylmalonyl CoA mutase activities leading to methylmalonic aciduria. These include mutations in the gene that encodes the enzyme can block its expression or result in expression of a defective protein, and other mutations that reduce the synthesis of its cofactor adenosylcobalamin. Individuals with these defects respond to vitamin B_{12} treatment.

- **Methylcobalamin** is the coenzyme for **methionine synthase**, which catalyzes the methylation of Hcy to regenerate methionine (MET). In this reaction, methylcobalamin serves as the methyl group carrier between the donor 5-methyltetrahydrofolate (5′-methyl-FH_4) and the acceptor Hcy. This reaction is a simple transfer of the single-C moiety. Because of diminished methionine synthase activity, vitamin B_{12}-deficient subjects show reduced availability of MET, which is essential for the synthesis of proteins and polyamines, and is the precursor of **S-adenosylmethionine (SAM)**. SAM is the primary donor of "labile" methyl groups for more than 100 enzymatic reactions with critical roles in metabolism.[49] SAM also serves as a key regulator of the transsulfuration and remethylation pathways, which involve the folate-dependent methylenetetrahydrofolate reductase. Losses of SAM lead to impairments in the synthesis of creatine, phospholipids, and the neurotransmitter acetylcholine, all of which have broad impacts on physiological function. Low vitamin B_{12} status, thus, results in the accumulation of both Hcy and 5′-methyl-FH_4 (via the methyl folate trap), the latter resulting in the loss of FH_4, the key functional form of folate. Methionine synthase can also catalyze the reduction of nitrous oxide to elemental nitrogen; in doing so, it generates a free radical that inactivates the enzyme. Methionine synthase expression is induced by vitamin B_{12} by the vitamin binding to a *trans*activating protein, inducing a conformational change that allows it a bind to an internal site on the methionine synthase mRNA, thus enhancing ribosomal recruitment and promoting translation.[50]

Genetic defects in methionine synthase and the production of its cofactor methylcobalamin result in homocysteinemia and, commonly, megaloblastic anemia. Individuals with these defects do not respond to vitamin B_{12} treatment, but their anemia can respond to folate supplementation. A 2756A>G polymorphism in methionine synthase has been described; women with the AG genotype have been found to experience double the risk of having a child with NTDs and a 3.5-fold increase in the risk of having a child with Down syndrome.[51] A 66A>G polymorphism of methionine synthase reductase, also involved in this functioning pathway, has been associated with similar effects.

Interrelationships With Folate

The major cycle of single-C flux in mammalian tissues is the serine hydroxymethyltransferase/5,10-methylene-FH_4 reductase/methionine synthase cycle. In this cycle, the committed step (5,10-methylene-FH_4 reductase) is feedback inhibited by SAM and product inhibited by 5-methyl-FH_4; but methionine synthase is rate limiting (Fig. 18.1). It depends on the transfer of labile methyl groups from 5-methyl-FH_4 to vitamin B_{12}. Methyl-B_{12} serves as the immediate methyl donor for converting Hcy to MET. Without adequate vitamin B_{12} to accept methyl groups from 5-methyl-FH_4, that metabolite accumulates at the expense of the other metabolically active folate pools. This is known as the "**methyl folate trap**" a blockade resulting in the accumulation of the intermediate FIGLU.[52] These interrelated

48. In bacteria, levels of adenosylcobalamin are controlled by regulating the genes responsible for its synthesis and import; this is effected by an adenosylcobalamin riboswitch, i.e., a regulatory segment of mRNA Johnson, J.E., Reyes, F.E., Polaski, J.T., et al., 2012. Nature 492, 133–137.
49. By loss of the flux of methyl groups via 5-methyl-FH_4:Hcy methyltransferase, folate deficiency causes a secondary hepatic choline deficiency.
50. Oltean, S. and Banerjee, R., 2005. J. Biol. Chem. 280, 32,662–32,668.
51. Doolin, M.T., Barbaux, S., McDonnel, M., et al., 2002. Am. J. Hum. Genet. 71, 1222–1226; Bosco, P., Guéant-Rodriguez, R.M., Anello, G., et al., 2003. Am. J. Med. Genet. 121A, 219–224.
52. Thus, elevated urinary FIGLU level after an oral histidine load is diagnostic of vitamin B_{12} deficiency.

metabolic pathways are affected by vitamin B_{12} deprivation in two ways:

- **Reduced MET regeneration** by the loss of methionine synthase activity, which results in a secondary folate deficiency due to the accumulation of 5-methyl-FH_4 by the "methyl folate trap" (the basis by which deficiency of either folate or vitamin B_{12} can cause macrocytic anemia) and the accumulation of Hcy manifest as homocysteinemia.
- **Reduced DNA methylation** by the reduced availability of single-C units. Hypomethylation of DNA cytosine bases and histone proteins alters chromatin structure in ways that affect transcription and can increase C→T transition mutation.[53] Thus, suboptimal status with respect to vitamin B_{12} (or folate) can affect gene expression and stability. Accordingly, chromosomal aberrations have been reported for some patients with pernicious anemia.[54] A cross-sectional study showed that the vitamin B_{12} levels of buccal cells were significantly lower in smokers and nonsmokers and that elevated levels of the vitamin were associated with reduced frequency of micronucleus formation.[55]

Physiological Functions

By participating in the regeneration of MET, vitamin B_{12} functions in the regulation of Hcy and, thus, the prevention of homocysteinemia, which can cause various adverse metabolic effects (discussed more extensively in Chapter 17). A prospective, community-based study found plasma Hcy to be weakly associated with plasma vitamin B_{12} concentration.[56] Circulating Hcy levels >13 μM have been associated with dysfunction that has been related specifically to vitamin B_{12} status:

- **Hematological development**. Vitamin B_{12} supports in hematopoietic cell division in the bone marrow by providing single-C units via the methionine synthase for the synthesis of thymidylate, which is required for normal DNA synthesis.
- **Neurological function**. Vitamin B_{12} has essential neurological functions including the synthesis of functional myelin sheaths and the synthesis of choline, the precursor of the neurotransmitter acetylcholine. These functions support both peripheral and cerebral–spinal aspects of neurological function, including cognition, sensation, and muscular coordination.
- **Fetal development**. It is likely that vitamin B_{12} has a role in supporting normal early fetal development. Studies have shown lower vitamin B_{12} levels in amnionic fluid from NTD pregnancies compared to healthy ones, even though the vitamin B_{12} contents of mothers' serum in both cases were in the normal range.[57] This suggests a limitation in the maternal capacity to provide the fetus with an adequate supply of the vitamin. Because women with NTD pregnancies are more likely to have the methionine synthase 66AG genotype, which presumably produces an aberrant enzyme, it is possible that compromised vitamin B_{12} function may be involved in the residual incidence of NTDs not prevented by folate supplementation.
- **Bone health**. An analysis of the NHANES 1999–2004 data found homocysteinemia to be associated with a two-fold increase is risk of lumbar spine osteoporosis.[58] This was found to be more prevalent among elderly Dutch women of marginal or deficient vitamin B_{12} status than those adequate with respect to the vitamin (Table 18.6). Studies have shown positive associations of serum vitamin B_{12} level and bone mineral density, markers of bone turnover, and risks of osteoporosis and hip fracture.[59] Most randomized trials using supplements of vitamin B_{12} and other B vitamins (folate, vitamin B_6) have found no effects on biomarkers of bone turnover,[60] although one found that prevention of homocysteinemia with a combined supplement of vitamin B_{12} and folate significantly reduced hip fracture risk.[61]

TABLE 18.6 Relationship of Vitamin B_{12} Status and Osteoporosis Risk Among Elderly Women

Plasma Vitamin B_{12} (pM)	n	Relative Risk
>320	34	1.0
210–320	43	4.8 (1.0–23.9)[a]
<210	35	9.5 (1.9–46.1)

[a] *95% confidence interval.*
Dhonukshe-Rutten, R.A.M., Lips, M., de Jong, N., et al., 2003. J. Nutr. 133, 801–807.

53. Methylated CpG sites appear to be at particularly high risk for C→T changes, the most common type of mutational change which are common in the p53 tumor suppressor gene.
54. Jensen, M.K., 1977. Mutat. Res. 45, 249–252.
55. Piyathilke, C.J., Macaluso, M., Hine, R.J., et al., 1995. Cancer Epid. Biomakers Prev. 4, 751–758.
56. Selhub, J., Jacques, P.F., Bostom, A.G., et al., 1996. J. Nutr. 126, 1258–1265S.
57. Ray, J.G. and Blom, H.J., 2003. Quart. J. Med. 96, 289–295.
58. Bailey, R.L., Looker, A.C., Lu, Z., et al., 2015. Am. J. Clin. Nutr. 102, 687–694.
59. Dhonukshe-Rutten, R.A., van Dusseldorp, M., Schneede, J., et al., 2005. Eur. J. Nutr. 44, 341–347; Tucker, K.L., Hannan, M.T., Qiao, N., et al., 2005. J. Bone Min. Res. 20, 152–158; Dhonukshe-Rutten, R.A., Pluijim, S.M., de Groot, L.C., et al., 2005. J. Bone Min. Res. 20, 921–927.
60. e.g., Green, T.J., McMahon, J.S., Skeaff, C.M., et al., 2007. Am. J. Clin. Nutr. 85, 460–464; van Wijgaarden, J.P., Swart, K.M.A., Enneman, A.W., et al., 2014. Am. J. Clin. Nutr. 100, 1578–1586.
61. Sato, Y., Honda, Y., Iwamoto, J., 2005. JAMA 293, 1082.

- **Hearing**. Low vitamin B_{12} status has been found in patients with tinnitus.[62,63] Such a relationship might indicate effects of homocysteinemia on the vascular or bone systems of the ear.

8. BIOMARKERS OF VITAMIN B_{12} STATUS

Vitamin B_{12} status can be assessed by analyses of blood:

- **Serum vitamin B_{12} concentration** is the most widely used tool to assess vitamin B_{12} status.[64] Normal values are in the 150–665 pM range; values <194 pM indicate deficiency as defined by significant risk of elevated serum MMA.[65] This biomarker is limited by the fact that it measures two different pools of the vitamin that turn over at different rates. Most (70–80%) serum vitamin B_{12} is bound to haptocorrin with a half-life of up to 10 days; whereas, the physiologically significant pool, i.e., vitamin B_{12} as holoTC, is smaller (10–20% of the serum total) with a much faster turnover (60–90 min). Therefore, short-term deprivation of vitamin B_{12} can reduce holoTC without affecting the haptocorrin-bound pool and, thus, not having a detectable effect on total vitamin B_{12} concentration. Therefore, the effects of short-term deprivation of the vitamin can easily be missed. Serum vitamin B_{12} level can be depressed in subjects expressing variants of the fucosyltransferase that influences the gastric secretion of IF,[66] and taking the antidiabetic drug metformin.[67] Markedly elevated serum vitamin B_{12} levels can occur in subjects with antibodies to holoTC given the vitamin intramuscularly and in patients with overproduction of haptocorrin due to myeloproliferative disease. Elevated serum vitamin B_{12} is among the diagnostic criteria for polycythemia vera[68] and hypereosinophilic syndrome.[69]
- **Serum holoTC concentration**[70] is a sensitive indicator of vitamin B_{12} recent absorption and status. It has been found to decline within a week after damage to the posterior ileum even though total serum concentrations of the vitamin were unaffected. Values <30 pM are considered indicative of deficiency; such levels have been found to be prevalent in many elderly subjects as well as in AIDS patients. HoloTC level does not, however, reflect declining vitamin B_{12} stores in subjects with adequate recent intakes of the vitamin.
- **Plasma/serum or urinary MMA**. Circulating and urinary MMA levels increase in vitamin B_{12} deficiency due to the diminished activity of MMA mutase. This can be tested after a meal of odd-chain fatty acids or a propionate load. Interpretation of MMA results necessitates evaluating renal function and diabetes status. Renal insufficiency can lead to elevated plasma MMA, which also occurs in patients with type 2 diabetes despite their normal serum vitamin B_{12} levels.[71]

Other biomarkers have been used to assess various aspects of vitamin B_{12} status or function but are not specific for vitamin B_{12}:

- **Plasma/serum Hcy concentration**. This is not a specific indicator of vitamin B_{12} status. While vitamin B_{12} deficiency can elevate plasma Hcy that outcome can also be produced by deficiencies of folate and/or MET, renal insufficiency, and some drugs. These factors must be considered in interpreting plasma Hcy level. Homocysteinemia (plasma Hcy >13 μM) is most likely to indicate suboptimal vitamin B_{12} status in populations exposed to folate-fortified foods.
- **Urinary FIGLU**. Deficiency of vitamin B_{12} increases urinary FIGLU excretion; but this is not a specific indicator of vitamin B_{12} status, as it is also caused by deprivation of folate.
- **Plasma/serum MET**. Serum MET levels are highly correlated with those of vitamin B_{12} in vitamin B_{12}-deficient subjects. About half of subjects with either vitamin B_{12} or folate deficiencies, and more than half of those with the combined deficiencies, show subnormal plasma MET concentrations.[72]

Distinguishing Deficiencies of Vitamin B_{12} and Folate

Macrocytic anemia and other clinical signs can result from deficiencies of either vitamin B_{12} or folate. This is based on their common participation in the regulation of the FH_4 pool. Deprivation of either vitamin will reduce that folate directly; vitamin B_{12} indirectly, via the methyl folate trap. In either case, the yield of 5,10-methylene-FH_4 is reduced.

62. i.e., The perception of sound within the ear in the absence of corresponding external sound.
63. Houston, D.K., Johnson, M.A., Nozza, R.J., et al., 1999. Am. J. Clin. Nutr. 69, 564–571; Shemesh, Z., Attias, J., Ornan, M., et al., 1993. Am. J. Otalaryngol. 2, 94–99.
64. Formerly determined by a microbiological growth assay (e.g., *Lactobacillus leichmannii*) or radioisotope dilution technique, this is now accomplished using enzyme-linking immunoassays.
65. Bailey, R.L., Durazo-Arvizu, R.A., Carmel, R., et al., 2013. Am. J. Clin. Nutr. 98, 460–467.
66. Chery, C., Hehn, A., Mrabet, N., et al., 2013. Biochimie 95, 995–1001.
67. Greib, E., Trolle, B., Bor, M.V., et al., 2013. Nutrients 5, 2475–2482.
68. A bone marrow disorder characterized by overproduction of erythrocytes.
69. A disease characterized by persistent, elevated eosinophil count (>1500/mm^3).
70. The ligand saturation of TC is not useful as a biomarker of vitamin B_{12} status, as apoTC (normally in 5- to 10-fold excess over holoTC) responds to inflammation and infection as an acute phase protein while holoTC does not.
71. Obeid, R., Jung, J., Falk, J., et al., 2013. Biochimie 95, 1056–1061.
72. Humans typically show plasma methionine concentrations in the range of 37–136 μM.

This limits the production of thymidylate and, thus, of DNA, resulting in impaired mitosis and being manifest as macrocytosis and anemia. Similarly, deficiencies of either folate or vitamin B_{12} increase urinary FIGLU excretion, as a diminished FH_4 pool reduces the capacity to degrade that metabolite by transfer of its formimino group to produce 5′-formimino-FH_4. The only way to distinguish deficiency of vitamin B_{12} from that of folate is on the basis of the urinary excretion of MMA. Lexus intake of folate can mask the anemia or FIGLU excretion associated with vitamin B_{12} deficiency by maintaining FH_4 in spite of the methyl folate trap. However, supplemental vitamin B_{12} does not affect the anemia or other signs of folate deficiency. However, **methylmalonic aciduria** occurs *only* in vitamin B_{12} deficiency, as the adenosylcobalamin-dependent methylmalonyl CoA mutase is not affected by folate status. Therefore, patients with macrocytic anemia, increased urinary FIGLU, and low blood folate levels can be diagnosed as vitamin B_{12} deficient if their urinary MMA levels are elevated, but as folate deficient if their urinary MMA is normal.

Distinguishing Vitamin B_{12} and Folate Deficiencies

Biomarker	Vitamin B_{12}-Deficient	Folate-Deficient
Urinary FIGLU	Elevated	Elevated
Urinary MMA	Elevated	Normal
Serum Hcy	Elevated	Elevated
Serum folate	Reduced	Reduced
Serum vitamin B_{12}	Normal-reduced	Normal

9. VITAMIN B_{12} DEFICIENCY

A study of elderly Americans found >40% to show elevations in urinary MMA levels; half also showed low serum vitamin B_{12} levels.[73] This levels have been observed in 10–15% of apparently healthy, elderly Americans with adequate vitamin B_{12} intakes, and in 60–70% of those with low vitamin B_{12} intakes.[74] The prevalence of low plasma vitamin B_{12} concentrations in all Central American age groups was found to be 35–90%.[75]

Vitamin B_{12} deficiency can have primary (privational) and secondary (nonprivational) causes. The major primary cause is the consumption of strict vegetarian diets.

Vegetarian Diets

Strict vegetarian diets, i.e., those containing no meats, fish, animal products, or vitamin B_{12} supplements, contain practically no vitamin B_{12} (Tables 18.7 and 18.8). Individuals consuming such vegan diets typically show very low circulating levels of the vitamin and elevated levels of Hcy.[76] Studies have found that >50% of vegetarians in India and the United States had low serum concentrations of vitamin B_{12}, i.e., <150pM. Yet, clinical signs among such individuals appear to be rare. Indeed, they may not be manifest for several years after starting a strict vegetarian dietary regimen. Serum vitamin B_{12} concentrations tend to vary inversely with the length of time of vegetarian practice, showing progressive declines for about 7 years—the time estimated to draw down hepatic stores of the vitamin (Fig. 18.3). Signs of vitamin B_{12} deficiency are more common among breast-fed infants of vegetarian mothers (Table 18.9). The vitamin B_{12} content of breast milk, like that of maternal serum, varies inversely with the length of maternal vegetarian practice.

TABLE 18.7 Vitamin B_{12} and Folate Status of Thai Vegetarians and Mixed Diet Eaters

Group	Vitamin B_{12} (pg/mL)	Folate (ng/mL)
Mixed Diet		
Men	490	5.7
Women	500	6.8
Vegetarian		
Men	117[a]	12.0[a]
Women	153[a]	12.6[a]

[a]$p > .05$.
Tungtrongchitr, V., Pongpaew, P., Prayurahong, B., et al., 1993. Int. J. Vit. Nutr. Res. 63, 201–207.

It should be remembered that not all vegetarians are strict vegans. Many **ovolactovegetarians** consume plant-based diets that also contain servings of dairy products, eggs, or fish to varying extents. Studies have shown that the occasional consumption of animal products (e.g., once per month) will support serum vitamin B_{12} levels comparable to those of people eating mixed diets (Table 18.10). In addition, vegetarians may include in their diets some seaweeds (*Nori* sp. and *Chlorella* sp.) and bacterially fermented foods that contain vitamin B_{12}. Some may be exposed to bacterial sources of the vitamin in contaminated foods or water. Many intentionally consume nutritional yeasts or nutritional supplements as sources of vitamin B_{12}.

The major secondary causes of vitamin B_{12} deficiency involve impaired absorption, metabolism, or metabolic function of the vitamin.

73. Norman, E.J., Morrison, J.A., 1993. Am. Med. J. 94, 589–594.
74. Carmel, R., Green, R., Jacobsen, D.W., et al., 1999, Am. J. Clin. Nutr. 70, 904–910; Carmel, R., 2000. Ann. Rev. Med. 51, 357–375.
75. Allen, L.H., 2004. Nutr. Rev. 62, S29–S33.
76. Obersby, D., Chappell, D.C., Dunnett, A., et al., 2013. Br. J. Nutr. 109, 785–794.

TABLE 18.8 Plasma Indicators of Vitamin B_{12} Status in Vegetarians and Nonvegetarians

		Lacto-/-Ovo-Vegetarians			
Plasma Analyte	Omnivorous Subjects	Vitamin Users	Vitamin Nonusers	Vitamin Users	Vegans Vitamin Nonusers
Vitamin B_{12} (pM)	287 (190–471)[a]	303 (146–771)	179 (124–330)	192 (125–299)	126 (92–267)
Transcobalamin (pM)	54 (16–122)	26 (30235)	23 (4–84)	14 (3–53)	4 (2–35)
Methylmalonic acid (nM)	161 (95–357)	230 (120–1344)	368 (141–2000)	708 (163–2651)	779 (222–3480)
Hcy (μM)	8.8 (5.5–16.1)	9.6 (5.5–19.4)	10.9 (6.4–27.7)	11.1 (5.3–25.9)	14.3 (6.5–52.1)
Folate (nM)	21.8 (14.5–51.5)	30 (14.8–119)	27.7 (16.0–76.9)	29.5 (18.8–71.8)	34.3 (20.7–72.7)

[a]*Mean, 95% confidence interval.*
Herrmann, W., Schorr, H., Obeid, R., et al., 2003. Am. J. Clin. Nutr. 78, 131–136.

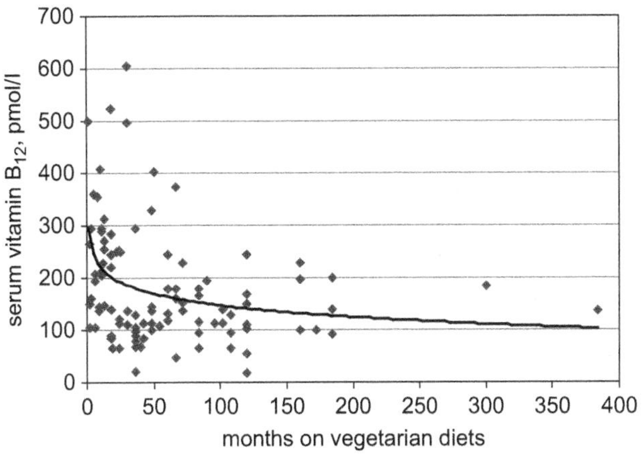

FIGURE 18.3 Inverse relationship of serum vitamin B_{12} concentrations and time following vegetarian eating practices in people in the northeastern United States. *From Miller, D.R., Specker, B.L., Ho, M.L., et al., 1991. Am. J. Clin. Nutr. 53, 524–529.*

TABLE 18.9 Ranges of Urinary Methylmalonic Acid (MMA) Excretion by Breast-Fed Infants of Vegetarian and Omnivorous Mothers

Group	MMA (μmol/mmol creatinine)
Vegetarian	2.6–791
Mixed-diet	1.7–21

Specker, B.L., Miller, D., Norman, E.J., et al., 1988. Am. J. Clin. Nutr. 47, 89–92.

TABLE 18.10 Impact of Occasional Consumption of Animal Products on Vitamin B_{12} Status in a Macrobiotic Community

Food	Consumed	Serum Vitamin B_{12} (pM)	Urine MMA (mmol/mol creatinine)
Dairy	Never	122	5.3
	≤1/week	183[a]	2.8[a]
	>1/week	179[a]	2.1[a]
Eggs	Never	139	4.8
	≤1/week	167	3.1
	>1/week	157	2.2
Sea foods[b]	Never	111	4.4
	≤1/week	145	5.3
	>1/week	161	2.6

[a]*p>.05.*
[b]*Includes various sea vegetables (e.g., wakame, kombu, hijiki, arame, nori, dulse).*
Miller, D.R., Specker, B.L., Ho, M.L., et al., 1991. Am. J. Clin. Nutr. 53, 524–529.

Malabsorption

Poor absorption of the vitamin is thought to account for at least one-third of cases of vitamin B_{12} deficiency. This can be caused by inadequate production of IF by gastric parietal cells and/or to defective functioning of ileal IF receptors.[77]

Loss of gastric parietal cell function. Vitamin B_{12} malabsorption occurs if IF production by gastric parietal cells

77. The **Schilling test** has been used to assess vitamin B_{12} absorption in clinical settings. It involves the oral administration of a tracer dose of ^{57}Co–vitamin B_{12} to a fasting subject, followed by the i.m. administration of a large dose of the vitamin to saturate plasma haptocorrin and TC. This allows the absorbed tracer to be cleared by the kidney and be quantified in the urine. Correction of low apparent absorption by orally administered IF in a stage II test indicates pernicious anemia.

is inadequate.[78] Such conditions can have causes of four general types:

- **Pernicious anemia** affects an estimated 2–3% of Americans, mostly women; although it is likely to be widely underdiagnosed. It is a disease of later life, 90% of cases are diagnosed in individuals >40 years of age. It presents as an autoimmune gastritis[79] involving destruction of the fundus and body of the stomach by antibodies to the parietal cells membrane H^+/K^+-ATPase. This causes progressive atrophy of those cells and loss of their production of acid[80] and IF, resulting in **hypochlorhydria** and vitamin B_{12} malabsorption ultimately (2–7 years) leading to macrocytic anemia.
- ***Heliobacter pylori*** infection affects an estimated 9–30% of Americans. It produces damage to the stomach referred to as type B chronic atrophic gastritis, which results in hypochlorhydria that limits the enteric absorption of vitamin B_{12}.
- **Other gastric diseases** can damage gastric parietal cells and, thus, reduce production of stomach acid and IF. Such damage can result in macrocytic anemia or, frequently, hypochromic anemia due to impaired iron absorption caused by the hypoacidic condition. These conditions can occur in patients with simple (nonautoimmune) atrophic gastritis as well as those undergoing gastrectomy. After bariatric surgery, 10–15% of patients develop vitamin B_{12} deficiency within a few years; all patients undergoing complete gastrectomy are placed in need of supplemental vitamin.
- **Chronic use of proton pump inhibitors** reduces parietal cell acid production, reducing the utilization of vitamin B_{12} from ingested food.
- **Hereditary disorders** comprise the most common vitamin B_{12} malabsorption in children. These include Imerslund–Gräsbeck syndrome and congenital IF deficiency.

Pancreatic insufficiency. The loss of pancreatic exocrine function can impair the utilization of vitamin B_{12}. For example, about one-half of all human patients with pancreatic insufficiency show abnormally low enteric absorption of the vitamin. This effect can be corrected by pancreatic enzyme replacement therapy, using oral pancreas powder or pancreatic proteases. Thus, the lesion appears to involve specifically the loss of proteolytic activity, resulting in the failure to digest intestinal R proteins, which thus retain vitamin B_{12} bound in the stomach instead of freeing it for binding by IF.

Intestinal disease. Disorders and removal of the terminal portion of the ileum, causing the loss of IF receptors, result in malabsorption of the vitamin. Such conditions include ileitis, inflammatory bowel disease (Crohn's disease) and tropical sprue.[81] In addition, intestinal parasites (e.g., the tapeworm *Diphyllobothrium latum*) and explosively growing bacterial floras can effectively compete with the host for uptake of the vitamin. Protozoal infections that cause chronic diarrhea (e.g., *Giardia lamblia*) can impair vitamin B_{12} absorption.

Chemical factors. Several factors can impair the utilization of vitamin B_{12}:

- **Xenobiotics** including biguanide antidiabetic agents, chronic alcohol consumption, and heavy smoking can damage the ileal epithelium causing loss of IF receptors.
- **Nitrous oxide (N_2O)** oxidizes cob(I)alamin to the inactive form cob(II)alamin, causing rapid inactivation of the methylcobalamin-dependent enzyme and the excretion of the vitamin. Repeated exposure to NO depletes the body of its vitamin B_{12} stores.[82]
- **Oral contraceptive steroid**[83] use has been associated with apparently asymptomatic reductions in plasma vitamin B_{12} concentrations independent of dietary intake of the vitamin.[84]

General Signs of Deficiency

Vitamin B_{12} deficiency causes **macrocytic anemia**. This type of anemia is caused by delay or failure of normal cell division in the bone marrow (it also occurs in the intestinal mucosa). The underlying biochemical lesion is arrested synthesis of DNA precursors due to diminished availability of single-C units as a result of decreased activity of the vitamin B_{12}-dependent methionine synthase.[85] This traps folate in the methyl folate trap and reduces the availability of 5,10-methylene-FH_4, which is needed for the synthesis of thymidylate and, thus, of DNA. This reduces mitotic rate results in a **megaloblastic transformation**, i.e., the formation of abnormally large, cytoplasm-rich cells. In the bone marrow, this results in a type of megaloblastic anemia referred to as macrocytic anemia.

Vitamin B_{12} deficiency also causes **neurologic abnormalities** in most species. These may also result from

78. Chronic atrophic gastritis can be a precancerous lesion, involving progressive metaplasia of the gastric mucosa leading to carcinoma.
79. It is also called type A chronic atrophic gastritis or gastric atrophy.
80. Gastric acid is needed to facilitate the dissociation of vitamin B_{12} from food proteins and to check the proliferation of enteric bacteria that compete for the vitamin. Gastric hypochlorhydria, therefore, reduces the bioavailability of vitamin B_{12} from foods.
81. Tropical sprue is endemic in south India, occurs epidemically in the Philippines and the Caribbean, and is frequently a source of vitamin B_{12} malabsorption experienced by tourists to those regions.
82. Much of the toxicity of N_2O may actually be due to impaired vitamin B_{12} function. Indeed, excessive dental use of *laughing gas* can lead to neurologic impairment.
83. i.e., Mixtures of estrogen and progestin.
84. McArthur, J.O., Tang, H.M., Petocz, P., et al., 2013. Nutrients 5, 3634–3645.
85. MET regeneration from Hcy can also be reduced by deficiencies of folate coenzymes (due to methyl folate "trapping"), which also reduce thymidylate synthesis—all leading to failed DNA replication.

TABLE 18.11 General Signs of Vitamin B_{12} Deficiency

Organ System	Signs
General	Reduced growth
Vital organs	Hepatic, cardiac, and renal steatosis
Fetus	Hemorrhage, myopathy, death
Circulatory	Anemia
Nervous	Peripheral neuropathy

TABLE 18.12 Recommended Vitamin B_{12} Intakes

US		FAO/WHO	
Age–Sex	RDA[a] (μg/day)	Age–Sex	RNI[b] (μg/day)
0–6 months	[0.4][c]	0–6 months	0.4
7–11 months	[0.5][c]	7–11 months	0.5
1–3 years	0.9	1–3 years	0.9
4–8 years	1.2	4–6 years	1.2
9–13 years	1.8	7–9 years	1.8
>13 years	2.4	>9 years	2.4
Pregnancy	2.6	Pregnancy	2.6
Lactation	2.8	Lactation	2.8

[a]*Food and Nutrition Board, 2000. Dietary Reference Intakes for Thiamin, Riboflavin, Niacin, Vitamin B_6, Folate, Vitamin B_{12}, Pantothenic Acid, Biotin and Choline. National Academy Press, Washington, DC, 564 pp.*
[b]*Recommended Nutrient Intakes, Joint WHO/FAO Expert Consultation, 2001. Human Vitamin and Mineral Requirements. WHO, Rome, 286 pp.*
[c]*RDAs have not been set; AIs are given instead.*

impaired MET biosynthesis; however, some investigators have proposed that they result from altered fatty acid metabolism due to the loss of MMA mutase activity. Neurological signs typically involve diffuse and progressive nerve demyelination, manifested as progressive neuropathy, often beginning in the peripheral nerves, and proceeding to the posterior and lateral columns of the spinal cord (Table 18.11). They tend to be manifested with relatively late onset due to the effective storage and conservation of the vitamin. Because folate can correct the anemia of vitamin B_{12} deficiency, lexus intakes of folate can mask that vitamin B_{12} deficiency, such that it may not be detected until possibly irreversible neurologic damage presents. Recommended dietary intakes for vitamin B_{12} have been established (Table 18.12).

Deficiency Signs in Humans

Vitamin B_{12} deficiency in humans produces hematologic and neurologic signs and symptoms. The hematological sign is **megaloblastic anemia**. Severely deficient infants present with feeding difficulties, developmental delay, and progressive neurological symptoms. In older children and adults, chronic deficiency can also produce progressive neurologic signs that are peripheral and/or cerebral in nature.[86] The earliest peripheral nervous symptoms are usually symmetrical paresthesia of the hands and feet, loss of proprioception and vibration sense of the ankles and toes, and ataxic gait. Rarely, patients also lose manual dexterity, taste, and smell and develop poor vision and orthostatic dizziness. Cerebral and psychiatric signs include memory impairment, depression, irritability, psychosis, and dementia.

Hematologic and neurologic signs do not necessarily manifest together in vitamin B_{12}-deficient subjects.[87] In most deficient subjects, either the anemia or neurologic signs predominate.[88] The metabolic basis for this phenomenon is not clear; nor is it clear why the neurologic signs in some subjects are predominantly peripheral nerve disorders, while in others they are predominantly cerebral disorders.

Low Vitamin B_{12} Status

Marginal deficiencies of vitamin B_{12} are estimated to be at least 10 times more prevalent than clinically overt deficiencies affecting apparently healthy people. An estimated 10–15% of people over the age of 60 have low serum vitamin B_{12} levels. However, that parameter underindicates the portion of marginal deficiencies involving metabolic changes marked by elevated circulating levels of FIGLU, Hcy, and MMA. Consideration of those parameters yields estimates of 30–40%. The prevalence of low vitamin B_{12} status is greatest in the elderly. That vitamin B_{12} status declines with age is thought to be related to declining intakes of the vitamin as well as increasing prevalence of atrophic gastritis and its associated hypochlorhydria, which can affect as much as half the geriatric population.

Homocysteinemia. Vitamin B_{12} deficiency may be the primary cause of homocysteinemia in many people; almost two-thirds of elderly subjects with homocysteinemia also show methylmalonic acidemia, indicative of vitamin B_{12} deficiency (Table 18.13). Still, less than one-third of individuals with low circulating vitamin B_{12} levels also show homocysteinemia. Epidemiologic studies have indicated associations of moderately elevated plasma Hcy and risks of coronary, peripheral and carotid arterial thrombosis and atherosclerosis, venous

86. McCaddon, A., 2012. Biochimie 95, 1066–1076.
87. Healton, E.B., Savage, D.G., Brust, J.C., et al., 1991. Medicine 70, 229–245.
88. Failure to appreciate this fact can lead to the underdiagnosis of vitamin B_{12} deficiency.

TABLE 18.13 Vitamin B_{12} and Folate Status of Elderly Subjects Showing Homocysteinemia

	Serum Hcy		Serum MMA (Methylmalonic Acid)	
Parameter	>3 SD	≤3 SD	>3 SD	≤3 SD
Serum vitamin B_{12} (pM)	197±77[a]	325±145	217±83[a]	332±146
Serum folate (nM)	12.7±8.2[a]	22.9±19.0	18.1±12.5[a]	22.7±19.5

SD, standard deviation.
[a]*p>.05.*
Lindenbaum, J., Rosenberg, I.H., Wilson, P.W., et al., 1994. Am. J. Clin. Nutr. 60, 2–11.

thrombosis, retinal vascular occlusion, carotid thickening, and hypertension.[89]

Neurological Effects

Insufficient vitamin B_{12} status is thought to lead to neurodegeneration as a result of abnormal incorporation of MMA into neuronal lipids including those in myelin sheaths, stimulation of the inflammatory cytokine tumor necrosis factor-α, and/or reduced synthesis of choline, the precursor of the neurotransmitter acetylcholine. Several aspects of neurological function are affected:

- **Cognition**. Serum Hcy level has been negatively correlated with the presentation of neuropsychiatric disorders in nonanemic subjects with low serum vitamin B_{12} levels.[90] Serum vitamin B_{12} levels <257 pM predicted cognitive decline in older subjects in the Framingham Heart Study.[91] These effects have been thought to be manifestations of white matter damage in the spinal cord and brain,[92] or atrophy of the brain,[93] both of which vary inversely with serum vitamin B_{12}. Recent studies have shown that poor memory performance by low serum vitamin B_{12} subjects is associated with damage to specific microstructural regions of the hippocampus.[94]

While one trial found high doses of cyanocobalamin to improve cognitive function in subjects with only mild impairment or with symptoms of recent onset (<6 months),[95] well-controlled, randomized clinical trials have found no benefits of vitamin B_{12} administration on low vitamin B_{12} subjects with cognitive impairment/dementia.[96] One trial found vitamin B_{12} therapy without effect on patients with dementia but to improve measures of verbal fluency in cognitively impaired patients.[97] A review of clinical experience in India suggested value of the vitamin in improving language function in patients.[98]

- **Alzheimer's disease (AD)**. Several studies have noted low concentrations of vitamin B_{12} in the serum and cerebrospinal fluid of nonanemic AD patients, who also tend to have homocysteinemia.[99] AD patients have also been found to have lower plasma levels of holoTC than nondemented elderly controls, despite having similar total plasma vitamin B_{12} levels.[100] In fact, holoTC level was inversely related to the subsequent risk of elderly subjects being diagnosed with AD over a 7-year period.[101]
- **Depression**. Low plasma levels of vitamin B_{12} have been reported in nearly one-third of patients with depression, who also tend to show homocysteinemia. Patients with high vitamin B_{12} status have been reported to have better treatment outcomes.[102]
- **Parkinson's disease (PD)**. PD patients have homocysteinemia,[103] which increases their risk for cognitive impairment. Some may also have elevated circulating MMA levels. Dietary supplementation with vitamin B_{12}

89. The low prevalence of coronary heart disease among South African blacks has been associated with their typically lower plasma Hcy levels and their demonstrably more effective Hcy clearance after methionine loading.
90. Lindenbaum, J., Healton, E.B., Savage, D.G., et al., 1988. N. Engl. J. Med. 318, 1720–1728.
91. Morris, M.S., Selhub, J., Jacques, P.F., 2012. J. Am. Geriatr. Soc. 60, 1457–1464.
92. de Lau, L.M., Smith, A.D., Refsum, H., et al., 2009. J. Neurol. Neurosurg. Psychiatry 80, 149–157.
93. Vogiatzoglou, A., Refsum, H., Johnson, C., et al., 2008. Neurology 71, 826–832.
94. Köbe, T., Witte, A.V., Schnelle, A., et al., 2016. Am. J. Clin. Nutr. 103, 1045–1054.
95. Martin, D.C., Francis, J. Protetch, et al., 1992. J. Am. Geriatr. Soc. 40, 168–172.
96. Carmel, R., Gott, P.S., Waters, C.H., et al., 1995. Eur. J. Haematol. 54, 245–253; Teunisse, S., Bollen, A.E., van Gool, V.A., et al., 1996. J. Neruol. 243, 522–529.
97. Eastley, R., Wilcock, G.K., Bucks, R.S., 2000. Psychiatry 15, 226–233.
98. Moretti, R., Torre, P., Antonello, R.M., et al., 2004. Neurol. India 52, 310–318.
99. McCaddon, A., 2013. Biochimie 95, 1066–1076.
100. Refsum, H. and Smith, A.D., 2003. J. Neurol. Neurosurg. Psychiatry 74, 959–961.
101. Hooshmand, B., Solomon, A., Kareholt, I., et al., 2010. Neurology 75, 1408–1414.
102. Levitt, A.J., Wesson, V.A., Joffe, R.T., 1998. Psychiat. Res. 79,123–129.
103. PD patients are treated with L-dopa, which can increase circulating Hcy levels.

(and folate) has been shown to reduce their plasma Hcy levels.[104]

- **Multiple sclerosis**. MS patients generally have elevated circulating levels of Hcy but relatively low levels of vitamin B_{12} and folate and are rarely anemic.[105] It has been suggested that low vitamin B_{12} status may exacerbate multiple sclerosis by enhancing the processes of inflammation and demyelination and by impairing those of myelin repair.
- **Hearing loss**. Serum vitamin B_{12} levels have been reported to be lower in subjects with tinnitus compared to normal hearing controls,[106] and vitamin B_{12} supplementation has been reported to lessen tinnitus in chronically affected subjects.[107]

Response to Treatment

Subclinical vitamin B_{12} deficiency, if diagnosed, is readily addressed. Biochemical indicators of vitamin B_{12} status, plasma/serum MMA and Hcy, fall within days of treatment with the vitamin. If correctly diagnosed in infants, intramuscular treatment with vitamin B_{12} can reverse both the biochemical indicators and clinical signs (regurgitations, delayed development of motor function) within a month.[108]

The hematological signs of vitamin B_{12} deficiency also respond quickly. Morphological abnormalities in the bone marrow are corrected within 2–3 days, reticulocyte numbers increase within 3–5 days, and this is followed by increases in erythrocyte numbers with the turnover of macrocytes.

Clinically overt vitamin B_{12} deficiency is more difficult to address, as it typically involves severe malabsorption. Also, neurological signs are corrected much slower, if at all. Muscular weakness and some psychiatric signs (e.g., irritability, confusion), particularly those of relatively recent onset (e.g., <3 months), may show improvement within weeks and may be completed corrected. Sensory signs may take longer and may never be completely corrected.

Deficiency Signs in Animals

Vitamin B_{12} deficiency in nonruminant animals is characterized most frequently by reductions in rates of growth and feed intake and by impairments in the efficiency of feed utilization. In a few species (e.g., swine) a mild macrocytic anemia develops. Swine may also develop rough skin and gastrointestinal disorders. Vitamin B_{12}-deficient gilts can have delayed estrus; pregnant gilts may deliver fewer progeny, which are of low birth weight. Growing chicks and turkey poults show impairments in growth and feed utilization, macrocytic anemia, neurologic signs, and defective feathering. They can also show perosis[109] as a secondary effect of reductions in MET and choline due to the reduced availability of labile methyl groups. Limited methyl group availability (for the synthesis of phosphatidylcholine) in poultry is also manifest as increased deposition of lipids in the liver, heart, and kidneys. For this reason, vitamin B_{12} is known as a **lipotrope** for poultry. Vitamin B_{12} deficiency in the chicken also causes embryonic death, with embryos showing myopathies of the muscles of the leg, hemorrhage, myocardial hypertrophy, as well as perosis. Laboratory rodent species typically show impaired growth and feed utilization; males can show impaired spermatogenesis. Monkeys show macrocytic anemia.

Synthesis of the vitamin in the rumen is dependent on the development of a rumen microbiome, which can take several weeks. Therefore, young ruminants, i.e., those less than 6 weeks old require a dietary source of vitamin B_{12}; else they develop anorexia, poor growth, and, sometimes, macrocytic anemia. Vitamin B_{12} supplements are required for calves and lambs fed diets containing no animal protein. That need disappears with the establishment of the rumen microbiome, which depends on an adequate supply of dietary Co for the microbial synthesis of the organic portion of the corrin nucleus. Ruminal production of vitamin B_{12} can also be affected by the composition of dietary fiber, the ratio of roughage to concentrate, and the level of dry matter intake. Most microbially produced vitamin B_{12} appears to be contained within rumen microbial cells; it is released for absorption only in the small intestine. Cattle are thought to be less susceptible than sheep to Co deficiency. Sheep fed a diet containing <70 μg Co per kilogram show signs of deficiency: anorexia, wasting, diarrhea, and watery lacrimation. Co deprivation reduces hepatic Co level and increases plasma methylmalonyl CoA expression but does not produce clinical signs. Therefore, most ruminants, and species with significant cecal microbiomes (e.g., horse, rabbits, some fish), do not have dietary needs for vitamin B_{12}.

104. Lamberti, P., Zoccolella, S., Armenise, E., et al., 2005. Eur. J. Neurol. 12, 365–368.

105. Moghaddsi, M., Mamarabadi, M., Mohevi, N., et al., 2013. Clin. Neurol. Neurosurg. 115, 1802–1805; Kocer, B., Enqur, S., Ak, F. et al., 2009. J. Clin. Neurosci. 16, 399–403.

106. Houston, D.K., Johnson, M.A., Nozza, R.J., et al., 1999. Am. J. Clin. Nutr. 69, 564–571; Shemesh, Z., Attias, J., Ornan, M., et al., 1993. Am. J. Otalaryngol. 2, 94–99.

107. Shemesh, Z., Attias, J., Ornan, M., et al., 1993. Am. J. Otolaryngol. 2, 94–99.

108. Torsvik, I., Ueland, P.M., Markestad, T., et al., 2013. Am. J. Clin. Nutr. 98, 1233–1240.

109. This is the anatomical condition sometimes called "slipped tendon" in which the gastrocnemius (achilles) tendon slips from the guiding condyles of the distal end of the tibia due to the twisting of that bone and widening of the tibiometatarsal joint. This impairs locomotion including access to food, typically resulting in reduced growth. Perosis can also be caused by deficiencies of manganese, choline, or niacin.

10. VITAMIN B_{12} IN HEALTH AND DISEASE

Anticarcinogenesis

It has been suggested that subclinical deficiencies of vitamin B_{12} may enhance carcinogenesis. That hypothesis is supported by the finding that low, asymptomatic vitamin B_{12} status can alter DNA base substitution and methylation in the rat model,[110] as well as observations in a prospective study of increased breast cancer risk in women ranking in the lowest quintile of plasma vitamin B_{12} concentration.[111] In contrast, another observational study found weak positive associations of serum vitamin B_{12} level and vitamin B_{12} intake and prostate cancer risk.[112] Large, placebo-controlled, randomized intervention trials have found combined treatment of vitamin B_{12}, vitamin B_6, and folate without effect on cancer risks of breast or total cancers.[113] Therefore, it is unlikely that vitamin B_{12} has a role in reducing cancer risk.

Cyanide binding. Cobalamins can bind cyanide to produce the nontoxic cyanocobalamin. For that reason, hydroxocobalamin is a well-recognized cyanide antidote. It has been proposed that vitamin B_{12} may have a role in the inactivation of the low levels of cyanide consumed in many fruits, beans, and nuts.

Antioxidant Activity

Vitamin B_{12} has antioxidant capacity. This is indicated by the findings that high levels of the vitamin can protect cells against in vitro exposure to hydrogen peroxide (H_2O_2).[114] This protection may involve stimulation of cellular methionine synthase activity, as well as direct reaction of the vitamin with the reactive oxygen species generated by H_2O_2. Any antioxidant effects in vivo are likely to occur only at pharmacological exposures to the vitamin or other cobalamins.

Protection against mineral toxicities. Nutritional levels of vitamin B_{12} have been effective in reducing intoxicating effects of selenium in Japanese quail[115] and cadmium in the rat.[116]

110. Choi, S.W., Friso, S., Ghandour, H., et al., 2004. J. Nutr. 134, 750–755.
111. Wu, K., Helzlsouer, K.J., Comstock, G.W., et al., 1999. Cancer Epidemiol. Biomarkers Prev. 8, 209–217.
112. Collin, S.M., Metcalfe, C., Refsum, H., et al., 2010. Cancer Epidemiol. Biomarkers Prev. 19, 1632–1642.
113. Zhang, Cook, N.R., Albert, C.M., et al., 2008. JAMA 300, 2012–2021; Andreeva, V.A., Touvier, M., Kesse-Guyot, E., 2012. Arch. Intern. Med. 172, 540–547.
114. Birch, C.S., Brasch, N.E., McCaddon, A., et al., 2009. Free Rad. Biol. Med. 47, 184–188.
115. Gad, M.A., El-Twab, S.M., 2009. Environ. Biol. Pharmacol. 27, 7–16.
116. Couce, M., Varela, J.M., Sánchez, A., et al., 1991. J. Inorg. Biochem. 41, 1–6.

11. VITAMIN B_{12} TOXICITY

Vitamin B_{12} has no appreciable toxicity. Upper tolerable intakes (ULs) for B12 have not been established. Results of studies with mice indicate that it is innocuous when administered parenterally in very high doses. Localized, injection-site, sclerodermoid reaction[117] secondary to vitamin B_{12} injection has been reported. Dietary levels of at least several hundred times the nutritional requirements are safe. High plasma levels of the vitamin are indicative of disease[118] rather than hypervitaminosis B_{12}.

12. CASE STUDY

Review the following case report, paying special attention to the diagnostic indicators on which the treatment was based. Next, answer the questions that follow.

Case

A 6-month-old boy was admitted in comatose condition. He had been born at term, weighing 3 kg, the first child of an apparently healthy 26-year-old *vegan*.[119] The mother had knowingly eaten no animal products for 8 years and took no supplemental vitamins. The infant was exclusively breast fed. He smiled at 1–2 months of age and appeared to be developing normally. At 4 months, his development began to regress; this was manifested by his loss of head control, decreased vocalization, lethargy, and increased irritability. Physical examination revealed a pale and flaccid infant who was completely unresponsive even to painful stimuli. His pulse was 136/min, respirations 22/min, and blood pressure 100 mmHg by palpation. His length was 65 cm (50th percentile for age) and his weight was 5.6 kg (<3rd percentile and at the 50th percentile for 3 months of age). His head circumference was 41 cm (3rd percentile). His optic disks[120] were pale. There were scattered ecchymoses[121] over his legs and buttocks. He had increased pigmentation over the dorsa of his hands and the feet, most prominently over the knuckles. He had no head control and a poor grasp. He showed no deep tendon reflexes. His liver edge was palpable 2 cm below the right costal margin.

117. Such reactions are not common but have been reported for various drugs and for vitamin K.
118. Elevated cobalamin levels are typical of myelogenous leukemia and promyelocytic leukemic and are used as diagnostic criteria for polycythemia vera and hypereosinophilic syndrome. Several liver diseases (acute hepatitis, cirrhosis, hepatocellular carcinoma, and metastatic liver disease) can cause similar increases, which are due to increased levels of TC_I.
119. A strict vegetarian.
120. Circular area of thinning of the sclera (the fibrous membrane forming the outer envelope of the eye) through which the fibers of the optic nerve pass.
121. Purple patches caused by extravasation of blood into the skin, differing from petechiae only in size (the latter being very small).

Laboratory Results

Parameter	Patient	Normal Range
Hemoglobin, g/dL	5.4	10.0–15.0
Hematocrit, %	17	36
Erythrocytes, $\times 10^6/\mu L$	1.63	3.9–5.3
White blood cells, $\times 10^3/\mu L$	3.8	6–17.5
Reticulocytes, %	0.1	<1
Platelets, $\times 10^3/\mu L$	45	200–480

A peripheral blood smear revealed mild macrocytosis[122] and some hypersegmentation of the neutrophils.[123] Bone marrow aspiration showed frank megaloblastic changes in both the myeloid[124] and the erythroid[125] series. Megakaryocytes[126] were decreased in number. The sedimentation rate, urinalysis, spinal fluid analysis, blood glucose, electrolytes, and tests of renal and liver function gave normal results. An electroencephalogram was markedly abnormal, as manifested by minimal background activity and epileptiform transients in both temporal regions. Analysis of the urine obtained on admission demonstrated a markedly elevated excretion of MMA, glycine, methylcitric acid, and Hcy. Shortly after admission, respiratory distress developed, and 5 mg of folic acid was given, followed by transfusion of 10 mL of packed erythrocytes per kilogram body weight. Four days later, a repeat bone marrow examination showed partial reversal of the megaloblastic abnormalities.

Other Laboratory Results

Parameter	Patient	Normal Range
Serum vitamin B_{12} (pg/mL)	20	150–1000
Serum folates (ng/mL)	10	3–15
Serum iron (μg/dL)	165	65–175
Serum iron-binding capacity (μg/dL)	177	250–410

Cyanocobalamin (1 mg/day) was administered for 4 days. The patient began to respond to stimuli after the transfusion; however, the response to vitamin B_{12} was dramatic. Four days after the initial dose he or she was alert, smiling, responding to visual stimuli, and maintaining his or her body temperature. As he or she responded, rhythmical twitching activity in the right hand and arm developed that persisted despite anticonvulsant therapy, and despite a concomitant resolution of electroencephalographic abnormalities. The mother showed a completely normal hemogram. Her serum vitamin B_{12} concentration was 160 pg/mL (normal, 150–1000 pg/mL), but she showed moderate methylmalonic aciduria. Her breast milk contained 75 pg of vitamin B_{12}/mL (normal, 1–3 ng/mL).

With vitamin B_{12} therapy, the infant's plasma vitamin B_{12} rose to 600 pg/mL and he or she continued to improve clinically. The abnormal urinary acids and homocystine disappeared by day 10; cystathionine persisted until day 20. On day 14, Hb was 14.4 g/dL, hematocrit was 41%, and the WBC was 5700/ml. The platelet count had become normal 20 days after admission. The unusual pigment on the extremities had improved considerably 2 weeks after he or she received the parenteral vitamin B_{12} and disappeared gradually over the next month. The liver was no longer palpable. The twitching of the hands disappeared within a month of therapy. Developmental assessment at 9 months of age revealed him or her to be functioning at the 5-month age level. A month later, he or she was sitting and taking steps with support. Head circumference had exhibited catch-up growth and at 44 cm was in the normal range for the first time since admission. His length was 70 cm (10th percentile) and weight 8.4 kg (10th percentile). By this time, the mother's serum vitamin B_{12} had dropped to 100 pg/mL, and she began taking supplemental vitamin B_{12}.

Case Questions

1. Which clinical findings suggested that two important coenzyme forms of vitamin B_{12} were deficient or defective in this infant? How do the clinical findings relate specifically to each coenzyme?
2. What findings allow the distinction of vitamin B_{12} deficiency from a possible folic acid-related disorder in this patient?
3. Offer a reasonable explanation for the fact that the mother, who had avoided vitamin B_{12}-containing foods for 8 years before her pregnancy, did not show overt signs of vitamin B_{12} deficiency.

13. STUDY QUESTIONS AND EXERCISES

1. Construct a decision tree for the diagnosis of vitamin B_{12} deficiency in humans or an animal species and, in particular, the distinction of this deficiency from that of folate.
2. What key feature of the chemistry of vitamin B_{12} relates to its coenzyme functions?

122. Occurrence of unusually large numbers of *macrocytes* (large erythrocytes) in the circulating blood; also called *megalocytosis*, *megalocythemia*, and *macrocythemia*.
123. A type of mature white blood cell in the granulocyte series.
124. Related to myocytes.
125. Related to erythrocytes.
126. An unusually large cell thought to be derived from the primitive mesenchymal tissue that differentiates from hematocytoblasts.

3. What parameters might you measure to assess vitamin B_{12} status of a human or animal?
4. What is the relationship of normal function of the stomach and pancreas with the utilization of dietary vitamin B_{12}?

RECOMMENDED READING

Alpers, D.H., Russell-Jones, G., 2013. Gastric intrinsic factor: the gastric and small intestinal stages of cobalamin absorption. A personal journey. Biochimie 95, 989–994.

Andrès, E., Serraj, K., Zhu, J., et al., 2013. The pathophysiology of elevated vitamin B_{12} in clinical practice. Q. J. Med. 106, 505–515.

Elmadfa, I., Singer, I., 2009. Vitamin B-12 and homocysteine status among vegetarians: a global perspective. Am. J. Clin. Nutr. 89, 1693S–1698S.

Froese, D.S., Gravel, R.A., 2010. Genetic disorders of vitamin B_{12} metabolism: eight complementation groups – eight genes. Exp. Rev. Molec. Med. 12, 1–20.

Gerashim, C., Lofgren, M., Banerjee, R., 2013. Navigating the B_{12} road: assimilation, delivery and disorders of cobalamin. J. Biol. Chem. 288, 13186–13193.

Gräsbeck, R., Tanner, S.M., 2011. Juvenile selective vitamin B_{12} malabsorption: 50 years after its description – 10 years of genetic testing. Pediatr. Res. 70, 222–228.

Guéant, J.L., Caillerez-Fofou, M., Battaglia-Hsu, S., et al., 2013. Molecular and cellular effects of vitamin B_{12} in brain, myocardium and liver through its role as a co-factor for methionine synthase. Biochimie 95, 1033–1040.

Green, R., Miller, J.W., 2014. Vitamin B_{12}. In: Zempleni, J., Suttie, J.W., Gregory, J.F., et al. (Eds.), Handbook of Vitamins, fifth ed. CRC Press, New York, pp. 447–489 (Chapter 12).

Hannibal, L., DiBello, P.M., Jacobsen, D.W., 2013. Proteomics of vitamin B_{12} processing. Clin. Chem. Lab. Med. 51, 477–488.

Kozyraki, R., Cases, O., 2013. Vitamin B_{12} absorption: mammalian physiology and acquired and inherited disorders. Biochemie 95, 1002–1007.

Li, F., Watkins, D., Rosenblatt, D.S., 2009. Vitamin B_{12} and birth defects. Molec. Cell Gen. Metab. 98, 166–172.

McCaddon, A., 2013. Vitamin B_{12} in neurology and ageing: clinical and genetic aspects. Biochimie 95, 1066–1076.

O'Leary, F., Samman, S., 2010. Vitamin B_{12} in health and disease. Nutrients 2, 299–316.

Smith, A.D., Refsum, H., 2009. Vitamin B-12 and cognition in the elderly. Am. J. Clin. Nutr. 89, 707S–711S.

Watanabe, F., Yabuta, Y., Bito, T., et al., 2014. Vitamin B_{12}-containing plant food sources for vegetarians. Nutrients 6, 1861–1873.

Chapter 19

Vitamin-Like Factors

Chapter Outline

Anchoring Concepts

1. The designation vitamin is specific for animal species, stage of development or production, and/or particular conditions of the physical environment and diet.
2. Each of the presently recognized vitamins was initially called an accessory factor or an unidentified growth factor, and these terms continue to be used to describe biologically active substances, particularly for species of lower orders.
3. To understand that choline and carnitine are vitamins for certain animal species.
4. To understand the metabolic functions of other conditionally essential nutrients: *myo*-inositol, pyrroloquinoline quinine, the ubiquinones, and orotic acid.
5. To understand why flavonoids, nonprovitamin A carotenoids, *p*-aminobenzoic acid, and lipoic acid are not called vitamins.

Have all the vitamins been discovered? From all indications in the extensive recent and current publications in the scientific literature dealing with the purification and effects of 'unidentified factors,' the answer appears to be 'no.' It is from such studies that new vitamins may be recognized and characterized

A.F. Wagner and K. Folkers[1]

LEARNING OBJECTIVES

1. To understand that the designation of a compound as a vitamin is biased in favor of avoiding deficiency in humans.
2. To understand that other substances have been proposed as vitamins.

1. Karl Folkers (1906–97) was an American biochemist who spent most of his career at Merck Pharmaceuticals where he made many contributions isolating, identifying, and synthesizing a wide variety of bioactive natural substances including vitamin B_{12}, pyidoxaine, pyridoxal, pyrodixamine, pantothenic acid, biotin, and ubiquinone. In 1990, he received the U.S. National Medal of Science. Among his colleagues at Merck was Arthur F. Wagner (1922–2010).

VOCABULARY

Acetylcholine
Acylcarnitine esters
Acylcarnitine translocase
Arachidonic acid
Beneficial dietary factor
Betaine
Betaine aldehyde dehydrogenase
Betaine:homocysteine methyltransferase
Branched-chain ketoacid dehydrogenase
γ-Butyrobetaine hydroxylase
Calcisomes
Canthaxanthin
Carnitine
Carnitine acyltransferases I and II (CATI, CATII)
Carnitine–acylcarnitine translocase (CACT)
Catechins
Ceramide
Choline
Choline acetyltransferase
Choline dehydrogenase

The Vitamins. http://dx.doi.org/10.1016/B978-0-12-802965-7.00019-8

Choline kinase
Choline oxidase
Choline phosphotransferase
Coenzyme Q10 (CoQ10)
Conditional essentiality
Cytidine diphosphatidylcholine (CDP–choline)
Dihydrolipoyl dehydrogenase
Dihydrolipoyl transacetylase
Dimethylglycine
Ethanolamine
Flavanols
Flavonols
Flavonoids
6-β-Galactinol
Glycerylphosphorylcholine
Glycerylphosphatidylcholine diesterase
Myo-inositol
Inositol 1,4,5-triphosphate (IP3)
Inositol phosphokinases
Isoflavones
α-Ketoglutarate dehydrogenase
Labile methyl groups
Lecithin
Lipoamide
Lipofuscin
Lipoic acid
Lipoic acid synthetase
Lipoyl transferase
Lycopene
Lysolecithin
Lysophosphatidylcholine
Lutein
Macula
Macular degeneration
Methionine
Mitochondrial fatty acid shuttle
Orotic acid
Phosphatidylcholine
Phosphatidylcholine–ceramide choline transferase
Phosphatidylcholine–glyceride choline transferase
Phosphatidylethanolamine
Phosphatidylethanolamine N-methyl transferase
Phosphatidylinositol (PI)
Phosphatidylinositol 4-phosphate (PIP)
Phosphatidylinositol 4,5-biphosphate (PIP2)
Phospholipases A1, A2, B, C, and D
Phosphorylcholine
Phytic acid
Phytoestrogens
Proanthocyanidins
Second messenger
Sphingomyelin
Tannic acid
Thioctic acid
Trimethylamine
ε-N-Trimethyllysine
Ubiquinones
Vitamin BT
Xanthophyll-binding protein (XBP)
Xanthophylls
Zeaxanthin

1. IS THE LIST OF VITAMINS COMPLETE?

The research that resulted in the recognition of the vitamins had both empirical and experimental phases. That is, initial associations between diet and health status generated hypotheses that could be tested experimentally. As is generally true in science, where hypotheses were clearly enunciated and adequate experimental approaches were available, insightful investigators were able to make remarkable progress in identifying these essential nutrients. Those endeavors, of course, also revealed some biologically active factors[2] to be identical or related to known nutrients,[3] some needed only by some species or under certain conditions, some to be biologically active but not essential in diets, and some not to be active at all.[4] The apparently irregular and often confusing array of informal vitamin names (Appendix A) reveals this history of discovery.

Over this history, the designation of bioactive factors as vitamins has been an informal process. It has tended to be anthropocentric in that it reflected, to a large extent, the nutritional needs of humans (Why is ascorbic acid called a vitamin, when it is synthesized by most species?). Designations have typically not been revised (Why is cholecalciferol still called a vitamin and not a hormone?). Most notably, it has been inconsistent in that several bioactives fit the traditional criteria for vitamin designation (*see* Chapter 1) under certain circumstances, e.g., for a particular species, genotype, stage of development, diet composition, nutritional status, or physical environment. Some of these have

2. In addition to the 40 or so recognized dietary essentials, foods contain an estimated 25,000+ biologically active compounds.
3. An example is vitamin T (also called "termitin," "penicin," "torutilin," "insectine," "hypomycin," "myocine," or "sesame seed factor"). This was extracted from yeast, sesame seeds, or insects and appeared to stimulate the growth of guppies, hamsters, baby pigs, chicks, mice, and insects; to promote wound healing in mice; and to improve certain human skin lesions. Ultimately, it was found to be a mixture containing folate, vitamin B_{12}, and amino acids. Similarly, vitamins M, B_c, B_{10}, T, and B_x were ultimately found to be forms of folate.
4. e.g., Early studies suggested that *p*-aminobenzoic acid, a bacterial metabolite in the biosynthesis of folate, increased growth in chicks and promoted lactation in rats, when those animals were fed marginal amounts of folate. For a time, PABA was called "vitamin B_x." Such responses were subsequently shown to be due to PABA being used by the intestinal microbiome for the synthesis of folate, which was made available to the host either directly in the gut or indirectly via the feces.

emerged as beneficial to health. Those dietary factors with vitamin-like properties fall into two categories:

- **Conditionally essential nutrients:** choline, carnitine, *myo*-inositol, ubiquinols, and lipoic acid.
- **Beneficial dietary bioactive:** nonprovitamin A carotenoids, flavonoids, lipoic acid, and orotic acid.

2. CHOLINE

Choline is a normal metabolite with essential functions in cells. Its metabolic derivatives are structural components of membranes, neurotransmitters, methyl group donors, and mediators of hepatic lipid metabolism. Choline can be biosynthesized, but there are circumstances in which that synthesis may not be sufficient for optimal physiologic functions, making it **conditionally essential**.

Recognition of a Role of Choline in Nutrition

The discovery of insulin by Banting and Best[5] in the mid-1920s led to studies with depancreatized dogs that showed dietary **lecithin (phosphatidylcholine, PC)** to be effective in mobilizing the excess lipids in the livers of insulin-deprived animals. Best and colleague Elinor Huntsman showed that the active component of lecithin is choline. Choline had been isolated by Strecker[6] in 1862, and its structure had been determined by Bayer[7] shortly after that. In 1940, Jukes showed that choline is required for normal growth and the prevention of the leg disorder called **perosis**[8] in turkeys; in fact, more choline was required to prevent perosis than to support normal growth. Jukes and Norris's group at Cornell then found that **betaine**, the metabolic precursor to choline, was not always effective in preventing choline-responsive perosis in turkeys and chicks. These findings made is clear that choline was more than a lipotrope.

5. Frederick Banting (1891–1949), a young surgeon working in the laboratory of John James Rickard Macleod at the University of Toronto, was assisted by a medical student, Charles Best (1899–1978). Banting and Macleod (1976–35) were awarded the 1923 Nobel Prize for medicine or physiology for the discovery and isolation of insulin. Banting shared his prize money with Best, and Macleod shared his with James Collip who had worked with the team on the purification of insulin.

6. Adolph Strecker (1822–1871) was then a Professor at the University of Tübingen; he coined the name "choline" after the Greek word "*chole*" for bile.

7. Carl Josef Bayer (1847–1904) was an Austrian chemist best known for inventing the eponymous process for extracting alumina from bauxite.

8. Perosis occurs in rapidly growing heavy-bodied poultry and involves the misalignment of the tibiotarsus causing slippage of the Achilles tendon. This impairs ambulation and can reduce feeding, consequently impairing growth. Perosis can also be caused by dietary deficiencies of niacin or manganese.

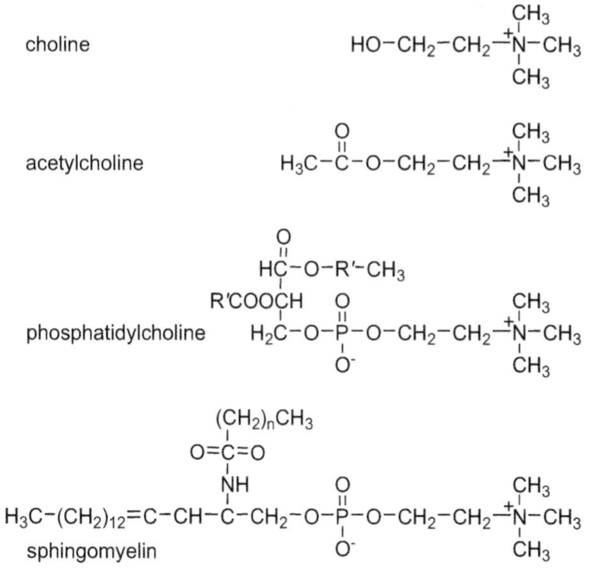

FIGURE 19.1 Choline and its functional metabolites.

Conditions of Need for Dietary Choline

For most individuals, endogenous biosynthesis of choline is sufficient to meet metabolic needs. However, that does not pertain in two situations:

- Insufficient biosynthesis of MET by 5-methyl-FH_4-dependent methylation of Hcy, e.g., due to suboptimal status with respect to folate, vitamin B_{12}, and/or preformed MET.
- Deficiencies in enzymes of choline or single-C metabolism.[9]

With low-choline intakes, these situations can, in principle, result in hepatic steatosis, hemorrhagic renal degeneration, and (in animals) depressed growth. Fatty infiltration of the liver reflects the need for PC for the synthesis of very low-density lipoprotein (VLDL), which is necessary for the export of triglycerides. Such manifestations have to do with nutritional status with respect to folate, vitamin B_{12}, and/or MET.

Conditions producing need for dietary choline are as follows:

- Deficiencies in single-C metabolism
- Deficiencies of choline-metabolizing enzymes

Chemical Properties of Choline

Choline is the trivial designation for the compound 2-hydroxy-*N,N,N*-trimethylethanaminium [also (β-hydroxyethyl)trimethylammonium] (Fig. 19.1). It is freely soluble in water

9. Zeisel, S.H., 2011. J. Nutr. 141, 531–534.

and ethanol but insoluble in organic solvents. It is a strong base and decomposes in alkaline solution with the release of trimethylamine. The prominent feature of its chemical structure is its triplet of methyl groups, which enables it to serve as a methyl donor.

Distribution of Choline in Foods and Feedstuffs

Americans consume some 440–630 mg/day.[10] All natural fats contain some choline (Table 19.1). The factor occurs naturally mostly in the form of **PC**[11] (also called **lecithin**), which, because it is a good emulsifying agent, is used as an ingredient or additive in many processed foods and food supplements. Some dietary choline (<10%) is present as the free base and some as **sphingomyelin**.[12] The richest sources in human diets are egg yolk, glandular meats (e.g., liver, kidney, brain), pork (meat, bacon), soybean products, wheat germ, and peanuts. Choline is added (as choline chloride and choline bitartrate) to infant formulas as a means of fortification.

The best sources of choline for animal feeding are the germs of cereals, legumes, and oilseed meals (e.g., soybean meal). Corn is notably low in choline (half the levels found in barley, oats, and wheat). Because wheat is rich in the choline-sparing factor betaine, the choline needs of livestock fed diets based on wheat are much lower than those of animals fed diets based on corn.

Bioavailability and stability. The bioavailability of choline in foods and feedstuffs appears to be generally good, i.e., >80%. It is dependent on those factors that affect the utilization of dietary fats. Naturally occurring choline, as well as the choline salts used as supplements, have good stability. The processing of foods/feedstuffs can enhance choline bioavailability, as mechanical disruption of plant cells by chopping and grinding, etc., can activate phospholipases to release choline in free form.

Absorption and Transport of Choline

Digestion. Choline is released from **PC**[13] by hydrolysis in the intestinal lumen through the action of phospholipases produced by the pancreas (**phospholipase A_2**, which cleaves the β-ester bond) and the intestinal mucosa

TABLE 19.1 Choline Contents (mg/100 g) of Common Foods

Food	Choline Equivalents[a]
Meats	
Beef	78.2
Beef liver	418.2
Chicken	65.8
Chicken liver	290.0
Pork	102.8
Bacon	124.9
Shrimp	70.6
Cod	83.6
Salmon	65.5
Vegetables	
Beans	13.5
Broccoli	40.1
Cabbage	15.5
Carrots	8.8
Corn	22.0
Cucumber	6.0
Lettuce	6.7
Mushrooms	16.9
Onions	6.1
Peas	27.5
Potatoes	14.4
Spinach	22.1
Soybeans	115.9
Tomatoes	6.7
Peanuts	52.5
Fruits	
Apples	3.4
Avocado	14.2
Blueberry	6.0
Banana	9.8
Cantaloupe	7.6
Grapefruit	7.5
Grapes	5.6
Oranges	8.4
Peaches	6.1

Continued

10. Choline intakes of Americans have been estimated to be 443±88 mg/day for women and 631±157 mg/day for men (Fischer, L.M., Scearce, J.A., Mar, M.H., et al., 2005. J. Nutr. 135, 826–829).

11. Phosphatidylcholine comprises 95% of total choline in eggs and 55–70% of total choline in meats and soy products.

12. i.e., Phosphatidylcholine analogues containing, instead of a fatty acid, sphingosine (2-amino-4-octadecene-1,3-diol) at the glycerol α-carbon.

13. Also called lecithin (lécithine), the name given by the French chemist Theodore Gobley (1811–76) to the factor he isolated from egg yolk (Greek, *lékith-os*).

TABLE 19.1 Choline Contents (mg/100 g) of Common Foods—cont'd

Food	Choline Equivalents[a]
Strawberries	5.7
Cereals	
Oats	7.4
Oat bran	58.6
Rice, polished	2.1
Rice, unpolished	9.2
Wheat	26.5
Wheat bran	74.4
Other	
Milk	14.3
Eggs	251.0

[a]*Includes free choline, phosphocholine, glycerophosphocholine, phosphatidylcholine, and sphingomyelin.*
Adapted from Zeisel, S.H., Mar, M.H., Howe, J.C., et al., 2003. J. Nutr. 133, 1302–1307.

(**phospholipases A_1** and **B**, both of which cleave the α-ester bond to yield **glycerylphosphatidylcholine, glyceryl-PC**). The mucosal enzymes are much less efficient than the pancreatic enzyme. Therefore, most of the PC that is ingested is absorbed as **lysophosphatidylcholine (lysoPC)**[14] (deacylated only in the α position), which is reacylated postabsorptively to yield PC. This reaction involves the dismutation of two molecules of lysolecithin to yield a molecule of glyceryl-PC and one of PC. Analogous reactions occur with sphingomyelin, which, unlike PC, is not degraded in the intestinal lumen but is taken up intact by the intestinal mucosa.

Absorption. LysoPC and sphingomyelin are partitioned into mixed micelles and are absorbed by micelle-dependent diffusion into enterocytes mainly in the duodenum and jejunum. Free choline is absorbed in the same region by a saturable, carrier-mediated process localized in the brush border, and efficient at low lumenal concentrations (<4 mM). It is also absorbed less efficiently by passive diffusion; at high intakes, the nonabsorbed major portion passes to the hindgut where it is catabolized by the intestinal microbiome to the end product **trimethylamine**[15] much of which is absorbed and excreted in the urine. Because PC, being better absorbed, is not subject to such extensive microbial metabolism, it produces less urinary trimethylamine. The oxidized metabolite of choline,

14. Also called lysolecithin.
15. The characteristic fishy odor of this product is identifiable after consumption of a choline supplement.

TABLE 19.2 Distribution of Phospholipids in Plasma Lipoproteins

Lipoprotein Class	Phospholipid Content, %
High-density lipoproteins (HDLs)	~30
Low-density lipoproteins (LDLs)	~22
Very low-density lipoproteins (VLDLs)	10–25
Chylomicra	3–15

betaine, is absorbed by way of a different carrier, the IMINO proline transporter.

Transport. Within the enterocyte, most recently absorbed lysoPC is esterified to form PC, some of which is thought to be incorporated into nascent high-density lipoproteins (HDLs). Those factors are transported into the lymphatic circulation (or the portal circulation in birds, fishes, and reptiles) bound to chylomicra, which are subject to clearance to the lipoproteins that circulate to the peripheral tissues. Thus, choline is transported to the tissues predominantly as phospholipids associated with the plasma lipoproteins (Table 19.2).

Free choline, a positively charged quaternary amine, does not cross biological membranes without a carrier. Three transport systems have been identified:

- **A high-affinity,**[16] **Na^+-dependent transporter CHT1** (SCL5A7) unique to cholinergic neurons of the brain, brain stem, and spinal column provide choline for the synthesis of acetylcholine. Genetic deletion of this transporter is lethal to mice within an hour of birth.[17]
- **Intermediate-affinity,**[18] **Na^+-dependent transporters** in the in the solute carrier family (SLC44) facilitate the transport of choline across plasma and mitochondrial membranes. Five choline transporter-like transporters have been identified. The predominant one, CTL1, is expressed in all human tissues. CTL2 is expressed mainly in lung, colon, spleen, and inner ear and to lesser extents in brain, liver, and kidney.
- **Low-affinity,**[19] **polyspecific transporters** in the solute carrier family transport choline nonspecifically. These include three organic cation transporters (OCTs) and two organic carnitine/cation transporters (OCTNs) and are expressed in many tissues.

16. K_m <10 μM.
17. Wurtman, R.J., Cansev, M., Ulus, I.H., 2009. Handbook of Neurochemistry and Molecular Neurobiology: Nerual Lipids. In: Lajitha, A., Tettamanti, G., Goracci, G. (Eds.). Springer, New York, p. 443–500.
18. K_m 10–30 μM.
19. K_m >30 μM.

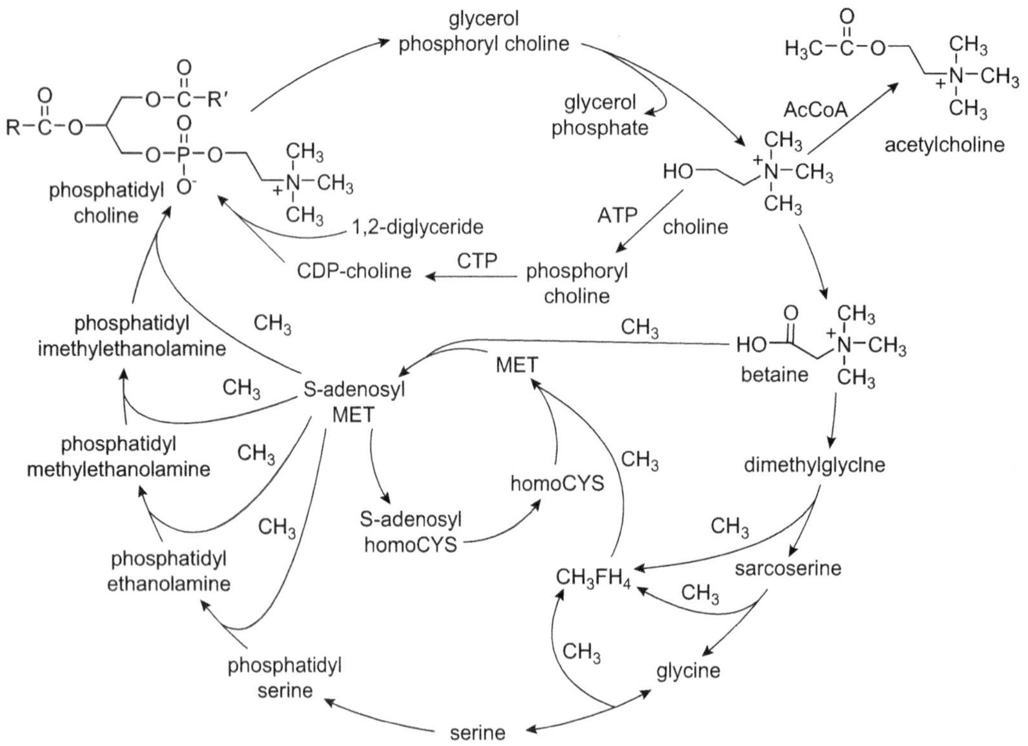

FIGURE 19.2 The biosynthesis and utilization of choline. *AcCoA*, acetyl-CoA; *CH3FH4*, 5-methyl tetrahydrofolate; *CTP*, cytidine triphosphate; *homo-Cys*, homocysteine; *MET*, methionine.

Tissue distribution. Choline is present in all tissues as an essential component of phospholipids in membranes of all types. Therefore, ~95% of the body pool of choline occurs as PC. It is not stored but occurs in the relatively great concentrations in the essential organs (e.g., brain, liver, kidney) mostly as PC and sphingomyelin.[20] Placental tissues are unique in that they accumulate large amounts of acetylcholine, presumably to meet fetal needs, which is otherwise present only in the parasympathetic nervous system.

Choline Metabolism

Biosynthesis. De novo biosynthesis of choline occurs by the sequential, *S*-adenosylmethionine (SAM)-dependent methylation of phosphatidylethanolamine by **phosphatidylethanolamine *N*-methyltransferase (PEMT)** to produce PC, which accounts for ~95% of the total choline pool (Fig. 19.2). This biosynthetic activity is due to two enzymes that use SAM as the methyl donor: (1) an inner plasma membrane enzyme adds the first methyl group and is rate limiting to choline synthesis; and (2) an outer membrane enzyme adds the second and third methyl groups. This pathway occurs in many tissues, but it greatest in liver in which it can account for as much as 40% of that organ's PC content. It is upregulated under conditions of dietary choline deficiency. PC produced by this pathway is composed mainly of long-chain polyunsaturated fatty acids.

Incorporation of choline into PC. Free choline is converted to PC in three steps. It is phosphorylated by cytosolic ATP-dependent **choline phosphotransferase** (also called **choline kinase**) in the rate-limiting step in PC synthesis. **Phosphorylcholine** is converted to **cytidine diphosphatidylcholine (CDP)** by **CDP:phosphocholine cytidyltransferase**. This step is upregulated by diacylglycerol and feedback-inhibited by PC. CDP is combined with diacylglycerol by **choline phosphotransferase** located on the endoplasmic reticulum. This pathway can account for some 70% of hepatic PC. PC produced by the CDP–choline pathway[21] is composed mainly of saturated fatty acids.

Mobilization of free choline. Cells mobilize choline by hydrolysis of PC (Fig. 19.2) and sphingomyelin. This can occur in two ways:

- The actions of **phospholipases A_2** and **B** yield lysoPC, which is converted to glyceryl-PC by **lysophospholipases**

20. Free choline comprises <1% of the total.

21. Also called the Kennedy pathway, after Eugene P. Kennedy (1919–2011) who first identified it in 1956 when he was on the faculty of the University of Chicago.

and, subsequently, to free choline by **glycerophosphodiester phosphodiesterases**. PC can also be metabolized by phospholipases C and D, which yield phosphorylcholine and phosphatidic acid, respectively. The former can be converted to free choline by **alkaline phosphatase**.

- By **phospholipid base exchange** of PC and serine, yielding choline and phosphatidylserine.

Choline oxidation. Most (60–90%) of the metabolism of free choline involves its oxidation, which frees its methyl groups to enter single-C metabolism. Choline oxidation is induced by dietary choline; it occurs in several tissues but is notably absent from brain, muscle, and blood. It is accomplished in two irreversible steps by a dual enzyme system collectively called **choline oxidase:**

- **Oxidation of choline** in the mitochondria by the FAD-linked enzyme **choline dehydrogenase** to yield betaine aldehyde and $FADH_2$.
- **Reduction of betaine aldehyde** to form **betaine** by the NAD-linked enzyme **betaine aldehyde dehydrogenase**, in both the mitochondria and cytosol, yielding NADH. This step occurs at a rate 10-fold greater than that of the incorporation of choline into phosphorylcholine.

The production of mitochondrial electron transport chain substrates ($FADH_2$ and NADH) by this process means that ~5 moles of ATP are generated for each mole of choline converted to betaine.[22]

Acetylcholine synthesis. Only a small fraction of choline is acetylated, but that amount provides the important neurotransmitter **acetylcholine**. Free choline can cross the blood–brain barrier and be taken up by presynaptic cholinergic neurons (by the low-affinity transporter) to be incorporated into PC and located in membranes. Choline can be mobilized by phospholipase D to provide free choline for the reaction with acetyl-CoA catalyzed by **choline acetyltransferase**. Because brain choline acetyltransferase is not saturated with either substrate, the availability of choline, and/or acetyl-CoA determines the rate of synthesis of acetylcholine.

Metabolic Functions of Choline

Choline and its derivative have several functions that are essential to normal metabolism and health.

- **Membrane structure**. PC is the major structural component of membranes; therefore, PC is required for membrane biogenesis in growth and cell division. With sphingomyelin it is localized in the outer leaflet of the lipid bilayer, contributing membrane asymmetry, which facilitates transmembrane signal transduction[23] and promotes lipid trafficking. In the lung, palmitate-rich PC is the active component of surfactant. PC is also a precursor to sphingomyelin,[24] the basic structure of membrane sphingolipids. Sphingomyelin comprises 10–20 mol% of the lipids in plasma membranes and is particularly abundant in the myelin sheaths of nerve cell axons. Sphingomyelin is formed by the addition of **ceramide**[25] by **PC–ceramide–phosphocholine transferase**.
- **Cell signaling**. Membrane PC also functions as a precursor of second messengers affecting cellular function. Hydrolysis by phospholipases produce diacylglycerol, an activator of **protein kinase C**, which phosphorylates a large number of target proteins in the cell; arachidonic acid (20:4), which is used in the synthesis of **eicosanoids**; phosphocholine, which is as a signaling molecule in cell division, including in tumorigenesis; free choline for incorporation into **platelet-activating factor**, which is important in clotting, inflammation,[26] uterine ovum implantation, fetal maturation, and induction of labor; free choline for incorporation into **plasmalogen**, which occurs at high levels in the sarcolemma and is important in myocardial function[27]; and free choline for the synthesis of **acetylcholine**, which is thought to act as a signaling molecule in immune cells and placenta.[28]
- **Lipoprotein synthesis**. PC is the major component of the lipid monolayer on the surface of VLDLs. The packaging of those particles occurs in the Golgi cisternae and is necessary for exporting lipids from the liver.
- **Methyl donor**. Choline is a key source of labile methyl groups via its oxidized metabolite betaine. A single methyl group is transferred from betaine to **homocysteine (Hcy)** to produce **dimethylglycine** and **methionine (MET)** by the Zn-containing enzyme **betaine:homocysteine *S*-methyltransferase (BHMT)** in the liver and kidney.[29] BHMT is downregulated by SAM and dimethylglycine. Additional methyl groups are rendered from dimethylglycine through its conversion to **sarcosine** by **dimethylg-**

22. Betaine is transported into cells by the betaine/γ-aminobutyric acid transporter.

23. Phospholipid-mediated signal transduction involves membrane phospholipases that trigger generation of inositol-1,4,5-triphosphate, which acts to release Ca^{2+} from stores in the endoplasmic reticulum.

24. Sphingomyelins can have either PC or phosphatidylethanolamine as the head group; they are found in cell membranes and myelin sheathes of nerve cell axons.

25. Ceramide is formed by adding a fatty acid to the amino group of sphingosine. Among the biological activities of ceramide are the stimulation of *apoptosis*, i.e., programmed cell death.

26. Overproduction of platelet-activating factor has been shown to produce a hyperresponsive condition, as occurs in asthma.

27. It is thought that the adverse effects of myocardial ischemia may involve the breakdown of plasmalogen.

28. Wessler, I., Kirkpatrick, C.J., 2008. Br. J. Pharmacol. 154, 1558–1571.

29. BHMT is among the more abundant proteins in liver, comprising as much as 2% of total soluble protein.

lycine dehydrogenase, and the subsequent conversion of sarcosine to **glycine** by **sarcosine dehydrogenase**. These transmethylation reactions link choline to folate metabolism. When the 5-methyl-FH_4 cannot meet intracellular demands for SAM, choline becomes an important dietary source of labile methyl groups for MET production.

- **Neurotransmission**. Acetylcholine is a key neurotransmitter in both the central and peripheral nervous systems. Through cholinergic neurons, it affects the functions of several organs (e.g., heart, lungs, gut, bladder, pancreas, and endocrine organs). Acetylcholine is also important for cognitive function and the development of the brain; choline deprivation of rat pups produce permanent memory impairment.
- **Bile**. Some 95% of biliary phosphor lipids[30] comprise a PC species that is particularly rich in palmitic (16:0) and linoleic (18:0) acids. These PCs are produced in the hepatic endoplasmic reticulum from circulating HDLs. Most (~95%) PC secreted with bile into the gut is reabsorbed; less than half of that amount returns to the liver, the major portion being used by extrahepatic tissues.
- **Osmoregulation**. The choline metabolites betaine and glyceryl-PC function as **organic osmolytes**. This function is particularly important in the renal cortex and medulla, which are exposed to high extracellular osmolarity as a consequence of concentrating urine for excretion.

These functions have several physiological effects:

- **Neurologic function**. The intake of choline can affect the concentrations of acetylcholine in the brain, suggesting that choline loading may be beneficial to patients with diseases involving deficiencies of cholinergic neurotransmission. Indeed, studies with animal models have shown choline supplementation during development to enhance cognitive performance, particularly, on more difficult tasks; to increase electrophysiological responsiveness; and to provide some protection against alcohol and other neurotoxic agents. A large cohort study found plasma choline level to be inversely associated with the incidence of symptoms of anxiety.[31] In humans, large doses (multiple gram quantities) of choline have been used to increase brain choline concentrations above normal levels, thereby stimulating the synthesis of acetylcholine in nerve terminals. Such supplementation has been found to help in the treatment of tardive dyskinesia, a movement disorder involving inadequate neurotransmission at striatal cholinergic interneurons.[32] Choline supplements have also been used with some success to improve free memory in subjects without dementia[33] and to diminish short-term memory losses associated with Alzheimer's disease,[34] a disorder involving deficiency of hippocampal cholinergic neurons. It has been suggested that autocannibalism of membrane PC may be an underlying defect in that disease; this is supported by the fact that patients treated with anticholinergic drugs develop short-term memory deficits resembling those associated with hippocampal lesions. PC has been reported to reduce manic episodes in patients, suggesting that it can be centrally active; however, such treatment has been found to exacerbate depression among tardive dyskinesia patients. A meta-analysis of 14 controlled trials found that supplementation with CDP–choline improved memory in elderly subjects with cognitive disorders.[35] Dietary choline intake was found to predict cognitive functioning in the Framingham Offspring cohort.[36]
- **Epigenetic effector**. Its contributions (via betaine) to the pool of labile methyl groups mean that the availability of choline can affect the metabolic functions of the single-C pool, particularly under circumstances that would limit that pool, e.g., when intakes of folate, vitamin B_{12}, and/or MET are limiting. Such circumstances can reduce the methylation of DNA and histones, thus modifying genetic expression. This phenomenon has been demonstrated in rodent models[37] and human volunteers.[38]

Dietary Choline in Health and Disease

Anticarcinogenesis. Choline deprivation in animal models promotes both the initiation and promotion phases of spontaneous and chemically induced

30. Phospholipids comprise ~3% of bile.
31. Bjelland, I., Tell, G.S., Vollset, S.E., 2009. Am. J. Clin. Nutr. 90, 1056–1060.
32. Tardive dyskinesia is prevalent among patients treated with neuroleptic drugs (affecting the autonomic nervous system) and is characterized by involuntary movements resembling both *chorea* (irregular and spasmodic) and *athetosis* (slow and writhing) of the face, extremities and, usually, the trunk.
33. Spiers, P., Myers, D., Hochanadel, G.S., et al., 1996. Arch. Neurol. 53, 441–448; Ladd, S.L., Sommer, S.A., LaBerge, S., et al., 1993. Clin. Neuropharmacol. 16, 540–549; Sitaram, N., Weingartner, H., Caine, E.D., et al., 1978. Life Sci. 22, 1555–1560.
34. Alvarez, X.A., Laredo, M., Corzo, D., et al., 1997. Methods Fund. Exp. Clin. Pharmacol. 19, 201–210.
35. Fioravanti, M., Yanagi, M., 2005. Cochrane Database Syst. Rev., CD000269. pub. 3.
36. Poly, C., Massar, J.M., Seshardi, S., et al., 2011. Am. J. Clin. Nutr. 94, 1584–1591.
37. Davison, J.M., Mellott, T.J., Kovacheva, V.P., et al., 2009. J. Biol. Chem. 284, 1982–1989.
38. Shin, W., Yan, J., Abratte, C.M., et al., 2010. J. Nutr. 140, 975–980; Jiang, X., Yan, J., West, A.A., et al., 2012. FASEB J. 26, 3563–3574.

hepatocarcinogenesis. These effects appear to result from decreases in tissue levels of SAM resulting in hypomethylation of DNA and, consequently, changes in gene transcription including modified expression of p53 protein. Epidemiological studies have found choline intake to be positively associated with risk to colorectal and prostate cancer[39]; although other studies have found no significant associations.[40]

Birth defects. Serum choline was inversely associated with risk of neural tube defect pregnancies.[41]

Choline Deficiency

Individuals with insufficient choline biosynthesis or deficiencies in choline or single-C metabolism can show hepatic steatosis, hemorrhagic renal degeneration, and (in animals) depressed growth.

Humans. Choline deficiency is not commonly observed in humans; this may reflect the adequate intakes of other methyl donors in the subjects most frequently studied. In a study in which healthy adults were fed a diet adequate in MET and folate, deprivation of choline produced indication of hepatic dysfunction (increased serum transaminase activities) that was corrected by feeding choline. In another study, chronic use of a choline-free parenteral feeding solution resulted in hepatic steatosis in adults.[42] Two cases have been identified in which hereditary enzyme deficiencies have produced needs for dietary choline: women with the 744C allele for PEMT, which yields that enzyme unresponsive to induction by estrogen[43]; and women with the 1958A allele for 5,10-methylenetetrahydrofolate dehydrogenase.[44] In 1998, the Food and Nutrition Board of the Institute of Medicine set recommended intakes for choline (Table 19.3).

Animals. Choline can be an indispensable dietary constituent for several species. This includes the chick, due to developmental deficiencies of PEMT, has an absolute need for dietary choline until about 13 weeks of age. Until that age, choline deprivation produces fatty liver and perosis, which signs are also see in older poultry fed diets deficient in methyl groups (e.g., MET-deficient). Clear needs for dietary choline have been demonstrated for fish,[45] signs of deficiency including impaired weight gain, reduced efficiency of feed utilization, and hepatic steatosis. It is generally assumed that most fishes cannot synthesize choline at levels sufficient to meet their physiological needs. Rodents require choline only if their capacity to methylate phosphatidylethanolamine is limited by the availability of methyl groups. In such cases, the feeding of methyl donors (MET, betaine) spares the need for choline. Rats fed a choline-deficient diet show 30–40% reductions in hepatic and brain levels of folate, resulting in a shift toward longer folylpolyglutamate metabolites and in the undermethylation of DNA. Pregnancy and lactation result in significant decreases in hepatic choline levels. Betaine can spare the need for choline to support growth and prevent fatty liver, apparently by providing single-C units.

TABLE 19.3 Recommended Choline Intakes

Age–Sex	AI[a], mg/day
0–6 months	125
7–12 months	150
1–3 years	200
4–8 years	250
9–13 years	375
>13 year males	550
14–18 year females	400
>18 year females	425
Pregnancy	450
Lactation	550

[a]*Adequate intakes are given, as RDAs have not been established.*
Food and Nutrition Board, 2000. Dietary Reference Intakes for Thiamin, Riboflavin, Niacin, Vitamin B_6, Folate, Vitamin B_{12}, Pantothenic Acid, Biotin and Choline. National Academy Press, Washington, DC, 564 pp.

Biomarkers of Choline Status

There are no satisfactory biomarkers of choline status. Plasma concentrations of free choline are regulated in the range of 6–13 μM[46] and are only minimally responsive to changes in choline intake. Urinary choline represents only 2% of the amount recently consumed.

39. Johansson, M., Van Guelpen, B., Vollset, S.E., et al., 2009. Cancer Epidemiol. Biomarkers Prev. 18, 1538–1543; Lee, J.E. Giovanussi, E., Fuchs, C.S., et al., 2010. Cancer Epidemiol. Biomarkers Prev. 19, 884–887.
40. Cho, E., Willett, W.C., Colditz, G.A., et al., 2007. J. Nat. Cancer. Inst. 99, 1224–1231; Cho, E., Homes, M.D., Hankinson, et al., 2010. Br. J. Cancer 102, 489–494.
41. Shaw, G.M., Finell, R.H., Blom, H.J., et al., 2009. Epidemiol. 20, 714–719.
42. Buchman, A.L., Ament, M.E., Sohel, M., et al., 2001. J. Parenteral Enteral Nutr. 25, 260–268.
43. da Costa, K.A., Kozyreva, O.G., Song, J., 2006. FASEB J. 20, 1336–1344.
44. Kohlmeier, M., da Costa K.A., Fischer, L.M., et al., 2005. Proc. Natl. Acad. Sci. U.S.A. 102, 16025–16030.
45. For example, red drum (*Sciaenops ocellatus*), striped bass (*Morone* spp.).
46. Abratte, C.M., Wang, W., Li, R., et al., 2009. J. Nutr. Biochem. 20, 62–69.

TABLE 19.4 Recommended Tolerable Upper Limits of Choline Intake

Age–Sex	UL[a], mg/day
0–12 months	–
1–8 years	1000
9–13 years	2000
>13 year males	3000
14–18 year females	3000
>18 year females	3500
Pregnancy	
<18	3000
≥18	3500
Lactation	
<18	3000
≥18	3500

[a]*Food and Nutrition Board, 2000. Dietary Reference Intakes for Thiamin, Riboflavin, Niacin, Vitamin B_6, Folate, Vitamin B_{12}, Pantothenic Acid, Biotin and Choline. National Academy Press, Washington, DC, 564 pp.*

Choline Toxicity

The toxicity of choline is low. However, deleterious effects have been reported for the salt choline chloride: growth depression, impaired utilization of vitamin B_6, and increased mortality. These effects may not relate to choline but, instead, to the perturbation of acid–base balance caused by the high level of chloride administered with large doses of that salt. In humans, high doses (e.g., 20 g) have produced dizziness, nausea, and diarrhea. Upper limits of choline intake have been established (Table 19.4).

3. CARNITINE

Carnitine is a normal metabolite that functions in the transport of long-chain fatty acids from the cytosol into the mitochondria for their oxidation as sources of energy. It is biosynthesized by most species; however, some species are not capable of its biosynthesis and some circumstances limit its biosynthesis for others, making it **conditionally essential**.

Recognition of a Nutritional Role of Carnitine

In the 1950s, Fraenkel[47] and colleagues found that the successful growth of the yellow mealworm (*Tenebrio molitor*) required the feeding of a natural substance, which they found present in milk, yeast, and many animal tissues. They purified the growth factor from whey solids and named it "**vitamin B_T**." Soon vitamin B_T was found to be **carnitine**, a known metabolite that had been identified in extracts of mammalian muscle five decades earlier. The first indication of a metabolic role came in the 1960s, when carnitine was found to stimulate the in vitro oxidation of long-chain fatty acids by subcellular fractions of heart muscle.[48] Research interest in carnitine was stimulated by the finding that carnitine, it is biosynthesized by mammals from the amino acid lysine,[49] which is limiting in the diets of many of the world's poor, and by the description of clinical syndromes (of apparently genetic origin) associated with carnitine deficiency.

Conditions of Need for Dietary Carnitine

Most species can synthesize carnitine at rates sufficient for their needs. In humans, this is indicated by the fact that subjects whose cereal-based diets provide very little preformed carnitine show plasma carnitine concentrations comparable to those of whose diets provide abundant amounts of the factor.[50] However, some circumstances can limit either the biosynthesis or utilization.

Humans. Neonates have compromised carnitine biosynthetic capacity due to very low hepatic γ-butyrobetaine hydroxylase activities. Their carnitine status is dependent on that of the mother, on the placental transfer of carnitine in utero, and on the availability of exogenous sources after birth.[51] Appreciable amounts of carnitine are found in the milk of several species.[52] Infants fed soy-based formulas, which contain little or no carnitine, are unable to maintain normal plasma carnitine levels; intravenous administration of L-carnitine allows them to do so.

Because carnitine is a key cofactor for energy metabolism, preterm infants can be at special risk. While their plasma carnitine levels tend to be nearly normal, their stores can be depleted rapidly during the course of intravenous

47. Gottfried S. Fraenkel (1901–84) was a German-born insect physiologist who, after fleeing Germany in 1933, spent most of his career as a Professor of Entomology at the University of Illinois. In 1968, he was named to the National Academy of Sciences.

48. Fritz, I.B., Yue, K.T.N., 1963. J. Lipid Res. 4, 279–288.

49. Tanphaichtr, V., Horne, D.W., Broquist, H.P., 1971. J. Biol. Chem. 246, 6364–6366.

50. Mean ± SD: men, 59 ± 12 μM (n = 40); women, 52 ± 12 μM (n = 45).

51. Examples are human milk, prepared infant formulas and milk replacers. It has been suggested that natural selection has resulted in mother's milk containing carnitine in proportion to the needs of the infant. In fact, the greatest concentrations of carnitine in human milk occur during the first 2–3 days of suckling. During the first 3 weeks of lactation, the carnitine content of human milk varies from 50 to 70 nmol/mL; after that time, it declines to about 35 nmol/mL by 6–8 weeks. Most milk-based infant formulas contain comparable or slightly greater amounts of carnitine; however, formulas based on soybean protein or casein and casein hydrolysate contain little or no carnitine. Lipid emulsions also contain no appreciable carnitine.

52. e.g., Human milk: 28–95 nmol/mL; cow's milk: 190–270 nmol/mL.

feeding with solutions unsupplemented with carnitine. One study found the plasma carnitine concentrations of preterm infants (gestational age <36 weeks) to drop from 29 to 13 μM during total parenteral feeding. The consequences of suboptimal carnitine status are direct, as newborns change from utilizing glucose to fats as the major fuel.[53] Thus, free fatty acids are the preferred metabolic fuels for the newborn, especially for the heart and skeletal muscle, which depend on the oxidation of fatty acids for more than half of their total energy metabolism, when glucose availability is limited. Neonates fed noncarnitine-fortified soy-based formula diets have shown, with 2 weeks, reduced hepatic carnitine concentrations with associated reductions in hepatic fatty acid oxidation and ketogenesis and hypertriglyceridemia. The long-term consequences of these reductions are unknown and no clinical signs of carnitine deficiency in infants have been described.

Individuals with low muscle and/or plasma carnitine levels typically show lipid accumulation in muscle with high risk of encephalopathy, progressive muscular weakness, and cardiomyopathy. Carnitine deficiency has also been recognized as a secondary feature of various other genetic disorders, e.g., organic acidurias[54] and Fanconi syndrome,[55] in which the renal tubular loss of total carnitine and of acylcarnitine esters in particular are elevated.

Other mammals and birds. While most species studied to date appear to synthesize carnitine at rates sufficient for their needs, the biosynthetic capacity of the rat can be impaired by deprivation of its metabolic precursors, lysine and/or methionine.[56] Carnitine biosynthetic capacities may also be limited in the fetus, as carnitine supplementation of pregnant and lactating sows has been found to enhance fetal glucose oxidation and result in higher birth weights and improved growth.[57] Neonatal rabbits fed a carnitine-free colostrum replacer or a carnitine-free weaning diet showed abnormally low-tissue and urinary carnitine levels, decreased plasma total and VLDL–cholesterol levels, and increased apolipoprotein levels. Carnitine supplementation of the diets of turkey hens was found to improve egg fertility, embryonic survival, and chick growth.[58]

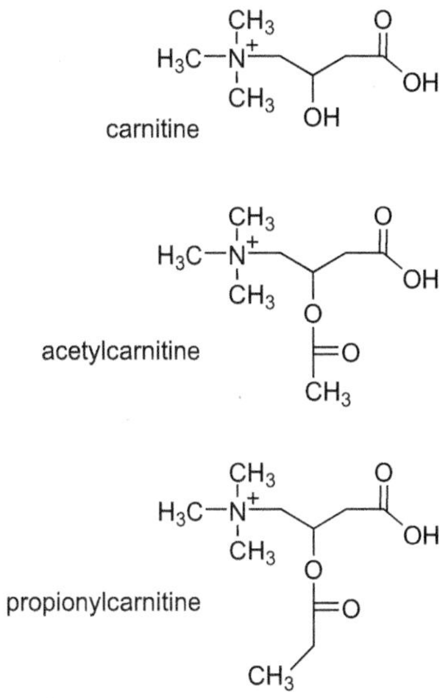

FIGURE 19.3 Carnitine and functional metabolites.

Insects. Carnitine is essential in the diets of at least some insect species. This includes beetles of the family *Tenebrionidae*,[59] the beetle *Oryzaephilus surinamensis*, and the fruit fly *Drosophila melanogaster*. It is presumed that carnitine plays the same essential role in the metabolism of fatty acids in insects that it does in mammals, and that their special dietary need is due to their inability to synthesize it from endogenous sources; however, that has not been investigated. For these species, carnitine is a vitamin.[60]

Conditions producing need for dietary carnitine are as follows:

- Limited biosynthesis in neonates
- Absent biosynthesis/transport in some insects.

Chemical Properties of Carnitine

Carnitine is the generic term for a number of compounds including L-carnitine (β[-]-β-hydroxy-γ-[*N,N,N*-trimethylaminobutyrate][61]) and its acetyl and propionyl esters (Fig. 19.3). Only the L-isomer is made by and is biologically active for eukaryotes. At physiological pH, carnitine

53. At birth, plasma-free fatty acids and β-hydroxybutyrate concentrations are rapidly elevated owing to the mobilization of fat from adipose tissue. These elevated levels are maintained by the utilization of high-fat diets such as human milk and many infant formulas, which typically contain more than 40% of total calories as lipid.
54. Examples include isovaleric, glutaric, propionic, and methylmalonic acidemias, which result from long- and medium-chain acyl-CoA dehydrogenase deficiencies.
55. Fanconi syndrome is a renal disease characterized by the excessive renal excretion of a number of metabolites that are normally reabsorbed (e.g., amino acids).
56. Some 0.1% of the lysine required by the rat is used to make carnitine; rats deprived of lysine develop mildly reduced tissue carnitine levels, depressed growth, and fatty liver (Eder, K., 2009. Br. J. Nutr. 102, 645–654).
57. Ramanau, A., Kluge, H., Eder, K., 2005. Br. J. Nutr. 93, 717–721; Eder, K., 2009. Br. J. Nutr. 102, 645–654.
58. Oso, A.O., Fafiolu, A.O., Adeleke, M.A., et al., 2014. J. Anim. Physiol. Anim. Nutr. 98, 766–774.
59. A family of mealworms.
60. Carnitine is an amino acid, but because it has no role in protein synthesis it meets the criteria of vitamin status (see Chapter 1).
61. Molecular weight, 161.5.

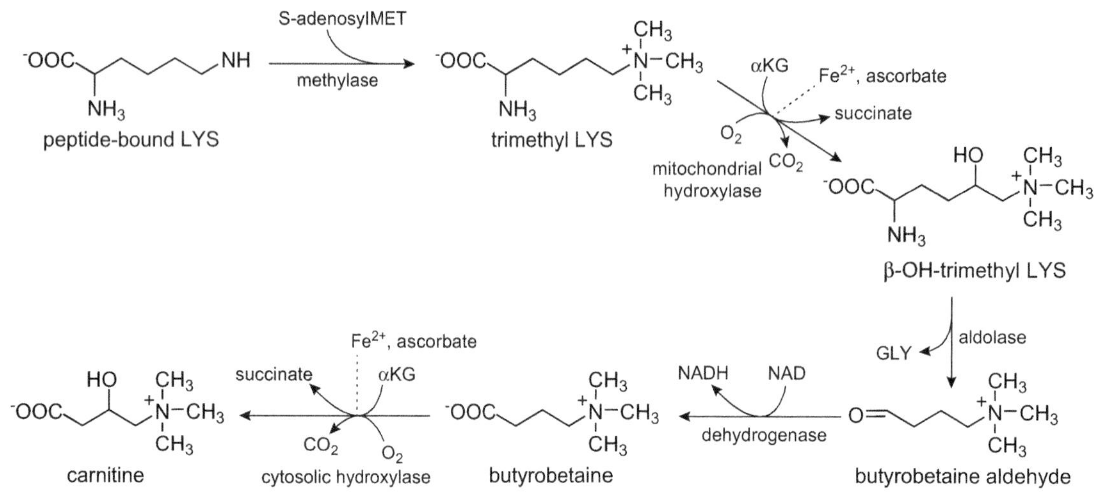

FIGURE 19.4 Biosynthesis of carnitine. *αKG*, α-ketoglutarate; *LYS*, lysine; *GLY*, glycine.

exists as a zwitterion,[62] with a positively charged quaternary amine and a negatively charged carboxyl. It forms esters with fatty acids by virtue of its hydroxyl.

Sources of Carnitine

Biosynthesis. Carnitine is synthesized in mammals. Humans are estimated to synthesize carnitine at ~1.2 μmol per kg body weight.[63] The synthesis starts with the post-translational modification of protein lysyl residues, some of which are thrice methylated to form ε-*N*-trimethyllysine using SAM as the methyl donor. This process is thought to occur in many proteins,[64] the turnover of which releases ε-*N*-trimethyllysine to be converted to L-carnitine through a sequence of reactions catalyzed by two Fe^{2+}- and ascorbate-dependent hydroxylases, a pyridoxal phosphate-dependent aldose, and an NAD^+-dependent dehydrogenase (Fig. 19.4). The first three of these enzymes occur in all tissues; but the last step, **γ-butyrobetaine hydroxylase**, occurs only in liver, kidney, and brain where it is present as multiple isoforms. It activity increases during development, peaking in the mid-teens.[65] This step does not, however, yield the quantitative conversion of γ-butyrobetaine to carnitine; therefore, the precursor is normally found in the urine. Tissue carnitine levels are depressed under conditions of vitamin B_6 deprivation and stimulated by the catabolic state (e.g., fasting), thyroid hormone, and the peroxisome proliferator clofibrate.[66]

62. i.e., A molecule with a positive and a negative electrical charge at different locations.

63. Rebouche, C.J., 1992. FASEB J. 6, 3379–3386.

64. e.g., Actin, myosin, ATP synthase, calmodulin, and histones.

65. For example, the hepatic γ-butyrobetaine hydroxylase activities of three infants and a 2.5-year-old boy were 12% and 30%, respectively, of the mean adult activity.

66. This effect appears to involve the peroxisome proliferator-activated nuclear receptor α (PPARα).

Dietary sources. The available data concerning the carnitine contents of foods are scant and must be considered suspect, owing to the use of nonstandard analytical methods. Nevertheless, it is apparent that materials of plant origin tend to be low in carnitine, whereas those derived from animals tend to be rich in the factor (Table 19.5). Red meats and dairy products are particularly rich sources. Typical mixed diets can be expected to provide 1–16 μg carnitine per day, the actual amount depending mainly on the intake of meats.

Absorption and Transport of Carnitine

Absorption. Carnitine is absorbed across the small intestine in two ways: by an active process dependent on Na^+ cotransport; and by a passive, diffusional process that may be important for the absorption of large doses of the factor. The efficiency of absorption appears to be high, ca. 55–95%, although high doses are absorbed at lower efficiencies (≤25%), with <1% appearing in the urine and very little appearing in the feces. The uptake of carnitine from the intestinal lumen into the mucosa is rapid, and about half of that taken up is acetylated in that tissue.

Transport. Carnitine is released slowly from tissues into the plasma in both the free and acetylated forms in simple solution. Plasma total carnitine concentrations in healthy adults are 30–89 μ*M*, with men typically showing slightly greater (by ~15%) concentrations than women. Carnitine is taken up against concentration gradients by peripheral tissues, most of which can also synthesize it.

Tissue distribution. Cellular uptake of carnitine and its short-chain acyl esters is facilitated by Na^+-dependent organic cation transporters (OCTNs). A high-affinity transporter (OCTN2) is expressed in kidney, skeletal muscle, heart, pancreas, testis, and placenta but is notably low in liver, brain, lung, and colon. The transporter in the colon

TABLE 19.5 Carnitine in Selected Foods and Feedstuffs

Food	Carnitine,[a] μg/100g
Vegetables	
Avocado	1.25
Cauliflower	0.13
Alfalfa	2.00
Peanut	0.76
Cereals	
Wheat	0.35–1.22
Bread	0.24
Meats	
Beef	59.8–67.4
Beef liver	2.6
Beef kidney	1.8
Beef heart	19.3
Chicken	4.6–9.1
Lamb, muscle	78.0
Other	
Cow's milk	0.53–3.91
Casein, acid washed	0.4
Torula yeast	1.60–3.29

[a]*None detected in cabbage, spinach, orange juice, barley, corn, egg.*
Adapted from Mitchell, M., 1978. Am. J. Clin. Nutr. 31, 293–306.

may serve in the uptake of any nonabsorbed dietary carnitine as well as β-butyrobetaine produced by the hindgut microbiome. Subjects with inborn errors in OCTN2 have been identified; they readily develop signs of carnitine deficiency unless they are given supplemental carnitine. A high-affinity, OTCN-related transporter, CT-2, is expressed only in testes. Short-chain carnitine derivatives are better utilized than the parent molecule by some tissues. Acetylcarnitine, which is structurally similar to acetylcholine, crosses the blood–brain barrier more readily than carnitine whereupon it is readily converted to carnitine. Propionylcarnitine, which is lipophilic, has high affinities for skeletal and cardiac muscles. Total body carnitine occurs in nonesterified (80–90%) and esterified (10–20%) forms. The greatest concentrations of carnitine in the human body are found in epididymal tissue, seminal plasma, and sperm.

Genetic disorders of carnitine transport. An autosomal, recessive disorder involving defective transport of carnitine has been described, **primary carnitine deficiency**.[67] In involves mutations in the *SLC22A5* gene that encodes OCTN2. Defects in the transporter result in reduced cellular uptake of carnitine and its consequent wastage in the urine. This impairs fatty acid oxidation, reduces ketogenesis, and increases cytosolic accumulation of long-chain fatty acids. The clinical manifestations can include hypoglycemia, hepatomegaly, weakness, and cardiomyopathy.[68]

Carnitine Metabolism

Turnover. Total body carnitine turns over every 66 days. However, that total is comprised of three kinetically distinct tissue pools.[69] In humans, 92–97% of body carnitine occurs in muscle, where it turns over relatively slowly (~191 h), but has relatively rapid flux (~427 μmol/h) due to the size of this pool. Liver and kidney contain 2–5% of body carnitine; this pool turns over in 12 h with a flux of ~277 μmol/h. Extracellular fluids contain 0.7–1.5% of body carnitine; this pool turns over quickly, e.g., 1.1 h. Carnitine is not catabolized in the tissues, which is released in the bile, it can be degraded by hindgut microbiome to γ-butyrobetaine, trimethylamine, and malic semialdehyde which appear in the feces. Trimethylamine can be absorbed across the colon to be oxidized to trimethylamine oxide by the liver.

Excretion. Carnitine is highly conserved by the human kidney, which reabsorbs >90% of filtered carnitine, being the dominant means of regulating plasma carnitine concentration. Renal reabsorption is facilitated by a brush border transporter, OTCN2, which recovers carnitine as well as its short-chain acyl esters. Renal tubular excretion of carnitine, its short-chain acyl esters, and γ-butyrobetaine adapt to circulating carnitine concentrations; some of this may come from the renal secretion of carnitine either in free form or as short-chain acylcarnitine esters. That urinary carnitine is typically comprised of a higher proportion of acylcarnitine esters than the general circulation suggests selective secretion of carnitine esters by renal tubular cells. Trimethylamine can be metabolized in the liver to yield trimethylamine oxide, which is also excreted in the urine.

Metabolic Functions of Carnitine

Mitochondrial fatty acid shuttle. Carnitine functions in the transport of long-chain fatty acids (fatty acyl-CoA) from the cytosol into the mitochondrial matrix for oxidation as sources of energy (Fig. 19.5). The mitochondrial inner membrane is impermeable to long-chain fatty acids and their CoA esters, which are therefore dependent on activation as carnitine esters for entry into that organelle. This transport process, referred to as the **carnitine transport**

67. Also called "carnitine uptake defect," the incidence is ~1:40,000 with a carrier rate of ~1%.

68. Fu, L., Huang, M., Chen, S., 2013. Korean Circ. 43, 785–792.

69. Rebouche, C.J., Engle, A.G., 1984. J. Clin. Invest. 73, 857–867.

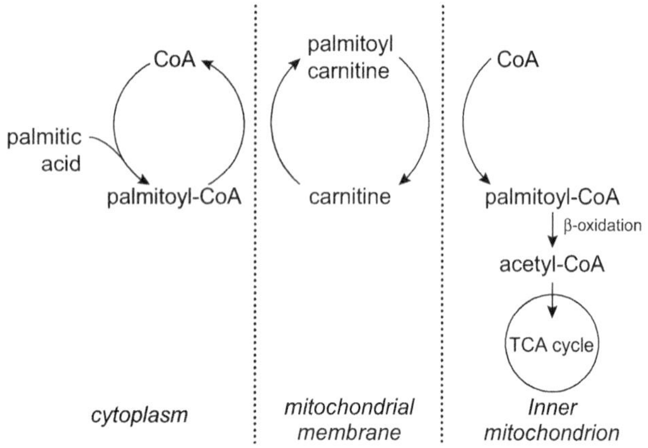

FIGURE 19.5 The mitochondrial fatty acid shuttle.

shuttle, is effected by two transesterifications involving fatty acyl-CoA esters and carnitine, and the action of three mitochondrial enzymes: **carnitine–acyltransferases I** and **II (CATI** and **CATII)** and **carnitine–acylcarnitine translocase (CACT)**.

CATI resides on the outer side of the inner mitochondrial membrane, while CATII is located on the matrix side and acylcarnitine translocase spans the inner membrane. These enzymes catalyze the formation and hydrolysis of fatty acylcarnitine esters.[70] The CATs provide an outlet for acetyl-CoA when the production of short-chain acyl-CoAs exceeds their rate of use by the TCA cycle, as in the transitions from fasting to feeding or from rest to vigorous physical activity, as well as in the overfed state.[71] Their activities in tissues are inhibited by accumulation of lipids, including palmitoyl-CoA, as in obesity.[72]

CACT catalyzes the exchange of carnitine and acylcarnitines produced by CATs across the membrane. The result of the concerted action of these enzymes is that long-chain fatty acids are brought into the mitochondrion by esterification to carnitine and transported as fatty acylcarnitine esters, after which carnitine is released and returned to the outer side of the membrane, thus, rendering the free fatty acid available for β-oxidation within the mitochondrion. It has been suggested that the carnitine transport shuttle may also function in the reverse direction by transporting acetyl groups back to the cytoplasm for fatty acid synthesis.[73]

Under normal metabolic conditions, it appears that short-chain acyl-CoAs are generated at rates comparable to the rates of their use, such that acylcarnitine does not accumulate. However, under conditions of propionic acidemia or methylmalonic academia/aciduria, which occur in vitamin B_{12} deficiency, the urinary excretion of acylcarnitine is enhanced owing to the increased formation of short-chain acylcarnitines.

The activity of the carnitine transport shuttle is typically low at birth but increases dramatically after birth. For example, the CATI activity in rat liver increases nearly five-fold within 24h of birth, peaking within 2–3days. Similar increases in the hepatic activities of the CATI and CATII have been observed in human infants. That these increases correspond to the development of fatty acid oxidation in the heart, liver, and adipose tissue suggests that the carnitine transport shuttle is rate limiting to that process.

Genetic disorders of carnitine metabolism. Genetic deficiencies have been described in both CATI and CATII. These deficiencies are rare; each involves impaired oxidation of long-chain fatty acids. CATI[74] deficiency is due to a single nucleotide variation in the enzyme. Clinical signs are typically seen in children, presenting as hepatic encephalopathy under conditions of increased energy demands are increased, e.g., in febrile illness.[75] At least 90 nucleotide changes have been identified in CATII; one variant accounts for some 60% of all mutant alleles. Those of CATII deficiency[76] present from infancy to adulthood in three multisystem diseases: a lethal neonatal form, a severe infantile hepatocardiomuscular form, and a mild myopathic form.[77]

Other functions. Carnitine and its esters have other biological actions:

- **Peroxisomal fatty acid shuttle**—Chain-shortened fatty acyl-CoAs are transesterified to carnitine within peroxisomes from which they are transported by OCTN3 into the cytoplasm for mitochondrial uptake.
- **Glucocorticoid-like actions**—The ability of carnitine to bind the glucocorticoid receptor-α has been related to its glucocorticoid-like effects in causing receptor-mediated release of cytokines and maintaining urea cycle activity to attenuate hyperammonemia.[78]
- **Phospholipid remodeling**—Carnitine-bound fatty acids are incorporated into erythrocyte membrane phospholipids during repair after oxidative attack, as well as into the lung surfactant dipalmitoylphosphatidylcholine.

70. The carnitine acyltransferases are actually a family of six related enzymes with different, but overlapping, chain length specificities that have been isolated from mitochondria (three each from the inner and outer sides of the inner membrane).

71. Muoio, D.M., Noland, R.C., Kovalik, J.P., et al., 2012. Cell Metab. 15, 764–777.

72. Seiler, S.E., Martin, O.J., Noland, R.C., et al., 2014. J. Lipid Res. 55, 635–644.

73. Even if such a reverse shuttle were to function, its contribution to fatty acid synthesis would be insignificant in comparison with that of the citrate shuttle, which transports acetyl-CoA to the cytoplasm by the action of a citrate cleavage enzyme.

74. Fewer than 60 cases are known, mostly in the native Alaskan population where the frequency of the variant is estimated to be 1.3/1000 live births.

75. Bennett, M.J., Santini, A.B., 2013. GeneReview NCI Bookshelf.

76. Wieser, T., 2014. GeneReview NCI Bookshelf.

77. These phenotypes have been described in 28, 28, and 300 families, respectively.

78. Manoli, I., De Martino, M.U., Kino, T., et al., 2004. Ann. N.Y. Acad. Sci. 1033, 147–157.

- **Insulin sensitivity**—Tissue carnitine levels, regulated through local expression of CATI and CACT, affect mitochondrial levels of acyl-CoA substrates control of which has been linked to insulin sensitivity.[79]
- **Cardiac function**—Carnitine stabilizes the cardiac mitochondrial membrane, protecting it from permeability transition in the presence of high-fatty acid β-oxidation.[80]
- **Thyroid function**—Carnitine appears to be a peripheral agonist of thyroid hormone. A randomized, controlled trial found administration of carnitine effective in reversing symptoms of hyperthyroidism.[81]

Dietary Carnitine in Health and Disease

Hepatic function. Hypocarnitinemia (plasma concentrations <55 μM) and tissue carnitine depletion appear to be common in patients with advanced cirrhosis, who not only tend to have marginal intakes of carnitine and its precursors but also have diminished capacity to synthesize carnitine. Carnitine supplementation has been found to protect against ammonia-induced encephalopathy in cirrhotics.[82]

Renal function. Patients with renal disease managed with chronic hemodialysis can be depleted of carnitine[83] due to losses in the dialysate that greatly exceed the amounts normally lost in the urine.[84] Tissue depletion of carnitine has been related to the complications of hemodialysis: hyperlipidemia, cardiomyopathy, skeletal muscle asthenia, and cramps. A meta-analysis of 49 randomized clinical trials found that carnitine treatment of dialysis patients with end-stage renal disease reduced serum concentrations of LDL and C-reactive protein (CRP)[85].

Cardiovascular health. Studies with animal models have shown that carnitine supplementation can benefit cardiac function.[86] Administration of propionylcarnitine improved cardiac function, as well as the functional recovery of the myocardium after ischemia.[87] Treatment with propionylcarnitine had a prostacyclin-like affect in countering vasoconstrictor activity, promoted endothelium-dependent arterial dilation in hypertensive individuals, and had antihyperlipidemic and antiatherosclerotic effects. Rats supplemented with carnitine showed prostaglandin responses associated with cardioprotection (i.e., reduced ratios of 6-keto-prostaglandin $F_{1\alpha}$ to thromboxane B_2 and leukotriene B_4) and reduced myocardial injury after ischemia/reperfusion.

Randomized, controlled trials with cardiac patients have found carnitine treatment to reduce hypertension, enhance vascular function, reduce left ventrical dilatation, and prevent ventricular remodeling. A multicenter, randomized trial found propionylcarnitine to enhance the exercise capacity of chronic heart failure patients with relatively intact myocardial function.[88]

Neurologic health. Studies have found that supplementation with carnitine can restore mitochondrial function in aging animals, which typically declines due to changes that include diminished expression of CACT.[89] Treatment with acetylcarnitine has been shown to reduce memory loss in old rats, the effect was associated with the release of acetylcholine in the striatum and hippocampus.

A multicenter, randomized, clinical intervention trial with Alzheimer's disease patients found attenuated progression for several parameters of behavior, disability, and cognitive performance.[90] Trials with Alzheimer's disease patients have found acetylcarnitine treatment to reduce deterioration in reaction time, reduce depressive symptoms, and improve cognitive performance.[91] A meta-analysis of 21 studies with patients with mild cognitive impairment or mild Alzheimer's disease concluded that treatment with acylcarnitine (1.5–2 g/day) improved both clinical and psychomotor assessments.[92] A randomized clinical trial found acylcarnitine to be as effective as, but better tolerated than, the antidepressive drug amisulpride in reducing depressive symptoms in dysthymia patients.[93] Studies have found carnitine supplementation effective in improving adaptive behavior and socialization skills, and reducing attention problems in subjects with attention deficient/hyperactivity

79. Noland, R.C., Koves, T.R., Seiler, S.E., et al., 2009. J. Biol. Chem. 284, 22840–22852.
80. Oyangi, E., Yano, H., Uchida, M., et al., 2011. Biochem. Biophys. Res. Commun. 412, 61–67.
81. Benvenga, S., Amato, A., Calvani, M., et al., 2004. Ann. N.Y. Acad. Sci. 1033, 158–167.
82. Malaguarnera, M., Pistone, G., Elvira, R., et al., 2005. World J. Gastroenterol. 11, 7197–7202.
83. In one study, the muscle carnitine concentrations of eight patients after hemodialysis was only 10% of that of healthy controls. It is of interest, however, that not all hemodialysis patients experience carnitine depletion. Some show chronic hypocarnitinemia, whereas others show a return of plasma carnitine concentrations to normal or higher than normal within about 6 h after dialysis. The recovery of the latter group is hastened (to about 2 h) if each patient is given 3 g of D,L-carnitine orally at the end of the dialysis period.
84. This can be prevented by adding carnitine to the dialysate, e.g., 65 μM.
85. Chen, Y., Abbate, M., Tang, L., et al., 2014. Am. J. Clin. Nutr. 99, 408–422.
86. Ferrari, R., Merli, E., Cicchitelli, G., et al., 2004. Ann. N.Y. Acad. Sci. 1033, 79–91.
87. Lango, R., Smoleński, R.T., Rogowski, J., et al., 2005. Cardiovasc. Drug Ther. 19, 267–275.
88. i.e., Ejection fractions of 30–40%; anonymous, 1999. Eur. Heart J. 20, 70–77.
89. Ames, B.N., Liu, J., 2004. Ann. N.Y. Acad. Sci. 1033, 108–116.
90. Spagnoli A., Lucca, U., Menasce, G., et al., 1991. Neurology 41, 1726–1732.
91. Rai, G., Wright, G. and Scott, L., et al., 1990. Curr. Med. Res. Opin. 11, 638–647.
92. Montgomery, S.A., Thal, L.J., Amrein, R., 2003. Int. Clin. Psychopharmacol. 18, 61–71.
93. Zanardi, R., Smeraldi, E., 2006. Eur. Neuropsychopharmacol. 16, 281–287.

disorder and fragile X syndrome.[94] A randomized trial found supplemental carnitine to reduce excessive daytime sleepiness in patients with narcolepsy.[95]

Exercise performance. It has been suggested that carnitine supplementation may improve exercise performance (i.e., acting as an ergogenic aid) and attenuate the deleterious effects of hypoxic training to hasten recovery from strenuous exercise. Implicit in such suggestions is the notion that carnitine supply may be rate limiting to fatty acid oxidation in muscle, a phenomenon demonstrated only when muscle carnitine drops to less than half normal levels due to the low K_m of CATI. A review of the published literature found no evidence that carnitine supplements can improve athletic performance.[96] A few studies, however, have reported modest improvements in exercise-induced muscle lactate accumulation in resistance-trained athletes,[97] and muscle soreness in nontrained subjects.[98]

Weight management. Some, but not all, studies with animals have found supplementation with carnitine or propionlycarnitine to enhance the loss of body fat under conditions of negative energy balance.[99] Controlled trials with humans, however, have not found such effects.

Diabetes. Reduction in the carnitine-dependent transport of fatty acids into the mitochondria results in cytosolic triglyceride accumulation and has been implicated in the development of insulin resistance. Prediabetic subjects and those with T2D show increased circulating levels of several acylcarnitines.[100] Studies with animal models have suggested that carnitine status can affect the control of glycolysis and/or gluconeogenesis; carnitine supplements have been found to stimulate the insulin-mediated disposal of glucose and confer protection of vascular function in insulin resistance.[101] Studies in the rat have shown experimentally induced diabetes to result in depletion of lens carnitine and acetylcarnitine, and carnitine supplementation to reduce the development of cataracts.

A clinical trial found carnitine to reduce fasting plasma glucose levels and increase fasting triglycerides in patients with T2D.[102] Intramuscular acetylcarnitine reduced pain in patients with diabetic neuropathy and in T2D patients with poorly controlled blood glucose.[103] Intravenous L-carnitine has been shown to improve insulin sensitivity,[104] and oral L-carnitine was found to reduce glycation end products in the skin of hemodialysis patients.[105]

Male reproductive health. Epididymal tissue and spermatozoa typically contain high amounts of carnitine. Studies in both rodent models and humans have related carnitine levels to sperm count, motility, and maturation, and have suggested that carnitine supplementation can improve sperm quality.[106] Two randomized trials found treatment with carnitine and/or acylcarnitine to improve sperm motility in low-fertility males. Propionylcarnitine was found to enhance the efficacy of the drug sildenafil in treating erectile dysfunction in diabetic patients and in postprostatectomy patients.[107]

Biomarkers of Carnitine Status

Carnitine status can be assessed on the basis of the concentration of carnitine in muscle or plasma/serum. Plasma carnitine levels are subject to several effects. They are reduced in individuals with a single-nucleotide polymorphism in medium-chain acyl-CoA dehydrogenase, a mitochondrial enzyme that catalyzes fatty acid β-oxidation,[108] protein malnutrition (Table 19.6), and have been noted in children with schistosomiasis and associated anemia. As Fe^{2+} is required in two steps in carnitine biosynthesis (ε-*N*-trimethyllysine hydroxylase, γ-butyrobetaine hydroxylase), it is possible that deficiencies of iron may reduce carnitine biosynthesis.

Carnitine Safety

A systematic review of published literature concluded that evidence of safety is strong for carnitine intakes as great

94. Torrioli, M.G., Vernacotola, S., Mariotti, P., et al., 1999. Am. J. Med. Genet. 87, 366–368; Van Oudheusden, L.J., Scholte, H.R., 2002. Prostaglandins Leukotrienes Essent. Fatty Acids 67, 33–38; Torrioli, M.G., Vernacotola, S., Peruzzi, L., et al., 2008. Am. J. Med. Genet. 146, 803–812.
95. Miyagawa, T., Kawamura, H., Obuchi, M., et al., 2013. PLOS One 8, e53707.
96. Brass, E.P., 2004. Ann. N.Y. Acad. Sci. 1033, 67–78.
97. Jacobs, P.L., Goldstein, E.R., Blackburn, W., et al., 2009. J. Int. Soc. Sports Nutr. 6, 9.
98. Spierling, B.A., Kraemer, W.J., Vingren, J.L., et al., 2007. J. Strength Cond. Res. 21, 259–264.; Ho, J.Y., Kraemer, W.J., Volek, J.S., et al., 2010. Metabolism 886, 223–230.
99. Heo, K., Odle, J., Han, I.K., et al., 2000. J. Nutr. 130, 1809–1814; Mingorance, C., del Pozo, M.G., Herra, M.D., et al., 2009. Br. J. Nutr. 102, 1145–1153; Schmengler, U., Ungru, J., Boston, R., et al., 2013. Livestock Sci. 155, 301–307.
100. Particularly, tetradecenoylcarnitine (C14:1), tetradecadienylcarnitine (C14:2), octadecenoylcarnitine (C18:1), and malonylcarnitine/hydroxylbutyrylcarnitine (C3DC+C4OH). Mai, M., Tönjes, A., Kovacs, P., et al., 2013. PLoS One 12, e382459.
101. Mingorance, C., del Pozo, M.G., Herra, M.D., et al., 2009. Br. J. Nutr. 102, 1145–1153.
102. Rahbar, A.R., Shakerhosseini, R., Saadat, N., et al., 2005. Eur. J. Clin. Nutr. 59, 592–596.
103. DeGrandis, D., Minardi, C., 2002. Drugs R. D. 3, 223; Sima, A.A., Calvani, M., Mehra, M., et al., 2005. Diabetes Care 28, 89–94.
104. Giancaterini, A., De Gaetano, A., Mingrone, G., et al., 2000. Metabolism 49, 704–708.
105. Fukami, K., Yamagishi, S.I., Sakai, K., et al., 2013. Rejuvenation Res. 16, 460–466.
106. Ng, C.M., Blackman, M.R., Wang, C., et al., 2004. Ann. N.Y. Acad. Sci. 1033, 177–188.
107. Gentile, V., Vicini, P., Prigiotti, G., et al., 2004. Curr. Med. Res. Opin. 20, 1377–1384; Cavalini, G., Modenini, F., Vitali, G., et al., 2005. Urology 66, 1080–1085.
108. Couce, M.L., Sánchez-Pintos, P., Diogo, L., et al., 2013. Orphanet J. Rare Dis. 8, 102–108.

TABLE 19.6 Apparent Carnitine Deficiency in Protein-Malnourished Children

Group	Plasma Carnitine, μmol/dL	Plasma Albumin, g/dL
Healthy controls	9.0±0.6 (8)[a]	3.5±0.1 (8)
Undernourished patients	6.4±0.9 (10)	2.7±0.2 (5)
Marasmus patients	3.7±0.5 (12)	2.7±0.2 (8)
Kwashiorkor patients	2.6±0.5 (13)	1.7±0.1 (9)

[a]*Mean±SD for (n) children.*
Adapted from Khan, L., Bamji, M.S., 1977. Clin. Chim. Acta 75, 163–166.

as 2000 mg/day on a chronic basis, although studies using higher levels have not observed adverse effects.[109]

4. *MYO*-INOSITOL

Myo-inositol is a normal metabolite that, as **phosphatidylinositol (PI)**, affects membrane structure and function, supporting the production of eicosanoids and cell signaling in the regulation of metabolism. *Myo*-inositol is biosynthesized by most species; however, some species are not capable of its biosynthesis, and dietary shortages of other vitamins may also limit its biosynthesis, making it **conditionally essential**.

Recognition of a Nutritional Role of *Myo*-Inositol

Although *myo*-inositol had been discovered in extracts of animal tissues almost 100 years earlier, interest in its potential nutritional role first occurred in 1940 when it was reported to be a new vitamin required for normal growth, hair and skin of the mouse, i.e., the "mouse antialopecia factor."[110] That original report was later questioned regarding the adequacy of the diet with respect to other known vitamins. Nevertheless, several groups found dietary supplements of *myo*-inositol to stimulate the growth of several species (chicks, turkeys, rats, mice) in ways dependent on shortages of other factors including pantothenic acid, biotin, and folate. Whether the observed effects were actually responses to a missing nutrient was debated. *Myo*-inositol was found to be essential for the growth of most cells in culture. However, it had been found earlier that the daily urinary excretion of *myo*-inositol by the rat exceeded the amount ingested, suggesting the factor was biosynthesized.[111] More recently, deprivation of *myo*-inositol has been shown to render hepatic triglyceride accumulation by the rat susceptible to the effects of the fatty acid composition of the diet, indicating a function resembling that of an essential nutrient. Further, Hegsted and colleagues found that the female Mongolian gerbil[112] develops intestinal lipodystrophy when deprived of the factor.[113] In fact, that group showed that the gerbil required a source of *myo*-inositol to prevent the disorder when fed a diet-containing adequate levels of all other known nutrients.

Conditions of Need for Dietary *Myo*-Inositol

A few species have overt dietary needs for *myo*-inositol. These include several fishes and the gerbil. Studies with fishes have shown dietary deprivation of *myo*-inositol to result in anorexia, fin degeneration, edema, anemia, decreased gastric emptying rate, reduced growth, and impaired efficiency of feed utilization. Studies with gerbils have shown *myo*-inositol deprivation to result in intestinal lipodystrophy, with associated hypocholesterolemia and reduced survival. These effects are observed only in females; males appear to have a sufficient testicular synthesis of *myo*-inositol. For these species, *myo*-inositol is a dietary essential.

Studies with nutritionally complete diets have failed to confirm early reports that suggested dietary needs for *myo*-inositol to prevent alopecia in rodents; fatty liver in rats and growth retardation in chicks, guinea pigs, and hamsters. It is possible that those earlier findings may have indicated beneficial effects of dietary sources of *myo*-inositol on the hindgut microbiome of those animals, i.e., that dietary *myo*-inositol might be beneficial in diets containing marginal amounts of such factors as choline and biotin, which are known to be synthesized by the microbiome. Support for this hypothesis comes from findings that supplemental *myo*-inositol reduced hepatic lipid accumulation in rats fed a choline-deficient diet, improved growth in rats fed a diet deficient in several vitamins, and reduced the incidence of fatty liver in and improved the growth of chicks fed a biotin-deficient diet containing an antibiotic.[114] Thus, it is plausible that a dietary source of *myo*-inositol could be useful under conditions that increase the need for myo-inositol for lipid transport (e.g., high-fat diets[115]) or disturb the

109. Hathcock, J.N., Shao, A., 2006. Regul. Toxicol. Pharmacol. 46, 23–28.
110. Woolley, D.W., 1940. Science 92, 384–385.
111. Needham, J., 1924. Biochem. J. 18, 891–904.
112. *Meriones unguiculatus*.
113. Hegsted, M, Hayes, K.C., Gallagher, A., et al., 1973. J. Nutr. 103, 302–307.
114. Antibiotics can reduce the numbers of microorganisms that normally produce *myo*-inositol and other nutrients.
115. High-fat diets may increase the needs for *myo*-inositol for lipid transport.

hindgut microbiome (e.g., antibiotic use, stressful physical environments).

Conditions producing need for dietary inositol are as follows:

- Limited biosynthesis in fish and gerbils
- Limited synthesis by the gut microbiome?

Chemical Nature

Myo-inositol is a water-soluble, hydroxylated, cyclic six-carbon compound (*cis*-1,2,3,5-*trans*-4,6-cyclohexanehexol) (Fig. 19.6). It is the only one of the nine possible stereoisomeric forms of cyclohexitol with biological activity.

Sources of *Myo*-Inositol

Biosynthesis. Most, if not all, mammals are capable of synthesizing *myo*-inositol de novo ultimately from glucose. Biosynthetic capacity has been found in the liver, kidney, brain, and testis of rats and rabbits, and in the kidney and other tissues in humans. Biosynthesis involves the cyclization of glucose-6-phosphate to inositol-1-phosphate by inositol-1-phosphate synthase, followed by a dephosphorylation by inositol-1-phosphatase. The human kidney produces several grams of *myo*-inositol daily. Renal synthesis of *myo*-inositol has been found to be about 4 g/day (~2 g/kidney/day).[116]

Dietary sources. *Myo*-inositol occurs in foods and feedstuffs in three forms: free *myo*-inositol, **phytic acid**,[117] and inositol-containing phospholipids. The richest sources of *myo*-inositol are the seeds of plants (e.g., beans, grains, and nuts) (Table 19.7). However, the predominant form occurring in plant materials is phytic acid (which can comprise most of the total phosphorus present in materials such as cereal

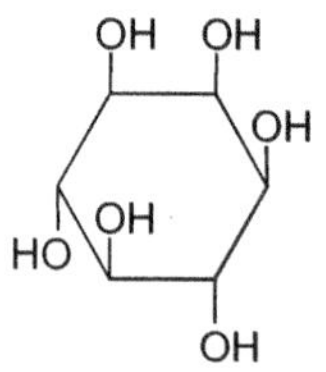

FIGURE 19.6 *Myo*-inositol.

grains[118]). Because most mammals have little or no intestinal phytase activity, phytic acid is poorly utilized as a source of either *myo*-inositol or phosphorus.[119,120] In animal products, *myo*-inositol occurs in free form as well as in inositol-containing phospholipids (primarily PI); free *myo*-inositol predominates in brain and kidney, whereas phospholipid inositol predominates in skeletal muscle, heart, liver, and pancreas. The richest animal sources of inositol are organ meats. Human milk is relatively rich in *myo*-inositol (colostrum, 200–500 mg/liter; mature milk, 100–200 mg/liter) in comparison with cow's milk (30–80 mg/liter). A disaccharide form of *myo*-inositol, 6-β-galactinol (6-*O*-β-D-galactopyranosyl-*myo*-inositol), comprises about one-sixth of the nonlipid *myo*-inositol in that material.

Myo-inositol is classified by the US FDA among the substances generally recognized as safe and, therefore, can be used in the formulation of foods without the demonstrations of safety and efficacy required by the Food, Drug and

---

116. Troyer, D.A., Schwertz, D.W., Kreisberg, J.I., et al., 1986. Ann. Rev. Physiol. 48, 51–71.

117. Inositol hexaphosphate.

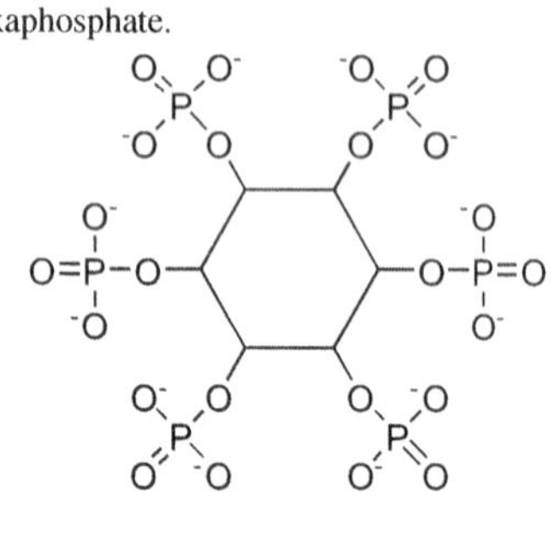

118. Of the total P present, phytic acid P comprises 48–73% for cereal grains (corn, barley, rye, wheat, rice, sorghum), 48–79% for brans (rice, wheat), 27–41% for legume seeds (soybeans, peas, broad beans), and 40–65% for oilseed meals (soybean meal, cottonseed meal, rapeseed meal).

119. The bioavailability of P from most plant sources is >50% for ruminants, which benefit from the phytase activities of their rumen microflora. Nonruminants, lacking intestinal phytase, derive much less P from plant phytic acid, depending on the phytase contributions of their intestinal microflora. For pigs and rats, such contributions appear to be significant, giving them moderate abilities (~40%) to utilize phytic acid P. In contrast, the chick, which, has only a sparse intestinal microflora, can use only ~8% of phytic acid P.

120. Phytic acid can also form a very stable chelation complex with zinc (Zn^{2+} is held by the negative charges on adjacent pyrophosphate groups), thus reducing its nutritional availability. For this reason, the bioavailability of zinc in such plant-derived foods as soybean is very low.

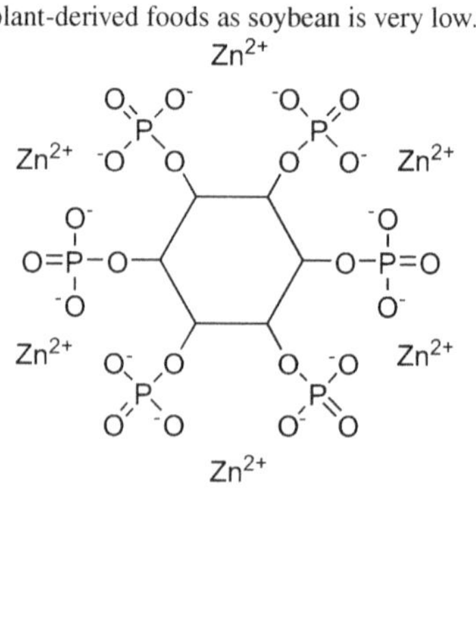

TABLE 19.7 Total *Myo*-Inositol Contents of Selected Foods

Food	Myo-inositol, mg/g
Vegetables	
Asparagus	0.29–0.68
Beans	
Green	0.55–1.93
White	2.83–4.40
Red	2.49
Broccoli	0.11–0.30
Cabbage	0.18–0.70
Carrot	0.52
Cauliflower	0.15–0.18
Celery	0.05
Okra	0.28–1.17
Pea	1.16–2.35
Potato	0.97
Spinach	0.06–0.25
Squash, yellow	0.25–0.32
Tomato	0.34–0.41
Fruits	
Apple	0.10–0.24
Cantaloupe	3.55
Grape	0.07–0.16
Grapefruit	1.17–1.99
Orange	3.07
Peach	0.19–0.58
Pear	0.46–0.73
Strawberry	0.13
Watermelon	0.48
Cereals	
Rice	0.15–0.30
Wheat	1.42–11.5
Meats	
Beef	0.09–0.37
Beef liver	0.64
Chicken	0.30–0.39
Chicken liver	1.31
Lamb	0.37
Pork	0.14–0.42

Continued

TABLE 19.7 Total *Myo*-Inositol Contents of Selected Foods—cont'd

Food	Myo-inositol, mg/g
Pork liver	0.17
Turkey	0.08–0.23
Trout	0.11
Tuna	0.11–0.15
Dairy Products and Eggs	
Milk	0.04
Ice cream	0.09
Cheese	0.01–0.09
Egg, whole	0.09
Egg yolk	0.34
Nuts	
Almond	2.78
Peanut	1.33–3.04

Adapted from Clements, Jr., S.R., Darnell, B., 1980. Am. J. Clin. Nutr. 3, 1954–1967.

Cosmetic Act. It is added to many prepared infant formulas (at about 0.1%). It is estimated that typical American diets provide adults with about 900 mg of *myo*-inositol per day, slightly over half of which is in phospholipid form.

Absorption and Transport of *Myo*-Inositol

Absorption. The enteric absorption of free *myo*-inositol occurs by active transport; uptake from the small intestine is virtually complete. The enteric absorption of phytic acid, however, depends on its digestion and the amounts of divalent cations in the diet/meal. Most animal species lack intestinal phytase activities and are, therefore, dependent on the presence of a gut microbiome with those enzymes. For species with foregut microbiomes (e.g., ruminants and long-gutted nonruminants), phytate is digestible, making it a useful source of *myo*-inositol. Dietary cations (particularly Ca^{2+}) can reduce the utilization of phytate by forming insoluble (and, thus, nondigestible and nonabsorbable) phytate **chelates**. Because a large portion of the total *myo*-inositol in mixed diets is typically in the form of phytic acid, its utilization from high-calcium diets can be <50% of that from diets containing low to moderate amounts of the mineral.[121]

121. For the same reason, the bioavailability of calcium is also low for high-phytate diets. This same effect occurs for the nutritionally important divalent cations Mn^{2+} and Zn^{2+}; the bioavailability of each is reduced by the presence of phytic acid in the diet.

Absorption of phospholipid *myo*-inositol is thought to be analogous to that of PC.[122]

Transport. The normal circulating concentration of *myo*-inositol in humans is about 30 μM. It circulates predominantly in the free form; a small amount occurs as PI associated with lipoproteins. Free *myo*-inositol is taken up by an active transport process in some tissues (kidney, brain) and by carrier-mediated diffusion in others (liver). The active process requires Na^+ and energy and is inhibited by high levels of glucose. Untreated diabetics show impaired tissue uptake and impaired urinary excretion of *myo*-inositol.

Tissue distribution. In tissues, *myo*-inositol is found as the free form and as the mono-, di-, and triphosphorylated metabolites. These PIs differ from other phospholipids by being enriched in **stearic acid** (predominantly at the 1-position) and **arachidonic acid** (predominantly at the 2-position). For example, the *myo*-inositol-containing phospholipids on the plasma membrane from human platelets contain about 42 mol% stearic acid and about 44 mol% arachidonic acid. The greatest concentrations of *myo*-inositol are found in neural and renal tissues.

Metabolism of *Myo*-Inositol

Conversion to PIs and soluble inositides. Free *myo*-inositol is converted to PI within cells by PI synthetase, which catalyzes its reacting with the liponucleotide cytidine diphosphate-diacylglycerol.[123] In turn, PI can be sequentially phosphorylated to the monophosphate (phosphatidylinositol 4-phosphate, PI_2) and diphosphate (phosphatidylinositol 4,5-diphosphate, PIP_2) forms by membrane **inositolphosphate kinases (IPKs)**.[124] Higher phosphorylated forms (PIP_4, PIP_5, PIP_6, and PIP_7) are formed by the action of **inositolpolyphosphate multikinase (IPMK)**. These PIs can be dephosphorylated by three phosphatases (PTEN, SHIP1, SHIP2[125]) that produce various soluble inositol phosphatides (e.g., IP, IP_2, IP_3, etc.) depending on the local expression of IPKs.

Turnover. In the presence of cytidine monophosphate, PI synthetase functions (in the reverse direction) to break down that form to yield CDP-diacylglycerol and *myo*-inositol. The kidney performs most of the further catabolism, clearing *myo*-inositol from the plasma, converting it to glucose, and then oxidizing it to CO_2 via the pentose phosphate shunt. The metabolism of *myo*-inositol appears to be relatively rapid; the rat can oxidize half of an ingested dose within 48 h.

Metabolic Functions of *Myo*-Inositol

The metabolically active forms of *myo*-inositol are PIs and IPs, which have several physiologically important roles [126,127]:

- **PIs affect membrane structure and function**. The PIs are membrane active due to their unique fatty acid compositions. For example, their polar head group and highly nonpolar fatty acyl chains facilitate specific electrostatic interactions while providing a hydrophobic microdomains in membranes that facilitate the recruitment of proteins and, ultimately, control the cytoskeleton and membrane dynamics during mitosis. This role also allows PIs to function as an activator of microsomal Na^+, K^+-ATPase, an essential constituent of acetyl-CoA carboxylase, a stimulator of tyrosine hydroxylase, cofactors of alkaline phosphatase and 5′-nucleotidase, a membrane anchor for acetylcholinesterase, a necessary factor for insulin sensitivity, and a regulator of endosome–lysosome trafficking and induction of pathogen killing and antigen-processing pathways in lymphoid cells.
- **PIs are ready sources of arachidonic acid**. PI serves as a source of releasable arachidonic acid for the formation of the **eicosanoids**[128] by cyclooxygenase and/or lipoxygenase. Although PI is less abundant in cells than other phospholipids (PC, phosphatidylethanolamine,

122. This would involve hydrolysis by pancreatic phospholipase A in the intestinal lumen to produce a lysophosphatidylinositol, which, on uptake by the enterocyte, would be reacylated by an acyltransferase or hydrolyzed further to yield glycerylphosphorylinositol.

123. This step is catalyzed by the microsomal enzyme CDP-diacylglycerol–inositol 3-phosphatidyltransferase (also called PI synthetase).

124. These are ATP:PI-4-phosphotransferase and ATP:PI-4-phosphate 5-phosphotransferase, respectively. They are located on the cytosolic surface of the erythrocyte membrane. There is no definitive evidence that *myo*-inositol can be isomerized or phosphorylated to the hexaphosphate level; however, such prospects would be of interest, as the isomer D-*chiro*-inositol has been shown to promote insulin function, and inositol hexaphosphate (phytic acid) has been found to be anticarcinogenic in different animal models.

125. Phosphatase and tensin homologue; PI-3,4,5-triphosphate 5-phosphatases 1 and 2.

126. Additional roles have been identified in lower organisms: transcriptional regulation in yeasts; stress tolerance in plants.

127. In addition, IP kinases have metabolic roles that appear unrelated to their catalytic activities. IPMK has been shown to mediate the effects of growth factors by stimulating the protein kinase Akt, which activates the central signaling regulator mTOR complex 1. This involves IPMK binding to mTORC1, promoting its stability and, thus, mediating the ability of amino acids to activate mTOR. IPMK similarly affects AMPK signaling, and a similar mechanism may be involved in the effect attributed to IP7 in maintaining exocytosis of insulin-containing secretory granules from mouse pancreatic β-cells.

128. The eicosanoids include prostaglandins, thromboxanes, and leukotrienes. The prostaglandins are hormone-like substances secreted for short-range action on neighboring tissues; they are involved in inflammation, in the regulation of blood pressure, in headaches, and in the induction of labor. The functions of the leukotrienes and thromboxanes are less well understood; they are thought to be involved in regulation of blood pressure and in the pathogenesis of some types of disease.

and phosphatidylserine), its enrichment in arachidonic acid renders it an effective source of that eicosanoid precursor.

- **IPs are second messengers of Ca^{2+} signaling**. IPs mediate the responses of target tissues to the signaling by cholinergic or β-adrenergic agonists (producing rapid responses) and mitogens (producing medium-term responses). This mediation involves its activation to the less abundant species, IP_2 to IP_3, which serves as a **second messenger** to activate the release of Ca^{2+} from intracellular stores (in discrete organelles called **calcisomes**) and stimulate the entry of Ca^{2+} into the cell across the plasma membrane.[129] Cell surface IP_3 receptors have been shown to effect primary control over the hydrolysis of IP_2 by regulating the activity of phospholipase C (phosphodiesterase) on the plasma membrane. Receptor occupancy, thus, activates the hydrolysis of PIP_2, which is favored at low intracellular concentrations of Ca^{2+} to produce IP_3 and, perhaps, other inositol polyphosphates. The former process involves a specific IP_3 receptor on the calcisomal membrane; IP_3 binding to this receptor opens an associated **Ca^{2+} channel**. IP_3-stimulated entry of Ca^{2+} into the cell requires an increase in the permeability of the plasma membrane signaled by the emptying of the IP_3-sensitive intracellular pool. Evidence suggests that 1,2-diacylglycerol, formed from receptor-stimulated metabolism of the *myo*-inositol-containing phospholipids, may signal the activation of protein kinase C for the phosphorylation of various proteins important to cell function. According to this hypothesis, 1,2-diacylglycerol functions with Ca^{2+} and phosphatidylserine, both of which are known to be involved in the activation of protein kinase C. The maintenance of balanced Ca^{2+} flux has also been shown to be important in left–right symmetrical development in the zebrafish; this is disrupted by genetic deletion of IPKs.[130]

Dietary *Myo*-Inositol in Health and Disease

Positive clinical findings have been reported for myo-inositol supplements in human patients:

- **Preterm infants**. Three randomized, clinical trials found *myo*-inositol supplementation to improve survival and reduce retinopathy of prematurity, bronchial dysplasia, and intraventricular hemorrhage.[131]
- **Obesity**. *Myo*-inositol supplementation has been found to reduce insulin resistance and gestational diabetes in two cohorts of obese pregnant women[132] and improve ovarian function in three cohorts of overweight/obese women with polycystic ovary syndrome.[133]
- **Psoriasis**. A small, randomized clinical trial found supplementation with *myo*-inositol to reduce the severity of symptoms of psoriasis patients.[134]
- **Neuropsychological responses**. Because mood stabilizers such as lithium, valproate, and carbamazepine function by stabilizing inositol signaling, it has been suggested that *myo*-inositol may have value in the treatment of depression and other psychiatric disorders. While *myo*-inositol supplementation has been reported to be helpful in treating bipolar depression and bulimia nervosa with binge eating, the limited findings in this area to date do not indicate clear benefits. A small randomized trial found *myo*-inositol to increase the efficacy of high-dose n-3 fatty acid therapy in reducing symptoms of mania and depression in children with mild to moderate bipolar spectrum disorders.[135] Another study found *myo*-inositol supplementation to improve symptoms in women with premenstrual dysphoric disorder.[136]

Biomarkers of *Myo*-Inositol Status

Plasma concentration of *myo*-inositol in humans are typically ~30 μM most of which is the free form. This parameter is not informative regarding the physiologically active PIs in cells.

Safety of Myo-Inositol

There are no reports of adverse effects of oral intakes of myo-inositol. It is presumed safe.

129. The Ca^{2+}-mobilizing activity of IP_3 is terminated by its dephosphorylation (via a 5-phosphatase) to the inactive inositol-1,4-bisphosphate, or by its phosphorylation (via a 3-kinase) to a product of uncertain activity, IP_4. There is some controversy concerning whether other inositol phosphates [e.g., inositol 1,3,4,5-tetraphosphate (IP_4), which is a product of a 3-kinase acting on IP_3] can also signal Ca^{2+} mobilization. Evidence suggests that, in at least some cells, IP_3 and IP_4 may have cooperative roles in Ca^{2+} signaling.
130. Tsui, M.M., York, J.D., 2010. Adv. Enzyme Regul. 50, 324–337.
131. Howlett, A., Ohlsson, A., 2003. Cochrane Database Syst. Rev. CD000366.
132. D'Anna, R., Di Benedetto, A., Scilipoti, A., et al., 2015. Obstet. Gyncecol. 126, 310–315; Matarelli, B., Vitacolonna, E., D'Angelo, M., et al., 2015. J. Matern. Fetal Neonatal. Med. 26, 967–972.
133. Kamenov, Z., Kolarov, G., Gatvba, A., et al., 2015. Gynecol. Endocrinol. 31, 131–135; Pizzo, A., Laganà, A.S., Barbaro, L., 2015. Gynecol. Endocrinol. 30, 205–208; Artini, P.G., Di Berardino, O.M., Papini, F., et al., 2015. Gynecol. Endocrinol. 29, 375–379.
134. Allan, S.J.R., Kavanagh, G.M., Herd, R.M., et al., 2004. Br. J. Dermatol. 150, 966–969.
135. Wozniak, J., Faraone, S.V., Chan, J., et al., 2015. J. Clin. Psychiatry 76, 1548–1555.
136. Gianfranco, C., Vittorio, U., Silvia, B., et al., 2011. Hum. Psychopharmacol. 26, 526–530.

5. UBIQUINONES

Ubiquinones are normal metabolites that are essential components of the mitochondrial electron transport chains of all cells. They are biosynthesized by most species; however, several conditions can limit their biosynthesis, making them **conditionally essential**.

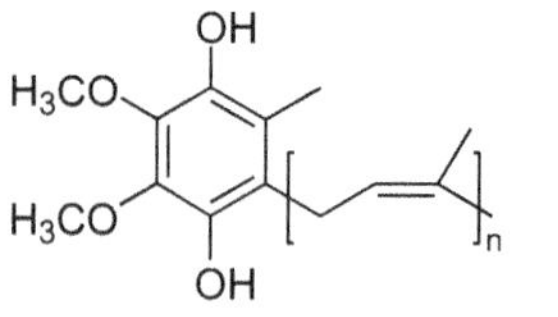

FIGURE 19.7 Structure ubiquinones (reduced form shown).

Recognition of Nutritional Roles of Ubiquinones

Isolated from the unsaponifiable fractions of the hepatic lipids from vitamin A-deficient rats, the principal species of the group, ubiquinone(50) was subsequently identified as an essential component of the mitochondrial electron transport chains of most prokaryotic and all eukaryotic cells. Folkers called this factor "vitamin Q." In subsequent decades, the term "coenzyme Q" came to be used as the general descriptor of this family of compounds.

Conditions of Need for Dietary Ubiquinones

Dietary sources of ubiquinones have been found to be beneficial under specific conditions: limited biosynthesis and increased need for mitochondrial antioxidant protection.

- Conditions limiting ubiquinone biosynthesis:
  Genetic disorders. Several genetic variants in enzymes in the ubiquinone biosynthetic pathway or mitochondrial electron transport chain have been found to reduce tissue CoQ_{10} losses markedly (33–97%).[137] Subject with these rare deficiencies that reduce plasma CoQ_{10} concentrations to more than two SD below the mean normal level show five major phenotypes: encephalopathy (4 cases reported), cerebellar ataxia (94 cases reported), an infantile multisystemic disease (17 cases reported), nephropathy and isolated myopathy (10 cases reported).[138]
  Aging. Deficient tissue levels of CoQ occur in older animals, suggesting that ubiquinone biosynthesis may decline with age. Tissue-specific deficiencies of CoQ have been identified in association with genetic deficiencies of enzymes involved in that ubiquinone biosynthesis. These deficiencies can be corrected with dietary supplements of CoQ_{10}.
  Statin therapy. Tissue CoQ levels can be reduced as a result of statin therapy, which impairs the synthesis of CoQ_{10} by inhibiting HMG-Co reductase to reduce farnesyl pyrophosphate, an intermediate in the pathway to CoQ. Statin treatment has been associated with myopathic conditions ranging from mild myalgia to fatal rhabdomyolysis as well as with subclinical cardiomyopathy. These conditions have been found to respond to CoQ_{10} supplementation or its synthetic analogue idebenone.[139]
- Conditions increasing antioxidative need are as follows:
  Vitamin E deficiency. Metabolic needs for CoQ are linked to vitamin E status, as the oxidized form of $CoQ_{10,}$ can support antioxidant functions *in lieu* of that vitamin. This has been demonstrated in the reduction of clinical signs of vitamin E deficiency in both rats[140] and rhesus monkeys.[141] In fact, responses to CoQ_{10} have been found to have more rapid onset than those to α-tocopherol.
  Disease-producing oxidative stress. Increased oxidative stress occurs in **hyperthyroidism**, which is associated with low circulating levels of CoQ_{10} and α-tocopherol.[142] **Parkinson's disease** patients face a higher than normal risk of CoQ deficiency,[143] which may be related to their excessive production of reactive oxygen species (ROS).

Conditions producing need for dietary ubiquinones are as follows:

- Limited biosynthesis: genetic disorders, aging, statin therapy
- Increased oxidative stress.

Chemical Nature of the Ubiquinones

The **ubiquinones** are a group of tetrasubstituted 1,4-benzoquinone derivatives with isoprenoid side chains of variable length (Fig. 19.7). The structure of the 6-chromanol moiety is remarkably similar to the oxidized form of vitamin E, tocopherylquinone, the difference being that ubiquinones have two methoxyl groups in ring positions where

137. Bentinger, M., Tekle, M., Dallner, G., 2010. Biochem. Biophys. Res. Commun. 396, 74–79.
138. Emmanuele, V., López, L.C., Berardo, A., et al., 2012. Arch. Neurol. 69, 978–983.
139. Littarru, G.P., Langsjoen, P., 2007. Mitochondrion 7S, S168–S174; Mancuso, M., Orsuci, D., Vopli, L., et al., 2010. Curr. Drug Targets 11, 111–121.
140. i.e., "Resorption–gestation" syndrome, *see* Chapter 8.
141. i.e., Anemia and the myopathy.
142. Mancini, A., Festa, R., Raimondo, S., et al., 2011. Int. J. Molec. Sci. 12, 9216–9225.
143. Mischley, L.K., Allen, J., Bradley, R., 2012. J. Neurol. Sci. 318, 72–75.

tocopherylquinone has methyl groups. The conventions of nomenclature for this group are similar to those for vitamin K. The terms ubiquinone and coenzyme Q (or CoQ) are synonymous, but their actual species, which are defined by the nature of their respective side chain, is indicated differently: the number of side chain carbons is indicated parenthetically for the ubiquinone designation and the number of side chain isoprenyl units is indicated in a subscript for the CoQ designation.[144]

Sources of Ubiquinones

Biosynthesis. CoQ_{10} is synthesized in all tissues from precursors available in inner mitochondrial membranes. The biosynthetic process derives the isoprenyl side chain is derived from mevalonate, the ring system from tyrosine, the hydroxyl groups from molecular oxygen, and the methyl groups from *S*-adenosylmethionine to produce a 50-C polyisoprene chain comprised of 10 isoprenoid units. A key enzyme in this pathway is hydroxymethylglutaryl-CoA reductase, which is feedback-inhibited by CoQ_{10}.[145] Endogenous biosynthesis appears sufficient to support membrane saturation levels.

Dietary sources. CoQ_{10}, localized in the mitochondria and other cellular membranes, is found in plant and animal tissues of high cellularity, particularly, those rich in mitochondria such as heart and muscle (Table 19.8).

Absorption and Transport of Ubiquinones

Absorption. Ubiquinones appear to be absorbed, transported, and taken up into cells by mechanisms analogous to those of the tocopherols. In the rat, the greatest absorption of CoQ_{10} has been found to occur in the duodenum, with demonstrable absorption also in the colon, ileum, and jejunum, suggesting the possibility of enterohepatic circulation.

Transport. Because all cell synthesizes apparently adequate amounts of ubiquinones, there is little redistribution via the circulation. The small amounts in plasma reflect CoQ recently absorbed from the diet and released by the liver in association with VLDL. Evidence indicates that the lipoprotein pool does not exchange with the tissues.

Tissue distribution. CoQ_{10} is normally present in membranes of all cells in the body as a result of ubiquitous biosynthesis. The total CoQ_{10} pool size in the human adult is estimated to be 0.5–1.5 g. Relatively great concentrations of CoQ_{10} are found in the liver, heart, spleen, kidney, pancreas, and adrenals.[146] Tissue ubiquinone levels increase under the influence of oxidative stress, cold acclimation, and thyroid hormone treatment and appear to decrease with cardiomyopathy, other muscle diseases, and aging. Dietary supplementation increases CoQ levels only in the liver and spleen.

Metabolism of Ubiquinones

The ubiquinones undergo reversible, two-electron redox reactions (Fig. 19.8).

Metabolic Functions of Ubiquinones

CoQ has several essential metabolic functions.

- **Mitochondrial respiratory chain**. CoQ is an electron acceptor for complexes I and II, passing electrons from flavoproteins (e.g., NADH or succinic dehydrogenases) to the cytochromes via cytochrome b_5. In this process, CoQ undergoes reversible reduction/oxidation to cycle between the 1,4-quinone (oxidized) and 1,4-dihydroxybenzene (reduced) species (Fig. 19.9).
- **Antioxidant**. CoQ_0 is a lipid-soluble antioxidant that protects α-tocopherol in membranes; protects LDL from oxidation; and generally participates in the prevention of the oxidation of lipids, proteins, and DNA. In LDLs, which are particularly susceptible to oxidation, CoQ_{10} is the primary antioxidant, being oxidized before α-tocopherol.[147]
- **Regulation of intracellular NAD^+/NADH balance**. CoQ is an essential cofactor for NADH oxidase in plasma membranes. This enzyme regulates intracellular NAD^+/NADH balance, which is important in cell growth and development.
- **Mitochondrial membrane pore regulation**. CoQ is a factor preventing the opening of mitochondrial membrane transition pores that otherwise admit large molecules capable of antagonizing the function of that organelle and promoting apoptosis.
- **Uncoupling of oxidative phosphorylation**. CoQ is required for the delivery of protons from fatty acids to uncoupling proteins. The uncoupling of the proton gradient from oxidative phosphorylation is used to produce heat.
- **Antiinflammatory effects**. CoQ stimulates lymphocytes to release factors signaling the expression of NF-κB-1-dependent genes to produce antiinflammatory cytokines.
- **Endothelial function**. CoQ stimulates endothelial release of nitric oxide.

144. e.g., Ubiquinone(50) is equivalent to CoQ10.

145. HMG-CoA reductase is upstream of the branch point in the mevalonate pathway whereby farnesyl pyrophosphate can be used for the synthesis of either CoQ or cholesterol. This fact has been exploited to reduced the cholesterol contents of eggs by supplementing the diets of laying hens with CoQ_{10} (Honda, K., Saneyasu, T., Motoki, T., et al., 2013. Biosci. Biotechnol. Biochem. 77, 1572–1574).

146. The contributions of foods and feedstuffs, many of which are now known to contain appreciable concentrations of ubiquinones, to these high tissue levels are unknown.

147. Stocker, R., Bowry, V.W., Frei, B., 1991. Proc. Natl. Acad. Sci. 88, 1646–1650.

TABLE 19.8 CoQ_{10} Contents of Foods

Food	CoQ_{10}, µg/g
Vegetables	
Asparagus	2
Avocado	10
Bean	2
Broccoli	6–9
Cabbage	1–5
Cauliflower	1–7
Eggplant	1–2
Onion	1
Parsley	8–26
Pepper, sweet	3
Potato	1
Soybean	7–19
Spinach	1–10
Sweet potato	3–4
Fruits	
Apple	1
Banana	1
Orange	1–2
Strawberry	1
Tomato	<1
Cereals	
Corn germ	7
Rice bran	5
Wheat germ	4–7
Meats	
Beef, muscle	16–37
Beef, liver	39–51
Beef, heart	113
Pork, muscle	14–45
Pork, liver	23–54
Pork, heart	118–282
Chicken, muscle	11–25
Chicken, liver	1116–132
Chicken, heart	92–192
Fish	3–130
Dairy products and eggs	
Milk	2
Butter	7
Eggs	1–4
Other	
Corn oil	13–139
Olive oil	4–160
Soybean oil	54–279

Adapted from Pravst, I., Zmitek, K., Zmitek, J., 2010. Crit. Rev. Food Sci. Nutr. 50, 269–280.

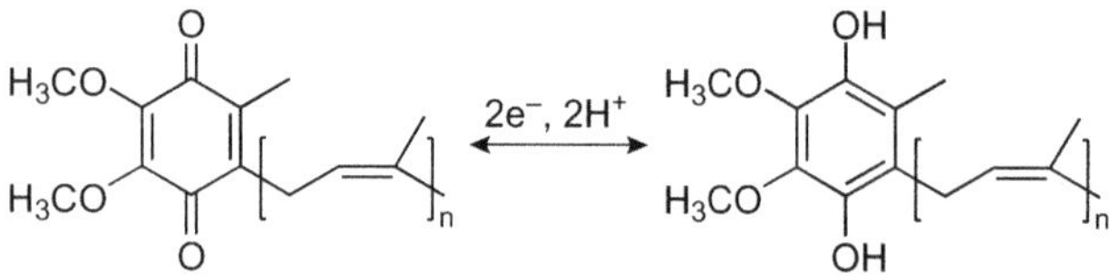

FIGURE 19.8 Redox function of ubiquinones.

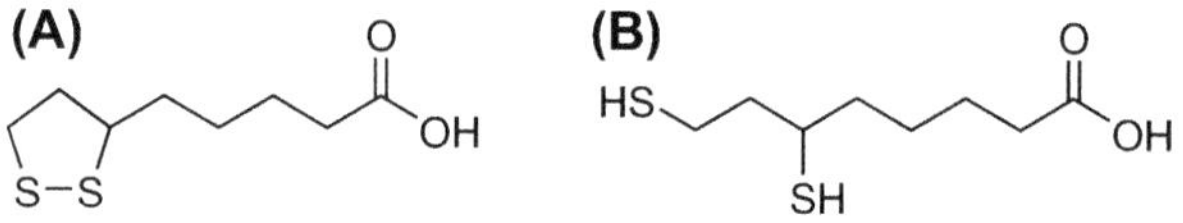

FIGURE 19.9 Lipoic acid (A) and dihydrolipoic acid (B).

Dietary Ubiquinones in Health and Disease

Cardiovascular health. Studies with animal models have shown that supplemental CoQ_{10} can help maintain the integrity of cardiac muscle under cardiomyopathic conditions.[148] Administration of CoQ_{10} has been found to protect against myocardial damage mediated by free-radical mechanisms (ischemia, drug toxicities) in animal models. Clinical trials with humans have indicated benefits of supplemental CoQ_{10} of several types:

- **Migraine**—A small, open-label trial found CoQ_{10} to reduce headache frequency.[149]

148. e.g., Cardiomyopathy induced in the rat by feeding a fructose-based, copper-deficient diet.
149. Sándor, P.S., Di Clemente, L., Coppola, G., et al., 2005. Neurol. 64, 713–715.

- **Congestive heart failure**—A meta-analysis of randomized controlled trials showed CoQ_{10} to reduce dyspnea, edema, and the frequency of hospitalization.[150]
- **Coronary artery disease**—Its antioxidant activity may be the basis of CoQ_{10} supplementation reducing inflammatory markers in patients with coronary artery disease.[151]
- **Hypertension**—A systematic review of eight randomized trials found that CoQ_{10} supplementation can decrease blood pressure (systolic: −16 mm Hg; diastolic: −10 mm Hg).[152]
- **Atherosclerosis**—A randomized trial found CoQ_{10} after myocardial infarction to reduce subsequent myocardial events and cardiac deaths.[153]
- **Endothelial dysfunction**—A randomized trial found CoQ_{10} to improve endothelial function of peripheral arteries of dyslipidemic patients with type 2 diabetes.[154]
- **Recovery from coronary artery bypass surgery**—In a randomized trial, patients preoperatively supplemented with CoQ_{10} had significantly fewer reperfusion arrhythmias and shorter hospitalizations than controls.[155]
- **Friedrich's ataxia**—High doses of CoQ_{10} administered with vitamin E improved cardiac and muscular function in patients with Friedrich's ataxia.[156,157]

Neurologic health. A clinical trial found CoQ_{10} supplementation to improve symptoms in children with autism.[158] Modest improvements in symptoms were reported from a small, randomized trial with Parkinson's disease patients.[159]

Other effects. CoQ_{10} supplementation has been reported to improve semen quality in men with idiopathic infertility, to reduce the risk of preeclampsia, to reduce ultraviolet light-induced skin wrinkling, to confer protection against cardiac or hepatic toxicity associated with cancer chemotherapy, and to improve muscular strength in patients with Duchenne muscular dystrophy. Studies in animal models have suggested that CoQ_{10} supplementation may improve insulin sensitivity.

150. Soja, A.M., Mortensen, S.A., 1997. Mol. Aspects Med. 18, S159–S168.
151. Lee, B.J., Huang, Y.C., Chen, S.J., et al., 2012. Nutrition 28, 767–772.
152. Rosenfeldt, F., Hilton, D., Pepe, S., et al., 2003. Biofactors 18, 91–100.
153. Singh, R.B., Neki, N.S., Kartikey, K., et al., 2003. Mol. Cell. Biochem. 246, 75–82.
154. Watts, G.F., Playford, D.A., Croft, K.D., et al., 2002. Diabetolgia 45, 420–426.
155. Makhija, N., Sendasgupta, C. Kiron, U., et al., 2008. J. Cardiothorac. Vasc. Anesth. 22, 832–839.
156. Cooper, J.M., Schapira, A.H.V., 2007. Mitochondrion 7S, S127–S135.
157. Friedreich's ataxia is an autosomal recessive condition involving deficient production of a mitochondrial protein, frataxin, thought to function in antioxidant regulation and/or iron metabolism. The condition manifests in adolescence as progressive limb and gait ataxias, and losses of deep tendon reflexes and position and vibration senses due to the loss of large sensory neurons in the dorsal root ganglia and deterioration of other cerebellar-spinal tracts.
158. Gvozdjáková, A., Kucharská, J., Ostatníková, D., et al., 2014. Oxid. Med. Cell. Long. Article ID:798957, 6 pp.
159. Muller, T., Büttner, T., Gholipour, A.F., et al., 2003. Neurosci. Lett. 341, 201–204.

Biomarkers of Ubiquinone Status

Assessments of CoQ in the plasma/serum reflect mostly the small amounts recently absorbed from the diet as well as that released by the liver in association with lipoproteins. In healthy individuals, plasma CoQ concentrations are typically ~1 μg/mL. While not informative regarding tissue CoQ status, plasma CoQ levels can be useful in evaluating the status of LDLs, which are particularly susceptible to oxidation. Tissue ubiquinone status is best assessed by direct measurement of CoQ in mitochondrial lipids.

Safety of Ubiquinones

Human studies have found CoQ to be safe and well tolerated as a dietary supplement. A chronic, high-level treatment (0.26% of diet) has been shown to exacerbate some cognitive and sensory impairments in older mice.[160]

6. LIPOIC ACID

Lipoic acid, also known as **thioctic acid**, is a normal metabolite first isolated from bovine liver in 1951.[161] It has subsequently been shown to function both as a coenzyme for mitochondrial enzymes and as an amphipathic antioxidant capable of quenching free radicals, regenerating other cellular antioxidants, and chelating prooxidative metal ions. While lipoic acid is biosynthesized in the mitochondria of apparently all animals, evidence suggests that its biosynthesis is likely to decline with age-related losses of mitochondrial mass, making lipoic acid **conditionally essential**. Its antioxidant properties appear to be the basis of beneficial health effects of lipoic acid supplements, making lipoic acid *also* a **beneficial dietary bioactive** for animals capable of its biosynthesis.

Benefits of dietary lipoic acid are as follows:

- Improved mitochondrial function in aging
- Increased antioxidant protection.

Chemical Nature of Lipoic Acid

Lipoic acid (1,2-dithiolane-3-pentanoic acid) contains two vicinal sulfur atoms that are subject to oxidation/reduction;

160. Sumien, N., Heinrich, K.R., Shetty, R.A., et al., 2009. J. Nutr. 139, 1926–1932.
161. Reed, L.J., DeBusk, B.G., Gansalus, I.C., et al., 1951. Ciences 114, 93–94.

the reduced form, dihydrolipoic acid, has two thiols. Lipoic acid contains one chiral center[162]; thus, the molecule has two possible optical isomers (*R*- and *S*-) (Fig. 19.9).

Sources of Lipoic Acid

Biosynthesis. Lipoic acid (the *R*-enantiomer) is synthesized in the mitochondria from unsaturated fatty acids and cysteine catalyzed by a **lipoyltransferase** and **lipoic acid synthetase**.[163] It is quickly reduced by **dihydrolipoyl dehydrogenase**. The amounts synthesized are limited,[164] but appear adequate for all species studied, e.g., no adverse effects have been observed in chicks, rats, or turkey poults fed lipoic acid-free diets.[165] However, because mitochondrial mass declines with age, lipoic acid biosynthesis may also decline in older individuals. This hypothesis is supported by findings that lipoic acid supplementation increased mitochondrial enzyme activities in older rats but not younger ones.[166] Genetic deficiencies have been identified in lipoic acid synthetase and lipoyltransferase I; these result in nonketotic hyperglycemia, defective energy metabolism, encephalopathy, and cardiomyopathy.

Dietary sources. Lipoic acid is present in a wide variety of foods, generally at low levels. The best sources are tissues rich in mitochondria (e.g., red meats, liver, heart, kidney) or chloroplasts (i.e., spinach, broccoli, Brussel sprouts, peas, tomatoes, potatoes, rice bran). In these foods, most lipoic acid is covalently bound to ε-amino groups of lysyl residues in proteins. Foods contain only the *R*-enantiomer; synthetic sources consist of equimolar mixtures of both the *R*- and *S*-enantiomers.

Absorption and Transport of Lipoic Acid

The enteric absorption of lipoic acid is thought to be facilitated by the monocarboxylate transporter and/or Na^+-dependent multivitamin transporter. *R*-lipoic acid is absorbed with moderate efficiency (~40%); *S*-lipoic acid is absorbed less well (~30%).[167] The bioavailability of food forms (lipoyllysine) remain unexamined, but are likely to be no more than 30–35%, depending on protein digestibility. Absorbed lipoic acid appears quickly in the plasma, apparently in free solution. It is cleared quickly, indicating ready tissue uptake. The same transporters are thought to facilitate the tissue uptake of lipoic acid. In tissues, lipoic acid is found in mitochondria.

Metabolism of Lipoic Acid

Lipoic acid is reduced to dihydrolipoic acid by the pyridine nucleotides, NADH or NADPH. Substrate stereospecificity varies; mitochondrial **NADH-dependent dihydrolipoamide dehydrogenase** shows a preference for *R*-lipoic acid; while cytosolic **NAPH-dependent glutathione reductase** shows a preference for *S*-lipoic acid. Lipoic acid is added to lipoic acid-dependent enzymes by covalently linking it to the ε-amino group of a lysyl residue through that action of **lipoyltransferase I**. Lipoic acid is catabolized in mitochondria by extensive β-oxidation of the carbon backbone, resulting in at least a dozen apparently inactive metabolites, including 3-ketolipoic acid; 2,4-bismethylmercaptobutanoic acid; and 4,6-bismethylmercaptohexanoic acid.[168]

Metabolic Functions of Lipoic Acid

Lipoic acid has two metabolic functions:

- **Coenzyme** for five multisubunit redox enzymes: two involved in energy metabolism (**α-ketoglutarate dehydrogenase, pyruvate dehydrogenase complex** [specifically, one of its three components, **dihydrolipoyl transacetylase**, also referred to E2[169]]); and three involved in amino acid metabolism (**branched-chain ketoacid dehydrogenase**, **2-oxoadipate dehydrogenase**) and the **glycine cleavage system**. In all cases, catalysis involves the amide form, **lipoamide**, undergoing reversible acylation/deacylation to transfer acyl groups to CoA as well as reversible redox ring opening/closing, coupled with the oxidation of an α-keto acid.
- **Antioxidant** involved in cellular antioxidant protection through both direct and indirect functions. This function is based on the redox cycling between the oxidized disulfide (lipoic acid) and the reduced sulfhydryl (dihydrolipoic acid) forms. This is a potent redox couple (0.32V reduction potential) capable of reducing oxidized forms of other cellular antioxidants, including glutathione, α-tocopherol, ascorbic acid, and CoQ_{10}, as well as quenching ROS and reactive nitrogen species. Lipoic acid can also chelate metal ions (iron, copper,

162. i.e., The lipoic acid molecule has a carbon atom to which four different moieties are bound, for which reason the molecule lacks an internal plane of symmetry.
163. Mayr, J.A., Feichtinger, R.G., Tort, F., et al., 2014. J. Inherit. Metab. Dis. 37, 553–563.
164. Carreau, J.P., 1979. Meth. Enzymol. 62, 152–158.
165. Exogenous sources of lipoic acid are, however, required by some species of bacteria (*Streptococcus faecalis*, *Lactobacillus casei*) and protozoa (*Tetrahymena geleii*).
166. Arivazhagan, P., Ramanathan, K., Panneerselavam, C., et al., 2001. Chem. Biol. Interact. 138, 189–198.
167. Hermann, R., Niebach, G., Borbe, H.O., et al., 1996. Eur. J. Pharm. Sci. 4, 167–174.
168. The dominant metabolites include bisnorlipoate, tetranorlipoate, β-hydroxy-bisnorlipoate, and the corresponding bis-methylated mercapto derivatives.
169. PDC also includes pyruvate dehydrogenase (E1) and dihydrolipoyl dehydrogenase (E3).

zinc, magnesium), which otherwise catalyze ROS-generating reactions. Its antioxidant properties underlie several metabolic functions of lipoic acid: increasing Nrf2, which induces expression of catalytic and regulatory subunits of γ-glutamylcysteine ligase, the rate-limiting enzyme in glutathione synthesis[170]; stimulating the insulin receptor and inducing phosphorylation of Akt to activate phosphoinositide-3 kinase; participating in the recruitment of the glucose transporter GLUT4 from its Golgi storage site in muscle; protection against bone loss by inhibiting osteoclastogenic ROS[171]; and preventing cysteine oxidation, which otherwise stimulates protein–tyrosine phosphatases.

Dietary Lipoic Acid in Health and Disease

The metabolic effects of dietary lipoic acid are related to age-related deficiencies in its biosynthesis and/or to its antioxidant properties, which over a wide range of exposure can affect cellular "redox tone."

Aging. Studies with animal models have shown lipoic acid supplementation to protect cardiac mitochondria from age-related dysfunction[172] and correct age-related deficits in the activities of mitochondrial enzymes of the TCA cycle and electron-transport chain affecting energy metabolism.[173] Other studies have found lipoic acid supplements and lipoic acid-rich foods to improve cognitive function and motor skills in aged animals,[174] deficits of which are thought to be related to mitochondrial dysfunction.

Antioxidant effects. Dietary lipoic acid has been shown to have value in preventing and/or treating conditions related to oxidative stress:

- **Diabetes**. Lipoic acid supplementation has been shown to enhance glucose utilization and improve glycemic control. Studies with animal models of diabetes have shown that supplemental lipoic acid can reduce glucose uptake by muscle, reduce exercise-induced lipid peroxidation, increase glucose disposal, reduce cataract formation, and improve motor neuron conductivity. Clinical trials with type 2 diabetic patients have found lipoic acid treatment to increase glucose clearance,[175] improve endothelium-dependent vasodilation,[176] and reduce blood glucose levels and lipid peroxidation products.[177] Meta-analyses of clinical trials concluded that lipoic acid treatment (300–600 mg/day) significantly improved diabetic neuropathies of the feet and lower limbs.[178] A clinical study found lipoic acid supplementation to improve visual contrast sensitivity caused by retinal microvascular damage in patients with either type 1 or type 2 diabetes.[179]
- **Cardiovascular health. A** small study found oral lipoic acid (300 mg/day for 4 weeks) to improve endothelial-dependent flow-mediated vasodilation in subjects with metabolic syndrome.[180]
- **Multiple sclerosis**—A clinical trial indicated that lipoic acid may be useful in treating patients by reducing serum matrix metalloproteinase-9 and reducing T cell migration into the central nervous system.[181]
- **Alzheimer's disease**—An uncontrolled clinical experiment with a small number of patients with probable Alzheimer's disease reported lipoic acid treatment to stabilize declining cognitive function.[182]
- **High-fat feeding**. Dietary supplementation with lipoic acid has been found to attenuate the oxidative stress, dyslipidemic and immunosuppressive effects of a high-fat diet for mice.[183]

Safety of Lipoic Acid

Lipoic acid is considered safe, having been widely used as in clinical therapy for several decades. Studies in dogs have indicated an LD_{50} of 400–500 mg/kg of body weight; however studies with rats have suggested LD_{50} values four to five times that range.[184] At very high doses, gastrointestinal signs (nausea, abdominal pain, vomiting, diarrhea), allergic skin reactions, and malodorous urine have been reported.

170. Suh, J.H., Shenvi, S.V., Dixon, B.M., et al., 2004. Proc. Natl. Acad. Sci. U.S.A. 101, 3381–3386.
171. Roberts, J.L., Moreau, R., 2015. Nutr. Rev. 73, 116–125.
172. Janson, M., 2006. Clin. Interv. Aging 13, 261–265.
173. Arivazhagan, P., Ramanathan, K., Panneerselavam, C., et al., 2001. Chem. Biol. Interact. 138, 189–198.
174. Stoll, S., Hartmann, H., Cohen, S.A., et al., 1993. Pharmacol. Biochem. Behav. 46, 799–805; Bickford, P.C., Gould, T., Briederick, L., et al., 2000. Brain Res. 866, 211–217; Roudebush, P., Ziccker, S.C., Cotman, C.W., et al., 2005. J. Am. Vet. Med. Assoc. 227, 722–728; Head, E., Nukala, V.N., Fenoglio, K.A., et al., 2009. Exp. Neurol. 220, 171–176.
175. Jacob, S., Henricksen, E.J., Tritschler, H.J., et al., 1996. Exp. Clin. Endcrinol. Diabetes 104, 284–288.
176. Heinsch, B.B., Francessconi, M., Mittermayer, F., et al., 2020. Eur. J. Clin. Invest. 40, 148–154.
177. Ziegler, D., Hanefeld, M., Ruhnau, K.J., et al., 1995. Diabetology 38, 1425–1433.
178. Ziegler, D., Nowack, H., Kempler, P., et al., 2004. Diabet. Med. 21, 114–1121; Han, T., Bai, J., Liu, W., et al., 2012. Eur. J. Enodccrinol. 167, 465–471.
179. Gębka, A., Serkies-Minuth, E., Raczyńska, D., 2014. Mediators Inflam. Article ID:131538, 7 pp.
180. Sola, S., Mir, M.Q., Cheema, F.A., et al., 2005. Circulation 111, 343–346.
181. Yadav, V., Marracci, G., Lovera, J., et al., 2005. Mult. Scler. 11, 159–165.
182. Hager, K., Marahens, A., Kenklies, M., et al., 2001. Arch. Gerotol. Geriatr. 32, 275–282.
183. Yang, R.L., Li, W., Shi, Y.H., et al., 2008. Nutrition 24, 582–588; Cui, J., Le, G., Yang, R., et al., 2009. Cell. Immunol. 260, 44–50.
184. Cremer, D.R., Rabeler, R., Roberts, A., 2006. Regul. Toxicol. Pharmacol. 46, 29–42.

7. NONPROVITAMIN A CAROTENOIDS

Of the hundreds of carotenoids that give orange and red colors to foods, a small number are absorbed, and an even smaller number are sequestered in certain tissues. These carotenoids lack β-ionone rings; hence, they cannot be metabolized to retinol. Nevertheless, a few are sequestered in the tissues where they exert important physiological effects, making them **beneficial dietary bioactives**.

Benefits of Nonprovitamin A Carotenoids

The significant fact of biology is that these pigments are accumulated in visual and nervous tissues where they appear to function as screening pigments and antioxidants. These functions would appear to underlie at least some of the demonstrated benefits of diets rich in fruits and vegetables. For some, this is grounds for establishing recommended intakes, although that has not been done.

Benefits of dietary nonprovitamin A carotenoids are as follows:
- Antioxidant protection
- Support of visual function
- Support of neurological function.

Chemical Properties of Nonprovitamin A Carotenoids

The carotenoids are polyisoprenoid compounds produced by plants for the purposes of harvesting light for photosynthesis and quenching free radicals, thereby, protecting plant tissues against oxidative stress. Both functions are due to the capabilities of the conjugated double bond systems of these compounds, which enable them to accept unpaired electrons by delocalizing that electronegativity across multiple carbons. Accordingly, carotenoids have potent antioxidant capabilities. This property allows some carotenoids to function in vision. However, most, i.e., those without the β-ionone head group necessary for that and other vitamin A functions (*see* Chapter 6), lack provitamin A activity. These include some carotenes and oxygenated analogues called **xanthophylls**. The most common nonprovitamin A carotenoids in human diets are **lycopene**, **lutein**, **zeaxanthin**, and **canthaxanthin** (Fig. 19.10).

The most common nonprovitamin A carotenoids in human diets are as follows:

lycopene, lutein, zeaxanthin, and canthoxanthin.

Sources of Nonprovitamin A Carotenoids

Most nonprovitamin A carotenoids are pigments, occurring in red-, yellow-, and orange-colored plant tissues (Table 19.9). The dominant form in US diets is lycopene, the nonaromatic, polyisoprenoid precursor to the biosynthesis of β-carotene in plants.[185] It is found in significant amounts, mainly as the all-*trans*-isomer, in such red-colored foods as tomatoes, watermelon, pink grapefruit, and guava. It is estimated that Americans consume an average of 3–11 mg lycopene daily; estimates from European studies have been similar. The xanthophyll lutein is present in significant concentrations in spinach, kale, corn, broccoli, collards, and eggs. Lutein and zeaxanthin in corn represent major sources of pigmentation for poultry diets.[186] The xanthophyll **astaxanthin** is the source of pink coloration of salmon.

Absorption and Transport of Nonprovitamin A Carotenoids

Nonprovitamin A carotenoids in foods are utilized in the same ways as their provitamin A counterparts (*see* Chapter 6). Their utilization as carotenoids depends on their not being cleaved by the carotene oxygenases.[187] The enteric absorption of nonprovitamin A carotenoids occurs by micelle-dependent diffusion, which depends on the presence of luminal fat and is subject to the antagonistic effects of binding to heat-labile food proteins. Hence, cooking or heat processing improves the bioavailability of lycopene from tomato products. Carotenoids enter the circulation as components of chylomicra from which they are moved to lipoproteins. Most are transported by HDL; however, lutein and zeaxanthin are equally distributed among HDL and LDL. Because the transfer of lipoprotein lipid contents depends on interactions with cell surface receptors, it is thought that the distribution of lutein and zeaxanthin to peripheral tissues, and the retinal capture of lutein and zeaxanthin in particular, depends on an individual's particular lipoprotein (particularly, apoE) profile. Serum levels of lycopene appear to decline with age, but that affect appears to be related to the lower consumption of fat and lycopene-rich, tomato-based foods by older adults (Fig. 19.11).[188]

Tissue distribution. Of the some three dozen carotenoids that have been identified in human serum, lycopene is preferentially accumulated by the prostate, and lutein and zeaxanthin are preferentially captured by the retinal pigment epithelium, the frontal lobe, and visual processing regions of the brain (Fig. 19.12). The human retina contains 25–200 ng of these pigments, zeaxanthin being concentrated in the center and lutein being distributed about the periphery. Greatest concentrations

185. Plants convert lycopene to β-carotene by forming to β-rings at its ends through the action of lycopene cyclase.
186. i.e., To promote coloration of egg yolks and broiler skin.
187. Carotenoid cleavage activity is highly variable between individuals; this is explained in part by single nucleotide polymorphisms in the carotene oxygenase-1 (BCO1) (Leitz, G., Lange, J., Rimbach, G., 2010. Arch. Biochem. Biophys. 502, 8–16.
188. Ganji, V., Kafai, M.R., 2005. J. Nutr. 135, 567–572.

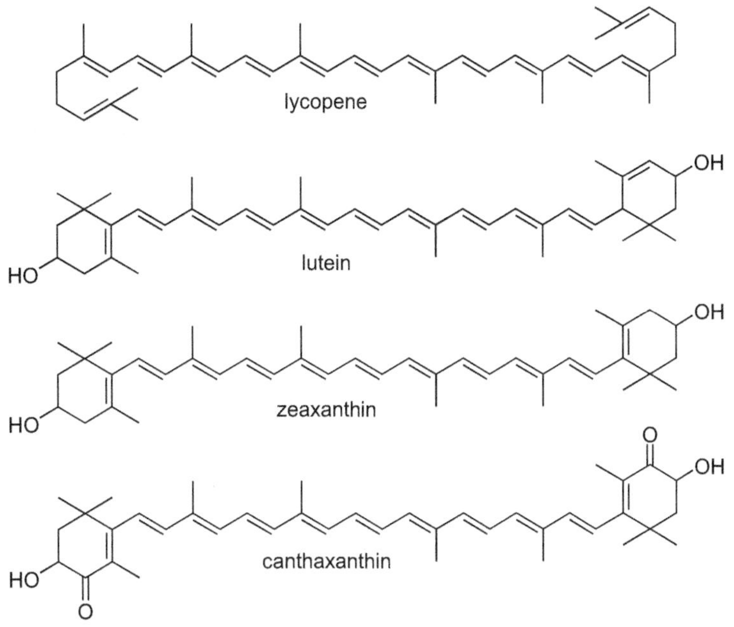

FIGURE 19.10 Structures of major nonprovitamin A carotenoids.

accumulate in the central region of the retina, known as the macula, giving that region, its characteristic yellow color.[189] Lutein and zeaxanthin are typically present in greatest amounts in the fovea (in a 2:1 molar ratio) but their concentrations decrease 100-fold in the periphery. The capture of the macular pigments is believed to be facilitated by a **xanthophyll-binding protein (XBP)**. Adipose tissue may also serve as a storage site for xanthophylls, as serum levels of lutein and zeaxanthin increase in response to weight loss, while macular pigment density is unaffected.[190] Retention of other flavonoids, notably quercetin in liver and naringenin in muscle, is effected by prenylation.[191]

Metabolism of Nonprovitamin A Carotenoids

Lycopene. The dominant forms of lycopene in the serum are its *cis*-isomers, which comprise 50–90% of the circulating carotenoid. This appears to be due to both the better absorption of that isomer than the dominant all-*trans*-isomer found in plant tissues, as well as the continuous *cis*-isomerization of the latter in the body.[192] In humans, lycopene has a half-life of 5 days; it is turned over by conversion to more polar metabolites. The enzyme **carotene oxygenase II** is thought to be involved in this metabolism, producing acylo-retinoids and carbonyl compounds, which are subject to autoxidation or radical-mediated oxidation. A large number of lycopene degradation products are possible.[193]

Lutein and zeaxanthin. While the macular pigments are exclusively of dietary origin, they include a zeaxanthin isomer, 3R,3′S-*meso*-zeaxanthin, that is *not* found in the diet. This isomer appears to be formed by the oxidation–reduction and double-bond isomerization of lutein (i.e., 3R,3′R,6′R-lutein). It is not found in other tissues, suggesting that it may be catalyzed photochemically in the retina or by enzymes expressed only in that tissue. Studies have shown that dietary supplementation with lutein and zeaxanthin increases the macular pigments in two ways: by direct capture and by stimulating the migration of retinal pigment epithelial cells to the macula.[194]

Metabolic Functions of Nonprovitamin A Carotenoids

Antioxidant. **Lycopene** is the most potent carotenoid antioxidant in vitro, being twice as effective (due to its extended conjugated diene system) than β-carotene in quenching singlet oxygen ($^1O_2$) and 10 times as effective as α-tocopherol. Whether this activity is the basis of its beneficial health effects is not clear, as tissue levels of lycopene tend to be much lower than those other antioxidants; such

189. Macular pigments occur in greater amounts in the central area, i.e., the **fovea**. Other mammals lack maculae; however, carotenoid-rich oil droplets have been found in the retinas of birds, reptiles, amphibians, and fish.

190. Kirby, M.L., Beatty, S., Stack, J., et al., 2011. Br. J. Nutr. 105, 1036–1046.

191. Terao, J., Mukai, R., 2014. Arch. Biochem. Biophys. 559, 12–16.

192. Unlu, N.Z., Bohn, T., Francis, D.M., 2007. Br. J. Nutr. 98, 140–146; Ross, A.B., Vuong, L.T., Ruckle, J., 2011. J. Nutr. 93, 1263–1273.

193. Several have been identified in humans or animals: 5,6-dihydroxy-5,6-dihydrolycopene; 2,6-cyclolycopene-1,5-diols; 5,6-dihydrolycopene; 5,6-dihydro-5-*cis*-lycopene; apo-8′-lycopenal; apo-10′-lycopenal; apo-10′-lycopenoic acid.

194. Leung, I.Y., Sandstrom, M.M., Zucker, C.L., et al., 2004. Invest. Ophthalmol. Vis. Sci. 45, 3244–3256.

TABLE 19.9 Food Contents of Nonprovitamin A Carotenoid Contents (μg/100 g or μg/100 mL)

Food	Lutein	Zeaxanthin	Lycopene
Fruits			
Apricot	123–188	0–39	54
Banana	86–192		0–254
Fig	80		320
Grapefruit, red			750
Guava			769–1816
Kiwi			<10
Mango			10–724
Nectarine, flesh			2–131
Papaya	93–318		0–7564
Pineapple			265–605
Watermelon, red			4,770–13,523
Watermelon, Yellow			56–287
Vegetables			
Avocado	213–361	8–18	
Basil	7050		
Bean, green	883		
Cabbage, white	450		
Carrot	254–510		
Cress	6510–7540		
Cucumber	459–840		
Dill weed	13,820		
Egg plant	170		
Endive	2060–6150		
Kale	4,800–11,470		
Leek	3680		
Lettuce	1000–4780		
Parsley	6,400–10,650		
Peas, green	1910		
Pepper, green	92–911	0–42	
Pepper, red	248–8506	593–1350	
Pepper, yellow	419–638		
Potato, sweet	50		
Pumpkin	630		500
Rhubarb			120
Sage	6350		
Spinach	5930–7900		

Continued

TABLE 19.9 Food Contents of Nonprovitamin A Carotenoid Contents (μg/100g or μg/100mL)—cont'd

Food	Lutein	Zeaxanthin	Lycopene
Tomato	46–213		850–12,700
Tomato ketchup			4,710–23,400
Tomato sauce			5,600–39,400
Grains			
Corn, flakes	0–52	102–297	
Durum lour	164		
Wheat flour	76–116		
Other			
Olive oil	350		
Butter	15–25	0–2	
Egg yolk	384–1320		
Milk, 4% fat	0.8–1.4	0–0.1	

Adapted from Maiani, G., Castón, M.J.P., Catasta, G., et al., 2009. Mol. Nutr. Food Res. 53, S194–S218.

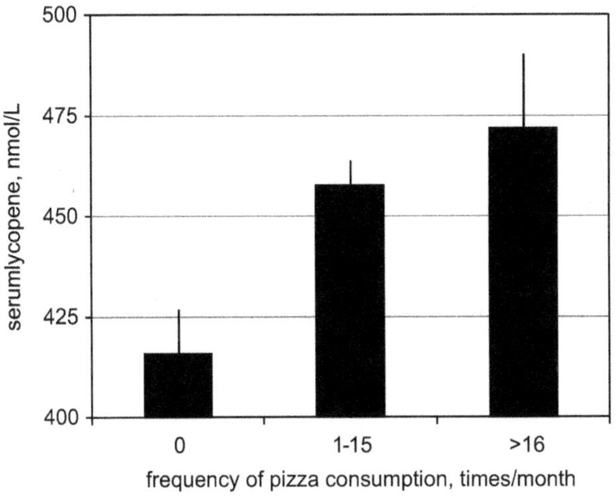

FIGURE 19.11 Relationship of serum lycopene concentration to pizza consumption in the NHANES 1988–94. *Ganji, V., Kafai, M.R., 2005. J. Nutr. 135, 567–572.*

effects may be mediated by lycopene metabolites (e.g., apo-10′-lycopenoid acid). These, and other carotinoids, have been shown to induce superoxide dismutase, glutathione-S-transferase, quinone reductase, and several phase I and II enzymes. Therefore, these responses appear to involve upregulation of **antioxidant response elements (ARE)**. Studies have found lycopene to have a very low affinity for the retinoid X receptor α (*see* Chapter 6) and, thus, to be a very weak transactivator of genes using the retinoic acid response element (RARE). The metabolite apo-10′-lycopenoid acid, however, has been shown to induce RARβ mRNA expression in some cell lines.[195] A meta-analysis of published epidemiological data found evidence that higher intakes of lutein being associated with lower risks of coronary heart disease and stroke.[196]

Vision. Lutein and zeaxanthin comprise the pigment of the posterior central region of the retina, i.e., the macula, where they are typically present in higher concentrations than in any other tissue. These carotenoids protect the macula from the damaging effects of blue-wavelength photons, as they absorb in the range of 420–480 nm, reducing by as much as 90% the incoming energy in this range from reaching macular photoreceptors. They accumulate selectively in the most vulnerable domains of membranes—those containing unsaturated phospholipids. Lutein and zeaxanthin can also scavenge ROS formed in photoreceptors, likely as a results of extramitochondrial oxidative phosphorylation in the rod outer segment. Light-catalyzed oxidative reactions also occur in the retina, as indicated by the accumulation of the autofluorescent pigment **lipofuscin**.[197] These xanthophylls support visual

195. Lian, F., Smith, D.E., Ernst, H., et al., 2007. Carcinogenesis 28, 1567–1574.

196. Leermakers, E.T.M., Darweesh, S.K.L., Baena, C.P., et al., 2016. Am. J. Clin. Nutr. 103, 481–494.

197. This appears to result from the condensation of two molecules of all-*trans*-retinal with one of phosphatidylethanolamine, which complex is taken up by the retinal pigment epithelium and converted to a stable pyridnium bis-retinoid that is cytotoxic and causes apoptosis and, hence, macular degeneration.

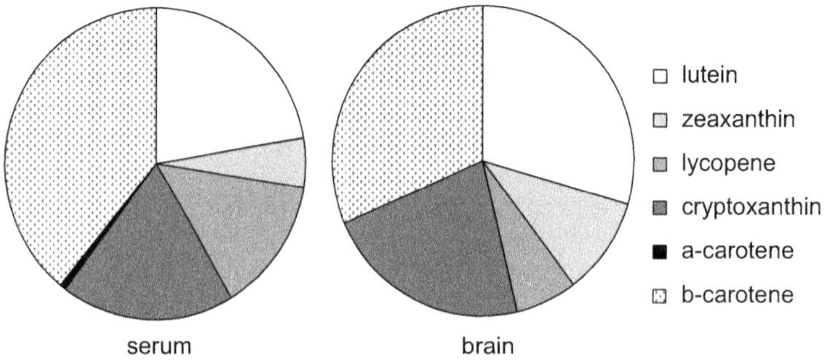

FIGURE 19.12 Distribution of carotenoids in serum and brain (average of cerebellum and frontal, occipital, and temporal cortices) of centenarian decedents. *After Johnson, E.J., Vishwanathan, R., Johnson, M.A., et al., 2013. J. Aging Res. Article ID:951786, 13 pp.*

development and protect against cataracts and age-related macular degeneration (AMD)[198]:

- **Visual function**. Evidence indicates that macular pigment carotenoids support visual acuity by reducing the effects of chromatic aberration or by preferentially absorbing short-wavelength dominant air light that produces veiling luminance in distance viewing (i.e., blue haze). Two well-controlled, randomized trials found lutein supplementation to improve central visual field and reduce loss of central field sensitivity.[199]
- **Cataracts**.[200] The human lens contains significant amounts of **lutein** and **zeaxanthin**, most of which is located in the epithelium and cortex. These antioxidant carotenoids are thought to be important in reducing the photoperoxidation of lens lipids both by quenching free radicals and by maintaining lens glutathione in the reduced state. Several cohort studies have produced strong evidence for relatively high intakes of lutein, zeaxanthin, and/or **lycopene** being associated with reduced risk for nuclear cataracts. One study found lutein supplementation (7 mg/day, i.e., the equivalent of ~100 g spinach) to improve visual acuity and reduce glare sensitivity in subjects with age-related cataracts.[201]
- **AMD**.[202] Evidence suggests that the macular pigments lutein and zeaxanthin may protect against AMD. A case–control study found that high consumption of carotenoid-rich foods was associated with a 43% reduction in risk for neovascular AMD.[203] The Age-Related Eye Disease Study (AREDS) found that individuals in the highest quintile of lutein and zeaxanthin consumption had risk reductions of 27–55% for various signs of AMD.[204] A randomized trial using lutein plus other antioxidants as the intervention agent in patients with progressive atrophic macular degeneration reported improvements in several aspects of visual acuity[205]; other trials found no protective effect of similar treatments.[206] These apparently discrepant findings may be due to different genetic susceptibilities to AMD in those cohorts, as intervention with lutein, zeaxanthin and other antioxidants was found effective in reducing AMD risk only in subjects with the greatest genetic risk.[207] AMD patients are advised to consume low-glycemic diets rich in green, leafy vegetables and to consume at least two

198. Vishwanathan, R., Johnson, E.J., 2012. In: Erdman, J.W., Macdonald, I.A., Zeisel, S.H. Eds.), Present Knowledge in Nutrition, tenth ed. Wiley-Blackwell, Ames, IA, pp. 939–981.

199. Bahrami, H., Melia, M., Dagnelie, G., 2006. BMC Ophthalmol. 6, 23; Berson, E.L., Rosner, B., Sanmdberg, M.A., et al., 2010. Arch. Ophthalmol. 122, 403–411.

200. Age-related cataract is a major cause of visual impairment and blindness, affecting as many as 50 M people worldwide and present in clinically significant ways in almost half of Americans 75–85 yrs, increasing from ~8% in those 52–64 yrs. Cataract can occur in different parts of the lens: center (nuclear, the most common), outer rim (cortical), or central posterior cortex (posterior subcapsular).

201. Olmedilla, B., Granado, F., Blanco, I., et al., 2003. Nutrition 19, 21–24.

202. AMD is a leading cause of severe vision loss in industrialized countries. It affects ~1.75 million Americans, including ~8% of adults 43–54 years. and ~30% of those >75 yrs with early signs (7% of the latter group with advanced disease). It affects the macula, which is responsible for high-acuity vision by virtue of its high density of concentration of cones. The fist sign of AMD is the accumulation of extracellular debris ("drusen") between the retinal pigment epithelium (RPE) and its basement membrane due to the phagocytic failure of RPE cells.

203. Seddon, J.M., Ajani, U.A., Sperduto, R.D., et al., 1994. JAMA 272, 1413–1420.

204. AREDS, 2007. Arch. Ophthamol. 125, 1225–1232.

205. Richer, S., Stiles, W., Statkute, L., et al., 2004. Optometry 75, 216–230.

206. Van den Langenberg, G., Mares-Perlman, J., Klein, R., et al., 1998. Am. J. Epidemiol. 148, 204–214; Taylor, H., Tikellis, G., Robman, L., et al., 2002. Br. Med. J. 325, 11–14; Moeller, M.S., Parekh, N., Tinker, L.L., et al., 2006. Arch. Ophthalmol. 124, 1151–1162; Bartlett, H.E., Eperjesi, F., 2007. Eur. J. Clin. Nutr. 61, 1121–1127.

207. Wang, J.J., Buitendijk, G.H.S., Rochtchina, E., et al., 2014. Ophthalmology 121, 667–675.

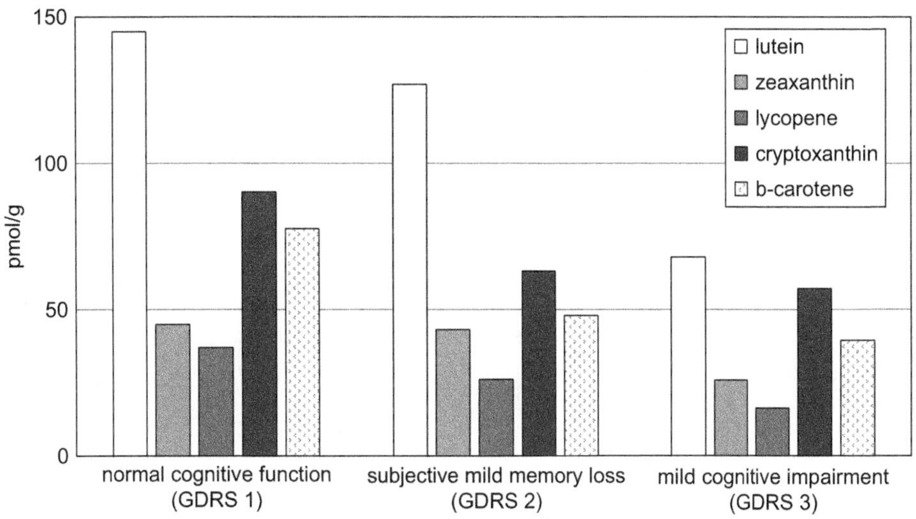

FIGURE 19.13 Carotenoids in brains (average of cerebellum and frontal, occipital and temporal cortices) of centenarian decedents according to premortem Global Deterioration Scores. *After Johnson, E.J., Vishwanathan, R., Johnson, M.A., et al., 2013. J. Aging Res. Article ID:951786, 13 pp.*

servings of fish weekly.[208] The use of an AREDS-based supplement (400 IU vitamin E, 500 mg vitamin C, 10 mg lutein, 2 mg zeaxanthin, 80 mg zinc, and 2 mg copper) is the most effective means of slowing the progression of atrophic AMD.[209]

Neurologic function. Lutein and zeaxanthin comprise more than two-thirds of the carotenoids in the frontal lobe and visual processing regions of the brain, with concentrations highly correlated with those of the retina. In vitro studies have shown that lutein can have a structural role in PUFA-rich neural membranes while also influencing interneuronal and neural–glial cell communications.[210] That such functions occur in vivo and affect visual processing is indicated by findings that the proxy variable macular pigment density was directly related to visuomotor ability in older subjects.[211] A study of octagenarians and centigenarians found that the serum lutein, zeaxanthin, and β-carotene were associated with better cognition, and that brain lutein and β-carotene concentrations determined postmortem were positively related to cognition (Fig. 19.13).[212] A double-blinded, randomized intervention trial found supplementation with lutein (12 mg/day) to improve verbal fluency, and supplementation with both lutein and docosahexaenoic acid (800 mg/day) to improve memory and learning rate.[213]

Dietary Nonprovitamin A Carotenoids in Health and Disease

Anticarcinogenesis. Epidemiological investigations have shown consumption of tomato-rich diets to be associated with reduced risk to cancers of the prostate and lung.[214] A meta-analysis of 31 prospective studies of breast cancer found β-carotene to be the only carotenoid intake of which was related (negatively) to breast cancer risk. A case–control analysis of 3004 women in the EPIC[215] cohort found significant inverse relationships of plasma concentrations of β- and α-carotenes and risk for estrogen receptor-negative (ER^-) breast cancer risk, but no relationship with ER^+ breast cancers.[216] One study found serum lutein concentration to be inversely related to breast cancer risk.[217] A meta-analysis of 11 case–control studies and 10 cohort or nested case–control studies found prostate cancer risk to be inversely related

208. Consumption of n-3 PUFAs has been shown to prevent and reduce progression of retinal lesions in animal models (Pinazo-Durán, M.D., Gómez-Ulla, F., Arias, L., et al., 2014. J. Ophthalmol. Article ID:901686).
209. Broadhead, G.K., Grigg, J.R., Chang, A.A., et al., 2015. Nutr. Rev. 73, 488–462.
210. Stahl, W., Sies, H., 2001. Biofactors 15, 95–98; Johnson, E.J., McDonald, K., Caldarella, S.M., et al., 2008. Nutr. Neurosci. 11, 75–98.
211. Hammond, B.R., Fletcher, L.M., 2012. Am. J. Clin. Nutr. 96, 1207S–1213S.
212. Johnson, E.J., Vishwanathan, R., Johnson, M.A., et al., 2013. J. Aging Res. Article ID:951786, 13 pp.
213. Johnson, E.J., McDonald, K., Caldarella, S,M., et al., 2008. Nutr. Neurosci. 11, 75–83.
214. Giovannucci, E., 2002. Exp. Biol. Med. 227, 852–859; Giovannucci, E., 2005. J. Nutr. 135, 2030S–2031S; Arab, L., Steck-Scott, S., Fleishaur, A.T., 2002. Exp. Biol. Med. 227, 894–899.
215. European Prospective Investigation into Cancer and Nutrition; this ongoing study includes 521,468 subjects managed in 23 centers in 10 countries.
216. Bakker, M.F., Peeters, P.H.M., Klaasen, V.M., et al., 2016. Am. J. Clin. Nutr. 103, 454–464.
217. A 25 μg/dL increment in serum lutein was associated with a 32% reduction in breast cancer risk. Aune, D., Chan, D.S.M., Vieira, A.R. et al., 2012. Am. J. Clin. Nutr. 96, 356–373.

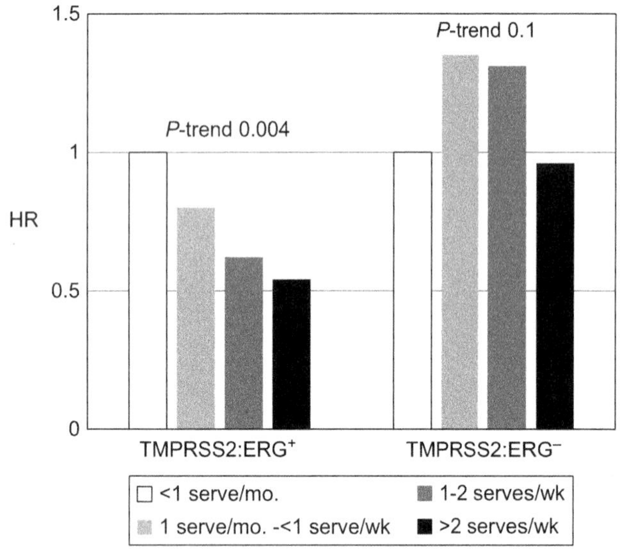

FIGURE 19.14 Relationship of tomato sauce intake and prostate cancer risk by TMPRSS2:ERG status. *Data are expressed as hazards ratios (HRs) comparing each intake group to the lowest one. After Graff, R.E., Pettersson, A., Lis, R.T., et al., 2016. Am. J. Clin. Nutr. 103, 851–860.*

to serum concentration of **lycopene**,[218] suggesting risk reduction of 25–30%. A more recent meta-analysis of 24 case–control and nested case–control studies found both dietary intake and plasma/serum levels of lycopene plus β-carotene (but not β-carotene alone) to be associated with reduced risk of prostate cancer.[219] A recent evaluation of the 23 years of follow-up of 884 man subset of the Health Professionals Follow-up Study found that the intake of tomato sauce was inversely related to risk of developing prostate cancer.[220] This effect was found to be driven by a strong protective effect for men expressing the transmembrane protease, TMPRSS2:ERG (Fig. 19.14).[221]

Studies with animal tumor models have shown lycopene treatment to affect molecular mechanisms associated with antitumorigenesis, e.g., reduced cell proliferation, increased apoptosis, reduced markers of oxidative stress. Two studies have found lycopene to reduce the incidence of spontaneous or chemically induced mammary cancers in the rat and of chemically induced lung cancers in male (but not female) mice. Lycopene has been found to cause cell cycle arrest and apoptosis in cultured prostate cancer cells, which is associated with upregulation of the expression of intercellular gap junction communication associated with decreased cell proliferation.[222] Only a few intervention studies have been conducted. One found lycopene supplementation to reduce cancer biomarkers and disease progression in patients with benign prostatic hyperplasia[223]; another found no effects on progression of high-grade prostatic intraepithelial neoplasias.[224]

Skin health. Epidemiological studies have pointed to a protective effect of tomato-rich diets against UV damage to the skin. A randomized trial found that consumption of lycopene-rich tomato paste conferred significant protection against UV damage in a small cohort of healthy women, increasing the median erythemal dose by 16%.[225]

Biomarkers of Nonprovitamin A Carotenoid Status

Carotenoid status can be assessed in two ways:

- **Plasma carotenoids** can be determined in plasma by visual absorbance after chromatographic separation. The responses of plasma lutein, lycopene, and β-carotene to meals containing those carotenoids are highly variable between individuals. The postprandial plasma lutein response has been found to be affected by at least 15 genes and single-nucleotide polymorphisms related to lutein metabolism and the chylomicron triacylglycerol response.[226] Serum lycopene levels have been similarly found to be affected by polymorphisms in at least three genes.[227]
- **Skin carotenoids** can be assessed in the skin optically by resonance Raman spectroscopy palmer scanning.[228] This method yields results that are correlated with plasma total carotenoid analyses (Fig. 19.15) and off the advances of a rapidly, noninvasive procedure.

Safety of Nonprovitamin A Carotenoids

Both **lycopene** and **lutein** are generally regarded as safe. A systematic risk review found no reports of adverse effects for either lutein (highest chronic exposures noted for

218. Emtinan, M., Takkouche, B., Caamano-Isoma, F., 2004. Cancer Epidemiol. Biomarkers Prev. 13, 340–345.
219. Wang, Y., Cui, R., Xiao, Y., et al., 2015. PLoS One 10, e0137427.
220. Graff, R.E., Pettersson, A., Lis, R.T., et al., 2016. Am. J. Clin. Nutr. 103, 851–860.
221. Serine 2:v-ets avian erythrocyte blastosis virus E26 oncogene homologue; this protein is expressed by ~50% of patients with prostate cancer.
222. i.e., Connexin 43; Heber, D., Lu, Q.Y., 2002. Exp. Biol Med. 227, 920–923.
223. Schwarz, S., Obermüller-Jevec, U.C., Hellmis, E., et al., 2008. J. Nutr. 138, 49–53.
224. Mariani, S., Lionetto, L., Cavallari, M., et al., 2014. Int. J. Mol. Sci. 15, 1433–1440.
225. Rizwan, M., Rodriguez-Blanco, I., Harbottle. A., et al., 2010. Br. J. Dermatol. 154, 154–162.
226. Borel, P., Desmarchelier, C., Nowicki, M., et al., 2014. Am. J. Clin. Nutr. 100, 168–175.
227. Zubair, N., Kooperberg, C., Liu, J., et al., 2015. J. Nutr. 145, 187–192.
228. Mayne, S.T., Cartmel, B., Scarmo, S., et al., 2013. Arch. Biochem. Biophys. 539, 163–170.

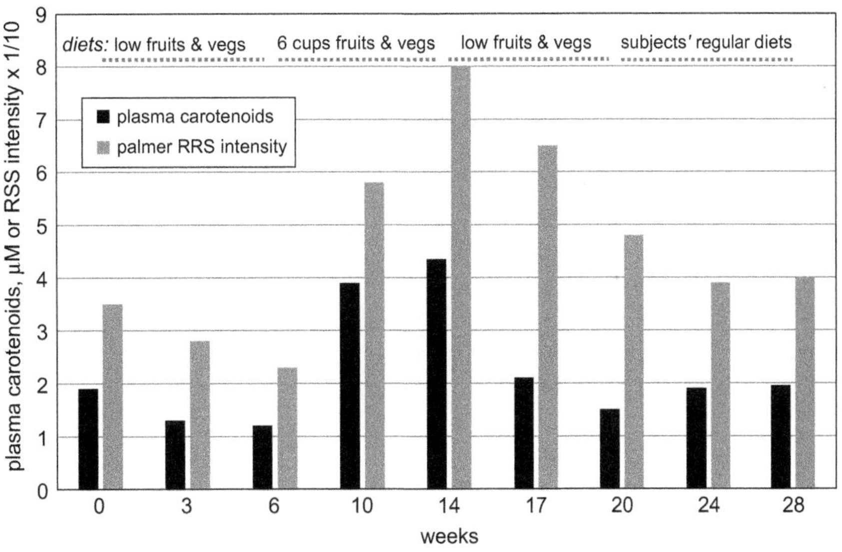

FIGURE 19.15 Responses of plasma carotenoids and palmar scanning by Raman resonance spectrometry (RRS) to changes in intakes of carotenoid-containing fruits and vegetables. In phases I (0–6 wks) and III (14–20 wks), subjects were asked to avoid carotenoid-containing fruits and vegetables. In phase II (6–14 wks), they were fed a diet that provided 1–2.5 cups of fruits and 3–4.5 cups of vegetables daily. In phase IV (20–28 wks) they returned to their regular self-selected diet. *After Jahns, L., Johnson, L.K., Mayne, S.T., et al., 2014. Am. J. Clin. Nutr. 100, 930–937.*

humans were 40 mg/day[229]) or lycopene (highest chronic exposures for humans were 150 mg/day).[230] Therefore, the observed safe levels (OSLs) for these carotenoids were set at ≤20 mg/day for lutein, ≤75 mg/day for lycopene.[231] However, because these OSLs were calculated without the benefit of no- or low-observed effect levels (NOAELs or LOAELs), they are necessarily cautious underestimates. Actual safe upper limits of exposure are likely to be at least several times greater.

8. FLAVONOIDS

Flavonoids are ubiquitous plant metabolites. More than 6000 different flavonoids have been identified, representing the major sources of red, blue, and yellow plant pigments other than the carotenoids. They have a wide variety of functions in plants: natural antibiotics,[232] predator feeding deterrents, photosensitizers, UV-screening agents, metabolic modulators. Several flavonoids and flavonoid-containing food have been found to have health benefits, making them **beneficial dietary factors**.

Recognition of Nutritional Roles of Flavonoids

The flavonoids were discovered by Szent–Györgyi as the component of lemon juice or red pepper that potentiated the antiscorbutic activities of those foods for the guinea pig. That factor was called by various groups "citrin," "vitamin P,"[233] and "vitamin C_2," but it was ultimately found to be a mixture of phenolic derivatives of 2-phenyl-1,4-benzopyrane, the flavane nucleus.

Benefits of Dietary Flavonoids

None of the flavonoids meet the criteria of essential nutrients. Still, it is clear that at least some of this diverse group of natural components of foods are beneficial to health. So, while there presently appears to be no situations of conditional need for flavonoids, there is clear wisdom in including sources of flavonoids in healthy diets. Diets containing such sources, particularly those that provide significant amounts of flavonols, flavones, flavanones, and anthocyanins, have been associated with greater likelihood of health and well-being in older age (Figs. 19.16 and 19.17).[234]

Benefits of dietary flavonoids are as follows:

- Protection against inflammation and related disorders
- Promotion of immune function
- Reduced risks for chronic disease

Chemical Properties of Flavonoids

Flavonoids are secondary metabolites of shikimic acid.[235] They are polyphenolic compounds containing two aromatic

229. Dagnelie, G., Zorge, I.S., McDonald, T.M., 2000. Optometry 71, 147–164.
230. Rao, A.V., Agarwal, S., 1998. Nutr. Cancer 31, 199–203.
231. Shao, A., Hathcock, J.N., 2006. Reg. Toxicol. Pharmacol. 45, 289–298.
232. e.g., Phytoalexins, antimicrobial plant metabolites which accumulate in infected tissues.
233. *P* indicated the *permeability* vitamin, because it improved capillary permeability.
234. Samieri, C., Sun, Q., Townsenmd, M.K., et al., 2014. Am. J. Clin. Nutr. 100, 1489–1497.
235. (3*R*,4*S*,5*R*)-3,4,5-trihydroxycyclohex-1-ene-1-carboxylic acid.

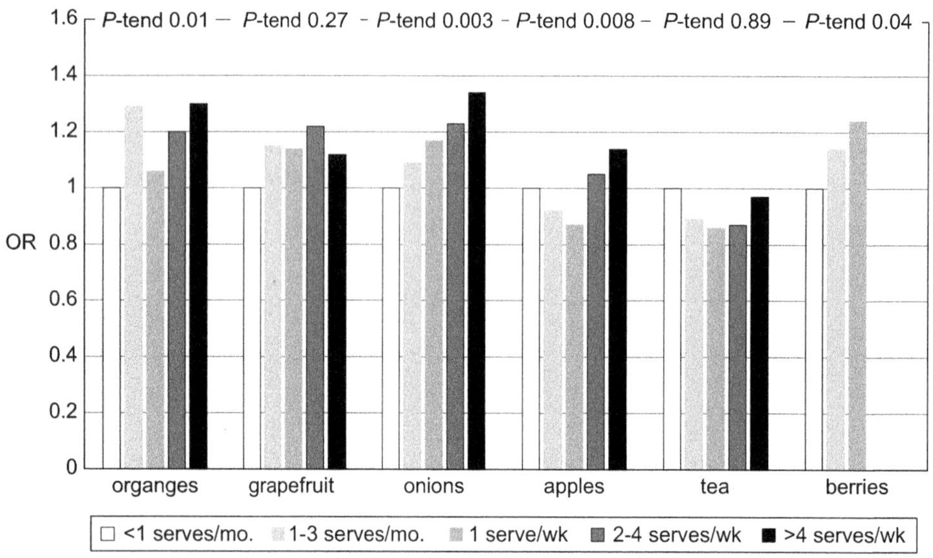

FIGURE 19.16 Relationship of intake of flavonoid-rich foods and healthy aging. Data from the Nurses' Health Study presented by quintiles of each class of flavonoid intake; relative healthy aging scores are presented as odds ratios (ORs) for each of the higher quintiles (Q2-Q5) compared to lowest quintile (Q1), n's 242–373. *After Samieri, C., Sun, Q., Townsend, M.K., et al., 2014. Am. J. Clin. Nutr. 100, 1489–1497.*

rings linked by an oxygen-containing, heterocyclic ring (Fig. 19.18). The hydroxyl groups of these polyphenols enable them to form glycosidic linkages with sugars, and most flavonoids occur naturally as glycosides.

There are six general classes of flavonoids, classified by their common ring substituents:

- **Flavonols** (R_3 hydroxy-, R_4 keto-derivatives) include quercetin, kaempferol, isorhamnetic and myricetin, the most abundant flavonoids in human diets. Flavonols are found in different fruits and vegetables, often as glycosides. Relatively high amounts (15–40 mg/100 g) are found in broccoli, kale, leeks, and onions.
- **Flavanols** (R_3 hydroxy derivatives), also called **catechins**, do not exist as glycosides (unlike other flavonoids). They include catechin, epicatechin, epigallocatechin, and their gallate derivatives found in apples, apricots, and red grapes (2–20 mg/100 g); green tea and dark chocolate are rich in catechins (40–65 mg/100 g).
- **Flavones** are a group of some 300 compounds that retain the basic flavane nucleus structure. It includes apigenin and luteolin, which occur in high concentrations (>600 mg/100 g) in parsley, and in lower but significant amounts in cereal grains, celery, and citrus rinds (as polymethoxylated forms).
- **Anthocyanins** (R_3 and R_4 reduced derivatives) exist as glycosides; their aglycones are called anthocyanidins of which there are several hundred, the most common being cyanidin, delphinidin, malvinidin, pelargonidin, peonidin, petunidin, and malvidin. Most are red or blue pigments. The richest sources (up to 600 mg/100 g) are raspberries, black berries, and blue berries; cherries, radishes, red cabbage, red skinned potato, red onions, and red wine are also good sources (50–150 mg/100 g). Anthocyanins have antioxidant properties. Unlike other flavonoids, anthocyanins are relatively unstable to cooking and high-temperature food processing.
- **Flavanones** (R_4 keto, "C" ring reduced derivatives) are found primarily in citrus pulp (15–50 mg/100 g) where they are also present as *O*- and *C*-glycosides and methoxylated derivatives. They include eriocitrin, neoericitrin, hesperetin, neohespiridin, naringin, narirutin, didymin, and poncirin.
- **Isoflavones** ("B" aromatic ring derivatives linked at R_3) are contained only in legumes, mostly as glycosides. They include daidzein, genistein, and glycitein, which are also referred to as **phytoestrogens** due to the affinities of their 7- and 4'-hydroxyl groups to binding mammalian estrogen receptors. Soy products (soy flour, tofu, tempeh) can contain 25–200 mg/100 g.
- **Tannins** are polymeric flavonoids present in all plants. Those conjoined by covalent, nonhydrolyzable C–C bonds are called **condensed tannins** or **proanthocyanidins**. Others containing hydrolysable nonaromatic polyol carbohydrate moieties include gallic and ellagic acids and have strong antioxidant properties in vitro.

Dietary Sources

Dietary intake of flavonoids varies widely according to dietary habits and preferences. Americans are estimated to consume ~190 mg/person/day, mostly as flavonols.[236] Similar estimates have been made for northern Europeans.

236. Chun, O.K., Chung, S.J., Song, W.O., 2007. J. Nutr. 137, 1244–1252.

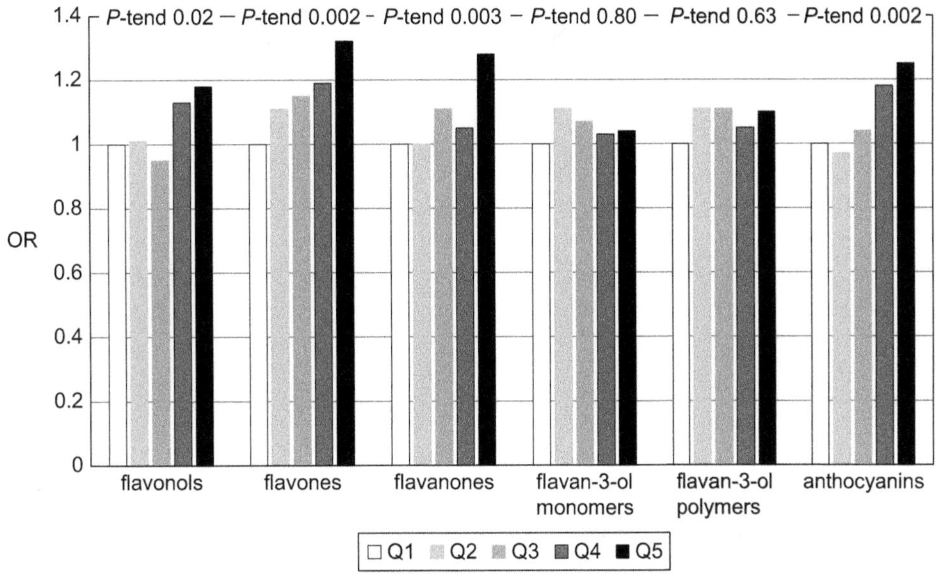

FIGURE 19.17 Relationship of flavonoid intake and healthy aging. Data from the Nurses' Health Study presented by quintiles of each class of flavonoids imputed from food intake data; relative healthy aging scores are presented as odds ratios (ORs) for each of the higher quintiles (Q2-Q5) compared to lowest quintile (Q1), n's 242–373. *After Samieri, C., Sun, Q., Townsend, M.K. et al., 2014. Am. J. Clin. Nutr. 100, 1489–1497.*

The greatest contributors of flavonoids in human diets are fruits and vegetables, fruit juices, green tea, and dark chocolate (Table 19.10). Most flavonoids tend to be concentrated in the outer layers of fruit and vegetable tissues (e.g., skin, peel). In general, the flavonoid contents of leafy vegetables are high; whereas, those of root vegetables are low, with the notable exception of red-skinned onions. The flavonoid contents of vegetables and fruits can vary between cultivars; for example, the quercetin contents of six commercial onion varieties were found to vary by 18-fold.[237] The greatest contributors to total flavonoid intakes tend to be tea, citrus fruits and juices, and wine. Flavonoid aglycones are stable during food processing and cooking; however, anthocyanidins are unstable to such conditions.

Absorption and Transport of Flavonoids

Absorption. Most flavonoids in foods occur as glycosides, which must be hydrolyzed by glycosidases in saliva and brush border of the intestine to be absorbed. The efficiency of these processes appears to be low, e.g., <10%, as many dietary flavonoids are esterified in nonhydrolyzable ways. Glucosylated flavonoids (e.g., quercetin, hesperetin) are hydrolyzed; their aglycone moieties are absorbed by the Na^+-dependent glucose transporter-1 (SGLT-1), as are nonglycosylated flavonoids (e.g., epicatechin, epigallocatechin).

The large portions of flavonoids that are not absorbed are metabolized by the hindgut microbiome. This includes anthocyanins, procyanidins, and flavonoids linked to glucose and rhamnose, e.g., hesperidin, naringin, and rutin.

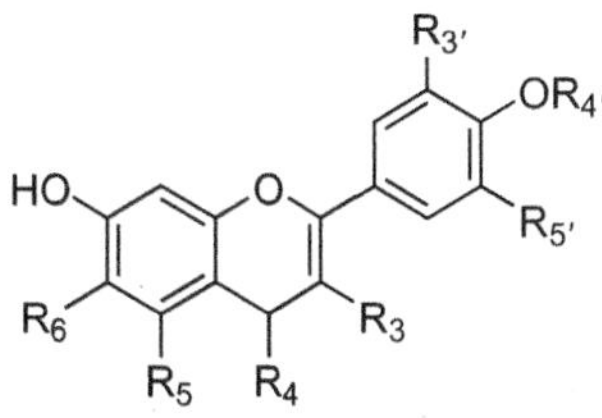

FIGURE 19.18 General structure of flavonoids.

Microbial metabolites of these flavonoids appear to be absorbed across the colon and are likely to be biologically active. Indeed, it has been suggested that the demonstrated benefits of black tea (and, perhaps, other) flavonoids in reducing inflammation, lowering blood pressure, and improving platelet and endothelial cell functions may be mediated by specific microbial metabolites.[238] The hindgut microbiome responds to the host's consumption of dietary flavonoids. This was demonstrated by the finding that pigs fed cocoa powder (a source of flavan-3-ols, epicatechin and catechin) showed increased abundance of *Lactobacillus* and *Bifidobacterium* species in their colonic microbiome.[239]

Transport. Most absorbed flavonoids are conjugated in the liver such that glucuronides, sulfates, and methylated derivatives comprise the dominant forms in the circulation. The notable exception is epigallocatechin gallate (EGCG), which circulates predominantly in unconjugated form.

237. Lee, J., Mitchell, A.E., 2011. J. Agr. Food Chem. 59, 857–863.

238. Van Duynhoven, J., Vaughan, E.E., van Dorsten, F., et al., 2013. Am. J. Clin. Nutr. 98, 1631S–1641S.

239. Jang, S., Sun, J., Chen, P., et al., 2016. J. Nutr. 146, 673–680.

TABLE 19.10 Dietary Sources of Flavonoids

Type	Flavonoid	Food sources
Flavonols	Quercetin, kaempferol, myricetin, isorhamnetin	*Rich sources:*[a] capers, dock, lovage leaves Apple, beans, black grapes, broccoli, buckwheat, carob, chokecherry, coriander, cranberry, elderberry, fennel leaves, ginger, Goji berry (wolfberry), juniper berries, kale, cress, mustard greens, New Zealand spinach, okra, onions, peppers (hot, sweet), plums, radicchio, radish leaves, sweet potato leaves
Flavanols	Epigallocatechin, epicatechin gallate, epigallocatechin gallate, isorhamnetin	Apples, arugula, asparagus, Chinese cabbage, chocolate, dill weed, grapes, parsley, sea buckthorn berries; service (Saskatoon) berries, watercress
Flavan-3-ols	Catechin, epicatechin, epicatechin 3-gallate	Bilberry, broad beans, chocolate (dark, milk), peaches, plums, soybeans, soy products, tea (brewed, black and green)
Flavones	Apigenin, luteolin	*Rich sources:*[a] celery seed, juniper berries, oregano, parsleyCelery, celery hearts, Chinese celery, kumquat, radicchio, thyme, vine spinach
Anthocyanidins	Cyanidin, delphinidin, malvidin, pelargonidin, peonidin, petunidin	*Rich sources:*[a] acai berries, apple skin, blackberries, blueberries, cabbage (red), Cedar bay cherries, chokecherries, currants, elderberries, grapes (red), Illawarra plums, pecans, radicchio, radishes, raspberries, service (Saskatoon) berries, Tasmanian peppers Apple skin, arctic bramble berry, bilberry, cherry (sweet, tart), black beans, cranberry, eggplant, lingonberry, strawberries, wines (red)
Flavanones	Hesperetin, naringenin, eriodictyol	*Rich sources:*[a] oregano Artichokes, lemons, limes, grapefruit, kumquat, oranges, pummelo, rosemary
Isoflavones	Daidzein, genistein, glycitein	Soybeans, soy products

[a]*Foods containing ≥100mg of respective flavonoids per 100g.*
Adapted from Foods containing ≥10mg of respective flavonoids per 100g.
From Bhagwat, S., Haytowitz, D.B., 2015. USDA Database for the Flavonoid Content of Selected Foods, Release 3.2 http://www.ars.usda.gov/nutrientdata/flav.

Metabolism of Flavonoids

Absorbed flavonoids are conjugated in the intestinal mucosa and are degraded to different phenolic compounds that are rapidly excreted in the bile and urine. Urinary flavonoids show highly variability, suggesting interindividual variation in flavonoid metabolism.

Significant flavonoid catabolism occurs in the hindgut **microbiome**. This includes cleavage of the heterocyclic ring, deesterification, and hydrolysis of sugars, resulting in the formation of various phenolic acids and their lactones. Some constituents of that microbiome, e.g., *Bacteroides* spp., have glycases and are, thus, able to metabolize polyphenylglycones. Their numbers are increased by consuming flavonoid-rich foods. The result is that ~80% of flavonoid metabolites are ultimately absorbed from the colon.[240]

Metabolic Effects of Flavonoids

Enzyme modulation. Flavonoids can interact with many enzymes, selectively affecting the activities of some. This includes induction of some (e.g., phase II enzymes) by binding to promoter regions of their respective genes, and inhibition of others (e.g., aldose reductase, phosphodiesterase, *O*-methyltransferase, and several serine and threonine kinases) by direct binding to the respective protein. For example, tea flavanols inhibit redox-sensitive transcription factors (NF-κB, AP-1) and prooxidative enzymes (lipoxygenases, cyclooxygenases, nitric oxide synthase, xanthine oxidase) but induce phase II and antioxidant enzymes (glutathione *S*-transferases, superoxide dismutases).[241] Flavanones can induce phase II enzymes and exert antiinflammatory effects; naringin, in particular, has been implicated in the effect of grapefruit juice in inhibiting cytochrome *P*450-dependent drug metabolism.[242] Flavonoids with B-ring catechols can

240. Stoupi, S., Williamson, G., Viton, F., et al., 2010. Drug Metab. Dispos. 38, 287–291.

241. Such effects have been cited as the basis of prospective antiinflammatory roles of flavonoids (Middleton, Jr., E., Kandaswami, D, Theoharides, T.C., 2000. Pharmacol. Rev. 52, 673–751).

242. This effect, which involves inhibition of the CYP3A4 isoform, may also involve other flavanones in grapefruit juice. The potency is evidenced by the fact that a single glass of grapefruit juice can affect the biological activity of drugs metabolized by this enzyme system, increasing the activities of some and decreasing the activities of others.

promote mitochondrial production of ROS by inhibiting succinoxidase. Others can cause uncoupling of mitochondrial oxidative phosphorylation and Ca^{+2} release by reducing membrane fluidity. Isoflavones can affect estrogen synthesis and the transactivation of estrogen receptors α and β to affect signal transduction pathways.

Antiestrogenic effects. That soy isoflavones can be ant-estrogenic was demonstrated by a study of Asian women whose intake of soy products was inversely associated with their circulating estrogen level.[243] This effect involves the binding of those isoflavones, genistein and daidzein to intranuclear type II estrogen receptors α and β, thus, affecting the estrogen-synthetic activity of 17β-steroid oxidoreductase and estrogen-dependent signal transduction pathways. This is believed to underlie epidemiological observations of inverse associations of soy products and loss of bone mineral density and symptoms of menopause or premenstrual syndrome. Some, but not all, clinical trials have found the consumption of soy products to reduce menopausal symptoms by as much as 50–60%.[244] One study found soy consumption effective in reducing premenstrual syndrome symptoms.[245] The flavonol quercetin is also a phytoestrogen, interacting with type II estrogen receptors.

As their estrogenic character might suggest, the consumption of soy isoflavones has been associated with higher bone mineral density in a limited number of epidemiological studies. Studies in animal models have shown maternal consumption of soy isoflavones to increase offspring bone mineral density. Such studies have shown that flavonoids modulate the expression of transcription factors that affect osteoblast function by affecting cellular signaling via mitogen-activated protein kinase (MAPK), bone morphogenic protein, estrogen receptor, and osteoprotegerin/receptor activator of NF-κB ligand.[246] Clinical trials conducted to test the hypothesis that soy isoflavones may be useful in improving bone mineralization for the prevention of osteoporosis have yielded inconsistent results.[247]

Putative antioxidant effects. It has been suggested that flavonoids must have metabolic value as antioxidants because of their ability to chelate divalent metal cations (e.g., Cu^{2+}, Fe^{2+}), which removes those catalysts of lipid peroxidation reactions and by scavenging radical

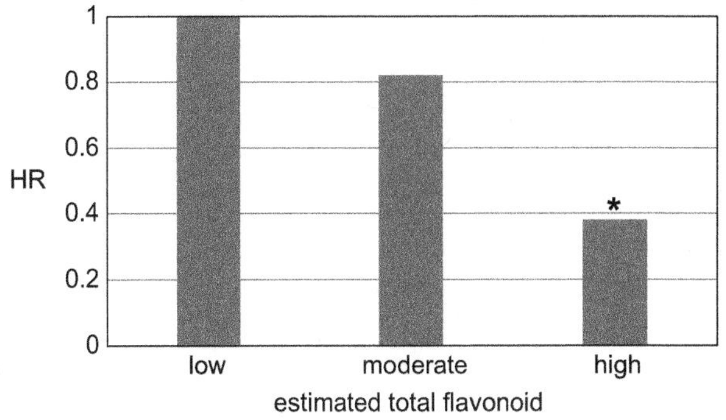

FIGURE 19.19 Risk of all-cause mortality by level of flavonoid intake among women. Data expressed as hazard ratio (HR). *After Ivey, K.L., Hodgson, J.M., Croft, K.D., et al., 2015. J. Nutr. 101, 1012–1020.*

intermediates.[248] Flavonols and some proanthocyanins have been shown to inhibit macrophage-mediated LDL oxidation in vitro, protecting LDL–α-tocopherol from oxidation. Quercetin, which has multiple phenolic hydroxyl groups (a carbonyl at C-4, and free hydroxyl groups at C-3 and C-5), can scavenge superoxide radical ions, hydroxyl radicals, and fatty acyl peroxyl radicals. This chemical property is the basis for in vitro chemical measurements of "total antioxidant capacity" in foods.[249] Such methods may be good indicators of total phenolic content; however, it is highly questionable whether they yield information of any physiologic relevance, as flavonoids are extensively metabolized after ingestion, so their circulating levels are low.[250] There is no direct evidence that health effects of flavonoids involve antioxidant functions in vivo.

Dietary Flavonoids in Health and Disease

Epidemiologic studies have demonstrated associations of diets high in flavonoids with increased longevity and reduced risks of cardiovascular diseases and cancer (Fig. 19.19). However, because such diets are typically rich in fruits and vegetables, it can be difficult to determine from these results whether the protective factor(s) are flavonoids or some other phytochemicals, vitamins, or minerals (e.g., β-carotene, ascorbic acid, fiber) also provided by those foods.

Antiinflammatory effects. Epidemiologic observations have found diets rich in fruits and vegetables to be associated with relatively low levels of inflammatory markers in the people consuming them. That flavonoids contribute to such antiinflammatory effects was suggested by an analysis of the NHANES 1999–2002 data,

243. Nagata, C., Takatsuka, N., Inaba, S., et al., 1998. J. Nat. Cancer Inst. 90, 1830–1835.
244. Albertazzi, P., Pansini, F., Bonaccorsi, G., et al., 1998. Obstet. Gynecol. 91, 6–11; Upmalis, D.H., Lobo, R., Bradley, L., et al., 2000. Menopause 7, 236–242.
245. Bryant, M., Cassidy, A., Hill, C., et al., 2005. Br. J. Nutr. 93, 731–739.
246. Trzeciakiewicz, A., Habauzit, V., Horcajada, M.N., 2009. Nutr. Res. Rev. 22, 68–81.
247. Messina, M., Ho, S., Alekel, D.L., 2004. Curr. Opin. Clin. Nutr. Metab. Care 7, 649–658.
248. Galleano, M., Verstraeten, S.V., Oteiza, P.I., et al., 2010. Arch. Biochem. Biophys. 510, 23–30.
249. Several systems have been used for this in vitro assessment: TRAP (total reactive antioxidant potential), TEAC (Trolox equivalent antioxidant capacity), ORAC (oxygen radical absorbance capacity).
250. Hollman, P.C.H., Cassidy, A., Comte, B., et al., 2011. J. Nutr. 141, 989S–1009S.

which showed that intakes of the flavonols, quercetin and kaempferol; the anthocyanins, malvidin and peonidin; and the isoflavone, genistein were each inversely related to serum concentrations of the inflammatory marker CRP.[251] Similarly, the Nurses' Health Study found that flavonol intake to be inversely related to circulating levels of soluble vascular adhesion molecule-1, and that intakes of flavonoid-rich foods to be associated with lower circulating levels of CRP and soluble tumor necrosis factor receptor-2 (sTNF-R2).[252] Studies in animal and cell models have pointed to several mechanisms underlying these antiinflammatory effects, including modulation of proinflammatory gene expression, inhibition of NF-κB activation, and inhibition of nuclear poly(ADP-ribose) polymerase-1 to reduce macrophage cytokine release.

Immunity. Quercetin has been found to affect aspects of immune cell function. This includes inhibition of induction and function of cytotoxic T lymphocytes, liposaccharide-induced production of NO and TNF-α by macrophages, and Ca^{+2} uptake and consequent histamine release by mast cells. Based on the latter effect, quercetin was administered in nasal spray containing other flavonols to small cohort of patients with allergic rhinitis and nasal congestion; results showed symptom relief within minutes and suppression of symptoms for several hours in most subjects.[253]

Cardiovascular health. Epidemiologic evidence points to cardiovascular disease mortality being inversely related to flavonoid intake, and particular flavanoids being protective against cardiovascular disease.[254] Diets high in the flavonol quercetin were associated with 21–53% reduced risks of cardiovascular disease prevalence.[255] An analysis of data from the Nurses' Health Study found that consumption of anthocyanins, some flavones, and some flavanols was associated with prevention of hypertension.[256] An analysis of a cohort of ~41,000 French women found individuals with high intakes of flavonols, anthocyanin and proanthocyanidins to be less likely to have hypertension.[257]

Several clinical intervention trials have found dark chocolate, green tea, or blueberry flavonols to improve vascular endothelial function,[258] coco and red grape flavanols to reduce platelet reactivity.[259] Intervention with quercetin was found to reduce blood pressure, particularly in hypertensive subjects.[260] Clinical trials with resveratrol and cardiovascular end points have yielded mixed results.[261]

Meta-analyses of clinical trials have found evidence for chocolate consumption increasing flow-mediated dilation and reducing blood pressure, for soy protein isolate consumption being associated with reductions in diastolic blood pressure and LDL cholesterol, and for green tea consumption being associated with reduced LDL cholesterol and stroke risk.[262]

Metabolic studies indicate several mechanisms for beneficial effects of flavonoids on vascular function. Quercetin has been found to inhibit the activity of angiotensin-converting enzyme and the activation of *c*-Jun N-terminal kinase in the modulation of angiotensin-induced hypertrophy of vascular smooth muscle cells. Various proanthocyanins have been shown to inhibit platelet activation; to inhibit the expression of interleukin-2; and to lower serum levels of glucose, triglyceride, and cholesterol. Flavonoid-rich foods (cocoa, grape juice, red wine) have been found to be antithrombotic by inhibiting platelet aggregation and promote vascular endothelial function by stimulating nitric oxide production. Both the flavanone hesperetin and the flavanol epigallocatechin have been found to block oxidized LDL-induced endothelial apoptosis,[263] a key process in atherosclerosis. Epicatechin and its metabolite methylepicatechin have been shown to inhibit NADPH oxidase and arginase, to modulate concentrations of nitric oxide in endothelial cells.[264] Flavonones, flavanols, and flavonols have been found to inhibit NF-κB activity.

251. Chun, O.K., Chung, S.J., Claycombe, K.J., et al., 2008. J. Nutr. 138, 753–760.
252. Landberg, R., Sun, Q., Rimm, E.B., et al., 2011. J. Nutr. 141, 618–625.
253. Remberg, P., Bjork, L., Hedner, T., et al., 2004. Phytomedicine 11, 36–42.
254. Knekt, P., Jarvinen, R., Renuanen, A., et al., 1996. Br. Med. J. 312, 478–481; Huxley, R.R., Neil, A.A., 2003. Eur. J. Clin. Nutr. 57, 904–980; Mink, P.J., Scraffod, C.G., Barraj, L.M., et al., 2007. Am. J. Clin. Nutr. 85, 895–909.
255. Arts, I.C.W., Hollman, P.C.H., 2005. Am. J. Clin. Nutr. 81, 317S–325S.
256. Cassidy, A., O'Reilly, E.J., Kay, C., et al., 2011. Am. J. Clin. Nutr. 93, 338–347.
257. Lajou, M., Rossignol, E., Fagherazzi, G., et al., 2016. Am. J. Clin. Nutr. 103, 1091–1098.
258. Fraga, C.G., Actis-Goretta, L., Ottaviani, J.L., et al., 2005. Clin. Dev. Immunol. 12, 11–17; Heiss, C., Kleinbongard, P. Dejam, A., et al., 2005. J. Am. Coll. Cardiol. 46, 1276–1283; Lorenz, M., Jochmann, N., von Krosigk, A., et al., 2007. Eur. Heart J. 28, 219–323; Grassi, D., Desideri, G., Necolzione, S., et al., 2008. J. Nutr. 138, 1671–1676; Dower, J.I., Geleijnse, J.M., Gijsbers, L., et al., 2015. Am. J. Clin. Nutr. 101, 914–921; Rodriguez-Mateos, A., Rendeiro, C., Bergillos-Meca, T., et al., 2013. Am. J. Clin. Nutr. 98, 1179–1191; Pereira, T., Maldonado, J., Laranjeiro, M., et al., 2014. Cardiol Res. Pract. Article ID:945951.
259. Keevil, J.G., Osman, H.E., Reed, J.D., et al., 2000. J. Nutr. 130, 53–56; Rein, et al., 2000; Hubbard, G.P., Wolfram, S., Lovegrve, J.A., et al., 2004. J. Thromb. Haemost. 2, 2138–2145.
260. Edwards, R.L., Lyon, T., Litwan, S.E., et al., 2007. J. Nutr. 137, 2405–2411; Edwards, R.L., Lyon, S.E., Rabovsky, A., et al., 2007. J. Nutr. 137, 2405–2411; Egert, S., Bosy-Westphal, A., Seiberl, J., et al., 2009. Br. J. Nutr. 102, 1065–1074.
261. Sahebkar, A., 2013. Nutr. Rev. 71, 822–835.
262. Hopper, L., Kroon, P.A., Rimm, E.B., et al., 2008. Am. J. Clin. Nutr. 88, 38–50; Reid, K., Sullivan, T., Fakler, P., et al., 2010. BMC Med. 8, 39; Arab, L., Liu, W., Elashoff, D., 2009. Stroke 40, 1786–1792; Arab, L., Khan, F., Lam, H., 2013. Am. J. Clin. Nutr. 98, 1651S–1659S.
263. Choi, J.S., Choi, Y.J., Shin, S.Y., et al., 2008. J. Nutr. 138, 983–990.
264. Steffen, Y., Schewe, T., Sies, H., 2007. Biochem. Biophys. Res. Commun. 359, 828–833; Schnorr, O., Brosssett, T., Momman, T.Y., et al., 2008. Arch. Biochem. Biophys. 476, 211–215.

Neurologic health. Clinical trials have found fruit flavonoids, flavanoid-rich foods (wine, tea, chocolate, grape juice, orange juice), and soy isoflavones to enhance cognitive function.[265] Studies in animal models have found extracts of flavonoid-rich foods (blueberry, spinach, strawberry) to reduce age-related declines in neuronal signal transduction and cognitive function.[266] Several mechanisms have been indicated in studies with animal models: increased expression of estrogen receptor-β, enhanced protein kinase, and lipid kinase signaling of transcription of factors involved in synaptic plasticity and cerebrovascular blood flow.

Obesity and diabetes. Flavonoids have been suggested as having antiobesity effects. A 14-year cohort study with 4280 men and women in the Netherlands, which found diets rich in flavonoids to be associated with lower increases in body mass index (BMI); the effect was significant only for women.[267] A study of ~1500 European adolescents found regular consumption of chocolate to be associated with lower adiposity.[268] Other trials have shown regular intake of flavonoid-rich fruits or green tea to reduce body weight. The green tea flavonol EGCG has been shown to inhibit obesity in the mouse model, but a meta-analysis of 15 studies found that green tea catechins were effective in producing modest reductions in BMI and body weight only in the presence of caffeine.[269] Such effects are thought to be due to increased thermogenesis and appetite suppression. The consumption of flavonoids has been shown to favor *Bacteroides* spp. in the hindgut. This may be a mechanism whereby flavonoids exert weight-lowering effects, as a *Bacteroides*-dominant hindgut microbiome has been associated with a relatively low yield of absorbable energy from fermentation.[270]

Flavonoids may be protective against the development of type 2 diabetes (T2D). Studies in three different cohorts found T2D risk to be inversely related to intakes of flavonoids (inputed from food frequency questionnaire data). In the ~340,000 subject EPIC[271] cohort, T2D risk was inversely related to intakes of flavan-3-ol and proanthocyanidin intakes.[272] In the 17-year follow-up of the Framingham Offspring cohort, a difference of 2.5-fold in flavonol intakes were associated with a 26% reduction in T2D risk.[273] In a 9-year follow-up of the ~18,200 subject Physicians Health Study cohort, T2D risk was inversely related to chocolate consumption.[274] A meta-analysis of clinical trials found interventions with cocoa/chocolate found those foods to reduce insulin resistance.[275] Another meta-analysis found resveratrol to improve glycemic control and insulin sensitivity in individuals with diabetes but not on nondiabetics.[276] A small, randomized clinical trial found intervention with anthocyanins to improve dyslipidemias and prevent insulin resistance in T2D patients.[277] Studies with rodent models have shown that flavonoids can improve glycemic control by enhancing insulin secretion and sensitivity.[278]

Anticarcinogenesis. Some (but not all) epidemiologic studies have found consumption of diets high in flavonoids to be associated with reduced risks of cancers of the lung and rectum (fruit catechins), and lung and prostate (soy isoflavones).[279] An analysis of the food intake data from the Nurses' Health Study cohorts (~172,000 women) found risk of developing ovarian cancer to be lowest among individuals in the highest quintile of tea consumption; the flavonoid intakes (imputed from food frequency data) suggested modest protective effects of flavonols and flavanones.[280]

Various flavonoids have been found to inhibit cell proliferation and angiogenesis in vitro and to inhibit phorbol ester-induced skin cancer in the mouse model. Several studies have shown that in vitro exposure to quercetin can synergize the effects of chemotherapeutic drugs on both resistant and nonresistant tumor cells. Underlying these effects may

265. Thorp, A.A., Sinn, N., Buckley, J.D., et al., 2009. Br. J. Nutr. 102, 1348–1354; Nurk, E., Refsum, H., Drevon, C.A., et al., 2009. J. Nutr. 139, 120–127; Lamport, D.J., Lawton, C.L., Merat, N., et al., 2016. Am. J. Clin. Nutr. 103, 775–783; Mastroiacovo, D., Kwik-Urike, C., Grassi, D., et al., 2015. Am. J. Clin. Nutr. 101, 538–548; Keen, R.J., Lamport, D.J., Dodd, G.F., et al., 2015. Am. J. Clin. Nutr. 101, 506–514.

266. Joseph, J., Shukitt-Hale, B., Denisova, N.A., et al., 1999. J. Neurosci. 19, 8114–8121.

267. Hughes, L.A.E., Arts, I.C.W., Amergen, T., et al., 2008. Am. J. Clin. Nutr. 88, 1341–1352.

268. Cuena-García, M., Ruiz, J.R., Ortega, F.B., et al., 2014. Nutrition 30, 236–239.

269. Phung, O.J., Baker, W.L., Matthews, L.J., et al., 2010. Am. J. Clin. Nutr. 91, 73–81.

270. Ley, R.E., Turnbargh, P.J., Klein, S., et al., 2006. Nature 444, 1022–1023.

271. European Prospective Investigation into Cancer and Nutrition, an 8-country, 26-center study.

272. Zamora-Ross, R., Forouhi, N., Sharp, S.J., et al., 2014. J. Nutr. 144, 335–343.

273. Jacques, P.F., Cassidy, A., Rogers, G., et al., 2013. J. Nutr. 143, 1474–1480.

274. Matsumotoa, C., Petrone, A.B., Sesso, H.D., et al., 2015. Am. J. Clin. Nutr. 101, 362–367.

275. Hooper, L., Kay, C., Abdelhamid, A., et al., 2012. Am. J. Clin. Nutr. 95, 740–751.

276. Liu, K., Zhou, R., Wang, B., et al., 2014. Am. J. Clin. Nut. 99, 1510–1519.

277. Li, D., Zhang, Y., Liu, Y., et al., 2015. J. Nutr. 145, 742–748.

278. Kwon, O., Eck, P., Chen, S., et al., 2007. FASEB J. 21, 366–377; Kobori, M., Masumoto, S., Akimoto, Y., et al., 2009. Mol. Nutr. Food Res. 53, 859–868; Takikawa, M., Inoue, S., HOrios, F., et al., 2010. J. Nutr. 140, 527–533; Youl, E., Bardy, G., Magous, R., et al., 2010. Br. J. Pharmacol. 161, 799–814; Babujanarthanam, R., Kavitha, P., Pandian, M.R., et al., 2010. Fundam. Clin. Pharmacol. 24, 357–364.

279. Knekt, P., Kumpulainen, J., Jävinen, R., et al., 2002. Am. J. Clin. Nutr. 76, 560–568; Arts, I.C., Jacobs, Jr., D.R., Gross, M., et al., 2002. Cancer Causes Control 13, 373–382; Nagata, Y., Sonoda, T., Mori, M., et al., 2007. J. Nutr. 137, 1974–1979; Shimazu, T., Inoue, M., Sasazuki, S., et al., 2010. Am. J. Clin. Nutr. 91, 722–728.

280. Cassidy, A., Huang, T., Rice, M.S., et al., 2014. Am. J. Clin. Nutr. 100, 1344–1351.

be any of several metabolic effects that have been demonstrated for various flavonoids[281]:

- inhibition of some enzymes (e.g., protein kinase C, inhibition of nuclear poly[ADP-ribose] polymerase-1, Akt to cause apoptosis, topoisomerase I, prolyl hydroxylase II to increase hypoxia-inducible factor-1α to impair cell proliferation);
- induction of other activities (e.g., estrogen receptor-β, p38 and the MAPK pathway, DNA repair, cytochrome *P*450-dependent carcinogen metabolism, phase 2 enzymes that conjugate carcinogens);
- preservation of intracellular NAD^+;
- alteration of DNA methylation patterns.

Other effects. It has been suggested that health benefits attributed to traditional, herbal medicaments may be due to bioactive flavonoids. Evidence supporting such hypotheses includes the findings that bilberry anthocyanins may reduce retinal hemorrhage in T2D patients. Some studies indicate that consumption of cranberry juice may reduce risk for recurrent urinary tract infections in women.[282] Such protection may be due to the effects of cranberry A-type proanthocyanidins, or their microbial metabolites, which were found capable of blocking bacterial binding to bladder epithelial cells.[283] Studies with animal models have shown effects of small magnitude on exercise performance associated with mitochondrial biogenesis; however, a meta-analysis of human trials concluded that any such effects are of minimal physiological significance.[284] Still, a study with multiple sclerosis patients found EGCG to increase working efficiency during moderate exercise, particularly in men.[285] High intakes of black tea, flavonols, and flavones have been associated with low risk to osetoporotic fracture.[286]

It was once thought that the effect of grapefruit juice consumption in increasing drug biopotency was due to its dominant flavonoid, the flavanone naringin (the content of which varies widely, 200–2000 µM), which was found to inhibit in vitro CYP3A, the cytochrome *P*450 enzyme that metabolizes more than 60% of commonly prescribed drugs.[287] However, further studies found that the administration of naringin or comparable amounts of grapefruit juice do *not* affect the pharmacokinetics of CYP3A metabolites.[288] This drug–food interaction is now thought to be due to another class of compounds in grapefruit juice, **furanocoumarins**, which are metabolized to reactive forms that irreversibly bind the CYP apoprotein and eliminate its enzymatic activity.[289]

Biomarkers of Flavonoid Status

Status with respect to some flavonoids can be assessed on the basis of **urinary flavonoid excretion**. That parameter can indicate the minimum amount of particular flavonoids that were absorbed. However, it is not useful for assessing quercetin status, as that flavonoid is excreted primarily in the bile.

Flavonoid Safety

While the toxicology of the flavonoids has not been investigated, those in foods are considered safe. There are, however, interactions that have been observed but not well explained. For example, a large, 10-year prospective study found women in the highest quintile of soy isoflavones intake (median 53.6 mg/day) to have small (17–26%) but significant increases in risk of ischemic stroke compared with those with lower intakes.[290]

9. OROTIC ACID

Orotic acid was isolated in the late 1940s from distillers' dried solubles.[291] For a while, it was called "*vitamin* B_{13}"; however, when studies failed to confirm vitamin activity, that designation was dropped. It is recognized as a normal metabolite; however, evidence shows that orotic acid supplementation can promote the synthesis of uracil nucleotides, and produce some health benefits, making it a **beneficial dietary bioactive**.

Benefits of Dietary Orotic Acid

Orotic acid supplementation of experimental animals has been found to increase the utilization of fatty acids by the heart,

281. Miles, S.L., McFarland, M., Niles, R.N., 2014. Nutr. Rev. 72, 720–734.
282. Maki, K.C., Kaspar, K.L., Khoo, C., et al., 2016. Am. J. Clin. Nutr. 103, 1434–1442.
283. Howell, A.B., Vorsa, N., Der Marerosian, A., et al., 1998. N. Engl. J. Med. 339, 1085–1086.
284. Pelletier, D.M., Lacerte, G., Goulet, E.D.B., 2013. Int. J. Prev. Med. 23, 73–82.
285. Mähler, A., Steiniger, J., Bock, M., et al., 2015. Am. J. Clin. Nutr. 101, 487–495.
286. Myers, G., Prince, R.L., Kerr, D.A., et al., 2015. Am. J. Clin. Nutr. 102, 958–965.
287. Owira, P.M., Ojewole, J.A., 2010. Cardiovasc. J. Africa 21, 280–285; Hanley, M.J., Cancalon, P., Widmer, W.W., et al., 2011. Expert Opin Drug Metab. Toxicol.7, 276–286.
288. Bailey, D.G., Arnold, J.M., Strong, H.A., et al., 1993. Clin. Pharmacol. Ther. 53, 589–594; Bailey, D.G., Arnold, J.M., Munoz, C., et al., 1993. Clin. Pharmacol. Ther. 53, 637–642; Rashin, J., McKinstray, C., Renwick, A.G., et al., 1993. Br. J. Clin. Pharmacol. 36, 460–463.
289. Lin, H.L., Kent, U.M., Hollenberg, P.F., 2005. J. Pharmacol. Exp. Ther. 313, 154–164.
290. Yu, D., Shu, X.O., Li, H., et al., 2015. Am. J. Clin. Nutr. 102, 680–686.
291. This feedstuff consists of the dried aqueous residue from the distillation of fermented corn. It is used mainly as a component of diets for poultry, swine, and dairy calves. It is rich in several B vitamins and has been valued as a source of UGFs, particularly for growing chicks and turkey poults.

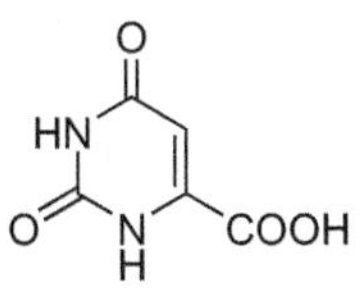

FIGURE 19.20 Orotic acid.

increase the activity of lipoprotein lipase, and increase expression of peroxisome proliferator-activated receptor α (PPARα) and its affected enzymes.[292] Supplemental orotic acid also increases hepatic levels of uracil nucleotides, presumably by increasing the flux through the pyrimidine pathway. Orotate treatment has been shown to be more effective than uracil in stimulating adaptive growth of the rat jejunum after massive small bowel resection.[293] Studies in cultured cells have found orotic acid to stimulate cell proliferation by downregulating AMPK, which activates mTORC1.[294] Magnesium orotate has been used to improve ventricular function and exercise tolerance in cardiac patients[295]; however, those effects would appear to be due to correction of magnesium depletion and not to the orotate moiety per se.

Benefits of dietary orotic acid are as follows:

- Increased pyrimidine production.

Nature of Orotic Acid

Orotic acid is a substituted pyrimidine: 1,2,3,6-tetrahydro-2,6-dioxo-4-pyrimidinecarboxylic acid (Fig. 19.20).

Sources of Orotic Acid

The most important dietary sources of orotic acid are milk, milk products, and root vegetables (beets, carrots). Notably, human milk lacks orotic acid.

Metabolism of Orotic Acid

Cellular uptake of orotic acid appears to be facilitated by an organic ion transporter OAT2 driven by glutamate antiport.[296] Orotic acid is an intermediate in the biosynthesis of **pyrimidines** (UTP, CTP, TTP). It is synthesized in the mitochondria from *N*-carbamylphosphate by dehydration (via dihydroorotase) and oxidation via **orotate reductase** to orotate, which is subsequently converted to UMP. An inborn error in UMP synthase, which catalyzes the last step, is characterized by orotic acid accumulating in the plasma and appearing in increased amounts in the urine. Orotic acid excretion is also increased in disorders of the urea cycle and has been proposed as an indicator of arginine depletion. It has been found to be increased in cases of subclinical mastitis in dairy cows.

Safety of Orotic Acid

Orotic acid supplements (0.1%) to the diets of rats have been found to reduce the conversion of tryptophan to niacin,[297] to induce hepatic steatosis and hepatomegaly.[298] The latter effects were associated with increases in the sterol regulatory element binding protein-1c, the target gene for which is involved in fatty acid synthesis. Orotic acid supplementation of the rat appeared to constitute an oxidative stress, as it reduced hepatic superoxide dismutase activity and increased the contents of conjugated dienes and protein carbonyls.[299]

10. UNIDENTIFIED FACTORS

Since the discovery of vitamin B_{12}, experimental nutritionists have observed beneficial effects, particularly stimulated growth, of natural materials added to purified diets. Many such responses have been found to involve interrelationships of known nutrients.[300] Some have involved diet palatability and, thus, the rate of food intake of experimental animals. One resulted in the discovery of an unrecognized essential nutrient, selenium. Other responses remain to be elucidated. For young monogastrics (particularly poultry), several feedstuffs that have such effects are regarded as having **"unidentified growth factor" (UGF)** activity (Table 19.11). Elucidating the nature of UGFs has been frustrated by the fact that growth is a nonspecific response and that the observed growth responses are often small and not reproducible, suggesting roles of the environment, the gut microbiome, etc.

The classical vocabulary of nutrition is not well suited to accommodate such cases. How should UGFs and other vitamin-like factors beneficial to health be described? What are the proper descriptors for

(1) Lycopene for reducing cancer risk?
(2) Xanthophylls for reducing risk to AMD?
(3) Carnitine for an individual with an OCTN2 deficit?
(4) Flavonoids or soluble fiber supporting colon health?[301]

---

292. Pôrto, L.C., de Castro, C.H., Savergnini, S.S., et al., 2012. Life Sci. 90, 476–483.
293. Evans, M.E., Tian, J., Gu, L.H., et al., 2005. J. Parenter. Enteral Nutr. 29, 315–320.
294. Jung, E.J., Lee, K.Y., Lee, B.H., 2012. J. Toxicol. Sci. 37, 813–821.
295. Classen, H.G., 2004. Rom. J. Intern. Med. 42, 491–501; Stepura, O.B., Martynow, A.I., 2009. Int. J. Cardiol. 134, 145–147.
296. Fork, C., Bsuer, T., Golz, S., et al., 2011. Biochem. J. 436, 305–312.
297. Fuluwatari, T., Morikawa, Y., Sugimoto, E., et al., 2002. Biosci. Biotechnol. Biochem. 66, 1196–1204.
298. Wang, Y.M., Hu, X.Q., Xue, Y., et al., 2011. Nutrition 27, 571–578; Shibata, K., Morita, N., Kawamura, T., et al., 2015. J. Nutr. Sci. Vitaminol. 61, 355–361.
299. Morifuji, M., Aoyama, Y., 2002. J. Nutr. Biochem. 13, 403–410.
300. e.g., At least part of the beneficial effect of including corn distillers' dried solubles in a soybean meal-based diet for chicks is due to its natural chelating activity, which increases the utilization of zinc.
301. Or soluble fiber, for that matter?

TABLE 19.11 Sources of UGF (unidentified growth factor) Activity for Poultry

Condensed fish solubles
Fish meal
Dried whey
Brewers' dried grains
Brewers' dried yeast
Corn distillers' dried solubles
Other fermentation residues

11. CASE STUDY

Instructions

Review each of the following case reports, paying special attention to the diagnostic indicators on which the treatments were based. Then answer the questions that follow.

In 1989, Killgore and colleagues demonstrated that mice fed a refined diet containing very low concentrations of a tricarboxylic acid with a fused quinone ring, pyrroloquinoline quinone (PQQ) (Fig. 19.21)[302] developed skin lesions that were prevented by supplementation with PQQ.[303] Not only did PQQ-fed mice grow faster, but one-fourth of the PQQ-deprived animals showed friable skin, mild alopecia, and a hunched posture, and one-fifth died by 8 weeks with aortic aneurysms or abdominal hemorrhages. The most frequent sign, friable skin, suggested an abnormality of collagen metabolism; PQQ-deprived animals showed abnormally low activities of lysyl oxidase and increased collagen solubility indicating reduced cross-linking.[304] PQQ is known to be a cofactor in several enzymes (now called quinoproteins) in bacteria, yeasts, and plants. It is ubiquitously present in common bacteria, soil, and plants and has been found in all foods examined. Mice fed the low-PQQ diet for 8–9 weeks produced either no litters, or litters in which the pups were immediately cannibalized at birth. Subsequent work showed deprivation of PQQ to alter mitogenic responses, reduce interleukin 2 levels, elevate plasma levels of glucose and several amino acids, and reduce mass of functional hepatic mitochondria.[305] PQQ has been shown to affect cell signaling.[306] Studies in cultured cells have demonstrated its effects in stimulating the activity of the ras oncogene in signal transduction pathways involved in growth and development. PQQ has also been found to affect the activity of the "Parkinson disease protein," DJ-1, a peptidase involved in androgen receptor-regulated transcription leading to mitochondrial biogenesis. PQQ has been shown to stimulate the activation, by phosphorylation, of the promoter of PPAR-γ-coactivator-1α (PGC-1α) in the stimulation of mitochondrial biogenesis.[307]

302. 4,5-Dihydro-4,5-dioxo-1*H*-pyrrolo[2,3-*f*]quinoline-2,7,9-tricarboxylic acid.
303. Killgore, J., Smidt, C., Duich, L., et al., 1989. Science 245, 850–852.
304. Steinberg, F., Stites, T.E., Anderson, P., et al., 2003. Exp. Biol. Med. 228, 160–162.
305. Stites, T., Storms, D., Bauerly, K., et al., 2006. J. Nutr. 136, 390–396.
306. Hiarakawa, A., Shimizu, K., Fukumitsu, H., et al., 2009. Biochem. Biophys. Res. Commun. 378, 308–312; Kamazawa, T., Hiwasa, T., Takiguchi, M., et al., 2007. Int. J. Mol. Med. 19, 765–770.

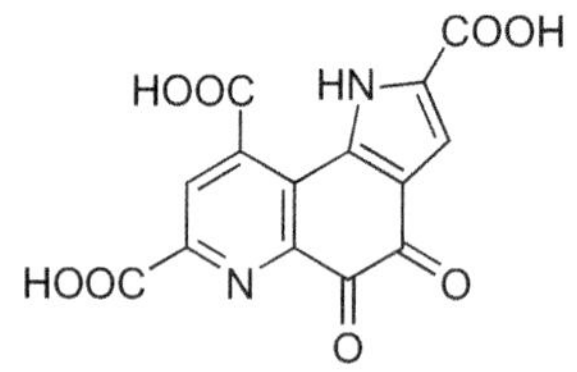

FIGURE 19.21 Pyrroloquinoline quinone.

PQQ has been found in several foods including all plants analyzed and at relatively high levels in fermented foods (60–800 μg/kg) and human milk (140–180 μg/kg). It has been estimated that most people consume 1–2 mg PQQ per day from foods.

The ability to redox cycle makes PQQ an antioxidant with capable of both one and two electron transfers. Studies in animal models have shown PQQ treatment to have antioxidant-like effects in protecting against carbon tetrachloride hepatotoxicity and inhibiting glucocorticoid-induced lenticular glutathione depletion and cataract.

Case Questions

1. Should PQQ be designated a vitamin? A conditionally essential nutrient? A beneficial dietary bioactive? Something else? Justify your answer.
2. Propose a series of experiments plan to clarify the status of PQQ as a putative nutrient.

12. STUDY QUESTIONS AND EXERCISES

1. List the questions that must be answered in determining the eligibility of a substance for vitamin status.
2. For each of the substances discussed in this chapter, list the available information that would support its designation as a vitamin, and that which would refute such a designation.
3. Outline the general approaches one would need to take to characterize the UGF activity of a natural material such as fish meal for the chick.
4. Prepare a concept map of the relationships of micronutrients and physiological function, including the specific relationships of the traditional vitamins, the quasivitamins, and ineffective factors.

307. Chowadnadisai, W., Bauerly, K.A., Tchaparian, E., et al., 2010. J. Biol. Chem. 285, 142–152.

RECOMMENDED READING

Carnitine

Flanagan, J.L., Simmons, P.A., Vehige, J., et al., 2010. Role of carnitine in disease. Nutr. Metab. 7, 30–37.

Fu, L., Huang, M., Chen, S., 2013. Primary carnitine deficiency and cardiomyopathy. Korean Circ. J. 43, 785–792.

Hathcock, J.N., Shao, A., 2006. Risk assessment for carnitine. Regul. Toxicol. Pharmacol. 46, 23–28.

Jones, L.L., McDonald, D.A., Borum, P.R., 2010. Acylcarnitines: role in brain. Prog. Lipid Res. 49, 61–75.

Marcovina, S.M., Sirtori, C., Peracino, A., et al., 2013. Translating the basic knowledge of mitochondrial functions to metabolic theory: the role of L-carnitine. Transl. Res. 161, 73–84.

Mingorance, C., Rodriguez-Rodriguez, R., Justo, M.L., et al., 2011. Pharmacological effects and clinical applications of propionyl-L-carnitine. Nutr. Rev. 69, 279–290.

Rebouche, G.J., 2012. Carnitine, Chapter 25. In: Erdman, J.W., Macdonald, I.A., Zeisel, S.H. (Eds.), Present Knowledge in Nutrition, tenth ed. Wiley-Blackwell, Ames, IA, pp. 391–404.

Schooneman, M.G., Vaz, F.M., HOuten, S.M., et al., 2013. Aceylcarnitines: reflecting or inflicting insulin resistance? Diabetes 62, 1–8.

Strijbis, K., Vaz, F.M., Distel, B., 2010. Enzymology of the carnitine biosynthesis pathway. IUBMB Life 62, 357–362.

Wall, B.T., Porter, G. (Eds.), 2014. Carnitine Metabolism and Human Nutrition. CRC Press, New York. 168 pp.

Zammit, V.A., Ramsay, R.R., Bonomini, M., et al., 2009. Carnitine, mitochondrial function and therapy. Adv. Drug Deliv. Rev. 61, 1353–1362.

Choline

Jiang, X., Yan, J., Caudill, M.A., 2014. Choline, Chapter 13. In: Zempleni, J., Suttie, J.W., Gregory, J.F., et al. (Eds.), Handbook of Vitamins, fifth ed. CRC Press, Washington, DC, pp. 491–513.

Mehedint, M.G., Zeisel, S.H., 2013. Choline's role in maintaining liver function: new evidence for epigenetic mechanisms. Curr. Opin. Clin. Nutr. Metab. Care 16, 339–345.

Ueland, P.M., 2011. Choline and betaine in health and disease. J. Inherit. Metab. Dis. 34, 3–15.

Zeisel, S.H., 2011. Nutritional genomics: defining the dietary requirement and effects of choline. J. Nutr. 141, 531–534.

Zeisel, S.H., Corbin, C.D., 2012. Choline, Chapter 26. In: Erdman, J.W., Macdonald, I.A., Zeisel, S.H. (Eds.), Present Knowledge in Nutrition, tenth ed. Wiley-Blackwell, Ames, IA, pp. 405–418.

Flavonoids

Blumberg, J.B., Camesano, T.A., Cassidy, A., et al., 2013. Cranberries and their bioactive constituents in human health. Adv. Nutr. 4, 618–632.

Bohn, T., 2014. Dietary factors affecting polyphenol bioavailability. Nutr. Rev. 72, 429–452.

Bondono, C.P., Croft, K.D., Ward, N., et al., 2015. Dietary flavonoids and nitrate: effect on nitric oxide vascular function. Nutr. Rev. 73, 216–235.

Cederroth, C.R., Nef, S., 2009. Soy, phyoestrogens and metabolism: a review. Mol. Cell. Endocrinol. 304, 30–42.

Clifford, M.N., van Hooft, J.J.J., Crozier, A., 2013. Am. J. Clin. Nutr. 98, 1619S–1630S.

Crozier, A., del Rio, D., Clifford, M.N., 2010. Bioavailability of dietary flavonoids and phenolic compounds. Mol. Asp. Med. 31, 446–467.

de Souza, P.L., Russel, P.J., Kearsley, J.H., et al., 2010. Clinical pharmacology of isoflavone and its relevance for potential prevention of prostate cancer. Nutr. Rev. 68, 542–555.

del Rio, D., Borges, G., Crozier, A., 2010. Berry flavonoids and phenolics: bioavailability and evidence of protective effects. Br. J. Nutr. 104, S67–S90.

Galleano, M., Oteiza, P.I., Fraga, C.G., 2009. Cocoa, chocolate and cardiovascular disease. J. Cardiovasc. Pharmacol. 54, 484–490.

González-Gellego, J., García-Mediavilla, M.V., Sánchez-Campos, S., et al., 2010. Fruit polyphenols, immunity and inflammation. Br. J. Nutr. 104, S15–S27.

Hodgson, J.M., Croft, K.D., 2010. Tea flavonoids and cardiovascular health. Mol. Asp. Med. 31, 495–502.

Kim, J., Kim, J., Shim, J., et al., 2014. Cocoa phytochemicals: recent advances in molecular mechanisms on health. Crit. Rev. Food Sci. Nutr. 54, 1458–1472.

Kumar, S., Pandey, A.K., 2013. Chemistry and biological activities of flavonoids: an overview. Sci. World J. Article ID:162750.

Mahmoud, A.M., Yan, W., Bosland, M.C., 2014. Soy isoflavoines and prostate cancer: a review of mechanisms. J. Steroid Biochem. Mol. Biol. 140, 116–132.

Miles, S.J., McFarland, M., Niles, R.M., 2014. Molecular and physiological actions of quercetin: need for clinical trials to assess its benefits in human disease. Nutr. Rev. 72, 720–734.

Sak, K., 2014. Site-specific anticancer effects of dietary flavonoid quercetin. Nutr. Cancer 66, 177–193.

Soni, M., Rahardjo, T.B.W., Soekardi, R., et al., 2014. Phytoestrogens and cognitive function: a review. Maturitas 77, 209–220.

van Duynhoven, J., Vaughn, E.E., van Dorsten, F., et al., 2013. Interactions of black tea polyphenols with human gut microbiota: implications for gut and cardiovascular health. Am. J. Clin. Nutr. 98, 1631S–1641S.

Wallace, T.C., Giusti, M.M. (Eds.), 2014. Anthocyanins in Health and Disease. CRC Press, New York. 355 pp.

Williamson, G., 2012. Dietary flavonoids, chapter 27. In: Erdman, J.W., Macdonald, I.A., Zeisel, S.H. (Eds.), Present Knowledge in Nutrition, tenth ed. Wiley-Blackwell, Ames, IA, pp. 419–433.

Myo-inositol

Campa, C.C., Martini, M., De Santis, M.C., et al., 2015. How PI3K-derived lipids control cell division. Front. Cell Dev. Biol. 3, 61.

Hawkins, P.T., Stephens, L.R., 2015. PI3K signaling in inflammation. Biochim. Biophys. Acta 1851, 882–897.

Kerr, W.G., Colucci, F., 2011. Inositol phospholipid signaling and the ebiology of natural killer cells. J. Innate Immun. 3, 249–257.

Lee, J.Y., Kim, Y.R., Park, J., et al., 2012. Insitol polyphosphate multikinase signaling in the regulation of metabolism. Ann. N.Y. Acad. Sci. 1271, 68–74.

Manna, P., Jain, S.K., 2015. Phosphatidylinositol-3,4,5-triphosphate and cellular signaling: implications for obesity and diabetes. Cell Physiol. Biochem. 35, 1253–1275.

Tsui, M.M., York, J.D., 2010. Roles of inositolf phosphates and inositol pyrolphosphates in development, cell signaling and nuclear processes. Adv. Enzyme Regul. 50, 324–337.

Lipoic Acid

Gorąca, A., Huk-Kolega, H., Piechota, A., et al., 2011. Lipoic acid – biological activity and therapeutic potential. Pharmacol. Rep. 63, 849–858.

Mayr, J.A., Feichtinger, R.G., Tort, F., et al., 2014. Lipoic acid biosynthesis defects. J. Inherit. Metab. Dis. 37, 553–563.

Nebbioso, M., Pranno, F., Pescosolido, N., 2013. Lipoic acid in animal models and clinical use in diabetic retinopathy. Expert Opin. Pharmacother. 14, 1829–1838.

Shay, K.P., Moreau, R.F., Smith, E.J., et al., 2009. Alpha-lopoic acid as a dietary supplement: molecular mechanisms and therapeutic potential. Biochim. Biophys. Acta 1790, 1149–1160.

Non-Provitamin A Carotenoids

Broadhead, G.K., Grigg, J.R., Chang, A.A., et al., 2015. Dietary modification and supplementation for the treatment of age-related macular degeneration. Nutr. Rev. 73, 448–462.

Hammond Jr., B.R., Fletcher, L.M., 2012. Influence of the dietary carotenoids lutein and zeaxanthin on visual performance: application to baseball. Am. J. Clin. Nutr. 96, 1207S–1213S.

Koushan, K., Rusovici, R., Wenhua, L., et al., 2013. The role of lutein in eye-related disease. Nutrients 5, 1823–1839.

Krishnadev, N., Meleth, A.D., Chew, E.Y., 2010. Nutritonal supplements for age-related macular degeneration. Curr. Opin. Ophthalmol. 21, 184–189.

Pinazo-Durán, M.D., Gómez-Ulla, F., Arias, L., et al., 2014. Do nutritional supplements have a role in age macular degeneration prevention? J. Ophthalmol. Article ID:901686.

Schleicher, M., Weikel, K., Garber, C., et al., 2013. Diminishing risk for age-related macular degeneration with nutrition: a curren view. Nutrients 5, 2405–2456.

Sommerburg, O., Siems, W., Kraemer, K. (Eds.), 2013. Carotenoids and Vitamin a in Translational Medicine. CRC Press, Boc Raton, FL. 405 pp.

van Breeman, R.B., Pajkovic, N., 2008. Multitarageted therapy of cancer by lycopene. Cancer Lett. 269, 339–351.

Wong, I.Y., Koo, S.C.Y., Chan, C.W.N., 2011. Prevention of age-related macular degeneration. Int. Ophthalmol. 31, 73–82.

Orotic Acid

Brosnan, M.E., Brosnan, J.T., 2007. Orotic acid excretion and arginine metabolism. J. Nutr. 137, 1656S–1661S.

Löffler, M., Carrey, E.A., Zameitat, E., 2015. Orotic acid, more than just an intermediate of pyrimidine de novo synthesis. J. Genet. Genomics 42, 207–219.

Wang, Y.M., Hu, X.Q., Xue, Y., 2011. Study of the possible mechanism of orotic acid-induced fatty liver in rats. Nutrition 27, 571–575.

Ubiquinones

Beatrycze, N., Kruk, J., 2010. Occurrence, biosynthesis and function of isoprenoid quinones. Biochim. Biophys. Acta 1797, 1587–1605.

Bentinger, M., Tekle, M., Dallner, G., 2010. Coenzyme Q – biosynthesis and functions. Biochem. Biophys. Res. Commun. 396, 74–79.

Littarru, G.P., Tiano, L., 2010. Clinical aspects of coenzyme Q_{10}: an update. Nutr 26, 250–254.

López-Lluch, G., Rodríguez-Aguilera, J.C., Santos-Ocana, C., et al., 2010. Is coenzyme Q a key factor in aging? Mech. Ageing Dev. 131, 225–235.

Quinzil, C.M., Hirano, M., 2010. Coenzyme Q and mitochondrial disease. Dev. Disabil. Res. Rev. 16, 183–188.

Wang, Y., Hekimi, S., 2013. Molecular genetics of ubiquinone biosynthesis in animals. Crit. Rev. Biochem. Mol. Biol. 48, 69–88.

Watmough, N.J., Frerman, F.E., 2010. The electron transfer flavoprotein: ubiquinone oxidoreductases. Biochim. Biophys. Acta 1797, 1910–1916.

Part III

Using Current Knowledge of the Vitamins

Chapter 20

Sources of the Vitamins

Chapter Outline

Anchoring Concepts

1. Estimates of vitamin contents of many foods and feedstuffs are available.
2. For some vitamins, only a portion of the total present in certain foods or feedstuffs is biologically available.
3. The total vitamin intake of an individual is the sum of the amounts of bioavailable vitamins in the various foods, feedstuffs, and supplements consumed.

The intakes of vitamins into the body calculated from standard tables are rarely accurate.

John H. Marks[1]

LEARNING OBJECTIVES

1. To understand the sources of error in estimates of vitamin contents of foods and feedstuffs
2. To understand the concept of a core food and to know the core foods for each of the vitamins
3. To understand the sources of potential losses of the vitamins from foods and feedstuffs
4. To understand which of the vitamins are most likely to be in insufficient supply in the diets of humans and livestock
5. To understand the means available for the supplementation of individual foods and total diets with vitamins.

VOCABULARY

Bioavailability
Biofortification
Core foods
Enrichment
Food labels
Fortification
Genetic engineering
Golden rice
National nutrient database
Nutrition Labeling and Education Act (NLEA)
Revitaminization
Selective breeding
Supplementation
Vitamin–mineral premix
Vitaminization

1. VITAMINS IN FOODS AND FEEDSTUFFS

Vitamin Content Data

Collation of best estimates of the nutrient composition of foods and feedstuffs has been an ongoing activity by several groups in the United States since the early 1900's.[2] Nutrient

1. ohn Henry Marks (b. 1925) is a prominent British physician who was a professor at Cambridge University. He is the author of well-written books on the vitamins, including "*The Vitamins: Their Role in Medical Practice*" (1985), Springer, New York, 224 pp.

2. The formal compilation of food composition data was initiated by the USDA food chemist W.O. Atwater in 1896. Since that time, developing information on the nutrient composition of foods has been an ongoing program of the USDA. Development of nutrient composition information for feedstuffs started in the United States in the early 1900s at several land grant colleges; in more recent times, those activities have passed largely into the private sector, being in the interest of corporate feed producers to have reliable data for contents in feedstuffs of those nutrients that most directly affect the cost of their formulations (e.g., metabolizable energy, protein, indispensable amino acids, calcium, and phosphorus).

The Vitamins. http://dx.doi.org/10.1016/B978-0-12-802965-7.00020-4

composition data for foods and feedstuffs are now available in many forms and from many sources. However, most compilations derive from relatively few primary sources. This is particularly true for the nutrient composition data for foods. For US foods, almost all current versions are renditions of the USDA **National Nutrient Database**[3] developed through an ongoing program of the U.S. Department of Agriculture. A similar database, the **Canadian Nutrient File** has been developed by that country.[4] Less extensive databases have been developed for other countries,[5] and efforts are being made to standardize the collection, compilation, and reporting of food nutrient composition data on a global basis.[6] Variances are to be expected in national food composition databases due to differences in analytical methodologies (a particular problem for folate) and data presentation, e.g., whether tocotrienols are included in the calculation of α-tocopherol equivalents.[7]

The nutrient composition of feedstuffs has, with few exceptions,[8] been developed less systematically and extensively. Data sets presently in the public domain have been compiled largely from original reports in the scientific literature. Therefore, effects of uncontrolled sampling, multiple and often old analytical methods, multiple analysts, unreported analytical precision, unreported sample variance, etc., are likely to be far greater for the nutrient databases for feedstuffs than for the corresponding databases for foods.

Use of any database for estimating vitamin intake is limited for reasons of accuracy and completeness of the data. Although the USDA National Nutrient Database is much more complete than most feed tables with respect to data for vitamins, it is least complete with respect to vitamins D, E, and K, as well as pantothenic acid.

Core Foods for Vitamins

Foods are the most important sources of vitamins in the daily diets of humans. However, the vitamins are unevenly distributed among the various foods that comprise human diets (Table 20.1). Therefore, evaluations of the degree of vitamin adequacy of diets and meal patterns are served by knowing which foods are likely to contribute significantly to the total intake of each particular vitamin, by virtue of the frequency and amounts of each food consumed as well as the probable concentrations of that vitamin in those foods. Identifying such **core foods** is difficult because both the voluntary intakes of foods by free-living people and the concentrations of vitamins in foods are difficult to estimate quantitatively. Nevertheless, attempts to do that have indicated a manageable number of core foods for each of the vitamins, e.g., for Americans, it has been estimated that 80% of the total intakes of several vitamins are provided by 50–200 foods.[9]

Vitamins in Staple Foods

Much of the world's poor, particularly those in resource-poor countries, rely on diets based largely on starchy foods that fail to provide adequate amounts of vitamins. This is evidenced by an analysis of the world's five leading staples (Table 20.2).[10] Therefore, individuals without access to diverse diets and reliant on staple foods are at risk for vitamin deficiencies.

3. This database is used for U.S. national food consumption surveys. Data are obtained from scientific publications, food processors, food industry groups, and USDA-contracted analyses. They are available for public use in the USDA National Nutrient Database for Standard Reference, Release 28, which can be accessed online (http://www.ars.usda.gov/ba/bhnrc/ndl). Additional data relevant to the vitamin status and intakes of Americans can be found in databases maintained by the USDA and the U.S. Food and Drug Administration (FDA).

- *USDA Nutrient Content of the U.S. Food Supply Series* (http://www.cnpp.usda.gov/USfoodsupply)—Historical data (since 1909) for amounts of nutrients available per capita per day, by major food group.
- *What We Eat in America* (http://www.ars.usda.gov/services/docs.htm?docid=13793)—the dietary intake component of the National Health and Nutrition Examination Survey (NHANES).
- *FDA Total Diet Studies* (http://www.fda.gov/Food/FoodScienceResearch/TotalDietStudy/)—ongoing monitoring of levels of various nutrients, pesticide residues, and contaminants in the US food supply.

4. http://www.hc-sc.gc.ca/fn-an/nutrition/fiche-nutri-data/cnf_downloads-telechargement_fcen-eng.php.

5. Europe (http://www.fao.org/infoods/infoods/tables-and-databases/europe/en/); Latin America (http://www.fao.org/infoods/infoods/tables-and-databases/latin-america/en/); Asia (http://www.fao.org/infoods/infoods/tables-and-databases/asia/en/); Africa (http://www.fao.org/infoods/infoods/tables-and-databases/africa/en/); Middle East (http://www.fao.org/infoods/infoods/tables-and-databases/middle-east/en/); Oceana (http://www.fao.org/infoods/infoods/tables-and-databases/oceania/en/); Canada, the Caribbean, and the US (http://www.fao.org/infoods/infoods/tables-and-databases/canada-caribbean-and-united-states/en/).

6. This is the purpose of the INFOODS project of the United Nations University Food and Nutrition Program (http://www.fao.org/infoods/).

7. Uusitaio, U., Kronberg-Kippila, C., Aronsson, C.A. et al., 2011. J. Food Compost. Anal. 24, 494–505.

8. The notable exception in the public domain was the program at the University of Maryland, which involved the ongoing analysis of feedstuffs commonly used in feeding poultry in the United States. That program focused on macronutrients. It was discontinued in the late 1970s; the last version of the data (*1979 Maryland Feed Composition Data*) was published as a supplement in the Proceedings of the Maryland Nutrition Conference in that year. Other widely used feed tables were derived in part from this source, e.g., Scott, M.L., Neshiem, M.C., Young, R.J., 1982. Nutrition of the Chicken. M.L. Scott Assoc., Ithaca, NY, p. 482.

9. Using data from the 1976 Nationwide Food Consumption Survey and the Continuing Survey of Food Intakes by Individuals, USDA nutritionists estimated that Americans obtained 80% of their total intakes of the following vitamins from the following numbers of foods: vitamin A, 60; vitamin E, 100; thiamin, 168; riboflavin, 165; niacin, 159; pyridoxine, 175; folate, 129; and vitamin B_{12}, 58.

10. The same situation pertains to other staples, such as plantain, the major component of diets in >50 countries.

TABLE 20.1 Core Foods for the Vitamins

Vitamin A	Vitamin D	Vitamin E	Vitamin K	Vitamin C	Thiamin	Riboflavin	Niacin	Vitamin B_6	Biotin	Pantothenic Acid	Folate	Vitamin B_{12}
As retinol	Milk[a]	Vegetable oils	Broccoli	Tomatoes	Meats	Eggs	Meats	Meats	Liver	Liver	Tomatoes	Liver
Milk (breast, animal[a])	Ghee		Asparagus	Potatoes	Potatoes	Liver	Eggs	Cabbage	Egg yolk	Milk	Beets	Fish
Butter, ghee, margarine[a]	Margarine[a]		Lettuce	Pumpkins	Whole grains	Meats	Fish	Potatoes	Cauliflower	Meats	Potatoes	Eggs
Liver	Cheese		Cauliflower	Citrus fruits	Some fish	Some fish	Whole grains	Liver	Kidney	Eggs	Wheat germ	Milk
Eggs	Chicken (skin)		Cabbage	Other fruits	Legumes	Asparagus	Legumes	Beans	Peanuts	Fish	Cabbage	
Small fish (eaten whole)	Liver		Brussels sprouts	Yams	Oilseeds	Milk	Milk	Whole grains	Soybeans	Grains	Eggs	
As carotene	Fatty fish		Turnip greens	Cassava	Milk	Whole grains	Liver		Peanuts	Wheat germ	Legumes	Meats
Red palm oil	Cod liver oil		Liver	Milk	Eggs	Green leaves	Peanuts		Soybeans	Oatmeal	Spinach	Peanuts
Dark/medium-green leaves	Egg yolk		Spinach	Legumes					Some fish	Carrots	Asparagus	Whole grains
Yellow/orange vegetables									Milk		Milk	Beans
Yellow/orange fruits												
Yellow maize												

[a] *The high vitamin content is due to fortification.*

TABLE 20.2 Vitamin Adequacy of Major Staple Foods

	% RDA (Recommended Dietary Allowance) Provided by Staple[b]				
Vitamin[a]	Wheat	Rice	Corn[c]	Potatoes	Cassava
Vitamin A	4	0	6	0	4
Vitamin D	0	0	0	0	0
Vitamin E	9	0	3	1	10
Vitamin K	1	0	0	43	21
Vitamin C	0	0	0	199	316
Thiamin	25	18	33	139	114
Riboflavin	17	12	10	23	30
Niacin	26	29	28	156	92
Vitamin B_6	8	31	28	275	44
Pantothenic acid	28	78	30	137	15
Folate	29	5	5	37	54
Vitamin B_{12}	0	0	0	0	0

[a]*This listing does not include biotin, which is not included in the USDA National Nutrient Database.*
[b]*Consumed to provide 80% of calories for lactating adult females.*
[c]*i.e., Maize.*
Fitzpatrick, T.B., Basset, G.J.C., Borel, P., et al., 2012. Plant Cell 24, 395–414.

Predicting Vitamin Contents of Foods and Feedstuffs

Data for the nutrient contents of foods and feedstuffs can be useful in judging the adequacy of food supplies and feedstuff inventories. However, estimates of the nutrient intakes of individuals as determined on the basis of these data are seldom accurate, owing to different factors that may alter the nutrient composition of a food or feedstuff before it is actually ingested. The errors associated with such estimates are particularly great for the vitamins. To accommodate these sources of error, most of which inflate estimates of nutrient intake, it is a common practice to discount by 10–25% the analytical values in the databases. It is likely that such modest discounts may still yield overestimates of intakes of at least some of the vitamins.

There are several sources of error in estimating vitamins in foods and feedstuffs:

- **Errors in sampling and analysis** contribute to inaccuracy of predicted values. Analytical errors are less likely to be problematic for vitamin E, thiamin, riboflavin, niacin, and pyridoxine, for which robust analytical techniques not prone to analyst effects are available.[11]
- **Variation among cultivars** in vitamin contents of plant foods can be as great as several orders of magnitude (Fig. 20.1). In some cases, these differences correspond to identifiable characteristics of the plant. For example, the ascorbic acid contents of lettuce, cabbage, and asparagus tend to be relatively high in the colored and darker green varieties; darker orange varieties of carrot tend to have greater provitamin A contents than do lighter-colored carrots. Still, vitamin contents are not necessarily related to such physical traits or to each other.
- **Variation among tissues** occurs, as most vitamins are not uniformly distributed among the various edible tissues of plants (Figs. 20.2–20.4). Gradients in the tissue contents of several nutrients correspond to the distribution of the phloem and xylem vascular network. Thus, relatively high concentrations have been observed for ascorbic acid in the stem end of oranges and pears; the top ends of pineapples; the apical ends of potatoes, the tuber end of sweet potatoes; both the lower and upper portions of carrots, the top ends of turnips; and the stem tips of asparagus, bamboo shoots, and cucumbers. In general, exposed tissues, i.e., skin/peel and outer leaves, tend to contain greater concentrations of vitamins, particularly ascorbic acid, which is found mostly in chloroplasts in tissues exposed to light.[12] In cereal grains, thiamin and niacin tend to be concentrated in tissues[13] that are removed in milling; therefore, breads made from refined wheat flour are much lower in those vitamins than are products made from maize, which is not milled. Mobilization of seed vitamin stores and, in some cases, biosynthesis of vitamins occur during germination, such that young seedlings (sprouts) tend to have relatively high vitamin contents.
- **Local agronomic factors and weather** conditions that affect growth rate and yield can also affect vitamin contents of plants (Tables 20.3A and 20.3B). For many plants, conditions that favor the production of lush vegetation will result in increased concentrations of several vitamins. Low temperatures have been shown to increase the ascorbic acid contents of beans and potatoes; the thiamin contents of broccoli and cabbage; the riboflavin contents of spinach, wheat, broccoli, and cabbage; and the niacin contents of spinach and wheat. However, low

11. Most nutrient analytical methods have been standardized by the Association of Official Analytical Chemists.
12. As much as 35–40% of the ascorbic acid in green plants may be present in chloroplasts, where its concentration can be as great as 50 mM. In the case of citrus fruits, three-quarters of the ascorbic acid is located in the peel.
13. That is, the scutellum and aleurone layer.

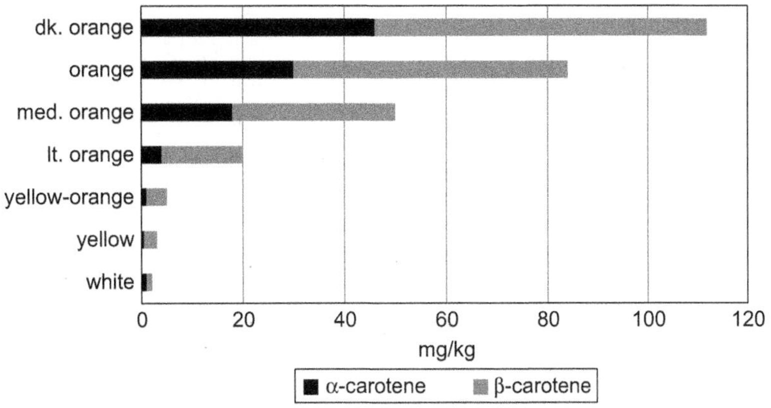

FIGURE 20.1 Variation in carotene contents among carrot cultivars. *After Leferriere, L., Gabelman, W.H., 1968. Proc. Am. Soc. Hortic. Sci. 93, 408–415.*

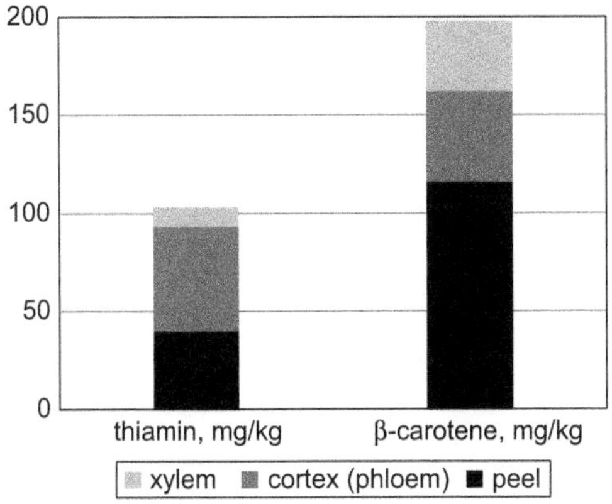

FIGURE 20.2 Distribution of vitamins in carrot tissues. *After Yamaguchi, Y., 1952. Proc. Am. Soc. Hortic. Sci. 60, 351–358.*

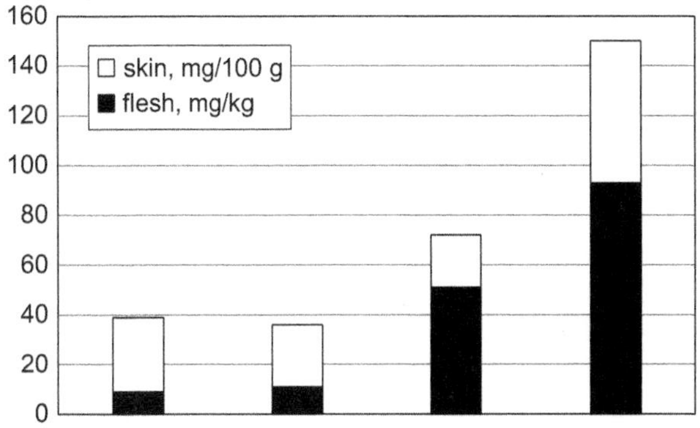

FIGURE 20.4 Distribution of ascorbic acid in tissues of four apple cultivars. *After Gross, E., 1943. Garterbauwissenschaften 17, 500–506.*

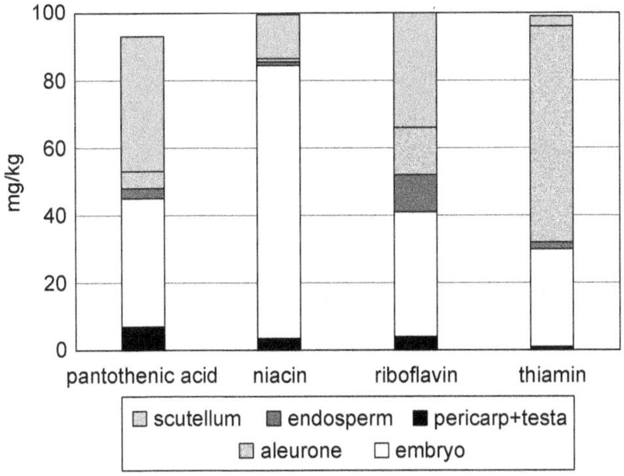

FIGURE 20.3 Distribution of vitamins in wheat tissues. *From Hinto, I., 1953. Nature (London) 173, 993–1000.*

temperatures decrease thiamin in beans and tomatoes; riboflavin in beans; niacin in tomatoes; and carotenoids in carrots, sweet potatoes, and papayas. Conditions at harvest can affect the vitamin content of some crops, e.g., the vitamin E content of fungus-infected corn grain can be less than half that of nonblighted corn. Legumes such as alfalfa and soybeans contain the enzyme lipoxygenase, which, if not inactivated (by drying) soon after harvest, catalyzes lipid oxidation reactions resulting in massive destruction of carotenoids and vitamin E. Accordingly, the vitamin contents of plant-based foods can be markedly different in different parts of the world, show both seasonal and annual changes. These fluctuations can be as great as 8-fold for the α-tocopherol content of alfalfa hay within a single season, and 11-fold for the ascorbic acid content of apples produced in different years.

TABLE 20.3A Fold Variations in Reported Vitamin Contents of Fruit and Vegetable Cultivars: β-Carotene, Ascorbic Acid, α-Tocopherol, Thiamin, and Riboflavin

Food	β-Carotene	Ascorbic acid	α-Tocopherol	Thiamin	Riboflavin
Apple		29		3.0	10
Apricot	2.9			1.5	1.3
Banana		9			
Barley				2.3	
Bean	2.3	2.9		2.7	3.7
Blueberry	17	3.0		1.3	1.8
Cabbage		3.8		2.5	2.8
Carrot	80	1.4		6.9	5.5
Cassava	113	1.9			
Cauliflower		1.7		1.4	1.4
Cherry	3.5	4.2		2.0	1.5
Collard	1.4	1.6			2.1
Cowpea				2.9	3.0
Grape		3.0		7.5	3.4
Grapefruit	9.3	1.3			
Guava		11			
Lemon		1.2			
Lemon		1.2			
Maize	24		2.0	1.8	
Mango	3.8	91			2.0
Muskmelon		20			
Nectarine	4.8	4.7		1.0	1.3
Oat				1.8	
Orange	6.8	1.5			
Palm, oil	5.1				
Papaya	5.7	2.7		1.3	1.5
Pea	4.3	3.4		5.2	1.7
Peach	6.0	4.2		2.0	1.7
Peanut				1.4	1.9
Pear		16		7.0	5.0
Pepper, green	1.3	1.8	18		
Pepper, chili	46	10			
Plum	3.2	1.5			1.2
Potato		5.1		2.5	6.2
Rapeseed, oil			3.4		
Raspberry		2.3			
Soybean		2.4	1.2		

TABLE 20.3A Fold Variations in Reported Vitamin Contents of Fruit and Vegetable Cultivars: β-Carotene, Ascorbic Acid, α-Tocopherol, Thiamin, and Riboflavin—cont'd

Food	β-Carotene	Ascorbic acid	α-Tocopherol	Thiamin	Riboflavin
Spinach		1.6			
Squash					
Summer		9.4			
Winter		3.5			
Strawberry		4.3			
Sunflower			2.7		
Soybean		2.4	1.2		
Sweet potato	89	3.1		2.9	3.1
Taro		3.2		4.9	2.5
Tomato	20	15		1.6	
Turnip, greens		1.1			
Watermelon	15				
Wheat			29	7.9	5.2
Yam		1.9		3.0	3.9

Mozafar, A., 1994. Plant Vitamins: Agronomic, Physiological and Nutritional Aspects. CRC Press, New York, p. 43.

- **Agronomic practices** can affect the vitamin contents of plant tissues. These relationships are complex, varying according to the soil type, plant species, and vitamin in question. In general, mineral fertilization can increase plant contents of ascorbic acid (P, K, Mn, B, Mo, Cu, Zn, Co); carotenes (N, Mg, Mn, Cu, Zn, B); thiamin (N, P, B); and riboflavin (N). However, nitrogen fertilization tends to decrease ascorbic acid concentrations despite increased yields. Organic fertilizers can increase the concentrations of some vitamins, particularly thiamin. These effects may be due to the lower nitrate contents in organic fertilizers compared with inorganic ones. Organic fertilizers may also contain vitamins in forms that plant roots can absorb. Practices that affect light exposure and plant growth rate can affect plant vitamin contents. This is especially true for ascorbic acid and the biosynthesis of which is related to plant carbohydrate metabolism. For example, field-grown tomatoes can have twice the ascorbic acid content of greenhouse-grown tomatoes, shaded fruits having less than those directly exposed to light.[14] Relatively high ascorbic acid contents have been found in peas, grapes, and tomatoes grown at lower planting densities; in lower yielding or smaller apples; and in field-ripened compared to artificially ripened apples.[15]
- **Conditions of feeding** can affect the vitamin contents of foods of animal origin. These can be highly variable according to country of origin, season of the year, age at slaughter, the composition of the diets used, etc. For example, the vitamin E content of poultry meat is greater from chickens fed supplements of the vitamin than from those that are not.[16]

Accounting for Variation in Vitamin Contents

Natural variation in the nutritional composition of foods and feedstuffs has generally been accommodated by the analysis of multiple representative samples of each material of interest. Nevertheless, most databases include only a single value, the mean of all analyses, and fail to indicate the variance around that mean. The practical necessity of using databases so constructed means that the nutritionist is faced with the dilemma of estimating vitamin intake through the use of data that are likely to be inaccurate but to an uncertain and unascertainable degree. Thus, if an average value of 150 mg/kg is used to represent the ascorbic acid concentration of potatoes, as is frequently the case, then it must be recognized that half of all samples will exceed that value (thus yielding an underestimate)

14. In fact, vitamin C levels can differ between exposed and shaded sides of the same fruit.
15. By exposure to ethylene either in storage or in transit to market.

16. A practical example of this comparison is the intensively managed commercial poultry flock fed formulated feeds in the United States *versus* the small courtyard flock largely subsisting on table scraps, insects, and grasses.

TABLE 20.3B Fold Variations in Reported Vitamin Contents of Fruit and Vegetable Cultivars: Niacin, Pyridoxine, Biotin, Pantothenic Acid, and Folate

Food	Niacin	Vitamin B_6	Biotin	Pantothenic Acid	Folate
Apple	2.0		1.1	4.0	
Apricot	1.3				
Avocado	1.5	1.6		13	
Barley	1.1			1.2	
Bean	3.8	2.2			4.6
Blueberry	1.7				
Cherry	1.5				
Cowpea	2.2	1.5	1.5	1.3	
Grape	2.4				
Maize	5.5			1.3	
Mango	18				
Nectarine	1.3				
Oat	1.4				
Papaya	2.3				
Pea	1.2	1.3			2.2
Peach	1.2				
Peanut	1.5				
Pear	4.0		1.1	2.5	
Pepper, green	1.2				
Plum	4.5				
Potato	2.7	3.2			
Rye	1.3				
Strawberry	1.3				
Sweet potato	3.4			2.2	
Taro	4.9				
Wheat	5.0	8.6		2.6	
Yam	2.7				

Mozafar, A., 1994. Plant Vitamins: Agronomic, Physiological and Nutritional Aspects. CRC Press, New York, p. 43.

while half will contain less than that value (thus, yielding an overestimate). In constructing databases for use in meal planning or feed formulation, a better way to accommodate such natural variation is to enter into the database values discounted by a multiple of the standard deviation that would yield an acceptably low probability of overestimating actual nutrient amounts.[17] That approach, however, requires a fairly extensive body of data from which to generate meaningful estimates of variance. Few, if any sets of food/feedstuff vitamin composition data are that extensive.

17. This approach was originated in the 1950s by G.F. Combs, Sr. at the University of Maryland in developing the Maryland Feed Composition Table. Those data were based on replicate analyses from multiple samples of each feedstuff and were expressed as the mean—0.9 SD units. That adjustment allowed a likelihood of overestimating actual nutrient concentration of $p=.20$.

2. VITAMIN BIOAVAILABILITY

Chemical analyses of vitamin contents may overestimate the amounts of vitamin that are bioavailable in a food or feedstuff (Table 20.4).[18] In the cases of niacin, biotin, pyridoxine, vitamin B_{12}, and choline, which in certain foods and

18. See Chapter 3 for a discussion of the concept of nutrient bioavailability.

TABLE 20.4 Foods and Feedstuffs With Low Vitamin Bioavailabilities

Vitamin	Form	Food/Feedstuff
Vitamin A	Provitamins A	Corn
Vitamin E	Nontocopherols	Corn oil, soybean oil
Ascorbic acid	Ascorbinogen	Cabbage
Niacin	Niacytin	Corn, potatoes, rice, sorghum grain, wheat
Pyridoxine	Pyridoxine 5′-β-glucoside	Corn, rice bran, unpolished rice, peanuts, soybeans, soybean meal, wheat bran, whole wheat
Biotin	Biocytin	Barley, fishmeal, oats, sorghum grain, wheat

feedstuffs can be poorly utilized, only the biologically available amounts have nutritional relevance. The bioavailability of a vitamin in a particular food depends on several factors extrinsic and intrinsic to the individual consuming that food:

Extrinsic factors

- **Chemical form**—can affects vitamin solubility, absorption, and/or metabolism.
- **Physical form** (including emulsifiers, coatings, etc.)—can affect vitamer interactions with other food components.
- **Concentration**—can affect vitamer solubility and absorption kinetics.
- **Food/diet composition**—can affect gastrointestinal transit time and vitamer digestion, absorption, and/or synthesis by gut microbiota.
- **Nonfood agonists** (e.g., cholestyramine, alcohol, drugs)—may impair vitamin absorption and/or metabolism.

Intrinsic factors

- **Age**—many older individuals have poor vitamin B_{12} absorption associated with loss of gastric production of intrinsic factor.
- **Health status**—impaired gastrointestinal function can reduce vitamin absorption and synthesis by gut microbiota.

For most vitamins, bioavailability must be determined using animal models. An in vitro method is available to measure niacin bioavailability; it involves comparing the amounts of niacin determined chemically before (free niacin) and after (total niacin) alkaline hydrolysis. The free niacin thus determined correlates with the bioavailable niacin determined using the growth response of niacin-deficient rats fed a low-tryptophan diet.

3. VITAMIN LOSSES IN FOODS

Every step in the handling of a food in the locally, nationally, and globally interrelated food system (Fig. 20.5) has the potential of contributing to losses of vitamins (Table 20.5). In theory, these losses can be modeled and, thus, predicted. However, in practice, the variation in the actual conditions of handling foods through each of these steps is so great that the only way to estimate vitamin intakes of people is to analyze the vitamin contents of foods as they are eaten.

Vitamin losses in foods can be of several types:

- **Storage losses** can occur due to postharvest oxidation and enzymatic decomposition. For example, the ascorbic acid contents of cold-stored apples and potatoes can drop by two-thirds and one-third, respectively, within 1–2 months. Those of some green vegetables can drop to 20–78% of original levels after a few days of storage at room temperature. Such losses can vary according to specific techniques of food processing and preservation.
- **Milling losses** can occur in removing the bran and germ portions of grains to produce flours.[19] Because those portions are typically rich in vitamin E and many of the water-soluble vitamins, highly refined flours are low in these vitamins (Fig. 20.6).
- **Processing losses** can occur during thermal processing in the preservation of foods. Blanching (mild heating to inactivate enzymes, reduce microbial numbers, and decrease interstitial gases) is usually minimally destructive, although it can result in the leaching of water-soluble vitamins from foods blanched in hot water. Otherwise, blanching usually improves vitamin stability. In contrast, canning and other forms of high-temperature treatment can accelerate reactions of vitamin degradation, depending on the chemical nature of the food (i.e., its pH, dissolved oxygen and moisture contents, presence of transition metals and/or other reactive compounds; Tables 20.5 and 20.6). Vitamins A, E, C, thiamin, and folate are sensitive to moist, high-temperature

19. The yield of flour obtained from the milling process is expressed as the extraction rate. A 100% extraction is whole meal flour containing all of the grain; removal of progressively more of the bran and germ produces whiter flours of lower extraction rates; 72% extraction is common white flour; "patent" flours comprised mostly of endosperm are of extraction rates of 30–50%.

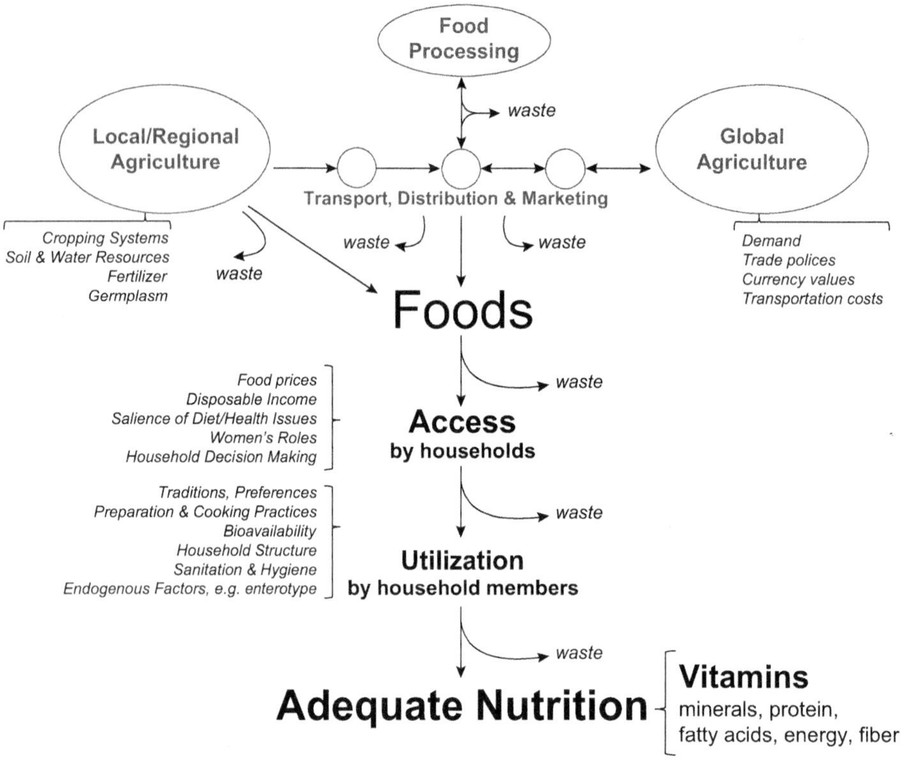

FIGURE 20.5 The food system.

TABLE 20.5 Effects of Food Processing Techniques on Vitamin Contents of Foods

Technique	Main Effects	Vitamins Destroyed[a]
Blanching	Partial removal of oxygen	Vitamin C (10–60%)[b,c]
	Partial heat inactivation of enzymes	Thiamin (2–30%), riboflavin (5–40%), niacin (15–50%), carotene (<5%)[c]
Pasteurization	Removal of oxygen[d]	Thiamin (10–15%)
	Inactivation of enzymes	Minor losses (1–5%) of niacin, vitamin B_6, riboflavin, and pantothenic acid
Canning	Exclusion of oxygen	Highly variable losses[e,f]
Freezing[h]	Inhibition of enzyme activity[g]	Very slight losses of most vitamins
Frozen storage[h]		Substantial losses of vitamin C and pantothenic acid; moderate losses of thiamin and riboflavin
Freeze drying	Removal of water	Very slight losses of most vitamins
Hot air drying	Removal of water	10–15% losses of vitamin C and thiamin
γ–Irradiation	Inactivation of enzymes	Some losses (about 10%) of vitamins C, E, K, and thiamin

[a]*Actual losses are variable, depending on exact conditions of time, temperature, etc.*
[b]*Loss of vitamin C is due to both oxidation and leaching.*
[c]*Losses of oxidizable vitamins can be reduced by rapid cooling after blanching.*
[d]*Vitamin losses are usually small, owing to the exclusion of oxygen during this process.*
[e]*Losses in addition to those associated with heat sterilization before canning.*
[f]*For example, 15% loss of vitamin C after 2 years at 10°C.*
[g]*While enzymatic decomposition is completely inhibited in frozen vegetables, reactivation occurs during thawing such that significant vitamin losses can occur. This is avoided by rapidly blanching before freezing.*
[h]*Thawing losses are associated with vitamin leaching into the syrup.*

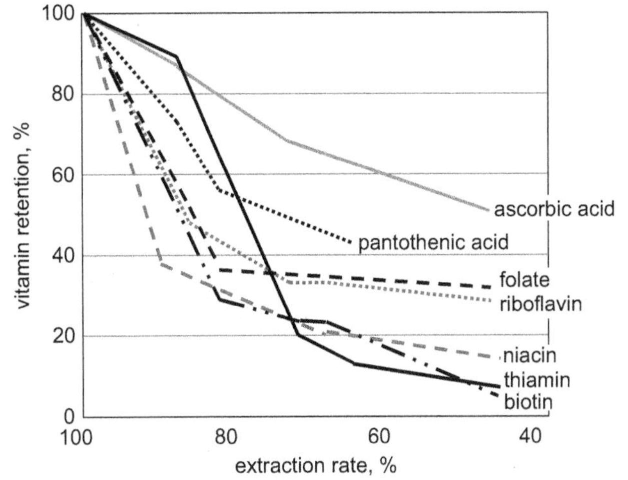

FIGURE 20.6 Loss of vitamins in the milling of wheat flour. Note: extraction rates of about 72% are typical for common bread flours. *From Moran, T., 1959. Nutr. Abstr. Rev. 29, 1–16.*

conditions such as those occurring during extrusion. Losses as great as 80% can occur, depending on the amount of water added to the food mixture and the temperature of the system. Freezing and drying usually result in only minor losses of most vitamins. Losses associated with ionizing (γ) irradiation vary according to the energy dose but are generally low (less than 10%).

- **Cooking losses** can result in further losses of vitamins from foods. However, methods used for cooking vary widely between different cultures and among different individuals, making vitamin losses associated with cooking highly variable. Washing fruits and vegetables in water before cooking can result in the extraction of water-soluble vitamins, particularly if they are soaked for long periods of time. Peeling fruits and vegetables can remove vitamins associated with the outer tissues.[20] Vitamin losses associated with cooking processes are also highly variable, but generally amount to about 50% for the less stable vitamins. The greatest losses are associated with long cooking times under conditions of exposure to air. Vitamin losses are less when food is cooked rapidly, as in a pressure cooker or a microwave oven, or by high-temperature stir frying. The baking of bread can reduce the thiamin content of flour by about 25% without affecting its contents of niacin or riboflavin. The susceptibilities of vitamins to processing and cooking losses are summarized in Table 20.7.
- **Waste** occurs throughout the food system due to such factors as spoilage, damage, pests, and discard. The United Nation has estimated that at least one-third of all food produced globally gets wasted.[21] In North America, food wastage has been estimated to reduce the food supplies by 30–40%. The food wastes of industrialized countries are estimated to be ~220M tons.

20. e.g., Peeling potatoes can substantially reduce their ascorbic acid content.

Vitamin losses from foods can be minimized using the following:

- fresh instead of stored food
- minimum amounts of water in preparing and cooking
- minimum cooking, using high temperatures for short times
- minimum storage

4. VITAMIN FORTIFICATION

Availability of Purified Vitamins

All of the vitamins are produced commercially in pure forms. Most are produced by chemical synthesis, but some are also isolated from natural sources (e.g., vitamin A from fish liver, vitamin D_3 from liver oil or irradiated yeast, vitamin E from soybean or corn oils, and vitamin K from fish meal) and some are produced microbiologically (e.g., thiamin, riboflavin, folate, pyridoxine, biotin, pantothenic acid, and vitamin B_{12}[22]). Before their commercial synthesis became feasible, which began only in the 1940s, vitamins were extracted from such natural sources as fish oils and rose hip syrup. Today, the production of vitamins is based predominantly on their chemical synthesis and/or microbiological production, the latter having been greatly impacted by the emergence of new techniques in biotechnology.[23] With the notable exception of the tocopherols,[24] there is no basis to the notion that biopotencies of vitamins prepared by chemical/microbiological synthesis are at least as great as those of vitamins isolated from natural sources. In some cases, synthetic vitamins may be appreciably more bioavailable than the vitamin from natural sources (e.g., purified niacin versus protein-bound niacytin; purified biotin versus protein-bound biocytin).

The use of purified vitamins offers obvious advantages for purposes of ensuring vitamin potency in a wide

21. Lipinski, B., Hanson, C., Lomax, J., et al., 2013. World Resources Institute Working Paper (http://www.wri.org/sites/default/files/reducing_food_loss_and_waste.pdf).
22. The commercial production of vitamin B_{12} is strictly from microorganisms.
23. The industrial production of the vitamins has been nicely reviewed: O'Leary, M.J., 1993. Industrial production. In: Ottaway, P.B. (Ed.), The Technology of Vitamins in Food. Chapman & Hall, London, p. 63.
24. Vitamers E produced by chemical synthesis vary in biopotency (*see* Chapters 3 and 8); the most potent is *RRR*-α-tocopherol.

TABLE 20.6 Typical Losses of Vitamins Through Canning (%)

Food	Vitamin A	Vitamin C	Thiamin	Riboflavin	Niacin	Vitamin B_6	Biotin	Pantothenic Acid	Folate
Asparagus	43	54	67	55	47	64	0		75
Lima bean	55	76	83	67	64	47		72	62
Green bean	52	79	62	64	40	50		60	57
Beet	50	70	67	60	75	9		33	80
Carrot	9	75	67	60	33	80	40	54	59
Corn	32	58	80	58	47	0	63	59	72
Mushroom		33	80	46	52		54	54	84
Green pea	30	67	74	64	69	69	78	80	59
Spinach	32	72	80	50	50	75	67	78	35
Tomato	0	26	17	25	0		55	30	54

Lund, D., 1988. Nutritional Evaluation of Food Processing. In: Karmas. E., Harris, R.S. (Eds.), third ed. Van Nostrand Reinhold, New York, 319 pp.

variety of formulated products, including fortificants for foods, premixed supplements for feeds, nutritional supplements, pharmaceuticals, and ingredients in cosmetics. Commercial vitamin production has grown steadily since the discovery of vitamin B_{12}. At that time, annual world vitamin production was estimated to be only <1500 metric tons; by 2005, it exceeded 20,000 metric tons. Vitamins are produced by at least 30 firms in some 17 countries; but a half-dozen companies presently dominate the world market.

Addition of Vitamins to Foods

The addition of vitamins to certain foods is a common practice in most countries. Vitamins are added for several purposes:

- **Fortification**—ensuring vitamin adequacy of populations, e.g., white flour (folate), milk (vitamins A and D), margarine (vitamin A), formula foods (multiple vitamins in infant formulas, liquid nutrient supplements, enteral formulas used for tube feeding, and parenteral formulas used for intravenous feeding).
- **Vitaminization**—making foods carriers of vitamins not normally present, e.g., many breakfast cereals (multiple vitamins), orange juice (vitamin D), wheat (vitamin A), table salt (multiple vitamins), and parental feeding solutions (multiple vitamins).
- **Revitaminization**—restoring the vitamin content to that originally present before processing, e.g., white flour (thiamin, riboflavin, niacin).
- **Enrichment**—increasing the amounts of vitamins already present.

These processes are subject to regulation by national food authorities. In the United States, the addition of nutrients to foods is regulated by the Food and Drug Administration (FDA), which has identified as candidates for addition to foods 22 nutrients including 12 vitamins (Table 20.8).[25] Fortification of wheat flour with folate has been mandatory in the United States since 1998 and in Ireland since 2006. Since 1966, the USDA and USAID[26] have also routinely fortified or enriched foods provided as foreign aid under Public Law 480 (Table 20.9).[27] In addition, many antixerophthalmia programs have used vitamin A fortification of such foods as dried milk, wheat flour, sugar, tea, margarine, and monosodium glutamate.[28] As a consequence, processed foods provide one-third of the vitamin D, two-thirds of the folate, and almost half of the vitamin B_{12} consumed by Americans.[29]

Stabilities of Vitamins Added to Foods

The stabilities and bioavailabilities of vitamins added to foods depend on the form of vitamin used, the composition of the food to which it is added, and the absorption status of the individual ingesting that food. The less stable vitamins can be lost from foods during storage, depending

25. In addition to these vitamins, other nutrients are approved: protein, calcium, phosphorus, magnesium, potassium, manganese, iron, copper, zinc, and iodine.
26. United States Agency for International Development.
27. The cost of this fortification is very low relative to the total value of the commodities. The ingredients (vitamins and minerals) used to enrich the processed and soy-fortified commodities cost less than 2.5% of the value of the product; those used to enrich the more expensive blended food supplements cost less than 5% of the product value.
28. MSG, used as a seasoning.
29. Weaver, C.M., Dwyer, J., Fulgoni, V.L., et al., 2014. Am. J. Clin. Nutr. 99, 1525–1542.

TABLE 20.7 General Susceptibilities of Vitamins to Processing and Cooking Losses

Vitamin	Conditions that Enhance Loss
Vitamin A	Highly variable but significant losses during storage and preparation
Vitamin D	(Stable to normal household procedures)
Vitamin E	Frying can result in losses of 70–90%; bleaching of flour destroys 100%; other losses in preparation or baking are small
Vitamin K	(Losses not significant due to synthesis by intestinal microflora)
Ascorbic acid	Readily lost by oxidation and/or extraction in many steps of food preparation, heat sterilization, drying, and cooking
Thiamin	Readily lost by leaching, by removal of thiamin-rich fractions from native foods (e.g., flour milling), and by heating; losses as great as 75% may occur in meats and 25–33% in breads
Riboflavin	Readily lost on exposure to light (90% in milk exposed to sun light for 2 h, 30% from milk exposed to room light for 1 day) but very stable when stored in dark; small losses (12–25%) on heating during cooking
Niacin	Leached during blanching of vegetables (≤40%) but very stable to cooking
Vitamin B_6	Leached during food preparation; pasteurization causes losses of 67%; roasting of beef causes losses of about 50%
Biotin	(Apparently very stable; limited data)
Pantothenic acid	Losses of 60% by milling of flour and of about 30% by cooking of meat; small losses in vegetable preparation
Folate	(Data not available)
Vitamin B_{12}	Only small losses on irradiation of milk by visible or ultraviolet light

on the conditions (time, temperature, and moisture) of that storage (Fig. 20.7).

5. BIOFORTIFICATION

The term "biofortification" was coined in the 1990s to describe the use of intrinsic metabolic capacities of plants to enhance their contents of micronutrients in major staple crops.[30] Increasing vitamin contents had not been an explicit goal of crop improvement, which had centered on traits directly related to yield and disease resistance. That attitude started to change with the recognition of the "**hidden hunger**"[31] of micronutrient malnutrition, i.e., the persistent, debilitating shortages of vitamins and essential minerals in the face of remarkable gains in the global production of total staple foods and total calories. That one-sixth of the world's population does not have access to the foods necessary for nutritionally balanced diets has made it impossible to overlook shortages of vitamin A, vitamin C, folate, vitamin B_{12}, iron, iodine, and zinc in the diets of the world's poor. Even in the industrialized world, where diet-related chronic diseases are substantial problems, opportunities exist to develop vitamin/mineral-rich fruits and vegetables and to use these aspects of specific, good nutrition as marketing "hooks." The international agricultural community has responded with a number of coordinated efforts using modern agricultural technologies to enhance the micronutrient contents of major staple foods eaten widely by the poor (rice, wheat, corn [maize], cassava, beans, sweet potato, and pearl millet).[32] These efforts, in effect, treat agriculture as an instrument of public health.

TABLE 20.8 Vitamins Approved by the FDA for Addition to Foods

Vitamin	Recommended Level of Addition (per 100 kcal)
Vitamin A	250 IU
Vitamin D	20 IU
Vitamin E	1.5 IU
Vitamin C	3 mg
Thiamin	75 μg
Riboflavin	85 μg
Niacin	1.0 mg
Vitamin B_6	0.1 mg
Biotin	15 μg
Pantothenic acid	0.5 mg
Folate	20 μg; wheat flour products: [a] 140 μg/100 g
Vitamin B_{12}	0.3 μg

[a] *Mandated in the United States for most enriched flour, breads, corn meals, rice, noodles, macaroni, and other grain products.*

30. The term "**field fortification**" has also been used.

31. "Hidden hunger" should in the lexicon of every person interested global health and well-being. It differs from other familiar terms: "**undernourishment**" and "**undernutrition**," which describe insufficient intakes of *energy and protein* to sustain growth and maintenance; and "**hunger**" and "**food insecurity**," which describe insufficient access to sufficient *quantities* of food.

32. The motive force for this effort has been the Consultative Group for International Agricultural Research (CGIAR) led by the International Food Policy Research Institute (IFPRI). With support from the Bill and Melinda Gates Foundation, IFPRI put together a global program, Harvest Plus (HarvestPlus@cgiar.org and www.HarvestPlus.org).

TABLE 20.9 Vitamins Added to P.L. 480 Title II Commodities Amount Added per 100 g[a]

Vitamin	Wheat–Soy Blend	Corn–Soy Blend	S-fortified Cereals	Nonfat Dry Milk	Others
Vitamin A (IU)	2314	2314	2204–2645	5000–7000	2204–2645
Vitamin D (IU)	198	198			
Vitamin E[b] (IU)	7.5	7.5			
Vitamin C (mg)	40.1	40.1			
Thiamin[c] (mg)	0.28	0.28	0.44–0.66		0.44–0.66
Riboflavin (mg)	0.39	0.39	0.26–0.40		0.26–0.40
Niacin (mg)	5.9	5.9	3.5–5.3		3.5–5.3
Vitamin B_6[d] (mg)	0.165	0.165			
Pantothenic acid (mg)	2.75	2.75			
Folate (μg)	198	198			
Vitamin B_{12} (μg)	3.97	3.97			

[a]*Processed blended foods are also fortified with Ca, P, Fe, Zn, I, and Na; Soy-fortified cereals and other processed foods are fortified with Ca and Fe.*
[b]*As all-rac-α-tocopheryl acetate.*
[c]*As thiamin mononitrate.*
[d]*As pyridoxine hydrochloride.*

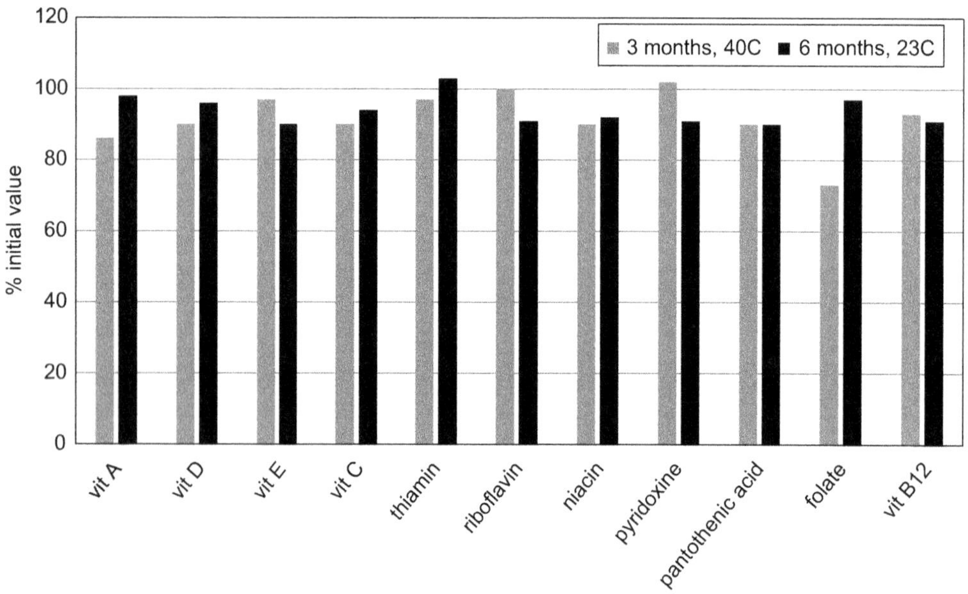

FIGURE 20.7 Stabilities of vitamins added to a breakfast cereal. *After Anderson, A.K., 1976. Food Technol. 30, 110–117.*

Biofortification involves the use of both conventional techniques of plant breeding as well as genetic modification of plant genomes to increase levels of key vitamins and essential minerals in crops. In cases of significant heterogeneity in tissue micronutrient content (provitamin A carotenoids in sweet potato, corn [maize], and cassava), conventional plant breeding techniques have been used to move high-micronutrient traits into agronomically superior varieties. In cases without such trait variability (provitamin A carotenoids and folate in rice), and for plants the breeding of which is prohibitively difficult or impossible (plantain, cassava, and potato), genetic engineering offers opportunities to improve tissue micronutrient contents. In both cases, improved germplasm can be moved into the respective crop breeding programs of local countries' national agricultural research systems to be incorporated into high-yielding cultivars well-suited to local agroecological conditions. This utilization of local agricultural systems to improve micronutrient nutrition amortizes the original costs of developing high-micronutrient traits over years-to-decades of crop production in nations of need.

Conventional selective breeding combined with marker-assisted selection has produced germplasm with enhanced vitamin content. Such efforts by collaborators in the HarvestPlus program exploited heterogeneity in the β-carotene contents among available varieties to produce new, high-β-carotene varieties:[33]

- **Orange-fleshed sweet potato**—Forty-six improved varieties with β-carotene contents as great as 24,900 μg/g (compared to 30–100 μg/g in common cultivars) have been released in throughout Africa, South America, and China.[34]
- **Yellow cassava**—Screening of germplasm revealed a range of β-carotene contents of 0–19 μg/g. Three hybrids with β-carotene contents of 6–8 μg/g have been released in Nigeria.[35]
- **High-β-carotene corn (maize)**—Screening revealed a range of β-carotene contents of 0–19 μg/g in existing lines. Genetic loci associated with β-carotene level were identified, and DNA markers were developed that facilitated breeding using marker-assisted selection to take advantage of rare genetic variation in the β-carotene hydroxylase-1 (*CRTRB1*) gene, and increase expression of *PSY1*, which encodes for phytoene synthase, the rate-limiting step in the carotenoid biosynthetic pathway.[36] Hybrids were produced with β-carotene contents of 6–8 μg/g; five have been released in Zambia, Nigeria, and Ghana.[37]
- **High-β-carotene plantain**[38]—Screening of >300 genotypes revealed a range of β-carotene contents of 1–345 μg/g (fresh weight). Five varieties with β-carotene contents of 17–106 μg/g have been identified for dissemination in Burundi and the Democratic Republic of the Congo.[39]

Others have used selective breeding to produce high-β-carotene carrots[40] and tomatoes[41] and high-anthocyanin carrots.[42] Proof-of-concept experiments have been done to increase other vitamins species used as models in plant research: tocopherols[43] and vitamin B_6[44] in *Arabidopsis thaliana*; tocopherols in *Brassica napus*.[45] Strain selection and optimization of culture conditions of baker's yeast (*Saccharomyces cerevisiae*) have been used to make three- to fivefold increases in the folate contents of breads.[46]

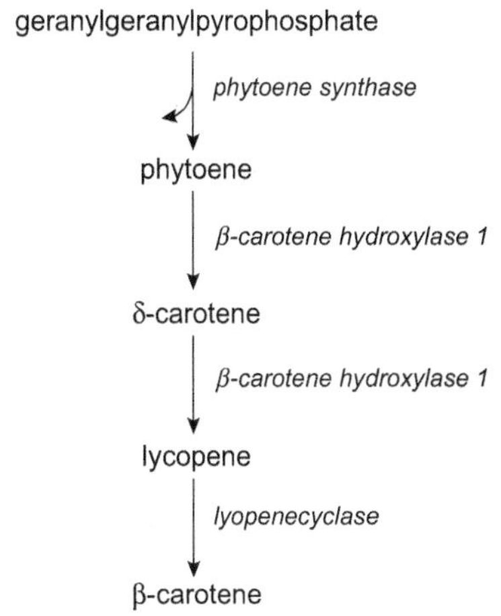

FIGURE 20.8 β-carotene biosynthetic pathway in plants.

Genetic engineering has been shown to be useful in adding vitamin biosynthetic capacity to plant species with incomplete biosynthetic pathways. This concept was demonstrated by the pioneering work of producing **"Golden Rice,"**[47] i.e., various rice containing ~35 μg/g β-carotene in the endosperm. This was achieved by inserting into the rice genome two missing genes needed for β-carotene synthesis (Fig. 20.8): *PSY* (encoding phytoene synthase), initially from daffodil (*Narcissus pseudonarcissus*) and subsequently from corn (maize); and *CRT1* (encoding β-carotene hydroxylase-1) from a soil bacterium (*Erwinia uredovora*). As the rice genome already contained the third gene required for this pathway, LYC (encoding *lycopene* cyclase), these additions completed β-carotene biosynthetic capacity.

Others have used genetic engineering to increase folate in lettuce[48] and rice.[49] Unlike conventional breeding, the

33. Increases in other micronutrients have also been achieved: high-iron pearl millet (India), bean (D.R. Congo, Rwanda), and high-zinc rice (Bangladesh, India) and wheat (India, Pakistan).
34. Andrade, M., 2014. Biofortification Progress Briefs, p. 13–15.
35. Kulakow, P., Parkes, E., 2014. Biofortification Progress Briefs, p.11–12.
36. Yan, J., Kandiani, C.B., Harjes, C.E., et al., 2010. Nature Genetics 42, 322–327; Toledo-Ortiz, G., Huq, E., Rodriguez-Concepcíon, M., 2010. PNAS 107, 11626–11631; Messias, R., Galli, V., Silva, S.D., et al., 2014. Nutrients 6, 546–563.
37. Dhliwayo, T., Palacios, N., Babu, R., et al., 2014. Biofortification Progress Briefs, p.9–10.
38. i.e., Cooking banana.
39. Ekesa, B., 2014. Biofortification Progress Briefs, p.15–16.
40. Mills, J.P., Simon, P.W., Tanumihardjo, S.A., 2008. J. Nutr. 138, 1692–1698.
41. Unlu, N.Z., Bohn, T., Francis, D., et al., 2007. J. Agr. Food Chem. 55, 1597–1603.
42. Simon, P.W., 1997. Hort. Sci. 32, 12–13.
43. Porfirova, S., Bergmüller, E., Tropf, S., et al., 2002. PNAS 99, 12495–12500.
44. Raschke, M., Boycheva, S., Crèvecoeur, M., et al., 2011. Plant J. 66, 414–432.
45. Raclaru, M. Gruber, J., Kumar, R., et al., 2006. Mol. Breed. 18, 93–107.
46. Hjortmo, S., Patring, J., Jastrebova, J., et al,. 2008. Int. J. Food Microbiol. 127, 32–36.
47. Ye, X. Al-Babili, S., Klöti, A., et al., 2000. Science 287, 303–305; Paine, J.A., Shipton, C.A., Chaggar, S., et al., 2005. Nat. Biotechnol. 23, 482–487.
48. Nunes, A.C.S., Kalkmann, D.C., Aragão, F.J.L., 2009. Transgenic Res. 18, 661–667.
49. Storozhenko, S., De Brouwer, V., Volkaert, M., et al., 2007. Nat. Biotechnol. 25, S212–S240.

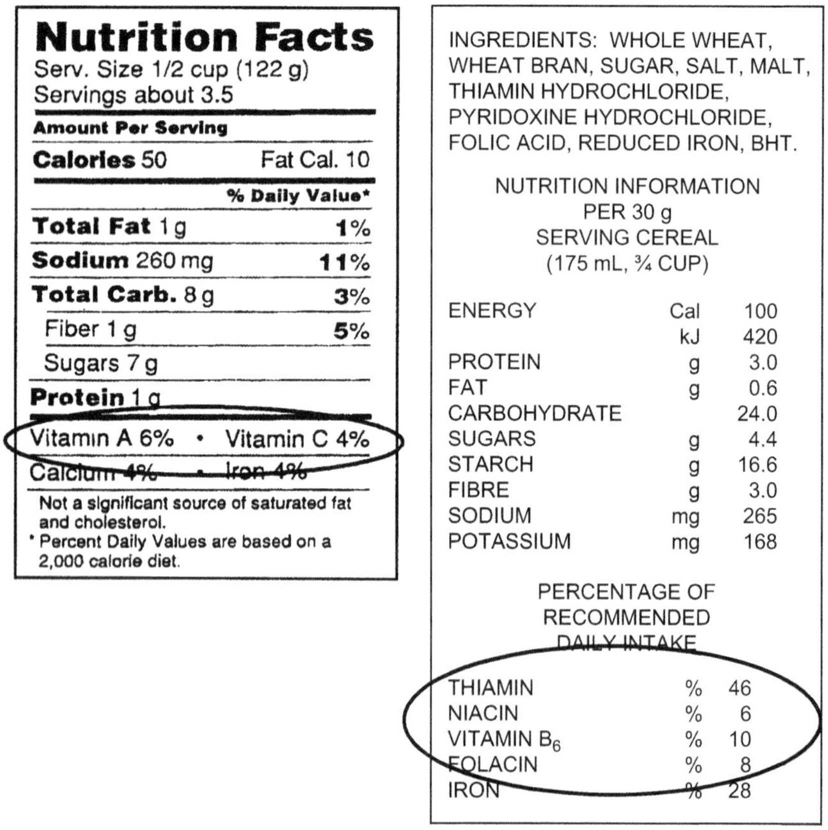

FIGURE 20.9 Nutrition information food labels, United States (left) and Canada (right). Note vitamin contents (*circled*).

products of genetic engineering are subject to regulatory legislation in each country. This has delayed the release of these enhanced varieties for use by farmers and, thus, to be available to consumers.[50]

6. VITAMIN LABELING OF FOODS

The labeling of nutrient contents of foods is a relatively new practice, having been instituted in the United States in 1972. The US regulations were respecified by the **Nutrition Labeling and Education Act** (**NLEA**) of 1990, the purpose of which was to provide, through a consistent food label format, useful information to consumers about the foods they eat in the context of their daily diets. The nutrition labeling of foods has the potential to influence consumer food use choices to the extent that the label information is accessible and can be acquired, processed, and used.

This US program involves compulsory labeling for most prepared and packaged foods.[51] It encourages voluntary labeling, either for individual products or at the point of purchase; for the most frequently consumed fresh fruits,[52] vegetables,[53] or seafood,[54] and at the point of purchase for fresh poultry and meats; and for prepared foods served in restaurants. In addition to information about the name of the product and its manufacturer and the measure/count of food contents, the act requires the food label to carry information about the ingredients, serving size and number of servings, and quantities of specified food components and nutrients (Fig. 20.9). Vitamin and mineral content information must be presented in comparison with a standard, the Reference Daily Intakes (RDIs) (Table 20.10).[55] For the information they present, nutrition labels draw on the USDA National Nutrient Data Bank, or an alternative data bank developed by the Produce Marketing Association.

50. Potrykus, I., 2010. New Biotechnol. 27, 466–472.

51. The act excludes foods containing few nutrients (e.g., plain coffee, tea, spices); foods produced by small businesses; and foods prepared and served by the same establishment.

52. Bananas, apples, watermelons, oranges, cantaloupe, grapes, grapefruits, strawberries, peaches, pears, nectarines, honeydew melons, plums, avocados, lemons, pineapples, tangerines, cherries, kiwi fruits, and limes.

53. Potatoes, iceberg lettuce, tomatoes, onions, carrots, celery, corn, broccoli, cucumbers, bell peppers, leaf lettuce, sweet potatoes, mushrooms, green onions, green beans, summer squash, and asparagus.

54. Shrimp, cod, pollock, catfish, scallops, salmon, flounder, sole, oysters, orange roughy, mackerel, ocean perch, rockfish, whiting, clams, haddock, blue crabs, rainbow trout, halibut, and lobster.

55. Most labels use the RDIs developed for adults and children 4 years of age or older; foods targeted to a certain age group must use the RDI developed for that group (see Chapter 5 for a discussion of DRIs).

TABLE 20.10 US RDAs (Recommended Dietary Allowances) Used in Food Labeling

Nutrient	Amount
Protein[a]	50 g
Minerals	
Calcium	1000 mg
Iron	18 mg
Iodine	150 μg
Copper	2 mg
Vitamins	
Vitamin A	5000 IU
Vitamin D	400 IU
Vitamin E	30 IU
Vitamin C	60 mg
Thiamin	1.5 mg
Riboflavin	1.7 mg
Niacin	20 mg
Vitamin B_6	2 mg
Pantothenic acid	10 mg
Biotin	300 μg
Folate	400 μg

The NLEA requires that information about vitamin A and vitamin C be carried on all food labels. It makes optional the disclosure of contents of other nutrients including vitamins for which RDAs (Recommended Dietary Allowances) have been established. In all cases, information must be presented according to the specified format.

7. VITAMINS IN HUMAN DIETS

Trends in US food intake patterns. Historical records of the American food supply show increases in the amounts of most of the vitamins available for consumption (Table 20.11).[56] Whether such increases have been reflected in the actual intakes of vitamins, or whether they have been distributed equitably across the American population is not indicated by such gross evaluations of the food supply. It is clear that people with low incomes tend to consume less food, although their food tends to have greater nutritional value per calorie than that consumed by people with greater incomes. While differences in diet quality due to income status appear to be small on average, variation in nutrient intake within groups of individuals appears to be very large. On average, at least, the vitamin intakes of Americans would appear to be generally adequate (Table 21.8). However, studies indicate that the vitamin intakes of many Americans may not meet the RDAs.

Despite an emerging picture of health benefits of diets richer in fruits and vegetables, surveys have shown that the regular intakes of fruits and vegetables of many Americans continue to fall short of the 5-A-Day goals. During the last decade, these intakes have increased by nearly 29% for vegetables and 38% for noncitrus fruits; however, the list of most frequently consumed fruits and vegetables continues to be short, with lowest consumption observed among lower socioeconomic groups and among individuals unaware of the health benefits attached to fruits and vegetables.

Vitamin Intakes From Foods

Food intake patterns are determined by many factors (e.g., tradition, taste, access, cost, and ease of preparation) but seldom nutrient content. In addition, patterns of food intake change. The most current estimates indicate that most Americans obtain most of their vitamins from their foods (Table 20.12); however, the intakes of vitamins A, D, E, K, B_6, B_{12}, thiamin, riboflavin, niacin, and folate were found to be generally lower among individuals living under 131% of poverty compared to other economic groups.[57] Individuals consuming strict vegetarian diets will not obtain adequate amounts of vitamins D and B_{12} through food sources alone.[58] Foods of both plant and animal origin provide vitamins in mixed diets for humans (Fig. 20.10; Table 20.13):

- **Meats and meat products**—Generally excellent sources of thiamin, riboflavin, niacin, pyridoxine, and vitamin B_{12}. Liver (including that from poultry or fish) is a very good source of vitamins A, D, E, and B_{12}, as well as folacin. Eggs are good sources of biotin. Animal products, however, are generally not good sources of vitamin C, K (except pork liver), or folate.
- **Beans, peas and lentils**—Generally good sources of thiamin, riboflavin, niacin, vitamin B_6, biotin, pantothenic acid, and folate.
- **Milk products**—Important sources of vitamins A and C, thiamin, riboflavin,[59] pyridoxine, and vitamin B_{12}. Because milk is widely enriched with irradiated ergosterol (vitamin D_2), it is also an important source of vitamin D.[60]

56. These result from increases in the availability of vegetables (+26%), fruits (+22%), grains (+44%), added fats and oils (+56%), and meats (+13%); and decreases in milk (−34%), eggs (−19%) over that same period of time (1970–2007) (Barnard, N.D., 2010. Am. J. Clin. Nutr. 91, 1530S–1536S).

57. USDA, Agricultural Research Service, 2010. What We Eat in American, NHANES 2007–2008 www.ars.usda.gov/ba/bhnrc/fsrg.

58. As well as calcium and long chain n-3 fatty acids.

59. In the United States, milk products supply an estimated 40% of the required riboflavin.

60. This practice has practically eliminated rickets in countries that use it.

TABLE 20.11 Vitamins Available for Consumption (per Person per Day) by Americans

Vitamin	1909–19	1920–29	1930–39	1940–49	1950–59	1960–69	1970–79	1980–89	1990–99	2000	2005
Vitamin A (IU)	1040	1090	1070	1210	1140	1150	1260	1230	1270	1260	1030
Vitamin E (mg)[a]	7.7	8.5	9.2	10.3	10.6	11.7	13.9	15.6	16.8	20.0	21.4
Vitamin C (mg)	95	100	104	112	98	93	112	119	127	130	115
Thiamin (mg)	1.5	1.5	1.4	1.9	1.8	1.9	2.3	2.6	3.0	3.0	2.9
Riboflavin (mg)	1.8	1.8	1.8	2.3	2.3	2.2	2.5	2.8	2.9	2.9	2.8
Niacin (mg)	18	17	16	20	20	20	25	29	32	33	33
Vitamin B_6 (mg)	2.1	2.0	1.9	2.0	1.8	1.8	2.0	2.2	2.4	2.5	2.5
Folate (μg)	309	305	309	325	297	284	326	356	449	706	682
Vitamin B_{12} (μg)	7.8	7.6	7.2	8.6	8.6	8.9	8.9	8.1	7.9	8.2	8.5

[a] *α-Tocopherol equivalents.*
Hiza, H.A.B., Bente, L., Fungwe, T., 2008. "Nutrient Content of the U.S. Food Supply", Home Economics Rept. 58, USDA, Washington, 72 pp.

TABLE 20.12 Average Daily Intakes of Vitamins From Foods by Americans, by Percentage of Usual Intake

Vitamin	Average Intake From Foods	% Consumed in				% Consumed Away From Home
		Breakfast	Lunch	Dinner	Snacks	
Vitamin A (μg)	607 ± 15	29	21	32	18	27
Vitamin D (μg)	4.6 ± 0.1	36	18	28	18	23
Vitamin E (mg)	7.2 ± 0.2	17	24	34	25	35
Vitamin K (mg)	88.9 ± 4.2	8	29	52	11	38
Vitamin C (mg)	84.2 ± 3.5	22	20	30	28	27
Thiamin (mg)	1.59 ± 0.03	25	24	34	17	31
Riboflavin (mg)	2.16 ± 0.04	30	20	28	22	30
Niacin (mg)	23.9 ± 0.34	19	25	39	17	34
Vitamin B_6 (mg)	1.91 ± 0.04	24	22	35	19	31
Pantothenic acid (mg)	–	–	–	–	–	–
Folate (μg)	527 ± 10	29	22	42	17	29
Vitamin B_{12} (μg)	5.19 ± 0.12	27	23	34	16	31

US Department of Agriculture, Agricultural Research Service, 2010. What We Eat in American, NHANES 2007–2008. www.ars.usda.gov/ba/bhnrc/fsrg.

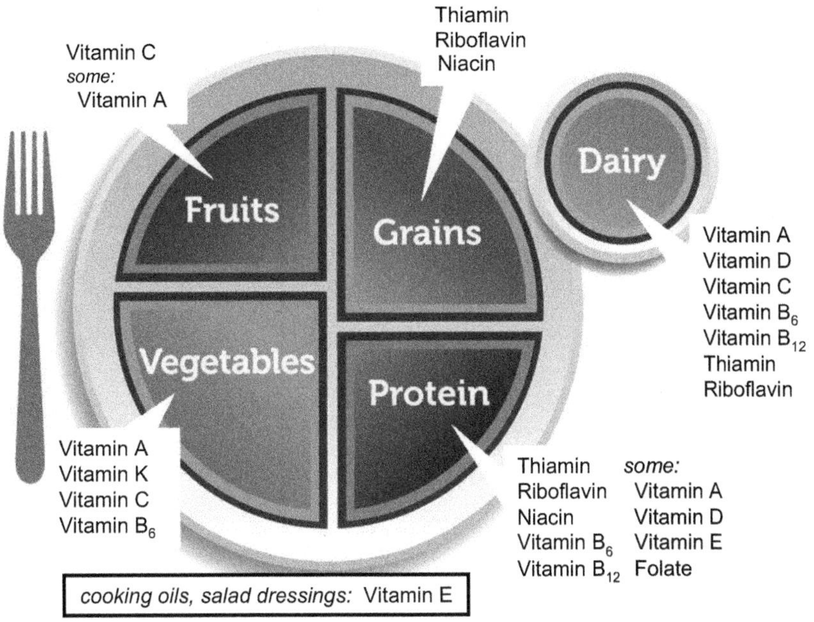

FIGURE 20.10 Vitamins provided by the major food groups, as depicted in ChooseMyPlate.gov.

- **Vegetables**—Generally good sources of vitamins A, K, and C and pyridoxine.
- **Fruits**—Generally good sources of vitamin C; some (e.g., mangoes) are also good sources of vitamin A.
- **Grain products**—Generally good sources of thiamin, riboflavin, and niacin.
- **Plant oils**—Generally good sources of vitamin E. Red palm oil is a particularly good source of vitamin A.

Dinner, typically the largest meal of the day, tends to be the most important for providing vitamins, but breakfast tends to be the most important in providing vitamin D likely due to the consumption of vitamin D-fortified milk. Snacking, now practiced by 97% of Americans,[61] provided nearly one-fifth of vitamin intake. An even greater amount, some 30%, was obtained from meals consumed away from the home, i.e., prepared by others. Studies show that Americans, on average, fall short of the daily recommendations for fruits (by 132%), vegetables (by 31%), whole grains (by 248%), and dairy products (by 66%).[62,63]

61. Piernas, C., Popkin, B.M., 2010. J. Nutr. 140, 325–332.

62. For an individual consuming 2000cal/day, the USDA Food Guide recommends the following daily servings: fruit, 2 cups; vegetables, 2.5 cups; whole grains, 3 oz.; dairy, 3 cups.

63. In fact, it is not clear whether the US agriculture could meet those recommended needs. In 2006, the USDA estimated that doing so would require doubling the US fruit production; increasing the production of nonstarchy vegetables by ~430% while reducing the production of starchy vegetables by 35%; and increasing dairy production by 66%. In contrast, increasing whole grain consumption at the expense of milled wheat would reduce total grain demand by 27% (Buxby, J.C., Wells, H.F., Vocke, G., 2006. Economic Res. Rept. No. 31, 29 pp.).

Vitamins in Breast Milk and Formula Foods

The vitamin contents of foods that are intended for use as the main or sole components of diets (e.g., human milk, infant formulas, and parenteral feeding solutions) are particularly important determinants of the vitamin status of individuals consuming them. The vitamin contents of human milk can vary, most being reduced under conditions of vitamin deprivation and responding to supplementation. For that reason, the contents of some vitamins in breast milk can vary,[64] even in well-nourished women. In general, the concentrations of all vitamins (except vitamin B_{12}) in human milk tend to increase during lactation. In comparison with cow's milk, human milk contains more of vitamins A, E, C, and niacin, but less vitamin K, thiamin, riboflavin, and pyridoxine (Table 20.14). Reference values for the contents of several vitamins in the breast milk of well-nourished women have been set for the purposes of establishing adequate intake values: thiamin, 0.21 mg/L; riboflavin, 0.35 mg/L; vitamin B_6, 0.13 mg/L, vitamin B_{12}, 0.42 µg/L; and choline, 160 mg/L.[65]

Because infant formulas and parenteral feeding solutions are carefully prepared and quality-controlled products, each is formulated largely from purified or partially refined ingredients to contain known amounts of the vitamins. For parenteral feeding solutions, however, some

64. e.g., Riboflavin concentrations in the breast milk of well-nourished women has been found in the range of 180–1800 µg/L (Roughead, Z.K., McCormick, D.B., 1990. Am. J. Clin. Nutr. 52, 854–857).

65. These values compare to those observed for breast milk fed to deficient infants: thiamin, 0.16 mg/L; riboflavin, 0.21 mg/L; vitamin B_6, 0.10 mg/L, vitamin B_{12}, <0.05 µg/L; choline, 90 mg/L (Allen, L.H., 2012. Adv. Nutr. 3, 362–369).

TABLE 20.13 Contributions of Food Groups to the Vitamin Contents of Mixed Diets for Americans

	Vitamin A	Vitamin E	Vitamin C	Thiamin	Riboflavin	Niacin	Vitamin B_6	Folate	Vitamin B_{12}
Per Capita Daily Availability									
	1030 μg	21.4 mg	115 mg	2.9 mg	2.8 mg	33 mg	2.5 mg	682 μg	8.5 μg
% Contributions by Food Group									
Vegetables	**27.1**[a]	5.9	**48.3**	8.6	5.7	9.8	20.1	12.3	0
Legumes, Nuts	0	4.8	0.1	4.5	1.6	4.0	3.6	9.3	0
Fruits	2.5	2.9	**40.5**	3.6	2.3	2.0	9.1	5.9	0
Grain products	5.3	3.9	4.7	**59.3**	**38.5**	**42.3**	18.3	**61.2**[b]	0.1
Meats, fish	**33.2**	3.8	2.3	18.0	17.9	**37.7**	**38.1**	3.8	**77.4**
Milk products	15.7	1.7	2.5	4.3	**25.0**	1.1	6.8	3.3	18.1
Eggs	6.6	1.8	0	0.7	6.3	0.1	1.9	2.5	4.2
Fats, oils	7.4	**74.6**	0	0	0.1	0	0	0	0.1
Other	2.2	0.6	1.7	0.9	1.9	3.2	1.9	1.7	0

[a]*Major contributing foods indicated in bold.*
[b]*Reflects mandatory fortification of wheat flour.*
Hiza, H.A.B., Bente, L., Fungwe, T., 2008. USDA Home Economics Rep. No. 58, 72 pp.

TABLE 20.14 Vitamin Contents of Human and Cow's Milk

Vitamin	Human Milk	Cow's Milk
Vitamin A (retinol) (mg/L)	0.60	0.31
Vitamin D_3 (μg/L)	0.3	0.2
Vitamin E (mg/L)	3.5	0.9
Vitamin K (mg/L)	0.15	0.6
Ascorbic acid (mg/L)	38	20
Thiamin (mg/L)	0.16	0.40
Riboflavin (mg/L)	0.30	1.90
Niacin (mg/L)	2.3	0.8
Pyridoxine (mg/L)	0.06	0.40
Biotin (μg/L)	7.6	20
Pantothenic acid (mg/L)	0.26	0.36
Folate (mg/L)	0.05	0.05
Vitamin B_{12} (μg/L)	<0.1	3

Porter, J.W.G., 1978. Proc. Nutr. Soc. 37, 225–230.

problems related to vitamin nutrition have occurred. One problem involved biotin, which was not added to such solutions before 1981 in the belief that intestinal synthesis of the vitamin was adequate for all patients except children with inborn metabolic errors or individuals ingesting large amounts of raw egg white. When it was found that children supported by total parenteral nutrition (TPN) frequently suffered from gastrointestinal abnormalities that responded to biotin,[66] the vitamin was added to TPN solutions.[67] Another problem with parenteral feeding solutions has been the loss of fat-soluble vitamins and riboflavin either by absorption to the plastic bags and tubing most frequently used, or by decomposition on exposure to light. Such effects can reduce the delivery of vitamins A, D, and E to the patient by two-thirds and to result in the loss of one-third of the riboflavin.

8. VITAMIN SUPPLEMENTATION

More than half of the US population consumes dietary supplements,[68] multivitamin/mineral supplements are the most popular and are used by 40% of American adults and more than 30% of children.[69] Dietary supplement use is greater among women than men and increases with age such that nearly three-quarters of Americans over 70 years. take a supplement. Studies have shown that supplement users tend to be more health conscious and have better diets, more education, and higher incomes than the general population. Supplement use is greater among health professionals, vegetarians, the elderly, and readers of health-focused magazines. Vitamin supplement use is most frequent among individuals who believe that diet affects disease, nondrinkers and lighter drinkers of alcohol, former smokers and individuals who never smoked, and individuals in the lowest three quartiles of BMI. Users of vitamin supplements have markedly greater vitamin intakes than nonusers (Table 20.15). The median levels of supplement use by Americans is one to two times the RDA for vitamins A, D, and B_6; niacin; pantothenic acid; and folate; and greater than twice the RDA for vitamins E, C, and B_{12}; thiamin; and riboflavin.

Studies in the US have shown that regular users of multivitamin/mineral supplements are more likely to have adequate vitamin intakes (Table 20.16), less likely to have suboptimal blood nutrient concentrations and more likely to have optimal levels of biomarkers of chronic disease.[70] However, it is not clear whether those users are actually at reduced chronic disease risk as a result of that practice. This is, in part, because users have generally healthy lifestyles, which independently reduce their low risk. Systematic reviews of the relevant clinical literature are limited by the paucity of rigorous studies conducted to date; most available results do not provide strong evidence for beneficial health effects of multivitamin/mineral supplements for most people.[71] However, there have been indications of some benefits: reduced fracture risks in postmenopausal women, marginal increases in cognitive performance in children.[72]

Interventions with multivitamin supplements have been useful in undernourished populations. For example, antenatal multiple-micronutrient supplementation reduced combined fetal loss and neonatal deaths by 11% and increased birth weight by 14% in Indonesia (particularly in undernourished and/or anemic mothers), and increased birth weight by in Nepal.[73] The use of vitamin supplements among peoples in developed countries has become great enough to make this means a significant contributor to the vitamin nutriture of

66. These patients appeared to have had altered gut microbiota as a result of antibiotic treatment.

67. Although there are no RDAs for biotin, it has been suggested that biotin supplements be given to individuals being fed parenterally (infants, 30 μg/kg/day; adults, 5 μg/kg/day). These levels are consistent with the provisional recommendations (infants, 10 μg/day; adults, 30–100 μg/day) of the National Research Council Food Nutrition Board (1989).

68. i.e., 53.4% of respondents in the NHANES 2003–06 (Bailey, R.L., Gahche, J.J., Lentino, C.V., et al., 2011. J. Nutr. 141, 261–266).

69. Rock, C.L., 2007. Am. J. Clin. Nutr. 85, 277S–279S; Gahche, J., Bailey, R., Burt, V., et al., 2011. NCHS Data Brief No. 61; Picciano, M.F., Dwyer, J.T., Radimer, K.L., et al., 2007. Arch. Pediatr. Adolesc. Med. 161, 978–985.

70. Block, G., Jensen, C.D., Norkus, E.P., et al., 2007. Nutr. J. 6, 30.

71. NIH State-of-the-Science Panel, 2007. Am. J. Clin. Nutr. 85, S257–S264; McCormick, D.B., 2010. Nutr. Rev. 68, 207–213.

72. Eilander, A., Gera, T., Sachdev, H.S., et al., 2010. Am. J. Clin. Nutr. 91, 115–130.

73. The Supplementation with Multiple Micronutrients Intervention Trial (SUMMIT) Study Group, 2008. Lancet 371, 215–222; Vaidya, A., Saville, N., Shresthra, B.P., et al., 2008. Lancet 371, 492–469.

TABLE 20.15 Contributions of Commonly Used Dietary Supplements to Vitamin Intakes of Americans

Vitamin	Sex	EAR[a]	Intake From Food	Intake From Supplement[a]	Total Intake
Vitamin A	Male	625 μg	656 μg	1050 μg	1706 μg
	Female	500 μg	564 μg	1050 μg	1614 μg
Vitamin E	Male	12 μg	8.2 μg	13.5 μg	21.7 μg
	Female	12 μg	6.3 μg	13.5 μg	19.8 μg
Vitamin C	Male	75 mg	105 mg	60 mg	165 mg
	Female	60 mg	84 mg	60 mg	144 mg

[a]*Vitamin content based on that of the most commonly consumed multivitamin/mineral supplement in the NHANES 2001–02.*
Dwyer, J.T., Holden, J., Andrews, K., et al., 2007. Anal. Bioanal. Chem. 389, 37–46.

TABLE 20.16 Effect of Multivitamin (MV) Use on Prevalence of Adequate Vitamin Intakes of Subjects in the Hawaii–Los Angeles Multiethnic Cohort

	Men			Women		
	Nonusers	Users		Nonusers	Users	
Vitamin	From Food	From Food	Total (Food + MV)	From Food	From Food	Total (Food + MV)
Vitamin A	59 ± 42[a]	61 ± 41	87 ± 29	69 ± 39	73 ± 39	89 ± 28
Vitamin E	27 ± 41	28 ± 41	68 ± 43	22 ± 38	23 ± 39	60 ± 45
Vitamin C	72 ± 42	76 ± 40	89 ± 29	82 ± 37	86 ± 33	93 ± 25
Thiamin	82 ± 35	84 ± 33	94 ± 23	79 ± 37	82 ± 35	92 ± 25
Riboflavin	87 ± 30	89 ± 28	95 ± 19	77 ± 20	79 ± 19	90 ± 16
Niacin	89 ± 27	90 ± 25	96 ± 18	84 ± 32	86 ± 30	94 ± 22
Vitamin B_6	79 ± 37	81 ± 36	93 ± 24	73 ± 41	77 ± 39	90 ± 28
Folate	93 ± 23	94 ± 21	97 ± 15	88 ± 30	90 ± 27	95 ± 20
Vitamin B_{12}	90 ± 28	90 ± 27	96 ± 18	83 ± 18	85 ± 34	94 ± 23

[a]*Mean ± S.E.*
Murphy, S.P., White, K.K., Park, S.Y., et al., 2007. Am. J. Clin. Nutr. 85, S280–S284.

those populations. Still, for most of the poor in those countries, access to multivitamin supplements remains limited by costs, making food-based approaches more sustainable for addressing prevalent multimicronutrient shortages.

Retail sales of vitamins and nutritional supplements in the United States are expected to exceed $36 billion by 2017, a doubling since 2001. The drivers of this trend have been identified as increasing awareness of consumers, increasing demand by the food industry facilitated by the GRAS[74] status of vitamins, increasing consumer demand for health-beneficial products, and increasing demand by livestock producers for performance enhancers. The United States is the largest consumer of vitamins, 40% of which is in multivitamin supplements most (>70%) of which are sold in drug stores, supermarkets, and health food stores. In the United States, dietary supplements are regulated by the FDA under the Dietary Supplement Health and Education Act of 1994.[75]

74. Generally Recognized As Safe.

75. This legislation charged the FDA to establish a framework for assuring safety, outline guidelines for literature displayed where supplements are sold, provide for use of claims and nutritional support statements, require ingredient and nutritional labeling, and establish good manufacturing practice regulations. The law changed previous legislation in that dietary supplements are no longer subject to the premarket safety evaluations required of other food ingredients.

Guidelines for Supplement Use

Healthy individuals can and should obtain adequate amounts of all nutrients, including vitamins, from a well-balanced diet based on different foods of good quality. Such an approach minimizes the risks of deficiencies as well as excesses of all nutrients. It also acknowledges that foods can provide health benefits that have yet to be fully elucidated.

For individuals with varied, balanced diets, the benefit of taking vitamin supplements is doubtful. However, certain circumstances may warrant the use of vitamin supplements:

- Folate for women, who may conceive, are pregnant or are lactating;[76]
- Multivitamins for individuals with very low caloric intakes (such that their consumption of total food is insufficient to provide all nutrients);
- Vitamin B_{12} for strict vegetarians, individuals with gastric achlorhydria, gastric resection, and most people over 50 years of age;
- Vitamin D for people living in the northern latitudes;
- Vitamin K (single dose) for newborn infants to prevent abnormal bleeding;
- Patients with diseases or medications that interfere with vitamin utilization.

9. VITAMINS IN LIVESTOCK FEEDING

Vitamins in Animal Feeds

The economic considerations in feeding livestock generally dictate the use of a relatively small number of feedstuffs with few (if any) day-to-day changes in diet formulation.[77] In livestock production, the continued use of the same or very similar diets has resulted in the empirical development of knowledge about the vitamin contents of feedstuffs (Table 20.17).

Unlike human diets, formulated diets for livestock generally do not provide adequate amounts of vitamins unless they are supplemented with either certain vitamin-rich feedstuffs or purified vitamins. In general, the relative vitamin adequacy of unsupplemented animal feeds depends on the relative complexity (the number of feedstuffs used in the mixture) of the diet. The vitamin contents of simple rations tend to be less than those of complex ones (Table 20.18). For example, complex diets such as those used for feeding poultry or swine in the 1950s[78] would be expected to contain in their constituent feedstuffs more than an adequate amount of vitamin B_{12}, and adequate (or nearly so) amounts of vitamin K, vitamin E, thiamin, riboflavin, niacin, pyridoxine, pantothenic acid, folate, and choline. In contrast, the simpler rations (based almost exclusively on corn and soybean meal) that are used today contain lower amounts, if any, of the more costly vitamin-rich feedstuffs previously used. Such simple rations can be expected to contain in constituent feedstuffs adequate levels only of vitamin E, thiamin, pyridoxine, and biotin (Table 20.19). The availability of stable, biologically available, and economical vitamins facilitated this change in complexity of animal feeds by replacing with inexpensive mixtures of vitamins the more costly vitamin-rich feedstuffs used previously.

Vitamins likely to be limiting in (unsupplemented) nonruminant diets:

- Vitamin A.
- Vitamin E
- Niacin
- Pantothenic acid
- Choline (chicks)
- *If raised indoors*—vitamin D
- *If raised on slatted or wire floors*—vitamin K, vitamin B_{12}

Losses of Vitamins From Feedstuffs and Finished Feeds

The vitamin contents of feedstuffs and finished feeds[79] are subject to destruction in ways very similar to those of foods. The storage losses that can occur in particular feedstuffs are dependent on the conditions of temperature and moisture during storage; heat and humidity enhance oxidation reactions of several of the vitamins (vitamins A and E, thiamin, riboflavin, and biotin). Vitamin losses are, therefore, minimized by drying feedstuffs quickly and storing them dry in weather-proof bins. Where the drying of a feedstuff is slow[80] or incomplete,[81] or where leaky bins are used for its storage, vitamin losses are greatest.

Vitamin losses from finished feeds are usually greater than those of individual feedstuffs. Finished feeds are supplemented with essential trace elements, some of which (Cu^{2+}, Fe^{3+}) can act as catalytic centers of oxidation reactions leading to vitamin destruction. Such effects are particularly important in high-energy feeds (e.g., broiler diets), which generally contain significant amounts of polyunsaturated fats. It is a common practice in many countries to

76. Pregnant and lactating women may also need supplements of iron and calcium.

77. For example, broiler chicks are typically fed the same diet from hatching to 3 weeks of age, and laying hens may be fed the same diet for 20 weeks before the formula is changed.

78. Those complex diets contained, in addition to a major grain and soybean meal, small amounts of alfalfa meal, corn distillers' dried solubles, fish meal, and meat-and-bone meal.

79. Complete, blended, ready-to-feed ratios.

80. For example, sun drying enhances the destruction of vitamin E in corn (although sun curing of cut hay is essential to provide vitamin D activity).

81. Where moisture is not reduced to less than about 15%.

TABLE 20.17 Feedstuffs Containing Significant Amounts of Vitamins

Vitamin A	Vitamin D	Vitamin E	Vitamin K	Vitamin C	Thiamin	Riboflavin	Niacin	Vitamin B_6	Biotin	Pantothenic Acid	Folate	Vitamin B_{12}	Choline
None[a]	None[a]	Alfalfa, dehydrated	Alfalfa, dehydrated	None[a,b]	None[a]	Dried skim milk	Barley	Sunflower seed meal	Corn germ meal	Molasses	Dried brewers' grains	Dried fish solubles	Liver/glandular meal
	Alfalfa, sun-cured		Alfalfa, sun-cured			Peanut meal	Cottonseed meal	Sesame meal	Brewers' yeast	Rice polishings	Alfalfa, dehydrated	Liver/glandular meal	Dried fish solubles
	Wheat germ meal					Brewers' yeast	Dried fish solubles	Meat/bone meal	Molasses	Sunflower seed meal	Brewers' yeast	Hydrolyzed feathers	Soybean meal
	Corn germ meal					Dried buttermilk	Rice bran, polishings		Torula yeast	Peanut meal	Soybean meal	Fish meals	Corn distillers' solubles
	Stabilized vegetable oils					Dried whey	Wheat bran		Hydrolyzed feathers	Torula yeast	Torula yeast	Crab meal	
						Torula yeast	Corn gluten feed		Safflower meal	Liver/glandular meal	Meat/bone meal	Dried skim milk	
						Corn distillers' solubles	Fish meals			Brewers' yeast	Corn distillers' solubles	Dried butter milk	
						Liver/glandular meal	Peanut meal				Alfalfa, sun-cured	Meat/bone meal	
							Torula yeast						
							Corn gluten meal						
							Corn distillers' solubles						
							Liver/glandular meal						
							Sunflower seed meal						
							Brewers' yeast						

[a] *Instability of the vitamin in most feedstuffs renders few, if any, predictable sources of appreciable amounts of it.*
[b] *Not required by livestock species.*

TABLE 20.18 Vitamins Provided by Constituent Feedstuffs in Older (Complex) and Modern (Simple) Chick Starter Diets

	1942 Diet[a] (%)	Modern Diet (%)
Ingredients		
Cornmeal	27.5	65.61
Oats	10.0	
Wheat bran	20.0	
Wheat middlings	10.0	
Soybean meal, 49% protein	10.0	19.08
Meat and bone meal	10.0	4.78
Poultry by-product meal		7.00
Dried whey	5.0	
Dehydrated alfalfa meal	5.0	
Blended fat		3.18
Limestone	2.0	
Salt	0.5	0.25
D,L-methionine (98%)		0.10
Trace minerals	+[b]	+[c]
Vitamins	+[d]	+[e]
Vitamins Provided by Feedstuffs		
Vitamin A (IU)	6000 (400)[f]	1360 (91)[f,g]
Vitamin E (IU)	27 (270)	20.5 (205)[g]
Vitamin K (mg)	0.73 (146)	0 (0)[g]
Thiamin (mg)	4.7 (261)	3.1 (172)
Riboflavin (mg)	5.4 (150)	2.3 (64)[g]
Niacin (mg)	69.9 (259)	28.3 (105)[g]
Pyridoxine (mg)	5.7 (190)	5.3 (177)
Biotin (g)	208 (139)	141 (94)
Pantothenic acid (mg)	15.7 (157)	7.4 (74)[g]
Folate (mg)	0.81 (145)	0.32 (58)
Vitamin B_{12} (g)	6.5 (72)	21.0 (233)[g]
Choline (mg)	1115 (86)	1395 (107)[g]

[a]*This was a state-of-the-art diet for starting chicks at Cornell University in 1942.*
[b]*$MnSO_4$, 125 mg/kg.*
[c]*Provides per kg of diet: ZnO, 66 mg; $MnSO_4$, 220 mg; Na_2SeO_3, 220 g.*
[d]*Vitamin D_3, 790 IU/kg.*
[e]*Provides per kg of diet: vitamin A, 4400 IU; vitamin D_3, 2200 IU; vitamin E, 5.5 IU; vitamin K_3, 2 mg; riboflavin, 4 mg; nicotinic acid, 33 mg; pantothenic acid, 11 mg; vitamin B_{12}, 1 g; choline, 220 mg.*
[f]*Numbers in parentheses give amounts of each vitamin as a percentage of current (1984) recommendations of the U.S. National Research Council.*
[g]*Included in the vitamin–mineral premix.*

TABLE 20.19 Insufficient Amounts of Vitamins in Turkey Starter Diet Feedstuffs

	Level From Feedstuffs, % NRC Requirement	
Vitamin	**Simple Feed[a]**	**Complex Feed[b]**
Vitamin A	20	40
Vitamin E	130	130
Thiamin	170	160
Riboflavin	60	90
Niacin	30	60
Vitamin B_6	130	100
Pantothenic acid	90	110
Biotin	120	140
Folate	50	50
Vitamin B_{12}	0	74
Choline	90	100

[a]*Corn, 40.5%; soybean meal, 51.2%; animal fat, 4%; $CaHPO_4$, 3%; limestone, 0.8%; salt, 0.3%; methionine, 0.15%; trace minerals, 0.05%.*
[b]*Milo, 20.5%; wheat, 20%; soybean meal, 33.9%; poultry meal, 6%; animal fat, 5%; meat and bone meal, 5%; fish meal, 4%; alfalfa meal, 2%; distillers' grains and solubles, 2%; limestone, 0.7%; $CaHPO_4$, 0.5%; salt, 0.3%; methionine, 0.13%; trace minerals, 0.05%.*
Anonymous, 1989. "Vitamin Nutrition for Poultry". Hoffman-La Roche, Inc., Nutley, New Jersey, pp.13–14.

compress many of these (and other) feeds into pellet form[82] by processes involving steam, heat, and pressure. Evidence suggests that pelleting can enhance the bioavailability of niacin and biotin, which occur in feedstuffs in bound forms; but it generally results in the destruction of vitamins A, D, E, K_3, C, and thiamin.

Vitamin Premixes for Animal Feeds

As purified sources of the vitamins have become available at low cost, it has become possible to use fewer feedstuffs in less complicated blends to produce diets of high quality that will support efficient and predictable animal performance. Thus, many feedstuffs formerly valuable as sources of vitamins (e.g., brewers' yeast, dried buttermilk, and green feeds[83]) are no longer economical to use in intensive animal management systems. This is most true in the economically developed parts of the world. In the developing world, such factors as the shortage of hard currency may make purified sources of vitamins too expensive to use in animal diets, making natural sources of the vitamins more valuable. Under such circumstances, it is prudent to exploit a wide variety of local feedstuffs, food wastes, and food by-products in the formulation of animal feeds that are adequate in terms of vitamins as well as all other known nutrients.

The use of purified vitamins as supplements to animal feeds has increased the economy of animal feeding by obviating the need to include relatively expensive vitamin-rich feedstuffs in favor of lower-priced feedstuffs that are lower in vitamin content but provide useful energy and protein. In modern practice, the addition of vitamins to animal feeds is accomplished by preparing a mixture of the specific vitamins required with a suitable carrier[84] to ensure homogeneous distribution in the feed as it is mixed. Such a preparation is referred to as **vitamin premix** (Tables 20.20 and 20.21) and is handled in much the same way as other feedstuffs in the blending of animal feeds. Typically, vitamin premixes are formulated to be blended into diets at rates of 0.5–1.0%.[85]

Premixes generally also contain synthetic antioxidants [e.g., ethoxyquin, butylated hydroxytoluene

82. There are many reasons for pelleting finished feeds. Pelleting prevents demixing of the feed during handling. By increasing bulk density, it can improve the economy of feed handling and improve the consumption of bulky, low-density feeds. It can improve the handling of feeds that are otherwise dusty and reduce wastage at the feeder. The metabolizable energy values of some feedstuffs may be improved by the steam treatment used in pelleting (e.g., soybean meal with significant residual antitryptic activity).

83. e.g., Fresh cabbage, grass.

84. e.g., Soybean meal, finely ground corn or wheat, corn gluten meal, wheat middlings.

85. The cost of the vitamin premix is typically <2% of the total cost of most finished feeds. Of that amount, two-thirds of the vitamin cost is accounted for by vitamin E, niacin, vitamin A, and riboflavin (roughly in that order).

TABLE 20.20 Vitamins Generally Included in Vitamin Premixes for Livestock Diets

Vitamin	Poultry	Piglets	Hogs	Calves	Cattle
Vitamin A	+	+	+	+	+
Vitamin D_3	+	+	+	+	+
Vitamin E	+	+	+	+	+[a]
Vitamin K	+	+	+[a]		
Ascorbic acid	+[b]+	+			
Thiamin	+[a]	+[a]			
Riboflavin	+	+	+	+[a]	
Niacin	+	+			
Vitamin B_6	+[a]	+	+[a]	+[a]	
Pantothenic acid	+	+	+	+[a]	
Biotin	+[a]	+[a]	+[a]		
Folate					
Vitamin B_{12}	+	+	+[a]		
Choline	+	+			

[a]*Sometimes added.*
[b]*Added in situations of stress.*

(BHT)] to enhance vitamin stability during storage.[86] In many cases, trace minerals are included in **vitamin–mineral premixes.**[87] It is a standard practice in the formulation of vitamin premixes to use amounts of vitamins that, when added to the expected amounts intrinsic to the component feedstuffs, will provide a comfortable excess above those levels found experimentally to be required to prevent overt deficiency signs. This is done in view of the many potential causes of increased vitamin needs; owing to the low cost of vitamin supplementation, this approach is considered a kind of low-cost nutrition insurance, as vitamin premixes usually account for only 1–2% of the total cost of feeds for nonruminant livestock. It should be remembered, however, that purified vitamins may not always be cheap, particularly in developing countries. Under those circumstances, the appropriate way to assess the value of using vitamin supplements is to compare their market prices with the estimated loss of production realized by not supplementing feeds that can be economically produced using locally available feedstuffs.

Stabilities of Vitamins in Feeds

Vitamins tend to be less stable in vitamin–mineral premixes used for livestock feeds owing to the redox reactions catalyzed by trace elements and physical abrasion of protective coatings (Table 20.22). Vitamin premixes that contain choline chloride typically show accelerated losses of vitamin B_6, which reacts with choline. During the storage of finished feeds, the migration of moisture to the shady, relatively cool side of a feed bin can result in the development of pockets of relatively high moisture, which can enhance both the chemical degradation of vitamins as well as support the growth of vitamin-consuming fungi. Feeds that are pelleted or extruded are also exposed to friction, pressure, heat, and humidity, all of which enhance vitamin loss.

The chemical stabilities of some vitamins can be improved by using a more stable chemical form or formulation. For example, the calcium salt of pantothenic acid is more stable than the free acid form, and esters of vitamins A and E (retinyl acetate, tocopheryl acetate) are much more resistant to

86. Loss of vitamin A from poultry feeds stored at moderate temperatures (ca. 15% in 30 days) can be reduced (to ca. 10%) by the addition of an antioxidant. Under conditions of high temperature and high humidity, vitamin A losses from finished feeds can be much greater (80–95%). Maximal protection by antioxidants is expected under conditions in which vitamin oxidation is moderate, e.g., short-term feed storage in hot, humid environments.

87. Owing to the presence of mineral catalysts in oxidative reactions, the stabilities of oxidant-sensitive vitamins in compound premixes can be expected to be less than in premixes of the vitamins alone.

TABLE 20.21 Examples of Vitamin Premixes for Animal Feeds

	Units/1000 kg		
Vitamin	**Practical Diet[a] for Chicks[c]**	**Diet Semipurified Diet[b] for Chicks[c]**	**Semipurified Diet[b] for Rats[d]**
Vitamin A[e] (IU)	8,800,000	50,000	40,000,000
Vitamin D_3 (IU)	2,200,000	4,500,000	1,000,000
Vitamin E (IU)[f]	5,500	50,000	50,000
Menadione $NaHSO_3$ (g)	2.2	1.5	50
Thiamin HCl (g)	15	6	
Riboflavin (g)	4.4	15	6
Niacin (g)	33	50	30
Pyridoxine HCl (g)	6	7	
d-Calcium pantothenate (g)	11	20	16
Biotin (mg)	0.6	0.2	
Folic acid (g)	6	2	
Vitamin B_{12} (mg)	10	20	10
Choline chloride (g)	220	2000	+[g]
Minerals	+[h]	+[i]	+[i]
Other Ingredients			
Antioxidant[j] (g)	125	100	100
Carrier (g)	To weight[k]	To weight[l]	To weight[l]

[a]*Composed of nonpurified natural feedstuffs (e.g., corn, soybean meal).*
[b]*Composed of purified/partially purified ingredients (e.g., isolated soy protein, casein, sucrose, and starch).*
[c]*From Scott, M.L., Nesheim, M.C., Young, R.J., 1982. "Nutrition of the Chicken", third ed. M.L. Scott & Assoc., Ithaca, New York.*
[d]*AIN-76 diet*
[e]*As all-trans-retinyl palmitate.*
[f]*As all-rac-tocopheryl acetate.*
[g]*Added as 0.2% choline bitartrate.*
[h]*Includes 66 g of ZnO, 220 g of $MnSO_4$, and 220 mg of Na_2SeO_3.*
[i]*Includes $CaHPO_4{\cdot}2H_2O$, $CaCO_3$, KH_2PO_4, $NaHCO_3$, $KHCO_3$, KCl, NaCl, $MnSO_4{\cdot}H_2O$, $FeSO_4{\cdot}7H_2O$, $MgCO_3$, $MgSO_4$, KIO_3, $CuO_4{\cdot}5H_2O$, $ZnCO_3$, $CoCl_2$, $NaMoO_4{\cdot}2H_2O$, and/or Na_2SeO_3 in amounts appropriate for the composition of the particular diet.*
[j]*e.g., Ethoxyquin, BHT (butylated hydroxytoluene).*
[k]*Corn meal.*
[l]*Sucrose.*

oxidation than the free alcohol forms. Vitamin preparations can also be coated or encapsulated[88] in ways that exclude oxygen and/or moisture, thus rendering them more stable. They are often spray-dried, spray-congealed, or prepared as adsorbates to improve their handling characteristics. Owing to such approaches, purified vitamins added to foods have been found to be as stable and bioavailable, if not more so, than the forms of the vitamins intrinsic to foods.

10. CASE STUDY

Cindy Stacey stepped into the center elevator the Longworth Building and punched "3." She was a new staffer for US Congressman Carl Rep. Pomerantz (ND), now in his fourth term. The congressman had the distinction of being a popular urban democrat in an overwhelmingly conservative, republican, agricultural state. His background in agricultural insurance had landed him a seat on the House Agriculture Committee where he was now the ranking minority member. His chief of staff hired Cindy during the intersession; she was just back from the Peace Corps where she taught English in Bangladesh.

Cindy liked the congressman but was less sure about his chief of staff Campbell Hurst. Hurst's job was to make sure that Mr Pomerantz had the information he needed when he needed it, which required him to be on top of the Congressman's political agenda. Campbell saw his own future as tied to Mr Pomerantz's political success. He tried

88. Gelatin, edible fats, starches, and sugars are used for this purpose.

TABLE 20.22 Typical Stabilities of Vitamins in a Broiler Feed

Vitamin	% Retained Activity			
	Premix Storage (2 months)	Pelleting/ Conditioning (93°C, 1 min)	Feed Storage (2 weeks)	Cumulative
Vitamin A[a]	98	90	92	81
Vitamin D_3	98	93	93	85
Vitamin E[b]	99	97	98	94
Vitamin K[c]	92	65	85	51
Thiamin[d]	99	89	98	86
Riboflavin	99	89	97	85
Niacin	99	90	93	83
Vitamin B_6[e]	99	87	95	82
Biotin	99	89	95	84
Pantothenic acid[f]	99	89	98	86
Folate	99	89	98	86
Vitamin B_{12}	100	96	98	86

[a]*all-trans-Retinyl acetate.*
[b]*all-rac-α-Tocopheryl acetate.*
[c]*Menadione sodium bisulfite complex.*
[d]*Thiamin mononitrate.*
[e]*Pyridoxine hydrochloride.*
[f]*Calcium pantothenate.*
BASF Keeping Current, No. 9138, 1992.

to steer the congressman into the most politically safe decisions whenever he could.

Two days ago, Campbell had given Cindy her first major assignment: review the needs and opportunities in agriculture-related international markets. This was to be used for Mr Pomerantz's work on the new House Agriculture bill. Mr Pomerantz had asked her specifically to investigate opportunities to address "hidden hunger," as he was interested in redirecting funds presently spent on the US surplus commodities. Campbell had added that she should look at activities of the USDA Agricultural Research Service related to market opportunities for the US commodities. She doubted that Campbell would want to hear much about micronutrients, as he saw political liabilities in the use of USDA budgetary support to address nondomestic problems.

Cindy also knew that Campbell had never seen those problems. Nor had she, until going to Bangladesh. In villages outside of the port city of Chittagong, she had been shocked at the blindness and bone deformities, the goiter, and diarrhea. She had been amazed at the stunted kids and the bent-over old women. They all seemed to have become part of the landscape—as though their infirmities were expected features of the lives of the poor. These conditions, she had learned, were caused by eating rice and little else. She did not like the term "hidden hunger" for something that was so obvious; but she was glad it was something Mr Pomerantz wanted to address.

Cindy suspected, though she had never discussed the matter with him, that the congressman's interest related to the fact that his own daughter had cerebral palsy. One could not miss the passion with which he talked about "hidden hunger" as "preventable sources of disability." Others in the office dated his interest in this area to an official trip he had made to Mozambique in his second term. Campbell had not had such experiences; for him, "hidden hunger" was something abstract.

So, for the last 2 days, Cindy had lived in the Library of Congress. She had poured over everything she could find publications from UNICEF and FAO and scores of research papers in plant breeding and nutritional surveillance in Africa and south Asia. She had called one of her old professors and had talked to a friend at USDA. During her investigations, she had come across "Golden Rice," a transgenic rice into which genes for the biosynthesis of beta-carotene had been added from other species. Rice that could be a source of vitamin A. This rice, still in the regulatory process in several countries, was being promoted as the solution to global vitamin A deficiency by its developers at the International Rice Research Institute in the Philippines.

Golden Rice fascinated Cindy. She saw opportunities to Bangladesh, where each day a child goes blind due to insufficient vitamin A. She saw opportunities to help millions of children in places where rice is a dietary staple. But she wondered about its politically sensitive GMO status, when it might be approved for *trans*national distribution, and whether it might carry any risks. She wondered what kind of political resistance there might be. And, if it were effective, she wondered whether that efficacy might be used as an excuse not to fund other international efforts related to food, agriculture, and market development. In short, she could see more questions than answers.

The elevator door opened on the third floor and Cindy stepped out, starting down the long corridor to the Pomerantz offices. Campbell had asked her to present her initial findings to Mr Pomerantz just before lunch, so she would spend the morning for preparing her recommendation.

Case question: How should Cindy present the Golden Rice case? Outline a presentation, highlighting the pros and cons of introducing genetically modified foods into developing nations with the intent of preventing the health and economic effects of "hidden hunger."

11. STUDY QUESTIONS AND EXERCISES

1. For a core food for any particular vitamin, construct a flow diagram showing all of the processes, from the growing of the food to the eating of it by a human, that might reduce the useful amount of that vitamin in the food.
2. In consideration of the core foods for the vitamins and your personal food habits, which vitamin(s) might you expect to have the lowest intakes from your diet? Which might you expect to be low in the typical American diet? Which might you expect to be low in vegetarian and low-meat diets?
3. Use a concept map to show the relationships of vitamin supplementation of animal feeds to the concepts of chemical stability, bioavailability, and physiological utilization.
4. Prepare a flow diagram to show the means by which you might first evaluate the dietary vitamin status of a specific population (e.g., in an institutional setting), and then improve it, if necessary.
5. What principles should be used in planning diets to ensure adequacy with respect to the vitamins (and other nutrients)?

RECOMMENDED READING

Allen, L., de Benoist, F., Dary, O., et al., 2006. Guidelines on Food Fortification with Micronutrients. FAO/WHO, Geneva. 376 pp.

Backstrand, J.R., 2002. The history and future of food fortification in the United States: a public health perspective. Nutr. Rev. 60, 15–26.

Briefel, R.R., McDowell, N.A., 2012. Nutrition monitoring in the United States, Chapter 64. In: Erdman, J.W., Macdonald, I., Zeisel, S.H. (Eds.), Present Knowledge in Nutrition, tenth ed. John Wiley & Sons, New York, pp. 1082–1109.

Burchi, F., Fanzo, J., Frison, E., 2011. The role of food and nutrition system approaches in tacking hidden hunger. Int. J. Environ. Res. Public Health 8, 358–373.

Dietary Guidelines Advisory Committee, 2015. Scientific Report of the 2015 Dietary Guidelines Advisory Committee. US Gov. Printing Office, Washington, DC. 571 pp.

Fitzpatrick, T.B., Basset, G.J.C., Borel, P., et al., 2012. Vitamin deficiencies in humans: can plant science help? Plant Cell 24, 395–414.

Food and Nutrition Board, 2003. Dietary Reference Intakes: Guiding Principles for Nutrition Labeling and Fortification. National Academy Press, Washington, DC. 205 pp.

Food and Nutrition Board, 2003. Dietary Reference Intakes: Applications in Dietary Planning. National Academy Press, Washington, DC. 237 pp.

Ottaway, P.B., 1993. The Technology of Vitamins in Foods. Chapman & Hall, London.

Yates, A.A., 2006. Which dietary reference intake is best suited to serve as the basis for nutrition labeling for daily values? J. Nutr. 136, 2457–2462.

Chapter 21

Assessing Vitamin Status

Chapter Outline

Anchoring Concepts

1. Detection of suboptimal vitamin status at early stages (before manifestation of overt deficiency disease) is desirable for the reason that vitamin deficiencies are most easily correctable in their early stages.
2. Vitamin status can be estimated by evaluating diets and food habits, but these methods are not precise.
3. Vitamin status can be determined by measuring the concentrations of vitamins and metabolites, and the activities of vitamin-dependent enzymes, in samples of tissues and urine.
4. Suboptimal status is more probable for some vitamins, and in some population demographics, than for others.

...the old idea, that the state of nutrition of a child could be at once established by mere cursory inspection by the doctor, has to be abandoned....[Such methods] gave us very little information about the occurrence of the milder degrees of deficiency, or of the earlier stages of their development.

L.J. Harris[1]

LEARNING OBJECTIVES

1. To understand the requirements of valid methods for assessing vitamin status.
2. To understand the methods available for assessing the vitamin status of humans and animals.
3. To be familiar with available information regarding the vitamin status of human populations.

1. Leslie, J., 1955. Harris was a professor of nutrition at Cambridge University; his "Vitamins in Theory and Practice", fourth ed. Cambridge Univ. Press, 366 pp., is a classic.

VOCABULARY

Anthropometric assessment
Biochemical assessment
Bioindicator
Biomarker
Clinical assessment
Dietary assessment
Dietary records
Food frequency questionnaires (FFQ)
Hidden hunger
Micronutrient malnutrition
Nutrient loading
Nutrition screening
Nutrition surveillance
Nutrition surveys
Nutritional assessment
Nutritional status
24-h recalls sociologic assessment

1. NUTRITIONAL ASSESSMENT

The need to understand and describe the health status of individuals, a basic tenet of medicine, spawned the development of methods to assess nutrition status as appreciation grew for the important relationship between nutrition and health. The first applications of nutritional assessment were in investigations of feed-related health and production problems of livestock and, later, in examinations of human populations in developing countries. Activities of the latter type, consisting mainly of organized nutrition surveys, resulted in the first efforts to standardize both the methods employed to collect such data and the ways in which those

The Vitamins. http://dx.doi.org/10.1016/B978-0-12-802965-7.00021-6

data are interpreted.[2] Ultimately, nutritional assessment became an essential part of the nutritional care of hospitalized patients and an important means of evaluating the impacts of public nutrition intervention programs.

Purposes of nutritional assessment are as follows:
- detection of deficiency states;
- evaluation of nutritional qualities of diets, food habits, and/or food supplies;
- prediction of health effects.

Approaches and Methods of Nutritional Assessment

Three nutritional assessment systems have been employed both in population-based studies and in the care of hospitalized patients.

- **Nutrition surveys** conducted to generate baseline nutritional data, to learn overall nutrition status, and to identify subgroups at nutritional risk. These are typically cross-sectional evaluations of selected population groups.
- **Nutrition surveillance** conducted to identify possible causes of malnutrition. These involve continuous monitoring of the nutritional status of selected population groups (e.g., at-risk groups) for extended periods of time.
- **Nutrition screening** conducted to identify malnourished individuals requiring nutritional interventions. These rely on the use of biomarkers of nutritional status.

Nutritional assessments employ any of five methodologies:

Dietary assessment—estimating nutrient intakes from evaluations of diets, food availability, and food habits. Several methods can be used for dietary assessment:

- **24-h recalls**—The USDA has improved interview-based methods with the development of a five-step computerized dietary recall instrument, the USDA Automated Multiple-Pass Method (AMPM)[3]; this is available as a web-based, self-administered instrument.
- **Dietary records**—Written records (food diaries) have been used; new approaches include smart phone-based methods with food image processing and voice and/or text input to capture eating episodes.[4] Studies have found these methods to yield useful results for vitamin intakes if conducted for several days.[5]
- **Food frequency questionnaires (FFQ)**—This approach depends on respondent memory and uses fixed lists of foods. FFQs yield information about the usual diet of the past at relatively low cost,[6] for which reasons they are the methods of choice for epidemiological studies. FFQs have been found to yield useful estimates of biomarkers of some vitamins.[7] A systematic review found that vitamin intake estimated from FFQs and 24-h recall methods showed correlation coefficients of 0.26–0.38, and that those from FFQs and dietary record methods showed correlation coefficients of 0.41–0.53.[8]

Due to the cumulative predictive uncertainties of quantitative food intakes and of vitamin contents of foods (see Chapter 20), these methods yield imprecise estimates of vitamin intakes. Nevertheless, these methods can identify features of diets and food habits that are likely to provide insufficient amounts of bioavailable vitamins:

- **monotonous diets** with little food variety,[9] particularly those based on milled cereal grains;
- **low caloric intakes**, i.e., low intakes of total food[10];
- **enteric malabsorption** due to deficiencies and/or imbalances, e.g., fat.

Biochemical assessment—estimating nutritional status from measurements of stores, functional forms, excreted forms, and/or metabolic functions of specific nutrients. Vitamin status refers to the functional reserve capacity provided by vitamins in tissue. This approach offers the best opportunity for detecting early-stage (i.e., subclinical) vitamin deficiencies. However, direct measurement of

2. In 1955, the US government organized the Interdepartmental Committee on Nutrition for National Defense (ICNND) to assist developing countries in assessing the nutritional status of their peoples, identifying problems of malnutrition, and developing practical ways of solving their nutrition-related problems. The ICNND teams conducted nutrition surveys in 24 countries. In 1963, the ICNND published the first comprehensive manual (ICNND, 1963. Manual for Nutrition Surveys, second ed. U.S. Government Printing Office, Washington, DC) in which analytical methods were described and interpretive guidelines were presented.
3. Thompson, F.E., Subar, A.F., Loria, C.M., et al., 2010. J. Am. Diet. Assoc. 110, 48–51.
4. Wharton, C.M., Johnston, C.S., Cunningham, B.K., et al., 2014. J. Nutr. Educ. Behav. 46, 440–444.
5. Presse, N., Payette, H., Shatenstein, B., et al., 2011. J. Nutr. 141, 341–346.
6. Block, G., Thompson, F.E., Hartman, A.M., et al., 1990. J. Am. Diet. Assoc. 92, 686–693.
7. Tangney, C.C., Bienias, J.L., Evans, D.A., et al., 2004. J. Nutr. 134, 927–934.
8. Serra-Majem, L., Anderson, F.L., Henríque-Sánchez, P., et al., 2009. Br. J. Nutr. 102, S3–S9.
9. This may include a strict vegetarian diet that does not include some source of vitamin B_{12}.
10. Reduced food intake is thought to be a major cause of subadequate nutrient intakes of the elderly. Studies have shown that Americans ≥71 years frequently have intakes less than the estimated average requirement: vitamin A, 50%; vitamin E, 75%; vitamin K, 49% of women and 34% of men; vitamin C, 40%; and folate, 40% of women and 16% of men (Marriott, B.P., Olsho, L., Hadden, L., et al., 2010. Crit. Rev. Food Sci. Nutr. 50, 228–258).

TABLE 21.1 Relevance of Nutritional Assessment Methodologies to the Stages of Vitamin Deficiencies

	Most Informative Methods				
Stage of Deficiency[a]	**Dietary**	**Biochemical**	**Clinical**	**Anthropometric**	**Sociologic**
1. Depletion of stores	+	+			
2. Cellular metabolic changes		+	+	+	
3. Clinical defects		+	+	+	
4. Morphological changes			+	+	
5. Behavioral signs					+

[a]*See Chapter 4.*

vitamin function is limited by the availability of functional biomarkers,[11] the existence of more than one metabolic function with different sensitivities to vitamin supply,[12] and/or the function of the vitamin in a loosely bound fashion unstable to methods of tissue preparation.[13]

Clinical assessment—estimating nutritional status from medical histories and physical examinations by a qualified observer to detect signs and symptoms associated with malnutrition. Diagnoses of vitamin deficiencies are generally most possible in the latter stages when physiologic dysfunction and/or morphological changes can be detected. Clinical evaluation can identify pathophysiological factors that may limit vitamin utilization:

- **enteric malabsorption** due to acquired or innate problems affecting the absorptive surface of the gut, e.g., enteritis, helminth infection, gastrointestinal surgery;
- **impaired vitamin retention/utilization** due to acquired or innate problems of hepatic or renal vitamin metabolism, e.g., hepatitis, nephritis.

Anthropometric assessment—estimating nutritional status from measurements of the physical dimensions and gross composition of an individual's body.

Sociologic assessment—collecting information on other variables known to affect or be related to nutritional status, e.g., socioeconomic status, food habits and beliefs, food prices and availability, food storage and cooking practices, drinking water quality, alcohol use, immunization records, incidence of low-birth weight infants, breast-feeding and weaning practices, age- and cause-specific mortality rates, birth order, family structure, etc.

2. BIOMARKERS OF VITAMIN STATUS

Risk of suboptimal vitamin status is determined largely by factors that limit access to a diet providing adequate amounts of vitamins and other essential nutrients, as well as factors that limit the body's ability to utilize them after ingestion. These factors can be best ascertained by assessing dietary practices, clinical status, and biochemical indicators (**biomarkers**) of vitamin status (Table 21.1).

Biomarker Definition

The term "biomarker" is used in the field of nutrition to describe a trait that can be measured objectively and used as an indicator of the status of normal or pathogenic biological processes, or responses to therapeutic interventions.[14] A useful biomarker of vitamin status must

- relate to the rate of vitamin intake, particularly within the nutritionally significant range, and respond to deprivation of the vitamin;
- relate to normal physiologic function and a meaningful period of time;
- be measurable in an accessible specimen, technically feasible, reproducible, and affordable; and
- have an available base of normative data.

11. For example, while vitamin E functions as a biological lipid antioxidant, measuring that function is not possible with any physiological relevance because all of the known products of lipid peroxidation (e.g., malonaldehyde, alkanes) are metabolized, making results difficult to interpret with respect simply to vitamin E status.

12. For example, pyridoxal phosphate (PalP) is an essential cofactor for each of two enzymes involved in the metabolic conversion of tryptophan to niacin: kynureninase and a transaminase. However, because kynureninase has a much greater affinity for PalP ($K_m = 10^{-3}$ M) than does the transaminase ($K_m = 10^{-8}$ M), under conditions of pyridoxine deprivation the transaminase activity can be reduced while kynureninase activity remains unaffected.

13. For example, the metabolically active forms of niacin, NAD(P)H, function as the cosubstrates of many redox enzymes. These enzyme–cosubstrate complexes are only transiently associated; therefore, dilution of biological specimens results in their dissociation and usually in the oxidation of the cosubstrate.

14. This distinguishes a biomarker from a "bioindicator," which is typically used to describe a measure of processes used to assess the quality of an individual's or community's the environment and its changes over time (Raiten, D.J., Combs Jr., G.F., 2015. Sight Life 29, 39–44).

TABLE 21.2 Tissues Accessible for Assessing Biomarkers of Vitamin Status

Tissue or Cell Type	Relevance
Blood	
Plasma/serum	Contains newly absorbed vitamins being transported to other tissues; therefore, tends to reflect recent vitamin intake; this effect can be reduced by collecting fasting blood
Erythrocytes	With a half-life of about 120 days, they tend to reflect chronic nutrient status; analyses can be technically difficult
Leukocytes	Have relatively short half-lives and, therefore, can be used to monitor short-term changes in nutrient status
Tissues	
Liver, adipose, muscle, marrow	Sampling is invasive, requiring research or clinical settings
Hair, nails	Easily collected and stored specimens offer advantages for studies of trace element status; not useful for assessing vitamin status
Skin, macula	Can be scanned noninvasively by resonance Raman spectroscopy for assessing carotenoids

Biomarkers of Vitamin Status

In some cases, it is possible to assess the current and stored metabolic function of a vitamin.[15] In many cases, however, direct measurement of vitamin function may not be possible, making other biomarkers useful: measurements of the vitamin, particular metabolites, and other enzymes in accessible tissues or urine (Tables 21.2 and 21.3). In this regard, biomarker data of informative value for individuals will call for combining it with genetic information regarding inborn metabolic errors. In addition, the emerging field of **metabolomics**, i.e., the measurement of multiple small molecular weight metabolites in biological specimens, is likely to offer useful new tools that will facilitate gaining insight into metabolic pathways affected by specific vitamins.[16] This approach has been used to identify metabolic profiles characteristic of general dietary patterns.[17]

15. e.g., Measuring prothrombin time to assess vitamin K status, and stimulation coefficients of erythrocyte transketolase and glutathione reductase to assess thiamin and riboflavin status, respectively.

16. Guertin, K.A., Moore, S.C., Sampson, J.N., et al., 2014. Am. J. Clin. Nutr. 100, 208–217.

17. e.g., High serum levels of acylcarnitines and acylalkylphosphatidylcholines were related to consumption of high amounts of butter; high serum hexose and phosphatidylcholines were related to consumption of red meat and fish with low amounts of whole grains and tea; high serum methionine and branched-chain amino acids were related to consumption of potatoes, dairy products, and cornflakes (Floegel, A., von Ruesten, A., Drogan, D., et al., 2013. Eur. J. Clin. Nutr. 67, 1100–1108).

Interpreting Biomarker Data

The relevance of biomarkers of vitamin status of individuals is not straightforward, owing to issues of intraindividual variation and confounding effects, which may be quantitatively more significant for individuals than for populations. For example, intraindividual variation is frequently noted in serum analytes. Therefore, a measurement of a single blood sample may not be appropriate for estimating the usual circulating level of the analyte of an individual, even though it may be useful in estimating the mean level of a population. Several factors can confound the interpretation of parameters of vitamin status: those affecting the response parameters directly, drugs that can increase vitamin needs, seasonal effects related to the physical environment[18] or food availability,[19] use of parenteral feeding solutions,[20] and use of vitamin supplements,[21] smoking,[22] etc. (Table 21.4). The guidelines originally developed by the ICNND remain useful for the interpretation of the results of biomarkers of vitamin status (Table 21.5). It is important to note, however, that those interpretive guidelines were developed for use in surveys of populations.

3. VITAMIN STATUS OF HUMAN POPULATIONS

Reserve Capacities of Vitamins

The reserve capacities of the vitamins vary. Each is affected by the history of vitamin intake, the metabolic needs for the vitamin, and the general health status of the individual. Typical reserve capacities of a healthy, adequately nourished human adult to meet normal metabolic needs are as follows:

- **4–10 days**—thiamin, biotin, and pantothenic acid
- **2–6 weeks**—vitamins D, E, K, and C; riboflavin, niacin, and vitamin B_6

18. For example, individuals living in northern latitudes typically show peak plasma levels of 25-OH-D_3 around September and low levels around February, with inverse patterns of plasma parathyroid hormone concentrations, owing to the seasonal variation in exposure to UV light.

19. For example, residents of Finland showed peak plasma ascorbic acid levels in August–September and lowest levels in November–January, owing to seasonal differences in the availability of vitamin C-rich fruits and vegetables.

20. Individuals supported by total parenteral nutrition (TPN) have frequently been found to be of low status with respect to biotin (due to their abnormal intestinal microflora) and the fat-soluble vitamins (due to absorption by the plastic bags and tubing and to destruction by UV light used to sterilize TPN solutions).

21. The NHANES I survey showed that more than 51% of Americans over 18 years of age used vitamin/mineral supplements, with 23.1% doing so on a daily basis. Multivitamins are the most commonly used supplements, followed by vitamin C, calcium, and vitamins E and A. The use of vitamin supplements has been found to have greater impact than that of vitamin-fortified food on both the mean and coefficient of variation (CV) of estimates of vitamin intake in free-living populations.

22. Smokers have been found to have abnormally low-plasma levels of ascorbic acid (with a corresponding increase in dehydroascorbic acid), pyridoxal, and pyridoxal phosphate.

TABLE 21.3 Biomarkers of Vitamin Status

Vitamin	Functional Parameters	Tissue Levels	Urinary Excretion
Vitamin A		Serum retinol[a]	
		Change in serum retinol after oral load[b]	
		Liver retinyl esters	
Vitamin D		Serum 25-$(OH)_2$-vitamin D_3[a]	
		Serum vitamin D_3	
		Serum 1,25-$(OH)_2$-D_3	
		Serum alkaline phosphatase	
Vitamin E	Erythrocyte hemolysis	Serum tocopherols[a]	
		Serum malondialdehyde	
		Serum 1,4-isoprostanes	
		Breath alkanes	
Vitamin K	Clotting time		
	Prothrombin time[a]		
Vitamin C		Serum ascorbic acid	Ascorbic acid
		Leukocyte ascorbic acid[a]	Ascorbic acid after load[c]
Thiamin	Erythrocyte transketolase	Blood thiamin	Thiamin (thiochrome)
	Stimulation[a]	Blood pyruvate	Thiamin after load[c]
Nacin		RBC NAD[a]	1-Methylnicotinamide
		RBC NAD:NADP ratio	1-Methyl-6-pyridone-3-carboxamide
		Plasma tryptophan	
Riboflavin	RBC glutathione reductase	Blood riboflavin	Riboflavin
	Stimulation[a]		Riboflavin after load[c]
Vitamin B_6	RBC transaminase	Plasma pyridoxal phosphate	Xanthurenic acid after tryptophan load[a,c]
		RBC transaminase stimulation	Quinolinic acid
		RBC pyridoxal phosphate	4-Pyridoxic acid
		Plasma pyridoxal	
Biotin		Blood biotin[a]	Biotin
Pantothenic acid		RBC sulfanilamide acetylase	Pantothenic acid
		RBC pantothenic acid	
		Blood pantothenic acid[a]	
Folate		Serum folates[a]	FIGLU[c] after histidine load[a,c]
		RBC folates[a]	
		Leukocyte folates	Urocanic acid after histidine load[c]
		Liver folates	
Vitamin B_{12}		Serum vitamin B_{12}[a]	FIGLU[d]
		RBC vitamin B_{12}	Methylmalonic acid[a]

[a]*Most useful parameter.*
[b]*Relative dose–response test.*
[c]*Single large oral dose.*
[d]*FIGLU, formiminoglutamic acid.*

- **3–4 months**—folate
- **1–2 years**—vitamin A
- **3–5 years**—vitamin B_{12}.

Differences in reserve capacities reflect differential abilities to retain and store vitamins. Such differences lead, therefore, to differential sensitivities to vitamin deprivation. For example, individuals with histories of generally adequate vitamin nutriture can be expected to sustain longer periods of deprivation of vitamins A or B_{12} than they could of thiamin, biotin, or pantothenic acid. Similarly, metabolic and physiologic lesions caused by deficiencies of thiamin, biotin, or pantothenic acid can be expected to appear much sooner than those of vitamins A or B_{12}, which may remain occult. Because nutritional intervention is typically most efficacious and cost-effective in earlier stages of vitamin

TABLE 21.4 Limitations of Some Biomarkers of Vitamin Status

Vitamin	Biomarker	Limitations
Vitamin A	Plasma[a] retinol	Reflects body vitamin A stores only at severely depleted or excessive levels; confounding effects of protein and zinc deficiencies and renal dysfunction
Vitamin D	Plasma[a] alkaline phosphatase	Affected by other disease states
Vitamin E	Plasma[a] tocopherol	Affected by blood lipid transport capacity
Thiamin	Plasma[a] thiamin	Low sensitivity to changes in thiamin intake
Riboflavin	Plasma[a] riboflavin	Low sensitivity to changes in riboflavin intake
Vitamin B_6	RBC glutamic–pyruvic	Genetic polymorphism transaminase
Folate	RBC folates	Also reduced in vitamin B_{12} deficiency
	Urinary FIGLU[b]	Also increased in vitamin B_{12} deficiency
Vitamin B_{12}	Urinary FIGLU[b]	Also increased in folate deficiency

[a]*Or serum.*
[b]*FIGLU, formiminoglutamic acid.*

TABLE 21.5 Interpreting Biomarkers of Vitamin Status

Vitamin	Parameter	Age Group	Values, by Category of Status[a]: Deficient (High Risk)	Low (Moderate Risk)	Acceptable (Low Risk)
Vitamin A	Plasma[b] retinol (μg/dL)	<5 months	<10	10–19	>20
		0.5–17 years	<20	20–29	>30
		Adult	<10	10–19	>20
Vitamin D	Plasma[b] 25-(OH)-D_3[c] (ng/mL)	All ages	<20[v]	20–29[v]	≥30[v]
	Plasma[b] alkaline phosphatase[c] (U/mL)	Infants	>390	298–390	99–298
		Adults	<40	40–56	57–99
Vitamin E	Plasma[b] α-tocopherol (mg/dL)	All ages	<0.35	0.35–0.80	>0.80
Vitamin K	Clotting time (min)	All ages	>10	~10	
	Prothrombin time (min)	—[d]	—[d]	—[d]	
Vitamin C	Plasma[b] ascorbic acid (mg/dL)	All ages	<0.20	0.20–0.30	>0.30
	Leukocyte ascorbic acid (mg/dL)	All ages	<8	8–15	>15
	Whole blood ascorbic acid (mg/dL)	All ages	<0.30	0.30–0.50	>0.50

TABLE 21.5 Interpreting Biomarkers of Vitamin Status—cont'd

Vitamin	Parameter	Age Group	Values, by Category of Status[a]		
			Deficient (High Risk)	Low (Moderate Risk)	Acceptable (Low Risk)
Thiamin	Urinary thiamin (μg/g creatinine)	1–3 years	<120	120–175	>175
		4–6 years	<85	85–120	>120
		7–9 years	<70	70–180	>180
		10–12 years	<60	60–180	>180
		13–15 years	<50	50–150	>150
		Adults	<27	27–65	>65
		Pregnant 2nd trim.	<23	23–55	>55
		Pregnant 3rd trim.	<21	21–50	>50
	Urinary thiamin				
	μg/24 h	Adults	<40	40–100	>100
	μg/6 h	Adults	<10	10–25	>25
	Urinary thiamin after load[e] (μg/4 h)	Adults	<20	20–80	>80
	RBC transketolase stimulated by TPP[f,g] (%)	Adults	>25	15–25	<15
Riboflavin	Urinary riboflavin (μg/g creatinine)	1–3 years	<150	150–500	>500
		4–6 years	<100	100–300	>300
		7–9 years	<85	85–270	>270
		10–15 years	<70	70–200	>200
		Adults	<27	27–80	>80
		Pregnant 2nd trim.	<39	39–120	>120
		Pregnant 3rd trim.	<30	30–90	>90
	Urinary riboflavin (μg/24 h)	Adults	<40	40–120	>120
	Urinary riboflavin (μg/6 h)	Adults	<10	10–30	>30
	Urinary riboflavin load[h] (μg/4 h)	Adults	<1000	1000–1400	>1400
	RBC riboflavin (μg/day)	Adults	<10.0	10.0–14.9	>14.9
	RBC glutathione reductase FAD[i] stimulation (%)	Adults	>40	20–40	<20
Niacin	Urinary *N′*-methylnicotinamide (μg/g creatinine)	Adults	<0.5	0.5–1.6	>1.6
		Pregnant 2nd trim.	<0.6	0.6–2.0	>2.0
		Pregnant 3rd trim.	<0.8	0.8–2.5	>2.5
	Urinary *N′*-methylnicotinamide (μg/6 h)	Adults	<0.2	0.2–0.6	>0.6
	Urinary 2-pyridone[j]:*N′*-methylnicotinamide	All ages	—[k]	<1.0	≥1.0
Vitamin B_6	Plasma PalP[l] (nM)	All ages	—[k]	<60[m]	≥60[m]
	Urinary vitamin B_6 (μg/g creatinine)	1–3 years	—[k]	<90[m]	≥90[m]
		4–6 years	—[k]	<75[m]	≥75[m]
		7–9 years	—[k]	<50[m]	≥50[m]
		10–12 years	—[k]	<40[m]	≥40[m]
		13–15 years	—[k]	<30[m]	≥30[m]
		Adults	—[k]	<20[m]	≥20[m]

Continued

TABLE 21.5 Interpreting Biomarkers of Vitamin Status—cont'd

Vitamin	Parameter	Age Group	Values, by Category of Status[a]		
			Deficient (High Risk)	Low (Moderate Risk)	Acceptable (Low Risk)
	Urinary 4-pyridoxic acid (mg/24 h)	Adults	<0.5[m]	0.5–0.8[m]	>0.8[m]
	Urinary xanthurenic acid after tryptophan load[h] (mg/24 h)	Adults	>50[m]	25–50[m]	<25[m]
	Urinary 3-OH-kynurenine after tryptophan load[h] (mg/24 h)	Adults	>50[m]	25–50[m]	<25[m]
	Urinary kynurenine after tryptophan load[h] (mg/24 h)	Adults	>50[m]	10–50[m]	<10[m]
	Quinolinic acid after tryptophan load[h] (mg/24 h) adults	Adults	>50[m]	25–50[m]	<25[m]
	Erythrocyte alanine aminotransferase stimulation by PalP[l] (%)	Adults	—[k]	>25[m]	≤25[m]
	Erythrocyte aspartate aminotrans-ferase stimulation by PalP[l] (%)	Adults	—[k]	>50[m]	≤50[m]
Biotin	Urinary biotin (μg/24 h)	Adults	<10[m]	10–25[m]	>25[m]
	Whole blood biotin (ng/mL)	Adults	<0.4[m]	0.4–0.8[m]	>0.8[m]
Pantothenic acid	Plasma[b] pantothenic acid (μg/dL)	Adults	—[k]	<6[m]	≥6[m]
	Blood pantothenic acid (μg/dL)	Adults	—[k]	<80[m]	≥80[m,n]
	Urinary pantothenic acid (mg/24 h)	Adults	—[k]	<1[m]	≥1[m,o]
Folate	Plasma[b] folates (ng/mL)	All ages	<3	3–6	>6
	RBC folates (ng/mL)	All ages	140	140–160	>160
	Leukocyte folates (ng/mL)	All ages	—[k]	<60	>60
	Urinary FIGLU[p] after histidine load[q] (mg/8 h)	Adults	>50[m]	5–50	<5[r]
Vitamin B_{12}	Plasma[b] vitamin B_{12} (pg/mL)	All ages	100	100–150	>150[s]
	Urinary methylmalonic acid after valine load[t] (mg/24 h)	Adults	≥300	2–300	≤2
	Urinary excretion of labeled B_{12} after a flushing dose[u] (%)	Adults	<3	3–8	>8

[a] *ICNND., 1963. Manual for Nutrition Surveys, second ed., U.S. Government Printing Office, Washington, DC; Sauberlich et al., 1974. Laboratory Tests for the Assessment of Nutritional Status. CRC Press, Cleveland, Ohio; Gibson, R.S., 1990. Principles of Nutritional Assessment. Oxford University Press, New York.*
[b] *Or serum.*
[c] *Subject to effects of season and sex.*
[d] *Results vary according to assay conditions; most assays are designed such that normal prothrombin times are 12–13 sec, with greater values indicating suboptimal vitamin K status.*
[e] *Single oral 2-mg dose.*
[f] *TPP, thiamin pyrophosphate.*
[g] *The TPP effect.*
[h] *Single oral 2-g dose.*
[i] *FAD, flavin adenine dinucleotide, reduced form, 1–3 μM.*
[j] *N′-methyl-2 pyridone-5-carboxamide.*
[k] *Database is insufficient to support a guideline.*
[l] *PalP, pyridoxal phosphate.*
[m] *These values have only a small database and, therefore, are considered as tentative.*
[n] *Normal values are about 100 μg/dL.*
[o] *Normal values are 2–4 mg/24 h.*
[p] *FIGLU, formiminoglutamic acid.*
[q] *Single oral 2- to 20-mg dose.*
[r] *Normal adults excrete 5–20 mg/8 h.*
[s] *Most healthy individuals show 200–900 pg/mL.*
[t] *Single oral 5- to 10-g dose.*
[u] *This is the Schilling test; it involves measurement of labeled vitamin B_{12} excreted from a 0.5- to 2-μg tracer dose after a large flushing dose (e.g., 1 mg) given 1 h after the tracer.*
[v] *Modified according to discussion in Chapter 5.*

deficiencies, the early detection of occult deficiencies is important for designing effective therapy and prophylaxis programs.

National Nutrition Surveillance

The United States has had a series of programs to track the nutritional adequacy of the food supply and/or the nutritional status of people. These efforts are now consolidated into the ongoing survey ***What We Eat in America***,[23] the dietary intake component of the National Health and Nutrition Examination Survey.[24] Other countries have conducted similar studies,[25] as well as regular food reporting.

Vitamins Status of Americans

The comprehensive review by the 2015 Dietary Guidelines Scientific Advisory Committee found that Americans widely underconsume vitamins D, E, and thiamin (Fig. 21.1), as well as calcium, magnesium, potassium, and fiber, and that adolescents and women also underconsume vitamin A, vitamin C, folate, and iron (Table 21.6).[26] This reflects the underconsumption of nutrient-rich foods; three-quarters of food energy comes from only half (16) of the subcategories of available foods. The vast majority of Americans does not consume the recommended amounts of fruit, vegetables, whole grains, and dairy foods, with lowest consumptions occurring in lower socioeconomic groups and the elderly.

It is estimated that at least one-fifth of heart disease and one-third of all cancers could be prevented by improving the American diet, specifically by increasing the consumption of fruits and vegetables and reducing intakes of saturated and total fat. Dietary patterns with the highest Healthy Eating Index (HEI) scores are generally those that emphasize seafood and plant proteins and dairy foods and have relatively few highly refined grains and "empty" calories. Vegetarian, but not fruitarian, diets can be nutritionally adequate if sensibly selected and appropriately supplemented (e.g., vitamin B_{12}).[27] Problems can arise in any type of diet if the variety of food is restricted and, particularly, if the consumption of dairy products is low. Therefore, emphases on increasing fruits and vegetables at the expense of meats (important sources of vitamins A, B_6, and B_{12}; thiamin; and niacin), and replacing vegetable oils (important sources of vitamin E) with reduced- and no-fat substitutes should be balanced to avoid reduced intakes of B vitamins (Table 21.7).

23. http://www.ars.usda.gov/services/docs.htm?docid=13793.
24. http://www.cdc.gov/nchs/nhanes/.
25. e.g., New Zealand National Nutrition Survey; Luxembourg Nutritional Surveillance System; United Kingdom Expenditure and Food Survey, National Food Survey and School Nutrition Dietary Assessment.
26. Dietary Guidelines Advisory Committee, 2015. Scientific Report of the 2015 Dietary Guidelines Advisory Committee. US Government Printing Office, 571 pp.
27. Craig, W.J., 2009. Am. J. Clin. Nutr. 89, S1627–S1633.

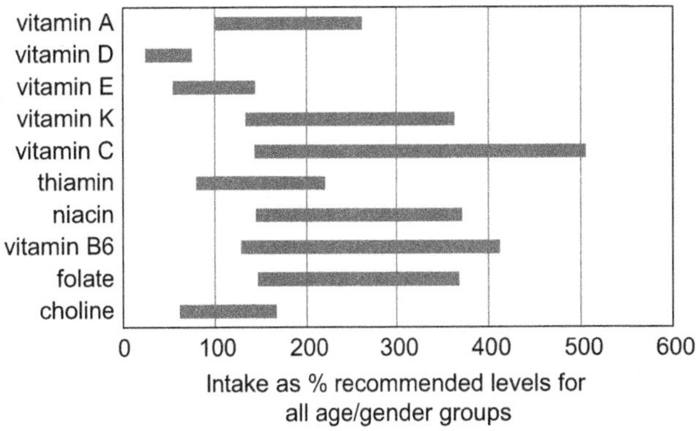

FIGURE 21.1 Estimated vitamins provided by American diets. *After Dietary Guidelines Advisory Committee, 2015. Scientific Report of the 2015 Dietary Guidelines Advisory Committee. US Government Printing Office, 571 pp.*

Nutritional Surveillance Reveals Vitamin Deficiencies

Data on food-nutrient supplies and apparent nutrient consumption are necessary for national food and health policy planning; but they yield no information useful in addressing questions of nutritional status of individuals within populations. Nutrition surveys were initiated to produce such data. The NHANES surveys have revealed the following[28]:

- **Vitamin A status**—Fewer than 1% of Americans had deficient levels of serum retinol (<20 µg/dL), but ~2% had levels indicative of risk to excess vitamin A (>100 µg/dL). Serum retinol concentrations increased with age. Serum β-carotene levels decreased throughout childhood and then increased, with females showing greater levels than males and non-Hispanic whites showing greater levels than other groups.
- **Vitamin E status**—About 2% of Americans had deficient levels of serum α-tocopherol (<500 µg/dL), although that portion was greater (2–4%) for adolescents. Serum α-tocopherol levels decreased throughout childhood and then increased with age. Serum γ-tocopherol levels were relatively stable throughout the life cycle.
- **Vitamin D status**—Results from 2003 to 2006 showed that 17% Americans had low-serum 25-OH-D_3 levels (<40 nM), with 8% being deficient (serum 25-OH-D_3 levels <30 nM) and <1% had excessive levels (>125 nM). Low-vitamin D status was more prevalent among females (10%) than males (6%) and among non-Hispanic blacks (31%) compared to other groups (non-Hispanic whites, 4%; Mexican Americans, 11%).
- **Vitamin C status**—Serum vitamin C concentrations showed a U-shaped distribution according to age, lowest

28. National Center for Health Statistics, 2012. Second National Report on Biochemical Indicators of Diet and Nutrition in the U.S. Population, 450 pp. http://www.cdc.gov/nutritionreport/report.html.

TABLE 21.6 Usual Intakes of American Women, 19–50 years, of Vitamins From Foods and Beverages Compared to Recommendations

Vitamin	Mean Intake[a]	EAR[b]	% Below EAR	UL[c]	% Above UL
Vitamin A, μg RE/d	549	500	48	3000	<3
Vitamin D, μg/d	3.9	10	>97	100	<3
Vitamin E, mg/d	6.9	12	95	–	–
Vitamin C, mg/d	76.6	60	45	2000	<3
Folate, μg/day	470	320	15	1000	<3

[a]*n* = 2957.
[b]*Estimated average requirement.*
[c]*Tolerable upper limit.*
Adapted from Dietary Guidelines Advisory Committee, 2015. Scientific Report of the 2015 Dietary Guidelines Advisory Committee. US Government Printing Office, 571 pp.

TABLE 21.7 Three Healthy Dietary Patterns

Food Group	Healthy US Style	Healthy Vegetarian	Healthy Mediterranean Style
Fruit	2 cups/day	2 cups/day	2.5 cups/day
Vegetables	2.5 cups/day	2.5 cups/day	2.5 cups/day
Legumes	1.5 cups/week	3 cups/week	1.5 cups/week
Whole grains	3 oz. equiv./day	3 oz. equiv./day	3 oz. equiv./day
Dairy	3 cups/day	3 cups/day	2 cups/day
Protein foods	5.5 oz. equiv./day	3.5 oz. equiv./day	6.5 oz. equiv./day
Meat	12.5 oz. equiv./day		12.5 oz. equiv./day
Poultry	10.5 oz. equiv./day		10.5 oz. equiv./day
Seafood	8 oz. equiv./day		15 oz. equiv./day
Eggs	3 oz. equiv./day	3 oz. equiv./day	3 oz. equiv./day
Nuts/seeds	4 oz. equiv./day	7 oz. equiv./day	4 oz. equiv./day
Processed soy	0.5 oz. equiv./day	8 oz. equiv./day	0.5 oz. equiv./day
Oils	27 g/day	27 g/day	27 g/day

Adapted from USDA Food Modeling Report, 2015.

levels occurring in persons 20–59 years. The prevalence of deficiency (serum vitamin C <11.4 μM) varied from 3% (Mexican Americans) to 7% (non-Hispanic whites) in 2003–06, with males, smokers, and individuals of low-socioeconomic status at elevated risk to being deficient.

- **Vitamin B_6 status**—The prevalence of low-vitamin B_6 status (plasma pyridoxal 5′-phosphate <20 nM) was 11% across age groups in 2005–06, with older individuals and women of childbearing age or using oral contraceptives and smokers being greatest risk of being deficient.
- **Folate status**—Since the introduction of folate fortification in 1998, serum folate concentrations of Americans have doubled and erythrocyte folate concentrations have increased by 50%. During the period of 1999–2006, fewer than 1% of Americans were of low-folate status (i.e., serum folate <2 ng/mL and erythrocyte folate <95 ng/mL), compared to 30 in 1988–94.[29] Serum and erythrocyte folate levels showed U-shaped distributions according to age, lowest levels occurring in adolescents and young adults.
- **Vitamin B_{12} status**—Results in 2003–06 showed <1–3% of children and 3–6% of adults to be of deficient

29. National Center for Health Statistics, 2012. Second National Report on Biochemical Indicators of Diet and Nutrition in the U.S. Population, 450 pp.

vitamin B_{12} status (serum B_{12} levels <200 pg/mL), with 20% showing marginal status.[30] The prevalence of low status was greater in older Americans (4%) and those with vegetarian (particularly vegan) dietary practices.[31] Vitamin B_{12} deficiency was indicated among subjects with low-serum vitamin B_{12} levels, by the fact that 17–19% of subjects also had elevated concentrations of methylmalonic acid and/or homocysteine, both of which variables increased with age.

- **Non-provitamin A carotenoids status**—Serum levels of α-carotene, lutein, zeaxanthin and cryptoxanthin decreased throughout childhood and then increased with age. Serum lycopene levels were greatest among young adults but decreased with age. Women showed greater α-carotene levels than men, but men showed greater lycopene levels than women.

4. GLOBAL UNDERNUTRITION

Under the auspices of national programs, bilateral programs, and international agencies, many nutrition surveys have been conducted in developing countries where malnutrition continues to be a problem. These have shown that, globally, more than 795 million people—one in nine of the world's people—do not to have access to enough food to meet their basic daily needs.[32] Malnutrition is an underlying cause of nearly half of all deaths and accounts for 11% of the global burden of disease.[33] The root causes of malnutrition, and underlying the food insecurity, are poverty and conflict.

Significant reductions have been made in global poverty. In fact, the first Millennium Development Goal (MDG) target, to cut the 1990 poverty rate in half by 2015, was accomplished in 2010, with the greatest reductions occurring in east Asia.[34] By 2015, the World Bank estimated the global prevalence of extreme poverty to be under 10% for the first time in recorded history. Still, on 2012, 896 M people (13% of the world's population) lived on no more than $1.90/day

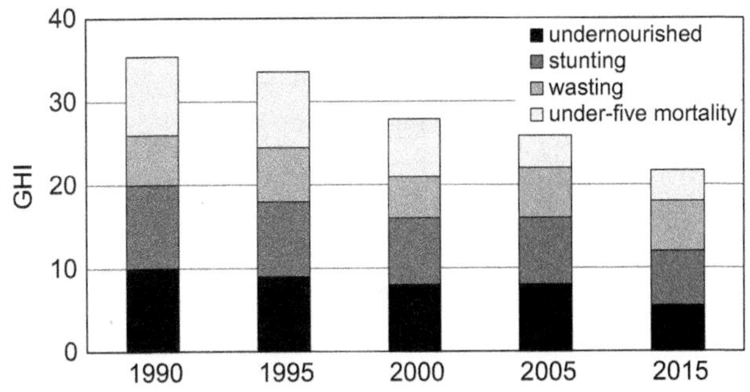

FIGURE 21.2 Trends in Global Hunger. The Global Hunger Index (GHI) was developed by the International Food Policy Institute (http://ghi.ifpri.org/results/).

and 2.1 B lived on less than $3.10/day. These changes have been accompanied by a 42% reduction in the prevalence of undernourished people; the MDG targets for developing countries, of cutting in half the prevalence of undernourished people by 2015 (i.e., from 23% in 1990–92 to 12% by 2015) has almost been met. The least progress has been made in sub-Saharan Africa where 23% are undernourished; despite progress south Asia has 276 M undernourished people.

Undernutrition has its most visible effects on children (Figs. 21.2 and 21.3). With diets inadequate in quantity and quality, 15–25% of the children in developing countries are stunted or wasted.[35] Undernourished children can experience as much as 160 days of illness in a year. At least half of the 11 million child deaths that occur each year are because of malnutrition and its potentiating effects on infectious disease[36]—the childhood mortality rate in developing countries is 10 times that of developed countries.

Hidden hunger. It is clear that the view of malnutrition resulting mainly from insufficient supplies of macronutrients (i.e., energy and protein) has led to gross underestimates of problems caused by deficiencies of critical micronutrients (i.e., vitamins and trace elements). The recognition of those problems due to micronutrient deficiencies generated the need for a new descriptor, **"hidden hunger**."[37] Two billion people live at risk of diseases resulting from deficiencies of vitamin A, iodine, and iron. Most are women and children living in the less-developed

30. Allen, L.H., 2009. J. Nutr. 89, S693–S696.
31. Elmadfa, I., Singer, I., 2009. Am. J. Clin. Nutr. 89, S1693–S1698.
32. Widespread malnutrition exists despite impressive gains in global agricultural production. In the last five decades, cereal yields have more than doubled and per capita supplies of food energy are at all-time high levels—exceeding present global needs. However, the high-yielding, "*green revolution*" varieties of major staple grains, being more profitable than traditional crops (including pulses), have displaced the latter and led to substantial reductions in the diversity of cropping systems. This has contributed to micronutrient malnutrition while increasing caloric output.
33. Malnutrition is considered the number 1 risk to health globally. Adults malnourished as children face 20% deficits in their earning potentials, reducing the gross domestic products of countries with prevalent malnutrition by 2–3%. It is estimated that each dollar spent on alleviating malnutrition yields $138 in benefits.
34. The World Bank estimated that 78% of the world's extremely poor live in south Asia (Pakistan, India, Bangladesh).
35. That is, more than two standard deviations below the median for height-for-age and weight-for-age, respectively. In 2010, the WHO estimated that 171 M children (98% in developing countries) were stunted. This trend shows improvement, particularly in Asia where the percentage has been cut in half since 1990. Still, stunting remains a major public health problem in many developing countries.
36. i.e., Diarrheal diseases (61% of deaths), malaria (57%), pneumonia (52%), and measles (45%) (Bryce, J., Boschi-Pinto, C., Shibuya, K., et al., 2005. Lancet 365, 1147–1152).
37. Kul Gautam, the former deputy executive director of UNICEF, described it well. He said, "*The 'hidden hunger' due to micronutrient deficiency does not produce hunger as we know it. You might not feel it in the belly, but it strikes at the core of your health and vitality.*"

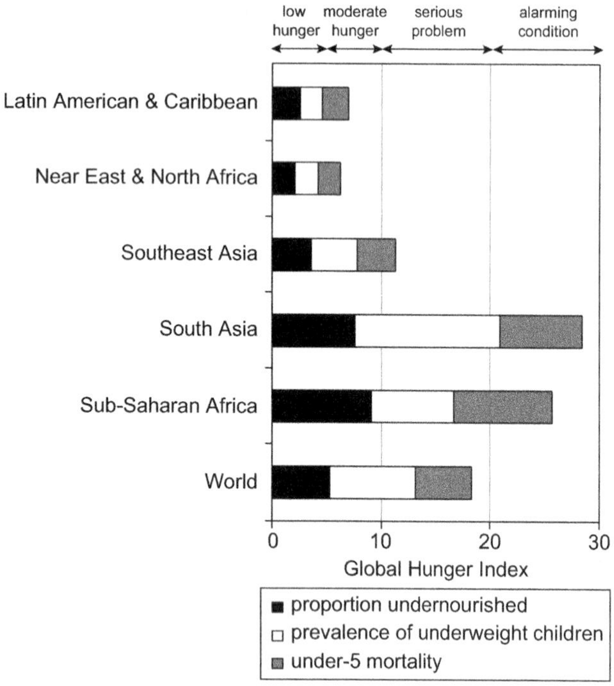

FIGURE 21.3 Regional rankings according to the 2010 Global Hunger Index. *After von Grebmer, K., Ruel, M.T., Menon, P., et al., 2010. 2010 Global Hunger Index, The Challenge of Hunger: Focus on the Crisis of Child Undernutrition. IFPRI, Washington.*

countries of sub-Saharan Africa, the eastern Mediterranean, southern and southeast Asia, Latin America, the Caribbean and the western Pacific.[38] The micronutrient deficiencies contributing most to hidden hunger are as follows:

- **Vitamin A deficiency**, affecting 190M preschool age children and 19M pregnant women, remains the single most important cause of childhood blindness in developing countries. **Frank vitamin A deficiency causes visual impairment and blindness**. Subclinical vitamin A deficiency **increases risk to infections including diarrhea and measles in children and** is associated with increased child mortality. Providing vitamin A can reduce child mortality by about 25% and birth-related, maternal mortality by 40%. One-third of vitamin A-deficient pregnant women are affected by night blindness.
- **Iodine deficiency**, affecting ~1.8B, causes brain damage in newborns and reduced mental capacity and goiter in adults
- **Iron deficiency**, affecting ~1.6B, causes anemia and reduced work capacity, impairing motor and cognitive development, and increasing maternal perinatal mortality and premature births.[39] Iron deficiency is thought to be responsible for at least half of the anemia affecting more than 40% of the world's women.[40,41]
- **Zinc deficiency**, affecting ~1.2B, causes stunting in children and a weakened immune system.
- **Other deficiencies** affect some groups who consume insufficient amounts of riboflavin, folate, vitamin B_{12}, and calcium.

38. An estimated 1600M people live in iodine-deficient areas. The most prevalent outcome of iodine deficiency is goiter, affecting some 200M people. In addition, some 6 million infants born annually to iodine-deficient mothers develop severe mental and neurological impairment known as *cretinism* (half of this number is in southern Asia). The deficiency also increases the rates of stillbirths, abortions, and infant deaths.

The Challenge

As global poverty is reduced and incomes increase, diets change. They tend to shift from cereals and tubers to meats, fats, and sugars, increasing total caloric consumption.[42] In addition, the forces of globalization appear to be leading to what has been called a "creeping homogenization"[43] in diets. These forces will put upward pressures on the prices of animal products and feed grains.

By 2050, the world population will exceed 9 billion, one-third more than today. Virtually all of this increase will occur in developing countries, most in their urban areas where it will be accompanied by continued increases in consumer purchasing power. Meeting this growing demand for food will call for increasing food production by ~70% above present levels. Experts believe that this can be achieved by increasing yields, reducing food wastes, and facilitating international trade. They agree that world has or can develop the requisite resources and technologies.

It is a practical and moral imperative that future demands for food be achieved in ways that will eliminate all types of hunger, including that of micronutrient malnutrition. This will require holistic approaches mindful of the inherent dependence of health on the balanced nutrition provided by access to diverse foods. The opportunities for food systems to advance public health are clear. They must inform the agendas not only of public health professionals but also of political leaders and other decision-makers in agriculture and trade.

39. It is estimated that one-fifth of maternal mortality is due to the direct (heart failure) or indirect (inability to tolerate hemorrhage) effects of anemia; severe anemia is responsible for nearly one-third of fatalities among children who are not given immediate transfusions.

40. This ranges from a high in southern Asia (64%) to the lowest but still surprisingly high rates (>20%) in industrialized countries.

41. Anemia can have multiple causes including malaria and intestinal parasitism.

42. An estimated 1.46B of the world's people are overweight/obese. In the developing world, that number increased from 250M in 1980 to 904M in 2008.

43. Keats, S., Wiggins, S., 2014. Future Diets Report, ODI, 116 pp.

5. STUDY QUESTIONS AND EXERCISES

1. Give an example of a situation wherein a particular biochemical test may be necessary for the diagnosis of a vitamin-related disorder detected by clinical examination.
2. Devise a system of biochemical measurements that could be performed on a 7-mL sample of fresh blood to yield as much information as possible about the vitamin status of the donor. (Assuming enzyme activities can be assayed using no more than 20 μL of plasma or erythrocyte lysate and other biochemical measurements can be made using no more than 100 μL each.)
3. In general terms, discuss the advantages and disadvantages of the various biomarkers for assessing vitamin status (e.g., functional tests, load tests, urinary excretion tests, circulating metabolite tests).

RECOMMENDED READING

Briefel, R.R., McDowell, M.A., 2012. Nutrition monitoring in the United States. In: Erdman, J.W., Macdonald, I.A., Zeisel, S.H. (Eds.), Present Knowledge in Nutrition, tenth ed. Wiley-Blackwell, New York, pp. 1082–1109. Chapter 62.

Combs Jr., G.F., Welch, R.M., Duxbury, J.M., Uphoff, N.T., Nesheim, M.C., 1996. Food-Based Approaches to Preventing Micronutrient Malnutrition: An International Research Agenda. Cornell University, Ithaca, New York. 68 pp.

Dietary Guidelines Advisory Committee, 2015. Scientific Report of the 2015 Dietary Guidelines Advisory Committee. US Government Printing Office. 571 pp.

Gibson, R.S., 1990. Principles of Nutritional Assessment. Oxford University Press, New York. 691 pp.

Hedrick, V.F., Dietrich, A.M., Estabrooks, P.A., et al., 2012. Dietary biomarkers: advances, limitations and future directions. Nutr. J. 11, 109–123.

Henríque-Sánchez, P., Sánchez-Villegas, A., Doreste-Alonso, J., Ortiz-Andrellucchi, A., Pfrimer, K., Serra-Majem, L., 2009. Dietary assessment methods for micronutrient intake: a systematic review on vitamins. Br. J. Nutr. 102, S10–S37.

Interdepartmental Committee on Nutrition for National Defense, 1963. Manual for Nutrition Surveys, second ed. U.S. Government Printing Office, Washington, DC. 327 pp.

Keats, S., Wiggins, S., 2014. Future Diets: Implications for Agriculture and Food Prices. ODI. 116 pp.

Pelletier, D.L., Olson, C.L., Frongillo, E.A., 2012. Food insecurity, hunger and undernutrition. In: Erdman, J.W., Macdonald, I.A., Zeisel, S.H. (Eds.), Present Knowledge in Nutrition, tenth ed. Wiley-Blackwell, New York, pp. 1165–1181. Chapter 68.

Román-Vinas, B., Serra-majem, L., Ribas-Barba, L., et al., 2009. Overview of methods used to evaluate the adequacy of nutrient intakes for individuals and populations. Br. J. Nutr. 101, S6–S11.

Sauberlich, H.E., 1999. Laboratory Tests for the Assessment of Nutritional Status, second ed. CRC Press, Cleveland, Ohio. 486 pp.

von Grebmer, K., Saltzman, A., Birol, E., et al., 2014. Global Hunger Index: The Challenge of Hidden Hunger. IFPR, Washington. 56 pp.

Appendix A

Current and Obsolete Designations of Vitamins (Bolded) and Other Vitamin-Like Factors

Name	Explanation
Aneurin	Infrequently used synonym for thiamin
A-N factor	Obsolete term for the *"antineuritic factor"* (thiamin)
Bios factors	Obsolete terms for yeast growth factors now known to include biotin
Citrovorum factor	Infrequently used term for a naturally occurring form of folic acid (N^5-formyl-5,6,7,8-tetrahydropteroylmonoglutamic acid), which is required for the growth of *Leuconostoc citrovorum*
Extrinsic factor	Obsolete term for the antianemic activity in liver, now called vitamin B_{12}
Factor U	Obsolete term for chick anti-anemic factor now known as a form of folate
Factor R	Obsolete term for chick antianemic factor now known as a form of folate
Factor X	Obsolete term used at various times to designate the rat fertility factor now called vitamin E and the rat growth factor now called vitamin B_{12}
Filtrate factor	Obsolete term for the antiblack tongue disease activity, now known to be niacin that could be isolated from the *"B_2 complex"* by filtration through fuller's earth; also used to describe the chick antidermatitis factor, now known to be pantothenic acid, isolated from acid solutions of the *"B_2 complex"* by filtration through fuller's earth
Flavin	Term originally used to describe the water-soluble fluorescent rat growth factors isolated from yeast and animal tissues; now, a general term for isoalloxazine derivatives including riboflavin and its active forms, FMN and FAD
Hepatoflavin	Obsolete term for the water-soluble rat growth factor, now known to be riboflavin, isolated from liver
Intrinsic factor	Accepted designation for the vitamin B_{12}-binding protein produced by gastric parietal cells and necessary for the enteric absorption of the cobalamins
Lactoflavin	Obsolete term for the water-soluble rat growth factor, now known to be riboflavin, isolated from whey
LLD factor	Obsolete term for the activity in liver that promoted the growth of *Lactobacillus lactis* Dorner, now known to be vitamin B_{12}
Norit eluate	Obsolete term for *Lactobacillus casei* growth promotant, factor now known as folic acid, that could be isolated from liver and yeasts by adsorption on norit
Ovoflavin	Obsolete term for the water-soluble rat growth factor, now known to be riboflavin, isolated from egg white
P–P factor	Obsolete term for the thermostable *"pellagra-preventive"* component, now known as niacin, of the *"water-soluble B"* activity of yeast

Continued

Name	Explanation
Rhizopterin	Obsolete synonym for the *"SLR factor"*, i.e., a factor from *Rhizobium* sp. fermentation that stimulated the growth of *Streptococcus lactis* R. (now called *Streptococcus faecalis*), which is now known to be a folate activity
SLR factor	Obsolete term for the *Streptococcus lactis* R. (now called *S. faecalis*) growth promotant later called *"rhizopterin"* and now known to be a folic acid activity
Streptogenin	A peptide present in liver and in enzymatic hydrolysates of casein and other proteins which promotes growth of mice and certain microorganisms (hemolytic streptococci and lactobacilli); not considered a vitamin
Vitamin A	Accepted designation of retinoids that prevent xerophthalmia and nyctalopia, and are essential for epithelial maintenance
Vitamin B	Original antiberiberi factor; now known to be a mixture of factors and designated as the vitamin B complex
Vitamin B complex	Term introduced when it became clear that *"water-soluble B"* contained more than one biologically active substance (such preparations were subsequently found to be mixtures of thiamin, niacin, riboflavin, pyridoxine, and pantothenic acid); the term has contemporary lay use as a nonspecific name for all of the B-designated vitamins
Vitamin B_1	Synonym for thiamin
Vitamin B_2	Synonym for riboflavin
Vitamin B_2 complex	Obsolete term for the thermostable "second nutritional factor" in yeast, which was found to be a mixture of niacin, riboflavin, pyridoxine, and pantothenic acid
Vitamin B_3	Infrequently used synonym for pantothenic acid; was also used for nicotinic acid
Vitamin B_4	Unconfirmed activity preventing muscular weakness in rats and chicks; believed to be a mixture of arginine, glycine, riboflavin, and pyridoxine
Vitamin B_5	Unconfirmed growth promotant for pigeons; probably niacin
Vitamin B_6	Synonym for pyridoxine
Vitamin B_7	Unconfirmed digestive promoter for pigeons; may be a mixture; also *"vitamin I"*
Vitamin B_8	Adenylic acid; no longer classified as a vitamin
Vitamin B_9	Unused designation
Vitamin B_{10}	Growth promotant for chicks; likely a mixture of folic acid and vitamin B_{12}
Vitamin B_{11}	Apparently the same as *"vitamin B_{10}"*
Vitamin B_{12}	Accepted designation of the cobalamins (cyano- and aquocobalamins) that prevent pernicious anemia and promote growth in animals
Vitamin B_{12a}	Synonym for aquacobalamin
Vitamin B_{12b}	Synonym for hydroxocobalamin
Vitamin B_{12c}	Synonym for nitritocobalamin
Vitamin B_{13}	Synonym for orotic acid, an intermediate of pyrimidine metabolism; not considered a vitamin
Vitamin B_{14}	Unconfirmed
Vitamin B_{15}	Synonym for *"pangamic acid"*; no proven biological value
Vitamin B_{17}	Synonym for laetrile, a cyanogenic glycoside with unsubstantiated claims of anticarcinogenic activity; not considered a vitamin
Vitamin B_c	Obsolete term for pteroylglutamic acid
Vitamin B_p	Activity preventing perosis in chicks; replaceable by choline and Mn
Vitamin B_t	Activity promoting insect growth; identified as carnitine
Vitamin B_x	Activity associated with pantothenic acid and *p*-aminobenzoic acid
Vitamin C	Accepted designation of the antiscorbutic factor, ascorbic acid
Vitamin C_2	Unconfirmed antipneumonia activity; also called *"vitamin J"*
Vitamin D	Accepted designation of the antirachitic factor (the calciferols)
Vitamin D_2	Accepted designation for ergocalciferol (a vitamin D-active substance derived from plant **sterols**)

Name	Explanation
Vitamin D_3	Accepted designation for cholecalciferol (a vitamin D-active substance derived from animal sterols)
Vitamin E	Accepted designation for tocopherols active in preventing myopathies and certain types of infertility in animals
Vitamin F	Obsolete term for essential fatty acids; also an abandoned term for thiamin activity
Vitamin G	Obsolete term for riboflavin activity; also an abandoned term for the *"pellagra-preventive factor"* (niacin)
Vitamin H	Obsolete term for biotin activity
Vitamin I	Mixture also formerly called *"vitamin B_7"*
Vitamin J	Postulated antipneumonia factor also formerly called *"vitamin C_2"*
Vitamin K	Accepted designation for activity preventing hypoprothrombinemic hemorrhage shared by related naphthoquinones
Vitamin K_1	Accepted designation for phylloquinones (vitamin K-active substances produced by plants)
Vitamin K_2	Accepted designation for prenylmenaquinones (vitamin K-active substances synthesized by microorganisms and produced from other vitamers K by animals)
Vitamin K_3	Accepted designation for menadione (synthetic vitamin K-active substance not found in nature)
Vitamin L_1	Unconfirmed liver filtrate activity, probably related to anthranilic acid, proposed as necessary for lactation
Vitamin L_2	Unconfirmed yeast filtrate activity, probably related to adenosine, proposed as necessary for lactation
Vitamin M	Obsolete term for antianemic factor in yeast now known to be pteroylglutamic acid
Vitamin N	Obsolete term for a mixture proposed to inhibit cancer
Vitamin O	Unused designation
Vitamin P	Activity reducing capillary fragility related to citrin, which is no longer classified as a vitamin
Vitamin Q	Unused designation (the letter was used to designate coenzyme Q)
Vitamin R	Obsolete term for folic acid; from Norris' chick antianemic *"factor R"*
Vitamin S	Chick growth activity related to the peptide *"streptogenin"*; the term was also applied to a bacterial growth activity probably related to biotin
Vitamin T	Unconfirmed group of activities isolated from termites, yeasts, or molds and reported to improve protein utilization in rats
Vitamin U	Unconfirmed activity from cabbage proposed to cure ulcers and promote bacterial growth; may have folic acid activity
Vitamin V	Tissue-derived activity promoting bacterial growth; probably related to NAD
Wills' factor	Obsolete term for the antianemic factor in yeast now known to be a form of folate
Zoopherin	Obsolete term for a rat growth factor now known as vitamin B_{12}

Appendix B

Original Reports for Case Studies

Chapter 6	Case 1 McLaren, D.S., Ahirajian, E., Tchalian, M., et al., 1965. Xerophthalmia in Jordan. Am. J. Clin. Nutr. 17, 117–130.
	Case 2 Wechsler, H.L., 1979. Vitamin A deficiency following small-bowel bypass surgery for obesity. Arch. Dermatol. 115, 73–75.
	Case 3 Sauberlich, H.E., Hodges, R.E., Wallace, D.L., et al., 1974. Vitamin A metabolism and requirements in the human studied with the use of labeled retinol. Vit. Horm. 32, 251–275.
Chapter 7	Marx, S.J., Spiegel, A.M., Brown, E.M., et al., 1978. A familial syndrome of decrease in sensitivity to 1,25-dihydroxyvitamin D. J. Clin. Endocrinol. Metab. 47, 1303–1310.
Chapter 8	Case 1 Boxer, L.A., Oliver, J.M., Spielberg, S.P., et al., 1979. Protection of Granulocytes by vitamin E in glutathione synthetase deficiency. N. Eng. J. Med. 301, 901–905.
	Case 2 Harding, A.E., Matthews, S., Jones, S., et al., 1985. Spinocerebellar degeneration associated with a selective defect in vitamin E absorption. N. Eng. J. Med. 313, 32–35.
Chapter 9	Case 1 Colvin, B.T., Lloyd, M.J., 1977. Severe coagulation defect due to a dietary deficiency of vitamin K. J. Clin. Pathol. 30, 1147–1148.
	Case 2 Corrigan, J. and Marcus, F.I., 1974. Coagulopathy associated with vitamin E ingestion. J. Am. Med. Assoc. 230, 1300–1301.
Chapter 10	Case 1 Hodges, R.E., Hood, J., Canham, J.E., et al., 1971. Clinical manifestations of ascorbic acid deficiency in man. Am. J. Clin. Nutr. 24, 432–443.
	Case 2 Dewhurst, K., 1954. A case of scurvy simulating a gastric neoplasm. Br. Med. J. 2, 1148–1150.
Chapter 11	Case 1 Burwell, C.S., Dexter, L., 1947. Beriberi heart disease. Trans. Assoc. Am. Physiol. 60, 59–64.
	Case 2 Blass, J.P., Gibson, G.E., 1977. Abnormality of a thiamin-requiring enzyme in patients with Wernicke-Korsakoff syndrome. New Eng. J. Med. 297, 1367–1370.
Chapter 12	Dutta, P., Gee, M., Rivlin, R.S., et al., 1988. Riboflavin deficiency and glutathione metabolism in rats: possible mechanisms underlying altered responses to hemolytic stimuli. J. Nutr. 118, 1149–1157.
Chapter 13	Vannucchi, H., Moreno, F.S., 1989. Interaction of niacin and zinc metabolism in patients with alcoholic pellagra. Am. J. Clin. Nutr. 50, 364–369.

Continued

Chapter 14	Case 1 Barber, G.W., Spaeth, G.L., 1969. The successful treatment of homocystinuria with pyridoxine. J. Pediatr. 75, 463–478.
	Case 2 Schaumberg, H., Kaplan, J., Windebank, A., et al., 1983. Sensory neuropathy from pyridoxine abuse. New Eng. J. Med. 309, 445–448.
Chapter 15	Mock, D.M., DeLorimer, A.A., Liebman, W.M., et al., 1981. Biotin deficiency: An unusual complication of parenteral alimentation. New Eng. J. Med. 304, 820–823.
Chapter 16	Lacroix, B., Didier, E., Grenier, J.F., 1988. Role of pantothenic and ascorbic acid in wound healing processes: in vitro study on fibroblasts. Int. J. Vit. Nutr. Res. 58, 407–413.
Chapter 17	Freeman, J.M., Finkelstein, J.D., Mudd, S.H., 1975. Folate-responsive homocystinuria and schizophrenia. A defect in methylation due to deficient 5,10-methylenetetrahydrofolic acid reductase activity. New Eng. J. Med. 292, 491–496.
Chapter 18	Higginbottom, M.C., Sweetman, L., Nyhan, W.L., 1978. A syndrome of methylmalonic aciduria, homocystinuria, megaloblastic anemia and neurological abnormalities of a vitamin B_{12}-deficient breast-fed infant of a strict vegetarian. New Eng. J. Med. 299, 317–323.
Chapter 19	Killgore, J., Smidt, C., Duich, L., et al., 1989. Nutritional Importance of Pyrroloquinoline Quinone, Science 245, 850–852.
	Rucker, R., Chowanadisai, W., Nakano, M., 2009. Potential physiological importance of Pyrroloquinoline Quinone. Altern. Med. Rev. 14, 268–277.

Appendix C

A Core of Current Vitamin Literature

The following tables list journals publishing original research and reviews (Table 1) about the vitamins, as well as website (Table 2) and reference texts (Table 3) useful to students, researchers, and clinicians.

TABLE C-1 Journals Presenting Original Research and Reviews

Title	Publisher/URL[a,b]
Advances in Nutrition	American Society for Nutrition *www.advances.nutrition.org*
American Journal of Clinical Nutrition	American Society for Nutrition *www.ajcn.nutrition.org*
American Journal of Epidemiology	Oxford Journals *www.aje.oxfordjournals.org*
American Journal of Medicine	Alliance for Academic Internal Medicine *www.www.amjmed.com*
American Journal of Public Health	American Public Health Association *www.ajph.aphypublications.org*
Annals of Internal Medicine	American College of Physicians *www.annals.org*
Annals of Nutrition and Metabolism	Karger *www.karger.com/Journal/223977*
Annual Review of Biochemistry	Annual Reviews *www.annualreviews.org/journal/biochem*
Annual Review of Medicine	Annual Reviews *www.annualreviews.org/loi/med*
Annual Review of Nutrition	Annual Reviews *www.annualreviews.org/journal/nutr*
Australian Journal of Nutrition and Dietetics	Dietetics Association of Australia *www.ajnd.org.au*
BBA (Biochemica et Biophysical Acta)	Elsevier *www.journals.elsevier.com/bba-general-subjects*
Biochemical and Biophysical Research Communications	Elsevier *www.journals.elsevier.com/biochemical-and-biophysical-research-communications*

Continued

TABLE C-1 Journals Presenting Original Research and Reviews—cont'd

Title	Publisher/URL[a,b]
Biochemistry	American Chemical Society *www.pubs.acs.org/journal/bichaw*
Biofactors	ILR Press, Oxford *www.onlinelibrary.wiley.com/journal/10.1002/(ISSN)1872-8081*
British Journal of Nutrition	Nutrition Society *www.journals.cambridge.org/action/displayJournal?jid=BJN*
British Medical Journal	BMJ Publishing Group *www.bmj.com*
Cancer Epidemiology, Biomarkers and Prevention	American Association for Cancer research *www.cebp.aacrjournals.org/*
Cell	Cell Press *www.cell.com*
Clinical Biochemistry	Canadian Society of Clinical Chemists *www.journals.elsevier.com/clinical-biochemistry*
Clinical Chemistry	American Association of Clinical Chemists *www.clinchem.org*
Critical Reviews in Biochemistry and Molecular Biology	Taylor & Francis *www.tandfonline.com/loi/ibmg20#.V2mIL7grlfo*
Critical Reviews in Food Science and Nutrition	Taylor & Francis Group *www.tandfonline.com/loi/bfsn20#.V2mIL7grlfo*
Current Nutrition and Food Science	Bentham Science Publishers *www.eurekaselect.com/612*
Current Opinion in Clinical Nutrition and Metabolic Care	Lippincott Williams & Wilkins *www.journals.lww.com/co-clinicalnutrition/Pages/default.aspx*
Epidemiology	Lippincott Williams & Wilkins *www.journals.lww.com/epidem/Pages/default.aspx*
European Journal of Biochemistry (FEBS Journal)	Federation of European Biochemical Societies *www.febs.onlinelibrary.wiley.com/hub/journal/10.1111/(ISSN)1742–4658/issues/*
European Journal of Clinical Nutrition	Stockton Press *www.nature.com/ejcn/index.html*
FASEB Journal	Federation of American Societies for Experimental Biology *www.fasebj.org*
FEBS Letters	Federation of European Biochemical Societies *www.febs.onlinelibrary.wiley.com/hub/journal/10.1002/(ISSN)1873–3468/*
Gastroenterology	American Gastroenterology Association Institute *www.gastrojournal.org*
International Journal of Epidemiology	Oxford Journals *www.ide.oxforedjournjals.org*
International Journal of Food Sciences and Nutrition	Taylor & Francis *www.tandfonline.com/loi/iijf20#.V2mMeLgrlf0*
International Journal for Vitamin and Nutrition Research	Hogrefe & Huber Publishers *www.econtent.hogrefe.com/loi/vit*
International Journal of Nutrition and Metabolism	Academic Journals *www.academicjournals.org/IJNAM*
International Journal of Obesity	Nature Publishing Group *www.nature.com/ijo/index.html*

TABLE C-1 Journals Presenting Original Research and Reviews—cont'd

Title	Publisher/URL[a,b]
Journal of Agricultural and Food Chemistry	American Chemical Society *www.pubs.acs.org/journal/jafcau*
Journal of the Academy of Nutrition and Dietetics	Academy of Nutrition and Dietetics *www.andjrnl.org*
JAMA (Journal of the American Medical Association)	American Medical Association www.jama.jamanetwork.com/ journal.aspx
JAMA Internal Medicine	American Medical Association *www.archinte.ama-assn.org*
Journal of Biological Chemistry	American Society of Biological Chemists *www.jbc.org*
Journal of Clinical Biochemistry and Nutrition	Institute of Applied Biochemistry *www.jstage.jst.go.jp/browse/jcbn*
Journal of Food Composition and Analysis	Elsevier *www.journals.elsevier.com/journal-of-food-composition-and-analysis*
Journal of Immunology	American Association of Immunologists *www.jimmunol.org*
Journal of Lipid Research	American Society for Biochemistry and Molecular Biology *www.jlr.org*
Journal of Nutrition	American Society for Nutrition *www.jn.nutrition.org*
Journal of Nutritional Biochemistry	Elsevier *www.jnutbio.com*
Journal of Nutritional Sciences and Vitaminology	The Vitamin Society of Japan and Japanese Society of Nutrition and Food Science *www.jsnfs.or.jp/english/english_jnsv.html*
Journal of Parenteral and Enteral Nutrition	American Society of Parenteral and Enteral Nutrition *www.pen.sagepub.com*
Journal of Pediatric Gastroenterology and Nutrition	Lippincott Williams & Wilkens *www.journals.lww.com/jpgn/pages.default.aspx*
Lipids	American Oil Chemists Society *www.link.springer.com/journal/11745*
New England Journal of Medicine	Massachusetts Medical Society *http://www.nejm.org*
Nutrition Abstracts and Reviews Series A (human, experimental)	CABI *www.cabi.org/publishing-products/online-information-resources/nutrition-abstracts-and-reviews-series-ahuman-and-experimental*
Nutrition Abstracts and Reviews Series B (feeds, feeding)	CABI *www.cabi.org/publishing-products/online-information-resources/nutrition-abstracts-and-reviews-series-b-livestock-feeds-and-feeding*
Nutrition and Food Science	MICS Publishing Group *www.omicsonline.org/nutrition-food-sciences.php*
Nutrition in Clinical Care	Wiley *www.onlinelibrary.wiley.com/journal/10.1111/(ISSN)1523-5408*
Nutrition	Elsevier *www.journals.elsevier.com/nutrition/*
Nutrition Journal	Cell & Bioscience *www.nutritionj.biomedcentral.com*

Continued

TABLE C-1 Journals Presenting Original Research and Reviews—cont'd

Title	Publisher/URL[a,b]
Nutrition Research	Elsevier *www.journals.elsevier.com/nutrition-research*
Nutrition Research Reviews	The Nutrition Society *www.journals.cambridge.org/action/displayJournal?jid=NRR*
Nutrition Reviews	Wiley-Blackwell *www.onlinelibrary.wiley.com/journal/10.1111/(ISSN)1753-4887*
Nutrition Today	Lippincott Williams & Wilkins *www.journals.lww.com/nutritiontodayonline/pages/default.aspx*
Obesity	The Obesity Society *www.obesity.org/publications/obesity-journal*
PNAS (Proceedings of the National Academy of Sciences)	National Academy of Sciences (US) *www.pnas.org*
Proceedings of the Nutrition Society	Nutrition Society *www.journals.cambridge.org/action/displayJournal?jid=PNS*
Proceedings of the Society for Experimental Biology and Medicine	Society for Experimental Biology and Medicine *www.sebm.org/journal*

[a]*URL, uniform resource locator.*
[b]*Sites accessed June 21, 2016.*

TABLE C-2 Some Useful Websites

Programs/Information	URL[a,b]
United Nations	
Food and Agricultural Organization (FAO)	*www.fao.org*
Agriculture and Consumer Protection	*www.fao.org/ag/portal/ag-home/en*
Codex Alimentarius[c]	*www.fao.org/fao-who-codexalimentarius/en*
Committee of World Food Security	*www.fao.org/cfs/en*
Food Composition—INFOODS[d] project	*www.fao.org/infoods/infoods/en*
Hunger	*www.fao.org/hunger/en*
Nutritional Assessment	*www.fao.org/nutrition/assessment/en*
Nutrition Country Profiles	*www.fao.org/ag/agn/nutrition/profiles_en.stm*
Nutrition Education and Consumer Awareness	*www.fao.org/ag/humannutrition/nutritioneducation/en*
Nutrition Requirements	*www.fao.org/nutrition/requirements/en*
Statistics	*www.fao.org/statistics/en*
Sustainable Development Goals	*http://www.fao.org/sustainable-development-goals/home/en*
World Food Situation	*www.fao.org/worldfoodsituation/en*
UN University	*www.unu.edu*
World Health Organization (WHO)	*www.who.int/en*
Child Growth Standards	*www.who.int/childgrowth/en*
Global Database on Child Growth and Malnutrition	*www.who.int/nutgrowthdb/en*
Global Health Library	*www.globalhealthlibrary.net/php/index.php*

TABLE C-2 Some Useful Websites—cont'd

Programs/Information	URL[a,b]
Growth Reference Database	*www.who.int/healthinfo/indicators/2015/en*
Health Data and Statistics	*www.who.int/healthinfo/statistics/en*
Nutrition for Health and Development	*www.who.int/nmh/about/nhd/en*
Vitamin and Mineral Information Systems (VMNIS)	*www.who.int/vmnis/en*
World Health Statistics	*www.who.int/gho/publications/world_health_statistics/2014/en*
United States Government	
Let's Move	*www.letsmove.gov*
Nutrition.gov	*www.nutrition.gov*
Department of Agriculture (USDA)	*www.usda.gov/wps/portal/usdahome*
Ag. Res. Service National Program in Human Nutrition	*www.ars.usda.gov/research/programs/programs.htm?NP_CODE=107*
Center for Nutrition Policy and Promotion	*www.cnpp.usda.gov*
Child Nutrition Programs	*www.fns.usda.gov/school-meals/child-nutrition-programs*
ChooseMyPlate.gov	*www.choosemyplate.gov*
Dietary Assessment Tools	*www.choosemyplate.gov/tools-supertracker*
Dietary Guidelines for Americans	*www.cnpp.usda.gov/dietary-guidelines*
Food Availability (per capita) Data System	*www.ers.usda.gov/data-products/food-availability-(per-capita)-data-system.aspx*
Food and Nutrition Information Center	*www.fnic.nal.usda.gov*
National Agricultural Library	*www.nalusda.gov*
National Institute for Food and Agriculture (NIFA)	*www.nifa.usda.gov*
National Nutrient Database for Standard Reference	*www.ndb.nal.usda.gov*
Department of Defense	
U.S. Army Research Institute of Environmental Medicine	www.usariem.army.mil/index.cfm/about
Department of Health and Human Services	*www.hhs.gov*
Center for Disease Control and Prevention	*www.cdc.gov*
National Center for Health Statistics	*www.cdc.gov/nchs*
National Health and Nutrition Examination Survey (NHANES)	*www.cdc.gov/nchs/nhanes*
Food and Drug Administration (FDA)	*www.fda.gov*
Center for Food Safety and Applied Nutrition (CFSAN)	*www.fda.gov/AboutFDA/CentersOffices/OfficeofFoods/CFSAN*
Dietary Supplements	*www.fda.gov/Food/DietarySupplements*
National Institutes of Health (NIH)	*www.nih.gov*
Eunice Kennedy Shriver National Institute of Child Health and Human Development	*www.nichd.nih.gov/Pages/index.aspx*
National Cancer Institute	*www.cancer.gov*
National Human Genome Research Institute	*www.genome.gov*
National Heart, Lung and Blood Institute	*www.nhlbi.nih.gov*
National Institute of Diabetes and Digestive and Kidney Diseases	*www.niddk.nih.gov/Pages/default.aspx*

Continued

TABLE C-2 Some Useful Websites—cont'd

Programs/Information	URL[a,b]
National Institute on Aging	*www.nia.nih.gov*
National Library of Medicine	*www.nlm.nih.gov*
Office of Dietary Supplements	*www.ods.od.nih.gov*
Department of State	
Agency for International Development Global Health Initiative	*www.usaid.gov/what-we-do/global-health/cross-cutting-areas/global-health-initiative*
Professional Societies	
Academy of Nutrition and Dietetics	*www.eatright.org*
American Society for Nutrition	*www.nutrition.org*
University On-Line Resources	
Cornell University: "Cornell NutritionWorks"	*www.nutritionworks.cornell.edu/home*
Harvard University: "The Nutrition Source"	*www.hsph.harvard.edu/nutritionsource*
Johns Hopkins Bloomberg School of Public Health:"Johns Hopkins Public Health"	*www.magazine.jhsph.edu*
Tufts University Friedman School: "Nutrition & Health Newsletter"	*www.nutritionletter.tufts.edu*
University of California—Berkeley: "Berkeley Wellness"	*www.berkeleywellness.com*

[a]*URL, uniform resource locator.*
[b]*Sites accessed June 21, 2016.*
[c]*Joint program of FAO and WHO.*
[d]*International Network of Food Data Systems.*

TABLE C-3 A Book Shelf of Useful References

Bales, C.W., Locher, J.L., Saltzman, E., eds. 2015. "Handbook of Clinical Nutrition and Aging", 3rd Edition, Springer, New York, pp. 442.
Ball, G.F.M., 2005. "Vitamins in Foods: Analysis, Bioavailability and Stability", CRC Press, New York, pp. 824.
Bender, D.A., 2009. "Nutritional Biochemistry of the Vitamins", 2nd Edition, Cambridge University Press, Cambridge, pp. 516.
Berdanier, C.D., Dwyer, J.T., Heber, D., eds. 2014. "Handbook of Nutrition and Food", CRC Press, Boa Raton, FL, pp. 1113.
Berdanier, C.D., Moustaid-Moussa, N., eds. 2004. "Genomics and Proteomics in Nutrition", Marcel Dekker, New York, pp. 528.
Bhagavan, N.V., Ha, C.E., 2011. "Essentials of Medical Biochemistry", Elsevier, New York, pp. 581.
Bray, G.A., Bouchard, C., 2004. "Handbook of Obesity" 2nd Edition, Marcel Dekker, New York, pp. 1046.
Brody, S., 1945. "Bioenergetics and Growth", Waverly Press, Baltimore, pp. 1022.
Bronner, F., ed. 1997. "Nutrition Policy in Public Health", Springer, New York, pp. 363.
Cheeke, P.R., Dierenfeld, E.A., 2010. "Comparative Animal Nutrition and Metabolism", CABI, New York, pp. 336.
Chernoff, R., 2014. "Geriatric Nutrition" 4th Edition, Jones & Bartlett, New York, pp. 581.
Eitenmiller, R.R., Landen, Jr., W.O., Ye, L., 2007. "Vitamin Analyses for the Health and Food Sciences" 2nd Edition, CRC Press, New York, pp. 664.
Erdman, J.W., Macdonald, I.A. and Zeisel, S.H., eds. (2012) "Present Knowledge in Nutrition", 10th Edition, ILSI Press, Washington, D.C., pp. 1035.
Escott-Stump, S., 2008. "Nutrition and Diagnosis-Related Care" 6th Edition, Lippincott Williams & Wilkins, New York, pp. 948.

TABLE C-3 A Book Shelf of Useful References—cont'd

Food and Nutrition Board 1997. "Dietary Reference Intakes for Calcium, Phosphorus, Magnesium, Vitamin D and Fluoride", National Academy Press, Washington, DC, pp. 207.
Food and Nutrition Board 1998. "Prevention of Micronutrient Deficiencies: Tools for Policymakers and Public Health Workers", National Academy Press, Washington, DC, pp. 432.
Food and Nutrition Board 2000. "Dietary Reference Intakes for Thiamin, Riboflavin, Niacin, Vitamin B6, Folate, Vitamin B12, Pantothenic Acid, Biotin and Choline", National Academy Press, Washington, DC, pp. 564.
Food and Nutrition Board 2000. "Dietary Reference Intakes for Vitamin C, Vitamin E, Selenium and Carotenoids", National Academy Press, Washington, DC, pp. 506.
Food and Nutrition Board 2001. "Dietary Reference Intakes for Vitamin A, Vitamin K, Arsenic, Boron, Chromium, Copper, Iodine, Iron, Manganese, Molybdenum, Nickel, Silicon, Vanadium and Zinc", National Academy Press, Washington, DC, pp. 773.
Food and Nutrition Board 2003. "Dietary Reference Intakes: Applications in Dietary Planning", National Academy Press, Washington, DC, pp. 237.
Food and Nutrition Board 2003. "Dietary Reference Intakes: Guiding Principles for Nutrition Labeling and Fortification", National Academy Press, Washington, DC, pp. 205.
Food and Nutrition Board 2010. "Dietary Reference Intakes: Calcium, Vitamin D", National Academy Press, Washington, DC, pp. 1105.
Gauch, Jr., H.G., 2003. "Scientific Method in Practice", Cambridge University Press, pp. 435.
Gibson, R.S., 1990. "Principles of Nutritional Assessment", Oxford University Press, New York, pp. 525.
Goldberg, G., ed. 2003. "Plants: Diet and Health", Blackwell, Oxford, pp. 349.
Ho, E., Domann, F., 2015. "Nutrition and Epigenetics", CRC Press, Boca Raton, FL, pp. 404.
Insel, P., Ross, D., McMahon, K. et al., 2014. "Nutrition" 5thrd Edition, Jones and Bartlett, Burlington, MA, pp. 960.
Kleiber, M., 1975. "The Fire of Life: an Introduction to Animal Energetics", Kreiger publishing, Huntington, NY, pp. 453
Kohlmeier, M., 2013. "Nutrigenetics: Applying the Science of Personal Nutrition", Elsevier, New York, pp. 384.
Leeson, S., Summers, J.D., 2001. "Scott's Nutrition of the Chicken", 4th Edition, University Press, Toronto, pp. 535.
Mahan, L.K., Escott-Stump, S., Raymond, J.L., 2012. "Krause's Food, Nutrition, & Diet Therapy", 13th Edition, W.B., Saunders, Philadelphia, pp. 1227.
Maulik, N., Maulik, G., 2011. "Nutrition, Epigenetic Mechanisms, and Human Disease", CRC Press, New York, pp. 426.
McDonald, P., Edwards, R.A., Greenhalgh, J.F. et al., 2010. "Animal Nutrition", 7th Edition, Benjamen-Cummings, New York, pp. 692.
McDowell, L.R., 1989. "vitamins in Animal Nutrition", Academic Press, New York, pp. 486.
Nelson, D.L., Cox, M.M., 2005. "Lehninger: Principles of Biochemistry" 4th Edition, Freeman & Co., New York, pp. 1119.
O'Neil, M., ed. 2013. "The Merck Index: an Encyclopedia of Chemicals, Drugs and Biologicals" 15th Edition, RSC publishing, New York, pp. 2708.
Ottaway, P.B., ed. 1999. "The Technology of Vitamins in Food", Aspen Publishers, Gaithersburg, Md., pp. 270.
Pond, W.G., Church, D.C., Pond, K.R., et al. 2005. "Basic Animal Nutrition and Feeding", 5th Edition, John Wiley & Sons, New York, pp. 580.
Ross, A.C., Caballero, B., Cousins, R.J. et al., eds. 2012. "Modern Nutrition in Health and Disease", 11th Edition, Lippincott Williams & Wilkins, New York, pp. 1648.
Sauberlich, H.E., 1999. "Laboratory Tests for the Assessment of Nutritional Status", 2nd Edition, CRC Press, New York, pp. 486.
Stein, N., 2015. "Public Health Nutrition: Principles and Practices in Community and Global Health", Jones & Bartlett, Burlington, MA, pp. 524.
Stipanuk, M.H. and Caudill, M.A., eds. 2014. "Biochemical and Physiological Aspects of Human Nutrition", 3rd Edition, W.B. Sanders, New York, pp. 948.
Villamena, F.A., 2013. "Molecular Basis of Oxidative Stress: Chemistry, Mechanisms and Disease Pathogenesis", Wiley, New York, pp. 420.
Whitney, E., Rolfes, S.R., 2012. "Understanding Nutrition" 13th Edition, Wadsworth publishing, New York, pp. 928.
Zemplini, J., Suttie, J.W., Gregory, J.F. et al., eds. 2014. "Handbook of Vitamins", 5th Edition, CRC Press, New York, pp. 591.

Appendix D

Vitamin Contents of Foods (units per 100 g Edible Portion)

Food, by Major Food Group	Vitamin A (IU)	Vitamin D (IU)	Vitamin E (mg)	Vitamin K (mg)	Vitamin C (mg)	Thiamin (mg)	Riboflavin (mg)	Niacin (mg)	Vitamin B6 (mg)	Pantothenic Acid (mg)	Folate (µg)	Vitamin B12 (mg)
Cereals												
Barley, pearled, cooked	7	0	0.01	0.08	0	0.083	0.062	2.063	0.115	0.135	16	0
Buckwheat groats, RSTD, cooked	0	0	0.09	1.9	0	0.04	0.039	0.94	0.077	0.359	14	0
Bulgur, cooked	2	0	0.01	0.5	0	0.057	0.028	1	0.083	0.344	18	0
Corn flour, whole grain, yellow	214	0	0.42	0.3	0	0.246	0.08	1.9	0.37	0.658	25	0
Cornmeal, whole grain, yellow	214	0	0.42	0.3	0	0.385	0.201	3.632	0.304	0.425	25	0
Couscous, cooked	0	0	0.013	0.1	0	0.063	0.027	0.983	0.051	0.371	15	0
Hominy, canned, white	1	0	0.05	0.2	0	0.003	0.006	0.033	0.005	0.154	1	0
Hominy, canned, yellow	110	0	NV	NV	0	0.003	0.006	0.033	0.005	0.154	1	0
Millet, cooked	3	0	0.02	0.3	0	0.106	0.082	1.33	0.108	0.171	19	0
Noodles, Chinese, chow mein	0	0	2.3	1.4	0	0.578	0.421	5.95	0.11	0.533	22	0
Noodles, egg, CKD, ENR	21	4	0.17	0	0	0.289	0.136	2.097	0.046	0.263	7	0.09
Noodles, egg, spinach, cooked, ENR	103	4	0.55	101	0	0.245	0.123	1.474	0.114	0.233	21	0.14
Noodles, Japanese, soba, CKD	0	0	NV	NV	0	0.094	0.026	0.51	0.04	0.235	7	0
Noodles, Japanese, somen, CKD	0	0	NV	NV	0	0.02	0.033	0.097	0.013	0.172	2	0
Oat bran, cooked	0	0	NV	NV	0	0.16	0.034	0.144	0.025	0.217	6	0
Oats	0	0	NV	NV	0	0.763	0.139	0.961	0.119	1.349	56	0
Rice, brown, long grain, CKD	0	0	0.17	0.2	0	0.178	0.069	2.561	0.123	0.38	9	0
Rice, white, glutinous, CKD	0	0	0.04	0	0	0.02	0.013	0.29	0.026	0.215	1	0
Rice, white, long grain, parboiled, CKD, ENR	0	0	0.01	0	0	0.212	0.019	2.309	0.156	0.323	3	0

Rice, white, long grain, REG, CKD	0	0	0.04	0	0	0.02	0.013	0.4	0.093	0.39	3	0
Rye flour, medium	0	0	1.43	5.9	0	0.287	0.114	1.727	0.268	0.492	19	0
Semolina, enriched	0	0	0.26	NV	0	0.811	0.571	5.99	0.103	0.58	72	0
Sorghum	0	0	0.5	NV	0	0.332	0.096	3.688	0.443	0.367	20	0
Pasta, CKD, ENR	0	0	0.06	0	0	0.274	0.136	1.689	0.049	0.112	7	0
Spaghetti, spinach, CKD	152	0	NV	NV	0	0.097	0.103	1.53	0.096	0.183	12	0
Pasta, whole wheat, CKD	4	0	0.23	0.6	0	0.156	0.099	3.126	0.093	0.268	21	0
Tapioca, pearl, dry	0	0	0	0	0	0.004	0	0	0.008	0.135	4	0
Wheat bran, crude	9	0	1.49	1.9	0	0.523	0.577	13.58	1.303	2.181	79	0
Wheat flour, whole grain	9	0	0.71	1.9	0	0.502	0.165	4.957	0.407	0.603	44	0
Wheat flour, white, all purpose, ENR, bleached	0	0	0.06	0.3	0	0.785	0.494	5.904	0.044	0.438	26	0
Wheat germ, crude	0	0	NV	NV	0	1.882	0.499	6.813	1.3	2.257	281	0
Wild rice, cooked	3	0	0.24	0.5	0	0.052	0.087	1.287	0.135	0.154	26	0
Breads, Cakes, and Pastries												
Bagels, plain, ENR	0	0	7.46	1.2	0	0.568	0.344	4.515	0.07	0.407	24	0
Biscuits, plain/ buttermilk	2	0	1.32	4.1	0	0.427	0.292	3.352	0.047	0.3	12	0.14
Bread, cornbread, w/ 2% milk	277	NV	NV	NV	0.3	0.291	0.294	2.254	0.113	0.339	19	0.15
Bread, cracked wheat	0	NV	NV	NV	0	0.358	0.24	3.671	0.304	0.512	39	0.03
Bread, French/Vienna/ sourdough	1	0	0.2	0.7	0	0.71	0.427	4.817	0.107	0.455	56	0
Bread, Irish soda	194	NV	NV	NV	0.8	0.298	0.269	2.405	0.083	0.25	10	0.05
Bread, Italian	1	0	0.29	1.2	0	0.473	0.292	4.381	0.048	0.378	30	0
Bread, mixed grain	0	0	0.37	1.4	0.1	0.279	0.131	4.042	0.263	0.336	75	0
Bread, oat bran	5	0	0.44	1.2	0	0.504	0.346	4.831	0.073	0.581	25	0
Bread, oatmeal	16	0	0.48	1.5	0	0.399	0.24	3.136	0.068	0.341	27	0.02
Bread, pita, white, enriched	0	0	0	0.2	0	0.599	0.327	4.632	0.034	0.397	24	0

Continued

Food, by Major Food Group	Vitamin A (IU)	Vitamin D (IU)	Vitamin E (mg)	Vitamin K (mg)	Vitamin C (mg)	Thiamin (mg)	Riboflavin (mg)	Niacin (mg)	Vitamin B6 (mg)	Pantothenic Acid (mg)	Folate (µg)	Vitamin B12 (mg)
Bread, pita, whole wheat	6	0	0.61	1.4	0	0.339	0.08	2.84	0.231	0.548	35	0
Bread, pumpernickel	0	0	0.42	0.8	0	0.327	0.305	3.091	0.126	0.404	34	0
Bread, raisin, enriched	0	0	0.28	1.7	0.1	0.339	0.398	3.466	0.069	0.387	34	0
Bread, rye	7	0	0.33	1.2	0.4	0.434	0.335	3.805	0.075	0.44	51	0
Bread, wheat bran	0	0	0.32	1.3	0	0.397	0.287	4.402	0.176	0.536	25	0
Bread, wheat germ	4	0	0.57	1.2	0.3	0.332	0.379	4.458	0.096	0.313	55	0.07
Bread, wheat	2	0	0.19	4.9	0.2	0.415	0.253	5.62	0.111	0.436	65	0
Bread, white	1	0	0.24	3.4	0	0.415	0.337	3.926	0.063	0.278	23	0.02
Bread, whole wheat	4	0	0.63	9	0	0.376	0.284	5.732	0.237	0.722	52	0
Cake, angel food	0	NV	NV	NV	0	0.102	0.491	0.883	0.031	0.198	3	0.06
Cake, Boston cream pie	82	5	0.15	3.1	0.2	0.408	0.27	0.191	0.026	0.301	8	0.16
Cake, fruitcake	22	0	0.9	1.5	0.5	0.05	0.099	0.791	0.046	0.226	3	0.01
Cake, gingerbread	48	NV	NV	NV	0.1	0.19	0.162	1.738	0.19	0.375	8	0.06
Cake, pound	241	34	0.65	1.7	0	0.173	0.249	1.105	0.036	0.485	28	0.36
Cake, shortcake, biscuit type	72	NV	NV	NV	0.2	0.311	0.272	2.573	0.03	0.248	10	0.07
Cake, sponge	154	9	0.24	0.2	0	0.243	0.269	1.932	0.052	0.478	13	0.24
Cake, white, w/o frosting	52	NV	0.12	5.1	0.2	0.186	0.242	1.533	0.021	0.184	7	0.08
Cake, yellow, w/o frosting	139	NV	NV	NV	0.2	0.183	0.233	1.456	0.036	0.31	10	0.16
Cheesecake	547	18	0.56	4.4	0.4	0.028	0.193	0.195	0.052	0.571	15	0.17
Cookies, animal crackers	0	0	0.12	5.9	0	0.35	0.326	3.47	0.022	0.376	14	0.05
Cookies, brownies	69	0	0.15	6.5	0	0.255	0.21	1.721	0.035	0.547	12	0.07
Cookies, butter, ENR	673	16	0.58	1.7	0	0.37	0.335	3.19	0.036	0.488	6	0.36
Cookies, choc chip, low fat	1	0	1.47	4.1	0	0.263	0.185	1.982	0.022	0.165	12	0
Cookies, choc sand-wich, w/ creme filling	5	0	1.58	11.2	0	0.095	0.204	1.373	0.044	0.178	7	0.05
Cookies, fig bars	33	0	0.65	5.8	0.3	0.158	0.217	1.874	0.075	0.364	10	0.09

Cookies, fortune	3	0	0.03	1.1	0	0.182	0.13	1.84	0.013	0.297	10	0.01
Cookies, gingersnaps	2	0	0.97	2.5	0	0.2	0.293	3.235	0.098	0.38	6	0
Cookies, graham crackers	0	0	1.51	14.3	0	0.265	0.317	4.439	0.001	0.005	19	0
Cookies, molasses	0	0	0.11	5.5	0	0.355	0.264	3.031	0.104	0.411	7	0
Cookies, oatmeal	5	0	0.26	8	0.5	0.267	0.23	2.227	0.066	0.386	7	0
Cookies, peanut butter	32	0	3.53	4.4	0	0.208	0.208	3.86	0.122	0.429	27	0.24
Cookies, raisin, soft type	5	0	2.24	3.9	0.4	0.216	0.206	1.967	0.052	0.244	9	0.03
Cookies, vanilla wafers	8	0	0.23	6	0	0.275	0.32	3.106	0.073	0.41	9	0.13
Crackers, cheese	156	1	2.19	9.4	0	0.562	0.338	6.113	0.17	0.472	25	0.34
Crackers, cheese, w/ peanut butter filling	2	0	2.37	12	0	0.552	0.294	5.831	0.151	0.485	25	0.28
Crackers, matzo	0	0	0.06	0.3	0	0.387	0.291	3.892	0.115	0.443	14	0
Crackers, melba toast	0	0	0.43	0.9	0	0.413	0.273	4.113	0.098	0.693	26	0
Crackers, rusk toast	41	NV	NV	NV	0	0.404	0.399	4.625	0.038	0.406	64	0.07
Crackers, rye, wafers	5	0	0.8	5.7	0.1	0.427	0.289	1.581	0.271	0.569	45	0
Crackers, rye	0	0	0.81	6	0	0.243	0.145	1.04	0.21	0.676	22	0
Crackers, saltiness	2	0	1.15	25.4	0	0.702	0.487	6.442	0.086	0.536	17	0.09
Crackers, wheat	0	0	1.55	14.2	0	0.284	0.146	4.022	0.244	0.577	36	0
Croissants, butter	206	0	0.84	1.8	0.2	0.388	0.241	2.188	0.058	0.861	28	1.6
Croutons, plain	0	NV	NV	NV	NV	0.623	0.272	5.439	0.026	0.429	22	0
Danish pastry, cheese	128	2	0.35	6.9	0.1	0.19	0.26	2	0.04	0.304	25	0.2
Danish pastry, fruit, ENR	51	0	0.34	5.3	3.9	0.263	0.22	1.992	0.043	0.634	16	0.09
Doughnuts, cake type, plain	16	0	2.02	7.9	1.3	0.163	0.125	1.59	0.024	0.126	28	0.1
Doughnuts, cake type, plain, sugared/glazed	10	NV	NV	NV	0.1	0.233	0.198	1.512	0.027	0.435	12	0.024
English muffins, plain, ENR	0	0	0.31	1.2	1.8	0.477	0.25	4.07	0.054	0.363	40	0.04
English muffins, wheat	1	0	0.45	0.8	0	0.431	0.292	3.356	0.09	0.444	39	0
French toast, w/ low-fat (2%) milk	503	NV	NV	NV	0.3	0.204	0.321	1.628	0.074	0.549	23	0.31

Continued

Food, by Major Food Group	Vitamin A (IU)	Vitamin D (IU)	Vitamin E (mg)	Vitamin K (mg)	Vitamin C (mg)	Thiamin (mg)	Riboflavin (mg)	Niacin (mg)	Vitamin B6 (mg)	Pantothenic Acid (mg)	Folate (μg)	Vitamin B12 (mg)
Hush puppies	186	0	1.26	23.6	0.2	0.352	0.332	2.782	0.102	0.357	20	0.19
Muffins, blueberry	73	4	1.63	39.2	0.9	0.168	0.163	1.418	0.04	0.47	12	0.16
Muffins, corn	208	0	0.8	2.3	0	0.273	0.326	2.037	0.084	0.444	34	0.09
Muffins, plain, w/ low-fat (2%) milk	0	NV	NV	NV	0.3	0.284	0.301	2.308	0.042	0.351	13	0.15
Pancakes, blueberry	199	NV	NV	NV	2.2	0.195	0.272	1.524	0.049	0.395	12	0.2
Pancakes, plain	196	NV	0	NV	0.3	0.201	0.281	1.567	0.046	0.405	12	0.22
Pie, apple, ENR FLR	124	0	1.52	3.5	3.2	0.028	0.027	0.263	0.038	0.119	4	0.001
Pie, banana cream	237	32	0.4	6.3	1.6	0.139	0.207	1.054	0.133	0.388	11	0.25
Pie, blueberry	140	0	1.04	10.5	2.7	0.01	0.03	0.3	0.037	0.136	4	0.01
Pie, cherry	237	0	0.76	7.6	0.9	0.023	0.029	0.2	0.041	0.319	8	0.01
Pie, chocolate crème	166	3	1.09	9.5	0	0.087	0.14	0.605	0.02	0.232	6	0.12
Pie, coconut crème	90	6	0.15	5.3	0	0.05	0.08	0.2	0.068	0.24	0.12	0.19
Pie, lemon meringue	173	7	0	2.1	3.2	0.062	0.209	0.649	0.03	0.793	8	0.17
Pie, peach	140	0	0.94	3.3	0.9	0.061	0.033	0.2	0.023	0.114	4	0
Pie, pecan	175	3	0.8	15.5	0	0.204	0.078	1.31	0.038	0.391	17	0.12
Pie, pumpkin	3434	2	0.76	13.2	0	0.177	0.124	1.107	0.063	0.452	10	0.35
Rolls, dinner, plain	5	0	0.28	10.6	0.2	0.526	0.374	5.367	0.097	0.452	30	0.13
Rolls, dinner, wheat	0	0	0.361	2.7	0	0.433	0.273	4.072	0.076	0.364	15	0
Rolls, French	0	0	0.3	1.8	0	0.523	0.3	4.352	0.039	0.452	33	0
Rolls, hamburger/hot-dog, plain	107	0	0.27	4.8	1.3	0.543	0.297	4.18	0.063	0.555	41	0.2
Rolls, hard (including kaiser)	0	0	0.42	0.6	0	0.478	0.336	4.239	0.035	0.41	15	0
Strudel, apple	30	0	1.42	2.9	1.7	0.04	0.025	0.33	0.046	0.27	6	0.22
Sweet rolls, cinnamon w/ raisins	214	0	1.99	4.4	2	0.324	0.265	2.384	0.107	0.406	24	0.14
Taco shells, baked	17	0	0.69	8.6	0	0.226	0.08	1.867	0.203	NV	6	0
Waffles, plain	228	NV	NV	NV	0.4	0.263	0.347	2.073	0.056	0.485	15	0.25
Wonton wrappers	14	NV	NV	NV	0	0.519	0.378	5.424	0.03	0.025	17	0.02

Breakfast Cereals											
All bran	1747	170	1.19	5.2	20	2.27	2.71	14.8	12	1.06	41	18.1
Corn flakes	3591	286	0.02	0	65	4.83	1.74	21.03	1.907	0.099	NV	5.36
Corn grits, CKD w/ water	0	0	0.03	0	0	0.086	0.058	0.79	0.046	0.046	14	0
Cream of rice, CKD w/ water	0	NV	0.02	NV	0	0.071	0.18	1.039	0.027	0.076	3	0
Cream of wheat, CKD w/ water	793	NV	NV	0.1	0	0.132	0.21	3.093	0.309	0.082	6	0
Farina, CKD w/ water	0	NV	0.04	0	0	0.126	0.065	1.493	0.096	0.256	17	0
Granola (homemade)	19	0	11.1	5.3	1.2	0.548	0.354	2.739	0.37	0.752	84	0
Oat bran	100	0	NV	NV	0	0.97	0.3	0.8	0.11	0.85	38	0
Oatmeal, instant regular	0	NV	0.47	NV	0	0.73	0.14	0.78	0.12	NV	32	0
Puffed rice	0	NV	NV	NV	0	2.6	1.8	35.3	0.075	0.32	19	0
Puffed wheat	0	NV	NV	NV	0	2.6	1.8	35.3	0.17	0.518	32	0
Raisin bran	1271	68	0.54	1.9	0.9	0.6	0.7	8.5	0.8	0.206	19	2.5
Rice cereal, crispy style	1515	121	0	0	18.2	1.13	1.28	15.14	1.51	NV	4	4.55
Rice cereal, check style	1852	148	0.35	1	22.2	1.39	1.6	18.5	1.85	NV	4	5.59
Shredded wheat	0	0	0	NV	0	0.226	0.105	5.24	0.256	NV	51	0
Wheat flakes	2586	138	46.35	1.4	207	5.17	5.86	69	6.9	34.5	19	21
Wheat germ, toasted	103	0	15.99	4	6	1.67	0.82	5.59	0.978	1.387	352	0
Vegetables												
Alfalfa seeds, sprouted, raw	155	0	0.02	30.5	8.2	0.076	0.126	0.481	0.034	0.563	36	0
Amarnath leaves, BLD, DRND	2770	0	NV	NV	41.1	0.02	0.134	0.559	0.177	0.062	57	0
Artichokes, BLD, DRND	13	0	0.19	14.8	7.4	0.5	0.89	1	0.081	0.24	89	0
Asparagus, BLD, DRND	1006	0	1.5	50.6	7.7	0.162	0.139	1.084	0.079	0.225	149	0
Balsam-pear (bitter gourd), tips, BLD, DRND	2416	0	1.45	163	55.6	0.147	0.282	0.995	0.76	0.06	88	0
Bamboo shoots, BLD, DRND	0	0	NV	NV	0	0.02	0.05	0.3	0.098	0.066	2	0
Beans, navy, sprouted, BLD, DRND	4	0	NV	NV	17.3	0.381	0.235	1.263	0.198	0.854	106.3	0

Continued

Food, by Major Food Group	Vitamin A (IU)	Vitamin D (IU)	Vitamin E (mg)	Vitamin K (mg)	Vitamin C (mg)	Thiamin (mg)	Riboflavin (mg)	Niacin (mg)	Vitamin B6 (mg)	Pantothenic Acid (mg)	Folate (μg)	Vitamin B12 (mg)
Beans, pinto, immature, FRZ, BLD, DRND	0	0	NV	NV	0.7	0.274	0.108	0.632	0.194	0.258	34	0
Beans, snap, green, BLD, DRND	633	0	0.46	47.9	9.7	0.074	0.097	0.614	0.056	0.074	33	0
Beans, snap, yellow, BLD, DRND	81	0	0.46	47.9	9.7	0.074	0.097	0.614	0.056	0.074	33	0
Beet greens, BLD, DRND	7654	0	1.81	484	24.9	0.117	0.289	0.499	0.132	0.329	14	0
Beets, BLD, DRND	35	0	0.04	0.2	3.6	0.027	0.04	0.331	0.067	0.145	80	0
Beets, pickled, CND, w/ liquid	49	0	0.06	0.3	2.3	0.01	0.048	0.251	0.05	0.137	27	0
Broadbeans, immature, BLD, DRND	270	0	NV	NV	19.8	0.128	0.09	1.2	0.029	0.066	58	0
Broccoli, BLD, DRND	1548	0	1.45	141	64.9	0.063	0.123	0.553	0.2	0.616	108	0
Broccoli, raw	623	0	0.78	102	89.2	0.071	0.117	0.639	0.175	0.573	63	0
Cabbage, Chinese (bak-choi), BLD, DRND	4249	0	0.09	34	26	0.032	0.063	0.428	0.166	0.079	41	0
Cabbage, BLD, DRND	80	0	0.14	109	37.5	0.061	0.038	0.248	0.112	0.174	30	0
Cabbage, raw	98	0	0.15	76	36.6	0.061	0.04	0.234	0.124	0.212	43	0
Cabbage, red, BLD, DRND	33	0	0.12	47.6	34.4	0.071	0.06	0.382	0.225	0.154	24	0
Cabbage, savoy, BLD, DRND	889	0	NV	NV	17	0.051	0.02	0.024	0.152	0.159	46	0
Carrots, baby, raw	13,796	0	NV	9.4	2.6	0.03	0.036	0.556	0.105	0.401	27	0
Carrots, BLD, DRND	17,033	0	0	13.7	3.6	0.066	0.036	0.645	0.153	0.232	14	0
Carrots, FRZ, BLD, DRND	16,928	0	1.01	13.6	2.3	0.03	0.037	0.416	0.084	0.174	11	0
Carrots, raw	16,706	0	0.66	13.2	5.9	0.066	0.058	0.983	0.138	0.273	19	0
Cassava, raw	13	0	0.19	1.9	20.6	0.087	0.048	0.854	0.088	0.107	27	0
Catsup	527	0	1.46	3	4.1	0.011	0.166	1.434	0.158	0.047	9	0
Cauliflower, BLD, DRND	12	0	0.07	13.8	44.3	0.042	0.052	0.41	0.173	0.508	44	0
Cauliflower, raw	0	0	0	15.5	48.2	0.05	0.06	0.507	0.184	0.667	57	0
Celery, raw	449	0	0.27	29.3	3.1	0.02	0.057	0.32	0.074	0.246	36	0

Chard, Swiss, BLD, DRND	6124	0	1.89	327	18	0.034	0.086	0.36	0.085	0.163	9	0
Chives, raw	4353	0	0.21	213	58.1	0.078	0.115	0.647	0.138	0.324	105	0
Collards, BLD, DRND	7600	0	0	407	18.2	0.04	0.106	0.575	0.128	0.218	16	0
Coriander, raw~	6748	0	0	310	27	0.067	0.162	1.114	0.149	0.57	62	0
Corn, sweet, yellow, BLD, DRND	263	0	0	0.4	5.5	0.093	0.057	1.683	0.139	0.792	23	0
Corn, sweet, yellow, raw	187	0	0	0.3	6.8	0.155	0.055	1.77	0.093	0.717	42	0
Corn, sweet, yellow, CND, w/ liquids	34	0	0	0	2.6	0.015	0.015	0.884	0.037	0.522	38	0
Cowpeas (blackeyes), BLD, DRND	75	0	0	36.8	2.6	0.26	0.064	0.728	0.095	0.213	141	0
Cucumber, w/ peel, raw	105	0	0	16.4	2.8	0.027	0.033	0.098	0.04	0.259	7	0
Dandelion greens, raw	10,161	0	3.44	778	35	0.19	0.26	0.806	0.251	0.084	27	0
Eggplant, BLD, DRND	37	0	0.41	2.9	1.3	0.076	0.02	0.6	0.086	0.075	14	0
Endive, raw	2167	0	0.44	231	6.5	0.08	0.075	0.4	0.02	0.9	142	0
Garlic, raw	9	0	0.08	1.7	31.2	0.2	0.11	0.7	1.235	0.596	3	0
Ginger root, raw	0	0	0.26	0.1	5	0.025	0.034	0.75	0.16	0.203	11	0
Gourd, calabash, BLD, DRND	0	0	NV	NV	8.5	0.029	0.022	0.39	0.038	0.144	4	0
Hearts of palm, canned	0	0	NV	NV	7.9	0.011	0.057	0.437	0.022	0.126	39	0
Kale, BLD, DRND	13,621	0	0.85	817	41	0.053	0.07	0.5	0.138	0.049	13	0
Kohlrabi, BLD, DRND	35	0	0	0.1	54	0.04	0.02	0.39	0.154	0.16	12	0
Leeks, BLD, DRND	812	0	0.5	25.4	4.2	0.026	0.02	0.2	0.113	0.072	24	0
Lemon grass (Citronella), raw	6	0	NV	NV	2.6	0.065	0.135	1.101	0.08	0.05	75	0
Lettuce, butterhead, raw	3312	0	0	102	3.7	0.057	0.062	0.357	0.082	0.15	73	0
Lettuce, cos/romaine, raw	8710	0	0.13	103	4	0.072	0.067	0.313	0.074	0.142	136	0
Lettuce, iceberg, raw	502	0	0	24.1	2.8	0.041	0.025	0.123	0.042	0.091	29	0
Lima beans, BLD, DRND	303	0	0.14	6.2	10.1	0.14	0.096	1.04	0.193	0.257	26	0
Lotus root, BLD, DRND	0	0	0.01	0.1	27.4	0.127	0.01	0.3	0.218	0.302	8	0
Mung beans, sprouted, stir-fried	31	0	NV	NV	16	0.14	0.18	1.2	0.13	0.559	70	0

Continued

Food, by Major Food Group	Vitamin A (IU)	Vitamin D (IU)	Vitamin E (mg)	Vitamin K (mg)	Vitamin C (mg)	Thiamin (mg)	Riboflavin (mg)	Niacin (mg)	Vitamin B6 (mg)	Pantothenic Acid (mg)	Folate (µg)	Vitamin B12 (mg)
Mushroom, cloud fungus, dried	0	0	NV	NV	0	0.015	0.844	6.267	0.112	0.481	38	0
Mushroom, oyster, raw	48	29	0	0	0	0.125	0.349	4.956	0.11	1.294	38	0
Mushrooms, CND, DRND	0	8	0.01	0	0	0.085	0.021	1.593	0.061	0.811	12	0
Mushrooms, raw	NV	18	NV	NV	NV	0.015	0.217	3.877	0.293	1.5	13	NV
Mushrooms, shitake, dried	0	154	0	0	3.5	0.3	1.27	14.1	0.965	21.879	163	0
Mushrooms, straw, CND, DRND	0	NV	NV	NV	0	0.013	0.07	0.224	0.014	0.412	38	0
Mustard greens, BLD, DRND	12,370	0	0	593	25.3	0.041	0.063	0.433	0.098	0.12	9	0
New zealand spinach, BLD, DRND	4400	0	1.42	337	16	0.03	0.107	0.39	0.237	0.256	8.3	0
Okra, BLD, DRND	575	0	0.69		30	0.04	0.13	0.5	0.304	0.312	15	0
Onions, BLD, DRND	283	0	0	40	16.3	0.132	0.055	0.871	0.187	0.213	46	0
Onions, raw	2	0	0	0.4	7.4	0.046	0.027	0.116	0.12	0.123	19	0
Parsley, raw	8424	0	0	1640	133	0.086	0.098	1.313	0.09	0.4	152	0
Parsnips, BLD, DRND	0	0	1	1	13	0.083	0.051	0.724	0.093	0.588	58	0
Peas, edible-pod type, BLD, DRND	1311	0	0.47	30.2	22	0.064	0.119	0.563	0.174	0.857	35	0
Peas, edible-pod type, raw	1087	0	0.39	25	60	0.15	0.08	0.6	0.16	0.75	42	0
Peas, green, raw	765	0	0.13	24.8	40	0.266	0.132	2.09	0.169	0.104	65	0
Peas, green, BLD, DRND	801	0	0.14	25.9	14.2	0.259	0.149	2.021	0.216	0.153	63	0
Pepper, banana, raw	340	0	0.69	9.5	82.7	0.081	0.054	1.242	0.357	0.265	29	0
Peppers, chili, green, CND	126	0	NV	NV	34.2	0.01	0.03	0.627	0.12	0.084	54	0
Peppers, Hungarian, raw	816	0	0.48	9.9	92.9	0.079	0.055	1.092	0.517	0.205	53	0
Peppers, Jalapeno, raw	1078	0	0	18.5	118.6	0.04	0.07	1.28	0.419	0.315	27	0

Peppers, sweet, green, raw	370	0	0.37	7.4	80.4	0.057	0.028	0.48	0.224	0.099	10	0
Peppers, sweet, red, raw	3131	0	0	4.9	127.7	0.054	0.085	0.979	0.291	0.317	46	0
Pickles, cucumber, sweet	764	0	0	47.1	0.7	0.025	0.03	0.115	0.024	0.051	1	0
Pickles, cucumber, dill	125	0	0	17.3	2.3	0.045	0.057	0.19	0.035	0.201	8	0
Pigeonpeas, BLD, DRND	3	0	NV	NV	0	0.146	0.059	0.781	0.05	0.319	111	0
Pimento, canned	2655	0	0.69	8.3	84.9	0.017	0.06	0.615	0.215	0.01	6	0
Potatoes, au gratin, w/ butter	264	0	NV	NV	9.9	0.064	0.116	0.993	0.174	0.387	8	0
Potatoes, BKD, flesh	0	0	0	0.3	12.8	0.105	0.021	1.395	0.301	0.555	9	0
Potatoes, CND, DRND	0	0	NV	NV	5.1	0.068	0.013	0.915	0.188	0.354	6	0
Potatoes, French fries, FRZ, oven heated	5	0	0.16	2.7	14.2	0.129	0.034	2.28	0.192	0.554	33	0
Potatoes, hashed brown	5	0	0	3.7	13	0.172	0.033	2.302	0.472	0.893	16	0
Potatoes, mashed, w/whole milk and margarine	187	7	0.42	6	10.5	0.092	0.42	1.174	0.247	0.476	9	0
Potatoes, microwaved in skin, flesh	0	0	NV	NV	15.1	0.129	0.025	1.625	0.319	0.597	12	0
Potatoes, scalloped, w/ butter	135	0	NV	NV	10.6	0.069	0.092	1.053	0.178	0.514	9	0
Pumpkin, CND	15,563	0	1.06	16	4.2	0.024	0.054	0.367	0.056	0.4	12	0
Radishes, raw	7	0	0	1.3	14.8	0.012	0.039	0.254	0.071	NA	25	0
Rutabagas, BLD, DRND	2	0	0.24	0.2	18.8	0.082	0.041	0.715	0.102	0.155	15	0
Sauerkraut, CND, w/ liquid	18	0	0	13	14.7	0.021	0.022	0.143	0.13	0.093	24	0
Shallots, raw	4	0	0.04	0.8	8	0.06	0.02	0.2	0.345	0.29	34	0
Soybeans, green, BLD, DRND	156	0	NV	NV	17	0.26	0.155	1.25	0.06	0.128	111	0
Spinach, BLD, DRND	10,481	0	2.08	494	9.8	0.095	0.236	0.49	0.242	0.145	146	0
Spinach, raw	9377	0	2.03	483	28.1	0.078	0.189	0.724	0.195	0.065	194	0
Squash, acorn, BKD	428	0	NV	NV	10.8	0.167	0.013	0.881	0.194	0.504	19	0
Squash, butternut, BKD	11,155	0	1.29	1	15.1	0.072	0.017	0.969	0.124	0.359	19	0
Squash, hubbard, BKD	6705	0	0.2	1.6	9.5	0.074	0.047	0.558	0.172	0.447	16	0

Continued

Food, by Major Food Group	Vitamin A (IU)	Vitamin D (IU)	Vitamin E (mg)	Vitamin K (mg)	Vitamin C (mg)	Thiamin (mg)	Riboflavin (mg)	Niacin (mg)	Vitamin B6 (mg)	Pantothenic Acid (mg)	Folate (µg)	Vitamin B12 (mg)
Squash, spaghetti, BLD, DRND/BKD	110	0	0.12	0.8	3.5	0.038	0.022	0.81	0.099	0.355	8	0
Squash, summer, BLD, DRND	85	0	0.12	3.5	10.8	0.051	0.025	0.464	0.085	0.079	21	0
Squash, zucchini, BLD, DRND	1117	0	0	4.2	12.9	0.035	0.024	0.51	0.08	0.288	28	0
Succotash (corn and lima beans), BLD, DRND	294	0	NV	NV	8.2	0.168	0.096	1.327	0.116	0.567	33	0
Sweetpotato leaves, STMD	2939	0	0.96	109	1.5	0.112	0.267	1.003	0.16	0.2	49	0
Sweetpotato, BKD in skin, flesh	19,218	0	0.71	2.3	19.6	0.107	0.106	1.487	0.286	0.884	6	0
Taro leaves, STMD	4238	0	NV	NV	35.5	0.139	0.38	1.267	0.072	0.044	48	0
Taro shoots, CKD	51	0	NV	NV	18.9	0.038	0.053	0.81	0.112	0.076	3	0
Taro, COOKED	84	0	2.93	1.2	5	0.107	0.028	0.51	0.331	0.336	19	0
Tomato juice, CND	450	0	0.32	2.3	70.1	0.1	0.078	0.673	0.07	NV	20	0
Tomato paste, CND	1525	0	0	11.4	21.9	0.06	0.153	3.076	0.216	0.142	12	0
Tomato sauce, CND	435	0	1.44	2.8	7	0.024	0.065	0.991	0.098	0.309	9	0
Tomatoes, green, raw	642	0	0.38	10.1	23.4	0.06	0.04	0.5	0.081	0.5	9	0
Tomatoes, red, ripe, CND, STWD	172	0	0.83	2.4	7.9	0.046	0.035	0.714	0.017	0.114	5	0
Tomatoes, red, ripe, raw	833	0	0.54	7.9	13.7	0.037	0.019	0.594	0.08	0.089	15	0
Turnip greens, BLD, DRND	8612	0	2.13	415	18.2	0.05	0.065	0.486	0.067	0.083	33	0
Turnips, BLD, DRND	0	0	0.03	0.06	11.6	0.027	0.023	0.299	0.067	0.142	9	0
Waterchestnuts, Chinese, CND	0	0	0.5	0.2	1.3	0.011	0.024	0.36	0.159	0.221	6	0
Watercress, raw	3191	0	1	250	43	0.09	0.12	0.2	0.129	0.31	9	0
Winged beans, BLD, DRND	0	0	NV	NV	0	0.295	0.129	0.83	0.047	0.156	10	0
Yam, BLD, DRND, BKD	122	0	0.34	2.3	12.1	0.095	0.028	0.552	0.228	0.311	16	0
Yardlong bean, BLD, DRND	450	0	NV	NV	16.2	0.085	0.099	0.63	0.024	0.051	45	0

Fruits and Fruit Juices												
Apple juice, CND/ BTLD, w/o added vitamin C	1	0	0.01	0	0.9	0.021	0.017	0.73	0.081	0.049	0	0
Apples, raw, w/skin	54	0	0.18	2.2	4.6	0.017	0.026	0.091	0.041	0.061	3	0
Applesauce, CND, w/o added vitamin C	29	0	0.16	0.5	21.2	0.026	0.03	0.084	0.027	0.041	3	0
Apricot nectar, CND, w/o added vitamin C	1316	0	0.13	1.2	0.6	0.009	0.014	0.26	0.022	0.096	1	0
Apricots, dehyd	12,669	0	NV	NV	9.5	0.043	0.148	3.58	0.52	1.067	4	0
Apricots, raw	1926	0	0	3.3	10	0.03	0.04	0.6	0.054	0.24	9	0
Avocados, raw	146	0	2.07	21	10	0.067	0.13	1.738	0.257	1.389	24	0
Bananas, raw	64	0	0	0.05	8.7	0.031	0.073	0.665	367	0.334	20	0
Blackberries, raw	214	0	0	19.8	21	0.02	0.026	0.646	0.03	0.276	25	0
Blueberries, raw	54	0	0.57	19.3	9.7	0.037	0.041	0.418	0.052	0.124	6	0
Cantaloupes, raw	3382	0	0.05	2.5	36.7	0.041	0.019	0.734	0.072	0.105	21	0
Casaba melons, raw	0	0	0.05	2.5	21.8	0.015	0.031	0.232	0.163	0.084	8	0
Cherries, sour, red, raw	1283	0	0.07	2.1	10	0.03	0.04	0.4	0.044	0.143	8	0
Cherries, sweet, raw	64	0	0.07	2.1	7	0.027	0.033	0.154	0.049	0.199	4	0
Crabapples, raw	40	NV	NV	NV	8	0.03	0.02	0.1	NV	NV	NV	0
Cranberries, raw	63	0	1.32	5	14	0.012	0.02	0.101	0.057	0.295	1	0
Cranberry sauce, CND	33	0	0.93	1.4	1	0.015	0.021	0.1	0.014	NV	1	0
Currants, European black, raw	230	NV	1	NV	181	0.05	0.05	0.3	0.066	0.398	NV	0
Custard apple (bullock's heart), raw	33	NV	NV	NV	19.2	0.08	0.1	0.5	0.221	0.135	NV	0
Elderberries, raw	600	NV	NV	NV	36	0.07	0.06	0.5	0.23	0.14	6	0
Figs, dried, uncooked	10	0	0.35	15.6	1.2	0.085	0.082	0.619	0.106	0.434	9	0
Figs, raw	142	0	0.11	4.7	2	0.06	0.05	0.4	0.113	0.3	6	0
Fruit cocktail, CND, water PK, w/liquids	250	0	0.4	2.6	2.1	0.016	0.011	0.363	0.052	0.062	3	0
Gooseberries, raw	290	NV	0.37	NV	27.7	0.04	0.03	0.3	0.08	NV	6	0
Grape juice, CND/ BTLD, w/o added vitamin C	8	0	0	0.4	0.1	0.017	0.015	0.133	0.032	0.048	0	0

Continued

Food, by Major Food Group	Vitamin A (IU)	Vitamin D (IU)	Vitamin E (mg)	Vitamin K (mg)	Vitamin C (mg)	Thiamin (mg)	Riboflavin (mg)	Niacin (mg)	Vitamin B6 (mg)	Pantothenic Acid (mg)	Folate (µg)	Vitamin B12 (mg)
Grapefruit juice, CND~	35	0	0.04	0	28.3	0.042	0.02	0.231	0.02	0.13	10	0
Grapefruit, raw, pink/red/white	927	0	0.13	0	34.4	0.036	0.02	0.25	0.042	0.283	10	0
Grapes, adherent skin type, raw	100	0	0.19	14.6	4	0.092	0.057	0.3	0.11	0.024	4	0
Guavas, raw	624	0	0.73	2.6	228.3	0.067	0.04	1.084	0.11	0.451	49	0
Honeydew melons, raw	50	0	0.02	2.9	18	0.038	0.012	0.418	0.088	0.155	19	0
Jackfruit, raw	110	NV	0.34	NV	13.7	0.105	0.055	0.92	0.329	0.235	24	0
Kiwi fruit, raw	87	0	1.46	40.3	92.7	0.027	0.025	0.341	0.063	0.183	25	0
Kumquats, raw	290	0	0.15	0	43.9	0.037	0.09	0.429	0.036	0.208	17	0
Lemon juice, CND/BTLD	33	0	0.23	0.1	14.3	0.021	0.017	0.18	0.037	0.08	9	0
Lemons, raw, w/o peel	22	0	0.15	0	53	0.04	0.02	0.1	0.08	0.19	11	0
Lime juice, CND/BTLD	16	0	0.12	0.5	6.4	0.033	0.003	0.163	0.027	0.066	8	0
Litchis, raw	0	0	0.7	0.4	71.5	0.011	0.065	0.603	0.1	NV	14	0
Mangos, raw	1082	0	0.9	4.2	36.4	0.028	0.038	0.669	0.119	0.197	43	0
Nectarines, raw	332	0	0.77	2.2	5.4	0.034	0.027	1.125	0.025	0.158	5	0
Olives, ripe, CND	403	0	1.65	1.4	0.9	0.003	0	0.037	0.009	0.015	0	0
Orange juice, including from concentrate	42	0	0.2	0	33.6	0.046	0.039	0.28	0.076	0.195	19	0
Oranges, raw	247	0	0.15	0	59.1	0.068	0.051	0.425	0.079	0.261	34	0
Papayas, raw	950	0	0.3	2.6	60.9	0.023	0.027	0.357	0.038	0.191	37	0
Passion fruit, purple, raw	717	NV	0.01	NV	29.8	0	0.131	1.46	0.05	NV	7	0
Peaches, CND, water PK, w/liquids	532	0	0.49	1.7	2.9	0.009	0.019	0.521	0.019	0.05	3	0
Peaches, raw	326	0	0.73	2.6	6.6	0.024	0.031	0.806	0.025	0.153	4	0
Pears, CND, water PK, w/liquids	0	0	0.08	0.3	1	0.008	0.01	0.054	0.014	0.022	1	0
Pears, raw	25	0	0.12	4.4	4.3	0.012	0.026	0.161	0.029	0.049	7	0
Pineapple, CND, water PK, w/liquids	38	0	0.1	0.3	7.7	0.093	0.026	0.298	0.074	0.1	5	0

Pineapple, raw	58	0	0.02	0.7	47.8	0.079	0.032	0.5	0.112	0.213	18	0
Plantains, CKD	909	0	0.13	0.7	10.9	0.046	0.052	0.756	0.24	0.233	26	0
Plums, raw	345	0	0	6.4	9.5	0.028	0.026	0.417	0.029	0.135	5	0
Pomegranates, raw	0	0	0.6	16.4	10.2	0.067	0.053	0.293	0.175	0.377	38	0
Prickly pears, raw	43	NV	NV	NV	14	0.014	0.06	0.46	0.06	NV	6	0
Prunes, CND, HVY syrup, w/liquids	797	0	NV	NV	2.8	0.034	0.122	0.866	0.203	0.1	0	0
Prunes, DEHYD, STWD	523	0	NV	NV	0	0.046	0.03	0.985	0.191	0.108	0	0
Prunes, DEHYD, UNCKD	1762	0	NV	NV	0	0.118	0.165	2.995	0.745	0.418	2	0
Quinces, raw	40	NV	NV	NV	15	0.02	0.03	0.2	0.04	0.081	3	0
Raisins, golden seedless	0	0	0.12	3.5	3.2	0.008	0.191	1.142	0.323	0.14	3	0
Raisins, seedless	0	0	0.12	3.5	2.3	0.106	0.125	0.766	0.174	0.095	5	0
Raspberries, raw	33	0	0.87	7.8	26.2	1.032	0.038	0.598	0.055	0.329	21	0
Rhubarb, FRZ, CKD	73	0	1.9	21.1	3.3	0.018	0.023	0.2	0.02	0.05	5	0
Strawberries, CND, heavy syrup, w/liquids	26	0	0.19	1.5	31.7	0.021	0.034	0.057	0.049	0.179	28	0
Strawberries, raw	12	0	0.29	2.2	58.8	0.024	0.022	0.386	0.047	0.125	24	0
Tangerines, raw	681	0	0.2	0	26.7	0.058	0.036	0.376	0.078	0.216	16	0
Watermelon, raw	569	0	0.05	0.1	8.1	0.033	0.021	0.178	0.045	0.221	3	0
Beans and Peas												
Black beans, BLD	6	0	0.87	3.3	0	0.244	0.059	0.505	0.069	0.242	149	0
Chickpeas, BLD	27	0	0.35	4	1.3	0.116	0.063	0.526	0.139	0.286	172	0
Cowpeas (blackeyes), BLD	791	0	0.22	26.6	2.2	0.101	0.148	1.403	0.065	0.154	127	0
Falafel	13	0	NV	NV	1.6	0.146	0.166	1.044	0.125	0.292	78	0
French beans, BLD	3	0	NV	NV	1.2	0.13	0.062	0.546	0.105	0.222	75	0
Great northern beans, BLD	1	0	NV	NV	1.3	0.158	0.059	0.681	0.117	0.266	102	0
Humus	30	0	NV	NV	0	0.18	0.064	0.502	0.2	0.132	83	0
Kidney beans, BLD	2	0	NV	NV	35.6	0.362	0.273	3.024	0.093	0.381	47	0
Lentils, BLD	8	0	0.11	1.7	1.5	0.169	0.073	1.06	0.178	0.638	181	0
Lima beans, BLD	0	0	0.18	2	0	0.161	0.055	0.421	0.161	0.422	83	0

Continued

Food, by Major Food Group	Vitamin A	Vitamin D	Vitamin E	Vitamin K	Vitamin C	Thiamin	Riboflavin	Niacin	Vitamin B6	Pantothenic Acid	Folate	Vitamin B12
	(IU)	(IU)	(mg)	(mg)	(mg)	(mg)	(mg)	(mg)	(mg)	(mg)	(μg)	(mg)
Lupins, BLD	7	0	NV	NV	1.1	0.134	0.053	0.495	0.009	0.188	59	0
Mung beans, BLD	24	0	0.51	2.7	1	0.164	0.061	0.577	0.067	0.41	159	0
Navy beans, BLD	0	0	0.01	0.6	0.9	0.237	0.066	0.649	0.138	NV	NV	0
Peanut butter, smooth style	0	0	9.1	0.3	0	0.15	0.192	13.11	0.441	1.137	87	0
Peanuts, BLD	0	0	4.1	0	0	0.259	0.063	5.259	0.152	0.825	75	0
Peanuts, dry-roasted	0	0	4.93	0	0	0.152	0.197	14.36	0.466	1.011	97	0
Peanuts, oil-roasted	0	0	6.94	0	0.8	0.085	0.089	13.83	0.461	1.202	120	0
Peas, split, BLD	7	0	0.03	5	0.4	0.19	0.056	0.89	0.048	0.595	65	0
Pigeon peas (red gram), BLD	3	0	NV	NV	0	0.146	0.059	0.781	0.05	0.319	111	0
Pinto beans, BLD	0	0	0.94	3.5	0.8	0.193	0.062	0.318	0.229	0.21	172	0
Refried beans, canned	NV	NV	0.09	NV	NV	0.04	0.015	0.37	0.107	0.175	NV	NV
Soy flour, full-fat, RSTD	122	0	1.98	71	0	0.412	0.941	3.286	0.351	1.209	227	0
Soybeans, RSTD	0	0	NV	NV	2.2	0.1	0.145	1.41	0.208	0.453	211	0
Tempeh	0	0	NV	NV	0	0.078	0.358	2.64	0.215	0.278	24	0.08
Tofu, raw	166	0	NV	NV	0.2	0.158	0.102	0.381	0.092	0.133	29	0
Winged beans, BLD	0	0	NV	NV	0	0.295	0.129	0.83	0.047	0.156	10	0
Yardlong beans, BLD	450	0	NV	NV	16.2	0.085	0.099	0.63	0.024	0.051	45	0
Nuts												
Acorns, dried	0	NV	NV	NV	0	0.149	0.154	2.406	0.695	0.94	115	0
Almonds, dry RSTD, unblanched	1	0	23.9	0	0	0.077	1.117	3.637	1.36.136	0.321	55	0
Brazilnuts, dried, unblanched	0	0	5.65	0	0.7	0.617	0.035	0.295	0.101	0.184	22	0
Butternuts, dried	124	0	NV	NV	3.2	0.383	0.148	1.045	0.56	0.633	66	0
Cashew nuts, dry RSTD	0	0	0.92	34.7	0	0.2	0.2	1.4	0.256	1.27	69	0
Chestnuts, European, RSTD	24	0	0.5	7.8	26	0.243	0.175	1.342	0.497	0.554	70	0
Coconut meat, dried, flaked	0	0	0	0	0	0.015	0.015	0.697	0.03	0.14	3	0
Coconut meat, raw	0	0	0.24	0.2	3.3	0.066	0.02	0.54	0.054	0.3	26	0

Coconut milk, CND	0	0	NV	NV	1	0.022	0	0.637	0.028	0.153	14	0
Coconut water	0	0	0	0	2.4	0.03	0.057	0.08	0.032	0.043	3	0
Filberts (hazelnuts), dry RSTD	61	0	15.28	NV	3.8	0.338	0.123	2.05	0.62	0.923	88	0
Pecans, dry RSTD	140	0	1.3	NV	0.7	0.45	0.107	1.167	0.187	0.703	16	0
Pine nuts, pinyon, dried	1	0	NV	NV	2	1.243	0.223	4.37	0.111	0.21	58	0
Pistachio nuts, dry RSTD	266	0	2.17	13.2	3	0.695	0.234	1.373	1.122	0.513	51	0
Pumpkin and squash seeds, WHL, RSTD	62	0	NV	NV	0.3	0.034	0.052	0.286	0.037	0.056	9	0
Sunflower kernels, dried	50	0	35.17	0	1.4	1.48	0.355	8.335	1.345	1.13	227	0
Tahini, from RSTD and TSTD sesame kernels	67	0	0.25	0	0	1.22	0.473	5.45	0.149	0.693	98	0
Walnuts, black, dried	40	0	2.08	2.7	1.7	0.057	0.13	0.47	0.583	1.66	31	0
Poultry												
Chicken, DK meat w/ skin, fried w/ batter	103	NV	NV	NV	0	0.117	0.218	5.607	0.25	0.953	9	0.27
Chicken, DK meat w/ skin, RSTD	201	NV	NV	NV	0	0.066	0.207	6.359	0.31	1.111	7	0.29
Chicken, giblets, simmered	9536	NV	NV	NV	5.5	0.09	1.047	4.971	0.41	2.971	367	9.48
Chicken, LT meat w/ skin, fried w/ batter	79	NV	NV	NV	0	0.113	0.147	9.156	0.39	0.794	6	0.28
Chicken, LT meat w/ skin, RSTD	110	NV	NV	NV	0	0.06	0.118	11.13	0.52	0.926	3	0.32
Duck, meat only, RSTD	77	4	0.7	3.8	0	0.26	0.47	5.1	0.25	1.5	10	0.4
Duck, meat w/ skin, RSTD	210	3	0.7	5.1	0	0.174	0.269	4.825	0.18	1.098	6	0.3
Goose, meat only, RSTD	40	NV	NV	NV	0	0.092	0.39	4.081	0.47	1.834	12	0.49
Turkey, breast w/ skin, RSTD	0	NV	NV	NV	0	0.057	0.131	6.365	0.48	0.634	6	0.36
Turkey, DK meat only, RSTD	18	10	0.07	0	0	0.06	0.375	6.685	0.438	1.015	9	1.65
Turkey, meat only, RSTD	14	10	0.06	0	0	0.047	0.28	9.5	0.643	NV	NV	0.94
Turkey, meat w/ skin, RSTD	39	15	0.07	0	0	0.045	0.281	9.573	0.616	0.948	9	1.02

Continued

Food, by Major Food Group	Vitamin A (IU)	Vitamin D (IU)	Vitamin E (mg)	Vitamin K (mg)	Vitamin C (mg)	Thiamin (mg)	Riboflavin (mg)	Niacin (mg)	Vitamin B6 (mg)	Pantothenic Acid (mg)	Folate (µg)	Vitamin B12 (mg)
Beef												
Brisket, 1/8″ fat, BRSD	0	16	0.19	1.9	0	0.06	0.2	3.27	0.26	0.31	7	2.4
Chuck (arm pot roast), 1/8″ fat, BRSD	0	13	0.53	1.9	0	0.061	0.178	4.273	0.295	0.594	9	2.22
Corned beef, CND	0	10	0.15	1.6	0	0.02	0.147	2.43	0.13	0.626	9	1.62
Dried beef, cured	0	1	0.38	1.3	0	0.065	0.163	5.164	0.386	0.566	10	1.59
Flank, BRSD	37	11	0.93	NV	0	0.022	0.085	3.41	0.118	0.129	NV	1.38
Ground, lean, BKD-MED	9	2	0.12	2.9	0	0.051	0.171	4.026	0.311	0.512	7	2.49
Ground, REG, BKD-MED	9	2	0.12	1.5	0	0.032	0.171	5.648	0.36	0.666	6	2.94
Ribs (10–12), 0″ fat, BRLD	0	NV	0.41	1.5	0	0.077	0.148	8.521	0.601	0.567	10	1.8
Round (eye of round), 0″ fat, RSTD	14	1	0.3	1.3	0	0.08	0.232	8.464	0.811	NV	NV	2.32
Round, full cut, 1/4″ fat, BRLD	0	5	0.14	1.5	0	0.1	0.22	4.26	0.4	0.41	10	3.17
Round, top round, 1/8″ fat, BRLD	0	6	0.43	1.5	0	0.07	0.163	5.36	0.403	0.605	10	1.71
Short loin (porterhouse steak), 0″ fat, BRLD	0	NV	0.18	NV	0	0.099	0.228	4.21	0.365	0.314	7	2.18
Short loin (top loin), 1/8″ fat, BRLD	0	NV	NV	NV	0	0.08	0.18	4.77	0.38	0.33	7	1.94
Tenderloin, 1/8″ fat, BRLD	0	NV	0.4	1.4	0	0.083	0.154	8.494	0.642	0.573	10	1.39
Top sirloin, 1/8″ fat, BRLD	0	9	0.44	1.6	0	0.073	0.13	7.176	0.564	0.531	8	1.59
Pork												
Bacon, Canadian style, pan fried	0	9	0.41	0.2	0	0.669	0.185	9.988	0.28	0.72	4	0.43
Bacon, BRLD/pan fried/RSTD	37	1	0.31	0.1	0	0.404	0.264	11.1	0.349	1.171	2	1.23
Cured ham, boneless, ex lean (5% fat), RSTD	0	32	0.25	0	0	0.754	0.202	4.023	0.4	0.403	3	0.65
Cured ham, boneless, REG (11% fat), RSTD	0	32	0.31	0	0	0.73	0.33	6.15	0.31	0.72	3	0.7

Cured ham, ex lean (4% fat),CND	0	93	0.17	0	0	0.836	0.23	5.302	0.45	0.492	6	0.82
Cured ham, REG (13% fat), CND	0	NV	NV	NV	14	0.82	0.26	5.3	0.3	0.73	5	1.06
Leg (ham), lean, RSTD	9	36	0.26	0	0.4	0.69	0.349	4.935	0.45	0.67	12	0.72
Loin, blade (chops), bone-in, BRSD	13	39	0.2	0	0	0.486	0.316	7.381	0.488	0.944	0	0.62
Loin (tenderloin), RSTD	0	10	0.08	0	0	0.95	0.387	7.432	0.739	1.012	0	0.57
Loin, top loin (chop), boneless, BRSD	11	38	0.24	0	0	0.526	0.238	9.905	0.537	1.03	0	0.67
Loin, top loin (roast), boneless, w/ fat, RSTD	4	20	0.11	0	0	0.547	0.232	7.104	0.692	0.674	0	0.58
Loin, lean, BRLD	7	37	0.27	0	0.7	0.923	0.338	5.243	0.492	0.729	6	0.72
Shoulder (arm picnic), lean, RSTD	0	35	0.26	0	0	0.727	0.226	4.798	0.37	0.654	4	1.11
Shoulder, lean, RSTD	0	34	0.26	0	0	0.68	0.254	5.02	0.47	0.498	4	0.7
Spareribs, w/ fat, BRSD	10	104	0.34	0	0	0.408	0.382	5.475	0.35	0.75	4	1.08
Sausages and Luncheon Meats												
Bologna, beef	90	28	0.56	2.4	15.2	0.03	0.065	2.321	0.157	0.385	3	1.19
Bologna, pork	0	56	0.26	0.3	0	0.523	0.157	3.9	0.27	0.72	5	0.93
Bologna, Turkey	32	26	0.45	0.3	13.3	0.049	0.095	2.607	0.243	0.466	9	0.23
Bratwurst, pork, CKD	6	44	0.26	3.4	0	0.459	0.307	4.795	0.327	0.666	3	0.73
Frankfurter, beef	0	38	0.2	1.8	NV	0.017	0.05	1.827	0.193	NV	10	1.44
Frankfurter, chicken	0	21	0.22	0	0	0.057	0.257	4.687	0.323	1.06	4	0.54
Frankfurter, Turkey	0	23	0.62	0	0	0.036	0.181	3.68	0.143	0.56	7	0.82
Ham, chopped, CND	0	24	0.25	0	2	0.535	0.165	3.2	0.32	0.28	1	0.7
Ham, sliced, REG (11% fat)	0	29	0.08	0	4	0.626	0.178	2.904	0.329	0.435	7	0.42
Italian sausage, pork, CKD	16	41	0.25	3.4	0.1	0.623	0.233	4.165	0.33	NV	5	1.3
Kielbasa, grilled	32	35	0.38	0	14.7	0.15	0.212	3.727	0.097	NV	NV	0.72
Knockwurst, pork and beef	0	44	0.57	1.6	0	0.342	0.14	2.734	0.17	0.32	2	1.18
Olive loaf, pork	200	44	0.25	3.4	0	0.295	0.26	1.835	0.23	0.77	2	1.26
Pastrami, Turkey	12	10	0.22	0	8.1	0.055	0.25	3.527	0.27	0.58	5	0.24

Continued

Food, by Major Food Group	Vitamin A (IU)	Vitamin D (IU)	Vitamin E (mg)	Vitamin K (mg)	Vitamin C (mg)	Thiamin (mg)	Riboflavin (mg)	Niacin (mg)	Vitamin B6 (mg)	Pantothenic Acid (mg)	Folate (µg)	Vitamin B12 (mg)
Polish sausage, pork	0	NV	NV	NV	1	0.502	0.148	3.443	0.19	0.45	2	0.98
Salami, beef	0	48	0.19	1.3	0	0.103	0.189	3.238	0.18	0.95	2	3.06
Salami, beef and pork	0	41	0.22	3.2	0	0.367	0.357	6.053	0.459	1.201	3	1.52
Smoked link sausage, pork	0	43	0.25	0	0	0.212	0.18	4.807	0.179	0.536	1	0.66
Smoked link sausage, pork and beef	74	44	0.13	0	0	0.192	0.106	2.94	0.163	0.525	2	0.58
Turkey breast meat	6	NV	NV	NV	0	0.058	0.115	5.2	0.48	0.621	7	0.42
Turkey ham	0	2	0.39	0	0	0.05	0.25	3.53	0.23	NV	6	0.26
Vienna sausage, beef and pork, CND	0	25	0.22	1.6	0	0.087	0.107	1.613	0.12	0.35	4	1.02
Fish and Seafood												
Abalone, fried~	5	NV	NV	NV	1.8	0.22	0.13	1.9	0.15	2.87	5	0.69
Anchovy, CND in oil, DRND	40	69	3.33	12.1	0	0.078	0.363	19.9	0.203	0.909	13	0.88
Carp, dry heat CKD	32	NV	NV	NV	1.6	0.14	0.07	2.1	0.219	NV	NV	1.47
Catfish, channel, breaded and fried	28	NV	NV	NV	0	0.073	0.133	2.282	0.19	0.73	17	1.9
Caviar, black/red	905	117	1.89	0.6	0	0.19	0.62	0.12	0.32	3.5	50	20
Clams, breaded and fried	302	NV	NV	NV	10	0.1	0.244	2.064	0.06	0.43	18	40.27
Cod, Atlantic, dry heat CKD	47	46	0.81	0.1	1	0.088	0.079	2.513	0.283	0.18	8	1.05
Cod, Atlantic, dried and salted	140	161	2.84	0.4	3.5	0.268	0.24	7.5	0.864	1.675	25	10
Crab, Alaska king, moist heat CKD	29	NV	NV	NV	7.6	0.053	0.055	1.34	0.18	0.4	51	11.5
Crab, blue, moist heat CKD	2	0	1.84	0.3	3.3	0.023	0.093	2.747	0.156	0.997	51	3.33
Crayfish, moist heat CKD	50	0	1.5	0.1	0.9	0.05	0.085	2.28	0.076	0.58	44	2.15
Eels, dry heat CKD	3787	NV	NV	NV	1.8	0.183	0.051	4.487	0.077	0.28	17	2.89
Flatfish (flounder/sole), dry heat CKD	37	139	0.77	0.1	0	0.026	0.025	1.278	0.115	0.227	6	1.31

Gefilte fish	89	NV	NV	NV	0.8	0.065	0.059	1	0.08	0.2	3	0.84
Haddock, dry heat CKD	62	23	0.55	0.1	0	0.023	0.069	4.119	0.327	0.494	13	2.13
Halibut, dry heat CKD	73	231	0.74	0	0	0.058	0.036	7.911	0.632	0.416	14	1.27
Herring, dry heat CKD	120	214	1.37	0.1	0.7	0.112	0.299	4.124	0.348	0.74	12	13.14
Herring, kippered	135	86	1.54	0.1	1	0.126	0.319	4.402	0.413	0.88	14	18.7
Herring, pickled	860	113	1.71	0.2	0	0.036	0.139	3.3	0.17	0.081	2	4.27
Lobster, Maine, moist heat CKD	4	1	1	0	0	0.023	0.017	1.83	0.119	1.667	11	1.43
Mackerel, dry heat CKD	180	NV	NV	NV	0.4	0.159	0.412	6.85	0.46	0.99	2	19
Mackerel, jack, CND, DRND	433	292	1.03	0.1	0.9	0.04	0.212	6.18	0.21	0.305	5	6.94
Mussel, blue, moist heat CKD	304	NV	NV	NV	13.6	0.3	0.42	3	0.1	0.95	76	24
Ocean perch, dry heat CKD	44	58	0.91	0.1	0	0.046	0.057	1.215	0.084	0.329	10	1.72
Oyster, breaded and fried	302	NV	NV	NV	3.8	0.15	0.202	1.65	0.064	0.27	14	15.63
Oyster, canned	300	1	0.85	0.1	5	0.15	0.166	1.244	0.095	0.18	9	19.13
Oyster, raw	270	NV	NV	NV	8	0.067	0.233	2.01	0.05	0.5	10	16
Perch, dry heat CKD	32	NV	NV	NV	1.7	0.08	0.12	1.9	0.14	0.87	6	2.2
Pike, northern, dry heat CKD	81	NV	NV	NV	3.8	0.067	0.077	2.8	0.135	0.87	17	2.3
Pollock, walleye, dry heat CKD	81	NV	NV	NV	0	0.312	0.195	2.801	0.138	0.865	17	2.31
Roe, mixed species, raw	299	484	7	0.2	16	0.24	0.74	1.8	0.16	1	80	10
Salmon, Chinook, smoked	87	685	1.35	0.1	0	0.023	0.101	4.72	0.278	0.87	2	3.26
Salmon, coho, wild, moist heat CKD	108	NV	NV	NV	1	0.115	0.159	7.779	0.556	0.834	9	4.48
Salmon, pink, CND, w/ bone and liquid	55	NV	NV	NV	0	0.023	0.186	6.536	0.3	0.55	15	4.4
Salmon, sockeye, dry heat CKD	193	670	0.99	0.1	0	0.157	0.246	10.12	0.827	1.274	7	4.47
Sardine, CND in oil, DRND	108	193	2.04	2.6	0	0.08	0.227	5.245	0.167	0.642	10	8.94
Scallops, breaded and fried	75	NV	NV	NV	2.3	0.042	0.11	1.505	0.14	0.2	18	1.32

Continued

Food, by Major Food Group	Vitamin A	Vitamin D	Vitamin E	Vitamin K	Vitamin C	Thiamin	Riboflavin	Niacin	Vitamin B6	Pantothenic Acid	Folate	Vitamin B12
	(IU)	(IU)	(mg)	(mg)	(mg)	(mg)	(mg)	(mg)	(mg)	(mg)	(µg)	(mg)
Sea bass, dry heat CKD	213	NV	NV	NV	0	0.13	0.15	1.9	0.46	0.87	6	0.3
Shark, battered and fried	180	NV	NV	NV	0	0.072	0.097	2.783	0.3	0.62	5	1.21
Shrimp, breaded and fried	189	5	1.3	1	1.5	0.129	0.136	3.07	0.098	0.35	23	1.87
Shrimp, moist heat CKD	301	4	2.2	0.4	0	0.032	0.024	2.678	0.242	0.519	24	1.66
Smelt, rainbow, dry heat CKD	58	NV	NV	NV	0	0.01	0.146	1.766	0.17	0.74	5	3.97
Snapper, dry heat CKD	115	NV	NV	NV	1.6	0.053	0.004	0.346	0.46	0.87	6	3.5
Squid, fried	35	NV	NV	NV	4.2	0.056	0.458	2.602	0.058	0.51	5	1.23
Surimi	67	NV	0.63	0.1	0	0.02	0.021	0.22	0.03	0.07	2	1.6
Swordfish, dry heat CKD	129	666	2.41	0.1	0	0.089	0.063	9.254	0.615	0.417	2	1.62
Trout, rainbow, dry heat CKD	50	NV	NV	NV	2	0.152	0.097	5.77	0.346	1.065	19	6.3
Tuna, bluefin, dry heat CKD	2520	NV	NV	NV	0	0.278	0.306	10.54	0.525	1.37	2	10.88
Tuna, light, CND in water, DRND	57	47	0.33	0.2	0	0.03	0.084	10.14	0.319	0.148	4	2.55
Whiting, dry heat CKD	100	57	0.3	0.1	0	0.056	0.046	1.3	0.156	0.326	13	2.3
Dairy Products												
Butter	2499	0	2.32	7	0	0.005	0.034	0.042	0.003	0.11	3	0.17
Cheese, blue	721	21	0.25	2.4	0	0.029	0.382	1.016	0.166	1.729	36	1.22
Cheese, Brie	592	20	0.24	2.3	0	0.07	0.52	0.38	0.235	0.69	65	1.65
Cheese, Camembert	820	0.4	0.21	2	0	0.028	0.488	0.63	0.227	1.364	62	1.3
Cheese, Cheddar	1242	24	0.71	2.4	0	0.029	0.428	0.059	0.066	0.41	27	1.1
Cheese, Colby	994	24	0.28	2.7	0	0.015	0.375	0.093	0.079	0.21	18	0.83
Cheese, cottage, 1% fat	41	0	0.01	0.1	0	0.021	0.165	0.128	0.068	0.215	12	0.63
Cheese, cottage, creamed	140	3	0.08	0	0	0.027	0.163	0.099	0.046	0.557	12	0.43
Cheese, cream	1111	0	0.86	2.1	0	0.023	0.23	0.091	0.056	0.517	9	0.22

Cheese, cream, fat free	8	0	0.01	0	0	0.023	0.226	0.044	0.016	0.446	9	0.46
Cheese, Edam	825	20	0.24	2.3	0	0.037	0.389	0.082	0.076	0.281	16	1.54
Cheese, feta	422	16	0.18	1.8	0	0.154	0.844	0.991	0.424	0.967	32	1.69
Cheese, Gouda	563	20	0.24	2.3	0	0.03	0.334	0.063	0.08	0.34	21	1.54
Cheese, Gruyere	984	24	0.28	2.7	0	0.06	0.279	0.106	0.081	0.562	10	1.6
Cheese, Monterey	769	22	0.26	2.5	0	0.015	0.39	0.093	0.079	0.21	18	0.83
Cheese, mozzarella, skim milk	481	12	0.14	1.6	0	0.018	0.303	0.105	0.07	0.079	9	0.82
Cheese, mozzarella, whole milk	676	16	0.19	2.3	0	0.03	0.283	0.104	0.037	0.141	7	2.28
Cheese, Muenster	1012	22	0.26	2.5	0	0.013	0.32	0.103	0.056	0.19	12	1.47
Cheese, Parmesan	974	21	0.53	1.7	0	0.026	0.358	0.08	0.081	0.45	6	1.4
Cheese, American	705	NV	NV	NV	0	0.03	0.446	0.074	0.014	0.977	5	1.28
Cheese, Provolone	880	20	0.23	2.2	0	0.019	0.321	0.156	0.073	0.476	10	1.46
Cheese, Ricotta, skim milk	384	6	0.07	0.7	0	0.021	0.185	0.078	0.02	0.242	13	0.29
Cheese, Ricotta, whole milk	445	10	0.11	1.1	0	0.013	0.195	0.104	0.043	0.213	12	0.34
Cheese, Swiss	1047	0	0.6	1.4	0	0.011	0.302	0.064	0.071	0.353	10	3.06
Cream, half and half	354	2	0.25	1.3	0.9	0.03	0.194	0.109	0.05	0.539	3	0.19
Cream, light, coffee/table	656	44	0.12	1.7	0.8	0.023	0.19	0.09	0.044	0.44	2	0.14
Cream, sour	447	0	0.38	1.5	0.9	0.02	0.168	0.093	0.041	0.472	6	0.21
milk, buttermilk, low fat	47	1	0.05	0.1	1	0.034	0.154	0.058	0.034	NV	NV	0.22
Milk, CND, Evap, whole, w/o added vitamin A	239	6	0.16	0.6	1.9	0.047	0.316	0.194	0.05	0.638	8	0.16
Milk, dry, skim, w/o added vitamin A	22	0	0	0.1	6.8	0.415	1.55	0.951	0.361	3.568	50	4.03
Milk, goat	198	51	0.07	0.3	1.3	0.048	0.138	0.277	0.046	0.31	1	0.07
Milk, human	212	3	0.08	0.3	5	0.014	0.036	0.177	0.011	0.223	5	0.05
Milk, 1% fat, w/ added vitamin A	204	40	NV	NV	1	0.04	0.173	0.09	0.045	0.336	5	0.38
Milk, 2% fat, w/ added vitamin A	75	NV	NV	NV	1.1	0.045	0.194	0.101	0.046	0.339	5	0.39

Continued

Food, by Major Food Group	Vitamin A (IU)	Vitamin D (IU)	Vitamin E (mg)	Vitamin K (mg)	Vitamin C (mg)	Thiamin (mg)	Riboflavin (mg)	Niacin (mg)	Vitamin B6 (mg)	Pantothenic Acid (mg)	Folate (µg)	Vitamin B12 (mg)
Milk, skim, w/ added vitamin A	204	47	0.01	0	0	0.045	0.182	0.094	0.037	0.357	5	0.5
Milk, whole (3.3% fat)	162	2	0.07	0.3	0	0.046	0.169	0.089	0.036	0.373	5	0.45
Yogurt, plain, whole milk	15	0	0.01	0	0	0.023	0.278	0.208	0.063	0.331	5	0.75
Eggs												
Egg, white, dried	0	0	0	0	0	0.005	2.53	0.865	0.036	0.775	18	0.18
Egg, white, raw	0	0	0	0	0	0.004	0.439	0.105	0.005	0.19	4	0.09
Egg, whole, hard-boiled	520	87	1.03	0.3	0	0.066	0.513	0.064	0.121	1.398	44	1.11
Egg, whole, fried	787	88	1.31	5.6	0	0.044	0.495	0.082	0.184	1.66	51	0.97
Egg, whole, raw	540	82	1.05	0.3	0	0.04	0.457	0.075	0.17	1.533	47	0.89
Egg, yolk, raw	1442	218	2.58	0.7	0	0.176	0.528	0.024	0.35	2.99	146	1.95
Fats and Oils												
Fat, chicken	0	191	2.7	0	0	0	0	0	0	0	0	0
Lard (pork fat)	0	102	0.6	0	0	0	0	0	0	0	0	0
Margarine, hard, corn/ soy/cottonseed (HYDR)	3577	NV	3.1	NV	0.2	0.01	0.037	0.023	0.009	0.084	1	0.1
Mayonnaise	0	0	11.79	24.7	0	0.01	0.06	0.01	0.01	NV	0	0
Oil, Canola	0	0	17.46	71.3	0	0	0	0	0	0	0	0
Oil, cocoa butter	0	NV	1.8	24.7	0	0	0	0	0	0	0	0
Oil, coconut	0	0	0.11	0.6	0	0	0	0	0	0	0	0
Oil, cod liver	100,000	10,000	NV	NV	0	0	0	0	0	0	0	0
Oil, corn	0	0	14.3	1.9	0	0	0	0	0	0	0	0
Oil, mustard	0	NV	NV	NV	0	0	0	0	0	0	0	0
Oil, olive	0	0	14.35	60.2	0	0	0	0	0	0	0	0
Oil, palm	0	NV	15.94	8	0	0	0	0	0	0	0	0
Oil, peanut	0	0	15.69	0.7	0	0	0	0	0	0	0	0
Oil, rice bran	0	NV	32.3	24.7	0	0	0	0	0	0	0	0
Oil, sesame	0	0	1.4	13.6	0	0	0	0	0	0	0	0
Oil, soybean	0	0	8.18	184	0	0	0	0	0	0	0	0

Oil, sunflower	0	NV	41.08	5.4	0	0	0	0	0	0	0	0
Oil, wheat germ	0	0	149.4	24.7	0	0	0	0	0	0	0	0
Salad dressing, 1000 island	213	0	4	69.1	0	1.445	0.058	0.418	0	0	0	0
Salad dressing, 1000 island, low fat	311	0	1	27.6	1.5	0.049	0.043	0.437	0	0	0	0
Salad dressing, French, low fat	541	0	1	17.8	4.8	0.024	0.052	0.467	0.055	0	2	0
Salad dressing, Italian, low fat	12	0	0	12.5	0	0.012	0.008	0.094	0.055	0	3	0
Salad dressing, Russian	577	0	3.32	53.7	6	0.029	0.046	0.594	0.097	0.4	5	0
Salad dressing, Russian, low fat	33	0	0.4	6.7	6	0.007	0.013	0.002	0.009	0.135	3	0.12
Shortening, soy bean and cottonseed oils (HYDR)	0	0	6.13	43	0	0	0	0	0	0	0	0
Tallow (beef fat)	0	28	2.7	0	0	0	0	0	0	0	0	0
Spices												
Allspice	540	0	NV	NV	39.2	0.101	0.063	2.86	0.21	NV	36	0
Anise seed	311	0	NV	NV	21	0.34	0.29	3.06	0.65	0.797	10	0
Basil	744	0	10.7	1714	0.8	0.08	1.2	4.9	1.34	0.838	310	0
Bay leaf	6185	0	NV	NV	46.5	0.009	0.421	2.005	1.74	NV	180	0
Caraway seed	363	0	2.5	0	21	0.383	0.379	3.606	0.36	NV	10	0
Cardamom	0	0	NV	NV	21	0.198	0.182	1.102	0.23	NV	NV	0
Celery seed	52	0	1.07	0	17.1	0.34	0.29	3.06	0.89	NV	10	0
Chili powder	29650	0	38.14	106	0.7	0.25	0.94	11.6	2.094	0.888	28	0
Cinnamon	295	0	2.32	31.2	3.8	0.022	0.041	1.332	0.158	0.358	6	0
Cloves	160	0	0	142	0.2	0.158	0.22	1.56	0.391	0.509	25	0
Coriander leaf, dried	5850	0	1.03	1359	566.7	1.252	1.5	10.71	0.61	NV	274	0
Coriander seed	0	0	NV	NV	21	0.239	0.29	2.13	NV	NV	0	0
Cumin seed	1270	0	3.33	5.4	7.7	0.628	0.327	4.579	0.435	NV	10	0
Curry powder	19	0	25.24	99.8	0.7	0.176	0.2	3.26	0.105	1.07	56	0
Dill seed	53	0	NV	NV	21	0.418	0.284	2.807	0.25	NV	10	0
Dill weed, dried	5850	0	NV	NV	50	0.418	0.284	2.807	1.71	NV	NV	0

Continued

Food, by Major Food Group	Vitamin A	Vitamin D	Vitamin E	Vitamin K	Vitamin C	Thiamin	Riboflavin	Niacin	Vitamin B6	Pantothenic Acid	Folate	Vitamin B12
	(IU)	(IU)	(mg)	(mg)	(mg)	(mg)	(mg)	(mg)	(mg)	(mg)	(µg)	(mg)
Fennel seed	135	0	NV	NV	21	0.408	0.353	6.05	0.47	NV	NV	0
Garlic powder	0	0	0.67	0.4	1.2	0.435	0.141	0.796	1.654	0.743	47	0
Ginger	30	0	0	0.8	0.7	0.046	0.17	9.62	0.626	0.477	13	0
Mace	800	0	NV	NV	21	0.312	0.448	1.35	0.16	NV	76	0
Marjoram, dried	8068	0	1.69	622	51.4	0.289	0.316	4.12	1.19	NV	274	0
Mustard seed, yellow	31	0	5.07	5.4	7.1	0.805	0.261	4.733	0.397	0.81	162	0
Nutmeg	102	0	0	0	3	0.346	0.057	1.299	0.16	NV	76	0
Oregano	1701	0	18.26	622	2.3	0.177	0.528	4.64	1.044	0.921	237	0
Paprika	42,954	0	29.1	80.3	0.9	0.33	1.23	10.06	2.141	2.51	49	0
Pepper, black	547	0	1.04	164	0	0.108	0.18	1.142	0.291	NV	NV	0
Pepper, red/cayenne	41,610	0	29.83	80.3	76.4	0.328	0.919	8.701	2.45	NV	106	0
Pepper, white	0	0	NV	NV	21	0.022	0.126	0.212	0.1	NV	10	0
Peppermint, fresh	4248	0	NV	NV	31.8	0.082	0.266	1.706	0.129	0.338	114	0
Poppy seed	0	0	1.77	0	1	0.854	0.1	0.896	0.247	0.324	82	0
Rosemary, dried	3128	0	NV	NV	61.2	0.514	0.428	1	1.74	NV	307	0
Saffron	530	0	NV	NV	80.8	0.115	0.267	1.46	1.01	NV	93	0
Sage	5900	0	7.48	17.1	32.4	0.754	0.336	5.72	2.69	NV	274	0
Savory	5130	0	NV	NV	50	0.366	NV	4.08	1.81	NV	NV	0
Spearmint, fresh	4054	0	NV	NV	13.3	0.078	0.175	0.948	0.158	0.25	105	0
Tarragon	4200	0	NV	NV	50	0.251	1.339	8.95	2.41	NV	274	0
Thyme	3800	0	7.48	1714	50	0.513	0.399	4.94	0.55	NV	274	0
Turmeric	0	0	4.43	13.4	0.7	0.058	0.15	1.35	0.107	0.542	20	0
Vanilla extract	0	0	0	0	0	0.011	0.095	0.425	0.026	0.035	0	0
Soups												
Bean w/ pork, CND	662	0	0.87	2.4	1.2	0.065	0.025	0.421	0.031	0.07	24	0.03
Black bean, CND	445	NV	0.36	1.8	0.2	0.042	0.039	0.41	0.07	0.16	66	0
Chicken broth, CND	0	0	0.01	0	0	0.006	0.046	2.23	0.02	0.04	4	0.2
Chicken gumbo, CND	98	NV	0.36	5.7	4	0.2	0.3	0.53	0.05	0.16	5	0.02
Chicken noodle, CND	410	0	0.05	0	0	0.079	0.07	1.222	0.042	0.107	5	0.12

Chicken w/ rice, CND	498	0	0.07	0.3	0	0.014	0.02	0.918	0.02	0.14	1	0.13
Clam chowder, Manhattan style, CND	762	NV	1.03	5.5	3.2	0.024	0.032	0.65	0.08	0.15	8	3.23
Clam chowder, New England style, CND	58	0	0.42	0.8	4.1	0.125	0.165	1.55	0.104	0.231	14	9.47
Cream of asparagus, CND	441	0	0.49	22	2.2	0.043	0.062	0.62	0.01	0.11	19	0.04
Cream of celery, CND	282	0	1.39	17.2	0.2	0.023	0.039	0.265	0.01	0.92	2	0.04
Cream of chicken, CND	182	0	0.54	4.1	0.1	0.013	0.046	0.392	0	0.192	2	0
Cream of mushroom, CND	8	9	0.5	19.6	0	0.012	0.016	0.345	0.015	0.148	2	0
Cream of onion, CND	113	0	0.43	2.2	1	0.04	0.06	0.4	0.02	0.24	6	0.04
Cream of potato, CND	68	0	0.07	1.1	0.2	0.028	0.029	0.43	0.03	0.7	2	0.04
Lentil w/ ham, CND	145	NV	NV	NV	1.7	0.07	0.045	0.545	0.09	0.14	20	0.12
Minestrone, CND	1708	0	0.46	7.8	0.9	0.044	0.036	0.77	0.08	0.28	13	0
Oyster stew, CND	58	NV	NV	NV	2.6	0.017	0.029	0.19	0.01	0.1	2	1.79
Pea, green, CND	25	0	0.18	0.4	1.3	0.082	0.052	0.943	0.04	0.1	1	0
Pea, split w/ ham, CND	331	NV	NV	NV	1.1	0.11	0.056	1.098	0.05	0.2	2	0.2
Tomato rice, CND	421	0	1.73	3	11.5	0.048	0.039	0.822	0.06	0.1	11	0
Tomato, CND	392	0	0.34	3.2	12.9	0.042	0.015	0.858	0.086	NV	0	0
Turkey noodle, CND	120	NV	NV	NV	0.1	0.03	0.026	0.572	0.015	0.07	1	0.06
Vegetable beef, CND	3104	0	0.48	5.6	1.9	0.029	0.039	0.823	0.06	0.28	8	0.25
Vegetarian vegetable, CND	2842	0	1.17	4.2	1.2	0.044	0.037	0.747	0.045	0.28	9	0
Beverages												
Beer, REG	0	0	0	0	0	0.005	0.025	0.513	0.046	0.041	6	0.02
Clam and tomato juice, CND	149	NV	0.11	NV	5	0.021	0.012	0.231	0.061	0.083	8	0.03
Cocoa mix, PDR	15	0	0.04	0.7	0	0.267	1.4	1.084	0.318	3.826	14	1.18
Coffee, brewed, espresso	0	0	0.01	0.1	0.2	0.001	0.177	5.207	0.002	0.028	1	0
Coffee, brewed, regular	0	NV	0	0	0	0	0.001	0.236	0	0.001	0	0
Distilled (Gin/rum/vodka/whiskey), 80 proof	0	0	0	0	0	0.006	0.004	0.013	0.001	0	0	0
Sodas (ginger ale/grape/orange)	0	0	0	0	0	0	0	0	0	0	0	0

Continued

Food, by Major Food Group	Vitamin A	Vitamin D	Vitamin E	Vitamin K	Vitamin C	Thiamin	Riboflavin	Niacin	Vitamin B6	Pantothenic Acid	Folate	Vitamin B12
	(IU)	(IU)	(mg)	(mg)	(mg)	(mg)	(mg)	(mg)	(mg)	(mg)	(μg)	(mg)
Sodas, lemon–lime	0	0	0	0	0	0	0	0.015	0	0	0	0
Tea, brewed	0	0	0	0	0	0	0	0	0	0	0	0
Wine, table	0	0	0	0	0	0.005	0.023	0.166	0.054	0.037	1	0
Snack Foods and Desserts												
Banana chips	83	0	0.24	1.3	6.3	0.085	0.017	0.71	0.26	0.62	14	0
Candies, caramels	42	0	0.46	1.8	0.4	0.103	0.256	0.148	0.056	0.62	4	0.3
Candies, gumdrops	0	0	0	0	0	0.006	0.013	0.01	0.005	0.012	0	0
Candies, hard	0	0	0	0	0	0.004	0.003	0.007	0.003	0.008	0	0
Candies, jellybeans	0	0	0	0	0	0.004	0.011	0.008	0.004	0.009	0	0
Chocolates, milk	195	0	0.51	5.7	0	0.112	0.298	0.386	0.036	0.472	11	0.75
Chocolates, semisweet	0	0	0.26	5.6	0	0.055	0.09	0.427	0.035	0.105	13	0
Frosting, vanilla, creamy	0	0	1.53	13	0	0.01	0.302	0.22	0	0.055	8	0
Gelatins, PREPD w/ water	0	0	0	0	0	0	0.006	0.001	0	0.002	1	0
Jams and preserves	0	0	0.12	0	8.8	0.016	0.076	0.036	0.02	0.02	11	0
Jellies	5	0	0	0.3	0.9	0.001	0.026	0.036	0.02	0.197	2	0
Marmalade, orange	62	0	0.06	0	4.8	0.005	0.025	0.052	0.019	0.015	9	0
Marshmallows	0	0	0	0	0	0.001	0.001	0.078	0.003	0.005	1	0
Molasses	0	0	0	0	0	0.041	0.002	0.93	0.67	0.804	0	0
Popcorn, air-popped	196	0	0.29	1.2	0	0.104	0.083	2.308	0.157	0.51	31	0
Popcorn, cakes	72	0	0.29	1.1	0	0.075	0.178	6.006	0.181	0.434	18	0
Popcorn, caramel-coated, w/ peanuts	78	0	0.85	3.9	0	0.051	0.126	1.99	0.185	0.23	16	0
Popcorn, oil-popped	11	NV	NV	NV	0.3	0.134	0.136	1.55	0.209	0.305	17	0
Potato chips, plain	0	0	10.45	22.1	21.6	0.213	0.088	4.762	0.531	0.956	29	0
Potato chips, barbecue	393	0	4.42	16.1	62.4	0.221	0.119	4.957	0.375	0.85	64	0
Pretzels, hard	0	0	0.47	2.8	2.1	0.424	0.332	5.27	0.074	0.322	166	0
Pudding, chocolate	46	0	0.31	0.6	0.3	0.024	0.072	0.123	0.018	0.244	3	0.09
Pudding, lemon, PREPD w/ 2% fat milk	170	33	NV	NV	0.8	0.033	0.137	0.072	0.036	0.267	4	0.3

Pudding, rice, PREPD w/ whole milk	115	34	NV	NV	0.7	0.074	0.138	0.441	0.034	0.283	4	0.24
Pudding, tapioca, PREPD w/ whole milk	108	34	NV	NV	0.7	0.03	0.14	0.073	0.033	0.272	4	0.25
Rice cakes, brown rice, plain	0	0	1.24	1.9	0	0.061	0.165	7.806	0.15	1	21	0
Sugar, brown	0	0	0	0	0	0	0	0.11	0.041	0.132	1	0
Sugar, granulated	0	0	0	0	0	0	0.019	0	0	0	0	0
Syrup, choc, fudge-type	2	0	2.63	2.5	0.2	0.034	0.091	0.25	0.02	0.119	4	0.06
Syrup, corn, light	0	0	0	0	0	0.059	0	0	0	0	0	0
Syrup, maple	0	0	0	0	0	0.066	1.27	0.081	0.002	0.036	0	0
Syrup, sorghum~	0	0	0	0	0	0.1	0.155	0.1	0.67	0.804	0	0
Tortilla chips, plain	4	0	3.53	20.9	0	0.14	0.07	0.838	0.179	0.297	12	0.36
Tortilla chips, taco-flavored	905	NV	NV	NV	0.9	0.242	0.204	1.999	0.297	0.29	21	0
American Fast Foods												
Biscuit, w/ egg, cheese and bacon	294	21	1.02	4.2	0	0.269	0.247	2.35	0.113	0.772	8	0.86
Biscuit, w/ ham	118	13	1.23	5.7	0.1	0.45	0.28	3.08	0.12	0.36	7	0.03
Burrito, w/ beans and cheese	68	0	0.37	7.6	0.3	0.287	0.143	2.48	0.085	0.403	58	0.08
Burrito, w/ beans and meat	212	NV	0.64	6.8	0.8	0.393	0.236	2.889	0.09	0.008	17	NV
Cheeseburger, single patty, w/ condiments	254	2	0.52	4.2	0	0.293	0.363	5.217	0.228	0.545	42	0.76
Chicken fillet sandwich, w/ cheese	32	5	2.41	8.5	0.8	0.23	0.3	7.72	0.383	1.2	35	0.13
Chicken, boneless pieces, breaded and fried	16	7	1.12	7	0.6	0.092	0.204	5.985	0.148	1.182	6	0.33
Chili con carne	467	1	0.5	4.6	0.2	0.019	0.029	0.782	0.105	0.297	21	0.28
Croissant, w/ egg, cheese and ham	529	31	0.76	2.1	0.1	0.34	0.2	2.1	0.15	0.82	22	0.66
Enchilada, w/ cheese	195	NV	0.89	NV	NV	0.065	0.22	0.78	0.125	0.37	NV	0.72
English muffin, w/ egg, cheese and Canadian bacon	347	26	0.91	0.8	1.1	0.378	0.293	3.187	0.071	NV	20	0.81

Continued

Food, by Major Food Group	Vitamin A	Vitamin D	Vitamin E	Vitamin K	Vitamin C	Thiamin	Riboflavin	Niacin	Vitamin B6	Pantothenic Acid	Folate	Vitamin B12
	(IU)	(IU)	(mg)	(mg)	(mg)	(mg)	(mg)	(mg)	(mg)	(mg)	(µg)	(mg)
French toast sticks	5	0	0.88	14.5	0	0.284	0.175	2.575	0.053	0.336	142	0
Hamburger, single patty, w/ condiments	57	2	0.07	4.9	0.3	0.349	0.186	4.061	0.108	0.37	17	1.2
Hotdog, beef, plain	0	38	0.2	1.8	NV	0.017	0.05	1.827	0.193	NV	10	1.44
Hotdog, w/ chili	51	NV	NV	NV	2.4	0.19	0.35	3.28	0.04	0.48	44	0.26
Hotdog, w/ corn flour coating (corn dog)	166	15	0.6	5.8	0.5	0.141	0.155	2.786	0.117	0.406	7	0.48
Hush puppies	1	0	1.06	3.4	0	0.17	0.07	2.13	0.11	NV	29	0
Nachos, w/ cheese	21	0	4.08	19.3	1.1	0.123	0.133	0.63	0.215	0.38	10	0.07
Onion rings, breaded and fried	0	0	NV	NV	4.6	0.1	0.079	0.693	0.103	0.24	19	0
Pizza w/ cheese~	358	0	0.83	6.7	1.4	0.39	0.195	3.825	0.08	NV	40	0.42
Pizza w/ cheese, meat and vegetables	339	0	1.13	8.2	3.4	0.216	0.233	2.379	0.149	0.332	0	0.62
Pizza w/ pepperoni~	365	0	0.87	6.4	0.9	0.43	0.22	4.14	0.08	NV	36	0.5
Potato salad	100	NV	NV	NV	1.1	0.07	0.11	0.27	0.15	0.37	25	0.12
Potato, mashed	205	2	0.29	2.4	0.1	0.085	0.088	1.02	0.109	0.405	6	0.22
Potatoes, hashed brown	NV	0	0.07	0.4	4.9	0.027	0.022	1.777	0.257	0.555	NV	0
Roast beef sandwich, plain	8	1	0.43	2.8	0	0.217	0.287	4.343	0.215	NV	39	0.96
Submarine sandwich, w/ cold cuts	170	7	0.41	4.5	7.5	0.343	0.267	4.323	0.203	0.555	9	0.22
Submarine sandwich, w/ roast beef	77	1	0.29	4.5	0	0.23	0.257	4.183	0.275	0.415	9	0.44
Submarine sandwich, w/ tuna salad	78	21	1.57	22.4	0	0.19	0.217	7.727	0.271	0.35	18	1.06
Taco	159	3	0.25	7.4	0.1	0.18	0.13	2.88	0.06	NV	16	0.84
Taco salad	297	NV	NV	NV	1.8	0.05	0.18	1.24	0.11	0.68	20	0.32

BKD, baked; *BLD*, boiled; *BTLD*, bottled; *BRSD*, braised; *CND*, canned; *CKD*, cooked; *DRND*, drained; *ENR*, enriched; *HYDR*, hydrogenated; *NV*, no value available; *PREPD*, prepared; *RSTD*, roasted; *STMD*, steamed; *STWD*, stewed.
Nutrient Data Laboratory, Agricultural Research Service, US Department of Agriculture, Release 28 (Database contains no biotin data.)

Appendix E

Vitamin Contents of Feedstuffs (units per kg)

Feedstuff	Vitamin E	Riboflavin	Niacin	Vitamin B6	Biotin	Pantothenic Acid	Folate	Vitamin B12	Choline
	IU	mg	mg	mg	mg	mg	mg	ug	mg
Alfalfa leaf meal, dehydrated	140	15	55	11	0.35	33	4	NV	1600
Alfalfa meal, dehydrated	120	13	46	10	0.33	27	3.5	NV	1600
Alfalfa meal, sun-cured	66	11	40	9	0.3	20	3.3	NV	1500
Bakery product, dehydrated	25	0.8	50	4.4	0.07	9	0.15	NV	660
Barley	36	2	57	2.9	15	6.6	0.5	NV	1100
Beans, field	1	1.8	24	0.3	0.11	3.1	1.3	NV	NV
Blood meal	0	4.2	29	0	NV	5.3	NV	NV	280
Brewers' dried grains	26	1.5	44	0.66	NV	8.8	9.7	NV	1600
Buckwheat	NV	11	18	NV	NV	5.9	NV	NV	13,000
Buttermilk, dried	6.3	30	9	2.4	0.3	30	0.4	20	1800
Casein, purified	NV	1.5	1.3	0.4	NV	2.6	0.4	NV	200
Citrus pulp, dried	NV	2.2	22	NV	NV	13	NV	NV	900
Coconut oil	35	0	0	0	0	0	0	0	0
Coconut oil meal (copra meal)	NV	3.5	24	4.4	NV	6.6	0.3	NV	1100
Corn and cob meal	20	1.1	20	5	0.05	5	0.3	NV	550
Corn germ meal	87	3.7	42	NV	3	3.3	0.7	NV	1540
Corn gluten feed	24	2.2	66	NV	0.3	0.5	0.2	NV	1100
Corn gluten meal	42	1.5	50	8	0.15	10	0.7	NV	330
Corn gluten meal, 60% protein	50	1.8	60	9.6	0.2	12	0.84	NV	400
Corn oil	280	0	0	0	0	0	0	0	0
Corn, dent, no. 2, yellow	22	1.3	22	7	0.06	5.7	0.36	NV	620
Cottonseed meal, dehulled	NV	5.7	51	7	0.1	15	1.1	NV	3300
Cottonseed meal, hydraulic/expeller	40	4	5	5.3	0.1	11	1	NV	2800
Cottonseed meal, solvent	15	5	44	6.4	0.1	13	1	NV	2900
Crab meal	NV	5.9	44	NV	NV	6.6	NV	330	2000
Distillers' dried grains (corn)	30	3.1	42	NV	0.7	5.9	NV	NV	1900
Distillers' dried grains w/sol's (corn)	40	8.6	66	NV	1.1	11	0.9	NV	2500
Distillers' dried solubles (corn)	55	17	115	10	1.5	22	2.2	NV	4800
Feathers, poultry, hydrolyzed	NV	2	24	NV	44	11	0.22	70	900

Fish meal, anchoveta	3.4	6.6	64	3.5	0.26	8.8	0.2	100	3700
Fish meal, herring	27	9	89	3.7	0.42	11	0.24	240	4000
Fish meal, menhaden	9	4.8	55	3.5	0.26	8.8	0.2	88	3500
Fish meal, pilchard	9	9.5	55	3.5	0.26	9	0.2	100	2200
Fish meal, redfish waste	6	NV	NV	3.3	0.08	NV	0.2	100	3500
Fish meal, whitefish waste	9	9	70	3.3	0.08	8.8	0.2	100	2200
Fish oils, stabilized	70	0	0	0	0	0	0	0	0
Fish solubles, dried	6	7.7	230	NV	0.26	45	NV	400	5300
Hominy feed, yellow	NV	2.2	44	11	0.13	7.7	0.28	NV	1000
Lard, stabilized	23	0	0	0	0	0	0	0	0
Liver and glandular meal	NV	40	160	5	0.8	105	4	440	10,500
Meat and bone meal, 45% protein	1	5.3	38	2.3	0.1	2.4	0.05	44	2000
Meat and bone meal, 50% protein	1	4.4	49	2.5	0.14	3.7	0.05	44	2200
Meat meal, 55% protein	1	5.3	57	3	0.26	4.8	0.05	44	2200
Milk, dried skim	NV	20	11	4.9	0.33	33	0.02	60	1400
Milo (grain sorghum)	12	1.2	40	4	0.18	11	0.24	NV	680
Molasses, beet	NV	0.4	40	5.4	88	66	0.2	NV	880
Molasses, cane	5	2.5	100	4.4	100	58	0.04	NV	880
Oat mill by-product	NV	1.5	10	NV	NV	3.3	NV	NV	440
Oatmeal feed	24	1.8	13	2.2	0.22	15	0.35	NV	1200
Oats, heavy	20	1.1	18	1.3	0.11	13	0.3	NV	1100
Olive oil	125	NV	NV	NV	NV	NV	NV	NV	NV
Peanut meal, dehulled, solvent	3	12	180	10	0.39	60	0.36	NV	2100
Peanut meal, solvent	3	11	170	10	0.39	53	0.36	NV	2000
Peanut oil	280	0	0	0	0	0	0	0	0
Peas, field dry	NV	1.8	37	1	0.18	10	0.36	NV	NV
Potato meal, white, dried	NV	0.7	33	14	0.1	20	0.6	NV	2600
Poultry by-product meal	2	11	40	NV	0.3	8.8	1	NV	6000
Poultry offal fat, stabilized	30	0	0	0	0	0	0	0	0
Rapeseed meal	19	3.7	155	NV	NV	9	NV	NV	6600
Rice bran oil	420	NV	NV	NV	NV	NV	NV	NV	NV
Rice bran	60	2.6	300	NV	0.42	23	NV	NV	1300

Continued

Feedstuff	Vitamin E	Riboflavin	Niacin	Vitamin B6	Biotin	Pantothenic Acid	Folate	Vitamin B12	Choline
	IU	mg	mg	mg	mg	mg	mg	ug	mg
Rice polishings	90	1.8	530	NV	0.62	57	NV	NV	1300
Rice, rough	14	0.5	37	NV	NV	3.3	0.25	NV	800
Rice, white, polished	3.6	0.6	14	0.4	NV	3.3	0.15	NV	900
Safflower meal	1	4	6	NV	1.4	4	0.44	NV	2600
Safflower oil	500	0	0	0	0	0	0	0	0
Sesame meal	NV	3.3	30	12.5	NV	6	NV	NV	1500
Sesame oil	250	0	0	0	0	0	0	0	0
Soybean meal	2	3.3	27	8	0.32	14.5	3.6	NV	2700
Soybean meal, dehulled	3.3	3.1	22	8	0.32	14.5	3.6	NV	2700
Soybean oil	280	0	0	0	0	0	0	0	0
Soybean, isolated protein	0	1.25	4.9	1.3	NV	0.63	NV	NV	NV
Soybeans, full-fat, processed	50	2.6	22	11	0.37	15	2.2	NV	2800
Sunflower oil	350	0	0	0	0	0	0	0	0
Sunflower seed meal, dehulled, solvent	20	7.2	106	16	NV	40	NV	NV	4200
Sunflower seed meal, solvent	11	6.4	91	16	NV	10	NV	NV	2900
Tallow, stabilized	13	0	0	0	0	0	0	0	0
Tomato pomace, dried	NV	6.2	NV	NV	NV	NV	NV	NV	NV
Wheat bran	17	3.1	200	10	0.48	29	0.78	NV	1000
Wheat germ meal	130	5	50	13	0.22	12	2.4	NV	3300
Wheat middlings	44	2	100	11	0.37	20	1.1	NV	1100
Wheat shorts	57	2	95	11	0.37	8	1.1	NV	930
Wheat, hard, northern US and Canada	11	1.1	60	4	0.11	13	0.4	NV	1000
Wheat, hard, south-central US	11	2	60	4	0.11	13	0.35	NV	1000
Wheat, soft	11	1.1	60	4	0.11	13	0.3	NV	1000
Whey product, dried	NV	40	15	3.2	0.28	60	0.8	40	260
Whey, dried	NV	30	11	2.5	0.25	47	0.58	0.3	200
Yeast, brewers', dried	0	35	450	3.3	1.3	110	12	NV	3900
Yeast, torula, dried	0	44	500	NV	2	83	21	NV	2900

NV, no value available. From Scott, M.L., Nesheim, M.C., Young, R.J., 1982. The Nutrition of the Chicken, third ed. Scott, M.L., & Assoc., Ithaca, NY, pp. 490–493; with permission.

Index

'*Note*: Page numbers followed by "f" indicate figures, "t" indicate tables and "b" indicates boxes.'

B

C

F

G

N

O

P

Q

R

S

U

V

W

X

Y

Z